Sourcebook of Models for Biomedical Research

Sourcebook of
Models for Biomedical Research

Edited by

P. Michael Conn, PhD

*Associate Director and Senior Scientist,
Oregon National Primate Research Center, Beaverton,
OR; Professor of Physiology and Pharmacology,
Professor of Cell and Developmental Biology,
Oregon Health and Science University, Portland, OR*

Humana Press
Totowa, New Jersey

999 Riverview Drive, Suite 208
Totowa, New Jersey 07512

humanapress.com

This publication is printed on acid-free paper. ∞

ANSI Z39.48-1984 (American Standards Institute) Permanence of Paper for Printed Library Materials.

Cover Illustration: We are indebted to Mr. Joel Ito, Medical Illustrator at the Oregon National Primate Research Center, Oregon Health & Sciences University for the design of this image that combines many of the elements of research models described in this book.

For additional copies, pricing for bulk purchases, and/or information about other Humana titles, contact Humana at the above address or at any of the following numbers: Tel.: 973-256-1699; Fax: 973-256-8341; or visit our Website: www.humanapress.com

10 9 8 7 6 5 4 3 2 1

e-ISBN: 978-1-59745-285-4

Library of Congress Control Number: 2007926774

Preface

The collection of systems represented in *Sourcebook of Models for Biomedical Research* is an effort to reflect the diversity and utility of models that are used in biomedicine. That utility is based on the consideration that observations made in particular organisms will provide insight into the workings of other, more complex, systems. Even the cell cycle in the simple yeast cell has similarities to that in humans and regulation with similar proteins occurs.

Some models have the advantage that the reproductive, mitotic, development or aging cycles are rapid compared with those in humans; others are utilized because individual proteins may be studied in an advantageous way and that have human homologs. Other organisms are facile to grow in laboratory settings or lend themselves to convenient analyses, have defined genomes or present especially good human models of human or animal disease.

We have made an effort not to be seduced into making the entire book homage to the remarkable success of the genomic programs, although this work is certainly well represented and indexed.

Some models have been omitted due to page limitations and we have encouraged the authors to use tables and figures to make comparisons of models so that observations not available in primary publications can become useful to the reader.

We thank Richard Lansing and the staff at Humana for guidance through the publication process.

As this book was entering production, we learned of the loss of Tom Lanigan, Sr. Tom was a leader and innovator in scientific publishing and a good friend and colleague to all in the exploratory enterprise. We dedicate this book to his memory. We will miss him greatly.

P. Michael Conn

Contents

Contributors

YUKSEL AGCA, DVM, PhD, *College of Veterinary Medicine, Comparative Medicine Center, University of Missouri-Columbia, Columbia, MO*

ADEBUSOLA A. ALAGBALA, BS, *Department of Pharmacology and Therapeutics, Roswell Park Cancer Institute, Buffalo, NY*

TIM ALLEN, MS, *US Department of Agriculture, National Agricultural Library, Animal Welfare Information Center, Beltsville, MD*

DAVID M. ANDERSON, DVM, *Washington National Primate Research Center, Department of Comparative Medicine, University of Washington, Seattle, WA*

PHILIP AVNER, PhD, *Departement de Biologie du Développement, Unite de Genetique Moleculaire Murine, Institut Pasteur, Paris, France*

LINA BADIMON, PhD, FAHA, FESC, *Centro de Investigacion Cardiovascular, CSIC-Institut Catala de Ciencies Cardiovasculars, Hosptial de la Santa Creu i Sant Pau, Barcelona, Cardiovascular Research Center, Pavello del Convent, Barcelona, Spain*

JANICE M. BAHR, PhD, *Department of Animal Sciences, University of Illinois, Urbana, IL*

VICTORIA L. BAUTCH, PhD, *Department of Biology, The University of North Carolina at Chapel Hill, Chapel Hill, NC*

BRYAN BELIKOFF, BS, MA, MS, *New England Inflammation and Tissue Protection Institute Consortium at Northeastern University; and Departments of Biology and Pharmaceutical Sciences, Boston, MA*

JANE BLOOD-SIEGFRIED, DNSc, CPNP, *Department of Nursing, Duke University School of Nursing, Durham, NC*

ERNESTO BOCKAMP, Dr. rer.nat., *Institute of Toxicology/Mouse Genetics, Johannes Gutenberg-Universitat Mainz, Mainz, Germany*

FLINT A. BOETTCHER, PhD, *Department of Communicative Sciences and Disorders, University of South Alabama, Mobile, AL*

NORKA RUIZ BRAVO, PhD, *Deputy Director for Extramural Research, National Institutes of Health, Department of Health and Human Services, Bethesda, MD*

ANDREW I. BROOKS, PhD, *Bionomics and Research Technology Center, Environmental and Occupational Health Science Institute, Piscataway, NJ*

JOSEPH L. BRYANT, DVM, *Animal Models Division, University of Maryland Biotechnology Institute, The Institute of Human Virology, Baltimore, MD*

MATTHEW J. BUECHNER, PhD, *Department of Molecular Biosciences, University of Kansas, Lawrence, KS*

ROCHELLE BUFFENSTEIN, *Department of Biology, City College of New York, The City University of New York, New York, NY*

JON A. BURAS, MD, PhD, *New England Inflammation and Tissue Protection Institute Consortium at Northeastern University; and Departments of Biology and Pharmaceutical Sciences, Boston, MA*

JAMES A. BYRNE, PhD, *Oregon National Primate Research Center, Oregon Health and Science University, Beaverton, OR*

NORMAN F. CAPRA, PhD, *Department of Biomedical Sciences, School of Dentistry, University of Maryland Baltimore, Baltimore, MD*

THOMAS CARLSSON, PhD student, *Brain Repair Group, School of Biosciences, Cardiff University, Cardiff, UK*

GREGORY W. CARTER, PhD, *Institute for Systems Biology, Seattle, WA*

LAURA CASANÍ, DVM, PhD, *Centro de Investigacion Cardiovascular, CSIC-Institut Catala de Ciences Cardiovasculars, Hospital de la Santa Creu I Sant Pau, Barcelona, Spain*

DAMAYANTI CHAKRABORTY, BS, *Integrated Graduate Program in Biomedical Sciences, University of Kansas Medical Center, Kansas, City, KS*

ANTHONY W.S. CHAN, DVM, PhD, *Division of Neuroscience, Yerkes National Primate Research Center, Department of Human Genetics, Emory University School of Medicine, Atlanta, GA*

LIHSIA CHEN, PhD, *Department of Genetics, Cell Biology and Development, University of Minnesota, Minneapolis MN*

M. AZHAR CHISHTI, DVM, PhD, *University of Toronto, Mississauga, ONT, Canada*

CHARLENE CHO, PhD, *Office of Extramural Research, National Institutes of Health, Bethesda, MD*

HAGIT COHEN, PhD, *Ministry of Health Mental Health Center, Anxiety and Stress Research Unit, Faculty of Health Sciences, Gen-Gurion University of the Negev, Beer-Sheva, Israel*

KRISTINE COLEMAN, PhD, *Oregon National Primate Research Center, Oregon Health and Science University, Beaverton, OR*

Antonio Conconi, PhD, *Departement de Microbiologie et d'Infectiologie, Faculte de Medecine, Universite de Sherbrooke, Sherbrooke, Québec, Canada*

P. Michael Conn, PhD, *Oregon National Primate Research Center, Beaverton, OR; Oregon Health and Science University, Portland, OR*

Victor A. Convertino, PhD, *US Army Institute of Surgical Research, Fort Sam Houston, TX*

Richard L. Crawford, DVM, *US Department of Agriculture, National Agricultural Library, Animal Welfare Information Center, Beltsville, MD*

Timothy A. Cudd, DVM, PhD, *Department of Veterinary Physiology and Pharmacology, College of Veterinary Medicine and Biomedical Sciences, The Texas A&M University System Health Science Center, College Station, TX*

Edwin Cuppen, PhD, *Hubrecht Laboratory, Utrecht, The Netherlands*

Chunsun Dai, MD, PhD, *Department of Pathology, University of Pittsburgh School of Medicine, Pittsburgh, PA*

Gionata De Vico, DVM, *Faculty of Mathematics, Physics, and Natural Sciences, Department of Biological Sciences, Naples, Italy*

Wanda A. Dorsett-Martin, DVM, *Division of Plastic Surgery, Department of Surgery, The University of Mississippi Medical Center, School of Medicine, Jackson, MS*

Stephen B. Dunnett, DSc, *Brain Repair Group, School of Biosciences, Cardiff University, Cardiff, UK*

Yael H. Edrey, *Department of Biology, City College of New York, The City University of New York, New York, NY*

Michelle M. Elekonich, PhD, *School of Life Sciences, University of Nevada Las Vegas, Las Vegas, NV*

Leonid Eshkind, PhD, *Institute of Toxicology/Mouse Genetics, Johannes Gutenberg-Universitat Mainz, Mainz, Germany*

Maria Luz Fernandez, PhD, *Department of Nutritional Sciences, University of Connecticut, Storrs, CT*

Xavier Fernàndez-Busquets, PhD, *Bioengineering Institute of Catalonia and Nanoscience and Nanotechnology Institute, Barcelona Science Park, University of Barcelona, Barcelona, Spain*

Barbara A. Foster, PhD, *Department of Pharmacology and Therapeutics, Roswell Park Cancer Institute, Buffalo, NY*

Timothy Galitski, PhD, *Institute for Systems Biology, Seattle, WA*

Robert T. Gerlai, PhD, *Department of Psychology, University of Toronto at Mississauga, Mississauga, Ontario, Canada*

Deborah J. Good, PhD, *Department of Human Nutrition, Foods and Exercise, Institute for Biomedical and Public Health Sciences, Virginia Polytechnic Institute and State University, Blacksburg, VA*

Michael N. Gould, PhD, *McArdle Lab for Cancer Research, University of Wisconsin-Madison, Madison, WI*

Matthew Gounis, PhD, *The University of Massachusetts, New England Center for Stroke Research, Boston, MA*

Paola Guglielmelli, MD, *Department of Hematology, University of Florence, Florence, Italy*

Geneviève Guillemin, PhD, *Laboratoire de Recherches Orthopediques, Université Denis Diderot, Paris, France*

Jill D. Haag, MS, *McArdle Lab for Cancer Research, University of Wisconsin-Madison, Madison, WI*

Nancy L. Haigwood, PhD, *Viral Vaccines Program, Seattle Biomedical Research Institute, Seattle, WA*

Mary Ann Handel, PhD, *The Jackson Laboratory, Bar Harbor, ME*

Lynette A. Hart, PhD, *UC Center for Animal Alternatives, School of Veterinary Medicine, University of California at Davis, Davis, CA*

Jann Hau, MD, *Chair of Comparative Medicine, Head of Department of Experimental Medicine, University of Copenhagen and Rigshospitalet, The Panum Institute, Copenhagen, N Denmark*

Eric A. Hendrickson, PhD, *Cancer Center Research Program, Genetic Mechanisms of Cancer, Department of Biochemistry, Molecular Biology and Biophysics, University of Minnesota Medical School, Minneapolis, MN*

Jan Hengstler, MD, *Center for Toxicology, Institute of Legal Medicine and Rudolf-Boehm Institute of Pharmacology and Toxicology, Leipzig, Germany*

Vanessa M. Hirsch, DVM, DSC, *Laboratory of Molecular Microbiology, National Institute of Allergy and Infectious Diseases, National Institutes of Health, Bethesda, MD*

Jana Hlucilova, DVM, *Institute of Animal Physiology and Genetics, The Academy of Sciences of the Czech Republic, Libechov, Czech Republic; and Center for Cell Therapy and Tissue Repair, Charles University, Praha, Czech Republic*

Junjiro Horiuchi, PhD, *Tokyo Metropolitan Institute for Neuroscience, Tokyo, Japan*

Shunji Ishiwata, *Laboratory of Molecular Pharmacogenomics, School of Pharmaceutical Sciences, Kinki University, Higashi-Osaka, Japan*

Takeshi Iwata, PhD, *National Institute of Sensory Organs, National Hospital Organzation Tokyo Medical Center, Tokyo, Japan*

Krishna Kandarpa, MD, PhD, *Radiologist-in-Chief, The University of Massachusetts Memorial Medical Center, Professor and Chairman, Department of Radiology, The University of Massachusetts, Worcester, MA*

Ronald E. Kettner, PhD, *Department of Physiology, Feinberg School of Medicine, Northwestern University, Chicago, IL*

J.P. Kim, JD, MBA, MSC, MALS, *Office of Extramural Research, National Institutes of Health, Bethesda, MD*

Deniz Kirik, MD, PhD, *CNS Disease Modeling Unit, Department of Experimental Medical Science, Lund University, Lund, Sweden*

Lawrence P. Kiss, MD, *Department of Pathology, University of Pittsburgh School of Medicine, Pittsburgh, PA*

Jiri Klima, PhD, *Institute of Animal Physiology and Genetics, The Academy of Sciences of the Czech Republic, Libechov, Czech Republic; and Center for Cell Therapy and Tissue Repair, Charles University, Praha, Czech Republic*

James J. Kohler, PhD, *Department of Pathology, Emory University School of Medicine, Atlanta, GA*

Atsushi Koike, *Division of Molecular Pharmacology and Pharmacogenomics, Department of Genome Sciences, Kobe University Graduate School of Medicine, Kobe, Japan*

Tetsuya Kominami, DSc, *Department of Biology, Faculty of Science, Ehime University, Matsuyama, Japan*

András Kónya, MD, PhD, *M.D. Anderson Cancer Center, Division of Diagnostic Imaging, Department of Diagnostic Radiology, Section of Interventional Radiology, University of Texas, Houston, TX*

Punita Koustubhan, bs, *The Forsyth Center for Regenerative and Developmental Biology and Harvard School of Dental Medicine, Boston, MA*

T. Rajendra Kumar, PhD, *Department of Molecular and Integrative Physiology, University of Kansas Medical Center, Kansas City, KS*

Takayoshi Kuno, *Division of Molecular Pharmacology and Pharmacogenomics, Department of Genome Sciences, Kobe University Graduate School of Medicine, Kobe, Japan*

Emma L. Lane, PhD, *Brain Repair Group, School of Biosciences, Cardiff University, Cardiff, UK*

Pamela L. Larsen, PhD, *Department of Cellular and Structural Biology, University of Texas Health Science Center at San Antonio, San Antonio, TX*

Michael S. Levin, PhD, *The Forsyth Center for Regenerative and Developmental Biology and Harvard School of Dental Medicine, Boston, MA*

William Lewis, MD, *Department of Pathology, Emory University School of Medicine, Atlanta, GA*

Brian A. Link, PhD, *Departments of Cell Biology, Neurobiology, and Anatomy, Medical College of Wisconsin, Milwaukee, WI*

Elwood A. Linney, PhD, *Department of Molecular Genetics and Microbiology, Duke University Medical Center, Durham, NC*

Youhua Liu, PhD, *Department of Pathology, University of Pittsburgh School of Medicine, Pittsburgh, PA*

Delphine Logeart-Avramoglou, PhD, *Laboratoire de Recherches Orthopediques, Université Denis Diderot, Paris, France*

M. Lysetskiy, PhD, *Department of Physiology and Interdepartmental Neuroscinece Program, Feinberg School of Medicine, Northwestern University, Chicago, IL*

Damodar K. Mahajan, MSc, PhD, *Department of Obstetrics and Gynecology, Louisiana State University Health Sciences Center, Shreveport, LA*

John B. Mailhes, PhD, *Department of Obstetrics and Gynecology, Louisiana State University Health Sciences Center, Shreveport, LA*

Paola Maiolino, DVM, *Associate Professor, Department of Pathology and Animal Health, Faculty of Veterinary Medicine, Naples University Federico II, Naples, Italy*

Willy J. Malaisse, MD, PhD, *Laboratory of Experimental Hormonology, Brussels Free University (CP 626), Brussels, Belgium*

Michael A. Matar, MD, *Ministry of Health Mental Health Center, Anxiety and Stress Research Unite, Faculty of Health Sciences, Ben-Gurion University of the Negev, Beer-Sheva, Israel*

Sean G. Megason, PhD, *Biological Imaging Center, Beckman Institute, California Institute of Technology, Pasadena, CA*

Marilyn Menotti-Raymond, PhD, *Laboratory of Genomic Diversity, National Cancer Institute-Frederick, Frederick, MD*

Anna Rita Migliaccio, PhD, *Department of Pathology, University of Illinois at Chicago, Chicago, IL; and Department of Hematology, Oncology and Molecular Medicine, Instituto Superiore di Sanita, Rome, Italy*

Yann S. Mineur, PhD, *Departments of Psychiatry, Neurology and Pharmacology, Yale School of Medicine, New Haven, CT*

Shoukhrat Mitalipov, PhD, *Oregon National Primate Research Center, Oregon Health and Science University, Beaverton, OR*

Jan Motlik, DVM, PhD, DSc, *Institute of Animal Physiology and Genetics, The Academy of Sciences of the Czech Republic, Libechov, Czech Republic; and Center for Cell Therapy and Tissue Repair, Charles University, Praha, Czech Republic*

Shaheen M. Nakeeb, DVM, PhD, *Carmel, IN*

Richard Nass, PhD, *Department of Pediatrics and Pharmacology, Vanderbilt University Medical Center, Nashville, TN*

Stephen J. O'Brien, PhD, *National Cancer Institute-Frederick, Fort Detrick, Frederick, MD*

Raymond R. O'Neill, PhD, *Division of Comparative Medicine, National Center for Research Resources, National Institutes of Health, Bethesda, MD*

Ricardo Ocaña-Riola, PhD, *Department of Statistics, Escuela Andaluza de Salud Publica, Campus Universitario Cartuja, Granada, Spain*

Tomas Olsson, MD, PhD, *Department of Clinical Neuroscience, Neuroimmunology Unit L8:04, Center of Molecular Medicine, Karolinska Hospital, Stockholm, Sweden*

Jitka Ourednik, PhD, *College of Veterinary Medicine, Iowa State University, Ames, IA*

Vaclav Ourednik, PhD, *College of Veterinary Medicine, Iowa State University, Ames, IA*

Maiolino Paola, DVM, *Department of Pathology and Animal Health, Faculty of Veterinary Medicine, Naples University, Naples, Italy*

James Parker, PhD, *Retired, Public Information Officer, Oregon National Primate Research Center, Portland, OR*

Hervé Petite, PhD, *Laboratoire de Recherches Orthopediques, Faculte de Medecine Lariboisiere Saint-Louis, Universite Denis Diderot Paris 7, Paris, France*

Radek Prochazka, DVM, PhD, *College of Veterinary Medicine, Iowa State University, Ames, IA*

Yureeda Qazi, MBBS, *Department of Biological and Biomedical Sciences, The Aga Khan University, Karachi, Pakistan*

Michael S. Rand, DVM, DACLAM, *Chief, Biotechnology Support Services, Department of Animal Care, University of Arizona, Tucson, AZ*

Caroline A. Rickards, PhD, *US Army Institute of Surgical Research, Fort Sam Houston, TX*

Jin Y. Ro, PhD, *Department of Biomedical Sciences, Baltimore College of Dental Surgery, University of Maryland Baltimore, Baltimore, MD*

Sara Rockwell, PhD, *Department of Therapeutic Radiology, Yale University School of Medicine, New Haven, CT*

Karen R. Rockwell, PhD, *Adj. Assistant Professor, Center for Pharmaceutical Biotechnology and Nanomedicine, School of Pharmacy, Northeastern University, Bouvé College of Health Sciences, Boston, MA*

Ute Christine Rogner, PhD, *Departement de Biologie du Développement, Unite de Genetique Moleculaire Murine, Institut Pasteur, Paris Cedex 15, France*

Dileep Kumar Rohra, MBBS, MPhil, PhD, *Department of Biological and Biomedical Sciences, The Aga Khan University, Karachi, Pakistan*

April E. Ronca, PhD, *Departments of Obstetrics and Gynecology and Neurobiology and Anatomy, Wake Forest University School of Medicine, Winston-Salem, NC*

JAMES D. ROSE, PhD, *Department of Zoology and Physiology, University of Wyoming, Laramie, WY*

NICOLE MARIE ROY, PhD, *Department of Molecular Genetics and Microbiology, Duke University Medical Center, Durham, NC*

ROBERT J. RUSSELL, DVM, *Director, Laboratory of Animal Medicine, Harlan Genetic Testing Services, Indianapolis, IN*

MINORU SAITOE, PhD, *Tokyo Metropolitan Institute for Neuroscience, Tokyo, Japan*

RICHARD SALVI, PhD, *Center for Hearing and Deafness, University at Buffalo, Buffalo, NY*

STEVEN J. SCHAPIRO, PhD, *The Michale E. Keeling Center for Comparative Medicine and Research, Department of Veterinary Sciences, The University of Texas M.D. Anderson Cancer Center, Bastrop, TX*

ABDULLAH SENER, PhD, *Laboratory of Experimental Hormonology, Brussels Free University, Brussels, Belgium*

AFONSO C. SILVA, PhD, *Cerebral Microcirculation Unit, Laboratory of Functional and Molecular Imaging, National Institute of Neurological Disorders and Stroke, National Institutes of Health, Bethesda, MD*

KRISHNA B. SINGH, MD, *Department of Obstetrics and Gynecology, LSU Health Sciences Center, Shreveport, LA*

ALISON C. SMITH, DVM, *Department of Comparative Medicine, Medical University of South Carolina, Charleston, SC*

BART M.G. SMITS, PhD, *McArdle Lab for Cancer Research, University of Wisconsin-Madison, Madison, WI*

DEBRA SOROCCO, ALM, *Department of Cytokine Biology, The Forsyth Institute, Boston, MA*

ROLF SPRENGEL, Dr. rer.nat., *Max Planck Institute for Medical Research, Heidelberg, Germany*

CRAIG W. STEVENS, PhD, *Professor of Pharmacology, Director of the Analgesia Research Laboratory, Oklahoma State University-Center for Health Sciences, College of Osteopathic Medicine, Tulsa, OK*

BOJANA STEFANOVIC, PhD, *Cerebral Microcirculation Unit, Laboratory of Functional and Molecular Imaging, National Institute of Neurological Disorders and Stroke, National Institutes of Health, Bethesda, MD*

REIKO SUGIURA, *Laboratory of Molecular Pharmacogenomics, School of Pharmaceutical Sciences, Kinki University Higashi-Osaka, Japan; and Division of Molecular Pharmacology and Pharmacogenomics, Department of Genome Sciences, Kobe University Graduate School of Medicine, Kobe, Japan*

M. SUH, PhD, *Department of Neurological Surgery, Weill Medical College of Cornell University, New Yorok, NY*

M. MICHAEL SWINDLE, DVM, *Department of Comparative Medicine, Thurmond/Gazes Research, Medical University of South Carolina, Charleston, SC*

HIROFUMI TAKADA, *Laboratory of Molecular Pharmacogenomics, School of Pharmaceutical Sciences, Kinki University, Higashi-Osaka, Japan*

HIROMI TAKATA, DSc, *Department of Biology, Faculty of Science, Ehime University, Matsuyama, Japan*

VESTEINN THORSSON, PhD, *Institute for Systems Biology, Seattle, WA*

STANISLAV TOMAREV, PhD, *National Eye Institute, National Institutes of Health, Bethesda, MD*

KARL THOMASETH, PhD, *National Research Council, Institute of Biomedical Engineering ISIB-CNR, Padova, Italy*

JOHN G. VANDENBERGH, PhD, *Professor, Zoology Department, North Carolina State University, Raleigh, NC*

ALESSANDRO M. VANNUCCHI, MD, *Division of Hematology, University of Florence, Florence, Italy*

VÉRONIQUE VIATEAU, DMV, PhD, *Unité Pédagogique de Pathologie Chirurgicale, Ecole Nationale Veterinaire d'Alfort, Maisons Alfort, France*

GEMMA VILAHUR, DVM, PhD, *Centro de Investigacion Cardiovascular, CSIC-Institut Catala de Ciences Cardiovasculares, Hospital de la Santa Creu I Sant Pau, Barcelona, Spain*

JEAN-LUC VILLEVAL, PhD, *INSERM, Universite Paris, Institut Gustave Roussy, Villejuif, France*

PETR VODICKA, MSc, *Department of Reproductive and Developmental Biology, Institute of Animal Physiology and Genetics, The Academy of Sciences of The Czech Republic, Libechov, Czech Republic*

ORIANNE WAGNER-BALLON, MD, *INSERM, Universite Paris, Institut Gustave Roussy, Villejuif, France; AP-HP, Laboratoire d'Hematologie, Hopital Henri Mondor, Creteil, France*

ERIK WALLSTRÖM, *Department of Neurology, Karolinska University Hospital, Karolinska Institute, Stockholm, Sweden*

C. JANE WELSH, PhD, *Department of Veterinary Integrative Biosciences, College of Veterinary Medicine and Biomedical Sciences, Texas A&M University, College Station, TX*

SHERYL J. WILDT, MS, *Laboratory of Animal Medicine, Harlan Genetic Testing Services, Indianapolis, IN*

DON P. WOLF, PhD, *Oregon National Primate Research Center, Oregon Health and Science University, Beaverton, OR*

MARY W. WOOD, MLS, *UC Center for Animal Alternatives, Carlson Health Sciences Library, University of California at Davis, Davis, CA*

RICHARD J. WOOD, PhD, *Department of Nutritional Sciences, University of Connecticut, Storrs, CT*

C. JEFFREY WOODBURY, PhD, *Department of Zoology and Physiology and Neuroscience Program, University of Wyoming, Laramie, WY*

KENNETH C. WRIGHT, PhD, *M.D. Anderson Cancer Center, Division of Diagnostic Imaging, Department of Diagnostic Radiology, Section of Interventional Radiology, University of Texas, Houston, TX*

ANNETTE B. WYSOCKI, PhD, *Division of Plastic Surgery, Department of Surgery, The University of Mississippi Medical Center, School of Medicine, Jackson, MS*

COLIN R. YOUNG, PhD, *Departments of Veterinary Integrative Biosciences and Psychology, College of Veterinary Medicine and Biomedical Sciences, Texas A&M University, College Station, TX*

JOSEPH ZOHAR, MD, *Department of Psychiatry, Chaim Sheba Medical Center, Sackler Medical School, Tel-Aviv University, Tel-Hashomer, Israel*

List of Color Plates

Color plates to follow page 240.

Figure 8–1. Example of enrichment items for laboratory mice. The hut is tinted such that the mouse cannot see out, but human caretakers can see in. (Photo by Jill Rawlins.)

Figure 13–4. Imaging in live zebrafish embryos. (A) Whole embryos can be quickly imaged on a fluorescent dissecting microscope in their chorions to sort positive transgenics (left) from wild-type siblings (right). These embryos were not treated with PTU so the melanocytes are visible. (B–E) Confocal microscopy permits much higher resolution imaging. (B) A quick method for labeling is to inject RNA encoding fluorescent proteins, in this case a histone2B-EGFP fusion, and a membrane localized mCherry was used to image all the cells of the inner ear. (C) GFP transgenics can be used to image neuronal projections from the trigeminal ganglion as they extend. (D) GFP transgenics can mark specific populations of cells, in this case rhombomeres 3 and 5. (E) GFP fusion proteins can reveal the subcellular localization pattern of proteins, in this case a cytoplasmic protein in the Rohon-Beard and motor neurons of the spinal cord. (Images form S.G. Megason, L.A. Trinh, and S.E. Fraser, unpublished.)

Figure 13–5. Creating genetic mosaics in zebrafish. (A) Donor cells are first lineage labeled with a tracer dye at the one-cell stage. Donor embryos can also be injected with morpholinos, RNA, or DNA. At the 1000-cell stage, totipotent blastula cells are transplanted into regions of the host embryo fated to give rise to specific structures. Donor and host embryos can be of either mutant or wild-type genotypes. Resultant chimeras are grown for subsequent analysis. (B) Fate-map of the pregastrula stage embryos. (Modified from Woo and Fraser, 1995.) (C) Mosaic embryo at 24 hpf showing bright-field (left) and fluorescent image (right) of rhodamine-dextran-labeled donor cells targeted to the eye and forebrain. (B.A. Link, unpublished.)

Figure 25–1. Feline chromosome maps (labeled at top) and homologous synteny blocks (HSBs) in the human (H) and dog (D) genomes. HSBs are shown to the right of each cat chromosome map (only the map scale is shown). The dark cross-marks on each cat chromosome correspond to 100-cR_{5000} intervals. The inferred centromere positions are shown by dark circles. HSBs are color coded by human or dog chromosome, defined by the key in the bottom right corner. (Reprinted from Murphy *et al.*[7] Copyright 2007, with permission from Elsevier.)

Figure 51–3. Changes in the CBV response to amphetamine (A) before and (B) 4 months after cessation of MPTP treatment in a cynomolgus monkey, showing an almost complete loss of amphetamine-induced CBV signal changes in dopaminergic regions. Parkinsonian primates had a prominent loss of response to amphetamine, with relative sparing of the nucleus accumbens and parafascicular thalamus. (Modified from Jenkins *et al.*[183] Copyright 2004 Society for Neurosciences.)

Figure 51–4. (A) Patterns of the fMRI response in the mouse main olfactory bulb (MOB) and accessory olfactory bulb (AOB, pink circle in slice 5) to the pheromone 2-heptanone, one of the urinary chemosignal compounds in mouse. The arrows point to two foci of activation suggestive of a pair of the nearly mirror projections of the receptor neuron subsets to the same MOB. Scale bar = 500 µm. (B) A flattened view of the olfactory bulb indicating the orientation of the odor maps shown in (C) and (D): A, anterior; D, dorsal; L, lateral; M, medial; P, posterior, V, ventral. (C, D) The odor maps of 2-heptanone (Hep) in two different mice show that this pheromone not only activates large regions of the MOB but also generates similar patterns across subjects. Interestingly, the odor map for amyl acetate (AA), a common odorant with an odor quality similar to that of 2-heptanone, was also similar to that of 2-heptanoneb. (Adapted from Xu *et al.*[217] Copyright 2005 Wiley-Liss, Inc.)

Figure 51–5. Signal changes in CBV-weighted fMRI obtained during stimulation of the cat visual cortex according to predetermined stimulus orientations. (A) Raw gray-scale functional map obtained by subtraction of images during 0° stimulation from prestimulus control. The center of each patch is marked with a green + sign. (B) Four different grating orientations (0°, 45°, 90°, and 135°) were presented in the study of one animal, enabling a composite angle map to be generated through pixel-by-pixel vector addition of the four single-condition maps. In the left hemisphere (marked by a white rectangular box), changes between pixels preferentially activated by a particular stimulus orientation were smoothed by a 3×3 Gaussian Kernel. (C) A composite angle map was generated with the region indicated by the white ROI in (B) "Pinwheel" structures indicated by small white dots were observed where domains for all orientations converge. (Adapted from Zhao *et al.*[114] Copyright 2005 Elsevier.)

Figure 51–6. Auditory activation in the songbird telencephalon shows (A) statistical maps illustrating the localization of significant signal intensity changes during auditory stimulation consisting of white noise (wn), a concerto of Bach (music), and a stimulation with song from a male starling (song). (B) The location of the activated areas (top), together with the average BOLD response amplitude for each stimulus (bottom). (Adapted from Van Meir *et al.*[231] Copyright 2005 Elsevier.)

Figure 61–1. The effect of single PSS exposure versus unexposed control on rat anxiety-like behavior and acoustic startle response and habituation. The representation of the data from both paradigms (EPM and ASR) shows two obvious and rather distinct features. First, it is clear that PSS exposure alters the response of the majority of individuals to at least some degree. Second, the cluster of individuals that forms in the upper left hand corner of the graph (i.e., had the more extreme responses to exposure) is quite distinct from the majority of individuals.[36,48]

Figure 68–2. Diabetic nephropathy induced by streptozotocin (STZ) in uninephrectomized mice. One week after uninephrectomy, male CD1 mice received an intravenous injection of STZ. (A–D) Representative micrographs demonstrate glomerular enlargement, mesangial expansion, and segmental glomerulosclerosis (periodic acid–Schiff staining; A, B) and glomerular collagen deposition (Masson–Trichrome staining; C, D) (A and C) Normal control; (B and D) diabetic mice. Arrows indicate injured glomeruli. (E) Albuminuria develops in diabetic mice in a time-dependent manner. Data are presented as means ± SEM. $*p < 0.05$. (F) Glomerular collagen deposition score in diabetic and normal mice. $*p < 0.05$. (Adapted and modified from Dai *et al.*[46])

Figure 68–3. Interstitial fibrosis in the mouse model of obstructive nephropathy. (A, B) Representative micrographs show the cross sections and gross morphology of the obstructed kidney at day 14 after ureteral ligation. (A) Sham control; (B) UUO. (C) Western blot analyses demonstrate a marked induction of α-smooth muscle actin (α-SMA), a molecular marker for myofibroblasts, in the obstructed kidney at day 14 after UUO. (D) Double immunofluorescence staining shows the α-SMA (red) and proximal tubular epithelial cell marker, fluorescein isothiocyanate (FITC)-conjugated lectin from *Tetragonolobus purpurea* (green). (E) Quantitative determination of total kidney collagen contents in sham and obstructed kidneys. Data are presented as means ± SEM. $*p < 0.01$. (Adapted and modified from Yang *et al.*[59])

Figure 73–4. A pathogenetic model for the cytokine-mediated stromal reaction observed in MMM.

INTRODUCTION

I

1 Animal Models for Human Diseases

An Overview

Jann Hau

ABSTRACT

This chapter provides an introduction to the concept of laboratory animal models, focusing on a general classification of animal models for the study of human diseases. Animal models can be grouped into one of the following five categories: (1) induced (experimental) models, (2) spontaneous (genetic, mutant) models, (3) genetically modified models, (4) negative models, and (5) orphan models. This is followed by a discussion of how knowledge concerning human biology and pathobiology can be extrapolation from results obtained from studies of animals. Finally the chapter discusses how the difference in body size and metabolic rate between small laboratory animals and humans has an impact on the calculation of relevant doses for animals used as models for humans in experimental studies.

Key Words: Animal model concept, Induced animal model, Spontaneous animal model, Transgenic animal model, Negative animal model, Orphan animal model, Body size—scaling, Extrapolation, Laboratory animal science—definition.

INTRODUCTION

Throughout history ethical and religious considerations as well as social prohibitions have prevented experimental studies of human biology and pathobiology. Even studies of human anatomy were for long periods of time in history a criminal offense and thus not possible. Although impressive anatomical teaching theaters were established in many old European universities, postmortem dissection was often restricted exclusively to criminals executed for their offenses. In the more quiet corners of Europe the teaching theaters remained unused for decades. Consequently, most of our present basic knowledge of human biology, physiology, endocrinology, and pharmacology has been derived from initial studies of mechanisms in animal models.[1] Throughout history scientists have performed experiments on animals to obtain knowledge of animal and human biological structure and function.[2,3] Often such studies have not been conducted, and are not possible to conduct, in humans. This may not only be due to ethical or religious considerations. Often practical, economic, and scientific reasons make initial studies in animals the best solution for studies of a biological phenomenon.

Laboratory animal science may be defined as the study of the scientific, ethical, and legal use of animals in biomedical research, i.e., a multidisciplinary field encompassing comparative biological and pathobiological specialties for the optimal scientific use of animals as models for human or other species. Basic laboratory animal science is concerned with the quality of animals as sentient tools in biomedical research. It encompasses the comparative biology of laboratory animals, aspects of breeding, housing, and husbandry, anesthesia, euthanasia, and experimental techniques. For animal welfare reasons as well as for scientific reasons it is vital that scientists using animal models in their research are competent and have a good knowledge of basic laboratory animal science.

This sourcebook provides a thorough introduction to the use of animal models for human diseases. High quality animals combined with first class animal care ensure the highest possible health and welfare status of the animals and are a prerequisite for good science and public acceptance of the use of animals in research.

THE ANIMAL MODEL CONCEPT

A laboratory animal model describes a biological phenomenon that the species has in common with the target species. A key word for understanding the concept of animal models is "analogy."[4] A model should not be considered a claim of identity with what is being modeled, but a convergent set of several kinds of analogies between the "target" phenomenon to be understood and the system that is being studied as a substitute for the target phenomenon. A more comprehensive definition has been given by Held on the basis of Wessler's original definition[5]: "a living organism in which normative biology or behavior can be studied, or in which a spontaneous or induced pathological process can be investigated, and in which the phenomenon in one or more respects resembles the same phenomenon in humans or other species of animal."

What is generally understood by the term animal model is modeling humans. It is not the image of the used animal that is the focus of research but the analogy of the physiological behavior of this animal to our own (or another) species. It would thus perhaps be more correct to refer to animals as "man models" in this context. Laboratory animal science, comparative medicine, and animal experiments are indeed much more about humans than about any other animal species.[6]

The practice of studying biological phenomena and diseases in laboratory animals and transferring the findings into solutions

From: *Sourcebook of Models for Biomedical Research* (P. M. Conn, ed.),

and treatments for improving human health and welfare has a long history and is well established in biomedical sciences. The significance and validity with respect to usefulness in terms of ”extrapolatability” of results generated in an animal model depend on the selection of a suitable animal model. A good knowledge of comparative anatomy and physiology is an obvious advantage when developing an animal model. Animal models may be found throughout the animal kingdom, and knowledge about human physiology has been obtained from species far removed from humans in terms of taxonomy. A good example is the importance of the fruit fly for the original studies of basic genetics. Animal models are used in most fields of biomedical research as reflected in the respective chapters of this book.

CLASSIFICATION OF ANIMAL MODELS

A plethora of animal models has been used and is being used and developed for studies of biological structure and function in humans. The models may be "exploratory," aiming to understand a biological mechanism whether this is a mechanism operative in fundamental normal biology or a mechanism associated with an abnormal biological function. Models may also be developed and applied as so-called "explanatory" models, aiming to understand a more or less complex biological problem. Explanatory models need not necessarily be reliant on the use of animals but may also be physical or mathematical model systems developed to unravel complex mechanisms. A third important group of animal models is "predictive" models. These models are used to discover and quantify the impact of a treatment, whether this is to cure a disease or to assess toxicity of a chemical compound. The anatomy or morphology of the model structure of relevance to the studies may be of importance in all three of these model systems. The extent of resemblance of the biological structure in the animal to the corresponding structure in humans has been termed fidelity. A high fidelity model with close resemblance to humans may seem an obvious advantage when developing certain models. What is often more important, however, is the discriminating ability of the models, in particular the predictive models. When using models to assess the carcinogenicity of a substance it is essential that at least one of the model species chosen responds in a manner that is predictive of the human response to this substance. Thus the similarity between human and model species with respect to relevant biological mechanisms is often more important than the fidelity of the model. Often the two go hand in hand and high fidelity models offer the best opportunity to study a particular biological function.

An animal model may be considered homologous if the symptoms shown by the animal and the course of the condition are identical to those of humans.[7] Models fulfilling these requirements are relatively few, but an example is well-defined lesion syndromes in, e.g., neuroscience.[8] An animal model is considered isomorphic if the animal symptoms are similar, but the cause of the symptoms differs in humans and the model. However, most models are neither homologous nor isomorphic but may rather be termed partial. These models do not mimic the entire human disease, but may be used to study certain aspects or treatments of the human disease.[8]

CLASSIFICATION OF DISEASE MODELS

The majority of laboratory animal models are developed and used to study the cause, nature, and cure of human disorders. Disease models may conveniently be categorized in one of the following five groups, of which the three first are the numerically most important[9]:

1. Induced (experimental) models
2. Spontaneous (genetic, mutant) models
3. Genetically modified models
4. Negative models
5. Orphan models

INDUCED (EXPERIMENTAL) MODELS As the name implies, induced models involve healthy animals in which the condition to be investigated is experimentally induced, e.g., the induction of diabetes mellitus with encephalomyocarditis virus,[10] allergy against cow's milk through immunization with minute doses of protein,[11] or partial hepatectomy to study liver regeneration.[12] The induced model group is the only category that theoretically allows a free choice of species. Although it might be presumed that extrapolation from an animal species to the human species is better the closer this species resembles humans (high fidelity), phylogenetic closeness, as fulfilled by primate models, is not a guarantee of validity of extrapolation, as the unsuccessful chimpanzee models in AIDS research have demonstrated.[13] It is just as decisive that the pathology and outcome of an induced disease or disorder in the model species resemble the respective lesions of the target species. Feline immunodeficiency virus (FIV) infection in cats may therefore for many studies be a better model for human AIDS than HIV infection in simians. Although mice and rats have many biological characteristics in common, they do not necessarily serve equally well as models of human disease. For example, schistosomiasis (mansoni) infection may be studied in experimentally infected mice, but not in rats whose immune system is able to fight the infection effectively.[14]

Most induced models are partial or isomorphic because the etiology of a disease experimentally induced in an animal is often different from that of the corresponding disease in humans. Few induced models completely mimic the etiology, course, and pathology of the target disease in humans.

SPONTANEOUS ANIMAL MODELS These models of human disease utilize naturally occurring genetic variants (mutants). Many hundreds of strains/stocks with inherited disorders modeling similar conditions in humans have been characterized and conserved (see, e.g., www.jax.org). The best known spontaneous models include the athymic nude mouse,[15] the use of which represented a turning point in the study of heterotransplanted tumors and enabled the first description of natural killer cells. Some of these mutants were discovered almost a century ago, like the famous spontaneous model Snell's Dwarf mouse without a functional pituitary[16] and the curly tail mouse in which fetuses develop a whole range of neural tube defects.[17] Many of the mutants are available in inbred strains with corresponding coisogenic or congenic strains. This is very useful because the influence of just one affected gene or locus may then be studied against a reference strain with a genetic background similar to the mutant.

An extensive literature is available on spontaneous models and the majority of these involve mice and rat models, although a wide range of mutants in many different species has been described. A good example of the amount of information available is the publication of Migaki,[18] referencing more than 200 diseases in animals exclusively caused by inborn errors of metabolism.

The spontaneous models are often isomorphic displaying phenotypic similarity between the disease in the animal and the corresponding disease in man, the so-called face validity, e.g., type I diabetes in humans and insulin requiring diabetes in the BB rat. This phenotypic similarity often extends to similar reactions to treatment in the model animal and the human patient, and spontaneous models have been important in the development of treatment regimens for human diseases.

However, if the object of a project is to study the genetic causes and etiology of a particular disease then comparable genomic segments involved in the etiology of the disorder—construct validity—is normally a requirement. It should be remembered, however, that an impaired gene or sequence of genes very often results in activation of other genes and mobilization of compensating metabolic processes. These compensatory mechanisms may of course differ between humans and the animal model species.

GENETICALLY MODIFIED MODELS The rapid developments in genetic engineering and embryo manipulation technology during the past decade have made transgenic disease models the most important category—in terms of numbers—of animal disease models. The technology and ability to genetically modify mice resulted in a substantial increase in the numbers of laboratory animals used worldwide, which is a trend that seems to continue. A multitude of animal models for important diseases have been developed since this technology became available in the 1980s, and the number of models has increased quickly. Mice are by far the most important animals for transgenic research purposes, but farm animals and fish are also receiving considerable interest.

Many physiological functions are polygenic and controlled by more than one gene, and it will require considerable research activities to identify the contribution of multiple genes to normal as well as abnormal biological mechanisms. The insertion of DNA into the genome of animals, or the deletion of specific genes, gives rise to sometimes unpredictable outcomes in terms of scientific results as well as animal well-being in the first generations of animals produced. Thereafter transgenic lines can be selected and bred or cloned to avoid or select for a specific genotype. It is not an accurate science, although the methodology is constantly improving, with the aim of eliminating unwanted effects. The embryo manipulation procedures in themselves do not appear to affect the welfare of offspring in the mouse,[19] and the large offspring syndrome common in farm animals has so far not been reported in the literature for rodents, although it has been observed (Johannes Wilbertz, Karolinska Institute, personal communication).

Mutations induced by the use of mutagens like ethylnitrosourea is another approach to the generation of new mutants, which may serve as models of human disorders. In many aspects these mutants may be similar to spontaneous mutants and to the ones generated by transgenic embryo manipulation. The maintenance of a line raises issues for chemically induced genetic mutants similar to those for animals genetically modified through embryo manipulation.

The recent completion of the maps of the genomes of mouse (and other model animals) and humans will increase the research activities in functional genomics and proteomics, and using high density microarray DNA chip technology in human patients as well as in animals will make it possible to investigate which genes are switched on or off in different diseases.[20]

Having both the human and the mouse genome maps available, this new technology is expected to rapidly increase our knowledge on the genetic background and etiology of important diseases. This paves the way for a range of new homologous animal models with homology between animal and humans (construct validity) for genotype as well as for phenotype. This development may result in a change in animal use from models for the identification of causative genes to models for studying the effects of changes in genetic pathways, gene–gene interactions, and gene–environment interactions.[21] The characterization and application of genetically modified mouse models are slowed down by difficulties in phenotyping the animals. There is a need for accurate and reliable behavioral assessment, biotechnology development for physiological assessment, and analysis of complex data as well as training scientists in whole-organism research.[22]

NEGATIVE MODELS Negative model is the term used for species, strains, or breeds in which a certain disease does not develop, e.g., gonococcal infection in rabbits following an experimental treatment that induces the disease in other animal(s). Models of infectious diseases are often restricted to a limited number of susceptible species and the remaining unresponsive species may be regarded as negative models for this particular human pathogenic organism. Negative models thus include animals demonstrating a lack of reactivity to a particular stimulus. Their main application is studies on the mechanism of resistance to gain insight into its physiological basis. Occasionally negative models give rise to the characterization of new spontaneous (mutant) models. Examples are found in studies of infectious diseases and carcinogenicity where individuals exhibiting resistance to a treatment may be developed into new spontaneous models.[23]

ORPHAN MODELS An orphan model disease is the term used to describe a functional disorder that occurs naturally in a nonhuman species but has not yet been described in humans, and which is recognized when a similar human disease is later identified. Examples include Marek's disease, papillomatosis, and bovine spongiform encephalopathy (BSE), Visna virus in sheep, and feline leukemia virus. When discovering that humans may suffer from a disease similar to one that has already been described in animals the literature generated in veterinary medicine may be very useful.

EXTRAPOLATION FROM ANIMALS TO HUMANS

When experimental results have been generated in an animal model they have to be validated with respect to their applicability to the target species, which normally is the human The term extrapolation is often used to describe how data obtained from animal studies reliably can be used to apply to humans. However, extrapolation is generally not performed in its mathematical sense where data fit a certain function that may be described graphically, and the graph extended beyond the highest or lowest sets of data to describe a situation outside the window of observation. Establishing toxicity data in animals and using these to determine safe levels of exposure for people is perhaps what comes closest to mathematical extrapolation in animal studies. However, most studies of animal structure and function are never extrapolated to be applicable to describing the corresponding features in humans; this is not relevant. What laboratory animal experimentation is about is very similar to other types of experiments. The scientists aim to obtain answers to specific questions. Hypotheses are being tested and the answers obtained, analyzed, and published. As an example of this, we might question the possible health hazards of a new synthetic steroid and ask a number of relevant questions to be answered in animal studies before deciding on its potential usefulness as a human hormonal contraceptive: Does it exist in the same form in humans and animals? How does it affect the estrus cycle in rodents? How does it affect endogenous hormone levels in rodents and other species? How soon after withdrawal do the animals revert to normal cyclicity? Does it interfere with pregnancy in rodents and primates? Does it affect fetal development in rodents and primates? Is the frequency of fetal malformations in mice affected? Are puberty, the ovarian cycle, and pregnancy in rodent and dog offspring of mothers treated with the substance affected? Analyzing the data from experiments of this nature would provide information on the potential of the new synthetic steroid as a hormonal contraceptive in humans.

A large multinational pharmaceutical company survey analyzed data compiled from 150 compounds for the concordance between adverse findings in human clinical data with data that had been generated in preclinical tests in animals.[24] The concordance rate was found to be 71% for rodent and nonrodent species, with nonrodents alone being predictive for 63% of human toxicity (HT) and rodents alone for 43%. High concordance rates were found, e.g., for cardiovascular HTs (80%), hematological HTs (91%), and gastrointestinal HTs (85%). Lower concordance rates were observed, e.g., for the neurological group, because it is difficult to identify symptoms such as headache and dizziness in the animals studied. The only gastrointestinal HT that did not correlate with animal studies was, not surprisingly, nausea. One of the conclusions reached in this study was that the choice of species used might be subject to more thoughtful consideration. By tradition studies are often carried out using rats and dogs, without an open-minded consideration of whether alternative species might be more appropriate for testing a specific compound.

Although the predictive value of animal studies may seem high if they are conducted thoroughly and have included several species, uncritical reliance on the results of animal tests can be dangerously misleading and has resulted in damage to human health in several cases, including drugs developed by large pharmaceutical companies. What is noxious or ineffective in nonhuman species can be innocuous or effective in humans and vice versa. For example, penicillin is fatal for guinea pigs but generally well tolerated by humans; aspirin is teratogenic in cats, dogs, guinea pigs, rats, mice, and monkeys but obviously not in pregnant women despite frequent consumption.[25] Thalidomide, which crippled 10,000 children, does not cause birth defects in rats[26] or many other species,[27] but does so in primates. Close phylogenetic relationship or anatomical similarity is not a guarantee of identical biochemical mechanisms and parallel physiological response, although this is the case in many instances.

The validity of extrapolation may be further complicated by the question of which humans. As desirable as it often is to obtain results from a genetically defined and uniform animal model, the humans to whom the results are extrapolated are genetically highly variable, with cultural, dietary, and environmental differences. This may be of minor importance for many disease models but can become significant for pharmacological and toxicological models.

It is not possible to provide reliable general rules for the validity of extrapolation from one species to another. This has to be assessed individually for each experiment and can often be verified only after first trials in the target species. An extensive and useful overview on the problem of predictive anthropomorphization, especially in the field of toxicology research, is *Principles of Animal Extrapolation* by Calabrese.[28] The rationale behind extrapolating results to other species is based on the extensive homology and evolutionary similarity between morphological structures and physiological processes among different animal species and between animals and humans.[29]

MODEL BODY SIZE AND SCALING

The use of laboratory animals as models for humans is often based on the premise that animals are more or less similar with respect to many biological characteristics and thus can be compared. However, there is one striking difference between mouse and human, and that is body size. In proportion to their body size mammals generally often have very similar organ sizes expressed as percentage of body weight. Take the heart, for instance, which often constitutes 5 or 6 g/kg body weight, or blood, which is often approximately 7% of total body weight.

It is well known that the metabolic rate of small animals is much higher than that of large animals. It has also been demonstrated that capillary density in animals smaller than rabbits increases dramatically with decreasing body weight.[30] However, considering that most animals are similar in having heart weights just above 0.5% of their body weight and a blood volume corresponding to 7% of their body weight, it becomes obvious that in order to supply the tissues of small animals with sufficient oxygen for their high metabolic rate it is not sufficient to increase the stroke volume. The stroke volume is limited by the size of the heart and heart frequency is the only parameter to increase, which results in heart rates well over 500 per minute in the smallest mammals. Other physiological variables, like respiration and food intake, are similarly affected by the high metabolic rate of small mammals.

This means that scaling must be an object for some consideration when calculating dosages of drugs and other compounds administered to animals in experiments.

If the object is to achieve equal concentrations of a substance in the body fluids of animals of different body size then the doses should be calculated in simple proportion to their body weights. If the object is to achieve a given concentration in a particular organ over a certain time period the calculation of dosage becomes more complicated and other factors including the physicochemical properties of the drug become important. Drugs and toxins exert their effect on an organism because of the way they are metabolized, the way they and their metabolites are distributed and bound in the body tissues, and how and when they are finally excreted.

However, metabolism or detoxification and excretion of a drug are not directly correlated with body size, but more accurately with the metabolic rate of the animal (see Schmidt-Nielsen[30,31] for more details). Kleiber[32] in 1932 was the first to demonstrate that in a log–log plot of mammalian body weight to metabolism the graph forms a straight line with a slope of 0.75.

The metabolic rate of an animal as expressed by oxygen consumption per gram body weight per hour is related to body weight in the following manner:

$$M = 3.8 \times \mathrm{BW}^{-0.25}$$

where M is the metabolic rate (oxygen consumption in milliliter per gram body weight per hour) and BW is body weight in grams. This equation may be used to calculate dosages for animals of different body weights if the dose for one animal (or man) is known.[33]

$$\text{Dose}_1/\text{Dose}_2 = \text{BW}_1^{-0.25}/\text{BW}_2^{-0.25}$$

$$\text{Dose}_1 = \text{Dose}_2 \times \text{BW}_1^{-0.25}/\text{BW}_2^{-0.25}$$

These equations should be considered as assistance for calculating dosages, but caution should be exerted with respect to too broad a generalization of their use, and the 0.50 power of body weight should be employed when dealing with animals having body weights below 100 g.[34] Some species react with particular sensitivity toward certain drugs and marked variations in the reaction of animals within a species occur with respect to strain, pigmentation, nutritional state, time of day, stress level, type of bedding, ambient temperature, etc.[33]

CONCLUSIONS

The selection of an animal model depends on a number of factors relating to the hypothesis to be tested, but often more practical aspects associated with the project and with project staff and experimental facilities play a significant role. The usefulness of a laboratory animal model should be judged on how well it answers the specific questions it is being used to answer, rather than how well it mimics the human disease.[35]

Often a number of different models may advantageously be used in order to scrutinize a biological phenomenon and for major human diseases such as diabetes, a whole range of well-described induced models are available as are spontaneous models in both mouse and rat strains.

Most of the regulating authorities require two species in toxicology screening, one of which has to be nonrodent. This does not imply that excessive numbers of animals will be used because an uncritical use of one-species models may mean that experimental data retrospectively turn out to be invalid for extrapolation, representing a waste of animals. The appropriateness of any laboratory animal model will eventually be judged by its capacity to explain and predict the observed effects in the target species.[36]

The free choice of species when developing animal models is more or less restricted to the induced models making use of clinically healthy animals, in which a condition deviating from normality is experimentally induced. Although all laboratory animal species are in principle available for model development, it has been a clear trend during the past 30 years that the most popular species, the house mouse and the Norwegian rat, are increasing in popularity at the expense of farm animal species and pet species, while the use of nonhuman primates seems to remain stable.[37,38] The completion of the map of the mouse genome and the dominating position of mice in transgenic research seem to indicate that the dominance of the mouse as the most popular model for humans will increase even more in the future.

REFERENCES

1. Coffey DS, Isaacs JT. Requirements for an idealized animal model in prostatic cancer. In: Murphy GP, Ed. *Models for Prostate Cancer.* New York: Alan R. Liss, 1980:379.
2. Held JR. Muhlbock memorial lecture: Considerations in the provision and characterization of animal models. In: Spiegel A, Erichsen S, Solleveld HA, Eds. *Animal Quality and Models in Biomedical Research, 7th ICLAS Symposium Utrecht 1979.* Stuttgart, Germany: Gustav Fisher Verlag, 1980.
3. Loew FM. Scholarship and clinical service; comparative and laboratory animal medicine. In: Nevalainen T, Hau J, Sarviharju M, Eds. *Frontiers in Laboratory Animal Science. Proceedings of Joint International Conference of ICLAS, Scand-LAS and FinLAS, Helsinki 1995. Scand J Lab Anim Sci* 1996;23(Suppl. 1):13.
4. Overmier JB. Animal models on human pathology: A bibliography of a quarter century of behavioral research, 1967–1992. In: Overmier JB, Burke PD, Eds. *Bibliographies in Psychology: Number 12.* Washington, DC: American Psychological Association, 1992:vii–xiv.
5. Wessler S. Introduction: What is a model? In: *Animal Models of Thrombosis and Hemorrhagic Diseases.* Bethesda, MD: National Institutes of Health, 1976:xi.
6. Salén J. Animal models, principles and problems. In: Svendsen P, Hau J, Eds. *Handbook of Laboratory Animal Science,* Vol. II, *Animal Models.* Boca Raton, FL: CRC Press, 1994:Chapter 1.
7. Kornetsky C. Animal models: Promises and problems. In: Hanin I, Usdin E, Eds. *Animal Models in Psychiatry and Neurology.* Oxford, UK: Pergamon Press, 1977:1.
8. Mogensen J, Holm S. Basic research and animal models in neuroscience—the necessity of "co-evolution." *Scand J Lab Anim Sci* 1989;16(Suppl. 1):51.
9. Hau J. Animal Models. In: Hau J, Van Hoosier G, Eds. *Handbook of Laboratory Animal Science,* Vol. II, *Animal Models.* Boca Raton, FL: CRC Press, 2003.
10. Hau J, Buschard K. Effect of encephalomyocarditis (EMC) virus on murine foetal and placental growth monitored by quantification of maternal plasma levels of pregnancy-associated murine protein-2 and alpha-fetoprotein. *Acta Pathol Microbiol Immunol Scand Sect B* 1986;94:339.
11. Poulsen OM, Hau J, Kollerup J. Effect of homogenization and pasteurization on the allergenicity of bovine milk analysed by a murine anaphylactic shock model. *Clin Allergy* 1987;17:449.
12. Hau J, Cervinkova Z, O'Brien D, Stodulski G, Simek J. Serum levels of selected liver proteins following partial hepatectomy in the female rat. *Lab Anim* 1995;29:185.
13. King NW. Simian models of acquired immunodeficiency syndrome (AIDS): A review. *Vet Pathol* 1986;23:345.
14. Farah IO, Kariuki TM, King CL, Hau J. An overview of animal models in experimental schistosomiasis and refinements in the use of non-human primates. *Lab Anim* 2001;35:205.
15. Pantelouris EM. Absence of thymus in a mouse mutant. *Nature* 1968;217:370.
16. Hau J, Poulsen OM, Dagnæs-Hansen NF. Induction of pregnancy-associated murine protein-1 (PAMP-1) in dwarf (dw) mice by growth hormone. *Lab Anim* 1990;24:183.
17. Jensen HE, Andersen LLI, Hau J. Fetal malformations and maternal alpha-fetoprotein levels in curly tail (ct) mice. *Int J Feto-Matern Med* 1991;4:205.
18. Migaki G. Compendium of inherited metabolic diseases in animals. In: *Animal Models for Inherited Metabolic Diseases.* New York: Alan R. Liss, 1982:473.
19. Van der Meer M, Baumans V, Olivier B, *et al.* Behavioral and physiological effects of biotechnology procedures used for gene targeting in mice. *Physiol Behav* 2001;73:719.
20. Lander ES. Array of hope. *Nat Genet* 1999;21(1 Suppl.):3.
21. Van Zutphen LFM. Is there a need for animal models of human genetic disorders in the post-genome era? *Comp Med* 2000;50:10.
22. Committee on New and Emerging Models in Biomedical and Behavioral Research. *Biomedical Models and Resources. Current Needs and Future Opportunities.* Washington, DC: National Academy Press, 1998.
23. Cui Z, Willingham MC, Hicks AM, Alexander-Miller MA, Howard TD, Hawkins GA, Miller MS, Weir HM, Du W, DeLong CJ. Spontaneous regression of advanced cancer: Identification of a unique genetically determined, age-dependent trait in mice. *Proc Natl Acad Sci USA* 2003;100(11):6682.
24. Olson H, Betton G, Robinson D, Thomas K, Monro A, Kolaja G, Lilly P, Sanders J, Sipes G, Bracken W, Dorato M, Van Deun K,

Smith P, Berger B, Heller A. Concordance of the toxicity of pharmaceuticals in humans and in animals. *Regul Toxicol Pharmacol* 2000;32:56.

25. Mann RD. *Modern Drug Use. An Enquiry on Historical Principles.* Lancaster: MTP, 1984.
26. Koppanyi T, Avery MA. Species differences and the clinical trial of new drugs: A review. *Clin Pharmacol Ther* 1966;7:250.
27. Lewis P. Animal tests for teratogenicity, their relevance to clinical practice. In: Hawkins, DF, Ed. *Drugs and Pregnancy: Human Teratogenesis and Related Problems.* Edinburgh: Churchill Livingstone, 1983;17.
28. Calabrese EJ. *Principles of Animal Extrapolation.* Chelsea, MI: Lewis Publishers Inc., 1991.
29. Beynen AC, Hau J. Animal models. In: Van Zutphen LFM, Baumans V, Beynen AC, Eds. *Principles of Laboratory Animal Science.* New York: Elsevier, 2001:Chapter 10.
30. Schmidt-Nielsen K. *How Animals Work.* London: Cambridge University Press, 1972.
31. Schmidt-Nielsen K. *Animal Physiology, Adaptation and Environment.* London: Cambridge University Press, 1975.
32. Kleiber M. Body size and metabolism. *Hilgardia* 1932;6:315.
33. Hau J, Poulsen OM. Doses for laboratory animals based on metabolic rate. *Scand J Lab Anim Sci* 1988;15:81.
34. Bartels H. Metabolic rates of mammals equals the 0.75 power of their body weight. *Exp Biol Med* 1982;7:1.
35. Snider GL, Lucey ED, Stone PJ. Animal models of emphysema. *Am Rev Respir Dis* 1986;133:149.
36. Frenkel JK. Choice of animal models for the study of disease processes in man. *Fed Proc* 1969;28:160.
37. Hagelin J, Hau J, Carlsson H.-E. The refining influence of ethics committees on animal experimentation in Sweden. *Lab Anim* 2003;37(1):10.
38. Carlsson HE, Schapiro SJ, Farah IO, Hau J. The use of primates in research: a global overview. *Am J Primatol* 2004;63(4):225.

2 Selection of Biomedical Animal Models

MICHAEL S. RAND

ABSTRACT

Laboratory animals play a crucial role in research discovery and technological advances, and they will continue to take part in improving the lives of people and other animals. It is incumbent upon the researcher to know his subject well in order to provide relevant information to the scientific world. In an effort to assist the biomedical researcher in gaining this knowledge, this chapter provides the following key elements: definition of types of animal models, legislative and legal requirements, criteria for choosing a model, extrapolation validity recommendations, and descriptive features for publication.

Key Words: Animal models, Laboratory animal(s), Animal model types, Animal use criteria, Choosing animal model, Animal factors, Extrapolation, Animal description.

INTRODUCTION

Over the last one and one-half centuries, almost all medical knowledge, treatment regimes, and medical device development have involved research using animals. The key factor in using animals in research is in its extrapolatability of results to humans. Animals in research have been and still are essential in developing treatments for asthma, HIV/AIDS, cancer, birth defects, bioterrorism medical countermeasures, vaccines, antibiotics, high blood pressure, and much more. Additionally, they have been vital in the development of antibiotics, vaccines, and organ transplantation techniques.[1] As the rise in emerging infectious diseases (e.g., West Nile virus and avian influenza) continues, animals will be key and essential in the development of preventive and treatment modalities.

THE HISTORY OF ANIMAL USE IN RESEARCH

The use of animals to study human physiology and anatomy can be traced back to the second century AD in which Galen was a Greek physician and philosopher. His research was based almost exclusively on studies using apes and pigs. Unfortunately, this initiated many errors based on his accepted authority and the prohibition by the Church of using human cadavers for research purposes. Galen was later blamed for using incorrect methods in research when in truth it would be more accurate to say that he drew wrong conclusions based on uncritical interspecies extrapolation of data. That is, he assumed that all extracted information derived from his use of animals could be directly applied to humans. It was not until the late sixteenth century that this error began to be recognized.

Modern research principles can be attributed to three physiologists from the 1860s. In 1865, Claude Bernard, a French physiologist, published *An Introduction to the Study of Experimental Medicine*.[2] This book was intended to provide physicians with guidance in experimental research. It proposed the use of chemical and physical induction of disease in animals, thus becoming the first published book to advocate creating "induced animal models" for biomedical research. His peers of the time were Louis Pasteur in France and Robert Koch in Germany. Louis Pasteur and Robert Koch introduced the concept of specificity into medicine and the "germ theory of disease." The turning of the century saw the development and use of animal models for infectious diseases and screening and the evaluation of new antibacterial drugs based upon the work of these three researchers.

During the first quarter of the nineteenth century, animal studies were crucial for less than one-third of the major advances that occurred. With the contributions of Claude Bernard, Louis Pasteur, and Robert Koch, animal studies contributed to more than half of the significant discoveries made thereafter. Since 1901, two-thirds and 7 of the last 10 Nobel Prizes in medicine have relied at least in part on animal research.[3] Today, researchers rely on the identification and development of animal models to explore all avenues of medical science to include assessment of pathogenic mechanisms, diagnostic and therapeutic procedures, nutrition and metabolic diseases, and the efficacy of novel drug development.

THE CONCEPT OF ANIMAL MODELS

WHAT IS AN ANIMAL? Etymologically, the word "animal" derives from the Latin *animal* meaning soul/spirit, thus describing living organisms that are *animated.*

WHAT IS A MODEL? A model is an *object of imitation,* something that accurately resembles something else, a person or thing that is the likeness or *image* of another. The *Holy Bible* tells us that God said, "Let us make man in our image, in our likeness, so God created man in his own image, in the image of God he created him; male and female he created them." God created man out of the dust of the ground and then breathed into his nostrils the breath of life to animate him. Thus, humans are "animal models" of God.

Consequently, combining the two definitions, an "animal model" is an *animated object of imitation,* an "image of Man" (or other species), used to investigate a physiological or pathological circumstance in question.[4]

From: *Sourcebook of Models for Biomedical Research*
(P. M. Conn, ed.), © 2008 Humana Press Inc., Totowa, NJ.

WHAT IS AN ANIMAL MODEL? The U.S. National Research Committee on Animal Models for Research on Aging attempted to define the term "laboratory animal model" as "an animal in which normative biology or behavior can be studied, or in which a spontaneous or induced pathological process can be investigated, and in which the phenomenon in one or more respects resembles the same phenomenon in humans or other species of animal."

Using the term animal model can be confusing because what is often meant by the term "animal model" is actually studying human conditions. In other words, it is not the image of the preferred animal that is the focus of research but the analogy of the physiological behavior of this animal to our own (or another) species. It would, thus, be more correct to speak of "human models" in this context. Indeed, although using animals in research can benefit other animals, it is much more focused on improving the human condition.

TYPES OF ANIMAL MODELS

When animals are used in research to study biological and functional systems in humans, they are broken down into the following categories:

1. **Exploratory.** Animals used in this category are used to gain an understanding of fundamental biological mechanisms, whether normal or abnormal. An example would be the use of a novel animal model of aging, particularly for identifying genes and biochemical pathways regulating longevity.
2. **Explanatory.** Animals used in this category are used to gain an understanding of complex biological problems. An example would be the use of cognitive and psychosocial animal models to provide an etiology for anorexia nervosa.[5]
3. **Predictive.** Animals used in this category are used to discover and quantify the impact of investigative treatments whether for diseases or chemical toxicities. Predictive animal testing models are important in improving the success of a drug or medical device in clinical trials and for generating new data in support of the ongoing marketing of existing products.

When animals are used in disease research, they are broken down into the following categories:

1. **Induced (or Experimental).** Induced models are ones in which normal animals are experimentally created either through surgical modifications, genetic modifications, or chemical injections. An example would be a myocardial infarction induced by coronary artery surgical ligation.
2. **Spontaneous.** Spontaneous models are genetic variants, which mimic the human condition. The variance occurs naturally through mutation and not by experimental induction. The *nu* mutation was first reported in 1966 in a closed stock of mice in a laboratory in Glasgow, Scotland. It was not until 1968, however, that it was discovered that the homozygous nude mouse also lacked a functional thymus, i.e., it was *athymic*. The mutation produces a hairless state, generating the name "nude." The other, unique defect of nude mice is the failure of the thymus to develop normally to maturity. The thymus remains rudimentary and produces reduced numbers of mature T cells. This means nude homozygotes (animals with identical mutant genes at corresponding chromosome loci) do not reject allografts and often do not reject xenografts (tissue from another species). The discovery that human neoplasms (tumors) could be grown in nude mice was immediately recognized as an important research tool. Thus, the spontaneous mutation of *nu* among laboratory mice was a serendipitous development that led to the nude mouse becoming the first animal model of a severe immunodeficiency. In the decades since, the nude mouse has been widely utilized by researchers studying factors regulating transplantable human tumor growth and cancer metastasis.[6]
3. **Transgenic.** Transgenic models are induced models in which DNA is inserted into or deleted (knockout) from the genome of the animal. The term "transgenic" was coined in 1981 by Gordon and Ruddle to describe an animal in which an exogenous gene was introduced into its genome. In the late 1980s, the term transgenic was extended to gene-targeting experimentation and the production of chimeric or "knockout" mice in which a gene (or genes) has been selectively removed from the host genome. Today, a transgenic animal can be defined as one having any specific, targeted genetic modification. Transgenic animals are most commonly produced through (1) germline modifications of gametes, (2) microinjection of DNA or gene constructs into zygotes (unicellular embryos), or (3) incorporating modified cells, including embryonic stem (ES) cells, into later stage embryos. After gamete or embryo modifications, the resultant embryos are matured to term in a recipient female.[7]
4. **Negative.** Negative models fail to react to a disease or chemical stimulus. Thus, their main use in biomedical research is for studies on the mechanism of disease resistance. A classic example is the failure of gonococcal infection to develop in rabbits after an experimental treatment that induces the disease in other animals. Negative animal models have become increasingly important with the advent of transgenic technology. For example, a novel transgenic mouse was created to study the lack of development of autoimmune thyroiditis with the injection of self-thyroglobulin. This strain of mice lacked certain surface epitopes to account for this negative reaction.[8]
5. **Orphan.** Orphan models are the opposite of negative models. Orphan models are animals in which a disease occurs but there is not a corresponding disease in humans. Orphan models may become induced models when a similar disease is recognized in humans later on. Historically, scrapie in sheep was such a model, but now is useful as a model for the human spongiform encephalopathies that are of so much concern (e.g., BSE, "mad cow disease," and CWD, chronic wasting disease in deer).

All categories above may be further subcategorized with the following divisions:

1. **Fidelity.** The extent a biological structure in an animal resembles that of a human. Thus, a high fidelity animal model gives a highly relevant biological closeness to the human structure. Model fidelity is best conceptualized as a continuous spectrum, ranging from low to high fidelity. Examples of low-fidelity models include bench models made of simple materials that often have little anatomical resemblance to reality. However, these models incorporate some of the key constructs of the simulated tasks. At the other end of the spectrum are high-fidelity models such as human or animal cadavers or the new array of virtual reality simulators. These simulators usually incorporate highly realistic visual and tactile cues in the midst of a highly interactive model. In between these two extremes, almost any kind of intermediate fidelity can exist.

2. **Homologous.** The symptoms shown in the animal are identical to those shown in the human. For instance, the recent discoveries of swine hepatitis E virus (HEV) from pigs and avian HEV from chickens afforded an opportunity to develop small homologous animal models for HEV.[9]

3. **Isomorphic.** The animal's symptoms or anatomy are similar to those in the human but the etiology or genetic character is different. For example, there is a set isomorphism between the human and mouse heart at the organ level and also at the organ part level: each species has a heart and a corresponding set of cardiac chambers (right and left atrium, right and left ventricle) and the wall of each chamber has a corresponding set of layers (epicardium, myocardium, endocardium).

4. **Partial.** These models do not mimic the entire human disease but enough similarities exist to allow their use in studying aspects of the disease or treatments. For instance, animal models of Alzheimer's disease can be created based on the accumulation of increased levels of amyloid-ß peptide in the brain and have many amyloid plaque deposits; however, they have only subtle behavioral and electrophysiological deficits, thus providing only a partial model of the human condition.[10]

5. **Face validity.** The degree to which there is a similar phenotypic display between the disease in the animal and the corresponding disease in the human. For example, it could be argued that the demonstration of drug effects in an animal model for depression after a period of chronic administration is important for establishing its face validity, but is not relevant to the model's predictiveness and therefore to its ability to serve as a screening test for treatments for the modeled disease.[11]

6. **Construct validity.** The degree to which there is a similar genetic display between the disease in the animal and the corresponding disease in the human. As an example of high construct validity, research was performed on three candidate dopaminergic genes (DRD2, DRD4, and DAT-1) that were sequenced in spontaneous hypertensive (SHR) and Wistar Kyoto (WKY) rats. No differences were found in DRD2 or DRD4 genes, but several variations were found in the DAT-1 gene that are of significance because several ADHD families show linkage to DAT-1. It also strengthened the validity of using WKY as a control for SHR, because their behavioral characteristics are similar to those of other rat strains.[12]

LEGISLATIVE AND LEGAL REQUIREMENTS FOR USING ANIMALS IN RESEARCH

Biomedical research is among the most regulated industries in the world. A comprehensive overview of global requirements can be found in the *Handbook of Laboratory Animal Science*, 2nd edition, Chapter 3.[13] Failure to comply with regulatory requirements can result in fines levied against the institution, suspension of authority to operate, permanent revocation of the facility's license, and withdrawal of public funding.

One newly regulated aspect of biomedical research not covered in this chapter occurred after the terrorist attack on September 11, 2001. The attack increased concerns in the United States for the possibility of bioterrorism using agents that would destroy human, animal, and plant life. This concern escalated the need for research that involved the development of therapeutic and preventive measures against such agents. In response, congress passed and President Bush signed into law the "Public Health Security and Bioterrorism Preparedness and Response Act of 2002" (Public Law 107–188) on June 12. The purpose of the act was to improve the capacity of the United States to prevent, prepare for, and respond to bioterrorism and other public health emergencies and to enhance the control of dangerous biological agents and toxins. The Centers for Disease Control and Prevention (CDC) is the agency with the primary responsibility for implementing the provisions of the Act with regard to human pathogens and toxins and the United States Department of Agriculture (USDA) with regard to animal and plant pathogens and toxins. The regulation provides for expanded regulatory oversight of select agents and toxins, and a process for limiting access to persons who have a legitimate need to possess, use, or store these agents. The regulation also establishes a requirement for a security risk assessment performed by the Federal Bureau of Investigation for those persons needing access to select agents and toxins. It also establishes and enforces safety and security procedures, including measures to ensure proper training and appropriate skills to handle agents and toxins; a requirement to designate an institutional Responsible Official to ensure compliance with the regulations; and a requirement to obtain a certificate of registration when there is a need to possess, use, or transfer select agents and toxins. Infectious agents labeled as "select" as determined by the CDC and USDA, registration forms, and other information concerning the Select Agents Program may be found at http://www.selectagents.gov.[14]

CHOOSING THE RIGHT MODEL

To quote the philosopher, Bernard Rollin, "The most brilliant design, the most elegant procedures, the purest reagents, along with investigator talent, public money, and animal life are all wasted if the choice of animal is incorrect." Once it has been determined that the use of laboratory animals is necessary, the most appropriate species, breed, and strain with the closest homology to humans must be chosen in order to give the research face and/or construct validity. Because new animal models are continually being identified and characterized and the field of biomedical research has become global in nature, the search for the appropriate animal model should start with a thorough literature search and a check of appropriate web sites (see Chapter 7). The Institute for Laboratory Animal Research maintains a very practical and useful search engine for this purpose.[15]

Selection of a species should not be based solely on availability, familiarity, or cost. Animals that meet these criteria may not provide the genetic, physiological, or psychological facets needed or wanted for the proposed project. It is almost impossible to give specific rules for the choice of the best animal model, because the many considerations that have to be made before an experiment can take place differ with each research project and its objectives. Nevertheless, some general rules can be given.

RESEARCH FACTORS

- Appropriateness as an analogue. Ensuring that the part or organ being studied has a function similar to the target species is vital in applying research-derived data from the chosen model.
- Transferability of information. The usual goal of research using an animal model is to define a process in a system with the hope of transferring the data gained to a more complex system. Traditionally, one-to-one modeling is

sought: modeling in one group of organisms that can be transferred to another group that has several analogous features of interest. This is especially helpful in modeling disease states. However, in modern research, many-to-many models are mainly used. This technique begins by analyzing the component parts of a process or disease, and then finding for each component analogous models in many taxa of living species.[16] This is especially helpful when a plurispecies approach is needed to gain approval for new medications or medical devices.

- Generalizability of the results. The ability to generalize results to the target species is important. Federal regulations prohibiting the unnecessary duplication of previous research highlights the importance of choosing an animal model in which testing results can be easily repeatable and verifiable on which to build new research. In May, 2006, the world was shocked when famed South Korean cloning scientist Dr. Hwang Woo-suk was charged with fraud and embezzlement when scientists could not verify his published data. In addition, if the ultimate target species is human, it is well known that this species is genetically highly variable, with cultural, dietary, and environmental differences. This may be of lesser importance in disease modeling since most diseases do not choose its victim based on genetic variability. However, this is now well known to be of importance in pharmacological and toxicological modeling and has opened up the new field of research in pharmacogenetics.
- Ethical implications. Certainly research must start with justification for using an animal at all. Federal regulations require the use of alternate methods if feasible. Alternate methods could consist of using cell lines, bacteria, computer models, or even human volunteers. The three Rs of Russell and Burch (replacement of existing experiments with animal-free alternatives, or reduction in the number of animals used, or refined methods to reduce animal suffering) help to meet the ethical concerns.[17]
- Numbers needed. Certainly consultation with a biostatistical analyst prior to submitting a proposal is highly recommended. Numbers needed to provide scientific validity, especially for publication, will impact many other factors such as cost and housing availability.
- Customary practice within a particular discipline. Caution must be displayed when using this criterion. Customary practices may not always mean that the most appropriate animal model has been used. The "customary" animal may not represent the most accurate genetic, microbiological, physiological, or psychological facets needed for the study. Historical evidence has revealed that using animal models just because others have has led to substandard results. However, customary practices when justified and supported by the other criteria listed can be a satisfactory and faster route of choosing the animal model needed.
- Existing body of knowledge of the problem under consideration. This criterion again emphasizes the need for a thorough literature search before forming the basis for the research project. The literature search will emphasize what is already known to prevent accidental duplication, but will also reveal what is not known. It will also make known published authorities in the discipline that may serve as a consultation source to prevent unnecessary and competitive research projects.
- Natural versus experimentally produced models. Unavailability of natural models will require the use of experimentally produced models. Depending on the objectives of the study, both may be needed.

ANIMAL CARE FACTORS

- Cost and availability. Certainly cost and availability are important factors when choosing an animal model, but they can be disastrous if the decision is based solely on cost and not the other listed factors. Cost also includes ongoing care not only for husbandry but also from experimental manipulations. Certainly, the best animal model can be in short supply as illustrated by the CNN news report on August 9, 2003. This report emphasized the increased demands in research due to public health crises such as AIDS and the threat of bioterrorism. The increased demands have led to a national shortage of rhesus macaques. In addition, the shortage has skyrocketed the cost per monkey.
- Housing availability. Another practical consideration in choosing the animal model is the accessibility of housing. Research animal housing requirements are stringent and may lessen the availability according to the species chosen. For instance, choosing a nonhuman primate may require the purchase of new caging and the hiring of additional personnel to provide specialized husbandry care, as opposed to choosing mice, which can be placed several to a cage and hundreds in a room.
- Husbandry expertise. Some models require not only special housing, but also special care.
- Stress factors. Stress sources from many different causes can affect the animal's physiology, biochemistry, and behavior. Sources of stress can be transportation, handling and manipulations, overcrowding, lack of environmental enrichment, and the research project itself.

PHYSICAL AND ENVIRONMENTAL FACTORS

- Ecological consequences. While the best animal model may be available only by capturing in the wild, ecological consequences must be considered in its removal. In addition, safety measures must be in place to prevent accidental escape from the research facility. A prime example is the *Xenopus* spp. frog. If it escapes, it can overrun local ponds and rivers endangering natural amphibian populations. Furthermore, care must be taken not to violate the Endangered Species Act (http://www.fws.gov/endangered/wildlife.html) or the Convention on International Trade in Endangered Species of Wild Fauna and Flora (http://www.cites.org).
- Hazardous components. Many research projects entail the use of chemicals, infectious agents, and radioisotopes. The uses of these components are highly regulated and the appropriate proper authority (Institutional Biosafety Committee, Radiation Safety Committee, Occupational Health and Safety Committee, Environmental Health and Safety Committee, and Institutional Animal Care and Use Committee) within each institution must give approval for

its use. In addition, all those who will be exposed, whether from the research side or the animal care side, must be notified.

- Environmental influences. Environmental aspects may be important to a particular animal species.[18] Environmental factors that fit into the broad categories of physical, chemical, biological, and social may impact the physiological and behavioral responses of animals. These factors include humidity, ventilation, light cycle and quality, noise, cage size and bedding materials, diet and water, and room temperature.[19] As an example, high temperature and humidity have been proven to impair memory in mice.[20]

ANIMAL-RELATED FACTORS

- Genetic aspects. Uniformity of organisms may be necessary where applicable. "This insidious evolution of the inbred genotype is known as genetic drift. It is capable of subverting the conclusions reached about comparable research results coming from different laboratories when each uses its own subline of the same inbred strain (Bailey DW, 1977)."[21] The importance and methods of preventing genetic drift in biomedical research can be found at http://jaxmice.jax.org/geneticquality/drift.html. In addition, it is important to remember that to be in compliance with the *National Institutes of Health's Guidelines*,[22] work with transgenic animals requires the approval of the Institutional Biosafety Committee as well as the Institutional Animal Care and Use Committee.
- Background knowledge of biological properties. Knowledge of biological properties such as generalized and specialized function of body components is needed in order to validate interspecies transfer of information. Certainly, a rat would not be the best choice in biliary studies due to the absence of a gallbladder. Knowing the biological properties also aids in the decision of whether the animal is a spontaneous model or must be experimentally induced.
- Ease of and adaptability to experimental manipulation. This is unquestionably a practical matter. Guinea pigs have highly inaccessible blood vessels and would be impractical in studies requiring repeated blood sampling. Prairie dogs and woodchucks can be vicious to handle; therefore, knowing the response to experimental manipulation may also influence the choice.
- Size of the animal. This item is important from several different aspects. The size of the animal impacts housing and husbandry availability. However, size is also important to consider when tissue sampling or blood collection is necessary. For instance, many proposals are rejected because the researcher failed to abide by published guidelines for removal of blood.[23] In addition, it is also important to incorporate the size of the animal into the decision-making apparatus when physiological or morphological properties such as joint strain or organ size must be identical to that of a human, especially when developing medical devices.
- Life span and age. Studies requiring components at different stages of life can certainly impact the species chosen. The average lifespan of a rat is 2.5–3.5 years, whereas it can be over 30 years for a rhesus monkey.
- Sex. The alternating cycle of hormonal production in the female gender and its influence on the data outcome must be considered when planning for the research project.
- Progeny needed. Female mice and rats can produce 5–10 progeny per month, whereas the rhesus monkeys only one or two per year. *Xenopus* sp. frogs produce thousands of ova during their lifetime, whereas mammals produce only dozens.
- Diseases or conditions that might complicate results. An excellent historical review on the struggle against pathogens in laboratory rodents can be found in Weisbroth.[24] The effects on research can be found in Baker.[25] Both publications emphasize the need for disease-free animals in research to prevent adverse effects on resultant data. Just as in human AIDS, the realm of disease-causing organisms changed with the advent of immune deficient models. Special caging and care procedures are fundamental in minimizing such infections.
- Special features of the animal such as unique responses or microflora. It is important to be familiar with unique anatomical or physiological features of the species you will be working with. The results could be quite unexpected otherwise. For example, in rabbits, the terminal portion of the ileum empties into an enlarged rounded viscus called the sacculus rotundus and not the colon as in humans. This unique feature of the rabbit is important to know when designing gastrointestinal studies.

Forming the above standards into a checklist will help to fulfill the criteria needed to choose the best model for the proposed research project. Model selection is the privilege of individual researchers, but they must be very cautious in their selection because in the end, it is up to them to convince the rest of the scientific community that they made the right choice.

Before choosing, consultations should occur with scientists who have already used the animal model. Just as with equipment purchase, communicating with previous users can be very helpful in learning unique features of the selected species, breed, and strain. Not all attributes (especially negative ones) are published, making it even more important to contact those who have experience with the animal model you choose.

Preparatory consultation should also occur with those who will be responsible for housing and maintaining the animals, as they will be the most familiar with the care of the animal and its physical and environmental needs. Preparatory consultation with the laboratory animal veterinary practitioner should also occur to discuss the animal-related factors.

EXTRAPOLATION FROM ANIMALS TO HUMANS

Extrapolation from animals to humans does not necessarily mean that biomedical research data obtained from using animals are then used to find a corollary in a human. Rather, a hypothesis is formed first based on human relevancy, and then tested on an animal. Answers are obtained, analyzed, and published based on the hypothesis. Although true in many cases, caution must be exerted in assuming that a close phylogenetic relationship or anatomical similarity guarantees an identical biochemical or physiological response in the animal. In addition, it must be realized that humans to whom the results are being extrapolated are genetically highly variable due to cultural, dietary, and environmental differences. This is of minor importance when developing disease models but is highly important for pharmacological and toxicological models.

So, how can the validity of extrapolation be verified? Complete reliability cannot be guaranteed; however, following the following vital requirements will help to avoid several of the mistakes of the past and overcome problems of the future:

- *Taking a plurispecies approach.* Most of the regulating authorities require two species in toxicology screening, one of which has to be nonrodent. This does not necessarily imply that excessive numbers of animals will be used. The uncritical use of one-species models can mean that experimental data retrospectively turn out to be invalid for extrapolation, representing real and complete waste of animals. Using more than one species is, of course, no guarantee for successful extrapolation either.
- *Metabolic patterns and speed and body size must match between species.* The use of laboratory animals as models for humans is often based on the premise that animals are more or less similar with respect to many biological characteristics and thus can be compared with humans. However, there is one striking difference between mouse and human, and that is body size. In proportion to their body size, mammals generally have very similar organ sizes expressed as percentage of body weight. Take the heart for instance, which often constitutes 5 or 6 g per kilogram of body weight, or blood, which is often approximately 7% of total body weight. It is well known that the metabolic rate of small animals is much higher than that of large animals and, thus, provisions must be made to adjust the study accordingly. Drugs and toxins exert their effect on an organism not per se but because of the way that they are metabolized, the way that they and their metabolites are distributed and bound in the body tissues, and how and when they are finally excreted. Adjusted doses should include the following provisions:
    - If the object is to achieve equal concentrations of a substance in the body fluids of animals of different body size, then the doses should be calculated in simple proportion to the animals' body weights.
    - If the object is to achieve a given concentration in a particular organ over a certain time period, the calculation of dosage becomes more complicated, and other factors, including the physicochemical properties of the drug, become important.
    - Metabolism or detoxification and excretion of a drug are not directly correlated with body size but, more accurately, to the metabolic rate of the animal.
    - Some species react with particular sensitivity toward certain drugs, and marked variations in the reaction of animals within a species occur with respect to strain, pigmentation, nutritional state, stress level, type of bedding, ambient temperature, age, sex, route or time of administration and sampling, diurnal variation, and season of the year. As much as possible, these items must be controlled.[4]
- *Experimental design and the life situation of the target species must correspond.* A model cannot be separated from the experimental design itself. If the design inadequately represents the "normal" life conditions of the target species, inaccurate conclusions may be drawn, regardless of the value of the model itself.

DESCRIPTION OF ANIMAL MODELS[16]

Unlike the old days when the researcher could write in the materials and methods section "black mice were used in the study," modern obligations require an exact description of the model. The description should include the following.

- Genetic strain and substrain using correct international nomenclature.[26,27]
- Special genetic features.
- Microbial status of the animal.
- Age.
- Housing standards.
- Maintenance procedures.
- Diet.
- If used in infectious disease studies, the description should also include
    - Strain of the organism.
    - Method of inoculum preparation.
    - Route of inoculation.

CONCLUSIONS

Laboratory animals play a crucial role in research discovery and technological advances, and they will continue to take part in improving the lives of people and other animals. It is incumbent upon the researcher to know the subject well in order to provide relevant information to the scientific world. The final judgment in the choice of the animal model will always be in its ability to elucidate and predict the observed effects in the target species.

REFERENCES

1. Americans for Medical Progress. Everyday wonders. (Accessed August 16, 2006, at http://www.amprogress.org.)
2. Bernard C. *Introduction to the Study of Experimental Medicine.* Mineola, NY: Dover Publications, 1957.
3. Foundation for Biomedical Research. Nobel prizes: The payoff from animal research. (Accessed on August 30, 2006, at http://www.fbresearch.org/education/nobels.htm.)
4. Hau J. Animal models. In: Hau J, Van Hoosier GL, Eds. *Handbook of Laboratory Animal Science*, 2nd ed., Vol. II. Boca Raton, FL: CRC Press, 2003:1–9.
5. Connan F, Campbell IC, Katzman M, Lightman SL, Treasure J. A neurodevelopmental model for anorexia nervosa. *Physiol Behav* 2003;79:13–24.
6. Immunodeficient rodents opening new doors for investigators. *Res Anim Rev* 1996;1(2):1–8.
7. Pinkert CA. The history and theory of transgenic animals. *Lab Anim* 1997;26:29–34.
8. Yan Y, Panos JC, McCormick DJ, *et al.* Characterization of a novel H2A(–)E+ transgenic model susceptible to heterologous but not self thyroglobulin in autoimmune thyroiditis: Thyroiditis transfer with Vbeta8+ T cells. *Cell Immunol* 2001;212(1):63–70.
9. Meng XJ. *Discoveries of Animal Strains of Hepatitis E Virus: Implications for Animal Models and Zoonosis.* Ithaca, NY: American College of Veterinary Pathologists and American Society for Veterinary Clinical Pathology, 2004. (Accessed at http://www.ivis.org.)
10. St. George-Hyslop PH, Westaway DA. Antibody clears senile plaque. *Nature* 1999;400:116–117.
11. Willner P. The validity of animal models of depression. *Psychopharmacology* 1984;83:1–16.

12. Sagvolden T, Russel VA, Aase H, Johansen EB, Farshbaf M. Rodent models of attention-deficit/hyperactivity disorder. *Biol Psychiatry* 2005;57:1239–1247.
13. Bayne K, deGreeve P. An overview of global legislation, regulation, and policies on the use of animals for scientific research, testing, or education. In: Hau J, Van Hoosier GL, Eds. *Handbook of Laboratory Animal Science*, 2nd ed., Vol. I. Boca Raton, FL: CRC Press, 2003:31–50.
14. Fleischauer AT, Ellis B, Ostorff S, Hemphill MS. The select agents and toxins interim final regulations: Applications to the small animal facility. *Contemp Top Lab Anim Sci* 2003;42(4):126–140.
15. Animal models and strains search engine. Washington, DC: Institute for Laboratory Animal Research. (Accessed August 16, 2006, at http://dels.nas.edu/ilar_n/ilarhome/search_amsst.shtml.)
16. Davidson MK, Lindsey JR, Davis JK. Requirements and selection of an animal model. *Isr J Med Sci* 1987;23:551–555.
17. Russell WMS, Burch RL. *The Principles of Humane Experimental Technique*. London: Methuen, 1959.
18. Environmental control for animals and plants. In: *ASHRAE Handbook*. Atlanta, GA: American Society of Heating, Refrigerating & Air Conditioning Engineers, Inc., 2005:Chapter 10.
19. Rose, MA. Environmental factors likely to impact on an animal's well-being-an overview. In: Baker RA, Jenkin G, Mellor DJ, Eds. *Improving the Well-Being of Animals in the Research Environment*. Adelaide, Australia: ANZCCART, 1994:284–301.
20. Inan SY, Aksu F. Amnesic effects of relative humidity and temperature in mice. *Lab Animal* 2002;31(2):40–48.
21. Bailey DW. Genetic drift: The problem and its possible solution by frozen-embryo storage. *Ciba Found Symp* 1977;291–303.
22. NIH guidelines for research involving recombinant DNA molecules. Bethesda, MD: National Institutes of Health, 2002.
23. Morton DB, Abbot D, Barclay R. Removal of blood from laboratory mammals and birds. *Lab Anim* 1993;27:1–22.
24. Weisbroth SH. Post-indigenous disease: Changing concepts of disease in laboratory rodents. *Lab Anim* 1996;25:25–33.
25. Baker DG. Natural pathogens of laboratory mice, rats, and rabbits and their effects on research. *Clin Microbiol Rev* 1998;11(2): 231–266.
26. Mouse Genome Informatics. Rules for nomenclature of mouse and rat strains, 2005. (Accessed August 30, 2006, at http://www.informatics.jax.org/mgihome/nomen/strains.shtml.)
27. Mouse Genome Informatics. Rules for nomenclature of genes, genetic markers, alleles, and mutations in mouse and rat, 2005. (Accessed August 30, 2006, at http://www.informatics.jax.org/mgihome/nomen/gene.shtml.)

3 Improved Models for Animal Research

Rolf Sprengel, Leonid Eshkind, Jan Hengstler, and Ernesto Bockamp

ABSTRACT

Experimental animal models are critical to understand gene function and human disease. Many rodent models are presently available providing avenues to elucidate gene function and/or to recapitulate specific pathological conditions. To a large extent, successful translation of clinical evidence or analytical data into appropriate mouse models is possible through progress in transgenic or gene deletion technology. Despite these significant improvements, major limitations still exist in manipulating the mouse genome. For this reason and to maximize success, the design and planning of mouse models need good knowledge concerning the requirements and limitations of commonly used strategies and emerging technologies. The purpose of this chapter is to provide a current overview of strategies for manipulating the mouse genome.

Key Words: Transgenic mice, Knockout mice, Conditional mouse models, Cre and Flp recombinase, Tetracycline system, RNAi, Knockdown mice, Functional genomics.

INTRODUCTION

One of the central issues facing biomedical research is the need to transform *in vitro* data into knowledge about gene function in mice or humans. In this respect genetically engineered laboratory mice are an excellent tool for modeling genetic disorders, assigning function to genes, evaluating the action of drugs and toxins, and answering fundamental questions in basic science. Animal models account for factors such as age, stress, cell-to-cell communication, pathogen–host interaction, physiology, immune response, brain function, and other key issues, providing an advantage over *in vitro* assays or computer models. For example, tumor initiation, progression, and spreading cannot be recapitulated *in vitro* but can be addressed with animal models. Nevertheless, animal models represent only an experimental surrogate and results obtained from mouse experiments do not necessarily recapitulate the human situation.

Furthermore, introducing genetic changes to the germ line of the mouse may indeed identify gene function, but also might result in severe developmental consequences, complicating or preventing analysis. Embryonic lethal phenotypes, frequently associated with null alleles or compensatory pleiotropic gene expression, induced through the absence or increased gene expression, are examples that can prevent the generation of a useful animal model. To overcome these limitations and more precisely control gene expression or gene deletion in a tissue- and time-specific fashion, conditional mouse models are used and are becoming increasingly popular. Therefore, these second-generation models may significantly improve our ability to examine gene function *in vivo*.

In this chapter, we will discuss different conditional transgenic and gene-targeting techniques, as well as provide a brief overview regarding conventional mouse transgenesis and germ line gene targeting. In addition, we will examine the emerging technology to knock down gene expression *in vivo* through small interfering RNA (siRNA) methods. While this chapter is not intended to be comprehensive or to provide specific technical details, we encourage the interested reader to explore more focused reviews on this topic.[1–6]

STANDARD MOUSE TRANSGENESIS

Since the pioneering work from Gordon and colleagues reporting the successful generation of transgenic mice by microinjection of DNA into the pronucleus of one-cell embryos, the genetic manipulation of the mouse embryo has been extremely useful for creating thousands of murine models for biomedical research.[7] Today, two different methods are routinely used for generating genetically modified mice.

TRANSGENIC MICE

First there is microinjection of recombinant DNA into the pronuclei of fertilized mouse eggs or infection of germ cells/early embryos with viral vectors carrying the foreign gene.[3,8] With this technique the introduced DNA is more or less randomly inserted in one or multiple copies into the mouse genome. Typically, the recombinant constructs used for pronucleus injection are composed of a selected DNA sequence linked to a promoter or promoter/enhancer combination that determines the expression pattern and expression level of the selected DNA sequence in a given tissue or developmental stage. Using this approach, it has been possible to characterize the function of well-defined transgenic gene products, dominant negative or constitutively active gene mutations, or specifically designed proteins. Furthermore, *in vivo* suppression of a particular endogenous gene can be achieved by transgenic expression of antisense mRNA, ribozymes, and small hairpin (shRNAs) or micro-RNAs (miRNAs).[2,9–15] Besides gene products, transcriptional control elements as enhancers,

From: *Sourcebook of Models for Biomedical Research*
(P. M. Conn, ed.), © 2008 Humana Press Inc., Totowa, NJ.

silencers, promoters, or complete locus control regions can be studied using transgenic mice.[16–18]

One recurrent problem observed in transgenic mouse models are variegational position effects often disturbing or masking the specific function of the used transcriptional control elements. Theoretically, it might be assumed that after stable integration into the genome the recombinant DNA should readily manifest its predicted mission. However, many experiments reveal that the genetic surrounding of the inserted transgenic construct is modulating the expression pattern of the transgene itself.[19] Obvious reasons for positional variegation are that the injected DNA (1) integrates into or near an endogenous gene locus with strong transcriptional control activity, affecting in *cis* the promoter of the transgenic construct or (2) integrates in a chromatin structure prone to be inactivated during development, which would lead to silencing of the transgenic construct. To avoid this recurring issue, single-copy integration of transgenes into a selected target locus or the use of so-called insulator elements has been reported.[20,21] Alternatively, large recombinant constructs like bacterial artificial chromosomes (BACs) or yeast artificial chromosomes (YACs) can be used for generating position-independent transgenic mouse lines.[22,23] As one of the consequences of the genome sequencing projects, a collection of well-characterized BAC and YAC clones that cover nearly the entire human and mouse genomes is now available (for example, http://www.rzpd.de/products/clones; http://bacpac.chori.org; http://www.sanger.ac.uk). In addition to that, convenient methods for site-specific modification of BACs and YACs have been established.[24–27] However, it has to be emphasized that by the random chromosomal insertion of BAC or YAC constructs endogenous gene loci may be destroyed and also the expression of neighboring genes might be modified.[28] In case these endogenous genes will be indispensable to life, this will become obvious when the transgenic mouse variant can be maintained only as a heterozygous line.

For the above reasons, a correlation between phenotype and transgene function is obvious only when at least two independent transgenic mouse lines with identical BAC, YAC, or transgenic constructs show the same phenotype. If only one transgenic line is analyzed, it can never be completely excluded that the observed phenotype is not linked to the transgene itself but reflects the compromised expression of neighboring endogenous genes.

CONVENTIONAL GENE TARGETING

The second widely used method for the generation of gene manipulated mice makes use of pluripotent mouse embryonic stem (ES) cells. Targeted gene modification is built on the finding that mammalian cells have the enzymatic machinery for exact homologous recombination between identical (homologous) DNA sequences.[29–31] Thus precise predefined modifications of an endogenous gene are possible in cultivated mammalian cells including mouse ES cells. Genetically modified ES cells can then be used to generate gene "knockout" or "knockin" mice that carry the planned DNA modification.[32] Normally, pluripotent ES cells are injected in early mouse embryos at the blastula stage or aggregated with morula stage embryos. During embryonic development the ES cells participate in the formation of different tissues including the germ cells and as a result the embryo is composed of two different genetic backgrounds, derived either from the wild-type cells or the genetically modified ES cells. However, in case ES cells have contributed to the germ cells of the animal, offspring from these chimeric mice will harbor the desired genetic modification in the germline (germline transmission). To generate mice with a defined background, inbreeding with animals of the same genetic background as the original ES cell will produce genetically identical homozygous offspring. However, mice crossed to commonly used ES cell mouse backgrounds often poorly reproduce and are therefore difficult to expand. For this reason, for most experiments chimeric mice are crossed to mouse strains with good breeding efficiencies.

Typically, the targeting construct for homologous recombination is composed of a central core region flanked by two regions—the so-called homology arms—that are identical in sequence to the nucleotide sequence of the target region in the genome. The homology arms are required for correct site-specific integration or replacement of the endogenous gene segment by the DNA of the targeting vector. The core region of the targeting construct incorporates the planned genetic modification together with a positive selection cassette, conferring resistance to ES cells containing the targeting construct. Most of the genomic insertions of the targeting vector are random, somewhere in the genome. In a few cases a replacement of the gene segment by the targeting vector takes place. Therefore, enrichment strategies for correctly recombined ES cell clones have been developed. These enrichment strategies make use of negative selection cassettes such as the herpes simplex type 1 thymidine kinase gene or the gene for the diphtheria toxin α-chain. The negative selection cassettes are placed outside the homology arms and are lost during homologous recombination, whereas during random integration the entire targeting vector including the negative selection marker is inserted into the genome of the ES cell. This permits a counterselection against randomly integrated clones and leads to an enrichment of ES cell clones with correct trageting.[33–35] For complex multistep genome manipulations including large chromosomal deletions or translocations convenient combinations of different selection markers have been described.[34]

Most of the published gene targeted mice are summarized in several electronically searchable databases in the internet (see http://www.bioscience.org/knockout/knochome.htm, http://www.nih.gov/science/models/mouse/index.html, and http://www.informatics.jax.org/imsr/index.jsp). In addition to the already existing gene targeted mice, academic institutes and commercial companies have generated gene trapped ES cell libraries that can be used as a source for generating knockout mice. Each individual ES cell clone of such a library harbors a single integration of a viral construct, which in turn serves as a signpost for identification of the trapped gene. Recently, the major gene trapping groups have centralized the access to all publicly available gene trap ES cell lines.[36] In this portal (www.genetrap.org) a collection of at the time 45,000 well-characterized ES cell clones is available on a noncollaborative basis. In addition, gene-trap-derived gene-specific knockout ES cell clones are commercially obtainable (http://www.lexicon-genetics.com/discovery/omnibank_ebiology.htm).

Importantly, studies involving gene targeted mice have to consider the genetic background. Depending on the mouse strain that was used the *in vivo* function or loss of function of a particular gene can be very different.[37,38] For example, γ-protein kinase C knockout mice showed a considerable difference in sensitivity to ethanol that was completely dependent on the genetic background.[39] This example illustrates that the effect of a single genetic alteration can be strongly affected by other genes and that the phenotype of a knockout mice is not a strict readout of the missing or destroyed gene but is the concerted action of the remaining genes. Therefore, in traditional genetics a phenotypic impairment of a gene knockout can be taken as just a hint for the function of a particular gene. The definitive proof is a rescue of the impaired phenotype of the knockout by, e.g., transgenic reexpression of the destroyed gene. However, the transgenic rescue of knockout mice is time consuming, expensive, and labor intensive. A simple shortcut strategy is usually applied to minimize the contribution of the individual genetic background. Knockout mice with the genetic origin of the 126SV mouse strain are crossed with C57BL6 or other inbred strains to generate F_1 hybrids that contain 50% 129SV and C57BL6 as genetic background. Next, the F_1 hybrids are intercrossed to generate knockout, heterozygous, and wild-type littermates. By analyzing knockout and control littermates from the F_1 hybrids, the influence of the genetic background is minimized since it will be very different from animal to animal. However, if this approach is used, the number of analyzed mice has to be high. As an alternative, the phenotype of the knockout can be analyzed in two very different genetic backgrounds, e.g., C57BL6 versus DBA. For this, C57BL6 and DBA congenic animals have to be generated. Congenic animals are genetically identical except for the modified region. However, 2–3 years of backcrossing to the defined mouse strain are necessary to achieve a statistically >99% homogeneous genetic background.[40] The Jackson Laboratory and the Center for Inherited Disease Research offer possible shortcuts for this time-consuming procedure that will approximately half the number of backcrosses (for more information about marker-assisted breeding see http://jaxmice.jax.org/services/speedcongenic.html and http://www.cidr.jhmi.edu/mouse/mouse_strp.html).

CONDITIONAL MOUSE MODELS

In conventional transgenic and gene knockout mice studies the clear-cut gene-function relation is hampered by several factors. Two of those factors were already discussed: the positional variegation and the genetic background. In addition, it has to be considered that the genetic manipulation is set in the early embryo at the one-cell stage. Therefore, the lack or malfunction of the manipulated gene can disturb the development. Consequently, the phenotype in adult mice might be caused by this abnormal development and not by the lack of proper gene function in adults. Likewise, the dysfunction of a particular gene can lead to the up- or downregulation of compensatory genes. Those pleiotropic compensatory effects would than mask a phenotype in adults.

A gene regulation system that permits the induction or deletion of a gene in a specific tissue or cell type at any given time would erase all the issues mentioned above. With a controlled gene switch it would be possible to study the function of a gene in a single animal before and after the gene is switched on or off. In an "off" and "on" state the genetic background and integration site are the same; the animal can develop normally since the gene switch can be turned after development and the time window for compensatory mechanism is narrow.

For this reason, it is obvious that the use of conditional mouse models is a much better choice than relying on conventional transgenic or knockout approaches. To decide which conditional animal model is most suitable, a detailed knowledge of the different established systems is essential. First, it is important to define the criteria that have to be met. The perfect regulatory system (1) should allow a tightly controlled regulation of the target gene without background activity, (2) should be reversible, (3) should be fast in induction kinetics, (4) should be effective in all target cells, and (5) should use a highly specific and nontoxic inducer.

Among the growing number of different conditional *in vivo* systems, the tetracycline (tet) regulated and the Cre-mediated site-specific recombination systems are most commonly used.[1,41,42] The scientific community has made a considerable effort to provide public resources comprising conditional mouse mutants (http://nagy.mshri.on.ca/PubLinks/indexmain.html and http://www.zmg.uni-mainz.de/tetmouse/index.htm). The activity of these initiatives provided the basis for sharing resources, valuable animal strains, and scientific information beyond the scientific scope of individual research projects. Taking a closer look at these resources prior the initiation of a conditional mouse project is therefore good practice and saves time and effort.

THE TET SYSTEM

Pioneering work from Gossen and Bujard established a general application of the *Escherichia coli* Tn10-derived tet resistance operon for regulating transgenic expression in mammalian cells.[43] In the TET system the tet repressor (TetR) is fused to the transcriptional transactivation domain VP16 of the herpes simplex virus, giving rise to the tet-controlled transactivator (tTA). In the absence of tetracycline [or other tetracycline derivatives like doxycycline (DOX), anhydrotetracycline, or 4-epidoxycycline], tTA specifically binds to the tTA-dependent promoter (Ptet) to initiate transcription of the Ptet-controlled transgene. Normally, Ptet is composed of seven copies of the tet operator consensus sequence (tetO) and a transcription initiation site (in most cases from the CMV immediate early promoter) that is placed close to the translational start codon of the Ptet transgene. If tetO7 is flanked by two transcription initiation sites (Ptet-bi), two genes can be regulated simultaneously by tTA.[44] Adding DOX to the system induces a conformational change of the transcriptional activator tTA, prevents its DNA binding, and at the Ptet promoter transcription is shut down. Thus the TET system requires two building blocks: as an effector the tet-dependant transactivator tTA and as responder a Ptet promoter-regulated gene.

The well-established pharmacological properties of tetracycline and its derivatives suggest that these compounds are safe and reliable regulators for tet-controlled gene function *in vivo*. The properties include predictable pharmacokinetics, good bioavailability and tissue distribution, known half-life times, lack of toxicity at the established working concentrations, and the ability to cross mammalian cell membranes, the blood–brain barrier, and the placenta. Until recently the only downside of using tetracycline or tetracycline derivatives in mice was a possible imbalance of the intestinal flora, which may result in diarrhea and to a less extent in colitis due to the antibiotic action of these compounds. In case unwanted antibiotic action of tetracycline or tetracycline derivatives poses a problem, 4-epidoxycycline, which is equally suitable for controlling tet on/off gene switches but lacks antibiotic activity, can be used instead.[45]

Almost immediately after the initial publication of the tTA "TET off" system in tissue culture experiments, the first report of a tet-responsive conditional mouse model provided evidence for the enormous potential of this system for reversibly controlling gene expression *in vivo*.[46] Although this first tet-based mouse model sparked much enthusiasm, certain limitations of the TET system became obvious. Major problems included residual background activity of transgene expression in the off state, cellular toxicity of the original transactivator, low sensitivity to DOX in certain tissues, internal cryptic splice acceptors, suboptimal eukaryotic codon usage, and slow *in vivo* induction kinetics. During the last decade these problems have been addressed and an exquisite toolbox for the generation of tet-based conditional mouse models is available now.

The first major improvement of the TET system was the development of the reverse "TET on" transactivator (rtTA), which is activated and induces transcription of Ptet responder genes upon DOX administration.[47] Using the rtTA system in transgenic mice increased the speed of transgene induction, reaching in some tissues complete Ptet–gene activation in 1 h as compared to the slow activation kinetics of up to 1 week for tTA.[48,49] This marked difference in responsiveness between the tTA and rtTA system *in vivo* is explained by the metabolic half-life time of DOX in different tissues, i.e., by the fact that in certain tissues it will take some time before DOX is completely eliminated. This difference in induction kinetics can be a major determinant when designing tet-based mouse models. Using the tTA "TET off" system might permit fast shut down and slow induction kinetics contrasting with the rtTA "TET on" system suitable for immediate induction and slow extinction of tet-controlled transgene expression.

An additional initial drawback of the TET system was the occasionally observed leakiness leading to unwanted Ptet–gene expression. The Ptet leakiness was addressed by developing antagonistic repressors that silence any residual Ptet–gene expression in the noninduced state. The antagonistic repressors are heterologous fusions of a tet-responsive DNA-binding domain with a strong transcriptional repressor, capable of silencing the residual activity of a Ptet[50,50a]. This strategy uses an antagonistic combination of tTA and the repressor that—depending on the administration of DOX and either tTA or the repressor—exclusively binds to tetO7. One way to avoid leakiness of Ptet is the generation of a tet repressor mouse line that ubiquitously expresses the tet-responsive repressor and the crossbreeding of this Ptet silencer line to the leaky tet-responder mouse[51] (unpublished data from our laboratory). Alternatively, it is also possible to use constructs that express both a transactivator and an antagonizing repressor.[51–54] Finally, engineered responder cassettes containing repositioned TetO elements have also been reported to minimize leakiness of the TET regulatory system.[55,56]

High tTA and rtTA expression can lead to cytotoxicity, most probably due to pleiotropic effects of high levels of the VP16 transactivation domain.[57] Therefore, a second generation of tTAs and rtTAs was developed.[58–60] Remodeling tTA and rtTA improved several features of the TET system including increased DOX sensitivity, faster activation kinetics, improved codon usage, abrogation of cryptic splice sites, and extended DNA-binding and dimerization properties.[56,61,62] These remarkable advances helped to optimize the design of conditional mouse models and to adjust the experimental setup according to the specific needs of the planned experiment.

Since the first report of a tet-controlled transgenic mouse model, the TET system has become increasingly popular. Today the huge number of tet-based mouse models represent an invaluable resource for combinations of tissue-specific or generalized tTA (rtTA) expressing and Ptet responder mice. However, the scope of this chapter does not allow for a detailed description of already generated "tet on/off" mouse models. The reader is encouraged to visit the specialized electronic databases as an additional resource (http://www.zmg.uni-mainz.de and http://www.tetsystems.com).

Besides its fundamental appeal as an established conditional regulatory system in mice, the "TET on/off" system has attracted considerable interest because of its unique potential of reversible gene regulation. Consequently, the TET system will be an excellent choice for the study of gene function relationships, for example, in behavioral neuroscience.

SITE-SPECIFIC RECOMBINASES

The first site-specific recombination system that was used in the mouse was the Cre/loxP system.[63,64] Initially discovered in the filamentous P1 phage, the Cre/loxP system has become a valuable tool for the induction of site-specific DNA recombination *in vivo*.[65] Based on the enzymatic activity of Cre (for "causes recombination") and depending upon the position and orientation of loxP sites (for "locus of crossover P1"), mice carrying gene deletions, duplications, inversions, or chromosomal translocations can be generated.[66,67] In a similar way the *Saccharomyces cerevisiae* recombinase Flp can be used for site-specific recombination between FRT (for "Flp recombination target") sites in mice.[58] Both the Cre and Flp recombinases have been applied for the induction of ubiquitous or cell-specific recombination in mice.

The success of any recombinase-mediated conditional *in vivo* experiment depends on the "DELETER" strains that express the Cre recombinase. However, some of Cre expressing mouse lines are "leaky," with widespread instead of specific Cre-mediated recombination. Consequently, recombination will not be restricted to the anticipated cell type but will also take place elsewhere. In most cases this effect is due to the early developmental activity of the promoter driving Cre expression. Drug-inducible Cre recombinases can bypass unwanted Cre expression during development and can be used as conditional gene switches *in vivo*. This can be achieved by tTA-controlled Cre expression in triple transgenic mice. These mice harbor a tissue-specific tTA transgene, a PtetO controlled Cre, and the floxed target gene.[69,70] Another option is Cre recombinase fusions with mutant estrogen receptor binding domains that selectively bind 17β-estradiol analogues like tamoxifen.[41,71] In this scenario Cre–estrogen receptor fusion proteins are retained in the cytoplasm but are translocated to the nucleus upon addition of the synthetic ligand.

As in the TET system, protein engineering provided better recombinases including codon improved Cre and Flp enzymes, Flp recombinases with better *in vivo* activities in mammals, and modified recombinases recognizing novel target sites.[68,72–75] Again, before starting experiments, writing grants, and planning the generation of novel transgenic or knockout mice, it will be rewarding to have a closer look at already established resources (http://nagy.mshri.on.ca/PubLinks/indexmain.html).

The application of mouse lines with low Cre activity revealed that an incomplete rearrangement of the target sequence can also depend on the accessibility of the loxP sites in the target gene.[76] In addition, the observed recombination efficiencies at any given Cre indicator locus can be taken as only a first hint for the usefulness of a particular recombination system. Therefore, it is good practice to analyze the efficiency of the Cre-mediated recombination at the target gene for each series of experiments.

RNA INTERFERENCE

RNA interference (RNAi) is a sequence-specific eukaryotic surveillance mechanism initiated by double-stranded RNA that ultimately will result in the repression of specific target genes.[77,78] Building on the observation that in mammalian cells double-stranded RNAs (dsRNAs) of less than 30 bp will induce specific degradation of a target mRNA but will not trigger a general translational block or interferon response, RNAi knockdown technology has become very popular.[79]

In 2002 Hasuwa and colleagues reported the generation of knockdown mice and rats, thus providing direct evidence that RNAi technology can be used for studying the functional consequences of gene inactivation in animal models.[80] The relative simplicity, speed, and cost-effectiveness of using RNAi for studying gene function *in vivo* have sparked great expectations and have led to the development of an increasing number of novel tools.[2,81–84] However, there are several important issues to be considered. First, *in vivo* knockdown strategies might significantly deplete the mRNA levels of the targeted gene but might not be sufficient to abolish gene function. Second, in different tissues the level of knockdown might be variable, also being determined by the targeted sequence, copy number of the inserted RNAi transgene, and its particular integration site[85] (unpublished observations from our laboratory). Nevertheless, a partial knockdown of a gene function can be of advantage since "mild" phenotypes might in some cases be more suitable for uncovering gene function as compared to a knockout.[86] Third, an additional concern is specificity. Indeed, RNAi-mediated knockdown may not be limited to the selected target gene but instead may also affect the function of other genes, an effect known as off-targeting.[87] This issue is further compounded by the fact that our present general understanding of the different mechanisms involving short RNAs, like micro-RNAs (miRNAs) or repeat-associated short interfering RNAs (rasiRNAs), remains sketchy. For this reason, the possibility that a particular small RNA might inadvertently bring on unwanted gene regulatory effects other than the sequence-specific knockdown of the target gene cannot be excluded. For this reason, it will be essential to implement workable and standardized controls before interpreting and validating RNAi data.[88] Finally, high-level shRNA or miRNA expression might compete for limiting cellular factors and thus severely interfere with endogenous siRNA regulatory pathways.[89]

Two alternative strategies have been used for gene selective knockdown in mice: cytoplasmic delivery of synthetic short siRNA oligonucleotides mimicking the active intermediate of an endogenous RNAi mechanism or the stable expression of recombinant shRNAs or miRNAs . The first approach uses small molecules, expression constructs, or viral particles that are injected or infused in a more or less transient fashion.[90–96] Possible applications for transient knockdown experiments in mice are the testing of siRNAs as gene therapeutic tools for humans or their use in antiviral treatments.[82,97,98]

For many applications, however, the use of germ-line transmitting knockdown mice might be the preferred option. In this case the knockdown can be either mediated through ubiquitous expression of shRNAs using RNA polymerase III or tRNA promoters or, alternatively, through miRNA expression from housekeeping or tissue-restricted RNA polymerase II promoters. During the past years several groups have reported the generation of germ-line-transmitting knockdown mouse lines.[99–101] The methodology used for the generation of these mice is similar to the approaches discussed above including standard transgenesis or gene targeting. Likewise, conditional activation of shRNA or miRNA expression in mice has been demonstrated.[13,14,102–104] With conditional RNAi systems knockdown-induced lethality can be bypassed and loss of gene function can be studied during any given time window and/or in specific tissues.

At present, possible avenues for implementing conditional knockdown in mice either allow the permanent induction/extinction of the knockdown (Cre/loxP-based systems) or facilitate the reversible induction of a knockdown (tet on/off-based systems).

The first strategy reported for inducing conditional knockdown phenotypes in mice relied on Cre-mediated activation or extinction of shRNA expression.[14,102,103] In this situation a transcriptional and translational stop element (loxP-STOP-loxP) is placed between the shRNA transcription start site and the upstream regulatory elements, needed for shRNA expression. Upon removal of the loxP-STOP-loxP element by Cre, transcription of the shRNA or miRNA is set off and the knockdown is initiated. Conversely, this strategy also allows for conditionally removing an actively transcribing shRNA expression element and thus permanently extinguishing the knockdown.[14]

The alternative approach for generating conditional knockdown mice is to use drug-controlled systems. In this respect TET systems were applied for regulated shRNA and miRNA expression in mice. In a recent publication Szulc and colleagues demonstrated reversible silencing of an RNA polymerase III-transcribed shRNA by virtue of a DOX-dependent KRAB repressor.[13] Although RNA polymerase III promoter-based conditional strategies are normally efficient in all cell types and provide good knockdown levels, a limitation of these systems is their lack of tissue specificity. The development of RNA polymerase II promoter-mediated miRNA-based shRNA TET on/off systems overcomes this limitation and initial *in vivo* experiments demonstrated an application of this approach.[105] Similarly, recent documentation of second-generation shRNA- or miRNA-based libraries that cover most if not all predicted mouse genes provides an excellent resource for constructing conditional RNAi mouse models.[106,107]

CONCLUDING REMARKS

Within the last few years, researchers have expanded the toolbox to manipulate the expression of mouse genes by developing several novel techniques, including RNAi knock-down, improved tools for conditional gene regulation *in vivo*, and the ability to engineer large BAC clones by "recombineering". In addition, searchable databases, providing detailed information on established mouse lines, together with available collections of pre-characterized ES cell clones and libraries, will facilitate the use of genetically modified mice. Although, this chapter only provides a general overview on the use and limitations of mouse models, we hope the here provided information will serve as a primer illustrating the exciting possibilities and exquisite tools available for *in vivo* investigating gene function in mice.

REFERENCES

1. Bockamp E, Maringer M, Spangenberg C, *et al.* Of mice and models: Improved animal models for biomedical research. *Physiol Genomics* 2002;11:115–132.
2. Prawitt D, Brixel L, Spangenberg C, *et al.* RNAi knock-down mice: An emerging technology for post-genomic functional genetics. *Cytogenet Genome Res* 2004;105:412–421.
3. Nagy A, Gertsenstein M, Vintersten R, Behringer M. *Manipulating the Mouse Embryo—A Laboratory Manual*, 3rd ed. Cold Spring Harbor, NY: Cold Spring Harbor Press, 2002.
4. Sung YH, Song J, Lee HW. Functional genomics approach using mice. *J Biochem Mol Biol* 2004;37:122–132.
5. Ristevski S. Making better transgenic models: Conditional, temporal, and spatial approaches. *Mol Biotechnol* 2005;29:153–163.
6. Glaser S, Anastassiadis K, Stewart AF. Current issues in mouse genome engineering. *Nat Genet* 2005;37:1187–1193.
7. Gordon JW, Scangos GA, Plotkin DJ, Barbosa JA, Ruddle FH. Genetic transformation of mouse embryos by microinjection of purified DNA. *Proc Natl Acad Sci USA* 1980;77:7380–7384.
8. Rulicke T. Pronuclear microinjection of mouse zygotes. *Methods Mol Biol* 2004;254:165–194.
9. Efrat S, Leiser M, Wu YJ, *et al.* Ribozyme-mediated attenuation of pancreatic beta-cell glucokinase expression in transgenic mice results in impaired glucose-induced insulin secretion. *Proc Natl Acad Sci USA* 1994;91:2051–2055.
10. Erickson RP. Antisense transgenics in animals. *Methods* 1999;18:304–310.
11. Kashani-Sabet M, Liu Y, Fong S, *et al.* Identification of gene function and functional pathways by systemic plasmid-based ribozyme targeting in adult mice. *Proc Natl Acad Sci USA* 2002;99:3878–3883.
12. Larsson S, Hotchkiss G, Andang M, *et al.* Reduced beta 2-microglobulin mRNA levels in transgenic mice expressing a designed hammerhead ribozyme. *Nucleic Acids Res* 1994;22:2242–2248.
13. Szulc J, Wiznerowicz M, Sauvain MO, Trono D, Aebischer P. A versatile tool for conditional gene expression and knockdown. *Nat Methods* 2006;3:109–116.
14. Ventura A, Meissner A, Dillon CP, *et al.* Cre-lox-regulated conditional RNA interference from transgenes. *Proc Natl Acad Sci USA* 2004;101:10380–10385.
15. Stegmeier F, Hu G, Rickles RJ, Hannon GJ, Elledge SJ. A lentiviral microRNA-based system for single-copy polymerase II-regulated RNA interference in mammalian cells. *Proc Natl Acad Sci USA* 2005;102:13212–13217.
16. Sanchez MJ, Bockamp EO, Miller J, Gambardella L, Green AR. Selective rescue of early haematopoietic progenitors in Scl(-/-) mice by expressing Scl under the control of a stem cell enhancer. *Development* 2001;128:4815–4827.
17. Navas PA, Swank RA, Yu M, Peterson KR, Stamatoyannopoulos G. Mutation of a transcriptional motif of a distant regulatory element reduces the expression of embryonic and fetal globin genes. *Hum Mol Genet* 2003;12:2941–2948.
18. Sinclair AM, Gottgens B, Barton LM, *et al.* Distinct 5′ SCL enhancers direct transcription to developing brain, spinal cord, and endothelium: Neural expression is mediated by GATA factor binding sites. *Dev Biol* 1999;209:128–142.
19. Wilson C, Bellen HJ, Gehring WJ. Position effects on eukaryotic gene expression. *Annu Rev Cell Biol* 1990;6:679–714.
20. Bronson SK, Plaehn EG, Kluckman KD, Hagaman JR, Maeda N, Smithies O. Single-copy transgenic mice with chosen-site integration. *Proc Natl Acad Sci USA* 1996;93:9067–9072.
21. Giraldo P, Rival-Gervier S, Houdebine LM, Montoliu L. The potential benefits of insulators on heterologous constructs in transgenic animals. *Transgenic Res* 2003;12:751–755.
22. Sparwasser T, Gong S, Li JY, Eberl G. General method for the modification of different BAC types and the rapid generation of BAC transgenic mice. *Genesis* 2004;38:39–50.
23. Giraldo P, Montoliu L. Size matters: Use of YACs, BACs and PACs in transgenic animals. *Transgenic Res* 2001;10:83–103.
24. Gong S, Yang XW, Li C, Heintz N. Highly efficient modification of bacterial artificial chromosomes (BACs) using novel shuttle vectors containing the R6Kgamma origin of replication. *Genome Res* 2002;12:1992–1998.
25. Lee EC, Yu D, Martinez de Velasco J, *et al.* A highly efficient *Escherichia coli*-based chromosome engineering system adapted for recombinogenic targeting and subcloning of BAC DNA. *Genomics* 2001;73:56–65.
26. Wang J, Sarov M, Rientjes J, *et al.* An improved recombineering approach by adding RecA to lambda Red recombination. *Mol Biotechnol* 2006;32:43–53.
27. Warming S, Costantino N, Court DL, Jenkins NA, Copeland NG. Simple and highly efficient BAC recombineering using galK selection. *Nucleic Acids Res* 2005;33:e36.
28. Rijkers T, Peetz A, Ruther U. Insertional mutagenesis in transgenic mice. *Transgenic Res* 1994;3:203–215.
29. Evans MJ, Kaufman MH. Establishment in culture of pluripotential cells from mouse embryos. *Nature* 1981;292:154–156.
30. Martin GR. Isolation of a pluripotent cell line from early mouse embryos cultured in medium conditioned by teratocarcinoma stem cells. *Proc Natl Acad Sci USA* 1981;78:7634–7638.
31. Smithies O, Gregg RG, Boggs SS, Koralewski MA, Kucherlapati RS. Insertion of DNA sequences into the human chromosomal beta-globin locus by homologous recombination. *Nature* 1985;317: 230–234.
32. Thomas KR, Capecchi MR. Site-directed mutagenesis by gene targeting in mouse embryo-derived stem cells. *Cell* 1987;51: 503–512.
33. Borrelli E, Heyman R, Hsi M, Evans RM. Targeting of an inducible toxic phenotype in animal cells. *Proc Natl Acad Sci USA* 1988;85: 7572–7576.
34. Chen YT, Bradley A. A new positive/negative selectable marker, puDeltatk, for use in embryonic stem cells. *Genesis* 2000;28: 31–35.
35. Yanagawa Y, Kobayashi T, Ohnishi M, *et al.* Enrichment and efficient screening of ES cells containing a targeted mutation: The use of DT-A gene with the polyadenylation signal as a negative selection maker. *Transgenic Res* 1999;8:215–221.
36. Nord AS, Chang PJ, Conklin BR, *et al.* The International Gene Trap Consortium Website: A portal to all publicly available gene trap cell lines in mouse. *Nucleic Acids Res* 2006;34:D642–648.
37. Phillips TJ, Hen R, Crabbe JC. Complications associated with genetic background effects in research using knockout mice. *Psychopharmacology (Berl)* 1999;147:5–7.
38. Sanford LP, Kallapur S, Ormsby I, Doetschman T. Influence of genetic background on knockout mouse phenotypes. *Methods Mol Biol* 2001;158:217–225.
39. Bowers BJ, Owen EH, Collins AC, Abeliovich A, Tonegawa S, Wehner JM. Decreased ethanol sensitivity and tolerance development in gamma-protein kinase C null mutant mice is dependent on genetic background. *Alcohol Clin Exp Res* 1999;23:387–397.
40. Sigmund CD. Viewpoint: Are studies in genetically altered mice out of control? *Arterioscler Thromb Vasc Biol* 2000;20:1425–1429.
41. Garcia-Otin AL, Guillou F. Mammalian genome targeting using site-specific recombinases. *Front Biosci* 2006;11:1108–1136.
42. Mallo M. Controlled gene activation and inactivation in the mouse. *Front Biosci* 2006;11:313–327.
43. Gossen M, Bujard H. Tight control of gene expression in mammalian cells by tetracycline-responsive promoters. *Proc Natl Acad Sci USA* 1992;89:5547–5551.
44. Baron U, Freundlieb S, Gossen M, Bujard H. Co-regulation of two gene activities by tetracycline via a bidirectional promoter. *Nucleic Acids Res* 1995;23:3605–3606.
45. Eger K, Hermes M, Uhlemann K, *et al.* 4-Epidoxycycline: An alternative to doxycycline to control gene expression in conditional mouse models. *Biochem Biophys Res Commun* 2004;323:979–986.
46. Furth PA, St Onge L, Boger H, *et al.* Temporal control of gene expression in transgenic mice by a tetracycline-responsive promoter. *Proc Natl Acad Sci USA* 1994;91:9302–9306.

47. Gossen M, Freundlieb S, Bender G, Muller G, Hillen W, Bujard H. Transcriptional activation by tetracyclines in mammalian cells. *Science* 1995;268:1766–1769.
48. Hasan MT, Schonig K, Berger S, Graewe W, Bujard H. Long-term, noninvasive imaging of regulated gene expression in living mice. *Genesis* 2001;29:116–122.
49. Kistner A, Gossen M, Zimmermann F, *et al.* Doxycycline-mediated quantitative and tissue-specific control of gene expression in transgenic mice. *Proc Natl Acad Sci USA* 1996;93:10933–10938.
50. Freundlieb S, Schirra-Muller C, Bujard H. A tetracycline controlled activation/repression system with increased potential for gene transfer into mammalian cells. *J Gene Med* 1999;1:4–12.
50a. Bockamp E, Christel C, Hameyer D, *et al.* Generation and characterization of tTS-H4: a novel transcriptional repressor that is compatible with the reverse tetracycline-controlled TET-ON system. *J Gene Med* 2007;9:308–318.
51. Zhu Z, Ma B, Homer RJ, Zheng T, Elias JA. Use of the tetracycline-controlled transcriptional silencer (tTS) to eliminate transgene leak in inducible overexpression transgenic mice. *J Biol Chem* 2001; 276:25222–25229.
52. Hayakawa T, Yusa K, Kouno M, Takeda J, Horie K. Bloom's syndrome gene-deficient phenotype in mouse primary cells induced by a modified tetracycline-controlled trans-silencer. *Gene* 2006; 369:80–89.
53. Forster K, Helbl V, Lederer T, Urlinger S, Wittenburg N, Hillen W. Tetracycline-inducible expression systems with reduced basal activity in mammalian cells. *Nucleic Acids Res* 1999;27:708–710.
54. Uchida S, Sakai S, Furuichi T, *et al.* Tight regulation of transgene expression by tetracycline-dependent activator and repressor in brain. *Genes Brain Behav* 2006;5:96–106.
55. Pluta K, Luce MJ, Bao L, Agha-Mohammadi S, Reiser J. Tight control of transgene expression by lentivirus vectors containing second-generation tetracycline-responsive promoters. *J Gene Med* 2005;7:803–817.
56. Krueger C, Danke C, Pfleiderer K, *et al.* A gene regulation system with four distinct expression levels. *J Gene Med* 2006;8(8): 1037–1047.
57. Izumi M, Gilbert DM. Homogeneous tetracycline-regulatable gene expression in mammalian fibroblasts. *J Cell Biochem* 1999; 76:280–289.
58. Akagi K, Kanai M, Saya H, Kozu T, Berns A. A novel tetracycline-dependent transactivator with E2F4 transcriptional activation domain. *Nucleic Acids Res* 2001;29:E23.
59. Baron U, Gossen M, Bujard H. Tetracycline-controlled transcription in eukaryotes: Novel transactivators with graded transactivation potential. *Nucleic Acids Res* 1997;25:2723–2729.
60. Urlinger S, Helbl V, Guthmann J, Pook E, Grimm S, Hillen W. The p65 domain from NF-kappaB is an efficient human activator in the tetracycline-regulatable gene expression system. *Gene* 2000;247: 103–110.
61. Krueger C, Berens C, Schmidt A, Schnappinger D, Hillen W. Single-chain Tet transregulators. *Nucleic Acids Res* 2003;31:3050–3056.
62. Urlinger S, Baron U, Thellmann M, Hasan MT, Bujard H, Hillen W. Exploring the sequence space for tetracycline-dependent transcriptional activators: Novel mutations yield expanded range and sensitivity. *Proc Natl Acad Sci USA* 2000;97:7963–7968.
63. Lakso M, Sauer B, Mosinger B Jr, *et al.* Targeted oncogene activation by site-specific recombination in transgenic mice. *Proc Natl Acad Sci USA* 1992;89:6232–6236.
64. Orban PC, Chui D, Marth JD. Tissue- and site-specific DNA recombination in transgenic mice. *Proc Natl Acad Sci USA* 1992;89: 6861–6865.
65. Hamilton DL, Abremski K. Site-specific recombination by the bacteriophage P1 lox-Cre system. Cre-mediated synapsis of two lox sites. *J Mol Biol* 1984;178:481–486.
66. van der Weyden L, Adams DJ, Bradley A. Tools for targeted manipulation of the mouse genome. *Physiol Genomics* 2002;11:133–164.
67. Branda CS, Dymecki SM. Talking about a revolution: The impact of site-specific recombinases on genetic analyses in mice. *Dev Cell* 2004;6:7–28.
68. Rodriguez CI, Buchholz F, Galloway J, *et al.* High-efficiency deleter mice show that FLPe is an alternative to Cre-loxP. *Nat Genet* 2000;25:139–140.
69. Shimshek DR, Jensen V, Celikel T, *et al.* Forebrain-specific glutamate receptor B deletion impairs spatial memory but not hippocampal field long-term potentiation. *J Neurosci* 2006;26: 8428–8440.
70. Schonig K, Schwenk F, Rajewsky K, Bujard H. Stringent doxycycline dependent control of CRE recombinase in vivo. *Nucleic Acids Res* 2002;30:e134.
71. Brocard J, Warot X, Wendling O, *et al.* Spatio-temporally controlled site-specific somatic mutagenesis in the mouse. *Proc Natl Acad Sci USA* 1997;94:14559–14563.
72. Shimshek DR, Kim J, Hubner MR, *et al.* Codon-improved Cre recombinase (iCre) expression in the mouse. *Genesis* 2002;32:19–26.
73. Buchholz F, Stewart AF. Alteration of Cre recombinase site specificity by substrate-linked protein evolution. *Nat Biotechnol* 2001;19: 1047–1052.
74. Koresawa Y, Miyagawa S, Ikawa M, *et al.* A new Cre recombinase gene based on optimal codon usage in mammals: A powerful material for organ-specific gene targeting. *Transplant Proc* 2000;32: 2516–2517.
75. Santoro SW, Schultz PG. Directed evolution of the site specificity of Cre recombinase. *Proc Natl Acad Sci USA* 2002;99:4185–4190.
76. Vooijs M, Jonkers J, Berns A. A highly efficient ligand-regulated Cre recombinase mouse line shows that LoxP recombination is position dependent. *EMBO Rep* 2001;2:292–297.
77. Meister G, Tuschl T. Mechanisms of gene silencing by double-stranded RNA. *Nature* 2004; 31:343–349.
78. Tomari Y, Zamore PD. Perspective: Machines for RNAi. *Genes Dev* 2005;19:517–529.
79. Elbashir SM, Harborth J, Lendeckel W, Yalcin A, Weber K, Tuschl T. Duplexes of 21-nucleotide RNAs mediate RNA interference in cultured mammalian cells. *Nature* 2001;411:494–498.
80. Hasuwa H, Kaseda K, Einarsdottir T, Okabe M. Small interfering RNA and gene silencing in transgenic mice and rats. *FEBS Lett* 2002;532:227–230.
81. Sandy P, Ventura A, Jacks T. Mammalian RNAi: A practical guide. *Biotechniques* 2005;39:215–224.
82. Lu PY, Xie F, Woodle MC. In vivo application of RNA interference: From functional genomics to therapeutics. *Adv Genet* 2005;54: 117–142.
83. Coumoul X, Deng CX. RNAi in mice: A promising approach to decipher gene functions in vivo. *Biochimie* 2006;88:637–643.
84. Wiznerowicz M, Szulc J, Trono D. Tuning silence: Conditional systems for RNA interference. *Nat Methods* 2006;3:682–688.
85. Hemann MT, Fridman JS, Zilfou JT, *et al.* An epi-allelic series of p53 hypomorphs created by stable RNAi produces distinct tumor phenotypes in vivo. *Nat Genet* 2003;33:396–400.
86. Rosenbauer F, Wagner K, Kutok JL, *et al.* Acute myeloid leukemia induced by graded reduction of a lineage-specific transcription factor, PU.1. *Nat Genet* 2004;36:624–630.
87. Jackson AL, Bartz SR, Schelter J, *et al.* Expression profiling reveals off-target gene regulation by RNAi. *Nat Biotechnol* 2003;21:635–637.
88. Huppi K, Martin SE, Caplen NJ. Defining and assaying RNAi in mammalian cells. *Mol Cell* 2005;17:1–10.
89. Grimm D, Streetz KL, Jopling CL, *et al.* Fatality in mice due to oversaturation of cellular microRNA/short hairpin RNA pathways. *Nature* 2006;441:537–541.
90. Matsuda T, Cepko CL. Electroporation and RNA interference in the rodent retina in vivo and in vitro. *Proc Natl Acad Sci USA* 2004;101:16–22.
91. Hommel JD, Sears RM, Georgescu D, Simmons DL, DiLeone RJ. Local gene knockdown in the brain using viral-mediated RNA interference. *Nat Med* 2003;9:1539–1544.

92. Xia H, Mao Q, Paulson HL, Davidson BL. siRNA-mediated gene silencing in vitro and in vivo. *Nat Biotechnol* 2002;20:1006–1010.
93. Kishida T, Asada H, Gojo S, *et al.* Sequence-specific gene silencing in murine muscle induced by electroporation-mediated transfer of short interfering RNA. *J Gene Med* 2004;6:105–110.
94. Song E, Lee SK, Wang J, *et al.* RNA interference targeting Fas protects mice from fulminant hepatitis. *Nat Med* 2003;9:347–351.
95. McCaffrey AP, Nakai H, Pandey K, *et al.* Inhibition of hepatitis B virus in mice by RNA interference. *Nat Biotechnol* 2003;21: 639–644.
96. Soutschek J, Akinc A, Bramlage B, *et al.* Therapeutic silencing of an endogenous gene by systemic administration of modified siRNAs. *Nature* 2004;432:173–178.
97. de Fougerolles A, Manoharan M, Meyers R, Vornlocher HP. RNA interference in vivo: Toward synthetic small inhibitory RNA-based therapeutics. *Methods Enzymol* 2005;392:278–296.
98. Leonard JN, Schaffer DV. Antiviral RNAi therapy: Emerging approaches for hitting a moving target. *Gene Ther* 2006;13: 532–540.
99. Kunath T, Gish G, Lickert H, Jones N, Pawson T, Rossant J. Transgenic RNA interference in ES cell-derived embryos recapitulates a genetic null phenotype. *Nat Biotechnol* 2003;21:559–561.
100. Rubinson DA, Dillon CP, Kwiatkowski AV, *et al.* A lentivirus-based system to functionally silence genes in primary mammalian cells, stem cells and transgenic mice by RNA interference. *Nat Genet* 2003;33:401–406.
101. Tiscornia G, Singer O, Ikawa M, Verma IM. A general method for gene knockdown in mice by using lentiviral vectors expressing small interfering RNA. *Proc Natl Acad Sci USA* 2003;100: 1844–1848.
102. Coumoul X, Shukla V, Li C, Wang RH, Deng CX. Conditional knockdown of Fgfr2 in mice using Cre-LoxP induced RNA interference. *Nucleic Acids Res* 2005;33:e102.
103. Chang HS, Lin CH, Chen YC, Yu WC. Using siRNA technique to generate transgenic animals with spatiotemporal and conditional gene knockdown. *Am J Pathol* 2004;165:1535–1541.
104. Xia XG, Zhou H, Samper E, Melov S, Xu Z. Pol II-expressed shRNA knocks down Sod2 gene expression and causes phenotypes of the gene knockout in mice. *PLoS Genet* 2006;2:e10.
105. Dickins RA, Hemann MT, Zilfou JT, *et al.* Probing tumor phenotypes using stable and regulated synthetic microRNA precursors. *Nat Genet* 2005;37:1289–1295.
106. Moffat J, Grueneberg DA, Yang X, *et al.* A lentiviral RNAi library for human and mouse genes applied to an arrayed viral high-content screen. *Cell* 2006;124:1283–1298.
107. Silva JM, Li MZ, Chang K, *et al.* Second-generation shRNA libraries covering the mouse and human genomes. *Nat Genet* 2005;37:1281–1288.

General Considerations

II

4 The Ethical Basis for Animal Use in Research

JAMES PARKER

ABSTRACT

The concern of biomedical researchers for the well-being of laboratory animals reflects a consensus that animals are conscious subjects. Partisans of the animal rights movement believe that researchers must go beyond the acknowledgment of consciousness and consequent attention to animal well-being. They should recognize just how similar human and animal consciousness are, and then address animal interests and animal rights. Consideration of these issues would challenge all uses, even painless uses, of animals. In any effort to identify the interests and rights of animals, even animal rights philosophers admit that the nature of consciousness and cognition—human and animal—matters. A theory of human consciousness and cognition developed by B.F.J. Lonergan is helpful in understanding and comparing human and animal consciousness. According to Lonergan, elemental wonder makes human consciousness and cognition a dynamic, self-assembling process moving from presentations given in experience through successive levels of understanding, judgment, and responsibility to the affirmation of values. Questioning sweeps humans across a divide between elementary knowing, which is shared with animals, and a type of knowing that is exclusively human. None of the animal behaviors catalogued by animal rights partisans reveals wonder and the drive to understand, affirm and decide. That drive is a single, restless activity generating successive levels of consciousness. If, as animal rights philosophers agree, animals are incapable of responsible behavior, it must also be that their cognitional feats are qualitatively different from those of humans. Absent questioning and the drive to understand, animals never emerge on the level of intelligent and rational, much less responsible, consciousness. Their cognitional achievements appear to fall within the province of elementary knowing, a realm in which they are accomplished associative learners—clever, to be sure, but not capable of the beginnings of full human knowing. Human consciousness and cognition differ enough from animal consciousness and cognition that humans claim rights while animals do not. So it is that even if we use animals in research, having vouched for their humane care, we use humans as experimental subjects only when they give informed consent. We safeguard the welfare of animals; we guarantee the rights of persons.

From: *Sourcebook of Models for Biomedical Research*
(P. M. Conn, ed.),

Key Words: Ethics, Consciousness, Animals, Biomedical, Research.

INTRODUCTION

If René Descartes (1596–1650) had not existed, animal rights philosophers would have had to invent him. They pillory this famous mathematician/philosopher for theorizing that animal operations are mechanical, not conscious—more like the movements of swans, propelled by water pressure, in the royal gardens of Versailles than like those of humans, directed by rational choices. Such a mechanistic understanding of animals, they assert, has provided justification for using them in biomedical research, even in painful experiments—something that we would never do to humans.

All philosophers today, however, are as certain as any pet owner that animals are endowed with consciousness. Except for viruses, bacteria, and, probably, insects and crustaceans, animals are aware of themselves and of the world around them that is made present by internal representations. Their movements and communication are more than the workings of mechanical stimulus and response.

Biomedical scientists are no less convinced about the existence of animal consciousness. In their professional organizations, in the laws and regulations they help draft, and in their laboratory deliberations they are keenly aware of issues about the humane treatment of research subjects that animal consciousness raises.[1] They follow and aid in advancing current "best practices" for ensuring animal well-being, keenly aware that in research institutions as well as in businesses or schools, there is always room for improving procedures.

On the topic of the existence of animal consciousness, Descartes has no contemporary disciples, philosophical or scientific. Instead, controversy swirls around the nature and degree of that consciousness. Several philosophers believe that we must go beyond an acknowledgment of consciousness and a concern for animal well-being and take into account the human-like nature of animal consciousness, address the issue of animal interests and animal rights, and challenge any use of animals, painless though it might be.

ANIMAL INTERESTS

The reputed father of the animal rights movement, Peter Singer, follows Jeremy Bentham (1748–1832), in positing that ethical reflection arises from our empathy with others in their pleasure or pain.[2] When ethics is based on calculations of pleasure and pain it includes responsibilities not just to rational humans, but also to any creatures that experience pleasure and pain. Singer quotes a Bentham proclamation that has become "The Great Sentence" of the animal rights movement: "The question is not, Can they reason? nor Can they talk? but, Can they suffer?"[3]

Bentham proposed that we align ourselves with a hypothetical observer who is impartial, benevolent, and capable of discerning every consequence of a given action. From such a position we would be able to choose actions that achieve the greatest utility—defined as the maximum pleasure and minimum pain—for the greatest number of individuals. In our calculations, or course, "Each [individual] is to count for one and none for more than one."

Singer extends Bentham's method to animals. He does not claim that humans and animals are equal or demand that we treat them equally. He alleges that most members of the chromosomal species *Homo sapiens* are persons—they possess rationality and self-consciousness—but most animals are not. Conversely, some chromosomal humans, such as the mentally disadvantaged and senile, are not persons, while some animals such as chimpanzees clearly are.[4] Persons or not, all suffer. Consequently, even though we need not treat each sentient creature equally—it makes no sense to extend the vote to dogs—we must consider the potential pleasures and pains of each equally as we ponder how to treat them. If we give special consideration to certain individuals solely on the basis of their membership in the human species we act as "speciesists." Singer concludes:

> If the experimenter is not prepared to use an orphaned human infant, then his readiness to use non-humans is simple discrimination, since adult apes, cats, mice, and other mammals are more aware of what is happening to them, more self-directing and, so far as we can tell, at least as sensitive to pain, as any human infant.[5]

Is it permissible, then, to experiment on and/or kill anesthetized subjects? To answer this question, Singer expands Bentham's principle of pleasure and pain. He counsels us to consider not just the pleasures and pains, but also the interests and harms of those affected by any action. We can best determine interests and harms by taking account of what would be individuals' preferences. No sentient subject would prefer being experimented on and killed, no matter how painless the procedures.

Because human consciousness is more complex—human pleasures are fuller and human futures richer—the harm and the wrong involved in overriding preferences are usually greater with humans than with animals. Still, the mental life and interests of some human infants are on par with that of many animals. Consequently, we must, if we intend to experiment on and kill animals, be prepared to do the same to those babies.[6]

Singer's arguments have provoked several lines of response.[7] One contends that measuring potential pleasures and pains, interests and harms that might derive from certain actions cannot serve as a reliable guide to moral decision-making. There is a limit to how well scientists can peer into the future. Seldom can they see clearly the connection between a given experiment and a new treatment, or forecast with accuracy the number of its potential beneficiaries. Scientists are explorers, not clairvoyants.[8]

Fellow traveler in the animal rights movement, philosopher Tom Regan, makes a second critique, faulting Singer for the "morally callous" idea that the ends justify the means. When Singer proposed that some human/animal sexual acts might not be wrong—after all, they could be mutually satisfactory—a piqued Regan responded:

> "The end does not justify the means" is a moral truth that applies beyond the boundaries of our species . . . As a utilitarian, he [Singer] believes that right and wrong depend on how much satisfaction results from our actions, an outlook that leads him to accept many practices that advocates of animal rights reject.[9]

Regan also notes that because utilitarian ethics aggregates interests and harms—the total of harms weighed against the sum of interests—it is a rickety defense for small numbers of laboratory animals, whose harms might be outweighed by the benefits achieved for many humans.[10]

Still another response asks if Singer doesn't undermine his utilitarian starting point by allowing that the harm of killing humans is usually greater than that of killing animals. He makes it appear, after all, that even for him the nature of consciousness and cognition matters very much in ethical theory.

ANIMAL CONSCIOUSNESS AND RIGHTS

Consciousness and cognition matter. Regan and animal rights lawyer Steven Wise base their arguments on this conviction.[11] They hold that animals are conscious in a way that requires not only humane treatment but also respect for the right of never being used for food, companionship, entertainment, or research.

Wise argues that common law should treat chimpanzees and possibly other animal species not as property but as persons. As persons they would enjoy at least some of the same rights as young children or profoundly retarded adults. Among those rights is that of not being caged and not being subjected to medical research.

Rights, Wise asserts, must rest on something objective, and that something is the conscious mind that makes us capable of giving responsible direction to our lives:

> cognition is a very big deal because the fundamental legal right of animals, the least porous barrier against oppression and abuse that humans have ever devised, depends on it.[12]

Wise describes several apparent similarities between human and animal consciousness and cognition. Then, counseling us to take three aspirin before we read on, he presents his distillation of the "ten top theories of consciousness."[13] Evidently, he believes that if philosophers and neuroscientists are stymied in head-splitting disagreement about the nature of human consciousness, then there is little reason to deny it—whatever it may be—to animals.

There is no need for such abject surrender. Regan, for one, goes to great lengths to define consciousness and cognition. By consciousness he means the awareness that mammals have of their environment, and by self-consciousness he means the awareness of oneself. Regan claims self-consciousness for many species of animals because they act on beliefs and in fulfillment of desires, which requires that they experience themselves as identities in time and know the world in the continuum of time and space rather than as something always new and singular.

By beliefs and desires, Regan means certain mental states underlying animal behavior. When my dog Fatigué, for example, sees me standing at the door with his leash, he wags his tail and paws my leg. According to Regan, Fatigué both believes that we are going for a walk and desires to get going. He has internal representations of the world around him, and can associate immediate representations (me at the door, leash in hand) to representations in his memory (previous walks). Moreover, his tail-wagging indicates that he has a welfare that matters to him. He is responding to more than a mindless physiological drive.

Fatigué's beliefs, Regan allows, are not as sophisticated as our concepts.[14] He may believe that a particular bone is good to gnaw on, whereas we can add the belief that the bone is part of a skeleton. Fatigué has few of the beliefs that make up a human concept. Still, he and other animals have beliefs, rudimentary as they may be.

Regan's contention requires that we inquire into animal consciousness and cognition. Is it so like ours that it can be said that animals possess rights and animal research is a moral wrong on par with experimentation on human subjects?

HUMAN CONSCIOUSNESS

The well-known philosopher of consciousness Daniel Dennett encourages us to "first devis[e] a theory [of consciousness] that concentrate[s] exclusively on *human* consciousness . . . and then look and see which features of that account apply to which animals, and why."[15] A good guide in the task of devising a theory of consciousness and cognition is the philosopher and theologian Bernard J.F. Lonergan.[16]

Lonergan begins by noting that conscious operations involve some sort of awareness that is missing in processes such as the growth of a beard. That awareness is immanent in cognitional and volitional acts: seeing, hearing, touching, feeling, smelling (all acts in the process of experiencing); imagining, thinking, conceiving (all acts in the process of understanding); reflecting, testing ideas, marshaling evidence, affirming and denying propositions (all acts in the process of judging); and taking stock of feelings, considering values, making decisions, committing ourselves (all acts in the process of willing). Unless we are in a dreamless sleep or coma, we are involved in these operations.

Each of these operations makes objects present to us. In other words, each is an act of knowing or willing objects, whether those objects are external stimuli or internal feelings. Because they make objects present (involving what Regan calls "consciousness"), Lonergan calls them intentional. Simultaneously and without any reflection or "bending back," each of these acts makes us present to ourselves. Because they make us present to ourselves (involving what Regan calls "self-consciousness"), Lonergan calls them conscious. It is because we—humans and animals alike—are present to ourselves (conscious) that objects can be present to us, or known.[17] Consciousness is the self-awareness that makes the knowledge and willing of objects possible.

Conscious and intentional operations (hereafter simply called conscious operations) are activities in a dynamic process that unfolds on four distinct but integrated levels. On the first level, we experience presentations that are given in acts of sensing and feeling—sounds and sights, odors, tastes, and feelings, both external and internal. All of these presentations, because they deliver objects, involve a type of knowing. We know the surroundings of sight, sound, smell, taste, and feel; of here, there, up, down, and over; of before and after; and of association ("A, then B"). While brewing the coffee, weaving in and out of traffic, or raking the leaves we know a taken-for-granted world.

It is obvious that animals share with humans this knowing that Lonergan calls elementary knowing. They, too, live in the taken-for-granted world, the world that is spontaneously and expected to be "out there" and real.[18] Elementary knowing allows animals to win sustenance, bear offspring, and stave off predators.

Due, however, to the spontaneous awakening of wonder, humans emerge from their habitat into the universe through a second type of knowing. What Lonergan terms full human knowing begins with questioning that transforms the presentations given in acts of sensing. Lonergan notes:

> if at moments I can slip into a lotus land in which mere presentations and representations are juxtaposed or successive, still that is not my normal state. The . . . world of mere impressions comes to me as a puzzle to be pieced together. I want to understand, to grasp intelligible unities and relations, to know what's up and where I stand.[19]

By virtue of a drive to find the cause and explain, solid, taken-for-granted surroundings are shot through with questions. We are puzzled by what might be causing that noise under the hood of the car, that ache in the back, that turmoil in the Middle East. Driven by the desire to understand we emerge on a second level of consciousness, the level that Lonergan designates as that of understanding. We continue attending to the objects perceived on the first level of consciousness, but we attend to them as questioners so that our consciousness is the awareness of a self that is puzzled, inquiring, and, at least sometimes, understanding. Figuring out a math problem, solving a crossword puzzle, or discovering clues in a murder mystery, we search for suggestive images and adopt a strategy of supposing ("What if . . . ?"). Then, suddenly—usually in a moment of rest after strenuous mental exertion but still with our suppositions—we generate a set of explanatory relationships, an understanding that is not present in sensations and feelings themselves. We relax in delight, "Ah, that's it!" Insight, or understanding, has occurred.

Notice that we can bend our native wonder back on itself in an effort to understand our own cognitive and emotional states. As we seek to understand the process of sensing, understanding, judging, and deciding, we make objects of these cognitional and volitional activities. All the while, of course, we remain aware of ourselves as inquiring, judging, and responsible subjects. We are introspectively conscious.

Insight, thrilling as it might be, is not the end point of human knowledge. The dynamic desire to know and decide—Lonergan calls it the eros of the human spirit—continues to unfold, impelling us next to check if our understanding is correct. On this third level of consciousness, what Lonergan calls rational consciousness, we ask, "Is this really so?" We marshal evidence for the concepts and hypotheses that formulate our insights. Scientists devise ways to test their hypotheses, and all of us draw up mental lists of conditions that have to be fulfilled before we can make ordinary judgments that something is true. Eventually our weighing of an idea leads to another insight when we discover that we have indeed fulfilled the conditions for saying, "This is so, this is true."

Most often we weigh conditions and come to insight about their fulfillment instantaneously—think of how quickly we are sure that we have gotten a joke. The rapidity of judgment can cause us to neglect how personal and significant is the act itself. In the act of judgment, as we satisfy ourselves that we are being neither rash nor indecisive, we take personal responsibility for the truth of an idea or the validity of a hypothesis. We realize that we are about to affirm or deny not only that something is true, but also and simultaneously that we are authentic knowers, in touch with what is.

Full human knowing, which implicitly includes affirmation by the knower that he or she is indeed a knower, comes to term in judgment. But the heart remains restless; persistent questioning now turns into deliberation. It opens a gap between what we have affirmed to be and what might and what ought to be. Again, the capacity for introspection becomes crucial. We take stock of feelings about what is, seek to understand those feelings, and assess how those feelings should guide decisions and actions. We ask, "Whence this passion? What is the nature of this fear? Does this friendship call for truth telling or dictate discreet silence?" Our self-awareness becomes the awareness of a subject in the throes of deliberation. Questioning launches us from the domain of fact into the uneasy realm of values. Consciousness acquires the name of conscience.

Values, of course, do not already exist in an "out there" reality. It is we who generate values in the process of judging our insights into the data of inner and outer experience. Whether we can say with confidence that something is really true and then truly valuable depends on whether we have paid good attention to the data, made sense of the data in insight, and stand prepared to take responsibility for a judgment of fact about what *is* and then for a judgment of value about what *ought to be*.[20]

The fact that we generate values and attempt to direct our lives in accord with them is the source of the claims that we make on each other. The inescapable burden of having to choose reveals both my freedom and my responsibility. I must decide what to do, and I must decide that in relationship to others who appear to possess the same consciousness. My choice is both a claim on others and a response to their claim on me. These claims are what we call rights.

This account of the dynamic desire to know what is, determine what should be, and commit ourselves to what is truly worthwhile reveals the crucial importance of wonder. In questions that express wonder, humans strive to understand, wrestle with the difference between the real and the apparent, and ponder what is to be done or not done. Questioning makes human consciousness and cognition a dynamic, self-assembling process moving from presentations given in experience through successive levels of understanding, judgment, and responsibility to the affirmation of values. Along the way, questioning sweeps us across a divide between elementary knowing, which we share with animals, and a second type of knowing that Lonergan calls full human knowing.

A final comment on Lonergan's analysis of conscious cognition: in its basic structure, it is not revisable. We cannot deny or revise our account of the process without employing the very same operations that we have outlined—operations of sensing, understanding, judging (affirming/denying), and deciding.

HUMAN AND ANIMAL CONSCIOUSNESS COMPARED

Following Dennett's counsel, my present task is to "look and see which features of [the above] account apply to which animals, and why." I propose to do that in reverse order, beginning with consciousness on the fourth level, the level of conscience. Neither Singer, nor Regan, nor Wise claim responsible consciousness for animals. Instead, they claim that its absence in animals is no more significant, morally and legally speaking, than its absence in infants, or in senile or mentally exceptional humans. Singer believes that treating such marginal human beings any differently that animals constitutes "speciesism," and Regan argues that these exceptional persons—he calls them moral patients in contrast to moral agents—are possessors of rights.[21]

Even animal rights philosopher Mary Ann Warren objects to the ethical relevance of occasional conditions or exceptional cases.[22] Infancy, senility, and mental incompetence are conditions that all of us have experienced or will experience in our lifetimes. And, just as we know in a rough and ready sort of way the difference between being conscious and unconscious, so we also know the difference between a smart animal and an exceptional child. Unlike Fatigué, whose lack of moral agency is something expected, a child with a dog's mental capacity is received into the world as an unexpected cause for grief and extra care.[23]

If animals do not function on the level of responsible consciousness, do they operate on the level of rational consciousness and full human knowing? We can surmise that none of our philosophers of animal rights would argue that animals take responsibility for the truth by affirming or denying that something is or is not the case.

They might contend, however, that animals operate on the second level of intelligent consciousness and have "aha" moments. In fact, it is here that Regan and Wise, who offer several examples of animal cunning, attempt to blur any distinction between humans and animals. I will return to those examples presently, after drawing attention once again to Lonergan's distinction between elementary knowing, on the one hand, and full human knowing, on the other. It is certain that animals are conscious. And, to be sure, they know their surroundings and act smartly in their habitat. Still, we must ask if they do so with wonder and its expression in that imperious desire to know and to decide that both initiates full human knowing and grounds human rights.

Regan, unfortunately, gives us no account of the characteristic human activity of questioning. Understandable as such an omission is—people today tend to think of knowing in terms of information rather than activity—it skews the picture. Paying no attention to the activity of asking questions, Regan fails to notice how significant the difference is between existing as a sentient subject of a life on the one hand and existing as a sentient, understanding, rational, and responsible subject of a life on the other. It is only with the full human knowing and willing generated by questioning that we enter into the realm of responsibility, morality, and law.

What goes on in strictly elementary knowing—that is, what goes on in Fatigué's mind—can be inferred from behavior. Much behavior, human and animal, appears to be coded in genes that evolution has selected for survival. Weeds mimic plants, viruses trick the immune system, birds build nests, predators stalk their prey, and people react to threats with "fight" or "flight" instincts. Such behavior involves no conscious knowing.

Other behavior, although selected by blind evolution, is conscious and based on associative learning. Humans and animals connect behavior "A" with consequence "B," a fact that makes it possible for humans to train children as well as pets. Fatigué's behavior of standing ready at the door, wagging his tail, can be explained by his having connected a past walk to a present image of me with his leash. That association suffices to create expectation; he does not question the image of me with the leash, and thus he will never say to himself, "He looks tired; maybe we ought to skip the walk tonight." As Stephen Budiansky comments:

> The intelligence of learned associations of this sort lies in part outside the animal's brain. It is not what is inside the head, but what the head is inside of—to use William Mace's felicitous phrase . . . the rationality of the world.[24]

Finally, some conscious behavior, still based on a capacity selected in evolution, entails not just associative learning, but also wonder and its products—insights, judgments, and values. Humans go beyond the givenness of presentations to develop theories, understand causation, reflect on thoughts and emotions, formulate values, and act to bring about our various visions of utopia.

Applying Occam's razor—one should choose the simplest adequate explanation of any phenomenon—we must ask if any of the many examples and stories of animal cunning require the insight, judgment, and responsible decision-making that are generated by questioning and characterize Lonergan's full human knowing. Fatigué's behavior tells us this much: in addition to undergoing vital processes such as the formation and nutrition of skeletal and muscular structures, he knows and desires things around him. Nevertheless, he probably does so in an elementary way. Curious as he may appear as he puts his nose to smells on his walk, he is not motivated by the drive to understand. When he has marked his territory, he goes on his way through a habitat that is anything but a world shot through with questions.

Is this supposition supported by observation? We can briefly review the fascinating literature of ethology by asking if any of four types of behavior manifests a type of knowing akin to full human knowing, which is initiated by questioning, or, rather, a type of knowing, which, however sophisticated, resembles elementary human knowing.

FEATS OF KNOWLEDGE Marian Dawkins explains that many vertebrates possess mental representations, carry around cognitive maps of food stores, retrieve hidden objects, sort and classify objects, count, and make and use tools.[25] Mice, to cite an example of tool making, use their excrement to plug vents carrying toxic cigarette smoke into their cages, probably acting on a learned association that we would express as "This (feces) here, then this (no smoke)." All such feats of knowledge and tool making remain within the scope of elementary knowing, in which animals associate "here" with "there," "before" to "after," "this means (a finger extended by a piece of straw) to that end (ants within the hill)."

SELF-REFLECTION Gordon Gallup reports that chimpanzees recognize themselves in mirrors and claims that this behavior proves their ability to reflect on themselves.[26] Dorothy Cheney and Robert Seyfarth, much more cautious, distinguish self-recognition—an ability to distinguish oneself from others that does not imply any awareness of doing so—from self-consciousness—a "meta-awareness" in which a subject can watch not just the mirror but himself looking into the mirror and thus have access to what is going on in his own mind. "There is ample evidence from studies of children . . . that many aspects of self-recognition do not require self-reflection."[27] Chimps not only know and reach for where they itch or feel pain but also recognize as part of themselves the hand they see in a mirror as it moves to the troubled site. Children rapidly acquire this same ability. In addition, however, and at about the time they begin asking questions, they begin to advance beyond such self-recognition to self-reflection as they turn the attention of inquiring intelligence upon their experience of such mental presentations. The different abilities of chimp and child correspond to the two types of knowledge, elementary and full human knowing.

ATTRIBUTION OF MENTAL LIFE The ability to attribute a mental life to others reveals the capacity for reflecting on one's own mental life. Humans come to such introspection by dint of questioning presentations of sense and feeling. If it were true that animals predict and attempt to change others' behavior because they understand that minds and intentions underlie that behavior, it would suggest that they can perform the introspection that we have claimed only for full human knowing.

Cheney and Seyfarth provide a careful analysis of the anecdotal and experimental evidence that chimpanzees possess a theory of mind. They conclude that even the most compelling examples of chimps acting as if they recognize that other individuals have beliefs—examples of apparent teaching and deceiving—can be explained in terms of animals' keen observation of behavior and their ability to follow associative ("A" then "B") rules to change that behavior.[28]

LANGUAGE To use a language is to employ signs or pictures or even words to fulfill a demand, obtain a reward, or achieve a desired end; to understand language is to employ the same instruments not just as means to ends, but also as conveyances of meaning. Computers can be programmed to use words. What they cannot do is understand themselves or understand the events of another's mind to which words point.[29]

Wise, who concedes that chimpanzees mostly employ signs as means of obtaining food or fulfilling human commands, is impressed that some chimps have demonstrated the use of more than 3000 words. But children at about the age of three years and without any specific training begin to go beyond the mere use of words to an understanding of thousands of words and the rules for joining them in an infinite number of novel combinations that express meaning in their minds and the minds of others.[30] For them, a word is a sign, or one thing standing for another, and also a symbol into which their minds and the minds of others can read significance.[31]

In summary, none of the animal behaviors catalogued by Wise reveals an underlying activity of asking questions or a basic capacity for wonder. All appear to fall within the province of elementary but not full human knowing. As chimpanzee researcher Daniel Povinelli notes, chimpanzees do not have an ambition to explain the world. Speculating about what would happen if he asked both a 3-year-old child and a chimpanzee why the chicken crossed the road, he concludes the following:

> For the three-year-old, the question is an opportunity to attribute feelings to the chicken and to make up a story—for example, the chicken must have missed its mommy and spotted her on the other side of the road. The chimpanzee has a much simpler answer: 'Yes.'[32]

It was Darwin who observed that the difference between human and animal consciousness is one of degree, not kind. It was also Darwin who noted that there is an enormous practical difference between humans and animals. Humans, he said, are the only "utopian animals."[33] It is the performance of asking questions that sweeps us into the realm of conscience and makes us "utopian animals." But that performance is a single, restless activity generating successive levels of consciousness. Just as it makes human behavior responsible, it also makes some acts of knowing acts of understanding. If nonhuman animals are incapable of moral behavior, then, most likely, their cognitional feats are qualitatively different from ours. In the absence of questioning, they probably never emerge on the levels of intelligent, rational, and responsible consciousness. They are clever associative learners in elementary knowing, but probably not capable of the beginnings of full human knowing.

CONCLUSION: THE USE OF ANIMALS IN RESEARCH

Two overarching issues frame ethical reflections on the use of animals in research. The first is the issue of pain and suffering. The ground of the contemporary concern for animal pain—ground on which philosophers, scientists, and the general public stand together—is an unchallenged consensus that most vertebrate species are conscious. Scientists are morally and, ever since the Animal Welfare Act of 1966, legally bound to ensure the well-being of laboratory animals.[34]

The second issue is that of animal use. If animal consciousness is no different than that of humans, then any use of animals—as pets, as a source of food and clothing, as subjects of research, even when painless and certainly when it involves their death—is as morally reprehensible as the use of humans for the same purposes. The argument of this chapter is that human consciousness and cognition are evidently different enough from animal consciousness and cognition that humans can claim rights while animals cannot. So even if we use animals in research, having vouched for their humane care, we use each other as experimental subjects only when we give informed consent. We safeguard the welfare of animals; we guarantee the rights of persons.

REFERENCES AND NOTES

1. Cf. American Association for the Advancement of Science: http://archives.aaas.org/docs/resolutions.php?doc_id=381; Federation of American Societies for Experimental Biology: http://opa.faseb.org/pages/PolicyIssues/animalresearch_restatement.htm; Society for Neuroscience: http://www.sfn.org/index.cfm?pagename=guidelinesPolicies_UseOfAnimalsandHumans; American Psychological Association: http://www.apa.org/science/anguide.html. Professional societies helped animal protectionists craft and then update the 1966 Animal Welfare Act that sets standards of animal care for research institutions. Cf. The National Research Council–Institute of Laboratory Animal Resources. *Guide for the Care and Use of Laboratory Animals.* Washington, DC: National Academy Press, 1996. Moreover, they endorse regulations that implement the Act—regulations that stipulate conditions for the housing, feeding, cleanliness, exercise, and medical needs of laboratory animals; prescribe anesthesia for potentially painful procedures and analgesic drugs during post-operative care; and require participation on Institutional Animal Care and Use Committees (IACUCs) to ensure that research animals are treated responsibly and compassionately.
2. Singer P. *Animal Liberation: A New Ethic for Our Treatment of Animals*, 2nd ed. New York: New York Review of Books, 1990 and Singer P. *Practical Ethics.* New York: Cambridge University Press, 1993.
3. Hearne V. What's wrong with animal rights. *Harpers* 1991;283: 1696:59–64. The citation is from Bentham J. *Introduction to the Principles of Morals and Legislation.* New York: Hafner Publishing Company, 1946:310–311.
4. See Singer (1990, 18–19),[2] and Singer (1993:117)[2].
5. See Singer (1990, 81–82).[2]
6. See Singer (1990, 21).[2]
7. Cf. Berkowitz P. Other people's mothers. *New Repub* 2000:4:434: 27–37; Russell SM, Nicoll CS. A dissection of the chapter "Tools for Research" in Peter Singer's *Animal Liberation. Proc Soc Exp Biol Med* 1996;211:109–139; Carruthers P. *The Animals Issue.* Cambridge, UK: Cambridge University Press, 1992.
8. Singer allows the use of animals when it is clear that small harm to few animals is outweighed by great benefits for many humans. Such cases, however, are very few. Biomedical research—especially basic research—does not proceed in a straight line from a handful of animals to certain specific persons. Ten years after declaring that the routine use of hundreds of thousands of laboratory animals in research "without the remotest prospect of significant benefits for human beings or any other animals" is ethically unacceptable, Singer heralded experiments in which experimentally damaged nervous systems of rats were repaired with stem cells derived from embryos. (Cf. Singer P. Sense and sentience: We might not need pig hearts if the ban on human embryo experiments were lifted. *Guardian* August 21, 1999:24.) Singer touted these experiments because they could lead to the practice of growing tissues for the repair or replacement of damaged organs, thus eliminating the need of temporary animal transplants for patients on human organ waiting lists. This research, however, depended on decades of basic research, much of it conducted with little understanding of its role in stem cell therapy and all of it involving the routine use of thousands of laboratory rats. None of it would have been approved if Singer's moral calculus had been used prospectively.
9. Regan T. Editorial. *News Observer*, Raleigh, NC. April 3, 2001: A11.
10. Regan T. The case for animal rights. In: Singer P, Ed. *In Defense of Animals.* New York: Harper & Row, 1985:20–21.
11. Regan T. *The Case for Animal Rights.* Berkeley, CA: University of California Press, 1983. Wise SM. *Rattling the Cage: Toward Legal Rights for Animals.* Cambridge, MA: Perseus Books, 2000.
12. See Wise SM (2000, 236).[11]
13. See Wise SM (2000, 125–131).[11]
14. See Regan T (1983, 56).[11]
15. Dennett DC. Animal consciousness: What matters and why. *Soc Res* 1995;62(3):700.
16. Lonergan BJF. *Insight.* New York: Longmans, Green and Todd, 1953. For readable summaries of Lonergan on consciousness and cognitional activity, cf. Lonergan BJF. Cognitional structure. *Continuum* 1964;2:230–242; and Novak M. Lonergan's starting place: The performance of asking questions. *Continuum* 1964;2:89–101.
17. People are inclined to say, "I am conscious of a bird flying by the window." Lonergan would say, "I have knowledge of the bird while being conscious or present to myself."
18. See Lonergan BJF (1953, 251).[16]
19. See Lonergan BJF (1953, 324).[16]
20. See Lonergan BJF. *Method in Theology.* New York: Herder and Herder, 1972:27–41.
21. See Regan T (1983, 279–281).[11]
22. Warren MA. The case for weak animal rights. In: Leone D, Leone B, Eds. *Animal Rights: Opposing Viewpoints.* San Diego, CA: Greenhaven Press, Inc., 1996:47.
23. Nelson JL. Animals, handicapped children and the tragedy of marginal cases. *J Med Ethics*1988;14:191–193. As Nelson reminds us, "The distinction is this: the birth of a 'marginal' human, or the reduction of a normal human to a marginal state, is a tragedy; the birth of say, a healthy collie pup, whose potentials are roughly on a par with the human's is not."
24. Budiansky S. *If a Lion Could Talk.* New York: The Free Press, 1998:27–28.
25. Dawkins MS. *Through Our Eyes Only? The Search for Animal Consciousness.* Oxford: W.H. Freeman Spektrum, 1993:24.
26. Gallup GG Jr. Chimpanzees: Self-recognition. *Science* 1970;167: 86–87.
27. Cheney D, Seyfarth RM. *How Monkeys See the World.* Chicago, IL: University of Chicago Press, 1990:240.
28. See Cheney D, Seyfarth RM.[27] Ever tentative, the two scientists leave open the possibility that chimpanzees possess a very limited theory

of mind. They may share another's feelings just as we spontaneously flinch at another's pain or grin at another's laugh, and they may be aware that feeling states such as pain underlie actions such as grimaces. Wise identifies such fellow-feeling as an "implicit theory of mind," while Lonergan names it the intersubjective and "potential meaning" of experience. Whatever it is called, it seems qualitatively different from an explicit theory of mind, which requires that one hold and compare two different states of mind, one's own and that of another. The ability to do that requires the introspection and active questioning of full human knowing.

29. So it is that the closest any language translation software has come to a meaningful alternative for "The spirit is willing, but the flesh is weak" is "The wine is good, but the meat is spoiled."
30. See Wise SM (2000, 221).[11] Cf. Trefil J. *Are We Unique?: A Scientist Explores the Unparalleled Intelligence of the Human Mind.* New York: John Wiley & Sons, Inc., 1997:55.
31. Wolfe A. *The Human Difference: Animals, Computers and the Necessity of Social Science.* Berkeley, CA: University of California Press, 1993:79.
32. Wheeler D. With ingenious experiments, a psychologist studies the mental lives of chimps. *Chron Higher Ed* June 25, 1999:B2.
33. Cited in Budiansky 1998, xv. Compare contemporary philosopher Charles Hartshorne, cited in Derr TS. Animal rights, human rights. *First Things* 1992;20:23–30, who, although sympathetic to animal rights issues, notes that the human/animal difference is "so vast that for many purposes one can safely forget that it is a matter of degree."
34. What constitutes animal well-being, we should note, is not immediately obvious. Counterintuitive as it seems, rats, for example, may be put on edge by being placed in larger cages. Researchers have an ethical obligation to advance knowledge about animal well-being, asking ever-new questions that lead to new and sometimes revised judgments and gradually round out the picture of humane animal care.

5 Bibliographic Searching Tools on Disease Models to Locate Alternatives for Animals in Research

A Website Companion

MARY W. WOOD AND LYNETTE A. HART

ABSTRACT

Nonvertebrate and vertebrate models are commonly used in research to study diseases that affect humans and animals and to clarify disease etiology and methods for disease prevention and treatment. The relevant information on models is published in a variety of resources and then indexed among free and proprietary databases. New searching tools specifically developed as a companion to this book can now be freely accessed at a designated website created as a companion for the *Sourcebook of Models for Biomedical Research*, where user-friendly searches can be executed at any time according to stored instructions, in a variety of databases pertaining to the relevant disease models. Some databases are freely available, such as PubMed, AGRICOLA, TOXLINE, and NCBI, as well as specialized resources such as FishBase and the JAX Mice Database. Other useful databases are proprietary, such as PsycInfo and Aquatic Sciences and Fisheries Abstracts. The website can be accessed: http://www.vetmed.ucdavis.edu/Animal_Alternatives/sourcebook.html.

Key Words: Animal models, Laboratory animal(s), Animal model types, Choosing animal model, Diseases, Alternatives, Disease models, Nonvertebrate, Vertebrate, Databases, Searching, Websites.

INTRODUCTION

Both nonvertebrate and vertebrate models are commonly used in research to study diseases that affect humans and animals and to clarify disease etiology and methods for disease prevention and treatment. A wide range of databases, both free and proprietary, are available to draw from when searching research literature on these topics.[1] Existing tools that use search filters in the PubMed database simplify identifying strains of mice as appropriate disease models.[2] In this chapter we describe other improved methods for effective literature searching for disease models, and present some new efficient strategies and tools for searching that are available in a new website companion for the *Sourcebook of Models for Biomedical Research.*

From: *Sourcebook of Models for Biomedical Research*
(P. M. Conn, ed.), © 2008 Humana Press Inc., Totowa, NJ.

SEARCHING FOR A RANGE OF POTENTIAL MODELS Anyone can quickly become frustrated trying to monitor and locate the relevant information on published research concerning disease models. Following the current literature, which is so widely dispersed, can seem an overwhelming task. The pertinent information is published in numerous journals and books and then organized and indexed among a variety of databases, some of which are freely available and others that are proprietary. New searching tools are now available in a website specifically developed as a companion to this book. The searching tools can be freely accessed at a designated website, where user-friendly searches can be executed at any time according to stored instructions,[3,4] in a variety of databases pertaining to the disease models described in this volume. The databases include those freely available and linked on the website, such as PubMed, TOXNET, AGRICOLA, TOXLINE, and NCBI, as well as specialized resources such as FishBase and the JAX Mice Database. Proprietary databases available by subscription are also listed, such as ISI Web of Science, PsycInfo, and Aquatic Sciences and Fisheries Abstracts (ASFA). We have previously developed comprehensive search grids to facilitate bibliographic searching on topics related to alternatives for veterinary medical education.[5–7]

WEBSITE ON MODELS AND DISEASES As described below and covered extensively in the interactive website, a working list of topics addressed by searching tools (using a variety of databases and resources) has been selected from the following, including nonvertebrate and vertebrate animal models by species: sea urchin embryo; budding yeast; fungus; social amoebae; round worm; *Drosophila*; bees; *Xenopus*, fish, and other aquatic animals; mice; rats, guinea pigs, hamsters, rabbits, and ferrets; large and farm animals; exotics; cats and dogs; and nonhuman primates. Searching tools are also organized by disease, including aging, AIDS, alcoholism, Alzheimer's disease, atherosclerosis, diabetes, eye diseases, hearing disorders, lung cancer, multiple sclerosis, neurodegenerative diseases, obesity, acute and chronic pain, Parkinson's disease, posttraumatic stress, renal disease, sepsis, thrombosis, and wound healing. The website was developed as a companion to the *Sourcebook of Models for Biomedical Research* and can be accessed: http://www.vetmed.ucdavis.edu/Animal_Alternatives/sourcebook.html.

The website offers direct links to particularly relevant databases and other resources, and indicates which ones are likely to be especially useful for particular models or diseases. In addition, the website includes some stored searches in complimentary databases, termed embedded search templates, that can be launched at will. With these templates, users can use point-and-click technology to avail themselves of the currently available literature on a particular topic, and then make the search instruction more specific to their needs. The search templates generally draw on the high-speed, comprehensive PubMed database, or TOXLINE or NCBI; all are freely available to anyone.

RELEVANT LEGISLATION AND REGULATIONS REGARDING ALTERNATIVES For each use of animals, scientists are required by the USDA's Animal Welfare Act[8] and the related USDA Policy 11[9] and Policy 12[10] to complete animal use protocols that include questions on possible alternatives to procedures causing more than momentary pain or distress, whether in teaching, research, or testing. While researchers seeking improved methods or completing animal-use protocols often find it difficult to locate the relevant information, searching the literature regularly could result in identifying an improved animal (or even yeast or fungus) model for a particular disease. The regulations require that principal investigators provide a written narrative in the protocol concerning the availability of alternatives, including the methods and sources used to determine the availability of refinements, reductions, and replacements. USDA animal care Policy 11 provides a detailed description of painful and distressful procedures. Policy 12 offers detailed guidance on the investigator's description of the database search for alternatives, including the names of the databases searched, the date the search was performed, the period covered by the search, and the key words and/or the search strategy used. Considering the ongoing creation of new animal models and the development of new technologies, frequent searching and consideration of alternatives are appropriate so as to identify and consider new methods and models. This chapter addresses a gap facing scientists by presenting user-friendly searching tools that are targeted toward (1) locating animal models and (2) conducting effective bibliographic searches on animal models and diseases, as required for animal use protocols. These tools simplify searching by providing streamlined access to the resources being sought. Facilitating efficient and effective searching by users can improve the research and simplify compliance with USDA requirements.

REVIEWING DATABASES BY ANIMAL MODEL OR SPECIES

Browsing among the databases pertinent to animal models can be a useful place to begin, to gain familiarity with the array of resources. We have sorted the databases, including 20 complimentary databases, 10 other resources, and 10 proprietary databases listed here, so that users can quickly explore the topics of greatest interest, whether they be animals or diseases in general, or particular animal models or diseases.

POTENTIAL ANIMAL MODELS As shown in Figure 5–1, a large number of complimentary databases are available that pertain to animal models. Some of them have more specialized purposes. For example, ECOTOX, INVITTOX, and TOXNET pertain specifically to toxicology testing, indicating substances that may play an adverse role in certain diseases. FishBase and oneFISH deal with literature on fish. AGRICOLA covers the general literature on veterinary medicine, and PubMed covers the general literature on human medicine. DTIC and DoD Biomedical Research are sites developed in response to military needs, but have general applications for psychological stress and wound healing.

A variety of other complimentary resources are available that are useful in finding animal models. Of particular note are the resources dealing with mice, including the International Mouse Strain Resource, the Mouse Models of Human Cancers Consortium, and the NCI Mouse Models. Ensembl Genome Browser taps into the expanding genomic developments.

Other databases are available as proprietary products. Many institutions subscribe to some of these, so users may have ready access to them at their worksites. One of the most useful of these, ISI's Web of Science, offers web access to citation information, allowing a search for all papers that have cited an earlier work. This feature, previously known as *Citation Index*, offers a feed-forward in tracking papers that have cited a classic work to follow significant lines of research. PsycInfo has good coverage of the human psychological literature, and sometimes addresses pain and stress topics in more detail than PubMed, with an added value of going back to very early literature. CAB and Biosis offer more comprehensive coverage of the European literature and the biology literature than either PubMed or AGRICOLA. Zoological Record covers a broad base of literature on animal behavior of many species and goes back to the earliest references. Fish & Fisheries Worldwide is another database that is essential for people working with fish. ERIC is a database specialized for accessing the literature pertaining to education.

NONVERTEBRATE MODELS Identifying useful models for a study not involving mammals, or perhaps not even vertebrates, offers a potential for reducing the pain or distress that may result from interventive procedures. A growing number of such models are proving useful. Studies using these can be easily accessed, as indicated in Figure 5–2.

Sea Urchin Embryo Search templates offer point-and-click instantaneous searches of some topics where sea urchin embryos have been useful, including cellular activity, the role of metals in early development, and transcription. Other search templates launch TOXLINE or HSDB, two resources on toxicity that are less well known, retrieving studies pertaining to cellular mechanisms and substances with adverse impacts on cellular activity.

Yeast Yeast has proven useful in studies of drug design, glucose, and DNA, topics for which search templates are provided. The DNA search retrieves studies on DNA damage from exposure to toxic substances.

Fungus The use of nonanimal models, such as fungi, can be reviewed in search templates that reveal studies of circadian oscillations in fungi, as compared with other organisms.

Social Amoebas Specialized databases for the social amoebas include DictyBase and Dicty Workbench. Embedded searches focus on cytokinesis, signal transduction, and development.

Round Worm The round worm has become a preferred model for studies of aging that relates to studies of human mitochondria. The database, WormBase, is an important special resource for studies of *Caenorhabditis elegans.*

Fruit Fly *Drosophila* has longstanding importance as an animal model for genetic studies, as reflected in the special database, Homophila. Search templates here pertain to using the fruit

Bibliographic Searching Tools on Disease Models:
Locating Alternatives for Animals in Research

by Species / Model
by Disease

A COMPANION TO: *SOURCEBOOK OF MODELS FOR BIOMEDICAL RESEARCH*

Databases by Species / Model : Identifying potential models

animal model selection	free databases & resources		proprietary databases *option if subscription available*
potential models	AGRICOLA AltBib CRIS CRISP DTIC DoD Biomedical Research ECVAM ECOTOX FishBase INVITTOX INVITRODERM ICCVAM JAX NCBI oneFish PrimateLit PubMed TOXNET WildPro ZFin	AltWeb Pain; Humane Endpoints ANZCCART Fact Sheet: Pain Ensembl Genome Browser ILAR Animal Models Intern'l Mouse Strain Resource Mouse Models of Human Cancers Consortium NCI Mouse Models NIH Model Organisms UKCCCR Guidelines NIGMS Trauma, Burn, Wound	ASFA Biosis CAB Embase ERIC Fish & Fisheries Worldwide ISI Web of Science PsycInfo SciFinder Zoological Record

Figure 5–1. Opening page of the website, showing selected resources suitable for browsing to identify potential disease models. Resources shown in shaded boxes are free to all users and are available via links on the website. The column on the right side lists useful proprietary databases available by subscription. http://www.vetmed.ucdavis.edu/Animal_Alternatives/sourcebook.html.

specific model	suggested databases		free embedded search examples
	free	proprietary	
sea urchin embryo	PubMed TOXNET	ASFA Biosis SciFinder	cellular activity metals and early development transcription *in* TOXLINE *in* HSDB
S. cerevisiae (budding yeast)	PubMed	Biosis ISI Web of Science	drug design glucose dna
Neurospora (filamentous fungus)	PubMed Resources	Biosis ISI Web of Science	model evolutionary
D. discoideum (social amoebae)	PubMed DictyBase Dicty Workbench	Biosis ISI Web of Science	cytokinesis signal transduction development
C. elegans (round worm)	PubMed WormBase	Biosis ISI Web of Science	aging neurological model
D. melanogaster (fruit fly)	PubMed Flybase Homophila	Biosis ISI Web of Science	hearing balance neurodegenerative model aging
bees	PubMed	Biosis ISI Web of Science	ethanol/alcohol

Figure 5–2. Nonvertebrate models featured on the website, highlighting free and proprietary databases in the middle two columns. The column on the right side indicates the use of the free database PubMed for stored embedded searches that are available for launching via links on the website. http://www.vetmed.ucdavis.edu/Animal_Alternatives/sourcebook.html.

specific model	suggested databases		free embedded search examples
	free	proprietary	
fish / frogs / aquatic	AGRICOLA PubMed TOXNET FishBase oneFISH Xenopus Resources Zebrafish Gene Collection ZFIN	ASFA Biosis Fish and Fisheries Worldwide Zoological Record	zfish as model zfish & obesity/diabetes frogs & oncogenesis frogs & neurogenesis
mice	AGRICOLA AltBib PubMed JAX Mice Database Mouse Biology Program	Biosis ISI Web of Science	as model alzheimers obesity
rats, guinea pigs, hamsters, rabbits, ferrets	AGRICOLA PubMed ANZCCART Fact Sheets	Biosis ISI Web of Science	ferrets & helicobacter gp & hyperlipidemia gp & osteoarthritis rats & wound healing rats & encephalomyelitis rabbits & thrombosis

Figure 5–3. Vertebrate models that are typically used in laboratories, featured on the website, highlighting free and proprietary databases in the middle two columns. The column on the right side indicates the use of the free database PubMed for stored embedded searches that are available for launching via links on the website. http://www.vetmed.ucdavis.edu/Animal_Alternatives/sourcebook.html.

fly for studies of hearing, balance, neurodegenerative models, and aging.

Bees Studies of alcohol effects are conducted on nonvertebrates, bees. These studies are easily accessed in an embedded search.

VERTEBRATE MODELS A number of vertebrate species are familiar in laboratory studies, primarily the aquatic vertebrates and the small mammals. The fish, frogs, and other aquatic animals, and the mice, rats, and other small mammals are readily accessed as shown in Figure 5–3. Less commonly studied, but also used as models, are large and other farm animals, exotic animals, dogs and cats, and nonhuman primates; searching strategies for these species are presented in Figure 5–4.

Fish, Frog, and Other Aquatic Animals The use of fish has expanded in recent years, in many cases as a vertebrate replacement for mammals. Specialized resources are most appropriate

specific model	suggested databases		free embedded search examples
	free	proprietary	
large / farm animals	AGRICOLA PubMed	Animal Health and Production Compendium Biosis CAB FASS Guide ISI Web of Science	cattle as model chickens as model
exotics	WildPro Information Network ANZCCART Fact Sheet	Biosis Wildlife & Ecology Studies Worldwide Zoological Record	as model
dogs, cats	AGRICOLA PubMed ANZCCART Fact Sheet: Dog, Cat	Biosis CAB ISI Web of Science	sleep apnea
non-human primates	PrimateLit PubMed Guidelines for the Care & Use of Mammals in Neuroscience & Behavioral Research	Biosis ISI Web of Science PsycInfo	as model HIV vaccine

Figure 5–4. Vertebrate models that are less typically used in laboratories, featured on the website, highlighting free and proprietary databases in the middle two columns. The column on the right side indicates the use of the free database PubMed for stored embedded searches that are available for launching via links on the website. http://www.vetmed.ucdavis.edu/Animal_Alternatives/sourcebook.html.

for searching on these species, including the complimentary Fish-Base, oneFISH, *Xenopus* Resources, Zebrafish Gene Collection, and ZFIN, and the proprietary ASFA and Fish and Fisheries Worldwide.

Mice Despite the growing number of nonvertebrate and fish models, mammals are essential for some studies. In mainstay models for studies in which a mammal is essential, the number of mice used far exceeds any other mammal. As examples, the search templates use mice as models for various diseases, for studies of Alzheimer's disease, and for studies of obesity. Specialized databases such as the JAX Mice Database and the Mouse Biology Program are important when searching for appropriate strains of mice. Specialized strategies for searching with search filters also can be employed.[2]

Rats, Guinea Pigs, Hamsters, Rabbits, and Ferrets Rats are frequently used as models in a wide range of studies where an animal larger than a mouse is needed. Other small mammals are used in specific research, such as ferrets in studies of *Helicobacter* and guinea pigs in hyperlipidemia. With these species, AGRICOLA provides coverage of some research that complements PubMed. The complimentary ANZCCART Fact Sheets provide practical information on basic parameters for working with small mammals.

Large and Other Farm Animals For agricultural animals, AGRICOLA invariably provides more comprehensive coverage of the research than PubMed. The FASS Guide is available at modest cost, providing another valuable resource on agricultural animals.

Exotic Animals Specialized databases are more effective when searching on exotic animals than PubMed, which is targeted specifically toward biomedical research and misses some important work utilizing exotic animals.

Dogs and Cats AGRICOLA is worthwhile for searching literature pertaining to dogs and cats, as is PubMed.

Nonhuman Primates Specialized databases are useful in locating literature on nonhuman primates, especially the complimentary PrimateLit, with research dating back to 1940.

REVIEWING DATABASES FOR ANIMAL MODELS BY DISEASE

Researchers face a dual challenge, to keep updated not only with regard to new animal models, but also concerning new work on the particular disease of interest.

DISEASES IN GENERAL As with animal models, a wide range of complimentary databases and resources is available for browsing, as indicated in Figure 5–5.

SPECIFIC DISEASES To provide useful examples of search strategies, four embedded search strategies are indicated in Figure 5–6 for each listed disease. For the embedded search templates stored concerning each of the specific conditions and diseases, two searches for each disease draw from PubMed, one focusing on a specific research strategy on a relevant topic for that disease and another addressing models for that disease. A third search for each disease draws from TOXLINE, the National Library of Medicine's bibliographic database for toxicology, covering the full range of effects of drugs and other chemicals, including biochemical, pharmacological, physiological, and toxicological effects. The fourth search for each disease uses NCBI (National Center for Biotechnology Information), offering an extended

Databases by Disease : Locating information based upon disease *and* model

free databases & resources		proprietary databases *options if subscriptions available*
AGRICOLA AltBib CRIS CRISP DTIC DoD Biomedical Research ECVAM ECOTOX FishBase INVITTOX INVITRODERM ICCVAM JAX NCBI oneFish PrimateLit PubMed TOXNET WildPro ZFin	AltWeb Pain; Humane Endpoints ANZCCART Fact Sheet: Pain ILAR Animal Models NCI Mouse Models NIH Model Organisms UKCCCR Guidelines NIGMS Trauma, Burn, Wound	ASFA Biosis CAB Embase Fish & Fisheries Worldwide ISI Web of Science PsycInfo RTECS Zoological Record

Figure 5–5. Selected resources on the website suitable for browsing by disease, to identify potential disease models. Resources shown in shaded boxes are free to all users and are available via links on the website. The column on the right side lists useful proprietary databases available by subscription. http://www.vetmed.ucdavis.edu/Animal_Alternatives/sourcebook.html.

disease model	free embedded searches *in* PubMed, TOXLINE, NCBI, DTIC
aging	stem cells ; models ; toxline ; ncbi
AIDS	HIV vaccine ; models ; toxline ; ncbi
alcoholism	neurogenesis ; models ; toxline ; ncbi ; dtic
Alzheimer's disease	vitamin E ; models ; toxline ; ncbi
atherosclerosis	helicobacter ; models ; toxline ; ncbi
diabetes	embryonic ; models ; toxline ; ncbi
eye diseases	macular degeneration ; models ; toxline ; ncbi
hearing disorders	hair cell regeneration ; models ; toxline ; ncbi
lung cancer	dna repair ; models ; toxline ; ncbi
multiple sclerosis	T-lymphocytes ; models ; toxline ; ncbi
neurodegenerative diseases	ALS ; models ; toxline ; ncbi
obesity	leptin ; models ; toxline ; ncbi
pain, acute & chronic	capsaicin ; models ; toxline ; ncbi
Parkinson's disease	LRRK2 ; models ; toxline ; ncbi
post-traumatic stress	neuropeptides ; models ; toxline ; ncbi ; dtic
renal disease	hepatocyte growth factor ; models ; toxline ; ncbi
sepsis	dexamethasone ; models ; toxline ; ncbi
thrombosis	atrial ; models ; toxline ; ncbi
wound healing	growth factor ; models ; toxline ; ncbi ; dtic

Figure 5–6. Search strategies on the website are presented as stored embedded searches by disease, drawing from the free databases PubMed, TOXLINE, NCBI, and DTIC. The stored searches are available for launching via the website. http://www.vetmed.ucdavis.edu/Animal_Alternatives/sourcebook.html.

searchable library of the life sciences literature, including PubMed, PubMed Central, books, Coffee Break (short reports on recent discoveries), Genes and Disease, and OMIM (catalog of human genes and genetic disorders). Thus, NCBI reflects the emerging emphasis on research pertaining to genetics interacting with other disciplines. We provide embedded searches for the 19 diseases and conditions listed below. Users can use the search templates as a starting point to adapt searches more specifically to their needs. Posttraumatic stress and wound healing are two topics that have broad application and benefit from the comprehensive military emphasis on databases, as found in DTIC and DoD Biomedical Research.

- Aging
- AIDS
- Alcoholism
- Alzheimer's disease
- Atherosclerosis
- Diabetes
- Eye diseases
- Hearing disorders
- Lung cancer
- Multiple sclerosis
- Neurodegenerative disorders
- Obesity
- Pain, acute and chronic
- Parkinson's disease
- Posttraumatic stress
- Renal disease
- Sepsis
- Thrombosis
- Wound healing

CONCLUSIONS

A website companion for the *Sourcebook of Models for Biomedical Research* developed by the UC Center for Animal Alternatives offers streamlined access to information concerning animal models for disease. The site offers expeditious links to relevant databases and provides embedded stored search examples for effective searching for models. The site can be used as an instructional tool to demonstrate how to search in these subject areas. The examples offer a starting point that a user can, with growing experience, use as a springboard for further searching.

ACKNOWLEDGMENTS

The extensive electronic library holdings of the University of California make it possible to explore, identify, and distinguish the unique features of the resources presented in these search grids. Hsin-Yi Weng creatively contributed to much of the developing technology offered in this chapter.

REFERENCES

1. Hart LA, Wood MW, Weng H-Y. Effective searching of the scientific literature for alternatives: Search grids for appropriate databases. *Anim Welf* 2005;14:287–289.
2. Wood MW, Hart LA. Search filters, a new tool in the search for alternatives: Locating mouse strains as disease models. *ATLA* 2004;32(Suppl. 1):599–602.
3. Wood MW, Hart LA. Considering animal alternatives and welfare via a comprehensive search of the scientific literature. *ALTEX* 2006;23(Spec. Issue):197–199.
4. Wood MW, Weng H-Y, Hart LA. Guide to bibliographic databases for alternatives searching: Database approach to a search. UC Center for Animal Alternatives, University of California, Davis, School of Veterinary Medicine. Accessed on September 7, 2006, at http://www.vetmed.ucdavis.edu/Animal_Alternatives/databaseapproach.html.
5. Wood MW, Hart LA, Weng H-Y. Effective bibliographic searching for animal alternatives in veterinary medical education: The UC Davis web site. *J Vet Med Ed* 2005;32:468–472.
6. Weng H-Y, Wood MW, Hart LA. Alternatives resources in education: Educational search grids. UC Center for Animal Alternatives, University of California, Davis, School of Veterinary Medicine. Accessed on September 7, 2006, at http://www.vetmed.ucdavis.edu/Animal_Alternatives/altsearch.htm.
7. Wood MW, Hart LA, Weng H-Y. Search grids for teaching and research protocols on veterinary medicine teaching alternatives. UC Center for Animal Alternatives, University of California, Davis, School of Veterinary Medicine. Accessed on September 7, 2006, at http://www.vetmed.ucdavis.edu/Animal_Alternatives/protocol.htm.
8. United States Animal Welfare Act 7USC 2131–2159, adopted 1966, amended 2002. Accessed September 7, 2006, at http://www.nal.usda.gov/awic/legislat/awa.htm.
9. United States Department of Agriculture, Animal and Plant Health Inspection Service Animal Care. USDA Animal and Plant Health Inspection Animal Care Policy Manual Policy #11—Painful Procedures—April 14, 1997. Accessed September 7, 2006, at http://www.aphis.usda.gov/ac/policy/policy11.pdf.
10. United States Department of Agriculture, Animal and Plant Health Inspection Service Animal Care. USDA Animal and Plant Health Inspection Animal Care Policy Manual Policy #12—Consideration of Alternatives to Painful/Distressful Procedures—June 21, 2000. Accessed September 7, 2006, at http://www.aphis.usda.gov/ac/policy/policy12.pdf.

6 NIH Policies on Sharing of Model Organisms and Related Research Resources

J.P. KIM, CHARLENE CHO, AND NORKA RUIZ BRAVO

ABSTRACT

The National Institutes of Health (NIH) has long recognized that the sharing of research materials and data is critically important to the progress of biomedical research. Effective sharing of research resources benefits the public by facilitating and promoting discovery across the full spectrum of the research community. In this spirit, the NIH Model Organism Sharing Policy builds upon extant NIH sharing policies to allow the NIH to continue to support and promote broad and enabling research effectively and efficiently. This chapter provides guidance, sample plans, approaches, and other resources for investigators considering how best to share model organisms and related research resources with the research community.

Key Words: Model, Organisms, Data, Federal, Funding, Grants, Sharing, Research, NIH, Policy.

INTRODUCTION

Model organisms have played a critical role in the history of biomedical research.[1] As tools of discovery and a means to demonstrate proof of concept, they have been pivotal in a broad range of seminal events, including among others, developing a vaccine against smallpox, testing the artificial heart, and discovering the genetic basis for lactose intolerance.[2] The sharing of research materials, including model organisms and associated data, has also been critically important in the progress of biomedical research.[3] Common use of model organisms by research laboratories has provided scientists with the opportunity to verify research results, explore previously unsuspected mechanisms of disease, and study complex biological questions using an integrative, holistic approach.[4]

As a public sponsor of biomedical research, the National Institutes of Health (NIH) is committed to supporting national and international efforts that encourage the sharing and dissemination of important research resources. This chapter will discuss the policies of the NIH on the sharing of model organisms and related research resources and its significance to biomedical research, as well as approaches and considerations of how such resources might most effectively be shared, consistent with NIH funding policies.

From: *Sourcebook of Models for Biomedical Research*
(P. M. Conn, ed.), © 2008 Humana Press Inc., Totowa, NJ.

EVOLUTION OF NIH SHARING POLICIES

FROM PUBLICATION TO SHARING OF RESEARCH TOOLS TO DATA AND ORGANISMS The NIH is a part of the U.S. Department of Health and Human Services and is the primary federal agency responsible for conducting and supporting biomedical research in the United States. With a history dating back to 1887, the NIH has played a prominent role not only in biomedical research, but also in the dissemination of research results and materials to the research community.

As a steward of federal funds and in keeping with its mission, the NIH has formulated policies regarding the sharing of research resources made possible as a result of public funding (see Table 6–1). The concept of sharing research results has remained a constant element of NIH grants policies, including those related to unique research resources. An early Public Health Service (PHS) Grants Policy Statement set the stage by stating a seminal principle for federally funded research: "Project directors and principal investigators are encouraged to make the results and accomplishments of their activities available to the public."[5] A May 1980 Special Announcement was published in the NIH Guide for Grants and Contracts regarding the "NIH Policy Relating to Reporting and Distribution of Hybridomas Produced Under Grants and Contracts."[6] This was followed by additional announcements highlighting the importance of sharing.[7] The Bayh-Dole Act, P.L. 96-517, passed into law on December 1980 to stimulate research and development of government-funded research, prompted additional reminders about the importance and value to scientific research of sharing research resources.[8]

THE NIH POLICY ON SHARING OF MODEL ORGANISMS FOR BIOMEDICAL RESEARCH

After consultation with the research community, the NIH announced its Policy on Sharing of Model Organisms for Biomedical Research (NOT-OD-04-042) on May 7 2004.[9] Under this policy, NIH applications that result or may result in new, genetically modified variants of model organisms are expected to include a plan for sharing those organisms and related research resources. The policy is based on existing NIH Grants Policy [NIH Grants Policy Statements (http://grants.nih.gov/grants/policy/#gps)], including the NIH Research Tools Policy, and essentially serves as implementation guidelines for earlier policies.[10,11] As such, this

Table 6–1
Highlights in the evolution of NIH sharing policies

Year	*Policy*	*Publication*
Pre-1980s	Policy encouraging making the results and accomplishments of NIH funded activities available to the public	PHS Grants Policy Statement, October 1976
1980	Policy on reporting and distribution of hybridomas produced under grants and contracts	NIH Guide Supplement, May 1980
1984	Policy on reporting and distribution of unique biological materials produced with NIH funding	NIH Guide Notice, March 1984
1996	Biological materials policy: "Handling Non-Election of Title to Patentable Biological Materials"	NIH Guide Notice, May 1996
1999	NIH research tools policy: "Sharing of Research Resources: Principles and Guidelines for Recipients of NIH Research Grants and Contracts on Obtaining and Disseminating Biomedical Research Resources"	Federal Register Notice, December 1999
2003	2003 NIH data sharing policy: "Final NIH Statement on Sharing Research Data"	NIH Guide Notice, February 2003
2004	NIH model organism sharing policy: "NIH Policy on Sharing of Model Organisms for Biomedical Research"	NIH Guide Notice, May 2004

policy applies to extramural investigators and institutions that are funded by a variety of mechanisms (see below). Although this policy does not apply to NIH intramural investigators, the NIH intramural research program has guidelines that are consistent with those for the extramural community (for example, see http://www1.od.nih.gov/oir/sourcebook/ethic-conduct/resources.htm). Accordingly, the NIH advocates that investigators share such resources across the entire research community. The policy assumes that sharing will take place in a manner consistent with accepted practice. For example, new, genetically modified model organisms developed with NIH funding may be shared as mature organisms, sperm, eggs, embryos, or even the vectors used to generate transgenic or knockout organisms.

RATIONALE FOR THE SHARING POLICY The NIH Model Organism Sharing Policy contributes to scientific progress by making new resources available more quickly to a large number of people and obviates the need to use limited resources and investigator time to reproduce previously developed model organisms. With the Model Organism Sharing Policy, the NIH expects that new, genetically modified model organisms and related resources generated with NIH funding will be distributed and shared with the scientific community in a timely manner (i.e., at least upon publication of the primary results announcing the development of the genetically modified model organisms).

This expectation serves to fulfill the mission of the NIH: "Science in pursuit of fundamental knowledge about the nature and behavior of living systems and the application of that knowledge to extend healthy life and to reduce the burdens of illness and disability." Sharing of model organisms and related research resources allows the biomedical community, including students, postdoctoral fellows, and investigators, to engage in new scientific research and training opportunities, which can extend the scope of inquiry beyond that envisioned by the original creator of the animal strain. As a result, sharing research resources is likely to result in more rapid breakthroughs for the diagnosis, prevention, and treatment of disease than a single investigator could achieve alone.

Broad access and use of research resources also contribute to the many goals of the NIH, including "promoting the highest level of scientific integrity, public accountability, and social responsibility in the conduct of science." Sharing of model organisms allows investigators a chance to verify research results, reduces needless and costly duplication of effort, and serves as a form of insurance against unforeseeable loss of research resources (e.g., disasters such as floods, fires, and power outages that have resulted in the loss of valuable mouse strains and other resources). Thus, everyone benefits from the sharing of research resources, including investigators, the scientific community, and the public.

Indeed, sharing is good practice.[12] Journals such as *Proceedings of the National Academy of Sciences USA, Cell, Nature, Neuron, Immunity, The Journal of Immunology, The Journal of Neuroscience*, and *Science* require investigators to make unique resources available so that research results can be verified and additional research promoted. Both the Society for Neuroscience, with a membership of 29,000 scientists, and the Federation of American Societies for Experimental Biology (FASEB), consisting of 19 societies and a membership of 66,000 scientists, have taken positions that are consistent with the NIH policy. For example, the policy of the Society for Neuroscience is that "unique and propagatable research materials used in studies being reported must be made available to qualified scientists for bona fide research purposes"(see http://www.sfn.org/index.cfm?pagename=responsibleConduct_authorsOfResearchManuscripts and http://www.sfn.org/skins/main/pdf/Guidelines/ResponsibleConduct.pdf) under section 1.8 of "Guidelines: Responsible Conduct Regarding Scientific Communication," Society for Neuroscience, 1998.[13]

MODEL ORGANISMS DEFINED For purposes of the NIH Model Organisms Sharing Policy, the NIH uses the term "model organism" to describe both mammalian (e.g., mouse and rat) and nonmammalian models (e.g., budding yeast, social amoebas, roundworm, *Arabidopsis*, and frog). Genetically modified model organisms are those in which spontaneous mutations have occurred and/or those in which mutations have been induced by chemicals, irradiation, transposons, or transgenesis (e.g., knockouts and injection of DNA into blastocysts). Congenic or consomic strains are also considered genetically modified model organisms. Detailed information about model organisms can be found on the NIH Model Organism for Biomedical Research website (http://

www.nih.gov/science/models/). Publications and electronic databases as well as websites of stock centers and repositories that list available strains also make the identification of model organisms relatively easy.

SCOPE OF THE POLICY The policy covers both model organisms and related resources. Related resources include materials and data necessary for the production and understanding of model organisms, such as vectors, nonhuman embryonic stem cells, established cell lines, protocols for genetic and phenotypic screens, mutagenesis protocols, and genetic and phenotypic data for all mutant strains. As a matter of practice, genetic variants of viruses, bacteria, and other prokaryotic organisms should be made widely available. At this time, however, the NIH is not expecting the submission of a sharing plan from investigators who intend to develop noneukaryotic organisms.

It should be noted that this NIH sharing policy applies only to nonhuman model organisms. For example, human cells that are not commercially available would not be subject to this policy. For information on regulations, policies, and guidelines for human subject specimens, see http://grants2.nih.gov/grants/policy/hs/specimens.htm.

Unlike the related, but distinct, 2003 NIH Data Sharing Policy (http://grants2.nih.gov/grants/guide/notice-files/NOT-OD-03-032.html), which applies to projects requesting $500,000 or more in direct costs in any given year,[14] the NIH Model Organism Sharing Policy applies to all projects that produce or may produce model organisms or related resources. In other words, the NIH expects that all investigators pursuing projects that would or could result in model organisms and related resources will include a sharing plan in their applications, regardless of the requested budget amount or final amount awarded.

FUNDING MECHANISMS AFFECTED With respect to NIH grant mechanisms, this policy applies to all research project grants (R mechanisms) and Centers (P mechanisms) as well as to Small Business Innovation Research (SBIR) and Small Business Technology Transfer (STTR) awards. The policy also applies to individual Career Development awards (Ks) if the scope and aims of the project include the development of model organisms. By contrast, institutional training grants (Ts) or individual fellowships (Fs) are not subject to the policy, but sharing would be applicable to such grantees as the production of model organisms presumably would be covered under a mentor's research grant. This policy also applies to cooperative agreements and to contracts from the NIH that will generate model organisms and related resources as described above.

It should also be noted that this policy applies to all tiers of a funding agreement, including international collaborations and foreign grants. For example, if the U.S. institution is the primary grantee, then it would be the responsible for its subgrantee or subcontract arrangements and would be expected to ensure that this policy is adequately addressed in the application. In some cases, different subprojects will have different sharing plans, or only one or a few of the subprojects will need sharing plans. Regardless of how many parties are involved or who is designated as the primary grantee, it is important to state clearly in the application which party [e.g., Principal Investigator (PI) or head of the subproject] will be responsible for the implementation of the proposed sharing plan(s).

DEVELOPING SHARING PLANS When submitting sharing plans to the NIH under this policy, such plans should provide information to ensure that sharing will be meaningful for other investigators in the community. For example, the following information should be addressed.

How novel strains will be made available to the scientific community, including the following:

- The form in which the investigator/institution will provide the organisms (e.g., adults, embryos, sperm).
- Related research resources and data that the investigator/institution will provide.
- A reasonable time frame for periodic deposition of material and associated data.
- Whether the investigator/institution will share under its own auspices or use a repository, and, if a repository, which one.
- For vertebrate animals and for other species for which pathogens or contaminants are potentially serious problems, how the investigator/institution will maintain the strains to minimize the risks of infection or contamination.

How technology transfer and intellectual property issues will be addressed, including the following:

- How the institution plans to make such organisms and resources widely available to the research community.
- How the institution plans to make certain any rights or obligations to third parties are consistent with the terms and conditions of the NIH award to ensure appropriate dissemination of model organisms or related research resources under the NIH award.
- A description of the mechanisms that will be used to distribute organisms and related research resources [e.g., material transfer agreements (MTA)].

Sharing plans may vary, depending on the organism, the nature of the resources that will be shared, the extent of intellectual property issues involved, and plans for distributing the resource(s), as well as the particular programmatic goals. The actual content and level of detail included in a sharing plan will be affected by several factors, such as the status of the development of a model organism, the potential impact of intellectual property rights on the availability of the model organism, and the method of sharing. Therefore, the NIH recommends that the investigator contact the funding Institute or Center about the level of detail expected.

Samples of sharing plans for model organisms and related resources may be found at http://grants.nih.gov/grants/policy/modelorganism. These three sample plans have been provided below to assist the applicant community in responding to this policy when submitting a funding application to the NIH:

- A *simple plan* that may be appropriate for a project, which has the goal of producing a model organism but has not produced one (Organism Sharing Simple) (http://grants.nih.gov/grants/policy/model_organism/model_organ_simple_plan.doc).
- A plan for *sharing mice* (Organism Sharing Mouse) (http://grants.nih.gov/grants/policy/model_organism/model_organ_mouse_plan.doc).
- An example of a *complex plan* for sharing mice with IP held by various parties (Organism Sharing Complex) (http://grants.nih.gov/grants/policy/model_organism/model_organ_complex_plan.doc).

Sharing plans may also take into account practical limitations. For example, if a large genetic screen is conducted and thousands of lines are generated, only a few of them may display a phenotype that is of interest to laboratories. Generally, the investigator in such a case would not be required to maintain and distribute all the lines. Similarly, if knowledge of their existence is limited to a publication in which they were described as being used in a cross and do not demonstrate a specific phenotype, there would be no need to maintain or distribute those lines. However, if the collection of lines was generated with NIH funding and with the purpose of serving as a community resource, then there would be an obligation to share the lines.

There are certain instances in which sharing of a resource may not be appropriate. For example, an investigator would not be expected simply to redistribute a resource acquired from another source. However, if that resource is used to breed, construct, or generate another resource using NIH funds, there would be an obligation to share the derived resource. In any case, if no sharing plan is included in the NIH application, the applicant would be obligated to state why sharing would be restricted or not possible. Failure to provide an adequate sharing plan may affect the NIH's decision to make an award. Equally important, failure to comply with the submitted and accepted sharing plan may be considered in future funding decisions for the investigator and the investigator's institution.

GENERAL RESOURCES AND CONSIDERATIONS FOR THE DISTRIBUTION AND SHARING OF MODEL ORGANISMS AND RELATED RESOURCES Sharing of model organisms and related resources is not intended to be challenging for the research community, and yet depending on circumstances (e.g., demand or supply), responding to requests for model organisms can be costly and labor intensive. Also, there could be other considerations that may need to be taken into account when transferring materials.

To minimize administrative burdens and to help streamline the processing of such requests, the NIH recommends that developers deposit their model organisms with existing repositories, which would provide for ready, long-term access by the biomedical research community. Repositories such as the Mouse Models for Human Cancer Consortium (http://emice.nci.nih.gov/), the Mutant Mouse Regional Resource Centers (http://www.mmrrc.org/), the Rat Resource and Research Center (http://www.nrrrc.missouri.edu/), and the Zebrafish Information Network (http://zfin.org/) are among some of the examples of repositories funded by the NIH and should be viewed as valuable resources for the sharing of model organisms.

If submitting the model organisms with a repository is not an option, the NIH also offers funds for research resource sharing. Applicants may request funds in their applications for expenses associated with sharing model organisms and related research resources, including administrative supplements for unanticipated expenses associated with sharing.

Whether it is through a repository or through the auspices of the developer, distributing model organisms and related research resources developed with NIH funding should occur in a manner consistent with the NIH Research Tools Policy. The majority of unpatented transfers to not-for-profit entities should occur under terms no more restrictive than the Uniform Biological Materials Transfer Agreement (UBMTA) (http://www.autm.net/aboutTT/aboutTT_umbta.cfm) or the Simple Letter Agreement (SLA) (http://www.nhlbi.nih.gov/tt/docs/sla_mta.pdf). The provider may specify whether the model organism can be distributed to other parties. If the materials are patented or licensed to an exclusive provider, other arrangements may be used, but terms providing for commercialization option rights, royalty reach-through, or product reach-through rights back to the provider are inconsistent with the NIH Research Tools Policy.

When NIH-funded materials are being transferred to for-profit entities that intend to restrict their use internally, recipients should ensure that the materials are transferred with the fewest encumbrances possible. The SLA may be expanded for use in this instance, and internal-use license agreements with execution or annual use fees may be appropriate, depending on circumstances.

Ascertaining which type of agreement to use in transferring materials can present quite a challenge. It is important to keep in mind that if no agreement is used in obtaining a material, there may be no applicable restrictions. Consequently, it is best to confer with the technology transfer office of the investigator's institution for guidance on such matters.

CONSIDERATIONS FOR DEVELOPERS AND PROVIDERS OF MODEL ORGANISMS The NIH Model Organism Sharing Policy states that the NIH expects the submission of sharing plans from investigators whose projects will or could yield new, genetically modified model organisms or related resources. Further, if a given project unexpectedly results in new, genetically modified model organisms or research resources, it would be appropriate to share the model organisms, consistent with NIH grants policy. Consulting with the NIH program official and including a sharing plan in the next noncompeting application or in a separate letter would also be considered appropriate in such a situation.

Since sharing plans can be highly complex and can involve issues pertinent to intellectual property, these plans should be consistent with NIH guidelines and in consultation with individuals who have expertise in this area. These experts could be staff from an institution's sponsored research office, technology transfer office, or office of general counsel. Such consultation often will clarify a given institution's policies regarding sharing and distribution of research resources such as mutant mice as well as its policies regarding intellectual property. Parties developing and/or providing research resources should ensure that research use rights are retained for noncommercial research purposes, which are consistent with practices supported by the Association of University Technology Managers (AUTM) and the Council on Governmental Relations (COGR).

If sharing plans for NIH awards are in conflict with obligations (i.e., title to the invention) to third parties (e.g., company X), NIH-funded parties will need to revise any such agreements so that they are consistent with the terms and conditions of the NIH funding award. For guidance regarding this issue, see the 2003 NIH Grants Policy Statement,[15] as well as "Developing Sponsored Research Agreements: Considerations for Recipients of NIH Research Grants and Contracts"[16] and "Intellectual Property Reporting for NIH Grantees That Also Have Involvement with the Veterans Administration."[17] Nonprofit institutions that wish to waive or assign title of inventions developed under an NIH funding award to another party must obtain approval from the NIH, except where such an assignment is made to an organization that has as one of its primary functions the management of inventions and are subject to the same provisions as the funding recipient.

As a term of award, it is the responsibility of the NIH-funded party to ensure that rights to inventions, such as model organisms and related research resources, arising out of NIH funding agreements are properly assigned to the institution. Failure to comply with NIH research resource sharing policies and the accepted plan may be considered by the NIH program staff in future funding decisions for the investigator and the investigator's institution.

An example of the Personal Benefits of Reserving Research Use Rights for the Research Community If the developer of a model organism reserves research use rights for at least noncommercial research purposes across the research community, this not only protects the nonprofit institutions, but it can also protect the developer of the mouse later on in his or her career. For example, in one scenario, a scientist is both an investigator at nonprofit institution A and the Chief Scientific Officer of for-profit X. While at nonprofit A, the investigator develops a model organism. The investigator does not want any competitors to have access to the resource, and therefore asks nonprofit A to license exclusively all rights for this resource to the investigator's for-profit X, while retaining a research use license only for nonprofit A. The investigator is terminated from both nonprofit A and for-profit X. The investigator moves to nonprofit B, hoping to continue the research, but can no longer access the resource because for-profit X refuses to provide access to the investigator and nonprofit A is not permitted to provide the resource to any third party. If a research use license had been retained for all nonprofit research institutions, the investigator would have been able to continue the research at nonprofit B unhindered.

CONSIDERATIONS FOR REQUESTORS AND RECIPIENTS OF MODEL ORGANISMS NIH sharing policies seek to ensure that research tools and results are made available to the biomedical community in a timely manner for furthering the research enterprise. Publications are the major medium by which the biomedical community learns of research results. When model organisms are reported in publications, users of the shared resource should acknowledge the source (repository and/or investigator) and reference the original citation describing the strains in their publications.

This citation helps identify the specific material used by identifying the source. Standard genetic nomenclature should be used to designate the strains. Scientific custom is to name providers of materials as authors if providers collaborate with recipients of organisms and contribute intellectually to the manuscript. Otherwise, it is appropriate for providers to be acknowledged as the source of research resources upon which the manuscript is based. This acknowledgment most appropriately appears in the Materials and Methods section of the publication.

As noted previously for developers and providers of model organisms, requestors and recipients of model organisms or related research resources produced with NIH funding should be mindful of the policies regarding inventions arising out of NIH funding agreements, including issues related to intellectual property. For example, if the recipient is employed by a not-for-profit institution (e.g., university) and conducts research with a model organism that results in a patentable invention, title to the invention must be properly assigned to the not-for-profit institution.

REVIEW AND EVALUATION OF SHARING PLANS When a sharing plan is submitted as part of the funding application, the Scientific Review Group, or study section, will comment on the adequacy of the sharing plan. Reviewers will be asked to describe their assessment of the sharing plan in an administrative note and, except in defined circumstances, will not include their assessment in the overall priority score. For some special initiatives, such as Funding Opportunity Announcements (FOAs) specifically directed to the development of model organisms, reviewers may be asked to integrate their evaluation of the sharing plan with other review criteria and factor their assessment into the overall evaluation of scientific merit. Any concerns must be resolved before an award can be made. Program staff, who will review sharing plans before making funding recommendations, should be the first point of contact for resolving sharing plan deficiencies.

Whether a sharing plan is reasonable can be determined by the reviewers on a case-by-case basis, taking into consideration the organism or related resource involved, the timeline, the applicant's decision to distribute the resource or deposit it in a repository, and other relevant considerations. For mechanisms other than special FOAs, the presence or adequacy of a plan should not enter into the scoring of the application. The sharing plan itself should be discussed after the application is scored and reviewer comments recorded in an administrative note. For special initiatives and research mechanisms, where the text of the announcement specifically includes the adequacy of a sharing plan as a review criterion, the adequacy of the plan will be discussed by the reviewers and will figure into the score. Finally, it should be noted that budget review would be part of the overall review consistent with usual NIH practice.

PROGRESS REPORTING AND MONITORING OF SHARING Grants management staff will ensure that an acceptable sharing plan is referenced in the Notice of Grant Award (NGA) either by the standard reference to the funded application or by specific reference to a revised plan. The NGA is a binding legal document. The terms and conditions incorporate, and the grantee accepts, the terms once funds are drawn to conduct the project. The Grants Management Officer works with the NIH program staff to complete this process and address enforcement if the plan is not followed. Recognizing that circumstances can change, grants management and program staff will work with grantees to renegotiate sharing plans as necessary. Each Institute or Center will determine an appropriate response, consistent with the NIH guidelines, if the terms and conditions of the award are violated.

Annual progress reports should report the number of sharing requests received and the number fulfilled when submitting a noncompeting renewal progress report, as well as the Final Progress Report.

As the expectations and tools available to facilitate model organism sharing continue to evolve, maximum flexibility is encouraged to allow for renegotiations during the project period at the request of either the Institute/Center or the funded institution in response to materially new and/or unforeseen information or developments. Sharing plans should be discussed with the appropriate NIH program contact who can work with the developer to modify the sharing plan accordingly.

It should also be noted that the sharing plan holds for the life of the grant. Accordingly, it should address how the institution plans to make the model organism and related research resources available after the grant has expired, such as through cost-reimbursable charges or donation to a repository. As noted earlier, repository use is highly encouraged to provide for easy long-term access.

CONCLUSIONS

The sharing of research materials and data is critically important in the progress of biomedical research. Along these lines, the NIH Model Organism Sharing Policy builds upon extant NIH sharing policies to continue to fund and promote broad and enabling research effectively and efficiently.

This chapter provides sample plans, approaches, guidance, and resources for investigators to consider in how to share model organisms and related research resources with the research community.

When done effectively, such sharing of research resources moves research forward and enables additional research to benefit public health.

APPENDIX: EXAMPLES OF REPOSITORIES AND ASSOCIATED RESOURCES FOR SHARING OF MODEL ORGANISMS AND RELATED RESOURCES AND DATA

NIH Model Organism for Biomedical Research website: http://www.nih.gov/science/models/.

Examples of rodent resources: NCRR Rodent Resources website: http://www.ncrr.nih.gov/ncrrprog/cmpdir/RODENT.asp.

Examples of mouse resources:

Trans-NIH Mouse Initiatives: Major resources including sequencing and mapping resources, mutant mouse resources, and databases and repositories: http://www.nih.gov/science/models/mouse/resources/index.html.

Mouse Models for Human Cancer Consortium: http://emice.nci.nih.gov.

Mutant Mouse Regional Resource Centers: http://www.mmrrc.org.

NIH Knockout Mouse Project (KOMP): http://www.nih.gov/science/models/mouse/knockout/index.html.

Partial listing of mouse repositories and databases: http://www.nih.gov/science/models/mouse/sharing/4.html.

Examples of rat resources:

Rat genomics and genetics: Major resources including rat databases and rat genomic tools: http://www.nih.gov/science/models/rat/resources/index.html.

Rat Resource and Research Center: http://www.nrrrc.missouri.edu.

NIH Autoimmune Rat Model Repository and Development Center: http://dvrnet.ors.od.nih.gov/ratcenter/index.html.

Examples of fish resources:

NCRR Fish Resources website: http://www.ncrr.nih.gov/ncrrprog/cmpdir/FISH.asp.

Zebrafish Information Network: http://zfin.org.

REFERENCES

1. Grieder FB, Strandberg JD. The contribution of laboratory animals to medical progress–past, present, and future. In: Hau J, Van Hoosier GL Jr, Eds. *Handbook of Laboratory Animal Science: Essential Principles and Practices*, 2nd ed., Vol. 1. Boca Raton, FL: CRC Press, 2002:1–11.
2. Hendrikson C. A short history of the use of animals in vaccine development and quality control. *Dev Biol Stand* 1996;86:3–10.
3. Malakoff D. The rise of the mouse, biomedicine's model mammal. *Science* 2000;288:248–253.
4. Rogers JR, Bradley A. The mouse genome sequence: Status and prospects. *Genomics* 2001;77:117–118.
5. Public Health Service Grants Policy Statement. Washington, DC: U.S. Department of Health, Education, and Welfare, 1976. [DHEW Publication No. (OS) 77-50,000.]
6. Special Announcement: NIH Policy Relating to Reporting and Distribution of Hybridomas Produced Under Grants and Contracts. The NIH Guide Supplement for Grants and Contracts, May 1, 1980. Bethesda, MD: National Institutes of Health (NIH), 1980.
7. Notice: NIH Policy Relating to Reporting and Distribution of Unique Biological Materials Produced with NIH Funding. The NIH Guide for Grants and Contracts. No. 13(4):1. Bethesda, MD: National Institutes of Health (NIH), 1984.
8. NIH Procedures for Handling Non-Election of Title to Patentable Biological Materials. The NIH Guide for Grants and Contracts. No. 25(16); PT 36:96–131. Bethesda, MD: National Institutes of Health (NIH), 1996.
9. NIH Policy on Sharing of Model Organisms for Biomedical Research. The NIH Guide for Grants and Contracts. NOT-OD-04-042. Bethesda, MD: National Institutes of Health (NIH), 2004. (Accessed September 1, 2006, at http://grants2.nih.gov/grants/guide/notice-files/NOT-OD-04-042.html.)
10. NIH Grants Policy Statements. Bethesda, MD: National Institutes of Health (NIH), 2003. (Accessed September 1, 2006, at http://grants.nih.gov/grants/policy/#gps.)
11. Principles and Guidelines for Recipients of NIH Research Grants and Contracts on Obtaining and Disseminating Biomedical Research Resources: Final Notice. U.S. Federal Register. No. 64(246):72090–72096. Bethesda, MD: National Institutes of Health (NIH), 1999.
12. National Research Council of the National Academies. *Sharing Publication-Related Data and Materials: Responsibilities of Authorship in the Life Sciences*. Washington, DC: The National Academies Press, 2003.
13. Society for Neuroscience. *Responsible Conduct Regarding Scientific Communication*, 1st ed. Washington, DC: Society for Neuroscience, 1998.
14. Final NIH Statement on Sharing Research Data. The NIH Guide for Grants and Contracts. NOT-OD-03-032. Bethesda, MD: National Institutes of Health (NIH), 2003. (Accessed September 1, 2006, at http://grants2.nih.gov/grants/guide/notice-files/NOT-OD-03-032.html.)
15. NIH Grants Policy Statement 12/03, 2003. Bethesda, MD: National Institutes of Health (NIH). (Accessed September 1, 2006, at http://grants.nih.gov/grants/policy/nihgps_2003/index.htm.)
16. Developing Sponsored Research Agreements: Considerations for Recipients of NIH Research Grants and Contracts. The NIH Guide for Grants and Contracts. No. 23(25); PT 34:94–213. Bethesda, MD: National Institutes of Health (NIH), 1994. (Accessed September 1, 2006, at http://grants2.nih.gov/grants/guide/notice-files/not94-213.html.)
17. Intellectual Property Reporting for NIH Grantees That Also Have Involvement with the Veterans Administration. The NIH Guide for Grants and Contracts. NOT-OD-01-033. Bethesda, MD: National Institutes of Health (NIH), 2001. (Accessed September 1, 2006, at http://grants.nih.gov/grants/guide/notice-files/NOT-OD-01-033.html.)

7 Databases for Biomedical Animal Resources

Richard L. Crawford and Tim Allen

ABSTRACT

Database searching for research involving warm-blooded animals is required by law in the United States and is also a requirement for performing animal research in many other countries. It is, therefore, necessary for the investigator/researcher to perform a reasonable database search before developing the research protocol.

Key Words: Databases, Searching, Websites, Animal models, Resources, Strategy, Protocol, Animal, Area of Study, Sources of Information.

INTRODUCTION

The United States (U.S.) Animal Welfare Act,[1] U.S. Department of Agriculture (USDA) animal welfare regulations,[2] and USDA-Animal Care Policy 12[3] require that the principal investigator, using warm-blooded animals in research, must ensure that pain and distress are minimized, that alternatives to the proposed procedure or protocol are not available, or if alternatives are available why they were not used, and that the research does not unnecessarily duplicate previous experiments. Also, many countries now have regulations that require researchers to document that they have conducted a search for possible alternatives when applying for permission to conduct animal research.

Database searching is the best method to find animal resources and to comply with the requirements of the Animal Welfare Act or similar regulations. It is important to become familiar with the database platform, the subjects covered in each database, and the sources of information for each database to perform an effective and reasonable search effort. An institution's librarian or information specialist can be of considerable help with database searching and should be consulted prior to developing the research protocol. Training of researchers in how to perform a proper database search is also recommended as the search strategy determines the type and quality of the information obtained.

From: *Sourcebook of Models for Biomedical Research*
(P. M. Conn, ed.), © 2008 Humana Press Inc., Totowa, NJ.

DATABASES

While there are many papers discussing the utility of various bibliographic and specialty databases and information centers in the development of animal research protocols,[4–18] it is still useful to point out major providers/suppliers/repositories of animal models for biomedical research.

In the United States, the National Institutes of Health (NIH) is the primary resource for the development and funding of biomedical animal models. Their website—Model Organisms for Biomedical Research (http://www.nih.gov/science/models)—is a good starting point for information on various projects under development by the NIH, its grantees, and other national and international laboratories. Among the organisms listed are mammalian models, especially rat and transgenic mouse models, yeast, fungi, round worm, fruit fly, and frogs. Another excellent resource is the NIH, National Center for Research Resources, Division for Comparative Medicine. (http://www.ncrr.nih.gov/comparative_med.asp). The Division for Comparative Medicine provides access to laboratory-bred nonhuman primates including macaques, baboons, chimpanzees, owl monkeys, squirrel monkeys, and others. The Division also maintains an aquatic animal models program (zebrafish, *Xiphophorus*, cephalopods, and *Aplysia*) and has initiatives for invertebrates (*Drosophila*, *Caenorhabditis*, *Aplysia*, *Tetrahymena*, and cephalopods). The Institute for Laboratory Animal Research at the U.S. National Academies of Science has a unique search engine available that searches the websites of suppliers and repositories that sell or distribute experimental animals. It is available at http://dels.nas.edu/ilar_n/ilarhome/models.shtml.

The European Union (EU) maintains the European Mouse Mutant Archive (EMMA) (http://www.emma.rm.cnr.it/), which preserves and distributes relevant strains collected from repositories and laboratories throughout the EU. The EMMA Strain Database is accessible through an easy to use search utility and can be found at http://andy.emma.cnr.it/jEmma/list.utf8.html. In Japan, the Riken Bioresource Center was established by the government to serve as a global resource for biomaterials. The Center serves as the Japanese repository for an extensive collection of mouse strains. The searchable catalog provides a detailed description of the model, strain information, the developer, and pertinent journal references. The English version is available at http://www.brc.riken.jp/lab/animal/en/. Another excellent inter-

Table 7–1
National and international sources of animals

Name	*Web address*
NIH Model Organisms for Biomedical Research	www.nih.gov/science/models
NIH Division for Comparative Medicine	www.ncrr.nih.gov/comparative_med.asp
Institute for Laboratory Animal Research Models search engine	dels.nas.edu/ilar_n/ilarhome/models.shtml
European Mouse Mutant Archive (EMMA)	www.emma.rm.cnr.it
EMMA Strain Database	Andy.emma.cnr.it/jEmma/list.utf8.html
Riken Bioresource Center	www.brc.riken.jp/lab/animal/en
Federation of International Mouse Resources	www.fimre.org
The Jackson Laboratory	www.jax.org
International Mouse Strain Resource	www.informatics.jax.org/imsr/index.jsp
Charles River Laboratories	www.criver.com
Taconic Farms/Lexicon Genetics	www.taconic.com
Harlan Sprague Dawley	www.harlan.com
Harlan Europe	www.harlaneurope.com
Lab Animal Suppliers Guide	guide.labanimal.com
Laboratory Animals Breeders and Suppliers	www.lal.org.uk/breedersandsupp1.html

national source of information on animal models and their availability is through the Federation of International Mouse Resources (FIMRe). This 11 nation collaboration provides researchers with access to a searchable mutant mouse catalog, searchable database of mouse genome informatics, nomenclature guidelines, and links to repository sites for contributing new strains. It is available at http://www.fimre.org. These websites also provide numerous links to university collections, national laboratories, and commercial vendors.

Not surprisingly, commercial vendors also maintain extensive databases of useful animal models. The Jackson Laboratory (http://www.jax.org), which operates as a nonprofit institution, is one of the world's foremost repositories for unique mouse models and maintains a searchable database of available strains. This database, the International Mouse Strain Resource, contains information on global stocks of inbred, mutant, and engineered mice. It can be found at http://www.informatics.jax.org/imsr/index.jsp. Charles River Laboratories (http://www.criver.com) is a commercial vendor of rodents, rabbits, chickens, and chicken eggs. It maintains a database of disease models and transgenic models and provides access to the Deltagen repository of 900 knockout mouse lines. Taconic Farms (http://www.taconic.com), with its partner Lexicon Genetics, maintains a searchable database of genetically modified mice and has numerous other rodent models listed with detailed descriptions of their utility. Harlan (http://www.harlan.com) and Harlan Europe (http://www.harlaneurope.com), as the world's largest suppliers of laboratory animals, provide information resources on a variety of animal models and maintain an extensive catalog of rats, mice, hamsters, guinea pigs, cotton rats, gerbils, and rabbits. More vendor information can be found at http://guide.labanimal.com and http://www.lal.org.uk/breedersandsupp1.html. Table 7–1 summarizes the information listed above.

Large bibliographic databases such as the National Library of Medicine's Pubmed (http://www.pubmed.gov), Elsevier's EmBase (http://www.embase.com), and Thomson Scientific's BIOSIS (http://www.biosis.org) are catalogs of the world's peer-reviewed scientific publishing. As such they contain a wealth of information on different animal models and provide background information on these models. It is important to remember that while there is some overlap in coverage, each database contains unique information not found in the others. Not only does subject coverage vary but the sources of information are varied as well. For example, a search of Medline should also include EmBase, which includes monographs, reports, and other useful sources not included in Medline, which covers only journals. Several core databases, such as Medline, EmBase, and Biosis, should be searched in order to conduct a comprehensive literature search. Elsevier has made this somewhat easier by combining their EmBase with unique records from Medline to create EmBase.com containing more than 17 million biological and pharmacological records. Choosing databases pertinent to the protocol will provide the researcher with better information and more accurate resources for the study and will better comply with the laws and regulations pertaining to animal research.

The following tables list selected bibliographic databases for animal resources (Table 7–2), the subjects covered (Table 7–3), and the sources of information for the selected databases (Table 7–4). Table 7–5 lists web addresses for database vendors or portals to a variety of online databases.

Table 7–2
Bibliographic databases, biomedical, veterinary, and biological

AGRICOLA—agricola.nal.usda.gov
CAB Abstracts—www.cabi.org
MEDLINE—www.ncbi.nlm.nih.gov/pubmed/
EMBASE—www.embase.com
BIOSIS—www.biosis.org
Zoological Record—www.biosis.org
PsycINFO—www.apa.org/psycinfo/
Scopus—www.scopus.com

Table 7–3
Subject coverage of selected databases

AGRICOLA	*BIOSIS*	*CAB*	*EMBASE*	*MEDLINE*	*Zoological Record*
General agriculture	General agriculture	General agriculture	Clinical medicine	Clinical medicine	Zoological information
Animal science	Aerospace biology	Animal science and production	Experimental medicine	Experimental medicine	Behavior and communication
Chemistry and biochemistry	Biochemistry and anatomy	Crop science	Pharmacology, drugs, potential drugs	Pharmacology	Physiology, immunology
Microbiology	Bacteriology (microbiology)	Forestry	Biochemistry	Microbiology	Biochemistry
Cytology	Cell biology	Pest control	Developmental biology	Administration	Diseases
Human and animal nutrition	Botany	Human nutrition	Forensic medicine	Nutrition	Evolution
Biotechnology	Anatomy	Biotechnology	Health economics	Nutrition	Genetics
Physiology	Physiology	Pesticides	Occupational health	Anatomy and physiology	Histology
Veterinary medicine	Clinical medicine	Veterinary medicine	Toxicology	Veterinary medicine	Taxonomy
Wildlife	Pathology	Machinery and buildings		Occupational medicine	Life cycles/ development
Zoology	Biophysics	Economics		Toxicology	Nomenclature
Entomology	Toxicology			Other medical topics	Paleontology

Source: United States Department of Agriculture, National Agricultural Library, Animal Welfare Information Center.

Table 7–4
Sources of information for selected databases

AGRICOLA 1970–present	*BIOSIS 1969–present*	*CAB 1972–present*	*EMBASE 1974–present*	*MEDLINE 1966–present*	*Zoological Record 1978–present*
>1000 journals	Approximately 9000 journals	>14,000 journals/ translations	Approximately 3700 journals	4300 journals	Approximately 6000 journals
Books, monographs	Books, monographs	Books, monographs	Books		Books, monographs
Proceedings	Proceedings/abstracts	Symposia	Conference proceedings, symposia, meetings		Conference proceedings, symposia, meetings
Research reports	Technical reports	Technical reports			Special reports
Theses	Nomenclature, rules	Theses, dissertations			Theses
Translations	Annual reviews	Review journals			
Bibliographies	Bibliographies	Bibliographies			
Electric documents	Patents 86–89	Patents			
Audio visuals	Letters, notes	Annual reports			
USDA publications	Guides	Guides			
Government documents	Research comments	Conferences			
Selected newsletters	Conferences	Meetings			
Manuals, SOPs	Symposia				
Toxicology protocols	Meetings				

Source: United States Department of Agriculture, National Agricultural Library, Animal Welfare Information Center.

Table 7–5
Databases, vendors, and portals

DIALOG—www.dialog.com
National Library of Medicine
Databases—www.nlm.nih.gov/databases
OVID—www.ovid.com
Primate Information Network—pin.primate.wisc.edu
AWIC—awic.nal.usda.gov
Databases—awic.nal.usda.gov/databases

Source: United States Department of Agriculture, National Agricultural Library, Animal Welfare Information Center.

Table 7–6
Additional databases available on the web

Scientific and Technical Information Network—stinet.dtic.mil (provides public access to unclassified military biomedical research)
Scirus—Scientific Information Search Engine developed by Elsevier Science—www.scirus.com/
PrimateLit—primatelit.library.wisc.edu/

WEB SEARCHING AND OTHER GENERAL DATABASES AVAILABLE ON THE WEB

Web searching is an information source that should not be ignored. There are many websites available and using the terminology "Animal Model" or "Animal Models" in any web search engine will bring up many websites concerning animal models. However, these will also include sites for animal replicas, sporting goods, top dog model, and other similar sites, and a lot of scrolling is required to locate pertinent and specific animal model sites. Consequently, it is necessary to choose your words wisely. Being specific will improve your results, i.e., use more than two or three words in your strategy. Remember, it is the World Wide Web, so consider international spelling, such as behavior, behaviour, tumor, tumour, estrogen, oestrogen, anesthetic, anasthetic, anaesthetic, and other words that may have several spellings. Enclosing a multiword phrase in quotation marks tells the search engine to list only sites that contain those words in that exact order. A plus (+) sign directly in front of a word indicates that the word or phrase must appear in the search results. A minus (−) sign indicates that the word or phrase should not be included in the search results.

By using proper searching strategies the results will be more accurate and meaningful to your protocol. In addition to the resources listed above, Table 7–6 lists other free resources that may provide useful information.

Because many countries require scientists using animals to consider alternative methods or to minimize pain and distress to those animals that are used, a listing of useful websites is provided below. Listed below are a few samples of websites for alternatives to animal use in research:

ALTERNATIVES

Alternatives to Skin Irritation & Corrosion Testing in Animals—www.invitroderm.com
Altweb—altweb.jhsph.edu
Animal Welfare Information Center—http://awic.nal.usda.gov
AVAR Alternatives in Education Database—www.avar.org
Center for Alternatives to Animal Testing (CAAT)—caat.jhsph.edu/
ECVAM: The European Centre for the Validation of Alternative Methods—ecvam.jrc.it/index.htm
Fund for the Replacement of Animals in Medical Experiments (FRAME)
www.frame.org.uk/index.htm
Humane Society of the United States (HSUS)—www.hsus.org
Interagency Coordinating Committee for the Validation of Alternative Methods (ICCVAM)—iccvam.niehs.nih.gov

Table 7–7
Comparison of alternative websites

	Altweb	*AVAR*	*AWIC*	*CAAT*	*FRAME*	*HSUS*	*ICC-VAM*	*Norwegian Reference Center*	*NORINA*	*Skin irritation*	*UC Center for Animal Alternative*
Articles	X		X	X	X	X	X	X			X
Abstracts	X	X	X						X	X	
Bibliographies			X			X				X	X
Conference proceedings	X		X	X	X		X				X
Dbase links	X		X		X	X		X			X
Document links	X		X	X	X	X		X	X		X
Grant information	X		X	X							
Grant program				X							
How to search			X		X						X
Legislation	X		X	X	X		X	X			
List of alternative methods	X	X	X		X	X	X		X	X	X
Organization links	X		X		X	X	X	X	X		X
Search engine	X		X			X		X	X	X	X

Source: United States Department of Agriculture, National Agricultural Library, Animal Welfare Information Center.

The Norwegian Reference Centre for Laboratory Animal Science & Alternatives—
oslovet.veths.no
NORINA (A Norwegian Inventory of Alternatives)—oslovet.veths.no/NORINA
University of California Center for Animal Alternatives—www.vetmed.usdavis.edu/Animal_Alternatives/main.html

Table 7–7 indicates the content and subject matter for the various web sites on alternatives to animal use.

SPECIFIC ANIMAL MODEL DATABASES

As previously indicated, the Internet has a lot of useful information available through general searches using broad terms such as "animal models of disease." There are animal model suppliers and vendors, books on animal models, journal articles on animal models and their use, lists of animal models available, and links to useful information on animal models. A good web search combined with a good multiple database search should provide useful information to the researcher that is pertinent to the research protocol. Listed below is a sample of databases to be found from various institutions.

- University of Michigan Transgenic Animal Model Core http://www.med.umich.edu/tamc
- Plant, Animal & Bacteria Cell Models http://www.cellsalive.com/cells/3dcell.htm
- Surgical Animal Models for Cardiovascular Diseases http://www.health.ufl.edu/anires/Models/index.htm
- Models for studying AIDS and AIDS associated cancers www.ihv.org/research/animal.html
- Animal Model of inherited human disorder www.vet.utk.edu/research/models.shtml
- Animal Models of Neuropsychiatric Diseases www.worldscibooks.com/medsci/p421.html
- Animal Models Center, Beth Israel Deaconess Medical Center http://cardiogenomics.med.harvard.edu/component-detail?project_id=233
- Animal Models of Diabetic Complications Consortium www.amdcc.org
- XENOGEN—Light Producing Animal Models www.xenogen.com/wt/page/animalmodel_list
- Zivic Laboratories Inc. www.zivic-miller.com/technical%20services%20department.htm
- Transgenic Animal Web www.med.umich.edu/tamc/links.html
- Canadian Council on Animal Care. Animal Models in Biomedical Research www.ccac.ca/en/CCAC_Programs/ETCC/Module05/04.html
- Zebrafish Animal Models for Human Hereditary Ocular Disease www.research-projects.unizh.ch/p5672.htm
- University of Arizona—Research Models & Disease Links www.ahsc.arizona.edu/uac/disease.shtml
- Animal Models Available from Baylor College of Medicine www.bcm.edu/blg/Animal/Mods.pdf

DATABASE SEARCH STRATEGY

Databases and web sites contain a great deal of useful research information, however, little pertinent information will be obtained without a proper search strategy to find the information desired. Listed below are pointers to keep in mind when preparing to do a database search.

1. What is the general area of study (drug testing, cardiology, toxicology, lipid metabolism, etc.)?
2. What is the type of protocol (research, teaching, testing, etc.)?
3. What is the proposed animal species?
4. Identify the systems or anatomy involved in the study.
5. List any drugs or compounds used in the procedure.
6. What are the objectives and end points of the protocol?
7. Use both American and European spelling.
8. List key words using terminology from your responses to the questions above.
9. If searching on a system that allows development of complex search strategies, keep your concepts separate and then combine them to obtain specific and pertinent information.
10. Choose databases that are appropriate to the area of study.

CONCLUSIONS

There is a tremendous amount of information available from databases and websites; however, it can be difficult to obtain the specific information desired if the proper databases are not searched and the terminology used is too broad or inappropriate to the protocol. Some databases, such as Agricola and Medline (PubMed), are available to search and download references for free. Other databases require subscriptions and the downloading references for a fee. If a facility is involved in a great deal of research, which necessitates database searching, a subscription to a database vendor would be helpful as multiple databases would be available for searching and those pertinent to the protocol or area of study could be utilized. Two, or more, databases should be searched and these databases should be pertinent to the protocol in both the subjects covered and the sources of information (not just journals). The search strategy, including databases used, key words used, and years searched, should be saved for use again when required or if changes are required in the protocol. Should the investigator not be familiar with database searching strategy, he or she should contact an institutional librarian or information specialist for assistance. Training in proper searching techniques is also very helpful to those who do not have librarians or information specialists available to them and are required to perform periodic database searches. Should neither of these options be available to the investigator, the Animal Welfare Information Center (U.S. Department of Agriculture, National Agricultural Library, Beltsville, MD; E-mail, awic@nal.usda.gov) is available for suggestions and assistance if desired.

ACKNOWLEDGMENTS

The use of trade, firm, or corporation names in this article is for the information and convenience of the reader. Such use does not constitute an official endorsement or approval by the United States Department of Agriculture or the Agricultural Research Service of any product or service to the exclusion of others that may be suitable.

REFERENCES

1. Animal Welfare Act. United States Code, Title 7, Chapter 54, Sections 2143(a)(3)(A)(B)(C).
2. Animal Welfare Regulations. Code of Federal Regulations, Title 9, Chapter 1, Subchapter A—Animal Welfare, Part 2, Subpart C, Sections 2.31 (d)(i)(ii)(iii).
3. Animal Care Policy 12, Written Narrative for Alternatives to Painful Procedures: June 21, 2000. www.aphis.usda.gov/ac/publications/policy/policy12.pdf.
4. Mattingly CJ, Rosenstein MC, Colby GT, Forrest JN, Boyer JL. The Comparative Toxicogenomics Database (CTD): A resource for comparative toxicological studies. *J Exp Zool A Comp Exp Biol* 2006;305(9):689–692.
5. Wiley JC, Prattipati M, Lin C-P, Ladiges W. Comparative Mouse Genomics Centers Consortium: The Mouse Genotype Database. *Mutat Res* 2006;595(1–2):137–144.
6. Sprague J, Bayraktaroglu L, Clements D, Conlin T, Fashena D, Frazer K, Haendel M, Howe DG, Mani P, Ramachandran S, Schaper K, Segerdell E, Song P, Sprunger B, Taylor S, Van Slyke CE, Westerfield M. The Zebrafish Information Network: The zebrafish model organism database. *Nucleic Acids Res* 2006;34(Database issue): D581–585.
7. Blake JA, Eppig JT, Bult CJ, Kadin JA, Richardson JE. The Mouse Genome Database (MGD): Updates and enhancements. *Nucleic Acids Res* 2006;34(Database issue):D562–567.
8. Smith AJ, Allen T. The use of databases, information centres, and guidelines when planning research that may involve animals. *Anim Welfare* 2005;14(4):347–359.
9. Krupke D, Naf D, Vincent M, Allio T, Mikaelian I, Sundberg J, Bult C, Eppig J. The Mouse Tumor Biology Database: Integrated access to mouse cancer. *Exp Lung Res* 2005;31(2):259–270. http://www.informatics.jax.org.
10. de la Cruz N, Bromberg S, Pasko D, Shimoyama M, Twigger S, *et al.* The Rat Genome Database (RGD): Developments towards a phenome database. *Nucleic Acids Res* 2005;33(Database issue): D485–491.
11. Weeks J, Hart RP. SCI-Base: An open-source spinal cord injury animal experimentation database. *Lab Anim (UK)* 2004;33(3):35–41.
12. Habermann B, Bebin A-G, Volkmer M, Herklotz S, Eckelt K, Pehlke K, Epperlein HH, Schackert HK, Wiebe G, Tanaka E. An Axolotl EST database: Gene information for studying regeneration. *Dev Biol* 2003;259(2):537.
13. FlyBase Consortium. The FlyBase database of the Drosophila genome projects and community literature. *Nucleic Acids Res* 2003;31(1):172–175. http://flybase.bio.indiana.edu.
14. Smith CM, Begley DA, Eppig JT, Finger JH, Hayamizu TF, Hill DP, Kadin JA, McCright IJ, Richardson JE, Ringwald M. The mouse Gene Expression Database (GXD): A resource for developmental biologists. *Dev Biol* 2003;259(2):492–493.
15. Santi PA, Beattie JW, Lukkes J, Nietfeld J. The Mouse Cochlea Database: Bibliographic Database Development. Society for Neuroscience Abstract Viewer and Itinerary Planner 2002, Abstract No. 162.3.
16. Anagnostopoulos AV (Reprint), Mobraaten LE, Sharp JJ, Davisson MT. Transgenic and knockout databases: Behavioral profiles of mouse mutants. *Physiol Behav* 2001;73(5):675–689.
17. Aleman CL, Mas RM, Rodeiro I, Noa M, Hernandez C, Menendez R, Gamez R. Reference database of the main physiological parameters in Sprague-Dawley rats from 6 to 32 months. *Lab Anim (UK)* 1998;32(4):457–466.
18. Allen T. On-line databases and the World Wide Web: What is available? What is missing? In: van Zutphen LFM, Balls M, Eds. *Animal Alternatives, Welfare, and Ethics. Developments in Animal and Veterinary Science,* Vol. 27. Amsterdam: Elsevier, 1997.

8 Psychological Enrichment for Animals in Captivity

KRISTINE COLEMAN

ABSTRACT

Animal care in biomedical facilities has undergone a dramatic transformation in the area of psychological well-being and enrichment over the past two decades. Today, attending to the behavioral needs of research animals is considered an integral part of animal care. Enrichment is defined as environmental stimuli provided to research animals in an effort to improve well-being by increasing species-specific and decreasing abnormal behaviors. Such environmental enhancement can help alleviate some of the stress associated with living in captivity, and can thus produce a better research model. This chapter discusses some of the issues surrounding the provision of enrichment and details some basic enrichment strategies, with emphasis on rodents, dogs, and nonhuman primates.

Key Words: Environmental enrichment, Psychological well-being, Animal welfare, Laboratory animals, Animal behavior, Animal husbandry.

INTRODUCTION

Animal care in biomedical facilities has undergone a dramatic transformation in the area of psychological well-being and enrichment over the past two decades. In the early part of the twentieth century, the emotional welfare of laboratory animals was largely disregarded. This situation changed as people realized that like humans, animals have certain basic behavioral needs, which, if not met, can adversely affect the behavior and physical health of the individual.

Today, providing for the psychological requirements of research animals is accepted as an integral part of animal care.[1] Most institutions devote substantial resources to ensure that animals receive enrichment (items provided to animals beyond their basic food and housing needs) of various sorts. Many facilities, particularly those housing nonhuman primates, have entire units dedicated to providing psychological enrichment for subjects.[2] Furthermore, enrichment is now provided not only to mammals, but also to birds,[3] snakes,[4] and fish.[5]

This chapter discusses the psychological enrichment of laboratory animals. While it focuses on rodents, dogs, and nonhuman primates, most of the concepts are universal, and can be applied to other laboratory species as well.

From: *Sourcebook of Models for Biomedical Research*
(P. M. Conn, ed.), © 2008 Humana Press Inc., Totowa, NJ.

WHAT IS PSYCHOLOGICAL ENRICHMENT? Environmental enrichment is "an animal husbandry principle that seeks to enhance the quality of captive animal care by identifying and providing the environmental stimuli necessary for optimal psychological and physiological well-being" (Shepherdson,[6] p. 1). In other words, it is a way to functionally simulate the natural environment of research animals, in an effort to increase the opportunity to express species-specific behaviors and decrease the occurrence of abnormal behaviors. The goals of providing psychological enrichment are to have animals that are in good physical condition, display a variety of species-typical behaviors, are resilient to stress, and easily recover (behaviorally and physiologically) from aversive stimuli.[7]

WHY PROVIDE ENRICHMENT?

ETHICAL CONCERNS The research community's views on the care of research animals reflect the significant change in attitude that has occurred in the general public over the past few decades. Much of this shift can be attributed to increases in scientific knowledge regarding the complex nature of animal behavior. Today, scientists and others using animals in research recognize that they have a responsibility to attend to the psychological well-being of their subjects.[8] In short, providing psychological enrichment, within experimental constraints, is generally considered the "right thing to do."

LEGAL OBLIGATIONS In most countries, including the United States, providing research animals with enrichment is not an option, but rather is required by various regulatory agencies. The Animal Welfare Act (AWA[9]) is the statute that safeguards laboratory animals in the United States. Nonhuman primates are afforded the greatest protection under the AWA, which mandates that research institutions housing nonhuman primates have in place a plan to ensure their psychological enrichment.[9] The AWA also calls for exercise for dogs used in research.[9] While the AWA does not contain provisions for mice, rats, or birds, scientists receiving federal funding from most U.S. funding agencies must adhere to the guidelines in the Public Health Service Policy On Humane Care And Use Of Laboratory Animals ["PHS Policy"; National Institutes of Health, Office of Laboratory Animal Welfare (OLAW)]. This policy covers all vertebrates and requires compliance with the AWA and the *Guide for the Care and Use of Laboratory Animals* ("*The Guide*"[8]). *The Guide* specifies that enrichment be provided for laboratory animals, including social housing when appropriate for the species and the use of objects and cage complexities that "increase the opportunities for the expression of species-specific postures and activities and enhance the animals' well-being" (Council,[8] p. 37). It also sets the standards for voluntary accreditation through the Association for Assessment and Accreditation of Laboratory Animal Care (AAALAC), an organization that promotes the humane treatment of animals used in science.

ENRICHMENT CAN IMPROVE THE ANIMAL MODEL One of the most compelling reasons to provide enrichment is to reduce stress and improve well-being for laboratory subjects, thereby improving the animal model. Stress can adversely affect many physiological parameters including the reproductive axis,[10] immunological parameters,[11,12] and brain function.[13] Stressed animals are at a higher risk for becoming ill[14] or developing maladaptive behaviors such as self-directed aggressive behavior,[15] which can affect a variety of research outcomes. Reducing this kind of stress can result in more reliable research subjects, and improve the interpretation of research results.

Providing social housing to rhesus macaques is an example of how enrichment can improve the animal model. Macaques and other nonhuman primates are social animals, and have evolved to live in complex societies. Socially housing monkeys affords them the opportunity to engage in species-specific behaviors and to develop social and cognitive skills necessary for group living.[16] It also reduces potential confounding factors, which can directly affect research results. For example, singly and group-housed monkeys differ with respect to various immunological parameters, such as natural killer cell activity, cytokine production, and proliferation responses to mitogens.[17–19] These differences can affect the results obtained in a variety of studies. Thus, it can be argued that housing monkeys in a more natural condition produces a better model for human conditions.

EFFECTS OF ENRICHMENT

Environmental enrichment in rodents has been associated with a variety of behavioral, neural, and physiological parameters, including increased exploratory behavior,[20–22] decreased anxiety on the elevated plus maze,[23] improved learning ability,[24,25] increased brain size and weight,[26] increased number of synapses per neuron,[27] increased neurogenesis,[28] and decreased cell death.[26,29–32] Mice in an enriched environment also have higher natural killer (NK) cell activity than nonenriched mice, suggesting better immune function.[23] On the other hand, enriched rodents tend to consume more food[33] and ethanol[22,34] compared to nonenriched rats, an outcome that needs to be carefully considered in the study design.

While enrichment has been shown to decrease stress (as measured by serum levels of corticosteroid) in nonhuman primates,[35] several studies have shown that it can increase basal corticosteroid (CORT) levels and cause larger adrenal weights in male rats.[23,36,37] However, these studies also reported that the CORT levels of the enriched rats were still within normal limits and were not associated with pathology.[23] Furthermore, enriched animals also showed less fear and exhibited less response to other stressors, such as predators or experimental stressors, suggesting that enrichment can attenuate the stress response to aversive events.[23,25,36,38] This mitigating effect of enrichment on stress response has been found in other species as well, including birds[39] and snakes.[4] This ability to recover from stressful experiences is considered an important aspect of well-being.[7,40]

ENRICHMENT AND EXPERIMENTAL VARIABILITY The findings described in the previous section underscore the importance of considering enrichment as a potential variable in studies. Even small changes in enrichment practices can have dramatic effects on many experimental outcomes, leading to the potential for variability among laboratories.[41] It is important to carefully document the type of enrichment provided, as well as when and how it is implemented, and to report this information in scientific publications.

However, the absence of enrichment can also increase variability between laboratories.[42] Lack of enrichment may increase an animal's vulnerability to the stressors that are being investigated.[43] Furthermore, there can be vast differences in how individuals respond to stress.[44–46] Using enrichment to decrease stress associated with experimental procedures may reduce variability and the development of stress-related problems, thereby increasing the validity of experimental results.

IMPLEMENTING AN ENRICHMENT PLAN

There are many resources available for implementing an enrichment plan for various species,[1,47–49] including a special issue of the *Institute for Laboratory Animal Research (ILAR) Journal* (2005, Volume 46). Below are some general guidelines.

ENRICHMENT SHOULD BE SPECIES-SPECIFIC Each species has evolved under different environmental conditions, resulting in unique behavioral patterns. Successful enrichment programs must take these specific behaviors into account in an effort to increase the biological functioning of the captive environment.[50] For instance, while most rodents are social, golden hamsters are not. Even within a species, enrichment needs may also vary by strain[51] or gender.[43] The Animal Welfare Information Center, part of the United States Department of Agriculture, maintains a website (http://www.nal.usda.gov/) containing useful information regarding the behavior of various laboratory animal species.

CONSISTENCY IS IMPORTANT Because enrichment can affect many research outcomes, care must be taken to ensure enrichment is given uniformly, and not haphazardly. Enrichment should be considered an integral part of animal care, and not as something extra to be done "when there is time."[1] It is also important for caretakers and researchers to avoid providing extra enrichment to "favorite" subjects. Not only can this be an experimental confound, but it can also affect the well-being of the animals. We have found that if only one or two monkeys in a room get an enrichment device, they are sometimes threatened by other, presumably more dominant, individuals (personal observation), which can lead to self-biting or other behavioral problems.

INTEGRATE BEHAVIORAL NEEDS WITH EXPERIMENTAL NEEDS Successful enrichment programs integrate the needs of the animal with experimental constraints. Studies on feeding behavior may prohibit the intake of extra calories, but other enrichment options can be used in lieu of food enrichment. Macaques can be given foraging devices containing toys that can be manipulated (e.g., small cat toys) instead of food.[52] New enrichment options should be tested prior to their implementation, to see if they will affect experimental outcomes.

ENRICHMENT SHOULD BE RESULTS-BASED An important, but difficult, part of a successful enrichment program is assessment. Enrichment offerings should be periodically evaluated, and provisions that do not have a demonstrable effect upon well-being of the animal should be reconsidered.[50] It is not always easy to decide whether or not a particular enrichment offering is effective. Parameters to be assessed should include behavior (enrichment items should increase species-typical and/or decrease abnormal behaviors) as well as physiological variables that indicate well-being (i.e., animals should be healthy, with few signs of chronic stress). It is also important to consider the cost and safety of enrichment items. Items that are difficult to clean may be fomites, increasing the risk of illness.[17]

Optimally, enrichment items directly affect the well-being of laboratory animals. However, there may be value in enrichment options that have only indirect effects on the animals for which they are designed. Televisions, for example, are often provided to nonhuman primate species, even though it is not part of the normal behavioral repertoire of wild monkeys. Still, because people like to watch television, they assume (rightly or wrongly) that animals would as well. In practice, most animals do not pay a lot of attention to TV enrichment, although some do. However, beyond the direct effects of TV enrichment on the animals (which may be small), it allows the caretakers to feel that they are doing something positive for the animals. This can increase their job satisfaction and improve the relationship between caretakers and the animals, which can, in turn, improve well-being for the primates.[53]

TYPES OF ENRICHMENT Enrichment can be broken down into five basic categories,[47,48] including social (housing gregarious species in groups or pairs), physical (perches, platforms, houses or platforms, toys, and other objects that can be explored or manipulated), food (novel food items, or objects that increase foraging time), sensory (televisions, radio), and occupational (physical and mental activity, including training). These enrichment options are not equally attractive to all species. Ideally, laboratory animals would receive multiple forms of enrichment.

ENRICHMENT FOR COMMONLY USED LABORATORY ANIMALS

RODENTS AND RABBITS Rodents and rabbits are among the most commonly used animals in biomedical research. Strikingly, although they have been bred in laboratories for centuries, they still maintain behaviors similar to wild counterparts,[54,55] including nest-building, burrowing, hiding, and foraging. Even knockout and genetically modified strains of mice and rats retain many behaviors of wild animals. A successful enrichment program should take species-typical behaviors into account.

Most rodents and rabbits are gregarious (although golden hamsters are not), and live in family groups.[56] Thus, social housing (in groups or pairs) should be one of the first considerations in implementing an enrichment program for these species. Such housing allows individuals to engage in many species-typical behaviors, including play and grooming.[55] Furthermore, having social companionship can buffer individuals against external stresses,[57] which, in turn, improves well-being.

However, social housing can lead to aggression in some species. Mice tend to be more aggressive than rats. Furthermore, both strain[58] and sex can affect aggression in mice. For example, female mice tend to fight less than males, particularly during the mating season. Providing additional enrichment can help ameliorate some of the negative consequences of social housing. Van Loo *et al.*[59] found that adding nesting material can reduce aggression in socially housed male mice.

In general, rodents and rabbits are prey species, which shapes much of their behavior. When confronted with an aversive or frightening stimulus, their first reaction is to hide or burrow. Providing the animals with shelter (e.g., tunnels or houses) or substrate in which to burrow affords them the ability to withdraw from aversive situations, which can alleviate fear and stress.[55] It also allows animals to exercise some control over their environment,[60–62] which itself reduces stress-related behaviors such as stereotypical behavior (defined as repetitive, invariant behavior patterns with no obvious goal or function).[37,57,63,64] These items, along with other forms of physical enrichment (e.g., platforms, nesting material, toys, and gnawing sticks), also increase exploration and manipulation and decrease abnormal behaviors in rodents and rabbits.[33,37,57,65–67] Many of these items (e.g., tunnels) can easily be made with commonly available materials, such as PVC tubing. Commercially available polycarbonate rodent huts provide shelter, but are tinted to allow observation of the animal by care staff (Figure 8–1).

In the wild, rodents and rabbits spend a great deal of their time foraging for food, and thus food enrichment is an important part of an enrichment program.[68] Indeed, rodents in the laboratory will work to obtain food, even when it is freely available.[69] Feeding enrichment can be as simple as hiding food in the substrate, which promotes foraging.

Exercise can also benefit rodents and should be included in an enrichment plan when possible, particularly if social housing is not an option. Exercise on a running wheel can mitigate some of the anxiety that occurs as a result of social isolation in mice.[70]

In addition to improving animal welfare, enrichment can also increase safety for humans. Providing enrichment can help to decrease the risk of bites and scratches, increasing safety for animal caretakers and researchers.[71] Van de Weerd *et al.*[23] found that mice from an enriched environment were easier to handle (e.g., showed more relaxed behavior) compared to mice in a standard environment.[33]

Providing enrichment is not without problems, however. The addition of objects can increase aggression among socially housed animals,[37] due to territoriality over particular items. Making more than one "favored item" available can help to reduce this aggression. Items must be carefully cleaned, to avoid causing illness.

Figure 8–1. Example of enrichment items for laboratory mice. The hut is tinted such that the mouse cannot see out, but human caretakers can see in. (See color insert.) (Photo by Jill Rawlins.)

And, care must be taken in deciding upon things to use. While nesting material is used without problem in many species, it can cause conjunctivitis in some genetically modified mice.[72] Therefore, enrichment options should be carefully researched prior to implementation, to ensure that they are appropriate for the species and strain being utilized.

DOGS Dogs are unique among laboratory animals in that they have evolved a well-developed bond with humans. Therefore, one of the best forms of enrichment for dogs is positive interactions with human handlers.[40] Socializing dogs with humans, particularly when they are puppies, allows them to learn appropriate social cues and behaviors.[73] Early socialization may also reduce stress for the dog and help to prevent behavior problems from starting.[74] Odendaal and Meintjes[75] demonstrated that simply petting a dog can increase affiliative neurotransmitters, including β-endorphins, for both the dog and the human. Working with socialized dogs is also safer for handlers. Dogs reared with human contact are more outgoing, easier to work with, and less likely to bite handlers compared to those without.[73,76]

Since the dog is a highly gregarious species, social housing, in pairs or groups, is important for their welfare. Social housing helps to increase species-specific behaviors, including play and exploration, and reduce abnormal behaviors such as stereotypies.[77] If social housing is precluded due to experimental constraints, dogs should be allowed to see conspecifics.[78]

Adding physical enrichment items such as beds, resting platforms, and toys can enhance the environment and improve the welfare of laboratory dogs.[73,77,79] It can also help increase learning in elderly dogs.[80] Providing dogs with a variety of objects from which to choose, including chew toys, can reduce boredom.[81]

Exercise is another important part of an enrichment plan for laboratory dogs. While exercise has not been shown to substantially improve welfare,[82] it is mandated by the Animal Welfare Act. Singly housed dogs must be given space to exercise.[9] However, simply giving the dog an area in which to play may not be enough to promote exercise. Dogs generally do not exercise unless they have stimulation from humans or other animals.[83] One way to ensure that dogs are getting exercise is to provide them with daily "play time" during which they can interact with other dogs. Another option is for handlers to take dogs for walks.

An effective way to improve welfare and refine common husbandry practices with dogs is to train them for various tasks using positive reinforcement training (PRT).[73] PRT techniques are a form of operant conditioning[84] in which the subject is presented with a stimulus (e.g., a verbal command), responds by performing a specific (i.e., target) behavior (e.g., present a body part for injection), and is provided with reinforcement (e.g., food treat). Training increases mental stimulation for the dogs.[73] It also makes dogs more obedient and easier to handle.[73] Bookstores are filled with resources on how to train dogs, including "Don't Shoot the Dog," [85] which provides an excellent introduction to PRT.

NONHUMAN PRIMATES Many species of nonhuman primates are used in biomedical research, although rhesus and other macaque species are among the most common. Unlike rodents or dogs, primates used in research have not been domesticated, and therefore have maintained even more of the behavioral repertoire of their wild counterparts than these other groups. Macaques and other nonhuman primates are social animals and live in complex societies, often with a strict dominance hierarchy. They form intricate social relationships in the wild, where they spend a great deal of time engaged in social behaviors such as grooming and huddling with other members of the troop. Thus, social housing is considered one of the best forms of enrichment for nonhuman primates.[16,86]

Options for social housing can vary from large outdoor enclosures to indoor cages, depending upon the constraints of the facility and the experimental design. For animals in cages, pair housing (i.e., housing two monkeys in a double cage with a removable slide) allows social contact between the partners, and as such is considered an acceptable form of social housing for several species of primates.[87,88] When full pair housing is not appropriate due to scientific reasons, the use of grooming contact slides[89] between adjacent cages should be considered. These slides are designed so that part of the slide is a solid panel while the other part consists of bars wide enough to permit grooming (Figure 8–2). They afford individuals privacy as well as increased social contact with the cage mate.[89]

Social housing increases the opportunity for animals to engage in many species-typical behaviors including play, feeding, and grooming.[90,91] It can also reduce abnormal behaviors such as stereotypical behavior and self-aggressive behavior,[92] although this is not always the case (personal observation). Social enrichment is, however, an important tool in preventing abnormal behaviors from occurring.[93] One of the main risk factors for the development of self-injurious behavior in macaques is having been reared without close social contact with conspecifics.[94] In addition, the presence of known companions can lessen the effects of various stressors,[95–97] and alter immune function,[18,19] which can improve well-being.

Social housing is not a panacea, however. It can result in injury if individuals are not compatible. Factors such as sex and age may affect how well individuals get along with each other. For macaques, males are generally harder to successfully pair house than females.[98] Temperament may also play a role in compatibility in pair-housed animals. In a study on female rhesus macaques (*Macaca mulatta*), we found that animals with similar temperaments (as measured by reaction to a novel object) were more likely to form successful pairs than those of different temperaments.[99]

Forming groups of monkeys is particularly challenging (see references 100–102 for strategies on group formation). Once formed, groups should remain stable. Introducing a new adult into an established group can result in injury or death to the newcomer. Even in the absence of overt aggression, social housing can be stressful, particularly in species with dominance hierarchies in which some animals have more access to resources than others. In these societies, individuals at the bottom of the hierarchy are more prone to stress and stress-related diseases than dominant individuals.[103,104]

The provision of physical enrichment is another important part of an enrichment plan for nonhuman primates. This type of enrichment, which includes toys, mirrors, wood, and other manipulatable objects, affords individuals the opportunity to express species-normative behaviors such as play and exploration.[35,105] Decisions regarding enrichment items should be based on the behavior of the species. For example, unlike macaques, owl monkeys (*Aotus* sp.) utilize nests in the wild, and should be provided with nest boxes in captivity. Macaques, on the other hand, utilize perches and other cage furniture.[86] Items such as toys should be rotated regularly to keep them novel and thus of interest

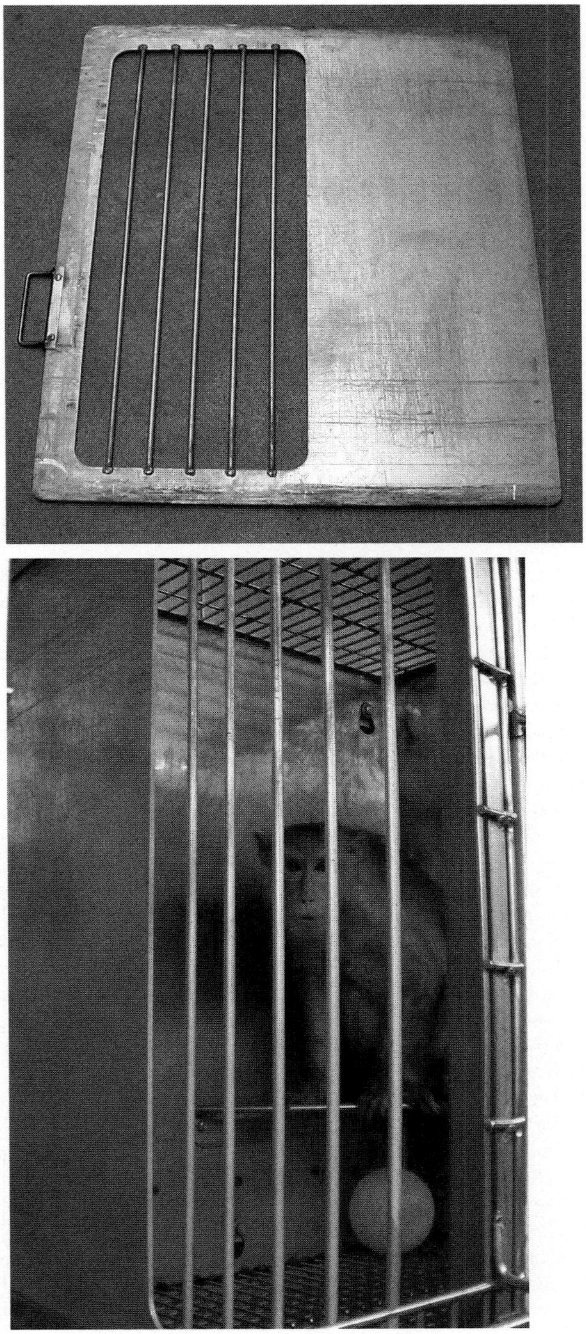

A

B

Figure 8–2. Grooming contact slide. The slide **(A)** is placed between two adjacent primate cages. The bars allow the partners to groom each other, while the solid part prevents full access to the other cage. **(B)** A slide within a cage (i.e., from the rhesus macaque's perspective). (Photos by Kevin Mueller.)

to the monkeys.[106–108] Specific enrichment devices, such as paint rollers or foraging boards covered with trail mix or other food items, are often provided to nonhuman primates in an effort to reduce the occurrence of abnormal behaviors such as stereotypy or overgrooming. There is debate as to the therapeutic value of such devices. While enrichment of this kind has been shown to decrease undesired behaviors in some studies,[105,109,110] others have found no discernable effect of the devices on abnormal behavior[111,112] or that the benefit was limited to the time the apparatus was present.[93]

Physical enrichment can increase the incidence of disease and injury for subjects. It is important to choose enrichment items that can be easily cleaned, to avoid spreading disease or illness.[17,113] Animals can injure themselves on certain types of hooks used to keep devices on the outside of cages (e.g., "S" hooks). Further, ropes and wire used to hang devices or swings should be covered with PVC tubing, to reduce the potential for strangulation.

Like rodents and rabbits, nonhuman primates spend much of their day foraging for food. This time is greatly reduced in a laboratory setting, where food is typically provided one or two times a day. Providing food enrichment can help functionally simulate the environment of nonhuman primate species.[50] Further, the USDA Draft Policy on Environmental Enhancement for Nonhuman Primates suggests that primates be provided with daily opportunities to forage. Devices have been developed to promote foraging and manipulation for the monkeys. These devices, which are typically put on the outside of cages and filled with trail mix and produce (depending on the species and dietary restrictions), provide monkeys with the challenge of obtaining food.[114] They have small holes through which the monkeys can forage for the treats. While these devices can be purchased commercially, they can also be made using materials such as PVC tubing (Figure 8–3). Handmade devices should be sturdy, so that they do not easily break. It is also important that devices can be thoroughly cleaned. Such enrichment devices have been found to decrease the occurrence of some undesirable behaviors, including stereotypies[112] and overgrooming,[105] when kept on the monkeys' cages continuously. However, these behaviors tend to return after time, likely due to increased boredom with devices that are kept on cages indefinitely. Rotating devices to keep them novel is recommended.[115] Adding a small toy to the device can also increase the amount of time they are utilized by the monkeys[52] (Figure 8–3).

Many facilities provide research primates with various forms of auditory and visual enrichment, including access to television, radio, brightly colored mobiles, or a window.[116] Computer screen savers can also be presented to the animals. Monkeys seem to watch the computer screens about the same amount as a television (personal observation). Television can be aversive if stressors, such as predators, appear on the screen. There is some evidence to suggest that sensory enrichment is beneficial to primates,[117] although there are contradictory results as well.[118] More work is needed to determine the benefits of this kind of enrichment.

Using positive reinforcement techniques to train monkeys to voluntarily cooperate with husbandry and/or research procedures is becoming recognized as an important tool in an effective enrichment program. PRT increases mental stimulation for subjects[119] and therefore is an effective form of psychological enrichment. Primates have been successfully trained to perform various tasks, such as presenting a body part for injection[120] or veterinary procedures,[121] taking oral medications,[122] and remaining stationary for blood samples.[120,123]

PRT can reduce the stress associated with common management procedures.[119,121,124] By allowing individuals to cooperate with the procedures, PRT may give primates an increased sense of control over their environment.[119] In addition, training can reduce abnormal behaviors in nonhuman primates. In a preliminary study,[125] we found that PRT reduced the occurrence of stereotypical behavior (e.g., pacing, rocking) in some, but not all, rhesus macaques. Furthermore, training primates to cooperate with research procedures can reduce the inherent risk of the

A

B

Figure 8–3. Foraging devices for nonhuman primates. **(A)** A device made by drilling extra holes into a wiffle ball. **(B)** A commercially available device (Crumble Disc Holder, Bio-Serv, Frenchtown, NJ), with a small cat toy added. (Photos by Jill Rawlins.)

procedures. It is clear that working with a cooperative animal is safer for caretakers and handlers than working with a highly stressed, and potentially reactive, animal.[126] While training is beneficial for the animal as well as the research, it can also involve a great deal of time. However, the time invested will likely be made up for by the decrease in time it takes to conduct procedures.[127]

CONCLUSIONS

In designing an enrichment program for an animal model it is essential to balance two equally important goals: the reduction of stress associated with the captive environment and the development of a model that (as closely as possible given the overriding need to care for the animal's physical and mental well-being) approximates the human condition. Life is as full of stressors for humans as it is for other animals.

Thus, it is not a realistic goal of any enrichment program to eliminate all stress. Not only is that goal unattainable, but it also is not necessarily in the animal's best interest. Stressors are not intrinsically detrimental,[7] and are a natural part of the behavioral repertoire of many animals.[16] Indeed, some can be adaptive, as they relieve boredom and help the animal learn to cope with environmental circumstances.[50] However, some stressors may not be adaptive, particularly if the animal has not evolved appropriate responses to them. These are the stressors that a successful enrichment program should address.

ACKNOWLEDGMENTS

I am grateful to Diana Gordon, Adriane Maier, Nicola DeBolt Robertson, Jill Rawlins, Greg Johnson, Kevin Mueller, and Victor Lucas for helpful comments on the manuscript and for their continued help and patience. I also thank Steve Schapiro for valuable conversations that have helped shape my thoughts on enrichment.

REFERENCES

1. Stewart K. Development of an environmental enrichment program utilizing simple strategies. *Anim Welf Inform Center Newsl* 2004;12:1–7.
2. Baker KC, Weed JL, Crockett CM, Bloomsmith MA. Survey of environmental enhancement programs for laboratory primates. *Am J Primatol* 2007;69(4):377–394.
3. Meehan CL, Garner JP, Mench JA. Environmental enrichment and development of cage stereotypy in Orange-winged Amazon parrots (*Amazona amazonica*). *Dev Psychobiol* 2004;44(4):209–218.
4. Almli LM, Burghardt GM. Environmental enrichment alters the behavioral profile of ratsnakes (elaphe). *J Appl Anim Welf Sci* 2006;9(2):85–109.
5. Kihslinger RL, Nevitt GA. Early rearing environment impacts cerebellar growth in juvenile salmon. *J Exp Biol* 2006;209 (Pt. 3):504–509.
6. Shepherdson DJ. Introduction: Tracing the path of environmental enrichment in zoos. In: Shepherdson D, Mellen J, Hutchins M, Eds. *Second Nature: Environmental Enrichment for Captive Animals.* Washington, DC: Smithsonian Institution Press, 1998:1–12.
7. Novak MA, Suomi SJ. Psychological well-being of primates in captivity. *Am Psychol* 1988;43(10):765–773.
8. Council NR. *Guide for the Care and Use of Laboratory Animals.* Washington, DC: National Academic Press, 1996.
9. Regulations AW. Animal Welfare Act. In: Federal Regulations. United States, 1991.
10. von Holst D. The concept of stress and its relevance for animal behavior. In: Moller AP, Milinski M, Slater PJB, Eds. *Stress and Behavior.* San Diego, CA: Academic Press, 1998:1–109.
11. Hickey MC, Drennan M, Earley B. The effect of abrupt weaning of suckler calves on the plasma concentrations of cortisol, catecholamines, leukocytes, acute-phase proteins and in vitro interferon-gamma production. *J Anim Sci* 2003;81(11):2847–2855.
12. Engler H, Dawils L, Hoves S, *et al.* Effects of social stress on blood leukocyte distribution: The role of alpha- and beta-adrenergic mechanisms. *J Neuroimmunol* 2004;156(1–2):153–162.
13. Conrad CD, McEwen BS. Acute stress increases neuropeptide Y mRNA within the arcuate nucleus and hilus of the dentate gyrus. *Brain Res Mol Brain Res* 2000;79(1–2):102–109.
14. Rogers CJ, Brissette-Storkus CS, Hayes LA, Cameron JL, Chambers WH. Selective reduction in CD2 expression on CD2bright/CD8+ lymphocytes from cynomolgus monkeys (*Macaca fascicularis*) in response to acute stress. *J Neuroimmunol* 1998;86(1):63–73.
15. Novak MA. Self-injurious behavior in rhesus monkeys: New insights into its etiology, physiology, and treatment. *Am J Primatol* 2003;59:3–19.
16. de Waal FBM. The social nature of primates. In: Novak MA, Petto AJ, Eds. *Through the Looking Glass.* Washington, DC: American Psychological Association, 1991:69–77.
17. Bayne K. Potential for unintended consequences of environmental enrichment for laboratory animals and research results. *ILAR J* 2005;46(2):129–139.

18. Lilly AA, Mehlman PT, Higley JD. Trait-like immunological and hematological measures in female rhesus across varied environmental conditions. *Am J Primatol* 1999;48(3):197–223.
19. Schapiro SJ, Nehete PN, Perlman JE, Sastry KJ. A comparison of cell-mediated immune responses in rhesus macaques housed singly, in pairs, or in groups. *Appl Anim Behav Sci* 2000;68(1):67–84.
20. Manosevitz M. Early environmental enrichment and mouse behavior. *J Comp Physiol Psychol* 1970;71(3):459–466.
21. Renner MJ, Rosenzweig MR. Social interactions among rats housed in grouped and enriched conditions. *Dev Psychobiol* 1986;19(4):303–314.
22. Fernandez-Teruel A, Driscoll P, Gil L, Aguilar R, Tobena A, Escorihuela RM. Enduring effects of environmental enrichment on novelty seeking, saccharin and ethanol intake in two rat lines (RHA/Verh and RLA/Verh) differing in incentive-seeking behavior. *Pharmacol Biochem Behav* 2002;73(1):225–231.
23. Benaroya-Milshtein N, Hollander N, Apter A, *et al.* Environmental enrichment in mice decreases anxiety, attenuates stress responses and enhances natural killer cell activity. *Eur J Neurosci* 2004;20(5):1341–1347.
24. Jankowsky JL, Melnikova T, Fadale DJ, *et al.* Environmental enrichment mitigates cognitive deficits in a mouse model of Alzheimer's disease. *J Neurosci* 2005;25(21):5217–5224.
25. Barbelivien A, Herbeaux K, Oberling P, Kelche C, Galani R, Majchrzak M. Environmental enrichment increases responding to contextual cues but decreases overall conditioned fear in the rat. *Behav Brain Res* 2006;169(2):231–238.
26. Greenough WT. Experiential modification of the developing brain. *Am Sci* 1975;63(1):37–46.
27. Benefiel AC, Greenough W. Effects of experience and environment on the developing and mature brain: Implications for laboratory animal housing. *ILAR J* 1998;39(1):5–11.
28. Kempermann G, Kuhn HG, Gage FH. More hippocampal neurons in adult mice living in an enriched environment. *Nature* 1997;386(6624):493–495.
29. Young D, Lawlor PA, Leone P, Dragunow M, During MJ. Environmental enrichment inhibits spontaneous apoptosis, prevents seizures and is neuroprotective. *Nat Med* 1999;5(4):448–453.
30. Walsh RN. Effects of environmental complexity and deprivation on brain anatomy and histology: A review. *Int J Neurosci* 1981;12(1):33–51.
31. Rosenzweig MR, Bennett EL. Psychobiology of plasticity: Effects of training and experience on brain and behavior. *Behav Brain Res* 1996;78(1):57–65.
32. van Praag H, Kempermann G, Gage FH. Neural consequences of environmental enrichment. *Nat Rev* 2000;1(3):191–198.
33. Van de Weerd HA, Aarsen EL, Mulder A, Kruitwagen CL, Hendriksen CF, Baumans V. Effects of environmental enrichment for mice: Variation in experimental results. *J Appl Anim Welf Sci* 2002;5(2):87–109.
34. Rockman GE, Gibson JE, Benarroch A. Effects of environmental enrichment on voluntary ethanol intake in rats. *Pharmacol Biochem Behav* 1989;34(3):487–490.
35. Boinski S, Swing SP, Gross TS, Davis JK. Environmental enrichment of brown capuchins (Cebus apella): Behavioral and plasma and fecal cortisol measures of effectiveness. *Am J Primatol* 1999;48(1):49–68.
36. Moncek F, Duncko R, Johansson BB, Jezova D. Effect of environmental enrichment on stress related systems in rats. *J Neuroendocrinol* 2004;16(5):423–431.
37. Marashi V, Barnekow A, Ossendorf E, Sachser N. Effects of different forms of environmental enrichment on behavioral, endocrinological, and immunological parameters in male mice. *Horm Behav* 2003;43(2):281–292.
38. Klein SL, Lambert KG, Durr D, Schaefer T, Waring RE. Influence of environmental enrichment and sex on predator stress response in rats. *Physiol Behav* 1994;56(2):291–297.
39. Reed HJ, Wilkins LJ, Austin SD, Gregory NG. The effect of environmental enrichment during rearing on fear reactions and depopulation trauma in adult caged hens. *Appl Anim Behav Sci* 1993;36(1):39–46.
40. Overall KL, Dyer D. Enrichment strategies for laboratory animals from the viewpoint of clinical veterinary behavioral medicine: Emphasis on cats and dogs. *ILAR J* 2005;46(2):202–215.
41. Benefiel AC, Dong WK, Greenough WT. Mandatory "enriched" housing of laboratory animals: The need for evidence-based evaluation. *ILAR J* 2005;46(2):95–105.
42. Weed JL, Raber JM. Balancing animal research with animal well-being: Establishment of goals and harmonization of approaches. *ILAR J* 2005;46(2):118–128.
43. Welberg L, Thrivikraman KV, Plotsky PM. Combined pre- and postnatal environmental enrichment programs the HPA axis differentially in male and female rats. *Psychoneuroendocrinology* 2006;31(5):553–564.
44. Meaney MJ, Viau V, Bhatnagar S, *et al.* Cellular mechanisms underlying the development and expression of individual differences in the hypothalamic-pituitary-adrenal stress response. *J Steroid Biochem Mol Biol* 1991;39(2):265–274.
45. De Kloet ER. Hormones and the stressed brain. *Ann NY Acad Sci* 2004;1018:1–15.
46. Bartolomucci A, Palanza P, Sacerdote P, *et al.* Social factors and individual vulnerability to chronic stress exposure. *Neurosci Biobehav Rev* 2005;29(1):67–81.
47. Bloomsmith MA, Brent LY, Schapiro SJ. Guidelines for developing and managing an environmental enrichment program for nonhuman primates. *Lab Anim Sci* 1991;41(4):372–377.
48. Keeling ME, Alford PL, Bloomsmith MA. Decision analysis for developing programs of psychological well-being: A bias-for-action approach. In: Novak MA, Petto AJ, Eds. *Through the Looking Glass*. Washington, DC: American Psychological Association, 1991:57–65.
49. Young RJ. *Environmental Enrichment for Captive Animals*. Oxford, UK: Blackwell Publishing, 2003.
50. Newberry RC. Environmental enrichment: Increasing the biological relevance of captive environments. *Appl Anim Behav Sci* 1995; 44(2–4):229–243.
51. Van de Weerd HA, Baumans V, Koolhaas JM, van Zutphen LF. Strain specific behavioural response to environmental enrichment in the mouse. *J Exp Anim Sci* 1994;36(4–5):117–127.
52. Rawlins JM, Poerstel S, Coleman K. Utilization of toy devices by rhesus macaques. *Contemp Top Lab Anim Sci* 2003;42:123.
53. Waitt C, Buchanan-Smith HM, Morris K. The effects of caretaker-primate relationships on primates in the laboratory. *J Appl Anim Welf Sci* 2002;5(4):309–319.
54. Boice R. Burrows of wild and albino rats: Effects of domestication, outdoor raising, age, experience, and maternal state. *J Comp Physiol Psychol* 1977;91(3):649–661.
55. Hutchinson E, Avery A, Vandewoude S. Environmental enrichment for laboratory rodents. *ILAR J* 2005;46(2):148–161.
56. Cowan DP. Group living in the European rabbit oryctolagus-cuniculus mutual benefit or resource localization. *J Anim Ecol* 1987;56(3):779–796.
57. Balcombe JP. Laboratory environments and rodents' behavioural needs: A review. *Lab Anim* 2006;40(3):217–235.
58. Crawley JN, Belknap JK, Collins A, *et al.* Behavioral phenotypes of inbred mouse strains: Implications and recommendations for molecular studies. *Psychopharmacology (Berl)* 1997;132(2):107–124.
59. Van Loo PLP, Kruitwagen CLJJ, Koolhaas JM, Van de Weerd HA, van Zutphen LFM, Baumans V. Influence of cage enrichment on aggressive behaviour and physiological parameters in male mice. *Appl Anim Behav Sci* 2002;76(1):65–81.
60. Joffe JM, Rawson RA, Mulick JA. Control of their environment reduces emotionality in rats. *Science* 1973;180(93):1383–1384.
61. Van de Weerd HA, Van Loo PLP, Van Zutphen LFM, Koolhaas JM, Baumans V. Preferences for nest boxes as environmental enrichment for laboratory mice. *Anim Welf* 1998;7(1):11–25.
62. Van de Weerd HA, Van Loo PLP, Van Zutphen LFM, Koolhaas JM, Baumans V. Strength of preference for nesting material as

environmental enrichment for laboratory mice. *Appl Anim Behav Sci* 1998;55(3–4):369–382.

63. Shepherdson D. Stereotypic behavior: What is it and how can it be eliminated or prevented? *J Assoc Br Wild Anim Keepers* 1993;16:100–105.
64. Wiepkema PR, Koolhaas JM. Stress and animal welfare. *Anim Welf* 1993;2:195–218.
65. Renner MJ, Rosenzweig MR. Social interactions among rats housed in grouped and enriched conditions. *Dev Psychobiol* 1986;19(4): 303–313.
66. Wemelsfelder F. Boredom and laboratory animal welfare. In: Rollin BE, Ed. The *Experimental Animal in Biomedical Research*, Vol. I, *A Survey of Scientific and Ethical Issues for Investigators*. Boca Raton, FL: CRC Press, Inc., 1990:243–272.
67. Chu L-R, Garner JP, Mench JA. A behavioral comparison of New Zealand White rabbits (Oryctolagus cuniculus) housed individually or in pairs in conventional laboratory cages. *Appl Anim Behav Sci* 2004;85(1–2):121–139.
68. Harris LD, Custer LB, Soranaka ET, Burge JR, Ruble GR. Evaluation of objects and food for environmental enrichment of NZW rabbits. *Contemp Top Lab Anim Sci* 2001;40(1):27–30.
69. Carder B, Berkowitz K. Rats' preference for earned in comparison with free food. *Science* 1970;167(922):1273–1274.
70. Pham TM, Brene S, Baumans V. Behavioral assessment of intermittent wheel running and individual housing in mice in the laboratory. *J Appl Anim Welf Sci* 2005;8(3):157–173.
71. Baumans V. Environmental enrichment for laboratory rodents and rabbits: Requirements of rodents, rabbits, and research. *ILAR J* 2005;46(2):162–170.
72. Bazille PG, Walden SD, Koniar BL, Gunther R. Commercial cotton nesting material as a predisposing factor for conjunctivitis in athymic nude mice. *Lab Animal* 2001;30(5):40–42.
73. Adams KM, Navarro AM, Hutchinson EK, Weed JL. A canine socialization and training program at the National Institutes of Health. *Lab Anim* 2004;33:32–36.
74. Hubrecht RC. Enrichment in puppyhood and its effects on later behavior of dogs. *Lab Anim Sci* 1995;45(1):70–75.
75. Odendaal JS, Meintjes RA. Neurophysiological correlates of affiliative behaviour between humans and dogs. *Vet J* 2003;165(3): 296–301.
76. Fox MW, Stelzner D. Behavioural effects of differential early experience in the dog. *Anim Behav* 1966;14(2):273–281.
77. Hubrecht RC, Serpell JA, Poole TB. Correlates of pen size and housing conditions on the behaviour of kennelled dogs. *Appl Anim Behav Sci* 1992;34(4):365–383.
78. Wells DL, Hepper PG. A note on the influence of visual conspecific contact on the behaviour of sheltered dogs. *Appl Anim Behav Sci* 1998;60(1):83–88.
79. Wells DL. A review of environmental enrichment for kennelled dogs, Canis familiaris. *Appl Anim Behav Sci* 2004;85(3–4):307–317.
80. Milgram NW, Head E, Zicker SC, *et al.* Learning ability in aged beagle dogs is preserved by behavioral enrichment and dietary fortification: A two-year longitudinal study. *Neurobiol Aging* 2005;26(1):77–90.
81. Hubrecht RC. A comparison of social and environmental enrichment methods for laboratory housed dogs. *Appl Anim Behav Sci* 1993;37(4):345–361.
82. Clark JD, Calpin JP, Armstrong RB. Influence of type of enclosure on exercise fitness of dogs. *Am J Vet Res* 1991;52(7):1024–1028.
83. Campbell SA, Hughes HC, Griffin HE, Landi MS, Mallon FM. Some effects of limited exercise on purpose-bred beagles. *Am J Vet Res* 1988;49(8):1298–1301.
84. Skinner BF. *The Behavior of Organisms*. New York: Appleton-Century-Crofts, 1938.
85. Pryor K. *Don't Shoot the Dog: The New Art of Teaching and Training*. New York: Simon & Schuster, 1999.
86. Crockett CM. Psychological well-being of captive non-human primates: Lessons from the laboratory studies. In: Shepherdson D, Mellen J, Hutchins M, Eds. *Second Nature: Environmental Enrichment for Captive Animals*. Washington, DC: Smithsonian Institution Press, 1998:129–152.
87. Reinhardt V. Pair-housing rather than single-housing for laboratory rhesus macaques. *J Med Primatol* 1994;23(8):426–431.
88. Majolo B, Buchanan-Smith HM, Morris K. Factors affecting the successful pairing of unfamiliar common marmoset (Callithrix jacchus) females: Preliminary results. *Anim Welf* 2003;12(3):327–337.
89. Crockett CM, Bellanca RU, Bowers CL, Bowden DM. Grooming-contact bars provide social contact for individually caged laboratory macaques. *Contemp Top Lab Anim Sci* 1997;36(6):53–60.
90. Schapiro SJ, Bloomsmith MA, Suarez SA, Porter LM. Effects of social and inanimate enrichment on the behavior of yearling rhesus monkeys. *Am J Primatol* 1996;40:247–260.
91. Spring SE, Clifford JO, Tomko DL. Effect of environmental enrichment devices on behaviors of single- and group-housed squirrel monkeys (Saimiri sciureus). *Contemp Top Lab Anim Sci* 1997;36(3): 72–75.
92. Reinhardt V. Pair-housing overcomes self-biting behavior in macaques. *Lab Primate News* 1999;38(1):4–5.
93. Lutz CK, Novak MA. Environmental enrichment for nonhuman primates: Theory and application. *ILAR J* 2005;46(2):178–191.
94. Novak MA. Self-injurious behavior in rhesus monkeys: New insights into its etiology, physiology, and treatment. *Am J Primatol* 2003;59(1):3–19.
95. Gust DA, Gordon TP, Brodie AR, McClure HM. Effect of a preferred companion in modulating stress in adult female rhesus monkeys. *Physiol Behav* 1994;55(4):681–684.
96. Gerber P, Schnell CR, Anzenberger G. Behavioral and cardiophysiological responses of common marmosets (Callithrix jacchus) to social and environmental changes. *Primates* 2002;43(3):201–216.
97. Epley SW. Reduction of the behavioral effects of aversive stimulation by the presence of companions. *Psychol Bull* 1974;81(5): 271–283.
98. Crockett CM, Bowers CL, Bowden DM, Sackett GP. Sex differences in compatibility of pair-housed adult longtailed macaques. *Am J Primatol* 1994;32(2):73–94.
99. McMillan J, Maier A, Tully L, Coleman K. The effects of temperament on pairing success in female rhesus macaques. *Am J Primatol* 2003;60:95.
100. Bernstein IS, Gordon TP, Rose RM. Aggression and social controls in rhesus monkey (Macaca mulatta) groups revealed in group formation studies. *Folia Primatol (Basel)* 1974;21(2):81–107.
101. Schapiro SJ, Lee-Parritz DE, Taylor LL, Watson L, Bloomsmith MA, Petto A. Behavioral management of specific pathogen-free rhesus macaques: Group formation, reproduction, and parental competence. *Lab Anim Sci* 1994;44(3):229–234.
102. Westergaard GC, Izard MK, Drake JH, Suomi SJ, Higley JD. Rhesus macaque (Macaca mulatta) group formation and housing: Wounding and reproduction in a specific pathogen free (SPF) colony. *Am J Primatol* 1999;49(4):339–347.
103. Coe CL. Is social housing of primates always the optimal choice? In: Novak MA, Petto AJ, Eds. *Through the Looking Glass*. Washington, DC: American Psychological Association, 1991:78–90.
104. Sapolsky RM. Stress in the wild. *Sci Am* 1990;262(1):116–123.
105. Schapiro SJ, Bloomsmith MA. Behavioral effects of enrichment on singly-housed yearling rhesus monkeys: An analysis including three enrichment conditions and a control group. *Am J Primatol* 1995;35:89–101.
106. Crockett CM, Bielitzki J, Carey A, Velez A. Kong toys as enrichment devices for singly-caged macaques. *Lab Primate News* 1989;28(2):21–22.
107. Bloomsmith MA, Finlay TW, Merhalski JJ, Maple TL. Rigid plastic balls as enrichment devices for captive chimpanzees. *Lab Anim Sci* 1990;40(3):319–322.
108. Wells DL. The influence of toys on the behaviour and welfare of kennelled dogs. *Anim Welf* 2004;13(3):367–373.
109. Kessel-Davenport A. Cage toys reduce abnormal behavior in individually housed pigtail macaques. *J Appl Anim Welf Sci* 1998;1(3):227–234.

110. Lam K, Rupniak NM, Iversen SD. Use of a grooming and foraging substrate to reduce cage stereotypies in macaques. *J Med Primatol* 1991;20(3):104–109.
111. Line SW, Morgan KN, Markowitz H. Simple toys do not alter the behavior of aged rhesus monkeys. *Zoo Biol* 1991;10:473–484.
112. Novak MA, Kinsey JH, Jorgensen MJ, Hazen TJ. Effects of puzzle feeders on pathological behavior in individually housed rhesus monkeys. *Am J Primatol* 1998;46:213–227.
113. Bayne KAL, Dexter SL, Strange GM, Hill EE. Kong toys for laboratory primates: Are they really an enrichment of just fomites? *Lab Anim Sci* 1993;43(1):78–85.
114. Holmes SN, Riley JM, Juneau P, Pyne D, Hofing GL. Short-term evaluation of a foraging device for non-human primates. *Lab Anim* 1995;29(4):364–369.
115. Rawlins JM, Johnson JG, Coleman K. The effect of novelty on device use in female rhesus macaques. *Contemp Top Lab Anim Sci* 2004;43:96.
116. Brent L, Stone AM. Long-term use of televisions, balls, and mirrors as enrichment for paired and singly caged chimpanzees. *Am J Primatol* 1996;39:139–145.
117. Bloomsmith MA, Lambeth SP. Videotapes as enrichment for captive chimpanzees (Pan troglodytes). *Zoo Biol* 2000;19:541–551.
118. Harris LD, Briand EJ, Orth R, Galbicka G. Assessing the value of television as environmental enrichment for individually housed rhesus monkeys: A behavioral economic approach. *Contemp Top Lab Anim Sci* 1999;38(2):48–53.
119. Laule GE, Bloomsmith MA, Schapiro SJ. The use of positive reinforcement training techniques to enhance the care, management, and welfare of primates in the laboratory. *J Appl Anim Welf Sci* 2003;6(3):163–173.
120. Priest GM. Training a diabetic drill (*Mandrillus leucophaeus*) to accept insulin injections and venipuncture. *Lab Prim News* 1991;30:1–4.
121. Schapiro SJ, Bloomsmith MA, Laule GE. Positive reinforcement training as a technique to alter nonhuman primate behavior Quantitative assessments of effectiveness. *J Appl Anim Welf Sci* 2003;6(3):175–187.
122. Klaiber-Schuh A, Welker C. Crab-eating monkeys can be trained to cooperate in non-invasive oral medication without stress. *Primate Rep* 1997;47:11–30.
123. Laule GE, Thurston RH, Alford PL, Bloomsmith MA. Training to reliably obtain blood and urine samples from a young, diabetic chimpanzee (*Pan troglodytes*). *Zoo Biol* 1996;15:587–591.
124. Bassett L, Buchanan-Smith HM, McKinley J, Smith TE. Effects of training on stress-related behavior of the common marmoset (Callithrix jacchus) in relation to coping with routine husbandry procedures. *J Appl Anim Welf Sci* 2003;6(3):221–233.
125. Maier A, Mcmillan JL, Coleman K. Effects of positive reinforcement training on stereotypic behaviour in rhesus macaques. *Folia Primatol* 2004;75:392–393.
126. Bloomsmith MA. Chimpanzee training and behavioral research: A symbiotic relationship. AAZPA/CAZPA 1992:403–410.
127. McKinley J, Buchanan-Smith HM, Bassett L, Morris K. Training common marmosets (Callithrix jacchus) to cooperate during routine laboratory procedures: Ease of training and time investment. *J Appl Anim Welf Sci* 2003;6(3):209–220.

Well-Established Models III

Yeast, Worms, Flies, Sea Animals, and Birds A

9 Integrated Network Modeling of Molecular and Genetic Interactions

Gregory W. Carter, Vesteinn Thorsson, and Timothy Galitski

ABSTRACT

In model organisms, systematic genetic perturbation and high-throughput data acquisition allow measurements of mutant phenotypes and detection of biomolecular interactions on a genome-wide scale. These data types are complementary in the sense that genetic interactions inferred from gene expression and other phenotype measurements can assign functional significance to the physical interactions between biomolecules. This chapter describes methods for the analysis of large-scale phenotype and expression data that employ genetic interactions to infer functional relationships between genes. Functional information is then used to direct the integration of physical interaction data to derive network models of cellular activity. These models, obtained from functional relationships but constrained by biochemistry, facilitate explicit and precise prediction of the effects of additional network perturbations.

Key Words: Genetic interaction, Network, Genomics, Cell differentiation.

INTRODUCTION

The identification of causal links between genetic variation and phenotype is central to the development of predictive, preventive, and personalized medicine. Many if not most phenotypes are determined by complex interactions among allelic variants and environmental factors. In biomedicine, these interacting elements are manifest as inherited and somatic genetic variants and polymorphisms, epigenetic effects on gene activity, environmental agents, and drug therapies including drug combinations. Development of new treatments for diseases with genetic susceptibilities will require not only the ability to genotype and classify patients on the basis of molecular fingerprints in tissues, but also an understanding of how genetic variants interact to affect clinical outcomes.

The large-scale study of how genotype contributes to the control of phenotype has been greatly aided by advances in methods to both systematically perturb genes and collect phenotype data on a genomic scale.[1–8] These high-throughput methods are designed to produce large quantitative data sets, requiring parallel development of computational and numerical modeling methods to interpret the output in terms of biological function. The primary objective of these methods is to identify functionally relevant genes and describe how they influence one another to produce cellular activity. In addition, this knowledge can be integrated with physical biochemistry to yield biomolecular models that carry out specific cellular functions. Such an integrated model facilitates validation of biological insights and identifies high-priority candidate genes for targeted therapeutic intervention.

One strategy to achieve this integration is to first infer functional interactions between trait genes, and then use these relationships to guide the selection of biochemical paths that are putatively responsible for the observed phenotypes. This provides a systematic framework for integrating two complementary data types: genetic interactions and physical interactions. Genetic interactions are inferred by observing how genetic perturbations interact to affect phenotypes, and encode functional relationships such as activation, repression, and information flow.[9] However, the biochemical interpretation of a genetic interaction is often ambiguous, frequently involving multiple molecular models and both direct and indirect mechanisms.[10,11] At the same time, the biochemical interactions plentifully generated through high-throughput methods are often lacking in functional interpretation.[12,13] Taking advantage of this complementarity by constraining the biochemical interpretation of genetic interactions by molecular wiring, we seek to construct biomolecular models that assign functional significance to physical interactions

This chapter outlines steps toward such an analysis. First, we outline a systematic analysis and classification of genetic interactions on a large scale. Second, we discuss methods to integrate genomic expression data and physical interaction data to generate molecular hypotheses of functional organization. Finally, we outline a systematic method to extend the two analyses in a predictive model for the control of a biological response in yeast.

GENETIC INTERACTION NETWORKS

A genetic interaction is the phenotypic manifestation of how two alleles combine to jointly affect cellular activity. This classical method of analysis is based on comparing the phenotypes of strains with specific allelic variations, and thus is most readily performed using targeted perturbations in inbred strains of model organisms. Primary examples are the use of the deletion mutant library for *Saccharomyces cerevisiae*,[1] RNAi technology in *Candida elegans*[14] and *Drosophila melanogaster*,[15] as well as

From: *Sourcebook of Models for Biomedical Research*
(P. M. Conn, ed.), © 2008 Humana Press Inc., Totowa, NJ.

curated data repositories such as FlyBase.[16] The properties and uses of genetic interactions learned with model organisms are of interest in the broader context of outbred populations that carry multiple alleles. The genetic basis of human disease susceptibility, in which polygenic interaction effects are often preeminent, is naturally the most urgent application.[17]

Genetic interactions are defined for a chosen phenotype in specific environmental conditions, so that the lone variables are the specific genetic perturbations (i.e., allele forms) carried by the strains being compared. The logic is as follows. Consider two genes, A and B, mutations of which yield observable phenotypes. Taking cell growth as an example phenotype, imagine a null mutation of A inhibits growth while a null mutation of B enhances growth. The growth rate of a double mutant, with null mutations on both A and B, is also observed to be inhibited. In this case the A mutant is said to be *epistatic* to the B mutant, as the A mutant phenotype does not change in the B-null background.* Such an interaction is often interpreted as evidence that A acts downstream of B, in that its knockout masks any modifications that occur further upstream of the phenotypic output.[9] So while this observation cannot, when taken alone, offer any biochemical explanation, it establishes a clear functional relationship between the two genes. Thus the genetic interaction can be considered as providing a comprehensive view of their coordinated activity, encompassing every step in cellular processing between genotype and phenotype.

Classical epistasis is one of many possible modes of genetic interaction. For instance, high-throughput screens in yeast have been developed to detect synthetic interactions,[2] which are manifest when a pair of genes that does not show phenotypic effects when mutated individually exhibits an altered phenotype when the mutations are combined in a single strain. With the proper experimental design, the modes of genetic interaction can be systematically analyzed and formally classified.[8] To this end, consider a genotype X, and its associated phenotype observation P_X. The phenotype could be a numerical measurement or any other observation that can be clearly compared across mutant genotypes (e.g., slow vs. standard vs. fast growth). If genotypes are labeled by mutant alleles, a set of four phenotype observations can be assembled that defines a genetic interaction: P_A and P_B for the A and B mutant alleles, P_{AB} for the AB double mutant, and P_{WT} for the wild type. The relationship between these four measurements defines a genetic interaction; for example, $P_{AB} = P_A < P_{WT} < P_B$ described an epistatic interaction while $P_{AB} < P_{WT} = P_A = P_B$ represents synthesis. It is immediately clear that some of these interactions are symmetric under an exchange of genes A and B (e.g., synthetic) while others are not (e.g., epistatic). There are a total of 75 distinct inequalities that can be constructed from four phenotypes.[8]

These inequalities were recently cataloged and classified as one of nine interaction modes, including epistasis and synthesis.[8] To test the applicability of this classification method over 1800 strains of systematic single and double gene perturbations were phenotyped for invasive growth, a cell differentiation response that occurs under specific environmental conditions. A software

*The term epistasis is sometimes used as a generic term meaning genetic interaction. Here we use it in the narrow sense of one mutant completely masking the effect of another.

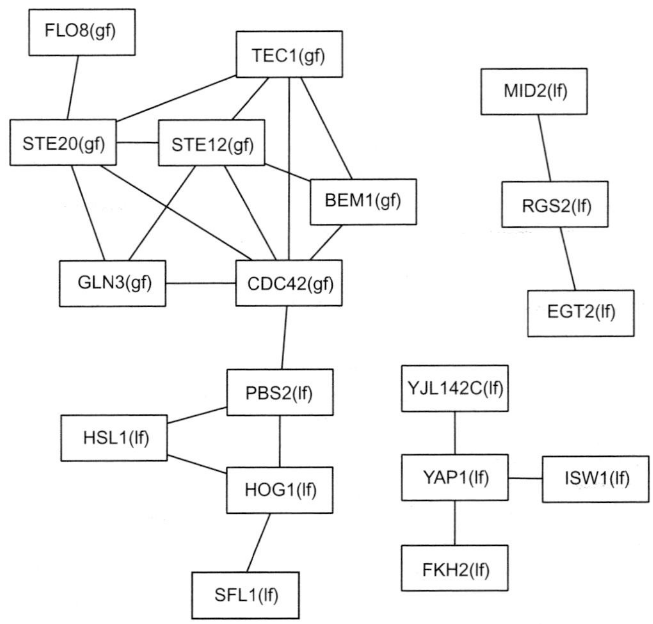

Figure 9–1. Networks of mutual information in patterns of genetic interaction show cliques. Nodes represent perturbed genes, gf indicates a gain-of-function allele, and lf indicates a loss-of-function allele. Edges connect gene pairs with significant mutual information in their patterns of genetic interaction. (Reprinted from Drees *et al.*[8])

application called *PhenotypeGenetics*, created as a plugin for the *Cytoscape* software platform,[18,19] was used to systematically construct a genetic interaction network from the high-throughput, quantitative phenotype data. All nine interaction modes were found in significant numbers, illustrating that a spectrum of genetic interaction effects is present in complex biological systems. Specific alleles were found to interact in a particular mode with neighbor genes of coherent biological function, leading to hypotheses on regulatory and pathway organization. Large-scale patterns of mutual information were also extracted from the data set, and groups of genes with significant mutual information between them formed network cliques corresponding to physical pathways (Figure 9–1). The genetic interaction patterns represent the biological system as a map of information flow from specific genetic perturbations to quantifiable phenotype effects.

DIRECTED DATA INTEGRATION

The invasive-growth network described in the previous section encodes functional information inferred from genetic interactions. We next consider a method to integrate these functional relationships with physical interactions in order to identify candidate molecular mechanisms that control cellular activity. In this section, we outline a strategy of data integration in which the analysis of genomic expression data directs the construction of regulatory networks.

By analyzing the effects of specific genetic perturbations on genomic expression, it is possible to identify specific genes that coordinate control of phenotypes at the transcriptional level. Moreover, gene expression can itself be viewed as a quantitative phenotype that is a suitable platform for studying complex phenomena such as genetic interactions.[20] Genomic expression measurements have many distinct advantages as a basis for quantitative analysis. DNA microarray technology provides a well-established platform for systematic data collection and many statistical methods now exist that identify groups of coexpressed genes. Sets of genes obtained can be systematically queried for common function in increasingly comprehensive annotation databases such as Gene Ontology.[21] Furthermore, the perturbations and measurements of genes introduce a concrete, molecular basis for further analysis, and large-scale interaction databases now make it possible to place these biomolecules in a network context.

We took advantage of these resources in a study of the Ras-cAMP signaling pathway in yeast.[22] This pathway has been implicated in filamentous growth,[23] cell proliferation, and glycolysis.[24–26] In yeast, the pathway centers on the activation of adenylate cyclase (Cyr1) by GTP-bound Ras2 and Ras1 proteins, which in turn facilitates the synthesis of cAMP. We perturbed key pathway elements and collected genomic expression data, with the intent of tracing the observed expression responses to the perturbation(s) that caused them. Specifically, we created strains with exogenously controlled levels of cAMP, dominant-active and dominant-negative alleles of Ras2, and genetic perturbations of the GTPase activating proteins Ira1 and Ira2. The intent was to infer a detailed map of the activity of these pathway elements, ranging from shared effects resulting from multiple perturbations to highly specific effects resulting from one or two perturbations. We wanted to identify and classify signals regardless of global magnitude, detecting not only quantitatively large signals but also fainter (yet coherent) expression effects. These faint effects may be of particular biological interest, possibly identifying functional differences between multiple perturbations of a single gene or two near-homologous genes. Expression patterns specifically related to phenotypes can also be quantitatively minor.[27,28]

To disentangle signals in the data, we performed singular value decomposition (SVD) on the global data set to identify overlapping sets of coexpressed genes, each weighted by global significance.[29,30] To test the presumption that these genes were cofunctional and coregulated, we queried gene sets for statistical enrichment of annotated functions and enrichment in targets of known transcription factors. Both were found for the majority of gene sets. With SVD we were also able to identify the group of experimental perturbations (i.e., microarray conditions) that specifically caused the expression pattern shared by each set of genes. This provided both molecular origins (the perturbations) and targets (the enriched transcription factors and hence the coexpressed genes) of regulatory signals. By querying physical interaction databases, we linked the former to the latter via molecular-interaction paths. The resulting series of linked subnetworks maps multiple pathways of information flow through a dense signaling network to targets throughout the genome (Figure 9–2). They are explicit hypotheses of biomolecular mechanisms for signal transduction, obtained through data integration directed by functional relationships.

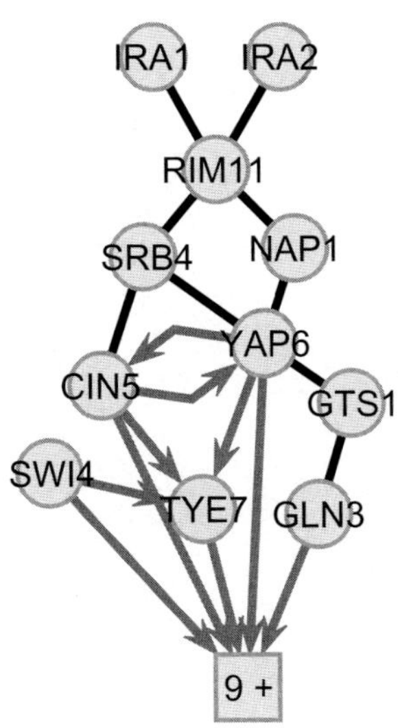

Figure 9–2. Regulatory network derived from perturbations of *IRA1* and *IRA2* genes, mapping putative biomolecular pathways for information flow to a set of genes sharing an expression component (box). The gene set, which is statistically enriched in Ty element transposons, is labeled "9+" to denote that it is the ninth SVD mode and the genes express the expression pattern with positive coefficients. Circular nodes represent individual proteins; black, nondirectional edges represent protein–protein interactions; gray, directional edges represent protein–DNA interactions. (Adapted from Carter *et al.*[22])

SYSTEM GENETICS

The functional information derived from genetic interaction analysis of phenotype is a powerful means of mapping relationships between genes in a pathway. However, many phenotypes are difficult to quantify and cannot yet be assayed with high throughput. This limits efforts to systematically exploit genetic interactions with quantitative methods. At the same time, we have demonstrated how analysis of microarray data for genetic perturbations can facilitate data integration and molecular network modeling. Thus we sought to combine genetic interactions analysis, numerical data decomposition, and functionally directed data integration to model the network control of gene expression and associated phenotypes.

We used a data-driven, linear decomposition of our expression data matrix to determine genetic influences from seed genes to genome-wide transcripts as well as cross-influences of the seed genes on one another's activity. Matrix decomposition methods such as singular value decomposition (SVD)[22,30] and generalized Network Component analysis (gNCA)[31] have proved successful in disentangling multiple overlapping quantitative signals in microarray data. The mathematical modeling was combined with physical interaction databases to infer (1) the degree to which certain transcription factors influence the expression level of every transcript in the genome, (2) the degree to which these transcription factors influence the regulatory activity of one another and hence cause genetic interactions, (3) the biochemical mechanisms that transmit these influences, and (4) how these biochemical mechanisms are organized into the biomolecular network that specifically controls the phenotype.

Our decomposition can be illustrated with the simplified case of two genes A and B, called seed genes since they are the starting point of the analysis, that influence the expression of two genes X and Y. For a strain background labeled with superscript *S*, consider a linear pair of equations for gene expression:

$$X^S = x_0 + x_A\, g_A^S + x_B\, g_B^S$$

$$Y^S = y_0 + y_A\, g_A^S + y_B\, g_B^S \qquad (9\text{–}1)$$

The parameters x_A, x_B, and x_0 represent contributions to the expression of X from the gene A, gene B, and the remainder of the genetic background (similarly for gene Y). These are independent of the genotype. The coefficients g_A^S and g_B^S are the inferred activity levels of the seed genes A and B in the strain background S, and are independent of the transcript measurements. Influences between seed genes (observed as genetic interactions) can be systematically inferred from changes in activity levels of one in a strain in which the other has been deleted, e.g., in a deletion strain of gene B, $g_A^B < g_A^{WT}$ would evince a positive influence from gene B on the activity of gene A. Rather than substituting transcript level data for these activities (as in regression methods) these model-derived parameters conceptually include all levels of gene control from initiation of transcription to protein localization, modification, and degradation.

The system of equations can be expanded to model an arbitrary number of expression values and seed genes by systematically adding parameters. They can be cast in matrix form and the solution can be represented as a network of influences, as illustrated in Table 9–1. These matrices can be expanded to model numerous observables (expression of genes X, Y, Z, . . .) under single and pairwise perturbations of multiple genes (A, B, C, . . .), simply by adding appropriate matrix rows and columns. For example, three seed genes would be written as

$$\begin{pmatrix} X^{WT} & X^{A} & X^{B} & X^{C} & X^{AB} & X^{AC} & X^{BC} \\ Y^{WT} & Y^{A} & Y^{B} & Y^{C} & Y^{AB} & Y^{AC} & Y^{BC} \\ Z^{WT} & Z^{A} & Z^{B} & Z^{C} & Z^{AB} & Z^{AC} & Z^{BC} \\ \vdots & \vdots & \vdots & \vdots & \vdots & \vdots & \vdots \end{pmatrix} = \begin{pmatrix} x_0 & x_A & x_B & x_C \\ y_0 & y_A & y_B & y_C \\ z_0 & z_A & z_B & z_C \\ \vdots & \vdots & \vdots & \vdots \end{pmatrix} \bullet \begin{pmatrix} 1 & 1 & 1 & 1 & 1 & 1 & 1 \\ g_A^{WT} & 0 & g_A^{B} & g_A^{C} & 0 & 0 & g_A^{BC} \\ g_B^{WT} & g_B^{A} & 0 & g_B^{C} & 0 & g_B^{AC} & 0 \\ g_C^{WT} & g_C^{A} & g_C^{B} & 0 & g_C^{AB} & 0 & 0 \end{pmatrix} \tag{9–2}$$

This procedure results in the decomposition of an expression data matrix **D** into two matrices: (1) an influence matrix, **X**, of coefficients for the influence of the seed genes on target genes and (2) a genotype matrix, **G**, of inferred activity levels for the

Table 9–1
Outline of mathematical modeling[a]

1. Model all possible influences (x_A, x_B, y_A, y_B) from genes A and B on the expression of genes X and Y	$X^{WT} = x_0 + x_A + x_B$ $Y^{WT} = y_0 + y_A + y_B$	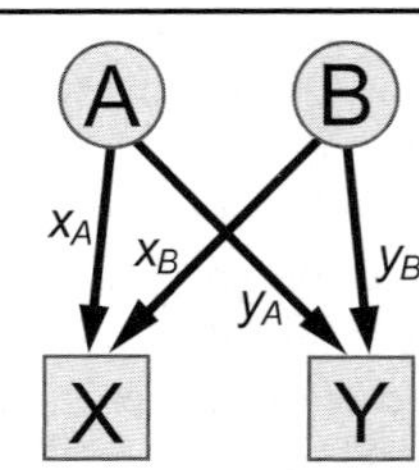
2. Add coefficients (g_A, g_B) representing possible effects of A on activity of B, and vice versa	$X^{WT} = x_0 + x_A\, g_A^{WT} + x_B\, g_B^{WT}$ $Y^{WT} = y_0 + y_A\, g_A^{WT} + y_B\, g_B^{WT}$	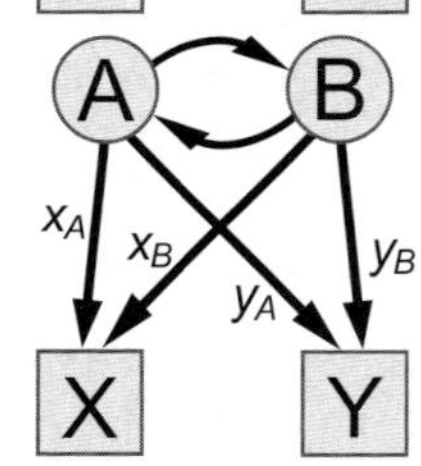
3. Model effects of a deletion of gene A (similar for deletion of B); g_A becomes 0; g_B^A represents the activity of gene B in the A deletion strain; faded lines represent possible influences lost	$X^{A} = x_0 + x_B\, g_B^{A}$ $Y^{A} = y_0 + y_B\, g_B^{A}$	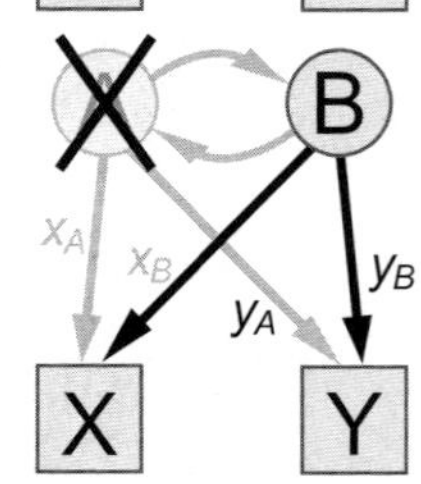
4. Model effects of double perturbation (deletion of genes A and B); faded lines represent possible influences lost	$X^{AB} = x_0$ $Y^{AB} = y_0$	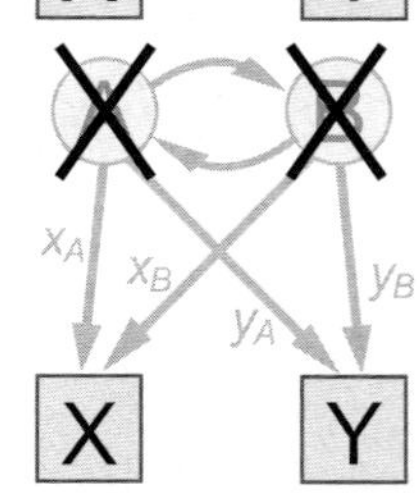
5. Rewrite equations as a matrix decomposition of the expression data	Expression data, Influence matrix, Genotype matrix: $\begin{pmatrix} X^{wt} & X^{A} & X^{B} & X^{AB} \\ Y^{wt} & Y^{A} & Y^{B} & Y^{AB} \end{pmatrix} = \begin{pmatrix} x_0 & x_A & x_B \\ y_0 & y_A & y_B \end{pmatrix} \bullet \begin{pmatrix} 1 & 1 & 1 & 1 \\ g_A^{WT} & 0 & g_A^{B} & 0 \\ g_B^{WT} & g_B^{A} & 0 & 0 \end{pmatrix}$	

[a]Modeling strategy illustrated for the simplified case of two seed genes (A and B) influencing the expression of two genes (X and Y).

seed genes in each genetic background. This is concisely written:

$$\mathbf{D} = \mathbf{X} \cdot \mathbf{G} \tag{9–3}$$

Thus, the more direct influences from the seed genes to target genes are separated quantitatively from the indirect effects that involve a second seed gene via a genetic interaction.

The structure of the genotype matrix [Eq. (9–2)] is fixed by the genetic composition of the laboratory strains. Activity levels of null alleles are fixed at zero and wild-type activity is set equal to one in all cases (gene-by-gene variation is encoded in the first column of the influence matrix). Other genotype matrix elements are unknown a priori. To set a scale for the inferred regulatory activities, we define the wild-type activities to be equal to one ($g_A = g_B = \ldots = 1$) and calculate the remaining activities as changes relative to wild type under perturbations of other seed genes (g_A^B, g_B^A, g_C^{AB}, etc.). The form of matrix **G** guarantees the existence of a unique best-fit solution due to the strict arrangement of ones and zeros required by the genotypes (i.e., the rows of matrix **G** are linearly independent and cannot be transformed to yield a similar format). In the decomposition of microarray data, the number of data points in Eq. (9–3) far exceeds the number of model parameters and a least-squares best-fit solution can be found.

From the genotype matrix **G** in Eq. (9–3), we can infer quantitative cross-influences between the seed genes that generate genetic interactions. These correspond to influences on inferred regulatory activity, rather than the gene expression influences encoded in the matrix **X**. They will allow us to make predictions for perturbations of path genes, with the modified influences experimentally manifest as altered genetic interactions. This is a further dimensional reduction of the genotype matrix **G**. Starting again with the case of two seed genes, A and B, we can write for their activity levels:

$$A = A_0 + m_{AB}\, B$$
$$B = B_0 + m_{BA}\, A \tag{9–4}$$

The variables A and B define generalized activity levels of the seed genes, and the parameters A_0 and B_0 represent basal input not directly due to genes A or B. The m_{ij} account for influences between A and B. Self-influences, such as m_{AA}, are not included since they cannot be numerically distinguished from the basal input. The model can be readily generalized to the case of N perturbed genes. For a vector of gene activities, we replace $\{A, B\}$ with $\boldsymbol{g} = \{g_1, g_2, \ldots, g_N\}$ and write

$$g_i = g_{0i} + m_{ij}\, g_j \tag{9–5}$$

where $\boldsymbol{g}_0$ is a vector of base activity and the m_{ij} form is an $N \times N$ matrix encoding the influence of the ith gene on the jth gene. Equation (9–5) can be solved for the activities and we find the vector solution

$$\boldsymbol{g}^{WT} = (\mathbf{1} - \mathbf{m})^{-1} \cdot \boldsymbol{g}_0 \tag{9–6}$$

where $\mathbf{1}$ is the $N \times N$ identity matrix. The vector $\{1, \boldsymbol{g}^{WT}\}$ is the first column of the genotype matrix **G**.

In this formulation, the deletion of a seed gene requires setting both its base activity and its influences on other seed genes to zero. This corresponds to replacing the appropriate entry in $\boldsymbol{g}_0$ with zero and the appropriate column in **m** with zeros. This can be achieved by rewriting Eq. (9–6) in terms of a diagonal base activity matrix, $\mathbf{G_0}$, formed by placing the elements of the vector $\boldsymbol{g}_0$ along the diagonal, and a scaled influence matrix with elements $M_{ij} = m_{ij}/(g_0)_i$. Defining the vector $\mathbf{1} = \{1, 1, \ldots, 1, 1\}$ of length N, we then have

$$\boldsymbol{g}^{WT} = [(\mathbf{G_0})^{-1} - \mathbf{M}]^{-1} \cdot \mathbf{1} \tag{9–7}$$

In this form, a deletion of gene A is modeled by taking the limit as $(G_0)_{AA} \rightarrow 0$, which means setting the basal activity of that gene to zero. The resulting $\boldsymbol{g}^A$ (with a 1 prepended) corresponds to the second column of the matrix **G**. Multiple deletions are modeled by taking multiple zero limits for entries of the diagonal matrix $\mathbf{G_0}$. This effectively removes all traces of the deleted genes from the system. Note that by after fixing $\boldsymbol{g}^{WT} = \{1, 1, 1, \ldots, 1\}$ we can find a solution for the matrix elements of **M** and the base activities $\mathbf{G_0}$ using the matrix elements of the solution for **G** as described above. Since this is a further dimensional reduction, reducing $N^2(N - 1)/2$ parameters to N^2, this generally requires a best fit solution.

To summarize, the linear influences decomposition assigns quantitative values to the influence of each seed gene on (1) the expression of every gene (encoded in **X**) and (2) the activity of the other seed genes (encoded in **M**). These influences amount to a global map of functional information flows of specified magnitude and sign (positive or negative).

A key use of the influence map is to serve as a template for integrating physical interaction data, in order to generate specific biomolecular hypotheses for regulatory activity. One strategy for this is to begin by selecting significant influences (i.e., the strongest ones). An example selection criterion is to use cross-validation to determine all influences that are highly unlikely to be nonzero.[28] Then, for each of these, a candidate molecular pathway is constructed that transmits the influence, using the integration of physical interaction data as previously described. In this way, networks can be constructed from physical interactions that were specifically selected based on the presence of numerical influences derived from the data. They propose specific biomolecular hypotheses of information flow in the network. Furthermore, perturbation of network elements predicts quantifiable changes in gene expression, since the corresponding numerical influence would be modified.

As a test of this modeling approach, we recently applied it to the control of filamentous growth in yeast. Certain strains of *S. cerevisiae* grow either as a round single-cell yeast form, or as a pathogenic, adhesive, invasive, filamentous form (also referred to as pseudohyphal development).[23] Many fungal pathogens of humans and plants exhibit similar dimorphism.[32] In response to environmental cues, the pathogenic chains of elongated cells adhere to host tissues, evade immune responses, and penetrate barriers. The invasive-growth phenotype previously discussed can be viewed as one part of this response. We chose five filamentation-related transcription factors as our seed genes and collected gene expression data for the 16 strains necessary to infer all genetic interactions between them: the wild type, 5 single knockouts, and 10 double knockouts. We then combined limited phenotype data with genomic expression data (thousands of measurements per strain) to infer a network that controls filamentous growth.

We identified a large set of genes with differential expression and performed the linear influences decomposition on the expression data (a 1863×16 data matrix). We then determined which

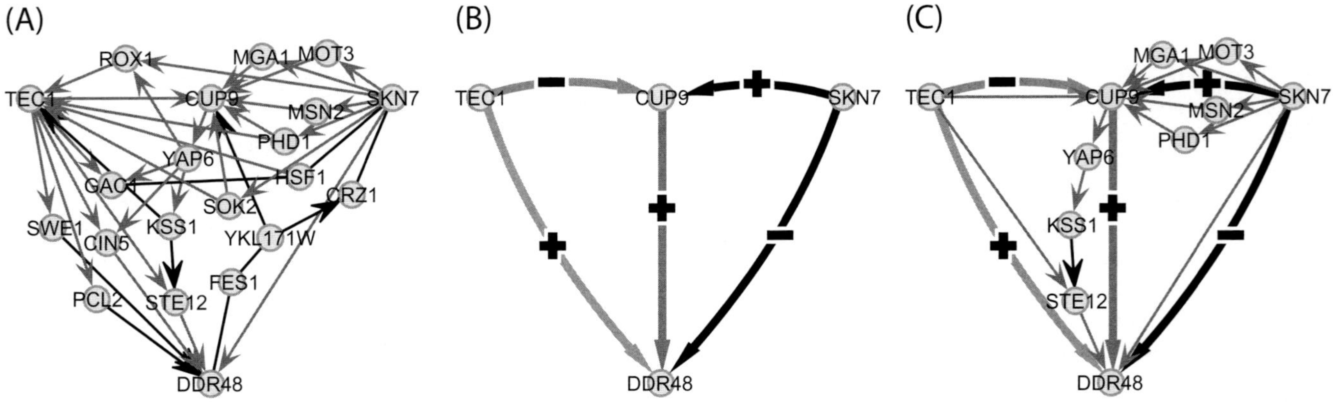

Figure 9–3. Data integration example: modeling how *TEC1*, *CUP9*, and *SKN7* control the expression of gene *DDR48*. (A) Network of physical interactions connecting the four genes from high-throughput data sets. This network is too dense and disorganized to identify functional pathways. Edges represent interactions as protein–protein (black, nondirectional), protein phosphorylation (black, directional), and protein–DNA (gray, directional). (B) Network of inferred influences using genetic influences decomposition. Bold edges indicate the direction of positive and negative influences (as labeled) and intensity indicates magnitude. (C) Integrated network constructed from the subset of physical pathways in (A) that are specific candidates for transmission of influences in (B). (Adapted from Carter *et al.*[28])

influences were significant and constructed putative biochemical networks controlling the expression of each gene. The strategy is illustrated in Figure 9–3 with a network for the transcriptional regulation of the gene *DDR48*, which encodes an ATPase involved in stress response, cell wall organization, and flocculation.[33]

We used the network model to predict genomic expression effects of additional genetic perturbations and tested these predictions by constructing additional combinatorial knockout strains and collecting microarray data. This served as a direct test of our genetic interaction modeling and data integration methods. We made predictions by removing (set to zero) numerical influences whenever the paths from influencer to influenced were broken under the downstream gene deletion. As a control, we made parallel predictions using a simpler linear model without genetic interactions. Our model provided a general improvement in fit across all genes. When assessed over thousands of expression measurements the improvement was highly significant, providing empirical evidence for both the importance of genetic interactions and the applicability of our modeling technique.[28]

In addition to microarray data, we also phenotyped each strain for filamentous growth. We then performed SVD on the genomic expression data[30] and identified a set of genes with an expression component correlated to the phenotype. The genes that strongly exhibit this expression component were a quantitative proxy for the phenotype, even though this component was quantitatively not the most dominant pattern in the data. These genes are statistically enriched in targets of eight transcription factors, of which six were known to have filamentous-growth-related phenotypes. Subsequently, we found that deletions of the other two have filamentation phenotypes.

We were able to determine which of our seed genes had strong expression influences on the phenotype-proxy gene set. By following the methods previously described to integrate physical interaction data, we connected these influencers to genes in the set in a biomolecular network. This constituted a candidate network for the regulation of filamentation, involving both known and novel regulators. Furthermore, from the topology of the network we were able to make systematic and precise predictions for the phenotypes of additional combinatorial perturbations. These predictions proved to be very accurate,[28] compared with predictions based a large training set of interactions.[8] We correctly predicted the phenotypes of all 13 of the novel double-knockout strains we constructed, compared to the training set expectation of seven correct predictions (significance of $p = 0.0002$). Furthermore, we were able to correctly predict the precise phenotype inequalities of six of the 13 strains, compared to an expectation of only two in the training set ($p = 0.009$).

DISCUSSION

The construction of predictive bionetworks has proved to be a substantial challenge due in part to the difficulty of extracting functionally relevant interactions from diverse high-throughput data types. With the increasing abundance of molecular interaction data, there are often numerous possible pathways of information flow between two genes.[34] Many recent works have sought to distill functionally important information by detecting systematic congruence in multiple large data sets.[10,35–39] These approaches have demonstrated substantial success in classifying genes according to functionality and identifying probable candidates for genetic interaction. In contrast, our goal was to infer functional relationships to drive network modeling. By using genetic influences decomposition of microarray data to guide data integration, our procedure assigned functional significance to specific physical interactions. This allowed us to make precise, testable predictions for novel combinatorial perturbations of specific genes.[28]

These models implicated new regulators and regulatory relationships. The network we obtained for the filamentous growth phenotype in yeast contained many genes known to be involved in filamentation, and the network also implicated new regulators of filamentation that were verified experimentally. In addition to implicating genes, our approach was often able to correctly infer relationships between the genes. The accuracy of the interaction network was evident in our broad success in predicting expression profiles and phenotypes for novel combinatorial perturbations.

A central concept in our approach was the assumption of linearity in quantitative relationships between genes. This allowed us to efficiently identify the effects of genetic interactions in the data and enabled systematic mapping of how the expression of each gene is influenced by a genetic perturbation. Although the linear approach was sufficient to predict the effects of novel pairwise gene deletions, we note that it has the potential to oversimplify complicated regulatory mechanisms. In particular, the nonlinearity inherent in many posttranscriptional and feedback mechanisms often leads to behavior too complex to be adequately modeled with static linear relationships. In such cases, the networks obtained from our approach might best be used as a starting point for dynamic modeling. With relatively few network elements connected by quantitative relationships, these networks are ideally suited for ordinary differential equation (ODE) and stochastic modeling.[40]

ACKNOWLEDGMENTS

We thank our collaborators for their contributions to this work: I. Avila-Campillo, B.L. Drees, G.R. Fink, B. Marzolf, C. Neou, M.Z. Raymond, A.W. Rives, S. Rupp, S. Prinz, P. Shannon, and J.P. Shelby. We also thank I. Shmulevich, R.J. Taylor, and D. Galas for discussions and P. Troisch for contributions. This work was supported in part by NIH Grant P50 GM076547. G.W.C. was supported in part by NSF Grant FIBR-0527023. T.G. is a recipient of a Burroughs Wellcome Fund Career Award in the Biomedical Sciences.

REFERENCES

1. Tong AH, Evangelista M, Parsons AB, Xu H, Bader GD, Page N, Robinson M, Raghibizadeh S, Hogue CW, Bussey H, *et al.* Systematic genetic analysis with ordered arrays of yeast deletion mutants. *Science* 2001;294(5550):2364–2368.
2. Tong AH, Lesage G, Bader GD, Ding H, Xu H, Xin X, Young J, Berriz GF, Brost RL, Chang M, *et al.* Global mapping of the yeast genetic interaction network. *Science* 2004;303(5659):808–813.
3. Fraser AG, Kamath RS, Zipperlen P, Martinez-Campos M, Sohrmann M, Ahringer J. Functional genomic analysis of C. elegans chromosome I by systematic RNA interference. *Nature* 2000;408(6810):325–330.
4. Tewari M, Hu PJ, Ahn JS, Ayivi-Guedehoussou N, Vidalain PO, Li S, Milstein S, Armstrong CM, Boxem M, Butler MD, *et al.* Systematic interactome mapping and genetic perturbation analysis of a *C. elegans* TGF-beta signaling network. *Mol Cell* 2004;13(4):469–482.
5. Winzeler EA, Shoemaker DD, Astromoff A, Liang H, Anderson K, Andre B, Bangham R, Benito R, Boeke JD, Bussey H, *et al.* Functional characterization of the S. cerevisiae genome by gene deletion and parallel analysis. *Science* 1999;285(5429):901–906.
6. Strausberg RL, Schreiber SL. From knowing to controlling: A path from genomics to drugs using small molecule probes. *Science* 2003;300(5617):294–295.
7. Swedlow JR, Goldberg I, Brauner E, Sorger PK. Informatics and quantitative analysis in biological imaging. *Science* 2003;300(5616):100–102.
8. Drees BL, Thorsson V, Carter GW, Rives AW, Raymond MZ, Avila-Campillo I, Shannon P, Galitski T. Phenotype and the interaction of genetic perturbations. *Genome Biol* 2004;6(4):R38.
9. Avery L, Wasserman S. Ordering gene function: The interpretation of epistasis in regulatory hierarchies. *Trends Genet* 1992;8(9):312–316.
10. Zhang LV, King OD, Wong SL, Goldberg DS, Tong AH, Lesage G, Andrews B, Bussey H, Boone C, Roth FP. Motifs, themes and thematic maps of an integrated Saccharomyces cerevisiae interaction network. *J Biol* 2005;4(2):6.
11. Kelley R, Ideker T. Systematic interpretation of genetic interactions using protein networks. *Nat Biotechnol* 2005;23(5):561–566.
12. Galitski T. Molecular networks in model systems. *Annu Rev Genomics Hum Genet* 2004;5:177–187.
13. Carter GW. Inferring network interactions within a cell. *Brief Bioinform* 2005;6(4):380–389.
14. Kamath RS, Fraser AG, Dong Y, Poulin G, Durbin R, Gotta M, Kanapin A, Le Bot N, Moreno S, Sohrmann M, Welchman DP, Zipperlen P, Ahringer J. Systematic functional analysis of the *Caenorhabditis elegans* genome using RNAi. *Nature* 2003;421:231–237.
15. Boutros M, Kiger AA, Armknecht S, Kerr K, Hild M, Koch B, Haas SA, Consortium HF, Paro R, Perrimon N. Genome-wide RNAi analysis of growth and viability in *Drosophila* cells. *Science* 2004;303:832–835.
16. Drysdale R. Phenotypic data in FlyBase. *Brief Bioinform* 2001;2:68–80.
17. Hartman JL IV, Garvik B, Hartwell L. Principles for the buffering of genetic variation. *Science* 2001;291:1001–1004.
18. Shannon P, Markiel A, Ozier O, Baliga NS, Wang JT, Ramage D, Amin N, Schwikowski B, Ideker T. Cytoscape: A software environment for integrated models of biomolecular interaction networks. *Genome Res* 2003;13:2498–2504.
19. Online at http://galitski.systemsbiology.net.
20. Van Driessche N, Demsar J, Booth EO, Hill P, Juvan P, Zupan B, Kuspa A, Shaulsky G. Epistasis analysis with global transcriptional phenotypes. *Nat Genet* 2005;37(5):471–477.
21. Online at http://www.geneontology.org.
22. Carter GW, Rupp S, Fink GR, Galitski T. Disentangling information flow in the Ras-cAMP signaling network. *Genome Res* 2006;16(4):520–526.
23. Gimeno CJ, Fink GR. The logic of cell division in the life cycle of yeast. *Science* 1992;257:626.
24. Thevelein JM. The RAS-adenylate cyclase pathway and cell cycle control in Saccharomyces cerevisiae. Antonie Van Leeuwenhoek 1992;62:109–130.
25. D'Souza CA, Heitman J. Conserved cAMP signaling cascades regulate fungal development and virulence. *FEMS Microbiol Rev* 2001;25:349–364.
26. Jones DL, Petty J, Hoyle DC, Hayes A, Ragni E, Popolo L, Oliver SG, Stateva LI. Transcriptome profiling of a Saccharomyces cerevisiae mutant with a constitutively activated Ras/cAMP pathway. *Physiol Genomics* 2003;16:107–118.
27. Mootha VK, Lindgren CM, Eriksson KF, Subramanian A, Sihag S, Lehar J, Puigserver P, Carlsson E, Ridderstrale M, Laurila E, *et al.* PGC-1alpha-responsive genes involved in oxidative phosphorylation are coordinately downregulated in human diabetes. *Nat Genet* 2003;34(3):267–273.
28. Carter GW, Prinz S, Neou C, Shelby JP, Marzolf B, Thorsson V, Galitski T. Prediction of phenotype and genomic expression for combinations of mutations. *Mol Syst Biol* 2007;3:96.
29. Weaver DC, Workman CT, Stormo GD. Modeling regulatory networks with weight matrices. *Pac Symp Biocomput* 1999;4(1):112–123.
30. Alter O, Brown PO, Botstein D. Singular value decomposition for genome-wide expression data processing and modeling. *Proc Natl Acad Sci USA* 2000;97:10101–10106.
31. Yang YL, Suen J, Brynildsen MP, Galbraith SJ, Liao JC. Inferring yeast cell cycle regulators and interactions using transcription factor activities. *BMC Genomics* 2005;6(1):90.
32. Lengeler KB, Davidson RC, D'Souza C, Harashima T, Shen WC, Wang P, Pan X, Waugh M, Heitman J. Signal transduction cascades regulating fungal development and virulence. *Microbiol Mol Biol Rev* 2000;64(4):746–785.
33. Tonouchi A, Fujita A, Kuhara S. Molecular cloning of the gene encoding a highly expressed protein in SFL1 gene-disrupted flocculating yeast. *J Biochem (Tokyo)* 1994;115(4):683–688.
34. Vidal M. Interactome modeling. *FEBS Lett* 2005;579(8):1834–1838.
35. Bader GD, Heilbut A, Andrews B, Tyers M, Hughes T, Boone C. Functional genoimcs and proteomics: Charting a multidimensional map of the yeast cell. *Trends Cell Biol* 2003;13:344–356.
36. Gunsalus KC, Ge H, Schetter AJ, Goldberg DS, Han JD, Hao T, Berriz GF, Bertin N, Huang J, Chuang LS, *et al.* Predictive models

of molecular machines involved in Caenorhabditis elegans early embryogenesis. *Nature* 2005;436(7052):861–865.

37. Sachs K, Perez O, Pe'er D, Lauffenburger DA, Nolan GP. Causal protein-signaling networks derived from multiparameter single-cell data. *Science* 2005;308(5721):523–529.
38. Zhong W, Sternberg PW. Genome-wide prediction of C. elegans genetic interactions. *Science* 2006;311(5766):1481–1484.
39. Workman CT, Mak HC, McCuine S, Tagne JB, Agarwal M, Ozier O, Begley TJ, Samson LD, Ideker T. A systems approach to mapping DNA damage response pathways. *Science* 2006;312(5776): 1054–1059.
40. Kaern M, Elston TC, Blake WJ, Collins JJ. Stochasticity in gene expression: From theories to phenotypes. *Nat Rev Genet* 2005;6:451–464.

10 The Sponge as a Model of Cellular Recognition

Xavier Fernàndez-Busquets

ABSTRACT

Sponges, the simplest extant Metazoans, have been traditionally used as models to study cell adhesion, since their abundant extracellular matrix allows a mild cell dissociation and the recovery of functionally active macromolecular structures. Dissociated sponge cells quickly reaggregate in a species-specific manner, differentiate, and reconstruct tissue, providing a simple model to investigate the molecular basis of animal intercellular recognition systems. Here, we review the application of sponges as biotools for the study of three cell recognition processes that are relevant for biomedical research. First, the species-specific association of sponge cells is discussed in the context of the role ascribed to carbohydrates in adhesion-related phenomena implicated, among other processes, in tumor metastasis. Second, we will summarize the current knowledge about a sponge proteoglycan that has been proposed as a model for the investigation of the mechanical properties of extracellular matrix proteoglycans in calcium-dependent cell adhesion events, whose impairment can lead to severe disorders such as Alzheimer's disease and Marfan syndrome. Finally, the self–nonself recognition reactions in sponges are presented as a window to the early period in the evolution of histocompatibility systems, and the corresponding molecular and cellular events will be compared with the processes and cells known to be involved in innate immunity and in placental implantation. Potential important applications of sponges in the search for new pharmaceuticals and in key biomedical areas such as stem cell research are just around the corner awaiting the development of sponge cell cultures.

Key Words: Cell adhesion, Extracellular matrix, Glycosaminoglycans, Invertebrate histocompatibility, Porifera, Proteoglycans, Self–nonself recognition, Single-molecule force spectroscopy.

SPECIES-SPECIFIC CELL ADHESION IN SPONGES AS A MODEL OF ORGAN-SPECIFIC CELL ADHESION

"Sponges grow spontaneously either attached to a rock or on sea-beaches, and they get their nutriment in slime: a proof of this statement is the fact that when they are first secured they are found to be full of slime." From Aristotle's History of Animals, 350 BC.[1]

From: *Sourcebook of Models for Biomedical Research*
(P. M. Conn, ed.), © 2008 Humana Press Inc., Totowa, NJ.

AGGREGATION ASSAYS WITH CELLS AND SYNTHETIC BEADS Sponges are the most primitive present-day multicellular animals. They lack organs, but their bodies consist of a few classes of cells that are responsible for specialized functions such as food intake, epithelium formation, and self–nonself recognition, among others. In 1907, Wilson[2] pioneered the use of sponges as model animals for the study of cell adhesion, describing the existence of species-specific reaggregation of marine sponge cells that had previously been mechanically dissociated by passing them through a fine cloth. After disaggregation the cells slowly settled down and started moving; upon contact they generally adhered and larger aggregates were formed as new encounters continued to occur, eventually reorganizing functional miniature sponges.[3] When dissociated cells of differently colored sponge species were mixed and allowed to aggregate, the cells of one species combined with each other, but not with those from a different species. This phenomenon is analogous to the organ-specific cell adhesion events that represent the basis of clinically important processes such as cancer metastasis and graft rejection. Sponge cells associate species specifically through multivalent interactions of carbohydrate structures on a type of extracellular proteoglycan termed aggregation factor (AF),[4–6] the *slime* Aristotle was referring to. Based on their molecular structure, AFs have been related to hyalectans[5]: large, extracellular aggregating modular proteoglycans. But unlike hyalectans, aggregation factors do not possess any of the main glycosaminoglycan (GAG) types described in higher animals; instead, they have complex and repetitive acidic carbohydrate motifs different from those found in classical proteoglycans and mucins,[7] which include novel acid-resistant and acid-labile carbohydrate domains, large and branched pyruvylated oligosaccharides,[8] and other previously unknown structures.[6,9–11] In the marine sponge *Microciona prolifera* the proteoglycan molecule, *Microciona* AF (MAF), binds cell membrane receptors via Ca^{2+}-independent interactions of small 6-kDa glycans (g6). Larger 200-kDa glycans (g-200) self-interact through calcium-dependent associations (Figure 10–1C and D). This bifunctional structure is common to all sponge AFs studied so far and represents one of the first specific cell adhesion systems that were necessary for the evolution of Metazoans.

In aggregation assays performed with synthetic beads coupled to AFs purified from different sponge species,[12,13] the beads sorted out and associated only with those carrying AFs from the same species, providing conclusive evidence that AFs are the molecular entity responsible for species-specific cell aggregation.

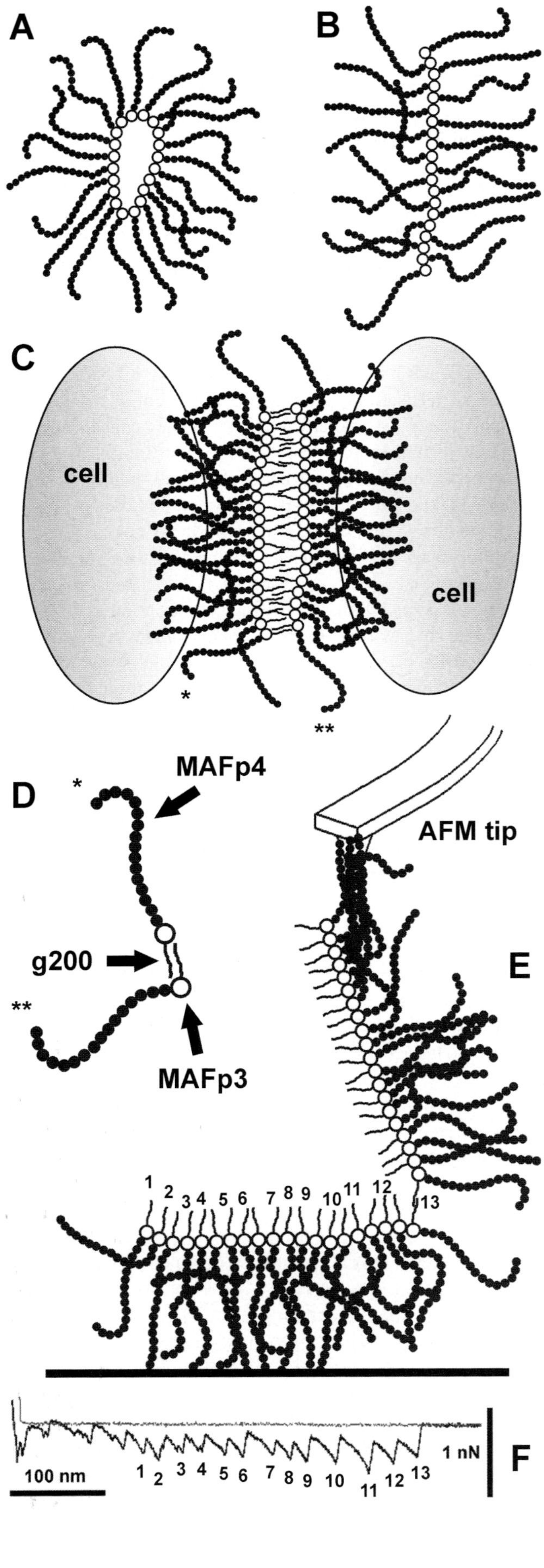

Figure 10–1. The aggregation factor of the marine sponge *M. prolifera* as a model for the study of proteoglycan structure–function relationships. (A) In its native form, MAF has the structure of a sunburst where the ring is formed by ~20 units of the MAFp3 protein (open circles), each noncovalently linked to a unit of the MAFp4 protein (an MAF "arm"). The solid circles forming MAFp4 represent repeated domains within the protein. (B) If the ring of MAF were open, the resulting structure is analogous to a classical proteoglycan, with MAFp3 and MAFp4 in place of link protein and proteoglycan monomer, respectively. (C) Model of the MAF interactions responsible for species specificity of cell adhesion: carbohydrates on MAFp4 bind receptors on the cell membrane, whereas the g-200 glycan on MAFp3 self-interacts in a calcium-dependent manner. For clarity, both MAF molecules are represented as linearized. (D) Detail of the interaction between the two subunits marked by asterisks (* and **) in (C). (E) Scheme of an SMFS experiment to study the adhesive interactions between two MAF molecules. For clarity, both molecules are represented as linearized. (F) Typical SMFS approach-retract curve obtained in experiments performed with native MAF, consistent with the representation from (E). The numbers from 1 to 13 correspond to the breaking of the individual g-200-g-200 interactions represented in (E) as the AFM tip is retracting from the surface. The 100-nm scale bar refers to parts (E) and (F) only. (Adapted from Garcia-Manyes *et al.*[18])

Glycoprotein subunits of MAF obtained through dissociative denaturing procedures bind *Microciona* cells, but with decreasing affinity correlating linearly with the decreasing size of the subunits,[14] suggesting polyvalency in the cell binding site. A small glycopeptide of 10 kDa obtained after trypsin digestion of MAF showed very little affinity for homotypic cells in its monomeric form, but reconstitution of binding affinity in the same order of magnitude as the native molecule could be obtained by polymerizing the glycopeptide fragment.[15] Eventually, both MAF–cell and MAF–MAF binding affinities were recovered upon chemical crosslinking of protein-free glycans isolated from the AF complex into large multivalent structures.[6,9,16] This indicated an active role for the carbohydrate moiety of MAF in the aggregation of sponge cells. In a final series of experiments it was shown that glycan-coated beads aggregated according to their species of origin, i.e., the same way as live sponge cells did.[17] Live cells also demonstrated species-selective binding to glycan-coated surfaces. These findings confirmed for the first time the existence of highly specific recognition between surface glycans, a process that may have significant implications in cellular interactions.

ROLE OF CARBOHYDRATES IN CELL ADHESION g-200 is not a common GAG, as suggested by its elevated fucose content and its resistance to usual GAG-digesting enzymes.[6] Preliminary nuclear magnetic resonance analyses performed with intact g-200 chains indicated such a structural diversity that the possible existence of a backbone containing a basic repetitive oligosaccharide unit could not be determined.[11] This complex structure of g-200, in contrast to the much simpler linear repetitive chains of chondroitin sulfate, heparan sulfate, or hyaluronan, can provide the basis for its highly selective adhesive properties in 10 mM Ca^{2+}, as illustrated by a strong intraspecies-specific binding,[18] weaker interspecies-specific interactions,[17] and very small adhesion forces with other unrelated GAGs.[18] Both the strength and the specificity of calcium-dependent carbohydrate–carbohydrate interactions might be guaranteed by polyvalency, by compositional and architectural differences between carbohydrates, and by the arrangement of the carbohydrate chain in a three-dimensional context, where calcium ions provide not only electrostatic forces but can contribute via coordinative forces to glycan superstructures.[19]

Specific carbohydrate–carbohydrate interactions are rarely reported in biologically relevant situations such as cell recognition. However, carbohydrate structures have immense structural diversity, a ubiquitous distribution in vertebrate and invertebrate tissues, and are associated with the cell surface, as required of cell recognition molecules.[20] Carbohydrate–carbohydrate interactions are characterized by relatively weak forces that, when multimerized, can be easily potentiated by orders of magnitude, representing a highly versatile form of cell adhesion given the extraordinary plasticity of their structures.[21] Among the few known examples of carbohydrate self-recognition proposed to be specific in biological processes are the multivalent binding of Lewis^x epitopes involved in the first steps of embryogenesis[22] and in cancer metastasis,[23,24] glycolipid–glycolipid interactions controlling cell adhesion, spreading, and motility,[25,26] and self-interactions of the glycan portion of sponge proteoglycan carbohydrates leading to species-specific cell adhesion.[17,19]

The species specificity of cell recognition mediated by sponge proteoglycans can be a useful model to study other adhesion processes at structural levels above individual cells. Sponges do not have organs, and therefore the molecular machinery regulating the recognition between cells from different sponge species might be based on the same principles that rule the interactions between cells from different organs in higher animals. Adhesion molecules are essential to motility by interacting with bound and soluble substrates within the extracellular matrix (ECM). Proteoglycans bound to the plasma membrane mediate the polyvalent interaction of the cell with ECM constituents and with molecules from neighboring cell surfaces. Given their direct involvement in cell–cell and cell–ECM interactions, proteoglycans are likely to be implicated in the regulation of cell movement, but to what degree the carbohydrate moiety affects this process is only partially understood.

Cadherins are a family of Ca^{2+}-dependent adhesion molecules that bind cells via homophilic interactions, and cells expressing distinct cadherins aggregate separately when mixed in culture.[27] Cadherins are responsible for the cell sorting that is necessary to distribute different cell types to their proper positions during development. During embryogenesis, the expression of different cadherins is spatiotemporally regulated, and correlates with events that involve cell aggregation or disaggregation. Alterations in cadherin expression or function occur frequently during carcinogenesis and tumoral angiogenesis.[28,29] Another group of cell adhesion molecules includes those related to the immunoglobulin (Ig) superfamily, which generally have Ca^{2+}-independent heterophilic binding.[30] AF-promoted cell adhesion in sponges is based on carbohydrate structures with both Ca^{2+}-dependent homophilic binding and Ca^{2+}-independent heterophilic interactions with cell surface receptors, suggesting that proteoglycan-linked glycans can have important functions in cell adhesion and recognition, comparables to those assigned to cadherins and to molecules of the Ig superfamily.

Metastasis, the process by which a primary malignancy establishes distant and discontiguous disease, clearly involves carbohydrate-mediated cell adhesion.[31] Not all tumors are capable of metastasis, and those that succeed in it rely on a series of steps that involve cell dissociation, motility, and subsequent readhesion.[32] CD44 is a widely distributed cell surface glycoprotein that binds to ECM components containing hyaluronic acid (HA).[33] This permits tumor attachment and migration toward degraded areas rich in HA. CD44 expression is associated with aggressive melanomas and it is an indicator of poor prognosis.[34] As we will see below, antibodies raised against CD44 specifically target a sponge cell type of elevated motility.[35] Study of the essential role that carbohydrates play in sponge cell recognition, motility, and specific adhesion might contribute to understand the participation of carbohydrate structures in the complex processes leading to cancer spreading.

SPONGE AGGREGATION FACTORS AS A MODEL FOR THE STUDY OF PROTEOGLYCAN STRUCTURE–FUNCTION RELATIONSHIPS

The biological roles assigned to proteoglycans are highly diversified, ranging from relatively straightforward mechanical functions to effects on more dynamic processes such as cell adhesion and motility, to complex and still poorly understood tasks in cell differentiation and development.[36–39] Angiogenesis, axon guidance and synapse development, metastasis, and patterning events are just some examples of processes that require finely tuned interactions between cells or between cells and the ECM where proteoglycans are involved.[40–44] Efforts aimed at a better understanding of proteoglycan structure and function are often hampered by the difficulties in their isolation from vertebrate tissues. The purification of proteoglycans is often complicated by (1) limited source quantities, (2) the necessity of chaotropic solvents, proteases, and high temperatures for efficient extraction, (3) their large molecular size, and (4) a lack of defined functions to enable purity to be assessed.[45] Marine sponge proteoglycans are relatively small and can be extracted in considerable amounts while preserving their biological activity by following a mild protocol that essentially consists of dissociating the sponge tissue in calcium-free sea water and spinning down the cells and other debris. The resulting supernatant contains aggregation factor as the main component in such quantities and purity that it can be directly used for cell aggregation assays and other functional experiments. For structural determinations, this preparation has to be centrifuged at high speed to pellet the AF that is finally purified in a cesium chloride density gradient. Concentrated AF solutions ($\geq$1 mg/ml) are highly resistant to proteolysis and can be stored at 4°C for months without significant loss in their specific cell aggregation capacity.

STRUCTURAL ANALYSES AND SINGLE MOLECULE FORCE SPECTROSCOPY STUDIES The supramolecular structure of MAF was elucidated by using immunochemical and electrophoretic procedures, combined with atomic force microscopy (AFM) imaging.[46] Twenty units from each of two N-glycosylated proteins, MAFp3 and MAFp4, form the central ring and radiating arms of MAF, respectively, stabilized by a hyaluronidase-sensitive component.[46] Each of the 20 arms is attached to one of the 20 globular structures in the ring in a 1:1 stoichiometry (Figure 10–1A), with an estimated molecular mass for the whole molecule of 2×10^7 Da.

Each MAFp3 ring unit carries on average one copy of the g-200 glycan involved in homologous self-interactions between aggregation factor molecules (Figure 10–1C and D), whereas each MAFp4 unit carries about 50 copies of the g-6 glycan that binds cell surface receptors.[46] MAFp3 is a highly polymorphic 35-kDa protein, as deduced from direct cDNA sequencing of several different allelic forms and from restriction fragment length polymorphism analysis of genomic DNA isolated from different individual sponges.[47,48] MAFp4 has an estimated molecular mass of ~400 kDa,[48] is also polymorphic, contains putative transmembrane regions, and has a stretch sharing 30% homology with the cytoplasmic domain of the Na^+–Ca^{2+} exchanger,[49] a membrane protein responsible for the maintenance of low intracellular Ca^{2+} levels.

Both MAFp3 and MAFp4 contain abundant Ca^{2+}-binding sites and glycosylation motifs. If the MAF circular core were open, the resulting linear structure would be remarkably similar to hyalectans[50] (Figure 10–1A and B), although the building blocks of both molecules lack sequence homologies. MAFp4 is analogous to the glycosylated core protein of the proteoglycan monomer, and carries the N-linked g-6 glycan instead of O-linked GAGs. MAFp3 is the sponge counterpart of the link protein, both molecules having very similar sizes. Like link protein, MAFp3 is N-glycosylated and contains eight cysteines involved in disulfide bonds. Unlike link protein, though, MAFp3 does not contain an Ig-like domain.

During the past decade, single molecule force spectroscopy (SMFS) has developed into a highly sensitive tool for the investigation of single biomolecule interactions.[51] Most SMFS experiments use either optical tweezers or AFM to measure dissociation forces of single ligand–receptor complexes in the piconewton range. AFM-based SMFS has the capacity to detect interactions as weak as individual hydrogen bonds. The molecular binding partners are attached to the nanoscale force sensor and to a sample holder by covalent chemistry. A critical step in the design of an SMFS experiment is the crosslinking of biomolecules to surfaces, which can introduce undesired reactive groups or steric constraints resulting in the alteration of the biological interaction under study. In the particular case of AFs, their abundant sulfate groups provide a very strong quasicovalent binding to gold surfaces, thus allowing immobilization in the absence of chemical modification, and without affecting the binding sites.[18]

When the AFM sensor and holder are brought into close contact, a specific bond between the individual molecules can form. By increasing the distance between the two surfaces again, the molecular bond is loaded under an external force until it finally breaks, yielding the molecular dissociation force. Upon systematical variation of the externally applied load and monitoring the mechanistic elasticity of the complex, information about the kinetic reaction rates, the mean lifetime, the equilibrium rate of dissociation, dissociation length, and the energy landscape of the interaction can be derived. SMFS studies performed with sponge-derived proteoglycans of their calcium-dependent interactions[18] have revealed functional intermolecular domains (Figure 10–1E) that can contribute to many adhesive and elastic ECM interactions. The multiplicity of individual binding sites with a mean force of ~250 pN for each adhesion event (Figure 10–1E and F) confers a high degree of modulability as required in most biological interactions, in contrast to the higher stability of a single, strong bond. This modular elongation mechanism, be it intramolecular or intermolecular, has been proposed to be a general strategy for conveying toughness to natural fibers and adhesives.[52] Sponge proteoglycan extension curves, then, derive from the existence of intermolecular adhesion domains that through the summation of multiple single binding sites provide strong adhesion forces for the resulting polymer.[18] The cooperativity of abundant, relatively weak intermolecular carbohydrate adhesion domains provides a molecular basis for some of the functions of modular proteoglycans.

IMPLICATION OF PROTEOGLYCANS IN EXTRACELLULAR MATRIX-RELATED PATHOLOGIES Proteoglycans are among the first ECM constituents to be produced during embryonic development and are often aberrantly expressed in a variety of inherited and acquired disorders. One of the major pathological features of Alzheimer's disease is the presence of plaques composed of β-structured fibrils made up of amyloid-β peptide (Aβ).[53] Nucleation of Aβ can occur by self-assembly or by heterogeneous nucleation resulting from seeding Aβ onto non-Aβ elements present in the ECM, such as proteoglycans.[54]

One of the most common nonenzymatic posttranslational protein modifications is the reaction of reducing sugars (or of other carbohydrates resulting from the fragmentation of diverse polysaccharides) with nucleophilic amino acid side chains such as those found in arginine and lysine. Subsequent restructuring, oxidations, and dehydrations generate a group of heterogeneous compounds termed advanced glycation endproducts (AGEs).[55] AGEs have been found in pathological protein deposits such as the senile plaques characteristic of Alzheimer's disease or the β_2-microglobulin deposits in hemodialysis patients.[56] Given their participation in the regulation of diverse growth factors and cytokines through receptor-mediated mechanisms, it has been proposed that AGE-modified proteins are involved in the pathology of several age-related diseases.[57] *In vitro*, AGE formation accelerates the growth of β-amyloid aggregates through covalent crosslinking of Aβ monomers.[58] The extracellular accumulation of AGEs can be caused by an accelerated oxidation of glycosylated proteins and proteoglycans.[59] Glycosylated proteins and, especially, proteoglycans (whose carbohydrate content can be higher than 50%) produce an amount of radicals almost 50 times higher than that generated by nonglycosylated proteins. Proteoglycans are associated with and are likely play a fundamental role in the deposition of amyloid fibrils in all amyloid diseases.[60] Fibril formation and the lateral aggregation of Aβ *in vitro* are increased by GAGs,[61] in a pH-dependent effect.[62] Therefore the interaction of GAGs with amyloid fibrils might be based on electrostatic interactions similar to those involved in the calcium-mediated aggregation of AFs. MAF has been shown to induce a structural transition in Aβ from random coil toward highly stable fibrillar β-sheet structures as detected by circular dichroism spectroscopy,[63] and to increase Aβ-induced toxicity of nerve growth factor-differentiated PC-12 cells in the absence of Ca^{2+}. The addition of Ca^{2+} to MAF-Aβ aggregates resulted in a moderate attenuation of toxicity possibly due to a reduction in Aβ–cell interactions caused by extensive lateral aggregation of the MAF–Aβ complexes. Such aggregation might be seeded by the self-association of MAF triggered by calcium. These results indicate that Aβ is generally susceptible to proteoglycan-mediated aggregation and fibril formation. Sponge proteoglycans, thus, represent a simple model system that can be used to examine potential drugs for the treatment of amyloid-related disorders.

Microfibrils are evolutionarily ancient macromolecular assemblies of the ECM formed by the protein fibrillin. They have unique extensible properties that endow vascular and other tissues with long-range elasticity. In humans, mutations in the principal structural component of microfibrils, fibrillin-1, are linked to the inheritable disease Marfan syndrome, which is characterized by severe cardiovascular, skeletal, and ocular defects.[64] Fibrillin-1 is a calcium-binding protein and it has been found that variations in its calcium-binding properties may be important for microfibril assembly and for the biomechanical properties of microfibrils within tissues.[65] The fibrillin-1 domain with the highest recorded affinity for calcium is located in the region of the protein where mutations (including those that specifically disrupt calcium binding) cause the most severe form of disease, neonatal Marfan syndrome. Proteoglycans have important, but still poorly understood structural interactions with microfibrils, and they may contribute to their integration into the ECM.[66]

ALLOGENEIC RECOGNITION IN SPONGES AS MODEL OF HISTOINCOMPATIBILITY REACTIONS

GRAFTING EXPERIMENTS, IMMUNOCYTOCHEMICAL ANALYSES, AND GENETIC STUDIES When tissues from different individuals of a given sponge species are brought into contact, they either fuse or reject through cellular events similar to those observed in vertebrate grafts.[67] There is, though, a peculiarity of the sponge self–nonself recognition system that so far has not been described in other phyla, which is the possible involvement of proteoglycans in histoincompatibility reactions.

In *M. prolifera*, sponge fragments can be easily grafted, thus allowing the design of tissue histocompatibility experiments that yield important information about the cell types involved in the process.[35] In a typical grafting experiment, 2-cm-long sponge papillae are pushed together on a 0.5-mm-thick stainless steel insect pin, keeping the tissues under seawater at all times. The pins are finally stuck into the underside of a Styrofoam rack and left to float on a tank with running seawater at 20°C. As soon as 2 h after the graft set-up, massive cell migration toward the zone of contact is evident. The study of the sponge cell types involved in this phenomenon has shed new light on our understanding of the evolution of early histocompatibility systems.

Polyclonal antibodies raised against MAFp3 are specific markers for archeocytes, the sponge stem cells.[35] Archeocytes are mobilized upon allogeneic contact (between tissues from two individuals of the same species), and they accumulate in the contact zone. A second type of cell that also migrates profusely to allograft interfaces, the gray cells, is specifically recognized by monoclonal antibodies raised against the HA receptor CD44.[35] In allograft contact areas an observed decrease in the intensity of archeocyte staining is accompanied by an increase in the number of gray cells, which are the most adhesive cell type in *M. prolifera*.[68] A higher adhesion would slow down cell movement and aggregation, contributing to the nonfusion characteristic of allografts. Besides this cell adhesion-related role of gray cells in allograft interfaces, other evidence suggests a more direct immunological function for this cell type, which has been proposed to be the sponge immunocyte.[69,70] In contrast, the contact zone of sponge isografts (grafts made between genetically identical tissue fragments) does not accumulate cells. Here, the absence of a massive gray cell migration would permit the cell movements and aggregation necessary for the final outcome of isografts, which invariably end up in the fusion of the interacting tissues. This differential behavior implies the existence of a genetically based system of individual-specific cell signaling molecules. Although the precise nature of these self–nonself recognition markers has not been established yet, genetic studies point at MAF or MAF-related entities as one of the elements of the sponge histocompatibility system.[47,48]

Due to the presence in sponge tissue of a potent nuclease activity, a special protocol had to be devised for the purification of high-quality sponge genomic DNA[49]: 0.1 ml from a pellet of freshly dissociated sponge cells is added to a 2-ml tube to 0.9 ml of a buffer containing 4 M guanidinium thiocyanate, 25 mM sodium citrate, pH 7, 0.5% sarcosyl, and 0.1 M 2-mercaptoethanol, and gently mixed by immersion until the viscous solution becomes homogeneous. The proteins are then extracted with a 1:1 mixture of phenol/chloroform equilibrated with TE buffer (10 mM Tris–HCl, pH 8, 1 mM EDTA). After a second extraction with chloroform/isoamyl alcohol 24:1, the DNA contained in the water phase is ethanol precipitated and stored at 4°C in TE, pH 7.5. Southern blot analysis of a significant number of sponges that were subjected to grafting experiments revealed that each genetically distinct sponge individual has a different set of genomic DNA coding sequences for MAFp3 and MAFp4.[47,48] Such inter-individual variability, also observed in the carbohydrate moiety of MAF,[18,47] matches the allelic complexity of the vertebrate major histocompatibility complex (MHC) and represents an early form of self-recognition. An evolutionary connection between cell adhesion and histocompatibility systems, however, has yet to be demonstrated. A prominent domain implicated in cell adhesion, the Ig domain, exists in several copies in the large basement membrane proteoglycan.[71] Similar domains are functionally important structures in most Ca^{2+}-independent adhesion molecules such as the neural cell adhesion molecule (NCAM),[72] the intercellular adhesion molecules (ICAMs),[73,74] and other neural or immune system-associated adhesion molecules. No obvious Ig-like domains have been identified in the cDNA-deduced MAF proteins, although the repeats found in MAFp4 exhibit interesting resemblances.[48]

SELF–NONSELF RECOGNITION IN SPONGES AS AN ANCESTOR OF IMMUNE SYSTEMS Natural killer (NK) cells are a subset of granular lymphocytes that express high levels of CD44 upon activation and mediate efficient MHC-unrestricted T cell receptor-independent lysis following binding of CD44.[33] The cross-reactivity of sponge gray cells with anti-CD44 antibodies, together with other morphological and functional similarities, suggests that deeper links between NK cells and sponge allogeneic reactions are possible. First, carbohydrates are crucial to the functions of NK cells[75]: NK cells can be identified and subdivided into functionally distinct subsets on the basis of the expression of lectin-like receptors.[76] Similarly, species-specific adhesion of sponge cells has a strict dependence on finely tuned carbohydrate interactions.[67] Second, NK cells are thought to be the remnants of a primeval immune system and are generally considered to be the evolutionary precursors of T cells.[77] In the human fetus, NK cells are the first lymphocyte lineage to appear,[78] followed by $\gamma\delta$T cells and αβT cells, in accordance with the controversial concept that ontogeny recapitulates phylogeny. NK-like cells have been suggested to exist in invertebrates like the snail,[79] a tunicate,[80] an earthworm,[81] and the leech.[82] Finally, in the allorecognition system of invertebrates no prior sensitization is necessary. The presence of self MHC class I molecules protects target cells from lysis mediated by NK cells,[75] verifying the "missing self" hypothesis that states that NK cell targets become susceptible to NK cell killing if they lack sufficient expression of self MHC molecules. Thus, unlike T cells, which recognize the presence of non-self, NK cells recognize the absence of self.[83] This observation prompted Burnet[84] to conclude that the primary mechanism of invertebrate allorecognition is based on self-recognition and any cell that lacks self antigens will be immediately recognized as foreign, a view that was later supported by investigations on the allogeneic responses of the colonial tunicate *Botryllus schlosseri*.[85] Some aspects of this invertebrate system appear to be retained in higher vertebrates.

There are many features of the immunology of allogeneic reactions in sponges that more closely resemble the immunological relationship between the animal fetus and its mother than between vertebrate allograft and host.[86] While contact with allogeneic cells in clinical transplantation is an artificial situation, in the crowded marine environment direct cellular contact among unrelated members of sessile invertebrate species occurs as naturally as the intermingling of allogeneic cells in placental implantation. In the latter case there is a well-defined boundary dividing invasive trophoblast and maternal decidua that mimics the collagen boundary separating two unrelated sponges, a situation very different from the extensive cell killing typical of vertebrate allograft rejection. In implantation, NK cells are the predominant cell type in deciduas and T cells are sparse; similarly, invertebrate allorecognition is mediated by NK-like cells, whereas vertebrate allorecognition is mediated by T and B cells. Finally, NK cells kill target cells deficient in MHC class I antigens, unlike vertebrate graft rejection where nonself antigens are detected; as mentioned above, current evidence suggests that sponge nonself recognition detects the absence of self antigens.[35] The development in early animals of the ability to discriminate self from nonself ensured the maintaining of individuality. It is possible that such a primitive form of allorecognition has been conserved to play a role in mammalian reproduction, which is the only natural situation in which allogeneic cells come into contact in vertebrates.

FUTURE PERSPECTIVES: DEVELOPMENT OF STABLE SPONGE CELL CULTURES, STEM CELL RESEARCH, AND BIODISCOVERY OF NEW PHARMACEUTICALS

Sponges possess totipotent cells, the archeocytes, that can differentiate into the rest of the cell types. Most important, other sponge cell types do spontaneously dedifferentiate into archeocytes, a phenomenon observed in naturally occurring processes such as tissue regression and remodeling, gemmule formation, and trauma induced by tissue removal.[87] Understanding the mechanism by which specialized sponge cells can return to an undifferentiated state will contribute to the quest for a reliable and ethically acceptable source of unlimited amounts of stem cells for therapeutic applications.

The main obstacle for the firm establishment of sponges as a generally accepted model laboratory animal is the current lack of stable sponge cell cultures. Interest in the development of sponge cell lines grew in parallel with the realization that sponges are an exceptional source of metabolites with potential pharmacological activities. Because of their longer evolutionary history, marine organisms have a greater molecular diversity than do their terrestrial counterparts. Among the numerous groups of marine invertebrates sponges rank first as the leading source of natural products in terms of the biogenetic diversity of secondary constituents and the sheer number of compounds isolated.[88] Furthermore, sponges have a high strike rate, especially for cytotoxic compounds: of the investigated marine sponge species, >10% has exhibited cytotoxic activity, suggesting production of potential medicinals. This percentage is considerably higher than that found for other marine animals (2%), terrestrial plants (<1%), or microorganisms (<1%).[89] Products with pharmacological activity identified in sponges include, among others, analgesics, antiinflammatory compounds, antitumorals, immunosuppressors, cardiovascular agents, neurosuppressors, muscle relaxants, antivirals, antimalarials, antibiotics, fungicides, and antifouling compounds.[90] Although many bioactives have been discovered in sponges, only a few of them have been commercialized.[91] Obtaining the necessary amounts of these metabolites requires quantities of sponge biomass that cannot be sustainably harvested from natural populations. Cultivation of sponge biomass in sea-based farms is feasible, but productivity is variable. Biomass production in controlled environments of aquariums has the potential to provide consistent yields, but many aspects of aquarium cultivation remain unknown for most sponges. In this context, the culture of sponge cells can become a future reliable source of metabolites.

However, sponge cell culture represents a unique purification challenge because, unlike higher Metazoans, there are no areas of a sponge from which a sterile primary culture can be obtained.[92] As an added complication, many sponges host endosymbionic microorganisms that may be released into an otherwise sterile culture if cells lyse. Control of microbial contamination is the most important obstacle to overcome in the establishment of primary sponge cell cultures. A combination of filtration, use of sterile media during cell dissociation, and selective sponge cell enrichment by density gradient separation has proven to be most effective. In addition, extended culture of dissociated sponge cells invariably requires the use of antibiotics to suppress bacterial contamination, especially in the rich media that are required. The antibiotic cocktails used do not inhibit fungal growth and proliferation of fungi is a recurring problem, and this at a cost that precludes any commercially sustainable metabolite production. As a result, current methodologies are at the stage of primary cultures without any significant progress toward a substantial increase in cell numbers or biomass due to cell proliferation.[93,94] The potential of sponges as model animals for the study of cellular recognition processes such as the ones described above will not be fully exploited until long-lasting cell cultures are successfully developed.

ACKNOWLEDGMENTS

This work was supported by Grant BIO2005-01591 from the Ministerio de Educación y Ciencia (MEC), Spain, that included FEDER funds. X.F.-B. holds a Ramón y Cajal tenure track position from the MEC.

REFERENCES

1. Aristotle. History of Animals, on line English translation by D'Arcy Wentworth Thompson (http://etext.library.adelaide.edu.au/a/aristotle/history/). Adelaide, Australia: The University of Adelaide Library, 350.
2. Wilson HV. On some phenomena of coalescence and regeneration in sponges. *J Exp Zool* 1907;5:245–258.
3. Galtsoff PS. Regeneration after dissociation (an experimental study on sponges). I. Behavior of dissociated cells of Microciona prolifera under normal and altered conditions. *J Exp Zool* 1925;42:183–221.
4. Humphreys S, Humphreys T, Sano J. Organization and polysaccharides of sponge aggregation factor. *J Supramol Struct* 1977;7(3–4):339–351.
5. Fernandez-Busquets X, Burger MM. Circular proteoglycans from sponges: First members of the spongican family. *Cell Mol Life Sci* 2003;60(1):88–112.
6. Misevic GN, Burger MM. Carbohydrate-carbohydrate interactions of a novel acidic glycan can mediate sponge cell adhesion. *J Biol Chem* 1993;268(7):4922–4929.
7. Misevic GN, Guerardel Y, Sumanovski LT, Slomianny MC, Demarty M, Ripoll C, *et al.* Molecular recognition between glyconectins as an adhesion self-assembly pathway to multicellularity. *J Biol Chem* 2004;279(15):15579–15590.
8. Guerardel Y, Czeszak X, Sumanovski LT, Karamanos Y, Popescu O, Strecker G, *et al.* Molecular fingerprinting of carbohydrate structure phenotypes of three porifera proteoglycan-like glyconectins. *J Biol Chem* 2004;279(15):15591–15603.
9. Misevic GN, Burger MM. The species-specific cell-binding site of the aggregation factor from the sponge Microciona prolifera is a highly repetitive novel glycan containing glucuronic acid, fucose, and mannose. *J Biol Chem* 1990;265(33):20577–20584.
10. Spillmann D, Thomas-Oates JE, van Kuik JA, Vliegenthart JF, Misevic G, Burger MM, *et al.* Characterization of a novel sulfated carbohydrate unit implicated in the carbohydrate-carbohydrate-mediated cell aggregation of the marine sponge Microciona prolifera. *J Biol Chem* 1995;270(10):5089–5097.
11. Spillmann D, Hard K, Thomas-Oates J, Vliegenthart JF, Misevic G, Burger MM, *et al.* Characterization of a novel pyruvylated carbohydrate unit implicated in the cell aggregation of the marine sponge Microciona prolifera. *J Biol Chem* 1993;268(18):13378–13387.
12. Popescu O, Misevic GN. Self-recognition by proteoglycans. *Nature* 1997;386(6622):231–232.
13. Jarchow J, Burger MM. Species-specific association of the cell-aggregation molecule mediates recognition in marine sponges. *Cell Adhes Commun* 1998;6(5):405–414.
14. Misevic GN, Jumblatt JE, Burger MM. Cell binding fragments from a sponge proteoglycan-like aggregation factor. *J Biol Chem* 1982;257(12):6931–6936.
15. Misevic GN, Burger MM. Reconstitution of high cell binding affinity of a marine sponge aggregation factor by cross-linking of small low affinity fragments into a large polyvalent polymer. *J Biol Chem* 1986;261(6):2853–2859.
16. Misevic GN, Finne J, Burger MM. Involvement of carbohydrates as multiple low affinity interaction sites in the self-association of the aggregation factor from the marine sponge Microciona prolifera. *J Biol Chem* 1987;262(12):5870–5877.
17. Bucior I, Scheuring S, Engel A, Burger MM. Carbohydrate-carbohydrate interaction provides adhesion force and specificity for cellular recognition. *J Cell Biol* 2004;165(4):529–537.
18. Garcia-Manyes S, Bucior I, Ros R, Anselmetti D, Sanz F, Burger MM, *et al.* Proteoglycan mechanics studied by single-molecule force spectroscopy of allotypic cell adhesion glycans. *J Biol Chem* 2006;281(9):5992–5999.
19. Bucior I, Burger MM. Carbohydrate-carbohydrate interaction as a major force initiating cell-cell recognition. *Glycoconj J* 2004;21(3–4):111–123.
20. Bucior I, Burger MM. Carbohydrate-carbohydrate interactions in cell recognition. *Curr Opin Struct Biol* 2004;14(5):631–637.
21. Spillmann D, Burger MM. Carbohydrate-carbohydrate interactions. In: *Carbohydrates in Chemistry and Biology. A Comprehensive Handbook.* Weinheim: Wiley-VCH Verlag GmbH, 2000:1061–1091.
22. Pincet F, Le Bouar T, Zhang Y, Esnault J, Mallet JM, Perez E, *et al.* Ultraweak sugar-sugar interactions for transient cell adhesion. *Biophys J* 2001;80(3):1354–1358.
23. Kawaguchi T. Cancer metastasis: Characterization and identification of the behavior of metastatic tumor cells and the cell adhesion molecules, including carbohydrates. *Curr Drug Targets Cardiovasc Haematol Disord* 2005;5(1):39–64.
24. Kannagi R. Molecular mechanism for cancer-associated induction of sialyl Lewis X and sialyl Lewis A expression–The Warburg effect revisited. *Glycoconj J* 2004;20(5):353–364.
25. Dicko A, Heng YM, Boggs JM. Interactions between glucosylceramide and galactosylceramide I(3) sulfate and microstructures formed. *Biochim Biophys Acta* 2003;1613(1–2):87–100.
26. Wang X, Sun P, Al Qamari A, Tai T, Kawashima I, Paller AS. Carbohydrate-carbohydrate binding of ganglioside to integrin alpha(5) modulates alpha(5)beta(1) function. *J Biol Chem* 2001; 276(11):8436–8444.
27. Leckband D, Prakasam A. Mechanism and dynamics of cadherin adhesion. *Annu Rev Biomed Eng* 2006;8:259–287.
28. Peinado H, Portillo F, Cano A. Transcriptional regulation of cadherins during development and carcinogenesis. *Int J Dev Biol* 2004;48(5–6): 365–375.
29. Cavallaro U, Liebner S, Dejana E. Endothelial cadherins and tumor angiogenesis. *Exp Cell Res* 2006;312(5):659–667.
30. Edelman GM, Crossin KL. Cell adhesion molecules: Implications for a molecular histology. *Annu Rev Biochem* 1991;60:155–190.
31. Kannagi R, Izawa M, Koike T, Miyazaki K, Kimura N. Carbohydrate-mediated cell adhesion in cancer metastasis and angiogenesis. *Cancer Sci* 2004;95(5):377–384.
32. Nguyen TH. Mechanisms of metastasis. *Clin Dermatol* 2004;22(3): 209–216.
33. Rafi-Janajreh AQ, Nagarkatti PS, Nagarkatti M. Role of CD44 in CTL and NK cell activity. *Front Biosci* 1998;3:d665–d671.
34. Pacifico MD, Grover R, Richman PI, Daley FM, Buffa F, Wilson GD. CD44v3 levels in primary cutaneous melanoma are predictive of prognosis: Assessment by the use of tissue microarray. *Int J Cancer* 2006;118(6):1460–1464.
35. Fernandez-Busquets X, Kuhns WJ, Simpson TL, Ho M, Gerosa D, Grob M, *et al.* Cell adhesion-related proteins as specific markers of sponge cell types involved in allogeneic recognition. *Dev Comp Immunol* 2002;26(4):313–323.
36. De Cat B, David G. Developmental roles of the glypicans. *Semin Cell Dev Biol* 2001;12(2):117–125.
37. Schwartz N. Biosynthesis and regulation of expression of proteoglycans. *Front Biosci* 2000;5:D649–D655.
38. Rapraeger AC. Molecular interactions of syndecans during development. *Semin Cell Dev Biol* 2001;12(2):107–116.
39. Kresse H, Schonherr E. Proteoglycans of the extracellular matrix and growth control. *J Cell Physiol* 2001;189(3):266–274.
40. Yamaguchi Y. Heparan sulfate proteoglycans in the nervous system: Their diverse roles in neurogenesis, axon guidance, and synaptogenesis. *Semin Cell Dev Biol* 2001;12(2):99–106.
41. Selleck SB. Genetic dissection of proteoglycan function in Drosophila and C. elegans. *Semin Cell Dev Biol* 2001;12(2):127–134.
42. Iozzo RV. Basement membrane proteoglycans: From cellar to ceiling. *Nat Rev Mol Cell Biol* 2005;6(8):646–656.
43. Hacker U, Nybakken K, Perrimon N. Heparan sulphate proteoglycans: The sweet side of development. *Nat Rev Mol Cell Biol* 2005;6(7):530–541.
44. Cattaruzza S, Perris R. Proteoglycan control of cell movement during wound healing and cancer spreading. *Matrix Biol* 2005;24(6): 400–417.
45. Fedarko NS. Isolation and purification of proteoglycans. *Experientia* 1993;49(5):369–383.

46. Jarchow J, Fritz J, Anselmetti D, Calabro A, Hascall VC, Gerosa D, *et al.* Supramolecular structure of a new family of circular proteoglycans mediating cell adhesion in sponges. *J Struct Biol* 2000;132(2): 95–105.
47. Fernandez-Busquets X, Burger MM. The main protein of the aggregation factor responsible for species-specific cell adhesion in the marine sponge Microciona prolifera is highly polymorphic. *J Biol Chem* 1997;272(44):27839–27847.
48. Fernandez-Busquets X, Gerosa D, Hess D, Burger MM. Accumulation in marine sponge grafts of the mRNA encoding the main proteins of the cell adhesion system. *J Biol Chem* 1998;273(45):29545–29553.
49. Fernandez-Busquets X, Kammerer RA, Burger MM. A 35-kDa protein is the basic unit of the core from the 2 × 10(4)-kDa aggregation factor responsible for species-specific cell adhesion in the marine sponge Microciona prolifera. *J Biol Chem* 1996;271(38):23558–23565.
50. Ng L, Grodzinsky AJ, Patwari P, Sandy J, Plaas A, Ortiz C. Individual cartilage aggrecan macromolecules and their constituent glycosaminoglycans visualized via atomic force microscopy. *J Struct Biol* 2003;143(3):242–257.
51. Zlatanova J, Lindsay SM, Leuba SH. Single molecule force spectroscopy in biology using the atomic force microscope. *Prog Biophys Mol Biol* 2000;74(1–2):37–61.
52. Smith BL, Schaffer TE, Viani M, Thompson JB, Frederick NA, Kindt J, *et al.* Molecular mechanistic origin of the toughness of natural adhesives, fibres and composites. *Nature* 1999;399(6738):761–763.
53. Arimon M, Diez-Perez I, Kogan MJ, Durany N, Giralt E, Sanz F, *et al.* Fine structure study of Abeta1–42 fibrillogenesis with atomic force microscopy. *FASEB J* 2005;19(10):1344–1346.
54. Yang DS, Serpell LC, Yip CM, McLaurin J, Chrishti MA, Horne P, *et al.* Assembly of Alzheimer's amyloid-beta fibrils and approaches for therapeutic intervention. *Amyloid* 2001;8(Suppl. 1):10–19.
55. DeGroot J. The AGE of the matrix: Chemistry, consequence and cure. *Curr Opin Pharmacol* 2004;4(3):301–305.
56. Richter T, Munch G, Luth HJ, Arendt T, Kientsch-Engel R, Stahl P, *et al.* Immunochemical crossreactivity of antibodies specific for "advanced glycation endproducts" with "advanced lipoxidation endproducts." *Neurobiol Aging* 2005;26(4):465–474.
57. Ahmed N. Advanced glycation endproducts–role in pathology of diabetic complications. *Diabetes Res Clin Pract* 2005;67(1):3–21.
58. Munch G, Kuhla B, Luth HJ, Arendt T, Robinson SR. Anti-AGEing defences against Alzheimer's disease. *Biochem Soc Trans* 2003;31(Pt. 6):1397–1399.
59. Snow AD, Wight TN, Nochlin D, Koike Y, Kimata K, DeArmond SJ, *et al.* Immunolocalization of heparan sulfate proteoglycans to the prion protein amyloid plaques of Gerstmann-Straussler syndrome, Creutzfeldt-Jakob disease and scrapie. *Lab Invest* 1990;63(5):601–611.
60. Alexandrescu AT. Amyloid accomplices and enforcers. *Protein Sci* 2005;14(1):1–12.
61. Fraser PE, Nguyen JT, Chin DT, Kirschner DA. Effects of sulfate ions on Alzheimer beta/A4 peptide assemblies: Implications for amyloid fibril-proteoglycan interactions. *J Neurochem* 1992;59(4): 1531–1540.
62. Brunden KR, Richter-Cook NJ, Chaturvedi N, Frederickson RC. pH-dependent binding of synthetic beta-amyloid peptides to glycosaminoglycans. *J Neurochem* 1993;61(6):2147–2154.
63. McLaurin J, Franklin T, Kuhns WJ, Fraser PE. A sulfated proteoglycan aggregation factor mediates amyloid-beta peptide fibril formation and neurotoxicity. *Amyloid* 1999;6(4):233–243.
64. Kielty CM, Wess TJ, Haston L, Ashworth JL, Sherratt MJ, Shuttleworth CA. Fibrillin-rich microfibrils: Elastic biopolymers of the extracellular matrix. *J Muscle Res Cell Motil* 2002;23(5–6):581–596.
65. Handford PA. Fibrillin-1, a calcium binding protein of extracellular matrix. *Biochim Biophys Acta* 2000;1498(2–3):84–90.
66. Isogai Z, Aspberg A, Keene DR, Ono RN, Reinhardt DP, Sakai LY. Versican interacts with fibrillin-1 and links extracellular microfibrils to other connective tissue networks. *J Biol Chem* 2002;277(6):4565–4572.
67. Fernandez-Busquets X, Burger MM. Cell adhesion and histocompatibility in sponges. *Microsc Res Tech* 1999;44(4):204–218.
68. Leith A. Role of aggregation factor and cell type in sponge cell adhesion. *Biol Bull* 1979;156(2):212–223.
69. Humphreys T. Rapid allogeneic recognition in the marine sponge Microciona prolifera. Implications for evolution of immune recognition. *Ann NY Acad Sci* 1994;712:342–345.
70. Yin C, Humphreys S. Acute cytotoxic allogeneic histoincompatibility reactions involving gray cells in the marine sponge, Callyspongia diffusa. *Biol Bull* 1996;191:159–167.
71. Kallunki P, Tryggvason K. Human basement membrane heparan sulfate proteoglycan core protein: A 467-kD protein containing multiple domains resembling elements of the low density lipoprotein receptor, laminin, neural cell adhesion molecules, and epidermal growth factor. *J Cell Biol* 1992;116(2):559–571.
72. Cunningham BA, Hemperly JJ, Murray BA, Prediger EA, Brackenbury R, Edelman GM. Neural cell adhesion molecule: Structure, immunoglobulin-like domains, cell surface modulation, and alternative RNA splicing. *Science* 1987;236(4803):799–806.
73. Simmons D, Makgoba MW, Seed B. ICAM, an adhesion ligand of LFA-1, is homologous to the neural cell adhesion molecule NCAM. *Nature* 1988;331(6157):624–627.
74. Staunton DE, Marlin SD, Stratowa C, Dustin ML, Springer TA. Primary structure of ICAM-1 demonstrates interaction between members of the immunoglobulin and integrin supergene families. *Cell* 1988;52(6):925–933.
75. Manilay JO, Sykes M. Natural killer cells and their role in graft rejection. *Curr Opin Immunol* 1998;10(5):532–538.
76. McCoy JP Jr, Chambers WH. Carbohydrates in the functions of natural killer cells. *Glycobiology* 1991;(4):321–328.
77. Janeway CA Jr. The immune system evolved to discriminate infectious nonself from noninfectious self. *Immunol Today* 1992;13(1):11–16.
78. Spits H, Lanier LL, Phillips JH. Development of human T and natural killer cells. *Blood* 1995;85(10):2654–2670.
79. Franceschi C, Cossarizza A, Monti D, Ottaviani E. Cytotoxicity and immunocyte markers in cells from the freshwater snail Planorbarius corneus (L.) (Gastropoda pulmonata): Implications for the evolution of natural killer cells. *Eur J Immunol* 1991;21(2):489–493.
80. Cammarata M, Arizza V, Parrinello N, Candore G, Caruso C. Phenoloxidase-dependent cytotoxic mechanism in ascidian (Styela plicata) hemocytes active against erythrocytes and K562 tumor cells. *Eur J Cell Biol* 1997;74(3):302–307.
81. Cooper EL, Cossarizza A, Suzuki MM, Salvioli S, Capri M, Quaglino D, *et al.* Autogeneic but not allogeneic earthworm effector coelomocytes kill the mammalian tumor cell target K562. *Cell Immunol* 1995;166(1):113–122.
82. de Eguileor M, Grimaldi A, Tettamanti G, Valvassori R, Cooper EL, Lanzavecchia G. Different types of response to foreign antigens by leech leukocytes. *Tissue Cell* 2000;32(1):40–48.
83. Ljunggren HG, Karre K. In search of the 'missing self': MHC molecules and NK cell recognition. *Immunol Today* 1990;11(7):237–244.
84. Burnet FM. "Self-recognition" in colonial marine forms and flowering plants in relation to the evolution of immunity. *Nature* 1971;232(5308):230–235.
85. Grosberg R, Quinn J. The genetic control and consequences of kin recognition by the larvae of a colonial marine invertebrate. *Nature* 1986;322:456–459.
86. Loke YW, King A. Immunology of human placental implantation: Clinical implications of our current understanding. *Mol Med Today* 1997;3(4):153–159.
87. Simpson TL. *The Cell Biology of Sponges*. New York: Springer-Verlag, 1984.
88. Faulkner DJ. Marine natural products. *Nat Prod Rep* 1993;10(5):497–539.
89. Garson MJ. The biosynthesis of sponge secondary metabolites: Why it is important. In: van Soest RWG, van Kempen MG, Braekman JC,

Eds. *Sponges in Time and Space*. Rotterdam: AA Balkema, 1994:427–440.

90. Sipkema D, Franssen MC, Osinga R, Tramper J, Wijffels RH. Marine sponges as pharmacy. *Mar Biotechnol (NY)* 2005;7(3):142–162.
91. Sipkema D, Osinga R, Schatton W, Mendola D, Tramper J, Wijffels RH. Large-scale production of pharmaceuticals by marine sponges: Sea, cell, or synthesis? *Biotechnol Bioeng* 2005;90(2):201–222.
92. Pomponi SA. Biology of the Porifera: Cell culture. *Can J Zool-Revue Can Zool* 2006;84(2):167–174.
93. Belarbi EH, Contreras GA, Chisti Y, Garcia CF, Molina GE. Producing drugs from marine sponges. *Biotechnol Adv* 2003;21(7): 585–598.
94. Rinkevich B. Marine invertebrate cell cultures: New millennium trends. *Mar Biotechnol (NY)* 2005;7(5):429–439.

11 Sea Urchin Embryo

A Model System for Analyzing Cellular Activities During Early Development

Tetsuya Kominami and Hiromi Takata

ABSTRACT

Sea urchin embryos have long been used as ideal experimental materials. In this chapter, we describe research areas to which sea urchin embryos have contributed, and the potential fields for which they may serve as a model system. The most valuable feature of sea urchin embryos is the availability of a large amount of homogeneous material, which facilitates biochemical and molecular biological approaches. The ease of gamete handling enables detailed analysis of the mechanism of fertilization, and the transparency and synchrony of fertilized eggs facilitate investigations on cell division and the cell cycle. Sea urchin embryos are an ideal model system for signal transduction, because the number of constituent cells is small. The simple organization of the embryo simplifies the analysis of morphogenetic movements. Both primary and secondary mesenchyme cells are interesting populations for studying cell movement. Sea urchin embryos will continue to contribute to the analysis of various unsolved problems.

Key Words: Sea urchin embryo, Fertilization, Cell division, Cell cycle, Signal transduction, Specification, Gastrulation, Invagination, Morphogenesis, Cell movement.

HANDLING OF ANIMALS AND EMBRYOS

Sea urchins belong to the phylum Echinoidermata (echinoderms), which consists of six classes (Crinoidea, Asteroidea, Concentricyclodea, Ophiuroidea, Echinoidea, and Holothuroidea). At present, nearly 900 species of sea urchins (class Echinoidea)—divided into nine orders—are known. Most live in intertidal zones or shallow seas, and their distributions extend from equatorial regions through the North and South Poles. Embryos of sea urchins are frequently used in biological and medical research including those of *Strongylocentrotus purpuratus*, *Lytechinus variegatus*, *Lytechinus pictus*, *Arbacia punctulata* (North America), *Paracentrotus lividus* (Mediterranean Sea), *Psammechinus miliaris* (North Sea), *Hemicentrotus pulcherrimus*, and *Anthocidaris crassispina* (Japan). The whole genome of *S. purpuratus*, a well known species in North America, has already been sequenced (http://sugp.caltech.edu/). In the United States, some species are available from commercial suppliers such as Susan Decker (Davie, FL) or Marinus Scientific (Long Beach, CA). Some marine laboratories also supply these materials in several countries. Adult sea urchins are reared in a small aquarium equipped with a simple filtering system and maintained without feeding for a couple of months.

During the breeding season, fertilizable oocytes are shed following intracoelomic injection of 0.5 M KCl. In some species, more than a million eggs are obtained from a single female. Usually, the testis is directly excised from adult males, and the leaked semen is diluted with seawater immediately before use. Fertilized eggs are obtained by simply adding the diluted sperm to an egg suspension.

For embryo culture, natural seawater should be filtered to remove debris and microorganisms. If fertilized eggs are deprived of the fertilization envelope for manipulations, it is preferable to add antibiotics (e.g., streptomycin or penicillin). Commercially supplied artificial seawater (e.g., Instant Ocean) is also available. At an appropriate density, fertilized eggs developed into pluteus larvae within 2–3 days without any special care. Detailed handling of animals and embryos has been described previously.[1]

FERTILIZATION

The finding of the acrosome reaction in starfish spermatozoa[2] led to detailed analysis of fertilization processes, which was accompanied by improved electron microscopic and biochemical techniques. Many important concepts of fertilization (cortical reaction, activation, Ca^{2+} release, acid release, change in membrane potential, block to polyspermy, etc.) have come from numerous studies of sea urchin fertilization. The jelly coat of *A. punctulata* contains a chemoattractant for spermatozoa,[3] although this has not yet been proved in other sea urchin species. When a sperm head attaches to the jelly coat (Figure11–1A), the acrosomal vesicle breaks down and the acrosomal process forms. Proteolytic enzymes contained in the acrosomal vesicle digest the jelly coat and allow the spermatozoon to approach the egg. The first contact between sperm and egg is mediated by bindin on the acrosomal process and its receptor on microvilli extruded from the oocyte.[4] The binding of bindin and its receptor shows species specificity, and leads to the membrane fusion of both gametes.

From: *Sourcebook of Models for Biomedical Research* (P. M. Conn, ed.), © 2008 Humana Press Inc., Totowa, NJ.

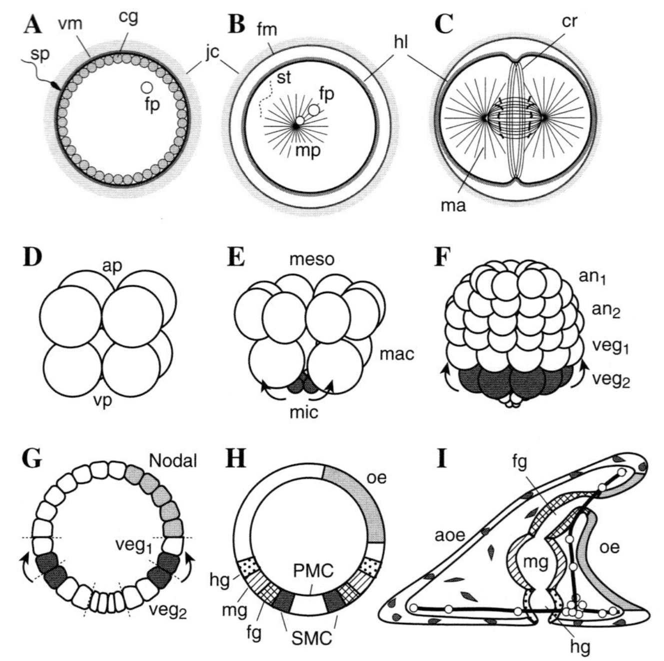

Figure 11–1. Early development of sea urchin embryos. (A) Unfertilized egg. (B) Fusion of male and female pronuclei. (C) First cleavage. (D) Eight-cell stage. (E) Sixteen-cell stage. (F) Sixty-cell stage. The four tiers of blastomeres are named an_1, an_2, veg_1, and veg_2 along the animal–vegetal axis. (G) Late blastula stage. (H) Fate map drawn in the median plane of the swimming blastula. (I) Pluteus larva (48 h after fertilization in *Hemicentrotus pulcherrimus*). (E–G) Positions of *Wnt8* and *Nodal* expressions are indicated in dark and light gray, respectively. (A–F) Frontal view. (G–I) Side view. Dotted lines in (G) indicate the boundaries of the veg_1 and veg_2 tiers. (H, I) Labels for SMCs (dark gray), foregut (reticulated), mid-gut (hatched), hind gut (dotted), and oral ectoderm (light gray) are common to Figure 11–2, which shows more details on stages between (H) and (I) of this figure. ap, animal pole; aoe, aboral ectoderm; cg, cortical granule; cr, contractile ring; fm, fertilization membrane; fp, female pronucleus; hl, hyaline layer; jc, jelly coat; ma, mitotic apparatus; mac, macromere; meso, mesomere; mic, micromere; mp, male pronucleus; oe, oral ectoderm; sp, spermatozoon; st, degenerating sperm tail; vm, vitelline membrane; vp, vegetal pole.

Following cell membrane fusion, the Na^+–K^+ ion channels of the egg act to elevate the intracellular Na^+ concentration and to depolarize the egg membrane. This electrical change is a rapid and transient block to polyspermy.[5] The influx of Ca^{2+} at the fusion site triggers the breakdown of cortical granules (cortical reaction), which is a typical Ca^{2+}-induced Ca^{2+} release reaction. The exocytosis of cortical granules propagates from the sperm entry point toward the opposite site concentrically. With cortical granule breakdown, the fertilization membrane is elevated (Figure 11–1B). This physical barrier is a late and permanent block to polyspermy. Molecules contained in the cortical granules cover the egg surface and form the hyaline layer (Figure 11–1B). Sea urchin fertilization is providing a model system for the analysis of intracellular signaling pathways for Ca^{2+} release from internal supplies.

After entry into the egg, the sperm head decondenses and becomes a male pronucleus. Centrioles brought into the egg with the sperm nucleus act as a microtubule-organizing center. The female pronucleus is pulled toward the centrioles along a ray of microtubules from the growing sperm monaster[6] and fuses with the male pronucleus. The transparency of the fertilized eggs enables us to observe these processes in detail.

CELL DIVISION AND THE CELL CYCLE

Fertilized sea urchin eggs have greatly contributed to the understanding of the mechanisms of cell division. This is due to the ease of handling and the transparency and synchrony of fertilized eggs. With differential interference contrast imaging or polarizing microscopy, dynamic changes in the mitotic apparatus—assembly and disassembly of microtubules—are observable (Figure 11–1C). Even the number of microtubules that constitutes the mitotic spindle can be estimated from the intensity of birefringence.[7]

By deforming fertilized eggs into various shapes, Rappaport has studied the spatial relationship between mitotic spindle and cleavage furrow.[8] It is now widely accepted that differences in the density of the microtubules that reach the egg cortex determine the position of the cleavage furrow (i.e., the position of the contractile ring), which exerts the main force for cytokinesis (Figure 11–1C). The existence of the contractile ring was first reported in fertilized sea urchin eggs.[9] To preserve this structure for electron microscopy, more than 500 combinations of buffers and fixatives were tested, but we can now detect microfilaments more easily using fluorescent phallacidins.

The fertilized eggs begin to cleave within 1.0–1.5 h after fertilization, and show radial and holoblastic cleavages. Up to the eight-cell stage, the cleavage pattern is typical (Figure 11–1D). In some species, blastomeres in the animal hemisphere are somewhat larger than those in the vegetal hemisphere. This is the first sign of the animal–vegetal axis. At the fourth cleavage, the animal blastomeres undergo meridional cleavage and give rise to eight blastomeres with the same volume (mesomeres). In the vegetal half, cleavage is horizontal and the cleavage plane shifts to the vegetal pole. As a result, four large blastomeres (macromeres) and small blastomeres (micromeres) are formed (Figure 11–1E). This formation of macromeres and micromeres is an excellent model for research of unequal cell division.

In most species, seven or eight rounds of synchronous cleavages occur at intervals of 0.5–1.0 h. The cell cycle then extends and the synchrony of division is lost. As is well known, the cell cycle is regulated by complexes of cyclin and Cdc(s) or Cdk(s). It is of note that cyclins were first found in sea urchin embryos.[10] By the stage of hatching, the fertilized eggs usually undergo 10 cleavages and develop into spherical and hollow blastulas (Figure 11–1H). The time required for development to the hatching stage is 10–12 h in most species. The hatched blastula develops into a pluteus larva through a series of morphogenetic movements and organogenesis within 2–3 days (Figure 11–1I). It is of note that each part of the larval body comprises a monolayered epithelium.

SIGNAL TRANSDUCTION AND GENE REGULATORY NETWORK

The double gradient theory has explained the results obtained from classical deletion and recombination experiments on sea urchin embryos.[11] Recent studies have revealed that micromeres and their descendants act as a signaling center for embryonic body patterning along the animal–vegetal axis. Briefly, β-catenin enters the nuclei of micromeres soon after their formation.[12] Then, *Wnt8* expression begins in micromeres[13] (Figure 11–1E). By the 60-cell stage (after the sixth cleavage), its expression in micromere descendants disappears. Instead, expression is noted in the veg_2 tier of blastomeres (Figure 11–1F). Furthermore, expression of *Wnt8* shifts to the descendants of the veg_1 tier of blastomeres as the cleavage progresses (Figure 11–1G).

The delta-Notch signaling pathway is also involved in the specification of mesodermal cells.[14] This pathway also participates in defining the boundary between mesoderm and endoderm.[15] These upward signal transductions seem to last until the mesenchyme blastula stage. In the vegetal plate of the mesenchyme blastula, a number of nonciliated cells (probably the progeny of small micromeres) are positioned at the center of the vegetal plate, and prospective mesodermal and endodermal cells are positioned concentrically.[16] The detailed gene regulatory network for the specification of the endodermal–mesodermal lineage has been intensively studied and demonstrated in Davidson's laboratory (http://www.its.caltech.edu/~mirsky/index.htm).

For specification of the dorsoventral (oral–aboral) axis, Nodal, which is a secreting protein, seems to play a crucial role.[17] Expression of Nodal begins at the mid-to-late blastula stage (Figure 11–1G). This is followed by the expression of *goosecoid* and *Brachyury*, which are the target genes of Nodal. However, the earlier phases of the oral–aboral axis specification remain unclear. Although unequal distribution of mitochondria in unfertilized eggs is one of the factors that leads to Nodal expression,[18] it is also possible that some molecules colocalizing with mitochondria affect the specification of the oral–aboral axis. Furthermore, it remains unclear whether sperm entry leads to the displacement of a cytoplasmic component that affects the oral–aboral specification.

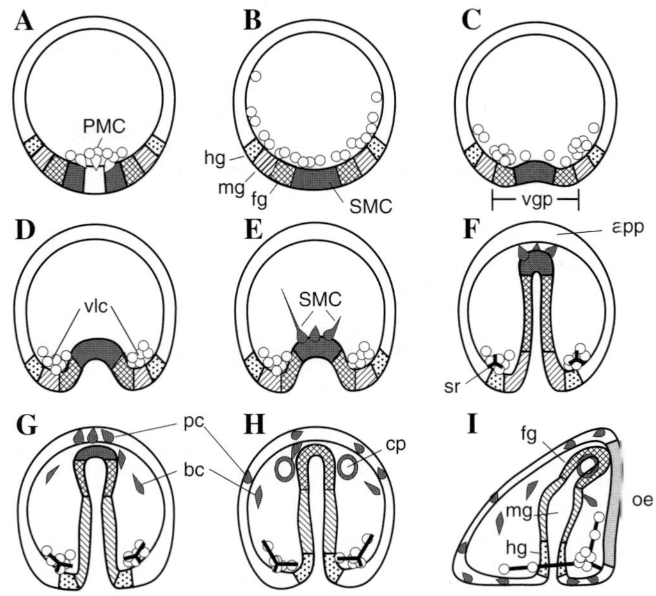

Figure 11–2. Process of gastrulation in *Hemicentrotus pulcherrimus.* (A) Early mesenchyme blastula stage (14h after fertilization). (B) Late mesenchyme blastula stage (16h). (C) Early phase of primary invagination (18h). (D) Early-to-mid gastrula (20h). (E) Mid-gastrula stage (22h). (F) Late gastrula stage (24h). (G) Early prism stage (28h). (H) Mid-prism stage (32h). (I) Late prism stage (36h). (A–H) Frontal view. (I) Side view. app, apical (animal) plate; bc, blastocoelar cell; cp, coelomic pouch; fg, foregut (esophagus); hg, hind gut (intestine); mg, mid-gut (stomach); pc, pigment cell; PMC, primary mesenchyme cell; SMC, secondary mesenchyme cell; sr, spicule rudiment; vlc, ventrolateral cluster of PMCs; vgp, vegetal plate.

MORPHOGENESIS OF A MONOLAYERED EPITHELIUM

Gastrulation is the most prominent morphogenetic event in early development, resulting in the formation of three germ layers: ectoderm, mesoderm, and endoderm. To determine the mechanism of gastrulation, sea urchin embryos have long been used as a model system owing to their transparency, simple organization, and synchrony. The process of sea urchin gastrulation is composed of five steps.[19]

Step 1: After hatching, primary mesenchyme cells (PMCs), the descendants of micromeres, begin to migrate into the blastocoel (Figure 11–2A). PMC precursor cells embedded in the blastocoel wall retract the anchoring microvilli from the hyaline layer, lose their cilia and adhesion to neighboring cells, disrupting the basal lamina that coats the inner surface of the PMCs.[20] As a result, the PMCs become rounded and slip from the vegetal plate into the blastocoel. This step is a model for the epithelium–mesenchyme transition.[21]

Step 2: Once invagination has started, the thickened vegetal plate bends inward and gives rise to a short, stub-like gut rudiment within a rather short period of time (Figure 11–2C and D). This step is called primary invagination,[22] and has been intensively studied to determine the mechanism of invagination of a monolayered epithelium. As in amphibian embryos,[23] apical constriction of bottle cells triggers the onset of invagination.[24] However, the cellular basis of primary invagination cannot solely be ascribed to bottle cells, since invagination occurs even in bottle cell-depleted embryos despite a significant delay in invagination.[25] Several models are proposed to explain primary invagination.[26] Each model partly explains the change in shape of the invaginating gut rudiment. However, the kinds of forces operating during this simple morphogenetic movement are not fully understood. Other factors such as the force exerted by noninvaginating epithelium and changes in the turgor pressure of the blastocoel should be considered for further understanding of the mechanism of primary invagination.[19]

Step 3: After the primary invagination, a pause (for 1–2h in most species) in the elongation of the gut rudiment is observed. During this period, secondary mesenchyme cells (SMCs) appear at the archenteron tip (Figure 11–2E). The SMCs extend long and thin filopodia toward the animal pole side, exploring the attachment site. With high-resolution Nomarski imaging, the presence of very thin filopodia (0.2–0.4µm diameter) extending from SMC have been shown.[27] These thin filopodia provide a means by which cells can contact others several cell-diameters away. Target recognition by the SMC filopodia seems to be mediated by localized carbohydrate chains, since the mannose-specific lectin *Lens culinaris* agglutinin blocks the anchoring of the SMC filopodia to the prospective oral opening.[28]

Step 4: After this pause in archenteron elongation, the gut rudiment rapidly elongates until its tip reaches the inner surface of the apical plate (secondary invagination), resulting in a slender archenteron (Figure 11–2F). Several lines of evidence suggest that the filopodia connecting the archenteron tip and the apical plate pull the gut rudiment upward.[29] Dozens of cells are initially observed on cross sections of the gut rudiment, but the number decreases to seven or eight after the completion of secondary invagination.[30] This explicitly indicates that the archenteron cells are rearranged during this morphogenetic process.

Step 5: It has long been thought that endodermal cells are exclusively derived from the veg_2 tier of blastomeres.[11] However, by labeling of veg_1 and veg_2 blastomeres with DiI, it was shown that the descendants of veg_1 blastomeres are also recruited into the archenteron.[31] Subsequent studies revealed that veg_1-derived cells are added to the archenteron after the completion of secondary invagination.[32] Cell trace experiments showed that cells that have been invaginated by the end of the secondary invagination differentiate into SMCs, esophagus, and the anterior half of stomach (Figure 11–2F). The posterior half of the stomach and the whole intestine are recruited into the archenteron during the prism stage (Figure 11–2G–I). This late invagination seems to be a typical convergence–extension movement.[33]

Thus, we can study various cellular activities involved in gastrulation using sea urchin embryos. The functions of various genes involved in the gastrulation process are becoming clearer.[34] With such knowledge, this model system may contribute to the further understanding of the morphogenesis of a monolayered epithelium.

CELL MOVEMENT

As described above, PMCs appear in the blastocoel after hatching. For a period of time, the PMCs stay at the bottom of the blastocoel (Figure 11–2A). Then they move about randomly on the inner surface (basal lamina) of the blastocoel wall (Figure 11–2B). Some PMCs move toward the animal pole side. After this phase, they begin to migrate toward the vegetal plate (Figure 11–2C) and form a ring pattern around the gut rudiment.[35] During these phases, the PMCs form very thin filopodia for locomotion and target recognition.[27] The migration pattern of PMCs seems to depend on changes in the composition of the basal lamina and the localization of some extracellular matrix (ECM) molecules.[36] After ring pattern formation, PMCs fuse with each other and form syncytial cytoplasmic cables in which skeletal spicules are deposited.[37] In transparent embryos with a large blastocoel, the behavior of individual PMCs can be traced with time-lapse video microscopy. In most species, a triradiate spicule rudiment appears in the ventrolateral cluster of PMCs at the end of secondary invagination (Figure 11–2F). Then, skeletal rods begin to elongate in three directions, and form the main frame of the larval skeleton (Figure 11–2G–I).

Micromeres formed in 16-cell embryos can be collected by centrifugation of dissociated blastomeres in a sucrose density gradient. The descendants derived from such isolated micromeres can differentiate autonomously into PMCs.[38] PMCs can also be isolated from mesenchyme blastulas using wheat germ agglutinin.[39] Using these materials, *in vitro* studies of the behavior of PMCs can be performed.

SMCs differentiate into four types of mesodermal cells: pigment cells, blastocoelar cells, coelomic pouch cells, and circum-esophageal muscles. Pigment cells may provide another model for analyzing cell movement.[40] In some species, pigment precursor cells leave the archenteron tip during gastrulation and enter the apical plate (Figure 11–2E-G). During this stage, they form long filopodia and move a considerable distance in the blastocoel. After entry into the apical plate, they begin to crawl in the ectodermal layer toward the vegetal pole side (Figure 11–2H). Thus, pigment cells change their manner of migration within a rather short time. Interestingly, pigment cells are distributed only in the aboral ectoderm (Figure 11–2I). Some ECM molecules, such as Ecto V expressed in the oral ectoderm[41] and arylsulfatase expressed in the aboral ectoderm,[42] are the candidates that cause this localization. However, the precise mechanism of pigment cell movement remains unknown. Pigment cells are another model for the analysis of cell–substrate interactions.

PHARMACOLOGY AND TOXICOLOGY

To now, the effects of various kinds of chemicals have been monitored using sea urchin embryos, because a large quantity of homogeneous material is available. With ryanodine receptor blockers, for example, it was found that sphingosine can activate the ryanodine receptor, resulting in transient Ca^{2+} release from thapsigargin-sensitive intracellular stores.[43] In this report, a homogenate of unfertilized sea urchin eggs was used as an assay system. Sea urchin fertilization is therefore an ideal model for the analysis of the intracellular signaling pathway of Ca^{2+} release, while other cellular activities are the targets of pharmacological investigations.

Toxicological effects are examined by immersing embryos in seawater containing the chemical to be tested. The effects are quickly and easily detected as delays in the development or malformation of embryos. For example, the interactive toxic effects of heavy metals at concentrations detected in mine effluents were examined. With various combinations of metals, it was found that zinc was one of the elements responsible for causing malformations (loss of the oral–aboral axis and diminution in size of the digestive tract), and its effects were intensified by the presence of other metals such as manganese, lead, iron, and copper.[44]

DISADVANTAGES IN BIOMEDICAL RESEARCH

Sea urchin embryos provide many models for the analysis of various cellular activities. However, they have some disadvantages. One is the limited breeding season. Full-grown and fertilizable oocytes that completed meiosis cannot be obtained from a single species throughout the year. Usually, two or three species of sea urchins with different breeding seasons are used in laboratories.

Creation of mutants is a powerful tool for analyzing the function of genes. Unfortunately, there is no established method for obtaining sea urchin mutants. This is due to difficulties in rearing metamorphosed juveniles. They change their source of nutrients as they grow. To our knowledge, there is no reported success in obtaining adult sea urchins under laboratory conditions. To analyze the function of genes, however, most of the methods used in other models (whole-mount *in situ* hybridization, overexpression of synthetic mRNAs, injection of dominant-negative mRNAs, or morpholino antisense oligonucleotides) are applicable to sea urchin embryos, although refined skill is necessary for the injection of synthetic mRNAs or nucleotides. There is no report of the use of RNAi in the analysis of gene functions in sea urchin embryos.

Micromanipulation is another powerful tool for analyzing the mechanisms that operate during early development. For this, the size of embryos is a critical factor. The diameter of sea urchin eggs, however, is around 100 µm in most available species. Sea urchin embryos are much more difficult to manipulate than amphibian embryos. While removal of a portion of embryonic tissue is possible at later stages, marking, deletion, recombination, and transplantation of blastomeres are primarily limited to embryos at the early cleavage stage.

Echinoderm is a member of the Deuterostomata as well as the Chordata, and is considered to be the ancestors of chordates. The tool kit for regulation of gene expression is almost common between echinoderms and chordates. However, the usage of some genes differs from that of chordates. For example, Nodal, which is indispensable for the differentiation of endoderms and mesoderms in chordates, is expressed in the prospective oral ectoderm in sea urchin embryos.[17] *Brachyury*, the expression of which is limited to the notochord in chordates, is expressed in cells near the blastopore, in addition to the prospective oral region.[45] Furthermore, some downstream genes such as *SM30*, which encode a spicule matrix protein, are specific to sea urchins.[46] Therefore, molecular biological and physiological results obtained in sea urchin embryos are not necessarily applicable to chordates including mammals.

Despite the disadvantages mentioned above, sea urchin embryos should continue to serve as models for research in various biomedical studies, since they have many features that are unavailable in other animal embryos.

REFERENCES

1. Hinegardner R. Care and handling of sea urchin eggs. In: Czihak G, Ed. *The Sea Urchin Embryos*. Berlin: Springer-Verlag, 1975:10–25.
2. Dan JC. Studies on the acrosome. II. Acrosome reaction in starfish spermatozoa. *Biol Bull* 1954;107:203–218.
3. Ward GE, Brokaw CJ, Garbers DL, Vacquier VD. Chemotaxis of *Arbacia punctulata* spermatozoa to react, a peptide from the jelly layer. *J Cell Biol* 1985;101:2324–2329.
4. Vacquire VD, Moy GW. Isolation of bindin: The protein responsible for adhesion of sperm to sea urchin eggs. *Proc Natl Acad Sci USA* 1977;74:2456–2460.
5. Jaffe LA. Fast block to polyspermy in sea urchin eggs is electrically mediated. *Nature* 1976;261:68–71.
6. Hamaguchi MS, Hiramoto Y. Fertilization process in the heart-urchin, *Clypeaster japonicus*, observed with a differential interference microscope. *Dev Growth Differ* 1980;22:517–530.
7. Hiramoto Y, Hamaguchi Y, Shoji Y, Schroeder TE, Shimoda S, Nakamura S. Quantitative studies on the polarization optical properties of living cells. II. The role of microtubules in birefringence of the spindle of the sea urchin egg. *J Cell Biol* 1981;89:121–130.
8. Rappaport R. Role of the mitotic apparatus in furrow initiation. *Ann NY Acad Sci* 1990;582:15–21.
9. Schroeder TE. The contractile ring. II. Determining its brief existence, volumetric changes, and vital role in cleaving *Arbacia* eggs. *J Cell Biol* 1972;53:419–434.
10. Evans T, Rosenthal ET, Youngblom J, Distel D, Hunt T. Cyclin: A protein specified by maternal mRNA in sea urchin eggs that is destroyed at each cleavage division. *Cell* 1983;33:389–396.
11. Hörstadius S. *Experimental Embryology of Echinoderms*. London: Oxford University Press, 1973.
12. Logan CY, Miller JR, Ferkowicz MJ, McClay DR. Nuclear beta-catenin is required to specify vegetal cell fates in the sea urchin embryo. *Development* 1999;126:345–357.
13. Wikramanayake AH, Peterson R, Chen J, Huang L, Bince JM, McClay DR, Klein WH. Nuclear beta-catenin-dependent Wnt8 signaling in vegetal cells of the early sea urchin embryo regulates gastrulation and differentiation of endoderm and mesodermal cell lineages. *Genesis* 2004;39:194–205.
14. Sherwood DR, McClay DR. LvNotch signaling mediates secondary mesenchyme specification in the sea urchin embryo. *Development* 1999;126:1703–1713.
15. Sherwood DR, McClay DR. LvNotch signaling plays a dual role in regulating the position of the ectoderm-endoderm boundary in the sea urchin embryo. *Development* 2001;128:2221–2232.
16. Ruffins SW, Ettensohn CA. A fate map of the vegetal plate of the sea urchin (*Lytechinus variegatus*) mesenchyme blastula. *Development* 1996:122:253–263.
17. Duboc, Lepage T. A conserved role for the Nodal signaling pathway in the establishment of dorso-ventral and left-right axes in deuterostomes. *J Exp Zool B Mol Dev Evol* 2006;306B:1–13.
18. Coffman JA, McCarthy JJ, Dickey-Sims C, Robertson AJ. Oral-aboral axis specification in the sea urchin embryo. II. Mitochondrial distribution and redox state contribute to establishing polarity in *Strongylocentrotus purpuratus*. *Dev Biol* 2004;273:160–171.
19. Kominami T, Takata H. Gastrulation in the sea urchin embryo: A model system for analyzing the morphogenesis of a monolayered epithelium. *Dev Growth Differ* 2004;46:309–326.
20. Fink RD, McClay DR. Three cell recognition changes accompany the ingression of sea urchin primary mesenchyme cells. *Dev Biol* 1985;107:66–74.
21. Rottinger E, Besnardeau L, Lepage T. A Raf/MEK/ERK signaling pathway is required for development of the sea urchin embryo micromere lineage through phosphorylation of the transcription factor Ets. *Development* 2004;131:1075–1087.
22. Okazaki K. Normal development to metamorphosis. In: Czihak G, Ed. *The Sea Urchin Embryos*. Berlin: Springer-Verlag, 1975:177–232.
23. Keller RE. An experimental analysis of the role of bottle cells and the deep marginal zone in the gastrulation of *Xenopus laevis*. *J Exp Zool* 1981;216:81–101.
24. Nakajima Y, Burke RD. The initial phase of gastrulation in sea urchins is accompanied by the formation of bottle cells. *Dev Biol* 1996;179:436–446.
25. Kimberly EL, Hardin J. Bottle cells are required for the initiation of primary invagination in the sea urchin embryo. *Dev Biol* 1998;204:235–250.
26. Davidson LA, Koehl MAR, Keller R, Oster GF. How do sea urchins invaginate: Using biomechanics to distinguish between mechanisms of primary invagination. *Development* 1995;121:2005–2018.
27. Miller J, Fraser SE, McClay DR. Dynamics of thin filopodia during sea urchin gastrulation. *Development* 1995;121:2501–2511.
28. Latham VH, Tully MJ, Oppenheimer SB. A putative role for carbohydrates in sea urchin gastrulation. *Acta Histochem* 1999;101:293–303.
29. Hardin JD. The role of secondary mesenchyme cells during sea urchin gastrulation studied by laser ablation. *Development* 1988;103:317–324.
30. Ettensohn CA. Gastrulation in the sea urchin embryos is accompanied by the rearrangement of invaginating epithelial cells. *Dev Biol* 1985;112:383–390.
31. Logan CY, McClay DR. The allocation of early blastomeres to the ectoderm and endoderm is variable in the sea urchin embryo. *Development* 1997;124:2213–2223.
32. Martins GG, Summers RG, Morrill JB. Cells are added to the archenteron during and following secondary invagination in the sea urchin *Lytechinus variegatus*. *Dev Biol* 1998;198:330–342.
33. Takata H, Kominami T. Shrinkage and expansion of blastocoel affect the degree of invagination in sea urchin embryos. *Zool Sci* 2001;18:1097–1105.
34. Beane WS, Gross JM, McClay DR. RhoA regulates initiation of invagination, but not convergent extension, during sea urchin gastrulation. *Dev Biol* 2006;292:213–225.
35. Ettensohn CA. The regulation of primary mesenchyme cell patterning. *Dev Biol* 1990;140:261–271.
36. Hodor PG, Illies MR, Broadley S, Ettensohn CA. Cell-substrate interactions during sea urchin gastrulation: Migrating primary mesenchyme cells interact with and align extracellular matrix fibers that contain ECM3, a molecule with NG2-like and multiple calcium-binding domains. *Dev Biol* 2000;222:181–194.

37. Hodor PG, Ettensohn CA. The dynamics and regulation of mesenchymal cell fusion in the sea urchin embryo. *Dev Biol* 1998;199:111–124.
38. Okazaki K. Spicule formation by isolated micromeres of the sea urchin embryo. *Am Zool* 1975;15:567–581.
39. Ettensohn CA, McClay DR. A new method for isolating primary mesenchyme cells of the sea urchin embryo. Panning on wheat germ agglutinin-coated dishes. *Exp Cell Res* 1987;168:431–438.
40. Kominami T, Takata H, Takaichi M. Behavior of pigment cells in gastrula-stage embryos of *Hemicentrotus pulcherrimus* and *Scaphechinus mirabilis*. *Dev Growth Differ* 2001;43:699–707.
41. Coffman JA, McClay DR. A hyaline layer protein that becomes localized to the oral ectoderm and foregut of sea urchin embryos. *Dev Biol* 1990:140:93–104.
42. Mitsunaga-Nakatsubo K, Akasaka K, Akimoto Y, Akiba E, Kitajima T, Tomita M, Hirano H, Shimada H. Arylsulfatase exists as non-enzymatic cell surface protein in sea urchin embryos. *J Exp Zool* 1998;280:220–230.
43. Floriddia EM, Pace D, Genazzani AA, Canonico PL, Condorelli F, Billington RA. Sphingosine releases Ca^{2+} from intracellular stores via the ryanodine receptor in sea urchin egg homogenates. *Biochem Biophys Res Commun* 2005;338:1316–1321.
44. Kobayashi N, Okamura H. Effects of heavy metals on sea urchin embryo development. Part 2. Interactive toxic effects of heavy metals in synthetic mine effluents. *Chemosphere* 2005;61:1198–1203.
45. Gross JM, McClay DR. The role of *Brachyury* (*T*) during gastrulation movements in the sea urchin *Lytechinus variegatus*. *Dev Biol* 2001;239:132–147.
46. George NC, Killian CE, Wilt FH. Characterization and expression of a gene encoding a 30.6-kDa *Strongylocentrotus purpuratus* spicule matrix protein. *Dev Biol* 1991;147:334–342.

12 *Caenorhabditis elegans* Models of Human Neurodegenerative Diseases

A Powerful Tool to Identify Molecular Mechanisms and Novel Therapeutic Targets

RICHARD NASS AND LIHSIA CHEN

ABSTRACT

Within the past decade it has become remarkably clear that the molecular components involved in basic cellular function and dysfunction are highly conserved across phyla from single cell organisms to humans. The high sequence similarities within eukaryotes and the advent of molecular technologies now allow us to utilize genetic models such as the nematode *C. elegans* to begin to elucidate the complex mechanisms involved in human neurodegenerative diseases. In this chapter we will briefly summarize recent advances using the nematode *C. elegans* to model a number of neurodegenerative diseases including Parkinson's disease, Alzheimer's disease, Huntington's disease, and hereditary spastic paraplegia, and describe how this relatively simple animal is being utilized to identify putative endogenous and exogenous molecules that may contribute to the disease. We also describe how this genetically tractable organism is amenable to high throughput technologies to identify novel drug targets and potential therapeutic leads to combat these devastating neurological disorders.

Key Words: Nematode, Genetics, *C. elegans*, Parkinson's Disease, Huntington's disease, Alzheimer's disease, Hereditary spastic paraplegia, 6-OHDA, High-throughput screens.

INTRODUCTION

The free-living soil nematode *Caenorhabditis elegans* was first introduced by Sydney Brenner in the mid-1960s as a model organism to dissect the genetic pathways that govern organ development, particularly of the nervous system, and behavior. Since then, the numerous advances resulting from the use of *C. elegans* have earned this simple animal a well-deserved reputation of a powerful genetic model system to elucidate cellular processes underlying normal development as well as disease. Indeed, the power of *C. elegans* was recognized some 35 years later when Sydney Brenner, Robert Horvitz, and John Sulston received the 2003 Nobel award for the establishment of the nematode *C. elegans* to explore organ development and programmed cell death.

It has become apparent in the past 15 years that the molecular pathways and mechanisms underlying basic biological processes are conserved between invertebrates and humans. Furthermore, the completion of the *C. elegans* genome and subsequently those of *Drosophila*, mouse, and human revealed genome conservation and revealed that the human genome contains only ~20% more genes than the *C. elegans* genome.[1,2] In addition to the genome, the worm's nervous system is highly conserved with mammals, and contains almost all of the known signaling and neurotransmitter systems found in the vertebrates.[3–7] These include enzymes and molecular pathways involved in the production of numerous neurotransmitters including acetylcholine, glutamate, α-aminobutyric acid, serotonin, and dopamine (DA), as well as neuropeptides and neurotransporters. Furthermore, most of the ion channels and components involved in mammalian synaptic neurotransmission are also present in the worm.[3,7] The similarity between worm and human nervous systems together with the conserved genomes and genetic pathways suggest that the knowledge and paradigms we gain by elucidating the molecular mechanisms involved in worm neurophysiology will have significant relevance to human physiology and disease.

In the present review, we discuss how *C. elegans* has allowed us to model several neurodegenerative diseases including Parkinson's, Alzheimer's, and Huntington's diseases, as well as hereditary spastic paraplegia, and how the opportunities presented by the worm will likely increase our understanding of the molecular basis of neurodegenerative disorders. We also describe how this genetically tractable system can be utilized to identify novel drug targets and potentially valuable therapeutic leads that can protect against cellular death.

CAENORHABDITIS ELEGANS AS A GENETIC MODEL

C. elegans is an anatomically simple organism containing under a thousand somatic cells. Its small size (1 mm long), large brood size (approximately 300–1000 progeny from a single hermaphrodite), short generation time (3.5 days), and ease of maintenance in the laboratory (tens of thousands can be grown on a 90-mm agar plate coated with bacteria) allow for inexpensive and rapid production of animals for experimental analysis.[8,9] In

From: *Sourcebook of Models for Biomedical Research*
(P. M. Conn, ed.), © 2008 Humana Press Inc., Totowa, NJ.

addition, the worm is transparent, allowing for the visualization of cells and developmental processes such as cell migration or apoptosis in a living animal. The worm can also easily be grown in liquid medium in standard 384 or 96-well microtiter plates, facilitating high or medium throughput screening (HTS) of animals for particular behavioral phenotypes or optical properties.[10,11] In addition to studies on an organismal level, large-scale primary cultures first developed by Laird Bloom in 1993 provide the opportunity to investigate biological processes on a cellular level.[12–16] For example, electrical properties of different cells can be determined and compared in both whole animals and primary cultures. These primary cultures also are amendable to cell sorting and allow for the development of tissue-specific cDNA libraries.[17]

While the aforementioned properties facilitate the use of *C. elegans* as a model system to study biological processes, the power of *C. elegans* in fact resides in its accessibility to genetic manipulation (see also Chapter 72, this volume).[8,18] Forward genetic screens using chemical mutagens or γ-irradiation are a common way to identify genes involved in a particular cellular process. An advantage to this approach is that no a priori knowledge about gene function is needed in order to determine whether the gene plays a role in the particular behavioral phenotype or cellular process.

C. elegans is also amendable to reverse genetic approaches that make it possible to determine of the function of a particular gene. As with forward genetic screens, chemical mutagens can be used to induce deletions in the gene of interest to produce genetic nulls or hypomorphs. Genetic knockout animals can be screened and identified within as little as a week.[19] A federally funded *C. elegans* gene knockout consortium as well as the Japanese National Bioresource Project are two resources in which deletion mutations in *C. elegans* genes are being generated. These mutants as well as over 8000 genetic strains isolated from different laboratories around the world and stored in the *C. elegans* Genetic Stock Center are available upon request (see *Caenorhabditis elegans* WWW Server: http://elegans.swmed.edu/).[20]

The relatively recent discovery of RNA-interference (RNAi) by Andrew Fire and Craig Mello, for which both received the 2006 Nobel award in physiology or medicine, has greatly altered the way genetic screens are performed in *C. elegans*. RNAi is the evolutionarily conserved process of sequence-specific degradation of mRNAs induced by the introduction of double-stranded RNA (dsRNA) containing homologous sequence.[20,21] RNAi can thus be used as a convenient way to knock down expression of any gene. Such targeted gene knockdown by RNAi can be obtained in most cell types *in vivo* by introducing gene-specific dsRNA into animals by injection, soaking animals in the dsRNA, or feeding animals bacteria that express the dsRNA.[22,23] The ability to introduce dsRNA by feeding has allowed genome-wide genetic screens to be performed on a rather routine basis. Indeed, genes participating in different processes such as synapse function or the pathophysiology of Huntington's disease (see below) have been identified in such screens using a commercially available RNAi-feeding library, which contains over 18,000 *Escherichia coli* strains each containing dsRNA for a specific gene within the *C. elegans* genome (representing approximately 86% of the *C. elegans* genome).[24–27]

In addition to mutagenesis or RNAi screens, the worm can be genetically manipulated with the introduction of transgenes. The ease and relatively short amount of time for generating transgenic animals have enabled functional studies to be performed routinely. These studies include the expression of green fluorescent protein (GFP) reporter or rescue constructs (about 4 days), which greatly facilitate spatial and temporal protein expression pattern studies as well as structure–functional analyses.[28–30]

The *C. elegans* nervous system has been reconstructed previously on the ultrastructural level. As such, the physical connections (axon targeting, as well as electrical and chemical synapses) of the 302 neurons that make up the *C. elegans* nervous system have been mapped.[31] Together with the accessible genetics and ease of high-throughput approaches, *C. elegans* provides a powerful tool to model neurodegenerative diseases to better dissect the molecular and cellular processes underlying the diseases.

MODELING PARKINSON'S DISEASE AND DOPAMINE NEURON CELL DEATH

Parkinson's disease (PD) is the second most prevalent neurodegenerative disease and results from the loss of over 80% of the DA neurons within the substantia nigra pars compacta (SNpc).[32,33] The disorder also results in the formation of protein aggregates called Lewy bodies (LBs) in a significant number of the surviving neurons.[34–36] The degradation of the DA neurons confers the classic triad of PD symptoms, including tremors, rigidity, and bradykinesia. Although the molecular basis of this disorder has not been identified, etiological and pathological data suggest that there is both a genetic and environmental component that causes oxidative damage, and proteosome and mitochondrial dysfunction.[33,37–39]

One of the most common mechanisms to model Parkinson's disease in vertebrates is through exposure of the animals to a DA neuron targeted toxin. Several weeks following exposure to the toxin, most of the animals display a Parkinsonian-like syndrome. The most common neurotoxins used are 6-hydroxydopamine (6-OHDA), 1-methyl-4-phenyl-1,2,3,6-tetrahydropyridine (MPTP), 1-methyl-4-phenylpyridinium ion [MPP^+ (the active metabolite of MPTP)], or the insecticide rotenone.[40–43] 6-OHDA and MPP^+ are transported into the cell by the high affinity dopamine transporter, DAT, which is also the target of drugs of abuse including cocaine and amphetamine.[41,43,44]

Genetic and epidemiological studies indicate that the etiology of PD likely involves specific molecular pathways involved in protein aggregation, degradation, oxidative stress, and mitochondria function. To date, at least seven genes have been independently identified that are associated with rare, familial forms of PD, including α-synuclein, parkin, DJ-1, UCH-1, NURR1, PINK-1, and LRRK2.[39,45–51] α-Synuclein, the first PD-associated gene identified, is a presynaptic protein that appears to interact with synaptic vesicles and could be involved in the regulation of both dopamine biosynthesis and dopamine transporter function.[52] α-Synuclein coding mutations A30P and A53T alter the structure of α-synuclein and affect DA neuron viability *in vitro* and *in vivo*.[34–36] Another mutation, α-synuclein K46E, has been recently identified and as with the other α-synuclein mutations, may interfere with the proteosomal degradation pathway.[51]

Toxin-induced and genetic *C. elegans* PD models that recapitulate many aspects of PD have been developed. We generated the first *C. elegans* PD model with a transgenic strain that expresses the GFP in all eight DA neurons that are thus visible under a fluorescent microscope.[53,54] When we briefly exposed (0.5–1 h) the animals to 6-OHDA, we find a time- and concentration-dependent

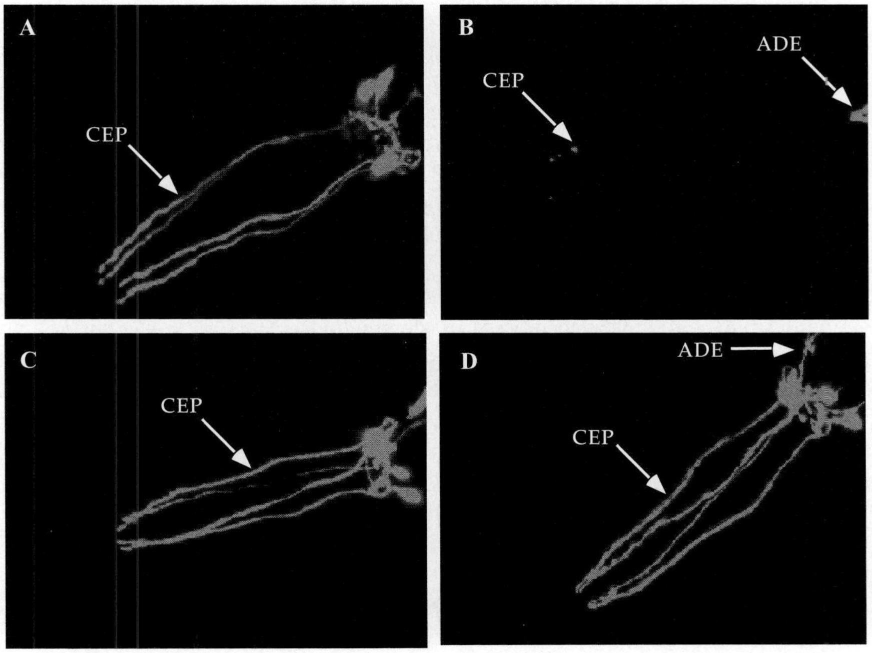

Figure 12–1. Suppression of 6-OHDA sensitivity of DA neurons in *C. elegans*. (A) $P_{dat\text{-}1}$::GFP animals exposed to vehicle; (B) $P_{dat\text{-}1}$::GFP worms exposed to 6-OHDA; (C) $P_{dat\text{-}1}$::GFP, *dat-1* (ΔDAT-1 worms) exposed to vehicle; (D) $P_{dat\text{-}1}$::GFP, dat-1 worms exposed to 6-OHDA. See Nass *et al.*[54] for details. (Reproduced with permission from Nass *et al.*[54])

loss of DA neuron integrity.[54] Consistent with the vertebrate PD models, the 6-OHDA-induced effects require the presence of the DAT, and can be completely blocked by coincubation with DAT agonist (e.g., amphetamine) or antagonist (e.g., cocaine) (Figure 12–1, data not shown).[54] MPP^+, the vertebrate mitochondrial complex I inhibitor, also causes loss of GFP in the DA neurons and likely DA neuron cell death, although no ultrastructural studies were performed to confirm degeneration (R. Nass, unpublished data).[55] We subsequently performed a forward genetic screen to identify suppressors of the 6-OHDA-induced cell death.[56] In our initial efforts we identified three novel *dat-1* alleles conferring complete tolerance to 6-OHDA, as well as additional mutants that confer partial tolerance. These studies provide proof of concept for the ability to identify specific toxin suppressors. Overall, our results as well as those from other groups mimic several significant aspects of vertebrate PD models. Thus the *C. elegans* PD model provides a powerful way to identify and characterize endogenous and exogenous molecules that may contribute to PD neurodegeneration.[32,55–57]

C. elegans does not contain an apparent homolog to the PD-associated protein α-synuclein, but the lack of an endogenous molecule can aid studies in dissecting potential interacting molecules, without concern for compensatory effects. In collaboration with Garry Wong at Kuopio University in Kuopio, Finland, we generated transgenic animals expressing either human wild-type (WT) or mutant A53T α-synuclein in the DA neurons within the worm.[58] We found that both synucleins confer DA neuron cell death. Motor deficits were also observed when α-synuclein was expressed pan-neuronally, and α-synuclein-containing inclusion bodies are seen in some of the DA neurons.[58] In a study by Kuwahara and colleagues, the expression of human A30P (and A53T) also appears to cause DA neurodegeneration, abnormal locomotion in response to food, and a loss of DA neuronal content, and importantly the addition of exogenous DA ameliorates the behavior defects.[59] Recently, Cooper and colleagues identified a Rab GTPase in a yeast plasmid overexpression screen that suppresses the α-synuclein toxicity in yeast.[60] Overexpression of the GTPase, which plays a role in Golgi vesicle docking and endoplasmic reticulum (ER) to Golgi vesicular transport, also rescues DA neuron cell death in both the fly and the worm, suggesting that PD-associated α-synuclein pathology may disrupt normal function of RabGTPase or the pathways involved in vesicle trafficking *in vivo*.[60] Furthermore, Wong and colleagues have recently shown significant gene expression changes in genes associated with the ubiquitin-proteasomal and mitochondrial systems in animals expressing α-synuclein, further supporting the role of these systems in PD-associated cell death.[61] The above studies suggest that the α-synuclein-induced toxicities observed in *C. elegans* may occur through pathways similar to those observed in vertebrate systems and that the worm could be a powerful model to explore human synucleopathies.

Two other genes associated with recessive PD have also been explored in *C. elegans*. *Parkin* encodes for an E3 ubiquitin ligase, one of a number of proteins that ubiquinate substrate proteins and mark them for degradation by the proteosome.[47] DJ, a protein of largely unknown function, may serve as a chaperone, and protect against neuronal oxidative stress and apoptosis.[48] *C. elegans* contains homologues for both parkin and DJ-1. Parkin mutations confer increased sensitivity to ER stress and aggregation, and result in lethality when coexpressed with human A53T α-synuclein in the worm.[62] Deletion of parkin or knockdown of DJ-1 increases the sensitivity to mitochondrial complex I inhibitors and can be partially rescued by compounds that improve mitochondrial function.[63] We have also found that the deletion of parkin renders the worms almost 2-fold more sensitive to 6-OHDA (R. Nass, unpublished data). Overall, these results support a strong role of the PD-associated proteins affecting proteosomal and mitochondrial function, and suggest that *C. elegans* is a viable model for PD-associated DA neuron dysfunction and cell death.

MODELING ALZHEIMER'S DISEASE

Alzheimer's disease (AD) is a heterogeneous neurodegenerative disorder that is the most common cause of dementia.[64] Patients exhibit cognitive decline such as impaired judgment and orientation, progressive memory loss, and personality changes. In the severe stages of the disease, patients lose their ability to function in daily activities and are generally bedridden and dependent on caregivers. Postmortem analysis of AD brains reveals extracellular senile plaques and intraneuronal neurofibrillary tangles (NFT), which are now considered hallmarks of AD. In addition to these structural changes, there is extensive neuron loss. The remaining neurons in the AD brain show morphological changes that include dystrophic neuritis and reduced synapses.

The major component of senile plaques is aggregated β-amyloid$_{1-42}$ (Aβ_{1-42}), a cleaved product of the amyloid precursor protein (APP).[65,66] Mutations in the APP gene as well as genes encoding presenilin 1 and 2 (PSEN1 and PSEN2) are associated with some familial forms of AD (FAD), in which an overproduction of Aβ_{1-42} is seen.[67–69] Presenilins are active components of the intramembranous protease complex, γ-secretase, which processes APP. These discoveries led to the amyloid cascade hypothesis, which holds that altered amyloid processing and aggregation are the key pathogenic causes of AD.[70]

Several transgenic animal models have been established to test the amyloid hypothesis. These include mouse, *Drosophila*, and *C. elegans* transgenic animals that overexpressed either Aβ_{1-42} or mutated APP that would result in increased levels of Aβ_{1-42}.[71] While none of these models recapitulates all aspects of AD, each presents facilities to dissect the molecular and cellular pathogenesis of AD with respect to the amyloid hypothesis.[72]

APP and presenilins are conserved in *C. elegans*. The worm APP homologue, encoded by the *apl-1* gene, interestingly lacks the Aβ_{1-42} sequence, as is the case for the *Drosophila* APP homologue.[73,74] RNAi knockdown of the worm APP homologue leads to an uncoordinated phenotype, while a genetic deletion in the gene results in lethality, indicating an essential function of the worm APP.[26,75] Genetic analysis of one of the worm presenilins, SEL-12, reveals a function in the notch signaling pathway in vulval development.[76] SEL-12 in the notch signaling can functionally be replaced by HOP-1, the second worm presenilin, and by WT human presenilins, indicating functional conservation of the homologues In contrast, human presenilins containing FAD mutations display reduced ability to replace *sel-12*.[77–79] The presence of these homologues in *C. elegans* not only makes it possible to determine the biological function of these proteins, but also to identify genetically interacting proteins in the pathway. Indeed, in a previous *C. elegans* screen for proteins involved in signaling of a notch-like protein, GLP-1, the *aph-2* gene was identified.[80] The human orthologue of APH-2, known as nicastrin, was independently identified as a protein that coimmunoprecipitates with presenilin 1 and has a role in APP processing.[81]

To create a *C. elegans* AD model, transgenic animals expressing human Aβ_{1-42} in body wall muscles were generated.[82–83] These transgenic animals exhibit progressive paralysis and amyloid-like plaques in the muscles and shortened life spans, phenotypes that parallel those observed in AD patients. Although this model system limits expression of human Aβ_{1-42} to body wall muscles, it has been useful in assaying for both molecular and cellular responses to the presence of human Aβ_{1-42}. For example, these *C. elegans* transgenic animals show signs of oxidation stress, as indicated by increased levels of carbonyl, a key marker for protein oxidation.[84] Similarly, AD brains and cultured hippocampal neurons exposed to exogenous synthetic Aβ_{1-42} show increased levels of carbonyl and signs of oxidation stress. The *C. elegans* AD model has also been useful in testing the toxicity of fibrillar deposits of Aβ_{1-42} and its role in AD progression. Using transgenic animals that had a temperature-inducible expression system of Aβ_{1-42}, it was determined that paralysis and increased levels of carbonyl were induced upon expression of Aβ_{1-42}, but prior to any detection of amyloid plaques.[85] This result indicates that fibrillar amyloid deposition does not correlate with gross pathology and protein oxidation in *C. elegans*.

Techniques to detect global changes in protein and gene expression as well as screens can easily be employed on the *C. elegans* AD model. For example, to better understand how Aβ_{1-42} induces protein oxidation, proteomic techniques were used to identify proteins that are specifically oxidated in the *C. elegans* transgenic AD animals.[86] In addition, microarray analysis of gene expression in the *C. elegans* transgenic AD animals was performed to identify global gene expression changes induced by Aβ_{1-42}. Identified in this screen was heat shock protein 16 (HSP16), which was also shown to colocalize with Aβ_{1-42}, in *C. elegans* AD model.[87] These screens have thus been valuable in identifying candidate amyloid-induced protiens that participate in AD. Drug screening and genetic modifier screens can also be performed easily on the *C. elegans* to identify potential therapeutic reagents and molecular pathogenesis of the Aβ_{1-42} in AD. For example, ginkgo biloba leaf extract (EGb761), was shown in clinical trials to be effective against AD forms of degenerative dementia.[88] Supporting these results is the recent discovery that EGb761 reduces the oxidative effects of Aβ_{1-42} in the *C. elegans* transgenic AD strains.[88,89]

Another feature of AD is NFTs, which are composed of fibrillar aggregates of hyperphosphorylated forms of tau, a microtubule-binding protein (MAP) that promotes microtubule assembly and stability.[64] It is unclear if or how NFTs and tau may be involved in AD progression. One hypothesis holds that tau aggregates are toxic to neurons. Another hypothesis posits hyperphosphorylated tau sequesters normal tau and other MAPs, leading to microtubule disassembly and thus disrupting axonal transport.

To dissect the molecular and cellular pathogenesis of NFTs and tau in AD in a simpler model system, a *C. elegans* model for tauopathy was established with the creation of transgenic animals expressing normal human tau in the nervous system.[90] These animals showed progressively uncoordinated locomotion that was accompanied by progressive accumulation of insoluble phosphorylated human tau, neurodegeneration, and reduced cholinergic neurotransmission; these phenotypes are much more severe with the expression of mutant tau containing pathogenic mutations that cause frontotemporal dementia and parkinsonism (FTDP-17).[91] Thus, these animals recapitulate many traits present in human tauopathies.

With a *C. elegans* model for tauopathy established, assays and screens similar to those done with the *C. elegans* model for amyloid-induced pathogenesis can be performed to better understand pathogenic mechanisms of tau and NFTs. In a recent genome-wide screen, RNAi on 16,757 genes was performed on the tau-expressing transgenic animals for altered uncoordinated phenotype as a means to identify players in this tau-induced phenotype.[92] Sixty genes that specifically enhanced the tau-induced phenotype were identified; these include stress-response proteins as well as kinases and phosphatases that could affect tau phosphorylation. Some of these genes, such as GSK3β, have previously been implicated in human tauopathies.[93]

The relative ease of culturing large numbers of animals, together with straightforward genetic manipulation, has allowed for screens and analysis on a global scale; this has led to the identification of participants in both amyloid- and tau-induced phenotypes, suggesting many advantages that *C. elegans* has to offer in elucidating the molecular basis of cell death in AD. With the establishment of *C. elegans* models for AD and tauopathies, it will be possible to study the relationship between amyloid and tau in the pathogenesis of AD, and identify novel therapeutic targets to inhibit neurodegeneration.

MODELING TRINUCLEOTIDE REPEATS OR HUNTINGTON'S DISEASE

Huntington's disease (HD) is a progressive autosomal dominant neurodegenerative disease that is characterized by abnormal movements, emotional disturbances, and cognitive decline. The disease is one of at least eight other diseases that are caused by the abnormal expansion of a CAG repeat encoding the poly-(Q) tract in a protein.[94] In HD the poly-Q tract is located within the N-terminus of the protein huntingtin.[95] Other poly-Q-related diseases include the X-linked motorneuron disease spinobulbar muscular atrophy (SBMA), which affects the androgen receptor, and spinocerebellar ataxia.[94] HD occurs when the poly-Q repeat is generally greater than 40 residues (individuals without the propensity to develop the disease have between 10 and 29 repeats), and there is an inverse relationship between the number of residues and the age of onset. The duration of the illness following onset is typically 15–20 years and the illness eventually results in death.

The molecular mechanisms underlying HD is not well understood, although current beliefs include both a gain- and loss-of-function of the huntingtin protein.[94,96,97] Huntingtin is ubiquitously expressed in the cytoplasm and is associated with vesicle membranes and vesicle trafficking proteins, and may contribute to neurogenesis.[98–100] Although the pathology is found throughout the central nervous system, the striatum appears to be particularly vulnerable. The expansion of the poly-Q causes insoluble granular and fibrous deposits and these aggregates may be responsible for the pathogenesis, including proteosomal impairment, transcriptional dysregulation, excitotoxicity, and loss of glutamate and dopamine receptors.[101–103] Mouse models indicate that HD-associated phenotypes become apparent before there is cell death. Furthermore, increases in the cleaved huntingtin fragment during pathogenesis causes a depletion of WT huntingtin, suggesting a loss of function, and may contribute to the oxidative stress and mitochondrial impairment seen in HD. Finally, the poly-Q-containing protein inclusions found in HD have also been proposed to be neuroprotective by decreasing the levels of mutant huntingtin in the cell, and limiting the amounts of the globular and protofibrillar intermediates that have been thought to contribute to neurotoxicity.[104]

Several *C. elegans* models of HD have been developed to explore the molecular basis of the toxicity. Overexpression of the expanded poly-Q fused with the huntingtin fragment in sensory and other neurons causes cellular dysfunction and protein aggregation.[105–107] As in vertebrate studies, the severity of the aggregation is dependent on the length of the poly-Q, and the aggregation is preceded by neuronal dysfunction, although the aggregation does not always lead to cell death. Poly-Q expression in muscle cells is also toxic, and the severity increases with the number of the glutamine repeats.[108,109] Furthermore, the aggregation threshold decreases as the animals aged. In an elegant follow-up study, Brignull, Morimoto, and colleagues show that poly-Q length correlates with apparent toxicity and neuronal dysfunction, but that the solubility and aggregate formation is neuron specific.[107] Furthermore, the development of protein inclusions does not always increase with animal age. Overall, these results suggest that aggregate formation is not the sole cause of cellular dysfunction and suggest an intrinsic property of the poly-Q expansion confers the significant pathology.

C. elegans models of HD have been developed to identify potential modifiers of the toxicity. Poly-Q aggregate formation and toxicity can be reversed by overexpression of the yeast chaperone Hsp104, providing further support that misfolded protein(s) may be the toxic moiety.[108] RNAi of several small endogenously expressed Hsps within the worm accelerated the onset of poly-Q aggregation, suggesting that these proteins may act to protect cells from aggregation-induced toxicities.[110] Tor-2, a putative chaperone and a protein found to colocalize with α-synuclein in PD Lewy bodies, also inhibits the formation of aggregates.[111] Although a homologue in humans has not been identified, overexpression of *pqe-1* suppresses poly-Q-mediated toxicity in worms, while loss-of-function mutations enhances it.[105] Recently Wang and colleagues have overexpressed the endogenous ubiquilin, a protein found to interact with presenilins in AD and found in neuropathological inclusions in HD and PD, in muscle cells in *C. elegans*. Overexpression prevents and rescues the motility defect associated with expression of the huntingtin poly-Q fusion in muscles. Furthermore, RNAi of the endogenous protein exacerbates the toxic effect. Although the molecular basis for the neuroprotection is not clear, it may modulate the ubiquitin–proteosome system or act as a molecular chaperone.[94]

Finally genome-wide RNAi screening has identified modulators of poly-Q aggregation. Nollen *et al.*[112] expressed poly-Q residues fused to the yellow fluorescent protein (YFP) in muscle cells and grew the worms on bacteria expressing dsRNA with the goal of identifying genes that could be involved in the aggregate formation. The investigators reported almost 200 proteins that result in the premature appearance of the aggregates, and include those involved in protein synthesis, folding and transport, and degradation. The screen also identified a number of proteins involved in RNA synthesis and processing, and although it is not clear how these genes could affect the propensity of poly-Q to aggregate, they may be involved in the regulation of the proteosome. This screen demonstrates the power of RNAi screening in *C. elegans* to identify novel proteins that could be involved in poly-Q aggregation.

HEREDITARY SPASTIC PARAPLEGIA

Hereditary spastic paraplegias (HSP) are heterogeneous neurodegenerative syndromes that are clinically marked by progressive spastic paralysis in the lower limbs, the onset of which can occur as early as infancy.[113] While teenage or later onset of the disease is typically accompanied by marked progression of limb weakness so that canes and wheelchairs are often required, there is generally little worsening of symptoms in early childhood or infancy onset of the disease. HSP syndromes can be classified clinically as uncomplicated if symptoms are limited to progressive spastic weakness in the legs, or complicated if the disorder is accompanied by other neurological abnormalities such as mental retardation. Histopathological studies of uncomplicated HSP generally show the terminal ends of axons associated with cortical motor neurons slowly degenerating in a "dying-back" fashion; i.e., degeneration begins at the distal end of the axon in a retrograde direction.[114]

Genetic mapping has placed HSP to at least 29 loci (SPG1–29), of which 11 genes have been identified.[115] The identified genes encode proteins that participate in extremely diverse processes. While much is still not known about many of these genes, current studies are just beginning to reveal how their loss or impaired function may result in HSP. These genes can be grouped into two classifications: (1) genes that participate in corticospinal tract development and (2) genes that are required to maintain the health of corticospinal tract axons. Genes of the latter classification can further be grouped into mitochondrial and axonal trafficking genes.[113,115]

L1 is a cell adhesion molecule (CAM) that belongs to the immunoglobulin (Ig) superfamily. Mutations in the SPG1 gene encoding L1 can result in X-linked HSP that is often complicated by mental retardation, adducted thumbs, and hydrocephalus.[116] Interestingly, manifestation of these symptoms is quite variable even with the same mutation in the L1 gene within the same family.[117] A striking observation in these patients is an absent or reduced size of the medullary pyramids, suggesting abnormal development of the corticospinal tract.[118] Mutations disrupting L1 function in mice produce phenotypes very similar to those observed in human patients. Indeed, these mice exhibit hind limb weakness and show hypoplasia of the corticospinal tract.[119] Together, these results suggest that abnormal development of the corticospinal tract may be responsible for the spastic paraplegia exhibited in the patients.

Because L1 is expressed in multiple tissues, including the nervous system, it is not surprising that L1 has both neuronal and nonneuronal functions.[117] In the nervous system, there is increasing evidence that L1 functions in axon guidance and fasciculation, efficient initiation and propagation of action potentials, and synapse function. Numerous studies have revealed multiple interacting proteins that help mediate L1 function; these include L1 itself, other Ig family proteins, integrins, neuropilin, and ankyrin. However, how impaired function of L1 results in corticospinal tract hypoplasia or the manifestation of the additional symptoms is not clear. Besides L1, the L1 CAM family consists of three additional genes that have apparent overlapping expression and function, complicating the dissection of L1 functions and mechanisms; they include NrCAM, neuroglian, and CHL1.

L1 CAMs are conserved in *C. elegans* in the form of two genes, *sax-7* and *lad-2*.[120,121] The SAX-7 protein is a single transmembrane protein that contains many hallmarks of the vertebrate L1 CAM family that include six Ig and five fibronectin type III repeats in the extracellular domain, a transmembrane domain, and a short but highly conserved cytoplasmic tail that contains an ankyrin-binding motif and a consensus PDZ-binding motif.[120] Like L1, SAX-7 is highly expressed in the nervous system as well as multiple nonneuronal tissues. The most striking phenotype exhibited in animals with reduced or loss of SAX-7 function is the position of neurons and their axons that are initially normally positioned become progressively displaced as the animals mature and age.[122,123] This displacement is reduced in animals that do not move as robustly as wild-type animals.[122] Together, these results indicate SAX-7 plays a maintenance role to ensure proper attachment of the neurons and axons against mechanical forces exerted by the environment and animals themselves.

LAD-2 is a noncanonical L1 CAM that is expressed only in a handful of neurons in *C. elegans* (X. Wang and L. Chen, unpublished data).[124] While the LAD-2 extracellular domain contains the conserved six Ig and five FNIII repeats, the LAD-2 cytoplasmic tail is much shorter and divergent and lacks the ankyrin-binding motif conserved in SAX-7 and vertebrate L1CAMs.[121] Reduced *lad-2* function results in axon guidance defects in the LAD-2-expressing axons (see Figure 12–2; X. Wang and L. Chen, unpublished data). This phenotype indicates that LAD-2 plays a developmental role to ensure proper axon pathfinding. Axon guidance defects have also been reported in L1 mouse knockouts and mutants of the sole *Drosophila* L1 homologue, neuroglian.[119,125]

These studies in *C. elegans* have revealed both a developmental and maintenance role in the nervous system for L1CAMs. Perhaps a combination of axon guidance defects and impaired positional maintenance of axons leads to hypoplasia of corticospinal tract, which leads to early onset of HSP in patients with impaired L1 function.

In neurons, axonal transport is heavily reliant on kinesins and dyneins, molecular motors that transport vesicles, membrane, and membranous organelles along microtubule tracks in an anterograde and retrograde fashion, respectively.[126] Several genes that are associated with HSP appear to participate in axonal transport. Indeed, SPG10, which is associated with dominant uncomplicated HSP, was identified as encoding KIF5A, a neuronal-specific kinesin heavy chain.[127] Pathogenic mutations are localized to the motor as well as the microtubule-binding domains. Mutations in the SPG4 gene encoding spastin are responsible for 40% of the autosomal dominant uncomplicated HSP.[128] Spastin is an AAA (ATPase associated with diverse cellular activities) ATPase that shares sequence homology with the microtubule

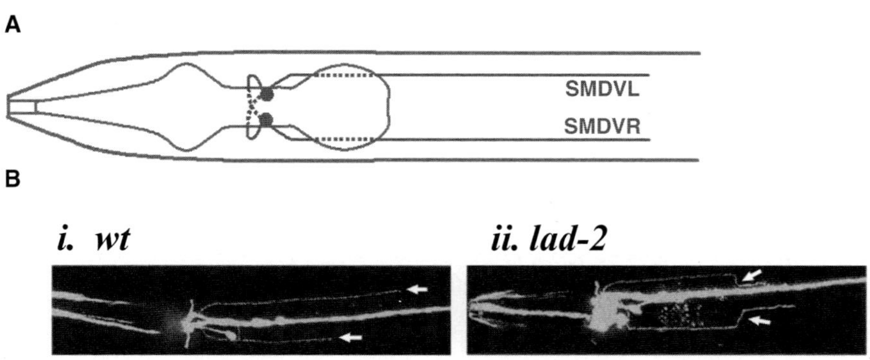

Figure 12–2. A schematic of the SMDL and SMDR axons is shown (A). Using the Pglr-1::GFP as a marker for SMDL and SMDR neurons and associated axons, *lad-2* mutant animals exhibit axon guidance defects for both neurons (arrows point to the point of deviation) (see Bii). The wild-type positions of both axons are shown (Bi).

severing protein, p60 katanin. Like many proteins that are involved in membrane trafficking, spastin contains a microtubule interacting and trafficking (MIT) domain. While microtubule binding and severing activity have recently been reported for spastin, its putative trafficking roles have yet to be confirmed.[129] A single spastin homologue is conserved in *C. elegans* and is encoded by the *spas-1* gene.[130] Like vertebrate spastin, the SPAS-1 protein is expressed in multiple tissues and can bind microtubules. Consistent with the broad expression of *spas-1*, *spas-1* genetic mutant animals exhibit pleiotropic phenotypes (Y. Matsushita and T. Ogura, personal communication). Detailed analysis of these phenotypes is necessary to dissect the mechanistic roles of SPAS-1 which in turn will reveal insight into how impaired or loss of spastin leads to HSP.

Consistent with a possible trafficking role for spastin is the recent discovery that spastin interacts with atlastin, a member of the dynamin family of GTPases that participates in vesicle trafficking and the dispersion of mitochondria. Atlastin, which is encoded by the SPG3A gene, is associated with 10% of autosomal dominant uncomplicated HSP.[131,132] Interestingly, a pathogenic mutation in atlastin has been shown to abolish interaction with spastin underscoring the functional importance of this protein interaction in healthy axons.[131]

While the function of KIF5A and its role in trafficking are uncontroversial, the functions of spastin and atlastin and their roles in the cellular pathogenesis of HSP are not clear. Based on the hypothesis that these proteins are indeed involved in trafficking, perhaps the less-than-efficient transport of proteins, membranes, and organelles such as the mitochondria, to the ends of the axons is detrimental to the health of the axon, resulting in the degeneration of axons that occurs in a distal–proximal fashion that is typical in HSP.[113]

In addition to trafficking proteins, there is increasing evidence that mitochondrial proteins are associated with HSP. Spartin, encoded by the SPG20 gene, is associated with recessive HSP complicated by distal muscle wasting.[133] Like spastin, spartin contains an MIT domain, suggesting microtubule-binding activity as well as intracellular trafficking.[134] Indeed spartin was recently shown to associate with microtubules and be localized to intracellular structures that include mitochondria and synaptic vesicles.[135,136] Moreover, a pathogenic mutation in spartin was shown to abolish its localization to the mitochondria, linking mitochondria to HSP.[135]

The SPG7 gene encodes for paraplegin, a nuclear-encoded mitochondrial metalloprotease belonging to the AAA family of proteins that has chaperon and proteolytic functions in the mitochondrial inner membrane. In addition to autosomal recessive HSP, patients with mutations in SPG7 exhibit structural and functional abnormalities in their muscles indicative of defects in mitochondrial oxidative phosphorylation.[137] *Spg7* knockout mice similarly exhibited degenerating axons that contain enlarged and structurally abnormal mitochondria clustered in synaptic terminals as well as inefficient retrograde axonal transport, suggesting that loss of paraplegin affects axonal transport.[138]

Like *SPG7* knockout mice, reduction of the sole *C. elegans* paraplegin homologue, SPG-7, by RNAi causes mitochondrial defects and premature lethality (T. Ogura, personal communication). The cause of this lethality has not been determined, but may be a result of defective mitochondria caused perhaps by an accumulation of abnormal proteins in the mitochondria. Indeed, *spg-7*(RNAi) results in increased levels of abnormally processed mitochondrial marker (GFP that contains a mitochondria import signal), indicating a role for SPG-7 in protein quality control. Furthermore, *spg-7*(RNAi) also activated the mitochondrial chaperone heat shock genes, *hsp-60* and *hsp-6*, homologues of the respective vertebrate HSP60 and HSP70 chaperones.[139] HSP60 is also associated with HSP although its role in the disease is not clear.[140] Together, these studies support a hypothesis that accumulation of abnormal mitochondrial proteins resulting from impaired paraplegin function leads to mitochondrial respiratory dysfunction, which in turn directly causes axon degeneration. Alternatively or additionally, the progressively abnormal mitochondria are less efficiently replaced, leading to a clogging of the transport system in the axon, which consequently degenerates due to inefficient axonal trafficking.[116]

NIP1A, a gene coding for unknown function, has also been implicated in causing juvenile onset HSP in at least six families.[141,142] Kindred studies suggest that the identified mutations within the putative integral membrane protein are pathogenic through a dominant negative mechanism. Expression studies involving the *C. elegans* NIP1A homologue show significant expression in the worm neurons, consistent with strong NIP1A expression in neuronal tissues within the human brain.[143] Overexpression of the wild-type NIP1A in *C. elegans* does not impair worm mobility; in contrast expression of NIP1A containing the corresponding pathogenic human mutations results in complete paralysis. This striking effect of pathogenic NIP1A in *C. elegans* reveals a valuable tool to dissect the functions and mechanisms of NIP1A to reveal participation of NIP1A in HSP. Taken together with the aforementioned *C. elegans* studies, the nematode is revealed as a powerful genetic model to dissect the molecular basis of the complex array of proteins involved in HSP and motor neuron dysfunction.

PERSPECTIVES AND FUTURE DIRECTIONS

The use of the nematode *C. elegans* in biomedical studies provides remarkable opportunities to identify and characterize potential human disease genes and proteins *in vivo*. The high similarities on the molecular level between the worm and humans suggest that the paradigms discovered using these systems are highly relevant to mammalian neurodegenerative diseases. Mammalian proteins that are associated with neurodegenerative diseases can complement and recapitulate many aspects of their pathology in worms. The ease of performing forward and reverse genetic screens will also very likely assist in the identification of novel molecules and pathways involved in neurodegeneration.

C. elegans is also easily amen able to chemical genetic screens to identify novel therapeutic drug leads, therapeutic targets, and the molecular pathway involved in the drug's efficacy. The worm is sensitive to a wide range of human neuroactive drugs, and has been utilized to screen for mutant proteins that have altered function in the presence of the drugs.[32,144,145,146] These screens also have the potential to identify molecular pathways not thought to play a role in the disease, or off-target sites of drugs that have not been previously identified.[5,147] For example, since *C. elegans* can easily be grown in liquid medium in 96-well microtiter plates, we could grow our DA neuron reporter line expressing α-synuclein in solution in the wells each containing a different compound from our small compound library. The plate can then be analyzed in a fluorescent plate reader, and those wells in which the DA neurons still fluoresce would contain a compound that has the potential to be neuroprotective in mammalian systems. Once a therapeutic has been identified, we can perform suppressor chemical screens to identify the molecules and interacting partners involved in the compound-mediated protection of the DA neurons. We would simply mutate the animals and screen for worms in which the DA neurons are *not* protected by the drug. Using standard *C. elegans* genetics and molecular biology, we could then quickly identify the drug targets and pathways involved in the drug's efficacy.

The high conservation of biological processes within the nervous system between *C. elegans* and humans and the strength and ease of incorporating forward and reverse genetics to identify and characterize novel components involved in cell dysfunction and death *in vivo* make this relatively simple animal a powerful and formidable tool in our arsenal in elucidating the molecules and their interactions that are involved in human neurodegenerative diseases. The strength and ease of incorporating the worm into high-throughput screening assays to identify novel compounds, therapeutic targets, and molecular pathways involved in neuroprotection are just beginning to be realized. It remains to be seen if the mighty worm will provide the therapeutic insight that has so far eluded us into combating the devastating effects of neurological diseases.

REFERENCES

1. The *C. elegans* Sequencing Consortium. Genome sequence of the nematode C. elegans: A platform for investigating biology. *Science* 1998;282:2012–2018.
2. Hiller LW, Coulson A, Murray JI, Bao Z, Sulston JE, Waterston RH. Genomics in *C. elegans*: So many genes, such a little worm. *Genome Res* 2006;15:1651–1660.
3. Rand JB, Nonet ML. Synaptic transmission. In: Riddle DL, Blumenthal T, Meyer BJ, Priess JR, Eds. *C. elegans II*. New York: Cold Spring Harbor Laboratory Press, 1997:611–643.
4. Thomas, JH, Lockery SH. Neurobiology. In: Hope IA, Ed. *C. elegans: A Practical Approach*. New York: Oxford University Press, 1999:143–179.
5. Bargmann CI. Neurobiology of the *Caenorhabditis elegans* genome. *Science* 1998;282:2028–2033.
6. Nonet ML. Studying mutants that affect neurotransmitter release in *C. elegans*. In: Bellen HJ, Ed. *Neurotransmitter Release*. New York: Oxford University Press. 1999.
7. Nass R, Blakely RB. The *Caenorhabditis elegans* dopaminergic system: Opportunities for insights into dopamine transport and neurodegeneration. *Annu Rev Pharmacol Toxicol* 2003;43:521–544.
8. Riddle DL, Blumenthal T, Meyer BJ, Priess JR, Eds. *C. elegans II*. New York: Cold Spring Harbor Laboratory Press, 1997.
9. Wood, WB. Introduction to *C. elegans*. In: WB Wood, Ed. *The Nematode Caenorhabditis elegans*. New York: Cold Spring Harbor Laboratory Press, 1988:1–16.
10. Link EM, Hardiman G, Sluder AE, Johnson CD, Liu LX. Therapeutic target discovery using *Caenorhabditis elegans*. *Pharmacogenomics* 2000;1:203–217.
11. Kaletta T, Hengartner MO. Finding function in novel targets: C. elegans as a model organism. *Nat Rev Drug Discov* 2006;5: 387–398.
12. Bloom L. Genetic and molecular analysis of genes required for axon outgrowth in *Caenorhabditis elegans*. PhD. Thesis, Massachusetts Institute of Technology, Cambridge, MA, 1993.
13. Buechner M, Hall DH, Bhatt H, Hedgecock EM. Cystic canal mutants in *Caenorhabtitis elegans* are defective in the apical membrane domain of the renal (excretory) cell. *Dev Biol* 1999;214:227–241.
14. Christensen M, Estevez A, Yin X, Fox R, Morrison R, McDonnell M, Gleason C, Miller DM, Strange K. A primary culture system for functional analysis of *C. elegans* neurons and muscle cells. *Neuron* 2002;33:503–514.
15. Carvelli L, McDonald PW, Blakely RD, Defelice LJ. Dopamine transporters depolarize neurons by a channel mechanism. *Proc Natl Acad Sci USA* 2004;101:16046–16051.
16. Bianchi L, Driscoll M. Culture of embryonic C. elegans cells for electrophysiological and pharmacological analyses. In: The *C. elegans* Research Community, Ed. Wormbook. http://www.wormbook.org, 2006.
17. Fox RM, Von Stetina SE, Barlow SJ, Shaffer C, Olszewski KL, Moore JH, Dupuy D, Vidal M, Miller, DM 3rd. A gene expression fingerprint of C. elegans embryonic motor neurons. *BMC Genomics* 2005;6:42.
18. Jorgensen EM, Mango SE. The art and design of genetic screens: Caenorhabditis elegans. *Nat Rev Genet* 2002;3:356–369.
19. Wicks SR, Yeh RT, Gish WR, Waterston RH, Plasterk RH. Rapid gene mapping in *Caenorhabditis elegans* using a high density polymorphism map. *Nat Genet* 2001;28:160–164.
20. See the *Caenorhabditis elegans* WWW Server at http://elegans.swmed.edu/.
21. Fire A, Xu S, Montgomery MK, Kostas SA, Driver SE, Mello CC. Potent and specific genetic interference by double-stranded RNA in Caenorhabditis elegans. *Nature* 1998;391:806–811.
22. Montgomery MK, Xu S, Fire A. RNA as a target of double-stranded RNA-mediated genetic interference in *Caenorhabditis elegans*. *Proc Natl Acad Sci USA* 1998;95:15502–15507.
23. Timmons L, Fire A. Specific interference by ingested dsRNA. *Nature* 1998;395:854.
24. Maeda I, Kohara Y, Yamamoto M, Sugimoto A. Large-scale analysis of gene function in Caenorhabditis elegans by high-throughput RNAi. *Curr Biol* 2001;11:171–176.
25. Kamath RS, Fraser AG, Dong Y, Poulin G, Durbin R, Gotta M, Kanapin A, Le Bot N, Moreno S, Sohrmann M, *et al.* Systematic functional analysis of the *Caenorhabditis elegans* genome using RNAi. *Nature* 2003;421:231–237.
26. Kamath RS, Ahringer J. Genome-wide RNAi screening in Caenorhabditis elegans. *Methods* 2003;30:313–321.
27. Buckingham SD, Esmaeili B, Wood M, Sattelle DB. RNA interference: From model organisms towards therapy for neural and neuromuscular disorders. *Hum Mol Genet* 2004;13(Spec. No. 2): R275–R288.
28. Mello C, Fire A. DNA transformation. *Methods Cell Biol* 1995;48:451–482.
29. Chalfie M, Tu Y, Euskirchen G, Ward WW, Prasher DC. Green fluorescent protein as a marker for gene expression. *Science* 1994;263:802–805.
30. Miller DM III, Desai NS, Hardin DC, Piston DW, Patterson GH, *et al.* Two color GFP expression for *C. elegans*. *BioTechniques* 1999;26:914–921.
31. White JG, Southgate E, Thompson JN, Brenner S. The structure of the nervous system of the nematode Caenorhabditis elegans. *Philos Trans R Soc London B Biol Sci* 1986;B314:1–340.
32. Mitchell SL, Kiely DK, Kiel DP, Lipsitz LA. The epidemiology, clinical characteristics, and natural history of older nursing home residents with a diagnosis of Parkinson's disease. *J Am Geriatr Soc* 1996;44:394:99.
33. Yahr MD, Bergman K, Eds. *Advances in Neurology: Parkinson's Disease*. New York: Raven, 1986.
34. Polymeropoulos MH, *et al.* Mutation in the alpha-synuclein gene identified in families with Parkinson's disease. *Science* 1997;276: 2045–2047.

35. Forloni G, Bertani I, Calella AM, Thaler F, Invernizzi R. Alpha-synuclein and Parkinson's disease: Selective neurodegenerative effect of alpha-synuclein fragment on dopaminergic neurons in vitro and in vivo. *Ann Neurol* 2000;47:632–640.
36. Masliah E, *et al.* Dopaminergic loss and inclusion body formation in alpha-synuclein mice: Implications for neurodegenerative disorders. *Science* 2000;287:1265–1269.
37. Jenner P. Oxidative mechanisms in nigral cell death in Parkinson's disease. *Mov Disord* 1998;13:24–34.
38. Witt SN, Flower TR. α-Synuclein, oxidative stress and apoptosis from the perspective of a yeast model of Parkinson's disease. *FEMS Yeast Res* 2006;6(8):1107–1110.
39. Marfin I, Vallejo J. Parkinson disease: From cellular and animal models to genomics *Curr Genomics* 2005;6:241–250.
40. Sauer H, Oertel WH. Progressive degeneration of nigrostriatal dopamine neurons following intrastriatal terminal lesions with 6-hydroxdopamine: A combined retrograde tracing and immunocytochemical study in the rat. *Neuroscience* 1994;59:401–415.
41. Glinka Y, Gassen M, Youdim MBH. Mechanism of 6-hydroxydopamine neurotoxicity. *J Neural Transm* 1997;50:55–66.
42. Hoglinger GU, Oerte WH, Hirsch EC. The rotenone model of parkinsonism—the five year inspection. *J Neural Transm Suppl* 2006;70:269–270.
43. Kitayama S, Shimada S, Uhl GR. Parkinsonism-inducing neurotoxin MPP+: Uptake and toxicity in nonneuronal COS cells expressing dopamine transporter cDNA. *Ann Neurol* 1992;32:109–111.
44. Gainetdinov RR, Jones SR, Fumagalli F, Wightman RM, Caron MG. Re-evaluation of the role of the dopamine transporter in dopamine system homeostasis. *Brain Res Brain ResRev* 1998;26: 148–153.
45. Kitada T, *et al.* Mutations in the parkin gene cause autosomal recessive juvenile parkinsonism. *Nature* 1998;392:605–608.
46. Leroy E, Anastasopoulos D, Konitsiotis S, Lavedan C, Polymeropoulos MH. Deletions in the parkin gene and genetic heterogeneity in a greek family with early onset Parkinsons disease. *Hum Genet* 1998;103:424–427.
47. Bonifati V, Oostra BA, Heutink P. Linking DJ-1 to neurodegeneration offers novel insights for understanding the pathogenesis of Parkinson's disease. *J Mol Med* 2004;82:163–174.
48. Gilks WP, Abou-Seiman PM, Gandhi S, Jain S, Singleton A, Lees AJ, Shaw K, Bhatia KP, Bonifati V, Quinn NP, Lynch J, Healy DG, Holton JL, Revesz T, Wood NW. A common LRRK2 mutation in idiopathic Parkinson's disease. *Lancet* 2005;365:415–416.
49. Valente EM, Abou-Sleiman PM, Caputo B, Muqit MM, Harvey K, Gispert S, Ali Z, Del Turco D, Bentivoglio AR, Healy DG, Albanese A, Nussbaum R, Gonzalez-Maldonado R, Deller T, Salvi S, Cortelli P, Gilks WP, Latchman DS, Harvey RJ, Dallapiccola B, Auburger G, Wood NW. Hereditary early-onset Parkinson's disease caused by mutations in pink-1. *Science* 2004;304:1158–1160.
50. Paisan-Ruiz C, Jain S, Evan EW, Gilks WP, Simon J, van der Brug M, Lopez de Munain A, Aparicio S, Gil AM, Khan N, Johnson J, Martinez JR, Nicholl D, Carrera IM, Pena AS, de Silva R, Lees A, Marti-Masso JF, Perez-Tur J, Wood NW, Singleton AB. Cloning of the gene containing mutations that cause *PARK8*-linked Parkinson's disease. *Neuron* 2004;44:595–600.
51. Greenbaum EA, Graves CL, Mishizen-Eberz AJ, Lupoli MA, Lynch DR, Englander SW., Axelsen PH, Giasson BI. The E46K mutation in alpha-synculein increases amyloid fibril formation. *J Biol Chem* 2005;280:7800–7807.
52. Lee FJ, Liu F, Pristupa ZB, Niznik HB. Direct binding and functional coupling of alpha-synuclein to the dopamine transporters accelerate dopamine-induced apoptosis. *FASEB J* 2001;15:916–926.
53. Nass R, Miller DM, Blakely RD. *C. elegans*: A novel pharmacogenetic model to study Parkinson's disease. *Parkinsonism Relat Disord* 2001;7:185–191.
54. Nass R, Hall DH, Miller DM 3rd, Blakely RD. Neurotoxin-induced degeneration of dopamine neurons in *Caenorhabditis elegans*. *Proc Natl Acad Sci USA* 2002;99:3264–3269.
55. Braungart E, Gerlach M, Riederer P, Baumeister R, Hoener MC. Caenorhabditis elegans MPP+ model of Parkinson's disease for high-throughput drug screenings. *Neurodegener Dis* 2004;1(4–5):175–183.
56. Wintle RF, Van Tol HHM. Dopamine signaling in *Caenorhabditis elegans*—potential for parkinsonism research. *Parkinsonism Relat Disord* 2001;7:177–183.
57. Westlund B, Stilwell G, Sluder A. Invertebrate disease models in neurotherapeutic discovery. *Curr Opin Drug Discov* 2004;7:169–178.
58. Lakso M, Vartianen S, Moilanen AM, Sirvio J, Thomoas JH, Nass R, Blakely RD, Wong G. Dopaminergic neuronal loss and motor deficits in *Caenorhabditis elegans* overexpressing human α-synuclein. *J Neurochem* 2003;86:165–172.
59. Kuwahara T, Koyama A, Gengyo-Ando K, Masuda M, Kowa H, Tsunoda M, Mitani S, Iwatsubo T. Familial Parkinson mutant α-synuclein causes dopamine neuron dysfunction in transgenic *Caenorhabditis elegans*. *J Biol Chem* 2006;281:334–340.
60. Cooper AA, Gitler AD, Casikar A, Haynes CM, Hill KJ, *et al.* α-Synuclein blocks ER-Golgi traffic and Rab1 rescues neuron loss in Parkinson's models. *Science* 2006;313:324–328.
61. Vartiainen S, Pehkonen P, Lakso M, Nass R, Wong G. Identification of gene expression changes in transgenic *C. elegans* overexpressing human α-synuclein. *Neurobiol Dis* 2006;22:477–486.
62. Springer W, Hoppe T, Schmidt E, Baumeister R. A Caenorhabditis elegans parkin mutant with altered solubility couples α-synuclein aggregation to proteotoxic stress. *Hum Mol Genet* 2005;14: 3407.
63. Ved R, Saha S, Westlund B, Perier C, Bumam L, Sluder A, Hoener M, Rodrigues CMP, Alfonso A, Steer C, Liu L, Przedborski S, Wolozin B. Similar patterns of mitochondrial vulnerability and rescue induced by genetic modification of a-synuclein, parkin, and DJ-1 in *Caenorhabditis elegans*. *J Biol Chem* 2005;280:42655–42668.
64. Blennow K, de Leon MJ, Zetterberg H. Alzheimer's disease. *Lancet* 2006;368:387–403.
65. Glenner GG, Wong CW. Alzheimer's disease and Down's syndrome: Sharing of a unique cerebrovascular amyloid fibril protein. *Biochem Biophys Res Commun* 1984;122:1131–1135.
66. Kang J, Lemaire HG, Unterbeck A, Salbaum JM, Masters CL, Greschik KH, *et al.* The precursor of Alzheimer's disease amyloid A4 protein resembles a cell-surface receptor. *Nature* 1987;325:733–736.
67. Goate A, Chartier-Harlin MC, Mullan M, *et al.* Segregation of a missense mutation in the amyloid precursor protein gene with familial Alzheimer's disease. *Nature* 349:704–776.
68. Sherrington R, Rogaev EI, Liang Y, *et al.* Cloning of a gene bearing missense mutations in early-onset familial Alzheimer's disease. *Nature* 1995;375:754–760.
69. Levy-Lahad E, Wasco W, Poorkaj P, *et al.* Candidate gene for the chromosome 1 familial Alzheimer's disease locus. *Science* 1995;269:973–977.
70. Hardy J, Selkoe DJ. The amyloid hypothesis of Alzheimer's disease: Progress and problems on the road to therapeutics. *Science* 2002;297:353–356.
71. Spires TL, Hyman BT. Transgenic models of Alzheimer's disease: Learning from animals. *NeuroRx* 2005;2:423–437.
72. Coulson EJ, Paliga K, Beyreuther K, Masters CL. What the evolution of the amyloid protein precursor supergene family tells us about its function. *Neurochem Int* 2000;36:175–184.
73. Daigle I, Li C. *apl-1*, a *Caenorhabditis elegans* gene encoding a protein related to the human beta-amyloid protein precursor. *Proc Natl Acad Sci USA* 1993;90:12045–12049.
74. Rosen DR, Martin-Morris L, Luo LQ, White K. A *Drosophila* gene encoding a protein resembling the human beta-amyloid protein precursor. *Proc Natl Acad Sci USA* 1989;86:2478–2482.
75. WormBase web site, http://www.wormbase.org, release WS163.
76. Levitan D, Greenwald I. Facilitation of lin-12-mediated signalling by *sel-12*, a *Caenorhabditis elegans* S182 Alzheimer's disease gene. *Nature* 1995;377:351–354.

77. Baumeister R, Leimer U, Zweckbronner I, Jakubek C, Grunberg J, Haass C. Human presenilin-1, but not familial Alzheimer's disease (FAD) mutants, facilitate *Caenorhabditis elegans* Notch signalling independently of proteolytic processing. *Genes Funct* 1997;1:149–159.
78. Levitan D, Doyle TG, Brousseau D, Lee MK, Thinakaran G, Slunt HH, Sisodia SS, Greenwald I. Assessment of normal and mutant human presenilin function in *Caenorhabditis elegans*. *Proc Natl Acad Sci USA* 1996;93:14940–14944.
79. Li X, Greenwald I. HOP-1, a *Caenorhabditis elegans* presenilin, appears to be functionally redundant with SEL-12 presenilin and to facilitate LIN-12 and GLP-1 signaling. *Proc Natl Acad Sci USA* 1997;94:12204–12209.
80. Goutte C, Hepler W, Mickey KM, Priess JR. *aph-2* encodes a novel extracellular protein required for GLP-1-mediated signaling. *Development* 2000;127:2481–2492.
81. Yu G, Nishimura M, Arawaka S, Levitan D, Zhang L, *et al.* Nicastrin modulates presenilin-mediated notch/glp-1 signal transduction and betaAPP processing. *Nature* 2000;407:48–54.
82. Link CD. Expression of human b-amyloid peptide in transgenic *Caenorhabditis elegans*. *Proc Natl Acad Sci USA* 1995;92:9368–9372.
83. Fay DS, Fluet A, Johnson CJ, Link CD. *In vivo* aggregation of beta-amyloid peptide variants. *J Neurochem* 1998;71:1616–1625.
84. Yatin SM, Varadarajan S, Link CD, Butterfield DA. *In vitro* and *in vivo* oxidative stress associated with Alzheimer's amyloid beta-peptide (1–42). *Neurobiol Aging* 1999;20:325–330; discussion 339–342.
85. Drake J, Link CD, Butterfield DA. Oxidative stress precedes fibrillar deposition of Alzheimer's disease amyloid beta-peptide (1–42) in a transgenic *Caenorhabditis elegans* model. *Neurobiol Aging* 1999;24:415–420.
86. Boyd-Kimball D, Poon HF, Lynn BC, Cai J, Pierce WM Jr, Klein JB, Ferguson J, Link CD, Butterfield DA. Proteomic identification of proteins specifically oxidized in *Caenorhabditis elegans* expressing human Abeta(1–42): Implications for Alzheimer's disease. *Neurobiol Aging* 2006;27:1239–1249.
87. Link CD, Taft A, Kapulkin V, Duke K, Kim S, Fei Q, Wood DE, Sahagan BG. Gene expression analysis in a transgenic Caenorhabditis elegans Alzheimer's disease model. *Neurobiol Aging* 2003;24:397–413.
88. Le Bars PL, Katz MM, Berman N, Itil TM, Freedman AM, Schatzberg AF. A placebo-controlled, double-blind, randomized trial of an extract of Ginkgo biloba for dementia. North American EGb Study Group. *JAMA* 1997;278:1327–1332.
89. Smith JV, Luo Y. Elevation of oxidative free radicals in Alzheimer's disease models can be attenuated by Ginkgo biloba extract EGb 761. *J Alzheimers Dis* 2003;5:287–300.
90. Kraemer BC, Zhang B, Leverenz JB, Thomas JH, Trojanowski JQ, Schellenberg GD. Neurodegeneration and defective neurotransmission in a Caenorhabditis elegans model of tauopathy. *Proc Natl Acad Sci USA* 2003;100:9980–9985.
91. Hutton M, Lendon CL, Rizzu P, Baker M, Froelich S, *et al.* Association of missense and 5'-splice-site mutations in tau with the inherited dementia FTDP-17. *Nature* 1998;393:702–705.
92. Kraemer BC, Burgess JK, Chen JH, Thomas JH, Schellenberg GD. Molecular pathways that influence human tau-induced pathology in *Caenorhabditis elegans*. *Hum Mol Genet* 2006;15:1483–1496.
93. Lucas JJ, Hernandez F, Gomez-Ramos P, Moran MA, Hen R, Avila J. Decreased nuclear beta-catenin, tau hyperphosphorylation and neurodegeneration in GSK-3beta conditional transgenic mice. *EMBO J* 2001;20:27–39.
94. Wang H, Lim PJ, Yin C, Rieckher M, Vogel BE, Monteiro MJ. Suppression of polyglutamine-induced toxicity in cell and animal models of Huntington's disease by ubiquilin. *Hum Mol Genet* 2006;15:1025–1041.
95. Zoghbi HY, Orr HT. Glutamine repeats and neruodegeneration. *Annu Rev Neurosci* 2000;23:217–247.
96. Arrasate M, Mitra S, Schweitzer ES, Sega MR, Finkbeiner S. Inclusion body formation reduces levels of mutant huntingtin and the risk of neuronal death. *Nature* 2004;431:805–810.
97. Bowman AB, Yoo SY, Dantuma NP, Zoghbi HY. Neuronal dysfunction in a polyglutamine disease model occurs in the absence of ubiquitin-proteosome system impairment and inversely correlates with the degree of nuclear inclusion formation. *Hum Mol Genet* 2005;14:679–691.
98. The Huntington Disease Collaborative Research Group. A novel gene containing a trinucleotide repeat that is expanded and unstable on Huntington's disease chromosomes. *Cell* 1993;72:971–983.
99. Velier J, Kim M, Schwarz C, Kim TW, Sapp E, Chase K, *et al.* Wild-type and mutant huntingtins function in vesicle trafficking in the secretory and endocytotic pathways. *Exp Neurol* 1998;152:34–40.
100. Nasir J, Floresco SB, O'Kusky JR, Diewert VM, Richmond JM, Zeisler J. Targeted disruption of the Huntington's disease gene results in embryonic lethality and behavioral and morphological changes in heterozygotes. *Cell* 1995;81:811–823.
101. Bence NF, Sampat RM, Kopita RR. Impairment of the ubiquitin-proteosome system by protein aggregation. *Science* 2001;292:1552–1555.
102. Waelter S, Boeddrich A, Lurz R, *et al.* Accumulation of mutant huntingtin fragments in aggresome-like inclusion bodies as a result of insufficient protein degradation. *Mol Biol Cell* 2001;12:1393–1407.
103. Cepeda C, Hurst RS, Calvert CR, *et al.* Transient and progressive electrophysiological alterations in the corticostriatal pathway in a mouse model of Huntington's disease. *J Neurosci* 2003;23:961–969.
104. Lee S, Kim M. Aging and neurodegeneration: Molecular mechanisms of neuronal loss in Huntington's disease. *Mech Ageing Dev* 2006;127:432–435.
105. Faber PW, Alter JR, MacDonald ME, Hart AC. Polyglutamine-mediated dysfunction and apoptotic death of a *Caenorhabditis elegans* sensory neuron. *Proc Natl Acad Sci USA* 1999;96:179–184.
106. Parker JA, Connolly JB, Wellington C, Hayden M, Dausset J, Neri C. Expanded polyglutamines in *Caenorhabditis elegans* cause axonal abnormalities and severe dysfunction of PLM mechanosensory neurons without cell death. *Proc Natl Acad Sci USA* 2001;98:13318–13323.
107. Brignull HR, Moore FE, Tang SJ, Morrimoto RI. Polyglutamine proteins at the pathogenic threshold display neuron-specific aggregation in a pan-neuronal *Caenorhabditis elegans* model *J Neurosci* 2006;26:7597–7606.
108. Satyal SH, Schmidt E, Kitagawa K, Sondheimer N, Lindquist S, Kramer JM, Morimoto RI. Polyglutamine aggregates alter protein folding homeostasis in *Caenorhabditis elegans*. *Proc. Natl Acad Sci USA* 2000;97:5750–5755.
109. Morley JF, Brignull HR, Weyers JJ, Morimoto RI. The threshold for polyglutamine-expansion protein aggregation and cellular toxicity is dynamic and influenced by aging in *Caenorhabditis elegans*. *Proc Natl Acad Sci USA* 2002;99:10417–10422.
110. Hsu AL, Murphy CT, Kenyon C. Regulation of aging and age-related disease by DAF-16 and heat-shock factor. *Science* 2003;300:1142–1145.
111. Caldwell GA, Songsong C, Sexton EG, Gelwix CC, Bevel JP, Caldwell KA. Suppression of polyglutamine-induced protein aggregation in *Caenorhabditis elegans* by torsin proteins. *Hum Mol Genet* 2003;12:307–319.
112. Nollen EAA, Garcia SM, van Haaften G, Kim S, Chavez A, Morimoto RI, Plasterk RHA. Genome-wide RNA interference screen identifies previously undescribed regulators of polyglutamine aggregation. *Proc Natl Acad Sci USA* 2004;101:6403–6408.
113. Fink JK. Hereditary spastic paraplegia: Nine genes and counting. *Arch Neurol* 2003;60:1045–1049.
114. Behan W, Maia M. Strumpell's familial spastic paraplegia: Genetics and neuropathology. *J Neurol Neurosurg Psychiatry* 1974;37:8–20.

115. Rugarli EI, Langer T. Translating m-AAA protease function in mitochondria to hereditary spastic paraplegia. *Trends Mol Med* 2006;12:262–269.
116. Kenwrick S, Watkins A, De Angelis E. Neural cell recognition molecule L1: Relating biological complexity to human disease mutations. *Hum Mol Genet* 2000;6:879–886.
117. Yamasaki M, Arita N, Hiraga S, Izumoto S, Morimoto K, Nakatani S, Fujitani K, Sato N, Hayakawa T. A clinical and neuroradiological study of X-linked hydrocephalus in Japan. *J Neurosurg* 1995;83:50–55.
118. Dahme M, Bartsch U, Martini R, Anliker B, Schachner M, Mantei N. Disruption of the mouse L1 gene leads to malformations of the nervous system. *Nat Genet* 1997;17:346–349.
119. Chen L, Ong B, Bennett V. LAD-1, the Caenorhabditis elegans L1CAM homologue, participates in embryonic and gonadal morphogenesis and is a substrate for fibroblast growth factor receptor pathway-dependent phosphotyrosine-based signaling. *J Cell Biol* 2001;154:841–855.
120. Hutter H, Vogel BE, Plenefisch JD, Norris CR, Proenca RB, Spieth J, Guo C, Mastwal S, Zhu X, Scheel J, *et al.* Conservation and novelty in the evolution of cell adhesion and extracellular matrix genes. *Science* 2000;287:989–994.
121. Sasakura H, Inada H, Kuhara A, Fusaoka E, Takemoto D, Takeuchi K, Mori I. Maintenance of neuronal positions in organized ganglia by SAX-7, a Caenorhabditis elegans homologue of L1. *EMBO J* 2005;24:1477–1488.
122. Wang X, Kweon J, Larson S, Chen L. A role for the C. elegans L1CAM homologue lad-1/sax-7 in maintaining tissue attachment. *Dev Biol* 2005;284:273–291.
123. Aurelio O, Hall DH, Hobert O. Immunoglobulin-domain proteins required for maintenance of ventral nerve cord organization *Science* 2002;295:686–690.
124. Fransen E, D'Hooge R, Van Camp G, Berhoye M, Sijbers J, Reyniers E, Soriano P, Hall SG, Bieber AJ. Mutations in the Drosophila neuroglian cell adhesion molecule affect motor neuron pathfinding and peripheral nervous system patterning. *J Neurobiol* 1997;32:325–340.
125. Hirokawa N, Takemura R. Molecular motors and mechanisms of directional transport in neurons. *Nat Rev Neurosci* 2005;6:201–214.
126. Reid E, Kloos M, Ashley-Koch A, Hughes L, Bevan S, Svenson IK, Graham FL, Gaskell PC, Dearlove A, Pericak-Vance MA, Rubinsztein DC, Marchuk DA. A kinesin heavy chain (KIF5A) mutation in hereditary spastic paraplegia (SPG10). *Am J Hum Genet* 2002;71: 1189–1194.
127. Hazan J, Fonknechten N, Mavel D, Paternotte C, Samson D, *et al.* Spastin, a new AAA protein, is altered in the most frequent form of autosomal dominant spastic paraplegia. *Nat Genet* 1999;23:296–303.
128. Evans KJ, Gomes ER, Reisenweber SM, Gundersen GG, Lauring BP. Linking axonal degeneration to microtubule remodeling by Spastin-mediated microtubule severing. *J Cell Biol* 2005;168:599–606.
129. Yakushiji Y, Yamanaka K, Ogura T. Identification of a cysteine residue important for the ATPase activity of C. elegans fidgetin homologue. *FEBS Lett* 2004;578:191–197.
130. Evans K, Keller C, Pavur K, Glasgow K, Conn B, Lauring B. Interaction of two hereditary spastic paraplegia gene products, spastin and atlastin, suggests a common pathway for axonal maintenance. *Proc Natl Acad Sci USA* 2006;103:10666–10671.
131. Sanderson CM, Connell JW, Edwards TL, Bright NA, Duley S, Thompson A, Luzio JP, Reid E. Spastin and atlastin, two proteins mutated in autosomal-dominant hereditary spastic paraplegia, are binding partners. *Hum Mol Genet* 2006;15:307–318.
132. Patel H, Cross H, Proukakis C, Hershberger R, Bork P, Ciccarelli FD, Patton MA, McKusick VA, Crosby AH. SPG20 is mutated in Troyer syndrome, an hereditary spastic paraplegia. *Nat Genet* 2002;31:347–348.
133. Ciccarelli FD, Proukakis C, Patel H, Cross H, Azam S, Patton MA, Bork P, Crosby AH. The identification of a conserved domain in both spartin and spastin, mutated in hereditary spastic paraplegia. *Genomics* 2003;81:437–441.
134. Lu J, Rashid F, Byrne PC. The hereditary spastic paraplegia protein spartin localises to mitochondria. *J Neurochem* 2006;98:1908–1919.
135. Robay D, Patel H, Simpson MA, Brown NA, Crosby AH. Endogenous spartin, mutated in hereditary spastic paraplegia, has a complex subcellular localization suggesting diverse roles in neurons. *Exp Cell Res* 2006;312:2764–2777.
136. Casari G, De Fusco M, Ciarmatori S, Zeviani M, Mora M, Fernandez P, De Michele G, Filla A, Cocozza S, Marconi R, Durr A, Fontaine B, Ballabio A. Spastic paraplegia and OXPHOS impairment caused by mutations in paraplegin, a nuclear-encoded mitochondrial metalloprotease. *Cell* 1998;93:973–983.
137. Ferreirinha F, Quattrini A, Pirozzi M, Valsecchi V, Dina G, Broccoli V, Auricchio A, Piemonte F, Tozzi G, Gaeta L, Casari G, Ballabio A, Rugarli EI. Axonal degeneration in paraplegin-deficient mice is associated with abnormal mitochondria and impairment of axonal transport. *J Clin Invest* 2004;113:231–242.
138. Yoneda T, Benedetti C, Urano F, Clark SG, Harding HP, Ron D. Compartment-specific perturbation of protein handling activates genes encoding mitochondrial chaperones. *J Cell Sci* 2004;117: 4055–4066.
139. Hansen JJ, Durr A, Cournu-Rebeix I, Georgopoulos C, Ang D, Nielsen MN, Davoine CS, Brice A, Fontaine B, Gregersen N, Bross P. Hereditary spastic paraplegia SPG13 is associated with a mutation in the gene encoding the mitochondrial chaperonin Hsp60. *Am J Hum Genet* 2002;70:1328–1332.
140. James PA, Talbot K. The molecular genetics of non-ALS motor neuron disease. *Biochim Biophys Acta* 2006;1762(11–12):986–1000.
141. Rainier S, Chai J, Tokarz D, Nicholls RD, Fink JK. NIPA1 gene mutations cause autosomal dominant hereditary spastic paraplegia (SPG6). *Am J Hum Genet* 2003;73:967–971.
142. Zhao J, Matthies DS, Nass R, Blakely RD, Hedera P. C. elegans model of hereditary spastic paraplegia (HSP) caused by mutations in NIPA1 gene homolog. Society for Neuroscience Meeting, Atlanta, GA, October 14–18, 2006.
143. Jouet M, Rosenthal A, Armstrong G, MacFarlane J, Stevenson R, Paterson J, Metzenberg A. X-linked spastic paraplegia (SPG1), MASA syndrome and X-linked hydrocephalus result from mutations in the L1 gene. *Nat Genet* 1994;7:402–407.
144. Rand JB, Johnson CD. Genetic pharmacology: Interactions betwen drugs and gene products in *Caenorhabditis elegans*. In: Epstein HF, Shakes DC, Eds. *Methods in Cell Biology: Caenorhabditis Elegans: Modern Biological Analysis of an Organism.* New York: Academic Press, 1995:187–204.
145. Link EM, Hardiman G, Sluder AE, Johnson CD, Liu LX. Therapeutic target discovery using *Caenorhabditis elegans*. *Pharmacogenomics* 2000;1:203–217.
146. Nass R, Hahn MK, Jessen T, McDonald PW, Carvelli L, Blakely RD. A genetic screen in Caenorhabditis elegans for dopamine neuron insensitivity to 6–hydroxydopamine identifies dopamine transporter mutants impacting transporter biosynthesis and trafficking. *J Neurochem* 2005;94:774–785.
147. Matthews DJ, Kopczynski J. Using model-system genetics for drug-based target discovery. *Drug Discov Today* 2001;6:141–149.

13 Zebrafish as a Model for Development

BRIAN A. LINK AND SEAN G. MEGASON

ABSTRACT

The zebrafish model has recently emerged as a significant experimental system for developmental biology studies. In this review, we provide an introduction to the zebrafish model system by overviewing set-up and husbandry requirements, the genome, and associated bioinformatics infrastructure. We then detail the utility of zebrafish for forward, reverse, and chemical genetic studies of development. We also discuss using zebrafish for live cell imaging experiments. Finally, we provide an overview of zebrafish community resources.

Key Words: Vertebrate, Genetics, Chemical-genetics, Time-lapse imaging, Organogenesis, Mutant analysis.

INTRODUCTION

Zebrafish, *Danio rerio*, are small freshwater teleosts that were insightfully selected as an experimental model based on several practical criteria as well as their genetic amenabilities. During the 1980s, the pioneering work of the late geneticist George Streisinger and his colleagues at the University of Oregon established the groundwork for utilizing zebrafish as an experimental system. This organism has now emerged as a significant model for developmental biology studies, acting as a segue between the genetic power of invertebrates and the relevance of rodents (Table 13–1).

Adult zebrafish are approximately 1.0–1.5 inches in length and can be housed inexpensively with modest space requirements (Figure 13–1). Sexual maturity is reached reliably at 3 months, but under optimal growth conditions this time can be reduced to 2 months. Females can produce 100–200 embryos per week. Fertilization occurs externally and initial development is completely transparent. Development through larval stages can also be rendered transparent with the use of melanin synthesis inhibitors.

Another attraction of zebrafish for developmental studies is the rapid ontogeny. Blastulas enter gastrulation at approximately 5 h postfertilization (hpf). Somitigenesis begins just before 10 hpf and by 24 hpf the major subdivisions within the nervous system are in place, as are the rudiments of the primary organs. For example, the heart primordium initiates rhythmic beating within the first day. Embryos begin to break from their protective chorion envelope around 60 hpf, although this nonessential shell can be manually removed at earlier times. By 72 hpf, organogenesis is well underway and embryos exhibit touch responses and swimming behaviors. Visually evoked behaviors are present by day 4. For the first 5 days of development, embryos feed endogenously from their yolk material. Specific stages of zebrafish embryonic development have been described in detail by Kimmel *et al.*[1] After day 5, larval zebrafish need to be fed externally and a brief discussion of zebrafish husbandry is provided next.

ZEBRAFISH HUSBANDY

In addition to the embryological advantages of zebrafish, their ease in care is another attraction as an experimental model. When starting a zebrafish colony, the typical researcher contacts one of several companies that specialize in the construction and installation of multitank systems. These systems can vary from accommodating 10 tanks to well over 1000 (Figure 13–1B). For laboratories that just need a small supply of embryos, individual 10-gallon tanks will suffice. Water quality is a key parameter in keeping and raising zebrafish. Zebrafish require slightly brackish water for proper osmoregulation (400–800 μS). Appropriate salinity is achieved by addition of balanced salt preparations, which are available commercially. The pH should be maintained around 7.0 and the temperature at approximately 28°C. One major concern with recirculating fish systems is the maintenance of the nitrogen cycle. Nitrogenous compounds are found in several forms as byproducts of fish metabolism: ammonia (NH_3) and nitrite (NO_2^-) are toxic to fish and need to be converted to the third nontoxic form of nitrate (NO_3^-). Larger zebrafish systems rely on biological (microbial) filtration to maintain the nitrogen cycle. Smaller individual tank systems can also be maintained with chemical filtration. Appropriate pH is maintained by addition of sodium bicarbonate or calcium carbonate to raise pH and hydrochloric acid to decrease pH. Additional water quality details can be obtained from the system supplier.

Circadian cycles are also import for healthy fish. Most researchers use a 14-h on/10-h off lighting cycle. Feedings should also occur at consistent times. In general, fish are fed two or three times daily with a vitamin-enriched dry food and freshly hatched brine shrimp. Live brine shrimp can easily be prepared in the laboratory by hatching purchased brine shrimp eggs or cysts using commercially available hatchers. To prevent fouling the tank with uneaten food, no more food should be given than can be consumed by the fish in the tank within 10 min.

Raising larvae to adulthood can be challenging for new fish researchers. The critical factors are having high-quality water slowly flowing through the tanks as soon as feeding begins, ensuring that the larvae cannot escape or get washed out of the tanks,

From: *Sourcebook of Models for Biomedical Research*
(P. M. Conn, ed.), © 2008 Humana Press Inc., Totowa, NJ.

and feeding a proper larval diet. Larvae can be fed a variety of foods including fresh *Paramecium*, rotifers, or dry food of the appropriate particle size. Live *Paramecium* can be ordered from the Zebrafish International Resource Center (Eugene, OR) and can then be propagated in the laboratory. High success rates in raising larvae can also be achieved with powdered foods designed especially for larvae (e.g., Hatchfry Encapsulon Grade I, Argent Labs). Larvae begin consuming live brine shrimp ~2 weeks postfertilization.

Embryos can be obtained by either pairwise or group crosses. Fish are netted from their housing tank and placed into breeding tanks the night before embryos are to be collected. Zebrafish typically breed at "dawn" when the lights turn on. Breeding tanks contain a porous insert sleeve, allowing the fertilized eggs to slip though to the bottom, where they are protected from cannibalization by the adult fish. Embryos can then easily be collected and used for microinjections or other experimentation and observation.

THE ZEBRAFISH GENOME

The zebrafish genome is diploid, is composed of 25 paired chromosomes, and is estimated at 1700 Mb, about half the size of the human genome.[2,3] A working draft of the assembled genome is available from the Sanger Centre (http://www.sanger.ac.uk/Projects/D_rerio/). Comparative mapping has shown extensive conservation in synteny (the ordering of genes along the chromosome) between zebrafish and humans. In fact, the local synteny between zebrafish and humans is even greater than between mouse and humans, as the mouse appears to have undergone substantially more chromosome rearrangements.[4] Analysis of the zebrafish genome, along with that from other bony fishes, has revealed an ancient genome duplication prior to the dramatic speciation of the teleosts followed by massive gene loss. In practical terms, this has resulted in the duplicate representation of approximately 30% of the zebrafish genes as compared to mammalian orthologs.[5] This has implications for both genome assemblies and gene functional studies. Having two loci for essentially the same gene can be viewed either as an advantage or a disadvantage, depending on the question at hand. For the majority of duplicated genes studied

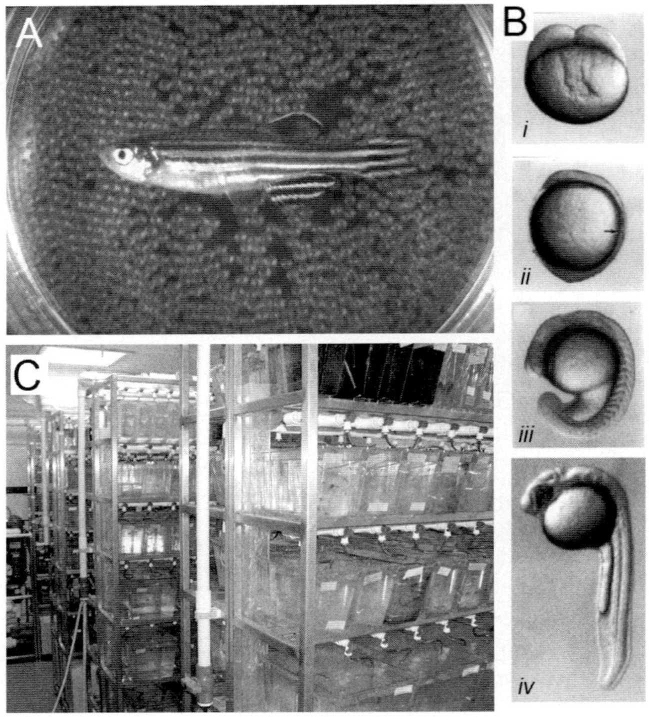

Figure 13–1. Adult and embryonic zebrafish. (A) Single female adult zebrafish in a 10-cm culture dish with newly fertilized embryos within their chorions. (B) Embryo morphology (chorions removed): (i) 2 cell (~45 min); (ii) 1 somite stage (arrow, ~10.5 h); (iii) 18 somite stage (18 h); (iv) 24-h embryo. (C) Typical multirack circulating system. In this system, each rack has 6 shelves of 6–12 tanks, depending on the size of the tanks. [Images in (B) modified from Kimmel *et al.*, 1995.]

in zebrafish, the divergence between the two genes is most dramatic within regions that control expression. The prevailing view is that while duplicated genes have subpartitioned their expression domains, they are often functionally equivalent. Of course this assumption warrants experimental validation. For forward genetic studies, gene duplication can be perceived as advantageous, as mutations potentially will yield more specific phenotypes than mutations in other vertebrate species. However, when both paired genes retain expression in any given domain, the disadvantages of genetic compensation can complicate experimental interpretation. Indeed, redundancy and compensation might prevent the isolation of mutations in particular genes because no measurable phenotype would be induced. From an experimental perspective, it is important to know whether the gene being investigated is duplicated. In the following sections, we next describe the most common techniques and utilities of zebrafish for studies in developmental biology.

Table 13–1
Advantages and disadvantages of popular animal models[a]

	Fish	*Frog*	*Worm*	*Fly*	*Mouse*
ENU screen	++	–	+++	+++	+
Insertional screen	+++	–	+++	+++	+
Small molecule screen	+++	++	++	+	–
DNA/RNA injection	+++	+++	+	+	+
Morpholinos	+++	+++	?	?	?
RNAi	?	?	+++	+++	+
Genetic misexpression	+++	–	++	+++	+++
Imaging	+++	+	++	+	+
Genetic mosaics	+++	–	++	++	++
Cell culture	+	+	+	+++	+++
Relevance to human health	++	++	+	+	+++

[a]This table is a rough guide for comparing the most commonly used model systems. It should serve as a general introduction as it summarizes complex comparisons that will not hold true for every situation. See the text for more details.

FORWARD GENETICS

MUTANT SCREENS Forward genetics refers to phenotype-based analysis followed by identification and characterization of the mutation in the causative gene (Figure 13–2). Mutant screens exemplify this phenotype-to-genotype approach. There is no doubt that forward genetics put zebrafish on the experimental systems map. Several large-scale mutagenesis screens have been conducted and others are currently underway. The first large-scale screens were done in tandem in the laboratories of C. Nusslien-

Volhard (Tübingen) and W. Driever (Boston/Freiburg) in the mid-1990s. Ethylnitrosourea (ENU) was used as the mutagen, which largely produces point mutations. The initial descriptions of these screens were published in a special issue of *Development* in 1996.[6] With chemically induced mutations, the causative gene mutation is usually isolated by combining positional cloning and candidate gene analysis techniques. Isolation of chemically induced mutations can be laborious, but the continued refinement of the zebrafish genome assembly greatly aids in such efforts.

Alternative approaches to ENU-induced mutagenesis are also possible. Shortly following the ENU screens, the Hopkins laboratory (MIT) conducted a retroviral-insertion screen.[7,8] The advantage of this strategy is that the phenotype-causing genes could be easily identified by simple inverse-polymerase chain reaction (PCR) assays. The disadvantage was that the mutagenesis was less efficient and the kinds of mutations created by large insertions, in principle, are more limited than point mutations. In addition, specialized expertise and facilities are required for the generation of high-titer pseudotyped viruses.

One variation on the more time and cost consuming *de novo* genetic screen is the shelf screen. Shelf screens refer to reanalysis of an existing collection of mutants for the phenotype of interest and several shelf screens have been successfully conducted.[9–12] For example, Neuhauss and colleagues screened for visual behavior deficits, identifying subtle ocular phenotypes in mutants that had been previously characterized for other defects.[10]

There are several important issues to consider with regard to mutational analysis in zebrafish. As with any genetic screen the assay for the phenotype of interest should be fast and robust. Most of the initial developmental genetic screens in zebrafish were morphology based. However, it is now more typical to screen for mutations that affect the expression of specific genes, antigens, or transgenes that express green fluorescent protein (GFP). Most screens for developmental anomalies require matings through several generations in order to bring recessive mutations to homozygosity (Figure 13–2). Dominant mutations can be selected in the F_1 generation, but the phenotype must be nonlethal in order to preserve the carrier. With zebrafish, when studying developmental phenotypes, the role of maternal mRNA and protein must be carefully considered. The large ooplasm of zebrafish contains very large supplies of maternally derived (wild-type) mature transcripts and proteins. This "start-up package" for the newly fertilized embryo can mask the early function of mutated genes. The degradation of specific mRNA and proteins is variable, but some maternal products have been shown to remain in zebrafish embryos for over 7 days, a relatively late developmental time point. The masking of early phenotypes by maternal products can be prevented through the use of maternal-zygotic (MZ) mutants in which both the mother and embryo are homozygous mutant. Excellent reviews on the logistics and further considerations of genetic screens in zebrafish are available.[6,13–15]

CHEMICAL GENETICS One other approach to phenotype-based study of developmental biology (as well as other biological phenomena) is termed chemical genetics. In 2000, R. Peterson and colleagues published a paper describing the use of combinatorial small molecule chemical libraries to induce phenotypes in developing zebrafish.[16] Once identified, small molecule compounds that disrupt developmental processes can be applied at different times. These agents, therefore, can be used similarly as conditional genetic alleles. In addition to allowing for time-specific disruptions, another advantage of chemical genetic screens is that founding generation embryos can be used. There is no need to progress through several generations of inbreeding as there is for genetic mutant screens. The major challenge with chemical genetics is target identification. What protein(s) and/or other cellular components does the compound affect? Traditional biochemical purification approaches are most common in target identification. More recently, chemical genetics has been used in tandem with genetic mutagenesis to identify potential small molecule targets as well as to conduct suppressor/enhancer screens. With this approach, mutagenized zebrafish embryos are screened in parallel with and without a phenotype-inducing small molecule. Those mutations that suppress or enhance the phenotype are strong candidates to disrupt genes that function in the same biological pathway as the small molecule. Excellent reviews on using zebrafish for chemical–genetic studies have been published.[17–19]

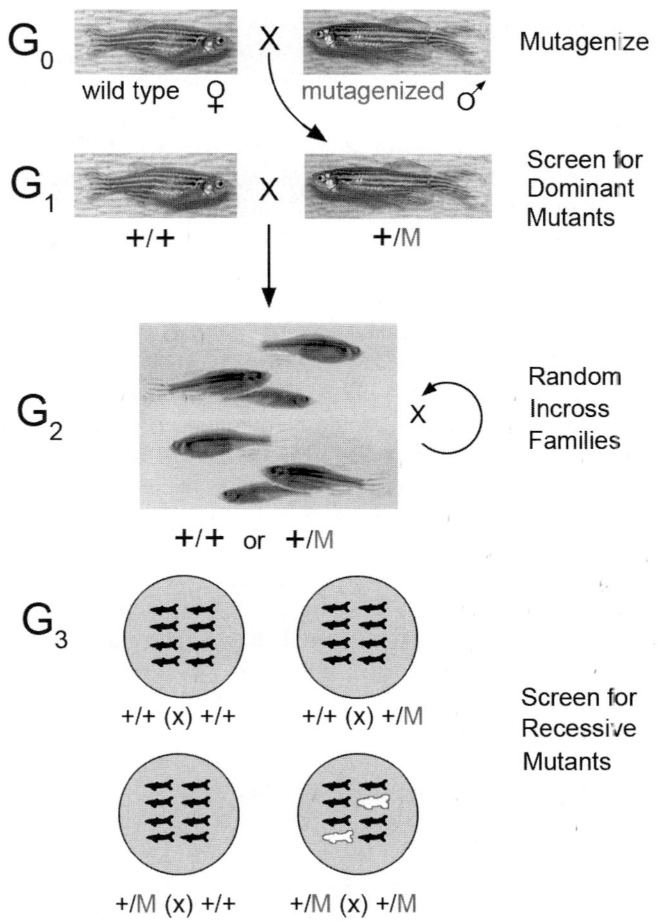

Figure 13–2. Classical three-generation screen for recessive mutants. In the founding G_0 generation, male spermatagonia are mutagenized. Resulting fish carry random mutations. This G_1 generation can be screened for nonlethal dominant mutations. To identify recessive mutations, G_1 fish are outcrossed again and the resulting G_2 families are incrossed. G_3 embryos will then show recessive phenotypes.

REVERSE GENETICS

In addition to being a superb system for phenotype-driven, forward genetic approaches, the zebrafish is also an excellent model for a variety of gene-driven techniques collectively referred to here as "reverse genetics." With reverse genetics, the investigator starts with a gene and tries to characterize its function as opposed to forward genetics in which the investigator starts with a function (i.e., phenotype) and tries to characterize the genes involved. For zebrafish, there are multiple and diverse techniques available for characterizing the function of a gene.

The primary advantage of zebrafish over other model systems is the ease with which reverse genetic approaches can be done and the wide array of available methods (see Table 13–1). The main reason that these techniques are easy to do in fish is that zebrafish eggs can be readily collected and injected in large numbers. In a typical experiment, a researcher would set up several pair matings the evening before, collect fertilized eggs from each pair in the morning (typically 50–250 eggs per pair), and inject 100–1000 eggs in a few hours. Injections are done under a dissecting microscope using a pressure injector and micromanipulator as detailed previously (http://zfin.org/zf_info/zfbook/chapt5/5.1.html). Most people learn how to be proficient at injections within a few sessions. It is best to do injections at the one-cell (zygote) stage to ensure an even distribution of the reagent. The initial divisions of zebrafish are meroblastic (incomplete cytokinesis), meaning that reagents can transfer between all the cells until the 16-cell stage.[20] Injections can also be done at later stages intentionally to achieve a mosaic distribution.

DNA/RNA INJECTIONS One of the most common ways to characterize a gene in zebrafish is simply to inject DNA or RNA based on the gene into eggs and assess the resulting phenotype of the embryos. The injected construct, for example, can be full length, dominant negative, or encode other mutant and truncated forms for structure–function analysis. Stabilized, capped RNA can be made using commercially available *in vitro* transcription kits from plasmid DNA templates. RNA can be injected into the yolk at the one- to four-cell stage and typically results in high-level, uniform expression. It is good practice to monitor the expression level and distribution of injected RNA by coinjecting RNA for a fluorescent protein [e.g., enhanced green fluorescent protein (EGFP)] or better yet using a fluorescent fusion protein construct. This allows properly injected embryos to be quickly sorted on a fluorescent dissecting microscope while still living. RNA typically starts expression within 4–5 h postfertilization and degrades within 2–4 days postfertilization, so injections are useful for studying developmental stages only within that window.

DNA injections should be done into the blastoderm rather than the yolk to improve expression, but they still typically result in mosaic expression even when done at the one-cell stage. Use of the Tol2 transposon system can substantially improve the distribution of expression.[21] The advantage of DNA injection over RNA is that DNA allows the use of enhancers to control when and where expression occurs and can result in more sustained expression. DNA injection is also how transgenic lines are made as described below.

MORPOLINO/SIRNA INJECTIONS DNA and RNA injections are typically used for gain-of-function studies. Injecting dominant-negative constructs can provide some insight into loss-of-function, but there are a number of caveats with this approach concerning the specificity and efficacy of the reagents. Traditionally the preferred method for doing loss-of-function studies was through the use of genetic null mutants, which with a reverse-genetic approach are often not available. Within the last several years, however, two methods, morpholinos and RNAi/siRNA, have become available to efficiently and specifically generate loss-of-function phenotypes in a number of organisms with morpholinos being much more popular for zebrafish.[22] Morpholinos are 18- to 25-bp synthetic oligonucleotides with a modified backbone that is resistant to enzymatic cleavage yet still base pairs with high affinity and specificity. They are designed to be antisense to the gene of interest such that they base pair near the start codon in the mature mRNA to block translation or across a splice junction in the pre-mRNA to block splicing. Both approaches can work well, but splice site morpholinos have the advantage that their efficacy can be assessed using reverse transcriptase (RT)-PCR. Morpholinos may not completely abolish expression, so the term "knockdown" rather than "knockout" is used to refer to their activity and "morphant" rather than "mutant" is used to refer to the affected embryos. When using morpholinos, it is essential to do controls to monitor the efficacy of knockdown (e.g., antibody staining) and specificity (e.g., rescue by coinjection of an RNA that lacks the target sequence). Morpholinos can be ordered through Gene Tools or Open Biosystems.

RNAi and the related technology of siRNA have worked well in a number of other systems, but in fish they have not yet been shown to work as robustly as morpholinos. Simple injection of dsRNA as is used for RNAi in worms and flies results in nonspecific effects in zebrafish.[23] siRNA is widely used in mammals and has the potential benefits of being cheaper and being able to be delivered genetically, but has received only minor use in zebrafish.[24,25]

TRANSGENIC MISEXPRESSION SYSTEMS The above methods take advantage of the ease of injecting zebrafish eggs to do transient assays. It is also useful to take advantage of the genetics of zebrafish to do transgenic-based studies to characterize a gene's function. This powerful ability to do both injection-based and stable transgenic-based approaches for reverse genetics is unique to fish (Table 13–1). Compared with injection, transgenic approaches are more labor intensive, but they allow more precise control.

A transgenic organism is an organism that contains an exogenous piece of DNA called a transgene. The term is most often used for stable transgenics in which the transgene is integrated into the genome and stably inherited from generation to generation. There are several ways to make transgenics in zebrafish.[26] The first approach is to simply inject DNA for a transgene into fertilized eggs, raise the embryos to adulthood, and screen for founders that pass on the transgene to their progeny. This method is not terribly efficient on a per embryo basis but since injecting and screening are easy in fish, transgenics are routinely made this way. Generally, around 5% of injected animals will become founders and they will pass the transgene on to around 5% of their progeny since the founders are mosaic. These numbers can vary depending on a number of factors including the construct, amount of DNA injected, and the skill of the person injecting. The transgene will follow normal Mendellian inheritance in subsequent generations. Since several hundred eggs can be injected in a few hour session, the injections do not require much work. The labor in generating transgenics is in raising the eggs to adulthood and screening to identify founders. Screening can be done visually for fluorescent protein (e.g., EGFP) transgenics but may require PCR for other constructs. Transgenics made by simple DNA injection

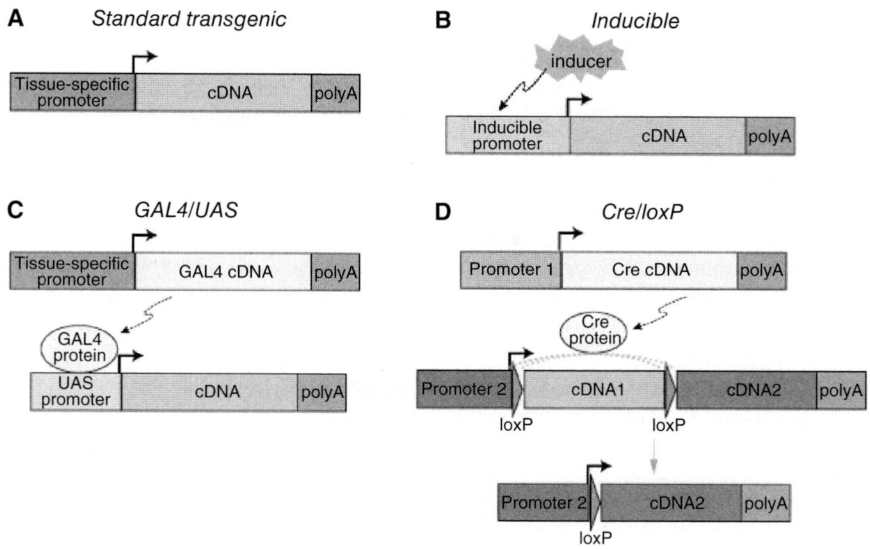

Figure 13–3. Transgenic misexpression systems. (A) For standard misexpression, a transgene contains an enhancer/promoter (typically tissue-specific) driving expression of a cDNA of interest. To ensure optimal expression it is important to include an intron (not shown) and a polyadenylation signal. (B) Inducible transgenics contain a single transgene containing a promoter that is activated in response to an outside factor such as heat or a drug. (C) In the GAL4/UAS system, two different transgenics are used that when separate have no effect. When combined (e.g., by crossing or coinjection), GAL4 made from one transgene will activate expression of the other transgene through binding the UAS sequence resulting in the cDNA being expressed in a domain determined by the tissue-specific promoter. (D) The Cre/loxP system also typically makes use of more than one transgene, although there are many variations on its use. In the example shown, Cre recombinase is expressed from one transgene and causes recombination between IoxP sites on a second transgene. The end result is that cDNA1 is expressed in cells that express promoter 2 and have *never* expressed promoter 2 and cDNA2 is expressed in cells that express promoter 2 and expressed promoter 1 at *any* time in their past. Such a system can be used for genetic fate mapping if visible markers are used or for controlled misexpression similar to GAL4, but with an important distinction caused by the irreversible nature of the loxP recombination.

often contain multiple tandem copies of the transgene. Multicopy transgenics can provide higher expression levels, but sometimes it is important to have single copy insertions (e.g., loxP reporters, gene trapping). Two transposon systems, Tol2[21] and Sleeping Beauty,[27] have recently been developed that generate single copy insertions and also integrate more efficiently than simple DNA injections. These transposons can be used for insertional screens as described above, but they can also be used to increase the efficiency of making specific transgenics. A final method for generating single copy insertions with improved efficiency is through the use of the meganuclease I-SceI.[28]

A common problem in using transgenics for studying the function of a gene is that the transgene can cause a dominant lethal phenotype, precluding the establishment of stable transgenic lines. There are several ways around this problem including inducible and binary systems. Inducible systems consist of a single transgene that is silent until expression is turned on by some outside factor. The promoter for the zebrafish heat shock protein (Hsp) 70 has only minimal activity in all tissues (except the lens) at 28.5°C, the normal temperature for raising embryos. Ubiquitous expression from Hsp70 can be induced by heat shocking the embryos for 1 h at 37°C (Figure 13–1A). Localized expression from Hsp70 can also be induced by local heating using a focused laser.[29] Inducible expression has also been achieved is zebrafish using the steroid glucocorticoid.[30] Binary systems use two separate transgenes to provide regulated expression. In the GAL4/UAS system, one transgenic fish expresses the yeast transcription factor GAL4 under a tissue-specific promoter.[30,31] A separate transgenic fish contains the gene of interest behind a minimal promoter containing UAS sites (Figure 13–1B). The UAS transgene is normally silent, but when crossed with the GAL4 line, it is turned on by GAL4 binding the UAS sequence and activating transcription in a pattern determined by the promoter regulating the GAL4 transgene. Once GAL4 and UAS lines have been established, a variety of experiments can be done simply by crossing different GAL4 and UAS lines. Another versatile binary system consists of the site-specific recombinase Cre and its recognition sequence loxP.[32] As shown in Figure 13–3, a transgene can be made to express one gene prior to recombination and a second gene after recombination. This transition can be controlled by combing the "loxP transgenic" with a second transgenic that expresses Cre tissue specifically. Numerous variations on this theme are possible using Cre/loxP and the analogous recombinase system FLP/FRT.

LOCUS-SPECIFIC MUTAGENESIS Although morpholinos can provide valuable loss-of-function data for a gene of interest, the cleanest and the traditional method is to have a genetic mutation for the gene. Until a few years ago obtaining mutants for any specific gene could be done with mouse only through the use of homologous recombination in embryonic stem cells. Recently, however, a new method called TILLING (Target-Induced Local Lesions IN Genomes) has been developed that uses a bank of mutagenized sperm and brute force molecular methods to detect sperm samples with mutations in the gene of interest.[33–35] These sperm samples are then used for *in vitro* fertilization to recover and analyze mutant animals. A review comparing various forward and reverse genetic techniques was recently published.[15]

Mutations found with TILLING are still random, point mutations and small deletions, but it may soon be possible to achieve the exquisite precision of genetic engineering possible with mouse

ES cells in zebrafish. Zebrafish have been produced from cultured cells through somatic cell nuclear transfer (i.e., cloning).[36] If these cells can be manipulated in culture through gene targeting as is done with mouse ES cells, then it may soon be possible to construct zebrafish mutants with any desired DNA sequence (e.g., conditional alleles, GFP fusions). Work is also being done to establish zebrafish ES cells. Germline chimeras have been made from ES-like cells after short-term culture[37] and these cells have been targeted using homologous recombination,[38] but to date these techniques have not been successfully combined.

IMAGING

ADVANTAGES One of the most compelling reasons for the use of zebrafish today and its choice by George Streisinger as a model organism over three decades ago is its superb suitability for imaging. Streisinger's choice is even more clairvoyant in light of the incredible technological revolutions in both fluorescent proteins and confocal/two-photon imaging during the past decade. A number of attributes make zebrafish ideal for imaging. Zebrafish are largely transparent for much of their development allowing excellent light penetration for optical microscopy. From fertilization to 24h postfertilization, zebrafish embryos are remarkably transparent. The embryo proper (as opposed to the yolk ball) has no color and extremely low autofluorescence. Shortly after fertilization (0–1 hpf) the yolk scatters moderately due to cytoplasmic streaming, but subsequently it is also transparent. Although transparent, it is best to avoid imaging through the yolk when possible, because it has a slight yellowish tinge and a higher refractive index. The yolk is also moderately autofluorescent. Zebrafish contain three types of pigment cells that begin to differentiate on the second day of development and can interfere with imaging. Melanocytes are the first and most obvious pigment cell (Figure 13–4A), but production of the dark melanin pigment from these cells can be easily blocked by adding phenylthiourea to the media. The other two cell types, xanthophores and iridophores, begin to appear around 48 hpf. Xanthophores produce a light yellow pigment and are slightly autofluorescent, while iridophores produce crystals of purine that are highly reflective but small. While these pigment cells make imaging some tissues difficult (such as the eye), most tissues in zebrafish can be readily imaged throughout their development in the embryonic and early larval stages. The egg shell or chorion of zebrafish embryos is also transparent. For routine imaging on a dissecting microscope, the chorion does not present a problem (Figure 13–4A), but for high-resolution imaging on a compound microscope, it is best to remove the chorion (Figure 13–4B–E).

Zebrafish embryos are externally fertilized and develop freely in an aquatic media meaning that they are accessible to imaging throughout their development and easy to keep alive during time-lapse imaging. Another important advantage for time-lapse imaging is the rapid development of zebrafish, which allows equivalent developmental processes to be imaged with shorter movies in fish and thus much easier. Zebrafish embryos are also a nice size for imaging. They are small enough to fit completely within the working distance of many high-end objectives yet are not so small that the details of their development are beyond the resolution available with optical microscopy. Zebrafish embryos are easy to label at high levels through injection of RNAs for fluorescent proteins (Figure 13–4B) or soaking in fluorescent dyes. Achieving high-level labeling is important in order to capture high signal-to-noise images quickly with minimal photodamage.

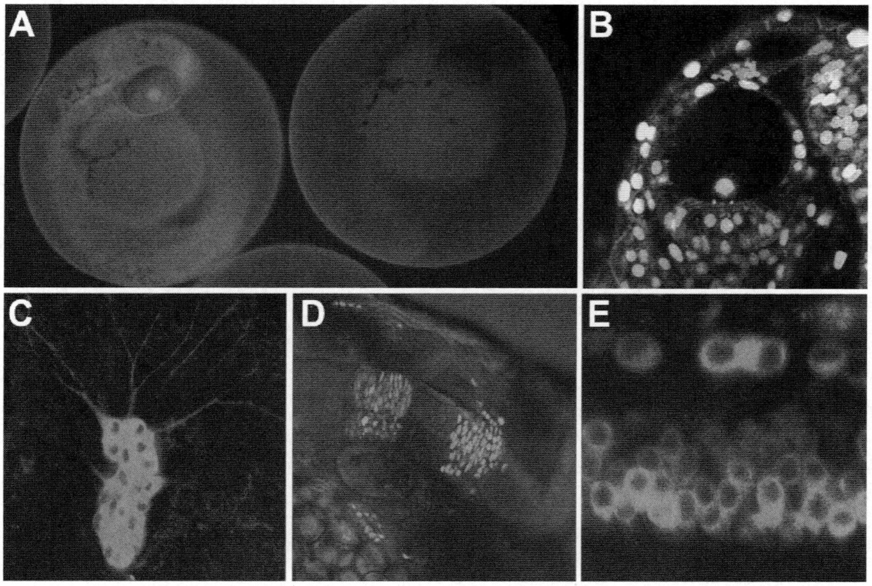

Figure 13–4. Imaging in live zebrafish embryos. (A) Whole embryos can be quickly imaged on a fluorescent dissecting microscope in their chorions to sort positive transgenics (left) from wild-type siblings (right). These embryos were not treated with PTU so the melanocytes are visible. (B–E) Confocal microscopy permits much higher resolution imaging. (B) A quick method for labeling is to inject RNA encoding fluorescent proteins, in this case a histone2B-EGFP fusion, and a membrane localized mCherry was used to image all the cells of the inner ear. (C) GFP transgenics can be used to image neuronal projections from the trigeminal ganglion as they extend. (D) GFP transgenics can mark specific populations of cells, in this case rhombomeres 3 and 5. (E) GFP fusion proteins can reveal the subcellular localization pattern of proteins, in this case a cytoplasmic protein in the Rohon-Beard and motor neurons of the spinal cord. (Images from S.G. Megason, L.A. Trinh, and S.E. Fraser, unpublished.) (See color insert.)

MICROSCOPES For quick, routine imaging, zebrafish embryos can be imaged on a dissecting or stereo microscope in a Petri dish full of egg water (Figure 13–2A). Most popular manufacturers now produce dissecting microscopes equipped for fluorescence that allow live zebrafish embryos expressing fluorescent proteins to be screened and sorted quickly. After 24 hpf, movement of the embryos can be a problem. Embryos can be anesthetized by adding a few milliliters of 0.1% tricaine to the Petri dish (0.010–0.015% final concentration). Embryos can be recovered from anesthesia with no ill effect by transferring to fresh egg water.

For high-resolution imaging a compound or better yet a confocal or two-photon microscope should be used (Figure 13–4B–E). Confocal and two-photon microscopy is advantageous because both approaches can eliminate out of focus light, which can be a problem unless only a small minority of cells is labeled. Tricaine should be used be used to prevent movement. At the proper concentration tricaine can be used to immobilize embryos for at least 48 h for time-lapse imaging while maintaining proper development (S.G. Megason and S.E. Fraser, unpublished observations). Careful mounting of embryos in the proper orientation and close to the coverslip is essential for high-resolution imaging. A fast and easy method for still images (not time-lapse) is to place the embryos in a drop of 2% methyl cellulose and cover with a coverslip supported by tape or vacuum grease. For time-lapse imaging it is essential to prevent the embryos from drifting during the time lapse. This can be accomplished by simply embedding the embryos in a dilute solution of low-melting temperature agarose. This method works well for imaging many parts of the body for moderate lengths of time (~12 h), but the rigidity of the agarose affects many aspects of morphogenesis such as tail extension. For more reproducible mounting and longer term time lapses of normal development, specially designed, micromachined templates can be used to cast agarose mounts to hold embryos in place for imaging (S.G. Megason and S.E. Fraser, unpublished observations).

LABELS For simple transmitted or reflected light microscopy, zebrafish can be stained using enzymatic markers with chromogenic substrates. This is routinely done with *in situ* hybridization to reveal RNA expression patterns and is also done using lacZ transgenics.[39] These techniques require the embryos to be fixed so time-lapse imaging cannot be done and are difficult to capture in three dimensions.

The true power of imaging zebrafish is through the use of fluorescent microscopy. Fluorescent imaging opens up two important dimensions (t and z): it allows molecular imaging to be done in real time on living embryos *and* it permits the use of confocal and two-photon microscopy to capture high-resolution volumetric images. Organic dyes can be used to label embryos such as bodipy ceramide to label membranes[40] or calcium indicators to monitor Ca^{2+} currents in neurons.[41] Fluorescent proteins such as GFP offer a number of advantages over organic dyes. They can be expressed in a spatial and temporal-specific manner through the use of transgenics.[26] Fluorescent proteins also come in a wide spectral range and are typically more photostable than organic dyes allowing for multicolor imaging (Figure 13–4B). Fluorescent proteins can also be engineered using standard techniques of molecular biology to act as reporters for a number of activities. At the simplest, they can serve as a coexpression marker—e.g., when making a transgenic to analyze the function of another gene, an internal ribosome entry site (IRES)-GFP cassette can be added to monitor the expression of the gene of interest. GFP transgenics can also be used to mark a particular cell type for analyzing mutant phenotypes or to simplify screening for new mutations involved in the development of that cell type (Figure 13–4C and D). Fluorescent protein transgenics can serve as markers for gene expression patterns. Once a transgenic line is available, it is often easier to use it to look at gene expression than performing *in situ* hybridization. GFP fusion proteins can also be used as "functional reporters." Fusions of GFP to a protein of interest can be used to monitor the subcellular localization of that protein, which often relates to its functional state (Figure 13–4E). For neuroscience, voltage, pH, and Ca^{2+}-sensitive versions of fluorescent proteins are available[42–44] as well as versions that can travel across synapses. For monitoring protein–protein interaction *in vivo*, fluorescent resonant energy transfer (FRET)[45] and bimolecular fluorescent complementation (BiFC)[46] can be used.

EMBRYOLOGICAL EXPERIMENTATION

While the genetic and imaging capabilities of zebrafish are the model's strengths, embryonic cell manipulation techniques have also been established for zebrafish. In this section we will introduce the most common "embryology" techniques used for zebrafish.

CELL AND TISSUE CULTURE To complement much of the *in vivo* analyses using zebrafish, there has been an increasing need to conduct cell biological experiments in culture. The use of zebrafish for this purpose is just in its beginning and no well-established zebrafish lines exist. However, protocols for primary cell culture are available for a variety of cell types.[47,48] In addition, cultures from blastula stage embryos have been shown to be mulitpotent and can be targeted for DNA transfections.[49,50] An area of active research and great promise is the use of these embryonic stem-like cells for targeted gene manipulation through homologous recombination.[38] The culture of whole tissue explants has also been developed and used for assessing inductive events as well as monitoring cell behaviors.[51,52]

LINEAGE LABELING AND FATE MAPPING One of the central questions in developmental biology is how progenitor cells distribute their progeny among various cell types. The ability to lineage label and fate map is essential for the utility of a model organism in developmental biology studies. In zebrafish, several strategies can be employed to label individual cells. The most common technique currently in use for labeling individual or small groups of cells relies on photolysis of caged fluorescein or other fluorescent labels.[53] With this technique, newly fertilized embryos are injected with the caged fluorescent compound. At later times, the individual cells can be labeled by "uncaging" the compound using a focused laser of the appropriate wavelength. The labeled cell is documented and its descendents can then be followed over time in a dynamic fashion or assessed at one particular time of interest. In a variation of this technique, fish with the *hsp70* promoter upstream of GFP (*hsp70*:GFP) can be used to photolabel cells in a similar manner.[29] In both techniques, the fluorescent label is diluted with each round of cell division. To permanently label cells, transgenic zebrafish have been established that contain *hsp70*:cre recombinase and lox{stop codon}lox-GFP.[54] The activation of cre recombinase will then catalyze the excision of the stop codon in front of the GFP transgene, permanently marking the heat-activated cells.

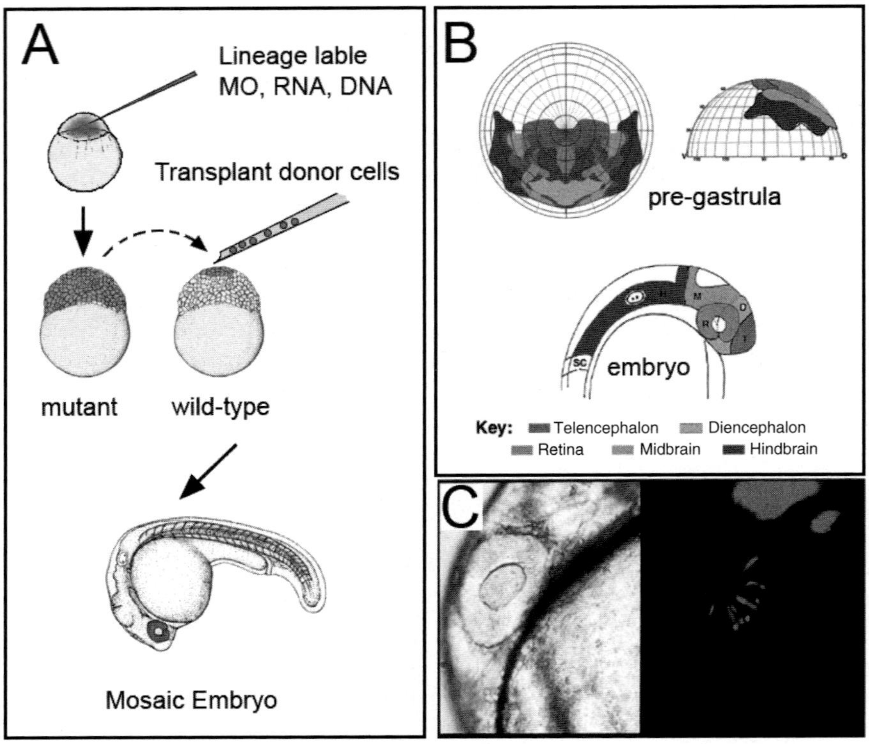

Figure 13–5. Creating genetic mosaics in zebrafish. (A) Donor cells are first lineage labeled with a tracer dye at the one-cell stage. Donor embryos can also be injected with morpholinos, RNA, or DNA. At the 1000-cell stage, totipotent blastula cells are transplanted into regions of the host embryo fated to give rise to specific structures. Donor and host embryos can be of either mutant or wild-type genotypes. Resultant chimeras are grown for subsequent analysis. (B) Fate-map of the pregastrula stage embryos. (Modified from Woo and Fraser, 1995.) (C) Mosaic embryo at 24 hpf showing bright-field (left) and fluorescent image (right) of rhodamine-dextran-labeled donor cells targeted to the eye and forebrain. (B.A. Link, unpublished.) (See color insert.)

In addition, focal labeling can be achieved through electroporation strategies. Protocols for labeling single cells[55] and larger fields of cells[56] have been established. With electroporation, localized electrical current carries small charged molecules across the plasma membrane into cells. For lineage marking, either labeled dextrans or plasmid DNA for GFP or other genetically encoded reporters can be delivered. Electroporation can also be used to manipulate gene expression as morpholinos or DNA for dominant negative/active proteins can be electroporated.

The described techniques have been used to establish several types of fate maps for the zebrafish embryo.[57–60] However, the need remains to assess cell fate determination with loss or gain of function of specific gene products, as well as at later times of development.

CELL AND TISSUE TRANSPLANTATION Zebrafish, like other *ex utero* organisms, are excellent specimens to conduct cell and tissue transplantation studies during development. The small size and tough epidermis of the zebrafish embryo, however, do pose problems for these types of experiments. In general, the approaches used for cell and tissue transplantation in other species such as chick and frog also apply to zebrafish.[61] Typically, the donor and host embryos are embedded adjacently in a low percent agarose gel or methyl cellulose. The fish are anesthetized using Tricane. Donor embryos are either labeled genetically with a fluorescent transgene or by injecting a tracer at the one-cell stage. For individual cell transplants, polished glass electrode pipettes are used to remove and expel cells. The positive and negative pressure needed to do this is created by a manual syringe driver. The details of this apparatus and technique have been described by J. Eisen (Westerfield, 1995; http://zfin.org/zf_info/zfbook/zfbk.html). For whole tissue transplants, sharpened tungsten needles or pulled glass electrodes are used to dissect and manipulate the tissue of interest. For ages over 1 day, the epidermis is thick but can be focally weakened by applying a small drop of mineral oil. Specific protocols for ocular tissue transplantation have been published.[62,63]

GENETIC MOSAIC ANALYSIS Genetic mosaic analysis is an elegant way to probe where and how specific genes function during development. Genetic mosaic embryos consist of cells of more than one genotype. In zebrafish, genetic mosaics are created by cell transplantation in early embryos, usually at the blastula or gastrula stages (Figure 13–5).[64,65] Because detailed fate maps exist for most tissue types in the early embryo, donor cells can be targeted to areas that will normally contribute to specific structures (Figure 13–5B and C). By transplanting genetically mutant cells into wild-type hosts and visa versa, the cellular autonomy of phenotypes associated with the mutation can be determined. As with other cell transplantation techniques, donor cells are labeled with a stable tracer at the one-cell stage or by utilizing transgenes. Excellent reviews of this technique and the analysis of genetic mosaics have been published.[66,67]

COMMUNITY RESOURCES

Although having grown tremendously in the past decade, the zebrafish community still has a well-organized infrastructure and strong community resources. The main website for all things zebrafish is ZFIN, the Zebrafish Information Network at http://zfin.org. It is maintained at the University of Oregon by Monte Westerfield and colleagues as the "official" NIH-supported, model organism database for zebrafish. ZFIN contains a wealth of information on genes, alleles (including mutants and transgenics), expression patterns, and anatomical atlases. ZFIN also contains an online copy of "The Zebrafish Book" by Monte Westerfield, which is essential reading for those new to using zebrafish in their research (http://zfin.org/zf_info/zfbook/zfbk.html). Other important resources for genomic information are NCBI GenBank (http://www.ncbi.nlm.nih.gov/) and the Sanger Institute (http://www.sanger.ac.uk/). The Sanger Institute is currently sequencing the zebrafish genome using a combined 10 × whole genome shotgun and BAC sequencing approach (http://www.sa nger.ac.uk/Projects/D_rerio/). As of September 2006, 64% of the genome is at finished quality and most of the remainder is covered as "unfinished" quality. The principal stock centers for zebrafish mutants and transgenics are ZIRC (the Zebrafish International Resource Center, http://zfin.org/zirc/) at the University of Oregon and the Tübingen Stock Center (http://www3.eb.tuebingen.mpg.de/core-facilities/zebrafish-stockcenter/tubingen-zebrafish-stockcenter) in Tübingen, Germany. A large variety of zebrafish strains can be requested from these stock centers for a nominal fee. The ZIRC also provides cDNAs/ESTs, anitibodies, and pathology services. The zebrafish community follows a standardized nomenclature for mutant and transgenic alleles. Each individual strain of fish is given an allele number such as b109. The initial letter designates the university or laboratory that identified the allele and the number is a unique number that specifically identifies that allele. A detailed description of the nomenclature guidelines can be found at ZFIN (http://zfin.org/zf_info/nomen.html). The most common companies for providing recirculating zebrafish systems are Aquatic Habitats, Aquaneering, and Marine Biotech in the United States and Aqua Schwarz in Germany. All of these companies as well as a number of other useful zebrafish-related companies are listed on the ZFIN company directory (http://mirror.zfin.org/cgi-bin/webdriver?MIval=aa-companyselect.apg). The zebrafish community holds an international conference on development and genetics every 2 years in the summer currently in Madison, WI. On alternating summers there is a large meeting in Europe, which rotates in its location. There are also a number of regional zebrafish meetings. Information on all these meetings is at ZFIN (http://zfin.org/zf_info/news/mtgs.html). There is a 2-week course on zebrafish at the Marine Biological Laboratory at Woods Hole, MA every summer for researchers new to fish (http://courses.mbl.edu/zebrafish/).

FUTURE PROSPECTS AND OTHER UTILITIES

To date zebrafish have principally been used as a model for developmental genetics, but they will likely find an even broader role in biomedical research in the future. This review has largely focused on embryonic analysis, but adult phenotypes can also be studied in zebrafish as well as adult-onset models of disease. This will be even truer as our ability to create inducible and conditional genetic alterations improves. Zebrafish are also an excellent model for neuroscience. Zebrafish neural activity can be monitored through imaging in whole, living larvae as they perform natural behaviors.[41] Zebrafish are also an excellent system for performing small-molecule/drug analysis. Small molecules can be screened in high-throughput with a chemical genetics strategy by simply soaking zebrafish embryos in an array of drugs using a 96-well plate-based format and screening for specific phenotypes.[18] Drugs can also be profiled rapidly for toxicological effects by embryo/larvae soaking.[68] Finally, zebrafish can be useful for identifying the target and mode of action of pharmaceuticals. Biochemical methods can be used to identify proteins that bind a drug as is normally done to identify targets, but the powerful genetic and genomic tools of zebrafish can also be applied. For example, it is possible to screen for interactions between genetic and small molecule-induced phenotypes.[18] Finally, since zebrafish can be analyzed very powerfully from a number of different angles (genetics, genomics, imaging, drug screening), they may also serve as an ideal model for integrative approaches such as systems biology.

REFERENCES

1. Kimmel CB, Ballard WW, Kimmel SR, Ullmann B, Schilling TF. Stages of embryonic development of the zebrafish. *Dev Dyn* 1995;203:253–310.
2. Fishman MC. Genomics. Zebrafish—the canonical vertebrate. *Science* 2001;294(5545):1290–1291.
3. Postlethwait J, Amores A, Force A, Yan Y. The zebrafish genome. *Methods Cell Biol* 1999;60:149–156.
4. Lyons LA, Raymond MM, O'Brien SJ. Comparative genomics: The next generation. *Anim Biotech* 1994;5:103–111.
5. Postlethwait J, Amores A, Cresko W, Singer A, Yan Y. Subfunction partitioning, the teleost radiation and the annotation of the human genome. *Trends Genet* 2004;20(10):481–490.
6. Haffter P, Odenthal J, Mullins MC, *et al.* Mutations affecting pigmentation and shape of the adult zebrafish. *Dev Genes Evol* 1996;206:260–276.
7. Golling G, Amsterdam A, Sun Z, *et al.* Insertional mutagenesis in zebrafish rapidly identifies genes essential for early vertebrate development. *Nat Genet* 2002;31(2):135–140.
8. Amsterdam A, Hopkins N. Retroviral-mediated insertional mutagenesis in zebrafish. *Methods Cell Biol* 2004;77:3–20.
9. Gross JM, Perkins BD, Amsterdam A, *et al.* Identification of zebrafish insertional mutants with defects in visual system development and function. *Genetics* 2005;170(1):245–261.
10. Neuhauss SC, Biehlmaier O, Seeliger MW, *et al.* Genetic disorders of vision revealed by a behavioral screen of 400 essential loci in zebrafish. *J Neurosci* 1999;19(19):8603–8615.
11. Sadler KC, Amsterdam A, Soroka C, Boyer J, Hopkins N. A genetic screen in zebrafish identifies the mutants vps18, nf2 and foie gras as models of liver disease. *Development* 2005;132(15):3561–3572.
12. Sun Z, Amsterdam A, Pazour GJ, Cole DG, Miller MS, Hopkins N. A genetic screen in zebrafish identifies cilia genes as a principal cause of cystic kidney. *Development* 2004;131(16):4085–4093.
13. Mullins MC, Nusslein-Volhard C. Mutational approaches to studying embryonic pattern formation in the zebrafish. *Curr Opin Genet Dev* 1993;3(4):648–654.
14. van Eeden FJ, Granato M, Odenthal J, Haffter P. Developmental mutant screens in the zebrafish. *Methods Cell Biol* 1999;60:21–41.
15. Amsterdam A. Insertional mutagenesis in zebrafish: Genes for development, genes for disease. *Brief Funct Genomic Proteomic* 2006;5(1):19–23.
16. Peterson R, Link B, Dowling J, Schreiber S. Small molecule developmental screens reveal the logic and timing of vertebrate development. *Proc Natl Acad Sci USA* 2000;97:12965–12969.
17. MacRae CA, Peterson RT. Zebrafish-based small molecule discovery. *Chem Biol* 2003;10(10):901–908.
18. Peterson RT, Fishman MC. Discovery and use of small molecules for probing biological processes in zebrafish. *Methods Cell Biol* 2004;76:569–591.
19. Yeh JR, Crews CM. Chemical genetics: Adding to the developmental biology toolbox. *Dev Cell* 2003;5(1):11–19.
20. Kimmel CB, Law RD. Cell lineage of zebrafish blastomeres. III. Clonal analyses of the blastula and gastrula stages. *Dev Biol* 1985;108(1):94–101.
21. Kawakami K. Transposon tools and methods in zebrafish. *Dev Dyn* 2005;234(2):244–254.
22. Nasevicius A, Ekker SC. Effective targeted gene 'knockdown' in zebrafish. *Nat Genet* 2000;26(2):216–220.
23. Oates AC, Bruce AE, Ho RK. Too much interference: Injection of double-stranded RNA has nonspecific effects in the zebrafish embryo. *Dev Biol* 2000;224(1):20–28.
24. Liu WY, Wang Y, Sun YH, *et al.* Efficient RNA interference in zebrafish embryos using siRNA synthesized with SP6 RNA polymerase. *Dev Growth Differ* 2005;47(5):323–331.

25. Kanungo J, Li BS, Goswami M, Zheng YL, Ramchandran R, Pant HC. Cloning and characterization of zebrafish (Danio rerio) cyclin-dependent kinase 5. *Neurosci Lett* 2006;412(3):233–238.
26. Megason S, Amsterdam A, Hopkins N, Lin S. Uses of GFP in transgenic vertebrates. *Methods Biochem Anal* 2006;47:285–303.
27. Ivics Z, Kaufman CD, Zayed H, Miskey C, Walisko O, Izsvak Z. The Sleeping Beauty transposable element: Evolution, regulation and genetic applications. *Curr Issues Mol Biol* 2004;6(1):43–55.
28. Grabher C, Joly JS, Wittbrodt J. Highly efficient zebrafish transgenesis mediated by the meganuclease I-SceI. *Methods Cell Biol* 2004;77:381–401.
29. Halloran MC, Sato-Maeda M, Warren JT, *et al.* Laser-induced gene expression in specific cells of transgenic zebrafish. *Development* 2000;127(9):1953–1960.
30. de Graaf M, Zivkovic D, Joore J. Hormone-inducible expression of secreted factors in zebrafish embryos. *Dev Growth Differ* 1998;40(6):577–582.
31. Scheer N, Campos-Ortega JA. Use of the Gal4-UAS technique for targeted gene expression in the zebrafish. *Mech Dev* 1999;80(2):153–158.
32. Dong J, Stuart GW. Transgene manipulation in zebrafish by using recombinases. *Methods Cell Biol* 2004;77:363–379.
33. Wienholds E, van Eeden F, Kosters M, Mudde J, Plasterk RH, Cuppen E. Efficient target-selected mutagenesis in zebrafish. *Genome Res* 2003;13(12):2700–2707.
34. Draper BW, McCallum CM, Stout JL, Slade AJ, Moens CB. A high-throughput method for identifying N-ethyl-N-nitrosourea (ENU)-induced point mutations in zebrafish. *Methods Cell Biol* 2004;77:91–112.
35. Sood R, English MA, Jones M, *et al.* Methods for reverse genetic screening in zebrafish by resequencing and TILLING. *Methods* 2006;39(3):220–227.
36. Lee KY, Huang H, Ju B, Yang Z, Lin S. Cloned zebrafish by nuclear transfer from long-term-cultured cells. *Nat Biotechnol* 2002;20(8):795–799.
37. Ma C, Fan L, Ganassin R, Bols N, Collodi P. Production of zebrafish germ-line chimeras from embryo cell cultures. *Proc Natl Acad Sci USA* 2001;98(5):2461–2466.
38. Fan L, Moon J, Crodian J, Collodi P. Homologous recombination in zebrafish ES cells. *Transgenic Res* 2006;15(1):21–30.
39. Bayer TA, Campos-Ortega JA. A transgene containing lacZ is expressed in primary sensory neurons in zebrafish. *Development* 1992;115(2):421–426.
40. Cooper MS, Szeto DP, Sommers-Herivel G, *et al.* Visualizing morphogenesis in transgenic zebrafish embryos using BODIPY TR methyl ester dye as a vital counterstain for GFP. *Dev Dyn* 2005;232(2):359–368.
41. O'Malley DM, Zhou Q, Gahtan E. Probing neural circuits in the zebrafish: A suite of optical techniques. *Methods* 2003;30(1):49–63.
42. Miesenbock G, De Angelis DA, Rothman JE. Visualizing secretion and synaptic transmission with pH-sensitive green fluorescent proteins. *Nature* 1998;394(6689):192–195.
43. Miyawaki A, Griesbeck O, Heim R, Tsien RY. Dynamic and quantitative Ca2+ measurements using improved cameleons. *Proc Natl Acad Sci USA* 1999;96(5):2135–2140.
44. Sakai R, Repunte-Canonigo V, Raj CD, Knopfel T. Design and characterization of a DNA-encoded, voltage-sensitive fluorescent protein. *Eur J Neurosci* 2001;13(12):2314–2318.
45. Siegel RM, Chan FK, Zacharias DA, *et al.* Measurement of molecular interactions in living cells by fluorescence resonance energy transfer between variants of the green fluorescent protein. *Sci STKE* 2000;2000(38):PL1.
46. Hu CD, Chinenov Y, Kerppola TK. Visualization of interactions among bZIP and Rel family proteins in living cells using bimolecular fluorescence complementation. *Mol Cell* 2002;9(4):789–798.
47. Collodi P, Kamei Y, Ernst T, Miranda C, Buhler DR, Barnes DW. Culture of cells from zebrafish (Brachydanio rerio) embryo and adult tissues. *Cell Biol Toxicol* 1992;8(1):43–61.
48. Helmrich A, Barnes D. Zebrafish embryonal cell culture. *Methods Cell Biol* 1999;59:29–37.
49. Sun L, Bradford CS, Ghosh C, Collodi P, Barnes DW. ES-like cell cultures derived from early zebrafish embryos. *Mol Mar Biol Biotechnol* 1995;4(3):193–199.
50. Fan L, Alestrom A, Alestrom P, Collodi P. Development of cell cultures with competency for contributing to the zebrafish germ line. *Crit Rev Eukaryot Gene Expr* 2004;14(1–2):43–51.
51. Grinblat Y, Lane ME, Sagerstrom C, Sive H. Analysis of zebrafish development using explant culture assays. In: Detrich HW, Westerfield M, Zon LI, Eds. *The Zebrafish: Biology*. San Diego, CA: Academic Press, 1999:128–155.
52. Langenberg T, Brand M, Cooper MS. Imaging brain development and organogenesis in zebrafish using immobilized embryonic explants. *Dev Dyn* 2003;228(3):464–474.
53. Kozlowski DJ, Murakami T, Ho RK, Weinberg ES. Regional cell movement and tissue patterning in the zebrafish embryo revealed by fate mapping with caged fluorescein. *Biochem Cell Biol* 1997;75(5):551–562.
54. Thummel R, Burket CT, Brewer JL, *et al.* Cre-mediated site-specific recombination in zebrafish embryos. *Dev Dyn* 2005;233(4):1366–1377.
55. Haas K, Sin WC, Javaherian A, Li Z, Cline HT. Single-cell electroporation for gene transfer in vivo. *Neuron* 2001;29(3):583–591.
56. Cerda GA, Thomas JE, Allende ML, Karlstrom RO, Palma V. Electroporation of DNA, RNA, and morpholinos into zebrafish embryos. *Methods* 2006;39(3):207–211.
57. Helde KA, Wilson ET, Cretekos CJ, Grunwald DJ. Contribution of early cells to the fate map of the zebrafish gastrula. *Science* 1994;265(5171):517–520.
58. Kimmel CB, Warga RM, Schilling TF. Origin and organization of the zebrafish fate map. *Development* 1990;108(4):581–594.
59. Woo K, Shih J, Fraser SE. Fate maps of the zebrafish embryo. *Curr Opin Genet Dev* 1995;5(4):439–443.
60. Stainier DY, Lee RK, Fishman MC. Cardiovascular development in the zebrafish. I. Myocardial fate map and heart tube formation. *Development* 1993;119(1):31–40.
61. Mizuno T, Shinya M, Takeda H. Cell and tissue transplantation in zebrafish embryos. *Methods Mol Biol* 1999;127:15–28.
62. Hutson LD, Campbell DS, Chien CB. Analyzing axon guidance in the zebrafish retinotectal system. *Methods Cell Biol* 2004;76:13–35.
63. Yamamoto Y, Jeffery WR. Probing teleost eye development by lens transplantation. *Methods* 2002;28(4):420–426.
64. Ho R, Kane D. Cell-autonomous action of zebrafish *spt-1* mutation in specific mesodermal precursors. *Nature* 1990;348:728–730.
65. Ho RK, Kimmel CB. Commitment of cell fate in the early zebrafish embryo. *Science* 1993;261(5117):109–111.
66. Moens CB, Fritz A. Techniques in neural development. In: Detrich HW, Westerfield M, Zon LI, Eds. *The Zebrafish: Biology*. San Diego, CA: Academic Press, 1999:133–147.
67. Rossant J, Spence A. Chimeras and mosaics in mouse mutant analysis. *Trends Genet* 1998;14:358–363.
68. Hill AJ, Teraoka H, Heideman W, Peterson RE. Zebrafish as a model vertebrate for investigating chemical toxicity. *Toxicol Sci* 2005;86(1):6–19.

14 Zebrafish as a Model for Studying Adult Effects of Challenges to the Embryonic Nervous System

NICOLE MARIE ROY AND ELWOOD A. LINNEY

ABSTRACT

Zebrafish is introduced as a model system to study environmental, chemical, and pharmaceutical challenges to the embryonic nervous system that can affect adult behavior/learning. The characteristics of the zebrafish system that make it possible to examine the developing nervous system in live embryos and larvae are presented. Gene discovery techniques, methodologies to generate fluorescent transgenic indicator embryos, as well as larval and adult zebrafish behavioral assays are described.

Key Words: Zebrafish, Transgenics, Neurodevelopment, Behavior, Microarrays, Neurotransmitter.

INTRODUCTION

In this chapter we will introduce the zebrafish model system and discuss how it can be used to investigate early challenges to the embryonic nervous system and the resulting impact of challenge on adult behavior. We hope to present (1) the importance of studying neurotoxins, embryonic nervous system challenge, adolescent and adult developmental disabilities, and societal impact; (2) the power of the zebrafish model system to study the nervous system, challenges to the developing nervous system, and associated learning, memory, and behavioral deficits; (3) the embryonic challenge hypothesis; and lastly (4) the methodologies and techniques used in zebrafish with which to study early embryonic challenge and adult learning/memory and behavior.

TOXINS AND DEVELOPMENTAL DISABILITIES Developmental, learning, and behavioral disabilities are of increasing concern among the population and pose a serious public health problem. It has recently been estimated that 12 million U.S. children under the age of 18 years are affected by developmental disabilities affecting cognitive function, language and learning ability, sensory and motor function, behavior, and emotion.[1] For example, attention deficit hyperactivity disorder (ADHD) affects conservatively 3–6% of all school aged children and treatment with Ritalin has doubled every 4–7 years since 1971.[2] Similarly, the incidence of autism has increased from a previously reported 0.5 case per thousand to 2.0 cases per thousand,[3] suggesting an increasing trend for these disorders over time. Children with developmental disabilities often require costly special educational and therapeutic treatments at significant financial cost to the family or state. Each year in the United States, between $81.5 and $167 billion is spent on neurodevelopmental deficits.[4] In addition, developmental disabilities are thought to affect only children and adolescents, but in reality, they pose lifelong challenges. Adults who suffered from childhood developmental disabilities have difficulties maintaining employment and learning new skills and techniques, are often socially alienated, and suffer from mood and anxiety disorders.[1] Thus, it is of significant importance to understand the etiology of these developmental disabilities and prevent their occurrence in future generations.

Normal neurological development is based on a choreographed sequence of cellular events. During development, brain cells are born, divide, migrate, differentiate, and establish synaptic connections in a highly organized series of events controlled by neurotransmitters and neurotrophic factors.[5] Disruption of this cascade, even subtle, can lead to serious long-term consequences. With the exception of single-gene disorders, heredity accounts for approximately half of the variances in cognitive, behavioral and personality traits.[5,6] The increase in the incidence of developmental disabilities has led to a considerable focus on environmental, pharmaceutical, or chemical factors that can influence the developing neural environment.

Studies with human fetal meconium have shown that in certain parts of the world human fetuses are being exposed to high levels of metals and pesticides[7] and several studies have demonstrated the neurotoxic effects of these heavy metals, pesticides and solvents. Lead is the most extensively studied developmental neurotoxicant and has been shown to cause deficits in language, learning, memory, attention, and motor coordination as well as hyperactivity, aggression, and even mental retardation in high doses.[1,8–11] These studies prompted the government to remove lead from gasoline and paint in the 1970s. Other heavy metals linked to developmental disabilities include cadmium, mercury, and manganese, which have also been linked to decreased IQs, attention deficit, hyperactivity, and motor dysfunction.[1,5] Pesticides are an important player in commercial food production and in lawn and garden weed and pest control and are used in quantities over one billion pounds per year. However, pesticides are toxic chemicals often designed to work specifically on the nervous system. For example, organophosphate pesticides have been linked to devel-

From: *Sourcebook of Models for Biomedical Research*
(P. M. Conn, ed.), © 2008 Humana Press Inc., Totowa, NJ.

opmental delays, hyperactivity, behavioral disorders, and motor dysfunction.[1,12,13] Other toxicants that pose a risk to the developing nervous system include the industrial solvents styrene, toluene, and xylene as well as the industrial lubricant and electrical insulator polychlorinated biphenyl (PCBs). These solvents are of particular concern as they contaminate soil and ground water affecting grazing animals used as beef cattle or as dairy animals.[14] Maternal ingestion of alcohol and nicotine also pose a risk to the developing brain.[11] Embryonic exposure to alcohol results in fetal alcohol syndrome (FAS), a condition hallmarked by craniofacial abnormalities, developmental delays, low brain weight, learning disabilities, memory impairment, and attention deficit.[15,16] Woman who smoke during pregnancy and expose the fetus to nicotine often have children who suffer from hyperactivity, learning disabilities, and developmental delays in cognitive function.[17–19] Lastly, as chemistry advances and pharmaceutical research progresses, more pharmacotherapeutics are routinely prescribed by medical practitioners. We are only in the infancy of understanding minute alterations these drugs may have on the developing brain or their long-term potential neurotoxicity.

Environmental, chemical, or pharmacological challenges to the nervous system often do not cause overt phenotypic changes to the developing brain, but rather cause minute, subtle changes that are not detectable with modern technologies. The complex nature of the human nervous system makes detecting these changes quite difficult to study. However, the zebrafish model system is an ideal organism with which to investigate early challenges to the developing nervous system and the subsequent impact on adult behavior and memory.

THE ZEBRAFISH MODEL ORGANISM Zebrafish are inexpensive, easy to maintain, have a short generation time, and individual females produce a large number of progeny. Unlike many other vertebrate models, zebrafish eggs develop *ex utero* and are transparent, meaning development is easily viewed under a simple dissection microscope. Zebrafish develop rapidly, are juveniles in a few weeks time, and are sexually mature adults by 3 months. More importantly, zebrafish are vertebrates, like humans, and follow the vertebrate path of embryonic development. In fact, there are several zebrafish models of human diseases including Alzheimer's disease, congenital heart disease, polycystic kidney disease, and cancer.[20] Accordingly, pharmaceutical companies are utilizing zebrafish in pharmacological trials to test drug candidates prior to investing in costly higher vertebrate models.

The zebrafish nervous system is highly amenable to investigation, particularly due to its transparency and the ability to clearly visualize fluorescent or stained neurons. The ability to identify small changes in neuronal architecture or molecular properties of neurons during development in response to environmental, chemical, or pharmaceutical challenge is of high importance in identifying potential neurotoxicants that cause early neural defects that can lead to adolescent and adult neurodevelopmental disorders. If minute changes can be detected, we can then link these early neuronal alterations with later staged behavioral changes and learning impairments. Luckily, work from several laboratories using calcium imaging, backfilling techniques, confocal imaging, *in situ* hybridization, immunohistochemistry, neuronal lesioning, and laser ablations has identified individual neurons and elucidated neuronal function and functional relationships between neuronal subtypes.[21–26] Furthermore, understanding the distribution patterns of subtypes of neurons is of importance to understanding neuronal circuitry. This is especially important in the spinal cord where neuronal circuits are essential for rhythmic movements and locomotor behavior. The development of the locomotor network in zebrafish has also been well characterized.[27,28] Full reviews of zebrafish movement and neuronal circuitry responsible for movement are available.[27–32]

Because of the relative long human life-span (and the shorter career span of scientists who might study such things), early exposure to any chemicals or pharmaceuticals that might produce an effect years later would be difficult to notice, let alone study. Therefore, developing an alternative model system that makes it possible to watch neuronal development and evaluate behavior within a reasonable amount of time could be a valuable means to examine early neural challenge and impact on adult behavior. The knowledge of the early developing nervous system and the imaging techniques available to study subsets of neurons and neuronal circuits coupled with rapid embryonic development into adulthood makes the zebrafish a particularly good candidate for studying early challenges to the nervous system and their impact on larval and adult learning, memory, and behavior.

CHALLENGING THE EMBRYONIC NERVOUS SYSTEM

THE EMBRYONIC CHALLENGE HYPOTHESIS This laboratory's work on challenging the zebrafish nervous system developed out of studies initiated through the Duke NIEHS funded Superfund Center. Following the lead of the Slotkin laboratory,[12] which utilized the rat model system to look at the effects of the organophosphate pesticide chlorpyrifos (Dursban) on embryonic signaling, we exposed zebrafish embryos to different concentrations of chlorpyrifos and examined its effects. High levels of chlorpyrifos exposure (500 ng/ml) resulted in muscle, notochord, and embryo morphology[33–35] defects, resembling the zebrafish acetylcholine esterase mutant phenotype. As we lowered the dose to one that allowed the fish to survive but still inhibited acetylcholine esterase activity, we found that adults that had been exposed to a low dose of chlorpyrifos demonstrated learning deficiencies.[36] From these studies we developed the following hypothesis: perturbing the neuronal signaling of the developing nervous system before it matures affects adult learning and/or behavior. Implicit in this hypothesis is the clear distinction that embryos are not just smaller adults in terms of their sensitivity to environmental insult and that any compound that has been used or tested on an adult may have a distinctly different effect and dose–response curve.

SENSING THE ENVIRONMENT AND REGULATORY CONSEQUENCES Looking back to the very early origins of molecular biology and the elegant experiments and interpretations of Jacob and Monod[37] we recall how their work generated basic concepts involving gene activation and repression. These were derived from experiments on how a simple bacterium senses the sugars in its environment and adjusts to this change with a regulation in gene expression. Obviously the cues that allow the nervous system to develop are somewhat pre-ordained in the developmental roadmap—proteins that allow some neuronal processes to move through the body to make connections that allow one part of the body to "speak" to another part. With industrialization and the development of the pesticide and pharmaceutical industries, we are being accidentally and purposely exposed to biological mimetics or cues to which a developing organism might respond. In a series of experiments that involved perturbing the developing *Xenopus* nervous system the Spitzer laboratory showed that perturbation could actually change the neurotransmitter produced by the perturbed neuron; this neurotransmitter "switching" was placed within the context of a homeostatic explanation or how the plasticity of the early nervous system might accommodate challenge by responding to an inhibitory influence with an excitatory influence.[38] While their study was focused upon the early embryonic nervous system and not the possible consequences to the adult, it does raise questions concerning how accommodating the nervous system can be to such challenges. As such, this laboratory has been approaching the developing nervous system in a more classical manner—with a focus on the decisions made by neurons to produce specific neurotransmitters rather than the more developmental focus of how regulators produce broader neuronal transitions. When the nervous system is viewed in this developmental context it provides an opportunity to address questions of neuronal decisions regarding the neurotransmitters that individual neurons produce. Similarly, since a presynaptic neuron's signal has to reach a postsynaptic receptor it raises the question of what neurotransmitter receptor subunits the postsynaptic neuron has available and how they become available. This level of questioning naturally leads to the use of a live fluorescent transgenic indicator approach to identify neurons producing, for example, acetylcholine rather than serotonin. This is one of our major technical approaches for addressing the question of early challenge of the nervous system.

TESTING THE HYPOTHESIS

CONCERNS FOR STUDYING THE MECHANISM: DIFFERENTIATION VERSUS DIRECT GENE REGULATION

One major concern in attempting to identify early events in developing embryos that might affect later behavior and learning, particularly from the standpoint of trying to dissect the mechanism, is that a compound that challenges embryos might influence differentiation in an inappropriate manner. Such an event might trigger an inappropriate repertoire of differentiation affecting many genes rather than a direct effect on a single gene regulator. This would make dissecting the mechanism by following gene expression extremely difficult. Therefore to investigate effects of early embryonic challenge on later events in life, it is necessary to consider several different approaches to try to distinguish the possible differences in mechanism. Below are described techniques others have used and techniques we employ.

TECHNIQUES USED BY OTHERS TO CREATE A KNOWLEDGE BASE ABOUT THE NERVOUS SYSTEM

Mutants The large scale genetic screens originally described in the December 1996 issue of *Development*[39] characterized many mutants by outward phenotype. Subsequent studies with these mutants and others have identified mutations affecting various forms of motility. There is now a battery of mutants affecting genes in neurotransmitter pathways (acetylcholine esterase,[33,34] glycine,[40,41] and other pathways), some of which have proven valuable in investigating early challenges to the nervous system. Furthermore, in those cases where genes have been linked with mutant phenotypes, the "morphant" phenotype can be generated utilizing antisense morpholino[42] technology. We have benefited from this since zebrafish exposed to a high concentration of organophosphate display phenotypic characteristics that overlap with the acetylcholine esterase mutant phenotype.

***In Situ* Techniques** Several laboratories have outlined the developing zebrafish nervous system by morphology using *in situ* hybridization and/or *in situ* immunolocalization or backfilling neurons with fluorescent probes. In a series of articles the Fetcho laboratory studied the development of the neural tube with reagents specific to GABA, glycine, glutamate,[26] serotonin, dopamine, and noradrenaline[43] producing neurons. While the specific anatomical localization and developmental time points were limited, they provide a very detailed and valuable picture of the developing zebrafish nervous system from embryogenesis through early larval development.

Recording Neuronal Activity The Drapeau laboratory[44] has pioneered techniques for recording electrical activity from the zebrafish nervous system including whole cell *in vivo* recording techniques via patch clamping. Others have used fluorophores for calcium imaging to evaluate neuronal activity.[25,45] Transgenic reporters for measuring ion movement can at least partially be used for the evaluation of changes in conductivity that reflect neuronal activity. This is currently being used in zebrafish,[46] but on a somewhat limited basis due to the weakness of fluorescent signal strength, a problem inherent with the physical characteristics of fluorescent molecules and/or the sensitivity of recording devices. Given the current limitations in recording neuronal activity as reported by fluorescent molecules, this technique is not a commonly used practice. Hopefully improvements in fluorescent reporter molecules will be made for recording neuronal activity via fluorescence directly.

Agonist/Antagonist Work The effects of compounds can be very complex and very dose specific. Therefore, in determining how some of their effects may impact the developing nervous system in more specific ways, it is sometimes useful to ask experimental questions using neurotransmitter-specific reagents such as receptor agonists, antagonists, and neurotransmitter specific reuptake inhibitors to investigate how disrupting a specific neurotransmitter pathway might lead to later effects. In this regard, reagents that affect various neurotransmitter pathways are available to us from various sources—toxins with specific receptor specificity (e.g., bungarotoxin with specific binding affinity to the acetylcholine receptor subunits, strychnine for glycine receptors), pharmaceuticals that impact upon neurotransmitter reuptake (e.g., the selective serotonin reuptake inhibitor Prozac), and a host of other selective reagents. Some of these display exquisite specificity and thus can be used in a selective manner, a benefit if antisense morpholino approaches are unsuccessful. In addition, antagonist/agonist strategies provide an option if the targeted gene product is something that is expressed later in development and hence not possible to effectively and selectively inhibit utilizing morpholinos.

TECHNIQUES WE HAVE BEEN USING

Transgenic Indicators This laboratory has a long history in the construction of transgenes and transgenic animals both in mice and in zebrafish. For example, we have made parallel transgenics in mice and zebrafish whose reporter activity is detected in at least a subset of embryonic tissue for which there is retinoic acid receptor activity.[47,48] These transgenics and their signal localization strongly support the homology in gene regulation via retinoic acid between mice and zebrafish. The transgenic lines also serve as inducible transgenes since exogenous retinoic acid exposure induces expression of the transgene ectopically in regions in which it is not necessarily expressed. In zebrafish, we have been working with several transgenic lines using fluorescent reporters [mainly green fluorescent protein (GFP) derived] that express selectively in the nervous system—lines we made and lines that we acquired from other laboratories.[35] However, as the organism develops it becomes increasingly complex to sort out the vast array of neurons in these lines. The early work that defined neurons present in the neural tube was based upon identifying neurons by morphology. However, since specific probes for neurons usually localize to one specific region of the neuron, an incomplete view of the neuron may be detected skewing interpretations. In our more recent work, we have used modified fluorescent reporter genes engineered with sequences that code for a farnesylation site. In doing so, the fluorescent reporter will be embedded into the neuronal membrane, yielding a more comprehensive visualization of the neuronal body and network. Currently, we have constructed transgenic lines containing a farnesylated GFP as well as a farnesylated mCherry, a monomeric red fluorescent protein (mCherry and several other different colored reporters were developed by the Tsien laboratory[49]).

As the organism develops, the neuronal network becomes quite vast and difficult to decipher. As such, the goal of our laboratory is to generate a series of transgenic lines that run from regulatory sequences of genes that identify neurons producing a specific neurotransmitter. Using this plan we hope to transgenically "dissect" the nervous system in fluorescent colors enabling us to visually examine different neurotransmitter pathways. It should be emphasized that while this program highlights the identification of presynaptic neurons that produce specific neurotransmitters, the production of transgenics that identify postsynaptic neurons that receive the neurotransmitter input is equally valuable. However, due to the complex diversity of subunit genes for neurotransmitter receptors, this is a task that we are not prepared to undertake at this time. However, once there is a better understanding of the expression specificity of various receptor subunits, transgenics can then be constructed. Overall, these transgenic lines will be used to identify the appearance of neurons producing specific neurotransmitters and for examining coexpression of different neurotransmitters produced by the same neuron. By utilizing these transgenic lines, we can visualize "neurotransmitter switching" in response to the challenge of a specific neurotransmitter pathway and possibly for following the development of the nervous system in a selective manner.

Transgenic Construction Our general procedures for making and characterizing transgenics have been detailed previously.[50] However, techniques evolve and these studies emphasize using much larger genomic regions for driving transgenes. There are two procedures for manipulating larger genomic regions into transgenic constructs. One, called recombineering, takes advantage of microbial recombination systems to homologously target a reporter gene in a genomic Bac or Pac clone so that the reporter gene is surrounded by the signals (known and unknown) that drive developmental expression of the endogenous gene.[51,52] This has been used in a limited way in zebrafish[53,54] and more commonly in mouse. While this is an elegant procedure, it does take considerable time since for most constructions two recombination events are necessary and the large clones have to be screened for the proper construction. An alternate, but less elegant procedure is to move large fragments into position to drive expression of the reporter gene. The larger genomic fragments, in most circumstances, will not have convenient restriction enzyme sites to allow one to recombine the proper construct. Thus, in order to recombine the construct, it is necessary to employ techniques for modifying "joints" of DNA so that they can ligate together: two common procedures are (1) to PCR change the sequence at the ends to create workable restriction sites for cloning and (2) to modify an existing restriction enzyme site with oligonucleotide adaptors to change an end of a DNA from one restriction site specificity to a more convenient one. We are currently using these nonrecombineering procedures for the design and construction of our transgenes.

Efficiency of Making Transgenics We find that if we prescreen transgenic injected embryos for expression, that approximately 10% of the adults with be germline for the transgene. However, we have found that in most cases, using our procedure, 2–50% of the resulting progeny are transgene positive, indicating that our procedures are producing founder transgenics that are chimeric for the transgene in their gonads. If we breed the transgene-positive progeny we obtain a normal Mendelian recovery of approximately 50% transgenic embryos. Occasionally we have transgenic founders with two separate integrations of the transgene resulting in transgene progeny percentages that are complicated until we can segregate the two transgenic integrants. Recently reports have come out indicating that the efficiency for producing germline transgenics can be enhanced by using the tol2 system.[55,56] By flanking their transgene with tol2 sequences and coinjecting the transgene with mRNA for the tol2 transposase, the efficiency of producing germline integrants is enhanced.

Microarray Analysis of Neurotransmitter Pathways For someone not familiar with using microarrays to investigate widespread gene expression changes a number of factors have to be considered: (1) the interpretation and analysis of the data are statistically based—many developmental biologists are not familiar with working with tab-delineated tables having thousands of rows. At best the information provides leads as to where to look and, thus, it is a technology that requires complementary or confirming techniques to develop an understanding of events; (2) in general the technology is expensive, and in going from "homemade arrays" that are arrayed at one's institutional microarray core to commercial arrays, the price increases; what adds to this cost is the need to have a number of biological repeats of the arrays to reach statistical significance; (3) the purpose of the array work: is it for gene discovery where one would want to work with an array having thousands of genes or is it to probe a defined pathway where making a smaller array would be more efficient and less costly; (4) for zebrafish in particular, the genome is still being assembled and many genes have yet to be defined and localized; some of the large commercial arrays are based not just on the whole gene sequence but on expressed sequence tags (ESTs) that can result in having several oligonucleotides on the array representing the same gene as the genome gets assembled; in our experience the Agilent 22k zebrafish array has more than several "genes" represented by two and sometimes three oligonucleotides; (5) because the assembly of the genome is still a work in progress, some of the arrays have little or no annotation for some of the oligonucleotides placed upon the array; this is improving with finer definition of the zebrafish genome.

Given all of the above, there are a variety of reagents available for zebrafish microarray work. One option is a Compugen library of ~16K oligonucleotides that can be purchased and arrayed. We have used this to study gene expression changes during the development of the neural tube[57] and to study transcript changes during early embryonic development.[58] However, in both cases the number of oligonucleotides that gave statistically significant data was proportionally small. This has caused us to use commercial microarrays whose oligonucleotides were based upon a newer assembly of the zebrafish genome. Both Affymetrix and Agilent produce larger zebrafish arrays (14.5k and 22k, respectively). While arrays of this kind provide a level of consistency that is usually not matched by "homemade" arrays, the flexibility of having oligonucleotides designed and synthesized for the genes of choice is a valuable factor when considering what to use. While we are developing a large (several hundred) Agilent 22k zebrafish array database, we find that the array is missing many neurotransmitter pathway genes, in part because many of these are at best algorithm predicted rather than defined by isolated cDNAs. That is why we are developing our own, much smaller array that will focus upon genes selectively expressed in neurons and genes associated with known neurotransmitter pathways.

Therefore, our purpose in this array will not be gene discovery but pathway exploration. In addition to microarrays for studying the steady-state level of transcripts, there are other microarrays for studying aspects of gene regulation beyond direct transcriptional control. There are now commercial designs of microarrays for using chromatin immunoprecipitation on microarrays (chIP-chip arrays) for zebrafish. These are oligonucleotide arrays that cover sequences near and 5 to mRNA start sites so that it is possible to examine which transcription factors are binding to which genes. The Agilent arrays cover over 11,000 zebrafish genes. Alternatively, it has become clear that microRNAs can play significant posttranscriptional roles in governing gene expression—either through hybridizing to mRNAs for their destruction or by inhibiting the translation of the mRNAs. In the second category, the homology between the microRNA and the mRNA does not have to be absolute, so some microRNAs can control the translation of many different mRNAs. Microarrays for examining the presence of microRNA have been constructed by researchers[59,60] and commercially.

In our use of the Agilent 22k zebrafish array, we have identified distinct changes in gene expression associated with a brief embryonic challenge to the glycine neurotransmitter pathway; this same brief embryonic challenge has been shown to have lasting behavioral effects on adults. The analysis of the data is complex, but we do see transcript level changes in genes associated with the glycine neurotransmitter pathway and the glutamate neurotransmitter pathway. However, as mentioned above, the coverage of genes in these pathways is not complete and is the major reason for our developing a new, smaller microarray.

For challenging the embryonic nervous system, these arrays can be used in combination with the challenges, i.e., comparing the transcript patterns of embryos and larvae with and without challenge. A chemical challenge via a pesticide exposure or a receptor agonist or antagonist makes it possible to limit the time of exposure; alternatively, but more specifically, injecting antisense morpholinos into one-cell embryos affords the possibility of knocking down expression of specific genes. The drawback to the morpholino approach is the lack of careful control over the time period by which the morpholino inhibits the synthesis of the target protein. We have and are using both of these approaches in our work.

Fluorescent Cell Sorting One of the benefits of the development by many laboratories of live, fluorescent reporter, zebrafish embryos is that it allows for a variety of different forms of visualization: fluorescent microscopy, confocal microscopy, multiphoton microscopy, and also the use of flow cytometers and fluorescent cell sorters to analyze populations. Some of the first fluorescent transgenic lines labeled blood cells and this made it possible to isolate fluorescent blood cells for transplantation into mutants defective in the hematopoietic system for subsequent real-time visualization of the reconstitution of a functional system.[61] As the number of reporter, fluorescent lines of zebrafish increases, each marking distinct subpopulations of cells in the embryo, it becomes possible to use embryo dissociation techniques in combination with fluorescent cell sorting to recover distinct subpopulations of cells. Dr. Ava Udvadia of the University of Wisconsin, Milwaukee has succeeding in doing this with a transgenic line driven by the gap43 promoter. This line marks embryonic neurons and she has successfully sorted fluorescent neurons and used them as a source of RNA to isolate cDNA clones of genes expressed in the neurons. We have recently collaborated with her to examine the expression of genes in these neurons using the Agilent 22k zebrafish microarray. We expect this procedure to be of value in the future in identifying transcript differences in different populations of neurons as well as a convenient alternative to laser microdissection techniques for examining gene expression in specific tissues of the embryo.

STRATEGIES FOR EXPLORING MECHANISMS We are suggesting that early challenge of the nervous system creates structural or regulatory effects that last beyond the exposure period and result in behavioral/learning effects in the adult. Obviously, explaining this mechanistically for something as complex as the brain is an immense goal. There are some possible clues regarding events, if not mechanisms, that might be affecting these changes. The work of the Spitzer laboratory[38] and some preliminary microarray work by this laboratory support the possibility of neurotransmitter switching. This could be tested in the organism with the transgenic lines we are generating to mark neurons producing specific neurotransmitters. Complementing this is the use of microarray analysis during the challenges to determine whether there are distinct changes in neurotransmitter pathway component expression. Introducing regions into the fluorescent reporters that allow them to be embedded into the membrane allows the outward morphology of the neurons to be delineated and we hope these lines will allow us to follow the development of the nervous system as the organism grows. Obviously, any challenge requires evidence that it does induce a behavioral change, and the definition of the vulnerable window of challenge should make it possible to narrow down the candidate changes. Both of these require analysis procedures for behavior and learning. Our progress with this is described below.

ADDRESSING BEHAVIORAL CHANGES

Swimming behavior of the larval zebrafish can be monitored with high-speed video and digital imaging software to study changes in burst and slow swims, tail bending amplitude, and fast, large-angle turns indicative of the escape response.[62] For example, video recording was used to differentiate wild-type and *accordion* mutant embryos as well as embryos chemically challenged with strychnine, to assess defects in touch response.[30,45,63] Zebrafish prey capture swimming behavior has also been analyzed by this method.[64] As the embryo matures, the swimming behavior can be monitored with more complex computerized video-tracking systems, e.g., Noldus-EthoVision, to track swimming and movement. Using the Noldus system, single zebrafish can be placed in multiwell plates and monitored simultaneously, greatly increasing the throughput of experiments. Once the fish are tracked, the software allows the investigator to dissect out various parameters of swimming movement including angular velocity (degrees/sec), meander (degrees/mm), time spent in defined zones, total distance moved, and time spent immobile, mobile, and strongly mobile. Analyzing these various parameters makes it possible to detect small changes in motor behavior in response to challenge. Assessment of learning and memory in adult zebrafish is still being developed, but researchers have devised ways to characterize simple learning, memory, and cognition by testing reflexes,[65–67] fear conditioning,[68] and spatial orientation.[69,70] For example, zebrafish can learn to swim to one side of the tank in response to a tap on the side that signals delivery of food.[71] In addition, a three-chambered shuttle maze can be used to test choice behavior and response latency. In this tank there is a central compartment and left and right choice compartments separated by sliding dividers. Zebrafish are trained to preferentially go to one side by placing the fish in the central compartment and allowing it to acclimate. After a period of acclimation, both side chambers are simultaneously opened. If the fish chooses the correct side chamber, the door closes and the fish is left alone. If the fish chooses the incorrect side chamber, the door is closed and a sliding partition is moved close to the wall placing the fish in a restrictive position. Within a few trials, the fish learn to avoid the chamber that will push them toward the wall.[70] Once trained, it is possible to measure the chamber choice accuracy and the time it takes for the fish to make a decision. Using this technique, it was demonstrated that early embryonic exposure to chlorpyrifos led to significant spatial discrimination impairments and response latency as adults.[36] Furthermore, utilizing this task, it was demonstrated that acute nicotine administration to adults causes a significant improvement in delayed spatial alternation at low nicotine doses, but impairs performance at high doses,[72] mimicking results seen in higher vertebrates.[70]

Recently, a novel study utilizing adult zebrafish to test diving and escape predation was performed.[73] When zebrafish are placed in a novel environment, they tend to dive to the bottom of the tank as a method to escape predation. Wild-type fish show a gradual decrease in the time spent in the bottom third of the tank and habituate to their new environment. Utilizing this test, Levin *et al.*[73] have demonstrated that acute nicotine treatment caused a significant decrease in diving throughout the test. The antianxiolitic effects of nicotine treatment were reversed by cotreatment with mecamylamine, a nicotinic receptor antagonist. These data suggest that the effect of nicotine on diving was due to net stimulation at nicotinic receptors, an effect that is blocked by mecamylamine.

Utilizing a variety of testing apparatuses and methodologies, researchers have demonstrated the use of adult zebrafish as a behavioral model for testing reflexes, habituation, Pavlovian conditioning, and operant conditioning. A review of the experiments and a full description of testing apparatuses can be found in Cerutti and Levin.[70] As behavioral biologists and psychologists continue to refine and create new testing parameters, we hope to elucidate more behavioral abnormalities and learning and memory deficits as a result of early challenge to the nervous system.

CONCLUSIONS

We have tried to outline how developing technologies and useful characteristics of the zebrafish model system can now allow us to investigate how embryonic exposures to the developing nervous system and concomitant adult effects can be studied. This combination of approaches and technologies will ultimately be used at various levels to study molecular, developmental, and behavioral effects relevant to the functions of the nervous system.

REFERENCES

1. Koger SM, Schettler T, Weiss B. Environmental toxicants and developmental disabilities: A challenge for psychologists. *Am Psychol* 2005;60(3):243–255.
2. Safer DJ, Zito JM, Fine EM. Increased methylphenidate usage for attention deficit disorder in the 1990s. *Pediatrics* 1996;98(6, Pt. 1):1084–1088.
3. Gillberg C, Wing L. Autism: Not an extremely rare disorder. *Acta Psychiatr Scand* 1999;99(6):399–406.
4. Muir T, Zegarac M. Societal costs of exposure to toxic substances: Economic and health costs of four case studies that are candidates for environmental causation. *Environ Health Perspect* 2001;109(Suppl. 6):885–903.
5. Schettler T. Toxic threats to neurologic development of children. *Environ Health Perspect* 2001;109(Suppl. 6):813–816.
6. Plomin R, Owen MJ, McGuffin P. The genetic basis of complex human behaviors. *Science* 1994;264(5166):1733–1739.
7. Ostrea EM, Morales V, Ngoumgna E, *et al.* Prevalence of fetal exposure to environmental toxins as determined by meconium analysis. *Neurotoxicology* 2002;23(3):329–339.
8. Bellinger DC, Stiles KM, Needleman HL. Low-level lead exposure, intelligence and academic achievement: A long-term follow-up study. *Pediatrics* 1992;90(6):855–861.
9. Lanphear BP, Dietrich K, Auinger P, Cox C. Cognitive deficits associated with blood lead concentrations <10 microg/dL in US children and adolescents. *Public Health Rep* 2000;115(6):521–529.
10. Needleman H. Lead poisoning. *Annu Rev Med* 2004;55:209–222.
11. Mendola P, Selevan SG, Gutter S, Rice D. Environmental factors associated with a spectrum of neurodevelopmental deficits. *Ment Retard Dev Disabil Res Rev* 2002;8(3):188–197.
12. Slotkin TA, Levin ED, Seidler FJ. Comparative developmental neurotoxicity of organophosphate insecticides: Effects on brain development are separable from systemic toxicity. *Environ Health Perspect* 2006;114(5):746–751.
13. Slotkin TA. Guidelines for developmental neurotoxicity and their impact on organophosphate pesticides: A personal view from an academic perspective. *Neurotoxicology* 2004;25(4):631–640.
14. Stein J, Schettler T, Wallinga D, Valenti M. In harm's way: Toxic threats to child development. *J Dev Behav Pediatr* 2002;23(1Suppl.):S13–22.
15. Welch-Carre E. The neurodevelopmental consequences of prenatal alcohol exposure. *Adv Neonatal Care* 2005;5(4):217–229.
16. Mattson SN, Schoenfeld AM, Riley EP. Teratogenic effects of alcohol on brain and behavior. *Alcohol Res Health* 2001;25(3):185–191.
17. Slikker W Jr, Xu ZA, Levin ED, Slotkin TA. Mode of action: Disruption of brain cell replication, second messenger, and neurotransmitter systems during development leading to cognitive dysfunction—developmental neurotoxicity of nicotine. *Crit Rev Toxicol* 2005; 35(8–9):703–711.
18. Eskenazi B, Castorina R. Association of prenatal maternal or postnatal child environmental tobacco smoke exposure and neurodevelopmental and behavioral problems in children. *Environ Health Perspect* 1999;107(12):991–1000.

19. Ferriero DM, Dempsey DA. Impact of addictive and harmful substances on fetal brain development. *Curr Opin Neurol* 1999; 12(2):161–166.
20. Ward AC, Lieschke GJ. The zebrafish as a model system for human disease. *Front Biosci* 2002;7:d827–833.
21. Bernhardt RR, Chitnis AB, Lindamer L, Kuwada JY. Identification of spinal neurons in the embryonic and larval zebrafish. *J Comp Neurol* 1990;302(3):603–616.
22. Bernhardt RR, Patel CK, Wilson SW, Kuwada JY. Axonal trajectories and distribution of GABAergic spinal neurons in wildtype and mutant zebrafish lacking floor plate cells. *J Comp Neurol* 1992;326(2): 263–272.
23. Ritter DA, Bhatt DH, Fetcho JR. In vivo imaging of zebrafish reveals differences in the spinal networks for escape and swimming movements. *J Neurosci* 2001;21(22):8956–8965.
24. Gahtan E, O'Malley DM. Rapid lesioning of large numbers of identified vertebrate neurons: Applications in zebrafish. *J Neurosci Methods* 2001;108(1):97–110.
25. Fetcho JR, O'Malley DM. Visualization of active neural circuitry in the spinal cord of intact zebrafish. *J Neurophysiol* 1995;73(1): 399–406.
26. Higashijima S, Mandel G, Fetcho JR. Distribution of prospective glutamatergic, glycinergic, and GABAergic neurons in embryonic and larval zebrafish. *J Comp Neurol* 2004;480(1):1–18.
27. Drapeau P, Saint-Amant L, Buss RR, Chong M, McDearmid JR, Brustein E. Development of the locomotor network in zebrafish. *Prog Neurobiol* 2002;68(2):85–111.
28. Brustein E, Saint-Amant L, Buss RR, Chong M, McDearmid JR, Drapeau P. Steps during the development of the zebrafish locomotor network. *J Physiol Paris* 2003;97(1):77–86.
29. Lewis KE, Eisen JS. From cells to circuits: Development of the zebrafish spinal cord. *Prog Neurobiol* 2003;69(6):419–449.
30. Saint-Amant L, Drapeau P. Synchronization of an embryonic network of identified spinal interneurons solely by electrical coupling. *Neuron* 2001;31(6):1035–1046.
31. Saint-Amant L, Drapeau P. Time course of the development of motor behaviors in the zebrafish embryo. *J Neurobiol* 1998;37(4):622–632.
32. Buss RR, Drapeau P. Synaptic drive to motoneurons during fictive swimming in the developing zebrafish. *J Neurophysiol* 2001;86(1): 197–210.
33. Behra M, Cousin X, Bertrand C, *et al.* Acetylcholinesterase is required for neuronal and muscular development in the zebrafish embryo. *Nat Neurosci* 2002;5(2):111–118.
34. Downes GB, Granato M. Acetylcholinesterase function is dispensable for sensory neurite growth but is critical for neuromuscular synapse stability. *Dev Biol* 2004;270(1):232–245.
35. Linney E, Upchurch L, Donerly S. Zebrafish as a neurotoxicological model. *Neurotoxicol Teratol* 2004;26(6):709–718.
36. Levin ED, Chrysanthis E, Yacisin K, Linney E. Chlorpyrifos exposure of developing zebrafish: Effects on survival and long-term effects on response latency and spatial discrimination. *Neurotoxicol Teratol* 2003;25(1):51–57.
37. Jacob F, Perrin D, Sanchez C, Monod J. [Operon: A group of genes with the expression coordinated by an operator.] *CR Hebd Seances Acad Sci* 1960;250:1727–1729.
38. Borodinsky LN, Root CM, Cronin JA, Sann SB, Gu X, Spitzer NC. Activity-dependent homeostatic specification of transmitter expression in embryonic neurons. *Nature* 2004;429(6991):523–530.
39. Zebrafish Issue. *Development* 1996;123.
40. Cui WW, Low SE, Hirata H, *et al.* The zebrafish shocked gene encodes a glycine transporter and is essential for the function of early neural circuits in the CNS. *J Neurosci* 2005;25(28):6610–6620.
41. Hirata H, Saint-Amant L, Downes GB, *et al.* Zebrafish bandoneon mutants display behavioral defects due to a mutation in the glycine receptor beta-subunit. *Proc Natl Acad Sci USA* 2005;102(23):8345–8350.
42. Nasevicius A, Ekker SC. Effective targeted gene "knockdown" in zebrafish. *Nat Genet* 2000;26(2):216–220.
43. McLean DL, Fetcho JR. Ontogeny and innervation patterns of dopaminergic, noradrenergic, and serotonergic neurons in larval zebrafish. *J Comp Neurol* 2004;480(1):38–56.
44. Drapeau P, Ali DW, Buss RR, Saint-Amant L. In vivo recording from identifiable neurons of the locomotor network in the developing zebrafish. *J Neurosci Methods* 1999;88(1):1–13.
45. Hirata H, Saint-Amant L, Waterbury J, *et al.* Accordion, a zebrafish behavioral mutant, has a muscle relaxation defect due to a mutation in the ATPase Ca2+ pump SERCA1. *Development* 2004;131(21): 5457–5468.
46. Li J, Mack JA, Souren M, *et al.* Early development of functional spatial maps in the zebrafish olfactory bulb. *J Neurosci* 2005;25(24): 5784–5795.
47. Balkan W, Colbert M, Bock C, Linney E. Transgenic indicator mice for studying activated retinoic acid receptors during development. *Proc Natl Acad Sci USA* 1992;89(8):3347–3351.
48. Perz-Edwards A, Hardison NL, Linney E. Retinoic acid-mediated gene expression in transgenic reporter zebrafish. *Dev Biol* 2001; 229(1):89–101.
49. Shaner NC, Campbell RE, Steinbach PA, Giepmans BN, Palmer AE, Tsien RY. Improved monomeric red, orange and yellow fluorescent proteins derived from Discosoma sp. red fluorescent protein. *Nat Biotechnol* 2004;22(12):1567–1572.
50. Linney E, Udvadia AJ. Construction and detection of fluorescent, germline transgenic zebrafish. *Methods Mol Biol* 2004;254:271–288.
51. Warming S, Costantino N, Court DL, Jenkins NA, Copeland NG. Simple and highly efficient BAC recombineering using galK selection. *Nucleic Acids Res* 2005;33(4):e36.
52. Misulovin Z, Yang XW, Yu W, Heintz N, Meffre E. A rapid method for targeted modification and screening of recombinant bacterial artificial chromosome. *J Immunol Methods* 2001;257(1–2):99–105.
53. Jessen JR, Willett CE, Lin S. Artificial chromosome transgenesis reveals long-distance negative regulation of rag1 in zebrafish. *Nat Genet* 1999;23(1):15–16.
54. Kucenas S, Soto F, Cox JA, Voigt MM. Selective labeling of central and peripheral sensory neurons in the developing zebrafish using P2X(3) receptor subunit transgenes. *Neuroscience* 2006;138(2):641–652.
55. Kawakami K, Shima A, Kawakami N. Identification of a functional transposase of the Tol2 element, an Ac-like element from the Japanese medaka fish, and its transposition in the zebrafish germ lineage. *Proc Natl Acad Sci USA* 2000;97(21):11403–11408.
56. Parinov S, Kondrichin I, Korzh V, Emelyanov A. Tol2 transposon-mediated enhancer trap to identify developmentally regulated zebrafish genes in vivo. *Dev Dyn* 2004;231(2):449–459.
57. Linney E, Dobbs-McAuliffe B, Sajadi H, Malek RL. Microarray gene expression profiling during the segmentation phase of zebrafish development. *Comp Biochem Physiol C Toxicol Pharmacol* 2004; 138(3):351–362.
58. Mathavan S, Lee SG, Mak A, *et al.* Transcriptome analysis of zebrafish embryogenesis using microarrays. *PLoS Genet* 2005;1(2): e29.
59. Miska EA, Alvarez-Saavedra E, Townsend M, *et al.* Microarray analysis of microRNA expression in the developing mammalian brain. *Genome Biol* 2004;5(9):R68.
60. Thomson JM, Parker J, Perou CM, Hammond SM. A custom microarray platform for analysis of microRNA gene expression. *Nat Methods* 2004;1(1):47–53.
61. Traver D, Paw BH, Poss KD, Penberthy WT, Lin S, Zon LI. Transplantation and in vivo imaging of multilineage engraftment in zebrafish bloodless mutants. *Nat Immunol* 2003;4(12):1238–1246.
62. Budick SA, O'Malley DM. Locomotor repertoire of the larval zebrafish: Swimming, turning and prey capture. *J Exp Biol* 2000;203(Pt. 17):2565–2579.
63. Granato M, van Eeden FJ, Schach U, *et al.* Genes controlling and mediating locomotion behavior of the zebrafish embryo and larva. *Development* 1996;123:399–413.

64. Borla MA, Palecek B, Budick S, O'Malley DM. Prey capture by larval zebrafish: Evidence for fine axial motor control. *Brain Behav Evol* 2002;60(4):207–229.
65. Carvan MJ, 3rd, Loucks E, Weber DN, Williams FE. Ethanol effects on the developing zebrafish: Neurobehavior and skeletal morphogenesis. *Neurotoxicol Teratol* 2004;26(6):757–768.
66. Loucks E, Carvan MJ 3rd. Strain-dependent effects of developmental ethanol exposure in zebrafish. *Neurotoxicol Teratol* 2004;26(6): 745–755.
67. Reimers MJ, Flockton AR, Tanguay RL. Ethanol- and acetaldehyde-mediated developmental toxicity in zebrafish. *Neurotoxicol Teratol* 2004;26(6):769–781.
68. Pradel G, Schmidt R, Schachner M. Involvement of L1.1 in memory consolidation after active avoidance conditioning in zebrafish. *J Neurobiol* 2000;43(4):389–403.
69. Arthur D, Levin ED. Spatial and non-spatial discrimination learning in zebrafish. *Anim Cogn* 2001;4:125–131.
70. Cerutti DT, Levin ED. *Cognitive Impairment Models Using Complementary Species*. Boca Raton, FL: CRS Press, 2006.
71. Williams FE, Messer WS. Memory function and muscarinic receptors in zebrafish. *Soc Neurosci Abstr* 1998;24:182.
72. Levin ED, Chen E. Nicotinic involvement in memory function in zebrafish. *Neurotoxicol Teratol* 2004;26(6):731–735.
73. Levin ED, Bencan Z, Cerutti, DT. Anxiolytic effects of nicotine in zebrafish. *Physiol Behav* 2007;90(1):54–58.

15 Modeling Cognitive and Neurodegenerative Disorders in *Drosophila melanogaster*

JUNJIRO HORIUCHI AND MINORU SAITOE

ABSTRACT

Significant conservation in cognitive and neurodevelopmental mechanisms exist between *Drosophila* and mammals. In addition, while mechanisms of aging are relatively well conserved, *Drosophila* age much faster than mammals. Thus *Drosophila* are an ideal organism in which to study cognitive and neurodegenerative diseases. Powerful genetic techniques exist that allow genes to be disrupted and overexpressed easily. Phenotypes observed from these mutants can then be used to screen for genetic modifiers to elucidate pathways involved in disease pathology. Here we describe recent advances obtained from *Drosophila* models for five diseases, neurofibromatosis 1, fragile X syndrome, Alzheimer's disease, tauopathies, and Parkinson's disease.

Key Words: Neurofibromatosis 1, Fragile X, Alzheimer's disease, Tauopathy, Parkinson's disease, *Drosophila melanogaster*.

INTRODUCTION

In choosing an organism for the study of a biological process, it is necessary to compare the relative advantages of using that particular organism versus the amount of conservation in the process between the organism in question and humans. *Drosophila melanogaster* has been used extensively as a model system in the study of development. *Drosophila* offers numerous advantages including powerful genetic techniques and a short life span. In this chapter, we will describe how *Drosophila* has been used in the study of several human cognitive and neurodegenerative diseases.

CONSERVATION OF PROCESSES BETWEEN HUMANS AND *DROSOPHILA*

How conserved are cognitive, behavioral, and neuronal processes between flies and mammals? *Drosophila* has been an essential organism in the study of behavior and memory. Two widely used, highly quantitative and reproducible behavioral assays, a Pavlovian olfactory association assay[1] and a courtship conditioning assay,[2] have been used to identify mechanisms of memory formation, storage, and recall. These mechanisms are conserved with those of mammals. Using the Pavlovian conditioning assay, a single cycle of training generates a short form of memory in flies that decays relatively quickly.[3] Similar to mammalian systems, multiple training trials generate longer lasting forms of memory and multiple training trials with rest intervals interspersed between trainings generate a protein-synthesis-dependent long-term memory.[4] In addition, the cyclic AMP/protein kinase A (cAMP/PKA) pathway is critical for memory formation in *Drosophila* and this is also the case in mammalian systems. Furthermore, activity of the CREB transcription factor has been shown to be essential for long-term memory formation in many organisms including *Drosophila* and mammals.[5]

For the study of synaptic morphology and neurotransmission, the *Drosophila* neuromuscular junction (NMJ) has been extensively used. NMJ preparations allow rapid screening for mutants that affect synaptic morphology.[6] Combined with temperature-sensitive paralytic mutants, they provide a critical tool for the identification of genes important in synaptic transmission.[7,8] Genes required for synapse formation and transmission are highly conserved between flies and mammals and genes involved in human diseases that affect synaptic structure or function often have similar functions at the *Drosophila* NMJ.[9–12]

An advantage that *Drosophila* has over mammals in the study of neurodegenerative diseases is a short life span of approximately 30–60 days after eclosion. This allows age-dependent changes that occur in *Drosophila* to be easily measured. Furthermore, *Drosophila* and mammals show much conservation in terms of mechanisms affecting aging.[13] Calorie restriction (CR) extends the life span of organisms ranging from yeast to mice.[14,15] CR seems to function by inactivating pathways that promote growth while activating pathways that increase protection from oxidative and other forms of stress. Thus, in many organisms including flies and mice, inhibition of the insulin/insulin-like-growth factor signaling (IIS) pathway extends the life span.[16] Fly homologs exist for the majority of genes involved in insulin signaling.

From: *Sourcebook of Models for Biomedical Research*
(P. M. Conn, ed.), © 2008 Humana Press Inc., Totowa, NJ.

ADVANTAGES OF STUDYING *DROSOPHILA*

A short generation time of approximately 10 days and the ease of maintaining large numbers of fly stocks in a laboratory setting have been invaluable in establishing *Drosophila* as a model genetic organism. *Drosophila* has a relatively small genome that has been completely sequenced and approximately 75% of the genes involved in human diseases have a counterpart in *Drosophila*.[17] Genetic techniques in *Drosophila* are highly developed with several being particularly advantageous in the study of disease. The commonly used GAL4/UAS system allows expression of a gene of interest in specific anatomical regions.[18] Briefly, a transgenic fly is generated expressing a gene of interest under control of the GAL4-UAS. Thus, the gene is expressed only in the presence of the heterologous yeast GAL4 activator. These flies can be crossed with various GAL4-enhancer trap lines, which express GAL4 in particular locations or cell types, to express the transgene at those specific regions. Newer versions of the GAL4/UAS system have been generated in which GAL4 has been mutated to be active only in the presence or absence of a drug, RU486. This allows gene expression to be regulated in both a spatial and a temporal manner.[19,20]

In forward genetic screens, enhancer-promoter (EP) lines are available where P-element transposons, carrying GAL4-regulated promoters, designed to express flanking genomic sequences, have been inserted throughout the genome.[21] If an EP element is inserted proximal to a gene in the appropriate orientation, that gene can often be expressed under GAL4 control. If the element is inserted in the opposite orientation, gene expression can often be inactivated. EP lines are particularly useful in genetic modifier screens when searching for genes whose altered expression will modify the phenotype of a particular mutant.

A particularly useful organ for genetic screening in flies is the compound eye.[22] GAL4 enhancer trap lines are available that specifically drive expression in the eye and expression of disease genes in this structure often causes a rough eye phenotype. Forward genetic screens using EP lines or random mutagenesis can be used to isolate genetic modifiers, either suppressors or enhancers, of this phenotype. In addition, reverse genetic techniques can be used to modify expression of specific candidate interacting genes to determine whether they have an effect on the rough eye phenotype. These techniques are used extensively to identify genetic interactions between genes of interest.

Finally, for pharmacological studies, similar to other organisms, flies can be injected with drugs. However, more usefully, flies are much more sensitive to ingested compounds compared to mammals. Thus, in contrast to mammals, in which active compounds may have to be injected directly into the brain, in flies, the compound can often be simply mixed with the food and fed to flies to produce effects.[23]

Taken together, we can conclude that *Drosophila* offers numerous advantages in the study of cognitive and age-related human diseases.

NEUROFIBROMATOSIS I

Neurofibromatosis type I (NF1) is a dominant genetic disorder caused by mutations in the NF1 gene, which encodes a Ras-specific GTPase-activating protein (Ras-GAP).[24] Symptoms include tumors of the nervous system, learning disabilities, defects in development, and a predisposition toward short stature.[25,26] Tumors result from the removal of NF1, which normally represses the growth-promoting functions of the Ras proteins. Until recently, it was also thought that the learning defects associated with the disease resulted from aberrant brain development due to abnormal Ras activity. However, results from fly models suggest an alternate function for NF1 in regulating intracellular cAMP levels which affect learning (Figure 15–1A).

The *Drosophila* homolog of NF1 is 60% identical to the human protein and viable NF1 null mutants have been generated.[25] Similar to phenotypes seen in human NF1 patients, NF1 null flies have learning defects and are physically smaller than their wild-type counterparts. While an increase in tumor incidence has not been reported in the fly, fly NF1 has ras-GAP activity similar to the human protein.[25]

Indications that *Drosophila* NF1 may be involved in regulating the cAMP pathway came from studies of the *Drosophila* NMJ in which the addition of the neuropeptide, pituitary adenylyl cyclase-activating polypeptide (PACAP), induces a strong enhancement in K^+ currents.[27] This enhancement is dependent on activation of both an adenylyl cyclase/cAMP/PKA pathway and a Ras-Raf kinase pathway. Interestingly, null mutants in NF1 are defective for this K^+ current enhancement.[28] However, neither mutations in GAP1, another Ras-specific GAP, nor overexpression of constitutively active Ras, blocked enhancement. Rather, a *rutabaga* (*rut*) mutant defective for an adenylyl cyclase had phenotypes similar to the NF1 null and cAMP analogs and adenylyl cyclase activators restored K^+ currents in NF1 mutants.[28] These results indicate that NF1 may function in a common pathway with *rut* to increase cAMP levels.

cAMP is known to be critical for learning and the *rut* mutant, which is defective for cAMP production, was initially identified as a learning mutant. This raises the possibility that the memory defect in NF1 mutants could be a result of defective cAMP rather than Ras signaling. In support of this idea, the learning defects of NF1 mutants are extremely similar to those of *rut* mutants and double *NF1;rut* mutants do not have phenotypes more severe than single mutants, demonstrating that both mutations affect the same pathway.[29] Furthermore, heat shock induction of a constitutively active PKA catalytic subunit complements the learning defects of NF1 mutants.

Interestingly, while acute expression of PKA seems to complement the learning defects of NF1 flies, developmental expression is required to complement the size defects. These results suggest three roles for NF1: an acute role in maintaining cAMP levels for memory formation, a developmental role in generating proper body size, and a Ras-GAP function in preventing tumor formation.

The role of NF1 in modulating K^+ currents[30] and in regulating cAMP levels[11,31] has been shown to be conserved in mammalian systems. If this conservation extends further, the learning defects of NF1 patients are likely to result from signaling rather than developmental defects, suggesting the hopeful idea that acute pharmacological interventions that increase cAMP signaling might be used to treat patients.

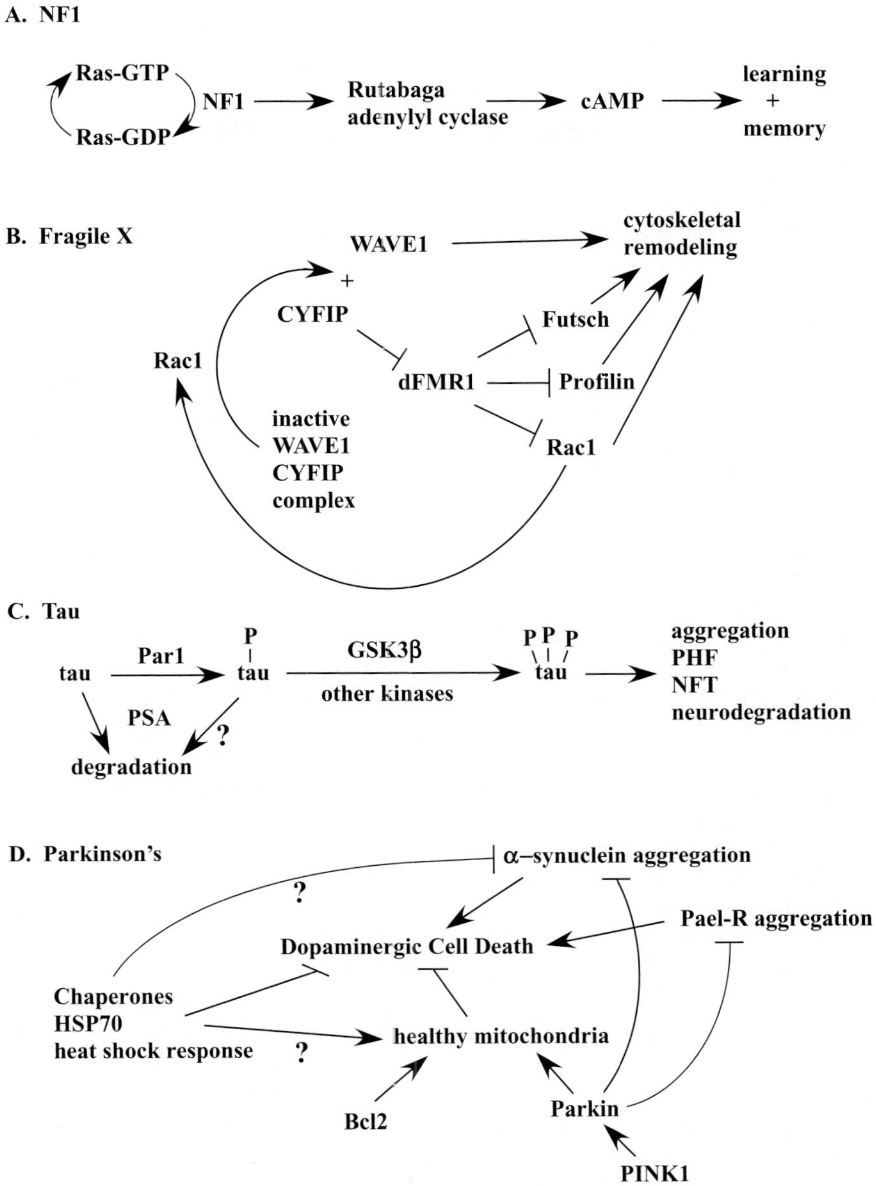

Figure 15–1. Selected disease pathways in which *Drosophila* models have contributed significantly (see text for references). (A) Although NF1 is well characterized as a Ras-GAP, *Drosophila* work has demonstrated that the learning defects associated with neurofibromatosis 1 may be due to a second role of NF1 in regulating cAMP levels through the rutabaga adenylyl cyclase. (B) Rac-GTPases control cytoskeletal components to regulate cell shape. Activation of Rac1 causes dissociation of an inactive complex containing WAVE1 and CYFIP as well as other proteins. Free WAVE1 stimulates cytoskeletal remodeling through actin nucleation while CYFIP may function by inhibiting FMR1. Inhibition of FMR1 allows excess futsch and profilin to be synthesized, stimulating microtubule and actin dynamics. In addition, Rac1 production is increased by inhibition of FMR1, suggesting a positive feedback mechanism. Fragile X syndrome, caused by loss of FMR1, may exert its effects through the increased activity of cytoskeletal remodeling components. (C) In tauopathies, increased accumulation and phosphorylation of tau lead to its misfolding and aggregation. Aggregated tau forms PHFs and NFTs leading to neurodegeneration. Phosphorylation follows an ordered sequence in which initial phosphorylation by the Par1 kinase primes tau for phosphorylation by GSK3β and other kinases. Increased expression of the peptidase PSA may be a way to treat tauopathies since it leads to degradation of tau. (D) Parkinson's disease results from progressive loss of dopaminergic neurons. Pathological phenotypes indicate that aggregation of proteins such as α-synuclein and Pael-R may be involved in cell death. Parkin, a ubiquitin E3 ligase, identified from familial forms of the disease, may be involved in preventing protein aggregation. In addition, parkin and PINK1 are required for normal mitochondrial function, suggesting that Parkinson's disease may result from mitochondrial damage. Increased activity of chaperones, heat shock response, and Bcl2 ameliorate toxicity in several *Drosophila* models of Parkinson's disease.

FRAGILE X SYNDROME

Fragile X syndrome, the most common inherited form of mental retardation, results from loss-of-function mutations in the FRM1 gene.[32] In addition to cognitive defects, symptoms include attention deficit disorders and hyperactivity, obsessive compulsive and autistic behaviors, and mild facial dysmorphologies. At the cellular level, loss of FMR1 leads to cytoskeletal defects, including increases in the density of synaptic spines, which tend to be longer, thinner and immature in appearance.[33,34] FMR1 encodes FMRP, a protein that is thought to function by binding mRNAs and repressing translation.[35,36] A fly homolog of FMR1, dFMR1, exists and mutations have been used to generate fly models of this disease.[37]

Although many potential RNA-binding targets have been identified in mammalian systems,[38,39] studies in *Drosophila* have yielded the most convincing data regarding the *in vivo* relevance of identified targets (Figure 15–1B). Similar to mammals, null mutations in dFMR1 cause altered cytoskeletal morphology at both NMJs and at central nervous system synapses. Reducing dFMR1 levels results in increased synaptic branching, increased numbers of boutons, and enlarged synaptic areas, while overexpression of dFMR1 reduces branching and growth and increases synaptic bouton size.[10,12] DFMR1 binds to the mRNAs of three proteins implicated in cytoskeletal structure, futsch, the *Drosophila* homolog of the mammalian microtubule-associated protein MAP1B,[10] profilin, a protein that stimulates both polymerization and depolymerization of filamentous actin,[40] and Rac1, a small rho-GTPase involved in cytoskeletal remodeling.[41] Both futsch and profilin protein levels are upregulated upon deletion of dFMR1 and futsch expression decreases when dFMR1 is overexpressed.[10,40] Furthermore, overexpression of futsch, profilin, or Rac1 causes increased growth and branching of neurites, similar to phenotypes of dFMR1 null mutants.[10,40,41] Reduction of futsch or Rac1 causes reduced branching and growth, similar to phenotypes observed when dFMR1 is overexpressed. Importantly, futsch, profilin, and Rac1 all show genetic interactions with dFMR1. *dFMR1;futsch* double mutants have morphologically normal synapses,[10] overexpressing dFMR1 suppresses the phenotypes caused by overexpressing profiling,[40] and overexpression of Rac1 suppresses the reduced branching phenotype caused by overexpressed dFMR1.[41] These data all support a model in which dFMR1 affects the synaptic structure by binding to the mRNAs and repressing expression of genes involved in cytoskeletal dynamics.

Interestingly, Rac1 may also function upstream of dFMR1. The cytoplasmic FMRP interacting proteins (CYFIP) 1 and 2 were initially identified in a two-hybrid screen for proteins that bind to human FMRP.[42] CYFIP1 also binds to Rac1 protein, indicating that FMRP is found in complexes with both Rac1 protein and Rac1 mRNA.[43] The genetic consequences of these interactions have been studied in *Drosophila*.[44] CYFIP null mutants are pupal lethal and display contracted or shortened synapses at NMJs. This phenotype is the opposite of the elongated synaptic phenotype seen in dFMR1 mutants. Furthermore, overexpression of dFMR1 in the *Drosophila* eye causes a rough eye phenotype that can be partially rescued by overexpression of CYFIP and exacerbated by reduction of CYFIP.[44] Thus CYFIP seems to work antagonistically to dFMR1. The Rho-GTPase pathways are thought to regulate cytoskeletal remodeling by causing dissociation of an inactive protein complex containing both WAVE1 and CYFIP.[45] Dissociated activated WAVE1 stimulates actin nucleation. The above results indicate that free CYFIP also has a function consisting of binding and inhibiting dFMR1. Inhibition of dFMR1 results in increased translation of proteins including futch, profilin, and Rac1, leading to changes in cytoskeletal structure.

Futsch, profiling, and Rac1 were identified as dFMR1 interactors by biochemical means (i.e., interaction of mRNAs with dFMR1 protein). Another gene, *Drosophila* lethal giant larvae (dlgl), was isolated in a genetic modifier screen as a dominant suppressor of the rough eye phenotype of flies overexpressing dFMR1.[46] Dlgl is a cytoskeleton-associated protein associated with the partitioning/atypical PKC (PAR) complex involved in cellular polarity.[47] Subsequent experiments have demonstrated genetic interactions between dFMR1 and members of the PAR complex,[46] indicating that dFMR1 functions in a variety of different complexes and demonstrating that the genetic approaches in *Drosophila* can be used to identify novel functional partners of dFMRP.

What are the electrophysiological phenotypes disrupting the FMR1 gene? Synaptic activity can cause alterations in connectivity between neurons resulting in long-lasting changes in synaptic strength. The best known of these changes are long-term potentiation (LTP), an activity-dependent strengthening of the synaptic connections, and LTD, an activity-dependent depression of synaptic connectivity.[48] LTP and LTD are not single phenomena, but rather consist of various types that utilize different mechanisms. LTD can be separated into a form that requires activation of postsynaptic *N*-methyl-D-aspartate (NMDA) receptors and one that requires activation of postsynaptic metabolic glutamate receptors (mGluRs).[49] mGluR-dependent LTD requires rapid translation of postsynaptic proteins. In addition, activation of mGluRs also induces translation of FMRP. Thus it seemed likely that mGluR-dependent LTD would be reduced in FMR1 mutants. Surprisingly though, LTD was enhanced in FMR1 knockout mice, suggesting that induction of FMRP functions to modulate LTD by inhibiting excess protein synthesis caused by mGluR activation.[50,51] If this is the case, inhibition of mGluR activity might be expected to suppress the phenotypes of FMR1 mutants.

Evidence supporting the mGluR theory of fragile X has recently been obtained from behavioral studies in *Drosophila*. Male dFMR1 null mutants court virgin females less actively than wild-type males.[52] This decreased courtship can be rescued by feeding flies mGluR inhibitors. Furthermore, when male flies are exposed to and attempt to court previously mated females, they are continually rejected and over time, decrease their attempts at courtship.[2] This is a learned behavior and both learning and memory can be quantified by measuring the number of courtship attempts that a male makes at various time points after courtship conditioning. Learning occurs normally in dFMR1 null flies such that nulls learn to reduce their attempts at courtship upon conditioning, similar to wild-type flies.[53] However, short forms of memory, within 1 h after conditioning, are defective in dFMR1 null flies and they resume courtship attempts much earlier than wild-type flies. Feeding flies mGluR inhibitors during adulthood rescues this memory defect, supporting the idea that increased mGluR activity may be responsible for the cognitive defects associated with fragile X syndrome.[53] It will be interesting to determine whether feeding flies mGluR inhibitors also decreases expression of futch, profilin, and Rac1 in dFMR1 mutants, possibly linking mGluR activity, cytoskeletal remodeling, and behavior. It is particularly hopeful to note that feeding flies mGluR inhibitors during adult stages improves their cognitive activity, suggesting that acute drug treatment may function in human patients as well.

ALZHEIMER'S DISEASE

Alzheimer's disease (AD), the most common cause of senile dementia in the human population, is characterized by two pathological phenotypes: accumulation of extracellular plaques consisting of aggregated amyloid β peptides, and accumulation of aggregated tau protein into intracellular neurofibrillary tangles.[54] The relationship between plaques and tangles is still unclear, but both are associated with progressive neurodegeneration.

β-AMYLOID MODELS Amyloid β (Aβ) peptides are generated by sequential cleavage of the amyloid precursor protein (APP), first by the β-site APP-cleaving enzyme (BACE) and subsequently by the γ-secretase enzyme.[55] In familial forms of Alzheimer's disease (FAD), mutations have been identified in the APP gene, and in the presenilin 1 and presenilin 2 genes, which encode components of the γ-secretase enzyme. Due to heterogeneity of γ-secretase cleavage, at least two peptides, Aβ40 and Aβ42, are generated, of which Aβ42 is considered more toxic.[56] FAD mutations are associated with increased levels of Aβ40 and Aβ42 or with an increased ratio of Aβ42 to total Aβ.

Several *Drosophila* transgenic lines useful for the study of Aβ-induced neurotoxicity have been developed. Lines overexpressing human APP, *Drosophila* presenilin with FAD-associated mutations, and human BACE are available. Using these lines, Greeve *et al.*[57] have shown that expression of APP in the *Drosophila* eye causes an age-dependent neurodegeneration in photoreceptor cells that is enhanced upon coexpression of presenilin and suppressed by reduction of endogenous presenilin. APP overexpression causes an age-dependent appearance of large globular deposits of Aβ, which can be accelerated by coexpression of presenilin. Also, expressing APP, BACE, and presenilin under actin promoter control results in semilethality that can be suppressed by raising flies on food supplemented with either BACE or γ-secretase inhibitors.[57]

Iijima *et al.*[58] used transgenic flies directly expressing secreted forms of Aβ40 and Aβ42. While expression of both peptides caused age-dependent learning defects, expression of Aβ42 but not Aβ40 caused age-dependent neurodegeneration, shortened life span, progressive locomoter defects, and diffuse amyloid deposits. Crowther *et al.*[59] extended these results by demonstrating that feeding Aβ42 flies MK-801 or Congo Red extends their life span and ameliorates neurodegeneration. Congo Red is an azo-dye that binds to Aβ42 *in vitro* and has been shown to inhibit aggregation. MK801 is an inhibitor of the NMDA glutamate receptor. It is unclear how MK801 extends the life span of Aβ42 flies, but it may function similarly to mamantine, a glutamate antagonist that has been shown to slow progression of Alzheimer's disease. Furthermore, Crowther *et al.*[59] demonstrate rough eye phenotypes in Aβ42 flies suggesting a quick screening method for isolating mutations and compounds that affect Aβ-mediated neurodegeneration.

TAUOPATHIES Alterations in the microtubule-associated protein tau have been implicated in a number of neurodegenerative diseases, collectively known as tauopathies, including AD, progressive supranuclear palsy, and frontotemporal dementia.[60] Tauopathies are characterized by the accumulation of abnormally phosphorylated forms of tau that aggregate into paired helical filaments (PHF) and cytoplasmic inclusions known as neurofibrillary tangles (NFTs). Formation of NFTs parallels the progression and anatomical distribution of neuronal degradation in tauopathies, suggesting that tau aggregation plays a role in neurodegeneration in these disorders.[61] A *Drosophila* model for tauopathy was developed by overexpressing wild-type human tau or tau mutated at sites associated with early onset dementia.[62] Overexpression of both wild-type and disease associated forms of tau cause an expression level-dependent shortening of the fly life span. Shortened life span is associated with neurodegeneration and increased tau phosphorylation, although NFTs are not observed in the fly models.[62]

Using this model system, modifiers of tauopathies were identified using both forward and reverse genetic techniques. Expression of tau specifically in *Drosophila* eyes results in a dose-dependent rough eye phenotype.[63,64] In a forward genetic screen for suppressors and enhancers of this phenotype, the major class of modifiers identified involved kinases and phosphatases, consistent with the idea that abnormal phosphorylation of tau leads to decreased association with microtubules and increased aggregation.[64] In a reverse genetic approach, a major kinase known to be involved in tau phosphorylation, GSK3β, encoded by the *Drosophila* gene *shaggy* (*sgg*), enhances the tau rough eye phenotype when overexpressed, and suppresses the phenotype when mutated.[63] Moreover, antibodies that specifically recognize a phosphorylated form of tau unique to PHFs show a dramatic increase in phosphorylated tau upon cooverexpression of *sgg*. While PHFs are not observed upon overexpression of tau alone, PHFs are observed when *sgg* is cooverexpressed with tau.[63] Inhibitors of apoptosis suppress tau-induced neurodegeneration.

Most substrates must be prephosphorylated by other kinases at nearby sites before they can become efficiently phosphorylated by GSK3β. Is there a priming phosphorylation required before tau is phosphorylated by GSK3β? Reverse genetic experiments by Nishimura *et al.*[65] indicate that Par-1, a homolog of the mammalian MARK kinase that regulates microtubule dynamics and neuronal differentiation, phosphorylates tau. Prior phosphorylation by Par-1 is required before downstream phosphorylation can occur by kinases including GSK3β (Figure 15–1C).

Although presenilins are generally thought to function in Alzheimer's disease by cleaving APP to produce cytotoxic Aβ peptide, a novel model proposes that presenilin may also function to regulate phosphorylation of tau. In support of this model, it has recently been shown that loss of presenilin drastically enhances the rough eye phenotype of flies overexpressing tau.[66]

In another approach, *Drosophila* has been used to identify the *in vivo* effects of genes identified in other organisms as possible tau interactors. Puromycin-sensitive aminopeptidase (PSA) was identified as a gene upregulated in the cerebellum following overexpression of a dominant mutated form of tau associated with frontal temporal dementia.[67] Since the cerebellum is particularly resistant to tau-dependent neurodegeneration, increased PSA expression was proposed to function in a protective manner. This was verified using *Drosophila* where it was shown that loss of function of PSA enhances the rough eye phenotype caused by overexpressed mutant tau.[67] Overexpression of PSA suppresses the phenotype. It was further determined that PSA functions to degrade tau *in vivo* and *in vitro*. This approach illustrates the efficacy of using *Drosophila* to test the *in vivo* phenotypes of biochemical connections identified in other species.

PARKINSON'S DISEASE

Parkinson's disease (PD), the second most prevalent human neurodegenerative disease, results in movement disorders, tremors, and muscle rigidity due to the selective loss of dopaminergic neurons. Pathological phenotypes associated with PD include the formation of Lewy bodies and Lewy neurites, protein inclusions containing α-synuclein, and ubiquitin.[68–70]

While PD is generally sporadic, mutations associated with familial forms have been characterized, leading to the identification of the *α-synuclein* gene.[71] Although an obvious *Drosophila* α-synuclein homolog has not been identified, a *Drosophila* model of PD was developed using transgenic flies expressing wild-type and mutant human α-synuclein.[72] Development and eclosion of transgenic flies occur normally. However, an age-dependent and specific loss of dopaminergic neurons was observed in transgenic flies, indicating that high expression of α-synuclein can cause PD. In addition, α-synuclein aggregates that strongly resemble human Lewy bodies are observed in aged transgenic flies.[72] Furthermore, an age-dependent loss of locomotor activity is greatly enhanced in transgenic flies. These phenotypes demonstrate significant similarities between the human disease and the fly model.

In addition to α-synuclein, at least five other genes have been linked to familial forms of PD and analysis of these genes suggests that two pathways may be involved in disease pathogenesis.[73] *Parkin*, which encodes an E3 ubiquitin ligase that directs proteins to be degraded by tagging them with ubiquitin, and ubiquitin carboxy-terminal hydrolase L1 are components of the ubiquitin proteosome system (UPS). DJ1 and PINK1, Pten-induced putative kinase 1, are involved in mitochondrial function. Although *parkin* loss-of-function mutants have been reported to have mitochondrial defects, direct interaction between the genes involved in the UPS and mitochondrial function have been lacking until recent results in flies.

Recently, several groups have generated *PINK1* mutant flies, either null mutants[74,75] or reduced activity mutants using RNAi.[76,77] *PINK1* mutants are, in general, viable, but have reduced longevity. They also suffer energy depletion, male sterility, and an age-dependent muscle degeneration associated with impaired mitochondrial function. In addition, an age-dependent loss of dopaminergic neurons is generally observed.[74,76,77] These phenotypes are similar to those observed in *parkin* mutants[78] and indeed reducing PINK1 expression results in reduced parkin levels.[76] Furthermore, overexpression of parkin dramatically restores *PINK1* loss-of-function phenotypes, *parkin;PINK1* double mutants do not have more severe phenotypes than single mutants, and overexpression of *PINK1* does not restore the defects of a *parkin* mutant.[74–76] These results indicate that parkin and PINK1 function in a common pathway with PINK1 functioning upstream to control parkin expression (Figure 15–1D). It will be of significant interest in the future to include the other four genes associated with familial PD in epistasis experiments in order to further elucidate the pathway or pathways involved in Parkinson's pathology.

Studies in *Drosophila* have facilitated the identification of possible therapeutic targets to combat PD. Auluck and collegues[79] have shown that overexpression of HSP70, a stress-response chaperone that functions in refolding misfolded proteins, suppresses the neurodegeneration seen in α-synuclein flies (Figure 15–1D). Furthermore, the drug geldanamycin, which blocks the activity of HSP90, a negative regulator of stress response genes, also prevents neurodegeneration.[80] Overexpressing Pael-R, a putative G-protein- coupled transmembrane protein, causes it to form insoluble aggregates and leads to endoplasmic reticulum stress and loss of dopaminergic neurons. Pael-R is degraded through ubiquitination by parkin.[81] Thus upregulation of stress response/chaperone pathways may be a useful therapeutic strategy for treatment of PD. Reducing PINK1 activity in the *Drosophila* compound eye results in ommatidial degeneration, which can be suppressed by expressing the human *SOD1* antioxidant gene or by feeding flies either vitamin E or SOD1 protein, both of which have antioxidant activity.[77] Also, overexpression of the fly *Bcl2* gene, which functions to inhibit apoptosis and is involved in protecting mitochondrial integrity, suppresses many of the phenotypes of *PINK1* mutants.[74] These results suggest that *Drosophila* models of PD have great potential in identifying factors or compounds that may be used in the treatment of the disease.

SUMMARY AND FUTURE DIRECTIONS

Drosophila have proven to be particularly well suited as a model organism for human cognitive and age-dependent neurodegenerative diseases as shown by the five disease models discussed in this chapter. This is likely due to several factors including the high degree of conservation between *Drosophila* and mammals with respect to memory formation, aging, neuronal development, and neurite guidance. In addition, these diseases all affect fundamental cellular processes such as cAMP/PKA signaling, translation, and protein misfolding and aggregation. As long as processes are this conserved, flies will continue to be extremely useful in elucidating pathways involved in disease progression and identifying therapeutic targets for drug screening.

Constructing *Drosophila* models of diseases is relatively straightforward. The *Drosophila* genome has been fully sequenced and, importantly, many P-element transposon lines and many genomic deletion lines exist. When a human disease is caused by mutations that result in reduced expression or loss of function of a gene, the *Drosophila* homolog can be readily identified and mutations in this homologous gene can be rapidly made by imprecise excision of nearby P-elements or by other mutagenic techniques. Reducing expression using RNAi is also commonly used. When disease pathology is caused by a dominant mutation or by increased accumulation of a gene product, transgenic *Drosophila* can be easily made that can be used to overexpress the gene of interest in a variety of tissues. This approach can even work when the gene of interest has no obvious homolog as seen in the case of overexpressing α-synuclein in a *Drosophila* Parkinson's disease model.

Once a disease model has been made, a major advantage of using *Drosophila* is that it is amenable to genetic screening for genes that can enhance or suppress the disease phenotype when either mutated or overexpressed. Large-scale unbiased screens of this nature are not practical in mammals. Even when genes encoding candidate interacting proteins are known, the large number of available mutants in *Drosophila* and the short generation time that facilitates the rapid construction of transgenic lines makes *Drosophila* an ideal choice in which to study genetic interactions. Furthermore, the ability to introduce many compounds into *Drosophila* through feeding allows for the rapid screening of large libraries of compounds for those that alter a disease phenotype. Of course, genes, genetic interactions, and compounds identified through *Drosophila* models will have to be confirmed using mammalian systems, but the use of *Drosophila* as an initial screen has the great advantage of rapidly identifying strong candidates.

Similarities between *Drosophila* and mammals suggest that in addition to the successes in studying specific diseases, *Drosophila* may be useful in the study of less well characterized aging effects. Recent advances in medicine and healthcare have led to increases in human life expectancy. A consequence of this has been an increase in the incidence of neurodegenerative diseases. However, normal age-related memory impairment (AMI), not associated with neurodegenerative diseases, is also becoming a major concern. What are the causes of AMI? Are the mechanisms involved in AMI the same as those involved in neurodegenerative diseases? How can AMI be treated? Recent results indicate that *Drosophila* also suffer a quantifiable and reproducible AMI.[82] Although large-scale screening of memory in aged-fly populations is challenging, identification of mutants or compounds that accelerate or suppress AMI should be feasible. In addition, compounds and mutations that suppress neurodegenerative diseases or extend the life span can be tested for their effects on AMI. Thus, *Drosophila* will continue to be a vital model system for the study and treatment of human diseases.

REFERENCES

1. Margulies C, Tully T, Dubnau J. Deconstructing memory in Drosophila. *Curr Biol* 2005;15:R700–713.
2. Mehren JE, Ejima A, Griffith LC. Unconventional sex: Fresh approaches to courtship learning. *Curr Opin Neurobiol* 2004;14:745–750.
3. Tully T, *et al.* A return to genetic dissection of memory in Drosophila. *Cold Spring Harb Symp Quant Biol* 1996;61:207–218.
4. Tully T, *et al.* Genetic dissection of consolidated memory in Drosophila. *Cell* 1994;79:35–47.
5. Yin JC, Tully T. CREB and the formation of long-term memory. *Curr Opin Neurobiol* 1996;6:264–268.
6. Liebl FL, *et al.* Genome-wide P-element screen for Drosophila synaptogenesis mutants. *J Neurobiol* 2006;66:332–347.
7. Vijayakrishnan N, Broadie K. Temperature-sensitive paralytic mutants: Insights into the synaptic vesicle cycle. *Biochem Soc Trans* 2006;34:81–87.
8. Huang FD, *et al.* Rolling blackout is required for synaptic vesicle exocytosis. *J Neurosci* 2006;26:2369–2379.
9. Kidokoro Y. Roles of SNARE proteins and synaptotagmin I in synaptic transmission: Studies at the Drosophila neuromuscular synapse. *Neurosignals* 2003;12:13–30.
10. Zhang YQ, *et al.* Drosophila fragile X-related gene regulates the MAP1B homolog Futsch to control synaptic structure and function. *Cell* 2001;107:591–603.
11. Tong J, *et al.* Neurofibromin regulates G protein-stimulated adenylyl cyclase activity. *Nat Neurosci* 2002;5:95–96.
12. Pan L, *et al.* The Drosophila fragile X gene negatively regulates neuronal elaboration and synaptic differentiation. *Curr Biol* 2004;14:1863–1870.
13. Tatar M, Bartke A, Antebi A. The endocrine regulation of aging by insulin-like signals. *Science* 2003;299:1346–1351.
14. Horiuchi J, Saitoe M. Can flies shed light on our own age-related memory impairment? *Ageing Res Rev* 2005;4:83–101.
15. Pletcher SD, Libert S, Skorupa D. Flies and their golden apples: The effect of dietary restriction on Drosophila aging and age-dependent gene expression. *Ageing Res Rev* 2005;4:451–480.
16. Longo VD, Finch CE. Evolutionary medicine: From dwarf model systems to healthy centenarians? *Science* 2003;299:1342–1346.
17. Fortini ME, *et al.* A survey of human disease gene counterparts in the Drosophila genome. *J Cell Biol* 2000;150:F23–30.
18. Brand AH, Perrimon N. Targeted gene expression as a means of altering cell fates and generating dominant phenotypes. *Development* 1993;118:401–415.
19. Mao Z, *et al.* Pharmacogenetic rescue in time and space of the rutabaga memory impairment by using Gene-Switch. *Proc Natl Acad Sci USA* 2004;101:198–203.
20. Osterwalder T, *et al.* A conditional tissue-specific transgene expression system using inducible GAL4. *Proc Natl Acad Sci USA* 2001;98:12596–12601.
21. Rorth P, *et al.* Systematic gain-of-function genetics in Drosophila. *Development* 1998;125:1049–1057.
22. Hay BA, Maile R, Rubin GM. P element insertion-dependent gene activation in the Drosophila eye. *Proc Natl Acad Sci USA* 1997;94:5195–5200.
23. Manev H, Dimitrijevic N, Dzitoyeva S. Techniques: Fruit flies as models for neuropharmacological research. *Trends Pharmacol Sci* 2003;24:41–43.
24. Xu GF, *et al.* The neurofibromatosis type 1 gene encodes a protein related to GAP. *Cell* 1990;62:599–608.
25. The I, *et al.* Rescue of a Drosophila NF1 mutant phenotype by protein kinase A. *Science* 1997;276:791–794.
26. Acosta MT, Gioia GA, Silva AJ. Neurofibromatosis type 1: New insights into neurocognitive issues. *Curr Neurol Neurosci Rep* 2006;6:136–143.
27. Zhong Y. Mediation of PACAP-like neuropeptide transmission by coactivation of Ras/Raf and cAMP signal transduction pathways in Drosophila. *Nature* 1995;375:588–592.
28. Guo HF, *et al.* Requirement of Drosophila NF1 for activation of adenylyl cyclase by PACAP38-like neuropeptides. *Science* 1997;276:795–798.
29. Guo HF, *et al.* A neurofibromatosis-1-regulated pathway is required for learning in Drosophila. *Nature* 2000;403:895–898.
30. Fieber LA. Ionic currents in normal and neurofibromatosis type 1-affected human Schwann cells: Induction of tumor cell K current in normal Schwann cells by cyclic AMP. *J Neurosci Res* 1998;54:495–506.
31. Dasgupta B, Dugan LL, Gutmann DH. The neurofibromatosis 1 gene product neurofibromin regulates pituitary adenylate cyclase-activating polypeptide-mediated signaling in astrocytes. *J Neurosci* 2003;23:8949–8954.
32. Zarnescu DC, *et al.* Come FLY with us: Toward understanding fragile X syndrome. *Genes Brain Behav* 2005;4:385–392.
33. Nimchinsky EA, Oberlander AM, Svoboda K. Abnormal development of dendritic spines in FMR1 knock-out mice. *J Neurosci* 2001;21:5139–5146.
34. Grossman AW, *et al.* Local protein synthesis and spine morphogenesis: Fragile X syndrome and beyond. *J Neurosci* 2006;26:7151–7155.
35. Darnell JC, Mostovetsky O, Darnell RB. FMRP RNA targets: Identification and validation. *Genes Brain Behav* 2005;4:341–349.
36. Ule J, Darnell RB. RNA binding proteins and the regulation of neuronal synaptic plasticity. *Curr Opin Neurobiol* 2006;16:102–110.
37. Wan L, *et al.* Characterization of dFMR1, a Drosophila melanogaster homolog of the fragile X mental retardation protein. *Mol Cell Biol* 2000;20:8536–8547.
38. Darnell JC, *et al.* Fragile X mental retardation protein targets G quartet mRNAs important for neuronal function. *Cell* 2001;107:489–499.
39. Brown V, *et al.* Microarray identification of FMRP-associated brain mRNAs and altered mRNA translational profiles in fragile X syndrome. *Cell* 2001;107:477–487.
40. Reeve SP, *et al.* The Drosophila fragile X mental retardation protein controls actin dynamics by directly regulating profilin in the brain. *Curr Biol* 2005;15:1156–1163.
41. Lee A, *et al.* Control of dendritic development by the Drosophila fragile X-related gene involves the small GTPase Rac1. *Development* 2003;130:5543–5552.
42. Schenck A, *et al.* A highly conserved protein family interacting with the fragile X mental retardation protein (FMRP) and displaying selective interactions with FMRP-related proteins FXR1P and FXR2P. *Proc Natl Acad Sci USA* 2001;98:8844–8849.
43. Kobayashi K, *et al.* p140Sra-1 (specifically Rac1-associated protein) is a novel specific target for Rac1 small GTPase. *J Biol Chem* 1998;273:291–295.

44. Schenck A, *et al.* CYFIP/Sra-1 controls neuronal connectivity in Drosophila and links the Rac1 GTPase pathway to the fragile X protein. *Neuron* 2003;38:887–898.
45. Eden S, *et al.* Mechanism of regulation of WAVE1-induced actin nucleation by Rac1 and Nck. *Nature* 2002;418:790–793.
46. Zarnescu DC, *et al.* Fragile X protein functions with lgl and the par complex in flies and mice. *Dev Cell* 2005;8:43–52.
47. Wirtz-Peitz F, Knoblich JA. Lethal giant larvae take on a life of their own. *Trends Cell Biol* 2006;16:234–241.
48. Malenka RC, Bear MF. LTP and LTD: An embarrassment of riches. *Neuron* 2004;44:5–21.
49. Huber KM, Roder JC, Bear MF. Chemical induction of mGluR5- and protein synthesis-dependent long-term depression in hippocampal area CA1. *J Neurophysiol* 2001;86:321–325.
50. Huber KM, *et al.* Altered synaptic plasticity in a mouse model of fragile X mental retardation. *Proc Natl Acad Sci USA* 2002;99:7746–7750.
51. Bear MF, Huber KM, Warren ST. The mGluR theory of fragile X mental retardation. *Trends Neurosci* 2004;27:370–377.
52. Dockendorff TC, *et al.* Drosophila lacking dfmr1 activity show defects in circadian output and fail to maintain courtship interest. *Neuron* 2002;34:973–984.
53. McBride SM, *et al.* Pharmacological rescue of synaptic plasticity, courtship behavior, and mushroom body defects in a Drosophila model of fragile X syndrome. *Neuron* 2005;45:753–764.
54. Blennow K, de Leon MJ, Zetterberg H. Alzheimer's disease. *Lancet* 2006;368:387–403.
55. Tanzi RE, Bertram L. Twenty years of the Alzheimer's disease amyloid hypothesis: A genetic perspective. *Cell* 2005;120:545–555.
56. Sisodia SS, St George-Hyslop PH. gamma-Secretase, Notch, Abeta and Alzheimer's disease: Where do the presenilins fit in? *Nat Rev Neurosci* 2002;3:281–290.
57. Greeve I, *et al.* Age-dependent neurodegeneration and Alzheimer-amyloid plaque formation in transgenic Drosophila. *J Neurosci* 2004;24:3899–3906.
58. Iijima K, *et al.* Dissecting the pathological effects of human Abeta40 and Abeta42 in Drosophila: A potential model for Alzheimer's disease. *Proc Natl Acad Sci USA* 2004;101:6623–6628.
59. Crowther DC, *et al.* Intraneuronal Abeta, non-amyloid aggregates and neurodegeneration in a Drosophila model of Alzheimer's disease. *Neuroscience* 2005;132:123–135.
60. Goedert M, Jakes R. Mutations causing neurodegenerative tauopathies. *Biochim Biophys Acta* 2005;1739:240–250.
61. Yen SH, *et al.* Alzheimer neurofibrillary lesions: Molecular nature and potential roles of different components. *Neurobiol Aging* 1995;16:381–387.
62. Wittmann CW, *et al.* Tauopathy in Drosophila: Neurodegeneration without neurofibrillary tangles. *Science* 2001;293:711–714.
63. Jackson GR, *et al.* Human wild-type tau interacts with wingless pathway components and produces neurofibrillary pathology in Drosophila. *Neuron* 2002;34:509–519.
64. Shulman JM, Feany MB. Genetic modifiers of tauopathy in Drosophila. *Genetics* 2003;165:1233–1242.
65. Nishimura I, Yang Y, Lu B. PAR-1 kinase plays an initiator role in a temporally ordered phosphorylation process that confers tau toxicity in Drosophila. *Cell* 2004;116:671–682.
66. Doglio LE, *et al.* gamma-cleavage-independent functions of presenilin, nicastrin, and Aph-1 regulate cell-junction organization and prevent tau toxicity in vivo. *Neuron* 2006;50:359–375.
67. Karsten SL, *et al.* A genomic screen for modifiers of tauopathy identifies puromycin-sensitive aminopeptidase as an inhibitor of tau-induced neurodegeneration. *Neuron* 2006;51:549–560.
68. Spillantini MG, *et al.* alpha-Synuclein in filamentous inclusions of Lewy bodies from Parkinson's disease and dementia with Lewy bodies. *Proc Natl Acad Sci USA* 1998;95:6469–6473.
69. Spillantini MG, Goedert M. The alpha-synucleinopathies: Parkinson's disease, dementia with Lewy bodies, and multiple system atrophy. *Ann NY Acad Sci* 2000;920:16–27.
70. Mezey E, *et al.* Alpha synuclein is present in Lewy bodies in sporadic Parkinson's disease. *Mol Psychiatry* 1998;3:493–499.
71. Polymeropoulos MH, *et al.* Mutation in the alpha-synuclein gene identified in families with Parkinson's disease. *Science* 1997;276: 2045–2047.
72. Feany MB, Bender WW. A Drosophila model of Parkinson's disease. *Nature* 2000;404:394–398.
73. Moore DJ, *et al.* Molecular pathophysiology of Parkinson's disease. *Annu Rev Neurosci* 2005;28:57–87.
74. Park J, *et al.* Mitochondrial dysfunction in Drosophila PINK1 mutants is complemented by parkin. *Nature* 2006;441:1157–1161.
75. Clark IE, *et al.* Drosophila pink1 is required for mitochondrial function and interacts genetically with parkin. *Nature* 2006;441: 1162–1166.
76. Yang Y, *et al.* Mitochondrial pathology and muscle and dopaminergic neuron degeneration caused by inactivation of Drosophila Pink1 is rescued by Parkin. *Proc Natl Acad Sci USA* 2006;103: 10793–10798.
77. Wang D, *et al.* Antioxidants protect PINK1-dependent dopaminergic neurons in Drosophila. *Proc Natl Acad Sci USA* 2006;103: 13520–13525.
78. Greene JC, *et al.* Mitochondrial pathology and apoptotic muscle degeneration in Drosophila parkin mutants. *Proc Natl Acad Sci USA* 2003;100:4078–4083.
79. Auluck PK, *et al.* Chaperone suppression of alpha-synuclein toxicity in a Drosophila model for Parkinson's disease. *Science* 2002; 295:865–868.
80. Auluck PK, Meulener MC, Bonini NM. Mechanisms of suppression of α-synuclein neurotoxicity by geldanamycin in Drosophila. *J Biol Chem* 2005;280:2873–2878.
81. Yang Y, *et al.* Parkin suppresses dopaminergic neuron-selective neurotoxicity induced by Pael-R in Drosophila. *Neuron* 2003;37: 911–924.
82. Tamura T, *et al.* Aging specifically impairs amnesiac-dependent memory in Drosophila. *Neuron* 2003;40:1003–1011.

16 Biomedical Research with Honey Bees

MICHELLE M. ELEKONICH

ABSTRACT

In addition to the economic benefits honey bees provide through pollination and honey production, this species is an important model system that provides insight into many areas of biomedical science. With a long history as a subject of studies of social behavior, learning and memory, and immunology (due to the allergens in their venom), the honey bee is emerging as a major model for phenotypic plasticity, development and aging, circadian rhythms, muscle metabolism and behavioral genomics. Sequencing of the honey bee genome revealed that honey bees share many genes and biochemical pathways with humans, thus opening new avenues of research. Biomedical research with honey bees benefits from this species' well known physiology, the tractability of its natural behaviors both in and out of the laboratory, a fully sequenced genome with accompanying suite of cellular and molecular tools, and a large interactive community of basic and applied researchers. Only *Drosophila* rivals the honey bee as an insect model system for biomedical research.

Key Words: *Apis mellifera*, Social behavior, Learning, Development, Aging, Muscle physiology, Circadian rhythms, Venom, Neurobiology, Olfaction, Genetics, Genomics, Beekeeping.

INTRODUCTION

Depicted in Spanish cave paintings dating from 6000 BC, the honey bee has long provided pollination of crops and numerous beneficial products including honey, wax, and propolis for human usage. Today the honey bee is also a well-established medical model for memory and learning, odor perception, behavioral plasticity, behavioral development, sociality, exercise metabolism, and aging. In addition, honey bee venom proteins have been the focus of many studies in immunology and protein biochemistry.

With a rich research literature containing more than 5000 references on honey bee evolution, behavior, physiology, and genetics dating back to the early 1800s, the honey bee model system is (1) the focus of a large interactive research community with basic and applied perspectives, (2) has well-described ecologies, natural behaviors, and phylogenies, (3) is well characterized with respect to natural genetic variation, associated phenotypic variation, and the evolutionary forces maintaining such variation, and (4) and has a sequenced experimentally tractable genome.[1] In fact, honey bees like *Drosophila* share many genes and biochemical pathways with humans,[2] suggesting many additional uses for this model system will soon emerge.

HONEY BEE NATURAL HISTORY

Honey bees are members of the insect order Hymenoptera, which includes approximately 100,000 species of sawflies, wasps, ants, and bees. The order Hymenoptera is one of 11 orders of holometabolous insects, those that undergo a metamorphic molt to reach their adult phenotype. Honey bees (*Apis mellifera*) live in large colonies usually containing one egg-laying queen and her progeny, some 20,000–40,000 nonreproductive female "workers," and 200–300 male "drones." Females, queens and workers, arise from fertilized (diploid) eggs laid by the queen. Diploid eggs become queens or workers depending on the cell in which they are laid and whether the resulting larvae are fed royal jelly or worker jelly by the bees performing brood care. In contrast, males, known as drones, arise from haploid unfertilized eggs. This system is known as haplodiploidy.[3] Male and female larvae undergo a series of larval stages followed by pupation and a full metamorphosis within a cell in the honeycomb. Emerging from the cell as fully formed adult bees, female workers undergo a form of behavioral development termed "temporal polyethism," moving through a series of behaviorally defined life history stages in an age-related fashion.[4]

BEEKEEPING STOCKS AND TOOLS

As of the year 2000, it was estimated that honey bee pollination adds as much as 14.6 billion dollars per year to crop productivity in the United States alone.[5] The important role of the honey bee in pollination means that beekeeping courses, equipment, and stocks of honey bees are easily obtained. To work with the bees a beekeeper will want a bee veil, bee gloves, and sturdy shoes at the minimum. Full bee suits with integrated veils and elastic at the ankles are a popular choice. A smoker to create cool smoke to control the bees, a hive tool, and a bee brush are also necessary. Each hive consists of a stand to keep the hive off the ground, two standard movable frame hive boxes (Langstroth hives), a bottom board, a queen excluder, smaller hive boxes (called supers) for the honey, and an inner cover, and a top cover. Individual beekeepers may vary in their use of the queen excluder and inner cover but they are standard parts of the hive. All pieces of equipment and hive parts are available premade from standard beekeeping suppliers (see the Appendix). Each colony will forage in an area a few hundred meters to 6 km around the hive. Thus, a beekeeper must ensure there is sufficient forage for a colony as well as a local water supply. Protection from wind and morning sun is

From: *Sourcebook of Models for Biomedical Research*
(P. M. Conn, ed.), © 2008 Humana Press Inc., Totowa, NJ.

also desirable. Finally, there are the bees themselves, which can be bought as a small hive (called a "split" or "nuc") or as a package (essentially a swarm in a container) to be placed in your equipment. Both of these can be bought with or without a queen, but will need to be picked up as neither the postal service nor any private carriers will deliver them. However, queens can currently be purchased and sent through the mail (see the Appendix).

There are numerous additional beekeeping tools that researchers sometimes find useful, although often for unintended uses. For example, the queen excluder can be used to modify the hive entrance. Then regular workers can exit but workers with attached raised tags that increase their size, known as "big-backed" workers, are confined to the hive. This manipulation can keep some members of a group of bees that are motivated and flight capable from flying to test the role of flight activity in physiology, neurobiology, or subsequent behavior.[6] Likewise, pollen traps that are meant to allow a beekeeper to collect bee gathered pollen, in the hands of the researcher, become tools to limit the amount of pollen coming into a colony. Because pollen is the only source of protein, this manipulation can be used to set up nutritional differences between colonies.[7] Similarly, cages typically used to introduce a new queen into a colony can be used to introduce groups of treated individuals.

Bees suffer from a number of parasites and diseases (bacterial, viral, and fungal). Beekeepers typically treat prophylactically for some of these and after the fact for others.[8] Many of the treatments are controlled as insecticides, so individual researchers will have to check to see which are legal for the state where they are located. The destructive force and spread of the varroa mite (*Varroa destructor*) has led to a renewed interest in integrated pest management and bee breeding. Various nonchemical solutions for varroa treatment such as rotating drone pupae and using screened bottom boards have been shown to decrease mite loads.[9,10] Although additional chemical treatments are available, beeswax retains many chemicals and the long-term hopes of the honey bee community to combat varroa, other parasites such as the small hive beetle, and diseases rest on the development of new strains of bees.

There are currently no isogenic lines of bees available. The behavioral and physiological qualities of a colony of bees are determined by the identity of the queen and the drones with which she mated. Virgin queens and male drones mate in the air at "drone congregation areas" that tend to be highly stable over time.[11–13] The price of a breeding queen depends on whether she was naturally mated, allowed to mate freely in nature with local drones, or artificially inseminated. Most beekeepers buy naturally mated queens, but if you require specific genotypes many queen breeders can produce custom inseminated queens for experiments. Instrumentally inseminated queens can cost up to 20 times as much as naturally mated queens, but give the researcher control over genotype. Courses in instrumental insemination are also available at Ohio State University (see the Appendix).

All commercially available honey bees in the United States are of the same species, *Apis mellifera,* mostly developed from the Italian *ligustica* strain, but scientists at the U.S. Department of Agricultrue (USDA), Universities, and commercial breeders have bred strains with various qualities in addition to the original *ligustica* qualities of fecundity, gentleness, and honey production. The Buckfast strain, one of the oldest and still available strains, was originally selected in part for resistance to tracheal mites. Other stocks available include the lighter colored Cordovan bees (also called Golden Italians). They are more golden and have brown rather than black bands. The gold color serves as a natural genetic marker. In contrast, Carniolan bees are darker colored and are bred for being winter hardy and producing lots of bees quickly with the onset of spring. Russian bees imported and tested at the USDA appear to have resistance to varroa mites, tracheal mites, and chalkbrood, a fungal disease.[14] Another strain developed at the USDA known as SMR (suppressed mite reproduction) and/or VSH (varroa sensitive hygiene)[15] and the strain developed at the University of Minnesota by Dr. Marla Spivak known as Minnesota Hygenic bees also exhibit increased resistance to the varroa mites. The Minnesota Hygenic bees are also well known for their resistance to foulbrood.[16,17] Other breeders have developed popular mite-resistant strains, but not all have scientifically tested the performance of those strains. In addition, some breeders offer queens resulting from combinations of these strains.

RESEARCH TOOLS AND MANIPULATIONS

A critical and widely used manipulation to discriminate age vs. behavioral effects during honey bee behavioral development is the use of single-cohort colonies (SCCs) to induce precocious foraging.[18–20] SCCs made entirely of newly emerged bees use skewed colony age demography to dissociate worker age and behavior. The lack of older workers in the hive causes about 10% of the bees in a single cohort colony to forage precociously, usually at 7–10 days of age.[19,21,22] In a typical colony older bees work outside the hive as foragers while younger bees would work inside the hive, while foragers and hive workers are the same age in an SCC. Critically, SCCs allow the observation and collection of same-aged bees performing different tasks. SCCs initially yield young (precocious) foragers and young (typical age) hive bees, and after a few weeks, old (typical age) foragers and old (overaged) hive bees. If all of the bees caring for brood in the hive are removed, a percentage of the members of the foraging force will revert to hive tasks including brood care (now a "reversion colony"). These individuals have been foragers and are the same age as the colony's foragers but now are behaviorally and physiologically at an earlier developmental stage.[23]

In addition to SCCs, honey bee researchers have invented a number of other manipulations that increase the usefulness of the honey bee as a model system. Of course the best known of these is the ability to train bees to specific feeding stations, individual markings, and observation hives as used by von Frisch[24] and his students in his Nobel Prize winning work on the honey bee waggle dance. Using cohorts of known age bees, as for an SCC described above, researchers can set up colonies with specific multiple-aged cohorts[25] to look at the role of age in socially mediated effects on behavior and physiology. As is the case for many other animals, honey bees can be cross-fostered[26] to separate the role of genotype from that of developmental environment. Colonies containing adult workers of mixed genotypes can also be made to investigate the genotypic influences on any number of individual and colony traits.[27,28]

Physical characters of bee colonies can be manipulated in many ways to facilitate data collection such as adding modified entrances to remove pollen,[29] collect dead bees,[30] lengthen the entrance to increase flight distance, or mark and/or capture bees individually as they leave or return.[31] The role of the queen and her pheromone secretions have been investigated using colonies where the queen's genotype differs from that of the workers or where the queen is removed and her presence simulated with pheromone.[32] Additionally, bee behavior and physiology have been manipulated with numerous pharmacological treatments[33] and exposure to pheromones[34] and other odors.[35]

The honey bee model system also has a number of neurobiological, physiological, and cellular tools. Honey bee brain neuroanatomy has been studied since the mid-1800s[36,37] and a bee brain atlas was recently completed.[38] Changes in the neuropil volume, dendritic arborization, and neurochemicals in both the mushroom bodies and antennal lobes of the honey bee brain occur in concert with behavioral development and its associated changes in physiology.[39–42]

In addition to anatomical studies of the honey bee brain, the functional role of individual types of cells, brain areas, receptors, and channels is increasingly being studied.[43,44] For example, the biogenic amine receptors, the cholinergic, GABAergic, and glutamatergic pathways in the central brain, and the subesophageal ganglion of the honeybee have all been well described.[34,45,46] Electrophysiological recordings and voltage clamp, *in situ* hybridization and immunocytochemistry are all well-established techniques for use in honey bees.[47–53]

Although there are many well-established insect cell lines derived from the fruit fly, house fly, and flesh fly,[54–57] honey bee cell lines are not yet available. Honey bee neurons have proven to be especially amenable to short-term culture[43,45,58–61] supporting studies of ion channels, receptors and neurotransmitters, and neurophysiology. Hemocytes (circulating immune system cells), antennal cells, and embryonic cells have also been cultured for short time periods.[62–64] Recently, embryonic cells taken from 36- to 40-h-old honey bee eggs have been cultured for as long as 3 months,[65] suggesting that the production of honey bee cell lines is not far off.

Transgenic bees have been created by incorporating DNA constructs into sperm used for instrumental insemination of queens. Transgenic lines produced in this way have been maintained for up to three generations. One drawback to this technique is that the construct did not appear to be integrated into the genomic DNA.[66] Chimeric bees have been created by transplanting embryonic cells into other embryos.[67,68] Although larval honey bees are typically reared in individual cells by the adults in the colony, *in vitro* rearing of embryonic and larval bees to adulthood is well established for toxicity studies on larvae and studies of caste determination.[69–76] Thus, with the ability to create chimeras in addition to the sequencing and annotation of the honey bee genome, the honey bee system is poised to pursue detailed studies of gene expression and biochemical pathways involved in phenotypic canalization and plasticity, development of genetic knockouts/knockins, and additional advances in honey bee breeding.

Several recent studies have effectively used RNA interference to disrupt gene expression. Injection of double-stranded RNA (dsRNA) into the neuropil of the antennal lobe disrupted expression of octopamine receptors at a level that decreased with distance from the injection site and length of time postinjection. Disruption of octopamine signaling decreased olfactory acquisition and recall in an operant conditioning task but did not disrupt olfactory discrimination.[77,78] Injection of dsRNA into the hemocoel via the intraabdominal segments decreased levels of vitellogenin, a storage protein, in the bees' hemolymph (blood). dsRNA injected into eggs also disrupted vitellogenin gene function across development.[79] Similarly, injection of the dsRNA encoding the E30 homeobox motif into preblastoderm stage embryos disrupted homeobox gene function producing embryonic phenotypes similar to those observed in the *Drosophila engrailed* mutants.[80] In another developmental study, Beye and colleagues disrupted sexual differentiation by suppressing *csd* expression in larvae that had developed from eggs injected with dsRNA for the part of the csd gene that is thought to be an SR-type protein.[81]

SOCIAL BEHAVIOR

Honey bees live in complex and cohesive societies characterized by "eusociality." Much like our own societies, generations overlap, there is cooperative care of young and importantly division of labor including reproduction.[3,82] Honey bee social behavior is studied both at the level of the individual and the level of the colony, often described as a "superorganism."[3,82,83] Studies of social behavior have focused on social control of behavior, individual response thresholds, physiology and group behavior, network dynamic models, communication, group decision making, the molecular basis of sociality, physiological control, colony need, and timing of response.[67,21,84–91] The following sections discuss some specific types of social behavior in more detail.

AGING AND BEHAVIORAL DEVELOPMENT

Worker bees perform several different tasks in the hive during the first 2–3 weeks of adult life, including brood care ("nursing") and hive maintenance, and then shift to foraging for nectar and pollen outside the hive for the remainder of their 5- to 7-week life.[4] Social interactions signaled by a pheromone pace the rate of adult worker bee behavioral development via the actions of juvenile hormone.[21,92–94] But juvenile hormone (JH) is not required for foraging behavior. Workers lacking JH following removal of the corpora allata (the sole source of JH) on the first day of adult life still become foragers but at older ages; this delay is eliminated with hormone replacement.[95] Adult behavioral development also varies with a colony's genetic background and is sensitive to factors such as weather, season, parasite infestation, and colony nutritional status.[18,22,27,96–99]

In times of colony need individual workers can develop more rapidly, in essence aging faster, to become precocious foragers. Similarly, if the colony loses many of its middle-aged workers, foragers can revert to earlier behaviors and their physiology also returns to more youthful states.[23,100] This socially mediated behavioral plasticity makes the honey bee a unique model for the physiology of aging.

As honey bees switch from in-hive tasks to foraging, they transition from constant exposure to the controlled homogeneous physical and sensory environment of the hive to prolonged periods in a far more heterogeneous environment outside the hive. Within the hive there is little light and the workers actively keep the temperature at 33–35°C by contracting their flight muscles and keep the humidity near 70%.[4] Foraging outside the hive occurs at air temperatures between 10 and 50°C[101] and exposes workers to wind, rain, and increased predation. The physiology of honey bees also changes as they age and move from nonflying tasks in the hive to flight-intensive foraging behaviors. For example, hypopharyngeal glands regress and produce enzymes for processing nectar instead of brood food, body mass decreases, body water content increases,and, as we describe in detail below, juvenile hormone levels and metabolic and flight capacity increase.[4,16,102–107]

In addition, more complex spatial and sensory information than that encountered in the hive must be integrated to successfully forage.[31,108]

Microarrays developed from a honey bee expressed sequence tag (EST) database[2] have revealed different patterns of gene expression in forager vs. nurse brains independent of age.[109] Among the genes differentially regulated between behavioral groups are those with strong sequence matches to annotated *Drosophila* genes important in axiogenesis, cell adhesion, and intracellular signaling. In the latter class is *foraging (for)*, a previously identified cGMP-dependent protein kinase whose pharmacological activation causes precocious positive phototaxis and precocious foraging.[33,110] Similarly, a recent macroarray study demonstrates higher expression levels for genes involved in signal transduction, ion channels, neurotransmitter transport, transcription factors, plasma membrane proteins, and most cell adhesion proteins in foragers as compared to newly emerged bees, suggesting plasticity and remodeling of neurocellular properties during aging and/or behavioral development in honey bees.[111]

Most adult honey bee workers live 3–6 weeks during spring and summer and up to 8 months if they emerge from metamorphosis in late fall and overwinter in the hive.[4,112,113] Bees that overwinter have higher levels of vitellogenin in the hemolymph, lower JH, and overall higher protein reserves.[112] Honey bee life span is determined by the onset of foraging; the earlier in life a worker becomes a forager, the sooner she dies.[114–117] As foragers age, the time required to collect food from a feeder at a known distance increases[118] and bees often return to the hive with only a partial load of nectar due perhaps to decreased flight ability.[119] Mechanistically, life span appears to be limited by foraging-induced damage to the flight apparatus. This damage includes physiological/biochemical impairment of the flight muscle and mechanical damage including the loss of wing surface area due to wing wear.[114,120,121] The amount of physiological and mechanical damage is directly related to the amount of flight, independent of age. As foragers age, they lose the ability to synthesize glycogen. Bees that make more flights per day exhaust their glycogen reserves and reach the point where glycogen synthesis breaks down sooner.[114,120,121]

In contrast, queen honey bees are quite long lived, typically living 1–3 years, but may live as long as 8 years.[4,113] Queen bees make one to several mating flights a few days after they emerge from the cell where they completed metamorphosis. The only other time a queen will fly during her entire lifetime is during colony reproduction or swarming.[4,113] Queens appear to age more gracefully than workers at the molecular level as well. Expression of most antioxidant genes decreases with age in queens but not workers.[122] The relatively short life span of worker bees in contrast to the longer life span of queens suggests that bees may be particularly well suited to investigations of the cellular mechanisms underlying individual differences in the rate of aging.[123]

CIRCADIAN RHYTHMS

The onset of foraging also marks a transition from relatively constant arrhythmic activity to diurnal activity patterns. As worker bees age, they develop a circadian rhythm,[124,125] but their behavioral rhythms are task dependent. Workers who are caring for brood, nurses, work around the clock while workers who forage work only during the day.[126] These behavioral rhythms are socially mediated—foragers that have reverted to brood care lose their behavioral rhythmicity.[100] Honey bee circadian rhythms are entrained by cycles of both light and temperature and act to coordinate foraging behavior with appropriate flight temperatures, sun compass navigation, and the timing of visits to specific floral sources with nectar availability.[127] Foraging age bees are positively phototactic. Phototaxis is modulated by the honey bee *foraging* gene (*Amfor*), which encodes a cGMP-dependent protein kinase (PKG). Increases in PKG activity cause precocious foraging and are correlated with increases in positive phototactic behavior.[9] As in other animals, honey bee circadian rhythms are influenced by *Period* gene expression in the brain.[125,128] However, the mechanism underlying the plasticity in behavioral rhythms and its relationship to the circadian rhythm are unknown.[100]

Circadian rhythms may also be important for the physiological support of foraging behavior. JH titers in foraging bees are consistently higher than those of bees working in the hive, even on their first foraging flight.[129,130] Increased JH titer causes degeneration of the hypopharyngeal glands and a shift from producing brood food to production of α-glucosidase, amylase, and glucose oxidase, the enzymes needed to process nectar into honey.[4,105,131] High JH titers also promote degradation of the fat body and consequent decreases in hemolymph vitellogenin, the most common storage protein in the hemolymph.[132] As brood food is high in protein, decreases in fat body and hemolymph vitellogenin levels further mark the switch from a bee well suited to caring for young to a bee well suited for flying, carrying heavy loads (in relation to body size), and processing nectar. In addition, there is a diurnal rhythm in JH titer in foragers that is not present in preadult bees.[129] Interestingly, the timing of the diurnal peaks in JH titer matches the timing of the peaks in *period* mRNA expression with the highest titers during the mid-subjective night.[125,129] However, it appears that JH titer does not influence diurnal locomotor activity as allactectomy has no effect on circadian locomotor rhythm, nor does treatment of young bees with methoprene, a JH analog, accelerate the onset of circadian behavior.[133]

It may be that the high titers and diurnal rhythms of JH support foraging physiology directly. In other insects, JH affects muscle properties directly influencing muscle maturation[134] and initiating muscle degeneration and ovarian development in postdispersing insects in preparation for reproduction.[135–137] JH also affects both cellular and whole animal metabolism. Allatectomized honey bee foragers have lower metabolic rates and are flight deficient compared to controls. However, this difference is abolished with treatment with the synthetic JH analog methoprene.[138]

LEARNING AND MEMORY

Experience-dependent changes in neural architecture and function may reflect changes in information processing and sensory thresholds in response to the greater complexity of foraging tasks relative to hive work. Honey bee foragers make use of visual, auditory, and olfactory information while locating flowers, assessing the sugar content of the nectar, communicating the quality and location of the food to other foragers, and repeatedly navigating between the hive and food sources.[4]

To support the complex task of foraging, olfactory sensitivity, learning ability, and volume of the neuropil of the mushroom bodies, associative brain areas involved in memory consolidation, increase as bees switch from in-hive to out-of-hive tasks.[40,108] Increased cholinergic signaling in the mushroom bodies is coincident with the foraging bees increased interaction with the world and increased mushroom body neuropil.[41] Field tests of honey bee learning make use of our ability to train honey bees to a food source, which the bees then communicate to new recruits via the waggle dance, thus in effect "teaching" other bees where to find food by reporting what has been learned.[24] A large body of literature suggests that foraging, recruitment via the waggle dance, and subsequent navigation to the food source require complex cognitive abilities including a compass sense, learning the local area and landmarks during orientation flights providing a detailed map, path learning, path integration, and higher order integration of these processes and redundancy of cues to enable foraging even in poor conditions.[108,139,140] These experiments continue to suggest that even with a small brain bees can be "smart."

In the laboratory, honey bee learning is investigated using the Proboscis Extension Reflex (PER) assay, a classical conditioning associative learning paradigm. Honey bees naturally extend their proboscis when receptors on the antennae are stimulated by sucrose, naturally found in nectar, to taste the food. In the PER assay restrained bees are taught to associate an odor with the food and learn to extend the proboscis in the presence of the odor.[141,142] These studies have shown that bees can learn simple associations, context, categorization, and discrimination, even performing successfully in reversal learning and negative patterning tasks.[58]

Because the bee is immobilized with the head, antennae, and mouthparts easily accessible, researchers can gain direct access to the brain to measure physiological correlates of learning during conditioning.[58,143] The octopaminergic VUM_{mx1} neuron of the subesophageal ganglion carries the information along the olfactory pathway from the antennal lobes to the mushroom bodies.[143] Likewise, downregulating VUM_{mx1} octopamine receptors using RNAi impairs learning.[77] In addition to the role of cholinergic signaling described above, olfactory memory processing in the mushroom bodies and antennal lobes relies on cAMP-dependent phosphorylation via protein kinase A (PKA) and the Ca^{2+}/phospholipid-dependent phosphorylation by protein kinase C (PKC), similar to other model systems.[33,144,145] Similar to vertebrate systems, glutamate signaling in the mushroom bodies facilitates memory stabilization.[146] Exposure to toxins such as pesticides can interrupt PER-based learning and may become an important tool for investigating sublethal effects of pesticides or environmental toxins.[72] The relative simplicity of the honey bee brain combined with powerful tools to study learning and physiology and the natural behavioral richness of the species probably owing to its high degree of sociality suggest that honey bees will continue to serve as a valuable model for studies of learning and memory.

MUSCLE METABOLISM

Honey bee flight muscles are very similar in structure and function to vertebrate skeletal muscle except that metabolism is completely aerobic, using only carbohydrates for fuel, and each muscle is composed of a sign fiber type.[147–151] In the honey bee, flight muscle metabolism accounts for over 90% of O_2 consumption during flight.[148] Honey bees possess an asynchronous flight muscle (AFM). Unlike synchronous muscles (which like typical striated muscles have a 1 : 1 ratio of neural stimuli to contractions with contraction initiated by intracellular calcium release and terminated by calcium uptake by the sarcoplasmic reticulum), AFMs show an approximately 1 : 10 ratio of neural stimuli to contractions. Neural stimulation in AFM releases intracellular calcium, which removes thin filament inhibition, but the cross-bridges themselves are activated by stretch and deactivated by sarcomere shortening. AFMs are stretched by thoracic deformation caused by contraction of antagonistic muscles, and this mechanical feedback keeps AFMs contracting over many cycles.[152] The large power-producing AFMs are controlled by a set of small synchronous muscles that produce little or no power but are capable of rapid and finely graded responses to neural stimuli.[153,154] Troponin-T is the tropomyosin-binding protein of the calcium-regulated troponin complex of striated muscle. As bees move from hive work to foraging, the troponin-T expression increases along with aldolase and the antioxidant protein superoxide dismutase.[115]

Adult honey bees are unable to fly during the first day following adult emergence, yet within 2 weeks generate spectacular rates of metabolism and aerodynamic power (up to $0.8\,W\,g^{-1}$ and $0.2\,W\,g^{-1}$, respectively)[155] that enable later work outside the hive, traveling up to 8 km from the hive and carrying loads equivalent to their body mass during foraging and undertaking. The development of flight ability generally occurs in two distinct periods, the first being the 3–4 days following adult emergence and the second typically at 14–21 days postemergence during the transition from hive work to foraging. Day-old bees that are physically agitated (a manipulation generally assumed to induce maximal metabolic capacity) can generate metabolic rates of only $0.1\,W\,g^{-1}$ and are isothermic with the surrounding air. Hovering 2-day-old bees have metabolic rates approaching $0.3\,W\,g^{-1}$ and are more endothermic; coincident with the increase in metabolic capacity of young bees are dramatic increases in thoracic pyruvate kinase and citrate synthase activities as well as thoracic glycogen levels.[103,114,156–158]

Such biochemical changes should in theory greatly increase flux capacity through the citric acid cycle and the rate of NADH and $FADH_2$ recycling. Flight metabolic rates, thoracic enzyme levels, and thoracic glycogen levels remain relatively constant over the 1- to 3-week period when the bees work within the hive. Then, at the onset of foraging (14–21 days postecolosion) there is an approximate 15% increase in agitated flight metabolic rate, coincident with an approximate doubling of thoracic glycogen levels.[156,157,159,160] In fact, flying honey bees have the highest aerobic metabolic rates ever measured in any animal, between 100 and 120 ml $O_2\,g^{-1}\,h^{-1}$. These values are 3-fold higher than hovering hummingbirds and 30-fold greater than human athletes undergoing maximal aerobic exercise.[161–164] Thus, in the honey bee mechanisms for resisting oxidative damage may be highly specialized and critically important. When combined with the ability to manipulate age, behavior, aerobic expenditure, oxidative stress, functional senescence, and life span independently, the exceedingly high rates of aerobic capacity suggest that the honey bee may be a superior exercise model.[151,165]

HONEY BEE VENOM

In the United States, 40–50 people die each year from systemic reactions to stings from Hymenopteran insects, most often due to anaphylactic reactions to stings from vespid wasps commonly known as yellow jackets.[166] Severe anaphylactic reactions can result in permanent hypoxic brain damage, myocardial infarction, and death.[167] After a systemic reaction, a patient diagnosed with an insect sting allergy is given immunotherapy over a period of months starting at low doses and building up to a maintenance dose, which is continued at 4–6 week intervals for as much as 5 years. Semirush protocols (giving multiple doses each week for several weeks) and rush protocols (where all doses are given on a single day) have been developed. These faster protocols carry some additional risk of adverse reaction relative to the conventional protocol.[102,168,169]

Cross-reactivity to different Hymenopteran venoms is observed in half of the patients with allergic reactions to wasp or bee stings.[101,170] In addition to species-specific proteins, Hymenopteran venoms contain a number of conserved cytolytic and cytotoxic compounds including phopholipases (A_1 in bees and A_2 in wasps), hyaluronidases, and acid phosphatases.[171,172] Cross-reactivity has been ascribed to immune responses to shared hyaluronidase epitopes and production of IgE specific to the carbohydrate side chains of the glycoproteins.[67,68,71,100] Melittin (Api m 4), Api m 6, and the recently identified icarpin are honey bee-specific allergenic proteins.[167,173,174]

Many of the components of honey bee venom have specific cellular targets and concomitantly stimulate the immune system. The bee-specific protein melittin acts by disrupting the cell membrane and activating phopholipases, calmodulin, and arachadonic acid release[175] and may also prove to be a useful molecule to target tumor cells, similar to phospholipase A_2.[176] Effects of bee venom may be somewhat cell specific. Bee venom appears to be more destructive to renal cells, inducing vasoconstriction, nephrotoxicity, and rhabdomyolysis, possibly resulting in renal failure.[177] Possible uses of venom components, individual differences in reactivity, reactivity of special groups such as beekeepers, test accuracy, and therapy safety and efficacy are ongoing foci of research in this area.[168,178]

Finally, there has been a renewed interest in apitherapy with therapeutic stinging touted to help everything from arthritis to multiple sclerosis. Most of the research in this area has occurred in the area of traditional Chinese medicine. In many of these studies bee stings or diluted bee venom are applied to traditional acupuncture points. Several studies suggest that bee venom therapy effectively reduces inflammation and induces analgesia in rodent models, but many of these studies were methodologically flawed.[179] However, two randomized controlled studies suggest that bee venom therapy can be useful for rheumatoid arthritis and osteoarthritis in the knee joint.[180–182] The outcomes for patients with multiple sclerosis are not so hopeful. To date, in the only randomized controlled study of the use of bee venom therapy to ameliorate symptoms of the disease, bee stings were not any more effective than no therapy at all.[183] There has also been a renewed interest in the wound-healing properties of honey and the antiinflammatory properties of propolis.[184] As the public's interest in complementary and alternative medicine continues to increase, this will likely continue to be an active area of research.

CONCLUSIONS

Research with insect models continues to suggest that many neural, hormonal, and molecular mechanisms are evolutionarily conserved across a broad range of taxa. A truly integrative approach to biomedical problems requires understanding the genetic, physiological, and social/ecological mechanisms and their interactions. With a combination of a sequenced genome, well-described natural history including ecology, social behavior that rivals our own in complexity and development, as well as a large, interactive community of researchers with a variety of tools to manipulate all aspects of honey bee biology, the honey bee will continue to be a highly tractable and interesting medical model system for a long time to come.

APPENDIX

BEEKEEPING EQUIPMENT

Beesource: includes lists of suppliers, plans for building equipment, and other useful information.
http://www.beesource.com/

A.I. Root, one of the oldest beekeeping families and supply companies in America, established in the 1800s and publisher of *Bee Culture.*
http://bee.airoot.com/beeculture/index.htm

Dadant & Sons, another beekeeping supply company established in the 1800s and publisher of *The Am Bee J.*
http://www.dadant.com/

Mann Lake Ltd: another large beekeeping supplier with a very informational website.
http://www.mannlakeltd.com/

Walter T. Kelly: beekeeping supplies, U.S. maker of honey extraction equipment.
http://www.kelleybees.com/

QUEEN BREEDERS For local breeders in your area, your best sources of information in the United States are the two major beekeeping magazines: *The Am Bee J* and *Bee Culture.*

Glenn Apiaries: sells instrumentally inseminated breeder queens.
http://members.aol.com/queenb95/catalog.html#anchor2391365

Ohio Queen Breeders: another source for instrumentally inseminated queens.
http://www.ohioqueenbreeders.com/

Rothenbuhler Honey Bee Lab at Ohio State University: information, equipment, and training in instrumental insemination from Sue Cobey.
http://www174.pair.com/birdland/Breeding/II.html

MORE INFORMATION

Cyberbee, a site with lots of practical information and links to bee research around the world, maintained by Dr. Zachary Huang at Michigan State University.
http://www.cyberbee.net/

Cornell University Master Beekeeper Program: information, news, master beekeeping courses, directions for making creamed honey.
http://www.masterbeekeeper.org/

State Apiarist Directory, contact information for each state's Apiary Inspector. http://www.mda.state.mn.us/ams/apiary/directory.htm

American Association for Professional Apiculturists, professional association and source of management information.
http://www.entomology.umn.edu/aapa/index.htm

USDA Beltsville, MD Beelab, includes information for submitting samples for testing. http://www.barc.usda.gov/psi/brl/brl-page.html

USDA Baton Rouge, LA Beelab, home of the SMRD breeding project.
http://msa.ars.usda.gov/la/btn/hbb
USDA Tuscon, AZ Beelab.
http://gears.tucson.ars.ag.gov/
USDA Welasco, TX Lab.
http://weslaco.ars.usda.gov/

REFERENCES

1. Feder ME, Mitchell-Olds T. Evolutionary and ecological functional genomics. *Nat Rev Genet* 2003;4:651–657.
2. Whitfield CW, Band MR, Bonaldo MF, *et al.* Annotated expressed sequence tags and cDNA microarrays for studies of brain and behavior in the honey bee. *Genome Res* 2002;12:555–566.
3. Queller DC, Strassmann JE. The many selves of social insects. *Science* 2002;296:311–313.
4. Winston ML. *The Biology of the Honey Bee*. Cambridge, MA: Harvard University Press, 1987:281.
5. Morse RA, Calderone NW. The value of honey bees as pollinators of U.S. crops in 2000. *Bee Cult* 2000;128:1–15.
6. Withers GS, Fahrbach SE, Robinson GE. Effects of experience and juvenile-hormone on the organization of the mushroom bodies of honey-bees. *J Neurobiol* 1995;26:130–144.
7. Hoover SER, Higo HA, Winston ML. Worker honey bee ovary development: Seasonal variation and the influence of larval and adult nutrition. *J Comp Physiol B Biochem System Environ Physiol* 2006;176:55–63.
8. Williams DL. A veterinary approach to the European honey bee (*Apis mellifera*). *Vet J* 2000;160:61–73.
9. Degrandi-Hoffman G, Curry R. The population dynamics of Varroa mites in honey bee colonies: Part III—How beekeeping practices could affect Varroa populations. *Am Bee J* 2005;145:709–710.
10. Delaplane KS, Berry JA, Skinner JA, Parkman JP, Hood WM. Integrated pest management against Varroa destructor reduces colony mite levels and delays treatment threshold. *J Apicult Res* 2005;44:157–162.
11. Koeniger N, Koeniger G, Gries M, Tingek S. Drone competition at drone congregation areas in four Apis species. *Apidologie* 2005;36:211–221.
12. Koeniger N, Koeniger G, Pechhacker H. The nearer the better? Drones (Apis mellifera) prefer nearer drone congregation areas. *Insect Soc* 2005;52:31–35.
13. Loper GM, Wolf WW, Taylor OR Jr. Honey bee drone flyways and congregation areas: Radar observations. *J Kans Entomol Soc* 1992;65:223–230.
14. Rinderer TE, De Guzman LI, Danka RG. A new phase begins for the USDA-ARS russian honey bee breeding program. *Am Bee J* 2005;145(7):579–582.
15. Harris JW, Harbo JR. The SMR trait explained by hygienic behavior of adult bees. *Am Bee J* 2005;145(5):430–431.
16. Ibrahim A, Spivak M. The relationship between hygienic behavior and suppression of mite reproduction as honey bee (*Apis mellifera*) mechanisms of resistance to Varroa destructor. *Apidologie* 2006;37:31–40.
17. Spivak M, Reuter GS. Honey bee hygienic behavior. *Am Bee J* 1998;138:283–286.
18. Giray T, Robinson GE. Effects of intracolony variability in behavioral-development on plasticity of division-of-labor in honey-bee colonies. *Behav Ecol Sociobiol* 1994;35:13–20.
19. Huang ZY, Robinson GE. Honeybee clony itegration—worker worker interactions mediate hormonally regulated plasticity in division-of-labor. *Proc Natl Acad Sci USA* 1992;89:11726–11729.
20. Robinson GE, Page RE, Strambi C, Strambi A. Hormonal and genetic-control of behavioral integration in honey bee colonies. *Science* 1989;246:109–111.
21. Huang ZY, Robinson GE. Regulation of honey bee division of labor by colony age demography. *Behav Ecol Sociobiol* 1996;39:147–158.
22. Schulz DJ, Huang ZY, Robinson GE. Effects of colony food shortage on behavioral development in honey bees. *Behav Ecol Sociobiol* 1998;42:295–303.
23. Robinson GE, Page RE, Strambi C, Strambi A. Colony integration in honey-bees—mechanisms of behavioral reversion. *Ethology* 1992;90:336–348.
24. von Frisch K. *The Dance Language and Orientation of Bees*. Cambridge, MA: Belknap Press of Harvard University Press, 1967:566.
25. Huang ZY, Robinson GE, Borst DW. Physiological correlates of division-of-labor among similarly aged honey-bees. *J Comp Physiol A Sensory Neural Behav Physiol* 1994;174:731–739.
26. Pankiw T, Tarpy DR, Page RE. Genotype and rearing environment affect honeybee perception and foraging behaviour. *Anim Behav* 2002;64:663–672.
27. Page RE, Robinson GE, Britton DS, Fondrk MK. Genotypic variability for rates of behavioral-development in worker honeybees (*Apis mellifera* L). *Behav Ecol* 1992;3:173–180.
28. Page RE, Fondrk MK, Hunt GJ, *et al.* Genetic dissection of honeybee (Apis mellifera L.) foraging behavior. *J Hered* 2000;91:474–479.
29. Todd FE, Bishop RK. Trapping honeybee-gathered pollen and factors affecting yields. *J Econ Entomol* 1940;33:866–870.
30. Gary NE, Lorenzen K. Improved trap to recover dead and abnormal honey bees *Apis mellifera* Hymenoptera: Apidae from hives. *Environ Entomol* 1984;13:718–723.
31. Capaldi EA, Robinson GE, Fahrbach SE. Neuroethology of spatial learning: The birds and the bees. *Annu Rev Psychol* 1999;50:651–682.
32. Slessor KN, Kaminski LA, Winston ML. The essence of royalty—queen honey bee pheromones. Abstracts of papers of the American Chemical Society 1988;195:55–AGRO.
33. Ben-Shahar Y, Leung HT, Pak WL, Sokolowski MB, Robinson GE. cGMP-dependent changes in phototaxis: A possible role for the foraging gene in honey bee division of labor. *J Exp Biol* 2003;206:2507–2515.
34. Grozinger CM, Sharabash NM, Whitfield CW, Robinson GE. Pheromone-mediated gene expression in the honey bee brain. *Proc Natl Acad Sci USA* 2003;100:14519–14525.
35. Wright GA, Smith BH. Different thresholds for detection and discrimination of odors in the honey bee (Apis mellifera). *Chem Sens* 2004;29:127–135.
36. Dujardin F. Memoire sur le syteme nerveux of insectes. *Ann Sci Nat Zool Biol Anim* 1850;14:195–206.
37. Kenyon FC. The brain of the bee: A preliminary contribution to the morphology of the nervous system of the Arthropoda. *J Comp Neurol* 1896;VI:134–205.
38. Brandt R, Rohlfing T, Rybak J, *et al.* Three-dimensional average-shape atlas of the honeybee brain and its applications. *J Comp Neurol* 2005;492:1–19.
39. Fahrbach SE, Farris SM, Sullivan JP, Robinson GE. Limits on volume changes in the mushroom bodies of the honey bee brain. *J Neurobiol* 2003;57:141–151.
40. Farris SM, Robinson GE, Fahrbach SE. Experience- and age-related outgrowth of intrinsic neurons in the mushroom bodies of the adult worker honeybee. *J Neurosci* 2001;21:6395–6404.
41. Ismail N, Robinson GE, Fahrbach SE. Stimulation of muscarinic receptors mimics experience-dependent plasticity in the honey bee brain. *Proc Natl Acad Sci USA* 2006;103:207–211.
42. Wang S, Zhang S, Sato K, Srinivasan MV. Maturation of odor representation in the honeybee antennal lobe. *J Insect Physiol* 2005;51:1244–1254.
43. Kloppenburg P, Kirchhof BS, Mercer AR. Voltage-activated currents from adult honeybee (*Apis mellifera*) antennal motor neurons recorded in vitro and in situ. *J Neurophysiol* 1999;81:39–48.
44. Wüstenberg DG, Boytcheva M, Grunewald B, Byrne JH, Menzel R, Baxter DA. Current- and voltage-clamp recordings and computer simulations of Kenyon cells in the honeybee. *J Neurophysiol* 2004;92:2589–2603.

45. Beggs KT, Hamilton IS, Kurshan PT, Mustard JA, Mercer AR. Characterization of a D2-like dopamine receptor (AmDOP3) in honey bee, Apis mellifera. *Insect Biochem Mol Biol* 2005;35: 873–882.
46. Blenau W, Baumann A. Molecular and pharmacological properties of insect biogenic amine receptors: Lessons from *Drosophila melanogaster* and *Apis mellifera. Arch Insect Biochem Physiol* 2001;48:13–38.
47. Mustard JA, Kurshan PT, Hamilton IS, Blenau W, Mercer AR. Developmental expression of a tyramine receptor gene in the brain of the honey bee, *Apis mellifera. J Comp Neurol* 2005;483:66–75.
48. Paul RK, Takeuchi H, Matsuo Y, Kubo T. Gene expression of ecdysteroid-regulated gene E74 of the honeybee in ovary and brain. *Insect Mol Biol* 2005;14:9–15.
49. Perk CG, Mercer AR. Dopamine modulation of honey bee (*Apis mellifera*) antennal-lobe neurons. *J Neurophysiol* 2006;95:1147–1157.
50. Sinakevitch I, Niwa M, Strausfeld NJ. Octopamine-like immunoreactivity in the honey bee and cockroach: Comparable organization in the brain and subesophageal ganglion. *J Comp Neurol* 2005;488:233–254.
51. Smirnov VB, Chesnokova EG, Lopatina NG, Voike E. Characteristics of neuron activity in the honey bee (*Apis Mellifera* L.) in conditions of kynurenine deficiency. *Neurosci Behav Physiol* 2006;V36:213–216.
52. Velarde RA, Sauer CD, Walden KKO, Fahrbach SE, Robertson HM. Pteropsin: A vertebrate-like non-visual opsin expressed in the honey bee brain. *Insect Biochem Mol Biol* 2005;35:1367–1377.
53. Zannat MT, Locatelli F, Rybak J, Menzel R, Leboulle G. Identification and localisation of the NR1 sub-unit homologue of the NMDA glutamate receptor in the honeybee brain. *Neurosci Lett* 2006;398:274–279.
54. Echalier G, Ohanessian A. *In vitro* culture of *Drosophila melanogaster* embryonic cells. *In vitro J Tissue Cult Assoc* 1970;6:162.
55. Eide PE. Establishment of a cell line from long-term primary embryonic house fly cell cultures. *J Insect Physiol* 1975;21:1431–1438.
56. Masakazu T, Mitsuhashi J, Ohtaki T. Establishment of a cell line from embryonic tissues of the fleshfly, Sarcophaga peregrina. *Dev Growth Differ* 1980;22:11–19.
57. Schneider I. Cell lines derived from late embryonic stages of Drosophila melanogaster. *J Embryol Exp Morphol* 1972;27:353.
58. Giurfa M. Cognitive neuroethology: Dissecting non-elemental learning in a honeybee brain. *Curr Opin Neurobiol* 2003;13:726–735.
59. Laurent S, Masson C, Jacob I. Whole-cell recording from honeybee olfactory receptor neurons: Ionic currents, membrane excitability and odorant response in developing worker bee and drone. *Eur J Neurosci* 1994;15:1139–1152.
60. Kreissl S, Bicker G. Dissociated neurons of the pupal honeybee brain in cell culture. *J Neurocytol* 1992;21:545–556.
61. Wüstenberg DG, Grünewald B. Pharmacology of the neuronal nicotinic acetylcholine receptor of cultured Kenyon cells of the honeybee, *Apis mellifera.* J Comp *Physiol* A: *Neuroethol Sens Neural, Behav Physiol* 2004;190:807–821.
62. Beisser K, Munz E, Reimann R, Renner-Müller ICE. Experimentelle Untersuchungen zur *in vitro*-Kultivierung von Zellen der Kärntner Honigbiene (*Apis mellifera carnica* Pollmann, 1879). *Zentralbl Veterinärmed B* 1990;37:509–519.
63. Gascuel J, Masson C, Bermudez I, Beadle DJ. Morphological analysis of honeybee antennal cells growing in primary cultures. *Tissue Cell* 1994;26:551–558.
64. Mitsuhashi J. Development of highly nutritive culture media. *In Vitro Cell Dev Biol* 2001;37:330–337.
65. Bergem M, Norberg K, Roseth A, Meuwissen T, Lien S, Asmodt RM. Chimeric honeybees (Apis mellifera) produced by transplantation of embryonic cells into pre-gastrula stage embryos and detection of chimerism by use of microsatellite markers. *Mol Reprod Dev* 2006;73:475–481.
66. Robinson KO, Ferguson HJ, Cobey S, Vaessin H, Smith BH. Sperm-mediated transformation of the honey bee, *Apis mellifera. Insect Mol Biol* 2000;9:625–634.
67. Beshers SN, Fewell JH. Models of division of labor in social insects. *Annu Rev Entomol* 2001;46:413–440.
68. Omholt SW, Rishovd S, Hagen A, Elmholdt O, Dalsgard B, Fromm S. Successful production of chimeric honeybee larvae. *J Exp Zool* 1995;272:410–412.
69. Aase ALTO, Amdam GV, Hagen A, Omholt SW. A new method for rearing genetically manipulated honey bee workers. *Apidologie* 2005;36:293–299.
70. Aupinel P, Fortini D, Dufour H, *et al.* Improvement of artificial feeding in a standard in vitro method for rearing *Apis mellifera* larvae. *J Insectol* 2005;58:107–111.
71. Czoppelt C, Rembold. H. Effect of parathion on honey bee larvae reared in vitro. *Anz Schadlingske* 1988;61:95–100.
72. Desneux N, Decourtye A, Delpuech J-M. The sublethal effects of pesticides on beneficial arthropods. *Annu Rev Entomol* 2007;52:81–106.
73. Peng YSC, Mussen E, Fong A, Montague MA, Tyler T. Effects of chlortetracycline of honey-bee worker larvae reared in vitro. *J Invertebr Pathol* 1992;60:127–133.
74. Rembold H, Lackner B. Rearing of honeybee larvae in vitro: Effect of yeast extract on queen differentiation. *J Apicult Res* 1981;20: 165–171.
75. Weaver N. Rearing of honeybee larvae on royal jelly in the laboratory. *Science* 1955;121:509–510.
76. Weaver N. Control of dimorphism in the female honeybee. 2: Methods of rearing larvae in the laboratory and of preserving royal jelly. *J Apicult Res* 1974;13:3–14.
77. Farooqui T, Robinson K, Vaessin H, Smith BH. Modulation of early olfactory processing by an octopaminergic reinforcement pathway in the honeybee. *J Neurosci* 2003;23:5370–5380.
78. Farooqui T, Vaessin H, Smith BH. Octopamine receptors in the honeybee (Apis mellifera) brain and their disruption by RNA-mediated interference. *J Insect Physiol* 2004;50:701–713.
79. Amdam GV, Simoes ZLP, Guidugli KR, Norberg K, Omholt SW. Disruption of vitellogenin gene function in adult honeybees by intra-abdominal injection of double-stranded RNA. *BMC Biotechnol* 2003;3:1–8.
80. Beye M, Hartel S, Hagen A, Hasselmann M, Omholt SW. Specific developmental gene silencing in the honey bee using a homeobox motif. *Insect Mol Biol* 2002;11:527–532.
81. Beye M, Hasselmann M, Fondrk MK, Page RE, Omholt SW. The gene csd is the primary signal for sexual development in the honeybee and encodes an SR-type protein. *Cell* 2003;114:419–429.
82. Wilson EO, Holldobler B. Eusociality: Origin and consequences. *Proc Natl Acad Sci USA* 2005;102:13367–13371.
83. Seeley TD. *Honeybee Ecology*. Princeton, NJ: Princeton University Press, 1985.
84. Fewell JH. Social insect networks. *Science* 2003;301:1867–1870.
85. Jeanson R, Kukuk PF, Fewell JH. Emergence of division of labour in halictine bees: Contributions of social interactions and behavioural variance. *Anim Behav* 2005;70:1183–1193.
86. Page RE, Erber J. Levels of behavioral organization and the evolution of division of labor. *Naturwissenschaften* 2002;89:91–106.
87. Robinson GE. Genomics and integrative analyses of division of labor in honeybee colonies. *Am Nat* 2002;160:S160–S172.
88. Robinson GE, Fahrbach SE, Winston ML. Insect societies and the molecular biology of social behavior. *BioEssays* 1997;19:1099–1108.
89. Robinson GE, Grozinger CM, Whitfield CW. Sociogenomics: Social life in molecular terms. *Nat Rev Genet* 2005;6:257–270.
90. Visscher PK. Colony integration and reproductive conflict in honey bees. *Apidologie* 1998;29:23–45.
91. Visscher PK. Group decision making in nest site selection among social insects. *Annu Rev Entomol* 2006;52:255–285.
92. Leoncini I, Crauser D, Robinson GE, Le Conte Y. Worker-worker inhibition of honey bee behavioural development independent of queen and brood. *Insect Soc* 2004;51:392–394.

93. Leoncini I, Le Conte Y, Costagliola G, *et al.* Regulation of behavioral maturation by a primer pheromone produced by adult worker honey bees. *Proc Natl Acad Sci USA* 2004;101:17559–17564.
94. Pankiw T, Roman R, Sagili RR, Zhu-Salzman K. Pheromone-modulated behavioral suites influence colony growth in the honey bee (Apis mellifera). *Naturwissenschaften* 2004;91:575–578.
95. Sullivan JP, Jassim O, Fahrbach SE, Robinson GE. Juvenile hormone paces behavioral development in the adult worker honey bee. *Horm Behav* 2000;37:1–14.
96. Giray T, Huang ZY, Guzman-Novoa E, Robinson GE. Physiological correlates of genetic variation for rate of behavioral development in the honeybee, *Apis mellifera. Behav Ecol Sociobiol* 1999;47:17–28.
97. Huang ZY, Robinson GE. Seasonal-changes in juvenile-hormone titers and rates of biosynthesis in honey-bees. *J Comp Physiol B Biochem Syst Environ Physiol* 1995;165:18–28.
98. Janmaat AF, Winston ML. Removal of *Varroa jacobsoni* infested brood in honey bee colonies with differing pollen stores. *Apidologie* 2000;31:377–385.
99. Kolmes SA, Winston ML. Division of labour among worker honey bees in demographically manipulated colonies. *Insect Soc* 1988; 35:262–270.
100. Bloch G, Robinson GE. Chronobiology—reversal of honeybee behavioural rhythms. *Nature* 2001;410:1048.
101. Hemmer W, Focke M, Kolarich D, Dalik I, Gotz M, Jarisch R. Identification by immunoblot of venom glycoproteins displaying immunoglobulin E-binding N-glycans as cross-reactive allergens in honeybee and yellow jacket venom. *Clin Exp Allergy* 2004;34: 460–469.
102. Frew AJ. Immunotherapy of allergic disease. *J Allergy Clin Immunol* 2003;111:712–719.
103. Harrison JF, Fewell JH. Environmental and genetic influences on flight metabolic rate in the honey bee, *Apis mellifera. Comp Biochem Physiol A Mol Integr Physiol* 2002;133:323–333.
104. Ohashi K, Sawata M, Takeuchi H, Natori S, Kubo T. Molecular cloning of cDNA and analysis of expression of the gene for alpha-glucosidase from the hypopharyngeal gland of the honeybee *Apis mellifera* L. *Biochem Biophys Res Commun* 1996;221:380–385.
105. Ohashi K, Natori S, Kubo T. Expression of amylase and glucose oxidase in the hypopharyngeal gland with an age-dependent role change of the worker honeybee (Apis mellifera L.). *Eur J Biochem* 1999;265:127–133.
106. Pontoh J, Low NH. Purification and characterization of beta-glucosidase from honey bees (*Apis mellifera*). *Insect Biochem Mol Biol* 2002;32:679–690.
107. Robinson GE, Vargo EL. Juvenile hormone in adult eusocial hymenoptera: Gonadotropin and behavioral pacemaker. *Arch Insect Biochem Physiol* 1997;35:559–583.
108. Capaldi EA, Smith AD, Osborne JL, *et al.* Ontogeny of orientation flight in the honeybee revealed by harmonic radar. *Nature* 2000;403:537–540.
109. Whitfield CW, Cziko AM, Robinson GE. Gene expression profiles in the brain predict behavior in individual honey bees. *Science* 2003;302:296–299.
110. Ben-Shahar Y, Robichon A, Sokolowski MB, Robinson GE. Influence of gene action across different time scales on behavior. *Science* 2002;296:741–744.
111. Tsuchimoto M, Aoki M, Takada M, *et al.* The changes of gene expression in honeybee (*Apis mellifera*) brains associated with ages. *Zool Sci* 2004;21:23–28.
112. Amdam GV, Omholt SW. The regulatory anatomy of honeybee lifespan. *J Theor Biol* 2002;216:209–228.
113. Page RE, Peng CYS. Aging and development in social insects with emphasis on the honey bee, *Apis mellifera* L. *Exp Gerontol* 2001;36:695–711.
114. Neukirch A. Dependence of the life span of the honeybee (*Apis mellifica*) upon flight performance and energy consumption. *J Comp Physiol B Biochem System Environ Physiol* 1982;146:35–40.
115. Schippers M-P, Dukas R, Smith RW, Wang J, Smolen K, McClelland GB. Lifetime performance in foraging honeybees: Behaviour and physiology. *J Exp Biol* 2006;209:3828–3836.
116. Schmid-Hempel P, Wolf T. Foraging effort and life span of workers in a social insect. *J Anim Ecol* 1988;57:500–521.
117. Tofilski A. Influence of age and polyethism on longevity of workers in social insects. *Behav Ecol Sociobiol* 2002;51:234–237.
118. Tofilski A. Senescence and learning in honeybee (*Apis mellifera*) workers. *Acta Neurobiol Exp* 2000;60:35–39.
119. Schmid-Hempel P, Kacelnik A, Houston AI. Honeybees maximize efficiency by not filling their crop. *Behav Ecol Sociobiol* 1985;17:61–66.
120. Higginson AD, Gilbert F. Paying for nectar with wingbeats: A new model of honeybee foraging. *Proc R Soc Lond B Biol Sci* 2004;271:2595–2603.
121. Higginson AD, Barnard CJ. Accumulating wing damage affects foraging decisions in honeybees (*Apis mellifera L.*). *Ecol Entomol* 2004;29:52–59.
122. Corona M, Hughes KA, Weaver DB, Robinson GE. Gene expression patterns associated with queen honey bee longevity. *Mech Ageing Dev* 2005;126:1230–1238.
123. Carey JR. Demographic mechanisms for the evolution of long life in social insects. *Exp Gerontol* 2001;36:713–722.
124. Stussi T, Harmelin ML. Recherche sur l'ontogenese du rythme circadien de la depense d'energie chez l'Abeille. *CR Acad Sci Hebd Seanc Acad Sci D* 1966;262:2066–2069.
125. Toma DP, Bloch G, Moore D, Robinson GE. Changes in period mRNA levels in the brain and division of labor in honey bee colonies. *Proc Natl Acad Sci USA* 2000;97:6914–6919.
126. Moore D, Angel JE, Cheeseman IM, Fahrbach SE, Robinson GE. Timekeeping in the honey bee colony: Integration of circadian rhythms and division of labor. *Behav Ecol Sociobiol* 1998;43: 147–160.
127. Moore D. Honey bee circadian clocks: Behavioral control from individual workers to whole-colony rhythms. *J Insect Physiol* 2001;47:843–857.
128. Bloch G, Solomon SM, Robinson GE, Fahrbach SE. Patterns of PERIOD and pigment-dispersing hormone immunoreactivity in the brain of the European honeybee (Apis mellifera): Age- and time-related plasticity. *J Comp Neurol* 2003;464:269–284.
129. Elekonich MM, Schulz DJ, Bloch G, Robinson GE. Juvenile hormone levels in honey bee (Apis mellifera L.) foragers: Foraging experience and diurnal variation. *J Insect Physiol* 2001;47:1119–1125.
130. Jassim O, Huang ZY, Robinson GE. Juvenile hormone profiles of worker honey bees, *Apis mellifera*, during normal and accelerated behavioural development. *J Insect Physiol* 2000;46:243–249.
131. Kubo T, Sasaki M, Nakamura J, *et al.* Change in the expression of hypopharyngeal-gland proteins of the worker honeybees (Apis mellifera L) with age and/or role. *J Biochem* 1996;119:291–295.
132. Fluri P, Lüsher M, Willie H, Gerig L. Changes in weight of the pharyngeal gland and haemolymph titers of juvenile hormone, protein and vitellogenin in worker honey bees. *J Insect Physiol* 1982;28:61–68.
133. Bloch G, Sullivan JP, Robinson GE. Juvenile hormone and circadian locomotor activity in the honey bee Apis mellifera. *J Insect Physiol* 2002;48:1123–1131.
134. Rose U, Ferber M, Hustert R. Maturation of muscle properties and its hormonal control in an adult insect. *J Exp Biol* 2001;204: 3531–3545.
135. Dingle H, Winchell R. Juvenile hormone as a mediator of plasticity in insect life histories. *Arch Insect Biochem Physiol* 1997;35:359–373.
136. Zera AJ, Cisper G. Genetic and diurnal variation in the juvenile hormone titer in a wing-polymorphic cricket: Implications for the evolution of life histories and dispersal. *Physiol Biochem Zool* 2001;74:293–306.
137. Zhao ZW, Zera AJ. The hemolymph JH titer exhibits a large-amplitude, morph-dependent, diurnal cycle in the wing-polymorphic cricket, Gryllus firmus. *J Insect Physiol* 2004;50:93–102.

138. Sullivan JP, Fahrbach SE, Harrison JF, Capaldi EA, Fewell JH, Robinson GE. Juvenile hormone and division of labor in honey bee colonies: Effects of allatectomy on flight behavior and metabolism. *J Exp Biol* 2003;206:2287–2296.
139. Dyer FC. The biology of the dance language. *Annu Rev Entomol* 2002;47:917–949.
140. Menzel R, Giurfa M. Dimensions of cognition in an insect, the honeybee. *Behav Cogn Neurosci Rev* 2006;5:24–40.
141. Bitterman ME, Menzel R, Fietz A, Schafer S. Classical conditioning of proboscis extension in honeybees (*Apis mellifera*). *J Comp Psychol* 1983;97:107–119.
142. Takeda K. Classical conditioned response in the honey bee. *J Insect Physiol* 1961;6:168–179.
143. Menzel R. Searching for the memory trace in a mini-brain, the honeybee. *Learn Mem* 2001;8:53–62.
144. Meller VH, Davis RL. Biochemistry of insect learning: Lessons from bees and flies. *Insect Biochem Mol Biol* 1996;26:327–335.
145. Müller U. Second messenger pathways in the honeybee brain: Immunohistochemistry of protein kinase A and protein kinase C. *Microsc Res Tech* 1999;45:165–173.
146. Locatelli F, Bundrock G, Muller U. Focal and temporal release of glutamate in the mushroom bodies improves olfactory memory in *Apis mellifera*. *J Neurosci* 2005;25:11614–11618.
147. Elder HY. Muscle structure. In: Usherwood PNR, Ed. *Insect Muscle*. London: Academic Press, 1975.
148. Rothe U, Natchtigall W. Flight of the honey bee IV: Respiratory quotients and metabolic rates during sitting, walking and flying. *J Comp Physiol B Biochem System Environ Physiol* 1989;158: 739–749.
149. Suarez RK. Energy metabolism during insect flight: Biochemical design and physiological performance. *Physiol Biochem Zool* 2000;73:765–771.
150. Stokes DR. Insect muscle innervated by single motorneurons: Structural and biochemical features. *Am Zool* 1987;27:1001–1010.
151. Wegener G. Flying insects: Model systems in exercise physiology. *Experientia* 1996;52:404–412.
152. Josephson RK, Malamud JG, Stokes DR. Asynchronous muscle: A primer. *J Exp Biol* 2000;203:2713–2722.
153. Dickinson MH, Tu MS. The function of Dipteran flight muscle. *Comp Biochem Physiol* 1997;116A:223–238.
154. Dickinson MH, Lehmann FO, Chan WP. The control of mechanical power in insect flight. *Am Zool* 1998;38:718–728.
155. Roberts SP, Harrison JF. Mechanisms of thermal stability during flight in the honeybee *Apis mellifera*. *J Exp Biol* 1999;202:1523–1533.
156. Fewell JH, Harrison JF. Variation in worker behavior of African and European honey bees. Proceedings of the Second International Congress on Africanized Bees and Bee Mites, Tucson, AZ, 2001.
157. Harrison JM. Caste-specific changes in honeybee flight capacity. *Physiol Zool* 1986;59:175–187.
158. Moritz RFA. Biochemical changes during honeybee flight muscle development. In: Nachtigall W, Ed. *The Flying Honeybee; Aspects of Energetics*. Stuttgart: Gustav Fischer, 1988:51–65.
159. Coelho JR, Mitten JB. Oxygen consumption during hovering is associated with genetic variation of enzymes in honey-bees. Funct Ecol 1988; 2:141–146.
160. Harrison JF, Fewell JH, Roberts SP, Hall HG. Achievement of thermal stability by varying metabolic heat production in flying honeybees. *Science* 1996;274:88–90.
161. Blomstrand E, Eckblom B, Newsholme E. Maximum activities for key glycolytic and oxidative enzymes in human muscle from differently trained individuals. *J Physiol* 1986;381:111–118.
162. Suarez RK, Staples JF, Lighton JRB. Turnover rates of mitochondrial respiratory chain enzymes in flying honey bees (Apis mellifera). *J Exp Zool* 1999;283:1–6.
163. Suarez RK, Lighton JRB, Moyes CD, Brown CS, Gass CL, Hochachka PW. Fuel selection in rufous hummingbirds: Ecological implications of metabolic biochemistry. *Proc Natl Acad Sci USA* 1990;87:9207–9210.
164. Suarez RK, Lighton JR, Joos B, Roberts SP, Harrison JF. Energy metabolism, enzymatic flux capacities and metabolic flux rates in flying honeybees. *Proc Natl Acad Sci USA* 1996;93:12616–12620.
165. Roberts SP, Elekonich MM. Muscle biochemistry and the ontogeny of flight capacity during behavioral development in the honey bee, *Apis mellifera*. *J Exp Biol* 2005;208:4193–4198.
166. Gupta P, Greenberger PA. 4. Stinging insect allergy and venom immunotherapy. *Allergy Asthma Proc* 2004;25:9–10.
167. Bilo BM, Rueff F, Mosbech H, Bonifazi F, Oude-Elberink JNG. Diagnosis of Hymenoptera venom allergy. *Allergy* 2005;60:1339–1349.
168. Hamilton RG. Diagnostic methods for insect sting allergy. *Curr Opin Allergy Clin Immunol* 2004;4:297–306.
169. Quercia O, Emiliani F, Pecora S, Burastero SE, Stefanini GF. Efficacy, safety, and modulation of immunologic markers by immunotherapy with honeybee venom: Comparison of standardized quality depot versus aqueous extract. *Allergy Asthma Proc* 2006;27: 151–158.
170. Hemmer W, Focke M, Kolarich D, *et al.* Antibody binding to venom carbohydrates is a frequent cause for double positivity to honeybee and yellow jacket venom in patients with stinging-insect allergy. *J Allergy Clin Immunol* 2001;108:1045–1052.
171. Rueff F, Przybilla B, Muller U, Mosbech H. The sting challenge test in Hymenoptera venom allergy. *Allergy* 1996;51:216–225.
172. Rueff F, Wolf H, Schnitker J, Ring J, Przybilla B. Specific immunotherapy in honeybee venom allergy: A comparative study using aqueous and aluminium hydroxide adsorbed preparations. *Allergy* 2004;59:589–595.
173. Kettner A, Hughes GJ, Frutiger S, *et al.* Api m 6: A new bee venom allergen. *J Allergy Clin Immunol* 2001;107:914–920.
174. Peiren N, de Graaf DC, Brunain M, *et al.* Molecular cloning and expression of icarapin, a novel IgE-binding bee venom protein. *FEBS Lett* 2006;580:4895–4899.
175. Pedersen SF, Poulsen KA, Lambert IH. Roles of phospholipase A2 isoforms in the swelling- and melittin-induced arachidonic acid release and taurine efflux in NIH3T3 fibroblasts. *Am J Physiol Cell Physiol* 2006;291(6):C1286–1296.
176. Putz T, Ramoner R, Gander H, Rahm A, Bartsch G, Thurnher M. Antitumor action and immune activation through cooperation of bee venom secretory phospholipase A2 and phosphatidylinositol-(3,4)-bisphosphate. *Cancer Immunol Immunother* 2006;55:1374–1383.
177. Grisotto LSD, Mendes GE, Castro I, *et al.* Mechanisms of bee venom-induced acute renal failure. *Toxicon* 2006;48:44–54.
178. Muller UR. Bee venom allergy in beekeepers and their family members. *Curr Opin Allergy Clin Immunol* 2005;5:343–347.
179. Lee JD, Park HJ, Chae Y, Lim S. An overview of bee venom acupuncture in the treatment of arthritis. *Evidence-Based Compl Alt Med* 2005;2:79–84.
180. Kang SS, Pak SC, Choi SH. The effect of whole bee venom on arthritis. *Am J Chin Med* 2002;30:73–80.
181. Lee SH, Hong SJ, Kim SY, *et al.* Randomized controlled double blind study of bee venom therapy on rheumatoid arthritis. *J Korean Acupunct Moxibustion Soc* 2003;20:80–83.
182. Wang OH, Ahn KB, Lim JK, Jang HS. Clinical study of the effectiveness of bee venom therapy on degenerative knee arthritis. *J Korean Acupunct Moxibustion Soc* 2001;18:35–47.
183. Wesselius T, Heersema DJ, Mostert JP, *et al.* A randomized crossover study of bee sting therapy for multiple sclerosis. *Neurology* 2005;65:1764–1768.
184. O'Connell N. It's all the buzz. *Nurs Stand* 2005;20:22–24.

17 Establishing and Maintaining a *Xenopus laevis* Colony for Research Laboratories

Punita Koustubhan, Debra Sorocco, and Michael S. Levin

ABSTRACT

This Chapter describes in detail the construction and maintenance of a *Xenopus laevis* colony. It covers *Xenopus* husbandry, facility design and construction, necessary maintenance protocols, and monitoring/disaster recovery procedures. Facility components and management protocols described within this chapter are based on our experience with the frog colony at The Forsyth Center for Regeneration and Developmental Biology as well as the information obtained from many other investigators. The system described in this chapter is AAALAC accredited and provides a steady supply of embryology-quality eggs and ensures excellent water quality and adult frog care. This information is a complete guide, allowing a new investigator to begin work with this powerful model species that has resulted in profound advances in developmental, cell, and cancer biology.

Key Words: *Xenopus laevis*, Frog, Developmental biology, Colony, Husbandry, Protocols.

INTRODUCTION

There are numerous advantages that steer investigators toward this South African clawed frog as a vertebrate model.[1] *Xenopus laevis* embryos develop externally allowing observation throughout all stages of development. Frog eggs are large and the developing embryo is visible making them an excellent research model. A tremendous number of cell, molecular, genetic, and developmental biology techniques have been developed for *Xenopus* and this model has been used in more than 33,000 papers in Medline as of 2006. *Xenopus* has been in practical use since at least the 1940s, once being a popular diagnostic tool for human pregnancy.[2–5] An excellent practical resource for embryonic work with *Xenopus* is Sive *et al.*,[6] and standard atlases provide extremely detailed information on embryonic development in this species.[7,8] The zoology of *Xenopus* tadpoles is covered in McDiarmid and Altig.[9] As a result, the *Xenopus* model system has contributed to both classical and modern breakthroughs in developmental and evolutionary biology, ethology, neurobiology, regeneration, endocrinology, toxicology, and cancer biology.

Xenopus is a convenient and powerful model for vertebrate biology. However, a wide variety of husbandry methods is used, with little standardization.[10–14] Current frog laboratories utilize techniques inherited from a previous mentor (and anecdotal suggestions abound in the community), but the task of designing, building, and maintaining a thriving frog colony is very daunting and time-consuming to the beginning investigator who needs to set up this model system from scratch. In the absence of such specific protocols and construction details, it is estimated that it took us over a year of hard work, from the opening of the new laboratory to the ability to routinely do embryo experiments. This chapter is presented in the hopes that we can radically shorten this time period for new investigators, and lower the barrier for others to become involved in the use of this model system.

This chapter provides most of the necessary information, protocols, considerations, and materials lists to run a facility. We give general advice where possible, but the specific details are taken directly from our running habitat at the Forsyth Center for Regenerative and Developmental Biology, where we have been using *Xenopus* to understand the biophysical control mechanisms of laterality, spinal cord regeneration, and eye development.[15–23] This chapter is intended to provide all of the information necessary to build and operate such a system. The lessons learned through the years of practical experience and discussions with the *Xenopus* community resulted in what we believe to be a very successful system. Our laboratory obtains large number of excellent embryos every week, with no seasonal variability.

Because our goal is to provide specific and detailed information on a working system, we do not give details on the construction of other types of habitats. A wide range exists in the community, and many people run successful colonies with standing tanks (in contrast to our flow-through system). We have found our system to be a good choice for a medium-sized laboratory, but there are many other ways to keep *Xenopus*.[24–26] The key principle is to build an environment conducive to very happy and healthy frogs, which is the best way to ensure high-quality eggs for any research program. In this chapter, we describe the optimal setup, but some aspects of it can be simplified if necessary to minimize cost.

FACILITY DESIGN AND COMPONENTS

PLANNING AND CONSIDERATIONS FOR DESIGN The first step is to make decisions on how the embryos will be used in the program. If *Xenopus* usage is low, the investment in a complex flow-through system may not be justified. In such cases, simple standing tanks or even purchasing females on an as-needed basis may be good alternatives. However, given the cost of good-quality *Xenopus* females (~$40), and the fact that a single female

From: *Sourcebook of Models for Biomedical Research* (P. M. Conn, ed.), © 2008 Humana Press Inc., Totowa, NJ.

Table 17–1
Estimate of facility design and installation costs

Item description	*Vendor*	*Part number*	*Price estimate*
Frog facility design and install	Marine Biotech, Inc.	X-MOD System	\$40,000 (2001)
Water chillers	Marine Biotech, Inc.	N/A	\$2,600 (2001)
Water polishing system	Hydro Service & Supplies	Picopure 2 UV Plus	\$3,400 (2001)
Supply cart, nets, buckets, ladder, digital thermometer, etc.	Aquatic Ecosystems	Varies	\$500 (2001)
Monitoring system	Electromechanica, Inc.	N/A	\$7,500 (2002)
Web camera	Veo Observer	V700000	\$160 (2002)
Total		~\$54,000	

should rest for 3–4 months between uses, a medium to large-sized laboratory will find it cost-effective to maintain the kind of facility described here. The expense should also be considered in terms of materials for the initial construction, supplies, and labor costs for maintenance. As a rough guideline, maintenance of this facility requires a minimum of 20 h per week. A dedicated technician who is available at least 30 h a week (including extended holidays and unpredictable emergency efforts) is ideal. The facility needs human attention daily, but when everything is working well, weekend attendance is not required. In general, the most crucial factor is the human element: it is essential to have a smart technician, preferably with experience in animal care and a dedication to their well-being, and to restrict access to the colony to as few individuals as possible. Establishing continuity of expert attention to the facility is the only way to ensure efficient and trouble-free operation. A commitment of at least 1 year is a minimum when choosing a technician. A skilled technician can learn to perform the day-to-day maintenance and monitoring in about 2 months. Becoming an expert in some of the rarer procedures and especially in handling crises (mechanical and biological) takes significantly longer.

The availability of the following nearby (especially making sure they are willing to make house calls on a time scale of 1–2 days) will prove very useful: an aquatic animal veterinarian, a company to do aquatic habitat plumbing, and a company providing water purification technology). The system we describe below services a laboratory in which five postdoctoral fellows utilize the frog embryo almost exclusively. It is sufficient to provide 16 frogs per week on a 365 days/year continuous cycle, which is enough for *Xenopus* experiments 4–5 days out of the week. Some of the costs associated with establishing this facility from scratch are shown in Tables 17–1–17–6. Also to be taken into account are space considerations; a room at least 600 ft^2 in area is needed, and if possible, all of the plumbing and pumps should be located in a separate, adjacent room to allow plumbers and other personnel to work without impacting the facility itself (also reducing the significant ambient noise for the technician maintaining the colony). Costs may be conserved by relying on ambient air temperature and water quality provided by the building instead of the water purifier and chillers we describe below. However, it has been our experience that in the long run, this is not cost effective because building air conditioning and water quality controls are rarely sufficient for a healthy colony, regardless of their official specifications.

OVERVIEW OF A FLOW-THROUGH X-MOD SYSTEM Good water quality is central to managing a frog colony. Any toxins or major fluctuations in water chemistry will cause physiological changes eventually leading to disease and reducing embryo quality even if adults do not seem obviously affected. This facility is designed to eliminate these sources of stress, and provide exceptional care for the frogs. As a rule, one frog per gallon (3.79 liters) is the optimum population density, however, there are other factors to consider when housing *X. laevis*. It is important that there is enough surface area available for the frogs to be able to rest on the bottom without touching any other frogs. Another important task is to calculate whether the biological filter and other components in the recirculating system, such as the carbon and particulate filters, are large enough to handle the amount of waste material generated by the frogs.

Table 17–2
Spare parts[a]

Item description	*Vendor*	*Part number*	*Price estimate*
Small UV (25 W)	Aqua Ultraviolet	A20025	\$58
Large UV (40 W)	Aqua Ultraviolet	A20040	\$68
UV sleeves	Aqua Ultraviolet	A10025, A10040	\$70, 80
Air pump	Hailea	LT24	\$170
Large water pump	March	TE-6T-MD	\$250
Reservoir pump	March	BC-3C-MD	\$190
Ultrameter pH/ORP probe	Myron-L	RPR	\$137
Total		\$1023	

[a]All items were purchased through Marine Biotech, Inc. circa 2001.

Table 17–3
Water chemistry start-up costs

Item	*Vendor*	*Part number*	*Price*
Water meter	Myron-L	Ultrameter 6P	$733
Calibration buffers	Myron-L	pH 4, 7, 10; conductivity KCl-1800 (1 qt); pH/ORP storage sol. (1 gallon)	$16/each $58
Alkalinity test kit	Aquarium Systems	TF-700[a]	$11.34
General and carbonate hardness test kit	Nutrafin (Hagen)	A-7830[a]	$4.50
Freshwater test kit (NH_3, NO_2)	Marine Enterprises	CA-10[a]	$11.98
Nitrate kit (NO_3)	Marine Enterprises	CA-11[a]	$9.35
Freshwater carbon (20 lb)	Kent Marine, Inc.	N/A	$70
Particulate filter	Spawarehouse.com	FC1460	$39
Prefilter pads 20 × 20 inches	Marine Biotech, Inc.	N/A	$3.75 × 12
Alkaline buffer (1 kg)	Seachem	CD508183[b]	$14.99
1 μm filter (polishing)	Marine Biotech, Inc.	FLTR0364	$26.25
Acid buffer (1 kg)	Seachem	CD923118[b]	$14.99
Sea salt (~50 lb)	Instant Ocean	16881[b]	$31.99
Cichlid lake salt (3.2 kg)	Seachem	CD922923[b]	$42.99
Equilibrium salts (600 g)	Seachem	CD900009[b]	$7.19
Biofilter bacteria	Fritz Pet Products	Fritz-Zyme Turbo-start 700[b]	$81.30 × 2
Sanaqua (16 oz)	Novalek	33246–16 fl oz[a]	$7
Total		~ $1350	

[a]Manufacturer part numbers, as distributed through Marine Biotech, Inc.
[b]Part numbers from DrsFosterSmith.com, the distributor from which we purchased these materials.

Table 17–4
Frog colony start-up costs[a]

Description	*Vendor*	*Catalog number*	*Quantity*	*Price*
Males	Nasco	LM00715M (male)	12 @ $21	$252
Wild-type females	Nasco	LM00531M (pig)	12 @ $31	$372
Albino females	Nasco	LM00531A (alb)	12 @ $36	$432
Frog chow (5 lb)	Nasco	SA05960 (LM)	1	$16
Oxytetracycline (50 g)	Sigma	O-5875	1	$102
Stresscoat (16 oz)	Xenopus Express	SC	1	$10
Levamisole HCl (10 g)	ICN[b]	15522810	1	$32.50
Total			$1200	

[a]We recommend increasing the population slowly, to allow the biofilter enough time to colonize, equalize, and handle increased waste loads.
[b]This item is ordered through Fisher Scientific as catalog number NC15522810.

Table 17–5
Operational costs and consumables

Description	*Cost*
Utilities (electricity/water/HVAC)	Varies with overhead
Water polishing system service contract	$3000/year
Water chemistry additives	$120/month
Cleaning agents (pipe cleaners, scrub mitts, solvents)	Varies
Filtration (prefilter, 1 μm, particulate, carbon)	$70/month
Frog brittle (for 200 frogs)	$16/month
Male frogs (36/year, $21/each)	$756/year
Medications (oxytetracycline, levamisole, stresscoat)	$144/year
Technician salary	$30,000/year
Total annual cost	$38,094

Table 17–6
Time commitment

Task	*Hours*
Daily water chemistry	2.5 h/week
Feeding/cleaning	5 h/week
System maintenance	4 h/week
Troubleshooting	1 h/week
Technical training period	~1 month
Annual technician hours	650 h/year

Described below is a recirculating housing system (Figure 17–1) known as an X-MOD (Marine Biotech), designed to hold many aquatic animals in a clean environment. One major feature of this facility is that if any of the electrical/mechanical components stop functioning, the frog tanks automatically stabilize as static containers with a constant water level. Thus, this system can handle emergencies lasting for about 2 days (after which wastes build up and algae growth becomes a problem). The design incorporates three racks. Each rack has two rows of clear, polycarbonate tanks. There are six 75.8-liter tanks on each rack. Every tank has a securable lid with holes drilled in it allowing airflow at the surface of the tank. At the center of a tank, there are inner and outer tubes that are designed to drain water and particles out of the tank while preventing the frogs from being pulled into the drain tubes. The water leaves the tank along a pipe that connects to the return pipe in the sump draining over a polyester filter pad into the sump. There are two sumps labeled "sump 1" and "sump 2." Both sumps are connected by an "equalization" pipe, which allows water flow from sump 1 to sump 2. There are three racks in the 10 × 16-foot room. The room has tiled walls and a tiled floor with two floor drains. Our design accommodates 275 frogs comfortably.

Fluorescent lights connected to an automatic timer regulate photoperiod. The lights are mounted on the ceiling in between the racks. The timer is set on a 12 : 12 day : night cycle and the frogs receive a consistent light cycle year round (the room contains no windows). Full spectrum lights may be used and are often suggested, but we have achieved excellent results with fluorescent lighting. It is questionable whether the additional wavelengths provided by full-spectrum bulbs can penetrate the water to any depth.

In their native habitat, South African clawed frogs rest on the bottom of dark ponds, under leaves or in logs, so naturally they do not like bright lights. Keeping the lights dim and providing the frogs' tank with large PVC pipes enable the frogs to have a welcome hiding place. Indirectly, it also discourages the growth of nuisance algae. Many frog facility managers are incorporating behavioral enhancement into animal protocols due to mounting evidence that environmental enrichment promotes well-being and health.[27,28] The tank walls are covered with black plastic sheets that are taped to the back and sides of each tank to provide a "cave-like" environment.

The logic of water flow in the system is shown in Figures 17–2 and 17–3A. Our tanks are maintained at a flow rate of 6 gallons per minute (GPM). This works out to a system turnover rate of about once every 22 min or approximately three full circulations per hour. These racks have flow meters measuring flow rate in GPM, and use simple ball valves to adjust the flow to the racks. Each tank has its own spigot to regulate water flow, which is used to turn off the water supply to an individual tank or control the water flow during feeding, cleaning, or quarantining frogs. Excessively strong water flow contributes to frog stress by irritating their sensitive lateral line system; water flow should be monitored daily.

A 0.5-horsepower pump drives water through the circulation loop. It is placed before the filtration series to maximize water pressure across the filters. All of the filters have an automated pressure gauge. The pressure gauge from each piece of equipment is also connected to an automated monitoring system (see below). Routine inspection and maintenance of filters and following a weekly cleaning protocol keep the water pressure constant throughout the system.

As water flows into each tank (Figure 17–3B), debris is swept toward the drain tubes; the outer tube allows water and debris to enter at the base, but creates a barrier between the frogs and the actual drain. Heavy debris accumulates at the bottom, while smaller particles are carried into the inner drain tube and down the drain. Two valves beneath each tank control the direction of tank water; one controls flow to the floor drain while the other recirculates the water. By default, we keep the system drain valve closed and the recirculation drain valve open. The only time to switch would be to temporarily switch the orientation of both valves simultaneously to clear debris, especially after feeding, or to close both drains and shut off the water flow when quarantining frogs. All of the tank wastewater is collected into two central wastewater pipes that empty onto prefilter assemblies over the sumps.

FILTRATION COMPONENTS OF AN X-MOD SYSTEM Filtration occurs at several levels in the frog facility. We will begin by looking at what happens in each frog tank, and then follow the flow of water through the recirculating system. As water flows into each tank, debris is swept toward the drain. The outer grated tube allows water and debris to enter at the base while creating a barrier between the frogs and the actual drain. Heavy debris accumulates at the bottom, while smaller particles are carried into the inner drain tube and down the drain. Two valves beneath each tank control whether the wastewater goes back into circulation (to the sump) or out the floor drain to a sewer. The system drain valve is kept closed and the recirculation drain valve is open unless the tank is being drained during quarantine or cleaning.

The prefilter assembly is composed of a polycarbonate housing that holds prefilter pads supported by a PVC grid underneath. Polyester prefilter pads provide the first level of mechanical filtration by capturing solid debris and allowing wastewater to drain through the grid to the biofilter in the sump below. Beneath the prefilter assembly is a sump with a separator for biofilter media (we use Kaldness™ bioballs). The biofilter (Figure 17–4) removes toxins released from decaying organic matter and metabolic by-products using a two-step biochemical reaction. The Kaldness™ media greatly increases the surface area available for nitrifying bacteria colonization. Nitrifying bacteria and seeding the sumps are discussed in the section on Water Quality. An air pump attached to a perforated PVC manifold provides bubbles to continuously move and thus aerate each biofilter.

The wastewater from each sump is diverted into the bead filter (Figure 17–5A), The X-MOD has a Hayward Pro™ series bead filter containing 1 ft^3 (55 lb) of Perm-Beads™. Not only does the

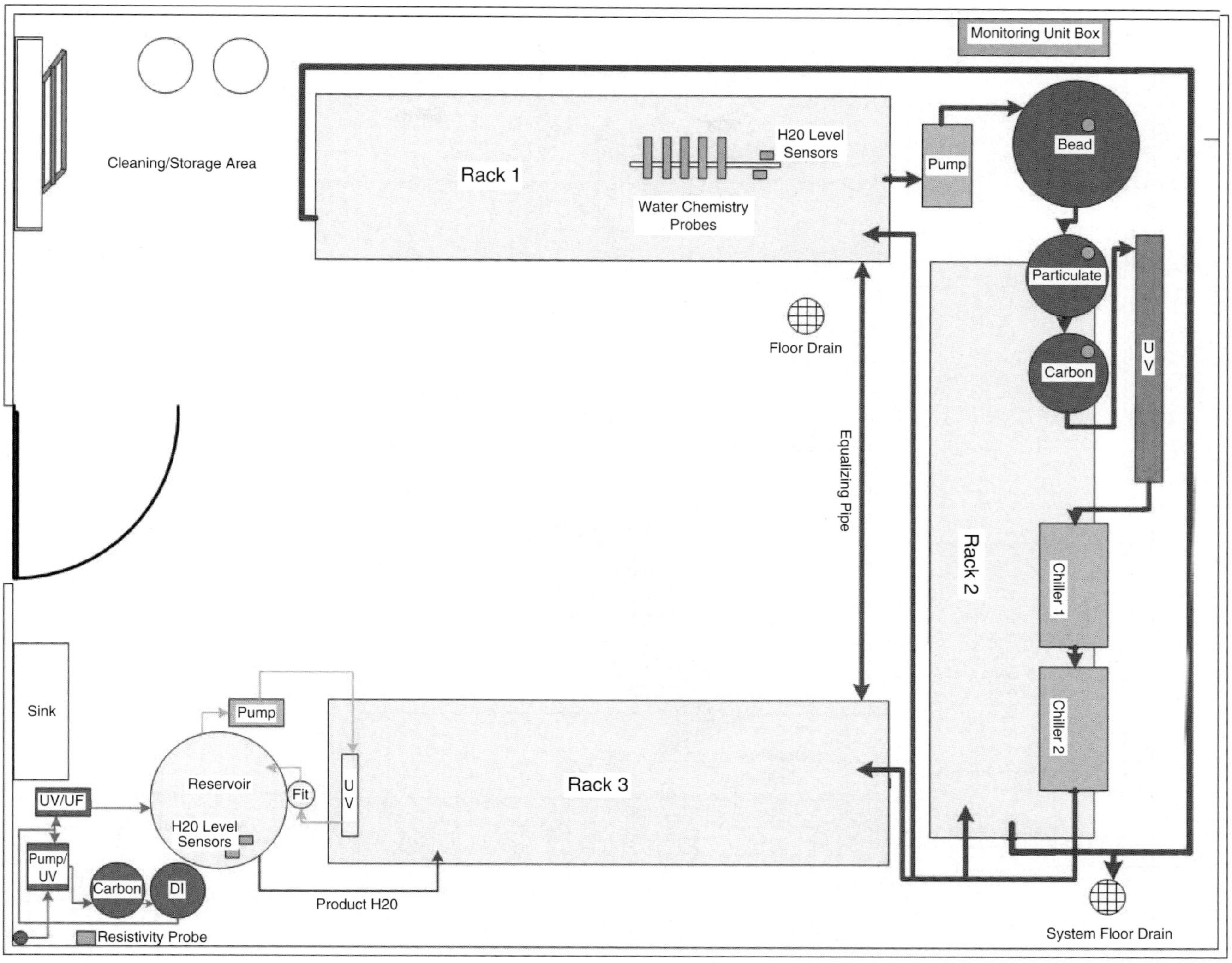

B

Figure 17–1. (A) A schematic of the FCRDB *Xenopus laevis* facility. Probes reporting to the online monitoring system are shown on Rack 1, Other components such as the RO-DI polishing system, housing racks, filters, pumps, and chillers are shown. The arrows indicate water flow through the system. (B) *Xenopus laevis* facility. A photograph of a three-rack system. Filters and chiller barrels are located below the middle rack against the back wall. Not shown are the holding tank and water purification equipment.

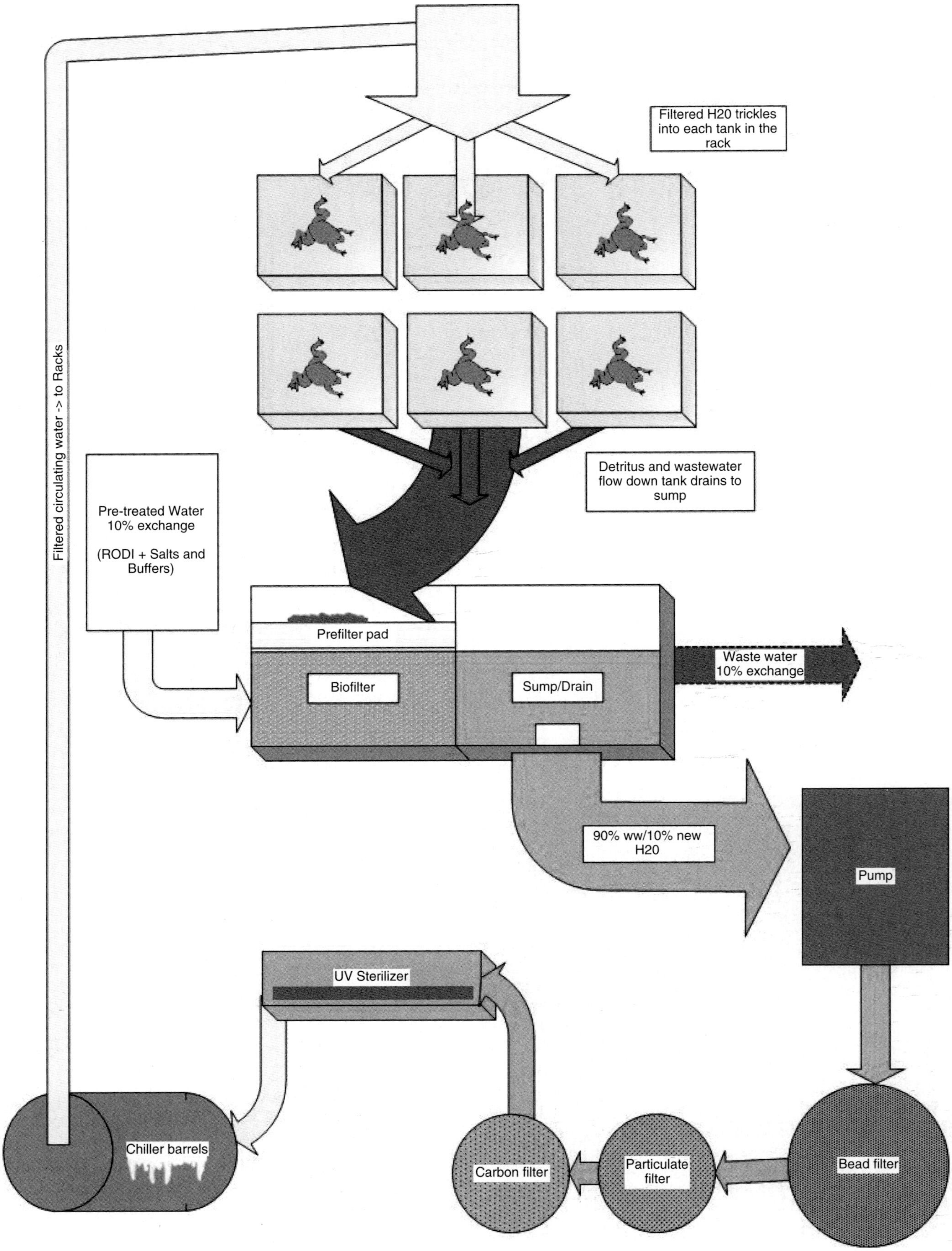

Figure 17–2. Schematic of water flow through the X-MOD system The water flow through the X-MOD system starts with the system pump, which drives water through the bead filter, the particulate filter, and the carbon filter, to thoroughly remove large particles and contaminants. The water passes through the UV sterilizer and the chiller barrels. The tanks are designed so that water is gently drawn toward the center of the tank pulling in debris and flushing out of the tanks down the pipe. The pipes are all connected and wastewater from the tanks ends up in the sump. A prefilter pad catches large particles.

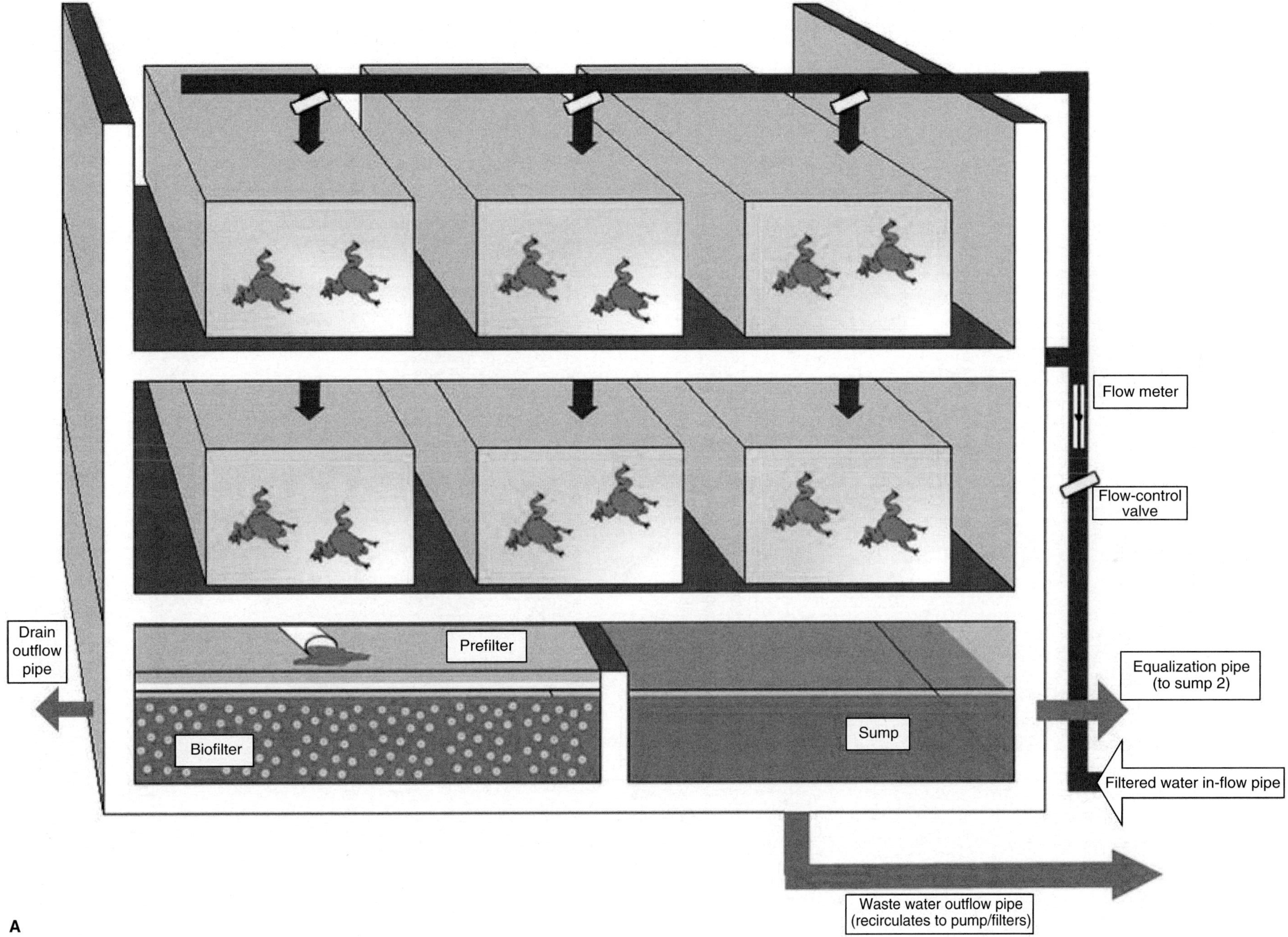

Figure 17–3. (A) Diagram of a single rack This schematic depicts how the racks are designed to hold frog tanks and circulate water. Water is pumped up the vertical pipe, into each tank, and then flows down into the sump, passing over the prefilter and biofilter. Wastewater flows back through filtration, out the drain overflow pipe, or to the other rack's sump to equalize the water level.

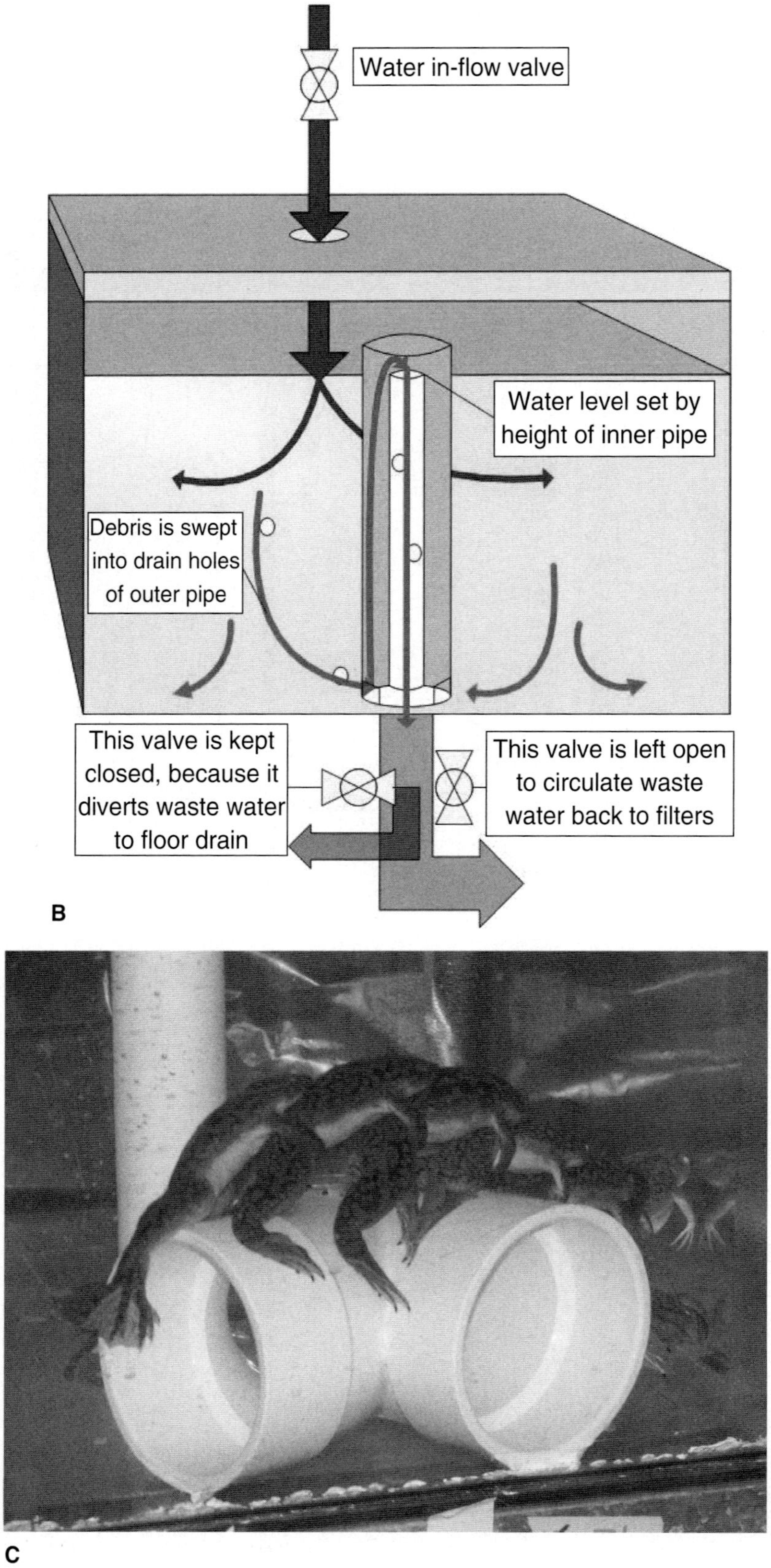

Figure 17–3. (continued) (B) Diagram of an individual tank (A). This diagram illustrates the design of each frog tank. Water pressure from the pump forces water into each tank, as controlled by the water in-flow valve. Excess water flows into the holes at the base of the outer pipe, and drips down the drain inside the inner pipe. The current created from the water flow draws wastewater and debris. (C) A chain of male frogs sitting on top of a PVC tube provided for enrichment.

Figure 17–4. Photo of a biofilter with Kaldness media The tank on the left contains Kaldness media or "bioballs" made of PVC. The bioballs provide adequate surface area for the growth of beneficial bacteria. The particles in this biofilter are continuously aerated by a powerful air pump. The tank on the right is one of the sumps, and the wires are connected to the WebDaq probes that continuously monitor water quality.

bead filter break up debris, assisting with the mechanical filtration component of the system, but it is also a major part of the biological filter. The Perma-Beads™ in the bead filter provide surface area for the bacteria to colonize and are almost indestructible. They have a unique oval shape that keeps water flowing and reduces clumping while breaking up debris and breaking down toxins. Most of the nitrification takes place in the bead filter. Often this bead filter will become loaded with frog waste material, so it is important to constantly monitor the pressure and backwash if it exceeds the proper pressure rating. In the absence of a pressure gauge, backwash after feeding or at least three times a week to keep the beads fluid and keep water flowing throughout all filter components. Follow the individual manufacturer's instructions if purchasing a bead filter from anyone other than Hayward Pro.

The particulate and carbon filters are the final step in mechanical filtration (Figure 17–5B). Together these filters remove smaller particles (particulate filter) and any colors or odors (activated carbon) as they exit the bead filter. These filters require changing at least once a month, or whenever there is a significant rise in pressure. A decreased flow rate through the frog tanks could indicate the need to change the particulate filter. When changing the particulate and carbon filters, turn off the pump and close the pump intake valve. Open the waste drain valve to drain the particulate and carbon filters. If using a CFR50 model, unscrew the valve on top releasing the pressure inside the chamber. Holding the yellow lever down, unscrew the large plastic band, firmly grip the lid, and pull it off. Once the chamber is open, slide the filter out to remove and replace it. When finished, reassemble the chamber, open the water valves, start the pump, and be sure to check for leaks. To service filters other than the Jacuzzi CFR50 free-standing carbon filter follow the manufacturer's instructions to properly remove and replace the filter.

The final filtration step is a 40-W sterilizing ultraviolet (UV) bulb, with an intensity of 100,000 μW·sec/cm^2. The UV light targets any microorganisms circulating in the system water. When servicing the unit, inspect the O-rings and seals for degradation and promptly replace worn parts. Follow the manufacturer's instructions concerning changing the bulb. In this facility, the bulbs are changed annually. Be sure to use gloves when handling

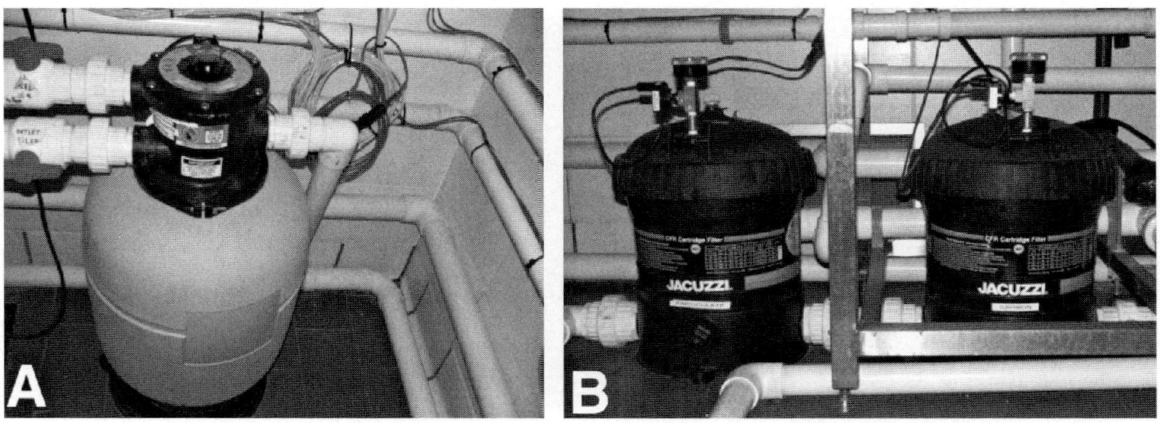

Figure 17–5. The filters. (A) Bead filter; (B) particulate filter and carbon filter with pressure gauges.

the UV bulb and quarts sleeve. Any residue or dirt left on the bulb will reduce the effectiveness of the UV sterilizer. Always have an extra supply of quarts sleeves and a UV bulb in stock in case of an emergency.

The water leaves the UV sterilizer and passes through the chiller barrels, cooling the water before entering the tanks. Temperature in our facility is maintained between 17.0°C and 18.5°C. Although these frogs can tolerate a wide range of temperatures, keeping the water at 18.0°C will ensure excellent egg quality[29] and healthy frogs. The X-MOD system is equipped with a chiller to maintain water temperature, which results in good egg quality throughout the year (avoiding the "Summer downtime" seen in many *Xenopus* laboratories). Our chiller system uses a combination of outdoor air supplementation as well as conventional refrigeration. An outdoor air-cooled chiller regulates water temperature in the winter and a mechanical water-cooled chiller controls it in the summer. This "hybrid" system saves energy and reduces wear on the chiller barrels. We set a 1°C temperature differential between the activation settings on each chiller. The primary chiller remains active at 1°C above desired temperature, and the backup chiller activates only if the temperature rises an additional degree or the primary chiller malfunctions.

WATER QUALITY

Water is the main determinant of health in the system. When building a facility, the main concern is to avoid distress and disease; investing in a polishing system exclusively for purifying incoming water for the frog facility is imperative.[30] Our institution provides the building with reverse osmosis deionized (RO-DI) water by passing municipal water through a series of sand, carbon, reverse-osmosis, and deionization membranes. The RO-DI water is held in a large storage tank on the roof of our building. Although this water is RO-DI and appears pure, there are many sources of contamination in the old pipes that can change the quality of the water needed for the facility. Public water systems in the United States use chlorine and/or chloramine as a disinfectant. Chlorine can be removed by aerating the water, but chloramines are persistent, requiring either chemical agents or carbon filtration for neutralization. Chloramine is a combination of ammonia and chlorine, and can directly enter the bloodstream of aquatic animals, slowly poisoning them. If you choose to use chemical neutralizing agents, you must monitor both ammonia and chlorine levels as each are released into the water after chloramine breakdown. There are many other dissolved substances with unknown effects on water chemistry that can impair the health of the colony. These substances include metals, minerals, dissolved gases, and organics, to name a few. Only a well-maintained water purification system will be able to effectively remove these harmful substances.

A resistivity meter displays the purity level of the water as it is drawn into the facility from the building's supply. The reading varies anywhere between 5 and 20 megohms (MΩ) and changes throughout the day. These readings are not consistent and indicate the variable presence of unknown dissolved ions and chemicals in the water flowing through the pipes to the facility. One major concern is that buildings' RO-DI systems are periodically flushed with harsh sterilization chemicals, which could contaminate the frog water and cause problems that are impossible to trace. The institutional RO-DI water is sterilized by the dedicated polishing system. This ensures that there are no contaminants introduced to the holding reservoir.

The frog facility's polishing system (Figure 17–6) consists of an activated carbon tank, a resin tank, two ultraviolet sterilizing lamps, and an ultrafilter cartridge. When the frog facility is not actively drawing water, the purifying system continuously recirculates in a loop so that there are no stagnant water pockets in the

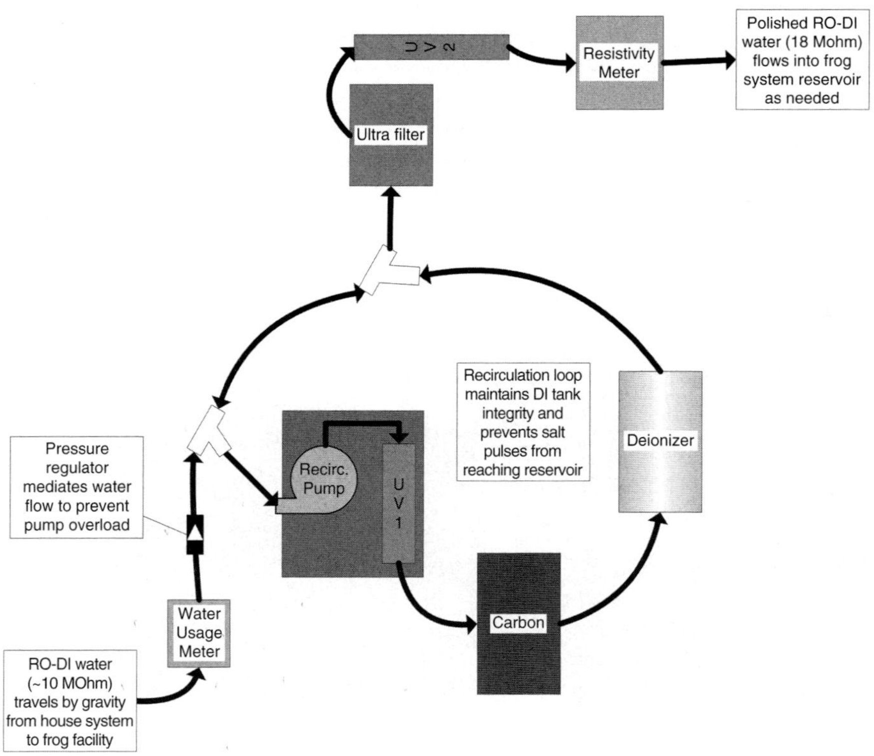

Figure 17–6. Schematic of a water polishing system The polishing system is necessary to purify water before it enters the holding reservoir. It is imperative to maintain the highest quality and standard of water. The water that is gravity fed to our facility from the building's water system on the roof passes through the water-usage meter, which records the amount of water entering the frog system in gallons. The water is driven by a small pump, through a UV sterilizer and a carbon tank, and then finally deionized as it passes through a resin tank. The water keeps recirculating through this loop until it is drawn into the frogwater reservoir as needed. As soon as the level in the holding tank drops (when it is drawn by the system for a daily 10% exchange), water is drawn through a UV sterilizer and "ultrafilter" trapping bacteria/particulates before filling. A resistivity meter was installed to monitor this system and to ensure the water is ~18.0 MΩ before entering the reservoir.

two tanks. When the water is ready to be drawn, it passes through the ultraviolet (UV) sterilizer and ultrafilter as a final step in purification. Measuring the resistivity is necessary to ensure consistent water quality. A meter measures the resistivity (a measure of purity, ideally 17.75 MΩ or above) at the last point before water enters our holding tank, ensuring ultrapure water for the system. The holding tank or reservoir has a KENT® float valve at the top that regulates the water level of the polishing system. When the reservoir is full, the float valve is horizontal and automatically stops filling until the level drops. The polishing system is always on, running to fill the reservoir on an as-needed basis.

After the incoming water passes through the polishing system, ultrapure water is drawn into the 90-gallon frog water holding reservoir that is manually dosed twice daily with salts and pH buffers. Since this reservoir is dosed by hand and not a sealed system, it is exposed to airborne materials and pathogens. To avoid contamination, the reservoir has its own circulating pump with a 1-μm particulate filter and UV sterilizer. We advise obtaining as large a tank as your space will allow: the more system water that can be replaced when necessary (e.g., when the building water supply is shut off), the easier it is to weather emergencies.

Individual components of the polishing system and their functions are as follows. (1) Activated carbon tank: with a porous surface for trapping organics and attracting charged articles, the activated carbon also removes chloramines. It is essential to protect the integrity of the RO membranes, which are susceptible to chlorine oxidation and clogging from organics. (2) Reverse osmosis deionization (RO-DI): RO-DI is the standard practice for producing reagent-grade water. Reverse osmosis is a process of forcing water against a tightly coiled semipermeable membrane at high pressure. The membrane allows only water to permeate its tiny pores, causing particles to be rejected when they cannot pass through the mesh. Deionization removes dissolved inorganics and gasses using a mixed-bed ion-exchange process. (3) Ultrafilter: the ultrafilter, the final filter on the polishing system, removes bacteria and pyrogens by trapping particles including microorganisms, their by-products, and organics from 10,000 MW to 100,000 MW. (4) Ultraviolet bulbs: a UV sterilizer at a wavelength of approximately 254 nm is sufficient to penetrate cells, damaging genetic material and proteins. Any bacteria or other pathogens present in the water will be destroyed or damaged by the UV filter. (5) Recirculation pump: the recirculation pump on the polishing system should always be running and maintaining a consistent pressure (PSI). Sometimes the PSI will drop when the polishing system is actively drawing water. Keeping the water circulating at specific pressure and rate maintains the integrity of the mixed bed DI tank and prevents contamination.

Service contracts are provided by the RO-DI system supplier. Service contracts must include routine maintenance of the pretreatment system (Table 17–7). A good service contract provides annual scheduled maintenance, including details on emergency repair costs. A list of equipment and replacement parts and maintenance information for the facility manager and technicians (monitoring and reporting system performance) should be reviewed before signing any contract. Since there are many unpredictable events that can happen in aquaculture, necessary steps need to be put in place to ensure that the polishing system is performing optimally. In our experience it is imperative to keep an eye on water system personnel working in the facility because many of them are not sufficiently experienced and/or sensitive to the needs of an animal habitat and can make mistakes that endanger the colony.

Table 17–7
Maintenance schedule

Daily checklist
Survey tanks for sick frogs
Flow meters at 6 gallons per minute
UVs on
Water meter pH/ORP compartment full and plugged
Air is being pumped through biofilter media
Water flowing into each tank
Polishing system resistivity is ≥16 MΩ/cm
Fill out system maintenance and daily water quality records
Adjust water quality as needed
Change prefilter pads if necessary
Tuesdays and Fridays
Feed frogs
Clean up after feeding
Check filter pressure gauges
Once a month
Clean particulate filter
Every other month
Change carbon
Every 3–4 months
Scrub out drainage tubes in each tank
Scrub tank interiors if dirty
Clean open areas in biofilter mixing tanks
Every 6 months
Clean quartz sleeve on UVs
Open bead filter and clear compacted beads/flush
Schedule semiannual maintenance on polishing system
Once a year
Change UV
Clean enclosed biofilter mixing tanks

Typically, UV lights, UV sleeves, the ultrafilter, RO-DI, and carbon tanks all require semiannual replacement. To keep the pump from running the UV filter dry, prevent any air from entering the RO-DI piping and make sure water levels do not fall below the intake of the pump. Another important way to safeguard against water contamination is by monitoring and recording the resistivity reading of the RO-DI water filling the reservoir on a daily basis. Any significant drop in resistivity indicates loss of purity and should be reported to the supplier immediately.

X. laevis has a life span of up to 20 years. Daily monitoring of water quality and proper conditioning will prolong the life and health of the frogs, keeping the facility thriving for many years. The main water testing parameters are temperature (air and water), pH, GH (general hardness), ammonia (NH_3), nitrate/nitrite, oxidation–reduction potential (ORP), and conductivity (μS). Although *X. laevis* can tolerate a wide range of salinity and pH, drastic changes in water chemistry are detrimental to its health, causing stress as well as bacterial and fungal attacks. Proactive measures should be taken to prevent fluctuations in water chemistry and provide frogs with excellent water quality standards that do not deviate from the designated parameters.

Table 17–8
Daily salt and buffer use and costs[a]

Add	*Concentration*	*10% exchange*	*Unit size*	*Cost*	*Ex/U*	*$/U*
Ocean	0.6 g/liter	90.6 g	11.3 kg	$34.99	124	0.28
CLS	0.4 g/liter	60.4 g	3.2 kg	$42.99	52	0.83
ES	0.067 g/liter	10.1 g	600 g	$7.19	59	0.12
Alk	0.325 g/liter	49.1 g	1 kg	$14.99	20	0.75
Ac	0.325 g/liter	48.9 g	1 kg	$14.49	20	0.72

[a]Cost per exchange = \$2.70. Cost per month = (Ex × 30) + (afternoon salts × 30) = \$81 + \$36.90 = \$117.90/month.

In the beginning, the additives that were used for the frog water in this facility were a combination of Marine Biotech's Sea Salt, along with $NaHCO_3$, to maintain the pH levels in the system water. This did not work well because it was difficult to strictly monitor parameters since we did not know what was in the Marine Biotech sea salt blend. It was also difficult to keep the pH in range because the sea salt that was used had additives that often increased the pH in the system and made it difficult to control since the sodium bicarbonate increases the pH. Instant Ocean salt components include calcium, magnesium, carbonate, and other trace elements needed by *X. laevis*, such as sodium chloride, iodide, bromide, and strontium. To increase GH and provide minerals, we supplement with Seachem Cichlid Lake Salt (CLS) and Seachem® Equilibrim salts to maintain hardness within 200–400 ppm. The protocol was changed to include SeaChem® buffers after reviewing the data in Godfrey and Sanders.[31] Their data showed that water hardness increases egg quality and is necessary for normal development of embryos to tailbud stages. Using Cichlid Lake Salts® in combination with the Instant Ocean® gives the proper salinity and trace minerals needed for *X. laevis* without affecting the pH. After changing our water chemistry protocol there was a noticeable improvement in water quality; the eggs were firm, and fewer incidents of illness were noted in the colony. Temperature, salinity, and water hardness (GH) all work together to maintain excellent water quality, health, and normal development of embryos. Guidelines for adjusting water chemistry are given in Table 17–8.

DAILY WATER CHANGES A 10% water change automatically occurs in the facility every day, at a designated time. An automatic timer controls a solenoid valve that is attached to the holding reservoir circulation loop. The solenoid valve is a two-way valve with an electromagnet that is triggered when the timer turns on, opening the valve. The valve stays open for 35 min draining water from the reservoir into sump 1, the approximate time needed for 30 gallons to exchange. The timer automatically shuts off, closing the valve. At 12:00 PM each day, the valve, which is operated by the timer, opens and clean frog water flows into sump 1. At the end of sump 2, there is a PVC pipe overflow drain that pushes the water out of the system when the water level rises. By adjusting the parameters of the water content in the holding tank, it is possible to ensure the gentle regulation of water quality and maintain strict water quality in the main facility, so that the frogs are always in good health. This helps discourage outbreaks that easily erupt if there is stress in the system. It has been observed and stated that wild *X. laevis* can withstand a wide range of salinity and pH; however, this is not necessarily true for laboratory frogs as they live in a clean environment where a certain range of salinity and pH should be maintained in the absence of unknown ecological components that may provide beneficial environmental buffering in the wild. Changes in pH, salinity, and even mineral deficiencies induce stress, making the animals susceptible to infections and parasites. Like all aquatic species, *X. laevis* is sensitive to changes in water quality and has an optimal range of conditions conducive to its survival and the survival of its offspring.

FROG FACILITY MONITORING

MANUAL MONITORING Trained technicians manually monitor the frog facility twice per day, checking frog health, water quality, and mechanical functionality of system components. System checks of pumps, UVs, filter pressures, temperatures, and water chemistry are necessary as a preventive measure (Table 17–9; Table 17–10 gives some sample readings in our facility over several days). The technicians dose the reservoir with appropriate salt and buffer ratios for the daily water exchange based on water readings in the holding tank and main system. Water chemistry is recorded for chlorine/chloramines in the reservoir periodically to make sure that the polishing system is performing properly. A consistent and disciplined monitoring schedule is essential to prevent disasters. The data should be kept and plotted in Microsoft Excel to show variations and provide records that can monitor long-term trends and other sources of potential problems.

TESTING EQUIPMENT AND PARAMETERS TO BE MONITORED We keep track of the following parameters, in both the

Table 17–9
Checklist for the facility monitoring

Temperature check
Air temperature
Water temperature
Pressure check: check for problems and leaks
Bead filter pressure (do backwash if pressure is greater than 20 psi)
Carbon pressure
Polishing system pump pressure
Water usage
How many gallons/liter per day? Record water usage
Tank water flow
Maintain 5 GPM?
Equilize individual tank flow: gentle water flow?
Check Air Pump
Bioballs being aerated?
Check UV sterilizers
Are they all ON?

Table 17–10
Sample daily water quality records of pH and salt measurements over a several-day period

Initials	*Date*	*Time*	*Tank*	*Conductivity*	*pH*	*ORP*	*Additives*
PK	1/2	10:00 AM	White Reservoir	1210 µS	6.01	214	90gIO 60gLS 39gAcd 50gAlk 12gEq
PK	1/2	10:00 AM	Tank 1	1830 µS	6.6	203	
PK	1/2	10:00 AM	Tank 2	1860 µS	6.62	205	
PK	1/2	4:00 PM	White Reservoir	1560 µS	6.24	235	45gIO 30gLS 39gAcd 50gAlk 12gEq
PK	1/2	4:00 PM	Tank 1	1787 µS	6.64	237	
PK	1/2	4:00 PM	Tank 2	1730 µS	6.97	233	
PK	1/3	10:00 AM	White Reservoir	1318 µS	6.11	213	90gIO 60gLS 39gAcd 50gAlk 12gEq
PK	1/3	10:00 AM	Tank 1	1727 µS	6.64	203	
PK	1/3	10:00 AM	Tank 2	1742 µS	6.69	207	
PK	1/3	4:00 PM	White Reservoir	1395 µS	6.18	181	90gIO 60gLS 39gAcd 50gAlk 12gEq
PK	1/3	4:00 PM	Tank 1	1740 µS	6.52	180	
PK	1/3	4:00 PM	Tank 2	1751 µS	6.53	182	
PK	1/4	10:00 AM	White Reservoir	1421 µS	6.1	175	45gIO 30gLS 39gAcd 50gAlk 12gEq
PK	1/4	10:00 AM	Tank 1	1770 µS	6.64	170	
PK	1/4	10:00 AM	Tank 2	1781 µS	6.69	173	
PK	1/4	4:00 PM	White Reservoir	1426 µS	6.04	250	45gIO 30gLS 39gAcd 50gAlk 12gEq
PK	1/4	4:00 PM	Tank 1	1805 µS	6.52	195	
PK	1/4	4:00 PM	Tank 2	1817 µS	6.53	196	
PK	1/5	10:00 AM	White Reservoir	1296 µS	6.01	176	90gIO 60gLS 39gAcd 50gAlk 12gEq
PK	1/5	10:00 AM	Tank 1	1827 µS	6.57	233	
PK	1/5	10:00 AM	Tank 2	1822 µS	6.67	238	
PK	1/5	4:00 PM	White Reservoir	1205 µS	6.1	188	90gIO 60gLS 39gAcd 50gAlk 12gEq
PK	1/5	4:00 PM	Tank 1	1800 µS	6.51	173	
PK	1/5	4:00 PM	Tank 2	1811 µS	6.56	180	
PK	1/6	10:00 AM	White Reservoir	1173 µS	6.17	187	90gIO 60gLS 39gAcd 50gAlk 12gEq
PK	1/6	10:00 AM	Tank 1	1844 µS	6.6	188	
PK	1/6	10:00 AM	Tank 2	1836 µS	6.61	190	

holding tank (to make decisions about how much salt and buffer to add) and the system itself. It is important to keep separate beakers for taking water samples for manual measurements from the holding tank and the main system, to avoid contaminating the holding reservoir with system biota. Resistivity (MΩ/cm) is a measure of opposition to electric current flow and is directly related to the purity of the water. Optimal RO-DI water should range between 16 and 18.2 MΩ/cm. The resistivity of the water entering the frog facility and polishing system product water should be monitored using in-line digital meters. Conductivity (µS) is the inverse of resistivity, measuring the water's ability to carry an electric current, and is used to measure the salinity of system water. System water in this facility is maintained within a pH range of 6.5 to 6.7. At a lower pH ammonia is less toxic and egg production and quality are improved. Godfrey and Sanders[31] observed that the *X. laevis* populations that live in acidic lakes of South Africa were thriving compared to the higher pH lakes.

Dissolved oxygen (DO, measured in mg/liter H_2O or percent saturation) is a critical parameter in the recirculation system.[32] Low DO (usually caused by decaying material) can stress out the animals as well as deplete aerobic bacteria from the biofilter causing nitrite and ammonia spikes, eventually killing the frogs. High DO can cause irreversible stress to the frogs, especially if they are exposed to supersaturated water. Safe ranges to maintain are between 30% and 70%. Levels over 90% will cause serious injury and death, as O_2 supersaturation causes gas bubble disease in the animals.[29] Supersaturation can be caused by an air leak before the intake of a pressurized pump. Even a hairline fracture can draw in enough air at an alarming rate by a Venturi effect. Air is drawn into the pump and at that pressure it dissolves into water and in minutes the system becomes supersaturated. Animals will appear listless, disoriented, and start to float. Many will start to die unless they are immediately transferred to new water. Air bubbles will enter their blood and their organs, potentially killing them before the problem is apparent. A DO meter or dissolved gas probe is an important investment to make for a facility.

ORP is a measure of the potential of a solution to either gain (oxidation) or lose (reduction) electrons. This is also a very important parameter to measure because it is an indicator of how well the water remains buffered (activity of ions) and of how well balanced the proportion of ions and minerals is in the system. Keeping the ORP in range keeps contaminants from fouling the system. Usually a rapid change in ORP is an indication of poor water chemistry. Normal ORP for a frog facility ranges between 100 and 300 units.

Temperature is maintained between 16.5 and 18°C for optimal egg production. We monitor the temperatures of the following components: the two chillers (with digital thermocouples), sump 1, recirculating water (with a digital thermometer), and air temperature in the room (with a digital thermometer). Variations in temperature related to seasonal changes in water and air temperature can drastically affect embryo quality and give rise to the well-known "Summer slowdown" experienced in some facilities.

Water level is topped off by KENT® float valves in sump 1 and the frog water reservoir. This water top off system is automatically controlled by changing the water level in the system. When the sump water level drops, the float valve position changes and starts to refill with water from the reservoir tank. The reservoir has the same float valve that is connected by a 0.5-inch pipe that draws water from the polishing system. As water evaporates or is lost from the system (during cleaning, etc.) it is automatically replaced. As an additional safety measure, float sensors placed in sump 2 indicate high or low water levels and are attached to the remote monitoring system (see below). In addition to this we have other float sensors that actually have audible and visual alarms that go off if the water gets low enough to allow air into the pump intake. Water pressure is monitored by pressure gauges on the bead, particulate, and carbon filters. These gauges are useful for diagnosing filter blockages and for alerting us of pump failure over the Internet through the automated monitoring system.

A Myron L ULTRAMETER® is used to measure all conductivity, pH, and ORP. Meter calibration should be done weekly to ensure accuracy. Follow the instructions for the pH calibration solutions from Myron L. In addition, we regularly use chemical Aquarium test kits to compare with the Myron L readings.

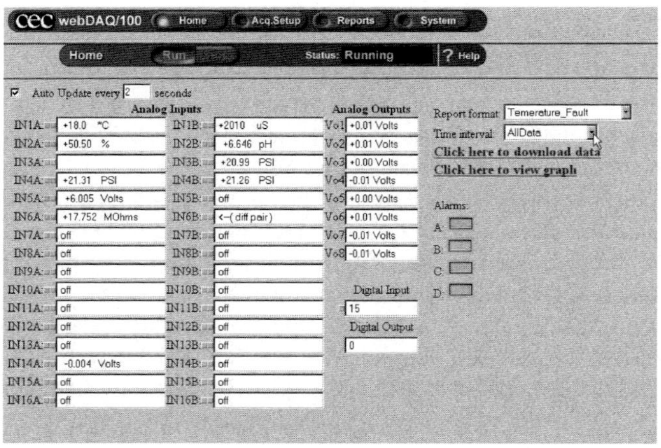

Figure 17–7. Screenshot of WebDaq monitoring system output This is a screenshot of real-time WebDaq information seen through a web browser. Various probes and gauges are connected to an embedded server that measures and records water quality parameters in the frog facility. The analog inputs correlate with probes in the frog facility. For example, "IN1A" is the analog input that is connected to the temperature probe in the sump and always shows the current water temperature. IN1B is connected to the conductivity probe, which is currently reporting the salinity in µS. Analog input "IN2B" is the pH probe. To visually keep track of what probes are linked to the inputs on the WebDAQ, we have included the units in the display. Notice the "click here to download data" and "click here to download graph" links. The download data link allows for instant downloading of text data of the desired report in a set number of time intervals. The download graph link displays the data, through Java Machine, in a customizable graph. These tools, and others of the WebDAQ program, ease and enhance remote monitoring of the facility.

Table 17–11
Daily water quality recording sheet

Daily water quality records								
Initials	*Date*	*Time*	*Tank*	*Conductivity (µS)*	*pH*	*ORP*	*Additives*	*Notes*
			Reservoir					
			Sump 1					
			Sump 2					
			Reservoir					
			Sump 1					
			Sump 2					

REMOTE MONITORING SYSTEM While a *Xenopus* facility requires attention and close monitoring throughout the day, it is usually not possible for most laboratories to maintain staffing in the overnight hours and weekends or holidays. To ensure that a situation does not get out of control and cause a major disaster, the FCRDB utilizes a *WebDAQ*© device to automate remote monitoring of the water quality (conductivity, pH, ORP, resistivity, temperature, and pressure). The *WebDAQ* is a self-contained wall-mounted system that contains inputs to which many different probes and sensors can be connected, and an embedded web server that provides these data in real time to any internet browser (Figure 17–7). *WebDAQ*'s monitoring capabilities include a variety of customizable reporting and alarm formats. The facility staff can be emailed any time a water quality parameter exceeds or falls below an acceptable range or when there is a change in pressure or water level. *WebDAQ* also stores these data so that trends and readings prior to a critical event can be examined later. Thus, *WebDAQ* provides two services: remote monitoring and recordkeeping. An uninterruptible power supply (UPS) is useful to ensure that the WebDAQ can send a message in the event of a power outage.

In addition to the *WebDAQ*, the FCRDB uses a webcam that allows visual monitoring of the facility at any time (the *WebDAQ* allows us to remotely turn on a light if it is necessary to see the facility when the overhead lights are off at night). If there is an emergency email, such as a low water alarm, it is important to be able to assess the situation remotely by checking the webcam to see if, for example, a blocked pipe is causing a tank to overflow or if the system lost a large volume of water. This has been a "lifesaver" for our facility and the investment and time taken to install these components have averted many calamities.

RECORDKEEPING Recordkeeping is crucial to tracking the performance of your frog facility as well as diagnosing problems and long-term trends. Below is a brief description of the parameters to document frog colony conditions (sample forms are given in Table 17–11 and 17–12, frog tank labels in Table 17–13, and frog identification systems in Table 17–14).

Water chemistry and facility maintenance: Water chemistry and support system conditions should be recorded daily: twice daily for water chemistry and once for other system parameters. These records are crucial for correlating environmental conditions with egg quality and frog health.

Quarantine/incoming frogs: New frogs are quarantined according to procedures described in the section on Disease. Quarantined animals are placed together in an isolated tank that is clearly labeled "Quarantine through mm/dd/yy." A treatment schedule is

Table 17–12
Daily system maintenance recording sheet

Daily system maintenance records

					Temperature (°C)				
			RODI polishing system		*Digital thermometer*		*Chiller barrels*		*Routine maintenance*
Initials	*Date*	*Time*	*Resistivity (MΩ)*	*Water meter (gallons)*	*Air*	*Tank*	*Air-cooled chiller*	*H_2O-cooled chiller*	*Backwash, filter changes, cleaning, etc.*

posted on the tank, and facility managers are required to initial the dates as they are completed.

Sick/euthanized frog log: Any frog that is isolated for treatment of illness or euthanization should be noted in the sick frog logbook. Documentation includes date of isolation, date of last use, frog size (body length), a description of the symptoms and condition, original tank number, and treatment type, dose, and dates. See the section on Disease for further details.

SYSTEM MAINTENANCE

Frog facility maintenance requires proper communication with facility staff and well-written service contracts with outside vendors. The following are periodic maintenance tasks that must be performed.

RESERVOIR STERILIZATION Despite the purity of the polished water supplied to your reservoir, you should periodically examine the reservoir for buildup or biofilm. A product called Minncare® Cold Sterilant, a peracetic acid solution that removes biofilms, is a widely used product to sterilize RO membranes. Follow the instructions for sterilizing RO units and be sure to rinse the reservoir several times. Keep the room well ventilated and follow the safety guidelines. This is a nontoxic and biodegradable product, but it emits fumes. Minncare® decomposes into oxygen, water, and acetic acid, and test strips are available to check for residue after rinsing.

Table 17–13
Frog tank labels

Date purchased	*Date used*	*Available after*	*Occupants (# WT/# Alb)*	*Notes*

Table 17–14
Frog identification systems

The following identification system resources were gathered from literature and suggested on the *X. tropicalis* listserv.[a]

We do not use any tagging procedure, so caution is strongly advised and no one system is advocated by the authors.

1. *Tattooing* involves the marking of frog skin with different colors and patterns of tattoo ink to identify individuals. Tattoos often fade within a few months.
2. Branding utilizes heat, cold, or chemicals to mark frog skin. A few respondents reported that cold-branding can last several years.
3. *Chip embedding systems* require skin incision or injection of a radio-signal microchip transponder that can be accessed by placing the tag reader over the animal.
   a. AVID labtrac system: http://www.avidid.com/special/index.html
   b. Biomark PIT system: http://www.biomark.com
4. *Tag embedding systems* utilize magnetized coded wire tags (CWT) or visual implant alphanumeric tags (VI-alpha) that are embedded in the tissue of the animal. One source is Northwest Marine Technologies CWT and VI-alpha systems: http://www.nmt.us/.
5. *Bead identification* involves the sewing of a small plastic thread behind the armpit, with different colors and combinations of beads.
6. *Skin grafting* uses transplantation of different shaped, colored, and sized skin grafts from the ventral to dorsal surface of the frog to uniquely identify it.
7. *Toe clipping* involves cutting off the tips of toes in codified combinations. This technique may not work effectively since *Xenopus* can regenerate the clipped toes.

[a]*X. tropicalis* listserv, post thread "frog identification system" from May 2004: http://faculty.virginia.edu/xtropicalis/newsgroup.html.

MAINTAINING PUMPS Electrical pumps generate heat during their operation, so it is important to keep airflow unobstructed. Inspect pump air intakes monthly for dust buildup to prevent overheating. The water-driving pumps (for the reservoir, recirculating system, and polishing system) can function only with an uninterrupted supply of water. If air enters the flow, or if a flow valve is closed before or after the pump, you may damage the motor. To reduce noise pollution, cushion the pumps on a bed of heat-resistant neoprene to dampen vibration.

MAINTAINING CLEAN FILTERS AND SUMPS Filters require regular monitoring and replacement to maintain clean water. Prefilter pads should be changed three times a week: Mondays and after feeding on Tuesdays and Fridays. If you neglect to change them debris will flow off the pad and into the biofilter. The biofilter should be cleaned only if a noticeable amount of debris has collected on the bottom of the sump. Once a year shut off the air pumps, remove the prefilter assembly, and carefully siphon out the debris making sure you do not remove any Kaldnes® beads in the process. The open portion of the sump can be cleaned every 3 to 4 months using a siphon. The bead filter requires regular backwashing to prevent the beads from being compacted with debris. In a healthy system the bead filter should be backwashed once a week. Every 3–6 months, open the bead filter to manually stir up the compacted beads with a thin PVC pipe to prevent buildup. The particulate filter must be changed at least once a month, or whenever there is a noticeable drop in pressure from the bead to the particulate. The carbon filter should be replaced with fresh carbon every time you change the particulate. Thoroughly rinse the new carbon to remove excess dust before placing the cartridge back in its housing.

To clean the biofilter tank, gently wipe down the sides with Python® mitts or any other kind of algae scraper. We use a gravel vacuum to suction all of the large debris from the bottom of the system sump. This process requires a 25% water change from the system. Carefully monitor and alter water chemistry as needed while the replacement water flows in.

ULTRAVIOLET STERILIZATION BULBS All of the UVs in the facility should be examined, with gloves, every 6 months for buildup on the quartz sleeves. If the sleeve is dirty, clean it with KimWipes® dipped in isopropyl alcohol. If there is residue on the bulb, or the bulb is discolored, it should be replaced. Generally, most manufacturers advise that UV bulbs should be replaced once a year. UV bulbs contain mercury and therefore legally must be disposed of as hazardous waste. When returning a bulb to its housing, check that the bulb and sleeve are completely dry and that the gasket is properly positioned to prevent leaks.

ROUTINE CLEANING OF EACH FROG TANK Once a month, remove both drain tubes and scrub them out, without using detergent. Never use any soaps or cleaners of any kind when cleaning the tanks or the pipes of the recirculation system. Invest in a few dozen bottle brushes of various sizes and use one brush per tank to avoid cross-contaminating with other tanks. Clean the outer tube first, then replace it and clean the inner tube, leaving one of the two tubes as a cover on the drain to prevent frogs from being sucked into the pipe drain. Frogs will instinctually swim with a current, so they will try to head down the tube and could get stuck. The vinyl inflow tubing that supplies water to each tank, and flow meters, can be scrubbed out whenever a visible biofilm accumulates. Once a year, if you are comfortable with manipulating the pipes, you can scrub out other elements of the system, including water intake pipes and sump drains. Be cautious when unscrewing the pipe fittings as PVC tends to become brittle over time. Never clean pipes or components of the system with detergents. Contaminating the system with chemical solvents and cleaning agents will poison the frogs and the beneficial bacteria.

After cleaning, soak all of the used items in a bucket with Sanaqua® overnight. Sanaqua® is a concentrated sanitizer and fungicide and should be used sparingly (add 5 drops per gallon until the water is slightly pink with suds). Let them soak overnight. Fill a new bucket with RO-DI water and transfer all of the items to the clean fresh water bucket. Wash each item individually making sure that there are no particles left on the nets or brushes. Wash in RO-DI water, making sure there are no suds or residue left on the equipment. Hang brushes, mitts, and other cleaning tools on hooks to dry.

RECOMMENDED CLEANING AGENTS We recommend avoiding harsh chemical cleaning agents to avoid frog stress and maintain proper water quality. The only chemicals used on a regular basis are Sanaqua® sanitizer and Anti-Chlor chlorine/chloramines removal solution. These solutions are used only for cleaning aquarium equipment such as nets and brushes in 5-gallon buckets. Never use any chemicals to clean any part of the recirculating system. Common cleaning applications include Sanaqua® for nets, sponges, mitts, or anything in contact with quarantined animals; Anti-Chlor for anything in contact with untreated tap water; and finally three rinses in RO-DI water to thoroughly clean off any residues.

FROG HUSBANDRY

Our facility currently houses 200–250 frogs. We have enough frogs to rotate through the months of the year while providing a rest period for each batch of frogs that are used per week. Our laboratory primes about 8–10 frogs every week allowing a "rest period" of 4 months between priming and hugging. The frogs are rotated to different tanks when they are returned to the colony after priming and hugging. The tanks are labeled according to the date at which the frogs are returned to the colony to "rest" for 4 months before they are primed again, so it is important to keep track of the frogs and label the tank appropriately to ensure an adequate rest period of at least 3 months.

Male frogs are kept separated from females in their own tank. When they arrive at the facility, they undergo a 3-week quarantine period in which they are treated for parasites and observed before putting them on the recirculation system to ensure that new parasites do not enter the system.

CYCLING THE BIOFILTER AND ADDING NEW FROGS Beneficial bacteria are responsible for breaking down and nitrifying animal wastes: *Nitrosomonas* oxidizes ammonia into nitrite, while *Nitrobacter* and *Nitrospira* oxidize the nitrite into a less harmful form.[33]

Our facility orders frogs from Nasco®, which is highly recommended over other suppliers. In the beginning, order 15–20 frogs and quarantine them for 6 weeks. During the quarantine, the frogs are treated with levamisole HCl (see the section on disease below). After treating the frogs with salt baths and deworming them with levamisole, check skin scrapings and examine them for scrapes or infection. When they are healthy, put them on the recirculation system. Simultaneously, seed the biofilter media with *Nitrosomonas* and *Nitrobacter* or *Nitrospira* when you add these frogs, and then carefully monitor the water quality twice a day. We add KoiZyme® at this time to allow a good combination of bacteria colonizing the system. Ammonia and nitrate testing should be conducted daily after adding bacteria to monitor the system biofilter growth. Ammonia, nitrates, and nitrites should be low (<0.1 ppm) in the beginning and then undetectable after a few days. It is a good indication that the nitrifying bacteria are doing their job if ammonia and nitrite test results are undetectable after a week. If the water quality parameters are in range, and there are no mysterious pH spikes, more frogs can be added gradually. Note that the bacteria levels cannot rise without frogs being present in the system (since they need the waste products to thrive); thus, animals and the bacterial colony should be ramped up in parallel.

FEEDING ADULT FROGS Frogs should be fed at least twice a week; some facilities feed frogs more often, however, these frogs have adjusted well to a 2 day per week feeding schedule. A number of diets have been used,[34,35] and in our experience commercial Nasco frog pellets work well. It is better to feed less than overfeed the frogs. *Xenopus* should not be handled after feeding because they will regurgitate their food. The frogs become nervous and stressed, so it is advisable to wait a few hours before cleaning the tank of debris and uneaten food particles. Females should be taken from the colony for priming only on days on which they have not been fed.

When feeding, turn the water supply off. Overfeeding creates excess debris and clogs up filters, so feed less food until it is known how much the frogs will consume in a 15-min time period. Most frogs eat approximately six or seven frog pellets each. Males eat about half the quantity of females and new arrivals usually take a few days to acclimate before they eat. Observe frog behavior during feeding to make sure every animal is eating and behaving normally. As soon as food is dropped into the tank the frogs lunge toward the food making a "shoveling" motion with their forelimbs. Sometimes when feeding, the whole group becomes competitive and can bite each other's limbs in a feeding frenzy. Usually there is no injury and the frog releases its victim after realizing they are missing out on food. In case of wounding, the frog should be removed and treated to avoid bacterial septicemia. After 15 min of feeding or whenever the food has been cleared in the tanks, return the water flow to the tanks.

CLEANING TUBES AND TANKS AFTER FEEDING After 3–4 h, uneaten food and fecal matter will accumulate in the drain tubes of every tank. The outer tube is used to avoid pulling large objects into the drain. The inner tube controls the water level and drain flow. Turn off the incoming water valve and clean the drain by carefully removing each tube (these tubes pop out of their slots) one at a time, blocking the drain hole to prevent curious frogs from escaping. Water drains rapidly when the tubes are removed, so the technician must act quickly. Remove the tubes and return the wider tube back to the drain. While the water drains, gently swirl the smaller tube above the drain to vortex the particles toward the center so that they get flushed down the pipe. Replace the inner tube and securely snap the lid on the tank. Finally, wipe down the PVC grid in the prefilter assembly and replace the prefilter pads. Check and balance the flow to the tanks. Check the pressure gauges and backwash filters if necessary.

QUARANTINE Disease prevention in the colony begins with effective quarantine procedures (Tables 17–15 and 17–16). New frogs entering the facility are generally the major sources of contamination. New frogs should be quarantined for a minimum of 3 weeks in a tank that is closed off from the recirculating system. This tank stays isolated from the rest of the system until the frogs have completed antihelmintic treatments.

REARING TADPOLES The day after fertilization, *X. laevis* tadpoles should be transferred to a clean Petri dish and cleaned with 0.1× MMR (Marc's Modified Ringer's[6]). Embryos should be incubated at 18°C in 0.1× MMR with a pH of 7.8. Like the tadpoles, the eggs need plenty of room for proper growth and development. Clean and observe the embryos every day, removing any dead or dying eggs. Transfer the eggs to new Petri dishes to avoid pathogenic attack.

The embryo will sustain itself off of the yolk for about 5 days, or until about stage 44, which is when they become free swimming (not clinging to the sides of the dish or laying on the bottom). Tadpoles will begin foraging for food with their head facing the bottom. *X. laevis* tadpoles are filter feeders at this stage of their development and will feed on 100- to 150-µm-sized algae and copepods. Split up the tadpoles so that there are 20 tadpoles per dish and start feeding *Spirulina*. Sera® Micron is a good choice because it contains a variety of microorganisms and *Spirulina,* which is an extremely nutritious type of algae. Feed twice or three times a day, making sure that the water is changed at the end of the day. Do not overfeed the tadpoles! Give them enough so that they clear the water in a few hours and then add more if they seem to be panning the bottom. The tadpoles hover at a 45° angle in the water column and feed off the bottom, so it is important to add the Sera® Micron and disperse it throughout the water, allowing it to settle on the bottom. Alternatively, mixing 1 mg/ml of water and then adding this mixture to the dish is an effective method for dispersing food.

Tadpoles can live in the dish for 2 days, at which point they need to be moved to a 4-quart Tupperware® for adequate spacing. Remember not to overcrowd the tadpoles. Cut holes in the lid and keep them covered. They should have at least 500 ml per tadpole and water should be twice as deep as the length of their body. At this point, an air-stone may be used to create circulation and oxygenate the water, but it is not essential since they use their lungs to gulp air at the surface. Slowly start adding NASCO® tadpole powder with the Sera® food at about stage 60[7]. NASCO® tadpole powder provides balanced nutrition, such as vitamin C, calcium, and magnesium, and it is a high protein diet providing the energy required during metamorphosis. During metamorphosis adding crushed pelleted food (NASCO® Frog Brittle) enhances their diet and growth rate. Soak the pellets in 0.1× MMR until they absorb and expand before adding them to the tank. The tadpoles cannot eat the dry powder/food if it floats on top of the water.

Metamorphosing tadpoles and froglets should be fed every other day for optimal growth. Nasco recommends feeding every day if time allows; otherwise it is sufficient to feed three times a week.

WATER QUALITY TESTING FOR TADPOLES It is important to maintain high general hardness (200 dH–400 dH) for the tadpoles because calcium and magnesium deficiencies will cause stunted growth and adversely affect metamorphosis. Again, water quality is the central theme because tadpoles generate large amounts of waste material that can alter the pH rapidly. Ammonia (NH_3, NH_4) is a silent killer that can wipe out all of the tadpoles in a few hours. Since there is no biological filter or biofilm established yet, it is necessary to change their water every day after the last feed. Although many husbandry manuals and

Table 17–15
Frog quarantine protocol

All new frogs must be quarantined for 6 weeks. Do not cross-contaminate with the rest of the colony. Wash your hands after handling quarantined frogs. Disinfect (Sanaqua) anything that comes in contact with quarantined frogs (net, sponge, etc.).

1. Label an empty tank "Quarantine Through __/__" with the date 6 weeks from arrival. Because of the life cycle of the nematode and to ensure that they are eradicated, we quarantine for 6 weeks in our facility.
2. Close off the water inlet and drain the tank to the upper notch (so drain tubes are above water level, about 4 inches). Close both drain valves.
3. Remove frogs from Nasco boxes and place in a "quarantine" labeled 5-gallon bucket filled with fresh frog water.
4. Use a net to remove excess peat moss floating in the bucket.
5. Place each frog, one by one, into the quarantine tank. Be sure to examine the frog for signs of infection or injury (if injured, keep isolated in a plastic container and see Table 17–18: Sick Frog Protocol).
6. Once all frogs have been placed in the tank, replace the tank cover and **do not turn on the water** because the tank will be closed off.
7. Treat the tank with 1.5 g levamisole hydrochloride (dosage 75 mg per frog or 20 mg/liter) overnight in standing water.
8. The next morning, drain the tank using the floor drain valve (the right-hand valve attached to the smaller drain pipe under the tank, **not the system valve on the left**), and refill with fresh frog water. It is beneficial to wipe away any biofilms that have accumulated in the tank(s).
9. Repeat this treatment for 4 days, allow a 1-week rest period, then resume 4-day levamisole hydrochloride soaks.
10. Additionally, frogs should be treated with a salt bath two or three times a week during quarantine. In the morning, weigh out extra "instant ocean." Increase the salinity to 3000 μS.
11. Keep the frogs in the salt bath for 30 min (unless you see signs of stress) and then drain and refill the tank as usual. Stress indicators include reddened skin and excessive movement (trying to escape).
12. Once the quarantine period is over, place frogs in a new tank and label it appropriately. Drain and scrub the quarantine tank (and drain tubes) with diluted Sanaqua and rinse thoroughly. Drain all wastewater into the floor drain (using the floor drain valve).
13. Reopen the system recirculating drain valve, close the floor drain valve, and refill the emptied tank.

Sample schedule

Sunday	*Monday*	*Tuesday*	*Wednesday*	*Thursday*	*Friday*
Week 1		Frogs arrive Levam PM	Salt bath AM Levam PM	Levam PM	Salt bath AM Levam PM
Week 2	Salt bath AM		Salt bath AM		Salt bath AM
Week 3	Salt bath AM	Levam PM	Salt bath AM Levam PM	Levam PM	Salt bath AM Levam PM

specialists advise against changing all the water, tadpoles in containers without any biofiltration are prone to stress caused by toxic levels of ammonia and nitrite. The same care and vigilance should be taken when raising tadpoles as with caring for adult frogs. Daily water testing, providing essential minerals, and quality nutrition should be standard practices when rearing tadpoles in the laboratory.

Metamorphosis rates differ within the tadpole population. Typically, after around 40 days[7] hind limbs grow and the tail begins to disappear into the body. Tadpoles become carnivorous at this stage and will devour their tank mates, especially the ones that have not morphed yet. Place the froglets into a separate container and continue to feed and check water quality on a daily basis.

TROUBLESHOOTING AND PREVENTING DISASTERS

A number of unexpected events can adversely impact the colony, leading to immediate or long-term delays in egg availability. Disasters are of two general types: specific and immediate electrical/mechanical/plumbing events, and steady-state problems in system parameters that affect frog health gradually.

Table 17–16
Quarantine schedule[a]

	Treat	*Date*	*Initial*	*Treat*	*Initials*	*Treat*	*Initials*	*Treat*	*Initials*
Week 1	Levam 1			Levam 2		Levam 3		Levam 4	
	Salt 1			Salt 2		Salt 3			
Week 2	Salt 1			Salt 2		Salt 3			
Week 3	Levam 1			Levam 2		Levam 3		Levam 4	
	Salt 1			Salt 2		Salt 3			

[a]Levamisole (Levam) HCl 20 mg/liter (1.5 g/20 gallons). Sea salt 200 g/20 gallons.

INTERRUPTED OR CONTAMINATED WATER SUPPLY Many problems can affect incoming water quality: bacterial outbreak, gas supersaturation, ammonia spike, etc. If there is a contaminant, the tanks can be closed off and water should be drained out of the system into the floor drain and replaced with fresh water from the reservoir. To save water, the tanks can be filled halfway or just enough so that the frogs have plenty of room to stretch without touching any other frogs. Ideally, the frog facility has a frog water reservoir large enough to be able to replace at least 50% of the water in the system at any given time. Our holding reservoir has enough water to do a 10% water change for 3 days.

PIPE LEAK/BROKEN SPIGOT/DAMAGED BALL VALVE Always have a backup of supplies. If pipes leak the system will need to be shut down until the leak is fixed. Usually, if it is more involved than resealing a joint, it will be necessary to have plumbing parts and tools, nontoxic aquarium cement, and hopefully someone with plumbing expertise to replace the pipe if it is broken. The spigots are easy to change and replacement parts should always be ordered and kept in the facility. If the main pump or any other pumps break or run dry, a disaster can occur if another pump is not on hand. Sometimes it can take weeks to order a replacement part. Always have a backup pump, as well as the tools and parts to replace the pump.

PH OUT OF RANGE Many factors can cause the pH to spike up or rapidly drop. In most cases, this might be caused by a break-down of the biofilter or wrong proportions of salts added to the system. Buffers are often made up of sodium bicarbonate that eventually disassociates after 12 h, causing the pH to crash. This can be adjusted by adding specific ratios of buffers (depending on whether you want to slowly raise or lower the pH) to the reservoir and manually exchanging the water until the pH is raised 0.1–0.2 units. **Caution: Never raise the pH more than 0.2 units a day** in an attempt to solve a pH problem. This could result in major stress that throws off the frogs' osmotic balance. The frogs can withstand a wide range of pH, but it is a rapid change in pH that degrades their slime coat, making them susceptible to infection.

LOSS OF ELECTRICAL POWER *X. laevis* can be left in standing water as long as there is enough reserve frog water to be able to do a partial water change daily. We strongly recommend having all of the key pumps, purifier systems, and monitoring probes on emergency backup power.

Table 17–17
Sick frog log sheets

Date	*Albino/WT*	*Size/description (cm from nose to cloaca)*	*Date of last use*	
Description of illness				

Date	*Appearance/ activity level*	*Location (isolation/ tank number)*	*Treatments used*	*Dosage*

FROG HEALTH PROBLEMS The diagnosis and treatment of medical conditions discussed in Tables 17–17 and 17–18[36,37] were carried out under the supervision of James G. Fox, D.V.M. One of the key features of a flow-through system is that diseases and infections can spread rapidly. Thus, it is imperative to keep a close watch on the animals and remove any that seem problematic (see Tables 17–19 and 17–20). If a treatment looks to be successful after a significant infection, frogs should not be returned to the facility since the infection can recur and spread to others. The purpose of treatment is mainly as a diagnostic for the problem (the success of the various treatments is informative as to what the illness was) and it is not worth the risk to return frogs that had a serious infection.

Frog skin is a crucial barrier to infection and skin quality is a major diagnostic for disease. *Xenopus* skin contains a number of

Table 17–18
Sick frog protocol and suggested treatments

1. *Nematode infestation*: Symptoms include grayish, sloughing skin that may feel tacky rather than slimy. Examine shed skin using a dissecting microscope for the presence of zigzagging tunnels in the skin, which indicate presence of nematodes. These parasites undergo a life cycle while embedded in the skin of the frog. Treatment involves a 3-week protocol of antihelmintic baths using levamisole HCl, 75 mg/frog (see Table 17–16 for the schedule).
2. *Protozoa*: We had one case of a frog infested with an Epistylis-like ciliate (Table 17–21: Tinsley, p. 237). Its symptoms included whitish, thread-like tufts growing off hind limb claws. We treated it with a series of 10 min 12–15 g/liter salt baths, but the infestation did not respond to treatment and led to secondary infections after 2 weeks, so the animal was euthanized.
3. *Red leg infection*: Symptoms include reddening of skin on the hind limbs, lower back, and ventral region. Reddening intensifies to dark open sores that quickly spread all over the body, and can lead to secondary infections. Frogs may begin to tremble at later stages. Red leg is caused by a few types of opportunistic bacteria (*Aeromonas*, *Pseudomonas*, etc.) that are normally associated with frog colonies but can overcome animals weakened by stress or injury. A few occasions of red leg have resulted from elevated pH (which may cause degradation of the protective slime coat), stress from excessive handling, and/or secondary to other infections (i.e., nematodes). These frogs are treated with oxytetracycline (0.2 mg/liter) until symptoms subside, but it is not 100% effective. In advanced cases the animal usually must be euthanized.
4. *Fungal infection*: Symptoms include long whitish threads growing off of skin and/or toes. We treat fungal infections with a commercially available product called fungus cure, at a dose of 1 ml per 4 liters, until symptoms subside. Again, this is not 100% effective, as in some cases secondary infections develop that are hard to treat.

Table 17–18
(continued)

5. *Bloating/edema*: Symptoms include gradual swelling just beneath the skin of hind limbs, back, etc. Frogs appear "full of water" under the skin. This may be caused by damage to the lymph system during injection of Chorulon, or general trauma from egg harvest procedures. I have not found a protocol that successfully treats this condition, despite recommendations for high salt solution baths.
6. *General ill health*: Symptoms can include undiagnosed skin discoloration (brownish or greenish), bruises (often "bite marks" from conspecifics during feeding), scratches, etc. We treat unknown but seemingly minor conditions with soaks in Melafix (1 ml/4 liters), a commercially available product consisting of tea tree oil, which is healing to the skin. If the frog appears infected, we may use a series of high salinity salt baths (see Table 17–15 for the quarantine protocol) and/or oxytetracycline (0.2 g/liter) to prevent bacterial infections.
7. *Dead frogs*: Dead frogs should be removed from the tank immediately. If discarding, follow your institution's animal disposal policy. You may wish to have the animal examined by a pathologist to determine the cause of death. If the frog is to be brought to a veterinarian or pathology laboratory for necropsy, cut a slit in the skin and connective tissues vertically along the underside of the frog (as if you were going to dissect it). Place the frog in either formalin or 10% formaldehyde. If none is available, DO NOT cut the frog open, but place it in a 4°C refrigerator until it can be brought to the veterinarian. It usually takes just over a week to receive the pathology results.

Table 17–19
Change in the appearance of frogs is indicative of health problems

Edema
Excess skin shedding
Skin discoloration–red, brown, gray, black
Growths on the skin–filamentous, tumors
Hemorrhages on skin
Loss of slime coat–sticky or rough skin
Anorexia
Eggs or shed skin sticking to the frog

antimicrobial and antifungal compounds,[38–42] and in the absence of handling and excessive contact, the regenerating slime will keep frogs resistant to the omnipresent background population of nematodes and bacteria.

Rough patchy skin and excessive shedding may indicate a *Pseudocapillaria xenopodi* infection, usually a secondary infection caused by stress from parasite load (usually involving nematodes). Examine shed skin under a microscope for holes and tunnels caused by exit wounds from the worms; sections of skin can reveal nematode cysts (Figure 17–8). There are usually many wavy lines throughout the skin samples. These parasites begin their life cycle in the subcutaneous skin layer. Treatment involves a 5-week protocol of antihelmintic treatment[43] using levamisole HCl (MP Biomedical), 100 mg/frog; one frog/gallon of system water has worked well in preventing the infection from returning.

Our facility has had a few cases of *Capillaria xenopodis* infections in the past 6 years. This nematode burrows in the skin and leaves cysts (embryos) that are not affected by the levamisole treatment when it is in this stage of the life cycle. These parasites cause severe skin irritation and make *Xenopus* prone to secondary bacterial infections, weakness, rapid weight loss, and finally death. Treating the colony is labor intensive for the facility staff but patience and careful observation will save the colony.

Table 17–20
Change in the behavior of frogs is indicative of health problems

Hanging at top of tank exclusively
Stiffness or lethargy
Loss of appetite
Vomiting
Trembling

Oxytetracycline treatment in conjunction with Stress Coat (aquarium pharmaceuticals) should be used for 1 day before and 1 day after hugging for egg harvesting.

CONCLUSION

Xenopus laevis and its genetically tractable cousin *Xenopus tropicalis*[44–48] are profoundly powerful vertebrate model systems in which physiological, developmental, and cell-biological mechanisms may be studied. They will be increasingly used by both basic biology and pharmaceutical efforts (as drug screening platforms), requiring stable, effective protocols for maintenance of adult frogs. The information presented in this chapter should be sufficient for any laboratory to design a useful system and maintain happy frogs that give high-quality eggs suitable for biochemical or molecular investigation of embryonic development (Table 17–21). The protocols for frog husbandry will surely continue to evolve, and we urge workers in *Xenopus* to make use of and contribute their experiences to the frog community.

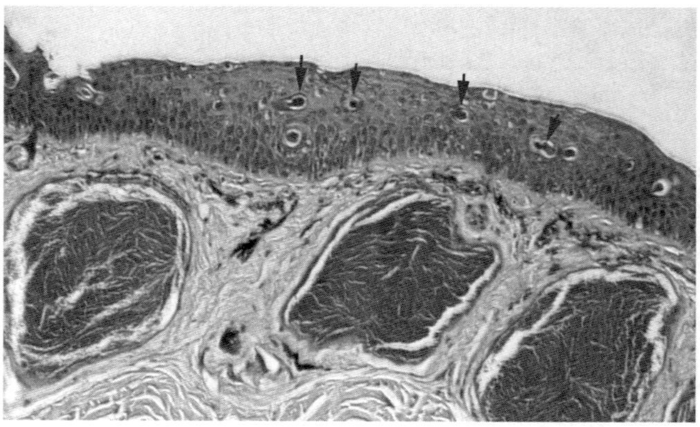

Figure 17–8. Photograph of skin parasites. *Capillaria* cysts just under the epidermis are shown at the sites of the arrows. These nematodes are at the larval stage and will be protected from levamisole treatment at this stage in their life cycle. This is a paraffin section of skin from a *Xenopus laevis* female from our facility that was infected by the nematode *Pseudocapillaria xenopodi*.

Table 17–21
Suggested reading/health references

1. Wu M, Gerhardt J. Raising Xenopus in the laboratory. *Methods Cell Biol* 1991;36:3–18.
2. National Research Council Report. "Amphibians: Guidelines for the breeding, care, and management of laboratory animals." Chapter 8, Records and Information Control. http://books.nap.edu/html/amphibian/eight.html.
3. O'Rourke Schaeffer D, Schultz TW. Biology and diseases of amphibians. In: Fox JG, Anderson LC, Loew FM, Quimby FW, Eds. *Laboratory Animal Medicine*, 2nd Ed. New York, Academic Press, 2002:793–823.
4. Chapter 13, Parasites of Xenopus. In: Tinsley RC, Kobel HR, Eds. *The Biology of Xenopus*. Oxford, UK: Clarendon Press, 1996:233–261.
5. Wright KM, Whitaker BR, Eds. AM*phibian Medicine and Captive Husbandry*. Malabar, FL: Krieger Publishing Company, 2001.

ACKNOWLEDGMENTS

We would like to thank Adam Crook for his invaluable assistance with the original construction of the colony, Katherine Gallant for help in frog room maintenance as well as proofreading and advice on the manuscript, and numerous members of the *Xenopus* community for useful advice. M.L. gratefully acknowledges the support of National Institute of Health grants GM067227 and GM07742, March of Dimes grant #6-FY04-65, the American Heart Association Established Investigator Grant 0740088N, and NHTSA grant DTNH22-06-G-00001.

REFERENCES

1. Deuchar EM. *Xenopus: The South African Clawed Frog*. New York: Wiley, 1975.
2. Schade H. [Possibility of definite pregnancy diagnosis with Xenopus laevis.] *Z Geburtshilfe Gynakol* 1951;134(3):300–303.
3. Vercammen-Grandjean PH. [Xenopus test in pregnancy diagnosis at the Costermansville laboratories.] *Ann Soc Belg Med Trop* 1950;30(2):313–321.
4. Manfredi P. [The dependability of the biological diagnosis of pregnancy by means of Xenopus laevis.] *Ann Ostet Ginecol* 1950;72(2): 153–164.
5. Harjola O, Toivonen S. Finnish experience of the care and use of the clawed frog (Xenopus laevis Daudin) in the diagnosis of pregnancy. *Ann Chir Gynaecol Fenn* 1949;38(Suppl. 3):68–82.
6. Sive HL, Grainger RM, Harland RM. *Early Development of Xenopus Laevis*. New York: Cold Spring Harbor Laboratory Press, 2000.
7. Nieuwkoop PD, Faber J. *Normal Table of Xenopus laevis (Daudin)*, 2nd ed. Amsterdam: North-Holland Publishing Company, 1967.
8. Bernardini G. *Atlas of Xenopus Development*. New York: Springer, 1999.
9. McDiarmid RW, Altig R. *Tadpoles: The Biology of Anuran Larvae*. Chicago, IL: University of Chicago Press, 1999.
10. Hilken G, Dimigen J, Iglauer F. Growth of Xenopus laevis under different laboratory rearing conditions. *Lab Anim* 1995;29(2): 152–162.
11. Dawson D, Schultz TW, Shroeder EC. Laboratory care and breeding of the African clawed frog. *Lab Anim* 1992;21(4):31–36.
12. Schultz TW, Dawson DA. Housing and husbandry of Xenopus for oocyte production. *Lab Anim* 2003;32(2):34–39.
13. Parker F, Robbins SL, Loveridge A. Breeding, rearing and care of the South African clawed frog. *Am Nat* 1997;81(796):38–49.
14. Major N, Wassersug RJ. Survey of current techniques in the care and maintenance of the African clawed frog (Xenopus laevis). *Contemp Top Lab Anim Sci* 1998;37(5):57–60.
15. Levin M, Mercola M. Expression of connexin 30 in Xenopus embryos and its involvement in hatching gland function. *Dev Dyn* 2000; 219(1):96–101.
16. Cheng SM, Chen I, Levin M. KATP channel activity is required for hatching in Xenopus embryos. *Dev Dyn* 2002;225(4):588–591.
17. Chen I, Levin M. The role of KATP channels in development of left-right asymmetry in Xenopus. *J Dent Res* 2004;83:A1340.
18. Bunney TD, De Boer AH, Levin M. Fusicoccin signaling reveals 14–3-3 protein function as a novel step in left-right patterning during amphibian embryogenesis. *Development* 2003;130:4847–4858.
19. Rutenberg J, Cheng SM, Levin M. Early embryonic expression of ion channels and pumps in chick and Xenopus development. *Dev Dyn* 2002;225(4):469–484.
20. Levin M, Thorlin T, Robinson KR, Nogi T, Mercola M. Asymmetries in H+/K+-ATPase and cell membrane potentials comprise a very early step in left-right patterning. *Cell* 2002;111(1):77–89.
21. Fukumoto T, Kema IP, Levin M. Serotonin signaling is a very early step in patterning of the left-right axis in chick and frog embryos. *Curr Biol* 2005;15(9):794–803.
22. Fukumoto T, Blakely R, Levin M. Serotonin transporter function is an early step in left-right patterning in chick and frog embryos. *Dev Neurosci* 2005;27(6):349–363.
23. Adams DS, Levin M. Strategies and techniques for investigation of biophysical signals in patterning. In: Whitman M, Sater AK, Eds. *Analysis of Growth Factor Signaling in Embryos*. London: Taylor and Francis Books, 2006:177–262.
24. Scharf SR, Rowning B, Wu M, Gerhart JC. Hyperdorsoanterior embryos from Xenopus eggs treated with D_2O. *Dev Biol* 1989; 134(1):175–188.
25. Hedberg E. [Breeding and use of Xenopus laevis in pregnancy tests.] *Nord Med* 1951;45(6):200–201.
26. Davys JS. The breeding of Xenopus laevis on a large scale in the laboratory. *Anim Technol J Instit Anim Technol* 1986;37(3): 217–223.
27. Major N, Wassersug RJ. Survey of current techniques in the care and maintenance of the African clawed frog (*Xenopus laevis*). *Cont Top Lab Anim Sci* 1998;37(5):57–60.
28. Kaplan ML. An enriched environment for the African clawed frog (Xenopus laevis). *Lab Anim* 1993;22(5):25–27.
29. Kobel HR, Tinsley RC. *The Biology of Xenopus*. Oxford: Published for the Zoological Society of London by Clarendon Press, 1996.
30. Hydroserve. *Hydro's Water Index: A Manual for Ultrapure Water System Design*. Research Triangle Park, NC: Hydro Service and Supplies Inc., 1999.
31. Godfrey EW, Sanders GE. Effect of water hardness on oocyte quality and embryo development in the African clawed frog (Xenopus laevis). *Comp Med* 2004;54(2):170–175.
32. Volunteers WA. *Dissolved Oxygen: Aquatic Life Depends on It*. Madison, WI: University of Wisconsin, 2003.
33. Carmingnani GM, Bennett JP. Rapid start-up of a biological filter in a closed aquaculture system. *Aquaculture* 1977;11:85–88.
34. Brown LE, Rosati RR. Effects of three different diets on survival and growth of larvae of the African clawed frog Xenopus laevis. *Progr Fish Cult* 1997;59(1):54–58.
35. Able DJ. An economical, balanced diet for Xenopus. *ILAR News* 1988;3:20–21.
36. Parker JM, Mikaelian I, Hahn N, Diggs HE. Clinical diagnosis and treatment of epidermal chytridiomycosis in African clawed frogs (Xenopus tropicalis). *Comp Med* 2002;52(3):265–268.
37. Hubbard GB. Aeromonas hydrophila infection in Xenopus laevis water borne bacillus. *Lab Anim Sci* 1981;31(3):297–300.
38. Reilly DS, Tomassini N, Zasloff M. Expression of magainin antimicrobial peptide genes in the developing granular glands of Xenopus skin and induction by thyroid hormone. *Dev Biol* 1994;162(1): 123–133.
39. James S, Gibbs BF, Toney K, Bennett HP. Purification of antimicrobial peptides from an extract of the skin of Xenopus laevis using

heparin-affinity HPLC: Characterization by ion-spray mass spectrometry. *Anal Biochem* 1994;217(1):84–90.

40. Moore KS, Bevins CL, Tomassini N, *et al.* A novel peptide-producing cell in Xenopus: Multinucleated gastric mucosal cell strikingly similar to the granular gland of the skin. *J Histochem Cytochem* 1992;40(3):367–378.
41. Moore KS, Bevins CL, Brasseur MM, *et al.* Antimicrobial peptides in the stomach of Xenopus laevis. *J Biol Chem* 1991;266(29): 19851–19857.
42. Soravia E, Martini G, Zasloff M. Antimicrobial properties of peptides from Xenopus granular gland secretions. *FEBS Lett* 1988;228(2): 337–340.
43. Iglauer F, Willmann F, Hilken G, Huisinga E, Dimigen J. Antihelmintic treatment to eradicate cutaneous capillariasis in a colony of South African clawed frogs (Xenopus laevis). *Lab Anim Sci* 1997;47(5):477–482.
44. Marchant JS, Parker I. Xenopus tropicalis oocytes as an advantageous model system for the study of intracellular Ca(2+) signalling. *Br J Pharmacol* 2001;132(7):1396–1410.
45. Kenwrick S, Amaya E, Papalopulu N. Pilot morpholino screen in Xenopus tropicalis identifies a novel gene involved in head development. *Dev Dyn* 2004;229(2):289–299.
46. Grammer TC, Khokha MK, Lane MA, Lam K, Harland RM. Identification of mutants in inbred Xenopus tropicalis. *Mech Dev* 2005;122(3):263–272.
47. Gilchrist MJ, Zorn AM, Voigt J, Smith JC, Papalopulu N, Amaya E. Defining a large set of full-length clones from a Xenopus tropicalis EST project. *Dev Biol* 2004;271(2):498–516.
48. Ishii Y, Asakawa S, Taguchi Y, Ishibashi S, Yagi T, Shimizu N. Construction of BAC library for the amphibian Xenopus tropicalis. *Genes Genet Syst* 2004;79(1):49–51.

18 The Chicken as a Model Organism

JANICE M. BAHR

ABSTRACT

Selection of appropriate animal models is essential to the advancement of basic and clinical research. This chapter provides an overview of the reproductive biology of the chicken (*Gallus domesticus*), also known as the domestic hen, and then introduces the chicken as a model for investigations of mechanisms controlling ovarian follicular growth and maturation, ovulation, and changes in ovarian function with aging. The chicken is currently the only animal model available to probe the etiology and progression of human ovarian cancer as well as to test chemotherapy agents. Finally the chicken is an ideal model for toxicology studies because of its sensitivity and rapid response to environmental toxicants and expression of external indicators, e.g., number of eggs laid, thickness of shell, as a mark of toxic exposure.

Key Words: Chicken, Ovary, Reproduction, Follicle, Ovulation, Ovarian cancer, Toxicology.

INTRODUCTION

Advances in basic and clinical sciences depend heavily on the successful use of appropriate animal models. Today there is widespread use of the mouse with its various genetic modifications, but there are also limitations to this model, particularly when an essential gene is deleted in all tissues. This systemic genetic modification of the entire animal's physiology may make it difficult if not impossible to draw valid conclusions regarding the function of a specific gene in a specific tissue. However, the choice of appropriate animal models is dependent on familiarity of scientists with different animal models, the availability of animal models, and the cost of the animal and housing requirements.

The chicken, also referred to as the domestic hen, has served science well. The chick embryo has been the basis for understanding the stages of early development and its control and is widely used in embryology classes. The young chick was the popular animal of choice for the discovery of steroid receptors, namely progesterone and estradiol receptors. The chick's oviduct, a rich source of these receptors following treatment with steroids, yielded large amounts of tissue for isolation, characterization, and cloning of the steroid receptors. A great deal of vitamin D research was done using the shell gland of the chicken. Awareness of the toxicity of some chemicals used in the environment, such as dichlorodyphenyltrichloroethane (DDT), came from observing that birds exposed to this chemical laid soft-shelled eggs.

From: *Sourcebook of Models for Biomedical Research* (P. M. Conn, ed.), © 2008 Humana Press Inc., Totowa, NJ.

The purpose of this chapter is to introduce the scientist to the chicken and indicate how the chicken is a valuable animal model for studies in basic reproductive biology, reproductive disease, and toxicology. First a brief overview of the basic reproductive physiology of the chicken will be given.

BRIEF OVERVIEW OF THE REPRODUCTIVE PHYSIOLOGY OF THE CHICKEN

HATCHING CHICKEN EGGS The chick egg has an incubation period of 21 days. Fertilized eggs are placed in an incubator that is maintained at a temperature of 100°F and has a humidified environment. To hatch a large number of eggs, it is advisable to have a commercial incubator. However, if only a small number of chicks is needed, e.g., 10–12, it is possible to incubate eggs in a laboratory incubator provided the eggs are rotated several times each day and humidity is high. Upon hatching, the chicks are placed in a warm environment maintained at 90°F for 3 weeks, after which the temperature can gradually be reduced to 70°F. At all times chicks are provided with commercial chick feed and water *ad libitum*. Chicks move to larger housing quarters as they increase in size. If chicks will not come in contact with other chickens or birds, it is not necessary to immunize the chicks against the common diseases of chickens. Whereas it is possible to raise adult chickens starting with the fertilized egg, it is easier and in some cases less expensive to purchase adult chickens from a producer.

BREEDS OF CHICKENS There are two commercial breeds of chickens, one which has been selected for egg laying and the other breed which has been selected for meat. Depending upon the research interest, either can be used. The laying breed grows slowly, reaching a weight of about 2.3 kg at maturity. In contrast, the meat breed grows rapidly, weighing 2.3 kg at 6 weeks of age, and if not feed restricted it will be obese at maturity. Generally the laying breed is selected for research because it does not have a weight problem, making it easier to manage. The most common strain is the white Leghorn, which is the strain used in most reproductive studies.

ONSET OF PUBERTY The egg laying Leghorn hen reaches sexual maturity at ~5 months of age and starts to lay eggs. Egg production is high during the first year of lay at a rate of 90–95%. As the hen ages, egg production will decrease slightly. In the commercial poultry production, hens are molted at the end of the first year of lay, which will result in increased egg production and also better egg quality. However, in a research environment, it is not necessary to molt hens after the first year of lay, albeit egg production will be less.

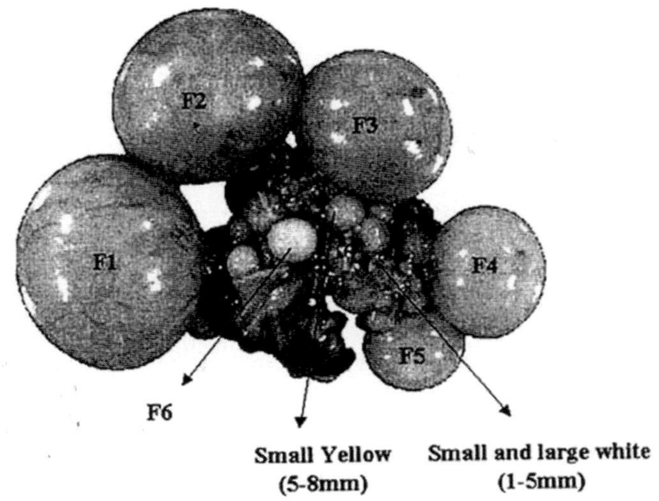

Figure 18–1. A chicken ovary. The preovulatory follicles of the hierarchy are identified according to size with the F1 follicle being the largest follicle and the next one to ovulate, followed by the F2 follicle, the second largest follicle, etc. Small follicles are classified according to size and color of yolk. The postovulatory follicle (not visible) is a saclike structure containing all the cell layers present in the preovulatory follicle.

LEFT OVARY DEVELOPS In the chicken only the left ovary and oviduct develop. The ovary contains five or six large yolk-filled follicles that are arranged in a hierarchy based on their size, F1–F6, with F1 being the largest follicle and the next follicle to ovulate (Figure 18–1). These preovulatory follicles are attached to the ovary by a stalk through which nerves and blood vessels pass. When the F1 follicle ovulates, the other preovulatory follicles move up one position in the hierarchy. At the same time, another follicle is selected from the small yellow follicles to enter the hierarchy. This follicular growth and maturation pattern works with almost clockwork precision in Leghorn chickens.

STRUCTURE OF THE FOLLICLE The follicle consists of an outer superficial epithelium single cell layer, a theca layer, a granulosa layer, a perivitelline membrane, a germinal disc, and a yolk (Figure 18–2). The theca layer, externa and interna, is a heterogeneous layer consisting of fibroblasts, steroidogenic cells, nerve cells, extracellular matrix, and blood vessels. In contrast, the granulosa layer is a single cell layer and is not innervated and has no blood vessels. The theca and granulosa layers are separated by a basement membrane, which allows an easy and clean separation of the two cell layers. The theca layer is the major site of testosterone and estrogen (estradiol-17β and estrone) synthesis, whereas the granulosa layer is the primary site of progesterone production.[1] Inhibin, a hormone that regulates follicle-stimulating hormone in a negative feedback manner, is produced mainly by the granulosa layer of the hierarchical follicles.[2]

OVULATORY CYCLE The ovulatory cycle, described as the interval from one ovulation to the next ovulation, is approximately 24–26 h in length depending upon the age of the chicken. Young hens, in their first year of lay, will ovulate at about 24 h intervals, whereas as the chicken ages this interval between ovulations can increase to ~26 h or even greater. Ovulation is preceded by a significant increase in progesterone, which induces the luteinizing hormone (LH) surge. Ovulation of the egg occurs as the stigma, a visible avascular strip on the follicles, ruptures (Figure 18–3). The egg, picked up by the oviduct, passes down the oviduct where albumen and shell membranes are secreted and then into the shell gland where the shell is deposited. It takes ~24 h for the egg to move down the reproductive tract. At the time of oviposition, the passage of the egg from the shell gland through the vagina is dependent upon prostaglandins $F_{2\alpha}$ and E_2 and arginine vasotocin. Upon the laying of the egg, ovulation of the next egg occurs 15–30 min after oviposition. The close temporal relationship between oviposition and ovulation makes it easy to determine the time of ovulation. For this reason, events during the ovulatory cycle are identified as hours before ovulation of the next (F1) egg. If the egg is to be fertilized, sperm stored in vagina-shell gland junction sperm storage glands are released and fertilize the egg as it enters the oviduct. For further information regarding the reproductive physiology of the chicken, see reviews by Bahr and Johnson[3] and Sharp.[4]

ENVIRONMENTAL CONDITIONS FOR MAXIMAL REPRODUCTIVE PERFORMANCE Chickens are easy to maintain and require only a few specific conditions. Commercial cages and watering systems are available and are relatively inexpensive. If possible, chicken cages should be suspended over a pit that has a stream of running water; however, this housing condition is not necessary. The water reduces the odor of the chickens and makes it easy to clean the pit by pulling the drain plug. Chickens should be housed in a room with adequate air exchange at an ambient temperature of 65°F with 14–17 h of light. Chickens are sensitive to wavelength, therefore warm fluorescent lights should be used and not the cool fluorescent lights. Chicken feed is available from commercial sources. Feed and water should be provided at all times.

MALE CHICKENS The rooster or cockerel reaches puberty at 4–5 months of age. The rooster has two testis that are located internally and attached to the dorsal wall of the body cavity. Spermatogenesis occurs at 41°C, the internal temperature of the rooster. There are no accessory reproductive glands, such as the

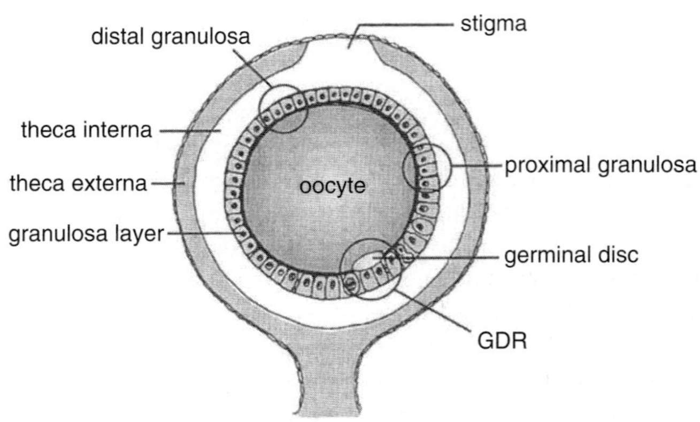

Figure 18–2. A cross section of a chicken preovulatory follicle. Note structural differences in the granulosa layer. The granulosa layer in contact with the germinal disc is called the germinal disc region (GDR). The columnar-shaped granulosa cells located proximal to the GDR (proximal granulosa) are proliferative and synthesize less progesterone. The cuboidal-shaped granulosa cells located distal to the GDR (distal granulosa) are differentiated and synthesize greater amounts of progesterone. (From Tischkau and Bahr,[5] reprinted with permission from the Society for the Study of Reproduction.)

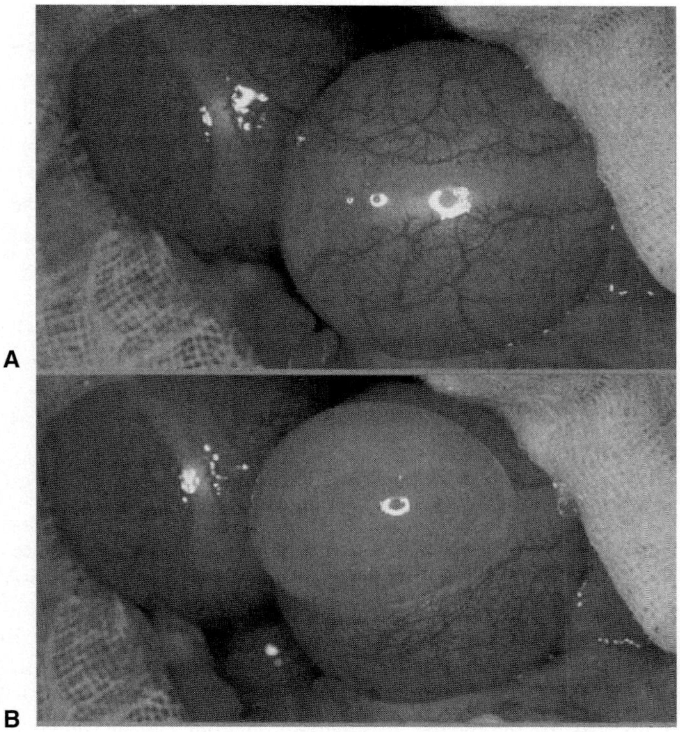

Figure 18–3. Ovulation of a chicken follicle. (A) A picture of the F1 follicle (largest) and the F2 follicle (smaller) several minutes before ovulation. The nonvascular stigma, the site of follicular rupture, is very obvious. (B) Ovulation of the F1 follicle as the stigma is ruptured. Note the size of the clear stigma in both the F1 follicle and the F2 follicle, which allows for easy isolation of the stigma and nonstigma parts of a follicle.

prostate and seminal vesicles. As a result, seminal plasma has the same constituents as the epididymal fluid. Roosters also lack an intromittent organ. Sperm, stored in the lower end of the ductus deferens, are transferred directly from the rooster's cloaca to that of the hen's cloaca at mating. For further information see the review by Sharp.[4]

VALUE OF THE CHICKEN AS A RESEARCH ANIMAL The benefits of using the chicken are many. Chickens are inexpensive to purchase, are healthy, and are easy to maintain. Specific pathogen-free eggs and chickens are available. Transgenic hens are produced commercially, though the supply is limited at this time. In the next section, examples of how the chicken is ideal for reproductive studies will be described.

THE CHICKEN AS A MODEL FOR BASIC REPRODUCTIVE STUDIES

FOLLICULAR DEVELOPMENT Its hierarchical arrangement of follicles, large preovulatory follicles, visible germinal disc, clean separation of the granulosa and theca layers, accurate prediction of ovulation, and change in follicular maturation and ovulation with aging are a few of the chicken's valuable qualifications as a model for ovarian studies. Hence the cellular changes and those associated with changes in gene expression that occur in the granulosa and theca layer as follicles mature and approach ovulation can be easily studied. In particular, the large number of granulosa cells (~5 million) easily and quickly obtained from the F1 follicle makes the chicken follicle an ideal model for investigating signaling, transcription factors, clock genes, etc. in the control of follicular development.

ROLE OF GERMINAL DISC For many years, it was thought that the oocyte was a passive inhabitant of the follicle. With the discovery of factors made by the oocyte that regulate the differentiation of the follicle, investigations into the role of the oocyte in follicular development has intensified. The chicken follicle is an excellent model to study the role of the germinal disc in follicle function. The germinal disc contains the genetic material and cellular organelles of the egg; thus the germinal disc is equivalent to the oocyte of the mammal. The germinal disc, attached to the overlying granulosa cells through cytoplasmic interdigitations, is called the germinal disc region (Figure 18–2). These granulosa cells have a higher rate of proliferation and are less differentiated than the granulosa cells distal to the germinal disc.[5] The visibility of the germinal disc *in vivo* is enhanced by injection of chickens the day before they are used with 0.8 ml Sudan black [10 mg/ml of 1:1 phosphate-buffered saline (PBS):100% ethanol, filtered through a 0.45-μm filter], which is taken up by the yolk.[6] To understand the role of the germinal disc, it can easily be destroyed *in vivo* by freezing it with dry ice for 20 sec.[7] The germinal disc along with its overlying granulosa cells can be obtained by isolating the granulosa layer.

OVULATION Identification of the biological changes, particularly expression of genes, associated with ovulation, has been a challenging and intriguing subject for many years. Whereas progress has been made in understanding the process of ovulation using the mammal, this animal model is limiting because of the inability to obtain adequate tissue to measure cellular changes in the stigma, the site of rupture, and in nonstigma tissue. The chicken's large preovulatory follicles overcome these difficulties. The hierarchical follicles have a distinct stigma area that can be easily isolated from the nonstigma area (see Figure 18–3). Furthermore comparison of the F1 follicle, the next one to ovulate, with the F2 follicle, the follicle to ovulate the next day, provides an excellent comparable tissue to measure the cellular changes associated with ovulation (Figure 18–4).[8] The short ovulatory cycle (24–25 h) and the accurate prediction of ovulation based on oviposition further bring superior precision to measuring the changes associated with ovulation. In the chicken follicle, unlike the mammalian follicle, there is no dilution of the measured endpoints due to the use of a large amount of ovarian tissue not associated with ovulation.

OVARIAN FUNCTION WITH AGING A longer interval between ovulations, hence a lengthening of the ovulatory cycle, occurs as the chicken ages. The follicle requires a longer time to gain competency to ovulate.[9] This delay in ovulation represents changes at the level of the follicle as well as at the level of the hypothalamus, which becomes less responsive to the positive feedback of progesterone to induce the luteinizing hormone surge. The accuracy with which the time of ovulation based on oviposition can be determined and the large amount of tissue available from individual follicles facilitate the investigation of changes in gene expression with age. The chicken is a particularly good model to study the role of clock genes at the level of the hypothalamus and the ovary in the aging process.

THE CHICKEN AS A MODEL FOR OVARIAN CANCER

Progress in eliminating a disease depends heavily upon the availability of an appropriate model. For many years scientists

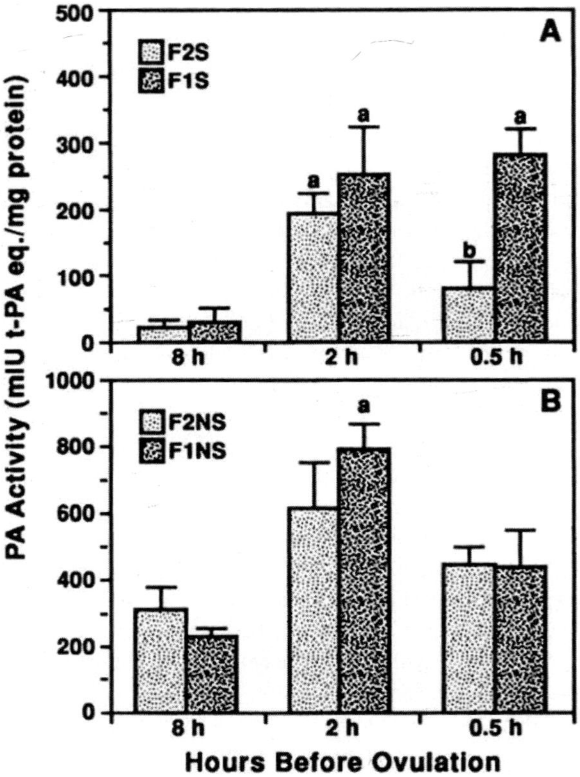

Figure 18–4. Plasminogen activity (PA) in (A) the stigma region (S) of the theca layer and (B) the nonstigma region (NS) of the theca layer of F1 and F2 preovulatory follicles, which will ovulate next and 26 h later, respectively. Follicles were obtained at 8 h (before the LH surge), 2 h (immediately after the LH surge), and 0.5 h immediately before ovulation. Each bar represents the mean ± SE of three to five experiments. a, significant differences in PA activity compared to 8 h before ovulation ($p < 0.05$); b, significant difference in PA activity in F1 versus F2 ($p < 0.05$). (From Jackson *et al.*,[8] reprinted with permission from the Society for the Study of Reproduction.)

Figure 18–5. Ovarian cancer of the hen. Gross appearance is characterized by firm, white nodules. Cystic or hemorrhagic follicles are also often present. (From Johnson and Giles,[10] reprinted with permission from the Poultry Science Association.)

have attempted to develop a mammalian model to investigate ovarian cancer in humans. Human ovarian cancer is the most common fatal gynecologic malignancy with a 5-year survival rate of 25–30%. One of the reasons for this low survival rate is the inability to detect ovarian cancer early. For example, the availability of a blood marker such as the prostate-specific antigen in males would be a significant advancement in early detection of ovarian cancer. Therefore a model that would make it possible to determine if there are markers in the blood for ovarian cancer and also the use of the presence of these markers to follow the effectiveness of chemotherapy would be ideal. To date, despite the use of specific hormone treatments of rat to induce ovarian cancer and the recent availability of a transgenic mouse model, a mammalian animal model that spontaneously develops ovarian cancer has not been developed.

In contrast, the chicken does develop ovarian cancer spontaneously and the types of cancers are similar to those that occur in the human.[10] The chicken seldom develops ovarian cancer before 2 years of age, but does develop it with increasing age. It has been hypothesized that women and chickens develop ovarian cancer because they are persistent ovulators.[11] The number of ovulations that occur in a woman's lifetime, which in most cases may be interrupted by one or two pregnancies, is similar to the number of ovulations a chicken will have during a 2-year period, with the chicken ovulating 80% of the time. In contrast, most mammals are either pregnant or nursing and therefore are not persistent ovulators.

Fredrickson[12] provides an early report of cancer in chickens. He biopsied 466 hens, ages 2–7 years, and found that 24% had malignant ovarian adenocarcinomas and that the incidence increased with age. He observed that the ovarian adenocarcinomas at their early stages are firm white cauliflower-like nodules that resemble atrectic follicles (Figure 18–5). Histological studies done by Fredrickson[12] indicated a similarity between the germinal epithelium and the ovarian adenocarcinoma cells. This observation supports the current hypothesis that the ovarian superficial epithelial (OSE) cells are the cellular source of the ovarian adenocarcinomas. Recent studies by Johnson and Giles[10] show that ovarian cancer in the chicken has a glandular growth pattern and simple columnar epithelium lines the glands (Figure 18–6). It is hypothesized that at the time of ovulation, cells located at the rupture site may have DNA damage; along with defective DNA repair mechanisms, this mutagenesis results in malignant transformation followed by clonal expansion and metastasis.[13] Another hypothesis that has a great deal of support is the epitheliomesenchymal conversion of OSE cells that occurs following ovulation when the OSE cells are displaced from the surface to the stroma.[11] This displacement can also occur in the aging ovary due to an invagination of the OSE. The epithelial cells that are converted to mesenchymal cells can be incorporated into the stroma. However, if the OSE remain epithelial, they can aggregate and form inclusion cysts that are the preferred sites of metaplastic and dysplastic changes that can lead to tumorigenesis. With malignant progression and depending upon the genetic predisposition of the woman, fewer of the OSE will become mesenchymal cells and be incorporated into the stroma.

The use of two chicken strains, C and K, that have a different incidence of ovarian cancer, different concentrations of estrogen

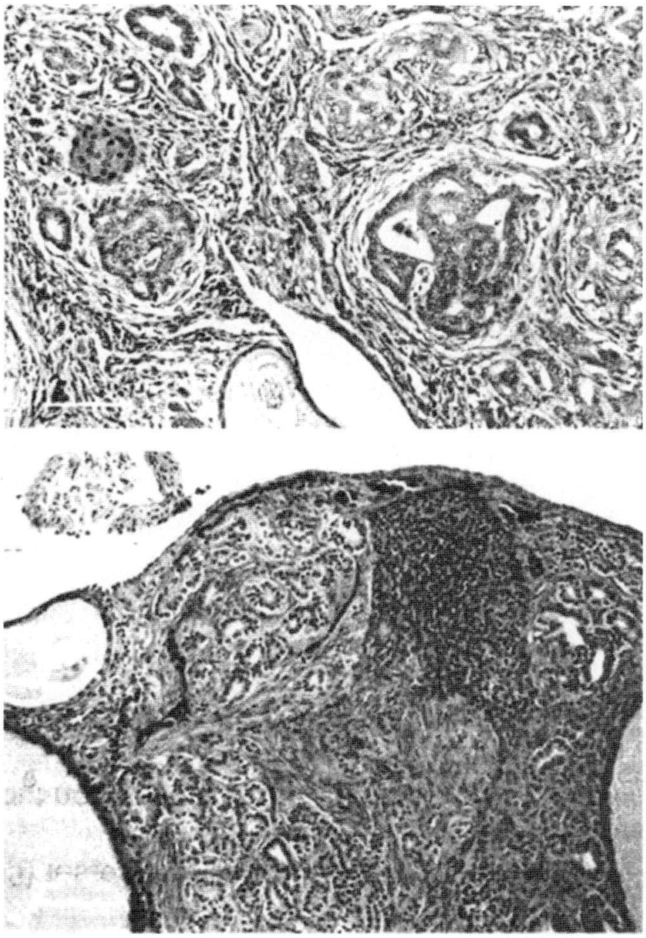

Figure 18–6. Hematoxylin and eosin photomicrographs of ovarian cancer in the hen. The cancer is characterized by a glandular growth pattern. The glands are generally lined by a simple columnar epithelium. Scale bar = 50 µm. (From Johnson and Giles,[10] reprinted with permission from the Poultry Science Association.)

in the blood, and a different amount of mRNA expression for the α-subunit of inhibin in the granulosa layer suggests that there may be a genetic basis for ovarian cancer.[10] The use of these two strains of chickens should be helpful in exploring the etiology of ovarian cancer.

Immunohistochemical expression of molecular markers associated with ovarian cancer in humans was performed on chicken ovarian cancer tissue.[14] A number of different antigens (cytokeratin AE1/AE3, pan cytokeratin, EGFR, Lewis Y, CEA, Tag 72, PCNA, p27, and TGF-α) were identified, which indicated the immunohistochemical similarity between human and chicken ovarian cancer tissue. Also this study demonstrated the potential utility of the avian model for the testing of chemoprevention agents.

THE CHICKEN AS A MODEL FOR TOXICOLOGY

Silent Spring, the book written by Rachel Carson in the late 1960s, indicates the value of birds to make us aware of dangerous toxicants in our environment. Rachel Carson called attention to the dangers of the pesticide, DDT. Birds were laying soft shelled eggs, which made it impossible for these eggs to hatch and produce the next generation of birds. Without these birds, spring would be silent. Just as the canary in the coal mine can detect dangerous gases before humans can smell them, wild birds can act as sentinels to alert us to dangerous chemicals in our environments. Whereas wild birds are not a good research animal for toxicology, the chicken is an excellent model. In fact, in many cases, the chicken is a more sensitive model with easily observable outcomes when compared to rodent models.

The list of xenobiotics and toxicants is long; however, a select few will be cited: DDT, polychlorinated biphenyl (PCB), and phenobarbital (PB). These lipophilic chemicals are metabolized by the mixed-function oxidases (MFO) present in the liver. The MFO metabolize substrates including drugs, lipophilic substances, and steroids. Hepatic cytochrome P450 (P450) is an indicator for the activity of this enzyme system. Exposure to these lipophilic substances results in an increase in the activity of the P450 in an effort to remove the toxic substances from the body. However, because these enzymes are not highly specific, an increase in their activity will increase steroid metabolism, thus lowering systemic concentrations of these steroids. Therefore physiological activities in the body that depend upon a certain blood concentration of steroids will be modified, which in the chicken would be the laying of soft shelled eggs or even complete cessation of egg laying.

The chicken, as an egg-producing species, is dependent upon a properly functioning ovary, liver, and shell gland to produce eggs. The ovary is the site of steroidogenesis and formation of yolky follicles. Ovarian estrogens act on the liver via an estrogen receptor to cause the synthesis of yolk proteins, which are then secreted into the blood and taken up by the ovarian follicles. Once the follicle has attained its proper size and is competent to respond to luteinizing hormone to ovulate, the egg is released and enters the oviduct where it obtains albumen and shell membranes. The egg then enters the shell gland where the shell, rich in calcium, is laid down. The mobilization of calcium from the medullary bone to the shell gland and its secretion into the shell gland are dependent primarily on estrogen. Therefore any chemical that changes the synthesis of estrogen by the ovary, its metabolism and ability to stimulate yolk synthesis in the liver, and the availability of estrogen to make available adequate calcium and its accretion for shell formation will affect the laying of eggs. Simply stated, the chicken has three specific tissues whose function can be altered by exposure to xenobiotics and toxicants. Furthermore, the response of chickens to these chemicals is rapid, occurring in 5–7 days, and can be easily determined by observing the number of eggs laid, measuring the thickness of the shell, and determining the absence of a hard shell.

Two toxicology studies done in the chicken[15,16] demonstrate the usefulness of the chicken for these types of studies. The objectives of these studies were to investigate the effect of xenobiotics on the induction or depression of the MFO, changes in hepatic cytochrome P450, blood concentration of estradiol-17β (E_2), plasma calcium (Ca), rate of egg lay, and thickness of egg shell in laying Leghorn hens.[15] They used PB and carbon tetrachloride (CCl_4) as the xenobiotics of choice. There was a dose–response decrease in E_2 concentrations in the blood (Figure 18–7) and an increase in the liver P450 concentration in response to feeding increasing amounts of PH (0–100 mg for 3 or 7 days; Figure 18–8). The response to CCl_4 for the concentration of P450 was

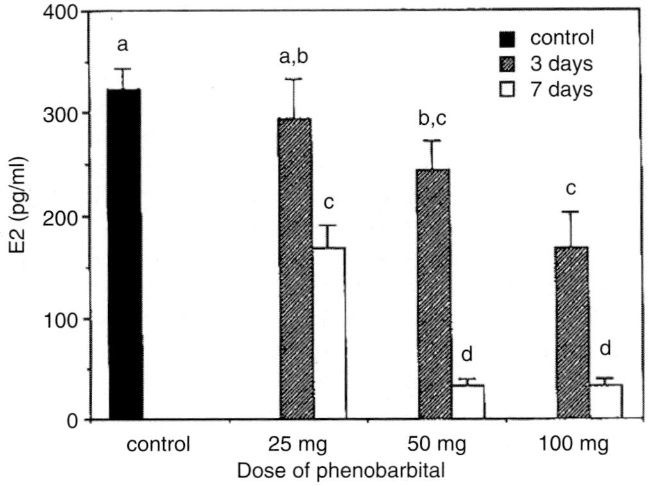

Figure 18–7. Concentrations (mean ± SE) of estradiol (E_2) in the plasma after the administration of phenobarbital for 3 or 7 days to laying hens. Means with the same letter are not different from each other ($p > 0.05$). Circulating E_2 was decreased by doses of phenobarbital in a linear manner. Three days: $R^2 = 0.413$ ($p < 0.001$). Seven days: $R^2 = 0.693$ ($p < 0.001$). (From Chen *et al.*,[15] reprinted with permission from the American Association of Animal Science.)

similar to the data obtained following phenobarbital. In the second study, Chen *et al.*[16] tested the effectiveness of four environmental toxicants that affect liver P450 concentrations. Three of the toxicants (aroclor 1254, a PCB, DDT, and benzo[α]pyrene) were hypothesized to induce different isoenymes of P450 and one toxicant, lead acetate, inhibited the induction of the P450. The PCB and DDT significantly increased P450 with a consequent depression of circulating E_2. Data from this study indicated that the environmental effects of these toxicants are an elevation of P450, which increases steroid metabolism by the liver and

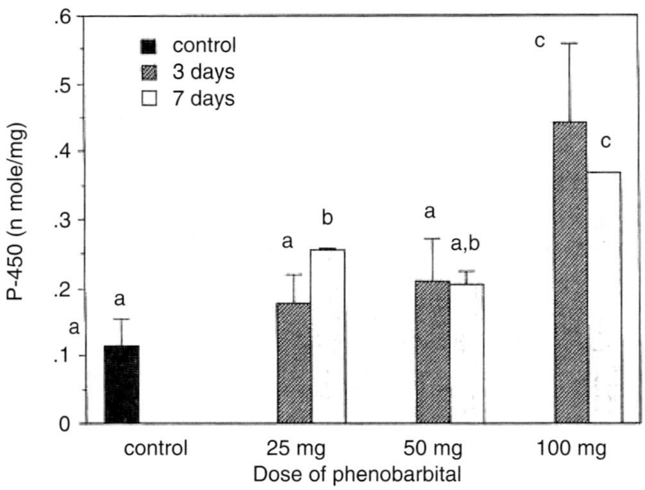

Figure 18–8. Concentrations (mean ± SE) of cytochrome P450 in the hepatic microsomal fraction in laying hens after the administration of phenobarbital for 3 or 7 days. Means with the same letter are not different from each other. The P450 was induced by doses of phenobarbital in a linear manner. Three days: $R^2 = 0.304$ ($p < 0.005$). Seven days: $R^2 = 0.389$ ($p < 0.001$). (From Chen *et al.*,[15] reprinted with permission from the American Society of Animal Science.)

decreases circulating concentrations of E_2 with negative effects on reproduction.

CONCLUSIONS

This chapter has provided a brief overview of the chicken's reproductive physiology to introduce scientists in biomedical research to a valuable model. Examples of three specific areas of research were presented, namely, basic reproductive biology of ovarian function and ovulation, ovarian cancer, and toxicology. The size of the follicles, availability of ample tissue, and accurate prediction of ovulation make the chicken an ideal model to probe basic questions about ovarian function. The fact that it is the only animal model that spontaneously develops ovarian cancer with histological changes and expression of genes similar to that observed in human ovarian cancer makes the chicken an extremely important model to investigate this serious gynecological disease in humans. Finally, with the increasing need to determine the toxicity of substances in our environment, the chicken has a number of attributes that makes it an invaluable model. Only a few select references regarding these studies in chickens were cited in this chapter; however, the interested scientist will find many more informative references in the published literature. The author of this chapter who has used chickens in research for 30 years is also a resource person to anyone needing help or having unanswered questions. She can be reached at jbahr@uiuc.edu.

REFERENCES

1. Bahr JM, Wang S-C, Huang M-Y, Calvo FO. Steroid concentrations in isolated theca and granulosa layers of preovulatory follicles during the ovulatory cycle of the domestic hen. *Biol Reprod* 1983;29:325–334.
2. Johnson P. Avian inhibin. *Poult Avian Biol Rev* 1997;8:21–31.
3. Bahr JM, Johnson PA. Reproduction in poultry. In Cupps PT, Ed. *Reproduction in Domestic Animals*, 4th ed. New York: Academic Press, 1991:555–575.
4. Sharp PJ. Chickens, control of reproduction. In: Knobil E, Neill JD, Eds. *Encyclopedia of Reproduction*, Vol. 1. San Diego: Academic Press, 1998:572–580.
5. Tischkau SA, Bahr JM. Avian germinal disc region secretes factors that stimulate proliferation and inhibit progesterone production by granulosa cells. *Biol Reprod* 1996;54:865–870.
6. Volentine K, Yao H-C, Bahr JM. Epidermal growth factor in the germinal disc and its potential role in follicular development in the chicken. *Biol Reprod* 1998;59:522–526.
7. Yoshimura Y, Tischkau SA, Bahr JM. Destruction of the germinal disc region of an immature preovulatory follicle suppresses follicular maturation and ovulation. *Biol Reprod* 1994;51:229–233.
8. Jackson JA, Zhang P, Bahr JM. Plasminogen activator activity in preovulatory follicles during the ovulatory cycle of the chicken. *Biol Reprod* 1993;49:1141–1146.
9. Johnson PA, Dickerman RW, Bahr JM. Decreased granulosa cell LH sensitivity and altered thecal estradiol concentration in the aged hen. *Biol Reprod* 1986;35:641–646.
10. Johnson PA, Giles JR. Use of genetic strains of chickens in studies of ovarian cancer. *Poult Sci* 2006;85:246–250.
11. Auersperg N, Wong AST, Choi K-C, Kang SK, Leung PCK. Ovarian surface epithelium: Biology, endocrinology and pathology. *Endocr Rev* 2001;22:255–288.
12. Fredrickson TN. Ovarian tumors of the hen. *Environ Health Perspect* 1987;73:35–51.
13. Murdock WJ, Martinchick JF. Oxidative damage to DNA of ovarian surface epithelial cells affected by ovulation: Carcinogenic implication and chemoprevention. *Exp Biol Med* 2004;229:546–552.

14. Rodriguez-Burford C, Barnes MN, Berry W, Partridge EE, Grizzle WE. Immunohistochemistry expression of molecular markers in an avian model: A potential model for preclinical evaluation of agents for ovarian cancer chemoprevention. *Gynecol Oncol* 2001;81:373–379.
15. Chen S-W, Francis BM, Dziuk PJ. Effect of concentration of mixed-function oxidase on concentration of estrogen, rate of egg lay, egg-shell thickness, and plasma calcium in laying hens. *J Anim Sci* 1993;71:2700–2707.
16. Chen S-W, Dziuk PJ, Francis BM. Effect of four environmental toxicants on plasma Ca and estradiol 17B and hepatic P450 in laying hens. *Environ Toxicol Chem* 1994;13:789–796.

Well-Established Models

III

Rodents

B

19 Rat Knockout and Mutant Models

BART M.G. SMITS, JILL D. HAAG, MICHAEL N. GOULD, AND EDWIN CUPPEN

ABSTRACT

The development of rat genetic models has rapidly progressed over the past decades and has become an important strategy to connect gene function to various aspects of health and disease. Reverse genetics or gene-driven genetics is a broadly accepted approach to study gene function. An indispensable reverse genetic tool is the generation and characterization of knockout and mutant animals for the genes of interest. Recently, the laboratories of Dr. Michael Gould (Madison, WI) and Dr. Edwin Cuppen (Utrecht, The Netherlands) have independently developed methods using *N*-ethyl-*N*-nitrosourea (ENU) mutagenesis to make rat knockouts and mutants in any gene of interest. This chapter describes experienced-based considerations from both laboratories that could serve as a guideline for initiating ENU mutagenesis-based screens in the rat.

Key Words: ENU mutagenesis, Point mutation, Reverse genetics, Inbred and outbred strain, Target gene, TILLING, CEL I, High-throughput resequencing, Yeast-based truncation assay.

INTRODUCTION

THE RAT AS A GENETIC MODEL ORGANISM Not long after rats were first captured for experimental purposes in the first half of the nineteenth century, rat genetic research (genetics of coat color) made its introduction.[1] Despite the early foundation, rat genetic research lost popularity, mainly since the early mammalian geneticists preferred the mouse for genetic experiments because of its smaller size and excellent reproductive characteristics. The rat with its larger size became the first choice for physiologists and nutritionists. Many inbred strains have been created to resemble specific aspects of human health and disease. The Rat Genome Database (http://rgd.mcw.edu/strains) currently contains over 500 rat strains. It is now the geneticist's turn to identify the genetic factors underlying the clinically relevant traits for which these strains have been bred. Comparative genomics by means of genetic mapping and positional cloning of quantitative trait loci (QTLs) in animal models has become a general strategy to genetically dissect these traits.[2,3] The possibility of rapidly creating congenic, consomic, and recombinant inbred strains has undoubtedly facilitated the annotation of genomic loci of importance to a phenotype. In addition to that, the availability of the rat genome,[4] a high-density radiation hybrid map,[5] and the ongoing generation of dense genetic marker sets[6–9] are completing the set of rat genomic resources.

Although these important advances in rat genetic research over the past decades have accelerated the identification and annotation of hundreds of QTLs, the next critical step in the process of coupling the clinically relevant trait to the genome is the identification of the causative genetic element(s).[10] Frequently, traditional positional cloning efforts result in the identification of a large locus containing many putative functional genetic elements, such as genes, enhancers, and noncoding RNAs. At the stage where continuing positional cloning becomes inefficient due to the lack of recombination events, the accessibility of genetic tools, such as knockout technology, is needed to evaluate the contribution of these functional genetic elements to the QTL.[10]

FUNCTIONAL GENOMICS Mammalian genomics in the years around the millennium change was characterized by the collecting of tremendous amounts of sequencing data culminating in the publication of the human,[11,12] mouse,[13] and rat[4] genomes. The "functional genomics" era aims for the identification and characterization of the functional genetic elements in these genome sequences, which will always require genetic approaches. Genetic approaches can roughly be divided into two categories: forward genetics and reverse genetics. Forward or classical genetics starts with a phenotype. To reveal the genetic variant that is causing the phenotype, genetic mapping and positional cloning are performed. Many model organisms have been subjected to large-scale phenotypic screening, including *Drosophila*,[14] zebrafish,[15,16] and mice.[17,18] These immense efforts have resulted in many interesting phenotypic mutants for which the underlying gene mutations have been identified.

On the other hand, reverse genetics is a gene-driven approach. For many genes the sequence is currently annotated or predicted. Hence, the gene of interest can be knocked out or mutated in the model organism of choice and the phenotype can be studied. For a long time, mammalian knockouts were composed solely of mice. The technology relied on homologous recombination in embryonic stem cells (ES cells) to replace an exon of interest with a selectable marker, thereby creating a nonfunctional allele of the target gene.[19] Rat embryonic stem cells capable of producing viable embryos have not been found thus far. This has hampered the production of rat knockouts for many decades, until recently. The first rat knockouts reported were generated by ENU mutagenesis-driven reverse genetics.[20–22]

From: *Sourcebook of Models for Biomedical Research* (P. M. Conn, ed.), © 2008 Humana Press Inc., Totowa, NJ.

N-ETHYL-N-NITROSOUREA MUTAGENESIS Mammalian *N*-ethyl-*N*-nitrosourea (ENU) mutagenesis was first developed in the mouse. Large-scale forward genetic screens could not rely on spontaneous mutation rates, since these are far too low to make this approach viable. Early work by Russell and colleagues[23] at Oak Ridge demonstrated that the mutation rate could be dramatically increased by germ-line mutagenesis. Several mutagenic agents, including chemical mutagens and X-irradiation, were examined using the specific locus test in mice. ENU was found to be the most favorable germ-line mutagen for three reasons.[24] First, ENU was the most potent mutagen in producing loss-of-function mutations. It is a directly acting alkylating agent not requiring metabolic activation to elicit its effect. Second, in contrast to X-irradiation, which produces larger deletions, ENU induces point mutations that merely affect single loci. Nonetheless, researchers should take into account that multiple background mutations could potentially distort phenotypic analysis. Third, ENU is considered a stem cell mutagen. When injected in males, DNA adduct formation occurs primarily in the spermatogonial stem cells,[25] which triggers misincorporation of bases during spermatogenesis, and ultimately results in fixed mutations.[26]

Additionally, repetitive application of lower doses of ENU gave a higher hit rate in the single locus test compared to a single high dose.[27,28] Above a certain threshold, the relationship between dose of ENU and hit rate in the specific locus test is linear, suggesting that at lower doses the DNA repair system in the spermatogonia is apparently capable of fixing the majority of ENU-induced DNA damage.[29] Finally, ENU-induced mutagenicity and toxicity are strongly strain dependent in the mouse.[24]

Recently, ENU mutagenesis has been utilized for gene-driven knockout/mutant production in several vertebrate species, such as zebrafish,[30] mice,[31] frogs,[32] and rats.[20–22] The universal methodology known as target-selected mutagenesis, gene-driven mutagenesis, or targeting local lesions in genomes (TILLING) is outlined in Figure 19–1. Male founder animals are mutagenized and mated with wild-type females to produce a large library of F_1 animals, carrying many random heterozygous point mutations across their genome. From each F_1 animal a tissue sample is taken and DNA or RNA is isolated and screened for induced point mutations in the genes of interest using a high-throughput mutation discovery platform. Mutations are randomly distributed across the genome, but only coding regions are screened. Missense mutations could potentially be very interesting, as amino acid changes could result in hypermorphic or hypomorphic proteins. Nonsense mutation that introduce a premature stop codon lead to an early truncated protein that is likely nonfunctional, thereby creating a knockout animal. This methodology allows for the generation of an allelic series, i.e., a nonsense allele, hypermorphic/hypomorphic alleles, and the wild-type allele, which is highly informative to study the function of the gene. Finally, the animals harboring interesting mutations are selected from the library and the mutation is outcrossed to reduce the amount of background mutations and incrossed to homozygosity.

PRODUCTION OF RAT KNOCKOUTS AND MUTANTS

Recently, Dr. Michael Gould's laboratory (Madison, WI) and Dr. Edwin Cuppen's laboratory (Utrecht, The Netherlands) have independently developed ENU mutagenesis protocols for several rat strains, resulting in the identification of the first rat knockouts and mutants worldwide. The Gould laboratory focused on genes related to breast cancer, whereas the Cuppen laboratory screened mainly genes involved in neurological processes and diseases. Although the technology seems rather uncomplicated, implementing it in the rat model system requires careful preparations. Which strain will be most suitable? Which doses of ENU and what mutation detection strategy could be used? After an interesting mutant has been identified, what follow-up will be performed? This section describes results and experiences from these two laboratories and can serve as a stepwise guideline for initiating an ENU mutagenesis-based screen in the rat.

STRAIN SELECTION Currently, over 500 rat strains have been developed to resemble a wide variety of aspects of human health and disease. It is often highly desirable to target candidate genes in the genetic background of interest, since numerous

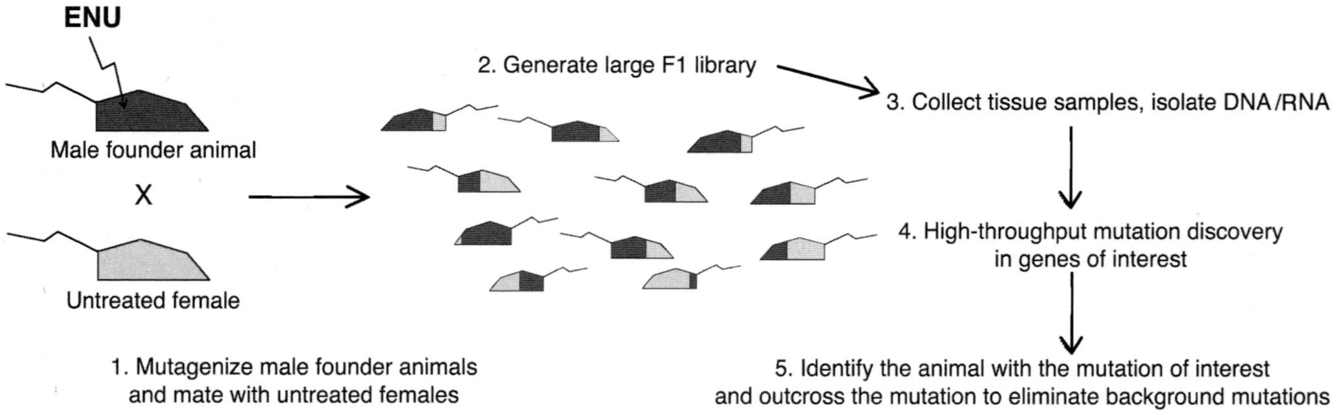

Figure 19–1. Schematic representation of a universal rat reverse genetic ENU mutagenesis screen. Mutagenized male founder animals are mated with untreated females to produce a large library of F_1 animals that contain random heterozygous point mutations in their genome. From each F_1 animal a tissue sample is collected from which DNA or RNA is extracted and screened using high-throughput mutation discovery technology for induced mutations in the genes of interest. Animals harboring interesting mutations are identified. The mutation is first outcrossed to the wild-type background and subsequently incrossed to homozygosity. Outcrossing for at least six generations will eliminate most (>98.5%) of the background mutations on other chromosomes and will on average leave <35% of the donor chromosome surrounding the mutation.

Table 19–1
Guidelines for choice of strain in *N*-ethyl-*N*-nitrosourea (ENU) mutagenesis experiments

Strain[a]	*Optimal dose*[b]	*Sterility dose*[b]	*Mutation rate*[c]	*Suitability(0–6)*[d] *RC*	*ET*	*MR*	*Total*	*Reference*
BN	3 × 20	3 × 40	1 in 2.91×10^6	0	1	1	2	Smits *et al.*[21]
F344/Crl	3 × 40	3 × 60	1 in 1.76×10^6	2	2	1	5	Smits *et al.*[21]
F344/NHsd	2 × 60	2 × 75	1 in 59	1	2	2	5	Zan *et al.*[22]
LEW	Unknown	3 × 20	Unknown	1	0	–	1	Smits *et al.*[21]
WF	Unknown	2 × 50	Unknown	0	0	–	0	Zan *et al.*[22]
WKy	Unknown	2 × 50	Unknown	0	0	–	0	Zan *et al.*[22]
SD	2 × 60	2 × 100	1 in 64	2	2	2	6	Zan *et al.*[22]
Wistar	3 × 40	3 × 60	1 in 1.24×10^6	2	2	2	6	Smits *et al.*[21]

[a]The first six strains are inbred; the last two strains are outbred.
[b]Dose of ENU in mg per kg bodyweight. 2× or 3× are weekly spaced injections. Optimal dose is determined as the dose at which at least half of the founder males are fertile at 10 weeks after the ENU treatment. Sterility dose is the dose at which none of the founder males is fertile at 10 weeks after the ENU treatment.
[c]Mutation rates BN, F344/Crl, and Wistar are molecular mutation rates determined by resequencing. The rates for F344/NHsd and SD are visible mutant phenotype rates, as determined by physical examination of the offspring. Nonmutagenized SD results in a visible mutant phenotype rate of 1 in 283.
[d]0 = not suitable, 6 = very suitable. Suitability is determined using three criteria: reproductive capabilities (RC), tolerability to ENU (ET), and induced mutation/mutant phenotype rate (MR). Each criterion is awarded 0, 1, or 2.

modifiers could influence phenotypic analysis dramatically. However, not every strain is equally suitable for ENU mutagenesis. Data from mouse ENU mutagenesis experiments have unambiguously shown a strong strain-dependent susceptibility toward the effects of ENU.[24] The Gould and Cuppen laboratories have independently established ENU mutagenesis protocols for a total of five inbred strains, Brown Norway (BN), Fischer (F344/Crl and F344/NHsd), Lewis (LEW), Wistar–Fürth (WF), and Wistar–Kyoto (WKy), as well as for two outbred strains, Sprague–Dawley (Hsd:SD; SD), and Wistar (Wistar/Crl). Table 19–1 shows the results of several titration experiments using split doses of ENU. The strains were awarded a "suitability score" based on three criteria: reproductive capabilities (RC), tolerability toward ENU (ET), and mutation/mutant phenotype rate (MR).

First, the reproductive capabilities of outbred strains are generally better than those of inbred strains. Among the inbred strains, F344 showed the best reproductive performance (http://rgd.mcw.edu/strains; on average 7–10 pups/litter in our hands, compared to 11–14 pups/litter for Wistar and SD). BN, WF, and WKy are known to be the moderate breeders (http://rgd.mcw.edu/strains), which make those strains less efficient for production of large libraries of mutant offspring.

The second criterion is the tolerability toward ENU (Table 19–1). The optimal split dose of ENU for the above-mentioned strains was determined. We define the optimal dose as the highest dose for which at least half of the males are fertile 10 weeks after the ENU treatment. A dose group in which less than half of the males are fertile will not be able to extensively contribute to the library. In general, the outbred strains Wistar and SD appear to tolerate higher doses of ENU compared to the inbred strains; however, inbred F344 animals appear equally tolerable toward ENU. The recommended doses for these strains are 2 × 60 (mg of ENU per kg body weight) or 3 × 40. The recommended doses for BN is 3 × 20. WF never reached a fertility percentage of over 50. The inbred strains LEW and WKy became completely sterile at the lowest split doses injected. The latter three strains are therefore considered less efficient for ENU mutagenesis screens. If a knockout or mutant in those genetic backgrounds is absolutely essential, two possibilities can be explored. First, the optimal dose for these strains could be determined by titrating a series of even lower split doses or single doses of ENU. The Gould laboratory obtained viable offspring from WF males injected with single doses of 25, 35, and 50 mg/kg, but only 33% or less of the males in all dose groups regained fertility (Table 19–2). However, it remains questionable if these low doses are capable of inducing mutations in these strains, knowing that ENU mutagenesis is efficient only above a certain threshold in mice.[29] Alternatively, the mutation of interest could be induced in a different genetic background and subsequently introgressed into the desired genetic background. Pitfalls of this procedure are discussed later on.

Finally, the suitability of using a certain strain in an ENU mutagenesis-based screen is dependent on the induced mutation rate. A higher mutation rate means that on average fewer animals need to be screened to find the desired mutant. The data from the Cuppen laboratory gives an accurate reflection of the molecular mutation frequencies in genes of interest in the F344/Crl and Wistar strains (Table 19–1). Given the optimal doses of ENU, the mutation frequencies were one mutation per 1.76 and 1.24 Mb, respectively.[21] The mutation frequency for the BN strain treated with its optimal dose of ENU was not reliable (only eight mutations with most of them derived from a single founder).

The Gould laboratory has estimated the mutagenicity of ENU to a strain from the rate of visible mutants produced.[22] A large-scale phenotypic screen in offspring of mutagenized SD (2 × 60) and F344/NHsd (2 × 60) animals revealed a visible mutant phenotype rate of 1 in 64 pups, respectively, compared to 1 in 283 abnormal phenotypes in untreated animals. To test the heritability of the induced mutant phenotypes, approximately half of the mutant phenotypes could be bred to the next generation.[22]

To summarize, based on these three criteria the two outbred strains were awarded the highest scores (Table 19–1). Among the

Table 19–2
Effects of *N*-ethyl-*N*-nitrosourea (ENU) treatment on male rat fertility[a]

Inbred strain	*Dose (mg/kg)*	*Fertile males*	*Fertile males (%)*	*Outbred strain*	*Dose (mg/kg)*	*Fertile males*	*Fertile males (%)*
BN	3 × 20	20 / 24	83%	SD	1 × 75	3 / 3	100%
	3 × 30	7 / 24	29%		1 × 100	4 / 5	80%
	3 × 40	0 / 24	0%		1 × 120	2 / 6	33%
F344/Crl	3 × 20	20 / 24	83%		1 × 150	0 / 3	0%
	3 × 30	19 / 24	21%		2 × 60	5 / 5	100
	3 × 40	17 / 24	71%		2 × 75	1 / 5	20%
	3 × 60	0 / 10	0%		2 × 100	0 / 3	0%
F344/NHsd	1 × 75	3 / 3	100%	Wistar	3 × 25	10 / 10	100%
	1 × 100	4 / 6	67%		3 × 30	8 / 10	80%
	1 × 120	0 / 6	0%		3 × 35	9 / 9	100%
	2 × 50	3 / 5	60%		3 × 40	6 / 10	60%
	2 × 60	2 / 5	40%		3 × 60	0 / 10	0%
	2 × 75	0 / 3	0%				
LEW	3 × 20	0 / 24	0%				
WF	1 × 25	3 / 10	30%				
	1 × 35	2 / 6	33%				
	1 × 50	3 / 12	25%				
	1 × 75	0 / 3	0%				
	2 × 15	1 / 6	17%				
	2 × 25	1 / 6	17%				
	2 × 50	0 / 3	0%				
WKy	2 × 50	0 / 10	0%				

[a]A founder is considered fertile if it produces more than one litter between 10 and 26 weeks after the ENU treatment.

inbred strains tested, F344 gave the highest scores. Hence, we recommend using Wistar, SD, or F344 for ENU mutagenesis-based genetic screens in the rat.

N-ETHYL-N-NITROSOUREA MUTAGENESIS Proper application of an exactly measured amount of ENU to the rats is critical for reproducibility in consecutive mutagenesis experiments. Hence, it is important to establish a standardized procedure for dissolving, measuring, and administering ENU. The Gould and Cuppen laboratories have used the same protocol with some minor differences.

ENU was always freshly dissolved less than 1 h prior to injections. ENU comes as a powder in a sealed vial (SIGMA; 1 g). It is predissolved by vigorous shaking in 95% (v/v) ethanol (Gould laboratory 10 ml; Cuppen laboratory 5 ml) and diluted with phosphate/citrate buffer (Gould laboratory: 90 ml of 0.2 M Na_2HPO_4, 0.1 M citric acid, pH 5.0; Cuppen laboratory: 95 ml of 0.1 M NaH_2PO_4, 0.05 M citric acid, pH 5.0). The concentration is determined by measuring the optical density (OD) of a 10 times dilution at 395 nm wavelength. Preferentially, vials with the same lot number are used for all injections. One OD unit equals a concentration of ~1 mg/ml. Final concentrations typically varied between 6 and 8.5 mg/ml, but were relatively constant for vials with the same lot number. The desired dose is injected intraperitonially in weekly intervals. The male germ-line should be fully developed (~8 weeks of age) at the first ENU treatment. The Gould laboratory started the first ENU treatment at 9 weeks of age, whereas the Cuppen laboratory started at 10 and 11 weeks of age, which in all cases resulted in successful mutagenesis.

The number of males to be mutagenized depends on the reproductive capabilities of the selected strain, as outlined above, on the fertility rate of the selected strain treated with the optimal dose of ENU, and on the mutation/mutant phenotype rate. Fertility rates of the rat strains tested by the Gould and Cuppen laboratories are listed in Table 19–2. For example, if researchers decide to mutagenize BN males with a dose of ENU of 3 × 20, they should envision that nearly 20% of the mutagenized males will become sterile and that BN has less than one-third of the reproductive capabilities of either the Wistar or SD strains. Additional males have to be mutagenized to compensate for the resulting moderate litter size.

LIBRARY PRODUCTION Using a mutation rate of roughly 1 in ~1.2 Mb, the Cuppen laboratory estimated that a knockout probability of 96% for an average-sized rat gene (~1300 bp of coding region) requires an F_1 library of ~50,000 mutant animals.[21] This estimation is based on a total coding region size of ~28.4 Mb, of which ~1.7 Mb could be mutated into a stop codon by the three most frequently occurring ENU-induced mutations (AT-TA, AT-GC, GC-AT).[21] Nonetheless, the Gould laboratory experimentally showed that the library sizes to find a truncation in breast cancer genes 1 and 2 (*Brca1* and *Brca2)* were 1965 and 788, respectively.[22] F_1 libraries of thousands of animals take considerable time to generate and require an adequate logistical plan. Most laboratories do not have sufficient housing capacity. Since successful rat sperm cryopreservation is still limited to a single laboratory worldwide[33] and is certainly not a routine procedure, we suggest a rolling-circle model for the library production, which means that preweanling pups are produced to be consistent with the screening capacity. Only the pups harboring interesting mutations are kept and all other pups are discarded shortly after screening to prevent accumulation in the animal facility.

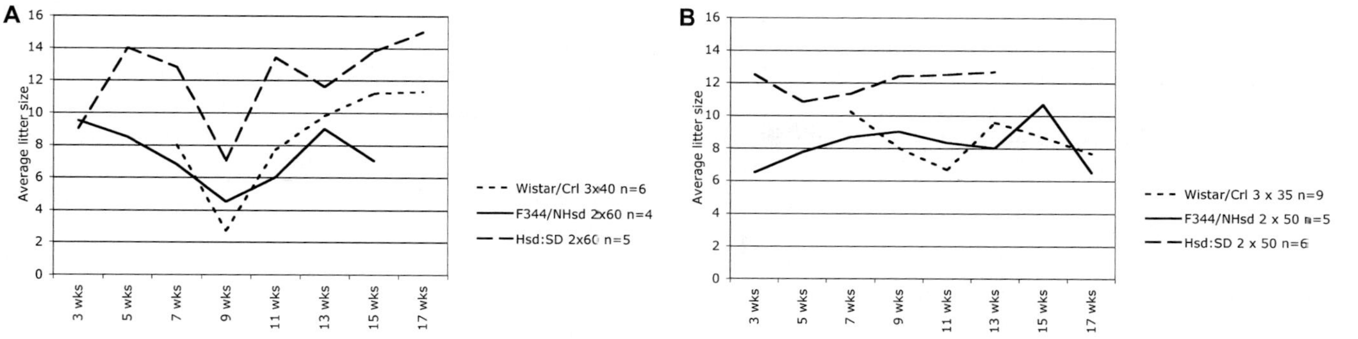

Figure 19–2. Average litter size for three different strains, Wistar/Crl, Sprague–Dawley (Hsd:SD), and Fischer (F344/NHsd), measured from 3 to 17 weeks (wks) after the ENU treatment (Hsd:SD, F344/Nhsd: wks after the first injection; Wistar/Crl: wks after the last injection). The dip in pup production manifests between 8 and 9 weeks after the last ENU injection. (A) Strains mutagenized with an optimal dose of ENU (in mg per kg body weight). (B) Strains mutagenized with a suboptimal dose of ENU (in mg per kg body weight).

The Gould and Cuppen laboratories set up matings with the mutagenized founder males every 2 or 3 weeks and each male was paired with one or two females. Using combinations of these two variable conditions, the pup production speed could be customized depending on the screening capacity. Mutagenized fertile males gave viable offspring up to 15 months of age, although some males experience a reduced life span due to tumor formation resulting from the mutagenesis.

To fix the mutations that are induced in the spermatogonial stem cells, it is important to let the mutagenized males undergo a full round of spermatogenesis (~60 days). However, to monitor fertility and to stimulate sperm production of the mutagenized founders, both laboratories have started matings as soon as 3 and 5 weeks after the ENU treatment. Pups born earlier than 10 weeks after the ENU treatment could be chimeric resulting from ENU-induced DNA adducts in spermatozoa and were subsequently discarded. These pups could be used to optimize the high-throughput mutation detection assay. Typically, the optimal dose of ENU for a certain strain is characterized by a dip in the pup production around 9 weeks after mutagenesis (Figure 19–2A), which is less pronounced or absent at suboptimal doses of ENU (Figure 19–2B). Unlike many mouse strains that might regain fertility up to 26 weeks after the ENU treatment,[24] ENU-treated male rats rarely regain fertility after a long period of sterility.

MUTATION DISCOVERY Due to the recent intensification of interest in genome-wide genetic variation, many high-throughput mutation discovery technologies became available. To date, three mutation discovery strategies have been applied to mutagenized rat libraries, namely CEL I-mediated heteroduplex cleavage,[20] high-throughput resequencing,[21] and a yeast-based screening assay.[22] This section will describe the most important pros and cons of each methodology (Table 19–3).

The Cuppen laboratory adopted the CEL I-mediated enzymatic heteroduplex cleavage method (CEL I) from the *Arabidopsis* field[34] and initially applied it to screening zebrafish in a high-throughput fashion.[35] The method relies on an exon-specific polymerase chain reaction (PCR) (~500–800 bp) with fluorescent primers that label both sides of the product with a different fluorophor. If a mutation was induced in the exon of interest, two alleles are present in that sample. Denaturing and reannealing of the samples will result in heteroduplexes in the mutant sample, which are specifically cut by the endonuclease CEL I. Denaturing polyacrylamide gel electrophoresis using an LI-COR DNA analyzer will detect the fluorescently labeled full-length and digested products. The mutations will appear as extra dots on a gel image, which are easily recognized by the eye. The corresponding sample is then resequenced to reveal the nature of the mutation.

An advantage of the method is that all reactions can be performed in an automated fashion, which makes the procedure easily scaleable (Table 19–3). In addition, after PCR amplification samples can be pooled up to four times to increase throughput. Investment costs are reasonably low. The assay requires a PCR machine, an LI-COR DNA analyzer, and depending upon the anticipated throughput, a liquid handling station for automation. The running costs are also low, especially if samples are pooled and PCRs are standardized using primers with universal adapters and universal fluorescent primers that fit these adapters. The disadvantages are that in our hands LI-COR gel images of variable quality (resolution, signal intensity) were produced. This was

Table 19–3
Overview of mutation discovery strategies applied to mutagenized rat libraries

	CEL I	*Resequencing*	*Yeast-based assay*
Automation[a]	+	++	––
High-throughput[b]	++	++	–
Scalability[c]	+	++	–
Flexibility[d]	–	+	–
Robustness[e]	–	++	+/–
Accuracy[f]	–	++	+/–
Investment costs	+	––	++
Running costs	+	+	++

[a]Automation: Can the method be run in a fully automated fashion, i.e., without manual operations?

[b]High-throughput: Does the method allow for the screening of large amounts of bases per day?

[c]Scalability: Can the amount of genomic DNA/RNA samples easily be scaled up or down?

[d]Flexibility: Does the method easily allow for the screening of additional amplicons/exons?

[e]Robustness: Is the quality of the data constant between amplicons, samples, or runs?

[f]Accuracy: Does the method produce false positives/negatives?

partially amplicon dependent and partially unknown, which makes the assay score lower in robustness. Additionally, researchers planning to use this technology should take into account that it is not 100% accurate. The CEL I method does miss some mutations and produces a considerable amount of false-positive dots that have to be followed up by an independent PCR and sequencing reaction.[35]

Dideoxy sequencing is considered the gold standard for the discovery of genetic variation.[36] The Cuppen laboratory has developed a cost-effective high-throughput automated resequencing platform that allows for the screening of literally millions of bases per day.[21] All reactions are performed in an automated fashion and data are analyzed using Polyphred software.[37] Polyphred awards a score to putative heterozygous positions, for which the original sequence trace was individually scanned by the eye. All potential mutations that pass this manual selection are repeated by an independent PCR and sequencing reaction for verification.

Resequencing is the most direct way of finding induced mutations and the method is extremely robust and scaleable (Table 19–3). Very few if any mutations are missed and the false-positive rate is low, making this method highly accurate. On the other hand, the investment costs are high, since the method requires a capillary sequencing machine, PCR machine, and liquid handling robot. As all reaction are standardized by the use of universal adapters and diluted to limited conditions, the running costs are surprisingly low.

The yeast-based screening assay employed by the Gould laboratory to generate knockouts in *Brca1* and *Brca2* allows for the identification of only truncations, caused by premature stop codons or frame shifts, induced by splice site mutations or insertions/deletions (INDELs). The advantage is that only those mutations that have the highest chance to alter gene function are found. However, potentially interesting functional missense mutations are neglected by this method. There are two versions of the method, namely the cDNA (copy DNA) version and the gDNA (genomic DNA) version. PCR-amplified fragments of the cDNA target or the large gDNA exon are transformed into competent yeast cells together with a linearized customized gap-repair vector containing sequences of the fragment of interest. Upon transformation, the amplified fragment is "cloned" *in vivo* into the gap-repair vector by homologous recombination, which places it behind the yeast *ADH1* promoter and in frame with the reporter gene *ADE2*. Yeast cells that produce the functional chimeric protein grow efficiently and form large white colonies on plates with selective medium. If the fusion protein is truncated due to a premature stop codon or frame shift mutation in the cloned fragment, the yeast cells will grow inefficiently and produce small red colonies. Since the original mutation is heterozygous, such a plate should theoretically contain half white and half red colonies. The corresponding sample is resequenced to reveal the nature of the mutation.

The yeast-based assay does involve substantial manual operation, especially the cDNA version, which includes isolation of total RNA and a reverse transcriptase reaction, and is not easily automated. In the final step, the transformed yeast cells have to be plated out manually and read by the eye. This also negatively affects the throughput and scalability of the method (Table 19–3). The method is not flexible, since it requires a different customized gap-repair vector containing a gDNA/cDNA product for every gene of interest. The method produces a background rate of red colonies, which is generally higher for the cDNA version (12.5–15%) compared to the gDNA version (0.5%).[22] Both versions do produce false positives, which are initially repeated in the same assay and subsequently followed up by resequencing. The main advantage of the yeast-based screening method is that the verified mutations will in almost all cases result in loss of gene function. Additionally, the cDNA version allows for the screening of larger fragments (up to the full-length cDNA). Finally, the investment costs are low, since only a PCR machine is needed. The running costs are also low; however, if the method is scaled up, personnel costs become a considerable issue due to many manual steps in the procedure.

These are three examples of mutation discovery platforms used in the rat, but many other methods not discussed here are available. Every mutation discovery platform has its own advantages and disadvantages. Depending on the purpose of the screening, the desired throughput, and the available personnel/funds, a method can be selected.

OUTCROSSING OF AN INTERESTING MUTATION All mutations identified in the F_1 library will be heterozygous. To obtain a homozygous mutant, the corresponding rat must first be outcrossed to an untreated rat. Progeny of the outcross are genotyped to ensure germ-line passage of the mutation in a Mendelian fashion. Although generating a homozygous mutant seems straightforward, a repeatedly posed question about the ENU mutagenesis-mediated knockout/mutant procedure is whether background mutations that are inevitably induced by ENU are a major concern. Given a mutation rate of 1 in ~1.2 Mb, a genome size of ~2.5 Gb, and a coding region size of ~28.4 Mb, of which ~1.7 Mb could be mutated into a stop codon by the three most common ENU-induced mutations, every mutagenized genome will on average contain ~2000 mutations, of which ~24 are located in coding regions, of which one or two will change an amino acid into premature stop codon. Theoretically, for every backcross the number of background mutations is reduced by on average 50%. Thus, after five backcrosses mutant animals will consist of ~97% of the wild-type genome and will contain approximately 1 mutation in coding regions and 63 mutations elsewhere in the genome. Although this may hold true for all unlinked background mutations, potentially confounding mutations that are linked to the mutation of interest will remain, as they are located nearby on the same chromosome. For the mouse, Keays *et al.*[38] modeled the decay of linked mutant DNA to a randomly selected position on a chromosome (e.g., a mutation) over the number of backcrosses and found that after 10 backcrosses there is still on average ~20% of mutant DNA linked to the selected site, which depending on the size of the chromosome corresponds to ~53.6 (chr *1*)–9.4 (chr *12*) Mb in the rat. These segments could still contain ~45–8 background mutations, which, depending on the proportion of coding or other functional regions surrounding the mutation of interest, could have a small chance of influencing phenotypic analysis.

In the case of introgressing an induced mutation into a different genetic background, the animal harboring the smallest portion of the mutant genome could be traced in each cross by marker-assisted selection,[38] thereby speeding up the backcrossing process. This procedure could more efficiently abandon confounding mutations and give insight in the amount of mutant DNA accompanying the mutation of interest; however, it could never eliminate all possibly linked mutations. Thus, while backcrossing to a wild type background is an effective strategy to eradicate unlinked mutations and to a certain extent linked mutations, the small chance of a closely linked confounding mutation contributing to the phenotype can be excluded only by functional experiments.

First, the phenotype–genotype relationship must always be 1 : 1, meaning that all homozygous mutant animals in every cross of the entire experiment must display the phenotype under study. If one homozygous animal does not display the phenotype, a closely linked mutation is most likely involved. This test becomes less penetrant if modifiers attenuate the phenotype. The gold standard to demonstrate the phenotype–genotype relationship is the generation of a second functional allele of the same gene that should give rise to a comparable phenotype. This is a rather common strategy in organisms with high induced mutation rates, like worms and zebrafish. However, for the rat, which allows for a fairly moderate mutation rate, it might not be feasible to find an additional functional allele, especially for below average-sized genes. Alternatively, a transgenic animal harboring a wild-type copy of the gene in the mutant background could be generated, which should rescue the phenotype. Technologies to produce transgenic rats are broadly established.[39,40] Finally, a phenocopy could be made by stable transgenic RNA interference-mediated gene silencing.[41]

RAT KNOCKOUT/MUTANT STATUS AND FUTURE PERSPECTIVES

Since the first publication of rat knockout technology in 2003, a total of 18 ENU-induced knockout rats have been reported by the Gould laboratory, the Cuppen laboratory, and the PhysGen program (http://pga.mcw.edu; Milwaukee, WI). In addition, 121 amino acid-replacing mutants have been reported. Several knockouts currently undergo detailed phenotypic characterization.[42] Recent publications on the rat knockouts for the adenomatous polyposis coli (*APC*) gene generated in a collaborative effort by the Gould and Dove laboratories (Madison, WI)[43] and the serotonin transporter (*SERT*) gene generated by the Cuppen laboratory[44] indicate that these models reflect human biology better than the corresponding mouse knockouts. The *Brca2* knockout, recently characterized by the Gould laboratory, does not provide an improved model for human *BRCA2*-related predisposition to breast cancer. However, it has a novel ocular phenotype (cataracts) never seen before in existing mouse *Brca2* knockouts.[45] In addition, homozygous knockout rats live longer compared to knockout mice with a truncation at a similar position in the *Brca2* gene. Finally, a rat knockout for *Myosin7a*, the gene affected in patients suffering from Usher syndrome Type 1B, was identified by an ENU mutagenesis-driven forward genetic approach.[46] In this case, the rat knockout model is phenotypically similar to existing mouse knockout models. However, due to the rat's greater size it provides better access to the affected organs, such as the inner ear, vestibular organ, and retina, and is also more suited for (micro)surgical manipulation.

This initial evidence suggests that rat knockouts could certainly complement other existing knockout models to study the genetics of specific aspects of human health and disease. Currently, the PhysGen program and the Cuppen laboratory are still pursuing ENU mutagenesis-based reverse genetics for genes involved in human cardiovascular processes, possibly resulting in novel interesting knockouts and mutants.

In addition to the reverse genetic approaches that have been discussed in this chapter, rat forward genetics could also lead to novel insights in gene function. Many rat strains have been thoroughly characterized in the past and are currently being systematically phenotyped (e.g., PhysGen program; The National Bio Resource Project in Kyoto, Japan; http://www.anim.med.kyoto-u.ac.jp/nbr). If these validated assays are converted into a streamlined high-throughput screening pipeline, they could be well suited for a large-scale systematic ENU mutagenesis-based screen. Previous ENU mutagenesis experiments have already produced a number of dominant and recessive mutant phenotypes, such as polydactyly, tail and eye aberrations, hip dysplasia, diphally, and circling behavior.[20,22,46] Only for the circling mutant (Tornado) has the underlying gene defect been found. This appeared to be a premature stop codon in *Myosin7a*, as mentioned above.

Despite the lack of embryonic stem cell-related technologies for the rat, the genetic toolbox has recently been filled rapidly. Historically, the rat has always been the model of choice for physiology rather than for genetics. However, because of these recent advances in ENU mutagenesis-based forward and reverse genetics, as well as the development of other genetic technologies for the rat, such as transgenesis,[40] and transient and permanent RNAi-mediated gene-silencing,[41] rat genetic and genomic approaches have recently gained momentum.[2,10]

ACKNOWLEDGMENTS

The work of the Gould laboratory has been supported by grants from the U.S. National Institutes of Health (CA28954, CA77494, and CA106216). The work of the Cuppen laboratory has been supported by the Dutch Ministry of Economic Affairs through the Innovative Oriented Research Program on Genomics and NV Organon.

REFERENCES

1. Lindsey JR. Historical foundations. In: Baker HJ, Lindsey JR, Weisbroth SH, Eds. *The Laboratory Rat*. New York: Academic Press, 1979:1–36.
2. Lazar J, Moreno C, Jacob HJ, Kwitek AE. Impact of genomics on research in the rat. *Genome Res* 2005;15(12):1717–1728.
3. Jacob HJ, Kwitek AE. Rat genetics: Attaching physiology and pharmacology to the genome. *Nat Rev Genet* 2002;3(1):33–42.
4. Gibbs RA, Weinstock GM, Metzker ML, *et al.* Genome sequence of the Brown Norway rat yields insights into mammalian evolution. *Nature* 2004;428(6982):493–521.
5. Kwitek AE, Gullings-Handley J, Yu J, *et al.* High-density rat radiation hybrid maps containing over 24,000 SSLPs, genes, and ESTs provide a direct link to the rat genome sequence. *Genome Res* 2004;14(4):750–757.
6. Guryev V, Berezikov E, Malik R, Plasterk RH, Cuppen E. Single nucleotide polymorphisms associated with rat expressed sequences. *Genome Res* 2004;14(7):1438–1443.
7. Mashimo T, Voigt B, Tsurumi T, *et al.* A set of highly informative rat simple sequence length polymorphism (SSLP) markers and genetically defined rat strains. *BMC Genet* 2006;7:19.
8. Moreno C, Kennedy K, Andrae JW, Jacob HJ. Genome-wide scanning with SSLPs in the rat. *Methods Mol Med* 2005;108:131–138.
9. Zimdahl H, Nyakatura G, Brandt P, *et al.* A SNP map of the rat genome generated from cDNA sequences. *Science* 2004;303(5659):807.
10. Smits BM, Cuppen E. Rat genetics: The next episode. *Trends Genet* 2006;22(4):232–240.

11. Lander ES, Linton LM, Birren B, *et al.* Initial sequencing and analysis of the human genome. *Nature* 2001;409(6822):860–921.
12. Venter JC, Adams MD, Myers EW, *et al.* The sequence of the human genome. *Science* 2001;291(5507):1304–1351.
13. Waterston RH, Lindblad-Toh K, Birney E, *et al.* Initial sequencing and comparative analysis of the mouse genome. *Nature* 2002;420(6915):520–562.
14. Nusslein-Volhard C, Wieschaus E. Mutations affecting segment number and polarity in Drosophila. *Nature* 1980;287(5785):795–801.
15. Driever W, Solnica-Krezel L, Schier AF, *et al.* A genetic screen for mutations affecting embryogenesis in zebrafish. *Development* 1996;123:37–46.
16. Haffter P, Granato M, Brand M, *et al.* The identification of genes with unique and essential functions in the development of the zebrafish, Danio rerio. *Development* 1996;123:1–36.
17. Hrabe de Angelis MH, Flaswinkel H, Fuchs H, *et al.* Genome-wide, large-scale production of mutant mice by ENU mutagenesis. *Nat Genet* 2000;25(4):444–447.
18. Nolan PM, Peters J, Strivens M, *et al.* A systematic, genome-wide, phenotype-driven mutagenesis programme for gene function studies in the mouse. *Nat Genet* 2000;25(4):440–443.
19. Capecchi MR. Gene targeting in mice: Functional analysis of the mammalian genome for the twenty-first century. *Nat Rev Genet* 2005;6(6):507–512.
20. Smits BM, Mudde J, Plasterk RH, Cuppen E. Target-selected mutagenesis of the rat. *Genomics* 2004;83(2):332–334.
21. Smits BM, Mudde JB, van de Belt J, *et al.* Generation of gene knockouts and mutant models in the laboratory rat by ENU-driven target-selected mutagenesis. *Pharmacogenet Genomics* 2006;16(3): 159–169.
22. Zan Y, Haag JD, Chen KS, *et al.* Production of knockout rats using ENU mutagenesis and a yeast-based screening assay. *Nat Biotechnol* 2003;21(6):645–651.
23. Russell WL, Kelly EM, Hunsicker PR, Bangham JW, Maddux SC, Phipps EL. Specific-locus test shows ethylnitrosourea to be the most potent mutagen in the mouse. *Proc Natl Acad Sci USA* 1979;76(11): 5818–5819.
24. Justice MJ, Carpenter DA, Favor J, *et al.* Effects of ENU dosage on mouse strains. *Mamm Genome* 2000;11(7):484–488.
25. van Zeeland AA, de Groot A, Neuhauser-Klaus A. DNA adduct formation in mouse testis by ethylating agents: A comparison with germ-cell mutagenesis. *Mutat Res* 1990;231(1):55–62.
26. Noveroske JK, Weber JS, Justice MJ. The mutagenic action of N-ethyl-N-nitrosourea in the mouse. *Mamm Genome* 2000;11(7): 478–483.
27. Hitotsumachi S, Carpenter DA, Russell WL. Dose-repetition increases the mutagenic effectiveness of N-ethyl-N-nitrosourea in mouse spermatogonia. *Proc Natl Acad Sci USA* 1985;82(19):6619–6621.
28. Russell WL, Hunsicker PR, Carpenter DA, Cornett CV, Guinn GM. Effect of dose fractionation on the ethylnitrosourea induction of specific-locus mutations in mouse spermatogonia. *Proc Natl Acad Sci USA* 1982;79(11):3592–3593.
29. Russell WL, Hunsicker PR, Raymer GD, Steele MH, Stelzner KF, Thompson HM. Dose–response curve for ethylnitrosourea-induced specific-locus mutations in mouse spermatogonia. *Proc Natl Acad Sci USA* 1982;79(11):3589–3591.
30. Wienholds E, Schulte-Merker S, Walderich B, Plasterk RH. Target-selected inactivation of the zebrafish rag1 gene. *Science* 2002;297(5578):99–102.
31. Coghill EL, Hugill A, Parkinson N, *et al.* A gene-driven approach to the identification of ENU mutants in the mouse. *Nat Genet* 2002;30(3):255–256.
32. Goda T, Abu-Daya A, Carruthers S, Clark MD, Stemple DL, Zimmerman LB. Genetic screens for mutations affecting development of Xenopus tropicalis. *PLoS Genet* 2006;2(6):e91.
33. Nakatsukasa E, Kashiwazaki N, Takizawa A, *et al.* Cryopreservation of spermatozoa from closed colonies, and inbred, spontaneous mutant, and transgenic strains of rats. *Comp Med* 2003;53(6):639–641.
34. Colbert T, Till BJ, Tompa R, *et al.* High-throughput screening for induced point mutations. *Plant Physiol* 2001;126(2):480–484.
35. Wienholds E, van Eeden F, Kosters M, Mudde J, Plasterk RH, Cuppen E. Efficient target-selected mutagenesis in zebrafish. *Genome Res* 2003;13(12):2700–2707.
36. Larsen LA, Christiansen M, Vuust J, Andersen PS. Recent developments in high-throughput mutation screening. *Pharmacogenomics* 2001;2(4):387–399.
37. Nickerson DA, Tobe VO, Taylor SL. PolyPhred: Automating the detection and genotyping of single nucleotide substitutions using fluorescence-based resequencing. *Nucleic Acids Res* 1997;25(14): 2745–2751.
38. Keays DA, Clark TG, Flint J. Estimating the number of coding mutations in genotypic- and phenotypic-driven N-ethyl-N-nitrosourea (ENU) screens. *Mamm Genome* 2006;17(3):230–238.
39. Mullins JJ, Peters J, Ganten D. Fulminant hypertension in transgenic rats harbouring the mouse Ren-2 gene. *Nature* 1990;344(6266): 541–544.
40. Tesson L, Cozzi J, Menoret S, *et al.* Transgenic modifications of the rat genome. *Transgenic Res* 2005;14(5):531–546.
41. Dann CT, Alvarado AL, Hammer RE, Garbers DL. Heritable and stable gene knockdown in rats. *Proc Natl Acad Sci USA* 2006;103(30): 11246–11251.
42. Smits BM, Cuppen E. Rats go genomic. *Genome Biol* 2006; 7(2):306.
43. Amos-Landgraf JM, Kwong LN, Kendziorski CM, *et al.* A target-selected Apcmutant rat kindred enhances the modeling of familial human colon cancer. *Proc Natl Acad Sci USA* 2007;104(10): 4036–41.
44. Homberg JR, Olivier JD, Smits BM, *et al.* Characterization of the serotonin transporter knockout rat: a selective change in the functioning of the serotonergic system. *Neuroscience* 2007;146(4): 1662–76.
45. Cotroneo MS, Haag JD, Zan Y, *et al.* Characterizing a rat Brca2 knockout model. *Oncogene* 2007;26(11):1626–1635.
46. Smits BM, Peters TA, Mul JD, *et al.* Identification of a rat model for usher syndrome type 1B by N-ethyl-N-nitrosourea mutagenesis-driven forward genetics. *Genetics* 2005;170(4):1887–1896.

20 Rodent Genetics, Models, and Genotyping Methods

SHERYL J. WILDT, ANDREW I. BROOKS, AND ROBERT J. RUSSELL

ABSTRACT

Laboratory mice (*Mus musculus*) and rats (*Rattus norvegicus*) are the most commonly used animals in biomedical and behavioral research. Today, there are many inbred, outbred, hybrid, congenic, and genetically modified mouse and rat models that have made essential contributions to advances in the diagnosis, prevention, and control of human disease. Routine genetic monitoring of these animals is essential to their successful use. Herein, we discuss the basic structure of the mouse and rat genome; common breeding protocols to preserve innate and exogenous genetic backgrounds; and the methods used for creating specific types of mutant model, research applications, and current genotyping methods.

Key Words: Genetics, Genotyping, Genetically modified, Microsatellite marker, Single nucleotide polymorphism, Polymerase chain reaction, Allelic discrimination.

MODEL USE AND HISTORY

Laboratory mice (*Mus musculus*) and rats (*Rattus norvegicus*) are the most commonly used animals in biomedical and behavioral research.[1] Mice and rats have made essential contributions to advances in the diagnosis, prevention, and control of human disease.[2–5] Mice and rats originated in Asia and then spread throughout the world as they became commensal with humans. *M. musculus* is most likely the result of a hybrid of several species of the species, *Mus*. The laboratory mouse is a subspecies of *M. musculus*; *M. musculus musculus*. In 1664, Robert Hooke was the first to use the mouse in the laboratory while studying the properties of air and oxygen. Albino mutant rats (*R. norvegicus*) were first brought into the laboratory in Germany in the 1870s, and thus, are the first mammal domesticated for experimental purposes.[4]

Many of the strains and stocks of mice and rats used today in research were initially started by coat color breeding fanciers (or pet breeders) in Asia (Chinese mouse fanciers are recorded as early as 307 AD), Europe, and the United States (Abby Lathrop). The first inbred mouse strain (DBA/1) was developed by C.C. Little in 1909. The Wistar Institute in Philadelphia is the original source of many of the currently used rat stocks and strains. Commercial and government supported breeding colonies of mice and rats were established throughout the world to supply a broad range of characterized genotypes for support of broad-based scientific studies. Standard nomenclature rules for mice and rats stocks and strains have been published.[6] Specific sources for mice and rats models are also available,[6] and genome databases have been established.[7–9]

Mice and rats have been widely used in the study of immunology, cancer, cardiovascular disorders, metabolic disorders, neurological disorders, organ transplantation, renal disease, genetics, toxicology, and behavior, just to name a few. The ever growing use of genetically modified mice and rats (transgenic, knockout, knockin, conditional mutants), in addition to the use of the standard stocks and strains of outbred, inbred, hybrid, congenic, consomic, and recombinant strains, has made major contributions to the advancement of genomic medical knowledge in the past 20 years.[10–14] The first transgenic mammal created by pronuclear injection was a mouse.[15] Recently, to ensure that unique strains will be available to future generations of scientists, a number of government sponsored repositories have been established in the United States, Europe, and Japan.[16–19] Improvements in genetic engineering technology and methodology are continuing for the development of inducible transgenics, conditional mutations, speed congenic production, transgenic and knockout rats, and chemical-induced and radiation-induced mutants. The need for evolving genetic testing services and technologies continues to be of utmost importance.[20,21]

BASIC RODENT GENETICS

THE MOUSE GENOME Scientific interest in the study of the mouse genome dates back to at least 1866 when Gregor Mendel began investigating coat color alleles in mice. Upon hearing of the monk's activities, a local bishop forbade Mendel to share his living quarters with any creature that engaged in sexual intercourse, and thus Mendel was forced to use pea plants for his now famous genetics research.[22] In 1902, Lucien Cuénot became inspired by reading Mendel's Laws of Independent Assortment and Segregation, and he began his own experiments, using the mouse to study segregation of the albino and yellow coat color alleles.[23] This work was continued by Cook and Little, who described nine color coat loci in mice.[24,25]

The mouse genome consists of 19 pairs of autosomes and the X and Y sex chromosomes. Each chromosome is acrocentric, meaning that the centromere, or "waist" of the chromosome, is positioned at one end. The short arm of acrocentric chromosomes have structures called "satellites," which are round, circular

From: *Sourcebook of Models for Biomedical Research*
(P. M. Conn, ed.), © 2008 Humana Press Inc., Totowa, NJ.

figures on the end of a stem that code for ribosomal DNA. The organization of the mouse genome consists of approximately 8% highly repetitive DNA, 76% repetitive DNA, 15% semirepetitive DNA, and 1% fold back DNA.[26,27]

THE RAT GENOME The scientific study of the rat dates back to as early as 1828. The first genetic experiments were performed on the rat by Crampe from 1877 to 1895 and focused on the inheritance of coat color alleles. In 1908 the first inbred rat (PA strain) was developed by King.[28]

In the early 1990s, many scientists argued that sequencing the rat genome would be redundant because the mouse and human genome had already been sequenced. It was assumed that the rat genome would be quite similar to the genome of the mouse. Despite scientific protest, investigators led by the Baylor College of Medicine Human Genome Sequencing Center sequenced the DNA from the Brown Norway rat (BN/RijHsd) obtained by the Medical College of Wisconsin from Harlan. The entire genome of the rat has not yet been sequenced; however, a high quality draft representing 90% of the DNA sequence is currently available. Despite the gaps in the DNA sequence, it is still quite evident that the organization and size of the rat genome were much different than first expected.[29]

The rat genome consists of 21 pairs of autosomes and the X and Y sex chromosomes. The size of the rat genome is approximately 2.75 gigabases (Gb), as compared to the mouse genome of 2.6 Gb and the human genome of 2.9 Gb.[30]

BREEDING METHODS

INBREEDING Each individual animal in an inbred strain is genetically uniform, thus eliminating experimental variability due to genetic effects. An inbred strain is defined as having at least 20 generations of brother × sister matings. An inbreeding program begins with breeding siblings produced from a first filial mating (F_1). Inbreeding is continued by performing full sibling matings at each successive generation. As the filial number increases, the level of heterogeneity in the colony decreases, and after 20 successive generations of brother × sister matings (F_{20}), a colony is considered to be fully inbred,[31] although many currently available strains are over 100 generations.

OUTBREEDING A rodent outbreeding program is the opposite of an inbreeding program with the goal being to retain maximum heterogeneity by avoiding matings between closely related individuals. Although outbred rodents (referred to as stocks) are not genetically defined, advantages for some investigators include increased body weight, larger size and frequency of litters, increased life expectancy, increased resistance to disease in some cases, and lower cost due to higher reproduction. Outbred stocks are useful in experiments in which the genetic background of the experimental model is not the major factor in the outcome of the study and they are used extensively in toxicology and safety assessment.[32]

F_1 HYBRIDS Fully inbred strains often display undesirable characteristics, including low fecundity, small litter sizes, reduced life span, and increased susceptibility to disease due to homozygosity. It is possible, however, to create a colony of rodents that is genetically identical without suffering the consequences of homozygosity. This is accomplished by crossing two inbred strains to produce an F_1 hybrid animal that is heterozygous at almost all loci. These animals display hybrid vigor for the characteristics mentioned above. F_1 hybrid strains also exceed the life span of either of the inbred parent strains.[33]

CONGENICS Congenics are created by backcrossing a "donor" strain that contains the gene of interest to an inbred "recipient" strain for 10 or more generations. With each generation, the allele of interest from the donor strain is maintained in a heterozygous state, while the rest of the genome is bred to homozygosity during the 10 generations. The resulting offspring are genetically identical to the recipient strain, except for a small chromosomal region that carries the gene of interest from the donor strain. Congenics are useful in studying multiple gene effects and also for studying a particular mutation of interest on various inbred strain or outbred stock genetic backgrounds.[32]

Marker-assisted breeding, or speed congenics, can reduce the time necessary to produce a transgenic by 50%. Speed congenics is performed by utilizing a genome wide screen at each generation to select breeders (typically males) for the next generation of backcrossing. Selected breeders not only carry the gene of interest, but are also the most genetically similar to the recipient strain.[34,35] Backcrossing using standard breeding methods and speed congenic techniques are commonly used today in most research institutions.

CONSOMICS Consomic strains are a variant on the congenic; they differ from the recipient strain by an entire chromosome instead of just a single gene of interest. They are created by performing complex long-term breeding schemes. These strains are useful in associating specific phenotypic traits with a particular chromosome, and also for studying and mapping complex genetic traits, such as quantitative trait loci (QTLs).[36]

SELECTED MUTANT MODELS

SPONTANEOUS MUTANTS A spontaneous mutation in the genome of an organism typically occurs due to errors made by the DNA replication and repair mechanisms in the cell, thus altering the gene nucleotide sequence. Chromosomal mutations arise due to a structural change or a change in the number of chromosomes in an organism. Scientists usually recognize mutations when they have some effect on the phenotype of the organism. Rodents that carry spontaneous mutations are very important in the study of basic biological processes, in studying gene relationships, and as human disease models.

A strain with a spontaneous mutation that has been extremely useful in the study of immunology, tumor cell growth, and transplantation studies is the C.B-17/Icr-*Prkdc*^{scid} inbred mouse. These mice accept grafts and tumors from other species of animals as well as mice. The *Prkdc*^{scid} mutation was first discovered when it was noted that several C.B-17 pups from a litter were deficient in IgM, IgG_1, and $IgG_{2\alpha}$.[37] The mutation, termed scid or severe combined immunodeficiency disease (SCID), was shown to be a recessive, single nucleotide polymorphism (SNP) located on mouse chromosome 16.[38,39] The *scid* mutation affects normal T and B cell development, and these mice are highly susceptible to opportunistic infections, similar to the severe combined immunodeficiency disease in humans.[37,40] The *scid* mouse is available on a variety of genetic backgrounds from several commercial breeders.

TRANSGENIC (MICROINJECTION) MODELS Transgenic models are created by modifying the genetic material by insertion of genetic material from another organism by nonbreeding methods. The most common technique used for creating a transgenic model is microinjection. This technique involves injection of foreign DNA directly into the male pronucleus of a fertilized oocyte.[41]

There are many examples of important rodent disease models that have been created by using microinjection techniques. One such model is the HLA-B27 transgenic rat. This model expresses human HLA-B27 and β_2M, and is useful in studying HLA-B27-related diseases in humans, including psoriasis, reactive arthritis, and inflammatory bowel disease.[42]

TARGETED MUTANTS Although microinjection is a powerful method, a major limitation is that it does not permit creation of targeted mutants. Several techniques, such as targeted mutagenesis and gene trapping, have now been developed to overcome this limitation. Through these techniques, investigators can create mutations at any locus of interest, and thus study the function of cloned genes.[43]

KNOCKOUT MUTANTS Knockout mice are generated by inactivating or "knocking out" the function of a single gene of choice without affecting any other gene in the genome. These techniques include homologous recombination and gene trapping.

Homologous recombination is performed by introducing a segment of DNA that has an identical sequence to the gene of interest, but also includes a mutation that renders it inactive, into an embryonic stem (ES) cell. ES cells from the 129 mouse strain are typically used. The cell's DNA replication and repair machinery recognizes the artificial DNA sequence and exchanges it with the target gene, thus "knocking out" the function of the gene of interest.[44]

The gene-targeted mouse ES cells are then microinjected into mouse blastocysts that are transferred to pseudopregnant female mice and permitted to develop to term. After the pups are born, some pups may be chimeric and they are usually identified by their coat color. The chimeric mice are bred to appropriate wild-type mice and the resulting pups are screened for germ-line transmission by identifying the heterozygous targeted allele. Homologous recombination is the most commonly used method for generating knockout mice.[45]

Gene trapping is a similar method, except that the artificial DNA sequence is designed to be inserted into any gene and disrupt the cell's RNA splicing mechanism. This technique prevents the gene that harbors the foreign DNA from producing a functional protein, and thus knocks out its function.

An example of an important knockout model is the *p53* knockout mouse. The *p53* gene codes for a protein that arrests cell division and thus halts tumor growth. Humans who have a mutated form of the *p53* gene have a much higher risk of developing many types of cancer at an early age, including breast cancer and leukemia. The *p53* knockout mouse has been an important tool for studying tumor growth and regulation of cell division.[46,47]

Rat ES cells are not available and homologous recombination and gene trapping are feasible only in mice. Until recently, the rat lagged behind the mouse in the development of genetic baseline information; however, the rat is now catching up.[7] Knockout rats have been developed. The recent advances in rat genetics will permit broader use of the rat in genomic studies with links to phenotypes and comparative mapping to both the mouse and the human.

A concerted effort to create a knockout for every gene in the mouse genome is currently being undertaken by major research institutions around the world, including Canada, Europe, the United States, and China. In the United States, the National Institutes of Health (NIH) has announced the Knockout Mouse Project (KOMP), which aims to create and identify approximately 10,000 knockout mouse lines for scientific research.[48]

KNOCKIN MUTATIONS Knockin mice are also created by microinjecting gene-targeted mouse ES cells into mouse blastocysts. The injected blastocysts are transferred to pseudopregnant female mice and allowed to develop to term. The chimeric mice are bred to appropriate wild-type mice and the resulting pups are screened for germ-line transmission.

One well-studied knockin mouse model is the *APOE2* mouse. The apolipoprotein E2 (*APOE2*) allele is recessive and is the cause of human Type III hyperlipidemia in humans. In the *APOE2* knockin mouse, the endogenous mouse *Apoe* gene has been replaced by the human *APOE2* gene. This mouse develops a severe diet-induced hyperlipidemia and shows marked cholesterol and triglyceride lowering upon treatment with fibrate, an agonist of peroxisome proliferator activator receptor α (PPAR-α).[49]

CONDITIONAL MUTANTS A variety of techniques are used to create conditional gain and loss of function mutations, but the most popular current method is the Cre-*loxP* recombinase system. The Cre recombinase enzyme functions like molecular scissors, cutting out the gene of interest that has been flanked by two *loxP* sites. Cre recombinase can be expressed in a spatial and temporal manner conveying cell type specificity, and allowing the targeted gene to be eliminated (or added) in specific cell types.[50,51]

The *BRCA1* conditional knockout mouse is an important model for the study of breast cancer tumorigenesis in women. Loss of function of the *BRCA1* tumor suppressor gene predisposes women to both breast and ovarian cancer. Knocking out the function of this gene in mouse tissue types results in embryonic lethality. To overcome this obstacle, a model was created by use of the Cre-*loxP* recombinase system, which knocked out the function of the gene in a tissue-specific manner.[52]

ETHYLNITROSOUREA MUTANTS Mutations in the DNA sequence of an organism can be induced by chemical mutagenesis. Ethylnitrosourea (ENU) is the most commonly used chemical mutagen, and it causes multiple untargeted point mutations. ENU mutants are created by exposing males to ENU and then breeding those males to normal females. Using high-throughput phenotyping methods, the offspring are screened for phenotypic traits of interest that originate from the point mutations inherited from their ENU-treated male parent.[53–55]

IRRADIATION MUTANTS By exposing an animal to ionizing radiation in the form of X-rays, gamma rays, or neutrons, multiple untargeted mutations in the DNA sequence can be induced.[56] Ionizing irradiation most often induces rearrangements, translocations, or deletions in the genome as much as 100 times higher than the rate of spontaneous mutation.[57] Although higher than the rate of spontaneous mutation, the frequency of mutation induced by ionizing radiation is still considered relatively low, and is practical only for laboratories with the means to screen large numbers of exposed animals using high-throughput phenotype screening programs.[28]

GENETIC MONITORING/GENOTYPING

HISTORICAL BACKGROUND The need for routine genetic monitoring of the thousands of rat and mouse strains used today in biomedical research is well established.[13,58–61]

Genetic quality control and monitoring of these animals are essential to their successful use. Continual change in the gene pool of inbred strains is unavoidable due to residual heterozygosity and spontaneous mutations that continually occur. In addition, genetic contamination due to mismating is a continuing threat in breeding colonies and must be prevented by appropriate husbandry techniques. The experimental use of nonauthentic or genetically contaminated rats and mice has been identified as the cause of erroneous research results.[62–69]

It is generally expected that genetically authentic or genetically modified animals will arrive as expected from the collaborating investigator, institutional source, or commercial supplier; however authenticity should be ascertained by the receiving laboratory. Information regarding the genetic background, genetic modification(s), and relationship to the background strain should be understood prior to research use. Genetic authenticity of strains should be ensured by appropriate testing and the results compared to those published for the reference strain. Increased transfer of animals between research institutions has resulted in an increased potential for genetic and microbiological contamination.

Institutional policies for the importation of mice and rats from outside sources should be developed and carefully followed to prevent potential genetic or health problems. Many strains and stocks (embryos, ovarian tissue, sperm) of mice and rats are now cryopreserved at repositories or individual institutions to protect their long-term genetic and microbiological quality.

Previously, biochemical markers;[13,59] strain-specific typing sera;[70,71] immunological methods using monoclonal and polyclonal antibodies to major histocompatibility haplotypes and lymphocyte antigens;[72–74] skin grafting;[63,75] phenotypic, morphological markers and characteristics (mandible shape);[62] coat color; behavior; and reproductive characteristics were used to monitor genetic quality. More recently, as our understanding of the mouse and rat genome has evolved, DNA from individual mice and rats is now used for genetic evaluation, and once the DNA has been extracted, it can be preserved indefinitely to perform a variety of analyses.

DNA fingerprinting,[76–79] using multilocus probes [also known as minisatellites or restriction length polymorphisms (RFLP)], microsatellite markers, and SNP markers [80,81] are now routinely used to replace the older methods that were primarily based on phenotypic expression.

DNA fingerprinting is useful for comparing different colonies of the same inbred strain. It is reassuring that the multiple bands observed using DNA fingerprints of inbred rat and mouse strains bred in different countries by the same supplier are similar, and that strains, known to differ at many biochemical loci as a result of past genetic contamination, have different DNA fingerprints. DNA fingerprinting is technically quite demanding even when nonisotopic fingerprinting protocols are used. The approach requires some time to become technically proficient in routinely producing high-quality Southern blot gels, preparing high-quality large-molecular-weight DNA, sample well loading, and membrane handling, not to mention data analysis and interpretation. A high mutation rate of visible bands implies that segregation of such bands within an inbred strain can be confused with genetic contamination. DNA fingerprinting also requires a longer period of time to obtain results than other techniques that are currently available.[82]

When microsatellite markers were introduced in the early 1990s, they greatly improved the efficiency of genetic evaluation. These tests are readily performed using polymerase chain reaction (PCR), and batteries of microsatellite markers, amplified using PCR, have been developed for both mice and rats. This method provides single locus DNA-based methods without the time limitation of fingerprinting and without having to amass large numbers of cloned probes. Microsatellites involve tandem repeat units, typically of one to five bases, and are detected using unique sequence DNA that flanks the tandem repeats. Only one or two bands are produced in each sample because DNA is amplified by unique sequence primers instead of the multiple primer sets used in DNA fingerprinting.

The availability of even greater numbers of SNPs provides the opportunity for greater resolution between DNA-based information. Over three million SNPs have been characterized in humans and one million SNPs in the mouse and rat. The resolution of SNPs in rodents is approximately 3 kb, allowing this information to be used in more detailed genetic mapping scenarios. SNP detection is currently becoming the method of choice for genetic evaluation of rats and mice.

MICROSATELLITE MARKERS The utilization of microsatellite markers has been a valuable tool and subsequently has been applied to many applications in genetics research laboratories. In the context of this review, the primary application being exploited is the use of microsatellite markers for monitoring genetic backgrounds in laboratory strains of mice and rats. Identifying well-defined polymorphic regions provides the information needed to distinguish the genetic background in most inbred strains. Although many public repositories exist to mine sequence information, the Center for Inherited Disease Research (CIDR) database (see Additional Web Links) is among the richest resource for selecting markers to characterize rodents. Once genetic regions of interest are identified, which include those that show the largest polymorphic differences between the strains being evaluated, a standard set of PCR conditions is often employed to "screen" animals. The sizing of the PCR products is aligned with reference controls in order to determine the genetic background of the animals of interest.[83]

The sizing of PCR products from microsatellite markers can be conducted in many ways. A variety of technologies can be employed depending on the resolution, throughput, and cost restraints of the experimentalist. Below is a review and description of the most commonly used techniques for sizing PCR products produced by microsatellite marker amplification.

Polymerase Chain Reaction The PCR is based on a molecular principal defined by Kary Mullis in the 1980s. The synthesis of DNA is used to make qualitative and, in some instances, quantitative measurements of nucleic acid. PCR, as defined by the 1989 editor of *Science*, Daniel Koshland Jr., is as follows:

> The starting material for PCR, the "target sequence," is a gene or segment of DNA. In a matter of hours, this target sequence can be amplified a million fold. The complementary strands of a double-stranded molecule of DNA are separated by heating. Two small pieces of synthetic DNA, each complementing a specific sequence at one end of the target sequence, serve as primers. Each primer binds to its complementary sequence.

Polymerases start at each primer and copy the sequence of that strand. Within a short time, exact replicas of the target sequence have been produced. In subsequent cycles, double-stranded molecules of both the original DNA and the copies are separated; primers bind again to complementary sequences and the polymerase replicates them. At the end of many cycles, the pool is greatly enriched in the small pieces of DNA that have the target sequences, and this amplified genetic information is then available for further analysis.

For most microsatellite marker PCR protocols, a standard set of reaction conditions is used to amplify products that range from 80 to 300 bp.

The main reason primers are designed with similar melting temperatures is to allow for many sequences to be amplified simultaneously. Following PCR, the analysis of PCR product sizes allows for the assessment of genetic contributions to segregate specific strains.

Visualization of DNA products: Capillary Electrophoresis, Traditional Gel Electrophoresis, and Lab-on-a-Chip Technologies A number of technologies exist to determine the size of DNA fragments produced during PCR. Although the importance of specificity and sensitivity of PCR is a given, the mechanism by which the results are measured is equally important. Until recently, gel electrophoresis in an agarose matrix was the technology of choice for measuring the size of DNA in a low-throughput manner. The resolution of detection is achieved by varying the percentage of agarose in the gel matrix. Lower concentrations of agarose lead to better resolution of larger DNA fragments (1 kb and higher), whereas higher concentrations of agarose lead to better resolution of smaller fragments (500 bp and lower) while resolution of even smaller fragments (100 bp and lower) requires a different matrix (i.e., polyacrylamide or PAGE gels).

Given that most microsatellite markers yield products that range from 80 to 250 bp, the resolution of PCR product sizes is of paramount importance when using absolute fragment sizes to compare PCR band sizes for the same marker between different inbred rodent strains because the strains typically differ by only a few base pairs. Traditional gel electrophoresis can be somewhat variable, leading to absolute differences between different markers, gels, and screens. Lastly, electrophoresis anomalies such as bowing, skidding, and smearing lead to variable interpretations that can confound important observations regarding the genetic integrity of a rodent colony. To this end, there are two new approaches that convey the sensitivity, specificity, and throughput to perform the sizing needed for this application.

The first is a capillary electrophoresis approach that is commonly employed by forensic laboratories for fragment analysis. Fragment analysis requires the use of a modified forward primer that results in a fluorescent PCR product that is resolved on a capillary electrophoresis DNA sequencer. In an optimal environment, markers can be multiplexed by using different fluorescently labeled oligonucleotides (primers) to distinguish marker-specific PCR products. Another slightly more cost effective option is Caliper's Lab-on-a-Chip technology.[84] Hundreds of samples can be run in 1 day, achieving resolution of less than 5 bp. This method allows for the most sensitive and efficient sizing of a wide variety of markers. Data are represented via an electropherogram, which depicts a sensitive array of all DNA products in a mixture. Either capillary electrophoresis or Lab-on-a-Chip technologies greatly increase the throughput and resolution to make accurate measurements that ensures reliable data interpretation.

SINGLE NUCLEOTIDE POLYMORPHISMS DETECTION TECHNOLOGIES SNPs are becoming a preferred method for monitoring the genetic basis of disease. In human studies, SNP profiling has become the primary source of genetic inquiry, surpassing the use of microsatellite markers for many clinical and forensic applications. As additional sequence information becomes available for different strains of rodents, the trend toward adaptation of SNP measurement to determine genetic purity is rapidly increasing. Currently, a large number of technologies can be employed to measure individual SNPs, thus enabling investigators to analyze over 500,000 SNPs at one time in a single sample.[85,86]

Low-, moderate-, and high-throughput techniques are available when using SNP analyses. Low-throughput analysis approaches are primarily PCR based and used for allelic discrimination applications as described later.

Moderate-throughput applications are capable of measuring up to several thousand SNPs in as many samples. Detection technologies range from high-order multiplex PCR (Applied Biosystems SNP-plex), single base extension technologies (Beckman Coulter SNPstream), and more recently, detection using mass spectrometry (Sequenom). All of these technologies provide robust data sets and use different detection technologies, which relate heavily to the cost associated with data generation.

High-throughput applications are currently microarray based. The Affymetrix GeneChip and Illumina bead-based arrays are the densest arrays available and allow for the simultaneous detection of over 500,000 genotypes in a single assay. Regardless of the mechanism for generating data, it is imperative to emphasize that the analytical tools used for analyzing genotyping data transcends the technology used for data acquisition. Since most panels for distinguishing inbred strains of rodents require at least several hundred SNPs for a concrete segregation, moderate-throughput technologies are typically employed for genetic monitoring and speed congenic applications.[87,88]

Allelic Discrimination Determination Allelic discrimination is an approach used to determine the zygosity of a spontaneous or genetically targeted single base pair mutation. One of the most commonly used technologies for this application is the Taqman fluorogenic 5′ nuclease assay. A single nucleotide polymorphism is identified through differential binding of two dual labeled probes designed to distinguish a single base pair mutation in the DNA sequence. The level of fluorescence generated by the cleaving of each probe (once bound to the template) allows for a sensitive and accurate detection of heterozygotes and homozygotes for the genetic mutation of interest. There are a number of real-time thermal cyclers capable of making measurements that range in capacity from 24 to 384 samples at a time.

Inbred Strain Verification and Speed Congenic Programs Animal models of disease are a critical component of most biomedical research programs. A major advantage of utilizing rodents in these paradigms is the ability to control most genetic contributions across generations. In addition, genetic diversity across inbred strains provides critical information regarding the genetic component of disease onset and progression. To this end, studying specific biological perturbations in the context of an a priori selected strain of rodent is of paramount importance.

Traditional backcrossing can require 3 years of breeding to transfer a gene of interest onto a genetic background different from the donor strain. Speed congenic or marker-assisted backcrossing methodologies reduce this time by 50% (or more), resulting in a new congenic strain in as few as five generations. Once a mutation has been characterized in a new genetic background, it is necessary to perform routine genetic monitoring to ensure the genetic integrity of the congenic strain. Both inbred strain verification and speed congenic programs employ microsatellite and SNP-based technologies.

The utilization of these approaches requires a fine balance between identifying the optimal resolution generating genetic information and the cost associated with data generation. Typically, chromosomal regions that require a finer resolution will benefit from the informatics and technology utilized in an SNP profiling panel. When using a panel that consists of several hundred SNPs, an investigator can cover all autosomes at approximately 15-cM intervals, to convey a level of specificity needed for most inbred strain and speed congenic applications. In addition to selecting a technology suitable for genetic monitoring, it is essential that a comprehensive husbandry strategy is employed to ensure the genetic integrity of any rodent colony.

GENE EXPRESSION ANALYSIS: DNA MICROARRAYS AND QUANTITATIVE POLYMERASE CHAIN REACTION Gene expression analysis has become a widely used approach for making predictions, characterizing biological perturbations, and defining the molecular mechanisms of complex diseases. Similar to SNP profiling, there are now a plethora of technologies that measure gene expression in a variety of tissue types.

Irrespective of the technology being used to measure gene expression, it is important to underscore the importance of RNA quality. Additionally, RNA is a labile and depletable resource, so many laboratories archive cDNA that is the stable template used for measuring gene expression. NuGEN Inc. has developed a technology (Ovation) that amplifies cDNA from limiting amounts of RNA in a regionally unbiased manner. This technology allows an investigator to use any downstream gene expression technologies and still archive a precious resource for future investigation. Assuming that the quality of RNA is acceptable, the use of DNA microarrays is widely accepted for high-throughput gene expression measurements. A number of array manufactures, including Affymetrix, Applied Biosystems, and Agilent Technologies, offer full genome expression arrays. These arrays simultaneously measure the entire transcriptome in a single assay. Many caveats associated with performing microarray studies are outside the scope of this review; nevertheless, the technology has now been accepted globally for a variety of profiling and mechanistic discovery programs.

The most commonly used validation technique is that of real-time quantitative (Q) PCR. There are a number of chemistries that can be used for QPCR that range from SYBR dyes that intercalate double-stranded DNA to probe-based technologies such as Taqman and Roche's Universal Probe Library. Quantitative real-time PCR has become a standard for measuring the most sensitive changes in gene expression when interrogating a handful of genes in large numbers of samples. It is necessary to proceed cautiously with gene expression experiments and be certain to create and execute a solid experimental design while confirming the quality of RNA before making any measurements. Lastly, the analysis of gene expression experiments requires significant bioinformatics tools and resources to maximize the use of these large data sets.[89,90]

CONCLUSIONS

Mice and rats are the animal models of choice in biomedical research due to the ability to understand, manipulate, and study their genome with relative ease not to mention the strong genetic relationship with humans allowing for the development of direct, translational models of human disease. Newer techniques for genotyping and genetic manipulation continue to evolve, ensuring that the mouse and rat will remain important in furthering our understanding of human disease, leading to earlier and more accurate diagnosis, the development of cutting edge therapeutics, and disease prevention through a better understanding of genetic predisposition and gene–environment interactions.[91]

REFERENCES

1. Festing MFW. *International Index of Laboratory Animals*, 6th ed. Carshalton, Surrey, UK: Lion Litho Ltd., 2003.
2. Crawley JN. What's Wrong with My Mouse? *Behavioral Phenotyping of Transgenic and Knockout Mice*. New York: Wiley-Liss, 2000.
3. Krinke GJ, Ed. *The Laboratory Rat*. New York: Academic Press, 2000.
4. Hedrich HJ, Bullock G, Eds. *The Laboratory Mouse*. New York: Elsevier Academic Press, 2004.
5. Suckow M, Weistbroth SH, Franklin CL, Eds. *The Laboratory Rat*. Burlington, MA: Elsevier Academic Press, 2006.
6. Institute for Laboratory Animal Research (ILAR). http://dels.nas.edu/ilar_n/ilarhome.
7. Rat Genome Database (RGI). http://rgd.mcw.edu.
8. Mouse Genome Informatics (MGI). http://www.informatics.jax.org.
9. Mammalian Genetics Unit, Harwell, UK. http://www.mgu.har.mrc.ac.uk.
10. Pinkert CA, Ed. *Transgenic Animal Technology*. New York: Academic Press, 1984.
11. Clarke AR, Ed. *Transgenesis Techniques,* 2nd ed. Totowa, NJ: Humana Press, 2002.
12. Kahl G, Ed. *The Dictionary of Gene Technology*, 3rd ed. Weinheim, Germany: Wiley-VCH Verlag GmbH & Co. DGaA., 2004.
13. Hedrich HJ, Ed. *Genetic Monitoring of Inbred Strains of Rats*. New York: Gustav Fischer Verlag, 1990.
14. Lyon MF, Rastan S, Brown SDM, Eds. *Genetic Variants and Strains of the Laboratory Mouse*, 3rd ed. Volumes I and II. New York: Oxford University Press, 1996.
15. Gordon JW, Scangos GA, Plotkin DJ, Barbosa JA, Ruddle FH. Genetic transformation of mouse embryos by microinjection of purified DNA. *Proc Natl Acad Sci USA* 1980;77:7380–7384.
16. European Mouse Mutant Mouse Archive. http://www.rm.cnr.it.
17. RIKEN Bioresource Center. http://www.brc.riken.jp.
18. Mutant Mouse Research Resource Center (MMRRC). http://www.mmrrc.org.
19. Rat Resource and Research Center (RRRC). http://nrrrc.missouri.edu.
20. Ince C, Ed. *The Physiological Genomics of the Critically Ill Mouse*. Boston, MA: Kluwer Academic Publishers, 2004.
21. Reece RJ. *Analysis of Genes and Genomes*. Chichester, UK: John Wiley & Sons, Ltd., 2005.
22. Eisen E. *The Mouse in Animal Genetics and Breeding Research*. Hackensack, NJ: World Scientific Publishing Company, 2005.
23. Cuénot, L. Notes et revues. *Arch Zool Exp Gen* 1902;XXVIII.
24. Castle WE, Little CC. The peculiar inheritance of pink eyes among colored mice. *Science* 1909;30:313–314.
25. Castle WE, Little CC. On the modified Mendelian ratio among yellow mice. *Science* 1910;32:868–870.
26. Ginelli E, Di Lernia R Corneo G. The organization of DNA sequences in the mouse genome. *Chromosoma* 1977;61(3):215–226.
27. Mouse Genome Sequencing Consortium. Initial sequencing and comparative analysis of the mouse genome. *Nature* 2002;420:520–562.

28. Baker HJ, Lindsey JR, Weisbroth SH, Eds. *The Laboratory Rat*. New York: Academic Press, 1979.
29. Gibbs RA, *et al.* Genome sequence of the Brown Norway rat yields insights into mammalian evolution. *Nature* 2004;428(6982):475–476.
30. Rat Genome Sequencing Project Consortium. Genome sequence of the Brown Norway rat yields insights into mammalian evolution. *Nature* 2004;428(6982):493–521.
31. Davidsson MT. Rules for nomenclature of inbred strains. In: *Genetic Variants and Strains of the Laboratory Mouse*, 3rd ed. Oxford, UK: Oxford University Press, 1996:1532–1536.
32. Silver LM. *Mouse Genetics: Concepts and Applications*. New York: Oxford University Press, 1995.
33. Green MC, Witham BA, Eds. *Handbook on Genetically Standardized JAX Mice*, 4th ed. Bar Harbor, ME: The Jackson Laboratory, 1991.
34. Markel P, Shu P, Ebeling C, Carlson GA, Nagle DL, Smutko JS, Moore KJ. Theoretical and empirical issues for marker-assisted breeding of congenic mouse strains. *Nat Genet* 1997;17:280–284.
35. Wakeland E, Morel L, Achey K, Yui M, Longmate J. Speed congenics: A classic technique in the fast lane (relatively speaking). *Immunol Today* 1997;18:472–477.
36. Nadeau JH, Singer JB, Matin A, Lander ES. Analyzing complex traits with chromosome substitution strains. *Nat Genet* 2000;24:221–225.
37. Bosma GC, Custer RP, Bosma MJ. A severe combined immunodeficiency mutation in the mouse. *Nature* 1983;301(5900):527–530.
38. Blunt T, Gell D, Fox M, Taccioli GE, Lehmann AR, Jackson SP, Jeggo PA. Identification of a nonsense mutation in the carboxyl-terminal region of DNA-dependent protein kinase catalytic subunit in the scid mouse. *Proc Natl Acad Sci USA* 1996;93(19):10285–10290.
39. Araki R, Fujimori A, Hamatani K, Mita K, Saito T, Mori M, Fukumura R, Morimyo M, Muto M, Itoh M, Tatsumi K, Abe M. Nonsense mutation at Tyr-4046 in the DNA-dependent protein kinase catalytic subunit of severe combined immune deficiency mice. *Proc Natl Acad Sci USA* 1997;94(6):2483–2493.
40. McKusick VA. *Mendelian Inheritance in Man. Catalogs of Autosomal Dominant, Autosomal Recessive, and X-Linked Phenotypes*. Baltimore, MD: The Johns Hopkins University Press, 1990.
41. Chan AW. Transgenic animals: Current and alternative strategies. *Cloning* 1999;1:25–46.
42. Hammer RE, Maika SD, Richardson JA, Tang JP, Taurog JD. Spontaneous inflammatory disease in transgenic rats expressing HLA-B27 and human b2M: An animal model of HLA-B27-associated human disorders. *Cell* 1990;63:1099–1112.
43. Smithies, O. Animal models of human genetic diseases. *Trends Genet* 1993;9:112–116.
44. Neale MJ, Keeney S. Clarifying the mechanics of DNA strand exchange in meiotic recombination. *Nature* 2006;442(7099):153–158.
45. Forrai A, Robb L. The gene trap resource: A treasure trove for hemopoiesis research. *Exp Hematol* 2005;8:845–856.
46. Tsukada T, Tomooka Y, Takai S, Ueda Y, Nishikawa S, Yagi T, Tokunaga T, Takeda N, Suda Y, Abe S. Enhanced proliferative potential in culture of cells from p53-deficient mice. *Oncogene* 1993;12:3313–3322.
47. Pritchard JB, French JE, Davis BJ, Haseman JK. The role of transgenic mouse models in carcinogen identification. *Environ Health Perspect* 2003;111:4.
48. Grimm D. Mouse genetics. A mouse for every gene. *Science* 2006;312:1863–1866.
49. Tsukamoto K, Maugeais C, Glick JM, Rader DJ. Markedly increased secretion of VLDL triglycerides induced by gene transfer of apolipoprotein E isoforms in apoE-deficient mice. *J Lipid Res* 2000;41(2):253–259.
50. Maddison K, Clarke AR. New approaches for modeling cancer mechanisms in the mouse. *J Pathol* 2005;205(2):181–193.
51. Mallo M. Controlled gene activation and inactivation in the mouse. *Front Biosci* 2006;11:313–327.
52. Deng CX, Xu X. Generation and analysis of brca1 conditional knockout mice. *Methods Mol Biol* 2004;280:185–200.
53. Russell WL, Kelly PR, Hunsicker PR, Bangham JW, Maddux SC, Phipps E L. Specific-locus test shows ethylnitrosourea to be the most potent mutagen in the mouse. *Proc Natl Acad Sci USA* 1979;76:5918–5922.
54. Popp RA, Bailiff EG, Skow LC, Johnson FM, Lewis SE. Analysis of a mouse alpha-globin gene mutation induced by ethylnitrosourea. *Genetics* 1983;105:157–167.
55. Russell LB, Hunsicker PR, Cacheiro NLA, Bangham JW, Russell WL, Shelby MD. Chlorambucil effectively induces deletion mutations in mouse germ cells. *Proc Natl Acad Sci USA* 1989;86:3704–3708.
56. Green, EL, Roderick TH. Radiation genetics. In: Green EL, Roderick TH, Eds. *Biology of the Laboratory Mouse*. New York: McGraw-Hill, 1966:165–185.
57. Rinchik EM. Chemical mutagenesis and fine-structure functional analysis of the mouse genome. *Trend Genet* 1991;7:15–21.
58. Festing MFW. *Inbred Strains in Biomedical Research*. New York: Macmillan Press, 1979.
59. Nomura T, Esaki K, Tomita T. *ICLAS Manual for Genetic Monitoring of Inbred Mice*. Tokyo: University of Tokyo Press, 1984.
60. Lyon MF, Searle AG. *Genetic Variants and Strains of the Laboratory Mouse,* 2nd ed. New York: Oxford University Press, 1989.
61. Russell RJ, Fernandez JL, Festing MWF. Genetic monitoring–mice and rats. *Lab Zhyvotnye* 1991;1:30–34.
62. Festing MFW. Genetic contamination of laboratory animal colonies: An increasingly serious problem. *ILAR News* 1982;25:6–10.
63. Gubbels ER, Keesom P, Hilgers J. Genetically-contaminated BALB/c nude mice. *Curr Top Microbiol Immunol* 1985;122:86–88.
64. Kahan BR, Auerbach BJ, *et al.* Histocompatibility and isoenzyme differences in commercially supplied "BALB/c" mice. *Science* 1982;217:379–381.
65. Kurtz TW Jr, Morris RC. Biological variability in Wistar-Kyoto rats: Implications for research with the spontaneously hypertensive rat. *Hypertension* 1987;10:127–131.
66. Kurtz TW, Montano M, Chan L, *et al.* Molecular evidence of genetic heterogeneity in Wistar-Kyoto rats: Implications for research with spontaneously hypertensive rats. *Hypertension* 1989;13:188–192.
67. Lewis JL, Russell RJ, Warnock DG. Analysis of the genetic contamination of salt-sensitive Dahl/Rapp rats. *Hypertension* 1994;24:255–259.
68. Lovell DP, Bigelow SW, *et al.* An investigation of genetic variation within a series of congenic strains of mice. *Lab Anim* 1984;18:291–297.
69 Nabika TY, Nara J, *et al.* Genetic heterogeneity of the spontaneously hypertensive rat. *Hypertension* 1991;18:12–16.
70. Festing MFW, Bender K. Genetic relationships between inbred strains of rats. An analysis based on genetic markers at 28 biochemical loci. *Genet Res* 1984;44:271–281.
71. Am J, Riordan SE, Pearson D, *et al.* Strain restricted typing sera (SRTS) for use in monitoring the genetic integrity of congenic strains. *J Immunol Methods* 1982;55:141–153.
72. Festing MFW, Totman P. Polyvalent strain specific alloantisera as tools for routine genetic quality control of inbred and congenic strains of rats and mice. *Lab Anim* 1980;14:173–177.
73. Fernandez JL, Weeks M. Genetic monitoring of inbred strains of mice using monoclonal antibodies to major histocompatibility haplotypes and lymphocyte antigens. *Lab Anim* 1986;20:293–297.
74. Fernandez JL, Weeks M. Genetic monitoring of rat strains using monoclonal antibodies specific for polymorphic class I and class II MHC antigens. *Lab Anim* 1988;22:235–239.
75. Bailey DW, Usama B. A rapid method of grafting skin on tails of mice. *Transplant Bull* 1960;(7):424.
76. Jeffreys AJ, Wilson V, Thein SW. Hypervariable "minisatellite" regions in human DNA. *Nature* 1985;314:67–72.
77. Jeffreys AJ, Wilson V, Kelly R, *et al.* Mouse DNA 'fingerprints': Analysis of chromosome localization and germ-line stability of hypervariable loci in recombinant inbred strains. *Nucleic Acids Res* 1987;15:2823.

78. Samani NJ, Swales JD, Jeffreys AJ, *et al.* DNA fingerprinting of spontaneously hypertensive and Wistar-Kyoto rats: Implications for hypertension research. *J Hypertens* 1989;7:809–816.
79. Russell RJ, Festing MWF, Deeny AA, Peters AG. DNA fingerprinting for genetic monitoring of inbred laboratory rats and mice. *Lab Anim Sci* 1993;435:460–465.
80. Petkov PP, *et al.* Development of a SNP genotyping panel for genetic monitoring of the laboratory mouse. *Genomics* 2003;83:902–911.
81. Petkov PP, *et al.* An efficient SNP system for mouse genomic scanning and elicidating strain relationships. *Genome Res* 2004;14:1806–1811.
82. Jeffreys AJ. Genetic fingerprinting. *Nat* Med 2005;11(10):1035–1039.
83. Brieger J, Kastner J, Gosepath J, Mann WJ. Evaluation of microsatellite amplifications at chromosomal locus 3q26 as surrogate marker for premalignant changes in mucosa surrounding head and neck squamous cell carcinoma. *Cancer Genet Cytogenet* 2006;167:26–31.
84. Hawtin P, Hardern I, Wittig R, Mollenhauer J, Poustka A, Salowsky R, Wulff T, Rizzo C, Wilson B. Utility of lab-on-a-chip technology for high-throughput nucleic acid and protein analysis. *Electrophoresis* 2005;19:3674–3681.
85. Smits BM, Guryev V, Zeegers D, Wedekind D, Hedrich HJ, Cuppen E. Efficient single nucleotide polymorphism discovery in laboratory rat strains using wild rat-derived SNP candidates. *BMC Genomics* 2005;29:170.
86. Frazer KA, Wade CM, Hinds DA, Patil N, Cox DR, Daly MJ. Segmental phylogenetic relationships of inbred mouse strains revealed by fine-scale analysis of sequence variation across 4.6 mb of mouse genome. *Genome Res* 2004;8:1493–1500.
87. Tsang S, Sun Z, Luke B, Stewart C, Lum N, Gregory M, Wu X, Subleski M, Jenkins NA, Copeland NG, Munroe DJ. A comprehensive SNP-based genetic analysis of inbred mouse strains. *Mamm Genome* 2005;7:476–480.
88. Eklund AC, Turner LR, Chen P, Jensen RV, DeFeo G, Kopf-Sill AR, Szallasi Z. Replacing cRNA targets with cDNA reduces microarray cross-hybridization. *Nat Biotechnol* 2006;24:1071–1073.
89. Kubista M, Andrade JM, Bengtsson M, Forootan A, Jonak J, Lind K, Sindelka R, Sjoback R, Sjogreen B, Strombom L, Stahlberg A, Zoric N. The real-time polymerase chain reaction. *Mol Aspects Med* 2006;2–3:95–125.
90. Arya M, Shergill IS, Williamson M, Gommersall L, Arya N, Patel HR. Basic principles of real-time quantitative PCR. *Expert Rev Mol Diagn* 2005;2:209–219.
91. Sundberg JP, Ichiki T, Eds. *Gentically Engineered Mice Handbook.* New York: CRC, Taylor and Francis Group, 2006.

ADDITIONAL WEB LINKS

Center for Inherited Disease Research: http://www.cidr.jhmi.edu/.
Applied Biosystems: http://www.appliedbiosystems.com/.
Caliper Life Sciences: http://www.caliperls.com/.
Beckman Coulter: http://www.beckmancoulter.com/Default.asp?bhfv=7.
Sequenom: http://www.sequenom.com/.
Affymetrix: http://www.affymetrix.com/index.affx.
Illumina: http://www.illumina.com/.
Agilent Technologies: http://www.home.agilent.com.
NuGEN Inc.: http://www.nugeninc.com/.
Roche: http://www.roche-diagnostics.com/.

21 The House Mouse in Biomedical Research

John G. Vandenbergh

ABSTRACT

The house mice has become the primary rodent model in much of biomedical research because of its ability to adapt to a variety of environments. They are omnivorous, adapt to the environment with varied social structures, show considerable genetic variability, and share many similar physiological functions with humans. Recently the mouse has become the species of choice in transgenic studies because of the relative ease to transfer genes from other species, including humans. This chapter reviews the origins of the laboratory mouse, the use and care of the mouse, and its social structure and adaptability to different environments. Furthermore, it discusses the influence of pheromones on puberty onset, ovarian synchrony, and social communication. By making appropriate genetic selections and genetic manipulations, and providing an appropriate environment for the laboratory mouse, it will continue to serve the biomedical research community.

Key Words: Laboratory mouse, Origins, Adaptability, Social structure, Reproduction, Pheromones, Puberty, Ovarian synchrony, Husbandry.

INTRODUCTION

The laboratory mouse is to biomedical research as Mickey Mouse is to Disneyland. Just as Mickey has captured the attention of children and adults around the world, the laboratory mouse has captured the attention of biomedical scientists. In fact, it is the most commonly used nonhuman mammal in biomedical research with over 32 million mice in use per year (estimate by J. Zurlo, ILAR). Rats, which are often associated with biomedical research by the public, are used less often, about 8 million per year. The ability to alter the genome in the mouse resulted in a sharp increase in their use and the numbers are likely to continue into the future. A detailed description of the use of genetically altered mice is given elsewhere in this volume. Here the focus will be on the natural biology of the mouse and how this relates to its life in the laboratory.

In the hands of cartoonists the mouse is cute and frisky. For researchers it is small, easily bred, often tame, and has proven to be a useful model of human physiology and pathology (see Table 21–1 for a list of attributes). In this chapter I will discuss the origins of the laboratory mouse, the natural history of the species that relates to its use as a laboratory animal, and how the adaptability of the mouse makes it useful as a mammalian model for biomedical research.

ORIGINS OF THE LABORATORY MOUSE

The Order Rodentia has a world-wide distribution and has been classified into over 2000 species grouped into 28 families.[1] The Family Muridae consists of approximately 1300 species. Three subspecies of this family in the complex of *Mus musculus* are pertinent to readers interested in the "laboratory mouse."[2]

The *M. musculus* complex originated in the Indian subcontinent and moved throughout Eurasia and into northern Africa before humans began serving as a vector for species distribution. The *M. musculus* complex consists of *M. m. domesticus* originally found in Europe, parts of Africa, and the near East, *M. m. castaneus* from Asia, and *M. m. molossinus* originally from the Japanese archipelago.[2] Mice of the *M. musculus* complex rarely hybridize, but where they are sympatric some genetic exchange occurs.[3,4] Thus, the taxonomy is complex with many authors using *M. musculus* and others using *M. m. domesticus* to identify the species being used. The latter is apparently utilized because much of the early research with the mouse occurred in Europe, the distribution of the *domesticus* subspecies. The majority of papers reporting on research involving the mouse simply use "mouse" or "house mouse" to identify the animal. The laboratory mouse is now available in many strains and with a number of genetic backgrounds,[5] so stain identity may be the most pertinent taxonomic designation for many studies. For more information on specific strains that have been developed see Chapter 22, 23, and 24.

The mouse has been associated with humans since the beginning of agriculture and the rise of cities. The earliest association between man and mouse is apparently in neolithic communities at around 6500 BC in what is now Turkey.[6] The mouse's association with agriculture may be reflected in the name for its genus, *Mus*. "Mus" came to Latin via Greek and before that from the ancient Sanscrit word "Mush," a word that means "to steal." [7] The mouse has been "stealing" food from humans for thousands of years. Its adaptability allowed it to disperse along with human hosts and, as the humans penetrated new environments, the mouse accompanied them. The house mouse, often seen as a pest species, is found throughout most of the world commensally with humans and as feral animals in a variety of natural habitats.[8]

The laboratory strains of mice used today derive from the activities of mouse breeders producing "fancy" mice, primarily

From: *Sourcebook of Models for Biomedical Research*
(P. M. Conn, ed.), © 2008 Humana Press Inc., Totowa, NJ.

Table 21–1
Laboratory-friendly attributes of the house mouse

Life history traits	*Reproduction*	*Genetic*
Small size	Breed year-round	Tolerates inbreeding well
Omnivorous	Large litters	Inbred strains available
Short generation time	Ease of cross-fostering	Genetic sequence described
Adapt to social grouping		Totipotent embryonic stem
Short life span		Cells can be grown *in vitro*

in Japan, England, and the United States.[9] Mice were bred for unusual pelage and for striking behaviors such as "waltzing." An important source of the laboratory mouse strains in the United States resulted from the activities of Abbie Lathrop. Miss Lathrop was a retired school teacher who bred fancy mice on her farm in Massachusetts at the turn of the twentieth century.[10] She obtained mice from dealers, other mouse fanciers, and from the wild. A few animals in her colony developed tumors and, in collaboration with Leo Loeb, she attempted to learn more about this condition. The work resulted in two early scientific reports on cancer in mice.[10,11] She further collaborated with William Castle and his student Clarence C. Little at Harvard. C.C. Little went on to found the Jackson Laboratory in Bar Harbor, Maine. The result of their collaboration is the founding of inbred strains, usually credited to C.C. Little.[11] The first named mouse strain emerged from Little's research: the *dba*,[12] with *dba* standing for dilute brown nonagouti. The strain is ancestral to the currently used DBA1 and DBA2 among others.[5] Little's line "C" was the origin of the C57/58 strains, which gave rise to the commonly used C57BL/6 and other strains (see Chapter 22 for more information on strains). The work of C.C. Little and his co-workers and successors resulted in the Jackson Laboratory growing into a primary source of mouse strains as well as a major research center.[11,13] Phylogenetic analyses of 144 separate loci confirm this historical record on the origin of strains and provide an accurate picture of the genealogical relationships among the mouse strains.[14]

Mice have been used as mammalian models to help explain biological phenomena from the earliest days of modern scientific research. Robert Hooke, for example, used mice in 1664 to determine the biological consequences of high air pressures.[8] With the mouse emerging over the past few decades as the primary mammalian genetic model, it is interesting to note that Gregor Mendel may have begun his selection studies to examine the inheritance of coat colors in the mouse.[13,15] His superiors apparently objected to the monk's use of mice. Their obvious interest in sex plus their odor probably made them unwelcome cooccupants of Mendel's cell. Under pressure to give up the mouse, Mendel switched to the pea as a model organism and, as they say, the rest is history.

ADAPTABILITY AND SOCIAL STRUCTURE

The adaptability of the house mouse allows it to live in the wild as a pest consuming or damaging agricultural products, or commensally in human habitations, or as a laboratory animal. The house mouse is considered to be the most adaptable of all mammals, especially for its reproductive traits.[16,17] It can adjust to the major factors that influence reproductive ability: diet, physical environment, and social environment, as well as the interactions among these factors.

The importance of diet was tested by creating bioenergetic stress on mice of both sexes through restricting caloric intake under different thermal conditions.[18] The surprising results showed that bioenergetic restriction affected reproduction only in the female. Both sexes showed a high tolerance to restriction, but the females showed delayed puberty and impaired reproduction in adulthood.[17] The males continued reproductive development even to the point of ceasing growth.[17] The genetic or hormonal mechanism for this remains unknown. Such a striking sex difference in the mouse should be considered in the design of biomedical studies on diet, environmental stress, and other bioenergetic issues.

Maintenance of laboratory mice in a relatively stable thermal neutral zone is important for their health and well being and may have an important effect in studies such as those of feeding behavior or with drugs having a thermoregulatory influence. The recommended dry bulb ambient temperature for housing mice is 18–26°C.[19] A review of studies on the laboratory mouse indicated that animals have a thermoneutral zone of about 26–34°C,[20] so it seems that the upper end of the recommended range is more appropriate for thermoneutrality. However, the use of covered cages, air flow-regulated racks, cage material, type and quantity of bedding used, and the number of mice housed in a cage can all influence the ambient temperature to which the mouse is exposed. The effects of different types of bedding and the depth of the bedding on core temperature, metabolic rate, and motor activity have been investigated.[21] Mice housed on deep wood shavings compared to shallow wood shavings or beta chips were able to maintain a significantly higher core temperature during the day, when they are relatively less active, than at night. The CD-1, an outbred strain, was used in this study.[21] Other strains or genetically modified mice may have specific thermal and metabolic requirements.

Another anomalous finding emerging from studies of mouse reproduction[21] is that reproduction in the house mouse does not seem to respond to seasonal changes in photoperiod. Seasonal breeding is strongly influenced by photoperiod in many mammals,[22] but not no in the mouse. Several stocks of mice have been maintained under constant day length for extended periods of time without apparent effect on reproduction. In addition, a study of wild-caught mice held under constant darkness for 75 days showed the same reproductive productivity as mice held under a 14:10 L:D schedule.[18] Though some seasonal variability occurs in feral populations with a peak in late spring/early summer, many breed throughout the year. In a meta-analysis of the breeding seasonality in feral and commensal (human structures) populations of mice in the temperate zone, Bronson[17] reported that four of five feral populations examined were seasonal breeders, whereas none of the four commensal populations were seasonal. The important variable regulating reproduction seems to be the continuous availability of food to females, rather than photoperiod. Despite the apparent absence of an effect of photoperiod on reproduction in the mouse, other physiological functions with a circadian cycle may be affected, so the usual laboratory light : dark cycle of 12:12 or 14:10 in laboratory colonies is probably warranted.

Mice are nocturnal rodents showing most activity in the dark; however, they will also show sporadic bursts of activity during daylight hours. Mice can control exposure to light and temperature by burrowing or building a nest if given an opportunity, which is seldom available to laboratory caged mice.

Under natural circumstances mice commonly live in a deme, a small group usually consisting of kin. As with reproduction, variability in social grouping is the rule. Individual feral male mice have been found to have home ranges of up to $400 m^2$.[23] At the other extreme, mice can be found in extremely dense populations as in a mouse plague of 100 mice per hectare.[24] Under conditions of high food availability such as in corn ricks in England, Southwick[25] found populations of over 2000 mice per corn rick or 15 adult mice per cubic meter. From these field studies of the mouse, it is apparent that the high degree of social flexibility available to the mouse allows it to adapt to and breed in a wide variety of captive conditions.

Although the importance of pheromonal communication is much greater in the mouse than in humans, attention to or manipulation of chemical signals can be useful in the laboratory setting. Mice use pheromones both to signal information to conspecifics and to prime several developmental and reproductive traits. For example, mice signal by marking continuously with urine as they move on a new substrate. Males show more intense marking behavior than females.[17] Such marking results in the deposition of pheromones that are important for territorial boundary marking among house mice. Signaling pheromones also provide information concerning an individual's sex, gonadal state, and social status.[26,27] The primary source of signaling pheromones seems to be urine, and that may help explain the commonly observed anogenital sniffing of strangers. However, mice also attend to the face area of a stranger, due perhaps to a sex identification pheromone that has recently been identified in lacrimal glands of the mouse.[28] It seems that tears can also convey olfactory information.

Pheromones can also induce developmental or physiological changes and are thus called "priming pheromones." Temporal regulation of ovulation in the female seems to be strongly influenced by such pheromones. The onset of puberty in the female is particularly sensitive to a chemical stimulus contained in male urine.[29,30] Female mice denied the presence of a male or his urine show a prolonged prepubertal period followed by a disorganized ovulatory cycle. The production of the puberty-accelerating pheromone is androgen dependent and as little as $10 \mu l$ of urine per day induces onset of puberty as early as 28 days of age. The active ingredient in urine that is responsible for puberty acceleration seems to be a protein-bound heptone.[31]

Puberty onset in the mouse is a consequence of both stimulatory and inhibitory pheromonal influences. Juvenile females housed in a group of four or more or those exposed to urine of grouped females show delayed onset of first ovulation.[32,33] The puberty-inhibiting pheromone has not been identified, but is known to result from grouping of the females and possibly from adrenal activation.[33]

One priming pheromonal effect that may have little relevance in wild populations[17] but may be useful in laboratory populations is the ovarian synchronicity effect. Inhibition of adult ovarian cycles occurs in all-female groups. In addition, if grouped females are exposed to an adult male or male urine, ovulation usually occurs 3 days later.[34] Thus, conception can be at least partly synchronized to a specific date by grouping females four or more per cage for 2 weeks prior to mating with a male.

CARE AND USE OF MICE IN THE LABORATORY

The *Guide for the Care and Use of Laboratory Animals*[19] contains the information needed and often required for maintaining mice used in biomedical research. Mice, rats, and poultry are not included in the Federal Animal Welfare Act of 1987 as amended, but many institutions have placed mice and rats under the same restrictions and oversight as other mammals covered in the Animal Welfare Act.[35] Details on housing conditions, diet, and use of mice can be found in the *Guide*.[19] To be useful as a laboratory model, mice should be maintained to allow as much normal behavior as possible in a cage. "Normal" is difficult to define in the case of the house mouse since it is such an adaptable animal as noted above. Certain behaviors, such as stereotypic actions expressed in confinement, may interfere with the usefulness of the animals as a model for biomedical research.[36] However, as a recent review suggests,[37] stereotypic behavior may also serve as a model for abnormal brain function and be useful in elucidating underlying causes of certain mental illnesses.

Avoidance or reduction of abnormal behaviors in experimental subjects is important in the majority of studies. Even though the house mouse is a highly adaptable species, it can show environmentally induced abnormal behaviors[38] that may interfere with the outcome of studies. An array of environmental enrichment procedures, both social and physical, includes housing animals in groups and providing nesting materials; providing some complexity to the cage environment may reduce or eliminate such behaviors.[39]

CONCLUSIONS

Both rats and mice are highly adaptable animals. As is true for humans, they are omnivorous, adapt to the environment with varied social structures, show considerable genetic variability, share many similar physiological functions, and show many of the same diseases. Perhaps for these reasons and others listed in Table 21–1 mice and rats, but especially mice, have become the model of choice for an array of biomedical studies.

By making appropriate genetic selections, genetic manipulations, and providing an appropriate environment for the laboratory mouse, it will continue to serve the research community just as the cartoon character, Mickey, continues to provide entertainment for the public. Humans must deal with the house mouse as a pest, but also should appreciate the role of the mouse in helping to understand the origins and cures of human and animal diseases. It seems that such an appreciation was attempted when C.C. Little tried to arrange a joint 25th birthday celebration at Walt Disney for both Mickey Mouse and the JAX mouse.[13] Unfortunately, Disney showed little interest in Little's proposal and Mickey got all the attention.

ACKNOWLEDGMENTS

I thank the W.M. Keck Center for Behavioral Biology at NCSU for support and the following colleagues for helpful suggestions: M.D. Engell, A.K. Hotchkiss, B.C. Ryan, D.J. Vandenbergh, and B.D. Vandenbergh.

REFERENCES

1. Hutchon D, Madsden O, Sibbald MJ, Ament K, Stanhope MJ, Catzeflis F, de Jong WW, Douzery EJP. Rodent phylogeny and a timescale for the evolution of Glires: Evidence from an extensive taxon sampling using three nuclear genes. *Mol Biol Evol* 2002; 19:1053–1065.
2. Guenet JL, Bonhomme F. Wild mice: An ever-increasing contribution to a popular mammalian model. *Trends Genet* 2003;19:24–31.
3. Boursot P, Auffray JC, Britton-Davidian J. The evolution of house mice. *Annu Rev Ecol Syst* 1993;24:119–152.
4. Yonekawa H, Moriwaki K, Gotoh O, Miyashita N, Matsushima Y, Shi LM, Cho WS, Zhen ZL, Tagashira Y. Hybrid origin of Japanese mice "Mus musculus molossinus": Evidence from restriction analysis of mitochondrial DNA. *Mol Biol Evol* 1988;5:63–78.
5. Linder CC, Davisson MT. Strains, stocks, and mutant mice. In: Hedrich HJ, Bullock G, Eds. *The Laboratory Mouse*. Amsterdam: Elsevier, 2004:25–46.
6. Brothwell D. The Pleistocene and Holocene archeology of the house mouse and related species. In: Berry RJ, Ed. *Symposium of the Zoological Society of London, The Biology of the House Mouse*, Vol. 47. London: Academic Press, 1981:1–13.
7. Silver LM. *Mouse Genetics*. New York: Oxford University Press, 1995.
8. Berry RJ. House mouse. *Biologist* 1987;34:177–186.
9. Morse HC III. *Origins of Inbred Mice*. New York: Academic Press, 1978.
10. Davisson MT, Linder CC. Historical foundations. In: Hedrich HJ, Bullock G. *The Laboratory Mouse*. Amsterdam: Elsevier, 2004:15–24.
11. Little CC. The dilute forms of yellow mice. *Science* 1911:33:896–897.
12. Rader KA. *Making Mice*. Princeton: Princeton University Press, 2004.
13. Paigen K. One hundred years of mouse genetics: An intellectual history. I. The classical period (1902–1980). *Genetics* 2003;163: 1–7.
14. Atchley WR, Fitch WM. Gene trees and the origins of inbred strains of mice. *Science* 1991:254:554–558.
15. Henig RM. *The Monk in the Garden: The Lost and Found Genetics of Gregor Mendel, the Founder of Genetics*. Boston, MA: Houghton Mifflin, 2000.
16. Bronson FH. The reproductive ecology of the house mouse. *Quart Rev Biol* 1979;54:265–299.
17. Bronson FH. The adaptability of the house mouse. *Sci Am* 1984;250:116–125.
18. Perrigo G, Bronson FH. Foraging effort, food intake, fat deposition, and puberty in female mice. *Biol Reprod* 1983;29:455–463.
19. *Guide for the Care and Use of Laboratory Animals*. ILAR, NAS/NRC, Washington, DC, 1996.
20. Gordon CJ. Effect of cage bedding on temperature regulation and metabolism of group-housed female mice. *Compara Med* 2004; 54:63–68.
21. Gordon CJ. *Temperature Regulation in Laboratory Rodents*. New York: Cambridge University Press, 1993.
22. Elliott JA. Circadian rhythms and photoperiodic time measurement in mammals. *Fed Proc* 1976;25:2339–2346.
23. Quadagno DM. Home range size in feral house mice. *J Mamm* 1968;49:140–151.
24. Pearson OP. History of two local outbreaks of feral house mice. *Ecology* 1963;44:540–549.
25. Southwick CH. Population characteristics of house mice living in English corn ricks: Density relationships. *Proc Zool Soc Lond* 1958;131:163–175.
26. Harrington JE. Recognition of the territorial boundaries by olfactory cues in mice. *Z Tierpsychol* 1976;41:295–306.
27. Jones RB, Nowell NW. The effect of urine on the investigatory behavior of male albino mice. *Physiol Behav* 1973;11:35–38.
28. Kimoto H, Haga S, Sato K, Touhara K. Sex-specific peptides from exocrine glands stimulate mouse vomeronasal sensory neurons. *Nature* 2005;347:898–901.
29. Vandenbergh JG. Effect of the presence of the male on the sexual maturation of female mice. *Endocrinology* 1967;81:345–348.
30. Vandenbergh JG. Acceleration and inhibition of puberty in female mice by pheromones. *J Reprod Fert* 1973;19:411–419.
31. Dulac C, Torello AT. Molecular detection of pheromone signals in mammals: From genes to behaviour. *Nat Rev Neurosci* 2003;4: 551–562.
32. Vandenbergh JG, Drickamer LC, Colby DR. Social and dietary factors in the sexual maturation of female mice. *J Reprod Fertil* 1972;28:397–405.
33. Drickamer LC. Sexual maturation of female house mice. *Dev Psychobiol* 1974;7:257–265.
34. Whitten WK. Modifications of the oestrus cycle of the mouse by external stimuli associated with the male. *J Endocrinol* 1956;13: 399–404.
35. VandeBerg JL, Williams-Blanchero S, Wolfle TL. US laws and norms related to laboratory animals. *ILAR J* 1999;40:15–21.
36. Garner JP. Stereotypies and other abnormal repetitive behaviors: Potential impact on validity, reliability, and replicability of scientific outcomes. *ILAR J* 2005;46:106–117.
37. Garner JP. Perservation and stereotypy–systems-level insights from clinical psychology. In: Rusen J, Mason G, Eds. *Stereotypic Animal Behaviour: Fundamentals and Applications to Welfare*. Wallingford, UK: CABI, 2007, Chapter 5.
38. Baumans V. Environmental enrichment for laboratory rodents and rabbits: Requirements of rodents, rabbits and research. *ILAR J* 2005;46:162–170.
39. Baumans V. The welfare of laboratory mice. In: Kaleste E, Ed. *The Welfare of Laboratory Animals*. Dordrecht: Kluwer Academic Publishers, 2004:119–152.

22 Mouse Model for Alzheimer's Disease

M. Azhar Chishti and Shaheen M. Nakeeb

ABSTRACT

Transgenic mouse models for Alzheimer's disease (AD) are now extensively used to understand the physiological and biochemical mechanisms of the disease as well as to discover new potential therapeutics. The identification of disease-causing mutations in proteins such as amyloid precursor protein (APP) and presenilins (PS1 and PS2) as well as pathogenic mutations in the tau protein has led to the creation of several transgenic mice including those expressing bigenic and trigenic constructs. Each model has unique pathologies that provide insight into disease mechanisms and features of neuropathologies and cognitive functions. Therapeutic hypotheses are now testable in these transgenic mice and the effectiveness of these therapeutics is testable in preclinical and clinical trials in AD patients. A number of therapeutic approaches including active and passive immunization, small molecule monoclonal antibodies, and nonviral DNA vaccines are available. It is hoped that in the near future a treatment for AD will be available as a result of these advancements.

Key Words: Transgenic mice, Amyloid plaques, Tau protein, Amyloid β peptide, β-APP, Pathology, Cognitive deficits, Immunization, Therapeutics.

INTRODUCTION

Alzheimer disease (AD), a progressive neurodegenerative disorder that is the most common cause of dementia, is characterized by a deterioration in mental functions leading to dementia and pathologically by two hallmark lesions: diffuse and neuritic amyloid plaques and neurofibrillary tangles composed of filamentous aggregates of hyperphosphorylated tau protein in the cerebral cortex and hippocampal region of the brain. The clinical manifestations of AD are mild cognitive impairments at the early stage leading to severe cognitive deficits, generalized behavioral impairments, and, ultimately, death. The typical course of the disease is 6–10 years.[1] AD may be characterized as early onset (30–60 years) or late onset (60 years and above). At least 10% of AD cases are of the early-onset type, which shows a strong genetic linkage and is transmitted as autosomal dominant traits with almost 100% penetrance. Since there is no naturally occurring rodent form of AD, there has been great interest in creating accurate transgenic facsimiles of AD. Such models have the potential to mirror the key pathological events and practical utility for testing interventions directed against the synthesis or deposition of the Aβ peptide or tau protein.

BIOCHEMICAL PATHWAY

AD amyloid β peptide is derived from two endoproteolytic "secretase" cleavages of the Alzheimer amyloid precursor protein (APP) encoded by the APP gene on chromosome 21. Familial Alzheimer's disease (FAD) mutations enhance proamyloidogenic cleavages at the β- and γ-secretase sites that release Aβ from APP,[2] while FAD mutations in presenilin 1 and 2 enhance γ-secretase cleavages (Figure 22–1).[3–6] The phenotypic ablation of presenilin 1 results in skeletal defects and accumulation of APP C-terminal fragments (CTFs) due to processing of notch signaling via a deficit in cleavage at the juxtamembrane "site3," further underscoring the connections between presenilins and endoproteolysis.[7–9] An anonymous FAD locus on chromosome 10 has been identified on the basis of presymptomatic elevations in plasma levels of Aβ peptide.[10] ApoE, a genetic modifier associated with the incidence of late onset AD,[11] also affects the biochemistry of Aβ peptide *in vitro*[12] and in transgenic mice.[13–14]

Thus FAD provides overwhelming evidence that an accumulation of Aβ peptide stands at the head of the pathogenic cascade that accounts for the hallmarks of the disease. This perspective adds two further pieces of information that help to clarify the roles of Aβ and tau in the cognitive deficits seen in AD.[15–17] First, this indicates that cognitive dysfunction is well correlated with the levels of soluble Aβ present within the parenchyma, and other studies have documented increased burdens of Aβ at earlier time points in the development of AD[18–22] and in Down syndrome patients (with trisomy 21 and hence three copies of the APP gene) prior to dementia.[23] The second concerns the pathogenic effect of mutant tau in frontotemporal dementia with Parkinsonism (FTPD17). A plethora of mutations in the tau gene indicate the pathogenic effects of an excess of "four repeat" tau, and in some cases missense mutations in the C-terminal region of tau.[24] These findings, as well as the generation of neurofibrillary tangle-like pathology and cell loss in transgenic mice expressing the P301L mutant form of human tau,[25] demonstrate a potential pathogenic role of tau. It is important to bear in mind that tau mutations do not exist within FAD kindred and that tau pathologies occur in many different diseases. In sum, penetrating FAD mutations causing AD indicate that Aβ must lie upstream of tau in a pathogenic pathway, with tau pathologies arising as a consequence of Aβ overaccumulation, perhaps contributing to the cognitive dysfunctions observed in the later stage of AD.

From: *Sourcebook of Models for Biomedical Research*
(P. M. Conn, ed.), © 2008 Humana Press Inc., Totowa, NJ.

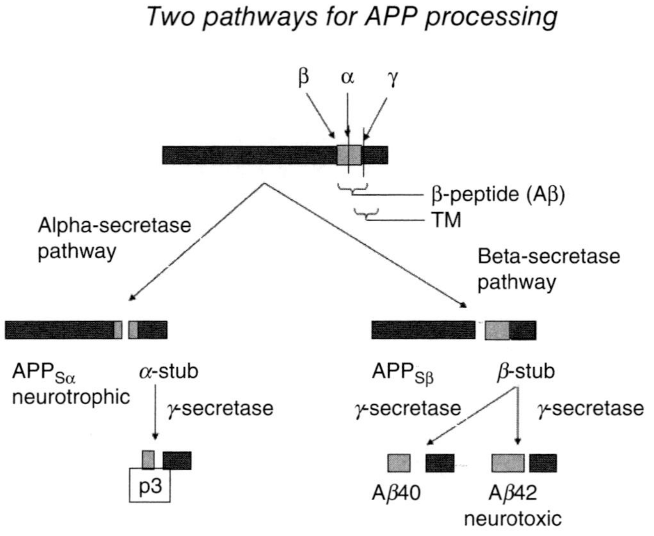

Figure 22–1. Schematic diagram showing that sites of secretase cleavage of amyloid precursor protein lead to generation of Aβ40 and neurotoxic Aβ42.

DIAGNOSIS OF ALZHEIMER'S DISEASE

The definitive diagnosis of AD is made on postmortem examination. The pathological hallmarks of this disease are extracellular amyloid plaques composed of amyloid β protein (Aβ), a metabolic product of amyloid precursor protein (APP), and intracellular neurofibrillary tangles composed of hyperphosphorylated tau protein.[26] This pathology shows a distinct spatial–temporal pattern, initially affecting areas of the temporal cortex, notably the transentorhinal region, before extending to the hippocampus and certain other cortical zones. At the latter stages, a more widespread pathology may be seen in certain brain areas, particularly cortical (frontal, temporal) and limbic regions, although other areas, such as the cerebellum, show no such signs.[27] Biochemical processes not completely understood, the accumulation of Aβ, and neurofibrillary tangles produce local neurodegeneration, which ultimately accounts for the clinical signs of this disease.

GENERATION OF TRANSGENIC MICE

Transgenic mice are produced by the introduction of a human gene sequence or mutated sequence into the mouse genome, resulting in the expression of a human protein. This is most commonly achieved by the microinjection of complementary DNA constructs of genomic fragments (with or without known altered mutations) complete with promoters, introns and exons, which have been used for the better expression of a single whole gene. Other methods are also used to manipulate endogenous genes to "knockin or knockout" a human sequence. Regardless, the challenge of creating a mouse model of AD involves recapitulating a chronic, progressive disease that takes decades to develop in humans, but develops in a short period of time, from 3 to 9 months, in a mouse that has a short life span of less than 2 years. Mice expressing pathogenic mutations of human genes have become a critical tool for biomedical research in testing hypotheses of AD pathogenesis and testing novel therapeutic strategies. The pathological and behavioral characteristics of AD transgenic mice are summarized in Table 22–1.

Table 22–1
Key features of transgenic mouse models for Alzheimer's disease[a]

Transgenic line	*Mutation*	*Promoter*	*Pathology*	*Behavior*	*Immunotherapy*	*Reference*
PDAPP	APPV717F	PDGF-β	Astrocytosis, microgliosis, Aβ deposits, and synaptic loss	Age-independent and age-related deficits in spatial learning	M266, more than 50% plaque reduction and cognitive improvement	28
Tg2576	APPSwedish	Hamster PrP	Aβ elevation and plaque deposition	Age-related deficits in spatial learning	Amyloid β 1–42, 50% plaque reduction, and cognitive improvement	29
TgAPP23	APPSwedish	Murine Thy-1	Amyloid plaques immunoreactive with tau hyperphosphorylation, angiopathy	Age-related deficits in spatial learning	Nonviral amyloid β DNA vaccine, 30–50% reduction	30
TgCRND-8	APPSwedish and V717F	Hamster PrP	Early onset amyloid plaques, neurotic pathology	Early impairment in learning acquisition	Amyloid β 1-42, 50% plaque reduction and cognitive improvement	33
PS/APP	APPSwedish and $PS1^{M146L}$	Hamster PrP/PDGF-β	Elevated levels of amyloid β 42 or 43	No information is available	Amyloid β 1-42, 50% plaque reduction, and cognitive improvement	35
JNPL3	Tau^{P301L} (4R,2-,3-)	Murine PrP	Development of neurofibrillary tangles	Age-related motor and behavioral deficits	N/A	41

Table 22–1 (continued)

Transgenic line	*Mutation*	*Promoter*	*Pathology*	*Behavior*	*Immunotherapy*	*Reference*
TgTau^{P301L}	Tau^{P301L}	Hamster PrP	Development of neurofibrillary tangles with gliosis, neuronal loss, and atrophy	Cognitive impairments	N/A	45
Triple transgenic	APPSwedish, PS1^{M146L}, and Tau^{P301L}	Thy-1.2	Development of amyloid plaques, neurofibrillary tangles, and synaptic dysfunction	Age-related cognitive deficits; LPT severely impaired at 6 months	Clearance of early hyperphosphorylated tau	46

[a]Behavioral and pathological features of different transgenic models of Alzheimer's disease (AD) and effects of immunotherapy in these AD transgenic mice.[52]

TRANSGENIC LINES ON THE BASE OF THE AMYLOID CASCADE

Most current animal models of AD are built upon early-onset genetic forms of the disease. However, familial AD cases number only about 10% of all cases of AD, with the sporadic form constituting the remaining 90%. The assumption that current models are reasonable facsimiles of nonfamilial forms of AD is based upon a variation in the theme that Aβ is the key pathogenic moiety in all forms of AD. Genetically engineered mice can be created by the addition of microinjected transgenes into unselected sites in the mouse genome that drive expression beyond the endogenous levels and accelerate the onset of neurodegenerative diseases. The main objective in creating an APP transgenic line is to obtain enough expression of human β APP in the central nervous system (CNS) to produce amyloid deposits without toxicity and mortality.

PDAPP MOUSE Games and co-workers[28] created PDAPP transgenic mice with a V717F mutation that showed many of the pathological features of AD including extensive deposition of extracellular amyloid plaques, astrocytosis, and neuritic dystrophy. A histological analysis of PDAPP mice revealed that plaques are first formed in the cingulate cortex and hippocampus followed by spreading to other subcortical areas except for the cerebellum, and that amyloid plaque loads increased as a function of age. At a later time point the amyloid plaque burden may reach 50% of the cortical areas, but clear neurodegeneration has not been demonstrated, except for some synaptic loss in the molecular layer of the dentate gyrus, and neurofibrillary tangles have not been identified. Behavioral studies with PDAPP mice describe early cognitive deficits before the amyloid plaques could really be detected. Modified Morris water maze testing showed memory impairments that were age dependent and correlated with amyloid plaque load. At a young age (6–9 months) PDAPP mice showed a level of acquisition similar to control mice, but middle-aged (13–15 months) and old (18–21 months) mice showed slow acquisition compared with age-matched controls. The common finding of PDAPP mice are both age-related (plaques) and age-unrelated cognitive changes. Test selection seems critical for their detection.

TG2576 MOUSE In 1996, Hsaio and colleagues[29] developed an APP695 transgenic line with a K670N/M671L (Swedish) mutation under the control of the prion protein promoter to ensure high CNS expression. Initially the development of this mouse was emphasized due to background strain effects in FVB/N or C57B/6 mice on pathology and early death, but changing the genetic background to C57B6XSJL resulted in better survival and deposits of thioflavin S-positive amyloid burden at age of onset of 9–12 months. The earliest deposition of amyloid plaques appears in the entorhinal and pyriform cortices and then the hippocampus, thus showing subtle differences from the PDAPP mouse possibly related to a different promoter. The Tg2576 mouse has since become the most widely used transgenic mouse in the AD field. With aging, Tg2576 mice exhibit a phenotype that includes learning and memory deficits, an abnormal pattern of glucose metabolism in the brain, and pathological changes including amyloid plaque deposits, elevated Aβ40 and Aβ42 levels, neuritic changes, phosphorylated tau epitopes, A-synuclein-positive dystrophic neuritis, gliosis, and inflammatory changes; however, these mice did not show any neurofibrial tangles and neuronal loss. A variety of inflammatory and oxidative markers have also been studied in these mice to examine the involvement of these processes in AD-like pathology.

APP23 MOUSE These transgenic mice were developed at Novartis[30] with an hAPP751 construct containing the Swedish mutation, under the control of the Thy-1 promoter for high neuronal expression. These mice were characterized extensively in terms of pathology and have features similar to PDAPP and Tg2576 mice such as neuritic plaques with local astrocytosis and microgliosis mainly in the neocortical and hippocampal regions. Interestingly, neurodegeneration has been reported in the CA1 region of the hippocampus and neocortex in 14- to 18-month-old mice that was not seen in the above lines. The other feature of the APP23 mouse is the development of cerebral amyloid angiopathy (CAA), which is significantly correlated clinically with hemorrhagic stroke in up to 90% of AD patients. Modest cholinergic deficits have also been reported in aged App23 mice.[31] Behavioral characterization of these mice showed no sensorimotor changes at 3, 18, and 25 months. However, visible and hidden platform acquisition performance was impaired at 3 and 25 months, suggesting transgene-induced neophobia and sensory changes.[32] Although soluble Aß levels did not change across the age range, subtle qualitative changes in the Aß form, such as conversion into small oligomers/protofibrils, may contribute to behavioral

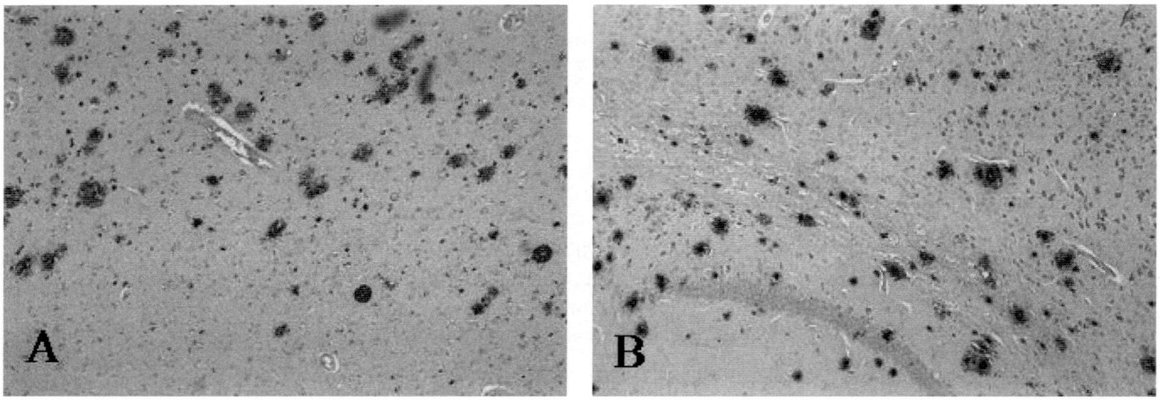

Figure 22–2. Amyloid plaque deposition is a pathological hallmark for Alzheimer's disease. (A) Human AD brain (70 years old): the deposition of amyloid-β plaques is visualized on cortical brain sections after immunolabeling with antiamyloid-β antibodies (4G8 monoclonal diluted 1:400). (B) Animal model (6 month old): transgenic CRND8 mice express human mutated amyloid precursor protein recapitulated pathological features of amyloidogenesis present in AD.

changes. The APP23 model has similarities to plaque-related changes in the PDAPP and Tg2576 models, but these are unrelated to cognitive changes.

CRND8 MOUSE The CRND8 mouse was created from a double mutated form of hAPP695 under the control of the hamster prion protein promoter. The combination of the Swedish K690N/M671L and V717F Indiana APP mutation results in extremely high levels Aß42, and a very early onset of thioflavin S-positive amyloid deposits becomes evident in all mice at 3 months of age.[33] These plaques are associated with dystrophic neurites and gliosis. A number of genetic backgrounds was used to overcome the early lethality evident in other lines. The most satisfactory is C57B6XC3H and 129 SvEv-Tac, with approximately 50% of CRND8 mice surviving by 12 months of age. The lethality may be associated with the increase in seizure liability, which is possibly related to the very high Aß levels attained in these mice. The amyloid burden in 6-month-old transgenic CRND8 mice is similar to that found in a human AD brain from a 70-year-old AD patient (Figure 22–2). In terms of cognitive status, at 11 weeks of age the CRND8 mice showed impaired acquisition of spatial learning in a Morris water maze test. Most recently, separate cohorts of 6- to 8-week old (preplaque) and 9- to 22-week old and 40-week old (postplaque) CRND8 mice showed spatial learning impairments increasing with age and a considerable burden of corticohippocampal plaques.[34]

PS/APP MOUSE Following the identification of pathogenic mutations in the presenilin (PS) genes in AD, transgenic animal models have been generated to address the role of PS1 and PS2 in APP processing *in vivo*. FAD mutant forms either alone (PS1) or in combination with APP and knockout of endogenous presenilin genes were tested. PS1 knockout mice are not viable and develop skeletal defects and CNS phenotype due to defects in the notch pathway. This lethal phenotype was rescued by expression of the FAD mutant PS1^{M146L}. Transgenic mouse lines overexpressing the *PS1* gene containing the M146L mutation were established in the hPS1^{M146L} mouse;[35] although these mice do not appear to develop plaques, they show an increase in extracellular Aβ1-42 levels. Subsequently, the crossing of the Tg2576 with the hPS1^{M146L} transgenic mouse resulted in a novel mouse line (PS/APP) that showed greatly accelerated plaque pathology and concomitant gliosis.[36] For example, in contrast to the 9- to 12-month onset of plaque deposition seen in the Tg2576 line, by 3 months of age thioflavin-S-positive plaques are invariably detected in the cingulate cortex of PS/APP mice. By 6–8 months of age, amyloid deposition and plaque size increased and extended to most cortical regions as well as the hippocampus. Beyond this time point, Aβ deposition appeared in the thalamus, striatum, and occasionally the cerebellum. Despite the extensive amyloid pathology by 12 months of age (an approximately 30% amyloid burden in the frontal cortex), no clear evidence for neurodegeneration in either frontal cortex or CA1 hippocampus was evident.[37] In common with all the aforementioned APP transgenic mouse models, neurofibrillary tangles have not been identified in the PS/APP mouse at any time. The behavioral phenotype of the PS/APP mouse has been described: a reduced Y-maze alternation in age-matched (12–14 week) PS/APP and Tg2576 mice compared to wild-type controls, a finding that clearly suggests no relationship to Aβ pathology. A subsequent study comparing hPS1^{M146L}, Tg2576, and PS/APP mice in cognitive and sensorimotor tests at 3, 6, and 9 months of age[38] showed Y-maze alternation was at chance levels in the PS/APP mice at each time point, and there was evidence of hyperactivity, since total arm entries were increased. Both of these behaviors were suggestive of impaired hippocampal function, although curiously, water maze performance (acquisition and 24 h probe test) was similar in all groups at the 6- and 9-month time points. In later studies, the clearest cognitive deficits emerged at 15–17 months of age, a time point at which there was intense amyloid deposition and signs of local oxidative stress and neuroinflammation.[39] This age of onset seems later than that described for the PDAPP and Tg2576 lines, which display less amyloid pathology, although an inverse correlation between cognitive performance and Aβ plaque load (frontal cortex and hippocampus) has been described.[40]

TRANSGENIC LINES ON THE BASES OF THE TAU CASCADE

In FAD, no mutations have been identified in the tau gene. However, exonic and intronic mutations were identified in the tau gene that were linked to frontotemporal dementia and

Parkinsonism (FTDP-17), a familial dementia related to AD. In comparison to the various lines of APP transgenic mice, relatively few tau-based transgenic mice have been reported to date. The transgene contains the most common tau mutation associated with the FTDP-17 P301L mutation under the control of the mouse or hamster prion protein promoter.

JNPL3 (TAU) MOUSE Hemizygous JNPL3 mice express the transgenic tau at levels equivalent to endogenous tau, while homozygous mice express double this amount. From 6 to 7 months, a neurological phenotype becomes manifest in hemizygous JNPL3 mice, consisting of a delayed righting reflex, hypolocomotion, and muscular weakness;[41] within 4 weeks of symptom onset, the mice become moribund. In terms of pathology, in contrast to APP transgenic mice, the JNPL3 mice develop neurofibrillary tangle (NFT) pathology, confirmed by tau immunostaining, silver staining, and electron microscopy (EM). No amyloid plaques have been identified. Astrogliosis also becomes evident in regions containing NFTs, which include the spinal cord and pons, and to a lesser extent the amygdala, cortex, and hippocampus.[41,42] Indeed, spinal cord pathology seemed to correlate with neurological phenotype. Not surprisingly, given the severity of this phenotype, no cognitive data have yet been described for the JNPL3 mouse. The TAPP mouse, a combination of mutant tau and APP, also supports the view of APP mismetabolism as the central event in the disease cascade. The JNPL3 mice develop relatively little NFT pathology in limbic areas; however, in age-matched TAPP mice, limbic NFTs are increased 7-fold.[42] Furthermore, Götz *et al.*[43] reported that discrete CNS microinjections of Aβ peptide into a mutant tau line[44] resulted in a 5-fold increase in NFTs in areas innervated by neurons originating from the injection site. Consequently, these studies support the view that APP/Aβ overexpression promotes NFT formation.

TGTAU^{P301L} MOUSE This transgenic line (TgTau^{P301L}) expresses a frontotemporal dementia with Parkinsonism linked to chromosome 17 (FTPD-17) mutations within the longest form of tau (2N, 4R) under the hamster prion protein promoter. TgTau^{P301L} mice developed florid pathology including neuronal pretangles, numerous Gallyas-Braak-positive neurofibrillary tangles, and glial fibrillary tangles in the frontotemporal areas of the cerebrum, in the brainstem, and to a lesser extent in the spinal cord.[45] These features were accompanied by gliosis, neuronal loss, and cerebral atrophy. Accumulated tau was hyperphosphorylated, conformationally changed, ubiquitinated, and Fn1 sarkosyl-insoluble, with EM demonstrating wavy filaments. Aged TgTau^{P301L} mice exhibited impairment in hippocampally dependent and independent behavioral paradigms, with impairments closely related to the presence of tau pathologies and levels of insoluble tau protein. These TgTau^{P301L} mice recreate the substantial phenotypic variation and spectrum of pathologies seen in FTDP-17 patients. Identification of genetic and/or environmental factors modifying the tau phenotype in these mice may shed light on factors modulating human tauopathies. These transgenic mice may aid therapeutic development for FTDP-17 and other diseases featuring accumulations of four-repeat tau, such as Alzheimer's disease, corticobasal degeneration, and progressive supranuclear palsy. These mice showed neurofibrillary tangles at 1 year of age that were similar to those seen in a 76-year-old human AD brain (Figure 22–3).

TRIPLE TRANSGENIC MOUSE To study the interaction between Aβ and tau and their effect on synaptic function, a triple transgenic model harboring the PS1^{M146V}, APP^{swe}, and Tau^{P301L} transgene was developed. Rather than crossing independent lines, two transgenes DNA was microinjected into single cell embryos from homozygous PS1^{M146V} knockin mice, resulting in a triple transgenic mouse with the same genetic background.[46] These triple transgenic mice progressively developed plaques and tangles. Intraneuronal Aβ immunoreactivity is first detected in neocortical regions and subsequently in CA1 pyramidal neurons. Extracellular amyloid plaques first become apparent at 6 months in the frontal cortex but there was no tau immunoreactivity, and were readily available by 12 months in the hippocampus and other regions. Synaptic dysfunction and phosphorylated and hyperphosphorylated tau were revealed in age-related progression between 12 to 15 months of age, but PHF1 did not become evident until the mice were 18 months of age. Tau-reactive dystrophic neurites were also evident surrounding globular structures in 18-month older brains. The triple transgenic mice develop a progressive and age-dependent Aβ and tau pathology, although Aβ deposits manifest prior to tangle formation. It is likely that Aβ pathology affects the development of tau pathology including LTP deficits in an age-related manner but before plaque and tangle pathology.

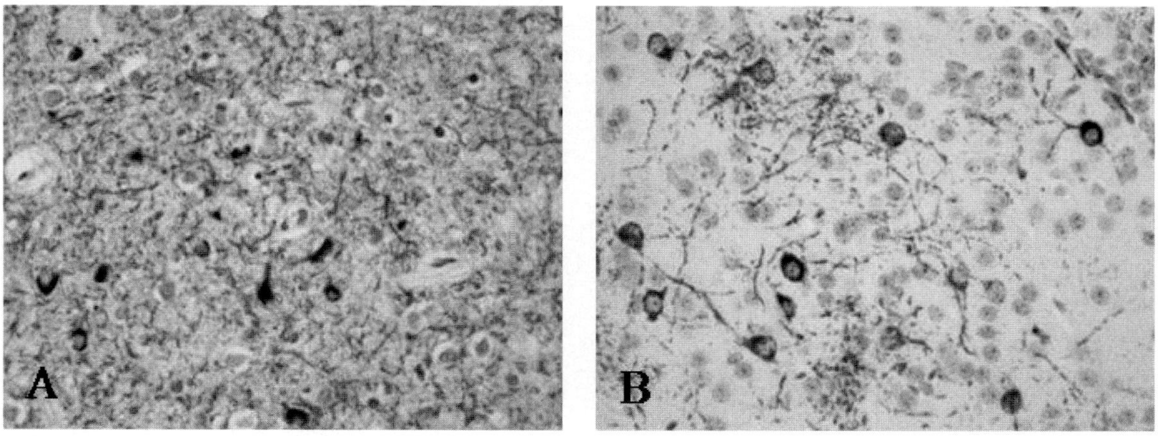

Figure 22–3. Alzheimer's disease brain reveals an intracellular accumulation of hyperphosphorylated tau. (A) Human AD brain (76 years old) cortical neurons showing neurofibrillary tangles. (B) A tau^{P301L} transgenic mouse (1 year old) showed similar neurofibrillary tangles in a cortical section immunolabeled with AT8 monoclonal antibodies.

TRANSGENIC MICE WEB SITES

The Alzheimer Research Forum website (http://www.alzaforum.org) contains important up-to-date information on the numerous strains of mice used in biomedical research related to Alzheimer's disease. Information is presented in chart format including the name or symbol, strain name, transgene or promoter and regulatory elements, genetic background, behavioral phenotype, neurological characteristics, availability, and primary citations for each of the mice strains related to APP, APOE, α-synuclein, Cox-2, PS1, PS2, tau, Double-Cross, and Triple-Cross. Furthermore, Jackson Laboratory (http://www.jax.org) data sheets on available strains contain essential information on the genetics, phenotypes, husbandry, and production; the same applies to Taconic (http://www.taconic.com), which provides data sheets on numerous available strains on AD. Transgenic animals require conventional care and do not need any special care.

THERAPEUTIC APPROACHES

The most widely replicated and thoroughly investigated approaches in these preclinical studies have involved the use of anti-Aβ immunotherapies[47] using AD transgenic mice. In 1999, the Schenk group described the preventive effects of AD neuropathology associated with immunizing the PDAPP transgenic mouse with Aβ peptide; this reduced amyloid plaques and preserved synaptophysin levels, important correlations to human AD.[48] Further studies in our laboratory and by other groups demonstrated a reduction of amyloid plaques and improvement of cognitive functions in a water maze test in hβAPP and PS1 transgenic mice. Passive immunization studies in which Aβ antibodies were given for short-term treatment resulted in positive cognitive changes in transgenic mice. The possible mechanism of action of these immunotherapies is still not clear, but antibodies may be capable of clearing plaques through both a monocyte-mediated and a direct effect on the clearance of soluble Aβ species that are neuroactive/toxic.[49] Immunization with Aβ1-42 peptide (AN-1792) was tested in human AD patients with noteworthy effects. These clinical trials were suspended due to the development of meningioencephalitis in a few cases.[50] These side effects highlight a limitation of animal models, which may not perfectly predict clinical outcomes due to limitations in the models' pathology as well as the more genetically varied human population. Interestingly, several cases from these studies have gone to autopsy and showed a strikingly similar reduction in AD neuropathologies as predicted in animal models. Recently, the cerebrospinal fluid biomarker tau showed an encouraging improvement of cognitive function and a robust reduction in pathology and passive immunization of Aβ monoclonal antibodies used in the original trials to avoid inflammatory responses. In a recent study, nonviral Aβ DNA vaccines against Alzheimer's disease were administered to model (APP23) mice and the reduction in Aβ burden was

Table 22–2
Therapeutic approaches to reduce the pathology of Alzheimer's disease (AD) in transgenic mouse models[a]

Test compound	*Mechanism of action*	*Dose and route*	*Duration*	*AD model*	*Age/amyloid burden at start of therapy*	*Reduction in plaque burden*	*Reference*
Curcumin	Amyloid binding antioxidant in curry powder	160 and 5000 ppm in chow	6 months	Tg2576	10 months with few plaques	At low dose more than 50%	53
iAβ5p	β-Sheet disrupting peptide	1.0 mg intravenously 3 times/week	8 weeks	APPV717I × PS1A246E	9 months (3 months after plaques start to appear)	45% in cortex and 29% in hippocampus	54
Lithium	GSK3ß kinase inhibition	2.4 g/kg in chow	7 months	PDAPP	1 month old before plaques	More than 75%	55
CP-113,818	ACAT inhibitor	7.2 mg/kg/day 2 month subcutaneous implant	2 months	APPSwedish + London	Plaques are just beginning	More than 90%	56
Clioquinol	Copper-zinc chelator	30 mg/kg/day orally	9 weeks	Tg2576	21 months of age; well after plaques start	49%	57
NCX-2216	NO-releasing NSAID; microglial activator	375 ppm in chow	5 months	APPSwedish + PS1^{M146L}	7 months; plaque formation is evident at this age	45%	58
Cyclohexanehexol	Amyloid peptide inhibitor	0.3–30 mg/kg/day oral gavage twice daily	6 months and 1 month	TgCRND8	6 weeks (just before plaques) to 6 months and 5 months full plaques	30–55%	59

[a]Therapeutic agents showed efficacy in reducing, at a minimum, amyloid plaque load after chronic treatment (more than 2 weeks). Although there are different mechanisms of action, all have reported efficacy.[52]

evaluated. Prophylactic treatments started before Aβ deposition reduced Aβ burden to 15.5% and 38.5% of that found in untreated mice at 7 and 18 months of age, respectively. Therapeutic treatment started after Aβ deposition reduced Aβ burden to approximately 50% at the age of 18 months. Importantly, this therapy induced neither neuroinflammation nor T cell responses to Aβ peptide in both APP23 and wild-type B6 mice, even after long-term vaccination. Although it is reported that other anti-Aβ therapies have pharmacological and/or technical difficulties, nonviral DNA vaccines are highly secure and easily controllable and are promising for the treatment of Alzheimer's disease.[51]

A wide range of therapies in addition to immunotherapy has been tested in several different AD transgenic models. Small molecules are orally bioavailable and have shown effectiveness by approaches ranging from natural products that bind to Aβ, metal chelation, lipid metabolism inhibitor, and inflammatory modulators. Preliminary results from these approaches involving more than 2 weeks of chronic treatment using transgenic models are summarized in Table 22–2.[52] The source of these compounds ranges from curry spice to drugs in clinical trials for other indications. These studies vary widely in the choice of model, duration of therapy, and, most importantly, the age/amyloid burden present when therapy was initiated. The great variability in paradigms excludes head-to-head efficacy assessments and predictions. Our experience with immunotherapy suggests that preventing amyloidosis may be easier than reducing existing burdens of older plaques. It is interesting to note in Table 22–2 that none of the tested compounds has the same proposed mechanisms of action or direct inhibition of secretases; this indicates that potential amyloid-directed therapies are quite diverse and effective therapies validated clinically will emerge in the near future.

ACKNOWLEDGMENTS

We thank Drs. Dora Games, Guy A. Higgins, and David Westaway for their kind permission to use tables and for support and encouragement in completing this chapter.

REFERENCES

1. Berg L, Morris JC. Diagnosis. In: Terry RD, Katzman R, Bick KL, Eds. *Alzheimer Disease*. New York: Raven Press, 1994:197–229.
2. Citron M, Oltersdorf T, Haass C, McConlogue C, Hung AY, Seubert P, Vigopelfrey C, Lieberburg I, Selkoe DJ. Mutation of the beta-amyloid precursor protein in familial Alzheimer's disease increases beta-protein production. *Nature* 1992;360:672–674.
3. Scheuner D, Eckman JM, Song X, Citron M, Suzuki N, Bird TD, Hardy J, Hutton M, Kukull W, Larson E, Levy-Hahad E, Viitanen M, Peskind E, Poorkaj P, Schllenberg G, Tanzi R, Wasco W, Lannfelt L, Selkoe D, Younkin S. Secreted amyloid beta-protein similar to that in the senile plaques of Alzheimer's disease is increased in vivo by the presenilin 1 and 2 and APP mutations linked to familial Alzheimer's disease. *Nat Med* 1996;2:864–870.
4. Duff, K, Eckman C, Zehr C, Yu X, Prada, CM, Perez-tur J, Hutton M, Buee L, Harigaya Y, Yager D, Morgan D, Gordon M, Holcomb L, Refolo L, Zenk B, Hardy J, Younkin S. Increased amyloid-beta42(43) in brains of mice expressing mutant presenilin 1. *Nature* 1996;383:710–713.
5. Borchelt DR, Thinakaran G, Eckman CB, Lee MK, Davenport F, Ratovitsky T, Parda CM, Kim G, Seekins S, Yager D, Slunt HH, Wang R, Seeger M, Levey AI, Gandy SE, Copeland NG, Jenkins NA, Price DL, Younkin SG, Sisodia SS. Familial Alzheimer's disease-linked presenilin 1 variants elevate Abeta1-42/1-40 ratio in vitro and in vivo. *Neuron* 1996;17:1005–1013.
6. Citron M, Westaway D, Zia W, Carlson GA, Diehl T, Levesque G, Johnson-Wood K, Lee M, Seubert P, Davis A, Kholodenko D, Motter R, Sherrington R, Perry B, Yao H, Strome R, Lieberburg I, Rommens J, Kim S, Schenk D, Fraser P, St George-Hyslop P, Selkoe D. Mutant presenilins of Alzheimer's disease increase production of 42-residue amyloid beta-protein in both transfected cells and transgenic mice. *Nat Med* 1997;3:67–72.
7. DeStrooper B, Saftig P, Craessaerts K, Vanderstichele H, Guhde G, Annaert W, Von Figura K, Van Leuven F. Deficiency of presenilin-1 inhibits the normal cleavage of amyloid precursor protein. *Nature* 1998;391:387–390.
8. DeStrooper B, Annaert W, Cupers P, Saftig P, Craessaerts K, Mumm JS, Schroeter EH, Schrijvers V, Wolfe MS, Ray WJ, Goate A, Kopan R. A presenilin-1-dependent gamma-secretase-like protease mediates release of Notch intracellular domain. *Nature* 1999;398(6767): 518–522.
9. Chen F, Yang DS, Petanceska S, Yang A, Tandon A, Yu G, Rozmahel R, Ghiso J, Nishimura M, Zhang DM, Kawarai T, Levesque G, Mills J, Levesque L, Song YQ, Rogaeva E, Westaway D, Mount H, Gandy S, St George-Hyslop P, Fraser PE. Carboxyl-terminal fragments of Alzheimer beta-amyloid precursor protein accumulate in restricted and unpredicted intracellular compartments in presenilin 1-deficient cells. *J Biol Chem* 2000;275(47):36794–36802.
10. Ertekin-Taner N, Graff-Radford N, Younkin LH, Eckman C, Baker M, Adamson J, Ronald J, Blangero J, Hutton M, Younkin SG. Linkage of plasma Abeta42 to a quantitative locus on chromosome 10 in late-onset Alzheimer's disease pedigrees. *Science* 2000;290(5500): 2303–2304.
11. Corder EH, Saunders AM, Stittmatter WJ, Schmechel D, Gaskell P, Small GW, Roses AD, Haines JL, Pericak-Vance MA. Gene dose of apolipoprotein E type 4 allele and the risk of Alzheimer's disease in late onset families. *Science* 1993;261:921–923.
12. Strittmatter WJ, Saunders AM, Schmechel D, Goldgaber D, Roses AD. Apolipoprotein E: High-avidity binding to beta-amyloid and increased frequency of type 4 allele in late-onset familial Alzheimer disease. *Proc Natl Acad Sci USA* 1993;90:1977–1981.
13. Holtzman DM, Bales KR, Tenkova T, Fagan AM, Parsadanian M, Sartorius LLJ, Mackey B, Olney J, McKeel D, Wozniak D, Paul SM. Apolipoprotein E isoform-dependent amyloid deposition and neuritic degeneration in a mouse model of Alzheimer's disease. *Proc Natl Acad Sci USA* 2000;97(6):2892–2897.
14. Poirier J. Apolipoprotein E and Alzheimer's disease. A role in amyloid catabolism. *Ann NY Acad Sci* 2000;924:81–90.
15. Hardy JA, Higgins GA. Alzheimer's disease: The amyloid cascade hypothesis. *Science* 1992;256:184–185.
16. Arriagada PV, Growdon JH, Hedley-Whyte ET, Hyman BT. Distribution of Alzheimer-type pathologic changes in nondemented elderly individuals matches the pattern in Alzheimer's disease. *Neurology* 1992;42:631–639.
17. Beach TG, Kuo YM, Spiegel K, Emmerling MR, Sue LI, Kokjohn K, Roher AE. The cholinergic deficit coincides with Abeta deposition at the earliest histopathologic stages of Alzheimer disease. *J Neuropathol Exp Neurol* 2000;59:308–313.
18. Jarrett JT, Berger EP, Lansbury PT. The carboxy terminus of the beta amyloid protein is critical for the seeding of amyloid formation: Implications for the pathogenesis of Alzheimer's disease. *Biochemistry* 1993;32:4693–4697.
19. Stine WB Jr, Snyder SW, Ladror US, Wade WS, Miller MF, Perun TJ, Holzman TF, Krafft GA. The nanometer-scale structure of amyloid-beta visualized by atomic force microscopy. *J Protein Chem* 1996;15:192–203.
20. McLean CA, Cherny RA, Fraser FW, Fuller SJ, Smith MJ, Beyreuther K, Bush AI, Masters CL. Soluble pool of Abeta amyloid as a determinant of severity of neurodegeneration in Alzheimer's disease. *Ann Neurol* 1999;46:860–866.
21. Wang J, Dickson DW, Trojanowski JQ, Lee VM. The levels of soluble versus insoluble brain Abeta distinguish Alzheimer's disease from normal and pathologic aging. *Exp Neurol* 1999;158:328–337.

22. Lue LF, Juo YM, Roher AE, Brachova L, Shen Y, Sue L, Beach T, Kurth JH, Ryder RE, Rogers J. Soluble amyloid beta peptide concentration as a predictor of synaptic change in Alzheimer's disease. *Am J Pathol* 1999;155:853–862.
23. Teller JK, Russo C, DeBusk LM, Angelini G, Zaccheo D, Dagna-Bricarelli F, Scartezzini P, Bertolini S, Mann DM, Tabaton M, Gambetti P. Presence of soluble amyloid beta-peptide precedes amyloid plaque formation in Down's syndrome. *Nat Med* 1996;2: 93–95.
24. Goedert M, Crowther RA, Spillantini MG. Tau mutations cause frontotemporal dementias. *Neuron* 1998;21:955–958.
25. Lewis J, McGowan E, Rockwood J, Melrose H, Nacharaju P, Van Slegtenhorst M, Gwinn-Hardy KP, Murphy M, Baker M, Yu X, Duff K, Hardy J, Corral A, Lin WL, Yen SH, Dickson DW, Davies P, Hutton M. Neurofibrillary tangles, amyotrophy and progressive motor disturbance in mice expressing mutant (P301L) tau protein. *Nat Genet* 2000;25:402–405.
26. Selkoe DJ. Alzheimer's disease: Genes, proteins, and therapy. *Physiol Rev* 2001;81:741–766.
27. Braak H, Braak E. Neuropathological stageing of Alzheimer-related changes. *Acta Neuropathol (Berl)* 1991;82:239–259.
28. Games D, Adams D, Alessandrini R, Barbous R, Berthelette P, Blackwell C, *et al.* Alzheimer-type neuropathology in transgenic mice overexpressing V717F [beta]-amyloid precursor protein. *Nature* 1995;373:523–527.
29. Hsaio K, Chapman P, Nilsen S, Eckman C, Harigaya Y, Younkin S, *et al.* Correlative memory deficits, A[beta] elevation, and amyloid plaques in transgenic mice. *Science* 1996;274:99–102.
30. Sturchler-Pierrat C, Abramowski D, Duke M, Wiederhold K-H, Mistl C, *et al.* Two amyloid precursor protein transgenic mouse models with Alzheimer disease-like pathology. *Proc Natl Acad Sci USA* 1997;94:13287–13292.
31. Boncristiano S, Calhoun ME, Kelly PH, Pfeifer M, Bondolfi L, Stadler M, *et al.* Cholinergic changes in the APP23 transgenic mouse model of cerebral amyloidosis. *J Neurosci* 2002;22:3234–3243.
32. Kelly PH, Bondolfi L, Hunziker D, Schlecht H-P, Carver K, Maguire E, *et al.* Progressive age-related impairment of cognitive behavior in APP23 transgenic mice. *Neurobiol Aging* 2003;24:365–378.
33. Chishti MA, Yang D-S, Janus C, Phinney AL, Horne P, Pearson J, *et al.* Early-onset amyloid deposition and cognitive deficits in transgenic mice expressing a double mutant form of amyloid precursor protein 695. *J Biol Chem* 2001;276:21562–21570.
34. Janus C, Pearson J, McLaurin J, Matthews PM, Jiang Y, Schmidt SD, *et al.* A[beta] peptide immunization reduces behavioural impairment and plaques in a model of Alzheimer's disease. *Nature* 2000; 408:979–982.
35. Duff K, Eckman C, Zehr C, Yu X, Prada C-H, Perez-Tur J, *et al.* Increased amyloid-[beta]42(43) in brains of mice expressing mutant presenilin 1. *Nature* 1996;383:710–713.
36. Holcomb L, Gordon MN, McGowan E, Yu X, Benkovic S, Jantzen P, *et al.* Accelerated Alzheimer-type phenotype in transgenic mice carrying both mutant amyloid precursor protein and presenilin 1 transgenes. *Nat Med* 1998;4:97–100.
37. Urbanc B, Cruz L, Le R, Sanders J, Hsaio Ashe K, *et al.* Neurotoxic effects of thioflavin S-positive amyloid deposits in transgenic mice and Alzheimer's disease. *Proc Natl Acad Sci USA* 2002;99:13990–13995.
38. Holcomb L, Gordon MN, Jantzen P, Hsaio K, Duff K, Morgan D. Behavioral changes in transgenic mice expressing both amyloid precursor protein and presenilin-1 mutations: Lack of association with amyloid deposits. *Behav Genet* 1999;29:177–185.
39. Jantzen PT, Connor KE, DiCarlo G, Wenk GL, Wallace JL, Rojiani AM, *et al.* Microglial activation and [beta]-amyloid deposit reduction caused by a nitric oxide-releasing non-steroidal anti-inflammatory drug in amyloid precursor protein plus presenilin-1 transgenic mice. *J Neurosci* 2002;22:2246–2254.
40. Gordon MN, King DL, Diamond DM, Jantzen PT, Boyett KV, Hope CE, *et al.* Correlation between cognitive deficits and A[beta] deposits in transgenic APP+PS1 mice. *Neurobiol Aging* 2001;22: 377–386.
41. Lewis J, McGowan E, Rockwood J, Melrose H, Nacharaju P, *et al.* Neurofibrillary tangles, amyotrophy and progressive motor disturbance in mice expressing mutant (P301L) tau protein. *Nat Genet* 2000;25:402–405.
42. Lewis J, Dickson DW, Lin W-L, Chisholm L, Corral A, Jones G, *et al.* Enhanced neurofibrillary degeneration in transgenic mice expressing mutant tau and APP. *Science* 2001;293:1487–1491.
43. Götz J, Chen F, Van Dorpe J, Nitsch RM. Formation of neurofibrillary tangles in P301L tau transgenic mice induced by Aβ42 fibrils. *Science* 2001;293:1491–1495.
44. Götz J, Chen F, Barmettler R, Nitsch RM. Tau filament formation in transgenic mice expressing P301L tau. *J Biol Chem* 2001;276: 529–534.
45. Murakami T, Paitel E, Kawarabayashi T, Ikeda M, Chishti MA, Janus C, Matsubara E, Sasaki A, Kawarai T, Phinney AL, Harigaya Y, Horne P, Egashira N, Mishima K, Hanna A, Yang J, Iwasaki K, Takahashi M, Fujiwara M, Ishiguro K, Bergeron C, Carlson GA, Abe K, Westaway D, St George-Hyslop P, Shoji M. Cortical neuronal and glial pathology in TgTauP301L transgenic mice: Neuronal degeneration, memory disturbance, and phenotypic variation. *Am J Pathol* 2006;169:1365–1375.
46. Oddo S, Caccamo A, Shepherd JD, Murphy MP, Golde TE, Kayed R, Metherate R, Mattson MP, Akbari Y, LaFerla FM. Triple-transgenic model of Alzheimer's disease with plaques and tangles: Intracellular Abeta and synaptic dysfunction. *Neuron* 2003;39:409–421.
47. Gelinas DS, DaSilva K, Fenili D, St George-Hyslop P, McLaurin J. Immunotherapy for Alzheimer's disease. *Proc Natl Acad Sci USA* 2004;101(Suppl. 2):14657–14662.
48. Schenk D, Barbour R, Dunn W, Gordon G, Grajeda H, Guido T, Hu K, Huang J, Johnson-Wood K, Khan K, Kholodenko D, Lee M, Liao Z, Lieberburg I, Motter R, Mutter L, Soriano F, Shopp G, Vasquez N, Vandevert C, Walker S, Wogulis M, Yednock T, Games D, Seubert P. Immunization with amyloid-beta attenuates Alzheimer-disease-like pathology in the PDAPP mouse. *Nature* 1999;8:173–177.
49. Lambert MP, Viola KL, Chromy BA, Chang L, Morgan TE, Yu J, Venton DL, Krafft GA, Finch CE, Klein WL. Vaccination with soluble Abeta oligomers generates toxicity-neutralizing antibodies. *J Neurochem* 2001;79:595–605.
50. Orgogozo JM, Gilman S, Dartigues JF, Laurent B, Puel M, Kirby LC, Jouanny P, Dubois B, Eisner L, Flitman S, Michel BF, Boada M, Frank A, Hock C. Subacute meningoencephalitis in a subset of patients with AD after Abeta42 immunization. *Neurology* 2003; 8:46–54.
51. Okura Y, Miyakoshi A, Kohyama K, Park IK, Staufenbiel M, Matsumoto Y. Nonviral Abeta DNA vaccine therapy against Alzheimer's disease: Long-term effects and safety. *Proc Natl Acad Sci USA* 2006;20:9619–9624.
52. Games D, Buttini M, Kobayashi D, Schenk D, Seubert P. Mice as models: Transgenic approaches and Alzheimer's disease. *J Alzheimers Dis* 2006;9:133–149.
53. Lim GP, Chu T, Yang F, Beech W, Frautschy SA, Cole GM. The curry spice curcumin reduces oxidative damage and amyloid pathology in an Alzheimer transgenic mouse. *J Neurosci* 2001;21:8370–8377.
54. Permanne B, Adessi C, Saborio GP, Fraga S, Frossard MJ, Van Dorpe J, Dewachter I, Banks WA, Van Leuven F, Soto C. Reduction of amyloid load and cerebral damage in a transgenic mouse model of Alzheimer's disease by treatment with a beta-sheet breaker peptide. *FASEB J* 2002;16:860–862.
55. Su Y, Ryder J, Li B, Wu X, Fox N, Solenberg P, Brune K, Paul S, Zhou Y, Liu F, Ni B. Lithium, a common drug for bipolar disorder treatment, regulates amyloid-beta precursor protein processing. *Biochemistry* 2004;43:6899–6908.
56. Hutter-Paier B, Huttunen HJ, Puglielli L, Eckman CB, Kim DY, Hofmeister A, Moir RD, Domnitz SB, Frosch MP, Windisch M, Kovacs DM. The ACAT inhibitor CP-113,818 markedly reduces amyloid pathology in a mouse model of Alzheimer's disease. *Neuron* 2004;44:227–238.
57. Cherny RA, Atwood CS, Xilinas ME, Gray DN, Jones WD, McLean CA, Barnham KJ, Volitakis I, Fraser FW, Kim Y, Huang X, Goldstein

LE, Moir RD, Lim JT, Beyreuther K, Zheng H, Tanzi RE, Masters CL, Bush AI. Treatment with a copper-zinc chelator markedly and rapidly inhibits beta-amyloid accumulation in Alzheimer's disease transgenic mice. *Neuron* 2001;30:665–676.

58. Jantzen PT, Connor KE, DiCarlo G, Wenk GL, Wallace JL, Rojiani AM, Coppola D, Morgan D, Gordon MN. Microglial activation and beta-amyloid deposit reduction caused by a nitric oxide-releasing nonsteroidal anti-inflammatory drug in amyloid precursor protein plus presenilin-1 transgenic mice. *J Neurosci* 2002;22:2246–2254.

59. McLaurin J, Kierstead ME, Brown ME, Hawkes CA, Lambermon MH, Phinney AL, Darabie AA, Cousins JE, French JE, Lan MF, Chen F, Wong SS, Mount HT, Fraser PE, Westaway D, St George-Hyslop P. Cyclohexanehexol inhibitors of Abeta aggregation prevent and reverse Alzheimer phenotype in a mouse model. *Nat Med* 2006; 12:801–808.

23 Guinea Pigs as Models for Human Cholesterol and Lipoprotein Metabolism

MARIA LUZ FERNANDEZ AND RICHARD J. WOOD

ABSTRACT

Lipoprotein anabolism and catabolism are highly complex when considering the multifaceted aspects that regulate the synthesis and removal of lipoproteins from plasma. In addition, lipoproteins undergo multiple modifications in the intravascular compartment leading to atherogenic or antiatherogenic subclasses. The use of an appropriate animal model mimicking human lipoprotein metabolism is critical to the understanding of how dietary interventions, drug treatment, sex, menopausal state, and exercise regulate these intricate processes. This chapter will discuss in detail the guinea pig as an appropriate animal model to determine the effects of different diets, lipid-lowering drugs, and life style modifications on hepatic and plasma cholesterol metabolism, and atherosclerosis. One of the most salient features of this animal model is that guinea pigs carry the majority of the cholesterol in low-density lipoprotein and possess cholesterol ester transfer protein and lipoprotein lipase activities, which result in reverse cholesterol transport and delipidation cascades equivalent to the human situation. Furthermore, the activities of hepatic enzymes involved in cholesterol metabolism and the mechanisms involved in secretion and removal of lipoproteins mimic those of humans. The similarity in plasma lipid responses to dietary interventions, drug treatment, and exercise between guinea pigs and humans further suggests the relevance of this model to elucidate the mechanisms involved.

Key Words: Guinea pigs, Animal models, Lipoprotein metabolism, Atherosclerosis, Dietary interventions, Drug therapy, Sex, Exercise.

INTRODUCTION

This chapter will provide a useful insight into the suitability of guinea pigs as animal models for cholesterol and lipoprotein metabolism. A discussion of the metabolic advantages plus the usefulness of this model to elucidate mechanisms by which dietary factors, drug therapy, exercise, gender, and menopause alter the regulatory pathways involved in the synthesis, catabolism, and intravascular processing of lipoproteins is presented.

There is some controversy as to whether guinea pigs (*Cavia porcellus*) should be classified as rodents.[1] While this issue is not relevant to the present discussion, there is one aspect of guinea pigs that makes them stand out from other rodents, the fact that they carry the majority of cholesterol in low-density lipoprotein (LDL).[2] This outstanding difference raises the question of whether guinea pigs present other similarities to humans in cholesterol and lipoprotein metabolism. Our laboratory and other investigators have found that indeed guinea pigs do have some analogies to human cholesterol metabolism that merit discussion.

USE OF ANIMAL MODELS FOR CHOLESTEROL AND LIPOPROTEIN METABOLISM

In addition to the guinea pigs, several animal models have been used to study lipoprotein metabolism and atherosclerosis including nonhuman primates, hamsters, mice, rats, and rabbits. Choosing the appropriate model to test a hypothesis is an important decision that should be based on the benefits and limitations of each particular model. In this section of the chapter we outline the advantages and disadvantages of several commonly used animal models in cholesterol and lipoprotein metabolism research. Advantages and disadvantages will be based on similarities and dissimilarities between the particular animal models and humans.

ADVANTAGES AND DISADVANTAGES OF OTHER MODELS

Nonhuman Primates Several types of nonhuman primates have been used in lipoprotein metabolism research including cebus monkeys,[3] squirrel monkeys,[4] baboons,[5] and owl monkeys.[6] One commonly used model for lipoprotein metabolism is the African green monkey, which is known to develop atherosclerosis in a fashion similar to humans with respect to morphology and cytology.[7] Like humans, African green monkeys express only apolipoprotein (apo) B 100 in the liver. A high cholesterol diet leads to a reduction in hepatic LDL receptor mRNA and down-regulation of cholesterol 7α-hydroxylase activity.[8] Feeding African green monkeys an atherogenic diet high in saturated fat led to the enrichment of LDL particles with cholesteryl oleate and predicted the magnitude of atherosclerotic development.[9,10] Atherosclerotic development appears to be less pronounced when these animals consume a diet composed largely of polyunsaturated fats (PUFA) when compared to monounsaturated fats or saturated fatty acids (SFA).[11] However, development of atherosclerosis in this animal model takes years (as opposed to weeks in other models), and these animals are expensive to obtain and maintain. Furthermore, African green monkeys tend to have higher plasma concentrations of HDL cholesterol (HDL-C) than LDL cholesterol (LDL-C), opposite to the human situation.[12]

From: *Sourcebook of Models for Biomedical Research*
(P. M. Conn, ed.), © 2008 Humana Press Inc., Totowa, NJ.

Rats There are numerous dissimilarities in lipoprotein metabolism between rats and humans. Wild-type rats do not have cholesterol ester transfer protein (CETP) activity,[13] which plays a major role in intravascular lipoprotein particle processing and lipid exchange in humans.[14] As such, CETP transgenic rats have been used.[15] Rats also have a very high activity of cholesterol 7α-hydroxylase (CYP7),[16] thus they readily eliminate excess body cholesterol. Furthermore, rats carry the majority of cholesterol in HDL particles,[17] which also differs from humans. LDL-C does not increase in rats fed saturated fat unless very high cholesterol (0.5–1.0% of diet) is fed in conjunction with cholic acid.[18] The densities of LDL and HDL particles overlap in rats, making isolation of these lipoproteins difficult. As a result of these characteristics of rat lipoprotein metabolism, rats are very resistant to atherosclerosis.

Mice Like rats, wild-type mice are not good models of human lipoprotein metabolism, thus several transgenic and knockout mice have been created. These genetic manipulations can affect the viability of these animals and increase the cost of the experiments. Wild-type mice carry the majority of cholesterol in HDL particles rather than in LDL particles.[17] Two commonly used genetically modified mouse models are the apo E-deficient and LDL receptor-deficient animals, which have been shown to respond to dietary and pharmacological interventions; however, the lipoprotein profile of these models is quite different from humans.[19] Like rats, mice respond to high dietary cholesterol by downregulating cholesterol synthesis and upregulating bile acid synthesis, which serves to maintain plasma cholesterol homeostasis. Mice also express apo B 48 in the liver,[20] which does not occur in humans.

Hamsters Significant research regarding LDL transport has been derived from the hamster model.[21] Compared to the rat, hamsters have moderate levels of expression of CYP7.[22] When fed a high cholesterol diet, hamsters develop hypertriglyceridemia, opposite to humans. Insulin resistance has been induced in hamsters that consumed a high fructose diet for 2 weeks, resulting in an increased mass of microsomal transfer protein (MTP) and increased very-low-density lipoprotein (VLDL) apo B secretion from the liver.[23] When fed a normal chow diet, hamsters carry cholesterol predominantly in HDL particles (56%) rather than in LDL particles (26%), though the majority of triglycerides circulate in VLDL.[24] The hamster is not a good model for dietary fiber since these animals have a forestomach where fiber is fermented prior to reaching the small intestine. Though fibrates function similarly in hamsters and humans,[24] hamsters are not a good model for 3-hydroxy-3-methylglutaryl coenzyme A (HMG-CoA) reductase inhibitors.[25]

Rabbits Both wild-type and transgenic rabbits are often used in cardiovascular disease research. Rabbits are susceptible to diet-induced atherosclerosis, perhaps due to their high activity of CETP. However, rabbits have almost no hepatic lipase. Like humans, the rabbit liver does not edit apolipoprotein B mRNA. Also, like hypertriglyceridemic patients, rabbits produce β-VLDL, which are large, triglyceride-laden VLDL particles characteristic of hypertriglyceridemia. Rabbits transgenic for lipoprotein (a), apolipoproteins AI, B, E2, and E3, hepatic lipase, and lecithin: cholesterol acyltransferase (LCAT) have all been used to study lipoprotein metabolism. The Watanabe Heritable Hyperlipidemic rabbits are a good model for familial hypercholesterolemia and St. Thomas' Hospital rabbits are a good model for familial combined hyperlipidemia.[26]

GUINEA PIGS

COMPARISONS OF GUINEA PIGS WITH HUMANS

Lipoprotein Metabolism In addition to carrying the majority of cholesterol in LDL, guinea pigs have other similarities in lipoprotein distribution and metabolism including a higher LDL to HDL ratio;[27] they possess plasma CETP,[28] LCAT,[29] and lipoprotein lipase (LPL)[30] activities for intravascular processing of plasma lipoproteins.

Females have higher HDL-C levels than males;[31] ovariectomized guinea pigs, a model for menopause, have elevated triglycerides (TG) and higher LDL-C similar to the plasma lipid profiles of postmenopausal women.[31] Guinea pigs during exercise lower their plasma TG and increase their plasma HDL-C.[32] Guinea pigs also respond to dietary interventions[33–35] and drug treatment[36–38] by lowering plasma LDL-C concentrations.

Hepatic Cholesterol Metabolism Compared to humans, guinea pigs have higher concentrations of free than cholesterol ester (CE) in the liver.[27,39] They also exhibit comparable moderate rates of hepatic cholesterol synthesis[27,40] and catabolism.[41,42] Similar to humans, the binding domain for the LDL receptor differentiates between normal and familial binding defective apo B-100.[43] Most importantly, apo B mRNA editing in liver is present in negligible amounts (<1%) compared to 18–70% in other species.[44] Table 23–1 illustrates the similarities of guinea pigs and humans compared to other animal models in several aspects of cholesterol and lipoprotein metabolism.

USE OF GUINEA PIGS FOR DIETARY INTERVENTIONS

DIETARY FIBER It has been well established that dietary soluble fiber is one of the major components in food, which has been associated with a decreased risk of coronary heart disease.[45] The hypocholesterolemic properties of dietary fiber are a result of the primary action of fiber in the intestinal lumen.[46] These mechanisms involve interruption of the enterohepatic circulation of bile acids and decreased absorption rate of lipids including cholesterol.[47] Based on this action of fiber, plasma LDL-C concentrations are reduced. Guinea pigs have been useful models to elucidate the secondary mechanisms associated with plasma cholesterol lowering.[48,49]

As a result of fiber's primary action in the small intestine, cholesterol delivered to the liver via the chylomicron remnant is reduced.[50] In addition, the interruption of bile acid cycling causes a major effect on hepatic cholesterol. Cholesterol hydroxylation at the alpha position is the initial and rate-limiting step in this process. This step is catalyzed by CYP7, the main regulatory enzyme in the bile acid synthesis pathway.[51] Due to these major alterations in hepatic cholesterol concentrations, CYP7 is upregulated leading to a greater depletion of the hepatic cholesterol pools. To maintain hepatic cholesterol homeostasis, the LDL receptor is upregulated to remove cholesterol from circulation leading to the decreases in LDL-C.

The expression of the LDL receptor gene in liver is regulated by a feedback mechanism involving hepatic cholesterol. When the demand for cholesterol increases, the liver cells express high levels of LDL receptor mRNA and when cholesterol accumulates

Table 23–1
Comparison of humans, guinea pigs, hamsters, mice, rabbits, and African green monkeys in parameters of lipoprotein metabolism[a]

Parameter	*Human*	*Guinea Pig*	*Hamster*	*Rat*	*Mouse*	*Rabbit*	*African Green Monkey*
LDL-C/HDL-C with no cholesterol challenge[2,8,17,22,26,57]	High	High	Moderate	Low	Low	Low	Moderate
Apo B-48 in liver[44]	No	No	ND[8]	Yes	Yes	No	No
Ratio of free to esterified cholesterol[8,22,27,39,57]	High	High	Low	Low	Low	Low	Moderate
Response to dietary cholesterol[8,22,26,62]	↑ LDL-C ↑ HDL-C	↑ LDL-C ↑ HDL-C	↑ LDL-C ↑ HDL-C ↑TG	↑ HDL-C ↑ LDL-C	↑ LDL-C ↑ HDL-C	↑ VLDL-C	↑ LDL-C
CETP activity[28]	Moderate	Moderate	Moderate	None	None	High	Low
Catabolism of cholesterol to bile acids[8,22,41,42]	Low	Low	Moderate	High	High	ND[1]	Low
Develops atherosclerosis[9,26,70]	Yes	Yes	Yes	No	No	Yes	Yes

[a]LDL-C, low-density lipoprotein C; HDL-C, high-density lipoprotein C; apo B, apolipoprotein B; CETP, cholesterol ester transfer protein; ND, not determined.

in the cell, the activity and expression of the LDL receptor are suppressed.[52] Depletion of hepatic cholesterol concentrations as a result of fiber intake leads to an upregulation of LDL receptors as demonstrated by *in vitro* measurements[50] of hepatic LDL receptor or *in vivo* determinations of LDL turnover.[46]

Dietary Fiber and Hepatic Cholesterol Metabolism In guinea pigs, significant reductions of hepatic cholesterol due to intake of different sources of soluble fiber including pectin, guar gum, psyllium,[53,54] or fiber from lime-treated corn husks have been reported.[55] Furthermore, increases in hepatic CYP7 activity and mRNA abundance have also been observed after pectin and psyllium intake,[56] which confirm the postulated mechanisms. Another important metabolic alteration is increased synthesis of cholesterol within the hepatocyte as demonstrated by higher activity of HMG-CoA reductase.[49] Such a compensatory increase in hepatic synthesis occurs when intestinal cholesterol absorption is impaired or when bile acid synthesis is stimulated. Under these conditions, acyl CoA cholesteryl:acyltransferase (ACAT) activity is downregulated[50] because the availability of free cholesterol for esterification is substantially reduced.[57,58]

Dietary Fiber and Lipoprotein Metabolism Lipoprotein metabolism is also altered by dietary fiber as a result of depleted hepatic pools. The decreases in hepatic cholesterol have been related to lower rates of hepatic apo B secretion and to a faster LDL turnover rate in guinea pigs.[46,47] In addition, the compositions of both the VLDL and LDL are significantly altered and these compositional modifications have metabolic implications.[59] Soluble fiber intake results in the formation of a large TG-enriched, CE-depleted VLDL particle, compositional changes associated with less conversion to IDL and LDL in the delipidation cascade.[6] It has also been postulated that large VLDLs are catabolized faster by the apo B/E receptor.[61] In addition, LDL fractional catabolic rates (FCR) *in vivo* are faster in guinea pigs fed soluble fiber.[2] Other important modifications that take place in the intravascular compartment are a significant decrease in CETP activity, which can be associated with the CE-depleted VLDL and LDL.[59] Lower CETP activity may contribute to the mechanisms of hypocholesterolemia attributed to soluble fiber.

The guinea pig model has been useful in elucidating the complex mechanisms by which soluble fiber lowers plasma LDL-C (Figure 23–1).

DIETARY CHOLESTEROL Guinea pigs, similar to humans, present different responses to dietary cholesterol[62] demonstrating

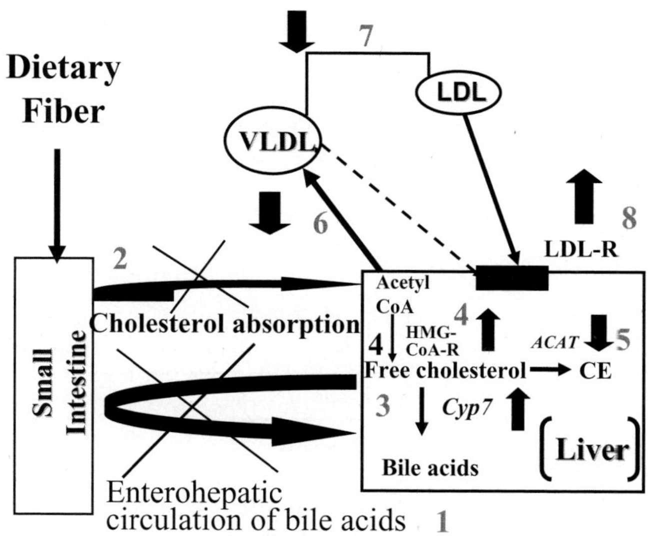

Figure 23–1. Mechanisms by which dietary fiber lowers plasma LDL cholesterol: (**1**) there is an interruption of the enterohepatic circulation of bile acids and (**2**) of cholesterol absorption in the intestinal lumen. As a result (**3**) cholesterol 7α-hydroxylase is upregulated to synthesize more bile acids. At the same time (**4**) HMG-CoA reductase activity is upregulated and (**5**) ACAT activity is reduced because of less substrate (free cholesterol). This leads to (**6**) decreased apo VLDL secretion. In addition, there is (**7**) less conversion of VLDL to LDL and (**8**) an upregulation of LDL receptors. All these mechanisms contribute to the decrease in LDL cholesterol.

that the responses are highly individualized in agreement with clinical studies.[63] Earlier reports on the deleterious effects of dietary cholesterol in guinea pigs, including hemolytic anemia characterized by accelerated destruction of the erythrocytes[64] and death before any considerable plaques have been developed, have precluded the use of guinea pigs for models of cholesterol and lipoprotein metabolism. In addition, these reports focused on the appearance of abnormal lipoproteins and changes in lipid metabolism, which were postulated to be species specific.[65,66] These studies failed to stress the fact that the amount of cholesterol provided to guinea pigs was 1–2%, which is equivalent to 7500–15,000 mg/day of cholesterol in the human situation. Thus the remarks concerning these earlier studies have to be taken with caution and the interpretation of the results viewed on the scope that these experiments cannot possibly have any clinical significance.

The effects of dietary cholesterol on plasma lipids, lipoprotein composition, hepatic LDL receptors, and HDL metabolism have been evaluated in guinea pigs fed 0.08, 0.17, or 0.33% dietary cholesterol, equivalent to 600–2500 mg cholesterol/day in the human situation.[62,67,68] These levels of dietary cholesterol correspond to an amount of absorbed cholesterol equal to one-half, one, and two times the endogenous cholesterol synthesis in guinea pigs.[38] Increasing the concentration of dietary cholesterol resulted in a dose-dependent increase in plasma cholesterol associated with the LDL fraction, which was independent of the dietary fat. HMG-CoA reductase activity was significantly downregulated with an absorbed amount equivalent to one-half times the endogenous cholesterol synthesis (0.08% dietary cholesterol) as a first compensatory mechanism. The amount of cholesterol in liver was also increased in a dose-dependent manner with higher levels of dietary cholesterol. These increases in hepatic cholesterol paralleled the increases in ACAT activity.[69] Hepatic LDL receptor numbers were measured by incubating increasing concentrations of ^{125}LDL with hepatic membranes and the number of receptors was decreased as the amount of dietary cholesterol increased.[67] These studies demonstrate the different compensatory mechanisms of the body when there is a challenge of dietary cholesterol. The effects of different levels of dietary cholesterol on major regulators of plasma cholesterol homeostasis are presented in Table 23–2.

In addition, our laboratory has also demonstrated that guinea pigs do present early atherosclerotic development and fatty streak accumulation after 12 weeks when challenged with amounts of dietary cholesterol in the range of 2000 mg/day human consumption.[70,71]

DIETARY FAT Effects of fatty acids varying in chain length and degree of saturation on cholesterol and lipoprotein metabolism have been studied in guinea pigs. Diets high in PUFA (corn oil) (CO) lower plasma LDL-C concentrations compared to highly SFA palm kernel (PK) rich in short-chain fatty acids or SFA lard rich in long-chain fatty acids.[33,72,73] Other metabolic alterations in lipoprotein assembly and secretion were also associated with the observed changes in plasma LDL-C. For instance, guinea pigs fed the PK diet had the highest apo B secretion rate.[73] The PK group also exhibited the slowest LDL FCR *in vivo*.[72] In contrast, guinea pigs fed the CO diet had the fastest LDL FCR and the highest number of hepatic LDL receptors.[72] Intake of CO also resulted in the smallest mature VLDL particle, as measured by electron microscopy.[74]

The effects of different types of fatty acids on hepatic enzymes involved in cholesterol homeostasis have been evaluated in guinea pigs.[75] Guinea pigs fed the PK diet had the highest plasma LDL-C followed by those fed palm oil rich in palmitic acid and CO intake resulted in the lowest plasma LDL-C concentrations. HMG-CoA reductase activity varied among groups and was independent of plasma LDL-C. In contrast, ACAT activity exhibited a positive correlation with plasma LDL-C concentrations. Two possibilities may account for these correlations: (1) ACAT activity determines rates of incorporation of CE into VLDL, which is subsequently converted into LDL, and (2) ACAT activity changes in response to rates of hepatic cholesterol influx. In support of the first theory, hypercholesterolemic diets are related to the production of a greater number of larger LDL particles in African green monkeys[9] and in guinea pigs.[33] The higher number of CE molecules, incorporated into newly secreted lipoproteins, has been correlated with

Table 23–2
Effects of different doses of dietary cholesterol on plasma and hepatic lipids and parameters of lipoprotein metabolism in guinea pigs[a,b]

Parameter	*0*	*0.08%*	*0.17%*	*0.33%*
LDL-C (mg/dl)[c]	40.0 ± 10.0	54.3 ± 20.9	124.7 ± 65.2	253.8 ± 71.1
HDL-C (mg/dl)[c]	14.0 ± 1.9	17.0 ± 3.1	20.2 ± 1.2	24.0 ± 1.2
Hepatic cholesterol (mg/g)[c]	1.51 ± 0.31	2.91 ± 0.31	3.53 ± 0.31	4.73 ± 0.70
HMG-CoA reductase (pmol/min·mg)[c]	41 ± 28	21 ± 9	11 ± 3	9 ± 1
ACAT (pmol/min·mg)[d]	16.1 ± 3.1	29.3 ± 10.0	132.3 ± 58.7	218.5 ± 43.7
LDL receptor B_{max} (μg/mg)[e]	2.04 ± 0.11	1.92 ± 0.18	1.47 ± 0.24	1.32 ± 0.07

[a]LDL-C, low-density lipoprotein C; HDL-C, high-density lipoprotein C; HMG-CoA, 3-hydroxy-3-methylglutaryl coenzyme A; ACAT, acyl CoA cholesteryl:acyltransferase.
[b]Values are expressed as mean ± SD for $n = 6$–8 guinea pigs. Values in the same row with different superscripts are significantly different as determined by one-way ANOVA and the Newman-Keules post hoc test ($p < 0.01$).
[c]Adapted from Lin *et al.*[62]
[d]Adapted from Sun *et al.*[67]
[e]Adapted from Lin *et al.*[69]

increased ACAT and with a higher number of atherosclerotic lesions in African green monkeys.[51] In support of the second theory, dietary fat saturation and chain length affect LDL absolute catabolic rate.[33,34] In a steady-state condition, LDL flux equals total catabolic rate and assuming 80% of LDL uptake by the liver, guinea pigs having the greatest influx (PK-fed animals) also had the highest ACAT activity. Similar results supporting both theories have been reported for guinea pigs fed low-fat (2.5%) or high-fat (25%) diets.[76]

These studies demonstrate that the mechanisms by which different fatty acids alter lipoprotein levels and composition and possibly atherosclerotic events are related to specific effects of these fatty acids on hepatic cholesterol homeostasis, VLDL secretion, alterations in the intravascular compartment, and LDL catabolic rates. In addition, the cholesterolemic response of guinea pigs to dietary fatty acids is similar to reports from clinical studies[77] and they are in agreement with proposed mechanisms reported in human studies.[78,79]

DIETARY CARBOHYDRATES The role of differing carbohydrate types (simple vs. complex) on lipoprotein metabolism in guinea pigs has been investigated.[80] Guinea pigs were fed isocaloric diets in which 52% of the energy from carbohydrates was derived from either sucrose or starch. Plasma cholesterol concentrations were not different between groups, but triglycerides and VLDL-C levels were significantly higher in the group consuming the sucrose-based diet. Also in the sucrose group, apolipoprotein B secretion was significantly higher, and VLDL contained less triglyceride per apolipoprotein B. This indicates that the higher plasma triglycerides were likely due to an increased quantity of VLDL particles. Guinea pigs consuming a high fat diet with complex carbohydrates display a more rapid removal of VLDL from plasma and less conversion of VLDL to LDL when compared to guinea pigs fed a high-fat diet with sucrose.[81] Similar results have been shown in humans where carbohydrate type affected plasma TG.

With cardiovascular diseases as the leading cause of death in the United States and most industrialized countries, risk management interventions often focus on dietary macronutrient composition. A varying macronutrient composition is known to affect lipoprotein metabolism. Low-fat diets appear to be particularly effective at lowering LDL-C, but may adversely affect TG and HDL-C. Restricting dietary carbohydrate can reduce TG and increase HDL-C, while the effects on LDL-C appear to vary between individuals.[82] We have recently shown that the response of dietary carbohydrate restriction in guinea pigs is similar to the response in humans.

Male Hartley guinea pigs were fed one of three diets for 12 weeks: (1) low cholesterol (0.04 g/100 g) with high carbohydrate (42% of energy), (2) high cholesterol (0.25 g/10 g) with high carbohydrate (42% of energy), or (3) high cholesterol (0.25 g/100 g) with low carbohydrate (11% of energy).[83] The groups consuming high cholesterol had significantly higher VLDL-C and LDL-C than the low-cholesterol group. Using nuclear magnetic resonance, lipoprotein particle size and subclass concentrations were determined. Interestingly, the quantity of total LDL particles and the quantity of medium and small LDL particles were significantly lower in the high-cholesterol, low-carbohydrate group than in the high-cholesterol, high-carbohydrate group. Similar results have been shown in humans[84] where a 12-week carbohydrate-restricted diet decreased LDL particle atherogenicity by reducing the quantity of total, medium, and small LDL particles while increasing mean LDL particle size. While future work is needed to investigate additional facets of lipoprotein metabolism affected by macronutrient distribution in guinea pigs, it appears that LDL metabolism is similarly impacted in humans and guinea pigs with carbohydrate restriction. In addition, as discussed below, carbohydrate restriction in guinea pigs results in improvements in inflammatory markers similar to those experienced by humans.

VITAMIN C DEFICIENCY Like humans, guinea pigs require an exogenous dietary source of vitamin C. Though vitamin C is produced in the liver or kidney of most animals, guinea pigs (like humans) lack the enzyme gulonolactone oxidase, which is required to synthesize it.

Epidemiological studies have shown a negative correlation between vitamin C intake and plasma cholesterol concentration.[85] Guinea pigs are an ideal model to investigate the effects of vitamin C status on hepatic lipid metabolism considering their similarity to humans with respect to ascorbic acid requirement and lipoprotein metabolism.[86] Consuming a diet with suboptimal vitamin C has been shown to negatively affect lipoprotein metabolism in guinea pigs.[87] Guinea pigs were fed diets composed of either high SFA or PUFA, while being fed either adequate (500 mg/kg diet) or suboptimal (50 mg/kg diet) amounts of vitamin C. Guinea pigs that consumed suboptimal vitamin C had 15% lower HMG-CoA reductase activity, a 25% reduction in LDL receptor quantity, 20% higher ACAT activity, a 28% greater hepatic TG and CE content, and a higher VLDL secretion rate in comparison to the vitamin C adequate group. Furthermore, the suboptimal vitamin C group displayed 45% lower plasma HDL-C concentrations, a 40% increase in VLDL-C, a 50% increase in LDL-C, and a 30% increase in CETP activity. Guinea pigs consuming suboptimal vitamin C and higher SFA showed the most pronounced deleterious effects on lipid metabolism. A potential mechanism for the elevations in TG is based on the role of vitamin C in carnitine synthesis. Lower levels of carnitine lead to a reduced transport of fatty acids to the mitochondria for oxidation, causing hepatic triglyceride accumulation.[88] Furthermore, inadequate vitamin C status is associated with reduced hepatic cytochrome P450[89] which results in decreased CYP7 activity and subsequent decreased bile acid formation.[90]

DRUG TREATMENT

Guinea pigs have been used to study drug effects on plasma cholesterol and TG and to clarify their hypocholesterolemic mechanisms. These drugs include probucol,[37] cholestyramine,[91] ACAT inhibitors,[92] HMG-CoA reductase inhibitors such as pravastatin,[93] simvastatin,[94,95] lovastatin,[36,96] and atorvastatin,[59,60,97] apical sodium-dependent bile acid transporter (ASBT) inhibitors,[98] MTP inhibitors,[38] and rapamycin, a drug used in organ transplant patients.[99]

CHOLESTYRAMINE Cholestyramine is very effective in reducing plasma LDL-C concentrations by 55–75% in guinea pigs.[58] Witztum *et al.*[91] measured cholestyramine effects of LDL size and composition and how these affected LDL FCR. Their findings suggest that LDL from cholestyramine-treated guinea pigs, which were smaller in size, had a slower turnover in plasma than LDL derived from control animals. The lowering of plasma LDL-C by cholestyramine was due to increases in LDL receptor in treated guinea pigs, suggesting that compositional changes in LDL have profound metabolic consequences.

3-HYDROXY-3-METHYLGLUTARYL COENZYME A REDUCTASE INHIBITORS Approximately 65% of total cholesterol in the body is endogenously synthesized in extrahepatic tissues.[100] Like humans, guinea pigs have moderate rates of hepatic cholesterol synthesis,[27,101] whereas the major site of cholesterol synthesis in the rat is the liver.[102] HMG-CoA reductase, the rate-limiting enzyme for cholesterol biosynthesis, is largely regulated by cholesterol via negative feedback regulation. Inhibition of HMG-CoA reductase activity results in reductions of plasma total and LDL-C concentrations.

In agreement with humans, guinea pigs treated with reductase inhibitors exhibit significant decreases in plasma LDL-C concentrations.[93–97] Most of these reductase inhibitors have shown that an increase in LDL receptor is a major mechanism for the observed hypocholesterolimia.[36,97] Lovastatin therapy has been shown to increase hepatic LDL receptor activity in guinea pigs and the drug induced compositional changes in the LDL particle, which had significant effects on removal of LDL from plasma.[36] LDL from lovastatin-treated guinea pigs had a slower LDL FCR compared to control LDL. When these studies were repeated in subjects with hyperlipidemia, results similar to those reported for guinea pigs were found, a decrease in particle affinity for the receptor and increases in receptor activity.[103] In our studies with atorvastatin, treatment with this reductase inhibitor resulted in decreased secretion of apo B in combination with increases in apo B/E receptors.[95,97] Our results are in total agreement with the reported mechanisms of LDL-C lowering in hyperlipidemic individuals treated with lovastatin,[103] confirming the suitability of guinea pigs in mimicking metabolic alterations in plasma lipoproteins induced by drug treatment.

APICAL SODIUM-DEPENDENT BILE ACID TRANSPORTER INHIBITORS Apical sodium-dependent bile acid transporter (ASBT) is a 348-amino acid protein localized on the apical surface of epithelial cells lining the ileum[104] and has been shown to mediate approximately 75% of the bile acids recycled in the human body.[105] Interruption of ileal bile acid reclamation leads to a reduction in plasma total and LDL-C levels via a significant increase in cholesterol catabolism.[50] SC-435 is a potent ASBT inhibitor (ASBTi) and has been studied in several animal models[106,107] including guinea pigs.[41,98,108]

West *et al.*[41] reported that guinea pigs treated with 13.4 mg/kg/day SC-435 had 44% lower LDL-C concentrations compared to controls. Hepatic CE were significantly reduced by 43, 56, and 70% in guinea pigs fed 0.8, 3.7, and 13.4 mg/kg/day of the ASBT inhibitor, respectively.[41] In addition, the highest dose of the inhibitor resulted in a 42% increase in the number of VLDL TG molecules and a larger VLDL diameter compared to controls. ACAT activity was 30% lower with the highest dose treatment while CYP7 was 30% higher with the highest ASBT inhibitor dose.[98] Cholesterol in the aortic arch was significantly reduced by 25 and 42% in guinea pigs treated with a monotherapy of 45 mg/day SC-435 and a coadministration of 17 mg/day SC-435 and 16 mg/day simvastatin , respectively.[98] In addition, SC-435 monotherapy and combination therapy also impact important regulatory enzymes of hepatic cholesterol homeostasis, which contribute to the lowering of plasma LDL-C.[98]

Taken together, these results indicate that the interruption of the enterohepatic circulation of bile acids by ASBTi and combination therapy may play a significant role in reducing cholesterol concentrations. This is a key finding since hypercholesterolemia in guinea pigs, like humans, results in an accumulation of cholesterol in the aortic arch, which ultimately leads to the development of atherosclerotic lesions

ACYL COA CHOLESTERYL:ACYLTRANSFERASE INHIBITORS ACAT is the rate-limiting enzyme in the formation of cholesteryl esters from cholesterol and long-chain fatty acyl coenzyme A. ACAT inhibitors have been shown to reduce plasma cholesterol levels by reducing cholesterol absorption[109] and decreasing the formation of VLDL particles. Furthermore, the inhibition of ACAT has been shown to prevent the conversion of macrophages to foam cells in the arterial wall.[110] The guinea pig has been used as a model to examine the effects of two different ACAT inhibitors, PD 321301-2[1] and CL 277,082[2]. At a lower dose (1 mg/kg), PD 321301-2 reduced plasma cholesterol by 21% and triglycerides by 25%. At a higher dose (10 mg/kg), VLDL-C was reduced by 87%, LDL-C by 64%, HDL-C by 75%, and triglycerides by 54%.[111] CL 277,082[2] was effective at lowering cholesterol only at a high dose (136 mg/kg). Junquero *et al.*[112] tested a systemic ACAT inhibitor (F 125113) using guinea pigs and found dose-dependent reductions in plasma cholesterol. These data suggest the potential utility of ACAT inhibitors in reducing plasma cholesterol, though additional work in humans is required. Furthermore, the effects of these drugs on HDL-C are an important consideration.

MICROSOMAL TRANSFER PROTEIN INHIBITORS MTP functions by transferring triglycerides and other lipids into the lumen of the endoplasmic reticulum. MTP is also a key in the assembly of chylomicrons. MTP inhibitors have been used to manage dyslipidemia, particularly because they target a different pathway of lipoprotein generation than HMG-CoA reductase inhibitors. The function of MTP inhibitors is to reduce packaging of lipids into chylomicrons and VLDL particles, decreasing plasma TG and reducing the formation of LDL particles.

An MTP inhibitor, CP-346086, has been administered to healthy humans. A single dose resulted in decreased plasma TG and VLDL cholesterol content, in a dose-dependent fashion. Following a 2-week treatment of 30 mg/day, total cholesterol was reduced by 47%, LDL-C by 72%, and TG by 75%.[113]

Though an MTP inhibitor can have profound effects on plasma lipids and lipoprotein particle lipid content, much concern still existed about hepatic lipid accumulation. JTT-130, an MTP inhibitor, was identified as specifically targeting the intestine. To investigate the effects of this JTT-30, guinea pigs were randomly assigned to consume either a control diet (no JTT-130), a low-dose diet, or a high-dose diet for 4 weeks.[38] In comparison to the control group, the low and high dose JTT-130 groups displayed significantly lower plasma total cholesterol, TG, and LDL-C. Perhaps noteworthy was that no differences between the control, low-dose, or high-dose groups were observed in terms of hepatic free cholesterol, CE, or TG content. It is believed that the JTT-130 reduced TG transfer to chylomicrons, leading to a reduced hepatic TG uptake via chylomicron remnants. Thus, the guinea pig model has been used to demonstrate that an MTP inhibitor specifically targeting the liver can reduce plasma LDL-C and TG without causing deleterious effects on hepatic lipid accumulation. Though further investigation in humans is necessary, the similarities between guinea pigs and humans with respect to lipoprotein metabolism indicate that treatment of humans with an

MTP inhibitor targeted to the intestine has the potential to be effective.

RAPAMYCIN Rapamycin is a natural compound produced by bacteria. Because of its immunosuppressive properties, rapamycin is approved by the FDA for use in preventing organ rejection in transplant patients. Rapamycin functions by binding to the immunophilin FK506 binding protein. Next, this complex interacts with the rapamycin binding domain, which results in the inactivation of a serine-threonine kinase known as the mammalian target of rapamycin (mTOR).[114] mTOR controls proteins that regulate mRNA translation initiation and G1 progression, and is targeted by numerous signals including insulin, growth factors, and amino acids. Through the action of these signals, regulatory cascades are adjusted to impact cell growth and proliferation.

Though these powerful immunosuppressive effects are clearly important in the prevention of transplant rejection, side effects of plasma lipids are known to occur. Specifically, dyslipidemia has been reported in 45% of liver transplant patients[115] and in 40% of kidney transplant patients.[116] Given the relationship between insulin and rapamycin, and the relationship between insulin resistance and dyslipidemia, the mechanisms behind the deleterious effects of rapamycin on lipoprotein metabolism provide insight concerning the risk for cardiovascular disease in transplant patients using this treatment.

The effects of rapamycin on plasma lipids are similar between guinea pigs and humans.[99] As such, the guinea pig model has provided considerable insight into the mechanisms behind the changes in plasma lipids and lipoprotein metabolism resulting from rapamycin prescription. To elucidate these mechanisms, 30 male guinea pigs consumed one of three treatments: control (no rapamycin), low dose, or high dose. The two groups treated with rapamycin displayed a 2-fold higher plasma triglyceride concentration in comparison to controls, with no difference between experimental groups. Guinea pigs treated with rapamycin also displayed large VLDL particle size, which contained a higher percentage of TG compared to the control group. Aortic cholesterol concentration and ACAT activity were significantly lower in the low-dose group as compared to the control and high-dose groups, and aortic triglyceride concentration was higher in the experimental than in the control groups. Also, plasma free-fatty acids and glucose were higher in the low-dose (65%) and high-dose (72%) groups in comparison to the control group. Furthermore, aortic tumor necrosis factor (TNF)-α concentrations were 3.6- and 10.4-fold greater in the low- and high-dose groups in comparison to the control groups.[99] These results indicate that the effects of rapamycin on plasma lipids and atherosclerosis may be due to alterations in the insulin signaling pathway. This appears to have caused an increased secretion of VLDL and VLDL TG, increasing TG accumulation in the aorta. These results also suggest that lower doses of rapamycin may reduce the unfavorable effects of higher doses of the drug.

EXERCISE AND LIPOPROTEIN METABOLISM

Exercise has been shown to favorably alter lipoprotein metabolism by decreasing plasma TG and increasing HDL-C.[117] Guinea pigs doing an exercise protocol consisting of running on a rodent treadmill at a rate of 33.3 revolutions/min for 30–40 min during 6 weeks exhibited lower plasma TG and higher HDL-C concentrations than sedentary animals.[32]

Prolonged exercise results in lowering of fasting plasma TG concentrations.[118] Exercise training increases LPL activity in plasma and in parenchymal cells, suggesting a role of exercise in increasing the capacity to clear TG from the circulation and suggesting that LPL is involved in the restoration of muscle TG stores reduced by exercise.[119] In addition, adipose tissue has increased LPL activity after exercise.[118] In agreement with human studies, exercised guinea pigs have been shown to have higher LPL activity both in heart and adipose tissue.[32] Patients who follow an exercise regimen in addition to dietary recommendations present significant regression and less progression of coronary atherosclerosis compared to usual-care control patients.[120] These results were correlated with significant decreases in plasma LDL-C and TG in the treated group. We have demonstrated that plasma lipid changes due to exercise in guinea pigs are similar to humans and that these animals develop atherosclerosis.[70,71] Based on these observations, the guinea pig could be an appropriate model to further evaluate the beneficial effects of exercise in the treatment of coronary heart disease.

GENDER AND HORMONAL STATUS

Results from the Framingham Heart Study have shown that higher plasma TG and low plasma HDL-C concentrations are highly associated with cardiovascular disease risk in women, while for men the major risk factor is elevated levels of plasma LDL-C.[121] Studies have shown that responses to dietary factors may differ between men and women.[122] In addition, postmenopausal women have significantly higher risk factors for the metabolic syndrome, insulin resistance, and increased plasma LDL-C, apo B, and TG concentrations.[123] Based on these observations, it is important to have an animal model with gender-associated responses to diet similar to humans. In addition, this model should present changes in lipoprotein profile in the absence of estrogen similar to postmenopausal women.

Female guinea pigs are more responsive to dietary cholesterol than males[124] in agreement with reported observations in humans.[125] In addition in females, fiber does not alter the regulatory enzymes of hepatic cholesterol metabolism, as is the case in males.[54] However, a slower apo B secretion rate and faster LDL FCR were observed in females compared to males,[53] indicating important differences in the gender response to dietary treatments.

We have also demonstrated the suitability of the ovariectomized guinea pig to mimic human menopause. Ovariectomized guinea pigs had higher levels of plasma TG than either males or females, higher concentrations of LDL-C, and overall a more detrimental lipoprotein profile.[31,126] In addition, ovariectomized guinea pigs exhibited higher susceptibility of LDL to oxidation compared to intact female animals. It has been shown that estrogen exerts a protective effect against free radical formation,[127] which might explain the higher susceptibility of LDL to oxidation in the ovariectomized guinea pigs. Gender and hormonal status effects on early atherosclerosis development were also evaluated in guinea pigs. Male guinea pigs had the greatest fatty streak accumulations in aortas, followed by ovariectomized animals, and females had the least lesion involvement ($p < 0.001$).[71] A comparison of plasma lipids, parameters of lipoprotein metabolism, and atherosclerosis among male, female, and ovariectomized guinea pigs is presented in Table 23–3.

Table 23–3

Gender/hormonal effects on plasma and hepatic lipids, lipoprotein metabolism, and atherosclerosis in male, female, and ovariectomized guinea pigs[a,b]

Parameter	*Male*	*Female*	*Ovariectomized*
LDL-C (mg/dl)[c]	69.5 ± 34.1	81.5 ± 53.1	123.4 ± 38.4
HDL-C (mg/dl)[c]	13.2 ± 5.4	18.6 ± 8.1	14.7 ± 4.7
Hepatic cholesterol (mg/g)[d]	2.2 ± 0.1	1.3 ± 0.5	2.3 ± 0.9
Hepatic triglycerides (mg/g)[d]	53.7 ± 25.1	24.0 ± 4.1	70.3 ± 35.5
TBARS (nmol MDA/non-HDL protein)[d]	1.4 ± 2.3	2.8 ± 3.4	8.2 ± 4.2
CYP7 activity (pmol/min·mg)[e]	1.99 ± 0.63	1.69 ± 1.03	0.98 ± 0.35
LDL apo B FCR (pools/h)[f,g]	0.08 ± 0.01	0.10 ± 0.02	ND
Aortic fatty streak (μm²/mm²)[d]	20705 ± 4752	14771 ± 2999	16576 ± 2094

[a]LDL-C, low-density lipoprotein C; HDL-C, high-density lipoprotein C; FCR, fractional catabolic rate.

[b]Values are expressed as mean ± SD for $n = 6–8$ guinea pigs. Values in the same row with different superscripts are significantly different as determined by one-way ANOVA and the Newman–Keules post hoc test ($p < 0.01$).

[c]Adapted from Roy *et al.*[31]

[d]Adapted from Cos *et al.*[71]

[e]Adapted from Roy *et al.*[56]

[f,g] Adapted from Fernandez[46] and Shen *et al.*[54]

GUINEA PIGS AS MODELS FOR ATHEROSCLEROSIS AND INFLAMMATION

Chronic dysfunctions of lipoprotein metabolism contribute to the formation of atherosclerotic plaque, and, potentially, a clinical event. Another important contributing factor to atherosclerosis is inflammation. Validation of an animal model with similarities to humans in terms of the end results of dysfunctional lipid metabolism and inflammation is a valuable resource in atherosclerosis research. Recent evidence suggests that the guinea pig is an appropriate model for the study of inflammation and atherosclerosis.

INFLAMMATION Atherosclerosis has been defined as a chronic, low-grade inflammatory condition. There are several causes of inflammation including vascular injury, lipid peroxidation, and infection.[128] Following such a stimulus, inflammatory cytokines, such as TNF-α, are released, which signal the release of messenger cytokines such as interleukin-6 (IL-6). Subsequently, systemic markers of inflammation like C-reactive protein (CRP) become elevated.[129] CRP has gained much attention recently in terms of cardiovascular disease risk since plasma CRP concentrations have strong predictive power for cardiovascular events.[130]

Our laboratory group has recently validated the guinea pig as a model for inflammation resulting from diet-induced atherosclerosis. Protein levels and mRNA expression of proinflammatory cytokines were measured for guinea pigs that consumed either high or low cholesterol diets that were either high or low in carbohydrate for 12 weeks. Compared with guinea pigs who consumed low-cholesterol diets, the high-cholesterol groups displayed significantly higher aortic cytokine proteins.[131] Another interesting finding was that the high-cholesterol, low-carbohydrate diet group displayed decreased protein and gene expression (assessed using real-time primers for guinea pigs) of interferon-γ, TNF-α, interleukin-1β, interleukin-8, and MCP-1. These results are similar to those reported for humans on a carbohydrate-restricted diet.[132] Furthermore, these results indicate that the guinea pig may be a promising model for the study of human inflammation resulting from diet-induced atherosclerosis.

ATHEROSCLEROSIS Atherosclerotic plaque formation is a complex process initiated via the oxidation of LDL particles. Once taken up by the arterial intima, LDL particles are susceptible to oxidation by free radicals. Oxidized LDLs enhance the adhesion and uptake of monocytes by the arterial intima. Monocytes become macrophages and engulf oxidized LDL particles to form the foam cells characteristic of atherosclerotic plaque.[133] LDL particles carried by guinea pigs are similar to those carried by humans,[134] and the LDL oxidation pattern in guinea pigs is more similar to that of humans (when compared to rats) with familial hypercholesterolemia.[135] Vitamin E has been shown to be protective of the integrity of the arterial intima in guinea pigs.[136] A vitamin C-deficient diet has been shown to increase the severity of atherosclerotic lesions, and the lesions were further exacerbated with high-cholesterol feeding.[137] Soluble fiber supplementation has been shown to reduce lesion development in guinea pigs.[71] Furthermore, guinea pigs have been shown to develop atherosclerosis, and the extent of atherosclerotic lesions is affected by gender and hormonal status.[71] Drug treatment[36] and grape polyphenols[70] can reduce the concentration of cholesterol in the aortic arch even in the presence of high dietary cholesterol.

CONCLUSIONS

This chapter presents compelling evidence regarding the suitability of guinea pigs as excellent models to evaluate the mechanisms by which diet interventions, drug treatments, and other life style factors alter plasma lipids and lipoprotein metabolism. Furthermore, we have shown that guinea pigs have an inflammatory response and develop atherosclerosis when challenged with a high cholesterol diet and that dietary interventions may prove to be beneficial in reducing the expression of inflammatory cytokines and the development of atherosclerosis. In addition, many of the mechanisms by which guinea pigs regulate cholesterol and lipoprotein metabolism as a response to diet or drug treatment are analogous to those reported in clinical experiments. These studies reinforce the importance of the use of guinea pigs rather than those more established and utilized animal models.

REFERENCES

1. Martignetti JA, Brosius J. Neural BCI RNA as an evolutionary marker: Guinea pig remains a rodent. *Proc Natl Acad Sci USA* 1993;90:9698–9702.
2. Fernandez ML, McNamara DJ. Dietary fat mediated changes in hepatic apo B/E receptors in the guinea pig. *Metabolism* 1989;38:1094–1102.
3. Khosla P, Hajri T, Pronzuc A, Hayes KC. Replacing dietary palmitic acid with elaidic acid (t-C18:1 delta9) depresses HDL and increases CETP activity in cebus monkeys. *J Nutr* 1997;127:531S-536S.
4. Morand OH, Aebi JD, Dehmlow H, Ji YH, Gains N, Lengsfeld H, Himber J. Ro 48-8.071, a new 2,3-oxidosqualene:lanosterol cyclase inhibitor lowering plasma cholesterol in hamsters, squirrel monkeys, and minipigs: Comparison to simvastatin. *J Lipid Res* 1997;38:373–390.
5. Kushwasha RS, Rosillo A, Rodriguez R, Chang J, VaderBerg JL. Expression levels of ACAT1 and ACAT2 genes in the liver and intestine of baboons with high and low lipemic responses to dietary lipids. *J Nutr Biochem* 2005;16:714–721.
6. Meydani, SN, Nicolosi RJ, Sehgal PK, Hayes KC. Altered lipoprotein metabolism in spontaneous vitamin E deficiency of owl monkeys. *Am J Clin Nutr* 1983;38:888–894.
7. Fincham JE, Benade AJ, Kruger M, Smuts CM, Gobregtz E, Chalton DO, Kritchevsky D. Atherosclerosis: Aortic lipid changes induced by diets suggest diffuse disease with focal severity in primates that model human atheromas. *Nutrition* 1998;14:17–22.
8. Rudel LL, Deckelm,an C, Wilson M, Scobey M, Anderson R. Dietary cholesterol and down-regulation of cholesterol 7α-hydroxylase and cholesterol absorption in African green monkeys. *J Clin Invest* 1994; 93:2463–2472.
9. Carr TP, Parks JS, Rudel LL. Hepatic ACAT activity in African green monkeys is highly correlated to plasma LDL cholesteryl ester enrichment and coronary artery atherosclerosis. *Arterioscler Thromb* 1992;12:1274–1283.
10. Wolfe MS, Sawyer JK, Morgan TM, Bullock LC, Rudel LL. Dietary polyunsaturated fat decreases coronary artery atherosclerosis in a pediatric-aged population of African green monkeys. *Arterioscler Thromb* 1994;14:587–597.
11. Rudel LL, Parks JS, Sawyer JK. Compared with dietary monounsaturated and saturated fat, polyunsaturated fat protects African green monkeys from coronary artery atherosclerosis. *Arterioscler Thromb Vasc Biol* 1995;15:2101–2110.
12. Huggins KW, Burleson ER, Sawyer JK, Kelly K, Rudel LL, Parks JS. Determination of the tissue sites responsible for the catabolism of large high density lipoprotein in the African green monkey. *J Lipid Res* 2000;41:384–394.
13. Speijer H, Groener JE, Van Ramshorts E, van Tol A. Different locations of cholesteryl ester transfer protein and phospholipid transfer protein activities in plasma. *Atherosclerosis* 1991;90:159–168.
14. Barter PJ, Brewer HB Jr, Chapman MJ, Hennekens CH, Rader DJ, Tall AR. Cholesteryl ester transfer protein: A novel target for raising HDL and inhibiting atherosclerosis. *Arterioscler Thromb Vasc Biol* 2003;23:160–167.
15. Zak Z, Gautier T, Dumont L, Masson D, Decker V, Duverruil L, Pais de Barros JP, Le Guern N, Schneider M, Moulin P, Bataillard A, Lagrost L. Effect of cholesteryl ester transfer protein (CETP) expression on diet-induced hyperlipidemias in transgenic rats. *Atherosclerosis* 2005;178:279–286.
16. Pandak WM, Vlahcevik ZR, Heuman DK, Hylemon PB. Regulation of bile acid synthesis. V. Inhibition of conversion of 7-dehydrocholesterol to cholesterol is associated with down-regulation of cholesterol 7 alpha-hydroxylase activity and inhibition of bile acid synthesis. *J Lipid Res* 1990;31:2149–2158.
17. Bergen WG, Mersmann HJ. Comparative aspects of lipid metabolism: Impact on contemporary research and use of animal models. *J Nutr* 2005;135:2499–2502.
18. Chiang M. Plasma lipoprotein cholesterol levels in rats fed a diet enriched in cholesterol and cholic acid. *Int J Vitam Nutr Res* 1998;68:328–334.
19. Wouters KT, Shiri-Sverdlof R, van Gorp PJ, van Milsen M, Hofker MH. Understanding hyperlipidemia and atherosclerosis: Lessons from genetically modified apoE and LDLr mice. *Clin Chem Lab Med* 2005;43:470–479.
20. Li X, Catalina F, Grundy SM, Patel S. Method to measure apolipoprotein B-48 and B-100 secretion rates in an individual mouse: Evidence for a very rapid turnover of VLDL and preferential removal of B-48- relative to B-100-containing lipoproteins. *J Lipid Res* 1996;37:210–220.
21. Woollett LA, Spady DK, Dietschy JM. Mechanisms by which saturated triacylglycerols elevate the plasma low density lipoprotein-cholesterol concentration in hamsters. Differential effects of fatty acid chain length. *J Clin Invest* 1989;84:119–128.
22. Horton JD, Cuthbert JA, Spady DK. Regulation of hepatic 7 alpha, hydroxylase expression in the rat and hamster. *J Biol Chem* 1995;270:5381–5387.
23. Taghibiligou C, Rashid-Kolvear F, Van Iderstine SC, Le-Tien H, Fantus IG, Lewis GF, Adeli K. Hepatic very low density lipoprotein-ApoB overproduction is associated with attenuated hepatic insulin signaling and overexpression of protein-tyrosine phosphatase 1B in a fructose-fed hamster model of insulin resistance. *J Biol Chem* 2002;277:793–803.
24. Plancke MO, Olivier P, Clavey V, Marzin D, Fruchart JC. Aspects of cholesterol metabolism in normal and hypercholesterolemic Syrian hamsters. Influence of fenofibrate. *Methods Find Exp Clin Pharmacol* 1988;10:575–579.
25. Sawada M, Matsuo M, Seki J. Inhibition of cholesterol synthesis causes both hypercholesterolemia and hypocholesterolemia in hamsters. *Biol Pharm Bull* 2002;25:1577–1582.
26. Brouseau ME, Hoeg JM. Transgenic rabbits as models for atherosclerosis research. *J Lipid Res* 1999;40(3):365–375.
27. Fernandez ML, Yount NY, McNamara DJ. Whole body cholesterol synthesis in the guinea pig. Effects of dietary fat quality. *Biochim Biophys Acta* 1990;1044:340–348.
28. Ha YC, Barter PA. Differences in plasma cholesteryl ester transfer protein activity in sixteen vertebrate species. *Comp Biochem Physiol* 1982;71:265–269.
29. Douglas G, Pownell JJ. Comparative specificity of plasma lecithin-cholesterol acyltransferase from ten animal species. *Lipids* 1991;26:416–420.
30. Olivecrona T, Bengsston-Olivecrona G. Lipoprotein lipase and hepatic lipase. *Curr Opin Lipidiol* 1993;4:187–196.
31. Roy S, Vega-Lopez S, Fernandez ML. Gender and hormonal status affect the hypolipidemic mechanisms of dietary soluble fiber in guinea pigs. *J Nutr* 2000;130:600–607.
32. Ensign W, McNamara DJ, Fernandez ML. Exercise improves plasma lipid profiles and modifies lipoprotein composition in guinea pigs. *J Nutr Biochem* 2002;12:747–753.
33. Fernandez ML, Lin ECK, McNamara DJ. Regulation of guinea pig plasma low density lipoprotein kinetics by dietary fat saturation. *J Lipid Res* 1992;33:97–109.
34. Fernandez ML, Lin ECK, McNamara DJ. Differential effects of saturated fatty acids on low density lipoprotein metabolism in the guinea pig. *J Lipid Res* 1992;33:1833–1842.
35. He L, Fernandez ML. Dietary carbohydrate type and fat saturation independently regulate hepatic cholesterol and LDL metabolism in guinea pigs. *J Nutr Biochem* 1998;9:37–46.
36. Berglund L, Sharkey ME, Elam RL, Witztum JL. Effects of lovastatin therapy on guinea pig low density lipoprotein composition and metabolism. *J Lipid Res* 1989;30:1591–1600.
37. Hikada K,Takada Y, Matsunaga A, Sasaki J, Arawaka K. Effects of probucol and low-density lipoprotein catabolism in guinea pigs. *Artery* 1992;19:162–176.
38. Aggarwal D, West KL, Zern TL, Shresthra S, Vergara-Jimenez M, Fernandez ML. JTT-130, a microsomal transfer protein inhibitor lowers plasma triglycerides and LDL cholesterol concentrations without increasing hepatic triglycerides in the guinea pig *BMC Card Dis* 2005;5:30.
39. Angelin B, Olivecrona H, Reihner H, Rudling E, Stahlberg D, Eriksson M, Ewerth S, Henricksson PK, Einarsson K. Hepatic

cholesterol metabolism in estrogen-treated men. *Gastroenterology* 1992;103:1657–1663.
40. Reihner E, Angelin B, Rudling M, Ewerth S, Bjorkhem I, Einarsson K. Regulation of hepatic cholesterol metabolism in humans: Stimulatory effects of cholestyramine on HMG-CoA reductase activity and low density lipoprotein receptor expression in gallstone patients. *J Lipid Res* 1990;31:2219–2226.
41. West KL, McGrane MM, Odon D, Kelley BL, Fernandez ML. SC-435, an ileal apical sodium-codependent bile acid transporter (ASBT) inhibitor alters mRNA levels and enzyme activities of selected genes involved in hepatic cholesterol and lipoprotein metabolism in guinea pigs. *J Nutr Biochem* 2005;16:722–728.
42. Reihner E, Angelin B, Bjorkhem I, Einarsson K. Hepatic cholesterol metabolism in cholesterol gallstone disease. *J Lipid Res* 1991;32: 469–475.
43. Corsini A, Mazzotti M, Villa A, Maggi FM, Bernini F, Romano L, Romano C, Fumagalli R, Catapano AL. Ability of the LDL receptor from several animal species to recognize the human apo B binding domain; studies with LDL from familial defective apo B-100. *Atherosclerosis* 1992;93:95–103.
44. Greeve J, Altkemper I, Dietrich J-H, Greten H, Windler E. Apolipoprotein mRNA editing in 12 diffeent mammalian species: Hepatic expression is reflected in low concentrations of apoB-containing plasma lipoproteins. *J Lipid Res* 1993;34:1367–1383.
45. Liu S, Stampfer, MJ, Hu FB, Giovannucci E, Rimm E, Manson JE, Hennekens CH, Willett WC. Whole grain-consumption and risk of coronary heart disease: Results from the Nurses' Health Study. *Am J Clin Nutr* 1999;70:412–419.
46. Fernandez ML. Distinct mechanisms of plasma LDL lowering by dietary soluble fiber. Specific effects of pectin, guar gum and psyllium. *J Lipid Res* 1995;36:2394–2404.
47. Fernandez ML. Soluble fiber and non-digestible carbohydrate effects on plasma lipids and cardiovascular risk. *Curr Opin Lipidiol* 2001;12:35–40.
48. Fernandez ML, Sun DM, Tosca M, McNamara DJ. Differential effects of guar gum on LDL and hepatic cholesterol metabolism in guinea pigs fed low and high cholesterol diets. A dose response study. *Am J Clin Nutr* 1995;61:127–134.
49. Fernandez ML, Sun D-M, Tosca M, McNamara DJ. Citrus pectin and cholesterol interact to regulate hepatic cholesterol homeostasis and lipoprotein metabolism. A dose response study in the guinea pig. *Am J Clin Nutr* 1994;59:869–878.
50. Fernandez ML, Ruiz LR, Conde AK, Sun D-M, Erickson S, McNamara DJ. Psyllium reduces plasma LDL in guinea pigs by altering hepatic cholesterol metabolism. *J Lipid Res* 1995;36:1128–1138.
51. Jelinek DF, Andersson S, Slaughter CA, Russel DW. Cloning and regulation of cholesterol 7α-hydroxylase, the rate limiting enzyme in bile acid biosynthesis. *J Biol Chem* 1990;265:8190–8197.
52. Brown MS, Goldstein JL. A receptor mediated pathway for cholesterol homeostasis. *Science* 1986;232:34–48.
53. Vergara-Jimenez M, Conde K, Erickson S, Fernandez ML. Hypolipidemic mechanisms of pectin and psyllium in guinea pigs fed high fat-sucrose diets: Alterations on hepatic cholesterol metabolism. *J Lipid Res* 1998;39:1455–1465.
54. Shen H, He L, Price RL, Fernandez ML. Dietary soluble fiber lowers plasma LDL cholesterol concentrations by altering lipoprotein metabolism in female guinea pigs. *J Nutr* 1998;128: 1434–1441.
55. Vidal-Quintanar RL, Hernandez L, Conde K. Vergara-Jimenez M, Fernandez ML. Lime-treated cornhusks lower plasma LDL cholesterol in guinea pigs by altering hepatic cholesterol metabolism. *J Nutr Biochem* 1997;8:479–486.
56. Roy S, Freake HC, Fernandez ML. Gender and hormonal status affect the regulation of hepatic cholesterol 7α-hydroxylase activity and mRNA abundance by dietary soluble fiber in the guinea pig. *Atherosclerosis* 2002;163:29–37.
57. Fernandez ML, Wilson TA, Conde K, Vergara-Jimenez M, Nicolosi RJ. Hamsters and guinea pigs differ in their plasma lipoprotein cholesterol distribution when fed diets varying in animal protein, soluble fiber or cholesterol content. *J Nutr* 1999;129:1323–1332.
58. Fernandez ML, Roy S, Vergara-Jimenez M. Resistant starch and cholestryramine have distinct effects on hepatic cholesterol metabolism in guinea pigs fed a hypercholesterolemic diet. *Nutr Res* 2000;20:837–849.
59. Fernandez ML, Conde K, Vergara-Jimenez M, Behr T, Abdel-Fattah G. Regulation of apolipoprotein B-containing lipoproteins by dietary soluble fiber in guinea pigs. *Am J Clin Nutr* 1997;65:814–822.
60. Ginsberg HN. Lipoprotein physiology and its relationship to atherogenesis. *Endocrinol Metab Clin North Am* 1990;19:211–222.
61. Nestel P, Billington T, Tada N, Nugent P, Fidge N. Heterogeneity of very-low-density lipoprotein metabolism in hyperlipidemic subjects. *Metabolism* 1983;32:810–817.
62. Lin ECK, Fernandez ML, McNamara DJ. Dietary fat type and cholesterol quantity interact to affect cholesterol metabolism in guinea pigs. *J Nutr* 1992;22:2019–2029.
63. McNamara DJ, Kolb R, Parker TS, Batwin H. Heterogeneity of cholesterol homeostasis in man. Response to changes in dietary fat quality and cholesterol quantity. *J Clin Invest* 1987;79:1729–1739.
64. Puppione DL, Sardet C, Yamanaka W, Ostwald R, Nichols AV. Plasma lipoproteins of cholesterol-fed guinea pigs. *Biochim Biophys Acta* 1971;231:295–301.
65. Guo LSS, Hamilton RL, Kane PJ, Fielding CJ, Chen GC. Characterization and quantitation of apolipoproteins A-I and E of normal and cholesterol-fed guinea pigs. *J Lipid Res* 1982;23:531–542.
66. Meng M, Guo L, Ostwald R. Isolation and partial characterization of a guinea pig serum apolipoprotein comigrating with apo E on sodium dodecyl sulphate-polyacrylamide electropherograms. *Biochim Biophys Acta* 1979;576:134–140.
67. Lin ECK, Fernandez ML, McNamara DJ. Dietary fat and cholesterol regulation of hepatic LDL in the guinea pig. *J Lipid Res* 1994;35:446–457.
68. Lin ECK, Fernandez ML, McNamara DJ. Dietary fat and cholesterol regulation of HDL metabolism: Hepatic HDL binding and in vivo apolipoprotein A-I catabolism in guinea pigs. *Atherosclerosis* 1995;112:161–175.
69. Sun D-M, Fernandez ML, Lin ECK, McNamara DJ. Regulation of guinea pig hepatic acyl-CoA:cholesterol acyltransferase activity by dietary fat saturation and cholesterol. *J Nutr Biochem* 1999;10:172–180.
70. Zern TL, West KL, Fernandez ML. Grape polyphenols decrease plasma triglycerides and reduce cholesterol accumulation in the aorta of ovariectomized guinea pigs. *J Nutr* 2003;133:2268–2272.
71. Cos E, Ramjiganesh T, Roy S, Yoganathan S, Nicolosi RJ, Fernandez ML. Soluble fiber and soybean protein reduce atherosclerotic lesions in guinea pigs. Sex and hormonal status determine lesion extension. *Lipids* 2001;11:1209–1216.
72. Fernandez ML, Abdel-Fattah G, McNamara DJ. Dietary fat saturation modifies the metabolism of LDL subfractions in guinea pigs. *Arterioscler Thromb* 1993;14:1418–1428.
73. Abdel-Fattah G, Fernandez ML, McNamara DJ. Regulation of guinea pig very low density lipoprotein secretion rates by dietary fat saturation. *J Lipid Res* 1995;36:1188–1198.
74. Abdel-Fattah G, Fernandez ML, McNamara DJ. Regulation of very low density lipoprotein apo B metabolism by dietary fat saturation and chain length in the guinea pig. *Lipids* 1998;33:23–31.
75. Fernandez ML, McNamara DJ. Dietary fat saturation and chain length modulate guinea pig hepatic cholesterol metabolism. *J Nutr* 1994;124:331–339.
76. Romero AL, Fernandez ML. Dietary fat amount and carbohydrate type regulate hepatic acyl CoA:cholesterol acyltransferase (ACAT) activity. Possible links between hepatic ACAT activity and plasma cholesterol levels. *Nutr Res* 1996;16:937–948.
77. Nicolosi RJ. Dietary fat saturation effects on low-density-lipoprotein concentrations and metabolism in various animal models. *Am J Clin Nutr* 1997;65(Suppl.):1617S-1627S.
78. Cortese C, Levy Y, Janus ED, Turner PR, Rao SN, Miller NE, Lewis B. Modes of action of lipid lowering diets in man: Studies of

apolipoprotein B kinetics in relation to fat consumption and dietary fatty acid composition. *Eur J Clin Invest* 1983;13:79–85.

79. Turner J, Lee N-A, Brown WV. Effects of changing dietary fat saturation on low-density lipoprotein metabolism in man. *Am J Physiol* 1981;241:E37–E65.
80. Fernandez ML, Conde K, Ruiz L, Montano C, McNamara DJ. Carbohydrate type and amount alter intravascular processing and catabolism of plasma lipoproteins in guinea pigs. *Lipids* 1995;30: 619–626.
81. Fernandez ML, Vergara-Jimenez M, Conde K, Abdel-Fattah G. Dietary carbohydrate type and fat amount alter VLDL and LDL metabolism in guinea pigs. *J Nutr* 1996;126:2494–2504.
82. Volek JS, Sharman M J, Forsythe CE. Modification of lipoproteins by very low-carbohydrate diets. *J Nutr* 2005;135:1339–1342.
83. Torres-Gonzalez M, Volek JS, Sharman M, Contois JH, Fernandez ML. Dietary carbohydrate and cholesterol influence the number of particles and distributions of lipoprotein subfractions in guinea pigs. *J Nutr Biochem* 2006;17(11):773–779.
84. Wood RJ, Volek JS, Liu Y, Shachter NS, Contois JH, Fernandez ML. Carbohydrate restriction alters lipoprotein metabolism by modifying VLDL, LDL and HDL subfraction distribution and size in overweight men. *J Nutr* 2005;136:384–389.
85. Cerna O, Ginter E. Blood lipids and vitamin-C status. *Lancet* 1978;1:1055–1056.
86. Fernandez ML. Guinea pigs as models for cholesterol and lipoprotein metabolism. *J Nutr* 2001;131:10–20.
87. Montano C, Fernandez ML, McNamara DJ. Regulation of apolipoprotein B-containing lipoproteins by vitamin C level and dietary fat saturation in guinea pigs. *Metabolism* 1998;47:883–891.
88. Hulse JD, Ellis SR, Hendersson LM. Carnitine biosynthesis. beta-Hydroxylation of trimethyllysine by an alpha-ketoglutarate-dependent mitochondrial dioxygenase. *J Biol Chem* 1978;253:1654–1659.
89. Greene YJ, Harwood HJ Jr, Stacpoole PW. Ascorbic acid regulation of 3-hydroxy-3-methylglutaryl coenzyme A reductase activity and cholesterol synthesis in guinea pig liver. *Biochim Biophys Acta* 1985;834:134–138.
90. Fernandez ML, Vega S, Ayala MT, Shen H, Conde K, Vergara-Jimenez M, Robbins A. Vitamin C level and dietary fat saturation alter hepatic cholesterol homeostasis and plasma LDL metabolism in guinea pigs. *J Nutr Biochem* 1997;8:414–424.
91. Witztum JL, Young SG, Elam RL, Carew TE, Fisher M. Cholestyramine-induced changes in low density lipoprotein composition and metabolism. I. Studies in the guinea pig. *J Lipid Res* 1985;26: 92–103.
92. Krause BR, Newton RS. Animal models for the evaluation of inhibitors of HMG-CoA reductase. *Adv Lipid Res* 1991;1:57–72.
93. Matsunaga A, Sasaki J, Takada Y, Hidaka K, Kazuko K, Arakawa K. Effect of simvastatin on receptor mediated metabolism of low density lipoprotein in guinea pigs. *Atherosclerosis* 1991;90:31–37.
94. Conde K, Roy S, Freake HC, Newton R, Fernandez ML. Atorvastatin and simvastatin have distinct effects on hydroxy methylglutaryl-CoA reductase activity and mRNA abundance in the guinea pig. *Lipids* 1999;34:1327–1332.
95. Conde K, Pineda G, Newton R, Fernandez ML. Hypocholesterolemic effects of 3-hydroxy-3-methylglutaryl coenzyme A (HMG-CoA) reductase inhibitors in the guinea pig. Atorvastatin versus simvastatin. *Biochem Pharmacol* 1999;58:1209–1219.
96. Krause BR, Bousley R, Kieft K, Robertson D, Stanfield R, Urda E, Newton RS. Comparison of lifibrol to other lipid-regulating agents in experimental animals. *Pharmacol Res* 1994;29:345–357.
97. Conde K, Vergara-Jimenez M, Crouse B, Newton R, Fernandez ML. Hypocholesterolemic actions of atorvastatin are associated with alterations on hepatic cholesterol metabolism and lipoprotein composition in the guinea pig. *J Lipid Res* 1996;37:2372–2382.
98. West KL, Ramjiganesh T, Roy S, Keller BT, Fernandez ML. SC-435, an ileal, apical sodium-dependent bile acid transporter inhibitor (ASBT) alters hepatic cholesterol metabolism and lowers plasma low-density-lipoprotein-cholesterol concentrations in guinea pigs. *J Pharmacol Exp Ther* 2002;303:291–299.
99. Aggarwal D, Fernandez ML, Soliman GA. Rapamycin, an mTOR inhibitor disrupts triglyceride metabolism in guinea pigs. *Metabolism* 2006;55:794–802.
100. Dietschy JM, Turley SD, Spady DK. Role of the liver in the maintenance of cholesterol and low density lipoprotein homeostasis in different animal species, including humans. *J Lipid Res* 1993; 34:1637–1659.
101. McNamara DJ. Cholesterol homeostasis in the guinea pig. The importance of quantitating net tissue accumulation of cholesterol in sterol balance studies. *Biochim Biophys Acta* 1984;796:51–54.
102. Swann A, Wiley MH, Sipperstein MD. Tissue distribution of cholesterol feedback control in the guinea pig. *J Lipid Res* 1975; 16:360–366.
103. Berglund L, Witztum JL, Galeano NF, Khuow AS, Ginsberg HN, Ramakrishnan R. Three-fold effect of lovastatin treatment on low density lipoprotein metabolism in subjects with hyperlipidemia: Increase in receptor activity, decrease in apo B production and decrease in particle affinity for the receptor. Results from a novel triple-tracer approach. *J Lipid Res* 1998;39:913–924.
104. Wong MH, Oelkers P, Dawson PA. Identification of a mutation in the ileal sodium-dependent bile acid transporter gene that abolishes transport activity. *J Biol Chem* 1995;270:27228–27234.
105. Chen F, Ma L, Al-Ansari N, Shneider B. The role of AP-1 in the transcriptional regulation of the rat apical sodium-dependent bile acid transporter. *J Biol Chem* 2001;276:38703–38714.
106. Huff MW, Telford DE, Edwards JY, Burnett JR, Barrett PH, Rapp SR, Nappawan L, Keller BT. Inhibition of both the apical sodium-dependent bile acid transporter and HMG-CoA reductase markedly enhances the clearance of LDL apoB. *J Lipid Res* 2003;44:943–952.
107. Batt BG, Rapp SR, Beaudry JA, Nappawan L, Butteiger DN, Hall KA, Null CL, Luo Y, Keller BT. Inhibition of ileal bile acid transport and reduced atherosclerosis in apoE-/- mice by SC-435. *J Lipid Res* 2003;44:1614–1621.
108. West KL, Zern TL, Butteinger D, Keller BT, Fernandez ML. SC-435, an ileal apical sodium co-dependent bile acid transporter (ASBT) inhibitor lowers plasma cholesterol and reduces atherosclerosis in guinea pigs. *Atherosclerosis* 2003;171:201–210.
109. Azuma A, Kawasaki T, Ikemoto K, Obata K, Ohno K, Yamada T, Yamasaki M, Nobuhara Y. Cholesterol-lowering effects of NTE-122, a novel acyl-CoA:cholesterol acyltransferase (ACAT) inhibitor, on cholesterol diet-fed rats and rabbits. *Jpn J Pharmacol* 1998;78:355–364.
110. Leon C, Hill JS, Wasan KM. Potential role of acyl-coenzyme A: cholesterol transferase (ACAT) inhibitors as hypolipidemic and antiatherosclerosis drugs. *Pharm Res* 2005;22:1578–1588.
111. Krause BR, Black A, Bousley R, Essenburg R, Cornicelli J, Holmes A, Homan R, Kieft K, Sekerke C, Shaw-Hess MK. Divergent pharmacologic activities of PD 132301-2 and CL 277,082, urea inhibitors of acyl-CoA:cholesterol acyltransferase. *J Pharmacol Exp Ther* 1993;267:734–743.
112. Junquero D, Oms P, Carilla-Durand E, Austin J, Tarayre J, Degryse A, Patoiseau J, Colpaert FC, Delhon A. Pharmacological profile of F 12511, (S)-2′,3′, 5′-trimethyl-4′-hydroxy-alpha-dodecylthioacetanilide a powerful and systemic acylcoenzyme A: Cholesterol acyltransferase inhibitor. *Biochem Pharmacol* 2001;61(1):97–108.
113. Chandler CE, Wilder DE, Pettine JL, Saviy YE, Pettras SF, Chang G, Vincent J, Harwood HJ Jr. CP-346086: An MTP inhibitor that lowers plasma cholesterol and triglycerides in experimental animals and in humans. *J Lipid Res* 2003;44:1887–1901.
114. Brown EJ, Albers MW, Shin TB, Ichikawa K, *et al.* A mammalian protein targeted by G1-arresting rapamycin-receptor complex. *Nature* 1994;369:756–758.
115. Mathis AS, Dave N, Knipp GT, Friedman GS. Drug-related dyslipidemia after renal transplantation. *Am J Health Syst Pharm* 2004;61:565–585.

116. Morrisett JD, Abdel-Fattah G, Hoogeven R, Mitchell E. Effects of sirolimus on plasma lipids, lipoprotein levels, and fatty acid metabolism in renal transplant patients. *J Lipid Res* 2002;43:1170–1180.
117. Kelley GA, Kelly KS, Franklin D. Aerobic exercise and lipids and lipoproteins in patients with cardiovascular disease: A meta-analysis of randomized controlled trials. *J Cardiopulm Rehabil* 2006; 26:131–139.
118. Savard R, Bouchard C. Genetic effects in the response to adipose tissue lipoprotein lipase activity to prolonged exercise. A twin study. *Int J Obesity* 1990;14:771–777.
119. Oscai LB, Essig DA, Palmer WK. Lipase regulation of muscle triglyceride hydrolysis. *J Appl Physiol* 1990;69:1571–1577.
120. Schlierf G, Schuler G, Hambrecht R, Niebauer J, Hauer H, Vogel G, Kubler W. Treatment of coronary heart disease by diet and exercise. *J Cardiovasc Pharmacol* 1995;25:S32–S34.
121. Castelli WP. Cholesterol and lipids in the risk of coronary heart disease. The Framingham Heart Study. *Can J Cardiol* 1988; 4:5A–10A.
122. Cobb M, Greenspan J, Timmons M, Teitelbaum H. Gender differences in lipoprotein responses to diet. *Ann Nutr Metab* 1993;37:225–236.
123. Mesch VR, Boero LV, Siseles NO, Royer M, Prada M, Sayegh F, Schreir L, Tenencia HJ, Berg GA. Metabolic syndrome throughout the menopausal transition: Influence of age and menopausal status. *Climacteric* 2006;9:40–48.
124. Fernandez ML,Vergara M, Romero AL, Erickson S, McNamara DJ. Gender differences in plasma and hepatic hypocholesterolemia induced by soluble fiber. Effects of pectin, guar gum and psyllium. *J Lipid Res* 1995;36:2191–2202.
125. Clifton PM, Abbey M, Noakes M, Beltrame S, Rumbelow N, Nestel PJ. Body fat distribution is a determinant of the high-density lipoprotein response to dietary fat and cholesterol in women. *Aterioscler Thromb Vasc Biol* 1995;15:1070–1078.
126. Fernandez ML, West KL, Roy S, Ramjiganesh T. Dietary fat saturation and gender/hormonal status modulate plasma lipids and lipoprotein composition. *J Nutr Biochem* 2001;12:703–710.
127. Sack MN, Rader DJ, Cannon RO II. Oestrogen and inhibition of oxidation of low-density lipoproteins in postmenopausal women. *Lancet* 1994;343:269–270.
128. Willerson JT, Ridker PM. Inflammation as a cardiovascular risk factor. *Circulation* 2004;109(Suppl. 1):II2–10.
129. Blake GJ, Ridker PM. Inflammatory mechanisms in atherosclerosis: From laboratory evidence to clinical application. *Ital Heart J* 2001;11:796–800.
130. Libby P, Willerson JY, Braunwald E. C-reactive protein and coronary heart disease. *N Engl J Med* 2004;351:295–298.
131. Fernandez ML, Volek JS. Guinea pigs: A suitable animal model to study lipoprotein metabolism atherosclerosis inflammation. *Nutr Metab (Lond)* 2006;3:17.
132. Sharman M, Volek JS. Weight loss leads to reductions in inflammatory biomarkers after a very-low-carbohydrate diet and a low-fat diet in overweight men. *Clin Sci (Lond)* 2004;107:365–369.
133. Parthsarathy S, Steinberg D, Witztum JL. The role of oxidized low-density lipoproteins in the pathogenesis of atherosclerosis. *Annu Rev Med* 1992;43:219–225.
134. Vazquez M, Merlos M, Adzet T, Laguna JC. Influence of lipid profile and fatty acid composition on the oxidation behavior of rat and guinea pig low density lipoprotein. *Comp Biochem Physiol* 1998;119B:311–316.
135. Vazquez M, Zambon D, Hernandez Y, Azdet T, Merlos M, Laguna JC. Lipoprotein composition and oxidative modification during therapy with gemfibrozil and lovastatin in patients with combined hyperlipidemia. *Br J Clin Pharmacol* 1998;45(3):265–269.
136. Barja G, Cadenas S, Rojas C, Perez-Campo R, Lopez-Torres M, Prat J, Pamplona R. Effect of vitamin E levels on fatty acid profiles and nonenzymatic lipid peroxidation in the guinea pig liver. *Lipids* 1996;31:963–970.
137. Sharma P, Pramod J, Sharma PK, Chaturvedi SK, Kothari LK. Effect of vitamin C administration on serum and aortic lipid profile of guinea pigs. *Indian J Med Res* 1988;87:283–287.

24 Reliability of Rodent Models

DILEEP KUMAR ROHRA AND YUREEDA QAZI

ABSTRACT

Despite the discrepancies observed between data generated from humans and animals, it is common practice to extrapolate data from animal models to humans. This chapter discusses the usefulness of animal models in research and development, but also stresses the need to exercise caution in interpreting the results. Complete reliance on the results from animal experiments can be dangerously misleading, resulting in damage to human health as well as loss of valuable time and resources. Furthermore, we will elaborate on the validity and reliability of rodent models, the role of confounding factors in using animal models, and precautionary advice when extrapolating data obtained from rodent models to a larger, heterogeneous human population.

Key Words: Rodent models, Reliability, Validity, Confounding factors, Transgenic mouse, Genetics.

INTRODUCTION

Animal models are important and essential tools in biomedical research. In 2006 alone, biomedical researchers have used rodent models to explore everything from the pathogenesis of pulmonary contusion to the beneficial effects of fruit extracts on neuronal function and behavior.[1,2] Recently, mouse models that faithfully recapitulate human head and neck small cell carcinoma (HNSCC) at the pathological and molecular levels have been developed, which will potentially allow the specific genes involved in HNSCC to be defined and reveal prospective therapeutic targets.[3]

ISSUE OF RELIABILITY OF RODENT MODELS

In spite of the fact that animal models have contributed considerably to our understanding of the pathological processes of various human diseases and their treatment, in many cases the issue of reliability has prevented results from animal models from being readily used to develop successful clinical treatments. At regular intervals, investigators have reported that animal models do not satisfactorily model the pathophysiology of complex diseases, cardiomyopathy being one of many examples.[4] To fully examine the reliability of such models, it is first necessary to briefly discuss the classification of such models and the means used to evaluate their validity. Using the concepts of validity, we will subsequently assess the reliability of a number of rodent models and indicate, where necessary, that caution should be used when extrapolating the results from such models to human patients.

From: *Sourcebook of Models for Biomedical Research*
(P. M. Conn, ed.), © 2008 Humana Press Inc., Totowa, NJ.

An animal model is an organism with a genetic, naturally obtained, or induced pathological process that closely resembles the same condition in humans.[5] Animal models bridge *in vitro* laboratory investigations and studies in humans.[6] Useful models are described as either correlational, isomorphic, or homologous.[7] A correlational model, for example, would show that compounds that ameliorate drug-induced seizures in mice are predicted to be anxiolytics in humans.[8] An isomorphic rodent model of a disease would share the signs and symptoms of the human disease but would differ in its etiology.[9] Snell's dwarf mouse model, with a nonfunctional pituitary, is an example of an isomorphic model. Generally, the most successful models have been homologous, where both the etiology and clinical features of a disease in an animal model mirror those found in the human patient, such as the mouse model for HNSCC.

To establish the validity of an animal model, three types of validity have been recognized: face validity, predictive validity, and construct validity.[8] Face validity indicates a degree of similarity between the clinical presentation of a condition seen in the patient and the signs and symptoms of that condition seen in the model. If a rodent model possesses predictive validity, it should correctly predict the results of a particular intervention when applied in a clinical setting. For example, if a mouse modeling hypertension is treated with a particular antihypertensive drug, its response should correctly predict the behavior of the drug when administered to a hypertensive human. Finally, a model possesses construct validity if both the pathogenesis and the clinical features of the disease expressed by the model mimic those of the clinical condition.[8]

Once a model has been established as being theoretically valid, it must also prove itself reliable. We would like to differentiate reliability into two types: experimental and extrapolation. An animal model is experimentally reliable if, under defined test conditions, the results attained can be reproduced by subsequent tests and by different experimenters following an identical protocol.[10] On the other hand, a model is extrapolation reliable if, under defined test conditions, the results produced by the model mirror the results produced by an identical test on a human subject. Refining our definition of "reliable" into two types, experimental and extrapolation, establishes two separate, overarching criteria under which models must be assessed. Depending on the test performed, a model that is experimentally reliable may or may not be extrapolation reliable. A rodent model's experimental reliability may be affected by differing strains of an animal, vendors, environmental conditions, diet, age, gender, and handling practices. However, if these factors are standardized, then a model's reliability becomes purely a function of its extrapolation validity, which is itself an expression of the model's conceptual validity. Most widely used rodent models such as the spontaneously hypertensive rats (SHR) have already established their experimental reliability. Therefore, when we discuss the reliability of certain findings based on a model like the SHR, it is, in fact, the extrapolation reliability that we wish to examine.

By examining rodent models such as the SHR model and transgenic models such as the APP23 model, we will examine the factors that call into question their general extrapolation reliability. Subsequently, we will examine some possible reasons for the discrepancy in findings between rodent models and the human disease states. Lastly, we will make some recommendations for working with rodent models.

ERRONEOUS EXTRAPOLATION

The SHR is a widely used model of human hypertension. It has been used extensively to study various aspects of hypertension including the complications associated with hypertension such as cardiac hypertrophy and stroke.[11] In recent studies, we showed that an acidic pH induces the contraction of aortas from hypertensive SHR and normotensive Wistar–Kyoto (WKY) rats.[9] Significantly, there was an exaggerated contractile response in the SHRs as compared to the WKY rats.[12] This increased aortic contractility in the SHR due to reduced pH hinted at a possible contributory mechanism for hypertension due to increased blood acidosis (not an uncommon finding where chronic renal failure and diabetes mellitus are comorbidities with hypertension). However, when similar experiments were performed on the precontracted internal mammary artery of normotensive and hypertensive human subjects, an acidic pH caused relaxation of blood vessels.[13] This contradictory finding requires us to define the nature of the extrapolation reliability of the SHR model. While the acidic pH-induced arterial contraction is reproducible between different SHR model organisms, it does not occur in human subjects. However, the SHR does model many other aspects of adult essential hypertension reasonably accurately.[11] It therefore becomes necessary to identify the parameters within which the SHR model remains valid.

SHR has good face validity with respect to hypertension since the symptoms and measurable variables (elevated average blood pressure) seen in the model reflect those seen in human subjects. However, the SHR can be thought to have poor construct validity as the mechanism by which the symptoms of hypertension appear to be established in the rat model do not occur in humans. For the researcher, this means that the SHR is an extrapolation reliable model when considering face valid facts. For example, SHR accurately predicts that humans suffering from uncontrolled hypertension are predisposed to developing strokes. However, we cannot make an a priori assumption that the mechanisms by which hypertension causes stroke in the SHR are the same as those in the human. Thus, while a model may readily acquire face validity, establishing construct validity, especially for models that attempt to represent complex disease states like hypertension, will require careful elucidation of each underlying mechanism and the confirmation that similar mechanisms exist in the human patient.

SHR has also been used to explore the possible genetic contribution to essential hypertension in humans. The SA gene was found to be overexpressed in the SHR kidneys.[14] Multiple F_2 crosses revealed that it was located in a chromosomal region linked to high blood pressure. The SA gene thus generated much interest as a potential candidate for hypertension genes in rats.[15] However, both linkage and association analyses performed on a white population turned out to be negative.[16] As postulated by Nabika, two hypotheses can explain the discrepancy between rat and human results: (1) the population tested lacked a mutation in the SA gene, and (2) the SA gene is a neutral marker and may link with hypertension gene(s) in rats but not in humans.[15] A similar genetic trail went cold after researchers discovered that while an altered atrial natriuretic peptide (ANP) gene was etiologically related to stroke in stroke-prone SHRs, no such etiology could be confirmed for stroke in humans.[17] Given our earlier observation that the SHR model is only face valid, such discrepancies are not unexpected. As we shall reiterate later, it is vital for researchers to understand the validity and extrapolation reliability of the model being used and thus remain wary of assuming that any data from their model are automatically relevant to human patients.

Our discussion of erroneous extrapolation has, until now, dealt with the more subtle case of models that are face valid but lack construct validity. However, there are a number of models whose face validity is questionable. One example is where rodent models are used to represent the human immune response, specifically, the hypersensitivity response seen in asthma.

Murine models for asthma gained widespread acceptance after it was shown that the hallmarks of asthma (epithelial damage, microvascular leakage and edema, and eosinophil activation) could be successfully created in inbred mice.[18–22] However, the extrapolation reliability of these models is called into question by several inconsistencies. In humans, bronchial asthma is well characterized by the presence of plasma exudate surrounding the inflamed bronchial mucosa.[23] In contrast, however, *in vivo* studies in mice show little exudation.[23,24] While eosinophil degranulation is a hallmark of human asthma, degranulation has not been demonstrated in murine models of asthma.[25] The role of mast cells in murine models of asthma further questions the model's validity. In humans, the release of inflammatory mediators from mast cells leads to bronchoconstriction and the influx of inflammatory cells. However, studies with both mast-cell-deficient and IgE-deficient mice have demonstrated that these animals show degrees of bronchial hyperresponsiveness, eosinophilia, and inflammation comparable to those of normal animals.[26] It would thus appear that IgE is not critical for the development of the murine "asthmatic" phenotype. The inconsistencies between the cellular profile and pathology associated with murine asthma as compared to human asthma mean that the murine model is rendered face invalid. Given that inbred mice and outbred humans inhabit different microenvironments and possess a greatly differing genetic background, it is not surprising that a number of researchers have questioned the use of a murine model for asthma.[22,27]

Inconsistencies between rodent models of disease and their human counterparts are especially prevalent when modeling central nervous system (CNS) function and dysfunction. One of the major drawbacks of the mouse model used to model Parkinson's disease (PD) is that it does not represent the progressive neurodegeneration and gradual intensification of motor disability that are the hallmarks of human PD.[28] By a stringent definition of the term, such a model could not even be classified as face valid. In a similar vein, the rodent models for multiple sclerosis (MS) and experimental allergic encephalomyelitis (EAE) led to false starts in the development of therapies for MS due to an apparent lack of construct validity. Several treatments, though successful in preclinical EAE trials, ended up being less successful when tested on patients or caused unexpected adverse events.[29] It is thought that key differences exist in the pathogenesis of EAE as compared with MS, which consequently limit the extrapolation reliability of data gained from trials with the EAE model.[24] Similarly, discrepancies between the results of rodent model studies and those of clinical trials have prevented the development of successful therapeutic interventions for the treatment of traumatic brain injury.[30] Where a model is unable to satisfactorily represent the symptoms and/or the mechanisms of the pathology, there can be no satisfactory extrapolation of the experimental results.

The study of atherosclerosis in mouse models has also been plagued with results that cannot be sensibly extrapolated to human patients. The inhibition of angiotensin converting enzyme (ACE) has been purported to have antiproliferative, antiinflammatory, and vasodilatory effects that modulate the development of atherosclerosis in experimental rodent models. Yet, no clinical benefit of ACE inhibition has been established and this awaits further clinical trials.[31] Other studies to examine the etiology of atherosclerosis using rodent models have produced counterintuitive results. In apolipoprotein E (apo $E^{-/-}$) mice, obese diabetic mice showed significantly decreased atherosclerosis when compared to lean, nondiabetic control mice.[32]

From our brief examination, it is readily apparent that there are shortcomings in a number of rodent models that prevent experimental results derived from these models from being extrapolated to human trials. The extrapolation reliability is primarily undermined by an incomplete or inaccurate representation of the disease state where the model lacks validity (either face validity or construct validity or both). One attempt to circumvent this problem and accurately model both the symptoms and the mechanism of a particular disease has been the development of transgenic mice.

THE TROUBLE WITH TRANSGENICS

The creation of transgenic rodent models holds out the promise of actually recreating human disease within a mouse's body. In theory, this would immediately render a model both face and construct valid and allow easy extrapolation of results from such models to humans. However, some caveats must be kept in mind.

Most commonly, transgenic rodents are created for the purpose of genotype transformation or for creating gene knockouts. Both approaches are based on the premise that there is congruence between mouse and human gene function and expression. As such, the results of any alteration in rodent gene function or expression by experimental manipulation should predict the results of altered gene function in humans as a result of pathology. With the exception of a few highly conserved genes, the reality is not quite as convenient.

While humans and mice shared a common ancestor at least 100 million years ago, the ensuing years of divergent evolution have left each species' genome significantly different from that of the other. Saliently, even highly conserved protein and regulatory encoding DNA sequences differ between species due to base substitutions in synonymous codons. In addition, human genomes contain a much larger number of introns separating functionally critical protein-encoding sequences.[33] Introns have recently been found to play an important role in gene splicing and transcriptional regulation with an average of 7.8 introns per gene.[34,35] The relative paucity of introns in the mouse genome and the resulting differences in the regulation of expression between human and mouse must be accounted for when extrapolating the results of transgenic studies to humans.

For a transgenic model to be extrapolation reliable, there must be careful confirmation that the symptomatology of a disease is corroborated by expected molecular changes. For example, the validity of the APP23 mouse as a model for Alzheimer's disease (AD) was increased by the finding that along with an AD-like histopathological and behavioral phenotype, the APP23 mice also showed disturbed levels of compounds in cholinergic, adrenergic, and serotonergic systems. However, the APP23 mice still showed discrepancies with postmortem findings on human brains of AD patients, indicating that the model did not yet represent the full spectrum of the disease.[36]

The creation of a more "human" representation of disease in rodent models has yielded promise for cancer therapy. The carcinoembryonic antigen (CEA), an example of a tumor-associated molecule, is one of many potential targets for novel cancer therapies like gene and immunotherapy. Rodents lack CEA and closely related family members, which compromises preclinical testing on such models. To overcome this problem, Chan and Stanners created a transgenic mouse that contained complete CEA genes, with similar spatiotemporal expression patterns.[37] Humanized rodent models will go some way to reducing the discrepancies between the results from transgenic models and those of clinical trials. However, it is likely that intrinsic species differences will remain a significant obstacle to achieving complete extrapolation reliability.

We have highlighted a number of issues with the extrapolation reliability of rodent models, including transgenic models. The discrepancies between rodent models and human trials may result from differences in pathogenesis, genotype, and confounding factors intrinsic to rodent use.

ROLE OF CONFOUNDING FACTORS

The creation of a rodent model often employs chemical or genetic manipulation to induce a disease state. For example, streptozocin (STZ) is used to create a model of type 1 diabetes. In the diabetic state, glucose causes oxidative glycation and free radical production to induce diabetes-related complications.[38] However, the aromatic hydroxylation observed in the STZ-induced diabetic mouse has been attributed, in part, to the action of STZ itself.[39] Thus, confounding factors in the experimental design of the STZ-induced diabetic mouse can undermine the inferences made from this model about the etiology of diabetes-related complications.

In contrast to their expression in rodent models, human disease states are often accompanied by comorbidities. Animal models are often free of comorbidities such as diabetes, hypertension, or cancer and have normal leukocyte counts.[9] In a model to study the pathophysiology of sepsis and test various treatment strategies, a number of confounding factors limited the clinical success rate of using protein C as a therapeutic option. Sepsis in patients with diabetes and atherosclerosis might exhibit lower potency of therapy because of downregulation of protein C and thrombomodulin receptors on endothelial vasculature, associated with these comorbid diseases.[40]

Variation in the etiology of disease between animal models and humans is one of the fundamental confounding factors in using rodent animal models. The induced sepsis and shock models used for drug development highlight the importance for models to be clinically relevant, especially in terms of etiology and mechanism of disease. These models can generally be classified as infection models, intravenous infusion models, and endotoxicosis models.[41,42] In the bolus intravascular bacterial or endotoxin models, short-term, large doses of bacteria are administered to previously healthy rodents, unlike the slowly seeding infective focus found in humans suffering from sepsis (41). This form of sepsis induction bears no clinical correlate.[41,43] To develop more humanized animal models, it is necessary to be cognizant of the pertinent role of the route of infection, the dose, and the virulence of the bacteria.[43] Researchers who use animal models to study cardiovascular, neurological, and immunological disorders must take into account differences in the genetic background, exposure to inciting pathogen, and temporal relation of exposure to pathogen as some of the factors that can affect the validity of the animal model and, consequently, the coherence of results.

Some of the disparities seen between the results from animals and human studies can be ascribed to a paucity of accurate animal models for chronic disease.[28] Such models are imperative to mimic the persistence of conditions such as asthma and the progressive neurodegeneration of human Parkinson's disease. However, headway has been made in improving the murine mouse model for asthma. A model closer to clinical reality, exhibiting airway-specific acute-on-chronic inflammation with cellular features seen in human asthma, has been developed.[44] This brings us closer to bridging the gap between chronic disease states seen in humans and rodents, allowing us to better understand the mechanisms of chronic diseases and investigate possible therapeutic interventions.

Another potential confounding factor is the use of anesthetics in experimental animals. The preparation of rodent models often involves anesthetization in order to provide immobility and analgesia, and reduce physical and mental stress. However, this can confound the results and outcome of drug trials in such animals.[45] When studying the effects of drugs or endogenous chemical compounds in anesthetized animals, it is necessary to consider the effects of anesthetic agents on cell membranes, which can consequently alter the drug's effects on physiological manifestations of the disease.[45] Rodents also have different clearance rates of drugs due to the reduced capillary density of lower animals, which, in turn, affects the free drug bioavailability in tissues.[46] This is a pertinent point to consider for the noncoherence of drug trials.

To produce data that can reliably be applied to humans, researchers must account for both confounders and bias factors. To ensure the validity and reproducibility of results from rodent models, variables such as substrains of rodents used, sources of diet for the rodents, and environmental conditions under which the animals are kept must also be considered.[47] All these precautions must be implemented before the extrapolation reliability of results from rodent models to humans can be assessed. To economize a scientist's time and resources, we propose confirming the preliminary results obtained from animals in humans during the early stages of the study. Failing that, researchers must critically evaluate the results obtained from animals based on a thorough understanding of the extrapolation reliability of the model.

ACKNOWLEDGMENTS

We gratefully acknowledge the significant intellectual input to this manuscript by Mr. Sheheryar Kabraji, a medical student of the University of Oxford, UK.

REFERENCES

1. Hoth JJ, Stitzel JD, Gayzik FS, Brownlee NA, Miller PR, Yoza BK, McCall CE, Meredith JW, Payne RM. The pathogenesis of pulmonary contusion: An open chest model in the rat. *J Trauma* 2006;61:32–45.
2. Shukitt-Hale B, Carey AN, Jenkins D, Rabin BM, Joseph JA. Beneficial effects of fruit extracts on neuronal function and behavior in a rodent model of accelerated aging. *Neurobiol Aging* 2007;28(8):1187–1194.
3. Lu SL, Herrington H, Wang XJ. Mouse models for human head and neck squamous cell carcinomas. *Head Neck* 2006;28(10):945–954.
4. Davidoff AJ, Gwathmey JK. Pathophysiology of cardiomyopathies: Part I. Animal models and humans. *Curr Opin Cardiol* 1994;9:357–368.
5. Hau J, Andersen LLI, Rye-Nielsen B, Poulsen OM. Laboratory animal models. *Scand J Lab Anim Sci* 1989;16:7–9.
6. Dawkins PA, Stockley RA. Animal models of chronic obstructive pulmonary disease. *Thorax* 2001;56:972–977.
7. Treit D. Animal models for the study of anti-anxiety agents: A review. *Neurosci Biobehav Rev* 1985;9:203–222.
8. Tkacs NC, Thompson HJ. From bedside to bench and back again: Research issues in animal models of human disease. *Biol Res Nurs* 2006;8:78–88.
9. Rohra DK, Jawaid A, Rehman T, Zaidi AH. Reliability of rodent animal models in biomedical research. *J Coll Physicians Surg Pak* 2005;15:809–812.
10. Willner P. Validity, reliability, and utility of the chronic mild stress model of depression: A 10-year review and evaluation. *Psychopharmacology (Berl)* 1997;134:319–329.
11. Ito H, Suzuki T. Pathophysiological overview of M-SHRSP. In: Saito H, Yamori Y, Minami M, Parvez SH, Eds. *Progress in Hypertension: New Advances in SHR Research-Pathophysiology and Pharmacology*, Vol. 3. Utrecht, the Netherlands: VSP BV, 1995:1–19.
12. Rohra DK, Saito S-Y, Ohizumi Y. Extracellular acidosis results in higher intracellular acidosis and greater contraction in spontaneously hypertensive rat aorta. *Eur J Pharmacol* 2003;465:141–144.
13. Rohra DK, Sharif HM, Zubairi HS, Sarfraz K, Ghayur MN, Gilani AH. Acidosis-induced relaxation of human internal mammary artery is due to activation of ATP-sensitive potassium channels. *Eur J Pharmacol* 2005;514:175–181.
14. Iwai N, Inagami T. Isolation of preferentially expressed genes in the kidneys of hypertensive rats. *Hypertension* 1991;17:161–169.
15. Nabika T. From animal models to humans. *Clin Exp Pharmacol Physiol* 1999;26:541–543.
16. Nabika T, Bonnardeaux A, James M, Julier C, Jeunemaitre X, Corvol P, Lathrop M, Soubrier F. Evaluation of the SA locus in human hypertension. *Hypertension* 1995;25:6–13.
17. Kato N, Ikeda K, Nabika T, Morita H, Sugiyama T, Gotoda T, Kurihara H, Kobayashi S, Yazaki Y, Yamori Y. Evaluation of the atrial natriuretic peptide gene in stroke. *Atherosclerosis* 2002;163:279–286.
18. Foster PS, Hogan SP, Ramsay AJ, Matthaei KI, Young IG. Interleukin 5 deficiency abolishes eosinophilia, airways hyperreactivity, and lung damage in a mouse asthma model. *J Exp Med* 1996;183:195–201.
19. Hamelmann E, Oshiba A, Loader J, Larsen GL, Gleich G, Lee J, Gelfand EW. Anti-interleukin 5 antibody prevents airway hyperresponsiveness in a murine model of airway sensitization. *Am J Respir Crit Care Med* 1997;155:819–825.
20. Hogan SP, Mould A, Kikutani H, Ramsay AJ, Foster PS. Aeroallergen-induced eosinophilic inflammation, lung damage, and airways hyperreactivity in mice can occur independently of IL-4 and allergen-specific immunoglobulins. *J Clin Invest* 1997;99:1329–1339.
21. Kaminuma O, Mori A, Ogawa K, Nakata A, Kikkawa H, Naito K, Suko M, Okudaira H. Successful transfer of late phase eosinophil

infiltration in the lung by infusion of helper T cell clones. *Am J Respir Cell Mol Biol* 1997;16:448–454.

22. Gordon J, Grafton G, Wood PM, Larche M, Armitage RJ. Modelling the human immune response: Can mice be trusted? *Curr Opin Pharmacol* 2001;1:431–435.
23. Persson CG, Erjefalt JS, Greiff L, Erjefalt I, Korsgren M, Linden M, Sundler F, Andersson M, Svensson C. Contribution of plasma-derived molecules to mucosal immune defence, disease and repair in the airways. *Scand J Immunol* 1998;47:302–313.
24. Persson CG, Erjefält JS, Korsgren M, Sundler F. The mouse trap. *Trends Pharmacol Sci* 1997;18:465–467.
25. Lefort J, Bachelet CM, Leduc D, Vargaftig BB. Effect of antigen provocation of IL-5 transgenic mice on eosinophil mobilization and bronchial hyperresponsiveness. *J Allergy Clin Immunol* 1996;97:788–799.
26. Mehlhop PD, van de Rijn M, Goldberg AB, Brewer JP, Kurup VP, Martin TR, Oettgen HC. Allergen-induced bronchial hyperreactivity and eosinophilic inflammation occur in the absence of IgE in a mouse model of asthma. *Proc Natl Acad Sci USA* 1997;94:1344–1349.
27. Persson CG. Con: Mice are not a good model of human airway disease. *Am J Respir Crit Care Med* 2002;166:6–7.
28. Bezard E, Imbert C, Gross CE. Experimental models of Parkinson's disease: From the static to the dynamic. *Rev Neurosci* 1998;9:71–90.
29. Friese MA, Montalban X, Willcox N, Bell JI, Martin R, Fugger L. The value of animal models for drug development in multiple sclerosis. *Brain* 2006;129:1940–1952.
30. Bullock MR, Lyeth BG, Muizelaar JP. Current status of neuroprotection trials for traumatic brain injury: Lessons from animal models and clinical studies. *Neurosurgery* 1999;45:207–217.
31. Curzen NP, Fox KM. Do ACE inhibitors modulate atherosclerosis? *Eur Heart J* 1997;18:1530–1535.
32. Lyngdorf LG, Gregersen S, Daugherty A, Falk E. Paradoxical reduction of atherosclerosis in apoE-deficient mice with obesity-related type 2 diabetes. *Cardiovasc Res* 2003;59:854–862.
33. Sakharkar MK, Chow VT, Kangueane P. Distributions of exons and introns in the human genome. *In Silico Biol* 2004;4:387–393.
34. Majewski J, Ott J. Distribution and characterization of regulatory elements in the human genome. *Genome Res* 2002;12:1827–1836.
35. Patthy L. Modular assembly of genes and the evolution of new functions. *Genetica* 2003;118:217–231.
36. Van Dam D, Marescau B, Engelborghs S, Cremers T, Mulder J, Staufenbiel M, De Deyn PP. Analysis of cholinergic markers, biogenic amines, and amino acids in the CNS of two APP overexpression mouse models. *Neurochem Int* 2005;46:409–422.
37. Chan CH, Stanners CP. Novel mouse model for carcinoembryonic antigen-based therapy. *Mol Ther* 2004;9:775–785.
38. Hunt JV, Wolff SP. Oxidative glycation and free radical production: A causal mechanism of diabetic complications. *Free Radic Res Commun* 1991;12–13:115–123.
39. Lubec B, Hermon M, Hoeger H, Lubec G. Aromatic hydroxylation in animal models of diabetes mellitus. *FASEB J* 1998;12:1581–1587.
40. Esmon CT. The roles of protein C and thrombomodulin in the regulation of blood coagulation. *J Biol Chem* 1989;264:4743–4746.
41. Wichterman KA, Baue AE, Chaudry IH. Sepsis and septic shock: A review of laboratory models and a proposal. *J Surg Res* 1980;29:189–201.
42. Fink MP, Heard SO. Laboratory models of sepsis and septic shock. *J Surg Res* 1990;49:186–196.
43. Deitch EA. Animal models of sepsis and shock: A review and lessons learned. *Shock* 1998;9:1–11.
44. Kumar RK, Foster PS. Modeling allergic asthma in mice: Pitfalls and opportunities. *Am J Respir Cell Mol Biol* 2002;27:267–72.
45. Bazin JE, Constantin JM, Gindre G. [Laboratory animal anesthesia: Influence of anesthetic protocols on experimental models.] *Ann Fr Anesth Reanim* 2004;23:811–818.
46. Greaves P, Williams A, Eve M. First dose of potential new medicines to humans: How animals help. *Nat Rev Drug Discov* 2004;3:226–236.
47. Anderson LE, Morris JE, Sasser LB, Loscher W. Effects of 50- or 60-hertz, 100 microT magnetic field exposure in the DMBA mammary cancer model in Sprague-Dawley rats: Possible explanations for different results from two laboratories. *Environ Health Perspect* 2000;108:797–802.

Well-Established Models III

Cats, Dogs, and Pigs C

25 The Domestic Cat, *Felis catus*, as a Model of Hereditary and Infectious Disease

MARILYN MENOTTI-RAYMOND AND STEPHEN J. O'BRIEN

ABSTRACT

The domestic cat, currently the most frequent of companion animals, has enjoyed a medical surveillance, as a nonprimate species, second only to the dog. With over 200 hereditary disease pathologies reported in the cat, the clinical and physiological study of these feline hereditary diseases provides a strong comparative medicine opportunity for prevention, diagnostics, and treatment studies in a laboratory setting. Causal mutations have been characterized in 19 felid genes, with the largest representation from lysosomal storage enzyme disorders. Corrective therapeutic strategies for several disorders have been proposed and examined in the cat, including enzyme replacement, heterologous bone marrow transplantation, and substrate reduction therapy. Genomics tools developed in the cat, including the recent completion of the 2-fold whole genome sequence of the cat and genome browser, radiation hybrid map of 1793 integrated coding and microsatellite loci, a 5-cM genetic linkage map, arrayed BAC libraries, and flow sorted chromosomes, are providing resources that are being utilized in mapping and characterization of genes of interest. A recent report of the mapping and characterization of a novel causative gene for feline spinal muscular atrophy marked the first identification of a disease gene purely from positional reasoning. With the development of genomic resources in the cat and the application of complementary comparative tools developed in other species, the domestic cat is emerging as a promising resource of phenotypically defined genetic variation of biomedical significance. Additionally, the cat has provided several useful models for infectious disease. These include feline leukemia and feline sarcoma virus, feline coronavirus, and Type C retroviruses that interact with cellular oncogenes to induce leukemia, lymphoma, and sarcoma.

Key Words: Domestic cat, *Felis catus*, Gene therapy, Whole genome sequence, Radiation hybrid map, Knockout model, FIV, SARS.

INTRODUCTION

Mankind has held a centuries-long fascination with the cat. The earliest arch eological records that have been linked to the domestication of *Felis catus* date to approximately 9500 years ago from Cyprus,[1] with recent molecular genetic analyses in our laboratory suggesting a Middle Eastern origin for domestication (C. Driscoll *et al.*, unpublished observations). Currently the most numerous of companion animals, numbering close to 90 million in households across the United States (http://www.appma.org/press_industrytrends.asp), the cat enjoys a medical surveillance second only to the dog and humankind. In this chapter we review the promise of the cat as an important model for the advancement of human hereditary and infectious disease and the genomic tools that have been developed for the identification, and characterization of genes of interest.

For many years we have sought to characterize genetic organization in the domestic cat and to develop genomic resources that establish *F. catus* as a useful animal model for human hereditary disease analogues, neoplasia, genetic factors associated with host response to infectious disease, and mammalian genome evolution.[2,3] To identify genes associated with inherited pathologies that mirror inherited human conditions and interesting pheno-types in the domestic cat, we have produced genetic maps of sufficient density to allow linkage or association-based mapping exercises.[4–11]

The first genetic map of the cat, a physical map generated from a somatic-cell hybrid panel, demonstrated the cat's high level of conserved synteny with the human genome, which offered much promise for the future application of comparative genomic inference in felid mapping and association exercises.[12] Several radiation hybrid (RH) and genetic linkage (GL) maps have since been published.[4–9,11,13,14]

THE DOMESTIC CAT RADIATION HYBRID MAP

Although previous versions of the cat gene map, based on somatic cell hybrid and ZOO FISH analysis,[15,16] revealed considerable conservation of synteny with the human genome, these maps provided no knowledge of gene order or intrachromosomal genome rearrangement between the two species, information that is critical to applying comparative map inference to gene discovery in gene-poor model systems. Radiation hybrid (RH) mapping has emerged as a powerful tool for constructing moderate- to high-density gene maps in vertebrates by obviating the need to identify interspecific polymorphisms critical for the generation of genetic linkage maps.[7]

The most recent RH map of the cat[8] includes 1793 markers: 662 coding loci, 335 selected markers derived from the cat 2X whole genome sequence targeted at breakpoints in conserved synteny between human and cat, and 797 short tandem repeat (STR) loci. The strategy used in developing the current RH map was to target gaps in the feline–human comparative map, and to provide more definition in breakpoints in regions of conserved synteny between cat and human. The 1793 markers cover the

From: *Sourcebook of Models for Biomedical Research*
(P. M. Conn, ed.), © 2008 Humana Press Inc., Totowa, NJ.

length of the 18 feline autosomes and the X chromosome at an average spacing of one marker every 1.5 Mb (megabase), with fairly uniform marker density.[8] An enhanced comparative map demonstrates that the current map provides 86% and 85% comparative coverage of the human and canine genomes, respectively.[8] Ninety-six percent of the 1793 cat markers have identifiable orthologues in the canine and human genome sequences, providing a rich comparative tool, which is critical in linkage mapping exercises for the identification of genes controlling feline phenotypes. Figure 25–1 presents a graphic display of each cat chromosome and blocks of conserved syntenic order with the human and canine genomes.[8] One hundred and fifty-two cat–human and 134 cat–dog homologous synteny blocks were identified. Alignment of cat, dog, and human chromosomes demonstrated different patterns of chromosomal rearrangement with a marked increase in interchromosomal rearrangements relative to human in the canid lineage (89% of all rearrangements), as opposed to the more frequent intrachromosomal rearrangements in the felid lineage (95% of all rearrangements) since divergence from a common carnivore ancestor 55 My ago.

With an average spacing of 1 marker every 1.5 Mb in the feline euchromatic sequence, the map provided a solid framework for the chromosomal assignment of feline contigs and scaffolds during assembly of the cat genome assembly,[17] and served as a comparative tool to aid in the identification of genes controlling feline phenotypes.

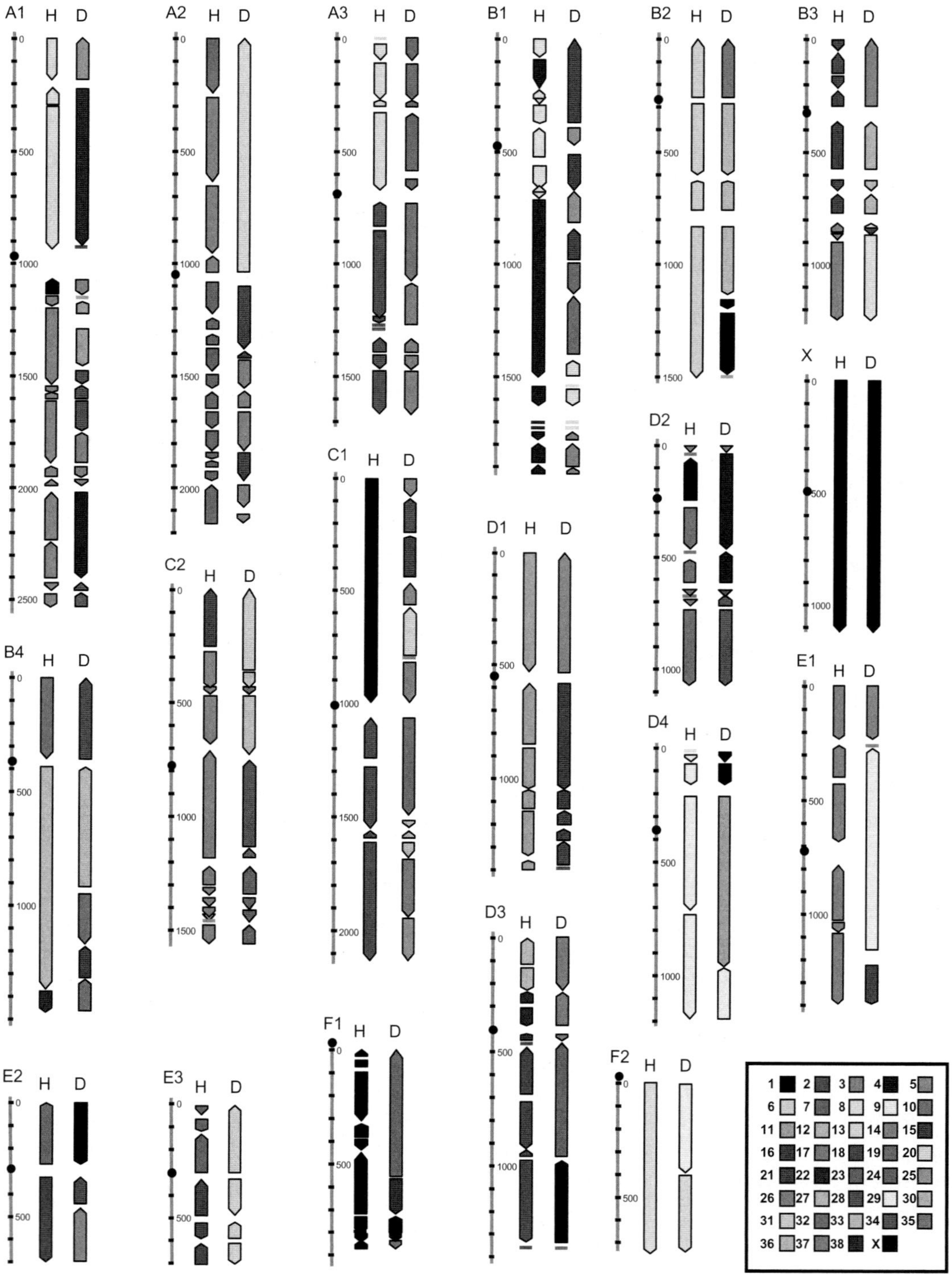

Figure 25–1. Feline chromosome maps (labeled at top) and homologous synteny blocks (HSBs) in the human (H) and dog (D) genomes. HSBs are shown to the right of each cat chromosome map (only the map scale is shown). The dark cross-marks on each cat chromosome correspond to 100-cR$_{5000}$ intervals. The inferred centromere positions are shown by dark circles. HSBs are color coded by human or dog chromosome, defined by the key in the bottom right corner. (Reprinted from Murphy *et al.*[7] Copyright 2007, with permission from Elsevier.) (See color insert.)

THE DOMESTIC CAT GENETIC LINKAGE MAP

As a complement to the RH map of the cat, a third generation linkage map of 625 STRs is currently nearing completion. The map has been generated in a large multigeneration domestic cat pedigree (n = 483 informative meioses).[18] Previous first- and second-generation linkage maps of the cat were generated in a multigeneration interspecies pedigree generated between the domestic cat and the Asian leopard cat, *Prionailurus bengalensis*,[7] to facilitate the mapping and integration of Type I (coding) and Type II (polymorphic STR) loci.[7] The current map, which spans all 18 autosomes with single linkage groups, has twice the STR density of previous maps, providing a 5-cM resolution. There is also greatly expanded coverage of the X chromosome, with some 75 STR loci. Marker order between the current generation RH and GL maps is highly concordant.[8]

Approximately 85% of the STRs are mapped in the most current RH map of the cat,[8] which provides reference and integration with Type I loci. Whereas the third-generation linkage map is composed entirely of STR loci, the sequence homology of extended genomic regions adjacent to the STR loci in the cat 2X whole genome sequence,[17] to the dog's homologous region,[19] has enabled us to obtain identifiable orthologues in the canine and human genome sequences for over 95% of the STRs. Thus, practically every STR acts as a "virtual" Type 1 locus, with both comparative anchoring and linkage map utility. Combined with the cat RH map, these genomic tools provide us with the comparative reference to other mammalian genomes critical for linkage and association mapping.

THE DOMESTIC CAT WHOLE GENOME (2X) SEQUENCE

The domestic cat is one of 26 mammalian species endorsed by the National Human Genome Research Institute (NHGRI) Human Genome Annotation committee for a "light" 2-fold whole genome sequence, largely to capture the pattern of genome variation and divergence that characterizes the mammalian radiations (http://www.hgsc.bcm.tmc.edu/projects/bovine/, http://www.broad.mit.edu/mammals/). Although light genome coverage provides limited sequence representation, (~80%),[20] one of the rationales for these light genome sequences included "enhancing opportunities for research on species providing human medical models." The 2-fold assembly of the domestic cat genome has recently been completed for a female Abyssinian cat, "Cinnamon," [17] and a 7X whole genome sequencing effort is planned in the near future.

A total of 9,161,674 reads were assembled to 817,956 contigs, covering 1.642 Gb with an N50 (i.e., half of the sequenced base pairs are in contigs <N50) contig length of 2378 bp. Assembled supercontigs (N = 217,790) had an N50 length of 117 kb[17] (http://hosted. abcc.ncifcrf.gov/cgi-gin/gbrowse/cat/). The estimated size of the genome was 2.7 Gb and the genome coverage was approximately 2-fold, predicting an average inclusion of 80–85% of the eukaryotic genome sequence.[21]

Feline coding genes were identified using a comparative approach based upon sequence homology and syntenic orthology of neighboring gene homologues in the genomes of six index mammal species (human, chimp, mouse, rat, cow, and dog). The results revealed nearly 21,080 feline genes plus 132,493 conserved sequence blocks (CSBs) used to build the gene map,[17] depending upon the framework RH map of 1794 ordered Type 1 markers.[8] The 2X feline genome sequence detects 83% of human genes, 89% of chimp or cattle genes, and 92% of dog genes based upon sequence identity to approximately 1000 bp of reciprocal base match[17] between the cat sequence and the genome sequence of the six index mammals.

A genome browser has been developed from the cat assembly, named Genome Annotation Region FIELD (GARFIELD), which provides a physical map of the 18 autosomes and the X chromosome, which can be inspected for sequence representation, including genes and the proportion of that gene available in the 2X cover, single nucleotide polymorphisms (SNPs), and STRs, which can be used in linkage and association mapping, and other genome features (http://ccr.cancer.gov/ labs/lab). Figure 25–2 illustrates a

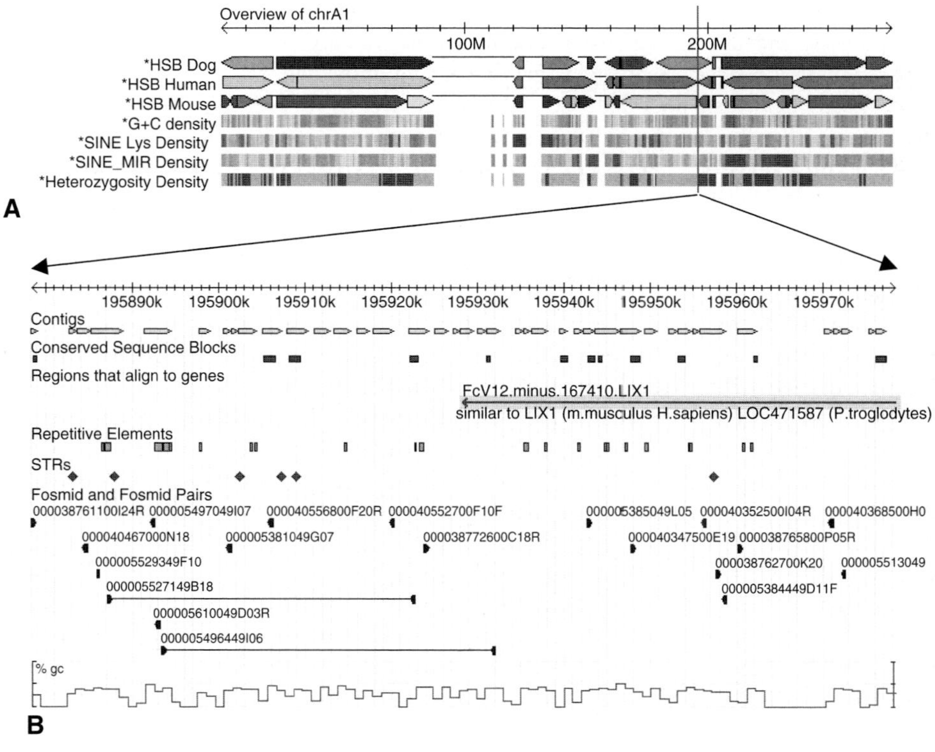

Figure 25–2. Gene annotation region fields (GARFIELD) displayed using Generic Genome Browser at two levels of resolution for the *LIX1* gene on chromosome A1. (A) Chromosome view showing homologous synteny blocks (HSB) for dog, human, and mouse, representation of G + C density, and of SINE, LINE, and SNPs. (B) A 200-kb view showing contigs of the region, conserved sequence blocks (CSB), regions that align to annotated genes in other mammalian genomes, regions masked by repeat masker, single tandem repeats, Fosmid reads with their partners, and a histogram of local GC content.

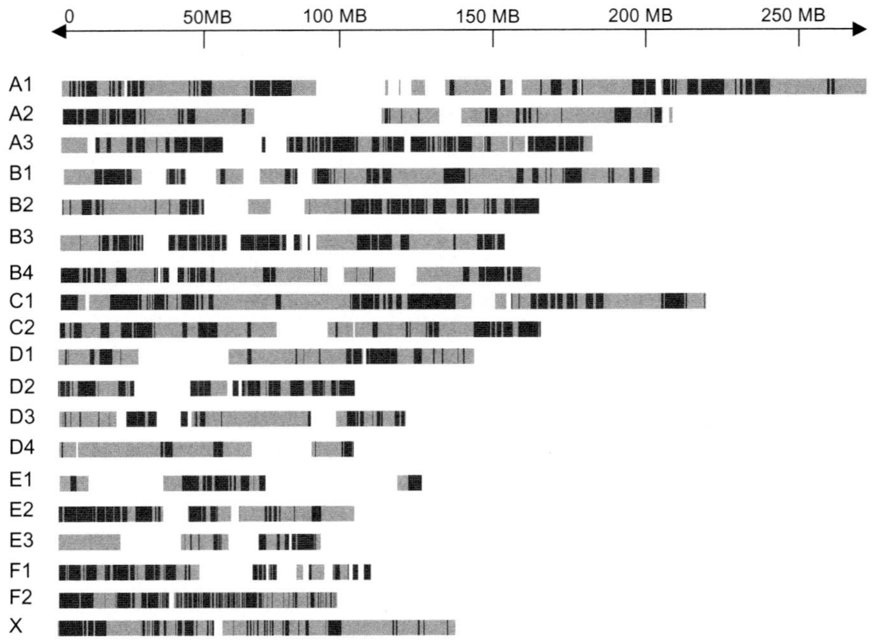

Figure 25–3. SNP profile across the 2X whole genome sequence of the domestic cat, Cinnamon. Heterozygosity across Cinnamon's chromosomes is represented in nonoverlapping windows of 100kb. Black represents regions with more than two SNPs per 100kb while gray represents the homozygous region (less than two SNP, 0/100kb). White represents gaps in the chromosome assembly.

representative view of GARFIELD demonstrating features for the *LIX1* gene on chromosome A1.[17]

SINGLE NUCLEOTIDE POLYMORPHISM ANALYSIS DEMONSTRATES POTENTIAL OF LINKAGE DISEQUILIBRIUM MAPPING IN CAT BREEDS

A total of 421,000 SNP variants were identified in Cinnamon's sequence, representing an incidence of 1/600bp.[17] Approximately 43% of Cinnamon's genome was heterozygous and 57% homozygous, which was not unexpected in a breed cat that is also the member of a highly inbred pedigree for retinal atrophy.[17,22] Long stretches of alternating homozygous and heterozygous segments were observed (Figure 25–3), which represent the consequences of close inbreeding during the domestication process, and the more recent generation of fancy breeds and inbred disease pedigree.[22] Similar patches of homozygous/heterozygous segments were observed in the recently released whole genome sequence of the dog.[19] The length of the segments is influenced by breed-specific history including effective population sizes, use of popular sires, and population bottlenecks.[23,24]

Linkage disequilibrium (LD) mapping has recently emerged as a powerful approach in humans for association mapping.[25] Long stretches of linkage disequilibrium in the target population greatly facilitate the success of the strategy and decrease the number of markers required for analysis.[25] The extended LDs observed in dog breeds, up to one hundred times the length observed in human populations,[26] is proving to be a powerful mapping strategy for identification of genes associated with breed-specific phenotypes,[27] including hereditary pathologies.[28,29]

The potential of this mapping approach in cat breeds was evaluated by examining breed-specific patterns of common segment homozygosity in 24 Cat Fancier Association (CFA) (http://www.cfainc.org) breeds.[17] The level of homozygosity reflected in a group of 665 SNPs was roughly half that seen in dogs,[26] likely reflecting a more extensive recent inbreeding within dog versus cat breeds.[30] This level of homozygosity was used to estimate[19] that some 45,000 equivalently spaced SNP variants would be required for a linkage disequilibrium/haplotype-based association genome search of a complex heritable disease within cat breeds.

Recently, tyrosinase-related protein 1 (*TYRP1*), one of the key enzymes in the melanogenic pathway, was linked to two coat color variants in the cat by association mapping in 38 cat breeds due to extensive LD.[31] Two DNA polymorphisms in *TYRP1*, an A3G substitution in the signal peptide and an in-frame insertion *TYRP1*–421ins17/18, were associated with the *chocolate* (b) allele. A premature UAG stop codon at position 100 of *TYRP1* was associated with a second allele of the B locus, *cinnamon* (bl).[32] SNP discovery is planned in the 7X whole genome sequencing of the cat through a resequencing strategy of selected genomic regions in several cat breeds as was recently performed in the 7X whole genome sequencing of the dog.[19]

THE MATURITY OF CURRENT FELID MAPPING RESOURCES DEMONSTRATED IN SUCCESSFUL WHOLE GENOME AND ASSOCIATION MAPPING EXERCISES

The majority of hereditary pathologies in the domestic cat for which the gene defect has been elucidated have resulted from the analysis of candidate genes (Table 25–1). However, with the availability of a detailed comparative map, and integration with developing GL and RH maps, and the cat 2X whole genome sequence, linkage and association-based mapping techniques have recently identified causative mutations for hereditary disease genes,[33,34] as well as several feline phenotypes (Table 25–1).[18,32,35–57]

Once a genomic region is implicated from association-based or linkage mapping exercises, fine mapping has been accomplished by development of new STRs or SNPs in the targeted region using the cat 2X whole genome sequence data accessed

Table 25–1
Feline genetic diseases/phenotypes characterized at a molecular level

Disease/phenotype	*Gene*	*Mutationa*	*Reference*
α-Mannosidosis	MAN2B1	1749_1752delCCAG leads to premature stop	38
Gangliosidosis G_{M1} (Sandhoff disease)	GLB1	R482P	39
Gangliosidosis G_{M2}	HEXB	39delC leads to premature stop or 1467_1491 inv; del exon12	40, 41
	GM2A	Del4bp in 3′ region leads to frame shift	42
Glycogenosis IV	GBE1	Gene rearrangement with insertion and large deletion in exon 12	43, 44
Hemophilia B	F8	R338X, C82Y	45
Hypertriglyceridemia (lipoprotein lipase deficiency)	LPL	G412R	46
Hypertrophic cardiomyopathy	MYBPC3	A31P	47
Mucolipidosis II (I-cell disease)	GNPTAB	C2655T	48
Mucopolysaccharoidosis Type I	IDUA	107_109 delCGA	49
Mucopolysaccharoidosis Type VI	ARSB	L476P (severe phenotype) D520N (mild phenotype)	50
Mucopolysaccharoidosis Type VII	GUSB	E351K	51
Muscular dystrophy, Duchenne type	DMD	Deletion in the dystrophin muscle promoter	52
Niemann–Pick disease, Type C	NPC1	G2864C	53
Oculocutaneous albinism (Type II)	TYR	R422Q	54
Polycystic kidney disease	PKD1	C3284X	33
Pyruvate kinase deficiency	PKLR	Splicing defect leads to 13 bp deletion in exon 6	55
Retinal degeneration in Abyssinian cats (rdAc)			
Spinal muscular atrophy	LIX1	~140 kb deletion	34
Albino	TYR	del975C leads to premature stop	36
Brown	TYRP1	A3G and 421_422 ins 18AA/19AA	32
Burmese	TYR	G227W	32, 37
Cinnamon	TYRP1	R100X	32
Dilute	MLPH	del83T leads to premature stop	35
Melanism	ASIP	123_124delCA leads to frame shift	18
Melanism (jaguar)	MC1R	301_315del	18
Melanism (jagaroundi)	MC1R	283_306del	18
Siamese	TYR	G301R	32, 37
Sweet taste receptor	TAS1R2	454_700del	56

[a]Mutation notation according to den Dunnen and Antonarakis.[53]

through the cat genome browser, GARFIELD.[17] A recent report of the mapping and characterization of a novel gene causative of feline spinal muscular atrophy[34] marked the first identification of a disease gene purely from positional reasoning.

Human spinal muscular atrophies (SMAs) are a genetically heterogeneous group of neuropathies that varies in clinical severity, from lethal in infancy to onset of mild weakness in adulthood, but all are characterized by neurogenic muscle atrophy due to degeneration of lower motor neurons of the spinal cord.[58] For approximately 97% of people affected with SMA, disease pathology is attributable to a mutation in the *SMN1* gene, on human chromosome 5q13, which is subject to a high frequency of deletions and gene conversion events with the divergent and only partially functional centromeric copy/copies of the duplicated *SMN2* locus.[59,60]

A domestic cat model of SMA has been described that is a model of autosomal recessive juvenile-onset SMA.[61] With the feline *SMN* gene excluded as the disease locus,[61] a full genome linkage scan was conducted in a pedigree segregating for SMA.[61] The disease phenotype was linked to chromosome A1q,[34] in a region of conserved synteny to human chromosome 5q15. Fine mapping was accomplished with development of new STRs and sequence tagged sites (STS), utilizing sequence information from the cat 2X whole genome sequencing effort, which ultimately identified an ~140 kb deletion and a novel gene candidate, *LIX1*[34] (Figure 25–4). Though the function of *LIX1* is unknown, the predicted secondary structure is compatible with a role in RNA metabolism. An exon sequence screen of 25 human SMA cases, not otherwise explicable by mutations at the *SMN1* locus, failed to identify comparable *LIX1* mutations.[34]

The *SMN1* gene product, SMN, is a ubiquitously expressed protein member of multiple ribonucleoprotein complexes with diverse roles in RNA metabolism, splicing, and transport in all cells.[62,63] A central focus of SMA research remains to discern the disease mechanism(s) and to understand why the primary disease pathology is localized to spinal lower motor neurons when all cells require SMN function. *LIX1* expression is largely restricted to the central nervous system (CNS), primarily in spinal motor neurons, thus offering an explanation of the tissue restriction of pathology in feline SMA. Determination of *LIX1* function may well provide fresh insight into the mechanisms of human SMA pathology, impetus for more targeted therapeutics, and answers to fundamental questions of motor neuron development, maintenance, and/or function.

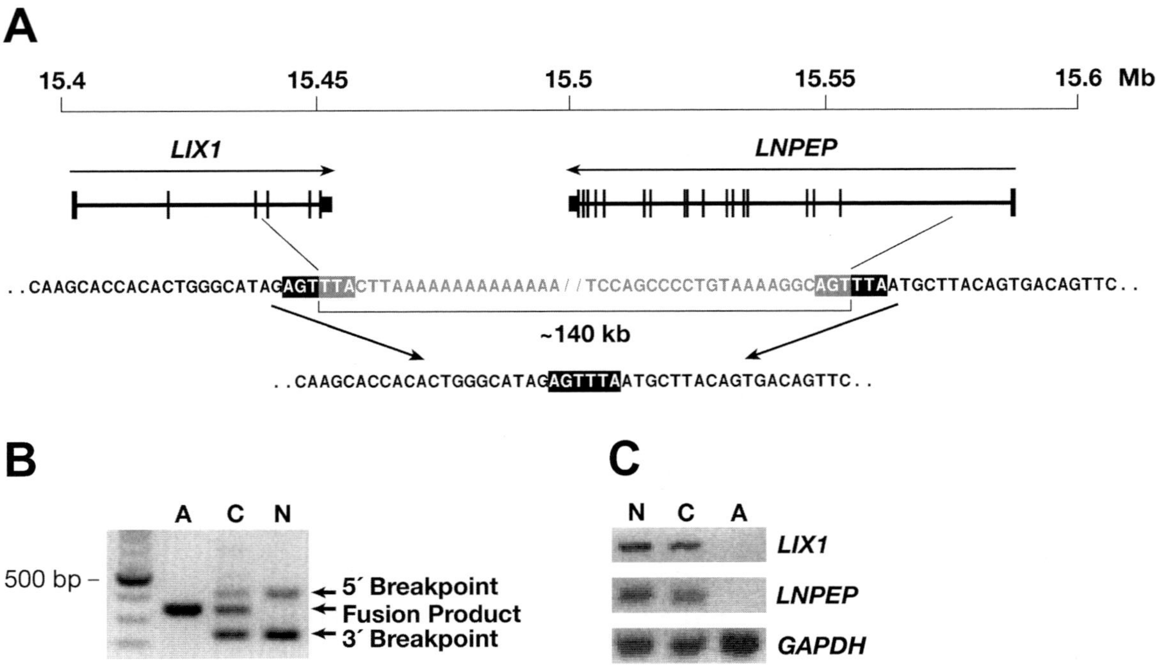

Figure 25–4. An ~140-kb deletion abrogates *LIX1* and *LNPEP* expression in SMA affected cats. (A) A schematic of the genomic organization of *LIX1* and *LNPEP* on FCA A1q. Arrows indicate the direction of transcription of each gene. The scale of chromosome coordinates is from the region of conserved synteny in the dog genome (CFA 3) (UCSC Genome Browser; July 2004 assembly). Sequences of the deleted and normal alleles are aligned showing the precise breakpoints *below* the schematic. The immediately flanking 6-bp sequence, AGTTTA, is in bold. (Note that the genome browser did not itself identify exon 1 of *LNPEP* by homology with RefSeq genes, but limited homology was identified by BLAT search of the dog genome with a cat sequence that was 82% identical over 300 bp of the human exon 1 and flanking 5′-UTR and intron 1 sequences.) (B) PCR products produced from genomic DNA in a multiplex reaction designed to amplify across both deletion breakpoints when the normal allele (lane N) was present. A third product corresponding to the affected cat sequence shown in (A) was amplified when the deleted allele was present, either in homozygous affected (lane A) or heterozygous carrier (lane C) cats. (C) RT-PCR products amplified from cervical spinal cord ventral horn RNA of a genetically normal cat (lane N), a clinically normal carrier (lane C), and an SMA affected cat (lane A). PCR primers for *LIX1* were from exons 1 and 4, *LNPEP* primers were from exons 14 and 15, and *GAPDH* primers were from exons 6 and 8. (Reprinted from Fyfe *et al.*[32] Copyright 2006, with permission from Cold Spring Harbor Laboratory Press.)

THE CAT AS ANIMAL MODEL FOR HUMAN HEREDITARY DISEASE

The world's veterinary schools produce thousands of practitioners each year, most of whom carefully document genetic and chronic diseases of our pets. The result is a comprehensive veterinary literature that has described over 200 feline hereditary pathologies[17] (http://www.angis.org.au/). The clinical and physiological study of these feline hereditary diseases provides a strong comparative medicine opportunity for prevention, diagnostics, and treatment studies in a laboratory setting. Additionally, large animal homologues are similar to humans in natural genetic diversity and offer the possibility of evaluating long-term effects of treatment.[64]

To date, causal mutations have been characterized in 19 felid genes that cause hereditary disease (Table 25–1). The largest representation comes from lysosomal storage enzyme disorders that arise from defects in genes playing a role in degradation of macromolecules targeted to the lysosomes. Many of the genes that cause these pathologies have been mapped in the cat.[65]

Corrective therapeutic strategies have been proposed and examined in the cat, including enzyme replacement, heterologous bone marrow transplantation, and substrate reduction therapy.[64,66] Limitations to these treatment strategies include high morbidity and mortality, limited positive outcomes, incomplete response to therapy, cost, and in some cases requirements for continuous lifelong therapy.[66] Gene therapy poses the most recent of intervention strategies. Feline models have been important in elucidating molecular pathogenesis and are now playing a critical role in evaluating and optimizing the range of therapeutic strategies prior to clinical trials in humans.

The mucopolysaccharidoses are disorders that result from the deficiency of lysosomal enzymes involved in the degradation of mucopolysaccharides. The cat offers homologous models for mucopolysaccharidosis Types I, VI, and VII.

Mucopolysaccharidosis Type I (MPS I), which results from a genetic deficient activity of the enzyme α-L-iduronidase (IDUA), can lead to mental retardation, growth abnormalities, and shortened life span in humans.[67] Naturally occurring models have been characterized in the cat[68,69] and dog.[70] Immune responses can nullify the effect of gene corrective therapy. It has been demonstrated that cats, but not dogs, mount a potent CTL response to canine IDUA after neonatal gene therapy, which can be prevented with transient CTLA4-Ig.[71] The efficacy of neonatal retroviral therapy has also been explored in the cat.[66] The cat model, additionally, provides an ideal system to study mechanisms of brain neurodegeneration and neural-directed strategies, especially given a large body of preexisting literature on cat neurology.

MPS VI or Marteaux–Lamy disease, deficient activity of arylsulfatase B (ARSB), is characterized in humans and cats with growth retardation, coarse facial features, corneal opacity, and skeletal deformities.[72–75] The feline model also exhibits abnormal lysosomal storage in occasional neurons and glia distributed throughout the cerebral cortex.[76] Fibroblast-mediated *in vitro* gene therapy has been examined in this cat model.[77] Recently, an adeno-associated vector containing feline *ARSB* has demonstrated gene therapy-based correction of corneal clouding in the MPS VI cat.[78]

MPS VII results from deficiency of β-glucuronidase (GUSB), which in humans manifests as cartilaginous and bony malformations, growth and mental retardation, abdominal organ enlargement, and corneal clouding.[79] Naturally occurring animal models have been described in mice,[80] dogs,[81] and the cat.[82] Enzymatic activity has been restored in fibroblasts and restored by retroviral gene transfer of rat GUSB cDNA. As GUSB is an essential housekeeping enzyme, this feline model is important for examination of exogenous genes and gene product delivery to a variety of tissue types, and could prove especially valuable due to extensive research conducted on the anatomy and physiology of the cat central nervous and visual systems. Three different serotypes of adeno-associated viral constructs of *GUSB* demonstrated the efficacy of this vector to achieve gene transfer in the normal cat brain, as a model for the efficacy of this construct in a large mammalian brain.[83]

Deficiency of lysosomal α-mannosidase leads to an accumulation of mannose-rich oligosaccharides,[84] which leads to mental retardation, recurrent infections, skeletal changes and hearing impairment.[85] This feline model was initially important in achieving bone marrow transplantation as corrective strategy for neuronal storage diseases of the CNS.[86,87] Recently, adeno-associated viral (AAV) constructs of feline α-mannosidase were used to demonstrate the efficacy of CNS gene therapy. Treated cats exhibited widespread improvement of neuropathology, showing the efficacy of this treatment in a large mammalian brain for CNS correction of human lysosomal enzyme deficiencies.[88]

Lipoprotein lipase (LPL) is a crucial enzyme involved in the regulation of lipoprotein and lipid metabolism ability to thrive.[89] Cats with LPL deficiency display a remarkably similar phenotype to humans, including severe pancreatitis, chylomicronemia, and failure to thrive.[46] There is currently no adequate treatment for this pathology in humans. Cats could prove to be the most valuable animal model of LPL deficiency, as of numerous animal model systems examined including the mouse, the cat most closely resembles the lipoprotein pattern and lipid transport system of humans. Recently, AAV-mediated transfer of a human LPL ^{S447X} variant into feline muscle cells demonstrated correction of the hypertriglyceridemia associated with feline pathology; this offers much promise for treatment of human LPL deficiency.[90]

A separate class of lysosomal storage disorder characterized in the cat is the gangliosidoses, G_{M1} and G_{M2}, which are heritable neurodegenerative diseases. A deficiency of lysosomal β-galactosidase results in the neuronal accumulation of the G_{M1} ganglioside, while the degradation of the G_{M2} ganglioside is initiated by coordinated action of at least three gene products, the α and β submits of β-*N*-acetylhexosaminidase and the G_{M2} activator (GM2A) protein.[42,91] Mutations in any of these enzymes result in an accumulation of gangliosides G_{M1} and G_{M2} in the lysosomes of affected neurons, resulting in progressive deterioration of the CNS. Feline models have been especially important in characterizing the pathobiology and molecular biology of these diseases. G_{M2}-gangliosidosis has been characterized in cat models deficient in the G_{M2} activator protein and HexB,[40,42,92,93] exhibiting remarkably similar pathology to human Sandhoff's disease.[91] Limited reduction in G_{M2} neuronal storage has been reported following bone marrow therapy.[86] Feline models will be important in the development of therapeutic strategies for these disorders.

Mucolipidosis II (I-cell disease) is caused by deficient activity of the enzyme *N*-acetylglucosamine-1-phosphotransferase, which leads to a failure to internalize enzymes into lysosomes. The cat is the only known animal model for this pathology.[94]

Congenital diseases of feline muscle and neuromuscular junction have been reviewed by Gashen *et al.*[95] Some pathologies have been observed in isolated breeds, including hypokalemic myopathy of Burmese cats,[96] glycogen storage disease type IV in Norwegian Forest cats,[44] and myopathy observed in the Devon Rex.[97] The cat is the only reported animal model for type IV glycogen storage disease. Myotonia congenital,[95,98] muscular dystrophy (dystrophin deficient),[52] and laminin α_2 deficiency[52,99] have also been reported in the cat.

X-linked muscular dystrophy in humans is characterized by progressive degeneration of skeletal and cardiac muscle. Mutations in humans lead to either an absence of or abnormality in the protein product dystrophin.[100,101] A deletion in the dystrophin muscle promoter characterized in the cat eliminates expression of muscle and Purkinje neuronal dystrophin isoforms.[52] Marked clinical heterogeneity is observed in these models, from severe disability exhibited in human and dog, to minor muscle fibrosis and an actual regenerative process leading to muscle hypertrophy in mouse and cat.[102–104] These different sequellae could be important in characterizing immediate and secondary consequences of the lack of dystrophin[105] and points out the importance of multiple animal models.

Hypertrophic cardiomyopathy is a clinically heterogeneous myocardial disease contributing to one of the most common causes of sudden cardiac death in young adults.[106] The cat represent the first spontaneous large animal model for this familial disease[47] and will prove to be valuable for examining pathophysiological processes and therapeutic interventions.

THE POTENTIAL FOR KNOCKOUT CATS

There is a continued demand for alternative mammalian models for studying human diseases. Compared to traditional murine models, the cat's more similar physiology, increased size, and longevity have made it ideal for testing the safety and efficacy of some therapeutic modalities in naturally occurring feline disease models.[107–110] Additionally, cats have a genetically heterogeneous background, similar to humans. A good example of a human disease in need of a better mammalian model is cystic fibrosis, as mice targeted with the most common human mutation (ΔF508) in the cystic fibrosis transmembrane receptor (*CFTR*) fail to spontaneously develop the same opportunistic lung infections that plague human patients with cystic fibrosis.[111,112] One group has been working to produce a ferret cystic fibrosis model through gene targeting of somatic cells for nuclear transfer.[113,114] While reproductive cloning has a consistently low success rate (1–4%) across mammalian species,[115] targeting genetic loci through homologous recombination in somatic cells is currently the only viable method for producing knockout models in all mammalian species other than the mouse, for which targeted embryonic stem cells are routinely used. Nuclear transfer of targeted fibroblasts has been successfully used to produce viable α-1,3-galactosyltransferase knockout pig and cow models for xenotransplantation studies.[116,117] With the successful reproductive cloning of cats by several groups,[118–120] the development of gene-targeted cat models through nuclear transfer is now feasible.[121] The imminent release of the annotated feline genome project,[17] integration of recombination and radiation hybrid maps,[5–10,122] and availability of multiple PAC, BAC,[123] flow-sorted autosome, and Y-chromosome libraries[124,125] (J. Pecon-Slattery *et al.*, unpublished observations) at the Laboratory of Genomic Diversity will facilitate efforts by researchers to develop new cat models of specific human genetic diseases.

COAT COLOR GENES IN THE DOMESTIC CAT

The coat color loci influence the development, maturation, and migration of melanocytes as well as the synthesis of melanin and the formation, transport, and transfer of melanosomes. Genes involved in these processes often have pleiotropic effects, which impact other important biochemical pathways. Coat color loci in the mouse have been known to be part of diverse cellular, developmental, and physiological processes and in some cases to be implicated in pathologies such as anemia, sterility, and neurological disorders.[126–128] The cat is an excellent model system with which to study coat color phenotypes. At least nine different coat color loci have been identified in the cat,[129] and several are now characterized on a molecular genetic level including, *a* (nonagouti) responsible for melanism,[18] *b* (Brown), which changes black pigmentation to brown or variants of brown,[32] *c* (color), causing the darker pigmentation at extremities (i.e., ears, tail), observed in Siamese and Burmese cats[32,37] and albinism,[36] and *d* (dilute), which causes dilution of expected color (i.e., black pigmentation appears gray).[35] The cat is also unique in mammalian species in exhibiting a variation in coat pattern, demonstrating agouti (*A*) (nonpatterned coat) and variants of the *T* (tabby) locus, which affect striping and spotting patterns.

VIRAL PATHOGENS OF THE DOMESTIC CAT

The cat has provided several useful models for infectious disease. These include feline leukemia and feline sarcoma virus, feline coronavirus, and Type C retroviruses that interact with cellular oncogenes to induce leukemia, lymphoma, and sarcoma.[130–132] Historically, many of the human oncogenes that define signal transduction pathways were originally discovered in the context of feline leukemia virus interaction in cat models. The cat provides the only naturally occurring model for human AIDS pathogenesis, in its endemic fatal transmissible feline immunodeficiency virus (FIV).[133,134] Similar to its close phylogenetic relative HIV, FIV induces CD4-T lymphocyte depletion in affected cats, an immune system collapse, and susceptibility to adventitious microbial agents as a prelude to wasting disease and death.[133,135] Interestingly, over 10 wildcat species (including lions, leopards, cheetahs, ocelots, pumas and other big cats) are endemic with their own species-specific strain of FIV[136–141]; however, unlike strains in domestic cats, the wildcat FIV strains do not appear to cause acute immunodeficiency in the wildcat species, perhaps a consequence of historic natural selection of host genetic resistance to the fatal virus.[139,142] Lion and puma-specific FIV strains have recently been demonstrated to utilize novel, more promiscuous mechanisms for cell entry than FIV, suggesting a divergent tropism and biological properties of these viruses.[143]

The World Health Organization reported a new human respiratory illness outbreak (severe acute respiratory syndrome, SARS) that emerged in Guangdong Province, China in 2003.[144,145] Sequence analysis demonstrated that the infectious agent was a previously unrecognized coronavirus.[146,147] An animal model demonstrating clinical symptoms and pathology of SARS-infected patients has not been reported.[148] Of interest, the recent report of a highly virulent feline coronavirus epidemic in captive African cheetahs, a disease model for human SARS, illustrates the critical role of ancestral population genetic variation.[132] In addition, cats injected with the SARS virus developed clinical symptoms, an important insight in implicating the virus in the SARS epidemic.[149–152]

The feline panleukopenia (feline distemper) virus has revealed a natural history parable in its abrupt transformation of the cat virus to an epidemic, fatal canine parvovirus, that emerged in the world's puppy population in 1978.[153] In contrast, the canine distemper virus, which is normally restricted to canid species, precipitously adapted to and decimated East African lions in 1994, killing one-third of the lions in the Serengeti ecosystem within a 9-month outbreak.[154] A clear involvement of host defense mecha-

Table 25–2
Feline genome project resources (September, 2006)

Resource	Reference
1. Somatic cell hybrid panel framework physical map > 100 Type I genes	12, 152
2. Interspecies backcross (ISB) genetic linkage maps	6, 7
3. Intraspecies Nestlé Purina pedigree genetic linkage map	Unpublished
4. 5000-rad radiation hybrid panel and maps	5, 6, 8–10, 122
5. Flow sorted feline chromosome libraries: reciprocal chromosome paint map	15, 125, 153
6. Arrayed BAC and PAC libraries	123
7. Tissue/cell line DNA repository of >10,000 exotic and domestic feline specimens	154, 155
8. Domestic cat breed forensic database of 38 breeds, 11 multiplexed optimized STRs	31
9. Domestic cat Y chromosome cosmid library	Unpublished
10. Completed sequence	
a. Whole genome sequence (2-fold coverage)	17
b. mtDNA sequence	156
c. Major histocompatibility complex	157
11. Cat genome browser (GARFIELD)	17

nisms in these and other infectious disease episodes renders the cats and their pathogens an excellent candidate species for characterizing the interaction of microbial adaptation and host disease gene defenses. Given the critical importance of infectious disease in scores of chronic and acute human disease, there are powerful research opportunities in the cat family.[142,155]

Finally, recent concern over the emergence of avian flu H5N1 has shown a strong susceptibility of cats, both domestic and large cats, again raising possibilities for pathogenesis and therapy development.[156,157]

With the development of genomic resources in the cat (Table 25–2) and the application of complementary comparative tools developed in other species, the domestic cat is emerging as a promising resource of phenotypically defined genetic variation of biomedical significance. Exploration of similar resources in other species, particularly the dog and mouse, has provided important insight into otherwise unexplained biomedical disorders.

REFERENCES

1. Vigne JD, Guilaine J, Debue K, Haye L, Gerard P. Early taming of the cat in Cyprus. *Science* 2004;304(5668):259.
2. O'Brien SJ, Menotti-Raymond, Murphy WJ, Yuhki N. The Feline Genome Project. *Annu Rev Genet* 2002;36:657–686.
3. O'Brien SJ. Cats. *Curr Biol* 2004;14(23):R988–999.
4. Sun S, Murphy WJ, Menotti-Raymond M, O'Brien SJ. Integration of the feline radiation hybrid and linkage maps. *Mamm Genome* 2001;12:436–441.
5. Menotti-Raymond M, David VA, Agarwala R, *et al.* Radiation hybrid mapping of 304 novel microsatellites in the domestic cat genome. *Cytogenet Genome Res* 2004;102(14):272–276.
6. Menotti-Raymond M, David VA, Lyons LA, *et al.* A genetic linkage map of microsatellites in the domestic cat (*Felis catus*). *Genomics* 1999;57:9–23.
7. Murphy WJ, Davis B, David VA, *et al.* A 1.5 megabase resolution radiation hybrid map of the cat genome and comparative analysis with the canine and human genomes. *Genomics* 2007;89(2): 189–196.
8. Menotti-Raymond M, David VA, Roelke ME, *et al.* Second-generation integrated genetic linkage/radiation hybrid maps of the domestic cat (Felis catus). *J Hered* 2003;94(1):95–106.
9. Murphy WJ, David VA, Schäffer AA, *et al.* A third-generation RH map of the domestic cat. *Cytogenet Genome Res* 2003.
10. Murphy WJ, Sun S, Chen ZQ, Pecon-Slattery J, O'Brien SJ. Extensive conservation of sex chromosome organization between cat and human revealed by parallel radiation hybrid mapping. *Genome Res* 1999;9(12):1223–1230.
11. O'Brien SJ, Nash WG. Genetic mapping in mammals: Chromosome map of domestic cat. *Science* 1982;216:257–265.
12. Murphy WJ, Sun S, Chen Z, *et al.* A radiation hybrid map of the cat genome: Implications for comparative mapping. *Genome Res* 2000;10(5):691–702.
13. Murphy WJ, Pearks Wilkerson AJ, Raudsepp T, *et al.* Novel gene acquisition on carnivore Y chromosomes. *PLoS Genet* 2006;2(3): e43.
14. O'Brien SJ, Wienberg J, Lyons LA. Comparative genomics: Lessons from cats. *Trends Genet* 1997;13:393–399.
15. O'Brien SJ, Cevario SJ, Martenson JS, *et al.* Comparative gene mapping in the domestic cat (Felis catus). *J Hered* 1997;88(5):408–414.
16. Pontius J, Mullikin J, Smith D, *et al.* The domestic cat genome sequence annotation and comparative inferences. 2006.
17. Lindblad-Toh K, Wade CM, Mikkelsen TS, *et al.* Genome sequence, comparative analysis and haplotype structure of the domestic dog. *Nature* 2005;438(7069):803–819.
18. Margulies EH, Blanchette M, Haussler D, Green ED. Identification and characterization of multi-species conserved sequences. *Genome Res* 2003;13(12):2507–2518.
19. Margulies EH, Vinson JP, Miller W, *et al.* An initial strategy for the systematic identification of functional elements in the human genome by low-redundancy comparative sequencing. *Proc Natl Acad Sci USA* 2005;102(13):4795–4800.
20. Narfström K. Hereditary progressive retinal atrophy in the Abyssinian cat. *J Hered* 1983;74:273–276.
21. Parker HG, Kim LV, Sutter NB, *et al.* Genetic structure of the purebred domestic dog. *Science* 2004;304(5674):1160–1164.
22. Sutter NB, Ostrander EA. Dog star rising: The canine genetic system. *Nat Rev Genet* 2004;5(12):900–910.
23. Morton NE. Linkage disequilibrium maps and association mapping. *J Clin Invest* 2005;115(6):1425–1430.
24. Sutter NB, Eberle MA, Parker HG, *et al.* Extensive and breed-specific linkage disequilibrium in Canis familiaris. *Genome Res* 2004;14(12):2388–2396.
25. Ostrander EA, Wayne RK. The canine genome. *Genome Res* 2005;15(12):1706–1716.
26. Chase K, Sargan D, Miller K, Ostrander EA, Lark KG. Understanding the genetics of autoimmune disease: Two loci that regulate late onset Addison's disease in Portuguese Water Dogs. *Int J Immunogenet* 2006;33(3):179–184.
27. Mellersh CS, Boursnell ME, Pettitt L, *et al.* Canine RPGRIP1 mutation establishes cone-rod dystrophy in miniature longhaired dachshunds as a homologue of human Leber congenital amaurosis. *Genomics* 2006;88(3):293–301.
28. Driscoll C, O'Brien SJ. Cat domestication age. *Science* 2006.
29. Menotti-Raymond MA, David VA, Wachter LL, Butler JM, O'Brien SJ. An STR forensic typing system for genetic individualization of domestic cat (Felis catus) samples. *J Forensic Sci* 2005;50(5): 1061–1070.
30. Schmidt-Kuntzel A, Eizirik E, O'Brien SJ, Menotti-Raymond M. Tyrosinase and tyrosinase related protein 1 alleles specify domestic cat coat color phenotypes of the albino and brown loci. *J Hered* 2005;96(4):289–301.
31. Lyons LA, Biller DS, Erdman CA, *et al.* Feline polycystic kidney disease mutation identified in PKD1. *J Am Soc Nephrol* 2004;15(10):2548–2555.
32. Fyfe JC, Menotti-Raymond M, David VA, *et al.* An ~140-kb deletion associated with feline spinal muscular atrophy implies an essential LIX1 function for motor neuron survival. *Genome Res* 2006;16(9):1084–1090.
33. Ishida Y, David VA, Eizirik E, *et al.* A homozygous single-base deletion in MLPH causes the dilute coat color phenotype in the domestic cat. *Genomics* 2006;88(6):698–705.
34. Imes DL, Geary LA, Grahn RA, Lyons LA. Albinism in the domestic cat (Felis catus) is associated with a tyrosinase (TYR) mutation. *Anim Genet* 2006;37(2):175–178.
35. Lyons LA, Imes DL, Rah HC, Grahn RA. Tyrosinase mutations associated with Siamese and Burmese patterns in the domestic cat (Felis catus). *Anim Genet* 2005;36(2):119–126.
36. Baker H, Smith BF, Martin DR, Foureman P. Molecular diagnosis of gangliosidosis: A model of inherited diseases in pure breeds. In: August JR, Ed. *Consultations in Feline Internal Medicine*. Philadelphia: W. B. Saunders Company, 2001:615–620.
37. Muldoon LL, Neuwelt EA, Pagel MA, Weiss DL. Characterization of the molecular defect in a feline model for type II GM2-gangliosidosis (Sandhoff disease). *Am J Pathol* 1994;144(5):1109–1118.
38. Martin DR, Krum BK, Varadarajan GS, Hathcock TL, Smith BF, Baker HJ. An inversion of 25 base pairs causes feline GM2 gangliosidosis variant. *Exp Neurol* 2004;187(1):30–37.
39. Martin DR, Cox NR, Morrison NE, *et al.* Mutation of the GM2 activator protein in a feline model of GM2 gangliosidosis. *Acta Neuropathol (Berl)* 2005;110(5):443–450.
40. Fyfe JC, Giger U, Winkle TJV, *et al.* Glycogen storage disease Type IV: Inherited deficiency of branching enzyme activity in cats. *Pediatr Res* 1992;32:719–725.
41. Fyfe JC, Kurzhals RL. Glycogen storage disease Type IV in Norwegian Forest Cats: Molecular detection of carriers. In: *First International Feline Genetic Disease Conference, June 25–28, 1998*. Philadelphia, PA: University of Pennsylvania, 1998.

42. Goree M, Catalfamo JL, Aber S, Boudreaux MK. Characterization of the mutations causing hemophilia B in 2 domestic cats. *J Vet Intern Med* 2005;19(2):200–204.
43. Ginzinger DG, Lewis MES, Ma YH, Jones BR, Liu GQ, Jones SD. A mutation in the lipoprotein lipase gene is the molecular basis of chylomicronemia in a colony of domestic cats. *J Clin Invest* 1996;97:1257–1266.
44. Meurs KM, Sanchez X, David RM, *et al.* A cardiac myosin binding protein C mutation in the Maine Coon cat with familial hypertrophic cardiomyopathy. *Hum Mol Genet* 2005;14(23):3587–3593.
45. Giger U, Tcherneva E, Caverly J, *et al.* A missense mutation in N-acetylglucosamine-1-phosphotransferase causes mucolipidosis II in domestic shorthair cats. *J Vet Intern Med* 2006;20:781.
46. Crawley AC, Yogalingam G, Muller VJ, Hopwood JJ. Two mutations within a feline mucopolysaccharidosis type VI colony cause three different clinical phenotypes. *J Clin Invest* 1998;101(1):109–119.
47. Fyfe JC, Kurzhals RL, Lassaline ME, *et al.* Molecular basis of feline beta-glucuronidase deficiency: An animal model of mucopolysaccharidosis VII. *Genomics* 1999;58(2):121–128.
48. Winand NJ, Edwards M, Pradhan D, Berian CA, Cooper BJ. Deletion of the dystrophin muscle promoter in feline muscular dystrophy. *Neuromuscul Disord* 1994;4(5–6):433–445.
49. Somers KL, Royals MA, Carstea ED, Rafi MA, Wenger DA, Thrall MA. Mutation analysis of feline Niemann-Pick C1 disease. *Mol Genet Metab* 2003;79(2):99–103.
50. Giebel LB, Tripathi RK, King RA, Spritz RA. A tyrosinase gene missense mutation in temperature-sensitive type I oculocutaneous albinism. A human homologue to the Siamese cat and the Himalayan mouse. *J Clin Invest* 1991;87(3):1119–1122.
51. Giger U, Rajpurohit Y, Wang P, *et al.* Molecular basis of erythrocyte pyruvate kinase (RPK) deficiency in cats. *Blood* 1997;90:5.
52. Li X, Li W, Wang H, *et al.* Cats lack a sweet taste receptor. *J Nutr* 2006;136(7Suppl.):1932S-1934S.
53. den Dunnen JT, Antonarakis SE. Nomenclature for the description of human sequence variations. *Hum Genet* 2001;109(1):121–124.
54. Talbot K, Davies KE. Spinal muscular atrophy. *Semin Neurol* 2001;21(2):189–197.
55. Lefebvre S, Burglen L, Reboullet S, *et al.* Identification and characterization of a spinal muscular atrophy-determining gene. *Cell* 1995;80(1):155–165.
56. Wirth B, Herz M, Wetter A, *et al.* Quantitative analysis of survival motor neuron copies: Identification of subtle SMN1 mutations in patients with spinal muscular atrophy, genotype-phenotype correlation, and implications for genetic counseling. *Am J Hum Genet* 1999;64(5):1340–1356.
57. He Q, Lowrie C, Shelton GD, *et al.* Inherited motor neuron disease in domestic cats: A model of spinal muscular atrophy. *Pediatr Res* 2005;57(3):324–330.
58. Meister G, Fischer U. Assisted RNP assembly: SMN and PRMT5 complexes cooperate in the formation of spliceosomal UsnRNPs. *EMBO J* 2002;21(21):5853–5863.
59. Haskins M, Casal M, Ellinwood NM, Melniczek J, Mazrier H, Giger U. Animal models for mucopolysaccharidoses and their clinical relevance. *Acta Paediatr Suppl* 2002;91(439):88–97.
60. Gilbert DA, O'Brien JS, O'Brien SJ. Chromosomal mapping of lysosomal enzyme structural genes in the domestic cat. *Genomics* 1988;2:329–336.
61. Ellinwood NM, Vite CH, Haskins ME. Gene therapy for lysosomal storage diseases: The lessons and promise of animal models. *J Gene Med* 2004;6(5):481–506.
62. Neufeld EF, Lim TW, Shapiro LJ. Inherited disorders of lysosomal metabolism. *Annu Rev Biochem* 1975;44:357–376.
63. Haskins ME, Jezyk PF, Desnick RJ, McDonough SK, Patterson DF. Alpha-L-iduronidase deficiency in a cat: A model of mucopolysaccharidosis I. *Pediatr Res* 1979;13(11):1294–1297.
64. Haskins ME, Jezyk PF, Desnick RJ, McDonough SK, Patterson DF. Mucopolysaccharidosis in a domestic short-haired cat–a disease distinct from that seen in the Siamese cat. *J Am Vet Med Assoc* 1979;175(4):384–387.
65. Spellacy E, Shull RM, Constantopoulos G, Neufeld EF. A canine model of human alpha-L-iduronidase deficiency. *Proc Natl Acad Sci USA* 1983;80(19):6091–6095.
66. Ponder KP, Wang B, Wang P, *et al.* Mucopolysaccharidosis I cats mount a cytotoxic T lymphocyte response after neonatal gene therapy that can be blocked with CTLA4-Ig. *Mol Ther* 2006;14(1):5–13.
67. Cowell KR, Jezyk PF, Haskins ME, Patterson DF. Mucopolysaccharidosis in a cat. *J Am Vet Med Assoc* 1976;169(3):334–339.
68. Jezyk PF, Haskins ME, Patterson DF, Mellman WJ, Greenstein M. Mucopolysaccharidosis in a cat with arylsulfatase B deficiency: A model of Maroteaux-Lamy syndrome. *Science* 1977;198(4319):834–836.
69. Haskins ME, Jezyk PF, Patterson DF. Mucopolysaccharide storage disease in three families of cats with arylsulfatase B deficiency: Leukocyte studies and carrier identification. *Pediatr Res* 1979;13(11):1203–1210.
70. Haskins ME, Jezyk PF, Desnick RJ, Patterson DF. Animal model of human disease: Mucopolysaccharidosis VI Maroteaux-Lamy syndrome, arylsulfatase B-deficient mucopolysaccharidosis in the Siamese cat. *Am J Pathol* 1981;105(2):191–193.
71. Walkley SU, Thrall MA, Haskins ME, *et al.* Abnormal neuronal metabolism and storage in mucopolysaccharidosis type VI (Maroteaux-Lamy) disease. *Neuropathol Appl Neurobiol* 2005;31(5):536–544.
72. Yogalingam G, Crawley A, Hopwood JJ, Anson DS. Evaluation of fibroblast-mediated gene therapy in a feline model of mucopolysaccharidosis type VI. *Biochim Biophys Acta* 1999;1453(2):284–296.
73. Ho TT, Maguire AM, Aguirre GD, *et al.* Phenotypic rescue after adeno-associated virus-mediated delivery of 4-sulfatase to the retinal pigment epithelium of feline mucopolysaccharidosis VI. *J Gene Med* 2002;4(6):613–621.
74. Neufeld EF, Muenzer J. The mucopolysaccharidoses. In: Scriver CR, Beaudet AL, Sly WS, Vallee D, Eds. *The Metabolic and Molecular Basis of Inherited Disease,* 6th ed. New York: McGraw-Hill Book Co., 1995:2465–2495.
75. Haskins ME, Desnick RJ, DiFerrante N, Jezyk PF, Patterson DF. Beta-glucuronidase deficiency in a dog: A model of human mucopolysaccharidosis VII. *Pediatr Res* 1984;18(10):980–984.
76. Gitzelmann R, Bosshard NU, Superti-Furga A, *et al.* Feline mucopolysaccharidosis VII due to beta-glucuronidase deficiency. *Vet Pathol* 1994;31(4):435–443.
77. Vite CH, Passini MA, Haskins ME, Wolfe JH. Adeno-associated virus vector-mediated transduction in the cat brain. *Gene Ther* 2003;10(22):1874–1881.
78. Thomas GH, Beaudet AL. Disorders of glycoprotein degradation: α-Mannosidosis, β-mannosidosis, sialidosis, aspartylglucosaminuria, and carbohydrate-deficient glycoprotein syndrome. In: Scriver CR, Beaudet AL, Sly WA, Valle D, Eds. The *Molecular and Metabolic Bases for Inherited Disease*, 6th ed. New York: McGraw-Hill Book Co., 1995:2529–2561.
79. Burditt LJ, Chotai K, Hirani S, Nugent PG, Winchester BG, Blakemore WF. Biochemical studies on a case of feline mannosidosis. *Biochem J* 1980;189(3):467–473.
80. Walkley SU, Thrall MA, Dobrenis K, *et al.* Bone marrow transplantation corrects the enzyme defect in neurons of the central nervous system in a lysosomal storage disease. *Proc Natl Acad Sci USA* 1994;91(8):2970–2974.
81. Haskins M, Abkowitz J, Aguirre G, *et al.* Bone marrow transplantation in animal models of lysosomal storage diseases. In: Ringden O, Hobbs J, Stewart C, Eds. *Correction of Genetic Diseases by Transplantation*. London: Cogent Press, 1997:1–11.
82. Vite CH, McGowan JC, Niogi SN, *et al.* Effective gene therapy for an inherited CNS disease in a large animal model. *Ann Neurol* 2005;57(3):355–364.
83. Brunzell JD. Familial lipoprotein lipase deficiency and other causes of the chylomicronemia syndrome. In: Scriver CR, Beaudet AL, Sly WS, Valle D, Eds. *Metabolic Basis of Inherited Disease*, 6th ed. New York: McGraw-Hill Book Co., 1995:1913–1932.

84. Ross CJ, Twisk J, Bakker AC, *et al.* Correction of feline lipoprotein lipase deficiency with adeno-associated virus serotype 1-mediated gene transfer of the lipoprotein lipase S447X beneficial mutation. *Hum Gene Ther* 2006;17(5):487–499.
85. Neuwelt EA, Johnson WG, Blank NK, *et al.* Characterization of a new model of GM2 gangliosidosis (Sandhoff's disease) in Korat cats. *J Clin Invest* 1985;76(2):482–490.
86. Gravel RA. The GM2 gangliosidoses. In: Scriver CR, Beaudet AL, Sly WS, Valle D, Eds. *The Metabolic and Molecular Bases of Inherited Diseases*, 6th ed. New York: McGraw-Hill Book Co., 1995:2839–2879.
87. Yamato O, Matsunaga S, Takata K, *et al.* GM2-gangliosidosis variant 0 (Sandhoff-like disease) in a family of Japanese domestic cats. *Vet Rec* 2004;155(23):739–744.
88. Mazrier H, Van Hoeven M, Wang P, *et al.* Inheritance, biochemical abnormalities, and clinical features of feline mucolipidosis II: The first animal model of human I-cell disease. *J Hered* 2003;94(5): 363–373.
89. Gaschen F, Jaggy A, Jones B. Congenital diseases of feline muscle and neuromuscular junction. *J Feline Med Surg* 2004;6(6):355–366.
90. Malik R, Mepstead K, Yang F, Harper C. Hereditary myopathy of Devon Rex cats. *J Small Anim Pract* 1993;34:539–546.
91. Hickford FH, Jones BR, Gething MA, Pack R, Alley MR. Congenital myotonia in related kittens. *J Small Anim Pract* 1998;39(6): 281–285.
92. Toll J, Cooper B, Altschul M. Congenital myotonia in 2 domestic cats. *J Vet Intern Med* 1998;12(2):116–119.
93. O'Brien DP, Johnson GC, Liu LA, *et al.* Laminin alpha 2 (merosin)-deficient muscular dystrophy and demyelinating neuropathy in two cats. *J Neurol Sci* 2001;189(1–2):37–43.
94. Hoffman EP, Brown RH Jr, Kunkel LM. Dystrophin: The protein product of the Duchenne muscular dystrophy locus. *Cell* 1987;51(6):919–928.
95. Gaschen FP, Hoffman EP, Gorospe JR, *et al.* Dystrophin deficiency causes lethal muscle hypertrophy in cats. *J Neurol Sci* 1992; 110(1–2):149–159.
96. Koenig M, Monaco AP, Kunkel LM. The complete sequence of dystrophin predicts a rod-shaped cytoskeletal protein. *Cell* 1988;53(2):219–226.
97. Anderson JE, Bressler BH, Ovalle WK. Functional regeneration in the hindlimb skeletal muscle of the mdx mouse. *J Muscle Res Cell Motil* 1988;9(6):499–515.
98. Carpenter JL, Hoffman EP, Romanul FC, *et al.* Feline muscular dystrophy with dystrophin deficiency. *Am J Pathol* 1989;135(5): 909–919.
99. Shelton GD. Muscular dystrophies: Expanding our knowledge in companion animals. *Vet J* 2004;168(1):6–8.
100. Marian AJ, Roberts R. The molecular genetic basis for hypertrophic cardiomyopathy. *J Mol Cell Cardiol* 2001;33(4):655–670.
101. Casal M, Haskins M. Large animal models and gene therapy. *Eur J Hum Genet* 2006;14(3):266–272.
102. Simonaro CM, Haskins ME, Abkowitz JL, *et al.* Autologous transplantation of retrovirally transduced bone marrow or neonatal blood cells into cats can lead to long-term engraftment in the absence of myeloablation. *Gene Ther* 1999;6(1):107–113.
103. Eckhorn R, Wilms M, Schanze T, *et al.* Visual resolution with retinal implants estimated from recordings in cat visual cortex. *Vision Res* 2006;46(17):2675–2690.
104. Ryugo DK, Kretzmer EA, Niparko JK. Restoration of auditory nerve synapses in cats by cochlear implants. *Science* 2005; 310(5753):1490–1492.
105. Colledge WH, Abella BS, Southern KW, *et al.* Generation and characterization of a delta F508 cystic fibrosis mouse model. *Nat Genet* 1995;10(4):445–452.
106. McCray PB Jr, Zabner J, Jia HP, Welsh MJ, Thorne PS. Efficient killing of inhaled bacteria in DeltaF508 mice: Role of airway surface liquid composition. *Am J Physiol* 1999;277(1Pt. 1):L183–190.
107. Li Z, Engelhardt JF. Progress toward generating a ferret model of cystic fibrosis by somatic cell nuclear transfer. *Reprod Biol Endocrinol* 2003;1:83.
108. Li Z, Sun X, Chen J, *et al.* Cloned ferrets produced by somatic cell nuclear transfer. *Dev Biol* 2006;293(2):439–448.
109. Wilmut I, Beaujean N, de Sousa PA, *et al.* Somatic cell nuclear transfer. *Nature* 2002;419(6907):583–586.
110. Lai L, Kolber-Simonds D, Park KW, *et al.* Production of alpha-1,3-galactosyltransferase knockout pigs by nuclear transfer cloning. *Science* 2002;295(5557):1089–1092.
111. Sendai Y, Sawada T, Urakawa M, *et al.* alpha1,3-Galactosyltransferase-gene knockout in cattle using a single targeting vector with loxP sequences and cre-expressing adenovirus. *Transplantation* 2006;81(5):760–766.
112. Shin T, Kraemer D, Pryor J, *et al.* A cat cloned by nuclear transplantation. *Nature* 2002;415(6874):859.
113. Gomez MC, Pope CE, Giraldo A, *et al.* Birth of African wildcat cloned kittens born from domestic cats. *Cloning Stem Cells* 2004;6(3):247–258.
114. Yin XJ, Lee HS, Lee YH, *et al.* Cats cloned from fetal and adult somatic cells by nuclear transfer. *Reproduction* 2005;129(2): 245–249.
115. Gomez MC, Pope CE, Dresser BL. Nuclear transfer in cats and its application. *Theriogenology* 2006;66(1):72–81.
116. Murphy WJ, Menotti-Raymond M, Lyons LA, Thompson ME, O'Brien SJ. Development of a feline whole-genome radiation hybrid panel and comparative mapping of human chromosome 12 and 22 loci. *Genomics* 1999;57:1–8.
117. Nash WG, Menninger JC, Wienberg J, Padilla-Nash HM, O'Brien SJ. The pattern of phylogenomic evolution of the Canidae. *Cytogenet Cell Genet* 2001;95:210–224.
118. Wienberg J, Stanyon R, Nash WG, *et al.* Conservation of human vs. feline genome organization revealed by reciprocal chromosome painting. *Cytogenet Cell Genet* 1997;77:211–217.
119. Silvers WK. *The Coal Colors of Mice: A Model for Mammalian Gene Action and Interaction*. New York: Springer-Verlag, 1979.
120. Fleischman RA. From white spots to stem cells: The role of the Kit receptor in mammalian development. *Trends Genet* 1993;9: 285–290.
121. Jackson IJ. Molecular and developmental genetics of mouse coat color. *Annu Rev Genet* 1994;28:189–217.
122. Vella CM, Robinson R. *Robinson's Genetics for Cat Breeders and Veterinarians*, 4th ed. Boston, MA: Butterworth-Heinemann, 1999.
123. Hardy WD, Essex M, McClelland AJ. *Feline Leukemia Virus*. New York: Elsevier, 1980.
124. Hardy WD. Feline oncoretroviruses. In: Levy JA, Ed. *Viruses: The Retroviridae*. New York: Plenum, 1993.
125. O'Brien SJ, Troyer J, Roelke M, Marker L, Pecon-Slattery J. Plagues and adaptation: Lessons from the Felidae models for SARS and AIDS. *Biol Conserv* 2006;131(2):255–267.
126. Pedersen NC. The feline immunodeficiency virus. In: Levy JA, Ed. *Viruses: The Retroviridae*. New York: Plenum, 1993:181–228.
127. Willett BJ, Flynn JN, Hosie MJ. FIV infection of the domestic cat: An animal model for AIDS. *Immunol Today* 1997;18(4):182–189.
128. Paillot R, Richard S, Bloas F, *et al.* Toward a detailed characterization of feline immunodeficiency virus-specific T cell immune responses and mediated immune disorders. *Vet Immunol Immunopathol* 2005;106(1–2):1–14.
129. Olmsted RA, Langley R, Roelke ME, *et al.* Worldwide prevalence of lentivirus infection in wild feline species: Epidemiologic and phylogenetic aspects. *J Virol* 1992;66(10):6008–6018.
130. Brown EW, Yuhki N, Packer C, O'Brien SJ. A lion lentivirus related to feline immunodeficiency virus: Epidemiologic and phylogenetic aspects. *J Virol* 1994;68(9):5953–5968.
131. Carpenter MA, O'Brien SJ. Coadaptation and immunodeficiency virus: Lessons from the Felidae. *Curr Opin Genet Dev* 1995;5(6):739–745.
132. Troyer JL, Pecon-Slattery J, Roelke ME, Black L, Packer C, O'Brien SJ. Patterns of feline immunodeficiency virus multiple infection and genome divergence in a free-ranging population of African lions. *J Virol* 2004;78(7):3777–3791.

133. Troyer JL, Pecon-Slattery J, Roelke ME, *et al.* Seroprevalence and genomic divergence of circulating strains of feline immunodeficiency virus among Felidae and Hyaenidae species. *J Virol* 2005;79(13):8282–8294.
134. O'Brien SJ. Genomic prospecting. *Nat Med* 1995;1(8):742–744.
135. Smirnova N, Troyer JL, Schissler J, Terwee J, Poss M, VandeWoude S. Feline lentiviruses demonstrate differences in receptor repertoire and envelope structural elements. *Virology* 2005;342(1):60–76.
136. Holmes KV. SARS-associated coronavirus. *N Engl J Med* 2003;348(20):1948–1951.
137. Rota PA, Oberste MS, Monroe SS, *et al.* Characterization of a novel coronavirus associated with severe acute respiratory syndrome. *Science* 2003;300(5624):1394–1399.
138. Snijder EJ, Bredenbeek PJ, Dobbe JC, *et al.* Unique and conserved features of genome and proteome of SARS-coronavirus, an early split-off from the coronavirus group 2 lineage. *J Mol Biol* 2003;331(5):991–1004.
139. Weiss SR, Navas-Martin S. Coronavirus pathogenesis and the emerging pathogen severe acute respiratory syndrome coronavirus. *Microbiol Mol Biol Rev* 2005;69(4):635–664.
140. Lun ZR, Qu LH. Animal-to-human SARS-associated coronavirus transmission? *Emerg Infect Dis* 2004;10(5):959.
141. Wu D, Tu C, Xin C, *et al.* Civets are equally susceptible to experimental infection by two different severe acute respiratory syndrome coronavirus isolates. *J Virol* 2005;79(4):2620–2625.
142. Abbott A. Pet theory comes to the fore in fight against SARS. *Nature* 2003;423(6940):576.
143. Martina BE, Haagmans BL, Kuiken T, *et al.* Virology: SARS virus infection of cats and ferrets. *Nature* 2003;425(6961):915.
144. Parrish CR. The emergence and evolution of canine parvovirus—an example of recent host range mutation. *Virology* 1994;5:121–132.
145. Roelke-Parker ME, Munson L, Packer C, *et al.* A canine distemper virus epidemic in Serengeti lions (Panthera leo). *Nature* 1996;379(6564):441–445.
146. Ewald PW. *Plague Time: How Stealth Infections Cause Cancer, Heart Disease, and Other Deadly Ailments*. New York: The Free Press, 2000.
147. Duke K. Germany says people in areas with bird flu should keep cats indoors. *Br Med J* 2006;332(7541):568.
148. van Riel D, Munster VJ, de Wit E, *et al.* H5N1 virus attachment to lower respiratory tract. *Science* 2006;312(5772):399.
149. Rimmelzwaan GF, van Riel D, Baars M, *et al.* Influenza A virus (H5N1) infection in cats causes systemic disease with potential novel routes of virus spread within and between hosts. *Am J Pathol* 2006;168(1):176–183.
150. Kuiken T, Rimmelzwaan G, van Riel D, *et al.* Avian H5N1 influenza in cats. *Science* 2004;306(5694):241.
151. Kuiken T, Fouchier R, Rimmelzwaan G, Osterhaus A, Roeder P. Feline friend or potential foe? *Nature* 2006;440(7085):741–742.
152. O'Brien SJ, Cevario SJ, Martenson JS, *et al.* Comparative gene mapping in the domestic cat (*Felis catus*). *J Hered* 1997;88:408–414.
153. Rettenberger G, Klett C, Zechner U, *et al.* ZOO-FISH analysis: Cat and human karyotypes closely resemble the putative ancestral mammalian karyotype. *Chromosome Res* 1995;3:479–486.
154. Pecon Slattery J, O'Brien SJ. Patterns of Y and X chromosome DNA sequence divergence during the Felidae radiation. *Genetics* 1998;148(3):1245–1255.
155. Johnson WE, O'Brien SJ. Phylogenetic reconstruction of the Felidae using 16S rRNA and NADH-5 mitochondrial genes. *J Mol Evol* 1997;44(Suppl. 1):S98–S116.
156. Lopez JV, Cevario S, O'Brien SJ. Complete nucleotide sequence of the domestic cat (*Felis catus*) mitochondrial genome and a transposed mtDNA tandem repeat (*Numt*) in the nuclear genome. *Genomics* 1996;33:229–246.
157. Yuhki N, Beck T, Stephens RM, Nishigaki Y, Newmann K, O'Brien SJ. Comparative genome organization of human, murine, and feline MHC class II region. *Genome Res* 2003;13(6A):1169–1179.

26 Swine in Biomedical Research

M. MICHAEL SWINDLE AND ALISON C. SMITH

ABSTRACT

This chapter provides an overview of the use of domestic and miniature breeds of swine in biomedical research. Differences between breeds of pigs and the most significant anatomic and physiological characteristics of the species related to their selection as biomedical research models are discussed. Husbandry and handling techniques that have proven to be beneficial for using these animals in research are detailed. Also included are the key references for developing and using swine and resources for additional training and refinement are listed.

Key Words: Swine, Miniature pig, Surgery, Animal models, Animal husbandry.

INTRODUCTION

Swine (*Sus scrofa domestica*) have been used extensively in biomedical research and have become a standard general surgical model for preclinical studies. Their usage continues to grow while other large animal models are decreasing in number. Since the 1980s there have been a number of conferences held to specifically identify uses of swine in research. There are detailed publications on the technical issues[1–6] involved in the use of this species and reference publications on biology[7–10] and models.[1,2,11–16] Detailed publications on anesthesia, analgesia, and perioperative care are available.[1–3,17–19]

In this chapter their common uses are described as well as some of the technical issues involved in their selection and use as biomedical models. Because of the large number of porcine models utilized, the anatomic and physiological parameters of the various systems that justify their use in biomedical research are detailed.

BREEDS AND SOURCES OF SWINE

Most breeds of swine are animals raised commercially for meat production. These animals are commonly referred to as domestic or farm breeds in the literature. There are hundreds of these breeds and common examples are Yorkshire, Landrace, Duroc, Hampshire, Pietrain, and Poland China. There are over 50 breeds of miniature pigs worldwide but only a few of them are important in biomedical research. Smaller miniature breeds are sometimes referred to as microswine. Some of the most common miniature swine used in research are the Yucatan, Hanford, Sinclair, Göttingen, Ossabaw, Banna, Vietnamese Potbellied, and Meishan.[1,2,4,8]

The most significant difference between miniature and domestic swine is the growth rate. Farm breeds grow from a birth weight of 0.5–1 kg to a weight of >100 kg in 4 months. This is an exponential growth rate that is an economic necessity when raising swine for food consumption. However, in a biomedical research facility this growth rate is a complication for research as well as a potential occupational health hazard. Large swine can be very difficult to handle and lifting anesthetized animals can lead to ergonomic injuries. In the same 4-month period of time miniature pigs grow from 0.25 kg to 7–20 kg depending upon the breed. Of the most commonly used minipigs the size ranges from smallest to largest are Göttingen, Sinclair, Yucatan, and Hanford. The Hanford achieves the body weight of an adult human in a year, while the smaller breeds weigh <30 kg at the same age.[1,2,4,8]

Consequently, using domestic breeds for long-term chronic experiments is discouraged. Unless growth is part of the study, projects using farm pigs for chronic experiments should be limited to 3–6 weeks depending on the size of the animal at the start of the experiment and the housing capabilities of the institution.

Selection of miniature pigs for an experiment is based upon the size and physiological characteristics of the particular breed. Piglets may be weaned from the sow between 3 and 6 weeks of age. Sexual maturity in swine generally is reached between 4 and 6 months of age. Consequently studying animals <4 months of age would be considered to be a study conducted in neonatal or juvenile animals. Depending upon the goals of the study this may be a complication.

In general all pigs of the same weight have similar sized organs and structures. The difference would be the physiological maturity of the organ of interest because the size of the pig is breed dependent. Most organs and systems are physiologically mature by sexual maturity. However, the range of weights between miniature and domestic breeds at sexual maturity may be as much as a 100 kg difference. Consequently, the size of an organ such as the heart may be substantially different at the same age between breeds.

The health status of the source herd is also important.[2,20,21] Pigs purchased from auctions or conventional farms may have complicating diseases and parasites. In fact, their use may be similar to using unconditioned pound source dogs. Large commercial suppliers and operations that specialize in supplying swine for

From: *Sourcebook of Models for Biomedical Research*
(P. M. Conn, ed.), © 2008 Humana Press Inc., Totowa, NJ.

research will generally have a program for prevention of diseases. In swine there is a designation of specific pathogen free (SPF), which has a proprietary connotation in the United States. These are animals that have been raised to prevent diseases that interfere with weight gain. SPF swine are not necessarily free of all diseases and it should be used as a starting point when evaluating a source of animals. Some pigs are available as Cesarian-derived and barrier-reared animals. Housing of these animals with conventional swine is not appropriate because they will become infected with organisms that may be subclinical in the other animals. The institutional veterinarian should be consulted to preapprove the source of animals after evaluation of the herd health program.

Shipping of swine between facilities may also be a substantial problem that must be overcome.[2,22] There are intense and specific regulations regarding the shipping of swine in interstate commerce within the United States, as regulated by the U.S. Department of Agriculture (USDA), and even more intense regulations need to be followed when shipping between countries. No exceptions are made for animals used in research. The institutional veterinarian should be knowledgeable about the regulations and shipping requirements for animals.

It is important to adequately describe the type of pig used in an experimental study in the materials and methods section of journal articles. This should include the breed, sex, weight, age, and health status of the animal. Hemodynamics should be indexed to the body surface area when they are reported. When making comparisons between studies these factors need to be taken into consideration.

GENETIC MODELS

There are only a handful of spontaneous mutations in swine that have been developed as genetically reproducible models. A colony of swine with ventricular septal defect (VSD) had a high incidence of perimembranous VSD that is similar to the most common form of VSD in humans. Affected animals also had a high incidence of patent foramen ovale. Pedigree studies indicated that the defect was inherited as a polygenic trait.[23] Pigs develop other congenital heart anomalies that occur in humans, but these have generally been described as single case reports and have not been genetically reproducible. Swine with mutations of the predominant bloodborne cholesterol transporter, low-density lipoprotein (LDL), have elevated plasma cholesterol levels and have been used to model familial hypercholesterolemia in humans.[24] Sinclair miniature swine spontaneously develop cutaneous melanoma lesions that are classified histologically as malignant melanomas. Features of the swine model shared with human melanomas are the genetic predisposition, spontaneous appearance, similar histological appearance, capability for malignant transformation, and pattern of metastasis. Unlike human melanomas, tumor development in swine is not related to exposure to ultraviolet radiation, spontaneous tumor regression occurs that is immunologically mediated in most pigs, and the melanomas develop only in darkly pigmented animals.[25] Currently, female Ossabaw swine fed a high fat/cholesterol chow are being developed as a model of the human metabolic syndrome, characterized by central obesity, insulin resistance, impaired glucose tolerance, dyslipidemia, and hypertension.[26]

Current emphasis is on developing transgenic and knockout swine for applications in the areas of xenotransplantation, improving agricultural production, pharmaceutical production, and the development of new animal models. The four primary methods for producing transgenic pigs include pronuclear injection, oocyte transduction, sperm-mediated gene transfer, and nuclear transfer (cloning). The various methods and the pigs that have been produced using these techniques have been summarized by Prather.[27] A transgenic swine model of retinitis pigmentosa has been produced that is characterized by early and severe loss of rod photoreceptors and a slower degeneration of cone photoreceptors, similar to the condition in humans.[28] A major goal in producing genetically modified swine has been focused on overcoming the immunological barriers to xenotransplantation. Cloning technology has been used to produce α-Gal-deficient pigs; however additional immunomodulating strategies will be necessary for xenografting to become clinically feasible.[29]

INDUCED MODELS

Most of the models using swine are induced, many by surgical means, rather than being spontaneous or genetic in origin. The most common use of swine is in cardiovascular research, including atherosclerosis.[2,11,13–15] Investigational surgical procedures of various types are also a very common area of usage. Swine have become the standard general surgical model, having replaced the dog and primates over the past two decades. Included in these invasive procedures would be endoscopy, laparoscopic surgery, interventional catheterization, transplantation, and xenographic technologies.[1–3,5,6,30] In addition to the cardiovascular system, swine have also been used extensively to study diseases of the digestive, urogenital, and integumentary systems.[1,2,13–16,31] Swine have been used to a lesser degree for other systems, but are emerging in a variety of new areas, such as safety pharmacology and toxicology.[2,12,32]

CARDIOVASCULAR Cardiovascular models are based upon shared anatomic and physiological characteristics with humans.[1,2,7,11,13–15] Sexually mature Hanford pigs have a similar sized heart and blood vessels, however, many of the smaller pigs can be used for cardiovascular experimentation if adult sized structures are not important. Anatomically the most significant difference from humans is the presence of a left azygous vein (hemiazygous) that drains the intercostal vessels into the coronary sinus of the heart. Therefore, coronary blood flow return in the pig is mixed systemic and myocardial blood. This vessel may be ligated if total coronary blood flow is required.

The blood supply to the conduction system is right side dominant as in 90% of the human population and significant collateral circulation in the coronary system is not present. If a coronary vessel is obstructed to create a myocardial infarction, a complete infarct is created. Infarcts may be created acutely by ligation of a coronary vessel or by injection or blocking agents such as microspheres. Chronic or gradual infarction may be created by surgically applying osmotic ameroid constrictors around the blood vessel or by creating atherosclerotic lesions in the vessel lumen.[1,2,7,11,13–15,33–35] Myocardial infarction is an intensive area of research in swine because of these characteristics.

None of the cardiac conduction systems in animals exactly matches that of the human. In the case of the pig the conduction tends to be neurogenic rather than myogenic; there are nerve fibers in the atrioventricular node and the bundle branches. The pig has large well-differentiated subendocardial Purkinje cells. However, the anatomy and parameters of the intracardial electrogram overlap the values in humans and the pig has been used in studies of cardiac system ablation and pacemaker testing.[36–38]

Pigs are susceptible to development of atherosclerosis in a manner similar to humans. Over time they develop complex lesions. In general, pigs can be fed a diet with 2–4% cholesterol and 40% fat to develop generalized lesions over a period of approximately 6 months. The process can be accelerated by damaging the endothelium with a balloon catheter. This offers the advantage of being able to localize the lesion and decreasing the time to production of a significant lesion to 3 months.[2,39,40]

Almost any model that can be created surgically in other large animal species can be developed in swine.[2] A major category in this area is the production of heart failure and cardiac hypertrophy. Concentric hypertrophy is produced by pressure overload of the ventricle. Pressure overload is created by placement of a Silastic band around one of the great vessels of the heart. The band can be placed without constriction in young animals and the band will produce constriction and overload as the animal grows. Alternatively, it can be placed tightly to produce a constriction and pressure gradient. In either case hypertrophy of the affected ventricle will develop over approximately 3 months.[2,41]

Volume overload can be produced by severance of the tendinous chordae producing mitral regurgitation. It can also be produced by creating an arteriovenous fistula or by creating a patent foramen ovale. The closer to the heart that the shunt is created, the faster is the production of volume overload heart failure and eccentric hypertrophy. Fistulas created in the leg will lead to hypertrophy and heart failure in approximately 3 months. Fistulas between the aorta and pulmonary artery may lead to heart failure within a week.[2]

Rapid epicardial pacing >180 bpm leads to the development of dilative cardiomyopathy and congestive heart failure in approximately 3 weeks. These animals tend to become compromised very quickly and frequently have to be maintained on diuretics because of the rapid development of disease.[42,43]

PULMONARY The pulmonary system consists of three lobes on either side plus an accessory lobe on the right lung. The pulmonary system of the pig is very friable and easily injured. The mediastinum is thin and friable and for practical purposes should be considered incomplete. The lungs are the shock organ for the pig, as in humans. For this reason, pigs have been used as endotoxic shock models. Pigs undergo the same phases of hypodynamic and hyperdynamic shock phases after being exposed to organisms, toxins, or lipopolysaccharides.[2]

Pigs have also been used in studies of adult respiratory distress, asthma, and oxidative stress. Predominately, they have been used in studies of surgical techniques and repair.[2]

INTERVENTIONAL CATHETER AND MINIMALLY INVASIVE SURGERY TECHNIQUES Swine have become the standard model for interventional catheterization, endoscopy, and robotic and laparoscopic surgical procedures.[1,2,34,35,37,38,44–53] At first they were utilized as models for training physicians in the techniques. This evolved into the replacement of dogs and primates as models for testing of these devices and techniques.

Anatomic and physiological characteristics of the vascular system contribute to the selection of this model for testing of intravascular devices.[39,44,46,50,51] The size of the blood vessels in larger minipigs is similar to that of humans. In vascular anastomosis procedures or when intravascular stents are implanted pigs can develop neointimal hyperplasia more readily than other animals at the procedural margin. Neonatal shunts, such as atrial septal defects (ASD) and patent ductus arteriosus (PDA), can be reopened using balloon catheter procedures. These shunts will remain open and can be used for the testing of closure devices. Other models such as aneurysms of various types can be created surgically. Growth of the cardiovascular system from birth until sexual maturity can be used as a model of growth of the human heart into the early third decade of life.[2]

Abdominal, thoracic, and central nervous center minimally invasive techniques are routinely performed in porcine models, mainly as feasibility and training procedures. Long-term testing of devices for preclinical trials are growing in number and the model has gained acceptance by the Food and Drug Administration (FDA).[2]

DIGESTIVE SYSTEM Physiologically the digestive system of the pig is similar to humans probably because pigs are true omnivores. However, the anatomy of the gastrointestinal tract has some significant differences from humans. The stomach has a muscular outpouching called the torus pyloricus that can cause delayed gastric emptying with some substances. The mesenteric vessels form vascular arcades in the subserosa rather than in the mesentery. The cecum and large intestine form a series of centrifugal and centripetal coils that adhere together to form the spiral colon in the left upper quadrant of the abdomen. The bile duct and the pancreatic duct enter the duodenum separately, rather than as a common duct as is found in humans.[2,16,54]

Swine are used in models involving the physiology of digestion and metabolism. They have special importance in the study of diabetes and metabolic syndrome. Pancreatitis and various aspects of islet cell transplantation have also been studied in porcine models. Endoscopic techniques, especially of the biliary system, are routinely studied in porcine models.[2,16,54–57]

UROGENITAL SYSTEM Swine have true multirenculate, multipapillate kidneys making the internal renal anatomy of the calyceal system analogous to humans. Swine have been used extensively in the study of renal function, renal hypertension, hydroureteronephrosis, intrarenal reflux, and the study of interventional devices for the correction of defects.[2,58–62]

The reproductive system of the female is a typical bicornuate type of uterus with epitheliochorial placentation. The fallopian tubes are long and torturous compared to other mammals. The male has the same types of accessory sex glands as the human, except that the dominant glands are the bulbourethral glands rather than the prostate, which is rudimentary. The male also has a corkscrew-shaped tip to the penis, a large preputial diverticulum, and a sigmoid flexure. All of these characteristics of the penis make it impossible to routinely catheterize the male bladder.[2,58–62]

INTEGUMENTARY SYSTEM/WOUND HEALING Swine have been standard models for wound healing and dermal toxicology because of the anatomic and physiological characteristics of the skin. Swine are relatively hairless with a skin that is tightly attached to the subcutaneous tissues as in humans. They have a comparable histology except that they have a paucity of eccrine sweat glands, a fatty subcutis, and a relatively higher pH value. There are similarities in cellular turnover time, permeability, transdermal absorption, and dermal metabolism. The microvascular anatomy is consistent with that of humans and the sizes of flaps, grafts, and dermal wounds have been standardized for comparison of therapeutic agents.[2,63–67]

OTHER SYSTEMS The systems described above represent the majority of the ones in which swine are used as a large animal model. Other organs and systems are utilized to a lesser degree and usually for a specific focus.[2]

For example, the musculoskeletal system is rarely utilized for studies because of the massive structure of the bones and muscle, as would be expected for an animal primarily used for food production. Also the epiphyseal closure may take as long as 4 years in domestic breeds, while the majority of the species are not maintained for that many years. However, some specialized efforts related to studies of bone and cartilage metabolism have been undertaken.[2]

The central nervous system (CNS) and sensory organs are emerging areas of using porcine models.[68–74] Pigs have been used in ophthalmic research, mainly because of characteristics of the retina and vitreous humor. They are being developed as stroke and CNS ischemia models.

The use of swine, miniature breeds in particular, for pharmacokinetic and toxicological studies has been increasing.[12,75,76] Like other species, swine have both similarities and differences from humans and multiple species are used for preclinical studies. In terms of the cytochrome P450 system pigs have CYP1A, CYP2E, and CYP3A activity similar to humans. There is either no information or low activity for CYP2B, CYP2C, and CYP2D. Pigs have high glucuronidation and acetylation activity but low sulfation activity. Pigs have 16 blood types (A–P) with weak blood group antigens. Factors V, VIII, and IX are higher compared to humans and pigs have faster clotting times. As a general rule in our laboratory when we are trying to use a new agent in swine, we use the human pediatric dosage.

Xenotransplantation studies are performed for both tissue and solid organ transplants. The research is mainly related to the transgenic technology issues discussed above. None of the various forms of rejection has been completely conquered. There also exists the issue of poor physiological function posttransplant, even if the rejection issues have been resolved. There is a major interest in protecting against xenozoonoses resulting from either mutation or a combination of viruses in human recipients. The reasons for using swine for this model are their availability as compared to nonhuman primates and because of the anatomic and physiological similarities.[2,77]

HUSBANDRY AND HANDLING

HUSBANDRY Swine are social animals and prefer to have contact with other swine. Laboratory swine can be housed individually or in small groups in pens. Individual housing is common in a research setting. If housed individually pigs should have visual, olfactory, and auditory contact with conspecifics to prevent stress associated with social deprivation.

Proper management practices and knowledge of normal swine behavior are required to successfully group house swine. Group sizes will be based on the size of the pens and the size of the pigs; however, group sizes should be limited to 10–15 animals to maintain a stable social hierarchy. Cage partitions should be provided to allow subordinant animals to avoid more dominant animals. Groups of animals should remain together as long as possible as dominance fighting occurs when new animals are added to the group. In a research setting, new groups should be established when a new shipment of pigs is received. Providing additional troughs or greater trough feeding space will allow all animals to eat simultaneously; however, differences in body weights may occur because of competition during feeding.[3]

In the United States, space recommendations are specified for swine used in biomedical research in the *Guide for the Care and Use of Laboratory Animals*.[69] The Council of Europe has recently revised their housing standards for research swine.[70]

Caging for swine should be sturdy since pigs are strong animals. Although pigs do not climb, they rub their sides along the sides of their cages with considerable force. As an expression of their rooting behavior, they will find and manipulate any loose items in their environment. Consequently, caging should be of solid construction and free of sharp edges and protrusions to prevent injury. Chain link fencing may be used for small swine. Other suitable materials include aluminum and stainless steel caging, which can be configured into vertical bars or solid panels. Regardless of cage material, the fencing should interface with the flooring securely and without gaps to prevent pigs from getting their hooves caught.

Swine require secure footing and can develop stress ulcers if flooring is slippery. Cage flooring can be either solid or on raised grids. If solid concrete or epoxy floors are used they should be textured to provide secure footing, and bedding of wood shavings or straw should be provided for rooting and nesting behavior. Grid floors increase the ease of sanitation since bedding is not usually provided. Spacing of the floor grids should be appropriate to prevent hoof damage. Plastic-coated diamond grids with five-eighths inch openings are suitable for housing swine of multiple sizes; however, swine will chew the plastic off the metal as soon as a tear in the plastic occurs. Fiberglass slatted floors with a slat width of 4.4 cm and a space of 0.64 cm between slats work well in the authors' experience. The slats can be coated with a medium grit to promote hoof wear. Regular hoof trimming at 3–6 month intervals is required for chronically housed animals if flooring does not maintain hoof wear.[2]

Cages that do not use bedding should be hosed daily to remove waste and minimize odors. Daily removal of soiled bedding is recommended when deep bedding is used. A complete change of bedding should be done one or two times a week. Cages should be sanitized no less than every 2 weeks by pressure washing for cleaning units in place or by cage washing modular units after they have been broken down. Pigs should be removed from their cages during routine cleaning to prevent contact with detergents and chilling from being wet. However, they readily accept baths and can be kept clean using a mild soap and warm water.

Swine readily use automatic watering systems, which are preferred to buckets. Swine require approximate 2.5 liters of water/kg of feed and can spill or soil water that is supplied in buckets. Feeders should be securely attached to cage sides to be able to withstand the excitement associated with feeding and the tendency of the pigs to root cage objects. Stainless steel or Teflon feed troughs are easily sanitized and are preferable to bucket feeders used in agricultural settings. Consideration should be given during cage design to the placement of the location of feeders and watering devices since swine will develop a dunging pattern and will defecate at the opposite end of the cage from where they are fed.

Laboratory swine should be provided with environmental enrichment to satisfy their intense need to chew and root, especially if bedding is not provided. We have successfully used large plastic balls, plastic dog toys, items that hang from chains, and nylon brushes attached to fencing. Hanging items satisfy the need to chew and rub while items provided on the floor are used for rooting. Enrichment items need to be included in routine cage sanitation since pigs will avoid them if they become soiled. Rotation of enrichment items and limiting access to them for several hours a day will help maintain their novelty.

HANDLING Laboratory swine are easily trained using gentle handling techniques and positive reinforcement. Food items such as canned dog food, cookies, apples, and carrots can be used for training. Agricultural methods of handling and restraint, such as snout snares and suspension by the rear legs, are aversive and stressful to swine and are inappropriate in a research setting. Swine adapt readily to daily routines, including special handling practices common to research settings. Small pigs may be carried in the handler's arms, similar to handling a large dog. Larger pigs can be gently restrained against the side of a cage using a swine board. If prolonged restraint is necessary animals can be trained to use a humane restraint sling following a period of acclimation.

Pigs should be approached quietly in a crouched position since bending toward them is perceived as threatening. Rubbing pigs gently on the abdomen has a calming effect and can be used to aid physical examination. Swine can be trained to move out of their cages and guided with the use of a pig board or they can be trained to walk using a harness and leash.[71]

Laboratory swine that are cared for and handled gently interact readily with people. In fact, human–pig interaction is an important part of their environmental enrichment and contributes to their well-being in a research setting.

CONCLUSIONS

Swine will continue to be developed as translational research models and it is expected that their use will continue to outpace the growth of other large animal species in this arena. We receive frequent inquiries to troubleshoot protocols from investigators outside our institution. In the authors' experiences most of the complications that occur in porcine protocols are related to anesthetic, perioperative care and surgical issues. Anesthetic and perioperative care protocols must be designed with specific attention to the physiological effects of the agents and their potential for interference with the experimental goals. Surgical complications are usually related to inadequate aseptic technique or improper wound closure techniques.

There are current courses, CD Rom training programs, and technical publications[2,3,5,6] to help ameliorate these issues. Many of the helpful techniques are located online in various websites, which are listed below:

1. Contains a swine literature database from the Animal Welfare Information Center:
   http://www.nal.usda.gov/awic/pubs/swine/swine.htm
2. Contains reviews of models and Sinclair, Hanford, and Yucatan information:
   http://www.sinclairresearch.com/
3. Tutorial on swine procedures in research: Laboratory Animal Training Association:
   http://www.latanet.com/online/onlinetr.htm
4. Biology and diseases of swine:
   http://www.ivis.org/advances/Reuter/swindle/chapter_frm.asp?LA=1
5. Basic information on swine:
   http://www.aphis.usda.gov/vs/ceah/cahm/Swine/swine.htm
   http://www.nal.usda.gov/awic/pubs/swinehousing/swinehousing2.htm
6. Göttingen minipig background information:
   http://minipigs.dk
7. CD Rom training series on husbandry, handling, injection techniques, anesthesia, analgesia, and perioperative care:
   http://www.latanet.com/desktop/drs.html
   http://www.latanet.com
8. National Swine Research Resource Center–Transgenic and Cryopreservation Technology:
   http://www.nsrrc.missouri.edu/
9. Medical University of SC-Swine Training Courses:
   http://research.musc.edu/dlar/Swine%20Training%20Courses.htm

REFERENCES

1. Swindle MM. *Surgery, Anesthesia and Experimental Techniques in Swine*. Ames, IA: Iowa State University Press, 1998.
2. Swindle MM. *Swine in the Laboratory: Surgery, Anesthesia, Imaging and Experimental Techniques*, 2nd ed. Boca Raton, FL: CRC Press, 2007.
3. Swindle MM. *Basic Surgical Exercises Using Swine*. Philadelphia, PA: Praeger Press, 1983.
4. Bollen PJA, Hansen AK, Rasmussen HJ, Suckow MA. *The Laboratory Swine*. Boca Raton, FL: CRC Press, 2000.
5. Swindle MM, Smith AC, Goodrich JG. Chronic cannulation and fistulation procedures in swine: A review and recommendations. *J Invest Surg* 1998;11(1):7–20.
6. Swindle MM, Nolan T, Jacobson A, Wolf P, Dalton MJ, Smith AC. Vascular access port (VAP) usage in large animal species. *Contemp Top Lab Anim Sci* 2005;44(3):7–17.
7. Swindle MM, Smith AC. Comparative anatomy and physiology of the pig. *Scand J Lab Anim Sci* 1998;25(Suppl. 1):1–10.
8. Pond WG, Mersmann HJ. *Biology of the Domestic Pig*. Ithaca, NY: Comstock Publishing Associates, 2001.
9. Sack WO. *Essentials of Pig Anatomy and Harowitz/Kramer Atlas of Musculoskeletal Anatomy of the Pig*. Ithaca, NY: Veterinary Textbooks, 1982.
10. Getty R, Ed. *Sisson and Grossman's The Anatomy of the Domestic Animal—Porcine*, Vol. 2. Philadelphia, PA: W.B. Saunders, 1975:1215–1422.
11. Stanton HC, Mersman HJ, Eds. *Swine in Cardiovascular Research*, Vol. 1–2. Boca Raton, FL: CRC Press, 1986.
12. Svendsen O. The minipig in toxicology: Proceedings of the Satellite Symposium to Eurotox 97. *Scand J Lab Anim Sci* 1998;25 (Suppl. 1).
13. Swindle MM, Ed. *Swine as Models in Biomedical Research*. Ames, IA: Iowa State University Press, 1992.
14. Tumbleson ME, Ed. *Swine in Biomedical Research*, Vol. 1–3. New York: Plenum Press, 1986.
15. Tumbleson ME, Schook LB, Eds. *Advances in Swine in Biomedical Research*, Vol. 1–2. New York: Plenum Press, 1996.
16. Lindberg JE, Ogle B. *Digestive Physiology of Pigs: Proceedings of the 8th Symposium*. New York: CABI Publishing, 2001.
17. Flecknell, PA. *Laboratory Animal Anaesthesia*, 2nd ed. New York: Academic Press, 1996.
18. Kohn DH, Wixson SK, White WJ, Benson GJ, Eds. *Anesthesia and Analgesia in Laboratory Animals*. New York: Academic Press, 1997.
19. Hawk CT, Leary SL, Morris TH. *Formulary for Laboratory Animals*, 3rd ed. Ames, IA: Blackwell Publishing, 2005.
20. Swindle MM, Laber K, Smith AC, Goodrich JA, Bingel SA. Biology and medicine of swine. In: Reuter JD, Suckow MA, Eds. *Laboratory*

Animal Medicine and Management, 2003. Ithaca, NY: International Veterinary Information Service, http://www.ivis.org/library.asp.
21. Straw BE, D'Allaire S, Mengeling WL, Taylor DJ. *Diseases of Swine*, 8th ed. Ames, IA: Iowa State University Press, 1999.
22. Laber KE, Swindle MM. Ethics and regulations for the care and use of laboratory animals. In:,_Wynek G, Bowlin G, Eds. *Encyclopedia of Biomaterials and Biomedical Engineering*. New York: Taylor and Francis, 2006. http://dekker.com/sdek/130097897–6529976/ issues+ db=enc+content=t713172959.
23. Swindle, MM, Thompson RP, Carabello BA, Smith AC, Green CT, Gillette, PC. Congenital cardiovascular disease. In: Tumbleson M, Schook L, Eds. *Advances in Swine in Biomedical Research*. New York: Plenum Press, 1996.
24. Attie, AD, Aiello, RJ, Checovich WJ. The spontaneously hypercholesterolemic pig as an animal model of human hypercholesterolemia. In: Tumbleson M, Schook L, Eds. *Advances in Swine in Biomedical Research*. New York: Plenum Press, 1996.
25. Johnson DK, Wisner ER, Griffey SM, Vessey AR, Haley PJ. Sinclair miniature swine melanoma as a model for evaluating novel lymphograpy contrast agents. In: Tumbleson M, Schook L, Eds. *Advances in Swine in Biomedical Research*. New York: Plenum Press, 1996.
26. Dyson M, Alloosh M, Vuchetich JP, Mokelke EA, Sturek M. Components of metabolic syndrome and coronary artery disease in female Ossabaw swine fed excess atherogenic diet. *Comp Med* 2006;56(1):35–45.
27. Prather RS. Transgenic approaches to developing swine models. In: *Swine in Biomedical Research, Update on Animal Models*. Memphis, TN: American Association for Laboratory Animal Science (AALAS), 2005.
28. Petters RM, Alexander CA, Wells KD. Genetically engineered large animal model for studying cone photoreceptor survival and degeneration in retinitis pigmentosa. *Nat Biotechnol* 1997;15(10):965–970.
29. Alisky JM. Xenografts are an achievable breakthrough. *Med Hypotheses* 2004;63(1):92–97.
30. Cramer DV, Podesta LG, Makawka L. *Handbook of Animal Models in Transplantation Research*. Boca Raton, FL: CRC Press, 1994.
31. Jensen SL, Gregersen H, Shokouh-Amiri MH, Moody FG. *Essentials of Experimental Surgery: Gastroenterology*. London: Harwood Academic Publishers, 1996.
32. Swindle MM, Wiest DB, Smith AC, Garner SS, Case CC, Thompson RP, Fyfe DA, Gillette PC. Fetal surgical protocols in Yucatan miniature swine. *Lab Anim Sci* 1996;46(1):90–95.
33. Bloor CM, White FC, Roth DM. The pig as a model of myocardial ischemia and gradual coronary occlusion. In: Swindle MM, Ed. *Swine as Models in Biomedical Research*. Ames, IA: Iowa State University Press, 1992:163–175.
34. Gootman PM. Cardiovascular system. In: Pond WG, Mersmann HJ, Eds. *Biology of the Domestic Pig*. Ithaca, NY: Cornell University Press, Comstock Publishing, 2001:533–559.
35. Reffelmann T, Sensebat O, Birnbaum Y, Stroemer E, Hanrath P, Uretsky BF, Schwarz ER. A novel minimal-invasive model of chronic myocardial infarction in swine. *Coron Artery Dis* 2004;15(1):7–12.
36. Gillette PC, Swindle MM, Thompson RP, Case CL. Transvenous cryoablation of the bundle of His. *Pacing Clinical Electrophysiol* 1991;14(4Pt. 1):504–510.
37. Hughes HC, Bowman TA. Intracardiac electrophysiology of swine for design and testing of cardiac pacemakers. In: Tumbleson ME, Ed. *Swine in Biomedical Research*, Vol. 1. New York: Plenum Press, 1986:327–331.
38. Smith AC, Knick B, Swindle MM, Gillette PC. A technique for conducting non-invasive cardiac electrophysiology studies in swine. *J Invest Surg* 1997;10(1–2):25–30.
39. Gal D, Isner JM. Atherosclerotic Yucatan microswine as a model for novel cardiovascular interventions and imaging, In: Swindle MM, Ed. *Swine as Models in Biomedical Research*. Ames, IA: Iowa State University Press, 1992:118–140.
40. Goodrich JA, Clarkson TB, Cline JM, Jenkins AJ, Del Signore MJ. Value of the micropig model of menopause in the assessment of benefits and risks of postmenopausal therapies for cardiovascular and reproductive tissues. *Fertil Steril* 2003;79(Suppl. 1):779–788.
41. Corno AF, Sekarski N, Gernath MA, Payot M, Tozzi P, von Segesser LK. Pulmonary artery banding: Long-term telemetric adjustment. *Eur J Cardiothorac Surg* 2003;23(3):317–322.
42. Capras SN, Clair MJ, Krombach RS, Hendrick JW, Houck WV, Kribbs SB, Mukherjee R, Tempel GE, Spinale FG. Brain blood flow patterns after the development of congestive heart failure: Effects of treadmill exercise. *Crit Care Med* 2000;28(1):209–214.
43. Eble DM, Spinale FG. Contractile and cytoskeletal content, structure, and mRNA levels with tachycardia-induced cardiomyopathy. *Am J Physiol* 1995;268(6Pt. 2):H2426–2439.
44. Dawson RC, Krisht AF, Barrow DL, Joseph GJ, Shengelaia GG, Bonner B. Treatment of experimental aneurysms using collagen-coated microcoils. *Neurosurgery* 1995;36(1):133–140.
45. Chitwood WR Jr, Wiley NL, Chapman WHH, Felger JE, Bailey BM, Ballint T, Mendleson KG, Kim VB, Young JA, Albrecht RA. Robotic surgical training in an academic institution. *Ann Surg* 2001; 234(4):475–486.
46. Fossum TW, Baltzer WI, Miller MW, Aguirre M, Whitlock D, Solter P, Makarski LA, McDonald MM, An MY, Humphrey JD. A novel coarctation model for studying hypertension in the pig. *J Invest Surg* 2003;16(1):35–44.
47. Kelly BS, Heffelfinger SC, Whiting JF, Miller MA, Reaves A, Armstrong J, Narayana A, Roy-Chaudhury P. Aggressive venous neointimal hyperplasia in a pig model of arteriovenous graft. *Kidney Int* 2002;62(7):2272–2280.
48. Maynar M, Qian Z, Hernandez J, Sun F, DeMiguel C, Crisostomo V, Uson J, Pineda LF, Espinoza CG, Castaneda WR. An animal model of abdominal aortic aneurysm created with peritoneal patch: Technique and initial results. *Cardiovasc Intervent Radiol* 2003;26(2): 168–176.
49. Mehran RJ, Ricci MA, Graham AM, Carter K, Smyes JF. Porcine model for vascular graft studies. *J Invest Surg* 1991;4(1):37–44.
50. Mitchell SE, Anderson JH, Swindle MM, Strandberg JD, Kan J. Atrial septostomy: Stationary angioplasty balloon technique. Experimental work and preliminary clinical applications. *Pediatr Cardiol* 1994;15(1):1–7.
51. Morrow WR, Smith VC, Ehler WJ, Van Dellen AF, Mullins CE. Balloon angioplasty with stent implantation in experimental coarctation of the aorta. *Circulation* 1994;89(6):2677–2683.
52. Gallegos RP, Nockel PJ, Rivard AL, Bianco RW. The current state of in-vivo pre-clinical animal models for heart valve evaluation. *J Heart Valve Dis* 2005;14(3):423–432.
53. Smerup M, Pedersen TF, Nyboe C, Funder JA, Christensen TD, Nielsen SL, Hjortdal V, Hasenkam JM. A long-term porcine model for evaluation of prosthetic heart valves. *Heart Surg Forum* 2004;7(4): E259–264.
54. Yen JT. Digestive system. In: Pond WG, Mersmann HJ, Eds. *Biology of the Domestic Pig*. Ithaca, NY: Cornell University Press, Comstock Publishing, 2001:399–453.
55. Kinnala PJ, Kuttila KT, Gronroos JM, Havia TV, Nevalainen TJ, Niinikoski JH. Pancreatic tissue perfusion in experimental acute pancreatitis. *Eur J Surg* 2001;167(9):689–694.
56. Nielsen TB, Yderstraede KB, Beck-Nielsen H. Isolation, transplantation, and functional studies of adult porcine islets of Langerhans. *Comp Med* 2002;52(2):127–135.
57. Larsen MO, Rolin B, Wilken M, Carr RD, Godtfredsen CF. Measurements of insulin secretory capacity and glucose tolerance to predict pancreatic b-cell mass in vivo in the nicotinamide/streptozotocin Göttingen minipig, a model of moderate insulin deficiency and diabetes. *Diabetes* 2003;52(1):118–123.
58. Dalmose AL, Hvistendahl JJ, Olsen LH, Eskild-Jensen A, Dhurhuus JC, Swindle MM. Surgically induced urologic models in swine. *J Invest Surg* 2000;13:133–145.
59. Olsen LH, Dalmose AL, Swindle MM, Jørgensen TM, Djurhuus JC. Male fetal pig lower urinary tract function in mid second and early third trimester of gestation. *J Urol* 2001;165(6Pt. 2):2331–2334.
60. Olsen LH, Dalmose AL, Swindle MM, Djurhuus JC, Jørgensen TM. Male fetal pig lower urinary tract function. Part II: Free voiding pattern close to term and in the newborn. *J Urol* 2004;171(6Pt.2): 2660–2663.

61. Swindle MM, Wiest DB, Smith AC, Garner SS, Case CC, Thompson RP, Fyfe DA, Gillette PC. Fetal surgical protocols in Yucatan miniature swine. *Lab Anim Sci* 1996;46(1):90–95.
62. Bazer FW, Ford JJ, Kensinger RS. Reproductive physiology. In: Pond WG, Mersmann HJ, Eds. *Biology of the Domestic Pig*. Ithaca, NY: Cornell University Press, Comstock Publishing, 2001:150–224.
63. Monteiro-Riviere NA. Integument. In: Pond WG, Mersmann HJ, Eds. *Biology of the Domestic Pig*. Ithaca, NY: Cornell University Press, Comstock Publishing, 2001:625–652.
64. Monteiro-Riviere NA, Riviere J. The pig as a model for human skin research. *Proceedings of the Swine in Biomedical Research, Update on Animal Models*, 2005, 56th American Association for Laboratory Animal Science (AALAS) Meeting, St. Louis, MO, November 2005.
65. Sullivan TP, Eaglstein WH, Davis SC, Mertz P. The pig as a model for human wound healing. *Wound Repair Regen* 2001;9(2):66–76.
66. Kerrigan CL, Zelt RG, Thomson JG, Diano E. The pig as an experimental animal in plastic surgery research for the study of skin flaps, myocutaneous flaps and fasciocutaneous flaps. *Lab Anim Sci* 1986;36(4):408–412.
67. Middelkoop E, van den Bogaerdt AJ, Lamme EN, Hoekstra MJ, Brandsma K, Ulrich MM. Porcine wound models for skin substitution and burn treatment. *Biomaterials* 2004;25(9):1559–1567.
68. Swindle KE, Hamilton PD, Ravi N. In situ formation of hydrogels as vitreous substitutes: Viscoelastic comparison to porcine vitreous. *Biomacromolecules* 2007.
69. National Research Council. *Guide for the Care and Use of Laboratory Animals*. Washington, DC; National Academy Press, 1996.
70. Council of Europe. 2006. European Convention for the protection of vertebrate animals used for experimental and other scientific purposes. ETS 123, Strasbourg. http://conventions.coe.int/Treaty/en/Treaties/Html/123.htm.
71. Smith AC. Swine behavior. In: *Swine in Biomedical Research, Update on Animal Models*. Memphis, TN: American Association for Laboratory Animal Science (AALAS), 2005.
72. Ghosh F, Wong F, Johansson K, Bruun A, Petters RM. Transplantation of full-thickness retina in the rhodopsin transgenic pig. *Retina* 2004;24(1):98–109.
73. Watanabe H, Andersen F, Simonsen CZ, Evans SM, Gjedde A, Cumming P, DaNeX Study Group. MR-based statistical atlas of the Göttingen minipigs brain. *Neuroimage* 2001;14:1089–1096.
74. Sakoh M, Røhl L, Gyldensted C, Gjedde A, østergaard L. Cerebral blood flow and blood volume measured by magnetic resonance imaging bolus tracking after acute stroke in pigs: Comparison with (^{15}O) H_2O positron emission tomography. *Stroke* 2000;31(8):1958–1964.
75. Wiest DB, Swindle MM, Garner SS, Smith AC, Gillette PC. Pregnant Yucatan miniature swine as a model for investigating fetal drug therapy. In: Tumbleson ME, Schook LB, Eds. *Advances in Swine in Biomedical Research*, Vol. 2. New York: Plenum Press, 1996:629–636.
76. Skaanild MT, Friis C. Cytochrome P450 differences in minipigs and conventional pigs. *Pharmacol Toxicol* 1999;85(3):174–180.
77. Swindle MM. Considerations of specific pathogen free (SPF) swine in xenotransplantation. *J Invest Surg* 1996;9(3):267–271.

Figure 8–1. Example of enrichment items for laboratory mice. The hut is tinted such that the mouse cannot see out, but human caretakers can see in. (Photo by Jill Rawlins.)

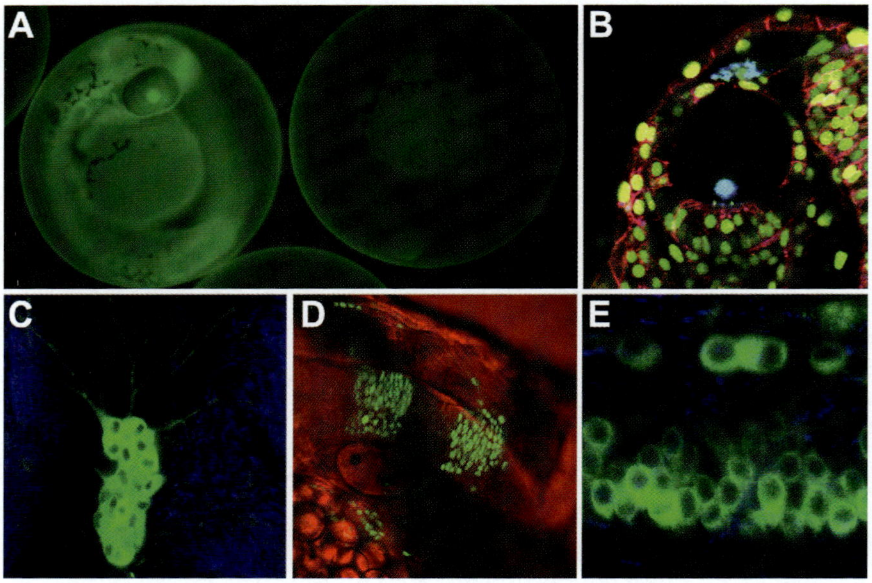

Figure 13–4. Imaging in live zebrafish embryos. (A) Whole embryos can be quickly imaged on a fluorescent dissecting microscope in their chorions to sort positive transgenics (left) from wild-type siblings (right). These embryos were not treated with PTU so the melanocytes are visible. (B–E) Confocal microscopy permits much higher resolution imaging. (B) A quick method for labeling is to inject RNA encoding fluorescent proteins, in this case a histone2B-EGFP fusion, and a membrane localized mCherry was used to image all the cells of the inner ear. (C) GFP transgenics can be used to image neuronal projections from the trigeminal ganglion as they extend. (D) GFP transgenics can mark specific populations of cells, in this case rhombomeres 3 and 5. (E) GFP fusion proteins can reveal the subcellular localization pattern of proteins, in this case a cytoplasmic protein in the Rohon-Beard and motor neurons of the spinal cord. (Images from S.G. Megason, L.A. Trinh, and S.E. Fraser, unpublished.)

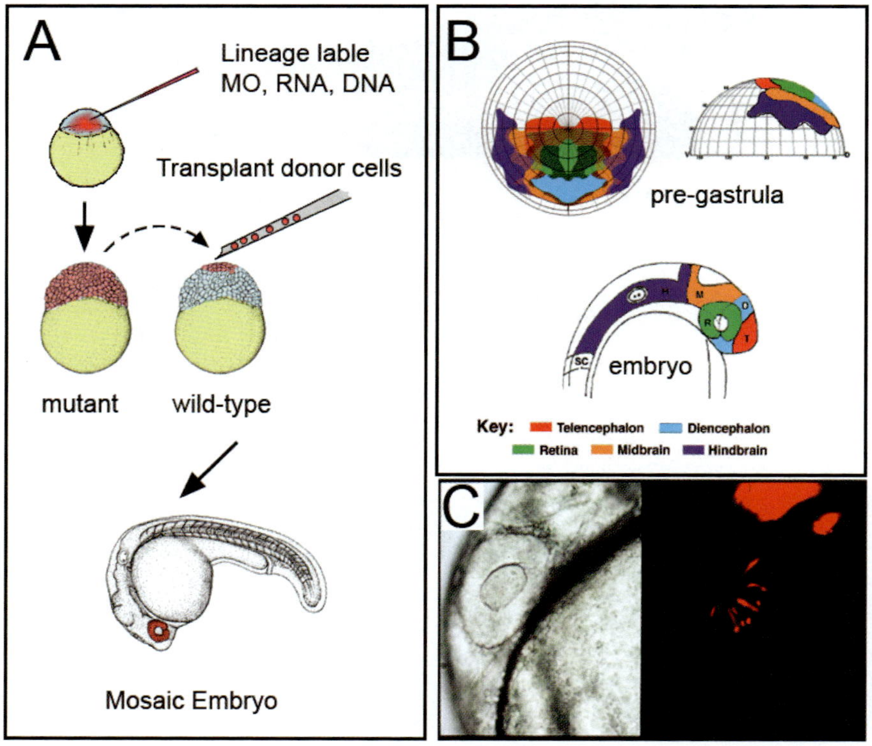

Figure 13–5. Creating genetic mosaics in zebrafish. (A) Donor cells are first lineage labeled with a tracer dye at the one-cell stage. Donor embryos can also be injected with morpholinos, RNA, or DNA. At the 1000-cell stage, totipotent blastula cells are transplanted into regions of the host embryo fated to give rise to specific structures. Donor and host embryos can be of either mutant or wild-type genotypes. Resultant chimeras are grown for subsequent analysis. (B) Fate-map of the pregastrula stage embryos. (Modified from Woo and Fraser, 1995.) (C) Mosaic embryo at 24 hpf showing bright-field (left) and fluorescent image (right) of rhodamine-dextran-labeled donor cells targeted to the eye and forebrain. (B.A. Link, unpublished.)

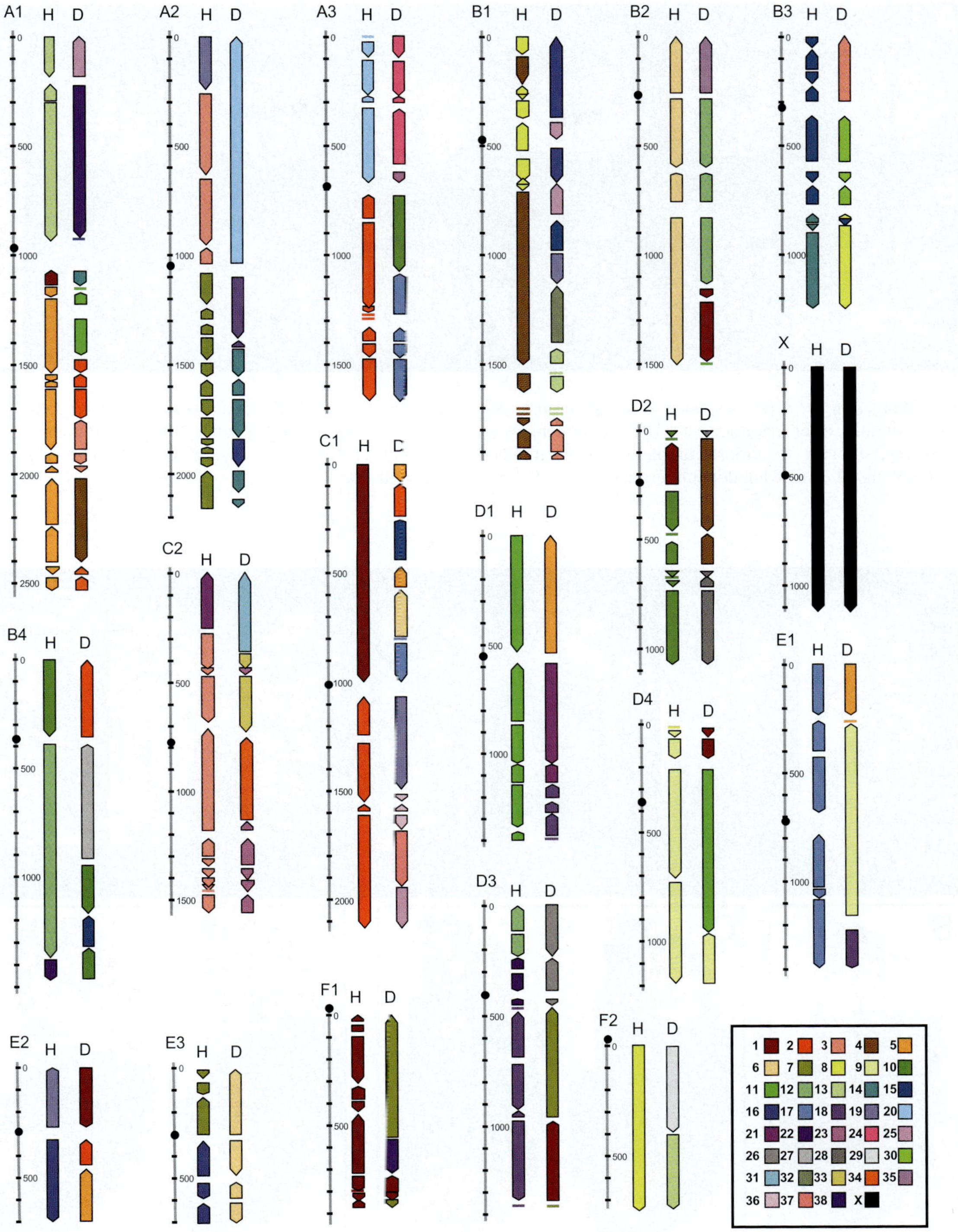

Figure 25–1. Feline chromosome maps (labeled at top) and homologous synteny blocks (HSBs) in the human (H) and dog (D) genomes. HSBs are shown to the right of each cat chromosome map (only the map scale is shown). The dark cross-marks on each cat chromosome correspond to 100-cR_{5000} intervals. The inferred centromere positions are shown by dark circles. HSBs are color coded by human or dog chromosome, defined by the key in the bottom right corner. (Reprinted from Murphy *et al.*[7] Copyright 2007, with permission from Elsevier.)

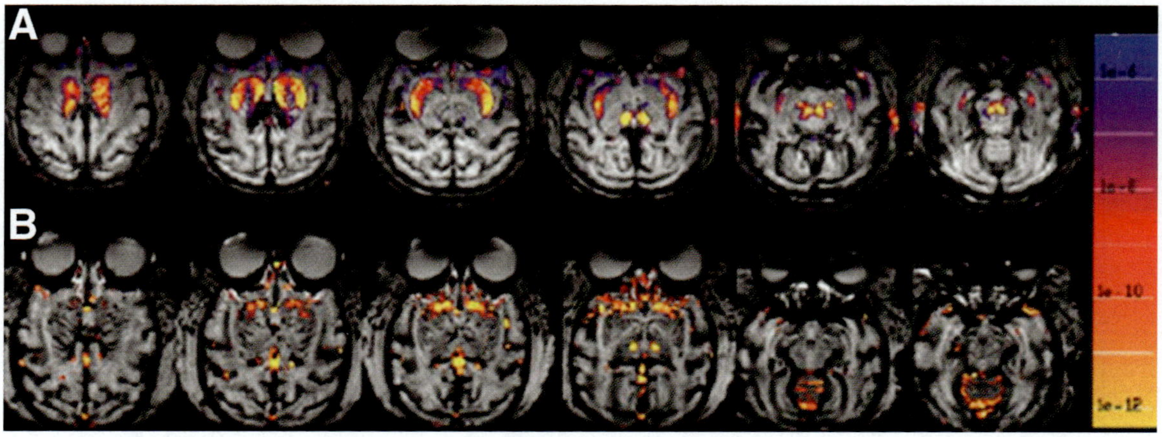

Figure 51–3. Changes in the CBV response to amphetamine (A) before and (B) 4 months after cessation of MPTP treatment in a cynomolgus monkey, showing an almost complete loss of amphetamine-induced CBV signal changes in dopaminergic regions. Parkinsonian primates had a prominent loss of response to amphetamine, with relative sparing of the nucleus accumbens and parafascicular thalamus. (Modified from Jenkins *et al.*[183] Copyright 2004 Society for Neurosciences.)

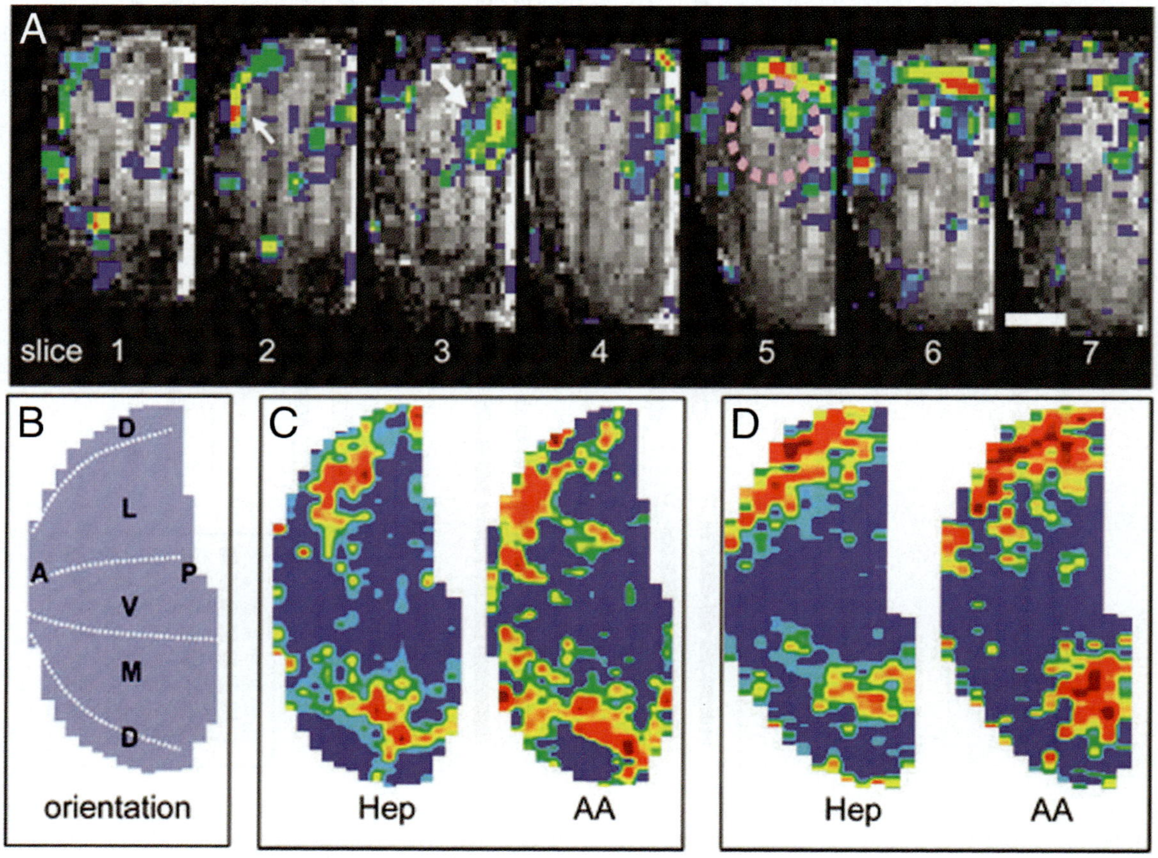

Figure 51–4. (A) Patterns of the fMRI response in the mouse main olfactory bulb (MOB) and accessory olfactory bulb (AOB, pink circle in slice 5) to the pheromone 2-heptanone, one of the urinary chemosignal compounds in mouse. The arrows point to two foci of activation suggestive of a pair of the nearly mirror projections of the receptor neuron subsets to the same MOB. Scale bar = 500 μm. (B) A flattened view of the olfactory bulb indicating the orientation of the odor maps shown in (C) and (D): A, anterior; D, dorsal; L, lateral; M, medial; P, posterior; V, ventral. (C, D) The odor maps of 2-heptanone (Hep) in two different mice show that this pheromone not only activates large regions of the MOB but also generates similar patterns across subjects. Interestingly, the odor map for amyl acetate (AA), a common odorant with an odor quality similar to that of 2-heptanone, was also similar to that of 2-heptanoneb. (Adapted from Xu *et al.*[217] Copyright 2005 Wiley-Liss, Inc.)

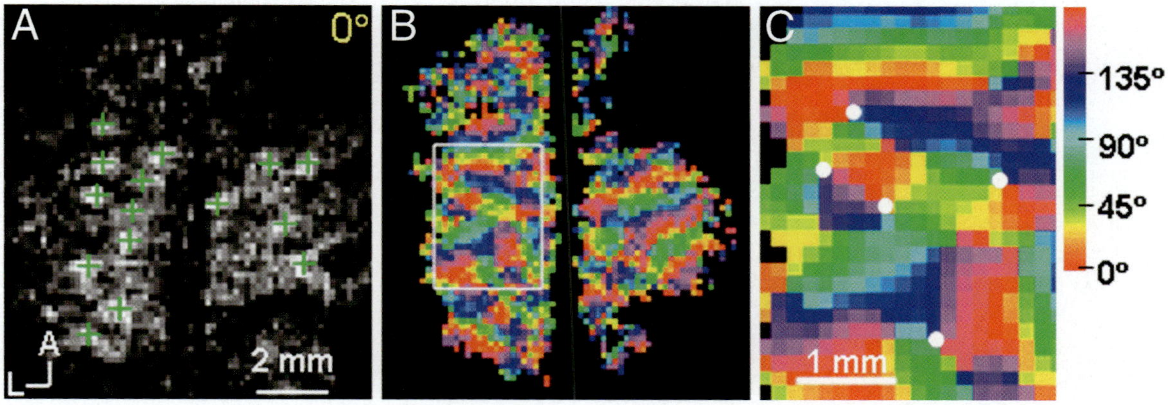

Figure 51–5. Signal changes in CBV-weighted fMRI obtained during stimulation of the cat visual cortex according to predetermined stimulus orientations. (A) Raw gray-scale functional map obtained by subtraction of images during 0° stimulation from prestimulus control. The center of each patch is marked with a green + sign. (B) Four different grating orientations (0°, 45°, 90°, and 135°) were presented in the study of one animal, enabling a composite angle map to be generated through pixel-by-pixel vector addition of the four single-condition maps. In the left hemisphere (marked by a white rectangular box), changes between pixels preferentially activated by a particular stimulus orientation were smoothed by a 3 × 3 Gaussian kernel. (C) A composite angle map was generated with the region indicated by the white ROI in (b). "Pinwheel" structures indicated by small white dots were observed where domains for all orientations converge. (Adapted from Zhao *et al.*[114] Copyright 2005 Elsevier.)

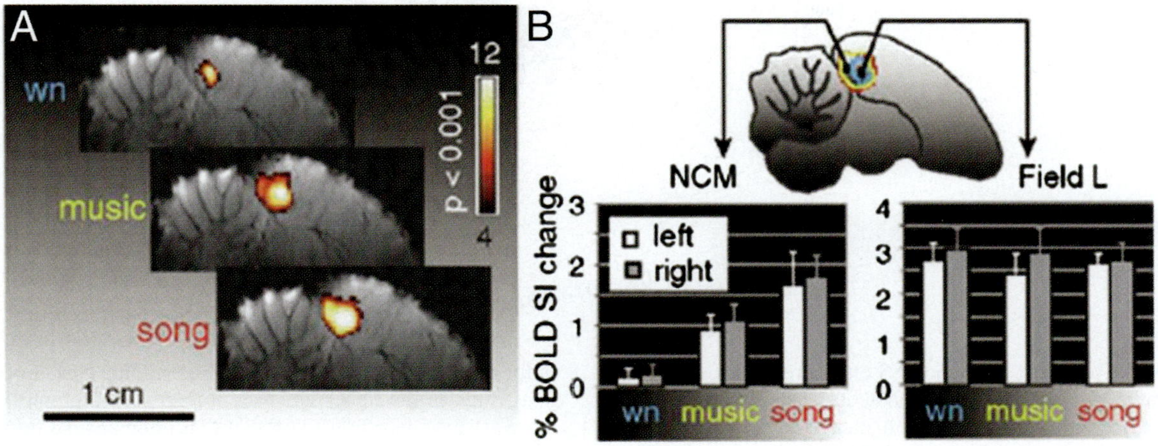

Figure 51–6. Auditory activation in the songbird telencephalon shows (A) statistical maps illustrating the localization of significant signal intensity changes during auditory stimulation consisting of white noise (wn), a concerto of Bach (music), and a stimulation with song from a male starling (song). (B) The location of the activated areas (top), together with the average BOLD response amplitude for each stimulus (bottom). (Adapted from Van Meir *et al.*[231] Copyright 2005 Elsevier.)

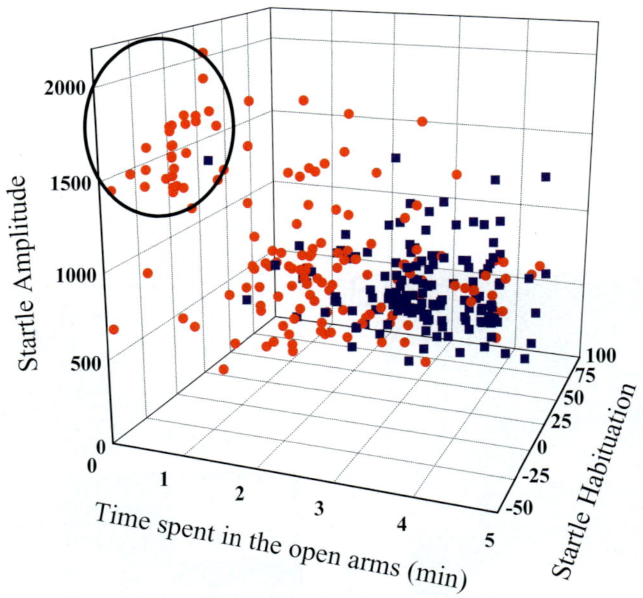

Figure 61–1. The effect of single PSS exposure versus unexposed control on rat anxiety-like behavior and acoustic startle response and habituation. The representation of the data from both paradigms (EPM and ASR) shows two obvious and rather distinct features. First, it is clear that PSS exposure alters the response of the majority of individuals to at least some degree. Second, the cluster of individuals that forms in the upper left hand corner of the graph (i.e., had the more extreme responses to exposure) is quite distinct from the majority of individuals.[36,48]

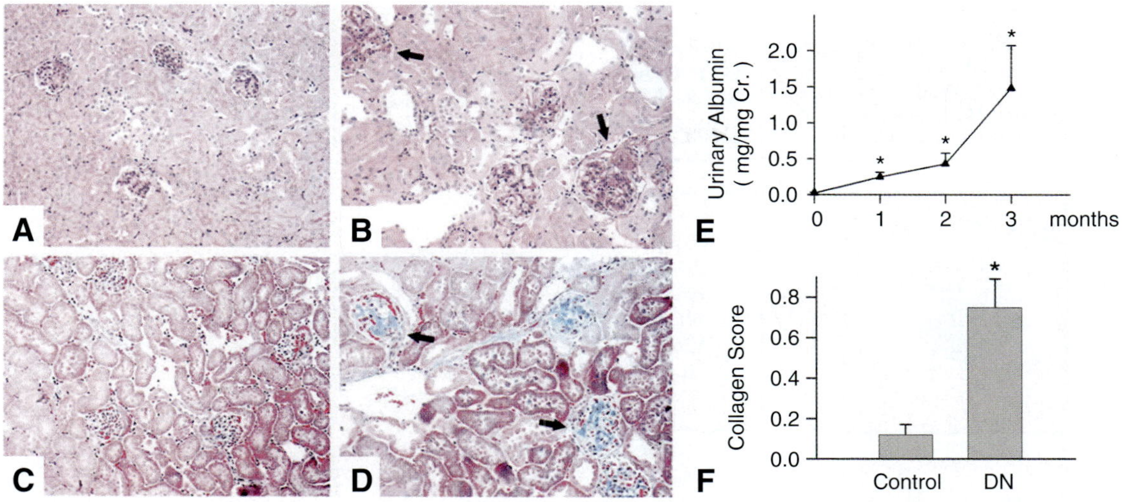

Figure 68–2. Diabetic nephropathy induced by streptozotocin (STZ) in uninephrectomized mice. One week after uninephrectomy, male CD1 mice received an intravenous injection of STZ. (A–D) Representative micrographs demonstrate glomerular enlargement, mesangial expansion, and segmental glomerulosclerosis (periodic acid–Schiff staining; A, B) and glomerular collagen deposition (Masson–Trichrome staining; C, D). (A and C) Normal control; (B and D) diabetic mice. Arrows indicate injured glomeruli. (E) Albuminuria develops in diabetic mice in a time-dependent manner. Data are presented as means ±SEM. $*p < 0.05$. (F) Glomerular collagen deposition score in diabetic and normal mice. $*p < 0.05$. (Adapted and modified from Dai *et al.*[46])

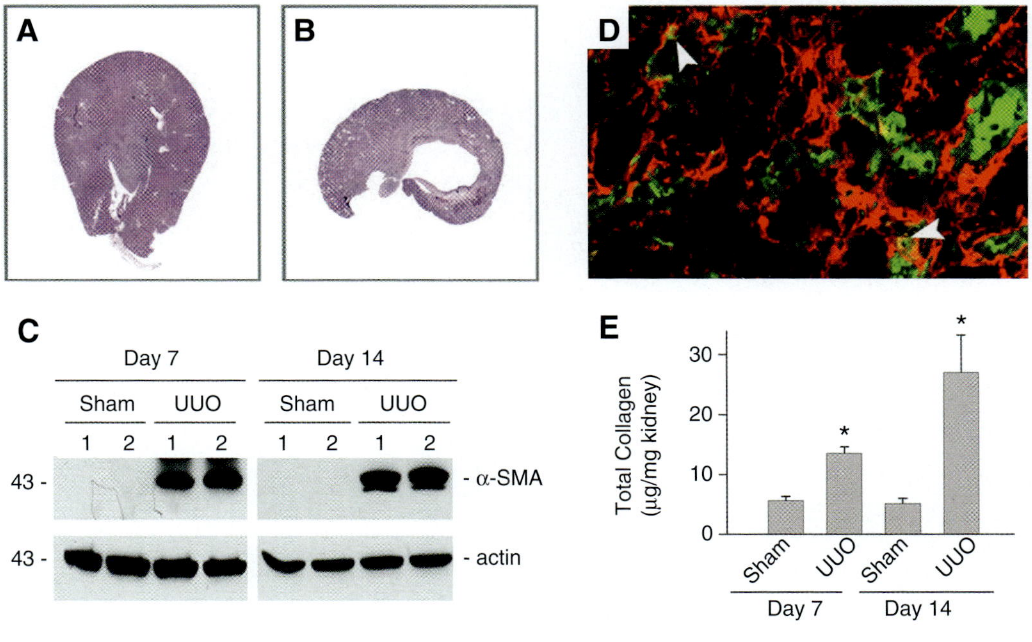

Figure 68–3. Interstitial fibrosis in the mouse model of obstructive nephropathy. (A, B) Representative micrographs show the cross sections and gross morphology of the obstructed kidneys at day 14 after ureteral ligation. (A) Sham control; (B) UUO. (C) Western blot analyses demonstrate a marked induction of α-smooth muscle actin (α-SMA), a molecular marker for myofibroblasts, in the obstructed kidney at day 14 after UUO. (D) Double immunofluorescence staining shows the α-SMA (red) and proximal tubular epithelial cell marker, fluorescein isothiocyanate (FITC)-conjugated lectin from *Tetragonolobus purpureas* (green). (E) Quantitative determination of total kidney collagen contents in sham and obstructed kidneys. Data are presented as means ±SEM. $*p < 0.01$. (Adapted and modified from Yang *et al.*[59]).

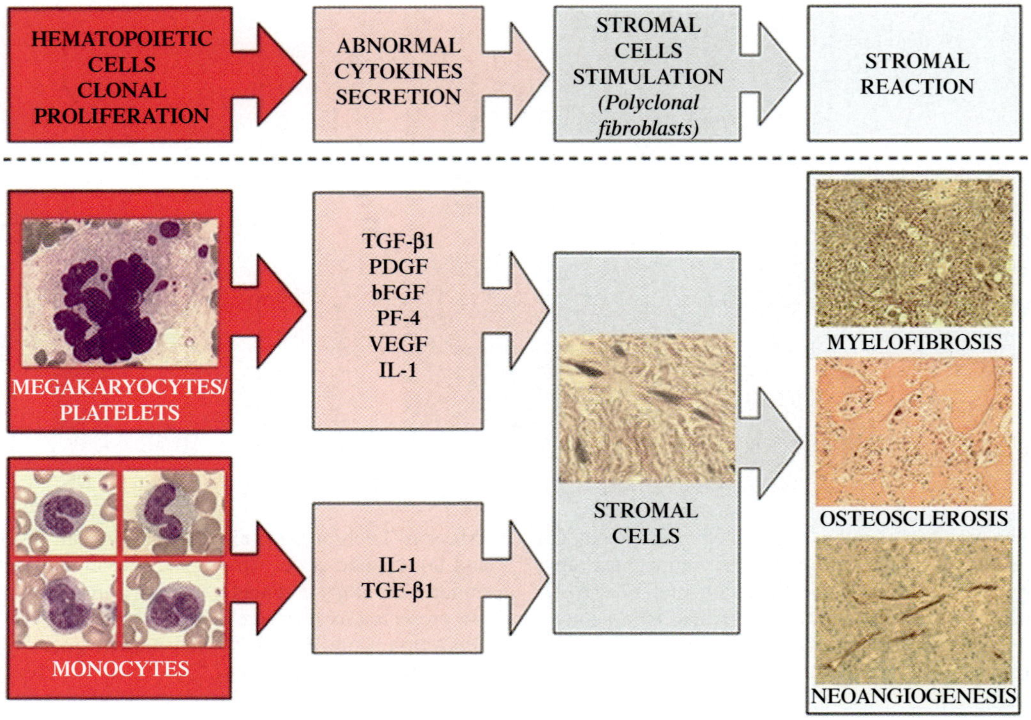

Figure 73–4. A pathogenetic model for the cytokine-mediated stromal reaction observed in MMM.

27 The Minipig as an Animal Model in Biomedical Stem Cell Research

Petr Vodicka, Jana Hlucilova, Jiri Klima, Radek Prochazka, Jitka Ourednik, Vaclav Ourednik, and Jan Motlik

ABSTRACT

Pigs and miniature pigs are steadily gaining importance as large animal models in the field of regenerative medicine, including stem cell research. With their size, organ capacity, and physiology resembling in several aspects that of humans, pigs are well suited for preclinical experiments and long-term safety studies. In this chapter, we summarize our experience with the isolation and culture of several somatic stem cell populations from fetal and adult pig tissue and briefly review their potential usefulness in future stem cell-based therapies. We also provide protocols for the isolation of fetal porcine neural stem cells (NSCs), adult bone marrow mesenchymal stem cells (MSCs), and epidermal progenitor cells (EPCs) from adult hair follicles.

Key Words: Animal model, Minipig, Swine, Neural stem cell, Mesenchymal stem cell, Epidermal stem cell, Cell transplantation, Biomedicine, Method(ology).

INTRODUCTION

In basic biomedical research, rodents still remain the first choice as animal models of disease due to their relatively low purchase and maintenance costs, short gestation time, large litter size, and the availability of a large number of genetically well-defined inbred and transgenic strains. However, to apply results obtained from rodent-based research at clinically more relevant levels, larger animals like dogs and nonhuman primates often become necessary. In that respect, pigs offer several advantages in comparison to these more traditional mammalian models:

1. Organ systems (e.g., heart and blood vessels, immune system, liver, kidney, and skin) resemble their human counterparts in size and physiology.
2. Life span is substantially longer than in rodents (10–15 years vs. ~3 years) allowing for long-term safety studies.
3. Easier handling, lower price, and less ethical controversy in comparison to nonhuman primates.

As immunological barriers to xenotransplantation are being overcome[1,2] and despite the recent debate over the risks of porcine endogenous retrovirus (PERV) transmission to humans,[3–5] pigs are considered the most suitable donors of organs and tissues for transplantation-based therapies.[6,7] Being so important in applied biomedical research, it is beyond the scope of this chapter to fully describe all the presently available pig disease models. Therefore, given the recent boom and belief in the promises of new stem cell-based therapeutic approaches, we decided to focus on novel methodologies specifically adapted for the isolation and propagation of several pig-derived stem cell populations and briefly review their possible applications. Although a few reports on the isolation of porcine embryonic stem and germ cells have been published,[8–10] optimal culture conditions for expanding these cells in their undifferentiated state have yet to be determined. Thus, we decided to cover the field of somatic stem cell isolation and culture, for which proven protocols are available.

The Institute of Animal Physiology and Genetics in Libechov imported the first miniature pigs in 1967 from the Hormel Institute, University of Minnesota (two boars and three sows) and from the Institute for Animal Breeding and Genetics, University of Göttingen, Germany (two boars and four sows). During the next 40 years of breeding, animal health and body shape were thoroughly controlled and outbreeding conditions were maintained by imports of several additional boars from Göttingen. The continuous selection made it possible to increase the average litter size (now about 8–10 piglets) and to fix the white color. The animals reach sexual maturity at about 4 months of age, when they weigh about 12–15 kg. While gradually including porcine somatic stem cells in our current research, we have also initiated a minipig inbreeding program. At present, we have two families with an inbreeding index of 0.6–0.7 allowing us, for example, to transplant neural stem cells without immunosuppression. All the experimental procedures described in this chapter were successfully performed on animals from this breed.

ISOLATION AND *IN VITRO* DIFFERENTIATION OF NEURAL STEM CELLS FROM FETAL PORCINE CENTRAL NERVOUS SYSTEM

Neural stem cells (NSCs) capable of self-renewal and differentiation into all major cell types of the neural lineage can be successfully isolated from the fetal and adult central nervous system (CNS) of a variety of mammals. These cells hold an enormous therapeutic potential for the cure of many devastating acquired and hereditary diseases, including spinal cord injuries and neurodegenerative disorders (Parkinson's, Huntington's, and Alzheimer's disease, or multiple sclerosis).[11]

From: *Sourcebook of Models for Biomedical Research* (P. M. Conn, ed.),

Parkinson's disease (PD), in which a specific population of dopaminergic neurons in the brainstem needs to be replenished and sufficient striatal dopamine levels reestablished to achieve a successful therapy, lends itself particularly well for a transplantation approach using fetal neural tissue or expanded neural progenitors. As fetal neural tissue of human origin is not available in sufficient amount and its use is highly controversial ethically, the ontogenesis of fetal porcine brain was studied with the intention to explore its usefulness as a possible source of dopaminergic cells.[12] It was concluded that the 28-day-old porcine fetus is optimal for derivation of brain tissue that could be used for xenotransplantation in PD treatment. The first transfer of mesencephalic tissue isolated from 25- to 28-day-old porcine fetuses occurred unilaterally into the striatum of 12 PD patients, a number of which experienced some improvement of symptoms without major adverse effects of the grafts.[13,14] Nevertheless, despite earlier assumptions that the brain is an absolutely immunoprivileged site, rejection of allogenic and xenogenic grafts can occur without immunosuppression.[15] Consequently, as neural precursors expanded *in vitro* appear to be less immunogenic than primary neural tissue and a recent study demonstrated the potential therapeutic usefulness of NSCs grafted into a rodent model of PD,[16] the survival of xenografted porcine neural precursors in a rat model of PD was investigated[17] and a pig model of MPTP-induced parkinsonism was successfully established for autotransplantation or allotransplantation studies.[18,19]

Huntington's disease (HD) represents another potential target for NSC-based therapy. Unfortunately, a first clinical study using fetal porcine striatal tissue for xenotransplantation into 12 HD patients[13] did not show any significant improvement of the patients' conditions. In 2000, Matsuyama *et al.*[20] identified the porcine huntingtin gene and 1 year later the derivation of the first miniature pig expressing the huntingtin transgene under the control of the neuron-specific enolase promoter was reported by Uchida *et al.*[21] Since no further reports on this topic have been published since, we still lack an effective HD model in the pig. Yet with the recent advances in transgenesis in swine, the establishment of a transgenic HD model in this animal is hopefully not too far away.

The protocol given below was optimized for the isolation of neural precursors from the telencephalon of 27-day-old (E27) porcine fetuses, but can be readily adapted for the isolation and expansion of cells from other parts of the fetal pig CNS and also for E40 neural progenitors. These cells can be significantly expanded in culture and used both for *in vitro* experiments and *in vivo* transplantation studies.

MATERIALS Unless stated otherwise, all media and reagents were purchased from Sigma. Where necessary, catalogue numbers are given in parentheses at the place of the first appearance in the text.

1. Common surgical instruments: forceps, scissors, scalpels, syringes, and needles.
2. Anesthesia: azaperonum (Stresnil, Janssen Animal Health), atropine (Atropin, Biotika), ketamine (Narkamon, Leciva), and halothane (Narcotan, Leciva).
3. Plasticware: 70-µm cell strainers (Falcon), 15-ml and 50-ml centrifuge tubes, plastic pipettes, 10-cm Petri dishes, 24-well culture plates (all TPP), 25-cm² low-adhesion culture flasks (Nunc), and cryogenic vials (Corning).
4. Glassware: sterile beakers, Pasteur pipettes, and 12-mm round coverslips.
5. P1000, P200, and P20 micropipettes.
6. Sterile phosphate-buffered saline (PBS), laminin (23017–015, Gibco), and fibronectin (F1141) for coating coverslips, Accutase (A6964) for cell dissociation, and fetal bovine serum (FBS, F7524).
7. NSC culture medium: 500 ml of DME/F12 with 15 mM HEPES (D6421), supplemented with 6.25 ml of 200 mM L-alanyl-L-glutamine (G8541), 0.5 ml of 50 mg/ml gentamicin (G1397), and 2.5 ml of heparin (H3149, 1 U/µl in DME/F12). Freezing medium: NSC medium with 20% dimethyl sulfoxide (DMSO, D2650).
8. Culture supplements and growth factors: B-27 supplement minus vitamin A (12587–010, Gibco), human recombinant basic fibroblast growth factor (bFGF) (100–18B, PeproTech), epidermal growth factor (EGF) (E9644), and all-*trans*-retinoic acid (R2625).
9. Trypan blue (T8154) and hemocytometer.

METHODS

Cell Isolation

1. Euthanize the timed pregnant sow by overdose anesthesia: after premedication with an im injection of 2.0 mg/kg body weight (bw) azaperonum and 1 mg/animal atropine, deeply anesthetize with 20 mg/ kg bw ketamine.
2. Remove the uterus by cesarean section and place on ice.
3. Rinse the uterus briefly with 70% ethanol, liberate the fetuses from their amniotic sacs, and collect them in a beaker with ice-cold phosphate-buffered saline (PBS).
4. From this time on, work in a laminar flow hood.
5. Place the fetus in a 10-cm dish and with sharp scissors or a surgical blade cut off a wedge-shaped head region containing most of the brain as indicated in Figure 27–1A.

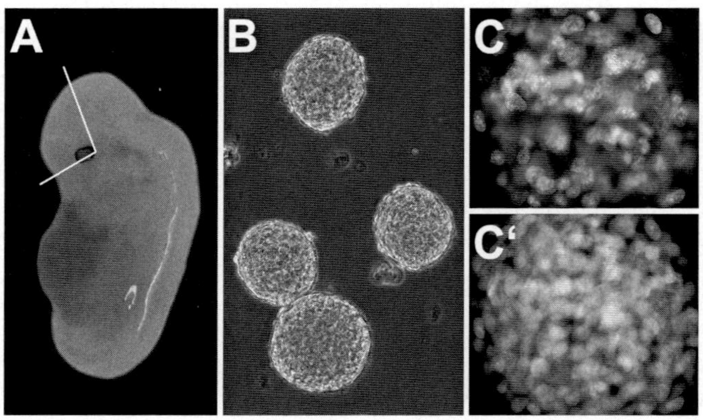

Figure 27–1. Preparation of neurospheres. (A) Example of a 27-day-old porcine fetus. The white line illustrates the cuts necessary to dissect the brain. (B) Phase contrast microphotography of neurospheres after 10 days in culture. (C) A secondary neurosphere 10 days after passage still contains many dividing cells, as indicated by immunolabeling for Ki-67 (C′). The nuclei of cells within the sphere are counterstained with propidium iodide.

6. Remove the brain from the membranous scull, wash with cold PBS, and, with sharp forceps, remove all meningeal membranes and blood vessels (pink color) from the rest of the tissue (white).
7. Collect the dissected brain tissue from all fetuses in ice-cold PBS and cut into small pieces with surgical scissors.
8. Centrifuge at $700 \times g$ for 3 min, resuspend in 5 ml of ice-cold NSC culture medium, and break up the tissue by trituration with a Pasteur pipette (bore size 2–3 mm).
9. Centrifuge at $700 \times g$ for 3 min, resuspend in 2 ml of ice-cold NSC culture medium, and make a final single-cell suspension by trituration using a P1000 micropipette tip.
10. Pass the cell suspension through a 70-µm Falcon cell strainer, wash the cell strainer with 10 ml of NSC medium, and collect a filtered, single-cell suspension in a 50-ml centrifuge tube.
11. Centrifuge at $700 \times g$ for 5 min to collect cells and wash with fresh medium. Repeat three times.
12. Resuspend the cells in NSC medium supplemented with B27 (20 µl/ml), EGF, and bFGF stocks (1 µl/ml, final concentration of 20 ng/ml each).
13. Count the cells using a hemocytometer and either plate them for neurosphere culture or cryopreserve for later use (see below).

Culture and Passaging

1. Plate the cells at a final density of $5–10 \times 10^3$ cells/cm^2 in low-adhesion 25-cm^2 culture flasks with 8 ml of complete NSC medium (with B27, EGF, and bFGF).

2. Culture at 37°C in a humidified 5% CO_2 incubator.

3. Supplement the cultures every third day with fresh growth factors (EGF and bFGF) and once a week replace half of the culture medium with fresh medium and supplement with B27.

4. Neurospheres should become visible under phase-contrast optics after 7–10 days (Figure 27–1B). They should be of a nearly perfect spherical shape with clearly defined edges and tightly packed cells.

5. Passage spheres, when they reach about 100–150 µm in diameter, usually after 12–14 days in culture, as follows.

6. Transfer the culture medium with neurospheres from the 25-cm^2 flask to a 15-ml centrifuge tube and spin at $700 \times g$ for 3 min. Aspirate the culture medium and save as conditioned medium for replating of the cells.

7. Resuspend the pelleted neurospheres in 1 ml of Accutase prewarmed to 37°C and incubate for 5 min at the same temperature.

8. Add 1 ml of fetal bovine serum (FBS) and 10 ml of cold NSC medium to stop the enzyme action. Centrifuge at $120 \times g$ for 3 min. Aspirate as much supernatant as possible to remove most of the FBS (can induce cell differentiation).

9. Add 400 µl of fresh NSC medium and finish the dissociation of the neurospheres to a single cell suspension by trituration through a P1000 micropipette tip (10–15 strokes).

10. Dilute a 10-µl aliquot of the cell suspension 1 : 1 with trypan blue and count the viable cells with a hemocytometer.

11. Replate the cells at a final concentration of $5–10 \times 10^3$ cells/cm^2 in a 1 : 1 mix of fresh and conditioned NSC medium and add fresh B27, EGF, and bFGF.

12. If a clonal origin of secondary neurospheres is desirable, plate the cells at a lower density (about 1000 cells/cm^2) in conditioned medium only, with fresh growth factors and B27 supplement.

Cryopreservation

1. Prepare freezing medium: 20% DMSO in complete NSC medium.

2. Prepare cryovials and label them with cell type, date, and cell passage number.

3. For cryopreservation of cultured cells, freezing of whole neurospheres when they reach a diameter of 50–100 µm usually gives the best results. All spheres from one 25-cm^2 culture flask are frozen in one cryovial.

4. Pellet the neurospheres by centrifugation at $700 \times g$ for 3 min, aspirate all but 0.5 ml of the culture medium, and gently resuspend the spheres.

5. Transfer the suspension to cryovial(s) and while gently swirling, add dropwise 0.5 ml of freezing medium.

6. Let the cell suspension equilibrate in the freezing medium for 10 min at room temperature.

7. Transfer the cryovials to a polystyrene box with ~2-cm-thick walls and leave it at −80°C overnight.

8. For long-term storage, transfer the cryovials to liquid nitrogen or a −180°C cryofreezer.

9. For freezing of cell suspensions from fresh isolations, dilute the cells to a final concentration of 2×10^7 cells/ml in complete NSC medium and aliquot 0.5 ml per cryovial. Continue from point 5 as with neurosphere freezing.

In Vitro Differentiation

1. Working in a laminar flow hood, place 12-mm round coverslips washed with ethanol into a 24-well culture plate and let them air dry.

2. Dilute the fibronectin stock with PBS to a final concentration of 10 µg/ml and pipette 0.4 ml of the solution per well. Incubate it for 1 h at room temperature.

3. Prepare a laminin solution by thawing the laminin stock slowly on ice and diluting it to 10 µg/ml in NSC medium. Aspirate the fibronectin solution from the wells and immediately add 0.4 ml of laminin to each well. Incubate it for 2–3 h at room temperature.

4. Aspirate the laminin solution and plate 6–10 neurospheres in 50 µl of NSC medium on each coverslip. Incubate the culture plate for 45 min at 37°C/5% CO_2 and let the spheres attach to the coverslips.

5. Add 1 ml of NSC medium with B27 supplement and 1 µM all-*trans*-retinoic acid without growth factors to each well.

6. Culture for 5–10 days.

7. Fix the cells on coverslips with 4% paraformaldehyde and process for standard immunolabeling. Several antibodies against NSCs and differentiated cells have been demonstrated to work on porcine cells; these are summarized in Table 27–1. An example of a whole neurosphere stained for the proliferation marker Ki-67 is shown in Figure 27–1C.

Table 27–1
Antibodies suitable for labeling of porcine neural stem cells and their differentiated progenies

Supplier/catalog number	*Antigen*	*Type*	*Recommended dilution*	*Labels*
Sigma, C5922	CNPase (2′,3′-cyclic nucleotide 3′-phosphodiesterase)	Mouse IgG_1	1:100–1:500	Oligodendrocytes, Schwann cells
Sigma, G9269	GFAP (glial fibrillary acidic protein)	Rabbit polyclonal	1:500–1:2000	Astrocytes
Sigma, C9205	GFAP (glial fibrillary acidic protein)	Mouse IgG_1, Cy3 conjugate	1:500–1:2000	Astrocytes
Dako, M7240	Ki–67	Mouse IgG_1	1:200–1:1000	Nuclei of proliferating cells
Endogen, MN–1070	MAP2 (microtubule-associated protein–2)	Mouse IgG_1	1:200–1:500	Neurons
BD, 93398	Nestin	Mouse IgG_1	1:400	Stem/progenitor cells
Chemicon, MAB377	NeuN (neuronal nuclei)	Mouse IgG_1	1:100–1:1000	Nuclei of postmitotic neurons
R&D, AF1638	Nucleostemin	Goat polyclonal, IgG	10 µg/ml	Nucleoli of stem cells
Sigma, T1299	Tyrosine hydroxylase	Mouse IgG_1	1:100	Catecholaminergic cells
Exbio, Tu–20	βIII–tubulin	Mouse IgG_1	1:500	Early neuronal differentiation

ISOLATION AND *IN VITRO* DIFFERENTIATION OF PORCINE MESENCHYMAL STEM CELLS

Mesenchymal stem cells (MSCs) represent a pluripotent cell population located in the stromal compartment of the bone marrow. These cells can differentiate into cells of mesenchymal lineage and form bone, cartilage, and adipose tissue.[22,23] Differentiation of MSCs into cardiomyocytes was also reported.[24] As MSCs can easily be isolated from a patient's own bone marrow and substantially expanded *in vitro*, they are a very attractive cell source for many cell transplantation-based therapies.

Infarcted myocardium is one of the potential targets for the therapeutic use of MSCs.[25] Many myocardium infarction models exist in the pig[26] and transplantation of allogenic porcine MSCs[27,28] into such models resulted in some improvement in heart function. Several methods of MSC delivery (intravenous, intracoronary, and endocardial) were studied by Freyman *et al.*[29]. Labeling of MSCs with superparamagnetic iron oxide (SPIO) allows targeted delivery and *in vivo* observation of such transplanted cells.[30–32]

MSCs readily differentiate into chondrocytes and osteoblasts and are regarded as an excellent source of cells for bone and cartilage reconstruction and tissue engineering. Thus, a reconstruction of bone by osteoinduction was recently demonstrated in a porcine model[33] where endogenous porcine MSCs were induced to differentiate into osteoblasts by acellular implants of recombinant human bone morphogenetic protein-2 (rhBMP-2) bound to an absorbable collagen sponge. Transplantation of chondrocytes was used for articular cartilage repair in a porcine model[34,35] but with limited amount of autologous donor cartilage available for chondrocyte harvest, MSCs represent an interesting alternative source.

In the following protocol, both isolation and *in vitro* expansion of porcine MSCs are described. *In vitro* differentiation into adipocytes, osteoblasts, and chondrocytes can be used to demonstrate the multilineage potential of isolated cells. A short-term chondrogenic induction protocol can be applied if MSCs are to be used for cartilage tissue therapy experiments.

MATERIALS Unless stated otherwise, all media and reagents were purchased from Sigma. Where necessary, catalogue numbers are given in parentheses at the place of the first appearance in the text.

1. Common surgical instruments: forceps, scissors, scalpels, syringes, and needles; Iliac-Crest 15 G/70-mm bioptic needle (Somatex).
2. Anesthesia: azaperonum (Stresnil, Janssen Animal Health), atropine (Atropin, Biotika), ketamine (Narkamon, Leciva), and halothane (Narcotan, Leciva).
3. Plasticware: injection syringes (Braun), 70-µm cell strainers (Falcon), 15-ml and 50-ml centrifuge tubes, plastic pipettes, 10-cm Petri dishes, 75-cm^2 culture flasks, and 24-well plates (all TPP).
4. Glassware: 12-mm round coverslips.
5. P1000, P200, and P20 micropipettes.
6. Ficoll-Paque™ Plus, density 1077 g/ml (Amersham), heparin solution 25,000 IU/ml (Zentiva), PBS, and FBS (F7524).
7. Trypan blue (T8154) and hemocytometer.
8. MSC culture medium: 500 ml of minimal essential medium (MEM)-α (22561–021, Gibco) with 0.55 ml of 50 mg/ml gentamicin (G1397) and 50 ml of FBS. Freezing medium: MSC medium with 20% DMSO (D2650).
9. Trypsin-EDTA solution: 0.25% trypsin (T4799) with 0.02% EDTA in Hanks' balanced salt solution.
10. Chondrogenic medium: MSC medium supplemented with 100 nM dexamethasone (D4902), 50 µg/ml ascorbic acid 2-phosphate, and 1% ITS (Insulin-Transferrin-Selenium, I3146); tumor growth factor (TGF)-β1 (240-B, R&D).
11. Osteogenic medium: MSC medium supplemented with 10 nM dexamethasone, 50 µg/ml ascorbic acid 2-phosphate, and

10 mM β-glycerophosphate; collagen type I 1 mg/ml solution (C8919) and vitronectin (V9881).

12. Adipogenic induction medium: MSC medium supplemented with 1% ITS, 10 mM sodium pyruvate, 0.5 mM IBMX (3-isobutyl-1-methyl-xanthine, I5879), 0.2 mM indomethacin (I7378), and 1 mM dexamethasone.

13. Adipogenic maintenance medium: MSC medium supplemented with 1% ITS and 10 mM sodium pyruvate.

METHODS

Isolation of Bone Marrow Mononuclear Cells

1. Premedicate a pig with an im injection of 1 mg/animal atropine, followed by 2.0 mg/kg bw of azaperonum 15 min later. As soon as sedation is induced, anesthetize the pig with 5 mg/kg bw of ketamine and maintain anesthesia with halothane.

2. Place the pig on its right side. Clean shave, clip, and aseptically prepare an operation field over the left coxal spine (tuber coxae allae ossis illi). Make a narrow scalpel incision through the skin and subcutis up to the coxal spine.

3. With a 15 G/70-mm bioptic needle, penetrate the bone and aspirate 10 ml of bone marrow from the coxal spine pulp chamber into a 20-ml syringe containing 10 ml of PBS with 2% FBS and 25 IU/ml heparin (Figure 27–2A).

4. From this time on, work in a laminar flow hood.

5. Let the Ficoll-Paque equilibrate to room temperature, mix well by inverting the bottle several times, and transfer 15 ml into a 50-ml centrifuge tube.

6. Carefully overlay 20 ml of bone marrow blood mixed 1 : 1 with PBS-FBS-heparin. **It is important not to mix the Ficoll-Paque with the diluted blood sample.**

7. Centrifuge at 400 × *g* for 30 min (20°C).

8. By centrifugation through the Ficoll-Paque density gradient, the cell suspension is separated, with erythrocytes and granulocytes at the bottom of the tube and the mononuclear cells located in a thin opalescent layer on the interface between Ficoll-Paque and blood plasma (Figure 27–2B).

9. Carefully aspirate and discard the blood plasma. Aspirate the opalescent layer of mononuclear cells and transfer into a 15-ml centrifuge tube containing 10 ml of MSC culture medium.

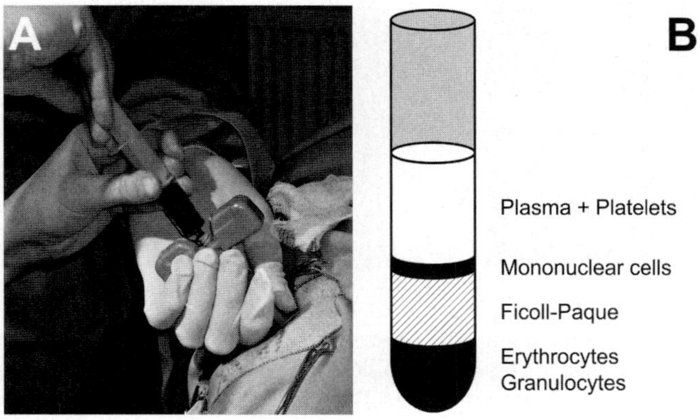

Figure 27–2. Isolation of mesenchymal stem cells. (A) Bone marrow is aspirated from the coxal spine pulp chamber with an iliac crest bioptic needle. (B) Distribution of cells separated by centrifugation through a Ficoll-Paque density gradient. Mesenchymal stem cells are present within the opalescent mononuclear cell layer.

10. Centrifuge at 250 × *g* for 10 min, discard the supernatant, and resuspend in 10 ml of fresh medium. Repeat the wash step once more.

11. Resuspend the cells in 5 ml of MSC medium, mix a 10-μl aliquot with 10 μl of trypan blue, and count the viable cells using a hemocytometer.

Cell Culture and Passaging

1. For initial selection of mesenchymal stem cells, plate isolated bone marrow mononuclear cells at a final density of 5×10^5 cells/cm^2 in MSC medium and culture at 37°C in a humidified 5% CO_2 incubator.
2. After 48 h of culture, aspirate and discard the culture medium with all the nonadherent cells, wash the culture flask with PBS, and feed the cells with fresh MSC medium.
3. Feed the cells with fresh medium every 3 days.
4. The first fibroblast-like colonies should appear in 4–5 days from the start of culture.
5. Passage cells when they reach ~80% confluency (usually 7–10 days from the start of the culture).
6. Aspirate the culture medium and wash the cell monolayer with PBS to remove all traces of FBS.
7. Add 5 ml of a trypsin-EDTA solution per 75-cm^2 culture flask and spread evenly over the cell layer. Leave this for 15–20 sec.
8. Aspirate the trypsin solution from the cells and incubate the flask for 3 min more at 37°C.
9. Add 10 ml of MSC medium and disperse the cells by repeated pipetting of the medium over the monolayer.
10. For routine subculture, split the cell suspension into a new flask 1 : 2 or 1 : 3.

Cryopreservation

1. Prepare freezing medium: 20% DMSO in MSC medium.
2. Prepare cryovials and label them with cell type, date, and cell passage number.
3. Detach the cells as described for passaging, resuspend them in 10 ml of MSC medium, and centrifuge at 700 × *g* for 3 min.
4. Resuspend the cells in 5 ml of MSC medium and count viable cell numbers by the trypan blue exclusion method using a hemocytometer.
5. Adjust the cell density to 1×10^7 cells/ml and aliquot 0.5 ml per cryovial. While gently swirling the cryovial, add dropwise 0.5 ml of freezing medium.
6. Let the aliquots equilibrate for 10 min at room temperature.
7. Transfer the cryovials to a polystyrene box with ~2-cm-thick walls and leave them at −80°C overnight.
8. For long-term storage, transfer the cryovials to liquid nitrogen or a −180°C cryofreezer.

In Vitro Chondrogenic Differentiation of Porcine Mesenchymal Stem Cells

1. For short-term induction of chondrogenic differentiation, resuspend 1×10^7 porcine MSCs in 2.5 ml of chondrogenic medium supplemented with 100 ng/ml of TGF-$ß_1$.
2. Incubate at 37°C for 30 min.
3. Centrifuge the cells at 700 × *g* for 3 min.

4. The cell suspension prepared in this way can be used for cell transplantation or embedding into a scaffold matrix.
5. For long-term *in vitro* chondrogenic differentiation, resuspend 2×10^5 porcine MSCs in 0.5 ml of chondrogenic medium containing 10 ng/ml of TGF-ß$_1$ in a 15-ml centrifuge tube.
6. Centrifuge at $700 \times g$ for 5 min.
7. Culture the pelleted cells in a 15-ml centrifuge tube with loosened cap to allow gas exchange at 37°C in a humidified 5% CO_2 incubator. After 24 h, the cells should form a single spherical cluster.
8. Culture the MSC cell cluster for 14 days and feed every 3 days with fresh chondrogenic MSC medium with 10 ng/ml of TGF-ß$_1$.

In Vitro Osteogenic Differentiation of Porcine Mesenchymal Stem Cells

1. Dissolve vitronectin in sterile ddH_2O to 0.5 mg/ml and mix with collagen in PBS to obtain a final concentration of 12 µg/ml for each ECM molecule. Add 0.5 ml of collagen/vitronectin mixture to each well in a 24-well culture plate and incubate overnight at 37°C in a humidified 5% CO_2 incubator. Just before use, aspirate the collagen/vitronectin solution and wash 1× with PBS.

2. Plate 6×10^4 MSCs in 1 ml of MSC medium per culture well.

3. Culture at 37°C/5% CO_2 until confluency.

4. Replace the MSC medium with osteogenic medium.

5. Continue the culture for 14–17 days. Every 3 days, feed the cells with fresh osteogenic medium.

In Vitro Adipogenic Differentiation of Porcine Mesenchymal Stem Cells

1. Plate MSCs at 1000 cells/cm^2 in adipogenic inducing medium and culture for 3 days at 37°C in a humidified 5% CO_2 incubator.
2. Exchange the medium for the adipogenic maintenance medium. Feed the cells every 3 days with fresh medium, alternating both media types, and continue the incubation for 28 days.

ISOLATION AND CULTURE OF PORCINE EPIDERMAL PROGENITOR CELLS

Pig skin has many similarities to its human counterpart and has been serving for some time as an excellent experimental model for wound healing, skin substitution, and burn treatment. Two extensive reviews addressing the pig as a model for skin wound healing were published recently by Sullivan *et al.*[36] and Middelkoop *et al.*[37]

Acellular porcine dermis can be used to support human epidermal sheet cultures[38] and autologous transplantation of human keratinocytes is currently the main approach for treatment of severe burns and chronic skin wounds. Unfortunately, the amount of autologous tissue that can be harvested for transplantation is limited. Hence, there is great interest in epithelial stem cells, which, after expansion *in vitro*, could produce sufficient numbers of keratinocytes. So far, skin epithelial stem cells and their niche have been studied mainly in the mouse model[39,40] and cells with stem cell characteristics were identified both in interfollicular epidermis and in the bulge region of hair follicles. Because porcine skin is a better model for human wound healing than rodent skin, the isolation of similar stem-like cells of porcine origin is likely to provide new insights and powerful treatment opportunities in the field of burn treatment and skin substitution.

The methodology described below is adapted from the article of Klima *et al.*[41] It allows the isolation of spherical clusters containing high proportions of undifferentiated porcine cells of epidermal origin.

MATERIALS Unless stated otherwise, all media and reagents were purchased from Sigma. Where necessary, catalogue numbers are given in parentheses at the place of the first appearance in the text.

1. Plasticware: 6-cm low adhesive culture dishes (Nunc), 6-cm tissue culture-treated culture dishes, and 15-ml centrifuge tubes (all TPP).

2. Surgical instruments: forceps, scissors, scalpels, and insulin syringe.

3. 70% ethanol, 10% Betadine (Egis Pharmaceuticals), PBS, crude collagenase (Sevapharma) solution: 0.4 mg/ml in Dulbecco's modified Eagle medium (DMEM), trypsin-EDTA solution: 0.25% trypsin (T4799) with 0.02% EDTA in Hanks' balanced salt solution.

4. EPC culture medium: 500 ml of DMEM (D6546) supplemented with 50 ml of FBS (F7524), 20 ml of 200 mM L-alanyl-L-glutamine (G8541), 0.5 ml of 50 mg/ml gentamicin (G1397), and 0.05 mM 2-mercaptoethanol.

5. Mitomycin-inactivated 3T3 fibroblast feeder layer.

METHODS

1. Wash large pieces of full thickness porcine skin briefly in 10% Betadine or 70% EtOH and rinse several times in sterile PBS.

2. From this time point, work in sterile conditions.

3. Using a stereomicroscope, dissect individual hair follicles while keeping the adjacent dermis (Figure 27–3A) and collect in PBS.

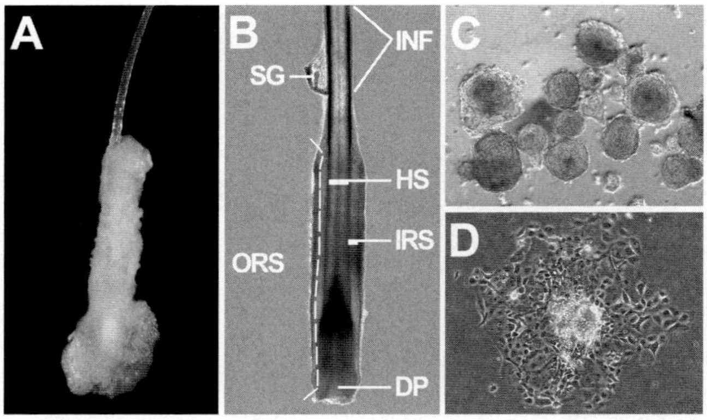

Figure 27–3. Isolation of epidermal progenitor cells from the hair follicle. (A) Photograph of a dissected porcine hair follicle with remnants of dermis. (B) A hair follicle after collagenase treatment and removal of dermal structures. INF, infundibulum; SG, sebaceous gland remnants; HS, hair shaft; ORS, outer root sheath; IRS, inner root sheath; DP, cavity formed by the removal of dermal papilla. The white dotted line demarks ORS, isolated from the hair follicle after trypsin treatment. (C) Phase contrast microphotography of spherical cell clusters formed by ORS cells after overnight culture. (D) Illustration of keratinocytes migrating out of the attached sphere.

4. Transfer approximately 20 dissected hair follicles into 1 ml of collagenase and incubate for 45–60 min at 37°C using a heat chamber.
5. Wash the hair follicles twice with EPC culture medium.
6. Using forceps and an insulin syringe needle, peel away the dermal remnants from the epidermal tissue of the follicle. Collect isolated hair follicles (Figure 27–3B) in PBS.
7. Wash once in PBS.
8. Transfer the follicles into 0.5 ml of trypsin. Incubate for 10–15 min at 37°C using a heat chamber.
9. Aspirate the trypsin solution and cover the hair follicles with EPC culture medium.
10. Using forceps and an insulin syringe needle, scrape the jelly-like outer root sheath (ORS in Figure 27–3B) from the hair follicles.
11. Gently transfer the resulting viscous suspension of the liberated ORS cells into a 15-ml centrifuge tube. Add 10 ml of EPC culture medium and spin down at 500 × *g* for 5 min.
12. Gently resuspend the resulting cell pellet in EPC culture medium and transfer into low-adhesion culture dishes.
13. Culture at 37°C in a humidified 5% CO_2 incubator. Spherical cell clusters, approximately 50–100 µm in diameter, should appear overnight (Figure 27–3C). The spheres of this size can be easily handled by a micropipette while using a stereomicroscope.
14. Seed the spheres onto an adhesive culture surface or a fibroblast feeder layer. They should adhere, and within 2–3 days keratinocytes should start to migrate out of the sphere (Figure 27–3D).

Note: Vigorous resuspending during steps 11 and 12 will result in more but profoundly smaller spherical colonies, which are not so easy to handle.

CELL-TYPE-SPECIFIC LABELING IN TRANSPLANTATION STUDIES

An important problem of cell transplantation-based therapies remains the unequivocal identification of grafted cells within the host tissue. In xenotransplantation experiments, species-specific antibodies can be used for immunohistochemical cell identification. In allotransplantation and autotransplantation experiments, cells should be prelabeled during *in vitro* culture. Several methods are being used at present for labeling of grafted cells, including the incorporation of fluorescent nuclear (Hoechst) or membrane-bound dyes (CM-DiI) or the DNA marker 5-bromo-2-deoxyuridine (BrdU). A recent publication of Coyne *et al.*,[42] however, shows the potential danger of the transfer of such donor labels into host tissue. Cells transgenic for the green fluorescent protein (GFP) represent an alternative and probably more reliable labeling approach. Recent advances in large animal transgenesis[43,44] allow cell isolation from GFP transgenic pigs and new lentiviral vectors coding for the marker can successfully transduce even nondividing cells right after isolation or during culture.[45] Another alternative approach to visualize donor cells uses SPIO particles as contrast agent for magnetic resonance imaging (MRI). Several types of commercially available SPIO varieties (e.g., Resovist, Endorem, Feridex) were shown to incorporate into stem cell endosomes without affecting cell viability and differentiation potential.[46] The advantage of SPIO-labeled cells is their traceability *in vivo* during and after transplantation[30–32] and their simultaneous detectability in histological sections, where these cells can be identified with a Prussian Blue stain.

A short protocol for MSC labeling is given below. The same protocol can be used for labeling of cells before short-time chondrogenic differentiation. In this case, cells are labeled before chondrogenic induction with TGF-ß$_1$.

MATERIALS AND METHODS

1. Add SPIO particles (Resovist, Schering) to cultured MSCs at 1 µl/ml of culture medium (final concentration of 25 µg Fe/ml medium) 48 h before transplantation and culture in a 37°C/5% CO_2 incubator.
2. Before harvesting of cells, wash the monolayer 1× with fresh culture medium and 3× with sterile PBS to remove nonincorporated particles.
3. Harvest the cells by standard trypsinization, spin down at 700 × *g* for 3 min, and resuspend in culture medium without FBS at the concentration required for transplantation.
4. Store on ice. Try to minimize the time between cell harvesting and transplantation, and gently resuspend the cells just before transplantation by trituration with a P1000 micropipette tip, as SPIO-labeled cells are sticky and quickly form cell clumps.

ACKNOWLEDGMENTS

This study was supported by the Centre for Cell Therapy and Tissue Repair (1M0538) and the Institutional Research Concept IAPG No. AV0Z50450515.

REFERENCES

1. Kolber-Simonds D, Lai L, Watt SR, Denaro M, Arn S, Augenstein ML, *et al.* Production of alpha-1,3-galactosyltransferase null pigs by means of nuclear transfer with fibroblasts bearing loss of heterozygosity mutations. *Proc Natl Acad Sci USA* 2004;101:7335–7340.
2. Tseng YL, Kuwaki K, Dor FJ, Shimizu A, Houser S, Hisashi Y, *et al.* alpha1,3-Galactosyltransferase gene-knockout pig heart transplantation in baboons with survival approaching 6 months. *Transplantation* 2005;80:1493–1500.
3. Chapman LE, Wilson CA. Implications of the advent of homozygous alpha 1,3-galactosyltransferase gene-deficient pigs on transmission of infectious agents. *Xenotransplantation* 2003;10:287–288.
4. Wood JC, Quinn G, Suling KM, Oldmixon BA, Van Tine BA, Cina R , *et al.* Identification of exogenous forms of human-tropic porcine endogenous retrovirus in miniature swine. *J Virol* 2004;78:2494–2501.
5. Quinn G, Wood JC, Ryan DJ, Suling KM, Moran KM, Kolber-Simonds DL, *et al.* Porcine endogenous retrovirus transmission characteristics of galactose alpha1–3 galactose-deficient pig cells. *J Virol* 2004;78:5805–5811.
6. Levy GA, Ghanekar A, Mendicino M, Phillips MJ, Grant DR. The present status of xenotransplantation. *Transplant Proc* 2001;33:3050–3052.
7. Kues WA, Niemann H. The contribution of farm animals to human health. *Trends Biotechnol* 2004;22:286–294.
8. Li M, Zhang D, Hou Y, Jiao L, Zheng X, Wang WH. Isolation and culture of embryonic stem cells from porcine blastocysts. *Mol Reprod Dev* 2003;65:429–434.
9. Li M, Ma W, Hou Y , Sun XF, Sun QY, Wang WH. Improved isolation and culture of embryonic stem cells from Chinese miniature pig. *J Reprod Dev* 2004;50:237–244.

10. Tsung HC, Du ZW, Rui R, Li XL, Bao LP, Wu J, *et al.* The culture and establishment of embryonic germ (EG) cell lines from Chinese mini swine. *Cell Res* 2003;13:195–202.
11. Park KI, Ourednik J, Ourednik V, Taylor RM, Aboody KS, Auguste KI, *et al.* Global gene and cell replacement strategies via stem cells. *Gene Ther* 2002;9:613–624.
12. Molenaar GJ, Hogenesch RI, Sprengers ME, Staal MJ. Ontogenesis of embryonic porcine ventral mesencephalon in the perspective of its potential use as a xenograft in Parkinson's disease. *J Comp Neurol* 1997;382:19–28.
13. Fink JS, Schumacher JM, Ellias SL, Palmer EP, Saint-Hilaire M, Shannon K, *et al.* Porcine xenografts in Parkinson's disease and Huntington's disease patients: Preliminary results. *Cell Transplant* 2000;9:273–278.
14. Schumacher JM, Ellias SA, Palmer EP, Kott HS, Dinsmore J, Dempsey PK, *et al.* Transplantation of embryonic porcine mesencephalic tissue in patients with PD. *Neurology* 2000;54:1042–1050.
15. Barker RA, Widner H. Immune problems in central nervous system cell therapy. *NeuroRx* 2004;1:472–481.
16. Ourednik J, Ourednik V, Lynch WP, Schachner M, Snyder EY. Neural stem cells display an inherent mechanism for rescuing dysfunctional neurons. *Nat Biotechnol* 2002;20:1103–1110.
17. Harrower TP, Tyers P, Hooks Y, Barker RA. Long-term survival and integration of porcine expanded neural precursor cell grafts in a rat model of Parkinson's disease. *Exp Neurol* 2006;197:56–69.
18. Mikkelsen M, Moller A, Jensen LH, Pedersen A, Harajehi JB, Pakkenberg H. MPTP-induced Parkinsonism in minipigs: A behavioral, biochemical, and histological study. *Neurotoxicol Teratol* 1999;21:169–175.
19. Cumming P, Gillings NM, Jensen SB, Bjarkam C, Gjedde A. Kinetics of the uptake and distribution of the dopamine D(2,3) agonist (R)-N-[1-(11)C]n-propylnorapomorphine in brain of healthy and MPTP-treated Gottingen miniature pigs. *Nucl Med Biol* 2003;30: 547–553.
20. Matsuyama N, Hadano S, Onoe K, Osuga H, Showguchi-Miyata J, Gondo Y, *et al.* Identification and characterization of the miniature pig Huntington's disease gene homolog: Evidence for conservation and polymorphism in the CAG triplet repeat. *Genomics* 2000;69: 72–85.
21. Uchida M, Shimatsu Y, Onoe K, Matsuyama N, Niki R, Ikeda JE, *et al.* Production of transgenic miniature pigs by pronuclear microinjection. *Transgenic Res* 2001;10:577–582.
22. Ringe J, Kaps C, Schmitt B, Buscher K, Bartel J, Smolian H, *et al.* Porcine mesenchymal stem cells. Induction of distinct mesenchymal cell lineages. *Cell Tissue Res* 2002;307:321–327.
23. Bosch P, Pratt SL, Stice SL. Isolation, characterization, gene modification, and nuclear reprogramming of porcine mesenchymal stem cells. *Biol Reprod* 2006;74:46–57.
24. Moscoso I, Centeno A, Lopez E, Rodriguez-Barbosa JI, Santamarina I, Filgueira P, *et al.* Differentiation "in vitro" of primary and immortalized porcine mesenchymal stem cells into cardiomyocytes for cell transplantation. *Transplant Proc* 2005;37:481–482.
25. Orlic D. BM stem cells and cardiac repair: Where do we stand in 2004? *Cytotherapy* 2005;7:3–15.
26. Dib N, Diethrich EB, Campbell A, Gahremanpour A, McGarry M, Opie SR. A percutaneous swine model of myocardial infarction. *J Pharmacol Toxicol Methods* 2006;53:256–263.
27. Makkar RR, Price MJ, Lill M, Frantzen M, Takizawa K, Kleisli T, *et al.* Intramyocardial injection of allogenic bone marrow-derived mesenchymal stem cells without immunosuppression preserves cardiac function in a porcine model of myocardial infarction. *J Cardiovasc Pharmacol Ther* 2005;10:225–233.
28. Kamihata H, Matsubara H, Nishiue T, Fujiyama S, Tsutsumi Y, Ozono R, *et al.* Implantation of bone marrow mononuclear cells into ischemic myocardium enhances collateral perfusion and regional function via side supply of angioblasts, angiogenic ligands, and cytokines. *Circulation* 2001;104:1046–1052.
29. Freyman T, Polin G, Osman H, Crary J, Lu M, Cheng L, *et al.* A quantitative, randomized study evaluating three methods of mesenchymal stem cell delivery following myocardial infarction. *Eur Heart J* 2006;27:1114–1122.
30. Dick AJ, Guttman MA, Raman VK, Peters DC, Pessanha BS, Hill JM, *et al.* Magnetic resonance fluoroscopy allows targeted delivery of mesenchymal stem cells to infarct borders in swine. *Circulation* 2003;108:2899–2904.
31. Hill JM, Dick AJ, Raman VK, Thompson RB, Yu ZX, Hinds KA, *et al.* Serial cardiac magnetic resonance imaging of injected mesenchymal stem cells. *Circulation* 2003;108:1009–1014.
32. Kraitchman DL, Heldman AW, Atalar E, Amado LC, Martin BJ, Pittenger MF, *et al.* In vivo magnetic resonance imaging of mesenchymal stem cells in myocardial infarction. *Circulation* 2003;107: 2290–2293.
33. Carstens MH, Chin M, Li XJ. In situ osteogenesis: Regeneration of 10-cm mandibular defect in porcine model using recombinant human bone morphogenetic protein-2 (rhBMP-2) and Helistat absorbable collagen sponge. *J Craniofac Surg* 2005;16:1033–1042.
34. Mainil-Varlet P, Rieser F, Grogan S, Mueller W, Saager C, Jakob RP. Articular cartilage repair using a tissue-engineered cartilage-like implant: An animal study. *Osteoarthritis Cartilage* 2001;9(Suppl A): S6–15.
35. Liu Y, Chen F, Liu W, Cui L, Shang Q, Xia W, *et al.* Repairing large porcine full-thickness defects of articular cartilage using autologous chondrocyte-engineered cartilage. *Tissue Eng* 2002;8:709–721.
36. Sullivan TP, Eaglstein WH, Davis SC, Mertz P. The pig as a model for human wound healing. *Wound Repair Regen* 2001;9:66–76.
37. Middelkoop E, van den Bogaerdt AJ, Lamme EN, Hoekstra MJ, Brandsma K, Ulrich MM. Porcine wound models for skin substitution and burn treatment. *Biomaterials* 2004;25:1559–1567.
38. Matouskova E, Vogtova D, Konigova R. A recombined skin composed of human keratinocytes cultured on cell-free pig dermis. *Burns* 1993;19:118–123.
39. Ito M, Liu Y, Yang Z, Nguyen J, Liang F, Morris RJ, *et al.* Stem cells in the hair follicle bulge contribute to wound repair but not to homeostasis of the epidermis. *Nat Med* 2005;11:1351–1354.
40. Tumbar T, Guasch G, Greco V, Blanpain C, Lowry WE, Rendl M, *et al.* Defining the epithelial stem cell niche in skin. *Science* 2004;303:359–363.
41. Klima J, Smetana K Jr, Motlik J, Plzakova Z, Liu FT, Stork J, *et al.* Comparative phenotypic characterization of keratinocytes originating from hair follicles. *J Mol Histol* 2005;36:89–96.
42. Coyne TM, Akiva JM, Woodbury D, Black IB. Marrow stromal cells transplanted to the adult brain are rejected by an inflammatory response and transfer donor labels to host neurons and glia. *Stem Cells* 2006;24(11):2483–2492.
43. Vodicka P, Smetana K Jr, Dvorankova B, Emerick T, Xu YZ, Ourednik J, *et al.* The miniature pig as an animal model in biomedical research. *Ann NY Acad Sci* 2005;1049:161–171.
44. Hofmann A, Kessler B, Ewerling S, Weppert M, Vogg B, Ludwig H, *et al.* Efficient transgenesis in farm animals by lentiviral vectors. *EMBO Rep* 2003;4:1054–1060.
45. Nguyen TH, Khakhoulina T, Simmons A, Morel P, Trono D. A simple and highly effective method for the stable transduction of uncultured porcine hepatocytes using lentiviral vector. *Cell Transplant* 2005;14:489–496.
46. Arbab AS, Yocum GT, Rad AM, Khakoo AY, Fellowes V, Read EJ, *et al.* Labeling of cells with ferumoxides-protamine sulfate complexes does not inhibit function or differentiation capacity of hematopoietic or mesenchymal stem cells. *NMR Biomed* 2005;18: 553–559.

Well-Established Models III

Nonhuman Primates D

28 The Nonhuman Primate as a Model for Biomedical Research

DAVID M. ANDERSON

ABSTRACT

This chapter provides a brief summary of issues surrounding the utilization of nonhuman primates in biomedical research. Although a relatively small proportion of the total number of animals utilized in biomedical research, nonhuman primates occupy a unique position as the species most closely related to humans, and thus have the potential to provide highly relevant information regarding human health issues. Nonhuman primates are utilized across a wide diversity of research topics and examples are provided including infectious disease, neuroscience, and genomics. Pertinent information relating to ethical issues, species selection, housing, and specific pathogen-free status are provided as an overview of relevant issues associated with selection of nonhuman primate models. Selected references are provided as a reference for more comprehensive information relating to these topics.

Key Words: Nonhuman primate, Animal model, Macaque, AIDS, Neuroscience, Genomics, Specific pathogen free.

INTRODUCTION

Animal models are an essential component of biomedical research. Over the past century, the overwhelming majority of Nobel prizes awarded in medicine and related sciences have involved significant work utilizing animal models. It is clear that information provided through the use of animal models, and in particular nonhuman primate models, will be essential for continued development of effective solutions to issues affecting the health and well-being of both human and animal populations.

Although nonhuman primates are a relatively small component of the total animal use in biomedical research, they are an extremely important resource for a wide range of investigations. Of over 1.1 million animals[1] (not including rodents, fish, or birds) used in biomedical research annually, less than 0.3% are nonhuman primates.[2] The foundation for this disproportionate significance lies in the quality of information available through nonhuman primate studies. Nonhuman primates are arguably the most applicable animal model to the human condition based on their anatomic, physiological, and genetic similarity to humans. Genetic relatedness with the chimpanzee [*Pan (Homo) troglodytes*], the most closely related nonhuman primate species, approaches 98%, suggesting a high likelihood of correlation for nonhuman primate studies[3,4] Other nonhuman primate species, while less related to humans, still demonstrate important similarities and differences, offering an incredible resource for studies spanning a broad spectrum ranging from infectious disease to evolution.

A wide variety of nonhuman primate species have been developed for biomedical studies.[2,5,6] These include several major families of primates spanning the spectrum from new to old world, apes to prosimians. Examples of commonly used species include the genera *Macaca, Pan (Homo), Papio, Aotus, Callithrix, Saimiri, Chlorocebus, Saquinus,* and *Cercocebus*. Again, species variations provide the basis for selective advantages between specific nonhuman primate species in certain research applications (Table 28–1).

CONSIDERATIONS FOR USE OF NONHUMAN PRIMATES

Ethical considerations associated with nonhuman primate use in biomedical research are a significant factor in model selection. Nonhuman primates occupy an exceptional position in the hierarchy of animal models due to their close relatedness to humans. As described in earlier chapters, both physical and cognitive needs must be met in order to guarantee appropriate use. Nonhuman primates are recognized as having elevated cognitive requirements in addition to standard physiological needs. Strict minimum housing space requirements as well as dietary and environmental regulations are in place through the Animal Welfare Act and subsequent amendments as well as the Guide for the Care and Use of Laboratory Animals.[7,8] These requirements are enforced through an extensive network of federal, state, and local regulations, scheduled and unscheduled site visits, and oversight by local Institutional Animal Care and Use Committees (IACUC). Use of nonhuman primates, as well as other commonly used laboratory animal species, requires prior justification and approval by the appropriate IACUC to safeguard the appropriate care and use of every animal.

As outlined in other chapters, the selection process utilized to evaluate the appropriateness of nonhuman primates as a model system involves a variety of factors. Of primary importance is the ability of the specific model system to address the biological question posed for each specific investigation. Nonhuman primates typically provide information very relevant to human application and therefore are highly recommended. Other factors, however, must also be considered in the process. If the biological question

From: *Sourcebook of Models for Biomedical Research*
(P. M. Conn, ed.), © 2008 Humana Press Inc., Totowa, NJ.

Table 28–1
Nonhuman primate species used in biomedical studies

Species name	*Common name*	*Applications*
Macaca nemestrina	Pigtail macaque	AIDS, reproduction, growth and development, behavior, neuroscience
Macaca mulatta	Rhesus macaque	AIDS, reproduction, growth and development, behavior, neuroscience
Macaca fascicularis	Long-tailed crab-eating macaque, cynomolgus	Infectious disease (AIDS, SARS), neuroscience
Papio anubis	Olive baboon	Reproduction, neuroscience, hematology, transplantation
Papio cynocephalus	Yellow baboon	Reproduction, neuroscience, hematology, transplantation
Cercocebus atys	Sooty mangabey	AIDS, leprosy, reproduction
Chlorocebus sabaeus	African green monkey	AIDS, neuroscience
Chlorocebus aethiops	Vervet	AIDS, neuroscience
Sanquinus oedipus	Cotton-top tamarin	Enteric disease, colitis, colorectal cancer
Callithirx jacchus	Common marmoset	Reproduction, endocrinology, vision, behavior
Aotus vociferans	Owl monkey	Vaccines (dengue, malaria), behavior, endocrinology, vision
Aotus nancymaae	Night monkey	Vaccines (dengue, malaria), behavior, endocrinology, vision
Saimiri sciureus	Squirrel monkey	Malaria, neuroscience
Pan (Homo) troglodytes	Common chimpanzee	Infectious disease (AIDS, hepatitis C), behavior

can be addressed through the use of lower species or nonanimal resources, current guidelines require use of these alternative means. In addition, issues such as housing requirements, species availability, and price can be important factors for consideration in the animal model selection process.

The relevancy of information developed in nonhuman primate models is well documented (see Table 28–1). Information developed in nonhuman primate models is critical to the design of subsequent human studies and the evaluation of new medical interventions into human health care. This correlation between human and nonhuman primate studies has application for both positive and negative results. Preliminary studies in nonhuman primate models of new lentiviral vaccines have closely mirrored subsequent results developed in human clinical studies.[9–16] Similarly, results in gene transfer studies have been generally disappointing in both human and nonhuman primates, reinforcing the similarity in response.

INFECTIOUS DISEASE

Nonhuman primates have been a critical resource in biomedical investigations related to infectious disease.[2,5,6] Based on the physiological similarities between species, nonhuman primates and humans share susceptibility to a wide variety of pathogens.[2,5,6] In addition, the balance of the host–pathogen relationship is largely dependent on both innate and adaptive immune responses, both of which are founded in the expression of the genetic code. Nonhuman primates possess a striking similarity to humans, a fact that has formed the basis for development of an extensive range of nonhuman primate models focused on transmission, pathogenesis, therapeutic intervention, and vaccine development for a variety of infectious pathogens.[2,5,6,9–24]

Macaque species are the most commonly utilized nonhuman primate species with the majority of public-funded studies directed toward infectious diseases. Macaques have a history of critical contributions to health concerns including the extensive utilization of rhesus monkeys (*Macaca mulatta*) for the development of the polio vaccine.[24] More recently, extensive use of a variety of species including, but not limited to, *M. mulatta*, *Macaca fascicularis*, and *Macaca nemestrina* has been closely integrated into the research on AIDS on a national and international basis.[10–16] Haigwood *et al.* have provided an excellent overview of this effort in Chapter 58 (this volume).

Of particular note is the expected utilization of nonhuman primate species for research related to "biodefense agents."[25] These agents have been identified by the Centers for Disease Control (CDC) and National Institutes of Health (NIH) as having potential use against the public health. To mitigate this potential, significant resources have been directed toward development of effective means for diagnosis, vaccination, and therapeutic intervention. It is highly likely that macaques will play a pivotal role in the development of these capabilities. This effort will require development and refinement of new nonhuman primate models to provide the necessary information required prior to implementation of human studies. This area of research emphasizes the highly relevant application of nonhuman primates as a precursor and substitute for human clinical studies. Serious ethical considerations accompany the concept of challenging humans with potentially deadly pathogens as part of the development process. Nonhuman primates offer the opportunity to evaluate the safety and efficacy of new vaccines and therapeutic agents with a high degree of confidence, yet without endangering human subjects.

New world nonhuman primate models, although less heavily utilized on a national basis, still provide critical contributions for specific diseases. From an international perspective, malaria is one of the most prevalent and devastating diseases, with a disproportionate impact in developing areas.[26–28] Development of effective malaria vaccines and therapeutics is a high priority for the national and international health care community.[26–30] *Saimiri* and *Aotus* species have proven to be excellent models for these investigations, contributing significant information on transmission, pathogenesis, and therapeutic and vaccine interventions.[22–24] Although availability and maintenance of these species are limiting factors, they will continue to be the most effective models for future translational research.

NEUROSCIENCE

Nonhuman primates have made significant contributions to neuroscience research through utilization of a variety of models over several decades. Nonhuman primates share more anatomic and physiological similarities to humans than any other species, and these factors have been used to provide important advances into the structure and function of the human brain.[31–35]

Nonhuman primates offer several advantages over other animals models, making them a unique resource for neurological investigation. The relatively close behavioral and cognitive abilities of nonhuman primates provide the opportunity to study the anatomic and physiological basis for these characteristics, a capability absent in lower species. In addition, nonhuman primates can be trained to perform a broad range of tasks, allowing development of models designed to investigate very specific neurological activities.[36–38]

As a correlate, many human neurological and cognitive tests have been adapted to nonhuman primates, allowing investigation of developmental aspects of cognition.[39–47] Developmental testing of reflex development, visual acuity, recognition memory, motor development, object concept development, and social interactions has been adapted to evaluate macaque species. These provide exceptional investigative tools and have been effectively utilized to identify important factors affecting cognitive and motor development.

A variety of nonhuman primate models have been developed to explore a broad spectrum of neurological structure and function. Extensive developmental work over several decades has resulted in well-characterized neurophysiological recording models allowing ongoing, active electrical recording from individual neurons within specific neurological networks. Characterization of interrelated neurophysiological function between specific neuronal groups has led to greater understanding of function within both motor and visual systems.[48–51]

Advances in imaging technology have provided exceptional opportunities for investigation of neurological function in nonhuman primate models. A wide variety of imaging modalities have been applied to nonhuman primate models including computed tomography (CT), magnetic resonance imaging (MRI), and positron emission tomography (PET). These noninvasive techniques offer the opportunity to perform multiple, detailed interrogations of specific neurological, anatomic, and physiological functions without physical disruption of neurological tissues. Additionally, the tractability of nonhuman primates has allowed awake models of neurological function using advanced imaging techniques including fMRI and fPET. These models provide a unique opportunity to investigate higher cognitive functions in species with a high degree of correlation to humans.

FUNCTIONAL GENOMICS

Perhaps the most important similarity between humans and nonhuman primates lies in the relative similarity of the genetic code.[52] In fact, the genomic similarities form the foundation for the physiological, morphological, and phenotypic likeness between species. Similarly, it appears that relatively small differences in either code or expression provide the basis for differences between species.[53–55] Investigation of the genetic code potentially provides an incredible resource for information across a wide spectrum of issues including evolution, cancer, basic physiology, immunology, cognition, and development.

Development of essential investigative resources for genomic studies has proceeded relatively quickly. Taking advantage of the research infrastructure established for sequencing the human genome, subsequent sequencing projects have provided initial full-length sequences for a number of nonhuman primate species including the chimpanzee, bonobo, and rhesus macaque, with sequencing projects underway for additional species.[54,55] Additionally, more limited information is available from a variety of other nonhuman primate species. Consistent with founding principles of the Genome Project, these data and others are available in public reference sites such as GenBank.[56] Microarray chips developed by both Agilent Technologies and Affymetrix have recently become available allowing directed investigation of specific genomic responses across a variety of biological studies. Advances in bioinformatics software capable of distilling complex microarray information into biologically relevant mechanistic pathways have provided a critical link between genome and phenotypic response. Taken together, these resources provide an exceptional opportunity to apply novel, highly sophisticated research methods to important biological issues.

Application of these research resources is rapidly increasing across a variety of biological fields. Evolutionary studies have derived significant benefit through comparative genomic studies of humans and nonhuman primates. Current genomic capabilities allow comparative studies of specific genetic composition and expression, providing important insights on evolutionary pressures underlying the relative divergence or convergence of primate species. In addition, genomic studies can provide significant comparative data for other investigations based on mitochondrial DNA and Y-chromosomal sequencing. Comparative genomic analyses provide the potential for extremely powerful genetic investigations defining the genetic basis for the morphological and phenotypic differences between species.

Infectious disease investigation is likely to be a primary beneficiary of genomic research capabilities. Both innate and adaptive immunity are critical components of the host–pathogen interaction. Recent microarray studies have clearly highlighted the vigorous host inflammatory response in animal models of pathogenic influenza infection.[57–59] It is clear that a more complete understanding of the host response to infection will provide critical information for the development of new generations of therapies and vaccines. As an example, the development of an effective AIDS vaccine has proven to be an exceptional challenge despite more than two decades of intense research. As yet, significant questions remain concerning the transmission events, viral pathogenesis, means of obstructing viral replication, and correlates of protection necessary for effective vaccination. Functional genomic analysis in appropriate nonhuman primate models has the potential to provide important evidence for the resolution of these issues.

In addition to functional genomics, proteomic studies in nonhuman primates have the potential to provide a key link between genome and phenotype. In that the proteome is the sum of expressed proteins based on the genetic composition of the organism and the current environment, the proteome is a critical link in achieving a more complete understanding of the ongoing relationship between genome, proteins, and function. Investigative tools are currently under development and promise to provide information necessary to fully characterize a wide variety of physiological responses. Efforts are currently underway to provide complementary experimental designs allowing for simultaneous investigation of genomic response, protein expression, and physiological phenotype in nonhuman primate model of infectious disease.[60,61]

HOUSING CONSIDERATIONS

Maintenance of nonhuman primates in laboratory environments presents significant challenges. A considerable commitment in expertise, resources, and effort is required to ensure appropriate care and support for nonhuman primates in research facilities. While outdoors, free-ranging native housing environments might be an ideal for nonhuman primate housing, it is likely that research protocol constraints, existing facilities, and geographic location would render this a nonviable option for most research institutions. Instead, current housing environments are designed to meet the environmental, nutritional, and psychological needs of each animal, allowing individual animals to express species-specific normative behaviors, albeit in a different setting.

While many components of an effective animal care program are similar among species used in biomedical research, nonhuman primate care programs typically must provide additional emphasis on provision for the psychological health of housed animals.[62] Specific elements of a comprehensive animal care program for nonhuman primates include effective health surveillance and veterinary care, husbandry, psychological enrichment/well-being, management of breeding activities, and oversight for research activities. Appropriate documentation of animal-related activities is a critical component of effective management as well as being necessary for regulatory compliance.

In particular, nonhuman primates have special requirements for environmental enrichment, social housing, and mitigation of experimental impacts, both physiological and cognitive. The perceptive and cognitive characteristics of nonhuman primates impose additional responsibility concerning their care and use in biomedical research. Behavioral assessment of nonhuman primates is considered an essential component of health and well-being monitoring programs. Assessment of behavior is considered an indicator of the psychological state of nonhuman primates, with the implication that abnormal behavior may be an outward manifestation of physiological or psychological stress.[63] Novak and Suomi[64] have provided recommendations for the assessment of nonhuman primate well-being, although additional measures may be considered.

The need for positive social interaction between conspecifics is recognized as a critical requirement for the psychological well-being of nonhuman primates. In natural settings, nonhuman primates are generally highly social animals maintaining a range of social relationships. These social interactions are influenced by age, sex, parentage, social standing, as well as a variety of other factors. Social housing of conspecific nonhuman primates is the accepted standard housing configuration unless precluded by specific research or medical or behavioral constraints.[62]

Social housing can take many forms and spans a spectrum of housing configurations. Basic elements of social housing include access to sensory cues from other nonhuman primates including sight, smell, hearing, and touch. The standard cage design for nonhuman primates typically employs spaced bars that restrict the animals to a desired space but also allow extensive interaction with environmental visual, olfactory, and auditory cues.

The highest level of social interaction is achieved through direct physical contact between conspecifics. Group housing configurations such as harem groups, family units, single-sex cohorts, age-matched cohorts, pair housing, or limited contact housing (grooming bars)[65] have all been utilized to provide effective socialization of nonhuman primates. Although a detailed examination of each housing configuration is beyond the scope of this chapter, each has the potential to provide the opportunity for species-specific behavioral interactions, thereby increasing environmental enrichment. Care must be taken, however, in the selection process for social housing partners. Significant differences in the success of social housing configurations can occur based on age, sex, and relatedness of individual animals. While direct social housing has the potential to provide significant environmental enrichment, unsuccessful social interactions can result in significant psychological and physical trauma, including death. Additionally, social interactions can precipitously change from stable, positive physical interactions to aggressive, potentially traumatic interactions without identification of a precipitating cause. Experience and ongoing close observation of socially housed nonhuman primates is necessary to establish and maintain positive social interactions.

Maintenance of health care for nonhuman primates is central to an effective animal care program. This point is substantiated in the Animal Welfare Act through placement of independent judgment and final decision authority in the Institutional Veterinarian. An effective health care program has several components including well-defined programs for preventive health care, health surveillance, and treatment of health issues. These components must be effectively combined into a seamless program that not only meets the spontaneous health needs of each animal but actively intercedes to avoid potential health care issues. Specific components are (1) preventive health care, including quarantine, isolation, annual physical examination, dentistry, and vaccination when appropriate; (2) health surveillance, including viral testing, microbiological testing, daily observations, behavioral assessment, and weight monitoring; and (3) treatment/veterinary care, including veterinary care and surgery support, treatment of spontaneous disease, and treatment of health issues associated with research activities.

CHARACTERIZATION

As the complexity of biomedical research increases, so too does the demand for better characterization of the nonhuman primate resource. Increasing sophistication of scientific investigation requires a more thorough knowledge of potentially confounding factors, including the health status and genetic profile of the experimental animals. Most recently, AIDS-related research in nonhuman primate models has provided the overriding impetus to characterize and segregate animals based on health status and genotype.[66–69] The heavy reliance on normal immune function in AIDS-related research studies requires well-characterized research animals capable of exhibiting a normal immune response to viral infection or vaccination.

Both ethical and financial considerations provide additional motivation for characterization of individual animals prior to recruitment into research projects. Ethical guidelines require the use of minimum numbers of animals to meet experimental objectives, typically the demonstration of the experimental hypothesis with statistically significant results.[70] Biostatistical consultation during the study design is a common part of the protocol development process. Additionally, nonhuman primate studies require a significant financial commitment for purchase and appropriate maintenance of experimental animals. These factors combine to produce an experimental design that includes the minimum number of animals necessary to produce statistically significant results. The dropout of individual animals due to unknown pathogen status can be the difference between experimental results that are statistically significant and results that are merely suggestive. Therefore, significant effort is directed toward the designation of specific health status prior to inclusion in research projects.

In addition to research requirements, significant motivation exists to eliminate agents with the potential for zoonotic transmission. Investigative and animal care staff typically work in close proximity to housed animals, providing the potential for transmission events between species. Several nonhuman primate pathogens have the potential to induce significant morbidity or mortality when transmitted into human hosts.[71–74] Although rare, a number of documented incidents of mortality associated with zoonotic transmission of herpes B virus from nonhuman primates to human hosts are available in the literature. Based on this potential, many facilities now require documented negative status of nonhuman primates for several infectious agents.

Genetic characterization of nonhuman primates has proven to be an important factor in studies involving immune response to infectious agents. Characterization of the MHC Class I alleles in the rhesus monkey has revealed a pattern of differential response to lentiviral infection based on the presence or absence of specific alleles.[66,67] It is therefore important to characterize the genetic composition of study animals prior to study onset either to avoid undesirable and unintended impacts to the study results or to provide important information for the interpretation of the results. As an important correlate, identification of these alleles may provide valuable information regarding relative susceptibility or resistance to disease progression following lentivirus infection.

Genetic characterization can extend beyond individual differences to important differences between subspecies. At one point in time, rhesus monkeys from either Indian or Chinese origin were largely considered interchangeable for infectious disease studies. Detailed comparative studies have revealed, however, that although both species appear to be equally susceptible to lentiviral infection, a significant difference exists in their ability to support ongoing, long-term viral replication.[75] Current recommendations therefore include segregation of subspecies for these studies, or equal distribution in experimental groups to avoid spurious distortion of experimental results.

Of note is the historically imprecise confirmation of the genetic origin of many domestic breeding colonies. Breeding records for many domestic nonhuman primate breeding colonies have depended on imprecise observational methods for documentation of parentage. Long-term productive management of breeding colonies requires accurate, ongoing genetic characterization to avoid undesirable inbreeding and to confirm the genotype for individual animals. Recent advances in genetic sequencing have provided more precise methods for confirmation of parentage and genetic origin. Techniques such as single nucleotide polymorphisms (SNPs) and microsatellite assays have made it possible to determine with relative certainty the parentage and general genetic stock and subspecies of individual animals.[76–79]

SPECIFIC PATHOGEN FREE

The characterization of the health status of nonhuman primates is critically important for a wide variety of experimental applications. In particular, studies involving exposure to infectious diseases or characterization of immunological response have the potential for unintended distortion due to coinfection with spurious pathogens. To avoid this potential confounding effect, significant efforts have been directed toward the development and characterization of nonhuman primates guaranteed negative for specific pathogens, designated specific pathogen free (SPF).

The SPF designation indicates negative status for one or more infectious agents. Although the SPF designation can technically be used for any pathogen, standard use refers to viral pathogens with a negative status for *Mycobacterium tuberculosis* assumed. Several different levels of SPF status exist; however, the most common designation refers to animals negative for simian retrovirus (SRV), simian T-lymphotrophic virus (STLV), simian immunodeficiency virus (SIV), and herpes B virus (B virus, CHV-1). This SPF designation has been established by the National Center for Research Resources (NCRR) and the Office of AIDS Research (OAR) through the development and support of SPF macaque colonies for AIDS-related research. This designation reflects the need to have animals free of indigenous retroviruses potentially capable of skewing immune function while at the same time providing protection for investigative and husbandry staffs against potential exposure to a significant zoonotic agent, herpes B virus.

Effective maintenance of SPF status requires ongoing surveillance, typically provided by strict viral testing programs. SPF status can be documented through a variety of techniques; however, several constraints exist for use of laboratory assays for colony surveillance screening. Sensitivity and specificity are key parameters for any laboratory assay and each should be quantitated and confirmed through a rigorous, ongoing quality control program. Additionally, colony surveillance assays must be relatively easy to perform and must be cost effective for efficient inclusion in a colony management program.

Under these considerations, a spectrum of screening tests has been developed to survey the viral status of macaques in breeding and research colonies. Serology is an industry standard with enzyme-linked immunosorbent assay (ELISA) and, more recently, multiplexed bead assays[80–86] utilized to identify antibody response to previous viral exposure. In addition, standard confirmatory tests such as Western blot, polymerase chain reaction (PCR), and virus isolation are utilized to confirm positive tests and resolve indeterminate results. While some institutions maintain proprietary diagnostic laboratories, commercial laboratories are available for support of viral surveillance programs.

Management of SPF colonies depends heavily on the surveillance of colony viral status. Typical management strategies include the means to test all animals with removal of individual animals testing positive or indeterminate for specified pathogens. Management protocols are available in the literature for the derivation of SPF nonhuman primate colonies from non-SPF founder stock as well as expansion of current SPF colonies.[87] These protocols rely heavily on rigorous adherence to surveillance testing schedules with exclusion of animals from the SPF colony until negative pathogen status is confirmed. Subsequent monitoring of the colony is maintained on a continuing basis to ensure identification and isolation of individual animals demonstrating conversion to positive status.

CONCLUSIONS

Nonhuman primates are a critical resource for biomedical research. The genetic, morphological, and physiological similarity to humans provides an invaluable resource for information directly applicable to both human and nonhuman primates. Although housing and maintenance of these species are challenging, well-defined programs are in place to meet the physiological and psychological needs of animals in research settings. Ongoing efforts continue to better characterize nonhuman primate genetics and pathogen status, thereby improving the application of these models to biomedical research questions. Given the current emphasis on translational research, it is highly likely that demand for these extremely relevant animal models will continue to increase in the foreseeable future.

REFERENCES

1. Animal Welfare Report; USDA/APHIS: http://www.aphis.usda.gov/ac/awreports/awreport2004.pdf.
2. Sibal LR, Samson KJ. Nonhuman primates: A critical role in current disease research. *ILAR J* 2001;42:74–84.
3. Wildman DE, Uddin M, Liu G, Grossman LI, Goodman M. Implications of natural selection in shaping 99.4% nonsynonymous DNA identity between humans and chimpanzees: Enlarging genus Homo. *Proc Natl Acad Sci USA* 2003;100:7181–7188.
4. Gagneuz P, Varki A. Genetic differences between humans and great apes. *Mol Phylogenet Evol* 2001;18:2–13.
5. Carlsson H-E, Schapiro SJ, Farah I, Hau J. Use of primates in research: A global overview. *Am J Primatol* 2004;63:223–237.
6. Bontrop RE. Nonhuman primates: Essential partners in biomedical research. *Immunol Rev* 2001;183:5–9.
7. Animal Welfare Act: http://www.aphis.usda.gov/ac/publications/AWA/AWAINDEX.html.
8. National Research Council. *Guide to the Care and Use of Laboratory Animals*, 1996.
9. Girard MP, Osmanov SA, Kieny MP. A review vaccine research and development: The human immunodeficiency virus (HIV). *Vaccine* 2006;24:4062–4081.
10. Letvin NL. Progress toward an HIV vaccine. *Annu Rev Med* 2005;56:213–223.
11. Letvin NL, Barouch DH, Montefiori DC. Prospects for vaccine protection against HIV-1 infection and AIDS. *Annu Rev Immunol* 2002;20:73–99.
12. Warren J. Preclinical AIDS vaccine research: Survey of SIV, SHIV and HIV challenge studies in vaccinated nonhuman primates. *J Med Primatol* 2002;31:237–256.
13. Smith SM. HIV vaccine development in nonhuman primate model of AIDS. *J Biomed Sci* 2002;9:100–111.
14. Duerr H, Wasserheit JN, Corey L. HIV vaccines: New frontiers in vaccine development. *Clin Infect Dis* 2006;43:500–511.
15. Kent SJ, Ada GL, Hayes E, Lewis IM. Determining the immune mechanisms of protection from AIDS: Correlates of immunity and the development of syngeneic macaques. *Immunol Rev* 2001;183:94–108.
16. Haigwood NL. Predictive value of primate models for AIDS. *AIDS Rev* 2004;6:187–198.
17. Haagmans BL, Osterhaus ADME. Nonhuman primate models for SARS. *PloS Med* 2006;3:e194.
18. Lawler JV, Endy TP, Hensley LE, Garrison A, Fritz EA, Lesar M, Baric RS, Kulesh DA, Norwood DA, Wasieloski LP, Ulrich MP, Slezak TR, Vitalis E, Huggins JW, Jahrling PB, Paragas J. Cynomolgus macaque as an animal model for severe acute respiratory syndrome. *PloS Med* 2006;3:e149.
19. Muchmore E. Chimpanzee models for human disease and immunobiology. *Immunol Rev* 2001;183:86–93.
20. Mansfield K. Marmoset models commonly used in biomedical research. *Comp Med* 2003;53:383–392.
21. Herrera S, Perlaza BL, Bonelo A, Arévalo-Herrera M. Aotus monkeys: Their great value for anti-malaria vaccines and drug testing. *Int J Parasitol* 2002;32:1625–1635.
22. Arévalo-Herrera M, Herrera S. Plasmodium vivax malaria vaccine development. *Mol Immunol* 2001;38:443–455.
23. Collins WE, Sullivan JS, Galland GG, Williams A, Nace D, Williams T, Barnwell JW. Plasmodium simium and Saimiri boliviensis as a model system for testing candidate vaccines against Plasmodium vivax. *Am J Trop Med Hyg* 2005;73:644–648.
24. Bayley MB. The story of the Salk anti-poliomyelitis vaccine, 1956 (online access: http://whale.to/vaccine/bayley.html).
25. Patterson JL, Carrion R Jr. Demand for nonhuman primate resources in the age of biodefense. *ILAR J* 2004;46:15–22.
26. Centers for Disease Control and Prevention: http://www.cdc.gov/malaria/facts.htm.
27. World Health Organization: http://www.who.int/topics/malaria/en.
28. National Institute of Allergy and Infectious Diseases, National Institutes of Health: http://www3.niaid.nih.gov/research/topics/malaria/.
29. Reed ZH, Friede M, Kien MP. Malaria vaccine development: Progress and challenges. *Curr Mol Med* 2006;6:231–245.
30. Guinovart C, Navia MM, Tanner M, Alonso PL. Malaria: Burden of disease. *Curr Mol Med* 2006;6:137–140.
31. Barbay S, Plautz EJ, Friel KM, Frost SB, Dancause N, Stowe AM, Nudo RJ. Behavioral and neurophysiological effects of delayed training following a small ischemic infarct in primary motor cortex of squirrel monkeys. *Exp Brain Res* 2006;169:106–116.
32. Roitberg BZ, Mangubat E, Chen E-Y, Sugaya K, Thulborn KR, Kordower JH, Pawar A, Konecny T, Emborg ME. Survival and early differentiation of human neural stem cells transplanted in a nonhuman primate model of stroke. *J Neurosurg* 2006;105:96–102.
33. Brok HPM, Bauer J, Jonker M, Blezer E, Amor S, Bontrop RE, Laman JD, Hart BA. Nonhuman primate models of multiple sclerosis. *Immunol Rev* 2001;183:173–185.
34. Genain CP, Hauser SL. Experimental allergic encephalomyelitis in the New World monkey, Callithrix jacchus. *Immunol Rev* 2001;183:159–172.
35. Nudo RJ, Larson D, Plautz EJ, Friel KM, Barbay S, Frost SB. A squirrel monkey model of poststroke motor recovery. *ILAR J* 2003;44:161–174.
36. Maier MA, Shupe LE, Fetz EE. Dynamic neural network models of the premotoneuronal circuitry controlling wrist movements in primates. *J Comput Neurosci* 2005;19:125–146.
37. Roitman JD, Shadlen MN. Response of neurons in the lateral intraparietal area during a combined visual discrimination reaction time task. *J Neurosci* 2002;22:9475–9489.
38. Fetz EE, Perlmutter SI, Prut Y, Seki K, Votaw S. Roles of primate spinal interneurons in preparation and execution of voluntary hand movement. *Brain Res Rev* 2002;40:53–65.
39. Apgar V. A proposal for new method of evaluation of the newborn infant. *Anesth Analg* 1953;32:260–267.
40. Brazelton TB. Brazelton neonatal behavioral assessment scale (National Spastics Monographs). In: *Clinics in Developmental Medicine #50*. London: William Heineman, 1973.
41. Fagan JF, Detterman DK. The Fagan test of infant intelligence: A technical summary. *J Appl Dev Psychol* 1992;13(2):173–193.
42. Burbacher TM, Grant KS, Mottet NK. Retarded object permanence development in methylmercury exposed Macaca fascicularis infants. *Dev Psychol* 1986;22:771–776.
43. Clarren SK, Astley SJ, Gunderson VM, Spellman D. Cognitive and behavioral deficits in nonhuman primates associated with very early embryonic binge exposures to ethanol. *J Pediatr* 1992;121(5 Pt 1):789–796.
44. Burbacher T, Grant KS, Shen D, Damian D, Ellis S, Liberato N. Reproductive and offspring developmental effects following mater-

nal inhalation exposure to methanol in nonhuman primates. Part II: Developmental effects in infants exposed prenatally to methanol. Health Effects Institute Report #89, 1999:69–117.
45. Diamond A, Goldman-Rakic P. Comparison of human infants and rhesus monkeys on Piaget's AB task: Evidence for dependence on dorsolateral prefrontal cortex. *Exp Brain Res* 1989;74:24–40.
46. Dobson V, McDonald M, Kohl P, Stern N, Samek M, Preston K. Visual acuity screening of infants and young children with the acuity card procedure. *J Am Optom Assoc* 1986;57:284–289.
47. Teller DY. The development of visual acuity in human and monkey infants. *Trends Neurosci* 1981;4:21–24.
48. Diller LC, Packer OS, Verweij J, McMahon MJ, Williams DR, Dacey DM. L- and M-cone contributions to the midget and parasol ganglion cell receptive fields of macaque monkey retina. *J Neurosci* 2004;24:1079–1088.
49. McMahon MJ, Packer OS, Dacey DM. The classical receptive field surround of primate parasol ganglion cells is mediated primarily by a non-GABAergic pathway. *J Neurosci* 2004;24:3736–3745.
50. Jagadeesh B, Chelazzi L, Mishkin M, Desimone R. Learning increases stimulus salience in anterior inferior temporal cortex of the macaque. *J Neurophysiol* 2001;86:290–303.
51. Leon MI, Shadlen MN. Representation of time by neurons in the posterior parietal cortex of the macaque. *Neuron* 2003;38:317–327.
52. Sikela JM. The jewels of our genome: The search for the genomic changes underlying the evolutionarily unique capacities of the human brain. *PloS Genet* 2006;2:e80.
53. Enard W, Khaitovich P, Klose J, Zollner S, Heissig F, Giavalisco P, Nieselt-Struwe K, Muchmore E, Varki A, Ravid R. Doxiadis GM, Bontrop RE, Pääbo S. Intra- and interspecific variation in primate gene expression patterns. *Science* 2002;296:340–343.
54. Magness CL, Fellin PC, Thomas MJ, Korth MJ, Agy MB, Proll SC, Fitzgibbon M, Scherer CA, Miner DG, Katze MG, Iadonato SP. Analysis of the *Macaca mulatta* transcriptome and the sequence divergence between *Macaca* and human. *Genome Biol* 2005;6 R60.
55. Barr CS, Newman TK, Becker ML, Parker CC, Champoux M, Lesch KP, Goldman D, Suomi SJ, Higley JD. The utility of the non-human primate model for studying gene by environment interactions in behavioral research. *Genes Brain Behav* 2003;2:336–340.
56. GenBank, National Center for Biotechnology Information, National Library of Medicine, National Institutes of Health: http://www.ncbi nlm.nih.gov/Genback.
57. Kash JC, Tumpey TM, Proll SC, Carter V, Perwirasari O, Thomas MJ, Basler CF, Palese P, Taubenberger JK, García-Sastre A, Swayne DE, Katze MG. Genomic analysis of increased host immune and cell death responses induced by 1918 influenza virus. *Nature* 2006;443:578–581.
58. Baskin CR, García-Sastre A, Tumpey TM, Bielefeldt-Ohmann H. Carter VS, Nistal-Villán E, Katze MG. Integration of clinical data, pathology, and cDNA microarrays in influenza virus-infected pigtailed macaques (*Macaca nemestrina*). *J Virol* 2004;78:10420–10432.
59. Baas T, Baskin CR, Diamond DL, Garcia-Sastre A, Bielefeldt-Ohmann H, Tumpey TM, Thomas MJ, Carter VS, Teal TH, Van Hoeven N, Proll SC, Jacobs JM, Caldwell ZR, Gritsenko MA, Hukkanen RR, Camp DG, Smith RD, Katze MG. An integrated molecular signature of disease: Analysis of influenza virus infected macaques through functional genomics and proteomics. *J Virol* 2006;80:10813–10828.
60. Wallace JC, Korth MJ, Diamond DL, Proll SC, Katze MG. Virology in the 21st century: Finding function with functional genomics. *Future Virol* 2006;1:47–53.
61. Spindel ER, Pauley MA, Jia Y, Bravett C, Thompson SL, Boyle NF, Ojeda SR, Norgren RB Jr. Leveraging human genomic information to identify nonhuman primate sequences for expression array development. *BMC Genom* 2005;6:160.
62. The Psychological Well-Being of Nonhuman Primates. Committee on Well-Being of Nonhuman Primate, Institute for Laboratory Animal Research, Commission on Life Sciences, National Research Council, 1998.
63. Honess PE, Marin CM. Enrichment and aggression in primates. *Neurosci Biobehav Rev* 2006;30:413–436.
64. Novak MA, Suomi SJ. Psychological well-being of primates in captivity. *Am Psychol* 1988;43:765–773.
65. Crockett CM, Bellanca RU, Bowers CL, Bowden DM. Grooming-contact bars provide social contact for individually caged laboratory macaques. *Contemp Top Lab Anim Sci* 1997;36:53–60.
66. Allen TM, Mothé BR, Sidney J, Jing P, Dzuris JL, Liebl ME, Vogel TU, O'Connor DH, Wang X, Wussow MC, Thomson JA, Altman JD, Watkins DI, Sette A. CD8+ lymphocytes from simian immunodeficiency virus-infected rhesus macaques recognize 14 different epitopes bound by the major histocompatibility complex class I molecule Mamu-A*01: Implications for vaccine design and testing. *J Virol* 2001;75:738–749.
67. Chen ZW, Li Y, Zeng X, Kurodea MJ, Schmitz JE, Shen Y, Lai Z, Shen L, Letvin NL. The TCR repertoire of an immunodominant CD8+ T lymphocyte population. *J Immunol* 2001;166:4525–4533.
68. Desrosires RC. The value of specific pathogen-free rhesus monkey breeding colonies for AIDS research. *AIDS Res Hum Retroviruses* 1997;13:5–6.
69. Cohen J. AIDS research: Vaccine studies stymied by shortage of animals. *Science* 2000;287:959–960.
70. Office of Laboratory Animal Welfare, National Institutes of Health. *Public Health Service Policy on Humane Care and Use of Laboratory Animals*, revised 2002.
71. Hankenson FC, Johnston NA, Weigler BJ, Di Giacomo RF. Zoönoses of occupational health importance in contemporary laboratory animal research. *Comp Med* 2003;53:579–601.
72. Weigler BJ, Di Giacomo RF. A national survey of laboratory animal workers concerning occupational risks for zoönotic diseases. *Comp Med* 2005;55:183–191.
73. Huff JL, Barry PA. B-virus (Cercopithicine herpesvirus 1) infection in humans and macaques: Potential for zoönotic disease. *Emerg Infect Dis* 2003;9:246–250.
74. Adams SR Jr, Muchmore E, Richardson JH, Renquist DM, Palmer AE, Broderson JR. Special Issue: Biohazards associated with natural and experimental diseases of nonhuman primates. *J Med Primatol* 1987;16(2).
75. Marthas ML, Lu D, Penedo MCT, Hendrickx AG, Miller CJ. Titration of an SIVmac stock by vaginal inoculation of Indian and Chinese origin rhesus macaques: Efficiency, viral loads and antibody responses. *AIDS Res Hum Retroviruses* 2001;17:1455–1466.
76. Carlson CS, Eberle MA, Rieder JM, Yi Q, Kruglyak L, Nickerson DA. Selecting a maximally informative set of single-nucleotide polmorphisms for association analysis using linkage disequilibrium. *Am J Hum Genet* 2004;74:106.
77. Kanthaswamy S, von Dollen A, Kurushima JD, Alminas O, Rogers J, Ferguson B, Lerche NW, Allen PC, Smith DG. Microsatellite markers for standardized genetic management of captive colonies of rhesus macaques (*Macaca mulatta*). *Am J Primatol* 2006;68:73.
78. Nair S, Ha JC, Rogers J. Nineteen new microsatellite DNA polymorphisms in pigtailed macaques (Macaca nemestrina). *Primates* 2000;41:343–350.
79. Penedo MC, Bontrop RE, Heijmans CM, Otting N, Noort R, Rouweler AJ, de Groot N, deGroot NG, Ward T, Doxiadis GG. Microsatellite typing of the rhesus macaque MHC region. *Immunogenetics* 2005;57:198.
80. Lairmore MD, Lerche NW, Schultz KT, Stone CM, Brown BG, Hermann LM, Jennings M. SIV, STLV-1, and type D retrovirus antibodies in captive rhesus macaques reactivity to SIV p27 in human and rhesus sera. *AIDS Res Hum Retroviruses* 1990;6:1233–1238.
81. National Institutes of Health, National Center for Research Resources. RFA RR-02-005: Establishment of specific pathogen free rhesus and pigtailed macaque colonies. Release date February 28, 2002.
82. Ward JA, Hilliard JK, Pearson S. Herpes B virus specific pathogen-free breeding colonies of macaques: Serology and the B-virus status of the macaque. *Contemp Top Lab Anim Sci* 2002;41:36–41.
83. Hilliard JK, Ward JA. B-virus specific-pathogen-free breeding colonies of macaques (Macaca mulatta) retrospective study of seven years of testing. *Lab Anim Sci* 1999;49:144–148.

84. Thouless ME, Wang Y, Welch MJ. Increased enzyme-linked immunosorbent assay specificity with solubilized simian retrovirus 2-infected cell membranes. *Lab Anim Sci* 1996;46:619–622.
85. Kuller L, Watanabe R, Anderson D, Grant R. Development of a whole-virus multiplex flow cytometric assay for antibody screening of a specific pathogen-free primate colony. *Diagnost Microbiol Infect Dis* 2005;53:185–193.
86. Khan IH, Mendoza S, Yee J, Deane M, Venkateswaran K, Zhou SS, Barry PA, Lerche NW, Liciw PA. Simultaneous detection of antibodies to six nonhuman-primate viruses by multiplex microbead immunoassay. *Clin Vaccine Immunol* 2006;13:45–52.
87. Mansfield K. Development of specific pathogen free nonhuman primate colonies. In: Coote SW, Ed. *The Laboratory Primate (Handbook of Experimental Animals)*. San Diego: Academic Press, 2005:229–239.

29 Primates as Models of Behavior in Biomedical Research

Steven J. Schapiro

ABSTRACT

Much can be learned about behaviors with biomedical relevance from behavioral studies of the seemingly uncomplicated prosimians to the seemingly complex great apes. This chapter will concentrate on providing insight into suitable, and perhaps even ideal, nonhuman primate (NHP) models that can be used, or that have been used, to attempt to answer specific behaviorally oriented biomedical research questions. Prosimians, monkeys, and apes are all organisms whose behavior has been investigated in the laboratory in an attempt to model, to simulate, to understand, or to change behavior. One of the primary biomedical uses of behavioral research with NHP is to identify, study, and treat behavioral abnormalities. A second important biomedical use of behavior is as an assay. Normal behavior can be assessed during a baseline condition and then an experimental manipulation can be implemented, with any significant changes in behavior being indicative of a meaningful effect of the manipulation. Nonhuman primate subjects can also be trained to perform specific tasks until they have reached target criteria and the effects of manipulations on performance can then be measured. For investigations focusing on the evolution of behavior, primate behavioral capabilities can be studied to better understand the evolution and/or development of human capabilities. Rhesus and cynomolgus macaques are the two most commonly utilized nonhuman primate models in studies of behavior. These medium-sized animals appear to represent a valuable compromise model, retaining many of the most attractive and useful characteristics of NHP as behavioral models, while simultaneously demonstrating many of the convenient and utilitarian characteristics required for successful laboratory animal models. Although specialized studies will require more specialized nonhuman primate models, rhesus and cynomolgus macaques appear to be exceptionally useful models for a variety of types of behavioral research.

Key Words: Animal models, Nonhuman primates, Behavior, Behavioral assays, Conditioning, Cynomolgus macaques, Rhesus macaques, Chimpanzees, Baboons.

INTRODUCTION

This chapter will focus on the use of nonhuman primates (NHP) as behavioral models in biomedical research. Other chapters within this volume will examine the behavior of other species[1] and other uses of nonhuman primates as models.[2,3] From studies of infants[4] to studies of the aged,[5] from studies of normal individuals[6] to studies of extremely abnormal subjects,[7] from studies of simple individual behaviors[8] to studies of extremely complex social behaviors,[9] nonhuman primates serve as important behavioral models in the biomedical research enterprise. Much can be learned about behaviors with biomedical relevance from studies of the seemingly uncomplicated prosimians to the seemingly complex great apes. Since the emphasis of this volume is on identifying appropriate models to address specific research questions, this chapter will concentrate on providing insight into suitable, and perhaps even ideal, NHP models that can be used, or that have been used, to attempt to answer specific behaviorally oriented biomedical research questions.

One of the primary biomedical uses of behavioral research with NHP is to identify, study, and treat behavioral abnormalities. Most behavioral studies of NHP have as their goal the exploration, explanation, prediction, and/or modification of behavior. Prosimians, monkeys, and apes are all organisms whose behavior has been investigated in the laboratory in an attempt to model, to simulate, to understand, or to change behavior. For example, Harlow demonstrated in considerable detail the behavioral abnormalities that result from inadequate mother–infant attachment.[4] Others continued this research and were able to identify methods to compensate for, and more importantly prevent, poor mother–infant relationships and their behavioral outcomes in the laboratory.[10–12] Addiction research shows that primates will work for "drugs," sometimes to the exclusion of food.[13] This work identifies the reinforcing components of drug use and abuse. Some studies of pharmaceuticals examine changes in abnormal behavior as a function of drug administration[14]—changes are good if the drug is having the desired effect and changes are bad if the drug is not supposed to affect behavior or is supposed to affect abnormal behavior differently. This chapter will highlight studies that have taken place in the laboratory setting, but will also include occasional mention of relevant work conducted with primates in natural settings (the field). This will be especially evident when examining the evolution of biomedically relevant behaviors.

A second important biomedical use of behavior is as an assay. There are two different, but related ways in which behavior can serve as an assay. First, normal behavior can be assessed during a baseline condition (or in control groups), and then an experimental manipulation can be implemented. Behavior can then be

From: *Sourcebook of Models for Biomedical Research*
(P. M. Conn, ed.), © 2008 Humana Press Inc., Totowa, NJ.

assessed again during the experimental condition, with any significant changes in behavior being indicative of a (hopefully biologically) meaningful effect of the manipulation. As an example, Kaplan, Shively, and their colleagues have studied the effects of a variety of manipulations (social stress, atherogenic diets, oral contraceptives) on the behavior of captive cynomolgus macaques,[15–20] primarily using between-subjects designs. Many studies of environmental enrichment involve multiple manipulations and control conditions and are most effectively conducted using within-subjects designs.[21,22]

Trained behaviors can also be used as assays. NHP subjects can be trained to perform specific tasks[23–26] until they have reached target criteria or have established stable patterns of responding. Once responding has stabilized, an experimental manipulation (i.e., administration of a pharmaceutical agent) can be implemented and changes in responding can be assessed. In many studies of psychoactive compounds, the effects of various doses of these compounds are measured on macaques or baboons that have been trained to reliably perform sophisticated operant tasks.[23] In addition to providing information about the behavioral effects of the manipulation, this experimental approach can also provide information about the sensory effects of the manipulation through the use of the techniques of psychophysics.[24,25]

A less biomedically oriented but still important use of studies of primate behavior involves investigations of the evolution of behavior. In this type of work, primate capabilities are studied to better understand the evolution and/or development of human capabilities. Studies of aggressive behaviors are one clear example[27] Cognitive behavior, especially the social cognition that is involved in the complex amalgamation of behaviors that yields "culture," is another example.[28] This chapter will stress behaviors that have considerable biomedical relevance. By doing so, we will be neglecting/ignoring a large, interesting, and useful component of the primate behavioral literature (e.g., studies of NHP: in space flight,[29] temperament/personality,[30,31] laterality,[32] and culture[28,33]). In a short chapter such as this one, it is inevitable that arbitrary decisions must be made and some subsets of the potential subject matter will not be treated in adequate detail.

One of the strongest justifications for using NHP as behavioral models in biomedical research is that it is possible to perform certain experimental manipulations with NHP subjects that cannot be performed with human subjects. This does not mean that potentially damaging or harmful research can be conducted without restrictions using NHP. All animal research, and especially primate research, is carefully regulated and approved by a variety of bodies with oversight responsibilities (i.e., IACUC, USDA, etc.). The similarity of NHP to humans makes them both desirable and problematic choices as animal models. In fact, NHP may be especially high fidelity models for behavioral investigations, as groups of subjects can be easily "randomized" and individuals are less likely to exhibit several potential confounding tendencies that may be unique to human subjects (noncompliance,[15] deception, biased behavior,[34] logically inconsistent preferences,[35] and spiteful behavior[36]). Of course, data from NHP are not as valid for generalizations to human behavior as are data from humans.

Although retrospective analyses can provide considerable insight into the understanding of behavior, prospective experimental investigations using appropriately selected,[37] conditioned,[38] desensitized,[39] and trained[40] primate models can more systematically, directly, and reliably address experimental hypotheses. Thus, assessments of the behavioral effects (natural or experimental) of glucocorticoids on the hippocampus in aged primate brains[41] are extremely relevant for studies of the behavioral effects of Alzheimer's disease in humans.

Inherent in any discussion of appropriate NHP models in biomedical research is the understanding that the care the animals receive while in captivity must be of the highest quality. In addition to outstanding husbandry and veterinary care, it is especially important for NHP maintained as models for behavioral research to be appropriately behaviorally managed.[42–44]

The different models and approaches used with nonhuman primates to study behaviors of biomedical relevance have been rather arbitrarily divided into several categories (see Table 29–1). These categories are intended to be neither exhaustive nor mutually exclusive. Since entire books and issues of journals have been devoted to topics relevant to this chapter, the reader is directed to additional sources, including the *ILAR Journal*, several recent books,[45–47] and an issue of the journal *Methods*[48] for more complete discussions of some of the topics treated in the present chapter.

NONHUMAN PRIMATE MODELS TO IDENTIFY AND TREAT BEHAVIORAL ABNORMALITIES

Nonhuman primate models have played a critical role in the study of abnormal behavior.[49–52] This applies to a variety of abnormal activities, including those behaviors that do not occur in natural circumstances, those behaviors that occur at abnormal frequencies under captive conditions, and those behaviors that are considered indicative of psychopathology. In all of these circumstances, NHP can and have been used to better understand, create, and treat abnormalities similar to those of clinical significance in humans.[50,51] Harlow's studies of poor/abnormal mother–infant attachment in rhesus monkeys (*Macaca mulatta*[4,53]) have probably made the largest contribution to our understanding of the development of a variety of abnormal behaviors, both in NHP and in humans. While Harlow is perhaps best known for his work that created rhesus monkeys that behaved abnormally, his research and the ongoing work of his academic descendants also demonstrate that it is possible (1) to mitigate abnormal behaviors once they have developed, but more importantly, (2) to prevent certain abnormalities from developing at all.[10–12] Much of what we know about proper mother–infant attachment and its lifelong consequences in NHP and humans comes from the laboratories of Harry Harlow and his colleagues[4,11,12,53,54] and the more human-focused John Bowlby.[55] In addition to the rhesus monkeys observed by Harlow, important contributions to the study of mother–infant attachment and abnormal behavior were made using squirrel monkey[56] and chimpanzee[57] models.

A number of abnormal behaviors may develop in laboratory primates that are housed singly for long periods of time, especially among those socially deprived when very young.[4,53,57] One of the most interesting and most relevant is self-injurious behavior (SIB).[7,58] As the name implies, SIB occurs when an individual injures itself, in the case of rhesus monkeys, typically through acts of self-directed biting. Aside from any parallels to human SIB, NHP SIB is of considerable importance in biomedical research because SIB can adversely affect the NHP and the study in which

Table 29–1
Research questions and appropriate model species

Research question	*Model species (references)*	*Special characteristics*
Assess and/or treat behavioral abnormalities		
Depression	Cynomolgus macaques[18–19]	Females and social stress
Self-injurious behavior	Rhesus[7,14,58]	Singly housed males
Abnormal development	Rhesus,[4,53] chimpanzees,[57] squirrel monkeys[56]	Socially isolated infants
Behavior as an assay		
Normal behaviors		
Effects of oral contraceptives	Cynomolgus macaques[15,17]	Females in captivity
Effects of diet	Cynomolgus macaques[17,20]	Laboratory setting
Environmental enrichment	Rhesus,[14,21,44] baboons[22]	Laboratory setting
Trained behaviors		
Drug effects	Rhesus,[8,3] baboons[23–25]	Laboratory trained to lever press
Brain manipulations	Rhesus,[26] squirrel monkeys[61]	Laboratory trained to lever press
Learning and memory	Many species[68,71–72,74]	Many circumstances
Control by thought	Rhesus[76,77]	Laboratory setting
Space flight	Rhesus,[29] chimpanzees[29]	Experimental settings
Study decrements to understand normal behavior		
Affiliative behavior	Cynomolgus,[63] rhesus[64]	Adults in unstable groups
AIDS susceptibility	Rhesus[64]	Subadult males in unstable groups
Alloparental	Marmosets,[86] Blue monkeys[87]	Naturalistic social groups
Response to capture	Vervets[66]	From wild to laboratory quarantine
Sexual behavior	Bonobos,[83–85] mandrills[81]	Naturalistic social groups
Study occurrence to understand evolution of behavior		
Aggression	Many species[27]	Laboratory and natural settings
Murder	Chimpanzees[79–80]	Natural setting
Rape	Orangutans[97]	Experimental settings
Cannibalism	Chimpanzees[98]	Natural setting
Infant abuse	Rhesus[10,53]	Socially isolated individuals
Social stress	Baboons[41,65,99]	Natural setting
	Cynomolgus macaques[16,19,59]	Laboratory setting
Eating	Rhesus[103–104]	Caloric restriction, obesity
	Cynomolgus macaques[105]	High cholesterol diets
Addiction	Baboons[23,25]	Laboratory setting
Infanticide	Proboscis monkeys[78]	Natural setting
Laterality	Chimpanzees[32]	All settings
Personality/temperament	Bushbabies[31]	Laboratory setting
Adoption	Capuchins[96]	Natural setting
Aging	Rhesus[5,106]	Laboratory settings
Culture	Chimpanzees[28,33]	Laboratory and natural settings
Self-medication	Chimpanzees[6]	Natural settings
Emerging techniques and new frontiers		
fMRI[a]	Rhesus,[108] marmosets[111]	Laboratory setting
PET[a] scans	Rhesus[54,109]	Laboratory setting
Gaze tracking	Ring-tailed lemurs[112]	Laboratory, but group-housed

[a]fMRI, functional magnetic resonance imaging; PET, positron emission tomography.

the animal is participating.[58] The focus of this volume is on identifying appropriate models for research questions. Therefore it should be obvious that singly caged male rhesus macaques are an appropriate model for studies of SIB and that SIB may make them inappropriate subjects for investigations requiring early and long-term social isolation. Because SIB is a condition that already exists among a subset of captive male rhesus,[7,58] there is little, if any, need to attempt to experimentally induce SIB.

Recent studies suggest that female cynomolgus macaques are particularly good models for investigations of depression,[19] demonstrating many useful similarities to human depression. These similarities appear to be relatively unique among animal models, and even NHP models, for depression. This research group has also profitably studied the effects of stress on behavior in this species.[59,60]

Nonhuman primates as models for abnormal behavior are probably the best known and reported use of NHP as behavioral biomedical models. Recent, excellent reviews of this topic are available from a number of sources.[50–52]

NONHUMAN PRIMATE BEHAVIOR AS AN ASSAY

As with many laboratory animal species, the behavior of NHP can be used as an assay to measure the effects of particular manipulations. These manipulations can be behavioral or nonbehavioral in nature. Typically, behavior during baseline periods is compared to behavior during experimental conditions or behavior in control groups of NHP is compared to behavior in experimental groups of NHP. Depending on the species and the experimental question, either natural behavior or trained behavior can be used as the assay.

CHANGES TO NORMAL NONHUMAN PRIMATE BEHAVIOR Perhaps the easiest way to view this approach to behavioral assays is to consider studies that are designed to establish relationships between particular brain structures and particular behaviors. Effects of a variety of brain manipulations (lesions, stimulation, pharmaceutical treatments, etc.) can be assessed by comparing behavior with and without the manipulation.[26,61] Macaques, primarily rhesus and cynomolgus macaques, are the most frequently utilized NHP models in neuroscience research.[62]

Unstable social settings have been shown to have a variety of adverse behavioral and physiological consequences on laboratory NHP, especially several macaque species *(M. mulatta* and *M. fascicularis*).[60,63,64] Similar findings exist for baboons in the wild[65] and for vervet monkeys recently captured from the wild and housed individually in captivity.[66] These studies are extremely relevant to biomedical research, as they compare stable to unstable social conditions and demonstrate that behavioral factors/manipulations can have profound physiological and behavioral consequences. In addition to opportunities to study these types of manipulations and their consequences as phenomena, it is important to be aware of factors such as potential confounds when developing and using NHP models in research. There are strong data describing the physiological and especially the immunological consequences of unstable social settings.[60,63,64] For example, rhesus macaques infected with simian immunodeficiency virus (SIV) get sicker faster when housed in unstable social groups than when housed in stable social groups.[64]

Cynomolgus macaques have also served as extremely important models in a long series of research projects that have examined the effects of a variety of behavioral and physiological manipulations, including social stress, atherogenic diets, alcohol consumption, ovariectomy, and oral contraceptives[15–20] on "normal" behavior. Most of these studies have taken advantage of between-subjects designs in which the behavior of the experimental group(s) is compared with the baseline behavior of the control group, which did not receive the atherogenic diet,[17] the oral contraceptives,[15] or the ovariectomy.[16]

USING TRAINED BEHAVIOR Nonhuman primates can be trained to perform a variety of simple and sophisticated operant responses.[8,13,23–26] Once trained, the response patterns of the NHP can be used to assess the effects of a variety of manipulations. Animals that respond in one way during control conditions but respond differently during experimental sessions provide important behavioral data to address research questions. For example, an NHP that has been trained to successfully avoid shock under baseline conditions, but fails to avoid shock while under the influence of alcohol,[67] provides a useful model for alterations in behavior as a function of alcohol consumption. The techniques of psychophysics can be applied to determine whether a subject's responses to target stimuli have changed, rather than its ability to perceive/sense the stimuli.[24,25]

Macaques, baboons, and chimpanzees can be trained to perform many different behaviors using food or fluid (water or juice) as the primary reinforcer.[68] Some of these experimental paradigms are conducted under conditions of food or water control, in which subjects receive much of their food or fluid as a function of performing the experimental task.[69] In these types of studies, food or water is often available only during, and shortly after, the experimental session, resulting in reasonably long periods of food or water restriction for the NHP.[23] These types of procedures are typically thought to increase the motivation of the subjects to perform the operant response, but may simultaneously adversely affect the validity of the experiment.[69,70]

NHP subjects can be trained to perform many different responses, including simple lever presses, making choices on touch screens, and using joysticks.[68,71,72] Similarly, the effects of a variety of manipulations can be examined using these behavioral responses, including the psychoactive effects of pharmaceutical compounds and the effects of brain lesions or brain stimulation.[8,13,23–26,61,67–69] NHP can also be trained to perform multiple operant responses with each response coupled to a different reinforcer in order to compare the value of the different reinforcers.[23,24] The most enlightening of these studies are ones that involve comparisons between food and psychoactive reinforcers (drugs or brain stimulation), providing substantial insight into the behavioral components of drug addiction.[23] The use of trained behaviors may be particularly valuable for assessing longitudinal effects of manipulations, including developmental, maturational, and aging processes.

Studies of Learning and Memory Trained behaviors are especially useful in studies of NHP learning and memory. Although memory is not really a behavior, learning is a constellation of behaviors with considerable biomedical relevance and learning and memory are so closely related that both learning and memory are important to include in the current discussion. Learning and memory have been studied in many primate species using naturalistic and artificial methods of assessment under both laboratory and field conditions.[73–75] The study of NHP learning and memory is likely to be limited, not by the capabilities of the animals, but by the ability of human experimenters to develop technology to adequately assess NHP learning and memory. Once again, macaques and baboons are the most frequently used NHP models for studies of learning and memory.[62] Chimpanzees, given their considerable cognitive abilities, are also attractive, well utilized, and productive models.[75]

Macaques have also been quite profitably utilized in research programs that have focused on training NHP to control body movements via thought (primarily hand movements).[76,77] Although neither a training manipulation in the strict operant conditioning sense nor a standard approach to examining learning and memory, these studies have successfully trained rhesus monkeys to control movements with thoughts. The data from these experiments are impressive and have many implications for therapeutic behavioral applications.

NONHUMAN PRIMATE MODELS OF THE EVOLUTION OF BEHAVIOR

When the behaviors of NHP are typically explained, evolutionary justifications for behavioral patterns are often provided. Although evolutionary explanations fit certain patterns of human and NHP behavior, explanations of human behavior are not limited solely to adaptations of NHP patterns. While male NHP may kill infants (infanticide) sired by rival males to increase their own likelihood of siring infants,[78] this type of explanation is unlikely to be invoked in human situations. On the other hand, male chimpanzees will systematically attack and kill other chimpanzees, a fact that some use to explain human tendencies toward murder and war.[79,80]

SEXUAL BEHAVIOR Ovulation is clearly signaled in some species of NHP[81] and concealed in others.[82,83] As reproductive strategies, both signaled and concealed ovulation impact patterns of sexual and paternal behavior. In NHP species with signaled ovulation,[81] males may be able to adopt behavioral strategies designed to monopolize females' breeding opportunities during fertile periods. If successful, males may then have some certainty of parenthood and should invest in the care of offspring they have "calculated" to be theirs. Alternatively, usurper males would know that they were not the father of any infants and might try to eliminate the offspring of a rival through infanticide. In NHP species with concealed ovulation, males may be more likely to invest in the infants of all females they mated with, since they cannot be completely certain that they sired any particular infant. NHP females may take advantage of this. Bonobos may be a particularly important model for studies of sexual behavior, since their ovulation appears to be concealed[83] and unlike many NHP species, sexual behavior is not strictly under "hormonal control."[84] In fact, bonobo sexual behavior appears to function both for reproductive and for nonreproductive purposes,[85] serving as a critical, human-like component of this species' repertoire of social behaviors.

CAREGIVING BEHAVIOR Nonhuman primates typically take very good care of their offspring, investing great quantities of time and resources in their young. As with most of the topics in this chapter, much has been learned about all types of caregiving behavior from studies of NHP caregiving.

Offspring of NHP are typically altricial, requiring considerable care from one or both parents (and sometimes additional individuals) during long developmental periods.[86,87] Mothers and fathers differ considerably in their initial investment in offspring, and much has been made of this discrepancy as an explanation for differences in parenting behavior.[88] In general, there is a positive correlation between an NHP male's certainty of paternity and his tendency to engage in paternal behavior.[89] As one example, males of species with typically monogamous mating systems (marmosets), on average, are considerably more paternal than males of species with typically polygamous mating systems (macaques[90]), providing a wide spectrum of potential NHP model species for studies of paternal behavior.

Nonhuman primate parents often rely on caregivers other than themselves to assist in the raising of offspring. Examples of these alloparenting situations include older siblings helping to care for younger siblings in several marmoset species,[91] and aunts and cousins caring for infants in many other species of nonhuman primates.[92–94] Grandparents are relatively infrequent alternate caregivers among NHP.[90] There are even examples of individuals purposely "adopting" and raising unrelated conspecific orphans (at significant costs to themselves and/or their own offspring[95]) or members of other species.[96]

AGGRESSIVE BEHAVIOR The vast array of aggressive, violent, and destructive behaviors of which humans are capable is at times unbelievable, but always of biomedical relevance. Clearly, examples of similar aggressive/violent tendencies exist among species of NHP.[79,80] Natural selection is typically posited as the cause of most aggression in wild populations of NHP, with animals competing with one another over the availability of resources and opportunities to pass their genes on to future generations. Competition among NHP may be aggressive and/or violent, but it is less frequently lethal, and according to evolutionarily based arguments, is rarely performed for "no apparent reason." It is of interest then that rape occurs among orangutans,[97] infanticide occurs in several species,[78,98] chimpanzees engage in war[79,80] and cannibalism,[98] and mother macaques will abuse their infants.[10] Clearly, rape, infanticide, war, cannibalism, and infant abuse are all behaviors that lend themselves to examination in a biomedical research context.

STRESS-RELATED BEHAVIOR AND DOMINANCE STATUS Several aspects of NHP agonistic interactions (dominance and subordinate behaviors) are affected by, and affect, multiple behavior patterns in many social circumstances and are thus relevant to biomedical investigations of behavior. Studies of chronic subordinate status in wild baboons serve as particularly relevant models in this regard. In addition to behaviors indicative of chronic stress, low-ranking male baboons suffer from persistently high levels of glucocorticoids that are frequently associated with a variety of physiological impairments with behavioral implications, including most importantly, death of neurons in the hippocampus.[65,99]

EATING BEHAVIOR Clearly, NHP eat in the wild[100] and in the laboratory.[101] Similarly, given the epidemic of obesity and diabetes that currently plagues many Western societies,[102] eating behavior is of great interest in biomedical research. The consumption of food by NHP has been studied in a variety of contexts, including normal foraging behavior and diet in wild NHP,[100] the effects of calorie restriction on longevity and behavior in captive NHP,[103] and the effects of different "human-like" diets on weight gain, behavior, and a variety of physiological parameters in captive NHP.[17,20,104] From the standpoint of this chapter in this volume, the most relevant body of work focuses on the effects of dietary manipulations (primarily the use of atherogenic diets) on the behavior of adult cynomolgus monkeys;[17,20,105] however there are many other useful NHP models for eating, obesity/diabetes,[104,106,107] and caloric restriction.[103]

EMERGING MODELS/QUESTIONS/FRONTIERS

Noninvasive imaging techniques are likely to increase in importance in studies of behavior in the biomedical research enterprise. Although few fully developed NHP models for use with these techniques currently exist, a number of promising models are currently under development, including rhesus monkeys for functional magnetic resonance imaging (fMRI)[108] and positron emission tomography (PET),[54,109] chimpanzees for MRI,[110] and marmosets for fMRI.[111]

Monitoring of the visual attention (gaze direction) of NHP is often an important dependent measure in biomedical research. In one of the few prosimian models in biomedicine, a system is currently under development to measure the visual orientation of socially living *Lemur catta*.[112]

CONCLUSIONS

This has been a very brief and very selective review that sought to identify some available and developing NHP behavioral models. The references in this chapter contain data from at least 20 different NHP taxa. A complete treatment of all NHP behavioral research that might be relevant to questions of biomedical interest is simply not possible in a short chapter. Many issues have not been treated, including extremely interesting topics such as NHP social cognition, language, culture, laterality, personality, and emotions,[28–33] and the reader is directed to several recent publications[45–48] for more detailed treatments of these and related topics.

For some behaviors, there is considerable justification for studying the behavior in NHP for its own sake, while for other behaviors, the primary, and perhaps only, justification for studying the behavior is to better understand human behavior. It should be obvious that there are numerous examples of behaviors in NHP that are useful for modeling, predicting, or understanding behaviors relevant to biomedical research investigations. In fact, it is probably difficult to think of a behavior that could not be studied with some face validity in biomedicine using an NHP model. Even many of the most sophisticated (language, learning, culture) and the most abnormal (self-aggression, infant abuse) patterns of behavior can be studied and/or induced in nonhuman primates.

Rhesus monkeys and cynomolgus macaques are the two most commonly utilized NHP models in studies of behavior.[62] While not all behavioral studies are biomedically oriented, the literature cited above suggests that these two species are among the most commonly utilized NHP models in biomedical behavioral research. These medium-sized animals appear to represent a valuable compromise model, retaining many of the most attractive and useful characteristics of NHP as behavioral models while simultaneously demonstrating many of the convenient and utilitarian characteristics required for successful laboratory animal models. While some specialized studies may require more specialized models like marmosets[86,91] or chimpanzees,[71,80,98,110] rhesus and cynomolgus macaques appear to be exceptionally useful models for a variety of types of behavioral research.

ACKNOWLEDGMENTS

This work was supported in part by the Department of Veterinary Sciences of The University of Texas M.D. Anderson Cancer Center.

REFERENCES

1. Zohar J. Animal models of posttraumatic stress disorder. In: Conn PM, Ed. *Sourcebook of Model Organisms in Biomedical Research.* New York: Humana Press, 2007.
2. Kettner RE. Primate cerebellum models. In: Conn PM, Ed. *Sourcebook of Model Organisms in Biomedical Research.* New York: Humana Press, 2007.
3. Wolf DP. Primate models for assisted reproductive technologies. In: Conn PM, Ed. *Sourcebook of Model Organisms in Biomedical Research.* New York: Humana Press, 2007.
4. Harlow HF, Harlow MK. Social deprivation in monkeys. *Sci Am* 1962;207:136–146.
5. Duce JA, Hollander W, Jaffe R, Abraham CR. Activation of early components of complement targets myelin and oligodendrocytes in the aged rhesus monkey brain. *Neurobiol Aging* 2006;27:633–644.
6. Huffman MA. Primate self-medication. In: Campbell CJ, Fuentes A, MacKinnon KC, Panger M, Bearder SK, Eds. *Primates in Perspective.* New York: Oxford University Press, 2007:677–690.
7. Tiefenbacher S, Novak MA, Lutz CK, Meyers JS. The physiology and neurochemistry of self-injurious behavior: A nonhuman primate model. *Front Biosci* 2005;10:1–11.
8. Foltin RW. Effects of sibutramine on the appetitive and consummatory aspects of feeding in non-human primates. *Physiol Behav* 2006;87:280–286.
9. Caldwell CA, Whiten A. Social learning in monkeys and apes: Cultural animals? In: Campbell CJ, Fuentes A, MacKinnon KC, Panger M, Bearder SK, Eds. *Primates in Perspective.* New York: Oxford University Press, 2007:652–664.
10. Maestripieri D, Lindell SG, Ayala A, Gold PW, Higley JD. Neurobiological characteristics of rhesus macaque abusive mothers and their relation to social and maternal behavior. *Neurosci Biobehav Rev* 2005;29:51–57.
11. Novak MA, Harlow HF. Social recovery of monkeys isolated for the first year of life: I. Rehabilitation and therapy. *Dev Psych* 1975;11:453–465.
12. Novak MA, Sackett GP. The effects of rearing experiences: The early years. In: Sackett GP, Ruppenthal GC, Elias K, Eds. *Nursery Rearing of Nonhuman Primates in the 21st Century.* New York: Springer, 2006:5–19.
13. Foltin RW. Effects of dietary and pharmacological manipulations on appetitive and consummatory aspects of feeding in non-human primates. *Appetite* 2005;45:110–120.
14. Taylor DK, Bass T, Flory GS, Hankenson FC. Use of low-dose chlorpromazine in conjunction with environmental enrichment to eliminate self-injurious behavior in a rhesus macaque (*Macaca mulatta*). *Comp Med* 2005;55:282–288.
15. Henderson JA, Shively CA. Triphasic oral contraceptive treatment alters the behavior and neurobiology of female cynomolgus monkeys. *Psychoneuroendocrinology* 2004;29:21–34.
16. Shively CA, Kaplan JR, Adams MR. Effects of ovariectomy, social instability and social status on female *Macaca fascicularis* social behavior. *Physiol Behav* 1986;36:1147–1153.
17. Shively CA, Kaplan JR, Clarkson TB. Carotid artery atherosclerosis in cholesterol-fed cynomolgus monkeys: The effects of oral contraceptive treatments, social factors and regional adiposity. *Arteriosclerosis* 1990;10:358–366.
18. Shively CA, Williams JK, Laber-Laird K, Anton RF. Depression and coronary artery atherosclerosis and reactivity in female cynomolgus monkeys. *Psychosom Med* 2002;64:699–706.
19. Shively CA, Register TC, Friedman DP, Morgan TM, Thompson J, Lanier T. Social stress-associated depression in adult female cynomolgus monkeys (*Macaca fascicularis*). *Biol Psychol* 2005;69:67–84.
20. Simon NG, Kaplan JR, Hu S, Register, TC, Adams MR. Increased aggressive behavior and decreased affiliative behavior in adult male monkeys after long-term consumption of diets rich in soy proteins and isoflavones. *Horm Behav* 2004;45:278–284.
21. Schapiro SJ, Suarez SA, Porter LM, Bloomsmith MA. The effects of different types of feeding enhancements on the behavior of single-caged, yearling rhesus macaques. *Anim Welf* 1996;5:129–138.
22. Bourgeois S, Brent L. Modifying the behavior of singly caged baboons: Evaluating the effectiveness of four enrichment techniques. *Anim Welf* 2005;14:71–81.

23. Weerts EM, Froestl W, Griffiths RR. Effects of GABAergic modulators on food and cocaine self-administration in baboons. *Drug Alcohol Depend* 2005;80:369–376.
24. Brady JV, Bradford LD, Hienz RW. Behavioral assessment of risk-taking and psychophysical functions in the baboon. *Neurobehav Toxicol* 1979;1(Suppl. 1):73–84.
25. Hienz RW, Weerts EM. Cocaine's effects on the perception of socially significant vocalizations in baboons. *Pharmacol Biochem Behav* 2005;81:440–450.
26. Rice DC. Effect of postnatal exposure to a PCB mixture in monkeys on multiple fixed interval-fixed ratio performance. *Neurotoxicol Teratol* 1997;19:429–434.
27. Bernstein IS. Social mechanisms in the control of primate aggression. In: Campbell CJ, Fuentes A, MacKinnon KC, Panger M, Bearder SK, Eds. *Primates in Perspective*. New York: Oxford University Press, 2007:562–571.
28. Whiten A, Horner V, de Waal FBM. Conformity to cultural norms of tool use in chimpanzees. *Nature* 2005;437(7059):737–740.
29. Hoban-Higgins TM, Robinson EL, Fuller CA. Primates in space flight. *Adv Space Biol Med* 2005;10:303–325.
30. Vazire S, Gosling SD, Dickey AS, Schapiro SJ. Measuring personality in nonhuman animals. In: Robins, RW, Fraley RC, Krueger, R, Eds. *Handbook of Research Methods in Personality Psychology*. New York: Guilford Press, 2007:190–208.
31. Watson SL, Ward JP. Temperament and problem-solving in the small-eared bushbaby (*Otolemur garnetti*). *J Comp Psych* 1996;110:377–385.
32. Hopkins WD, Wesley MJ, Russell JL, Schapiro SJ. Parental and perinatal factors influencing the development of handedness in captive chimpanzees. *Dev Psychobiol* 2006;48:428–435.
33. Hopper L, Spiteri A, Lambeth SP, Schapiro SJ, Horner V, Whiten A. Experimental studies of traditions and underlying transmission processes in chimpanzees. *Anim Behav* 2007;73:1021–2032.
34. Tversky A, Kahneman D. The framing of decisions and the psychology of choice. *Science* 1981;211:453–458.
35. Thaler R, Shefrin HM. An economic theory of self-control. *J Pol Econ* 1981;89:392–406.
36. Werner G, Schmittberger R, Schwartze, B. An experimental analysis of ultimatum bargaining. *J Econ Behav Org* 1982;3:367–388.
37. Capitanio JP, Kyes RC, Fairbanks LA. Considerations in the selection and conditioning of Old World monkeys for laboratory research: Animals from domestic sources. *ILAR J* 2006;47:294–306.
38. Tardif S, Bales K, Williams L, Moeller EL, Abbott D, Schultz-Darken N, Mendoza S, Mason W, Bourgeois S, Ruiz J. Preparing new world monkeys for laboratory research. *ILAR J* 2006;47: 307–315.
39. Schapiro SJ, Everitt JI. Preparation of animals for use in the laboratory: Issues and challenges for the Institutional Animal Care and Use Committee (IACUC). *ILAR J* 2006;47:370–375.
40. Bloomsmith MA, Schapiro SJ, Strobert EA. Preparing chimpanzees for laboratory research. *ILAR J* 2006;47:316–325.
41. Sapolsky RM, Finch CE. Alzheimer's disease and some speculations about the evolution of its modifiers. *Ann NY Acad Sci* 2000;924:99–103.
42. Bloomsmith MA, Else JG. Behavioral management of chimpanzees in biomedical research: The state of the science. *ILAR J* 2005;46:192–201.
43. Lutz CK, Novak MA. Primate natural history and social behavior: Implications for laboratory housing. In: Wolfe-Coote S, Ed. *The Laboratory Primate*. San Diego, CA: Elsevier, 2005:133–142.
44. Schapiro SJ. Effects of social manipulations and environmental enrichment on behavior and cell-mediated immune responses in rhesus macaques. *Pharmacol Biochem Behav* 2002;73:271–278.
45. Wolfe-Coote S. *The Laboratory Primate*. San Diego: Elsevier, 2005.
46. Campbell CJ, Fuentes A, MacKinnon KC, Panger M, Bearder SK. *Primates in Perspective*. New York: Oxford University Press, 2007.
47. Maestripieri D. *Primate Psychology*. Cambridge, MA: Harvard University Press, 2003.
48. Coleman K. Introduction: Models for primate behavior. *Methods* 2006;38:161.
49. Capitanio JP. Behavioral pathology. In: Erwin J, Mitchell G, Eds. *Comparative Primate Biology*, Vol. 2, Part A: *Behavior, Conservation and Ecology*. New York: Alan R. Liss, 1986:411–454.
50. Murison R. Animal models for psychological disorders. In: Hau J, Van Hoosier GL Jr, Eds. *Handbook of Laboratory Animal Science*, Vol. II, *Animal Models*. New York: CRC Press, 2003:111–125.
51. Troisi A. Psychopathology. In: Maestripieri D, Ed. *Primate Psychology*. Cambridge, MA: Harvard University Press, 2003:451–470.
52. Maestripieri D, Wallen K. Nonhuman primate models of developmental psychopathology: Problems and prospects. In: Cicchetti D, Walker EF, Eds. *Neurodevelopmental Mechanisms in Psychopathology*. New York: Cambridge University Press, 2003:187–214.
53. Mitchell GD, Raymond EJ, Ruppenthal GC, Harlow HF. Long-term effects of total social isolation upon behavior of rhesus monkeys. *Psych Rep* 1966;18:567–580.
54. Ichise M, Vines DC, Gura T, Anderson GM, Suomi SJ, Higley JD, Innis RB. Effects of early stress on [11C]DASB positron emission tomography imaging of serotonin transporters in adolescent peer- and mother-reared rhesus monkeys. *J Neurosci* 2006;26: 4638–4643.
55. Bowlby J. Making and breaking of affectional bonds: 1. Etiology and psychopathology in light of attachment theory. *Br J Psych* 1977;130:201–210.
56. Vogt J, Levine S. Response of mother and infant squirrel monkeys to separation and disturbance. *Physiol Behav* 1980;24: 829–832.
57. Menzel EW Jr. The effects of cumulative experience on responses to novel objects in young isolation-reared chimpanzees. *Behavior* 1963;21:1–12.
58. Novak MA. Self-injurious behavior in rhesus monkeys: New insights into its etiology, physiology and treatment. *Am J Primatol* 2003;59:3–19.
59. Kaplan JR, Manuck SB. Ovarian dysfunction, stress, and disease: A primate continuum. *ILAR J* 2004;45:89–115.
60. Cohen S, Kaplan JR, Cunnick JE, Manuck SB, Rabin BS. Chronic social stress, affiliation, and cellular immune response in nonhuman primates. *Psychol Sci* 1992;3:301–304.
61. Dressnandt J, Juergens U. Brain stimulation-induced changes of phonation in the squirrel monkey. *Exp Brain Res* 1992;89: 549–559.
62. Carlsson HE, Schapiro SJ, Farah I, Hau J. Use of primates in research: A global overview. *Am J Primatol* 2004;63:225–237.
63. Kaplan JR, Heise ER, Manuck SB, Shively CA, Cohen S, Rabin BS, Kasprowicz AL. The relationship of agonistic and affiliative behavior patterns to cellular immune function among cynomolgus monkeys (*Macaca fascicularis*) living in unstable social groups. *Am J Primatol* 1991;25:157–173.
64. Capitanio JP, Mendoza SP, Lerche NW, Mason WA. Social stress results in altered glucocorticoid regulation and shorter survival in simian acquired immune deficiency syndrome. *Proc Natl Acad Sci USA* 1998;95:4714–4719.
65. Sapolsky RM. The physiology of dominance in stable versus unstable dominance hierarchies. In: Mason WA, Mendoza SP, Eds. *Primate Social Conflict*. Albany, NY: SUNY Press, 1993:171–204.
66. Suleman MA, Wango E, Farah IO, Hau J. Adrenal cortex and stomach lesions associated with stress in wild male African green monkeys (*Cercopithecus aethiops*) in the post-capture period. *J Med Primatol* 2000;29:338–342.
67. Barrett JE. Effects of alcohol, chlordiazepoxide, cocaine and pentobarbital on responding maintained under fixed-interval schedules of food or shock presentation. *J Pharmacol Exp Ther* 1976;196: 605–615.
68. Weed MR, Taffe MA, Polis I, Roberts AC, Robbins TW, Koob GF, Bloom FE, Gold LH. Performance norms for a rhesus monkey neuropsychological testing battery: Acquisition and long-term performance. *Cogn Brain Res* 1999;8:95.

69. Taffe MA. Effects of parametric feeding manipulations on behavioral performance in macaques. *Physiol Behav* 2004;81:59–70.
70. Toth LA, Gardiner TW. Food and water restriction protocols: Physiological and behavioral considerations. *Contemp Top Lab Anim Sci* 2000;39:9–17.
71. Matsuno T, Kawai N, Matsuzawa T. Color classification by chimpanzees (*Pan troglodytes*) in a matching-to-sample task. *Behav Brain Res* 2004;148:157–165.
72. Leighty KA, Fragaszy DM. Primates in cyberspace: Using interactive computer tasks to study perception and action in nonhuman animals. *Anim Cogn* 2003;6:137–139.
73. Weller RE, LeDoux MS, Toll LM, Gould MK, Hicks RA, Cox JE. Subdivisions of inferior temporal cortex in squirrel monkeys make dissociable contributions to visual learning and memory. *Behav Neurosci* 2006;120:423–446.
74. Wich SA, de Vries H. Male monkeys remember which group members have given alarm calls. *Proc Royal Soc Lond B Biol Sci* 2006;273:735–740.
75. Matsuzawa T, Tomonaga M, Tanaka M. *Cognitive Development in Chimpanzees*. New York: Springer, 2006.
76. Santhanam G, Ryu SI, Yu BM, Afshar A, Shenoy KV. A high performance brain-computer interface. *Nature* 2006;442:195–198.
77. Musallam S, Corneil BD, Greger B, Scherberger H, Andersen RA. Cognitive control signals for neural prosthetics. *Science* 2004;305:258–262.
78. Agoramoorthy G, Hsu MJ. Occurrence of infanticide among wild proboscis monkeys (*Nasalis larvatus*) in Sabah, Northern Borneo. *Folia Primatol* 2005;76:177–179.
79. Wrangham R. Why apes and humans kill. In: Jones M, Fabian AC, Eds. *Conflict*. New York: Cambridge University Press, 2006:43–62.
80. Wrangham RW, Wilson ML, Muller MN. Comparative rates of violence in chimpanzees and humans. *Primates* 2006;47:14–26.
81. Setchell JM Wickings EJ. Sexual swelling in mandrills (*Mandrillus sphinx*): A test of the reliable indicator hypothesis. *Behav Ecol* 2004;15:438–445.
82. Ostner J, Chalise MK, Koenig A, Lornhardt K, Nikolei J, Podzuweit D, Borries C. What Hanuman langur males know about female reproductive status. *Am J Primatol* 2006;68:701–712.
83. Reichert KE, Heistermann M, Hodges JK, Boesch C, Hohmann G. What females tell males about their reproductive status: Are morphological and behavioral cues reliable signals of ovulation in bonobos (*Pan paniscus*). *Ethology* 2002;108:583–600.
84. Manson JH, Perry S, Parish AR. Nonconceptive sexual behavior in bonobos and capuchins. *Int J Primatol* 1997;18:767–786.
85. Parish AR, de Waal FBM. The other "closest living relative." How bonobos (*Pan paniscus*) challenge traditional assumptions about females, dominance, intra- and intersexual interactions, and hominid evolution. *Ann NY Acad Sci* 2007;907:97–113.
86. Tardif SD. The bioenergetics of parental behavior and the evolution of allomaternal care in marmosets and tamarins. In: Solomon NG, French JA, Eds. *Cooperative Breeding in Mammals*. New York: Cambridge University Press, 1997:11–33.
87. Foerster S, Cords M. Socialization of infant blue monkeys (*Cercopithecus mitis stuhlmanni*): Allomaternal interactions and sex differences. *Behavior* 2005;142:869–896.
88. Trivers RL. Parental investment and sexual selection. In: Campbell B, Ed. *Sexual Selection and the Descent of Man*. Chicago: Aldine, 1972:136–140.
89. Werren JH, Gross MR, Shine R. Paternity and the evolution of male parental care. *J Theor Biol* 1980;82:619–631.
90. Fairbanks LA. Parenting. In: Maestripieri D, Ed. *Primate Psychology*. Cambridge, MA: Harvard University Press, 2003:144–170.
91. Goldizen AW. Tamarins and marmosets: Communal care of offspring. In: Smuts BB, Cheney DL, Seyfarth RM, Wrangham RW, Struhsaker TT, Eds. *Primate Societies*. Chicago: University of Chicago Press, 1987:34–43.
92. Hrdy SB. The care and exploitation of nonhuman primate infants by conspecifics other than the mother. In: Rosenblatt J, Hinde R, Shaw E, Beer C, Eds. *Advances in the Study of Behavior*, Vol. 6. New York: Academic Press, 1976:101–158.
93. Nishida T. Alloparental behavior in wild chimpanzees of the Mahale Mountains of Tanzania. *Folia Primatol* 1983;41:1–33.
94. Williams L, Gibson S, McDaniel M, Bazzel J, Barnes S, Abee C. Allomaternal interactions in the Bolivian squirrel monkey (*Saimiri boliviensis boliviensis*). *Am J Primatol* 1994;34:145–156.
95. Thierry B, Anderson JR. Adoption in anthropoid primates. *Int J Primatol* 1986;7:191–216.
96. Izar P, Verderane MP, Visalberghi E, Ottoni EB, Gomes De Oliveira M, Shirley J, Fragaszy D. Cross-genus adoption of a marmoset (*Callithrix jacchus*) by wild capuchin monkeys (*Cebus libidinosus*): Case report. *Am J Primatol* 2006;68:692–700.
97. Knott CD, Kahlenberg SM. Orangutans in perspective: Forced copulations and female mating resistance. In: Campbell CJ, Fuentes A, MacKinnon KC, Panger M, Bearder SK, Eds. *Primates in Perspective*. New York: Oxford University Press, 2007:290–305.
98. Watts DP, Mitani JC, Sherrow HM. New cases of inter-community infanticide by male chimpanzees at Ngogo, Kibale National Park, Uganda. *Primates* 2002;43:263–270.
99. Sapolsky RM, Alberts SC, Altmann J. Hypercortisolism associated with social subordinance or social isolation among wild baboons. *Arch Gen Psychiatry* 1997;54:1137–1143.
100. Lambert JE. Primate nutritional ecology: Feeding biology and diet at ecological and evolutionary scales. In: Campbell CJ, Fuentes A, MacKinnon KC, Panger M, Bearder SK, Eds. *Primates in Perspective*. New York: Oxford University Press, 2007:482–495.
101. National Academy of Sciences. *Nutritional Requirements of Nonhuman Primates*, 2nd ed. Washington, DC: National Academy of Sciences, 2003.
102. Kopelman PG. Obesity as a medical problem. *Nature* 2000;**404:** 635–643.
103. Mattison JA, Lane MA, Roth GS, Ingram DK. Calorie restriction in rhesus monkeys. *Exp Gerontol* 2003;38:35–46.
104. Grove KL, Grayson BE, Glavas MM, Xiao XQ, Smith MS. Development of metabolic systems. *Physiol Behav* 2005;86:646–660.
105. Kaplan JR, Fontenot MB, Manuck SB, Muldoon MF. Influence of dietary lipids on agonistic and affiliative behavior in *Macaca fascicularis*. *Am J Primatol* 1996;38:333–347.
106. Tigno XT, Erwin JM, Hansen BC. Nonhuman primate models of human aging. In: Wolfe-Coote L, Ed. *The Laboratory Primate*. San Diego, CA: Elsevier Academic Press, 2005:449–466.
107. Wagner JD, Kavanagh K, Ward GM, Auerbach BJ, Harwood HJ Jr, Kaplan JR. Old World nonhuman primate models of type 2 diabetes mellitus. *ILAR J* 2006;47:259–271.
108. Gamlin PD, Ward MK, Bolding MS, Grossmann JK, Twieg DB. Developing functional magnetic resonance imaging techniques for alert macaque monkeys. *Methods* 2006;38:210–220.
109. Gil-da-Costa R, Braun A, Martin A. Using PET H2015 brain imaging to study the functional-anatomical correlates of non-human primate communication. *Methods* 2006;38:221–226.
110. Dadda M, Cantalupo C, Hopkins WD. Further evidence of an association between handedness and neuroanatomical asymmetries in the primary motor cortex of chimpanzees (*Pan troglodytes*). *Neuropsychologia* 2006;44:2583–2586.
111. Brevard ME, Meyer JS, Harder JA, Ferris CF. Imaging brain activity in conscious monkeys following oral MDMA ("ecstasy"). *Magn Reson Imaging* 2006;24:707–714.
112. Shepherd SV, Platt ML. Noninvasive telemetric gaze tracking in freely moving socially housed prosimian primates. *Methods* 2006;38:185–194.

30 Primate Models for Understanding Brain Mechanisms of Cognitive Behavior

Ronald E. Kettner, M. Lysetskiy, and M. Suh

ABSTRACT

An approach to understanding cognitive brain function is described for behaviors that are best studied in primates. The approach begins with the development of an experimental paradigm that exhibits important aspects of the cognitive behavior, can be learned by a monkey, and allows quantification and experimental control. Behavioral and neural responses during task performance are then recorded and analyzed. Finally, biologically realistic neural network models of specific brain regions are created that generate appropriate behavioral and neural outputs. This approach is ambitious and depends on the combined efforts of many laboratories, but has a high payoff in providing a mechanistic explanation of cognitive function. To illustrate this approach, we discuss work that seeks to understand the neural basis of predictive control during smooth eye pursuit of visible, as well as imagined, targets in monkeys. This is a cognitive process that facilitates eye tracking based on the prediction of upcoming target motions from memory. Neural recording takes place in the flocculus/paraflocculus regions of the cerebellum and the frontal eye fields of the frontal lobe, regions that are involved in predictive eye pursuit. Mathematical modeling of neural/behavioral processing is done at several levels. Single-neuron-firing models are used to quantify single-neuron responses during task performance in terms of sensory, motor, and cognitive variables. Local-circuit models describe how these neural firing patterns might be processed using local synaptic and intracellular mechanisms. Finally, a neural network model is presented that describes how predictive eye tracking could be generated in the cerebellum.

Key Words: Primate, Monkey, Model, Pursuit, Eye, Neuron, Behavior, Prediction, Cognition.

INTRODUCTION

A primary goal of neuroscience research is the understanding of how brain systems generate cognitive behaviors such as memory, perception, prediction, dance, music, and rational thought at the single-neuron level. In this chapter we describe a multilevel approach to understanding cognitive brain function using a combination of experimental and mathematical modeling techniques. The goal is a mechanistic description of the brain processes that create cognitive behavior. Mechanistic descriptions go beyond simple correlative descriptions that merely point to the involvement of particular brain areas in mental function. They make it possible to understand at a very fundamental level how specific brain areas might produce cognitive performance, how disruptions in this processing might produce mental dysfunction, and how processing can be changed to improve cognition. Mechanistic explanations of how biological systems might produce cognitive performance have also proven useful to computer scientists and engineers who want to create robots and intelligent computer systems with cognitive abilities.

To study a specific cognitive behavior, it is necessary first to develop an experimental paradigm that allows the control and quantification needed for scientific study while still retaining the most important aspects of the behavior. Experimental paradigms are also called animal models, behavioral models, or model systems (see other chapters in this book). In this chapter, which explores behavioral, neural recording, and mathematical approaches to understanding cognition, "model" will refer to mathematical models and "experimental paradigm" will refer to the animal preparation and associated experimental tasks used to study a specific behavior.

We have been interested in the cognitive abilities that underlie predictive behaviors. Real world examples include the quarterback who throws a football to the predicted future location of his receiver, the dancer who moves elegantly and in perfect synchrony with the beat of the music, and the Formula 1 driver who skillfully anticipates a turn in the road before it becomes visible. At a more mechanistic level, predictive motor control allows language production based on a rich variety of internal and external inputs that are not driven by auditory feedback control. Finally, predictive eye movements keep the eye on a target when it moves behind a tree or when it moves along an erratic but remembered trajectory.

Our experimental paradigm for the study of predictive control utilizes macaque monkeys as experimental subjects and experimental tasks that examine eye pursuit along predictable and nonpredictable visual target trajectories. We first train the monkeys to perform pursuit tasks and then characterize this behavior quantitatively using behavioral models. Once trained, single-neuron responses are recorded during task performance.

Our modeling approach strives for biological realism at each processing stage without abandoning the primary goal of developing models of cognitive brain function. We do not follow a strict "bottom-up" approach that requires a complete knowledge of

From: *Sourcebook of Models for Biomedical Research*
(P. M. Conn, ed.), © 2008 Humana Press Inc., Totowa, NJ.

low-level function at the cellular level before creating theories of higher-level function at the cognitive level. Nor do we adhere to a strict "top-down" philosophy that builds models of higher-level function with minimal anchoring in empirical data. Instead we use a combination of top-down and bottom-up approaches. Single-neuron and behavioral data provide measures of input, output, and internal signaling that guide the construction and evaluation of our modeling work. The architecture of the neural network model used to generate high-level performance is based on neuroanatomical data. Finally, and perhaps most importantly, the processing performed by the simulated neurons and local neural circuits within the model are biologically reasonable. For example, individual neurons can sum inputs from other neurons but cannot perform a Fourier or Laplace transform on input signals. This construction of higher-level performance from large numbers of units with simple input–output transformations is central to the spirit of the neural network modeling approach to understanding brain function.

THE USE OF MONKEYS TO STUDY COGNITIVE BRAIN PROCESSING

Like many other investigators of cognitive function, our empirical studies utilize macaque monkeys as experimental subjects because they are easily trained to perform a variety of interesting cognitive behaviors that cannot be studied in other animals. Primate investigations also tend to study important motor behaviors that can be studied only in primates. For example, studies of hand and finger movement require primate subjects. Our own work focuses on smooth pursuit eye movements, a class of behaviors unique to primates.

Monkeys also have several methodological advantages for experiments that correlate behavioral performance with single-neuron recording. Monkeys enjoy doing complex tasks for juice reward and once trained perform reliably for several hours while sitting comfortably in a primate chair that maintains their head in a fixed position. This quiet pattern of behavior coupled with neurons that are relatively large in size makes it possible to record from a single neuron for about 1 h while the monkey performs under a variety of test and control conditions.

Although there are obvious concerns about the expense of using monkey subjects, this cost is offset by the fact that a great deal of information can be obtained from each monkey. From 100 to 200 neurons are typically studied in a single monkey during a series of recording sessions lasting a month or more. This means that detailed analyses of a specific brain region generally require only two or three monkeys. This small number of subjects contrasts with other animal studies that use many more subjects. Perhaps most importantly for cognitive studies, the ability to collect large amounts of data from each monkey makes it possible to study cognitive tasks that require months of training. It would not be feasible to train an animal for 3 months and then obtain data during a single test day. Thus, for many types of experiment, primates are the species of choice in terms of cost and efficiency.

Repeated recording sessions across days are made possible by experimental procedures that have become standard in the field. During sterile surgery conducted under deep anesthesia, a recording chamber is surgically implanted over a skull opening that provides access to the brain region of interest. The exposed brain is protected from infection and damage between recording sessions by preserving the membrane that encloses and protects the brain, flooding the chamber with an antibiotic solution, and sealing the chamber with a cap. On the day of an experiment, the chamber cap is removed and a microdrive system is attached to the chamber. This allows accurate positioning of a microelectrode at a specified chamber location. The microelectrode is then slowly lowered into the brain until a stable, single neuron is encountered. At this time, the monkey is allowed to perform in one or more tasks while action potential events are recorded from the isolated neuron. This process is repeated for other neurons until the monkey indicates that he wants to go back to his cage. This generally occurs when the monkey has had his fill of juice.

BEHAVIORAL PARADIGMS FOR STUDYING PREDICTIVE EYE PURSUIT

In the laboratory setting, several experimental paradigms have been developed to study predictive eye movements.[1–3] Perhaps the simplest requires the subject to track a target that moves back and forth at a fixed sinusoidal amplitude and frequency. Subjects quickly lock onto the target so that the lag between eye and target is essentially zero. This result cannot be explained by direct visual control because the brain processes visual information with a delay of ~100 msec. That is, if direct visual control were the only resource available to the primate there would be a 100-msec delay between eye and target during sinusoidal pursuit, a delay that is not observed. To produce the performance that is actually observed, the brain must utilize cognitive processes that predict changes in target motion well before information about these changes can be processed by the visual system.

Our first behavioral studies[4,5] demonstrated that predictive control was more complex than previously thought: it could also be used to generate accurate eye pursuit along more complex sum-of-sines trajectories. Figure 30–1A illustrates this result using a sum-of-sines task that required tracking of a trajectory created by combining two sinusoidal motions with different frequencies along horizontal and vertical axes. Notice that repeated cycles of eye pursuit (thin lines) were relatively accurate, albeit with some variability about the target trajectory (thick lines). In particular, changes in eye direction were well timed with little overshoot following relatively abrupt changes in target motion. The use of predictive control was indicated by quantitative measurements that showed a delay between eye and target of essentially zero, far better than would have been possible using visual inputs processed with a 100-msec delay on this and a variety of other complex trajectories.

Our next studies[5–7] utilized half-circle trajectories that are less complex than sum-of-sines trajectories but more demanding in requiring right angle changes in the pursuit direction. The abrupt changes in target direction also facilitate the viewing and measurement of pursuit delays. For example, Figure 30–1B shows eye tracking performance during the random-half-circle task where the half-circle trajectory is presented randomly on 25% of cycles intermixed with circle trajectories on the remaining 75% of cycles. As expected, a consistent overshoot was observed on the cycles that contained the half-circle perturbation because its occurrence was unpredictable. The system had to rely on visual information about changes in target direction that was processed with a 100-msec delay. This idea was supported by quantitative analyses that indicated an average lag of 95 msec following the top right angle transition.

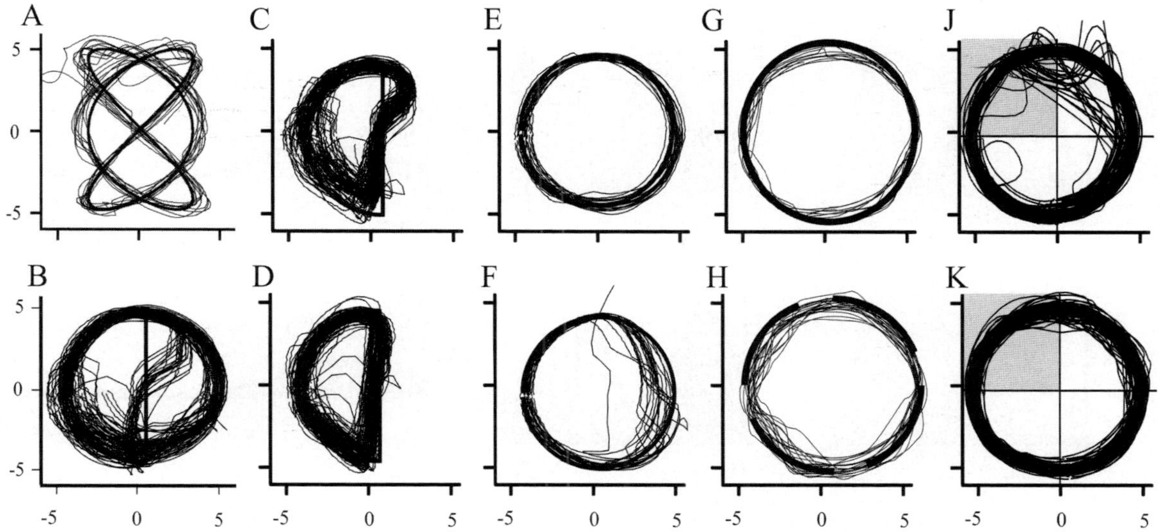

Figure 30–1. Behavioral results during (A) the sum-of-sines task, (B) the random-half-circle task, (C, D) the repeated-half-circle task before and after training, (E, F) the circle task before and after training in the repeated-half-circle task, (G, H) the circle task versus the circle-with-four-gaps task, and (J, K) the circle-with-random-gaps task before and after training. Thin lines represent multiple cycles of eye pursuit along a target trajectory indicated by thicker lines. Axes indicate eye position in degrees.

To see whether predictive control could be used to improve performance on this demanding half-circle trajectory, its occurrence was made highly predictable by presenting it on every cycle in the repeated-half-circle task.[6,7] As expected, overshoot errors were initially observed (Figure 30–1C). However, after several hundred repeated presentations a highly accurate pursuit profile developed (Figure 30–1D). This repeated presentation schedule allowed the monkey to utilize predictive control that appropriately decelerated his eye in anticipation of the top change in target direction. Deceleration occurred well before visual information about the change in target direction became available. Interestingly, the ability to use predictive control during the repeated-half-circle task develops rather slowly across hundreds of repeated cycles. This suggests that predictive control in this task results from permanent alterations in the brain pursuit systems. This idea was verified by the presence of aftereffect errors in comparisons of circle tracking before (Figure 30–1E) and after (Figure 30–1F) training. Notice that after training the eye consistently decelerated at the top to the circle in anticipation of the expected right angle change in direction. These aftereffect errors continued for hundreds of cycles.

Other behavioral paradigms developed for the study of nonvisual predictive control turn the target off briefly during ongoing pursuit.[8–13] In these "gap" tasks the subject is required to maintain pursuit when the target goes off. It is argued that any pursuit during target gaps cannot result from direct visual drive, but must instead result from predictive signals that maintain target motion in the absence of a visual target. In short 100-msec target gaps during linear pursuit, the eye continues to move with little reduction in ongoing velocity.[8] Here the level of required predictive control is relatively simple: the maintenance of ongoing eye speed and direction. We extended this result using a circle-with-four-gaps task that introduces four gaps (of either 100 or 200 msec) at constant positions along a repeated circle trajectory.[14] The monkey subjects showed good pursuit continuation during circle trajectories both with (Figure 30–1G) and without (Figure 30–1H) target gaps under conditions that required the prediction of a target that was continually changing direction.

When the duration of a target gap is increased beyond 100 msec during linear target pursuit declines in pursuit velocity are observed at about 200 msec, although some residual eye velocity is maintained.[9] This result is based on task conditions designed to make the time and duration of gap events unpredictable. We have extended these studies using a circle-with-random-gaps task that turned the target off on random cycles of circle pursuit but always during the same 312-msec period in the upper left quadrant of the circle trajectory.[15,16] Our goal was to study pursuit during longer gap periods for a trajectory that constantly changed direction and thereby put higher demands on a predictive controller. In these experiments, the monkeys initially did quite badly (Figure 30–1J) showing severe declines in pursuit ~100 msec after the target light was switched off. When the target reappeared, a large catch-up saccade was then generated after ~200 msec that pulled the eye back on target. Both results are what would be expected based on past experiments. A new finding (Figure 30–1K) was that performance during these longer gaps improved markedly with extended training, indicating the presence of predictive pursuit during these longer gap periods.

NEURAL PARADIGMS FOR STUDYING PREDICTIVE EYE PURSUIT

Studies of smooth eye pursuit have focused on three brain regions: the flocculus and ventral paraflocculus of the cerebellum, the frontal eye field of the frontal neocortex, and the superior colliculus. Their involvement in smooth eye pursuit has been established by a number of recording, stimulation, and lesion studies.[17–34] Other brain sites are also likely to be involved and deserve further study including the middle-temporal, middle-superior-temporal, lateral-intraparietal, and supplementary-eye-field regions of the neocortex.[1,8,17] The contribution of brain systems closer to motor output or sensory input are less likely involved in higher-level predictive processing, but their

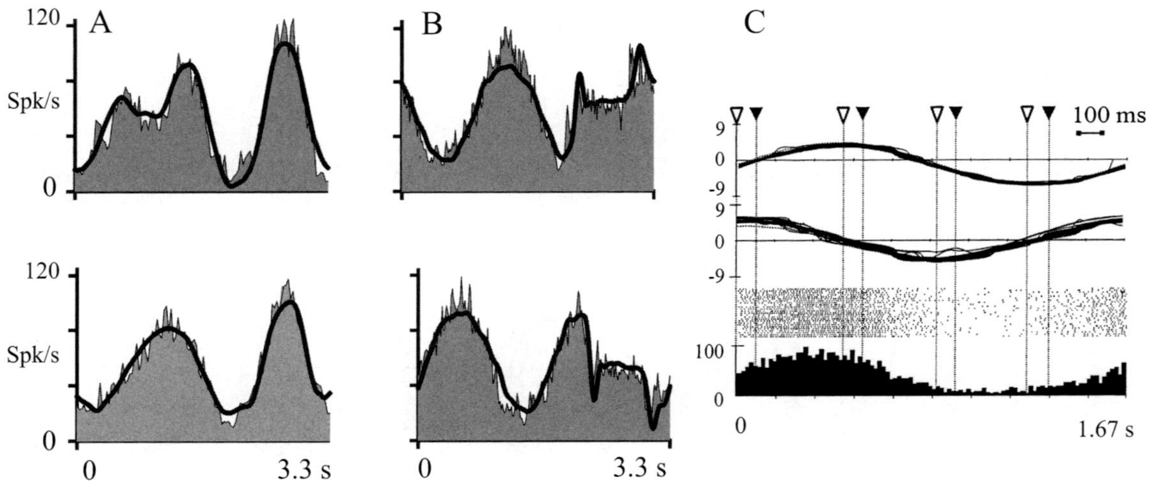

Figure 30–2. Cerebellar neural responses during (A) the sum-of-sines task, (B) the random-half-circle task, and (C) the circle-with-four-gaps task. In (A) and (B), the gray histogram represents averaged neural responses and the dark line indicates fits of the eye-motion model to the data. (C) Horizontal and vertical eye positions are shown at the top and neural raster and average neural response data at the bottom. Vertical lines indicate the start and stop of the 100-msec gap periods.

contribution needs to be studied to understand the complete system.

One approach that we have used to understand the brain mechanisms of predictive pursuit is to record from single neurons while a monkey performs behavioral tasks designed to contrast predictive versus nonpredictive control. For example, in studies of the cerebellum[14] we recorded activity from single neurons during performance in both the sum-of-sines task (see Figure 30–2A) and in the random-half-circle tasks (see Figure 30–2B) to contrast responses related to predictive and nonpredictive pursuit, respectively. Our analyses produced two important results. First, these cerebellar regions are most directly involved in predictive control and only secondarily involved in nonpredictive control. Second, predictive control is derived from nonvisual signals.

Evidence for predictive control in the cerebellar flocculus and ventral paraflocculus was based on a timing analysis that compared changes in neural firing to changes in eye movement. Put briefly, we found that average neural responses led eye motion by 3 msec during predictive pursuit in the sum-of-sines task, but lagged eye motion by 7 msec during nonpredictive pursuit in the random-half-circle task. Because activity in brain areas responsible for generating pursuit must lead eye motion while those playing a follow-up role will show response lags, we concluded that the cerebellar flocculus and ventral paraflocculus were more strongly involved in generating predictive pursuit than in generating nonpredictive pursuit.

Our second finding was that predictive processing was based on nonvisual input. During performance in the circle-with-four-gaps task designed specifically to test for the presence of visual input, no modulation in firing rate was observed when the target was turned on and off (see Figure 30–2C). Any system involved in direct visual control should have shown some response to these dramatic on–off changes in visual input. An absence of visual drive was also shown using the modeling analyses described in the next section. These analyses showed strong correlations with eye movement and negligible influence from retinal-slip inputs during the sum-of-sines, random-half-circle, and circle-with-four-gaps tasks. This result is particularly compelling for data obtained from the random-half-circle task that generated reliable periods of visual input when the target image on the retina was moved away from the fovea following unexpected right angle changes in target direction.

MATHEMATICAL MODELING OF SINGLE-NEURON FIRING PATTERNS

Our modeling work begins with the quantitative modeling of single-neuron responses. This is important for our later models that must be supplied with appropriate inputs. A standard approach to modeling neural responses with multilinear equations is used.[25,27,35,36] We model two types of neural response: the first related to eye motion input and the second related to visual input variables.

The eye-motion firing-rate model defines the neural firing rate due to eye motion, $R_M(t)$, by the following equation:

$$R_M(t) = b_M + \mathbf{e}_P \cdot \mathbf{E}_P(t) + \mathbf{e}_V \cdot \mathbf{E}_V(t) \qquad (30\text{–}1)$$

It states that at a given time, t, the firing rate of a neuron can be determined by adding a baseline firing rate, b_M, to weighted influences due to eye position, $\mathbf{E}_P(t)$, and eye velocity, $\mathbf{E}_V(t)$. The weights are constants that define sensitivities to eye position, $\mathbf{e}_P$, and eye velocity, $\mathbf{e}_V$. In all instances, bold letters indicate vectors and their multiplication is by the dot product. In this vector formulation the length and direction of the sensitivity vector define the magnitude and direction of the variable it specifies. This equation is often reexpressed in terms of x and y components. For example, $\mathbf{e}_P \cdot \mathbf{E}_P(t) = e_{PX} E_{PX} + e_{PY} E_{PY}$ based on the dot product operation on $\mathbf{e}_P = (e_{PX}, e_{PY})$ and $\mathbf{E}_P = (E_{PX}, E_{PY})$.

The dark lines in Figure 30–2A and B show the excellent fits provided by the eye-motion model for cerebellar responses during performance in the sum-of-sines and random-half-circle tasks. Further support is provided in Figure 30–3A and B by model descriptions of frontal-eye-field neurons during a circles task that alternates six cycles of CW (clockwise) pursuit with six cycles of CCW (counterclockwise) pursuit.[34] Interestingly, the neuron shown in Figure 30–3B fired more strongly during CCW cycles, a result obtained when position and velocity sensitivity vectors

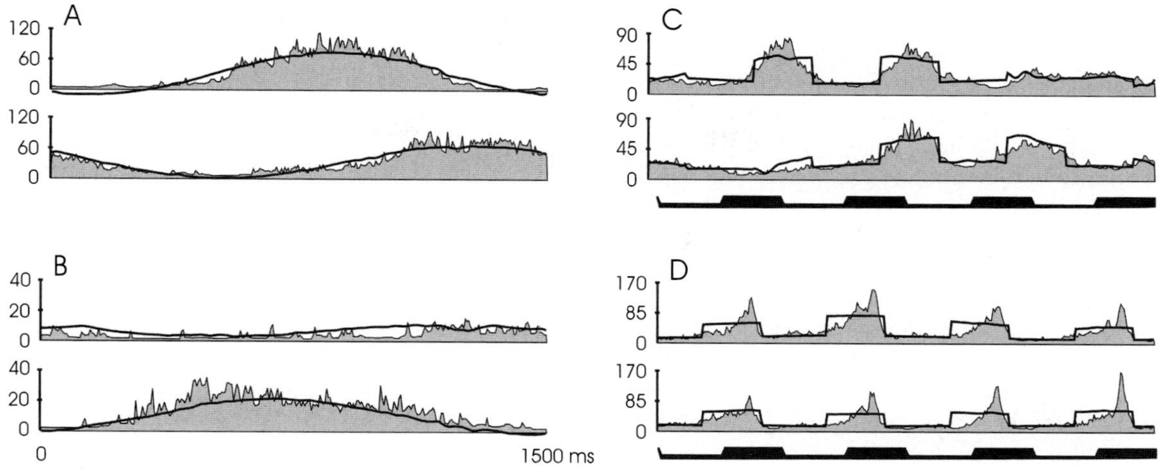

Figure 30–3. Comparisons between model fits (thick lines) and average firing rate (thin lines) for four single neurons recorded from the frontal eye field. (A, B) Responses during the circle task. (C, D) Responses during the circle-with-four-gaps task. Target on times are indicated by bars. In each panel, the top and bottom plots show CW and CCW pursuit along a circle of radius 5°, a period of 1500 msec, and a tangential velocity of 20.9°/sec.

point in different directions. Rotation-specific neurons were also observed in our cerebellar studies.[27]

The visual-input firing-rate model defines the neural firing rate due to visual input, $R_V(t)$, by the following equation:

$$R_V(t) = L(t)\,[b_V + R_S(t) + R_G(t)] \qquad (30\text{–}2)$$

Because visual input is defined only when the target light is on, its contribution to firing rate is modulated by the variable, $L(t)$, which depends on the intensity of the target light. Here, it is defined to be one when the light is on and zero otherwise. Thus, the baseline value, b_V, indicates the average increase in firing rate due to the presence to the target light. Variations in neural firing rate result from two factors.

The first factor, $R_S(t)$, defines changes in visual input that result from movement of the target image on the retina:

$$R_S(t) = \mathbf{s}_P \cdot \mathbf{S}_P(t) + \mathbf{s}_V \cdot \mathbf{S}_V(t) \qquad (30\text{–}3)$$

Although slip position, $\mathbf{S}_P(t)$, cannot be measured directly, it can be computed by subtracting eye position from target position, $\mathbf{T}_P(t)$, using the equation $\mathbf{S}_P(t) = \mathbf{E}_p(t) - \mathbf{T}_P(t)$. Slip velocity, $\mathbf{S}_V(t)$, is defined similarly.

A second factor, $R_G(t)$, defines gain-field changes in visual input due to eye motion:

$$R_G(t) = \mathbf{g}_P \cdot \mathbf{E}_P(t) + \mathbf{g}_V \cdot \mathbf{E}_V(t) \qquad (30\text{–}4)$$

Several investigators have reported visual responses that vary with the position of the eye and have described these results in terms of gain fields. In our work, we have found it useful to extend this idea to include an additional dependence on eye velocity.

The addition of the eye-motion and visual-input firing-rate models produces the total firing-rate model that defines the total firing rate, $R_{Total}(t)$, using the following equation:

$$R_{Total}(t) = R_M(t + \Delta_M) + R_V(t + \Delta_V) \qquad (30\text{–}5)$$

Here we have introduced two delay constants: a movement delay, Δ_M, and a visual delay, Δ_V. Estimates of the movement delay allow us to determine whether neural activity leads or lags eye motion. As noted in the previous section, leads and lags are important criteria for determining whether a brain area is creating a behavior (response lead) or simply monitoring a behavior after it has been created elsewhere in the brain (response lag). The visual delay is needed to compensate for observed delays in visual processing of ~100 msec during visual pursuit tasks. Both delay terms can be used to identify cognitive events that precede motor and visual signals and therefore cannot be explained by these signals.

The total firing-rate model makes it possible to quantify the relative contributions of movement and visual variables in terms of both magnitude and preferred direction. Figure 30–3C and D shows fits of the total firing-rate model for two frontal-eye-field neurons during the circle-with-four-gaps task.[34] One neuron (Figure 30–3C) shows responses related to a combination of eye-motion and visual-input variables as indicated by modulation during both target-on and target-off periods. Responses from another neuron (Figure 30–3D) show a stronger reliance on visual input with minimal variation in firing when the target was off. Interestingly, the responses related to visual input began before the target turned on. Thus, they reflect cognitive processes that anticipate the appearance of the target in the absence of direct visual drive and that could be useful in predictive control.

LOCAL-CIRCUIT MODELS

Sometimes information key to the creation of a systems-level model is not available. For example, it has proven difficult to record from cerebellar granule cells in performing monkeys because of their small size. Neuroscientists have also made slow progress in studying the synaptic mechanisms of learning because intracellular recordings are difficult to maintain during the course of learning in behaving animals. Because of these difficulties, modelers sometimes make hypotheses about the nature of missing information. This is a useful approach that produces testable model predictions that can be evaluated experimentally.

A more direct approach is to use local-circuit models to generate the needed information from data that are available. For example, we use local-circuit models to generate granule cell responses that are difficult to study in behaving monkeys using the mossy-fiber data that are available. Another local-circuit model is used to describe changes in the efficacy of parallel-fiber-to-Purkinje-cell synapses during learning. Local-circuit models are based on cellular-level information that describes the anatomical, electrical, and chemical characteristics of a small group of interconnected neurons. Their implementation is facilitated by computer programs (e.g., NEURON, GENESIS) that provide a standard framework to which the cellular modeler can add electrophysiological and neurochemical response data specific to the local circuit being modeled. The primary use of this type of modeling has been to explain intracellular responses to simple electrical or chemical inputs. However, there is increasing interest in using cellular models to study responses to more realistic input streams.

Thus, local-circuit models allow systems-level models to benefit from the very large base of information available from "reduced preparations." For instance, the "slice" preparation has revolutionized neuroscience by providing the access and stability required for intracellular recording using patch-clamp, sharp-electrode, and imaging techniques, as well as the controlled study of responses elicited by intracellular current injection, extracellular stimulation of axonal and dendritic inputs, and the application of chemical agents. Of course, slice experiments necessarily eliminate much of the brain circuitry and input at play in intact organisms. Thus, they have been most useful for the analysis of specific channels and/or receptor subtypes. However, the use of this information to simulate natural conditions using local-circuit models provides an important bridge between these highly specialized cellular experiments and systems-level analyses. From a philosophical point of view, the use of local-circuit models to facilitate systems-level modeling represents an elegant combination of top-down and bottom-up approaches to understanding and modeling brain function.

CEREBELLAR LOCAL-CIRCUIT MODELS

To illustrate the use of local-circuit models we first describe a model of the cerebellar glomerulus that has been developed by Dr. Lysetskiy and its use in creating a local-circuit model that implements onset detection in our network model. His analyses indicate that local processing within the glomerulus produces granule cell activation only during the initial few milliseconds of a sustained mossy-fiber input. This onset-response transformation greatly reduces the fraction of active granule cells at any given time relative to the fraction of active mossy-fiber inputs. At the systems level, the result is a sparser coding of eye motion information within the granule cell population as compared with the highly overlapping coding of eye motion information observed experimentally for the mossy-fiber population. The importance of this sparse recoding will be discussed in the context of our neural network model of the cerebellum.

The glomerular model is shown schematically in Figure 30–4. It models synaptic interactions among the triad of neural processes within each cerebellar glomerulus: mossy-fiber input, Golgi-dendrite input, and granule-dendrite output. The model was developed in the NEURON modeling environment by constructing the glomerular triad and inserting into each neural process the large number of membrane-channel and receptor properties that have been described in the cellular literature (e.g., D'Angelo *et al.*[37]). A primary finding was the production of a phasic burst of granule cell action potentials (upper-left graph) in response to a

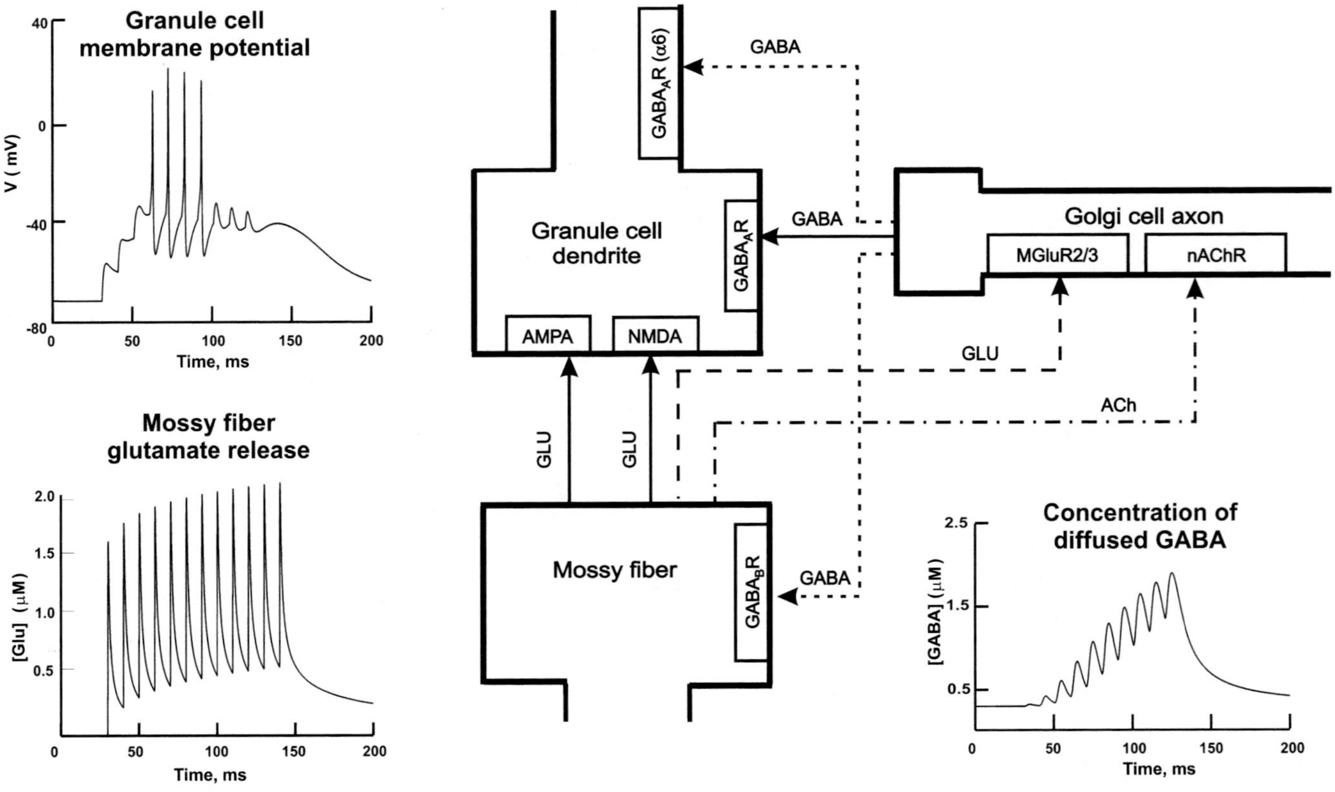

Figure 30–4. Local-circuit model of the cerebellar glomerulus. See the text for additional details.

sustained pattern of mossy-fiber glutamate release (lower-left graph). The on phase of the response was generated by the expected direct activation of sodium channels via activation of glutamate receptors. The off phase was more surprising. It resulted from several factors that acted in combination including sustained inactivation of spike-generating sodium currents, membrane hyperpolarization induced by slow (I_M like) potassium currents, and increased inhibition due to accumulated GABA (lower-right graph). The idea that granule cells might produce phasic responses also has initial support from empirical studies that show burst responses under some conditions.[37]

Another important local-circuit model implements a coincidence-detection mechanism that further refines the mossy-fiber input signal at the granule cell soma. Put simply, more than one dendritic input to a granule cell must be active simultaneously before an action-potential output is generated. This model is motivated by slice experiments[38] that indicate that granule cells show no spontaneous firing and require more than one input pulse in close temporal proximity before an action potential is generated. Because glomerular processing causes individual dendrites to generate a single onset burst in response to mossy-fiber input, it is likely that more than one dendrite must be active at any given time before an action potential is generated at the granule cell soma. This coincidence detection is implemented by a local-circuit model that limits the maximum contribution of each dendritic input to the membrane potential at the granule cell soma so that more than one input must be active before action-potential generation can occur. At the system level, this coincidence detection mechanism further enhances sparse coding and the computational capacity of our neural network model.

A final local-circuit model specifies the synaptic learning rule that we use to modify the influence of parallel-fiber axons on Purkinje cell dendrites. Granule cell axons ascend and then bifurcate to form parallel fibers that innervate Purkinje cell dendrites. Plasticity at the parallel-fiber-to-Purkinje-dendrite synapse has been studied extensively in slice preparations.[39–43] Initial studies focused on conditions that cause a long-term depression (LTD) of synaptic strength, but more recent studies have established additional conditions that produce long-term potentiation (LTP) of the same synapse. This dual mechanism is important for the training of our neural network. Without the ability to drive synaptic strength both up and down, a synapse that becomes too strong cannot be pulled back to an appropriate level. A simple learning rule based on the combined occurrence of recent parallel-fiber activation and climbing-fiber input allows our cerebellar network model to learn.

BIOLOGICALLY REALISTIC NEURAL NETWORK MODELS

Several central concepts distinguish biologically realistic neural network models from other brain models. First, neural networks contain a large number of computational units. The original idea was to mimic the large number of neurons observed experimentally in brain systems. For example, our cerebellar models utilize from 10,000 to 40,000 computational units depending on the pursuit accuracy that is required. Over time, network research has indicated the advantages associated with processors with large numbers of units computing in parallel: they are faster than serial processors and are better able to handle the large number of sensory and feedback signals received by real brain systems. Thus, evolution has increased the size of the primate brain for important reasons.

A second guiding idea used in the construction of neural network models is that each unit performs a relatively simple computation. Again, the original idea was to mimic biological neurons that were then assumed to perform relatively simple computations. This computation has most frequently been the weighted summation of inputs coupled with a "squashing function" that smoothly limits output levels. Later, slightly more complex operations involving multiplication and spiking events have also been used. The simplicity of the unit computations in network models is in sharp contrast to the complexity of unit computations in models containing only a handful of interconnected units. The units in these models are sometimes labeled "black boxes" to indicate the complexity of the operations they perform and a lack of interest in understanding the mechanisms that create this complexity. For those interested in brain mechanisms, these black box models are unappealing. However, they can prove useful for dissecting a complex process into subprocesses.

A third guiding principal is that the computational unit is often the synapse. This principle acknowledges the fact that simple operations beyond the simple multiplication of input by a synaptic weight can sometimes occur. For example, the glomerular transformation at the mossy-to-granule synapse results in the conversion of a sustained input into a phasic output. This principle implies that relatively complex operations are performed by neurons that receive thousands of synaptic inputs and that less complex operations are performed by neurons with only a few synaptic contacts. In the cerebellar model described below, two Purkinje units perform predictive control by driving horizontal and vertical eye muscles respectively. These complex operations arise from the convergence on each Purkinje cell of more than 5000 synaptic inputs from parallel fibers. In contrast, each granule cell performs a relatively simple coincidence detection operation based on input from four dendrites.

A final defining concept is that a biologically realistic neural network model should be constrained by neural data. This is not the same as saying that such a model needs to be a simple, bottom-up reconstruction of a brain system. This is a very tedious exercise that produces a result that is already known: that the complete system can generate responses observed experimentally. Rather, the goal of the modeler is to help us understand how a brain system works by creating a simpler version of the system that retains biologically realistic connectivity and processing. One simplifying idea is that brain systems consist of thousands of nearly identical modules each consisting of a few cells. It is then possible to understand the larger system by understanding processing within a module. Another useful approach is the deletion of neural elements that are not needed for the generation of a particular function. Here it is presumed that these discarded elements are used only to refine the processing of the simple model. It is also possible that excluded elements play their role in other functions performed by the network.

Of course there are differences in opinion regarding how schematic a model can become before it stops being useful to biologists. For this reason there has been a split between those interested in creating biologically realistic neural network models and others interested in computational networks without the constraints required for realistic biological modeling. The result is a healthy and synergistic interaction between two perspectives that facilitates a common goal of understanding complex, highly parallel, computational systems in multiple contexts. That said, we place ourselves in the group interested in using neural networks to understand brain function.

A NEURAL NETWORK MODEL OF CEREBELLAR PREDICTIVE CONTROL

A neural network model will now be described that puts these pieces together to provide a mechanistic explanation of how the cerebellum could generate predictive smooth eye pursuit. It benefits from a variety of other approaches to cerebellar modeling.[44–48] A schematic diagram of the model's architecture is shown in Figure 30–5. It follows well-established anatomical connections between the primary neural elements in the model, although the precise anatomical positioning of these neural elements has been altered. For example, the exit direction for Purkinje cell axons has been reversed to allow a more orderly left-to-right presentation of information flow through the model. The number of neural elements is also underrepresented: the model contains 240 mossy fibers and 7200 granule cells that provide input to two Purkinje cells and one Golgi cell.

Mossy fibers entering the cerebellum at the left of the diagram provide the primary input to the model. Their signals are then processed within cerebellar glomeruli before activating multiple granule cells. Each granule cell, in turn, receives input from multiple glomeruli. This produces an expansive recoding of the mossy-fiber population into a much larger granule cell population with a more diverse pattern of activation. Granule cell axons then travel to the molecular layer where they bifurcate at right angles to form the parallel fibers. The Golgi cell also receives input from parallel fibers and in turn influences granule cell responses within the glomeruli.

Purkinje cells provide the sole output for the cerebellar model. They receive their primary input from parallel fibers and climbing fibers that arise from the inferior olivary nucleus. The climbing fibers provide a training signal[24,49,50] that is used to modify the strength of parallel-to-Purkinje synapses using the biologically based learning rule described above. The Purkinje cells provide predictive output that activates a well-established system of brainstem nuclei that further process the signal before activating eye muscles. A rudimentary model of saccadic generation and a visual-pursuit model based on existing pursuit models[51–53] have been included to complete the model environment. They are needed in the initial training of the cerebellar network and to provide contrasts between what is expected from visual control alone and what can be achieved with the addition of predictive signals.

Figure 30–6A and C shows the excellent performance of the model during sum-of-sines and circle pursuit. Figure 30–6B and D shows the much poorer levels of performance observed without predictive control using only delayed visual control. The jagged appearance of the pursuit results from a saccade generator that pulls the eye back on target at intermittent times that never exceed the saccade refractory time of ~200 msec. Between saccades, the eye continues the pursuit, but in the wrong direction, based on information about the direction of target motion that is delayed by 100 msec. Figure 30–6E shows how the sparse activation of the granule cell population provides a good representation of eye position at the top of the circle trajectory. Figure 30–6F shows the much broader pattern of activation across the granule cell population that would have been observed without the local-circuit refinements that produce sparse recoding. Without this sparse representation the model is unable to learn predictive tracking.

Put simply, the model creates predictive pursuit at a given point along a predictable trajectory by using the currently active subpopulation of granule cells to drive Purkinje cells in a fashion that generates correct eye motion at that position. This drive varies as first one subpopulation of granule cells and then another subpopulation is activated along the trajectory so that a changing drive signal is created across the entire trajectory. This process requires the sparse coding of trajectory location by the granule cell population. Otherwise it would be impossible to assign the unique drive pattern required at each point along the trajectory. Put another way, the network model creates predictive eye pursuit by using the active granule cell subpopulation to retrieve

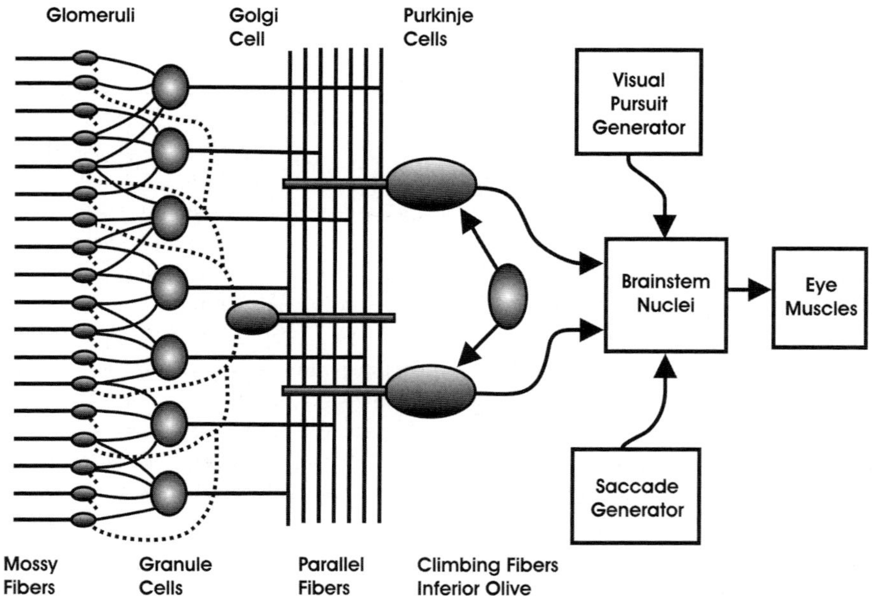

Figure 30–5. Architecture of the neural network model of the cerebellum described. Although Golgi cell axons innervate each glomerulus, these connections are indicated in complete detail for only a subset of glomeruli. See the text for additional details.

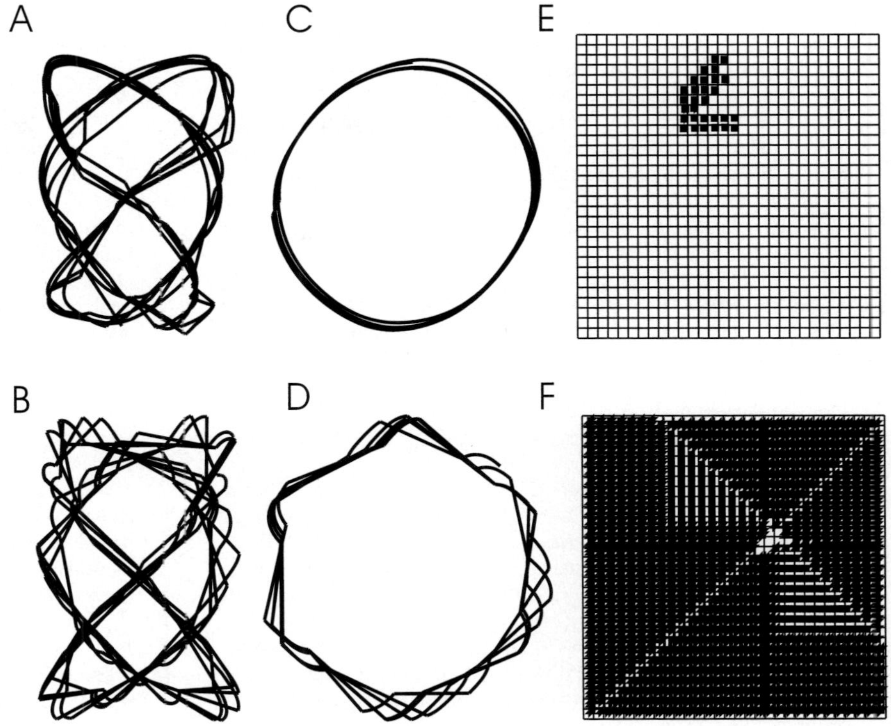

Figure 30–6. Neural network processing. (A, B) Eye movements during sum-of-sines pursuit with and without predictive control. (C, D) Eye movements during circular pursuit with and without predictive control. (E, F) Activation of the granule cell population of 7200 units with and without sparse coding. Each of the 900 boxes corresponds to a specific eye position and contains eight granule cells with different preferred directions. In the top panel, cells in black indicate the granule cell subpopulation active when the eye is at the top of the circle trajectory. The bottom panel shows that most of the population is active when the glomerular onset-detection mechanism is turned off at the same top eye position.

remembered control information stored by granule-to-Purkinje synapses. Importantly, the model does not rely on visual input because the mossy-fiber population is activated exclusively by signals related to eye motion.

CONCLUSIONS

This chapter describes how monkeys are used to study the neural mechanisms of cognitive processing. The approach is multipronged using information from behavioral studies that test the limits of cognitive performance in primates, neural recording studies that examine brain responses in performing monkeys, and modeling studies that attempt to pull this information together in a biologically realistic fashion. Examples of this approach are presented based on our own work that seeks to understand the cognitive processing that results in predictive control. We first present evidence that predictive processing is used by monkey subjects during eye pursuit in a variety of experimental tasks. Single-neuron recording experiments indicate that the cerebellum and frontal eye fields show responses correlated with predictive control. Finally, a biologically realistic neural network model is presented that generates predictive control and explains the experimental data.

ACKNOWLEDGMENTS

We wish to thank the NIH (R01EY12984, R01EY014606) and the NSF (IBN-9723846) for financial support.

REFERENCES

1. Leigh RJ, Zee DS. *The Neurology of Eye Movements*, 3rd ed. New York: Oxford University Press, 1999.
2. Kowler E, Steinman RM. The effect of expectations on slow oculomotor control. I. Periodic target steps. *Vision Res* 1979;19:619–632.
3. Barnes GR, Donnelly SF, Eason RD. Predictive velocity estimation in the pursuit reflex response to pseudo-random and step displacement stimuli in man. *J Physiol* 1987;389:111–136.
4. Kettner RE, Leung H-C, Peterson BW. Predictive smooth pursuit of complex two-dimensional trajectories in monkey: Component interactions. *Exp Brain Res* 1996;108:221–235.
5. Leung H-C, Kettner RE. Predictive smooth pursuit of complex two-dimensional trajectories demonstrated by perturbation responses in monkeys. *Vision Res* 1997;37:1347–1354.
6. Kettner RE, Suh M, Davis D, Leung H-C. Modeling cerebellar flocculus and paraflocculus involvement in complex predictive smooth eye pursuit in monkeys. *Ann NY Acad Sci* 2002;978:455–467.
7. Kettner RE, Suh M, Davis D, Leung H-C. Cerebellar control of complex predictive pursuit: A model system for cerebellar studies of skilled movement. *Arch Ital Biol* 2002;140:331–340.
8. Newsome WT, Wurtz RH, Komatsu H. Relation of cortical areas MT and MST to pursuit eye movements. II. Differentiation of retinal from extraretinal inputs. *J Neurophysiol* 1988;60:604–620.
9. Becker W, Fuchs AF. Prediction in the oculomotor system: Smooth pursuit during transient disappearance of a visual target. *Exp Brain Res* 1985;57:562–575.
10. Barnes GR, Asselman PT. The mechanisms of prediction in human smooth pursuit eye movements. *J Physiol* 1991;439:439–461.
11. Barnes GR, Asselman PT. Pursuit of intermittently illuminated moving targets in the human. *J Physiol* 1992;445:617–637.
12. Churchland MM, Chou IH, Lisberger SG. Evidence for object permanence in the smooth-pursuit eye movements of monkeys. *J Neurophysiol* 2003;90:2205–2218.
13. Madelain L, Krauzlis RJ. Effects of learning on smooth pursuit during transient disappearance of a visual target. *J Neurophysiol* 2003;90:972–982.
14. Suh M, Leung H-C, Kettner RE. Cerebellar flocculus and ventral paraflocculus Purkinje cell activity during predictive and visually driven pursuit in monkey. *J Neurophysiol* 2000;84:1835–1850.
15. Suh M, Davis D, Kettner RE. Smooth eye movements improve with training during random target gaps in monkey. *Soc Neurosci Abstr* 2000;26:1716.
16. Kettner RE, Suh M, Grimes K. Learned predictive saccades in the absence of a visible target allow improved performance during tracking of a circular motion with target gaps in monkey. *Soc Neurosci Abstr* 2003;441.10.
17. Krauzlis RJ. Recasting the smooth pursuit eye movement system. *J Neurophysiol* 2004;91:591–603.

18. Lisberger SG, Fuchs AF. Role of primate flocculus during rapid behavioral modification of vestibuloocular reflex. I. Purkinje cell activity during visually guided horizontal smooth-pursuit eye movements and passive head rotation. *J Neurophysiol* 1978;41:733–763.
19. Lisberger SG, Fuchs AF. Role of primate flocculus during rapid behavioral modification of vestibuloocular reflex. II. Mossy fiber firing patterns during horizontal head rotation and eye movement. *J Neurophysiol* 1978;41:764–777.
20. Noda HD, Suzuki A. The role of the flocculus of the monkey in fixation and smooth pursuit eye movements. *J Physiol (Lond)* 1979;294:335–348.
21. Miles FA, Fuller JH, Braitman DJ, Dow BM. Long-term adaptive changes in primate vestibuloocular reflex. III. Electrophysiological observations in flocculus of normal monkeys. *J Neurophysiol* 1980;43:1437–1476.
22. Zee DS, Atsumi Y, Butler PH, Güçer G. Effects of ablation of flocculus and paraflocculus on eye movements in primate. *J Neurophysiol* 1987;46:878–899.
23. Stone LS, Lisberger SG. Visual responses of Purkinje cells in the cerebellar flocculus during smooth pursuit eye movements in monkeys. I. Simple spikes. *J Neurophysiol* 1990;63:1241–1261.
24. Stone LS, Lisberger SG. Visual responses of Purkinje cells in the cerebellar flocculus during smooth pursuit eye movements in monkeys. II. Complex spikes. *J Neurophysiol* 1990;63:1262–1275.
25. Gomi H, Shidara M, Takemura A, Inoue Y, Kawano K, Kawato M. Temporal firing patterns of Purkinje cells in the cerebellar ventral paraflocculus during ocular following responses in monkeys. I. Simple spikes. *J Neurophysiol* 1998;80:818–831.
26. Kobayashi Y, Kawano K, Takemura A, Inoue Y, Kitama T, Gomi H, Kawato M. Temporal firing patterns of Purkinje cells in the cerebellar ventral paraflocculus during ocular following responses in monkeys. II. Complex spikes. *J Neurophysiol* 1998;80:832–848.
27. Leung HC, Suh M, Kettner RE. Cerebellar flocculus and paraflocculus Purkinje cell activity during circular pursuit in monkey. *J Neurophysiol* 2000;83:13–30.
28. MacAvoy MG, Gottlieb JP, Bruce CJ. Smooth-pursuit eye movement representation in the primate frontal eye field. *Cereb Cortex* 1991; 1:95–102.
29. Gottlieb JP, MacAvoy MG, Bruce CJ. Neural responses related to smooth-pursuit eye movements and their correspondence with electrically elicited smooth eye movements in the primate frontal eye field. *J Neurophysiol* 1994;72:1634–1653.
30. Shi D, Friedman HR, Bruce CJ. Deficits in smooth-pursuit eye movements after muscimol inactivation within the primate's frontal eye field. *J Neurophysiol* 1998;80:458–464.
31. Keating EG, Pierre A, Chopra S. Ablation of the pursuit area in the frontal cortex of the primate degrades foveal but not optokinetic smooth eye movements. *J Neurophysiol* 1996;76:637–641.
32. Tanaka M, Lisberger SG. Regulation of the gain of visually guided smooth-pursuit eye movements by frontal cortex. *Nature* 2001; 409:191–194.
33. Tanaka M, Lisberger SG. Role of arcuate frontal cortex of monkeys in smooth pursuit eye movements. I. Basic response properties to retinal image motion and position. *J Neurophysiol* 2002;87:2684–2699.
34. Lysetskiy M, Lugtu KR, Segraves MA, Kettner RE. Frontal eye field responses during circle and circle-with-gap pursuit produce both simple and complex activation patterns in monkey. *Soc Neurosci Abstr* 2006;138.5.
35. Shidara M, Kawano K, Gomi H, Kawato M. Inverse-dynamics model eye movement control by Purkinje cells in the cerebellum. *Nature* 1993;365:50–52.
36. Hirata Y, Highstein SM. Acute adaptation of the vestibuloocular reflex: Signal processing by floccular and ventral parafloccular Purkinje cells. *J Neurophysiol* 2001;85:2267–2288.
37. D'Angelo E, Nieus T, Maffei A, Armano S, Rossi P, Taglietti V, Fontana A, Naldi G. Theta-frequency bursting and resonance in cerebellar granule cells: Experimental evidence and modeling of a slow k+-dependent mechanism. *J Neurosci* 2001;21:759–770.
38. Chadderton P, Margrie TW, Hausser M. Integration of quanta in cerebellar granule cells during sensory processing. *Nature* 2004;428: 856–860.
39. Linden DJ, Connor JA. Long-term synaptic depression. *Annu Rev Neurosci* 1995;18:319–357.
40. Hartell NA. Parallel fiber plasticity. *Cerebellum* 2002;1:3–18.
41. Ito M. Cerebellar long-term depression: Characterization, signal transduction, and functional roles. *Physiol Rev* 2001;81:1143–1195.
42. Chen C, Thompson RF. Temporal specificity of long-term depression in parallel fiber-Purkinje synapses in rat cerebellar slice. *Learn Mem* 1995;2:85–98.
43. Schreurs BG, Oh MM, Alkon DL. Pairing-specific long-term depression of Purkinje cell excitatory postsynaptic potentials results from a classical conditioning procedure in the rabbit cerebellar slice. *J Neurophysiol* 1996;75:1051–1060.
44. Marr D. A theory of cerebellar cortex. *J Physiol* 1969;202:437–470.
45. Albus JS. A theory of cerebellar function. *Math Biosci* 1971;10:25–61.
46. Fujita M. Adaptive filter model of the cerebellum. *Biol Cybern* 1982;45:195–206.
47. Tyrrell T, Willshaw DJ. Cerebellar cortex: Its simulation and the relevance of Marr's theory. *Proc R Soc Lond B Biol Sci* 1992;336:239–257.
48. Kettner RE, Mahamud S, Leung H-C, Sitkoff N, Houk J, Peterson BW, Barto AG. Prediction of complex two-dimensional trajectories by a cerebellar model of smooth pursuit eye movement. *J Neurophysiol* 1997;77:2115–2130.
49. Simpson JI, Alley KE. Visual climbing fiber input to rabbit vestibulocerebellum: A source of direction-specific information. *Brain Res* 1974;82:302–308.
50. Fushiki H, Sato Y, Miura A, Kawasaki T. Climbing fiber responses of Purkinje cells to retinal image movement in cat cerebellar flocculus. *J Neurophysiol* 1994;71:1336–1350.
51. Robinson DA, Gordon JL, Gordon SE. A model of the smooth pursuit eye movement system. *Biol Cybern* 1986;55:43–57.
52. Krauzlis RJ, Lisberger SG. A model of visually-guided smooth pursuit eye movements based on behavioral observations. *J Comput Neurosci* 1994;1:265–283.
53. Ringach DL. A 'tachometer' feedback model of smooth pursuit eye movements. *Biol Cybern* 1995;73:561–568.

Models for Specific Purposes IV

Visual and Auditory Disease A

31 Animal Models for Eye Diseases and Therapeutics

TAKESHI IWATA AND STANISLAV TOMAREV

ABSTRACT

It is believed that more than 80% of the information our brain receives comes from the visual system. Dysfunction of the visual system can significantly lower the quality of life. The most prevalent causes of visual impairment are cataracts, glaucoma, and age-related macular degeneration (AMD), which are responsible for 69% of blindness globally. In spite of the high incidence of AMD and glaucoma, a limited amount of information is available on the underlying pathological mechanisms causing these diseases. Because experimental studies of AMD and glaucoma are limited in humans, the availability of animal models is very valuable to investigate molecular mechanisms and to test new therapeutic interventions. Appropriate animal models, such as monkey, mouse, rat, and zebrafish, facilitate the identification of new genes involved in the pathology and elucidate the genetic relationships between causative and modifier genes. In this chapter the advantages and difficulties of using animal models for vision research will be discussed. Several animal models including a primate model with defined macula for AMD research and genetically modified mice models for glaucoma research will be introduced.

Key Words: Vision, Age-related macular degeneration, Retina, Macula, Drusen, Glaucoma, Retinal ganglion cells, Optic nerve.

VISUAL IMPAIRMENT AND IMPORTANCE OF ANIMAL MODELS FOR EYE DISEASES

It is believed that more than 80% of the information our brain receives comes from the visual system. Dysfunction of the visual system can alter the normal human life style and significantly lower the quality of life. The causes of visual impairments and blindness vary among ethnic groups and the global regions where they live. There are many causes of visual impairments including diabetic complications, infections, and trauma; however the most prevalent causes of visual impairment are cataracts, glaucoma, and AMD. According to the World Health Organization, there were more than 161 million visually impaired individuals in 2002; 124 million of this group had low vision and 37 million were blind (http://www.who.int/mediacentre/factsheets/fs282/en/index.html) (Figure 31–1).

Cataract, glaucoma, and AMD are responsible for 69% of blindness globally. Although cataracts are the leading cause of blindness worldwide, recent advances in cataract surgery have significantly reduced the visual impairments caused by cataracts, especially in developed countries. Glaucoma, an optic neuropathy, is often associated with elevated intraocular pressure and is responsible for blindness in 6.7 million people across the world. Glaucoma is more common in individuals of African ancestry, and the incidence of glaucoma increases with age.

The most prevalent eye disease for elderly Europeans and Americans is AMD. This degenerative disease progresses from retinal deposits called drusen to neovascularization and retinal hemorrhages resulting in irreversible loss of central vision.

In spite of the high incidence of AMD and glaucoma, a limited amount of information is available on the underlying pathological mechanisms causing these diseases. Obtaining tissues for any disease is often difficult, and even when obtained, they may not be informative because the tissues are usually collected many hours or even days after death. Because experimental studies of AMD and glaucoma are limited in humans, the availability of animal models is very valuable because they can be used to investigate the molecular mechanisms causing these diseases and to test new therapeutic interventions. Animal models, compared to other experimental methods, e.g., cell and organ cultures or postmortem models, allow the study of different pathological factors and therapeutic treatments under *in vivo* conditions, i.e., with the visual and other systems of the body intact. Appropriate animal models, e.g., monkey, mouse, and zebrafish, also facilitate the identification of new genes involved in the pathology as well as elucidate the genetic relationships between causative and modifier genes. Equally important, specific genes can be altered in these models. Thus, it is possible to induce mutations in animals, and then search for specific phenotypes, e.g., abnormal intraocular pressure (IOP) and retinal ganglion cell degeneration. Subsequently, the affected genes can be identified by standard genetic procedures.

Many animal models of AMD and glaucoma have been developed in different mammalian and nonmammalian species. None of these models is a perfect reproduction of the human disease, and when choosing the animal model for research, the investigator should evaluate the following: (1) the similarity of the visual system of the model to that of humans, especially the eye; (2) the similarity of the time course of pathological changes in the model and in human eyes; (3) the ability to perform genetic manipulations; (4) the training required to produce affected animals; (5) the size of the eye; (6) the availability and difficulties in the

From: *Sourcebook of Models for Biomedical Research*
(P. M. Conn, ed.), © 2008 Humana Press Inc., Totowa, NJ.

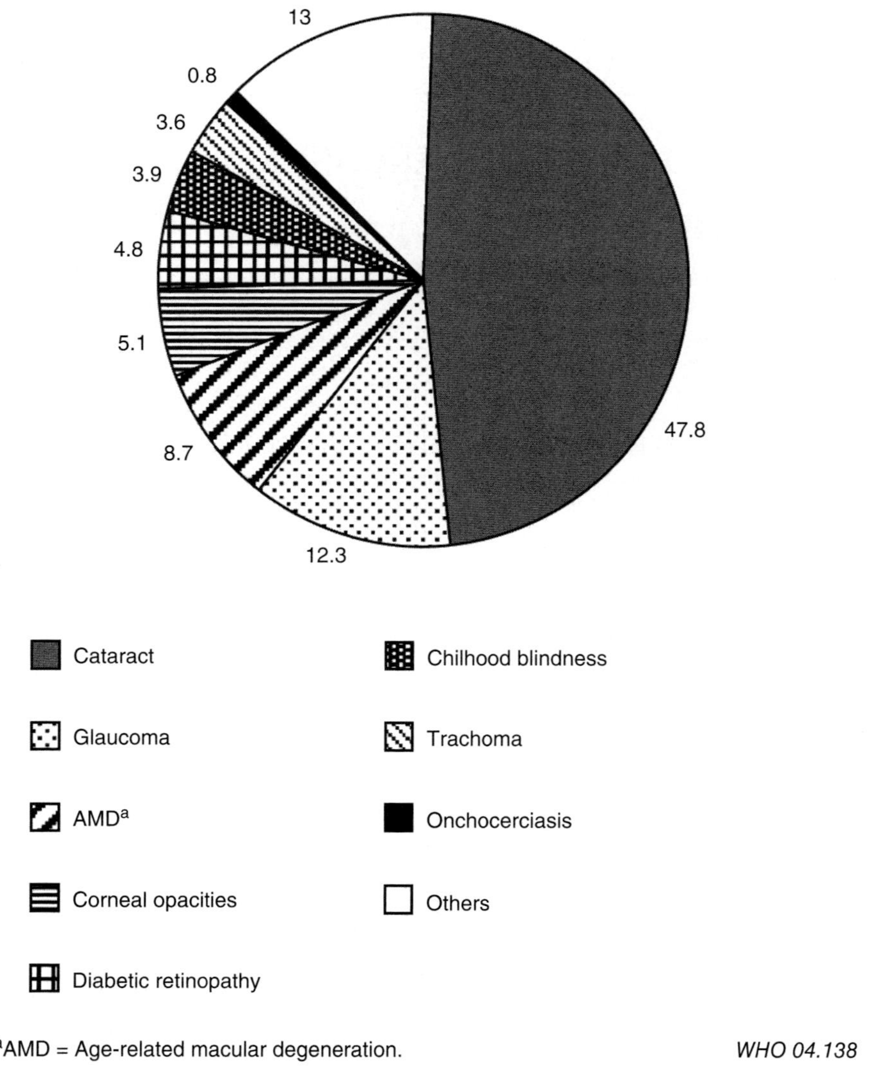

Figure 31–1. Global causes of blindness as a proportion of total blindness in the year 2002 (WHO). AMD, age-related macular degeneration.

methods of analysis; (7) the availability of animals; and (8) the cost.

AGE-RELATED MACULAR DEGENERATION

INTRODUCTION The retina is composed of nine layers of neural and glial cells that are arranged concentrically at the posterior pole of the eye. Incoming light is focused on the central area of the retina called the fovea, which is located in the center of the macular area (Figure 31–2). In humans, the size of the macula is approximately 6 mm in diameter (Figure 31–3). The outer (posterior) surface of the retina is covered by a monolayer of retinal pigment epithelial (RPE) cells that forms a diffusion barrier between the neural retina and the choroidal blood supply. The RPE regulates the transport of proteins to the retina and controls the hydration and ionic composition of the subretinal space. The physiological condition of the RPE is closely associated with the pathogenesis of AMD.

AMD is a blinding disorder characterized by a marked decrease in central vision associated with RPE atrophy with or without choroidal neovascularization (CNV). Many factors, including genetic, behavioral, and environmental, are involved in this disease. AMD is characterized by the degeneration of cone photoreceptors in the foveal region of the retina resulting in a decrease of central visual acuity. The progressive impairment of the RPE cells and damage to Bruch's membrane and choriocapillaris results in retinal atrophy and photoreceptor dysfunction. In some cases, CNV develops, and the new vessels penetrate Bruch's membrane and pass into the subretinal space.

Two types of AMD are recognized: the nonneovascular type is called the dry-type AMD and includes more than 80% of the cases; the neovascular type is called the wet-type AMD and it is progressive with a higher probability of blindness. The prevalence of AMD differs considerably among the different ethnic groups, but the incidence increases with age in all groups. A lower prevalence of AMD has been reported in individuals of African ancestry than of Anglo-Saxon ancestry. Other risk factors for AMD are cigarette smoking, obesity, hypertension, and atherosclerosis.

EPIDEMIOLOGY AND GENETICS Extensive epidemiological studies have shown a genetic component for AMD. Thus, twin studies have shown a higher concordance for AMD in monozygotic twins than in dizygotic twins.[1–3] In addition, first-degree relatives of individuals with AMD have a 2- to 4-fold higher incidence of AMD than individuals without a family history of AMD. Genetic segregation studies have also shown a genetic effect that accounts for approximately 60% of AMD with a single major gene accounting for about 55% of the risk of developing

AMD. Overall, the data have suggested that the etiology of AMD has a significant genetic component. Only a small proportion of the families with AMD shows Mendelian inheritance, and the majority of the individuals inherit AMD in a complex multigene pattern.

There have been a number of attempts to identify the genes that cause AMD. With the help of the haplotype marker project (HapMap Project), genome-wide scanning has identified at least 13 loci linked to AMD on different chromosomes.[4–6]

Recently, a polymorphism of complement factor H gene (*Y402H*) was shown to be associated with an increased risk for AMD.[7–10] These results were confirmed in many of the countries with large white populations but not in Japan.[11,12] This gene is located on chromosome 1q25-31 where one of the candidate loci was identified by linkage studies. Another recent study reported that a haplotype association of tandemly located complement 2 and factor B was protective for AMD.[13]

PATHOLOGY AND BIOCHEMISTRY The early stage of the dry-type AMD is characterized by a thickening of Bruch's membrane, aggregation of pigment granules, and increasing numbers of drusen. The thickening of Bruch's membrane obstructs its function as a "barrier" between the choroid and the RPE that protects the neural retina from the choriocapillary. Drusen are small yellowish-white deposits that are composed of lipids, proteins, glycoproteins, and glycosaminoglycans. They accumulate in the extracellular space and the inner aspects of Bruch's membrane (Figure 31–4). Drusen are not directly associated with visual loss but represent a risk factor for both the dry-type and wet-type AMD. The classification of hard and soft drusen is based on their size, shape, and color; hard drusen are yellowish with diameters <50 μm and are found in eyes that are less likely to progress to advanced stages of the disease, while soft drusen are darker yellow and larger in size, and are found in eyes more likely to progress to more advanced stages of AMD. A small percentage of dry-type AMD patients progress to the late stage of the wet-type AMD that is characterized by geographic atrophy or detachment of RPE and the development of CNV in the macular region. The presence of a CNV is the factor that most damages the neural retina because the newly developed vessels grow from the choriocapillaris through Bruch's membrane and extend laterally through the RPE cell layer (classic CNV) or extend between the inner Bruch's membrane and RPE (occult CNV). In advanced stages of AMD, the CNV and fluid leaked into the subretinal or intraretinal regions lead to cell death and retinal detachment.

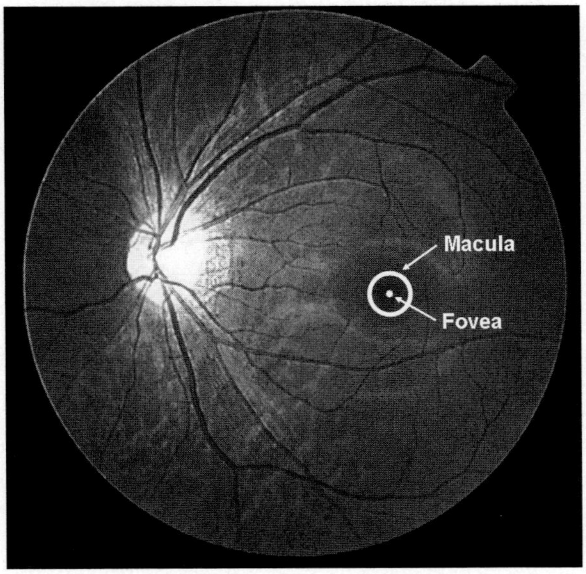

Figure 31–3. Fundus photograph of a normal human retina showing the location of the macula and the fovea.

Recent analyses of the progression of drusen have provided important clues that help understand the molecular pathology of AMD. Using both immunohistochemistry and proteomic techniques, the materials in drusen were found to be composed of molecules that mediate inflammatory and immune processes.[14,15] These molecules include components of the complement pathway and modulators of complement activation, viz. vitronectin, clusterin, membrane cofactor protein, and complement receptor-1. In addition, molecules triggering inflammation, viz. amyloid P component, α_1-antitrypsin, and apolipoprotein E, were identified in drusen. Cellular debris from macrophages, RPE cells, and choroidal dendritic cells has also been identified in drusen. On the other hand, crystallins, EEFMP1, and amyloid-β have been found at

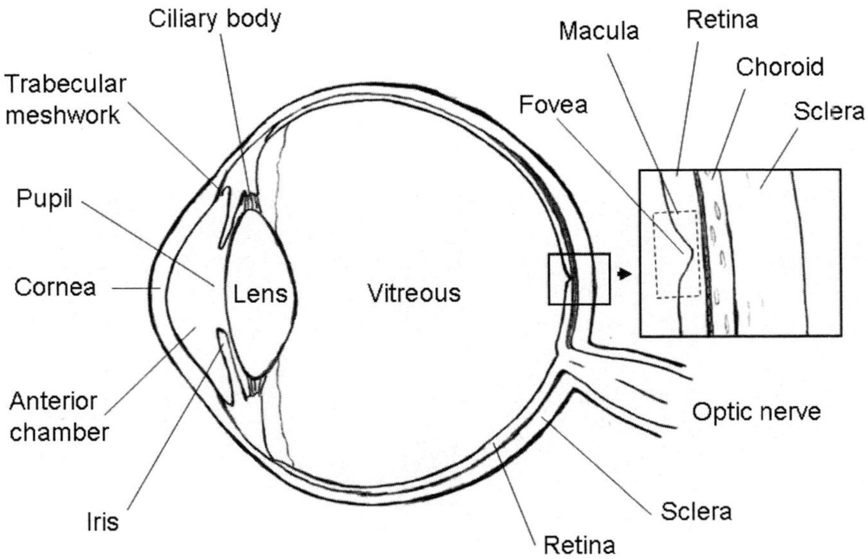

Figure 31–2. Schematic diagram of the human eye.

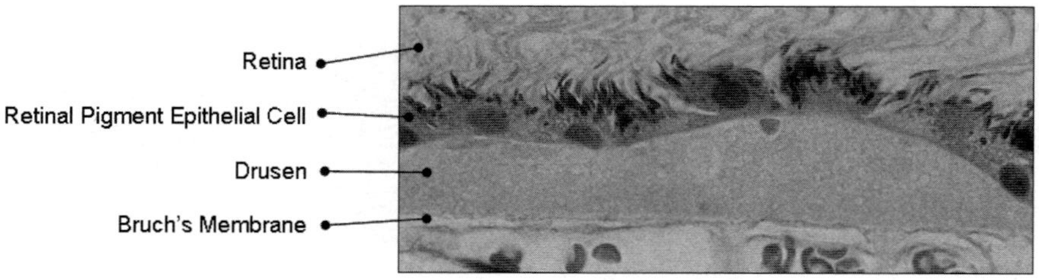

Figure 31–4. Retinal histological section showing the accumulation of drusen between the retinal pigment epithelium and Bruch's membrane in a primate model of AMD.

higher levels in drusen from individuals unaffected by AMD. The presence of immunoreactive proteins and the oxidative modifications of many proteins in drusen imply that both oxidation and immune functions are involved in the pathogenesis of AMD.

All of these findings suggest that complement activation triggers innate immune responses in the subretinal space. The codistribution of immunoglobulin G (IgG) and terminal complement complexes in drusen indicate that immune responses that directly target antigens in retinal cells might also be occurring. Antiretinal autoantibodies have been reported in a number of ocular disorders, e.g., macular degeneration in an aged monkey model.

ANIMAL MODELS Access to appropriate biological materials from affected donors at different stages of a disease is an absolute necessity for the study of mechanisms underlying the disease process. However, because it is nearly impossible to obtain retinal tissues from patients or controls, the development of animal models becomes crucial for investigating the biological pathways involved in the progression of the disease and for the development of therapeutic strategies.

Over the past few years, genetic engineering techniques have generated a number of animal models of AMD in mice, rats, rabbits, pigs, and dogs.[16] However, in mammals, a well-defined fovea is found only in primates (humans and monkeys), and a search for a monkey line affected with macular degeneration has been persistent for a long time.

A monkey with macular degeneration was first described by Stafford *et al.* in 1974. They reported that 6.6% of the elderly monkeys they examined showed pigmentary disorders and drusen-like spots.[17] El-Mofty *et al.* reported that the incidence of maculopathy was 50% in a colony of rhesus monkeys at the Caribbean Primate Research Center of the University of Puerto Rico.[18] At the Tsukuba Primate Research Center (Tsukuba City, Japan), Suzuki *et al.* found a single cynomolgus monkey (*Macaca fascicularis*) in 1986 with a large number of small drusen around the macular region (Figure 31–5).[19–21] This single affected monkey has multiplied to a large pedigree of more than 65 affected and 210 unaffected monkeys. Drusen were observed in the macular region as early as 1 year after birth, and the numbers increased and spread toward the peripheral retina throughout life. No histological abnormalities have been found in the retina, retinal vessels, or choroidal vasculatures of the eyes with drusen. Immunohistochemical and proteomic analyses of the drusen from these monkeys showed that the drusen were very similar to those in other monkeys with aged macular degeneration sporadically found in older monkeys and also with human drusen.[22,23] These observations by Umeda *et al.* have shown that the Tsukuba monkeys produce drusen that are biochemically similar to those in human AMD patients, but the development of the drusen occurs at an accelerated rate of over 25 times. Currently, 240 loci of the cynomolgus monkey are being investigated to try to identify the disease-causing gene and to understand the biological pathways leading to complement activation.

The eyes of monkey are structurally similar to human eyes, which make them extremely valuable for macular degeneration studies. However, there are limitations in using this species over other laboratory animals. Monkeys have a relatively longer life span, have a longer gestation period, have lower birth numbers resulting in a slower rate of expanding the pedigree, and are more difficult to genetically manipulate; in addition, the cost of maintenance is high. In other laboratory animals, the differences in the eye structure, lack of a fovea, and a low cone-to-rod ratio compared to humans have been considered to be a disadvantage for using them as AMD models. However, they are easier to manipulate genetically and easier and less expensive to maintain. This has made the development of a mouse model of AMD very attractive, and a number of mouse AMD models have been reported recently.

The mouse model described by Ambati *et al.* is deficient either in monocyte chemoattractant protein-1 or its cognate C–C chemokine receptor-2. These mice were found to develop the cardinal

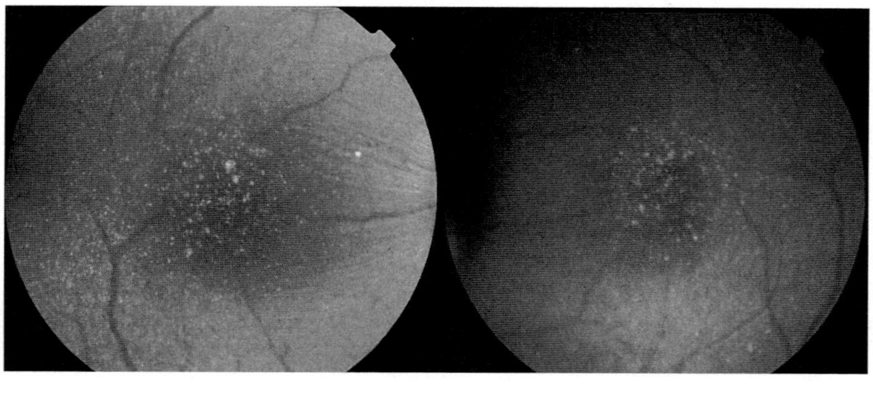

Figure 31–5. Photographs of the fundus of two monkeys with AMD in the Tsukuba Primate Research Center showing drusen.

features of AMD including accumulation of lipofuscin in drusen beneath the RPE, photoreceptor atrophy, and CNV.[24] An impairment of macrophage recruitment allowed the accumulation of C5a and IgG, which leads to the production of vascular endothelial growth factor by the RPE cells and the development of CNVs.

Another mouse model that has three known AMD risk factors—age, high-fat cholesterol-rich diet, and expression of human apolipoprotein E (apoE2, apoE3, andapoE4)—has been developed.[25] ApoE4-deficient mice are severely affected showing diffuse subretinal pigment epithelial deposits, drusen, thickened Bruch's membrane, and atrophy, hypopigmentation, and hyperpigmentation of the RPE.

Oxidative stress has long been linked to the pathogenesis of AMD. Imamura, *et al.* reported Cu, Zn-superoxide dismutase (SOD1)-deficient mice that had features typical of AMD in humans. Senescent $Sod1^{-/-}$ mice had drusen, thickened Bruch's membrane, and choroidal neovascularization.[26] The number of drusen increased with age and also after exposure of young $Sod1^{-/-}$ mice to excess light. The retinal pigment epithelial cells of $Sod1^{-/-}$ mice showed oxidative damage, and their β-catenin-mediated cellular integrity was disrupted. These findings suggested that oxidative stress may affect the junctional proteins necessary for the barrier integrity of the RPE. These observations strongly suggested that oxidative stress may play a major role in AMD.

The complement components, C3a and C5a, are present in drusen, and were observed in Bruch's membrane of a laser-induced CNV mice model. Neutralization of C3a or C5a by antibody or by blockade of their receptors by a complement inhibitor significantly reduced the CNV. These observations revealed a role of immunological mechanisms for angiogenesis and provided evidence for future therapeutic strategies for AMD.

Although the pathology of AMD is pronounced in the macula area, it is not confined to this region. Characteristics of human AMD such as thickening of Bruch's membrane, accumulation of drusen, and CNV have been observed in mouse models. Nevertheless, the primate model will still be the choice for AMD studies, especially at the stage when new therapeutic methods are tested and evaluated for the first time. However, it would be wise and more productive to study both primate and mouse models in AMD research. This will be necessary to determine the mechanisms underlying the disease and to identify clinical and molecular markers for the early stages of AMD. The findings from these studies will provide critical information needed to develop therapies for AMD.

GLAUCOMA

OVERVIEW Glaucoma is a heterogeneous group of complex neurodegenerative disorders that is characterized by the constriction of the visual field, death of retinal ganglion cells (RGCs), and a pathognomonic deformation of the optic nerve head (ONH) known as glaucomatous cupping. Glaucomas are classified into three main types: open-angle, closed-angle, and congenital glaucoma. Each of these types is subdivided into primary and secondary types.

EPIDEMIOLOGY AND GENETICS Primary open-angle glaucoma (POAG) is the most common form of glaucoma; it occurs in about 4.5 million people worldwide and accounts for 12% of all global blindness. By the year 2020, over 11 million people will be blind from primary glaucoma.[28] POAG is often, but not always, associated with elevated IOP, which is one of the main risk factors in glaucoma. However, about a third of all patients with POAG develop the disease without an IOP elevation, and in these patients, the IOP is continuously below 21 mm Hg. This form of POAG is called normal tension or low tension glaucoma (NTG). A reduction of the IOP, even in cases of NTG, is the main, clinically proven, treatment for glaucoma.

ANIMAL MODELS Among the different animal models of glaucoma, the monkey models are superior because of the anatomical similarity of the monkey eyes to human eyes and the phylogenetic similarities of these two species. At the same time, monkeys are extremely expensive and experiments on them require a highly skilled team of investigators.

Most of the existing animal models of POAG, including the monkey models, are based on the elevation of the IOP. An elevation of the IOP develops from an imbalance between aqueous humor production and outflow. Aqueous humor, a fluid produced by the ciliary body of the eye, drains out of the eye and into the blood circulatory system. The eye's outflow system consists of a series of endothelial cell-lined structures that include the trabecular meshwork (TM), Schlemm's canal (SC), which serves as a collector vessel, and the episcleral venous system (Figure 31–6). In most glaucoma models, the IOP is elevated as a result of a reduction or blockage of the aqueous humor outflow. In monkeys, an elevation of IOP is commonly induced by laser photocoagulation of the TM.[29,30] Several days after the laser treatment, the IOP increases and this elevation may last for more than a year, although more than one laser session is usually required to achieve a sustained elevation of the IOP. The IOP in treated eyes is usually between 25 and 60 mm Hg.

Other methods to elevate the IOP in monkey eyes include the anterior chamber injection of ghost red cells,[31] latex microspheres,[32] cross-linked polyacrylamide gels,[33] and enzymes.[34] Topical steroids have also been shown to elevate the IOP.[35] These latter treatments produce less consistent elevations than laser photocoagulation.[36]

Monkey glaucoma models have been shown to have changes in the optic disk, optic nerve, RGC, and nerve fiber layers similar to those observed in glaucomatous human eyes. Apoptosis was shown to be the cause of the RGCs death in a monkey photocoagulation model,[37] and apoptosis was later confirmed to be the cause in other animal models and in humans with glaucoma. Monkey glaucoma models have also been successfully used to study changes in the retinal gene expression pattern 30 days after laser photocoagulation of the TM,[38] and to test the effectiveness of new classes of hypotensive drugs.[36]

Rodent became the animal of choice when large numbers of animals were required, e.g., when examining the mechanism of RGC degeneration and neuroprotection. Several rat models of elevated pressure-induced optic nerve damage have been developed during the past decade, and they have been used to study changes in the retina and the optic nerve. Rats are easy to handle and the relatively large size of their eyes allows multiple, awake measurements of the IOP with commercially available equipment.[39] This latter is important because it is well documented that general anesthesia induces a rapid decrease in the IOP. Although there are certain differences in the structure of the rat and human eyes, all of the eye structures affected in glaucomatous human eyes exist in the rat eye.

In rat models, the IOP elevation is achieved by injection of concentrated saline solution into the episcleral veins,[40] laser

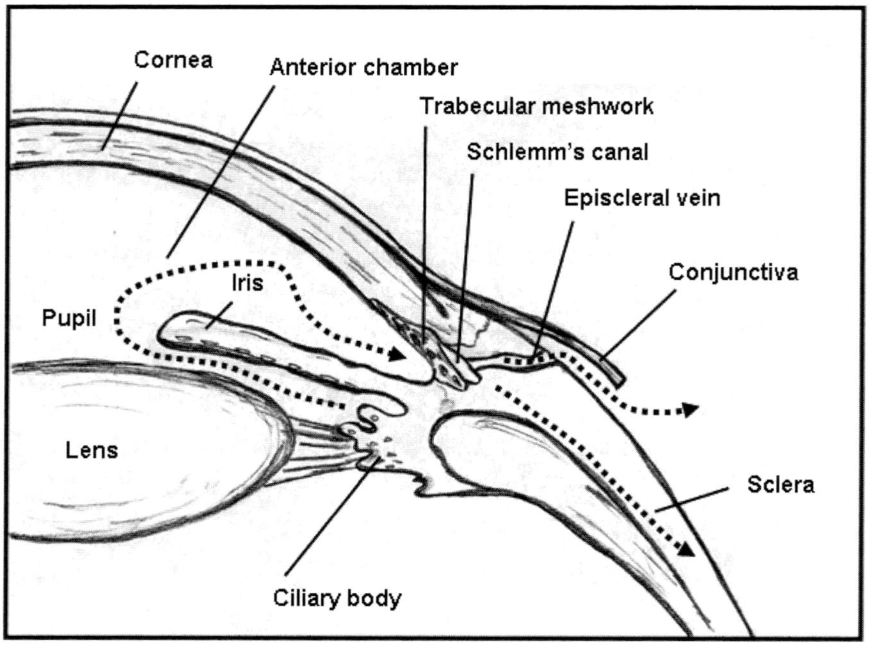

Figure 31–6. Schematic diagram of the anterior segment of the eye showing the trabecular meshwork and uveoscleral outflow pathways of aqueous humor.

photocoagulation of the TM after an injection of Indian ink into the anterior chamber,[41] laser photocoagulation of the TM,[42] and laser cauterization of episcleral veins.[43] All of these methods that lead to an elevation of the IOP require special training of the investigators.

Successful treatment of the eye leads to a rapid elevation of the IOP, although the level of elevation varies from eye to eye. Saline injection generally produces a wide range of IOP elevation from a very minimal rise to a 2-fold increase over the IOP in control eyes. The elevation of the IOP generally lasts for several weeks, and a second laser treatment is often required in the photocoagulation method to maintain an elevated IOP for more than 3 weeks.

A chronic elevation of the IOP in rats leads to apoptosis of the RGCs, degeneration of the optic nerve fibers, and remodeling of the ONH similar to those observed in human glaucomatous eyes.[40,44–45] Rat models of glaucoma have been used to study the effects of elevated IOP on the electroretinogram,[46] neuroprotective drugs,[47] and molecular changes in the retina and optic nerve using the candidate gene approach and array hybridization.[48]

A mutant rat strain was reported to have unilateral or bilateral enlargement of the eyes with an IOP ranging from 25 to 45 mm Hg. In this strain of rat, cupping of the optic nerve head was detected by funduscopic examination, and the cupping was more pronounced in older animals. The number of RGCs also declined with age.[49] Unfortunately, this strain was obtained from the Royal College of Surgeons colony that has a mutation in the receptor tyrosine kinase gene, leading to degeneration of the photoreceptors. Therefore, this strain can hardly be considered a good glaucoma model.

The construction of mouse models of glaucoma has lagged behind rat glaucoma models for a long time despite the advantages of mice over rats and other mammalian species for cost-effective genetic manipulations, availability of a wide spectrum of methods, and the existence of many genetically modified strains. However, it should be remembered that mouse and human eyes have certain important differences including the arterial blood supply to the optic nerve head and the absence of a lamina cribrosa.[50] The lamina cribrosa, a collagenous scaffold supporting the optic nerve, plays a critical role in the damage/protection of the human optic nerve.

One of the main difficulties working with mice is that their eyes are much smaller than the eyes of humans and rats, and new methods had to be developed to measure the IOPs in mice. To date, several invasive and noninvasive methods of IOP measurements have been developed for mice. The first remains one of the most reliable and accurate methods and does not depend upon the mechanical properties of the cornea. It involves the insertion of a glass microneedle connected to a pressure transducer into the anterior chamber of the eye. Using this method, it was shown that common mouse strains have different IOP between 10 and 20 mm Hg.[51] Other methods of IOP measurements in mice were later developed including noninvasive techniques. Noninvasive methods allow multiple IOP measurements to be completed in a short period of time, but the results of these measurements may depend upon mechanical properties of the cornea. To obtain reliable IOP readings, all of the described techniques require training.

Transgenic and gene-targeted knockout approaches have been used to develop several mouse models of glaucoma. The main advantage of these models is that the animals with the mutated gene provide a more uniform elevation of the IOP and damage to the retina and optic nerve similar to that found with surgically induced elevated IOPs. A large number of animals can be produced, and once a mutant mouse line is obtained, no special training is needed to produce more affected mice.

Several lines of transgenic mice have been developed that contained BAC DNAs with a mouse Tyr423His point mutation and Tyr437His point mutation of the human myocilin (Myoc) genes. The Tyr437His mutation in the Myoc gene leads to severe glaucoma in humans, and the mouse Tyr423His mutation corresponds to this human mutation. Expression of mutated mouse Myoc in the ocular drainage structures led to moderate (about 2 mm Hg during the day and 4 mm Hg at night) elevation of the

IOP and progressive degenerative changes in the peripheral RGC layer and optic nerve that resembled glaucomatous changes in human eyes.[52] In 1-year-old animals, the peripheral retina of transgenic mice had approximately 20% fewer RGCs than the peripheral retina of control littermates.

Transgenic mice with a targeted mutation in the gene for the α_1-subunit of collagen type I have also been constructed. This mutation blocks the cleavage of collagen by matrix metalloproteinase-1. Transgenic mice expressing mutated collagen had elevated IOPs. The difference in the IOP between control and transgenic mice gradually increased to a maximum of 4.8 mm Hg at 36 weeks. Because these mice had progressive optic nerve axon loss with normal organization of the drainage structures, it has been suggested that they may be used as a mouse model of POAG.[53]

Recent data have demonstrated that transgenic mice expressing mutated optineurin under the control of the ubiquitous ROSA26 promoter develop optic nerve cupping and death of the RGCs without elevation of the IOP.[54] These transgenic mice may represent the first animal model of NTG.

The surgical methods used to produce rat glaucoma models have also been used in mice. However, performing surgery on the mouse eye is even more challenging than on rat eyes because of the difference in size. A significant elevation of the IOP was found in the eyes of C57BL/6J mice that had an injection of indocyanine green dye into the anterior chamber and diode laser treatment of the TM and episcleral vein region.[55] At 10 days after the surgery, the mean IOP in the operated eyes was 33.6 ± 1.5 mm Hg versus 15.2 ± 0.6 mm Hg in the control eyes. However, the IOP returned to normal 60 days after the surgery. Histological examination of the treated eyes 65 day after the surgery revealed anterior synechia, a decrease in the number of RGCs, thinning of all retinal layers, and damage to the optic nerve structures without evidence of prominent cupping.[55] A reduction in the function of the outer retinal layers, confirmed by electroretinographic studies, may indicate that this model produces more extensive changes in the retina compared to the glaucoma in humans.

Similar to the above model, an elevation of IOP was induced by argon laser photocoagulation of the episcleral and limbal veins in C57BL/6J mouse eyes[56] or by cauterization of three episcleral veins in CD1 mouse eyes.[57] During the first 4 weeks following laser treatment, the mean IOP in the treated eyes was about 1.5 times higher than in control eyes. The number of RGCs had decreased by 22.4 ± 7.5% of that in the controls at 4 weeks after treatment. Most of the TUNEL-positive apoptotic cells were detected in the peripheral retina.[56]

Cauterization of the episcleral veins led to a maximum IOP elevation within 2–9 days, and the IOP decreased progressively thereafter to reach more or less normal values after 24–33 days. There was a 20% decrease in the number of RGCs 2 weeks after the surgery.[57]

The DBA/2J strain has high IOP and has become a popular mouse model to study secondary angle-closure glaucoma. This mouse strain has mutations in two genes, *Tyrp1* and *Gpnmb*.[58] DBA/2J mice develop pigment dispersion, iris transillumination, iris atrophy, and anterior synechia. At the age of 9 months, the IOP was elevated in most mice and the elevation was accompanied by the death of the RGCs, optic nerve atrophy, and optic nerve cupping. Although no group of the RGCs was especially vulnerable or resistant to degeneration, fan-shaped sectors of dead or surviving RGC radiated from the ONH.[59] It has been suggested that axon damage at the ONH might be a primary lesion in this model.[59]

Several important observations were made from the studies on the DBA/2J model. It was shown that the proapoptotic protein BAX is required for the survival of RGCs but not for RGC axon degeneration, suggesting that BAX may be a candidate human glaucoma susceptibility gene.[60] Unexpectedly, a high dose of γ-irradiation accompanied by syngeneic bone marrow transfer protected the RGCs in DBA/2J mice.[61] Similar to the results obtained with rat and monkey models, genes involved in the glial activation and immune response were activated in DBA/2J retina as shown by array hybridization.[62]

Complement component, 1Q, was upregulated in the retina of several animal models of glaucoma as it is in human glaucoma with the timing suggesting that complement activation plays a significant role in the pathogenesis of glaucoma.[63]

Taken together, these findings confirm that animal models might be used to look for a molecular mechanism involved in glaucoma in humans.

The modulation of the activity of genes involved in the development of the anterior segment of the eye may lead to relatively rare developmental glaucomas that account for less than 1% of all glaucoma cases. Several genes have been implicated in congenital glaucoma and anterior segment disgenesis. They include *Cyp1b1*, *Foxc1*, *Foxc2*, *Pitx2*, *Lmx1b*, and *Pax6*. Several lines of mice with defects in these genes have been studied with glaucoma in mind (see Gould *et al.*[64] for a review). For example, mutation in the *CYP1B1* gene (cytochrome P450, family 1, subfamily b, polypeptide 1) may lead to primary congenital glaucoma (PCG) in humans. Although Cyp1b1 knockout mice did not develop elevated IOP, they had ocular abnormalities similar to the defects in humans with PCG, viz. small or absent Schlemm's canal, defects in the TM, and attachment of the iris to the TM and anterior synechia.

Mutations in the *FOXC1* gene, which encodes a transcription factor with a forkhead-winged-helix DNA binding domain, cause a range of eye abnormalities associated with glaucoma, e.g., iris hypoplasia, Axenfeld and Rieger anomaly, and Rieger syndrome. Foxc1$^{-/-}$ mice die at birth, while Foxc1$^{+/-}$ animals are viable but have defects in the eye drainage structures without changes in IOP. Similar eye defects were observed in Foxc2$^{+/-}$ mice. It has been suggested that *Foxc1*$^{+/-}$ and *Foxc2*$^{+/-}$ mice are useful models for studying anterior segment development and anomalies, and may allow the identification of genes that interact with *Foxc1* and *Foxc2* (or *FKHL7* and *FKHL14*) to produce a phenotype with elevated IOP and glaucoma.

Other animals, including rabbit, pig, and bovine, have also been used to develop animal models of glaucoma, but none of them is widely used for different reasons. Zebrafish became a powerful model for advanced genetic studies in vertebrates, especially in the case of complex diseases, and was proposed as a model for identification of modifier genes for glaucoma.[65]

CONCLUSIONS

In summary, animal models of glaucoma, including the most widely used rodent and monkey models, have already provided interesting new information about mechanisms of glaucoma in humans. However, it should be remembered that even in monkey models the time course of changes in the glaucomatous eyes may be significantly accelerated compared to that in human glaucomatous eyes, and all discussed models are indeed just models of glaucoma in humans. Results obtained with these models should not be automatically applied to human glaucoma and should be confirmed by testing in humans whenever possible. It has become clear that reaction to the same insult, e.g., elevated IOP, may be somewhat different in different animal models. Glaucoma studies in animals may help us identify the molecular mechanisms involved in the development of glaucoma in each particular model. By comparing these mechanisms, it may be possible to find some common mechanism that might be involved in glaucoma formation in humans. This will be extremely valuable for the development of new therapeutic approaches for the treatment and prevention of glaucoma in humans.

REFERENCES

1. Heiba IM, Elston RC, Klein BE, Klein R. Sibling correlations and segregation analysis of age-related maculopathy: The Beaver Dam Eye Study. *Genet Epidemiol* 1994;11:51–67.
2. Seddon JM, Ajani UA, Mitchell BD. Familial aggregation of age-related maculopathy. *Am J Ophthalmol* 1997;123:199–206.
3. Hammond CJ, Webster AR, Snieder H, Bird AC, Gilbert CE, Spector TD. Genetic influence on early age-related maculopathy: A twin study. *Ophthalmology* 2002;109:730–736.
4. Iyengar SK, Song D, Klein BE, Klein R, Schick JH, Humphrey J, Millard C, Liptak R, Russo K, Jun G, Lee KE, Fijal B, Elston RC. Dissection of genomewide-scan data in extended families reveals a major locus and oligogenic susceptibility for age-related macular degeneration. *Am J Hum Genet* 2004;74:20–39.
5. Schick JH, Iyengar SK, Klein BE, Klein R, Reading K, Liptak R, Millard C, Lee KE, Tomany SC, Moore EL, Fijal BA, Elston RC. A whole-genome screen of a quantitative trait of age-related maculopathy in sibships from the Beaver Dam Eye Study. *Am J Hum Genet* 2003;72:1412–1424.
6. Majewski J, Schultz DW, Weleber RG, Schain MB, Edwards AO, Matise TC, Acott TS, Ott J, Klein ML. Age-related macular degeneration—a genome scan in extended families. *Am J Hum Genet* 2003;73:540–550.
7. Klein RJ, Zeiss C, Chew EY, Tsai JY, Sackler RS, Haynes C, Henning AK, SanGiovanni JP, Mane SM, Mayne ST, Bracken MB, Ferris FL, Ott J, Barnstable C, Hoh J. Complement factor H polymorphism in age-related macular degeneration. *Science* 2005;308:385–389.
8. Edwards AO, Ritter R 3rd, Abel KJ, Manning A, Panhuysen C, Farrer LA. Complement factor H polymorphism and age-related macular degeneration. *Science* 2005;308:421–424.
9. Haines JL, Hauser MA, Schmidt S, Scott WK, Olson LM, Gallins P, Spencer KL, Kwan SY, Noureddine M, Gilbert JR, Schnetz-Boutaud N, Agarwal A, Postel EA, Pericak-Vance MA. Complement factor H variant increases the risk of age-related macular degeneration. *Science* 2005;308:419–421.
10. Hageman GS, Anderson DH, Johnson LV, Hancox LS, Taiber AJ, Hardisty LI, Hageman JL, Stockman HA, Borchardt JD, Gehrs KM, Smith RJ, Silvestri G, Russell SR, Klaver CC, Barbazetto I, Chang S, Yannuzzi LA, Barile GR, Merriam JC, Smith RT, Olsh AK, Bergeron J, Zernant J, Merriam JE, Gold B, Dean M, Allikmets R. A common haplotype in the complement regulatory gene factor H (HF1/CFH) predisposes individuals to age-related macular degeneration. *Proc Natl Acad Sci USA* 2005;102:7227–7232.
11. Okamoto H, Umeda S, Obazawa M, Minami M, Noda T, Mizota A, Honda M, Tanaka M, Koyama R, Takagi I, Sakamoto Y, Saito Y, Miyake Y, Iwata T. Complement factor H polymorphisms in Japanese population with age-related macular degeneration. *Mol Vis* 2006;12:156–158.
12. Gotoh N, Yamada R, Hiratani H, Renault V, Kuroiwa S, Monet M, Toyoda S, Chida S, Mandai M, Otani A, Yoshimura N, Matsuda F. No association between complement factor H gene polymorphism and exudative age-related macular degeneration in Japanese. *Hum Genet* 2006;120:139–143.
13. Gold B, Merriam JE, Zernant J, Hancox LS, Taiber AJ, Gehrs K, Cramer K, Neel J, Bergeron J, Barile GR, Smith RT; AMD Genetics Clinical Study Group; Hageman GS, Dean M, Allikmets R. Variation in factor B (BF) and complement component 2 (C2) genes is associated with age-related macular degeneration. *Nat Genet* 2006;38:458–462.
14. Russell SR, Mullins RF, Schneider BL, Hageman GS. Location, substructure, and composition of basal laminar drusen compared with drusen associated with aging and age-related macular degeneration. *Am J Ophthalmol* 2000;129:205–214.
15. Mullins RF, Russell SR, Anderson DH, Hageman GS. Drusen associated with aging and age-related macular degeneration contain proteins common to extracellular deposits associated with atherosclerosis, elastosis, amyloidosis, and dense deposit disease. *FASEB J* 2000;14:835–846.
16. Chader GJ. Animal models in research on retinal degenerations: Past progress and future hope. *Vision Res* 2002;42:393–399.
17. Stafford TJ, Anness SH, Fine BS. Spontaneous degenerative maculopathy in the monkey. *Ophthalmology* 1984;91:513–521.
18. El-Mofty A, Gouras P, Eisner G, Balazs EA. Macular degeneration in rhesus monkey (Macaca mulatta). *Exp Eye Res* 1978;27:499–502.
19. Nicolas MG, Fujiki K, Murayama K, Suzuki MT, Mineki R, Hayakawa M, Yoshikawa Y, Cho F, Kanai A. Studies on the mechanism of early onset macular degeneration in cynomolgus (Macaca fascicularis) monkeys. I. Abnormal concentrations of two proteins in the retina. *Exp Eye Res* 1996;62:211–219.
20. Nicolas MG, Fujiki K, Murayama K, Suzuki MT, Shindo N, Hotta Y, Iwata F, Fujimura T, Yoshikawa Y, Cho F, Kanai A. Studies on the mechanism of early onset macular degeneration in cynomolgus monkeys. II. Suppression of metallothionein synthesis in the retina in oxidative stress. *Exp Eye Res* 1996;62:399–408.
21. Suzuki MT, Terao K, Yoshikawa Y. Familial early onset macular degeneration in cynomolgus monkeys (*Macaca fascicularis*). *Primates* 2003;44:291–294.
22. Umeda S, Ayyagari R, Allikmets R, Suzuki MT, Karoukis AJ, Ambasudhan R, Zernant J, Okamoto H, Ono F, Terao K, Mizota A, Yoshikawa Y, Tanaka Y, Iwata T. Early-onset macular degeneration with drusen in a cynomolgus monkey (*Macaca fascicularis*) pedigree: Exclusion of 13 candidate genes and loci. *Invest Ophthalmol Vis Sci* 2005;46:683–691.
23. Umeda S, Suzuki MT, Okamoto H, Ono F, Mizota A, Terao K, Yoshikawa Y, Tanaka Y, Iwata T. Molecular composition of drusen and possible involvement of anti-retinal autoimmunity in two different forms of macular degeneration in cynomolgus monkey (Macaca fascicularis). *FASEB J* 2005;19:1683–1685.
24. Ambati J, Anand A, Fernandez S, Sakurai E, Lynn BC, Kuziel WA, Rollins BJ, Ambati BK. An animal model of age-related macular degeneration in senescent Ccl-2- or Ccr-2-deficient mice. *Nat Med* 2003;9:1390–1397.
25. Malek G, Johnson LV, Mace BE, Saloupis P, Schmechel DE, Rickman DW, Toth CA, Sullivan PM, Bowes Rickman C. Apolipoprotein E allele-dependent pathogenesis: A model for age-related retinal degeneration. *Proc Natl Acad Sci USA* 2005;102:11900–11905.
26. Imamura Y, Noda S, Hashizume K, Shinoda K, Yamaguchi M, Uchiyama S, Shimizu T, Mizushima Y, Shirasawa T, Tsubota K. Drusen, choroidal neovascularization, and retinal pigment epithelium dysfunction in SOD1-deficient mice: A model of age-related macular degeneration. *Proc Natl Acad Sci USA* 2006;103:11282–11287.
27. Nozaki M, Raisler BJ, Sakurai E, Sarma JV, Barnum SR, Lambris JD, Chen Y, Zhang K, Ambati BK, Baffi JZ, Ambati J. Drusen complement components C3a and C5a promote choroidal neovascularization. *Proc Natl Acad Sci USA* 2006;103:2328–2333.
28. Quigley HA, Broman AT. The number of people with glaucoma worldwide in 2010 and 2020. *Br J Ophthalmol* 2006;90:262–267.
29. Gaasterland D, Kupfer C. Experimental glaucoma in the rhesus monkey. *Invest Ophthalmol Vis Sci* 1974;13:455–457.

30. Quigley HA, Hohman RM. Laser energy levels for trabecular meshwork damage in the primate eye. *Invest Ophthalmol Vis Sci* 1983;24:1305–1307.
31. Quigley HA, Addicks EM. Chronic experimental glaucoma in primates. I. Production of elevated intraocular pressure by anterior chamber injection of autologous ghost red blood cells. *Invest Ophthalmol Vis Sci* 1980;19:126–136.
32. Weber AJ, Zelenak D. Experimental glaucoma in the primate induced by latex microspheres. *J Neurosci Methods* 2001;111:39–48.
33. Kaufman PL, Lutjen-Drecoll E, Hubbard WC, Erickson KA. Obstruction of aqueous humor outflow by cross-linked polyacrylamide microgels in bovine, monkey, and human eyes. *Ophthalmology* 1994;101:1672–1679.
34. Zimmerman LE, De Venecia G, Hamasaki DI. Pathology of the optic nerve in experimental acute glaucoma. *Invest Ophthalmol* 1967;6:109–125.
35. Amfrmaly MF. Aqueous outflow facility in monkeys and the effect of topical corticoids. *Invest Ophthalmol* 1964;3:534–538.
36. Rasmussen CA, Kaufman PL. Primate glaucoma models. *J Glaucoma* 2005;14:311–314.
37. Quigley HA, Nickells RW, Kerrigan LA, Pease ME, Thibault DJ, Zack DJ. Retinal ganglion cell death in experimental glaucoma and after axotomy occurs by apoptosis. *Invest Ophthalmol Vis Sci* 1995;36:774–786.
38. Miyahara T, Kikuchi T, Akimoto M, Kurokawa T, Shibuki H, Yoshimura N. Gene microarray analysis of experimental glaucomatous retina from cynomologous monkey. *Invest Ophthalmol Vis Sci* 2003;44:4347–4356.
39. Moore CG, Milne ST, Morrison JC. Noninvasive measurement of rat intraocular pressure with the Tono-Pen. *Invest Ophthalmol Vis Sci* 1993;34:363–369.
40. Morrison JC, Moore CG, Deppmeier LM, Gold BG, Meshul CK, Johnson EC. A rat model of chronic pressure-induced optic nerve damage. *Exp Eye Res* 1997;64:85–96.
41. Ueda J, Sawaguchi S, Hanyu T, Yaoeda K, Fukuchi T, Abe H, *et al.* Experimental glaucoma model in the rat induced by laser trabecular photocoagulation after an intracameral injection of India ink. *Jpn J Ophthalmol* 1998;42:337–344.
42. Levkovitch-Verbin H, Quigley HA, Martin KR, Valenta D, Baumrind LA, Pease ME. Translimbal laser photocoagulation to the trabecular meshwork as a model of glaucoma in rats. *Invest Ophthalmol Vis Sci* 2002;43:402–410.
43. Shareef SR, Garcia-Valenzuela E, Salierno A, Walsh J, Sharma SC. Chronic ocular hypertension following episcleral venous occlusion in rats [letter]. *Exp Eye Res* 1995;61:379–382.
44. Garcia-Valenzuela E, Shareef S, Walsh J, Sharma SC. Programmed cell death of retinal ganglion cells during experimental glaucoma. *Exp Eye Res* 1995;61:33–44.
45. Johnson EC, Morrison JC, Farrell S, Deppmeier L, Moore CG, McGinty MR. The effect of chronically elevated intraocular pressure on the rat optic nerve head extracellular matrix. *Exp Eye Res* 1996;62:663–674.
46. Bayer AU, Danias J, Brodie S, Maag KP, Chen B, Shen F, *et al.* Electroretinographic abnormalities in a rat glaucoma model with chronic elevated intraocular pressure. *Exp Eye Res* 2001;72:667–677.
47. McKinnon SJ, Lehman DM, Tahzib NG, Ransom NL, Reitsamer HA, Liston P, *et al.* Baculoviral IAP repeat-containing-4 protects optic nerve axons in a rat glaucoma model. *Mol Ther* 2002;5:780–787.
48. Ahmed F, Brown KM, Stephan DA, Morrison JC, Johnson EC, Tomarev SI. Microarray analysis of changes in mRNA levels in the rat retina after experimental elevation of intraocular pressure. *Invest Ophthalmol Vis Sci* 2004;45:1247–1258.
49. Thanos S, Naskar R. Correlation between retinal ganglion cell death and chronically developing inherited glaucoma in a new rat mutant. *Exp Eye Res* 2004;79:119–129.
50. May CA, Lutjen-Drecoll E. Morphology of the murine optic nerve. *Invest Ophthalmol Vis Sci* 2002;43:2206–2212.
51. Savinova OV, Sugiyama F, Martin JE, Tomarev SI, Paigen BJ, Smith RS, *et al.* Intraocular pressure in genetically distinct mice: An update and strain survey. *BMC Genet* 2001;2:12.
52. Senatorov VV, Malyukova I, Fariss R, Wawrousek E, Swaminathan S, Sharan SK, Tomarev S. Expression of mutated mouse myocilin induces open-angle glaucoma in transgenic mice. *J Neurosci* 2006;26(46):11903–11914.
53. Mabuchi F, Lindsey JD, Aihara M, Mackey MR, Weinreb RN. Optic nerve damage in mice with a targeted type I collagen mutation. *Invest Ophthalmol Vis Sci* 2004;45:1841–1845.
54. Akahori M, Obazawa M, Minami M, Noda T, Noda S, Nakaya N, Tomarev S, Sasaoka M, Shimazaki A, Kawase K, Yamamoto T, Miyake Y, Iwata T. Expression of mutated optineurin leads to normal tension glaucoma and disruption of optineurin-Rab8 interaction in mice. Submitted.
55. Grozdanic SD, Betts DM, Sakaguchi DS, Allbaugh RA, Kwon YH, Kardon RH. Laser-induced mouse model of chronic ocular hypertension. *Invest Ophthalmol Vis Sci* 2003;44:4337–4346.
56. Gross RL, Ji J, Chang P, Pennesi ME, Yang Z, Zhang J, *et al.* A mouse model of elevated intraocular pressure: Retina and optic nerve findings. *Trans Am Ophthalmol Soc* 2003;101:163–169.
57. Ruiz-Ederra J, Verkman AS. Mouse model of sustained elevation in intraocular pressure produced by episcleral vein occlusion. *Exp Eye Res* 2006;82:879–884.
58. Anderson MG, Smith RS, Hawes NL, Zabaleta A, Chang B, Wiggs JL, *et al.* Mutations in genes encoding melanosomal proteins cause pigmentary glaucoma in DBA/2J mice. *Nat Genet* 2002;30: 81–85.
59. Jakobs TC, Libby RT, Ben Y, John SW, Masland RH. Retinal ganglion cell degeneration is topological but not cell type specific in DBA/2J mice. *J Cell Biol* 2005;171:313–325.
60. Libby RT, Li Y, Savinova OV, Barter J, Smith RS, Nickells RW, *et al.* Susceptibility to neurodegeneration in a glaucoma is modified by Bax gene dosage. *PLoS Genet* 2005;1:17–26.
61. Anderson MG, Libby RT, Gould DB, Smith RS, John SW. High-dose radiation with bone marrow transfer prevents neurodegeneration in an inherited glaucoma. *Proc Natl Acad Sci USA* 2005;102: 4566–4571.
62. Steele MR, Inman DM, Calkins DJ, Horner PJ, Vetter ML. Microarray analysis of retinal gene expression in the DBA/2J model of glaucoma. *Invest Ophthalmol Vis Sci* 2006;47:977–985.
63. Stasi K, Nagel D, Yang X, Wang RF, Ren L, Podos SM, *et al.* Complement component 1Q (C1Q) upregulation in retina of murine, primate, and human glaucomatous eyes. *Invest Ophthalmol Vis Sci* 2006;47:1024–1029.
64. Gould DB, Smith RS, John SW. Anterior segment development relevant to glaucoma. *Int J Dev Biol* 2004;48:1015–1029.
65. McMahon C, Semina EV, Link BA. Using zebrafish to study the complex genetics of glaucoma. *Comp Biochem Physiol C Toxicol Pharmacol* 2004;138:343–350.

32 Animal Models of Noise-Induced Hearing Loss

RICHARD SALVI AND FLINT A. BOETTCHER

ABSTRACT

Animal models have played a major role in understanding noise-induced hearing loss, a major cause of hearing loss in industrialized societies. Behavioral models using positive and negative reinforcement techniques have played an important role in delineating the relationship between the physical characteristic of the noise, namely intensity, spectral content, and duration, and the magnitude of the resulting hearing loss. An important contribution from these studies was the discovery that the hearing loss reaches an asymptotic value of threshold shift for exposures lasting 24 h or more. Animal models of noise-induced hearing loss have revealed a significant loss of frequency sensitivity, making it difficult to discriminate signals in noise, and a breakdown in temporal resolution that makes it difficult to perceive the temporal information contained in complex signals such as speech. Animal models have shown that prolonged exposure to high-level noise mainly damages the sensory hair cells in the inner ear. The stereocilia bundles on the sensory hair cells, which contain the mechanically gated ion channels responsible for converting sound into neural activity, are especially vulnerable to acoustic overstimulation. High-level noise can obliterate or damage the stereocilia. More severe sound exposures can result in degeneration of the hair cell body, beginning with the outer hair cells followed by the inner hair cells. The functional status of the outer hair cells can be evaluated in noise-exposed animals using distortion product otoacoustic emissions. When healthy ears are stimulated with two primary tones, f1 and f2, the inner ear generates a distortion tone, the most prominent of which occurs at 2f1–f2. Acoustic overstimulation that primarily damages the outer hair cells and leads to mild to moderate hearing loss almost completely abolishes the distortion product otoacoustic emission. The type I spiral ganglion neurons, which make synaptic contact with single inner hair cells, can be used to assess the functional status of a very narrow region of the cochlea. The functional status of small populations of type I neurons can be evaluated using tone bursts to elicit the compound action potential. Narrow band noise exposures that damage specific regions of the cochlea cause an increase in the compound action potential threshold at frequencies associated with the region of cochlear pathology. Microelectrode recordings from single type I auditory nerve fibers can be used to delineate the pathophysiology at specific sites along the cochlea. Sound exposures that selectively damage the outer hair cells lead to an elevation of threshold and loss of tuning. In cases in which acoustic overstimulation destroys both the outer hair cells and inner hair cells, nerve fibers associated with the damaged region are unresponsive to sound resulting in a dead region. The biological mechanisms that lead to noise-induced degeneration of hair cells is not fully understood, but there is growing awareness that hair cell death may be initiated by oxidative stress that leads to hair cell death by necrosis or apoptosis.

Key Words: Chinchilla, Gerbil, Mice, Noise exposure, Hearing loss, Tuning, Oxidative stress, Hair cells, Auditory nerve.

NOISE-INDUCED HEARING LOSS

In industrialized countries where recreational and work place noise are increasing, noise-induced hearing loss (NIHL) has become the leading cause of hearing loss among those under 50 years of age.[1] The early studies of NIHL in humans between 1940 and late 1960 helped to establish a basic understanding of the magnitude of the hearing loss resulting from noise exposures of a given intensity, duration, and spectral content; however, because of ethical constraints and the difficulty of carrying out long-term noise exposure with humans, most of the early human studies focused on sound exposures that resulted in temporary threshold shifts (TTS). With the growing concern that TTS-inducing sound exposures might cause permanent damage to the sensory cells, supporting cells, and spiral ganglion neurons in the inner ear, more recent work on NIHL has turned to animal models, in particular, studies aimed at understanding the basic neuroanatomical, neurophysiological, and biochemical mechanisms underlying NIHL.

Which animal model is the most useful for investigating NIHL and to what extent is it possible to extrapolate from animal models to humans? The answer to the question is complex and probably depends on the specific issue or problem being investigated. Most biological studies of sensorineural hearing loss have been conducted on cat, chinchilla, gerbil, guinea pig, mouse, and rat. Because cats are particularly hardy animals, they have proved especially useful for long-term electrophysiological recording.[2] Chinchillas have often been used because their range of hearing is nearly identical to that of humans;[3] however, they may be more susceptible to acoustic trauma than humans.[4] The upper range of hearing in guinea pigs and gerbils is slightly

From: *Sourcebook of Models for Biomedical Research* (P. M. Conn, ed.), © 2008 Humana Press Inc., Totowa, NJ.

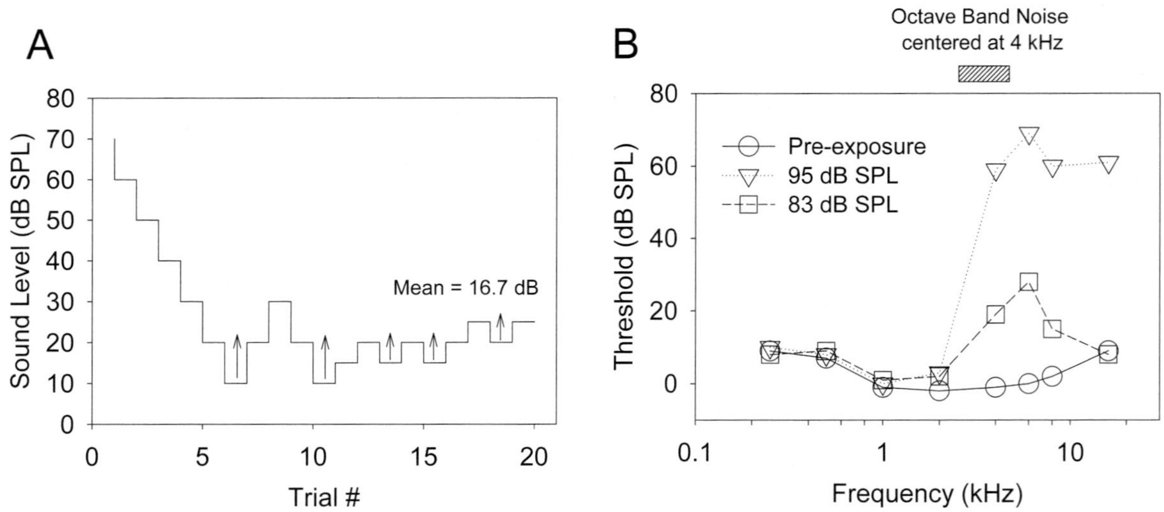

Figure 32–1. (A) Behavioral threshold tracking algorithm in which the sound level is initially decreased in 10-dB steps after a response is made on a tone trial (hit) and increased in 10-dB steps if no response is made on a tone trial (miss). The step size is reduced to 5 dB after two 10-dB reversals. The threshold is calculated as the mean of the last three reversals with a 5-dB step size. (B) Behavioral thresholds between 0.25 and 16 kHz preexposure and after long-term noise exposure to an octave band noise centered at 4 kHz and presented at 83 and 95 dB SPL. Note the large increase in high-frequency behavioral thresholds and normal low-frequency thresholds after the exposure. (Modified from Salvi *et al.*[9])

higher than in humans, but their cochlea is accessible for physiological recordings and anatomical analysis.[5,6] With the rapid developments in molecular biology and genomics, research has shifted toward the use of rats and mice, in particular, inbred strains in which the effects of genetic variability are greatly reduced.[7] A wide range of techniques has been used to assess the effects of NIHL in animal models. The following sections will review some of the most commonly used techniques that have provided fundamental insights into questions such as the following: How much noise is too much noise? What anatomical structures in the inner ear are damaged by noise? What physiological changes are associated with NIHL? Are there any therapeutic treatments for NIHL?

BEHAVIORAL ASSESSMENT OF NOISE-INDUCED HEARING LOSS Behavioral measures, although time consuming, allow for the most comprehensive assessment of the full range of hearing deficits associated with NIHL. Animal psychophysical techniques for assessing hearing fall into two broad categories: those involving positive reinforcement and those utilizing negative reinforcement.[8,9] Regardless of the technique, sufficient training must occur so that the animal's behavior is under stimulus control. The psychophysical procedures used to define threshold in quiet typically involve the method of limits, method of constant stimuli, or a tracking algorithm in which the sound level is decreased or increased by a specified dB step size after a correct or incorrect response, respectively (Figure 32–1). For example, chinchillas have been trained to press and hold down a key and wait for a variable time interval until a tone is presented. If the tone is detected, the animal releases the bar and receives a food pellet (hit). Failure to release the bar on tone trials is scored as a miss resulting in a time out indicted by turning off the house lights. Blank (no tone) trials are presented on 50% of the trials. If the bar hold is maintained throughout the trial, a food pellet is administered and the trial is scored as a correct rejection. Release of the bar on a no-tone trial is scored as a false alarm. To more accurately define threshold, the step size can be reduced after a certain number of threshold reversals has occurred. In the example shown in Figure 32–1A, threshold was defined as the mean of the last three reversals, 16.7 dB sound pressure level (SPL), with a step size of 5 dB. Threshold can be measured over a broad range of frequencies to obtain a plot of threshold versus frequency, or the audiogram. Figure 32–1B illustrates the audiogram of the chinchilla, an animal model that has been used extensively in behavioral studies of NIHL. Thresholds in quiet are lowest in the 2–4 kHz range and begin to increase near the lower and upper range of hearing for this species.[9]

Procedures similar to those used to measure the audiogram have also been used to assess suprathreshold measures of hearing such as frequency discrimination, frequency tuning curves, temporal resolution, intensity discrimination, masking patterns, and the critical band.[10–13]

BEHAVIORAL ASSESSMENT OF NOISE-INDUCED HEARING LOSS Chinchillas have been used extensively to map out the relationship between the three main parameters of the noise exposure (intensity, frequency, and exposure frequency) and degree of TTS and/or permanent threshold shifts (PTS).[14–18] Figure 32–1B illustrates the pattern of TTS that would result from exposure to an octave band of noise (OBN) centered at 4 kHz at a low level (83 dB SPL) and a high level (95 dB SPL). The hearing loss is typically greatest one-half to one octave above the center frequency of the noise exposure, a phenomenon referred to as the one-half octave shift. At low exposure levels, the hearing loss drops off rapidly above and below the exposure frequency; consequently, there is little or no threshold shift at 16 or 2 kHz. However, at high exposure levels, the hearing loss spreads extensively toward the high frequencies, but comparatively little toward the low frequencies. The spread of TTS and PTS toward the high frequency results mainly from the traveling wave vibration pattern that spreads from the point of maximum vibration toward the basal high-frequency region of the cochlea.[19]

EXPOSURE LEVEL AND DURATION A major advantage of animal models is that it is possible to use long-term, well-controlled noise exposures that extend over days, weeks, months, or years, something that is impossible to do in human laboratory studies. Prior to the 1970s, it was not clear if the degree of hearing loss would continue to increase with exposure duration or if it would reach an asymptotic value. When the experiments were finally carried out, a major revelation was that the degree of hearing loss increased with exposure duration, but reached an asymptotic value after 18–24 h. Figure 32–2A illustrates the growth of TTS at 715 Hz in the chinchilla resulting from a 7-day exposure to an octave band of noise centered at 500 Hz.[20] The level of the octave band noise was set at 75, 85, 95, or 105 dB SPL. Hearing loss was measured approximately 4 min after the animal was removed from the noise. When hearing loss is plotted on a logarithmic time scale, TTS increased approximately linearly with exposure duration over the first 24 h and then reached a plateau, or asymptotic threshold shift (ATS). This plateau remains stable for exposure durations as long as 3 years.[21] The ATS value increased according to the following equation: $TTS_{4\,min} = m(OBL - k)$, where OBL is the octave band noise level in dB, k is the exposure level that must be exceeded for a hearing loss to occur, and m is the rate of growth of hearing loss. For an octave band noise centered at 500 Hz, $TTS_{4\,min} = 1.6(OBL - 65)$. That is, the ATS value will increase at 1.6 dB/dB once the OBN level exceeds 65 dB SPL. The variables m and k vary with species and/or frequency. It was originally assumed that if the noise exposure went on indefinitely, ATS would go on to become PTS. If this assumption were correct, then this would be theoretically important because ATS could be used to predict the maximum PTS that could result from a given noise exposure. However, 5–20 dB of recovery has been observed even after a 3-year period of ATS. Thus, ATS is a combination of TTS and PTS where the PTS component increases with exposure duration and level.[21]

RECOVERY Following the end of a noise exposure, thresholds typically recover over a period of approximately 10–15 days. Thresholds can fully or partially recover depending on the exposure level and exposure duration. Figure 32–2B shows the pattern of threshold recovery at 5.7 kHz following an intermittent (6 h/day for 9 days) exposure to an octave band noise centered at 4 kHz.[15] The exposure resulted in an ATS that ranged from 10 dB with an exposure level of 52 dB SPL to approximately 70 dB when the exposure level increased to 92 dB SPL. On a log-recovery time scale, three phases are seen: an early phase during the first 200 min (~3 h) where there is little recovery, a rapid recovery phase extending from approximately 200 to 7000 min (~5 days), and finally a very slow recovery phase followed by a plateau. In general, small ATS values produced by low exposure levels are associated with a shorter recovery time and little or no PTS, whereas large ATS values produced by high exposure levels resulted in longer recovery times and PTS.

FREQUENCY SELECTIVITY If NIHL simply resulted in a loss in hearing sensitivity, then correcting the problem would simply entail turning up the volume on your hearing aid or stereo amplifier. Unfortunately, NIHL also disrupts a listener's frequency selectivity, that is, the ability to extract out and hear the signal of interest from nearby frequencies in the background noise.[12,22] Behavioral psychophysical tuning curves can be measured by having animals respond when they hear a fixed-frequency, short-duration probe tone burst, for example, a 20-msec tone burst at 2000 Hz presented 10–15 dB above the threshold. The low-intensity probe tone presumably excites a limited group of neurons that is tuned to the frequency of the probe signal. Next, a continuous pure tone masker is added to the background. The frequency of the masker is set to a frequency far below the probe-tone frequency, for example, 500 Hz, and then the level of the 500-Hz continuous masker is increased until the animal can no longer hear the 2000-Hz probe tone. The intensity of the 500-Hz

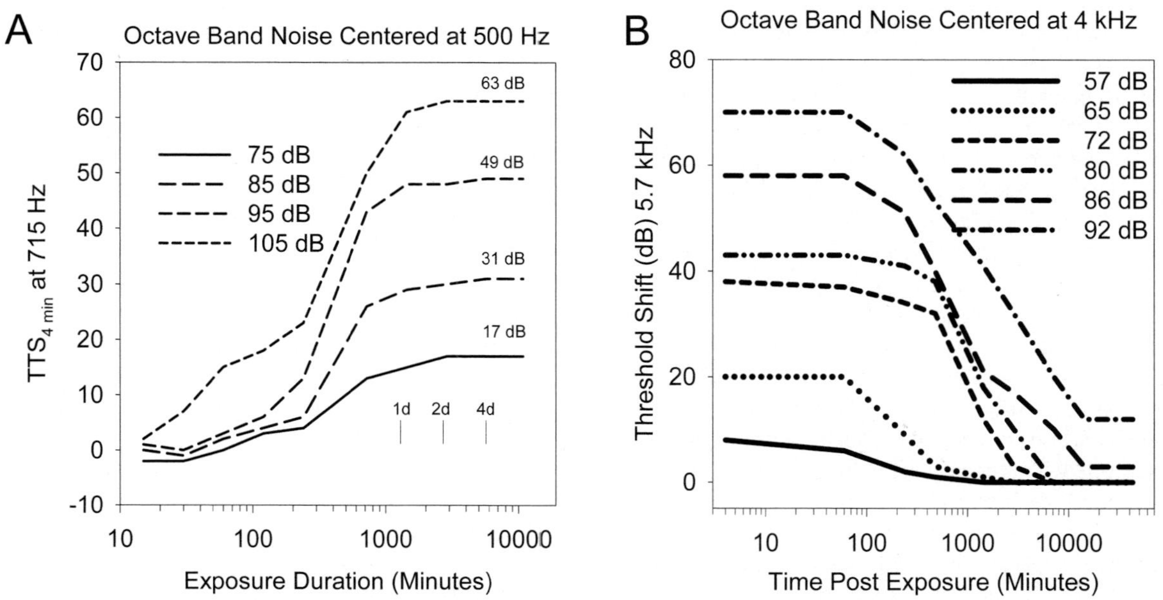

Figure 32–2. (A) The amount of TTS at 715 Hz that occurred 4 min after chinchillas are removed from an octave band noise centered at 500 Hz. Exposure levels were 75, 85, 95, and 105 dB SPL; the ATS values associated with these exposure values were 17, 31, 49, and 63 dB. (B) The recovery of threshold as a function of log recovery time following exposure to an octave band noise centered at 4 kHz. Exposure levels ranged from 57 to 102 dB SPL as shown in the inset. (Modified from Clark.[21])

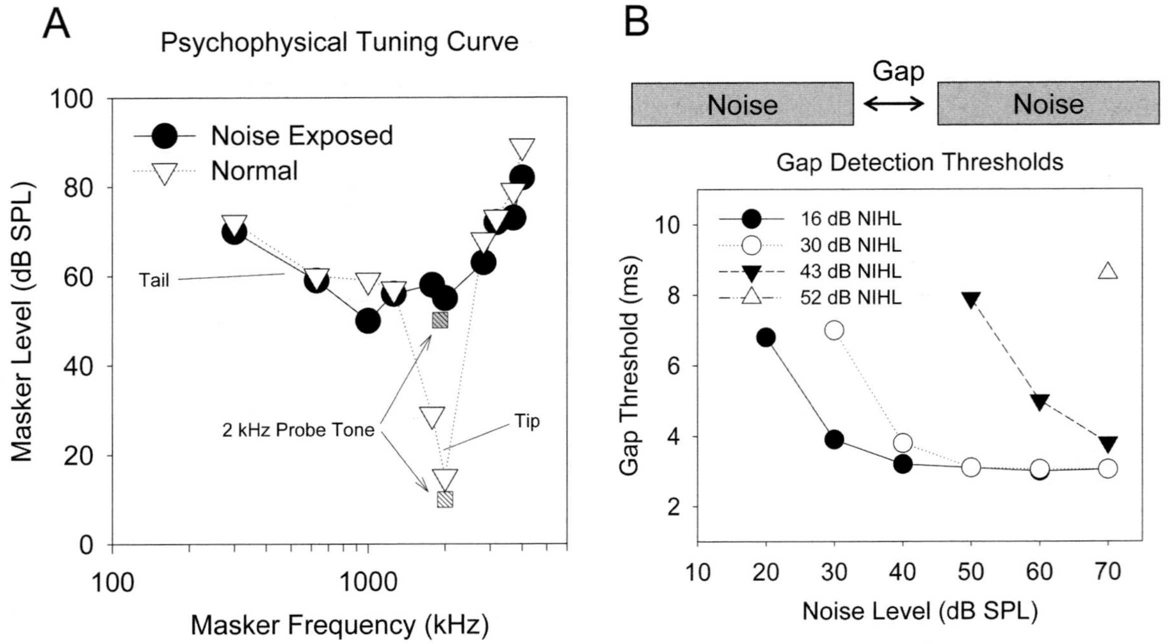

Figure 32–3. (A) Psychophysical tuning curves from a chinchilla using a 2-kHz probe tone presented approximately 10 dB above threshold before and after a noise exposure that resulted in approximately 30 dB of permanent NIHL. The normal psychophysical tuning curve has a narrow, V-shaped, low-level tip near the 2-kHz probe tone frequency and a high-level, low-frequency tail that extends toward the low frequencies. After a noise exposure, the level of the 2-kHz probe tone must be increased so that it is again 10 dB above threshold and the psychophysical tuning curve is remeasured. In animals with NIHL, the psychophysical tuning curve has a broad U-shaped or blunt W-shaped tip due to the increased masked threshold near the probe frequency. Masked thresholds in the tail of the psychophysical tuning curve show little change or sometimes become hypersensitive. (Modified from Salvi *et al.*[47]) (B) Changes in gap thresholds as a function of stimulus level in chinchillas with a flat NIHL of approximately 16, 30, 43 or 52 dB. (Modified from Giraudi-Perry *et al.*[13])

masker that prevents the detection of the 2000 Hz tone bursts represents the masked threshold at 500 Hz. The frequency of the masker is change to another frequency, and the masked threshold is remeasured to obtain a second point on the psychophysical tuning curve. When the masked threshold is measured over a wide range of frequencies below, near, and above the frequency of the probe tone, the entire psychophysical tuning curve can be mapped out. Figure 32–3A shows a normal psychophysical tuning curve measured in a chinchilla using a 2-kHz probe tone presented approximately 10 dB above threshold. The masked thresholds are lowest near the probe tone frequency; masked thresholds increase rapidly as masker frequency increases resulting in a steep high-frequency edge to the tuning curve. Masked thresholds also increase rapidly as the masker frequency decreases; however, at about one octave below the probe frequency, masked thresholds reach a plateau. Thus, normal psychophysical tuning curves consist of two distinct regions, a narrowly tuned V-shaped tip centered at the probe frequency and a high-threshold, broadly tuned tail region. The shape of the psychophysical tuning curve changes significantly in subjects with NIHL as shown in Figure 32–3A. In subjects with approximately 30 dB of NIHL, the probe tone intensity must be increased so that it is 10 dB above the threshold in the impaired ear. When the psychophysical tuning curve is measured in an ear with NIHL, the masked thresholds in the vicinity of the tip are found to increase (hyposensitive), whereas masked thresholds in the tail of the tuning curve remain the same or sometime decrease (hypersensitive). Thus, psychophysical tuning curves in animals with NIHL have a high-threshold, broadly tuned tip connected to a high-threshold tail region, which give rise to a blunt W- or U-shaped tuning curve. Consequently, sounds far below the probe tone, which normally would have little effect on its detection, become almost as effective at masking the signal as those located near the probe signal.

TEMPORAL RESOLUTION The information content of sounds such as speech and music is largely contained in the temporal fluctuations in sound frequency and intensity. Thus, it is possible to ask whether NIHL has any impact on the ability of listeners to extract out and respond to the temporal fluctuations present in sounds. One simple task that can be used to assess the temporal resolving power is the ability to hear a brief silent gap in an otherwise continuous background noise (Figure 32–3B).[13,23] An animal is placed in a continuous background noise and trained to make a behavioral response when a silent gap appears in the ongoing background noise. As shown in Figure 32–3B, the shortest silent gap that can be detected in broadband noise is on the order of 3 msec at moderate to high sound levels. When the background noise intensity falls below 40 dB SPL, the gap threshold decreases to around 7 msec when the background noise is just above the threshold of hearing.

When chinchillas were exposed for many days to a broadband noise that induced relatively flat hearing loss across all frequencies, gap thresholds systematically increased as the noise-induced ATS increased.[13] With a 16-dB ATS, gap thresholds were longer than normal only when the gap thresholds were measured in low-level background noise. With higher levels of background noise gap thresholds were essentially normal. However, as the noise-induced ATS increased from 30 to 43 and then 52 dB, the gap threshold became longer than normal and could be measured at high levels of background noise only when the stimulus was above threshold. Since it is possible to systematically manipulate the degree of NIHL in animal models and assess the deterioration in hearing performance, animal models have been especially useful in determining exactly how auditory temporal resolution changes as the degree of NIHL increases.

HISTOPATHOLOGIES ASSOCIATED WITH NOISE-INDUCED HEARING LOSS

The major histopathologies associated with NIHL occur in the inner ear (cochlea). The spiral-shaped sensory epithelium is encased in a fluid-filled, bony labyrinth located in the temporal bone (Figure 32–4A). Sounds that are collected by the external ear are channeled toward the tympanic membrane that is coupled to the middle ear ossicles, which consist of a chain of three small bones, the malleus, incus, and stapes. Sound-induced vibrations of the tympanic membrane are transmitted through the middle ear to the stapes, a piston-like bone that is loosely coupled by ligaments to the oval window, the point of entry of sounds into the fluid-filled cochlea. Complex sounds entering the cochlea are spatially filtered and set up a traveling wave such that high-frequency sounds produce maximal vibration in the base of the cochlea; lower frequencies produce maximal vibration further toward the apex.[19] The spiral tube of bone that forms the cochlea is subdivided along its entire length by three parallel, fluid-filled compartments, scala tympani, scala vestibuli, and scala media, as shown in Figure 32–3B. The cochlea spirals around the modiolus, a centrally located, hollow bony core that contains the spiral ganglion neurons whose axons give rise to the auditory nerve fibers. The sensory cells, which transduce the incoming sounds, rest on supporting cells of the basilar membrane. The basilar membrane is attached medially to the bony modiolus and laterally near the spiral limbus along the lateral bony wall of the cochlea. The outer hair cells (OHC) form three rows and the inner hair cells (IHC) form a single row that spiral along the length of the cochlea from base to apex. Although there are roughly three times as many OHC as IHC, the majority, 95%, of auditory nerve fibers make synaptic contact with the IHC. Only about 5% of auditory nerve fibers contact the OHC.[24] The sensory hair cells can be subdivided into three main regions, the cell body, a cuticular plate containing a stereocilia bundle on the cell's apical surface facing the endolymph, and a basal pole where the afferent dendrites from the spiral ganglion neurons make synaptic contact. The stereocilia bundles that protrude from the apical pole of the hair cells are most easily visualized using scanning electron microscopy (SEM).

SCANNING ELECTRON MICROSCOPY AND STEREOCILIA DAMAGE Our procedure for visualizing the stereocilia

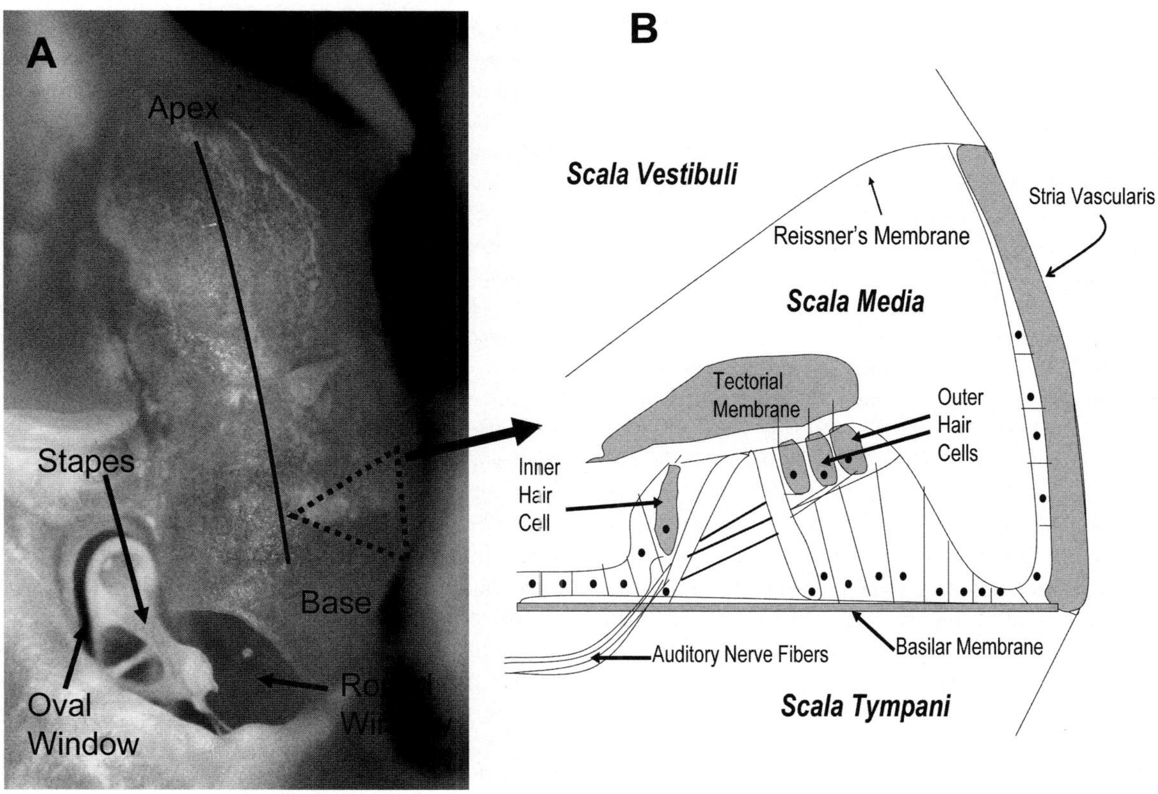

Figure 32–4. (A) Photomicrograph of the chinchilla cochlea showing the stapes plugged into the oval window in the base of the cochlea next to the round window. The cochlea spirals from base to apex around the modiolus, a central bony core containing the spiral ganglion neurons whose fibers give rise to the auditory nerve. The solid line depicts the axis of the modiolus within the cochlea; the dotted triangle depicts a cross section through the basal turn of the cochlea. (B) Schematic cross section of one turn of the cochlea showing the three fluid-filled compartments, the scala media, scala vestibuli, and scala tympani. The scala media is separated from the scala vestibuli by Reissner's membrane; the stria vascularis lies along the lateral wall of the cochlea. The organ of Corti containing the sensory hair cells and supporting cells rests on the basilar membrane. The tectorial membrane, an acellular structure, lies above the outer hair cells and inner hair cells. Approximately 95% of the auditory nerve fibers contact the IHC; the remaining 5% contact the OHC.

Figure 32–5. (A) The apical surface of the sensory epithelium from a normal control animal with three rows of OHC and a single row of IHC. The stereocilia bundle on each OHC is arranged in a W-shaped pattern with the bottom of the W oriented toward the lateral wall and the interior of the W facing the modiolus. The IHC stereocilia bundle is arranged in a gently curving arc. (B) High-level sound exposures primarily damage the sensory hair cells. In some cases, the entire stereocilia bundle is destroyed leaving behind the W-shaped imprint of the stereocilia rootlets on the apical surface of the OHC (white arrowheads). In other cases, the stereocilia bundle can be partially obliterated, in disarray, or the stereocilia can be fused together or bent over (black arrowheads).

with SEM involves euthanizing the animal; afterward, the cochlea is quickly removed, the stapes and oval windows are opened, and fixative (e.g., 2.5% glutaraldehyde) is slowly perfused through the round window.[25] The cochlea is immersed in fixative for 24 h, postfixed in 1% osmium tetroxide, and dehydrated with ethyl alcohol (ETOH). Afterward, the bony shell surrounding the sensory epithelium is dissected away along with the stria vascularis. Segments of the cochlea are removed, rehydrated, immersed in thiocarbohydrazide, dehydrated, rinsed in acetone, critically point dried in CO_2, mounted on aluminum stubs coated with conductive graphite, sputter coated with gold (25 nm), and examined on an SEM at an accelerating voltage between 10 and 25 kV. Figure 32–5A shows the apical surface of the sensory epithelium from a normal control animal with three rows of OHC and a single row of IHC. The stereocilia bundle on each OHC is arranged in a W-shaped pattern with the bottom of the W oriented toward the lateral wall and the interior of the W facing the modiolus. The IHC stereocilia bundle is arranged in a gently curving arc.

High-level sound exposures primarily damage the sensory hair cells, although secondary damage has been seen in auditory nerve fibers, stria vascularis, or fibrocytes in the sensory epithelium.[26,27] With low-intensity or short-duration sound exposures that primarily lead to TTS or a low level of PTS, damage appears to be largely confined to the stereocilia bundles on the OHC.[28,29] In some cases, the damage is confined to the tip links that contain the mechanically gated ion channels.[30] As the intensity and/or duration of the noise increases, the amount of stereocilia damage increases.[25] In some cases, the entire stereocilia bundle is destroyed leaving behind the W-shaped imprint of the stereocilia rootlets on the apical surface of the OHC (Figure 32–5B, white arrowheads). In other cases, the stereocilia bundle can be partially obliterated, in disarray, or the stereocilia can be fused together or bent over (Figure 32–5B, black arrowhead). Since the transduction channels on the stereocilia play an essential role in the mechanical-to-electrical transduction process that leads to depolarization of the hair cells, significant damage to the stereocilia bundle could result in nonfunctional, but surviving hair cells. With careful sample preparation that reduces the risk of preparation artifacts, it is possible to quantify the amount of stereocilia damage following acoustic overstimulation;[25] however, the technique can be difficult and time consuming, especially if it is necessary to evaluate the condition of the stereocilia over the entire length of the cochlea.

SURFACE PREPARATION TECHNIQUE AND COCHLEOGRAM One of the most useful techniques for surveying the degree of hair cell damage along the entire length of the cochlea is the surface preparation technique. Detailed methods for preparing and staining surface preparations of the cochlea can be found in Ding *et al.*[31] Briefly, the cochlea is removed, stained, and/or fixed using a procedure similar to those described above. Afterward, the bony shell and lateral wall surrounding the cochlea are carefully dissected away and then the sensory epithelium containing the organ of Corti with the IHC and OHC is carefully removed as a surface preparation. Sections of the cochlea are typically removed in half turns. The half turns are mounted in glycerin or other medium on glass slides in sequential order, coverslipped, and then observed through a light microscope typically at 400× magnification. Two of the staining techniques we have found to be especially useful are Harris's hematoxylin and succinate dehydrogenase histochemistry (SDH). For Harris's hematoxylin staining, the cochleas are first fixed with 10% formalin for approximately 4 h and then immersed in hematoxylin staining solution for approximately 5 min. Afterward, specimens are dipped in 0.3% hydrochloric acid for 3 min to remove excess stain followed by immersion for a few seconds in ammonium aluminum sulfate until the tissue turns blue. This is an inexpensive, reliable, and convenient method that intensely labels the nuclei of hair cells and supporting cells. SDH staining is a particularly useful method for selectively staining just the IHC and OHC since SDH and other dehydrogenase enzymes are heavily expressed in the hair cells, but not in the surrounding supporting cells. The SDH technique utilizes colorless tetrazolium salts that are reduced to colored formazans in the presence of hydrogen donors. Briefly, after removing the cochlea, the round and oval windows are opened and a small hole is made in the apex. Then SDH staining solution is gently perfused through the round window followed by immersion in SDH for approximately 45 min at 37°C. Afterward, the cochlea is fixed with 10% formalin, dissected out as a flat surface preparation, and mounted on glass slides for analysis.

Figure 32–6A shows two segments of surface preparations from SDH-stained cochleas, one from a leupeptin-protected ear and the other from the opposite unprotected ear of a chinchilla that had been exposed for 48 h to a 105 dB SPL OBN centered at 4 kHz. The IHC and OHC in the protected ear (Figure 32–6A,

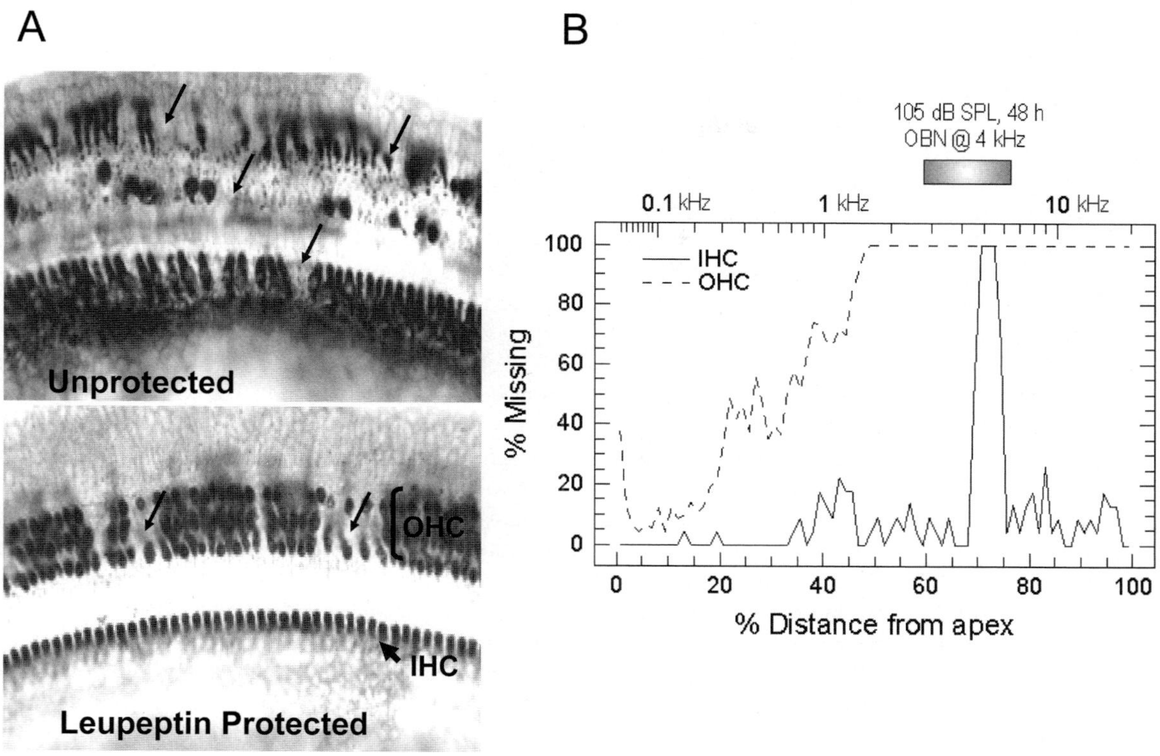

Figure 32–6. (A) Segment of a surface preparation of chinchilla cochlea stained with SDH. Surface preparations were obtained from a chinchilla that had been exposed for 48 h to an octave band noise centered at 4 kHz at 105 dB SPL. One ear (bottom, left panel) had been infused with a protective agent, leupeptin; the opposite ear (upper, left panel) was unprotected (no treatment). The protected ear shows intense dark purple staining of the single row of IHC and three rows of OHC; all of the IHC are present in the protected ear whereas a few OHC are missing (arrows). Many OHC are missing (arrows) in the unprotected ear as well as some IHC. (B) Cochleogram showing the percentage of missing IHC and OHC as a function of percent distance from the apex of the cochlea from a chinchilla that had been exposed to a 105-dB SPL octave band noise (OBN) centered at 4 kHz for 48 h. The position along the length of the cochlea is related to frequency using a frequency–position map for the chinchilla.[32] The horizontal bar at the top shows the approximate bandwidth of the OBN on the frequency–place map. A large IHC loss was present near the 4-kHz place in the cochlea. Note the large widespread loss of OHC extending from the base of the cochlea up to 30% distance from the apex.

bottom, left) are darkly stained while the surrounding supporting cells are unstained, making it especially easy to identify the hair cells. All of the IHC are present in the protected ear and most of the OHC are present in three parallel rows, except for the scattered loss of a few OHC (arrows). In the unprotected ear from the opposite cochlea (Figure 32–6A, upper left), many OHC are missing as well as a few missing IHC.

The degree of IHC and OHC loss can be quantified by counting the number of hair cells that are present or absent at each cochlear location from the apex and proceeding toward the base. For ease of comparison, hair cell loss can be converted to percentage loss; however, this requires that the hair cell density be determined for each cochlear location in the species being studied.[31] Figure 32–6B is a cochleogram from a noise-exposed chinchilla (105 dB SPL, OBN at 4 kHz, 48 h) showing the percentage of missing OHC and IHC as a function of the percent distance from the apex (lower abscissa) of the chinchilla cochlea. Using the frequency-position map developed for the chinchilla, percent distance from the apex of the cochlea can be related to frequency as shown in the upper abscissa.[32] The cochleogram in Figure 32–6B shows a large, discrete lesion located at a distance of 70–75% from the apex; the location of the IHC lesion lies near the 4 kHz region on the chinchilla frequency-place map. This chinchilla sustained a 100% loss of OHC extending from the base of the cochlea down to the middle of the cochlea (~1.8 kHz) followed by a decline of the OHC lesion in the apical half of the cochlea.

PHYSIOLOGICAL MEASURES OF NOISE-INDUCED HEARING LOSS

While behavioral measures have been extremely useful in delineating the perceptual changes that are associated with NIHL, the assessment technique is time consuming and provides little insight regarding the functional changes that are occurring in the cochlea. Consequently, researchers have developed a number of useful physiological tools for evaluating the functional status of the hair cells in ears with NIHL. Three of the most useful tools are discussed below.

OTOACOUSTIC EMISSIONS Otoacoustic emissions, discovered by Kemp in the late 1970s, refer to sounds generated by an active process within the inner ear. Otoacoustic emissions can be recorded by a sensitive microphone placed in the ear canal.[33,34] The biological source of the spontaneous and sound-evoked otoacoustic emissions remained a mystery until it was discovery that OHC possess an electromotile response. Changes in the membrane potential of the OHC cause extremely rapid, cycle-by-cycle longitudinal movements of the OHC body.[35,36] OHC elongate during membrane hyperpolarization and contract during depolarization. The electromotile response is asymmetric with contraction being substantially larger than elongation. The electromotile response arises from the voltage-sensitive motor protein prestin, which is heavily expressed in the lateral wall of the OHC.[37] Knockout mice lacking prestin fail to produce otoacoustic emissions, even though the OHC appear normal, and show hearing loss on the order of 50 dB.[38] Furthermore, destruction of the OHC, but not the IHC, leads to substantial or complete loss of otoacoustic emissions.[39,40] Thus, otoacoustic emissions provide a sensitive and selective tool for assessing the status of the OHC in NIHL.

Distortion Product Otoacoustic Emissions Distortion product otoacoustic emissions (DPOAE) can be evoked from a normal, healthy ear by presenting two primary tones, f1 (lower frequency) and f2 (higher frequency), through a pair of loudspeakers. DPOAE amplitudes are largest when the f2/f1 ratio is on the order of 1.2 and the intensity level of the lower frequency, L1, is slightly greater than the intensity level of the higher tone, L2. When the primary tones, f1 and f2, are delivered to the ear canal, they propagate in the forward direction into the inner ear where each tone sets up a traveling wave with a peak at its characteristic place along the cochlear (Figure 32–7A). The two primary tones produce maximum activation of OHC in the region of the cochlear where the peaks of the two traveling waves overlap. Since the mechanical response of the cochlear partition vibrates nonlinearly at its characteristic place, presumably due to the nonlinear electromotile response of the OHC, distortion tones, the most prominent of which is 2f1–f2, are generated in the vicinity of the primaries. The distortion tones generated in the cochlea propagate in two directions, in the forward direction to the characteristic place associated with 2f1–f2 and in the reverse direction, back through the middle ear and into the ear canal. As shown in Figure 32–7B, the two primary tones together with the distortion tone, 2f1–f2, can be recorded with a sensitive microphone in the ear canal and analyzed with a spectrum analyzer to identify the amplitude of f1, f2, and 2f1–f2. When the OHC are healthy, then a robust 2f1–f2 DPOAE can be recorded as schematized in Figure 32–7B. The functional status of OHC located at high, mid, or low frequencies can be assessed simply by sweeping the f1, f2 pairs from high to low frequencies and measuring the amplitude of the DPOAE at each frequency location. A plot of DPOAE amplitude versus the geometric mean of each f1, f2 pair is sometimes referred to as a DPgram, somewhat akin to a behavioral audiogram. When the OHC are damaged or destroyed, DPOAE amplitude is greatly reduced or absent.

Distortion Product Otoacoustic Emission Amplitude and Noise-Induced Hearing Loss The amplitude of the DPOAE can be evaluated at each f1, f2 pair as the level of the two primary

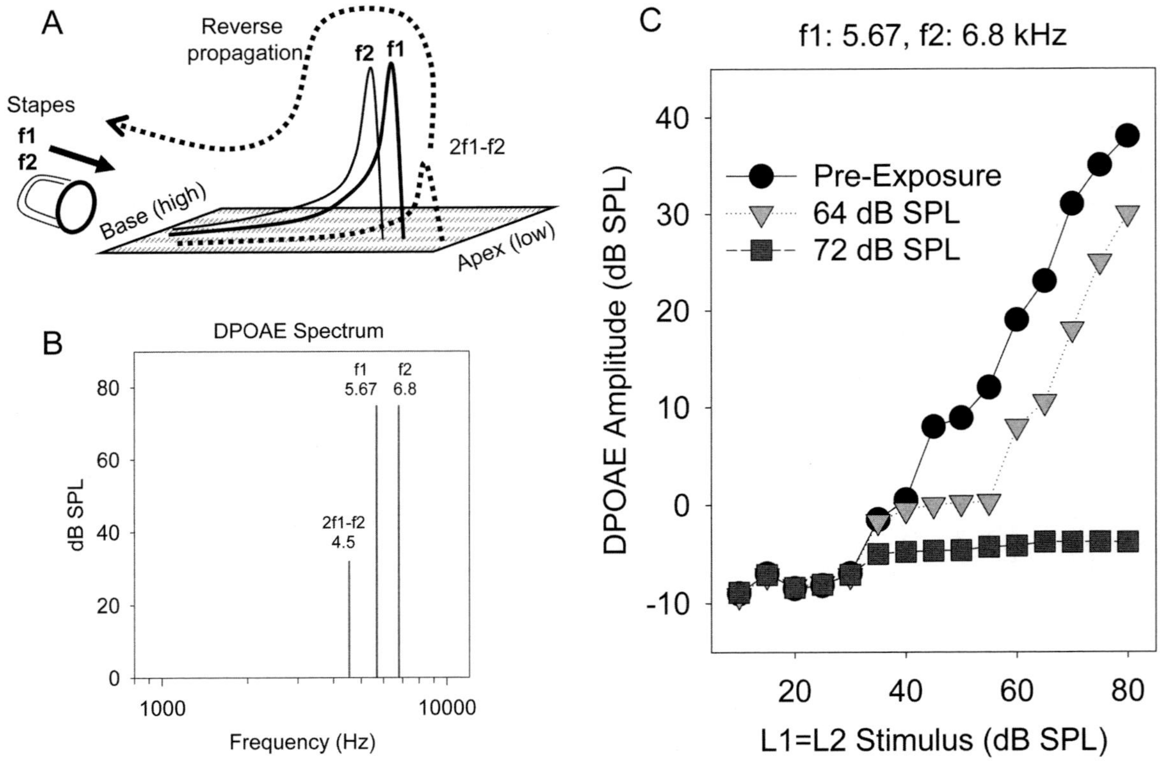

Figure 32–7. (A) Basilar membrane vibration pattern (forward transduction) to f1 (lower) and f2 (higher) primary input tones with an f2/f1 ratio of 1.2. Primary tones generate a distortion signal, 2f1–f2, that propagates in the forward direction from the region of the primaries to its characteristic place in the cochlea; the distortion tone also propagates in the reverse direction back toward the stapes, through the middle ear, and into the external ear canal, where it is detected with a sensitive microphone. The high-frequency base of the cochlea is shown on the left and the low-frequency apex of the cochlea is shown on the right. (B) Sound spectrum recorded with a sensitive microphone in the ear canal of a healthy ear. Primary tones f1 (5.67 kHz) and f2 (6.8 kHz) were presented at 75 dB SPL. Note the distortion production otoacoustic emission (DPOAE) signal, 2f1–f2, at 4.5 kHz at an intensity of approximately 35 dB SPL. (C) DPOAE input/output functions measured preexposure and after a 2–3 day exposure to an octave band of noise centered at 4 kHz presented at 64 dB SPL and then 72 dB SPL. (Redrawn from Eddins *et al.*[41])

tones, L1 and L2 (L1 = L2), are increased. DPOAE input/output functions can be used to assess the functional status of the OHC in the vicinity of f1 (5.67 kHz) and f2 (6.8 kHz) before and after a noise-induced ATS using an octave band noise centered at 4 kHz (see Figure 32–9C). Prior to the noise exposure, DPOAE amplitude begins to rise above the measurement noise floor when the stimulus level exceeds 30 dB SPL. Between 40 and 80 dB SPL, DPOAE amplitude increases from approximately 0 to nearly 40 dB, that is, a growth rate of 1 dB output per dB of input. When chinchillas are exposed to a long-duration (several days) noise that leads to ATS, DPOAE amplitude drops precipitously as the level of the octave band noise centered at 4 kHz increases.[41] Exposure to an octave band noise of 64 dB SPL, which induces ~30 dB of ATS, results in roughly a 10 dB decrease in DPOAE amplitude at nearly all stimulus levels. When the level of the octave band noise was increased to 72 dB SPL, an exposure level that induces 40 dB of ATS, the DPOAE was almost completely abolished. Importantly, the DPOAE remained essentially normal at low frequencies where there was little or no hearing loss from the exposure. A set of six to eight DPOAE input/output functions between 1000, and 12,000 Hz can be measured in about 30 min or less. In some species such as the chinchilla, DPOAE can be measured while the animals are awake and lightly restrained avoiding potential confounding effects associated with anesthesia. These results indicate the DPOAE are an extremely useful tool for assessing NIHL and the functional status of the OHC; however, otoacoustic emissions have a number of limitations. First, DPOAE are almost completely abolished once the hearing loss exceeds 30–35 dB; thus otoacoustic emissions cannot differentiate between large NIHL in the range of 35 and 100 dB. Second, DPOAE cannot assess the functional status of the IHC and the auditory nerve, which provide the main conduit for transmitting information to the central auditory system. Therefore two additional electrophysiological techniques, the compound action potential (CAP) and single fiber recordings from the auditory nerve, provide important insight on the output transmitted through the IHC and into the central auditory system.

Compound Action Potential and Noise-Induced Hearing Loss The CAP represents the local field potential that originates from the synchronous response of the type I auditory nerve fibers that make synaptic contact with the IHC. The CAP onset response can be elicited with a click or short duration (1–2 msec) tone bursts with a rapid rise–fall time (1 msec). CAP recordings can be obtained relatively easily by making a small opening in the middle ear space and placing a silver ball wire electrode on or near the round window (Figure 32–4A) and a ground electrode at some remote location. In most cases, the CAP is recorded from acute surgical preparations; however, it is possible to chronically implant the recording electrode on the round window, seal the opening in the middle ear space with dental cement, and then thread the output of the electrode to a connector permanently mounted on the skull.[25]

The output of an electrode contains the cochlear microphonic (CM) potential, mainly arising from the OHC, the summating potential (SP) mainly coming from the IHC, and the CAP from the type I neurons that contact the IHC. The CAP response can be isolated by filtering (100–3000 Hz) to eliminate the SP and reversing the phase of the stimulus 180° on every other stimulus presentation to cancel the CM.[42,43] Figure 32–8A shows the main features of the CAP waveform recorded from the round window. The CAP consists of two negative peaks, N1 and N2, separated by approximately 1 msec. The latency between stimulus onset and the N1 peak is approximately 1.5 msec; however, latency increases as the intensity and frequency of the stimulus decrease. The amplitude of the N1 component (baseline to negative peak) can be measured as a function of stimulus intensity to obtain the CAP input/output function to a click stimulus or tone burst. Figure 32–8B shows the CAP input/output function for a tone burst under normal conditions and after an NIHL. As stimulus intensity is increased, the amplitude of the N1 response rises above the noise floor of the recording system. In normal animals, the visual detection threshold for the CAP is on the order of 20–25 dB. In animals with NIHL, the CAP amplitude is reduced and the intensity needed to produce a just noticeable response is shifted toward higher intensities to reflect the degree of NIHL. In the example shown in Figure 32–8B, the noise exposure resulted in a threshold shift of the CAP of approximately 25 dB and a roughly 50% reduction in the maximum amplitude.

The tone burst-evoked CAP is useful for assessing the degree of NIHL because the tone burst frequency can be varied to determine the CAP threshold in different frequency regions thereby obtaining a CAP "audiogram." Figure 32–8C shows what the CAP thresholds might look like in a normal chinchilla and one that had been exposed for 24 h to an OBN centered at 4 kHz and an intensity of 86 dB SPL. The CAP thresholds in the normal control are lowest in the mid-frequencies (~20–25 dB) and increase to around 40 dB at the highest and lowest frequencies. Animals that have been exposed to the OBN centered at 4 kHz show elevated CAP thresholds above 2 kHz and normal thresholds at the low frequencies. The CAP threshold shifts are on the order of 50 dB at 4 and 8 kHz and slightly less at 16 kHz. Thus, the CAP can provide a global look at the degree of hearing loss.

Single Auditory Nerve Fibers and Noise-Induced Hearing Loss The auditory nerve forms the only pathway by which acoustic information transduced by the hair cells can be relayed to the central auditory system. The auditory nerve is composed of type I and type II nerve fibers arising from the spiral ganglion neurons that lie within the modiolus. Dye-labeling studies indicate that the type I fibers, which comprise 95% of the total population, appear to be the only fibers that respond to sound stimuli.[44,45] Since each type I nerve fiber makes one-to-one synaptic contact with a single IHC at some point along the length of the cochlea, the response properties recorded from a type I fiber reflect the combined output of the OHC, IHC, and supporting cells from a discrete region of the cochlea. The fibers are tonotopically organized; those innervating hair cells in the apex of the cochlea are most sensitive to low frequencies and those located progressively closer to the base of the cochlea respond best to progressively higher frequencies.

Glass micropipettes with diameters less than 1 μm are typically used to record the all or none spike discharges from single fibers in the stump of the auditory nerve as it passes out of the cochlea through the internal auditory meatus and into the brainstem where the fibers synapse on neurons in the cochlear nuclei.[46,47] Figure 32–9A shows the spike discharge pattern that would occur when a tone burst is presented at a frequency and intensity sufficient to activate the neuron. The number of spike discharges is highest during the onset of the tone burst and gradually declines to a steady-state level; after the tone burst is turned off the spontaneous spike rate is low and gradually returns to the baseline

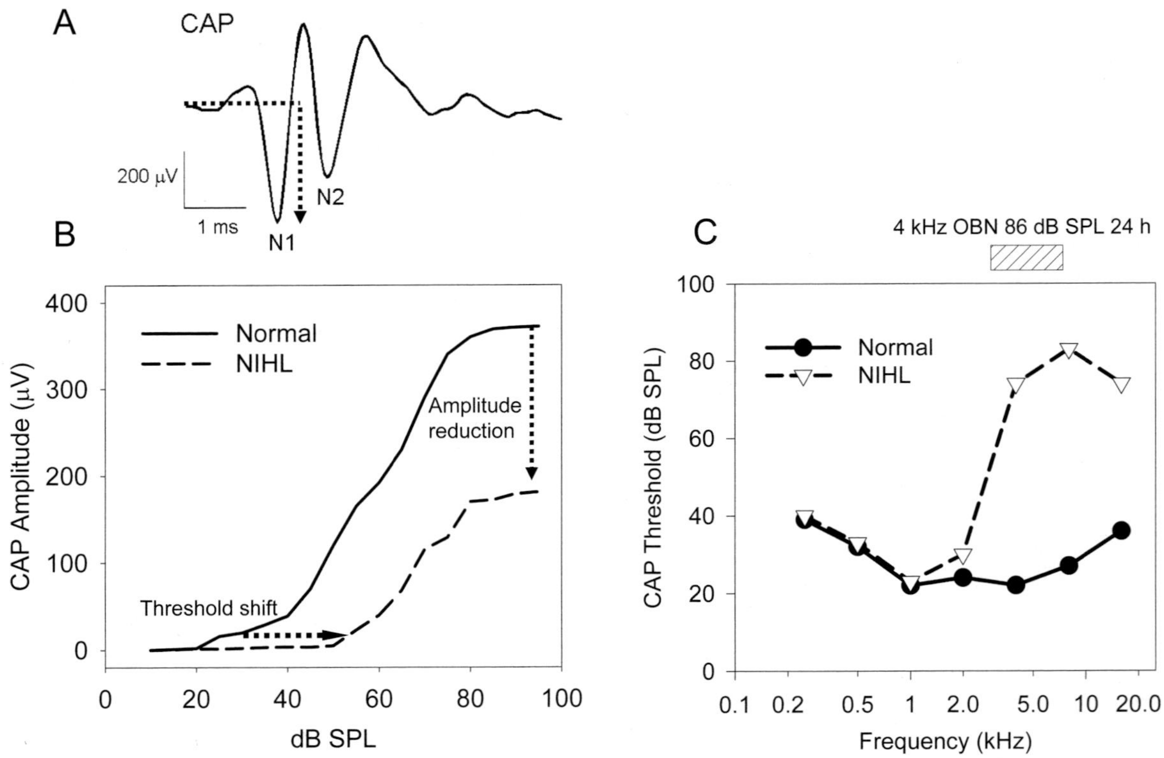

Figure 32–8. (A) CAP waveform with N1 and N2 components. (B) CAP input/output function before and after NIHL. Note the reduction in CAP amplitude and the increase in CAP threshold resulting in threshold shift. (C) CAP threshold versus tone burst frequency in the normal control group and in the group exposed for 24 h to octave band noise centered at 4 kHz at 86 dB SPL.

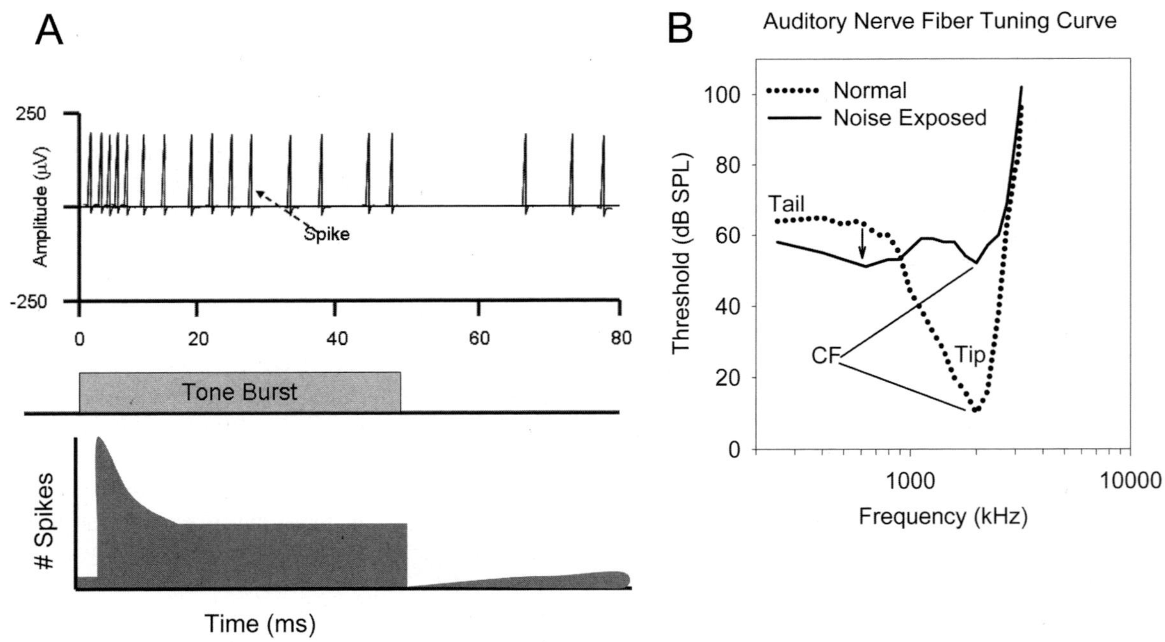

Figure 32–9. (A) Action potentials from an auditory nerve fiber during the presentation of a tone burst at a frequency–intensity combination that caused an increase in spike rate during the presentation of the tone burst. The lower panel is a poststimulus time histogram showing the number of spike discharges occurring during and after the tone burst. (B) Frequency–threshold tuning curves from an auditory nerve fiber in a normal chinchilla (dotted line) and one from a noise-exposed animal (solid line). The normal tuning curve has a low-threshold, sharply tuned tip with a characteristic frequency (CF, frequency with lowest threshold) near 2 kHz; the normal tuning curve has a high-threshold, broadly tuned tail beginning roughly one octave below CF. The tuning curve from the noise-exposed chinchilla has elevated thresholds and broad tuning at the CF region located along the high-frequency side of the tuning curve. Thresholds in the low-frequency tail of the tuning curve are somewhat lower than normal (hyposensitive). Below the high-frequency cutoff, the threshold is nearly constant across the frequency resulting in poor frequency selectivity. (Redrawn from Salvi *et al.*[47])

spontaneous rate. The average spike discharge pattern during and following the tone burst can be measured by presenting the tone burst many times and counting the number of spikes that occur at specific time bins during and following the stimulus; this type of plot is referred to as a poststimulus time histogram and is shown below the tone burst in Figure 32–9A.

Tone burst frequency and intensity can be varied to determine which frequency–intensity combinations are just able to produce a significant increase in discharge rate above the spontaneous rate; this information is used to construct a frequency–threshold tuning curve for each fiber. The dotted line in Figure 32–9B shows the frequency–threshold tuning curve of an auditory nerve fiber in the chinchilla that has a characteristic frequency (CF) of 2 kHz; the CF is defined as the frequency with the lowest threshold. This fiber presumably innervates an IHC located near the middle of the chinchilla cochlea. The auditory nerve fiber tuning curves are similar in shape to psychophysical tuning curves (Figure 32–3A). Physiological tuning curves have a low-threshold, narrowly tuned tip at CF, a steep high-frequency slope, and a threshold plateau in the low-frequency tail. If recordings are obtained from many neurons, auditory nerve fiber tuning curves can be found with CFs spanning the entire range of hearing of the animal.[47] That is, the auditory nerve contains fibers tuned to a frequency range covering the entire range of hearing.

Auditory nerve fiber tuning curves have been evaluated in animals with varying degrees of NIHL and varying amounts of OHC and IHC damage. When animals are exposed to high-level noise that causes 40–50 dB of hearing loss and primarily damages the OHC, then fiber thresholds increase significantly around the CF and the tip of the tuning curve (Figure 32–9B). In addition, thresholds in the tail of the tuning curve sometimes decrease (hyposensitive). Thus, the fiber not only becomes less sensitive, but more broadly tuned responding with almost equal sensitivity to most frequencies from the tail up to the high-frequency cutoff of the tuning curve. These results indicate that OHC are largely responsible for giving each fiber its extraordinary sensitivity and exquisite tuning. In cases in which both the OHC and IHC are destroyed, acoustic information can no longer be relayed to the auditory nerve fibers innervating the region of total hair cell damage. In cases in which there is total loss of OHC and IHC, a gap appears in the CF distribution.[47] Psychoacousticians often refer to regions with complete loss of IHC and OHC as cochlear "dead regions" since sound-induced vibrations from these regions fail to produce any neural activity.[48,49]

BIOLOGICAL BASIS OF NOISE-INDUCED HEARING LOSS

Over the past 30 years, much has been learned about how intense sounds affect hearing, the anatomical structures in the ear, and the physiological properties of the cochlea. In the past few years, research has focused on understanding the cellular basis of NIHL with the long-term goal of developing intervention strategies that might protect the hair cells from noise-induced damage and reduce the risk of NIHL.

CELL DEATH, OXIDATIVE STRESS, AND NOISE-INDUCED HEARING LOSS As with many other forms of tissue injury, noise-induced hair cell loss can occur through necrosis or apoptosis.[50,51] Necrotic cell death is characterized by swelling of the cell body and lyses of the cell membrane leading to leakage of the cell's content into the surrounding intercellular space, which can generate an inflammatory response. In cases of extremely high levels of noise, the mechanical vibrations in the inner ear may rip apart sections of the sensory epithelium leading to the rupture of cell membranes and necrotic cell death.[52–54] Apoptosis, or programmed cell death, is characterized morphologically by chromatin condensation, nuclear condensation and fragmentation, membrane blebbing, and formation of apoptotic bodies. Hair cell death associated with lower level noise exposures that do not lead to direct mechanical injury to the hair cells may arise from the overproduction of reactive oxygen species arising from metabolic stress.[55] Reactive oxygen species implicated in NIHL include the superoxide radical, hydrogen peroxide, the hydroxyl radical, and peroxynitrite. The metabolic stress associated with acoustic overstimulation has been shown to lead to the overproduction of reactive oxygen species.[56,57] Since acoustic overstimulation can decrease blood flow to the cochlea, the subsequent reperfusion of blood may be a major source of reactive oxygen species.[57]

PHARMACOLOGICAL STRATEGIES FOR PREVENTING NOISE-INDUCED HEARING LOSS Armed with the fact that noise may induce the overproduction of reactive oxygen species in the cochlea, numerous animal studies have been undertaken to identify pharmacological strategies that can prevent NIHL and block hair cell death by suppressing the production of reactive oxygen species. For example, *R*-phenylisopropyladenosine (*R*-PIA), which upregulates glutathione and superoxide dismutase, endogenous antioxidants, reduces NIHL and hair cell loss.[58] Similarly, acetyl-L-carnitine, which enhances the function of mitochondria during oxidative stress, and D-methionine, which enhances the production of the endogenous antioxidant, glutathione, reduce NIHL and hair cell loss.[59] Likewise, when glutathione monoethyl ester, a precursor of glutathione, is applied to the round window, it reduced the amount of NIHL and hair cell loss.[60]

CONCLUSIONS

Animal models have played an important role in determining how the intensity, duration, and spectrum of a noise affects the amount of hearing loss and how hearing perception changes when the OHC and IHC are damaged. High-level noise exposures tend to damage OHC before IHC. The first signs of damage are often seen on the stereocilia bundles that contain the mechanical-to-electrical transduction apparatus. Noise-induced destruction of the OHC results in a 40–50 threshold elevation plus a loss of frequency selectivity and temporal resolution. OHC damage leads to a significant reduction or complete loss of DPOAE, whereas auditory nerve fibers show a selective elevation of threshold near CF and significant loss of tuning. When the noise exposure is severe enough to destroy both the IHC and OHC, neural activity can no longer be relayed to the central nervous system resulting in a cochlear "dead region." Since the hair cells in the mammalian inner ear do not spontaneously regenerate, preventing NIHL hearing loss from occurring has become a major research focus. Since high-level noise exposures appear to generate toxic reactive oxygen species such as the superoxide radical, hydrogen peroxide, and the hydroxyl radical, a promising approach to preventing NIHL is to enhance the antioxidant capacity of the inner ear using endogenous or exogenous therapeutic agents. Positive results obtained in animal models of NIHL are therefore likely to lead to translational research aimed at preventing NIHL in humans who are routinely exposed to high-level noise.

REFERENCES

1. Ward WD. Presbycusis, sociocusis and occupational noise-induced hearing loss. *Proc R Soc Med* 1971;64(2):200–203.
2. Liberman MC, Kiang NYS. Acoustic trauma in cats: Cochlear pathology and auditory nerve activity. *Acta Otolaryngol Suppl* 1978;358:5–63.
3. Miller JD. Audibility curve of the chinchilla. *J Acoust Soc Am* 1970;48:513–523.
4. DeCory L, Dancer AL, Aran J-M. Species differences and mechanisms of damage. In: Dancer AL, Henderson D, Salvi R, Hamernik RP, Eds. *Noise-Induced Hearing Loss*. St. Louis, MO: Mosby Year Book, 1992:73–88.
5. Mills JH, Schmiedt RA, Kulish LF. Age-related changes in auditory potentials of Mongolian gerbil. *Hear Res* 1990;46(3):201–210.
6. Puel JL, Saffiedine S, Gervais d'Aldin C, Eybalin M, Pujol R. Synaptic regeneration and functional recovery after excitotoxic injury in the guinea pig cochlea. *CR Acad Sci III* 1995;318(1):67–75.
7. Davis RR, Kozel P, Erway LC. Genetic influences in individual susceptibility to noise: A review. *Noise Health* 2003;5(20):19–28.
8. Clark WW, Clark CS, Moody DB, Stebbins WC. Noise-induced hearing loss in the chinchilla, as determined by a positive-reinforcement technique. *J Acoust Soc Am* 1974;56(4):1202–1209.
9. Salvi RJ, Hamernik RP, Henderson D. Discharge patterns in the cochlear nucleus of the chinchilla following noise induced asymptotic threshold shift. *Exp Brain Res* 1978;32(3):301–320.
10. Long GR, Miller JD. Tone-on-tone masking in the chinchilla. *Hear Res* 1981;4:279–286.
11. Henderson D, Salvi R, Pavek G, Hamernik R. Amplitude modulation thresholds in chinchillas with high-frequency hearing loss. *J Acoust Soc Am* 1984;75(4):1177–1183.
12. Salvi RJ, Perry J, Hamernik RP, Henderson D. Relationship between cochlear pathologies and auditory nerve and behavioral responses following acoustic trauma. In: Hamernik RP, Henderson D, Salvi RJ, Eds. *New Perspectives on Noise-Induced Hearing Loss*. New York: Raven Press, 1982:165–188.
13. Giraudi-Perry DM, Salvi RJ, Perry D. Gap detection in hearing-impaired chinchillas. *J Acoust Soc Am* 1982;72:1387–1393.
14. Mills JH. Threshold shifts produced by a 90-day exposure to noise. In: Henderson D, Hamernik RP, Mills JH, Dosanjh D, Eds. *Effects of Noise on Hearing*. New York: Raven Press, 1976:265–275.
15. Saunders JC, Mills JH, Miller JD. Threshold shift in the chinchilla from daily exposure to noise for six hours. *J Acoust Soc Am* 1977;61(2):558–570.
16. Mills JH. Temporary and permanent threshold shifts produced by nine-day exposures to noise. *J Speech Hear Res* 1973;16(3):426–438.
17. Mills JH, Talo SA, Gordon GS. Decay of temporary threshold shift in noise. *J Speech Hear Res* 1973;16(2):267–270.
18. Mills JH, Talo SA. Temporary threshold shifts produced by exposure to high-frequency noise. *J Speech Hear Res* 1972;15(3):624–631.
19. Rhode WS. Cochlear partition vibration–recent views. *J Acoust Soc Am* 1980;67(5):1696–1703.
20. Carder HM, Miller JD. Temporary threshold shifts from prolonged exposure to noise. *J Speech Hear Res* 1972;15(3):603–623.
21. Clark WW. Recent studies of temporary threshold shift (TTS) and permanent threshold shift (PTS) in animals. *J Acoust Soc Am* 1991; 90(1):155–163.
22. Wightman FL. Psychoacoustic correlates of hearing loss. In: Hamernik RP, Henderson D, Salvi RJ, Eds. *New Perspectives on Noise-Induced Hearing Loss*. New York: Raven Press, 1982:375–394.
23. Giraudi D, Salvi R, Henderson D, Hamernik R. Gap detection by the chinchilla. *J Acoust Soc Am* 1980;68(3):802–806.
24. Spoendlin H. The afferent innervation of cochlea. In: Naunton R, Fernandez C, Eds. *Evoked Electrical activity in the Auditory Nervous System*. New York: Academic Press, 1978:21–38.
25. Boettcher FA, Spongr VP, Salvi RJ. Physiological and histological changes associated with the reduction in threshold shift during interrupted noise exposure. *Hear Res* 1992;62(2):217–236.
26. Puel JL, Ruel J, Gervais d'Aldin C, Pujol R. Excitotoxicity and repair of cochlear synapses after noise-trauma induced hearing loss. *Neuroreport* 1998;9(9):2109–2114.
27. Hirose K, Liberman MC. Lateral wall histopathology and endocochlear potential in the noise-damaged mouse cochlea. *J Assoc Res Otolaryngol* 2003;4(3):339–352.
28. Liberman MC, Dodds LW. Single-neuron labeling and chronic cochlear pathology, III: Stereocilia damage and alterations in threshold tuning curves. *Hear Res* 1984;16:55–74.
29. Pickles JO, Osborne MP, Comis SD. Vulnerability of tip links between stereocilia to acoustic trauma in the guinea pig. *Hear Res* 1987;25(2–3):173–183.
30. Hudspeth AJ, Gillespie PG. Pulling springs to tune transduction: Adaptation by hair cells. *Neuron* 1994;12(1):1–9.
31. Ding D, McFadden S, Salvi RJ. Cochlear hair cell densities and inner ear staining techniques. In: Willott J, Ed. *The Auditory Psychobiology of the Mouse*. Boca Raton, FL: CRC Press, 2001:189–204.
32. Greenwood DD. A cochlear frequency-position function for several species–29 years later. *J Acoust Soc Am* 1990;87(6):2592–2604.
33. Kemp D. Stimulated acoustic emissions from within the human auditory system. *J Acoust Soc Am* 1978;64:1386–1391.
34. Kemp DT. Towards a model for the origin of cochlear echoes. *Hear Res* 1980;2(3–4):533–548.
35. Brownell WE, Bader CR, Bertrand D, de Ribaupierre Y. Evoked mechanical responses of isolated cochlear outer hair cells. *Science* 1985;227:194–196.
36. Ashmore JF. A fast motile response in guinea-pig outer hair cells: The cellular basis of the cochlear amplifier. *J Physiol (Lond)* 1987; 388:323–347.
37. Zheng J, Shen W, He DZ, Long KB, Madison LD, Dallos P. Prestin is the motor protein of cochlear outer hair cells. *Nature* 2000; 405(6783):149–155.
38. Liberman MC, Gao J, He DZ, Wu X, Jia S, Zuo J. Prestin is required for electromotility of the outer hair cell and for the cochlear amplifier. *Nature* 2002;419(6904):300–304.
39. Hofstetter P, Ding D, Powers N, Salvi RJ. Quantitative relationship of carboplatin dose to magnitude of inner and outer hair cell loss and the reduction in distortion product otoacoustic emission amplitude in chinchillas. *Hear Res* 1997;112(1–2):199–215.
40. Brown AM. Acoustic distortion from rodent ears: A comparison of responses from rats, guinea pigs, and gerbils. *Hear Res* 1987;31:25–38.
41. Eddins AC, Zuskov M, Salvi RJ. Changes in distortion product otoacoustic emissions during prolonged noise exposure. *Hear Res* 1999;127(1–2):119–128.
42. Salvi RJ, Hamernik RP, Henderson D. Auditory nerve activity and cochlear morphology after noise exposure. *Arch Oto-Rhino-Laryngol* 1979;224(1–2):111–116.
43. Salvi R, Henderson D, Hamernik RP, Parkins C. VIII nerve response to click stimuli in normal and pathological cochleas. *Hear Res* 1980;2(3–4):335–342.
44. Liberman MC. Single-neuron labeling and chronic cochlear pathology. I. Threshold shift and characteristic-frequency shift. *Hear Res* 1984;16(1):33–41.
45. Robertson D. Horseradish peroxidase injection of physiologically characterized afferent and efferent neurones in the guinea pig spiral ganglion. *Hear Res* 1984;15:113–121.
46. Salvi R, Hamernik RP, Henderson D. Auditory nerve fiber activity and cochlear morphology after noise trauma. *Acta Otolaryngol* 1979;224:111–116.
47. Salvi R, Hamernik RP, Henderson D. Response patterns of auditory nerve fibers during temporary threshold shift. *Hear Res* 1982;10: 37–67.
48. Huss M, Moore BC. Dead regions and pitch perception. *J Acoust Soc Am* 2005;117(6):3841–3852.
49. Moore BC. Dead regions in the cochlea: Conceptual foundations, diagnosis, and clinical applications. *Ear Hear* 2004;25(2):98–116.
50. Samali A, Gorman AM, Cotter TG. Apoptosis–the story so far. *Experientia* 1996;52(10–11):933–941.

51. Snider BJ, Gottron FJ, Choi DW. Apoptosis and necrosis in cerebrovascular disease. *Ann NY Acad Sci* 1999;893:243–253.
52. Wang J, Dib M, Lenoir M, *et al.* Riluzole rescues cochlear sensory cells from acoustic trauma in the guinea-pig. *Neuroscience* 2002; 111(3):635–648.
53. Thorne PR, Gavin JB, Herdson PB. A quantitative study of the sequence of topographical changes in the organ of Corti following acoustic trauma. *Acta Otolaryngol* 1984;97(1–2):69–81.
54. Hu BH, Henderson D, Nicotera TM. Extremely rapid induction of outer hair cell apoptosis in the chinchilla cochlea following exposure to impulse noise. *Hear Res* 2006;211(1–2):16–25.
55. Henderson D, Bielefeld EC, Harris KC, Hu BH. The role of oxidative stress in noise-induced hearing loss. *Ear Hear* 2006;27(1):1–19.
56. Yamane H, Nakai Y, Takayama M, *et al.* The emergence of free radicals after acoustic trauma and strial blood flow. *Acta Otolaryngol Suppl* 1995;519:87–92.
57. Ohlemiller KK, Wright JS, Dugan LL. Early elevation of cochlear reactive oxygen species following noise exposure. *Audiol Neurootol* 1999;4(5):229–236.
58. Hu BH, Zheng XY, McFadden SL, Kopke RD, Henderson D. R-phenylisopropyladenosine attenuates noise-induced hearing loss in the chinchilla. *Hear Res* 1997;113(1–2):198–206.
59. Kopke RD, Coleman JK, Liu J, Campbell KC, Riffenburgh RH. Candidate's thesis: Enhancing intrinsic cochlear stress defenses to reduce noise-induced hearing loss. *Laryngoscope* 2002;112(9):1515–1532.
60. Hight NG, McFadden SL, Henderson D, Burkard RF, Nicotera T. Noise-induced hearing loss in chinchillas pre-treated with glutathione monoethylester and R-PIA. *Hear Res* 2003;179(1–2):21–32.

Models for Specific Purposes IV

Trauma, Pain, and Neurology B

33 Human and Animal Models for the Study of Muscle Pain

Norman F. Capra and Jin Y. Ro

ABSTRACT

Muscle pain typically results from engaging in novel or unaccustomed levels of muscle activity. It may also result from a sharp blow or other direct trauma. Typically this type of pain resolves over a few days. However, several clinical conditions are associated with muscle pain that may range in severity from an intermittent but recurring distraction, to a debilitating disorder lasting for months that is accompanied by severe stress and in some cases significant motor impairment. The cost of medical treatment and lost work productivity runs in the billions of dollars, yet little is known about the etiology of chronic muscle pain. This chapter will explore how experimental approaches have been used in an effort to model clinical pain in humans and in experimental animals, and provide practical information regarding the implementation, interpretation, and limitations of these methods. Parallels between human and animal experiments will be emphasized.

Key Words: Chronic muscle pain, Hyperalgesia, Allodynia, Experimental models, Intramuscular injection, Exercise, Ischemic contraction, Induction, Assessment.

INTRODUCTION

The costs of medical treatment, lost work productivity, and suffering resulting from chronic musculoskeletal pain conditions run in the billions of dollars, yet relatively little is known about the underlying mechanisms of chronic muscle pain. Chronic muscular pain conditions may be regional [e.g., temporomandibular disorders (TMD) and low back pain] or widespread, involving multiple tender points distributed over most of the body [e.g., fibromyalgia syndrome (FMS)]. Progress in understanding the pathophysiological basis for these disorders has been limited by the lack of adequate human and animal models to study chronic muscle pain. However, renewed interest in the subject has led to refinements to early methods and the development of promising new approaches for the study of chronic muscle pain.

Several strategies have been used to experimentally induce muscle pain including ischemic contraction, novel or prolonged exercising, mechanical stimulation, electrical stimulation, and intramuscular injection of algesic substances (i.e., chemical stimulation) in humans. Each of these methods has also been adapted to study peripheral and central nociceptive mechanisms in experimental animals. The judicious implementation of experimental tools that are not feasible in humans and the development of novel methods for the assessment of nociceptive processing in animal models have provided insight into the mechanisms of persistent pain in humans. This chapter will consider several experimental approaches in which parallels between the results of human and animal experiments have enhanced the understanding of the mechanisms of chronic muscle pain.

Muscle pain produces characteristic sensory and sensorimotor manifestations. Precise assessment of these characteristic features is necessary to evaluate therapeutic efficacy and to study underlying mechanisms. Pain of muscle origin is generally more diffuse than cutaneous pain, it spreads to adjacent muscles, and it is commonly referred to remote sites. Assessing the distribution of pain in patients with chronic muscle pain is clinically useful. The quality of pain originating in muscles differs from superficial pain and is often described by patients as aching, tender, and throbbing. Tenderness, a characteristic symptom of persistent muscle pain, results from sensitization of peripheral afferents and central neurons. Sensitization results in lowering of the mechanical threshold of muscle afferent receptors to noxious and innocuous mechanical stimuli leading to hyperalgesia and allodynia, respectively. In addition, complex emotional signs and autonomic symptoms[1] often accompany chronic muscle pain.

Pain arising in muscle and associated deep tissues (e.g., investing fascia, blood vessels) begins with the activation of small-diameter group III and group IV afferent nerve fibers. The morphology and distribution of muscle afferents contribute to the characteristic properties of muscle pain. They are diffusely distributed and terminate in unmyelinated (free) nerve endings. Many of these afferents are polymodal, responding to a wide range of stimuli. The neurobiology of muscle pain, a glossary of pain terminology, and a wealth of information on clinical muscle pain conditions are provided by Mense and colleagues.[1–3]

EXPERIMENTAL MUSCLE PAIN: INDUCTION

In the 1930s, Kellgren introduced the technique of making intramuscular injections with hypertonic saline (HS; 5%) to model deep pain.[4] This method is recognized as a safe, reliable, and reproducible experimental procedure. Most importantly, subjects report that pain resulting from a single intramuscular HS injection is similar in quality and intensity to clinical pain of muscle origin Each of these conditions—safety, reliability, reproducibility, and clinical relevance—is an essential ingredient for the study of experimentally induced muscle pain in humans.

From: *Sourcebook of Models for Biomedical Research*
(P. M. Conn, ed.), © 2008 Humana Press Inc., Totowa, NJ.

Also in the 1930s, Lewis introduced the technique of ischemic contraction to produce muscle pain.[5,6] This method has been used successfully to address certain multidimensional aspects of deep tissue pain[7] and is appropriate for the study of clinical pain disorders that appear to be primarily related to vascular insufficiency.

Exercise-based models, which produce pain in response to unaccustomed muscle activity, have been widely used by sports physiologists to produce delayed onset muscle soreness (DOMS).[8] The cause, the onset, and the time course for resolution of DOMS are well characterized. As more is learned about the mechanism of DOMS, it is becoming increasingly clear that it is connected to inflammatory processes similar to those associated with chronic inflammatory pain.[9,10] As a result, the potential of this method to address clinically relevant questions is being recognized by experimentalists.

INTRAMUSCULAR INJECTION Of the methods described above, intramuscular injection of algesic and inflammatory agents is the most common method of induction. This method involves either injecting a discrete bolus of a sterile solution containing an algogen (pain-producing agent) or the continuous infusion of an algogen (usually HS) over a period of minutes into the muscle of interest. Pain assessments are typically made before, during, and after algesic administration.

Algogens In addition to the advantages described above, HS may be injected several times in the same muscle or it may be preceded or followed by injection with isotonic saline, which serves as a vehicle to control nonspecific effects. This is possible because single injections of HS do not produce sensitization or "tenderness," a characteristic feature of clinical muscle pain. To study sensitization, an alternate algogen or injection strategy must be considered.

A number of *endogenous* and *exogenous* algogens have been used for making intramuscular injections in human and animal experiments (Table 33–1).[11–16] Endogenous algogens are substances found naturally in the body such as the metabolites of exercising muscle or substances that accumulate in the interstitial tissues subsequent to injury or with inflammation (i.e., myositis). Most algogens produce measurable sensitization (e.g., a reduction in mechanical threshold) following a single injection and when combined, endogenous algogens may synergize to produce greater pain and sensitization than injection of either substance alone.[17] In addition to algogens tested in humans, exogenous agents not considered safe for human use have been used to study nociceptive mechanisms in animals. For example, complete Freund's adjuvant (CFA) produces significant and prolonged inflammation and is used to model persistent muscle pain.

Many algogens operate through specific receptor-mediated mechanisms. Capsaicin, which activates C fiber (group IV) nociceptors, binds to the TRPV1 receptor and is present in a subpopulation of muscle afferents. Capsaicin is one of the few exogenous algogens that have been used to induce muscle pain in human studies.[18] Sequential intramuscular injection strategies, involving injection with specific receptor agonists and antagonists prior to algesic injections, have been exploited to study the involvement of specific membrane receptors in peripheral aspects of muscle pain. A list of algogens is provided in Table 33–1 with references to selected literature describing how they were used to study muscle pain.

Bolus Injections

Injection Volume In most cases, injection volume has been empirically determined based on the size and morphology of the muscle under study. For animal experiments, adequacy of the injection volume and intramuscular location may be confirmed by including an inert dye in the injection solution and performing careful postmortem dissection. Following injection of algogens that produce significant tissue edema, Evan's Blue may be administered near the end of the experiment. Localization of the injection site can be inferred from the distribution of Evans Blue and the extent of edema can be determined by measures of plasma extravasation.[19,20] Cairns *et al.*[21] used magnetic resonance imaging to determine edema volume in the masseter muscle following injection with glutamate.

Injection Parameters Consideration must also be given to the duration, timing, and sequencing of injections. Injections are usually made slowly to limit effects due to increases in local tissue pressure. If multiple injections of algogen are to be performed, sufficient time must be allowed for the effects of each injection to clear. If drugs are injected prior to algogen the timing is critical to maximize the drug effect. By carefully timing the injection of antagonists prior to injections with glutamate, Ro and others have provided compelling evidence that glutamate has important actions at peripheral receptors in muscle nociception.[22–25] Algesic injections are evaluated against isovolemic injections with vehicle and observations are made with the experimenter blind to the injection condition.

The initial effect and the duration of algesic action are also important considerations. For example, the primary effects of injecting a bolus of HS (e.g., a pain report in humans or increased neuronal activity in animals) dissipate within several minutes and vehicle or another dose of HS can be reinjected into the same site in the same subject. In contrast to HS, glutamate produces post-injection sensitization and edema. In such cases, reinjection into the same muscle is not practical, so experiments must involve separate experimental and control groups.

Table 33–1
Algogens used for intramuscular injection[a]

Substance	*Site of action*
Endogenous	
Acid saline (pH)[26,27]	ASICS receptors
Adenosine[11]	P2X receptors
Bradykinin[17]	B1 and B2 receptors
Potassium chloride[12,13]	NS
Glutamate[22,23]	NMDA, AMPA, MGluR receptors
Hypertonic saline[28,35,37]	NS
NGF[14]	Trk A receptor
PGE_2[9]	EP 1–4 receptors
Serotonin[17]	5-HT receptors
Substance P[17]	NK-1 receptors
TNF-α [15]	TNFA R1 and RII
Exogenous	
Capsaicin[18]	TRPV1
Carrageenan[44,56]	NS
Formalin[54]	NS
Complete Freund's adjuvant[50]	NS
Mustard oil[16,22,23]	TRPA1

[a]NS, nonselective; NMDA, *N*-methyl-D-aspartate; AMPA, α-amino-3-hydroxy-5-methyl-4-isoxazolepropionate; NGF, nerve growth factor; PGE_2, prostaglandin E_2; 5-HT, 5-hydroxytryptamine; TNF, tumor necrosis factor.

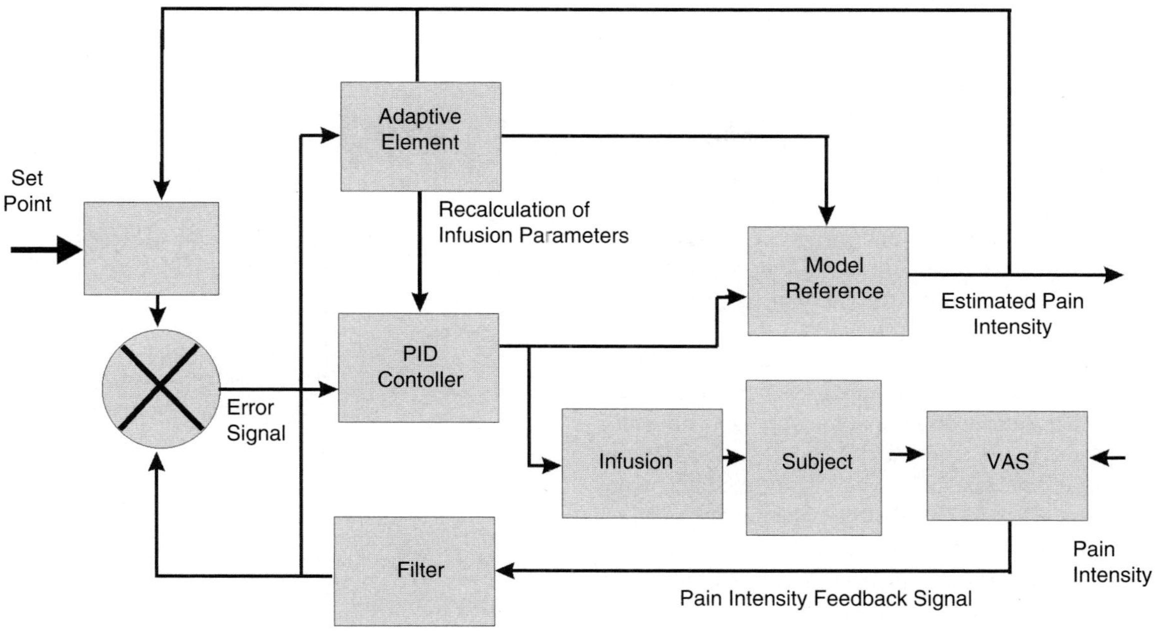

Figure 33–1. Block diagram of the closed loop system designed to induce and maintain pain intensity. (Adapted from Stohler.[31])

An interesting approach developed by Sluka and co-workers[26,27] involved making repeated intramuscular injections of acid saline buffered to pH 5.0. In contrast to HS in a "normal" pH range, repeated injections with acid saline produced a prolonged reduction in withdrawal threshold to cutaneous mechanical stimuli and direct muscle pressure, with forceps attached to a force transducer. The long-term behavioral changes observed in animals after repeated acid injection may be equated with allodynia and hyperalgesia in humans.

Intramuscular Infusion Zhang and colleagues[28] developed an infusion system based on closed-loop drug delivery systems to model chronic masseter muscle pain, a common complaint of patients with TMD. Feedback from continuously rated visual analog pain ratings (a measure of pain intensity) was used to adjust HS infusion rates and maintain a constant pain level for as long as 20 minutes.[29,30] The major elements of the system include a microprocessor-controlled infusion pump interfaced with a PC, a proportional integral derivative (PID) controller, a strain gage pressure transducer, an electronic visual analog scale (VAS), and a tubing delivery system with meticulously defined mechanical properties to account for factors such as compliance and dead space (Figure 33–1). Complete methodological details are provided in the original publications.[28]

For the experiment, an infusion needle was placed in a standard location (i.e., in the "middle" of the masseter muscle). Once any pain produced by placing the needle subsided, subjects received an initial standard algesic injection and pain intensity ratings were obtained every 15 sec until it dissipated. The overall magnitude and time to peak pain for the initial response were recorded (i.e., the open loop response) and compared to a preestablished group mean response. A control algorithm used these values to calculate an initial infusion rate. The infusion was started and updated pain intensity values were used as feedback to adjust and maintain a constant pain level (VAS score).

Sustained infusion of HS produced complex patterns of pain-related states including primary and secondary muscle hyperalgesia and altered cutaneous sensibility. Importantly, this approach also evoked emotional and autonomic responses similar to those observed in chronic pain conditions. This method has provided insight into opioid mechanisms in the sensory and affective dimensions of pain,[30] revealed sex differences in opioid sensitivity,[32] and provided evidence that genetic factors underlie differential muscle pain sensitivity.[33] Based on the system described above, others have developed a standard set of infusion protocols referenced to the subject's initial pain report,[34–36] and further characterized the sensory and sensorimotor manifestations of persistent muscle pain.

To the best of our knowledge only one study employed HS infusion in an animal model.[37] This experiment clearly showed an enhanced activation of central nociceptive neurons compared to bolus injections. This response was clearly attenuated in a dose-dependent manner by preinjection into the muscle with MK-801, a noncompetitive *N*-methyl-D-aspartate (NMDA) receptor antagonist.

ISCHEMIC CONTRACTION Originally described by Lewis,[5,6] ischemic contraction was the first method developed to systematically produce deep tissue pain in humans. A modification known as the submaximum effort tourniquet technique was introduced by Smith and colleagues.[38] Following venous exsanguination of the forearm with an Esmarch bandage, a pressure cuff was inflated to occlude arterial flow. Subjects were instructed to perform a sequence of repeated contractions (squeeze a hand exerciser) at approximately half of their predetermined maximum voluntary contraction force and then to rest until they registered a maximum pain rating or reached a predetermined time limit before deflating the cuff.

This procedure is very effective in producing muscle pain. However, Graven-Neilsen *et al.*[39] reported that the quality of pain produced by ischemic contraction differed from pain reports following HS or adenosine injections and suggested that these methods activated different subpopulations of muscle nociceptors. Rainville *et al.*[7] compared four different types of phasic and tonic pain and found that in addition to producing intense pain, ischemic pain tended to be more "unpleasant" than other types of experimental pain.

A potential disadvantage of the ischemic contraction method is that ischemia is not limited to one muscle or muscle group so that there are concomitant changes in the afferent input from other nonmuscle deep tissues and from cutaneous sources. This may, in part, contribute to the observation that the pressure pain threshold[39] increases over the ischemic muscle, whereas it is generally reduced over the site of intramuscular algesic injection. Although the method lacks the selectivity of intramuscular injection, to model conditions with muscle pain in the face of reduced blood flow (e.g., compartment syndromes or intermittent claudication) this may be the best approach.

Ischemic contraction has not been used widely to study muscle pain in animals, but this method was important in the initial electrophysiological characterization of muscle nociceptive afferents. Bessou and Laporte[40] used a contraction model to demonstrate that group IV muscle afferents were specifically activated under ischemic conditions. Chabel applied a tourniquet to the hindlimb of rats and recorded nerve fiber activity from the proximal sciatic nerve.[41] While this study was not specifically directed toward the identification of muscle afferents, it revealed the gradual onset of spontaneous activity in slowly conducting afferents during the ischemia.

DELAYED ONSET MUSCLE SORENESS Exercise protocols used to produce delayed-onset muscle soreness (DOMS) typically involve a significant amount of eccentric contraction (i.e., muscle contraction during lengthening movements of the muscle) along with concentric contraction (shortening contractions). Onset of pain generally develops within 24–72 h after exercise and is associated with loss of strength, reduced range of motion, and elevated creatinine kinase levels.[8]

Although many strategies can be implemented to produce DOMS, the technique is well illustrated by the work of Tedeger and colleagues[9] who utilized a simple "stepping" protocol to produce DOMS. Subjects were instructed to stand with one foot on a 45° sloped step, lifted on the tip of the toes (concentric contraction), and then to slowly lower the heel until the sole of the foot made full contact on the sloped surface of the step (eccentric contraction). The tips of the toes were lifted again and contractions were repeated until subjects completed two sets of 50 combined concentric/eccentric contractions separated by a 5-min rest

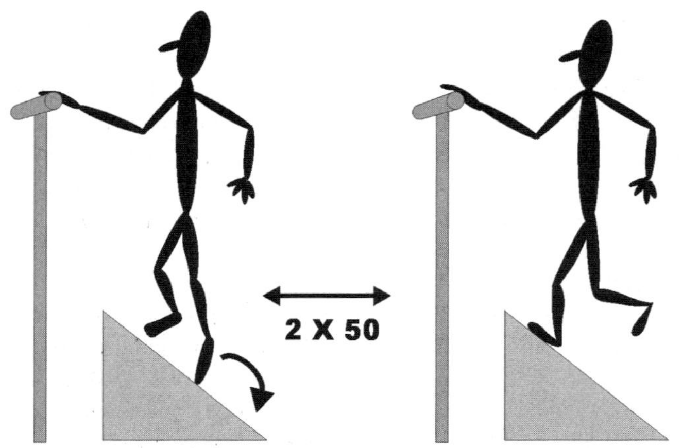

Figure 33–2. Illustration of the exercise protocol used by Tegeder *et al.*[9] to produce delayed onset of muscle soreness. (Adapted from Tegeder *et al.*,[9] with permission.)

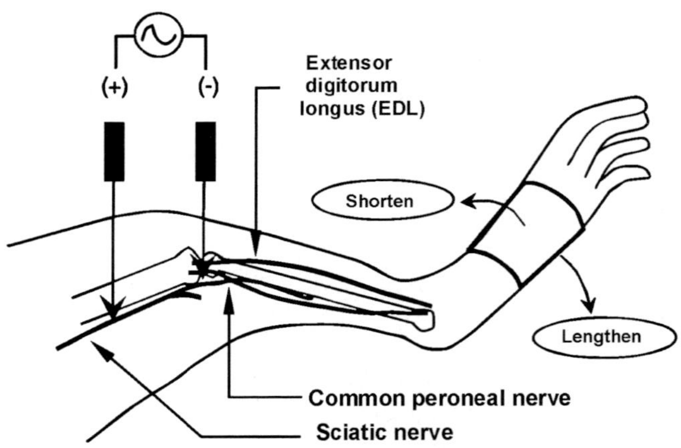

Figure 33–3. Schematic illustrating the experimental setup used by Taguchi *et al.*[43] to produce eccentric contraction in an animal model. (Adapted from Taguchi *et al.*,[43] with permission.)

(Figure 33–2). Pain intensity was measured prior to the exercise, immediately after, and 24 h after the exercise. Pain assessments were made in the DOMS and the control leg was assessed at rest, walking, going downstairs, and tiptoeing.

There is an impressive literature utilizing "exercise" or controlled muscle contraction (concentric and eccentric) in animals, but relatively few animal studies have used this strategy to explore mechanisms of deep tissue nociception. Taguchi and co-workers[42,43] have shown that repeated bouts of muscle contractions, which include an eccentric component, yield evidence for primary muscle hyperalgesia and sensitization in a rat model. In their model, the hindpaw was fixed to a bar, which in turn was connected to a linear motor. Transcutaneous needle electrodes were inserted near the sciatic nerve and the common peroneal nerve, respectively (Figure 33–3). Repetitive contractions of the muscle were induced by cyclic electrical stimulation of the nerve at three times the threshold for twitch contraction. The stimulus frequency was set at 50 Hz to produce tetanic contractions. Stimulus trains (1 sec duration) alternated with 3 sec of rest. Timing of the lengthening movements were synchronized with the electrical stimuli to produce contractions during lengthening. The control group was subjected to stretching without contraction. This procedure reliably resulted in reduction of the mechanical threshold to noxious and innocuous stimuli. Kehl and Fairbanks[44] ran rats on a treadmill with a 30° downhill incline to produce eccentric contraction of the limb muscles. This method produced a measurable reduction in grip force (an index of hyperalgesia, see below) compared to animals running on a level treadmill.

EXPERIMENTAL MUSCLE PAIN ASSESSMENT

HUMANS The sensory manifestations of muscle pain are commonly assessed with psychophysical tools that measure the intensity, the quality, and the distribution or location of the pain. Pain intensity is commonly assessed with magnitude rating scales. An ordinal rating scale numbered from 0 to 10, with 0 representing no pain and 10 being the worst pain imaginable, allows subjects to score their pain. Alternately, a subject or patient may rate pain by extending a line along a fixed distance (e.g., 0–10 cm) with a pencil on paper or on a computer monitor using a "mouse" or other pointing device. Computerized or electronic measures may be time stamped and the ratings may be saved in a spreadsheet or database during the course of an experiment. Although there is a great deal of intersubject variability, magnitude rating scales have been shown to be very reliable indicators of within-subject muscle pain.[45]

The McGill Pain Questionnaire (MPQ) and comparable instruments that involve verbal descriptors are used to assess the quality of muscle pain. Melzak and colleagues demonstrated that it was possible to correctly classify 91% of patients with facial pain conditions using seven verbal descriptors.[46] The discriminative capacity of the MPQ can be improved by the addition of terms uniquely suited for the type of pain condition. This is a particularly valuable suggestion when attempting to characterize pain of deep tissue origin. Discrimination of muscle pain would benefit from inclusion of both temporal and spatial descriptors. For example, it is important to ask whether the pain is constant or intermittent (temporal) or if it is local or diffuse (spatial).

Since muscle pain is commonly referred to other deep tissue sites, it is important to identify the distribution of the pain. This may be done with simple body maps in which the patient or subject is asked to draw or otherwise indicate the location or distribution of spontaneous or experimentally evoked pain. In addition to identification of spontaneous sites of muscle pain, palpation is commonly used to demonstrate areas of reduced mechanical threshold. The diagnosis of FMS depends upon the ability to demonstrate a minimum of 11 out of 18 "tender" points as described by Wolfe and others.[47] Although palpation can be performed manually to identify these areas, there are increasing efforts to obtain a quantitative measure of palpation pressure with algometry.

Hand-held algometers are widely used to assess muscle sensitivity. Several types are available ranging from a simple "analog" version to more sophisticated digital instruments. Most have a circular footplate with a diameter of 1 cm^2. Regardless of the design, all algometers provide a numerical value of the force (e.g., kg/cm^2) necessary for the subject to report pain. This value is referred to as the pressure pain threshold or PPT. Several caveats must be considered with the use of algometry. The PPT is strongly influenced by the method of application (most instruments are designed to be applied perpendicularly over the test site) and the composition of the tissue at the test site (e.g., the amount of body fat). Physical considerations such as the size of the footplate and the rate of application are also critical variables that influence PPT values. Last, but not least, the skill and experience of the examiner are essential in order to obtain reproducible results.

Evaluation of the sensorimotor response to pain has been the subject of much research and has led to a better understanding of the relationships between muscle pain and motor function. During voluntary motor activity muscle pain is generally accompanied by reduced motor activity when the sore or injured muscle is working as an agonist. This is often accompanied by increased activity in the antagonist muscles. This response, referred to as pain adaptation,[48] serves to protect the injured muscle. In contrast, myofascial pain is associated with focal regions of increased muscle activity (i.e., trigger points) characterized by the demonstration of "taut bands." It is important to distinguish between these two quite distinct entities when attempting to generalize about the relationships of muscle pain and motor function. This topic has been thoroughly reviewed by Mense and Simons.[1]

ANIMAL MODELS Although animals cannot report the subjective experience that we refer to as pain, methods for the induction of muscle pain in humans characteristically produce activation of nociceptors, recruitment of central nociceptive pathways, and predictable nocifensive behaviors in animals. In addition to the general ethical issues associated with the use of animals in research, the study of nociception in animal models requires special consideration. Extreme care must be taken to minimize exposure of animals to nociceptive stimuli, to recognize symptoms of unacceptable morbidity, and whenever possible to provide an "escape" from prolonged intense noxious stimuli. These and other important considerations are highlighted in *Definition of Pain and Distress and Reporting Requirements for Laboratory Animals* 42 http://www.nap.edu/catalog/10035.html.

Assessment of Muscle Nociception in Anesthetized Animals Many insights regarding muscle nociceptive mechanisms can be obtained from studies in anesthetized animals. Nociceptive afferents and their central target show changes in firing to graded noxious stimulation and are the electrophysiological correlate of intensity. Changes in the receptive field size of nociceptive neurons in response to experimental manipulations (e.g., intramuscular algesic injection) provide an assessment of spatial changes related to noxious stimuli. An anatomical approach to assess the intensity and extent of nociceptive activation is to study the pattern of expression of the immediate-early gene *c-fos* after noxious stimulation. Fos protein expression shows predictable increases, under carefully controlled conditions, and has been widely used as a measure of nociceptive activation in muscle.[23,37,50,51]

Behavioral Assessment of Muscle Pain in a Lightly Anesthetized Model A useful strategy to minimize discomfort in animals is to study nocifensive behaviors that are elicited under conditions of light anesthesia. Monitoring electromyogram (EMG) activity following intramuscular injection[52] has been used successfully as a method to document nociceptive activation in orofacial muscles following injection of the masseter muscle with algesic chemicals in halothane-anesthetized animals. Lightly anesthetized animals are often employed to assess changes in the magnitude of muscle reflexes, and the production of nocifensive behavior in response to noxious deep tissue stimulation. However, nocifensive reflexes and behaviors documented under light anesthesia must be carefully assessed and validated. Responses obtained with one anesthetic agent may be qualitatively and quantitatively different from those obtained with another agent in the same species.

Examination of nocifensive behaviors in lightly anesthetized animals has been used successfully by Ro and colleagues to make inferences about the intensity and the duration of algesic stimuli injected into the masseter muscle.[53] Algesic injections made into the masseter muscle of lightly anesthetized rats elicited a brisk repetitive hindpaw shaking that appeared to reflect an attempt to rub the site of injection while injection with vehicle produced minimal hindpaw shaking behavior. This behavior is similar to previously described orofacial pain-induced grooming behavior in awake rats.[54] The anesthetic level was maintained by infusing a constant rate of sodium pentobarbital (Nembutal; 3–5 mg/h). Animals in the light anesthetic state typically responded to noxious tail pinch with an abdominal contraction and noxious hindpaw pinch with a calibrated clip (600 g) produced a brisk single withdrawal response. Core temperature and heart rate were monitored continuously to ensure the animals were within a normal physiological range throughout the experiment.

Once the animals were stable, selected algogens were injected into the masseter muscle. The behavior was recorded with a digital camera and stored on compact disc for subsequent analysis. Hindpaw shakes were counted by an individual, blinded to the experimental condition. Counts were distributed into 30-sec bins for the duration of the behavior. The mean peak count, which usually occurred in the first 30 sec, and area under the curve were calculated from the resulting histograms. Both parameters varied in a concentration-dependent manner following injection with mustard oil. Systemic administration of morphine sulfate (3 and 0.3 mg/kg), ip, dose-dependently attenuated the shaking behavior. Lidocaine injected locally 5 min prior to mustard oil injection also significantly decreased the behavior. Similar results have been obtained with injection of glutamate, capsaicin, acid saline, and complete Freund's adjuvant. Regardless of the algogen used, this method produced a relatively finite stereotypical hindpaw shaking response that lasted no more than 4 or 5 min following injection. These experiments validate this model for assessing the intensity and duration of acute nociceptive activation and may also be a useful test of novel pharmacological interventions.

Postinjection Sensitization in a Lightly Anesthetized Model It is possible to rekindle the hindpaw shaking behavior by mechanical stimulation with von Frey filaments or an electronic von Frey in lightly anesthetized rats. Approximately 5 min after the hindpaw behavior subsides, application of von Frey monofilaments of relatively high bending forces generates a few cycles of shaking behavior in animals injected with sensitizing agents (e.g., glutamate, capsaicin, acid saline). Monofilaments are applied perpendicular to the injection site in a stepwise fashion and the smallest diameter von Frey capable of "rekindling" the hindpaw shaking behavior in response to four consecutive applications at bending force was considered threshold. Repeating the procedure periodically (e.g., every 5–10 min) resulted in the demonstration of a treatment-dependent reduction in threshold that persisted for at least an hour after injection.

Behavioral Assessment in Awake Animals It is possible to condition animals to sit quietly for measurements in response to stimulation with manual or electronic von Frey filaments to determine changes in mechanical threshold.[20,55] This approach has been used to study changes in mechanical threshold following intramuscular injection with algesic chemicals.[20] Sensorimotor manifestations of muscle inflammation and nociceptive stimulation may be assessed by indirect approaches such as monitoring locomotor behavior or studying feeding behavior. Recently, however, behavioral assays based on the predictions of the pain adaptation model have been used to assess muscle hyperalgesia and allodynia in animals.

Grip Force as a Model for Muscle Hyperalgesia Rats and mice exhibit an innate grasping behavior when suspended over a surface like a cage top or wire grid. Kehl *et al.*[56] took advantage of this behavior to develop a rodent model for assessment of muscle hyperalgesia. They postulated that injection of carrageenan into the triceps muscles would produce an "anatomically" specific reduction of grip force with respect to preinjection values. During testing, each animal was held by the tail, just above wire mesh grids attached to force transducers and gently pulled in a rostrocaudal direction. Output from the force transducers was digitized and led to a computer that display force versus time traces. Initially, three experimental groups were examined. One group received bilateral carrageenan injections (3%, 75 μl), a second received bilateral vehicle injections [phosphate-buffered saline (PBS), 75 μl], and a third received carrageenan in one triceps and vehicle (PBS) in the contralateral triceps. Carrageenan injections produced a significant time- and treatment-dependent

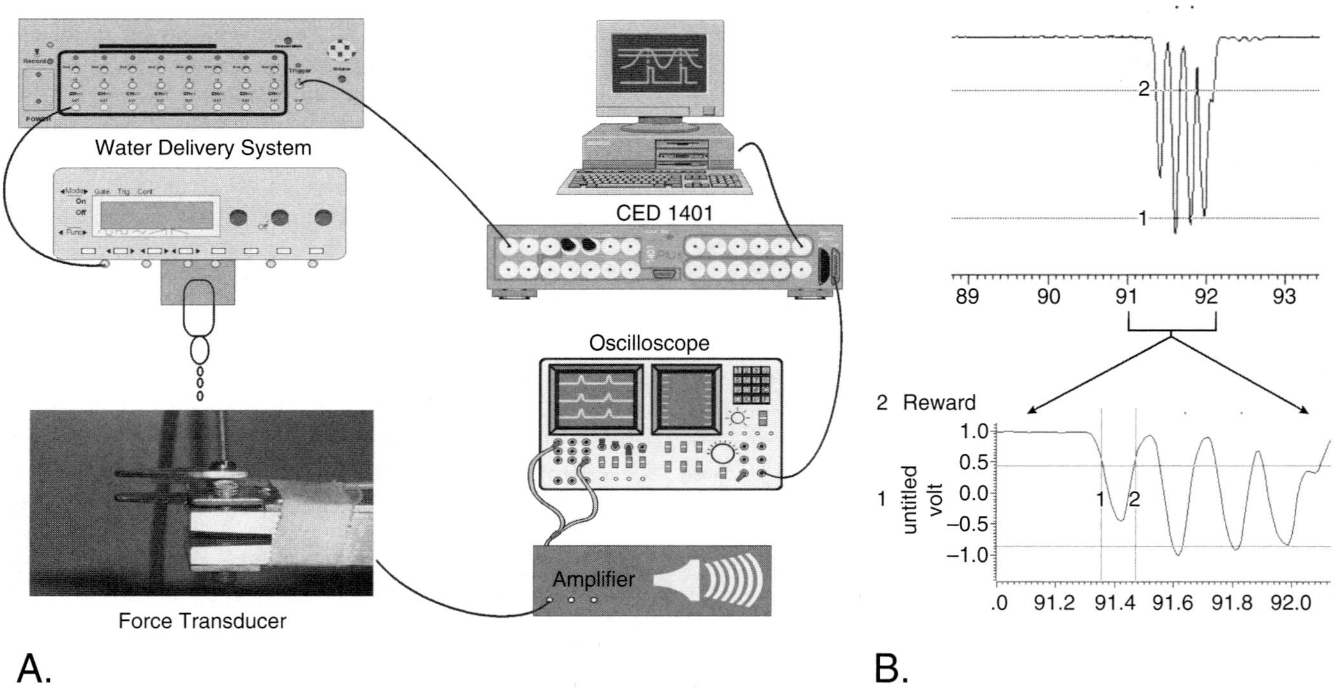

Figure 33–4. Schematic illustrating the bite force apparatus. In addition to the description in the text, full methodological descriptions are provided in Nies and Ro.[57] (A) The signal from the bite plates is led to a device to provide water and the output of the transducer is led to an A/D to record the bite force. (B) A software program is used to set horizontal and vertical (1 and 2) cursors to measure the bite force and duration.

reduction in gripping force that peaked at 12 h but remained significantly reduced at 36 h. Treatment with six different concentrations (0–6 mg) produced dose-dependent reductions in the grasp strength measured at 24 h after injection.

The demonstration that significant recovery of grip force was obtained by treatment with levorphanol (a μ-opioid selective agonist) provided compelling evidence that the reduction of threshold was a specific response. The reduction in grip strength was also attenuated in animals with significantly reduced unmyelinated afferents (produced by neonatal capsaicin treatment). These observations, which were consistent with the predictions of the pain adaptation model (i.e., reduced activity to protect the injured muscle), were interpreted as behavioral indications of reduced mechanical threshold and therefore a measure of muscle hyperalgesia. Additional details about the grasp-force device, manufacturer information, and source of drugs are provided in the original manuscript and a subsequent review of methodology for study of muscle pain in animals.[44]

Bite Force as a Measure for Craniofacial Muscle Hyperalgesia Nies and Ro[57] developed an operant conditioning protocol in which rats are trained to generate higher than normal bite force in order to obtain water reinforcement. On the basis of the pain adaptation model, they hypothesized that animals with sore or inflamed jaw muscles would not be able to generate the same bite force as normal animals.[24] Sprague–Dawley rats were positioned so they could bite on a pair of brass plates attached to a force transducer with the incisors. A water tube placed between the bite plates was connected to a Schmitt trigger-activated delivery system to provide water reinforcement for successful biting trials. The output of the transducer was led to a CED 1401 laboratory interface and data were stored on the hard drive for offline analysis (Figure 33–4A).

Training was initiated after determining the average weight and water consumption for each animal. During the training and the experimental phase, water intake was limited but was sufficient to maintain body weight within 80% of normal. After some exploratory behavior, rats gradually learned to associate biting with water reinforcement. Once this association was made they were rewarded for bite forces 400 g or greater and only the rewarded trials were saved for analysis (bites below this "cutoff value" were difficult to distinguish as discrete bites).

Once rats successfully learned the task, they were trained to generate gradually increasing target forces by rewarding only bites whose force exceeded the target. Animals were trained until a stable success rate (>70%) was attained at the final target force (>1.3 kg). Although there were individual differences in biting pattern (e.g., multiple bites versus single bites), the learning curves were similar and rats generally reached target force within 3 weeks.

Bite force data for each session were analyzed using Bite Analysis, a "script" program designed for use with Spike 2. The program is interactive so that users can manipulate cursors to specify cutoff and target forces and select the duration of the data sample (Figure 33–4B). The program automatically computes parameters such as the success rate and maximum and mean bite force and stores these values for subsequent analysis.

CONCLUSIONS

Lewis proposed that pain produced by ischemic contraction resulted from a "factor P". This factor, which acted in the interstitial (tissue) spaces under ischemic conditions, was dependent upon processes occurring within the muscle fiber as a result of contraction. An interesting aspect of the methods considered in this review is that each involves the introduction, the production, or the accumulation of algesic metabolites at some time throughout the course of the experiment. Many of the endogenous algogens have been shown to produce a significant level of muscle pain and nociceptive activation. Beyond the initial insult it is clear that both peripheral and central factors are likely players in the maintenance of persistent pain and there is substantial evidence that both are involved. Through the further evolution of models that specifically address the problem of muscle pain, we may one day unravel the complexly timed chain of biological events from the initial injury to the chronic pain condition and use this information to develop meaningful treatment strategies.

REFERENCES

1. Mense S, Simons DG. *Muscle Pain: Understanding Its Nature, Diagnosis, and Treatment*. Philadelphia, PA: Lippincott Williams & Wilkins, 2001.
2. Mense S. Nociception from skeletal muscle in relation to clinical muscle pain. *Pain* 1993;54(3):241–289.
3. Simons DG, Mense S. Understanding and measurement of muscle tone as related to clinical muscle pain. *Pain* 1998;75(1):1–17.
4. Kellgren JH. Observations on referred pain arising from muscle. *Clin Sci* 1938;3:175.
5. Lewis T. Pain in muscular ischemia. *Arch Int Med* 1932;49:713.
6. Lewis T, Pickering GW, Rothschild P. Observations upon muscular pain in intermittent claudication. *Heart* 1931;15:359.
7. Rainville P, Feine JS, Bushnell MC, Duncan GH. A psychophysical comparison of sensory and affective responses to four modalities of experimental pain. *Somatosens Mot Res* 1992;9(4):265–277.
8. Miles MP, Clarkson PM. Exercise-induced muscle pain, soreness, and cramps. *J Sports Med Phys Fitness* 1994;34(3):203–216.
9. Tegeder L, Zimmermann J, Meller ST, Geisslinger G. Release of algesic substances in human experimental muscle pain. *Inflamm Res* 2002;51(8):393–402.
10. Pyne DB. Exercise-induced muscle damage and inflammation: A review. *Aust J Sci Med Sport* 1994;26(3–4):49–58.
11. Ambalavanar R, Moritani M, Dessem D. Trigeminal P2X3 receptor expression differs from dorsal root ganglion and is modulated by deep tissue inflammation. *Pain* 2005;117(3):280–291.
12. Jensen K, Norup M. Experimental pain in human temporal muscle induced by hypertonic saline, potassium and acidity. *Cephalalgia* 1992;12(2):101–106.
13. Mense S, Schmidt RF. Activation of group IV afferent units from muscle by algesic agents. *Brain Res* 1974;72(2):305–310.
14. Svensson P, Cairns BE, Wang K, Arendt-Nielsen L. Injection of nerve growth factor into human masseter muscle evokes long-lasting mechanical allodynia and hyperalgesia. *Pain* 2003;104(1–2):241–247.
15. Schafers M, Sorkin LS, Sommer C. Intramuscular injection of tumor necrosis factor-alpha induces muscle hyperalgesia in rats. *Pain* 2003;104(3):579–588.
16. Jordt SE, McKemy DD, Julius D. Lessons from peppers and peppermint: The molecular logic of thermosensation. *Curr Opin Neurobiol* 2003;13(4):487–492.
17. Babenko V, Svensson P, Graven-Nielsen T, Drewes AM, Jensen TS, Arendt-Nielsen L. Duration and distribution of experimental muscle hyperalgesia in humans following combined infusions of serotonin and bradykinin. *Brain Res* 2000;853(2):275–281.
18. Witting N, Svensson P, Gottrup H, Arendt-Nielsen L, Jensen TS. Intramuscular and intradermal injection of capsaicin: A comparison of local and referred pain. *Pain* 2000;84(2–3):407–412.
19. Jancso-Gabor A, Szolcsanyi J, Jansco N. A simple method for measuring the amount of azovan blue exuded into the skin in response to an inflammatory stimulus. *J Pharm Pharmacol* 1967;19(7):486–487.

20. Ambalavanar R, Moritani M, Moutanni A, Gangula P, Yallampalli C, Dessem D. Deep tissue inflammation upregulates neuropeptides and evokes nociceptive behaviors which are modulated by a neuropeptide antagonist. *Pain* 2006;120(1–2):53–68.
21. Cairns BE, Gambarota G, Svensson P, Arendt-Nielsen L, Berde CB. Glutamate-induced sensitization of rat masseter muscle fibers. *Neuroscience* 2002;109(2):389–399.
22. Ro JY. Contribution of peripheral NMDA receptors in craniofacial muscle nociception and edema formation. *Brain Res* 2003;979(1–2): 78–84.
23. Ro JY, Capra NF, Masri R. Contribution of peripheral N-methyl-D-aspartate receptors to c-fos expression in the trigeminal spinal nucleus following acute masseteric inflammation. *Neuroscience* 2004;123(1): 213–219.
24. Ro JY, Nies M, Zhang Y. The role of peripheral N-methyl-D-aspartate receptors in muscle hyperalgesia. *Neuroreport* 2005;16(5): 485–489.
25. Cairns BE, Svensson P, Wang K, *et al.* Activation of peripheral NMDA receptors contributes to human pain and rat afferent discharges evoked by injection of glutamate into the masseter muscle. *J Neurophysiol* 2003;90(4):2098–2105.
26. Sluka KA, Rohlwing JJ, Bussey RA, Eikenberry SA, Wilken JM. Chronic muscle pain induced by repeated acid injection is reversed by spinally administered mu- and delta-, but not kappa-, opioid receptor agonists. *J Pharmacol Exp Ther* 2002;302(3):1146–1150.
27. Sluka KA, Kalra A, Moore SA. Unilateral intramuscular injections of acidic saline produce a bilateral, long-lasting hyperalgesia. *Muscle Nerve* 2001;24(1):37–46.
28. Zhang X, Ashton-Miller JA, Stohler CS. A closed-loop system for maintaining constant experimental muscle pain in man. *IEEE Trans Biomed Eng* 1993;40(4):344–352.
29. Stohler CS, Zhang X, Lund JP. The effect of experimental jaw muscle pain on postural muscle activity. *Pain* 1996;66(2–3):215–221.
30. Zubieta JK, Smith YR, Bueller JA, *et al.* Regional mu opioid receptor regulation of sensory and affective dimensions of pain. *Science* 2001;293(5528):311–315.
31. Stohler C. Human experimental pain models. In: Max MB, Lynne J, Eds. *Interactive Textbook on Clinical Symptom Research*, Chapter 21, Human Experimental Pain Models. http://symptomresearch.nih.gov/index.htm.http://symptomresearch.nih.gov/index.htm.
32. Zubieta JK, Smith YR, Bueller JA, *et al.* Mu-opioid receptor-mediated antinociceptive responses differ in men and women. *J Neurosci* 2002;22(12):5100–5107.
33. Zubieta JK, Heitzeg MM, Smith YR, *et al.* COMT val158met genotype affects mu-opioid neurotransmitter responses to a pain stressor. *Science* 2003;299(5610):1240–1243.
34. Graven-Nielsen T, Fenger-Gron LS, Svensson P, Steengaard-Pedersen K, Arendt-Nielsen L, Staehelin Jensen T. Quantification of deep and superficial sensibility in saline-induced muscle pain—a psychophysical study. *Somatosens Mot Res* 1998;15(1):46–53.
35. Svensson P, Arendt-Nielsen L, Houe L. Muscle pain modulates mastication: An experimental study in humans. *J Orofac Pain* 1998;12(1):7–16.
36. Svensson P, Graven-Nielsen T, Arendt-Nielsen L. Mechanical hyperesthesia of human facial skin induced by tonic painful stimulation of jaw muscles. *Pain* 1998;74(1):93–100.
37. Ro JY, Capra NF, Lee JS, Masri R, Chun YH. Hypertonic saline-induced muscle nociception and c-fos activation are partially mediated by peripheral NMDA receptors. *Eur J Pain* 2006;11(4): 398–405.
38. Smith GM, Egbert LD, Markowitz RA, Mosteller F, Beecher HK. An experimental pain method sensitive to morphine in man: The submaximum effort tourniquet technique. *J Pharmacol Exp Ther* 1966;154(2):324–332.
39. Graven-Nielsen T, Jansson Y, Segerdahl M, *et al.* Experimental pain by ischaemic contractions compared with pain by intramuscular infusions of adenosine and hypertonic saline. *Eur J Pain* 2003;7(1): 93–102.
40. Bessou P, Laporte Y. [Activation of non-medullated afferent fibers of muscular origin.] *C R Seances Soc Biol Fil* 1958;152(11): 1587–1590.
41. Chabel C, Russell LC, Lee R. Tourniquet-induced limb ischemia: A neurophysiologic animal model. *Anesthesiology* 1990;72(6):1038–1044.
42. Taguchi T, Matsuda T, Tamura R, Sato J, Mizumura K. Muscular mechanical hyperalgesia revealed by behavioural pain test and c-Fos expression in the spinal dorsal horn after eccentric contraction in rats. *J Physiol* 2005;564(Pt. 1):259–268.
43. Taguchi T, Sato J, Mizumura K. Augmented mechanical response of muscle thin-fiber sensory receptors recorded from rat muscle-nerve preparations in vitro after eccentric contraction. *J Neurophysiol* 2005;94(4):2822–2831.
44. Kehl LJ, Fairbanks CA. Experimental animal models of muscle pain and analgesia. *Exerc Sport Sci Rev* 2003;31(4):188–194.
45. Graven-Nielsen T, Arendt-Nielsen L, Svensson P, Jensen TS. Experimental muscle pain: A quantitative study of local and referred pain in humans following injection of hypertonic saline. *J Musculoskeletal Pain* 1997;5(1):49.
46. Melzack R, Terrence C, Fromm G, Amsel R. Trigeminal neuralgia and atypical facial pain: Use of the McGill Pain Questionnaire for discrimination and diagnosis. *Pain* 1986;27(3):297–302.
47. Wolfe F, Smythe HA, Yunus MB, *et al.* The American College of Rheumatology 1990 Criteria for the Classification of Fibromyalgia. Report of the Multicenter Criteria Committee. *Arthritis Rheum* 1990;33(2):160–172.
48. Lund JP, Donga R, Widmer CG, Stohler CS. The pain-adaptation model: A discussion of the relationship between chronic musculoskeletal pain and motor activity. *Can J Physiol Pharmacol* 1991;69(5):683–694.
49. Definition of Pain and Distress and Reporting Requirements for Laboratory Animals, 2000. http://www.nap.edu/catalog/10035.html.
50. Ikeda T, Terayama R, Jue SS, Sugiyo S, Dubner R, Ren K. Differential rostral projections of caudal brainstem neurons receiving trigeminal input after masseter inflammation. *J Comp Neurol* 2003;465(2):220–233.
51. Ohtori S, Takahashi K, Chiba T, *et al.* Fos expression in the rat brain and spinal cord evoked by noxious stimulation to low back muscle and skin. *Spine* 2000;25(19):2425–2430.
52. Lam DK, Sessle BJ, Cairns BE, Hu JW. Peripheral NMDA receptor modulation of jaw muscle electromyographic activity induced by capsaicin injection into the temporomandibular joint of rats. *Brain Res* 2005;1046(1–2):68–76.
53. Ro JY, Capra N, Masri R. Development of a behavioral assessment of craniofacial muscle pain in lightly anesthetized rats. *Pain* 2003;104(1–2):179–185.
54. Clavelou P, Pajot J, Dallel R, Raboisson P. Application of the formalin test to the study of orofacial pain in the rat. *Neurosci Lett* 1989;103(3):349–353.
55. Ren K. An improved method for assessing mechanical allodynia in the rat. *Physiol Behav* 1999;67(5):711–716.
56. Kehl LJ, Trempe TM, Hargreaves KM. A new animal model for assessing mechanisms and management of muscle hyperalgesia. *Pain* 2000;85(3):333–343.
57. Nies M, Ro JY. Bite force measurement in awake rats. *Brain Res Brain Res Protoc* 2004;12(3):180–185.

34 Animal Models of Parkinson's Disease

Emma L. Lane, Thomas Carlsson, Deniz Kirik, and Stephen B. Dunnett

ABSTRACT

Parkinson's disease is a movement disorder, caused predominantly by the degeneration of the dopaminergic nigrostriatal pathway. Various animal models have been developed in both rodents and nonhuman primates that have been used successfully in the evaluation of pharmacotherapeutics. The most commonly used animal models are generated through the administration of toxins; those most extensively characterized with a view to analyzing motor deficits are the 6-hydroxydopamine (6-OHDA) lesioned rat and the 1-methyl-4-phenyl-1,2,3,6-tetrahydropyridine (MPTP)-treated nonhuman primate. We review how these models are developed and the most common behavioral tests used to discern motor impairments and the improvements afforded by both pharmacological and surgical interventions. A major problem in the treatment of Parkinson's disease is the generation of dyskinesia, or abnormal involuntary movements following the long-term administration of L-DOPA. Until recently, research on dyskinesia was restricted to the use of the MPTP-treated nonhuman primate, but the development of a rodent model has improved understanding of the mechanisms responsible and expedited screening of drugs for their dyskinesiogenic potential. Although both rodent and nonhuman primate models have significantly furthered our understanding of the mechanisms behind Parkinson's disease and the associated motor complications, both still have significant limitations. Therefore, the search continues for a model that can reproduce not just the neuropathology and symptoms of the disease, but also recreate more faithfully the progressive pathogenic aspects of the disease.

Key Words: Parkinson's disease, MPTP-treated nonhuman primate, 6-hydroxydopamine lesioned rat, Abnormal involuntary movements.

INTRODUCTION

Parkinson's disease (PD) is a neurodegenerative motor disorder that presents as a loss or paucity of movement (akinesia/bradykinesia), postural instability, resting tremor, rigidity, and altered gait. The primary cause of these symptoms is the continuous deterioration of dopaminergic (DAergic) neurons of the substantia nigra pars compacta (SNc) that project to the striatum via the nigrostriatal pathway. To study the pathogenic mechanisms and evaluate potential therapies of PD, a range of animal models has been used, based on the administration of toxins, either locally (6-hydroxydopamine, 6-OHDA; 1-methyl-4-phenyl-1,2,3,6-tetrahydropyridine, MPTP) or systemically (reserpine, MPTP, rotenone, proteosomal inhibitors). These models are used to test potential therapeutic strategies, such as drugs, cell replacement, and gene therapy, and it is predominantly their overt behavioral phenotypes, correlated to aspects of the human disorder, that make them so useful. The primary aim of this chapter is to detail the animal models that are most widely used to evaluate potential therapies to improve motor function in PD. Thus, transgenic mouse models based on specific mutations identified in a small number of patients, which have not yet been used effectively in therapeutic development, will not be covered (although some of the mouse behavioral tests described can also be applied in that context[1]).

This chapter will focus on the use of two classical toxin-based models of PD, the unilateral 6-OHDA-lesioned rodent and the MPTP-treated primate. The use of these toxins in alternative species, i.e., the 6-OHDA-lesioned primate and the MPTP-treated mouse, will be considered and we also include a brief description of the other toxin-based rodent models. These are less well established and are still to be fully characterized, but are contributing a new dimension to toxin-based models of PD.

6-HYDROXYDOPAMINE-LESIONED RAT

The first use of 6-OHDA to obliterate a catecholaminergic pathway in the search for a model of Parkinson's disease was described in 1968 by Ungerstedt.[2] 6-OHDA is taken up through the catecholamine transporter on DAergic and noradrenergic neurons where it accumulates in the cytosol. Here it inhibits mitochondrial complexes in the respiratory chain and also produces reactive hydroxyl radicals that contribute to the induction of a nonapoptotic cell death.[3] As it does not cross the blood–brain barrier, it is administered directly into the required area (in this case along the nigrostriatal tract) using stereotaxic procedures. Animals are placed in a stereotaxic frame (Figure 34–1) and using a sagittal cut the skull and bregma are exposed. The location of the appropriate target track is determined on the surface of the skull in coordinates from the interaural line and bregma using a rat brain atlas, and a small burr hole is drilled. A cannula is then lowered to the correct depth and the toxin is slowly infused either manually or using an automated pump. The cannulas are typically made of fine stainless-steel tubing (0.2–0.4 mm o.d.), but alternatively glass capillaries can be pulled to much finer tips (~50 μm)

From: *Sourcebook of Models for Biomedical Research* (P. M. Conn, ed.), © 2008 Humana Press Inc., Totowa, NJ.

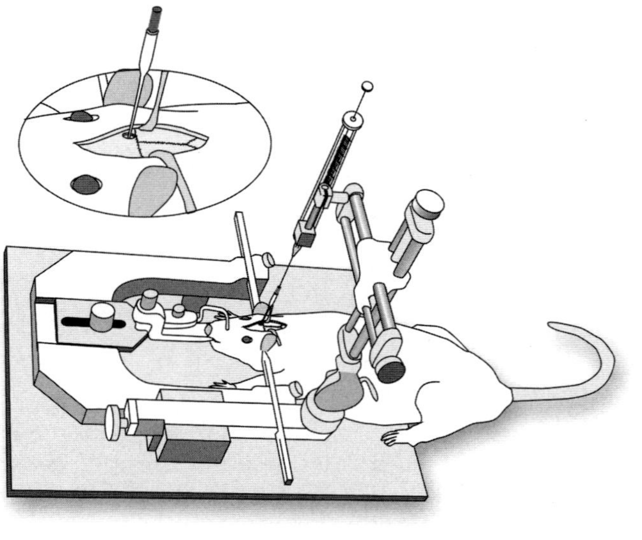

Figure 34–1. Schematic representation of a rat in a stereotaxic frame. The head is secured in position with ear and tooth bars. This frame holds a Hamilton syringe with a glass capillary attached through which the toxin is injected; alternatively, metal cannulas can be used connected to a syringe placed in a syringe pump.

to minimize tissue damage, especially where multiple injection sites are used. After the injection, the cannula should remain in place for a short time to allow the toxin to diffuse into the surrounding tissue before it is retracted and the wound sutured. When using 6-OHDA it is important to remember that it oxidizes easily and should be dissolved in saline containing 0.02% ascorbic acid as an antioxidant. The powder and solution should be kept cold and in the dark as far as is practically possible, with fresh solutions made up at regular intervals (every 1–2 h), which should be discarded as soon as a brown or pink coloration is observed. The cannula should be loaded just prior to each injection to minimize exposure to light.

The extent of the lesion produced by 6-OHDA is dependent on the location at which the toxin is applied and the concentration used. The most common lesion model is the complete destruction of the nigrostriatal pathway in which 6-OHDA is injected into the median forebrain bundle (MFB). This protocol produces a >97% depletion of the DA neurons in the SNc and up to 80% in the ventral tegmental area (VTA). A more restricted, progressive, partial lesion of the nigrostriatal pathway can also be achieved by injecting the 6-OHDA into the striatum.[4] DAergic terminals take up the toxin and retrogradely transport it to the cell bodies in the SNc. The extent of the partial lesion can range from a 30% to 90% loss of DA neurons of the SNc, dependent upon the number of striatal injection sites and the quantity of 6-OHDA administered.[6] A four-site striatal lesion gives rise to about an 80–90% loss of the DA neurons in the SNc, but only 15% of VTA neurons are affected.[4] Partial lesions can also be achieved by targeting the MFB with lower concentrations of the toxin,[5] although this strategy is typically more variable. In addition, 6-OHDA has also been directly injected in the SNc, resulting in substantial degeneration of the nigrostriatal pathway, but this lesion method is less commonly used.

BEHAVIORAL TESTS FOR UNILATERAL 6-HYDROXYDOPAMINE LESIONS The unilateral denervation produces a number of deficits on the side contralateral to the lesion, displayed as spontaneous circling toward the (ipsilateral) lesioned side, reduced use of the contralateral forelimb, and a neglect of the contralateral side. A very wide range of behavioral tests has been described in the nearly four decades since 6-OHDA came into widespread use. We describe below a selected range of behavioral tests in widespread use, which have been found by ourselves and others to be sensitive, objective, and easily quantified. The list is of course not exhaustive, and other tests can be found elsewhere.[1,6–8]

Rotational Behavior Circling behavior is the most frequently used behavioral test, largely due to its simplicity.[9] It is affected differentially by the administration of dopaminomimetics, the direction and nature of the behavior being related to their respective modes of action. The number of rotations can be counted manually, although commonly this is done with the use of automated rotometers. The output is the total number of rotations in either direction over a predetermined period of time, which can also be expressed as net rotations. Some days after receiving a unilateral lesion, rats exhibit a tendency to turn during spontaneous locomotion to the (ipsilateral) side of the lesion. This ipsilateral bias is markedly enhanced by activating the animal, for example, by exposure to mildly stressful stimuli or injection with a stimulant drug. Thus, amphetamine will increase DA release on the intact side to a greater extent than the DA-depleted lesioned side, and will enhance marked turning toward the side ipsilateral to the lesion. Significant asymmetry can be induced by amphetamine with more than a 30% loss of the DA neurons in the SN or a 40–50% depletion of the DA content in the striatum. In contrast, direct acting dopamine receptor agonists will induce a contralateral circling behavior by acting predominantly on the DA receptors on the postsynaptic striatal neurons, made supersensitive by the deafferenting lesion. More extensive lesions are required to induce high levels of rotation with apomorphine. Many studies have reported a close correlation between the rates of amphetamine and apomorphine-induced rotation and the extent of denervation, such that rotation scores can be used to determine the extent of lesions on an animal-by-animal basis, allowing exclusion from further study of animals with inadequate lesions and allocation of the remainder to matched groups prior to subsequent investigations. Generally, 14 days postlesion, in excess of five ipsilateral rotations per minute induced by *d*-amphetamine sulfate (2–5 mg/kg ip or sc in saline) measured over 90 min is used to indicate >90% denervation. Similarly, over 90% dopamine depletion is assumed from >5 contralateral rotations per minute following the administration of apomorphine (0.5 mg/kg sc in 0.02% ascorbic acid) over 40–60 min. It should be noted that apomorphine can sensitize the circling response and should be avoided if the animals are intended for use in studies of striatal sensitivity; in addition, compared to amphetamine, apomorphine is less responsive to dopamine replacement measures.

Compounds inducing contralateral rotations, such as direct acting DA agonists and L-DOPA, are predicted to have antiparkinsonian effects, while a reduction in amphetamine-induced ipsilateral rotations is frequently used to demonstrate improved DAergic function on the lesioned side following cell transplantation or treatment with viral vectors.[10,11]

Spontaneous Forelimb Test (Cylinder Test) The cylinder test evaluates forelimb akinesia by determining the animals' ability to support their body weight against the wall of a glass cylinder during explorative behavior. The cylinder test was first described by Schallert and Lindner[12] and further modified by Schallert and Tillerson.[7] The animals are placed in a transparent glass cylinder (approximately 20 cm in diameter for rats and 10 cm for mice) and videotaped; two mirrors are placed at a 90° angle behind the cylinder to allow visualization of the whole surface. An observer blinded to the treatment of the animals counts the number of independent weight-bearing touches made with each forelimb, usually expressed as percent touches with the lesioned paw of the total performed. A normal animal will, on average, achieve a score of 50% (Figure 34–2A, "normal"), while animals with DA depletion of the nigrostriatal fibers will exhibit a marked reduction in the use of the contralateral paw (Figure 34–2A, "lesion"). To improve the sensitivity of this test Schallert and Tillerson[7] describe a more detailed analysis where the first touches during rearing and landing and differential counting of simultaneous versus lateralized paw use are separately scored.

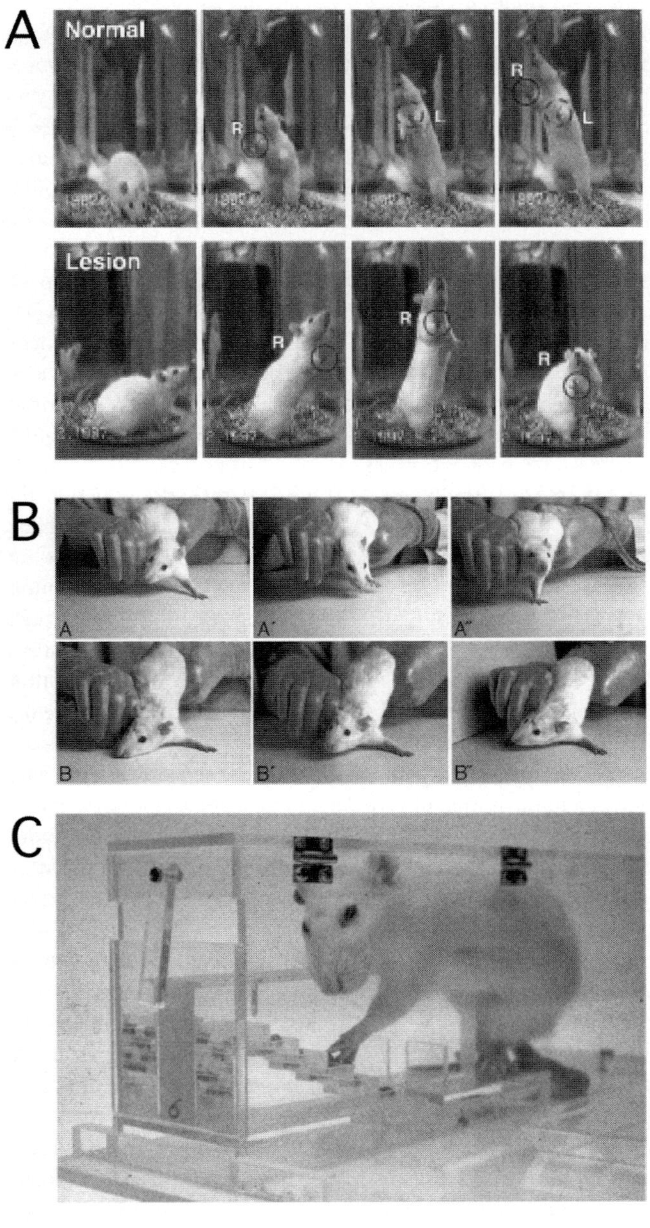

Figure 34–2. Photographic demonstration of (A) the cylinder test of forepaw asymmetry. A "normal" rat will use both forepaws to explore the cylinder, but a 6-OHDA "lesion" will cause a predominant use of the paw on the ipsilateral side to the lesion. (B) The stepping test, in which normal rats (A, A′, A″) will make adjusting steps with the weight-bearing forepaw as it is moved across a surface, but a lesioned animal (B, B′, B″) will not. (C) A rat in the paw-reaching "staircase" apparatus.

There are several advantages of the cylinder test, in that no drug administration is required, food restriction is not necessary, and there is no or limited handling by the experimenter, reducing the influence of external parameters. However, especially where repeat testing is necessary, the animals can become habituated and more reticent to explore the cylinder. To encourage a good performance, the test should be carried out in the latter part of the light cycle, and the following can also be applied: (1) turning the light off and/or switching the light on and off two or three times (requires a video camera with infrared recording capability), (2) mildly shaking the cylinder for a few seconds, (3) removing the animals from the cylinder for 10–30 sec and returning them, and (4) exposing the animals to female/male odors or mild stressors.

Stepping Test The stepping test forces the use of individual forelimbs and determines the rats' ability to perform adjusting steps when moved sideways. The design was first described by Schallert *et al.*[13] and further defined by Olsson *et al.*[14] The experimenter supports the animal lightly round the waist, restraining the hind limbs and the forelimb that is not being tested. The animal is lowered to the table surface and the free forepaw is allowed to provide placing support. The animal is then moved sideways over a 90 cm surface, from one edge to the other, over about 5 sec, during which time the tested paw is touching the surface (Figure 34–2B). The number of steps, while the animal is moved over the surface, is counted. The test is repeated twice per day for three consecutive days for each forepaw in both a forehand and backhand direction. Prior to testing, a habituation trial period of 3–7 days is required to ensure that the animals perform the test properly and to give consistent results. A normal rat will perform about 9–13 adjusting steps in the forehand direction and a few more, 12–16, in the backhand direction. Rats with extensive 6-OHDA lesions of the nigrostriatal pathway (>80%) show a marked reduction in adjusting steps for the contralateral forepaw to only 0–2[4,14,15] (Figure 34–2B, B′). Stepping in the backhand direction tends to be much less affected than stepping in the forehand direction, and may be only marginally impaired in rats with less completed intrastriatal 6-OHDA lesions. Intermediate 6-OHDA lesions, 50–80%, will make intermittent forelimb steps, while a lesion of less than 40% will not significantly alter forelimb stepping.[4,15] In conducting this test it should be noted that a subjective element is introduced when there is direct interaction between experimenter and animal. This makes comparing the results between different experimenters and studies difficult and the same experimenter should conduct all the testing within a study. Other factors such as strain used, testing environment, and stress can also influence the rats' performance.

Skilled Forelimb Use in the Staircase Test This test, also called the *paw reaching test*, was introduced by Montoya *et al.*[16] to independently evaluate individual forepaws in a skilled test and is based on the animals' ability to successfully retrieve and eat sugar pellets with different levels of difficulty. The apparatus is generally a transparent Plexiglas box with a central platform separating two staircases such that the animal can reach down into the staircase on each side only with the forepaw on the same side (Figure 34–2C). This configuration allows the animal's performance to be assessed separately for the two forelimbs without having to restrain the animal. Each staircase has a well on each step in which a predefined number of sugar pellets is placed (between 2 and 10 are used in different laboratories). The animals are placed in the box and allowed to collect and eat the pellets for 10–15 min.[4,16] The number of pellets taken from each well and the number of displaced ("missed") pellets are counted, giving the total number of successfully eaten pellets for each forepaw. For the animals to perform well in this task they should be food deprived prior to starting the test (1–3 days) and be maintained at around 90% of their body weight during the testing period; sugar pellets can be introduced as food during this time. On the first day of testing, the animals generally explore the test box and begin to associate the task with the sugar pellets, and with repeated sessions on subsequent days begin to learn the task. The time required for training varies between groups of animals and can be up to 2 weeks in older rats before a stable performance is reached. When more than 50% of DA is lost in the SNc and greater than 80% in the striatum[4,17,18] the animals start to show significant deficits. Complete unilateral 6-OHDA lesions reduce performance in this test by about 50–70% taken pellets and 30–50% eaten pellets compared to intact rats.[14,16,17,19,20] It should be noted that lesioned animal usually improve on their intact paw and retrieve up to 10% more pellets than intact control animals. The test can also be used as a forced paw choice test, where sugar pellets are placed only on one side of the staircase, forcing the use of the paw on that side.[21]

Other less widely used but validated tests include the rotarod and forelimb-placing test. The rotarod is a rotating cylinder on which the rats (or mice[22]) are trained to walk. There are two main approaches: the first uses a fixed-speed protocol in which rats perform on set speeds (increments between 4 and 40 rpm) and their time to stay on the drum at different speeds is measured. In the second approach accelerating paradigms have the rod increasing in speed (4–44 rpm) at a fixed rate (increasing over 90–300 sec)[23–25]; the time at which the rat falls off the drum is recorded and can be directly related to a speed. A comparative study between the two protocols found that the fixed-speed assay was more predictive of therapeutic efficacy, but accelerating protocols could be used more efficiently and easily to test lesion extent.[23] Again, animals do require some training on this apparatus to reach a stable performance in the task before any testing can be carried out, but this can be as little as one or two training sessions.[25]

These tests were primarily developed in rats, but have been adapted for other species, most notably mice and primates. Thus, unilateral nigrostriatal lesions induce similar postural and rotation asymmetries in mice, although their size typically makes it easier to monitor rotation by direct observation or using automated video tracking devices than mechanical rotometers.[22] Similarly, mice show performances akin to rats in staircase and other paw-reaching tasks with the apparatus scaled down to be suitable for the smaller species.[26] Whereas the effects of DAergic lesions in large primates have been more extensively explored using peripheral administration of MPTP (see below), the brains of small New World marmoset monkeys are sufficiently consistent to allow stereotaxic positioning using external reference landmarks to produce reliable nigrostriatal lesions with 6-OHDA, which, as in rats, produce a range of lateralized motor and sensorimotor impairments. Rotation is less consistent because these monkeys typically move in a three-dimensional space, but contralateral postural and head-turning biases are prominent and easy to measure by video recording.[27] Moreover, Marshall, Ridley, and Kendall have developed primate analogues of the staircase test in which the monkeys reach through slots behind transparent vertical baffles that constrain them to use just one arm in reaching to retrieve palatable foods from staircase wells behind the barrier.[28,29] In each of these tasks, marmosets exhibit lateralized changes in response to unilateral nigrostriatal lesions and to pharmacological and cellular therapeutic interventions directly comparable to rats.[30,31]

1-METHYL-4-PHENYL-1,2,3,6-TETRAHYDROPYRIDINE

A byproduct of the illicit manufacture of a pethidine analogue, MPTP was identified as the cause of rapidly developing parkinsonism in a group of drug addicts in the United States in the early 1980s.[32] Postmortem analysis revealed that MPTP had induced a self-perpetuating cell death of DAergic neurons in the SNc, with few other anatomical changes. It was found to have similar effects in nonhuman primates;[33–35] this has developed into what is currently considered the best available model for PD, providing both a similar behavioral profile and neuropathology. MPTP has also been used in a wide range of other species from the mouse to the worm and zebra fish. In a series of enzyme steps, non-DAergic cells such as glia and serotonergic neurons convert MPTP into the active metabolite MPP^+. It is then released into the synaptic space and subsequently taken up by DAergic neurons through the DA transporter. Although the mechanism through which it induces selective DAergic cell death has not been fully elucidated, it is known to act as an inhibitor of complex I of the mitochondrial respiratory chain, producing a decrease in tissue ATP content and an increase in oxidative stress.

1-METHYL-4-PHENYL-1,2,3,6-TETRAHYDROPYRIDINE-TREATED PRIMATE The behavioral effects of MPTP have been well characterized in various nonhuman primates including the squirrel, rhesus, and cynomolgus monkey, and the common marmoset.[33–35] The progression of the degeneration can be controlled by the paradigm of administration. Most protocols use a relatively acute regime of administration for convenience, but a more prolonged administration of regular small doses of weeks or months is required to reproduce the slow progressive deterioration and the absence of spontaneous recovery seen in patients. The mode of administration to give a bilateral lesion can be subcutaneous, intraperitoneal, intravenous, or intramuscular. For example, a common acute bilateral paradigm in marmosets is the subcutaneous administration of MPTP at 2 mg/kg over 5 days.[34] With this regime the nigrostriatal DAergic cell loss is upward of 90%, representing late stage PD. After a few days of treatment the primates show akinesia, limb rigidity, reduced vocalization in the marmosets, and reduced grooming. Although resting tremor is not common, it can occasionally be seen as a whole body tremor. A regime of 1 mg/kg for 3 days has also been described recently in the marmoset, in which a subthreshold DA loss produces more limited behavioral deficits that more closely represent presymptomatic to early stage disease states.[36] The more chronic treatment protocols that have been described use between 0.2 and 4.5 mg/kg MPTP repeatedly over weeks or months.[37–39] However, there appears to be an optimum treatment regime to create a stable cell death, as recovery of the DAergic losses has been described with protocols of 29 days,[37] while with 15 days a reportedly nonrecoverable DAergic deficiency is induced.[38]

MPTP is highly toxic, and the literature reports a number of cases of permanent neurological impairment in laboratory personnel working with it.[40] Supreme care should therefore be taken when working with MPTP, typically to category III standard, to avoid self-administration or any contact with animal products during the administration period and for a washout period of a few days afterward. The animals require careful observation over several weeks following administration and hand feeding is advisable until the animals can eat unaided and sustain their body weight. Occasionally antiparkinsonian medication such as L-DOPA is given to increase their ability for self-care.

The bilateral MPTP-treated primate model is commonly used to screen pharmaceuticals for their antiparkinsonian potential. Following treatment, animals are observed either in their home cages or in specialized locomotor cages to independently determine their activity levels, while subjective rating scales for activity and disability are also used. Several different scales are used by different research groups, each with its own criteria and grading system, often based on clinical rating scales such as the UPDRS (United Parkinson's Disease Rating Scale). Many of these scales include criteria such as bradykinesia, freezing, rigidity, tremor, alertness, reaction to external stimuli/interaction, posture, balance/coordination, and vocalization (for comparisons of different scales see Imbert *et al.*[41]).

In Old World primates such as the macaque and rhesus monkeys, the toxin may be administered though the intracarotid artery, giving a predominantly unilateral syndrome. This surgery is more invasive but, as with bilateral versus unilateral lesion with 6-OHDA, the unilateral lesion results in improved survival and requires less posttreatment care. The behaviors are evaluated using rating scales similar to the bilateral model but there are individual scores for the gross motor skills for each limb. The fine motor skills of each forelimb can also be evaluated in food retrieval tasks.[42,43]

1-METHYL-4-PHENYL-1,2,3,6-TETRAHYDROPYRIDINE-TREATED MOUSE Rats have a relative insensitivity to MPTP that may be the result of a species-specific metabolism or sequestration of MPP^+.[44] Very high doses are required to induce DAergic cell death compared to mice and the administration of guanethidine is required to prevent peripheral catecholamine release and mortality.[45] Mice, however, are susceptible to the effect of MPTP and have provided a useful small animal model of MPTP-induced parkinsonism. There are caveats to this model, as mice are not as sensitive as primates to the toxin and susceptibility is also hugely variable between mouse strains. Furthermore, with some treatment regimes, cell death is not complete and the mice can show a spontaneous functional recovery. This is more often seen with acute paradigms, while longer-term treatment may provide a more stable DAergic loss. The toxin is usually injected subcutaneously (although it can be given via other routes) typically in the range of 2–45 mg/kg/day for up to 10 days (for a full summary see Seledis *et al.*[46]). As in the primate, this produces a bilateral lesion and therefore the behavioral tests that are applied are distinct from those used for the unilateral 6-OHDA lesioned rodents.

BEHAVIOR TESTS WITH THE 1-METHYL-4-PHENYL-1,2,3,6-TETRAHYDROPYRIDINE-TREATED MOUSE

ACTIVITY Depending on the paradigm of administration, MPTP-treated mice may develop a hypokinesia that can be assessed using simple tests of locomotor activity and catalepsy. The simplest test of akinesia involves placing the animal of a flat surface and timing how long it takes to move all four limbs (akinesia), while for catalepsy the animals are placed with their fore or hind limbs supporting them on an elevated block or a raised bar, and the latency to restore normal posture is recorded.[1] More generally, activity can be monitored in a large square open field (typically 100 cm × 100 cm). In the manual version of the task, the field is divided into squares by lines on the floor and the number of squares entered in a fixed time interval is recorded by direct observation. Frequently, the total counts are subdivided into central squares and squares adjacent to the wall of the open field, as an index of exploratory activity (more anxious animals will spend more time at the edge). The open field test can be readily automated by mounting an array of infrared beams to traverse the open field or a standard cage. The number of beam breaks is used to calculate the time spent locomoting and the total distance traveled. In addition, the location and pattern of beam breaks can also be interpreted, i.e., whether the central beams are broken or just in the peripheral areas of the cage, or if single beams are repeatedly broken, indicating stereotypic behaviors, and with a decline of counts over time taken as an index of "habituation." Higher beams can also be used to measure vertical rearing activity, as a second measure of exploration. As mice are nocturnal, better results are achieved if they are tested during the dark cycle, as their spontaneous activity is higher. Placing a novel object in a center square of the open field and recording the rise in the time spent and number of entries in the squares immediately adjacent to the object can further assay exploration more directly.

POLE TEST The pole test[47] is very sensitive to changes in nigrostriatal function as it requires maneuvering and grip function. The mouse is placed head up near the top of a vertical pole (1 cm in diameter, approximately 50 cm in height) and the time to rotate 180° and time to descend are used as measures of motor function. Similarly, tests of grip strength are also sensitive to changes in nigrostriatal DA.

The rotarod, as described above, in the 6-OHDA rat model has been very widely used for assessing motor abnormalities in mutant mice.[1,48] However, MPTP has been reported to have variable effects on the rotarod,[46] and the use of different protocols in the administration of the toxin as well as in the ways the rotarod test is conducted makes comparisons difficult. Other tests used in the evaluation of motor function in the MPTP-treated mouse include swim tests (set in a narrow tank of water at 22–25°C for 30 min) assessing active swimming versus passive floating over a period of time both in the presence and absence of an escape platform, grip strength, and apomorphine-induced climbing.

DYSKINESIA IN ANIMAL MODELS OF PARKINSON'S DISEASE

The gold standard for pharmacological dopamine replacement in PD is L-DOPA, the metabolic precursor to DA. This is a highly effective treatment, but within 10 years of initiating L-DOPA therapy abnormal involuntary movements (AIMs), known as dyskinesia, develop in most patients limiting its use.[49] Thus, a major area of research using animal models of PD has involved identifying the cause of dyskinesia and evaluating the potential of new compounds to elicit, exacerbate, or reduce this side effect.

DYSKINESIA TESTING IN THE 6-HYDROXYDOPAMINE-LESIONED RAT The ability of drugs to induce contralateral rotations in the 6-OHDA-lesioned rat is taken as an indicator of

Table 34–1
The amplitude ratings for the abnormal involuntary movement scale in rodents given in addition to scores of duration[a]

Score	*Axial*	*Limb*	*Orolingual*
1	Persistent deviation of the head to a 30° angle	Small movements of the paw and distal forelimb around a fixed position	Chewing and repeated jaw movements
2	Lateral deviation of the head and neck at an angle between 30° and 60°	Low-amplitude movements of the distal and proximal forelimb	With tongue protrusion
3	Lateral deviation of greater than 60° without loss of balance	Translocation of the whole limb with the shoulder muscle	N/A
4	Twisting of the head and neck such that balance is lost	Vigorous limb and shoulder movements to their maximum extent	N/A

[a]Axial and limb movements are scored 1–4 and orolingual 1–2.

their antiparkinsonian potential, but with repeated administration the circling in response to L-DOPA changes in profile.[50] This sensitivity of the locomotor response was considered an indicator of dyskinesiogenic potential. However, this model has been further developed by the extensive characterization of the simultaneous development of abnormal involuntary movement of the torso, fore and hind limbs, and orolingual area; this is considered to be rodent correlates of the L-DOPA-induced dyskinesia seen in patients.[51,52] Rating scales have been devised for these behaviors based on their duration and the magnitude to which they are expressed. To evaluate these behaviors, rats are individually observed in sawdust-free, transparent cages without food or water. Each animal is injected with the compound of interest at 1 min intervals and then observed at regular intervals (10–30 min) for 1–2 min each and rated. Some researchers prefer to take fewer time points and video the behaviors, permitting scores of additional parameters such as head nodding, and also to distinguish between choreic and dystonic movements of the forelimb. However, care should be taken to ensure that enough time points are evaluated as differences between single time points could be attributed to shifts in the time course, rather than a change in the severity of the response. There is a three-tier 1–4 point rating scale based on the duration, class, and amplitude of dyskinetic behaviors, as follows for the duration of behaviors of each class:

1. The behavior is present for less than half the observation time.
2. The behavior is present for more than half but not all the time.
3. The behavior is continuously present but interruptible by external stimuli, e.g., a sharp rap on the outside of the cage.
4. The behavior is uninterruptible by external stimuli.

Scores for amplitude in each category are listed in Table 34–1.[20] Duration and amplitude ratings are combined to give an overall score of severity for each category at each time point, which were accumulated to give a cumulative severity score. The same methodology and rating scales can also be applied to 6-OHDA-lesioned mice.[53]

DYSKINESIA TESTING THE 1-METHYL-4-PHENYL-1,2,3,6-TETRAHYDROPYRIDINE-TREATED MARMOSET The MPTP-treated primate has been extensively used to determine the neural mechanisms underlying the development of L-DOPA-induced dyskinesia and to assess the dyskinesiogenic potential of new compounds in addition to the antiparkinsonian activity. The dyskinesia takes the form of choreic movements of the fore and hind limbs, athetosis, and dystonic posturing and a separate dyskinesia rating scale based on the clinical UPDRS is used. The scoring occurs at repeated intervals (generally every 30 min), often for the duration of drug action, on several days during the treatment regime. A rater blind to the treatment received by each animal, as these are subjective scales, should carry these out.

CHOOSING THE RIGHT MODEL

UNILATERAL VERSUS BILATERAL LESIONS 6-OHDA and to a certain extent MPTP can produce either unilateral or bilateral lesions. PD generally affects patients on both sides, although the symptoms are generally worse on one side. In the early stages of disease, PD is often observed to be asymmetric with greater symptoms on one side, but the degree of lateralization observed in extensive unilateral lesions in animals is seldom seen, with both sides affected significantly in advanced disease. Therefore the bilateral models can be seen as more closely representative of the human disease. However, generally the bilateral lesions are more problematic in terms of the health of the animal. Extensive bilateral lesions in rats, whether induced by intraventricular or intracerebral injections of 6-OHDA, typically result in profound inanition, with the animals failing to eat, drink, or groom spontaneously, requiring prolonged and intensive care to maintain basic health status.[54] By contrast, both primates and rodents with unilateral lesions can feed and drink adequately to sustain their own body weight and maintain basic health, although MPTP-treated primates with severe bilateral lesions will often require feeding by hand for the first few weeks following treatment and may require antiparkinsonian medication. Partial, bilateral, striatal, and nigral 6-OHDA lesions have been used successfully and the rats are then able to sustain themselves,[55,56] but because the deficit is then bilateral, a different combination of behavior tests is then used including open field and operant tests such as the reaction time task (this is described in more detail elsewhere).[8,56]

In terms of the behavioral tasks that can be used to assess motor function and recovery, in the bilateral primate models especially, qualitative observational parameters are predominantly used to evaluate changes in motor function. In contrast, the tests that utilize the asymmetry of the unilateral models generally provide a more quantitative assessment. A further advantage to the unilateral model is that it enables comparisons of the physiological events in damaged and intact hemispheres, allowing functional deficits to be related to physiological changes. The intact side may therefore be used as an internal control, but caution must be exercised, as there are some changes to the intact side as a result of the lesion.

6-HYDROXYDOPAMINE VERSUS 1-METHYL-4-PHENYL-1,2,3,6-TETRAHYDROPYRIDINE MPTP is a highly dangerous, lipophilic toxin and special care has to be taken in all aspects of handling and injecting it and when cleaning out the cages during and shortly after the toxin-administration period. 6-OHDA should also be handled with care, but the main concern, as detailed above, is in keeping the toxin fresh. MPTP (in the bilateral model) can be given as a simple systemic injection and in the unilateral model by peripheral intracarotid injection, but 6-OHDA must be centrally injected by stereotactic surgery, which requires a greater level of skill and equipment, is more time consuming, and because of the accuracy of stereotactic placement adds an additional degree of variability. Ultimately, however, central injection of either toxin is required to achieve extensive unilateral lesion in the absence of any depletion on the contralateral side.

Using the modes of administration described above, both 6-OHDA and MPTP are relatively selective for the DAergic nigrostriatal pathway. Indeed, although 6-OHDA can also be taken up by and can kill noradrenergic neurons, there is only sparse noradrenergic innervation of the striatum such that these lesions should have little impact. If this is a concern, desipramine can be administered 30 min prior to surgery, which will block the noradrenaline transporter, thereby increasing the selectivity for DAergic neurons. The actions of MPTP and 6-OHDA on the serotonin system have been debated, and both toxins have been claimed to leave the serotonin system unaffected, induce serotonergic sprouting, or decrease levels of serotonin in the striatum following DAergic denervation.[57–60] This may depend on the mode of administration in the case of 6-OHDA (striatal vs. mfb) or the time course of treatment for MPTP.[37,61,62]

One of the pathological hallmarks of PD is the presence of intracellular inclusions termed Lewy bodies, now known to be attributable to the abnormal aggregation of the α-synuclein protein.[63] The administration of 6-OHDA does not result in the development of inclusions. However, MPTP has been reported to induce protein accumulation in the SN of aged primates that resembles some features of Lewy bodies.[64–66]

RODENT VERSUS PRIMATE The main considerations in deciding between rodent and primate models are both ethical and practical. Both legislative and welfare considerations require use of the least advanced species to address the specific experimental hypotheses. Moreover, primates are more expensive and have limited availability, and the level of care is higher than for rodents as they require larger and more specialized housing facilities.[67] In addition, larger equipment for recording activity is also required, while rodents are easier to handle, house, and care for. Nevertheless, some experimental questions can be addressed only in the larger primate brain. A major advantage of the use of nonhuman primates, especially in the testing of pharmaceuticals, is their phylogenetic similarity to humans and the expression of PD pathology. Rodents do not have a distinguishable caudate and putamen, which are instead combined in a singular structure, the neostriatum, and thus in this regard nonhuman primates have a closer structural arrangement to humans. Aside from the pathology, the expression of parkinsonian symptoms in primates is more similar to the movement disorder in humans, and so has greater face validity in preclinical studies. Indeed many of the movement disorder rating scales used in primates are adapted from those used clinically to assess patients. By contrast, while many of the behaviors observed in rodents are less clearly recognizable as "parkinsonism," there are multiple behavioral tasks that can be applied in order to probe different components of the underlying motor deficits such as neglect versus motor function. When comparing rodent species, rats are larger, more docile, and more amenable to both surgical manipulation and behavioral analysis than are mice. Nevertheless, there has been greater interest in recent years to developing mouse models also, because of the greater numbers of mutant strains and the opportunities provided by new transgenic and knockout technologies to probe specific genetic and molecular influences on degeneration and behavior.

OTHER TOXIN MODELS One of the earliest toxins to be used in PD research is reserpine, a toxin that induces the release of catecholamines from their stores in the nerve terminals. This induces a hypokinetic state in most rodents and drugs can then be administered systemically or into selected brain regions through implanted cannulas. Reduced akinesia and increased activity and grooming can be observed when drugs are administered systemically, or the induction of a locomotor side bias when drugs are injected unilaterally. However, unlike the disease state, the depletion of the catecholamine stores and thus the resulting behavioral deficits are only temporary, and the stores are replenished and locomotor activity restored over a few days. It is therefore difficult to investigate experimental deficits such as dyskinesia that accumulate over time, or to evaluate chronic therapeutic treatments against a stable baseline. Furthermore, there are no morphological effects on the DAergic neurons, no permanent cell death or damage is observed, and no inclusions are documented. There is no selectivity for nigrostriatal as opposed to mesolimbic DAergic neurons, and other monoamine neurotransmitters are also severely depleted. Despite these failings, this model has been successfully used to evaluate pharmacotherapeutics and was fundamental in demonstrating the efficacy of L-DOPA in restoring motor function following DA depletion.[68]

In the past few years, models have been proposed based on rotenone and paraquat, two agricultural chemicals that have both been associated with an increased incidence of PD. However, these models produce a high mortality and decidedly variable results, and the induced degeneration is not specific to the SNc.[69–72] More recently, abnormalities of protein processing by the proteosome have been identified as a possible mechanism for inducing PD and the accumulation of proteins such as α-synuclein. McNaught *et al.*[73] reported that chronic administration of PSI, a synthetic peptide aldehyde that inhibits proteasomal function, reproduced both the motor and pathological consequences of PD in rats. They describe specific degeneration of dopaminergic neurons, proteinous deposits akin to Lewy bodies, in addition to bradykinesia, tremor, and rigidity. However, although similar results have been reported by other researchers,[74] equally many have not been able to reproduce these findings.[75,76] Each of these models still has to be fully and reliably characterized before being used to evaluate therapeutic or neuroprotective approaches to the treatment of PD.

CONCLUSIONS

6-OHDA and MPTP are the tools most frequently used to induce selective and effective degeneration of the DAergic nigrostriatal pathway in experimental animals, and thereby model the primary pathology and symptoms of PD in the laboratory. In deciding which model to use, the focus should be on first selecting the most appropriate model and, more importantly, the behavioral tests most appropriate for that model for the specific experimental hypotheses or preclinical strategies to be evaluated. The alternatives will be weighed according to a range of criteria involving dimensions of practicality, efficiency, validity, and animal welfare. Generally, several tests are used to determine the generality of results and the extent and limitations of their application. Treatments may not necessarily improve function in all of them. For example, transplants of DAergic tissue produce improved forepaw asymmetry in the cylinder test but not in the more complex paw-reaching task. However, while the 6-OHDA-lesioned rat and MPTP-treated primate are well-established models of PD that have provided considerable insight into the disease, it should be remembered that both still have significant limitations and the search continues for a model that can not just reproduce the neuropathology and symptoms of the disease, but also recreate more faithfully the progressive pathogenic aspects of the disease.

ACKNOWLEDGMENT

The authors would like to thank Bengt Mattson for providing the figures.

REFERENCES

1. Meredith GE, Kang UJ. Behavioral models of Parkinson's disease in rodents: A new look at an old problem. *Mov Disord* 2006;21(10): 1595–1606.
2. Ungerstedt U. 6-Hydroxy-dopamine induced degeneration of central monoamine neurons. *Eur J Pharmacol* 1968;5(1):107–110.
3. Glinka YY, Youdim MB. Inhibition of mitochondrial complexes I and IV by 6-hydroxydopamine. *Eur J Pharmacol* 1995;292(3–4): 329–332.
4. Kirik D, Rosenblad C, Bjorklund A. Characterization of behavioral and neurodegenerative changes following partial lesions of the nigrostriatal dopamine system induced by intrastriatal 6-hydroxydopamine in the rat. *Exp Neurol* 1998;152(2):259–277.
5. Visanji NP, O'Neill MJ, Duty S. Nicotine, but neither alpha4beta2 nor alpha7 nAChR subtype selective agonists, protects against a partial 6-hydroxydopamine lesion of the rat nigrostriatal tract. *Neuropharmacology* 2006;51(3):506–516.
6. Schwarting RK, Huston JP. The unilateral 6-hydroxydopamine lesion model in behavioral brain research. Analysis of functional deficits, recovery and treatments. *Prog Neurobiol* 1996;50(2–3):275–331.
7. Schallert T, Tillerson JL. Intervention strategies for degeneration of dopamine neurons in parkinsonism: Optimising behavioural assessment of outcome. In: *Innovative Models of CNS Disease: From Molecule to Therapy*. Clifton, NJ: Humana Press, 1999:131–151.
8. Deumens R, Blokland A, Prickaerts J. Modeling Parkinson's disease in rats: An evaluation of 6-OHDA lesions of the nigrostriatal pathway. *Exp Neurol* 2002;175(2):303–317.
9. Ungerstedt U, Arbuthnott GW. Quantitative recording of rotational behaviour in rats after 6-hydroxy-dopamine lesions of the nigrostriatal dopamine system. *Brain Res* 1970;24(3):485–493.
10. Kirik D, Georgievska B, Burger C, Winkler C, Muzyczka N, Mandel RJ, Bjorklund A. Reversal of motor impairments in parkinsonian rats by continuous intrastriatal delivery of L-dopa using rAAV-mediated gene transfer. *Proc Natl Acad Sci USA* 2002;99(7):4708–4713.
11. Dunnett SB, Bjorklund A, Stenevi U, Iversen SD. Behavioural recovery following transplantation of substantia nigra in rats subjected to 6-OHDA lesions of the nigrostriatal pathway. II. Bilateral lesions. *Brain Res* 1981;229(2):457–470.
12. Schallert T, Lindner MD. Rescuing neurons from trans-synaptic degeneration after brain damage: Helpful, harmful, or neutral in recovery of function? *Can J Psychol* 1990;44(2):276–292.
13. Schallert T, De Ryck M, Whishaw IQ, Ramirez VD, Teitelbaum P. Excessive bracing reactions and their control by atropine and L-DOPA in an animal analog of parkinsonism. *Exp Neurol* 1979;64(1): 33–43.
14. Olsson M, Nikkhah G, Bentlage C, Bjorklund A. Forelimb akinesia in the rat Parkinson model: Differential effects of dopamine agonists and nigral transplants as assessed by a new stepping test. *J Neurosci* 1995;15(5Pt. 2):3863–3875.
15. Chang JW, Wachtel SR, Young D, Kang UJ. Biochemical and anatomical characterization of forepaw adjusting steps in rat models of Parkinson's disease: Studies on medial forebrain bundle and striatal lesions. *Neuroscience* 1999;88(2):617–628.
16. Montoya CP, Campbell-Hope LJ, Pemberton KD, Dunnet SB. The "staircase test": A measure of independent forelimb reaching and grasping abilities in rats. *J Neurosci Methods* 1991;36(2–3):219–228.
17. Barneoud P, Parmentier S, Mazadier M, Miquet JM, Boireau A, Dubedat P, Blanchard JC. Effects of complete and partial lesions of the dopaminergic mesotelencephalic system on skilled forelimb use in the rat. *Neuroscience* 1995;67(4):837–848.
18. Lee CS, Sauer H, Bjorklund A. Dopaminergic neuronal degeneration and motor impairments following axon terminal lesion by instrastriatal 6-hydroxydopamine in the rat. *Neuroscience* 1996;72(3): 641–653.
19. Nikkhah G, Duan WM, Knappe U, Jodicke A, Bjorklund A. Restoration of complex sensorimotor behavior and skilled forelimb use by a modified nigral cell suspension transplantation approach in the rat Parkinson model. *Neuroscience* 1993;56(1):33–43.
20. Winkler C, Bentlage C, Nikkhah G, Samii M, Bjorklund A. Intranigral transplants of GABA-rich striatal tissue induce behavioral recovery in the rat Parkinson model and promote the effects obtained by intrastriatal dopaminergic transplants. *Exp Neurol* 1999;155(2): 165–186.
21. Dobrossy MD, Dunnett SB. Training specificity, graft development and graft-mediated functional recovery in a rodent model of Huntington's disease. *Neuroscience* 2005;132(3):543–552.
22. Iancu R, Mohapel P, Brundin P, Paul G. Behavioral characterization of a unilateral 6-OHDA-lesion model of Parkinson's disease in mice. *Behav Brain Res* 2005;162(1):1–10.
23. Monville C, Torres EM, Dunnett SB. Comparison of incremental and accelerating protocols of the rotarod test for the assessment of motor deficits in the 6-OHDA model. *J Neurosci Methods* 2006;158(2): 219–223.
24. Lundblad M, Vaudano E, Cenci MA. Cellular and behavioural effects of the adenosine A2a receptor antagonist KW-6002 in a rat model of l-DOPA-induced dyskinesia. *J Neurochem* 2003;84(6):1398–1410.
25. Rozas G, Guerra MJ, Labandeira-Garcia JL. An automated rotarod method for quantitative drug-free evaluation of overall motor deficits in rat models of parkinsonism. *Brain Res Brain Res Protoc* 1997;2(1):75–84.
26. Baird AL, Meldrum A, Dunnett SB. The staircase test of skilled reaching in mice. *Brain Res Bull* 2001;54(2):243–250.
27. Annett LE, Rogers DC, Hernandez TD, Dunnett SB. Behavioural analysis of unilateral monoamine depletion in the marmoset. *Brain* 1992;115(Pt. 3):825–856.
28. Kendall AL, Rayment FD, Torres EM, Baker HF, Ridley RM, Dunnett SB. Functional integration of striatal allografts in a primate model of Huntington's disease. *Nat Med* 1998;4(6):727–729.
29. Marshall JW, Ridley RM. Assessment of functional impairment following permanent middle cerebral artery occlusion in a non-human primate species. *Neurodegeneration* 1996;5(3):275–286.
30. Annett LE, Martel FL, Rogers DC, Ridley RM, Baker HF, Dunnett SB. Behavioral assessment of the effects of embryonic nigral grafts in marmosets with unilateral 6-OHDA lesions of the nigrostriatal pathway. *Exp Neurol* 1994;125(2):228–246.
31. Eslamboli A, Cummings RM, Ridley RM, Baker HF, Muzyczka N, Burger C, Mandel RJ, Kirik D, Annett LE. Recombinant adeno-associated viral vector (rAAV) delivery of GDNF provides protection against 6-OHDA lesion in the common marmoset monkey (Callithrix jacchus). *Exp Neurol* 2003;184(1):536–548.

32. Langston JW, Ballard P, Tetrud JW, Irwin I. Chronic parkinsonism in humans due to a product of meperidine-analog synthesis. *Science* 1983;219(4587):979–980.
33. Burns RS, Chiueh CC, Markey SP, Ebert MH, Jacobowitz DM. Kopin IJ. A primate model of parkinsonism: Selective destruction of dopaminergic neurons in the pars compacta of the substantia nigra by N-methyl-4-phenyl-1,2,3,6-tetrahydropyridine. *Proc Natl Acad Sci USA* 1983;80(14):4546–4550.
34. Jenner P, Rupniak NM, Rose S, Kelly E, Kilpatrick G, Lees A, Marsden CD. 1-Methyl-4-phenyl-1,2,3,6-tetrahydropyridine-induced parkinsonism in the common marmoset. *Neurosci Lett* 1984;50(1–3): 85–90.
35. Langston JW, Forno LS, Rebert CS, Irwin I. Selective nigral toxicity after systemic administration of 1-methyl-4-phenyl-1,2,5,6-tetrahydropyridine (MPTP) in the squirrel monkey. *Brain Res* 1984;292(2): 390–394.
36. Iravani MM, Syed E, Jackson MJ, Johnston LC, Smith LA, Jenner P. A modified MPTP treatment regime produces reproducible partial nigrostriatal lesions in common marmosets. *Eur J Neurosci* 2005; 21(4):841–854.
37. Russ H, Mihatsch W, Gerlach M, Riederer P, Przuntek H. Neurochemical and behavioural features induced by chronic low dose treatment with 1-methyl-4-phenyl-1,2,3,6-tetrahydropyridine (MPTP) in the common marmoset: Implications for Parkinson's disease? *Neurosci Lett* 1991;123(1):115–118.
38. Bezard E, Dovero S, Prunier C, Ravenscroft P, Chalon S, Guilloteau D, Crossman AR, Bioulac B, Brotchie JM, Gross CE. Relationship between the appearance of symptoms and the level of nigrostriatal degeneration in a progressive 1-methyl-4-phenyl-1,2,3,6-tetrahydropyridine-lesioned macaque model of Parkinson's disease. *J Neurosci* 2001;21(17):6853–6861.
39. Colosimo C, Granata R, Del Zompo M, Piccardi MP, Perretta G, Albanese A. Chronic administration of MPTP to monkeys: Behavioural morphological and biochemical correlates. *Neurochem Int* 1992;20(Suppl.):279S–285S.
40. Langston JW, Ballard PA Jr. Parkinson's disease in a chemist working with 1-methyl-4-phenyl-1,2,5,6-tetrahydropyridine. *N Engl J Med* 1983;309(5):310.
41. Imbert C, Bezard E, Guitraud S, Boraud T, Gross CE. Comparison of eight clinical rating scales used for the assessment of MPTP-induced parkinsonism in the Macaque monkey. *J Neurosci Methods* 2000;96(1):71–76.
42. Emborg ME, Ma SY, Mufson EJ, Levey AI, Taylor MD, Brown WD, Holden JE, Kordower JH. Age-related declines in nigral neuronal function correlate with motor impairments in rhesus monkeys. *J Comp Neurol* 1998;401(2):253–265.
43. Collier TJ, Steece-Collier K, Kordower JH. Primate models of Parkinson's disease. *Exp Neurol* 2003;183(2):258–262.
44. Sundstrom E, Samuelsson EB. Comparison of key steps in 1-methyl-4-phenyl-1,2,3,6-tetrahydropyridine (MPTP) neurotoxicity in rodents. *Pharmacol Toxicol* 1997;81(5):226–231.
45. Giovanni A, Sonsalla PK, Heikkila RE. Studies on species sensitivity to the dopaminergic neurotoxin 1-methyl-4-phenyl-1,2,3,6-tetrahydropyridine. Part 2: Central administration of 1-methyl-4-phenylpyridinium. *J Pharmacol Exp Ther* 1994;270(3):1008–1014.
46. Sedelis M, Schwarting RK, Huston JP. Behavioral phenotyping of the MPTP mouse model of Parkinson's disease. *Behav Brain Res* 2001;125(1–2):109–125.
47. Ogawa N, Hirose Y, Ohara S, Ono T, Watanabe Y. A simple quantitative bradykinesia test in MPTP-treated mice. *Res Commun Chem Pathol Pharmacol* 1985;50(3):435–441.
48. Crawley JN. Behavioral phenotyping of transgenic and knockout mice: Experimental design and evaluation of general health, sensory functions, motor abilities, and specific behavioral tests. *Brain Res* 1999;835(1):18–26.
49. Ahlskog JE, Muenter MD. Frequency of levodopa-related dyskinesias and motor fluctuations as estimated from the cumulative literature. *Mov Disord* 2001;16(3):448–458.
50. Papa SM, Engber TM, Kask AM, Chase TN. Motor fluctuations in levodopa treated parkinsonian rats: Relation to lesion extent and treatment duration. *Brain Res* 1994;662(1–2):69–74.
51. Lundblad M, Andersson M, Winkler C, Kirik D, Wierup N, Cenci MA. Pharmacological validation of behavioural measures of akinesia and dyskinesia in a rat model of Parkinson's disease. *Eur J Neurosci* 2002;15(1):120–132.
52. Cenci MA, Tranberg A, Andersson M, Hilbertson A. Changes in the regional and compartmental distribution of FosB- and JunB-like immunoreactivity induced in the dopamine-denervated rat striatum by acute or chronic L-dopa treatment. *Neuroscience* 1999;94(2): 515–527.
53. Lundblad M, Usiello A, Carta M, Hakansson K, Fisone G, Cenci MA. Pharmacological validation of a mouse model of l-DOPA-induced dyskinesia. *Exp Neurol* 2005;194(1):66–75.
54. Zigmond MJ, Stricker EM. Proceedings: The effects on ingestive behaviors of damage to central dopamine-containing neurons. *Psychopharmacol Bull* 1974;10(3):40–41.
55. Sakai K, Gash DM. Effect of bilateral 6-OHDA lesions of the substantia nigra on locomotor activity in the rat. *Brain Res* 1994;633(1–2):144–150.
56. Amalric M, Moukhles H, Nieoullon A, Daszuta A. Complex deficits on reaction time performance following bilateral intrastriatal 6-OHDA infusion in the rat. *Eur J Neurosci* 1995;7(5):972–980.
57. Karstaedt PJ, Kerasidis H, Pincus JH, Meloni R, Graham J, Gale K. Unilateral destruction of dopamine pathways increases ipsilateral striatal serotonin turnover in rats. *Exp Neurol* 1994;126(1):25–30.
58. Balcioglu A, Zhang K, Tarazi FI. Dopamine depletion abolishes apomorphine- and amphetamine-induced increases in extracellular serotonin levels in the striatum of conscious rats: A microdialysis study. *Neuroscience* 2003;119(4):1045–1053.
59. Takeuchi Y, Sawada T, Blunt S, Jenner P, Marsden CD. Effects of 6-hydroxydopamine lesions of the nigrostriatal pathway on striatal serotonin innervation in adult rats. *Brain Res* 1991;562(2):301–305.
60. Guerra MJ, Liste I, Labandeira-Garcia JL. Effects of lesions of the nigrostriatal pathway and of nigral grafts on striatal serotonergic innervation in adult rats. *Neuroreport* 1997;8(16):3485–3488.
61. Perez-Otano I, Herrero MT, Oset C, De Ceballos ML, Luquin MR, Obeso JA, Del Rio J. Extensive loss of brain dopamine and serotonin induced by chronic administration of MPTP in the marmoset. *Brain Res* 1991;567(1):127–132.
62. Pifl C, Schingnitz G, Hornykiewicz O. Effect of 1-methyl-4-phenyl-1,2,3,6-tetrahydropyridine on the regional distribution of brain monoamines in the rhesus monkey. *Neuroscience* 1991; 44(3):591–605.
63. Spillantini MG, Schmidt ML, Lee VM, Trojanowski JQ, Jakes R, Goedert M. Alpha-synuclein in Lewy bodies. *Nature* 1997;388(6645): 839–840.
64. Kowall NW, Hantraye P, Brouillet E, Beal MF, McKee AC, Ferrante RJ. MPTP induces alpha-synuclein aggregation in the substantia nigra of baboons. *Neuroreport* 2000;11(1):211–213.
65. Gibb WR, Terruli M, Lees AJ, Jenner P, Marsden CD. The evolution and distribution of morphological changes in the nervous system of the common marmoset following the acute administration of 1-methyl-4-phenyl-1,2,3,6-tetrahydropyridine. *Mov Disord* 1989;4(1): 53–74.
66. Forno LS, DeLanney LE, Irwin I, Langston JW. Similarities and differences between MPTP-induced parkinsonism and Parkinson's disease. Neuropathologic considerations. *Adv Neurol* 1993;60:600–608.
67. Eslamboli A. Marmoset monkey models of Parkinson's disease: Which model, when and why? *Brain Res Bull* 2005;68(3):140–149.
68. Utley JD, Carlsson A. Relative effects of L-DOPA and its methyl ester given orally or intraperitoneally to reserpine-treated mice. *Acta Pharmacol Toxicol (Copenh)* 1965;23(2):189–193.
69. Brooks AI, Chadwick CA, Gelbard HA, Cory-Slechta DA, Federoff HJ. Paraquat elicited neurobehavioral syndrome caused by dopaminergic neuron loss. *Brain Res* 1999;823(1–2):1–10.
70. Corasaniti MT, Strongoli MC, Rotiroti D, Bagetta G, Nistico G. Paraquat: A useful tool for the in vivo study of mechanisms of neuronal cell death. *Pharmacol Toxicol* 1998;83(1):1–7.
71. Höglinger GU, Oertel WH, Hirsch EC. The rotenone model of parkinsonism—the five years inspection. *J Neural Transm* 2006; (Suppl.)70:269–272.

72. Betarbet R, Sherer TB, MacKenzie G, Garcia-Osuna M, Panov AV, Greenamyre JT. Chronic systemic pesticide exposure reproduces features of Parkinson's disease. *Nat Neurosci* 2000;3(12):1301–1306.
73. McNaught KS, Perl DP, Brownell AL, Olanow CW. Systemic exposure to proteasome inhibitors causes a progressive model of Parkinson's disease. *Ann Neurol* 2004;56(1):149–162.
74. Schapira AH, Cleeter MW, Muddle JR, Workman JM, Cooper JM, King RH. Proteasomal inhibition causes loss of nigral tyrosine hydroxylase neurons. *Ann Neurol* 2006;60(2):253–255.
75. Manning-Bog AB, Reaney SH, Chou VP, Johnston LC, McCormack AL, Johnston J, Langston JW, Di Monte DA. Lack of nigrostriatal pathology in a rat model of proteasome inhibition. *Ann Neurol* 2006;60(2):256–260.
76. Kordower JH, Kanaan NM, Chu Y, Suresh Babu R, Stansell J 3rd, Terpstra BT, Sortwell CE, Steece-Collier K, Collier TJ. Failure of proteasome inhibitor administration to provide a model of Parkinson's disease in rats and monkeys. *Ann Neurol* 2006;60(2): 264–268.

35 Transgenic Animal Models of Neurodegenerative Diseases

ANTHONY W.S. CHAN AND YUKSEL AGCA

ABSTRACT

Neurodegenerative diseases are complex disorders involving neuropathological and psychiatric alterations. Transgenic mice expressing mutant human Alzheimer precursor protein (βAPP), α-synuclein, and mutant huntingtin genes have been developed for Alzheimer's, Parkinson's, and Huntington's disease (AD, PD, and HD), respectively. However, none of the existing mouse models completely represents the pathology of these diseases, including neuronal loss, cerebral atrophy, widespread neurofibrillary tangles, and neuropil threads in AD, formation of Lewy bodies and the extensive loss of dopaminergic neurons in PD, and neuronal loss and aggregate formation in HD. Although rat models have been shown to be superior to mouse models because of the brain size and the availability of more sophisticated behavioral testing, a common disadvantage in rodent models is the lack of neurodegeneration. Furthermore, rodents have relatively smaller brains and thus provide a limited behavioral repertoire for assessing transgene-associated structural and cognitive deficits. Nonhuman primates are ideally suited for modeling the neuropathology of neurodegenerative diseases because of their biological proximity to humans, size, and age-related development of lesions. Thus the development of a transgenic nonhuman primate model of neurodegenerative diseases holds greater promise for success in the discovery of diagnoses, treatments, and cures than approaches using other animal species. Nevertheless, it is undeniable that rodent models contribute significantly to the understanding of neurodegenerative diseases and the development of cures. A transgenic nonhuman primate model will be able to mimic human conditions not only physiologically but also genetically. However, the development of transgenic rodent and nonhuman primate models of neurodegenerative diseases is equally important for unraveling the mystery of neurodegeneration and the development of early diagnosis and cures.

Key Words: Neurodegenerative disease, Neurodegeneration, Animal model, Transgenic rodent, Transgenic nonhuman primates.

INTRODUCTION

Animal models of neurodegenerative diseases (NDs) have been developed quickly in the past decades because of the advancement in molecular genetics and the latest development of gene transfer technology. Parkinson's disease (PD), Alzheimer's disease (AD), and Huntington's disease (HD) are the most common NDs and affect more than six million people in the United States. Compared to HD, which is caused by the extension of the polyglutamine [poly(Q)] tract at the 5′ region of the huntingtin (htt) gene, PD and AD are more complex.[1–6] Extensive effort has been made to determine the genetic and environmental components that lead to complex genetic disorders such as PD and AD.[7,8] Candidate genes are often identified in a small number of families that are scattered around the globe.[3,7,9] Although candidate genes have been identified, no single gene is unanimously linked to PD and AD. The latest discovery of the leucine-rich repeat kinase 2 (LRRK2) gene has the greatest link to familial parkinsonism (FP), which contributes to only 2–10% of these cases.[1,2,9,10] Several mutations have also been identified in the Alzheimer precursor protein (APP) gene that are linked to the development of AD.[3,11–13] However, no single mutation leads to neuropathology and cognitive behavioral declines in animal models equivalent to that seen in human patients.[3,10]

Transgenic animal technology has revolutionized animal modeling, which becomes a very useful tool for elucidating the link between genetics and human diseases.[3,10,14–18] The development of a new animal model is not only important for a better understanding of human diseases, but also important for the development of diagnostic tools and treatments.[18,19] Unfortunately, animals that carry genetic defects similar to those found in patients do not develop identical patterns.[20–21] This suggests fundamental differences between human and other mammalian species such as rodents and nonhuman primates (NHPs).[19] Physiological differences include life span, brain complexity, cellular metabolism, and endocrine and reproductive function, which could be the result of genetic redundancy or variations in biochemical pathways.[22–24] Hence, a good animal model is not only dependent on the genetic constitution of the transgene but also the genetic and physiological background of the animal species.[25] It is particularly important for NDs because of the complexity in genetic composition, the unknown neuropathological mechanisms, and the gradual decline in cognitive and behavioral function. For example, most of the transgenic HD mice do not develop identical neuropathology. Although brain size was reduced by 20%, specific neuronal cell loss was not found in the striatal region or cortex.[26] A similar phenomenon was also observed in AD and PD mice.

Due to physiological differences between species, the rat has certain advantages over the mouse because of brain

From: *Sourcebook of Models for Biomedical Research*
(P. M. Conn, ed.), © 2008 Humana Press Inc., Totowa, NJ.

size, neurocircuitry, and availability of cognitive–behavioral testing.[12,13,27–31] Additionally, NHPs are considered the best animal models, especially for NDs, because of the complex correlation between aging and progressive neurodegeneration, which leads to deleterious alterations in behavior and psychiatric status associated with diseases such as AD, PD, and HD in humans.[19,32,33] NHPs are the only animal model providing accessibility to a wide range of cognitive–behavioral testing including the possibility for high-resolution brain imaging such as functional magnetic resonance imaging (fMRI) and positron emission tomography (PET).[19,32] Thus, a transgenic animal with a genetic alteration that leads to human inherited NDs might best mimic patient conditions; this could increase our understanding of disease development, thereby paving the way for the development of cures.[32]

Besides the selection of target genes and animal species, promoter is also an important factor for the development of a transgenic model. Promoter regulates the expression of disease-related genes, which affects the etiological pattern of the animals.[3,10,16] The R6/2 mouse model of HD has good representation in neuropathology and motor deficits; however, mutant *htt* expressed under the control of the human *htt* promoter results in an excessive decline in health and sudden death with no known cause.[34–36] On the other hand, transgenic mice that carry mutant *htt* with 82Q under the control of the prion protein promoter have neuron-specific pathology with distinctive stages of disease development, which mimics the course of HD progression.[37] These studies suggest that promoter is important in determining disease development, neuropathology, and cognitive–behavioral patterns in transgenic animal models.

Here we will focus on animal model development of AD, PD, and HD as a representative of NDs. Our goal is to provide a brief overview of currently available animal models and new methodology for creating transgenic rat and NHP models.

METHODS FOR CREATING TRANSGENIC ANIMALS

The method for creating transgenic animals has a significant impact on the application of the disease model.[14,18] In general, transgenic animals can be categorized by the expression pattern, which includes (1) overexpression, (2) gene targeting: knockin, knockout, and knockdown, (3) conditional expression, and (4) inducible expression. Although the expression pattern could be manipulated by strategies such as the tetracycline-inducible system, the Cre-lox conditional expression system, and artificial chromosome (AC), an exogenous gene construct has to be inserted into the target cell genome in order to achieve long-term expression. Additionally, there are three fundamental criteria determining the success of a transgenic animal: (1) germline transmission, (2) stable expression, and (3) all body cells must carry the transgene.[38] Two exceptions are chimeric mice generated by blastocyst injection of genetically modified embryonic stem cells and local gene transfer such as stereotaxic viral injection in a selected brain region.[33,39–41]

Several methods have been used to generate transgenic animals. Among these methods, pronuclear microinjection (PI) is the most commonly used method in creating transgenic mice and rats. A similar method has also been used to generate transgenic livestock; however, somatic cell nuclear transplantation (SCNT; cloning), a recent development, is slowly replacing PI in livestock because more versatile genetic manipulation could be achieved.

Table 35–1
Focal transgenesis versus germline transgenesis

	Focal transgenesis	*Germline transgenesis*
Number of transgenic cells	Regional and limited	All body and germ cells
Expression pattern	Regional	Whole body (dependent on promoter)
Inheritance	No	Yes
Timeline	1–6 months	6–24 months (dependent on species)

SCNT will not be discussed in detail because it is not commonly used in ND models. Compared to PI, viral vectors such as retrovirus and lentivirus have been developed with high efficiency in generating transgenic animals.[14,42,43] Retroviral gene transfer was first described by Jaenisch and colleagues in 1976. However, retrovirus was not widely used for the production of transgenic animals because of biosafety concerns. Not until the mid-1990s was a replication-defective pseudotyped retroviral vector developed.[44,45] Pseudotyped retroviral vectors have the advantage of achieving a high titer and broad host cell range, which are the two major limitations of conventional retroviral vector systems. Chan and colleagues have successfully generated transgenic cattle and the first transgenic monkey using pseudotyped retrovirus at high efficiency.[14,42] However, due to the life cycle of retroviruses, efficient gene transfer could be achieved only in actively dividing cells.[14,42]

The recent development of the lentiviral vector has led to a new era of *in vivo* genetic manipulation.[46,47] Nondividing cells could be transfected efficiently using pseudotyped lentivirus.[46,48] Replication-defective lentivirus is derived from HIV-1 with most of the viral structural genes removed. Vesicular stomatitis virus envelope glycoprotein protein G (VSVG) pseudotyped lentivirus is widely used in neurological research because gene transfer in neuronal cell types that are not actively dividing could be achieved at high efficiency.[49] Stereotaxic injection of lentivirus at selected brain regions provides an efficient[46,48] and effective method for studying gene function in a relatively short time compared to germline transgenesis[43,47,50] (Table 35–1). There are significant differences between focal and germline transgenesis. Focal transgenesis is limited to a small region in the brain and is able to transfect only a limited number of neuronal cell types in an individual. Therefore, each model has to be generated independently and variation between animals is expected. This technique offers the opportunity for an *in vivo* gene function study in species that are highly relevant to humans but for which efficient gene transfer techniques are not available for the creation of transgenic models such as rat and NHPs.[14,19] The development of transgenic rat and NHP models of NDs will also be discussed.

TRANSGENIC MOUSE MODEL OF NEURODEGENERATIVE DISEASES

The transgenic mouse is the most commonly used and diversified animal model in NDs.[3,4,10,15,16,36,37] Extensive information on the mouse genetic background is important when developing an

ND model. Mice with different genetic backgrounds have been shown to be a crucial factor for the future development of a cognitive–behavioral pattern in a transgenic mouse model.[25] Additionally, well-established techniques in handling gametes, embryos, and surrogates, inexpensive and relatively limitless supplies, and short generation times allow a more versatile application of the transgenic mouse model than models of other species.[19] Therefore, transgenic rodents, developed with genetic defects similar to those seen in patients, are widely used in biomedical research.

Various genetic manipulation strategies are possible only in mice but not in species such as rats or NHPs.[19] Besides the overexpression of a transgene, downregulation by RNAi or microRNA, or conditional expression of a target gene, disease-related genes could be altered in embryonic stem cells by recombination techniques followed by blactocyst injection. However, the resulting chimeric mice require an additional breeding procedure in order to establish a genetically defined transgenic line. Although mouse cloning technology has not been widely used in ND, the prospect of an identical mouse model is expected to have significant impact. Transgenic mouse models of AD, PD, and HD have also been developed with various promoters, which include tissue- and cell type-specific promoters such as the prion protein (PrP) and neuron-specific enolase (NSE) promoter, the thymus-1 minigene (thy-1), a viral and housekeeping gene promoter such as cytomegalovirus (CMV), ubiquitin, platelet-derived growth factor (PDGF) promoters, and endogenous promoters such as the htt promoter.

AD mouse models are generated primarily by the overexpression of wild-type APP or APP with mutations at positions K670N and M671L (Swedish), E639Q (Dutch), D694N (Iowa), and V717F (London) that favors Aβ processing.[3,10,51] These mutants are important factors for achieving Aβ levels sufficient to lead to amyloid deposition.[3] Additionally, presenillin 1 (PS1) and 2 (PS2) affect the activity of -secretase, which results in the alteration of the Aβ42 level.[10,13] Several transgenic mice with PS1 mutations have no significant effect on Aβ deposition. However, the coexpression of APP Swedish or London mutations and the PS1 mutant, a significant increase of Aβ42, and amyloidosis at an early age were observed compared to mice with APP mutations alone.[3] A double mutant mouse (PS1 M146L/APP K670N/M671L) has plaque formation by 3 months of age compared to 9–12 months in an APP K670N/M671L mouse.[3] Besides APP and PS1, overexpression of wild-type or mutant tau protein in mice has led to various degrees of neurofibrillary pathology, which is highly related to the number of tau repeats.[3,10] A triple transgenic mouse model with the combination of mutant APP, PS1, and tau has resulted in an early onset at 3 months of age of plaque formation, positive for phosphorylated tau and neurofibrillary tangles.[52,53] A transgenic mouse with triple APP mutations (Swe/Dut/Iowa) has also been generated with early-onset and robust amyloid accumulation.[51]

Animal model targeting modifiers of amyloid pathology have also been developed. These include overexpression and knockout models of apolipoprotein E (ApoE) and BACE1.[54–57] An increase in the level of ApoE results in enhancement of amyloid deposition,[55] whereas knockout of ApoE results in a decrease of deposition in transgenic mice.[54] A similar result was observed in BACE1. Consistently, a mouse deficient in low-density lipoprotein receptor-related protein (LRP), an ApoE receptor, has an increase in amyloid deposition.[10] However, transgenic mice overexpressing TGF-β1 have an increase in vascular deposition and a decrease in plaque deposition. These animal models suggest how complicated the neuropathology in AD is and these modifiers play an influential role in amyloid pathology in APP mice. In fact, ApoE is considered one of the risk factors for AD.

Several candidate genes including PARK 1 (α-synuclein), 2 (Parkin), 5 (UCH-L1), 6 (PINK1), 7 (DJ1), and 8 (LRRK2) have been identified and are related to PD.[1,7,9] Among these genes, α-synuclein is the first linked to PD. Several mutations of α-synuclein have been identified and are sporadically distributed at low percentages in familial PD cases. Overexpression of both wild-type and mutant α-synuclein has resulted in a decrease of dopamine and dopamine transporter with behavioral impairment.[4,27] However, no significant difference was found when different promoters were used, which includes tyrosine hydroxylase (TH), PDGF, and the Thy-1 promoter.[4] On the other hand, the parkin gene is related to familial autosomal recessive juvenile PD.[1,7,9,27] Loss of parkin function results in an abnormal accumulation of its substrate such as glycosylated α-synuclein.[15] A variation in neuropathology was also found in the deletion of exons 2, 3, and 7 of the parkin gene. Some common features include no loss of nigrastriatal DA neurons and behavioral impairments that were observed in exon 3 and 7 deletions but were undetectable in exon 2-deleted mice.[15] Mouse models with DJ-1 knockout, reduced Nurr1, and the loss of Pitx3-aphakia have also been generated recently. Variations in neuropathology and behavioral impairments suggest that Nurr1 and Pitx3-aphakia mice are well suited for studying later stages of AD because of the loss of nigrastriatal dopaminergic neurons, whereas DJ-1 mice may be best for the study of early-onset parkinsonism.

Genetically HD is less complex than AD and PD.[2,7,30] The extended glutamine tract at the 5′ region of exon 1 of htt has been identified and linked to HD.[58–60] Variations in neuropathology and behavioral decline are highly correlated with the length of the poly(Q) in patients as well as in HD mice.[58,60–62] In addition to the poly(Q) tract, promoter, which regulates the expression of mutant htt, is also important in the development of animal models. Although mice are not the perfect model for humans, HD mice present features that are highly relevant to HD and are great tools in understanding the etiology and the development of cures.[19,32] Although the function of *htt* has not been clearly determined, *htt* knockout is embryonic lethal and knockin mice do not have neuropathology identical to that seen in HD patients. Systemic overexpression of mutant *htt* is lethal with a board spectrum of HD phenotypes including motor impairment and neuropathology. However, the cause of death in R6/2 mice, one of the most popular HD mouse models,[61] is not directly linked to the deteriorating effect of HD but to complications that lead to physiological dysfunction.

Although most of the PD, AD, and HD mice present certain features of cognitive–behavioral decline and motor deficit, which lead to further insight on the molecular basis of neurodegenerative diseases, highly sensitive testing for in-depth analysis in mice is very limited. Therefore, the need for an animal model that allows longitudinal studies such as noninvasive imaging analysis as well as access to more accurate testing methodology for evaluating cognitive–behavioral and motor impairment are important in unraveling the etiology of the disease and developing diagnostic tools and cures.

TRANSGENIC RAT MODEL OF NEURODEGENERATIVE DISEASES

For over a century the rat has been the model of choice for studies of many human diseases. Several hundred rats lines including inbred strains, congenics, and mutant rats exist. Unfortunately, germline transgenic rats are less available to biomedical researchers. In the context of neurological science, the rats have served as important animals for the study of neural development and disease, neurodegeneration, addiction, and behavior. There are several reasons for the wide use of rats in biomedicine: (1) the availability of extensive baseline data that make the interpretation of the experimental results more meaningful, (2) their larger size makes many live operations feasible, and (3) certain physiological similarities of the rat to humans.

Although transgenic mice have proven to be important models for human diseases, there are many cases in which rats are superior models due to metabolic similarities and larger body size including larger brain for the development of neurosurgical interventions. In some instances, transgenic rats provided a more appropriate phenotype when compared to mice bearing the same transgene.[63] For example, rats are a superior animal model for studying autoimmune diseases such as rheumatoid arthritis.[64] Transgenic rats have also been very useful in transplantation research, in which microsurgery is an essential procedure. However, there are currently only a handful of transgenic rat lines compared to transgenic mouse lines. This is mainly due to inefficient gene transfer methods and limited understanding of gamete and embryo manipulation skills in the rat.[65] In recent years, little advancement has been made in rat germline modification techniques; further improvement in gene transfer technology will ensure the wider application of transgenic rats as human disease models.

Fortunately, genomic resources for the better utilization of rat models of human disease are growing rapidly.[66] With the completion of the Human Genome Project, there is an unprecedented need for animal models to determine gene functions and to test new therapeutic strategies aimed at curing human disease. There is no doubt that the recent completion of the rat genome sequence further improves the utility of the rat as a model organism. Therefore, genetically modified rats will play a pivotal role in those advances since they offer many unique advantages as models of human disease. Despite the availability of remarkable new genomic tools, the development of genetically modified rats is still in its early stages and complex genetic manipulation such as gene targeting is not currently available.

PRONUCLEAR MICROINJECTION METHOD FOR THE GENERATION OF TRANSGENIC RATS

The generation of transgenic rats by PI often requires a large number of embryo donor and recipient animals. The procedure is labor intensive, and requires significant technical skill. Moreover, the production of transgenic rats by PI is inefficient due, in part, to difficulties associated with observing the pronuclei and a high postinjection lysis rate (30–35%). Rat zygotes are often not synchronized as they are in mice and thus finding the appropriate timing for visualization and injection of the pronucleus is timely and labor intensive. To generate five or six transgenic founder animals, about 800–1000 embryos have to be injected if outbred strains such as Sprague–Dawley and Wistar (see Table 35–2) are used. These shortcomings are magnified when creating transgenic rats in an inbred background such as Fischer 344 and Lewis, primarily due to their suboptimal response to superovulation, higher sensitivity to microinjection, and lower gene transfer efficiency. Due to low efficiencies many investigators initially create transgenic rats using outbred stocks, i.e., Sprague–Dawley, and subsequently backcross them to the desired inbred genetic background using a congenic approach. This approach is not only expensive due to animal housing, but also time consuming, requiring at least 10 generations of backcrossing.[67,68] Since the significance of genetic background in animal disease models cannot be discounted, the availability of a simple and reliable methodology to create transgenic rats with desired genetic background is urgently needed.

GENERATION OF TRANSGENIC ANIMALS USING VIRAL VECTORS

Since the production of transgenic rats by PI is inefficient, the development of alternative gene delivery method(s) is critical for the advancement of transgenic rat models of NDs. An alternative approach to generating germline transgenic animals is the microinjection of viral vectors into the perivitelline space (PVS) of zygotes or MII oocytes. This method appears to yield a much higher transgenic rate (Table 35–2) compared to the PI method and overcomes many of the problems associated with the highly inefficient PI method in the rat. In addition, lentivirus-mediated delivery of small interfering RNA (siRNA) into cycling, noncycling mammalian cells, embryonic stem cells, zygotes, and their derivatives has been successfully demonstrated.[69,70] The combination of lentiviral gene transfer and siRNA technology has further accelerated the need for rat disease models that require the suppression or the loss of gene function.

Table 35–2
Transgenic efficiency using pronuclear microinjection of plasmid DNA or perivitelline space microinjection of lentiviral vector in the rat

Method of injection	*Number of zygotes injected*	*Postinjection survival (%)*	*Number of embryo transfers*	*Number of transgenic founder*	*Overall (number injected/ number transgenic (%))*
Pronuclear microinjection[a]	800–1000	65–70	25–27	5–6	0.6
Perivitelline space microinjection[b]	30–40	90	2–3	8–10	25–30

[a]Based on Charreau *et al.*[63]
[b]Unpublished data.

An alternative to the transgenic rat model, lentiviral gene transfer, has led to the possibility of focal transgenesis. Lentivirus is injected into a selected region of the brain and gene function could then be studied. Focal transgenesis has been widely used in ND research because it is fast and effective in determining gene function. Functional studies of α-synuclein and parkin in PD, APP, and tau in AD rat models have been reported, which accelerate the understanding of gene function and the role of genes in disease development.[11,27,33,41,71]

TRANSGENIC RAT MODEL OF NEURODEGENERATIVE DISEASES

Only limited numbers of transgenic rat models of NDs were developed in recent years because of the low efficiency. Most of these transgenic rats were generated by the traditional PI method. A transgenic rat model of NDs was first reported in 1998.[13] The PS1 gene was introduced into rats and a significant increase of apoptosis in primary cortical neurons was observed.[13] Transgenic rat models that carry various combinations of APP and PS1 mutations have also been developed. However, no senile plaque of aggregated Aβ was observed in these rats.[29] In addition to an AD model, a truncated rat htt cDNA with 52 CAG repeats under the control of the rat htt promoter was microinjected into the pronucleus of rat zygote.[31] These rats exhibit adult-onset HD with typical histopathological alterations.[31] *In vivo* imaging including MRI and PET was also performed in these rats and longitudinal studies could be accomplished. Although several rat models of HD were developed afterward, the development of ND rat model remains slow.[30,71] So far, no transgenic rat model of PD has yet been reported and the search for candidate genes remains controversial.

Only recently has lentivirus been shown to be highly efficient in generating transgenic animals.[43,47,50] Compared to PI, lentiviral gene transfer results in a much higher transgenic rate (Table 35–2). Although transgenic rat lines remain primarily generated by PI, the versatile application of lentivirus has thrived in recent years not only in focal gene transfer but also in the creation of a transgenic animal model.[33,41,47,49,50,72,73] We have generated several transgenic rat lines at a high rate as summarized in Table 35–2. A brief but precise protocol will be discussed in the following section.

GENERATION OF TRANSGENIC RAT USING LENTIVIRUS

RAT SUPEROVULATION Female rats (28–30 days old) are superovulated using SC implantation of 8 units of follicle-stimulating hormone (FSH), introduced via Alzet miniosmotic pumps, and followed by an intraperitoneal (ip) injection of 15 units of luteinizing hormone (LH) approximately 50–52 h after FSH implantation.[74] To obtain zygotes, donors will be mated with male rats right after LH injection.

RAT EMBRYO COLLECTION Superovulated female rats are euthanized by CO_2 inhalation 20–24 h post-LH injection. The oviducts are excised and placed into a culture dish containing TL-HEPES supplemented with 1 mg/ml hyaluronidase (Sigma Chemical Co., St. Louis, MO) and 4 mg/ml bovine serum albumin (BSA) (Sigma Chemical Co., St. Louis, MO). The extended, translucent oviductal ampulla is dissected to release the clutch of zygotes into the solution in a culture dish. Zygotes are then incubated with hyaluronidase for about 5 min to enable dissociation and removal of the cumulus cells. An inverted microscope is used to evaluate zygotes for the presence of polar bodies, pronuclei, and sperm tail to verify fertilization. The cumulus-free zygotes are washed by transferring them through three drops of hyaluronidase-free TL-HEPES before being placed into 40 µl culture drops of KSOM under oil in culture dishes, and placed in an incubator (humidified 5% CO_2/95% N_2) at 37°C. The zygotes are subjected to lentiviral injection via PVS 1–2 h after collection.

SURGICAL EMBRYO TRANSFER To obtain pseudopregnant embryo recipient rats, 8- to 10-week-old Sprague–Dawley female rats (HSD: Sprague Dawley® SD®) are synchronized by ip injection of 40 µg gonadotropin-releasing hormone (GnRH) analog des-Gly[10] [D-Ala[6]] ethylamide (Sigma L4513) and with mature, vasectomized Sprague–Dawley males 4 days after hormone injection. For verification of mating and hence pseudopregnancy, the female rats are removed the following morning and checked for the presence of a vaginal plug.

For embryo transfer (ET), the recipient rats are anesthetized by injecting the ketamine:xylazine mixture intramuscularly (im) into the biceps femoris muscle. Approximately 5 min after the injection, the dorsal and lateral skin is clipped and the surgery site prepared with three alternate scrubs of betadine and alcohol. The skin of the dorsum is sagittally incised (10–15 mm in length) on the midline at the level of the paralumbar fossa. The skin incision is then rolled laterally to superimpose the paralumbar fossa. The muscle layers and the peritoneal wall are opened via incision and spread to create a window about 8–12 mm in length that permits retraction of the ovarian fat pad. The ovary is gently grasped with a tissue forceps and retracted/extracted until the ovary, the ovarian bursa, and the oviduct are exposed and visualized using a stereomicroscope at 10–15×. A few drops (~25 µl) of epinephrine (VEDCO, St. Joeseph, MO; 1 mg/ml) are sprinkled over the bursa to reduce bleeding. The bursa is gently dissected using watchmaker forceps to gain access to the infundibulum. PVS-injected embryos are loaded into a small column consisting of a few microliters of phosphate-buffered saline (PBS) within a 150-µm-diameter transfer pipette. The transfer pipette is inserted into the infundibulum, 8–10 injected zygotes are discharged into the oviduct, and then the ovary is carefully replaced into the peritoneal cavity. After repeating the procedure on the contralateral side, the skin is closed using wound clips (Fisher Scientific, Pittsburgh, PA).

LENTIVIRAL VECTOR CONSTRUCTION The gene of interest is inserted into a lentiviral vector backbone such as "pFUW"[43] [F, flap; U, ubiquitin promoter; G, green fluorescent protein; W, woodchuck hepatitis virus posttranscriptional regulatory element (WRE)]. In brief, pFUW is a self-inactivating vector that is composed of the WRE in order to increase the transcription level and minimize the position effect. Additionally, an HIV flap element was inserted between the 5′ LTR and the internal promoter to increase the titer of the virus.

COLLECTION OF AND CONCENTRATION OF VECTOR 293FT cells are plated into a 10-cm tissue culture plate the day before the cotransfection with pΔ8.9 and pVSV-G into a 293FT packaging cell (Invitrogen, Inc.). Culture medium is collected at 48 h posttransfection for 3 consecutive days at 24-h intervals. The supernatant is centrifuged at 25,000 × *g* for 90 min. The vector pellet is resuspended and aliquoted in 50 µl of 0.1× Hanks' balanced salt solution and kept frozen at −80°C. All procedures were performed inside a Biosafety level II cabinet. All of the plasticware and instruments that have contacted the vector are decontaminated with 10% bleach or autoclaved before disposal.[38]

LENTIVIRAL INJECTION The procedure for lentiviral injection in the PVS is performed as described for NHP oocytes in the next section.

GENOTYPING Two weeks after delivery, tailsnips are taken from all potential founder rat pups, DNA extracted, and genotyped using an appropriate primer set by polymerase chain reaction (PCR). Furthermore, Southern blot analysis is performed to determine copy numbers of the transgene of interest.

CURRENT STATUS OF THE TRANSGENIC NONHUMAN PRIMATE MODEL OF NEURODEGENERATIVE DISEASES

NDs are complex disorders in which declines in neuropathology and psychiatric alterations are highly related to aging. Although transgenic mouse and rat models of neurodegenerative diseases such as AD, PD, and HD with certain clinical features have been created, a limited neuropathological pattern such as neurodegeneration is the major drawback of these animals. Chemical induction of HD and PD in NHPs has also been reported with similar neuropathological damage and psychiatric alteration; however, the role of genetic defects cannot be identified in the course of disease development. Due to the close relationship between human and NHPs, the development of transgenic NHP models will be able to mimic human conditions not only physiologically but also in terms of the genetic defects that lead to AD, PD, or HD. Therefore, disease onset and development could then be elucidated with the aid of a transgenic NHP model. Additionally, high-resolution brain imaging technology such as MRI and PET and cognitive–behavioral testing can be carried out in the NHP simultaneously, which is not possible in other animal models. A high degree of similarity in motor repertoire between NHPs and humans allows direct comparison of the neurological phenotype between the NHP model and patients. Because of this, a transgenic NHP is expected to have a significant impact on ND research. However, due to the difficulty in creating a transgenic NHP, an alternative strategy has been developed to bypass the creation of a germline transgenic monkey and to study gene function at a selected brain region. Focal transgenesis (stereotaxic delivery of a viral vector) is one of the most popular techniques in neuroscience because of the latest development of a lentiviral vector that allows efficient gene transfer in inactive cells such as neurons.[33,41] Although focal transgenesis is fast and efficient, the role of the gene defect will be restricted to a selected brain region and a selected group of neurons, and its role beyond neurological functions may be easily overlooked. Therefore, focal transgenesis is not the solution for a germline transgenic animal that carries the genetic defect in all body cells, which best mimics patient conditions. Although both strategies are unique and are very useful in ND research, careful interpretation of results should be taken when using these animal models.

Focal transgenesis by the injection of lentivirus has been widely used to study disease-related gene function in rodents and NHPs.[33,41] However, there is only one report of the creation of a transgenic NHP.[14] Although a transgenic NHP is not commonly used in the ND model, the potential impact of the transgenic NHP model in ND research could be considerable because of the great similarity to humans including motor repertoire, brain structure and function, and genome constitution. Here we have a brief protocol for the generation of a transgenic NHP model of NDs. Preparation of a lentiviral vector was previously described when the transgenic rat model was discussed.

GENERATION OF TRANSGENIC NONHUMAN PRIMATES

FOLLICLE STIMULATION Hyperstimulation of female rhesus monkeys exhibiting regular menstrual cycles is induced with exogenous gonadotropins. In brief, subcutaneous injection of a GnRH antagonist (Antide; Serono Inc.) and twice daily injection of r-hFSH (Serono Inc.) for 6 days is followed by twice daily injections of r-hFSH and r-hLH (Serono Inc.). On day 7 of the stimulation, ultrasonography is performed and r-hCG is administered for induction of ovulation when follicles of 3–4 mm in diameter are observed.

FOLLICULAR ASPIRATION BY LAPAROSCOPY Follicular aspiration is performed at 27 h post-hCG. Oocytes are aspirated from follicles using a needle suction device lined with Teflon tubing. Multiple individual follicles are aspirated via continuous vacuum, and collection tubes are immediately transported to a dedicated primate oocyte/zygote laboratory for oocyte recovery and evaluation of the maturation stage.

COLLECTION AND EVALUATION OF RHESUS OOCYTES The contents of each collection tube are diluted in TALP-HEPES supplemented with 2 mg/ml hyaluronidase. Oocytes are rinsed and then transferred to preequilibrated CMRL medium prior to the evaluation of maturation. MII-arrested oocytes, exhibiting expanded cumulus cells, a distinct PVS, and a first polar body, are selected for PVS injection and subsequent fertilization.

***IN VITRO* MATURATION** Oocytes are matured in modified CMRL-1066 containing 20% heat-inactivated fetal bovine serum (FBS; Hyclone Laboratories Inc., Logan, UT) supplemented with 5 μg/ml ovine FSH and 10 ng/ml ovine LH for up to 36 h in 50-μl drops of medium under saline-equilibrated silicone oil at 37°C in a humidified atmosphere of 5% CO_2 in air.

INFECTION OF MATURE RHESUS OOCYTES/ZYGOTES WITH THE VSVG PSEUDOTYPE MII-arrested oocytes and fertilized oocytes are selected for PVS injection followed by *in vitro* fertilization in oocytes. VSVG pseudotype solution is loaded into the injection needle by micropipette, mounted onto a micromanipulator, and connected to a microinjector. After microinjection, the oocytes are returned to the maturation drop and fertilized by intracytoplasmic sperm injection (ICSI), whereas the zygote is returned to the culture medium.

COLLECTION, PREPARATION, AND HANDLING OF RHESUS SPERM Rhesus males of proven fertility are trained to routinely produce semen samples by penile electroejaculation. After serial washes, the sperm samples are counted and diluted to a concentration of 20×10^6 sperm/ml. Sperm suspensions are incubated at 37°C under 5% CO_2 in air for 6 h, at which point 1 mM caffeine and 1 mM dibutyryl cyclic adenosine monophosphate (dbcAMP) is added for the final hour to stimulate hyperactivation.

FERTILIZATION BY INTRACYTOPLASMIC SPERM INJECTION A single sperm is aspirated tail-first into an injection needle from the sperm-polyvinylpyrrolidone (PVP) to the oocyte-containing drop. Oocytes are immobilized with the polar body at the 12 o'clock position and the injection needle is inserted through the zona into the cytoplasm at the 3 o'clock position. The oolemma is breached by gentle cytoplasmic aspiration and the sperm will then be released into the oocyte.

ASSESSMENT OF SUCCESSFUL FERTILIZATION AND *IN VITRO* EMBRYO CULTURE Fertilization is assessed within 3–6 h by detection of the second polar body using RC optics. Pronuclei are assessed between 12 and 16 h post-ICSI. After completion of the first cleavage division (24–28 h post-ICSI), two-cell embryos are cocultured in CMRL + 10% FBS (Hyclone Laboratories, Inc., Logan, UT) on a Buffalo rat liver cell (BRL 1442; ATCC, Rockville, MD) in 100 µl drops overlaid with oil. Embryos are selected at the 4- to 16-cell stage for transfer into selected surrogates.

SELECTION AND PREPARATION OF SURROGATE FEMALES Rhesus females with normal menstrual cycles and natural cycles similar to those of the egg donor are screened as potential surrogate mothers. Blood samples, beginning on day 8 of the menstrual cycle (day 1 is the first day of menses), are collected and analyzed daily for serum progesterone and estrogen. Surgical ET is performed on day 2 or 3 following ovulation by transferring two four- to eight-cell embryos into the oviduct of the recipient.

RHESUS EMBRYO TRANSFER BY LAPAROTOMY Surgical ET is performed by mid-ventral laparotomy. The oviduct is cannulated using a Tomcat catheter containing two four- to eight-cell stage embryos in HEPES-buffered TALP, containing 3 mg/ml BSA. Embryos are expelled from the catheter in about 0.05 ml of medium while the catheter is withdrawn. The catheter is flushed with medium after removal from the female to ensure that the embryos are successfully transferred.

CONFIRMATION OF PREGNANCY To confirm embryonic implantation, blood samples are collected daily and analyzed for serum estrogen and progesterone concentrations. If hormone levels indicate a possible pregnancy, the pregnancy is confirmed by a transabdominal ultrasound on day 35 posttransfer.

INFANT CARE The pregnant animal will be monitored daily for the 150 days of pregnancy because rhesus monkeys occasionally experience preeclampsia, and newborns can be lost due to complications at birth. Therefore, cesarean sections may be performed. Babies are weighed and measured, and their head circumference recorded; they are kept in an incubator and raised in the nursery facility if needed.

REFERENCES

1. Morris HR. Trends in molecular medicine: Genetics in Parkinson's disease. *Ann Med* 2005;37:86–89.
2. Jain S, Wood NW, Healy DG. Molecular genetic pathways in Parkinson's disease: A review. *Clin Sci* 2005;109:355–364.
3. Higgins GA, Jacobsen H. Transgenic mouse models of Alzheimer's disease: Phenotype and application. *Behav Pharmacol* 2003;14: 419–438.
4. Fernagut PO, Chesselet MF. Alpha-synuclein and transgenic mouse models. *Neurobiol Dis* 2004;17:123–130.
5. Huntington's Disease Collaborative Research Group. A novel gene containing a trinucleotide repeat that is expanded and unstable on Huntington's disease chromosomes. *Cell* 1993;72:971–983.
6. Yu ZX, Li SH, Evans J, Pillarisetti A, Li H, Li XJ. Mutant huntingtin causes context-dependent neurodegeneration in mice with Huntington's disease. *J Neurosci* 2003;23(6):2193–2202.
7. Abeliovich A, Beal MF. Parkinsonism genes: Culprits and clues. *J Neurochem* 2006;99(4):1062–1072.
8. Pardo LM, van Duijn CM. In search of genes involved in neurodegenerative disorders. *Mutat Res* 2005;592:89–101.
9. Gasser T. Genetics of Parkinson's disease. *Curr Opin Neurol* 2005;18:363–369.
10. Spires TL, Hyman BT. Transgenic models of Alzheimer's disease: Learning from animals. *NeuroRx* 2005;2:423–437.
11. Bass B, Olanow CW, Kordower JH. Gene transfer of trophic factors and stem cell grafting as treatments for Parkinson's disease. *Neurology* 2006;66:S89–S103.
12. Vercauteren FGG, Clerens S, Roy L, Hamel N, Arckens L, Vandesande F, Alhonen L, Janne J, Szyf M, Cuello AC. Early disregulation of hippocampal proteins in transgenic rats with Alzheimer's disease-like mutations in amyloid precursor protein and presenilin 1. *Mol Brain Res* 2004;132:241–259.
13. Czech C, Lesort M, Tremp G, Terro F, Blanchard V, Schombert B, Carpentier N, Dreisler S, Bonici B, Takashima A, Moussaoui S, Hugon J, Pradier L. Characterization of human presenilin 1 transgenic rats: Increased sensitivity to apoptosis in primary neuronal cultures. *Neuroscience* 1998;87(2):325–336.
14. Chan AW, Chong KY, Martinovich C, Simerly C, Schatten G. Transgenic monkeys produced by retroviral gene transfer into mature oocytes. *Science* 2001;291(5502):309–312.
15. Fleming SM, Fernagut PO, Chesselet MF. Genetic mouse models of Parkinsonism: Strengths and limitations. *NeuroRx* 2005;2:495–503.
16. Levine MS, Cepeda C, Hickey MA, Fleming SM, Chesselet MF. Genetic mouse models of Huntington's and Parkinson's diseases: Illuminating but imperfect. *Trends Neurosci* 2004;27(11):691–697.
17. Petters RM, Sommer JP. Transgenic animals as model for human disease. *Transgenic Res* 2000;9:347–351.
18. Vaitukaitis JL. Animal models of human disease for the 21st century. *Lab Anim Sci* 1998;48(6):562–564.
19. Chan AWS, Chong KY, Schatten G. Production of transgenic non-human primates. In: Pinkert CA, Ed. *Transgenic Animal Technology: A Laboratory Handbook*, 2nd ed. San Diego, CA: Academic Press, 2002:359–382.
20. Reddy S, Smith DB, Rich MM, Leferovich JM, Reilly P, *et al.* Mice lacking the myotonic dystrophy protein kinase develop a late onset progressive myopathy. *Nat Genet* 1996;13(3):325–335.
21. Jansen G, Groenen PJ, Bachner D, Jap PH, Coerwinkel M, *et al.* Abnormal myotonic dystrophy protein kinase levels produce only mild myopathy in mice. *Nat Genet* 1996;13(3):316–324.
22. Polymeropoulos MH, Lavedan C, Leroy E, Ide SE, Dehejia A, *et al.* Mutation in the alpha-synuclein gene identified in families with Parkinson's disease. *Science* 1997;276(5321):2045–2047.
23. Sango K, Yamanaka S, Hoffmann A, Okuda Y, Grinberg A, *et al.* Mouse models of Tay-Sachs and Sandhoff diseases differ in neurologic phenotype and ganglioside metabolism. *Nat Genet* 1995; 11(2):170–176.
24. Yamanaka S, Johnson MD, Grinberg A, Westphal H, Crawley JN, *et al.* Targeted disruption of the Hexa gene results in mice with biochemical and pathologic features of Tay-Sachs disease. *Proc Natl Acad Sci USA* 1994;91(21):9975–9979.
25. Adriaan Bouwknecht J, van der Gugten J, Groenink L, Olivier B, Paylor RE. Behavioral and physiological mouse models for anxiety: Effects of flesinoxan in 129S6/SvEvTac and C57BL/6J mice. *Eur J Pharmacol* 2004;494(1):45–53.
26. Brouillet E, Conde F, Beal MF, Hantraye P. Replicating Huntington's disease phenotypes in experimental animals. *Prog Neurobiol* 1999;59:427–468.
27. Yamada M, Mizuno Y, Mochizuki H. Parkin gene therapy for alpha-synucleinopathy: A rat model of Parkinson's disease. *Hum Gene Ther* 2005;16(2):262–270.
28. Lopez EM, Bell KFS, Ribeiro-da-Silva A, Cuello AC. Early changes in neurons of the hippocampus and neurocortex in transgenic rats expressing intracellular human A–β. *J Alzheimers Dis* 2004;6(4): 421–431.
29. Echeverria V, Ducatenzeiler A, Alhonen L, Janne J, Grant SM, Wandosell F, Muro A, Baralle F, Li H, Duff K, Szyf M, Cuello AC. Rat transgenic models with a phenotype of intracellular Aβ accumulation in hippocampus and cortex. *J Alzheimers Dis* 2004;6:209–219.

30. Cao C, Temel Y, Blokland A, Ozen H, Steinbusch HWM, Vlamings R, Nguyen HP, von Horsten S, Schmitz C, Visser-Vandewalle V. Progressive deterioration of reaction time performance and choreiform symptoms in a new Huntington's disease transgenic rat model. *Behav Brain Res* 2006;170:257–261.
31. van Horsten S, Schmitt I, Nguyen HP, Holzmann C, Schmidt T, Walter T, Bader M, Pabst R, Kobbe P, Krotova J, Stiller D, Kask A, Vaarmann A, Rathke-Hartlieb S, Schulz JB, Grasshoff U, Bauer I, Vieira-Saecker AMM, Paul M, Jones L, Lindenberg KS, Landwehrmeyer B, Bauer A, Li XJ, Riess O. Transgenic rat model of Huntington's disease. *Hum Mol Genet* 2003;12(6):617–624.
32. Chan AW. Transgenic nonhuman primates for neurodegenerative diseases. *Reprod Biol Endocrinol* 2004;2:39.
33. Kordower JH, Emborg ME, Bloch J, Ma SY, Chu Y, Leventhal L, McBride J, Chen EY, Palfi S, Roitberg BZ, Brown WD, Holden JE, Pyzalski R, Taylor MD, Carvey P, Ling ZD, Trono D, Hantraye P, Deglon N, Aebischer P. Neurodegeneration prevented by lentiviral vector delivery of GDNF in primate models of Parkinson's disease. *Science* 2000;290:767–773.
34. Yamamoto A, Lucas JJ, Hen R. Reversal of neuropathology and motor dysfunction in a conditional model of Huntington's disease. *Cell* 2000;101(1):57–66.
35. Davies SW, Turmaine M, Cozens BA, DiFiglia M, Sharp AH, *et al.* Formation of neuronal intranuclear inclusions underlies the neurological dysfunction in mice transgenic for the HD mutation. *Cell* 1997;90(3):537–548.
36. Bates GP, Mangiarini L, Mahal A, Davies SW. Transgenic models of Huntington's disease. *Hum Mol Genet* 1997;6(10):1633–1637.
37. Schilling G, Becher MW, Sharp AH, Jinnah HA, Duan K, *et al.* Intranuclear inclusions and neuritic aggregates in transgenic mice expressing a mutant N-terminal fragment of huntingtin. *Hum Mol Genet* 1999;8(3):397–407.
38. Chan AWS. Transgenic animals: Current and alternative strategies. *Cloning* 1999;1(1):25–46.
39. Kordower JH, Bloch J, Ma SY, Chu Y, Palfi S, *et al.* Lentiviral gene transfer to the nonhuman primate brain. *Exp Neurol* 1999; 160(1):1–16.
40. Palfi S, Leventhal L, Chu Y, Ma SY, Emborg M, *et al.* Lentivirally delivered glial cell line-derived neurotrophic factor increases the number of striatal dopaminergic neurons in primate models of nigrostriatal degeneration. *J Neurosci* 2002;22(12):4942–4954.
41. de Almeida LP, Rodd CA, Zala D, Aebischer P, Deglon N. Lentiviral-mediated delivery of mutant huntingtin in the striatum of rats induces a selective neuropathology modulated by polyglutamine repeat size, huntingtin expression levels, and protein length. *J Neurosci* 2002;22(9):3473–3483.
42. Chan AWS, Homan EJ, Ballou LU, Burns JC, Bremel RD. Transgenic cattle produced by reverse-transcribed gene transfer in oocytes. *Proc Natl Acad Sci USA* 1998;95(24):14028–14033.
43. Lois C, Hong EJ, Pease S, Brown EJ, Baltimore D. Germline transmission and tissue-specific expression of transgenes delivered by lentivirus vectors. *Science* 2002;295(5556):868–872.
44. Burns JC, Friedmann T, Driever W, Burrascano M, Yee JK. Vesicular stomatitis virus G glycoprotein pseudotyped retroviral vectors: Concentration to very high titer and efficient gene transfer into mammalian and nonmammalian cells. *Proc Natl Acad Sci USA* 1993;90(17):8033–8037.
45. Yee JK, Miyanohara A, LaPorte P, Bouic K, Burns JC, Friedmann T. A general method for the generation of high-titer, pantropic retroviral vectors: Highly efficient infection of primary hepatocytes. *Proc Natl Acad Sci USA* 1994;91(20):9564–9568.
46. Miyoshi H, Blomer U, Takahashi M, Gage FH, Verma IM. Development of a self-inactivating lentivirus vector. *J Virol* 1998;72(10): 8150–8157.
47. Pfeifer A, Ikawa M, Dayn Y, Verma IM. Transgenesis by lentiviral vectors: Lack of gene silencing in mammalian embryonic stem cells and preimplantation. *Proc Natl Acad Sci USA* 2001;99(4):2140–2145.
48. Miyoshi H, Takahashi M, Gage FH, Verma IM. Stable and efficient gene transfer into the retina using an HIV-based lentiviral vector. *Proc Natl Acad Sci USA* 1997;94(19):10319–10323.
49. Naldini L, Blomer U, Gage FH, Trono D, Verma IM. Efficient transfer, integration, and sustained long-term expression of the transgene in adult rat brains injected with a lentiviral vector. *Proc Natl Acad Sci USA* 1996;93(21):11382–11388.
50. Tiscornia G, Singer O, Ikawa M, Verma IM. A general method for gene knockdown in mice by using lentiviral vectors expressing small interfering RNA. *Proc Natl Acad Sci USA* 2003;100(4):1844–1848.
51. Davis J, Xu F, Deane R, Romanov G, Previti ML, Zeigler K, Zlokovic BV, Van Nostrand WE. Early-onset and robust cerebral microvascular accumulation of amyloid β-protein in transgenic mice expressing low levels of a vasculotrophic Dutch/Iowa mutant form of amyloid β-protein precursor. *J Biol Chem* 2004;279(19): 20296–20306.
52. Oddo S, Caccamo A, Kitazawa M, Tseng BP, LaFerla FM. Amyloid deposition precedes tangle formation in a triple transgenic model of Alzheimer's disease. *Neurobiol Aging* 2003;24(8):1063–1070.
53. Oddo S, Caccamo A, Shepherd JD, Murphy MP, Golde TE, *et al.* Triple-transgenic model of Alzheimer's disease with plaques and tangles: Intracellular Abeta and synaptic dysfunction. *Neuron* 2003;39(3):409–421.
54. Bales KR, Verina T, Dodel RC, Du Y, Altstiel L, *et al.* Lack of apolipoprotein E dramatically reduces amyloid beta-peptide deposition. *Nat Genet* 1997;17(3):263–264.
55. Holtzman DM, Bales KR, Tenkova T, Fagan AM, Parsadanian M, *et al.* Apolipoprotein E isoform-dependent amyloid deposition and neuritic degeneration in a mouse model of Alzheimer's disease. *Proc Natl Acad Sci USA* 2000;97(6):2892–2897.
56. Luo Y, Bolon B, Damore MA, Fitzpatrick D, Liu H, *et al.* BACE1 (beta-secretase) knockout mice do not acquire compensatory gene expression changes or develop neural lesions over time. *Neurobiol Dis* 2003;14(1):81–88.
57. Mohajeri MH, Saini KD, Nitsch RM. Transgenic BACE expression in mouse neurons accelerates amyloid plaque pathology. *J Neural Transm* 2004;111(3):413–425.
58. Rubinsztein DC. Lessons from animal models of Huntington's disease. *Trends Genet* 2002;18(4):202–209.
59. Davies S, Ramsden DB. Huntington's disease. *J Clin Pathol Mol Pathol* 2001;54:409–413.
60. Cummings CJ, Zoghbi HY. Trinucleotide repeats: Mechanisms and pathophysiology. *Annu Rev Genomics Hum Genet* 2000;1:281–328.
61. Mangiarini L, Sathasivam K, Seller M, Cozens B, Harper A, *et al.* Exon 1 of the *HD* gene with an expanded CAG repeat is sufficient to cause a progressive neurological phenotype in transgenic mice. *Cell* 1996;87:493–506.
62. Menalled LB, Chesselet MF. Mouse models of Huntington's disease. *Trends Pharmacol Sci* 2002;23(1):32–39.
63. Charreau B, Tesson L, Soulillou JP, Pourcel C, Anegon I. Transgenesis in rats: Technical aspects and models. *Transgenic Res* 1996;5:223–234.
64. Taurog JD, Maika SD, Satumtira N, Dorris ML, McLean IL, *et al.* Inflammatory disease in HLA-B27 transgenic rats. *Immunol Rev* 1999;169:209–223.
65. Agca Y, Critser JK. Assisted reproductive technologies and genetic engineering in rats. In: Suckow MA, Weisbroth SH, Franklin CL, Eds. *The Laboratory Rats*, 2nd ed. San Diego, CA: Academic Press, 2005:165–190.
66. Jacob HJ. Functional genomics and rat models. *Genome Res* 1999;9(11):1013–1016.
67. Kuramoto T, Yamasaki K, Kondo A, Nakajima K, Yamada M, Serikawa T. Production of WTC.ZI-zi rat congenic strain and its pathological and genetic analyses. *Exp Anim* 1998;47:75–81.
68. Kose H, Moralejo DH, Ogino T, Mizuno A, Yamada T, Matsumoto K. Examination of OLETF-derived non-insulin-dependent diabetes mellitus QTL by construction of a series of congenic rats. *Mamm Genome* 2002;13:558–562.

69. Pfeifer A, Ikawa M, Dayn Y, Verma IM. Transgenesis by lentiviral vectors: Lack of gene silencing in mammalian embryonic stem cells and preimplantation embryos. *Proc Natl Acad Sci USA* 2002;99:2140–2145
70. Rubinson DA, Dillon CP, Kwiatkowski AV, Sievers C, Yang L, Kopinja J, *et al.* A lentivirus-based system to functionally silence genes in primary mammalian cells, stem cells and transgenic mice by RNA interference. *Nat Genet* 2003;33:401–406.
71. Kantor O, Temel Y, Holzmann C, Raber K, Nguyen HP, Cao C, Turkoglu HO, Rutten BPF, Visser-Vandewalle V, Steinbusch HWM, Blokland A, Korr H, Riess O, van Horsten S, Schmitz C. Selective striatal neuron loss and alterations in behavior correlate with impaired striatal function in Huntington's disease transgenic rats. *Neurobiol Dis* 2006;22:538–547.
72. Takahashi M, Miyoshi H, Verma IM, Gage FH. Rescue from photoreceptor degeneration in the rd mouse by human immunodeficiency virus vector-mediated gene transfer. *J Virol* 1999;73(9):7812–7816.
73. Gallichan WS, Kafri T, Krahl T, Verma IM, Sarvetnick N. Lentivirus-mediated transduction of islet grafts with interleukin 4 results in sustained gene expression and protection from insulins. *Hum Genet Ther* 1998;9(18):2717–2726.
74. Armstrong DT, Opavsky MA. Superovulation of immature rats by continuous infusion of follicle-stimulating hormone. *Biol Reprod* 1988;39(3):511–518.

36 Animal Models of Nociception and Pain

JAMES D. ROSE AND C. JEFFREY WOODBURY

ABSTRACT

Research on the neurobiological bases of nociception and pain and related investigations of potential therapies require great reliance on animal models. There are unique challenges in the development of well-validated models in this field because of the distinction between nociception, the processing and response to potentially pain producing stimuli by lower levels of the nervous system, and pain, the conscious result of nociceptive stimulus processing by the cerebral cortex. The most frequently used models actually represent tests of nociception only and are appropriate for investigating diverse pathophysiological processes that cause nociceptive activity in peripheral tissues, nerves, the spinal cord, or subcortical regions. However, because human pain is a complex end result of nociception and consciousness-dependent processes, models intended to address pain must be validated for this purpose. Models assessing processes related to pain are relatively rare and more difficult to validate and use than those relevant only for nociception. A failure to recognize the pain–nociception distinction has significant practical consequences for successful extrapolation of results from laboratory to clinical practice.

Key Words: Nociception–pain dichotomy, Construct validity, Neocortex, Nocifensive behavior, Neural substrate.

THE NEED FOR ANIMAL MODELS OF NOCICEPTION AND PAIN

Pain research with human subjects is productive on many fronts, as shown by the large and diverse literature surveyed in the most recent edition of Wall and Melzack's *Textbook of Pain.*[1] A particularly prominent area of progress is in the use of brain imaging methods such as positron emission tomography and functional magnetic resonance imaging to advance our understanding of the higher brain processes that underlie pain.[2] However, there are great limitations on the use of humans in experimental studies of pain and animal subjects continue to be vital. *In vivo* models are particularly important because pain and its underlying mechanisms are emergent processes of a whole nervous system; these processes cannot be fully simulated in highly reduced cell or tissue systems. In addition to bettering our understanding of nociception and pain, these models are valuable out of welfare concerns for achieving a better understanding of pain–nociception processes in animals. In spite of great recent progress in deciphering the neurobiological basis of nociception and pain, this knowledge has yet to see large-scale translation into effective pain therapies. A limiting factor has been the often unsuccessful extrapolation from animal models to human clinical practice, as exemplified by recent attempts to develop new pharmacological treatments for migraine headache.[3–5] Beneficial applications from animal models could be fostered by rigorous examination of the validity and limits of these models.[3] This goal also hinges on a better understanding of the similarities and differences between nociception and pain in humans and in the animal models, an understanding that would make model selection and interpretation more valid for human applications.

This chapter's principal objectives are to clarify distinctions between nociception and pain to improve the interpretation and validity of animal models and to discuss and evaluate commonly used and important models. There are several recent reviews, some highly detailed, on animal models in nociception–pain research that should be consulted by readers wishing further information.[3,6–8]

DEFINING PAIN IN HUMANS: IMPLICATIONS FOR ANIMAL MODELS

A valid working definition of pain is vital for efforts to explain its underlying mechanisms or develop therapeutic interventions. To this end, the International Association for the Study of Pain[9] defines human pain as follows: (1) pain is an unpleasant sensory and emotional experience associated with actual or potential tissue damage, or described in terms of such damage; (2) pain is always subjective; and (3) pain is sometimes reported in the absence of tissue damage and the definition of pain should avoid tying pain to an external eliciting stimulus. One of the most critical conceptual advances in the understanding of pain is the distinction between nociception and pain. As Wall[10] emphasized, nociception that is "activity induced in the nociceptor and nociceptive pathways by a noxious stimulus is not pain, which is always a psychological state." It is also critical to understand that the pain experience requires conscious awareness.[11,12] In the usual course of events, tissue-damaging forms of stimuli excite nociceptors and this activity is conducted through peripheral nerves, the spinal cord, and subcortical brain structures to the cerebral cortex. If a person is conscious when nociception-related activity arrives in the cortex, further processing by extensive cortical regions results in pain.[11–14]

From: *Sourcebook of Models for Biomedical Research*
(P. M. Conn, ed.), © 2008 Humana Press Inc., Totowa, NJ.

The separateness of pain and nociception is seen in diverse ways. First, nociceptive processes do not always lead to pain. People can sustain severe injuries in warfare, sports, or everyday life and either not report pain or report it differently than the extent of an injury would suggest.[15–17] Second, people with "functional" pain syndromes experience chronic, disturbing pain without any demonstrable tissue damage or pathology. Third, pain can be reduced by psychological manipulations such as a visual illusion[18] or hypnotic suggestion.[19] Fourth, pain has a strong social learning component and depends greatly on prior experience with it and interpersonal interactions that accompany this experience.[13]

THE NEUROLOGY OF NOCICEPTION AND PAIN There has been great progress in identifying the functional neuroanatomy underlying nociception and pain. We will present a brief account of the neural structures and systems implicated in nociception and pain as it pertains to the use and interpretation of animal models. Numerous excellent reviews provide more detailed information.[1,2,20–22]

Studies of diverse mammals have shown that nociceptive stimuli activate two types of nociceptive receptors in body tissues: those that appear to respond exclusively to noxious mechanical, thermal, or chemical stimuli and those that respond to combinations of these stimuli (polymodal nociceptors). Activity is conducted from these receptors to the spinal cord through both myelinated (Aβ and Aδ) and unmyelinated (C) axons. These axons synapse on dorsal horn spinal neurons, principally in the more superficial laminae, where extensive processing occurs. Ascending projections arise from neurons in diverse dorsal horn laminae, principally lamina I and V, and travel through the contralateral lateral and ventral spinal pathways.

The ascending pathways have synaptic terminations in diverse regions of the brainstem, mainly catecholamine neuron groups, parabrachial nuclei of the pons, midbrain periaqueductal gray, and diverse sites through the brainstem reticular formation. The thalamus receives multiple direct spinal projections to topographically organized lateral and nontopographically organized medial nuclei. There are significant differences between mammalian species at this level, particularly in the existence of a posterior ventromedial nucleus that seems to exist only in primates and is greatly enlarged in humans.[21] Functional imaging studies in humans have consistently shown a diverse array of cortical structures to be activated specifically in association with perceived pain, including the first and second somatosensory areas, anterior cingulate gyrus, insula, and prefrontal cortex. Diverse evidence indicates that the somatosensory cortical zones are critical for the sensory-discriminative dimension of pain, that pain intensity is related to activation of multiple zones, especially involving both hemispheres, and that the emotional-evaluative (suffering) component depends on the anterior cingulate gyrus, insula, and prefrontal cortex. In addition, it is now well established that in humans, pain experience is absolutely dependent on the functioning of these neocortical and limbic cortical areas.[22,23] The dependence of pain on these cortical regions makes sense also when it is considered that pain depends on the concurrent existence of another function: consciousness. Extensive evidence shows that the cortical regions known to be essential for pain greatly overlap with those vital to the existence of consciousness.[11,14]

In addition to the ascending pathways is a network of descending modulatory controls, centered in the periaqueductal gray and rostral ventromedial medulla, that exerts both antinociceptive and pronociceptive actions on ascending nociceptive signaling.[24]

THE ADAPTIVENESS OF NOCIFENSIVE BEHAVIORS Nociceptors form a common underlying thread throughout the evolutionary history of multicellular organisms. Nociceptors have been observed in all bilaterally symmetrical multicellular organisms that have been examined, with the notable exception of elasmobranch fishes.[14] Even the leech has nociceptive neurons,[25] many of which display close similarity to the polymodal nociceptor population that has been so well characterized in mammals.

Although nociceptors are common in the animal kingdom, the existence of nocifensive behaviors, the unconscious protective responses to noxious stimuli, is even more widespread and not specifically tied to possession of nociceptors.[26] The single-celled paramecium (absent any possibility of a nervous system) exhibits protective responses to adverse environmental stimuli. Likewise, sponges with no nervous system and jellyfish with simple nerve nets have simple, but functional nocifensive behaviors. In these and more advanced bilaterally symmetrical invertebrate organisms as well as vertebrates, the nociceptive system and nocifensive behaviors constitute an essential component to survival. Importantly, the suite of responses to nociceptive stimuli does not end simply with withdrawal reflexes, but in advanced multicellular organisms also includes complex arrays of endocrine and autonomic responses that help prepare the organism for a defense of disturbed homeostasis. Species with nociceptors showing properties in common with those of mammals could serve as useful models for investigating peripheral nociception.

DISTINGUISHING NOCICEPTION FROM PAIN: WHY IT MATTERS As explained above, nociception and pain are distinctly different things, with differing underlying mechanisms. Unless one is studying the processes that specifically mediate the conscious experience of pain or a behavioral response that is specific to such processes, nociception is being studied and the term pain should not be used. Unfortunately, these terms are frequently used in ambiguous or inconsistent ways, with significant practical costs in the use and interpretation of animal models and theoretical costs in understanding of mechanisms.

The Nociception–Pain Dichotomy in Clinical Neurology The behavioral separateness of nociception and nocifensive responses from pain is commonly seen in humans with severe neurological injury such as a spinal cord transection. Noxious stimulation of a limb below the level of the transection excites nociceptive sensory receptors and nociceptive pathways within the spinal cord. This spinal activity produces a nocifensive limb withdrawal response, but because nociceptive pathways are interrupted between the spinal cord and the cortex, no pain is felt. The pain–nociception distinction does not stop at this level.

Humans with massive damage or dysfunction of the cerebral cortex are unconscious, but can be awake and show grimacing, vocalization, and organized avoidance reactions in response to a nociceptive stimulus.[12] The importance of understanding the nociception–pain distinction was shown by the confusion and contentiousness surrounding the recent tragic case of Terri Shaivo, who in 1990 experienced a prolonged period of anoxia. Although examining neurologists agreed that she was unconscious, in an irreversible, persistent vegetative state, Mrs. Shaivo was awake, was quite reactive to noxious stimuli, and exhibited nocifensive and emotion-like behaviors, which led to claims by some that Ms. Schaivo's behaviors had to be consciously mediated. After a protracted legal dispute and interventions by government officials, her feeding tube was removed. Subsequent autopsy confirmed that her cerebral cortex had massively degenerated, a fact consistent with the diagnosis of unconsciousness.[28] The important points to be gleaned from this case are that a striking array of nocifensive behaviors can be unconsciously expressed in a human; that these are behaviors mediated by the brainstem and spinal systems rather than conscious, pain-related responses; and there is a widespread public unawareness of such things with associated societal consequences.

Valid Selection and Interpretation of Models As stated previously, effective therapies for pain have lagged behind advances in basic research. A practical example will serve to illustrate this point. Substance P was described in 1931 and its role in nociceptive neurotransmission has been known for decades.[29] Immense resources have been devoted to development of substance P antagonists, principally NK_1 receptor antagonists for pain relief, especially migraine headache. In spite of their effectiveness in animal models of dural inflammation, these drugs were ineffective in relieving human migraine pain. This negative result was not a matter of a receptor affinity mismatch because high level NK_1 receptor blockade has been achieved in the brains of humans.[30] This costly failure has been attributed to many things, including recognition that there is more to human migraine headache pain than events in the dura mater.[4,31] Here, the animal model response measure was at the most peripheral level of the complex hierarchy of nociception–pain process, but the human clinical response measure, the pain report, was at the end stage of all intervening processing.

WHAT WOULD THE PAIN OF ANIMALS BE LIKE? A most important question regarding animal models for pain is which species are capable of pain experience, or at least a pain experience that meaningfully resembles that of a human. The foregoing explanation should make it clear that most species in the animal kingdom can detect and respond to noxious stimuli but not all can experience conscious pain. There has been interest in using invertebrates and some nonmammals such as amphibians[32] as animal models for nociception on the assumption that they were more humane models than mammals. Examination of the collective neurological and behavioral evidence has led to the conclusion that fishes and amphibians are very unlikely to have a capacity for conscious pain experience, at least anything resembling that of humans.[14,32] However, there are differing views.[33,34] It is clear, though, that structural organization of the forebrain differs dramatically between mammals and nonmammalian vertebrates. Most notable is the unique presence of a neocortical component of the forebrain pallium in mammals,[35,36] with its suite of structures, and especially interconnections between them, essential to pain experience. Nonetheless, the forebrain complexity of reptiles and birds is considerable[36,37] and the presence of functional homologs of pain or consciousness-related cerebral structures, especially in birds, remains to be investigated.

The majority of animals used in nociception–pain research are mammals and the species used all have, to varying degrees, the cortical regions that correspond to those as essential to human pain experience.[14,36] However, large expanses of higher–order nonsensory, nonmotor cortex are part of the pain-related mosaic and in nonprimate mammals, only a small proportion of cerebral cortex, less than 10% in rats and mice,[38] fits this structural designation. As mentioned previously, the subcortical pathways that distribute nociception-related information to pain-related cortical zones also differ substantially between primates (especially humans) and nonprimate mammals. Clearly, significant functional differences may exist between species of mammals in general and humans in particular in terms of the how pain processing might occur.

WHEN IS AN ANIMAL MODEL APPROPRIATE FOR THE INVESTIGATION OF PAIN?

Much current research employs models specifically addressed at particular types of human pain conditions. Some models are also used to address fundamental mechanistic questions concerning nociceptive processing or actual pain experience. In either case, where an animal model is being used to investigate some aspect of human nociception and pain, it is vital to know that the model system is actually valid for the purpose. In particular, pain-related response measures must be distinguished from purely nociceptive–nocifensive response measures because pain is a psychological process that is not directly observable and must be measured indirectly through behavior.

As shown above, clinical neurology provides human examples of the pain–nociception distinction, but clear examples have been in the animal literature for many years. Responses to noxious stimuli have been studied in several mammalian species following decerebration where all of the forebrain above the midbrain is removed. Chronically decerebrate rats[39] react strongly to the insertion of a feeding tube, struggling, pushing at it with the forepaws, and vocalizing. When receiving an injection, these rats react indistinguishably from a normal rat: vocalizing, attempting to bite the syringe or the experimenter's hand and lick the injection site. Since a large body of evidence indicates that consciousness (and accordingly pain experience) depends on the neocortex, it must be concluded that these reactions are nocifensive and unconscious rather than expressions of conscious pain. It can be concluded that nocifensive behaviors can be far more complex than "simple reflexes" and even ostensibly purposive, a fact that makes the behavioral distinction between nociception and pain difficult. In fact, many assumptions about indications of pain have been based on behaviors that are sustained, organized, or directed to the site of nociceptive stimulation,[7,8,40] the type of responses fully within the capacity of decerebrate rats.

Identification of pain in humans usually depends on a verbal report, but verbal reports and other pain-related behaviors are not always reliably interpretable.[19] So, validated rating scales or other tools adopted from cognitive psychology provide sufficiently reliable means of measuring pain in humans. Correspondingly, there is a long history in experimental psychology of using nonverbal behavioral methods to assess the internal "psychological" state of an animal. It is quite possible to assess the aversiveness of a stimulus in terms of whether the animal will learn to avoid the stimulus, escape from it, or perform some behavior to escape that reflects the aversiveness of the nociceptive stimulus. An example of the last case is that a rat will leave a dark chamber and enter a brightly illuminated chamber (normally aversive to a rat) in order to escape a hot plate or electric shock.[8] Learned avoidance, or conditioned emotional responses to nociceptive stimuli, however, does not prove the existence of conscious pain, because associative learning of these types is believed to be unconsciously mediated.[41]

Another effort to establish the existence of pain in an animal, as opposed to measurement of pain intensity, was Bateson's[40] eight criteria for animal pain, which have been of interest to welfare biologists. However, when viewed in terms of what is currently known about pain[14] these criteria are flawed because they fail to distinguish nociception and nocifensive behaviors from conscious pain.[34]

EXAMPLES OF COMMONLY USED ANIMAL MODELS AND AN EVALUATION OF THEIR INTERPRETATION

Although overlap occurs in the use of some models, they can be broadly divided into those primarily used for investigation of acute nociception–pain and those for study of chronic nociception–pain. Because of the great clinical importance of chronic pain, more diverse and specialized models are used to represent particular human pathological processes such as neuropathic, bone cancer, or arthritis pain. In using waking animals for studies of nociception–pain, there is the ever-present issue of humane treatment.[6,8] Investigators are bound by ethics and statute to minimize the exposure of animals to presumably painful procedures, within the objectives of a study. This requirement has shaped the development of the animal models for nociception–pain research toward the use of relatively low-intensity, brief-duration stimuli. The welfare consideration has been an impetus to develop *in vitro* models or use species (such as amphibians; see Chapter 37 by Stevens in this volume) viewed as less likely to have a capacity for pain experience. Accordingly, as addressed below, these animal models must be validated for their intended application.

CONSTRUCTS AND TERMINOLOGY Nociception (and frequently pain) is a normal consequence of noxious stimulation, of course, but modifications of normal nociception–pain are more commonly of clinical importance. These include allodynia, hyperalgesia, analgesia, dyesthesia, paresthesia, causalgia, and neuropathic pain (Table 36–1). These terms represent human clinical complaints that animal models are intended to represent, but because some of these complaints, like paresthesia or causalgia, are identified by subjective reports, it is impossible to know how well the animal model represents them.

There are two main ways to categorize nociception/pain models[1]: (1) by stimulus duration, such as short duration (acute) stimuli and longer duration (chronic) stimuli, or (2) by level of the nervous system presumed to mediate the responses (Table 36–2[8]). Ideally, a model should meet the following criteria: specificity, sensitivity, construct validity, predictive validity, and reliability (Table 36–1).

The Commonly Used Models Contemporary models vary greatly in sophistication and validity but the predominantly used behavioral models employ rats or mice and are based on spinal reflexes such as tail flick and paw withdrawal and entail, variously, measures of response threshold and/or latency. Additional measures associated with paw withdrawal are paw lifting, flinching, guarding, and licking. The most frequently used tests of nociception have been the tail flick, hot plate, paw pressure, writhing, and formalin tests.[6,8] The tail flick test entails radiant heat application to a localized tail region or immersion in preheated water. Tail flick is known to be a spinal reflex, but is likely modulated by descending influences from the brainstem.[6,8,24] A variation on the tail flick test is the Hargreave's test in which heat is applied to the plantar surface of a foot and withdrawal latency is measured. In the hot plate test the animal is placed on a heated metallic plate and latencies for paw licking and jumping responses are recorded. These behaviors are mediated through a combination of spinal and brainstem processes (Table 36–2). In the plantar pressure test, increasing force is applied on the plantar surface of a foot with a von Frey-type filament and threshold pressure for paw withdrawal is monitored. The latter two responses involve supraspinal, probably brainstem control (Table 36–2).

Electrical stimulation of the tail or paw has also been frequently used to elicit a hierarchy of responses, including twitching, escape behavior, vocalization, and biting the electrodes, which are thought to reflect progressively more complex neural mediation. Dental pulp stimulation has also been used for the purpose of simulating trigeminal nociception–pain states and also

Table 36–1
Constructs and terminology

Allodynia: pain caused by a normally innocuous stimulus.

Analgesia: a consequence of a manipulation, such as drug administration, that causes a previously pain-provoking stimulus to become nonpainful.

Causalgia: an abnormal spontaneous or stimulus-evoked burning sensation, generally due to neuropathic conditions.

Dyesthesia: an unpleasant abnormal sensation.

Hyperalgesia: intensification of the painfulness of a normally pain-producing stimulus.

Neuropathic pain: diverse pain experiences, including allodynia, hyperalgesia, and spontaneous pain due to a nervous system lesion or disease.

Paresthesia: an abnormal spontaneous or stimulus-evoked sensation.

Specificity: the stimuli used should be genuinely nociceptive. In practice, test stimuli may activate combinations of nociceptive and nonnociceptive sensory afferents. The behavioral response model should distinguish nonnociceptive from nociceptive stimuli.

Sensitivity: it should be possible to detect a range of responsiveness across a range of stimulus magnitudes, from below to above the nocifensive response threshold. Ideally, the model should also be sensitive to pharmacological or other manipulations that might modify responsiveness.

Construct validity: a most critical but often overlooked criterion is that the model should actually be an indication of nociception or pain. It is particularly important to know if the model is actually assessing pain as opposed to nociception alone. Put another way, the model should really be assessing the process or variable that it is thought to assess. If bone cancer pain in humans is of fundamental interest, the model must realistically replicate key pathophysiological and psychological attributes of human bone cancer pain. This issue emerges in many contexts, but is particularly salient in consideration of models that do not employ behavioral measures, such as histological or gene expression measures.

Predictive validity: effects of pharmacological or other therapies on the animal model should predict the effects of these treatments on human pain.

Reliability: The model must give the same results each time it is used, within and between laboratories, to make the results obtained adequately generalizable.

Table 36–2
Common nociception testing paradigms,[a] observed behavioral outcomes, and their presumed neural substrates

Stimulus modality	*Testing paradigms[b]*	*Quantified behaviors*	*Minimal central substrates[c]*
Mechanical	Hargreave's (von Frey)	Withdrawal	Segmental (spinal)
Thermal	Tail immersion (hot or cold), tail flick, hot plate, cold plate, Hargreave's (radiant heat)	Withdrawal, jumping, escape, licking, guarding	Segmental and suprasegmental (brainstem)
Chemical	Formalin, acid, capsaicin, taxol	Licking, biting, guarding, altered posture, writhing, vocalizing	Segmental and suprasegmental

[a]Employing natural stimuli; nonnatural stimuli (e.g., electrical stimulation) are not included; see Le Bars *et al.*[6]
[b]Reviewed in Mogil *et al.*[48]
[c]Reviewed in Vierck.[8]

on the (questionable[6]) assumption that all the associated afferents are nociceptors.

The commonly used formalin and writhing tests exemplify tests eliciting long-duration nociceptive stimulation by irritant injection. Intradermal formalin injections are most common, but hypertonic saline, complete Freund's adjuvant, capsaicin, bee venoms, and other agents have been used. When injected into the dorsal surface of a rat's forepaw, a 0.5–15% formalin solution evokes a constellation of quantifiable behaviors, including reduced weight bearing on the paw, paw lifting, licking, nibbling, shaking, or biting. Irritant injection into the footpad of rats or mice, in contrast, produces a biphasic response, with an initial phase of about 3 min latency, a subsequent period during which responses dissipate, and a later (20–30 min latency) reappearance of behavioral reactions, which appears to stem from predominant activation of C fiber nociceptors.[8]

The writhing test involves intraperitoneal injection of any of several irritants, including phenylbenzoquinone, acetylcholine, dilute hydrochloric or acetic acid, and bradykinin. These substances irritate serous membranes of the peritoneal cavity and in rats and mice evoke abdominal contractions, large body movements, including the hind paws, asymmetric contraction of dorsal abdominal muscles, and reduced locomotion. Although intended to be a model for visceral pain, it is likely that somatic afferents are also activated by the algogenic substances.

Models using stimulation of hollow organs, particularly gastrointestinal or urogenital structures, have taken various forms: injection of formalin or other algogenic substances into the colon, capsaicin injection into the bladder, distention of organs like the colon and rectum with an inflatable balloon, or solid material introduction into a ureter. The behaviors produced depend to some extent on the specific type and location of the nociceptive stimulus, but include various types of abdominal reflexive contractions, stretching, and abdominal licking.

Models specifically designed to investigate chronic nociception–pain often include some of the same behavioral measures and nociceptive treatments previously described for investigations of acute nociception–pain. For example, the threshold nociceptive pressure required for paw withdrawal can be assessed in a normal rat by pressure with a filament (the von Frey test), but this method can test hyperalgesia after injection of an irritant like formalin into a paw. Likewise, this test can be used to assess allodynia associated with neuropathic pain caused by direct nerve injury or chronic nerve constriction.

Interpretations of the Commonly Used Behavioral Models The above-described tests are widely used because they appear to have face validity, that is, they appear to produce states that would be comparable to those that would be associated with pain in humans. In addition, the most widely used of these, like the hot plate and intradermal formalin tests, are relatively easy to perform and score on large numbers of animals, thereby lending themselves to dose–effect studies of potentially analgesic drugs or other antinociceptive manipulations. The seeming simplicity of such tests is illusory. Detailed critiques of the limitations in these models have been presented by Le Bars *et al.*,[6] Blackburn-Munro,[3] and Vierck,[8] and the interested reader is urged to read these informative and thoughtful reviews.

A detailed discussion of the technical and theoretical limitations of the various nociception–pain models is not possible here, but we will outline the major concerns. The first such as noxious temperature paradigms (e.g., tail flick, hot plate, Hargreave's) or paw withdrawal tests, there are many poorly controlled and variable aspects to nociceptive stimulus application. For example, a rat's tail is a thermal exchange organ, which places it in a very different functional category from other tissues regarding effects of thermal stimuli.[6] Furthermore, in the hot plate test, the animal's movements make thermal stimulus application highly variable. Although nociceptive stimuli are often administered in ways intended to avoid or minimize tissue trauma, it may result in variable degrees, thereby introducing the dimension of allodynia or hyperalgesia in a test where these effects are not intended. Effects of injected algogenic stimuli will differ depending on the specific agent used (formalin, capsaicin, mustard oil, etc.) and route of administration (intraperitoneal, in the bladder) presenting a complex stimulus, to say the least. In addition, nonnociceptive afferent activation is often a confounding variable in that those afferents may contribute to the elicitation of responses that are presumed to be purely nocifensive.

Response measures themselves also present difficulties in the foregoing test paradigms.[6,8] Investigator judgments are frequently a major source of variation in quantifying responses. Repeated testing of individual animals in some of the procedures, like the hot plate test, can be associated with learned aversion behaviors that modify stimulus application as well as the character of the responses. Sometimes tests of threshold become confounded with response latency, especially with longer duration stimuli. In addition, activation of nonnociceptive afferents in testing paradigms (above) may elicit flexion reflexes, confounding the interpretation of the nociceptive nature of the stimulus and response. Genetic variability and strain differences can also be a major source of confounding variability in a number of models.[6,8]

In most cases, the responses examined in the most frequently used tests such as tail flick, hot plate, paw withdrawal, and their variations could be entirely mediated by spinal reflexes or brainstem–spinal motor programs, thus constituting unconscious nocifensive responses (Table 36–2[8]). Nonetheless, some higher brain influence is probably operating in an intact, awake animal, but its presence and nature are hard to separate from subcortical processes and it is also likely to be unconsciously mediated. Consequently, none of these tests can be legitimately viewed as tests of pain, since they do not depend on a consciously mediated response. In some cases, investigators are aware of this constraint and strictly adhere to the term nociception rather than pain in interpreting their results. Unfortunately this is far from a universal practice and erroneous language and inference are common. Frequently "pain processing" or "pain transmission" is used to describe what is clearly nociceptive processing at the receptor, spinal, or subcortical level. Statements that can be found in the literature, such as pain in labor originates in the cervix or that there is inflammatory pain in the bladder, are mechanistically incorrect by implying that the conscious, psychological process of pain is somehow produced and experienced by peripheral tissues.

Development of well-validated models for pain, as opposed to nociception, is one of the most significant challenges to this field. To this end, some investigators have utilized diverse paradigms requiring behaviors clearly beyond the realm of reflexes or complex, but probably unconsciously mediated motor programs. Behaviors that are more complex than stimulus-bound flexion reflexes and are directed at the site of nociceptive stimulation, or sustained after stimulus termination, such as the vocalization afterdischarge, are sometimes taken to reflect pain. But recall that even decerebrate rats are capable of such behaviors. Measures involving learned behaviors, tasks where the animal can regulate nociceptive stimulus intensity, and assessments of disrupted feeding, sleep, or social behavior may have more validity as reflecting pain.[3,8] Presumably, these types of behaviors reflect more complex, higher-order processing in the brain, but in the absence of rigorous validation, their acceptance as indications of pain should be considered tentative.

Interpretation of Nonbehavioral Measures of Nociception/Pain Numerous reduced preparations, even *in vitro* models, are being used for studies of nociception–pain. Tissue response measures, such as neuronal expression of the early oncogene c-*fos*[42] and other markers of gene expression,[43,44] are potentially associated with processes of importance for nociception, but their significance must be qualified because they are quite removed from pain or even the dynamic, functional context of nociception. Expression of c-*fos* is commonly interpreted as an indication of neuronal response to sensory, including nociceptive stimuli. But unlike c-*fos* expression, neuronal responses to nociceptive stimuli are not all or none, not always excitatory, and not invariant. Rather, they may entail changes in pattern or inhibition, all of which are mechanistically important, but undetectable with markers such as c-*fos*. In addition, the proportion of neurons that expresses c-*fos* is often much smaller than the population that actually responds electrophysiologically[42,45] and neurons in some neurophysiologically responsive brain regions do not express c-*fos*.[42,45] There are additional problems with false-positive responses and dissociation between behavioral and c-*fos* responses to nociceptive stimuli.[42] Clearly, these indirect measures, like behavioral measures, require careful validation with respect to what they do or do not mean. Likewise, interpretations from response measures such as reflexes or neurophysiological recordings obtained under anesthesia are also subject to qualification. Measures of brain neuronal activity in animals or measures used in humans, such as functional magnetic resonance imaging, alone are never proof of pain. The relationship of such indicators to pain can be assessed only after pain has independently been shown to coexist with and be essential for putative pain-related properties of such neurophysiological correlates. Furthermore, nothing short of an intact, fully functioning animal, particularly a mammal, is a potentially suitable model for investigations of human-like pain. Highly reduced preparations are primarily valuable for understanding some elements of nociceptive signaling or peripheral pathophysiological processes that might initiate nociceptive signaling.

RECOMMENDATIONS FOR MODEL SELECTION AND INTERPRETATION

This chapter has stressed the importance of critically examining the validity of animal models to foster progress in understanding mechanisms of pain and the development of more effective treatments. To this end, we propose that particular consideration be given to the following questions: (1) is the problem or phenomenon of interest expressed at the level of nociceptive processing and nocifensive responses or the higher-order, end stage of pain; (2) is the response measure valid and appropriate for the question under investigation; (3) how valid are assumptions concerning the equivalence between the animal model and the human pain condition; and (4) is peripheral nociception of greatest importance, or is it essential to evaluate pain specifically?

The difficulty of knowing how well the putative pain experience of an animal such as a rat or mouse translates to a human-like experience is likely to remain a limiting factor in development of models that validly assess the psychological experience of pain. Future success in development of pain therapies may often hinge on more exacting attention to this issue. There is a good deal of disagreement between investigators of animal psychology on which species might manifest consciousness, a prerequisite to pain experience as we know it,[34,46] but there is a different approach to the question of pain. The dependence of the suffering dimension of human pain on cortical functioning, especially the cingulate gyrus, insula, and prefrontal cortex, is now well established.[2,13,14,22,23] It is also commonly and probably safely assumed that similar cortical regions, where present, work in at least roughly similar ways across mammalian species. On this basis, it would be possible to provide a preliminary validation of a putative animal model for pain by showing that behaviors allegedly reflecting it depend on the functional integrity of these pain-mediating cortical zones. There would still be potential for misinterpretation by confusing nocifensive behaviors with pain-dependent behaviors, but by placing the control of the response measure at the same cortical regions known to be essential to pain experience in humans, the potential for examining common mechanisms would be greatly facilitated. There is currently evidence that investigators of pain are using this approach.[8,47]

REFERENCES

1. McMahon SB, Koltzenburg M, Eds. *Wall and Melzack's Textbook of Pain*, 5th ed. Philadelphia: Elsevier Churchill Livingstone, 2006.

2. Bushnell CM, Apkarian AV. Representation of pain in the brain. In: McMahon SB, Koltzenburg M, Eds. *Wall and Melzack's Textbook of Pain*, 5th ed. Philadelphia: Elsevier Churchill Livingstone, 2006:107–124.
3. Blackburn-Munro G. Pain-like behaviours in animals–how human are they? *Trends Pharmacol Sci* 2004;25:299–305.
4. Hill R. NK_1 (substance P) receptor antagonists–why are they not analgesic in humans? *Trends Pharmacol Sci* 2000;21:244–246.
5. Villanueva L. Is there a gap between preclinical and clinical studies of analgesia? *Trends Pharmacol Sci* 2000;21:461–462.
6. Le Bars D, Gozariu M, Cadden SM. Animal models of nociception. *Physiol Rev* 2001;53:597–652.
7. Hogan Q. Animal pain models. *Reg Anesth Pain Med* 2002;27: 385–401.
8. Vierck CJ. Animal models of pain. In: McMahon SB, Koltzenburg M, Eds. *Wall and Melzack's Textbook of Pain*, 5th ed. Philadelphia: Elsevier Churchill Livingstone, 2006:175–186.
9. Bond MR. Psychiatric disorders and pain. In: McMahon SB, Koltzenburg M, Eds. *Wall and Melzack's Textbook of Pain*, 5th ed. Philadelphia: Elsevier Churchill Livingstone, 2006:259–266.
10. Wall PD. Pain: Neurophysiological mechanisms. In: Adelman G, Smith B, Eds. *Encyclopedia of Neuroscience*, 2nd ed. Amsterdam: Elsevier, 1999:1565–1567.
11. Laureys S, Goldman S, Phillips C, Van Bogaert P, Aerts J, Luxen A, Franck G, Maquet P. Impaired cortical connectivity in vegetative state: Preliminary investigation using PET. *Neuroimage* 1999;9:377–382.
12. Young GB, Ropp AH, Bolton CF. *Coma and Impaired Consciousness*. New York: McGraw-Hill, 1998.
13. Derbyshire SWG. Locating the beginnings of pain. *Bioethics* 1999;13:1–31.
14. Rose JD. The neurobehavioral nature of fishes and the question of awareness and pain. *Rev Fisheries Sci* 2002;10:1–38.
15. Beecher HK. *Measurement of Subjective Responses*. New York: Oxford University Press, 1959.
16. Carlen PL, Wall PD, Nadvrona H, Steinbach T. Phantom limbs and related phenomena in recent traumatic amputations. *Neurology* 1978;28:211–217.
17. Melzack R, Wall PD, Ty TC. Acute pain in an emergency clinic: Latency of onset and descriptor patterns. *Pain* 1982;14:33–43.
18. Ramachandran VS, Rogers-Ramachandran D. Synaesthesia in phantom limbs induced with mirrors. *Proc Biol Sci* 1996;263:377–386.
19. Price DD, Rainville P. Hypnotic analgesia. In: McMahon SB, Koltzenburg M, Eds. *Wall and Melzack's Textbook of Pain*, 5th ed. Philadelphia: Elsevier Churchill Livingstone, 2006:329–338.
20. Craig AD. How do you feel? Interoception: The sense of the physiological condition of the body. *Nat Rev Neurosci* 2002;3:655–666.
21. Dostrovsky JO, Craig AD. Ascending projection systems. In: McMahon SB, Koltzenburg M, Eds. *Wall and Melzack's Textbook of Pain*, 5th ed. Philadelphia: Elsevier Churchill Livingstone, 2006:187–204.
22. Price DD. *Psychological Mechanisms of Pain and Analgesia*. Seattle: International Association for the Study of Pain, 1999.
23. Treede RD, Kenshalo DR, Jones AKP. The cortical representation of pain. *Pain* 1999;79:105–111.
24. Fields HL, Basbaum AI, Heinricher HH. Central nervous system mechanisms of pain modulation. In: McMahon SB, Koltzenburg M, Eds. *Wall and Melzack's Textbook of Pain*, 5th ed. Philadelphia: Elsevier Churchill Livingstone, 2006:125–142.
25. Pastor J, Soria B, Belmonte C. Properties of nociceptive neurons of leech segmental ganglion. *J Neurophysiol* 1996;75:2268–2279.
26. Dewsbury DA, Rethlingshafer DA. *Comparative Psychology, a Modern Survey*. New York: McGraw-Hill, 1973.
27. Jouvet M. Coma and other disorders of consciousness. In: Vinken PJ, Bruyn GW, Eds. *Handbook of Clinical Neurology*. New York: Elsevier, 1969;3:62–79.
28. Thogmartin JR. Report of autopsy #5050439, Theresa Schaivo, Pasco and Pinellas Counties, Florida, 2005.
29. Wahlestedt C. Neuropharmacology: Reward for persistence in substance P research. *Science* 1998;281:1624–1625.
30. Keller M, Montgomery S, Ball W, Morrison M, Snavely D, Guanghan L, Hargreaves R, Hietala J, Lines C, Beebe K, Reines S. Lack of efficacy of the substance P ($neurokinin_1$ receptor) antagonist aprepitant in the treatment of major depressive disorder. *Biol Psychiatry* 2006;59:216–223.
31. Ramadan NM. Acute treatments: Some blind alleys. *Curr Med Res Opin* 2001;17(Suppl. 1):s71–80.
32. Stevens CW. Opioid research in amphibians: An alternative pain model yielding insights on the evolution of opioid receptors. *Brain Res Rev* 2004;46:204–215.
33. Chandroo KP, Duncan IJH, Moccia RD. Can fish suffer?: Perspectives on sentience, pain, fear and stress. *Appl Anim Behav Sci* 2004;86:225–250.
34. Rose JD. Anthropomorphism and "mental welfare" of fishes. *Dis Aquat Org* 2007;75(2):139–154.
35. Northcutt RG, Kaas JH. The emergence and evolution of mammalian neocortex. *Trends Neurosci* 1995;18:373–379.
36. Nieuwenhuys R, ten Donkelaar HJ, Nicholson C. *The Central Nervous System of Vertebrates*. Berlin: Springer, 1998.
37. Avian Brain Nomenclature Consortium. Avian brains and a new understanding of vertebrate brain evolution. *Nat Rev Neurosci* 2005;6:151–159.
38. Mountcastle VB. *Perceptual Neuroscience*. Cambridge, MA: Harvard University Press, 1998.
39. Woods JW. Behavior of chronically decerebrate rats. *J Neurophysiol* 1964;27:635–644.
40. Bateson P. Do animals feel pain? *New Sci* 1992;134:30–33.
41. Macphail EA. *The Evolution of Consciousness*. New York: Oxford University Press, 1998.
42. Harris J. Using c-fos as a neural marker of pain. *Brain Res Bull* 1998;45:1–8.
43. Pace MC, Mazzariello L, Passavanti MB, Sansone P, Barabarisi M, Aurilio C. Neurobiology of pain. *J Cell Physiol* 2006;209:8–12.
44. Lacroix-Fralish ML, Tawfik VL, Tanga FY, Spratt KF, DeLeo JA. Differential spinal cord gene expression in rodent models of radicular and neuropathic pain. *Anesthesiology* 2006;104:1283–1292.
45. Alexander BM, Rose JD, Stellflug JN, Fitzgerald JA, Moss GE. Fos-like immunoreactivity in brain regions of domestic rams following exposure to rams or ewes. *Physiol Behav* 2001;73:75–80.
46. Wynne CDL. The perils of anthropomorphism. *Nature* 2004;428: 606.
47. Johansen JP, Fields HL, Manning BH. The affective component of pain in rodents: Direct evidence for a contribution of the anterior cingulate cortex. *Proc Natl Acad Sci USA* 2001;98:8077–8082.
48. Mogil JS, Wilson SG, Bon K, Lee SE, Chung K, Raber P, Pieper JO, Hain HS, Belnap JK, Hubert L, Elmer GI, Chung JM, Devor M. Heritability of nociception I: Responses of 11 inbred mouse strains on 12 measures of nociception. *Pain* 1999;80:67–82.

37 Nonmammalian Models for the Study of Pain

CRAIG W. STEVENS

ABSTRACT

Processing nociceptive information is a consistent feature of the nervous system in all vertebrate species. This chapter reviews the nonmammalian models developed for pain and analgesia research, with a special emphasis on models that were used for detecting opioid antinociception. Nociceptive pathways in nonmammalian vertebrates are reviewed and endogenous opioid systems are described. Compared to mammalian models for pain research, there are relatively few models in each vertebrate class. Details are provided for the methods and the results of using an amphibian model for the testing of opioid analgesics. The chapter ends with a case study of the comparative approach in pain and analgesia research applied to the molecular evolution of opioid receptor proteins.

Key Words: Opioids, Pain, Analgesia, Animal models, Amphibians, Evolution.

INTRODUCTION: ALTERNATIVES TO THE USE OF MAMMALS FOR PAIN RESEARCH

Pain research is the study of neural pathways that convey noxious stimuli and the investigation of substances or manipulations that alter the neurotransmission of nociceptive information. As the goal of pain research is to generate safer and more effective treatments for patients in pain, this field of research is better described as "analgesia research," but this more accurate term is not widely used.

Unlike other types of biomedical research employing nonhuman animals, pain and analgesia research has an added ethical component due to the testing of analgesic agents (or manipulations) by first inducing a "painful" state in the animal subject by the application of a noxious stimulus. The effectiveness of the analgesic treatment is determined by the difference of treatment values from baseline values in the duration or intensity of the noxious stimulus at the time the animal exhibits a behavior (a nocifensive behavior) that signals nociception. Though most exposures to the noxious stimuli in the experimental setting are brief (like the hot plate and tail-flick algesiometric tests in rodents), a number of persistent pain models are used to more closely mimic the clinical scenario of chronic pain patients. The core of the ethical concern in pain and analgesia research, like many ethical issues, comes from a basic lack of knowledge: the capacity of nonhuman animals to experience nociceptive transmission as humans do, i.e., "pain," is not known. It is unknowable because "pain" is a conscious perception and it is not known if nonhuman animals have the capacity for consciousness that may be needed for the "pain" perception to exist. To think otherwise is not science, but science fiction.

But science can guide rational thought and based on the comparative neurology of central nervous system (CNS) pathways and brain structures shown to be important for nociceptive transmission and supraspinal appreciation of pain in mammalian studies, it was proposed that the potential for pain, if it exists in nonhuman animals, is less in earlier-evolved vertebrates than in mammals.[1] This idea arose from considerations of an amphibian pain model,[2–4] but similar conclusions were made in the discussion of pain and awareness in fish.[5] Thus, a continuum of the potential for pain likely exists among vertebrate classes, perhaps correlated with the capacity for consciousness, and the relative difference in this potential for pain between, say, a frog and a rat may justify developing an alternative or adjunct* nonmammalian model for pain and analgesia research on ethical grounds.[6]

The scientific basis for the use of adjunct models for analgesia research depends on the value of a comparative approach to address the question at hand. For example, the value of investigating mechanisms of opioid analgesia in adjunct models using nonmammalian vertebrates has enriched the understanding of opioid receptors and the selectivity of opioid drugs, as highlighted at the end of this chapter. While there are a great number of pharmacological and nonpharmacological means to modulate nociceptive transmission and therefore many avenues of investigation in pain research, this chapter will focus on nonmammalian models developed primarily to investigate the role of opioids in producing antinociception in behavioral models. An initial review of current knowledge of the nociceptive pathways and endogenous opioid systems in nonmammalian vertebrates is followed by a review of pain and analgesia studies by vertebrate class. A more detailed examination of the methodology and results using an amphibian model follows next. The chapter closes with a "case study" of a comparative approach to investigate the binding of opioid analgesics that leads to an increased understanding of the evolution of vertebrate opioid receptors.

From: *Sourcebook of Models for Biomedical Research*
(P. M. Conn, ed.), © 2008 Humana Press Inc., Totowa, NJ.

*The use of "adjunct model" avoids any connotation present in the term "alternative model."

NOCICEPTIVE AND ENDOGENOUS OPIOID SYSTEMS IN NONMAMMALIAN VERTEBRATES

This section presents a summary of the existing scientific literature on nociceptive processing and endogenous antinociceptive systems in fish, amphibians, reptiles, and birds. It should be noted that there are relatively few data available on nociception in nonmammalian vertebrates, and as shown in the next section, even fewer data on the action of opioid analgesic agents. However, the basic plan of the nervous system is consistent in all vertebrates, with the major difference being a lack of neocortical brain development in earlier-evolved vertebrate species. Thus, nociceptive transmission at the lower levels in the nervous system (nociceptive primary afferent fibers and spinal cord pathways) is homologous in all vertebrates. In contrast, pain processing at higher, supraspinal centers arising later in vertebrate phylogeny may not be present in fish, amphibians, reptiles, and birds, which do not share homologous supraspinal structures.

NOCICEPTIVE PATHWAYS IN EARLIER-EVOLVED VERTEBRATES Noxious information in mammals is parallel processed throughout the nervous system, with separate pathways for the sensory-discriminative and motivational-affective aspects of pain perception.[7] To some extent, this duality of pain perception begins with the primary afferent fiber type: thinly myelinated Aδ fibers initiate the sensory-discriminative pathway and the unmyelinated C fibers may signal the motivational-affective component of pain. In mammals, the highest level termination of the sensory pathways is the sensory neocortex and the destination of motivational pathways is the limbic neocortex.[8,9]

Electrophysiology experiments using dissected nerves and characterizing single nociceptive primary afferent fibers with defined nociceptive receptive fields are not routinely done using nonmammalian species. However, there are a number of studies characterizing nociceptive pathways in the rainbow trout, *Oncorhynchus mykiss*. Anatomical data and electrophysiological examination of single units recorded from trigeminal ganglia cells identified nociceptors and nociceptive afferents that were not significantly different from those found in other vertebrates.[10,11] Notable differences include more Aδ fibers and less C fibers in fish trigeminal nerve than in mammals.[12] There are few studies of the central termination of nociceptive afferents in fish spinal cord, although substance P, a marker of nociceptive primary afferent fibers, was localized densely in the superficial aspect of the substantia gelatinosa and scattered throughout the nucleus proprius, intermediate zone, and the ventral horn.[13]

Investigations of amphibian primary afferent fibers, including nociceptive afferents, were among the first studies characterizing sensory fibers in vertebrates.[14–16] In contrast to numerous specialized end organs on mammalian afferent fibers, only three groups of afferent nerve endings were identified in amphibians. These include an "expanded tip" morphology ending at the superficial level of the dermis and two types of free nerve endings: one type terminating in the epidermis and the other ending in the deeper layers of the dermis.[17–19] Based on early electrophysiological work, the detection of all noxious sensory stimuli, including chemical, was attributed exclusively to the free nerve endings associated with thermal sensitivity in amphibians.[16] Amphibians possess both myelinated and unmyelinated afferent fibers running concurrently in mixed-fiber peripheral sensory nerves. Investigators delineate these fibers into three general classes based on morphology, conduction velocities, latency of response, and action potential characteristics: large heavily myelinated A fibers, thinly myelinated B fibers, and small unmyelinated C fibers. Their characteristics correlate well with those observed in corresponding classes of somatic sensory afferents in mammals: Aβ, Aδ, and C fibers, respectively. In general, the cutaneous nociceptors in frog skin (those responding to noxious thermal, mechanical, and chemical stimuli) arise from peripheral terminations of the thinly myelinated C fibers. In studies that separated fibers both by conduction velocity and fiber diameter in amphibians, small slowly conducting fibers transmitted the majority of all impulses induced by noxious heat, pinching, pin pricks, and the application of dilute acid to the skin.[16,20–22]

There is controversy over the central terminations of primary afferent fibers within the amphibian spinal gray,[23] although most studies indicate that sensory afferents from the skin terminate in areas of the dorsal spinal cord that correspond to laminae I–IV. The termination of small-caliber fibers associated with nociception was found exclusively in the superficial dorsal laminae (the substantia gelatinosa), while only larger fibers associated with transmission of nonnoxious information penetrate to the deeper laminae.[24,25] Electrophysiological evidence supports the existence of primary afferent synapses on motoneuron dendrites within the amphibian dorsal horn.

Little research has been done on the nociceptive pathways of reptiles. Reptiles also appear to possess only simple nonencapsulated corpuscular endings and free nerve endings, both of which are distributed throughout the dermis and epidermis.[26,27] The separation of receptor structures into categories of noxious and nonnoxious sensory function appears less discreet in the reptile and, thus far, no stimulus-specific receptor structures have been identified.[28,29] Likewise, there are remarkably few data available concerning peripheral fiber types in reptiles. In contrast to reports on amphibians, researchers report only a unimodal distribution of sensory fiber sizes, which extends over the sizes of the B and C classes of amphibian fibers.[28] Furthermore, all stimulus modalities elicit responses throughout both ranges of conduction velocity and fiber size. Additionally, the dorsal roots containing peripheral sensory afferents entering the spinal cord do not clearly segregate according to large and small fibers as in higher vertebrates.[30] Thus it is difficult to classify any specific subgroup of reptilian sensory afferents as primarily nociceptive. However, at least in the turtle, a separation of peripheral fiber types according to their sites of termination in the spinal cord was shown.[31] Smaller type A fibers associated with cutaneous sensation and nociception arborized primarily in the superficial laminae (I–III) as compared to muscle spindle and joint proprioceptive fibers that penetrated to the deeper laminae (IV–VII) where they formed monosynaptic reflex pathways with the dendritic endings of motor neurons.

Investigation of nociceptive pathways in birds utilized the domestic chicken, *Gallus gallus*. Electrophysiological recordings of single afferent fibers from teased-apart leg nerves identified both mechanical and thermal nociceptors.[32] The majority of responses were classified as C fibers responding to both mechanical and thermal noxious stimuli, but Aδ fibers were also noted. Nociceptive afferents innervating chicken skeletal muscle respond to noxious mechanical stimulation (muscle compression) and also to the injection of acetic acid into the muscle tissue.[33] Further studies in chickens showed that nasal trigeminal nociceptors are activated by ammonia vapor,[34] that joint nociceptors are activated

Table 37–1
μ, δ, and κ opioid receptor proteins cloned and sequenced in nonmammalian vertebrates

Class	*Common name*	*Genus species*	*Access number*[a]			*References*
			MOR	*DOR*	*KOR*	
Pisces	Zebrafish	*Danio rerio*	AF132813	AJ001596	AF285173	138–140
Amphibia	Grass frog	*Rana pipiens*	AF530571	AF530572	AF530573	4, 149
Amphibia	Newt[a]	*Taricha granulosa*	AY751784	AY751785	AY725197	146, 147

[a]Access numbers for the nucleotide sequences deposited in GenBank at http://www.ncbi.nlm.nih.gov/Genbank/index.html.

by the intraarticular injection of urate crystals,[35,36] and that beak amputation stimulates the firing of nociceptive afferents in the trigeminal nerve.[37] The spinal cord projections of ankle joint primary afferents were mapped to the upper layers of the chicken spinal cord dorsal horn[38] and substance P-containing fibers terminate in laminae I–II.[39,40]

SUPRASPINAL PATHWAYS OF NOCICEPTION There is scant literature on the supraspinal pathways that may process nociceptive information in nonmammalian vertebrates. Functional studies of evoked responses in supraspinal sites after specific noxious stimulation of peripheral nociceptive fibers are lacking; however, electrical stimulation of the sciatic nerve produced evoked potentials in posterior thalamic nuclei and primordial hippocampal structures in frogs.[41] Single unit recordings in trout and goldfish (*Carassius auratus*) obtained from electrodes placed in the spinal cord, cerebellum, tectum, and telencephalon after timed noxious stimuli revealed supraspinal nociceptive pathways, but more precise mapping by histological confirmation of electrode placement was not done.[42]

The sensory-discriminative pathway originating from spinal neurons makes direct connections with neuron groups farther toward the front of the brain as we ascend the phylogeny of vertebrates.[43] Therefore, fish and amphibians have direct spinal connections to the brainstem, reptiles and birds to the brainstem and the dorsal thalamus in the midbrain, and mammals to the brainstem, thalamus, and primary cortex. Pathways that appear to contribute to the motivational-affective dimension of pain follow a similar evolutionary pattern, except that more medial target sites are contacted in the brain: the medial thalamic nuclei in reptiles and the limbic cortex in mammals. Throughout phylogeny, all the target sites of spinal pathways in the brain increase in complexity, specialization, and number of neurons, suggesting that nociceptive messages to the thalamus in a reptile and mammal may not be comparable.[44]

ENDOGENOUS OPIOID PEPTIDES Nociceptive information is carried from the peripheral nociceptors to the spinal cord by the primary afferent nerve fibers in all vertebrates. These fibers terminate and synapse on second-order neurons in the dorsal horn, releasing neurotransmitters such as substance P, calcitonin gene-related peptide (CGRP), and glutamate. Using immunohistochemical techniques, all of these substances are readily identified in abundance in the spinal dorsal horn of fish, amphibians, reptiles, and birds.[45–48] Substance P and glutamate excite second-order neurons that have their cell bodies in the dorsal horn and send long fibers upward to form the ascending nociceptive pathways. Such second-order neurons within the dorsal horn that receive direct dorsal root input have not yet been identified in nonmammalian vertebrates.[49]

Also present in the spinal cord are intrinsic neurons that release met-enkephalin, an endogenous opioid peptide. Met-enkephalin inhibits the release of substance P from the central terminations of nociceptive primary afferents and decreases the firing of the second-order pain neurons. Met-enkephalin is present in the spinal cord of all vertebrate species examined.[2,39,48,50–57] In amphibians, immobilization stress produces antinociception that is blocked by the pretreatment of animals with the opioid antagonist naloxone.[58,59] This finding demonstrates that enkephalinergic neurons modulate nociceptive threshold in earlier-evolved vertebrates as they do in mammals. The molecular evolution of opioid peptides and their precursors is the subject of a number of recent studies.[60–64]

OPIOID RECEPTORS IN NONMAMMALIAN VERTEBRATES Opioid receptors are membrane proteins belonging to the superfamily of G-protein-coupled receptors (GPCR) and are the products of four distinct genes in mammalian genomes.[65] μ, δ, and κ opioid receptors, abbreviated MOR, DOR, and KOR, mediate the analgesic effects of opioids, while the role of the fourth type of opioid receptor, the nociceptin or orphanin FQ receptor (ORL), is less clear.[66] Before the advent of the genomic era, binding studies using tissue homogenates demonstrated the presence of opioid-binding sites in fish,[67–70] amphibian,[71–76] bird,[77–80] and reptile[81–83] CNS. More definite evidence for the expression of opioid receptor proteins in nonmammalian vertebrates comes from the cloning and sequencing of opioid receptor mRNA. To date, μ, δ, and κ opioid receptors were cloned and sequenced in three nonmammalian vertebrates: the zebrafish, the Northern grass frog, and the rough-skinned newt (see Table 37–1). These sequences represents two classes of nonmammalian vertebrates (Pisces and Amphibia) with no opioid receptor sequences yet available for species from the Reptilia and Aves classes. As shown at the end of the chapter, an analysis of opioid receptor sequences from nonmammalian and mammalian vertebrates demonstrates the utility of a comparative approach to investigate opioid analgesic action.

PAIN AND ANALGESIA RESEARCH USING NONMAMMALIAN VERTEBRATES

Key factors in the development of a nonmammalian model for pain and analgesia research are the identification of a nocifensive behavior that is unique to a noxious stimulus and the demonstration that the same noxious stimulus solely activates nociceptive primary afferents. In an amphibian model called the acetic acid test, the wiping response is the nocifensive behavior and a drop of dilute acetic acid is the noxious stimulus that activates nociceptors and nociceptive primary afferent fibers exclusively.[20,22,84] For other nonmammalian models, integration of neurophysiological data with behavioral data in most cases are still lacking. A review of these adjunct models for analgesia research by vertebrate class is presented next, followed by a detailed examination of an amphibian model for analgesia research. The adjunct models for behavioral assessment of opioid antinociception are listed in Table 37–2.

PISCES Fish are not easily amenable to behavioral assays of antinociception. A study aimed primarily at the metabolism of morphine in goldfish (*C. auratus*) also described a behavioral assay using an electric prod applied caudal to the dorsal fin as the noxious stimulus.[85] The nocifensive behavior that was observed was called the agitated swim response. The voltage threshold for eliciting the swim response was consistent for each animal and with repeated testing. Morphine sulfate added to the tank water produced a dose-dependent increase in the electrical voltage needed to elicit the agitated swim response. It was also shown that goldfish became tolerant to the antinociceptive effects of morphine with repeated administration.[85] There was no attempt to attain a measure of morphine's potency (i.e., no ED_{50} value of morphine was reported) nor was opioid receptor involvement verified by pretreatment with an opioid antagonist such as naloxone. Further studies using the same model in goldfish showed that morphine administered via the intracranial route (into a space above the optic tectum) produced a dose-dependent increase in the voltage needed to elicit the swim response.[86] In this study, opioid receptor involvement was shown by naloxone antagonism of morphine's effect.

The antinociceptive effects of dermorphin, a potent and selective MOR opioid peptide originally isolated from the skin of *Phyllomedusa* frogs,[87] was tested using electrodes implanted in

Table 37–2
Behavioral models for testing opioid antinociception in nonmammalian vertebrate species

Class	*Animal/(genus species)*	*Model*	*Stimulus*	*Response*	*Notes*
Pisces	Goldfish (*Carasius auratus*)	Caudal electroshock	Electric current by handheld prod	Agitated swim response	Acute model; measured agitated swim response (ASR); morphine added to tank water or intracranial; blocked by naloxone [85,86]
Pisces	Codfish (*Gadus orhuamarisalbi*)	Caudal electroshock	Electric current by implanted electrode	Agitated swim response	Acute model; ASR quantified by force-transducer; dermorphin given by intranasal application; β-casomorphin given by IP and IM injection[88,90]
Pisces	Rainbow trout (*Oncorhynchus mykiss*)	Caudal electroshock	Electric current by implanted electrode	Agitated swim response	Acute model; ASR quantified by force-transducer; dermorphin given by intranasal application[89]
Pisces	Carp (*Cyprinus carpio*)	Caudal electroshock	Electric current by implanted electrode	Agitated swim response	Acute model; ASR quantified by force-transducer; tramadol given by IM injection; blocked by naloxone[91]
Pisces	Rainbow trout (*Oncorhynchus mykiss*)	Lip inflammation	Acetic acid/bee venom lip injection	Pain-related behaviors	Chronic model; pain-related behaviors quantified; morphine effects noted[92–95]
Amphibia	European frog (*Rana esculenta*)	Electrified grid	Electric current in floor grid	Jumping	Acute model; no antinociceptive effect of morphine[96]
Amphibia	European frog (*Rana esculenta*)	Hot plate test	Heated surface	Jumping	Acute model; no antinociceptive effect of morphine[97]
Amphibia	Northern grass frog (*Rana pipiens*)	Acetic acid test	Dilute acetic acid	Wiping response	Acute model; well-developed and validated adjunct model; sensitive to opioid and nonopioid analgesics, stress- and hypothermia-induced analgesia[58,59,98,99,121–131,136]
Amphibia	Japanese firebelly newt (*Cynops pyrrhogaster*)	Tail-flick test	Heat from lamp	Tail-flick	Acute model; effects of met-enkephalin and RFamide blocked by naloxone[100]

Table 37–2
(continued)

Class	*Animal/(genus species)*	*Model*	*Stimulus*	*Response*	*Notes*
Reptilia	Green anole lizard (*Anolis carolinensis*)	Tail-flick test	Heat from lamp	Tail-flick	Acute model; antinociceptive effects of morphine[101]
Reptilia	Nile crocodile (*Crocodylus niloticus*)	Hot plate test	Heated surface	Leg-lift or escape response	Acute model; antinociceptive effects of morphine and meperidine (pethidine)[103]
Reptilia	Green iguana (*Iguana iguana*)	Tail-flick test	Heat from thermal electrode	Tail-flick	Acute model; preliminary investigation of butorphanol[104]
Aves	Chicken (*Gallus gallus*)	Toe-pinch	Noxious mechanical	Limb withdrawal	Acute model; no antinociceptive effect of morphine[105]
Aves	Chicken (*Gallus gallus*)	Wing-shock	Electric current to underside of wing	Flight response	Acute model; morphine antinociception[106]
Aves	Chicken (*Gallus gallus*)	Hot plate test	Noxious thermal to feet	Jumping	Acute model; morphine, codeine hyperalgesia; developmental stage specific[106–111]
Aves	Chicken (*Gallus gallus*)	Formalin test	Noxious chemical in foot	Leg-lifting	Chronic model; morphine hyperalgesia, strain dependent[112,113]
Aves	Chicken (*Gallus gallus*)	Carrageenan inflammation	Inflammation-induced thermal hyperalgesia	Foot withdrawal from thermal stimulus	Chronic model; morphine produced dose-dependent antinociception[114]
Aves	Chicken (*Gallus gallus*)	Monoarthritis model	Urate crystals in joint capsule	Pain-related behaviors	Chronic model; no effect of morphine, fentanyl, buprenorphine[115–117]

the caudal region of Derjugin codfish, *Gadus morhuamarisalbi*, as the noxious stimulus.[88] The nocifensive response was similar to the agitated swim response mentioned above, but the fish was held in a flow-through chamber and the swim response was quantified by a force transducer. Dermorphin was administered into the intranasal passages (olfactory sacs) of the codfish and produced a dose-dependent antinociceptive effect. Further studies using rainbow trout, *O. mykiss*, with the same sophisticated behavioral apparatus, mapped the most sensitive areas of noxious stimulation on the trout and confirmed the antinociceptive effect of intranasal dermorphin in this species.[89] In codfish, intraperitoneal (IP) and intramuscular (IM) administration of the MOR selective opioid peptide, β-casomorphin, produced antinociceptive effects.[90] The opioid analgesic, tramadol, an MOR opioid agonist and norepinephrine/serotonin reuptake inhibitor, produced dose-dependent antinociception in the carp, *Cyprinus carpio*.[91] In this last study, pretreatment with the opioid antagonist naloxone blocked the antinociception produced by tramadol.

A chronic pain model was developed in fish by injecting acetic acid or bee venom into the lip region of trout.[92,93] These animals develop pain-related behaviors that can be quantified and antinociception determined by the reduction in pain-related behaviors. Morphine had inhibiting effects on nociceptive behaviors per se and in the presence of novel objects.[94,95]

AMPHIBIA Early studies employing a hot plate test and electrified floor grid in amphibians did not detect an antinociceptive effect of morphine.[96,97] It is likely that this was due to an inappropriate experimental design, as previously noted.[98] A behavioral assay using acetic acid as the noxious stimulus and the wiping response as the nocifensive behavior was developed using the Northern grass frog, *Rana pipiens*, and is discussed in greater detail in the next section. Other studies using the acetic acid test in amphibians demonstrated that regional hypothermia is antinociceptive and is mediated in part by endogenous opioid peptides.[99] A tail-flick test was also used to assess the antinociceptive effects of opioids in the Japanese firebelly newt, *Cynops pyrrhogaster*.[100] Intraperitoneal injection of met-enkephalin and the putative opioid peptide RFamide produced a mild antinociceptive effect that was blocked by the concurrent administration of the opioid antagonist naloxone.

REPTILIA To test the antinociceptive effects of opioids in reptiles, a tail-flick apparatus was used in the green anole lizard, *Anolis carolinensis*.[101] Intraperitoneal injection of morphine (5 mg/kg) produced a weak, but significant increase in the latency of the lizard to remove its tail from a noxious heat stimulus.

A number of nociceptive assays were characterized in the Nile crocodile, *Crocodylus niloticus*.[102] Crocodiles placed on a hot plate attempt to guard their foot pads by lifting them alternately as well as attempting to escape the heated surface. A dilute capsaicin solution instilled in the eye of the crocodile produces behavioral responses including blinking, rubbing, and head shaking. Injection of formalin in the crocodile forepaw results in the development of paw guarding behaviors. Morphine or meperidine (pethidine) produced antinociception as measured by the increased latency for crocodiles to lift their leg or escape on the hot plate test, which was the most suitable of three assays.[103] However, these studies did not include dose–response analysis and did not assess the opioid nature of the antinociceptive effect by pretreatment with an opioid antagonist, such as naloxone or naltrexone. As mentioned in a clinical review of analgesia for exotic species,[104] there are preliminary data that the partial opioid agonist butorphanol produced antinociception in the green iguana, *Iguana iguana*, using a thermal tail-flick test.

AVES An early attempt to measure the antinociceptive effect of opioids in chicks using toe-pinch as the noxious stimulus and limb withdrawal as the nocifensive behavior did not find any antinociceptive effect of morphine up to a dose of 200 mg/kg.[105] Using an electroshock to the underside of the wing as the noxious stimulus and an elicited flight response as the nocifensive behavior in chickens, morphine at a single dose of 30 mg/kg produced measurable antinociception in this model.[106]

Using a heated floor grid, other studies found that morphine produced a *decrease* in the latency for chickens to jump.[107] This hyperalgesic effect was dependent on the age of the chicken; morphine administered to young cockerels less than 14 days old did not produce hyperalgesia.[108] The hyperalgesic effect of morphine was dose dependent and exhibited a U-shaped time course curve.[109] Using selective opioid antagonists, the hyperalgesic effect of morphine in chickens was shown to be predominantly mediated by μ opioid receptors.[110] Codeine produced both hypoalgesia (antinociception) and hyperalgesia in this model.[111]

A formalin algesiometric assay was developed in chickens, quantified by the pain-related behaviors observed after injection of formalin into the foot.[112] There was a strain difference in the hyperalgesic effect of morphine, perhaps due to differences in descending monoamine pathways.[113] Carrageenan, a chemical irritant like formalin, provided the noxious stimulus in a chronic inflammatory model using the Hargreaves method.[114] In this assay, chickens are measured for the latency to withdrawal of an inflamed foot from a lamp placed under the floor grid. Morphine produced a dose-dependent antinociception but had no effect on the edema produced by carrageenan. A chronic arthritis model was developed in chickens using an intraarticular injection of urate crystals, which models gouty arthritis.[115,116] However, intraarticular injection of morphine, fentanyl, or buprenorphine did not have any effect in reducing pain-related behaviors due to the experimentally induced arthritis.[117]

AN AMPHIBIAN MODEL FOR PAIN RESEARCH

THE ACETIC ACID TEST The acetic acid test (AAT) to determine the nociceptive threshold (NT) in frogs consists of 11 concentrations of acetic acid serially diluted from glacial acetic acid (two parts acid : one part water).[98] The concentrations are given a code number from 0 to 10 with the lowest code number equal to the lowest concentration of acetic acid. The actual molar concentration of acetic acid per each solution is given by the following equation:

$$\text{Log}[M] = 0.1716 \times (\text{Code \#}) - 0.5849$$

Nociceptive testing is done by placing, with a Pasteur pipette, a single drop of acid on the dorsal surface of the frog's thigh. Testing begins with the lowest concentration and proceeds with increasing concentrations until the NT is reached. The NT is defined as the code number of the lowest concentration of acid that causes the frog to vigorously wipe the treated leg with either hindlimb. To prevent tissue damage, the acetic acid is immediately wiped off with a gentle stream of distilled water once the animal responds or after 5 sec if the animal fails to respond. If the animal fails to respond, testing continues on the opposite hindlimb. An animal that fails to respond to the highest concentration (#10) is assigned the cutoff value of 11. The wiping response in frogs, like the tail-flick in rodents, remains intact after a high spinal transection, demonstrating sufficient circuitry in the spinal cord to mediate this behavior.[118] The wiping response has not been observed in the laboratory in the absence of noxious stimuli and appears specific for assessing nociception. As noted above, acetic acid was shown to activate nociceptive afferents in amphibians. The wiping response is also the basis for much research on the motor systems of the amphibian spinal cord.[119]

ADMINISTRATION OF ANALGESICS TO AMPHIBIANS The technique of systemic administration (SC) of drugs in frogs is similar to subcutaneous administration in rodent species. The frog is gently grasped and a tuberculin syringe fitted with a 25-gauge needle is slipped under the skin on the dorsal side of the animal. The initial location of the injection is the mid-back region, which overlies the dorsal lymph sac in amphibians (see Figure 37–1, top). The needle is gently inserted into the lymph sac (felt as a slight resistance as if injecting into a sponge) and the drug is delivered in a volume of 10 µl/g body weight. Each animal is used only once for injections to prevent possible confounding effects of repeated injections in the same animal. A saline-injected control group is run with each experiment.

The technique of intraspinal (IS) administration of drugs in frogs is remarkably similar to that described for mice.[120] Frogs are gently held with the nondominant hand and using a tapered 26-gauge needle fitted to a microsyringe, the needle is introduced into the spinal cord slightly lateral to the neural spine at the articulation between the seventh and eighth vertebrae. This articulation is felt by probing with the needle and is located by triangulation of the dorsal protuberances of the iliac crest of the frog (see Figure 37–1, middle). The region of intraspinal injection overlies the area of the lumbar enlargement. Penetration of the needle sometimes produces a mild hindlimb extension or trembling, which discontinues after a 2–3 sec delivery of the drug. Previous experiments employing radiolabeled morphine demonstrated that over 98% of intraspinal morphine remained within two or three segments of the site of injection throughout the time course of the experiment.[121]

Using a modification of the methods for intracerebroventricular (ICV) administration in rodents, animals are injected with 3 µl of drug or saline vehicle into the third ventricle. Briefly, animals are administered a local anesthetic (0.5 ml, 2% lidocaine) suffused just under the skin of the skull region. A midline incision is made and the top of the skull is exposed by retraction. A hand-held microsyringe, fitted with an 8 cm length of PE-10 and a 1.5-inch, 29-gauge needle cannula is used to deliver the drugs. Placement of the needle is made through the skull at a depth of 2.5 mm, midline and just anterior to the optic tectum, a site shown in pilot studies to fill the brain ventricles with dye (see Figure 37–1, bottom). Visualization of the brain structures through the thin

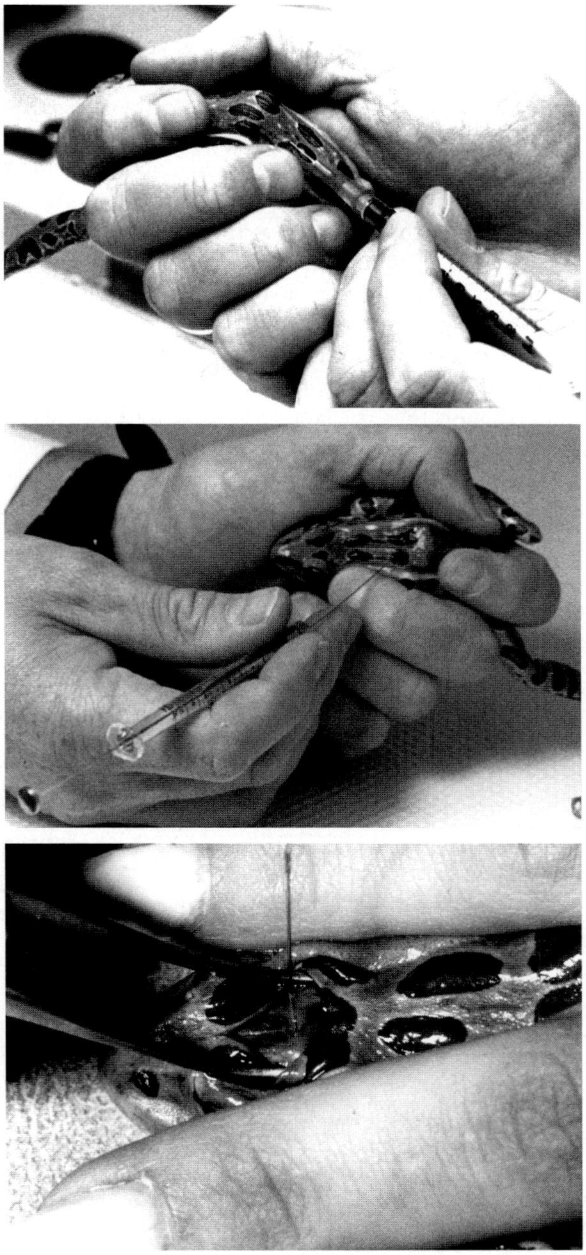

Figure 37–1. Routes of drug administration in frogs. Drugs can be given by the systemic route (top), the intraspinal route (middle), and the intracerebroventricular route (bottom). See text for details.

cartilaginous skull of amphibians also aids in the consistency of injection sites.

EFFECTS OF OPIOID ANALGESICS IN AMPHIBIANS Initial studies of the analgesic effects of opioid administration in amphibians were conducted using nonselective opioid agonists, endogenous opioid peptides, and antagonists.[121–124] These studies showed that both exogenous opioid agonists and endogenous opioid peptides could raise the nociceptive threshold in amphibians by an action at opioid receptors. Tolerance to the analgesic effects of daily morphine administration was documented[125] and stress-induced release of endogenous opioids was shown to produce analgesia, which was potentiated by enkephalinase inhibitors.[59] Other behavioral studies include an investigation of the effects of opioids on noxious and nonnoxious sensory modalities,[84,126] and an examination of agents acting on α_2-adrenergic receptors after systemic and spinal administration.[127,128]

Later results of systematic studies examining the antinociception of selective μ, δ, or κ opioid agonists administered by different routes yielded an important finding: the relative antinociceptive potency of μ, δ, or κ opioid agonists after systemic, intraspinal, or intracerebroventricular administration in amphibians was highly correlated to that observed in typical mammalian models and to the relative analgesic potency of opioid analgesics in human clinical studies.[129–131] These data established the amphibian model as a robust and predictive adjunct model for the testing of opioid analgesics.[1,6,132] Table 37–3 provides a summary of the opioid agonists used in behavioral studies of opioid

Table 37–3
Antinociceptive potency of opioid agonists administered to *Rana pipiens*

Opioid agonist	*Receptor*[a]	*Route*[b]	*ED_{50} value*[c]
Fentanyl	*MOR*	SC	1.4
CI-977 (enadoline)	*KOR*	SC	5.8
Remifentanil	*MOR*	SC	7.1
Levorphanol	*MOR*	SC	7.5
U50488H	*KOR*	SC	8.5
Methadone	*MOR*	SC	19.9
Bremazocine	*KOR*	SC	44.4
Morphine	*MOR*	SC	86.3
Buprenorphine	*MOR*	SC	99.1
Meperidine	*MOR*	SC	128.1
Codeine	*MOR*	SC	140.3
Nalorphine	*KOR*	SC	320.9
Dermorphin	*MOR*	IS	0.04
DAMGO	*MOR*	IS	0.13
DSLET	*DOR*	IS	0.13
DADLE	*MOR*	IS	0.14
Fentanyl	*MOR*	IS	0.94
Morphine	*MOR*	IS	2.26
Remifentanil	*MOR*	IS	3.20
DPDPE	*DOR*	IS	3.29
Deltorphin	*DOR*	IS	13.50
CI-977 (enadoline)	*KOR*	IS	13.51
Bremazocine	*KOR*	IS	22.89
U50488H	*KOR*	IS	36.82
Nalorphine	*KOR*	IS	43.13
DADLE	*MOR*	ICV	1.3
Morphine	*MOR*	ICV	2.0
DPDPE	*DOR*	ICV	8.3
Fentanyl	*MOR*	ICV	23.0
CI-977 (enadoline)	*KOR*	ICV	35.5
U50488H	*KOR*	ICV	61.5

[a]Selectivity of the listed opioid for MOR, DOR, or KOR (μ, δ, or κ opioid receptor).
[b]Route of administration: SC, subcutaneous; IS, intraspinal; ICV, intracerebroventricular.
[c]Dose that gives 50% antinociceptive effects; in nanomoles/gram for SC, and in nanomoles/frog for IS and ICV administration. Data are from references 129, 130, 131, and 148.

antinociception following systemic, spinal and supraspinal routes of administration.

However, emerging results from both behavioral and binding studies in *R. pipiens* suggested that the selectivity of amphibian opioid receptors was different from the selectivity of mammalian opioid receptors. The MOR, DOR, and KOR type selective opioid antagonists, β-FNA, naltrindole, and nor-BNI,[133–135] did not show type selectivity in blocking the antinociceptive effects of selective opioid agonists in amphibians.[136] In radioligand binding studies, β-FNA, naltrindole, and nor-BNI competed with [^{3}H]naloxone binding from brain and spinal cord homogenates with equal apparent affinity, but did show type-selective competition against selective opioid agonist radioligands.[74–76] Still lacking was the ultimate identification of the types of opioid receptor proteins expressed in *R. pipiens* from molecular cloning studies.

AN EXAMPLE OF THE COMPARATIVE APPROACH: THE EVOLUTION OF VERTEBRATE OPIOID RECEPTORS

The number of full-length cDNA sequences for opioid receptors in nonmammalian vertebrates deposited in GenBank is surprisingly small. At present, there are no databank sequences for any type of opioid receptor from any species of Reptilia or Aves. From class Pisces there was an initial entry of μ opioid receptor sequences expressed in the white suckerfish, *Castomus commersoni*,[137] and more recently, the triad of μ, δ, and κ opioid receptor cDNAs was cloned from the zebrafish, *Danio rerio*.[138–140] The cloning of full-length opioid receptors in *R. pipiens* was particularly needed given that there is a corresponding dataset of opioid behavioral and binding data. Other nonmammalian species do not have a corresponding dataset of opioid binding and behavioral studies to integrate molecular cloning and *in vitro* results with whole animal, behavioral data on opioid antinociception.

CLONING AND SEQUENCING OF AMPHIBIAN OPIOID RECEPTORS Three full-length clones of opioid receptors were obtained by reverse transcriptase polymerase chain reaction (RT-PCR) amplification of *R. pipiens* brain and spinal cord cDNA. On the basis of high homologies to existing vertebrate sequences, these clones were named *rpMOR* (GenBank number AF530571), *rpDOR* (AF530572), and *rpKOR* (AF530573) as novel orthologs to mammalian μ, δ, and κ opioid receptors. A fourth type of opioid receptor protein, *rpORL* (AY434690), was recently sequenced and was homologous to existing *ORL* proteins.

The percent identity of amino acids in *rpMOR* compared to the zebrafish μ opioid receptor was 79%, and ranges from 83 to 84% compared to mammalian μ opioid receptors. The frog δ opioid receptor, *rpDOR*, had 77% identity with the zebrafish ortholog and 73% identity compared to mammalian δ opioid receptors. Mammalian κ opioid receptors ranged from 70 to 71% identical to *rpKOR* and the zebrafish κ opioid receptor was 64% identical to *rpKOR*.

BIOINFORMATICS OF VERTEBRATE OPIOID-LIKE RECEPTORS The bioinformatic analysis of the amphibian opioid receptor used existing matched datasets of μ, δ, and κ opioid receptors from other vertebrate species. Somewhat surprising, there are only four vertebrate species, excluding the amphibian data, that have full-length sequences for all three types of opioid receptor cDNA deposited in GenBank. These species are the zebrafish (*Danio rerio*), mouse (M*s* μ*scularis*), rat (*Rattus norvigecus*), and human (*Homo sapiens*). Phylogenetic dendrograms (trees) were constructed using existing alignment (ClustalW) and molecular evolution software. The nearest-neighbor joining trees for each vertebrate opioid receptor type gave the accepted evolutionary relationship, with the fourth type of opioid receptors (ORL) used as an outgroup to root the three classical types of opioid receptors (see Figure 37–2). As shown, the three groups of opioid receptor types did not arise from a single, common ancestral sequence but rather the group of κ opioid receptors shared a common ancestral sequence with the branch bifurcating later to the μ and δ opioid receptor groups.

As pharmacological selectivity is correlated with similarity of amino acid sequences at the GPCR family level (e.g., opioid receptors vs. muscarinic receptors) it is reasonable to assume that the type selectivity within family members (e.g., *MOR*, *DOR*, and

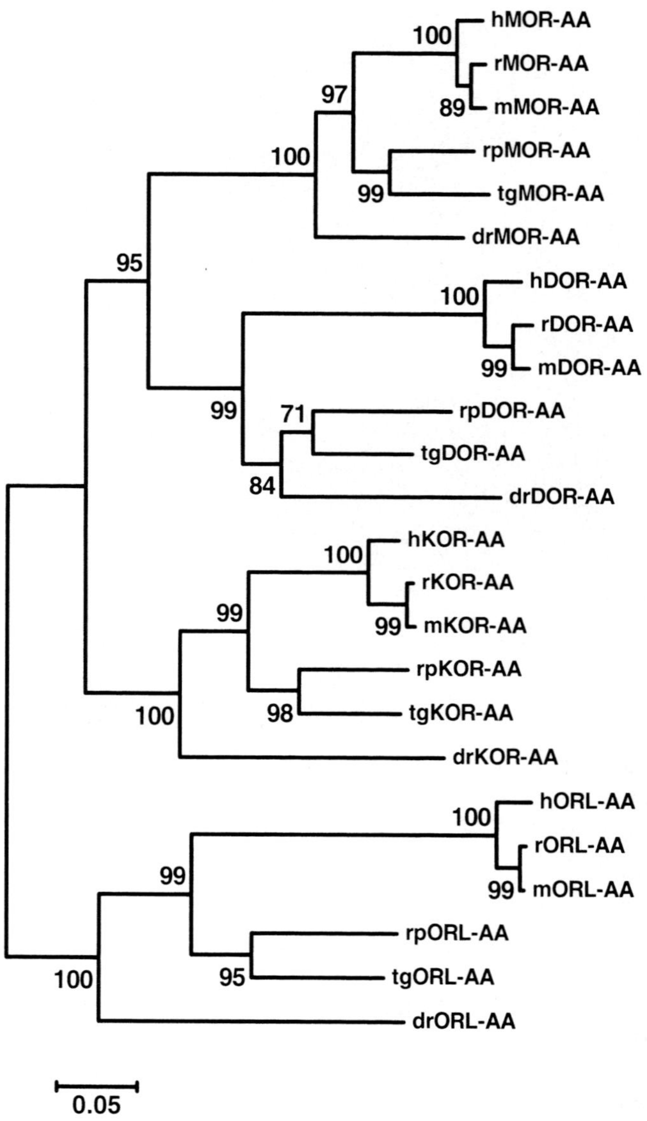

Figure 37–2. Phylogenetic analysis of the vertebrate *OpR* proteins. The neighbor-joining algorithm was used (h, human; r, rat; m, mouse; rp, frog; tg, newt; and dr, fish). The corresponding vertebrate *ORL* proteins were used as the outgroup. Numbers provide the bootstrap value (percent reliability out of 1000 replicates) and the scale bar is the proportional difference (e.g., 0.05 = 5% divergence in sequence) for a given unit of branch length.

Table 37–4
Homology of *MOR, DOR,* and *KOR* receptors within different vertebrate species[a]

	MOR *vs.* DOR	MOR *vs.* KOR	DOR *vs.* KOR	*Mean (SEM)[b]*
Identity				
Zebrafish	68	62	68	66.0 (2.0)
Frog	73	65	63	67.0 (3.1)
Rat	66	61	61	62.7 (1.7)
Mouse	61	60	61	60.7 (0.3)
Human	62	60	59	60.3 (0.9)
Similarity				
Zebrafish	80	78	82	80.0 (1.2)[c]
Frog	85	81	77	81.0 (2.3)[c]
Rat	76	75	74	75.0 (0.6)
Mouse	71	75	73	73.0 (1.2)
Human	73	73	72	72.7 (0.3)

[a]Percent identity and similarity determined by BLAST-P. BLAST-P settings: matrix = Blossum62, gap open = 11, gap extension = 1, x-drop-off = 50, expect = 10.00, wordsize = 3, and filter off.

[b]Standard error of the mean.

[c]Denotes significantly different from rat, mouse, and human mean values ($p < 0.05$, one-way ANOVA followed by post hoc Newman–Kuels test).

KOR) is also correlated with percent identity or similarity. The results of pair-wise BLAST analysis[141] for the three types of opioid receptor sequences within each species yielded a rank order of divergence such that in earlier-evolved vertebrates, μ, δ, and κ opioid sequences were more closely related to each other than in humans and other mammals. For the percent similarity comparison, these values reached statistical difference such that the mean values of similarity of μ, δ, and κ opioid receptors in nonmammals were greater than the similarity of μ, δ, and κ opioid receptors in mammals (Table 37–4). Insofar as divergence of molecular sequence is related to the greater type selectivity of opioid receptors, this finding gave rise to the novel hypothesis that opioid receptors are more type selective in mammals than in nonmammalian species. The hypothesis that opioid receptors in earlier-evolved vertebrate are less selective is supported by behavioral and binding studies in amphibians[74,136] and by binding studies on zebrafish opioid receptors expressed in Chinese hamster ovary (CHO) cells.[138,142]

Further bioinformatic analysis of the vertebrate opioid receptor sequence dataset leads to identification of individual amino acids that may be crucial for determining type selectivity of opioid analgesics. Using this comparative approach, novel hypotheses were generated that lead to a greater understanding of the molecular evolution of vertebrate opioid receptors and the mechanism of the opioid analgesic selectivity.[4,149]

CONCLUSIONS

The comparative approach to investigations of pain and analgesia has great promise for contributing to a greater understanding of analgesic mechanisms in humans. Additionally, much research using adjunct nonmammalian models for the assessment of antinociception is needed to make sound clinical decisions in veterinary treatment of exotic animals (i.e., not dogs or cats, but fish, amphibians, reptiles, and birds).[143] There has been some transfer of knowledge from basic studies of antinociceptive effects in *R. pipiens* to the clinical application of analgesics for use during oocyte harvesting in *Xenopus laevis*.[144] However, for both improved clinical treatment of humans and animals,[145] additional studies of nociception in nonmammalian species and the development of adjunct models for analgesia are critically needed.

REFERENCES

1. Stevens CW. An amphibian model for pain research. *Lab Anim* 1995;24:32–36.
2. Stevens CW. Opioid antinociception in amphibians. *Brain Res Bull* 1988;21:959–962.
3. Stevens CW. An amphibian model for the assessment of opioid analgesia: Systemic and spinal studies. *Analgesia* 1995;1:683–686.
4. Stevens CW. Opioid research in amphibians: An alternative pain model yielding insights on the evolution of opioid receptors. *Brain Res Rev* 2004;46:204–215.
5. Rose JD. The neurobehavioral nature of fishes and the question of awareness and pain. *Rev Fish Sci* 2002;10:1–38.
6. Stevens CW. Alternatives to the use of mammals for pain research. *Life Sci* 1992;50:901–912.
7. Hammond DL. New insights regarding organization of spinal cord pain pathways. *News Physiol Sci* 1989;4:98–101.
8. Willis WD, Westlund KN. Neuroanatomy of the pain system and of the pathways that modulate pain. *J Clin Neurophys* 1997; 14:2–31.
9. Wilkinson SV, Neary MT, Jones RO, Sunshein KF. The neuroanatomy of pain. *Pain Manag* 1994;11:1–13.
10. Sneddon LU. Anatomical and electrophysiological analysis of the trigeminal nerve in a teleost fish, *Oncorhynchus mykiss*. *Neurosci Lett* 2002;319(3):167–171.
11. Sneddon LU. Trigeminal somatosensory innervation of the head of a teleost fish with particular reference to nociception. *Brain Res* 2003;972(1–2):44–52.
12. Sasaki Y, Matsui M, Taguchi M, *et al.* D-ARG2-dermorphin tetrapeptide analogs: A potent and long-lasting analgesic activity after subcutaneous administration. *Biochem Biophys Res Commun* 1984;120:214–218.
13. Ritchie TC, Leonard RB. Immunohistochemical studies on the distribution and origin of candidate peptidergic primary afferent neurotransmitters in the spinal cord of an elasmobranch fish, the atlantic stingray (*Dasyatis sabina*). *J Comp Neurol* 1983;213:414–425.

14. Adrian ED. The impulses produced by sensory nerve endings. Part 4: Impulses from pain receptors. *J Physiol* 1926;62:33–51.
15. Adrian ED. Impulses in sympathetic fibres and in slow afferent fibres. *Proc Physiol Soc* 1930;20–21.
16. Adrian ED, Cattell M, Hoagland H. Sensory discharges in single cutaneous nerve fibres. *J Physiol* 1931;72:377–391.
17. Roberts A, Hayes BP. The anatomy and function of free nerve endings in an amphibian skin sensory system. *Proc Biol Sci* 1977;196:415–420.
18. Fox H, Whitear M. Observations of Merkel cells in amphibians. *Biol Cell* 1978;32:223–232.
19. Spray DC. Cutaneous receptors: Pain and temperature receptors of anurans. In: Llinas R, Precht W, Eds. *Frog Neurobiology: A Handbook*. Berlin: Springer-Verlag, 1976:607–628.
20. Hamamoto DT, Forkey MW, Davis WL, Kajander KC. The role of pH and osmolarity in evoking the acetic acid-induced wiping response in a model of nociception in frogs. *Brain Res* 2000; 862:217–229.
21. Maruhashi J, Mizuguchi K, Tasaki I. Action currents in single afferent nerve fibres elicited by stimulation of the skin of the toad and the cat. *J Physiol* 1952;117:129–151.
22. Hamamoto DT, Simone DA. Characterization of cutaneous primary afferent fibers excited by acetic acid in a model of nociception in frogs. *J Neurophysiol* 2003;90:566–577.
23. Nikundiwe AM, De Boer-van Huizen R, Ten Donkelaar HJ. Dorsal root projections in the clawed toad (*Xenopus laevis*) as demonstrated by anterograde labeling with horseradish peroxidase. *Neuroscience* 1982;7:2089–2103.
24. Szekely G. The morphology of motoneurons and dorsal root fibers in the frog's spinal cord. *Brain Res* 1976;103:275–290.
25. Szekely G, Matesz K, Baker RE, Antal M. The termination of cutaneous nerves in the dorsal horn of the spinal cord in normal and in skin-rotated frogs. *Exp Brain Res* 1982;45:19–28.
26. Landmann L, Halata Z. Merkel cells and nerve endings in the labial epidermis of a lizard. *Cell Tissue Res* 1980;210:353–357.
27. Pac L. Nerve endings in the lizard skin (*Lacerta viridis*). *Z Mikrosk Anat Forsch* 1984;98:939–950.
28. Kenton B, Kruger L, Woo M. Two classes of slowly adapting mechanoreceptor fibres in reptile cutaneous nerve. *J Physiol* 1971;212:21–44.
29. Kenton B, Kruger L. Information transmission in slowly adapting mechanoreceptor fibers. *Exp Neurol* 1971;31:114–139.
30. Kusuma A, Ten Donkelaar HJ. Dorsal root projections in various types of reptiles. *Brain Behav Evol* 1980;17:291–309.
31. Ruigrok TJ, Crowe A, Ten Donkelaar HJ. Morphology of primary afferents to the spinal cord of the turtle Pseudemys scripta elegans. *Anat Embryol (Berl)* 1985;171:75–81.
32. Gentle MJ, Tilston V, McKeegan DE. Mechanothermal nociceptors in the scaly skin of the chicken leg. *Neuroscience* 2001;106:643–652.
33. Sandercock DA. Putative nociceptor responses to mechanical and chemical stimulation in skeletal muscles of the chicken leg. *Brain Res Brain Res* Rev 2004;46:155–162.
34. McKeegan DE, Demmers TG, Wathes CM, Jones RB, Gentle MJ. Response characteristics of nasal trigeminal nociceptors in *Gallus domesticus*. *NeuroReport* 2002;13:1033–1035.
35. Gentle MJ, Thorp BH. Sensory properties of ankle joint capsule mechanoreceptors in acute monoarthritic chickens. *Pain* 1994; 57:361–374.
36. Gentle MJ. Sodium urate arthritis: Effects on the sensory properties of articular afferents in the chicken. *Pain* 1997;70:245–251.
37. Gentle MJ. The acute effects of amputation on peripheral trigeminal afferents in *Gallus gallus var domesticus*. *Pain* 1991;46:97–103.
38. Gentle MJ, Hunter LN, Sterling RJ. Projections of ankle joint afferents to the spinal cord and brainstem of the chicken (*Gallus g. domesticus*). *J Comp Neurol* 1995;361:669–680.
39. Atsumi S, Sakamoto H, Kawate T, Zhai XY. Substance P-containing primary afferent receives inhibitory modulation directly from enkephalin- and GABA-containing interneurons in the dorsal horn of the chicken. *Regul Pept* 1993;46:410–412.
40. Zhai XY, Atsumi S. Large dorsal horn neurons which receive inputs from numerous substance P-like immunoreactive axon terminals in the laminae I and II of the chicken spinal cord. *Neurosci Res* 1997;28:147–154.
41. Vesselkin NP, Agayan AL, Nomokonova LM. A study of thalamo-telencephalic afferent systems in frogs. *Brain Behav Evol* 1971; 4:295–306.
42. Dunlop R, Laming P. Mechanoreceptive and nociceptive responses in the central nervous system of goldfish (*Carassius auratus*) and trout (*Oncorhynchus mykiss*). *J Pain* 2005;6:561–568.
43. Sarnat HB, Netsky MG. *Evolution of the Nervous System*, 2nd ed. New York: Oxford University Press, 1981.
44. Northcutt RG. Evolution of the vertebrate central nervous system: Patterns and processes. *Am Zool* 1984;24:701–716.
45. Inagaki S, Senba E, Shiosaka S. Regional distribution of substance P-like immunoreactivity in the frog brain and spinal cord: Immunohistochemical analysis. *J Comp Neurol* 1981;201:243–254.
46. Lorez HP, Kemali M. Substance P, met-enkephalin and somatostatin-like immunoreactivity distribution in the frog spinal cord. *Neurosci Lett* 1981;26:119–124.
47. Wouters W, Ermes-Busio MCE, Van Den Bercken J. Potentiation by naloxone of reflex activity in the isolated spinal cord of *Xenopus*. *Eur J Pharmacol* 1979;55:431–433.
48. Luthman J, Fernandez A, Radmilovich M, Trujillo-Cenoz O. Immunohistochemical studies on the spinal dorsal horn of the turtle *Chrysemys d'orbigny*. *Tissue Cell* 1991;23:515–523.
49. Simpson JI. Functional Synaptology of the Spinal Cord. In: Llinas R, Precht W, Eds. *Frog Neurobiology: A Handbook*. Berlin: Springer-Verlag, 1976:728–749.
50. Naik DR, Sar M, Stumpf WE. Immunohistochemical localization of enkephalin in the central nervous system and pituitary of the lizard, *Anolis carolinensis*. *J Comp Neurol* 1981;198:583–601.
51. Snow PJ, Renshaw GM, Hamlin KE. Localization of enkephalin immunoreactivity in the spinal cord of the long-tailed ray *Himantura fai*. *J Comp Neurol* 1996;367:264–273.
52. Vecino E, Perez MT, Ekstrom P. Localization of enkephalinergic neurons in the central nervous system of the salmon (*Salmo salar L.*) by in situ hybridization and immunocytochemistry. *J Chem Neuroanat* 1995;9:81–97.
53. Vecino E, Pinuela C, Arevalo R, Lara J, Alonso JR, Aijon J. Distribution of enkephalin-like immunoreactivity in the central nervous system of the rainbow trout: An immunocytochemical study. *J Anat* 1992;180:435–453.
54. Cameron AA, Plenderleith MB, Snow PJ. Organization of the spinal cord in four species of elasmobranch fish: Cytoarchitecture and distribution of serotonin and selected neuropeptides. *J Comp Neurol* 1990;297:201–218.
55. Kawate T, Sakamoto H, Yang C, Li Y, Shimada O, Atsumi S. Immunohistochemical study of delta and mu opioid receptors on synaptic glomeruli with substance P-positive central terminals in chicken dorsal horn. *Neurosci Res* 2005;53:279–287.
56. Hamada S, Ogawa M, Okado N. Immunohistochemical examination of intraspinal serotonin neurons and fibers in the chicken lumbar spinal cord and coexistence with leu-enkephalin. *Cell Tissue Res* 1995;282:387–397.
57. Du F, Dubois PM. Development and distribution of enkephalin-immunoreactive elements in the chicken spinal cord. *Neuroscience* 1988;27:251–266.
58. Pezalla PD, Dicig M. Stress-induced analgesia in frogs: Evidence for the involvement of an opioid system. *Brain Res* 1984; 296:356–360.
59. Stevens CW, Sangha S, Ogg BG. Analgesia produced by immobilization stress and an enkephalinase-inhibitor in amphibians. *Pharmacol Biochem Behav* 1995;51:675–680.
60. Lecaude S, Alrubaian J, Sollars C, Propper C, Danielson P, Dores RM. Organization of proenkephalin in amphibians: Cloning of a proenkephalin cDNA from the brain of the anuran amphibian, *Spea* µ*ltiplicatus*. *Peptides* 2000;21:339–344.
61. Danielson PB, Hoversten MT, Fitzpatrick M, Schreck C, Akil H, Dores RM. Sturgeon orphanin, a molecular "fossil" that bridges the

gap between the opioids and orphanin FQ/nociceptin. *J Biol Chem* 2001;276:22114–22119.

62. Dores RM, Lecaude S, Bauer D, Danielson PB. Analyzing the evolution of the opioid/orphanin gene family. *Mass Spectrom Rev* 2002;21(4):220–243.
63. Dores RM, Lecaude S. Trends in the evolution of the proopiomelanocortin gene. *Gen Comp Endocrinol* 2005;142:81–93.
64. Alrubaian J, Lecaude S, Barba J, Szynskie L, Jacobs N, Bauer D, Brown C, Kaminer I, Bagrosky B, Dores RM. Trends in the evolution of the preprodynorphin gene in teleosts: Cloning of eel and tilapia prodynorphin cDNAs. *Peptides* 2006;27:797–804.
65. Fredriksson R, Lagerstrom MC, Lundin LG, Schioth HB. The G-protein-coupled receptors in the human genome form five main families: Phylogenetic analysis, paralogon groups, and fingerprints. *Mol Pharmacol* 2003;63:1256–1272.
66. Mogil JS, Pasternak G. The molecular and behavioral pharmacology of the orphanin FQ/nociceptin peptide and receptor family. *Pharmacol Rev* 2001;53:381–415.
67. Gonzalez-Nunez V, Barrallo A, Traynor JR, Rodriguez RE. Characterization of opioid binding sites in zebrafish brain. *J Pharmacol Exp Ther* 2006;316:900–904.
68. Brooks AI, Standifer KM, Cheng J, Ciszewska GR, Pasternak GW. Opioid binding in giant toad and goldfish brain. *Receptor* 1994;4:55–62.
69. Bird DJ, Jackson M, Baker BI, Buckingham JC. Opioid binding sites in the fish brain: An autoradiographic study. *Gen Comp Endocrinol* 1988;70:49–62.
70. Buatti MC, Pasternak GW. Multiple opiate receptors: Phylogenetic differences. *Brain Res* 1981;218:400–405.
71. Simon EJ, Hiller JM, Groth J, Itzhak Y, Holland MJ, Beck SG. The nature of opiate receptors in toad brain. *Life Sci* 1982;31:1367–1370.
72. Benyhe S, Varga E, Hepp J, Magyar A, Borsodi A, Wollemann M. Characterization of kappa1 and kappa2 opioid binding sites in frog (*Rana esculenta*) brain membrane. *Neurochem Res* 1990;15:899–904.
73. Simon J, Szucs M, Benyhe S, Borsodi A, Zemlan FP, Wollemann M. Solubilization and characterization of opioid binding sites from frog (*Rana esculenta*) brain. *J Neurochem* 1984;43:957–963.
74. Newman LC, Wallace DR, Stevens CW. Selective opioid agonist and antagonist displacement of [^{3}H]-naloxone binding in amphibian brain. *Eur J Pharmacol* 2000;397:255–262.
75. Newman LC, Wallace DR, Stevens CW. Selective opioid receptor agonist and antagonist displacement of [^{3}H]-naloxone binding in amphibian spinal cord. *Brain Res* 2000;884:184–191.
76. Newman LC, Sands SS, Wallace DR, Stevens CW. Characterization of μ, κ, and δ opioid binding in amphibian whole brain tissue homogenates. *J Pharmacol Exp Ther* 2002;301:364–370.
77. Kawashima M, Imai S, Takahashi T, Kamiyoshi M, Tanaka K. An opiate receptor in the neurohypophysis of laying hens. *Poult Sci* 1995;74:716–722.
78. Csillag A, Stewart MG, Szekely AD, Magloczky Z, Bourne RC, Steele RJ. Quantitative autoradiographic demonstration of changes in binding to delta opioid, but not mu or kappa receptors, in chick forebrain 30 minutes after passive avoidance training. *Brain Res* 1993;613:96–105.
79. Martin R, McGregor GP, Halbinger G, Falke N, Voigt KH. Methionine5-enkephalin and opiate binding sites in the neurohypophysis of the bird, *Gallus domesticus*. *Regul Pept* 1992;38:33–44.
80. Csillag A, Bourne RC, Stewart MG. Distribution of mu, delta, and kappa opioid receptor binding sites in the brain of the one-day-old domestic chick (*Gallus domesticus*): An in vitro quantitative autoradiographic study. *J Comp Neurol* 1990;302:543–551.
81. Pert CB, Aposhian D, Snyder SH. Phylogenetic distribution of opiate binding. *Brain Res* 1974;75:356–361.
82. Xia Y, Haddad GG. Major difference in the expression of δ- and μ-opioid receptors between turtle and rat brain. *J Comp Neurol* 2001;436:202–210.
83. Bakalkin GY, Pivovarov AS, Kobylyansky AG, Nesterenko PN, Yarygin KN. Lateralization of opioid receptors in turtle visual cortex. *Brain Res* 1989;480:268–276.
84. Willenbring S, Stevens CW. Thermal, mechanical and chemical peripheral sensation in amphibians: Opioid and adrenergic effects. *Life Sci* 1996;58:125–133.
85. Jansen GA, Greene NM. Morphine metabolism and morphine tolerance in goldfish. *Anesthesiology* 1970;32:231–235.
86. Ehrensing RH, Michell GF, Kastin AJ. Similar antagonism of morphine analgesia by MIF-1 and naloxone in Carassius auratus. *Pharmacol Biochem Behav* 1982;17:757–761.
87. Stevens CW, Yaksh TL. Spinal action of dermorphin, an extremely potent opioid peptide from frog skin. *Brain Res* 1986;385:300–304.
88. Chervova LS, Kamenskii AA, Malyukhina GA, Baturina EY, Deigin VI, Yaroav EP. Study of dermorphin's intranasal action mechanism in fish and mammal species. *Zh Evolyutsionnoi Biokhim Fiziol* 1992;28:45–48.
89. Chervova LS, Lapshin DN, Kamenskii AA. Pain sensitivity of trout and analgesia induced by intranasal administration of dermorphin. *Doklady Biol Sci* 1994;338:424–425.
90. Chervova LS. Pain sensitivity and behavior of fishes. *J Ichthyol* 1997;37:98–102.
91. Chervova LS, Lapshin DN. Opioid modulation of pain threshold in fish. *Doklady Biol Sci* 2000;375:703–704.
92. Sneddon LU, Braithwaite VA, Gentle MJ. Do fishes have nociceptors? Evidence for the evolution of a vertebrate sensory system. *Proc Biol Sci* 2003;270:1115–1121.
93. Sneddon LU. Evolution of nociception in vertebrates: Comparative analysis of lower vertebrates. *Brain Res Brain Res Rev* 2004;46:123–130.
94. Sneddon LU, Braithwaite VA, Gentle MJ. Novel object test: Examining nociception and fear in the rainbow trout. *J Pain* 2003;4:431–440.
95. Sneddon LU. The evidence for pain in fish: The use of morphine as an analgesic. *Appl Anim Behav Sci* 2003;83:153–162.
96. Nistri A, Pepeu G, Cammelli E, Spina L, De Bellis AM. Effects of morphine on brain and spinal acetylcholine levels and nociceptive threshold in the frog. *Brain Res* 1974;80:199–209.
97. Nistri A, Pepeu G. Increase in brain and spinal acetylcholine levels without antinociceptive actions following morphine administration in the frog. *Br J Pharmacol* 1973;47:650P-651P.
98. Pezalla PD. Morphine-induced analgesia and explosive motor behavior in an amphibian. *Brain Res* 1983;273:297–305.
99. Suckow M, Terril L, Grigdesby C, March P. Evaluation of hypothermia-induced analgesia and influence of opioid antagonists in leopard frogs (*Rana pipiens*). *Pharmacol Biochem Behav* 1999;63:39–43.
100. Kanetoh T, Sugikawa T, Sasaki I, *et al.* Identification of a novel frog RFamide and its effect on the latency of the tail-flick response of the newt. *Comp Biochem Physiol C Toxicol Pharmacol* 2003;134:259–266.
101. Mauk MD, Olson RD, LaHoste GJ, Olson GA. Tonic immobility produces hyperalgesia and antagonizes morphine analgesia. *Science* 1981;213:353–354.
102. Kanui TI, Hole K, Miaron JO. Nociception in crocodiles: Capsaicin instillation, formalin and hot plate tests. *Zool Sci* 1990;7:537–540.
103. Kanui TI, Hole K. Morphine and pethidine antinociception in the crocodile. *J Vet Pharmacol Ther* 1992;15:101–103.
104. Hawkins MG. The use of analgesics in birds, reptiles, and small exotic mammals. *J Exot Pet Med* 2006;15:177–192.
105. Schneider C. Effects of morphine-like drugs in chicks. *Nature* 1961;191:607–608.
106. Bardo MT, Hughes RA. Shock-elicited flight response in chickens as an index of morphine analgesia. *Pharmacol Biochem Behav* 1978;9:147–149.
107. Hughes RA, Sufka KJ. The ontogeny of thermal nociception in domestic fowl: Thermal stimulus intensity and isolation effects. *Dev Psychobiol* 1990;23:129–140.
108. Hughes RA, Bowes M, Sufka KJ. Morphine hyperalgesic effects on developmental changes in thermal nociception and respiration in domestic fowl (*Gallus gallus*). *Pharmacol Biochem Behav* 1992;42:535–539.

109. Sufka KJ, Hughes RA. Dose and temporal parameters of morphine-induced hyperalgesia in domestic fowl. *Physiol Behav* 1990;47:385–387.
110. Sufka KJ, Hughes RA, Giordano J. Effects of selective opiate antagonists on morphine-induced hyperalgesia in domestic fowl. *Pharmacol Biochem Behav* 1991;38:49–54.
111. Sufka KJ, Hughes RA. Time-dependent codeine hypoalgesia and hyperalgesia in domestic fowl. *Pharmacol Biochem Behav* 1992;41:349–353.
112. Hughes RA, Sufka KJ. Morphine hyperalgesic effects on the formalin test in domestic fowl (*Gallus gallus*). *Pharmacol Biochem Behav* 1991;38:247–251.
113. Sufka KJ, Hoganson DA, Hughes RA. Central monoaminergic changes induced by morphine in hypoalgesic and hyperalgesic strains of domestic fowl. *Pharmacol Biochem Behav* 1992;42:781–785.
114. Roach JT, Sufka KJ. Characterization of the chick carrageenan response. *Brain Res* 2003;994:216–225.
115. Gentle MJ, Corr SA. Endogenous analgesia in the chicken. *Neurosci Lett* 1995;201:211–214.
116. Gentle MJ. Pain-related behaviour following sodium urate arthritis is expressed in decerebrate chickens. *Physiol Behav* 1997;62:581–584.
117. Gentle MJ, Hocking PM, Bernard R, Dunn LN. Evaluation of intraarticular opioid analgesia for the relief of articular pain in the domestic fowl. *Pharmacol Biochem Behav* 1999;63:339–343.
118. Fukson OI, Berkinblit MB, Feldman AG. The spinal frog takes into account the scheme of its body during the wiping reflex. Science 1980;209:1261–1263.
119. Giszter SF, McIntyre J, Bizzi E. Kinematic strategies and sensorimotor transformations in the wiping movements of frogs. *J Neurophysiol* 1989;62:750–767.
120. Hylden JLK, Wilcox GL. Intrathecal morphine in mice: A new technique. *Eur J Pharmacol* 1980;67:313–316.
121. Stevens CW, Pezalla PD. A spinal site mediates opiate analgesia in frogs. *Life Sci* 1983;33:2097–2103.
122. Stevens CW, Pezalla PD. Naloxone blocks the analgesic action of levorphanol but not of dextrorphan in the leopard frog. *Brain Res* 1984;301:171–174.
123. Pezalla PD, Stevens CW. Behavioral effects of morphine, levorphanol, dextrorphan and naloxone in the frog *Rana pipiens*. *Pharmacol Biochem Behav* 1984;21:213–217.
124. Stevens CW, Pezalla PD, Yaksh TL. Spinal antinociceptive action of three representative opioid peptides in frogs. *Brain Res* 1987;402:201–203.
125. Stevens CW, Kirkendall K. Time course and magnitude of tolerance to the analgesic effects of systemic morphine in amphibians. *Life Sci* 1993;52:PL111–116.
126. Willenbring S, Stevens CW. Spinal μ, δ and κ opioids alter chemical, mechanical and thermal sensitivities in amphibians. *Life Sci* 1997;61:2167–2176.
127. Brenner GM, Klopp AJ, Deason LL, Stevens CW. Analgesic potency of alpha adrenergic agents after systemic administration in amphibians. *J Pharmacol Exp Ther* 1994;270:540–545.
128. Stevens CW, Brenner GM. Spinal administration of adrenergic agents produces analgesia in amphibians. *Eur J Pharmacol* 1996;316:205–210.
129. Stevens CW, Klopp AJ, Facello JA. Analgesic potency of μ and κ opioids after systemic administration in amphibians. *J Pharmacol Exp Ther* 1994;269:1086–1093.
130. Stevens CW. Relative analgesic potency of μ, δ and κ opioids after spinal administration in amphibians. *J Pharmacol Exp Ther* 1996;276:440–448.
131. Stevens CW, Rothe KS. Supraspinal administration of opioids with selectivity for μ-, δ-, and κ-opioid receptors produces analgesia in amphibians. *Eur J Pharmacol* 1997;331:15–21.
132. Stevens CW. Opioid research in amphibians: A unique perspective on mechanisms of opioid analgesia and the evolution of opioid receptors. *Rev Analgesia* 2003;7:122–136.
133. Portoghese PS, Sultana M, Takemori AE. Naltrindole, a highly selective and potent non-peptide δ opioid receptor antagonist. *Eur J Pharm* 1988;146:185–186.
134. Takemori AE, Ho BY, Naeseth JS, Portoghese PS. Nor-binaltorphimine, a highly selective κ-opioid antagonist in analgesic and receptor binding assays. *J Pharmacol Exp Ther* 1988;246:255–258.
135. Ward SJ, Portoghese PS, Takemori AE. Pharmacological characterization in vivo of the novel opiate, *beta*-funaltrexamine. *J Pharmacol Exp Ther* 1982;220:494–498.
136. Stevens CW, Newman LC. Spinal administration of selective opioid antagonists in amphibians: Evidence for an opioid unireceptor. *Life Sci* 1999;64:PL125–PL130.
137. Darlison MG, Greten FR, Harvey RJ, *et al.* Opioid receptors from a lower vertebrate (*Castomus commersoni*): Sequence, pharmacology, coupling to a G-protein-gated inward-rectifying potassium channel (GIRK1), and evolution. *Proc Natl Acad Sci USA* 1997;94:8214–8219.
138. Barrallo A, González-Sarmiento R, Alvar F, Rodríguez RE. ZFOR2, a new opioid receptor-like gene from the teleost zebrafish (*Danio rerio*). *Mol Brain Res* 2000;84:1–6.
139. Rodríguez RE, Barrallo A, Garcia-Malvar F, McFadyen IJ, González-Sarmiento R, Traynor JR. Characterization of ZFOR1, a putative δ-opioid receptor from the teleost zebrafish. *Neurosci Lett* 2000;288:207–210.
140. Alvarez FA, Rodríguez-Martin I, Gonzalez-Nunez V, de Velasco EM, Gonzalez SR, Rodríguez RE. New κ opioid receptor from zebrafish Danio rerio. *Neurosci Lett* 2006;405:94–99.
141. Altschul SF, Gish W, Miller W, Myers EW, Lipman DJ. Basic local alignment search tool. *J Mol Biol* 1990;215:403–410.
142. McFadyen IJ, Metzger TG, Paterlini MG, Ferguson DM. Exploring the unique pharmacology of a novel opioid receptor, ZFOR1, using molecular modeling and the 'message-address' concept. *Protein Eng* 2001;14:953–960.
143. Read MR. Evaluation of the use of anesthesia and analgesia in reptiles. *J Am Vet Med Assoc* 2004;224:547–552.
144. Green SL. Postoperative analgesics in South African clawed frogs (*Xenopus laevis*) after surgical harvest of oocytes. *Comp Med* 2003;53:244–247.
145. Paul-Murphy J, Ludders JW, Robertson SA, Gaynor JS, Hellyer PW, Wong PL. The need for a cross-species approach to the study of pain in animals. *J Am Vet Med Assoc* 2004;224:692–697.
146. Bradford CS, Walthers EA, Searcy BT, Moore FL. Cloning, heterologous expression and pharmacological characterization of a kappa opioid receptor from the brain of the rough-skinned newt, *Taricha granulosa*. *J Mol Endocrinol* 2005;34:809–823.
147. Bradford CS, Walthers EA, Stanley DJ, Baugh MM, Moore FL. Δ and μ opioid receptors from the brain of a urodele amphibian, the rough-skinned newt, *Taricha granulosa*: Cloning, heterologous expression, and pharmacological characterization. *Gen Comp Endocrinol* 2006;146:275–290.
148. Mohan SK, Stevens CW. Systemic and spinal administration of the *mu* opioid, remifentanil, produces antinociception in amphibians. *Eur J Pharmacol* 2006;534:89–94.
149. Stevens CW, Brasel CM, Mohan SK. Cloning and bioinformatics of amphibian *mu*, *delta*, *kappa*, and nocieptin opioid receptors expressed in brain tissue: evidence for opioid receptor divergence in mammals. *Neurosci Lett* 2007;419:189–194.

Models of Behavior V

Cardiovascular A

38 Animal Models of Vascular Development and Endothelial Cell Biology

Victoria L. Bautch

ABSTRACT

There have been great advances in the past 15 years in our understanding of how blood vessels form and function. Many of these advances result from the availability of several animal models that allow for embryological, molecular, and genetic manipulations of the developing vasculature. Here I describe four vertebrate animal models used in studies of vascular development. The frog, zebrafish, avian, and mouse embryos are compared historically and currently. Recent advances using each model are highlighted. Finally, I describe the use of animal derivatives, the chick chorioallantoic membrane and mouse embryonic stem cells, for studies of blood vessel formation.

Key Words: Vascular development, Angiogenesis, Animal models, *Xenopus* embryos, Zebrafish embryos, Avian embryos, Mouse embryos, Transgenesis, Genetic analysis.

OVERVIEW OF ANIMAL MODELS OF VESSEL DEVELOPMENT

Numerous animal models, both historically and currently, have helped us understand blood vessel formation. Important steps include the differentiation of angioblasts to endothelial cells, the initial assembly of endothelial cells to form vessels in the embryo, the expansion of the vessel network, and the remodeling of vessels to accommodate the physiological needs of the organism. These processes were first described, then perturbed using physical agents, pharmacological compounds, and genetic tools. The first known description of embryonic blood vessels was provided by Aristotle in 350 BC, when he cracked open chicken eggs and followed the stages of avian development![1]

The animal models of vascular development revolve around vertebrate organisms, since invertebrates do not have a closed circulatory system. However, it might be argued that the fruitfly *Drosophila melanogaster* has provided a model of blood vessel formation, because a tracheal system develops that opens to the environment and allows diffusion of oxygen into the organism.[2] Elegant genetic dissection of tracheal development in the fly has led to a model of tube formation that seems to hold for vertebrate vessels in its essential points. Specifically, both systems appear to develop according to a "hard-wired" genetic program that starts the process, while later physiological inputs from the organism are translated into molecular cues that modify and modulate the tube system to maximize efficient function. Moreover, it was recently shown that the formation of lumens in developing vessels proceeds via vesicle fusion within the endothelial cells, in a manner very similar to lumen formation in the fly tracheal system.[3]

This chapter will focus on four major vertebrate models of blood vessel formation: the frog (*Xenopus laevis* and *tropicalis*), the zebrafish (*Danio rerio*), the avian chick (*Gallus gallus*) and the Japanese quail (*Coturnix japonica*), and the rodent mouse (*Mus musculus*) and a stem cell model derived from mouse cells. Historically the frog and avian embryos were the tools of choice, for reasons outlined below. However, in recent years the relatively new field of zebrafish development has blossomed, and this vertebrate is particularly useful for aspects of vessel development. The mouse has also gained in popularity in recent years, with the advent of genetic tools and more sophisticated technology for analysis of mammalian embryos. Finally, we and others have utilized mouse embryonic stem cells, which can differentiate to produce endothelial cells and primitive blood vessels, as a parallel model to address questions of mammalian vascular development.

XENOPUS AS A MODEL OF VASCULAR DEVELOPMENT

The historical advantages of the frog *X. laevis* are that the embryos are large and development occurs rapidly outside the mother, allowing for manipulation and visualization. Thus fate mapping studies showed that the endothelial cells that form blood vessels arise from mesoderm in the ventral part of the embryo.[4,5] The frog was also used by Cleaver and Krieg in ground-breaking studies showing that migration of precursor cells (angioblasts) from lateral areas to the midline in response to a vascular endothelial growth factor (VEGF) signal initiates formation of the dorsal aorta, a large and important vessel for all vertebrates.[6,7] More recently, the injection of a compound that binds endothelial cells, DiI-Ac-LDL, was used to image the patent vasculature at successive stages of frog development.[8]

Despite these and other important studies, until recently the use of *Xenopus* has been limited because much of the frog vascular development is not amenable to perturbation by morpholinos, the method of choice to manipulate gene expression in *Xenopus* embryos. This is because morpholinos, which block gene expression by blocking splicing or translation, are injected early and diluted out with cell division. Thus they are ineffective for

From: *Sourcebook of Models for Biomedical Research*
(P. M. Conn, ed.), © 2008 Humana Press Inc., Totowa, NJ.

probing many aspects of vascular development in the frog that occur at later stages. However, in recent years a second species, *X. tropicalis*, has gained in popularity because it is possible to do forward genetics in this model. That is to say, the generation time is such that it is feasible to mutagenize a male, and derive offspring that carry mutations that can be screened and analyzed.[9] This is not possible in the original *X. laevis*. Thus we can look forward to genetic dissection of aspects of vascular development in the frog to complement forward genetics in zebrafish. This will be useful since the frog eventually develops lungs and breathes air, while the fish has gills and gets oxygen from the water. Moreover, it was recently shown that the frog develops a lymphatic vasculature that resembles that of mammals,[10] so the development of this parallel vascular system to move lymph fluid can be dissected using the frog model.

ZEBRAFISH AS A MODEL OF VASCULAR DEVELOPMENT

HISTORY OF ZEBRAFISH AS A MODEL ANIMAL As mentioned above, the zebrafish has only recently gained popularity as an animal model for development and vascular development. In the 1970s, George Streisinger discovered that zebrafish embryos could be physically manipulated to develop as homozygous diploids, thus allowing for easy genetic analysis.[11] This technology proved difficult—more developmental defects were introduced by the process than by the mutagen used to generate mutations! However, he and others realized that the zebrafish had a generation time that allowed for forward genetic screens, with many of the physical advantages of the frog embryo. The zebrafish was firmly established when several major investigators conducted genetic screens for developmental defects and began to analyze these mutations.[12,13]

ADVANTAGES OF ZEBRAFISH FOR VASCULAR DEVELOPMENT The zebrafish embryo presents several advantages for the study of vascular development. Similar to the frog embryo, it is fairly large, development occurs outside the mother, and the embryos are clear, allowing for easy visualization of developmental processes. Once blood flow commences, patent vessels can be easily seen through the movement of the blood cells. Additionally, it is possible to use microangiography, and several years ago Weinstein and colleagues used this technology to produce a complete atlas of vascular development in the zebrafish.[14] Even more recently, several groups have generated transgenic zebrafish that express green fluorescent protein (GFP) in the developing vasculature,[15,16] thus allowing for identification of subtle vascular defects.

Additionally, it is possible to perturb cardiac or vascular development in the zebrafish embryo without these perturbations inducing immediate lethality, because the embryo obtains oxygen from the water. Thus cardiovascular mutations and perturbations can be dissected *in vivo*, and the effects of cardiac function on vessel development can be assessed. For example, a mutation named *silent heart* fails to produce a beating heart, yet the vasculature continues to develop for some time, indicating that a heartbeat or blood flow is not essential for these stages of vascular development.[17] These advantages have led several groups to invest in the development of highly sophisticated imaging technologies that allow for live image analysis of vascular development in zebrafish, and lineage tracing of progenitor cells.[18,19]

The major advantage of the zebrafish model is that it is a vertebrate embryo with a developing vasculature that is amenable to forward genetic screens. Thus a series of mutations can be isolated, characterized, and cloned to further our understanding of novel pathways that may be involved in vascular development, as well as novel relationships among players and pathways already identified.[20] Investigators have combined tools for visualization of vascular development described above with forward genetic screens.[21] It is also possible to do chemical screens to identify compounds that perturb wild-type vascular development or rescue vascular mutations, since embryos will develop for a time in 96-well formats.[22] This aspect of zebrafish biology is especially attractive to pharmaceutical companies that have access to chemical libraries and would like to uncover new therapeutic targets or drugs.

RECENT ADVANCES USING ZEBRAFISH TO STUDY VASCULAR DEVELOPMENT Among the recent outpouring of studies using the zebrafish to study blood vessel formation, the studies that elucidated a genetic pathway for specification of arteries vs. veins are particularly compelling. It was known that both genetic cues[23] and blood flow[24] influenced the decision of vessels to become arteries or veins, but the mechanism was not known. Weinstein and colleagues used a combination of available mutants, morpholino or RNA injection, and chemical inhibitors to define a molecular pathway that determines artery fate in zebrafish. In genetic terms, *hedgehog* signaling is upstream of *vegf* signaling, which in turn is upstream of *Notch* signaling, to determine arteries.[25,26] A mutant screen uncovered an additional mutation that proved to be a lesion in the phospholipase C γ gene, and this gene is downstream of *vegf* signaling in arterial specification.[27]

AVIAN MODELS OF VASCULAR DEVELOPMENT

HISTORY OF AVIAN MODELS OF VASCULAR DEVELOPMENT As mentioned above, the chick embryo was used to visualize vascular development historically. The chick, along with its cousin the quail, has continued to be used in recent times, primarily because the embryos are accessible to manipulation *in ovo*, and further development is possible. The identification of an epitope on quail angioblasts and blood vessels via production of a specific monoclonal antibody called QH1[28] led to the first complete description of blood vessel formation in a model animal.[29] Moreover, the QH1 antibody has allowed for distinction of quail vessels (QH1 is not found on chick vessels) in transplants of quail tissue into chick hosts. These quail–chick chimeras were used to show that most mesoderm contains cells that can differentiate into endothelial cells and form vessels,[30] and that angioblasts from different mesoderm sources have different potentials. Specifically, mesoderm near endoderm is thought to produce angioblasts that can give rise to "hemogenic endothelium," endothelium that can differentiate into blood cells, while mesoderm near ectoderm does not have this capacity.[31,32] The avian models have been overshadowed in recent times because they are not amenable to genetic analysis, but as we shall see below, new technologies can overcome that deficit to some extent.

ADVANTAGES OF AVIANS AS A MODEL OF VASCULAR DEVELOPMENT The accessibility of the avian embryo has been combined with gene delivery techniques to successfully manipulate gene expression in some areas of the embryo. Some studies have used viral vectors that encode genes for overexpression studies.[33,34] However, recently embryo electroporation has been used to deliver gene expression to embryonic sites. The most popular site is the neural tube, since it is possible to deliver the DNA to the central cavity prior to electroporation.[35,36] This has so far limited the use of this technology for widespread studies of vascular development in the avian, but technically any tissue can be targeted if DNA can be provided to it. The DNA usually encodes proteins that lead to overexpression of the gene of interest, or dominant negative constructs that block function. However, recently modified vectors to produce small hairpin RNAs that lead to local loss of function have been reported, and it is likely that viral delivery of interfering RNA will become a popular method to produce loss-of-function phenotypes in the avian vasculature.[37] We and others have adapted the protocol to deliver genes to quail embryos, thus allowing for subsequent analysis of vessel pattern with the QH1 antibody (J. James and V.L. Bautch, unpublished results).

RECENT ADVANCES USING AVIAN MODELS OF VASCULAR DEVELOPMENT

How Does Blood Flow Influence Vessel Identity? Eichmann and colleagues[24] used the avian yolk sac to manipulate flow parameters in an elegant study showing that blood flow is required to maintain arterial identity once it is established. They took advantage of the accessibility of the embryo and yolk sac and ligated the right vitelline artery in the yolk sac. The arterial marker neuropilin 1 was immediately downregulated, and with time venous markers such as neuropilin 2 and Tie-2 were upregulated. Moreover, reperfusion led to upregulation of the arterial markers, showing that in the avian yolk sac blood flow is a determinant of vessel identity.

Imaging Vascular Development Little and colleagues have produced an elegant analysis of vascular development in the avian embryo using dynamic imaging techniques.[38,39] They used the avian embryo because the early stages of development are relatively flat and accessible compared to the mouse. They labeled the QH1 antibody with a fluoror, and injected it into living embryos. The labeled antibody binds the developing vessels and allows for imaging of early events. Using this live imaging technology they have described migration of endothelial cells over preexisting vessels.

How Does the Neural Tube Pattern Blood Vessels? We took advantage of the avian model to develop mouse–quail chimeras that carried grafts of mouse presomitic mesoderm, a source of angioblasts. These grafts and embryonic stem (ES) cell-derived embryoid body grafts were shown to contribute mouse endothelial cells to the vascular plexus that forms around the developing avian neural tube (perineural vascular plexus, PNVP).[40,41] Further embryo manipulations placed a mouse neural tube ectopically into the avian embryo, and showed that the neural tube was the source of a signal leading to PNVP formation.[42] Studies using explanted tissues placed in collagen showed that VEGF-A is a critical component of the signal.[42] Recent work in our laboratory has utilized electroporation technology to manipulate gene expression locally in the developing neural tube, with a goal of further defining the role of VEGF-A in vessel patterning around the neural tube (J. James and V.L. Bautch, unpublished results).

AVIAN DERIVATIVES AS MODELS OF VASCULAR DEVELOPMENT It is noteworthy that alongside the avian embryo, the developing vasculature of an extraembryonic tissue called the chorioallantoic membrane (CAM) has been used extensively to study aspects of angiogenesis. The CAM was first used to show that tumor tissue could induce new blood vessel formation,[43] and many cancer researchers have tested both pro- and antiangiogenic compounds using the CAM. Briefly, a day 3 chick embryo is opened, and a paper or bead containing a compound is placed on top of the CAM. After 3–4 days of incubation the CAM is harvested and the amount of vasculature determined by quantitative measures.

RODENT MODELS OF VASCULAR DEVELOPMENT

HISTORY OF RODENT MODELS OF VASCULAR DEVELOPMENT Mouse and rat are the primary experimental models for mammalian development because of their size, ability to reproduce, and a rich history of genetics for the mouse. However, their utility for studies of vascular development was hampered by the fact that development occurs *in utero* and is difficult to visualize. Moreover, most perturbations to the developing vasculature quickly lead to embryonic death, and historically there was not a good *in situ* marker for immature mammalian vessels. However, the finding that PECAM (platelet endothelial cell adhesion molecule) is expressed on mouse angioblasts and early vessels[44–46] and the development of a high-quality monoclonal antibody to mouse PECAM[47] have provided a powerful tool for visualization of mouse embryonic vessels. This and advances in genetic manipulation discussed below have made the mouse the premier model for studies of mammalian vascular development in recent years. Of the models discussed here, the processes and mechanisms of vascular development in the mouse most closely mimic those of human vascular development, so there is great interest in using and refining this model of vascular development.

ADVANTAGES OF THE MOUSE AS A MODEL OF VASCULAR DEVELOPMENT The primary advantage of the mouse is that it is possible to manipulate the genome by introducing mutations into specific genes.[48] This results from several technologies developed in the 1980s and 1990s. First, ES cells that can produce descendants contributing to the germline of the animal have been isolated. These ES cells can be manipulated *in vitro* prior to reintroduction to a mouse embryo. Second, it is possible to target a specific gene locus in ES cells via homologous recombination. Thus it is possible to mutate or delete one of the two alleles of a gene in ES cells, select for the rare homologous recombination ("targeting") events, and then reintroduce the cells to embryos, eventually leading to animals with the mutation in both alleles. In this way numerous genes have been deleted, and the effects on mammalian vascular development analyzed. Examples are the two high-affinity receptors for VEGF-A, flk-1 (VEGFR-2) and flt-1 (VEGFR-1), that when deleted lead to perturbed vascular development and death.[49,50] However, the phenotypes of the mutant embryos were very different; deletion of flk-1 led to reduced vessel development, while deletion of flt-1 led to vessel overgrowth. This was among the first evidence that the receptors did not function in the same way in vascular development.

In recent years it has become possible to selectively delete genes in the developing vasculature.[51,52] This is done by introducing recombination sites (loxP) into one allele of a gene in ES cells, and then generating an animal with sites in both alleles. These mice are then mated with mice that express the relevant recombinase (Cre) in only the developing blood vessels.[53–56] This modification has allowed investigators to analyze the function of genes in vessels that may have required roles in other places of the embryo earlier in development. Even more recently, an inducible form of the recombinase has been used, and in this scenario excision of the allele occurs in vascular tissues only when the inducer is present.[57,58]

Along with these techniques to manipulate gene function, the ability to analyze vascular phenotypes in the mouse embryo has grown quite sophisticated. There are now numerous markers of endothelial cells: markers that distinguish arteries from veins and markers that distinguish blood-carrying vessels from lymphatic vessels.[59] These markers are of three types: antibodies that recognize proteins or epitopes that are vessel specific (see PECAM description above), antisense RNA *in situ* hybridization probes that recognize vessel-specific mRNAs, and reporter genes that are placed into a gene locus, then are expressed when gene expression is activated. The most popular of the reporter genes is bacterial *lacZ*. The *lacZ* gene encodes the β-galactosidase protein that turns a substrate blue only in the cells that express *lacZ*, so vessels or vessel subsets are stained blue after the chemical reaction. Recently GFP has also been used as a reporter gene because it allows for live imaging.

RECENT ADVANCES USING MOUSE MODELS OF VASCULAR DEVELOPMENT

What Regulates Blood Vessel Branching? Analysis of the pattern of blood vessels in mid-gestation mouse embryos carrying mutations has implicated several molecular signaling pathways in the regulation of blood vessel branching. Among these branching regulators are pathways that regulate axon guidance in the developing nervous system, including the semaphorin (Sema) pathway and the netrin pathway. The Sema pathway is complicated, because some Semas interact with neuropilins, and the neuropilin coreceptors can complex with either vascular-expressed flk-1 or neuronal or vascular-expressed plexins. However, recently it was shown that plexin D1 is expressed primarily in blood vessels, and a knockout mutation led to vascular defects.[60] The netrin pathway was shown to be involved in branching by analysis of a mutation in a netrin receptor, UNC5B.[61] It is not clear whether this pathway promotes or decreases branching,[62,63] but it is clearly involved in the processes.

Does Vasculogenesis Occur after Birth? Until the mid-1990s the dogma in the field was that vasculogenesis, the process whereby endothelial progenitor cells form a vessel while differentiating, occurred only during the early stages of development. Subsequent to those stages, all vessel formation and expansion were thought to occur via angiogenesis—the migration, division, and fusion of endothelial cells already in vessels. However, evidence suggested that perhaps endothelial cells could be formed *de novo* in some situations, and evidence for this model was provided through the use of the mouse. Mice were irradiated so their bone marrow was destroyed, and then bone marrow from a "marked" donor was grafted into the host. Once recovered, the chimeric mouse now had a "marked" bone marrow, and it was subjected to various situations in which new vessels form. In ischemic injury recovery and tumor angiogenesis, marked cells were seen in the perivascular area, suggesting that bone marrow-derived cells could contribute to neoangiogenesis.[64] While there is controversy about the lineage and ultimate fate of the bone marrow cells, most investigators think that an endothelial progenitor (EPC) resides in the bone marrow, and it can be mobilized to sites of new vessel formation and perhaps differentiate into an endothelial cell. Thus vasculogenesis may not be confined to embryonic stages, but may occur alongside angiogenesis in the adult as well.

MOUSE DERIVATIVES AS MODELS OF VASCULAR DEVELOPMENT The major mouse derivative used for studies of vascular development is the ES cell that was discussed above in the context of targeted mutations. ES cells can be aggregated and differentiate *in vitro* to form embryoid bodies, which can then cavitate to become cystic.[65,66] This programmed differentiation leads to the formation of endoderm and mesoderm, and subsequently derivatives of these lineages.[67,68] Amazingly, endoderm-derived factors such as VEGF-A provide appropriate signals for the differentiation of endothelial precursors, vasculogenesis leads to the formation of primitive vessels, and then angiogenesis results in the expansion of a primitive vascular plexus in this model.[69,70] We and others realized that if the embryoid bodies are reattached prior to cavitation, vascular development proceeds as described above.[67,69] Thus embryoid bodies or ES cell cultures can be manipulated both genetically and pharmacologically to address questions of vascular development. Simon and colleagues have shown a requirement for hypoxia-driven gene expression in vascular development using the ES cell model.[71] We recently developed ES cell lines that express vascular-restricted GFP, and this allowed us to image the dynamic processes of vascular development.[68] Using this tool we showed that the VEGF-A receptor flt-1 (VEGFR-1) positively modulates branching morphogenesis of developing vessels, probably via modulation of the amount and/or spatial context of signaling through the flk-1 (VEGFR-2) receptor.[72]

Keller and colleagues[73,74] used the ES cell model to elucidate lineage relationships. Specifically, evidence for a bipotential common precursor of both blood and endothelial cells, called the hemangioblast, was presented. Mouse ES cells were partially differentiated as described above, then dissociated to single cells and plated in conditions conducive to differentiation of both blood cells and endothelial cells. The finding that single cells could give rise to both endothelial and blood cells was consistent with the presence of a hemangioblast among the cells. Subsequent studies *in vivo* corroborated this finding,[75] but these studies did not distinguish between a true bipotential cell and a multipotential progenitor that could also differentiate into other mesodermal lineages. Recently an elegant study using zebrafish provided evidence that a bipotential hemangioblast does exist, although many endothelial cells do not seem to go through this intermediate.[19] The latter study points out the utility of having multiple models of vascular development available for analysis. The cells were tracked by injecting a compound that was then uncaged in a single progenitor cell using a laser. This was technically challenging in the zebrafish embryo and is not currently feasible in the other models.

CONCLUSIONS

Our understanding of blood vessel formation has benefited enormously from the availability of several animals models of vascular development, and there have been great strides in our understanding of vascular development at the cell and molecular level in the past several years. I predict that in the near future much will be added to this knowledge base through the creative use of animal models of vascular development such as those described here. More to the point, even more advances will likely come from the integrated use of multiple animal models of vascular development. For example, it is possible to envision a forward genetic screen in zebrafish to uncover a novel gene important in vascular development, then manipulation of the gene through RNA and morpholinos in both the fish and frog, followed by vascular-specific inducible deletion in the mouse. In this way a comprehensive analysis of the role of a particular gene in vascular development should be elucidated. The sequencing of the relevant genomes and reagents for microarray analysis will no doubt augment future efforts to further understand the mechanisms regulating the development of blood vessels.

ACKNOWLEDGMENTS

This work was supported by grants from the NIH (R01HL43174 and R01HL83262) to V.L.B.

REFERENCES

1. Gilbert S. *Developmental Biology*. Sunderland, MA: Sinauer Associates, Inc., 2003:838.
2. Ghabrial A, Luschnig S, Metzstein MM, Krasnow MA. Branching morphogenesis of the Drosophila tracheal system. *Annu Rev Cell Dev Biol* 2003;19:623–647.
3. Kamei M, Brian Saunders W, Bayless KJ, Dye L, Davis GE, Weinstein BM. Endothelial tubes assemble from intracellular vacuoles in vivo. *Nature* 2006;442:453–456.
4. Rovainen C. Labeling of developing vascular endothelium after injections of rhodamine-dextran into blastomeres of Xenopus laevis. *J Exp Zool* 1991;259:209–221.
5. Mills KR, Kruep D, Saha MS. Elucidating the origins of the vascular system: A fate map of the vascular endothelial and red blood cell lineages in Xenopus laevis. *Dev Biol* 1999;209:352–368.
6. Cleaver O, Tonissen KF, Saha MS, Krieg PA. Neovascularization of the Xenopus embryo. *Dev Dyn* 1997;210:66–77.
7. Cleaver O, Krieg PA. Molecular mechanisms of vascular development. In: Harvey RP, Rosenthal N, Eds. *Heart Development*. New York: Academic Press, 1999:221–252.
8. Levine AJ, Munoz-Sanjuan I, Bell E, North AJ, Brivanlou AH. Fluorescent labeling of endothelial cells allows in vivo, continuous characterization of the vascular development of Xenopus laevis. *Dev Biol* 2003;254:50–67.
9. Carruthers S, Stemple DL. Genetic and genomic prospects for Xenopus tropicalis research. *Sem Cell Dev Biol* 2006;17:146–153.
10. Ny A, Koch M, Schneider M, Neven E, Tong RT, Maity S, Fischer C, Plaisance S, Lambrechts D, Heligon C, Terclavers S, Ciesiolka M, Kalin R, Man WY, Senn I, Wyns S, Lupu F, Brandli A, Vleminckx K, Collen D, Dewerchin M, Conway EM, Moons L, Jain RK, Carmeliet P. A genetic Xenopus laevis tadpole model to study lymphangiogenesis. *Nat Med* 2005;11:998–1004.
11. Streisinger G, Walker C, Dower N, Knauber D, Singer F. Production of clones of homozygous diploid zebra fish (Brachydanio rerio). *Nature* 1981;291:293–296.
12. Mullins MC, Hammerschmidt M, Haffter P, Nusslein-Volhard C. Large-scale mutagenesis in the zebrafish: In search of genes controlling development in a vertebrate. *Curr Biol* 1994;4:189–202.
13. Amsterdam A, Burgess S, Golling G, Chen W, Sun Z, Townsend K, Farrington S, Haldi M, Hopkins N. A large-scale insertional mutagenesis screen in zebrafish. *Genes Dev* 1999;13:2713–2724.
14. Isogai S, Horiguchi M, Weinstein BM. The vascular anatomy of the developing zebrafish: An atlas of embryonic and early larval development. *Dev Biol* 2001;230:278–301.
15. Lawson ND, Weinstein BM. In vivo imaging of embryonic vascular development using transgenic zebrafish. *Dev Biol* 2002;248:307–318.
16. Jin S-W, Beis D, Mitchell T, Chen J-N, Stainier DYR. Cellular and molecular analyses of vascular tube and lumen formation in zebrafish. *Development* 2005;132:5199–5209.
17. Isogai S, Lawson ND, Torrealday S, Horiguchi M, Weinstein BM. Angiogenic network formation in the developing vertebrate trunk. *Development* 2003;130:5281–5290.
18. Weinstein BM. Vascular cell biology in vivo: A new piscine paradigm? *Trends Cell Biol* 2002;12:439–445.
19. Vogeli KM, Jin S-W, Martin GR, Stainier DYR. A common progenitor for haematopoietic and endothelial lineages in the zebrafish gastrula. *Nature* 2006;443:337–339.
20. Stainier DYR, Fouquet B, Chen J, Warren K, Weinstein BM, Meiler S, Mohi M, Neuhauss S, Solnica-Krezel L, Schier A, Zwartkruis F, Stemple DL, Driever W, Fishman MC. Mutations affecting the formation and function of the cardiovascular system in the zebrafish embryo. *Development* 1996;123:285–292.
21. Weinstein BM. Plumbing the mysteries of vascular development using the zebrafish. *Semin Cell Dev Biol* 2002;13:515–522.
22. Peterson RT, Shaw SY, Peterson TA, Milan DJ, Zhong TP, Schreiber SL, MacRae CA, Fishman MC. Chemical suppression of a genetic mutation in a zebrafish model of aortic coarctation. *Nat Biotechnol* 2004;22:595–599.
23. Mukouyama YS, Shin D, Britsch S, Taniguchi M, Anderson DJ. Sensory nerves determine the pattern of arterial differentiation and blood vessel branching in the skin. *Cell* 2002;109:693–705.
24. le Noble F, Moyon D, Pardanaud L, Yuan L, Djonov V, Matthijsen R, Breant C, Fleury V, Eichmann A. Flow regulates arterial-venous differentiation in the chick embryo yolk sac. *Development* 2004;131:361–375.
25. Lawson ND, Scheer N, Pham VN, Kim CH, Chitnis AB, Campos-Ortega JA, Weinstein BM. Notch signaling is required for arterial-venous differentiation during embryonic vascular development. *Development* 2001;128:3675–3683.
26. Lawson ND, Vogel AM, Weinstein BM. Sonic hedgehog and vascular endothelial growth factor act upstream of the notch pathway during arterial endothelial differentiation. *Dev Cell* 2002;3:127–136.
27. Lawson ND, Mugford JW, Diamond BA, Weinstein BM. Phospholipase C gamma-1 is required downstream of vascular endothelial growth factor during arterial development. *Genes Dev* 2003;17:1346–1351.
28. Pardanaud L, Altmann C, Kitos P, Dieterlen-Lievre F, Buck CA. Vasculogenesis in the early quail blastodisc as studied with a monoclonal antibody recognizing endothelial cells. *Development* 1987;100:339–349.
29. Coffin JD, Poole TJ. Embryonic vascular development: Immunohistochemical identification of the origin and subsequent morphogenesis of the major vessel primordia in quail embryos. *Development* 1988;102:735–748.
30. Noden DM. Embryonic origins and assembly of blood vessels. *Am Rev Respir Dis* 1989;140:1097–1103.
31. Pardanaud L, Yassine F, Dieterlen-Lievre F. Relationship between vasculogenesis, angiogenesis and haemopoiesis during avian ontogeny. *Development* 1989;105:473–485.
32. Pardanaud L, Luton D, Prigent M, Bourcheix LM, Catala M, Dieterlen-Lievre F. Two distinct endothelial lineages in ontogeny, one of them related to hemopoiesis. *Development* 1996;122:1363–1371.
33. Iba H. Gene transfer into chicken embryos by retrovirus vectors. *Dev Growth Differ* 2000;42:213–218.
34. Ishii Y, Reese D, Mikawa T. Somatic transgenesis using retroviral vectors in the chicken embryo. *Dev Dyn* 2004;229:630–642.
35. Ogura T. In vivo electroporation: A new frontier for gene delivery and embryology. *Differentiation* 2002;70:163–171.
36. Swartz M, Eberhart J, Mastick GS, Krull CE. Sparking new frontiers: Using in vivo electroporation for genetic manipulations. *Dev Biol* 2001;233:13–21.
37. Bomberg-White J, Webb C, Patacsil V, Miranti C, Williams B, Holmen S. Delivery of short hairpin RNA sequences by using a replication-competent avian retroviral vector. *J Virol* 2004;78:4914–4916.
38. Rupp PA, Czirok A, Little CD. Novel approaches for the study of vascular assembly and morphogenesis in avian embryos. *Trends Cardiovasc Med* 2003;13:283–288.

39. Rupp PA, Czirok A, Little CD. $\alpha v\beta 3$ integrin-dependent endothelial cell dynamics in vivo. *Development* 2004;131:2887–2897.
40. Ambler CA, Nowicki JL, Burke AC, Bautch VL. Assembly of trunk and limb blood vessels involves extensive migration and vasculogenesis of somite-derived angioblasts. *Dev Biol* 2001;234:352–364.
41. Ambler CA, Schmunk GA, Bautch VL. Stem cell-derived endothelial cells/progenitors migrate and pattern in the embryo using the VEGF signaling pathway. *Dev Biol* 2003;257:205–219.
42. Hogan KA, Ambler CA, Chapman DL, Bautch VL. The neural tube patterns vessels developmentally using the VEGF signaling pathway. *Development* 2004;131:1503–1513.
43. Norrby K. In vivo models of angiogenesis. *J Cell Mol Med* 2006;10:588–612.
44. Baldwin H, Shen H, Yan H, DeLisser H, Chung A, Mickanin C, Trask T, Kirschbaum N, Newman P, Albelda S. Platelet endothelial cell adhesion molecule-1 (PECAM-1/CD31): Alternatively spliced, functionally distinct isoforms expressed during mammalian cardiovascular development. *Development* 1994;120:2539–2553.
45. DeLisser H, Christofidou-Solomidou M, Strieter R, Burdick M, Robinsion C, Wexler R, Kerr J, Garlanda C, Merwin J, Madri J, Albelda S. Involvement of endothelial PECAM-1/CD31 in angiogenesis. *Am J Pathol* 1997;151:671–677.
46. Redick SD, Bautch VL. Developmental platelet endothelial cell adhesion molecule expression suggests multiple roles for a vascular adhesion molecule. *Am J Pathol* 1999;154:1137–1147.
47. Nourshargh S, Krombach F, Dejana E. The role of JAM-A and PECAM-1 in modulating leukocyte infiltration in inflamed and ischemic tissues. *J Leukoc Biol* 2006;80:714–718.
48. Downing GJ, Battey JF Jr. Technical assessment of the first 20 years of research using mouse embryonic stem cell lines. *Stem Cells* 2004;22:1168–1180.
49. Shalaby F, Ho J, Stanford WL, Fischer KD, Schuh AC, Schwartz L, Bernstein A, Rossant J. A requirement for Flk1 in primitive and definitive hematopoiesis and vasculogenesis. *Cell* 1997;89:981–990.
50. Fong GH, Rossant J, Gertsenstein M, Breitman M. Role of the Flt-1 receptor tyrosine kinase in regulating the assembly of vascular endothelium. *Nature* 1995;376:66–70.
51. Metzger D, Chambon P. Site- and time-specific gene targeting in the mouse. *Methods* 2001;24:71–80.
52. Sauer B. Inducible gene targeting in mice using the cre/lox system. *Methods* 1998;14:381–392.
53. Gustafsson E, Brakebusch C, Hietanen K, Fassler R. Tie-1-directed expression of Cre recombinase in endothelial cells of embryoid bodies and transgenic mice. *J Cell Sci* 2001;114:671–676.
54. Kisanuki YY, Hammer RE, Miyazaki J-i, Williams SC, Richardson JA, Yanagisawa M. Tie2-Cre transgenic mice: A new model for endothelial cell-lineage analysis in vivo. *Dev Biol* 2001;230: 230–242.
55. Motoike T, Markham D, Rossant J, Sato T. Evidence for novel fate of Flk1 progenitor: Contribution to muscle lineage. *Genesis* 2003;35:153–159.
56. Alva JA, Zovein A, Monvoisin A, Murphy T, Salazar A, Harvey N, Carmeliet P, Iruela-Arispe M. VE-Cadherin-Cre-recombinase transgenic mouse: A tool for lineage analysis and gene deletion in endothelial cells. *Dev Dyn* 2006;235:759–767.
57. Forde A, Constien R, Grone H, Hammerling G, Arnold B. Temporal Cre-mediated recombination exclusively in endothelial cells using Tie2 regulatory elements. *Genesis* 2002;33:191–197.
58. Gothert JR, Gustin SE, Hall MA, Green AR, Gottgens B, Izon DJ, Begley CG. In vivo fate-tracing studies using the Scl stem cell enhancer: Embryonic hematopoietic stem cells significantly contribute to adult hematopoiesis. *Blood* 2005;105:2724–2732.
59. Eichmann A, Yuan L, Moyon D, le Noble F, Pardanaud L, Breant C. Vascular development: From precursor cells to branched arterial and venous networks. *Int J Dev Biol* 2005;49:259–267.
60. Gitler AD, Lu MM, Epstein JA. PlexinD1 and semaphorin signaling are required in endothelial cells for cardiovascular development. *Dev Cell* 2004;7:107–116.
61. Lu X, le Noble F, Yuan L, Jiang Q, de Lafarge B, Sugiyama D, Breant C, Claes F, De Smet F, Thomas J-L, Autiero M, Carmeliet P, Tessier-Lavigne M, Eichmann A. The netrin receptor UNC5B mediates guidance events controlling morphogenesis of the vascular system. *Nature* 2004;432:179–186.
62. Park KW, Crouse D, Lee M, Karnik SK, Sorensen LK, Murphy KJ, Kuo CJ, Li DY. The axonal attractant Netrin-1 is an angiogenic factor. *Proc Natl Acad Sci USA* 2004;101:16210–16215.
63. Wilson BD, Ii M, Park KW, Suli A, Sorensen LK, Larrieu-Lahargue F, Urness LD, Suh W, Asai J, Kock GAH, Thorne T, Silver M, Thomas KR, Chien C-B, Losordo DW, Li DY. Netrins promote developmental and therapeutic angiogenesis. *Science* 2006;313: 640–644.
64. Barber CL, Iruela-Arispe ML. The ever-elusive endothelial progenitor cell: Identities, functions and clinical implications. *Pediatr Res* 2006;59:26R-32R.
65. Doetschman TC, Eistetter H, Katz M, Schmidt W, Kemler R. The *in vitro* development of blastocyst-derived embryonic stem cell lines: Formation of visceral yolk sac, blood islands and myocardium. *J Embryol Exp Morphol* 1985;87:27–45.
66. Risau W, Sariola H, Zerwes HG, Sasse J, Ekblom P, Kemler R, Doetschman T. Vasculogenesis and angiogenesis in embryonic-stem-cell-derived embryoid bodies. *Development* 1988;102:471–478.
67. Bautch VL. Embryonic stem cell differentiation and the vascular lineage. In: Turksen K, Ed. *Methods in Molecular Biology*, Vol. 185, *Embryonic Stem Cells: Methods and Protocols*. Totowa, NJ: Humana Press, 2001:117–125.
68. Kearney JB, Bautch VL. In vitro differentiation of mouse ES cells: Hematopoietic and vascular development. *Methods Enzymol* 2003;365:83–98.
69. Bautch VL, Stanford WL, Rapoport R, Russell S, Byrum RS, Futch TA. Blood island formation in attached cultures of murine embryonic stem cells. *Dev Dyn* 1996;205:1–12.
70. Bautch VL, Redick SD, Scalia A, Harmaty M, Carmeliet P, Rapoport R. Characterization of the vasculogenic block in the absence of vascular endothelial growth factor-A. *Blood* 2000;95:1979–1987.
71. Ramirez-Bergeron DL, Runge A, Dahl KDC, Fehling HJ, Keller G, Simon MC. Hypoxia affects mesoderm and enhances hemangioblast specification during early development. *Development* 2004;131: 4623–4634.
72. Kearney JB, Kappas NC, Ellerstrom C, DiPaola FW, Bautch VL. The VEGF receptor flt-1 (VEGFR-1) is a positive modulator of vascular sprout formation and branching morphogenesis. *Blood* 2004;103: 4527–4535.
73. Choi K, Kennedy M, Kazarov A, Papadimitriou J, Keller G. A common precursor for hematopoietic and endothelial cells. *Development* 1998;125:725–732.
74. Park C, Ma YD, Choi K. Evidence for the hemangioblast. *Exp Hematol* 2005;33:965–970.
75. Huber TL, Kouskoff V, Joerg Fehling H, Palis J, Keller G. Haemangioblast commitment is initiated in the primitive streak of the mouse embryo. *Nature* 2004;432:625–630.

39 Models of Behavior

Cardiovascular

LINA BADIMON, LAURA CASANÍ, AND GEMMA VILAHUR

ABSTRACT

Experimental models of thrombosis are important research resources that have helped to increase our understanding and knowledge of the pathogenesis of the atherothrombotic process as well as to develop novel therapeutic approaches. In fact, different animal models and procedures have been implemented in thrombosis research in order to provide new insights into the mechanisms that have already been outlined in isolated cells and protein studies. Moreover, animal preclinical studies have rendered a solid rationale for conducting prospective randomized trials in patients with coronary artery disease.

INTRODUCTION

This chapter reviews the more common animal models used in the study of thrombotic disease, focusing primarily on the management, particularities, and topics related to the porcine animal model because of its human resemblance, particularly in the cardiovascular system.[1–7]

THE USE OF ANIMAL MODELS TO STUDY ARTERIAL THROMBOSIS

Autopsy and angiographic studies have pointed out the importance of plaque rupture with subsequent exposure of atherosclerotic components as a key step of thrombus formation and the subsequent clinical events. However, "the paradigm of plaque rupture" presents several contradictions that must be considered. As such, autopsies have shown that a substantial minority of victims of sudden coronary death shows superficial plaque erosion without plaque rupture,[8,9] and although plaque disruption accounts for up to two-thirds of patients in whom unstable angina or acute myocardial infarction develops, in most cases plaque disruptions are asymptomatic, although they may contribute to the growth of the atherosclerotic plaque.[10,11] Despite this, there is no doubt that a defect in the surface of the atherosclerotic plaque is almost always associated with mural thrombosis, which, at least, leads to plaque progression.

From: *Sourcebook of Models for Biomedical Research*
(P. M. Conn, ed.), © 2008 Humana Press Inc., Totowa, NJ.

The study of thrombosis and its inhibition is based on *in vitro*, *ex vivo*, and *in vivo* approaches that combined have helped to widen the knowledge of the thrombotic process itself. Indeed, the availability of transgenic and knockout animals makes it possible to pinpoint the relative functional importance of single changes in specific gene products and hence facilitates the formulation of new strategies for cardiovascular protection and the prevention and treatment of thrombosis. However, on the other hand, domestic swines have been widely used as biomedical research models for human diseases due to their similarities in cardiovascular anatomy and physiology as well their ability to develop spontaneous or diet-induced atherosclerosis. Nevertheless, the objectives of a study should mandate the type of animal model to use in order to avoid repetitive and unneeded experiments. For instance, although swine are up to 90% similar to humans, there are some molecular differences in the mechanisms involved in arterial thrombosis that must be taken into account in order to properly evaluate the data.

ANIMAL SIZE MATTERS?

Some of the reasons for the frequent use of small animal models in research include (1) low cost, (2) ready availability, (3) reduced ethical concern compared to large animals, especially primates, and (4) small size that limits the quantities of new agents required for *in vivo* screening. These characteristics have permitted rapid evaluation of new agents in sufficiently large populations to perform meaningful statistical analyses. In addition to these practical indications for their use, small animal models have the added advantage of well-defined genetic characterization, and, in the case of mice, the availability of transgenic and gene knockout animals. Furthermore, small animals have figured prominently in the study of the contributions of thrombosis to luminal narrowing following arterial injury.[12] In contrast, large animal models such as pigs have the advantage of being more comparable to humans, especially because of their susceptibility to develop atherosclerosis.[1,7,8,13] Furthermore, their large size has permitted the evaluation of their coronary arteries, rather than the aortas studied in small animals because of its relatively larger size. In addition, the ability to induce atheromatous disease, which simulates the lesions observed in humans, is a strong attribute of the pig as a large animal model. Conversely, the disadvantages of large animal models are primarily the reverse of the advantages of small animal models. These include greatly increased cost, heightened ethical concerns, less precise genetic characterization, and a scarcity of transgenic models and antibodies.

THE PIG AS A COMMONLY ACCEPTED RESEARCH ANIMAL MODEL

Previously published articles mainly present three types of atherothrombotic-related experiments, sometimes combined, performed in pigs: (1) feeding animals with fat-rich diets, (2) intravascular interventions to induce damage (balloon, grafting, etc.), and (3) extracorporeal arteriovenous shunts into which flow chambers of varying patterns are coupled. However, before thoroughly reviewing these topics several pig-related features deserve to be acknowledged to better understand the swine model of atherothrombosis.

SWINES NUTRITIVE REQUIREMENTS The majority of swines used in research weight between 10 and 80 kg and require approximately 600–2800 g of food per day.[14] Experimental animals can be fed with the commercial chow used for feeding farm animals, although it has been specially formulated in order to obtain a rapid growth rate. However, if there is a research requirement to use animals between 9 and 18 kg (40–60 days old) they should be fed with a nutrient-enriched diet (i.e., postweaning chow). Diet transition should be performed by mixing increasing parts of growing chow with the postweaning chow in order to reduce the presence of food-related diarrheas. For long time period studies, it is desirable to feed the animal with a restricted diet (3.5% of their total weight) to avoid handling difficulties related to animal overweight. This food restriction (also used in farm animals) covers all the animal nutritive and metabolic requirements as published by the National Research Council (NRC).[15] Commercial diets for farm pigs usually contain antibiotics, promoters, and other supplements to increase weight gain. If these components can somehow interfere with the study objectives, free-additive diets can be especially formulated and obtained from local venders. Finally, oral drug administration in swines is easy to perform, especially after overnight fasting. The suggestion is to mix or dissolve the drug within a small amount of chow before offering them the calculated ration. Swine approximately require 2.5 liters of water per kg of consumed food per day. Water should be provided by an automated watering system to avoid water restriction-related neurological syndromes such as "salt poisoning" or "sodium toxicosis."[16,17]

HEART ANATOMY AND VASCULARIZATION The heart of the pig is anatomically similar to humans except for the presence of the left azygous (hemiazygous) vein, which drains the intercostal system into the coronary sinus,[18] and for the size, which tends to be a bit smaller. The heart blood supply is mostly right side dominant since it originates from the posterior septal artery[19] and both anatomy and function of the pig coronary system as well as the histological anatomy of the aorta are comparable to humans. However, on the other hand, pig blood vessels are more friable and prone to vasospasm during manipulation, and thus require careful handling during blood withdrawals. Although animal models generally have the advantage of a shorter time to lesion development, pig atherosclerosis is quite slow and occurs both spontaneously and by experimental induction (i.e., intake of a high atherogenic diet). Moreover, if allowed to develop over time, mild atherosclerotic lesions first appear in coronary arteries and both atherosclerotic plaque distribution and composition (lipid, fibrinogen, smooth muscle cells, and macrophage content)[20] follow a pattern comparable to that of humans.[21–23] Human-like pig lipoprotein metabolisms may help to explain in part the above mentioned similarities. In relation to pig hemodynamics, either physiological cardiac function or mechanically induced myocardial infarction and the subsequent arrhythmogenic activity in reperfusion are also analogous to humans as well as the wound-healing process.[24] Actually, although wound healing in the myocardium has typically used both swine and sheep models, ruminants' healing somehow differs from humans since it is mainly characterized by the formation of collagenous scars.[25] Finally, unlike other animal models, after gradual occlusion of the coronary vessels induced by both balloon injury and an atherogenic diet, swines may develop coronary restenosis syndrome.[19,21,22,24,26–28]

HUSBANDRY AND MANAGEMENT Recommendations for housing laboratory swine are regulated by the European Directive 2003/65/CE, by local governmental animal care laws, and in the United States by the National Institutse of Health, *Guide for the Care and Use of Laboratory Animals*.[29] Plastic slats (PVC, vinyl) are preferred in biomedical research facilities because they are easy to clean and better isolate animals from humidity. If possible, heating plates should be installed in all pens (Figure 39–1). Swine reared for meat received standard treatments for endoparasites and ectoparasites and vaccinations against common swine diseases at appropriate ages and intervals. However, according to our previous experience, animals that are going to be part of a long-term project should undergo preventive treatment with antibiotics and antiparasites for 3 days as soon as they arrive at the animal facilities to avoid the presence of infectious diseases that may modify physiological pig parameters.

Restraint methods should be discouraged in biomedical research because of the stress they induce.[30] Subsequently, short-term chemical restraint agents and anesthetics may be used in order to obtain blood samples.

ANESTHESIA, ANALGESIA, AND SURGERY Animal pain and suffering are clinically important conditions that adversely affect the quality of life of the animals and as a result may influence the experimental data. Drugs, techniques, or husbandry methods used to prevent and control pain must be tailored to individual animals and should be based, in part, on the species, breed, age, procedure performed, degree of tissue trauma, individual behavioral characteristics, degree of pain, and health status. Preventive analgesia before major surgery must be performed with agents that do not interfere with platelets or the coagulation system. Accordingly, the use of opioids, particularly buprenorphine and fentanyl, is highly recommended. In our group, we use different agents to induce swine anesthesia depending on the amount of sample required and on the nature of the objectives we want to accomplish. For instance, tranquilization with 2.2 mg/kg of azaperone is more than enough to obtain a small blood sample (less than 10 ml) from the auricular vein or to cannulate a superficial vein (auricular or mammary vein). Once tranquilized, animals can be induced to anesthesia with an intravenous bolus of pentobarbital (5 mg/kg) to either obtain larger

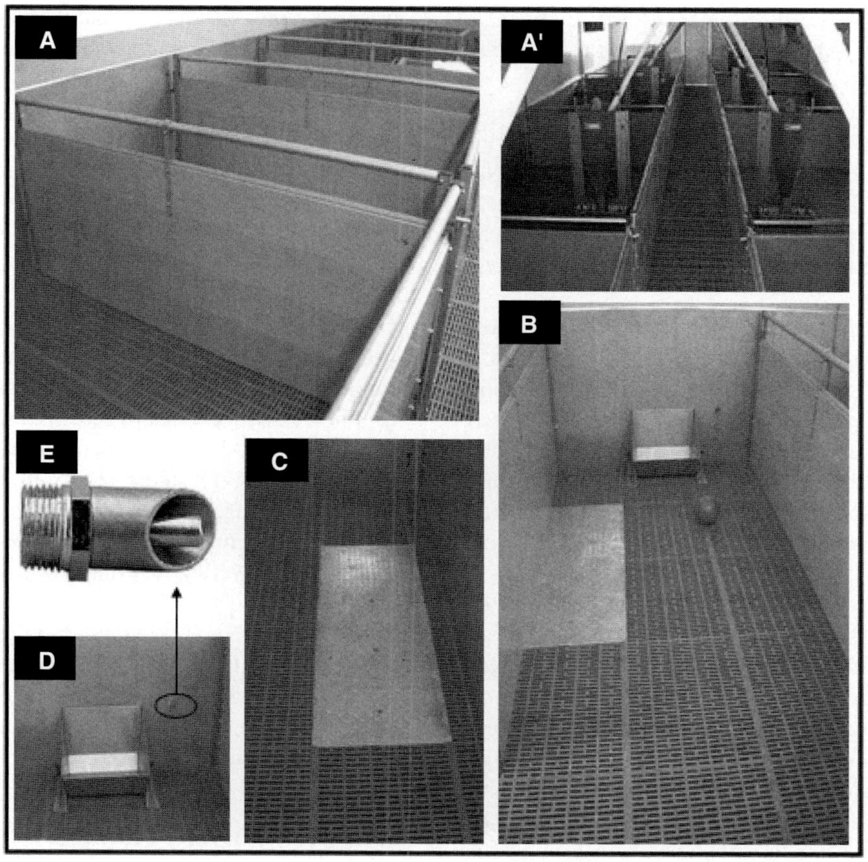

Figure 39–1. Swine PVC pens in a biomedical research facility. (A, A′) General view of pens. The walls in contact with the animals are covered with stainless steel. These are 1-m height plastic separations between pens and slip-free fiberglass slat flooring for greater hygiene. (A) The size of the pens allows a resting and a feeding area; toys can be as simple as balls. (C–E) Details of integrated aluminum heating plates, stainless-steel feeders (easily fixed to the floor with a mobile screw), and nipple drinkers placed at a standard height of 30 cm.

amounts of blood via the femoral vein/artery or to perform small surgical procedures. An intramuscular combination of tiletamine/zolazepam and medetomidine induces a fast anesthetic status sufficient to perform small surgery such as central vessel catheterization.

To achieve general deep anesthesia a continuous infusion of pentobarbital-saline (infusion rate of 2 mg/min) should be used, rather than repeated pentobarbital bolus injections. Deep anesthesia can also be induced by pentobarbital and maintained with volatile anesthesia (i.e., isofluorane). Within this context, it is important to emphasize that higher doses of pentobarbital (>30 mg/kg) have previously been described to reduce platelet reactivity[31] and an overdose can cause fatal respiratory and cardiac depression.

THE PIG AS A TOOL FOR EVALUATING ATHEROTHROMBOTIC DISEASE

DIET-INDUCED ATHEROSCLEROSIS Swine can develop hypercholesterolemia and atherosclerotic lesions by diet induction (high cholesterol content diets), reaching plasma cholesterol levels similar to those in humans. Indeed, after a 50-day period with a standard hypercholesterolemic diet, we have previously observed that almost all pigs developed early atherosclerotic lesions (fatty streaks) localized in the abdominal aorta and to a lesser extent in the coronary arteries. In such cases, lesion composition was similar to early stage human atherosclerosis.[20] As expected, increasing the diet induction period to 100 and 150 days was associated with a higher degree of lesion severity made evident when suitable fat stain was applied (Sudan IV). Notwithstanding, animals employed in these long period studies tend to be within the prepuberty period and as a consequence their lesions are not as severe as those developed in adult humans over the years. Older animals (over 6 months) can be used to obtain atherosclerotic models of greater human resemblance,[32] but management will be even more difficult because of their considerable weight (up to 200 kg). To overcome size-related problems, minipigs are becoming increasingly used. Indeed, miniature swine are preferable to commercial swine as animal models because of their small size and small growth rate, which allow them to maintain weight and size throughout adulthood. They are docile and easily handled, and there are several miniature strains available to better fit the scientific purpose. Miniature pigs have been used in several fields of biomedical research studies such as cardiovascular disease (models of thrombosis and restenosis), diabetes, transplantation, and dermal applications.

Miniature pigs can be allocated in the same pens as larger pigs (increasing the number of animals per pen) and feeders and drinkers can be adapted to their size. However, the principal disadvantages of using miniature pigs as a substitute for common pigs are the maintenance requirements and the high cost of the animals. As a consequence, miniature pig breeders are not extensively distributed in all countries.

***EX VIVO* PORCINE MODEL OF THROMBOSIS** The *ex vivo* porcine model of thrombus formation (e.g., flow chambers coupled to extracorporeal shunts) has become essential for testing the effect of blood elements and rheology as well as atherosclerotic vessel components in thrombus formation in a controlled manner.[33–40] At the same time it is reproducible and easy to manipulate and thus allows the evaluation of the thrombogenic effect of different plaque constituents (e.g., collagen, fatty streaks, smooth muscle cells),[33–35] the different degrees of shear stress (to mimic different degrees of stenosis),[36,37] and the antiplatelet effects of new antithrombotic compounds.[20,35,37,40–43]

Shear stress measures the difference in blood velocity between the center of the vessel and along the vessel wall. Atherosclerosis preferentially occurs in areas of turbulent blood flow and low fluid shear stress, whereas laminar blood flow and high shear stress are atheroprotective. The Badimon perfusion chamber allows control of the blood flow as well as the degree of stenosis and thus evaluates moderate to high-risk thrombotic situations. A positive correlation between vessel stenosis and platelet deposition has previously been described.[44] Furthermore, platelet axial distribution analysis indicates that the apex (not the flow recirculation zone distal to the apex) is the segment with the greatest platelet accumulation, suggesting the crucial role of sudden changes in the degree of stenosis after plaque rupture, that is, shear stress changes in acute platelet response.[36] The Badimon perfusion chamber also makes it possible to analyze the thrombogenicity of different degrees of vascular injury as occurs during atherothrombogenesis. As such, in situations in which a relatively mild injury to the vessel wall may occur, the thrombogenic stimulus is relatively small, and the resultant thrombus is labile and may then be incorporated into the plaque, as may occur in the episodic asymptomatic progression of atherosclerotic plaques as well as in patients with unstable angina and transient ischemia. In contrast, deep vessel injury, particularly in the presence of severe stenosis, leads to relatively persistent thrombotic occlusion as may occur in myocardial infarction.

INSIGHTS FROM VON WILLEBRAND DISEASE ANIMAL MODELS

von Willebrand disease (vWD) is a genetic bleeding disorder that arises from the abnormalities in the vW factor (vWF). It is the most common inherited bleeding disorder in humans, and over the past years several animal species have also been described as suffering from this disease, whether through spontaneous mutation (pigs and dogs) or through a genetically engineered mutation (mice).[45] All these different animal models have been extremely useful in exploring the characteristics of vWD and in testing new treatments. Indeed, the interaction of platelets with vWF is crucial in the initiation and development of any thrombotic process since it enables platelets, via its surface glycoprotein receptors, to adhere to exposed subendothelium and to respond to blood shear stress.[46–48] In fact, Fuster and Badimon[1,49] were among the first to show a striking difference in the atherosclerotic lesions in the aorta between normal pigs and vWD pigs, both in spontaneous and in diet-induced atherosclerosis, demonstrating protection against atherosclerosis in V vWD pigs. The pig is a good model for studying vWD since vWF localization in endothelial cells and platelets mimics that of humans as do clotting and platelet characteristics. Additionally, in normal pigs the level of vWF is close to the human level whereas in cow, sheep, or goat it is 6- to 12-fold higher.[45] Canine vWD is also similar to humans. Over the years, many dog breeds have been identified as suffering from this disease, making vWD the most common inherited bleeding disorder in dogs. However, under the term "canine vWD" there seems to be a very heterogeneous group of bleeding disorders with different subtypes and mode of inheritance.[50] Actually, canine vWD can vary in genetic transmission, clinical severity, and diagnostic laboratory findings. Thus, in contrast to pigs, vWD dogs have not been used extensively for research purposes.[51] Finally, it deserves to be mentioned that vWD has also been reported in other animal species such as murine,[52,53] rabbits,[54] and cats.[55] However, extrapolation of the results obtained from these animals to humans remains to be completely elucidated.

ANIMAL MODELS OF PLAQUE RUPTURE

By its very nature, rupture of an atherosclerotic plaque is difficult to study in humans. Moreover, since rupture of an atherosclerotic plaque occurs in a stochastic fashion, it is also difficult to identify triggering factors and equally hard to investigate treatments addressed toward plaque stabilization. Thus, a good animal model not only may help to better understand the mechanisms behind its rupture but also to test treatments to prevent it from happening. Spontaneous hemorrhage and rupture are considered extremely rare events in large animal models and thus have been found only in the coronary arteries of pigs with inherited hyper-low-density lipoprotein cholesterolemia or in cholesterol-fed pigs with streptozotocin-induced diabetes.[56,57] In contrast, many interesting rodent studies related to plaque rupture have been published in recent years. Some authors have described the presence of blood-filled channels within the advanced coronary lesions as well as plaque ruptures and thrombi in the aortic origin of old apolipoprotein (apo) $E^{-/-}$ mice[58] whereas other studies have reported the presence of luminal thrombi in ruptured plaques of spontaneously dead apo$E^{-/-}$ mice.[59] On the other hand, other investigators have evaluated the harmful effect of combining several risk factors. As such, it has been reported that either in double knockout mice with homozygous null mutations in the apoE and the high-density lipoprotein (HDL) receptor, scavenger receptor class B, or a combination of hypertension and dyslipemia (hypertensive rats transgenic from human cholesteryl ester transfer protein) increases plaque vulnerability and thrombi formation. Despite all these, the most simplistic approach to induce plaque rupture and subsequent thrombus formation is by direct mechanical injury of the vessel of interest. Table 39–1 describes different approaches already established in order to induce plaque rupture and the subsequent thrombi formation in different animal models.[60–70]

Table 39–1
Previously applied methods to induce plaque rupture and/or thrombus formation

Reference	*Method applied to induce plaque rupture*	*Animal model and/or thrombus formation*
Reddick *et al.*[60]	Forceps squeezing of the aorta	$ApoE^{-/-}$mice
Gertz *et al.*,[61] Rekhter *et al.*[62]	Combination of double balloon injury and hypercholesterolemia and further angioplasty-induced plaque rupture	Rabbits
Eitzman *et al.*[63]	Photochemical reaction to previously formed atherosclerotic plaques	$ApoE^{-/-}$ mice
Constantinides *et al.*,[64] Abela *et al.*,[65] Nakamura *et al.*[66]	Intraperitoneal injection of Russell's viper venom followed by intravenous injection of histamine (vasopressor), hypercholesterolemic serotonin, or angiotensin II	Rabbits
Rekhter *et al.*[67]	Intracerebroventricular injection of corticotropin-releasing factors ("stress hormone")	$ApoE^{-/-}$ mice
Heras *et al.*[68]	Carotid angioplasty	Pig
Badimon *et al.*,[34] Lassila *et al.*[44]	Thrombogenic effect of shear stress and atherosclerotic vessel components on the extracorporeal perfusion system	
Fuster *et al.*,[49,69] Badimon *et al.*[1]	von Willebrand factor deficiency	Pigs with von Willebrand disease

RABBIT MODELS OF THROMBOSIS

The use of rabbits has been widely extended because with the appropriate precaution, they are easy to handle and cost-effective. There are three breeds of rabbits commonly used in biomedical research: New Zealand White, Dutch Belted, and Flemish Giant. All of them are mainly used as animal models for the study of atherogenesis,[2,61] plaque instability and rupture,[62,65] and myocardial infarction,[71] and to evaluate the efficacy of new antithrombotic and antiatherosclerotic drugs.[72]

Rabbits do not develop spontaneous atherosclerosis as they are vegetarian. However, rabbits can rapidly induce vascular lesions after the administration of a rich atherosclerotic diet (0.5–4% cholesterol content) during 8–16 weeks.[2] Moreover, transgenic rabbits have also been developed to obtain models of hereditary hyperlipidemia (WHHL-rabbit) Yet, rabbit atherosclerotic lesions differ from human atheroma since their lipid and macrophage content is much higher[10] as is their hypercholesterolemic index. Accelerated atherosclerotic lesions can also be developed in rabbits by a combination of balloon injury and hypercholesterolemia feeding.[61]

MICE MODELS OF THROMBOSIS

Mice do not develop atherosclerosis without genetic manipulation since they have a lipid physiology that is radically different from that in humans, most of the cholesterol being transported in HDL-like particles. On the other hand, the mouse has not been as widely used as the pig as a model to test the possible therapeutic usefulness of antithrombotic agents.[8] The reasons for this may be related to the small size of the murine arteries, which limits the manipulation of isolated vessel segments, the lack of standardization of the model, and mostly the differences from the human model, which decidedly limit the extrapolation of mouse findings to clinical relevance in humans. Thus, an accurate determination of the mouse hemostatic system with regard to coagulation and fibrinolytic systems, platelet structure, and platelet receptor/enzyme system is required and differences between the two species must be acknowledged and appreciated in the interpretation of the data. As an example, platelet counts in mice on average are four times those of humans and platelets are only approximately one-half the volume of human platelets. An excellent and comprehensive review of the murine hemostatic system has been provided by Tsakiris *et al.*,[72] and important aspects of murine platelet receptor function have recently been reviewed by Ware.[73] However, despite these species-related differences, the use of murine models is particularly advantageous. On one hand, it makes it possible to easily perform several techniques of vascular injury as described in Table 39–2. On the other hand, the murine model not only offers numerous immunological approaches (e.g., monoclonal antibodies), but also provides genetically modified animal strains with defined defects in platelet function that open new ways to identify the individual roles and the interplay of platelet proteins in thrombus formation.[74–81]

Finally, the usefulness of intravital microscopy in the mice model of thrombosis deserves to be mentioned since it allows testing the *in vivo* relevance of the *in vitro* observations. Indeed, the study of the microcirculation by intravital microscopy represents a sophisticated research tool to examine complex biological interactions and disease mechanisms as well as to develop and test novel approaches aimed at the prevention or attenuation of disease-associated microvascular disorders and cellular dysfunction such as the ones that occur in atherosclerosis and thrombosis pathogenesis. In addition, using the intravital microscopic technique, circulatory and cellular disorders in surgical diseases and procedures, such as ischemia/reperfusion, trauma, sepsis, and inflammation, as well as wound healing, may be analyzed.

Table 39–2
Different models to induce vascular injury

Vessel wall injury models	*Method*
Photochemical injury	Vessel irradiation in the presence of rose Bengal or sodium fluorescein[77]
Mechanical injury	Pinching arteries with needle forceps (Folts model)[78]
	Denuding the endothelium via balloon catheters
	Infusion of a portion of a vessel with air or saline
Electrical injury	Application of an electrical current through the vessel with an electromicrocoagulator
Chemical injury	*Ex vivo* application of ferric chloride; applicable only in guinea pigs, rats, and mice[79–81]
Laser injury	Argon laser: irradiation followed by visualization of the injury with intravital microscopy[82]

ANIMAL MODELS OF VENOUS THROMBOSIS

Thrombogenesis in veins is mainly attributed to some combination of hypercoagulability, stasis, and vascular injury. All these triggering factors induce tissue thromboplastin (tissue factor) release to the flowing blood forming thrombin and fibrin that trap red blood cells. Thus, in contrast to arterial thrombotic lesions (platelet-rich thrombus, "white thrombus"), venous thrombosis is predominantly "red" and fibrin-rich. Animal models for venous thrombosis appear to be particularly useful for studying the pathophysiology of blood coagulation *in vivo* and the pathogenesis of venous thrombosis. For instance, this model has helped to define the role of activated protein C, the interplay between inflammatory and procoagulant mediators, and the regulatory role of PAI-1 in thrombolysis in addition to therapeutic approaches.[82] In fact, Dorffler-Melly[83] and Levi[82] have previously published interesting literature searches in MEDLINE and EMBASE databases offering an in-depth review of small and large animal models of venous thrombosis, pointing out their advantages, usefulness, and limitations.

CONCLUSIONS

Clinical observations provide the substrate to build up pathophysiological hypotheses, but for obvious ethical reasons our ability to test these hypotheses in humans is very limited. Cell biology-related studies have helped to answer mechanistic questions, but lack the complexity of a real disease, thus limiting the scope of testable hypotheses. On the other hand, studies using rodent or large animal models have proved to be essential for *proof of concept* since they yield *in vivo* approaches to confirm critical hypotheses previously evaluated in relevant *in vitro* models. Moreover, DNA technology made it possible to create a number of knockout and transgenic rodent models that have enhanced our understanding of cardiovascular disease and expedited the development of new gene therapies. However, while small animal models provide experimental convenience and ease in manipulation, more clinically relevant models are necessary to study the mechanisms surrounding human atherothrombogenesis. Large animal models, although associated with higher cost and handling-related difficulties, have shown an atherothrombotic pattern more comparable to that of humans. To date, considering all models, the porcine model is one of the most useful currently available atherothrombotic models. Indeed, pig animal models have been shown to address specific questions related to blood and atherosclerotic vessel mechanisms involved in thrombus formation that have been eventually translated to clinical situations. Despite all of this, efforts in developing the "ideal" animal model for atherothrombotic evaluation must continue. Furthermore, for this purpose, it may be critical to construct animal preclinical trials that resemble human trials in order to make clear what data and important conclusions can be justifiably extracted from animal models.

REFERENCES

1. Badimon L, Steele P, Badimon JJ, Bowie EJ, Fuster V. Aortic atherosclerosis in pigs with heterozygous von Willebrand disease. Comparison with homozygous von Willebrand and normal pigs. *Arteriosclerosis* 1985;5:366–370.
2. Badimon L, Fuster V, Chesebro JH, Dewanjee MK. New "ex vivo" radioisotopic method of quantitation of platelet deposition—studies in four animal species. *Thromb Haemost* 1983;50:639–644.
3. Steele PM, Chesebro JH, Stanson AW, *et al.* Balloon angioplasty. Natural history of the pathophysiological response to injury in a pig model. *Circ Res* 1985;57:105–112.
4. Fuster V, Badimon L, Badimon JJT, V, Lie J, Bowie E. Experimental approach to vascular disease in swine with von Willebrand's disease: In: Tumbleson ME, Ed. *Swine in Biomedical Research.* New York: Plenum Publishing Corp., 1986:1527–1542.
5. Galvez A, Badimon L, Badimon JJ, Fuster V. Electrical aggregometry in whole blood from human, pig and rabbit. *Thromb Haemost* 1986;56:128–132.
6. Badimon L, Badimon JJ, Turitto VT, Fuster V. Thrombosis: Studies under flow conditions. *Ann NY Acad Sci* 1987;516:527–540.
7. Fuster V, Lie JT, Badimon L, Rosemark JA, Badimon JJ, Bowie EJ. Spontaneous and diet-induced coronary atherosclerosis in normal swine and swine with von Willebrand disease. *Arteriosclerosis* 1985;5:67–73.
8. Farb A, Burke AP, Tang AL, *et al.* Coronary plaque erosion without rupture into a lipid core. A frequent cause of coronary thrombosis in sudden coronary death. *Circulation* 1996;93:1354–1363.
9. van der Wal AC, Becker AE, van der Loos CM, Das PK. Site of intimal rupture or erosion of thrombosed coronary atherosclerotic plaques is characterized by an inflammatory process irrespective of the dominant plaque morphology. *Circulation* 1994;89:36–44.
10. Badimon L. Atherosclerosis and thrombosis: Lessons from animal models. *Thromb Haemost* 2001;86:356–365.
11. Burke AP, Kolodgie FD, Farb A, *et al.* Healed plaque ruptures and sudden coronary death: Evidence that subclinical rupture has a role in plaque progression. *Circulation* 2001;103:934–940.
12. Handley DA. Experimental therapeutics and clinical studies in (re)stenosis. *Micron* 1995;26:51–68.
13. Johnson GJ, Griggs TR, Badimon L. The utility of animal models in the preclinical study of interventions to prevent human coronary artery restenosis: Analysis and recommendations. On behalf of the Subcommittee on Animal, Cellular and Molecular Models of Thrombosis and Haemostasis of the Scientific and Standardization Committee of the International Society on Thrombosis and Haemostasis. *Thromb Haemost* 1999;81:835–843.
14. Lewis A, Southern L. *Swine Nutrition*, 2nd ed. Boca Raton, FL: CRC Press Inc., 2000.
15. Committee on Animal Nutrition, National Research Council. *Nutrient Requirements of Swine*, 10th revised ed. Washington, DC: National Academies Press Publications, 1998.

16. Swindle M. Defining appropriate health status and management programs for specific-pathogen-free swine for xenotransplantation. *Ann NY Acad Sci* 1998;862:111–120.
17. Laber K, Whary M, Bingel S, *et al.* Biology diseases of swine in laboratory Medicine In: Fox JG, Anderson LC, Loew FM, Quimby FW, Eds. *Laboratory Animal Medicine*, 2nd ed. San Diego, CA: Academic Press, 2002.
18. Swindle MM, Horneffer PJ, Gardner TJ, *et al.* Anatomic and anesthetic considerations in experimental cardiopulmonary surgery in swine. *Lab Anim Sci* 1986;36:357–361.
19. Gardner T, Johnson D. Cardiovascular system. In: Swindle MM, Adams RJ, Eds. *Experimental Surgery and Physiology: Induced Animal Models of Human Disease*. Philadelphia, PA: Lippincott, Williams & Wilkins, 1988:74–124.
20. Casani L, Sanchez-Gomez S, Vilahur G, Badimon L. Pravastatin reduces thrombogenicity by mechanisms beyond plasma cholesterol lowering. *Thromb Haemost* 2005;94:1035–1041.
21. Gal D, Chokshi SK, Mosseri M, Clarke RH, Isner JM. Percutaneous delivery of low-level laser energy reverses histamine-induced spasm in atherosclerotic Yucatan microswine. *Circulation* 1992;85: 756–768.
22. White FC, Carroll SM, Magnet A, Bloor CM. Coronary collateral development in swine after coronary artery occlusion. *Circ Res* 1992;71:1490–1500.
23. White FC, Bloor CM. Coronary vascular remodeling and coronary resistance during chronic ischemia. *Am J Cardiovasc Pathol* 1992;4:193–202.
24. White C, Ramee S, Banks A, Wiktor D, Price H. The Yucatan miniature swine: An atherogenic model to assess the early potency rates of an endovascular stent. In: Swindle MM, Ed. *Swine as Models in Biomedical Research*. Ames, IA: Iowa State University Press, 1992:156–162.
25. Mehran R, Ricci M, Graham A, Carter K, Smyes J. Porcine model for vascular graft studies. *J Invest Surg* 1991;4(l):37–44.
26. Swindle MM. Defining appropriate health status and management programs for specific-pathogen-free swine for xenotransplantation. *Ann NY Acad Sci* 1998;862:111–120.
27. Bloor C, White F, Roth D. The pig as a model of myocardial ischemia and gradual coronary artery occlusion. In: Swindle MM, Ed. *Swine as Models in Biomedical Research*. Ames, IA: Iowa State University Press, 1992:163–175.
28. Stanton H, Mersmann H. *Swine in Cardiovascular Research*, Vols. 1 and 2, Boca Raton, FL: CRC Press, Inc., 1986.
29. Institute of Laboratory Animal Resources Commission on Life Sciences, National Research Council. *Guide for the Care and Use of Laboratory Animals*. Washington, DC: National Academies Press Publications, 1996.
30. Swindle M, Smith A, Laber-Laird K, Dungan L. Swine in biomedical research: Management and models. *ILAR News* 1994;36(l):1–5.
31. O'Rourke S, Folts J, Albrecht R. Studies on the inhibition of canine platelet aggregation by barbiturates. *J Lab Clin Med* 1986; 108:206–212.
32. Jorgensen L. The role of platelets in the initial stages of atherosclerosis. *J Thromb Haemost* 2006;4:1443–1449.
33. Fernandez-Ortiz A, Badimon JJ, Falk E, *et al.* Characterization of the relative thrombogenicity of atherosclerotic plaque components: Implications for consequences of plaque rupture. *J Am Coll Cardiol* 1994;23:1562–1569.
34. Badimon L, Badimon JJ, Galvez A, Chesebro JH, Fuster V. Influence of arterial damage and wall shear rate on platelet deposition. Ex vivo study in a swine model. *Arteriosclerosis* 1986;6:312–320.
35. Vilahur G, Duran X, Juan-Babot O, Casani L, Badimon L. Antithrombotic effects of saratin on human atherosclerotic plaques. *Thromb Haemost* 2004;92:191–200.
36. Badimon L, Badimon JJ. Mechanisms of arterial thrombosis in nonparallel streamlines: Platelet thrombi grow on the apex of stenotic severely injured vessel wall. Experimental study in the pig model. *J Clin Invest* 1989;84:1134–1144.
37. Vilahur G, Segales E, Salas E, Badimon L. Effects of a novel platelet nitric oxide donor (LA816), aspirin, clopidogrel, and combined therapy in inhibiting flow- and lesion-dependent thrombosis in the porcine ex vivo model. *Circulation* 2004;110:1686–1693.
38. Roussi J, Berge N, Bal dit Sollier C, *et al.* Clopidogrel-induced qualitative changes in thrombus formation correlate with stent patency in injured pig cervical arteries. *Thromb Res* 2002;105: 209–216.
39. Mailhac A, Badimon JJ, Fallon JT, *et al.* Effect of an eccentric severe stenosis on fibrin(ogen) deposition on severely damaged vessel wall in arterial thrombosis. Relative contribution of fibrin(ogen) and platelets. *Circulation* 1994;90:988–996.
40. Roque M, Rauch U, Reis ED, Chesebro JH, Fuster V, Badimon JJ. Comparative study of antithrombotic effect of a low molecular weight heparin and unfractionated heparin in an ex vivo model of deep arterial injury. *Thromb Res* 2000;98:499–505.
41. Badimon JJ, Lettino M, Toschi V, *et al.* Local inhibition of tissue factor reduces the thrombogenicity of disrupted human atherosclerotic plaques: Effects of tissue factor pathway inhibitor on plaque thrombogenicity under flow conditions. *Circulation* 1999;99: 1780–1787.
42. Vilahur G, Segales E, Casani L, Badimon L. A novel anti-ischemic nitric oxide donor inhibits thrombosis without modifying haemodynamic parameters. *Thromb Haemost* 2004;91:1035–1043.
43. Osende JI, Shimbo D, Fuster V, Dubar M, Badimon JJ. Antithrombotic effects of S 18886, a novel orally active thromboxane A2 receptor antagonist. *J Thromb Haemost* 2004;2:492–498.
44. Lassila R, Badimon JJ, Vallabhajosula S, Badimon L. Dynamic monitoring of platelet deposition on severely damaged vessel wall in flowing blood. Effects of different stenoses on thrombus growth. *Arteriosclerosis* 1990;10:306–315.
45. Denis CV, Wagner DD. Insights from von Willebrand disease animal models. *Cell Mol Life Sci* 1999;56:977–990.
46. Ruggeri ZM, Orje JN, Habermann R, Federici AB, Reininger AJ. Activation-independent platelet adhesion and aggregation under elevated shear stress. *Blood* 2006;108:1903–1910.
47. Reininger AJ, Heijnen HF, Schumann H, Specht HM, Schramm W, Ruggeri ZM. Mechanism of platelet adhesion to von Willebrand factor and microparticle formation under high shear stress. *Blood* 2006;107:3537–3545.
48. Donadelli R, Orje JN, Capoferri C, Remuzzi G, Ruggeri ZM. Size regulation of von Willebrand factor-mediated platelet thrombi by ADAMTS13 in flowing blood. *Blood* 2006;107:1943–1950.
49. Fuster W, Bowie EJ, Lewis JC, Fass DN, Owen CA Jr, Brown AL. Resistance to arteriosclerosis in pigs with von Willebrand's disease. Spontaneous and high cholesterol diet-induced arteriosclerosis. *J Clin Invest* 1978;61:722–730.
50. Johnson GS, Turrentine MA, Kraus KH. Canine von Willebrand's disease. A heterogeneous group of bleeding disorders. *Vet Clin North Am Small Anim Pract* 1988;18:195–229.
51. Nichols TC, Bellinger DA, Reddick RL, *et al.* The roles of von Willebrand factor and factor VIII in arterial thrombosis: Studies in canine von Willebrand disease and hemophilia A. *Blood* 1993;81: 2644–2651.
52. Sweeney JD, Novak EK, Reddington M, Takeuchi KH, Swank RT. The RIIIS/J inbred mouse strain as a model for von Willebrand disease. *Blood* 1990;76:2258–2265.
53. Denis C, Methia N, Frenette PS, *et al.* A mouse model of severe von Willebrand disease: Defects in hemostasis and thrombosis. *Proc Natl Acad Sci USA* 1998;95:9524–9529.
54. Benson R, Dodds W. Autosomal factor VIII deficiency in rabbits: Size variations of rabbit factor VIII. *Thromb Haemost* 1977;38:380.
55. French TW, Fox LE, Randolph JF, Dodds WJ. A bleeding disorder (von Willebrand's disease) in a Himalayan cat. *J Am Vet Med Assoc* 1987;190:437–439.
56. Prescott MF, McBride CH, Hasler-Rapacz J, Von Linden J, Rapacz J. Development of complex atherosclerotic lesions in pigs with inherited hyper-LDL cholesterolemia bearing mutant alleles for apolipoprotein B. *Am J Pathol* 1991;139:139–147.
57. Gerrity RG, Natarajan R, Nadler JL, Kimsey T. Diabetes-induced accelerated atherosclerosis in swine. *Diabetes* 2001;50:1654–1665.

58. Rosenfeld ME, Polinsky P, Virmani R, Kauser K, Rubanyi G, Schwartz SM. Advanced atherosclerotic lesions in the innominate artery of the apoE knockout mouse. *Arterioscler Thromb Vasc Biol* 2000;20:2587–2592.
59. Johnson JL, Jackson CL. Atherosclerotic plaque rupture in the apolipoprotein E knockout mouse. *Atherosclerosis* 2001;154:399–406.
60. Reddick RL, Zhang SH, Maeda N. Aortic atherosclerotic plaque injury in apolipoprotein E deficient mice. *Atherosclerosis* 1998; 140:297–305.
61. Gertz SD, Fallon JT, Gallo R, *et al.* Hirudin reduces tissue factor expression in neointima after balloon injury in rabbit femoral and porcine coronary arteries. *Circulation* 1998;98:580–587.
62. Rekhter MD, Hicks GW, Brammer DW, *et al.* Animal model that mimics atherosclerotic plaque rupture. *Circ Res* 1998;83:705–713.
63. Eitzman DT, Westrick RJ, Xu Z, Tyson J, Ginsburg D. Hyperlipidemia promotes thrombosis after injury to atherosclerotic vessels in apolipoprotein E-deficient mice. *Arterioscler Thromb Vasc Biol* 2000;20:1831–1834.
64. Constantinides P, Chakravarti RN. Rabbit arterial thrombosis production by systemic procedures. *Arch Pathol* 1961;72:197–208.
65. Abela GS, Picon PD, Friedl SE, *et al.* Triggering of plaque disruption and arterial thrombosis in an atherosclerotic rabbit model. *Circulation* 1995;91:776–784.
66. Nakamura M, Abe S, Kinukawa N. Aortic medial necrosis with or without thrombosis in rabbits treated with Russell's viper venom and angiotensin II. *Atherosclerosis* 1997;128:149–156.
67. Rekhter MD, Hicks GW, Brammer DW, *et al.* Hypercholesterolemia causes mechanical weakening of rabbit atheroma: Local collagen loss as a prerequisite of plaque rupture. *Circ Res* 2000;86:101–108.
68. Heras M, Chesebro JH, Penny WJ, Bailey KR, Badimon L, Fuster V. Effects of thrombin inhibition on the development of acute platelet-thrombus deposition during angioplasty in pigs. Heparin versus recombinant hirudin, a specific thrombin inhibitor. *Circulation* 1989;79:657–665.
69. Fuster V, Fass DN, Kaye MP, Josa M, Zinsmeister AR, Bowie EJ. Arteriosclerosis in normal and von Willebrand pigs: Long-term prospective study and aortic transplantation study. *Circ Res* 1982;51: 587–593.
70. Shiomi M, Ito T, Yamada S, Kawashima S, Fan J. Correlation of vulnerable coronary plaques to sudden cardiac events. Lessons from a myocardial infarction-prone animal model (the WHHLMI rabbit). *J Atheroscler Thromb* 2004;11:184–189.
71. Alfon J, Pueyo Palazon C, Royo T, Badimon L. Effects of statins in thrombosis and aortic lesion development in a dyslipemic rabbit model. *Thromb Haemost* 1999;81:822–827.
72. Tsakiris DA, Scudder L, Hodivala-Dilke K, Hynes RO, Coller BS. Hemostasis in the mouse (Mus musculus): A review. *Thromb Haemost* 1999;81:177–188.
73. Ware J. Dysfunctional platelet membrane receptors: From humans to mice. *Thromb Haemost* 2004;92:478–485.
74. Nieswandt B, Aktas B, Moers A, Sachs UJ. Platelets in atherothrombosis: Lessons from mouse models. J *Thromb Haemost* 2005;3: 1725–1736.
75. Carmeliet P, Moons L, Collen D. Mouse models of angiogenesis, arterial stenosis, atherosclerosis and hemostasis. *Cardiovasc Res* 1998;39:8–33.
76. Rosenblum WI, El-Sabban F. Platelet aggregation in the cerebral microcirculation: Effect of aspirin and other agents. *Circ Res* 1977;40:320–328.
77. Gallagher KP, Osakada G, Kemper WS, Ross J Jr. Cyclical coronary flow reductions in conscious dogs equipped with ameroid constrictors to produce severe coronary narrowing. *Basic Res Cardiol* 1985;80: 100–106.
78. Aoki T, Cox D, Senzaki K, *et al.* Comparison of the antithrombotic effects of FK633, GPIIb/IIIa antagonist, and aspirin in a guinea pig thrombosis model. *Thromb Res* 1998;89:129–136.
79. Kurz KD, Main BW, Sandusky GE. Rat model of arterial thrombosis induced by ferric chloride. *Thromb Res* 1990;60:269–280.
80. Fay WP, Parker AC, Ansari MN, Zheng X, Ginsburg D. Vitronectin inhibits the thrombotic response to arterial injury in mice. *Blood* 1999;93:1825–1830.
81. Rosen ED, Raymond S, Zollman A, *et al.* Laser-induced noninvasive vascular injury models in mice generate platelet- and coagulation-dependent thrombi. *Am J Pathol* 2001;158:1613–1622.
82. Levi M, Dorffle-Melly J, Johnson GJ, Drouet L, Badimon L. Usefulness and limitations of animal models of venous thrombosis. *Thromb Haemost* 2001;86:1331–1333.
83. Dorffler-Melly J, Schwarte LA, Ince C, Levi M. Mouse models of focal arterial and venous thrombosis. *Basic Res Cardiol* 2000;95: 503–509.

40 Animal Models for Atherosclerosis, Restenosis, and Endovascular Aneurysm Repair

ANDRÁS KÓNYA, KENNETH C. WRIGHT, MATTHEW GOUNIS, AND KRISHNA KANDARPA

ABSTRACT

Animal models have significantly advanced the understanding of mechanisms involved in atherosclerosis and restenosis and have allowed the evaluation of therapeutic options. The current focus of research is to develop strategies to prevent restenosis. These include pharmacological and biological interventions directed primarily against smooth muscle cell proliferation, endovascular devices for recanalization and/or drug delivery, and an integrated approach using both devices and pharmacobiological agents. Devices aimed at the percutaneous endoluminal exclusion of aortic aneurysms have also generated a great deal of interest. Experience over many decades with animal models for vascular research has established that a single, ideal model for atherosclerosis, restenosis, or aneurysm formation does not exist, although recent development of transgenic variants of currently available models has widened the available options. Nevertheless, an appreciation of the individual features of natural or stimulated disease in each species is of the utmost importance for the proper design and execution of relevant experiments.

Key Words: Animal models, Atherosclerosis, Arteriosclerosis, Restenosis, Angioplasty, Stenting, Drug-eluting stents, Endovascular aneurysm repair, Endovascular grafts, Transcatheter embolization.

INTRODUCTION

Almost all medical knowledge and treatment, especially in the last century, has involved work with laboratory animal models. Since the early part of the twentieth century, research using animal models has contributed to more than half of the significant medical discoveries and during some periods, accounted for more than 75% of the major advances. Two-thirds of the Nobel prizes awarded since 1901 have been for discoveries involving the use of laboratory animal models.

The selection of the laboratory species, breed, and strain to be used as a model for human disease is one of the most important decisions to be made by an investigator. The goals of the research play an important role in the selection of the laboratory animal model to be used. The validity of any information derived from the animal is dependent on the appropriateness of the model. Thus, a detailed consideration of species and model features that are favorable and applicable to the specific purpose of the research is important. Since new animal models are continually being identified and characterized, a thorough literature search should always be conducted to determine what models are available and which are the most relevant.

From: *Sourcebook of Models for Biomedical Research* (P. M. Conn, ed.), © 2008 Humana Press Inc., Totowa, NJ.

ATHEROSCLEROSIS

BACKGROUND Atherosclerosis is a focal disease that localizes mainly in areas of disturbed blood flow[1] and bears many features of chronic inflammation.[2] The lesion of plaque consists of endothelial cells, monocytes/macrophages, smooth muscle cells (SMC), and a matrix of collagen, proteoglycans, and calcification. Lesion formation begins with the influx of monocytes/macrophages, which become loaded with cholesteryl ester to create the so-called "fatty streak." These lesions may progress by the influx of additional monocytes and T cells, whose secretory products, along with those of endothelial cells, result in migration of SMC, where they proliferate and synthesize matrix components. Some cells in the plaque die, leaving a necrotic core and extracellular cholesterol clefts, which are hallmarks of advanced lesions. The luminal face of the lesion often has a fibrous cap, generated by SMC stabilizing the plaque.[3]

Previously, atherosclerosis has been viewed as a disease of uncontrolled plaque growth eventually leading to total occlusion.[4] New clinical findings suggest that acute coronary events are triggered by rupture and hemorrhage of vulnerable atherosclerotic plaques.[5] The plaque consists of a highly thrombogenic core and a slender and fragile fibrous cap that separates the core from blood coagulation factors, thereby precluding formation of thrombus on the surface of the intact plaque.[4] Disturbance of cap integrity leads to exposure of the plaque's thrombogenic interior and subsequent thrombosis. Such ruptures of the fibrous cap may recur over many years without causing complete occlusion of the vessel, and thereby may often remain clinically silent.

The reasons for the thinning and rupture of the fibrous cap are not known. One theory hypothesizes that the rupture is a result of

Table 40–1
Characteristics of an ideal animal model of arteriosclerosis[11]

1. Inexpensive	
2. Easy to maintain and handle	
3. Proper size	
4. Reproducible in a laboratory setting	
5. Well-defined genetic characteristics	
6. Sharing important aspects of human disease	
7. Lesion development:	Develop naturally on a normal diet
	Progress slowly over the animals' lifetime
	Cause clinical sequelae at middle to old age
8. Lesion pathogenesis:	Complete range from fatty streak to atheromatous plaques
9. Lesion complications:	Calcification
	Ulceration
	Hemorrhage
	Stenosis
	Thrombosis

loss of SMC, thought to be the main producers of cap-stabilizing collagen. This is caused by the loss of cells from apoptosis that may be mediated by the interaction between SMC and monocytes/macrophages.[6,7] According to a second hypothesis, plaque rupture is the result of an imbalance between the production of plaque-stabilizing collagen and the action of corrosive enzymes.[8] An immunohistochemical study has shown that macrophages accumulated in the plaque express high levels of matrix metalloproteinases and tissue factors[9] that are thought to make the fibrous plaque fragile by digesting the extracellular matrix. Therefore, the search for an animal model of atherosclerotic plaque rupture should be seen against this background.

Arteriosclerosis is a disease that really affects only humans and one or two other species, such as the pig and certain nonhuman primates, and because of this researchers must resort to genetically modified models to even partially reproduce the condition.[10] It has never been easy to find a good animal model of the atherosclerotic process itself. The ideal animal model of human arteriosclerosis should posses several important characteristics[11] (Table 40–1).

ANIMAL MODELS OF ATHEROGENESIS

Diet-Induced Nonmurine Models Historically, relatively large animals, such as nonhuman primates, swine, and rabbits, have been predominantly used for arteriosclerosis-related research. Rats and dogs are not good models because they do not develop spontaneous lesions and require heavy modifications of diet to produce a vascular lesion. In recent years, however, some transgenic rat models have been produced that develop lesions resembling human atherosclerosis.[12,13]

Rabbits are useful because they are highly responsive to cholesterol manipulation and develop lesions in a fairly short time.[14] The lesions, however, are much more fatty and macrophage-rich (i.e., more inflammatory) than human lesions and plasma cholesterol levels are extraordinarily high.[15] The Watanabe heritable hyperlipidemic rabbits (WHHL) are genetic variants that mimic familial hypercholesterolemia.[16] This strain of rabbit was produced by inbreeding from a mutant discovered in 1973. The rabbits consistently showed spontaneous development of aortic arteriosclerosis over 5 months of age. The existence of rabbit strains with congenital abnormalities in lipid metabolism has contributed to the feasibility of creating particularly informative transgenic rabbits.

The guinea pig provides valuable models for studying early arteriosclerosis development.[17] Guinea pigs carry the majority of their plasma cholesterol in low-density lipoprotein (LDL), and present other striking similarities to humans in terms of cholesterol and lipoprotein metabolism making them a unique animal model. Many of the mechanisms by which guinea pigs regulate cholesterol and lipoprotein metabolism as a response to diet or drug treatment are analogous to those reported in clinical human studies.

Nonhuman primates including chimpanzees and various monkeys fed a high-cholesterol diet develop a form of atherosclerosis that is very similar to that of humans. However, the cost of nonhuman primates is prohibitive and many of these species are protected. Therefore, arteriosclerosis research in primates is today generally confined to the study of complex issues such as the effects of psychological stress.[10]

The domestic pig is one of the most useful currently available large animal models of arteriosclerosis. Over time, pigs develop atherosclerosis even on a normal porcine diet.[18] When fed a high-cholesterol diet, they develop elevated plasma cholesterol levels and atherosclerotic lesions that are similar to those seen in humans. However, the maintenance of pigs can be expensive and they require special facilities and handling.

Miniature pigs (e.g., Yucatan miniswine) have been used in several studies to gain insight into early-stage arteriosclerotic disease. One study concluded that early pathological changes (i.e., increased intima-media thickness and foam cell accumulation) induced by a high fat cholesterol diet precede decreases in endothelium-dependent relaxation (endothelial dysfunction) in femoral arteries.[19] This model may be an important tool to evaluate the effectiveness of interventions in modifying or blocking the development of diet-induced vascular disease and the role of the vascular endothelium in these processes. Yucatan minipigs have been used to study material property alterations in femoral and carotid arteries with atheroma induced by endothelial denudation and a high cholesterol diet.[20,21] The data demonstrated a striking difference between the two arteries. While lesion progression resulted in stiffening vascular wall elastic modulus in carotid arteries, the elastic modulus in the femoral arteries decreased first with early stages of atheroma development and increased as lesions progressed.

The use of pig models of atherosclerosis initially revealed that monocyte infiltration was one of the primary cellular events in the atherogenic process.[22] Studies in monkeys and rabbits were pivotal in defining the cellular events in the initiation and development of atherosclerotic lesions.[23,24]

Murine Models In recent years there has been an explosion in the number of *in vivo* studies that is largely attributable to the use of murine models to study atherogenic mechanisms.[25] Mice are highly resistant to atherosclerosis. The only exception is the C57BL/6 strain. A special diet (30% fat, 5% cholesterol, 2% cholic acid) led to arteriosclerosis in C57BL/6 mice, but the lesions differed from the human condition in their histological nature and location. Additionally, this diet was toxic and resulted in weight loss and respiratory infections. Paigen *et al.* modified the diet (15% fat, 1.25% cholesterol, 0.5% cholic acid), but the lesions were very small (range of 200–1000 μm^2) in the aortic root and usually did not develop beyond the early foam-cell, fatty streak stage.[26]

It is now possible to add exogenous transgenes into mice as well as many other species. However, uniquely in mice, it is also possible to knock out or replace endogenous genes, which is one of the main advantages of working with the mouse.[25] Knockout murine models can be used to identify arteriosclerosis susceptibility modifying genes, the role of various cell types in atherogenesis, and environmental factors affecting atherogenesis and to assess therapies that block atherogenesis or lesion progression.[25]

Although genetic modifications render mice susceptible to atherogenesis, lipid-enriched diets are indispensable to induce or accelerate the rate of development of atherosclerotic lesions. It appears that the induction of persistent hypercholesterolemia to levels >300 mg/dl is required for the development of experimental atherosclerosis in the mouse where dietary cholesterol is the major proatherogenic component.[27]

Apolipoprotein E-Deficient Mice Apolipoprotein E (apoE) is a glycoprotein that is a structural component of all lipoprotein particles other than low-density lipoprotein (LDL). One of its most important functions is to serve as a high-affinity ligand for the apoB- and apoE-LDL receptor and for the chylomicron-remnant receptor, thereby allowing the specific uptake of apoE-containing particles by the liver.[28]

In 1992, apoE-deficient mice were generated by selectively inactivating the apoE gene.[29] The apoE-knockout mice exhibited increased total plasma cholesterol levels, which were five times those of normal littermates. Mice naturally have high levels of high-density lipoprotein (HDL) and low levels of LDL, in contrast to humans who have high LDL and low HDL. The mice lack the cholesteryl ester transferase enzyme that converts HDL to very low-density lipoprotein (VLDL) to LDL. The sequential events involved in the formation of atherosclerotic lesions in apoE-deficient mice are strikingly similar to those in humans. Initial fatty streak lesions rapidly progress to advanced lesions with a necrotic core surrounded by proliferating SMC and varying amounts of extracellular matrix. These lesions have well-formed fibrous caps. It is not uncommon for the inflammatory lesion to erode deep into the medial wall of the aorta, sometimes resulting in the formation of aortic aneurysms.

When apoE-deficient mice are fed a "Western type" diet (21% fat, 0.15% cholesterol, no cholic acid),[30] they exhibit over a 3-fold elevation in plasma cholesterol (to 2000 mg/dl), while wild-type mice show a 2-fold elevation. On this diet, formation of atherosclerotic lesions is greatly accelerated and lesion size is increased. Lesions in the apoE-deficient mouse, as in humans, tend to develop at vascular branch points and progress from a foam cell stage to a fibroproliferative stage with well-defined fibrous caps and necrotic lipid cores.

ApoE-deficient mice are considered to be one of the most relevant models for arteriosclerosis since they are hypercholesterolemic and develop spontaneous arterial lesions.[31] The apoE-deficient mouse contains the entire spectrum of lesions observed during atherogenesis and was the first mouse model to develop lesions similar to those in humans.

In 1995, Kashyap *et al.* described the successful correction of apoE deficiency in apoE-deficient mice by using an approach involving systemic delivery of recombinant adenoviral vectors expressing human apoE to the liver. Thus the single genetic change causing the absence of apoE and severe hypercholesterolemia is sufficient to convert the mouse from a species that is highly resistant to atherosclerosis to one that is highly susceptible.[32]

Low-Density Lipoprotein Receptor-Deficient Mice Gene targeting in embryonic stem cells has recently been used to create LDL receptor-knockout mice, a model of familial hypercholesterolemia.[33] The lipoprotein abnormality in these mice is more modest than that in the apoE-deficient mice, featuring increases in LDL and VLDL with plasma cholesterol of about 250 mg/dl on a normal chow diet without developing arteriosclerosis. However, after being fed the Paigen diet, their plasma cholesterol level soars to 1500 mg/dl and large arteriosclerotic lesions form.[34] The less toxic Western type diet also leads to the development of large lesions in these mice with a plasma cholesterol level of 400 mg/dl.

Other Murine Models Overexpression of human apoA-I in apoE-deficient mice increased HDL cholesterol levels 2-fold and substantially decreased fatty streaks and advanced fibroproliferative lesion formation.[35,36] All but 3–5% of apoE-deficient mice had detectable fatty streaks by 4 months of age, and had lesions that were highly organized and that occluded on an average 25% of the aortic lumen by 8 months of age. In contrast, more than 50% of apoE-deficient mice that overexpressed human apoA-I had no lesions by 4 months of age and by 8 months of age the lesions that occurred occluded on average only 5% of the aortic lumen. These data suggest that overexpression of apoA-I can diminish lesion size and slow the initiation of fatty streak formation.

A human apoB transgenic murine model has also been created. These mice have mildly increased LDL and total cholesterol levels with a lipoprotein profile showing a distinct LDL peak, which differs from the wild-type mice that have only a distinct HDL peak.[37,38] This model also shows a strong diet responsiveness; while a normal chow diet is not atherogenic, arteriosclerotic lesions develop upon feeding the Paigen and/or the Western diet.

ApoE/LDL receptor (LDLR) double knockout mice have recently been created.[34] These animals develop severe hyperlipidemia and arteriosclerosis[39] and have been reported to develop more marked arteriosclerotic lesions with accelerated progression than the apoE-deficient mice even when fed a regular chow diet.[40] Thus, the double knockout model is suitable for studying antiarteriosclerotic compounds without having to feed the animals an atherogenic diet.

More recently, Veniant *et al.* successfully elevated blood cholesterol levels in chow-fed apoE-deficient and LDLR-deficient mice by making both mouse models homozygous for the apolipoprotein B-100 allele.[41] That change resulted in alleviating the hypercholesterolemia in the setting of apoE deficiency, but worsening it in the setting of LDLR deficiency. Furthermore, the LDLR-deficient apoB100/100 mice developed extensive arteriosclerosis even on a normal chow diet.

Conclusions In summary, the apoE-deficient mice are the most widely used experimental model of atherosclerosis because they spontaneously produce lesions that simulate complex human atherosclerotic lesions.[42] It is generally believed that apoE-deficient animals, even when maintained on regular chow, exhibit more severe or complex lesions than do LDLR-deficient mice fed a Western type diet despite higher plasma cholesterol levels. The apoE-deficient mouse model, however, is not an ideal model for most forms of human arteriosclerosis, which often involve an elevation in LDL. In that regard, the LDLR-deficient mouse coupled with apoB-editing deficiency (double knockout mice) seems to be more promising as an LDL model that develops arteriosclerosis on a low-fat diet.[41]

Transgenic Rabbit Models By filling the gap between the laboratory mouse and larger domesticated mammals, transgenic rabbits expressing human genes have become a relatively large mammalian model to explore the pathophysiological processes of arteriosclerosis. To date, a wide variety of human transgenes [e.g., apo(a), apoA-I, apoB-100, apoE-2, and apoE-3] have been expressed in rabbits. These models have helped to elucidate the metabolic roles of selected proteins in atherogenesis.

Elevated plasma levels of lipoprotein(a) [Lp(a)] constitute an independent risk factor for coronary heart disease, stroke, and restenosis.[43] However, apo(a), a unique component of Lp(a), is naturally present exclusively in Old World monkeys, humans, and hedgehogs. Therefore, there are no convenient experimental animal models for studying Lp(a). Using the apo(a) transgenic rabbit model, no arteriosclerotic lesions were found on a regular chow diet, suggesting that lower plasma apo(a) is not atherogenic.[44] When transgenic rabbits were fed a 0.3% cholesterol diet for 16 weeks they exhibited more extensive atherosclerotic lesions than nontransgenic rabbits, although cholesterol levels were similarly elevated. Compared to the lesions in nontransgenic control animals, the areas of atherosclerotic lesions in the transgenic rabbits were increased in the aorta as well as the iliac, carotid, and coronary arteries.

Transgenic Rabbits versus Transgenic Mice Rabbit strains have even more diverse genetic backgrounds than inbred and outbred mouse strains. On the one hand, this might be advantageous for studying complex disease models such as arteriosclerosis, but it is obviously a disadvantage for defining the effects of gain or loss of target gene function.

Rabbits like humans are LDL mammals and have features of lipoprotein metabolism that differ from mice that are HDL mammals. Therefore, the two species show different phenotypes even when the same gene is introduced. For instance, overexpression of apoE in mice caused inhibition of atherosclerosis,[45] but led to increased plasma LDL and spontaneous arteriosclerosis in transgenic rabbits.[46]

ANIMAL MODELS OF PLAQUE VULNERABILITY Prevention of heart attack and stroke depends on the detection of vulnerable plaques and the development of plaque-stabilizing therapies.[47] Vulnerable plaque is one of the toughest cases in animal model design. Plaque rupture is a complication of an already complex atherosclerotic process, and precise mechanisms of this complication remain hypothetical.[48] The current paradigm of plaque rupture may be described as a combination of four overlapping components: mechanical, thrombogenic, inflammatory, and lipid.[4]

Plaque rupture occurs as a result of interactions between external and internal mechanical triggers and vulnerable (i.e., weak) regions of the plaque when forces acting on the plaque exceed its tensile strength. Disruption of the protective fibrous cap allows the thrombogenic (lipid/necrotic) core to come into contact with the blood coagulation system, which leads to the formation of thrombus. Plaques with a thin fibrous cap and large core are considered vulnerable. Hence, the cap thickness, core size, and cap/core ratio are the most basic established "vulnerability endpoints."

Models of Plaque Rupture and Plaque-Associated Thrombosis

Spontaneous Plaque Rupture/Thrombosis Spontaneous hemorrhage and rupture were observed in the coronary arteries of 39- to 54-month-old pigs with inherited hyper-LDL cholesterolemia bearing mutant alleles for apolipoprotein B.[49]

In apoE-deficient mice aged 42–54 weeks, a high frequency (up to 75%) of intraplaque hemorrhage and fibrotic conversion of necrotic zones was found with the loss of the fibrous cap in the innominate (brachiocephalic) artery.[42] Surprisingly, intraplaque hemorrhages were not associated with mural thrombosis. These lesions had been previously overlooked because the majority of studies concentrated on the aortic root and the aortic arch. These observations indicate that the character of atherosclerotic lesions is site dependent.

In another study, 37- to 59-week-old apoE-deficient mice fed the Western type diet died spontaneously. Necropsy revealed luminal thrombi associated with ruptured plaques in the brachiocephalic artery.[50] Recently, these observations were further extended. Of 98 mice, 51 had acutely ruptured plaque in the same locations. The major finding was a significant increase in the number of buried caps in ruptured versus intact lesions, probably indicating previous ruptures in the same plaque.[51]

Induced Plaque Rupture/Thrombosis The use of direct mechanical injury to induce plaque rupture and thrombosis ignores the complexity of clinically relevant trigger mechanisms. Platelet and fibrin-rich thrombi have been induced in association with atherosclerotic plaques in apoE-deficient mice after squeezing the aorta between forceps.[52] Similarly, accelerated atherosclerotic lesions were developed in rabbits by combining balloon injury and hypercholesterolemia, and subsequently mechanically disrupted by inflation of an angioplasty balloon in the arterial lumen.[53] In another labor-intensive rabbit model, arteriosclerotic plaque could be ruptured at will after an inflatable balloon was embedded into the plaque.[54] This model could also be used for induction of thrombi associated with plaque rupture where the pressure needed to inflate the plaque-covered balloon is an indicator of the overall mechanical strength of the plaque.

A more recent study described a pharmacological model of plaque rupture in which plaque was first made vulnerable and then challenged by a systemic stressor.[55] First, accelerated arteriosclerosis was initiated in apoE-deficient mice by placement of a perivascular collar. The resulting plaques were transfected with the p53 gene, which led to an increase in cap cell apoptosis, and in return resulted in a marked decrease in the cellular and extracellular content of the cap. Whereas spontaneous plaque rupture was rare, it was found in 40% of cases after treatment with the vasopressor compound phenylephrine.

Myocardial Infarction-Prone Rabbit Model Originally, the Watanabe heritable hyperlipidemic (WHHL) rabbit was established as a model of human familial hypercholesteremia due to an LDL receptor deficiency.[16] Based on the WHHL rabbit strain, a special rabbit model was developed, designated as WHHL-MI, in which myocardial infarction (MI) occurred spontaneously.[56] This model possesses several types of coronary plaques that are possibly correlated to sudden cardiac events. Although many of the coronary plaques in WHHL-MI rabbits appeared histologically to be rupture prone in nature, no occlusive thrombus was detected in any rabbits, suggesting that some additional stimuli play a definitive role in causing disruption of the rupture-prone plaques and thrombosis. The most common feature of WHHL-MI rabbits is plaques that occlude most of the arterial lumen due to macrophage accumulation, suggesting that they are responsible for sudden cardiac events. This model could be useful for studying the mechanisms of plaque rupture and thrombogenesis if plaque rupture/thrombosis could be induced by additional triggering factors.

Mouse Models Simulating Plaque Vulnerability The currently used mouse models are plaque-centric and do not take into consideration the processes that lead to occlusive thrombus formation after plaque rupture. The current modeling of vulnerable plaques emphasizes the acute nature of plaque rupture. In reality, several layers of spontaneous plaque fissures are frequently observed in coronary arteries and aortas indicating a series of disruptions of the protective cap in which only the last one becomes a clinical event.[48] Even more disturbingly, emerging clinical data suggest that coronary thrombi can arise without plaque rupture and without signs of severe local inflammation.[57]

Any model is based upon a limited set of assumptions and therefore by default has its own limitations. A model often exaggerates only one feature of a complex process to make analysis more convenient and more accurate.[48]

Researchers have largely focused on mouse models for the study of arteriosclerosis because of their obvious advantages (e.g., easy handling, extensive knowledge of mouse physiology, large amount of available genetic information). The mouse models, however, have fundamental limitations.[10] Mice have a lipid physiology that is radically different from that in humans (HDL vs. LDL dominance) and therefore they do not develop arteriosclerosis without genetic manipulation. Furthermore, mice weigh about 25–30 g, some 3000 times less than the average human. Because mouse cells are about the same size as human cells, a section of coronary artery in the mouse contains about 3000 times fewer cells than an equivalent section of human coronary artery. Consequently, the endothelial layer in mice lies directly on the internal elastic lamina and the media consist of only a few layers of SMC. In contrast to humans, arteriosclerotic lesions in the mouse coronary artery often extend beyond the elastic lamina. Also, it is difficult in mice to make a distinction between plaque erosion, as defined by endothelial denudation, and complete rupture of the fibrous cap.[58] Although classic eccentric atheromas with a single fibrous cap exist in lesion-prone murine models, multiple necrotic core areas with or without separate fibrous caps are the norm.[31,59,60] Disruption of these lesions may not mimic plaque rupture in humans, placing a fundamental limit on the applicability of mouse models for the investigation of rupture mechanisms.[58]

In addition to the difficulties arising from the differences between mouse and human biology, problems also occur when two different genetic models of a particular illness are used to investigate the effect of a third genetic manipulation. Another major problem with the genetically manipulated mouse strains is the genetic heterogeneity between the strains used to generate transgenic and knockout mice. This may lead to a situation in which animals containing the same genetic manipulation exhibit profoundly different phenotypes when present on diverse genetic backgrounds.[10] The ideal features of an animal model of plaque rupture are summarized in Table 40–2.

Table 40–2
Ideal features of an animal model of plaque rupture[10]

1. The atherosclerotic process in the animal model should be histologically identical to that in humans
2. The atherosclerotic plaque in the animal model should show the same vulnerability to rupture as its human counterpart
3. The events leading to vulnerability and rupture (i.e., cap/core ratio, cellular composition, collagen production, and breakdown) should as far as possible be identical to those in humans
4. Plaque rupture in the animal model should occur without the need for manipulations such as cuffing of the artery or insertion of balloon catheters
5. Plaque rupture should at least in some cases be accompanied by the formation of platelet-rich fibrin thrombi
6. Response to treatment in the animal model should have the potential for duplication in humans
7. The species should be available for research, easy and economical to maintain, and plaque rupture should occur under reproducible conditions within a reasonable time frame

CONCLUSIONS Arteriosclerosis in humans is a multifactorial condition that develops over many years and modeling it in animals is a notoriously tricky business.[10] Reproducing the events leading up to occlusive thrombosis, the event that actually kills and disables humans, is even more challenging. Existing animal models are useful for studying certain features of arteriosclerosis in regard to the function of specific genes. However, because of the complexity of the processes involved in plaque rupture and thrombosis, it is not certain whether a good model of these complications will ever be developed.

RESTENOSIS

BACKGROUND Restenosis in the form of accelerated arteriosclerosis and intimal hyperplasia constitutes the greatest threat to prolonged vessel patency following treatment of vascular occlusive diseases by percutaneous endovascular revascularization procedures. Although such procedures can achieve high initial success rates (>95%), a significant proportion of the treated vessels (20–50%) develops restenosis within 3–6 months.[61–64]

Animal models have been used successfully to study specific aspects of the early phase of the restenotic process (e.g., smooth muscle cell proliferation). Ideally, restenosis research should be performed on arteries with advanced arteriosclerotic disease. However, animal models with complex arteriosclerotic obstructive plaques are difficult to produce and require expensive, lengthy experiments. Complex lesions similar to those clinically significant in humans have been produced experimentally only in hyperlipemic pigs[49] and in nonhuman primates.[65]

Although arteriosclerosis progresses over a lifetime in humans, experimental settings try to reproduce atherosclerotic lesions in weeks or months based on diet modifications, and then following vascular injury the restenosis process is studied over a period of weeks or months. Furthermore, there are significant interspecies differences relevant to vascular disease. While elastic arteries (aorta and its primary branches) from various species have been reported to demonstrate remarkable structural similarity, muscular arteries (coronary and femoral arteries) exhibit major differences between species with respect to their thickness, medial elastin content, and intimal cellularity.[66] In addition, the type, duration, and number of successive injuries as well as the associated forces (e.g., shear stress) cause a proportional injury-related vascular response in various species, which is measured by an intima-to-media area ratio or luminal cross-sectional area narrowing by neointima. However, the relationships vary among species; the injury-neointimal thickness regression slope is steep in pigs, intermediate in baboons and rabbits, and almost flat in dogs.[67–69]

VASCULAR INJURY TECHNIQUES Localized arterial endothelial and mural injuries are primarily used for studying restenosis (morphology, mechanisms, and therapeutic strategies). Vascular injury accelerates lesion development and can be produced in animal models using a variety of methods including dietary manipulation. Some of the nondietary methods for inducing vascular injury include the Baumgartner balloon catheter technique,[70] nylon and wire loop techniques,[71,72] air desiccation,[73] chemical irrigation,[74] thermal drying,[75] and irradiation.[76] As the names imply, mechanical devices (balloon catheters or wire loops) or agents (chemical, electrical, thermal, or radiation) are used to create site-specific injury in the arterial tree, and lesions develop shortly thereafter. Restenosis can be studied on either normal or previously injured arteries. In the latter model, the animal may be placed on a high fat/high cholesterol (HFHC) diet after focal injury to enhance lesion formation.

The most commonly employed vascular injury technique is focal balloon catheter injury. This technique involves inflating a Fogarty balloon at the desired site within the artery and then withdrawing the dilated balloon along a specific length of the vessel to create adequate shear stress on the wall to produce a significant injury. Nondietary methods such as this can be used either alone or in combination with an HFHC diet.

The decision to use a particular method or a combined approach to induce vascular injury depends on the type of lesion that is required, which in turn depends on the purpose of the study. When combinations of techniques of disease induction are utilized, lesions develop at an accelerated pace and with greater similarity to the human prototype.[77] Such manipulations also allow greater control over the type of lesion that can be created. The utilization of combined techniques has been more successful in creating a range of morphological lesions of desired severity.[77–84] For instance, a combination of an atherogenic HFHC diet and balloon-mediated endothelial injury in Yucatan microswine produces consistent high-grade arterial stenoses, and this model has been proposed for the study of cardiovascular interventional techniques.[78]

The creation of fatty streaks and plaques with minimal luminal narrowing may be sufficient for assessing the cellular and molecular roles in the histopathogenesis of restenosis,[82] whereas a lesion of intermediate magnitude with moderate narrowing of the luminal diameter is needed for pharmacological therapy research.[83] The investigation of endovascular recanalization devices requires consistent high-grade stenoses.[84] Injury of the media is critical to the formation of such lesions and an integrated approach to their creation may be useful.[85]

ANIMAL MODELS Arterial injury models simulating vascular injury in humans resulting from revascularization procedures are created in animals with the aim of determining morphology, pathogenesis, and preventive as well as therapeutic strategies. However, there is no single ideal model. The rat, rabbit, swine, and nonhuman primate have all been used as models of restenosis, with the last two affording the best approximation to humans. The rat carotid artery balloon injury model developed in the 1960s has been well characterized. However, the rat as well as the rabbit are small animals and do not serve all the demands of restenosis models. In this regard, the larger size of the Flemish Giant rabbit (6–7 kg) offers an advantage over the common New Zealand white rabbit (3–4 kg), and this strain of rabbits has been used to study the effects of percutaneous transluminal angioplasty (PTA) on the arterial wall.[86] As for testing coronary stent designs, the rabbit iliac artery offers a reliable model.[87,88]

Studying restenosis following stent placement in peripheral vessels requires large animal models such as the sheep, dog, and pig.[89–91] These models allow evaluation of postplacement healing, short- and mid-term stent patency, as well as mechanical integrity of the stent.[92] They also allow determination of the mechanical factors dictated by the stent design that may play a significant role in the development of intimal hyperplasia.

In large animal models, arterial size allows easy targeting of the selected vessel (e.g., coronary arteries), and this is one reason the porcine coronary artery model has become a common model for coronary restenosis studies.[93,94] In addition, the neointima that forms in this model is identical to human restenotic neointima, and the amount of neointimal thickening is directly proportional to the degree of injury. Therefore, this model permits the creation of an injury-response regression relationship that can be used to quantitate the response to potential therapeutic strategies.[95]

Previously, research activities related to restenosis following balloon angioplasty focused on intimal hyperplasia and the smooth muscle cell as crucial players in its pathogenesis. Dissenting opinions, however, have emerged. The main reason for this has been that despite favorable results with numerous pharmacological strategies in many animal models, clinical trials in humans did not match the experimental success.[96] This discrepancy led to a rethinking of the pathogenesis of restenosis, and arterial wall remodeling is now being considered as a major contributor to the underlying mechanisms of restenosis.[97]

Endovascular procedures in humans are carried out on atherosclerotic arteries with large plaques, whereas in the majority of animal experiments, procedures are used to create injury in the normal disease-free vascular wall. Therefore, factors contributing to the development of restenosis may be different in animal models, which would explain the lack of success of therapeutic strategies in humans.

Angioplasty Until recently, PTA has been the most commonly used revascularization procedure in clinical practice. PTA alleviates luminal stenosis by affecting plaque fractures and remodeling with overstretching of the disease-free portion of the arterial wall.[98,99] Acute restenosis is often the result of wall spasm, *in situ* thrombosis, and/or elastic recoil.

Constrictive vascular remodeling, which is defined as a loss of vessel area within the external elastic lamina, is the major determinant of chronic restenosis in humans after balloon angioplasty. Excessive neointimal formation, including proliferation of smooth muscle cells and synthesis of the extracellular matrix, has traditionally been considered the major cause of restenosis. However, a third process, the formation of neoadventitia, has been suggested to play an active role in both vascular remodeling[100] and neointimal formation.[101,102] It was hypothesized that neoadventitia could act in a way similar to scar contraction leading to "strangulation" of the vessel.[100] It was also shown that adventitial fibroblasts migrate toward the lumen and take on a myofibroblastic phenotype to form the neointima.[101–103]

A recent porcine model has provided more accurate insight into the pathophysiology of constrictive vascular remodeling.[97] After 30% overdilation of the porcine coronary arteries, loss of lumen, constrictive remodeling, and formation of neointima occurred between days 7 and 28. These processes were preceded by the formation of neoadventitia and adventitial neovascularization occurring during the first 4 days after angioplasty. Neointimal formation began just after day 7; therefore, the constrictive remodeling coincided with the formation of neointima rather than the formation of neoadventitia. At late follow-up, loss of lumen was found to be correlated with remodeling, not neointimal formation. However, the neointimal area was not correlated with either late loss of lumen or vascular remodeling. These results indicate that qualitative rather than quantitative changes within the vessel wall neointima, media, and adventitia may play an important role in late luminal narrowing.

In conclusion, the time course of events after PTA is as follows: (1) platelet adhesion/aggregation and *in situ* thrombosis occur minutes after a type II/III arterial injury,[63,104] (2) formation of neoadventitia including adventitial neovascularization occurs within 4 days, and (3) late loss of lumen consisting of constrictive remodeling and formation of neointima occurs between 7 and 28 days.[97] Late loss of lumen is more dependent on vascular remodeling than on the formation of neointima. The time course of remodeling, however, coincides with the growth of neointima rather than with changes in the adventitia, indicating that constrictive vascular remodeling might depend on the formation of neointima. Thus, the present knowledge of restenosis after balloon angioplasty indicates that in humans and pigs late loss of lumen after balloon angioplasty is caused more by constrictive remodeling than by neointimal formation.

Stenting and Drug-Eluting Stents PTA has been replaced by arterial stenting as the most common revascularization procedure in certain instances. This is especially true for the coronary artery and the majority of the peripheral vascular territories where primary stenting has become the treatment of choice.

Compared to percutaneous transluminal coronary angioplasty (PTCA), the introduction of intracoronary stents reduced the 3- to 6-month restenosis rate to 20–30%.[105] To further reduce the "in-stent" restenosis rate, drug-eluting stents (DESs) have recently been introduced into clinical practice for coronary artery application. This change in clinical practice was made possible by extensive preclinical studies that required relevant changes in the animal models used to evaluate restenosis after stenting.

The pathological responses to balloon dilatation (PTA) and stenting differ. Regional constrictive remodeling predominates after balloon angioplasty, whereas after stent implantation, restenosis is due solely to neointima thickening. Early studies in the porcine coronary injury model suggested that the deeper the arterial injury, the greater the neointimal thickening.[106] This proportionality in the pig model resulted in the development of stent designs that reduce arterial injury. Early balloon expandable stents caused substantial injury when they were overexpanded, whereas the currently used self-expanding slotted tubular stent designs produce less injury.

Overstretch injury to pig coronary arteries holds important lessons for neointimal response to injury. Simple overstretch without stent implantation causes medial fracture and laceration with frequent dissections.[107] A typical balloon/artery ratio is 1.2/1 or 1.3/1, which creates enough injury for acceptable neointimal thickening without the risk of large dissections. Larger balloon/artery ratios yield the likelihood of severe dissection with resulting thrombosis and occlusion. When stents are implanted, dissections are usually controlled except at the stent margins. However, stent/artery ratios of >1.3/1 often result in chronic vessel injury dominated by a high degree of inflammation. This inflammation is undesirable because drug elution cannot overcome such severe injury, making assessment of stent/drug efficacy impossible. Therefore, for preclinical DES testing a stent/artery ratio of 1.1/1 is usually employed so as to produce only a minimal to mild degree of neointima formation, which in turn makes it possible to determine the efficacy of the drug used for coating the stent.

The porcine coronary model suggests a temporal comparability for healing after stenting that can be predicted for human application. Stent healing occurs approximately six times faster in pigs compared to humans.[107] This proven time relationship has established standardized times for porcine studies (1, 3, 6, and 12 months). Times points after 1 month are used principally to evaluate safety issues only because few changes occur in the pig model beyond this time with the exception of the slow healing process.

Self-expanding stents induce a neointima composed of cells migrating from the media. Conversely, balloon overstretch injury resulting in complete rupture of the media and exposure of the adventitia to blood components can lead to SMC stimulation in addition to activation and proliferation of adventitial fibroblasts followed by their inward migration and participation in the formation of neointima. The different SMC markers characteristic of these two different pathomechanisms of neointimal formation were demonstrated in the established porcine coronary artery model.[108] The lessons learned from animal restenosis models are summarized in Table 40–3.

It is essential to rapidly and effectively screen candidate compounds for DES before large, expensive, and time-consuming preclinical trials take place. Moreover, it is necessary to evaluate

Table 40–3
Lessons learned from animal restenosis models[107]

1. Arterial injuries are major determinants of neointimal thickening; the optimization of stent design should limit arterial injury as much as possible
2. Neointimal formation on drug-eluting stents (DESs) develops the same as on bare-metal stents
3. Although DESs limit neointimal formation, they may also delay or cause incomplete healing to a greater degree than bare-metal stents
4. The porcine coronary and rabbit iliac artery models provide useful information regarding stent thrombosis risk, and thus can be used to measure safety
5. Significant lumen loss in animal models resulting from medial cell death, inflammation, and neointimal thickening should be taken into consideration when initiating clinical trials
6. Efficacy testing in currently used preclinical models (rabbit, porcine) has proven difficult to establish
7. Innovative preclinical model enhancements are required to achieve more rapid turnaround time of the results of experimental interventions

safety-related pathological features of the existing antirestenotic compounds to properly define the therapeutic window of candidate compounds for DES.[109] Therefore, it is important that the animal model used possesses the underlying atherosclerotic lesion. DES may not only affect the formation of in-stent restenosis, but could also affect the formation or progression of the underlying atherosclerotic stented lesions (stabilization vs. destabilization of the stented plaque).

While the porcine coronary stent model appears predictive for stent thrombosis and adverse neointimal formation, this model is simply too expensive and time consuming to be used for screening drugs suitable for DES technology. Furthermore, the lack of arteriosclerosis in this native-vessel model is a major limitation to the proper evaluation of the efficacy and safety of DES strategies.

Well-established genetically modified murine models appear to be helpful for the purpose of accurately screening candidate compounds for use on DES. Pires *et al.* recently described an inventive new model using an established transgenic mouse strain to screen possible drugs for DES.[110] First, a nonconstrictive polyethylene perivascular cuff is surgically placed around the femoral artery of diet-driven hypercholesterolemic apoE*-Leiden transgenic mice, which results in the development of atherosclerotic-like lesions in 2 weeks. Then, the first cuff is replaced by a second drug-eluting cuff constructed from a poly(ε-caprolactone) polymer matrix that is loaded with the drug of interest. This novel mouse model enables cost-effective screening of candidate drugs for DES and the establishment of dose–response relationships with respect to both therapeutic and harmful effects on the vessel wall. The candidate drugs with reasonable potential for DES can subsequently be tested in large animal models. Evaluating DES in the presence of atherosclerosis results in new insights into restenosis and elucidates the effects of DES on atherosclerotic lesions.

CONCLUSIONS Animal models have been used successfully to study specific aspects of the restenotic process. These arterial models are created to simulate the vascular injury seen in humans following revascularization techniques (i.e., balloon angioplasty and stenting) with the aim of determining morphology and pathogenesis as well as preventive and therapeutic strategies. However, none of the described methods for inducing restenosis produces a single ideal model.

ENDOVASCULAR ANEURYSM REPAIR

BACKGROUND Among the animal models that have been created for studying aneurysm formation and repair, abdominal aortic aneurysm (AAA) models have dominated the field. The ideal animal model for AAA should possess all of the features of the human disease. While the manifestations of AAA may vary considerably depending on the stage of the disease, consistent features of the dilated segment are medial degeneration and adventitial inflammation. There are several small animal models (mouse, rat) in which biochemical or genetic perturbations have been used to develop AAAs. The interested reader is referred to detailed articles on the basic mechanical and clinical implications as well as small animal models.[111–114]

In the past decade, the use of stent grafts has become the therapeutic alternative of choice for the endovascular repair of AAAs, while minimally invasive transcatheter embolization techniques have become the treatment of choice for endovascular treatment of intracerebral aneurysms. Recent developments in these two fields of endovascular aneurysm repair have been based on extensive experimental studies using a variety of animal models. An overview of these models follows.

AORTIC ANEURYSM MODELS One of the most exciting developments in interventional imaging guided minimally invasive techniques in recent years is endovascular stent grafts for the endoluminal exclusion of aneurysms. Animal testing of endovascular grafts, and of the techniques employed in their deployment, is an essential step for the successful clinical application of these devices. Such experimental trials require the creation of an aneurysm in an animal model for effective testing. Since 1986, when Balko *et al.*[115] reported the first endoluminal exclusion of an experimentally created AAA in sheep, several large animal aortic aneurysm models have been used for the study of endovascular grafts. These have included models of dissecting, saccular, fusiform, and acutely rupturing aneurysms. The choice of research animal and the choice of aneurysm model to be used for the study of endovascular grafts are two of the most important decisions to be made by an investigator. Animal models can serve as an adequate test for endovascular prostheses only to the extent that they simulate the human condition.

Animals The choice of an experimental animal is especially important because of species differences relative to humans. Appreciable differences in the coagulation and fibrinolytic systems between humans and the experimental animal used to evaluate an endovascular prosthesis could significantly bias the interpretation of test results. Comparative studies have shown that the clotting and fibrinolytic systems of the calf and nonhuman primates are more similar to the human than to those of dogs or pigs.[116] Interspecies differences also exist regarding neointimal formation. Apart from specific situations (subsequently described), all models reported to date have employed normotensive, nonatherosclerotic animals.

In 1993, the Ad Hoc Committee of the Joint Councils of the Society for Vascular Surgery and the International Society for Cardiovascular Surgery, North American chapter, recommended that the canine be used for preclinical testing of arterial grafts in most instances.[117] According to the committee's report, the dog has two characteristics that make it a suitable animal model. One is the lack of significant spontaneous endothelialization of prosthetic surfaces, which is similar to humans. The other is a variable and relatively unpredictable tendency toward hypercoagulability, which presents a challenging test of device thrombogenicity. Other advantages of the dog include large peripheral arteries, ease of handling, and tolerance of prolonged anesthesia. However, the availability of dogs is being restricted by animal-rights activists, and as a result, the cost of using dogs is rising sharply. Furthermore, the dog has an extremely potent fibrinolytic system.[116]

Pigs and sheep also have been used for the study of endovascular grafts, although not as frequently as dogs. The arterial morphology of the pig is most similar to humans; they exhibit a tendency toward hypercoagulability, and their fibrinolytic system is not as active as the canine. When compared with dogs, however, they are not as easy to handle, are less tolerant of anesthesia, and their arteries are smaller, making it difficult to use large introducer systems and prostheses. Sheep, however, are more similar in size to the human and their arteries are large enough to accommodate larger diameter prostheses (18 mm). In addition, their coagulation system is closer to the human than either dogs or pigs. The sheep is an easy animal to manage in the laboratory and to use for testing of vascular devices. However, sheep can harbor a rickettsial infection (Q fever) that is transmissible to humans.

Experimental Models and Techniques Only a few published reports have dealt with the evaluation of endovascular grafts in animal models of dissecting and acutely rupturing aortic aneurysms. The standard technique for the creation of dissection aortic aneurysms continues to be the Blanton technique.[118] The aorta is surgically exposed, cross-clamped, and a 120° transverse aortotomy incision is performed. An intramural pocket is formed by dissecting a free edge of the vessel caudally, and the outer wall of the free edge is sutured back to the proximal edge of the aortotomy site creating a dissecting aneurysm. Williams *et al.*[119] reported the use of covered stents in a model of acutely rupturing thoracic aortic aneurysm. The model was created by puncturing the aortic wall with a Colapinto needle and then balloon dilating the tract. However, immediate treatment and homologous blood transfusions were needed to prevent rapid exsanguination.

To date, endovascular grafts have been studied primarily for the purpose of intraluminal exclusion of saccular and fusiform aneurysms of the infrarenal abdominal aorta. Since the first published studies in the mid-1980s several experimental AAA models have been used to test the delivery, biocompatibility, and efficacy of endovascular grafts.

Anterior Patch Model The anterior patch AAA model has been one of the most popular animal models for the study of endovascular grafts. Saccular AAAs have been created in dogs, pigs, and sheep using various patch materials including Dacron,[115,120] rectus fascia without[120,121] and with[122] peritoneum, jejunum,[123,124] iliac vein,[125] and double layer peritoneum.[126] The advantage of using autologous tissue as a patch material is that it can progressively enlarge and rupture similar to a human AAA. This has been shown to occur with fascial and jejunal patches. Although iliac vein patches enlarge, they do not rupture because scar tissue forms around the aneurysm. The double layer peritoneum model created in swine has been shown to have the closest behavior to human AAA in regard to gradual expansion and tendency for rupture. Its rupture rates (70% in long and 20% in short aneurysms within 2 weeks) are higher than the rate associated with rectus fascia with peritoneum (11% in 5 months)[122] and with iliac vein (0% in 4 ± 2 months),[125] much lower than that with untreated jejunum (100% within 42 h),[123] and comparable to that with glutaraldehyde-treated jejunum (66% at 11 days).[123]

During the creation of the anterior patch aneurysm model, the aorta is isolated, the animal is heparinized, the aorta is cross-clamped below the renal arteries and above the iliac bifurcation, and collaterals are temporarily controlled with vessel loops or hemoclips that are removed upon the completion of the aortic patch suture. A longitudinal aortotomy is created in the anterior aspect of the aorta and an elliptical-shaped patch is sutured into the incision. No mural thrombus has been reported within fascial or iliac vein patch aneurysms and only trace amounts of thrombus have been found in Dacron patch aneurysms prior to graft implantation. It is unclear whether this is a limitation of these models because the effect that thrombus deposition within an aneurysm has on the early healing of the graft is unknown. Aneurysms formed by a jejunal patch have been found to contain thrombus prior to grafting, but none has been completely thrombosed. Approximately 95% of the aortic collateral arteries associated with anterior patch AAAs have remained patent prior to graft replacement. AAA models without patent lumbar and inferior mesenteric arteries are unrealistic. The absence of collateral branches prevents the study of their role in maintaining patency of the aneurysmal sac after endovascular grafting and the efficacy of any particular device in preventing aneurysm rupture. At 30 and 60 days after graft placement, approximately 80% of the lumbar arteries bridged by the prosthesis have reportedly thrombosed. Collateral flow has remained in some instances in which blood flow into and out of the excluded aneurysmal sac occurred via two separate vessels. Unfortunately, with the anterior patch model, lumbar arteries are usually flush with the endovascular graft and are more likely to thrombose than if they were more separated from the prosthesis, as may be the case in human aneurysms.

An "endoleak" AAA model based on a modified anterior patch technique was created to yield a large aneurysmal sac after stent-graft placement for testing of techniques for its percutaneous occlusion.[127] Infrarenal AAAs were created in dogs by anastomosis of an isolated segment of the inferior vena cava (IVC) to the right side of the abdominal aorta in combination with a large anterior patch from the external jugular vein. During surgery, the aorta and IVC were isolated from the renal arteries to the bifurcation, and both were cross-clamped at these levels. The aorta and IVC were opened along their apposing surfaces, anterior to the origin of the lumbar arteries and veins. The medial aspects of the aorta and IVC were joined over a distance of approximately 3 cm in a side-to-side anastomosis. At least one or two lumbar arteries and one or two lumbar veins were incorporated into the AAA sac formed. The anterior wall of the aorta was then partially resected and replaced with a large vein patch. The patch was sewn to the aorta and to the anterior aspect of the IVC incision to create the anterior aneurysmal pouch. The IVC was then ligated immediately below and above the level of the side-to-side anastomosis, thereby transferring the anastomosed segment of the IVC and its lumbar veins to the aorta. One hour later, the animals underwent percutaneous implantation of polytetrafluoroethylene-covered Z stent endografts with three 3-mm-diameter holes through the fabric. Immediately after endograft placement, all nine animals had artificial type III endoleaks with angiographic filling of lumbar arteries and veins. One aneurysm ruptured at 1 week. Of the remaining eight endografts, six remained patent for 1 week (n = 1), 1 month (n = 1), 2 months (n = 2), 3 months (n = 1), and 6 months (n = 1), and two were occluded at 3 and 6 months, respectively. The aneurysm sac had enlarged by approximately 50% in seven animals during the follow-up period. The model proved to be technically feasible, but was not found to be reliable for chronic evaluation because of rapid progressive thrombosis in most aneurysm sacs and occasional complete thrombosis of the AAA and endograft. The model was later modified by Hiraki *et al.* to eliminate the use of the jugular vein patch and thereby shorten the aortic clamping time to about 30 min.[128]

Mural-Stripping Model Economou *et al.*[129] described a method for creating saccular aortic aneurysms in dogs that involved removal of the adventitia and the outer 60–70% of the media, while leaving the remaining media, internal elastic membrane, and subendothelial layer intact. After dissection of the adventitia, the media was cut through in layers with gentle strokes of the scalpel. By not occluding the aorta during dissection, the pressure within the vessel produced a bulge at the dissected area, which acted as a guide to the depth of cutting necessary. Once the desired depth of the cut was achieved, the media was peeled along a cleavage plane at the depth of the dissection. An ellipse of media was then cut away. Optimal aneurysm formation occurred when approximately 70% of the media had been excised. A lesser dissection failed to produce an aneurysm, whereas removal of more than 70% of the media resulted in acute rupture of the resultant aneurysm. All aneurysms persisted for up to 13 weeks and no evidence of spontaneous repair was noted microscopically. Moreover, production of hypertension before and after aneurysm creation did not result in larger aneurysms at the time of death. This method was used by Mirich *et al.*[130] to create AAAs in dogs for the evaluation of an aortic stent graft. Although cross-clamping of the aorta and its side branches was not required, surgical stripping of the adventitia and the media was slow, uncertain, and difficult to perform. Furthermore, no definite aneurysm was identified in any dog 2 weeks after surgery and angioplasty was required to achieve mild aortic dilation. The mean aortic diameters after the procedures increased by only an average of 46% (range, 36–58%).

Interposition Graft Model Parodi *et al.*[131] described a fusiform AAA model for the study of endovascular grafts that involved replacement of a portion of infrarenal aorta with a large crimped artificial graft. To date, this model has been created in dogs, pigs, and sheep with interposition grafts to mimic an aneurysm made from crimped woven Dacron,[131–135] biomedical grade elastomeric polyurethane,[133] and segments of glutaraldehyde-treated bovine internal jugular vein.[136]

After isolation of the aorta, the animal is heparinized, the aorta is cross-clamped below the renal arteries and above the bifurcation, and collaterals are ligated and divided. A segment of infrarenal aorta 2–6 cm in length is excised. The artificial aneurysm is anastomosed end to end with the aorta as an interposition graft.

All aneurysms created with an artificial conduit have remained patent prior to endovascular graft placement. A decrease in aneurysmal diameter has been reported, but it has been difficult to determine if this decrease was associated with the formation of mural thrombi or with fibrosis and shrinkage of the graft material. In some studies, nonsignificant mural thrombus formation was noted in the aneurysmal sac by the time of endovascular graft placement. A moderate to severe fibrous reaction was present around the artificial aneurysm in all animals.

All aneurysms created with glutaraldehyde-treated bovine internal jugular vein were immediately pulsatile and remained so until endovascular exclusion.[136] No significant change in maximum aneurysm diameter was reported, although 1 of 12 aneurysms ruptured. High-flow jets at the proximal and distal anastomoses with turbulent flow in the sac itself were noted prior to stent placement. This was due to fibrointimal hyperplasia at the anastomoses. No mural thrombus developed in the sac prior to stent placement, and in all cases the sac wall demonstrated a florid inflammatory reaction with foreign body giant cells.

The interposition graft AAA model is useful for the evaluation of delivery systems and of graft deployment techniques; advantages of this model include an aneurysmal size and configuration that more closely resemble human AAAs. However, it is not appropriate for studying either the role of the biological aneurysmal wall in the healing process around the prosthetic bypass or the effects of the patent aortic side branches. Furthermore, scarring of the prosthetic aortic system occurs, which is an experimental variable not likely to be present in human AAA.

Elastase Model Martin *et al.*[137] reported the creation of AAA in dogs by temporary exposure to the elastolytic action of pancreatic elastase. Segments of the abdominal aorta measuring 3 cm in length were exposed to 0.1% trypsin or 0.1% elastase for a period of 90 min while blood flow was arrested. All animals underwent thyroidectomy and were given an HFHC diet with a large dose of vitamin D. None of the dogs exposed to trypsin developed aneurysms, whereas 50% of the dogs exposed to elastase developed aneurysms of the exposed aortic segment 2 months after the procedure. The elastin content of the aorta in the exposed area averaged 51% of that in the unexposed area. The technique, however, proved to be inconsistent, and when successful resulted in the formation of small aneurysms.

Boudghène *et al.*[138] reported the use of elastase to create an AAA model in dogs for evaluation of endovascular grafting. All infrarenal lumbar arteries, except for one lumbar branch at the caudal end of the aortic segment, were ligated. After heparinization, a 4-cm length of infrarenal aorta was isolated between two clamps. This isolated segment was catheterized via the preserved lumbar artery, which was then tied to the catheter. The blood was removed from the isolated segment and the segment was flushed several times with saline. A solution of porcine type I elastase was then infused at a rate of 60 ml/h for 40 min. The dose of elastase was found to be critical. When low doses of elastase (<2000 U) were used, no aneurysm formation occurred, and when high doses of elastase (>3000 U) were used, the aneurysms ruptured within 24–72 h. All animals treated with 2800 U of elastase developed a 4-cm-long aneurysm of the infrarenal aorta. Aneurysmal diameter was approximately twice the size of the normal aorta and aortic wall thickness decreased by an average of 50%. Parietal thrombus was present in three of eight animals. Histological examination showed intense inflammation around the induced aneurysm. In addition, the elastic laminae were frayed and broken in all sections and had totally disappeared in some areas. These lesions were not uniform throughout the length or circumference of the perfused segment and are similar to acquired aneurysms in patients that result from the progressive degeneration of the arterial wall with diffuse loss of elastic tissue. Ligation of the lumbar arteries is a technical disadvantage in the assessment of their role in maintaining AAA patency after grafting and in the evaluation of the efficacy of the graft in preventing aneurysm rupture.

Transluminally Created Model Hallisey[139] reported the first transluminal creation of a fusiform AAA in dogs. After systemic heparinization, a Palmaz stent was deployed in the infrarenal aorta. The stent and aorta were then dilated to twice the diameter of the normal aorta with an appropriately sized angioplasty balloon. The animals were maintained on oral aspirin daily for the entire 30-day follow-up period. All lumbar arteries remained patent and the AAAs remained unchanged. No gross evidence of rupture or retroperitoneal hematoma was noted in any of the dogs. A less than 1-cm-long nonstented but dilated segment was seen at either end of the aneurysm. The normal mural architecture was present, although hidden by stent compression. Some degeneration of the smooth muscle and replacement by fibrous connective tissue were seen. The luminal surface of the AAA was smooth and there was no significant inflammation around the stent wires, even when they entered the tunica media of the aortic wall.

This AAA model is easily created transluminally without the need for retroperitoneal dissection or major surgery. The aneurysms are short in length, with a normal neck of abdominal aorta above and below. Lumbar arteries and the lumen of the aneurysm remain patent. However, the model lacks true elastin breakdown, inflammatory cell reaction, and arterial wall degradation, all of which are found in human AAAs. The Palmaz stent also prevents any reduction in the size of the aneurysmal sac that might occur after endovascular graft placement.

The model originally described by Hallisey was modified by Schoder *et al.* who created an infrarenal AAA sheep model.[140] First, the aorta was overdilated with a transluminally placed Palmaz stent similar to the Hallisey model. Then an approximately 20% undersized endograft suspended between two stent-graft adapters was used to bridge the aneurysm in such a manner that two pairs of lumbar arteries remained patent within the residual aneurysm sac. The stent-graft adapters were large-vessel occluders consisting of a 20-mm-square stent with four barbs and a small intestinal submucosa (SIS) cover with a 6-mm-long central slit. The specific goal of this modification was to percutaneously create an AAA model that duplicates the anatomic and physiological events associated with endoleaks after endografting.

CEREBRAL ANEURYSM MODELS The selection of an experimental *in vivo* cerebral aneurysm model (CAM) for evaluation of medical devices for repair requires two critical decisions: the first is the selection of the specific animal species and the second is the methodology of CAM creation. For example, device safety is tested in a swine model due to its hypercoagulative and hyperproliferative nature. However, the efficacy of the bioimplant is difficult to assess in pigs since aneurysms in these animals readily thrombose. In contrast, a device successful for aneurysm treatment in humans will more likely be effective following evaluation in canine models because of the dogs' strong fibrolytic profile. High patency rates in canine sidewall CAMs have been demonstrated by angiography in excess of 6 months.[141] Thus, the selection criteria of an experimental *in vivo* CAM are strongly dependent upon the study objective.

Experimental Models and Techniques Numerous experimental *in vivo* CAMs have been reported in the literature, and the methods of creation may be divided into the categories of mechanical vessel wall trauma, chemical injury, enzymatic elastolysis, surgical anastomosis of a venous pouch, and the combination of induced renal hypertension, unilateral carotid ligation, and lathyrism. In the latter model, high hemodynamic stress and the administration of neurotoxic amino acids leading to angiolathyrism create an environment for the eventual formation of aneurysms. This model produced in nonhuman primates is the most representative model of the pathogenesis of human cerebral aneurysms; however, unpredictable aneurysm formation, high cost, and the length of time for aneurysmal dilations make this model unattractive.[142] This model may be reliably produced in rats or mice, but the small size of the vessels makes evaluation of endovascular repair impractical.[143–145] Only enzymatic digestion and surgical venous pouch construction have been shown to reliably produce practical CAMs.

One of the oldest and most frequently reported ways to create a CAM involves an arteriotomy with a vein segment sutured to the margins of the incision.[146] Many variations of this technique have been employed in multiple animal species including the rabbit, canine, swine, and rat. Furthermore, a myriad of configurations of both the venous pouch anastomosis as well as the parent artery have evolved to produce a variety of aneurysm types including fusiform, saccular sidewall, bifurcation, and terminal bifurcation geometries. Stehbens was the first to describe a bifurcation venous pouch CAM made in the rabbit at aortic bifurcation.[147] Strother *et al.* developed a method of constructing a bifurcation CAM by performing an end-to-side anastomosis of the left common carotid artery (CCA) to the right CCA and subsequently suturing a vein segment to the apex of this juncture.[148] One key advantage of the lateral wall configuration of the venous pouch model is the ability to create similar CAMs bilaterally in the same animal allowing for the evaluation of the safety profile of a repair device compared with a control device in the contralateral CCA. However, the lateral wall model does not produce realistic intraaneurysmal flow characteristics, which may be simulated with the more complex bifurcation models.

Although the surgical techniques used to create the bifurcation models allow excellent manipulation of the aneurysm-parent vessel geometry and the ability to create various aneurysm types in many different animal species allowing precise control of the local hemodynamics,[148,149] the venous pouch CAMs are not histologically similar to the human aneurysms that they are intended to represent.[150,151] One striking dissimilarity is the average aneurysmal wall thickness, which was found to be more than four times greater in the venous pouch CAM model compared with human aneurysms.[150] Moreover, the attenuation or absence of the tunica media and the internal elastic lamina at the dome of the human aneurysm is not reproduced in the experimental CAM, which might explain the infrequent rupture of venous pouch CAMs (other than acutely at the suture line of the neck). Within 2 h of creation, the experimental CAM is extensively infiltrated by inflammatory leukocytes. Rapid, spontaneous, and progressive fibromuscular neointimal proliferation occurs shortly after the venous pouch is sutured to the arterial segment and has been shown to continue for as long as 2 years.[152] It has been postulated that the disruption of the endothelium by suturing a venous patch to the margins of an arteriotomy causes the release of platelet-derived growth factor, which leads to fibrotic scarring.[153,154] The presence of fibrotic scarring in the model, which is frequently absent in human aneurysms, could complicate treatment efficacy studies[154] if they are not conducted over sufficient time, since mural thickening occurs over several months.

More recently, a method of experimental CAM construction has been proposed wherein the wall of the aneurysm is arterial and not venous.[153,154,155–158] The fundamental concept of the technique is to create an arterial "stump" in the CCA of the rabbit either by surgical ligation or by the use of an endovascular balloon, and then to incubate within the stump pancreatic elastase for a period between 10 and 45 min. This technique has been employed to create CAMs that resemble human cerebral aneurysms for testing of endovascular treatment modalities.[141,156]

The elastase-induced rabbit CAM has been extensively studied over the past decade, and the results indicate that this model offers certain advantages over the experimental vein pouch CAM. The technique is simple, the model has been shown to remain patent for more than 1 year, and depending on the artery used to construct the CAM, both lateral aneurysms located on the outside aspect of a curved parent vessel and bifurcation-type aneurysms may be induced that mimic the geometry typical of human intracranial aneurysms. Furthermore, the elastase CAM is histologically representative of human cerebral aneurysms, with a similar wall thickness, attenuation or absence of the internal elastic lamina beyond the neck, the absence of intramural inflammation, and only a moderate presence of cellular elements within the aneurysmal sac.[150] The migration of fibroblasts and cellular elements into the sac of the venous pouch CAM due to the disruption of the basement membrane by the surgical wound is avoided by the elastase technique. Thus, it has been proposed that the elastase-induced CAM shares more common morphological and histological characteristics with human cerebral aneurysms than the vein pouch model.[150] However, control of the angioarchitecture of the elastase CAM is not reliable, as it has been discovered that hemodynamic factors determine the final dimensions of the CAM.[159] Another point to keep in mind is that if this model is created in the CCA, the proximity of the model to the thorax creates vascular motion due to respiration. This motion can potentially affect the evaluation of treatment strategies since it is not present within the intracranial vessels.

CONCLUSIONS In conclusion, judicious selection of an aortic or cerebral aneurysm animal model allows specific scientific hypotheses to be addressed. Therefore, the choice of the animal species and the technique employed to create the aneurysm must be based on the desired characteristics of the model that are compatible with the objectives of the experiment.

ACKNOWLEDGMENT

This updated work was substantially based on the following prior review article: Narayanaswamy M, Wright KC, Kandarpa K. Animal Models for Atherosclerosis, Restenosis, and Endovascular Graft Research. JVIR 2000;11:5–17.

REFERENCES

1. Gimbrone MA Jr. Vascular endothelium, hemodynamic forces, and atherogenesis. *Am J Pathol* 1999;155:1–5.
2. Ross R. Atherosclerosis—an inflammatory disease. *N Engl J Med* 1999;340:115–126.
3. Reardon CA, Getz GS. Mouse models of atherosclerosis. *Curr Opin Lipidol* 2001;12:167–173.
4. Rekhter M. Vulnerable atherosclerotic plaque: Emerging challenge for animal models. *Curr Opin Cardiol* 2002;17:626–632.
5. Libby P. Current concepts of the pathogenesis of the acute coronary syndromes. *Circulation* 2001;104:365–372.
6. Bennett M, Macdonald K, Chan SW, Luzio JP, Simari R, Weisberg P. Cell surface trafficking of Fas: A rapid mechanism of p53-mediated apoptosis. *Science* 1998;282:290–293.
7. Boyle JJ, Bowyer DE, Weissberg PL, Bennett MR. Human blood-derived macrophages induce apoptosis in human plaque-derived vascular smooth muscle cells by Fas-ligand/Fas interactions. *Arterioscler Thromb Vasc Biol* 2001;21:1402–1407.
8. Libby P, Geng YJ, Aikawa M, Schoenbeck U, Mach F, Clinton SK, Sukhova GK, Lee RT. Macrophages and atherosclerotic plaque stability. *Curr Opin Lipidol* 1996;7:330–335.
9. Aikawa M, Rabkin E, Sugiyama S, Voglic SJ, Fukumoto Y, Furukawa Y, Shiomi M, Schoen FJ, Libby P. An HMG-CoA reductase inhibitor, cerivastatin, suppresses growth of macrophages expressing matrix metalloproteinases and tissue factor in vivo and in vitro. *Circulation* 2001;103:276–283.
10. Cullen P, Baetta R, Bellosta S, Bernini F, Chinetti G, Cignarella A, von Eckardstein A, Exley A, Goddard M, Hofker M, Hurt-Camejo E, Kanters E, Kovanen P, Lorkowski S, McPheat W, Pentikainen M, Rauterberg J, Ritchie A, Staels B, Weitkamp B, deWinther M; MAFAPS Consortium. Rupture of the atherosclerotic plaque: Does a good animal model exist? *Arterioscler Thromb Vasc Biol* 2003;23:535–542.
11. Fan J, Watanabe T. Cholesterol-fed and transgenic rabbits for the study of atherosclerosis. *J Atheroscler Thromb* 2000;7:26–32.
12. Herrera VL, Makrides SC, Xie HX, Adari H, Krauss RM, Ryan US, Ruiz-Opazo N. Spontaneous combined hyperlipidemia, coronary heart disease and decreased survival in Dahl salt-sensitive hypertensive rats transgenic for human cholesteryl ester transfer protein. *Nat Med* 1999;5:1383–1389.
13. Russel JC, Graham SE, Richardson M. Cardiovascular disease in the JCR:LA-cp rat. *Mol Cell Biochem* 1998;188:113–126.
14. Drobnik J, Dabrowski R, Szczepanowska A, Giernat L, Lorenc J. Response of aorta connective tissue matrix to injury caused by vasopressin-induced hypertension or hypercholesterolemia. *J Physiol Pharmacol* 2000;51:521–533.
15. Badimon L. Atherosclerosis and thrombosis: Lessons from animal models. *Thromb Haemost* 2001;86:356–365.
16. Watanabe Y. Serial inbreeding of rabbits with hereditary hyperlipidemia (WHHL-rabbit). *Atherosclerosis* 1980;36:261–268.
17. Feranandez ML. Guinea pigs as models for cholesterol and lipoprotein metabolism. *J Nutr* 2001;131:10–20.
18. Fuster V, Lie JT, Badimon L, Rosemark JA, Badimon JJ, Bowie EJ. Spontaneous and diet-induced coronary atherosclerosis in normal swine and swine with von Willebrand disease. *Arteriosclerosis* 1985;5:67–73.
19. Turk JR, Henderson KK, Vanvickle GD, Watkins J, Laughlin MH. Arterial endothelial function in a porcine model of early stage atherosclerotic vascular disease. *Int J Exp Pathol* 2005;86:335–345.
20. Hamilton AJ, Kim H, Nagaraj A, Mun JH, Yan LL, Roth SI, McPherson DD, Chandran KB. Regional material property alterations in porcine femoral arteries with atheroma development. *J Biomech* 2005;38:2354–2364.
21. Nagaraj A, Kim H, Hamilton AJ, Mun JH, Smulevitz B, Kane BJ, Yan LL, Roth SI, McPherson DD, Chandran KB. Porcine carotid arterial material property alterations with induced atheroma: An in vivo study. *Med Eng Phys* 2005;27:147–156.
22. Gerrity RG. The role of the monocyte in atherogenesis: I. Transition of blood-borne monocytes into foam cells in fatty lesions. *Am J Pathol* 1981;103:181–190.
23. Fagiotto A, Ross R, Harker L. Studies of hypercholesterolemia in the nonhuman primate. I. Changes that lead to fatty streak formation. *Arteriosclerosis* 1984;4:323–340.
24. Rosenfeld ME, Tsukada T, Gown AM, Ross R. Fatty streak initiation in Watanabe heritable hyperlipemic and comparably hypercholesterolemic fat-fed rabbits. *Arteriosclerosis* 1987;7:9–23.
25. Jawien J, Nastalek P, Korbut R. Mouse models of experimental atherosclerosis. *J Physiol Pharmacol* 2004;55:503–517.
26. Paigen B, Morrow A, Holmes PA, Mitchell D, Williams RA. Quantitative assessment of atherosclerotic lesions in mice. *Atherosclerosis* 1987;68:231–240.
27. Getz GS, Reardon CA. Diet and murine atherosclerosis. *Arterioscler Thromb Vasc Biol* 2006;26:242–249.
28. Zhang SH, Reddick RL, Piedrahita JA, Meada N. Spontaneous hypercholesterolemia and arterial lesions in mice lacking apolipoprotein E. *Science* 1992;258:468–471.
29. Piedrahita JA, Zhang SH, Hagaman JR, Oliver PM, Maeda N. Generation of mice carrying a mutant apolipoprotein E gene inactivated by gene targeting in embryonic stem cells. *Proc Natl Acad Sci USA* 1992;89:4471–4475.
30. Plump AS, Smith JD, Hayek T, Aalto-Setala K, Walsh A, Verstuyft JG, Rubin EM, Breslow JL. Severe hypercholesterolemia and atherosclerosis in apolipoprotein E-deficient mice created by homologous recombination in ES cells. *Cell* 1992;71:343–353.
31. Nakashima Y, Plump AS, Raines EW, Breslow JL, Ross R. ApoE-deficient mice develop lesions of all phases of atherosclerosis throughout the arterial tree. *Arterioscler Thromb* 1994;14: 133–140.
32. Kashyap VS, Santamaria-Fojo S, Brown DR, Parrott CL, Applebaum-Bowden D, Meyn S, Talley G, Paigeb B, Maeda N, Brewer HB Jr. Apolipoprotein E deficiency in mice: Gene replacement and prevention of atherosclerosis using adenovirus vectors. *J Clin Invest* 1995;96:1612–1620.
33. Ishibashi S, Brown MS, Goldstein JL, Gerard RD, Hammer RE, Herz J. Hypercholesterolemia in low density lipoprotein receptor

knockout mice and its reversal by adenovirus-mediated gene delivery. *J Clin Invest* 1993;92:883–893.
34. Ishibashi S, Herz J, Maeda N, Goldstein JL, Brown MS. The two-receptor model of lipoprotein clearance: Tests of the hypothesis in "knockout" mice lacking the low density lipoprotein receptor, apolipoprotein E, or both proteins. *Proc Natl Acad Sci USA* 1994;91:4431–4435.
35. Paszty C, Maeda N, Verstuyft J, Rubin EM. Apolipoprotein AI transgene corrects apolipoprotein E deficiency-induced atherosclerosis in mice. *J Clin Invest* 1994;94:899–903.
36. Plump AS, Scott CJ, Breslow JL. Human apolipoprotein A-I gene expression increases high density lipoprotein and suppresses atherosclerosis in the apolipoprotein E-deficient mouse. *Proc Natl Acad Sci USA* 1994;91:9607–9611.
37. Linton MF, Farese RV Jr, Chiesa G, Grass DS, Chin P, Hammer RE, Hobbs HH, Young SG. Transgenic mice expressing high plasma concentrations of human apolipoprotein B100 and lipoprotein(a). *J Clin Invest* 1993;92:3029–3037.
38. Callow MJ, Stoltzfus LJ, Lawn RM, Rubin EM. Expression of human apolipoprotein B and assembly of lipoprotein(a) in transgenic mice. *Proc Natl Acad Sci USA* 1994;91:2130–2134.
39. Bonthu S, Heistad DD, Chappel DA, Lamping KG, Faraci FM. Atherosclerosis, vascular remodeling, and impairment of endothelium-dependent relaxation in genetically altered hyperlipidemic mice. *Arterioscler Thromb Vasc Biol* 1997;17:2333–2340.
40. Witting PK, Petterson K, Ostlund-Lindqvist AM, Westerlund C, Eriksson AW, Stocker R. Inhibition by a coantioxidant of aortic lipoprotein lipid peroxidation and atherosclerosis in apolipoprotein E and low density lipoprotein receptor gene double knockout mice. *FASEB J* 1999;13:667–675.
41. Veniant MM, Sullivan MA, Kim SK, Ambroziak P, Chu A, Wilson MD, Hellerstein MK, Rudel LL, Walzem RL, Young SG. Defining the atherogenicity of large and small lipoproteins containing apolipoprotein B100. *J Clin Invest* 2000;106:1501–1510.
42. Rosenfeld ME, Polinsky P, Virmani R, Kauser K, Rubanyi G, Schwartz SM. Advanced atherosclerotic lesions in the innominate artery of the apoE knockout mouse. *Arterioscler Thromb Vasc Biol* 2000;20:2587–2592.
43. Ishibashi S. Lipoprotein(a) and atherosclerosis. *Arterioscler Thromb Vasc Biol* 2001;21:1–2.
44. Fan J, Shimoyamada H, Sun H, Marcovina S, Honda K, Watanabe T. Transgenic rabbits expressing human apolipoprotein(a) develop more extensive atherosclerotic lesions in response to a cholesterol-rich diet. *Arterioscler Thromb Vasc Biol* 2001;21:88–94.
45. Shimano H, Yamada N, Katsuki M, Yamamoto K, Gotoda T, Harada K, Shimada M, Yazaki Y. Plasma lipoprotein metabolism in transgenic mice overexpressing apolipoprotein E. Accelerated clearance of lipoproteins containing apolipoprotein B. *J Clin Invest* 1992;90:2084–2091.
46. Fan J, Ji ZS, Huang Y, deSilva H, Sanan D, Mahley RW, Innerarity TL, Taylor JM. Increased expression of apolipoprotein E in transgenic rabbits results in reduced levels of very low density lipoproteins and an accumulation of low density lipoproteins in plasma. *J Clin Invest* 1998;101:2151–2164.
47. Libby P. Changing concepts of atherogenesis. *J Intern Med* 2000;247:349–358.
48. Rekhter MD. How to evaluate plaque vulnerability in animal models of atherosclerosis? *Cardiovasc Res* 2002;54:36–41.
49. Prescott MF, McBride CH, Hasler-Rapacz J, Von Linden J, Rapacz J. Development of complex atherosclerotic lesions in pigs with inherited hyper-LDL cholesterolemia bearing mutant alleles for apolipoprotein B. *Am J Pathol* 1991;139:139–147.
50. Johnson JL, Jackson CL. Atherosclerotic plaque rupture in the apolipoprotein E knockout mouse. *Atherosclerosis* 2001;154:399–406.
51. Williams H, Johnson JL, Carson KG, Jackson CL. Characteristics of intact and ruptured atherosclerotic plaques in brachiocephalic arteries of apolipoprotein E knockout mice. *Arterioscler Thromb Vasc Biol* 2002;22:788–792.
52. Reddick RL, Zhang SH, Maeda N. Aortic atherosclerotic plaque injury in apolipoprotein deficient mice. *Atherosclerosis* 1998;140: 297–305.
53. Gertz SD, Fallon JT, Gallo R, Taubman MB, Banai S, Barry WL, Gimple LW, Nemerson Y, Thiruvikraman S, Naidu SS, Chesebro JH, Fuster V, Sarembock IJ, Badimon JJ. Hirudin reduces tissue factor expression in neointima after balloon injury in rabbit femoral and porcine coronary arteries. *Circulation* 1998;98:580–587.
54. Rekhter MD, Hicks GW, Brammer DW, Work CW, Kim JS, Gordon D, Keiser JA, Ryan MJ. Animal model that mimics atherosclerotic plaque rupture. *Circ Res* 1998;83:705–713.
55. von der Thusen JH, van Vlijmen BJ, Hoeben RC, Kockx MM, Havekes LM, van Beerkel TJ, Biessen EA. Induction of atherosclerotic plaque rupture in apolipoprotein E-/- mice after adenovirus-mediated transfer of p 53. *Circulation* 2002;105:2064–2070.
56. Shiomi M, Ito T, Hasegawa M, Yoshida K, Gould KL. Correlation of vulnerable coronary plaques to sudden cardiac events. Lessons from a myocardial infarction-prone animal model (the WHHLMI rabbit). *J Atheroscler Thromb* 2004;11:184–189.
57. Virmani R, Kolodgie FD, Burke AP, Farb A, Schwartz SM. Lessons from sudden coronary death: A comprehensive morphological classification scheme for atherosclerotic lesions. *Arterioscler Thromb Vasc Biol* 2000;20:1262–1275.
58. Calara F, Silvestre M, Casanada F, Yuan N, Napoli C, Palinski W. Spontaneous plaque rupture and secondary thrombosis in apolipoprotein E-deficient and LDL receptor-deficient mice. *J Pathol* 2001;195:257–263.
59. Reddick RL, Zhang SH, Maeda N. Atherosclerosis in mice lacking apo E. Evaluation of lesional development and progression. *Arterioscler Thromb* 1994;14:141–147.
60. Palinski W, Ord VA, Plump AS, Breslow JL, Steinberg D. ApoE-deficient mice are a model of lipoprotein oxidation in atherogenesis. Demonstration of oxidation-specific epitopes in lesions and high titers of autoantibodies to malondialdehyde-lysine in serum. *Arterioscler Thromb* 1994;14:605–616.
61. Roubin GS, Douglas JS Jr, King SB III, Lin SF, Hutchison N, Thomas RG, Gruentzig AR. Influence of balloon size on initial success, acute complications and restenosis after percutaneous transluminal angioplasty: A prospective randomized study. *Circulation* 1988;78:557–565.
62. Mabin TA, Holmes DR Jr, Smith HC, Vlietstra RE, Reeder GS, Bresnahan JF, Bove AA, Hammes LN, Elveback LR, Orszulak TA. Follow-up clinical results in patients undergoing percutaneous transluminal angioplasty. *Circulation* 1985;71:754–760.
63. Phillips-Hughes J, Kandarpa K. Restenosis: Pathophysiology and preventive strategies. *J Vasc Interv Radiol* 1996;7:321–333.
64. Serruys PW, de Jasegere P, Kiemeneij F, Macaya C, Rutsch W, Heyndrickx G, Emanuelsson H, Marco J, Legrand V, Materne P, Belardi J, Sigwart U, Colombo A, Goy JJ, van den Heuvel P, Delcan J, Morel M-A. A comparison of balloon-expandable stent implantation with balloon angioplasty in patients with coronary artery disease. *N Engl J Med* 1994:331:489–495.
65. Vesselinovitch D. Animal models and the study of atherosclerosis. *Arch Pathol Lab Med* 1988;112:1011–1017.
66. Sims FH. A comparison of structural features of the walls of coronary arteries from 10 different species. *Pathology* 1989;21:115–124.
67. Gertz SD, Gimple LW, Ragosta M, Roberts WC, Haber HL, Powers ER, Perez LS, Sarembock IJ. Response of femoral arteries of cholesterol-fed rabbits to balloon angioplasty with or without laser: Emphasis on the distribution of foam cells. *Exp Mol Pathol* 1993;59:225–243.
68. Schwartz RS, Edwards WD, Bailey KR, Camrud AR, Jorgenson MA, Holmes DR Jr. Differential neointimal response to coronary artery injury in pigs and dogs. Implications for restenosis models. *Arterioscler Thromb* 1994;14:395–400.
69. Schwartz RS. Neointima and arterial injury: Dogs, rats, pigs, and more. *Lab Invest* 1994;71:789–791.

70. Baumgartner HR. The role of blood flow in platelet adhesion, fibrin deposition and formation of mural thrombi. *Microvasc Res* 1973;5:167–179.
71. Reidy MA, Schwartz SM. Endothelial regeneration. III. Time course of intimal changes after small defined injury to rat aortic endothelium. *Lab Invest* 1981;44:301–308.
72. Walker LN, Ramsay MM, Bower DE. Endothelial healing following defined injury to rabbit aorta: Depth of injury and mode of repair. *Atherosclerosis* 1983;47:123–130.
73. Fishman JA, Ryan GB, Karnovsky MJ. Endothelial regeneration in the rat carotid artery and the significance of endothelial denudation in the pathogenesis of myointimal thickening. *Lab Invest* 1975; 32:339–351.
74. Hardin NJ, Minick CR, Murphy GE. Experimental induction of atherosclerosis by the synergy of allergic injury to arteries and lipid rich diet. 3. The role of earlier acquired fibromuscular intimal thickening in the pathology of later developing atherosclerosis. *Am J Pathol* 1973;73:301–326.
75. Douek PC, Correa R, Neville R, Unger EF, Shou M, Banai S, Ferrans VJ, Epstein SE, Leon MB, Bonner RF. Dose-dependent smooth muscle cell proliferation induced the thermal injury with pulsed infrared lasers. *Circulation* 1992;86:1249–1256.
76. Fajardo LF, Berthrong M. Vascular lesions following radiation. *Pathol Annu* 1988;23:297–330.
77. Fritz KE, Daoud AS, Augustyn JM, Jarmolych J. Morphological and biochemical differences among grossly defined types of swine aortic atherosclerotic lesions induced by a combination of injury and atherogenic diet. *Exp Mol Pathol* 1980;32:61–72.
78. Gal D, Rongione AJ, Slovenkai GA, DeJesus ST, Lucas A, Fields CD, Isner JM. Atherosclerotic Yucatan microswine: An animal model with high-grade fibrocalcific, nonfatty lesions suitable for testing catheter-based interventions. *Am Heart J* 1990;119:291–300.
79. Faxon DP, Weber VJ, Haudenschild C, Gottsman SB, McGovern WA, Ryan TJ. Acute effects of transluminal angioplasty in three experimental models of atherosclerosis. *Arteriosclerosis* 1982;2:125–133.
80. Lee WM, Lee KT. Advanced coronary atherosclerosis in swine produced by combination of balloon catheter injury and cholesterol feeding. *Exp Mol Pathol* 1975;23:491–499.
81. Lee KT, Lee WM, Han J, Jarmolych J, Bishop MB, Goel BG. Experimental model for study of "sudden death" from ventricular fibrillation or asystole. *Am J Cardiol* 1973;32:62–73.
82. Clowes AW, Reidy MA, Clowes MM. Mechanisms of stenosis after arterial injury. *Lab Invest* 1983;49:208–215.
83. Muller DW, Ellis SG, Topol EJ. Experimental models of coronary artery restenosis. *J Am Coll Cardiol* 1992;19:418–432.
84. Schatz RA, Palmaz J, Tio FO, Garcia F, Garcia O, Reuther SR. Balloon expandable intracoronary stents in the adult dog. *Circulation* 1987;76:450–457.
85. Fischell TA, Grant G, Johnson DE. Determinants of smooth muscle injury during balloon angioplasty. *Circulation* 1990;82:2170–2184.
86. LeVeen RF, Wolf GL, Villanueva TG. New rabbit atherosclerosis model for the investigation of transluminal angioplasty. *Invest Radiol* 1982;17:470–475.
87. Garasic JM, Edelman ER, Squire JC, Seifert P, Williams MS, Rogers C. Swine and primates best approximate the requirements of a model of restenosis and demonstrate lesions that are markedly similar in physiology and morphology to their human counterpart. Circulation 2000;101:812–818.
88. LaDisa F Jr, Olson LE, Molthen RC, Hettrick DA, Pratt PF, Hardel MD, Kersten JR, Warltier DC, Pagel PS. Alterations in wall shear stress predict sites of neointimal hyperplasia after stent implantation in rabbit iliac arteries. *Am J Physiol Heart Circ Physiol* 2005;288: H2465–2475.
89. Cejna M, Virmani R, Jones R, Bergmeister H, Loewe C, Schoder M, Grgurin M, Lammer J. Biocompatibility and performance of the Wallstent and the Wallgraft, Jostent, and Hemobahn stent-grafts in a sheep model. *J Vasc Interv Radiol* 2002;13:823–830.
90. Barth KH, Virmani R, Froelich J, Takeda T, Lossef SV, Newsome J, Jones R, Lindisch D. Paired comparison of vascular wall reactions to Palmaz stents, Strecker tantalum stents, and Wallstents in canine iliac and femoral arteries. *Circulation* 1996;93:2161–2169.
91. Fontaine AB, Spigos DG, Eaton G, Das Passos S, Christoforidis G, Khabiri H, Jung S. Stent-induced intimal hyperplasia: Are there fundamental differences between flexible and rigid stent designs? *J Vasc Interv Radiol* 1994;5:739–744.
92. Andrews RT, Venbrux AC, Magee CA, Bova DA. Placement of a flexible endovascular stent across the femoral joint: An in vivo study in the swine model. *J Vasc Interv Radiol* 1999;10:1219–1228.
93. Wanibuchi H, Dingemans KP, Becker AE, Ueda M, Naruko T, Tanizawa S, Nakamura K. Is the Watanabe heritable hyperlipidemic rabbit a suitable experimental model for percutaneous transluminal coronary angioplasty in humans? A light microscopic, immunohistochemical and ultrastructural study. *J Am Coll Cardiol* 1993;21:1490–1496.
94. Schwartz RS, Murphy JG, Edwards WD, Camrud AR, Vliestra RE, Holmes DR. Restenosis after balloon angioplasty: A practical proliferative model in porcine coronary arteries. *Circulation* 1990;82:2190–2200.
95. Schwartz RS, Topol EJ, Serrujs PW, Sangiorgi G, Holmes DR Jr. Artery size, neointima, and remodeling: Time for standards. *J Am Coll Cardiol* 1998;32:2087–2094.
96. O'Brien ERM, deBlois D, Schwartz SM. A critical examination of animal models of restenosis following angioplasty. In: Dobrin P, Ed. *Intimal Hyperplasia*. Austin, TX: R. G. Landes, 1994:229–256.
97. Mæng M, Olesen PG, Emmertsen NC, Thorwest M, Nielsen TT, Kristensen BO, Falk E, Andersen HR. Time course of vascular remodeling, formation of neointima and formation of neoadventitia after angioplasty in a porcine model. *Coron Artery Dis* 2001;12: 285–293.
98. Kinney TB, Chin AK, Rurik GW, Finn JC, Shoor PM, Hayden WG, Fogarty TJ. Transluminal angioplasty: A mechanical-pathophysiological correlation of its physical mechanisms. *Radiology* 1984;153:85–89.
99. Castenada-Zuniga WR, Formanek A, Tadavarthy M, Vlodaver Z, Edwards JE, Zollikofer C, Amplatz K. The mechanism of balloon angioplasty. *Radiology* 1980;135:565–571.
100. Andersen HR, Mæng M, Thorwest M, Falk E. Remodeling rather than neointimal formation explains luminal narrowing after deep vessel wall injury. Insights from a porcine coronary restenosis model. *Circulation* 1996;93:1716–1724.
101. Shi Y, O'Brien JE, Fard A, Mannion JD, Wang D, Zalewski A. Adventitial myofibroblasts contribute to neointimal formation in injured porcine coronary arteries. *Circulation* 1996;94:1655–1664.
102. Scott NA, Cipolla GD, Ross CE, Dunn B, Martin FH, Simonet L, Wilcox JN. Identification of a potential role for the adventitia in vascular lesion formation after balloon overstretch injury of porcine coronary arteries. *Circulation* 1996;93:2178–2187.
103. Siow RC, Mallawaarachchi CM, Weisberg PL. Migration of adventitial myofibroblasts following vascular balloon injury: Insights from in vivo gene transfer to rat carotid arteries. *Cardiovasc Res* 2003;59:212–221.
104. Wilentz JR, Sanborn TA, Haudenschild CC, Valeri CR, Ryan TJ, Faxon DP. Platelet accumulation in experimental angioplasty: Time course and relation to vascular injury. *Circulation* 1987;75:636–642.
105. Fischman DL, Leon MB, Baim DS, Schatz RA, Savage MP, Penn I, Detre K, Veltri L, Ricci D, Nobuyoshi M, Cleman M, Heuser R, Almond D, Teirstein PS, Fish RD, Colombo A, Brinker J, Moses J, Shaknovich A, Hirshfeld J, Bailey S, Ellis S, Rake R, Goldberg S for The Restenosis Study Investigators. A randomized comparison of coronary-stent placement and balloon angioplasty in the treatment of coronary artery disease. Stent Restenosis Study Investigators. *N Engl J Med* 1994;331:496–501.
106. Schwartz RS, Huber KC, Murphy JG, Edwards WD, Camrud AR, Vliestra RE, Holmes DR. Restenosis and the proportional neointi-

mal response to coronary artery injury: Results in a porcine model. *J Am Coll Cardiol* 1992;19:267–274.
107. Schwartz RS, Chronos NA, Virmani R. Preclinical restenosis models and drug-eluting stents. Still important, still much to learn. *J Am Coll Cardiol* 2004;44:1373–1385.
108. Christen T, Verin V, Bochaton-Piallat ML, Popowski Y, Ramaekers F, Debruyne P, Camenzind E, van Eys G, Gabbiani G. Mechanisms of neointima formation and remodeling in the porcine coronary artery. *Circulation* 2001;103:882–888.
109. Pires NMM, Jukema JW, Daemen JAP, Quax PHA. Drug-eluting stents studies in mice: Do we need atherosclerosis to study restenosis? *Vascul Pharmacol* 2006;44:257–264.
110. Pires NMM, van der Hoeven BL, de Vries MR, Havekes LM, van Vlijmen BJ, Hennink WE, Quax PHA, Jukema JW. Local perivascular delivery of anti-restenotic agents from a drug-eluting poly(epsilon-caprolactone) stent cuff. *Biomaterials* 2005;26:5386–5394.
111. Thompson RW, Geraghty PJ, Lee JK. Abdominal aortic aneurysms: Basic mechanisms and clinical implications. *Curr Probl Surg* 2002;39:110–230.
112. Carell TWG, Smith A, Burnand KG. Experimental techniques and models in the study of the development and treatment of abdominal aortic aneurysm. *Br J Surg* 1999;86:305–312.
113. Manning MW, Cassis LA, Huang J, Szilvassy SJ, Daugherty A. Abdominal aortic aneurysms: Fresh insights from a novel animal model of the disease. *Vasc Med* 2002;7:45–54.
114. Daugherty A, Cassis LA. Mouse models of abdominal aortic aneurysms. *Arterioscler Thromb Vasc Biol* 2004;24:429–434.
115. Balko A, Piasecki GJ, Shah DM, Carney WI, Hopkins RW, Jackson BT. Transfemoral placement of intraluminal polyurethane prosthesis for abdominal aortic aneurysm. *J Surg Res* 1986;40:305–309.
116. Mason RG, Read MS. Some species differences in fibrinolysis and blood coagulation. *J Biomed Mater Res* 1971;5:121–128.
117. Abbott WM, Callow A, Moore W, Rutherford R, Veith F, Weinberg S. Evaluation and performance standards for arterial prosthesis. *J Vasc Surg* 1993;17:746–756.
118. Blanton FS Jr, Muller WH Jr, Warren WD. Experimental production of dissecting aneurysms of the aorta. *Surgery* 1959;45:81–90.
119. Williams DM, Andrews JC, Chee SS, Marx MV, Abrams GD. Canine model of acute aortic rupture: Treatment with percutaneous delivery of a covered Z stent–work in progress. *J Vasc Interv Radiol* 1994;5:797–803.
120. Verbin C, Donayre C, Kopchok G, Scoccianti M, White RA. Anterior patch aortic aneurysm model for the study of endoluminal grafts. *J Invest Surg* 1995;8:381–388.
121. Ruiz CE, Zhang HP, Douglas JT, Zuppan CW, Kean CJ. A novel method for the treatment of abdominal aortic aneurysms using percutaneous implantation of a newly designed endovascular device. *Circulation* 1995;91:2470–2477.
122. Palmaz JC, Tio FO, Laborde JC, Clem M, Rivera FJ, Murphy KD, Encarnacion CE. Use of stents covered with polytetrafluoroethylene in experimental abdominal aortic aneurysms. *J Vasc Interv Radiol* 1995;6:879–885.
123. Criado E, Marston WA, Woosley JT, Ligush J, Chuter TA, Baird C, Suggs CA, Mauro MA, Keagy BA. An aortic aneurysm model for the evaluation of endovascular exclusion prostheses. *J Vasc Surg* 1995;22:306–315.
124. Marston WA, Criado E, Baird CA, Keagy BA. Reduction of aneurysm pressure and wall stress after endovascular repair of abdominal aortic aneurysm in a canine model. *Ann Vasc Surg* 1996;10:166–173.
125. Eton D, Warner D, Owens C, McClenic B, Cava R, Ofek B, Borhani M, Baraniewski H, Schuler JJ. Results of endoluminal grafting in an experimental aortic aneurysm model. *J Vasc Surg* 1996;23:819–831.
126. Maynar M, Qian Z, Hernandez J, Sun F, DeMiguel C, Crisostomo V, Usón J, Pineda L-F, Espinoza CG, Castañeda WR. An animal model of abdominal aortic aneurysm created with peritoneal patch: Technique and initial results. *Cardiovasc Intervent Radiol* 2003;26:168–176.
127. Pavcnik D, Andrews RT, Yin Q, Uchida BT, Timmermans HA, Corless C, Toyota N, Nakata M, Kaufman J, MD, Keller FS, Rösch J. A canine model for studying endoleak after endovascular aneurysm repair. *J Vasc Interv Radiol* 2003;14:1303–1310.
128. Hiraki T, Pavcnik D, Uchida BT, Timmermans HA, Yin Q, Wu R-H, Niyyati M, Keller FS, Rösch J. Prophylactic residual aneurysmal sac embolization with expandable hydrogel embolic devices for endoleak prevention: Preliminary study in dogs. *Cardiovasc Intervent Radiol* 2005;28:459–466.
129. Economou SG, Taylor CB, Beattie EJ Jr, Davis CB Jr. Persistent experimental aortic aneurysms in dogs. *Surgery* 1960;47:21–28.
130. Mirich D, Wright KC, Wallace S, Yoshioka T, Lawrence DD Jr, Charnsangavej C, Gianturco C. Percutaneously placed endovascular grafts for aortic aneurysms: Feasibility study. *Radiology* 1989;170:1033–1037.
131. Parodi JC, Palmaz JC, Barone HD. Transfemoral intraluminal graft implantation for abdominal aortic aneurysms. *Ann Vasc Surg* 1991;5:491–499.
132. Laborde JC, Parodi JC, Clem MF, Tio FO, Barone HD, Rivera FJ, Encarnacion CE, Palmaz JC. Intraluminal bypass of abdominal aortic aneurysm: Feasibility study. *Radiology* 1992;184:185–190.
133. Hagen B, Harnoss BM, Trabhardt S, Ladeburg M, Fuhrmann H, Franck C. Self-expandable macroporous nitinol stents for transfemoral exclusion of aortic aneurysms in dogs: Preliminary results. *Cardiovasc Intervent Radiol* 1993;16:339–342.
134. Piquet P, Rolland PH, Bartoli JM, Tranier P, Moulin G, Mercier C. Tantalum-dacron coknit stent for endovascular treatment of aortic aneurysms: A preliminary experimental study. *J Vasc Surg* 1994;19:698–706.
135. Gorin DR, Arbid EJ, D'Agostino R, Yucel EK, Solovay KS, La Morte WW, Quist WC, Mulligan N, Menzoian JO. A new generation endovascular graft for the treatment or repair of abdominal aortic aneurysms. *Am J Surg* 1997;173:159–164.
136. Whitbread T, Birch P, Rogers S, Majeed A, Rochester J, Beard JD, Gaines P. A new animal model for abdominal aortic aneurysms: Initial results using a multiple-wire stent. *Eur J Vasc Endovasc Surg* 1996;11:90–97.
137. Martin DE III, Nasbeth DC, Rowe MI. Production of experimental aneurysms with pancreatic elastase. *Surg Forum* 1962;8:237–239.
138. Boudghene F, Anidjar S, Allaire E, Osborne-Pellegrin M, Bigot JM, Michel JB. Endovascular grafting in elastase-induced experimental aortic aneurysms in dogs: Feasibility and preliminary results. *J Vasc Interv Radiol* 1993;4:497–504.
139. Hallisey MJ. A transluminally created abdominal aortic aneurysm model. *J Vasc Interv Radiol* 1997;8:305–312.
140. Schoder M, Pavcnik D, Uchida BT, Corless C, Timmermans HA, Yin Q, Brountzos E, Nakata M, Hiraki T, Niyyati M, Kaufman JA, Keller FS, Rösch J. Small intestinal submucosa aneurysm sac embolization for endoleak prevention after abdominal aortic aneurysm endografting: A pilot study in sheep. *J Vasc Interv Radiol* 2004;15:69–83.
141. Kallmes DF, Altes TA, Vincent DA, Cloft HJ, Do HM, Jensen ME. Experimental side-wall aneurysms: A natural history study. *Neuroradiology* 1999;41:338–341.
142. Hashimoto N, Kim C, Kikuchi H, Kojima M, Kang Y, Hazama F. Experimental induction of cerebral aneurysms in monkeys. *J Neurosurg* 1987;67:903–905.
143. Hashimoto N, Handa H, Nagata I, Hazama F. Experimentally induced cerebral aneurysms in rats: Part V: Relation of hemodynamics in the circle of Willis to formation of aneurysms. *Surg Neurol* 1980;13:41–45.
144. Moriwaki T, Takagi Y, Sadamasa N, Aoki T, Nozaki K, Hashimoto N. Impaired progression of cerebral aneurysms in interleukin-1-B-deficient mice. *Stroke* 2006;37:900–905.
145. Nagata I, Handa H, Hashimoto N, Hazama F. Experimentally induced cerebral aneurysms in rats: Part VI. Hypertension. *Surg Neurol* 1980;14:477–479.
146. German W, Black S. Experimental production of carotid aneurysms. *N Engl J Med* 1954;250:104–106.

147. Stehbens WE. Experimental production of aneurysms by microvascular surgery in rabbits. *Vasc Surg* 1973;7:165–175.
148. Strother CM, Graves VB, Rappe A. Aneurysm hemodynamics: An experimental study. *Am J Neuroradiol* 1992;13:1089–1095.
149. Wakhloo AK, Schellhammer F, de Vries J, Haberstroh J, Schumacher M. Self-expanding and balloon-expandable stents in the treatment of carotid aneurysms: An experimental study in a canine model. *Am J Neuroradiol* 1994;5:493–502.
150. Abruzzo T, Shengelaia R, Dawson RC, Owens DS, Cawley CM, Gravanis MB. Histologic and morphologic comparison of experimental aneurysms with human intracranial aneurysms. *Am J Neuroradiol* 1998;19:1309–1314.
151. Miskolczi L, Guterman LR, Flaherty JD, Szikora I, Hopkins LN. Rapid saccular aneurysm induction by elastase application in vitro. *Neurosurgery* 1997;41:220–229.
152. Stehbens WE. Chronic changes in experimental saccular and fusiform aneurysms in rabbits. *Arch Pathol Lab Med* 1981;105: 603–607
153. Cloft HJ, Altes TA, Marx WF, Raible RJ, Hudson SB, Helm GA, Mandell JW, Jensen ME, Dion JE, Kallmes DF. Endovascular creation of an in vivo bifurcation aneurysm model in rabbits. *Radiology* 1999;213:223–238.
154. Crawley CM, Dawson RC, Shengelaia G, Bonner G, Barrow DL, Colohan AR. Arterial saccular aneurysm model in the rabbit. *Am J Neuroradiol* 1996;17:1761–1766.
155. Fujiwara NH, Cloft HJ, Marx WF, Short JG, Jensen ME, Kallmes DF. Serial angiography in an elastase-induced aneurysm model in rabbits: Evidence for progressive aneurysm enlargement after creation. *Am J Neuroradiol* 2001;22:698–703.
156. Marx WF, Cloft HJ, Helm GA, Short JG, Do HM, Jensen ME, Kallmes DF. Endovascular treatment of experimental aneurysms by use of biologically modified embolic devices: Coil-mediated intraaneurysmal delivery of fibroblast tissue allografts. *Am J Neuroradiol* 2001;22:323–333.
157. Miskolczi L, Guterman LR, Flaherty JD, Szikora I, Hopkins LN. Rapid saccular aneurysm induction by elastase application in vitro. *Neurosurgery* 1997;41:220–229.
158. Miskolczi L, Guterman LR, Flaherty JD, Hopkins LN. Saccular aneurysm induction by elastase digestion of the arterial wall: A new animal model. *Neurosurgery* 1998;43:595–600.
159. Onizuka M, Miskolczi L, Gounis MJ, Seong J, Lieber BB, Wakhloo AK. Elastase-induced aneurysms in rabbits—effect of post-construction geometry on final size. *Am J Neuroradiol* 2006;27: 1129–1131.

41 Transgenic Mouse Models of HIV-1/AIDS and Cardiac Performance

JAMES J. KOHLER AND WILLIAM LEWIS

ABSTRACT

Cardiomyopathy (CM) associated with HIV/AIDS is important clinically and increasingly may impact the epidemic of AIDS as survival with HIV/AIDS increases due to effective antiretroviral therapies. Animal models offer sophisticated *in vivo* biological systems to recapitulate many direct and indirect effects of HIV-1 infection or the effects of therapeutics on the host. Few small animal models are available to investigate cardiovascular parameters associated with HIV/AIDS or highly active antiretroviral therapy (HAART). Unlike nonhuman primates, murine models of AIDS offer a safe, cost-effective approach to evaluate cardiac structure and function in AIDS CM. Experimentally, the mouse serves as perhaps the foremost laboratory mammal for genetic and immunological research. Several models utilize the well-established approach of transgenesis for general expression or specific, cardiac-targeted expression of viral genes. Gene targeting to generate gain or loss of function mutations in the mouse has yielded remarkable advances in understanding the roles played by specific gene products. This chapter focuses on mechanisms of CM in AIDS through the use of transgenic murine (TG) models as an important biological tool to study the impact of HIV-1, its gene products, and antiretroviral therapy on the pathophysiology of CM.

Key Words: Transgenic mouse, HIV/AIDS, Cardiomyopathy, Antiretrovirals, Mitochondria.

INTRODUCTION

Few small animal models are available to investigate cardiovascular parameters associated with HIV/AIDS. Animal models ideally completely recapitulate HIV-1 infection, its systemic or cardiovascular effects, or the effects of therapeutics. The use of nonhuman primates, for example, offers a model that closely reflects many virological and immunological effects of human HIV-1 but with limited availability, high cost, and intrinsic complexity that make cardiac studies in this model more difficult.[1,2] Murine models of AIDS, on the other hand, offer a safe, cost-effective approach to the evaluation of cardiac structure and function in AIDS cardiomyopathy (CM), particularly if they capture features of the disease and/or its therapy. Therefore, this chapter focuses on mechanisms of CM in AIDS through the use of rodent and transgenic murine (TG) models.

NUCLEOSIDE REVERSE TRANSCRIPTASE INHIBITOR–HIGHLY ACTIVE ANTIRETROVIRAL THERAPY TOXICITY AND SMALL RODENT MODEL

Nucleoside analog reverse transcriptase inhibitors (NRTI) are a class of antiretrovirals commonly used as part of the combination known as highly active antiretroviral therapy (HAART). Compelling experimental and clinical evidence suggests that combined HAART is a formidable way to combat HIV-1 infection and to treat AIDS.[3,4] Despite the benefits of HAART, adverse effects associated with antiretroviral therapies include clinical symptoms such as myopathy, CM, polyneuropathy, lactic acidosis, steatosis, pancreatitis, and lipodystrophy.[5–9] The central pathogenetic mechanism in these complications is hypothesized to relate to mitochondrial dysfunction with morphologically damaged mitochondria, decreased mitochondrial DNA (mtDNA), and depletion of mitochondrial-encoded enzymes.[6,10,11] Of note, these are all findings in clinical mitochondrial genetic diseases.[12,13]

A supply of nucleotides and balanced nucleotide pools is required for mtDNA replication.[14–27] Even in postmitotic cells, mtDNA replication is continuous.[28] Phosphorylated nucleotides within mitochondria comprise a distinct pool.[28,29] The clinical impact of NRTI therapy on this pool is largely unknown, despite extensive therapeutic use of NRTIs. Some NRTIs exhibit cardiac toxicity, and may be etiologically linked to AIDS CM, the latter of which is a focus of our laboratory studies. Our group has identified deleterious cardiac effects from NRTIs (including zidovudine, fialuridine, stavudine, and others; see Table 41–1), which cause cardiac mitochondrial dysfunction and related effects in other tissues.[10,11,20,30–35]

Because we focus sharply on NRTI toxicity in the heart, an overview of mechanisms of NRTI toxicity is presented here. A number of mechanisms may underlie NRTI mitochondrial toxicity. mtDNA polymerase-gamma (pol-γ), for example, is the eukaryotic mtDNA replication enzyme and is thought to play a key role in NRTI MT.[36] "The DNA pol-γ hypothesis" was proposed over a decade ago.[5] It postulates that inhibition of pol-γ leads to the depletion of mtDNA and thereby causes mitochondrial dysfunction. Specifically, possible mechanisms of mitochondrial toxicity include direct inhibition of mtDNA pol-γ without

From: *Sourcebook of Models for Biomedical Research* (P. M. Conn, ed.), © 2008 Humana Press Inc., Totowa, NJ.

Table 41–1
Nucleoside Reverse Transcriptase Inhibitors with mitochondrial toxicity

Zidovudine (AZT)[14,15]
Tenofovir (TDF)[16,17]
Zalcitabine (ddC)[18]
Fialuridine (FIAU)[19–21]
Stavudine (d4T)[22,23]
Lodenosine (FddA)[24]
Adefovir (Adf)[26,27]

NRTI incorporation, chain termination by incorporation of NRTI triphosphate into mtDNA by DNA pol-γ, persistence of incorporated analogs in mtDNA because of inefficient excision by DNA pol-γ and resultant defective template mtDNA, or a combination of these mechanisms.[30]

Additionally, the hypothesis addresses the impact of NRTIs on pools of native nucleotides in mitochondria. With respect to intramitochondrial nucleotide pools, four potential pathways exist from which phosphorylated nucleotides arise within mitochondria. First, phosphorylated nucleotides could be synthesized in the cytosol and taken up by specific transport systems.[37] Second, cytosolic phosphorylated nucleotides could be dephosphorylated and then the corresponding deoxyribonucleoside diphosphates (dNDPs) taken up by the deoxynucleotide transporter (DNC)[38] and converted to triphosphorylated nucleotides intramitochondrially. Third, ribonucleoside diphosphates could be imported by DNC, then reduced within the mitochondrion by mitochondrial ribonucleotide reductase,[39] but evidence to support the latter moiety is sketchy. Last, mitochondria could directly import deoxyribonucleosides using the equilibrative nucleoside transporter (ENT1),[40–43] although evidence is emerging for species-related differences.[44]

USE OF SPECIFIC MARKERS IN GENETICALLY MANIPULATED MOUSE

Experimentally, the mouse serves as perhaps the foremost laboratory mammal for genetic and immunological research.[45,46] Biochemical screening of genetically manipulated mice offers the distinct advantage of a scientifically panoramic, *integrative* view of the general health of the living, genetically manipulated mouse while focusing on a parametric evaluation of cardiovascular dysfunction. The approach offers the advantages of broad screening for a variety of analytes, both murine general health and disease (cardiac)-related screening. Analogous screenings are routinely performed for human patients for the presence of health and disease, and serve as important tools for the presence of cardiovascular diseases, their prognosis, and efficacy of therapy. It follows that use of mouse models to elucidate mechanisms of AIDS CM in humans includes such screening methods and data analysis of markers in mice. These screening tests offer the investigator an opportunity to explore pathophysiological events in the cardiovascular system while maintaining its integrity *in vivo* in a mammalian system. Rapid nonlethal analyses are performed while simultaneous examination of the general well being of the genetically manipulated mouse is monitored. It should be understood that such analyses would likely not fall under routine monitoring procedures and animal care, and specialized breeding, husbandry, and care of the mice are required.

With respect to heart failure studies, one clinically relevant approach is to evaluate the presence of heart failure through the screening of plasma or serum analyses and immunoassay. This approach offers the advantage of defining illness early in the genetically manipulated mouse where sequential blood samples may be obtained. Cytokines, in particular, can be measured. Circulating or locally acting cytokines including endothelin (ET-1) and tumor necrosis factor-α (TNF-α) play key roles in cardiac dysfunction, cardiomyopathy, myocarditis, and other cardiovascular diseases.[47–50] TNF-α, interleukin (IL)-1β, and interferon (IFN)-γ can lead to increased levels of inducible nitric oxide syntheses (iNOS), increased nitric oxide (NO) synthesis, and myocyte death.[51]

Cytokines do not cause myocyte dysfunction or necrosis directly. However, cytokines may alter myocyte function, through the β_1-adrenoceptor-G-protein-adenylyl cyclase axis.[52–54] From *in vitro* studies, long-term treatment of cardiomyocytes with immune cell supernatants containing IL-1 and TNF-α reduced contractility and cyclic adenosine monophosphate (cAMP) accumulation by inhibition of β-adrenergic responsiveness.[55] Furthermore, TNF-α depressant effects on cardiomyocytes have been demonstrated in culture and lead to left ventricle (LV) dysfunction *in vivo*.[56,57]

Plasma ET-1 concentrations are increased in patients with cardiac heart failure (CHF)[58–60] and suggest poor prognosis in patients with New York Heart Association (NYHA) class III or IV CHF. ET receptors are expressed on several cell types in the myocardium, including myocytes, fibroblasts, and endothelial cells. Stimulation with the ET-1 ligand, endothelin, stimulates myocyte hypertrophy[61,62] and impacts the synthesis and degradation of the extracellular matrix.[63]

The progressive nature of cardiac dysfunction encompasses a complex of molecular and cellular events termed "remodeling" that leads to changes in the structure, function, and phenotype of myocardium.[64] These changes include hypertrophy and death of myocytes, regression to a molecular phenotype characterized by the expression of fetal genes and proteins, and alterations in the quantity and composition of the extracellular matrix. NF-κB substantially increases the progression of cardiac remodeling.[65] LV expression of atrial natriuretic factor (ANF), a fetal gene product, is a marker of LV hypertrophy (LVH) and cardiac dysfunction.[66] Thus, ANF serum and/or plasma levels offer affective markers for consideration. ANF is a 28-amino-acid peptide that is normally synthesized in the atria and to a lesser extent in the ventricles and is released into the circulation during atrial distention.[67] In patients with CHF, plasma ANF concentrations rise as atrial pressure increase. Increased secretion of atrial natriuretic peptide in early CHF may attenuate or delay systemic and renal arterial vasoconstriction, venoconstriction with increased cardiac preload, and renal sodium retention. In animals with CHF, administration of antibodies to ANF peptide or antagonists to ANF receptors decreases renal blood flow; increases right atrial pressure, plasma rennin activity, and sodium retention; and worsens diastolic dysfunction by impairing myocardial relaxation.[68]

Plasma ANF and brain natriuretic peptide (BNP) levels increase in patients with heart failure with progression of clinical symptoms.[69,70] Therefore, monitoring plasma BNP may be useful as a noninvasive biomarker for identifying possible cardiac dysfunction. Echocardiography (ECHO) is a noninvasive, highly accurate and quick assessment of the overall health of the heart through generation of a two-dimensional ultrasound image of the heart, providing a quick assessment of the heart valves and degree of heart muscle contraction. Together with echocardiography, the routine assay of plasma BNP offers important markers to monitor cardiac function in mouse models.

Increased local and circulating concentrations of norepinephrine (NE) may also contribute to myocyte hypertrophy, but are difficult to measure in the mouse. Their action can occur directly through stimulation of α_1- and β-adrenergic receptors or secondarily by activating renin–angiotensin–aldosterone.[71,72] NE's toxicity to myocardial cells is mediated through calcium overload and/or the induction of apoptosis.[48] Blocking β-adrenergic receptors can prevent NE-induced myocyte death. Patients with plasma NE concentrations >800 pg/ml (4.7 nmol/liter) have a 1 year survival rate of less than 40%. Thus, a simple method of screening of plasma NE levels may be effective for identification of CHF in mice, but has not been widely used.

As mentioned, renin–angiotensin–aldosterone system activity is also increased in CHF. Together with plasma NE and ANF, measurement of plasma renin activity provides a prognostic index in patients.[73] If such measurements were possible in microvolume samples, this could prove a useful screen of cardiovascular disease in mice.

TRANSGENIC MURINE STRATEGIES

GENERAL EXPRESSION TGs are well-established, efficient models of living systems that are useful to evaluate viral gene function and pathogenesis in AIDS and other viral illnesses.[74–76] Transgenesis, particularly in the models with HIV, is predicated on the cellular incorporation of viral sequences into genetic material. This technique eliminates possible problems associated with viral administration since the viral sequences are contained within every target cell. The extensive use of TG in cardiovascular research and HIV/AIDS is established.[46,77] The power of transgenesis provides a unique experimental system to dissect complex aspects of AIDS CM.

For the purposes of studying HIV-1 effects in the heart, the challenge has been to develop a TG harboring a defective provirus that results in expression in the heart. An early model developed by Iwakura *et al.*,[78] while expressed primarily in the eye, had no viral mRNA detected in the heart, kidney, liver, or bone marrow. Alternatively, a hemizygous NL4-3Δ *gag/pol* TG mouse, which contains an internal deletion that eliminates most of the gag/pol coding sequence, was developed,[79] which has become an important generalized model of AIDS that can be used to evaluate CM in AIDS. No founder mice carrying the transgene developed disease during their life span. Three founders (Tg22, Tg25, and Tg26) produced progeny that developed renal disease, and one founder (Tg26) developed myopathy/myositis.[80] Variable tissue distribution and level of viral transgene expression occurred among the three lines. Three mRNA species were detected (full-length, single and double spliced messages, respectively) with widespread expression but varied tissue distribution (high levels in skin and muscle, followed by moderate levels in thymus, gastrointestinal tract, kidney, eye, brain and spleen, but low levels in heart). More importantly, NL4-3Δ *gag/pol* TG demonstrate key features of HIV/AIDS including AIDS nephropathy, wasting, and skin diseases that phenotypically resemble clinical counterparts in AIDS. Specific viral protein expression patterns, as determined by immunoblotting, were tissue specific. Altered Rev functions in different tissues may explain the apparent disparate protein expression patterns.

GENE-SPECIFIC EXPRESSION TGs expressing HIV Tat have also been generated.[81,82] In this model, a single gene from HIV (*Tat*) is driven by the viral long terminal repeat (LTR) and by the β-actin promoter. Tat TG mice develop hepatocellular carcinomas, endothelial proliferation, and skin lesions that resemble Kaposi's sarcoma.[83–85] Some Tat TG lines appear to potentiate zidovudine (AZT)-induced toxicity and oxidative damage.[86] Other Tat-TG murine models have demonstrated a molecular mechanism of reduced glutathione levels and induction of B-lymphomas.[87,88]

Another replication-defective HIV-1 TG created by Gilbert Jay is based on a construct with a *gag/pol/env* deletion. Viral mRNA expression occurred in all tissues with a perivascular inflammation as a prominent phenotypic feature.[77] These TGs had minimal cardiac physiological changes, compared to NL4-3Δ *gag/pol* TG, which have striking defects in contractility and relaxation.[33]

Lastly, a series of HIV-1 TG lines was generated by Paul Jolicoeur's group that expresses coding sequences of HIV-1 regulated by the CD4 gene.[89–92] Thus, expression of HIV-1 genes occurs in the same subset of immune cells that is infected by HIV-1 ($CD4^+CD8^+$ immature thymic T cells, circulating $CD4^+CD8^-$ T cells, and $CD4^+$ monocytes/macrophages). Interestingly, these TG mice spontaneously develop severe cardiac inflammatory lesions. Thus, this TG line may constitute an animal model of inflammatory cardiomyopathy, a subset of the clinically relevant heart diseases found in human AIDS.

CARDIAC-SPECIFIC TARGETING OF INDIVIDUAL GENES Targeting transgenic expression exclusively to a specific tissue target is a more focused, organ-relevant approach. Although plausible, a complete infectious HIV-1 genome is technically possible, but problematic. Such a system employing an intact human provirus in rodents would warrant serious concerns for laboratory safety. Mice expressing intact HIV-1 may constitute animal reservoirs for virus and potentially be infectious to humans.[93]

The fundamental approach taken to develop TG CM models is based on the fact that a specific protein relates to cardiac function and dysfunction.[94] Modification of the cardiac protein repertoire in a defined manner through stable transgenesis facilitates expression of engineered proteins in the heart. The most widely used and successful promoter candidate for these purposes is the α-myosin heavy chain promoter (α-MyHC), although it is not the only approach available. This promoter meets the criteria of the key features desired, including a high degree of cardiac specificity, high levels of expression in adult atria and ventricles, the presence of sequences necessary for copy-number dependent and position-independent expression, and homogeneous expression across the cardiac myocyte population. Then, cardiac contractility and structure can be examined exclusively.[94] We have utilized Robbins' approach[95] to achieve cardiac targeted transgenesis. In principle, key features included selection of an ideal promoter for cardiac expression that (1) would elicit a high level of cardiac-specific gene expression at appropriate times of development with a minimum of position-sensitive effects of chromosomal context and (2) would be copy number dependent and (3) position independent.

Table 41–2
HIV-1 viral proteins used in transgenics (in the Lewis laboratory)

Protein	*Function*
Tat	Transactivation[98–101]
Nef	Pleiotropic, can increase or decrease virus[102,103]
Vpr	Helps in virus replication[104,105]
Vpu	Helps in virus release; disrupts gp160–CD4 complexes; expression regulated by Vpr[106]

Applying this TG approach, we targeted selected HIV-1 transgenes to cardiac ventricular myocytes in mice. Targeting single-gene TG with cardiac myocyte-specific expression of individual HIV-1 genes has proven to be a unique and powerful model system to define structural and functional effects of HIV-1 gene products on cardiac performance and myocyte structure. Granted the disadvantages of the single-gene-targeted transgenesis may be that the interaction of multiple viral proteins may be required to induce the observed clinical pathological or physiological (phenotypic) changes. Moreover, isolation of a gene from its genomic context may alter or eliminate critical *cis*-acting control elements required for effective, native gene expression.

Despite some potential limitations of single-gene-targeted transgenesis, we successfully generated several TG lines expressing HIV-1 gene products using this cardiac-targeted approach (see Table 41–2). TG AIDS lines that express HIV-1 *Tat* in ventricular cardiac myocytes with different levels of *Tat* expression were created. Progeny of founders demonstrate exclusive expression of HIV-1 *Tat* mRNA in cardiac myocytes. Hemizygotes demonstrate left ventricle hypertrophy pathologically, and ultrastructural defects in mitochondrial cristae with enlarged, abnormal mitochondria.[96] Six TG lines targeting HIV-1 viral protein R (Vpr) to cardiomyocytes have been developed; each develops congestive heart failure (~8 weeks of age), abnormal cardiomyocyte nuclei, and mitoses.[97] In addition, Vpr TGs have induced four-chamber dilation, defective contraction, and atrial masses. We developed TG lines each expressing HIV-1 *Nef* exclusively targeted in ventricular cardiac myocytes. Phenotypic analysis of mice up to 12 months of age reveals no gross pathological changes in hemizygous *Nef* TG.[98–106]

ORGANELLE-SPECIFIC TARGETING As mentioned above, because our focus in AIDS has been to examine the effects of antiretroviral agents on mtDNA replication, other models of mitochondrial biogenesis may be important tools. A number of other transgenic and knockout mouse models have been examined to delineate the role of specific, related signaling pathways on cardiac performance and CM.[107] Sah *et al.*[108] generated transgenic mice in which constitutively activated RhoA or wild-type (WT) RhoA were expressed in a cardiac-specific manner through the α-MHC promoter to determine the role of Rho family GTPases on cardiac performance. Results demonstrated evidence of heart failure seen in RhoA transgenics, with severe edema, ventricular chamber dilation, increased cardiac fibrosis, atrial enlargement, and decreased fractional shortening. Unlike Rho, Rac1 proved to be more challenging by use of conventional gene targeting approaches as deletion results in embryonic lethality.[109] Instead, genetic manipulation of Rac1 to study the consequences of increased myocardial Rac1 activity was accomplished by cardiac-specific transgenesis with an unusual dichotomy: (1) rapid-onset, high-level postnatal expression that led to lethal dilated CM or (2) slow-onset postnatal expression leading to transient hypertrophy in juvenile mice that resolved with age.[110]

Transgenic mice have also been created to investigate the role of cardiac regulatory proteins, protein kinase C (PKC) and troponin I (cTNI).[111] Pioneering work in the laboratory of Chris Vlahos[112] has shown that PKC is upregulated in the human heart in end-stage failure. To demonstrate causality of PKC in heart failure, transgenic mice have been created in which the PKC isoform is overexpressed in the heart, often in a constitutively active form.[113–116] Collectively, studies with these TG show that chronic PKC overexpression can lead to features of cardiac failure including hypertrophy, reexpression of fetal genes, wall thinning, and left ventricular dysfunction.

Mouse transgenic models also may alter expression of genes involved in cardiac uptake and metabolism of either lipid or carbohydrate.[117] Disruption of the insulin-sensitive GLUT4, for example, by various transgenic techniques results in cardiac hypertrophy.[118,119] Studies by Yang *et al.* utilized TG murine models to investigate the role of myosin-binding protein C and molecular regulation of lipotoxicity in the heart.[120,121] Results from these and many other TG models demonstrate that alterations in cardiac uptake and metabolism may precede both the functional and morphological alterations that accompany diabetic cardiomyopathy and heart failure.

Understanding the critical role of NO signaling in myocardial function has also advanced with targeted overexpression of endothelial NOS (NOS3).[122] In the hearts of three lines of mice carrying the αMHC-NOS3 TG, pressure development of the LV was inversely correlated with NOS3 transgene expression.[123,124] On the other hand, the contractile response to norepinephrine and acetylcholine was not affected by TG expression. Janssens *et al.*[125] and Champion *et al.*[126] confirmed overexpressed NOS3 localized to caveolae by confocal microscopy and immunoprecipitation studies. In short, the availability of mice in which NOS3 is overexpressed has enabled investigators to characterize the function of this enzyme in mouse models of cardiovascular disease.

CONDITIONAL EXPRESSION Gene targeting to generate gain or loss of function mutations in the mouse has yielded remarkable advances in understanding the roles played by specific gene products.[127] Not all biological processes, however, can be accessed and studied by gene-inactivation or transgene-expression strategies. Insights have been gained into both embryogenesis and later gene function; however obstacles exist that include indirect systemic defects that can complicate transgenic constructs in the germ line to elicit the role of genes in a specific pathophysiological context. For example, many genes have several roles during embryogenesis and adulthood for which their ablation results in an embryonic-lethal phenotype, precluding an analysis of its function in later development. To circumvent these limitations, mouse geneticists developed strategies that allow genes to be activated or silenced where and when the investigator chooses.

The ideal genetic "switch" results in low or zero basal gene expression when "off" and high levels of gene expression when "on." Furthermore, the switch ideally is reversible and specific for the target gene, and should not interfere with other cellular components or with general metabolism. Early attempts included use of promoters that could be induced by heavy metals, heat shock, interferon, or steroids. Unfortunately, these strategies fell short due to high basal activity in the absence of induction.[128] More recently, success was reached with binary transgenic systems. Gene expression is controlled by the interaction of two components: an "effector" transgene and a "target" transgene.[45]

The most widely used binary transcription transactivation systems are the tetracycline-dependent regulatory systems developed by Mandfred Gossen and Hermann Bujard.[129] The effector is a fusion of sequences that encodes the VP16 transactivation domain and the *Escherichia coli* tetracycline repressor (TetR) protein, which specifically binds both tetracycline and the 19-bp operator sequences (*tetO*) of the *tet* operon in the target transgene, resulting in transcription. Two versions exist: (1) the tetracycline controlled transactivator (tTA) protein cannot bind DNA when the inducer is present ("tet-off"), and (2) the "reverse tTA" (rtTA) binds DNA only when the inducer is present ("tet-on"). Conditional expression with this *tetO* promoter system was used in a recent study to induce cardiac fibrosis and heart failure using an antisense mRNA of mineralocorticoid receptor targeted in cardiomyocytes.[130]

The first nonviral binary system to be used in mice was the Gal4/upstream activator sequence (UAS) system. Gal4, an isolated transcriptional activator of *Saccharomyces cerevisiae*, directs the transcription of Gal4-responsive genes by binding to (UASs). Two members of the integrase family of site-specific recombinases, *Cre* from bacteriophage P1[131] and *Flp* from *S. cerevisiae*[132] are currently used to conditionally control gene expression by site-specific DNA recombination.

Somatic, tissue-specific recombination has gained appeal more recently. One strategy for this exploits the *cre-loxP* system derived from bacteriophage P1.[133,134] The 38-kDa Cre protein functions as a site-specific recombinase, splicing DNA between specific 34-bp sequences known as *loxP* sites. Mating mice that contain a *loxP*-flanked gene with mice expressing Cre under the control of a cell specific-promoter result in Cre-mediated excision selectively in targeted cells. The cre–loxP system is entirely contingent on the exogenous recombines, so that recombination can be regulated via the timing and tissue specificity of *Cre* transgene expression.[135,136] A strategy has been generated for cell type-specific plus temporal control by tissue-restricted expression of a conditionally functional chimeric Cre protein, comprising a fusion between Cre and the mutated ligand binding domain (LBD) of a steroid hormone receptor.[135–140] These chimeric proteins become selectively active on binding a synthetic ligand, in preference to endogenous progesterone and estrogen, respectively.

PHARMACOLOGICAL MANIPULATION OF AIDS TRANSGENIC MURINE MODELS

We extensively used the NL4-3Δ *gag/pol* TG and the WT littermates (FVB/n mice) in both AZT and d4T monotherapy and combination therapy studies. TG and WT received water *ad libitum* with and without AZT, for example, for 21 or 35 days. After 21 days, molecular indicators of cardiac dysfunction were identified. Abundances of mRNA for cardiac sarcoplasmic reticulum calcium ATPase (SERCA2), sodium calcium exchanger (NCX1), and ANF were determined individually. Depressed SERCA2 and increased ANF mRNA abundance were found in LV from AZT-treated TG. NCX1 abundance remained unchanged. Eccentric LV hypertrophy was determined echocardiographically. After 35 days, cardiac dysfunction was worse in TG, with or without AZT treatment. Decreases in the first derivative of the maximal change in LV systolic pressure with respect to time (*+dP/dt*) occurred at baseline in TG. The increased half-time of relaxation and ventricular relaxation (*−dP/dt*) occurred in TG as well, independent of treatment. However, increased time to peak pressure was found only in AZT-treated TG. Histopathological and ultrastructural (transmission electron microscopy) changes were also identified. Ultrastructurally, mitochondrial destruction was most pronounced in AZT-treated TG, but was also found in AZT treated WT. In short, TG mice that express HIV-1 demonstrate cardiac dysfunction. AZT-treatment of WT causes mitochondrial ultrastructural alterations and *−dP/dt* changes. Therefore, data collectively suggest that HIV-1 and AZT each contributes to cardiac dysfunction in this transgenic model of AIDS CM.[33]

Additional studies have been performed using a combination HAART [AZT, lamivudine (3TC), and indinavir] or vehicle control treatment for 10 or 35 days in NL4-3Δ *gag/pol* TG and FVB/n WT mice. This "2 × 2" protocol provides a systematic determination of the effect(s) of TG and/or treatment. At the termination of the experiments (day 35), mice underwent echocardiography, quantitation of ANF, SERCA2, and determination of plasma lactate.[141] Myocardial histological features were also analyzed semiquantitatively and results were confirmed by transmission electron microscopy. Echocardiographic analysis revealed a 160% increase in LV mass in the TG + HAART cohort. Molecularly, ANF mRNA increased 250% and SERCA2 mRNA decreased 57%. Biochemically, plasma lactate was elevated (8.5 ± 2.0 mM). Pathologically, granular cytoplasmic changes were found in cardiac myocytes indicating enlarged, damaged mitochondria. Findings were confirmed ultrastructurally. Surprisingly, no changes occurred in the other cohorts. Cumulative HAART causes mitochondrial CM in AIDS TG mice.

GENERAL HEALTH OF THE GENETICALLY MANIPULATED MOUSE

Mouse genetic research has provided an important biological tool resulting in remarkable advances in the understanding of disease, pathogenesis, and treatment strategies. Important concerns remain, however, regarding the specificity of genetic manipulations of the cardiovascular system and other physiological systems. Concerns that the resultant changes yield an authentic model of human disease and that the degree of accurate phenotypic characteristics of the disease the genetic manipulation displays have been raised by clinicians and researchers alike. More importantly is the concern that nontargeted organ systems remain intact and functional. This is the rationale for a complete evaluation for general health parameters in genetically manipulated mice.

Establishing a mouse "strain-specific baseline" for routine parameters is the key to identifying and ensuring that phenotypic characteristics of the particular strain remain, overall, intact when evaluating the phenotypic changes that occur as a result of a particular genetic manipulation. Phenotypic parameters of size, weight-to-age ratios, for example, can offer basic, observational, noninvasive assessments when comparing a native (WT) strain to a genetically manipulated mouse line. Overall appetite, feeding and activity patterns can also serve as observational comparisons.

Table 41–3
Relevant murine cardiac analytes[a]

Atrial natriuretic peptide
(ANF)[142] Brain natriuretic peptide
(BNP)[143]
Troponin I[144]
Troponin T[145]
Lactic acid[146]
Lactate dehydrogenase (LDH)[147]
Creatinine kinase[148]

[a]All of the above assays are based on enzymatic colorimetric or immunoturbidometric assays and have been tested for feasibility.

Commercially available assays exist to address the general health of domestic animals. A slight modification of these assays could prove specifically useful in the mouse model, for various strains. Blood sampling, for example, is ideal for repetitive sampling as a function of age and disease development or in response to treatment of disease while preserving the living organism. Previous experience with obtaining blood from FVB/n and our own genetically manipulated mice indicates that retroorbital bleeding was rapid, safe, and effective. Collection of 200 µl of blood into heparinized capillary tubes yields approximately 100–120 µl of plasma. From this total volume, ~90 µl is usable for the analyte panel and ~10–30 µl plasma for other analyses.

The panel of analytes shown in Table 41–3 may serve as a marker for cardiovascular disease to evaluate the genetically manipulated mouse and the controls.[142–148]

CONCLUSIONS

Cardiomyopathy associated with HIV-1/AIDS is important clinically, and increasingly may impact the epidemic of AIDS as survival with HIV-1 increases due to effective antiretroviral therapies. The mechanisms of AIDS CM are not yet completely understood. TG models, however, particularly murine models, provide useful biological tools to study the impact of HIV-1, its gene products, as well as side effects of NRTI-HAART therapy on the pathophysiology and pathogenesis of CM.

REFERENCES

1. Shannon RP. SIV cardiomyopathy in non-human primates. *Trends Cardiovasc Med* 2001;11(6):242–246.
2. Yearley JH, Pearson C, Carville A, Shannon RP, Mansfield KG. SIV-associated myocarditis: Viral and cellular correlates of inflammation severity. *AIDS Res Hum Retroviruses* 2006;22(6):529–540.
3. Hammer SM, Saag MS, Schechter M, *et al.* Treatment for adult HIV infection: 2006 recommendations of the International AIDS Society-USA panel. *JAMA* 2006;296(7):827–843.
4. Resino S, Resino R, Maria Bellon J, *et al.* Clinical outcomes improve with highly active antiretroviral therapy in vertically HIV type-1-infected children. *Clin Infect Dis* 2006;43(2):243–252.
5. Lewis W, Dalakas MC. Mitochondrial toxicity of antiviral drugs. *Nat Med* 1995;1(5):417–422.
6. Brinkman K, ter Hofstede HJ, Burger DM, Smeitink JA, Koopmans PP. Adverse effects of reverse transcriptase inhibitors: Mitochondrial toxicity as common pathway. *AIDS* 1998;12(14):1735–1744.
7. Moyle G. Clinical manifestations and management of antiretroviral nucleoside analog-related mitochondrial toxicity. *Clin Ther* 2000; 22(8):911–936; discussion 898.
8. Villarroya F, Domingo P, Giralt M. Lipodystrophy associated with highly active anti-retroviral therapy for HIV infection: The adipocyte as a target of anti-retroviral-induced mitochondrial toxicity. *Trends Pharmacol Sci* 2005;26(2):88–93.
9. Sharma PL, Nurpeisov V, Hernandez-Santiago B, Beltran T, Schinazi RF. Nucleoside inhibitors of human immunodeficiency virus type 1 reverse transcriptase. *Curr Top Med Chem* 2004;4(9): 895–919.
10. Lewis W, Day BJ, Copeland WC. Mitochondrial toxicity of NRTI antiviral drugs: An integrated cellular perspective. *Nat Rev Drug Discov* 2003;2(10):812–822.
11. Lewis W, Copeland WC, Day BJ. Mitochondrial DNA depletion, oxidative stress, and mutation: Mechanisms of dysfunction from nucleoside reverse transcriptase inhibitors. *Lab Invest* 2001;81(6): 777–790.
12. Sukernik RI, Derbeneva OA, Starikovskaia EB, *et al.* [The mitochondrial genome and human mitochondrial diseases]. *Genetika* 2002;38(2):161–170.
13. Wallace DC. The mitochondrial genome in human adaptive radiation and disease: On the road to therapeutics and performance enhancement. *Gene* 2005;354:169–180.
14. Arnaudo E, Dalakas M, Shanske S, Moraes CT, DiMauro S, Schon EA. Depletion of muscle mitochondrial DNA in AIDS patients with zidovudine-induced myopathy. *Lancet* 1991;337(8740):508–510.
15. Dalakas MC, Illa I, Pezeshkpour GH, Laukaitis JP, Cohen B, Griffin JL. Mitochondrial myopathy caused by long-term zidovudine therapy. *N Engl J Med* 1990;322(16):1098–1105.
16. Murphy MD, O'Hearn M, Chou S. Fatal lactic acidosis and acute renal failure after addition of tenofovir to an antiretroviral regimen containing didanosine. *Clin Infect Dis* 2003;36(8):1082–1085.
17. De Clercq E. New approaches toward anti-HIV chemotherapy. *J Med Chem* 2005;48(5):1297–1313.
18. Dubinsky RM, Yarchoan R, Dalakas M, Broder S. Reversible axonal neuropathy from the treatment of AIDS and related disorders with 2′,3′-dideoxycytidine (ddC). *Muscle Nerve* 1989;12(10): 856–860.
19. Colacino JM. Mechanisms for the anti-hepatitis B virus activity and mitochondrial toxicity of fialuridine (FIAU). *Antiviral Res* 1996;29(2–3):125–139.
20. Lewis W, Griniuviene B, Tankersley KO, *et al.* Depletion of mitochondrial DNA, destruction of mitochondria, and accumulation of lipid droplets result from fialuridine treatment in woodchucks (Marmota monax). *Lab Invest* 1997;76(1):77–87.
21. Lewis W, Meyer RR, Simpson JF, Colacino JM, Perrino FW. Mammalian DNA polymerases alpha, beta, gamma, delta, and epsilon incorporate fialuridine (FIAU) monophosphate into DNA and are inhibited competitively by FIAU triphosphate. *Biochemistry* 1994;33(48):14620–14624.
22. Dalakas MC, Semino-Mora C, Leon-Monzon M. Mitochondrial alterations with mitochondrial DNA depletion in the nerves of AIDS patients with peripheral neuropathy induced by 2′3′-dideoxycytidine (ddC). *Lab Invest* 2001;81(11):1537–1544.
23. Lim SE, Ponamarev MV, Longley MJ, Copeland WC. Structural determinants in human DNA polymerase gamma account for mitochondrial toxicity from nucleoside analogs. *J Mol Biol* 2003; 329(1):45–57.
24. Highleyman L. Lodenosine trials stopped due to safety concerns. *BETA* 1999;12(4):4.
25. Moreno A, Quereda C, Moreno L, *et al.* High rate of didanosine-related mitochondrial toxicity in HIV/HCV-coinfected patients receiving ribavirin. *Antiviral Ther* 2004;9(1):133–138.
26. Baker R. Kidney dysfunction: A safety update on adefovir (preveon). *BETA* 1998;Jul:9–10.
27. Ho ES, Lin DC, Mendel DB, Cihlar T. Cytotoxicity of antiviral nucleotides adefovir and cidofovir is induced by the expression of human renal organic anion transporter 1. *J Am Soc Nephrol* 2000;11(3):383–393.
28. Song S, Wheeler LJ, Mathews CK. Deoxyribonucleotide pool imbalance stimulates deletions in HeLa cell mitochondrial DNA. *J Biol Chem* 2003;278(45):43893–43896.
29. Pontarin G, Gallinaro L, Ferraro P, Reichard P, Bianchi V. Origins of mitochondrial thymidine triphosphate: Dynamic relations to

cytosolic pools. *Proc Natl Acad Sci USA* 2003;100(21): 12159–12164.
30. Lewis W. Mitochondrial dysfunction and nucleoside reverse transcriptase inhibitor therapy: Experimental clarifications and persistent clinical questions. *Antiviral Res* 2003;58(3):189–197.
31. Lewis W. Mitochondrial DNA replication, nucleoside reverse-transcriptase inhibitors, and AIDS cardiomyopathy. *Prog Cardiovasc Dis* 2003;45(4):305–318.
32. Lewis W, Gonzalez B, Chomyn A, Papoian T. Zidovudine induces molecular, biochemical, and ultrastructural changes in rat skeletal muscle mitochondria. *J Clin Invest* 1992;89(4):1354–1360.
33. Lewis W, Grupp IL, Grupp G, *et al.* Cardiac dysfunction occurs in the HIV-1 transgenic mouse treated with zidovudine. *Lab Invest* 2000;80(2):187–197.
34. Lewis W, Papoian T, Gonzalez B, *et al.* Mitochondrial ultrastructural and molecular changes induced by zidovudine in rat hearts. *Lab Invest* 1991;65(2):228–236.
35. Lewis W, Simpson JF, Meyer RR. Cardiac mitochondrial DNA polymerase-gamma is inhibited competitively and noncompetitively by phosphorylated zidovudine. *Circ Res* 1994;74(2):344–348.
36. Kaguni LS. DNA polymerase gamma, the mitochondrial replicase. *Annu Rev Biochem* 2004;73:293–320.
37. Bridges EG, Jiang Z, Cheng YC. Characterization of a dCTP transport activity reconstituted from human mitochondria. *J Biol Chem* 1999;274(8):4620–4625.
38. Dolce V, Fiermonte G, Runswick MJ, Palmieri F, Walker JE. The human mitochondrial deoxynucleotide carrier and its role in the toxicity of nucleoside antivirals. *Proc Natl Acad Sci USA* 2001;98(5):2284–2288.
39. Young P, Leeds JM, Slabaugh MB, Mathews CK. Ribonucleotide reductase: Evidence for specific association with HeLa cell mitochondria. *Biochem Biophys Res Commun* 1994;203(1):46–52.
40. Chang C, Swaan PW, Ngo LY, Lum PY, Patil SD, Unadkat JD. Molecular requirements of the human nucleoside transporters hCNT1, hCNT2, and hENT1. *Mol Pharmacol* 2004;65(3): 558–570.
41. Lai Y, Bakken AH, Unadkat JD. Simultaneous expression of hCNT1-CFP and hENT1-YFP in Madin-Darby canine kidney cells. Localization and vectorial transport studies. *J Biol Chem* 2002; 277(40):37711–37717.
42. Lai Y, Tse CM, Unadkat JD. Mitochondrial expression of the human equilibrative nucleoside transporter 1 (hENT1) results in enhanced mitochondrial toxicity of antiviral drugs. *J Biol Chem* 2004;279(6):4490–4497.
43. SenGupta DJ, Lum PY, Lai Y, *et al.* A single glycine mutation in the equilibrative nucleoside transporter gene, hENT1, alters nucleoside transport activity and sensitivity to nitrobenzylthioinosine. *Biochemistry* 2002;41(5):1512–1519.
44. Lee EW, Lai Y, Zhang H, Unadkat JD. Identification of the mitochondrial targeting signal of the human equilibrative nucleoside transporter 1 (hENT1): Implications for interspecies differences in mitochondrial toxicity of fialuridine. *J Biol Chem* 2006;281(24): 16700–16706.
45. Lewandoski M. Conditional control of gene expression in the mouse. *Nat Rev Genet* 2001;2(10):743–755.
46. Klotman PE, Notkins AL. Transgenic models of human immunodeficiency virus type-1. *Curr Top Microbiol Immunol* 1996;206:197–222.
47. Bristow MR. Tumor necrosis factor-alpha and cardiomyopathy. *Circulation* 1998;97(14):1340–1341.
48. Mann DL. Stress-activated cytokines and the heart: From adaptation to maladaptation. *Annu Rev Physiol* 2003;65:81–101.
49. Welch S, Plank D, Witt S, *et al.* Cardiac-specific IGF-1 expression attenuates dilated cardiomyopathy in tropomodulin-overexpressing transgenic mice. *Circ Res* 2002;90(6):641–648.
50. Dorn GW 2nd. Physiologic growth and pathologic genes in cardiac development and cardiomyopathy. *Trends Cardiovasc Med* 2005;15(5):185–189.
51. Pinsky DJ, Cai B, Yang X, Rodriguez C, Sciacca RR, Cannon PJ. The lethal effects of cytokine-induced nitric oxide on cardiac myocytes are blocked by nitric oxide synthase antagonism or transforming growth factor beta. *J Clin Invest* 1995;95(2):677–685.
52. Levine B, Kalman J, Mayer L, Fillit HM, Packer M. Elevated circulating levels of tumor necrosis factor in severe chronic heart failure. *N Engl J Med* 1990;323(4):236–241.
53. Barry WH. Mechanisms of immune-mediated myocyte injury. *Circulation* 1994;89(5):2421–2432.
54. Lange LG, Schreiner GF. Immune mechanisms of cardiac disease. *N Engl J Med* 1994;330(16):1129–1135.
55. Gulick T, Chung MK, Pieper SJ, Lange LG, Schreiner GF. Interleukin 1 and tumor necrosis factor inhibit cardiac myocyte beta-adrenergic responsiveness. *Proc Natl Acad Sci USA* 1989;86(17): 6753–6757.
56. Pagani FD, Baker LS, Hsi C, Knox M, Fink MP, Visner MS. Left ventricular systolic and diastolic dysfunction after infusion of tumor necrosis factor-alpha in conscious dogs. *J Clin Invest* 1992;90(2): 389–398.
57. Pagani FD, Baker LS, Knox MA, Cheng H, Fink MP, Visner MS. Load-insensitive assessment of myocardial performance after tumor necrosis factor-alpha in dogs. *Surgery* 1992;111(6):683–693.
58. Pacher R, Stanek B. Prostaglandin E1 in heart failure: The Vienna experience. *Wien Klin Wochenschr* 1996;108(16):491–495.
59. Pacher R, Stanek B, Hulsmann M, *et al.* Prognostic impact of big endothelin-1 plasma concentrations compared with invasive hemodynamic evaluation in severe heart failure. *J Am Coll Cardiol* 1996;27(3):633–641.
60. Cody RJ, Haas GJ, Binkley PF, Capers Q, Kelley R. Plasma endothelin correlates with the extent of pulmonary hypertension in patients with chronic congestive heart failure. *Circulation* 1992;85(2):504–509.
61. Shubeita HE, McDonough PM, Harris AN, *et al.* Endothelin induction of inositol phospholipid hydrolysis, sarcomere assembly, and cardiac gene expression in ventricular myocytes. A paracrine mechanism for myocardial cell hypertrophy. *J Biol Chem* 1990;265(33): 20555–20562.
62. Wei CM, Lerman A, Rodeheffer RJ, *et al.* Endothelin in human congestive heart failure. *Circulation* 1994;89(4):1580–1586.
63. Weber KT. Extracellular matrix remodeling in heart failure: A role for de novo angiotensin II generation. *Circulation* 1997;96(11): 4065–4082.
64. Colucci WS. Molecular and cellular mechanisms of myocardial failure. *Am J Cardiol* 1997;80(11A):15L–25L.
65. Onai Y, Suzuki J, Maejima Y, *et al.* Inhibition of NF-{kappa}B improves left ventricular remodeling and cardiac dysfunction after myocardial infarction. *Am J Physiol Heart Circ Physiol* 2006;292(1): H530–538.
66. Rosenzweig A, Seidman CE. Atrial natriuretic factor and related peptide hormones. *Annu Rev Biochem* 1991;60:229–255.
67. Feringa HH, Schouten O, Dunkelgrun M, *et al.* Plasma N-terminal pro-B-type natriuretic peptide as long-term prognostic marker after major vascular surgery. *Heart* 2006;93(2):226–231.
68. Drexler H, Hirth C, Stasch HP, Lu W, Neuser D, Just H. Vasodilatory action of endogenous atrial natriuretic factor in a rat model of chronic heart failure as determined by monoclonal ANF antibody. *Circ Res* 1990;66(5):1371–1380.
69. Clerico A, Iervasi G, Del Chicca MG, *et al.* Circulating levels of cardiac natriuretic peptides (ANP and BNP) measured by highly sensitive and specific immunoradiometric assays in normal subjects and in patients with different degrees of heart failure. *J Endocrinol Invest* 1998;21(3):170–179.
70. Date T, Shinozaki T, Yamakawa M, *et al.* Elevated plasma brain natriuretic peptide level in cardiac sarcoidosis patients with preserved ejection fraction. *Cardiology* 2006;107(4):277–280.
71. Bristow MR, Abraham WT, Yoshikawa T, *et al.* Second- and third-generation beta-blocking drugs in chronic heart failure. *Cardiovasc Drugs Ther* 1997;11(Suppl. 1):291–296.
72. Abraham WT, Raynolds MV, Gottschall B, *et al.* Importance of angiotensin-converting enzyme in pulmonary hypertension. *Cardiology* 1995;86(Suppl. 1):9–15.

73. Francis GS, Benedict C, Johnstone DE, *et al.* Comparison of neuroendocrine activation in patients with left ventricular dysfunction with and without congestive heart failure. A substudy of the Studies of Left Ventricular Dysfunction (SOLVD). *Circulation* 1990;82(5):1724–1729.
74. Small JA, Scangos GA, Cork L, Jay G, Khoury G. The early region of human papovavirus JC induces dysmyelination in transgenic mice. *Cell* 1986;46(1):13–18.
75. Hinrichs SH, Nerenberg M, Reynolds RK, Khoury G, Jay G. A transgenic mouse model for human neurofibromatosis. *Science* 1987;237(4820):1340–1343.
76. Kim CM, Koike K, Saito I, Miyamura T, Jay G. HBx gene of hepatitis B virus induces liver cancer in transgenic mice. *Nature* 1991;351(6324):317–320.
77. Tinkle BT, Ngo L, Luciw PA, Maciag T, Jay G. Human immunodeficiency virus-associated vasculopathy in transgenic mice. *J Virol* 1997;71(6):4809–4814.
78. Iwakura Y, Shioda T, Tosu M, *et al.* The induction of cataracts by HIV-1 in transgenic mice. AIDS 1992;6(10):1069–1075.
79. Dickie P, Felser J, Eckhaus M, *et al.* HIV-associated nephropathy in transgenic mice expressing HIV-1 genes. *Virology* 1991;185(1):109–119.
80. Kopp JB, Klotman ME, Adler SH, *et al.* Progressive glomerulosclerosis and enhanced renal accumulation of basement membrane components in mice transgenic for human immunodeficiency virus type 1 genes. *Proc Natl Acad Sci USA* 1992;89(5):1577–1581.
81. Cai SR, Xu G, Becker-Hapak M, Ma M, Dowdy SF, McLeod HL. The kinetics and tissue distribution of protein transduction in mice. *Eur J Pharm Sci* 2006;27(4):311–319.
82. Gibellini D, Vitone F, Schiavone P, Re MC. HIV-1 tat protein and cell proliferation and survival: A brief review. *New Microbiol* 2005;28(2):95–109.
83. Vogel J, Hinrichs SH, Napolitano LA, Ngo L, Jay G. Liver cancer in transgenic mice carrying the human immunodeficiency virus tat gene. *Cancer Res* 1991;51(24):6686–6690.
84. Corallini A, Altavilla G, Pozzi L, *et al.* Systemic expression of HIV-1 tat gene in transgenic mice induces endothelial proliferation and tumors of different histotypes. *Cancer Res* 1993;53(22):5569–5575.
85. Barbanti-Brodano G, Sampaolesi R, Campioni D, *et al.* HIV-1 tat acts as a growth factor and induces angiogenic activity in BK virus/tat transgenic mice. *Antibiot Chemother* 1994;46:88–101.
86. Prakash O, Teng S, Ali M, *et al.* The human immunodeficiency virus type 1 Tat protein potentiates zidovudine-induced cellular toxicity in transgenic mice. *Arch Biochem Biophys* 1997;343(2):173–180.
87. Choi J, Liu RM, Kundu RK, *et al.* Molecular mechanism of decreased glutathione content in human immunodeficiency virus type 1 Tat-transgenic mice. *J Biol Chem* 2000;275(5):3693–3698.
88. Kundu RK, Sangiorgi F, Wu LY, *et al.* Expression of the human immunodeficiency virus-Tat gene in lymphoid tissues of transgenic mice is associated with B-cell lymphoma. *Blood* 1999;94(1):275–282.
89. Hanna Z, Kay DG, Rebai N, Guimond A, Jothy S, Jolicoeur P. Nef harbors a major determinant of pathogenicity for an AIDS-like disease induced by HIV-1 in transgenic mice. *Cell* 1998;95(2):163–175.
90. Hanna Z, Priceputu E, Hu C, Vincent P, Jolicoeur P. HIV-1 Nef mutations abrogating downregulation of CD4 affect other Nef functions and show reduced pathogenicity in transgenic mice. *Virology* 2006;346(1):40–52.
91. Jolicoeur P, Kay DG, Cool M, Jothy S, Rebai N, Hanna Z. A novel mouse model of HIV-1 disease. *Leukemia* 1999;13(Suppl. 1):S78–80.
92. Zhong J, Zuo Y, Ma J, *et al.* Expression of HIV-1 genes in podocytes alone can lead to the full spectrum of HIV-1-associated nephropathy. *Kidney Int* 2005;68(3):1048–1060.
93. Locardi C, Puddu P, Ferrantini M, *et al.* Persistent infection of normal mice with human immunodeficiency virus. *J Virol* 1992;66(3):1649–1654.
94. Robbins J. Remodeling the cardiac sarcomere using transgenesis. *Annu Rev Physiol* 2000;62:261–287.
95. Subramaniam A, Jones WK, Gulick J, Wert S, Neumann J, Robbins J. Tissue-specific regulation of the alpha-myosin heavy chain gene promoter in transgenic mice. *J Biol Chem* 1991;266(36):24613–24620.
96. Raidel SM, Haase C, Jansen NR, *et al.* Targeted myocardial transgenic expression of HIV Tat causes cardiomyopathy and mitochondrial damage. *Am J Physiol Heart Circ Physiol* 2002;282(5):H1672–1678.
97. Lewis W, Miller YK, Haase CP, *et al.* HIV viral protein R causes atrial cardiomyocyte mitosis, mesenchymal tumor, dysrhythmia, and heart failure. *Lab Invest* 2005;85(2):182–192.
98. Karn J, Dingwall C, Finch JT, Heaphy S, Gait MJ. RNA binding by the tat and rev proteins of HIV-1. *Biochimie* 1991;73(1):9–16.
99. Kim YS, Panganiban AT. The full-length Tat protein is required for TAR-independent, posttranscriptional trans activation of human immunodeficiency virus type 1 env gene expression. *J Virol* 1993;67(7):3739–3747.
100. Marciniak RA, Calnan BJ, Frankel AD, Sharp PA. HIV-1 Tat protein trans-activates transcription in vitro. *Cell* 1990;63(4):791–802.
101. Vives E, Charneau P, van Rietschoten J, Rochat H, Bahraoui E. Effects of the Tat basic domain on human immunodeficiency virus type 1 transactivation, using chemically synthesized Tat protein and Tat peptides. *J Virol* 1994;68(5):3343–3353.
102. Aiken C, Trono D. Nef stimulates human immunodeficiency virus type 1 proviral DNA synthesis. *J Virol* 1995;69(8):5048–5056.
103. Romero IA, Teixeira A, Strosberg AD, Cazaubon S, Couraud PO. The HIV-1 nef protein inhibits extracellular signal-regulated kinase-dependent DNA synthesis in a human astrocytic cell line. *J Neurochem* 1998;70(2):778–785.
104. Elder RT, Benko Z, Zhao Y. HIV-1 VPR modulates cell cycle G2/M transition through an alternative cellular mechanism other than the classic mitotic checkpoints. *Front Biosci* 2002;7:d349–357.
105. Subbramanian RA, Kessous-Elbaz A, Lodge R, *et al.* Human immunodeficiency virus type 1 Vpr is a positive regulator of viral transcription and infectivity in primary human macrophages. *J Exp Med* 1998;187(7):1103–1111.
106. Willey RL, Maldarelli F, Martin MA, Strebel K. Human immunodeficiency virus type 1 Vpu protein regulates the formation of intracellular gp160-CD4 complexes. *J Virol* 1992;66(1):226–234.
107. Brown JH, Del Re DP, Sussman MA. The Rac and Rho hall of fame: A decade of hypertrophic signaling hits. *Circ Res* 2006;98(6):730–742.
108. Sah VP, Minamisawa S, Tam SP, *et al.* Cardiac-specific overexpression of RhoA results in sinus and atrioventricular nodal dysfunction and contractile failure. *J Clin Invest* 1999;103(12):1627–1634.
109. Sugihara K, Nakatsuji N, Nakamura K, *et al.* Rac1 is required for the formation of three germ layers during gastrulation. *Oncogene* 1998;17(26):3427–3433.
110. Sussman MA, Welch S, Walker A, *et al.* Altered focal adhesion regulation correlates with cardiomyopathy in mice expressing constitutively active rac1. *J Clin Invest* 2000;105(7):875–886.
111. Walker JW. Protein kinase C, troponin I and heart failure: Overexpressed, hyperphosphorylated and underappreciated? *J Mol Cell Cardiol* 2006;40(4):446–450.
112. Vlahos CJ, McDowell SA, Clerk A. Kinases as therapeutic targets for heart failure. *Nat Rev Drug Discov* 2003;2(2):99–113.
113. Goldspink PH, Montgomery DE, Walker LA, *et al.* Protein kinase Cepsilon overexpression alters myofilament properties and composition during the progression of heart failure. *Circ Res* 2004;95(4):424–432.
114. Huang L, Wolska BM, Montgomery DE, Burkart EM, Buttrick PM, Solaro RJ. Increased contractility and altered Ca(2+) transients of mouse heart myocytes conditionally expressing PKCbeta. *Am J Physiol Cell Physiol* 2001;280(5):C1114–1120.
115. Takeishi Y, Chu G, Kirkpatrick DM, *et al.* In vivo phosphorylation of cardiac troponin I by protein kinase Cbeta2 decreases cardiomyo-

cyte calcium responsiveness and contractility in transgenic mouse hearts. *J Clin Invest* 1998;102(1):72–78.
116. Wakasaki H, Koya D, Schoen FJ, *et al.* Targeted overexpression of protein kinase C beta2 isoform in myocardium causes cardiomyopathy. *Proc Natl Acad Sci USA* 1997;94(17):9320–9325.
117. Hartil K, Charron MJ. Genetic modification of the heart: Transgenic modification of cardiac lipid and carbohydrate utilization. *J Mol Cell Cardiol* 2005;39(4):581–593.
118. Abel ED, Kaulbach HC, Tian R, *et al.* Cardiac hypertrophy with preserved contractile function after selective deletion of GLUT4 from the heart. *J Clin Invest* 1999;104(12):1703–1714.
119. Katz EB, Stenbit AE, Hatton K, DePinho R, Charron MJ. Cardiac and adipose tissue abnormalities but not diabetes in mice deficient in GLUT4. *Nature* 1995;377(6545):151–155.
120. Yang Q, Sanbe A, Osinska H, Hewett TE, Klevitsky R, Robbins J. A mouse model of myosin binding protein C human familial hypertrophic cardiomyopathy. *J Clin Invest* 1998;102(7):1292–1300.
121. Yang Q, Hewett TE, Klevitsky R, Sanbe A, Wang X, Robbins J. PKA-dependent phosphorylation of cardiac myosin binding protein C in transgenic mice. *Cardiovasc Res* 2001;51(1):80–88.
122. Bloch KD, Janssens S. Cardiomyocyte-specific overexpression of nitric oxide synthase 3: Impact on left ventricular function and myocardial infarction. *Trends Cardiovasc Med* 2005;15(7):249–253.
123. Brunner F, Andrew P, Wolkart G, Zechner R, Mayer B. Myocardial contractile function and heart rate in mice with myocyte-specific overexpression of endothelial nitric oxide synthase. *Circulation* 2001;104(25):3097–3102.
124. Brunner F, Maier R, Andrew P, Wolkart G, Zechner R, Mayer B. Attenuation of myocardial ischemia/reperfusion injury in mice with myocyte-specific overexpression of endothelial nitric oxide synthase. *Cardiovasc Res* 2003;57(1):55–62.
125. Janssens S, Pokreisz P, Schoonjans L, *et al.* Cardiomyocyte-specific overexpression of nitric oxide synthase 3 improves left ventricular performance and reduces compensatory hypertrophy after myocardial infarction. *Circ Res* 2004;94(9):1256–1262.
126. Champion HC, Georgakopoulos D, Takimoto E, Isoda T, Wang Y, Kass DA. Modulation of in vivo cardiac function by myocyte-specific nitric oxide synthase-3. *Circ Res* 2004;94(5):657–663.
127. Rajewsky K, Gu H, Kuhn R, *et al.* Conditional gene targeting. *J Clin Invest* 1996;98(3):600–603.
128. Yarranton GT. Inducible vectors for expression in mammalian cells. *Curr Opin Biotechnol* 1992;3(5):506–511.
129. Gossen M, Bujard H. Tight control of gene expression in mammalian cells by tetracycline-responsive promoters. *Proc Natl Acad Sci USA* 1992;89(12):5547–5551.
130. Beggah AT, Escoubet B, Puttini S, *et al.* Reversible cardiac fibrosis and heart failure induced by conditional expression of an antisense mRNA of the mineralocorticoid receptor in cardiomyocytes. *Proc Natl Acad Sci USA* 2002;99(10):7160–7165.
131. Sauer B, Henderson N. Cre-stimulated recombination at loxP-containing DNA sequences placed into the mammalian genome. *Nucleic Acids Res* 1989;17(1):147–161.
132. O'Gorman S, Fox DT, Wahl GM. Recombinase-mediated gene activation and site-specific integration in mammalian cells. *Science* 1991;251(4999):1351–1355.
133. Lakso M, Sauer B, Mosinger B Jr, *et al.* Targeted oncogene activation by site-specific recombination in transgenic mice. *Proc Natl Acad Sci USA* 1992;89(14):6232–6236.
134. Sauer B, Henderson N. Site-specific DNA recombination in mammalian cells by the Cre recombinase of bacteriophage P1. *Proc Natl Acad Sci USA* 1988;85(14):5166–5170.
135. Kellendonk C, Tronche F, Monaghan AP, Angrand PO, Stewart F, Schutz G. Regulation of Cre recombinase activity by the synthetic steroid RU 486. *Nucleic Acids Res* 1996;24(8):1404–1411.
136. Kellendonk C, Tronche F, Casanova E, Anlag K, Opherk C, Schutz G. Inducible site-specific recombination in the brain. *J Mol Biol* 1999;285(1):175–182.
137. Reichardt HM, Kellendonk C, Tronche F, Schutz G. The Cre/loxP system—a versatile tool to study glucocorticoid signalling in mice. *Biochem Soc Trans* 1999;27(2):78–83.
138. Tsujita M, Mori H, Watanabe M, Suzuki M, Miyazaki J, Mishina M. Cerebellar granule cell-specific and inducible expression of Cre recombinase in the mouse. *J Neurosci* 1999;19(23):10318–10323.
139. Indra AK, Warot X, Brocard J, *et al.* Temporally-controlled site-specific mutagenesis in the basal layer of the epidermis: Comparison of the recombinase activity of the tamoxifen-inducible Cre-ER(T) and Cre-ER(T2) recombinases. *Nucleic Acids Res* 1999;27(22):4324–4327.
140. Brocard J, Warot X, Wendling O, *et al.* Spatio-temporally controlled site-specific somatic mutagenesis in the mouse. *Proc Natl Acad Sci USA* 1997;94(26):14559–14563.
141. Lewis W, Haase CP, Raidel SM, *et al.* Combined antiretroviral therapy causes cardiomyopathy and elevates plasma lactate in transgenic AIDS mice. *Lab Invest* 2001;81(11):1527–1536.
142. Milne R, Gutkowska J, Thibault G, *et al.* A murine monoclonal antibody against rat atrial natriuretic factor (ANF) which cross-reacts with mouse ANF. *Mol Immunol* 1987;24(2):127–132.
143. Nakagawa M, Tanaka I, Suga S, *et al.* Preparation of a monoclonal antibody against mouse brain natriuretic peptide (BNP) and tissue distribution of BNP in mice. *Clin Exp Pharmacol Physiol Suppl* 1995;22(1):S186–187.
144. O'Brien PJ, Dameron GW, Beck ML, Brandt M. Differential reactivity of cardiac and skeletal muscle from various species in two generations of cardiac troponin-T immunoassays. *Res Vet Sci* 1998;65(2):135–137.
145. Lim SS, Tu ZH, Lemanski LF. Anti-troponin-T monoclonal antibody crossreacts with all muscle types. *J Muscle Res Cell Motil* 1984;5(5):515–526.
146. Kromenaker SJ, Srienc F. Effect of lactic acid on the kinetics of growth and antibody production in a murine hybridoma: Secretion patterns during the cell cycle. *J Biotechnol* 1994;34(1):13–34.
147. Sigal LJ, Berens S, Wylie D. A lactate dehydrogenase (LDH)-based immunoassay for detection of cell surface antigens and its application to the study of MHC class I-binding peptides. *J Immunol Methods* 1994;177(1–2):261–268.
148. Laurino JP, Fischberg-Bender E, Galligan S, Chang J. An immunochemical mass assay for the direct measurement of creatine kinase MB2. *Ann Clin Lab Sci* 1995;25(3):252–263.

Models of Behavior V

Reproduction B

42 Primate Models for the Assisted Reproductive Technologies and Embryonic Stem Cell Biology

Don P. Wolf, Shoukhrat Mitalipov, and James A. Byrne

ABSTRACT

Nonhuman primates (NHPs) represent clinically relevant animal models used in studies on the etiology and treatment of human diseases. In the context of reproduction, NHP models are relevant to research interests as diverse as the etiology and treatment of infertility and contraceptive development to an evaluation of cell or tissue-based therapies of disease employing embryonic stem cell-derived phenotypes. The assisted reproductive technologies (ARTs) are used in the production of animals carrying desired MHC alleles for HIV vaccine development and could be used in the production of genetically identical animals by somatic cell nuclear transfer (SCNT). SCNT in turn is relevant to rescuing the unique genetics of aging or even deceased animals, to testing the efficacy of therapeutic cloning, and in combination with gene targeting, to creating monkey models of genetically based, neurodegenerative diseases. Since most NHP experience in applying the ARTs and in deriving embryonic stem cells (ESCs) involves Old World monkeys, principally rhesus and cynomolgus macaques and baboons, studies with these species will be featured in this chapter.

Key Words: Nonhuman primates, Assisted reproductive technology, Embryonic stem cells, Cloning.

GENERAL CHARACTERISTICS

Nonhuman primates (NHPs) are among the least available and most costly of animal models employed in biomedical research and their use is carefully regulated and should be exhaustively justified. While there are approximately 200 species in the order Primates, representatives from seven genera are most commonly used in research including both Old World and New World primates (see Chapter 28, this volume, on nonhuman primates). Among the great apes, the chimpanzee (*Pan troglodytes*) has played a unique role in biomedical research predicated on its susceptibility to human infectious diseases including HIV.[1] Of obvious relevance to their role in biomedical research, NHPs resemble human beings much more closely in anatomy and physiology than do other commonly employed animals such as rodents. A case in point, because NHP brains more closely resemble those of humans in size and complexity, they exhibit complex behavior and cognitive skills that are difficult to discern in lower mammals. This feature is particularly important in assessing whether treatments designed to ameliorate subtle cognitive defects are safe and effective. Other unique characteristics of NHPs include their susceptibility to human infectious agents and their responses to experimentally induced diseases.[1] They also serve as relevant models for stress or nutrition-related research. In the more specific context of assisted reproductive technologies (ARTs) and stem cell biology, NHPs models are relevant to research interests as diverse as the etiology and treatment of infertility and contraceptive development to evaluations of cell or tissue-based treatments of disease using embryonic stem cell (ESC)-derived phenotypes. The ARTs are used in the production of desired genotypes for HIV vaccine development and could be used to produce genetically identical animals for nature versus nurture comparisons. When also considering somatic cell nuclear transfer (SCNT) or reproductive cloning, the rescue of the unique genetics of aging or even deceased animals is possible as is the creation of genetically identical animals, testing the efficacy of therapeutic cloning and in combination with gene targeting, producing monkey models of genetically based diseases.

ADVANTAGES OF THE MODEL Rhesus monkeys are recognized as an important preclinical model. For instance, the methyl-4-phenyl-1,2,3,6-tetrahydropyridine (MPTP)-treated monkey is the best animal model of Parkinson's disease (see Chapter 35, this volume, on transgenic models of neurodegenerative diseases). With regard to the reproductive and embryonic stem cell technologies, monkeys show menstrual cycles that are similar to humans in length and characteristics. This analogy is also true for spermatogenesis, early preimplantation development/embryology, and placentation. ESCs derived from NHP embryos are similar to those from humans and markedly different from murine ESCs. Relevant areas of research include intracytoplasmic sperm injection (ICSI) or spermatid injection,[2] *in vitro* oocyte maturation,[3–7] optimizing embryo culture conditions, therapeutic and reproductive cloning, and models of contraception and implantation. NHP ESCs could also be employed in preclinical studies involving cell or tissue-based treatments of human degenerative diseases, in the prospective systematic evaluation of health consequences of the ARTs, and even in the development of female fertility preservation strategies.[8] The availability of NHP ESCs and the ability to use federal monies to generate addi-

From: *Sourcebook of Models for Biomedical Research*
(P. M. Conn, ed.), © 2008 Humana Press Inc., Totowa, NJ.

tional lines represent advantages over current efforts with human ESCs. Applications of the ARTs in conjunction with SCNT should eventually allow the creation of NHP models, for instance, of genetically based diseases.

DISADVANTAGES ASSOCIATED WITH NONHUMAN PRIMATE RESEARCH Features of NHP research include high costs in animal lease and per diem fees and limited access or availability. Obviously, unique facilities are required in terms of animal husbandry, caging, and surgical facilities. There is also substantial public concern over the use of such highly sentient animals in biomedical research. As is true for all animal models employed in research there are and should be restrictions on use, that is, when rodent models are not available or when the unique features of the NHP necessitate it. Other restrictions that impact the use of the model include limits on the number and types of surgeries to which animals can be exposed and the large number of animals required in animal production using the ARTs—egg donors and embryo transfer recipients—or in studies involving infrequent events, for instance, investigations into intracytoplasmic sperm injection-related abnormalities in newborns.

HUSBANDRY AND AVAILABLE STOCKS

As significant increases in our knowledge of the biology and care of NHPs have been realized, the number of NHPs imported has declined (the 2002 estimation of the number of NHPs imported annually is 50–60,000) because of bans on exportation from countries of origin (case in point, Indian-origin rhesus macaques), the impact of conservation movements, animal activist group activities, and more efficient animal use.[1] Domestic breeding programs initiated in the 1970s have become the major supply source of rhesus monkeys, baboons, chimpanzees and squirrel monkeys (consult Chapter 28, this volume, on NHPs and psic@bart.rprc.washington.edu for an Annual Resource Guide concerning animal suppliers). The National Primate Research Center system is a major supplier of animals for National Institutes of Health (NIH)-sponsored research with infrastructural support from National Center for Research Resources (NCRR). There are currently eight Centers in this system with a major emphasis on the propagation of rhesus and cynomolgus macaques and baboons. Specific pathogen-free production colonies are currently under development with NCRR sponsorship that might, parenthetically, benefit from the use of the ARTs, and should eventually become the standard resource. For ARTs and ESC biology, only a few NHP models have been developed sufficiently to allow itemization of reproducible protocols. In several contexts, the human serves as the model for the NHP as clinical applications of the ARTs are now responsible for the production of over 1% of the births in Western nations accounting, cumulatively, for millions of children. Furthermore, federally sponsored research on human ESCs dwarfs funded efforts employing NHP ESCs.

The great apes can be eliminated from consideration simply based on economic, practical, and ethical arguments. Indeed, in this case, it is more likely that the ARTs may ultimately be used to propagate endangered animals rather than serve as useful models in translational research. As for the more commonly available New World monkeys, such as owl or squirrel monkeys, extensive efforts to establish routine protocols for the ARTs have met with limited success, allocating this group into species currently requiring protocol development.[9] In a few applications, specific advantages may be compelling in the selection of a species for study, for instance in polyovulatory marmosets where dizygotic twinning is the rule[1] or baboons where external signs, such as sex skin changes, can be used to conveniently monitor the menstrual cycle or where seasonal anovulation may not be limiting as it is in rhesus macaques.

Based on the extensive experience and success achieved in the past two decades, Old World macaques, rhesus and cynomolgus monkeys, will be featured in this chapter. The baboon is still limited by suboptimal *in vitro* embryo culture conditions and while proof of principle experimentation has been reported recently in *Macaca nemenstrina*, the experience is limited.[10] As an example of the robust experience now available with the rhesus monkey, a highly unique and valuable resource exists in the ART Core at Oregon National Primate Research Center (ONPRC) (http://www.onprc.ohsu.edu/) established in 1988 under the co-direction of Drs. Don Wolf and Richard Stouffer. The focus of the core has been and continues to be on utilizing the rhesus macaque as a NHP model for research on gametogenesis, fertilization, embryogenesis, and implantation. This facility is responsible for 11 of the first 15 test tube rhesus monkeys produced in the world. Parenthetically, Dr. Barry Bavister's group at the University of Wisconsin is responsible for the other four including Petri, the first success.[11] The ART Core was, in part, responsible for the first monkey produced following the transfer of an embryo that had been frozen, cryobanked, and thawed after successful *in vitro* fertilization,[12] the first twin pregnancies,[12] the first infants produced by nuclear transfer of embryonic cells,[13] the first infant born following the transfer of an ICSI-produced blastocyst employing a nonsurgical procedure,[14] the first live birth resulting from the transfer of a demiembryo created by blastomere separation at the two-cell stage,[15] the first animal produced from transplanted ovarian tissue,[8] and the first pregnancy to result from embryo transfer (ET) during a controlled ovarian stimulation (COS) cycle (unpublished result). The program has contributed to improvements in COS[16] and in its more recent configuration has added an ESC component to its activities. Currently the ART/ESC Core program provides oocytes, sperm, embryos, follicular fluid, cumulus, and granulosa cells and distributes rhesus monkey ESC lines along with mouse embryonic fibroblasts, media, protocols, and expertise to the global scientific community.

It is worth noting that a major limitation in developing the ARTs or ESC technology in NHPs is the focus by NIH scientific reviewers on basic or applied research as opposed to protocol development. One approach to overcoming this limitation has been to couple protocol development with the production of a highly desirable but rare product such as monkeys with specific major histocompatibility alleles or with efforts to produce specific pathogen-free colonies, in both cases reflecting the unmet needs of the HIV vaccine development community. However, despite these limitations remarkable progress has been made in the rhesus monkey with implantation rates from ART-produced embryos comparable to those obtained in humans at approximately one in six embryos transferred. Of relevance to investigators interested in gene expression is the recently released gene chip with probe sets interrogating over 47,000 rhesus monkey transcripts (www.affymetrix.com/products/arrays/specific/rhesus_macaque.affx).

METHODS

ASSISTED REPRODUCTIVE TECHNOLOGIES A general pitfall in the application of ARTs in NHPs is the realization that it is not necessarily possible to extrapolate technologies from one species to another. Typically there are advantages and disadvantages associated with each species, for instance, in their availability, health, size, relevance to a specific project, and availability of established disease models. However, the trump card at present in selecting an NHP animal model has to be the existence of established protocols and relevant experience in that species. Fortunately, as noted above, much of the methodology involved in the routine application of the ARTs in rhesus monkeys has been worked out and published in detail.[17–19] A comprehensive itemization of references available in NHPs on gamete collection, fertilization, embryo culture, and transfer is included in Table 42–1.[20–72] A few specific recommendations or potential concerns are discussed below.

Generally sperm collection involves penile or rectal electroejaculation or needle biopsy of the caudal epididymas or vas deferens. Rhesus monkey sperm collected by penile electroejaculation is notable for its uniform nature and high motility, for example, at ONPRC, only animals with sperm concentrations greater that 60 million/ml and with more that 70% motile cells with normal morphology are acceptable. Inducing sperm capacitation in the rhesus monkey is more that simply washing sperm as is the case in humans. Typically, washed sperm must be exposed to an intracellular cAMP enhancer such as caffeine or the cell-permeable dibutyryl derivative.[37] This exposure is associated with a transition in sperm motility and a tendency of sperm to agglutinate, which can create real problems when *in vitro* insemination is involved. While there are several reports on the cryostorage of NHP sperm, a notably loss of motility in the rhesus macaque makes the prospect of using frozen/thawed sperm for artificial insemination by donor (AID) or *in vitro* fertilization (IVF) from a wide spectrum of donors questionable. In situations in which IVF is involved, ICSI can be employed to overcome sperm motility limitations (see below).

Egg collection for ARTs in rhesus macaques now typically involves subjecting females to COS using recombinant human follicle-stimulating hormone, luteinizing hormone, and chorionic gonadotropin. COS does not always result in the achievement of preestablished thresholds in animal response and the percentage of "nonresponders" varies by season showing an increase during the summer months, reaching over 35% in June and July. During summer, despite housing in controlled environments, many females also become anovulatory and it is impractical to attempt COS. Females can be recycled for COSs, however, the response to recombinant human gonadotropins is gradually decreased with increasing numbers of stimulations, apparently due to an immune reaction.[16,73] Practically, up to three stimulations on average can be performed per female with the recovery of a reasonable number of high-quality oocytes. The availability of monkey recombinant gonadotropins may allow the more efficient and extended use of females. This immune response, although greatly reduced compared to when animal or human urinary preparations were employed, still limits animal responses and the number of repeat cycles that are possible. The egg yield is highly variable in rhesus monkeys from a few to over 100 using such fixed stimulation protocols and, not surprisingly, oocyte quality is also variable.[4] The recovery of high oocyte numbers, in our hands, is associated with poor performance as measured by fertilization and/or *in vitro* developmental potential. Fertilization by IVF or ICSI is feasible with the latter conveniently used with frozen-thawed sperm or whenever poor motility compromises IVF. We have previously reported that fertilization rates by ICSI with cryopreserved sperm injected shortly after thaw are significantly lower than with fresh sperm.[19] However, preincubation times of more than 3 h after thaw were associated with ICSI fertilizing capacity similar to that seen with fresh sperm. Piercing the plasma membrane during the ICSI procedure is critical and often difficult to assess since the needle can grossly invaginate the membrane without breaking through. For that reason, the injection pipette should be inserted through the zona pellucida and "into" the oocyte across approximately one-third of the egg's diameter. Cytoplasm is slowly aspirated into the ICSI pipette (as far back as the needle junction with the zona pellucida) until the plasma membrane breaks. A "pop" or sudden movement of cytoplasm into the pipette indicates the release of membrane tension. Once the membrane is penetrated, the sperm can be expelled into the cytoplasm.

Table 42–1
Gamete collection, fertilization, embryo culture, and transfer in nonhuman primates[a]

	ART procedures				
Species	*Semen recovery and cryostorage*	*COS and egg collection*	*Fertilization*	*Embryo culture*	*Embryo transfer; pregnancy outcome*
Old World macaques					
Rhesus	11, 20–25	20, 26, 27	21, 22, 28	20, 22, 29–38	13, 20–22, 30, 39, 40
Cynomolgus	41–46	41, 47	41, 48, 49	48	41, 50
Pig tailed	10, 51, 52	10, 51	10, 51	10	10
Baboon	53	54, 55	53, 54	56	57
Vervet	58–60	60, 61	60	60	60
New World macaques					
Marmoset	62, 63	63	62, 63	63	62, 64
Squirrel	65, 66	67, 68	65, 66		
Great apes					
Chimpanzee	69				
Gorilla	70–72	72	70–72		70

[a]ART, assisted reproductive technology; COS, controlled ovarian stimulation.

Embryo culture has progressed from BRL cell coculture to reliance on chemically defined medium such as HECM-9 or -10 fortified with a protein source, usually fetal bovine serum (FBS), at later stages in preimplantation development. Development of the primate embryo as in other mammals includes morphologically distinct developmental stages from the formation of the zygote after fertilization to cleavage, morula formation, compaction, and finally cavitation of the blastocyst. In HECM-9aa + FBS, blastocysts containing 100 or so cells can be seen by day 6 or 7 *in vitro* at a control rate of approximately 50% in the rhesus monkey.

Embryo transfer in the context of producing animals can be accomplished by laparoscopy or by minilaparotomy with direct deposition into the oviduct. A nonsurgical transcervical route, although evaluated, has proven troublesome in the rhesus monkey secondary to difficulties in passing a cannula through the tortuous cervix in this species.[14,19] For information on optimal timing of embryo transfer see Wolf *et al.*[19] Pregnancy outcomes in NHPs following ET of ART-produced embryos have been evaluated; however, a significant experience is available only in the rhesus macaque where approximately 100 infants have been born following IVF/ICSI.[17–19]

PRODUCTION OF IDENTICAL ANIMALS The ARTs can, in theory, be employed for the production of genetically identical animals by twinning, embryonic cell nuclear transfer (ECNT), or SCNT. This would be a highly significant accomplishment as there is a need among the research community for monkeys with specific genotypes that cannot be satisfied by the importation of animals from the wild or by the identification and propagation of founder animals using conventional breeding. Moreover, spontaneous or naturally occurring NHP models of human genetic diseases are exceedingly rare and not currently available in useful numbers. If a technology for preserving, propagating, and indeed creating NHP models was available, it could be a key to understanding factors involved in disease progression, and it would also provide a relevant platform for testing therapies.

Embryo twinning alternatives, as perfected in domestic animal species, have been evaluated in efforts to create genetically identical monkeys. The limitation of this approach is that sets of only two or possibly four identical animals could be produced by embryo separation at early developmental stages or by blastocyst splitting. Monkeys have been produced following the transfer of demiembryos or split embryos[15,18,74] into a single recipient; however, the outcomes have been disappointing with only a few twin pregnancies resulting and none progressing to term. Nuclear transfer using embryonic blastomeres is an alternative option for the production of identical animals, but it is also restricted theoretically by a small clone size.[13,17–19,75] Protocols for ECNT in the rhesus monkey, including oocyte and embryo production, micromanipulation, and embryo transfer, are described in detail in the *Methods in Molecular Biology Series*.[76]

With success in producing identical animals by SCNT in the past decade, attention has focused on the use of this technology in nonhuman primates. Several advantages accrue to SCNT including the potential to introduce precise genetic modifications into cultured cells, and then use these cells as nuclear donors for the production of transgenic or loss-of-function monkey models of human diseases. SCNT embryos could also provide a resource for the derivation of autologous ESCs and the subsequent therapeutic use of these cells in preclinical trials of therapeutic cloning. Initial efforts at SCNT by conventional protocols in the rhesus monkey were disappointing because SCNT embryos seldom progressed beyond the eight-cell stage *in vitro* when fetal fibroblasts were used as the source of donor nuclei.[75] While this failure to develop was likely related to incomplete or erroneous reprogramming, the depletion of essential microtubule and centrosomal proteins during oocyte enucleation has also been implicated,[75] to the extent that reproductive cloning in nonhuman primates was thought to be impossible.[77] However, these obstacles were subsequently overcome, to some degree, by technique alteration[78] with normal spindles present after the introduction of somatic nuclei into enucleated cytoplasts.[79,80] An encouraging outcome was achieved in a prospective study comparing SCNT outcome in the monkey with fibroblasts as the nuclear donor source using a conventional protocol versus a one-step method. SCNT-derived embryos developed to the blastocyst stage at a rate in excess of 20% using the one-step method.[80] In the latest rendition of this technique, we incorporated several modifications.[81] First, to prevent premature degradation of cytoplasmic reprogramming factors, the manipulation medium was modified to exclude calcium and magnesium. As before, the electrofusion step was replaced by direct donor cell injection. The latter was performed with a Piezo drill similar to procedures described for mouse SCNT. Cultured donor cells were dispersed and a blunt transfer pipette (5–7 μm outer diameter) was used to disrupt plasmalemma integrity by aspiration before a lysed cell with intact nucleus was injected into a cytoplast. A second modification was implemented to avoid possible deleterious effects of Hoechst staining and ultraviolet (UV) exposure on cytoplast integrity. Visualization of the metaphase spindle during enucleation was accomplished using Oosight Imaging System (CRI, Inc., Woburn, MA) that allowed noninvasive, polarized light imaging and detection of the spindle based on birefringence. Using this innovative approach, we were able to locate and quickly enucleate the spindle, real-time with 100% efficiency. Several fetal and adult somatic cell types supported significant blastocyst development and these achievements represent breakthroughs that allow, for the first time, the *in vivo* and *in vitro* characterization of SCNT blastocysts in this species.

EMBRYONIC STEM CELLS ESCs can proliferate indefinitely, maintain an undifferentiated pluripotent state, and differentiate into any cell type. These pliable embryonic cells are derived from the inner cell mass (ICM) of mammalian blastocyst-stage embryos. Embryonic stem cells—or at least ES-like cells—have been derived from the embryos of multiple species, most notably mice,[82] rhesus monkeys,[83] marmoset monkeys,[84] cynomolgus monkeys,[85] and humans.[86] The ability of ESCs to differentiate into any cell type means we may be able to utilize these cells to cure or alleviate the symptoms of many degenerative diseases. However, unresolved issues regarding cellular survival, maintenance of functionality, and tumor formation mean a prudent approach should be taken toward advancing ESCs into human clinical trials. The short-term and long-term efficacy and safety of ESCs must be thoroughly examined. Due to practical and ethical considerations these studies cannot be performed *in vivo* in humans and clinically relevant nonhuman models are therefore essential. ESCs derived from NHP or human blastocysts demonstrate extensive similarities not observed in murine ESCs.[87,88]

There are currently 42 NHP ESC lines available: 26 in the rhesus monkey, of which eight were derived from *in vivo* produced embryos,[83,89] and 18 from *in vitro* efforts,[90] five cynomolgus monkey ESC lines, of which four were derived from *in vitro* fertilized embryos[85] and one from a parthenogenetically generated embryo,[91] and 11 common marmoset ESC lines that were all derived from *in vivo* produced embryos.[84,92] A summary of the ESC lines currently available in the nonhuman primate is presented in Table 42–2.

To produce these lines, blastocysts were either obtained following *in vivo* fertilization and nonsurgical uterine flushing[83] or though *in vitro* production[90] including conventional *in vitro* fertilization,[85] ICSI,[90] and parthenogenesis.[91,93] Once the blastocyst is obtained, the same basic technique is generally used to derive an ESC line. Zonae pellucidae are removed by a brief exposure to pronase. The blastocysts are then exposed to an animal serum containing anti-rhesus antibodies that can bind to the trophectoderm cells that surround the embryo's ICM. Next, the embryos are moved into serum complement. The complement serum proteins respond to the trophectoderm-bound antibodies and induce lysis. The lysed trophectodermal cells are gently removed with a small bore pipette and the isolated ICM is plated onto dishes containing a monolayer of mitotically inactivated mouse embryonic fibroblast (MEF) feeder cells in ESC culture media. The ICMs attach to the feeder layer and produce small outgrowths after a few days that are manually dissociated into small cell clumps and replated onto fresh dishes of MEFs. Colonies with an ESC-like morphology (flat monolayer colony with distinctive cobbled stem cell morphology, prominent nucleoli, and a high nucleocytoplasmic ratio) are manually selected for further propagation, characterization, and low temperature storage. Medium is changed daily and ESC colonies are split every few days onto dishes with fresh MEF feeder layers. The nondividing MEF feeder cells provide various growth factors that help maintain the NHP ESCs in an undifferentiated state and encourage proliferation. It should be noted that NHP ESC culture is technically difficult. The majority of novice groups supplied with NHP ESCs have had severe difficulty in maintaining their colonies in an undifferentiated state. It is therefore highly recommended that investigators interested in NHP ESCs seek out and attend a relevant training workshop such as an NIH-sponsored workshop on human ESCs or an NHP ESC workshop provided by an active laboratory in this field.

Manual passaging of ESCs is, of course, labor intensive and is incompatible with the scaled up production numbers required for *in vivo* transplantation. Theoretically, combining enzymatic passage to produce multiple subclones with continuous passage-by-passage karyotypic analysis could be used to discard aneuploid subclones and select for an enzymatically passaged karyotypically stable NHP ESC line that would be ideal for scaled up production. This type of NHP ESC line would be an important addition to any NHP ESC bank. NHP ESC lines have, for the most part, been maintained on mouse feeder layers that may contaminate the primate cells with murine pathogens and/or xenoepitopes, either of which could induce a severe immune response following transplantation. Possible solutions to this problem would be to grow NHP ESCs either on NHP feeders[94] or in feeder-free conditions.[95,96] Our research has suggested that NHP ESCs are best cultured on MEFs in DMEM/F12 medium supplemented with 15% FBS, 0.1 mM 2-mercaptoethanol, 1% nonessential amino acids, 2 mM L-glutamine, 4 ng/ml fibroblast growth factor 2 (FGF2) (pH adjusted to 7.2 with NaOH) at 37°C in 3% CO_2, 5% O_2, and 92% N_2.[90]

NHP ESC lines derived from *in vivo*-produced embryos have been characterized by a normal karyotype; they express the same stem cell markers as human ESCs, such as OCT-4, SSEA-3, SSEA-4, TRA-1–60, and TRA-1–81, and possess the ability to differentiate into multiple cell types *in vitro* and into teratomas *in vivo*.[83,84,92] NHP ESC lines derived from *in vitro*-produced embryos also express the aforementioned stem cell markers and can differentiate into multiple cell types both *in vitro* and *in vivo*.[85,90–92,97] However, abnormal karyotypes were discovered in three of the six *in vitro*-derived rhesus monkey ESC lines that were enzymatically passaged during their initial derivation. Primate ESCs appear to be particularly prone to karyotypic aneuploidy during enzymatic passaging as none of the 12 *in vitro*-derived rhesus monkey ESC lines that were manually passaged demonstrated any aneuploidy.[90] Most of the other biological characteristics of NHP ESCs have been elucidated using rhesus monkey ESCs. The cell cycle of rhesus monkey ESC is characterized by an extremely rapid transition though G_1, which is not arrested by serum starvation, pharmacological inhibition, or gamma irradiation, reflecting an absence of a G_1 checkpoint similar to murine ESCs.[98]

NHP ESCs can be either differentiated *in vitro* or *in vivo*. For *in vitro* differentiation, embryoid bodies (EBs) are formed and then placed into various culture conditions until the desired cell types are obtained. For *in vivo* differentiation, the ESCs are injected into a mouse and allowed to spontaneously form a tumor. To obtain embryoid bodies, entire ESC colonies are detached from the feeder cells and manually transferred into feeder-free ultralow adhesion dishes and cultured in suspension in ESC medium without FGF for several days. The NHP ESCs initially clump up into a ball and then cavitate to form a blastocyst-like embryoid body. To induce spontaneous differentiation, the embryoid bodies are attached to gelatin-coated culture dishes and cultured for 1 month in FGF-free ESC culture media. This allows

Table 42–2
A summary of the NHP ESC lines currently available[a]

Primate species	*Blastocyst origin*	*ESC lines (number)*	*Reference*
Rhesus monkey (*Macaca mulata*)	*In vivo*	8	83, 89
Rhesus monkey (*Macaca mulata*)	*In vitro*	18	90
Cynomolgus monkey (*Macaca fascicularis*)	*In vitro*	4	85
Cynomolgus monkey (*Macaca fascicularis*)	*In vitro*	1 (parthenote derived)	91, 93
Common marmoset (*Callithrix jacchus*)	*In vivo*	11	84, 92

[a]NHP, nonhuman primate; ESC, embryonic stem cell.

the ESCs to differentiate down all possible pathways and produces a wide spectrum of cell types. To form teratomas, large numbers of NHP ESCs (usually 10^6–10^7) are injected into the leg muscle of a severe combined immune deficient (SCID) mouse. These mice have no functional immune systems, so the NHP ESCs are tolerated and survive. Over 2 months the NHP ESC differentiate into a tumor composed of many different cell types and containing representatives of all three germ layers. This type of tumor is referred to as a teratoma.

Finally a few of the limitations associated with NHP ESCs will be noted. ESC cloning strategies have proved successful using human ESCs,[99] but are not yet available in the NHP. However, the initial difficulties in genetically modifying NHP ESCs have been largely solved using lentiviral vectors for transfection, resulting in high-level sustained transgene expression in both NHP and human ESCs.[100,101] This lentiviral-based ability to produce gene knockout and transgenic NHP ESCs has significant implications for the use of NHP ESCs in developmental biology, disease pathology, and gene therapy. In the mouse, genetically modified ESCs are inserted into tetraploid host blastocysts and the implanted chimeras produce transgenic and gene knockout animals.[102] If these chimera techniques eventually proved transferable to NHP ESCs then it would permit the generation of gene knockout and transgenic primates. Chimera research in the NHP is still in the preliminary stages. We have demonstrated that membrane marked NHP ESCs—when transferred to a host embryo at the four- to eight-cell stage—have the capacity to proliferate and contribute to the ICM and trophectoderm of the resulting NHP chimeras[90] and when green fluorescent protein (GFP)-marked NHP ESCs have been transferred into NHPs at the fetal stages they have successfully integrated into the host tissues.[103]

Recent evidence has demonstrated that multiple NHP ESC lines demonstrate a loss of imprinting.[104] Aberrant imprinting has also been observed in mouse ESCs[105] and in high passage human ESCs,[106] although it should be noted that low-passage human ESCs appear to demonstrate normal imprinting.[106] How this loss of imprinting will affect the therapeutic potential of ESCs is still unknown, but as the loss of imprinting has been associated with germ cell tumors[107] this question becomes yet another reason for extensive research into the safety of ESC-based therapy in the NHP before human clinical trials are initiated.

REFERENCES

1. Bernacky BJ, Gibson SV, Keeling ME, Abee CR. Nonhuman primates. In: Fox JG, Anderson LC, Loew FM, Quimby FW, Eds. *Laboratory Animal Medicine*, 2nd ed. London: Academic Press, 2002:676–791.
2. Hewitson L. Primate models for assisted reproductive technologies. *Reproduction* 2004;128:293–299.
3. Schramm RD, Bavister BD. A macaque model for studying mechanisms controlling oocyte development and maturation in human and non-human primates. *Hum Reprod* 1999;14:2544–2555.
4. Schramm RD, Paprocki AM. Birth of rhesus monkey infant after transfer of embryos derived from in-vitro matured oocytes: Short communication. *Hum Reprod* 2000;15:2411–2414.
5. Zheng P, Si W, Bavister BD, Yang J, Ding CH, Ji W. 17β-Estradiol and progesterone improve in-vitro cytoplasmic maturation of oocytes from unstimulated prepubertal and adult rhesus monkeys. *Hum Reprod* 2003;18:2137–2144.
6. Yin H, Duffy DM, Gosden RG. Comparative maturation of cynomolgus monkey oocytes in vivo and in vitro. *Reprod Biol Endocrinol* 2006;4:14.
7. Chen C, Liow S-L, Abdullah RB, Embong WKW, Yip W-Y, Tan L-G, Tong G-Q, Ng SC. Developmental competence of transported in vitro matured macaque oocytes. *Reprod Biol Endocrinol* 2006;23:50–59.
8. Lee DM, Yeoman RR, Battaglia DE, Stouffer RL, Zelinski-Wooten MB, Fanton JW, Wolf DP. Live birth after ovarian tissue transplant. *Nature* 2004;428:137–138.
9. Dukelow WR, Pierce DL, Roudebush WE, Jarosz SJ, Sengoku K. In vitro fertilization in nonhuman primates. *J Med Primatol* 1990;19:627–639.
10. Kubisch HM, Gagliardi C, Williams VM, Ribka EP, Ratterree MS. In vitro fertilization in the pigtailed macaque (Macaca nemestrina). *Theriogenology* 2006;66(4):749–754.
11. Bavister BD, Boatman DE, Collins K, Dierschke DJ, Eisele SG. Birth of rhesus monkey infant after in vitro fertilization and nonsurgical embryo transfer. *Proc Natl Acad Sci USA* 1984;81:2218–2222.
12. Wolf DP, Thomson JA, Zelinski-Wooten MB, Stouffer RL. In vitro fertilization-embryo transfer in nonhuman primates: The technique and its applications. *Mol Reprod Dev* 1990;27:261–280.
13. Meng L, Ely JJ, Stouffer RL, Wolf DP. Rhesus monkeys produced by nuclear transfer. *Biol Reprod* 1997;57:454–459.
14. Nusser KD, Mitalipov S, Widmann A, Gerami-Naini B, Yeoman RR, Wolf DP. Developmental competence of oocytes after ICSI in the rhesus monkey. *Hum Reprod* 2001;16:130–137.
15. Mitalipov SM, Yeoman RR, Kuo HC, Wolf DP. Monozygotic twinning in rhesus monkeys by manipulation of in vitro-derived embryos. *Biol Reprod* 2002;66:1449–1455.
16. Stouffer RL, Zelinski-Wooten M, Chandrasekher YA, Wolf DP. Stimulation of follicle and oocyte development in macaques for IVF procedures. In: Wolf DP, Stouffer RL, Brenner RM, Eds. *In Vitro Fertilization and Embryo Transfer in Primates*. New York: Springer-Verlag, 1992:124–141.
17. Ouhibi N, Zelinski-Wooten M, Thomson JA, Wolf DP. Assisted fertilization and nuclear transfer in non human primates. In: Wolf DP, Zelinski-Wooten M, Eds. *Assisted Fertilization and Nuclear Transfer in Mammals*. Totowa, NJ: Humana Press, 2001:253–284.
18. Wolf DP. Assisted reproductive technologies in rhesus macaques. *Reprod Biol Endocrinol* 2004;2:37.
19. Wolf DP, Thormahlen S, Ramsey C, Fanton J, Mitalipov S. Use of assisted reproductive technologies in the propagation of rhesus macaque offspring. *Biol Reprod* 2004;71:486–493.
20. Wolf DP, Vandevoort CA, Meyer-Haas GR, Zelinski-Wooten MB, Hess DL, Baughman WL, Stouffer RL. In vitro fertilization and embryo transfer in the rhesus monkey. *Biol Reprod* 1989;41:335–346.
21. Lanzendorf SE, Zelinski-Wooten MB, Stouffer RL, Wolf DP. Maturity at collection and the developmental potential of rhesus monkey oocytes. *Biol Reprod* 1990;42:703–711.
22. Boatman DE. *In vitro* growth of nonhuman primate pre- and peri-implantation embryos. In: Bavister BD, Ed. T*he Mammalian Preimplantation Embryo: Regulation of Growth and Differentiation In Vitro*. New York: Plenum, 1987:273–308.
23. Roussel JD, Austin CR. Preservation of primate spermatozoa by freezing. *J Reprod Fertil* 1967;13:333–335.
24. Leverage WE, Valerio DA, Schultz AP, Kingsbury E, Dorey C. Comparative study on the freeze preservation of spermatozoa. Primate, bovine, and human. *Lab Anim Sci* 1972;22:882–889.
25. Yeoman RR, Mitalipov S, Gerami-Naini B, Nusser KD, Wolf DP. Low temperature storage of rhesus monkey spermatozoa and fertility evaluation by intracytoplasmic injection. *Theriogenology* 2005;63:2356–2371.
26. VandeVoort CA, Baughman WL, Stouffer RL. Comparison of different regimens of human gonadotropins for superovulation of rhesus monkeys: Ovulatory response and subsequent luteal function. *J In Vitro Fert Embryo Transfer* 1989;6:85–91.
27. Abassi R, Kenigsberg D, Danforth D, Falk RJ, Hodgen HG. Cumulative ovulation rate in human menopausal/human chorionic gonadotropin-treated monkeys: "Step-up" versus "step-down" dose regimens. *Fertil Steril* 1987;47:1019–1024.

28. Boatman DE, Morgan PM, Bavister BD. Variables affecting the yield and developmental potential of embryos following superstimulation and in vitro fertilization in rhesus monkeys. *Gamete Res* 1986;13:327–338.
29. Zhang L, Weston AM, Denniston RS, Goodeaux LL, Godke RA, Wolf DP. Developmental potential of rhesus monkey embryos produced by in vitro fertilization. *Biol Reprod* 1994;51:433–440.
30. Weston AM, Zelinski-Wooten MB, Hutchison JS, Stouffer RL, Wolf DP. Developmental potential of embryos produced by in-vitro fertilization from gonadotrophin-releasing hormone antagonist-treated macaques stimulated with recombinant human follicle stimulating hormone alone or in combination with luteinizing hormone. *Hum Reprod* 1996;11:608–613.
31. Seshagiri PB, Hearn JP. In-vitro development of in-vivo produced rhesus monkey morulae and blastocysts to hatched, attached, and post-attached blastocyst stages: Morphology and early secretion of chorionic gonadotrophin. *Hum Reprod* 1993;8:279–287.
32. Weston AM, Wolf DP. Differential preimplantation development of rhesus monkey embryos in serum-supplemented media. *Mol Reprod Dev* 1996;44:88–92.
33. Schramm RD, Bavister BD. Development of in-vitro-fertilized primate embryos into blastocysts in a chemically defined, protein-free culture medium. *Hum Reprod* 1996;11:1690–1697.
34. Alak BM, Wolf DP. Rhesus monkey oocyte maturation and fertilization in vitro: Roles of the menstrual cycle phase and of exogenous gonadotropins. *Biol Reprod* 1994;51:879–887.
35. Goodeaux L, Thibodeaux JK, Voelkel S, Anzalone CA, Roussel JD, Cohen JC, Menezo Y. Collection, co-culture and transfer of rhesus preimplantation embryos. *ARTA* 1990;1:370–379.
36. Seshagiri PB, Dierschke DJ, Eisele SG, Scheffler J, Hearn JP. Recovery of rhesus monkey preimplantation embryos by non-surgical uterine flushing. *Biol Reprod* 1991;44(Suppl. 1):157.
37. Boatman DE, Bavister BD. Stimulation of rhesus monkey sperm capacitation by cyclic nucleotide mediators. *J Reprod Fertil* 1984;71:357–366.
38. Morgan PM, Boatman DE, Bavister BD. Relationships between follicular fluid steroid hormone concentrations, oocyte maturity, in vitro fertilization and embryonic development in the rhesus monkey. *Mol Reprod Dev* 1990;27:145–151.
39. Goodeaux LL, Anzalone CA, Thibodeaux JK, Menezo Y, Roussel JD, Voelkel SA. Successful non-surgical collection of *Macaca mulatta* embryos. *Theriogenology* 1990;34:1159–1167.
40. Hewitson L, Dominko T, Takahashi D, Martinovich C, Ramalho-Santos J, Sutovsky P, Fanton J, Jacob D, Monteith D, Neuringer M, Battaglia D, Simerly C, Schatten G. Unique checkpoints during the first cell cycle of fertilization after intracytoplasmic sperm injection in rhesus monkeys. *Nat Med* 1999;5:431–433.
41. Balmaceda JP, Pool TB, Arana JB, Heitman TS, Asch RH. Successful *in vitro* fertilization and embryo transfer in cynomolgus monkeys. *Fertil Steril* 1984;42:791–795.
42. Kreitman O, Lynch A, Nixon WE, G.D. H. Ovum collection, induced luteal dysfunction, *in vitro* fertilization, embryo development and low tubal ovum transfer in primates. In: Hafez ESE, Semm K, Eds. *In Vitro Fertilization and Embryo Transfer*. Lancaster, UK: MTP Press Ltd., 1982:303–324.
43. Cho F, Honjo S, Makita T. Fertility of frozen-preserved spermatozoa of cynomolgus monkeys. In: Kondo S, Kawai M, Ehara A, Eds. *Contemporary Primatology*. Basel: S. Karger, 1975:125–133.
44. Cho F, Honjo S. A simplified method for collecting and preserving cynomolgus macaque semen. *Jpn J Med Sci Biol* 1973;26:261–268.
45. Mahone JP, Dukelow WR. Semen preservation in *Macaca fascicularis*. *Lab Anim Sci* 1978;28:556–561.
46. Tollner TL, VandeVoort CA, Overstreet JW, Drobnis EZ. Cryopreservation of spermatozoa from cynomolgus monkeys (Macaca fascicularis). *J Reprod Fertil* 1990;90:347–352.
47. Schenken RS, Williams RF, Hodgen GD. Ovulation induction using "pure" follicle-stimulating hormone in monkeys. *Fertil Steril* 1984;41:629–634.
48. Fujisaki M, Suzuki M, Kohno M, Cho F, Honjo S. Early embryonal culture of the cynomolgus monkey (*Macaca fascicularis*). *Am J Primatol* 1989;18:303–313.
49. Balmaceda JP, Gastaldi C, Ord T, Borrero C, Asch RH. Tubal embryo transfer in cynomolgus monkeys: Effects of hyperstimulation and synchrony. *Hum Reprod* 1988;3:441–443.
50. Balmaceda JP, Heitman TO, Garcia MR, Pauerstein CJ, Pool TB. Embryo cryopreservation in cynomolgus monkeys. *Fertil Steril* 1986;45:403–406.
51. Cranfield MR, Schaffer N, Bavister BD, Berger NG, Boatman DE, Kempske SE, *et al.* Assessment of oocytes retrieved from stimulated and unstimulated ovaries of pig-tailed macaques (Macaca nemestrina) as a model to enhance the genetic diversity of captive lion-tailed macaques (Macaca silenus). *Zool Biol* 1989;Suppl. 1:33–46.
52. Cranfield MR, Bavister BD, Boatman DE, Berger NG, Schaffer N, Kempske SE, *et al.* Assisted reproduction in the propagation management of the endangered lion-tailed macaque (*Macaca silenus*). In: Wolf DP, Stouffer RL, Brenner RM, Eds. *In vitro Fertilization and Embryo Transfer in Primates*. New York: Springer-Verlag, 1993:331–348.
53. Clayton O, Kuehl TJ. The first successful *in vitro* fertilization and embryo transfer in a nonhuman primate. *Theriogenology* 1984; 21:228.
54. Fourie FR, Snyman E, van der Merwe JV, Grace A. Primate *in vitro* fertilization research: Preliminary results on the folliculogenic effects of three different ovulatory induction agents on the chacma baboon, *Papio ursinus*. *Comp Biochem Physiol* 1987;87:889–893.
55. Kraemer DC, Vera Cruz NC. Collection, gross characteristics and freezing of baboon semen. *J Reprod Fertil* 1969;20:345–348.
56. Pope V, Pope E, Beck L. *In vitro* development of the primate embryo. In: Brans YW, Kuehl TJ, Eds. *Nonhuman Primates in Perinatal Research*. New York: John Wiley & Sons; 1988: 161–174.
57. Pope C, Pope V, Beck LR. Live birth following cryopreservation and transfer of a baboon embryo. *Fertil Steril* 1984;42:143–145.
58. Hess DL, Hendrickx AG, Stabenfeldt GH. Reproductive and hormonal patterns in the African green monkey (*Cercopithecus aethiops*). *J Med Primatol* 1979;8:273–281.
59. Van Der Horst G, Seier JV, Spinks AC, Hendricks S. The maturation of sperm motility in the epididymis and vas deferens of the vervet monkey, *Cercopithecus aethiops*. *Int J Androl* 1999;22:197–207.
60. Sparman ML, Ramsey C, Thomas CM, Mitalipov S, Fanton J, Maginnis GM, Stouffer RL, Wolf DP. Evaluation of the vervet (*Cercopithecus aethiops*) as a model for the assisted reproductive technologies. *Am J Primatol* 2007;69:1–13.
61. Sankai T, Cho F, Yoshikawa Y. In vitro fertilization and preimplantation embryo development of African green monkeys (*Cercopithecus aethiops*). *Am J Primatol* 1997;43:43–50.
62. Lopata A, Summers PM, Hearn JP. Births following the transfer of cultured embryos obtained by in vitro and in vivo fertilization in the marmoset monkey (Callithrix jacchus). *Fertil Steril* 1988;50: 503–509.
63. Wilton LJ, Marshall VS, Piercy EC, Moore HD. In vitro fertilization and embryo development in the marmoset monkey (Callithrix jacchus). *J Reprod Fertil* 1993;97:481–486.
64. Marshall VS, Kalishman J, Thomson JA. Nonsurgical embryo transfer in the common marmoset monkey. *J Med Primatol* 1997; 26:241–247.
65. Gould KG, Cline EM, Williams WL. Observations on the induction of ovulation and fertilization *in vitro* in the squirrel monkey (Saimiri sciureus). *Fertil Steril* 1973;24:260–268.
66. Kuehl TJ, Dukelow WR. Time relations of squirrel monkey (Saimiri sciureus) sperm capacitation and ovum maturation in an *in vitro* fertilization system. *J Reprod Fertil* 1982;64:135–137.
67. Dukelow WR. Human chorionic gonadotropin: Induction of ovulation in the squirrel monkey. *Science* 1979;206:234–235.
68. Denis LT, Poindexter AN, Ritter MB, Seager SWJ, Deter RL. Freeze preservation of squirrel monkey sperm for use in timed fertilization studies. *Fertil Steril* 1976;27:723–729.

69. Gould KG. Ovum recovery and *in vitro* fertilization in the chimpanzee. *Fertil Steril* 1983;40:378–383.
70. Pope CE, Dresser BL, Chin NW, Liu JH, Loskutoff NM, Behnke EJ, Brown C, McRae MA, Sinoway CE, Campbell MK, Cameron KN, Owens OM, Johnson CA, Evans RR, Cedars MI. Birth of a western lowland gorilla (Gorilla gorilla gorilla) following in vitro fertilization and embryo transfer. *Am J Primatol* 1997;41: 247–260.
71. Lanzendorf SE, Holmgren WJ, Schaffer N, Hatasaka H, Wentz AC. *In vitro* fertilization and gamete micromanipulation in lowland gorilla. *J Assist Reprod Genet* 1992;9:358–364.
72. Huntress SL, Loskutoff NM, Raphael BL, Yee B, Bowsher TR, Putman JM, Kraemer DC. Pronucleus formation following *in vitro* fertilization of oocytes recovered from a gorilla (*Gorilla gorilla gorilla*) with unilateral endometroid adenocarcinoma of the ovary. *Am J Primatol* 1989;18:259–266.
73. Iliff SA, Molskness TA, Stouffer RL. Anti-human gonadotropin antibodies generated during in vitro fertilization (IVF)-related cycles: Effect on fertility of rhesus macaques. *J Med Primatol* 1995;24:7–11.
74. Chan AW, Dominko T, Luetjens CM, Neuber E, Martinovich C, Hewitson L, Simerly CR, Schatten GP. Clonal propagation of primate offspring by embryo splitting. *Science* 2000;287:317–319.
75. Mitalipov SM, Yeoman RR, Nusser KD, Wolf DP. Rhesus monkey embryos produced by nuclear transfer from embryonic blastomeres or somatic cells. *Biol Reprod* 2002;66:1367–1373.
76. Mitalipov S, Wolf DP. Nuclear transfer in non-human primates. In: Verna P, Trounson A, Eds. *Nuclear Transfer Protocols: Cell Reprogramming and Transgenesis*. Totowa, NJ: Humana Press, 2006.
77. Simerly C, Dominko T, Navara C, Payne C, Capuano S, Gosman G, Chong KY, Takahashi D, Chace C, Compton D, Hewitson L, Schatten G. Molecular correlates of primate nuclear transfer failures. *Science* 2003;300:297.
78. Simerly C, Navara C, Hyun SH, Lee BC, Kang SK, Capuano S, Gosman G, Dominko T, Chong KY, Compton D, Hwang WS, Schatten G. Embryogenesis and blastocyst development after somatic cell nuclear transfer in nonhuman primates: Overcoming defects caused by meiotic spindle extraction. *Dev Biol* 2004;276:237–252.
79. Ng SC, Chen N, Yip WY, Liow SL, Tong GQ, Martelli B, Tan LG, Martelli P. The first cell cycle after transfer of somatic cell nuclei in a non-human primate. *Development* 2004;131:2475–2484.
80. Zhou Q, Yang SH, Ding CH, He XC, Xie YH, Hildebrandt TB, Mitalipov SM, Tang XH, Wolf DP, Ji WZ. A comparative approach to somatic cell nuclear transfer in the rhesus monkey. *Hum Reprod* 2006;22:22.
81. Mitalipov SM, Zhou Q, Byrne J, Ji WZ, Wolf DP. Reprogramming events and developmental competence of rhesus monkey embryos produced by somatic cell nuclear transfer. Annual Conference of the International Embryo Transfer Society, Orlando, FL. 7–11 January 2006, 139.
82. Evans MJ, Kaufman MH. Establishment in culture of pluripotential cells from mouse embryos. *Nature* 1981;292:154–156.
83. Thomson JA, Kalishman J, Golos TG, Durning M, Harris CP, Becker RA, Hearn JP. Isolation of a primate embryonic stem cell line. *Proc Natl Acad Sci USA* 1995;92:7844–7848.
84. Thomson JA, Kalishman J, Golos TG, Durning M, Harris CP, Hearn JP. Pluripotent cell lines derived from common marmoset (Callithrix jacchus) blastocysts. *Biol Reprod* 1996;55:254–259.
85. Suemori H, Tada T, Torii R, Hosoi Y, Kobayashi K, Imahie H, Kondo Y, Iritani A, Nakatsuji N. Establishment of embryonic stem cell lines from cynomolgus monkey blastocysts produced by IVF or ICSI. *Dev Dyn* 2001;222:273–279.
86. Thomson JA, Itskovitz-Eldor J, Shapiro SS, Waknitz MA, Swiergiel JJ, Marshall VS, Jones JM. Embryonic stem cell lines derived from human blastocysts. *Science* 1998;282:1145–1147.
87. Donovan PJ, Gearhart J. The end of the beginning for pluripotent stem cells. *Nature* 2001;414:92–97.
88. Ginis I, Luo Y, Miura T, Thies S, Brandenberger R, Gerecht-Nir S, Amit M, Hoke A, Carpenter MK, Itskovitz-Eldor J, Rao MS. Differences between human and mouse embryonic stem cells. *Dev Biol* 2004;269:360–380.
89. Marshall VS, Waknitz MA, Thomson JA. Isolation and maintenance of primate embryonic stem cells. *Methods Mol Biol* 2001;158:11–18.
90. Mitalipov S, Kuo HC, Byrne J, Clepper L, Meisner L, Johnson J, Zeier R, Wolf D. Isolation and characterization of novel rhesus monkey embryonic stem cell lines. *Stem Cells* 2006;1:1.
91. Cibelli JB, Grant KA, Chapman KB, Cunniff K, Worst T, Green HL, Walker SJ, Gutin PH, Vilner L, Tabar V, Dominko T, Kane J, Wettstein PJ, Lanza RP, Studer L, Vrana KE, West MD. Parthenogenetic stem cells in nonhuman primates. *Science* 2002;295:819.
92. Sasaki E, Hanazawa K, Kurita R, Akatsuka A, Yoshizaki T, Ishii H, Tanioka Y, Ohnishi Y, Suemizu H, Sugawara A, Tamaoki N, Izawa K, Nakazaki Y, Hamada H, Suemori H, Asano S, Nakatsuji N, Okano H, Tani K. Establishment of novel embryonic stem cell lines derived from the common marmoset (Callithrix jacchus). *Stem Cells* 2005;23:1304–1313.
93. Vrana KE, Hipp JD, Goss AM, McCool BA, Riddle DR, Walker SJ, Wettstein PJ, Studer LP, Tabar V, Cunniff K, Chapman K, Vilner L, West MD, Grant KA, Cibelli JB. Nonhuman primate parthenogenetic stem cells. *Proc Natl Acad Sci USA* 2003;100: 11911–11916.
94. Li T, Wang S, Xie Y, Lu Y, Zhang X, Wang L, Yang S, Wolf D, Zhou Q, Ji W. Homologous feeder cells support undifferentiated growth and pluripotency in monkey embryonic stem cells. *Stem Cells* 2005;23:1192–1199.
95. Xu C, Inokuma MS, Denham J, Golds K, Kundu P, Gold JD, Carpenter MK. Feeder-free growth of undifferentiated human embryonic stem cells. *Nat Biotechnol* 2001;19:971–974.
96. Levenstein ME, Ludwig TE, Xu RH, Llanas RA, VanDenHeuvel-Kramer K, Manning D, Thomson JA. Basic fibroblast growth factor support of human embryonic stem cell self-renewal. *Stem Cells* 2006;24:568–574.
97. Byrne J, Mitalipov S, Wolf D. Current progress with primate embryonic stem cells. *Curr Stem Cell Res Ther* 2006;1:127–138.
98. Fluckiger AC, Marcy G, Marchand M, Negre D, Cosset FL, Mitalipov S, Wolf D, Savatier P, Dehay C. Cell cycle features of primate embryonic stem cells. *Stem Cells* 2006;24:547–556.
99. Amit M, Carpenter MK, Inokuma MS, Chiu CP, Harris CP, Waknitz MA, Itskovitz-Eldor J, Thomson JA. Clonally derived human embryonic stem cell lines maintain pluripotency and proliferative potential for prolonged periods of culture. *Dev Biol* 2000;227: 271–278.
100. Takada T, Suzuki Y, Kondo Y, Kadota N, Kobayashi K, Nito S, Kimura H, Torii R. Monkey embryonic stem cell lines expressing green fluorescent protein. *Cell Transplant* 2002;11:631–635.
101. Ma J, Ramezani A, Lewis R, Hawley RG, Thomson JA. High-level sustained transgene expression in human embryonic stem cells using lentiviral vectors. *Stem Cells* 2003;21:111–117.
102. Wang Z, Jaenisch R. At most three ES cells contribute to the somatic lineages of chimeric mice and of mice produced by ES-tetraploid complementation. *Dev Biol* 2004;275:192–201.
103. Asano T, Ageyama N, Takeuchi K, Momoeda M, Kitano Y, Sasaki K, Ueda Y, Suzuki Y, Kondo Y, Torii R, Hasegawa M, Ookawara S, Harii K, Terao K, Ozawa K, Hanazono Y. Engraftment and tumor formation after allogeneic in utero transplantation of primate embryonic stem cells. *Transplantation* 2003;76:1061–1067.
104. Fujimoto A, Mitalipov SM, Kuo HC, Wolf DP. Aberrant genomic imprinting in rhesus monkey embryonic stem cells. *Stem Cells* 2006;24:595–603.
105. Dean W, Bowden L, Aitchison A, Klose J, Moore T, Meneses JJ, Reik W, Feil R. Altered imprinted gene methylation and expression in completely ES cell-derived mouse fetuses: Association with aberrant phenotypes. *Development* 1998;125:2273–2282.
106. Rugg-Gunn PJ, Ferguson-Smith AC, Pedersen RA. Epigenetic status of human embryonic stem cells. *Nat Genet* 2005;37: 585–587.
107. Nonomura N, Nishimura K, Miki T, Kanno N, Kojima Y, Yokoyama M, Okuyama A. Loss of imprinting of the insulin-like growth factor II gene in renal cell carcinoma. *Cancer Res* 1997;57:2575–2577.

43 Rat Models of Polycystic Ovary Syndrome

KRISHNA B. SINGH

ABSTRACT

Polycystic ovary (PCO) syndrome is a common endocrine disorder of unknown etiology. This condition is a major cause of menstrual irregularity and infertility in women with chronic anovulation. This chapter reviews the literature on methods of producing experimental PCO models in the laboratory rat, and the findings that support their resemblance to the human condition. The principal rat PCO models that have been validated include constant light exposure, hypothalamic lesions, sex steroid-induced models, and the mifepristone (RU486) model. The biological systems involved in developing chronic anovulation and PCO in these models include the hypothalamus, the pituitary gland, the adrenals, and the ovary with its paracrine, autocrine, and intracrine regulators. Animal models are central to making the transition from scientific concepts to understanding the reality of a human disease. Validated animal models can be used for therapeutic screens, in preclinical trials, and for basic research in reproductive biology. However, animal models of chronic anovulation and PCO may not fully reproduce the reproductive events seen in the human syndrome.

Key Words: Animal models, Persistent estrus, Polycystic ovary, Polycystic ovary syndrome, Chronic anovulation.

INTRODUCTION

Polycystic ovary (PCO) syndrome is a common but complex endocrine disorder among women of reproductive age. Women with this disorder present with menstrual irregularity, hirsutism, chronic anovulation, and infertility. Despite many clinical and basic animal studies reported in the literature, the etiology and pathogenesis of chronic anovulation and PCO syndrome remain enigmatic.[1,2]

The first systematic studies on a rat PCO model were reported in the 1960s.[3–5] Since then newer rat PCO models have been developed and validated. These studies provide unique concepts in understanding the nature of chronic anovulation and polycystic ovary.[6–11]

This chapter first provides background information on the estrous cycle in rats and explains the terminology frequently used in basic research. It then reviews the literature on various methods of producing experimental PCO models in the laboratory rat, and the findings that support their resemblance to the human condition. The focus of the review will be on morphological and hormonal data on the validated rat PCO models. In the final section, the key issues pertaining to animal models in PCO research are discussed.

ESTROUS CYCLE

The short estrous cycle (4–5 days) in rats and mice makes them ideal for exploring many physiological changes that occur during a reproductive cycle.[12] The ovarian cycle once initiated at puberty is characterized by repeating patterns of cellular growth, differentiation, and transformation of ovarian tissues. The dynamic events include follicular development and ovulation with subsequent formation and regression of the corpora lutea. The sequence of these events is orchestrated by several interacting ovarian factors and the cyclic production of hormones from the anterior pituitary and hypothalamus.

Estrogen exerts a critical regulatory influence upon the biosynthetic and secretory activity of the gonadotropin-releasing hormone (GnRH) neurons. It seems likely that estrogen regulates the behavior of the GnRH neuron through multiple transsynaptic, neuronal–glial, and direct membrane modes of action. Preovulatory GnRH and luteinizing hormone (LH) surges depend on activation of estrogen (E_2)-inducible progesterone (P) receptors in the preoptic area. Surges do not occur in males or in perinatally androgenized females (see below).

TERMINOLOGY

The term persistent estrus means failure of two or more consecutive estrous cycles in adult rats. Persistent estrus is documented by precornified and cornified vaginal smears lasting for more than 10 days. The term PCO disease means the presence of persistent estrus, chronic anovulation, and polycystic or polyfollicular ovaries.[4,5,7,11] The term polycystic ovary syndrome refers to the clinical disorder or syndrome in women of reproductive age.

The following section describes studies in female rats according to the methods for induction of persistent estrus, chronic anovulation, and polycystic ovaries.

CONSTANT LIGHT

The constant-light rat model is a simple, reversible, inexpensive, and reproducible model to study the pathogenesis of chronic anovulation and its reproductive outcomes. The exposure of young female Sprague–Dawley rats to continuous light gradually

From: *Sourcebook of Models for Biomedical Research*
(P. M. Conn, ed.), © 2008 Humana Press Inc., Totowa, NJ.

causes failure of the estrous cycle and infertility because of chronic anovulation associated with persistent estrus. During the early stages of constant light exposure the rats show regular estrous cycles, but the cycles become irregular and persistent estrus develops with increased duration of constant light. On changing the light schedule from constant light to light and dark, persistent estrus was interrupted in 32 of 34 (94.1%) rats after an average of 7.5 days. The ovaries of rats exposed to constant light for 56 days, then to light and dark for 28 days, were normal in appearance.[3–5] When placed in darkness for 10h, 80% of the animal ovulated 46h later.[13]

The onset of persistent estrus in female rats depends on the intensity, duration, and spectral characteristics of the environmental light. The absence of 24-h time cues contributes to the induction of persistent estrus when the animals are housed under constant light.[14] Constant light exposure of female rats for as little as 3 days causes a shift in phase of the LH surge related to decreased sensitivity to E_2/P involved in GnRH release at the hypothalamus.[15] The exposure of constant light disrupts the estrous cycle and decreases fertility in young female rats by inducing chronic anovulation associated with persistent estrus. However, light–dark cyclicity is not critical for postpartum ovulation in constant-light rats.[16]

Histological examinations of the ovary in constant-light rats show occasional primordial or secondary follicles, several cystic follicles, and the absence of corpora lutea.[4–6] Electron microscopic studies have further revealed abnormal findings in the tunica albuginea, granulosa cells, and theca cells compared with age-matched controls.[17]

The hormonal findings reported in constant-light rats have been consistent among published studies by various authors. Compared with control values, serum LH levels are normal, follicle-stimulating hormone (FSH) levels are low, E_2 levels are high and P levels are low.[17–20] With high E_2/P ratios the constant-light rat model represents an estrogen-dominant condition.

In our laboratory, the constant-light PCO rat models have been studied for several medical and surgical methods of ovulation induction, and for mating and pregnancy outcomes. Results show the rats exposed to constant light resume estrous cycles when exposed to the light and dark laboratory environment. The constant-light rats ovulate following mating with healthy male partners, and when treated with injections of P and human chorionic gonadotropin (hCG) or unilateral ovariectomy. Pregnancy outcomes are normal, and the pups raised in the constant-light environment do not develop polycystic ovaries until they reach puberty.[3–5,21,22]

The mechanism of chronic anovulation in constant-light rats is not fully understood. The GnRH sensitivity of the pituitary gland for releasing LH is decreased in constant-light rats with persistent estrus.[3] Giving injections of naloxone elevates serum FSH and LH during ovulation and significantly increases the hypothalamic levels of β-endorphin compared with saline-injected controls. Blockade of opiate receptors increases β-endorphin production and uptake and/or decreases its release from the hypothalamus.[23]

There is evidence that anovulation in constant-light rats may result from activation of the sympathetic nervous system that results in adrenal gland hypertrophy during light exposure.[4,5] In animals exposed to chronic lighting, the absolute and relative volumes of zona fasciculata are significantly increased. The serum concentration of corticosterone is also significantly increased in these rats in comparison with controls. These findings suggest that exposure of female rats to constant light increases growth and secretory activity of zona fasciculata cells, which may be related to chronic stress.[24]

Chronic anovulation during continuous light exposure in rats may also be caused by decreased pineal gland activity and suppression of melatonin production. Treatment with melatonin induces ovulation in about 70% of constant-light rats.[25] Feeding precooked maize devoid of tryptophan to light-exposed rats suppresses persistent estrus and significantly reduces estrus–proestrus frequency.[26]

LIMBIC SYSTEM

The limbic system of the brain consists of the hippocampus, hypothalamus, thalamus, amygdala, and the pineal gland. The neural elements obligatory for GnRH release in female rats are found within and or immediately close to the medial preoptic nucleus of the hypothalamus. Preovulatory GnRH and LH surges depend on activation of estrogen-inducible P receptors in the preoptic area of the hypothalamus. Estrogen signaling to GnRH neurons is critical for coordinating the preovulatory surge release of LH that leads to full follicular maturation and ovulation. The preovulatory surges of GnRH and LH are activated by increased E_2 concentrations in blood.[27] The mechanisms for the blockade of the GnRH surge in persistent estrus rats are poorly understood.

Several experimental approaches to manipulate the limbic system have been taken to induce persistent estrus in female rats. Electrolytic lesions using 5 and 10 μA of direct current into the anterior medial preoptic nucleus or the suprachiasmatic nucleus of female rats induced persistent estrus or irregular estrous cycles about 5 weeks after lesions. The preovulatory surges of LH and prolactin (PRL) were eliminated in both groups of persistent estrus rats.[28]

Amygdala kindling in adult female rats causes persistent estrus and development of PCOs associated with high serum E_2 and increased pituitary weight. The ovary shows many cystic follicles, as well as follicles in various stages of growth and atresia. Treatment with P restored cyclicity in the persistent estrus kindled rats. The amygdaloid-lesioned rats provide powerful tools to investigate the association of PCO and epilepsy.[29,30]

Pinealectomy in female rats induces persistent estrus with morphological features of PCO as seen in the constant light rat model.[31]

ANDROGENS

Several studies have shown that giving an androgen such as testosterone propionate (TP) to female rats during early postnatal life results in persistent estrus and polyfollicular anovulatory ovaries during adulthood.[5–7,11] In one study Wistar female rats were injected with testosterone (T) or 5β-dihydrotestosterone (5β-DHT) for the first 5 days of life. Neonatal treatment with 1 mg of T resulted in persistent estrus in all of the animals; however, among the females injected with 1 mg of 5ß-DHT, 74% of rats became sterile at 120 days of age. When the daily dose of 5β-DHT was reduced to 0.1 mg, only 33% of the rats became sterile.[32] In another study, the androgenized rats showed hyperprolactinemia 40 days after they were given 1.25 mg TP at 5 days of age. The polycystic ovaries showed hyperthecosis, but no thickening of the tunica albuginea.[33]

The serum LH, FSH, T, and E_2 values in TP androgenized rats were not different from control values, and treatment with FSH did not increase serum E_2 concentrations or the ovarian weights. The androgenized rats showed a similar postunilateral ovariectomy increase in serum FSH, and compensatory ovarian hypertrophy was not noted compared with controls. Because of these findings it was suggested that the androgenized rat is not a proper model for the human PCO syndrome.[34]

Daily treatment with 4 or 20 mg of danazol in the TP androgenized rats caused regression of cystic follicles. Serum levels of LH, T, and E_2 were significantly decreased, and the vaginal smears changed to diestrus suggesting inhibition of estrogenic activity.[35] Giving an antiandrogen flutamide caused partial normalization of the hormonal and morphological characteristics and recovery of fertility in the animals.[36] Long-term persistence of hormone imbalances in androgen-sterilized rats can induce atypical hyperplasia and adenocarcinoma of the uterus after 500 days of age.[37]

An androgen-sterilized rat (ASR) model that showed obesity was established by subcutaneous injections of 1.25 mg TP to Sprague–Dawley female rats at the age of 9 days. The rats sacrificed around 112 days of age showed polycystic ovaries, anovulation, high food intake, increased body weight, and obesity. High levels of serum E_2 and leptin were noted, but serum FSH and LH levels were reduced significantly compared with normal rats. After feeding herbal tea to the animals, the endocrine changes were normalized, ovulation was induced, and the animals also lost weight.[38] Another ASR model of obesity and polycystic ovaries was developed in Wistar-Hannover rats that showed a significant increase in body weight starting at the fifth week of life compared with control rats.[39]

Implanting androstenedione (ADO) Silastic capsules that released 63.2 ± 4.4 μg/24 h ADO into cyclic rats resulted in significant prolongation of the estrous cycles; most rats showed 5-day cycles after the first normal cycle. There was a decline in ovulation rate with low serum LH levels and normal FSH on the morning of estrus.[40]

Giving dehydroepiandrosterone (DHEA) to neonatal female rats results in persistent estrus and polycystic ovaries with or without corpora lutea. Serum FSH and LH levels showed an ovulatory surge on day 30 at the time of precocious ovulation. Serum FSH levels were comparable to control rats, LH levels were low, and serum PRL was raised significantly. The withdrawal of DHEA treatment resulted in restoration of cyclicity in this model.[8,9,41,42] In DHEA-induced PCO rats, serum FSH, LH, and PRL levels did not differ significantly compared with controls, and the levels of serum DHEA, T, and ADO were high.[43] In another study, the DHEA-treated rats showed normal serum LH and FSH levels, increased DHEAS, ADO, T, and E_2, and reduced serum P level.[44]

With pharmacological doses of androgens given to female rats, the acyclic pattern of gonadotropin secretion is associated with irreversible changes in the hypothalamic cyclic centers normally responsive to steroid feedback. Dose–response studies using various androgens in the neonatal pups to induce persistent estrus during adulthood are limited. According to one study, the majority (99%) of the rats treated with androgens other than DHEA were in persistent estrus by 90 days of age.[45]

In another dose–response study, female rats were androgenized by giving a single injection of 10 or 100 μg TP at 5 days of age. Treatment with 100 μg TP resulted in persistent estrus while treatment with 10 μg TP resulted in regular estrous cycles followed by persistent estrus vaginal smears.[46] Dose–response studies of the reproductive and developmental effects of exposure to prenatal TP and other sex steroids provide unique androgenized rat models.[47]

ESTROGENS

Like androgens, giving estrogens to female rats during early postnatal life results in persistent estrus and polyfollicular anovulatory ovaries during adulthood.[5–7,11] Female rats injected with 2 mg estradiol valerate (EV) develop persistent estrus and small polyfollicular ovaries within 2 months. Plasma E_2 concentrations are high while LH and FSH concentrations are within the high and low normal range, respectively. This model has been used to study the pathogenesis of chronic anovulation in female rats.[48–53]

Increased sympathetic activity plays a role in forming polyfollicular ovarian cysts in EV-treated rats. This theory was tested using a combined cold and restraint stress that induced an increase in the ovarian sympathetic nerves.[54] The activation of the sympathetic nervous system precedes the induction of polycystic ovaries in EV-treated rats. The mechanism involves both direct and neurogenic components.[55,56] There is evidence that persistent estrus in estradiol benzoate (EB)-treated rats is associated with delayed maturation of the tonic GnRH systems and that it affects estrogen receptor synthesis in the uterine tissues during adulthood.[57,58]

ANTIPROGESTERONE

Mifepristone (RU486) is an 11ß-dimethylaminophenyl derivative of norethindrone with a high affinity for P and glucocorticoid receptors. Subcutaneous administration of 4 mg of RU486 to cyclic rats over 8 consecutive days induces persistent estrus and anovulatory cystic ovaries. In this model, serum LH/FSH and T/E_2 ratios are high.[59] Giving RU486 to cyclic rats on proestrus advances the preovulatory surge of LH, which results in 3-day estrous cycles.[60]

The mechanism of inducing chronic anovulation in the RU486 rat model is not fully understood. Several studies in RU486 rats have been conducted to elucidate the pathogenesis of chronic anovulation and to show the validity of the model.[61] Like the constant light and DHEA rat models, the RU486 rat PCO model is also reversible.

AROMATASE INHIBITORS

Giving a third-generation aromatase inhibitor letrozole daily for 21 days induces polycystic ovaries in cyclic female rats. The ovaries a showed high incidence of subcapsular polyfollicular cysts, capsular thickening, incomplete luteinization, and a decreased number of corpora lutea. Serum E_2 and P levels were reduced in a dose-dependent manner and serum LH and T levels were high. Serum levels of FSH were increased at higher doses of letrozole (0.5 and 1.0 mg/kg), contrary to a low dose of letrozole (0.1 mg/kg) at which a slight decrease was observed. The low-dose letrozole persistent estrus rat model shows morphological and hormonal findings that mimic the human PCO syndrome.[62]

DISCUSSION

This chapter reviews various methods of producing persistent estrus, chronic anovulation, and PCO in the laboratory rat. The principal validated models are constant light exposure, hypothalamic lesions, androgen- and estrogen-induced models, and the mifepristone (RU486) model. The vast literature on other animal models has been reviewed by the author at http://obg.lsuhsc-s.edu/pco/index.html.

The main biological systems involved in the development of chronic anovulation and PCO in animal models include the hypothalamus, the pituitary gland, the adrenals, and the ovary with its paracrine, autocrine, and intracrine regulators. The arrested follicular growth, lack of dominant follicles, and thecal and stromal hyperplasias in the polycystic ovaries may represent expressions of several pathogenic pathways including the ovarian paracrine, autocrine, and intracrine regulators.[63]

The pathogenesis of chronic anovulation, the hallmark feature of the PCO syndrome, has been recently studied using animal models such as the prenatally androgenized sheep and rhesus monkey.[1,2] The laboratory rats and mice have key advantages over the androgenized sheep and monkey models to study chronic anovulation and PCO because of their small size, high reproductive index, genetic diversity, and low maintenance costs.

The newer animal models and their implications for future research should be considered in the context for replicating the human disorders. For example, the transgenic mice models have been investigated to determine how excess LH leads to cyst formation in the ovary associated with infertility. Targeted overexpression of LH in transgenic mice leads to obesity, infertility, polycystic ovaries, and the formation of ovarian tumors.[64,65] The prenatally androgenized transgenic mice show irregular estrous cycles and elevated testosterone and luteinizing hormone levels, suggesting altered hypothalamic–pituitary–gonadal axis function.[66] Hormone receptors and knockout mouse models may also provide new insights into the interacting roles of hormones, receptors, and enzymes in chronic anovulation.[67–69]

Environmental factors that cause significant stress can also trigger chronic anovulation and PCO both in animal models and in women with this syndrome.[70–72] Studies in our laboratory have shown, for example, that exposure of adult female Sprague–Dawley rats to continuous noise for more than a month produces persistent estrus and PCO much like the constant-light rats. The ovaries of the noise-exposed rats after 150 days of high-intensity noise showed morphological features similar to the constant-light rats.[73]

Polycystic ovary syndrome is a complex and heterogeneous disorder characterized by hyperandrogenemia, hyperinsulinemia, insulin resistance, and chronic anovulation. Currently PCO syndrome is considered a polygenic trait that might result from the interactions of susceptible and protective genomic variants and environmental factors during either prenatal or postnatal life. Prenatal androgenization of the female fetus induced by genetic and environmental factors, or the interaction of both, may program differentiating target tissues toward the development of the PCO syndrome phenotype in adult life.[1,2,74] Computer models and *in vitro* cell culture studies have enhanced our understanding of physiologically important phenomena in reproductive biology.[75]

Animal models are central to making the transition from scientific concepts to understanding the reality of a human disease. Validated animal models can be used for therapeutic screens, in preclinical trials, and for basic research in reproductive biology. However, animal models of chronic anovulation and PCO may not fully reproduce the reproductive events seen in the human syndrome.

REFERENCES

1. Tsilchorozidou T, Overton C, Conway GS. The pathophysiology of polycystic ovary syndrome. *Clin Endocrinol (Oxf)* 2004;60:1–17.
2. Abbott DH, Dumesic DA, Franks S. Developmental origin of polycystic ovary syndrome—a hypothesis. *J Endocrinol* 2002;174:1–5.
3. Singh KB, Greenwald GS. Effects of continuous light on the reproductive cycle of the female rat: Induction of ovulation and pituitary gonadotrophins during persistent oestrus. *J Endocrinol* 1967;38: 389–394.
4. Singh KB. Induction of polycystic ovarian disease in rats by continuous light. I. The reproductive cycle, organ weights, and histology of the ovaries. *Am J Obstet Gynecol* 1969;103:1078–1083.
5. Singh KB. Persistent estrus: An experimental model of the polycystic ovary syndrome. *Obstet Gynecol Surv* 1969;24:2–17.
6. Baldissera SF, Motta LD, Almeida, MC, Antunes-Rodrigues J. Proposal of an experimental model for the study of polycystic ovaries. *Braz J Med Biol Res* 1991;24:747–751.
7. Mahajan DK. Polycystic ovarian disease: Animal models. *Endocrinol Metab Clin North Am* 1988;17:705–732.
8. Mahesh VB, Costoff A, Mills TM, Bagnell CA. The polycystic ovary syndrome and experimental models for the study of its pathogenesis. *Prog Clin Biol Res* 1982;112:301–313.
9. Mahesh VB, Mills TM, Bagnell CA, Conway BA. Animal models for study of polycystic ovary and ovarian atresia. *Adv Exp Med Biol* 1987;219:237–257.
10. Szukiewicz D, Uilenbroek JT. Polycystic ovary syndrome—searching for an animal model. *J Med* 1998;29:259–275.
11. Singh KB. Persistent estrus rat models of polycystic ovary disease: An update. *Fertil Steril* 2005;84:1228–1234.
12. Marcondes FK, Bianchi FJ, Tanno AP. Determination of the estrous cycle phases of rats: Some helpful considerations. *Braz J Biol* 2002;62:609–617.
13. Takahashi M, Ford JJ, Yoshinaga K, Greep RO. Ovulation in light-estrous rats induced by darkness. *Endocrinol Jpn* 1977;24:89–96.
14. Weber AL, Adler NT. Delay of constant light-induced persistent vaginal estrus by 24-hour time cues in rats. *Science* 1979;204: 323–325.
15. Watts AG, Fink G. Effects of short-term constant light on the proestrous luteinizing hormone surge and pituitary responsiveness in the female rat. *Neuroendocrinology* 1981;33:176–180.
16. Rowland DL, van der Schoot P. Effect of constant light on parturition and postpartum reproduction in the rat. *Physiol Behav* 1995;58: 567–572.
17. Singh KB, Mahajan DK. Ultrastructural basis for continued steroidogenesis in the rat polycystic ovary. *J Reprod Med* 1990;35: 222–228.
18. Singh KB, Mahajan DK, Tewari RP. Hormonal modulation of the vaginal bacterial flora in experimental polycystic ovarian disease. *J Clin Lab Anal* 1996;10:233–238.
19. Takeo Y. Influence of continuous illumination on estrous cycle of rats: Time course of changes in levels of gonadotropins and ovarian steroids until occurrence of persistent estrus. *Neuroendocrinology* 1984;39:97–104.
20. Takeo Y, Kohno J, Hokano M. Ultrastructural evidence for estradiol synthesis in the ovary of persistent-estrous rats exposed to continuous illumination. *Acta Anat* 1986;127:161–170.
21. Singh KB. Induction of polycystic ovarian disease in rats by continuous light. II. Observations on mating, pregnancy, and the postpartum period. *Am J Obstet Gynecol* 1969;104:1004–1007.
22. Singh KB. Induction of polycystic ovarian disease in rats by continuous light. III. Mechanism of ovarian compensatory hypertrophy and ovulation after unilateral oophorectomy. *Am J Obstet Gynecol* 1969;104:1008–1011.
23. Hulse GK, Coleman GJ, Copolov DL, Lee VW. The role of endogenous opioid peptides in the effects of constant illumination on reproductive function in the rat. *Pharmacol Biochem Behav* 1985;23:535–539.
24. Milosevic V, Trifunovic S, Sekulic M, Sosic-Jurjevic B, Filipovic B, Negic N, *et al.* Chronic exposure to constant light affects morphology

and secretion of adrenal zona fasciculata cells in female rats. *Gen Physiol Biophys* 2005;24:299–309.

25. Trentini GP, Mess B, De Gaetani CF, Ruzsas C. Effect of melatonin on induction of ovulation in the light-induced constant estrous-anovulatory syndrome and possible role of the brain serotoninergic system. *J Endocrinol Invest* 1978;1:305–310.
26. Giammanco S, Ernandes M, La Guardia M. Effects of environmental lighting and tryptophan devoid diet on the rat vaginal cycle. *Arch Physiol Biochem* 1997;105:445–449.
27. Petersen SL, Ottem EN, Carpenter CD. Direct and indirect regulation of gonadotropin-releasing hormone neurons by estradiol. *Biol Reprod* 2003;69:1771–1778.
28. Ma YJ, Kelly MJ, Ronnekleiv OK. Pro-gonadotropin-releasing hormone (ProGnRH) and GnRH content in the preoptic area and the basal hypothalamus of anterior medial preoptic nucleus/suprachiasmatic nucleus-lesioned persistent estrous rats. *Endocrinology* 1990;127:2654–2664.
29. Edwards HE, Burnham WM, Ng MM, Asa S, MacLusky NJ. Limbic seizures alter reproductive function in the female rat. *Epilepsia* 1999;40:1370–1377.
30. Edwards HE, MacLusky NJ, Burnham, WM. The effect of seizures and kindling on reproductive hormones in the rat. *Neurosci Biobehav Rev* 2000;24:753–762.
31. Prata Lima MF, Baracat EC, Simoes MJ. Effects of melatonin on the ovarian response to pinealectomy or continuous light in female rats: Similarity with polycystic ovary syndrome. *Braz J Med Biol Res* 2004;37:987–995.
32. Arai Y, Yamanouchi K, Mizukami S, Yanai R, Shibata K, Nagasawa H. Induction of anovulatory sterility by neonatal treatment with 5 beta-dihydrotestosterone in female rats. *Acta Endocrinol (Copenh)* 1981;96:439–443.
33. Ota H, Fukushima M, Maki M. Endocrinological and histological aspects of the process of polycystic ovary formation in the rat treated with testosterone propionate. *Tohoku J Exp Med* 1983;140:121–131.
34. Jones HM, Vernon MW, Rush ME. Systematic studies invalidate the neonatally androgenized rat as a model for polycystic ovary disease. *Biol Reprod* 1987;36:1253–1265.
35. Raj SG, Raj MH, Talbert LM, Dy RC. Structural and functional regression of polycystic ovaries by danazol. *Fertil Steril* 1981;36:392–395.
36. Reznikov AG, Sinitsyn PV, Tarasenko LV, Polyakova LI. Neuroendocrine mechanisms of development of experimental hyperandrogen-induced anovulation. *Neurosci Behav Physiol* 2003;33:773–776.
37. Morikawa S, Sekiya S, Naitoh M, Iwasawa H, Takeda B, Takamizawa H. Spontaneous occurrence of atypical hyperplasia and adenocarcinoma of the uterus in androgen-sterilized SD rats. *J Natl Cancer Inst* 1982;69:95–101.
38. Sun F, Yu J. The effect of a special herbal tea on obesity and anovulation in androgen-sterilized rats. *Proc Soc Exp Biol Med* 2000; 223:295–301.
39. Borges da Silva B, Rocha Gontijo JA, Crete AR, de Jesus Simoes M. Evaluation of body weight in androgenized female rats. *Clin Exp Obstet Gynecol* 2002;29:97–99.
40. Gnodde,HP, van Dieten JA, Van Look PF. Effect of androstenedione on the estrous cycle of the rat. *J Reprod Fertil* 1979;56:675–678.
41. Knudsen JF, Costoff A, Mahesh VB. Dehydroepiandrosterone-induced polycystic ovaries and acyclicity in the rat. *Fertil Steril* 1975;26:807–817.
42. Ward RC, Costoff A, Mahesh VB. The induction of polycystic ovaries in mature cycling rats by the administration of dehydroepiandrosterone (DHA). *Biol Reprod* 1978;18:614–623.
43. Anderson E, Lee GY, O'Brien K. Polycystic ovarian condition in the dehydroepiandrosterone-treated rat model: Hyperandrogenism and the resumption of meiosis are major initial events associated with cystogenesis of antral follicles. *Anat Rec* 1997;249:44–53.
44. Henmi H, Endo T, Nagasawa K, Hayashi T, Chida M, Akutagawa N, *et al.* Lysyl oxidase and MMP-2 expression in dehydroepiandrosterone-induced polycystic ovary in rats. *Biol Reprod* 2001;64: 157–162.
45. Mallampati RS, Johnson DC. Gonadotropins in female rats androgenized by various treatments: Prolactin as an index to hypothalamic damage. *Neuroendocrinology* 1974;15:255–266.
46. Handa RJ, Gorski RA. Alterations in the onset of ovulatory failure and gonadotropin secretion following steroid administration to lightly androgenized female rats. *Biol Reprod* 1985;32:248–256.
47. Wolf CJ, Hotchkiss A, Ostby JS, LeBlanc GA, Gray LE Jr. Effects of prenatal testosterone propionate on the sexual development of male and female rats: A dose-response study. *Toxicol Sci* 2002; 65:71–86.
48. Brawer JR, Naftolin F, Martin J, Sonnenschein C. Effects of a single injection of estradiol valerate on the hypothalamic arcuate nucleus and on reproductive function in the female rat. *Endocrinology* 1978;103:501–512.
49. Brawer J, Schipper H, Naftolin F. Ovary-dependent degeneration in the hypothalamic arcuate nucleus. *Endocrinology* 1980;107:274–279.
50. Brawer JR., Munoz M, Farookhi R. Development of the polycystic ovarian condition (PCO) in the estradiol valerate-treated rat. *Biol Reprod* 1986;35:647–655.
51. Grosser PM, McCarthy GF, Robaire B, Farookhi R, Brawer JR. Plasma patterns of LH, FSH and prolactin in rats with a polycystic ovarian condition induced by oestradiol valerate. *J Endocrinol* 1987;114:33–39.
52. McCarthy GF, Farookhi R, Brawer JR. Plasma gonadotropin patterns characterizing the development of polycystic ovaries in the estradiol valerate treated rat. *Can J Physiol Pharmacol* 1990;68:28–33.
53. McCarthy GF, Brawer JR. Induction of Stein-Leventhal-like polycystic ovaries (PCO) in the rat: A new model for cystic ovarian disease. *Anat Rec* 1990;228:137–144.
54. Paredes A, Galvez A, Leyton V, Aravena G, Fiedler JL, Bustamante D, *et al.* Stress promotes development of ovarian cysts in rats: The possible role of sympathetic nerve activation. *Endocrine* 1998;8:309–315.
55. Lara HE, Dorfman M, Venegas M, Luza SM, Luna SL, Mayerhofer A, *et al.* Changes in sympathetic nerve activity of the mammalian ovary during a normal estrous cycle and in polycystic ovary syndrome: Studies on norepinephrine release. *Microsc Res Tech* 2002; 59:495–502.
56. Stener-Victorin E, Lindholm C. Immunity and beta-endorphin concentrations in hypothalamus and plasma in rats with steroid-induced polycystic ovaries: Effect of low-frequency electroacupuncture. *Biol Reprod* 2004;70:329–333.
57. Aihara M, Hayashi S. Induction of persistent diestrus followed by persistent estrus is indicative of delayed maturation of tonic gonadotropin-releasing systems in rats. *Biol Reprod* 1989;40:96–101.
58. Aihara M, Kobayashi H, Kimura T, Hayashi S, Kato J. Changes in uterine estrogen receptor concentrations in persistent estrous and persistent diestrous rats. *Endocrinol Jpn* 1988;35:57–70.
59. Sanchez-Criado JE, Sanchez A, Ruiz A, Gaytan F. Endocrine and morphological features of cystic ovarian condition in antiprogesterone RU486-treated rats. *Acta Endocrinol (Copenh)* 1993;129: 237–245.
60. Tebar M, Ruiz A, Gonzalez D, Hernandez G, Alonso R, Sanchez-Criado JE. Effect of RU486 injected on proestrous morning on LHRH, LH and 17beta-estradiol secretion during the estrous cycle in rat. *J Physiol Biochem* 1998;54:91–97.
61. Ruiz A, Aguilar R, Tebar M, Gaytan F, Sanchez-Criado JE. RU486-treated rats show endocrine and morphological responses to therapies analogous to responses of women with polycystic ovary syndrome treated with similar therapies. *Biol Reprod* 1996;55:1284–1291.
62. Kafali H, Iriadam M, Ozardali I, Demir N. Letrozole-induced polycystic ovaries in the rat: A new model for cystic ovarian disease. *Arch Med Res* 2004;35:103–108.
63. Dye RB, Rabinovici J, Jaffe RB. Inhibin and activin in reproductive biology. *Obstet Gynecol Surv* 1992;47:173–185.
64. Mann RJ, Keri RA, Nilson JH. Consequences of elevated luteinizing hormone on diverse physiological systems: Use of the LHbetaCTP transgenic mouse as a model of ovarian hyperstimulation-induced pathophysiology. *Recent Prog Horm Res* 2003;58:343–375.

65. Risma KA, Clay CM, Nett TM, Wagner T, Yun J, Nilson JH. Targeted overexpression of luteinizing hormone in transgenic mice leads to infertility, polycystic ovaries, and ovarian tumors. *Proc Natl Acad Sci USA* 1995;92:1322–1326.
66. Sullivan SD, Moenter SM. Prenatal androgens alter GABAergic drive to gonadotropin-releasing hormone neurons: Implications for a common fertility disorder. *Proc Natl Acad Sci USA* 2004;101: 7129–7134.
67. Couse JF, Yates MM, Sanford R, Nyska A, Nilson JH, Korach KS. Formation of cystic ovarian follicles associated with elevated luteinizing hormone requires estrogen receptor-beta. *Endocrinology* 2004;145:4693–4702.
68. Rosenfeld CS, Roberts RM, Lubahn DB. Estrogen receptor- and aromatase-deficient mice provide insight into the roles of estrogen within the ovary and uterus. *Mol Reprod Dev* 2001;59:336–346.
69. Britt KL, Drummond AE, Dyson M, Wreford NG, Jones ME, Simpson ER, Findlay JK. The ovarian phenotype of the aromatase knockout (ArKO) mouse. *J Steroid Biochem Mol Biol* 2001;79: 181–185.
70. Greiner M, Paredes A, Araya V, Lara HE. Role of stress and sympathetic innervation in the development of polycystic ovary syndrome. *Endocrine* 2005;28:319–324.
71. Singh KB. Menstrual disorders in college students. *Am J Obstet Gynecol* 1981;140:299–302.
72. Singh KB: Effects of sound on the female reproductive system. *Am J Obstet Gynecol* 1972;112:981–991.
73. Singh KB, Rao PS. Studies on the polycystic ovaries of rats under continuous auditory stress. *Am J Obstet Gynecol* 1970;108: 557–564.
74. Xita N, Tsatsoulis A. Review: Fetal programming of polycystic ovary syndrome by androgen excess: Evidence from experimental, clinical, and genetic association studies. *J Clin Endocrinol Metab* 2006;91:1660–1666.
75. Washington TM, Blum JJ, Reed MC, Conn PM. A mathematical model for LH release in response to continuous and pulsatile exposure of gonadotrophs to GnRH. *Theor Biol Med Model* 2004;1:9. Available at http://www.tbiomed.com/content/1/1/9. Accessed May 30, 2006.

44 Murine Models for Reproduction

Damayanti Chakraborty and T. Rajendra Kumar

ABSTRACT

Recent developments in mouse manipulation techniques have resulted in the generation of various single gene knockout mouse models. It is estimated that approximately 300 novel strains of mice with mutations in loci directly or indirectly affecting reproductive development have already been generated. Furthermore, the ability to spatiotemporally inactivate or activate gene expression *in vivo* using the "Cre-lox" technology has recently emerged as a powerful approach to understand various developmental processes involved in reproduction in a physiological setting. In parallel with these gene-targeting approaches, large-scale mutagenesis programs have been initiated with the goal of uncovering novel genes important for reproduction. These and other related transgenic approaches are offering unlimited opportunities to systematically analyze the complex process of reproduction. The principles of conventional and conditional gene targeting technologies, the recent advances in functional genomic approaches, and their applications to reproductive biology will be discussed.

Key Words: Mice, Transgene, Knockout, Hypothalamus, Pituitary, Gonads.

INTRODUCTION

Reproduction is a complex process that involves multiple interactions. The mammalian reproductive axis consists of three major centers, namely, the hypothalamus, pituitary, and gonads. Defects in regulation of the reproductive axis may lead to cancers, hyperstimulation syndromes, and infertility.[1–3] The majority of the genes and the corresponding proteins they encode are evolutionarily conserved across various mammalian species.[1,4]

In recent years, the mouse has emerged as a useful model to manipulate the reproductive axis and several murine models of reproduction have been generated.[5–9] There are several advantages of using mice to study basic aspects of reproduction and model human reproductive diseases. First, mice are relatively cheap, easy to maintain in a laboratory setting, and have a short gestation time and life span. Second, an exhaustive amount of data exist with regard to their genetic make-up and linkage of genes. Third, techniques to manipulate the mouse genome have been well established.[10] Finally, mouse genome sequencing has been completed and the information is disseminated into public databases. There is a high degree of sequence similarity between mice and humans in many of the loci, important for reproduction.[4]

Both gain- and loss-of-function mouse models have been generated that have defects ranging from primordial germ cell specification, migration, sex determination, somatic and germ cell development and function, fertilization, implantation, and embryo development. Furthermore, models with defects in pituitary gonadotrope and hypothalamus development and function and their reproductive consequences have also been characterized. Studies with these random or targeted transgenic and knockout mouse models have also been complemented with analysis of naturally occurring mutant strains. Here, we will present a summary of various ways to manipulate the mouse genome and provide lists of representative models that have reproductive phenotypes.

TRANSGENIC APPROACHES

Nearly 25 years ago, transgenic mice expressing human growth hormone were successfully developed.[11] This work revolutionized various fields of biology including reproductive biology and provided a basis for many gain-of-function experiments. Several hundreds of transgenic mice related to reproduction have since been produced. In this approach, foreign genes or cDNAs are first injected into fertilized one-cell mouse embryos usually obtained following a superovulation protocol.[10] Next, these transgene-injected embryos are transferred into oviducts of pseudopregnant foster females.[10] Subsequently, pups produced are screened for the presence and number of copies of the injected transgene. Independent lines will then be established by breeding the transgene-carrying founder mice. Because the transgene is injected into the mouse genome prior to the first round of replication, the foreign DNA is randomly integrated into multiple tissues.[10]

Most often, the transgenes are expressed in a tissue/cell-specific manner using either homologous (i.e., mouse) or heterologous (for example, rat, cow, sheep, human, pig) gene regulatory sequences.[12,13] Mouse genes or cDNAs appropriately marked with random oligos or engineered with heterologous downstream polyadenylation DNA sequences are also used.[12,13] The effects of overexpression of such transgenes on reproductive function are then monitored. In some instances, ectopic expression of the transgene is also achieved by purposefully directing its expression into tissues/cells different from those in which the corresponding mouse gene is normally expressed.[12,13]

From: *Sourcebook of Models for Biomedical Research*
(P. M. Conn, ed.), © 2008 Humana Press Inc., Totowa, NJ.

MAPPING REGULATORY REGIONS OF GENES THAT CONFER TISSUE/CELL-SPECIFIC EXPRESSION Promoter or other regulatory elements that confer hormone responsiveness and cell-specific expression and dictate combinatorial binding of various transcription factors are routinely mapped using cell culture models.[14–20] However, many times these data may not correlate well with the *in vivo* expression data. In other instances, well-established cell lines may not be available for fine mapping studies of regulatory regions of some genes.[21] In these cases, transgenic mice can be used as a powerful *in vivo* expression system. First, a series of deletion constructs are engineered from a larger piece of the gene that is known to confer cell/specific expression.[21] Subsequently, the truncated transgenes are microinjected to produce transgenic mice and as an endpoint, its expression in the selected cell type is monitored at the RNA and/or protein level.[21] Similar strategies have also been used in which a known transcription factor-binding site is mutated on a given promoter driving the expression of a transgene and its functional consequence tested *in vivo*.[22–26]

TRANSGENIC MICE AND QUANTIFIABLE PROMOTER ACTIVITY If the promoter elements of a given gene are well mapped, then these can be used to direct the expression of useful reporters that can be assayed quantitatively. Strength or the effects of positive or negative factors that regulate gene expression or mutations in the promoter region that affect transcription factor binding can be quantified by this approach. The most commonly used quantifiable reporters include lacZ from *Escherichia coli*, chloramphenicol acetyltransferase, and the firefly luciferase.[10] Several promoter regions of genes involved in reproduction have been tested using these strategies (Table 44–1).

CELL LINEAGE MARKING AND PURIFICATION OF DESIRED CELL TYPES Developmental expression of many genes that have important reproductive function can be tracked using lineage marking and cell fate mapping. Depending on the

Table 44–1
A representative list of various transgenic mouse models related to reproduction

Models	*Expression/use*		*Reference*
Mouse models for mapping regulatory regions of genes			
Gene	**Tissue of expression**		
Oxytocin	Hypothalamus		119–121
FSH-β	Pituitary		21
Oct-4	Early germ cells		122
SF-1	Hypothalamus, pituitary, gonads		123–125
Gdf-9	Ovary, testis		126
SP-10	Testis		127–129
Models for quantifiable promoter activity			
Promoter	**Reporter used**		
Bovine α-GSU	LacZ, CAT		22, 130
Rat LH-β	Luciferase		131
Ovine FSH-β	Luciferase		132, 133
Mouse Gdf-9	Luciferase		126
Models for lineage marking			
Promoter	**Reporter used**		
Mouse SF-1	GFP		134, 135
Mouse RARE	LacZ		136
Mouse Oct-4	GFP		137, 138
Mouse GnRH	GFP		139, 140
Models for cell ablation			
Promoter	**Toxin used**		
Human α-GSU	Diphtheria toxin		23
Mouse inhibin-α	Herpes simplex virus thymidine kinase		32
Rat FSH-β	Herpes simplex virus thymidine kinase		30
Models for cell-specific immortalization			
Promoter	**Oncogene used**	**Cell immortalized**	
Rat GnRH	SV40 TAg	Hypothalamic neurons	141, 142
Human α-GSU	SV40 TAg	Gonadotrope	39
Rat LH-β	SV40 TAg	Gonadotrope	38
Mouse GnRH-R	SV40 TAg	Gonadotrope	143
Human FSH-β	SV40 ts TAg	Gonadotrope	36
Ovine FSH-β	SV40 TAg	Gonadotrope	37
Human anti-Mùllerian hormone	SV40 TAg	Sertoli cell	144
Mouse inhibin-α	SV40 TAg	Leydig cell	145, 146
Mouse inhibin-α	SV40 TAg	Granulosa cell	145, 146

specificity and expressivity of the gene regulatory sequences and the earliest time at which these are activated during embryogenesis, expression of either lacZ, alkaline phosphatase, or various fluorescent reporters (for example, GFP, CFP, YFP or dsRed) can be targeted to specific cell types.[10] Tissues/cells are then harvested and the activity of lacZ (formation of a blue product), alkaline phosphatase (formation of a blue or red product), or the visualization of distinct colors under ultraviolet illumination is monitored starting from embryonic stages.[10] Tracking through distinct developmental stages will provide a novel way to study the linage specification and differentiation of desired cell types.

Because cells expressing fluorescent reporters can be sorted by fluorescence-activated cell sorting, these transgenic mice will provide novel resources to purify desired cell types from a tissue consisting of heterogeneous populations of various other cells.[27–29] These can be further used for gene/protein expression profiling under normal physiological conditions.

CELL ABLATION BY TARGETED EXPRESSION OF TOXINS Identification of cell/specific regulatory elements has also been useful for selectively ablating cells at desired times and studying the consequences of the loss of hormones secreted from these cells.[10] This has been achieved by expressing diphtheria toxin,[23] or herpes simplex virus thymidine kinase,[30] or viral-specific ion channels.[31] In the latter two cases, either an appropriate substrate (gancyclovir) or an ionophore (calcium or sodium channel activator or blocker) is used to produce either a cell-toxic product or changes in ion flux that affect hormone secretion. These approaches have been used to study the consequences of ablation of gonadotropes on gonadal development and reproduction.[30] More recently, ablation of Sertoli cells has been achieved to study the consequences on germ cell and consequently male reproduction.[32]

CELL-SPECIFIC IMMORTALIZATION Targeted expression of viral oncogenes in transgenic mice is a powerful way to immortalize desired cell types *in vivo*.[33] This strategy permits immortalization of rare cell types that are otherwise difficult to obtain in large numbers and good purity for routine cell transfection analyses.[34–39] Moreover, novel cell lines are derived from these tumors and established as useful *in vitro* tools for various studies. Since many cell types within the reproductive axis are postmitotic, this approach has been particularly useful for immortalizing these cell types and establishing novel cell lines.[34–39] Many of these cell lines have been used to investigate specific signal transduction pathways, and transcriptional regulation.[34–39] In some cases, these tumor-prone mouse models also phenocopy known human cancers and thus have tremendous potential to help in understanding the pathobiology of the human disease.[36] Furthermore, these models can also be a useful resource for identifying novel cancer biomarkers.

HIGH BASAL LEVELS AND INDUCIBLE EXPRESSION OF TRANSGENES Although expression of a given transgene is dependent on the promoter that drives its expression, the site of the integration, and the copy number, some promoters exhibit high basal activity in many tissues.[40,41] This feature has been exploited to ectopically express various hormones in multiple tissues at high basal levels; the consequences of this on reproduction have been analyzed.[42–44] Some of these promoters are also inducible, for example, metallothionein-1 is induced by heavy metals such as zinc and cadmium.[40,41]

An alternate strategy to temporally induce transgene expression has been developed. In this approach, the desired protein encoding cDNA or gene is engineered downstream of a conditionally activated "gene switch."[45,46] This "gene switch" consists of DNA sequences that encode a bacterial repressor protein and an operator region that is linked to the gene of interest. The repressor binds to the adjacent operator DNA sequence that is linked to the gene of interest and keeps it inactive. This entire "gene switch" cassette along with the gene of interest is under the regulation of a tissue/cell-specific promoter. When a drug that binds the repressor and prevents it from binding to the operator region is administered to transgenic mice, it allows the transcription of the desired gene at specific times. Based on this principle and using tet^{on},[47] ecdysone,[45,48] RU486,[49] and tamoxifen[50] as inducible gene expression models, transgenic mice have been developed.

Several representative models illustrating the application of the above strategies to reproduction research are listed in Table 44–1. Many transgenic approaches have also been successfully used in gene targeting experiments to genetically rescue mutant mice, spatiotemporally inactivate genes, cell fate mapping, and gene/protein expression profiling. These gene knockout models will be described in the following sections.

GENE KNOCKOUT APPROACHES

Loss-of-function mouse models are commonly produced through site-specific mutations at desired loci in embryonic stem (ES) cells by homologous recombination. ES cells are derived from the inner cell mass of blastocysts at embryonic day 3.5.[10] These are propagated *in vitro* on fibroblast cell feeder layers and in an undifferentiated state by providing selected growth factors. Desired mutations are engineered using considerable length of homologous gene sequences flanking the deletion and an appropriate drug selectable marker.[51–54] These engineered DNA sequences are called the targeting vectors and are usually electroporated into ES cells derived from agouti coat color mice.[10] Following appropriate drug selection, enrichment and expansion into individual clones, and identification of the correct gene-targeting event (i.e., one allele disrupted), the mutant ES cells are injected into host blastocysts derived from black coat color mice and transferred to uterine horns of pseudopregnant foster mice.[10] The injected mutant ES cells mix with the inner cell mass of the host and contribute to all parts of the body, including the germ line, and the resulting chimeric mice have patches of both agouti and black coat colors.[10,51] The chimeric mice, mostly males, are bred to normal females and germ line transmission of the mutant allele is achieved resulting in production of heterozygous mice. If viable and fertile, the heterozygous male and female mice are intercrossed to generate the homozygous gene knockout mice.[10,51] If the mutation in the homozygous recessive condition does not cause embryonic lethality, then 25% of knockout mice are produced according to Mendelian inheritance. The majority of the murine models of reproduction have been produced by this standard approach (Tables 44–2 to 44–4). More recently, application of several variations of this basic theme have resulted in production of numerous gene knockout models (Table 44–5). These approaches are described in the following sections.

TISSUE/CELL-RESTRICTED AND TEMPORALLY REGULATED CONDITIONAL GENE INACTIVATION MODELS Many proteins are expressed throughout the reproductive axis or,

Table 44–2
Examples of knockout mice with reproductive defects only in males

Mouse model	*Major reproductive phenotypes*	*Reference*
Acrosin	Delayed fertility	147
AMH or Amh-R	Infertility in males due to obstruction by uteri development	96, 148
Bax	Premeiotic stage block in spermatogenesis; infertility	149
Bclw	Infertile; progressive loss of Sertoli and germ cells	150
Bmp8a	Germ cell degeneration, spermiogenesis defects, epididymis degenerate and progressive infertility	151
Bmp8b	Infertility; defects in primordial germ cell proliferation	152
Calmegin	Infertility due to impaired sperm binding to zona pellucida of eggs	153, 154
CREM	Infertility; block at first stage of spermiogenesis	155, 156
Dhh	Defective germ, Sertoli, and Leydig cell development; infertility	157–159
Hoxd13	Defects in male accessory sex glands	160
HR6B	Infertility; defects in histone polyubiquitination and degradation during spermatogenesis	161
Hsp70	Infertility; increased spermatocyte apoptosis	162, 163
KLHL10	Haploinsufficiency causes asynchronous spermatid development; infertility	164
PC4	Impaired fertilizing ability of spermatozoa; infertility	165
RARα	Seminiferous tubule degeneration leading to male infertility	166
RARβ	Germ cell maturation defects and infertility	167
RARγ	Squamous metaplasia of seminal vesicles and prostate and male infertility	168
TEX14	Infertility; failure to form intracellular bridges after spermatogonia differentiate	169

additionally, in other tissues during distinct developmental stages.[55–57] A deletion of the corresponding gene throughout the body may often result in embryonic lethality.[56,57] Thus, it precludes analyzing the roles in adult physiology, for example, in gonad development and function. This commonly encountered problem is circumvented by a conditional or tissue/cell-specific gene deletion approach. In this approach, first, within the gene of interest, typically in introns flanking the exon to be deleted, two loxP (a 34-bp sequence that consists of two 13-bp inverted repeats and an 8-base pair asymmetric spacer region) sites, one on each side, are engineered in the same orientation.[58–60] A "floxed" mouse is generated in a heterozygous or homozygous condition by standard gene targeting approaches, and this introduction of loxP sites does not interfere with the normal expression of the gene of interest. Second, a transgenic line of mice is produced in which cre, a phage P1 recombinase, is selectively expressed in desired tissue/cells.[58–60] When the flox mice are intercrossed with the cre line, in desired cells, the DNA between the loxP sites is deleted by cre-mediated recombination, leaving one loxP site intact at the locus.[58–60]

Table 44–3
Examples of knockout mice with reproductive defects only in females

Mouse model	*Major reproductive phenotypes*	*Reference*
Activin/inhibin βB	Large litters but delayed parturition and nursing defects	170
Caspase-2	Excess number of germ cells in ovaries; oocytes resistant to cell death following drug treatment	171
C/EBPβ	Infertility; defects in ovulation and corpora lutea formation	172
c-mos	Ovarian cysts and teratoma formation, decreased fertility	173, 174
Connexin 37	Infertility; defects in late stage ovarian folliculogenesis	175
Connexin 43	Defects in primordial germ cells, block at the primary follicle stage	176
Cox2	Infertility; defects in ovulation and corpora lutea formation	177
Foxo3a	Oocyte death, early depletion of functional ovarian follicles, and secondary infertility	178
Gdf9	Defects at one layered follicle stage; infertility	179
IL-11 R	Infertility; implantation defects	180
NGFI-A	Infertility; suppressed luteinizing hormone and ovulatory defects	181, 182
Nobox	Infertility; postnatal oocyte loss, arrest in folliculogenesis at primordial stage	183
Npm2	Nucleolar defects in oocytes and preimplantation embryos, subfertility	184
PR	Infertility; defects in ovary, pituitary, and hypothalamus	185
PRL	Infertility, irregular estrus cycles	186
Prostaglandin F-R	Infertility; lack of induction of oxytocin receptor	187
Ptx3	Subfertility due to defects in the integrity of the cumulus cell–oocyte complexes	188
Srd5a1	Reduced litter size; parturition defects	189
Stat5a/5b	Infertility; absence of corpora lutea and implantation failure	190
Zar1	Infertility; developmental arrest of one-cell embryos	191
ZP3	Infertility; no zona pellucida formation	192, 193

Table 44–4
Examples of knockout mice with reproductive defects in both males and females

Mouse model	*Major reproductive phenotypes*	*Reference*
α-Inhibin	Infertility in both sexes, granulosa/Sertoli cell tumors	194
Acvr2a	Hypogonadism in both sexes, infertility in females, subfertility and male sexual behavioral defects in males	55, 195
α-GSU	Infertile; hypogonadism and hypothyroidism	196
Atm	Male and female infertility; absence of germ cells	197, 198
A-myb	Male infertility; pachytene arrest of spermatogenesis; nursing defects in females due to impaired mammary gland development	199
β1,4-Galactosyltransferase	Male and female infertility due to abnormal glycoprotein hormone glycosylation	200
Cyclin D2	Females infertile; block in folliculogenesis at the preantral stage; hypogonadism in males	201
Dazla	Male and female infertility; loss of germ cells	202
Emx2	Accelerated degeneration of Wolffian duct and mesonephric tubules without formation of Mùlllerian duct	203
ERα	Uterine/ovarian defects in females; small testes, reduced sperm number	204, 205
FSH-β	Hypogonadism, males fertile with reduced sperm number and motility; females infertile, preantral stage block in folliculogenesis	206
IGF-1	Hypogonadal and infertile; preantral block in folliculogenesis; defects in Leydig cell development	207, 208
LH/LH-R	Hypogonadism and infertility in both sexes; spermiogenesis block, reduced testosterone levels, reduced estrogen, and an ovulation defect in females	209–212
Mlh1	Male and female infertility; pachytene stage defects in males; meiosis II defects in females	213, 214
Nhlh2	Males infertile; females fertile only in the presence of males; hypothalamic defects	215
$p27^{kip1}$	Female infertility; corpus luteum defects; males exhibit macroorchidism	216–218
PRL-R	Females infertile due to irregular estrus cycles and implantation defects, idiopathic infertility in males	219–221

Table 44–5
Examples of various mutagenesis approaches

Approach	*Mouse model*	*Reference*
Knockin		
	Activin βB knockin to the βA locus	69
	Cre knockin to the *Amhr* locus	222, 223
	lacZ knockin to the ROSA locus	224, 225
Cre/lox		
	Sertoli cell-specific deletion of androgen receptor	226–228
	Granulosa cell-specific deletions of follistatin, SMAD4	223, 229
	Gonadotrope-specific deletions of GATA2, SF-1	230, 231
	Hypothalamus-specific deletion of SF-1	232, 233
Long-range deletions		
	Down's syndrome	70
	Padre–Willi syndrome	70
Large-scale mutagenesis		
Gene trap	LRG 4-R	234, 235
ENU	GnRH-R	236
Multiple mutations		
	Inhibin/AMH	237
	Acvr2a/FSH-β	92
	Inhibin/FSH-β	43
	ERα/ERβ	90
	Protamines 1 and 2	238
	Transition proteins 1 and 2	239
	Cyclins D1, D2, and D3	240
Genetic rescue		
	FSH-β null mice with human FSH-β transgenes	97
	Dax1 null mice with a human DAX1 cDNA	241, 242
	SF-1 null mice with a rat YAC containing SF-1 genomic sequences	124, 243

Although the above approach provides spatially restricted gene deletion, temporal control is not possible with this cre/lox-based system. Recently, inducible strategies similar to those previously described have been developed in which cre enzyme expression is induced in desired cells at distinct times.[58,61–68] A combination of spatially and temporally regulated gene inactivation offers limitless opportunities to study cell-specific gene expression and function throughout the course of development without affecting the same gene in other tissues.

GENE KNOCKIN STRATEGIES A gene targeting approach can eliminate many of the drawbacks with transgene-based approaches. For example, reporters can be knocked-in to the desired loci (lacZ or GFP or Cre); thus random integration or effects of site of integration can be avoided.[10] In this way, transgenes can be expressed as a single copy from the same endogenous locus. Moreover, bicistronic cassettes can be introduced into endogenous loci and thus gene disruption along with expression of two markers from the same disrupted locus can be achieved.[10] If a known gene is already disrupted in ES cells, variants or mutant forms of the same gene can be knocked-in at the same locus and the effects monitored directly *in vivo*.[10] The other advantage of the knockin strategy is that genes encoding related family members with overlapping patterns of expression are replaced with one another.[69] Thus while disruption of one member is achieved, the other member is transcribed from the disrupted locus using the same regulatory sequences.[69] Most importantly, knockin of cre gene sequences into desired loci and subsequent generation of mice provide a powerful genetic tool for conditional gene deletion purposes.[58,61–68]

LONG-RANGE DELETIONS AND CHROMOSOME ENGINEERING The advent of cre-lox-mediated recombination has led to the emergence of another genetic approach, called chromosome engineering.[70] In this strategy, long-range deletions of desired DNA sequences are achieved in ES cells, although at a low frequency, by selectively engineering loxP sites over megabase intervals on a given chromosome.[70] By altering the orientation of the loxP sites, various chromosomal events including duplication and inversion are also achieved, similar to those originally pioneered in classical *Drosophila* genetics.[70–72] Mice bearing these global chromosomal rearrangements are often maintained as balancer stocks.[70–72] These models coupled with other genetic approaches provide an *in vivo* platform for systematic functional analysis of all the genes that lie within the long intervals on chromosomes.

LARGE-SCALE RANDOM MUTAGENESIS SCHEMES TO GENERATE MULTIPLE DOMINANT OR RECESSIVE MUTANT MODELS It is estimated that ~30,000 genes are expressed in the genome, and about a third of these are believed to be important for reproductive function.[4] Conventional gene targeting approaches are relatively time consuming and mostly aim at mutations in single genes. To rapidly and randomly generate hundreds of mutations in ES cells, several approaches have been used. In one approach, special vectors designated gene trap vectors are randomly introduced into ES cells, and individual vector integrated ES cells are identified and used to produce mutant mice.[73] These vectors are usually designed to disrupt either exons or promoter regions and carry a lacZ reporter expressed from the inserted locus. This facilitates tracking the expression of the gene and eventually cloning the gene itself.[73]

In another approach, *N*-ethyl-*N*-nitrosourea (ENU), an alkylating agent, is injected into wild-type male mice. ENU is a powerful mutagen in mouse spermatogonial stem cells and the goal is to saturate the mouse chromosomes with point mutations.[74–79] The mutagenized males are bred to propagate scores of these mutations into subsequent generations.[74–79] By designing appropriate phenotypic screens, mutants with reproductive defects are further characterized. With the aid of additional genetic tools, the transmission of the mutant allele can be followed visually by coat color and the gene of interest cloned.[74–79] One advantage of this random mutagenesis scheme is that it is possible to recover many types of mutations: null, hypomorphic, and dominant or recessive. Thus, an allelic series of mutations in a given locus can be generated by this approach.[74–79] ENU mutagenesis in ES cells has also been achieved, and methodologies have been developed to identify the mutant alleles in these mutagenized ES cells. The ES cells with desired point mutations are subsequently used to derive mutant strains of mice.[80,81]

Finally, transposon-mediated recombination is also being used to create random mutations in mice.[82–85] Transposons are mobile genetic elements that can integrate into DNA, disrupt the gene, and move over to another region within the same locus or genome-wide.[82–85] This genetic event is facilitated by transposase, the enzyme that recognizes specific sequences to carry out this process, called transposition. Initially, two strains of transgenic mice are generated, one that carries the transposon vector along with a green fluorescent marker between the transposon recognition sequences and the other, transposase. These two lines are bred together and double transgene-positive mice are produced. They are then mated to wild-type mice and several progeny are screened for transposition events that took place in the germ line. By this approach, various mutations clustered within the 3–4 megabase range and outside this region have been produced.[86,87] Although considered not as efficient as the ENU method, transposon-mediated mutagenesis appears mostly useful for region-specific mutagenesis schemes.

COMBINATION OF MULTIPLE MUTATIONS The availability of mutations in desired loci permits a simple genetic intercross strategy to combining these mutations. This is possible if the targeted mutations are on distinct chromosomes or sufficiently apart on the same chromosome.[88–96] Multiple mutations in different loci have been combined for various studies. This approach is useful to analyze the synergistic and/or redundant roles of related family members in a given pathway, or to understand the effect of several modifier factors that regulate a tumorigenesis pathway.[88–96] Similarly, targeted or ectopic expression of transgenes in the null genetic background results in a genetic rescue.[97] If the rescue can be achieved conditionally, then it will provide a useful tool to developmentally follow the effect of the transgene at desired times in the absence of the corresponding endogenous mouse gene.

RESOURCES

Although we have summarized various principles of generating mouse mutations, the reader is encouraged to browse through excellent mouse and mouse genome resources available. The Jackson Labs in Barr Harbor, ME and the mouse genome informatics are great resources for all the mouse genome-based information. This is tightly linked to the National Center for Biotechnology Information (NCBI) that provides various mouse genomics tools. Similarly, the ENSEMBL website maintained by

Table 44–6
Web links to some useful resources

International Gene Trap Consortium	http://www.genetrap.org/
Baygenomics	http://baygenomics.ucsf.edu/
The Jackson Laboratory (General site)	http://jax.org/
Federation of International Mouse Resources	http://www.fimre.org/
European Mouse Mutant Archive	http://www.emma.rm.cnr.it/
Ensembl Mouse	http://www.ensembl.org/ Mus_musculus/index.html
Mouse Genome Informatics	http://www.informatics.jax.org/
Trans-NIH Mouse Initiative	http://nih.gov/science/models/mouse/resources/index.html
Lexicon Genetics	http://www.lexicon-genetics.com/index.php
Endocrine Reviews	edrv.endojournals.org/cgi
Baylor College of Medicine (ENU project)	http://www.mouse-genome.bcm.tmc.edu/
Cornell University (ENU project)	http://www.vertebrategenomics.cornell.edu/
The Jackson Laboratory (infertility website)	http://jaxmice.jax.org/library/notes/496c.html
The Sloan-Kettering Mouse Project	http://mouse.ski.mskcc.org/

the Sanger Institute in Cambridge, UK has numerous genetic tools related to mouse and human as well as other species. Two authoritative reference collections of knockout mouse models are available through The American Endocrine Society and The BioMed Central. A number of institution-based websites list and periodically update mutations obtained with their ENU schemes or provide information particularly focused on reproductive phenotypes. Some of these useful sites are listed in Table 44–6. Finally, ready-to-use practical manuals are also available that describe details of transgenic and ES cell technologies.[10,98,99]

CONCLUSIONS AND FUTURE DIRECTIONS

There are now several mouse manipulation approaches available to study the *in vivo* roles of genes at a whole organism level. These include global or conditional (cell-specific or temporally inducible) gain- or loss-of-function mouse models. These models are powerful genetic tools and when used in combination with genomics and proteomics, they offer limitless opportunities to study reproduction.[100–103] New methods of rapid germline modification can be anticipated using spermatogonial stem cells instead of ES cells.[104–106] Recent advances in RNA-interference, and micro-RNA-regulated pathways are readily feasible *in vivo* using transgenic/ES cell-based techniques.[107–114] These novel *in vivo* approaches will enable us to achieve more tightly controlled regulation of gene expression *in vivo*. Large-scale mutagenesis programs coupled with long-range deletions can be used to mimic microdeletions on the Y-chromosome that commonly occur in several cases of male infertility.[115,116] Moreover, the genetic tools will facilitate identification of candidate genes responsible for many naturally occurring mouse mutations that have not yet been characterized.[117] Finally, several new targets for either fertility control or enhancement may also be identified by these genetic approaches.[100,118]

ACKNOWLEDGMENTS

We wish to emphasize that for simplicity's sake, in preparing this chapter, we have chosen only representative examples to illustrate the principles and applications of transgenic and knockout mice. We acknowledge the financial support received from the National Institutes of Health, The Moran Foundation, Houston, TX, and The Hall Family Foundation, Kansas City, MO. D.C. is a graduate student in and is supported by the Integrated Graduate Program in Biomedical Sciences, University of Kansas Medical Center.

REFERENCES

1. Nishimori K, Matzuk MM. Transgenic mice in the analysis of reproductive development and function. *Rev Reprod* 1996;1:203–212.
2. Seminara SB, Crowley WF Jr. Perspective: The importance of genetic defects in humans in elucidating the complexities of the hypothalamic-pituitary-gonadal axis. *Endocrinology* 2001;142:2173–2177.
3. Seminara SB, Crowley WF Jr. Genetic approaches to unraveling reproductive disorders: Examples of bedside to bench research in the genomic era. *Endocr Rev* 2002;23:382–392.
4. Matzuk MM, Lamb DJ. Genetic dissection of mammalian fertility pathways. *Nat Cell Biol* 2002;4(Suppl.):s41–49.
5. Burns KH, Matzuk MM. Minireview: Genetic models for the study of gonadotropin actions. *Endocrinology* 2002;143:2823–2835.
6. Furnes B, Schimenti JC. Fast forward to new genes in mammalian reproduction. *J Physiol* 2006;578(Pt. 1):25–32.
7. Jorgez CJ, Lin YN, Matzuk MM. Genetic manipulations to study reproduction. *Mol Cell Endocrinol* 2005;234:127–135.
8. Roy A, Matzuk MM. Deconstructing mammalian reproduction: Using knockouts to define fertility pathways. *Reproduction* 2006;131:207–219.
9. Wang H, Dey SK. Roadmap to embryo implantation: Clues from mouse models. *Nat Rev Genet* 2006;7:185–199.

10. Nagy A, Gertsenstein M, Vintersten K, Behringer R. *Manipulating the Mouse Embryo: A Laboratory Manual*, 3rd ed. Woodbury, NY: Cold Spring Harbor Laboratory Press, 2003.
11. Palmiter RD, Brinster RL, Hammer RE, Trumbauer ME, Rosenfeld MG, Birnberg NC, Evans RM. Dramatic growth of mice that develop from eggs microinjected with metallothionein-growth hormone fusion genes. *Nature* 1982;300:611–615.
12. Brinster RL, Palmiter RD. Introduction of genes into the germ line of animals. *Harvey Lect* 1984;80:1–38.
13. Palmiter RD, Brinster RL. Germ-line transformation of mice. *Annu Rev Genet* 1986;20:465–499.
14. Bailey JS, Rave-Harel N, McGillivray SM, Coss D, Mellon PL. Activin regulation of the follicle-stimulating hormone beta-subunit gene involves Smads and the TALE homeodomain proteins Pbx1 and Prep1. *Mol Endocrinol* 2004;18:1158–1170.
15. Barnhart KM, Mellon PL. The orphan nuclear receptor, steroidogenic factor-1, regulates the glycoprotein hormone alpha-subunit gene in pituitary gonadotropes. *Mol Endocrinol* 1994;8:878–885.
16. Harris AN, Mellon PL. The basic helix-loop-helix, leucine zipper transcription factor, USF (upstream stimulatory factor), is a key regulator of SF-1 (steroidogenic factor-1) gene expression in pituitary gonadotrope and steroidogenic cells. *Mol Endocrinol* 1998;12: 714–726.
17. Liu F, Austin DA, Mellon PL, Olefsky JM, Webster NJ. GnRH activates ERK1/2 leading to the induction of c-fos and LHbeta protein expression in LbetaT2 cells. *Mol Endocrinol* 2002;16: 419–434.
18. Schoderbek WE, Kim KE, Ridgway EC, Mellon PL, Maurer RA. Analysis of DNA sequences required for pituitary-specific expression of the glycoprotein hormone alpha-subunit gene. *Mol Endocrinol* 1992;6:893–903.
19. Vasilyev VV, Lawson MA, Dipaolo D, Webster NJ, Mellon PL. Different signaling pathways control acute induction versus long-term repression of LHbeta transcription by GnRH. *Endocrinology* 2002;143:3414–3426.
20. Vasilyev VV, Pernasetti F, Rosenberg SB, Barsoum MJ, Austin DA, Webster NJ, Mellon PL. Transcriptional activation of the ovine follicle-stimulating hormone-beta gene by gonadotropin-releasing hormone involves multiple signal transduction pathways. *Endocrinology* 2002;143:1651–1659.
21. Kumar TR, Schuff KG, Nusser KD, Low MJ. Gonadotroph-specific expression of the human follicle stimulating hormone beta gene in transgenic mice. *Mol Cell Endocrinol* 2006;247:103–115.
22. Hamernik DL, Keri RA, Clay CM, Clay JN, Sherman GB, Sawyer HR Jr, Nett TM, Nilson JH. Gonadotrope- and thyrotrope-specific expression of the human and bovine glycoprotein hormone alpha-subunit genes is regulated by distinct cis-acting elements. *Mol Endocrinol* 1992;6:1745–1755.
23. Kendall SK, Saunders TL, Jin L, Lloyd RV, Glode LM, Nett TM, Keri RA, Nilson JH, Camper SA. Targeted ablation of pituitary gonadotropes in transgenic mice. *Mol Endocrinol* 1991;5:2025–2036.
24. Keri RA, Bachmann DJ, Behrooz A, Herr BD, Ameduri RK, Quirk CC, Nilson JH. An NF-Y binding site is important for basal, but not gonadotropin-releasing hormone-stimulated, expression of the luteinizing hormone beta subunit gene. *J Biol Chem* 2000;275: 13082–13088.
25. Keri RA, Nilson JH. A steroidogenic factor-1 binding site is required for activity of the luteinizing hormone beta subunit promoter in gonadotropes of transgenic mice. *J Biol Chem* 1996;271:10782–10785.
26. Quirk CC, Lozada KL, Keri RA, Nilson JH. A single Pitx1 binding site is essential for activity of the LHbeta promoter in transgenic mice. *Mol Endocrinol* 2001;15:734–746.
27. Balthasar N, Mery PF, Magoulas CB, Mathers KE, Martin A, Mollard P, Robinson IC. Growth hormone-releasing hormone (GHRH) neurons in GHRH-enhanced green fluorescent protein transgenic mice: A ventral hypothalamic network. *Endocrinology* 2003;144:2728–2740.
28. Bonnefont X, Lacampagne A, Sanchez-Hormigo A, Fino E, Creff A, Mathieu MN, Smallwood S, Carmignac D, Fontanaud P, Travo P, Alonso G, Courtois-Coutry N, Pincus SM, Robinson IC, Mollard P. Revealing the large-scale network organization of growth hormone-secreting cells. *Proc Natl Acad Sci USA* 2005;102: 16880–16885.
29. Magoulas C, McGuinness L, Balthasar N, Carmignac DF, Sesay AK, Mathers KE, Christian H, Candeil L, Bonnefont X, Mollard P, Robinson IC. A secreted fluorescent reporter targeted to pituitary growth hormone cells in transgenic mice. *Endocrinology* 2000;141:4681–4689.
30. Markkula M, Kananen K, Klemi P, Huhtaniemi I. Pituitary and ovarian expression of the endogenous follicle-stimulating hormone (FSH) subunit genes and an FSH beta-subunit promoter-driven herpes simplex virus thymidine kinase gene in transgenic mice; specific partial ablation of FSH-producing cells by antiherpes treatment. *J Endocrinol* 1996;150:265–273.
31. Le Tissier PR, Carmignac DF, Lilley S, Sesay AK, Phelps CJ, Houston P, Mathers K, Magoulas C, Ogden D, Robinson IC. Hypothalamic growth hormone-releasing hormone (GHRH) deficiency: Targeted ablation of GHRH neurons in mice using a viral ion channel transgene. *Mol Endocrinol* 2005;19:1251–1262.
32. Ahtiainen M, Toppari J, Poutanen M, Huhtaniemi I. Indirect Sertoli cell-mediated ablation of germ cells in mice expressing the inhibin-alpha promoter/herpes simplex virus thymidine kinase transgene. *Biol Reprod* 2004;71:1545–1550.
33. Hanahan D. Heritable formation of pancreatic beta-cell tumours in transgenic mice expressing recombinant insulin/simian virus 40 oncogenes. *Nature* 1985;315:115–122.
34. Alarid ET, Holley S, Hayakawa M, Mellon PL. Discrete stages of anterior pituitary differentiation recapitulated in immortalized cell lines. *Mol Cell Endocrinol* 1998;140:25–30.
35. Alarid ET, Windle JJ, Whyte DB, Mellon PL. Immortalization of pituitary cells at discrete stages of development by directed oncogenesis in transgenic mice. *Development* 1996;122:3319–3329.
36. Kumar TR, Graham KE, Asa SL, Low MJ. Simian virus 40 T antigen-induced gonadotroph adenomas: A model of human null cell adenomas. *Endocrinology* 1998;139:3342–3351.
37. Pernasetti F, Spady TJ, Hall SB, Rosenberg SB, Givens ML, Anderson S, Paulus M, Miller WL, Mellon PL. Pituitary tumorigenesis targeted by the ovine follicle-stimulating hormone beta-subunit gene regulatory region in transgenic mice. *Mol Cell Endocrinol* 2003;203:169–183.
38. Thomas P, Mellon PL, Turgeon J, Waring DW. The L beta T2 clonal gonadotrope: A model for single cell studies of endocrine cell secretion. *Endocrinology* 1996;137:2979–2989.
39. Windle JJ, Weiner RI, Mellon PL. Cell lines of the pituitary gonadotrope lineage derived by targeted oncogenesis in transgenic mice. *Mol Endocrinol* 1990;4:597–603.
40. Palmiter RD. Molecular biology of metallothionein gene expression. *Experientia Suppl* 1987;52:63–80.
41. Palmiter RD, Norstedt G, Gelinas RE, Hammer RE, Brinster RL. Metallothionein-human GH fusion genes stimulate growth of mice. *Science* 1983;222:809–814.
42. Guo Q, Kumar TR, Woodruff T, Hadsell LA, DeMayo FJ, Matzuk MM. Overexpression of mouse follistatin causes reproductive defects in transgenic mice. *Mol Endocrinol* 1998;12:96–106.
43. Kumar TR, Palapattu G, Wang P, Woodruff TK, Boime I, Byrne MC, Matzuk MM. Transgenic models to study gonadotropin function: The role of follicle-stimulating hormone in gonadal growth and tumorigenesis. *Mol Endocrinol* 1999;13:851–865.
44. Matzuk MM, DeMayo FJ, Hadsell LA, Kumar TR. Overexpression of human chorionic gonadotropin causes multiple reproductive defects in transgenic mice. *Biol Reprod* 2003;69:338–346.
45. Ryding AD, Sharp MG, Mullins JJ. Conditional transgenic technologies. *J Endocrinol* 2001;171:1–14.
46. Schnutgen F, Doerflinger N, Calleja C, Wendling O, Chambon P, Ghyselinck NB. A directional strategy for monitoring Cre-mediated recombination at the cellular level in the mouse. *Nat Biotechnol* 2003;21:562–565.

47. Zhu Z, Zheng T, Lee CG, Homer RJ, Elias JA. Tetracycline-controlled transcriptional regulation systems: Advances and application in transgenic animal modeling. *Semin Cell Dev Biol* 2002;13: 121–128.
48. Karzenowski D, Potter DW, Padidam M. Inducible control of transgene expression with ecdysone receptor: Gene switches with high sensitivity, robust expression, and reduced size. *Biotechniques* 2005;39:191–192, 194, 196 passim.
49. Pierson TM, Wang Y, DeMayo FJ, Matzuk MM, Tsai SY, Omalley BW. Regulable expression of inhibin A in wild-type and inhibin alpha null mice. *Mol Endocrinol* 2000;14:1075–1085.
50. Albanese C, Hulit J, Sakamaki T, Pestell RG. Recent advances in inducible expression in transgenic mice. *Semin Cell Dev Biol* 2002;13:129–141.
51. Bradley A, Hasty P, Davis A, Ramirez-Solis R. Modifying the mouse: Design and desire. *Biotechnology (NY)* 1992;10:534–539.
52. Hasty P, Rivera-Perez J, Bradley A. The length of homology required for gene targeting in embryonic stem cells. *Mol Cell Biol* 1991;11:5586–5591.
53. Hasty P, Rivera-Perez J, Chang C, Bradley A. Target frequency and integration pattern for insertion and replacement vectors in embryonic stem cells. *Mol Cell Biol* 1991;11:4509–4517.
54. Zhang H, Hasty P, Bradley A. Targeting frequency for deletion vectors in embryonic stem cells. *Mol Cell Biol* 1994;14:2404–2410.
55. Matzuk MM, Kumar TR, Bradley A. Different phenotypes for mice deficient in either activins or activin receptor type II. *Nature* 1995;374:356–360.
56. Matzuk MM, Kumar TR, Vassalli A, Bickenbach JR, Roop DR, Jaenisch R, Bradley A. Functional analysis of activins during mammalian development. *Nature* 1995;374:354–356.
57. Matzuk MM, Lu N, Vogel H, Sellheyer K, Roop DR, Bradley A. Multiple defects and perinatal death in mice deficient in follistatin. *Nature* 1995;374:360–363.
58. Le Y, Sauer B. Conditional gene knockout using Cre recombinase. *Mol Biotechnol* 2001;17:269–275.
59. Sauer B, Henderson N. Cre-stimulated recombination at loxP-containing DNA sequences placed into the mammalian genome. *Nucleic Acids Res* 1989;17:147–161.
60. Sauer B, Henderson N. Targeted insertion of exogenous DNA into the eukaryotic genome by the Cre recombinase. *New Biol* 1990; 2:441–449.
61. Belteki G, Haigh J, Kabacs N, Haigh K, Sison K, Costantini F, Whitsett J, Quaggin SE, Nagy A. Conditional and inducible transgene expression in mice through the combinatorial use of Cre-mediated recombination and tetracycline induction. *Nucleic Acids Res* 2005;33:e51.
62. Buch T, Heppner FL, Tertilt C, Heinen TJ, Kremer M, Wunderlich FT, Jung S, Waisman A. A Cre-inducible diphtheria toxin receptor mediates cell lineage ablation after toxin administration. *Nat Methods* 2005;2:419–426.
63. Chow LM, Tian Y, Weber T, Corbett M, Zuo J, Baker SJ. Inducible Cre recombinase activity in mouse cerebellar granule cell precursors and inner ear hair cells. *Dev Dyn* 2006;235:2991–2998.
64. Hayashi S, McMahon AP. Efficient recombination in diverse tissues by a tamoxifen-inducible form of Cre: A tool for temporally regulated gene activation/inactivation in the mouse. *Dev Biol* 2002; 244:305–318.
65. Hunter NL, Awatramani RB, Farley FW, Dymecki SM. Ligand-activated Flpe for temporally regulated gene modifications. *Genesis* 2005;41:99–109.
66. Tannour-Louet M, Porteu A, Vaulont S, Kahn A, Vasseur-Cognet M. A tamoxifen-inducible chimeric Cre recombinase specifically effective in the fetal and adult mouse liver. *Hepatology* 2002;35: 1072–1081.
67. Weber P, Schuler M, Gerard C, Mark M, Metzger D, Chambon P. Temporally controlled site-specific mutagenesis in the germ cell lineage of the mouse testis. *Biol Reprod* 2003;68:553–559.
68. Zhao J, Nassar MA, Gavazzi I, Wood JN. Tamoxifen-inducible NaV1.8-CreERT2 recombinase activity in nociceptive neurons of dorsal root ganglia. *Genesis* 2006;44:364–371.
69. Brown CW, Houston-Hawkins DE, Woodruff TK, Matzuk MM. Insertion of Inhbb into the Inhba locus rescues the Inhba-null phenotype and reveals new activin functions. *Nat Genet* 2000;25: 453–457.
70. Bradley A, van der Weyden L. Mouse: Chromosome engineering for modeling human disease. *Annu Rev Genomics Hum Genet* 2006;7 [Epub ahead of print].
71. Zheng B, Mills AA, Bradley A. Introducing defined chromosomal rearrangements into the mouse genome. *Methods* 2001;24:81–94.
72. Zheng B, Sage M, Cai WW, Thompson DM, Tavsanli BC, Cheah YC, Bradley A. Engineering a mouse balancer chromosome. *Nat Genet* 1999;22:375–378.
73. Chen YT, Liu P, Bradley A. Inducible gene trapping with drug-selectable markers and Cre/loxP to identify developmentally regulated genes. *Mol Cell Biol* 2004;24:9930–9941.
74. Clark AT, Goldowitz D, Takahashi JS, Vitaterna MH, Siepka SM, Peters LL, Frankel WN, Carlson GA, Rossant J, Nadeau JH, Justice MJ. Implementing large-scale ENU mutagenesis screens in North America. *Genetica* 2004;122:51–64.
75. Herron BJ, Lu W, Rao C, Liu S, Peters H, Bronson RT, Justice MJ, McDonald JD, Beier DR. Efficient generation and mapping of recessive developmental mutations using ENU mutagenesis. *Nat Genet* 2002;30:185–189.
76. Justice MJ. From the atomic age to the Genome Project. *Genetica* 2004;122:3–7.
77. Justice MJ, Carpenter DA, Favor J, Neuhauser-Klaus A, Hrabe de Angelis M, Soewarto D, Moser A, Cordes S, Miller D, Chapman V, Weber JS, Rinchik EM, Hunsicker PR, Russell WL, Bode VC. Effects of ENU dosage on mouse strains. *Mamm Genome* 2000;11:484–488.
78. Kile BT, Hentges KE, Clark AT, Nakamura H, Salinger AP, Liu B, Box N, Stockton DW, Johnson RL, Behringer RR, Bradley A, Justice MJ. Functional genetic analysis of mouse chromosome 11. *Nature* 2003;425:81–86.
79. Noveroske JK, Weber JS, Justice MJ. The mutagenic action of N-ethyl-N-nitrosourea in the mouse. *Mamm Genome* 2000;11: 478–483.
80. Chen Y, Vivian JL, Magnuson T. Gene-based chemical mutagenesis in mouse embryonic stem cells. *Methods Enzymol* 2003;365: 406–415.
81. Vivian JL, Chen Y, Yee D, Schneider E, Magnuson T. An allelic series of mutations in Smad2 and Smad4 identified in a genotype-based screen of N-ethyl-N-nitrosourea-mutagenized mouse embryonic stem cells. *Proc Natl Acad Sci USA* 2002;99:15542–15547.
82. Luo G, Ivics Z, Izsvak Z, Bradley A. Chromosomal transposition of a Tc1/mariner-like element in mouse embryonic stem cells. *Proc Natl Acad Sci USA* 1998;95:10769–10773.
83. Score PR, Belur LR, Frandsen JL, Guerts JL, Yamaguchi T, Somia NV, Hackett PB, Largaespada DA, McIvor RS. Sleeping Beauty-mediated transposition and long-term expression in vivo: Use of the LoxP/Cre recombinase system to distinguish transposition-specific expression. *Mol Ther* 2006;13:617–624.
84. Collier LS, Largaespada DA. Hopping around the tumor genome: Transposons for cancer gene discovery. *Cancer Res* 2005;65: 9607–9610.
85. Gueguen E, Rousseau P, Duval-Valentin G, Chandler M. The transpososome: Control of transposition at the level of catalysis. *Trends Microbiol* 2005;13:543–549.
86. Yae K, Keng VW, Koike M, Yusa K, Kouno M, Uno Y, Kondoh G, Gotow T, Uchiyama Y, Horie K, Takeda J. Sleeping beauty transposon-based phenotypic analysis of mice: Lack of Arpc3 results in defective trophoblast outgrowth. *Mol Cell Biol* 2006;26:6185–6196.
87. Yant SR, Wu X, Huang Y, Garrison B, Burgess SM, Kay MA. High-resolution genome-wide mapping of transposon integration in mammals. *Mol Cell Biol* 2005;25:2085–2094.
88. Su YQ, Wu X, O'Brien MJ, Pendola FL, Denegre JN, Matzuk MM, Eppig JJ. Synergistic roles of BMP15 and GDF9 in the development and function of the oocyte-cumulus cell complex in mice: Genetic

evidence for an oocyte-granulosa cell regulatory loop. *Dev Biol* 2004;276:64–73.
89. Kumar TR, Agno J, Janovick JA, Conn PM, Matzuk MM. Regulation of FSHbeta and GnRH receptor gene expression in activin receptor II knockout male mice. *Mol Cell Endocrinol* 2003;212: 19–27.
90. Burns KH, Agno JE, Chen L, Haupt B, Ogbonna SC, Korach KS, Matzuk MM. Sexually dimorphic roles of steroid hormone receptor signaling in gonadal tumorigenesis. *Mol Endocrinol* 2003;17: 2039–2052.
91. Burns KH, Agno JE, Sicinski P, Matzuk MM. Cyclin D2 and p27 are tissue-specific regulators of tumorigenesis in inhibin alpha knockout mice. *Mol Endocrinol* 2003;17:2053–2069.
92. Kumar TR, Varani S, Wreford NG, Telfer NM, de Kretser DM, Matzuk MM. Male reproductive phenotypes in double mutant mice lacking both FSHbeta and activin receptor IIA. *Endocrinology* 2001;142:3512–3518.
93. Cipriano SC, Chen L, Burns KH, Koff A, Matzuk MM. Inhibin and p27 interact to regulate gonadal tumorigenesis. *Mol Endocrinol* 2001;15:985–996.
94. Yan C, Wang P, DeMayo J, DeMayo FJ, Elvin JA, Carino C, Prasad SV, Skinner SS, Dunbar BS, Dube JL, Celeste AJ, Matzuk MM. Synergistic roles of bone morphogenetic protein 15 and growth differentiation factor 9 in ovarian function. *Mol Endocrinol* 2001;15:854–866.
95. Lau AL, Kumar TR, Nishimori K, Bonadio J, Matzuk MM. Activin betaC and betaE genes are not essential for mouse liver growth, differentiation, and regeneration. *Mol Cell Biol* 2000;20:6127–6137.
96. Mishina Y, Rey R, Finegold MJ, Matzuk MM, Josso N, Cate RL, Behringer RR. Genetic analysis of the Mullerian-inhibiting substance signal transduction pathway in mammalian sexual differentiation. *Genes Dev* 1996;10:2577–2587.
97. Kumar TR, Low MJ, Matzuk MM. Genetic rescue of follicle-stimulating hormone beta-deficient mice. *Endocrinology* 1998;139: 3289–3295.
98. Joyner AL. *Gene Targeting: A Practical Approach*, 2nd ed. New York: Oxford University Press, 2001.
99. Hofker MH, Van Deursen J. *Transgenic Mouse: Methods and Protocols, Methods in Molecular Biology*, Vol. 209. Totowa, NJ: Humana Press, 2003.
100. Chu DS, Liu H, Nix P, Wu TF, Ralston EJ, Yates JR 3rd, Meyer BJ. Sperm chromatin proteomics identifies evolutionarily conserved fertility factors. *Nature* 2006;443:101–105.
101. Lin YN, Matzuk MM. High-throughput discovery of germ-cell-specific genes. *Semin Reprod Med* 2005;23:201–212.
102. Rockett JC. Genomic and proteomic techniques applied to reproductive biology. *Genome Biol* 2001;2:REPORTS4020.
103. Sirard MA, Dufort I, Coenen K, Tremblay K, Massicotte L, Robert C. The use of genomics and proteomics to understand oocyte and early embryo functions in farm animals. *Reprod Suppl* 2003;61:117–129.
104. Shinohara T, Kato M, Takehashi M, Lee J, Chuma S, Nakatsuji N, Kanatsu-Shinohara M, Hirabayashi M. Rats produced by interspecies spermatogonial transplantation in mice and in vitro microinsemination. *Proc Natl Acad Sci USA* 2006;103:13624–13628.
105. Kanatsu-Shinohara M, Ikawa M, Takehashi M, Ogonuki N, Miki H, Inoue K, Kazuki Y, Lee J, Toyokuni S, Oshimura M, Ogura A, Shinohara T. Production of knockout mice by random or targeted mutagenesis in spermatogonial stem cells. *Proc Natl Acad Sci USA* 2006;103:8018–8023.
106. Kanatsu-Shinohara M, Toyokuni S, Shinohara T. Transgenic mice produced by retroviral transduction of male germ line stem cells in vivo. *Biol Reprod* 2004;71:1202–1207.
107. Glaser S, Anastassiadis K, Stewart AF. Current issues in mouse genome engineering. *Nat Genet* 2005;37:1187–1193.
108. Lin SL, Chang DC, Ying SY. Isolation and identification of gene-specific microRNAs. *Methods Mol Biol* 2006;342:313–320.
109. Lin SL, Chang SJ, Ying SY. Transgene-like animal models using intronic microRNAs. *Methods Mol Biol* 2006;342:321–334.
110. Lin SL, Ying SY. Gene silencing in vitro and in vivo using intronic microRNAs. *Methods Mol Biol* 2006;342:295–312.
111. Muljo SA, Kanellopoulou C. Mouse embryonic stem cells as a model genetic system to dissect and exploit the RNA interference machinery. *Methods Mol Biol* 2006;342:57–72.
112. Prawitt D, Brixel L, Spangenberg C, Eshkind L, Heck R, Oesch F, Zabel B, Bockamp E. RNAi knock-down mice: An emerging technology for post-genomic functional genetics. *Cytogenet Genome Res* 2004;105:412–421.
113. Ristevski S. Making better transgenic models: Conditional, temporal, and spatial approaches. *Mol Biotechnol* 2005;29:153–163.
114. Ying SY, Chang DC, Miller JD, Lin SL. The microRNA: Overview of the RNA gene that modulates gene functions. *Methods Mol Biol* 2006;342:1–18.
115. Affara NA, Mitchell MJ. The role of human and mouse Y chromosome genes in male infertility. *J Endocrinol Invest* 2000;23: 630–645.
116. Burgoyne PS. The role of Y-encoded genes in mammalian spermatogenesis. *Semin Cell Dev Biol* 1998;9:423–432.
117. Lyon MF, Rastan S, Brown SDM, Eds. *Genetic Variants and Strains of the Laboratory Mouse*, 3rd ed. New York: Oxford University Press, 1995.
118. Heck S, Qian X, Velleca M. Genetically engineered mouse models for drug discovery: New chemical genetic approaches. *Curr Drug Discov Technol* 2004;1:13–26.
119. Gainer H. Cell-specific gene expression in oxytocin and vasopressin magnocellular neurons. *Adv Exp Med Biol* 1998;449:15–27.
120. Jeong SW, Castel M, Zhang BJ, Fields RL, Paras P, Arnheiter H, Chin H, Gainer H. Cell-specific expression and subcellular localization of neurophysin-CAT-fusion proteins expressed from oxytocin and vasopressin gene promoter-driven constructs in transgenic mice. *Exp Neurol* 2001;171:255–271.
121. Ratty AK, Jeong SW, Nagle JW, Chin H, Gainer H, Murphy D, Venkatesh B. A systematic survey of the intergenic region between the murine oxytocin- and vasopressin-encoding genes. *Gene* 1996;174:71–78.
122. Yeom YI, Fuhrmann G, Ovitt CE, Brehm A, Ohbo K, Gross M, Hubner K, Scholer HR. Germline regulatory element of Oct-4 specific for the totipotent cycle of embryonal cells. *Development* 1996;122:881–894.
123. Daggett MA, Rice DA, Heckert LL. Expression of steroidogenic factor 1 in the testis requires an E box and CCAAT box in its promoter proximal region. *Biol Reprod* 2000;62:670–679.
124. Karpova T, Presley J, Manimaran RR, Scherrer SP, Tejada L, Peterson KR, Heckert LL. A FTZ-F1-containing yeast artificial chromosome recapitulates expression of steroidogenic factor 1 in vivo. *Mol Endocrinol* 2005;19:2549–2563.
125. Scherrer SP, Rice DA, Heckert LL. Expression of steroidogenic factor 1 in the testis requires an interactive array of elements within its proximal promoter. *Biol Reprod* 2002;67:1509–1521.
126. Yan C, Elvin JA, Lin YN, Hadsell LA, Wang J, DeMayo FJ, Matzuk MM. Regulation of growth differentiation factor 9 expression in oocytes in vivo: A key role of the E-box. *Biol Reprod* 2006;74:999–1006.
127. Acharya KK, Govind CK, Shore AN, Stoler MH, Reddi PP. Cis-requirement for the maintenance of round spermatid-specific transcription. *Dev Biol* 2006;295:781–790.
128. Reddi PP, Flickinger CJ, Herr JC. Round spermatid-specific transcription of the mouse SP-10 gene is mediated by a 294-base pair proximal promoter. *Biol Reprod* 1999;61:1256–1266.
129. Reddi PP, Shore AN, Shapiro JA, Anderson A, Stoler MH, Acharya KK. Spermatid-specific promoter of the SP-10 gene functions as an insulator in somatic cells. *Dev Biol* 2003;262:173–182.
130. Keri RA, Andersen B, Kennedy GC, Hamernik DL, Clay CM, Brace AD, Nett TM, Notides AC, Nilson JH. Estradiol inhibits transcription of the human glycoprotein hormone alpha-subunit gene despite the absence of a high affinity binding site for estrogen receptor. *Mol Endocrinol* 1991;5:725–733.
131. Fallest PC, Trader GL, Darrow JM, Shupnik MA. Regulation of rat luteinizing hormone beta gene expression in transgenic mice by

steroids and a gonadotropin-releasing hormone antagonist. *Biol Reprod* 1995;53:103–109.
132. Huang HJ, Sebastian J, Strahl BD, Wu JC, Miller WL. Transcriptional regulation of the ovine follicle-stimulating hormone-beta gene by activin and gonadotropin-releasing hormone (GnRH): Involvement of two proximal activator protein-1 sites for GnRH stimulation. *Endocrinology* 2001;142:2267–2274.
133. Huang HJ, Sebastian J, Strahl BD, Wu JC, Miller WL. The promoter for the ovine follicle-stimulating hormone-beta gene (FSHbeta) confers FSHbeta-like expression on luciferase in transgenic mice: Regulatory studies in vivo and in vitro. *Endocrinology* 2001;142: 2260–2266.
134. Stallings NR, Hanley NA, Majdic G, Zhao L, Bakke M, Parker KL. Development of a transgenic green fluorescent protein lineage marker for steroidogenic factor 1. *Endocr Res* 2002;28:497–504.
135. Stallings NR, Hanley NA, Majdic G, Zhao L, Bakke M, Parker KL. Development of a transgenic green fluorescent protein lineage marker for steroidogenic factor 1. *Mol Endocrinol* 2002;16: 2360–2370.
136. Bowles J, Knight D, Smith C, Wilhelm D, Richman J, Mamiya S, Yashiro K, Chawengsaksophak K, Wilson MJ, Rossant J, Hamada H, Koopman P. Retinoid signaling determines germ cell fate in mice. *Science* 2006;312:596–600.
137. Boiani M, Kehler J, Scholer HR. Activity of the germline-specific Oct4-GFP transgene in normal and clone mouse embryos. *Methods Mol Biol* 2004;254:1–34.
138. Keiser JT, Jobst PM, Garst AS, Boone JT, Geyer CB, Phelps C, Ayares DL, Page RL. Preimplantation screening for transgenesis using an embryonic specific promoter and green fluorescent protein. *Cloning* 2001;3:23–30.
139. Suter KJ, Song WJ, Sampson TL, Wuarin JP, Saunders JT, Dudek FE, Moenter SM. Genetic targeting of green fluorescent protein to gonadotropin-releasing hormone neurons: Characterization of whole-cell electrophysiological properties and morphology. *Endocrinology* 2000;141:412–419.
140. Suter KJ, Wuarin JP, Smith BN, Dudek FE, Moenter SM. Whole-cell recordings from preoptic/hypothalamic slices reveal burst firing in gonadotropin-releasing hormone neurons identified with green fluorescent protein in transgenic mice. *Endocrinology* 2000;141: 3731–3736.
141. Mellon PL, Windle JJ, Goldsmith PC, Padula CA, Roberts JL, Weiner RI. Immortalization of hypothalamic GnRH neurons by genetically targeted tumorigenesis. *Neuron* 1990;5:1–10.
142. Weiner RI, Wetsel W, Goldsmith P, Martinez de la Escalera G, Windle J, Padula C, Choi A, Negro-Vilar A, Mellon P. Gonadotropin-releasing hormone neuronal cell lines. *Front Neuroendocrinol* 1992;13:95–119.
143. Albarracin CT, Frosch MP, Chin WW. The gonadotropin-releasing hormone receptor gene promoter directs pituitary-specific oncogene expression in transgenic mice. *Endocrinology* 1999;140:2415–2421.
144. Peschon JJ, Behringer RR, Cate RL, Harwood KA, Idzerda RL, Brinster RL, Palmiter RD. Directed expression of an oncogene to Sertoli cells in transgenic mice using Mullerian inhibiting substance regulatory sequences. *Mol Endocrinol* 1992;6:1403–1411.
145. Kananen K, Markkula M, el-Hefnawy T, Zhang FP, Paukku T, Su JG, Hsueh AJ, Huhtaniemi I. The mouse inhibin alpha-subunit promoter directs SV40 T-antigen to Leydig cells in transgenic mice. *Mol Cell Endocrinol* 1996;119:135–146.
146. Rahman NA, Kananen Rilianawati K, Paukku T, Mikola M, Markkula M, Hamalainen T, Huhtaniemi IT. Transgenic mouse models for gonadal tumorigenesis. *Mol Cell Endocrinol* 1998;145: 167–174.
147. Adham IM, Nayernia K, Engel W. Spermatozoa lacking acrosin protein show delayed fertilization. *Mol Reprod Dev* 1997;46: 370–376.
148. Jamin SP, Arango NA, Mishina Y, Hanks MC, Behringer RR. Genetic studies of the AMH/MIS signaling pathway for Mullerian duct regression. *Mol Cell Endocrinol* 2003;211:15–19.
149. Knudson CM, Tung KS, Tourtellotte WG, Brown GA, Korsmeyer SJ. Bax-deficient mice with lymphoid hyperplasia and male germ cell death. *Science* 1995;270:96–99.
150. Ross AJ, Waymire KG, Moss JE, Parlow AF, Skinner MK, Russell LD, MacGregor GR. Testicular degeneration in Bclw-deficient mice. *Nat Genet* 1998;18:251–256.
151. Zhao GQ, Liaw L, Hogan BL. Bone morphogenetic protein 8A plays a role in the maintenance of spermatogenesis and the integrity of the epididymis. *Development* 1998;125:1103–1112.
152. Zhao GQ, Deng K, Labosky PA, Liaw L, Hogan BL. The gene encoding bone morphogenetic protein 8B is required for the initiation and maintenance of spermatogenesis in the mouse. *Genes Dev* 1996;10:1657–1669.
153. Ikawa M, Nakanishi T, Yamada S, Wada I, Kominami K, Tanaka H, Nozaki M, Nishimune Y, Okabe M. Calmegin is required for fertilin alpha/beta heterodimerization and sperm fertility. *Dev Biol* 2001;240:254–261.
154. Ikawa M, Wada I, Kominami K, Watanabe D, Toshimori K, Nishimune Y, Okabe M. The putative chaperone calmegin is required for sperm fertility. *Nature* 1997;387:607–611.
155. Blendy JA, Kaestner KH, Weinbauer GF, Nieschlag E, Schutz G. Severe impairment of spermatogenesis in mice lacking the CREM gene. *Nature* 1996;380:162–165.
156. Nantel F, Monaco L, Foulkes NS, Masquilier D, LeMeur M, Henriksen K, Dierich A, Parvinen M, Sassone-Corsi P. Spermiogenesis deficiency and germ-cell apoptosis in CREM-mutant mice. *Nature* 1996;380:159–162.
157. Bitgood MJ, Shen L, McMahon AP. Sertoli cell signaling by Desert hedgehog regulates the male germline. *Curr Biol* 1996;6:298–304.
158. Clark AM, Garland KK, Russell LD. Desert hedgehog (Dhh) gene is required in the mouse testis for formation of adult-type Leydig cells and normal development of peritubular cells and seminiferous tubules. *Biol Reprod* 2000;63:1825–1838.
159. Pierucci-Alves F, Clark AM, Russell LD. A developmental study of the Desert hedgehog-null mouse testis. *Biol Reprod* 2001;65: 1392–1402.
160. Podlasek CA, Duboule D, Bushman W. Male accessory sex organ morphogenesis is altered by loss of function of Hoxd-13. *Dev Dyn* 1997;208:454–465.
161. Roest HP, van Klaveren J, de Wit J, van Gurp CG, Koken MH, Vermey M, van Roijen JH, Hoogerbrugge JW, Vreeburg JT, Baarends WM, Bootsma D, Grootegoed JA, Hoeijmakers JH. Inactivation of the HR6B ubiquitin-conjugating DNA repair enzyme in mice causes male sterility associated with chromatin modification. *Cell* 1996;86:799–810.
162. Dix DJ, Allen JW, Collins BW, Mori C, Nakamura N, Poorman-Allen P, Goulding EH, Eddy EM. Targeted gene disruption of Hsp70–2 results in failed meiosis, germ cell apoptosis, and male infertility. *Proc Natl Acad Sci USA* 1996;93:3264–3268.
163. Dix DJ, Allen JW, Collins BW, Poorman-Allen P, Mori C, Blizard DR, Brown PR, Goulding EH, Strong BD, Eddy EM. HSP70–2 is required for desynapsis of synaptonemal complexes during meiotic prophase in juvenile and adult mouse spermatocytes. *Development* 1997;124:4595–4603.
164. Yan W, Ma L, Burns KH, Matzuk MM. Haploinsufficiency of kelch-like protein homolog 10 causes infertility in male mice. *Proc Natl Acad Sci USA* 2004;101:7793–7798.
165. Mbikay M, Tadros H, Ishida N, Lerner CP, De Lamirande E, Chen A, El-Alfy M, Clermont Y, Seidah NG, Chretien M, Gagnon C, Simpson EM. Impaired fertility in mice deficient for the testicular germ-cell protease PC4. *Proc Natl Acad Sci USA* 1997;94: 6842–6846.
166. Lufkin T, Lohnes D, Mark M, Dierich A, Gorry P, Gaub MP, LeMeur M, Chambon P. High postnatal lethality and testis degeneration in retinoic acid receptor alpha mutant mice. *Proc Natl Acad Sci USA* 1993;90:7225–7229.
167. Kastner P, Mark M, Leid M, Gansmuller A, Chin W, Grondona JM, Decimo D, Krezel W, Dierich A, Chambon P. Abnormal spermatogenesis in RXR beta mutant mice. *Genes Dev* 1996;10:80–92.

168. Lohnes D, Kastner P, Dierich A, Mark M, LeMeur M, Chambon P. Function of retinoic acid receptor gamma in the mouse. *Cell* 1993;73:643–658.
169. Greenbaum MP, Yan W, Wu MH, Lin YN, Agno JE, Sharma M, Braun RE, Rajkovic A, Matzuk MM. TEX14 is essential for intercellular bridges and fertility in male mice. *Proc Natl Acad Sci USA* 2006;103:4982–4987.
170. Vassalli A, Matzuk MM, Gardner HA, Lee KF, Jaenisch R. Activin/inhibin beta B subunit gene disruption leads to defects in eyelid development and female reproduction. *Genes Dev* 1994;8: 414–427.
171. Bergeron L, Perez GI, Macdonald G, Shi L, Sun Y, Jurisicova A, Varmuza S, Latham KE, Flaws JA, Salter JC, Hara H, Moskowitz MA, Li E, Greenberg A, Tilly JL, Yuan J. Defects in regulation of apoptosis in caspase-2-deficient mice. *Genes Dev* 1998;12: 1304–1314.
172. Sterneck E, Tessarollo L, Johnson PF. An essential role for C/EBPbeta in female reproduction. *Genes Dev* 1997;11:2153–2162.
173. Furuta Y, Shigetani Y, Takeda N, Iwasaki K, Ikawa Y, Aizawa S. Ovarian teratomas in mice lacking the protooncogene c-mos. *Jpn J Cancer Res* 1995;86:540–545.
174. Hashimoto N, Watanabe N, Furuta Y, Tamemoto H, Sagata N, Yokoyama M, Okazaki K, Nagayoshi M, Takeda N, Ikawa Y, *et al.* Parthenogenetic activation of oocytes in c-mos-deficient mice. *Nature* 1994;370:68–71.
175. Simon AM, Goodenough DA, Li E, Paul DL. Female infertility in mice lacking connexin 37. *Nature* 1997;385:525–529.
176. Juneja SC, Barr KJ, Enders GC, Kidder GM. Defects in the germ line and gonads of mice lacking connexin43. *Biol Reprod* 1999;60:1263–1270.
177. Lim H, Paria BC, Das SK, Dinchuk JE, Langenbach R, Trzaskos JM, Dey SK. Multiple female reproductive failures in cyclooxygenase 2-deficient mice. *Cell* 1997;91:197–208.
178. Castrillon DH, Miao L, Kollipara R, Horner JW, DePinho RA. Suppression of ovarian follicle activation in mice by the transcription factor Foxo3a. *Science* 2003;301:215–218.
179. Dong J, Albertini DF, Nishimori K, Kumar TR, Lu N, Matzuk MM. Growth differentiation factor-9 is required during early ovarian folliculogenesis. *Nature* 1996;383:531–535.
180. Robb L, Li R, Hartley L, Nandurkar HH, Koentgen F, Begley CG. Infertility in female mice lacking the receptor for interleukin 11 is due to a defective uterine response to implantation. *Nat Med* 1998;4:303–308.
181. Lee SL, Sadovsky Y, Swirnoff AH, Polish JA, Goda P, Gavrilina G, Milbrandt J. Luteinizing hormone deficiency and female infertility in mice lacking the transcription factor NGFI-A (Egr-1). *Science* 1996;273:1219–1221.
182. Topilko P, Schneider-Maunoury S, Levi G, Trembleau A, Gourdji D, Driancourt MA, Rao CV, Charnay P. Multiple pituitary and ovarian defects in Krox-24 (NGFI-A, Egr-1)-targeted mice. *Mol Endocrinol* 1998;12:107–122.
183. Rajkovic A, Pangas SA, Ballow D, Suzumori N, Matzuk MM. NOBOX deficiency disrupts early folliculogenesis and oocyte-specific gene expression. *Science* 2004;305:1157–1159.
184. Burns KH, Viveiros MM, Ren Y, Wang P, DeMayo FJ, Frail DE, Eppig JJ, Matzuk MM. Roles of NPM2 in chromatin and nucleolar organization in oocytes and embryos. *Science* 2003;300:633–636.
185. Lydon JP, DeMayo FJ, Funk CR, Mani SK, Hughes AR, Montgomery CA Jr, Shyamala G, Conneely OM, O'Malley BW. Mice lacking progesterone receptor exhibit pleiotropic reproductive abnormalities. *Genes Dev* 1995;9:2266–2278.
186. Horseman ND, Zhao W, Montecino-Rodriguez E, Tanaka M, Nakashima K, Engle SJ, Smith F, Markoff E, Dorshkind K. Defective mammopoiesis, but normal hematopoiesis, in mice with a targeted disruption of the prolactin gene. *EMBO J* 1997;16: 6926–6935.
187. Sugimoto Y, Yamasaki A, Segi E, Tsuboi K, Aze Y, Nishimura T, Oida H, Yoshida N, Tanaka T, Katsuyama M, Hasumoto K, Murata T, Hirata M, Ushikubi F, Negishi M, Ichikawa A, Narumiya S. Failure of parturition in mice lacking the prostaglandin F receptor. *Science* 1997;277:681–683.
188. Varani S, Elvin JA, Yan C, DeMayo J, DeMayo FJ, Horton HF, Byrne MC, Matzuk MM. Knockout of pentraxin 3, a downstream target of growth differentiation factor-9, causes female subfertility. *Mol Endocrinol* 2002;16:1154–1167.
189. Mahendroo MS, Cala KM, Russell DW. 5 Alpha-reduced androgens play a key role in murine parturition. *Mol Endocrinol* 1996;10: 380–392.
190. Teglund S, McKay C, Schuetz E, van Deursen JM, Stravopodis D, Wang D, Brown M, Bodner S, Grosveld G, Ihle JN. Stat5a and Stat5b proteins have essential and nonessential, or redundant, roles in cytokine responses. *Cell* 1998;93:841–850.
191. Wu X, Viveiros MM, Eppig JJ, Bai Y, Fitzpatrick SL, Matzuk MM. Zygote arrest 1 (Zar1) is a novel maternal-effect gene critical for the oocyte-to-embryo transition. *Nat Genet* 2003;33:187–191.
192. Dean J. Oocyte-specific genes regulate follicle formation, fertility and early mouse development. *J Reprod Immunol* 2002;53: 171–180.
193. Rankin TL, O'Brien M, Lee E, Wigglesworth K, Eppig J, Dean J. Defective zonae pellucidae in Zp2-null mice disrupt folliculogenesis, fertility and development. *Development* 2001;128:1119–1126.
194. Matzuk MM, Finegold MJ, Su JG, Hsueh AJ, Bradley A. Alpha-inhibin is a tumour-suppressor gene with gonadal specificity in mice. *Nature* 1992;360:313–319.
195. Ma X, Reyna A, Mani SK, Matzuk MM, Kumar TR. Impaired male sexual behavior in activin receptor type II knockout mice. *Biol Reprod* 2005;73:1182–1190.
196. Kendall SK, Samuelson LC, Saunders TL, Wood RI, Camper SA. Targeted disruption of the pituitary glycoprotein hormone alpha-subunit produces hypogonadal and hypothyroid mice. *Genes Dev* 1995;9:2007–2019.
197. Barlow C, Hirotsune S, Paylor R, Liyanage M, Eckhaus M, Collins F, Shiloh Y, Crawley JN, Ried T, Tagle D, Wynshaw-Boris A. Atm-deficient mice: A paradigm of ataxia telangiectasia. *Cell* 1996;86:159–171.
198. Xu Y, Ashley T, Brainerd EE, Bronson RT, Meyn MS, Baltimore D. Targeted disruption of ATM leads to growth retardation, chromosomal fragmentation during meiosis, immune defects, and thymic lymphoma. *Genes Dev* 1996;10:2411–2422.
199. Toscani A, Mettus RV, Coupland R, Simpkins H, Litvin J, Orth J, Hatton KS, Reddy EP. Arrest of spermatogenesis and defective breast development in mice lacking A-myb. *Nature* 1997;386: 713–717.
200. Lu Q, Hasty P, Shur BD. Targeted mutation in beta1,4-galactosyltransferase leads to pituitary insufficiency and neonatal lethality. *Dev Biol* 1997;181:257–267.
201. Sicinski P, Donaher JL, Geng Y, Parker SB, Gardner H, Park MY, Robker RL, Richards JS, McGinnis LK, Biggers JD, Eppig JJ, Bronson RT, Elledge SJ, Weinberg RA. Cyclin D2 is an FSH-responsive gene involved in gonadal cell proliferation and oncogenesis. *Nature* 1996;384:470–474.
202. Ruggiu M, Speed R, Taggart M, McKay SJ, Kilanowski F, Saunders P, Dorin J, Cooke HJ. The mouse Dazla gene encodes a cytoplasmic protein essential for gametogenesis. *Nature* 1997;389:73–77.
203. Miyamoto N, Yoshida M, Kuratani S, Matsuo I, Aizawa S. Defects of urogenital development in mice lacking Emx2. *Development* 1997;124:1653–1664.
204. Krege JH, Hodgin JB, Couse JF, Enmark E, Warner M, Mahler JF, Sar M, Korach KS, Gustafsson JA, Smithies O. Generation and reproductive phenotypes of mice lacking estrogen receptor beta. *Proc Natl Acad Sci USA* 1998;95:15677–15682.
205. Lubahn DB, Moyer JS, Golding TS, Couse JF, Korach KS, Smithies O. Alteration of reproductive function but not prenatal sexual development after insertional disruption of the mouse estrogen receptor gene. *Proc Natl Acad Sci USA* 1993;90:11162–11166.
206. Kumar TR, Wang Y, Lu N, Matzuk MM. Follicle stimulating hormone is required for ovarian follicle maturation but not male fertility. *Nat Genet* 1997;15:201–204.

207. Baker J, Hardy MP, Zhou J, Bondy C, Lupu F, Bellve AR, Efstratiadis A. Effects of an Igf1 gene null mutation on mouse reproduction. *Mol Endocrinol* 1996;10:903–918.
208. Baker J, Liu JP, Robertson EJ, Efstratiadis A. Role of insulin-like growth factors in embryonic and postnatal growth. *Cell* 1993; 75:73–82.
209. Lei ZM, Mishra S, Zou W, Xu B, Foltz M, Li X, Rao CV. Targeted disruption of luteinizing hormone/human chorionic gonadotropin receptor gene. *Mol Endocrinol* 2001;15:184–200.
210. Ma X, Dong Y, Matzuk MM, Kumar TR. Targeted disruption of luteinizing hormone beta-subunit leads to hypogonadism, defects in gonadal steroidogenesis, and infertility. *Proc Natl Acad Sci USA* 2004;101:17294–17299.
211. Rao CV, Lei ZM. Consequences of targeted inactivation of LH receptors. *Mol Cell Endocrinol* 2002;187:57–67.
212. Zhang FP, Poutanen M, Wilbertz J, Huhtaniemi I. Normal prenatal but arrested postnatal sexual development of luteinizing hormone receptor knockout (LuRKO) mice. *Mol Endocrinol* 2001;15:172–183.
213. Edelmann W, Cohen PE, Kane M, Lau K, Morrow B, Bennett S, Umar A, Kunkel T, Cattoretti G, Chaganti R, Pollard JW, Kolodner RD, Kucherlapati R. Meiotic pachytene arrest in MLH1-deficient mice. *Cell* 1996;85:1125–1134.
214. Edelmann W, Cohen PE, Kneitz B, Winand N, Lia M, Heyer J, Kolodner R, Pollard JW, Kucherlapati R. Mammalian MutS homologue 5 is required for chromosome pairing in meiosis. *Nat Genet* 1999;21:123–127.
215. Good DJ, Porter FD, Mahon KA, Parlow AF, Westphal H, Kirsch IR. Hypogonadism and obesity in mice with a targeted deletion of the Nhlh2 gene. *Nat Genet* 1997;15:397–401.
216. Fero ML, Rivkin M, Tasch M, Porter P, Carow CE, Firpo E, Polyak K, Tsai LH, Broudy V, Perlmutter RM, Kaushansky K, Roberts JM. A syndrome of multiorgan hyperplasia with features of gigantism, tumorigenesis, and female sterility in p27(Kip1)-deficient mice. *Cell* 1996;85:733–744.
217. Kiyokawa H, Kineman RD, Manova-Todorova KO, Soares VC, Hoffman ES, Ono M, Khanam D, Hayday AC, Frohman LA, Koff A. Enhanced growth of mice lacking the cyclin-dependent kinase inhibitor function of p27(Kip1). *Cell* 1996;85:721–732.
218. Nakayama K, Ishida N, Shirane M, Inomata A, Inoue T, Shishido N, Horii I, Loh DY, Nakayama K. Mice lacking p27(Kip1) display increased body size, multiple organ hyperplasia, retinal dysplasia, and pituitary tumors. *Cell* 1996;85:707–720.
219. Goffin V, Binart N, Clement-Lacroix P, Bouchard B, Bole-Feysot C, Edery M, Lucas BK, Touraine P, Pezet A, Maaskant R, Pichard C, Helloco C, Baran N, Favre H, Bernichtein S, Allamando A, Ormandy C, Kelly PA. From the molecular biology of prolactin and its receptor to the lessons learned from knockout mice models. *Genet Anal* 1999;15:189–201.
220. Grosdemouge I, Bachelot A, Lucas A, Baran N, Kelly PA, Binart N. Effects of deletion of the prolactin receptor on ovarian gene expression. *Reprod Biol Endocrinol* 2003;1:12.
221. Kelly PA, Binart N, Lucas B, Bouchard B, Goffin V. Implications of multiple phenotypes observed in prolactin receptor knockout mice. *Front Neuroendocrinol* 2001;22:140–145.
222. Jeyasuria P, Ikeda Y, Jamin SP, Zhao L, De Rooij DG, Themmen AP, Behringer RR, Parker KL. Cell-specific knockout of steroidogenic factor 1 reveals its essential roles in gonadal function. *Mol Endocrinol* 2004;18:1610–1619.
223. Jorgez CJ, Klysik M, Jamin SP, Behringer RR, Matzuk MM. Granulosa cell-specific inactivation of follistatin causes female fertility defects. *Mol Endocrinol* 2004;18:953–967.
224. Soriano P. Generalized lacZ expression with the ROSA26 Cre reporter strain. *Nat Genet* 1999;21:70–71.
225. Zambrowicz BP, Imamoto A, Fiering S, Herzenberg LA, Kerr WG, Soriano P. Disruption of overlapping transcripts in the ROSA beta geo 26 gene trap strain leads to widespread expression of beta-galactosidase in mouse embryos and hematopoietic cells. *Proc Natl Acad Sci USA* 1997;94:3789–3794.
226. Chang C, Chen YT, Yeh SD, Xu Q, Wang RS, Guillou F, Lardy H, Yeh S. Infertility with defective spermatogenesis and hypotestosteronemia in male mice lacking the androgen receptor in Sertoli cells. *Proc Natl Acad Sci USA* 2004;101:6876–6881.
227. De Gendt K, Swinnen JV, Saunders PT, Schoonjans L, Dewerchin M, Devos A, Tan K, Atanassova N, Claessens F, Lecureuil C, Heyns W, Carmeliet P, Guillou F, Sharpe RM, Verhoeven G. A Sertoli cell-selective knockout of the androgen receptor causes spermatogenic arrest in meiosis. *Proc Natl Acad Sci USA* 2004;101: 1327–1332.
228. Holdcraft RW, Braun RE. Androgen receptor function is required in Sertoli cells for the terminal differentiation of haploid spermatids. *Development* 2004;131:459–467.
229. Pangas SA, Li X, Robertson EJ, Matzuk MM. Premature luteinization and cumulus cell defects in ovarian-specific Smad4 knockout mice. *Mol Endocrinol* 2006;20:1406–1422.
230. Charles MA, Saunders TL, Wood WM, Owens K, Parlow AF, Camper SA, Ridgway EC, Gordon DF. Pituitary-specific Gata2 knockout: Effects on gonadotrope and thyrotrope function. *Mol Endocrinol* 2006;20:1366–1377.
231. Zhao L, Bakke M, Krimkevich Y, Cushman LJ, Parlow AF, Camper SA, Parker KL. Steroidogenic factor 1 (SF1) is essential for pituitary gonadotrope function. *Development* 2001;128:147–154.
232. Davis AM, Seney ML, Stallings NR, Zhao L, Parker KL, Tobet SA. Loss of steroidogenic factor 1 alters cellular topography in the mouse ventromedial nucleus of the hypothalamus. *J Neurobiol* 2004;60:424–436.
233. Dellovade TL, Young M, Ross EP, Henderson R, Caron K, Parker K, Tobet SA. Disruption of the gene encoding SF-1 alters the distribution of hypothalamic neuronal phenotypes. *J Comp Neurol* 2000;423:579–589.
234. Nord AS, Chang PJ, Conklin BR, Cox AV, Harper CA, Hicks GG, Huang CC, Johns SJ, Kawamoto M, Liu S, Meng EC, Morris JH, Rossant J, Ruiz P, Skarnes WC, Soriano P, Stanford WL, Stryke D, von Melchner H, Wurst W, Yamamura K, Young SG, Babbitt PC, Ferrin TE. The International Gene Trap Consortium Website: A portal to all publicly available gene trap cell lines in mouse. *Nucleic Acids Res* 2006;34:D642–648.
235. Skarnes WC, von Melchner H, Wurst W, Hicks G, Nord AS, Cox T, Young SG, Ruiz P, Soriano P, Tessier-Lavigne M, Conklin BR, Stanford WL, Rossant J. A public gene trap resource for mouse functional genomics. *Nat Genet* 2004;36:543–544.
236. Pask AJ, Kanasaki H, Kaiser UB, Conn PM, Janovick JA, Stockton DW, Hess DL, Justice MJ, Behringer RR. A novel mouse model of hypogonadotrophic hypogonadism: N-ethyl-N-nitrosourea-induced gonadotropin-releasing hormone receptor gene mutation. *Mol Endocrinol* 2005;19:972–981.
237. Matzuk MM, Finegold MJ, Mishina Y, Bradley A, Behringer RR. Synergistic effects of inhibins and Mullerian-inhibiting substance on testicular tumorigenesis. *Mol Endocrinol* 1995;9:1337–1345.
238. Cho C, Willis WD, Goulding EH, Jung-Ha H, Choi YC, Hecht NB, Eddy EM. Haploinsufficiency of protamine-1 or -2 causes infertility in mice. *Nat Genet* 2001;28:82–86.
239. Shirley CR, Hayashi S, Mounsey S, Yanagimachi R, Meistrich ML. Abnormalities and reduced reproductive potential of sperm from Tnp1- and Tnp2-null double mutant mice. *Biol Reprod* 2004;71:1220–1229.
240. Kozar K, Ciemerych MA, Rebel VI, Shigematsu H, Zagozdzon A, Sicinska E, Geng Y, Yu Q, Bhattacharya S, Bronson RT, Akashi K, Sicinski P. Mouse development and cell proliferation in the absence of D-cyclins. *Cell* 2004;118:477–491.
241. Jeffs B, Ito M, Yu RN, Martinson FA, Wang ZJ, Doglio LT, Jameson JL. Sertoli cell-specific rescue of fertility, but not testicular pathology, in Dax1 (Ahch)-deficient male mice. *Endocrinology* 2001;142:2481–2488.
242. Meeks JJ, Russell TA, Jeffs B, Huhtaniemi I, Weiss J, Jameson JL. Leydig cell-specific expression of DAX1 improves fertility of the Dax1-deficient mouse. *Biol Reprod* 2003;69:154–160.
243. Karpova T, Maran RR, Presley J, Scherrer SP, Tejada L, Heckert LL. Transgenic rescue of SF-1-null mice. *Ann NY Acad Sci* 2005;1061:55–64.

45 Pig Model to Study Dynamics of Steroids During Ovarian Follicular Growth and Maturation

DAMODAR K. MAHAJAN

ABSTRACT

The pig is a polyovular farm animal and its prolificacy differs in different breeds. The female pig has an estrous cycle of 20–21 days with a short follicular phase of only 3 days and a long luteal phase. The luteal phase starts at day 1 after ovulation to day 15 of the estrous cycle and it overlaps the early follicular phase on day 16–17. The follicular phase lasts only 3–4 days. The early follicular growth is initiated while a high concentration of progesterone (P) is secreted by active corpora lutea, circulating it in the peripheral blood. In this study, P, androstenedione (A), testosterone (T), estrone (E_1), and estradiol-17β (E_2) were measured in peripheral blood, ovarian venous blood, follicular fluid, and perifollicular blood at various stages of the pig estrous cycle. Concentrations of P in peripheral blood, ovarian venous blood, follicular fluid, and perifollicular blood (blood from the vasculature surrounding the follicle) during the active luteal phase of gilts were 19.0 ± 2.4, 59.2 ± 12.7, and 3937 ± 710 ng/ml, respectively. Interestingly, the high concentration of P in perifollicular blood did not equilibrate with the follicular fluid and it appears that there is a blood barrier between the two. This level of P was significantly lower in the late luteal, follicular, and estrous stages. The follicular fluid in the follicular phase contained the highest concentrations of A (163 ± 37.9 ng/ml) of all the stages of follicular growth. During this stage the perifollicular blood also contained a concentration of A higher than that observed at any other stage of the estrous cycle. On the day of estrus the amounts of both A and T in the follicular fluid as well as in ovarian venous blood were at their lowest. Concentrations of T in perifollicular blood were higher in the luteal phase than in the follicular phase. The high levels of E_1 (22.94 ± 2.95 ng/ml) and E_2 (18.96 ± 1.80 ng/ml) in the follicular fluid during the follicular phase may be due to aromatization of A and T, which are also present in high concentrations in follicular fluid and are in contact with granulosa cells. On the day of estrus E_1 and E_2 in the follicular fluid were 7.42 ± 1.45 and 6.93 ng/ml, respectively, which may precede the onset of the preovulatory surge. The lower levels of E_1 and E_2 in perifollicular blood than in follicular fluid suggest that the main site of aromatization could be the granulosa cells in the follicles.

From: *Sourcebook of Models for Biomedical Research* (P. M. Conn, ed.), © 2008 Humana Press Inc., Totowa, NJ.

Key Words: Gilts, Steroids, Ovary, Follicle, Follicular fluid, Perifollicular fluid, Follicular growth, Progesterone, Androgens, Estrogens.

INTRODUCTION

The pig is an ideal animal model to study reproductive endocrinology and, in particular, to understand ovarian physiology. The ovarian morphology, tissue differentiation, and actions of individual cell types have been extensively studied. The ease of obtaining ovaries from abattoir facilitated the initial basic investigation of ovarian function, though the background of the gilts or sows were in question. However, availability of defined breeds, knowledge of their ages, nutrition, and any hormonal treatment and assessment of the estrous period made it easy to study carefully defined protocols of investigation. The pig is polyovular and the important limiting factor affecting sow productivity is prolificacy, which is defined as the number of viable piglets born per litter. Chinese Meishan farrow have an average of three to five more piglets per litter than the European breed White Composite. Other mixed breeds vary with a minimum of 12 to 20 piglets.[1,2] The ovulation rate is a major determinant of reproductive efficacy, which is genetically controlled within a particular breed. After recruitment of the primordial follicles, the preovulatory follicles are selected, which then become dominant and progress toward ovulation. Growth, maturation, and ovulation are dependent on endocrine and paracrine control. The sequential effect of follicle-stimulating hormone (FSH) and leutinizing hormone (LH) on the development of the follicles is controlled by a hypothalamic–pituitary–ovarian feedback mechanism. Binding of insulin to various follicular cells as well as the action of insulin-like growth factors IGF-I and IGF-II and their binding proteins,[3,4] such as inhibin and activin,[5] produced within the follicle exert metabolic actions directly or indirectly on the follicular cells and overall follicular growth. Follicular angiogenesis is promoted by vascular endothelial growth factor (VEGF), which is produced by the granulosa cells within the follicle. Steroids produced by interstitial, theca, and granulosa cells are distributed within and out of the ovary through the basement membrane and the follicular vasculature of the follicle to reach various organs to exert or promote metabolic actions.

Synthesis of various steroids in the ovary and their secretion into the circulation are a complex process because of the ovarian–

pituitary feedback system. The two-cell type theory of steroid synthesis in ovarian tissue postulating that the theca interna cells of the follicle produce mainly androgens and estrogens, while the luteinized granulosa cells secrete progestins[6] was suggested by Short in 1962, and it was based on steroid concentrations in the follicular fluid and luteal tissue. In the following year it was reported that equine corpora lutea have a low level of 17α-hydroxylase ($P450_{17\alpha}$) and 17α-hydroxyprogesterone-20-lyase ($P450_{c17}$) enzyme activity.[7] Thus, the large amount of progesterone (P) produced by the corpus luteum was not converted to C_{19} steroids within the luteal tissue. The theca cells of the Graafian follicle are able to produce P and contain all the necessary enzymes to convert P to androgens and estrogens.[7–9] The granulosa cells appear to have lower $P450_{c17}$ activity than theca cells but possess very active aromatase ($P450_{arom}$), which converts Δ^4-androstenedione (A) and testosterone (T) to estrogens.[10–12] When steroid synthesis by theca and granulosa cells in culture was studied, it was observed that the combined cell culture more efficiently synthesized estradiol-17β (E_2) than either of the cell types alone.[13–15]

From these *in vitro* studies and the data on steroid concentrations in follicular fluid at various stages of follicular growth, it is postulated that P, A, T, estrone (E_1), and E_2 are synthesized in different compartments of the ovary and that the ability of these different tissues to produce steroids changes during various stages of follicular growth. These compartments do not appear to be isolated and there appears to be an active exchange of steroids between different types of steroid-producing cells. Since the corpus luteum, stroma, and theca are vascularized, steroids produced by these cells would be readily mixed into the follicular circulation and may accumulate in the follicular fluid. There is sufficient evidence to indicate that substances in the follicular fluid constantly equilibrate with the ovarian blood,[6,16] although no correlation was found between the concentration of steroids in follicular fluid and the corresponding steroid level in peripheral plasma of the same pig during the estrous cycle.[17] Based on the observation that very high amounts of ovarian steroids were present in the follicular fluid of preovulatory follicles, it was theorized that follicular fluid serves as a storage for steroids, particularly E_2 and P, which are required by the uterus for implantation. In the present investigation P, A, T, E_1, and E_2 were measured in peripheral blood, ovarian venous blood, follicular fluid, and perifollicular blood during the entire estrous cycle of gilts; there is speculation as to whether they correlate and whether the active corpus luteum could be a source of progesterone to theca and granulosa cells for conversion into C_{19} and C_{18} steroids and serve as one of the regulators of steroidogenesis in the pig.

The quantitative determination of steroidogenic enzyme mRNAs and certain steroids in follicular tissue and follicular fluid was carried out during either the regumate or altrenogest-synchronized preovulatory period of follicular growth of the gilts.[18–20] Day 1 was considered the prefollicular phase, day 3 the early follicular phase, day 5 the mid-follicular phase, and day 7 the late follicular phase. It was concluded that the $P450_{arom}$ enzyme mRNA increased successively from day 1 to day 5, dropped significantly on day 7 to a level less than day 1, and correlated only partially with the E_2 in follicular fluid. A similar poor correlation was true for $P450_{17\alpha}$ mRNA to testosterone and cholesterol side chain cleavage ($P450_{scc}$) to progesterone. In animal husbandry when the pigs are grown for the meat industry, all animals are treated with progestational compound or progestational compound plus estrogenic compounds to arrest ovulation or continuation of cycling and also to increase their weight. By rules and regulations, steroid treatment is stopped at least 2 weeks before they are brought to abattoirs. Therefore ovaries collected at the abattoir have small, immature follicles of 2–5 mm or so, and the results of studies conducted on these ovaries could be different from studies conducted on naturally cycling gilts. In the cycling gilts after ovulation, days 3–5 are early luteal phase with active corpora lutea producing a large amount of progesterone. At this stage the surface of the ovary is covered with corpora lutea and the small immature follicles are situated among the corpora lutea in close proximity. It is reported that follicular fluid contains a large amount of progesterone,[21] which may not be produced by the follicular tissue, but may be accumulated from the vasculature of the corpora lutea. This progesterone may inhibit some enzyme activity, specifically $P450_{17\alpha}$ and $P450_{c17}$,[7] and at the same time it may act as a precursor for steroid-producing enzymes. Hence, in such situations the follicular steroid enzymatic mRNA may not correlate with steroids present in the follicular fluid of these small follicles, some of which may be destined to become atretic follicles.

VARIOUS BLOOD SAMPLES AND FOLLICULAR FLUID COLLECTION

Ovarian follicular fluid, perifollicular blood, and peripheral blood samples were collected from 40 gilts at the swine research farm at Ohio Agricultural Research and Development Center, Wooster, OH. Gilts 1–2 years of age were observed for regularity of their estrous cycles according to a method previously described[22] and only those having three consistent consecutive estrous cycles of 19–21 days were chosen for the study. The first day of estrous behavior was designated day 0 of the cycle. For each gilt on the appropriate day of the estrous cycle, surgical anesthesia was induced with 1.5–2 g of sodium thiamylal (10% solution) intravenously via an ear vein, and surgical anesthesia was maintained by methoxyflurane and oxygen inhalation with a closed circuit anesthesia machine. First, a 20-ml peripheral blood sample was withdrawn from the anterior vena cava and allowed to clot. Laparotomy was performed and the ovaries were exposed. Follicular fluid from an individual "healthy-looking" follicle (a follicle full of fluid, reddish in color, and transparent without white deposit in it) was withdrawn with a tuberculin syringe fitted with a 26-gauge hypodermic needle until the follicle had completely collapsed. Then, within a few seconds, the empty follicle was filled with blood, which was then withdrawn over a 1-min period with 1-min intervals between samples. These samples were considered to be perifollicular blood.[17,23] Five such consecutive samples of perifollicular blood were collected. Sometimes external pressure was applied on the surrounding side of the follicle to fill the empty antrum with blood. Follicular fluid and perifollicular blood samples from two or three follicles were collected from both ovaries. Blood samples less than 25 μl were discarded. The remaining follicles on each ovary were aspirated, and that fluid was pooled for the right or left ovary. The volume of the follicular fluid or blood sample was measured immediately by a capillary pipette or by a narrow graduated tube. The individual blood volume varied from 30 μl to 600 μl and steroids were measured in the whole blood of perifollicular blood samples. Fluid from nonovulated follicular cysts of the previous estrous cycle were not included in the pool of follicular fluid of an individual ovary, but were processed separately. The follicles up to day 5 of the estrous cycle were too small in size (<2 mm in diameter) to permit follicular fluid collection as individual samples from a single follicle or perifollicular blood, hence only samples from day 5 to day 21 were included in these studies.

STEROID QUANTITATION

The various steroids were measured by specific radioimmunoassay (RIA). The following radioactive steroids were obtained from New England Nuclear, Boston, MA, as reference steroids and their purity was tested by paper chromatography before use: [1,2,6,7-^{3}H]progesterone (114 Ci/mmol), Δ^4-[1,2-^{3}H]androstenedione (60 Ci/mmol), [1,2,6,7-^{3}H]testosterone (93.9 Ci/mmol), [2,4,6,7-^{3}H]estrone (95.5 Ci/mmol), and [2,4,6,7-^{3}H]estradiol-17β (90 Ci/mmol). Steroids with 1% or more impurities were further purified by thin-layer chromatography. The radio inert steroids for references were obtained from Steraloids (Wilten, NH) and Sigma Chemical Company (St. Louis, MO), and were crystallized from organic solvents before being used as standards for the radioimmunoassays; their purity was checked by melting point determination and/or absorption spectra. Stock standard solutions were made by dissolving weighed amounts of steroids in methanol and the concentrations were confirmed by ultraviolet (UV) absorption.

The steroids were measured using the double antibody procedures for steroid RIA.[24,25] Antibodies used to measure P, A, and E_2 were highly specific. For example, the highest cross-reactivity of 20α-hydroxyprogesterone (OHP) with antiprogesterone antibody was only <0.001% and 17α-OHP did not cross-react with this antiserum. The cross-reaction of E_1 with the anti-E_2 antibody was <1.0% and estriol cross-reacted at 0.1%. The anti-A antibody was very specific and cross-reacted at <0.01% with T. The anti-T antibody cross-reacted at 2% with dihydrotestosterone (DHT) and 0.3% and 0.2% with androstan-3β,17-diol (3β-A-diol) and androstan-3α,17-diol (3α-A-diol), respectively. The E_1 antibody cross-reacted about 1% with E_2 compared to E_1 and did not cross-react with estriol. The follicular fluid samples were extracted with diethyl ether and P, A, and E_2 were fractionated by paper chromatography.[25] The intraassay coefficient of variation ranged between 5% and 14% depending on the concentration of steroids in the samples. The interassay coefficients of variation for all the various steroids were >12.5%.

Student's *t* test was used for statistical comparison of values of different steroids for different sera and follicular fluid at various stages of the estrous cycle. Two-tailed *p* was calculated using Graphpad Instate software to compare differences between the groups; differences were considered statistically significant at $p < 0.05$. The results are expressed as mean ± standard error (SE).

STEROIDS IN PERIPHERAL BLOOD, OVARIAN VENOUS BLOOD, FOLLICULAR FLUID, AND PERIFOLLICULAR BLOOD DURING THE ESTROUS CYCLE OF GILTS

PROGESTERONE Most breeds of female pigs have an estrous cycle of 20–21 days with a short follicular phase of 3 days and a long luteal phase. The luteal phase is from day 1 to day 15 of the estrous cycle and the late luteal phase overlaps the early follicular phase on day 16–17. The follicular phase lasts for only 3–4 days and estrus is on day 20 or 21.[22] The concentration of P in peripheral blood serum, ovarian venous blood serum, follicular fluid, and perifollicular blood on various days of the estrous cycle is shown in Figure 45–1A. The secretory activity of the corpora lutea of the ovary could be judged from the concentration of P in ovarian venous blood and in peripheral blood. In the active luteal phase (days 5–12) the peripheral blood serum contained 19.0 ± 2.4 ng/ml, which decreased to 7.9 ± 1.3 ng/ml in the late luteal phase and was further significantly reduced to 2.5 ± 0.7 ng/ml ($p < 0.01$) during the follicular phase. A similar pattern of P concentration was observed in ovarian venous blood. Ovarian secretion of P was higher during the luteal phase, 97.8 ± 12.0 ng/ml, but significantly declined to 4.0 ± 0.7 ng/ml ($p < 0.001$) during the follicular phase, then increased to 10.3 ± 2.0 on the day of estrus, which suggests that the granulosa cells could have started luteinization prior to ovulation. Interestingly, follicular fluid on days 5–12 of the estrous cycle contained high amounts of P, though the follicles were small (2–3 mm) in size. In the active luteal phase, between 5 and 12 days of the estrous cycle, after the removal of follicular fluid from the follicle antrum, in the first minute collection of perifollicular blood, the concentration of P was 3937.2 ± 710.6 ng/ml and did not vary significantly in the subsequent four blood samples collected at 1 min intervals over a period of 8 min (Figure 45–1B). It appears that such a high amount of the progesterone is produced in the adjacent active corpora lutea, mixes with the venous blood, and enters the follicular antrum. On days 14–15 of the estrous cycle the concentration of perifollicular blood P was lower (848.0 ± 205.7 ng/ml) and did not vary significantly in the subsequent four blood samples obtained at 1 min intervals over a period of 8 min. In the late luteal phase, as well as in the follicular phase, the P content further declined. On the day of estrus, when the follicle size was about 0.8–1.2 cm, though the follicular fluid contained a high concentration of P, the perifollicular blood level of P was at its lowest (Figure 45–1B).

Δ^4-ADROSTENEDIONE The concentrations of A in peripheral blood serum, ovarian venous blood serum, follicular fluid, and perifollicular blood samples are given in Figure 45–2A. The amount of circulating A during days 5–12, i.e., during the active luteal phase, was 2.5 ± 0.3 ng/ml, which was similar to that in the late luteal phase. It decreased slightly during the follicular phase and on the day of estrus. In the ovarian venous blood serum, the concentration of A in the early luteal phase was 6.0 ± 0.7 ng/ml and rose to 16.1 ± 3.4 ng/ml at the late luteal–early follicular phase; this venous concentration of A did not change during the short follicular phase and significantly declined at estrus. Follicular fluid from small follicles during the active luteal phase contained an average concentration of 81.0 ± 11.0 ng/ml of A, which further declined in the late luteal or early follicular phase to 29.2 ± 3.4 ng/ml. During the active follicular phase when the follicles are large in size (0.8 ± 1.2 cm), the concentration of A in the follicular fluid rose to 163.6 ± 37.9 ng/ml and varied considerably among animals. The high concentration may be due to the presence of a large number of active theca cells surrounding the follicle, and as estrus approached, greater amounts of A were converted to E_1 and E_2. Therefore, the amount of A in the follicular fluid was reduced significantly on the day of estrus.

During the active luteal phase when the follicles were small and corpora lutea were proliferative and highly vascularized, the concentration of A in the perifollicular blood was 88.5 ± 16.3 ng/ml, but successively declined in the subsequent four 1 min blood

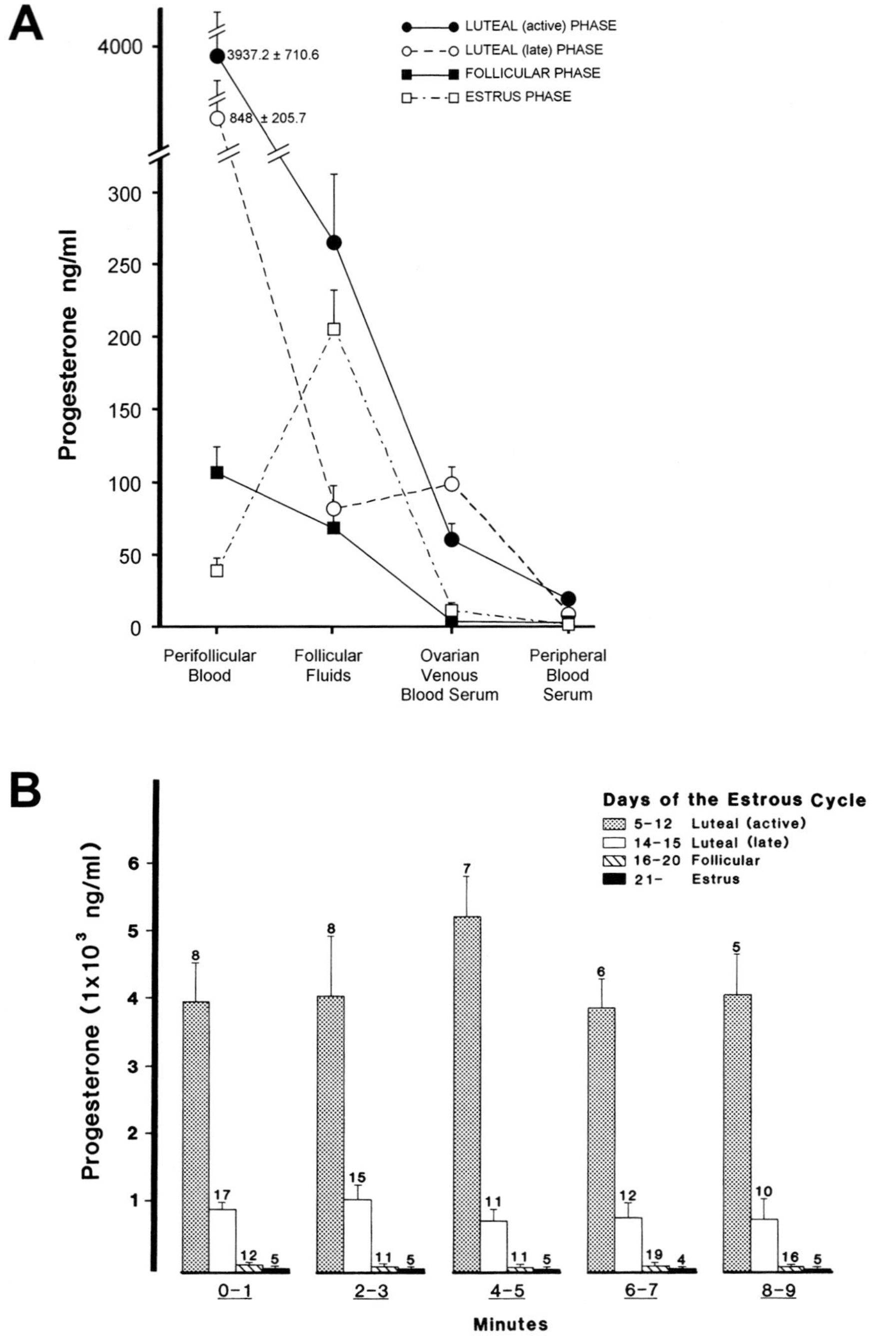

Figure 45–1. Progesterone. (A) Concentration of P in the peripheral blood, ovarian venous blood, follicular fluid, and perifollicular blood in various phases of the estrous cycle of pig. In the luteal (active) and luteal (late) phases the concentration of P in perifollicular blood was 3937.2 ± 710.6 (SE) and 848 ± 205.7 (SE) ng/ml, indicating that the source was the corpora lutea. Even in follicular fluid in the active luteal phase it was 264.5 ± 48.0 (SE) ng/ml, which appears to be from the corpora lutea. At this stage, the follicles are 2–3 mm in size and contain a small number of theca and granulosa cells. In the follicular phase, the concentration of P in the follicular fluid was 67.9 ± 16.9 (SE) ng/ml, which could be synthesized within the follicle. In ovarian venous blood, the concentration of P was then <97.8 ± 12.0 (SE) ng/ml in the luteal phase and peripheral blood contained <10.3 ± 2.0 (SE) ng/ml. (B) The perifollicular blood samples were drawn from the follicle continuously for 1 min at 1 min intervals. In the active luteal phase, the concentration of P was 3937.2 ± 710.6 (SE) ng/ml and remained at high levels in all subsequent four samples. The concentration of P was reduced to 848.0 ± 205.7 (SE) ng/ml and remained constant in all the remaining samples. The concentration of P decreased to <100 mg/ml in the follicular phase and was only <28.4 ± 7.2 (SE) ng/ml during estrus. The number above each column indicates the number of animals used to collect the perifollicular blood samples.

samples. However, when the follicular size increased during the late luteal phase the A content of perifollicular blood did not change significantly in five 1 min samples drawn consecutively (Figure 45–2B). However, on the day of estrus the concentration of A in the perifollicular blood was the lowest. This phenomenon could be due to the beginning of the process of luteinization of the theca cells surrounding the follicle.

TESTOSTERONE The concentration of T in the peripheral blood of mature gilts was approximately one-tenth the concentration of A, the highest being at the time of estrus and the lowest during the late luteal phase. During the estrous cycle, the overall concentration of T in peripheral blood serum was 0.10 ± 0.01 ng/ml on days 14–15, and on the day of estrus it was 0.51 ± 0.13 ng/ml. The concentration of T in ovarian venous blood did not vary significantly during the entire estrous cycle; however, a higher level of T was observed during the luteal phase (5–12 and 14–15 days of the estrous cycle). Similar differences in the concentration of T were also observed in the follicular fluid during the estrous cycle, the highest concentration (17.81 ± 3.2 ng/ml), occurring during days 5–12 and the lowest (6.78 ± 0.64 ng/ml) at the time of estrus (Figure 45–3A). The concentration of T in all five samples of perifollicular blood was approximately the same on days 5–17 of the estrous cycle and decreased at estrus. The pattern of T concentrations was similar to that of A except for the follicular fluid (Figure 45–3B).

ESTRONE The concentration of E_1 in peripheral blood serum in the active luteal phase of the estrous cycle was 0.018 ± 0.006 ng/ml, but increased through the late luteal and early follicular phase

Figure 45–2. Δ^4-Androstenedione. (A) Concentration of A in the perifollicular blood, follicular fluid, ovarian venous blood serum, and peripheral blood serum during the estrous cycle of pig. Follicular fluid in the follicular phase contained 163.6 ± 37.9 (SE) ng/ml A, which did not equilibrate with ovarian venous blood; its concentration in ovarian venous blood serum was only 16.0 ± 6.8 (SE) ng/ml. The concentration in the peripheral blood serum was <1.8 ± 0.3 (SE) ng/ml. The presence of A in perifollicular blood drawn from the follicle after removal of follicular fluid suggests that the perifollicular blood drawn from the follicle may secrete A, which is contaminated in perifollicular blood, or it may be partially secreted by the active luteal tissue. (B) Concentration of A in perifollicular blood during different days of the estrous cycle in pig. The concentration of A increased gradually from 84.3 ± 4.0 (SE) ng/ml to the fifth sample of the perifollicular blood reaching 159.6 ± 33.1 (SE) ng/ml, suggesting that in the follicular phase the perifollicular blood mainly comes from the follicular vasculature. Since the perifollicular blood in both active and late luteal phases contained A, A must also be the product of the corpora lutea. The number above each column indicates the number of animals used to collect the perifollicular blood samples.

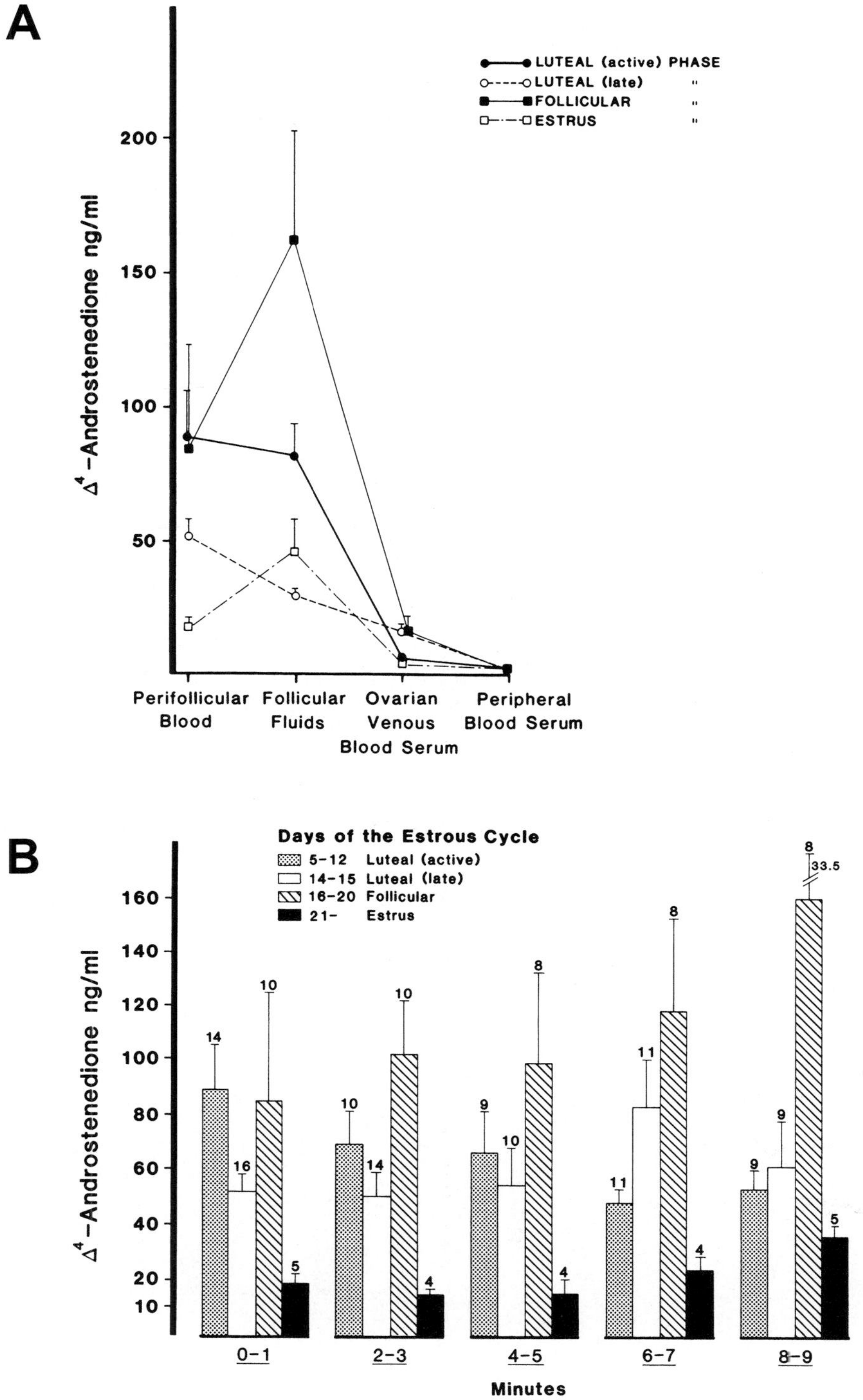

(Figure 45–4A). The amount of E_1 in the ovarian venous blood serum was about 10 times higher than in peripheral blood and did not significantly change during the luteal or early follicular phases of the estrous cycle. On the day of estrus just prior to ovulation, the concentration of E_1, similar to E_2, was low in the ovarian venous blood, follicular fluid, and perifollicular blood (Figure 45–4A and B). The concentration of E_1 in the follicular fluid remained constant during the luteal and follicular phases. The E_1 content in perifollicular blood was lower during the active luteal phase (days 5–12) than during the late luteal or early follicular phases and decreased to its lowest level on the day of estrus (Figure 45–4B).

ESTRADIOL-17β The overall change in the pattern of E_2 concentration in peripheral blood serum did not correspond to the changes of E_2 levels in ovarian venous blood serum or the E_2 content of follicular fluid. The peripheral blood concentration of E_2 varied between 0.09 ± 0.01 ng/ml and 0.31 ± 18 ng/ml throughout the estrous cycle of the gilts, and was higher than E_1. The concentration of E_2 in ovarian venous blood serum in the early follicular phase was higher than that in the luteal phase and fell to its lowest level (0.28 ± 0.1 ng/ml) on the day of estrus (Figure 45–5A). Similarly, the concentration of E_2 in the follicular fluid varied between 13.04 ± 3.07 and 21.29 ± 4.18 ng/ml during the estrous cycle and significantly decreased to 6.93 ng/ml on the day

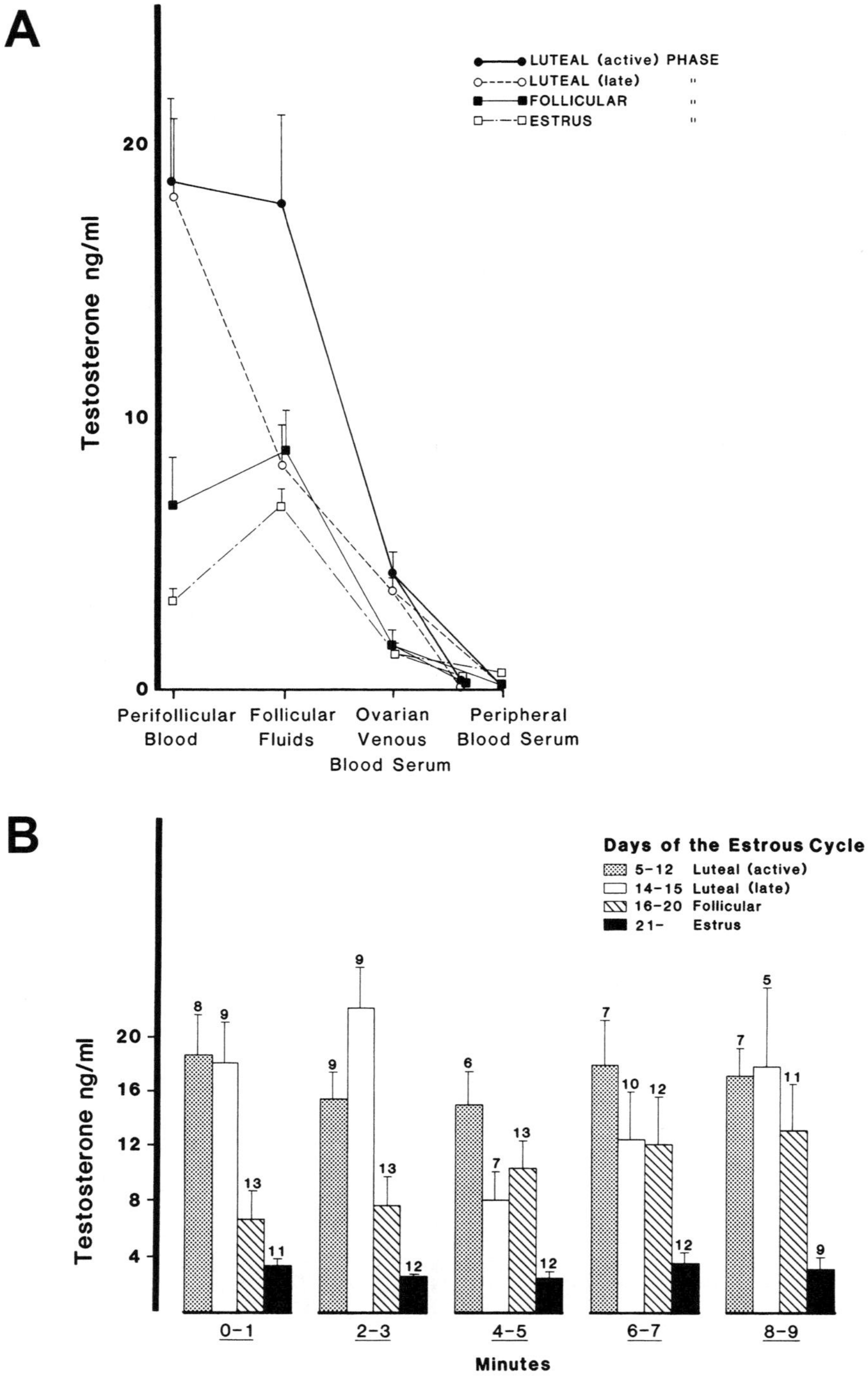

Figure 45–3. Testosterone. (A) Concentration of T in the perifollicular blood, follicular fluid, ovarian venous blood serum, and peripheral blood serum. The overall concentration of T in the perifollicular blood, follicular fluid, and ovarian venous blood serum is approximately 10% of the concentration of A in these fluids. This suggests that A is a preferred substrate for estrogen synthesis in both follicle and luteal tissue. However, the presence of T in the perifollicular blood in amounts such as 18.63 ± 3.03 (SE) ng/ml and 18.10 ± 2.84 (SE) ng/ml in luteal active and late phases indicates that T is also synthesized in luteal tissue, which may serve as a substrate for the $P450_{arom}$ enzyme. (B) Concentration of T in the perifollicular blood in various stages of the estrous cycle of pig. The amount of T in the first minute samples of perifollicular blood varied from 18.63 ± 3.03 (SE) ng/ml in the early luteal phase to 6.78 ± 0.64 (SE) ng/ml in the estrous phase, and it continued to be present in all the second, third, fourth, and fifth samples of perifollicular blood. This suggests that the source of testosterone is from both luteal tissue and the follicular vasculature. In follicular and estrous phases, its main source must be from the follicular vasculature, probably from theca cells and interstitial tissue. The number above each column indicates the number of animals used to collect the perifollicular blood samples.

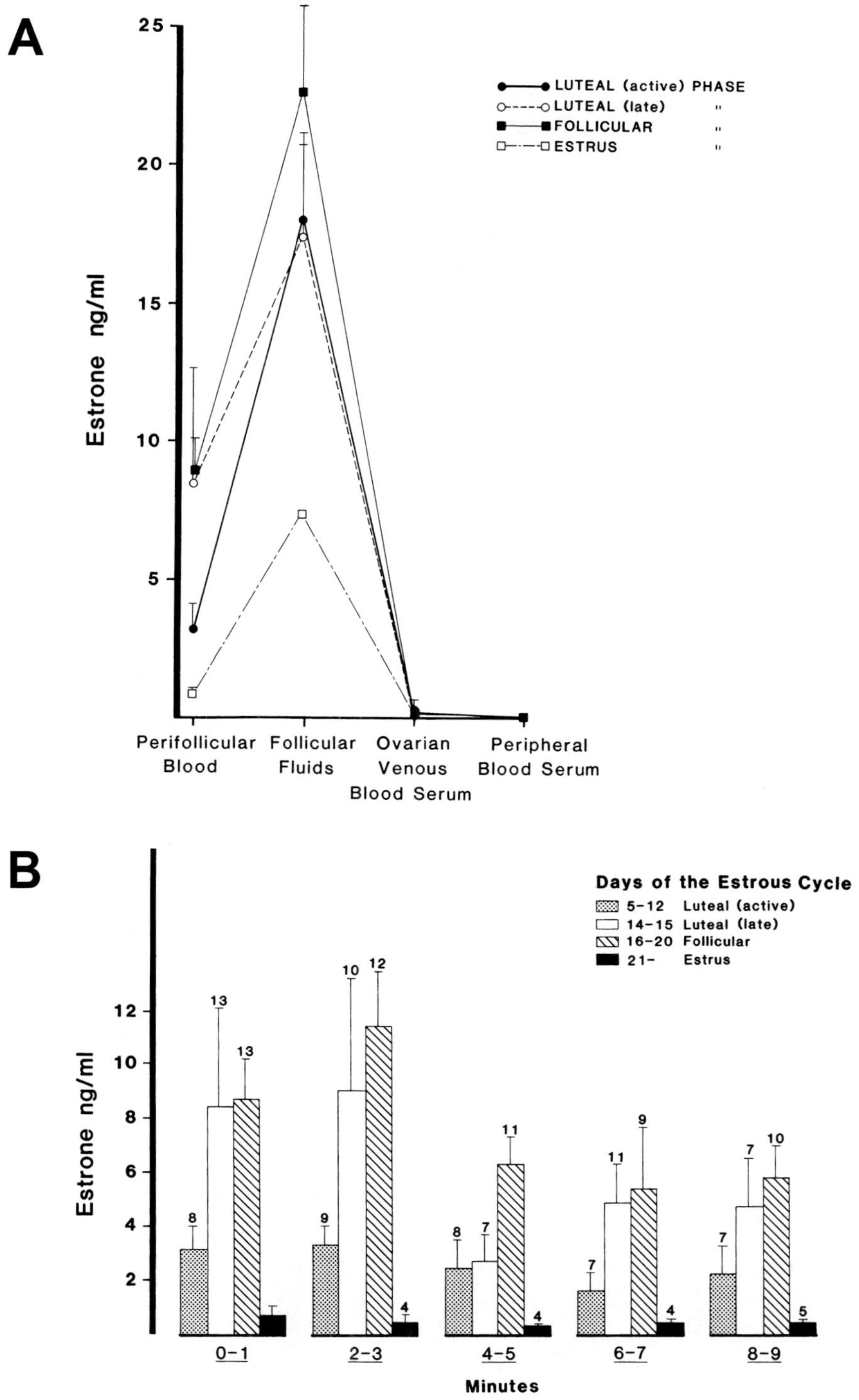

Figure 45–4. Estrone. (A) Concentration of E_1 in the perifollicular blood, follicular fluid, ovarian venous blood serum, and peripheral blood serum during the various stages of the estrous cycle in pig. Since E_1 is synthesized mainly by the granulosa and theca cells, the highest concentrations were observed in the follicular fluid of all the phases of the estrous cycle. The amount of E_1 in the follicular phase was 22.94 ± 3.0 (SE) ng/ml and in the ovarian venous blood serum it was only 0.23 ± 0.11(SE) ng/ml. In the peripheral blood serum of all the estrous cycle phases the concentrations of E_1 were <0.28 ± 0.01 (SE) ng/ml. Its presence in the perifollicular blood during the late luteal and follicular phases, 8.44 ± 4.21 (SE) and 8.68 ± 1.50 (SE) ng/ml, respectively, suggests that its origin is mainly from the follicular vasculature. (B) Concentrations of E_1 in five consecutive perifollicular blood samples collected for 1 min at 1 min intervals during each phase of the estrous cycle. High concentrations of E_1 were present in the perifollicular blood of the late luteal and follicular phases, indicating that E_1 is synthesized by both luteal and follicular granulosa and theca cells. The number above each column indicates the number of animals used to collect the perifollicular blood samples.

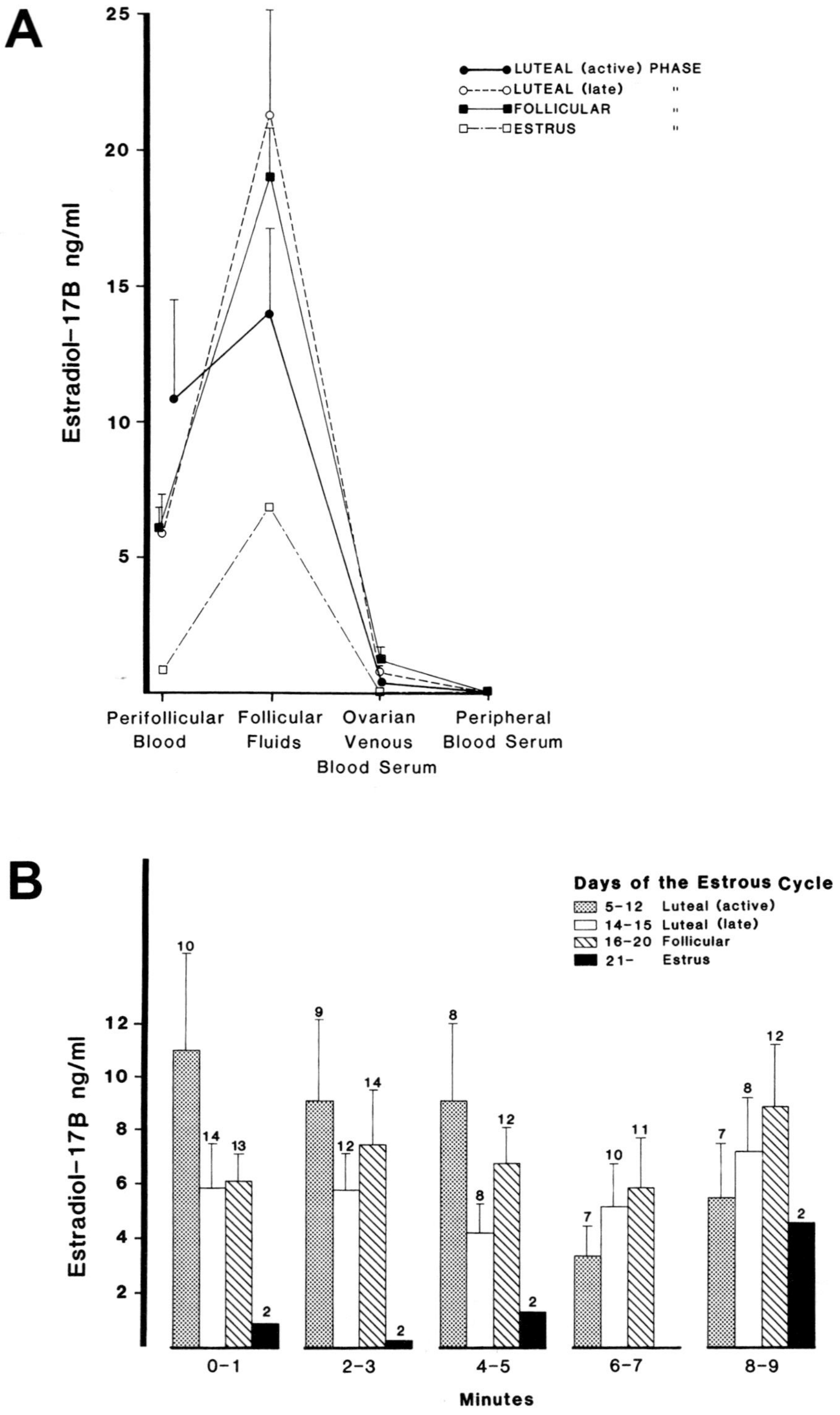

Figure 45–5. Estradiol-17β. (A) Concentration of E_2 in the perifollicular blood, follicular fluid, ovarian venous blood serum, and peripheral blood serum. The amount of E_2 [21.29 ± 4.18 (SE), n = 24] in the follicular fluid of the late luteal phase was not significantly different from the amount of E_2 [18.96 ± 1.80 (SE), n = 21] in the follicular fluid of the follicular phase ($p = 0.63$). This may be due to the growing follicles in the late luteal phase, in which both the theca and granulosa cells are actively synthesizing E_2. The concentrations of E_2 in the perifollicular blood of both the late luteal and follicular phases were nearly equal, suggesting the perifollicular blood may be mainly from the follicular vasculature. The amount of E_2 in ovarian venous blood was <1.2 ng/ml in all phases of the estrous cycle. (B) E_2 was determined in perifollicular blood samples drawn for 1 min with 1 min intervals in all the four phases, as described for the estrous cycle of pig. The first perifollicular blood sample of the active luteal phase had 10.88 ± 3.6 (SE) ng/ml of E_2, indicating that a part of these values may be contributed by the active luteal tissue in addition to theca and granulosa cells of the follicle. During the estrous phase, the values of E_2 were <1.0 ng/ml, except for the fifth minute sample which was 4.52 ng/ml. The number above each column indicates the number of animals used to collect the perifollicular blood samples.

of estrus. In perifollicular blood E_2 concentrations during the luteal and early follicular phases varied between 3.31 ± 1.12 ng/ml and 10.88 ± 3.59 ng/ml and there was no significant change in the concentration of E_2 in the samples taken at 1 min intervals; however, at days 5–12 of the estrous cycle a sequential reduction from 10.88 ± 3.59 to 3.31 ± 1.12 ng/ml was observed from the first to the fourth blood sample (Figure 45–5B). Such a significant variation was not observed during the early follicular phase. The concentration of E_2 in perifollicular blood was significantly reduced on the day of estrus and did not vary during the time of collection of five consecutive samples of perifollicular blood (Figure 45–5B).

DISCUSSION

The presence of various steroids including progestins, androgens, and estrogens in follicular fluid has been demonstrated by a number of investigators.[26] In these studies the follicles were obtained from mares, cows, sows, and ewes, and most of the time they were collected at slaughterhouses. It has been shown that the data obtained from slaughterhouse material do not agree with those obtained with follicular fluid at the time of laparotomy of sows with well-monitored estrous cycles.[17] Usually, these studies measured fewer steroids than measured in the present report. It is still not known if the various steroids secreted by the ovary are always synthesized by individual cell types or whether a particular precursor is synthesized in one compartment and transferred to another for further metabolism.

Is P present in the follicular fluid in the luteal phase synthesized by the granulosa and theca cells surrounding the follicular fluid, or is it also a fraction of a large quantity of P synthesized by the corpus luteum? The present data reveal that the follicular fluid collected from follicles at days 5–12 of the estrous cycle contained higher concentrations of P than did follicular fluid collected on days 14–15, days 16–20, or on the day of estrus. The peripheral blood and ovarian venous blood levels of P at days 5–12 were also higher than at any time during the estrous cycle. Interestingly, all five of the consecutive perifollicular blood samples collected during the active luteal phase when the corpora lutea are proliferative and highly vascularized (5–12 days of the estrous cycle) contained very high concentrations of P, which varied between 3.9 and 5.2 µg/ml. The concentration of P in perifollicular blood did not decrease during the 8-min withdrawal of perifollicular blood, suggesting that P may be synthesized continuously at a high rate and secreted into the blood. It has been shown that in the newly formed corpus luteum of the ewe P values reach about 2 µg /ml of luteal fluid.[27] The P values in perifollicular blood of the gilt in the active luteal phase (days 5–12) appear to be slightly higher than those in the luteal fluid of the ewe. These high values would suggest that in perifollicular blood P is produced by the corpora lutea. If the ovarian blood flow in the luteal phase is assumed to be similar to the ewe, approximately 20 ml/min,[28] the production rate of P by the ovary at the peak activity of the corpora lutea could be approximately 4.6 mg/h. However, this value may be far from accurate as the blood volume withdrawn per minute in the present study varied greatly, and the first of the five consecutive samples of perifollicular blood drawn is most likely contaminated with follicular fluid.

Second, if the steroids in sample 4 or 5 had been lower than in earlier samples, it could be assumed that the steroids in the blood had been from intracellular spaces and not from an active secretory process. The concentrations of P in all five consecutive samples were similar, which suggested that P in the perifollicular blood is likely to be a continuous secretory process (Figure 45–1B.) The concentration of P observed during the active stage of the corpus luteum fell by about 20% on days 14 and 15 of late luteal phase of the estrous cycle, indicating that the corpus luteum at this stage was less active in the synthesis of P; it appears that P in perifollicular blood was still from the luteal tissue. The level in perifollicular blood decreased during the follicular phase when the luteal tissue was degenerated and most of the P must have been secreted by theca and mainly by the granulosa cells. Ovarian venous blood, peripheral blood, and follicular fluid during the follicular phase also contained very low values compared to the concentration of P in these fluids during the luteal phase. This suggests that theca cells are capable of producing and secreting low levels of P. It is interesting that the P content in the follicular fluid on the day of estrus had sharply increased from its lowest level, but this was not true for P in the perifollicular blood. All the ovarian steroid concentrations declined to their lowest values just prior to ovulation. It is not known whether this phenomenon is due to a sudden increase in the follicular fluid volume so that the contents of the follicular fluid are diluted, or whether the cellular composition of the follicle is in the process of changing to new types of cells with decreased steroid output during the transitory phase, or if there is a temporal downregulated response of the cells to elevated, cyclic surges of gonadotropins. The luteal tissue has lower $P450_{17\alpha}$ and $P450_{c17}$ activity compared to theca cells.[11] These enzymatic changes could explain the lower concentration of A, T, E_1, and E_2, but the drop in concentration of P in estrus remains to be explained.

The role of P in the control of ovarian follicular growth and development remains unclear. It was observed that exogenously administered P to 27-day-old rats increased the human chorionic gonadotropin (hCG) response to E_2 accumulation and aromatase activity in small antral follicles and can stimulate the growth of these follicles to the preovulatory stage in intact prepubertal rats.[29] Since the theca and granulosa cells of antral follicles have not developed the receptors for LH and FSH in sufficient number, P may serve as a precursor for C_{19} and C_{18} steroids and may not act as a direct stimulant to steroidogenesis or follicular growth. On the other hand, it has been shown that P inhibits $P450_{c17}$ enzymes in rat testicular homogenates, but may not greatly reduce the formation of C_{19} because P itself is metabolized to form C_{19} steroids by the action of the $P450_{17\alpha}$ and $P450_{c17}$ enzyme system without accumulating 17α-OHP.[11] The synthetic P analog (R5020) inhibited the FSH binding to granulosa cells and estrogen synthesis was reduced.[30] It was demonstrated that the reduction in E_2 synthesis due to R5020 was not the result of an inhibitory effect of R5020 on the aromatase enzyme but the result of an inhibition of the action of FSH on granulosa cells. It was suggested that R5020 and P inhibited the binding of FSH to granulosa cells by impairing the receptor sites for FSH. The actual mechanism of the reduction of FSH receptor sites is not explained. Furthermore, it was observed that due to the reduction of FSH receptor sites on granulosa cells, induction of the receptors for LH/hCG was inhibited. This effect is specific to P because 17α-OHP and 20α-OHP had no effect on FSH binding to granulosa cells or induction of the LH/hCG receptor.[31] This may be why estrogen synthesis is reduced in the antral follicle due to a high concentration of P that seeps from an adjacent corpus luteum resulting in inhibition of growth.

Attempts have been made to explain the inhibitory effect of P on folliculogenesis in the monkey by luteectomy in one ovary, thereby raising FSH secretion, and by supplying exogenous P. It was observed that the monkeys that did not receive P had an expected LH surge at day 14 ± 0.8 after luteectomy. In contrast, in monkeys implanted with Silastic capsules containing P for 10 days, the LH surge occurred on day 25 ± 2.7 after luteectomy and when these capsules were removed, they ovulated on day 14 ± 2.7. These observations suggest that P has a regulatory action on folliculogenesis and on the maturation of follicles.[32] Exogenous administration of 17α-OHP after luteectomy did not have any effect on the day of LH surge or ovulation, indicating that the delay of LH surge or follicular growth was specifically due to P. The average concentration of A in peripheral blood ranged between 1.6 and 2.5 ng/ml and did not show significant changes during the estrous cycle, though a trend for slightly higher values is shown during the luteal phase (Figure 45–3A). The ovarian venous blood and peripheral blood levels of A do not show a close correlation. This could be due to a significant amount of A secretion by the adrenal cortex. Such findings were also previously reported.[17]

The follicular fluid concentration of A showed two significant peaks, one during the active luteal phase (days 5–12) and the other at the follicular phase (days 16–20), and the level of A was much higher during the follicular than the luteal phase. The values of A were lower during the late luteal phase and also on the day of estrus (Figure 45–2). In another study it was observed that in the cycling pig during early and late estrus the range of concentration of E_2 was 6–848 ng/ml, of A was 7–1001 ng/ml, of T was 1–272 ng/ml, and of P was 69–612 ng/ml in the follicular fluid, which suggested that the concentrations of E_2, A, and T in the follicular fluid were correlated significantly with the concentration of LH and FSH in circulating blood serum, but P was negatively correlated with other steroids.[33] The decrease in these steroid concentrations in the follicular fluid was associated with the reduced expression of $P450_{c17}$ in theca cells and reduced expression of $P450_{arom}$ in theca and granulosa cells of the follicle.

During the follicular phase, if it extends from day 16 to day 21, the concentrations of E_2 and T were higher on day 20 than on day 21, but the concentration of P was 497.2 ± 175.8 ng/ml on day 21. This high concentration of P could be the result of the theca and granulosa cells having started luteinization just prior to ovulation. The mean $P450_{arom}$ activity (ng E_2/2 h/follicle) was higher on day 16 (7.3) and day 18 (6.2) than on day 20 (2.6) and day 21 (0.2), suggesting that the granulosa cells were becoming luteinized by day 21.[34] In another study, ovaries were collected in a local abattoir and assigned to functional stages of the estrous cycle by checking for corpora hemorrhagica, corpora lutea, corpora albicantia, and developing follicles and by determining their size, color, and number. The luteal and follicular phases were each subdivided into early, mid, and late stages. It was not determined whether the sows were cycling normally. Steroids were measured in the combined follicular fluid of two follicles of the same ovary. Androgens and estrogens were at their lowest level during mid and late luteal phases, whereas progesterone was about 250 ng/ml of follicular fluid. All the steroids were at their highest level during late follicular phase; P was about 550 ng/ml, A was about 375 ng/ml, T was about 100 ng/ml, and E_1 and E_2 were about 80 and 200 ng/ml, respectively. These steroids decreased to 10% or less during the follicular ovulatory stage.[35]

When P, A, and E_2 were measured in the follicular fluid of cycling pigs in nonatretic follicles during the luteal phase, P was about 200 ng/ml in the preovulatory follicle on day 7, which decreased gradually to 100 ng/ml on day 13. The A and E_2 values were also high on day 7 and decreased to their lowest values on day 13. The high concentration of P in the follicular fluid of nonatretic follicles on day 7 may be difficult to attribute to steroid-producing theca, interstitial, and granulosa cells.[36] Furthermore, in altrenogest-treated noncycling pigs, the 3β-hydroxysteroid dehydrogenase in the follicular tissue at day 7 was below detectable levels.[20] These studies indicate that P must be entering the follicle from the perifollicular blood of the surrounding vasculature. The higher levels of A and E_2 in the follicular fluid on day 7 could be synthesized by the active theca and granulosa cells. The levels of A in the follicular fluid of women vary during the menstrual cycle.[37] The high levels of A in the follicular fluid during these stages of follicular growth might serve as a precursor for synthesis of E_1 and E_2. The perifollicular concentration of A was significantly higher during the follicular phase than during other stages of the estrous cycle. At this period the vascularized theca cells are active in producing A, which appears in perifollicular blood and in follicular fluid. Thus, it reaches the granulosa cell where aromatization to E_2 can occur. The high concentration of A in perifollicular blood also suggests that A is a more important C_{19} steroid precursor than T for the synthesis of estrogens. Diffusion of A from the vasculature of theca and interstitial tissue of the follicle to perifollicular blood may also take place; however, granulosa cells have a lower $P450_{c17}$ enzyme activity than do theca cells.[11,38]

Although conclusive data regarding the rate of transmission of steroids from blood to follicle and from follicular fluid into circulating blood are not available, based on the transfer of radioactive steroids injected into the follicle, the passage of A from the follicle into the circulating blood is slower than the passage of E_2 from the follicle into the circulation.[25] The amount of T in the follicular fluid was higher in the active luteal phase than in the late luteal or follicular phases. It was lowest on the day of estrus. The concentration of T in the follicular fluid corresponded approximately to the amount of T in perifollicular blood. The overall level of T during the luteal phase was about 20% of the level of A in the ovarian fluids, and similar to the concentration of A in the follicular fluid and perifollicular blood, it did not peak during the follicular phase (days 16–20 of the estrous cycle), suggesting that A may be the principal androgen secreted by thecal and stromal tissue and may serve as a precursor for estrogen synthesis. Furthermore, thecal tissue from healthy human follicles[13,39] or from porcine ovarian follicles[15] cultured *in vitro* produced higher amounts of A than T. The low values of T in perifollicular blood may be caused by enzymatic changes occurring in theca cells, which become luteinized after ovulation. In ovarian venous blood, the level of T was much lower than that in perifollicular blood and tended to be higher during the luteal phase than the follicular phase. The concentration of T in peripheral blood did not change significantly during the estrous cycle. The concentration of T in perifollicular blood was significantly higher in the luteal phase than the concentration in follicular fluid. However, levels of E_1 and E_2 were higher in the follicles of the follicular phase, which is also true in humans[37,40] and in monkey ovarian follicles.[41] Although granulosa cells have a very active $P450_{arom}$ enzyme system, large amounts of estrogens are produced *in vitro* only when the granulosa cells are mixed with theca cells.[14] The concentrations of E_1 and E_2 were similar in the follicular fluid at all stages of the estrous cycle of the gilts; however, the level of E_2 in perifollicular blood was two to three times higher during the active luteal stages. Luteal tissue has an active $P450_{arom}$, which could convert A and T into E_2. In addition to luteal tissue the vascularized theca cells could also contribute to E_2 synthesis and release into perifollicular blood. On the day of estrus levels of both E_1 and E_2 were very low, similar to that of P and androgens in ovarian venous blood, follicular fluid, and perifollicular blood. The high concentrations of E_2 and other steroids such as P and A in follicular antral fluid probably play an important role locally in regulating follicular development. Estrogens are essential for inducing the receptor sites for FSH first and then E_2; FSH increases the LH receptor sites on granulosa cells,[42,43] thereby increasing the steroid-producing enzyme activity of granulosa cells.[44,45]

The follicular fluid and perifollicular blood are in very close proximity. Substances injected in the general circulation were found in the follicular fluid. Peptides such as insulin, FSH, and LH are secreted elsewhere, but they enter the follicle and bind to the receptors of granulosa, theca, and interstitial cells. Although the follicular compartment containing the granulosa cell layer and cumulus is avascularized, the majority of the constituents of the follicular antral fluid are derived from serum. The barrier that separates the antrum from the circulation has properties that admit most blood proteins into the fluid.[45] However, when the ratio of the quantities of steroids in the follicular fluid to the quantities of steroids in the perifollicular blood was calculated, none of the steroids showed a ratio of 1 (unity) at various phases of the estrous cycle (Figure 45–6). These ratios indicate that the steroids in the follicle are not at equilibrium with the steroids of the perifollicular blood. It appears that at the follicular–blood barrier there is a selective process of exchanging steroids from the surrounding vasculature. E_1 and E_2 have positive ratios in all phases of the estrous cycle while P, A, and T have negative ratios during luteal phases. That is, during all the estrous phases the quantities of E_1 and E_2 are higher in the follicular fluid and the quantities of P, A, and T are lower in the follicular fluid than in the perifollicular blood.

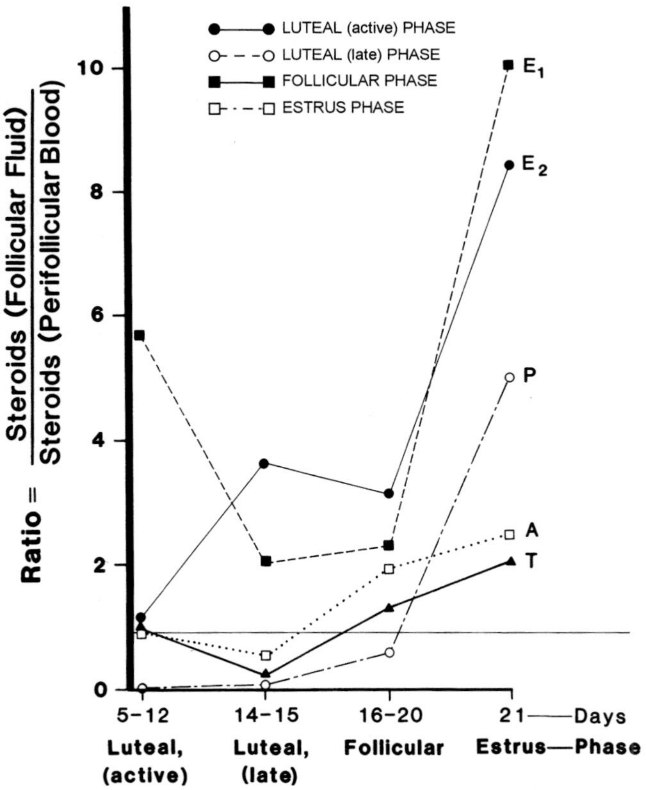

Figure 45–6. The ratios of steroids in the follicular fluid to the corresponding steroids in the first 1 min collection of perifollicular blood. During active follicular growth such as in the late luteal stage of the estrous cycle, the ratio did not reach 1. This suggests that none of the steroid was in an equilibration state. It demonstrates that at the follicular–blood barrier a selective process of exchange exists. E_1 and E_2 have a positive ratio at all the stages of the estrous cycle, indicating high $P450_{arom}$ activity in the follicle, whereas P, A, and T have negative ratios except during estrus.

CONCLUSIONS

This study was undertaken to observe steroid dynamics in the ovary during the estrous cycle of regularly cycling gilts. Forty cycling gilts were examined for three consecutive cycles to select regularly cycling gilts for this study. On the appropriate day of the cycle needed for the study laparotomy was done and the peripheral blood, ovarian venous blood, follicular fluid, and five samples of perifollicular blood were collected at 1 min intervals. P, A, T, E_1, and E_2 were measured in the blood serum samples, follicular fluid, and perifollicular blood using specific RIA methods.

Concentrations of P in peripheral blood, ovarian venous blood, follicular fluid, and perifollicular blood (blood from the vasculature surrounding the follicle) during the active luteal phase of gilts were 19.0 ± 2.4, 59.2 ± 12.7, and 3937 ± 710 ng/ml, respectively. These values were significantly lower in the follicular phase. The concentration of P in perifollicular blood in the late luteal phase was 848.0 ± 205.7 ng/ml, which was reduced to approximately one-tenth of this value in the follicular phase. Conversely, concentrations of A in the follicular fluid from the follicles in the follicular phase contained the highest concentration (153 ± 37.9 ng/ml) of all the stages of follicular growth. During this stage the perifollicular blood also contained a higher concentration of A than that observed at any other stage of the estrous cycle. On the day of estrus, the level of A in the follicular fluid as well as in ovarian venous blood was at its lowest. During the follicular phase the follicular fluid contained 8.74 ± 1.15 ng/ml of T, which did not significantly vary in the late luteal phase and was about twice as much as that in the follicular phase. Higher concentrations of T were also found in perifollicular blood in the luteal phase than in the follicular phase.

The high level of E_1 (22.94 ± 2.95 ng/ml) and E_2 (18.96 ± 1.80 ng/ml) in the follicular fluid during the follicular phase may be due to aromatization of A, which is also present in high concentrations in the follicular fluid and is in contact with granulosa cells. On the day of estrus E_1 and E_2 in the follicular fluid were 7.42 ± 1.45 and 6.93 ng/ml, respectively, which may be prior to the onset of the preovulatory surge. The lower levels of E_1 and E_2 in perifollicular blood than in follicular fluid suggest that the main site of aromatization could be the granulosa cells, with the product then partially equilibrating with the perifollicular blood. The concentration of A and T appeared to be higher or at least similar in these two ovarian fluids indicating that theca may be the main source of A and T. Interestingly, very high concentrations of P in the perifollicular blood suggest its origin in the corpus luteum, which may act as a precursor for C_{19} and C_{18} steroids in the follicle and may also regulate the sequential events that occur during follicular growth. The follicular fluid and perifollicular blood are in very close proximity, but the steroids present in them are not in perfect equilibrium.

ACKNOWLEDGMENTS

Dr. Finnie A. Murray, Jr., Ph.D., Associate Professor, Department of Animal Science, Ohio Agriculture Research and Development Center, Ohio State University, Wooster, Ohio, deserves profound acknowledgment for his professional supervision on cycling and skillful surgery of laparotomy of the gilts. Thanks also to Dr. David S. Knight, Ph.D., Associate Professor, Department of Anatomy and Cell Biology, L.S.U.H.S.C., Shreveport, LA, for reading through the manuscript.

REFERENCES

1. McCoard SA, Wise TH, Ford JJ. Germ cell development in Meisham and White composite gilts. *Anim Reprod Sci* 2003;77(1–2):85–105.
2. Hunter MG, Robinson RS, Mann GE, Webb R. Endocrine and paracrine control of follicular development and ovulation rate in farm species. *Anim Reprod Sci* 2004;82–83:461–477.
3. Adashi EY, Besnick CE, D'Ercole J, Svoboda ME, Van Wyk JJ. Insulin-like growth factors as intraovarian regulators of granulose cell growth and function. *Endocr Rev* 1985;6:400–420.
4. Hammond JM, Mondschein JS, Samaras SE, Smith SA, Hagen DR. The ovarian insulin-like growth factor system. *J Reprod Fertil Suppl* 1991;43:199–208.
5. Hiller SG, Miro F. Inhibin, activin and follistatin, potential roles in ovarian physiology. *Ann NY Acad Sci* 1992;687:29–38.
6. Short RV. Steroids in follicular fluid with the corpus luteum of the mare. A two cell type theory of ovarian steroid synthesis. *J Endocrinol* 1962;24:59–63.
7. Mahajan DK, Samuels LT. Biosynthesis of steroids by different ovarian tissues. *Fed Proc* 1963;22:531.
8. Ryan KJ, Petro Z, Kaizer J. Steroid formation by isolated and recombined ovarian granulosa and theca cells. *J Clin Endocrinol Metab* 1968;28:355–358.
9. Fortune JE, Armstrong DT. Hormonal control of 17β-estradiol biosynthesis in proestrus rat follicles: Estradiol production by isolated theca versus granulosa. *Endocrinology* 1978;102:227–235.
10. Ryan KJ, Short RV. Formation of estradiol by granulosa and theca cells of the equine ovarian follicle. *Endocrinology* 1965;76: 108–114.
11. Mahajan DK, Samuels LT. Inhibition of 17,20(17-hydroxyprogesterone)-lyase by progesterone. *Steroids* 1974;25:217–228.
12. Tsang BK, Armstrong DT, Whitfield JF. Steroid biosynthesis by isolated human ovarian follicular cells in vitro. *J Clin Endocrinol Metab* 1980;51:1407–1411.
13. McNatty KP, Makris A, DeGrazia C, Osathanondh R, Ryan KJ. Steroidogenesis by recombined follicular cells from the human ovary in vitro. *J Clin Endocrinol Metab* 1980;51:1286–1292.
14. Batta SK, Wentz AC, Channing CP. Steroidogenesis by human ovarian cell types in culture: Influence of mixing cell types and effect of added testosterone. *J Clin Endocrinol Metab* 1980;50:274–279.
15. Evans G, Dobias M, Kinmg GJ, Armstrong DT. Estrogen, androgen, and progesterone biosynthesis by theca and granulosa of preovulatory follicles in the pig. *Biol Reprod* 1981;25:673–682.
16. Mancini RE, Vilar O, Hendrich JJ, Davidson OW, Alvarex B. Transference of circulating labeled serum proteins to the follicle of the rat ovary. *J Histochem Cytochem* 1963;11:80–88.
17. Eiler H, Nalbandov AV. Sex steroid in follicular fluid and blood plasma during the estrous cycle of pigs. *Endocrinology* 1977; 100:331–338.
18. Guthrie HD, Barbars JA, Leighton JK, Hammond JM. Steroidogenic cytochrome P450 enzyme messenger ribonucleic acid and follicular fluid steroids in individual follicles during preovulatory maturation in the pig. *Biol Reprod* 1994;51:465–471.
19. Driancourt MA, Terqui M. Follicular growth and maturation in hyperprolific and large white sows. *J Anim Sci* 1996;74: 2231–2238.
20. Garrett WM, Guthrie HD. Steroidogenic enzyme expression during preovulatory follicle maturation in pigs. *Biol Reprod* 1997;56: 1424–1431.
21. Mahajan DK, Murray FA Jr, Singh KB. Does perifollicular progesterone influence the secretion of sex-steroids by ovarian follicles in the gild? 68th Annual Meeting, Endocrine Society, Anaheim, CA, 1986.
22. Hafez ESE. Sexual behavior in farm animals. In: Hafez ESE, Ed. *Reproduction in Farm Animals*. Philadelphia, PA: Lea & Febiger, 1962:161.
23. Mahajan DK, Murray FA Jr, Little B. Concentration of steroids in peripheral, ovarian venous and perifollicular bloods and follicular fluid of gild. Second International Conference on Pig Reproduction, Columbia, MO, 1985.
24. Mahajan OK, Wahlen JD, Tyler FH, West CD. Plasma 11-deoxycortisol radioimmunoassay for metyrapone tests. *Steroids* 1972;20:609–620.
25. West CD, Mahajan DK, Chavre VJ, Nabors CJ, Tyler FH. Simultaneous measurement of multiple plasma steroids by radio immunoassay demonstrating episodic secretion. *J Clin Endocrinol Metab* 1973;36: 1230–1236.
26. Edwards RG. Follicular fluid. *J Reprod Fertil* 1974;37:189–219.
27. McNatty KP, Gibb M, Dobson C, Thurlye DC, Findley JK. Changes in the concentration of gonadotrophic and steroidal hormones in the antral fluid of ovarian follicles throughout the estrous cycle of the sheep. *Aust J Biol Sci* 1981;34:67–80.
28. Niswender GD, Reimers TJ, Diekman MA, Nett TM. Blood flow: A mediator of ovarian function. *Biol Reprod* 1976;14:64–81.
29. Richard JS, Bogovich K. Effect of human chorionic gonadotropin and progesterone on follicular development in the immature rat. *Endocrinology* 1982;111:1429–1438.
30. Schreiber JR, Nakamura K, Erickson G. Progesterone inhibits FSH stimulated steroidogenesis in cultured rat granulosa cells. *Mol Cell Endocrinol* 1980;19:165–173.
31. Schreiber JR, Nakamura K, Truscello AM, Erickson GF. Progesterone inhibits FSH-induced function of LH receptors in cultured rat granulosa cells. *Mol Cell Endocrinol* 1982;25:113–124.
32. Goodman AL, Hodgen GD. Antifolliculogenic action of progesterone despite hyper secretion of FSH in monkeys. *Am J Physiol* 1982;243:E387–E397.
33. Conley AJ, Howard HJ, Slanger WD, Ford JJ. Steroidogenesis in preovulatory porcine follicle. *Biol Reprod* 1994;51:655–661.
34. Grant SA, Hunter, MG, Foxcroft GR. Morphological and biochemical characteristics during ovarian follicular development in the pig. *J Reprod Fertil* 1989;86:171–183.
35. Babalola AO, Shapizo BH. Correlation of follicular steroid hormone profiles with ovarian fertility in sows. *J Reprod Fertil* 1988;84: 79–87.
36. Guthrie HD, Cooper BS. Follicular atresia, follicular fluid hormones and circulating hormones during the multiluteal phase of the estrus cycle in pigs. *Biol Reprod* 1996;55:543–547.
37. McNatty KP, Baird DT, Bolton A, Chambers P, Corker CS, McLean H. Concentration of estrogens and androgens in human ovarian venous plasma and follicular fluid throughout the menstrual cycle. *J Endocrinol* 1976;71:77–85.
38. Short RV. Ovarian steroid synthesis and secretions in vivo. *Recent Prog Horm Res* 1964;20:303–332.
39. McNatty KP, Makris A, DeGrazia C, Osathanondh R, Ryan KJ. The production of progesterone, androgens and estrogens by granulosa cells, thecal tissue and stromal tissue from human ovaries in vitro. *J Clin Endocrinol Metab* 1979;49:687–699.
40. Sanyal MK, Berger MJ, Thompson, IE, Taymor ML, Hrone HW. Development of Graafian follicles in adult human ovary. I: Correlation of estrogen and progesterone concentration in antral fluid with growth of follicles. *J Clin Endocrinol Metab* 1974;380:828–835.
41. Channing CP, Coudert SP. Contribution of granulosa cells and follicular fluid to ovarian estrogen secretion in the rhesus monkey in vivo. *Endocrinology* 1976;98:590–596.
42. Richards VS, Midgley AR. Protein hormone action: A key to understanding ovarian follicular and luteal cell development. *Biol Reprod* 1976;14:82–94.
43. Channing CP. Progesterone and estrogen secretion by cultured monkey ovarian cell type: Influence of follicular size, serum luteinizing hormone levels and follicular fluid estrogen levels. *Endocrinology* 1980;10Z:342–352.
44. McNatty KP, Sawers RS. Relationship between the endocrine environment within the Graafian follicle and the subsequent rate of progesterone secretion by human granulosa cells in vitro. *J Endocrinol* 1975;66:391–400.
45. Shalgi R, Kraicer P, Pinto M, Soferman N. Proteins in human follicular fluid: The blood-follicular barrier. *Fertil Steril* 1973;24: 429–434.

Models of Behavior — V

Drug Development and Research Models — C

46 Molecular Genetic Approach to Identify Inhibitors of Signal Transduction Pathways

Fission Yeast as a Model System for Drug Discovery

SHUNJI ISHIWATA, TAKAYOSHI KUNO, HIROFUMI TAKADA, ATSUSHI KOIKE, AND REIKO SUGIURA

ABSTRACT

The fission yeast *Schizosaccharomyces pombe* (*S. pombe*) has become a valuable model system to elucidate the mechanisms of basic cellular functions of higher eukaryotes, including cell cycle control, membrane trafficking, and signal transduction. Having the smallest genome size among eukaryotes and with its powerful genetics, this organism is also an excellent model system for drug discovery. In addition, many signaling molecules targeted by the drug or the homologues of disease-linked human genes are highly conserved. We have been studying the signal transduction pathway in fission yeast with special emphasis on calcineurin phosphatase and mitogen-activated protein kinase (MAPK) signal transduction pathways. Our molecular genetic approach, which utilizes a cross-talk between calcineurin and MAPK signaling, has identified several regulators of Pmk1 MAPK, which is a homologue of extracellular signal-regulated kinase (ERK) in mammals. As MAPK signal transduction pathways are one of the most attractive targets for cancer therapy, inhibitors that target this signaling appear to be promising drug candidates for the treatment of cancer. Here, we first give an overview of the use of yeast as a model system for drug discovery and then we introduce our molecular genetic strategy to identify regulators of MAPK signaling and the application of this approach to drug discovery.

Key Words: MAP kinase, Fission yeast, Drug discovery, Pharmacogenomics.

INTRODUCTION

Recent progress in molecular sciences has enabled the identification of molecular alterations in human tumor cells, which frequently involve the conditional or constitutive activation of signaling pathways. Such cellular signaling pathways can operate to promote cancer development. The conservation of several components of these cancer-related signaling pathways with higher eukaryotes combined with the emerging genomics technologies and their sophisticated genetic approaches makes yeast an excellent model to study the molecular mechanisms of tumorigenesis as well as a valuable tool for drug discovery for cancer therapy.

THE CHARACTERISTICS OF THE RESEARCH USING YEAST

ADVANTAGES OF YEAST AS A MODEL FOR DRUG DISCOVERY Many of the genes that are frequently altered in human tumors have structural or functional homologues in model genetic systems, including the yeasts, *Saccharomyces cerevisiae* and *Schizosaccharomyces pombe*, the nematode *Caenorhabditis elegans*, and the fruit fly *Drosophila melanogaster*. The genetic methods have been established in these model organisms. Since fission yeast has the smallest genome size, the study using this yeast allows us to elucidate the complex signaling pathways and provides insight into the molecular mechanisms of fundamental cellular processes.[1]

The characteristic features of the yeast as the model in drug research are summarized as follows.

1. The yeast has essential functions similar to those observed in eukaryotic cell and many of its proteins are conserved in higher eukaryotes. The gene expression profiles in nematode, as a representative of multicellular organisms, and yeast, as a representative of monocellular organisms, are similar in that both have the essential genes observed in eukaryotic cells. They are different from each other in that they do not have the genes observed only in multicellular organisms. This suggests that the functions of the nervous and vascular systems, which are observed only in multicellular organisms, were formed by the arrangement and evolution of the genes in monocellular organisms. Therefore, yeast is an excellent model organism that possesses a minimum and essential genome set, especially useful in drug research.

2. Powerful genetic techniques are available in yeast. Genetic methods such as gene disruption and replacement can easily be

From: *Sourcebook of Models for Biomedical Research*
(P. M. Conn, ed.), © 2008 Humana Press Inc., Totowa, NJ.

carried out by homologous recombination in yeast. Most importantly, genetic strategies such as "synthetic lethality" and "multicopy suppressor" as described in the following sections can also be used. These genetic approaches are difficult to be utilized in other eukaryotes and allow us to identify unexpected genes and to reveal functional interactions between signaling pathways.

3. Abundant genetic resources are available in yeast. Since yeast research played a pioneering role in genetic research, almost all yeast genes have been cloned and analyzed. These accumulated findings can be advantageous, especially when identifying the genes influenced by drug administration. The databases on yeast genome are abundant as well. For budding yeast, SGD (*Saccharomyces* Genome Database) maintained by Stanford University and CYGD (the MIPS comprehensive yeast genome database) provide abundant information on the genome such as sequences, localizations, expression levels, and their related subjects. For fission yeast, the *S. pombe* gene database is maintained by the Wellcome Trust Sanger Institute. The *S. pombe* postgenome database has been recently constructed by Riken, which provides information on fission yeast ORFeome cloning as well as basic information on genes and proteins.[2] A bioknowledge library maintained by Incyte Corporation is also available. The strains and plasmids can be made available from National-Bio-Resource-Project Yeast (http://yeast.lab.nig.ac.jp/nig/english/index.html) and other public resources.

4. Rapid growth of yeast enables accelerated research. Yeast divides once in 2–3 h and this accelerates the speed of experiments. The methods of culture and storage are easily maintained and the costs of experiments are low.

BUDDING YEAST AND FISSION YEAST

BUDDING YEAST The analysis of the whole genome structure of the budding yeast has been the first to be completed among eukaryotes.[3] The budding yeast has 16 chromosomes and its whole length is 12 Mb, which is approximately 1/300 of those of humans. Approximately 6300 genes have been identified to date. The budding yeast usually grows as a diploid cell and a haploid cell can be readily obtained by induction of meiosis. The haploid cell can grow stably, and more than 6000 sets of gene disruption mutants are established and available.[4] The high-throughput screening aiming to identify the targets of the drugs can be planned using these gene disruption sets. The sensitivity to the drug is enhanced in haploid cells, since it has only one copy of the gene. However, the excessive sensitivity to the drugs can induce cell death, and this will make it difficult to analyze the target molecules encoded by essential genes. These drug targets can be studied using the deletion sets of diploid cells. The combination of these gene disruption sets and the drugs acting on signaling molecules makes it possible to clarify the complex signal transduction pathways affected by the drug.

FISSION YEAST The fission yeast is a monocellular organism and usually grows as a haploid cell. It has three chromosomes and its whole length is approximately 14 Mb, which has more than 4800 genes.[1] The genome size of the fission yeast is the smallest among the eukaryotes to date. However, its genome structures with regulatory mechanisms such as splicing resemble those of higher eukaryotes compared to those of budding yeast. More than 100 types of genes related to human disease, such as Down's syndrome, are conserved in yeast, which suggests the applicability of the fission yeast as a model organism.

GENETIC STRATEGIES

As mentioned above, genetic strategies such as "multicopy suppressor" and "synthetic lethality" are utilized in yeast to allow the identification of unexpected genes and to reveal functional interactions between signaling pathways.

HIGH GENE-DOSAGE SUPPRESSION OF THE DRUG EFFECT AND ITS APPLICATION TO THE IDENTIFICATION OF TARGET MOLECULES If a drug exerts its action by inhibiting the function of a target protein thereby inducing a specific phenotype including cell growth arrest, we can identify the unknown target molecules by utilizing yeast genetics of high-dosage suppressors. Moreover, it is also helpful to detect and reveal novel gene products that functionally interact with the drug.[5] It is also possible to identify the human genes related to drug resistance using a human cDNA library instead of a yeast genome library. For example, from genetic studies in *S. cerevisiae*, the TOR1 and TOR2 gene products were found to be targets of rapamycin, and from *S. pombe* studies, the cellular target of leptomycin B was identified as Crm1.

SYNTHETIC LETHALITY Synthetic lethality has been extensively used both to characterize interactions between genes previously identified as likely to be involved in similar processes as well as to uncover new interactions. The term "synthetic lethality" was first used by Dobzhansky to describe the phenomenon in which alleles of different genes are separately viable, but inviable when combined in a double mutant. The possibility that a certain agent can be a candidate anticancer drug was tested by using the mutant of DNA repair genes, in which the sensitivity for the agent was compared among the mutants.[6]

The budding yeast has approximately 6000 genes; of these approximately 1000 genes are essential and the deletion mutants remaining genes are viable. Parson *et al.* analyzed the sensitivity of 12 drugs on approximately 4700 deletion mutants and isolated 647 hypersensitive mutants.[7] They made a chemical genetic-interaction profile in which the genes are plotted on the horizontal axis and the drugs are plotted on the vertical axis. The results showed that the drug and its suggested interactive genes form a cluster. The chemical-genetic profile of benomyl, the inhibitor of polymerization of microtubules, forms a cluster consisting of the genes related to cell structure, chromosome, and mitosis. Cyclosporin A and FK506, the inhibitors of calcineurin, were reported to form similar clusters that include the genes containing cell wall integrity and the function of the vacuole.

We also utilized this "synthetic lethality" screening method to identify genes that are functionally interacting with calcineurin, the Ca^{2+}/calmodulin-dependent protein phosphatase in fission yeast. Calcineurin is an *in vivo* target of FK506 in fission yeast, and inhibition of calcineurin activity by gene disruption or by the addition of FK506 to the media does not affect the vegetative growth of fission yeast. This suggests that other regulatory pathways in the cell exist, and that they perform overlapping functions with calcineurin. Thus, mutants that require calcineurin activity for vegetative growth would not grow in the presence of FK506. A genetic screen to search for the mutations that show sensitivity to the immunosuppressive drug FK506 has been performed, and eight complementation groups (*its1-8* for immunosuppressant and temperature sensitive) were identified.[8–13] The gene products identified in our screen include various genes involved in membrane trafficking, such as Ypt3 (Rab11 homologue), Ypt3 (Rab6 homologue), Apm1 (μ1 subunit of the clathrin-adaptor complex-1), and Gid1 (GDI1), as well as signaling molecules such as PI4P5K. These genetic approaches contribute to the identification of previously unknown genetic and functional interactions between signaling molecules that are affected by the drug and also contribute to revealing the unexpected action of the drug.

GENOMIC TECHNOLOGY: DNA MICROARRAY The cDNA chip for the all genes of budding yeast is used in the research on the effects of the drug on genomic expression.[14] Comparing the gene expression profiles between the deletion mutants and the yeast cells under drug treatment on the cDNA chip, the targets of the drugs will be identified. Moreover, an analysis of a profile change of genes induced by a certain drug in the cells in which the primary target of the drug is disrupted revealed secondary target of the genes, thereby leading to the discovery of a new drug and an unknown pathway of the drug.

Examples Marton *et al.* examined the change of the gene expression profile in the cell treated with cyclosporine A and FK506, and they showed that both drugs targeted calcineurin similar to mammals.[15] Moreover, the correlation coefficient between the difference in FK506-treated yeast and the wild type and the difference in the deletion mutant of the calcineurin gene and the wild type is not more than 0.80, suggesting an alternative pathway not through calcineurin. An analysis of the deletion mutant of Gcn4, a transcriptional factor, suggests that the pathway not via calcineurin is via Gcn4.

THE DISCOVERY OF MITOGEN-ACTIVATED PROTEIN KINASE INHIBITORS USING A FISSION YEAST GENETIC APPROACH In this section, we will introduce our genetic strategy to identify regulators of the mitogen-activated protein kinase (MAPK) signaling pathway based on our discovery that the Pmk1 MAPK and calcineurin phosphatase act antagonistically in the Cl^- homeostasis in fission yeast and its application for drug discovery.

Isolation of Multicopy Suppressors of Calcineurin Deletion Calcineurin (CN) is a highly conserved Ca^{2+}/calmodulin-dependent protein phosphatase and two important immunosuppressive drugs, cyclosporine A and tacrolimus (FK506), are potent and specific inhibitors of calcineurin activity in animal cells as well as in fission yeast. We demonstrated that knockout of the fission yeast calcineurin gene *ppb1*$^+$ or inhibition of calcineurin activity by immunosuppressants results in hypersensitivity to Cl^-, and that calcineurin and Pmk1 MAPK, a homologue of extracellular signal-related kinase (ERK)/MAPK in mammals, play antagonistic roles in Cl^- homeostasis[16] (Figure 46–1).

Based on this genetic interaction between calcineurin and Pmk1 MAPK, we screened for multicopy suppressors of the Cl^--hypersensitive phenotype of the calcineurin knockout. Notably, all the genes we identified turned out to negatively regulate Pmk1 MAPK signaling, thereby suppressing the Cl^- hypersensitivity of the calcineurin knockout. The *first* gene we identified encodes the MAPK phosphatase Pmp1. Pmp1 is a homologue of the mammalian dual-specificity phosphatase CL100/MKP-1, and dephosphorylates the MAPK Pmk1, thereby inhibiting Pmk1 signaling (Figure 46–1). The *second* gene identified was *pek1*$^+$, encoding MAPK kinase (MAPKK), which phosphorylates and activates Pmk1 in its phosphorylated form, thus acting as an upstream

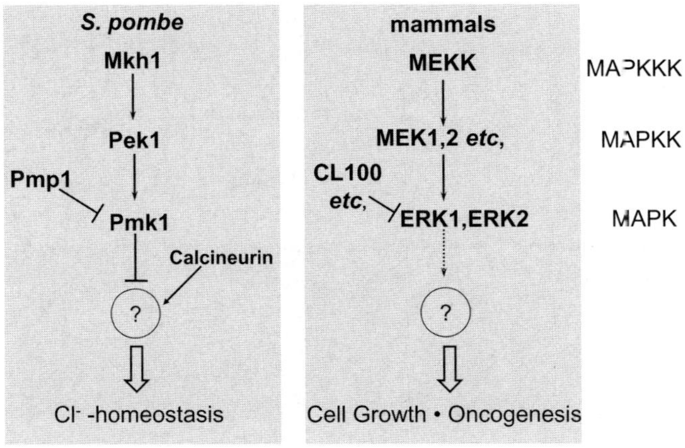

Figure 46–1. MAPK signaling pathway is conserved from *S. pombe* to mammals.

MAPKK for Pmk1 (Figure 46–1). Surprisingly, Pek1 in its unphosphorylated state binds phosphorylated Pmk1, thereby inhibiting Pmk1 MAPK signaling, and hence Pek1 acts as a phosphorylation-dependent molecular switch. The *third* gene was *rnc1*$^+$, encoding a novel KH-type RNA-binding protein. Rnc1 was shown to bind to and stabilize the inherently unstable mRNA of Pmp1, an MAPK phosphatase for Pmk1. Consistently, Pmp1 mRNA was dramatically destabilized and Pmp1 protein was barely detectable in Rnc1 knockout cells. Moreover, Pmk1 MAPK directly phosphorylates Rnc1 *in vivo* and *in vitro*, and this phosphorylation by Pmk1 MAPK regulates the RNA-binding activity of Rnc1. Thus, the identification of Rnc1 as a target of the MAPK signaling cascade has revealed a novel feedback mechanism for the posttranscriptional regulation of MAPK phosphatase gene expression.

Altogether, our screen for multicopy suppressors of calcineurin deletion efficiently identified negative regulators of Pmk1 MAPK signaling.

Isolation of *vic* (Viable in the Presence of Immunosuppressant and Chloride Ion) Mutants We recently developed a novel genetic screen to identify new components of the Pmk1 MAPK pathway. As shown above, knockout of the components of the Pmk1 MAPK pathway, such as Pmk1 (MAPK), Pek1 (MAPKK), Mkh1 (MAPKKK), or Pck2 (protein kinase C), complemented the Cl^--hypersensitive phenotype of calcineurin knockout. Notably, these knockout strains are viable in the presence of immunosuppressant FK506 and high concentrations of $MgCl_2$, while the wild-type cells are inviable in the same condition[16,17] (Figure 46–2). We therefore hypothesized that if we isolate mutants that can grow in the presence of FK506 and 0.15 M $MgCl_2$, the genes responsible for the mutation are expected to function in protein kinase C–Pmk1 MAPK signaling (Figure 46–2). These may include mutations in the unknown factor downstream of Pmk1, or in novel factors required for the activation or function of Pmk1 MAPK. We then isolated the mutants that are viable in the presence of immunosuppressant and chloride ion, and hence named these *vic* (viable in the presence of immunosuppressant and chloride ion) mutants.

Based on this rationale, we identified the *vic1-1/cpp1-v1* mutant, in addition to the mutation allele in the known components of the Pmk1 MAPK pathway including *pmk1*$^+$, *pek1*$^+$, *mkh1*$^+$, as well as *pck2*$^+$.[13] The *cpp1*$^+$ gene encodes a β subunit of the

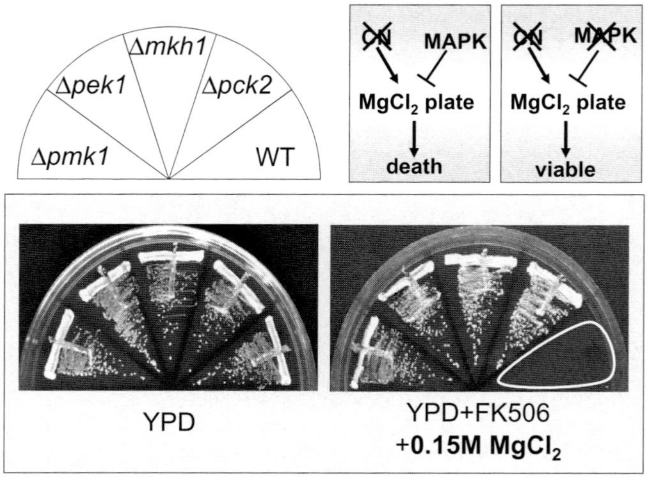

Figure 46–2. Calcineurin and Pmk1 MAPK play antagonistic roles in Cl^- homeostasis.

farnesyltransferase (FTase) that is highly conserved through evolution. We further showed that a small GTPase Rho2 is a novel target of Cpp1. Moreover, Cpp1 and Rho2 act upstream of the Pck2–Pmk1 MAPK signaling pathway, thereby resulting in the *vic* phenotype upon their mutations. Thus, our genetic screen of *vic* mutants utilizing the antagonistic interaction between calcineurin and Pmk1 signaling successfully identified Cpp1 as a novel upstream component of Pck2–Pmk1 signaling.[13]

A Genetic Approach to Screen for Inhibitors of Mitogen-Activated Protein Kinase Signaling We next wished to apply this powerful genetic strategy to screen for compounds that target MAPK/PCK signaling and we established a unique method based on the following rationale. MAPK/PCK signaling plays an important role in cell growth as well as in oncogenesis, and compounds that target MAPK signaling may help contribute to the discovery of anticancer drugs. As shown in Figure 46–2, there was a complementation of the Cl^--hypersensitive phenotype of calcineurin knockout cells upon the inhibition of Pmk1 MAPK signaling. Thus, treatment with a chemical compound that allows calcineurin knockout cells to be viable in the presence of $MgCl_2$ suggests that the compound might act as an inhibitor of MAPK/PKC signaling (Figure 46–3).

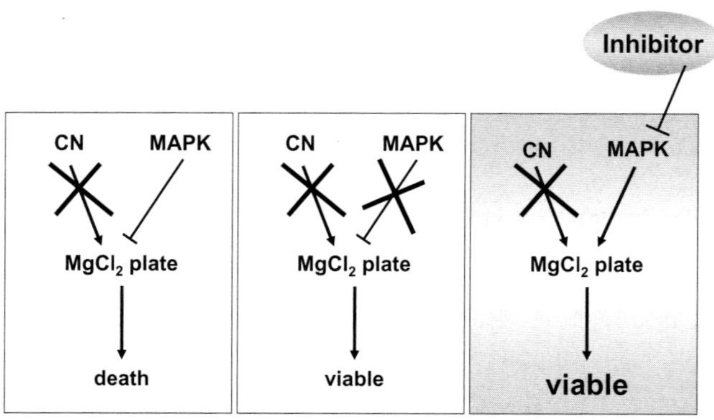

Figure 46–3. A strategy for screening for inhibitors of MAPK/PCK signaling.

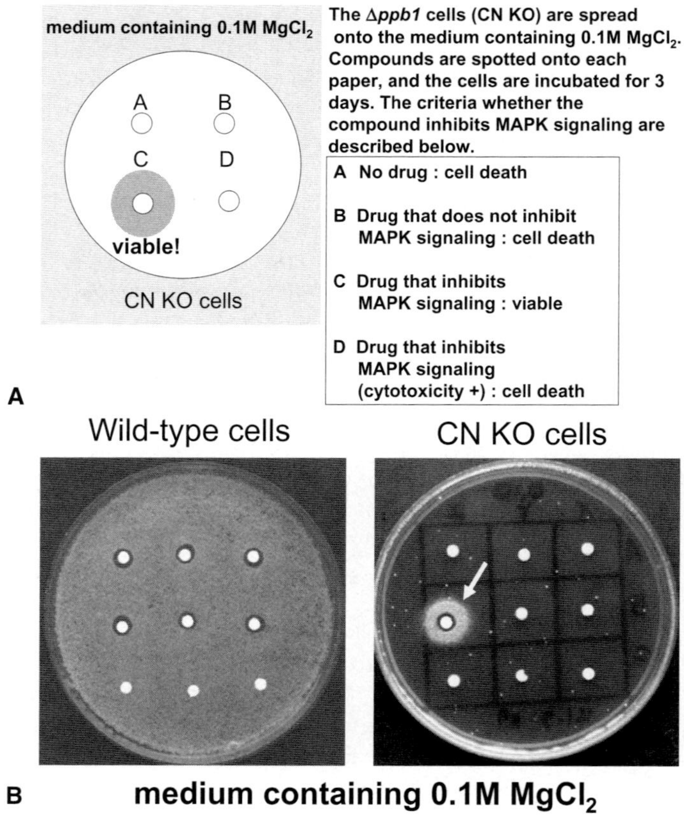

Figure 46–4. (A) A scheme to identify inhibitors of MAPK signaling. (B) Hit compound that may target fission yeast MAPK signaling.

To identify compounds that target MAPK/PKC signaling, the screen was performed as described in Figure 46–4A. First, spread calcineurin knockout cells onto the medium containing 0.1 M $MgCl_2$ and then put a filter paper disc in each section. Next, add the chemical compound onto each disc then grow the cells at 27°C for 3 days. Notably, the addition of a known human MAPK inhibitor makes calcineurin knockout cells viable in the presence of $MgCl_2$, indicating that the human MAPK inhibitor inhibits MAPK Pmk1 signaling in fission yeast (our unpublished results). Screening a chemical library of more than 8000 compounds resulted in the identification of three hit compounds that may target fission yeast MAPK signaling; one of the hit compounds is shown in Figure 46–4B. The three hit compounds that target fission yeast MAPK signaling may inhibit human MAPK signaling.

This method has several advantages. First, this screen is unique in that the ability to inhibit MAPK signaling can be assayed by the growth of calcineurin knockout cells in the presence of $MgCl_2$. Therefore, the compound with a cytotoxic effect will not be isolated in our screen. In a screen for an inhibitor wherein the inhibitory effect is evaluated by the inhibition of cell growth, it is difficult to discriminate between an inhibitory effect on the signaling and a toxicity of the compound. The cytotoxic effect of the compound is also examined by the addition of compounds to the wild-type cells (Figure 46–4B). Second, in our assay, the compound was added onto a paper disc, which enables the diffusion of the compound into the medium, thereby establishing a gradient concentration. Therefore, the optimum concentration of the compound is efficiently achieved in our assay, which allows a high-throughput screening of a chemical compound library. And third, our assay evaluates the activity of the compound on Pmk1 MAPK signaling by the *in vivo* parameter of hypersensitivity of calcineurin deletion. Therefore, the compound that has *in vitro* activity but is unable to pass through the plasma membrane is automatically eliminated as a drug candidate, resulting in a more cost-effective and time-saving method.

CONCLUSIONS

In summary, our genetic approach has effectively identified regulators of Pmk1 MAPK signaling, including negative regulators, upstream activating factors, as well as candidate inhibitors of the signaling. Given that MAPK/PCK signaling plays an important role in cell growth as well as in oncogenesis, the application of these powerful genetic approaches to screen for compounds that target MAPK/PCK signaling emphasizes the potential contribution of these compounds to the discovery of anticancer drugs. Also, the targets of these drugs may be one of the genes that will be identified by our screening of the *vic* mutants.

ACKNOWLEDGMENTS

We thank Dr. Takashi Toda, Dr. Mitsuhiro Yanagida, Dr. Chikashi Shimoda, and the National Bio-Resource Project for providing strains and plasmids, Susie O. Sio for critical reading of the manuscript, and Fujisawa JAPAN Inc. for gifts of FK506. This work was supported by the 21st Century COE Program, the Asahi Glass Foundation, the Uehara Memorial Foundation, and research grants from the Ministry of Education, Culture, Sports, Science and Technology of Japan.

REFERENCES

1. Wood V, Gwilliam R, Rajandream MA, *et al.* The genome sequence of *Schizosaccharomyces pombe*. *Nature* 2002;415:871–880.
2. Matsuyama A, Arai R, Yashiroda Y, *et al.* ORFeome cloning and global analysis of protein localization in the fission yeast *Schizosaccharomyces pombe*. *Nat Biotechnol* 2006;24:841–847.
3. Goffeau A, Barrell BG, Bussey H, *et al.* Life with 6000 genes. *Science* 1996;274:546, 563–567.
4. Tong AH, Lesage G, Bader GD, *et al.* Global mapping of the yeast genetic interaction network. *Science* 2004;303:808–813.
5. Giaever G, Flaherty P, Kumm J, *et al.* Chemogenomic profiling: Identifying the functional interactions of small molecules in yeast. *Proc Natl Acad Sci USA* 2004;101:793–798.
6. Hartwell LH, Szankasi P, Roberts CJ, Murray AW, Friend SH. Integrating genetic approaches into the discovery of anticancer drugs. *Science* 1997;278:1064–1068.
7. Parsons AB, Brost RL, Ding H, *et al.* Integration of chemical-genetic and genetic interaction data links bioactive compounds to cellular target pathways. *Nat Biotechnol* 2004;22:62–69.
8. Zhang Y, Sugiura R, Lu Y, *et al.* Phosphatidylinositol 4-phosphate 5-kinase Its3 and calcineurin Ppb1 coordinately regulate cytokinesis in fission yeast. *J Biol Chem* 2000;275:35600–35606.
9. Yada T, Sugiura R, Kita A, *et al.* Its8, a fission yeast homolog of Mcd4 and Pig-n, is involved in GPI anchor synthesis and shares an essential function with calcineurin in cytokinesis. *J Biol Chem* 2001;276:13579–13586.
10. Cheng H, Sugiura R, Wu W, *et al.* Role of the Rab GTP-binding protein Ypt3 in the fission yeast exocytic pathway and its connection to calcineurin function. *Mol Biol Cell* 2002;13:2963–2976.
11. Kita A, Sugiura R, Shoji H, *et al.* Loss of Apm1, the micro1 subunit of the clathrin-associated adaptor-protein-1 complex, causes distinct phenotypes and synthetic lethality with calcineurin deletion in fission yeast. *Mol Biol Cell* 2004;15:2920–2931.
12. He Y, Sugiura R, Ma Y, *et al.* Genetic and functional interaction between Ryh1 and Ypt3: Two Rab GTPases that function in *S. pombe* secretory pathway. *Genes Cells* 2006;11:207–221.
13. Ma Y, Kuno T, Kita A, Asayama Y, Sugiura R. Rho2 is a target of the farnesyltransferase cpp1 and acts upstream of pmk1 mitogen-activated protein kinase signaling in fission yeast. *Mol Biol Cell* 2006;17:5028–5037.
14. Lockhart DJ. Mutant yeast on drugs. *Nat Med* 1998;4:1235–1236.
15. Marton MJ, DeRisi JL, Bennett HA, *et al.* Drug target validation and identification of secondary drug target effects using DNA microarrays. *Nat Med* 1998;4:1293–1301.
16. Sugiura R, Toda T, Shuntoh H, Yanagida M, Kuno T. *pmp1*$^{+}$, a suppressor of calcineurin deficiency, encodes a novel MAP kinase phosphatase in fission yeast. *EMBO J* 1998;17:140–148.
17. Sugiura R, Toda T, Dhut S, Shuntoh H, Kuno T. The MAPK kinase Pek1 acts as a phosphorylation-dependent molecular switch. *Nature* 1999;399:479–483.

47 Yeast as a Model System to Study DNA Damage and DNA Repair

ANTONIO CONCONI

ABSTRACT

This chapter provides procedures for determining cell survival and the extent of DNA damage and repair after treatment of yeast cultures with DNA-damaging agents such as bleomycin, methyl methanesulfonate, dimethyl sulfate, and ultraviolet light. Comprehensive protocols for basic methods in yeast biology such as how to prepare different media, how to grow and store yeast cells, how to isolate nucleic acids, how to transform and prepare yeast mutant strains, and how to use Internet-based yeast resources are available.

Key Words: Benzo[*a*]pyrenediol, Bleomycin, CPD, Dimethyl sulfate, Double-stranded breaks, Methyl methanesulfonate, Nucleotide excision repair, Single-stranded breaks, 2,3,5-Triphenyltetrazolium chloride.

INTRODUCTION

The budding yeast *Saccharomyces cerevisiae* is classified as a safe microorganism and has been used for many years as a biological model.[1–3] It has provided important contributions to our current understanding of basic eukaryotic cellular functions. Furthermore, yeast is employed as a model to investigate mammalian diseases. For instance, at least 31% of yeast proteins have human homologues and nearly 50% of human genes implicated in heritable diseases have yeast homologues.[4] Although its nuclear genome is only 3.5 times larger than that of *Escherichia coli*, the yeast genome displays most of the hallmarks of higher eukaryotic genomes. The genome of haploid yeast cells is about 1.2×10^7 bp and is organized into 16 chromosomes, ranging in size from 200 to 2200 kb.[5] Finally, yeast DNA is found within a nucleus and its nucleosomal organization of chromatin is similar to that present in higher eukaryotes, except that there is no histone H1.[6,7]

The powerful yeast genetics and its relatively high rate of recombination facilitate the insertion of DNA sequences at specific locations in the genome, allowing the creation of yeast strains carrying gene knockouts,[8] the tagging of proteins with different epitopes,[9] and the engineering of regulated promoters upstream of a given open reading frame (ORF).[10] In yeast there are 6466 ORFs.[11,12] Their functions, information on the corresponding proteins, and their cellular localization are available through databases such as the *Saccharomyces* Genome Database (SGD; www.yeastgenome.org), Yeast Protein Database (YPD; www.proteome.com), Munich Information Center for Protein Sequences (MIPS) Comprehensive Yeast Genome Database (CYGD; mips.gsf.de/genre/proj/yeast/index.jsp), the Yeast Resource Center (depts.washington.edu/~yeastrc), and yeast green fluorescent protein (GFP) fusion localization database (yeastgfp.ucsf.edu). Furthermore, 5916 ORFs were deleted to construct a heterozygous diploid collection that contains deletions in both essential and nonessential genes.[13] About 4800 nonessential deletions were generated as haploid MATa and MATα strains or as homozygous diploids, and these collections can be obtained from Research Genetics (ResGen) (www.resgen.com/products/) or the Euroscarf consortium (web.uni-frankfurt.de/fb15/mikro/euroscarf/col_index.html) in readily usable forms.[14] Among numerous applications of the ORF deletions libraries, they have been employed to understand the genetics that determine cell sensitivity and resistance to drugs, and some multidrug resistance genes have already been identified.[15] Also, the toxicity of chemicals is currently tested in yeast that present metabolic defects similar to those observed in human cancers,[16] and high-throughput screenings are employed to determine the mode of action of DNA-damaging agents such as those employed in chemotherapy.[17] Finally, studies on cytotoxic compounds such as methyl methanesulfonate (MMS) and hydroxyurea,[18] bleomycin (BLEO),[19] and cisplatin,[20] on ionizing radiation,[21] and on ultraviolet light (UV) radiation[22] have revealed genes that are involved in DNA repair, cell cycle checkpoints, DNA recombination, and DNA replication. The results have highlighted the importance of the nucleotide excision repair (NER) mechanism together with homologous recombination, postreplication repair, and cellular responses to DNA damage.

MMS and DMS are DNA methylating agents that produce a variety of damaged bases, of which N^7-methylguanines and N^3-methyladenine constitute about 80% and 10% of the lesions, respectively.[23] In yeast, *N*-methylpurines are enzymatically removed by the base excision repair pathway (BER), and the sensitivity to MMS increases in cells lacking genes involved in BER. However, homologous recombination (HR) may also participate in the repair of methylpurines. Bleomycin is a radiomimetic drug that in the presence of cofactors creates single strand breaks (SSB) and double strand breaks (DSB) by a concerted free radical attack on the sugar moieties of the DNA backbone.[23] DNA strand breaks block DNA replication, and the difficulty of repairing DSBs is the major source of BLEO's cytotoxicity. The

From: *Sourcebook of Models for Biomedical Research*
(P. M. Conn, ed.), © 2008 Humana Press Inc., Totowa, NJ.

recombination and postreplication DNA repair pathways are involved in the repair of BLEO-induced DNA lesions. UV light induces the formation of two major photoproducts: *cis-sin* cyclobutane pyrimidine dimers (CPD) and (6-4) pyrimidine–pyrimidone dimers [(6-4)PD].[23] These lesions block gene transcription and DNA replication, and are repaired by the NER pathway. Although UV photoproducts induced by UV-C serve as the paradigm to study the mechanisms of NER, this ubiquitous repair pathway operates on a large variety of chemical adducts to DNA.[23]

NER is characterized by the incision of the damaged DNA strand on both sides of the lesion, resulting in the removal of an oligonucleotide fragment (25~30 nucleotides long) containing the damage. This is followed by DNA repair synthesis and ligation. The yeast genes involved in NER can be grouped into two classes; the first consists of *RAD1, 2, 3, 4, 10, 14,* and *25* genes, and the second consists of *RAD7, 16,* and *23* and the *MMS19* genes. Mutations in the genes of class 1 cause a very high degree of UV sensitivity, whereas mutations in the genes of class 2 cause moderate UV sensitivity.[24] An additional factor encoded by the *RAD26* gene is involved in transcription-coupled repair (TCR), the subpathway of NER that is responsible for removing DNA lesions from the transcribed strand of active genes.[25]

Photoreactivation is an alternative mechanism to NER that specifically removes pyrimidine dimers. This process is catalyzed by one enzyme (photolyase) and is light dependent (blue light). Photolyases are widespread in nature, though there is no photoreactivation of DNA damage in human cells.[23] To inhibit the activity of photolyase, all the experiments done with UV-irradiated yeast cultures are performed in a dark room, under yellow light (e.g., Rapid Start Fluorescent lamps, Reprographic-Gold 40 W 2250 K°, Standard Products Inc.; catalog 6050RGO).

GROWING YEAST CELLS AND INDUCTION OF DNA DAMAGE

Culturing yeast is simple, economical, and rapid. They grow in liquid and on solid media, forming a bud that pinches off to originate daughter cells. During exponential growth (or log-phase) in rich media, yeast have a short life cycle of about 90 min. As cell density increases, nutrients drop and the rate of cell division slows. The log-phase is divided into three stages: early-log phase ($<10^7$ cells/ml), mid-log phase (between 1 and 5×10^7 cells/ml), and late-log phase (between 5×10^7 and 2×10^8 cells/ml). At a density of 2×10^8 cells/ml, yeast enter the stationary (G_0) phase. The growth curve of newly inoculated cells (e.g., at 2×10^5/ml) is characterized by a lag phase of two to three cell divisions (~5 h), exponential growth for about six cell divisions (~9 h), and the stationary phase (during this last phase, cells undergo about two more divisions). The optimal growth conditions can vary depending on the yeast genetic background but, in general, yeast grow very rapidly at 28–30°C in the most commonly used YEPD medium. The recipes for rich, minimal, and complete minimal dropout media and their uses have been reported.[1–3]

Below is described how we grow yeast cultures and incubate them with DNA-damaging agents and how we monitor DNA repair. However, optimal growth conditions and treatments with genotoxic compounds could vary from strain to strain. Thus it is advisable to carry out pilot experiments to follow the dose survival curves. Also, while small cultures provide enough cells for the survival tests, large quantities of cells are needed for DNA repair analyses.

GROWTH OF LARGE YEAST CULTURES AND PREPARATION OF CELLS FOR TREATMENT WITH DNA-DAMAGING AGENTS

1. For liquid cultures, add the YEPD medium (10 g yeast extract, 20 g peptone, 20 g glucose, and distilled water to 1 liter) to a flask that holds two to three times the required volume. For solid plates, prior to autoclaving add 2% (w/v) agar to the YEPD in a flask containing a stir bar. Autoclave at 121°C and 15 psi (1 atm) for 20 min. Gently mix the agar-containing medium on a stirrer and let it cool down to ~50–55°C before pouring the plates (20–25 ml/Petri dish). Allow the plates to solidify and dry at room temperature (RT) for 2 days; store them at 4°C in plastic bags.
2. Streak out the cells on solid YEPD and incubate the plate at 28–30°C until colonies form (2–3 days). Prepare a preculture as follows: use one colony to inoculate ~7 ml YEPD in a sterile culture tube (e.g., borosilicate glass 16 × 150 mm, Fisher Scientific catalog no. 14-961-31). Usually, cells grow to a stationary phase when incubated overnight at 28–30°C with continuous rotation or shaking.
3. DNA repair experiments are done with exponentially growing cultures (2×10^7 cells/ml). One 2-liter flask containing 500 ml of YEPD is inoculated with 300–400 μl of preculture cells (final concentration ~10^5 cells/ml). The culture is incubated at 30°C with continuous shaking (~300 rpm) overnight. To monitor cell growth blank the spectrophotometer with sterile medium, remove an aliquot, and measure the absorbance of the cells at 600 nm [optical density $(OD)_{600}$]. Use dilutions to make sure that measurements are done in the linear range of the spectrophotometer, and calculate the number of cells/ml considering that OD_{600} ~ 0.1 is approximately 2×10^6 cells/ml. [*Note*: The correlation OD vs. number of cells varies not only from strain to strain but also for a given strain grown in different media. Therefore it is important to determine the cell number per OD unit using a counting chamber, and plotting the growth curves to determine the times of cell division.]
4. Cells are harvested when they are in mid-exponential phase (1 to 2×10^7 cells/ml) in a benchtop centrifuge (2000–3000 g/5 min), washed in ice cold phosphate-buffered saline (PBS) (8 g NaCl, 0.2 g KCl, 1.44 g Na_2HPO_4, 0.24 g of KH_2PO_4, pH 7.4 per liter), and reharvested to prepare them for treatment with DNA-damaging agents.

TREATMENT OF SMALL CULTURES OF CELLS WITH METHYL METHANESULFONATE AND BLEOMYCIN TO DETERMINE CELL SURVIVAL

1. Yeast strains (10 ml cultures) are grown exponentially (10^7 to 2×10^7 cells/ml), collected by centrifugation, washed with 10 ml of sterile water, and resuspended in 5 ml of water to 0.8×10^7 cells/ml.
2. MMS (Sigma, catalog no. M 4016) is used as the stock solution. BLEO (from MeadJohnson or Sigma) is dissolved in sterile water to 5 U/ml, and stored in aliquots at −20°C.
3. Chronic exposures of yeast cultures to MMS and BLEO are done in triplicate: 5 μl of cell suspensions (0.8×10^7 cells/ml in sterile water) is used to inoculate the wells of a microplate containing 95 μl of YEPD media and various concentrations of MMS (0 as control, 0.005, 0.01, and 0.02%; v/v) or BLEO (0 as control,

6 and 10 mU/ml). The microplates are incubated as described in the section on Measuring Yeast Sensitivity to Genotoxic Agents in Liquid Cultures. Alternatively, 10-fold serial dilutions of cells are prepared in sterile water and 5 µl of each dilution is spotted on YEPD plates containing different concentrations of MMS or BLEO (see above). The plates are prepared and incubated as described in the section on Measuring Yeast Sensitivity to Genotoxic Agents by Colony Spot Testing.

4. Acute exposures of yeast cultures to MMS are done in triplicate: 1 ml of yeast cultures in sterile water (prepared as in step 1) is collected by centrifugation, resuspended in 1 ml of fresh YEPD (0.8×10^7 cells/ml) containing different concentrations of MMS (0 as control, 0.01, 0.025, and 0.05%; v/v), and incubated at 30°C for 90 min under continuous rotation. Thereafter, cells are harvested by centrifugation, washed twice with sterile water, and resuspended in 1 ml of water. As described above, 5 µl of cell suspensions (4×10^4 cells) is used to inoculate the wells of a microplate containing 95 µl of fresh YEPD (without DNA-damaging agents; MMS). The microplates are incubated as described in the section on Measuring Yeast Sensitivity to Genotoxic Agents in Liquid Cultures. Alternatively, 10-fold serial dilutions of washed cells are prepared in sterile water and 5 µl of each dilution is spotted on YEPD plates (without DNA-damaging agents). The plates are incubated as described in the section on Measuring Yeast Sensitivity to Genotoxic Agents by Colony Spot Testing.

ULTRAVIOLET LIGHT IRRADIATION OF SMALL CULTURES OF CELLS TO DETERMINE CELL SURVIVAL

1. In sterile water 350 µl of 0.8×10^7 cells/ml is spotted on an autoclaved glass plate and irradiated (254 nm) with UV doses of 20, 80, and 150 J/m^2, as measured with a UVX radiometer (Ultra-Violet Products, Upland, CA). How to monitor UV intensities and convert them to UV doses is described in the section on Ultraviolet Light Irradiation of Large Cell Cultures (step 3).

2. Triplicates of 5 µl of UV irradiated, or mock treated (control), cell suspensions are used to inoculate the wells of a microplate containing 95 µl of YEPD media. The microplates are incubated in the dark to avoid photoreversal of DNA damage by photolyase as described in the section on Measuring Yeast Sensitivity to Genotoxic Agents in Liquid Cultures. Alternatively, 5 µl of 10-fold serial dilutions of irradiated and nonirradiated cells are spotted on YEPD plates. The plates are incubated in the dark to avoid photoreversal of DNA damage by photolyase as described in the section on Measuring Yeast Sensitivity to Genotoxic Agents by Colony Spot Testing.

TREATMENT OF LARGE CULTURES OF CELLS WITH DIMETHYL SULFATE

1. Yeast cells are grown at 30°C in YEPD to mid-log phase ($OD_{600} \sim 0.4$) as described in the section on Growth of Large Yeast Cultures. An aliquot of $1–2 \times 10^9$ cells is kept as control [no treatment with dimethyl sulfate (DMS)].

2. The rest of the culture is mixed with DMS (Sigma, undiluted solution) to give a final concentration of 0.03% (v/v), and incubated at RT for 2 min.

3. Cells are washed twice with PBS, resuspended in 200 ml of prewarmed YEPD at 30°C, poured into a 500-ml Erlenmeyer flask, and incubated at 30°C with continuous shaking (~300 rpm) for different repair incubation times (e.g., 0, 1, 2, and 4 h). Then 50-ml aliquots ($1–2 \times 10^9$ cells) are collected on ice and cells are harvested by centrifugation. [*Note:* If needed, to prevent DNA replication during repair incubation, hydroxyurea is added to a final concentration of 100 mM (e.g., 25 and 26 and references therein).]

4. Total genomic DNA is isolated as described in the section on Isolation of Genomic DNA from Yeast Cells.

TREATMENT OF LARGE CULTURES OF CELLS WITH BLEOMYCIN

1. Yeast cells are grown at 30°C in YEPD to mid-log phase ($OD_{600} \sim 0.4$) as described in the section on Growth of Large Yeast Cultures.

2. After washing twice with sterile ice-cold water, cells are resuspended in 50 mM NaCl, 2 mM $MgCl_2$, and 0.02% glucose to a cell density of 1 to 2×10^9 cells/ml.

3. Digitonin (Sigma, 10% stock) and $Fe(NH_4)_2(SO_4)_2$ (10 mM, freshly dissolved in H_2O) are added to the cell suspension to final concentrations of 0.05% and 50 µM, respectively.[26]

4. The cells suspension is divided into five aliquots and one aliquot is kept as the negative control (no treatment with BLEO). BLEO (Sigma, 20 units/ml stock) is added to final concentrations of 12.5, 25, 50, and 100 mU/ml, followed by a 12-min incubation at 30°C.

5. To stop the reaction, the five cell suspensions are mixed with 100 volumes of ice-cold YEPD and the cells are collected by centrifugation. [*Note:* BLEO can aggregate at cell plasma membranes and even in the presence of ethylenediaminetetraacetic acid (EDTA), DNA can be damaged during the extraction process. BLEO molecules are efficiently removed from mammalian cell membranes by trypsin,[27] and from yeast cells walls by washing the cells with yeast media.[26]]

6. Cells are washed a second time in ice-cold YEPD, collected by centrifugation and prepared for total genomic DNA extraction as described in the section on Isolation of Genomic DNA from Yeast Cells.

ULTRAVIOLET LIGHT IRRADIATION OF LARGE CELL CULTURES

1. Yeast cells are grown at 30°C in YEPD to mid-log phase ($OD_{600} \sim 0.4$) as described in the section on Growth of Large Yeast Cultures. The cell pellet from a 500-ml culture (total 5×10^9 cells) is resuspended in 300 ml PBS (~1.67×10^7 cells/ml). [*Note:* Cells are irradiated in PBS, as YEPD filters out 254 nm UV light.]

2. An aliquot of 50 ml is kept as the control (–UV) and the remaining cells are poured into trays (21.5 × 14 cm of mat gray plastic from Rubbermaid) and irradiated with primary 254 nm UV (UVC) light. [*Note:* For optimal and equal UV irradiation the thickness of yeast suspensions should not be above 2 mm, the maximal cell density should be kept below 3×10^7 cells/ml, and the UV lamps should be turned on about 5 min before use.]

3. Cell suspensions are irradiated with 254-nm emitting germicidal lamps (15 W; General Electric). The UV intensity is measured with a radiometer (UVX from Ultra-Violet Products, Upland, CA) and converted to UV dose with the following equation:

$$\text{Dose (J/m}^2) = \text{Time of irradiation (seconds)} \times \text{Intensity } (\mu\text{W/cm}^2)/100$$

For instance, to study DNA repair of chromatin in yeast, we routinely irradiate cells at ~150 J/m.[2]

4. Soon after irradiation, cells of a 50-ml aliquot are harvested by centrifugation and used as control (+UV, no DNA repair). The remaining 200 ml of the cell suspension is collected as previously described (see the section on Growth of Large Yeast Cultures, step 4), resuspended in 200 ml of 30°C prewarmed YEPD, poured into a 500-ml Erlenmeyer flask, and incubated at 30°C with continuous shaking (~300 rpm) to allow repair. At different time points (e.g., 30 min, 1 h, 2 h, and 4 h after UV irradiation) 50-ml aliquots are collected and cells are harvested by centrifugation. Isolation of total genomic DNA is described in the section on Isolation of Genomic DNA from Yeast Cells.

ASSESSING YEAST SURVIVAL RATES

To test cytotoxic compounds on yeast from the ORF deletions libraries, cells carrying deleted genes are plated onto solid media containing the drugs, and the strains that are sensitive to the drugs are recognized by their slow growth. By this assay, sets of multidrug resistance genes have already been identified.[15] However, analyses of cell growth in liquid media are better suited for the detection of weak sensitivities, and are thought to be more representative than the plating assays.[15,28] Finally, methods that measure cell metabolism are also used to determine the fraction of surviving cells.[29] One such test is the reduction of 2,3,5-triphenyltetrazolium chloride (TTC) to formazan by the mitochondria.

MEASURING YEAST SENSITIVITY TO GENOTOXIC AGENTS BY COLONY SPOT TESTING

1. The yeast strains are grown exponentially in YEPD at 30°C under continuous rotation. Cells are collected by centrifugation, resuspended in sterile water, and treated with the DNA-damaging agents (see the section on Treatment of Small Cultures of Cells and the section on Ultraviolet Light Irradiation).

2a. For chemical agents 10-fold serial dilutions of yeast cultures (0.8×10^7, 0.8×10^6, 0.8×10^5, 0.8×10^4, and 0.8×10^3 cells/ml) are prepared in sterile water and 5 µl of each dilution (corresponding, respectively, to 4×10^4, 4×10^3, 4×10^2, 4×10^1, and 4×10^0 cells) is spotted on YEPD plates containing different concentrations of DNA-damaging agents. The plates are incubated at 30°C for 48 h, and cell survival is scored by visual inspection such as comparing the size of the grown spots of yeast colonies.

2b. For UV irradiation 5 µl of 10-fold serial dilutions of irradiated and nonirradiated cells (0.8×10^7, 0.8×10^6, 0.8×10^5, 0.8×10^4, and 0.8×10^3 cells/ml) are spotted on YEPD plates. After UV irradiation, the plates are incubated at 30°C for 48 h in the dark (to avoid photo-reversal), and cell survival is scored by visual inspection of the plates.

SURVIVAL CURVES OF ULTRAVIOLET-IRRADIATED CELLS BY COLONY COUNTING

1. Yeast stock cultures are grown overnight in 5 ml YEPD at 30°C and under continuous rotation to early stationary phase.

2. Serial dilutions are prepared as follows:

- 10^{-4} to 10^{-6} dilutions of the stock cultures are made for the controls (no UV) and the WT (or *RAD*$^+$) strains irradiated at UV doses up to 150 J/m^2.
- 10^{-3} to 10^{-5} dilutions of the stock cultures are made for class I of *rad* mutant strains (see Introduction) when irradiated at doses between 5 J/m^2 and 10 J/m^2. For higher irradiations (up to 30 J/m^2) prepare 10^{-2} to 10^0 dilutions.

3. Of each yeast strain 100 µl is plated in triplicate for each UV dose and dilution. The plates are allowed to dry for 10 min at RT before irradiation (without a lid). After irradiation, the plates are incubated at 30°C for 48 h in the dark. The colonies are counted and plotted as percent survival versus UV doses.

MEASURING YEAST SENSITIVITY TO GENOTOXIC AGENTS IN LIQUID CULTURES

1. Yeast strains (10 ml cultures) are grown exponentially (10^7 to 2×10^7 cells/ml) in YEPD, collected by centrifugation, washed with 10 ml, and resuspended in 5 ml of sterile water to final concentrations of 0.8×10^7 cells/ml.

2. For a control, the OD_{600nm} of the 5-ml cell suspensions can be remeasured.

3. Yeast are treated with DNA-damaging agents as described in the section on Treatment of Small Cultures of Cells and the section on Ultraviolet Light Irradiation.

4. Prepare a 96-well plate (flat bottom, nontreated, low evaporation lid, polystyrene, sterile; Costar, catalog no. 3370) with 95 µl of sterile YEPD in each well with or without DNA-damaging agent.

5. Triplicates of 5-µl yeast samples (corresponding to 4×10^4 cells, treated or mock treated with DNA-damaging agents) are deposited in each well and mixed with the YEPD by pipetting up and down (it is convenient to use a multichannel pipette).

6. To avoid evaporation during the incubation period, the 96-well plates are sealed with Parafilm.

7. Cell growth is monitored with a PowerWave microplate scanning spectrophotometer (Bio-Tek). The OD of cell growth is recorded using KC4 microplate data analysis software (Bio-Tek), programmed as follows: running time, 48 h; OD reading interval, 10 min; OD wavelength, 660 nm; shaking, intensity 2 and duration of 595 sec before every reading; temperature, 30°C.

8. The data are exported to Excel (Microsoft) for plotting.[28]

MEASURING YEAST SENSITIVITY TO GENOTOXIC AGENTS BY THE REDUCTION OF 2,3,5-TRIPHENYLTETRAZOLIUM CHLORIDE The water-soluble and colorless tetrazolium salts are reduced in a variety of cells to red, water-insoluble formazan compounds. The reaction occurs in the mitochondria where these compounds are reduced at different sites along the electron-transport chain, with the exact site of reduction depending on their chemical substitution. Among these, TTC is one of the most commonly used. Its reduction to red formazan occurs at the site of the cytochrome oxidase complex in place of molecular oxygen reduction.[29] Consequently, measurements of cell metabolism by TTC reduction can be used to determine cell survival.

1. Yeast cultures are grown to mid-log phase, treated with DNA-damaging agents, and incubated for different times to allow cells to repair the DNA.

2. For each time point, 35 ml of cell culture (~7×10^8 cells) is pelleted, resuspended in 1 ml sodium phosphate buffer (0.05 M, pH 7.5), transferred to an Eppendorf tube containing 350 µl of TTC stock solution (0.5% TTC in 0.05 M phosphate buffer, Sigma), and mixed well.

3. Incubation is done for 20 h at RT in the dark (without stirring).

4. Cells are pelleted by centrifugation in a microcentrifuge for 2 min, washed in sterile water, and pelleted again.

5. An equal volume of glass beads (425–600 μm, Sigma) is added to the cell pellet, together with 500 μl ethanol : acetone (1 : 1). The suspension is vortexed vigorously to break open the cells.

6. Cell debris and beads are pelleted in a microfuge for 2 min and the supernatant is transferred into a 15-ml tube.

7. Steps 5 and 6 are repeated four times, followed by two extractions with 500 μl of 100% acetone.

8. For each DNA repair time point, all the supernatants are collected and adjusted to 3.5 ml final volumes with ethanol : acetone (1 : 1), followed by 5 min centrifugation at maximum speed in a benchtop centrifuge to pellet cell debris.

9. Absorbance of the combined washing extractions is read at 485 nm and measurements are conveniently taken in 1-ml plastic cuvettes. [*Note:* Dilute the sample if OD_{485} is above 2.]

10. OD_{485} measurements of mock-treated cells are considered as 100%, and the percents of OD_{485} are plotted versus the doses of DNA-damaging agents employed.

MEASUREMENTS OF DNA DAMAGE AND DNA REPAIR RATES

The quantification of the formation and removal of lesions is crucial to understanding the processes of DNA damage and repair, and the preparation of good-quality, high-molecular-weight DNA is an essential step of these analyses. Several methods are used to break the yeast wall, which are either based on mechanical or enzymatic assays (Zymolyase, Lyticase, Glusulase).[1–3] Because cells could rapidly remove some DNA lesions, it is important that the initial steps of the DNA isolation be carried out rapidly. In our laboratory, the method of choice is to break the cells by vigorous shaking in the presence of glass beads.

Currently, the most common techniques that are used to quantify the number of photoproducts in total genomic DNA are either immunological methods[30] or are based on the T4endoV assay. Monoclonal antibodies specific to CPDs and to (6.4)PDs [or to other types of DNA lesions that are not induced by UV light, such as benzo[*a*]pyrenediol (BPDE)] are commercially available, and antibody binding can be measured even with small quantities of antigens. Alternatively, the T4endoV assay developed in the Hanawalt laboratory[31] combines T4 enzymatic activity with Southern blotting. Namely, CPD formation and CPD removal are detected with CPD-specific endonuclease T4endoV, which makes single-DNA strand cuts at CPD sites. Southern blotting of denaturing (alkaline) agarose gels and hybridization with specific radioactive probes are used to quantify the frequency of CPDs in DNA. This technique also makes it possible to determine levels of DNA repair in small fractions of the genome such as single copy genes.

ISOLATION OF GENOMIC DNA FROM YEAST CELLS To follow NER of UV-induced photoproducts, the following steps are carried out under yellow lamps (peak at ~550–650 nm):

1. The cells are irradiated and 50-ml aliquots are collected by centrifugation as described in the section on Ultraviolet Light Irradiation of Large Cell Cultures. The cell pellets are resuspended in ice-cold 1.5 ml NIB (nuclei isolation buffer: 17% glycerol, 50 mM MOPS, 150 mM K-acetate, 2 mM $MgCl_2$, 0.5 mM spermine, and 0.15 mM spermidine, pH 8.0) and transferred to 15-ml tubes containing 1.5-ml glass beads (Sigma; ~425–600 μm) prewashed twice in NIB and prepared on ice.

2. The cells are vortexed vigorously 10–15 times for 30 sec each time with 30-sec gaps on ice.

3. The cells lysates are transferred to new 15-ml tubes on ice and the glass beads are rinsed twice with 1 ml NIB. The combined washing extractions are spun in a benchtop centrifuge at 3500 rpm (~2000 × *g*) for 2.5 min.

4. The supernatants (nuclei suspensions) are divided into two 2-ml Eppendorf tubes and spun in a microcentrifuge for 5 min at 4°C.

5. The pellets are resuspended in 500 μl TE total volume and collected into one 2-ml Eppendorf tube. After addition of 225 μl 3 M Na-acetate (pH 6.5), mixing, and 35 μl 10% sodium dodecyl sulfate (SDS), mixing (final volume 770 μl), the samples are extracted twice with phenol : chloroform : isoamyl alcohol (25 : 24 : 1) (equilibrated with Tris to pH ~ neutral) and once with chloroform (avoid taking the interphase). Finally, the nucleic acids are precipitated with one volume isopropanol at −20°C from 1 h to overnight.

6. After centrifugation for 30 min at 4°C, the white pellets are resuspended (without drying) in 500 μl TE; 10 μl RNase A (10 mg/ml) is added to each tube and the samples are incubated at 37°C for 30 min. Thereafter, 225 μl 3 M Na-acetate (pH 6.5) and 35 μl 10% SDS are added as described in step 5, followed by one extraction with phenol : chloroform : isoamyl alcohol (25 : 24 : 1) and one extraction with chloroform.

8. The samples are precipitated in isopropanol as described above. After centrifugation, the pellets are rinsed twice in 70% (v/v) ethanol, dried, resuspended in 200 μl TE, and stored at −20°C.

IMMUNOASSAY FOR THE DETECTION OF ULTRAVIOLET-INDUCED PHOTOPRODUCTS IN TOTAL GENOMIC DNA

1. Total genomic DNA is digested with two restriction enzymes to reduce its average size and resuspended in TE buffer. The samples are denatured at 95°C for 15 min and chilled on ice. [*Note:* the amounts of DNA must be determined empirically, e.g., 50–500 ng per slot.]

2. The nitrocellulose membrane (Protran, Schleicher & Schuell) is prewetted in ddH_2O and then equilibrated in 20 × SSPE (174 g NaCl, 27.6 g $NaH_2PO_4{\cdot}H_2O$, 7.4 g EDTA, pH 7.4 per liter) for ~30 min. Using a slot-blot apparatus, the DNA samples are spotted in triplicate onto the membrane as follows:

- While applying a constant vacuum add 500 μl 20 × SSPE to each well.
- When all the liquid is adsorbed add the DNA samples in 250 μl.
- When all the liquid is adsorbed, wait an additional 5 min, keeping the vacuum constant. Then rinse the slots twice with 250 μl 20 × SSPE.
- When all the liquid is adsorbed, wait an additional 5 min, dismantle the slot-blotter, and place the membrane (DNA-side up) on 3MM papers and let it dry for 30 min.

3. After baking the membrane for 2 h at 80°C to fix the DNA, the membrane is rehydrated in PBS.

4. Blocking of the membrane is done in 5% nonfat dry milk in PBS-T (0.05% Tween-20 in PBS), at RT for 1 h.

5. The blocking solution is discarded and the membrane is briefly rinsed twice with PBS-T followed by four washes in PBS-T, 10 min each.

6. Incubation with antibodies for CPDs and (6,4)PDs (e.g., from Abcam, Affitech, Sigma-Aldrich, Trevigen) is done in 1% milk in PBS (~0.15 ml/cm^2 of membrane surface) for 3 h at RT or overnight at 4°C. [*Note:* The concentration of the antibodies is predetermined empirically.]

7. Washing of the membrane is done as described in step 5.

8. Secondary antibodies (Amersham) are prepared in 1% milk in PBS and the incubation is done for 1 h at RT, followed by the washing step as described above. [*Note:* The concentration of antibodies is predetermined empirically.]

9. Detection is done with chemiluminescence reagents (ECL; Amersham) according to the manufacturer's instruction, and the membranes are exposed to X-ray film or a PhosphorImager.

10. To normalize the amount of DNA in each sample and slot, DNA is spotted onto a parallel membrane (steps 1–3). Prehybridization and hybridization with ^{32}P-labeled DNA are done following standard protocols.

11. To determine the number of photoproducts and the DNA repair rates, normalize the signals obtained with the antibodies to the respective amounts of DNA in each slot. Subtract the signals generated from nonradiated samples (–UV, controls), which correspond to the nonspecific binding of the antibodies to DNA, and determine the percent repair (0 time repair = 100% photoproducts remaining). [*Note:* The same technique can be used to detect BPDE, the ultimate carcinogenic metabolite of benzo(*a*)pyrene.]

THE ALKALINE AGAROSE GEL ELECTROPHORESIS AND SOUTHERN BLOT ASSAY FOR THE DETECTION OF DNA STRAND BREAKS AND ULTRAVIOLET-INDUCED PHOTOPRODUCTS IN TOTAL GENOMIC DNA The strand breaks, for example, induced by T4endoV cleavage at CPD sites, cause a shift of the migration of DNA in denaturing agarose gel electrophoresis (Figure 47–1A; compare – and + T4endoV, – and + UV). The shift can be measured by separation of the DNA in alkaline agarose gels, transfer to nylon membranes, hybridization with ^{32}P-labeled genomic DNA (e.g., by random priming), and detection with a PhosphorImager.[32] The storage phosphor technology is very sensitive and has a linear response range over five orders of magnitude.

Discrete-step scans of images (Figure 47–1B) are recorded using ImageQuant software (Molecular Dynamics, Sunnyvale, CA). The half-area DNA migration points, or average length of the population of DNA molecules, are calculated on an Excel spreadsheet (Microsoft, Redmond, WA) from the migration profile of the population of DNA molecules, relative to DNA size standards. Finally, the number of CPDs/kb (or DNA strand breaks/kb) is calculated according to Freeman *et al.*[33]

1. The isolated DNA is treated with T4endoV (Figure 47–1A) as described in the section on The T4endoV Assay.

2. Alkaline agarose gels (1%) are prepared as previously described[34] and electrophoresis is run for about 13 h at 25 V (0.7 V/cm) with buffer circulation. Thereafter, gels are first neutralized for 1 h in 1.5 M NaCl, 1 M Tris–HCl (pH 7.5), then soaked in 0.25 N HCl for 30 min, and finally treated in 0.4 N NaOH for 15 min. The DNA is transferred in 0.4 M NaOH to a Hybond-XL membrane (Amersham) following the alkaline Southern blotting protocol.

3. The radioactive probe is generated from yeast total genomic DNA digested with *Eco*RI (to obtain shorter fragments) using random priming (Amersham). Prehybridization, hybridization, and washing are done as described in standard protocols, and the membranes are exposed to PhosphorImager screens.

4. Determination of the average DNA lengths (L^{av}) is done as follows:

- The image of one lane is overlaid with an array of duplicated rectangles, with the height that corresponds to 1 mm distance on the gel [Figure 47–1B; (–)T4endoV]. This array (1 to *n*) is copied onto the other lanes [Figure 47–1B; (+)T4endoV] and onto an empty lane that is used for background correction (Figure 47–1B; N).
- The image density (or volume) within each rectangle (1 to *n*) is measured with ImageQuant and the data are transferred to an Excel spreadsheet (Microsoft). A chart is generated showing the DNA profile of each lane (Figure 47–1C). [*Note*: Figure 47–1C is a schematic representation of a typical DNA profile of UV-irradiated genomic DNA, treated or mock treated with T4endoV.]
- Three backgrounds are subtracted from the image density data shown in Figure 47–1C. (a) The background generated from nonspecific binding of radiolabeled probe to the nylon membrane is removed by subtracting the density values obtained from an empty lane (Figure 47–1B, N) to the corresponding rectangles (1 to *n*) in the (–) and (+) T4endoV lanes and (b) the background due to random nicks generated during DNA isolation [visible in the lanes (–)T4endoV] is removed. (b1) First, the signal (smear) below the broad band corresponding to high-molecular-weight DNA [Figure 47–1B; (–)T4endoV, from 5 to *n*) or region "b1" (Figure 47–1D): each position in the (–)T4endoV lane is subtracted from itself and from its corresponding position in the (+)T4endoV lane. (b2) Second, region "b2" (Figure 47–1D), which corresponds to nicked high-molecular-weight DNA and is described as the area under the straight line ($y = ax$), which is generated from the origin of the high-molecular-weight DNA and the first data point of the smear (rectangle 5); the *y*-values are subtracted from each corresponding data point in the (–) and (+)T4endoV lanes. The results obtained after subtraction of the three backgrounds are shown in Figure 47–1E.
- The total areas encompassed by the corrected scans are calculated, and the data points corresponding to half of these areas are determined (Figure 47–1E; X_{med}). By using the standard curve generated from the migration of the DNA size markers (Figure 47–1A; M), X_{med} is converted to the median length L_{med} (in bases). The average molecular length (L^{av}) is then calculated using the following equation[35]:

$$L^{av} = 0.6 \times L_{med}$$

assuming a Poisson distribution of DNA fragments.

5a. For UV-irradiated DNA, the number of CPDs/kb is calculated using the following equation[33]:

$$\text{CPDs/kb} = (L^{av})^{-1}_{(+)\text{T4endoV}} - (L^{av})^{-1}_{(-)\text{T4endoV}}$$

where $(L^{av})_{(+)\text{T4endoV}}$ and $(L^{av})_{(-)\text{T4endoV}}$ are the L^{av} values for samples treated or mock treated with T4endoV, respectively.

5b. For DNA SSBs, for example, induced by BLEO, the number of SSBs/kb is calculated using the following equation[32]:

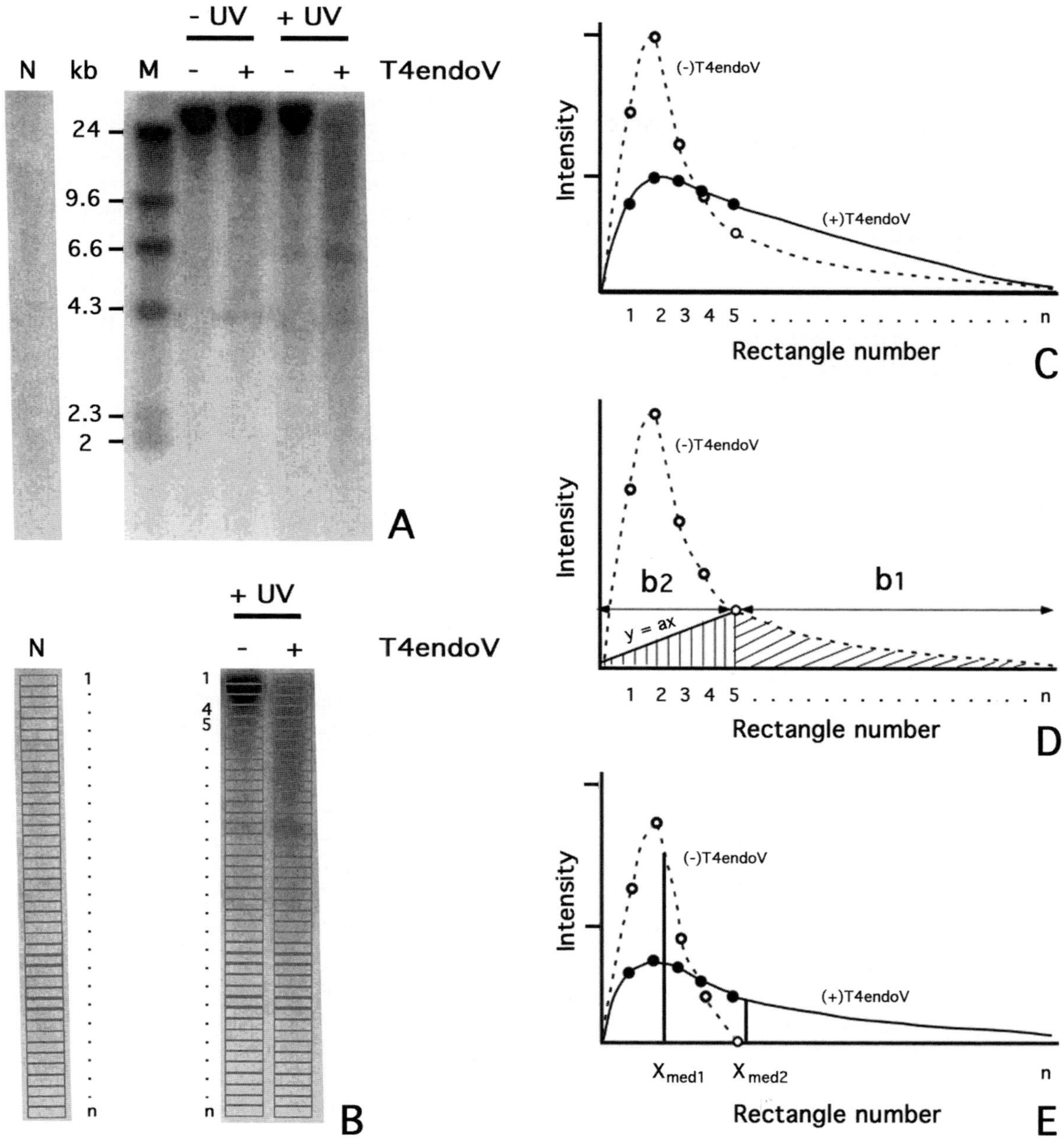

Figure 47–1. Measurements of DNA strand breaks (induced by T4endoV cleavage at CPD sites) by alkaline agarose gel electrophoresis and Southern blot.

$$SSB/kb = (L^{av})^{-1}$$

6. For repair experiments, CPDs/kb (or SSB/kb) at various repair times are expressed as percent of the damage measured at time of 0 repair (see the section on The T4endoV Assay).

THE T4ENDOV ASSAY TO MEASURE NUCLEOTIDE EXCISION REPAIR OF ULTRAVIOLET-INDUCED PHOTOPRODUCTS IN SPECIFIC DNA SEQUENCES

1. DNA is isolated before UV irradiation (−UV), soon after UV irradiation (+UV; 0 repair), and after various repair times (+UV; 1/2, 1, 2, and 4 h) (Figure 47–2A). About 6 µg of DNA for each time point is digested with the suitable restriction enzyme. [For example, in Figure 47–2A the DNA was digested with *Eco*RI, which releases an ~2.9-kb DNA fragment when blots are hybridized with a probe corresponding to the central portion of ribosomal genes (rDNA).]

2. After extraction with phenol:chloroform:isoamyl alcohol and chloroform, the DNA is precipitated in ethanol and resuspended in 32 µl of T4endoV buffer (50 mM Tris–HCl, pH 7.5, and 5 mM EDTA).

3. On ice, dilute 1 µl of T4endoV enzyme (e.g., from Epicentre) 1 : 50 in T4endoV buffer, and prepare the reactions in 0.5-ml Eppendorf tubes as follows:

- To the mock-treated samples [(−)T4endoV] add 2 µl of T4endoV buffer.
- To the test samples (+)T4endoV] add 2 µl of the diluted T4endoV enzyme.

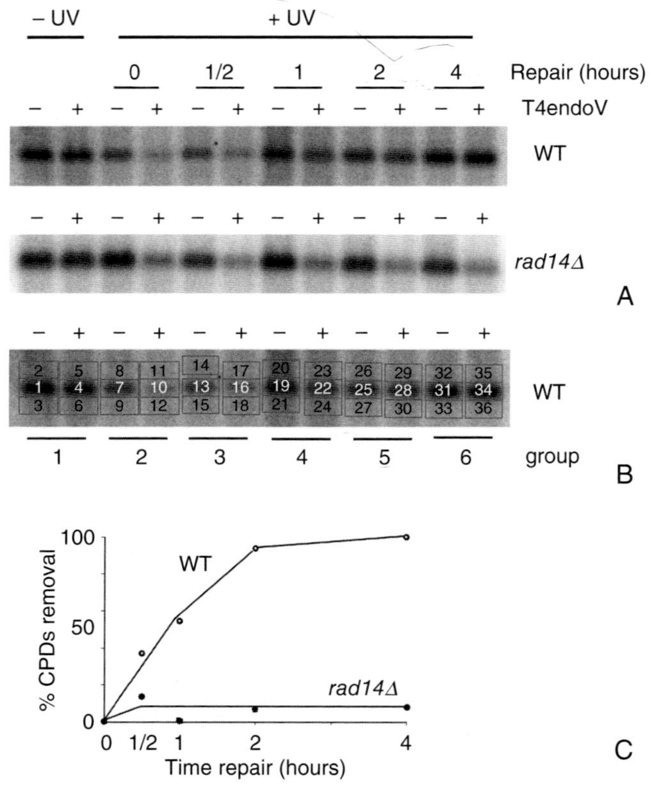

Figure 47–2. Measurements of repair of CPDs in defined DNA sequences by the T4endoV assay.

- Add 15 µl of the DNA suspensions (out of the 32 µl total) to the test samples and to the mock-treated controls, and incubate the reaction for 45 min at 37°C.

4. To each sample add 3 µl of 6× alkaline loading buffer [300 mM NaOH, 6 mM EDTA, 18% (w/v) Ficoll (type 400), 0.15% (w/v) bromocresol green, 0.25% (w/v) xylene cyanol], which brings the final volumes to 20 µl, and load the samples onto 1% alkaline agarose gels that are prepared as previously described.[34] The gel electrophoresis is run for about 13 h at 25 V (0.7 V/cm) with buffer circulation, and the DNA is transferred in 0.4 M NaOH to a Hybond-XL membrane (Amersham) following the alkaline Southern blotting protocol. Thereafter, the membrane is hybridized with ^{32}P-labeled specific DNA sequences. [In Figure 47–2A, the blots carrying DNA from the wild-type (WT) and from the NER-deficient mutants (*rad14Δ*) were hybridized with a rDNA probe.] Changes in bands intensities, compare lanes (–) and (+), reflect the number of CPDs. Whereas DNA repair is observed in WT, no removal of CPDs is found in *rad14Δ* cells.

5. Quantification of the image scans is done with ImageQuant software (Molecular Dynamics, Sunnyvale, CA) as follows:

- A rectangle is drawn tightly around the signal corresponding to –UV (–)T4endoV (Figure 47–2B; group 1, rectangle 1), taking care not to touch the band. Five duplicated rectangles are overlaid on the (+)T4endoV band (rectangle 4) and on the regions adjacent to the bands that are used for background correction (rectangles 2, 3, 5, and 6). This procedure is repeated for each of the six groups.
- The densities (or volumes, *V*) of each rectangle (1 to 36) are measured (ImageQuant) and the data are transferred to an Excel spreadsheet (Microsoft).
- The volume (*V*) of each band is corrected by subtracting its average background, which is generated from the nonspecific binding of radiolabeled probe to the membrane (rectangle above the band) and from random nicks generated during DNA isolation (rectangle below the band), and is called $\bar{V}$ [e.g., $\bar{V}_{19} = V_{19}\,(V_{20} + V_{21})/2$].
- The number of CPDs per DNA fragment (in Figure 47–2A; the ~2.9-kb *Eco*RI rDNA fragment) is calculated with the following equation:

$$\text{CPDs/DNA fragment} = \ln(\bar{V}_{(-)\text{T4endoV}} / \bar{V}_{(+)\text{T4endoV}})$$

6. The resistance of DNA cutting by T4endoV is monitored as a measure of repair, and the percent of CPD removal is plotted over the repair incubation time (Figure 47–2C). (In the rDNA of WT cells ~50% of CPDs is removed after 2 h and ~100% repair is achieved after 4 h. No repair is measured for the same DNA fragment in *rad14Δ* cells.)

ACKNOWLEDGMENTS

I would like to thank Dr. K. Kobryn at the Université de Sherbooke and D. Fahy at Washington State University for critical reading of the manuscript. Also, I would like to thank the students in my laboratory, M. Toussaint, M. Tremblay, G. Levasseur and J. Doyon for providing me with the most up to date protocols. The Natural Sciences and Engineering Research Council of Canada (NSERC) supported this project.

REFERENCES

1. Ausubel FM, Brent R, Kingston RE, Moore DD, Seidman JG, Smith JA, Struhl K. *Current Protocols in Molecular Biology*. New York: John Wiley & Sons, Inc., 2004.
2. Curran BPG, Bugeja V. Basic Investigations in *Saccharomyces cerevisiae*. In: Wei X, Ed. *Yeast Protocols, Methods in Molecular Biology* 313. Totowa, NJ: Humana Press, 2006.
3. Guthrie C, Fink GR. Guide to yeast genetics and molecular biology. In: *Methods in Enzymology*, Vol. 194. San Diego: Academic Press, 1991.
4. Hartwell LH. Yeast and cancer. *Biosci Rep* 2004;24:523–544.
5. Cherry JM, Ball C, Weng S, Juvik G, Schmidt R, Adler C, Dunn B, Dwight S, *et al.* Genetic and physical maps of *Saccharomyces cerevisiae*. *Nature* 1997;387:67–73.
6. Bash R, Lohr D. Yeast chromatin structure and regulation of *GAL* gene expression. *Progr Nucleic Acids Res Mol Biol* 2001; 65:197–259.
7. Conconi A, Wellinger RJ. A new link for a linker histone. *Mol Cell* 2003;11:1421–1423.
8. Wach A, Brachat A, Pohlmann R, Philippsen P. New heterologous modules for classic or PCR-based gene disruptions in *Saccharomyces cerevisiae*. *Yeast* 1994;10:1793–1808.
9. De Antoni A, Gallwitz D. A novel multi-purpose cassette for repeated integrative epitope tagging of genes in *Saccharomyces cerevisiae*. *Gene* 2000;246:179–185.
10. Mnaimneh S, Davierwala AP, Haynes J, Moffat J, Peng WT, Zhang W, Yang X, Pootoolal J, *et al.* Exploration of essential gene functions via titrable promoter alleles. *Cell* 2004;118:31–44.
11. Kumar A, Harrison PM, Cheung KH, Lan N, Echols N, Bertone P, Miller P, Gerstein MB, *et al.* An integrated approach for finding overlooked genes in yeast. *Nat Biotechnol* 2002;20:58–63.
12. Mewes HW, Frishman D, Guldener U, Mannhaupt G, Mayer K, Mokrejs M, Morgenstern B, Munsterkotter M, *et al.* MIPS: A database for genomes and protein sequences. *Nucleic Acids Res* 2002;30:31–34.
13. Giaever G, Chu AM, Ni L, Connelly C, Riles L, Veronneau S, Dow S, Lucau-Danila A, *et al.* Functional profiling of the *Saccharomyces cerevisiae* genome. *Nature* 2002;418:387–391.

14. Hudson JR, Dawson EP, Rushing KL, Jackson CH, Lockshon D, Conover D, Lanciault C, Harris JR, *et al.* The complete set of predicted genes from *Saccharomyces cerevisiae* in a readily usable form. *Genome Res* 1997;7:1169–1173.
15. Suter B, Auerbach D, Stagljar I. Yeast-based functional genomics and proteomics technologies: The first 15 years and beyond. *Biotechniques* 2006;40:625–644.
16. Hartwell LH, Szankasi P, Roberts CJ, Murray AW, Friend SH. Integrating genetic approaches into the discovery of anticancer drugs. *Science* 1997;278:1064–1068.
17. Lee W, St Onge RP, Proctor M, Flaherty P, Jordan MI, Arkin AP, Davis RW, Nislow C, *et al.* Genome-wide requirements for resistance to functionally distinct DNA-damaging agents. *PLoS Genet* 2005;1: e24.
18. Chang M, Bellaoui M, Boone C, Brown GW. A genome-wide screen for methyl methanesulfonate-sensitive mutants reveals genes required for S phase progression in the presence of DNA damage. *Proc Natl Acad Sci USA* 2002;99:16934–16939.
19. Aouida M, Page N, Leduc A, Peter M, Ramotar D. A genome-wide screen in Saccharomyces cerevisiae reveals altered transport as a mechanism of resistance to the anticancer drug bleomycin. *Cancer Res* 2004;64:1102–1109.
20. Huang RY, Eddy M, Vujcic M, Kowalski D. Genome-wide screen identifies genes whose inactivation confer resistance to cisplatin in *Saccharomyces cerevisiae. Cancer Res* 2005;65:5890–5897.
21. Game JC, Birrel GW, Brown JA, Shibata T, Baccari C, Chu AM, Williamson MS, Brown JM. Use of a genome-wide approach to identify new genes that control resistance of *Saccharomyces cerevisiae* to ionizing radiation. *Radiat Res* 2003;160:14–24.
22. Birrell GW, Giaever G, Chu AM, Davis RW, Brown JM. A genome-wide screen in *Saccharomyces cerevisiae* for genes affecting UV radiation sensitivity. *Proc Natl Acad Sci USA* 2001;98: 12608–12613.
23. Friedberg EC, Walker GC, Siede W, Wood RD, Schultz RA, Ellenberger T. *DNA Repair and Mutagenesis,* 2nd ed. Washington, DC: ASM Press, 2005.
24. Prakash S, Prakash L. Nucleotide excision repair in yeast. *Mutat Res* 2000;451:13–24.
25. Conconi A, Vyacheslav BA, Smerdon MJ. Transcription-coupled repair in RNA polymerase I-transcribed genes of yeast. *Proc Natl Acad Sci USA* 2002;99:649–654.
26. Li S, Smerdon MJ. Nucleosome structure and repair of N-methylpurines in the GAL1-10 genes of *Saccharomyces cerevisiae. J Biol Chem* 2002;277:44651–44659.
27. Solomon LR, Beerelli RD, Moseley PL. Bleomycin-iron can degrade DNA in the presence of excess ethylenediaminetetraacetic acid in vitro. *Biochemistry* 1989;28:9932–9937.
28. Toussaint M, Levasseur G, Gervais-Bird J, Wellinger RJ, Abou Elela S, Conconi A. A high-throughput method to measure the sensitivity of yeast cells to genotoxic agents in liquid cultures. *Mutat Res* 2006;606:92–105.
29. Conconi A, Jager-Vottero P, Zhang X, Beard BC, Smerdon MJ. Mitotic viability and metabolic competence in UV-irradiated yeast cells. *Mutat Res* 2000;459:55–64.
30. DNA Repair Protocols. In: Henderson DS, Ed. *Methods in Molecular Biology*, Vol. 113. Totowa, NJ: Humana Press, 1999.
31. Bohr VA, Smith CA, Okumoto DS, Hanawalt PC. DNA repair in an active gene: Removal of pyrimidine dimers from the DHFR gene of CHO cells is much more efficient than in the genome overall. *Cell* 1985;40:359–369.
32. Bespalov VA, Conconi A, Zhang X, Fahy D, Smerdon MJ. Improved method for measuring the ensemble average of strand breaks in genomic DNA. *Environ Molec Mutagen* 2001;38:166–174.
33. Freeman SE, Blackett AD, Monteleone DC, Setlow RB, Sutherland BM, Sutherland JC. Quantitation of radiation-, chemical-, or enzyme-induced single strand breaks in nonradioactive DNA by alkaline gel electrophoresis: Application to pyrimidine dimers. *Anal Biochem* 1986;158:119–129.
34. Maniatis T, Fritsch EF, Sambrook J. *Molecular Cloning, A Laboratory Manual.* Cold Spring Harbor, NY: Cold Spring Harbor Laboratory, 1982.
35. Veatch W, Okada S. Radiation-induced breaks of DNA in cultured mammalian cells. *Biophys J* 1969;9:330–346.

MODELS OF BEHAVIOR

V

Physiology

D

48 Human Models of Space Physiology

VICTOR A. CONVERTINO AND CAROLINE A. RICKARDS

ABSTRACT

Maintaining the health and fitness of astronauts is essential to the successful exploration of space. Consequently, the development of methods (i.e., countermeasures) designed to minimize potential adverse clinical effects of space flight depends on an understanding of the physiological mechanisms that underlie human adaptation to microgravity. However, uncontrolled conditions and mission logistics associated with spaceflight present significant limitations to the scientific study of human physiology in the space environment. Although no one model precisely simulates the actual space environment, the quantitative as well as qualitative comparisons of space physiology with the 6° head-down bed rest (HDBR) model are striking. Selective comparisons presented in this chapter demonstrate distinct qualitative and quantitative similarities in the underlying physiology of body fluids, cardiovascular and autonomic functions, muscle, bone, and metabolism in humans between HDBR and space. Because of these similarities, the use of the HDBR model has provided critical direction for the investigation of space physiology that otherwise could not be addressed adequately in the space environment. HDBR has proven to be one of the most effective and valuable models for assessing the effects of prolonged exposure to microgravity on human physiological functions by inducing physical and physiological changes similar to those that occur when humans are exposed to the actual environment of space.

Key Words: Head-down bed rest, Blood volume, Cardiovascular functions, Autonomic nervous function, Bone, Muscle, Metabolism, Orthostatic intolerance, Physical exercise.

INTRODUCTION

Since the beginning of human spaceflight, the value of understanding mechanisms of physiological adaptation to microgravity became apparent to life scientists who were interested in maintaining crew health and developing countermeasures against the adverse effects of a mission. However, several characteristics associated with the logistics of spaceflight presented significant limitations to the scientific study of human adaptation to microgravity. Because space missions are so infrequent and involve minimal numbers of crewmembers, meaningful statistical analyses of data are limited. The reproducibility of results from experiments conducted in space is difficult to assess since there are few repeated space missions involving the same crewmembers. Since the emphasis of space missions is placed on operations, human physiological experiments are often compromised by numerous confounding and uncontrolled factors (e.g., time, diet, physical activities, circadian rhythms) that can impact measured responses. With the limited time available to collect data in space, crewmembers are required to participate in numerous simultaneous experiments proposed by multiple investigators that can adversely affect the outcomes of each independent experiment. The technology and ability to measure physiological functions necessary to test specific hypotheses can also be severely limited by physical space and power constraints of the space environment. Finally, technical and logistical aspects of space missions such as launch delays, extended missions, and operational emergencies can significantly compromise the timing and control of experiments. These limitations have stimulated scientists to develop ground-based models of microgravity in an effort to investigate the effects of space on physiological function in a controlled experimental setting.

TRADITIONAL MODELS FOR SPACE PHYSIOLOGY

The term *microgravity* describes a condition in which gravitational forces acting on the long axis of the body are minimized in space. The key to effective simulation of microgravity on Earth is to remove this gravitational stimulus as much as possible. There are several methods that have been attempted to achieve this requirement in the simulation of microgravity on humans.

FREE FALL AND PARABOLIC FLIGHT Free fall occurs when an object plunges downward toward Earth at the rate of acceleration due to gravity (980 cm/sec^2). One model often used to investigate space physiology is a special aircraft that flies repeated parabolic arcs that result in a controlled, 45° nose-down, free-fall dive. The passengers inside the aircraft experience an immediate transition into free fall (weightlessness) from the top of the arc that persists during the downward portion of the parabola. Unfortunately, the average duration of weightlessness with this technique is only 20–25 sec, limiting the collection of data and the interpretation of physiological responses to a very acute

From: *Sourcebook of Models for Biomedical Research* (P. M. Conn, ed.), © 2008 Humana Press Inc., Totowa, NJ.

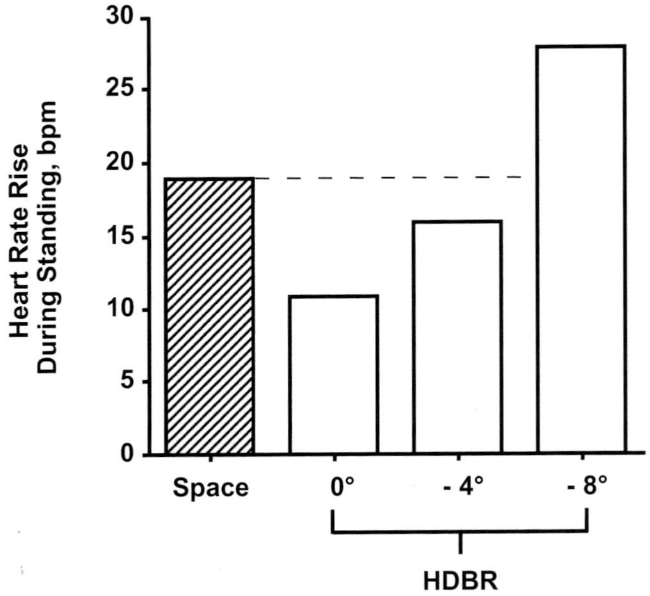

Figure 48–1. Elevation in heart rate during standing after exposure to space (closed bar) and 0°, 4°, and 8° HDBR (open bars). (Data modified from Kakurin *et al.*[1])

time course. Additionally, parabolic flight often induces symptoms of motion sickness such as nausea and vomiting that can potentially confound the measured outcome variables.

WATER IMMERSION Placement of human subjects in a small pool of thermoneutral (~34.5°C) water with their head just above water is used to simulate microgravity, as the buoyancy of the surrounding water removes the full effect of Earth's gravity. A major difficulty experienced in such studies is the high rate of heat transfer between the subject and the surrounding water, and prolonged exposure of the skin to the immersion fluid. To avoid such problems, models of "dry" immersion have been developed that enclose subjects in a waterproof material. This model of microgravity has obvious limitations for performing experimental investigations that require physical access to subjects, and there are also limitations on the amount of time a subject can be exposed to this environment.

SUSPENSION AND IMMOBILIZATION Musculoskeletal unloading similar to space can be produced by suspending subjects above the ground with springs, or by supporting them from below with air jets (air-bearing floors). Such methods create regions of tissue compression that can be tolerated for only short periods of time. Limb immobilization or casting has been used to study regional changes in muscle, but fixing the limb may not represent the conditions of spaceflight where astronauts move their limbs freely. Models of unilateral lower limb suspension have been developed that involve a support strap or high platform shoe allowing the subject to ambulate with crutches while the unloaded leg hangs freely. This model is primarily limited to the study of alterations in muscle and bone associated with space travel without inducing the headward fluid shifts associated with exposure to space.

HEAD-DOWN BED REST Bed rest has become the most popular model since it can be applied for prolonged periods of time (weeks to months) and affects most of the physiological systems of the body. The work of Kakurin and co-workers[1] represents one of the first comparisons of bed rest data to space data. These investigators determined that the magnitude of the heart rate response to a stand test following return from space was between the responses produced by exposure of human subjects to 4° and 8° head-down bed rest (HDBR) (Figure 48–1). As a result of this investigation, 6° HDBR has been adopted as the universal human model for the study of space physiology. Thus, the remainder of this chapter will focus on the comparison of select human physiological data collected from HDBR experiments with those obtained from space experiments.

BODY FLUIDS

One of the earliest effects on the human body upon entry into the microgravity of space is a headward fluid shift and enlargement of the heart.[2] A fundamental adaptation to this headward fluid shift is a reduction in plasma and blood volume.[3] The time course of the change in circulating vascular volume is characterized by a rapid reduction within 24–48 h followed by a newly stabilized hypovolemic state (Figure 48–2). Figure 48–2 demonstrates the striking similarity in magnitude as well as time course of plasma volume reductions during exposure to space and HDBR. Interestingly, the average magnitude of plasma volume reduction (13%) from the subsequent SLS-1, SLS-2, and D-2 NASA space missions after 9–14 days of flight is nearly identical to that predicted by the HDBR graph generated in Figure 48–2.[3] The alteration in body fluids with exposure to space is also reflected in similar reductions in body weight observed following similar durations of spaceflight (3%) and HDBR (4%).[4,5] Thus, the similarity in time course and magnitude between space and HDBR data suggests that the latter provides an effective model to investigate the mechanisms associated with reduced plasma volume and body fluid redistribution that cannot be readily studied in space.

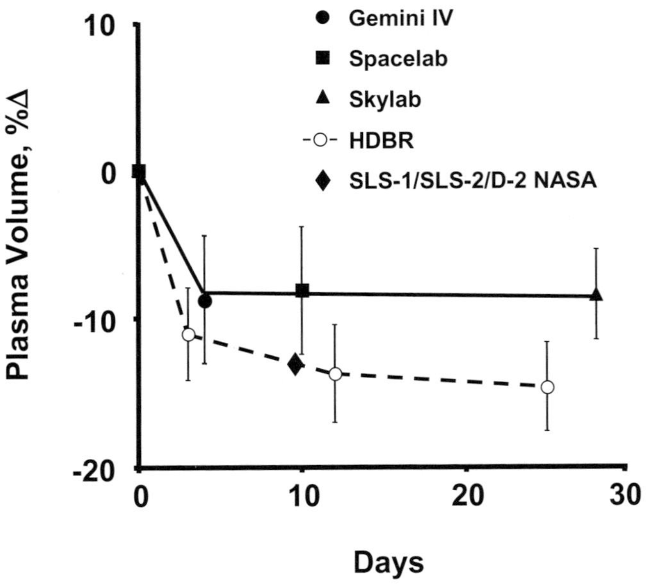

Figure 48–2. Comparison of time courses of plasma volume changes (%Δ) during adaptation to space (closed symbols and solid line) and HDBR (open circles and broken line). (Modified from Convertino.[3])

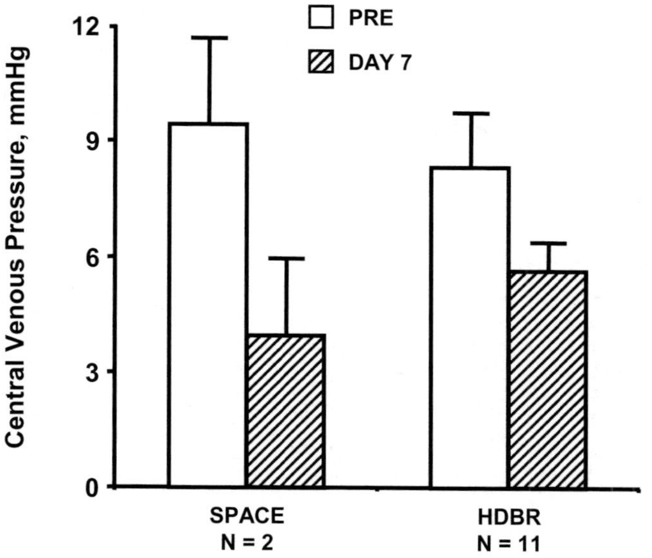

Figure 48–3. Comparison of estimated central venous pressure (CVP) before (open bars) and on day 7 of (closed bars) exposure to space and HDBR. Values are mean ± 1 SEM (lines). (Modified from Convertino.[8])

CARDIOVASCULAR

There is a vast collection of data on the physiology of the cardiovascular system and its adaptation to space, perhaps because of its important contribution to astronaut health and performance. In conjunction with the decrease in plasma volume, the general cardiovascular adaptation to space and HDBR includes a reduction in stroke volume with compensatory elevations in peripheral arterial vasoconstriction and heart rate leading to the maintenance of cardiac output and arterial blood pressures.[2,5,6] Magnetic resonance imaging measurements conducted on four astronauts who participated in the D-2 NASA space mission revealed an average 14% reduction in left ventricular mass, which is comparable to the 8% decrease in cardiac mass following HDBR.[7] Data obtained from peripheral catheters suggested a reduced central venous pressure (CVP) of similar magnitude after 7 days of exposure to both space and HDBR (Figure 48–3). This reduction in CVP was later verified with direct measurements from indwelling catheters placed in the right atrium of five astronauts while in space.[2] Venous compliance of the calf is also decreased in HDBR and in space.[8] These comparisons support the idea that HDBR represents an excellent model for the investigation of cardiovascular responses associated with the physiological adaptation to space.

AUTONOMIC NERVOUS SYSTEM

It is almost impossible to interpret responses of the sympathetic nervous system (SNS) and the parasympathetic nervous system (PNS) while in space because of wide variations in conditions such as physical activity, gender, energy balance, and diet. However, direct and indirect indices of SNS and PNS responses measured before and after exposure to space and HDBR are qualitatively similar. PNS activity, indicated by the high-frequency spectral power of R-R intervals calculated from an electrocardiogram, is reduced following adaptation to space and HDBR.[9,10] Conversely, direct measurements from muscle sympathetic nerves indicate that SNS activity is dramatically elevated following exposure to space and HDBR.[11] Autonomically mediated baroreflexes also appear to be affected in a similar fashion by HDBR and space.[8] A shift of the carotid–cardiac baroreflex stimulus–response relationship downward and to the right indicates that for a given reduction in arterial pressure. there will be a smaller compensatory increase in heart rate (i.e., less sensitivity of the baroreflex) after adaptation to microgravity or HDBR (Figure 48–4). The similarity in these data supports the notion that HDBR provides a valid model to investigate the effects of space on mechanisms of autonomic nervous function and regulation.

MUSCLE

Exposure to space and HDBR causes unloading of skeletal muscles, especially those of the lower weight-bearing extremities. The reduction in leg volume is similar in individuals exposed to space and HDBR and has been used as an indirect indication of muscle atrophy (Table 48–1). Actual muscle biopsies taken from

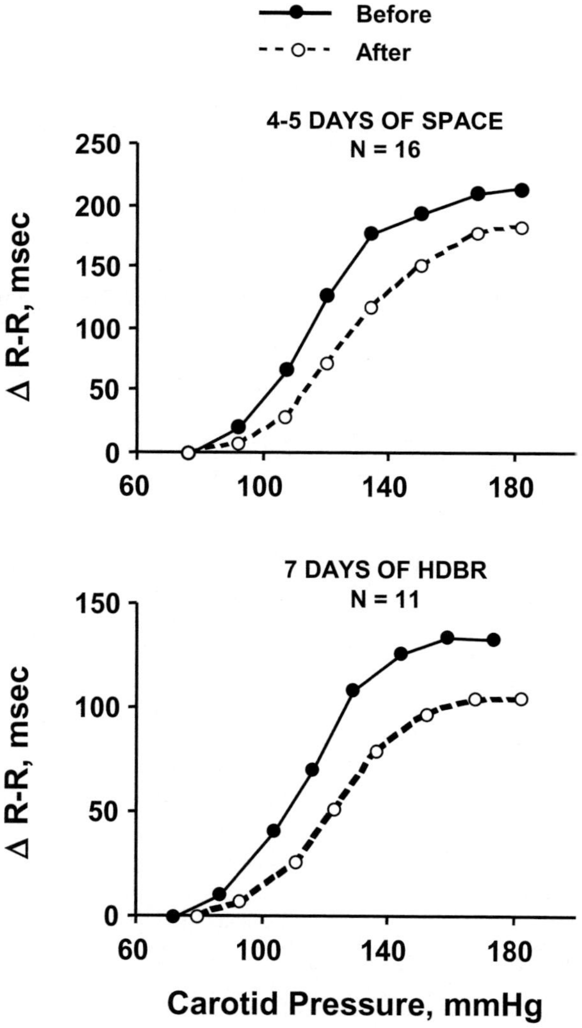

Figure 48–4. Average carotid–cardiac baroreflex stimulus–response relationships before (closed circles and solid line) and after (open circles and broken line) exposure to space (top panel) and HDBR (bottom panel). (Modified from Convertino *et al.*[8])

Table 48–1
Comparison of changes reported following adaptation to space and head-down bed rest (HDBR)[a]

Physiological variable	*References*	*Microgravity analog*	*Days of exposure*	*N*	*Condition*	%Δ
Body weight	Convertino[4*]	Spaceflight	28	3	Rest	3%↓
Body weight	Convertino[4*]	6° HDBR	30	8	Rest	4%↓
Venous pressure	Convertino[8*]	Spaceflight	7	2	Rest	58%↓
Venous pressure	Convertino[8*]	6° HDBR	7	11	Rest	32%↓
Baroreflex	Convertino[8*]	Spaceflight	4–5	16	Rest	28%↓
Baroreflex	Convertino[8*]	6° HDBR	12	11	Rest	31%↓
Plasma volume	Fischer *et al.*[20]	Spaceflight	4	2	Rest	9%↓
Plasma volume	Convertino[3*]	6° HDBR	3	11	Rest	12%↓
Leg volume	Convertino[4*]	Spaceflight	28	3	Rest	10%↓
Leg volume	Convertino[4*]	6° HDBR	30	8	Rest	10%↓
Knee flexors	Convertino[4*]	Spaceflight	28	3	Maximum torque	8%↓
Knee flexors	Convertino[4*]	6° HDBR	30	8	Maximum torque	10%↓
Knee extensors	Convertino[4*]	Spaceflight	28	3	Maximum torque	21%↓
Knee extensors	Convertino[4*]	6° HDBR	30	8	Maximum torque	21%↓
Oxygen uptake	Kakurin *et al.*[1]	Spaceflight	5	7	700 Kgm/min	4%↑
Oxygen uptake	Convertino[8*]	6° HDBR	7	5	600 Kgm/min	5%↑
Heart rate	Kakurin *et al.*[1]	Spaceflight	5	7	700 Kgm/min	13%↑
Heart rate	Convertino[8*]	6° HDBR	7	5	600 Kgm/min	12%↑
Oxygen pulse	Kakurin *et al.*[1]	Spaceflight	5	7	700 Kgm/min	9%↑
Oxygen pulse	Convertino[8*]	6° HDBR	7	5	600 Kgm/min	8%↑
Stroke volume	Convertino[19*]	Spaceflight	237	2	765 Kgm/min	31%↓
Stroke volume	Convertino[19*]	0° HDBR	10	12	835 Kgm/min	28%↓
Ejection fraction	Convertino[19*]	Spaceflight	237	2	765 Kgm/min	13%↑
Ejection fraction	Convertino[19*]	0° HDBR	10	12	825 Kgm/min	20%↑
Cardiac mass	Perhonen *et al.*[7]	Spaceflight	10	4	Supine	14%↓
Cardiac mass	Perhonen *et al.*[7]	6° HDBR	42	5	Supine	8%↓
Heart rate	Convertino[8*]	Spaceflight	7	9	Standing	47%↑
Heart rate	Convertino[8*]	6° HDBR	30	10	Standing	55%↑
Fiber XS area	Edgerton *et al.*[12]	Spaceflight	14	5	Rest fast twitch	23%↓
Fiber XS area	Convertino[4*]	6° HDBR	30	7	Rest fast twitch	18%↓
Fiber XS area	Edgerton *et al.*[12]	Spaceflight	14	5	Rest slow twitch	16%↓
Fiber XS area	Convertino[4*]	6° HDBR	30	7	Rest slow twitch	11%↓
Capillary-to-fiber ratio	Edgerton *et al.*[12]	Spaceflight	14	5	Rest slow twitch	20%↓
Capillary-to-fiber ratio	Convertino[4*]	6° HDBR	30	7	Rest slow twitch	37%↓
Succinate dehydrogenase	Edgerton *et al.*[12]	Spaceflight	14	5	Rest slow twitch	13%↓
β-Hydroxyocyl-CoA DH	Convertino[4*]	6° HDBR	30	7	Rest slow twitch	18%↓
Bone loss—calcaneus	Collet *et al.*[21]	Spaceflight	182	1	Rest	13%↓
Bone loss—calcaneus	LeBlanc *et al.*[22]	6° HDBR	119	6	Rest	10%↓
Bone mineral density—tibia	Vico *et al.*[23]	Spaceflight	182	11	Rest	5%↓
Bone mineral density—trochanter	Zerwekh *et al.*[24]	0° HDBR	84	11	Rest	4%↓
Bone resorption—[hydroxyproline]	Smith *et al.*[25*]	Spaceflight	59	6	Rest	33%↑
Bone resorption—[hydroxyproline]	Zerwekh *et al.*[24]	0° HDBR	84	11	Rest	45%↑
Urinary calcium	Smith *et al.*[25*]	Spaceflight	84	3	Rest	47%↑
Urinary calcium	LeBlanc *et al.*[26]	0° HDBR	119	8	Rest	39%↑
Serum calcium	Smith *et al.*[25*]	Spaceflight	115	3	Rest	2%↑
Serum calcium	Zerwekh *et al.*[24]	0° HDBR	84	11	Rest	2%↑
Vitamin D	Smith *et al.*[25*]	Spaceflight	115	3	Rest	25%↓
Vitamin D	LeBlanc *et al.*[26]	0° HDBR	119	8	Rest	17%↓

[a]An asterisk (*) indicates data obtained from a review article rather than a primary source.

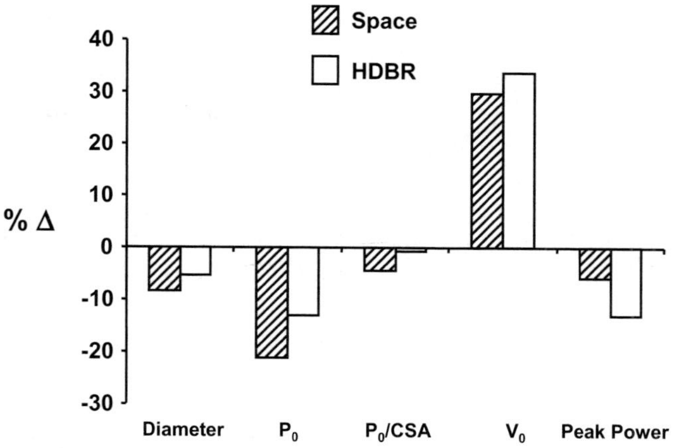

Figure 48–5. Average relative changes (%Δ) in single muscle fiber diameter, peak force (P_0), normalized peak force corrected for fiber size (P_0/CSA), unloaded shortening velocity (V_0), and peak power obtained from human muscle biopsies of the soleus (calf) muscle before and after exposure to HDBR (open bars) and space (closed bars). (Modified from Widrick *et al.*[13])

subjects before and after exposure to space and HDBR have confirmed significant muscle atrophy under both conditions as evidenced by similar reductions in cross-sectional areas of both slow-twitch and fast-twitch muscle fibers of the vastus lateralis (thigh) muscle.[4,12] Muscle atrophy was accompanied by similar reductions in the capillary-to-fiber ratio in slow-twitch fibers after exposure to HDBR (11%) and space (16%).[4,12] In a study specifically designed to compare the HDBR model to space exposure, measurements of muscle structure and function were performed on single soleus (calf) muscle fibers obtained from biopsies of four astronauts and eight subjects before and after 17 days of muscle unloading.[13] Measurements included muscle fiber structural (i.e., diameter) and contractile (i.e., peak force, unloaded shortening velocity, peak power) properties. Muscle fiber diameter, peak force, and peak power were reduced while unloaded shortening velocity increased (Figure 48–5). The comparisons revealed alterations of similar magnitude and direction for space and HDBR. Clearly, HDBR provides an effective model to investigate the unloading of muscle caused by space.

METABOLISM

In addition to changes in muscle morphology and function, exposure to microgravity has been associated with histochemical and biochemical alterations in muscle that can impact cellular metabolism. Although there were no significant changes in the activities of glycolytic enzymes (e.g., ATPase, lactate dehydrogenase, phosphofructokinase) in the soleus and vastus lateralis muscles following HDBR or space, the activities of enzymes associated with aerobic metabolic pathways (e.g., succinate dehydrogenase, citrate synthase) were reduced in both slow- and fast-twitch muscle fibers.[4,12] Measurements of energy intake and expenditure were performed using the doubly labeled water method on four astronauts and eight subjects before and after 17 days of exposure to space and HDBR.[14] Energy intake was 22% lower and energy expenditure was 32% higher in space compared with HDBR (Figure 48–6). As a result, astronauts experienced a significant negative energy balance and weight loss in space that were not observed in the HDBR subjects. Despite an experiment designed to match physical exercise regimens, the differences in energy exchange between space and HDBR are not unexpected given the additional energy requirements of spacecraft movement and extravehicular activities. It is clear from these data that adding more physical activity or decreasing dietary intake during HDBR would provide an improved model for the study of metabolism during space.

BONE

As with the muscular system, the absence of gravity has a profound impact on both the structure and function of bone, specifically the weight-bearing bones of the lower body. Generally, unloading of bone in space and bed rest has been associated with reduced bone formation, increased bone resorption (breakdown), and changes in the distribution of bone, ultimately leading to reductions in absolute bone mass and in bone mineral density.[15] Skeletal unloading leads to an altered regulation and metabolism of calcium as indicated by increased calcium excretion (urinary and blood), negative calcium balance, and alterations in bone-regulating hormones such as vitamin D. The magnitudes of these changes in bed rest are similar to those reported in space (Table 48–1) and are proportional to the time of exposure. Thus, bed rest provides a valid model to investigate the effects of space on the mechanisms of bone unloading.

OPERATIONAL FUNCTIONS

ORTHOSTATIC PERFORMANCE Syncopal events (i.e., fainting) reported in 28–65% of mission specialists studied during stand or tilt tests after returning from space are comparable with the rate of syncope (~40%) observed in subjects during a stand test following HDBR.[8] In orthostatic tests without symptomatic episodes, the magnitude of the elevation in heart rate with stable blood pressure is similar after exposure to both HDBR and space (Figure 48–7). However, subjects demonstrated elevated peripheral vascular resistance during orthostatic tests following

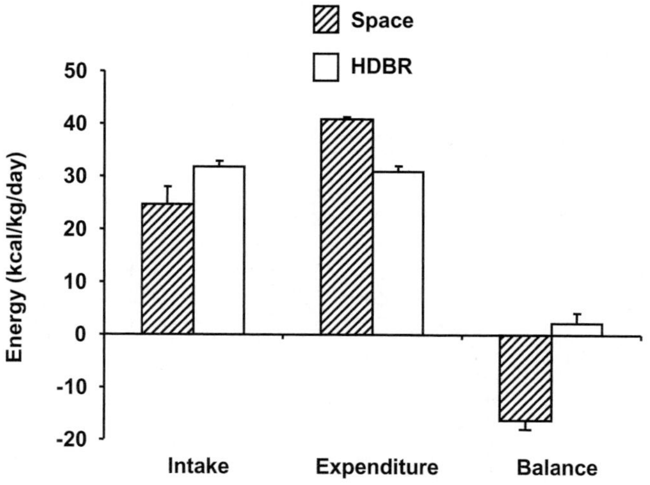

Figure 48–6. Energy intake, expenditure, and balance during 17 days of exposure to space (closed bars) and HDBR (open bars). Values are mean ± 1 SEM (lines). (Data from Stein *et al.*[14])

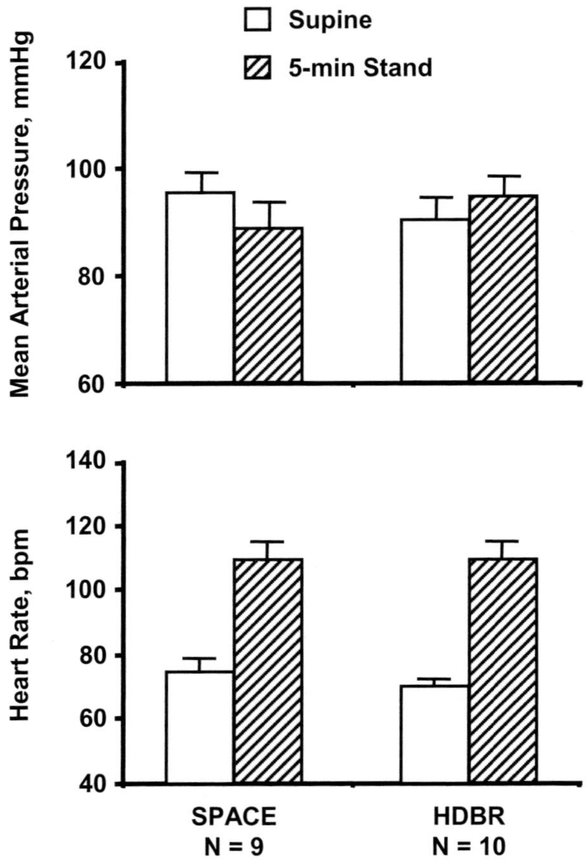

Figure 48–7. Comparison of heart rate and blood pressure responses during 5 min of standing after exposure to space and HDBR. Values are mean ± 1 SEM (lines). (Modified from Convertino.[8])

exposure to both space and HDBR.[6,16] Thus, similarities in cardiovascular responses during orthostatic challenges support the use of the HDBR as an appropriate model for the study of mechanisms that underlie orthostatic intolerance following return from space and for the development of effective countermeasures.

RESPONSES TO PHYSICAL WORK Exercise tests have proven to be one of the most effective methods of assessing the integrated effects of space on multiple physiological systems. Generally, cardiovascular responses to exercise in space and with HDBR are marked by decreased cardiac output and oxygen (O_2) pulse despite compensatory elevations in heart rate at a given intensity of physical work.[1,5,8,17] Exposure to space and HDBR results in similar reductions of stroke volume and increases in cardiac ejection fraction at a given intensity of physical work compared with prespace/HDBR measurements (Figure 48–8). The metabolic response to exercise during HDBR is also similar to space. The capacity of the body to utilize oxygen is decreased in space and HDBR.[5,18] In a study specifically designed to compare the HDBR model to space, cardiorespiratory and metabolic measurements were performed on four astronauts (VO_2 max decreased 10%) and eight subjects (VO_2 max decreased 7%) before and after 17 days of exposure to space and HDBR.[18] Figure 48–9 demonstrates that the relative change (%Δ) in oxygen uptake, heart rate, ventilatory volume (V_E), O_2 pulse, and ventilatory equivalent (V_E/VO_2) with maximal exercise was similar in space and HDBR. From these investigations, HDBR has become a primary model for simulating the effects of prolonged exposure to space on cardiorespiratory and metabolic functions.

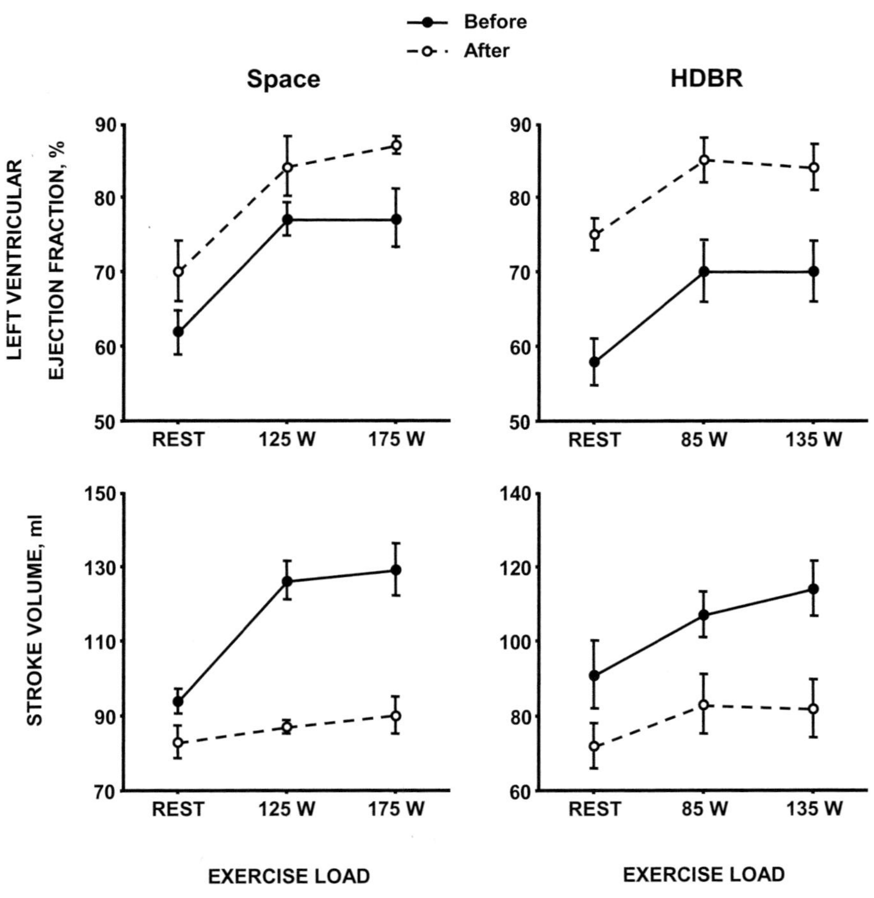

Figure 48–8. Left ventricular ejection fraction and stroke volume during rest and graded exercise before (closed circles and solid lines) and after (open circles and broken lines) space (left panels) and HDBR (right panels). Values are mean ± 1 SEM (lines). (Modified from Convertino and Cooke.[19])

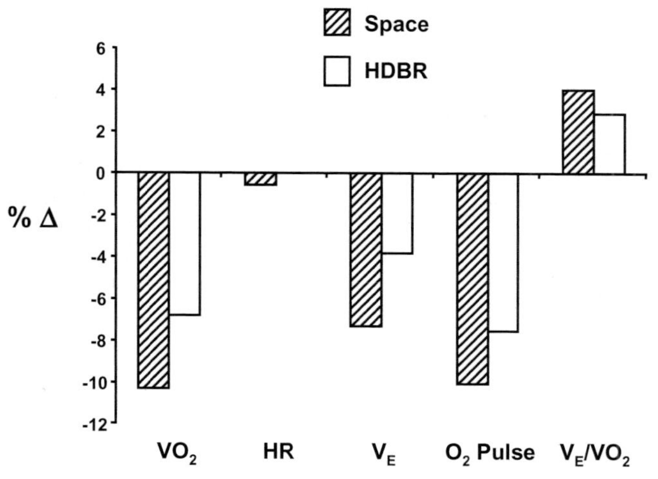

Figure 48–9. Average relative change (%Δ) in oxygen uptake (VO_2), heart rate (HR), ventilatory volume (V_E), oxygen (O_2) pulse, and ventilatory equivalent (V_E/VO_2) during maximal exercise following 17 days in space (closed bars) and HDBR (open bars). (Data from Trappe *et al.*[18])

CONCLUSIONS

No one model precisely simulates the actual space environment. In an attempt to assess the validity of various models for the simulation of microgravity effects on human physiological functions, difficulties arise with replicating the uncontrolled experimental conditions surrounding investigations conducted in space. In addition, space experiments suffer from the lack of true ground controls. Despite the extreme variability in experimental conditions and mission activities associated with space, the quantitative as well as qualitative comparisons of space physiology with the 6° HDBR model are striking (Table 48–1). Selective comparisons presented in this chapter demonstrate distinct qualitative and quantitative similarities in the underlying physiology of body fluids, cardiovascular and autonomic functions, muscle, bone, and metabolism in humans between HDBR and space. Because of these similarities, the use of the HDBR model has provided critical direction for the investigation of space physiology that otherwise could not be addressed adequately in the space environment. HDBR has proven to be one of the most effective and valuable models for assessing the effects of prolonged exposure to microgravity on function in humans by inducing physical and physiological changes similar to those that occur when humans are exposed to the actual environment of space.

REFERENCES

1. Kakurin LI, Lobachik VI, Mikhailov VM, Yu AS. Antiorthostatic hypokinesia as a method of weightlessness simulation. *Aviat Space Environ Med* 1976;47:1083–1086.
2. Buckey JC Jr, Gaffney FA, Lane LD, Levine BD, Watenpaugh DE, Wright SJ, Yancy CW, Meyer D, Blomqvist CG. Central venous pressure in space. *J Appl Physiol* 1996;81:19–25.
3. Convertino VA. Clinical aspects of the control of plasma volume at microgravity and during return to one gravity. *Med Sci Sports Exerc* 1996;28:45–52.
4. Convertino VA. Neuromuscular aspects in development of exercise countermeasures. *Physiologist* 1991;34:S125–S128.
5. Convertino VA. Exercise and adaptation to microgravity environments. In: Fregly MJ, Blatteis CM, Eds. *Handbook of Physiology: Environmental Physiology. III. The Gravitational Environment*, Vol. 1. New York: Oxford University Press, 1995:815–843.
6. Convertino VA, Doerr DF, Ludwig DA, Vernikos J. Effect of simulated microgravity on cardiopulmonary baroreflex control of forearm vascular resistance. *Am J Physiol Regul Integr Comp Physiol* 1994;266:R1962–R1969.
7. Perhonen MA, Franco F, Lane LD, Buckey JC, Blomqvist CG, Zerwekh JE, Peshock RM, Weatherall PT, Levine BD. Cardiac atrophy after bed rest and spaceflight. *J Appl Physiol* 2001;91:645–653.
8. Convertino VA. Insight into mechanisms of reduced orthostatic performance after exposure to microgravity: Comparison of ground-based and space flight data. *J Gravit Physiol* 1998;5:P87–P90.
9. Fritsch-Yelle JM, Charles JB, Jones MM, Beightol LA, Eckberg DL. Spaceflight alters autonomic regulation of arterial pressure in humans. *J Appl Physiol* 1994;77:1776–1783.
10. Crandall CG, Engelke KA, Pawelczyk JA, Raven PB, Convertino VA. Power spectral analysis and time based analysis of heart rate variability following 15 days of simulated microgravity exposure in humans. *Aviat Space Environ Med* 1994;65:1105–1109.
11. Cooke WH, Convertino VA. Sympathetic nervous system and space flight. *Acta Astronaut* 2007;60:223–233.
12. Edgerton VR, Zhou M-Y, Ohira Y, Klitgaard H, Jiang B, Bell G, Harris B, Saltin B, Gollnick PD, Roy RR, Day MK, Greenisen M. Human fiber size and enzymatic properties after 5 and 11 days of spaceflight. *J Appl Physiol* 1995;78:1733–1739.
13. Widrick JJ, Knuth ST, Norenberg KM, Romatowski JG, Bain JL W, Riley DA, Karhanek M, Trappe SW, Trappe TA, Costill DL, Fitts RH. Effect of a 17 day spaceflight on contractile properties of human soleus muscle fibres. *J Physiol (Lond)* 1999;516 915–930.
14. Stein TP, Leskiw MJ, Schluter MD, Hoyt RW, Lane HW, Gretebeck RE, LeBlanc AD. Energy expenditure and balance during spaceflight on the space shuttle. *Am J Physiol Regul Integr Comp Physiol* 1999;45:R1739–R1748.
15. Giangregorio L, Blimkie CJR. Skeletal adaptations to alterations in weight-bearing activity. A comparison of models of disuse osteoporosis. *Sports Med* 2002;32:459–476.
16. Buckey JC Jr, Lane LD, Levine BD, Watenpaugh DE, Wright SJ, Moore WE, Gaffney FA, Blomqvist CG. Orthostatic tolerance after spaceflight. *J Appl Physiol* 1996;81:7–18.
17. Convertino VA, Bisson R, Bates R, Goldwater D, Sandler H. Effects of antiorthostatic bedrest on the cardiorespiratory responses to exercise. *Aviat Space Environ Med* 1981;52:251–255.
18. Trappe T, Trappe S, Lee G, Widrick J, Fitts R, Costill D. Cardiorespiratory responses to physical work during and following 17 days of bed rest and spaceflight. *J Appl Physiol* 2006;100:951–957.
19. Convertino VA, Cooke WH. Evaluation of cardiovascular risks of spaceflight does not support the NASA bioastronautics critical path roadmap. *Aviat Space Environ Med* 2005;76:869–876.
20. Fischer CL, Johnson PC, Berry CA. Red blood cell and plasma volume changes in manned spaceflight. *JAMA* 1967;200:579–583.
21. Collet P, Uebelhart D, Vico L, Moro L, Hartmann D, Roth M, Alexandre C. Effects of 1- and 6-month spaceflight on bone mass and biochemistry in two humans. *Bone* 1997;20:547–551.
22. LeBlanc AD, Schneider VS, Evans HJ, Engelbretson DA, Krebs JM. Bone mineral loss and recovery after 17 weeks of bed rest. *J Bone Miner Res* 1990;5:843–850.
23. Vico L, Collet P, Guignandon A, Lafage-Proust M-H, Thomas T, Rehailia M, Alexandre C. Effects of long-term microgravity exposure on cancellous and cortical weight-bearing bones of cosmonauts. *Lancet* 2000;355:1607–1611.
24. Zerwekh JE, Rumi LA, Gottschalk F, Pak CYC. The effects of twelve weeks of bed rest on bone histology, biochemical markers of bone turnover, and calcium homeostasis in eleven normal subjects. *J Bone Miner Res* 1998;13:1594–1601.

25. Smith SM, Wastney ME, Morukov BV, Larina IM, Nyquist LE, Abrams SA, Taran EN, Shih C-Y, Nillen JL, Davis-Street JE, Rice BL, Lane HW. Calcium metabolism before, during, and after a 3-mo spaceflight: Kinetic and biochemical changes. *Am J Physiol Regul Integr Comp Physiol* 1999;277:R1–R10.

26. LeBlanc A, Schneider V, Spector E, Evans H, Rowe R, Lane H, Demers L, Lipton A. Calcium absorption, endogenous excretion, and endocrine changes during and after long-term bed rest. *Bone* 1995;16:301S–304S.

49 Developmental Space Biology of Mammals

Concepts and Methods of Study

April E. Ronca

ABSTRACT

In the past decade, emerging concepts and methods for studying mammalian development under altered gravitational fields have led to a new generation of space biology research. Studies of the Norway rat (*Rattus norvegicus*) provide unequivocal evidence that mammalian reproduction and ontogenesis can proceed under gravitational forces that deviate considerably from the constant force of the Earth's $1 \times g$. Mammalian birth has not yet been attempted within the microgravity of space, however, pregnant and lactating rat dams and their progeny have flown onboard the NASA Space Shuttle. Collectively, spaceflight and ground-based research have led to identification of important requirements for supporting developing mammals in altered gravitational fields, particularly the necessity of preserving the coordinated interactions between mothers and offspring, the ontogenetic context within which development occurs. Developmental studies utilizing gravitational manipulations are yielding new information relevant to health and disease on Earth, especially forces that shape and maintain the vestibular, balance, and motor systems and contribute to our understanding of body-weight regulation, prenatal stress, and adaptation to environmental change. Major concepts, approaches, and methods for studying the developmental space biology of mammals and prospective contributions of this nascent yet fertile area of scientific query to contemporary concerns in biomedicine are described.

Key Words: Prenatal, Postnatal, Maternal, Fetus, Infant, Pregnancy, Lactation, Microgravity, Hypergravity, Centrifugation.

INTRODUCTION

A major goal of space biology research is to broaden scientific knowledge of the Earth's constant gravitational force ($1 \times g$) on living organisms. To this end, studying the life cycles of mammals in space and in other altered gravitational environments promises to uncover exciting new insights into how gravity shaped life on Earth. The life cycles of different vertebrates are characterized by vastly differing adaptations that vary in number and complexity, yet no vertebrate has yet been raised from conception to sexual maturity in the absence of gravity.[1–4] A number of pioneering experiments with fish, amphibians, birds, and rodents flown at various phases during their reproductive and developmental cycles have shown that vertebrate reproduction and development can proceed in the microgravity of space.[5] For example, female frogs (*Xenopus laevis*) flown on the NASA Space Shuttle underwent ovulation and their eggs yielded larvae.[6] Medaka killfish (*Oryzias latipes*) mated and successfully fertilized then hatched fry in space.[7] Female Norway rats (*Rattus norvegicus*) spent the latter half of their pregnancies in space and gave birth to healthy offspring within 48 h of returning to Earth.[8] These and other space biology studies, many relying on established ground-based models,[9,10] have been instrumental in validating new approaches and methods for studying reproduction and development in altered gravity. Collectively, this research has opened the door to a new era of space and ground-based experimentation with vertebrates. Historically, the rat has served as the primary model for mammalian developmental space biology experimentation, and therefore will be the primary focus of this chapter.

WHY STUDY DEVELOPING MAMMALS?

Mammalian reproduction and development are comprised of an intricate and complex series of biological events. Internal fertilization, pregnancy, placentation, embryogenesis, fetal growth and development, labor and birth, lactation, parental care, and postnatal development, including sexual maturation, are major steps involved in the reproductive cycles of mammals. Throughout development, rapid and continuous change occurs within the context of correlated environmental transformations.[11] The fetus begins life within a warm uterine enclave with physiological resources automatically supported. At birth, profound neural and biochemical changes foster the emergence of new behaviors, including postpartum breathing and feeding.[12] The mother's caretaking behavior helps direct and regulate postnatal development.

These and other patterns that characterize development and the context in which it occurs are unique to and universal among mammals. Commonalities among developmental processes represent the shared ancestry of contemporary mammalian species.[11] Because mammals share many biological processes throughout the life cycle, the Norway rat has become a predominant model

From: *Sourcebook of Models for Biomedical Research*
(P. M. Conn, ed.), © 2008 Humana Press Inc., Totowa, NJ.

for human function and disease in both space flight and ground-based experiments. Developmental analysis provides an especially useful approach. Whereas organisms change and adapt throughout their life span, changes that occur during development are far more rapid and are expressed more dramatically ("magnified") as compared to slow, subtle changes in adulthood (for further discussion, see Alberts and Ronca[11]). The developmental approach, which provides an important resource for space studies, can be used in various ways.

DEVELOPMENTAL HISTORY One strategy used for understanding biological processes in adult forms is to study the developmental history of a physiological process or behavior. Determining the sequential organization of a biological phenomenon during maturation may provide clues about how that system is organized and its underlying mechanisms. For example, the appearance of particular morphological changes during development may correspond to specific functional changes. Some complex functions may have multiple, intertwined components with separate developmental histories. Of course, in the infant, the process may differ from the adult or may not be fully developed. This developmental strategy has been used extensively to study the organization of ingestive behavior and has revealed multiple and separate regulations.[13] Applied to space biology, this approach would entail perturbing the system during development by altering the gravitational milieu, then determining how the specific changes observed correlate with the response of individual components in the adult system. In this way, it is possible to learn how the system is built or assembled.

DEVELOPMENTALLY DISTINCTIVE BIOLOGICAL PROCESSES There is a vast array of biological processes and behaviors specific to particular developmental phases. Suckling for milk in early life, for example, is a behavior that appears to be under separate physiological regulation than is adult or infant feeding.[13] The developing animal has a specific ecological and physiological niche, and there are many illuminating examples of the adaptive strategies that young use to thrive in their unique ontogenetic setting.[11] Studies of processes specific to development, including those that are gravity sensitive, can reveal how the environment shapes early behavior and function. Thus, placing a newborn rat on its back on a surface quickly elicits a body turn *against* the gravity vector to achieve the prone position.[14] However, during nipple attachment the neonate performs a similar motor sequence to orient itself *toward* the gravity vector and attain a supine posture.[15]

DEVELOPMENT AS A MODEL SYSTEM Another perspective on studying developing animals is that the immature animal can provide a model system for analyzing relationships between environmental change and biological function. The use of maturing animals can provide information that cannot readily be obtained in adult animals—not because young animals are simpler than adult animals, because they are likely to be more complex. As we begin to better characterize responses and adaptations of animals of different ages to altered gravity, we can relate their patterns of response to age-specific changes in physiological and behavioral systems.

FUNDAMENTAL CONCEPTS IN SPACE AND GRAVITATIONAL BIOLOGY

A basic appreciation for the space environment and the physical stimulus of gravity is a prerequisite to space biology research.

THE SPACE ENVIRONMENT Spacecraft orbiting the Earth are in continuous "free-fall," balanced by equal and opposite forces toward (gravitational) and away from (centripedal) the Earth. The term "microgravity" refers to the small amount of gravity present on an Earth-orbiting spacecraft that is estimated to be approximately one-millionth of its value on Earth.[16]

Microgravity is a major component of the space environment. Radiation, arising from charged particles, neutrons, or ionizing photons, is a significant factor as well. Radiation levels in space exceed both background exposure at the Earth's surface and occupational exposure levels for radiation and health workers. Accurate dosimetric profiles of the space environment, including information on fluence, charge, velocity, specific energy and time course of dose deposition, and potential synergistic effects of microgravity and radiation require detailed analyses within the space environment and cannot be answered by ground-based studies.[16]

Acceleration, vibration, and noise encountered during launch and reentry are significant variables in spaceflight experiments that are difficult to reliably duplicate in ground-based simulations. For these reasons, studies incorporating $1 \times g$ controls in the space environment must eventually be performed. The centrifuge accommodation module planned for the International Space Station (ISS) is essential for validating space biology studies and answering critical questions about the space environment.

GRAVITY AS A CONTINUUM Gravity (g) is the universal force of attraction between bodies. It is a continuum beginning at zero and continuing through infinity. Table 49–1 shows gravitational loads associated with different environments. Models used to study gravity's effects on humans and animals rely on removing or adding gravity, usually within about $1 \times g$ of the Earth's gravity in rats.

Early space life science investigators viewed the application of gravitation force as a continuum with applications of varying magnitude producing graded biological responses. Unfortunately, they did not have the opportunity to test their ideas using spaceflight.[10] Although there is a lack of definitive data obtained in space, growing evidence suggests that dose–response relationships exist across gravity levels exceeding $1 \times g$.[10,17] Thus, some biological systems respond to increased and decreased gravity with responses that are opposite in directionality. For example, antigravity muscles undergo hypotrophy in weightlessness, but *hyper*trophy in response to gravitational loading.[18] We recently

Table 49–1
Range of g-loads associated with different gravitational environments

Spaceflight	*Earth's moon*	*Mars*	*Earth*	*Centrifugation*	*Ultracentrifugation*
10^{-5}–$10^{-2} \times g$	$0.17 \times g$	$0.3 \times g$	$1 \times g$	$>1 \times g$	10^3–$10^5 \times g$

observed a strong ($R^2 = 0.98$) linear relationship between mammary metabolic activity and gravity loading at 0, 1.25, 1.5, 1.75, and $2 \times g$.[19] In other cases, responses appear in an "all or none" fashion at and above a certain gravity threshold, with no response observed below that threshold. Still other systems may show the same response to deviations from $1 \times g$ in opposite directions. In this case, a biological system that evolved for maximal efficiency at $1 \times g$ may show a degraded efficiency at g loads that are sufficiently greater or less than the Earth's $1 \times g$.

METHODS OF STUDY

In addition to reducing the risks of manned spaceflight, a major objective of space life science research is to use the space environment to broaden scientific knowledge about the influence of gravity on living systems. While spaceflight studies are critical to this venture, it is both difficult and costly to conduct research in space. For these reasons, ground-based approaches have been developed for simulating some of the major physiological effects of microgravity ($0 \times g$). These models include bed rest, water immersion, and hindlimb unloading. To increase gravitational loading (induce hypergravity), centrifugation is used. Both spaceflight and ground-based approaches require specialized methodologies and considerations, particularly when applied to reproducing and developing animals.

RATS IN SPACE The first adult rodents were flown in space in the 1950s.[16] Studies of mating, pregnant, and lactating rats and their developing offspring were begun in 1979 with a total of six missions launched through 1998. (Pregnant mice were flown on the last of these flights.) Table 49–2 shows the year, mission, flight duration, and subject characteristics corresponding to each spaceflight. Collectively, these studies (1) established the feasibility of spaceflight experimentation on reproducing and developing rodents, (2) permitted identification of critical requirements for proper husbandry and support of animals flown at sensitive developmental phases, and (3) led to new scientific discoveries, thereby advancing our understanding of spaceflight and gravitational influences on reproductive and developmental processes of mammals.

Characteristic of spaceflight studies, these efforts involved multiple analyses by different investigative teams, small numbers of animals, and detailed and systematic data collection techniques. With few exceptions, the experiments were designed to maximize the quantity and quality of the data obtained. In some cases, unexpected problems during flight afforded less interpretable data but revealed important "lessons learned."[20,21] The methods that were developed and the fundamental observations that emerged from these different spaceflights are reviewed here. Detailed review of the experimental results can be found elsewhere.[3,11]

Space Habitats A variety of different habitats have been used to fly reproducing and developing rats in space. Figure 49–1 illustrates caging configurations that have been used. These different systems all provide for automated food and water delivery, ventilation, waste control, and light timers to regulate the circadian cycle. For each depiction, the rats are shown in the $1 \times g$ configuration.

Figure 49–1 (upper left) shows the BIOS-Vivarium cage. In the only mammalian mating experiment attempted in space (Cosmos-1129), male and female rats were launched in separate compartments and a partition was removed during flight to enable copulation. As for most Cosmos flights, animals were fed a formulated paste diet. For unknown reasons, neither experimental nor control females showed clear evidence of conception or pregnancies.

The BIOS-Vivarium cage was also used on the unmanned Cosmos-1514 mission. Although the flight dams weighed 18% less than identically housed synchronous $1 \times g$ control dams, this important study demonstrated the feasibility of flying pregnant rats in space.

NASA's STS-66 mission carried 10 pregnant rats into space as part of a joint NASA–NIH venture, NIH.Rodent (R)1. Considerably longer in duration than Cosmos-1514, NIH.R1 and its sister experiment NIH.R2 were flown 2 years later on STS-70. R1 and R2 collectively yielded valuable new information on the biological and behavioral responses of mothers and offspring to prenatal spaceflight. For each mission, rats were housed in the NASA Animal Enclosure Module (AEM), depicted in the upper right of Figure 49–1. The AEM has a habitable volume of $863\,cm^2$, thus five rats can be housed in each unit. Food bars line the walls of the AEM and continuous water access is provided by a central lixit system.

The growth and well being of suckling rat pups in space were evaluated on the STS-72 mission. Family units (mothers and their litters) were pair-housed in the AEM-Nursing Facility (NF), separated by a bisecting partition (Figure 49–1, lower left). A small compartment fitted with a "nursing wedge" afforded physical support to the dam during nursing. This was critical for the young

Table 49–2
Spaceflight missions using rats (*Rattus norvegicus*) and mice (*Mus musculus*)[a]

Year	*Mission number*	*Duration (days)*	*Subjects characteristics*
1979	Cosmos-1129	18.5	Adult male and female rats
1982	Cosmos-1514	4.5	Pregnant and fetal rats (G[b]13–18)
1994	STS[c]-66	11	Pregnant and fetal rats (G13–18)
1995	STS-70	9	Pregnant and fetal rats (G13–18)
1996	STS-72	9	Nursing rat litters (P[d]5, 8, 14)
1998	STS-90	16	Nursing rat litters (P8, 14) Pregnant mice (G3, 5, 8)

[a]Data associated with each flight are reviewed by Ronca.[3]
[b]G, gestational day; conception, gestational day 1; birth occurs on gestational day 22/23.
[c]STS, Space Transport System.
[d]P = postnatal day where P0 = day of birth.

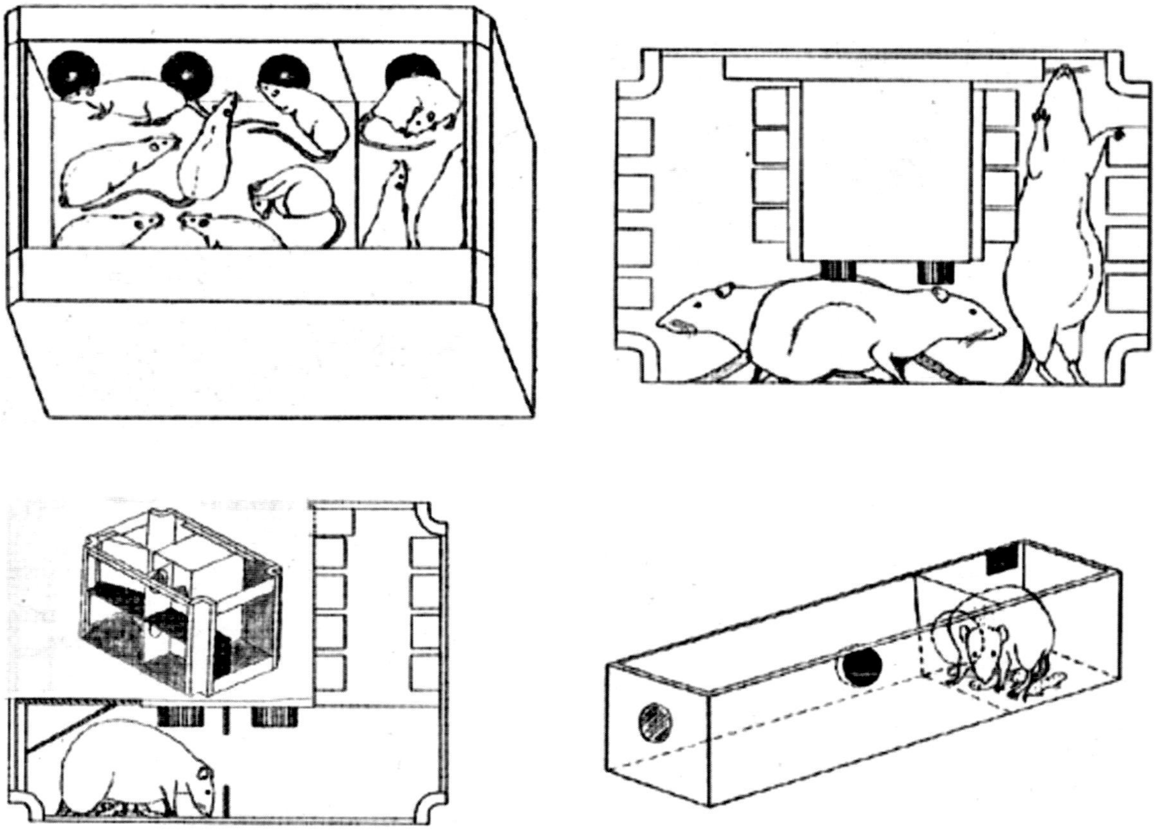

Figure 49–1. Flight housing used for studies of reproducing and developing rats in space. For each image, rats are shown in the 1 × *g* configuration. Upper left: The Russian Bios-Vivarium cage was used for the rodent mating experiment flown on Cosmos-1129 and Cosmos-1514. For Cosmos-1129, a partition (demarcated by the solid interior lines) separated male and female rats prior to launch. (Redrawn from Gazenko, 1993.) Upper right: The Animal Enclosure Module (AEM) carried mid-pregnant rats on the NASA Space Shuttle for the NIH.Rodent (R)1 and NIH.R2 experiments. Lower left: The Animal Enclosure Module-Nursing Facility (AEM-NF, a modified AEM) carried the first lactating dams and litters into space (NIH.R3); an experiment to house older (P14) litters on the Neurolab mission. Two dams and their litters were housed in each AEM-NF. The inset shows a perspective drawing of the AEM-NF with the darker shading depicting the partition between litters and the lighter shading illustrating the habitable areas of a single compartment. Lower right: The Research Animal Holding Facility (RAHF) was used to house rat litters that were P9 at launch on the Neurolab mission. (Drawings by Kenneth E. Johnson. Reprinted from Ronca,[3] with permission from Elsevier.)

ages tested since rat pups do not wean from the mother until around P21. NIH.R3 was a prelude to the Neurolab SpaceLab Mission and provided a "proof-of-concept" that suckling rats could survive and thrive in space. In contrast to the R3 mission, 9-day-old Neurolab litters were housed in the Research Animal Holding Facility (RAHF), a rectangular design with a habitable volume of approximately 1100 cm^3 (Figure 49–1, lower right). Pups of similar age flown on the R3 and Neurolab missions had vastly different outcomes, as many of the Neurolab pups died in space.[20,21] The results of these studies led to important insights regarding habitat requirements for nursing rat litters in space.[3,11,20,21]

Methodological Considerations in Spaceflight Studies Space comprises a unique laboratory environment. Spaceflight studies are highly specialized, extremely costly, labor intensive, and typically will not be repeated. For these reasons, it is imperative to utilize methods and controls that will optimize interpretation of the data.

Habitat Controls Spaceflight studies require custom designed caging (see Figure 49–1), necessitating the use of multiple 1 × *g* control conditions. One of these is the Vivarium Control in which animals are housed in a standard colony environment in vivarium caging. Synchronous Controls are also needed. In this condition, subjects are housed in flight caging and treated identically to animals in the flight condition, except that they are not exposed to spaceflight. In NASA studies (e.g., NIH.R1 and R2), in-flight temperature, humidity, and lighting have been simulated in the 1 × *g* Synchronous Control AEMs using data down linked from the Space Shuttle to the Kennedy Space Center (KSC).

Spaceflight is known to induce atrophy in most skeletal muscles; however, the flight cage configuration alone is sufficient to cause hypertrophy in some skeletal muscles.[22] A second consideration in interpreting spaceflight data is that animals are weightless during flight and can therefore utilize the full expanse of cage volume, whereas Synchronous Controls are Earth bound and may be relatively more crowded.

Video Footage of Rats in Space There are very few experiments in which video footage of the animal subjects has been collected in flight. In the NIH.R1 and R2 experiments, the

Figure 49–2. Video image of the Animal Enclosure Module (AEM) and five weightless rats onboard the space shuttle during the 11-day NIH.R1 mission.

astronauts collected daily video segments of the weightless pregnant rats through the lexan wall of each AEM (Figure 49–2). This was invaluable for assessing the rats' overall well being, behavior, and social interactions. Preflight, the rats' fur was colored with a nontoxic hair dye, which proved useful in relating the individual dams' in-flight behaviors to the pups' postflight vestibular reactions.[23] In the NIH.R3 experiment, video footage of lactating dams and pups provided evidence for maternal nursing and maternal care (pup retrieving and licking) during the flight.[24] These behaviors persisted in flight despite the lack of overall coherence and drastically changed dynamics with the litters.

Surgical Manipulation and Treatment of Pregnant Dams Sixteen NIH.R1 and R2 science teams collected data from pregnant and lactating rats dams and their fetal and neonatal offspring.[3] Dams were selected for inclusion in the study based on surgical laparotomies performed at 1 week postconception that enabled us to visualize fertilization sites preflight, thereby ensuring adequate numbers of specimens. At landing, unilateral hysterectomies provided both prenatal and postnatal subjects from each dam.[8,25] Excision of one of each dams' paired uterine horns yielded fetal specimens. Births of pups from the intact uterine horn occurred 48–72 h later (at the expected time, G22/23). R2 dams were not surgically manipulated following flight, affording a rare opportunity to validate the findings across two spaceflights and differing methodologies.[8]

Following birth, the dams were euthanized and their tissues harvested for study. Pups from all conditions were fostered to nonflight dams for extended postnatal study.[26,27]

GROUND-BASED MODELS Ground-based models continue to provide valuable information on effects of gravity on developing animals, and lend important insights into the design and interpretation of spaceflight experiments. Centrifugation is a commonly used technique for studying responses of reproducing and developing animals to gravitational loading.[28–32] In contrast, the Morey–Holton hindlimb suspension (HLS) model[9] is used to simulate certain features of microgravity exposure. In this model, the hindquarters are elevated 30° to produce head-down tilt with resulting cephalad fluid shift. Although used in just a few reproductive and developmental studies, HLS has produced some interesting findings.[33–35] Highly detailed information on HLS can be found elsewhere,[9] therefore discussion of ground-based methods in this chapter will focus on methods for centrifugation.

Centrifugation Centrifugation induces hypergravity by summing the normal gravity of Earth with acceleration forces that are produced by changing the direction or linear rate of motion.[10] While centrifugation can be used to simulate the relatively brief accelerations associated with launch and landing, most developmental gravitational biology studies involving centrifuges are designed to induce long duration gravity loading.

The use of centrifugation in developmental gravitational biology studies has grown immensely in recent years. Using this technique, fractional increments in g-load exceeding $1 \times g$ can be continuously applied to biological specimens for extended periods and dose–response relationships established.[10,17]

Figure 49–3 shows an image of the 24-foot-diameter centrifuge of the NASA Ames Research Center for Gravitational Biology Research, an apparatus designed to create a hypergravity environment spanning >1 to $3.5 \times g$ for small animal research.

The centrifuge has 10 radial arms with two opaque enclosures situated on each arm. Each enclosure holds four standard rat vivarium cages. Each arm has multiple enclosure mounting locations at different distances from a central vertical shaft spindle, thereby allowing various combinations of gravity levels to be introduced to different groups of rats in a single experiment. In quadripeds, gravitational forces are applied from the back through the feet.[10] To accomplish this, the animal housing units on the centrifuge must be gimbaled such that the resultant gravitational force is applied through the floor.

Investigators have used centrifugation to study acute responses and adaptation to increased gravity environments.[10] Responses of pregnant and lactating rats to centrifugation are generally similar to those observed in nonreproducing adult animals, namely, an initial decline in feeding, drinking, body mass, physical activity, and temperature, with a return, following 4–6 days of acclimation, to an approximate 8–15% reduction in body mass relative to controls.[36] Increased mortality may be observed in postnatal animals, depending upon their age at the onset of centrifugation

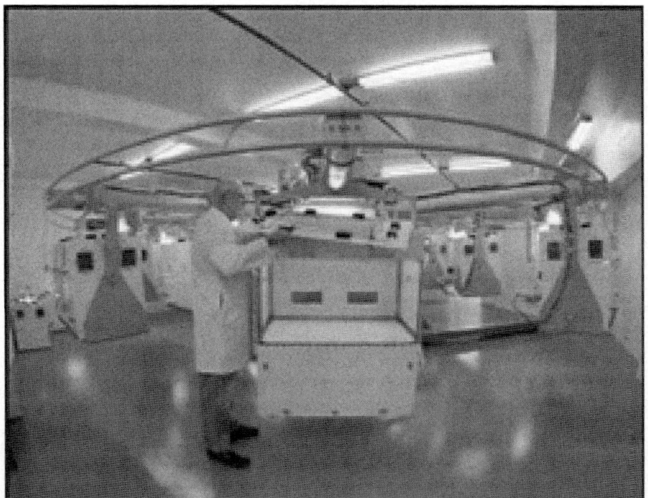

Figure 49–3. The 24-foot small animal centrifuge at the Center for Gravitational Biology Research (CGBR), NASA Ames Research Center, Moffett Field, CA. (Photo by Thomas Trouwer. Reprinted courtesy of NASA Ames Research Center.)

and gravity-induced changes in the mother.[37,38] (See the section on Novel Challenges in the Study of Space Biology in Developing Mammals.)

Methodological Considerations in Centrifugation Studies Necessary control groups in centrifugation experiments involve placing subjects at or near the center of rotation to control for rotational stimulation, and stationary controls. Numerous factors need to be considered in the design and interpretation of centrifuge studies.[10] These include (1) scaling (the larger the mass of an animal, the greater the impact of gravitational loading), (2) rotation (more apparent at shorter diameters from the center of the centrifuge), (3) Coriolis forces (produced when the orientation of the animal's body varies in relation to the direction of acceleration), (4) duration of exposure (acute versus chronic responses), and (5) resultant level of gravity. The interested reader is referred to Wade[10] for a detailed review of the attendant issues.

NOVEL CHALLENGES IN THE STUDY OF SPACE BIOLOGY IN DEVELOPING MAMMALS

A key feature of mammalian development is the presence of the mother. The biological and behavioral interactions that exist between mothers and their young evolved under the constant force of the Earth's $1 \times g$ gravitational field. It is therefore reasonable to predict that maternal–offspring interactions will be altered in the weightless environment of space and the hyperweighted environment of the centrifuge. The mother's role in development also makes it more challenging to accurately determine the source of gravitational effects on the offspring. Finally, the young are dependent upon nursing, licking, and retrieving by the dam, behavioral responses that may change in altered gravity environments.[3,11,38]

THE MOTHER–OFFSPRING SYSTEM Development of the young progresses within the context of an intact, integrated system composed of a mother and her young.[3,11] This family system, as a unit, is characterized by coordinated changes within and between a mother and her young. As the offspring are developing within the womb and nest, the maternal–offspring system follows its own ontogenetic trajectory. Mammalian development involves bidirectional linkages between mothers and infants that are important for normal growth and development.

SOURCES OF GRAVITATIONAL INFLUENCE A major challenge in studies of the developing mammal is separating effects of altered gravity on mothers and offspring. Due to the fact that the maternal–offspring system is so highly intertwined in mammals, it is difficult to differentiate among "direct" and "indirect" effects of gravity on the young. "Direct" effects of gravity are those that operate through a primary relation with the recipient organism, tissue, or cell.[3,11] In contrast, "indirect" effects of gravity are those expressed through avenues of the mammalian system. For example, if mothers cannot stabilize their bodies in the nursing posture or if pups cannot retain metabolic heat because their huddling behavior is disrupted, the altered growth effect would clearly be an indirect consequence of weightlessness.

MATERNAL CARE AND REPRODUCTIVE EXPERIENCE Altricial species, such as the Norway rat, are born in a highly immature state with relatively undeveloped sensory and motoric capabilities. The mammalian mother affords vital resources that lead to proper growth and development. She does this through maternal behavior, a constituent set of behavioral activities including nursing, touching, and carrying the young.[39] The neonate relies on stimulation by the mother for induction and maintenance of normal behavior and physiology.[11–13] Because the mother plays a major role in the regulation of developmental processes in her young, the quality and integrity of maternal care are vital to making accurate assessments of direct and indirect effects of altered gravity on the pups.

Prior birth experience facilitates the appearance of maternal care patterns when pups are present.[39] Reproductive experience, or parity, is an important consideration in developmental space biology studies.[38] Second-time (bigravid) dams exposed to $1.5 \times g$ hypergravity gave birth to more live newborn pups as compared to first-time (primigravid) dams (94% versus 82% survival), and pup mortality was strongly associated with changes in the dams' maternal behavior. Under unusual or stressful conditions, parity can be an important factor in promoting the health and longevity of the young.

CONSEQUENCES OF DISRUPTING MATERNAL–OFFSPRING INTERACTIONS Factors that disrupt either the mother or her offspring are liable to disrupt the maternal–offspring system.[3,11] If this occurs, the organizing influence of the system on its constituent parts is likely to change and the ontogenetic processes within the individuals can be disrupted. In spaceflight studies, upsetting the important interrelations between mothers and their young can pose major interpretive problems by obscuring meaningful results or creating false impressions of how microgravity affects development. Maternal behavior and physical relationships between mothers and pups are apt to be altered in the weightless space environment. The mother's maneuverability of pups and her proximity to them may be impeded during attempts to huddle and nurse; pup licking may be diminished. Retrieving behavior, important for maintaining pups within a coherent nest, may not be possible if pups continuously float away from the nest. Pup milk intake, warmth, and tactile stimulation normally provided by the mother may be diminished and compromise the pups.

CONTRIBUTIONS TO BIOMEDICINE

Developmental space biology research affords unique perspectives on biomedical concerns on Earth and in space. Gravity is a fundamental factor in the external environment, stimulating vestibular organs and exerting major forces on the body. The tools of space biology, namely, spaceflight, centrifugation, and other ground-based methods, can be applied to induce identifiable and meaningful changes in physiological systems, brain areas, nerve pathways, and gravity sensors. Developmental analysis is especially useful for studying gravity-induced changes in structure and function. It is widely recognized that activity-dependent processes and experience play important roles in the establishment of neural architecture and function. Researchers are utilizing developmental space biology approaches in new studies of vestibular development, equilibrium, and locomotion. The population of the United States is living longer than ever before, with a nearly 30-year increase in life span in the past century. It is predicted that the number of people over 85 will double by the year 2020 and that by 2050, over 20% of the population will be over 65. Vestibular and equilibrium disorders are frequent symptoms of aging, related to "multisensorial decay," age-related conditions, Parkinson's disease, and other degenerative conditions. Due to vestibular and proprioceptive dysfunction, the elderly suffer disastrous falls from which many never recover. Fetuses with vestibular disorders are prone to breech births. Babies born after maternal bedrest during pregnancy appear to crave vestibular simulation and require more rocking to be quieted. Altered gravity is a useful method for studying the development of body-weight regulation and metabolism. Changes in body mass and energy balance induced by gravitational loading may provide clues to developmental factors that contribute to obesity. Recent gravitational biology research is addressing prenatal contributions to the establishment of body-weight regulation in later life, and the effects of chronic exposure to stress prior to birth. These are central issues in the field of prenatal programming of adult disease. Adaptation, a prominent feature of living organisms in general, and developing organisms in particular, can be analyzed in altered gravity. Adaptation to extreme environments is relevant to a range of biomedical concerns, including shifts experienced in high-altitude and oceanic environments on Earth. In addition to major contributions to the concerns of astronauts facing long-duration missions, developmental space biology can provide immense benefits on Earth, and will undoubtedly contribute to the foundation of knowledge needed for the development of effective therapies and cures in contemporary biomedicine.

ACKNOWLEDGMENTS

Supported by NIH Grants MH46485 and HD50201 and NASA Grant NNA04CK83.

REFERENCES

1. Miquel J, Souza KA. Gravity effects on reproduction, development, and aging. *Adv Space Biol Med* 1991;1:71–97.
2. Moody SA, Golden C. Developmental biology research in space: Issues and directions in the era of the international space station. *Dev Biol* 2000;228:1–5.
3. Ronca AE. Mammalian development in space. In: Marthy HJ, Ed. *Advances in Space Biology and Medicine*, Vol. 9, *Development in Space*. Amsterdam, The Netherlands: Elsevier, 2003:217–251.
4. Wassersug RJ. Vertebrate biology in microgravity. *Am Sci* 2001;89:46–52.
5. Marthy HJ, Ed. *Advances in Space Biology and Medicine*, Vol. 9, *Development in Space*. Amsterdam, The Netherlands: Elsevier, 2003.
6. Fejtek M, Souza K, Neff A, Wassersug R. Swimming kinematics and respiratory behaviour of *Xenopus laevis* larvae raised in altered gravity. *J Exp Biol* 1998;201:1917–1926.
7. Ijiri K. Life-cycle experiments of medaka fish aboard the international space station. In: Marthy HJ, Ed. *Advances in Space Biology and Medicine*, Vol. 9, *Development in Space*. Amsterdam, The Netherlands: Elsevier, 2003:201–216.
8. Ronca AE, Alberts JR. Physiology of a Microgravity Environment Selected Contribution: Effects of spaceflight during pregnancy on labor and birth at 1G. *J Appl Physiol* 2000;89:849–854.
9. Morey-Holton ER, Globus RK. Hindlimb unloading rodent model: Technical aspects. *J Appl Physiol* 2002;9:1367–1377.
10. Wade CE. Responses across the gravity continuum: Hypergravity to microgravity. In: Sonnenfeld G, Ed. *Advances in Space Biology and Medicine*, Vol. 10, *Experimentation with Animal Models in Space*. Amsterdam, The Netherlands: Elsevier, 2005:225–245.
11. Alberts JR, Ronca AE. Development as adaptation: A paradigm for gravitational and space biology. In: Sonnenfeld G, Ed. *Advances in Space Biology and Medicine*, Vol. 10, *Experimentation with Animal Models in Space*. Amsterdam, The Netherlands: Elsevier, 2005:175–207.
12. Ronca AE, Abel RA, Alberts JR. Perinatal stimulation and adaptation of the neonate. *Acta Pediatr* 1996;415:8–15.
13. Phifer CB. The study of early feeding and drinking behaviors. In: Shair HN, Barr GA, Hofer MA, Eds. *Developmental Psychobiology: New Methods and Changing Concepts*. New York: Oxford University Press, 1991:189–205.
14. Pellis VC, Pellis SM, Teitelbaum P. A descriptive analysis of the postnatal development of contact righting in rats *(Rattus norvegicus)*. *Dev Psychobiol* 1991;24:236–267.
15. Eilam D, Smotherman WP. How the neonatal rat gets to the nipple: Common motor modules and their involvement in the expression of early motor behavior. *Dev Psychobiol* 1998;32:57–66.
16. Souza K, Hogan R, Ballard R. Life into space: Space life sciences experiments, NASA Ames Research Center, 1965–1990. NASA Reference Publication 1372, 1995.
17. Phillips R. Gravity: It's the law. *J Gravit Physiol* 2002;9: P15–16.
18. Vasques M, Lang C, Grindeland RE, Roy RR, Daunton N, Bigbee AJ, Wade CE. Comparison of hyper- and microgravity on rat muscle, organ weights and selected plasma constituents. *Aviat Space Environ Med* 1998;69:2–8.
19. Plaut K, Maple R, Baer L, Wade C, Ronca A. Mammary to metabolic function: Is gravity a continuum? *J Appl Physiol* 2003;95: 2350–2354.
20. Maese AC, Ostrach LH. Neurolab: Final report for the Ames Research Center payload. NASA 2002;TM-211841.
21. Reichhardt T. Animal deaths turn shuttle into 'necrolab'. *Nature* 1998;393:4.
22. Fejtek M, Wassersug R. Effects of laparotomy, cage type, gestation period and spaceflight on abdominal muscles of pregnant rodents. *J Exp Zool* 1999;284:252–264.
23. Ronca AE, Fritzsch B, Alberts JR, Bruce LL. Effects of microgravity on vestibular development and function in rats: Genetics and environment. *Kor J Biol Sci* 2000;4:215–221.
24. Daly ME, Ronca AE. Lactating rats retain nursing behavior and maternal care in space. *Dev Psychobiol* 2002;39:75.
25. Alberts JR, Burden HW, Hawes N, Ronca AE. Sampling pre- and postnatal offspring from individual rat dams enhances animal use without compromising development. *Contemp Top Lab Anim Sci* 1996;35:61–65.
26. Hoban-Higgins TM, Murakami DM, Tang IH, Fuller PM, Fuller CA. Development of circadian rhythms in rat pups exposed to microgravity during gestation. *J Gravit Physiol* 1999;6:71–79.
27. Wong AM, DeSantis M. Rat gestation during space flight: Outcomes for dams and their offspring born after return to Earth. *Integr Physiol Behav Sci* 1997;32:322–342.
28. Baer LA, Rushing L, Wade CE, Ronca AE. Prenatal centrifugation: A model for developmental programming of adult body weight? *J Gravit Physiol* 2005;12:P181–182.
29. Bouet V, Borel L, Harlay F, Gahery Y, Lacour M. Kinematics of treadmill locomotion in rats conceived, born, and reared in a hypergravity field (2*g*). Adaptation to 1*g*. *Behav Brain Res* 2004; 150:207–216.
30. Brocard F, Clarac F, Vinay L. Gravity influences the development of inputs from the brain to lumbar motoneurons in the rat. *Neuroreport* 2003;14:1697–1700.
31. Ronca, AE, Rushing L, Tou J, Wade CE, Baer LA. Centrifugation effects on estrous cycle, mating success and pregnancy outcome. *J Gravit Physiol* 2005;12:P183–184.
32. Wubbels R, Bouet V, de Jong H, Gramsbergen A. Development of sensory motor reflexes in 2G exposed rats. *J Gravit Physiol* 2004;11: P21–22.
33. Gharbi N, El Fazaa S, Fagette S, Gauquelin G, Gharib C, Kamcun A. Cortico-adrenal function under simulated weightlessness during gestation in the rat—effects on fetal development. *J Gravit Physiol* 1996;3:63–68.
34. Huckstorf BL, Slocum GR, Bain JL, Reiser PM, Sedlak FR, Wong-Riley MT, Riley DA. Effects of hindlimb unloading on neuromuscular development of neonatal rats. *Dev Brain Res* 2000;119: 169–178.
35. Serova LV. Simulation models of weightlessness in mammalian's developmental program. *J Gravit Physiol* 1998;5:P127–128.
36. Ronca AE, Baer LA, Mills NA, Sajdel-Sulkowska EM, Wade CE. Body mass, food and water intake, and activity of pregnant and

lactating rat dams during 1.5-g centrifugation. *J Gravit Physiol* 2000;7:P131–P132.

37. Baer LA, Ronca AE, Wade CM. Survival and growth of developing rats during centrifugation at 2.0 g. *J Gravit Physiol* 2000;7:17–22.
38. Ronca AE, Baer LA, Daunton NG, Wade CM. Maternal reproductive experience enhances birth outcome and early postnatal survival following gestation and birth of rats in hypergravity. *Biol Reprod* 2001;65:805–813.
39. Bridges RS. Endocrine regulation of parental behavior in rodents. In: Krasnegor NA, Bridges RS, Eds. *Mammalian Parenting: Biochemical, Neurobiological, and Behavioral Determinants*. New York: Oxford University Press, 1990.

50 A Practical Approach to Animal Models of Sepsis

BRYAN BELIKOFF AND JON A. BURAS

ABSTRACT

The underlying mechanisms active during the pathogenesis of sepsis are not clearly defined; however, several animal models of sepsis have been used in an attempt to understand these complex cellular and molecular interactions that result in disease. There are three general categories of sepsis models, the host barrier disruption model, the chemical shock model, and the exogenous infection model. Host barrier disruption sepsis models initiate infection with the release of endogenous bacteria into normally sterile compartments. The chemical shock model employs intravenous or intraperitoneal administration of a toll-like receptor (TLR) agent, such as lipopolysaccharide (LPS or endotoxin) or zymosan, to initiate a state of proinflammatory cytokine-induced shock. The exogenous infection model of sepsis utilizes administration of an exogenous viable pathogen, typically bacteria, directly (by an intravenous or intraperitoneal route) into the host. Each of these models has particular strengths and weaknesses with respect to their ability to mimic the clinical progression of sepsis in human patients.

Key Words: Sepsis, Animal models, Peritonitis, Inflammation, Infection, Septic shock, Cecal ligation and puncture (CLP), Endotoxemia.

SEPSIS OVERVIEW

Sepsis is defined as the dysregulation of the immune and inflammatory responses following infection resulting in an immunocompromised state with microvascular dysfunction and organ damage that eventually leads to death of the host.[1–3] The immune system serves as the host's defense against pathogens from invading and disrupting normal cellular function. Paradoxically, immune dysfunction and the subsequent excess of inflammation can progress to extensive, irreversible cell damage in inflamed or infected tissue, as well as to the surrounding tissues. These processes eventually result in impaired microvascular and organ function, ultimately resulting in host death if the infection is not successfully contained. Sepsis is a complex state of immunoinflammatory dysfunction that is typically a consequence of unsuccessful infection containment.[4]

The inflammatory response during sepsis may be divided into two overlapping, nonexclusive stages: the systemic inflammatory response syndrome (SIRS) and the compensatory antiinflammatory response syndrome (CARS).[5] SIRS is characteristic of the early stage of sepsis and is thought to result from innate immune cell activation caused by increases in the circulating proinflammatory cytokines interleukin-1 (IL-1), IL-6, tumor necrosis factor-α (TNF-α), and IL-12, and chemokines such as IL-8.[4,5] The overwhelming proinflammatory response results in epithelial and endothelial cell dysfunction, dissemination of the inciting pathogen, and ultimately may lead to multiple organ dysfunction (MOD) and septic shock. Normally, the host immune system is able to effectively regulate the proinflammatory response through increased production of antiinflammatory cytokines IL-4, IL-10, IL-13, IL-1 receptor antagonist, and tumor growth factor-β (TGF-β), expressed during the CARS response.[4,5] Sepsis progression remains unchecked when the host immunoinflammatory response is unable to modulate the SIRS–CARS system appropriately, resulting in a dysfunctional immune response and an immunocompromised state with a reduced ability to contain infection.

The central goal of sepsis research is to understand the mechanisms of sepsis pathogenesis in order to develop effective therapeutic modalities for the treatment of human sepsis. Human sepsis pathogenesis develops in gradual, sequential stages that have been difficult to characterize due to a lack of adequate staging markers.[6,7] In the most simplistic characterization, the progression of sepsis is thought to involve two generalized components of the inflammatory response.[5] The initial (early) inflammatory stage of human sepsis is proinflammatory and characterized by SIRS. The early physiological response to the septic insult involves physiological alterations characterized by high cardiac output (CO), increased heart rate (HR), and low total peripheral (vascular) resistance (TPR), and is collectively referred to as the hyperdynamic phase of sepsis.[8–12] In this phase, the drop in TPR is balanced by an increase in CO, resulting in maintenance of a normal blood pressure. Subsequent to the hyperdynamic phase, blood pressure drops in the hypodynamic (late) phase sepsis, as a result of impaired cardiac function (i.e., decreased HR and low CO with sustained low TPR).[9] The hypodynamic phase reflects the condition of septic shock. The late, hypodynamic phase of sepsis is associated with the development of host immunocompromise.[13] This state of immune dysfunction results in part from upregulation of antiinflammatory cytokine production during the CARS response, which normally assists in termination of the initial early SIRS.[5]

From: *Sourcebook of Models for Biomedical Research*
(P. M. Conn, ed.), © 2008 Humana Press Inc., Totowa, NJ.

INDIVIDUAL SEPSIS MODELS

There are three categories of animal sepsis models including exogenous toxin administration models, exogenous bacterial infection models, and host barrier disruption models.[14] Review of the advantages and disadvantages of each model class suggests that there is no perfect model in the study of sepsis; however, understanding the limitations of each model assists in correlating experimental results with the human disease state. Detailed discussion of animal sepsis models may be found in several reviews.[14–18]

ENDOTOXEMIA MODEL In the early 1940s, Andre Boivin first isolated endotoxins from Gram-negative bacteria; Bordon and Hall[19] later suggested that they induce human endotoxic shock. Today, lipopolysaccharide (LPS) is the more common endotoxin used to induce endotoxicosis in sepsis models; however, other toll-like receptor (TLR) ligands have been used to illicit the shock state, including the TLR2 synthetic lipopeptide agonists and the TLR9 agonist CpG-DNA.[20–22] LPS/mCD14 binding and recruitment of TLRs on monocytes and macrophages trigger the release of IL-1α, IL-1β, IL-6, and TNF-α. The release of these proinflammatory cytokines is capable of inducing physiological hemodynamic changes that are in some ways similar to those observed in septic shock.

Endotoxemia sepsis models represent an easily controlled single variable model in both humans and in animals, making it an attractive experimental model. Unfortunately there are significant differences with the endotoxemia model and true sepsis. Similar physiological effects of LPS may be observed in both humans and rodents; however, higher doses of LPS are required to achieve the same increase in serum TNF-α in rodents.[23] Furthermore, despite the general similarities of the LPS response to those seen in septic shock, the hemodynamic alterations after a bolus injection of LPS in human patients are not identical to alterations observed in sepsis patients.[24] The hemodynamic changes in murine endotoxemia models do not completely mimic hemodynamic changes in sepsis patients, as there are no distinct hyperdynamic and hypodynamic phases.[16,17] A bolus injection of LPS induces a rapid reduction in HR and mean arterial blood pressure (MAP) that may be transient and not lethal (Figure 50–1), or persistent and lethal (Figure 50–2), depending on the dose administered. Shortcomings of the endotoxemia model likely contributed to a number of failed sepsis clinical trials that were based on false assumptions extrapolated from the use of this model system.[25] Despite these shortcomings, the endotoxemia model overall has allowed significant contributions to sepsis research by providing knowledge about the pathways activated by pathogen TLR agonists during the host response to infection.

EXOGENOUS INFECTION MODELS Infection with a load of exogenous bacteria has been used to study mechanisms of sepsis.[15–17] Unfortunately, the doses of bacteria required to elicit mortality do not mimic the responses of typical host infection as these bacteria do not colonize and replicate due to complement-mediated lysis.[26] Furthermore, significant differences are noted in the host cytokine response depending on the exact bacterial strain.[27,28] Also, the compartment of infection (blood versus peritoneal) may affect the host cytokine response.[29,30] Based on these findings, exogenous infection has been utilized more as a system to study the host response to a particular pathogen infecting a particular compartment than as a generalized sepsis model. Based on these facts, this class of model system is not considered further in this chapter, which focuses on general sepsis models.

HOST BARRIER DISRUPTION MODELS Host barrier disruption models utilize the host's normal mixed bacterial flora as the infecting insult and allow passage of these organisms into a previously sterile body compartment through an existing protective barrier such as skin or intestinal wall. The most commonly used methods involve disruption of the intestinal barrier and include the cecal ligation and puncture (CLP) and colon ascendens stent peritonitis (CASP) models.

Cecal Ligation and Puncture Model In 1980, Irshad Chaudry developed the CLP procedure from a previous rat sepsis model consisting only of ligation, which was thought to allow bacteria to spill out of the necrotic cecum into the peritoneum.[15,31]

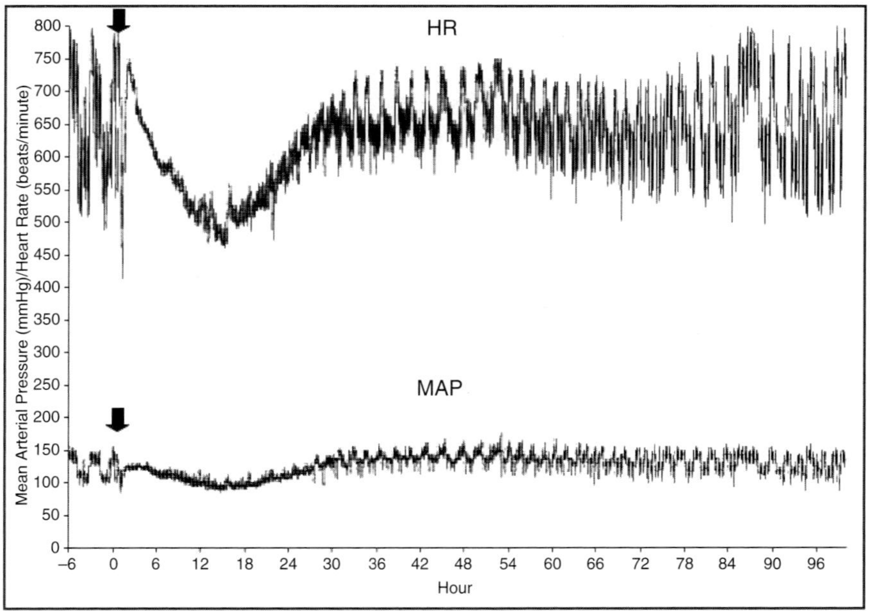

Figure 50–1. Telemetry analysis of physiological responses to sublethal LPS exposure. An example of a single mouse implanted with a wireless PA-C10 catheter (Data Scientific International) in the carotid artery with baseline measurements of heart rate (HR) and mean arterial pressure (MAP) beginning at −6h from LPS injection. Normal variability of HR and MAP is observed in this time period. At time 0, the mouse received an intravenous LPS injection of 5 mg/kg (black arrow). Note the rapid onset of bradycardia and reduction in MAP occurring within 3h of LPS injection. LPS-induced hemodynamic changes are transient and return to baseline approximately 28h after injection.

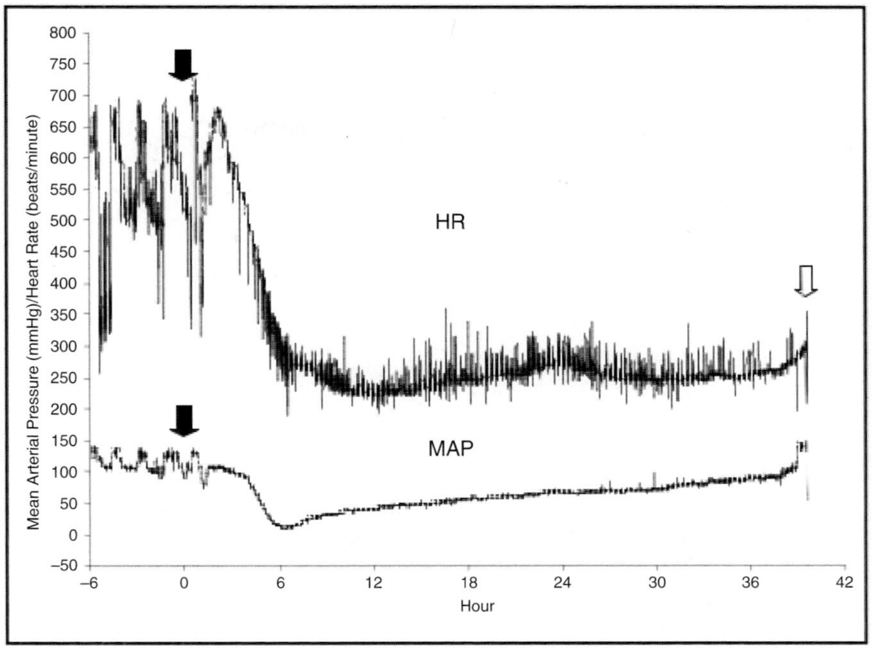

Figure 50–2. Telemetry analysis of physiological responses to lethal LPS exposure. An example of a single mouse implanted with a wireless PA-C10 catheter (Data Scientific International) in the carotid artery with baseline measurements of heart rate (HR) and mean arterial pressure (MAP) beginning at –6h from LPS injection. Normal variability of HR and MAP is observed in this time period. At time 0, the mouse received an intravenous LPS injection of 10mg/kg (black arrow). Note the rapid onset of bradycardia and reduction in MAP occurring within 3h of LPS injection. Altered hemodynamics persist until death of the mouse at 38h (white arrow).

Chaudry was unable to induce septic shock via ligation alone and modified the existing procedure by puncturing the ligated cecum. Cecal puncture allowed controlled bacteria spillage into the abdominal cavity resulting in a simple reproducible model of sepsis.[15] The CLP sepsis model mimics the human clinical disease of perforated appendicitis or diverticulitis. The CLP model has evolved into the "gold standard" and has been used to characterize the inflammatory stages of early sepsis.[32] It is an eloquent model as it is a simple procedure, and survival rates are controllable and reproducible. Also, sepsis physiology in rodents shares similarities with the biphasic hemodynamic patterns observed during the clinical course of human sepsis is an early hyperdynamic phase followed by a hypodynamic phase (Figure 50–3). Furthermore, similar metabolic, immunological, and apoptotic responses are observed in the CLP model as in human disease, strengthening the validity of this model.[15,33,34]

Advantages of the Cecal Ligation and Puncture Model There are many features of the CLP model that contribute to its effectiveness as a model in sepsis research. The model is highly reproducible, cost effective, and can be performed on small and large animals. Telemetry data using a wireless hemodynamic transmitter implanted in the carotid artery (Data Scientific International; PA-C10) demonstrates that CLP mice progress through a hyperdynamic phase from 12 to 24h and subsequently to a hypodynamic phase of sepsis mimicking the clinical progression of human sepsis (Figure 50–3). This is in clear distinction to the physiological response to bolus administration of LPS in either a sublethal or lethal dose (Figures 50–1 and 50–2). As compared to endotoxemia, CLP generates live circulating polymicrobial organisms in the blood that are capable of distant organ colonization. The dissemination of multiple strains of bacteria through the bloodstream allows evaluation of distant organ bacterial seeding mechanisms that the endotoxemia model cannot provide.

Another appealing aspect of the CLP model is the ability to control the rate and degree of mortality for a given experimental group.[35] This is accomplished by varying the size of the needle used for cecal puncture (Figure 50–4). It is important to realize, however, that altering the mortality rate by changing needle size is a relative variable and different results may be observed between laboratories.[36,37] Another factor affecting the severity of the CLP model is the percentage of cecum ligated, with a larger amount of necrotic, stool-filled tissue creating a greater infectious and inflammatory insult.[38] The amount of cecum ligated should be reproducibly maintained between experiments and reported in subsequent publications. The exact needle size and percentage of cecum ligated must be established empirically in each investigator's laboratory and for each operator.

The CLP model has been criticized as not being representative of human disease since perforated appendicitis is treated surgically. The CLP model does have the advantage that surgical resection of the cecum may be performed after the initial CLP procedure.[39–41] Furthermore, it appears that there is a critical point in the host response when sepsis has progressed into a stage at which cecal resection is no longer capable of rescuing the animal from mortality.[40,41] This time may vary with the severity of the initial CLP insult and needs to be established empirically within the investigator's laboratory.

Disadvantages of the Cecal Ligation and Puncture Model Host response to local infection promotes abscess formation around the septic focus, to prevent dissemination of the inciting pathogen. A disadvantage of the CLP model of sepsis is the host's ability to form an abscess and successfully contain the infection, abrogating the development of sepsis. While abscess formation is the desired host response to the local infection source, it may interfere with the interpretation of experimental results regarding sepsis therapies. It is possible that an experimental intervention may enhance abscess formation rather than preserve some other systemic pathway unrelated to direct infection containment. In this case, a therapy might be erroneously identified as protective during sepsis in a general sense, whereas in reality it may merely improve a local response to infection. The problem of abscess formation can be circumvented in part through invasive

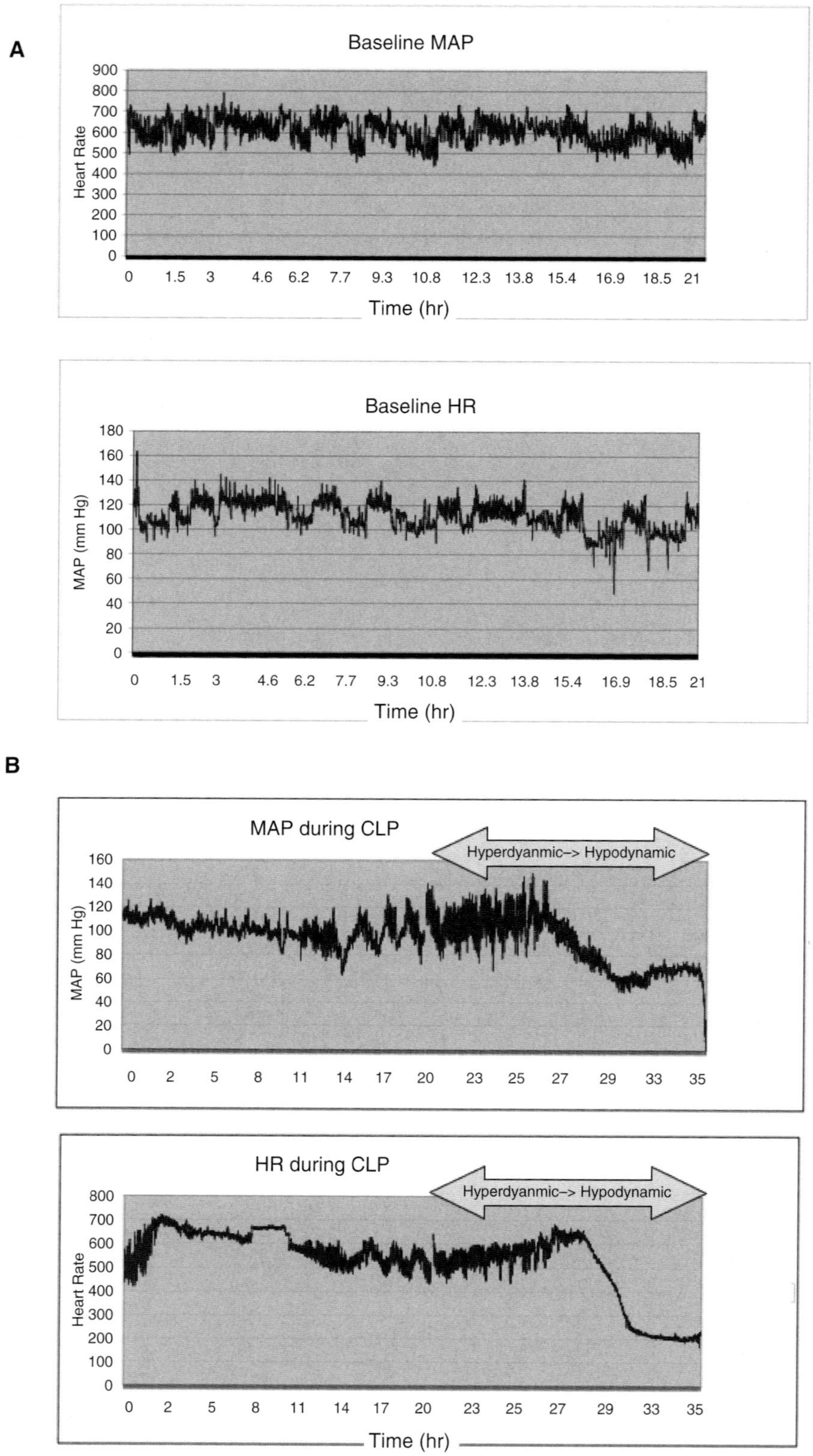

Figure 50–3. Telemetry analysis of physiological responses following CLP. (A) Continuous baseline HR and MAP data are shown over a 21-h period following wireless PA-C10 catheter monitor implantation (Data Scientific International) in the carotid artery. (B) Telemetry analysis following induction of CLP sepsis. After a 3-day recovery period, sepsis was induced via CLP with a 21-gauge needle at time 0. Blood pressure and heart rate were recorded continuously over a 35-h period until the death of the animal. The figure demonstrates the reduction of MAP and initial compensatory increase in HR, followed by the decline in both MAP and HR. This demarcates the hyperdynamic and hypodynamic phases of sepsis (depicted by arrowhead).

telemetry monitoring, biomarker evaluation, and necropsy evaluation of abscess formation. By monitoring these parameters, the investigator should be able to determine whether the animal had progressed to a stage of critical illness.

Colon Ascendens Stent Peritonitis Model Fewer data exist on the CASP model, since it represents the newest animal sepsis model.[42,43] The CASP model of abdominal sepsis was developed to complement the CLP model. The CASP technique is performed by implanting a stent with its origin in the cecum and an exit point in the normally sterile peritoneum. The CASP model induces diffuse peritonitis through the continual release of bacteria from the implanted stent, and is less prone to abscess formation as compared to the CLP model. Changing the stent diameter controls the severity of sepsis. A smaller gauge size stent correlates with an increase in bacterial peritoneal load and sepsis severity. The hemodynamic responses to CASP are not well

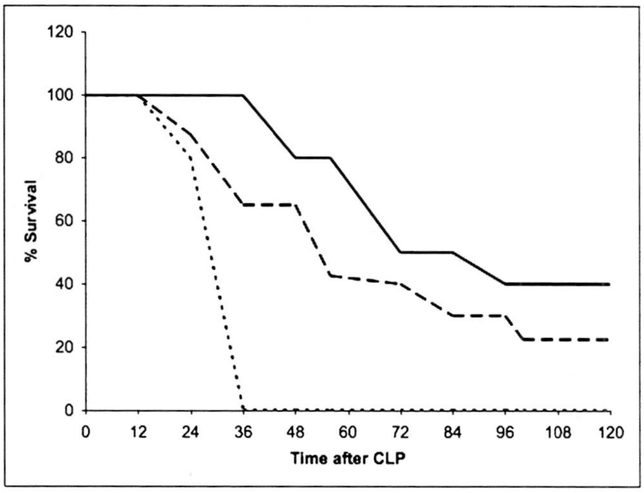

Figure 50–4. Altering cecal puncture and ligation alter CLP sepsis model severity. CLP was performed in male C57/B6 mice using a 20-gauge needle through and through (double) puncture with 40–50% cecal ligation (dashed line, *n* = 10), a 21-gauge needle single puncture (only one hole is made in the cecum) with 30–35% cecal ligation (broken line, *n* = 40), and a 27-gauge needle through and through puncture with 30–35% cecal ligation (solid line, *n* = 10). Both 20-gauge (0 survivors by 36h) and 21-gauge (approximately 20% survivors by 120h) CLP are considered severe models of sepsis, with the 20-gauge model being the most severe. The 27-gauge CLP is considered a moderate model of sepsis (40% survivors by 120h). In severe sepsis the majority of mortality occurs between 24 and 36h, whereas the majority of mortality occurs between 56 and 72h in moderate sepsis.

characterized and the disadvantages of this model are relatively unknown.[14] Further studies characterizing the pathophysiology of the CASP model are required to validate this promising sepsis model.

Comparison between Cecal Ligation and Puncture and Colon Ascendens Stent Peritonitis Models The host-barrier disruption sepsis models, CLP and CASP, use endogenous gut flora as a source of infection and retain the advantage of evaluating bacterial clearance by the host, as opposed to endotoxemia models of sepsis that lack a replicating pathogen. A closer look at these two models reveals subtle yet significant differences, indicating that they represent independent sepsis models. Although both are peritonitis models, the CASP model reflects a diffuse peritonitis versus the peritonitis model of CLP.[43] Arecent study compared CLP and CASP.[43] Unlike CLP mice, bacteremia in CASP mice constantly increased significantly at regular time intervals, over 18h, suggesting continual bacterial leakage from the stent. At 18h, CASP mice demonstrated a more diffuse peritonitis through an increased number of bacteria found in vital organs, particularly in the lungs, liver, and kidney, compared with CLP mice.[43] These results suggest that CASP may represent a better diffuse peritonitis sepsis model, complimenting the CLP model of intraabdominal abscess formation.

METHODS FOR SEPSIS MODELS

The following represents a generalized description of techniques for the CLP, CASP, and endotoxemia models. The focus of this chapter is primarily the CLP method, which represents the current "gold standard" of sepsis models because it most closely approximates the physiological and immunological responses observed in human sepsis.[14]

CECAL LIGATION AND PUNCTURE METHOD

1. All procedures must follow the Institutional Animal Care and Utilization Committee (IACUC) or an institutionally approved animal care protocol. Animals are maintained on a standard *ad libitum* diet including water throughout the experiment. Methods in the early literature describe fasting animals 12h prior to CLP surgery.[15] Fasting animals helps to promote uniform stool consistency, leading to a subsequent similar flow of fecal contents into the peritoneum between experimental animals. Initial rodent experiments in our laboratory followed this procedure. We have determined in our laboratory that there was no benefit to fasting animals before CLP. In a rat CLP model, Singleton and Wischmeyer found fasted animals varied in individual cecum size length and thus cecal stool content, introducing heterogeneity in animals.[38] In the unlikely event of problems with heterogeneous stool consistency, fasting animals may assist in promoting stool homogeneity. Otherwise, fasting animals any time before CLP is not recommended or required for reproducible results.

2. Anesthetize the animal using a mixture of isoflurane and 100% oxygen. Other agents may be administered for animal anesthesia, including intraperitoneal ketamine (80mg/kg) and xylazine (10mg/kg); however, we favor inhalational agents delivered with a precision vaporizor via a tight-fitting nosecone as it allows for rapid performance of procedures with a reproducible depth of anesthesia and a rapid recovery time.

3. Prepare the animal for surgery by shaving and thoroughly disinfecting the abdominal area with an appropriate agent such as 7.5% povidone iodine cleansing solution and sterile gauze. Note that a heating pad or other source should be utilized in an attempt to ensure normothermia throughout the procedure. Meticulous attention to detail, organization of materials, and excellent surgical technique represent the cornerstone of successful reproducibility in this procedure.

4. Perform a 1.5-cm midline laparotomy to expose the cecum. The peritoneal cavity is protected by two layers: skin and the muscular abdominal wall. Once an incision is made to expose the underlying abdominal musculature, the second incision will be made through the translucent linea alba. The linea alba is a vertical midline of connective tissue running from the breast bone to the inguinal region that can be determined through appearance. An incision made in the linea alba will not bleed since the connective tissue lacks blood vessels, in contrast to the vascularized abdominal muscles. Use the contrast of the pink color of the abdominal muscle to find the translucent linea alba.

5. Locate and carefully exteriorize the cecum, making sure not to disrupt any vessels while untwisting the cecum (Figure 50–5A). Initially, locating the cecum can be time consuming. It is best to develop a systematic approach to this method. If the abdominal cavity is divided into four regions, the cecum is typically located in the upper left abdominal quadrant in the region of the spleen or it is immediately visible following laparotomy. A very effective method for locating the cecum developed in our laboratory is the "reverse 'N'" technique and is used if the cecum is not immediately found. With the mouse in a supine position, start at the traced end of the "N," the upper left abdominal compartment equivalent. Inspect each compartment systematically by reverse

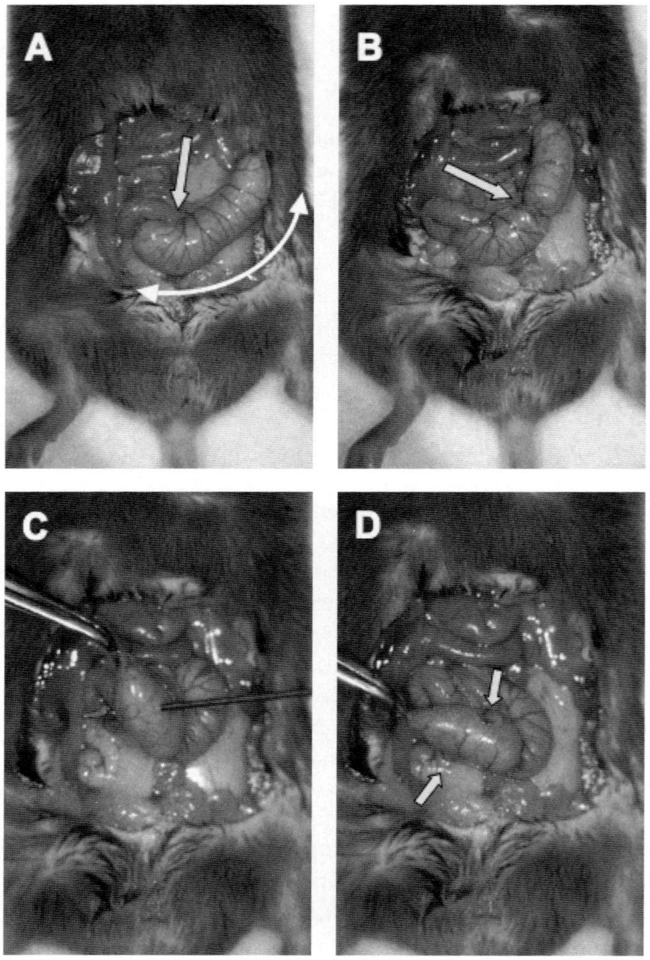

Figure 50–5. CLP Methods. (A) Normal cecal anatomy. The ileocecal junction (arrow) is exposed following midline laparotomy. The white double arrow demonstrates the entire length of the cecum, corresponding to 100% of the cecal length. (B) Cecal ligation. Cecal ligation is performed at approximately 50% of the length of the cecum in this case (the arrow denotes the site of cecal ligation). Care is taken to avoid damage to arterial branches during ligation and ligature is placed through the mesentery inferiorly to arterial branches. (C) Cecal puncture. A through and through (double) puncture is made while carefully immobilizing the ligated distal end of the cecum with a 21-gauge needle. (D) Milking of stool from cecum. Both arrows point to a 2-mm-diameter collection of stool milked from the cecum to the exterior to ensure puncture site patency.

tracing the "N," subsequently, moving to its lower left, skipping through the middle to the upper right, and finally to the lower right compartment until the cecum is located. This method of detection is a quick and efficient way to search the entire abdominal cavity. Remember that the cecum is the only blind-ended part of the intestine in rodents and is ultimately used for its positive identification. Care should be taken not to damage the cecum or intestines during handling. The operator should keep a body-temperature bowl of sterile saline nearby to keep gloves moist, which will assist in easier manipulation of the intestines.

6. Now that the cecum has been removed from the abdominal cavity, milk the stool contents within the cecum distally starting below the ileocecal valve to fill the blind-ended pouch that will subsequently be ligated.

7. Place 3.0 cm of 6.0 vicryl suture underneath the cecal artery and ligate the cecum (Figure 50–5B). To insert the 3.0-cm vicryl suture underneath the cecal artery, first carefully blunt dissect the connective tissue, connecting the cecal artery to the cecum with curved tweezers. Once this is complete, leave the tweezers in place and grab the suture with a second pair of tweezers. Transfer the suture to the first pair of curved tweezers and pull it through, underneath the cecal artery, subjacent to the cecum. Care must be taken to avoid trauma to the cecal artery and its branches.

8. Fold the cecum over the suture, remilk the cecum, and double tie the ligature loop tightly to completely ligate the cecum (Figure 50–5B). Gas bubbles may introduce variability within the model and consequently alter mortality. Before ligating the cecum, look for the presence of gas bubbles while milking the stool distally to fill the blind ended pouch patent with stool. Make sure that there are no gas bubbles present in the section distal to the ligation site. It is also important to make note of the stool consistency; many unpublished observations indicate that suprahydrated stool increases mortality. It is plausible that watery stool could be the result of an unknown gastrointestinal infection and may alter the severity of the CLP model. It is recommended that mice with either severely dehydrated stool or overly hydrated stool not be used for experimental purposes as the quality of the stool may limit or enhance the spread of infection, respectively.

An important step in the CLP procedure is the ligation of the cecum; remaining consistent with the distance of cecum ligated between animals as a higher percentage of cecum ligated correlates with increased sepsis severity.[38] The mouse's anatomy is an easily used reference control. Branching off from the main cecal artery are several perpendicular small arteries that can be used as a good point of reference for ligation. Starting distally, count the number of branch points up to the location chosen for ligation. The branch point number can be used on subsequent mice to approximate the total cecum length that was ligated in the previous procedure(s) to ensure consistent measurements. When publishing an article, the percentage of cecum ligated should be reported in the Methods section, as it represents important and frequently omitted information.

9. Puncture the ligated cecum as desired (Figure 50–5C). More severe models use needle gauge sizes ranging from 18- to 21- gauge and less severe models use needle gauge sizes ranging from 21- to 27-gauge depending on the number of puncture(s), percent of ligated cecum, and amount of feces manually extruded from the ligated cecum (see step 10).

10. Manually extrude 1–2 mm of feces from the ligated cecum (Figure 50–5D). With the same gauge size needle, the model could be made more or less severe as a sepsis model by adding or taking away approximately 1.0 mm of extruded feces. To improve the control over the amount of extruded feces, hold the ligated cecum parallel with straight tweezers while inserting the needle for puncture; this permits greater pressure control to extrude the feces from the cecum. With the tweezers, add a slight amount of pressure to the cecum before completely withdrawing the needle from it. Continue to exert pressure while retracting the needle from the cecum. The feces will flow out as a continuum with the retracting needle, mitigating the chances of the puncture hole closing off, which can lead to difficulties in controlling the amount of stool removed from the ligated cecum.

11. Carefully place the cecum back inside the abdominal cavity. Holding the abdominal muscles rather than holding the skin with tweezers facilitates replacement of the cecum back within the abdominal cavity. When there is an excessive loss of extruded feces from the cecum prior to placing it back into the abdominal cavity, the severity of the model may be diminished.

12. Close the abdominal cavity with 6.0 vicryl suture.

13. Administer 1 ml of saline fluid resuscitation subcutaneously using a 25-gauge needle. Fluid resuscitation replaces fluids lost during the course of surgery that may cause a premature decrease in blood pressure and early mortality.

14. Administer an analgesic, such as buprenorphine (1.6 mg/kg), for pain control every 12 h for 3 days.

15. Continually warm the animal on a heating pad for 5–10 min or until normal body temperature is reached and the animal has recovered from anesthesia by demonstrating self-righting ability.

16. Place the animals in separate cages. The animals should be housed in separate cages to prevent them from chewing another mouse's surgical incision. To increase the accessibility of food, gel food may be substituted for normal chow and placed directly into the cage along with the recovering animals during periods of reduced mobility.

17. Administer antibiotics if desired. When including antibiotics, a potent broad-spectrum antibiotic, such as imipenem (25 mg/kg), should be administered by intraperitoneal injection every 12 h. Additionally, antibiotic therapy should be initiated more than 2 h postprocedurally to better mimic the human clinical situation of delayed presentation and antibiotic administration. Note that when using imipenem for antibiotic therapy, the selective dihydropeptidase inhibitor cilastatin must be coadministered to prevent degradation of imipenem.

GENERAL FEATURES OF THE CECAL LIGATION AND PUNCTURE SURVIVAL STUDY The central goal of any animal model is to reproduce the clinical scenario of a disease as closely as possible. The time of deaths that occur during a survival CLP study is an important determinant of whether the model is correct in its severity. The severity of the CLP model ranges from sublethal to lethal with differences in the rate of mortality as well as the absolute percentage of mortality in the experimental group. In general, the first death should occur 24 h after the procedure to rule out iatrogenic effects, such as hemorrhage, that may also cause death. In the more severe CLP model, most deaths will occur between 24 and 36 h after CLP with mortality typically greater than 70% within 5 days (Figure 50–4). In the less severe CLP model most deaths occur between 48 and 72 h after CLP with a 30–40% range of mortality at the end of 5 days (Figure 50–4). These rates and percentages are presented as a guideline and the actual characteristics will vary based on the technique and experimental needs.

COLON ASCENDENS STENT PERITONITIS METHOD Before surgery remove the needle from the desired size venous catheter (14, 16, or 18 gauge).

1. Follow steps 1–4 as for the CLP procedure.

2. Locate the antimesenteric side of the ascending colon, and make an incision approximately 1 cm from the ileocecal valve where the stent will be surgically inserted. The ascending colon is connected to the small bowel at the ileocecal valve junction.

3. With 7-0 Ethicon thread stitch the stent into the lumen of the ascending colon.

4. Anchor the stent in place with two stitches.

5. Remove the inner needle of the stent, followed by cutting the stent at the prepared site.

6. To ensure the proper placement of the stent into the lumen of the ascending colon, squeeze a small amount of feces out of the stent.

7. Replace the intestinal content back into the abdominal cavity.

8. Follow steps 12–16 of the CLP procedure.

ENDOTOXEMIA METHOD

1. Weigh the animal.

2. Intraperitoneally inject (1 ml) LPS at the desired milligram/kilogram dose from the desired *Escherichia coli* serotype, or other chemical agent (e.g., zymosan or CpG DNA). Do not stir the LPS solution vigorously to prevent aggregates from forming. Sonication of the LPS solution for 2 min will help mitigate the presence of aggregates. Anesthetize the animal using isofluorane. When delivering an intraperitoneal injection, carefully insert the needle shallowly into the abdomen to avoid damage to underlying blood vessels and organs by tenting the abdominal wall. For intravenous injection, ensure that the LPS solution does not extravasate from the vein and infiltrate surrounding tissue.

3. Monitor the animals to ensure proper recovery from anesthesia.

IMPORTANT GENERAL CONSIDERATIONS FOR SEPSIS MODELS

HUMANE ENDPOINTS IN SEPSIS RESEARCH The use of fair and ethical animal treatment throughout sepsis research and any other experiments that cause pain or distress to animals must be strictly maintained at all times. This often presents challenges to scientists, as early termination may lead to inaccurate results, especially in experimental models involving complex longer-term pathogenesis such as the CLP model.[44] Organizations that advocate more humane endpoints in research animals and create guidelines for research animal care include Public Health Services (PHS), the National Research Council, and the United States Department of Agriculture (USDA) (www.grants.nih.gov/grants/olaw/references/phspol.htm, www.nap.edu, *Fed Reg.* 54). It is the responsibility of all researchers to be familiar with these guidelines and understand how they should be applied to their model(s). Guidance should also be sought from local experienced investigators and the institutional veterinarian and IACUC.

An outstanding review of the issue of humane endpoints in sepsis research has recently been published by Nemzek *et al.*[44] Death as an endpoint should be avoided whenever possible and any animal in a perimoribund state should be euthanized. However, as mentioned above, the limitations of current mortality predictors prevent development of any "gold standard" recommendation. Each investigator must determine the degree of illness and moribund state dictating euthanasia for the model in question. Clearly, implementing early termination of any experiment increases the risk of data collection errors that may lead to erroneous conclusions. Consequently, fastidious decisions regarding early endpoints must be taken to prevent these types of errors. The decision to terminate the experiment is a complex one. To prevent hasty

Table 50–1
Morbidity findings during development of sepsis

Objective	*Subjective*
Abnormal respiration	Aggression
Absence of bedding preparation	Decreased activity
Ataxia	Decreased movement in response to external stimuli
Reduced food/water consumption	
Body temperature reduction	
Body weight reduction	
Diarrhea/constipation	
Hunched posture	
Orbital exudates	
Ruffled hair/fur	

decisions multiple parameters should be evaluated before euthanizing an experimental animal. For example, in the murine CLP model physiological parameters, such as body temperature, blood pressure, and IL-6 levels, may represent predictors of CLP mortality.[44] Defining these parameters, in conjunction with the physical signs of CLP morbidity (Table 50–1), can potentially be used to successfully predict CLP mortality in mice.[44] However, the use of such predictors must be established empirically within the investigator's laboratory and be tested for reproducibility.

MONITORING As a general rule, animals should be closely monitored every 4h for the first 48h postprocedure. Depending on the severity of the model, particular timeframes following surgery are critical and require more attention. For example, in a severe sepsis model the majority of deaths occur between 24 and 36h. Within this critical time frame, it is important to frequently assess for animal pain and distress caused by illness, in addition to checking for mortalities to ensure proper animal care and data collection, respectively. Analgesia with an agent such as buprenorphine may be administered at an increased frequency during this period. In the event that death occurs overnight, an average time of death can be calculated by taking the average of the last hour the animal(s) were found alive and the hour of death. When the animals have passed the critical period when most succumb to sepsis and exhibit normal behavior (survivors), observation times may be less frequent. A thorough understanding of these timeframes during the course of animal sepsis is beneficial to both the animals and the investigator. This knowledge serves to ensure that proper animal care is given at all times, and will reduce valuable labor while maintaining a high standard of animal care.

REPRODUCIBILITY Successful development of a functional sepsis model requires a significant amount of experience with the chosen technique. The level of experience is directly proportional to the complexity of the procedure in question. The simplest model involves a bolus injection of endotoxin and the most surgically complex and demanding model is the CASP technique. Regardless of which model is chosen, the investigator must perform several pilot experiments to define the timing and survival rate to ensure reproducibility. It is also highly recommended that experiments be performed in a randomized and blinded manner whenever possible. This becomes an issue as the control group for each study must be run with every experiment and this increases the number of animals required for completion of the study. However, historical controls unfortunately do not suffice as there may be variability in the experiments based on subclinical infections, alteration of food or bedding lots, and operator dependence. When possible, it is optimal to retain the same operator for the experiment, as subtle differences in surgical style may significantly alter the timing of illness progression and survival rates.

Attention to detail and repetition are paramount in the successful development and utilization of sepsis models. Animal survival remains the primary outcome measurement in study of sepsis mechanisms and in preclinical therapy studies because there are limited markers for predicting mortality or defining sepsis stages, and also because the use of *in vivo* telemetry is complex. To ensure that the operator has enough experience to initiate the end experiment, several independent control experiments should be conducted. For example, using the CLP model, our experience dictates an expected profile of mortality for a given needle gauge used (Figure 50–4). An inexperienced operator, after becoming generally familiar with the CLP technique, typically must perform three to five independent experiments with a sufficient number of animals (typically 10 per experimental group) to demonstrate that the operator can reproduce the expected profile and rate of mortality. The concept of generating reproducible illness is critical when conducting experiments that do not use the mortality endpoint, as in these shorter studies there is no other way to ensure that the proper level of illness has been established. Hopefully, advances in our understanding of sepsis staging markers will assist with this issue and make mortality studies obsolete.

BIOMARKERS IN SEPSIS The complexity of the dysregulated immune response in sepsis makes it difficult to predict clinical outcomes in patients and animal models. Over the years, immunological studies performed to unravel this enigmatic process have also unveiled potential biomarkers predictive of sepsis mortality in attempts to classify the development of sepsis.[45,46]

Translation of therapeutic agents successful in animal model studies to the clinic represents the current stumbling block of sepsis therapy developments.[14] Biomarkers are particularly useful in discriminating between populations of patients who may benefit from particular therapies and for animal model purposes in discriminating sepsis stage and severity when designing sepsis therapies.[14] Many studies have demonstrated a strong association between interleukin 6 (IL-6) levels and sepsis severity in both humans and animal models of sepsis, suggesting that serum IL-6 level may represent a possible biomarker of sepsis mortality.[32,40,47,48] In fact, endogenous IL-6 was utilized successfully to stratify patients into groups of sepsis severity in a recent clinical trial evaluating the efficacy of the antitumor necrosis factor antibody $F(ab')_2$ fragment afelimomab.[49]

The lethal predictive threshold level of circulating IL-6 has been shown to vary with genetic background in murine studies, and it is not associated with disease severity and antibiotic administration.[50] In less severe CLP sepsis with 77% survival after 3 days, serum IL-6 levels were measured 6h after the CLP procedure and ranged from 83pg/ml to 10,000pg/ml.[32] Female BALB/c mice used in this experiment with serum IL-6 levels greater than 2000pg/ml had significantly more rapid onset of mortality and greater mortality rates compared with those mice with IL-6 levels less than 2000pg/ml.[32] In the study by Turnbull *et al.,* male ND4 mice demonstrated that the "lethal predictive threshold" of serum IL-6 levels was at 14,000pg/ml.[50] The difference in the "lethal predictive threshold" of IL-6 described in these two studies was attributed to the consequences of strain differences.

It is likely that the gender, age, strain, or all variables in combination account for differences in lethal predictive threshold IL-6 levels, as there is clear variability within the CLP model depending on the age, gender, and animal strain employed.[14] The effect of age is difficult to assess between the above two studies due to the lack of consistency in age descriptions. The ages for BALB/c mice were ambiguously described as "adult," as opposed to the more accurately identified 6- to 8-week-old ND4 mice. Conclusions drawn from these experiments emphasize the importance of careful documentation of the characteristics of animal subjects so that potential confounding variables may be identified when comparing the interpretation of results among studies.

ANTIBIOTICS The successful control of a systemic infection is an important component of the body's response to infection and involves both innate and adaptive immune responses. Broad-spectrum antibiotics are routinely prescribed for human sepsis patients once the infecting agent is identified, to help in the destruction of bacterial pathogens.[51] To more closely mimic the clinical scenario of sepsis, antibiotics have been used in a variety of sepsis models, including CLP, CASP, and exogenous bacterial implantation models. Currently, imipenem (25 mg/kg) represents a broad-spectrum antibiotic and is commonly used in sepsis models. Dosing of antibiotic also affects agent selection as less frequent administration reduces the number of injections each animal will receive. Imipenem is attractive in this regard as it may be dosed at 12 h intervals.

It has long been a concern that the administration of antibiotics results in the acceleration of cytokine-induced septic shock due to the release of endotoxin bacterial products after bacterial killing.[52] Antibiotic administration may increase the circulating levels of endotoxin in murine sepsis models, and may theoretically lead to a more robust host inflammatory response. However, further studies have demonstrated that the effects of increased serum endotoxin levels resulting from antibiotic administration have no effect on inflammation and prognosis in the CLP sepsis model suggesting that antibiotics will not artifactually alter mortality.[53] In a murine sepsis lung injury model, antibiotics delayed mortality time points but did not improve overall survival.[54] In conclusion, administration of antibiotics appears to decrease the severity of sepsis but does not necessarily decrease mortality in polymicrobial sepsis models and does not appear to increase model severity due to endotoxin release.

CONCLUSIONS

Successful development and use of sepsis models in research require a thorough theoretical understanding of the advantages and disadvantages of each model system, as well as a meticulous technical approach to establishing a reproducible model system. With such understanding and attention to detail, experiments utilizing these complex models can generate important data that will assist in further defining sepsis pathophysiology and identify novel sepsis therapies.

REFERENCES

1. Perl M, Chung CS, Garber M, Huang X, Ayala A. Contribution of anti-inflammatory/immune suppressive processes to the pathology of sepsis. *Front Biosci* 2006;11:272–299.
2. Vincent JL, De Backer D. Microvascular dysfunction as a cause of organ dysfunction in severe sepsis. *Crit Care* 2005;9(Suppl. 4): S9–12.
3. Gullo A, Iscra F, Di Capua G, *et al.* Sepsis and organ dysfunction: An ongoing challenge. *Minerva Anestesiol* 2005;71(11):671–699.
4. Hoesel LM, Gao H, Ward PA. New insights into cellular mechanisms during sepsis. *Immunol Res* 2006;34(2):133–141.
5. Bone RC. Sir Isaac Newton, sepsis, SIRS, and CARS. *Crit Care Med* 1996;24(7):1125–1128.
6. Levy MM, Fink MP, Marshall JC, *et al.* 2001 SCCM/ESICM/ACCP/ATS/SIS International Sepsis Definitions Conference. *Crit Care Med* 2003;31(4):1250–1256.
7. Ulloa L, Tracey KJ. The "cytokine profile": A code for sepsis. *Trends Mol Med* 2005;11(2):56–63.
8. Parker MM, Parrillo JE. Septic shock. Hemodynamics and pathogenesis. *JAMA* 1983;250(24):3324–3327.
9. Abraham E, Shoemaker WC, Bland RD, Cobo JC. Sequential cardiorespiratory patterns in septic shock. *Crit Care Med* 1983;11(10): 799–803.
10. Groeneveld AB, Bronsveld W, Thijs LG. Hemodynamic determinants of mortality in human septic shock. *Surgery* 1986;99(2): 140–153.
11. Parker MM, Shelhamer JH, Bacharach SL, *et al.* Profound but reversible myocardial depression in patients with septic shock. *Ann Intern Med* 1984;100(4):483–490.
12. Parker MM, Shelhamer JH, Natanson C, Alling DW, Parrillo JE. Serial cardiovascular variables in survivors and nonsurvivors of human septic shock: Heart rate as an early predictor of prognosis. *Crit Care Med* 1987;15(10):923–929.
13. Shelley O, Murphy T, Paterson H, Mannick JA, Lederer JA. Interaction between the innate and adaptive immune systems is required to survive sepsis and control inflammation after injury. *Shock* 2003;20(2):123–129.
14. Buras JA, Holzmann B, Sitkovsky M. Animal models of sepsis: Setting the stage. *Nat Rev Drug Discov* 2005;4(10):854–865.
15. Wichterman KA, Baue AE, Chaudry IH. Sepsis and septic shock—a review of laboratory models and a proposal. *J Surg Res* 1980;29(2): 189–201.
16. Fink MP, Heard SO. Laboratory models of sepsis and septic shock. *J Surg Res* 1990;49:186–196.
17. Deitch EA. Animal models of sepsis and shock: A review and lessons learned. *Shock* 1998;9(1):1–11.
18. Esmon CT. Why do animal models (sometimes) fail to mimic human sepsis? *Crit Care Med* 2004;32(5 Suppl.):S219–222.
19. Borden CW, Hall WH. Fatal transfusion reactions from massive bacterial contamination of blood. *N Engl J Med* 1951;245(20): 760–765.
20. Meng G, Rutz M, Schiemann M, *et al.* Antagonistic antibody prevents toll-like receptor 2-driven lethal shock-like syndromes. *J Clin Invest* 2004;113(10):1473–1481.
21. Sparwasser T, Miethke T, Lipford G, *et al.* Bacterial DNA causes septic shock. *Nature* 1997;386(6623):336–337.
22. Hemmi H, Takeuchi O, Kawai T, *et al.* A toll-like receptor recognizes bacterial DNA. *Nature* 2000;408(6813):740–745.
23. Reinhart K, Karzai W. Anti-tumor necrosis factor therapy in sepsis: Update on clinical trials and lessons learned. *Crit Care Med* 2001;29(7 Suppl.):S121–125.
24. Riedemann NC, Guo RF, Ward PA. The enigma of sepsis. *J Clin Invest* 2003;112(4):460–467.
25. Zeni F, Freeman B, Natanson C. Anti-inflammatory therapies to treat sepsis and septic shock: A reassessment. *Crit Care Med* 1997;25(7):1095–1100.
26. Cross AS, Opal SM, Sadoff JC, Gemski P. Choice of bacteria in animal models of sepsis. *Infect Immun* 1993;61(7):2741–2747.
27. Greenberger MJ, Strieter RM, Kunkel SL, Danforth JM, Goodman RE, Standiford TJ. Neutralization of IL-10 increases survival in a murine model of Klebsiella pneumonia. *J Immunol* 1995;155(2): 722–729.
28. van der Poll T, Marchant A, Keogh CV, Goldman M, Lowry SF. Interleukin-10 impairs host defense in murine pneumococcal pneumonia. *J Infect Dis* 1996;174(5):994–1000.
29. Evans GF, Snyder YM, Butler LD, Zuckerman SH. Differential expression of interleukin-1 and tumor necrosis factor in murine septic shock models. *Circ Shock* 1989;29(4):279–290.

30. Zanetti G, Heumann D, Gerain J, *et al.* Cytokine production after intravenous or peritoneal gram-negative bacterial challenge in mice. Comparative protective efficacy of antibodies to tumor necrosis factor-alpha and to lipopolysaccharide. *J Immunol* 1992;148(6): 1890–1897.
31. Ryan NT, Blackburn GL, Clowes HA Jr. Differential tissue sensitivity to elevated endogenous insulin levels during experimental peritonitis in rats. *Metabolism* 1974;23(11):1081–1089.
32. Remick DG, Bolgos GR, Siddiqui J, Shin J, Nemzek JA. Six at six: Interleukin-6 measured 6h after the initiation of sepsis predicts mortality over 3 days. *Shock* 2002;17(6):463–467.
33. Ayala A, Chaudry IH. Immune dysfunction in murine polymicrobial sepsis: Mediators, macrophages, lymphocytes and apoptosis. *Shock* 1996;6(Suppl. 1):S27–38.
34. Hotchkiss RS, Tinsley KW, Karl IE. Role of apoptotic cell death in sepsis. *Scand J Infect Dis* 2003;35(9):585–592.
35. Baker CC, Chaudry IH, Gaines HO, Baue AE. Evaluation of factors affecting mortality rate after sepsis in a murine cecal ligation and puncture model. *Surgery* 1983;94(2):331–335.
36. van der Poll T, Marchant A, Buurman WA, *et al.* Endogenous IL-10 protects mice from death during septic peritonitis. *J Immunol* 1995;155(11):5397–5401.
37. Oberholzer A, Oberholzer C, Bahjat KS, *et al.* Increased survival in sepsis by in vivo adenovirus-induced expression of IL-10 in dendritic cells. *J Immunol* 2002;168(7):3412–3418.
38. Singleton KD, Wischmeyer PE. Distance of cecum ligated influences mortality, tumor necrosis factor-alpha and interleukin-6 expression following cecal ligation and puncture in the rat. *Eur Surg Res* 2003;35(6):486–491.
39. Yang S, Zhou M, Koo DJ, Chaudry IH, Wang P. Pentoxifylline prevents the transition from the hyperdynamic to hypodynamic response during sepsis. *Am J Physiol Heart Circ Physiol* 1999;277(3Pt. 2): H1036–1044.
40. Latifi SQ, O'Riordan MA, Levine AD. Interleukin-10 controls the onset of irreversible septic shock. *Infect Immun* 2002;70(8): 4441–4446.
41. Manley MO, O'Riordan M A, Levine AD, Latifi SQ. Interleukin 10 extends the effectiveness of standard therapy during late sepsis with serum interleukin 6 levels predicting outcome. *Shock* 2005; 23(6):521–526.
42. Zantl N, Uebe A, Neumann B, *et al.* Essential role of gamma interferon in survival of colon ascendens stent peritonitis, a novel murine model of abdominal sepsis. *Infect Immun* 1998;66(5):2300–2309.
43. Maier S, Traeger T, Entleutner M, *et al.* Cecal ligation and puncture versus colon ascendens stent peritonitis: Two distinct animal models for polymicrobial sepsis. *Shock* 2004;21(6):505–511.
44. Nemzek JA, Xiao HY, Minard AE, Bolgos GL, Remick DG. Humane endpoints in shock research. *Shock* 2004;21(1):17–25.
45. Marshall JC. Biomarkers of sepsis. *Curr Infect Dis Rep* 2006;8(5): 351–357.
46. Spapen HD, Hachimi-Idrissi S, Corne L, Huyghens LP. Diagnostic markers of sepsis in the emergency department. *Acta Clin Belg* 2006;61(3):138–142.
47. Spittler A, Razenberger M, Kupper H, *et al.* Relationship between interleukin-6 plasma concentration in patients with sepsis, monocyte phenotype, monocyte phagocytic properties, and cytokine production. *Clin Infect Dis* 2000;31(6):1338–1342.
48. Groeneveld AB, Tacx AN, Bossink AW, van Mierlo GJ, Hack CE. Circulating inflammatory mediators predict shock and mortality in febrile patients with microbial infection. *Clin Immunol* 2003;106(2): 106–115.
49. Panacek EA, Marshall JC, Albertson TE, *et al.* Efficacy and safety of the monoclonal anti-tumor necrosis factor antibody F(ab')2 fragment afelimomab in patients with severe sepsis and elevated interleukin-6 levels. *Crit Care Med* 2004;32(11):2173–2182.
50. Turnbull IR, Javadi P, Buchman TG, Hotchkiss RS, Karl IE, Coopersmith CM. Antibiotics improve survival in sepsis independent of injury severity but do not change mortality in mice with markedly elevated interleukin 6 levels. *Shock* 2004;21(2):121–125.
51. Mazuski JE, Sawyer RG, Nathens AB, *et al.* The Surgical Infection Society guidelines on antimicrobial therapy for intra-abdominal infections: An executive summary. *Surg Infect (Larchmt)* 2002;3(3): 161–173.
52. Hopkin BDA. A nasty shock from antibiotics. *Lancet* 1985;2(8455): 594.
53. Vianna RC, Gomes RN, Bozza FA, *et al.* Antibiotic treatment in a murine model of sepsis: Impact on cytokines and endotoxin release. *Shock* 2004;21(2):115–120.
54. Doerschug KC, Powers LS, Monick MM, Thorne PS, Hunninghake GW. Antibiotics delay but do not prevent bacteremia and lung injury in murine sepsis. *Crit Care Med* 2004;32(2):489–494.

51 Animal Models in Functional Magnetic Resonance Imaging

AFONSO C. SILVA AND BOJANA STEFANOVIC

ABSTRACT

Functional magnetic resonance imaging (fMRI) has significantly advanced brain research, both in basic sciences as well as in translational and clinical studies. Animal models have been at the forefront of research into the mechanisms of magnetic resonance imaging (MRI) and fMRI. The use of animal models in fMRI has been particularly advantageous in preclinical and translational studies of various models of brain disease, and fundamental in the exploration of basic neuroscience questions, such as the mechanisms of perception, behavior, and cognition. This chapter describes the practical aspects of the use of animal models in fMRI studies of the brain.

Key Words: Brain, BOLD, Cat, Cerebral blood flow, Cerebral blood volume, Cerebrovascular coupling, Nonhuman primate, Rodent, Songbird.

INTRODUCTION

Since its inception in the early 1990s, functional magnetic resonance imaging (fMRI) has had a profound impact on brain function research, particularly in cognitive neuroscience[1] and in translational and clinical studies.[2] Like positron emission tomography (PET) and diffuse optical tomography, fMRI relies on the cerebrovascular coupling, a tight relationship between changes in neural activity and local cerebral blood flow (CBF), cerebral blood volume (CBV), and oxygen consumption ($CMRO_2$).[3] fMRI provides excellent contrast to noise ratio, submillimeter spatial resolution, coverage of the whole brain, and relative ease of implementation. On the other hand, the underlying fMRI signal mechanism and its functional specificity are still subjects of research. Moreover, the temporal resolution of fMRI is relatively low (particularly with respect to the time scale of neuronal events), and the magnetic resonance imaging (MRI) environment poses stringent restrictions on subject positioning and movement.

While the brain is a particularly difficult target for translational studies, critical insight into its functioning has been gained through animal models. Indeed, the use of animal models has been of fundamental importance to the development of fMRI techniques. The first study employing blood oxygenation level-dependent (BOLD) contrast was performed in rats[4,5] as were the first CBF measurements using exogenous MRI tracers, such as deuterium,[6,7] fluorine,[8,9] and gadolinium chelates,[10,11] which allowed estimation of CBV in addition to CBF. The advantages of using an endogenous tracer led to the development of arterial spin labeling (ASL) techniques, with the first demonstrations occurring in rodents.[12,13]

Animal models of functional activation of the brain have been widely employed to address issues related to spatial localization of the functional signals, the magnitude of signal changes as a function of stimulation parameters, as well as temporal aspects of the hemodynamic response. Furthermore, the use of animal models in fMRI has been particularly advantageous in preclinical and translational studies of various models of brain disease. This chapter is aimed at describing the practical aspects of the use of animal models in fMRI studies of the brain.

BASICS OF FUNCTIONAL MAGNETIC RESONANCE IMAGING

ACTIVATION PHYSIOLOGY At the heart of brain function lies the generation and integration of electrical and chemical signals. Electrical signals arise via transient changes in the neuronal membrane potential, with >~15 mV depolarization[14] leading to action potential generation or neuronal firing. In most instances, this depolarization arises from temporal and spatial summation of the graded potentials generated by the synaptic input to the neuron, with coincidental activation of hundreds of the thousands of synapses of a typical neuron. While only a miniscule amount of charge is transported across the membrane in the course of the 1-msec activation potential,[14] the long-term preservation of the ionic gradients[15] and hence the excitability of the neuron is ensured by the continued operation of the Na,K-ATPase pump.

Under normal steady-state conditions, almost all of the brain's global energetic demands (in the form of ATP, the primary energy currency in the body) are supplied by oxidative metabolism of glucose (CMRGlc), the O_2-to-glucose ratio being ~5.5 mol/mol.[16] Whereas glucose consumption is closely correlated with local functional activity,[17–20] only ~50% of the glucose extracted by tissue from blood is actually catabolized by neurons and glia.[21] Indeed, it is frequently surmised that under normal physiological conditions blood flow upregulation following functional activation is not subserving the increased demand for glucose.[22]

Increases in local $CMRO_2$ following functional activation are typically much smaller than the accompanying rises in CBF, translating into a higher blood oxygenation level at activation relative to baseline. The customarily quoted reason is that the efficiency of oxygen extraction from the capillary network

From: *Sourcebook of Models for Biomedical Research*
(P. M. Conn, ed.), © 2008 Humana Press Inc., Totowa, NJ.

[i.e. oxygen extraction fraction (OEF)] decreases with increases in flow,[23] so that a much larger CBF increase is required to support a given rise in $CMRO_2$. In other words, given low tissue PO_2 yet constant parenchymal geometry, the O_2 concentration gradient and hence O_2 diffusion are increased through a rise in the capillary PO_2 achieved by a drop in the O_2 extraction.[22]

In addition to increased metabolism, a focal rise in neuronal activity is accompanied by an increase in the local CBF. One or more vasodilatory substances[24] thus act on the smooth muscle in the arteriolar wall, causing a drop in the arteriolar vascular resistance following functional activation. On the level of capillaries, the focal rise in CBF is thought to be predominantly mediated by an increase in capillary blood velocity rather than capillary dilatation or recruitment.[25] The ensuing rise in the pressure seen at the veins causes their passive dilatation and results in a significant increase in the total blood volume. Consistent with this view of CBV changes being the result rather than the cause of the blood flow increase is the observation of delayed CBV relative to CBF response in rodent somatosensory cortex following forepaw stimulation.[26]

BOLD FUNCTIONAL MAGNETIC RESONANCE IMAGING CONTRAST The magnetic properties of the heme iron of a hemoglobin molecule change following oxygenation,[27] constituting the source of contrast for BOLD fMRI. Specifically, the deoxygenated heme iron ion is paramagnetic, i.e., it has a small positive magnetic susceptibility, ($\chi > 0$). Upon binding with oxygen, the iron ion changes to a low-spin state and becomes diamagnetic, exhibiting a very small, yet negative magnetic susceptibility.[22,28] Overall, the magnetic susceptibility of blood changes linearly with its oxygen saturation.[29] Upon an increase in local neuronal activity, the blood oxygenation level rises (i.e., assuming normal conditions, the fraction of deoxygenated heme iron atoms decreases), so that the magnetic field profile across a vascularized tissue element becomes more homogeneous.

The signal measured in a typical BOLD fMRI study reflects a nonequilibrium value of coherent transverse water magnetization, generated by radiofrequency (RF) excitation. In particular, the timing parameters of BOLD fMRI experiments are chosen to produce a weighting of the signal that emphasizes the activation-induced changes in the time constant of the roughly exponential decay of spin phase coherence in the plane orthogonal to the direction of the external static magnetic field. Such activation-induced differences in the apparent transverse relaxation rate ($R2^*$) arise from the changes in the magnetic flux density (B) experienced by a spin ensemble under investigation following focal functional activation. A higher BOLD fMRI signal will be measured at activation, relative to baseline, as a result of the slower decay (smaller $R2^*$) produced by the increased homogeneity in the focal microscopic magnetic field profile, arising from the activation-induced increase in the blood oxygenation levels.

OTHER FUNCTIONAL MAGNETIC RESONANCE IMAGING CONTRASTS In addition to BOLD fMRI, fMRI measurements of blood flow and blood volume have also been frequently employed to track brain function.

Cerebral Blood Flow fMRI affords noninvasive means of direct CBF measurements using arterial water as a perfusion tracer.[30–32] The general principle behind ASL techniques is to differentiate the net magnetization of endogenous arterial water flowing proximally to the organ of interest from the net magnetization of tissue. As arterial blood perfuses the tissue, water exchange occurs, effectively changing the net magnetization of tissue proportionally to the blood flow rate. Therefore, CBF can be quantitatively related from the difference of two images acquired consecutively: one with and one without spin labeling. ASL MRI techniques can be implemented with either pulsed labeling or continuous labeling.[30,32,33] The pulsed ASL (PASL) methods use single or multiple RF pulses to label arterial blood water spins. The continuous ASL (CASL) technique uses a long RF pulse in the presence of a longitudinal field gradient to label the arterial spins according to the principles of adiabatic fast passage. ASL techniques have major advantages over other techniques based on the administration of exogenous tracers. They are completely noninvasive and because of the very short half-life of the labeled spins, repeated measurements of CBF can be performed freely. Moreover, the techniques preserve the high spatial resolution of ^1H MRI, enabling a direct and precise anatomical localization of CBF changes. On the other hand, proper perfusion contrast is achieved only when enough time is allowed for the labeled arterial spins to travel into the region of interest and exchange with tissue spins, making it difficult to detect changes in CBF with a temporal resolution greater than the decay time of the label.

Arterial spin labeling affords absolute quantification of a uniquely defined physiological parameter; it is specific to tissue signal change, better localized to the focus of the increased electrical activity (as it is sensitive to the arterial side of the vascular tree, in particular to capillaries), and it involves a shorter delay following the neuronal activity increase when compared to the BOLD technique. It is also less affected by venous drainage and is largely insensitive to scanner drift and local susceptibility gradients, in contrast to BOLD. Overall, then, it offers a more sensitive and specific measure of the neuronal activity, with respect to the assessment of its magnitude, spatial extent, and temporal evolution. It is thus the tool of choice for longitudinal functional imaging and has been used in a variety of clinical applications.[34] Nonetheless, ASL has a significantly lower contrast-to-noise ratio and temporal resolution than BOLD and is arguably harder to implement.

Cerebral Blood Volume Most of the existing MR data on CBV come from contrast-enhanced MRI studies. Indeed, the first human fMRI brain maps, obtained by Belliveau *et al.* in 1991,[35,36] were of CBV changes accompanying visual stimulation. The passage of a bolus of gadolinium-DTPA (Gd-DTPA, an intravascular MR contrast agent) was rapidly imaged using $T2^*$-weighted acquisitions.[36,37] The effect of Gd-DTPA is to produce microscopic magnetic field inhomogeneities, thus enhancing the rate of spin–spin relaxation. The analysis relies on the application of the indicator dilution theory to establish the relationship between the contrast agent and CBV in addition to the observed linear dependence of the tissue transverse relaxation rate on the contrast agent concentration.[35,36] Specifically, as the contrast agent passes through the brain, signal attenuation directly proportional to the concentration of the contrast agent is observed. Integrating the area under the concentration–time curve provides a measure of CBV. By repeating this experiment in baseline and activation conditions and subtracting the calculated CBV images, a functional activation map can be produced. The major shortcomings of the method, however, are poor temporal resolution and the limited number of functional measurements that can be performed, due to the administration of the contrast agent.

Unlike dynamic bolus methods, steady-state techniques require a contrast agent with a long intravascular half-life (e.g., ultrasmall superparamagnetic particles) that alleviates the requirement for a bolus injection and affords continuous monitoring of CBV changes. One such agent is MION (monocrystalline iron oxide nanocolloid), which has been used in anesthetized rats to provide a 5-fold increase in signal-to-noise ratio (SNR) over BOLD contrast at 2 T[26] and in awake, behaving macaques, resulting in a three-fold increase in sensitivity over BOLD at 3 T.[38] While MION concentrations used in these experiments are considerably higher than those approved for humans, initial testing of stable blood-pool contrast agents in humans is underway.[39] This approach provides high SNR and temporal resolution, though as with Gd-DTPA bolus methods, toxicity issues limit the number and frequency of studies that can be performed on an individual.

PERFORMING FUNCTIONAL MAGNETIC RESONANCE IMAGING IN ANIMALS

THE CHOICE OF THE ANIMAL MODEL The use of animals in biomedical research is governed by the scientific rationale that a living organism provides a whole, integrated biological system that is the best surrogate for humans in the laboratory. Animal research has led to a greater understanding of living systems and has generated knowledge that can be generalized to humans, facilitating the development of effective therapies, cure, and/or prevention of diseases. When applied to neuroscience, the use of animal models has afforded a better grasp of the biophysical mechanisms of neuroimaging techniques, a greater understanding of neurodegenerative diseases and the underlying mechanisms of neuronal cell damage and death, aided in the development of effective pharmacological therapies and treatments, and facilitated the study of the mechanisms of sensory perception, cognition, and emotion.

There are several factors that influence the choice of the animal model to be used in biomedical research. First, the investigator must consider not only the appropriateness of a particular species, but also whether the information obtained from the research can be extrapolated to human beings. Second, the investigator must be equipped with the experimental setup necessary to collect physiologically meaningful data. This is particularly important when performing fMRI experiments in animal models, when the type of instrument poses constraints on data acquisition, such as spatial and temporal resolution, spatial coverage, SNR, etc. Broadly, the fMRI experiments employing animal models focused on three major areas: the study of biophysical models of fMRI contrast; the investigation of neurological disorders, such as stroke, Alzheimer's disease, Parkinson's disease, and epilepsy; and the exploration of basic neuroscience questions, such as the mechanisms of perception, behavior, and cognition. Examples of each of these applications are described below.

Use of Animal Models to Understand the Biophysical Mechanisms of Functional Magnetic Resonance Imaging Animal models have been at the forefront of basic science research into the mechanisms of MRI and fMRI. From the early days of MRI, mostly rodents, but also dogs and cats, provided a stable and useful platform for the development and characterization of several intravascular contrast agents, such as Gd-DTPA[10,40–43] and iron-oxide particles.[44] These studies led to the use of such intravascular agents for the measurement of CBF and CBV,[10,11,37,45] angiography,[46,47] detection of brain tumors,[48–53] brain infarct,[54–56] and of defects in the blood–brain barrier.[57–59] At the same time, MRI techniques sensitive to blood flow were developed in rodents[6,7,60] and cats[9,61–68] using first exogenous agents, and later endogenous arterial water as a perfusion tracer.[12,13] It was long known that changes in CBF and CBV accompanied changes in neural activity due to brain stimulation.[69] Thus, MRI techniques sensitive to blood flow and blood volume were employed to probe brain function. The first experiments used intravascular contrast agents and were performed in dogs[35,70] and then in humans.[36,71,72] In fact, the discovery of the BOLD effect (see Figure 51–1) was made in rodents (rats and mice).[4,5,73] Thus, the use of animal models had a profound impact on the development of fMRI techniques.

The rodent model has been the workhorse for the investigation of theoretical issues and development of experimental approaches aimed at elucidating quantification of regional CBF and CBV, as well as for the studies of BOLD signal mechanism. In arterial spin labeling, for example, the rodent model has been of fundamental importance in the study of issues such as the effects of magnetization transfer,[74,75] the transit-time from the labeling plane to the detection voxel,[76,77] the perfusion territory of specific arterial vessels,[78] and the exchange of arterial water with tissue water.[75,79–82]

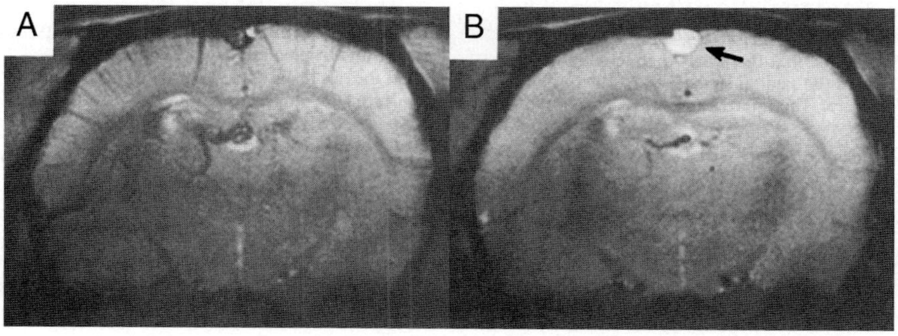

Figure 51–1. A demonstration of the effect of inspired CO_2 on BOLD contrast in a urethane-anesthetized rat. (A) T_2*-weighted coronal image obtained when the animal breathed 100% O_2. Venous blood vessels appear dark due to a short T_2* associated with a reduced venous blood oxygenation level *Y*. (B) The same image obtained when the animal breathed a mixture of 90% O_2/10% CO_2. In response to hypercapnia, CBF increases, elevating *Y* and thus increasing the MRI signal in the veins, such as the sagittal sinus (arrow). (Modified from Ogawa *et al.*[4] Copyright 1990 National Academy of Sciences USA.)

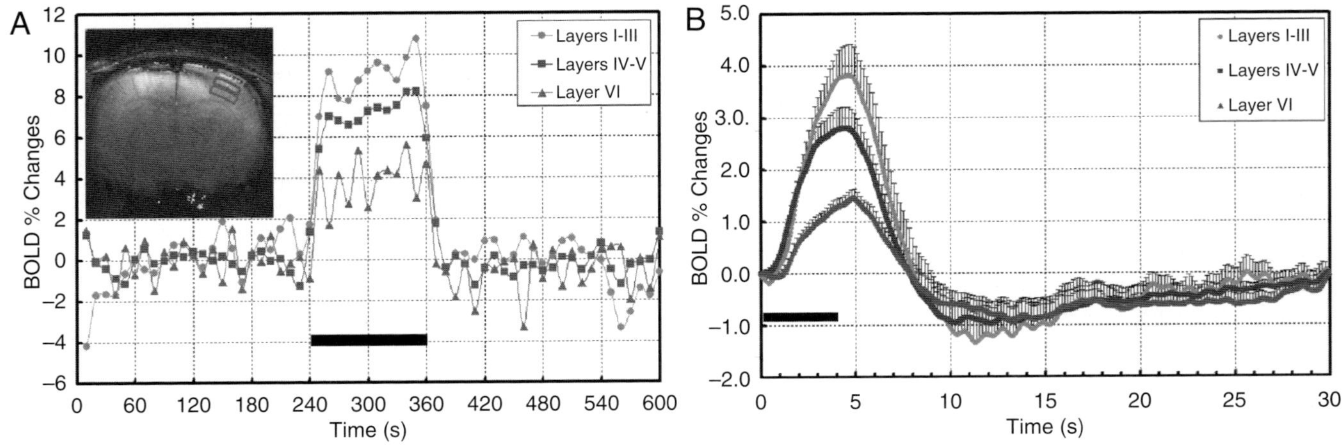

Figure 51–2. (A) Laminar-dependent spatial heterogeneity of BOLD signal changes to electrical stimulation of the rat forepaw, obtained at 50 × 50 × 2000 μm^3. The BOLD response in the supragranular layers was 44% larger than the one in the intermediate layers and 144% larger than the one in the bottom layer. (B) BOLD time courses obtained at 40-msec temporal resolution averaged across nine animals. Layers IV–V present the fastest onset time to stimulation, followed by layers I–III and VI. The inset shows the difference in onset times. (Modified from Silva and Koretsky.[110] Copyright 2002 National Academy of Sciences USA.)

These studies were aimed at improving the accuracy of quantification of CBF and its changes during functional brain stimulation. The use of intravascular contrast agents to quantify CBV proliferated in the 1990s in both rodents[83–88] and cats.[56,89–91]

Subsequent rodent studies focused on comprehending how changes in blood oxygenation related to BOLD signal changes[92–96] and on the use of sensorimotor functional paradigms as robust models of brain functional activation.[26,97–105] More recently, animal models of functional activation of the brain have been employed to investigate the spatial localization of fMRI techniques,[99,101,105–115] the dependence of signal magnitudes on stimulation parameters,[86,116–120] as well as the temporal aspects of the functional time-courses.[110,121–126] Recently, in a rat model of somatosensory stimulation, we were able to show that the BOLD signal changes have distinct amplitude and temporal characteristics that vary spatially across cortical layers, providing strong evidence that functional hemodynamic signals encode and preserve the temporal evolution of neuronal events (Figure 51–2).

Use of Animal Models to Study Neurological Disorders A second major area of application of animal models in fMRI is the study of neurological disorders, such as stroke, epilepsy, Parkinson's disease, and Alzheimer's disease, which involve a disruption in cerebrovascular and cerebrometabolic coupling. Animal models were extensively used to establish experimental, translational, and preclinical protocols aimed at understanding human diseases. Moreover, fMRI provided an invaluable tool for disease characterization, as well as the assessment of the functional recovery and reorganization of the brain.

Several experimental models of brain ischemia have been developed, involving global and focal reductions in CBF.[127,128] Rodent[129–135] and, in particular, feline[56,63,136–142] models of brain ischemia helped establish robust experimental platforms for preclinical and translational therapeutic approaches,[143–145] although their relevance to human disease merits further consideration.[146–148] The use of fMRI techniques in experimental animal models of stroke has provided an important tool in the evaluation of brain plasticity and the recovery of function of the infarcted brain tissue. In a rodent model of stroke, there was no functional response in the ischemic side following either electrical stimulation of the forepaw or bicuculline administration[149,150] from 24 to 216 h following the permanent occlusion of the middle cerebral artery (MCA).[151] However, two other studies using a permanent[152] and a temporary occlusion model[153] have shown a functional reorganization of the cortex following unilateral stroke, which induced a large ipsilateral infarct and acute dysfunction of the contralateral forelimb, accompanied by loss of stimulus-induced activation in the ipsilesional sensorimotor cortex in spite of preservation of cerebrovascular reactivity as measured by CO_2 challenges. At 3 days following stroke, activation-induced responses were detected in the contralesional hemisphere; at 14 days, significant responses were observed in the infarction periphery. The authors thus suggested that limb dysfunction was related to loss of brain activation in the ipsilesional sensorimotor cortex and that restoration of function was associated with biphasic recruitment of perilesional and contralesional functional fields in the brain.

Reese *et al.* also used a transient MCA occlusion model to study local changes in CBV elicited by systemic infusion of the GABA(A) antagonist bicuculline, and found that the CBV response was negatively correlated with the duration of ischemia.[154] More recently, Shen *et al.* combined functional, perfusion, and diffusion MRI to study the effects of permanent and transient focal ischemic brain injury in rats during the acute phase, and concluded that with permanent ischemia, forepaw stimulation BOLD fMRI response in the primary somatosensory cortices was lost, while with temporary ischemia, vascular coupling and forepaw BOLD fMRI response remained intact.[155] On the other hand, Kim *et al.* showed that 2 weeks after stroke, animals displayed variable degrees of fMRI activation in ipsilesional cortex that did not correlate with structural damages as measured using apparent diffusion coefficient, fractional anisotropy, blood volume, and vessel size index.[156]

Animal models have been of fundamental importance in the study of focal seizures and in the preclinical evaluation of antiepileptic treatments.[157] There are several different models of epilepsy available, spanning a wide range of etiologies, acute or chronic presentations of partial seizures, and models for complex partial seizures.[157] Animal models of generalized seizures include rodents, cats, chicken, and nonhuman primates.[157,158] Because animal models that present spontaneous seizure activity can yield invaluable information on the molecular epileptic mechanisms, transgenic rodents are widely used,[159–163] although cats present epileptic symptoms more akin to the ones seen in humans.[164] MRI and fMRI of epilepsy have focused on mapping changes in the apparent water diffusion coefficient during seizure discharges,[165] the changes in local CBF and $CMRO_2$ during spontaneous spike-wave seizures,[166,167] and the detection of tissue damage in kainic acid-induced epilepsy.[168]

Parkinson's disease (PD) is characterized by a progressive loss of motor control caused by degeneration of the nigrostriatal dopaminergic (DA) pathway. There are two major categories of experimental animal models of PD[169–171]: one based on depletion of dopamine by a reversible pharmacological treatment with either reserpine[172] or amphetamine and the other involving permanent lesioning with neurotoxins such as intranigral injection of 6-hydroxydopamine (6-OHDA)[173–176] or systemic administration of 1-methyl-4-phenyl-1,2,3,6-tetrahydropyridine (MPTP).[177–180] Both rodent and nonhuman primate models of PD have been particularly useful in providing data on the disease progression and the effects of therapy.[169,171,179,181–183] While most of the work in animal models of PD is done via PET, MRI and fMRI techniques are increasingly providing complimentary information on the changes in functional brain activity at the onset of the disease,[174,176] or the regional alterations in CBV in response to an amphetamine challenge,[183,184] as shown in Figure 51–3. MRI has also been used to follow the differentiation of mouse embryonic stem cells into DA neurons in the rat striatum,[185] a gradual and sustained behavioral restoration of dopamine-mediated motor asymmetry suggesting stem cell-based therapy can restore cerebral function and behavior.

Alzheimer's disease (AD) is the most common brain dementia afflicting the aging population. Thus, there has been great effort in creating animal models of AD, most notably using transgenic mice models.[186–193] The major role that MRI has played in the study of AD is to provide noninvasive, longitudinal structural information[194] about the neuronal death and myelin breakdown typical of the disease,[195] as well as the development and evolution of amyloid plaques.[195–206] In addition, the combination of MRI and angiographic techniques has been particularly useful in revealing cerebrovascular abnormalities and CBF disturbances associated with AD.[207,208] More recently, fMRI has been combined with behavioral tests to investigate functional deficits in a rat model of AD.[209]

Use of Animal Models to Study Basic Neuroscience Questions Several different animal models have been used to investigate brain function, including rodents, cats, songbirds, and nonhuman primates. Examples of contributions of each animal model to basic neuroscientific work are given below.

Rodents The most common use of rodents in fMRI experiments has been in the study of the somatosensory system. Complimentary to a vast literature based on electrophysiology, histological techniques, and autoradiography, fMRI allows a noninvasive and repetitive view of the entire somatosensory pathway in rodents. The first fMRI experiments in the rodent somatosensory cortex established electrical stimulation of the forepaw as a robust functional paradigm for producing BOLD,[97,98] CBV,[26] and CBF[94,105] contrast in the forelimb area of the primary somatosensory cortex. In the forepaw rat model, BOLD and CBF tuning curves were obtained[94,105] showing maximum responses at low frequencies (1–3 Hz) and no responses above 9 Hz. However, the tuning curve can be greatly affected either by the stimulus parameters[210] or by anesthesia,[211] which may shift the peak response to the 8–12 Hz range (alpha band), which is the natural tuning frequency of the whisker barrel cortex, another functional paradigm used in fMRI of rodents.[99,101,212,213] In addition to distinguishing responses as a function of the stimulus frequency, fMRI has also been used to reveal the functional representation of the entire somatosensory pathway,[112,113,214,215] enabling the investigation of the changes that occur in the activated network during learning or after injury.[152,153,156] The ability of fMRI to reveal functionally interconnected brain regions has been exploited in a recent study of the visual system in the rat,[210] whereby flashing light displayed at different frequencies evoked maximum BOLD response at low frequencies (1–5 Hz) in the visual cortex (VC) and in the flocculus-paraflocculus (FL-PFL) of the cerebellum, and at high frequencies (8–12 Hz) in the superior colliculus (SC). These observed correlations between frequency-dependent responses of different visual areas suggested a functional relationship between the VC and FL-PFL rather than between the SC and FL-PFL.

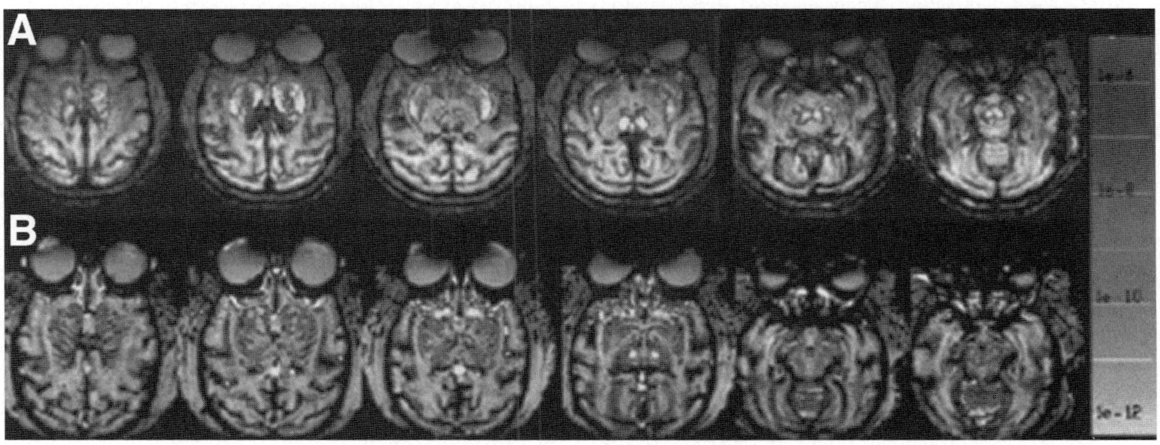

Figure 51–3. Changes in the CBV response to amphetamine (A) before and (B) 4 months after cessation of MPTP treatment in a cynomolgus monkey, showing an almost complete loss of amphetamine-induced CBV signal changes in dopaminergic regions. Parkinsonian primates had a prominent loss of response to amphetamine, with relative sparing of the nucleus accumbens and parafascicular thalamus. (Modified from Jenkins *et al.*[183] Copyright 2004 Society for Neurosciences.) (See color insert.)

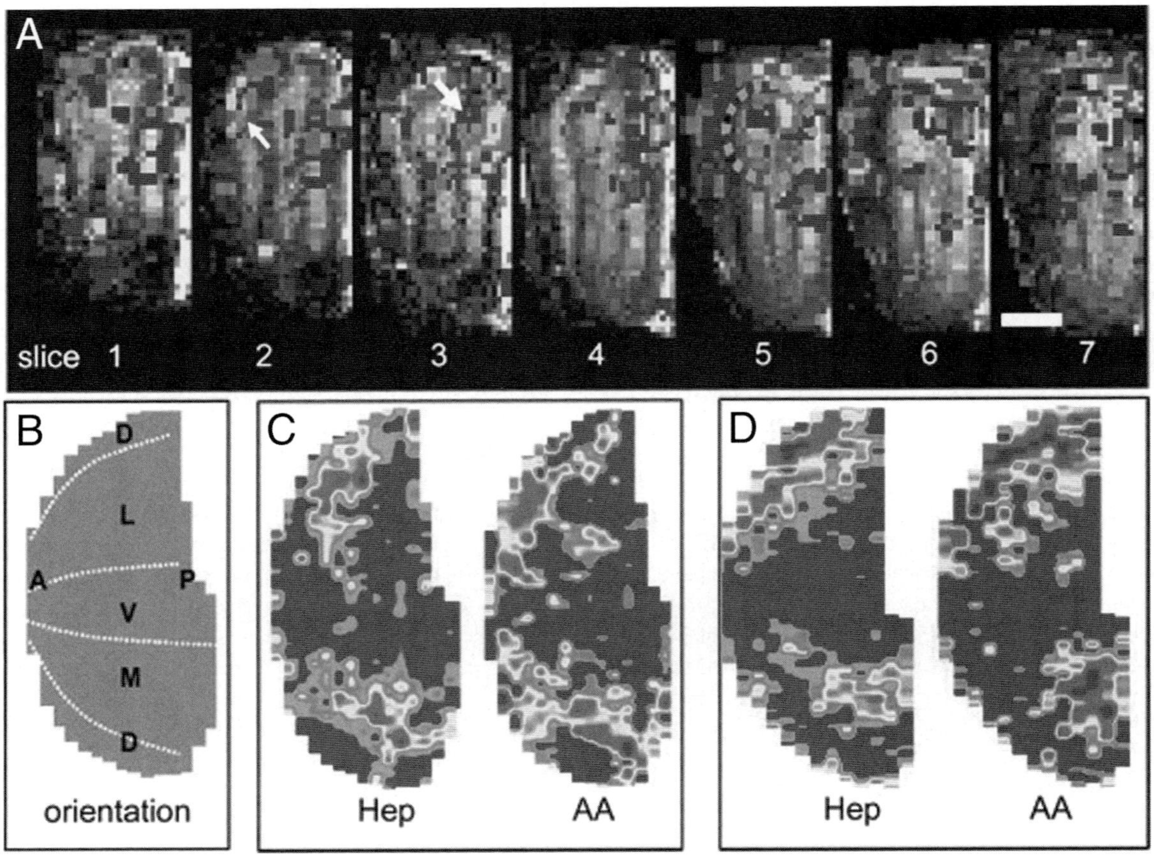

Figure 51–4. (A) Patterns of the fMRI response in the mouse main olfactory bulb (MOB) and accessory olfactory bulb (AOB, pink circle in slice 5) to the pheromone 2-heptanone, one of the urinary chemosignal compounds in mouse. The arrows point to two foci of activation suggestive of a pair of the nearly mirror projections of the receptor neuron subsets to the same MOB. Scale bar = 500 μm. (B) A flattened view of the olfactory bulb indicating the orientation of the odor maps shown in (C) and (D): A, anterior; D, dorsal; L, lateral; M, medial; P, posterior; V, ventral. (C, D) The odor maps of 2-heptanone (Hep) in two different mice show that this pheromone not only activates large regions of the MOB but also generates similar patterns across subjects. Interestingly, the odor map for amyl acetate (AA), a common odorant with an odor quality similar to that of 2-heptanone, was also similar to that of 2-heptanoneb. (Adapted from Xu *et al.*[217] Copyright 2005 Wiley-Liss, Inc.) (See color insert.)

Another prominent rodent sensory system amenable to fMRI investigation is olfaction.[103,109,111,216–218] The spatial resolution of BOLD fMRI allows discrimination of laminae in the olfactory bulb.[103,109] BOLD fMRI has hence been used to describe the spatial activation patterns in the glomerular layer in response to different odors,[111,216] and to reveal the responses in the main olfactory bulb (MOB) and accessory olfactory bulb (AOB) to odors and pheromones[217] (Figure 51–4), and map their ability to adapt to the presence of background odors.[218]

Cats Cats have been most frequently employed in studies of the visual system. The first fMRI experiment in cat visual cortex targeted the establishment of a robust stimulus protocol consisting of a high-contrast drifting grating, whose speed and spatial frequency were optimized for area 18 (V2).[219] Since then, experiments in cat visual cortex have focused on detection of isoorientation columns in the primary visual cortex using the early negative BOLD signal,[220] which was reportedly better suited to resolve the isoorientation columns than the early positive BOLD response.[221] The cat model has been used frequently to determine whether fMRI can provide high enough spatial specificity to resolve subcortical domains, such as functional columns (Figure 51–5) or layers.[114,222–229]

Songbirds Songbirds have evolved a vocal communication system responsible for the production of songs of various complexities. Songbirds make a good model of neuroplasticity in the auditory system since they can learn to produce specific vocalizations. The song control nuclei control the learning and vocalization of songs with seasonal periodicity, providing an ideal system to study brain plasticity and the interactions between neuroplasticity and learning. The group of Annemarie Van der Linden has been studying the song control system in starlings, canaries, and zebra finches using manganese-enhanced MRI (MEMRI),[230] which allows mapping of the song control nuclei and changes in their volume and connectivity as a function of seasonal and hormonal influences. More recently, BOLD fMRI has been employed to observe functional activation in the auditory regions of the telencephalon in starling birds.[231] Distinct spatiotemporal properties of the BOLD response to white noise, to music, and to a conspecific song of a male starling bird recorded during the breeding season were found (Figure 51–6). Only primary auditory regions responded to white noise, but both complex music and the socially relevant conspecific song elicited activation in the descending auditory pathway, indicating a more complex neuronal processing of complex sounds.

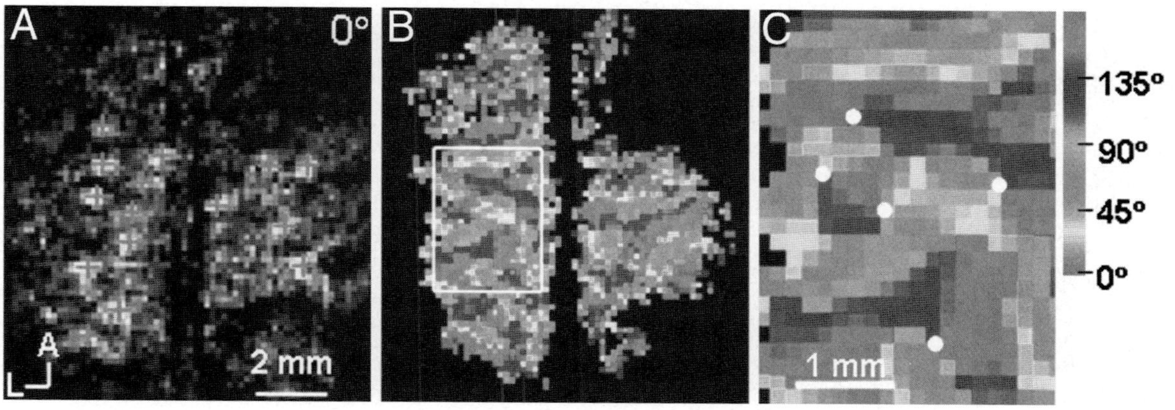

Figure 51–5. Signal changes in CBV-weighted fMRI obtained during stimulation of the cat visual cortex according to predetermined stimulus orientations. (A) Raw gray-scale functional map obtained by subtraction of images during 0° stimulation from prestimulus control. The center of each patch is marked with a green + sign. (B) Four different grating orientations (0°, 45°, 90°, and 135°) were presented in the study of one animal, enabling a composite angle map to be generated through pixel-by-pixel vector addition of the four single-condition maps. In the left hemisphere (marked by a white rectangular box), changes between pixels preferentially activated by a particular stimulus orientation were smoothed by a 3 × 3 Gaussian kernel. (C) A composite angle map was generated with the region indicated by the white ROI in (B). "Pinwheel" structures indicated by small white dots were observed where domains for all orientations converge. (Adapted from Zhao *et al.*[114] Copyright 2005 Elsevier.) (See color insert.)

Nonhuman Primates Since they are phylogenetically closer to humans, nonhuman primates have been the main model used in studies of perceptual, motor, and cognitive brain processes. Until the mid-1990s, the major experimental techniques in neuroscience-oriented studies of the monkey brain were invasive electrophysiology, histological neuronal tract tracing, and PET. In the late 1990s, the first fMRI experiments were performed in anesthetized or awake behaving monkeys during visual stimulation[232–234] or somatosensory stimulation.[235]

In practice, a number of methodological issues have to be overcome to enable fMRI experiments in awake, behaving monkeys. First, all components of the head and body restraining devices must be made of nonferrous materials, so they can be safely used in the magnetic fields. Second, to mimic the actual scanning environment, the monkeys must be trained to perform behavioral tasks in an extremely noisy and thus potentially distracting environment. A controversial issue related to the training of the monkeys lies in the orientation of the magnetic bore. Electrophysiological experiments are typically performed with the animal sitting erect. However, the majority of MRI magnets are oriented horizontally. While several research groups have successfully obtained results from animals trained and restrained in a horizontal position,[232,233,236] the group of Nikos Logothetis strongly advocated a vertical magnet.[234,237] Results obtained from such technical "tour de force" studies have been remarkable,[238] so that there are now multiple nonhuman primate vertical magnets in operation. However, it is still unclear whether the benefits justify the substantial additional costs associated with acquisition, site planning, and operation of such systems.

In isoflurane anesthetized monkeys, Disbrow *et al.* used fMRI to demonstrate the mediolateral topographic representation of the foot, hand, and face in primary somatosensory cortex,[235] and found about 55% concordance between the fMRI maps and maps obtained in the same animals with microelectrode recording

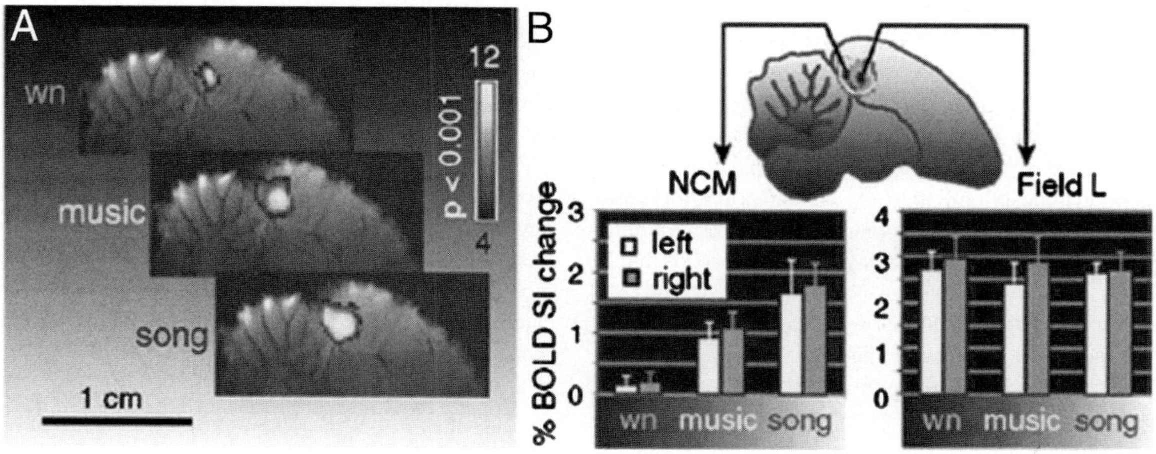

Figure 51–6. Auditory activation in the songbird telencephalon shows (A) statistical maps illustrating the localization of significant signal intensity changes during auditory stimulation consisting of white noise (wn), a concerto of Bach (music), and a stimulation with song from a male starling (song). (B) The location of the activated areas (top), together with the average BOLD response amplitude for each stimulus (bottom). (Adapted from Van Meir *et al.*[231] Copyright 2005 Elsevier.) (See color insert.)

techniques.[239] They concluded that most of the discrepancies were due to the sensitivity of fMRI to "draining veins." Another report demonstrated the separation of primary and secondary somatosensory areas of hand and face.[240] A recent fMRI study of the hand representation in macaques has revealed an ipsilateral hand input in area 3b of the primary somatosensory cortex that could not be distinguished from the fMRI response of the contralateral hand responses.[241] Comparison of the fMRI data to electrophysiology showed that the colocalized ipsilateral response was influenced by inhibition, suggesting that ipsilateral inputs may have modulatory effects on contralateral input processing.[241]

Most of the fMRI studies performed in nonhuman primates have investigated the visual system. In a seminal paper, Logothetis *et al.* described the presentation of checkerboard patterns or pictures to anesthetized monkeys during high spatial resolution (0.5 μl) fMRI experiments at a magnetic field of 4.7 T. Even under the influence of anesthesia, visual stimulation produced robust, reproducible, focal activation of the lateral geniculate nucleus (LGN), the primary visual area (V1), and a number of extrastriate visual areas, including areas in the superior temporal sulcus.[234] Using local RF coils implanted in the skull, the same group was able to push the spatial resolution down to 0.0113 μl and show laminae-specific activation in V1 and V2 by comparing the activity elicited by moving and flickering visual stimuli,[237] while moving to 7 T improved the image SNR.[242] Subsequent reports by the same group emphasized the effect of noise on the BOLD responses to visual stimulation,[243] the processing of motion by the visual pathway,[244] and the three-dimensional representation of shape in the "what" and the "where" pathways.[245] Meanwhile, other groups reported the use of intravascular agents as a way to enhance contrast in fMRI experiments of nonhuman primates,[38,236,246] which allowed definition of motion-sensitive areas in the visual cortex.[236,247] Other fMRI studies in macaques included mapping of the retinotopic organization throughout the visual cortex,[248] and comparing the monkey to human responses in the integration of local image features into global shapes,[249] with respect to the relative spatial extent and magnitude of activation in prefrontal cortex,[250] in higher visual areas,[251] in frontal and parietal eye fields,[252] and in object adaptation in shape-sensitive regions.[253]

THE USE OF ANESTHESIA The use of anesthetics offers important advantages of ensuring compliance, minimizing movement, and alleviating stress in addition to obviating the need for extensive training to both the MRI scanning environment and the functional paradigm of interest. Stress can be a significant confound in the study of brain function, particularly when hemodynamic variables are used as surrogate markers of neural activity. It is difficult to dynamically monitor the level of stress in the awake animals and almost impossible to establish its absence even with extensive training: indeed, continued research on training protocols and evaluation of stress indicators are subjects of much research.[254]

Urethane[124,255,256] and α-chloralose[94,97,98,102,105,113,257,258] are among the most widely used in studies of the somatosensory cortex. However, both substances are toxic and thus not adequate for longitudinal use.

Indeed, an additional challenge in the use of anesthesia is encountered in longitudinal studies of brain function. The inhalational compounds halothane[258] and isoflurane[211,259–263] are safer to use in repetitive studies in the same animal, but these anesthetics suppress neuronal activity and thus greatly influence the cerebrovascular coupling, making it necessary to redefine the proper stimulus parameters to produce robust activation.[211] Propofol is an injectable anesthetic with a rapid mechanism of action. It is increasingly used in fMRI experiments in animal models,[259,264–267] as the depth of anesthesia can be readily adjusted by varying the rate of infusion, and the animals quickly recovered at the end of the experiment, thus facilitating longitudinal studies. Another agent that has been recently proposed as suitable for repetitive studies is the α_2-adrenoreceptor agonist medetomidine hydrochloride,[68,269] which has been shown to allow robust fMRI responses[269] and whose sedative and analgesic effects can be quickly reversed with the application of atipamezole hydrochloride.

The influence of anesthetics on the neuronal responses and their coupling to hemodynamics severely restricts the choice of the functional paradigm and has been the subject of active research.[257,258,261,265,269–275] In addition to affecting the response magnitude and its evolution, the anesthetic frequently interacts with pharmacological agents administered.[276] It thus often represents another complex confound in the interpretation of the results and, most importantly, in the translation of findings from animal to human models.

ANIMAL PREPARATION AND PHYSIOLOGICAL MONITORING The use of anesthesia is predicated on some degree of preparatory surgical procedures: these are usually necessary to allow administration of anesthesia over the course of the experiment and/or facilitate physiological monitoring so as to ensure stable physiology under the influence of the anesthetic. The well-established anesthetics for functional experiments, namely urethane, propofol, and α-chloralose, are all injectable. Whereas urethane is usually given in a few intraperitoneal boluses and propofol and α-chloralose are administered as continuous infusion, either intraperitoneally or intravenously. Both urethane and α-chloralose typically require an additional anesthetic to be used during the preparatory surgery. The halogenated anesthetics (halothane, sevoflurane, and isoflurane) are the customary choices for induction of anesthesia prior to the commencement of rodent surgery. In contrast, an intramuscular bolus of ketamine (either alone or in combination with its potentiators xylazine, acepromazine, or medetomidine hydrochloride) is frequently used for induction in both cats and nonhuman primates.

Typically, following the induction of anesthesia, the animals are typically orally intubated or tracheostomized and mechanically ventilated to allow dynamic adjustments of respiratory parameters so as to ensure stable physiology—under the effect of the anesthetic—throughout the experiment. The importance of the maintenance of stable physiological conditions cannot be overstated, as the physiological condition of the animal exerts profound effects on its functional responses. Monitoring the physiological state of the animal typically includes continuous recording of the temperature (rectal or cutaneous), mean arterial blood pressure (via a pressure cuff or an indwelling arterial catheter), end-tidal CO_2 pressure (via capnography), and arterial blood oxygen saturation (via pulse oximetry). Periodic sampling of arterial blood is also performed to yield information on arterial blood gases, hematocrit, and electrolytes and is achieved through cannulation of a major artery, typically the femoral or the iliac. Data collected during unstable physiology are best discarded. Furthermore, large deviations in the arterial blood gases often call for administration of corrective pharmacological agents. The normal ranges on blood gas values are given in Table 51-1.

Table 51–1
Arterial blood gas values for animals breathing air[287]

Parameter	*Normal range*
PCO_2	28–40 mmHg
PO_2	82–94 mmHg
pH	7.35–7.45

MRI poses stringent requirements on immobilization. After surgery, the animals are thus placed in a stereotaxic-like head holder and strapped to a cradle/chair. The design of the cradle/chair requires considerable engineering effort.[232–234] A muscular relaxant (e.g., pancuronium bromide or tubocarine) may also be periodically administered intravenously to aid with immobilization.

FUNCTIONAL PARADIGM As with most other modalities for studying brain function, the environment of the MRI scanner and the statistical nature of the fMRI analysis restrict the nature of the functional paradigm employed. Functional paradigms need to be carefully designed so as to isolate the responses evoked by the stimuli/task from the "resting" activity. The definition of the control condition is a particularly difficult one and the evoked responses are arguably significantly modulated (and, at times, dictated) by the so called "baseline." Broadly, there are two categories of functional paradigms employed in fMRI research: the slow frequency "block" paradigms and the pseudorandom, brief on-period "event-related" paradigms. The choice of the two is influenced by the issues related to the neuroscience question at hand in addition to the signal detection versus estimation considerations.[22]

PHARMACOLOGICAL PERTURBATIONS Animal models have been widely employed in pharmacological MRI (phMRI) studies to address both basic science and clinical questions.[277] The latter have been frequently exploited to provide pharmacodynamic assays and identify the potential therapeutic agents, thereby guiding later research in human subjects.[278]

The spatial and temporal characterizations of neurotransmitter interactions have thus been undertaken in the rodent[184,279] and primate models.[183] For example, to shed further light on the progression of Parkinson's disease, BOLD responses to DA stimulation have been studied in both rats[280,184] and rhesus monkeys.[281] Ketamine has been administered to produce schizophrenia-like behaviors in anesthetized rodents.[276] Finally, pharmacological studies in animals have yielded important insight into addiction.[282,283]

COMBINATION WITH OTHER MODALITIES Whereas fMRI affords relatively straightforward noninvasive assessment of function, the complex mapping from hemodynamics to neuronal state has hindered the interpretation of many BOLD fMRI studies. To address these issues, a number of investigators have combined BOLD fMRI with other invasive modalities to provide validation and refine the interpretation of the BOLD fMRI measurements. Animal models are particularly well suited to such invasive, multimodal investigations. BOLD fMRI has thus been combined with optical imaging, histology and behavioral assessments,[281] immunoreactivity,[284] microdialysis,[285] and PET.[286]

CONCLUSIONS

The use of animal models in functional MRI investigations has had a profound affect on the development of fMRI techniques and the understanding of the biophysical processes underlying the various fMRI contrasts. It has provided much valuable data into basic neuroscience questions and aided the investigation of a plethora of neurological diseases and disorders. In light of the ever increasing demand on the sensitivity, specificity, and quantification of brain function, animal models will certainly continue to play a central role in the refinement of the existing techniques as well as in the development of novel contrast agents, new endogenous contrast sources, and more realistic models of various brain pathologies.

REFERENCES

1. Nyberg L. Functional neuroimaging of cognition: State-of-the-art. *Scand J Psychol* 2001;42(3):163–165.
2. Matthews PM, Honey GD, Bullmore ET. Applications of fMRI in translational medicine and clinical practice. *Nat Rev Neurosci* 2006;7(9):732–744.
3. Attwell D, Iadecola C. The neural basis of functional brain imaging signals. *Trends Neurosci* 2002;25(12):621–625.
4. Ogawa S, Lee TM, Kay AR, Tank DW. Brain magnetic resonance imaging with contrast dependent on blood oxygenation. *Proc Natl Acad Sci USA* 1990;87(24):9868–9872.
5. Ogawa S, Lee TM. Magnetic resonance imaging of blood vessels at high fields: In vivo and in vitro measurements and image simulation. *Magn Reson Med* 1990;16(1):9–18.
6. Kim SG, Ackerman JJ. Quantification of regional blood flow by monitoring of exogenous tracer via nuclear magnetic resonance spectroscopy. *Magn Reson Med* 1990;14(2):266–282.
7. Detre JA, Eskey CJ, Koretsky AP. Measurement of cerebral blood flow in rat brain by 19F-NMR detection of trifluoromethane washout. *Magn Reson Med* 1990;15(1):45–57.
8. Detre JA, Williams DS, Koretsky AP. Nuclear magnetic resonance determination of flow, lactate, and phosphate metabolites during amphetamine stimulation of the rat brain. *NMR Biomed* 1990;3(6): 272–278.
9. Barranco D, Sutton LN, Florin S, Greenberg J, Sinnwell T, Ligeti L, *et al.* Use of 19F NMR spectroscopy for measurement of cerebral blood flow: A comparative study using microspheres. *J Cereb Blood Flow Metab* 1989;9(6):886–891.
10. Villringer A, Rosen BR, Belliveau JW, Ackerman JL, Lauffer RB, Buxton RB, *et al.* Dynamic imaging with lanthanide chelates in normal brain: Contrast due to magnetic susceptibility effects. *Magn Reson Med* 1988;6(2):164–174.
11. Rosen BR, Belliveau JW, Buchbinder BR, McKinstry RC, Porkka LM, Kennedy DN, *et al.* Contrast agents and cerebral hemodynamics. *Magn Reson Med* 1991;19(2):285–292.
12. Williams DS, Detre JA, Leigh JS, Koretsky AP. Magnetic resonance imaging of perfusion using spin inversion of arterial water. *Proc Natl Acad Sci USA* 1992;89(1):212–216.
13. Detre JA, Leigh JS, Williams DS, Koretsky AP. Perfusion imaging. *Magn Reson Med* 1992;23(1):37–45.
14. Widmaier EP, Raff H, Strang KT. *Vander, Sherman, & Luciano's Human Physiology: The Mechanisms of Body Function*, 7th ed. New York: McGraw-Hill, 1998.
15. Attwell D, Laughlin SB. An energy budget for signaling in the grey matter of the brain. *J Cereb Blood Flow Metab* 2001;21(10): 1133–1145.
16. Sokoloff L. Energy metabolism in neural tissues in vivo at rest and in functionally altered states. In: Shulman RG, Rothman DL, Eds. *Brain Energetics and Neuronal Activity Applications to fMRI and Medicine*. West Sussex: John Wiley & Sons, Ltd., 2004:11–30.
17. Kennedy C, Des Rosiers MH, Sakurada O, Shinohara M, Reivich M, Jehle JW, *et al.* Metabolic mapping of the primary visual system

of the monkey by means of the autoradiographic [14C]deoxyglucose technique. *Proc Natl Acad Sci USA* 1976;73(11):4230–4234.
18. Schwartz WJ, Smith CB, Davidsen L, Savaki H, Sokoloff L, Mata M, *et al.* Metabolic mapping of functional activity in the hypothalamo-neurohypophysial system of the rat. *Science* 1979;205(4407): 723–725.
19. Sokoloff L. The F. O. Schmitt Lecture in Neuroscience 1980. The relationship between function and energy metabolism: Its use in the localization of functional activity in the nervous system. *Neurosci Res Program Bull* 1981;19(2):159–207.
20. Phelps ME, Mazziotta JC. Positron emission tomography: Human brain function and biochemistry. *Science* 1985;228(4701): 799–809.
21. Gjedde A. Does deoxyglucose uptake in the brain reflect energy metabolism? *Biochem Pharmacol* 1987;36(12):1853–1861.
22. Buxton RB. *Introduction to Functional Magnetic Resonance Imaging: Principles and Techniques*, 1st ed. Cambridge, UK: Cambridge University Press, 2002.
23. Buxton RB, Frank LR. A model for the coupling between cerebral blood flow and oxygen metabolism during neural stimulation. *J Cereb Blood Flow Metab* 1997;17(1):64–72.
24. Lauritzen M. Reading vascular changes in brain imaging: Is dendritic calcium the key? *Nat Rev Neurosci* 2005;6(1):77–85.
25. Kuschinsky W, Paulson OB. Capillary circulation in the brain. *Cerebrovasc Brain Metab Rev* 1992;4(3):261–286.
26. Mandeville JB, Marota JJ, Kosofsky BE, Keltner JR, Weissleder R, Rosen BR, *et al.* Dynamic functional imaging of relative cerebral blood volume during rat forepaw stimulation. *Magn Reson Med* 1998;39(4):615–624.
27. Pauling L, Coryell CD. The magnetic properties and structure of hemoglobin, oxyhemoglobin and carbonmonoxyhemoglobin. *Proc Natl Acad Sci USA* 1936;22(4):210–216.
28. Cho ZH, Ro YM, Lim TH. NMR venography using the susceptibility effect produced by deoxyhemoglobin. *Magn Reson Med* 1992;28(1):25–38.
29. Weisskoff RM, Kiihne S. MRI susceptometry: Image-based measurement of absolute susceptibility of MR contrast agents and human blood. *Magn Reson Med* 1992;24(2):375–383.
30. Barbier EL, Lamalle L, Decorps M. Methodology of brain perfusion imaging. *J Magn Reson Imaging* 2001;13(4):496–520.
31. Silva AC, Kim SG. Perfusion-based functional magnetic resonance imaging. *Concepts Magn Reson Part A* 2003;16A(1):16–27.
32. Golay X, Hendrikse J, Lim TC. Perfusion imaging using arterial spin labeling. *Top Magn Reson Imaging* 2004;15(1):10–27.
33. Calamante F, Thomas DL, Pell GS, Wiersma J, Turner R. Measuring cerebral blood flow using magnetic resonance imaging techniques. *J Cereb Blood Flow Metab* 1999;19(7):701–735.
34. Detre JA, Wang J. Technical aspects and utility of fMRI using BOLD and ASL. *Clin Neurophysiol* 2002;113(5):621–634.
35. Belliveau JW, Rosen BR, Kantor HL, Rzedzian RR, Kennedy DN, McKinstry RC, *et al.* Functional cerebral imaging by susceptibility-contrast NMR. *Magn Reson Med* 1990;14(3):538–546.
36. Belliveau JW, Kennedy DN Jr, McKinstry RC, Buchbinder BR, Weisskoff RM, Cohen MS, *et al.* Functional mapping of the human visual cortex by magnetic resonance imaging. *Science* 1991;254(5032):716–719.
37. Rosen BR, Belliveau JW, Vevea JM, Brady TJ. Perfusion imaging with NMR contrast agents. *Magn Reson Med* 1990;14(2):249–265.
38. Leite FP, Tsao D, Vanduffel W, Fize D, Sasaki Y, Wald LL, *et al.* Repeated fMRI using iron oxide contrast agent in awake, behaving macaques at 3 Tesla. *Neuroimage* 2002;16(2):283–294.
39. Scheffler K, Seifritz E, Haselhorst R, Bilecen D. Titration of the BOLD effect: Separation and quantitation of blood volume and oxygenation changes in the human cerebral cortex during neuronal activation and ferumoxide infusion. *Magn Reson Med* 1999;42(5): 829–836.
40. Brasch RC, Weinmann HJ, Wesbey GE. Contrast-enhanced NMR imaging: Animal studies using gadolinium-DTPA complex. *AJR Am J Roentgenol* 1984;142(3):625–630.
41. Schmiedl U, Ogan MD, Moseley ME, Brasch RC. Comparison of the contrast-enhancing properties of albumin-(Gd-DTPA) and Gd-DTPA at 2.0 T: An experimental study in rats. *AJR Am J Roentgenol* 1986;147(6):1263–1270.
42. Tauber U, Weinmann HJ, Panzer M, Acksteiner B, Vollert B, Schulze PE. Whole-body autoradiographic studies in rats with gadolinium-diethylenetriaminepentaacetic acid, a new contrast agent for magnetic resonance imaging. *Arzneimittelforschung* 1986;36(7): 1089–1091.
43. Schmiedl U, Ogan M, Paajanen H, Marotti M, Crooks LE, Brito AC, *et al.* Albumin labeled with Gd-DTPA as an intravascular, blood pool-enhancing agent for MR imaging: Biodistribution and imaging studies. *Radiology* 1987;162(1Pt. 1):205–210.
44. Weissleder R, Elizondo G, Wittenberg J, Rabito CA, Bengele HH, Josephson L. Ultrasmall superparamagnetic iron oxide: Characterization of a new class of contrast agents for MR imaging. *Radiology* 1990;175(2):489–493.
45. Rudin M, Sauter A. Noninvasive determination of regional cerebral blood flow in rats using dynamic imaging with Gd(DTPA). *Magn Reson Med* 1991;22(1):32–46.
46. Moseley ME, White DL, Wang SC, Wikstrom MG, Dupon JW, Gobbel G, *et al.* Vascular mapping using albumin-(Gd-DTPA), an intravascular MR contrast agent, and projection MR imaging. *J Comput Assist Tomogr* 1989;13(2):215–221.
47. Benderbous S, Bonnemain B. Superparamagnetic nanoparticles as blood-pool contrast agents. Contribution to MRI preclinical investigations. *Radiologe* 1995;35(11 Suppl. 2):S248–S252.
48. Carr DH. The use of proton relaxation enhancers in magnetic resonance imaging. *Magn Reson Imaging* 1985;3(1):17–25.
49. Norman AB, Thomas SR, Pratt RG, Samaratunga RC, Sanberg PR. A magnetic resonance imaging contrast agent differentiates between the vascular properties of fetal striatal tissue transplants and gliomas in rat brain in vivo. *Brain Res* 1989;503(1):156–159.
50. Baba T, Moriguchi M, Natori Y, Katsuki C, Inoue T, Fukui M. Magnetic resonance imaging of experimental rat brain tumors: Histopathological evaluation. *Surg Neurol* 1990;34(6):378–382.
51. Kornguth S, Anderson M, Turski P, Sorenson J, Robins HI, Cohen J, *et al.* Glioblastoma multiforme: MR imaging at 1.5 and 9.4 T after injection of polylysine-DTPA-Gd in rats. *AJNR Am J Neuroradiol* 1990;11(2):313–318.
52. Rajan SS, Rosa L, Francisco J, Muraki A, Carvlin M, Tuturea E. MRI characterization of 9L-glioma in rat brain at 4.7 Tesla. *Magn Reson Imaging* 1990;8(2):185–190.
53. Schmiedl UP, Kenney J, Maravilla KR. MRI of blood-brain barrier permeability in astrocytic gliomas: Application of small and large molecular weight contrast media. *Magn Reson Med* 1991;22(2): 288–292.
54. Hayakawa K, Yamashita K, Matsuda T, Miyake T, Ito H. MR imaging of acute cerebral infarction: Experimental study with Gd-DTPA. *Radiat Med* 1989;7(2):58–65.
55. Kent TA, Quast MJ, Kaplan BJ, Najafi A, Amparo EG, Gevedon RM, *et al.* Cerebral blood volume in a rat model of ischemia by MR imaging at 4.7 T. *AJNR Am J Neuroradiol* 1989;10(2):335–338.
56. Maeda M, Itoh S, Ide H, Matsuda T, Kobayashi H, Kubota T, *et al.* Acute stroke in cats: Comparison of dynamic susceptibility-contrast MR imaging with T2- and diffusion-weighted MR imaging. *Radiology* 1993;189(1):227–232.
57. Hawkins CP, Munro PM, MacKenzie F, Kesselring J, Tofts PS, du Boulay EP, *et al.* Duration and selectivity of blood-brain barrier breakdown in chronic relapsing experimental allergic encephalomyelitis studied by gadolinium-DTPA and protein markers. *Brain* 1990;113(Pt. 2):365–378.
58. Prato FS, Frappier JR, Shivers RR, Kavaliers M, Zabel P, Drost D, *et al.* Magnetic resonance imaging increases the blood-brain barrier permeability to 153-gadolinium diethylenetriaminepentaacetic acid in rats. *Brain Res* 1990;523(2):301–304.
59. Norman AB, Bertram KJ, Thomas SR, Pratt RG, Samaratunga RC, Sanberg PR. Magnetic resonance imaging of rat brain following in vivo disruption of the cerebral vasculature. *Brain Res Bull* 1991;26(4):593–597.

60. Rudin M, Sauter A. Non-invasive determination of cerebral blood flow changes by 19F NMR spectroscopy. *NMR Biomed* 1989;2(3): 98–103.
61. Ewing JR, Branch CA, Helpern JA, Smith MB, Butt SM, Welch KM. Cerebral blood flow measured by NMR indicator dilution in cats. *Stroke* 1989;20(2):259–267.
62. Detre JA, Subramanian VH, Mitchell MD, Smith DS, Kobayashi A, Zaman A, *et al.* Measurement of regional cerebral blood flow in cat brain using intracarotid 2H2O and 2H NMR imaging. *Magn Reson Med* 1990;14(2):389–395.
63. Brunetti A, Nagashima G, Bizzi A, DesPres DJ, Alger JR. Cerebral blood flow in experimental ischemia assessed by 19F magnetic resonance spectroscopy in cats. *Stroke* 1990;21(10):1439–1444.
64. van Zijl PC, Ligeti L, Sinnwell T, Alger JR, Chesnick AS, Moonen CT, *et al.* Measurement of cerebral blood flow by volume-selective 19F NMR spectroscopy. *Magn Reson Med* 1990;16(3):489–495.
65. Pekar J, Ligeti L, Ruttner Z, Lyon RC, Sinnwell TM, van Gelderen P, *et al.* In vivo measurement of cerebral oxygen consumption and blood flow using 17O magnetic resonance imaging. *Magn Reson Med* 1991;21(2):313–319.
66. Branch CA, Ewing JR, Helpern JA, Ordidge RJ, Butt S, Welch KM. Atraumatic quantitation of cerebral perfusion in cats by 19F magnetic resonance imaging. *Magn Reson Med* 1992;28(1):39–53.
67. Fiat D, Kang S. Determination of the rate of cerebral oxygen consumption and regional cerebral blood flow by non-invasive 17O in vivo NMR spectroscopy and magnetic resonance imaging: Part 1. Theory and data analysis methods. *Neurol Res* 1992;14(4): 303–311.
68. Fiat D, Kang S. Determination of the rate of cerebral oxygen consumption and regional cerebral blood flow by non-invasive 17O in vivo NMR spectroscopy and magnetic resonance imaging. Part 2. Determination of CMRO2 for the rat by 17O NMR, and CMRO2, rCBF and the partition coefficient for the cat by 17O MRI. *Neurol Res* 1993;15(1):7–22.
69. Roy CS, Sherrington CS. On the regulation of the blood supply of the brain. *J Physiol* 1890;11:85–108.
70. Kwong KK, Hopkins AL, Belliveau JW, Chesler DA, Porkka LM, McKinstry RC, *et al.* Proton NMR imaging of cerebral blood flow using H2(17)O. *Magn Reson Med* 1991;22(1):154–158.
71. Belliveau JW, Cohen MS, Weisskoff RM, Buchbinder BR, Rosen BR. Functional studies of the human brain using high-speed magnetic resonance imaging. *J Neuroimaging* 1991;1(1):36–41.
72. Belliveau JW, Kwong KK, Kennedy DN, Baker JR, Stern CE, Benson R, *et al.* Magnetic resonance imaging mapping of brain function. Human visual cortex. *Invest Radiol* 1992;27(Suppl. 2): S59–S65.
73. Ogawa S, Lee TM, Nayak AS, Glynn P. Oxygenation-sensitive contrast in magnetic resonance image of rodent brain at high magnetic fields. *Magn Reson Med* 1990;14(1):68–78.
74. Zhang W, Silva AC, Williams DS, Koretsky AP. NMR measurement of perfusion using arterial spin labeling without saturation of macromolecular spins. *Magn Reson Med* 1995;33(3):370–376.
75. Silva AC, Zhang W, Williams DS, Koretsky AP. Estimation of water extraction fractions in rat brain using magnetic resonance measurement of perfusion with arterial spin labeling. *Magn Reson Med* 1997;37(1):58–68.
76. Zhang W, Williams DS, Detre JA, Koretsky AP. Measurement of brain perfusion by volume-localized NMR spectroscopy using inversion of arterial water spins: Accounting for transit time and cross-relaxation. *Magn Reson Med* 1992;25(2):362–371.
77. Barbier EL, Silva AC, Kim SG, Koretsky AP. Perfusion imaging using dynamic arterial spin labeling (DASL). *Magn Reson Med* 2001;45(6):1021–1029.
78. Detre JA, Zhang W, Roberts DA, Silva AC, Williams DS, Grandis DJ, *et al.* Tissue specific perfusion imaging using arterial spin labeling. *NMR Biomed* 1994;7(1–2):75–82.
79. Silva AC, Williams DS, Koretsky AP. Evidence for the exchange of arterial spin-labeled water with tissue water in rat brain from diffusion-sensitized measurements of perfusion. *Magn Reson Med* 1997;38(2):232–237.
80. Ewing JR, Cao Y, Fenstermacher J. Single-coil arterial spin-tagging for estimating cerebral blood flow as viewed from the capillary: Relative contributions of intra- and extravascular signal. *Magn Reson Med* 2001;46(3):465–475.
81. St Lawrence KS, Lee TY. An adiabatic approximation to the tissue homogeneity model for water exchange in the brain: I. Theoretical derivation. *J Cereb Blood Flow Metab* 1998;18(12):1365–1377.
82. St Lawrence KS, Lee TY. An adiabatic approximation to the tissue homogeneity model for water exchange in the brain: II. Experimental validation. *J Cereb Blood Flow Metab* 1998;18(12):1378–1385.
83. Donahue KM, Weisskoff RM, Parmelee DJ, Callahan RJ, Wilkinson RA, Mandeville JB, *et al.* Dynamic Gd-DTPA enhanced MRI measurement of tissue cell volume fraction. *Magn Reson Med* 1995;34(3):423–432.
84. Berry I, Benderbous S, Ranjeva JP, Gracia-Meavilla D, Manelfe C, Le Bihan D. Contribution of Sinerem used as blood-pool contrast agent: Detection of cerebral blood volume changes during apnea in the rabbit. *Magn Reson Med* 1996;36(3):415–419.
85. Donahue KM, Weisskoff RM, Chesler DA, Kwong KK, Bogdanov AA Jr, Mandeville JB, *et al.* Improving MR quantification of regional blood volume with intravascular T1 contrast agents: Accuracy, precision, and water exchange. *Magn Reson Med* 1996;36(6): 858–867.
86. Lin W, Paczynski RP, Kuppusamy K, Hsu CY, Haacke EM. Quantitative measurements of regional cerebral blood volume using MRI in rats: Effects of arterial carbon dioxide tension and mannitol. *Magn Reson Med* 1997;38(3):4–8.
87. Schwarzbauer C, Morrissey SP, Deichmann R, Hillenbrand C, Syha J, Adolf H, *et al.* Quantitative magnetic resonance imaging of capillary water permeability and regional blood volume with an intravascular MR contrast agent. *Magn Reson Med* 1997;37(5):769–777.
88. Caramia F, Yoshida T, Hamberg LM, Huang Z, Hunter G, Wanke I, *et al.* Measurement of changes in cerebral blood volume in spontaneously hypertensive rats following L-arginine infusion using dynamic susceptibility contrast MRI. *Magn Reson Med* 1998;39(1): 160–163.
89. Moseley ME, Chew WM, White DL, Kucharczyk J, Litt L, Derugin N, *et al.* Hypercarbia-induced changes in cerebral blood volume in the cat: A 1H MRI and intravascular contrast agent study. *Magn Reson Med* 1992;23(1):21–30.
90. White DL, Aicher KP, Tzika AA, Kucharczyk J, Engelstad BL, Moseley ME. Iron-dextran as a magnetic susceptibility contrast agent: Flow-related contrast effects in the T2-weighted spin-echo MRI of normal rat and cat brain. *Magn Reson Med* 1992;24(1): 14–28.
91. Hamberg LM, Macfarlane R, Tasdemiroglu E, Boccalini P, Hunter GJ, Belliveau JW, *et al.* Measurement of cerebrovascular changes in cats after transient ischemia using dynamic magnetic resonance imaging. *Stroke* 1993;24(3):444–450.
92. Ogawa S, Lee TM, Barrere B. The sensitivity of magnetic resonance image signals of a rat brain to changes in the cerebral venous blood oxygenation. *Magn Reson Med* 1993;29(2):205–210.
93. Jones RA, Muller TB, Haraldseth O, Baptista AM, Oksendal AN. Cerebrovascular changes in rats during ischemia and reperfusion: A comparison of BOLD and first pass bolus tracking techniques. *Magn Reson Med* 1996;35(4):489–496.
94. Kerskens CM, Hoehn-Berlage M, Schmitz B, Busch E, Bock C, Gyngell ML, *et al.* Ultrafast perfusion-weighted MRI of functional brain activation in rats during forepaw stimulation: Comparison with T2-weighted MRI. *NMR Biomed* 1996;9(1):20–23.
95. Hyder F, Shulman RG, Rothman DL. A model for the regulation of cerebral oxygen delivery. *J Appl Physiol* 1998;85(2):554–564.
96. Kennan RP, Scanley BE, Innis RB, Gore JC. Physiological basis for BOLD MR signal changes due to neuronal stimulation: Separation of blood volume and magnetic susceptibility effects. *Magn Reson Med* 1998;40(6):840–846.
97. Hyder F, Behar KL, Martin MA, Blamire AM, Shulman RG. Dynamic magnetic resonance imaging of the rat brain during

forepaw stimulation. *J Cereb Blood Flow Metab* 1994;14(4):649–655.

98. Gyngell ML, Bock C, Schmitz B, Hoehn-Berlage M, Hossmann KA. Variation of functional MRI signal in response to frequency of somatosensory stimulation in alpha-chloralose anesthetized rats. *Magn Reson Med* 1996;36(1):13–15.
99. Yang X, Hyder F, Shulman RG. Activation of single whisker barrel in rat brain localized by functional magnetic resonance imaging. *Proc Natl Acad Sci USA* 1996;93(1):475–478.
100. Hyder F, Rothman DL, Mason GF, Rangarajan A, Behar KL, Shulman RG. Oxidative glucose metabolism in rat brain during single forepaw stimulation: A spatially localized 1H[13C] nuclear magnetic resonance study. *J Cereb Blood Flow Metab* 1997; 17(10):1040–1047.
101. Yang X, Hyder F, Shulman RG. Functional MRI BOLD signal coincides with electrical activity in the rat whisker barrels. *Magn Reson Med* 1997;38(6):874–877.
102. Bock C, Krep H, Brinker G, Hoehn-Berlage M. Brainmapping of alpha-chloralose anesthetized rats with T2*-weighted imaging: Distinction between the representation of the forepaw and hindpaw in the somatosensory cortex. *NMR Biomed* 1998;11(3): 115–119.
103. Yang X, Renken R, Hyder F, Siddeek M, Greer CA, Shepherd GM, *et al.* Dynamic mapping at the laminar level of odor-elicited responses in rat olfactory bulb by functional MRI. *Proc Natl Acad Sci USA* 1998;95(13):7715–7720.
104. Brinker G, Bock C, Busch E, Krep H, Hossmann KA, Hoehn-Berlage M. Simultaneous recording of evoked potentials and T2*-weighted MR images during somatosensory stimulation of rat. *Magn Reson Med* 1999;41(3):469–473.
105. Silva AC, Lee SP, Yang G, Iadecola C, Kim SG. Simultaneous blood oxygenation level-dependent and cerebral blood flow functional magnetic resonance imaging during forepaw stimulation in the rat. *J Cereb Blood Flow Metab* 1999;19(8):871–879.
106. Lee SP, Silva AC, Ugurbil K, Kim SG. Diffusion-weighted spin-echo fMRI at 9.4 T: Microvascular/tissue contribution to BOLD signal changes. *Magn Reson Med* 1999;42(5):919–928.
107. Mandeville JB, Marota JJ. Vascular filters of functional MRI: Spatial localization using BOLD and CBV contrast. *Magn Reson Med* 1999;42(3):591–598.
108. Duong TQ, Silva AC, Lee SP, Kim SG. Functional MRI of calcium-dependent synaptic activity: Cross correlation with CBF and BOLD measurements. *Magn Reson Med* 2000;43(3):383–392.
109. Kida I, Xu F, Shulman RG, Hyder F. Mapping at glomerular resolution: fMRI of rat olfactory bulb. *Magn Reson Med* 2002;48(3): 570–576.
110. Silva AC, Koretsky AP. Laminar specificity of functional MRI onset times during somatosensory stimulation in rat. *Proc Natl Acad Sci USA* 2002;99(23):15182–15187.
111. Xu F, Liu N, Kida I, Rothman DL, Hyder F, Shepherd GM. Odor maps of aldehydes and esters revealed by functional MRI in the glomerular layer of the mouse olfactory bulb. *Proc Natl Acad Sci USA* 2003;100(19):11029–11034.
112. Keilholz SD, Silva AC, Raman M, Merkle H, Koretsky AP. Functional MRI of the rodent somatosensory pathway using multislice echo planar imaging. *Magn Reson Med* 2004;52(1):89–99.
113. Keilholz SD, Silva AC, Raman M, Merkle H, Koretsky AP. BOLD and CBV-weighted functional magnetic resonance imaging of the rat somatosensory system. *Magn Reson Med* 2006;55(2):316–324.
114. Zhao F, Wang P, Hendrich K, Kim SG. Spatial specificity of cerebral blood volume-weighted fMRI responses at columnar resolution. *Neuroimage* 2005;27(2):416–424.
115. Goense JB, Logothetis NK. Laminar specificity in monkey V1 using high-resolution SE-fMRI. *Magn Reson Imaging* 2006;24(4): 381–392.
116. van Bruggen N, Busch E, Palmer JT, Williams SP, de Crespigny AJ. High-resolution functional magnetic resonance imaging of the rat brain: Mapping changes in cerebral blood volume using iron oxide contrast media. *J Cereb Blood Flow Metab* 1998;18(11): 1178–1183.
117. Kida I, Hyder F, Kennan RP, Behar KL. Toward absolute quantitation of bold functional MRI. *Adv Exp Med Biol* 1999;471:681–689.
118. Hyder F, Renken R, Kennan RP, Rothman DL. Quantitative multimodal functional MRI with blood oxygenation level dependent exponential decays adjusted for flow attenuated inversion recovery (BOLDED AFFAIR). *Magn Reson Imaging* 2000;18(3):227–235.
119. Hyder F, Kida I, Behar KL, Kennan RP, Maciejewski PK, Rothman DL. Quantitative functional imaging of the brain: Towards mapping neuronal activity by BOLD fMRI. *NMR Biomed* 2001;14(7–8): 413–431.
120. Hyder F. Neuroimaging with calibrated FMRI. *Stroke* 2004;35(11 Suppl. 1):2635–2641.
121. Marota JJ, Ayata C, Moskowitz MA, Weisskoff RM, Rosen BR, Mandeville JB. Investigation of the early response to rat forepaw stimulation. *Magn Reson Med* 1999;41(2):247–252.
122. Silva AC, Kim SG. Pseudo-continuous arterial spin labeling technique for measuring CBF dynamics with high temporal resolution. *Magn Reson Med* 1999;42(3):425–429.
123. Silva AC, Lee SP, Iadecola C, Kim SG. Early temporal characteristics of cerebral blood flow and deoxyhemoglobin changes during somatosensory stimulation. *J Cereb Blood Flow Metab* 2000;20(1): 201–206.
124. Wu G, Luo F, Li Z, Zhao X, Li SJ. Transient relationships among BOLD, CBV, and CBF changes in rat brain as detected by functional MRI. *Magn Reson Med* 2002;48(6):987–993.
125. Kida I, Maciejewski PK, Hyder F. Dynamic imaging of perfusion and oxygenation by functional magnetic resonance imaging. *J Cereb Blood Flow Metab* 2004;24(12):1369–1381.
126. Lu H, Soltysik DA, Ward BD, Hyde JS. Temporal evolution of the CBV-fMRI signal to rat whisker stimulation of variable duration and intensity: A linearity analysis. *Neuroimage* 2005;26(2):432–440.
127. Lythgoe MF, Sibson NR, Harris NG. Neuroimaging of animal models of brain disease. *Br Med Bull* 2003;65:235–257.
128. van der Weerd L, Thomas DL, Thornton JS, Lythgoe MF. MRI of animal models of brain disease. *Methods Enzymol* 2004;386: 149–177.
129. Sauter A, Rudin M. Experimental studies with isradipine in stroke. *Drugs* 1990;40(Suppl. 2):44–51.
130. Sauter A, Rudin M. Prevention of stroke and brain damage with calcium antagonists in animals. *Am J Hypertens* 1991;4(2Pt. 2):121S–127S.
131. de Crespigny AJ, Tsuura M, Moseley ME, Kucharczyk J. Perfusion and diffusion MR imaging of thromboembolic stroke. *J Magn Reson Imaging* 1993;3(5):746–754.
132. Sauter A, Rudin M. Strain-dependent drug effects in rat middle cerebral artery occlusion model of stroke. *J Pharmacol Exp Ther* 1995;274(2):1008–1013.
133. Quast MJ, Wei J, Huang NC, Brunder DG, Sell SL, Gonzalez JM, *et al.* Perfusion deficit parallels exacerbation of cerebral ischemia/ reperfusion injury in hyperglycemic rats. *J Cereb Blood Flow Metab* 1997;17(5):553–559.
134. Zhang Z, Zhang RL, Jiang Q, Raman SB, Cantwell L, Chopp M. A new rat model of thrombotic focal cerebral ischemia. *J Cereb Blood Flow Metab* 1997;17(2):123–135.
135. Pierce AR, Lo EH, Mandeville JB, Gonzalez RG, Rosen BR, Wolf GL. MRI measurements of water diffusion and cerebral perfusion: Their relationship in a rat model of focal cerebral ischemia. *J Cereb Blood Flow Metab* 1997;17(2):183–190.
136. Moseley ME, Cohen Y, Mintorovitch J, Chileuitt L, Shimizu H, Kucharczyk J, *et al.* Early detection of regional cerebral ischemia in cats: Comparison of diffusion- and T2-weighted MRI and spectroscopy. *Magn Reson Med* 1990;14(2):330–346.
137. Moonen CTW, Pekar J, de Vleeschouwer MH, van Gelderen P, van ZP, DesPres DJ. Restricted and anisotropic displacement of water in healthy cat brain and in stroke studied by NMR diffusion imaging. *Magn Reson Med* 1991;19(2):327–332.
138. Kucharczyk J, Mintorovitch J, Asgari HS, Moseley ME. Diffusion/ perfusion MR imaging of acute cerebral ischemia. *Magn Reson Med* 1991;19(2):311–315.

139. van Gelderen P, de Vleeschouwer MH, DesPres D, Pekar J, van Zijl PC, Moonen CT. Water diffusion and acute stroke. *Magn Reson Med* 1994;31(2):154–163.
140. Kobayashi H, Ide H, Kodera T, Handa Y, Kabuto M, Kubota T, *et al.* Effect of mannitol on focal cerebral ischemia evaluated by magnetic resonance imaging. *Acta Neurochir Suppl (Wien)* 1994; 60:228–230.
141. Davis D, Ulatowski J, Eleff SM, Izuta M, Mori S, Shungu D, *et al.* Rapid monitoring of changes in water diffusion coefficients during reversible ischemia in cat and rat brain. *Magn Reson Med* 1994; 31:454–460.
142. Kobayashi H, Ide H, Kabuto M, Handa Y, Kubota T, Ishii Y. Effect of mannitol on focal cerebral ischemia evaluated by somatosensory-evoked potentials and magnetic resonance imaging. *Surg Neurol* 1995;44(1):55–61.
143. Hossmann KA. Experimental models for the investigation of brain ischemia. *Cardiovasc Res* 1998;39(1):106–120.
144. Hoehn M, Nicolay K, Franke C, van der SB. Application of magnetic resonance to animal models of cerebral ischemia. *J Magn Reson Imaging* 2001;14(5):491–509.
145. Weber R, Ramos-Cabrer P, Hoehn M. Present status of magnetic resonance imaging and spectroscopy in animal stroke models. *J Cereb Blood Flow Metab* 2006;26(5):591–604.
146. Zivin JA, Grotta JC. Animal stroke models. They are relevant to human disease. *Stroke* 1990;21(7):981–983.
147. Wiebers DO, Adams HP Jr, Whisnant JP. Animal models of stroke: Are they relevant to human disease? *Stroke* 1990;21(1): 1–3.
148. Rosenblum WI. Criteria for valid preclinical trials using animal stroke models. *Stroke* 1993;24(10):1601–1602.
149. Sauter A, Reese T, Porszasz R, Baumann D, Rausch M, Rudin M. Recovery of function in cytoprotected cerebral cortex in rat stroke model assessed by functional MRI. *Magn Reson Med* 2002; 47(4):759–765.
150. Reese T, Porszasz R, Baumann D, Bochelen D, Boumezbeur F, McAllister KH, *et al.* Cytoprotection does not preserve brain functionality in rats during the acute post-stroke phase despite evidence of non-infarction provided by MRI. *NMR Biomed* 2000;13(6): 361–370.
151. Rudin M, Baumann D, Ekatodramis D, Stirnimann R, McAllister KH, Sauter A. MRI analysis of the changes in apparent water diffusion coefficient, T(2) relaxation time, and cerebral blood flow and volume in the temporal evolution of cerebral infarction following permanent middle cerebral artery occlusion in rats. *Exp Neurol* 2001;169(1):56–63.
152. Dijkhuizen RM, Ren J, Mandeville JB, Wu O, Ozdag FM, Moskowitz MA, *et al.* Functional magnetic resonance imaging of reorganization in rat brain after stroke. *Proc Natl Acad Sci USA* 2001;98(22):12766–12771.
153. Dijkhuizen RM, Singhal AB, Mandeville JB, Wu O, Halpern EF, Finklestein SP, *et al.* Correlation between brain reorganization, ischemic damage, and neurologic status after transient focal cerebral ischemia in rats: A functional magnetic resonance imaging study. *J Neurosci* 2003;23(2):510–517.
154. Reese T, Bochelen D, Baumann D, Rausch M, Sauter A, Rudin M. Impaired functionality of reperfused brain tissue following short transient focal ischemia in rats. *Magn Reson Imaging* 2002; 20(6):447–454.
155. Shen Q, Ren H, Cheng H, Fisher M, Duong TQ. Functional, perfusion and diffusion MRI of acute focal ischemic brain injury. *J Cereb Blood Flow Metab* 2005;25(10):1265–1279.
156. Kim YR, Huang IJ, Lee SR, Tejima E, Mandeville JB, van Meer MP, *et al.* Measurements of BOLD/CBV ratio show altered fMRI hemodynamics during stroke recovery in rats. *J Cereb Blood Flow Metab* 2005;25(7):820–829.
157. Fisher RS. Animal models of the epilepsies. *Brain Res Brain Res Rev* 1989;14(3):245–278.
158. Engel J Jr. Experimental animal models of epilepsy: Classification and relevance to human epileptic phenomena. *Epilepsy Res Suppl* 1992;8:9–20.
159. Buchhalter JR. Animal models of inherited epilepsy. *Epilepsia* 1993;34(Suppl. 3):S31–S41.
160. Hosford DA. Models of primary generalized epilepsy. *Curr Opin Neurol* 1995;8(2):121–125.
161. Lason W. Genetic animal models of epilepsy. *Pol J Pharmacol* 1998;50(1):77–79.
162. Toth M, Tecott L. Transgenic approaches to epilepsy. *Adv Neurol* 1999;79:291–296.
163. Coenen AM, Van Luijtelaar EL. Genetic animal models for absence epilepsy: A review of the WAG/Rij strain of rats. *Behav Genet* 2003;33(6):635–655.
164. Fariello RG. Critical review of the animal models of generalized epilepsies. *Ital J Neurol Sci* 1995;16(1–2):69–72.
165. Zhong J, Petroff OA, Prichard JW, Gore JC. Barbiturate-reversible reduction of water diffusion coefficient in flurothyl-induced status epilepticus in rats. *Magn Reson Med* 1995;33(2): 253–256.
166. Nersesyan H, Hyder F, Rothman DL, Blumenfeld H. Dynamic fMRI and EEG recordings during spike-wave seizures and generalized tonic-clonic seizures in WAG/Rij rats. *J Cereb Blood Flow Metab* 2004;24(6):589–599.
167. Nersesyan H, Herman P, Erdogan E, Hyder F, Blumenfeld H. Relative changes in cerebral blood flow and neuronal activity in local microdomains during generalized seizures. *J Cereb Blood Flow Metab* 2004;24(9):1057–1068.
168. Fabene PF, Sbarbati A. In vivo MRI in different models of experimental epilepsy. *Curr Drug Targets* 2004;5(7):629–636.
169. Kaakkola S, Teravainen H. Animal models of parkinsonism. *Pharmacol Toxicol* 1990;67(2):95–100.
170. Tolwani RJ, Jakowec MW, Petzinger GM, Green S, Waggie K. Experimental models of Parkinson's disease: Insights from many models. *Lab Anim Sci* 1999;49(4):363–371.
171. Betarbet R, Sherer TB, Greenamyre JT. Animal models of Parkinson's disease. *Bioessays* 2002;24(4):308–318.
172. Colpaert FC. Pharmacological characteristics of tremor, rigidity and hypokinesia induced by reserpine in rat. *Neuropharmacology* 1987;26(9):1431–1440.
173. Hall S, Rutledge JN, Schallert T. MRI, brain iron and experimental Parkinson's disease. *J Neurol Sci* 1992;113(2):198–208.
174. Pelled G, Bergman H, Goelman G. Bilateral overactivation of the sensorimotor cortex in the unilateral rodent model of Parkinson's disease—a functional magnetic resonance imaging study. *Eur J Neurosci* 2002;15(2):389–394.
175. Kondoh T, Bannai M, Nishino H, Torii K. 6-Hydroxydopamine-induced lesions in a rat model of hemi-Parkinson's disease monitored by magnetic resonance imaging. *Exp Neurol* 2005;192(1):194–202.
176. Pelled G, Bergman H, Ben Hur T, Goelman G. Reduced basal activity and increased functional homogeneity in sensorimotor and striatum of a Parkinson's disease rat model: A functional MRI study. *Eur J Neurosci* 2005;21(8):2227–2232.
177. Brownell AL, Jenkins BG, Isacson O. Dopamine imaging markers and predictive mathematical models for progressive degeneration in Parkinson's disease. *Biomed Pharmacother* 1999; 53(3):131–140.
178. Zhang Z, Zhang M, Ai Y, Avison C, Gash DM. MPTP-induced pallidal lesions in rhesus monkeys. *Exp Neurol* 1999;155(1):140–149.
179. Brownell AL, Canales K, Chen YI, Jenkins BG, Owen C, Livni E, *et al.* Mapping of brain function after MPTP-induced neurotoxicity in a primate Parkinson's disease model. *Neuroimage* 2003;20(2): 1064–1075.
180. Podell M, Hadjiconstantinou M, Smith MA, Neff NH. Proton magnetic resonance imaging and spectroscopy identify metabolic changes in the striatum in the MPTP feline model of parkinsonism. *Exp Neurol* 2003;179(2):159–166.
181. Chen Q, Andersen AH, Zhang Z, Ovadia A, Gash DM, Avison MJ. Mapping drug-induced changes in cerebral R2* by multiple gradient recalled echo functional MRI. *Magn Reson Imaging* 1996; 14(5):469–476.

182. Chen Q, Andersen AH, Zhang Z, Ovadia A, Cass WA, Gash DM, *et al.* Functional MRI of basal ganglia responsiveness to levodopa in parkinsonian rhesus monkeys. *Exp Neurol* 1999;158(1):63–75.
183. Jenkins BG, Sanchez-Pernaute R, Brownell AL, Chen YC, Isacson O. Mapping dopamine function in primates using pharmacologic magnetic resonance imaging. *J Neurosci* 2004;24(43):9553–9560.
184. Chen YC, Choi JK, Andersen SL, Rosen BR, Jenkins BG. Mapping dopamine D2/D3 receptor function using pharmacological magnetic resonance imaging. *Psychopharmacology (Berl)* 2005;180:705–715.
185. Bjorklund LM, Sanchez-Pernaute R, Chung S, Andersson T, Chen IY, McNaught KS, *et al.* Embryonic stem cells develop into functional dopaminergic neurons after transplantation in a Parkinson rat model. *Proc Natl Acad Sci USA* 2002;99(4):2344–2349.
186. Sturchler-Pierrat C, Sommer B. Transgenic animals in Alzheimer's disease research. *Rev Neurosci* 1999;10(1):15–24.
187. Yamada K, Nabeshima T. Animal models of Alzheimer's disease and evaluation of anti-dementia drugs. *Pharmacol Ther* 2000;88(2):93–113.
188. Bornemann KD, Staufenbiel M. Transgenic mouse models of Alzheimer's disease. *Ann NY Acad Sci* 2000;908:260–266.
189. Duff K. Transgenic mouse models of Alzheimer's disease: Phenotype and mechanisms of pathogenesis. *Biochem Soc Symp* 2001;(67):195–202.
190. Phinney AL, Horne P, Yang J, Janus C, Bergeron C, Westaway D. Mouse models of Alzheimer's disease: The long and filamentous road. *Neurol Res* 2003;25(6):590–600.
191. Higgins GA, Jacobsen H. Transgenic mouse models of Alzheimer's disease: Phenotype and application. *Behav Pharmacol* 2003; 14(5–6):419–438.
192. Sankaranarayanan S. Genetically modified mice models for Alzheimer's disease. *Curr Top Med Chem* 2006;6(6):609–627.
193. McGowan E, Eriksen J, Hutton M. A decade of modeling Alzheimer's disease in transgenic mice. *Trends Genet* 2006;22(5): 281–289.
194. Dhenain M, Michot JL, Volk A, Picq JL, Boller F. T2-weighted MRI studies of mouse lemurs: A primate model of brain aging. *Neurobiol Aging* 1997;18(5):517–521.
195. Coimbra A, Williams DS, Hostetler ED. The role of MRI and PET/SPECT in Alzheimer's disease. *Curr Top Med Chem* 2006;6(6): 629–647.
196. Helpern JA, Jensen J, Lee SP, Falangola MF. Quantitative MRI assessment of Alzheimer's disease. *J Mol Neurosci* 2004;24(1): 45–48.
197. Zhang J, Yarowsky P, Gordon MN, Di Carlo G, Munireddy S, van Zijl PC, *et al.* Detection of amyloid plaques in mouse models of Alzheimer's disease by magnetic resonance imaging. *Magn Reson Med* 2004;51(3):452–457.
198. Lee SP, Falangola MF, Nixon RA, Duff K, Helpern JA. Visualization of beta-amyloid plaques in a transgenic mouse model of Alzheimer's disease using MR microscopy without contrast reagents. *Magn Reson Med* 2004;52(3):538–544.
199. Jack CR Jr, Garwood M, Wengenack TM, Borowski B, Curran GL, Lin J, *et al.* In vivo visualization of Alzheimer's amyloid plaques by magnetic resonance imaging in transgenic mice without a contrast agent. *Magn Reson Med* 2004;52(6):1263–1271.
200. Wadghiri YZ, Sigurdsson EM, Wisniewski T, Turnbull DH. Magnetic resonance imaging of amyloid plaques in transgenic mice. *Methods Mol Biol* 2005;299:365–379.
201. Jack CR Jr, Wengenack TM, Reyes DA, Garwood M, Curran GL, Borowski BJ, *et al.* In vivo magnetic resonance microimaging of individual amyloid plaques in Alzheimer's transgenic mice. *J Neurosci* 2005;25(43):10041–10048.
202. Falangola MF, Lee SP, Nixon RA, Duff K, Helpern JA. Histological co-localization of iron in Abeta plaques of PS/APP transgenic mice. *Neurochem Res* 2005;30(2):201–205.
203. Vanhoutte G, Dewachter I, Borghgraef P, Van Leuven F, Van der LA. Noninvasive in vivo MRI detection of neuritic plaques associated with iron in APP[V717I] transgenic mice, a model for Alzheimer's disease. *Magn Reson Med* 2005;53(3):607–613.
204. Bartlett S. MRI for in vivo detection of amyloid plaques. *Lancet Neurol* 2005;4(5):276.
205. Harms MP, Kotyk JJ, Merchant KM. Evaluation of white matter integrity in ex vivo brains of amyloid plaque-bearing APPsw transgenic mice using magnetic resonance diffusion tensor imaging. *Exp Neurol* 2006;199(2):408–415.
206. Dhenain M, Delatour B, Walczak C, Volk A. Passive staining: A novel ex vivo MRI protocol to detect amyloid deposits in mouse models of Alzheimer's disease. *Magn Reson Med* 2006;55(3): 687–693.
207. Beckmann N, Schuler A, Mueggler T, Meyer EP, Wiederhold KH, Staufenbiel M, *et al.* Age-dependent cerebrovascular abnormalities and blood flow disturbances in APP23 mice modeling Alzheimer's disease. *J Neurosci* 2003;23(24):8453–8459.
208. Krucker T, Schuler A, Meyer EP, Staufenbiel M, Beckmann N. Magnetic resonance angiography and vascular corrosion casting as tools in biomedical research: Application to transgenic mice modeling Alzheimer's disease. *Neurol Res* 2004;26(5):507–516.
209. Hu ZH, Wang XC, Li LY, Liu ML, Liu R, Ling Z, *et al.* Correla-tion of behavior changes and BOLD signal in Alzheimer-like rat model. *Acta Biochim Biophys Sin (Shanghai)* 2004;36:803–810.
210. Van Camp N, Verhoye M, De Zeeuw CI, Van der LA. Light stimulus frequency dependence of activity in the rat visual system as studied with high-resolution BOLD fMRI. *J Neurophysiol* 2006;95(5):3164–3170.
211. Masamoto K, Kim T, Fukuda M, Wang P, Kim SG. Relationship between neural, vascular, and BOLD signals in isoflurane-anesthetized rat somatosensory cortex. *Cereb Cortex* 2007;17(4):942–950.
212. Lu H, Patel S, Luo F, Li SJ, Hillard CJ, Ward BD, *et al.* Spatial correlations of laminar BOLD and CBV responses to rat whisker stimulation with neuronal activity localized by Fos expression. *Magn Reson Med* 2004;52(5):1060–1068.
213. Kennerley AJ, Berwick J, Martindale J, Johnston D, Papadakis N, Mayhew JE. Concurrent fMRI and optical measures for the investigation of the hemodynamic response function. *Magn Reson Med* 2005;54(2):354–365.
214. Peeters RR, Verhoye M, Vos BP, Van Dyck D, Van der LA, De Schutter E. A patchy horizontal organization of the somatosensory activation of the rat cerebellum demonstrated by functional MRI. *Eur J Neurosci* 1999;11(8):27–30.
215. Van Camp N, Peeters RR, Van der LA. A comparison between blood oxygenation level-dependent and cerebral blood volume contrast in the rat cerebral and cerebellar somatosensoric cortex during electrical paw stimulation. *J Magn Reson Imaging* 2005;22(4): 483–491.
216. Xu F, Kida I, Hyder F, Shulman RG. Assessment and discrimination of odor stimuli in rat olfactory bulb by dynamic functional MRI. *Proc Natl Acad Sci USA* 2000;97(19):10601–10606.
217. Xu F, Schaefer M, Kida I, Schafer J, Liu N, Rothman DL, *et al.* Simultaneous activation of mouse main and accessory olfactory bulbs by odors or pheromones. *J Comp Neurol* 2005;489(4):491–500.
218. Schafer JR, Kida I, Rothman DL, Hyder F, Xu F. Adaptation in the rodent olfactory bulb measured by fMRI. *Magn Reson Med* 2005;54(2):443–448.
219. Jezzard P, Rauschecker JP, Malonek D. An in vivo model for functional MRI in cat visual cortex. *Magn Reson Med* 1997;38(5): 699–705.
220. Kim DS, Duong TQ, Kim SG. High-resolution mapping of isoorientation columns by fMRI. *Nat Neurosci* 2000;3(2): 164–169.
221. Duong TQ, Kim DS, Ugurbil K, Kim SG. Spatiotemporal dynamics of the BOLD fMRI signals: Toward mapping submillimeter cortical columns using the early negative response. *Magn Reson Med* 2000;44(2):231–242.
222. Duong TQ, Kim DS, Ugurbil K, Kim SG. Localized cerebral blood flow response at submillimeter columnar resolution. *Proc Natl Acad Sci USA* 2001;98(19):10904–10909.

223. Harel N, Lee SP, Nagaoka T, Kim DS, Kim SG. Origin of negative blood oxygenation level-dependent fMRI signals. *J Cereb Blood Flow Metab* 2002;22(8):908–917.
224. Kim DS, Kim M, Ronen I, Formisano E, Kim KH, Ugurbil K, *et al.* In vivo mapping of functional domains and axonal connectivity in cat visual cortex using magnetic resonance imaging. *Magn Reson Imaging* 2003;21(10):1131–1140.
225. Zhao F, Wang P, Kim SG. Cortical depth-dependent gradient-echo and spin-echo BOLD fMRI at 9.4T. *Magn Reson Med* 2004; 51(3):518–524.
226. Kim DS, Ronen I, Olman C, Kim SG, Ugurbil K, Toth LJ. Spatial relationship between neuronal activity and BOLD functional MRI. *Neuroimage* 2004;21(3):876–885.
227. Nagaoka T, Zhao F, Wang P, Harel N, Kennan RP, Ogawa S, *et al.* Increases in oxygen consumption without cerebral blood volume change during visual stimulation under hypotension condition. *J Cereb Blood Flow Metab* 2006;26(8):1043–1051.
228. Zhao F, Wang P, Hendrich K, Ugurbil K, Kim SG. Cortical layer-dependent BOLD and CBV responses measured by spin-echo and gradient-echo fMRI: Insights into hemodynamic regulation. *Neuroimage* 2006;30(4):1149–1160.
229. Bolan PJ, Yacoub E, Garwood M, Ugurbil K, Harel N. In vivo micro-MRI of intracortical neurovasculature. *Neuroimage* 2006; 32(1):62–69.
230. Van der Linden A, Van Meir V, Tindemans I, Verhoye M, Balthazart J. Applications of manganese-enhanced magnetic resonance imaging (MEMRI) to image brain plasticity in song birds. *NMR Biomed* 2004;17(8):602–612.
231. Van Meir V, Boumans T, De Groof G, Van Audekerke J, Smolders A, Scheunders P, *et al.* Spatiotemporal properties of the BOLD response in the songbirds' auditory circuit during a variety of listening tasks. *Neuroimage* 2005;25(4):1242–1255.
232. Stefanacci L, Reber P, Costanza J, Wong E, Buxton R, Zola S, *et al.* fMRI of monkey visual cortex. *Neuron* 1998;20(6):1051–1057.
233. Dubowitz DJ, Chen DY, Atkinson DJ, Grieve KL, Gillikin B, Bradley WG Jr, *et al.* Functional magnetic resonance imaging in macaque cortex. *Neuroreport* 1998;9(10):2213–2218.
234. Logothetis NK, Guggenberger H, Peled S, Pauls J. Functional imaging of the monkey brain. *Nat Neurosci* 1999;2(6):555–562.
235. Disbrow E, Roberts TP, Slutsky D, Krubitzer L. The use of fMRI for determining the topographic organization of cortical fields in human and nonhuman primates. *Brain Res* 1999;829(1–2):167–173.
236. Vanduffel W, Fize D, Mandeville JB, Nelissen K, Van Hecke P, Rosen BR, *et al.* Visual motion processing investigated using contrast agent-enhanced fMRI in awake behaving monkeys. *Neuron* 2001;32(4):565–577.
237. Logothetis N, Merkle H, Augath M, Trinath T, Ugurbil K. Ultra high-resolution fMRI in monkeys with implanted RF coils. *Neuron* 2002;35(2):227–242.
238. Barinaga M. fMRI provides new view of monkey brains. *Science* 1998;282(5393):1397.
239. Disbrow EA, Slutsky DA, Roberts TP, Krubitzer LA. Functional MRI at 1.5 Tesla: A comparison of the blood oxygenation level-dependent signal and electrophysiology. *Proc Natl Acad Sci USA* 2000;97(17):9718–9723.
240. Hayashi T, Konishi S, Hasegawa I, Miyashita Y. Short communication: Mapping of somatosensory cortices with functional magnetic resonance imaging in anaesthetized macaque monkeys. *Eur J Neurosci* 1999;11(12):4451–4456.
241. Lipton ML, Fu KM, Branch CA, Schroeder CE. Ipsilateral hand input to area 3b revealed by converging hemodynamic and electrophysiological analyses in macaque monkeys. *J Neurosci* 2006;26(1): 180–185.
242. Pfeuffer J, Merkle H, Beyerlein M, Steudel T, Logothetis NK. Anatomical and functional MR imaging in the macaque monkey using a vertical large-bore 7 Tesla setup. *Magn Reson Imaging* 2004;22(10):1343–1359.
243. Rainer G, Augath M, Trinath T, Logothetis NK. Nonmonotonic noise tuning of BOLD fMRI signal to natural images in the visual cortex of the anesthetized monkey. *Curr Biol* 2001;11(11):846–854.
244. Tolias AS, Smirnakis SM, Augath MA, Trinath T, Logothetis NK. Motion processing in the macaque: Revisited with functional magnetic resonance imaging. *J Neurosci* 2001;21(21):8594–8601.
245. Sereno ME, Trinath T, Augath M, Logothetis NK. Three-dimensional shape representation in monkey cortex. *Neuron* 2002;33(4): 635–652.
246. Dubowitz DJ, Bernheim KA, Chen DY, Bradley WG Jr, Andersen RA. Enhancing fMRI contrast in awake-behaving primates using intravascular magnetite dextran nanoparticles. *Neuroreport* 2001; 12(11):2335–2340.
247. Vanduffel W, Fize D, Peuskens H, Denys K, Sunaert S, Todd JT, *et al.* Extracting 3D from motion: Differences in human and monkey intraparietal cortex. *Science* 2002;298(5592):413–415.
248. Fize D, Vanduffel W, Nelissen K, Denys K, Chef dC, Faugeras O, *et al.* The retinotopic organization of primate dorsal V4 and surrounding areas: A functional magnetic resonance imaging study in awake monkeys. *J Neurosci* 2003;23(19):7395–7406.
249. Kourtzi Z, Tolias AS, Altmann CF, Augath M, Logothetis NK. Integration of local features into global shapes: Monkey and human FMRI studies. *Neuron* 2003;37(2):333–346.
250. Denys K, Vanduffel W, Fize D, Nelissen K, Sawamura H, Georgieva S, *et al.* Visual activation in prefrontal cortex is stronger in monkeys than in humans. *J Cogn Neurosci* 2004;16(9):1505–1516.
251. Orban GA, Van Essen D, Vanduffel W. Comparative mapping of higher visual areas in monkeys and humans. *Trends Cogn Sci* 2004;8(7):315–324.
252. Koyama M, Hasegawa I, Osada T, Adachi Y, Nakahara K, Miyashita Y. Functional magnetic resonance imaging of macaque monkeys performing visually guided saccade tasks: Comparison of cortical eye fields with humans. *Neuron* 2004;41(5):795–807.
253. Sawamura H, Georgieva S, Vogels R, Vanduffel W, Orban GA. Using functional magnetic resonance imaging to assess adaptation and size invariance of shape processing by humans and monkeys. *J Neurosci* 2005;25(17):4294–4306.
254. King JA, Garelick TS, Brevard ME, Chen W, Messenger TL, Duong TQ, *et al.* Procedure for minimizing stress for fMRI studies in conscious rats. *J Neurosci Methods* 2005;148(2):154–160.
255. Kannurpatti SS, Biswal BB. Effect of anesthesia on CBF, MAP and fMRI-BOLD signal in response to apnea. *Brain Res* 2004;1011(2): 141–147.
256. Martin C, Martindale J, Berwick J, Mayhew J. Investigating neural-hemodynamic coupling and the hemodynamic response function in the awake rat. *Neuroimage* 2006;32(1):33–48.
257. Peeters RR, Tindemans I, De Schutter E, Van der Linden A. Comparing BOLD fMRI signal changes in the awake and anesthetized rat during electrical forepaw stimulation. *Magn Reson Imaging* 2001;19(6):821–826.
258. Austin VC, Blamire AM, Allers KA, Sharp T, Styles P, Matthews PM, *et al.* Confounding effects of anesthesia on functional activation in rodent brain: A study of halothane and alpha-chloralose anesthesia. *Neuroimage* 2005;24(1):92–100.
259. Willis CK, Quinn RP, McDonell WM, Gati J, Parent J, Nicolle D. Functional MRI as a tool to assess vision in dogs: The optimal anesthetic. *Vet Ophthalmol* 2001;4(4):243–253.
260. Heinke W, Schwarzbauer C. Subanesthetic isoflurane affects task-induced brain activation in a highly specific manner: A functional magnetic resonance imaging study. *Anesthesiology* 2001;94(6): 973–981.
261. Liu ZM, Schmidt KF, Sicard KM, Duong TQ. Imaging oxygen consumption in forepaw somatosensory stimulation in rats under isoflurane anesthesia. *Magn Reson Med* 2004;52(2):277–285.
262. Abo M, Suzuki M, Senoo A, Miyano S, Yamauchi H, Yonemoto K, *et al.* Influence of isoflurane concentration and hypoxia on functional magnetic resonance imaging for the detection of bicuculline-induced neuronal activation. *Neurosignals* 2004;13(3):144–149.

263. Dashti M, Geso M, Williams J. The effects of anaesthesia on cortical stimulation in rats: A functional MRI study. *Australas Phys Eng Sci Med* 2005;28(1):21–25.
264. Scanley BE, Kennan RP, Cannan S, Skudlarski P, Innis RB, Gore JC. Functional magnetic resonance imaging of median nerve stimulation in rats at 2.0 T. *Magn Reson Med* 1997;37(6):969–972.
265. Lahti KM, Ferris CF, Li F, Sotak CH, King JA. Comparison of evoked cortical activity in conscious and propofol-anesthetized rats using functional MRI. *Magn Reson Med* 1999;41(2):412–416.
266. Kalisch R, Elbel GK, Gossl C, Czisch M, Auer DP. Blood pressure changes induced by arterial blood withdrawal influence bold signal in anesthesized rats at 7 Tesla: Implications for pharmacologic MRI. *Neuroimage* 2001;14(4):891–898.
267. Makiranta MJ, Lehtinen S, Jauhiainen JP, Oikarinen JT, Pyhtinen J, Tervonen O. MR perfusion, diffusion and BOLD imaging of methotrexate-exposed swine brain. *J Magn Reson Imaging* 2002;15(5):511–519.
268. Ferris CF, Snowdon CT, King JA, Duong TQ, Ziegler TE, Ugurbil K, *et al.* Functional imaging of brain activity in conscious monkeys responding to sexually arousing cues. *Neuroreport* 2001;12(10):2231–2236.
269. Weber R, Ramos-Cabrer P, Wiedermann D, Van Camp N, Hoehn M. A fully noninvasive and robust experimental protocol for longitudinal fMRI studies in the rat. *Neuroimage* 2006;29(4):1303–1310.
270. Crosby G, Crane AM, Sokoloff L. Local changes in cerebral glucose utilization during ketamine anesthesia. *Anesthesiology* 1982;56(6):437–443.
271. Crosby G, Crane AM, Jehle J, Sokoloff L. The local metabolic effects of somatosensory stimulation in the central nervous system of rats given pentobarbital or nitrous oxide. *Anesthesiology* 1983;58(1):38–43.
272. Crosby G, Crane AM, Sokoloff L. A comparison of local rates of glucose utilization in spinal cord and brain in conscious and nitrous oxide- or pentobarbital-treated rats. *Anesthesiology* 1984;61(4):434–438.
273. Ueki M, Mies G, Hossmann KA. Effect of alpha-chloralose, halothane, pentobarbital and nitrous oxide anesthesia on metabolic coupling in somatosensory cortex of rat. *Acta Anaesthesiol Scand* 1992;36(4):318–322.
274. Nakao Y, Itoh Y, Kuang TY, Cook M, Jehle J, Sokoloff L. Effects of anesthesia on functional activation of cerebral blood flow and metabolism. *Proc Natl Acad Sci USA* 2001;98(13):7593–7598.
275. Rojas MJ, Navas JA, Rector DM. Evoked response potential markers for anesthetic and behavioral states. *Am J Physiol Regul Integr Comp Physiol* 2006;291(1):R189–R196.
276. Littlewood CL, Cash D, Dixon AL, Dix SL, White CT, O'neill MJ, *et al.* Using the BOLD MR signal to differentiate the stereoisomers of ketamine in the rat. *Neuroimage* 2006;32(4):1733–1746.
277. Steward CA, Marsden CA, Prior MJ, Morris PG, Shah YB. Methodological considerations in rat brain BOLD contrast pharmacological MRI. *Psychopharmacology (Berl)* 2005;180:687–704.
278. Wise RG, Tracey I. The role of fMRI in drug discovery. *J Magn Reson Imaging* 2006;23(6):862–876.
279. Schwarz A, Gozzi A, Reese T, Bertani S, Crestan V, Hagan J, *et al.* Selective dopamine D(3) receptor antagonist SB-277011-A potentiates phMRI response to acute amphetamine challenge in the rat brain. *Synapse* 2004;54(1):1–10.
280. Ireland MD, Lowe AS, Reavill C, James MF, Leslie RA, Williams SC. Mapping the effects of the selective dopamine D2/D3 receptor agonist quinelorane using pharmacological magnetic resonance imaging. *Neuroscience* 2005;133(1):315–326.
281. Zhang Z, Andersen AH, Ai Y, Loveland A, Hardy PA, Gerhardt GA, *et al.* Assessing nigrostriatal dysfunctions by pharmacological MRI in parkinsonian rhesus macaques. *Neuroimage* 2006;33(2):636–643.
282. Choi JK, Mandeville JB, Chen YI, Kim YR, Jenkins BG. High resolution spatial mapping of nicotine action using pharmacologic magnetic resonance imaging. *Synapse* 2006;60(2):152–157.
283. Gozzi A, Schwarz A, Reese T, Bertani S, Crestan V, Bifone A. Region-specific effects of nicotine on brain activity: A pharmacological MRI study in the drug-naive rat. *Neuropsychopharmacology* 2006;31(8):1690–1703.
284. Kalisch R, Salome N, Platzer S, Wigger A, Czisch M, Sommer W, *et al.* High trait anxiety and hyporeactivity to stress of the dorsomedial prefrontal cortex: A combined phMRI and Fos study in rats. *Neuroimage* 2004;23(1):382–391.
285. Schwarz AJ, Zocchi A, Reese T, Gozzi A, Garzotti M, Varnier G, *et al.* Concurrent pharmacological MRI and in situ microdialysis of cocaine reveal a complex relationship between the central hemodynamic response and local dopamine concentration. *Neuroimage* 2004;23(1):296–304.
286. Chen YC, Galpern WR, Brownell AL, Matthews RT, Bogdanov M, Isacson O, *et al.* Detection of dopaminergic neurotransmitter activity using pharmacologic MRI: Correlation with PET, microdialysis, and behavioral data. *Magn Reson Med* 1997;38(3):389–398.
287. Flecknell PA. Laboratory Animal Anesthesia, 2nd ed. London: Academic Press, 1996:91.

52 Animal Models in Aging Research

A Critical Examination

ROCHELLE BUFFENSTEIN, YAEL H. EDREY, AND PAMELA L. LARSEN

ABSTRACT

Understanding why different organisms show diverse rates of aging may provide useful insights into basic aging processes. Biogerontologists have converged on a few model organisms that represent only a minute fraction of the animal kingdom, but nevertheless span a considerable distance in animal evolution. Shared features of these evolutionary divergent animals have highlighted some conserved regulatory processes in animal aging. However, these traditional models are all short-lived and may have unintentionally constrained research to focus on only those areas in which their use is most appropriate. Surprisingly few studies focus on slow-aging organisms, or nontraditional model organisms that may be better suited to address successful aging and issues more relevant to long-living humans. This chapter critically assesses both traditional and nontraditional animal models used in aging research, and emphasizes the importance and judicious use of the comparative method to test the ubiquity of aging theories, mechanisms, and their potential translation for human application.

Key Words: Longevity, Slow-aging, Comparative approach, Life span extension, Oxidative damage, Phylogenetically independent contrasts, Disposable soma theory of aging.

INTRODUCTION

Humans are inherently fascinated by the aging process and fearful of growing old. The quest for immortality is a frequent theme in mythology and historical accounts, such that many people, both before and after the famed Spanish explorer Juan Ponce de Leon (1513), have searched in vain for the "Fountain of Youth," for to drink its water may bestow eternal youth. Despite this elusiveness, both human life expectancy and quality of life at advanced ages have increased dramatically since that time. These demographic patterns are a product of improved public hygiene, better nutrition, and biomedical advancements rather than the result of any retardation of the mechanisms of aging. Aging itself is often considered life-limiting, and is the major risk factor for the predominant causes of death (cardiovascular disease and cancer). To continue to enhance the longevity and quality of life, we need to convert the enigma of aging into an understanding of the mechanisms involved in this ubiquitous process.

Animal models have long facilitated the exploration of biological mechanisms, including research on aging. "Salacious animals and those abounding in seed age quickly," claimed Greek philosopher Aristotle in his 350 BCE treatise *On Longevity and Shortness of Life*. Aristotle cited the mule (typically a sterile hybrid) as living "longer than either the horse or the ass from which it sprang." Similar claims were made in "the rate of living theory" popularized as "live fast, die young," proposed by Pearl.[1] Although these theories are inherently logical, they were refuted following careful scrutiny based on numerous animal studies. Whether results supported or provided an exception to these theories depended in part on the species studied. Each animal model affords different strengths for physiological, biochemical, and genetic research. Since the aim is to uncover the full assortment of mechanisms that are important components of aging, the discussion here will focus on how to gain the most knowledge from the use of different animal models.

August Krogh, the Nobel laureate of 1920 and a founding father of comparative biology, was renowned not only for his research on regulatory processes in capillaries, but also for his insightful approach to setting the modalities of this field. His profound statement—"for many biological problems, there is an animal on which it can be most conveniently studied"—is often cited when animal models are chosen.[2] Biogerontologists have converged on a few model organisms (single-celled yeast, *Saccharomyces cerevisiae*, the nematode *Caenorhabditis elegans*, the fruitfly *Drosophila melanogaster,* and the laboratory mouse *Mus musculus*). These four species represent a minute fraction of the animal kingdom, but nevertheless span a considerable distance in animal evolution (Figure 52–1). Shared features of these evolutionary divergent animals strongly indicate the presence of some conserved processes. However, there is still considerable debate concerning the extent of generality of aging mechanisms. These traditional models have provided valuable insight into putative regulators of longevity including the genetic, molecular, and biochemical basis, and have contributed immensely to current understanding of aging processes.[3–6] Unintentionally though, they may have constrained research to focus on only those areas in which the use of these species is most appropriate.

Careful scrutiny of why these classical model organisms are selected suggests that as in many other areas of biomedical research, they are primarily selected for expediency, rather than for any specific biological properties that make them ideally suited for studying aging processes relevant to humans.[7,8] Regardless of the research field, they are used because they have many advantages in a laboratory setting: they are small, easy to care for,

From: *Sourcebook of Models for Biomedical Research*
(P. M. Conn, ed.), © 2008 Humana Press Inc., Totowa, NJ.

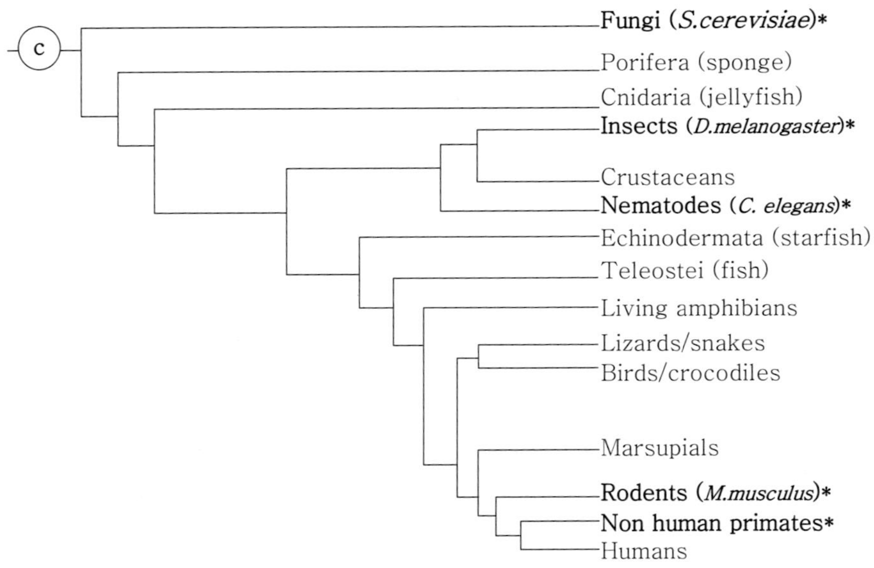

Figure 52–1. Phylogenetic relationships of model organisms used in aging research. "C" represents a common ancestor. Animals in bold are traditional models used in aging research. (Modified from the American Museum of Natural History website.)

and breed well in captivity, such that the cost of maintaining statistically significant numbers is not prohibitive.[9] Furthermore, their basic biology and genome are known. This, together with appropriate new genetic and molecular tools, has facilitated considerable advances in scientific endeavor.

A shared weakness of traditional aging models is that they do not show well-developed mechanisms that protect them from aging and they are all short-lived.[7] There are examples of slow-aging animals, which exhibit only slight age-related declines in physiological capacity, reproductive rate, and/or disease resistance. Surprisingly, few studies involve these animals, even though they display characteristics that we try to emulate in short-lived models. A complementary approach would be to assess how slow-aging animals attain their prolonged longevity. There is also a need to understand mechanisms of aging in species phylogenetically closer to humans (e.g., primates, Figure 52–1), as well as in vertebrates with life spans that approach the life span of humans. Research on aging could benefit from a comparative approach using traditional models and wise addition of nontraditional model organisms, especially those better suited to addressing aging issues more relevant to those related to humans.

AGING

Aging can be defined as a *progressive, irreversible, endogenous*, and *deleterious* process[10] that occurs postmaturation. This turns young healthy adults into older, more frail adults, increasingly susceptible to environmental challenges (such as extreme temperature or disease-inducing infectious agents), and with a concomitant greater risk of death. Its progressive character suggests that causes of aging are present throughout life. Thus, it may be discerned at all ages and is most pronounced in postmitotic tissues in which cells that are irreversibly damaged or lost cannot be replaced by mitosis. Irreversible damage (from reactive oxygen species (ROS), advanced glycation end products (AGEs), and DNA mutations) may accrue at different rates in different cell types. Exogenous factors (environmental stressors, e.g., toxic chemicals) may interact with genotype-dependent factors, either enhancing or diminishing their effects, and thereby determining rates of aging. These combined effects may explain why different species or strains, maintained under similar conditions, age at diverse rates and exhibit unequal organ-specific vulnerability. Regardless of these dissimilarities, the prominence of age-related declines makes it possible to distinguish young from old throughout the animal kingdom.

Certain strains or species are more prone to specific age-related changes and associated pathologies. For instance humans, and naked mole-rats have a far lower incidence of cancer than do mice at any age.[11–13] In contrast, some mouse strains (e.g., AKR/J) are so susceptible to neoplasia that if they survive more than 24 months, at death 100% of the individuals will present with a strain-specific tumor (AKR/J ~ thymic lymphoma), even if is not the determined cause of death.[14] Aging research in these animals thus may be compounded with disease effects.

Of critical concern is that a reductionist approach has been used in aging research. This limits the number of variables under study, and usually ignores synergistic interactions, thereby oversimplifying the process. For example, if higher levels of oxidative damage are found in an older cohort, a reductionist would conclude that this damage causes aging. However, this may potentially be a consequence of upstream mechanisms. Numerous similar features have been cataloged in young and old representatives within a species, without providing real insight into the aging process. Clearly, to make significant advances in aging research, an *in vivo* broad integrative approach needs to be taken and experimental data from the plethora of studies need to be collated, and thereafter used in a systems-based predictive modality.

ANIMAL MODELS

There are three primary criteria for employment of animal models: (1) their use must be feasible, (2) the specific question under investigation can be appropriately addressed with this model, and (3) the findings can be generalized.

MODEL SYSTEM SELECTION

Feasibility of Use Choice of animal model depends to a large extent on logistical considerations such as its ease of acquisition, housing and husbandry requirements, and costs of care in captivity. As such, it is highly unlikely that elephants or rockfish, despite exhibiting a life span that overlaps the human range, will ever be considered an "ideal" animal model for aging research. There are many other species that would not present as logistical a problem and from which aging research could benefit. Their use would to a large extent depend upon prior knowledge of the basic biology and the nutritional and housing requirements of the selected species. For instance, guinea pigs, like humans, cannot synthesize vitamin C and require daily supplements to maintain good health, while other laboratory rodents can produce their own supply.[15] Furthermore, guinea pigs, like many wild-derived species, are particularly sensitive to light availability. Inadequate lighting may alter their endocrine profiles and physiological function, and markedly affect research outcome. This may contribute to the lack of interlaboratory reproducibility of findings. Similarly certain animals, in particular primates, do not function optimally when housed without enrichment activities, and may present with abnormal responses in cognitive tests and altered neuronal function and hormone profiles.[16] A thorough knowledge of species-specific needs makes it possible to better interpret findings and ensure the appropriate species are used for the hypothesis under investigation.[17]

Specificity of Question The organism of choice would depend upon the specifics of the study. For instance, an organism whose somatic cells are all postmitotic would not be suitable for assessing changes in rates of cell turnover. However, this organism with a postmitotic adult soma could be a powerful animal model with which to assess damage accrual, for new cells are not recruited to confound the interpretation.

General Application A key criterion in aging research is the ease with which the information gleaned can be generalized and applied to a wide range of species, and in particular to humans. To achieve this goal, a specific question may be explored in a wide variety of phylogenetically unrelated organisms. Alternate ways of addressing these issues include assessment of both phylogenetically distinct and closely related species that show disparate longevity, and/or in similar-sized organisms with markedly different life spans. Regardless of methodology, useful insights may be obtained from both data that converge and diverge.

PRINCIPLES OF USE OF ANIMAL MODELS There are certain basic principles for use of animal models that should be adhered to for biogerontological research. These are briefly summarized here, but for an in depth review see Miller and Nadon.[8] Regardless of age, animals should be healthy, pathogen and disease free, and have no signs of tumors or lesions. It is imperative to carefully choose the age cohorts such that animals used are neither too young nor too old. Animals that are nearing the end of their lives may already be riddled with age-associated diseases, whereas those that are too young may still be undergoing the complex process of development and maturation. If a particular trait differs from young controls only in the oldest cohort, it is hard to ascertain if this is specifically due to aging, maturation of the young cohort, or age-associated pathologies (Figure 52–2). As such it is far better in initial surveys of novel model organisms to use more age cohorts and avoid the risk of misinterpreting data, or missing the important age-related changes in those variables (as highlighted in Figure 52–2).

STRAIN SPECIFICITY When planning a study, the genetic background of the model organism is a critical design decision. In many cases, particular strains are used simply because they have been previously used by other scientists and are readily available.[18] Indeed, more than 70% of the mice requested from

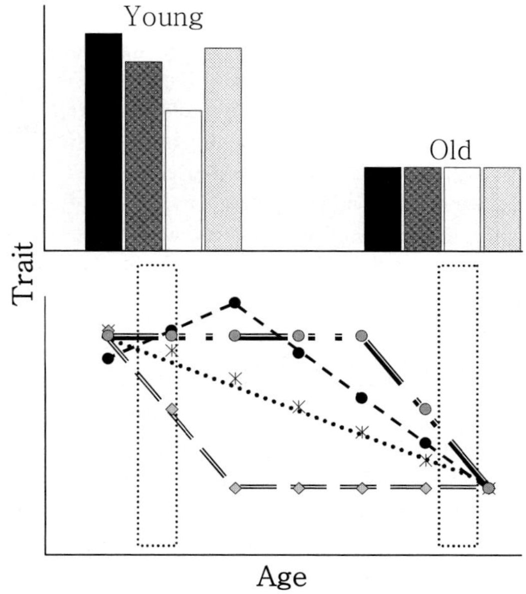

Figure 52–2. Age-related changes in four hypothetical species in which interspecies trait values are similar within each age cohort, although patterns of aging vary dramatically, and this cannot be duduced from the data provided in the two age cohorts.

the U.S. National Institute of Aging are of the C57BL/6 strain.[19] This means that rodent aging research is primarily based upon a single genotype.

Inbred strains have the advantage of low genetic heterogeneity so that some traits show only slight intrastrain variation. Thus, statistical significance can be acquired using small sample sizes within a group. Use of highly inbred strains, however, is almost equivalent to repeated measures on a single individual, thereby limiting informed inferences about the generality of findings. For instance, C57BL/6 mice are extremely susceptible to atherosclerosis.[20] Any experimental intervention that reduces mortality from this strain-specific pathology may spuriously be assumed to affect normal aging and longevity. This is considered one of the main drawbacks of using any mouse strain as a model for human aging. F_1 hybrids, four way crosses, or outbred heterogeneous strains tend to be hardier and exhibit longer life spans than their parental stock.[13,21] While costs of experiments based upon genetically diverse populations are higher (since larger samples sizes are needed to overcome the greater variability in measured traits), the outcome is influenced less by any particular genotype, and there is more confidence in broadly applying the conclusions gleaned from these studies.

Aging studies should be undertaken in both genders, for differences in genotype may be evident. For example, old male CBA mice have a stronger likelihood of developing hepatocellular tumors than females.[13] Similarly, quantitative trait locus (QTL) mapping of *Drosophila* genes has identified gender specific loci that differentially alter longevity.[22]

ANIMAL HUSBANDRY If studies are to be reproducible, animals ought to be housed under identical conditions and have standardized diets. Housing and dietary requirements for worms, flies, and mice have been well established to provide optimal living conditions in the laboratory. For this reason interlaboratory comparisons and experimental interventions can be undertaken facilitating greater confidence in data obtained.

In novel species, housing animals under simulated natural conditions and providing a varied diet may ameliorate captivity-associated stresses. For instance, naked mole-rats are highly social species that naturally live in underground burrows. In captivity these animals are housed communally in a series of interconnected plexiglass tubes. Animals housed in this way thrive, whereas those housed under standard rodent conditions do not, and develop dry skin and associated maladies (R. Buffenstein, personal observations). Thus, the health of the animal may outweigh the scientific desire for uniformity in diet and environment.

TRADITIONAL MODELS OF AGING

The three traditional multicellular model species for aging research (worms, flies, and mice) have been thoroughly studied. Breakdown in the maintenance of genomic stability, stochastic damage to DNA, and inadequate repair processes, as well as oxidative damage and impaired protein processing and folding have been widely implicated in their aging.[23,24] Caloric or dietary restriction (DR) appears to prolong life span in all classical models.[25] Similarly, single-gene mutations that alter life span have been identified in yeast,[26,27] worms,[28] flies,[29] and mice.[30] Furthermore, a similar genetic mechanism and a concomitant biochemical pathway have been found across the phyla.[31]

EXPERIMENTAL MANIPULATION DR without nutrient deficiency induces life span extension in all four evolutionary divergent traditional model species. In all cases metabolism is attenuated initially and reproduction is inhibited.[25] The ubiquity of this process in other organisms, however, is equivocal.[32–34] DR may exert its life-extending effects by acting as a low-level stressor, inducing reduced ROS formation and less oxidative damage accrual,[35] lowering glucose and insulin levels,[25] and/or decreasing insulin-like growth factor-1 (IGF-1).[36] DR appears to attenuate the rate of aging and extend longevity through a variety of mechanisms.[37] While the life-extending effects of DR on human longevity are unknown, a National Institute on Aging (NIA) study to address this was initiated in 2002.[38]

Although most studies of DR involve energy reduction without a concomitant nutrient deficit, it has been shown that limiting protein intake, or more specifically methionine, may lead to similar, though not as pronounced life-extending effects.[37] Methionine residues on proteins are readily prone to ROS attack and activity of a broad range of proteins is reportedly altered by its oxidation.[39] Limiting dietary methionine content lowers ROS production,[40] and also increases oxidative stress resistance.[17] In addition, methionine-restricted diets reduce glucose and insulin levels and lower IGF-1 concentrations in mice.[17] It therefore appears that methionine restriction may act on the same pathways as DR when exerting its longevity-extending effects.[37]

Kenyon[31] suggests that the conserved responses to DR provide further evidence that there are regulatory genes for longevity that may retard aging by delaying reproduction. She proposes that by linking reproduction and aging, these genes may be selected for; animals that do not show a decline in reproductive output when nutrients are scarce would become extinct.

GENETIC MANIPULATION Genetics has proved to be an incredibly powerful tool with which to probe proximate causes of aging. It substantiates the existence of genes that control life span.[4,31,41,42] Mutations in worms, flies, and mice that extend the life span are all associated with loss of function[42,43] and for the majority a concomitant decline in reproduction. This implies that genes enhancing reproductive fitness have been selected in the laboratory setting and possibly also from an evolutionarily standpoint. Research on guppies in the wild has shown that evolutionary tradeoffs between longevity and reproduction can manifest within as few as 60 generations, such that populations in predation-prone zones show earlier sexual maturity, greater reproductive fitness, and shorter life spans than those in areas with low predator density.[44] These genes, promoting early life reproductive fitness, appear to be detrimental later in life, and are a good example of selection by antagonistic pleiotropy.[45]

Many of the genetic manipulations that extend longevity reduce the size of the organism.[42] However, while IGFs have been implicated as life span determinants in all traditional models,[46] such that animals with low levels of IGF-1 live longer, small body size does not appear to be a requirement for this life span extension.[47,48] Heterozygous chico null/+ flies are normal in size yet live 36% longer than wild-type flies.[49] Similarly, the p66^{shc} knockout mouse shows no reduction in size yet nevertheless shows a 30% increase in longevity,[50] an increment comparable to that of Snell dwarf mice.[51]

Genetic studies in invertebrates and mammals have elucidated mechanisms associated with prolonged longevity by either genetic or environmental manipulations. DR of dwarf mice (lacking pituitary functions) further extends longevity in an additive manner.[42] These data and the observation that DR is DAF-16/FOXO independent suggest that the longevity mechanisms are independent of each other.[52,53] Long-lived mutant *C. elegans*[53] and *Drosophila*[54] and p66^{shc} mice[50] all show increased resistance to oxidative stressors. Similarly, oxidative stress studies using fibroblasts from dwarf mice also show that cell survival following exposure to oxidative insults correlates with life span.[55] Other studies, however, report contrary findings inconsistent with the mechanism of oxidative stress resistance causing longevity. Chico mutant flies show resistance to oxidative stressors similar to wild-type animals, despite being longer-lived.[49] In mice, Hauck *et al.*[56] reported that long-lived GHR/BP$^{-/-}$ mice are extremely susceptible to oxidative insults, while Harper *et al.*[17] observed extreme sensitivity to hepatotoxins in long-living dw/dw and GHR-KO mutant mice. Also, the increased oxidative stress and damage in Mn SOD-deficient mice are not accompanied by a concomitant decline in life span.[57] Thus these data suggest that while resistance to oxidative stress and other toxins is a common characteristic of experimental life span extension, it is not an essential determinant of longevity.

NONHUMAN PRIMATES AS MODELS FOR AGING RESEARCH

Although considerable information has been gleaned from traditional aging models, expanding the species under study to nonhuman primates is important given their phylogenetic (evolutionary) proximity to humans. Rhesus monkeys (*Macaca mulatta*), the most commonly used primate model in aging research,[13] share approximately 90% of their genome with humans[58] and have many more traits in common with humans than traditional aging models.[13] Rhesus monkeys have regular monthly menstrual cycles and show cessation thereof with aging with menopausal hormone profiles similar to that of humans.[59] Thus, they may be particularly useful for studies addressing reproductive senescence.[60] Age-related changes in neurological function and memory of monkeys also parallel those of humans.[61,62]

While the advantages of primate use in aging biology are obvious, to date their use in aging research has been rather limited. This appears to be primarily due to relative costs compared to other animal models,[13] fear of disruption by animal rights groups, as well as logistical issues associated with their comparatively long life spans. These issues may be potentially overcome through the use of small, rat-sized primates (e.g., common marmoset or mouse lemur) that have both shorter maturation times and life spans than rhesus monkeys; their research potential has been only scantly (i.e., Zeng *et al.*[63]) exploited to date.

NONTRADITIONAL MODELS FOR AGING RESEARCH AND THE COMPARATIVE APPROACH

Experimental manipulation in traditional models has led to 20–400% increases in mean and maximum life span (MLS). A major enticement to employ a comparative method is the more than 40,000-fold natural difference in life span among animals. Just within mammals, maximum life span varies by at least two orders of magnitude (Figure 52–3). Generally MLS lengthens in a predictable manner (i.e., MLS is proportional to $\text{mass}^{0.2}$, see Speakman[64]) as species increase in size such that for every doubling of body mass in mammals, there is on average a 16% increase in MLS. Body mass, however, can explain only ~35% of the variation in MLS of mammals. For our size *Homo sapiens* is a very long-living mammal, with an MLS (122.5 years) greater than that of elephants (MLS 80 years) and even much larger blue whales (105 years),[3] and that is 5 times[65] to 10 times[66] longer than predicted by mass (Figure 52–4). To date we do not know why humans, and a few other long-living mammals, lie more than two standard deviations away from this allometric relationship, nor do we have a good understanding of the mechanisms that facilitate slow aging, even though this clearly is highly relevant to biogerontology.

ADVANTAGES AND DISADVANTAGES OF THE COMPARATIVE APPROACH Comparative differences among species with disparate longevity provide a powerful source of putative mechanistic traits that can be further explored in depth; however, uncritical application may lead to numerous erroneous generalizations. This is especially true when comparisons are made between organisms with different complexities of biological organization and where specialized functional adaptations have evolved.

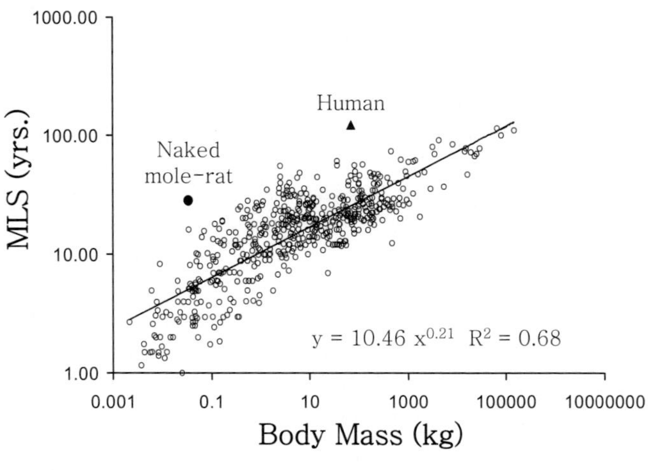

Figure 52–3. The allometric relationship between body size and maximum life span (MLS) for 500 nonvolant mammals. Data from the various orders generally fall close to the descriptor ($\text{MLS} = \text{M}^{0.21}$); notable exceptions are humans (triangles) and naked mole-rats (circles).

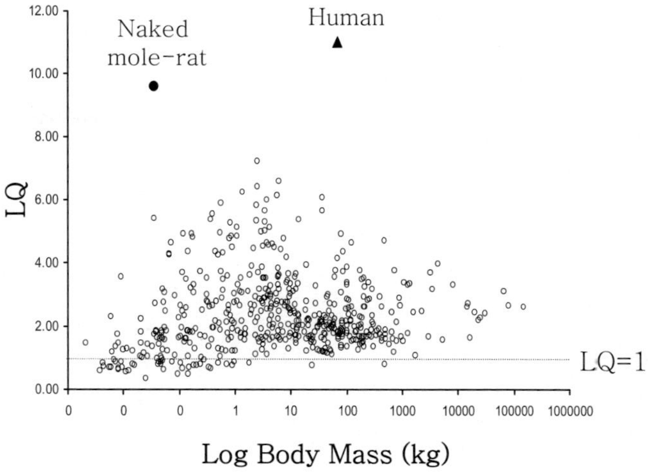

Figure 52–4. The longevity quotient (LQ; the ratio of observed maximum life span to that predicted by body mass using the equation of Prothero and Jurgens[66]; $y = 5.3 \times \text{mass}^{0.174}$). Both naked mole-rats and humans have exceptionally high LQs.

Two problems routinely plague comparative studies: first, the covariation of a specific trait with body mass is frequently ignored when interpreting findings, and second, phylogenetic relatedness between species may affect the statistical independence of data suites. Indeed, it is an axiom of comparative biology that species with a close evolutionary relationship share more traits than more distantly related species and comparative studies ignoring phylogeny may lead to spurious conclusions. Speakman[64] highlighted this problem in a hypothetical study outlined below (Figure 52–5) comparing a trait that appears to be well correlated with both maximum life span and body mass in nine species. Although data appear well correlated and even residual analyses show significant relationships, careful scrutiny reveals that data from species within the same order cluster together. In effect this hypothetical study pseudoreplicates data, such that although nine species were examined, in actuality it is only three orders with mutation events that clearly are not shared between orders. This problem of overestimating the number of degrees of freedom has long been recognized in comparative biology, and is overcome with sophisticated statistical methods (phylogenetically independent contrasts). There are, however, several situations in which it is more appropriate to use nonphylogenetic analyses such as when traits imply ancestral and nonevolved conditions.[67] While modern biological research occurs in an implicitly evolutionary context, it nevertheless is still possible to conduct comparative studies without taking the phylogeny into account, provided the aims of the question and the extrapolations thereof are carefully constrained. When traits are shared across taxa and classes, and these appear to be modified in tandem by some experimental intervention, with greater confidence it can be concluded that this is an important and conserved pathway, such as is evidenced with DR and reduced insulin/IGF-1 signaling in worms, flies, and mice.

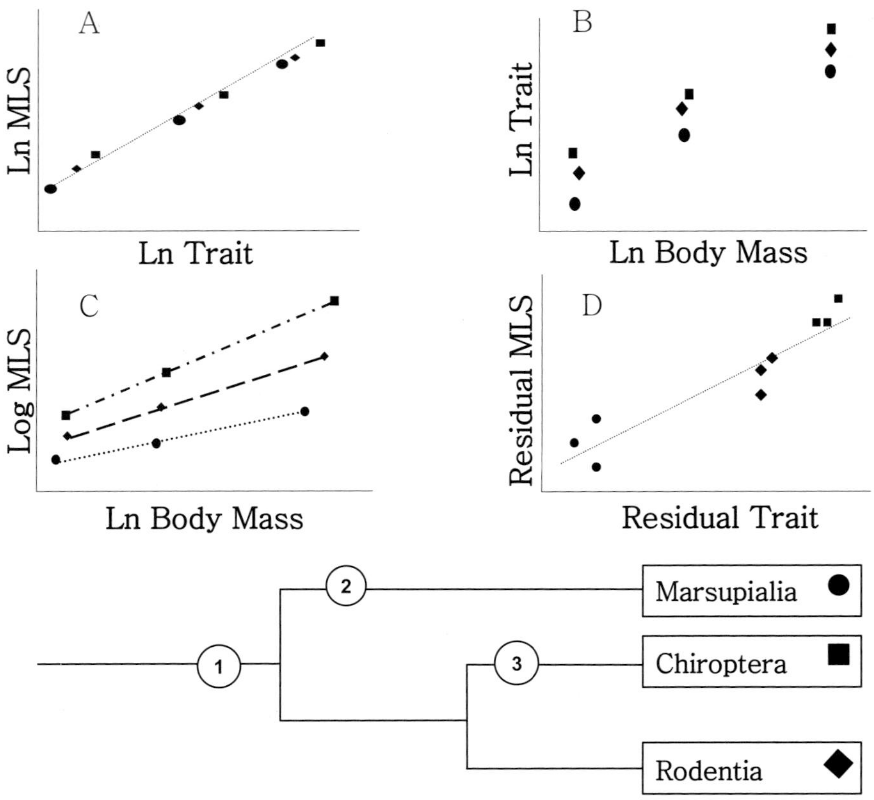

Figure 52–5. Hypothetical relationship based on nine species [three bats (squares), three rodents (diamonds), and three marsupials (circles)], modified from Speakman.[64] (A) The relationship between log MLS and log mass; (B) the relationship of a trait thought to be involved in aging as a function of body mass; (C) the relationship between log MLS and that trait, such that these raw data show a significant relationship between the trait and MLS. (D) The residual variation of the trait in (B) plotted against the residual MLS. This figure reveals a significant relationship between the trait and MLS; however, close scrutiny reveals that data are clustered by phylogeny. (E) The hypothetical phylogenetic tree of the nine species. Mutations (1, 2, and 3) affect all species downstream from that point; clearly sampling points downstream of the mutation points do not give phylogenetically independent samples and may lead to spurious conclusions.

WHICH ANIMALS TO STUDY? Organisms that live at least twice as long as expected from allometry (Figure 52–4), such as bats and mole-rats, as well as birds, fish, and social insects, are good candidates for discerning mechanisms employed in slow aging.[9] Less can be gleaned from animals with longevity quotients (the ratio of reported MLS to that predicted by mass) less than one, for it is extremely difficult to discriminate the effects due to accelerated aging from altered susceptibility to disease and inadequate husbandry of captive exotic species.

The evolutionary theory of aging posits that aging results from the declining power of natural selection to favor advantageous alleles, or eliminate deleterious ones at successive ages after sexual maturity.[43] Thus, animals that experience high extrinsic mortality due to living in challenging environments (e.g., climatic conditions and/or high predation risk) will evolve life history traits that facilitate early reproduction, often at the expense of somatic maintenance, and will have short life spans, and vice versa.[43] It can be predicted that animals that can avoid predation, such as volant species (e.g., birds and bats), or those living in protected areas (e.g., subterranean mole-rats, social insects) will have low extrinsic mortality and evolve mechanisms facilitating extended longevity. Captive animals provide an opportunity to observe intrinsic mortality away from confounding problems associated with extrinsic mortality. Studies based upon comparative differences among species with rapid and retarded rates of aging may elucidate mechanistic differences.

INTERESTING EXCEPTIONS TO FAVORED HYPOTHESES IN NONTRADITIONAL MODELS FOR AGING RESEARCH The disposable soma theory posits that aging is ultimately due to the tradeoff between partitioning energy into somatic maintenance and reproduction.[68] In favor of reproduction, somatic maintenance is compromised and random molecular damage is inadequately repaired and accrues, leading to a concomitant decline in organismic function. Surprisingly, both the eusocial insects[69] and rodents[12] do not support this theory. In both insect and naked mole-rat colonies, the breeding animals live at least as long, if not considerably longer, than their nonbreeding relatives.[12,69] Similarly, many fish, amphibians, and reptiles fail to show signs of aging and maintain reproductive potential throughout their long lives.[70,71]

The free radical theory asserts that oxidative damage results from an imbalance between endogenous ROS and their neutralization by antioxidants, as well as removal of any incurred damage by organismal repair systems.[72] Some comparative studies have shown that both the generation of oxidative damage[73] and accrual are negatively correlated with species longevity.[74,75] However, naked mole-rat data show similar rates of ROS generation and paradoxically higher levels of accrued oxidative damage even at a young age than shorter-living rodents.[76] Another exception is that in eusocial insects antioxidant defenses are not superior in "queens," compared to "workers," and oxidative damage may accrue at similar rates.[77,78] Honeybee queens produce greater amounts of vitellogenin, a yolk precursor protein that reportedly has strong antioxidative properties,[79] and this may be linked to queen longevity.[80] Finally, Hamilton and collaborators[81] have reported higher oxidative damage in long-living birds than in shorter-living rodents. Consequently, while the oxidative stress theory may play an important role in homeostatic imbalance with aging, there are likely other factors that modulate both it and rates of aging.

CONCLUSIONS

The use of animal models in aging research has advanced the field considerably and continues to hold great promise for further advancing our knowledge of the aging process, as well as finding mechanisms to retard aging. However, much more work is needed to identify novel model organisms for aging research that may provide new insights into successful aging and compare the ubiquity of findings thereof with traditional models. There is a critical need for appropriate use of the comparative method and the development of better cross-species databases that make it possible to test the ubiquity of theories, mechanisms, and their potential translation for human application.

ACKNOWLEDGMENTS

Funding from the NIH/NIA to R. Buffenstein (NIH/NIA AG 022891) is gratefully acknowledged. Blazej Andziak is sincerely thanked for his assistance.

REFERENCES

1. Pearl R. *The Rate of Living*. New York: A. A Knopf, 1928.
2. Krogh A. The progress of physiology. *Science* 1929;70:200–204.
3. Ricklefs RE, Finch CE. *Aging: A Natural History*. New York: Scientific American Library, HPHLP, 1995.
4. Tatar M, Khazaeli AA, Curtsinger JW. Chaperoning extended life. *Nature* 1997;390:30.
5. Guarente L, Kenyon C. Genetic pathways that regulate ageing in model organisms. *Nature* 2000;408:255–262.
6. Liang H, Masoro EJ, Nelson JF, Strong R, McMahan CA, Richardson A. Genetic mouse models of extended lifespan. *Exp Gerontol* 2003;38:1353–1364.
7. Finch CE. *Longevity, Senescence and the Genome*. Chicago, IL: Chicago University Press, 1990.
8. Miller RA, Nadon NL. Principles of animal use for gerontological research. *J Gerontol A Biol Sci Med Sci* 2000;55:B117–123.
9. Austad SN. An experimental paradigm for the study of slowly aging organisms. *Exp Gerontol* 2001;36:599–605.
10. Strehler BL. *Time, Cells and Aging*. New York: Academic Press, 1962.
11. deBoer J, Hoeijmakers JH. Cancer from the outside, aging from the inside: Mouse models to study the consequences of defective nucleotide excision repair. *Biochimie* 1999;81:127–137.
12. Buffenstein R. The naked mole-rat: A new long-living model for human aging research? *J Gerontol* 2005;60:1369–1377.
13. Nadon NL. Of mice and monkeys: National Institute on Aging resources supporting the use of animal models in biogerontology research. *J Gerontol A Biol Sci Med Sci* 2006;61:813–815.
14. Vaage J, Smith GH, Asch BB, Teramoto Y. Mammary tumorigenesis and tumor morphology in four C3H sublines with or without exogenous mammary tumor virus. *Cancer Res* 1986;46:2096–2100.
15. Gershoff SN. Vitamin C (ascorbic acid): New roles, new requirements. *Nutr Rev* 1993;51:313–326.
16. Washburn DA, Rumbaugh DM. Investigations of rhesus monkey video-task performance: Evidence for enrichment. *Contemp Top Lab Anim Sci* 1992;31:6–10.
17. Miller RA, Buehner G, Chang Y, Harper JM, Sigler R, Smith-Wheelock M. Methionine deficient diet extends mouse lifespan, slows immune and lens aging, alters glucose, T4, IGFI and insulin levels, and increases hepatocyte MIF levels and stress resistance. *Aging Cell* 2005;4:119–125.
18. Weindruch R, Masoro EJ. Concerns about rodent models for aging research. *J Gerontol* 1991;46:B87–88.
19. Austad SN. Introduction to animal models. *Exp Gerontol* 2003;38:1327–1328.
20. Nishina PM, Lowe S, Verstuyft J, Naggert JK, Kuypers FA, Paigen B. Effects of dietary fats from animal and plant sources on diet induced fatty streak lesions in C57BL/6J mice. *J Lipid Res* 1993;34:1413–1422.
21. Miller RA, Austad SN, Burke D, *et al.* Exotic mice as models for aging research: Polemic and prospectus. *Neurobiol Aging* 1999;20: 217–231.
22. Nuzhdin SV, Mackay TF. Direct determination of retrotransposon transposition rates in *Drosophila melanogaster*. *Genet Res* 1994;63: 139–144.
23. Hasty P, Campisi J, Hoeijmakers J, van Steeg H, Vijg J. Aging and genome maintenance: Lessons from the mouse? *Science* 2003;299: 1355–1359.
24. Martin GM. Genetic engineering of mice to test the oxidative damage theory of aging. *Ann NY Acad Sci* 2005;1055:26–34.
25. Masoro EJ. *Caloric Restriction: A Key to Understanding and Modulating Aging*. Amsterdam, The Netherlands: Elsevier, 2002.
26. Jazwinski SM. The RAS genes: A homeostatic device in *Saccharomyces cerevisiae* longevity. *Neurobiol Aging* 1999;20:471–478.
27. Karberlein M, McVey M, Guarente L. The SIR2/3/4 complex and SIR2 alone promote longevity in *Saccharomyces cerevisiae* by two different mechanisms. *Genes Dev* 1999;13:2570–2580.
28. Friedman DB, Johnson TE. A mutation in the *age-1* gene in *C. elegans* lengthens life and reduces hermaphrodite fertility. *Genetics* 1988;118:75–86.
29. Arking R. Successful selection for increased longevity in *Drosophila*: Analysis of the survival data and presentation of a hypothesis on the genetic regulation of longevity. *Exp Gerontol* 1987;22:199–220.
30. Brown-Borg HM, Borg KE, Meliska CJ, Bartke A. Dwarf mice and the ageing process. *Nature* 1996;384:33.
31. Kenyon C. A conserved regulatory system for aging. *Cell* 2001;105: 165–168.
32. Kirk L. Dietary restriction and aging: Comparative tests of evolutionary hypotheses. *J Gerontol A Biol Sci Med Sci* 2001;56: B123–129.
33. Carey JR, Liedo P, Harshman L, *et al.* Life history response of Mediterranean fruit flies to dietary restriction. *Aging Cell* 2002;1: 140–148.
34. Mair W, Piper MD, Partridge L. Calories do not explain extension of lifespan by dietary restriction in *Drosophila*. *PloS Biol* 2005;3: e223.
35. Sohal RS, Weindruch R. Oxidative stress, caloric restriction, and aging. *Science* 1996;273:59–63.
36. Breese CR, Ingram RL, Sonntag WE. Influence of age and long-term dietary restriction on plasma insulin like growth factor 1 (IGF1), IGF1 gene expression, and IGF1 binding proteins. *J Gerontol* 1991;46:B180–B187.
37. Zimmerman JA, Malloy V, Krajcik R, Orentreich N. Nutritional control of aging. *Exp Gerontol* 2003;38:47–52.
38. Heilbronn LK, Ravussin E. Calorie restriction and aging: Review of the literature and implications for studies in humans. *Am J Clin Nutr* 2003;78:361–369.
39. Levine RL, Moskovitz J, Stadtman ER. Oxidation of methionine in proteins: Roles in antioxidant defense and cellular regulation. *IUBMB Life* 2000;50:301–307.
40. Pamplona R, Barja G. Mitochondrial oxidative stress, aging and caloric restriction: The protein and methionine connection. *Biochim Biophys Acta* 2006;1757:496–508.
41. Larsen PL, Albert PS, Riddle DL. Genes that regulate both development and longevity in *Caenorhabditis elegans*. *Genetics* 1995;139: 1567–1583.
42. Bartke A, Wright CJ, Mattison JA, Ingram DK, Miller RA, Roth GS. Extending the lifespan of long lived mice. *Nature* 2001;414: 412.
43. Kirkwood TBL, Austad SN. Why do we age? *Nature* 2000;408: 233–238.
44. Reznick D, Buckwalter G, Groff J, Elder D. The evolution of senescence in natural populations of guppies (*Poecilia reticulata*): A comparative approach. *Exp Gerontol* 2001;36:791–812.
45. Medawar PB. *An Unsolved Problem of Biology*. London: H. K. Lewis, 1952.

46. Rincon M, Muzumdar R, Atzmon G, Barzilai N. The paradox of the insulin/IGF1 signaling pathway in longevity. *Mech Ageing Dev* 2004;125:397–403.
47. McCulloch D, Gems D. Body size, insulin/IGF signaling and aging in the nematode. *Exp Gerontol* 2003;38:129–136.
48. Houthoofd K, Fidalgo MA, Hoogewijs D, *et al.* Metabolism, physiology and stress defense in three aging Ins/IGF1 mutants of the nematode *Caenorhabditis elegans*. *Aging Cell* 2005;4:87–95.
49. Clancy DJ, Gems D, Harshman LG, *et al.* Extension of lifespan by loss of CHICO, a *Drosophila* insulin receptor substrate protein. *Science* 2001;292:104–106.
50. Migliaccio E, Giorgio M, Mele S, *et al.* The p66shc adaptor protein controls oxidative stress response and life span in mammals. *Nature* 1999;402:309–313.
51. Flurkey K, Papaconstantinou J, Harrison DE. The Snell dwarf mutation Pit1(dw) can increase lifespan in mice. *Mech Ageing Dev* 2002; 123:121–130.
52. Walker GA, Lithgow G. Lifespan extension in *C. elegans* by molecular chaperone dependent upon insulin-like signals. *Aging Cells* 2003;2:131–139.
53. Johnson TE. Genes, phenes, and dreams of immortality: The 2003 Kleemeier Award lecture. *J Gerontol A Biol Sci Med Sci* 2005;60: 680–687.
54. Partridge L, Gems D. Beyond the evolutionary theory of ageing, from functional genomics to evo-gero. *Trends Ecol Evol* 2006;21: 334–340.
55. Salmon AB, Murakami S, Bartke A, Kopchick J, Yasumura K, Miller RA. Fibroblast cell lines from young adult mice of long-lived mutant strains are resistant to multiple forms of stress. *Am J Physiol Endocrinol Metab* 2005;289:E23–29.
56. Hauck SJ, Aaron JM, Wright C, Kopchick JJ, Bartke A. Antioxidant enzymes, free radical damage, and response to paraquat in liver and kidney of long-living growth hormone receptor/binding protein gene disrupted mice. *Horm Metab Res* 2002;34:481–486.
57. Van Remmen H, Ikeno Y, Hamilton M, *et al.* Life long reduction in MnSOD activity results in increased DNA damage and higher incidence of cancer but does not accelerate aging. *Physiol Genomics* 2003;16:29–37.
58. Sibley CG, Ahlquist JE. DNA hybridization evidence of hominoid phylogeny: Results from an expanded data set. *J Mol Evol* 1987;26: 99–121.
59. Roth GS, Mattison JA, Ottinger MA, Chachich ME, Lane MA, Ingram DK. Aging in rhesus monkeys: Relevance to human health interventions. *Science* 2004;305:1423–1426.
60. Rapp PR, Morrison JH, Roberts JA. Cyclic estrogen replacement improves cognitive function in aged ovariectomized rhesus monkeys. *J Neurosci* 2003;23:5708–5714.
61. Small SA, Chawla MK, Buonocore M, Rapp PR, Barnes CA. Imaging correlates of brain function in monkeys and rats isolates a hippocampal subregion differentially vulnerable to aging. *Proc Natl Acad Sci USA* 2004;101:7181–7186.
62. Smith DE, Rapp PR, McKay HM, Roberts JA, Tuszynski MH. Memory impairment in aged primates is associated with focal death of cortical neurons and atrophy of subcortical neurons. *J Neurosci* 2004;24:4373–4381.
63. Zeng BY, Medhurst AD, Jackson M, Rose S, Jenner P. Proteasomal activity in brain differs between species and brain regions and changes with age. *Mech Ageing Dev* 2005;126:760–766.
64. Speakman JR. Body size, energy metabolism and lifespan. *J Exp Biol* 2005;208:1717–1730.
65. Austad SN, Fischer KE. Mammalian aging, metabolism, and ecology: Evidence from the bats and marsupials. *J Gerontol* 1991;46: B47–B53.
66. Prothero J, Jurgens KD. Scaling of maximal lifespan in mammals: A review. *Basic Life Sci* 1987;42:49–74.
67. Freckleton RP, Harvey PH, Pagel M. Phylogenetic analysis and comparative data: A test and review of evidence. *Am Nat* 2002;160:712–726.
68. Kirkwood TB. Evolution of ageing. *Nature* 1977;24:301–304.
69. Keller L, Jemielty S. Social insects as a model to study the molecular basis of ageing. *Exp Gerontol* 2006;41:553–556.
70. DeMagalhaes JP, Toussaint O. The evolution of mammalian aging. *Exp Gerontol* 2002;37:769–775.
71. Reznick D, Ghalambor C, Nunney L. The evolution of senescence in fish. *Mech Ageing Dev* 2002;123:773–789.
72. Beckman KB, Ames BN. The free radical theory of aging matures. *Physiol Rev* 1998;78:547–581.
73. Perez-Campo R, Lopez-Torres M, Rojas C, Cadenas S, Barja G. Longevity and antioxidant enzymes, non-enzymatic antioxidants and oxidative stress in the vertebrate lung: A comparative study. *J Comp Physiol B* 1994;163:682–689.
74. Barja G, Herrero A. Oxidative damage to mitochondria DNA is inversely related to maximum lifespan in the heart and brain of mammals. *FASEB J* 2000;14:312–318.
75. Pamplona R, Portero-Otin M, Riba D, Requena JR, Thorpe SR, Lopez-Torres M, Barja G. Low fatty acid unsaturation: A mechanism for lowered lipoperoxidative modification of tissue proteins in mammalian species with long life spans. *J Gerontol* 2000;55A: B286–B291.
76. Andziak B, O'Connor TP, Qi W, DeWaal EM, Pierce A, Chaudhuri AR, Van Remmen H, Buffenstein R. High oxidative damage levels in the longest-living rodent, the naked mole-rat. *Aging Cell* 2006; 5(6):463–471.
77. Parker JD, Parker KM, Sohal BH, Sohal RS, Keller L. Decreased expression of CuZn superoxide dismutase1 in ants with extreme lifespan. *Proc Natl Acad Sci USA* 2004;101:3486–3489.
78. Corona M, Hughes KA, Weaver DB, Robinson GE. Gene expression patterns associated with queen honey bee longevity. *Mech Ageing Dev* 2005;126:1230–1238.
79. Amdam GV, Omholt SW. The regulatory anatomy of honeybee lifespan. *J Theor Biol* 2002;216:209–228.
80. Seehuus SC, Norberg K, Gimsa U, Krekling T, Amdam GV. Reproductive protein protects functionally sterile honey bee workers from oxidative stress. *Proc Natl Acad Sci USA* 2006;103:962–967.
81. Hamilton ML, Guo Z, Fuller CD, Van Remmen H, Ward WF, Austad SN, Troyer DA, Thompson I, Richardson A. Are liable assessment of 8-oxo-2-deoxyguanosine levels in nuclear and mitochondrial DNA using the sodium iodide method to isolate DNA. *Nucleic Acids Res* 2001;29:2117–2126.

Models of Behavior

V

Genetics

E

53 Gene Targeting in Human Somatic Cells

ERIC A. HENDRICKSON

ABSTRACT

Somatic gene targeting in human cells has two general applications of importance and wide interest. One is the inactivation of genes ("knockouts"), a process utilized to delineate the loss-of-function phenotype(s) of a particular gene. The second application is the process of gene therapy, which involves correcting a preexisting mutated allele(s) of a gene back to wild-type in order to ameliorate some pathological phenotype associated with the mutation. Both of these processes require a form of DNA double-strand break repair known as homologous recombination. Although bacteria and lower eukaryotes utilize homologous recombination almost exclusively, a competing process, known as nonhomologous end joining, predominates in higher eukaryotes and was presumed to prevent the use of gene targeting in human somatic cells in culture. A series of molecular and technical advances developed in the 1990s disproved this notion, but still resulted in a process that was cumbersome, labor intensive, highly inefficient, and slow. Within the past 5 years, the use of new gene delivery vectors such as recombinant adeno-associated virus and the identification of cell lines such as Nalm-6 that appear to undergo high rates of gene targeting have significantly brightened the outlook for this field and resulted in a gene delivery system that facilitates both gene knockouts and gene therapy modifications at robust levels. Thus, gene targeting in human somatic cells in culture has become not only feasible, but also relatively facile, and it harbingers a golden age for directed mutagenesis.

Key Words: Gene targeting, Gene knockouts, Gene therapy, DNA double-strand breaks, Nonhomologous end joining, Homologous recombination, Recombinant adeno-associated virus, HCT116 cells, NALM-6 cells.

INTRODUCTION

Somatic gene targeting is the intentional modification of a genetic locus in a living cell.[1,2] This technology has two general applications of importance and wide interest. One is the inactivation of genes ("knockouts"), a process in which the two wild-type alleles of a gene are mutated in order to delineate the loss-of-function phenotype(s) of that particular gene.[3] The second application is the clinically more relevant process of gene therapy, which, in its strictest sense, involves correcting a preexisting mutated allele(s) of a gene back to wild-type in order to ameliorate some pathological phenotype associated with the mutation.[4] Importantly, although these applications are—at the DNA level—reciprocal opposites of one another, they are mechanistically identical and utilize the same four basic steps: (1) a search for homologous sequences between the incoming donor DNA and the chromosomal DNA, (2) breakage [usually in the form of double-stranded breaks (DSBs)] of the DNA at the site of targeting, (3) exchange of DNA/genetic information between the donor DNA and the chromosomal DNA, and (4) ligation of the broken chromosome to restore its structural integrity. Together, these four steps define a process referred to as homologous recombination (HR), which is absolutely required for gene targeting to occur.[5,6] Although HR was known to predominate in bacteria and lower eukaryotes, the competing process of nonhomologous end joining (NHEJ), in which the incoming donor DNA is randomly integrated within the genome, predominates in higher eukaryotes and was presumed to prevent the use of gene targeting in human somatic cells in culture.[7] A series of molecular and technical advances disproved this notion, but still resulted in a process that was cumbersome, labor intensive, highly inefficient, and slow.[8] Within the past 5 years, the use of new gene delivery vectors such as recombinant adeno-associated virus (rAAV)[9] and the identification of cell lines such as Nalm-6 that appear to undergo high rates of gene targeting[10] have significantly brightened the outlook for this field and resulted in a gene delivery system that facilitates both gene knockouts and gene therapy modifications at robust levels. Thus, gene targeting in human somatic cells in culture has become not only feasible, but relatively facile, and it harbingers a golden age for directed evolution.

BACKGROUND

NONHOMOLOGOUS END JOINING A double-stranded piece of linear DNA introduced into a cell can be incorporated into that cell's genome by either HR or NHEJ.[11] Bacteria and lower eukaryotes almost exclusively utilize HR for the uptake of foreign DNA. In higher eukaryotes, however, integration proceeds more frequently by a process that does not require extended regions of homology. Specifically, mammalian cells—and humans in particular—have evolved a highly efficient ability to join nonhomologous DNA molecules together.[7,12,13] In their seminal work on gene targeting, Capecchi and co-workers showed that although somatic mammalian cells can integrate a linear duplex DNA into corresponding homologous chromosomal sequences using HR,

From: *Sourcebook of Models for Biomedical Research*
(P. M. Conn, ed.), © 2008 Humana Press Inc., Totowa, NJ.

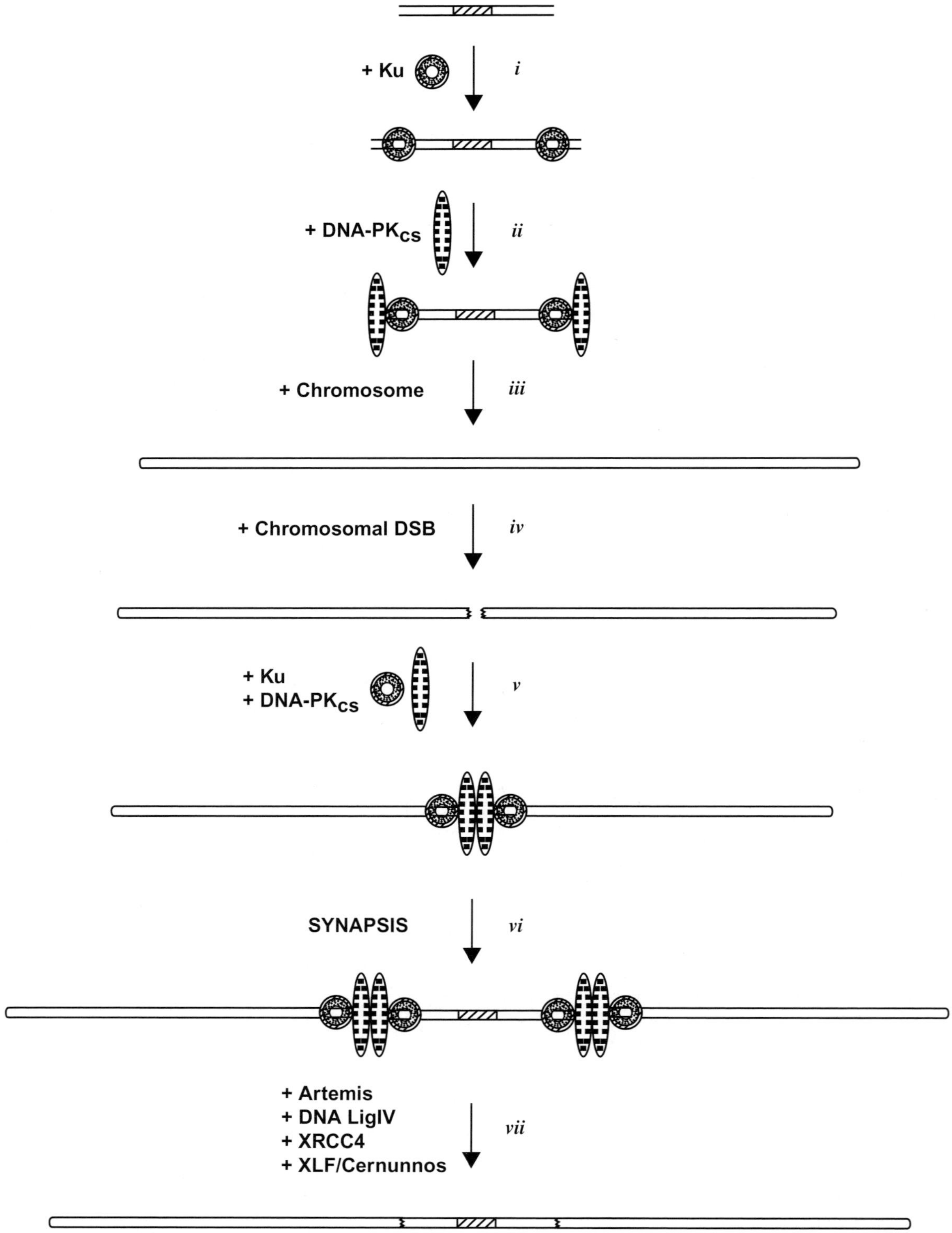

Figure 53–1. A hypothetical pathway for the role of NHEJ factors in random integration. The double line represents a transfected donor dsDNA that has homology to some location within the recipient cell's genome and the hatched box represents a positive drug selection marker. (*i*) The Ku heterodimer (stippled circle) binds onto the ends of the donor DNA. (*ii*) Ku recruits DNA-PK$_{cs}$ (hatched oval) to the ends of the donor DNA. (*iii*) The donor DNA, complexed with DNA-PK, now interacts with a cellular chromosome (long double line with hairpinned ends). (*iv*) By an unknown process, a DSB is introduced into the chromosome (jagged slash). (*v*) DNA-PK complex components then assemble on the ends of the chromosomal DNA. (*vi*) The donor DNA, complexed with DNA-PK, synapses with the chromosomal DNA, also complexed with DNA-PK, in a process that probably involves DNA-PK$_{cs}$: DNA-PK$_{cs}$ homotypic interactions. (*vii*) DNA-PK activates the downstream effectors Artemis, DNA LIGIV, XRCC4, and XLF/Cernunnos, which act in concert to repair the chromosomal DSB (jagged slashes), resulting in the random introduction of the donor DNA into the chromosome.

the frequency with which recombination into nonhomologous sequences occurred was at least 1000-fold greater.[14] This NHEJ pathway appears to be predominantly active during the G_1/early S phase of the cell cycle.[15,16] Given that NHEJ DSB repair is generally error prone, an attribute that bacteria and lower eukaryotes can ill afford, the increased percentage of noncoding DNA in higher eukaryotes may have facilitated the evolution of this pathway.

While the many details of NHEJ remain to be worked out, a reasonable model is that following the introduction of a linear double-stranded DNA (dsDNA) into a cell, the Ku86 : Ku70 heterodimer[17] binds to the broken DNA ends to prevent unnecessary DNA degradation[18–20] (Figure 53–1, *i*). The binding of Ku to the free DNA ends recruits and activates the DNA-dependent protein kinase complex catalytic subunit[21,22] (DNA-PK_{cs}) (Figure 53–1, *ii*). This DNA is then brought into contact with a chromosome into which a DSB is introduced (Figure 53–1, *iii* and *iv*). This step(s) is poorly understood. The chromosomal DSB ends are almost certainly also occupied by Ku and DNA-PK_{cs} (Figure 53–1, *v*) and this probably facilitates the formation of a synaptic complex with the donor DNA (Figure 53–1, *vi*). DNA-PK_{cs} : DNA-PK_{cs} homotypic interactions are the critical feature required for synapsis.[23–25] Once DNA-PK_{cs} is properly assembled at the broken ends it, in turn, recruits and activates a nuclease, Artemis[26–28] (Figure 53–1, *vii*), to help trim the damaged DNA ends. The rejoining of the DNA DSB (Figure 53–1, *vii*) requires the recruitment[29] of DNA ligase IV[30,31] (DNA LigIV) and its two accessory factors: X-ray cross-complementing group 4[32,33] (XRCC4) and XRCC4-like factor (XLF)/Cernunnos.[34,35] In summary, mammals are different from bacteria and lower eukaryotes in that DSB repair proceeds primarily through an NHEJ recombinational pathway. Moreover, NHEJ must be overcome in order to facilitate gene targeting and this can occur when the incoming DNA is shunted into the HR pathway.

HOMOLOGOUS RECOMBINATION In HR,[6] the DNA ends of the incoming DNA are likely resected to yield 3′-single-stranded DNA overhangs (Figure 53–2, *i*). Despite intense inves-

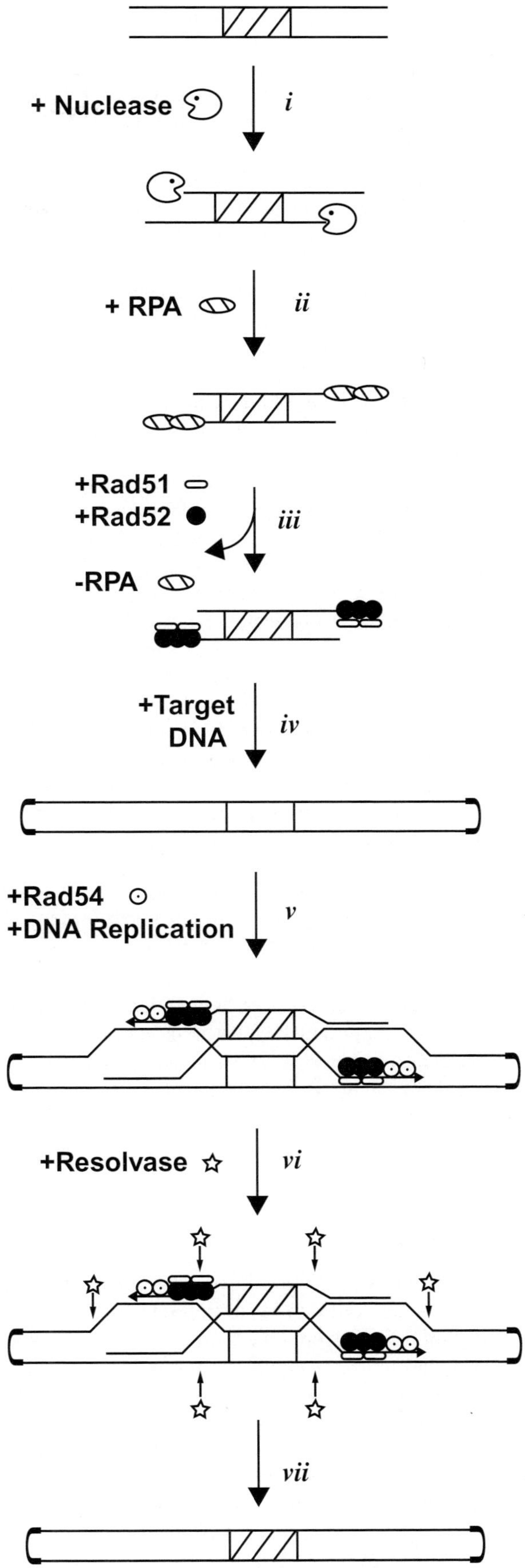

Figure 53–2. A hypothetical pathway for the role of HR factors in gene targeting. The double line represents a transfected donor dsDNA that has homology to some location within the recipient cell's genome and the hatched box represents a positive drug selection marker or a section of DNA containing the researcher's desired modification. (*i*) An unknown nuclease (PacMan™) resects the ends of the donor DNA. (*ii*) RPA (hatched oval circle) then coats the ssDNA ends. (*iii*) Rad51 (empty ellipse) and Rad52 (filled circle) then bind onto the ssDNA ends, displacing RPA in the process. (*iv*) The donor DNA complexed with Rad51 and Rad52 then associates with a chromosome (long double line with hairpinned ends) containing homologous sequences (open box). (*v*) With the assistance of Rad54 (open circle with dot) and DNA replication, the resected ends invade the donor DNA and set up a double Holliday junction. *Note:* The donor DNA shown in (*v*) is rotated with respect to the donor DNA shown in (*iv*) simply for the sake of presentation. (*vi*) The repeated action of a resolvase (star) is then required to complete the recombination process. (*vii*) At the end, the donor DNA has been precisely integrated into a homologous region on an endogenous chromosome. In all panels, the vertical arrows are drawn implying a temporal order to each process, although in many cases the precise sequence of events is not known. In all panels, the italicized Roman numerals refer to steps in each pathway that are described in the text.

tigation, the identity of this nuclease(s) is still undetermined, although the MRN complex [Mre11/Rad50/Nbs1[36] and ExoI (exonuclease I)] has been repeatedly implicated as the likely culprit(s).[37] The resulting overhangs are then coated by replication protein A (RPA), a heterotrimeric single-stranded DNA-binding protein, which removes the secondary structures from the overhangs.[38] RPA subsequently helps to recruit radiation-sensitive 51 (Rad51) and radiation-sensitive 52 (Rad52) to the overhangs, although it is itself displaced in the process (Figure 53–2, *iii*). Rad51 is the key strand-exchange protein in homologous recombination.[39] It is essential for the homology searches on the target DNA, i.e., the entire human genome (Figure 53–2, *iv*), that are required to localize the incoming DNA to its specific, cognate chromosomal counterpart.[40] In humans, there are at least seven Rad51 family members and almost all of them have been implicated in some aspect of HR and also in human disease.[41] Rad52 is an essential accessory factor for Rad51 and it facilitates strand exchange, probably by overcoming the inhibitory role of RPA.[42] Strand invasion into the homologous chromosomal sequence requires radiation-sensitive 54 (Rad54) and DNA replication (Figure 53–2, *v*). Rad54 is a double-stranded DNA-dependent ATPase that can remodel chromatin, and it probably plays critical roles at several steps in the recombination process.[43] In particular, Rad54 is critical for stabilizing the Rad51-dependent joint molecule formation (Figure 53–2, *v*) as well as for promoting the disassembly of Rad51 following exchange.[44] Gene targeting generates a complex structure (Figure 53–2, *v*) that is essentially identical to the linearized plasmid "ends-out" recombination intermediates that have been extensively defined in yeast.[45] Ultimately, the resolution of this structure probably requires the participation of helicases of the recombination defective Q (RecQ) family[46] and certainly requires the action of a resolvase(s) to repeatedly nick the strands (Figure 53–2, *vi*). Although the identity of the human resolvase(s) is still debated, radiation-sensitive 51C (Rad51C) and X-ray cross-complementing 3 (XRCC3) have emerged as the most likely candidates.[47,48] The resolution of the cross-stranded intermediates with crossovers generates a modified chromosome in which the original chromosomal sequences have been precisely replaced by the sequences present on the incoming donor DNA (Figure 53–2, *vii*). In summary, human somatic cells express all of the gene products needed to carry out gene targeting. These events occur, however, at very low frequency due to the preferred usage of NHEJ.

METHODOLOGY

SELECTIONS AND VECTORS Since correct gene targeting is a rare event, strategies needed to be developed in order to detect them. Almost all of these rely on some form of selection.[1,2] Some of the earliest studies utilized naturally occurring loci and selections. Thus, the hypoxanthine phosphoribosyltransferase (HPRT) gene has been a popular choice for study since mutation of this gene can be negatively selected for by the acquisition of resistance to 6-thioguanine (6-TG), a nucleoside analogue that HPRT would otherwise metabolically activate to a toxic compound. Moreover, HPRT is an X-linked gene and male-derived hemizygous cell lines require only one round of targeting to generate the null (6-TG resistance) phenotype.[49–56] However, since most genes of interest do not provide such natural and accommodating selection strategies, the use of positive selection schemes was required. Some esoteric and specialized positive selections have been described, such as relying upon the gain of a cell surface epitope expression and fluorescence-activated cell sorting (FACS) analysis to obtain the correctly targeted cell line.[57] This and similar strategies,[58] however, were not widely applicable and thus they did not come into general use. Instead, the expression of antibiotic resistance genes that can be positively selected for have gained wide usage. One of the first—and by far still most popular and useful—marker selections is the neomycin resistance (NEO) gene. NEO encodes an enzyme that in bacteria confers resistance to the neomycin family of antibiotics, which kill by inhibiting protein translation. In mammals, the expression of NEO provides resistance to a similar compound called gentamycin 418 (G418). Thus, the inclusion of NEO on the donor DNA can be used to select for those cells that have stably taken up the DNA based upon their conversion to G418 resistance (Figure 53–3A). Similarly, other antibiotic resistance genes have been pirated for use in human gene targeting, including hygromycin (HYG), puromycin (PUR), histidinol (HIS), and blasticidin (BSD). Since two rounds of gene targeting are necessary to disrupt a diploid locus, investigators generally incorporate two of these markers into otherwise identical gene-targeting vectors to permit sequential selection for the modification of both alleles.[59–62]

Even with strong positive selection schemes, the majority of clones resulting from such an experiment are randomly targeted. This results in a time-consuming and laborious second step in which all of the drug-resistant colonies must be screened at the molecular level [usually either by polymerase chain reaction (PCR) or by Southern blotting analysis] in order to identify the rare correctly targeted clone. To enrich for the desired recombination reactions, gene targeting methodology was subsequently improved by the use of positive : negative selection (PNS) vectors.[1,2] These vectors provided the same positive selection schemes that the earlier vectors possessed, but they also incorporated a negative selection against nonhomologous integrations. This method consists of attaching a genetic element that can be selected against onto the 5′- and/or 3′-flank of the targeting vector. If this vector is randomly integrated into the host cell using NHEJ then the negative marker will be retained and this can be used to eliminate these cells from the analysis by a simple drug selection. If the vector integrates via HR, then the negative marker, being nonhomologous to the targeted locus, will be lost during the integration event (Figure 53–3B). The most commonly used negative marker is herpes simplex virus thymidine kinase (HSV TK). Although TK genes are widely conserved and ubiquitously expressed among all species, the viral TK genes have a slightly altered substrate specificity and this difference has been experimental exploited. Thus, the antiviral drug gangcyclovir (GANC) can be metabolized to a toxic analogue by the viral TK, but not by the cellular enzyme. Therefore, cells containing the viral TK gene (at randomly targeted sites) will be killed by GANC exposure, whereas cells expressing only the cellular TK gene (homologously targeted clones) will survive.[63,64] While HSV TK is still widely in use and appears to be adequate for most gene-targeting strategies, even stronger negative selections, such as using the diphtheria toxin (DT) gene, can significantly enhance the targeting frequency.[64]

Even with these improvements, the isolation of correctly targeted clones was a daunting affair and generally was accomplished only by the most obstinate of researchers. A huge improvement (by one to two orders of magnitude) came with the development of promoterless selection. These vectors are almost identical to the original positive selection with the important exception that the promoter driving the expression of the selectable marker has been deleted.[1,2] As a consequence of this, the selectable marker can be expressed only if it correctly integrates and/or if it is inserted randomly next to or within an actively transcribing gene. Since this requirement is not often met, the total number of drug-resistant clones obtained is significantly reduced and therefore so is the work required to characterize them. To facilitate the expression of the selectable marker, a slice acceptor sequence is often inserted 5′ of the marker gene (Figure 53–3C). This allows the selectable marker, if it is in frame, to be expressed as a fusion protein with the preceding endogenous exons (Figure 53–3C). Alternatively, an internal ribosome entry site (IRES) can be included 5′ to the splice acceptor to facilitate independent expression of the selectable marker. This is an important consideration, especially when exons at the 3'-end of a locus are being targeted.[65] Lastly, empirical observation suggests that it is probably most efficacious to combine the positive promoterless selection strategy with a negative selection (Figure 53–3D).[64,66] This is especially true for hard-to-transfect cells, such as human embryonic stem cell lines.[54]

SITE-SPECIFIC CRE : LOXP RECOMBINATION Although the promoterless vectors utilizing antibiotic drug resistance selections described above represent the current state of the art, they are still somewhat limited in their utility. For example, bold targeting schemes in which multiple loci are to be knocked out or modified would not be possible since the investigator would ultimately run out of selectable markers. In addition, essential genes cannot be directly knocked out since successful targeting would result in dead cells. These technical and biological difficulties, respectively, have been overcome through the use of the site-specific recombination system, cyclization recombination : locus of crossing over (x), P1 (Cre : LoxP).[67,68] Cre is a member of the integrase family of recombinases and it is required in its native state for a productive infection by the bacteriophage, P1.[69] Cre binds and recombines LoxP target sites, which are 34 bp long, that contain an asymmetric 8-bp core flanked by 13-bp inverted repeats. When the LoxP sites are artificially arranged in direct orientation on a chromosome, the ensuing Cre-mediated recombination results in two products: the chromosome with a single

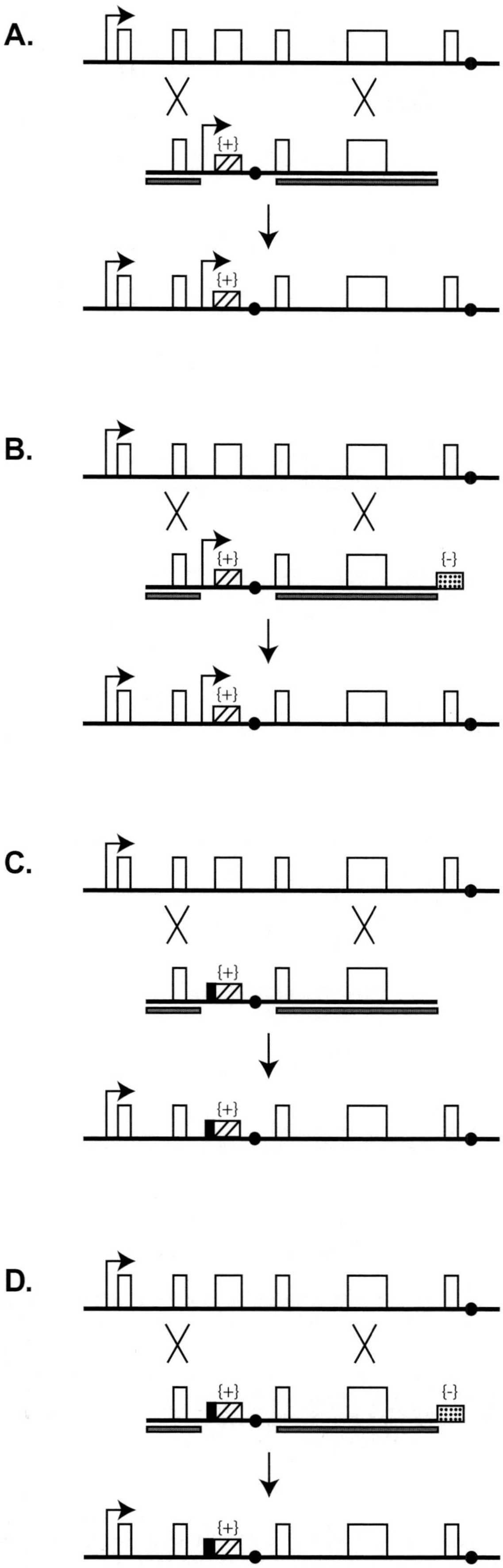

Figure 53–3. Different selection strategies utilized during classical gene targeting. (A) Positive selection; (B) positive/negative selection; (C) promoterless positive selection; (D) promoterless positive/negative selection. The solid line represents chromosomal DNA. The open rectangles represent exons. The 90° bent arrow represents a promoter. The small, solid circle represents a polyadenylation sequence. The large Xs represent crossover recombination reactions. The hatched rectangle with a {+} over it represents a marker for which a positive selection exists. The shaded bars represent regions of homology between the donor DNA and the recipient chromosome. The stippled rectangle with a {−} over it represents a marker for which a negative selection exists. The solid rectangle represents a splice acceptor sequence. In all panels, the vertical arrows are drawn implying a temporal order to each process, although in many cases the precise sequence of events is not known.

residual LoxP site and a circle containing the other LoxP site plus all the genomic DNA that used to lie in between the two loxP sites (Figure 53–4A). This basic recombination reaction can be used to the researcher's advantage to remove the selection marker from an integrated knockout allele. Thus, if loxP sites are built onto the targeting vector within the flanking regions of both sides of the positive selection marker, the marker is said to be "flox"ed (flanking loxP sites). After the marker is used (positive selection) to identify the correct homologous recombination event, it can be removed by the expression of Cre (Figure 53–4B). Even after

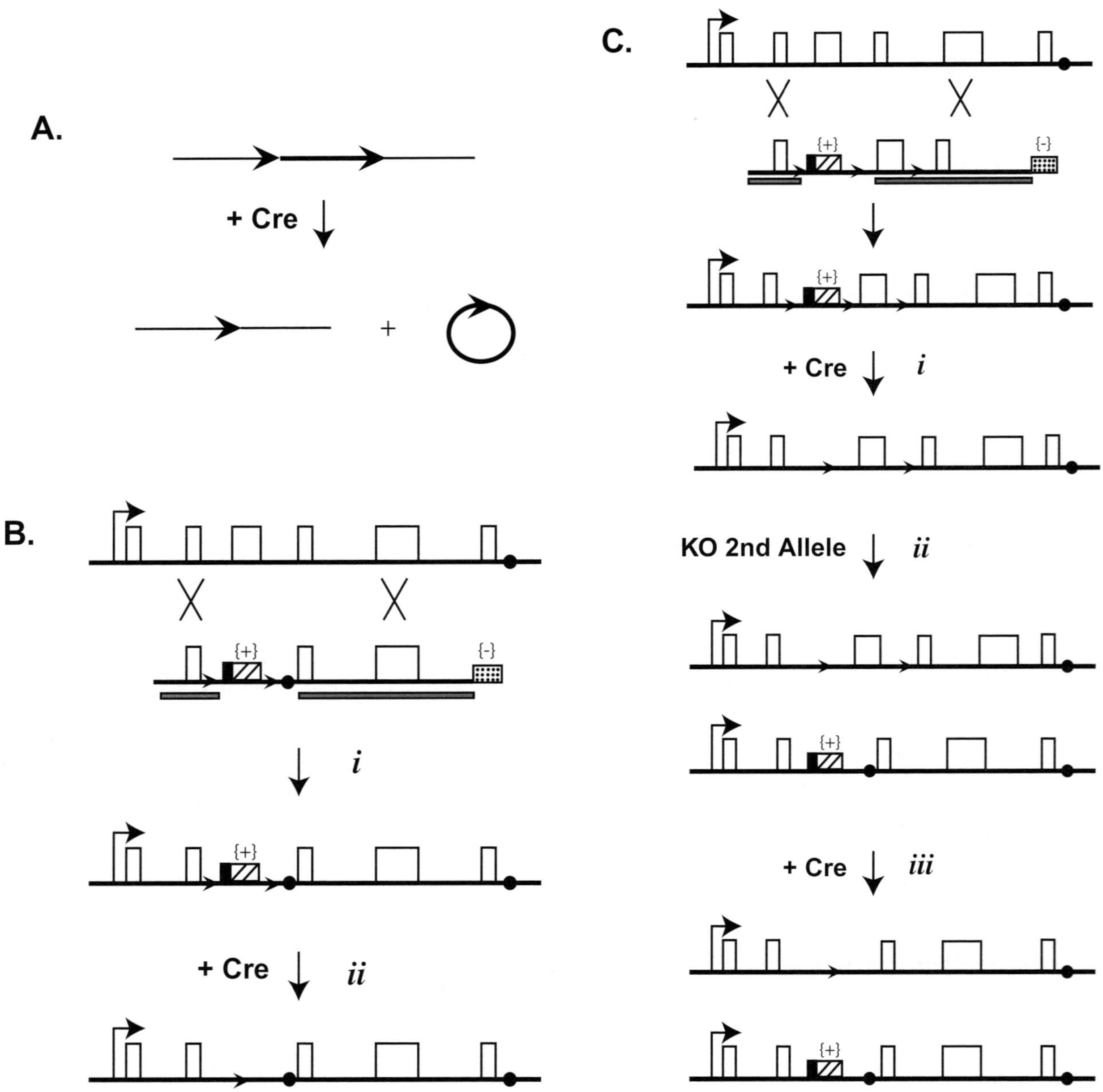

Figure 53–4. The basic Cre recombination reaction and its utility in gene targeting. (A) Cre deletional recombination. The line represents chromosomal DNA and the triangles represent loxP sites. Upon the addition of the Cre recombinase (+ Cre), recombination ensues resulting in a deleted chromosome containing a single loxP sequence and a circle, containing a single loxP sequence and all of the DNA (darker line) that used to exist between the two loxP sequences. (B) Selection marker removal. All symbols, with the exception of the triangles, which represent loxP sites, are as described in the legend to Figure 53–3. (*i*) A promoterless positive/negative gene-targeting event, in which the positive selection marker has been floxed, is shown. The cell harboring this construct is resistant to the drug specified by the marker. (*ii*) The addition of Cre (+ Cre) results in the deletional excision of the selectable marker. The cell harboring this construct is now sensitive to the drug specified by the marker. Note, however, that the allele is nonetheless still nonfunctional as an exon is missing and a polyadenylation sequence has been introduced into the locus. (C) Construction of a conditionally null allele. As in (B), a promoterless positive/negative selection-targeting event is shown. In this instance, the targeting vector contained three loxP sites, two flanking the positive selection marker and two flanking an exon to be deleted. The middle loxP site is shared by both pairs. (*i*) Upon the addition of Cre (+ Cre), deletional recombination between a pair of loxP sites can occur. Although three recombination events are possible, the indicated clone, in which the selection marker has been removed and the exon remains floxed, can be identified by molecular analysis. (*ii*) This cell line is subjected to a second round of gene targeting to create a heterozygous cell line in which the only functional allele is the floxed one. (*iii*) Upon a second exposure to Cre (+ Cre), deletion of the floxed exon occurs and results in a null cell line.

removal of the positive marker, the resulting allele is nonetheless inactive, since the exon(s) removed in the initial targeting event is still missing. Most importantly, however, the cell in which this event has occurred now becomes sensitive to G418. Thus, the original targeting vector and selection scheme can be reused to target the second allele. Using this approach, a positive selectable marker can be reused over and over again to permit a theoretically limitless number of gene targeting events within the same cell line.[64] Thus, in a fashion analogous to murine model systems, in which animal husbandry is utilized to construct strains of mice containing two or more knockout alleles,[70,71] human somatic cell lines can be constructed containing multiple gene disruptions.[10,52,61,62,72]

In a comparable and often utilized scheme, three loxP sites can be built into the targeting vector, such that both the positive selection marker and the exon that ultimately is to be deleted are floxed (Figure 53–4C). After the selection marker is used to identify the correctly targeted clone, the cells are transiently exposed to Cre. A screening step is then utilized to identify those cells in which the selectable marker, but not the exon, was excised by Cre.[63] The resulting cell line is again sensitive to the antibiotic utilized in the positive selection and contains a floxed exon (Figure 53–4C, *i*). These cells can then be subjected to a second round of gene targeting in which the remaining wild-type allele is inactivated in a direct knockout scheme (Figure 53–4C, *ii*), resulting in a "conditionally null" cell line. This cell line is hemizygous for and expresses the gene of interest from an allele that is floxed. Thus, at the researcher's discretion, Cre can be reintroduced into this cell line. This deletes the remaining functional allele and results in a cell line that is null for the gene of interest (Figure 53–4C, *iii*). In this manner, the mechanism of action of essential genes can be studied.[73]

In summary, the bacteriophage P1 Cre : LoxP recombination system has been conscripted by molecular biologists to provide experimental schemes for the reuse of selectable markers and as a means to study essential genes in human somatic cells. This system can also be used to introduce subtle, hypomorphic point mutations into genes[2] and to create complex chromosomal rearrangements[3] for the mouse, but the strategies are identical for human cells.

CELL LINES AND TARGETING FREQUENCIES The choice of cell line to use is of obvious importance when considering a gene targeting approach. Fortunately, a large number of researchers working in disparate fields have, due to the biological questions that interested them, attempted to functionally alter genes in a wide variety of cell lines. For the ease of discussion, these cell lines can be grouped into either (1) the HCT116 cell line, (2) the NALM-6 cell line, and (3) all other cell lines (non-HCT116, non-NALM-6 cell lines). Another important parameter to consider is the expected frequency of targeting. Here, the historical variability is great, ranging from multiple instances of total failure (0% efficiency[59,61,64,74,75]) to a recent report of achieving correct targeting in the only clone analyzed (100% efficiency[10]). These two issues, choice of cell line and targeting efficiency, are discussed in more depth below.

Non-HCT116, Non-NALM-6 Cell Lines The first published report of gene targeting in human cells occurred just over two decades ago. Smithies *et al.* attempted to disrupt the β-globin locus in the human EJ bladder carcinoma cell line.[74] No correctly targeted cell lines were obtained, but the same authors were subsequently able to successfully target the human β-globin locus in a human : mouse hybrid cell line (Table 53–1). Several years later, Jasin *et al.*[57] reported the first isogenic targeting of an endogenous locus (the CD4 gene) in a completely human (the T-lymphocyte JM) cell line.[57] Since that time, a total of 17 different genes (albeit some of them multiple times) have been disrupted, in what amounts to a rather pedestrian ~1 gene per year ratio (Table 53–1). The genes that have been disrupted represent rather diverse areas of biology and include transcription factors (p53), oncogenes (Ki-Ras), cell surface receptors (CD43), and checkpoint effectors (p21), in addition to genes with clear clinical relevance [e.g., Bloom's syndrome (BLM) and cystic fibrosis transmembrane receptor (CFTR)].

As importantly, these 17 genes were disrupted in 17 different cell lines (Table 53–1), reinforcing the belief that almost all human somatic cells adequately express the HR machinery required for this process. In theory, the most relevant cell line to use for gene targeting—such that it might be an appropriate model system for human health—would be a "normal," nonimmortalized, nontransformed cell line. This has, in fact, been accomplished, both for the inactivation of a single allele of the β_2-microglobulin[76] and p53[77] genes as well as for the technically impressive double knockout of the p21 gene.[78] Unfortunately, nonimmortalized cell lines usually have proliferative potential for only 30–40 population doublings before they undergo senescence. Since 30–40 population doublings are also required to successfully carry out two rounds of gene targeting, the resulting cell lines have limited research utility, unless the area of investigation is specifically senescence.[78] Consequently, most researchers have turned to immortalized (and often transformed) cell lines in a "damned if you do and damned if you don't" scenario, begrudgingly losing some clinical and/or general relevance for the sake of obtaining a cell line with which they can actually do experiments. The most popular cell line has been the HT1080 line, which has independently been used 11 different times (Table 53–1). This tumorigenic cell line was derived from a fibrosarcoma biopsy from a 35-year-old male[79] and it contains an activated Ras oncogene.[80] The popularity of this cell line probably has more to do with its practical attributes of a rapid doubling time, good cloning efficiency, and good transfectability than with its developmental origin. Other human cell lines, however, that are routinely found in the laboratory have also been used for gene-targeting study including fibroblastic, epithelial, adherent, and nonadherent lines representing many terminally differentiated states. Of special note was the successful use of human embryonic stem (ES) cells.[54] The ability to genetically modify genes in ES cells is an obligatory step for many gene therapy approaches.[4] While many of the moral and ethical issues of this subject still need to be resolved, this study clearly showed that technically, such experiments are possible.

Overall, 188 correctly targeted gene disruption events have been reported for the 8296 drug-resistant clones carefully examined (Table 53–1). This represents a relatively robust targeting frequency of 2.27%. At this frequency, the amount of work required to disrupt a gene is minimal and can easily be carried out even by a small, graduate student-driven laboratory. However, great variability in the targeting frequency clearly exists. Thus, reports of targeting frequencies of 30–50%[51,54] must be contrasted with reports in which the frequency was orders of magnitude lower.[81,82] Similarly, there is little consistency within a cell line,

Table 53–1
Gene targeting studies in non-HCT116, non-Nalm-6 cell lines

Gene	+[a]	#[b]	*Frequency*[c]	*Geneotype*[d]	*Miscellaneous*[e]	*Reference*
6-16			~0.0007	+/+ > +/–	HeLa cells	125
6-16			~0.0007	+/+ > +/–	HT1080 cells	125
6-16			~0.0007	+/+ > +/–	HT1080 cells	126
6-16	6	49	0.1224	+/– > –/–	HT1080 cells	127
6-16	32	114	0.2807	+/+ > +/–	HT1080 cells	51
6-16	66	117	0.5641	+/+ > +/–	HT1080 cells + RAD51 o/e	51
ADA				/– > –/+	JB and LB cells + oligos	128
ALKP				Transgene	HT1080 cells	53
ApoB	2	400	0.0050	+/+/+ > +/+/–	HepG2 cells	81
B-Raf				+/+ > +/–	HEC1A cells	129
B-Raf				+/– > –/–	HEC1A cells	129
β-Globin	0	~3,000	0.0000	+/+ > +/–	EJ cells	74
β-Globin	1	~700	~0.0014	+/+ > +/–	Hu11 cells	74
β2MG				+/+ > +/–	RPE/RPK cells	76
BLM				+/+ > +/–	HT1080 cells	56
CD4	1	~900	~0.0011	+/+ > +/–	JM cells	57
CD43	1	~450	~0.0022	+/+ > +/–	CEM cells	58
CDC2	3	150	0.0200	+/+ > +/–	HT1080 cells	82
CDC2	3	700	0.0043	+/– > –/–	HT1080 cells + CDC2 o/e	82
CFTR				+/– > +/+	CFE	130
CFTR	0	28	0.0000	+/+/+/+ > +/+/+/–	HT29-18-C1	75
CFTR	0	296	0.0000	+/+/+/+ > +/+/+/–	HT29-18-C1	75
CFTR	1	11	0.0909	+/+/+/+ > +/+/+/–	HT29-18-C1	75
GP91	3	726	0.0041	+/O > –/O	PLB-985 cells	131
HPRT			~0.0002	+/O > –/O	HT1080 cells	53
HPRT			0.0000	+/O > –/O	HT1080 cells	53
HPRT			~0.0001	+/O > –/O	HT1080 cells	53
HPRT				+/O > –/O	HT1080 cells	56
HPRT				–/O > +/O	HT1080 cells ± BLM RNAi	56
HPRT				+/O > –/O	HT1080 cells	50
HPRT			~0.0100	+/O > –/O	HT1080 cells	51
HPRT			~0.0200	+/O > –/O	HT1080 cells + RAD51 o/e	51
HPRT				+/O > –/O	HT1080 cells	49
Ki-Ras	7	24	0.2917	+/–* > +/–	DLD-1 cells	83
Ki-Ras				+/–* > –/–*	DLD-1 cells	83
Ki-Ras				+/–* > +/–	HEC1A cells	129
Ki-Ras				+/–* > –/–*	HEC1A cells	129
Neo	1	78	0.0128	Transgene	EJ cells	132
Neo				Transgene	HT1080 cells	133
p21	3	20	0.1500	+/+ > +/–	LF1 cells	78
p21	1	24	0.0417	+/– > –/–	LF1 cells	78
p53			~0.0500	+/+ > +/–	LF1 cells	77
p53			~0.0600	+/+ > +/–	LF1 cells	77
p53			~0.3100	+/+ > +/–	LF1 cells	77
p53				+/– > +/+	SNB-19 cells + oligos	128
POU5F1	50	159	0.3145	+/+ > +/–	ES cells	54
Totals	188	8,296	0.0227			

[a]+: Number of correctly targeted clones identified.

[b]#: Total number of clones analyzed.

[c]Frequency: Number of correctly targeted clones (+) divided by the total number of clones (#) analyzed.

[d]Genotype: (+) designates wild type; (–) designates mutated; (–*) designates a preexisting mutated allele; (O) designates a hemizygous configuration.

[e]Miscellaneous: (+RAD51 o/e) designates that the experiment was performed in cells overexpressing hRAD51; (+oligos) designates a study done solely by oligonucleotide transfections without a targeting vector; (+CDC2 o/e) designates that the experiment was performed in cells overexpressing hCDC2; (±BLM RNAi) designates a study done in the presence and absence, respectively, of RNAi directed against hBLM1.

with targeting frequencies as high as 28%[51] and as low as 0.01%[53] having been reported for HT1080 cells. Parameters such as chromatin or accessibility differences, frequency of repetitive DNA within a given locus, variable expression of HR gene products between cell lines, use or nonuse of "isogenic" DNA,[8] choice of delivery for the vector, length of homology on the donor DNA, to mention just a few, have all been invoked to explain differences in targeting frequency. Unfortunately, it is not known how important any of these parameters really are. Thus, there currently does not exist an a priori method to determine whether a particular gene can be correctly targeted and, if so, at what frequency. A comprehensive investigation of why some genes "target better" than other genes would greatly benefit the field.

The HCT116 Cell Line In the early 1990s, Shirasawa *et al.* described successful gene targeting of the Ki-Ras gene in two human colon carcinoma cell lines, DLD-1 and HCT116.[83] The laboratories of Drs. Bert Vogelstein and Kenneth Kinzler, which already had an established record in the investigation of colon cancer and genetic instability,[84,85] apparently took this report to heart and initiated their own series of studies in which they disrupted genes thought to be involved in *DNA Repair*/chromosome stability in the HCT116 cell line.[59,77,86–89] The impressive success of these studies caught the attention of the *DNA Repair* community and many other laboratories have subsequently followed in their footsteps. Overall, 22 genes have been disrupted in this cell line in the intervening 13 years (Table 53–2). Some of these genes

Table 53–2
Gene targeting studies in the HCT116 cell line

Gene	+[a]	#[b]	*Frequency*[c]	*Genotype*[d]	*Miscellaneous*[e]	*Reference*
14-3-3σ				+/+ > +/–		86
14-3-3σ				+/– > –/–		86
ATR				+/+ > +/–		73
ATR				+/– > –/–		73
Bax	2	~5,000	~0.0004	+/– > –/–		98
β-Catenin	26	216	0.1204	+/–* > +/–		134
β-Catenin				+/–* > –/–*		134
β-Catenin				+/–* > +/–		99
β-Catenin				+/–* > –/–*		99
β-Catenin	5	576	0.0087	+/–* > +/–		65
β-Catenin				+/–* > –/–*		65
BLM				+/+ > +/–		72
BLM				+/– > –/–		72
DNMT1				+/+ > +/–		87
DNMT1				+/– > –/–		87
DNMT3b				+/+ > +/–		89
DNMT3b				+/– > –/–		89
DNMT3b	1	190	0.0053	+/+ > +/–		72
DNMT3b	12	106	0.1132	+/+ > +/–	In a $BLM^{-/-}$ bkg	72
EME1	2	5,250	0.0004	+/+ > +/–		135
HPRT	4	9,700	0.0004	+/O > –/O	In a $p53^{-/-}$ bkg	52
HPRT	4	12,000	0.0003	+/O > –/O		52
Ki-Ras	9	93	0.0968	+/–* > +/–		83
Ki-Ras				+/–* > –/–*		83
Ku86	0	500	0.0000	+/+ > +/–	HSV TK selection	64
Ku86	2	354	0.0056	+/+ > +/–	DT selection	64
Ku86	3	487	0.0062	+/– > +/–	Retargeting	64
Ku86	2	487	0.0041	–/+ > –/–		64
Ku86	2	229	0.0087	+/+ > +/–	In a $p53^{-/-}$ bkg	62
Ku86	1	54	0.0185	+/+ > +/–		62
Ku86	1	279	0.0036	+/– > +/–	Retargeting	62
Mus81	2	744	0.0027	+/+ > +/–		135
Mus81	2	2,380	0.0008	+/– > –/–		135
ORC2			~0.0250	+/+ > +/–		136
ORC2			~0.0880	+/– > –/–		136
p21	0	232	0.0000	+/+ > +/–	Promoter +	59
p21	37	100	0.3700	+/+ > +/–	Promoterless	59
p21	5	20	0.2500	+/+ > +/–		59
p53	1	600	0.0017	+/+ > +/–		77
p53	1	940	0.0011	+/– > –/–		77
p53	1	1,141	0.0009	+/+ > +/–		72
p53	8	527	0.0152	+/+ > +/–	In a $BLM^{-/-}$ bkg	72

continued

Table 53–2
(continued)

Gene	$+^a$	$\#^b$	*Frequency*[c]	*Genotype*[d]	*Miscellaneous*[e]	*Reference*
RAD51C	5	1,930	0.0026	+/+ > +/–		90
RAD51C	3	1,361	0.0022	+/+ > +/–	In a $XRCC3^{-/-}$ bkg	90
RAD51L2	6	7,197	0.0008	+/+ > +/–	In a $RAD54B^{+/+/+}$ bkg	61
RAD51L2	6	5,948	0.0010	+/+ > +/–	In a $RAD54B^{+/+/-}$ bkg	61
RAD51L2	7	6,509	0.0011	+/+ > +/–	In a $RAD54B^{+/-/-}$ bkg	61
RAD51L2	1	10,382	0.0001	+/+ > +/–	In a $RAD54B^{-/-/-}$ bkg	61
RAD51L2	0	7,398	0.0000	+/+ > +/–	In a $RAD54B^{-/-/-}$ bkg	61
RAD54B	2	25	0.0800	+/+/+ > +/+/–		61
RAD54B	9	61	0.1475	+/+/– > +/–/–		61
RAD54B	2	78	0.0256	+/–/– > –/–/–		61
RAD54B	7	283	0.0247	+/+/+ > +/+/–		90
RAD54B	6	349	0.0172	+/+/+ > +/+/–	In a $XRCC3^{-/-}$ bkg	90
Securin				+/+ > +/–		88
Securin				+/– > –/–		88
Smad4	4	300	0.0133	+/+ > +/–		137
Smad4	2	254	0.0079	+/– > –/–		137
XRCC2	6	94	0.0638	+/+ > +/–	In a $RAD54B^{+/+/+}$ bkg	61
XRCC2	7	93	0.0753	+/+ > +/–	In a $RAD54B^{+/+/-}$ bkg	61
XRCC2	14	166	0.0843	+/+ > +/–	In a $RAD54B^{+/-/-}$ bkg	61
XRCC2	1	173	0.0058	+/+ > +/–	In a $RAD54B^{-/-/-}$ bkg	61
XRCC2	0	120	0.0058	+/+ > +/–	In a $RAD54B^{-/-/-}$ bkg	61
XRCC3				+/+ > +/–		90
XRCC3	5	193	0.0259	+/– > –/–		90
Totals	226	85,119	0.0027			

[a–d]As defined in Table 53–1.
[e]Miscellaneous: (In a $BLM^{-/-}$, $p53^{-/-}$, $XRCC3^{-/-}$, $RAD54^{+/+/+}$, $RAD54^{+/+/-}$, $RAD54^{+/-/-}$, $RAD54^{-/-/-}$, bkg) designates studies in which the gene targeting was done in the indicated respective backgrounds; (HSV TK selection) designates a study using herpes simplex virus thymidine kinase gene as the negative selection; (DT selection) designates a study using the diphtheria toxin gene as the negative selection; (Retargeting) designates events in which the targeting vector has integrated in the previously disrupted allele; (Promoter +) designates a study done with a vector using a selectable marker driven by a promoter; (Promoterless) designates a study done with a vector using a promoterless selectable marker.

are identical to those that have been disrupted in other cell lines (e.g., p21, p53, and HPRT; compare Table 53–1 with Table 53–2), but many of them are novel. In particular, the HCT116 studies are noteworthy because of the large number of times that two rounds of successful gene targeting have been carried out, allowing for an assessment of the null phenotype (Table 53–2). Moreover, HCT116 is the only cell line, with the recent exception of NALM-6 cells (see below), in which compound knockout strategies—in which two or more genes are disrupted to investigate their genetic interaction with one another— have succeeded.[52,61,62,72,90] Overall, 226 correctly targeted integrations have been reported for 85,119 drug-resistant clones analyzed (Table 53–2). This represents a rather pathetic targeting frequency of 0.27%, which is an order of magnitude lower than that reported for the non-HCT116, non-NALM-6 cell lines (Table 53–1). At this frequency, the work required to identify a correctly targeted clone becomes a significant issue and is thus usually undertaken only by laboratories with sufficient manpower. This frequency does, however, include a study in which the RAD54B gene was first disrupted in HCT116 cells and then (mostly unsuccessful) attempts were made to target the RAD51L2 and XRCC2 loci, thus validating the requirement for RAD54 in HR/gene targeting (Figure 53–2). Even if this large, negative study (two positive clones from 17,973 clones analyzed) and other studies in which the HCT116 genetic background was altered prior to use for gene targeting are omitted from analysis, the targeting frequency in a wild-type HCT116 background is still a paltry 0.39%. This frequency is low enough to daunt all but the most committed of researchers. Still, it should be noted that there is great variability in the targeting frequency in HCT116 between loci, with frequencies as high as 37% being reported for the p21 locus[59] and many instances of targeting frequencies ranging from 1 to 10% (Table 53–2). Again, unfortunately, there is no a priori way to know ahead of time how well a particular gene may or may not target.

In summary, for the past 5–10 years, the HCT116 cell line has been the gold standard for performing gene knockouts in human somatic cells. Although it is an immortalized, transformed colorectal carcinoma cell line, its utility stems from its relatively stable karyotype[91–93] and because it is diploid and wild-type for almost all of the relevant *DNA Repair*, DNA checkpoint, and chromosome stability genes[59,77,88] with the exception of meiotic recombination defective 11[94] (Mre11). Nonetheless, HCT116 is not a "normal" cell line. (1) It is mutated in the Mut L homolog 1 (MLH1) gene, which makes it a mismatch repair[95,96] as well as a transcription-coupled repair[97] defective cell line. As such, it has an overall microsatellite instability phenotype. In addition, genes that have homopolymeric tracts within them, such as Bax and Mre11, often spontaneously mutate one or both alleles.[94,98] (2) HCT116 cells have gains of parts of chromosomes 8q (accounting for the triploid nature of the RAD54B gene[61]), 10q, 16q, and 17q and also contain two derivative chromosomes carrying two translocations, der(16)t(8;16) and der(18)t(17;18).[91,92] (3) Finally,

Table 53–3
Gene targeting studies in the NALM-6 cell line

Gene	*+*[a]	*#*[b]	*Frequency*[c]	*Genotype*[d]	*Miscellaneous*[e]	*Reference*
APRT	5	134	0.0373	+/+ > +/–	±BLM RNAi	56
BLM	14	168	0.0833	+/+ > +/–		101
BLM	5	274	0.0182	+/– > –/–		101
HPRT				–/O > +/O	Rescue ± BLM RNAi	56
Ku70	3	80	0.0375	+/+ > +/–		100
Ku86	7	84	0.0833	+/+ > +/–		100
LIGIV	2	124	0.0161	+/+ > +/–		60
LIGIV	1	20	0.0500	+/– > –/–		60
LIGIV			~0.0340	+/+ > +/–		66
LIGIV			~0.0085	+/– > –/–		66
p53	1	9	0.1111	+/+ > +/–		10
p53	9	116	0.0776	+/– > –/–		10
p53	1	1	1.0000	+/+ > +/–	In a BLM$^{-/-}$ bkg	10
p53	4	42	0.0952	+/– > –/–	In a BLM$^{-/-}$ bkg	10
TDP1	2	7	0.2857	+/+ > +/–		10
XRCC3	1	6	0.1667	+/+ > +/–		10
XRCC3	1	6	0.1667	+/+ > +/–	In a LIGIV$^{-/-}$ bkg	10
Totals	56	1,071	0.0523			

[a–d]As defined in Table 53–1.

[e]Miscellaneous: (±BLM RNAi) designates a study in which the gene targeting was carried out in the presence (+) and absence (–), respectively, of RNAi directed against the Bloom's syndrome helicase; (Rescue ± BLM RNAi) designates a study in which the correcting gene targeting a knockin experiment was carried out in the presence (+) and absence (–), respectively, of RNAi directed against the Bloom's syndrome helicase; (In a BLM$^{-/-}$ bkg and LIGIV$^{-/-}$ bkg) designates studies in which the gene targeting was done in the indicated respective genetic backgrounds.

HCT116 cells also contain a dominant gain-of-function mutation in the β-catenin gene, a potent transcription factor that drives c-myc expression.[99] As a word of caution, it should be noted that the aberrancies described above are the ones that are known and reported. It is quite likely that other genes are mutated or misexpressed relative to a normal, nonimmortalized, nontransformed cell in HCT116 cells and the impact of such differences on the phenotypes of a knockout of your favorite gene is hard to assess. In the same cautionary vein, recent reports of phenotypic differences between a knockout of the Ku86 gene in HCT116 cells[64,93] and NALM-6 cells[100] suggest that investigators need to take care in generalizing their results to other cell lines and other systems.

The NALM-6 Cell Line In 1998, Grawunder *et al.* described the successful targeting of both alleles of the DNA LIGIV gene in the human pre-B acute lymphoblastic leukemia (ALL) cell line, NALM-6.[60] A total of three correctly targeted clones were obtained from 144 analyzed for a relatively robust 2.08% targeting frequency. This cell line then lay (scientifically) dormant for the next 5 years until a report appeared in 2004 from the laboratory of Dr. Koyama describing the disruption of the BLM gene.[101] In the subsequent 2 years, the Koyama laboratory published a series of papers[10,56,66,100] describing the disruption of an additional eight genes in this cell line (Table 53–3). Impressively, 56 correct targeting events from a total of 1071 drug-resistant clones analyzed have been reported, for an overall frequency of 5.23% (Table 53–3). Thus, this cell line appears to carry out gene targeting at least twice as well as most other human cell lines (Table 53–1) and 20 times better than HCT116 cells (Table 53–2). If these frequencies can be reproduced in other laboratories, it is likely that the NALM-6 cell line will supplant HCT116 as a model system for human somatic cell knockouts.

It must be reiterated, however, that NALM-6, like HCT116, cells are not "normal." This line was established in culture some three decades ago from the peripheral blood of a 19-year-old man suffering from ALL.[102] Although the cell line is diploid, it carries at least one derivative chromosome containing a translocation [der t(5;12)(q33.2;p13.2)].[103] This translocation has fused the platelet-derived growth factor receptor-β (PDGFRB) gene (a tyrosine kinase involved in cell proliferation) to the translocation-ETS-leukemia (TEL) transcription factor gene,[104] resulting in a dominant gain-of-function chimeric kinase. Coincidentally, NALM-6, like HCT116, is also defective in mismatch repair. In the case of NALM-6 neither mutS homolog 2 (MSH2) nor mutS homolog 6 (MSH6) is expressed.[104] Thus, although it has not been directly demonstrated, it seems highly likely that NALM-6 cells will show microsatelite instability and high mutability of genes that have homopolymeric tracts within them. In addition, it can be anticipated that the more popular this cell line becomes, the greater is the likelihood that other genes that are mutated or misexpressed relative to a normal, nonimmortalized, nontransformed cell will be uncovered. Lastly, the same cautionary note concerning the recent reports of phenotypic differences between a knockout of the Ku86 gene in HCT116 cells[64,93] and NALM-6 cells[100] suggests that investigators utilizing NALM-6 cells—just like those utilizing HCT116 cells—need to take care in generalizing their results to other cell lines and other systems.

THE FUTURE OF GENE TARGETING: RECOMBINANT ADENO-ASSOCIATED VIRUS

Adeno-associated virus (AAV) is a human parvovirus that has both latent and lytic life cycles.[9,105,106] During the latent infection, AAV integrates stably into the human genome, preferentially, but not exclusively, at a site on chromosome 19.[107,108] In an ironic twist, it was this characteristic of site-specific integration that first drew the attention of the gene therapy field to AAV, since the use of AAV as a gene delivery vector would not have the serious problem of random insertional mutagenesis attendant with almost every other viral vector delivery system. While this has turned out to be true[109] and clinical trials using rAAV are underway,[110] the use of AAV as a gene targeting vector has far outpaced its use as a gene therapy delivery vector. The ability to undergo a latent infection is a necessary part of the AAV life cycle since AAV requires the presence of a helper virus (generally adenovirus) for a productive lytic infection. Thus, AAV has been detected only in individuals undergoing the symptoms of an adenoviral[105,111] or herpes simplex virus[112] infection. Approximately 85% of the human population is seropositive for AAV, although there is no pathology or disease that has been attributed to an AAV infection.[106]

AAV is a single-stranded DNA virus of 4.68 kb. Either strand of the integrated AAV DNA can be packaged[113] and both are equally infectious.[114] Multiple serotypes of the virus are known, but only the AAV-2 serotype is routinely utilized in the laboratory. AAV encodes two genes, Rep (replication) and Cap (capsid). Rep is required for viral replication, needed for viral integration, and can potentiate viral gene expression.[106] Cap encodes the viral capsid proteins, of which three isoforms, VP1, VP2, and VP3, exist.[115,116] These genes are flanked by identical inverted terminal repeats (ITRs) of 145 bp that form T-shaped hairpins, due to three palindromic sequences residing within the ITRs.[117] The ITRs are essentially the only *cis*-acting sequences necessary for viral replication,[118] packaging, and integration.[119] This fact greatly facilitated the development of rAAV as a gene delivery and gene-targeting vector. Thus, if Rep and Cap are provided in *trans*, then the internal 4.39 kb of AAV can consist of recombinant DNA of the investigator's choosing.[120] Similarly, the adenoviral gene products E1A, E1B, E4, E2A, and VA are required for AAV replication.[106] Originally, these were provided by coinfecting with live adenovirus, which, of course, precluded the use of the resulting AAV stock for gene therapy experiments. This significant detriment was overcome once again simply by providing recombinant adenoviral functions in *trans*.[121] In summary, AAV has been known to the research community for some four decades and the life cycle of the virus is quite well understood. The virus was subsequently conscripted in the 1980s by investigators who were interested in viral vectors for use in gene therapy. Recombinant molecular biological methodologies were then used to develop an rAAV system for the production of virus stocks that could productively infect every human (and many other mammalian) cell line tested to date.

This story took a very unexpected and surprising turn in 1998, when Russell and Hirata demonstrated that rAAV vectors could be used for gene targeting.[122] Impressively, they used rAAV not just to inactivate genes, but to introduce defined modifications into homologous chromosomal sequences at very high frequency. They demonstrated this first with a Neo reporter construct, but also showed that they could modify an endogenous locus (HPRT) with similar efficiency.[122] In the intervening 8 years, the Russell laboratory and many others have confirmed and extended these original observations. Thus, 15 endogenous loci have been inactivated in a wide variety of human cell lines, ranging from investigator favorites such as HCT116 and NALM-6 to more biologically relevant normal diploid fibroblastic or stem cell lines (Table 53–4). The most spectacular use of rAAV to date was the correction of a dominant negative mutation in the COL1A1 gene in stem cells derived from patients afflicted with osteogenesis imperfecta.[123] Overall, 234 correctly targeted events have been recorded from a total of 6305 drug-resistant colonies examined (Table 53–4). This is an overall targeting frequency of 3.71%, which is much better than the traditional transfection-based approaches using HCT116 cells (compare Table 53–2 to Table 53–4) and comparable to the targeting frequencies reported for non-HCT116, non-NALM-6 (compare Table 53–1 to Table 53–4), and NALM-6 (Compare Table 53–3 to Table 53–4) cells.

Why rAAV works so well for gene targeting is not clear. A priori, the virus, because of its ITRs, has at least 145 nt of non-homology at either end of the donor DNA and this is generally a deterrent to homologous integration. Moreover, because the drug selection cassettes that are usually present on a targeting vector are 2 to 2.5 kb in length, this restricts the left and right homology arms to 0.9 to 1.0 kb each, such that the resulting recombinant virus can be packaged properly. This length of homology for the arms is extremely short in comparison to traditional approaches using transfection of linear dsDNA, where arms of 3–8 kb are standard. That rAAV gene targeting does work, however, is indisputable. Since invasion of a single-stranded donor DNA into the recipient chromosome is an essential feature of gene targeting (Figure 53–2), it is probable that rAAV—an ssDNA virus—readily becomes an HR intermediate.[9] Needless to say, other factors, such as the likely higher concentration of DNA in the nucleus provided by a viral infection as compared to DNA transfection, probably also work in its favor. Research into other parameters, such as multiplicity of infection, DNA sequence of the arms, cell cycle effects, and promoterless selection schemes, are currently all under investigation and it is likely that the rAAV gene targeting frequency can be substantially optimized.

What must be emphasized, however, is the ease of working with rAAV in comparison to the traditional methodology. Thus, historically, an investigator needed to spend 6 months or more assembling a targeting vector (although see Iiizumi *et al.*[66] for a faster alternative) and then spend another 6 months to a year laboriously transfecting their targeting construct into cells and screening the resulting drug-resistant clones by Southern blot analysis. With rAAV methodology, the entire experiment starting from the idea, "I'd like to knockout gene X," to having a cell line in hand can take as little as 2 months (B. Ruis, S. Oh, and E.A. Hendrickson, unpublished data). Importantly, Kohli *et al.* did the field a major service by developing a protocol, driven almost exclusively by PCR, for the development of targeting vectors and viral stocks.[124] Since the viral vectors and helper plasmids (expressing the necessary AAV and adenoviral gene products) are all commercially available (Stratagene), anyone can adopt this methodology into their laboratory. Additionally, in an ironic twist, because the rAAV targeting arms are so short, it becomes technically feasible to screen the resulting clones not by Southern blot analysis, but by PCR, thus once again expediting the targeting process.[124] It should be pointed out that almost all of the rAAV studies carried out to date have used a simple positive selection, in which the rAAV vector carries a drug selection marker constitutively expressed using a strong viral promoter. Recently, a promoterless version of an rAAV vector has been described and targeting frequencies of ~20% were reported.[65] If the use of this

Table 53–4
Gene targeting studies using rAAV vectors

Gene	+[a]	#[b]	*Frequency*[c]	*Genotype*[d]	*Miscellaneous*[e]	*Reference*
ALKP				Transgene	MHF2 cells	138
ALKP				Transgene	MHF2 cells	139
β-Catenin	14	60	0.2333	+/–* > +/– or –/–*	HCT116 cells	65
β-Catenin	3	741	0.0400	+/–* > –/–*	HCT116 cells; knockin	65
β-Catenin	3	170	0.0176	+/–* > –/–*	HCT116 cells; promoterless	65
β-Catenin	20	109	0.1835	+/+ > +/–	DLD1 cells; knockin	65
CCR5	12	173	0.0694	+/+ > +/–	hTERT-RPE cells	124
CCR5	13	244	0.0533	+/+ > +/–	HCT116 cells	124
CDX2				+/+ > +/–	MKN5 cells	140
CDX2				+/– > –/–	MKN5 cells	140
CDX2				+/+ > +/–	LOVO cells	141
CDX2			~0.0100	+/– > –/–	LOVO cells	141
CDX2				+/+ > +/–	SW48 cells	141
CDX2	67	2,000	0.0335	+/– > –/–	SW48 cells	141
COL1A1	35	52	0.6731	+/–* > +/– or –/–*	Stem cells; knockin	123
COL1A1	19	27	0.7037	+/+ > +/–	MHF2 cells	123
DNA-PK_{cs}	2	55	0.0364	+/+ > +/–	HCT116 cells	BR and EAH, unpublished[f]
FHIT	2	92	0.0217	+/+ > +/–	HCT116 cells	124
GFP				Transgene	293 cells; ±IsceI	142
GFP			~0.0100	Transgene	293 cells	143
HAUSP				+/+ > +/–	HCT116 cells	144
HAUSP				+/– > –/–	HCT116 cells	144
HAUSP				+/+ > +/–	SW48 cells	144
HAUSP				+/– > +/–	SW48 cells; retargeting	144
HAUSP				+/+ > +/–	hTERT-RPE cells	144
HAUSP				+/– > +/–	hTERT-RPE cells; retargeting	144
HPRT				+/O > –/O	HT1080 cells	122
HPRT				+/O > –/O	MHF1 cells; knockin	122
HPRT				+/O > –/O	MHF2 cells; knockin	122
HPRT				+/O > –/O	MHF3 cells; knockin	122
HPRT				+/O > –/O	MHF2 cells	139
HPRT				+/O > –/O	MHF2 cells	145
Ku70	3	437	0.0069	+/+ > +/–	HCT116 cells	RF and EAH, unpublished[f]
Ku70	2	121	0.0165	+/+ > +/–	NALM-6 cells	RF and EAH, unpublished[f]
Ku86	2	90	0.0222	+/+ > +/–	HCT116 cells	BR and EAH, unpublished[f]
Ku86	2	379	0.0053	+/+ > +/–	NALM-6 cells	BR and EAH, unpublished[f]
LIGIV	2	176	0.0114	+/+ > +/–	HCT116 cells	SO and EAH, unpublished[f]
NEO			~0.0014	Transgene	HT1080 cells	146
NEO			~0.0008	Transgene	MHF2 cells ± hTERT	146
NEO			~0.0010	Transgene	HeLa cells	146
NEO				Transgene	HeLa cells	122
p53	4	95	0.0421	+/+ > +/–	HCT116 cells	65
p53	26	94	0.2766	+/+ > +/–	HCT116 cells; promoterless	65
SCO2	3	1,200	0.0025	+/+ > +/–	HCT116 cells	147
XIAP				+/+ > +/–	HCT116 cells	148
XIAP				+/– > –/–	HCT116 cells	148
XIAP				+/+ > +/–	DLD-1 cells	148
XIAP				+/– > –/–	DLD-1 cells	148
Totals	234	6,305	0.0371			

[a–d]As defined in Table 53–1.

[e]Miscellaneous: For each study, the cell line in which the gene targeting was performed is indicated; (knockin) designates a study in which a subtle mutation was introduced into an exon as opposed to the complete ablation of the exon; (Promoterless) designates a gene targeting study in which the positive selection marker relied upon an endogenous promoter for expression; (±IScel) designates a study in which the gene targeting was carried out in the presence (+) and absence (–), respectively, of the endonuclease, I(ntron)-*Saccharomyces cerevisiae* I (I-Scel); (Retargeting) designates a study in which the gene targeting event occurred on a previously targeted allele; (±hTERT) designates a study in which the gene targeting was done in a cell line in the presence (+) and absence (–), respectively, of h(uman) telomeric reverse transcriptase (hTERT).

[f]EAH, Eric A. Hendrickson; BR, Brian Ruis; RF, Riaz Fattah; SO, Sehyun OH.

vector becomes generally applicable, there will be few obstacles to targeting any gene of interest. Lastly, it should be noted that one obvious improvement might involve using rAAV methodology with the NALM-6 cell line, which appears to be so proficient in standard gene targeting. However, in our unpublished studies, rAAV seems to target no better in NALM-6 than in HCT116 cells (Table 53–4).[125–148]

CONCLUSIONS

In the 21 years that have elapsed since the first report of a gene-targeting study in human somatic cells,[74] a total 46 endogenous genes have been disrupted or modified. While modestly impressive in its own right, this number pales in comparison to the number of knockout mice strains that have been generated in the same time frame. Some of this difference is simply a reflection of the choice of model systems and the biological questions being addressed. Thus, for example, questions of development cannot be addressed in somatic cells in culture. Some of the difference, however, has been due to the fact that carrying out knockouts in human cells in culture is inherently more difficult than in murine ES cell lines. All in all, a total of 704 correctly targeted events in human somatic cells have been recorded from 100,791 clones examined, for a less than robust 0.7% success rate. Importantly, however, the identification of new cell lines such as NALM-6 and, most importantly, the development of the novel rAAV technology portend great improvements in this success rate in the foreseeable future. Moreover, since the rAAV methodology is still relatively in its infancy, we can expect that modifications will be developed to improve this already promising technique. While it is unlikely that gene targeting in human somatic cells will ever be as facile as gene replacement in yeast, it is clear that powerful methodology already exists to make such studies practical. Moreover, if the promise of the rAAV system comes to fruition, researchers working on human somatic cells in culture will have a powerful and facile weapon at their scientific disposal.

ACKNOWLEDGMENTS

I thank Dr. Anja-Katrin Bielinsky (University of Minnesota Medical School) for her comments and helpful discussions. I thank the past and current members of my laboratory who have contributed to the development of this methodology. This work has been supported in part by Grants HL079559 and GM069576 from the NIH to E.A.H.

REFERENCES

1. Sedivy JM, Joyner A. *Gene Targeting*. New York: W.H. Freeman and Company, 1992.
2. Sorrell DA, Kolb AF. Targeted modification of mammalian genomes. *Biotechnol Adv* 2005;23:431–469.
3. Yu Y, Bradley A. Engineering chromosomal rearrangements in mice. *Nat Rev Genet* 2001;2:780–790.
4. Park F, Gow KW. Gene therapy: Future or flop. *Pediatr Clin North Am* 2006;53:621–638.
5. Cahill D, Connor B, Carney JP. Mechanisms of eukaryotic DNA double strand break repair. *Front Biosci* 2006;11:1958–1976.
6. van Veelen L, Wesoly J, Kanaar R. Biochemical and cellular aspects of homologous recombination. In: Seide W, Kow YW, Doetsch P, Eds. *DNA Damage Recognition*. New York: Taylor & Francis Group, 2006:581–607.
7. Hefferin ML, Tomkinson AE. Mechanism of DNA double-strand break repair by nonhomologous end joining. *DNA Repair* 2005;4:639–648.
8. Sedivy JM, Vogelstein B, Liber HL, Hendrickson EA, Rosmarin A. Gene targeting in human cells without isogenic DNA. *Science* 1999;283:9a.
9. Vasileva A, Jessberger R. Precise hit: Adeno-associated virus in gene targeting. *Nat Rev Microbiol* 2005;3:837–847.
10. Adachi N, So S, Iiizumi S, *et al.* The human pre-B cell line Nalm-6 is highly proficient in gene targeting by homologous recombination. *DNA Cell Biol* 2006;25:19–24.
11. Sonoda E, Hochegger H, Saberi A, Taniguchi Y, Takeda S. Differential usage of nonhomologous end-joining and homologous recombination in double strand break repair. *DNA Repair* 2006;5: 1021–1029.
12. Roth DB, Wilson JH. Relative rates of homologous and nonhomologous recombination in transfected DNA. *Proc Natl Acad Sci USA* 1985;82:3355–3359.
13. Lieber MR, Ma Y, Kefei Y, Pannicke U, Schwarz K. The mechanism of vertebrate nonhomologous DNA end joining and its role in immune system gene rearrangements. In: Seide W, Kow YW, Doetsch P, Eds. *DNA Damage Recognition*. New York: Taylor & Francis Group, 2006:609–627.
14. Thomas KR, Capecchi MR. Site-directed mutagenesis by gene targeting in mouse embryo-derived stem cells. *Cell* 1987;51:503–512.
15. Lee SE, Mitchell RA, Cheng A, Hendrickson EA. Evidence for DNA-PK-dependent and -independent DNA double-strand break repair pathways in mammalian cells as a function of the cell cycle. *Mol Cell Biol* 1997;17:1425–1433.
16. Takata M, Sasaki MS, Sonoda E, *et al.* Homologous recombination and non-homologous end-joining pathways of DNA double-strand break repair have overlapping roles in the maintenance of chromosomal integrity in vertebrate cells. *EMBO J* 1998;17:5497–5508.
17. Hendrickson EA, Huffman JL, Tainer JA. Structural aspects of Ku and the DNA-dependent protein kinase complex. In: Seide W, Kow YW, Doetsch P, Eds. *DNA Damage Recognition*. New York: Taylor & Francis Group, 2006:629–684.
18. Getts RC, Stamato TD. Absence of a Ku-like DNA end binding activity in the *xrs* double-strand *DNA Repair*-deficient mutant. *J Biol Chem* 1994;269:15981–15984.
19. Liang F, Romanienko PJ, Weaver DT, Jeggo PA, Jasin M. Chromosomal double-strand break repair in Ku80-deficient cells. *Proc Natl Acad Sci USA* 1996;93:8929–8933.
20. Inamdar KV, Yu Y, Povirk LF. Resistance of 3'-phosphoglycolate DNA ends to digestion by mammalian DNase III. *Radiat Res* 2002;157:306–311.
21. Gottlieb TM, Jackson SP. The DNA-dependent protein kinase: Requirement for DNA ends and association with Ku antigen. *Cell* 1993;72:131–142.
22. Suwa A, Hirakata M, Takeda Y, Jesch SA, Mimori T, Hardin JA. DNA-dependent protein kinase (Ku protein-p350 complex) assembles on double-stranded DNA. *Proc Natl Acad Sci USA* 1994;91:6904–6908.
23. DeFazio LG, Stansel RM, Griffith JD, Chu G. Synapsis of DNA ends by DNA-dependent protein kinase. *EMBO J* 2002;21:3192–3200.
24. Weterings E, Verkaik NS, Bruggenwirth HT, Hoeijmakers JH, van Gent DC. The role of DNA dependent protein kinase in synapsis of DNA ends. *Nucleic Acids Res* 2003;31:7238–7246.
25. Spagnolo L, Rivera-Calzada A, Pearl LH, Llorca O. Three-dimensional structure of the human DNA-PKcs/Ku70/Ku80 complex assembled on DNA and its implications for DNA DSB repair. *Mol Cell* 2006;22(4):511–519.
26. Moshous D, Callebaut I, de Chasseval R, *et al.* Artemis, a novel DNA double-strand break repair/V(D)J recombination protein, is mutated in human severe combined immune deficiency. *Cell* 2001;105:177–186.
27. Ma Y, Pannicke U, Schwarz K, Lieber MR. Hairpin opening and overhang processing by an Artemis/DNA-dependent protein kinase complex in nonhomologous end joining and V(D)J recombination. *Cell* 2002;108:781–794.

28. Moshous D, Callebaut I, de Chasseval R, *et al.* The V(D)J recombination/DNA Repair factor Artemis belongs to the metallo-beta-lactamase family and constitutes a critical developmental checkpoint of the lymphoid system. *Ann NY Acad Sci* 2003;987:150–157.
29. Critchlow SE, Bowater RP, Jackson SP. Mammalian DNA double-strand break repair protein XRCC4 interacts with DNA ligase IV. *Curr Biol* 1997;7:588–598.
30. McElhinny SAN, Snowden CM, McCarville J, Ramsden DA. Ku recruits the XRCC4 ligase IV complex to DNA ends. *Mol Cell Biol* 2000;20:2996–3003.
31. Teo SH, Jackson SP. Lif1p targets the DNA ligase Lig4p to sites of DNA double-strand breaks. *Curr Biol* 2000;10:165–168.
32. Li Z, Otevrel T, Gao Y, *et al.* The XRCC-4 gene encodes a novel protein involved in DNA double-strand break repair and V(D)J recombination. *Cell* 1995;83:1079–1089.
33. Gao Y, Frank KM, Dikkes P, *et al.* A critical role for DNA end-joining proteins in both lymphogenesis and neurogenesis. *Cell* 1998;95:891–902.
34. Ahnesorg P, Smith P, Jackson SP. XLF interacts with the XRCC4-DNA ligase IV complex to promote DNA nonhomologous end-joining. *Cell* 2006;124:301–313.
35. Buck D, Malivert L, de Chasseval R, *et al.* Cernunnos, a novel nonhomologous end-joining factor, is mutated in human immunodeficiency with microcephaly. *Cell* 2006;124:287–299.
36. Hopfner K-P. The Mre11/Rad50/Nbs1 complex. In: Seide W, Kow YW, Doetsch P, Eds. *DNA Damage Recognition*. New York: Taylor & Francis Group, 2006:705–721.
37. Moreau S, Morgan EA, Symington LS. Overlapping functions of the Saccharomyces cerevisiae Mre11, Exo1 and Rad27 nucleases in DNA metabolism. *Genetics* 2001;159:1423–1433.
38. Iftode C, Daniely Y, Borowiec JA. Replication protein A (RPA): The eukaryotic SSB. *Crit Rev Biochem Mol Biol* 1999;34:141–180.
39. Kawabata M, Kawabata T, Nishibori M. Role of recA/RAD51 family proteins in mammals. *Acta Med Okayama* 2005;59:1–9.
40. Sung P, Trujillo KM, Van Komen S. Recombination factors of Saccharomyces cerevisiae. *Mutat Res* 2000;451:257–275.
41. Thacker J. The RAD51 gene family, genetic instability and cancer. *Cancer Lett* 2005;219:125–135.
42. Sugiyama T, Kowalczykowski SC. Rad52 protein associates with replication protein A (RPA)-single-stranded DNA to accelerate Rad51-mediated displacement of RPA and presynaptic complex formation. *J Biol Chem* 2002;277:31663–31672.
43. Heyer WD, Li X, Rolfsmeier M, Zhang XP. Rad54: The Swiss Army knife of homologous recombination? *Nucleic Acids Res* 2006;34:4115–4125.
44. Solinger JA, Kiianitsa K, Heyer WD. Rad54, a Swi2/Snf2-like recombinational repair protein, disassembles Rad51 : dsDNA filaments. *Mol Cell* 2002;10:1175–1188.
45. Hastings PJ, McGill C, Shafer B, Strathern JN. Ends-in vs. ends-out recombination in yeast. *Genetics* 1993;135:973–980.
46. Sharma S, Doherty KM, Brosh RM Jr. Mechanisms of RecQ helicases in pathways of DNA metabolism and maintenance of genomic stability. *Biochem J* 2006;398:319–337.
47. Liu Y, Masson JY, Shah R, O'Regan P, West SC. RAD51C is required for Holliday junction processing in mammalian cells. *Science* 2004;303:243–246.
48. Liu Y, West SC. Happy Hollidays: 40th anniversary of the Holliday junction. *Nat Rev Mol Cell Biol* 2004;5:937–944.
49. Zheng H, Hasty P, Brenneman MA, *et al.* Fidelity of targeted recombination in human fibroblasts and murine embryonic stem cells. *Proc Natl Acad Sci USA* 1991;88:8067–8071.
50. Yanez RJ, Porter AC. Influence of DNA delivery method on gene targeting frequencies in human cells. *Somat Cell Mol Genet* 1999;25:27–31.
51. Yanez RJ, Porter AC. Gene targeting is enhanced in human cells overexpressing hRAD51. *Gene Ther* 1999;6:1282–1290.
52. Bunz F, Fauth C, Speicher MR, *et al.* Targeted inactivation of p53 in human cells does not result in aneuploidy. *Cancer Res* 2002;62:1129–1133.
53. Hendrie PC, Hirata RK, Russell DW. Chromosomal integration and homologous gene targeting by replication-incompetent vectors based on the autonomous parvovirus minute virus of mice. *J Virol* 2003;77:13136–13145.
54. Zwaka TP, Thomson JA. Homologous recombination in human embryonic stem cells. *Nat Biotechnol* 2003;21:319–321.
55. Trobridge G, Hirata RK, Russell DW. Gene targeting by adeno-associated virus vectors is cell-cycle dependent. *Hum Gene Ther* 2005;16:522–526.
56. So S, Nomura Y, Adachi N, *et al.* Enhanced gene targeting efficiency by siRNA that silences the expression of the Bloom syndrome gene in human cells. *Genes Cells* 2006;11:363–371.
57. Jasin M, Elledge SJ, Davis RW, Berg P. Gene targeting at the human CD4 locus by epitope addition. *Genes Dev* 1990;4:157–166.
58. Manjunath N, Johnson RS, Staunton DE, Pasqualini R, Ardman B. Targeted disruption of CD43 gene enhances T lymphocyte adhesion. *J Immunol* 1993;151:1528–1534.
59. Waldman T, Kinzler KW, Vogelstein B. p21 is necessary for the p53-mediated G1 arrest in human cancer cells. *Cancer Res* 1995;55:5187–5190.
60. Grawunder U, Zimmer D, Fugmann S, Schwarz K, Lieber MR. DNA ligase IV is essential for V(D)J recombination and DNA double-strand break repair in human precursor lymphocytes. *Mol Cell* 1998;2:477–484.
61. Miyagawa K, Tsuruga T, Kinomura A, *et al.* A role for RAD54B in homologous recombination in human cells. *EMBO J* 2002;21:175–180.
62. Ghosh G, Li G, Myung K, Hendrickson EA. The lethality of Ku86 loss-of-function mutations in human cells is p53-independent. *Radiat Res* 2007;167(1):66–79.
63. Gu H, Marth JD, Orban PC, Mossmann H, Rajewsky K. Deletion of a DNA polymerase beta gene segment in T cells using cell type-specific gene targeting. *Science* 1994;265:103–106.
64. Li G, Nelsen C, Hendrickson EA. Ku86 is essential in human somatic cells. *Proc Natl Acad Sci USA* 2002;99:832–837.
65. Topaloglu O, Hurley PJ, Yildirim O, Civin CI, Bunz F. Improved methods for the generation of human gene knockout and knockin cell lines. *Nucleic Acids Res* 2005;33:e158.
66. Iiizumi S, Nomura Y, So S, *et al.* Simple one-week method to construct gene-targeting vectors: Application to production of human knockout cell lines. *Biotechniques* 2006;41:311–316.
67. Sauer B, Henderson N. Site-specific DNA recombination in mammalian cells by the Cre recombinase of bacteriophage P1. *Proc Natl Acad Sci USA* 1988;85:5166–5170.
68. Kolb AF. Genome engineering using site-specific recombinases. *Cloning Stem Cells* 2002;4:65–80.
69. Austin S, Ziese M, Sternberg N. A novel role for site-specific recombination in maintenance of bacterial replicons. *Cell* 1981;25:729–736.
70. Couedel C, Mills KD, Barchi M, *et al.* Collaboration of homologous recombination and nonhomologous end-joining factors for the survival and integrity of mice and cells. *Genes Dev* 2004;18:1293–1304.
71. Celli GB, Denchi EL, de Lange T. Ku70 stimulates fusion of dysfunctional telomeres yet protects chromosome ends from homologous recombination. *Nat Cell Biol* 2006;8:885–890.
72. Traverso G, Bettegowda C, Kraus J, *et al.* Hyper-combination and genetic instability in BLM-deficient epithelial cells. *Cancer Res* 2003;63:8578–8581.
73. Cortez D, Guntuku S, Qin J, Elledge SJ. ATR and ATRIP: Partners in checkpoint signaling. *Science* 2001;294:1713–1716.
74. Smithies O, Gregg RG, Boggs SS, Koralewski MA, Kucherlapati RS. Insertion of DNA sequences into the human chromosomal beta-globin locus by homologous recombination. *Nature* 1985;317:230–234.
75. Montrose-Rafizadeh C, Kole J, Bartkowski LM, *et al.* Gene targeting of a CFTR allele in HT29 human epithelial cells. *J Cell Physiol* 1997;170:299–308.

76. Williams SR, Ousley FC, Vitez LJ, DuBridge RB. Rapid detection of homologous recombinants in nontransformed human cells. *Proc Natl Acad Sci USA* 1994;91:11943–11947.
77. Bunz F, Dutriaux A, Lengauer C, *et al.* Requirement for p53 and p21 to sustain G2 arrest after DNA damage. *Science* 1998;282: 1497–1501.
78. Brown JP, Wei W, Sedivy JM. Bypass of senescence after disruption of p21CIP1/WAF1 gene in normal diploid human fibroblasts. *Science* 1997;277:831–834.
79. Rasheed S, Nelson-Rees WA, Toth EM, Arnstein P, Gardner MB. Characterization of a newly derived human sarcoma cell line (HT-1080). *Cancer* 1974;33:1027–1033.
80. Geiser AG, Anderson MJ, Stanbridge EJ. Suppression of tumorigenicity in human cell hybrids derived from cell lines expressing different activated ras oncogenes. *Cancer Res* 1989;49:1572–1577.
81. Farese RV Jr, Flynn LM, Young SG. Modification of the apolipoprotein B gene in HepG2 cells by gene targeting. *J Clin Invest* 1992;90:256–261.
82. Itzhaki JE, Gilbert CS, Porter AC. Construction by gene targeting in human cells of a "conditional" CDC2 mutant that rereplicates its DNA. *Nat Genet* 1997;15:258–265.
83. Shirasawa S, Furuse M, Yokoyama N, Sasazuki T. Altered growth of human colon cancer cell lines disrupted at activated Ki-ras. *Science* 1993;260:85–88.
84. Powell SM, Zilz N, Beazer-Barclay Y, *et al.* APC mutations occur early during colorectal tumorigenesis. *Nature* 1992;359:235–237.
85. Vogelstein B, Kinzler KW. The multistep nature of cancer. *Trends Genet* 1993;9:138–141.
86. Chan TA, Hermeking H, Lengauer C, Kinzler KW, Vogelstein B. 14–3-3s is required to prevent mitotic catastrophe after DNA damage. *Nature* 1999;401:616–620.
87. Rhee I, Jair KW, Yen RW, *et al.* CpG methylation is maintained in human cancer cells lacking DNMT1. *Nature* 2000;404:1003–1007.
88. Jallepalli PV, Waizenegger IC, Bunz F, *et al.* Securin is required for chromosomal stability in human cells. *Cell* 2001;105:445–457.
89. Rhee I, Bachman KE, Park BH, *et al.* DNMT1 and DNMT3b cooperate to silence genes in human cancer cells. *Nature* 2002;416: 552–556.
90. Yoshihara T, Ishida M, Kinomura A, *et al.* XRCC3 deficiency results in a defect in recombination and increased endoreduplication in human cells. *EMBO J* 2004;23:670–680.
91. Masramon L, Ribas M, Cifuentes P, *et al.* Cytogenetic characterization of two colon cell lines by using conventional G-banding, comparative genomic hybridization, and whole chromosome painting. *Cancer Genet Cytogenet* 2000;121:17–21.
92. Abdel-Rahman WM, Katsura K, Rens W, *et al.* Spectral karyotyping suggests additional subsets of colorectal cancers characterized by pattern of chromosome rearrangement. *Proc Natl Acad Sci USA* 2001;98:2538–2543.
93. Myung K, Ghosh G, Fattah FJ, *et al.* Regulation of telomere length and suppression of genomic instability in human somatic cells by Ku86. *Mol Cell Biol* 2004;24:5050–5059.
94. Giannini G, Ristori E, Cerignoli F, *et al.* Human MRE11 is inactivated in mismatch repair-deficient cancers. *EMBO Rep* 2002;3:248–254.
95. Papadopoulos N, Nicolaides NC, Wei YF, *et al.* Mutation of a mutL homolog in hereditary colon cancer. *Science* 1994;263:1625–1629.
96. Liu B, Nicolaides NC, Markowitz S, *et al.* Mismatch repair gene defects in sporadic colorectal cancers with microsatellite instability. *Nat Genet* 1995;9:48–55.
97. Mellon I, Rajpal DK, Koi M, Boland CR, Champe GN. Transcription-coupled repair deficiency and mutations in human mismatch repair genes. *Science* 1996;272:557–560.
98. Zhang L, Yu J, Park BH, Kinzler KW, Vogelstein B. Role of BAX in the apoptotic response to anticancer agents. *Science* 2000;290: 989–992.
99. Chan TA, Wang Z, Dang LH, Vogelstein B, Kinzler KW. Targeted inactivation of CTNNB1 reveals unexpected effects of b-catenin mutation. *Proc Natl Acad Sci USA* 2002;99:8265–8270.
100. Uegaki K, Adachi N, So S, Iiizumi S, Koyama H. Heterozygous inactivation of human Ku70/Ku86 heterodimer does not affect cell growth, double-strand break repair, or genome integrity. *DNA Repair* 2006;5:303–311.
101. So S, Adachi N, Lieber MR, Koyama H. Genetic interactions between BLM and DNA ligase IV in human cells. *J Biol Chem* 2004;279:55433–55442.
102. Hurwitz R, Hozier J, LeBien T, *et al.* Characterization of a leukemic cell line of the pre-B phenotype. *Int J Cancer* 1979;23:174–180.
103. Wlodarska I, Aventin A, Ingles-Esteve J, *et al.* A new subtype of pre-B acute lymphoblastic leukemia with t(5;12)(q31q33;p12), molecularly and cytogenetically distinct from t(5;12) in chronic myelomonocytic leukemia. *Blood* 1997;89:1716–1722.
104. Matheson EC, Hall AG. Assessment of mismatch repair function in leukaemic cell lines and blasts from children with acute lymphoblastic leukaemia. *Carcinogenesis* 2003;24:31–38.
105. Atchison RW, Casto BC, Hammon WM. Adenovirus-associated defective virus particles. *Science* 1965;149:754–756.
106. Muzyczka N. Use of adeno-associated virus as a general transduction vector for mammalian cells. *Curr Top Microbiol Immunol* 1992;158:97–129.
107. Kotin RM, Siniscalco M, Samulski RJ, *et al.* Site-specific integration by adeno-associated virus. *Proc Natl Acad Sci USA* 1990; 87:2211–2215.
108. Samulski RJ, Zhu X, Xiao X, *et al.* Targeted integration of adeno-associated virus (AAV) into human chromosome 19. *EMBO J* 1991;10:3941–3950.
109. Lu Y. Recombinant adeno-associated virus as delivery vector for gene therapy—a review. *Stem Cells Dev* 2004;13:133–145.
110. Grieger JC, Samulski RJ. Adeno-associated virus as a gene therapy vector: Vector development, production and clinical applications. *Adv Biochem Eng Biotechnol* 2005;99:119–145.
111. Blacklow NR, Hoggan MD, Rowe WP. Serologic evidence for human infection with adenovirus-associated viruses. *J Natl Cancer Inst* 1968;40:319–327.
112. Buller RM, Janik JE, Sebring ED, Rose JA. Herpes simplex virus types 1 and 2 completely help adenovirus-associated virus replication. *J Virol* 1981;40:241–247.
113. Rose JA, Berns KI, Hoggan MD, Koczot FJ. Evidence for a single-stranded adenovirus-associated virus genome: Formation of a DNA density hybrid on release of viral DNA. *Proc Natl Acad Sci USA* 1969;64:863–869.
114. Samulski RJ, Chang LS, Shenk T. A recombinant plasmid from which an infectious adeno-associated virus genome can be excised *in vitro* and its use to study viral replication. *J Virol* 1987;61: 3096–3101.
115. Johnson FB, Ozer HL, Hoggan MD. Structural proteins of adeno-virus-associated virus type 3. *J Virol* 1971;8:860–863.
116. Rose JA, Maizel JV Jr, Inman JK, Shatkin AJ. Structural proteins of adenovirus-associated viruses. *J Virol* 1971;8:766–770.
117. Lusby E, Fife KH, Berns KI. Nucleotide sequence of the inverted terminal repetition in adeno-associated virus DNA. *J Virol* 1980;34: 402–409.
118. Xiao X, Xiao W, Li J, Samulski RJ. A novel 165-base-pair terminal repeat sequence is the sole cis requirement for the adeno-associated virus life cycle. *J Virol* 1997;71:941–948.
119. Samulski RJ, Chang LS, Shenk T. Helper-free stocks of recombinant adeno-associated viruses: Normal integration does not require viral gene expression. *J Virol* 1989;63:3822–3888.
120. Ferrari FK, Xiao X, McCarty D, Samulski RJ. New developments in the generation of Ad-free, high-titer rAAV gene therapy vectors. *Nat Med* 1997;3:1295–1297.
121. Xiao X, Li J, Samulski RJ. Production of high-titer recombinant adeno-associated virus vectors in the absence of helper adenovirus. *J Virol* 1998;72:2224–2232.
122. Russwll DW, Hirata RK. Human gene targeting by viral vectors. *Nat Genet* 1998;18:325–330.

123. Chamberlain JR, Schwarze U, Wang PR, *et al.* Gene targeting in stem cells from individuals with osteogenesis imperfecta. *Science* 2004;303:1198–1201.
124. Kohli M, Rago C, Lengauer C, Kinzler KW, Vogelstein B. Facile methods for generating human somatic cell gene knockouts using recombinant adeno-associated viruses. *Nucleic Acids Res* 2004;32: e3.
125. Itzhaki JE, Porter AC. Targeted disruption of a human interferon-inducible gene detected by secretion of human growth hormone. *Nucleic Acids Res* 1991;19:3835–3842.
126. Itzhaki JE, Barnett MA, MacCarthy AB, Buckle VJ, Brown WR, Porter AC. Targeted breakage of a human chromosome mediated by cloned human telomeric DNA. *Nat Genet* 1992;2:283–287.
127. Porter AC, Itzhaki JE. Gene targeting in human somatic cells. Complete inactivation of an interferon-inducible gene. *Eur J Biochem* 1993;218:273–281.
128. Culver KW, Hsieh WT, Huyen Y, *et al.* Correction of chromosomal point mutations in human cells with bifunctional oligonucleotides. *Nat Biotechnol* 1999;17:989–993.
129. Kim JS, Lee C, Foxworth A, Waldman T. B-Raf is dispensable for K-Ras-mediated oncogenesis in human cancer cells. *Cancer Res* 2004;64:1932–1937.
130. Kunzelmann K, Legendre JY, Knoell DL, Escobar LC, Xu Z, Gruenert DC. Gene targeting of CFTR DNA in CF epithelial cells. *Gene Ther* 1996;3:859–867.
131. Zhen L, King AA, Xiao Y, Chanock SJ, Orkin SH, Dinauer MC. Gene targeting of X chromosome-linked chronic granulomatous disease locus in a human myeloid leukemia cell line and rescue by expression of recombinant gp91phox. *Proc Natl Acad Sci USA* 1993;90:9832–9836.
132. Song KY, Schwartz F, Maeda N, Smithies O, Kucherlapati R. Accurate modification of a chromosomal plasmid by homologous recombination in human cells. *Proc Natl Acad Sci USA* 1987;84: 6820–6824.
133. Thyagarajan B, Johnson BL, Campbell C. The effect of target site transcription on gene targeting in human cells *in vitro*. *Nucleic Acids Res* 1995;23:2784–2790.
134. Kim JS, Crooks H, Dracheva T, *et al.* Oncogenic beta-catenin is required for bone morphogenetic protein 4 expression in human cancer cells. *Cancer Res* 2002;62:2744–2748.
135. Hiyama T, Katsura M, Yoshihara T, *et al.* Haploinsufficiency of the Mus81-Eme1 endonuclease activates the intra-S-phase and G2/M checkpoints and promotes rereplication in human cells. *Nucleic Acids Res* 2006;34:880–892.
136. Dhar SK, Yoshida K, Machida Y, *et al.* Replication from oriP of Epstein-Barr virus requires human ORC and is inhibited by geminin. *Cell* 2001;106:287–296.
137. Zhou S, Buckhaults P, Zawel L, *et al.* Targeted deletion of Smad4 shows it is required for transforming growth factor b and activin signaling in colorectal cancer cells. *Proc Natl Acad Sci USA* 1998;95:2412–2416.
138. Hirata RK, Russell DW. Design and packaging of adeno-associated virus gene targeting vectors. *J Virol* 2000;74:4612–4620.
139. Inoue N, Dong R, Hirata RK, Russell DW. Introduction of single base substitutions at homologous chromosomal sequences by adeno-associated virus vectors. *Mol Ther* 2001;3:526–530.
140. Dang LH, Chen F, Knock SA, *et al.* CDX2 does not suppress tumorigenicity in the human gastric cancer cell line MKN45. *Oncogene* 2006;25:2048–2059.
141. Dang LH, Chen F, Ying C, *et al.* CDX2 has tumorigenic potential in the human colon cancer cell lines LOVO and SW48. *Oncogene* 2006;25:2264–2272.
142. Porteus MH, Cathomen T, Weitzman MD, Baltimore D. Efficient gene targeting mediated by adeno-associated virus and DNA double-strand breaks. *Mol Cell Biol* 2003;23:3558–3565.
143. Liu X, Yan Z, Luo M, *et al.* Targeted correction of single-base-pair mutations with adeno-associated virus vectors under nonselective conditions. *J Virol* 2004;78:4165–4175.
144. Cummins JM, Rago C, Kohli M, Kinzler KW, Lengauer C, Vogelstein B. Tumour suppression: Disruption of HAUSP gene stabilizes p53. *Nature* 2004;428:1.
145. Hirata R, Chamberlain J, Dong R, Russell DW. Targeted transgene insertion into human chromosomes by adeno-associated virus vectors. *Nat Biotechnol* 2002;20:735–738.
146. Inoue N, Hirata RK, Russell DW. High-fidelity correction of mutations at multiple chromosomal positions by adeno-associated virus vectors. *J Virol* 1999;73:7376–7380.
147. Matoba S, Kang JG, Patino WD, *et al.* p53 regulates mitochondrial respiration. *Science* 2006;312:1650–1653.
148. Cummins JM, Kohli M, Rago C, Kinzler KW, Vogelstein B, Bunz F. X-linked inhibitor of apoptosis protein (XIAP) is a nonredundant modulator of tumor necrosis factor-related apoptosis-inducing ligand (TRAIL)-mediated apoptosis in human cancer cells. *Cancer Res* 2004;64:3006–3008.

54 Animal Models for Investigating the Causes and Mechanisms of Mammalian Germ Cell Aneuploidy

JOHN B. MAILHES AND MARY ANN HANDEL

ABSTRACT

Aneuploidy is a condition whereby the chromosome number of a cell or an organism is not an exact multiple of the haploid number for the species. It represents the most prevalent genetic abnormality of mankind and is associated with failed reproduction, mental and physical retardation, and cancer. Although research involving various organisms is providing information about the interactions between the biochemical pathways and the organelles responsible for chromosome segregation, we are still far removed from understanding the numerous potential molecular mechanisms of aneuploidy, especially in mammalian germ cells. Furthermore, the mechanisms of aneuploidy need to be studied in both male and female germ cells because of the inherent differences between male and female gametogenesis. The mouse aneuploid models presented here include those relevant to unique human disorders (Down syndrome) and those involving knockdown and knockout of genes needed for chromosome segregation. Also, models resulting from RNA interference technology and the exposure of cells to chemicals that specifically disrupt specific biochemical pathways or cellular organelles during gametogenesis are presented.

Key Words: Animal models, Aneuploidy, Female germ cells, Male germ cells, Molecular mechanisms.

SIGNIFICANCE OF ANEUPLOIDY FOR HUMAN HEALTH AND REPRODUCTION

Aneuploidy represents the greatest genetic affliction of humans. However, little is known about the causes and even less about the underlying molecular mechanisms of aneuploidy, especially in mammalian germ cells, where most aneuploidy is thought to arise. Human aneuploidy is linked with embryonic loss, mental and physical anomalies, and cancer. Approximately 10–30% of fertilized human eggs,[1] 10% of all human conceptuses,[2,3] 50% of spontaneous abortuses,[4,5] and 0.31% (204/64887) of human newborns[6] have an abnormal number of chromosomes. Even though numerous hypotheses have been proposed for the etiology of human aneuploidy, the only consistent findings are its positive correlation with maternal age[4,7,8] and its more frequent occurrence during female meiosis I for certain chromosomes.[9–12] Although the relationship between human aneuploidy and maternal age remains elusive, data suggest that such an association is linked to decreased frequencies of chiasmata formation,[13] abnormally placed chiasmata,[14] progressive losses of cohesion proteins,[15,16] and spindle checkpoint proteins.[17–19] It is now generally accepted that both nondisjunction and premature centromere separation (PCS) represent major events leading to human germ cell aneuploidy,[15,20,21] and that variation exists among chromosomes for the likelihood of missegregation.[22,23] Due the multitude of potential mechanisms, it is not surprising that a unique precept about the etiology of aneuploidy remains elusive. Recent information about the molecular mechanisms of chromosome segregation in various species is providing a foundation for research designed to investigate the causes and mechanisms of human aneuploidy.

While animal models for aneuploidy (e.g., trisomy models) have been useful for studying the consequences of aneuploidy, most animal models for the genesis of aneuploidy have not been specifically evaluated for their relevance to humans. Such comparisons are difficult because of ethical and sample size limitations and the biological variability inherent among human gametes. Thus, animal models are the only approach to obtain fundamental information about potential genetic hazards when humans are exposed to certain chemicals and other aneuploidy-inducing agents. It is important not only to identify those agents that significantly increase the incidence of aneuploidy in germ cells, but also to know if aneuploid germ cells are transmitted to progeny. However, the value of animal models goes far beyond the fact that they provide an experimentally malleable system for testing various agents. In fact, their greatest value to research on the etiology of aneuploidy may lie in the numerous genetic models that are available. These models enable determination of the roles of specific molecules in processes of meiotic chromosome dynamics and chromosome segregation that lead to gametic aneuploidy.

Here we review first, the basic processes of meiosis that, when they go awry, can lead to gametic aneuploidy, second, a sampling of the wealth of genetic models available for dissecting processes of meiosis and the genesis of gametic aneuploidy, and finally, what we have learned from analysis of environmental agents that cause aneuploidy. Our focus is on the laboratory mouse.

From: *Sourcebook of Models for Biomedical Research* (P. M. Conn, ed.), © 2008 Humana Press Inc., Totowa, NJ.

OVERVIEW OF MAMMALIAN GAMETOGENESIS

One interesting feature about human aneuploidy is its sexually dimorphic aspects, namely, it is highly correlated with maternal age and occurs more frequently in maternal meiosis for some chromosomes. Thus, much of the value of animal models lies in the exploitation of the sexual dimorphism of gametogenesis, which we highlight here in this overview of gametogenic processes and mechanisms, the context in which the origins of aneuploidy must be considered. Fundamental to the origins of aneuploidy is the process of meiosis. This specialized cell division process is defining of gametogenesis, as it occurs only in germ cells. In brief, the stages of meiosis include a premeiotic S-phase (often temporally longer than a mitotic S-phase), an extended prophase of meiosis I, during which chromosome pairing, synapsis, and recombination occur, and, finally, the two meiotic divisions that establish germ cell haploidy. For convenience, the extended meiosis I prophase is described in substages: during leptonema, chromosomes condense and begin to pair; during zygonema, chromosomes are paired and their intimate synapsis is initiated; during the pachytene stage, synapsis is completed and recombination (crossing over) occurs; and during the diplotene stage, desynapsis occurs. The intimate chromosome synapsis required for recombination is mediated by the synaptonemal complex (SC). The crossovers, visually apparent as chiasmata, serve to ensure the stable bipolar orientation of homologous pairs of chromosomes, a prerequisite to their correct segregation during the division process. Thus, although the primary defect leading to gametic aneuploidy may lie in prophase events, the nondisjunction that gives rise to incorrect chromosome segregation occurs in the meiotic division phases. The first meiotic division is a reductional one in which the homologous chromosomes are segregated into daughter cells; during the second, equational, division, sister chromatids are separated.

OOGENESIS Mammalian meiosis begins in the fetal ovary and is later arrested postpartum at the diplotene (dictyate) stage of meiosis I. Unless oocytes undergo atresia, they remain in diplotene until meiosis resumes in response to the luteinizing hormone (LH) surge prompting ovulation. Oocyte maturation (OM) is critical in setting up a correct meiotic division and encompasses the transition from diplotene to metaphase II (MII) and involves both nuclear and cytoplasmic remodeling.[24–27] In most mammals, metaphase II (MII) oocytes are ovulated and primed for fertilization, which initiates anaphase II.

Although differences in the biochemical pathways of OM exist among marine invertebrates, amphibians, fish, and mammals, the initiation and orderly temporal sequence of events during OM are influenced by species-dependent protein modifications by kinases and phosphatases.[28] In mammals, the intraoocyte titer of cyclic adenosine monophosphate (cAMP) plays a major role in initiating OM. Elevated levels of cAMP favor cAMP-dependent kinase activity and retention of oocytes in the diplotene stage of meiotic prophase. Conversely, low cAMP levels shift the equilibrium toward cAMP-dependent phosphatase activity, which is needed for activating maturation promoting factor (MPF).[24–27,29,30] MPF is composed of a 34-kDa catalytic subunit (CDC2A, sometimes referred to as $p34^{cdc2}$) that exhibits serine-threonine kinase activity and a 45-kDa cyclin B regulatory subunit (CCNB1). MPF is activated when CDC2A is dephosphorylated at the tyrosine 15 residue and coupled with cyclin B1, whereas, it becomes deactivated when tyrosine is phosphorylated.[31–33] MPF activity oscillates; it is highest during metaphase, but decreases at anaphase[34,35] and following fertilization[36–39] or parthenogenetic activation.[38,39]

Besides MPF, other kinases and phosphatases are also involved in the nuclear and cytoplasmic changes that occur during OM. Mitogen-activated protein kinases (MAPKs) are serine-threonine protein kinases that phosphorylate many of the same sites as active MPF.[40,41] These kinases mediate intracellular signal transmission in response to external stimuli, participate in assembling the first meiotic spindle, and prevent rodent oocytes from entering interphase during the interval between meiosis I and II.[42,43] Unlike MPF, MAPK activity remains high throughout OM.

Another kinase that has important functions during OM is the *Mos* protooncogene product MOS. This serine-threonine kinase is detected throughout OM[44–46] and exerts numerous species-dependent functions during OM including activation of the MAPK pathway.[47] MOS functions as a cytostatic factor by preventing premature exit from MII in mouse oocytes.[48,49] Oocytes from *Mos*-deficient mice are capable of completing OM, but fail to arrest at MII and subsequently undergo spontaneous parthenogenetic activation.[49–51] In addition to their roles during OM, the kinases MPF, MAPKs, and MOS also have important roles in the spindle assembly checkpoint (SAC), anaphase-promoting complex (APC) activation, and initiating the metaphase–anaphase transition (MAT).[40,47,52–54]

Upon completing OM, mouse oocytes remain in MII for a limited time period until fertilization, spontaneous activation, or atresia. Fertilization or parthenogenesis leads to elevated intraoocyte calcium levels that enable the formation of a calcium–calmodulin complex and calcium-dependent protein kinase II activation.[55] This kinase then initiates a cascade of kinase-phosphatase reactions that ultimately leads to unique dephosphorylation events, proteolysis of specific proteins, and inactivation of MPF, MAPKs, and MOS.

Considering that the synchronous interaction between unique kinases and phosphatases with their target compounds is needed for the normal temporal progression of OM and the MAT, it seems reasonable to suspect that alteration of their normal activities may lead to downstream effects manifested by chromosome missegregation. Based on their antagonistic effects relative to the degree of tyrosine CDC2A phosphorylation,[56] certain kinase and phosphatase inhibitors can alter the rate of OM and induce spindle abnormalities and aneuploidy in rodent oocytes. Okadaic acid (OA) specifically inhibits the protein phosphatases 1 (PPP1) and 2A (PPP2A) that dephosphorylate serine and threonine residues.[57,58] When mouse oocytes were treated with OA, MPF activation was inhibited[59] and abnormalities involving spindle fibers, multipolar spindles, kinetochores, and chromosome alignment were detected.[60–62] Furthermore, elevated frequencies of PCS and aneuploidy were found in mouse oocytes exposed to OA.[63] Some of the above observations may be related to the finding that OA treatment resulted in hyperphosphorylation of both microtubule-organizing centers and microtubule-associated proteins (MAPs)[59,60] and that hyperphosphorylated MAPs appear to have a reduced affinity for microtubules.[61]

The kinase inhibitor 6-dimethylaminopurine (6-DMAP) can disrupt CDC2A kinase and MAPK activities and reversibly block mouse oocytes at various stages during OM.[64–65] When mouse dictyate oocytes were exposed to 6-DMAP during *in vitro* culture, both protein phosphorylation and germinal vesicle breakdown

(GVBD) were inhibited, and when 6-DMAP was added after GVBD, first polar body formation was prevented.[64] Other data showed that 6-DMAP inhibited protein phosphorylation in activated mouse MII oocytes, which resulted in the premature disappearance of phosphorylated proteins and abnormalities involving polar body extrusion and pronuclei formation.[65] A relevant point about either *in vivo* or *in vitro* oocyte aging is that the pattern of protein dephosphorylation events appears to be correlated with increased frequencies of spontaneous oocyte activation and PCS.[15,66–71]

SPERMATOGENESIS The tempo and timetable of spermatogenesis differ markedly from that of oogenesis. Germ cells in fetal rodent testes enter into a period of mitotic arrest until shortly after birth. During the same period, the germ cells, positioned in the center of the seminiferous tubules, initiate ameboid activity and move between the Sertoli cells to establish themselves on the basal lamina of the tubules, a niche favorable for stem cell activity.[72] During the second week of life, large numbers of differentiated spermatogonia enter meiotic prophase, and by the end of the third week, the meiotic divisions occur and postmeiotic spermatid differentiation, or spermiogenesis, is initiated. Thus, the first wave of spermatogenesis is more or less synchronous, but subsequently, in adult life, these distinct processes of mitotic proliferation, meiosis, and spermiogenesis occur simultaneously in the testis and different generations of germ cells are present simultaneously. Because of the complexity of the adult testis, the first wave of spermatogenesis is frequently studied experimentally to determine temporal relationships and expression patterns, especially for meiotic events.

Unlike meiosis in the female germ cell, there is no meiotic arrest during spermatogenesis. After the extended prophase (more than a week long in mice), germ cells proceed through diplonema and the two meiotic divisions rapidly, in a period of 24 h or so. This rapid and uninterrupted timing has important consequences that may be informative about the relative incidence of gametic aneuploidy in males versus females. First, the recombination-based crossovers, manifest in chiasmata, are not "old" at the time of the meiotic division. Similarly, other "binder" proteins, such as chromosomal cohesins, are relatively "new" in spermatocytes at the time of the meiotic divisions. Together or separately, these could contribute to greater fidelity of metaphase chromosome alignment and subsequent segregation in male gametogenesis. In contrast, in females, both chiasmata and binder proteins are established during prophase, in fetal life, and thus they could be quite "old" at the time of the meiotic divisions, although nothing is known about the potential turnover and replacement of these proteins in oocytes.

Experimental analysis of chromosome segregation during male meiosis is hampered by the fact that very little is known about the regulation of the onset of the meiotic division phase in spermatocytes.[73,74] Although oocytes spontaneously enter the meiotic division phase upon removal from the follicle, removal of mid-prophase spermatocytes from their surrounding Sertoli cells does not prompt the onset of the division phase. Thus, the signal for the onset of the division phase is likely to be a positive one, either autonomous to the germ cell or emanating from the Sertoli cells. Although genetic evidence is lacking, the G_2/MI transition, as the progress from late meiosis I prophase to the first meiotic division is known, is likely to be mediated, as in oocytes, by the universal cell cycle regulator MPF.[75] Experimentally, the G_2/MI transition can be induced by treating pachytene spermatocytes with OA, but the exposed cells do not complete division, hampering analysis of chromosome segregation. Another experimental approach to the analysis of MI chromosome segregation is the injection of pachytene spermatocytes into oocytes, but in this case, chromosome segregation is not normal,[76] thus perhaps providing a model system for analysis of the genesis of gametic aneuploidy.

GENETIC MODELS FOR ANEUPLOIDY

The quintessential value of the mouse for the study of aneuploidy is the plethora of genetic models available. These include models of aneuploidy (e.g., mouse models for Down syndrome) for analysis of the consequences of aneuploidy, as well as single gene mutations that can be used to decipher the intricacies of chromosome segregation, contributing to the analysis of causes and mechanisms of aneuploidy. Prior to the development of such mutant gene models (mostly by genetic knockout technologies), studies on aneuploidy had been limited to various chemicals, mainly those that damage microtubules, that induce aneuploidy in mammalian germ cells,[77–79] which can be transmitted to progeny.[23,80–82] These studies are useful (and are covered in the section on Environmentally Induced Aneuploidy in Germ Cells), but lack the precision and power of genetic models. Gene knockout and knockdown models, RNA interference technology, and mutant genetic strains are now being employed for studying aneuploidy as summarized below. It should be noted that the gene models represent a selective list (Table 54–1) of those proving most useful for analysis of the genesis of aneuploidy, and also, almost by definition, involve only one event in a complex, and yet not completely understood, series of biochemical reactions and cellular organelles that orchestrate chromosome segregation. Most of these models directly relate to aneuploidy, but some involve events that predispose cells to aneuploidy. Importantly, the full potential of these models has not yet been realized; for example, they may provide "sensitized" models for testing of potential environmental disruptors of meiotic chromosome segregation.

Note: In the section below and in Table 54–1, standard gene and protein nomenclature, approved by international commissions on nomenclature, is used.

GENETIC MODELS FOR ANEUPLOID CONDITIONS The most studied human autosomal aneuploidy is trisomy 21 (Ts21) or Down syndrome. Several mouse models for Ts21 have become available.[83–88] The Ts(17^{16})65Dn, commonly known as Ts65Dn, mouse shows many features characteristic of Down syndrome[85,89] and has been particularly informative for teasing apart gene–phenotype relationships for the relevant chromosomal regions in attempts to understand the complexity of Down syndrome manifest in humans. In addition to these models for autosomal aneuploidy, models for sex chromosome aneuploidy have also been exploited to determine the consequences for physiology and meiotic pairing of extra (or deficient) sex chromosomes.[90–92]

ANEUPLOIDY IN MODELS FOR EVENTS OF CHROMOSOME PAIRING, SYNAPSIS, AND RECOMBINATION Chromosome pairing is mediated by cohesins, proteins that promote sister-chromatid cohesion, and by proteins of the SC. The synaptonemal complex protein 3, SYCP3 (encoded by the *Sycp3* gene in mice and by the *SYCP3* gene in humans), is a meiosis-specific structural protein needed for chromosomal structural integrity,

Table 54–1
Mammalian genetic models for chromosome missegregation

Gene	*Cell type*	*Cytogenetic event*	*Reference*
Sycp3[a]	Mouse oocyte	Aneuploidy	97
Smc1b[b]	Mouse oocyte	Univalents, single chromatids	101
Smc1b[b]	Mouse oocyte	Abnormal homolog synapsis, premature chromatid separation	100
Smc1b[b]	Mouse spermatocyte	Meiotic arrest	100
Bub1b[c]	Mouse fibroblasts	Aneuploidy	114
Mad2l1[d]	Mouse blastocyst	Chromosome missegregation	115
Mad2l1[d]	Human/mouse somatic cells	Premature chromatid separation	116
Mad2l1[d]	Mouse oocyte	Aneuploidy	19
Atrx[e]	Mouse oocyte	Abnormal chromosome alignment	109
Top3b[f]	Mouse spermatocytes	Aneuploidy	111
Mei1[g]	Mouse spermatocytes	Abnormal homolog synapsis	117
Mei1[g]	Mouse oocytes	Homolog asynapsis, abnormal chromosome alignment	117
Mlh1[h]	Mouse oocytes	Abnormal homolog synapsis	103, 104, 106
PL/J mice	Mouse spermatocytes	Homolog asynapsis, aneuploidy	119

[a]Synaptonemal complex protein.
[b]Structural maintenance of chromosomes protein.
[c]BubR1 spindle checkpoint protein.
[d]Mad2 spindle checkpoint protein.
[e]Centromeric binding protein.
[f]Encodes DNA topoisomerase IIIb.
[g]Meiosis defective 1.
[h]DNA mismatch repair protein.

chiasmata formation, and chromosome synapsis. Expression of SYCP3 protein is restricted to meiotic prophase I and disappears during diakinesis in mouse oocytes.[93] In meiotic prophase chromosome preparations, SYCP3 is found in the SC during meiotic prophase,[94] where it colocalizes with the RAD21 (also known as SCC1) cohesin subunit.[95] In meiosis division-phase spermatocytes, SYCP3 localizes to the centromeres by metaphase I and subsequently disappears by telophase I.[95] Spermatocytes in homozygous *Sycp3*$^{-/-}$ male mice exhibit synaptic failure and meiotic arrest during zygotene, and the male mice are infertile.[96] The females, however, are not infertile and oocytes from *Sycp3*$^{-/-}$ females display incomplete synapsis and alignment of homologs during pachytene and abnormal position and reduced frequency of chiasmata; chromosomes are found as univalents in oocytes and aneuploid zygotes are produced,[97] apparently by oocytes that evade checkpoint control.[98] This sexually dimorphic phenotype is not clear; it may in some way reflect the fact that SC proteins remain associated with centromeres until meiotic division phase in males, whereas they disappear prior to this in females.[93]

The "structural maintenance of chromosomes" (SMC) family of proteins includes both cohesins and condensins. SMC1B (also known as SMC1 beta, encoded by the mouse *Smc1b* gene) is a meiosis-specific isoform of a cohesin subunit.[99] Genetic knockout analysis reveals that SMC1B has a role in preserving chiasmata and maintaining chromosome cohesion until anaphase.[100,101] *Smc1b*-deficient mice also display sexual dimorphism; spermatocytes arrest during meiotic prophase, whereas oocytes progress to MII.[93,100] The pachytene arrest noted in males was attributed to incomplete synapsis of homologs. In oocytes, reduced recombination and premature loss of sister chromatid cohesion during meiotic prophase I suggest that SMC1B helps stabilize chiasmata until anaphase, and, interestingly, this may be a factor in the age-related increase in oocyte aneuploidy.[101]

The MLH1 DNA mismatch repair protein promotes meiotic recombination in yeast.[102] Mouse genetic models are available, as the mouse *Mlh1* gene has been knocked out.[103,104] In spermatocytes of *Mlh1*-null mice, chromosome pairs fall apart, presenting as univalents at metaphase I,[103] the condensed univalents do not assemble onto the spindle apparatus, and spermatocytes die, perhaps in response to a recombination or chiasmata checkpoint.[105] This was also found to be the case for female mutants,[106] where there is failure to exclude the second polar body and zygotic death.[104] These observations suggest that the MLH1 protein is essential for the maintenance of chiasmata. In a robust validation of the utility of such models for basic insights, it was subsequently found that MLH1 localizes to sites of crossing over that become the chiasmata.[107] Likewise, deficiency for the related MLH3 mismatch repair protein causes meiotic arrest and aneuploidy phenotypes.[108]

Other chromosomal proteins have also been found to play a role in chromosome dynamics that lead to correct meiotic chromosome segregation. ATRX (α-thalassemia/mental retardation syndrome X-linked homolog) is a centromeric-binding protein of the SNF2 family of helicase-ATPases. When ATRX activity was inhibited by antibody microinjection or RNA interference in mouse oocytes, chromosome alignment was disrupted during metaphase II.[109] TOP3B (DNA topoisomerase III beta) is a type IA DNA topoisomerase.[110] Targeted disruption of the *Top3b* gene in mice causes a homozygous phenotype of infertility with a high incidence of aneuploidy in male germ cells.[111] Topoisomerases are required for meiotic chromosome condensation and division in yeast and mammals,[112] but much remains to be learned about their precise role.

ANEUPLOIDY IN MODELS FOR CHECKPOINTS THAT MONITOR GENETIC INTEGRITY BUB1B (also known as BUB1R) comprises one of several spindle checkpoint proteins.[113]

Generation of *Bub1b* mutant mice resulted from a gene-targeting strategy creating a hypomorphic allele in mice embryonic stem cells.[114] Progressive reduction of BUB1B expression in mouse embryonic fibroblasts, spermatocytes, and oocytes resulted in aneuploidy. Also, a correlation between chronological aging of wild-type mice with diminished expression of BUB1B in somatic and germinal cells was noted. This relationship suggested a role for spindle checkpoint protein activity and the maternal age effect of aneuploidy.[114]

Another spindle checkpoint protein well studied in yeast is MAD2, known in the mouse as MAD2L1. Morpholino-based gene silencing of *Mad2l1* transcripts in mouse oocytes during meiosis I resulted in premature proteolysis of both cyclin B and securin, attenuating meiosis I and causing an elevated incidence of aneuploidy. Conversely, overexpression of a human *MAD2L1*-GFP construct in oocytes led to homolog nondisjunction.[19] Insufficiency of MAD2 L in mouse,[115] as well as partial knockout of *Mad2l1* in human and mouse mitotic cells,[116] has been shown to result in an accelerated rate of meiosis, PCS, and aneuploidy. Additional data from real-time polymerase chain reaction (PCR) experiments have demonstrated that *MAD2L1* transcripts in human oocytes decrease as a function of maternal age[17] as well as postovulatory age in mouse oocytes.[18] Together, these data indicate an important role for MAD2L1, a spindle checkpoint, in monitoring chromosome segregation.

OTHER GENETIC MODELS OF GAMETIC ANEUPLOIDY Meiosis-defective 1 (*Mei1*$^{-/-}$) mice were derived from chemically mutagenized embryonic stem cells.[117] The *Mei1* gene encodes a protein, MEI1, of unknown function.[118] Spermatocytes from homozygous *Mei1*$^{-/-}$ mice exhibited zygotene arrest with abnormal synapsis of homologs; but some oocytes progressed to metaphase I and displayed asynapsis and abnormal chromosome alignment.[117] Thus, when revealed, the function of the MEI1 protein may shed light on processes involved in setting up accurate meiotic chromosome segregation.

Males of the PL/J mouse strain are characterized by a high frequency of morphologically abnormal sperm, an elevated frequency of aneuploid sperm, as well as spermatocytes exhibiting chromosome asynapsis and other meiotic prophase abnormalities.[119] Genetic analysis revealed that these features are complex traits, and thus far the underlying genes have not yet been identified.[119] Also, senescence-accelerated mice (SAM) exhibit age-related defects in chromosome alignment during oocyte meiosis I and II.[120] This phenotype is reminiscent of that of the *Smc1b*-null mice, and it was speculated that cohesion defects in aged mice may be associated with misalignment and PCS.

ENVIRONMENTALLY INDUCED ANEUPLOIDY IN GERM CELLS

As mentioned previously, many studies have shown that various chemicals, mainly those that damage microtubules, can induce aneuploidy in mammalian germ cells.[77–79] Moreover, several studies have demonstrated that aneuploid germ cells can be transmitted to progeny.[23,80–82]

Animal models for aneuploidy have not been specifically evaluated for their relevance to human aneuploidy. Such comparisons are not possible due to ethical and sample size limitations plus the biological variability inherent among human oocytes. Because of this, animal models are the only experimental approach to obtain information about potential genetic hazards when humans are exposed to certain chemicals and other aneuploidy-inducing agents. As mentioned earlier, it is not only important to identify those agents that can significantly increase the incidence of aneuploidy in germ cells, but also to know if such aneuploid germ cells can be transmitted to offspring. Several relevant agents are described below and in Table 54–2.

FEMALE GERM CELLS Several review papers list compounds (aneugens) reported to induce aneuploidy in mammalian female germ cells.[23,78,79,121,122] Most of the chemicals reported in the earlier reviews dealt with spindle poisons such as colchicine, vinblastine sulfate, and benomyl, whereas, more recent articles have involved compounds that inhibit unique biochemical[63,123] or cellular organelle activities.[22,109] Thus, it is well-recognized that various types of compounds are capable of inducing aneuploidy in dividing cells. Based on the numerous unique cellular and biochemical reactions involved with chromosome segregation, the list of aneugens can be expected to increase. For example, it has recently been shown that exposure to bisphenol A can increase meiotic nondisjunction in mouse oocytes.[124] Furthermore, environmental agents can be expected to act not only on individual gametes, but also on the conceptus produced by their union. For example, conditions associated with fertilization *in vitro* may increase mitotic nondisjunction in early cleavage,[125] and the mouse is an excellent model for study of this phenomenon. Although identification of aneugens has merit, it is also of importance to describe their mode of action. Agents may act proximally, on mediators of cell division, but may also act distally and upstream, perhaps on endocrine receptors, to create a "susceptible" cellular environment. Thus, for each cell type, the molecular mechanisms of normal chromosome segregation need to be known as well as the effects of a particular perturbation during the chain of events comprising mitosis and meiosis. How such information

Table 54–2
Environmental agents affecting mammalian germ cell chromosome segregation

Agent	*Cell type*	*Cytogenetic event*	*Reference*
MG132[a]	Mouse oocytes	Meiotic arrest, aneuploidy, premature chromatid separation	167
Vanadate[b]	Mouse somatic cells	Aneuploidy, premature chromatid separation	168
	Mouse oocytes	Premature anaphase	
Okadaic acid[c]	Mouse oocytes	Aneuploidy, premature chromatid separation	63
Monastrol[d]	Mouse oocytes	Aneuploidy, premature chromatid separation	124

[a]Proteasome inhibitor.
[b]Tyrosine phosphatase inhibitor.
[c]Protein phosphatase 1 and 2A inhibitor.
[d]Kinesin EG5 inhibitor.

can be utilized for reducing the incidence of human aneuploidy still remains the ultimate task.

MALE GERM CELLS Experimental analysis of the effect of environmental agents on genesis of germ cell aneuploidy is facilitated in the male by the relative accessibility of the critical germ cell stages. Two primary germ cell stages are used for analysis of environmental agent effects directly on male germ cells: primary spermatocytes, for assessment of impairment of the SC and/or pairing and recombination, and sperm, for assessment of effects on chromosome segregation. In spite of the availability of these germ cell stages, the effect of environmental agents on transmissible aneuploidy can be detected only at the level of the zygote or offspring.

A number of agents can affect the synaptonemal complex structure of mouse spermatocytes, introducing breaks or discontinuities, which are detected after silver staining or with immunofluorescent labeling of surface-spread chromatin. These include colchicine,[126,127] cyclophosphamide,[128,129] and etoposide. The latter is the most thoroughly studied agent with respect to both the full spectrum of effects and species, and is an excellent example of the power of combined genetic, cellular, and cytogenetic analyses. First found in 1998 to produce peak mutagenicity in primary spermatocytes of the mouse,[130] etoposide was subsequently found to reduce crossing over[131] and to produce heritable chromosome abnormalities and aneuploidy.[132,133] Together, these studies illustrate the spectrum of techniques that can be brought to bear on an analysis of germ cell effects of environmental agents. The first discovery of germ cell stage sensitivity came from analysis of the mutagenicity of etoposide in a genetic-specific locus test,[134] relying on detection of specific mutations in offspring. This was followed by careful analysis of SC and chromosome morphology in surface spread chromatin coupled with genetic recombination analysis (of offspring) to show abnormalities in meiotic protein localization and reduced recombination.[131] The discovery that etoposide causes nondisjunction and gametic aneuploidy required scoring of offspring and, especially, the use of sperm FISH. This technique, involving hybridization of chromosome-specific probes to denatured sperm DNA *in situ*,[135] is among the most useful for direct determination of gametic aneuploidy. FISH analysis of sperm from etoposide-treated mice revealed increases in aneuploid sperm.[133,136,137] Interestingly, FISH data on sperm from treated human males (undergoing chemotherapy) also revealed an increase in the frequency of aneuploid sperm.[138] Etoposide and diazepam[139] are the only chemicals known for which both mouse and human sperm aneuploidy data are available. Even with the knowledge that sperm are aneuploid, transmissible aneuploidy can be detected only by examination of offspring, which has been a limiting factor in these studies. However, the technique of PAINT/DAPI has facilitated rapid and accurate scoring of chromosomes in zygotic first-cleavage metaphase[140]; application of this to studies of environmental agents could provide new insights.[141] Nonetheless, although these studies highlight the potential utility of rodent models for environmental induction of gametic aneuploidy, we still do not know the exact mechanism(s) by which etoposide or other environmental agents act in promoting nondisjunction and heritable gametic aneuploidy.

ETIOLOGY OF ANEUPLOIDY

Although the incidence of aneuploidy has not been routinely studied in mammals other than humans and rodents, the available data show that it occurs more frequently in humans than in rodents, and more frequently in females than in males. Is there a unique aspect of human female reproduction that renders it more vulnerable to chromosome missegregation? One feature is the absence of a defined estrus period during the menstrual/estrous cycle. Most domestic mammals ovulate during or shortly after the estrus period of their estrous cycle in order to ensure that freshly ovulated oocytes are fertilized.[142] Since this situation does not occur in humans, the possibility exists that fertilization of postovulatory aged oocytes might occur. Indeed, it has been proposed that fertilization of postovulatory aged oocytes may contribute to the higher incidence of aneuploidy in humans relative to other species.[6,143–147] Furthermore, two epidemiological studies offered support for an association between fertilization of aged human oocytes and early embryonic failure[148] and trisomic offspring.[143]

Postovulatory and *in vitro* oocyte aging involves a progressive, functional deterioration of the biochemical and cellular organelles essential for accurate chromosome segregation, normal fertilization, and embryonic development.[120,148] During this aging process, a narrow window of time exists for the optimal expression of gamete physiology. The fertilizable life span of mammalian oocytes ranges from 12 to 24 h and only 6 to 8 h in mares.[142] Although the fertilizable mean life span for both induced and naturally ovulated mouse oocytes is approximately 15 h postovulation, their optimal time for fertilization ranges from 4 to 6 h postovulation.[149–151] After this period, time-dependent intraoocyte changes take place among certain cellular organelles and biochemical profiles that can lead to apoptosis.[152–155]

In addition to postovulatory aging, other hypotheses include reduced and abnormally positioned chiasmata,[13,14] biological aging that is associated with reduced oocyte pools and reproductive hormones as maternal age increases,[156] compromised gonadal environment,[91,157] and premature centromere separation.[15,16,20] Recent data suggest that maternal age is positively correlated with progressive deterioration of cohesion between bivalents,[21,97] progressive loss of cohesion proteins,[95,101] and spindle checkpoint proteins.[17] Experimental data have shown that chronological aging of female mice deficient for the SMC1B chiasmata-binding protein led to deterioration of both chiasmata and arm cohesion as well as elevated levels of univalents and single chromatids in oocytes.[101] Recognizing that many of the reported aneuploid hypotheses are not necessarily mutually exclusive, it now appears that both preovulatory (maternal aging) and postovulatory-related alterations in the biochemical pathway regulating chromosome segregation can predispose oocytes to aneuploidy. Furthermore, these same cellular features governing chromosome segregation may also contribute to male gametic aneuploidy. Considering the numerous intracellular biochemical and organelle changes that can alter normal chromosome segregation, it seems reasonable to state that various molecular approaches will be needed to better understand the intricate details of aneuploidy.

FUTURE DIRECTIONS AND CONCLUSIONS

Since many potential mechanisms may lead to aneuploidy, specific investigations into the molecular mechanisms of aneuploidy that alter one event, or even several events, of the chromosome segregation pathway represent only a small component of the overall format underlying chromosome segregation. It has been estimated that approximately 500 yeast genes directly or indirectly control chromosome segregation.[158] Accurate chromosome segregation requires the temporally coordinated interaction among unique biochemical and cellular organelles. Some of these events include protein kinases and phosphatases, topoisomerases, microtubule kinetics, kinetochores and their proteins, centrosomes, motor proteins, passenger proteins, spindle checkpoint proteins, anaphase promoting complex, proteasomes, and securin, cohesin, and separin proteins. Abnormalities involving microtubule capture by kinetochores, removal of catenations between sister chromatids, chromosome condensation and biorientation, kinetochore–microtubule tension, and the spindle checkpoint have the potential for predisposing cells to aneuploidy.

Technologies such as DNA microarrays[159] and RNA-mediated interference of genes and protein kinases involved with cell cycle progression[160,161] offer avenues for deciphering cell cycle progression and chromosome segregation pathways. RNAi silencing and gene knockdown technologies have been used to alter the expression of the MAD2L1 protein in mouse oocytes[19,162] and BUB1B protein in mouse germ and somatic cells.[114] Reduced expression of these SAC proteins was associated with aneuploidy. Additionally, large-scale knockout strategies for genes upregulated during yeast meiosis found that deletion of specific genes required for maintaining centromeric cohesion during anaphase I resulted in chromosome missegregation.[163,164] Furthermore, data derived from genomic and proteomic analyses are expected to provide clues about the fundamental processes associated with cell division. Gene expression analyses of cancer cells showed that a subset of genes is universally activated in most cancers,[165] and that overexpression of cell division regulatory genes was linked with chromosome aberrations and neoplastic progression.[166] Continued research involving the chronological sequence of molecular events comprising chromosome segregation during mitosis and meiosis coupled with cytogenetic analyses is required to further our knowledge about the origin of gametic aneuploidy.[167,168]

ACKNOWLEDGMENTS

We are indebted to Alicia Valenzuela and Drs. Muriel Davisson and Fengyun Sun for helpful comments on the manuscript. M.A.H.'s contributions to this article were supported by a grant from the NIH, HD48998.

REFERENCES

1. Hassold T, Hunt PA. To err (meiotically) is human: The genesis of human aneuploidy. *Nat Rev Genet* 2001;2:280–291.
2. Ford JH. Nondisjunction. In: Burgio GR, Fraccaro M, Tiepolo L, Wolf U, Eds. *Trisomy 21*. Berlin: Springer-Verlag,1981:103–143.
3. Hansmann I. Factors and mechanisms involved in nondisjunction and X-chromosome loss. In: Sandberg AA, Ed. *Cytogenetics of the Mammalian X Chromosome, Part A: Progress and Topics in Cytogenetics*, Vol. 3A. New York: Alan R. Liss, 1983:131–170.
4. Bond DJ, Chandley AC. *Aneuploidy*. Oxford Monographs on Medical Genetics, No. 11. Oxford, UK: Oxford University Press, 1983.
5. Hook EB. The impact of aneuploidy upon public health: Mortality and morbidity associated with human chromosome abnormalities. In: Dellarco VL, Voytek PE, Hollaender A, Eds. *Aneuploidy: Etiology and Mechanisms*. New York: Plenum Press, 1985:7–33.
6. Hecht F, Hecht BK. Aneuploidy in humans: Dimensions, demography, and dangers of abnormal numbers of chromosomes. In: Vig BK, Sandberg AA, Eds. *Aneuploidy, Part A: Incidence and Etiology*. New York: Alan R. Liss, 1987:9–49.
7. Hook EB. Maternal age, paternal age, and human chromosome abnormality: Nature, magnitude, etiology, and mechanisms of effects. In: Dellarco VL, Voytek PE, Hollaender A, Eds. *Aneuploidy: Etiology and Mechanisms*. New York: Plenum Press, 1985: 117–132.
8. Chandley AC. Aneuploidy: An overview. In: Vig BK, Sandberg AA, Eds. *Aneuploidy: Part A: Incidence and Etiology*. New York: Alan R. Liss, 1987:1–8.
9. Lamson SH, Hook EB. A simple function for maternal age-specific rates of Down syndrome in the 20-to-48 year age range and its biological implications. *Am J Hum Genet* 1980;32:743–753.
10. Hassold T, Chiu D, Yamane JA. Parental origin of autosomal trisomies. *Ann Hum Genet* 1984;48:129–144.
11. Hassold TJ. The origin of aneuploidy in humans. In: Dellarco VL, Voytek PE, Hollaender A, Eds. *Aneuploidy: Etiology and Mechanisms*. New York: Plenum Press, 1985:103–115.
12. Nicolaidis P, Petersen MB. Origin and mechanisms of non-disjunction in human autosomal trisomies. *Hum Reprod* 1998;13: 313–319.
13. Henderson SA, Edwards RG. Chiasma frequency and maternal age in mammals. *Nature* 1968;218:22–28.
14. Lamb NE, Freeman SB, Savage-Austin A, Pettay D, Taft L, Hershey J, Gu YC, Shen J, Saker D, May KM, Avramopoulos D, Petersen MB, Hallberg A, Mikkelsen M, Hassold TJ, Sherman SL. Susceptible chiasmate configurations of chromosome 21 predispose to non-disjunction in both maternal meiosis I and meiosis II. *Nat Genet* 1996;14:400–405.
15. Angell RR, Xian J, Keith J, Ledger W, Baird DT. First meiotic division abnormalities in human oocytes: Mechanism of trisomy formation. *Cytogenet Cell Genet* 1994;65:194–202.
16. Angell RR. First-meiotic division nondisjunction in human oocytes. *Am J Hum Genet* 1997;61:23–32.
17. Steuerwald N, Cohen J, Herrera RJ, Sandalinas M, Brenner CA. Association between spindle assembly checkpoint expression and maternal age in human oocytes. *Mol Hum Reprod* 2001;7:49–55.
18. Steuerwald N, Steuerwald MD, Mailhes JB. Postovulatory aging of mouse oocytes leads to decreased MAD2 transcripts and increased frequencies of premature centromere separation and anaphase. *Mol Hum Reprod* 2005;11:623–630.
19. Homer HA, Mcdougall A, Levasseur M, Yallop K, Murdoch AP, Herbert M. Mad2 prevents aneuploidy and premature proteolysis of cyclin B and securin during meiosis I in mouse oocytes. *Genes Dev* 2005;19:202–207.
20. Angell RR. Predivision in human oocytes at meiosis I: A mechanism for trisomy formation in man. *Hum Genet* 1991;86:383–387.
21. Wolstenholme J, Angell RR. Maternal age and trisomy–a unifying mechanism of formation. *Chromosoma* 2000;109:435–438.
22. Sun FY, Schmid TE, Schmid E, Baumgartner A, Adler I-D. Trichlorfon induces spindle disturbances in V79 cells and aneuploidy in male mouse germ cells. *Mutagenesis* 2000;15:17–24.
23. Eichenlaub-Ritter U. Aneuploidy in aging oocytes and after toxic insult. In: Trounson A, Gosden RG, Eds. *Biology and Pathology of the Oocyte; Its Role in Fertility and Reproductive Medicine*. Cambridge, UK: Cambridge University Press, 2003:220–257.
24. Dekel N. Regulation of oocyte maturation: The role of cAMP. *Ann NY Acad Sci* 1988;541:211–216.
25. Schultz RM, Montgomery RR, Belanoff JR. Regulation of mouse oocyte maturation: Implication of a decrease in oocyte cAMP and protein dephosphorylation in commitment to resume meiosis. *Dev Biol* 1983;797:264–273.
26. Schultz RM. Regulatory functions of protein phosphorylation in meiotic maturation of mouse oocytes in vitro. In: Haseltine FP, First NL, Eds. *Meiotic Inhibition: Molecular Control of Meiosis*. New York: Alan R. Liss, 1988:137–151.
27. Racowsky C. Somatic control of meiotic status in mammalian oocytes. In: Haseltine FP, Heyner S. Eds. *Meiosis II, Contemporary Approaches to the Study of Meiosis*. Washington, DC: American Association for the Advancement of Science Press, 1993:107–116.
28. Yamasita M, Mita K, Yoshida N, Kondo T. Molecular mechanisms of the initiation of oocyte maturation: General and species-specific aspects. *Prog Cell Cycle Res* 2000;4:115–129.

29. Boernslaeger EA, Mattei P, Schultz RM. Involvement of cAMP-dependent protein kinase and protein phosphorylation in regulation of mouse oocyte maturation. *Dev Biol* 1986;114:453–462.
30. Downs SM, Daniel SAJ, Bornslaeger EA, Hoppe PC, Eppig JJ. Maintenance of meiotic arrest in mouse oocytes by purines: Modulation of cAMP levels and cAMP phosphodiesterase activity. *Gamete Res* 1989;23:323–334.
31. Dunphy WG, Kumagai A. The cdc25 protein contains an intrinsic phosphatase activity. *Cell* 1991;67:189–196.
32. Gautier J, Solomon MJ, Booher RN, Bazan JF, Kirschner ME. cdc25 is a specific tyrosine phosphatase that directly activates $p34^{cdc2}$. *Cell* 1991;67:197–211.
33. Strausfeld U, Labbe JC, Fesquet D, Cavadore JC, Picard A, Sadhu K, Russell P, Doree M. Dephosphorylation and activation of a $p34^{cdc2}$/cyclin B complex in vitro by human CDC25 protein. *Nature* 1991;351:242–245.
34. Arion D, Meifer L, Brizuela L, Beach D. cdc2 is a component of the M phase-specific histone H1 kinase: Evidence for identity with MPF. *Cell* 1988;55:371–378.
35. Draetta G, Beach D. Activation of cdc2 protein kinase during mitosis in human cells: Cell cycle dependent phosphorylation and subunit rearrangement. *Cell* 1988;54:17–26.
36. Choi T, Aoki F, Mori M, Yamashita M, Nagahama Y, Kohmoto K. Activation of $p34^{cdc2}$ protein kinase activity in meiotic and mitotic cell cycles in mouse oocytes and embryos. *Development* 1991; 113:789–795.
37. Fulka J Jr, Jung T, Moor RM. The fall of biological maturation promoting factor (MPF) and histone H1 kinase activity during anaphase and telophase in mouse oocytes. *Mol Reprod Dev* 1992;32:378–382.
38. Collas P, Sullivan EJ, Barnes FL. Histone H1 kinase activity in bovine oocytes following calcium stimulation. *Mol Reprod Dev* 1993;34:224–231.
39. Kikuchi K, Izaike Y, Noguchi J, Furukawa T, Daen FP, Naito K, Toyoda Y. Decrease of histone H1 kinase activity in relation to parthenogenetic activation of pig follicular oocytes matured and aged in vitro. *J Reprod Fertil* 1995;105:325–330.
40. Murray AW. MAP kinases in meiosis. *Cell* 1998;92:157–159.
41. Takenaka K, Moriguchi T, Nishida E. Activation of the protein kinase p38 in the spindle assembly checkpoint and mitotic arrest. *Science* 1998;280:599–602.
42. Sobajima T, Aoki F, Kohomoto K. Activation of mitogen-activated protein kinase during meiotic maturation in mouse oocytes. *J Reprod Fertil* 1993;97:389–394.
43. Verlhac MH, Kubiak JZ, Clarke HJ, Maro BH. Microtubule and chromatin behavior follow MAP kinase activity but not MPF activity during meiosis in mouse oocytes. *Development* 1994;120: 1017–1025.
44. Paules RS, Buccione R, Moschel RC, Vande Woude GF, Eppig JJ. Mouse mos protooncogene product is present and functions during oogenesis. *Proc Natl Acad Sci USA* 1989;86:5395–5399.
45. Sagata N. What does Mos do in oocytes and somatic cells? *BioEssays* 1997;19:13–21.
46. Singh B, Arlinghaus RB. Mos and the cell cycle. *Prog Cell Cycle Res* 1997;3:251–259.
47. Dekel N. Protein phosphorylation-dephosphorylation in the meiotic cell cycle of mammalian oocytes. *Rev Reprod* 1996;1:82–88.
48. Sagata N. Meiotic metaphase arrest in animal oocytes: Its mechanisms and biological significance. *Trends Cell Biol* 1996;6: 22–28.
49. Hashimoto N. Role of c-mos proto-oncogene product in the regulation of mouse oocyte maturation. *Horm Res* 1996;46:11–14.
50. Colledge WH, Carlton MBL, Udy GB, Evans MJ. Disruption of *c-mos* causes parthenogenetic development of unfertilized mouse eggs. *Nature* 1994;370:65–68.
51. Hashimoto N, Watanabe N, Furuta Y, Tamemoto H, Sagata N, Yokoyama M, Okazaki K, Nagayoshi M, Takeda N, Ikawa Y, *et al.* Parthenogenetic activation of oocytes in c-mos deficient mice. *Nature* 1994;370:68–71.
52. Hyman AA, Mitchison TJ. Two different microtubule-based motor activities with opposite polarities in kinetochores. *Nature* 1991;351: 206–211.
53. Karsenti E. Mitotic spindle morphogenesis in animal cells. *Semin Cell Biol* 1991;4:251–260.
54. Dorée M, Le Peuch C, Morin N. Onset of chromosome segregation at the metaphase to anaphase transition of the cell cycle. *Prog Cell Cycle Res* 1995;1:309–318.
55. Santella L. The role of calcium in the cell cycle: Facts and hypotheses. *Biochem Biophys Res Commun* 1998;244:317–324.
56. Jesus C, Rime H, Haccard O, Van Lint J, Goris J, Merlevede W, Ozon R. Tyrosine phosphorylation of $p34^{cdc2}$ and p42 during meiotic maturation of *Xenopus* oocyte: Antagonistic action of okadaic acid and 6-DMAP. *Development* 1991;111:813–820.
57. Cohen P, Holmes CF, Tsukitani Y. Okadaic acid: A new probe for the study of cellular regulation. *Trends Biochem Sci* 1990;15: 98–102.
58. Schönthal S. Okadaic acid—a valuable new tool for the study of signal transduction and cell cycle regulation? *New Biol* 1992;4:16–21.
59. Schwartz DA, Schultz RM. Stimulatory effect of okadaic acid, an inhibitor of protein phosphorylation in mouse oocytes and one-cell zygotes. *Dev Biol* 1991;145:119–127.
60. Vandre DD, Willis VL. Inhibition of mitosis by okadaic acid: Possible involvement of a protein phosphatase 2A in the transition from metaphase to anaphase. *J Cell Sci* 1992;101:79–91.
61. Zernicka-Goetz M, Kubiak JZ, Antony C, Maro B. Cytoskeletal organization of rat oocytes during metaphase II arrest and following abortive activation: A study by confocal laser scanning microscopy. *Mol Reprod Dev* 1993;35:165–175.
62. De Pennart H, Helene-Verlhac M, Cibert C, Santa Maria A, Maro B. Okadaic acid induces spindle lengthening and disrupts the interaction of microtubules with the kinetochores in metaphase II-arrested mouse oocytes. *Dev Biol* 1993;157:170–181.
63. Mailhes JB, Hilliard C, Fuseler JW, London SN. Okadaic acid, an inhibitor of phosphatase 1 and 2A, induces premature separation of sister chromatids during meiosis I and aneuploidy in mouse oocytes *in vitro*. *Chromosome Res* 2003;11:619–631.
64. Rime H, Neant I, Guerrier P, Ozon R. 6-Dimethylaminopurine (6-DMAP), a reversible inhibitor of the transition to metaphase during the first meiotic cell division of the mouse oocyte. *Dev Biol* 1989;133:169–179.
65. Szollosi MS, Kubiak JZ, Debey P, de Pennart H, Szollosi D, Maro B. Inhibition of protein kinases by 6-dimethylaminopurine accelerates the transition to interphase in activated mouse oocytes. *J Cell Sci* 1993;104:861–872.
66. Spielmann H, Krüger C, Stauber M, Vogel R. Abnormal chromosome behavior in human oocytes which remained unfertilized during human in vitro fertilization. *J In Vitro Fertil Embryo Transf* 1985;2:138–142.
67. Saito H, Koike K, Saito T, Nohara M, Kawagoe S, Hiroi M. Aging changes in the alignment of chromosomes after human chorionic gonadotropin stimulation may be a possible cause of decreased fertility in mice. *Horm Res* 1993;39:28–31.
68. Sakurada K, Ishikawa H, Endo A. Cytogenetic effects of advanced maternal age and delayed fertilization on first-cleavage mouse embryos. *Cytogenet Cell Genet* 1996;72:46–49.
69. Dailey T, Dale B, Cohen J, Nunné S. Association between non-disjunction and maternal age in meiosis II oocytes. *Am J Hum Genet* 1996;59:176–184.
70. Mailhes JB, Young D, London SN. 1,2-Propanediol-induced premature centromere separation in mouse oocytes and aneuploidy in one-cell zygotes. *Biol Reprod* 1997;57:92–98.
71. Mailhes JB, Young D, London SN. Postovulatory ageing of mouse oocytes in vivo and premature centromere separation and aneuploidy. *Biol Reprod* 1998;58:1206–1210.
72. Orth JM, Qiu JP, Jester WF, Pilder S. Expression of the c-kit gene is critical for migration of neonatal rat gonocytes in vitro. *Biol Reprod* 1997;57:676–683.

73. Handel MA, Cobb J, Eaker S. What are the spermatocyte's requirements for successful meiotic division? *J Exp Zool* 1999;285: 243–250.
74. Inselman A, Eaker S, Handel MA. Temporal expression of cell cycle-related proteins during spermatogenesis: Establishing a timeline for the onset of the meiotic divisions. *Cytogenet Genome Res* 2003;103:277–284.
75. Wiltshire T, Park C, Caldwell KA, Handel MA. Induced premature G2/M-phase transition in pachytene spermatocytes includes events unique to meiosis. *Dev Biol* 1995;169:557–567.
76. Kimura Y, Tateno H, Handel MA, Yanagimachi R. Factors affecting meiotic and developmental competence of primary spermatocyte nuclei injected into mouse oocytes. *Biol Reprod* 1998;59: 871–877.
77. Allen JW, Liang JC, Carrano AV, Preston RJ. Review of literature on chemical-induced aneuploidy in mammalian male germ cells. *Mutat Res* 1986;167:123–137.
78. Mailhes JB, Preston RJ, Lavappa KS. Mammalian in vivo assays for aneuploidy in female germ cells. *Mutat Res* 1986;167: 139–148.
79. Mailhes JB, Marchetti F. Chemically-induced aneuploidy in mammalian oocytes. *Mutat Res* 1994;320:87–111.
80. Mailhes JB, Preston RJ, Yuan ZP, Payne HS. Analysis of mouse metaphase II oocytes as an assay for chemically induced aneuploidy. *Mutat Res* 1988;198:145–152.
81. Mailhes JB, Yuan ZP, Aardema MJ. Cytogenetic analysis of mouse oocytes and one-cell zygotes as a potential assay for heritable germ cell aneuploidy. *Mutat Res* 1990;242:89–100.
82. Adler I-D, Schmid TE, Baumgartner A. Induction of aneuploidy in male mouse germ cells detected by the sperm-FISH assay: A review of the present data base. *Mutat Res* 2002;504:173–182.
83. Akeson EC, Lambert JP, Narayanswami S, Gardiner K, Bechtel LJ, Davisson MT. Ts65Dn–localization of the translocation breakpoint and trisomic gene content in a mouse model for Down syndrome. *Cytogenet Cell Genet* 2001;93:270–276.
84. Davisson MT. Mouse models of Down syndrome. *Drug Discov Today: Dis Models* 2005;2:103–109.
85. Davisson MT, Costa ACS. Mouse models of Down syndrome. In: Popko B, Ed. *Mouse Models in the Study of Genetic Neurological Disorders. Advances in Neurochemistry,* Vol. 9. New York: Kluwer Academic/Plenum, 1999.
86. Sago H, Carlson EJ, Smith DJ, Kilbridge J, Rubin EM, Mobley WC, Epstein CJ, Huang TT. Ts1Cje, a partial trisomy 16 mouse model for Down syndrome, exhibits learning and behavioral abnormalities. *Proc Natl Acad Sci USA* 1998;95:6256–6261.
87. O'Doherty A, Ruf S, Mulligan C, Hildreth V, Errington ML, Sesay A, Modino S, Vanes L, Hernandez D, Linehan JM, Sharpe PT, Brandner S, Bliss TV, Henderson DJ, Nizetic D, *et al.* An aneuploid mouse strain carrying human chromosome 21 with Down syndrome phenotypes. *Science* 2005;309(5743):2033–2037.
88. Villar AJ, Belichenko PV, Gillespie AM, Kozy HM, Mobley WC, Epstein CJ. Identification and characterization of a new Down syndrome model, Ts[Rb(12.1716)]2Cje, resulting from a spontaneous Robertsonian fusion between T(171)65Dn and mouse chromosome 12. *Mamm Genome* 2005;16:79–90.
89. Reeves RH, Irving NG, Moran T, Wohn A, Sisodia SS, Schmidt C, Davisson MT. A mouse model for Down syndrome exhibits learning and behavior deficits. *Nat Genet* 1995;11:177–183.
90. Hunt PA, Worthman C, Levinson H, Stallings J, LeMaire R, Mroz K, Park C, Handel MA. Germ cell loss in the XXY male mouse: Altered X chromosome dosage affects prenatal development. *Mol Reprod Dev* 1998;49:101–111.
91. Mroz K, Hassold TJ, Hunt PA. Meiotic aneuploidy in the XXY mouse: Evidence that a compromised testicular environment increases the incidence of meiotic errors. *Hum Reprod* 1999;14: 1151–1156.
92. Burgoyne PS, Ojarikre OA, Tuner JMA. Evidence that postnatal growth retardation in XO mice is due to haploinsufficiency for a non-PAR X gene. *Cytogenet Genome Res* 2002;99:252–256.
93. Hodges CA, LeMaire-Adkins R, Hunt PA. Coordinating the segregation of sister chromatids during the first meiotic division: Evidence for sexual dimorphism. *J Cell Sci* 2001;114:2417–2426.
94. Dobson MJ, Pearlman RE, Karaiskakis A, Spyropoulos B, Moens PB. Synaptonemal complex proteins: Occurrence, epitope mapping and chromosome disjunction. *J Cell Sci* 1994;107:2749–2760.
95. Parra MT, Viera A, Gomez R, Page J, Benavente R, Santos JL, Rufas JS, Suja JA. Involvement of the cohesion Rad21 and Scp3 in monopolar attachment of sister kinetochores during mouse meiosis I. *J Cell Sci* 2004;117:1221–1234.
96. Yuan L, Liu JG, Zhao J, Brundell E, Daneholt B, Hoog C. The murine SCP3 gene is required for synaptonemal complex assembly, chromosome synapsis, and male fertility. *Mol Cell* 2000;5:73–83.
97. Yuan L, Liu JG, Hoja MR, Wilbertz J, Nordqvist K, Hoog C. Female germ cell aneuploidy and embryo death in mice lacking the meiosis-specific protein SCP3. *Science* 2002;296:1115–1118.
98. Wang H, Hoog C. Structural damage to meiotic chromosomes impairs DNA recombination and checkpoint control in mammalian oocytes. *J Cell Biol* 2006;173:485–495.
99. Revenkova E, Eijpe M, Heyting C, Gross B, Jessberger R. Novel meiosis-specific isoform of mammalian SMC1. *Mol Cell Biol* 2001;21:6984–6998.
100. Revenkova E, Eifpe M, Heyting C, Hodges CA, Hunt PA, Liebe B, Scherthan H, Jessberger R. Cohesin SMC1beta is required for meiotic chromosome dynamics, sister chromatid cohesion and DNA recombination. *Nat Cell Biol* 2004;6:555–562.
101. Hodges CA, Revenkova E, Jessberger R, Hassold T, Hunt PA. SMC1beta-deficient female mice provide evidence that cohesions are a missing link in age-related nondisjunction. *Nat Genet* 2005;37:1351–1355.
102. Hunter N, Borts RH. Mlh1 is unique among mismatch repair proteins in its ability to promote crossing-over during meiosis. *Genes Dev* 1997;11:1573–1582.
103. Baker S, Plug A, Prolla T, Bronner C, Harris A, Yao X, Christie DM, Monell C, Arnheim N, Bradley A, *et al.* Involvement of mouse MLH1 in DNA mismatch repair and meiotic crossing over. *Nat Genet* 1996;13:336–342.
104. Edelmann W, Cohen PE, Kane M, Lau K, Morrow B, Bennett S, Umar A, Kunkel T, Cattoretti G, Chaganti R, *et al.* Meiotic pachytene arrest in MLH1-deficient mice. *Cell* 1996;85:1125–1134.
105. Eaker S, Cobb J, Pyle A, Handel MA. Meiotic prophase abnormalities and metaphase cell death in MLH1-deficient mouse spermatocytes: Insights into regulation of spermatogenic progress. *Dev Biol* 2002;249:85–95.
106. Woods LM, Hodges CA, Baart E, Baker SM, Liskay M, Hunt PA. Chromosomal influence on meiotic spindle assembly: Abnormal meiosis I in female *MLh1* mutant mice. *J Cell Biol* 1999;145:1395–1406.
107. Anderson LK, Reeves A, Webb LM, Ashley T. Distribution of crossing over on mouse synaptonemal complexes using immunofluorescent localization of MLH1 protein. *Genetics* 1999;151:1569–1579.
108. Lipkin SM, Moens PB, Wang V, Lenzi M, Shanmugarajah D, Gilgeous A, Thomas J, Cheng J, Touchman JW, Green ED, Schwartzberg P, Collins FS, Cohen PE. Meiotic arrest and aneuploidy in MLH3-deficient mice. *Nat Genet* 2002;31:385–390.
109. De La Fuente R, Viveiros MM, Wigglesworth K, Eppig JJ. ATRX, a member of the SNF2 family of helicase/ATPases, is required for chromosome alignment and meiotic spindle organization in metaphase II stage mouse oocytes. *Dev Biol* 2004;272:1–14.
110. Champoux JJ. DNA topoisomerases: Structure, function, and mechanism. *Annu Rev Biochem* 2001;70:369–413.
111. Kwan KY, Moens PB, Wang JC. Infertility and aneuploidy in mice lacking a type IA DNA topoisomerase IIIbeta. *Proc Natl Acad Sci USA* 2003;100:2526–2531.
112. Cobb J, Reddy RK, Park C, Handel MA. Analysis of expression and function of topoisomerase I and II during meiosis in male mice. *Mol Reprod Dev* 1997;46:489–498.

113. Nasmyth K. How do so few control so many? *Cell* 2005;120: 739–746.
114. Baker DJ, Jeganathan KB, Cameron JD, Thompson M, Juneja S, Kopecka A, Kumar R, Jenkins RB, de Groen PC, Roche P, van Deursen JM. BubR1 insufficiency causes early onset of aging-associated phenotypes and infertility in mice. *Nat Genet* 2004; 36:744–749.
115. Dobles M, Liberal V, Scott ML, Benerza R, Sorger PK. Chromosome missegregation and apoptosis in mice lacking the mitotic checkpoint protein Mad2. *Cell* 2000;101:635–645.
116. Michel LS, Liberal V, Chatterjee A, Kirchwegger R, Pasche B, Gerald W, Dobles M, Lorger PK, Murty VVVS, Benerza R. MAD2 haplo-insufficiency causes premature anaphase and chromosome instability in mammalian cells. *Nature* 2001;409:355–359.
117. Libby BJ, DeLaFuente R, O'Brien MJ, Wigglesworth K, Cobb J, Inselman A, Eaker S, Handel MA, Eppig JJ, Schimenti JC. The mouse meiotic mutation *mei1* disrupts chromosome synapsis with sexually dimorphic consequences for meiotic progression. *Dev Biol* 2002;242:174–187.
118. Bannister LA, Reinholdt LG, Munroe RJ, Schimenti JC. Positional cloning and characterization of mouse mei8, a disrupted allele of the meiotic cohesin Rec8. *Genesis* 2004;40:184–194.
119. Pyle A, Handel MA. Meiosis in male PL/J mice: A genetic model for gametic aneuploidy. *Mol Reprod Dev* 2003;64:471–481.
120. Liu L, Keefe DL. Aging-associated aberration in meiosis of oocytes from senescence-accelerated mice. *Hum Reprod* 2002;17:2678–2685.
121. Pacchierotti F. Chemically induced aneuploidy in germ cells of mouse. In: Vig BK, Sandberg AA, Eds. *Aneuploidy, Part B: Induction and Test Systems, Progress and Topics in Cytogenetics*, Vol. 7b. New York: Alan R. Liss, 1988:123–139.
122. Mailhes JB, Marchetti F. Mechanisms and chemical induction of aneuploidy in rodent germ cells. *Cytogenet Genome Res* 2005;111: 384–391.
123. Mailhes JB, Mastromatteo C, Fuseler JW. Transient exposure to the EG5 kinesin inhibitor monastrol leads to syntelic orientation of chromosomes and aneuploidy in mouse oocytes. *Mutat Res* 2004;559:153–167.
124. Hunt PA, Koehler KE, Susiarjo M, Hodges CA, Ilagan A, Voight RC, Thomas S, Thomas BF, Hassold TJ. Bisphenol a exposure causes meiotic aneuploidy in the female mouse. *Curr Biol* 2003;13: 546–553.
125. Bean CJ, Hassold TJ, Judis L, Hunt PA. Fertilization in vitro increases non-disjunction during early cleavage divisions in a mouse model system. *Hum Reprod* 2002;17:2362–2367.
126. Tepperberg JH, Moses MJ, Nath J. Colchicine effects on meiosis in the male mouse .1. Meiotic prophase: Synaptic arrest, univalents, loss of damaged spermatocytes and a possible checkpoint at pachytene. *Chromosoma* 1997;106:183–192.
127. Tepperberg JH, Moses MJ, Nath J. Colchicine effects on meiosis in the male mouse—II. Inhibition of synapsis and induction of nondisjunction. *Mutat Res* 1999;429:93–105.
128. Allen JW, Gibson JB, Poorman PA, Backer LC, Moses MJ. Synaptonemal complex damage induced by clastogenic and anti-mitotic chemicals: Implications for non-disjunction and aneuploidy. *Mutat Res* 1988;201:313–324.
129. Backer LC, Gibson JB, Moses MJ, Allen JW. Synaptonemal complex damage in relation to meiotic chromosome aberrations after exposure of male mice to cyclophosphamide. *Mutat Res* 1988;203:317–330.
130. Russell LB, Hunsicker PR, Johnson DK, Shelby MD. Unlike other chemicals, etoposide (a topoisomerase-II inhibitor) produces peak mutagenicity in primary spermatocytes of the mouse. *Mutat Res* 1998;400:279–286.
131. Russell LB, Hunsicker PR, Hack AM, Ashley T. Effect of the topoisomerase-II inhibitor etoposide on meiotic recombination in male mice. *Mutat Res* 2000;464:201–212.
132. Marchetti F, Bishop JB, Lowe X, Generoso WM, Hozier J, Wyrobek AJ. Etoposide induces heritable chromosomal aberrations and aneuploidy during male meiosis in the mouse. *Proc Natl Acad Sci USA* 2001;98:3952–3957.
133. Russell LB, Hunsicker PR, Kerley M, Pyle A, Saxton AM. Etoposide exposure during male mouse pachytene has complex effects on crossing-over and causes nondisjunction. *Mutat Res* 2004;565: 61–77.
134. Russell LB. Effects of male germ-cell stage on the frequency, nature and spectrum of induced specific-locus mutations in the mouse. *Genetica* 2004;122:25–36.
135. Davisson MT, Handel MA. Cytogenetics. In: Smith AG, Ed. *The Mouse in Biomedical Research*. Amsterdam, The Netherlands: Elsevier, 2006:146–164.
136. Attia SM, Schmid TE, Badary OA, Hamada FM, Adler ID. Molecular cytogenetic analysis in mouse sperm of chemically induced aneuploidy: Studies with topoisomerase II inhibitors. *Mutat Res* 2002;520:1–13.
137. Marchetti F, Pearson FS, Bishop JB, Wyrobek AJ. Etoposide induces chromosomal abnormalities in mouse spermatocytes and stem cell spermatogonia. *Hum Reprod* 2006;21:888–895.
138. De Mas P, Daudin M, Vincent MC, *et al.* Increased aneuploidy in spermatozoa from testicular tumour patients after chemotherapy with cisplatin, etoposide and bleomycin. *Hum Reprod* 2001;16:1204–1208.
139. Baumgartner A, Schmid TE, Schuetz CG, Adler ID. Detection of aneuploidy in rodent and human sperm by multicolor FISH after chronic exposure to diazepam. *Mutat Res* 2001;490:11–19.
140. Marchetti F, Lowe X, Moore DI, Bishop J, Wyrobek AJ. Paternally inherited chromosomal structural aberrations detected in mouse first-cleavage zygote metaphases by multicolor fluorescence in situ hybridization painting. *Chromosome Res* 1996;4:604–613.
141. Marchetti F, Bishop JB, Cosentino L, Moore D, 2nd, Wyrobek AJ. Paternally transmitted chromosomal aberrations in mouse zygotes determine their embryonic fate. *Biol Reprod* 2004;70: 616–624.
142. Hafez ESE. *Reproduction in Farm Animals*, 6th ed. Philadelphia, PA: Lea & Febiger, 1993:144–164.
143. Juberg RC. Origin of chromosomal abnormalities: Evidence for delayed fertilization in meiotic nondisjunction. *Hum Genet* 1983;64:122–127.
144. Blazak WF. Incidence of aneuploidy in farm animals. In: Vig BK, Sandberg AA, Eds. *Aneuploidy, Part A: Incidence and Etiology*. New York: Alan R. Liss, 1987:103–116.
145. Mailhes JB. Incidence of aneuploidy in rodents. In: Vig BK, Sandberg AA, Eds. *Aneuploidy, Part A: Incidence and Etiology*. New York: Alan R. Liss,1987:67–101.
146. Pellestor F. Frequency and distribution of aneuploidy in human female gametes. *Hum Genet* 1991;86:283–288.
147. Zenzes MT, Casper RF. Cytogenetics of human oocytes, zygotes, and embryos after in vitro fertilization. *Hum Genet* 1992;88: 367–375.
148. Wilcox AJ, Weinberg CR, Baird DD. Post-ovulatory ageing of the human oocyte and embryo failure. *Hum Reprod* 1998;13:394–397.
149. Lewis WH, Wright ES. On the early development of the mouse egg. *Carnegie Inst Contrib Embryol* 1935;25:113–143.
150. Edwards RG, Gates AH. Timing of the stages of the maturation divisions, ovulation, fertilization and the first cleavage of eggs of adult mice treated with gonadotrophins. *J Endocrinol* 1959;18:292–304.
151. Marston JH, Chang MC. The fertilizable life of ova and their morphology following delayed insemination in mature and immature mice. *J Exp Zool* 1964;155:237–252.
152. Exley GE, Tang C, McElhinny AS, Warner CM. Expression of caspase and BCL-2 apoptotic family members in mouse preimplantation embryos. *Biol Reprod* 1999;61:231–239.
153. Morita Y, Tilly JL. Oocyte apoptosis: Like sand through an hourglass. *Dev Biol* 1999;213:1–17.
154. Perez GI, Tao XJ, Tilly JL. Fragmentation and death (a.k.a. apoptosis) of ovulated oocytes. *Mol Human Reprod* 1999;5:414–420.
155. Gordo AC, Rodrigues P, Kurokawa M, Jellerette T, Exley GE, Warner C, Fissore R. Intracellular calcium oscillations signal

apoptosis rather than activation in in vitro aged mouse eggs. *Biol Reprod* 2002;66:1828–1837.
156. Warburton D. Biological aging and the etiology of aneuploidy. *Cytogenet Genome Res* 2005;111:266–272.
157. Hodges CA, Ilagan A, Jennings D, Keri R, Nilson J, Hunt PA. Experimental evidence that changes in oocyte growth influence meiotic chromosome segregation. *Hum Reprod* 2002;17:1171–1180.
158. Yanagida M. Basic mechanisms of eukaryotic chromosome segregation. *Phil Trans R Soc B* 2005;360:609–621.
159. Schlecht U, Primig M. Mining meiosis and gametogenesis with DNA microarrays. *Reproduction* 2003;125:447–456.
160. Stein P, Svoboda P, Schultz RM. Transgenic RNAi in mouse oocytes: A simple and fast approach to study gene function *Dev Biol* 2003;256:187–193.
161. Bettencourt-Dias M, Giet R, Sinka R, Mazumdar A, Lock WG, Balloux PJ, Safiropoulos PJ, Yamaguchi S, Winter S, Carthew RW, Cooper M, Jones D, Frenz L, Glover DM. Genome-wide survey of protein kinases required for cell cycle progression. *Nature* 2004;432:980–987.
162. Prawitt D, Brixel L, Spangenberg C. Eshkind L, Heck R, Oesch F, Zabel B, Bockamp E. RNAi knock-down mice: An emerging technology for post-genomic functional genetics. *Cytogenet Genome Res* 2004;105:412–421.
163. Gregan J, Rabitsch PK, Sakem B, Csutak O, Latypov V, Lehmann E, Kohli J, Nasmygh K. Novel genes required for meiotic chromosome segregation are identified by a high-throughput knockout screen in fission yeast. *Curr Biol* 2005;15:1663–1669.
164. Marston AL, Tham WH, Shah H, Amon A. A genome-wide screen identifies genes required for centromeric cohesion. *Science* 2004;303:1367–1370.
165. Rhodes DR, Yu J, Shanker K, Deshpande N, Varambally R, Ghosh D, Barrette T, Pandey A, Chinnaiyan AM. Large-scale meta-analysis of cancer microarray data identifies common transcriptional profiles of neoplastic transformation and progression. *Proc Natl Acad Sci USA* 2004;101:9309–9314.
166. Rajagopalan H, Lengauer C. Aneuploidy and cancer. *Nature* 2004; 432:338–341.
167. Mailhes JB, Hilliard C, Lowery M, London SN. MG-132, an inhibitor of proteasomes and calpains, induced inhibition of oocyte maturation and aneuploidy in mouse oocytes. *Cell Chromosome* 2002;1:2.
168. Mailhes JB, Hilliard C, Fuseler JW, London SN. Vanadate, an inhibitor of tyrosine phosphatases, induced premature anaphase in oocytes and aneuploidy and polyploidy in mouse bone marrow cells. *Mutat Res* 2003;538:101–107.

55 Genetic Models of Alzheimer's Disease

YANN S. MINEUR

ABSTRACT

This chapter presents and discusses different strategies to dissect genetically the molecular mechanisms of Alzheimer's disease. First, a short preamble sketches the common phenotypical features of Alzheimer's disease. Second, a basic introduction defines the characteristics of a good animal model along with the different molecular strategies commonly used to develop transgenic animals. Third, an exclusive census of genetically modified mouse models describes some of the remarkable characteristics observed in these animals and tries to weigh the pros and cons of each of them. Finally, a brief note highlights gene–environment interactions relevant to the field of genetic mouse models and Alzheimer's disease. All sections are illustrated with examples focusing on behavioral phenotypes with a specific interest on learning and memory.

Key Words: Alzheimer's disease, Transgenic, Mouse models, Neuroanatomy, Behavior, Learning, Memory, Amyloid, Secretases, Tau.

INTRODUCTION

Alzheimer's disease (AD) is expected to affect 16 million individuals in the United States by the year 2050. While the onset of AD-associated dementia can be subtle, the inexorable evolution of this disease ultimately leads to global cognitive deficits including memory, orientation, reasoning, and judgment. Unfortunately, besides limited cholinergic therapies aimed at improving cognitive decline in patients, no effective therapies are currently available.

Extracellular neuritic plaques and intraneuronal fibrillary tangles (NFTs) are the two classical hallmarks of AD pathology in the brain.[1,2] Neuritic plaques are typically composed of β-amyloid oligomers (Aβ) surrounding dystrophic neurites. These plaques will precipitate and accumulate (primarily in the cortex and the limbic system),[3–5] leading to neuronal death, which is thought to be responsible for dementia observed in AD patients[6] (see Figure 55–1).

Tau, a microtubule-associated protein, is the primary component of NFTs.[7] In AD, tau is hyperphosphorylated, resulting in disruption of microtubule dynamics, impaired axonal transport, and intraneuronal polymerization of the cytoskeleton. These molecular events will also result in neuronal death.

For a pathological diagnosis of AD, neuritic plaques and NFTs are required.

It is now commonly believed that abnormal production and aggregation of Aβ (especially the more fibrillogenic and insoluble $A\beta_{42}$ isoform) are the primary cause of the plaques found in AD.

ANIMAL MODELS OF ALZHEIMER'S DISEASE?

Investigating the molecular mechanisms of AD in human subjects is particularly difficult because only postmortem tissue can be studied, once the disease has reached its most advanced stages. Neuronal tissue harvested from human brain is generally fragile and very sensitive, affected by aging, medications, and other possible neurological defects. In addition, the complexity is increased by broad individual variability due to genetic and environmental factors. Moreover, the availability of brain tissue is limited. Most of these difficulties can be overcome by using animal models.

Given the wealth of data available on models of AD, this chapter will principally focus on the mouse. Mice are the most popular animal models (for AD), primarily because their genome can be manipulated easily. This advantage has led to the generation of a variety of genetic mouse models, although rat models do exist as well (see, for instance, Hu *et al.*[8]).

Nevertheless, invertebrate models such as *Drosophila melanogaster* and *Caenorhabditis elegans* should not be thrust aside too hastily. In spite of their clear limitations—mainly the lack of mammalian brain structures and greater genetic distance from human than mice—these model organisms can aid in the study of AD. These "limitations" can also be an advantage: simpler organisms are generally easier for mapping of neuronal networks and for investigating molecular interactions. From a practical point of view, these animals are also smaller, cheaper to house and breed, and have shorter generation times than any mammal. Thus, many generations can be produced in a short time span. As a result, transgenic fly and worm AD models have been developed for both Aβ and tau expression.[9–11]

HOW DO WE DEFINE A "GOOD" MODEL OF ALZHEIMER'S DISEASE?

Modeling a disease in an animal can be challenging. A good AD model should, as much as feasible, present the following features:

- Reproducibility: a good model should yield consistent phenotypic features across generations and time, and

From: *Sourcebook of Models for Biomedical Research* (P. M. Conn, ed.), © 2008 Humana Press Inc., Totowa, NJ.

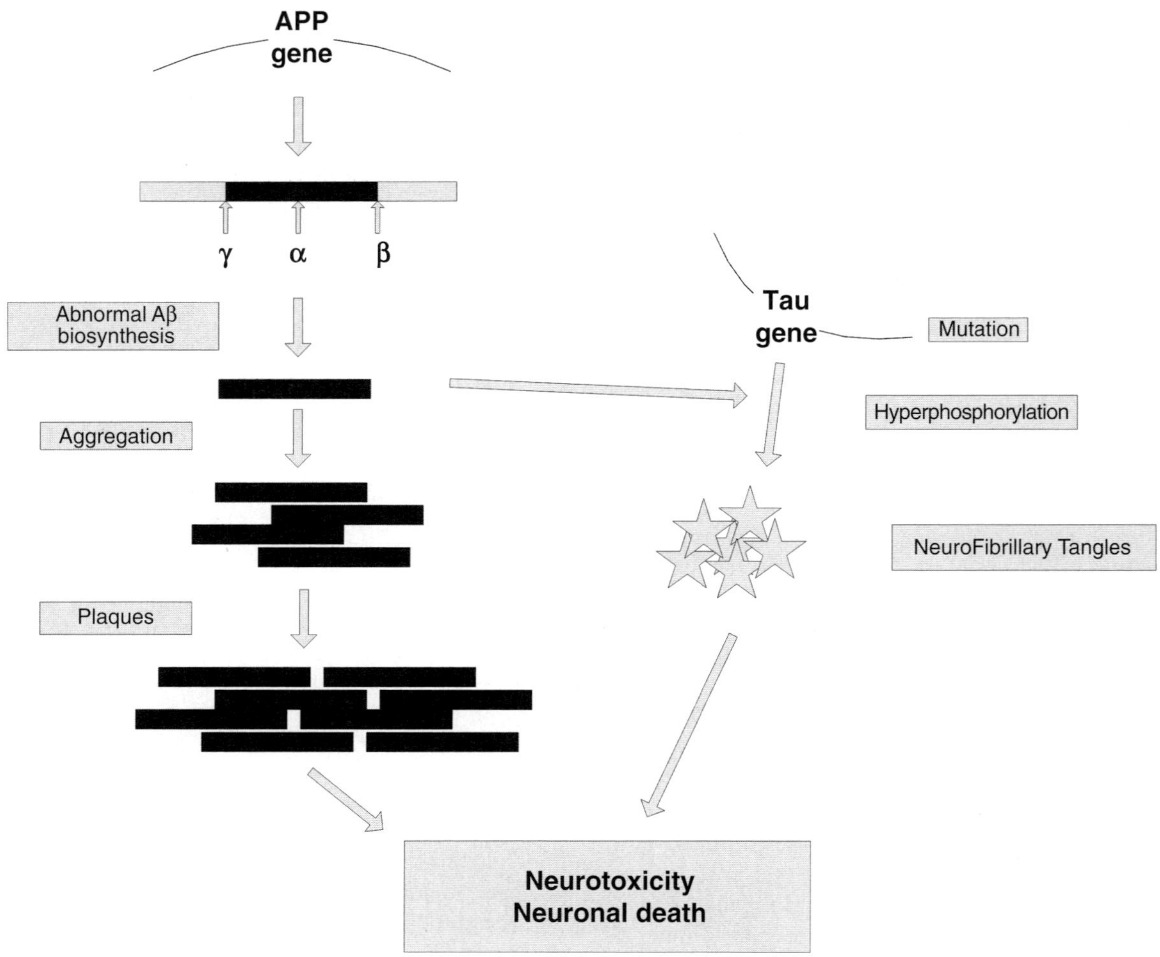

Figure 55–1. Simplified overview of the amyloid cascade hypothesis of Alzheimer's disease. See text for details.

possibly across investigators and laboratories. However, the latter can be quite difficult to achieve despite the best efforts to do so.[12]

- Face validity: the phenotypic features should "look" the same as those observed in the human disease. Of course, this can be problematic for phenotypic traits related to higher cognitive function and/or for human characteristics impossible to assess and model in mice. Nevertheless, a plethora of behavioral assays performed in mice permits evaluation of memory processing and cognitive functions, for instance. In addition, physiological and biochemical analyses allow the identification of amyloid plaques and subsequent neuronal damage similar to that observed in AD.
- Construct validity: the model either relies on, or reveals, the *same* basic *underlying mechanisms* as AD, including the accumulation of Aß peptides, hyperphosphorylation of tau, and subsequent neuronal death. By extension, genetic validity is part of this category when a disease or its risk factors engage genetic components similar to the ones in human subjects.
- Predictive validity: the observed effects of particular disease-triggering elements should be similar between the animal model and human subjects. For instance, β amyloid-induced plaques with neuronal losses would constitute good predictive validity of an animal model of AD. A good predictive validity can also help in defining how useful and/or safe a drug/treatment can be.

These three latter categories are not mutually exclusive one with another. Phenotypical factors can belong to one, two, or all of the categories.

GENETIC MOUSE MODELS OF ALZHEIMER'S DISEASE

It is crucial to realize that "natural" development of plaques and tangles does not occur in mice (or other species commonly used in scientific research). Thus, engineering a model for AD requires experimental manipulation, either genetic and/or pharmacological, the latter generally consisting of the exogenous administration of different Aβ peptides into the normal rodent brain. Both techniques have advantages and disadvantages, but are essentially complementary. This chapter will focus on genetic models of AD. For a review of mouse models of AD that use exogenous Aβ administration and their comparison to transgenic models, the reader is referred to Stephan and Phillips.[13]

Genes encoding proteins shown to play a role in the onset of AD pathology would be the prime candidates for manipulations. Consequently, a large number of genetically modified mouse lines have been generated using several of the following common techniques:

- Gene deletion or knockout (KO): in these models, defined gene(s) expression is absent throughout the life of the animal, from the earliest stage of development. Refinements have been added to this techniques resulting in conditional KO mice, in which the expression of a gene is abolished but only in restricted brain area(s)/neuronal types, and/or at a specific time during development of the animal. The deletion of a gene can also be reversible.
- Transgenic overexpression: in this case, a foreign gene, e.g., human APP (hAPP), is introduced into the genome and its expression (level and region) is driven by specific promoters.
- Targeted mutation or knockin (KI): a specific region of a gene is mutated, leading to loss or alteration of the proteins encoded by the targeted gene.

Combinations of these genetic techniques are possible and are becoming common in mouse models of AD. The following section will describe and compare several AD mouse models generated to date.

APP TRANSGENICS The first successful genetically engineered mouse models of AD were transgenic mice, characterized by increased Aβ levels induced by overexpression of its precursor (APP). Commonly, hAPP—and usually a mutant form linked to the early-onset familial form of AD—is introduced into the mouse genome. In turn, this gene transfer leads to the expression of the transgene. To date, numerous hAPP models have been generated including PDAPP, Tg2567, APP23, TgCRNDN8, and J20. Each model uses either different APP mutations, different promoters, or a different genetic background of the host mouse strain (see Table 55–1 for more details and http://www.alzforum.org/res/com/tra/ for a comprehensive list of transgenic models). Not surprisingly, the phenotypes are therefore dissimilar between these strains, characterized by different qualitative and quantitative levels of neuroanatomical defects.

These strains have all been assessed behaviorally, some extensively. PDAPP and Tg2567 have been studied in great detail and high APP levels induced by the transgene were shown to be responsible for cognitive abnormalities in both lines. Both lines are considered to be reliable AD models, because they exhibit clear learning deficits over time and across laboratories. Moreover, the cognitive decline observed in these animals was observed in many cognitive tasks. For instance, Tg2576 mice, developed by Hsiao *et al.*,[14] show a progressive impairment in various forms of the water navigation task,[15] deficits in a version of the Barnes maze,[16] and poorer performance in a T-maze alternation task and contextual fear conditioning.[17]

SECRETASE TRANSGENICS The β- and γ-secretases catalyze the processing of APP into the various forms of Aβ, whereas α-secretase is part of the nonamyloidogenic pathway (see Figure 55–1). Thus, genes encoding these secretases are candidate genes of interest for AD.

β-Secretase BACE1 is the predominant neuronal β-secretase. Both transgenic mice overexpressing BACE1 and knockout mice lacking BACE1 have been generated and, intriguingly, at least one of their behavioral phenotypes is opposite: BACE1 KO mice exhibit more anxiety-like behaviors than controls, whereas transgenics showed less anxiety-like behaviors. However, cognitive deficits have not been observed.

The development of double transgenic mice generated by crossbreeding mice overexpressing hAPP with those either lacking or overexpressing BACE1 has also helped identify the role of BACE1 in AD. As expected, double transgenic mice (hBACE/Tg2576) have increased amyloid pathology, high concentrations of both total Aβ and $A\beta_{42}$, and greater numbers of plaques than hAPP mice alone. However, elimination of BACE1 in the presence of hAPP rescues specific cognitive deficits associated with Aβ deposits.[18]

γ-Secretase The secondary cleavage of APP leading to $A\beta_{42}$ involves the activity of γ-secretase. Functional γ-secretase is a complex holoenzyme made of several individual enzymes, including presenilin (PS)1, PS2, Nicastrin, Aph-1, and Pen-2.[19,20] The presenilins (PS1, PS2) have been thoroughly investigated since human subjects with mutations in PS1 show early-onset AD. Many genetically engineered mouse models have been developed including KOs, transgenics, and KIs, double and even triple transgenics. While PS1 KO mice are not viable,[21] the development of conditional KOs (cKOs), in which the loss of the gene was limited to the postnatal forebrain, circumvented this lethality. PS1 cKO mice show mild cognitive impairments in long-term spatial reference memory and retention.[22]

Overexpression of a human form of PS1 (hPS1) in mice, however, induced only slight behavioral perturbations. KIs generated by a targeted missense mutation in murine PS1 overproduce $A\beta_{42}$; however, this does not lead to the aggregation of amyloid plaques. Nonetheless, mice perform poorly in an object recognition task, but are similar to controls in a water navigation test. These results suggest that changes in PS1 activity could influence

Table 55–1
Examples of different APP transgenic mice

Common abbreviations	*Transgene*	*Promoter*	*Genetic background*
PDAPP	Human APP with Val-717 substituted by Phe (Indiana mutation)	Human platelet-derived growth factor β (PDGF- β) chain gene promoter	Swiss × C57BL/6 × DBA2
Tg2567	Human APP with mutations Lys-670 substituted to Asn and Met-671 to Leu (Swedish mutation)	Hamster prion promoter	C57BL/6 × SJL
APP23	Human APP751 with Swedish mutation	Murine Thy-1.2 gene	C57BL/6
TgCRNDN8	APP695 with Swedish and Indiana mutations	Hamster prion promoter	C3H/He × C57BL/6J
J20	APP695 with Swedish and Indiana mutations	Human platelet-derived growth factor β (PDGF- β) chain gene promoter	C57BL/6 × DBA2

hippocampus-dependent memory pathways.[23,24] Mice lacking PS2 do not exhibit obvious neurological or behavioral abnormalities, and can reproduce successfully. In contrast, PS2 overexpressing transgenic mice and KIs with mutations in PS1 exhibit poor performance in a water navigation task. cPS1/APP double transgenic mice have also been generated (cPS1 carries a neuron-specific conditional KO of the PS1 gene). In this line, accumulation of Aβ is faster and Aβ levels are higher. Behaviorally, cPS1/APP double transgenic mice performed poorly in several cognitive tasks (for details, see Kobayashi and Chen[18]). As was seen in hAPP/PS double transgenics, the cPS1/APP double transgenic mice do not develop amyloid plaque; behavioral abnormalities in an object recognition task were rescued in the double transgenic mice.[25] This reinforces the conclusion that inhibition of secretase could constitute a promising target for AD treatment or prevention.

α-Secretase α-secretase cleaves about 90% of APP. However, α-secretase induces production of the P3 peptide, which does not appear to be associated with amyloid neurotoxicity. Although P3 can be detected in amyloid plaques, very few reports suggest deleterious effects on neurons. Hence, the α-secretase-processing pathway is referred to as nonamyloidogenic.[26,27]

Numerous enzymes have α-secretase activity including the ADAM (a disintegrin and metalloproteinase) family. These proteins all have α-secretase cleavage activity, but differential patterns of expression.[28–30]

In "normal" mice, overexpression of ADAM10 does not appear harmful; however, in transgenic models of AD, neuronal function can be restored by ADAM10 overexpression. For instance, Postina *et al.*[31] demonstrated that increased α-secretase expression prevents plaque deposits in aged animals overexpressing hAPP.

In contrast, overexpression of inactivated ADAM10 (mADAM10) in APP mice worsens amyloid aggregation. In a water navigation task, mADAM10/APP double transgenic animals showed deficits in the acquisition phase of place learning and in the probe trial, whereas the ADAM10/APP double transgenics performed similarly to control animals. Long-term potentiation (LTP) was also improved in the double transgenic mice as compared to APP transgenics, suggesting a rescue of synaptic plasticity by increased α-secretase activity. Previous findings from Moechars *et al.*[32] corroborate these results: in transgenic animals carrying a modified hAPP in which the α-secretase cleavage site has been altered (APP/RK), similar results were observed. The APP/RK transgenic animals showed increased APP expression with a shift toward β-site cleavage amyloid peptides. Additionally, these animals had shorter life spans than control mice, were more aggressive and hyperactive, and displayed abnormalities in the amygdala, cortex, and hippocampus.[33]

TAU In addition to the "amyloid-related" transgenics, complementary mouse models have been developed that overexpress human tau in its normal and/or mutated forms (including a form associated with frontotemporal dementia and Parkinson's disease, FDPD).

Human tau proteins are encoded by a gene on chromosome 17. Six brains isoforms resulting from alternative splicing have been described and can be divided into two classes: proteins with three C-terminal repeat domains (3R) and proteins with four domains (4R). Most transgenic models have been generated with one of the mutated 4R forms, varying from "regular" 4R tau to mono-mutated forms (P301L, P301S, V337M) and multiple-mutated 4R forms.

Although tau-related pathologies have been detected in the brains of these transgenics,[9,10] few studies have characterized these animals behaviorally.

In the elevated plus-maze, 11-month-old transgenic mice overexpressing the V337 mutant tau gene displayed less anxiety-like behavior compared to their wild-type littermates. In contrast, no differences were detected in a water navigation test. These animals showed impaired habituation in an open-field exploration task. Additionally, these tau transgenic mice showed an irregular neuronal-shaped hippocampus and altered neuronal activity.

Another related line carries the R406W tau mutation (found in human subjects with a tauopathy resembling AD). Observations made in aged animals (ranging from 16 to 23 months of age) revealed a blunted fear response to cues, but not to context, after 2 days of conditioning. The same animals showed lower levels of contextual fear after 15 days of conditioning, suggesting that long-term memory loss may be more dramatic in aged R406W tau transgenic mice. These data suggest that there is a deficit in associative memory in mice overexpressing a mutated tau gene. Neuroanatomically, transgenic mice develop tau inclusions in the hippocampus, amygdala, and neocortex, structures known to be involved in memory formation.[34]

Subtle differences and specific impairments were also observed in transgenic mice overexpressing the P301L mutation of tau.[35] Aggregation and NFTs were observed in many brain areas, including the amygdala and the hippocampus, brain regions that are also severely affected by NFTs in AD patients. Because these mice show accelerated extinction in a conditioned taste aversion test, the authors evaluated their behaviors in tasks requiring an intact hippocampus, including the Morris Water maze and the Y maze. Differences were detected only in spatial working memory, with mice showing greater impairment with age, starting at 6 months. However, the transgenic mice were more active and slightly less prone to novelty-induced anxiety, which suggests that the impairment observed in these mice is not related to affective behavior unrelated to the hippocampus.

POLYTRANSGENICS Because Aβ and tau are two primary, and complementary, factors in the development of AD, and because the presenilins are clearly genetic susceptibility loci for the disease,[36] the interactions of these proteins were tested by engineering a triple-transgenic mouse model. To generate these mice, two transgenes (Tg2576, an APP mutant, and tau P301L) were transferred onto a PS1-heterozygous knockin mutation background. Triple-transgenic (3xTg) mice developed progressive age-dependent neuropathology, including plaques and tangles. Remarkably, the pattern of evolution of these abnormalities closely mimics some of those described in AD. In these mice, presynaptic dysfunction and LTP deficits precede the accumulation of extracellular Aβ. This suggests a significant role for intracellular Aβ in AD-dependent pathology, even in the absence of structural changes. Plaques first appeared in cortical regions, then in the hippocampus, and later in the amygdala, whereas tangles followed the reversed pattern, beginning in the amygdala. This sequence of events is in line with the amyloid cascade hypothesis of AD (cf. Figure 55–1). 3xTg mice also demonstrate age-dependent cognitive deficits.[37] The earliest cognitive impairments were recorded at 4 months of age, and were characterized by a deficit in long-term memory correlated with the intraneuronal accumulation of Aβ in the hippocampus and amygdala. No plaques or tangles could be detected at this age, suggesting that they contribute to cognitive dysfunction only at older ages. Immunotherapy treatments eliminated intraneuronal Aβ deposits and, consequently, rescued the deficits observed in hippocampal-dependent cognitive tasks. The reemergence of Aβ pathology after immunotherapy led to the return of the cognitive deficits observed before treatment. Given these promising results and the very integrative nature of this model, triple transgenic mice have now been studied further and offer promising insights on the sequence of molecular and cellular events leading to AD.[38]

A NOTE ON EPIGENETIC AND ENVIRONMENTAL FACTORS IN ALZHEIMER'S DISEASE GENETICALLY MODIFIED MOUSE MODELS

Because familial cases of AD represent only 15% of AD and because genetic factors are essentially risk factors, an important field of research aims to determine the interaction between epigenetic factors and specific genetic alteration in an effort to identify gene–environment interactions. Two studies by Lazarov *et al.*[39] and Jankowsky *et al.*[40] have investigated the effect of long-term exposure to an enriched environment on double transgenic mice coexpressing APPswe and PS1 polypeptide variants. Although there are some discrepancies in the results (see Lazarov *et al.*[39] for discussion), the results unequivocally demonstrated that environmental factors can influence amyloid deposition in a gene-dependent fashion. Identifying other environmental factors that modulate disease progression could be very valuable in the treatment or prevention of AD.

CONCLUSIONS

Genetic mouse models have enhanced our understanding of the pathogenesis of AD and its underlying mechanisms. While inherent differences between mice and humans will always limit our ability to model this disease in animals, recent studies demonstrate that it is possible to generate plaques and tangles in a mouse, similar to human neuropathology. The evolution of both quantitative and qualitative genetic mouse models of AD has provided invaluable tools to better understand this pathology, along with the possibility of developing new therapeutics. When looking at the progress accomplished between the first transgenic models of AD and the current development of neuron-specific, conditional and polytransgenic models, it is clear that genetic models of AD will play an important role in finding a cure for this disease.

ACKNOWLEDGMENTS

This chapter was based on a review by Mineur *et al.*[41] I would like to greatly thank Marina Picciotto (Yale University, New Haven, CT) for her helpful comments and suggestions.

REFERENCES

1. Lovestone S, McLoughlin DM. Protein aggregates and dementia: Is there a common toxicity? *J Neurol Neurosurg Psychiatry* 2002;72: 152–161.
2. Selkoe DJ. Cell biology of protein misfolding: The examples of Alzheimer's and Parkinson's diseases. *Nat Cell Biol* 2004;6: 1054–1061.
3. Glenner GG, Wong CW. Alzheimer's disease and Down's syndrome: Sharing of a unique cerebrovascular amyloid fibril protein. *Biochem Biophys Res Commun* 1984;122:1131–1135.
4. Glenner GG, Wong CW. Alzheimer's disease: Initial report of the purification and characterization of a novel cerebrovascular amyloid protein. *Biochem Biophys Res Commun* 1984;120:885–890.
5. Glenner GG, Wong CW, Quaranta V, Eanes ED. The amyloid deposits in Alzheimer's disease: Their nature and pathogenesis. *Appl Pathol* 1984;2:357–369.
6. Wilquet V, De Strooper B. Amyloid-beta precursor protein processing in neurodegeneration. *Curr Opin Neurobiol* 2004;14: 582–588.
7. Stoothoff WH, Johnson GV. Tau phosphorylation: Physiological and pathological consequences. *Biochim Biophys Acta* 2005;1739: 280–297.
8. Hu ZH, Wang XC, Li LY, *et al.* Correlation of behavior changes and BOLD signal in Alzheimer-like rat model. *Acta Biochim Biophys Sin (Shanghai)* 2004;36:803–810.
9. Brandt R, Hundelt M, Shahani N. Tau alteration and neuronal degeneration in tauopathies: Mechanisms and models. *Biochim Biophys Acta* 2005;1739:331–354.
10. Lee VM, Kenyon TK, Trojanowski JQ. Transgenic animal models of tauopathies. *Biochim Biophys Acta* 2005;1739:251–259.
11. Link CD. Invertebrate models of Alzheimer's disease. *Genes Brain Behav* 2005;4:147–156.
12. Crabbe JC. Animal model in neurobehavioral genetics: Methods for estimating genetic correlation. In: Jones C, Mormede P, Eds. *Neurobehavioral Genetics: Methods and Applications*. Boca Raton, FL: CRC Press, 1999.
13. Stephan A, Phillips AG. A case for a non-transgenic animal model of Alzheimer's disease. *Genes Brain Behav* 2005;4:157–172.
14. Hsiao K, Chapman P, Nilsen S, *et al.* Correlative memory deficits, Abeta elevation, and amyloid plaques in transgenic mice. *Science* 1996;274:99–102.
15. Westerman MA, Cooper-Blacketer D, Mariash A, *et al.* The relationship between Abeta and memory in the Tg2576 mouse model of Alzheimer's disease. *J Neurosci* 2002;22:1858–1867.
16. Pompl PN, Mullan MJ, Bjugstad K, Arendash GW. Adaptation of the circular platform spatial memory task for mice: Use in detecting cognitive impairment in the APP(SW) transgenic mouse model for Alzheimer's disease. *J Neurosci Methods* 1999;87:87–95.
17. Corcoran KA, Lu Y, Turner RS, Maren S. Overexpression of hAPPswe impairs rewarded alternation and contextual fear conditioning in a transgenic mouse model of Alzheimer's disease. *Learn Mem* 2002;9:243–252.
18. Kobayashi DT, Chen KS. Behavioral phenotypes of amyloid-based genetically modified mouse models of Alzheimer's disease. *Genes Brain Behav* 2005;4:173–196.
19. De Strooper B, Woodgett J. Alzheimer's disease: Mental plaque removal. *Nature* 2003;423:392–393.
20. Francis R, McGrath G, Zhang J, *et al.* aph-1 and pen-2 are required for Notch pathway signaling, gamma-secretase cleavage of betaAPP, and presenilin protein accumulation. *Dev Cell* 2002;3:85–97.
21. Shen J, Bronson RT, Chen DF, Xia W, Selkoe DJ, Tonegawa S. Skeletal and CNS defects in presenilin-1-deficient mice. *Cell* 1997;89:629–639.
22. Yu H, Saura CA, Choi SY, *et al.* APP processing and synaptic plasticity in presenilin-1 conditional knockout mice. *Neuron* 2001;31: 713–726.
23. Huang XG, Yee BK, Nag S, Chan ST, Tang F. Behavioral and neurochemical characterization of transgenic mice carrying the human presenilin-1 gene with or without the leucine-to-proline mutation at codon 235. *Exp Neurol* 2003;183:673–681.
24. Janus C, D'Amelio S, Amitay O, *et al.* Spatial learning in transgenic mice expressing human presenilin 1 (PS1) transgenes. *Neurobiol Aging* 2000;21:541–549.
25. Dewachter I, Reverse D, Caluwaerts N, *et al.* Neuronal deficiency of presenilin 1 inhibits amyloid plaque formation and corrects hippocampal long-term potentiation but not a cognitive defect of amyloid precursor protein [V717I] transgenic mice. *J Neurosci* 2002;22:3445–3453.
26. Naslund J, Jensen M, Tjernberg LO, Thyberg J, Terenius L, Nordstedt C. The metabolic pathway generating p3, an A beta-peptide

fragment, is probably non-amyloidogenic. *Biochem Biophys Res Commun* 1994;204:780–787.

27. Wei Q, Holzer M, Brueckner MK, Liu Y, Arendt T. Dephosphorylation of tau protein by calcineurin triturated into neural living cells. *Cell Mol Neurobiol* 2002;22:13–24.
28. Buxbaum JD, Liu KN, Luo Y, *et al.* Evidence that tumor necrosis factor alpha converting enzyme is involved in regulated alpha-secretase cleavage of the Alzheimer amyloid protein precursor. *J Biol Chem* 1998;273:27765–27767.
29. Karkkainen I, Rybnikova E, Pelto-Huikko M, Huovila AP. Metalloprotease-disintegrin (ADAM) genes are widely and differentially expressed in the adult CNS. *Mol Cell Neurosci* 2000;15: 547–560.
30. Lammich S, Kojro E, Postina R, *et al.* Constitutive and regulated alpha-secretase cleavage of Alzheimer's amyloid precursor protein by a disintegrin metalloprotease. *Proc Natl Acad Sci USA* 1999; 96:3922–3927.
31. Postina R, Schroeder A, Dewachter I, *et al.* A disintegrin-metalloproteinase prevents amyloid plaque formation and hippocampal defects in an Alzheimer disease mouse model. *J Clin Invest* 2004;113:1456–1464.
32. Moechars D, Lorent K, De Strooper B, Dewachter I, Van Leuven F. Expression in brain of amyloid precursor protein mutated in the alpha-secretase site causes disturbed behavior, neuronal degeneration and premature death in transgenic mice. *EMBO J* 1996;15:1265–1274.
33. Moechars D, Lorent K, Van Leuven F. Premature death in transgenic mice that overexpress a mutant amyloid precursor protein is preceded by severe neurodegeneration and apoptosis. *Neuroscience* 1999;91: 819–830.
34. Tatebayashi Y, Miyasaka T, Chui DH, *et al.* Tau filament formation and associative memory deficit in aged mice expressing mutant (R406W) human tau. *Proc Natl Acad Sci USA* 2002;99:13896–13901.
35. Pennanen L, Wolfer DP, Nitsch RM, Gotz J. Impaired spatial reference memory and increased exploratory behavior in P301L tau transgenic mice. *Genes Brain Behav* 2006;5:369–379.
36. Oddo S, Caccamo A, Kitazawa M, Tseng BP, LaFerla FM. Amyloid deposition precedes tangle formation in a triple transgenic model of Alzheimer's disease. *Neurobiol Aging* 2003;24:1063–1070.
37. Billings LM, Oddo S, Green KN, McGaugh JL, Laferla FM. Intraneuronal Abeta causes the onset of early Alzheimer's disease-related cognitive deficits in transgenic mice. *Neuron* 2005;45:675–688.
38. Laferla FM, Oddo S. Alzheimer's disease: Abeta, tau and synaptic dysfunction. *Trends Mol Med* 2005;11:170–176.
39. Lazarov O, Robinson J, Tang YP, *et al.* Environmental enrichment reduces Abeta levels and amyloid deposition in transgenic mice. *Cell* 2005;120:701–713.
40. Jankowsky JL, Xu G, Fromholt D, Gonzales V, Borchelt DR. Environmental enrichment exacerbates amyloid plaque formation in a transgenic mouse model of Alzheimer disease. *J Neuropathol Exp Neurol* 2003;62:1220–1227.
41. Mineur YS, McLoughlin D, Crusio WE, Sluyter F. Genetic mouse models of Alzheimer's disease. *Neural Plast* 2005;12:299–310.

Models of Behavior

V

Immunology and Virology

F

56 Rat Models of Experimental Autoimmune Encephalomyelitis

ERIK WALLSTRÖM AND TOMAS OLSSON

ABSTRACT

Multiple sclerosis (MS) is a chronic inflammatory disease of the central nervous system (CNS) leading to neurological deficits. The relative inaccessibility of the CNS for sampling and the problems of experimental manipulations in humans make proper experimental models indispensable for progress in the basic understanding of the disease as well as for testing of therapy. There is no single experimental model mimicking all aspects of MS and a model should be chosen based on the scientific question being investigated. We describe our experience with antigen-induced models in the rat species. The clinical course of disease, monophasic versus relapsing/protracted, and the histopathological features depend on the strain of rats used, their MHC and non-MHC genes, as well as the myelin autoantigen used for immunization. Other factors such as age, weight, and gender also influence the outcome. Use of myelin oligodendrocyte glycoprotein in DA rats results in an MS-like relapsing disease with plaques of demyelination. Use of myelin basic protein in LEW rats results in an acute monophasic inflammatory disease with little demyelination. Several disease-determining factors and practical schemes for MS models in rats are presented.

Key Words: Multiple sclerosis, Autoimmunity, Myelin, Myelin basic protein, Myelin oligodendrocyte glycoprotein, Proteolipid protein, T cells, B cells, Antibodies, Major histocompatibility complex, Genetics, Demyelination, Inflammation, Axons, Neurons, Oligodendrocytes, Astrocytes, Macrophages, Microglia.

GENERAL CHARACTERISTICS

WHY MODEL MULTIPLE SCLEROSIS? Multiple sclerosis (MS) is an inflammatory disease of the central nervous system (CNS) with foci of inflammation, demyelination, and damage to axons. It is typically a disease of the Western world with prevalence there varying from 0.5 to 2 per thousand. This distribution most likely depends primarily on a higher genetic predisposition in these countries, but differences in environmental factors may also play a role. Despite its relatively low incidence and prevalence, it has a major impact on the afflicted individual and on society as a whole; its impact is also economic, since the disease starts early in life and persists for decades. In addition, drug companies have realized that MS is a valuable target, since currently approved drugs either have only modest effects or can result in rare but fatal adverse events.

Since there is a great medical need for better and safer treatments, in turn requiring better knowledge of disease mechanisms, MS continues to be a strong focus for research both in academia and in the pharmaceutical industry; disease-appropriate animal models are therefore indispensable for further progress. This is particularly so in MS, because the CNS is relatively inaccessible for sampling, observations in humans in most cases are descriptive, and experimental manipulation in humans is rarely possible.

In principle, a rodent model for MS can be used for studies of disease mechanisms, either by classical hypothesis-driven methods or by unbiased genetic tools, or for studies of new therapeutic strategies. There is no single model of experimental autoimmune encephalomyelitis (EAE) that covers all aspects of human MS, and the choice of a particular EAE variant should be carefully thought through depending on the scientific question.

To enable such a selection of a model, in depth knowledge of the natural course of the disease and its basic immunopathology is necessary. We refer the reader to several recent reviews on these topics, and give a summary of the most critical factors in the following.

MULTIPLE SCLEROSIS: COURSE AND SUBPHENOTYPES MS rarely starts before the age of 15 or later than 50 years. The average age of onset is 28 years, with a 2/1 female/male ratio. In about 85% of all cases of MS the disease follows a remitting relapsing course, with relapses of new neurological deficits on average every second year. Recovery can be complete or incomplete. Interestingly, the introduction of magnetic resonance imaging (MRI) into the study of MS has shown that there are up to 10 subclinical inflammatory events in the CNS on each clinical recognizable relapse. With time, most patients enter a secondary chronic progressive course with steadily increasing neurological defects. The mean time from onset to this phase is around 15 years. Current dogma suggests that the mechanisms during this progressive phase differ from those during the relapsing phase. None of the currently approved drugs for MS has had any effect on the progressive phase of the disease. Of MS patients, 10–15% start with a steady progression, categorized as primary progressive MS. Of note, few rodent MS models can mimic these very chronic aspects of MS.

The symptoms and signs of MS vary depending on where in the CNS the inflammatory lesions occur, and almost all

From: *Sourcebook of Models for Biomedical Research*
(P. M. Conn, ed.), © 2008 Humana Press Inc., Totowa, NJ.

neurological functions can ultimately be affected. These are focal neurological disturbances. In addition, MS is accompanied by general symptoms and signs.

To enhance the understanding of the MS models described below we provide a short summary of common signs.

Afferent neurological functions are often affected. Patients may develop regional hypoesthesias/paresthesias and defects in deep sensation. Optic neuritis, with loss of vision in one eye, is common. In most cases vision recovers over weeks. Brainstem lesions may lead to signs such as dysfunction of ocular motility and central vertigo. Cerebellar lesions can lead to dystaxia. Motor tracts are often affected, resulting in monoparesis of an extremity, most often in legs, hemiparesis, or paraparesis. In the chronic progressive phase a slowly increasing paraparesis folowed by tetraparesis is common.

General signs include fatigue, which most patients refer to as their worst manifestation of MS. Depression is common. A mild or modest decrease of cognitive function occurs in up to 50% of MS patients. In quite rare instances full blown dementia may develop.

Just as there are scoring scales for the EAE models described below, there are rating scales for MS, used in clinical follow-up or during clinical drug trials. The best established is the expanded disability scale (EDSS), with which different neurological functions are assessed and rated in a neuroexamination. These are added together to produce a score from 0 to 10 : 0 is no neurological dysfunction, 6 reflects decreased motor function with the patient able to walk 100 m with unilateral support, and 10 is death due to MS. The multiple sclerosis functional composite (MSFC) is a more recently introduced scale in which performance on a simple cognitive test, motor function in the upper extremity, and walking ability are added together. There are numerous other scales, some based on patient self-report and some measuring quality of life.

MULTIPLE SCLEROSIS: HISTOPATHOLOGY MS histopathology is characterized by foci of inflammation, demyelination, and damage to neurons/axons. Initial periventricular lesions are composed of mononuclear cells, mainly macrophages, as well as an abundance of lymphocytes: $CD4^+$, $CD8^+$, T cells, and B cells. Demyelination ensues through macrophage attack. In addition, axons are damaged in this acute inflammatory attack.[1] Further damage to axons also occurs more chronically.

Classical demyelinated plaques varying in size from millimeters to centimeters ensue; these also contain astroglial scarring. To a variable extent demyelinated axons may become remyelinated by oligodendroglial precursors that are recruited.

PRESUMED IMMUNE PATHOGENESIS The formal proof for any autoimmune or alternative viral etiology is lacking. However, there is strong circumstantial evidence suggesting an autoimmune pathogenesis based on the following: (1) Increased numbers of myelin autoreactive T and B cells enriched to the cerebrospinal fluid can be detected in persons with MS. (2) Certain HLA class II alleles, gene products being restriction elements for $CD4^+$ T cells, dispose for MS. (3) Genomic regions disposing for MS overlap regions defined in rodent models known to be autoimmune.[2] (4) The organ specificity of the inflammatory attack in MS is most easily reconciled in the context of the immune system, perhaps the only system able to discriminate between such a broad variety of molecular targets. Finally, (5) some of the induced EAE models described in this chapter closely mimic both the clinical course and histopathological features of MS.

Based on immense numbers of studies in EAE and descriptive observations in human MS, a condensed sequential immunopathogenesis of MS can be described as follows.

For unknown reasons CNS autoantigen-reactive T cells become activated in the periphery (lymphoid organs/systemic circulation). Upon reaching the endothelium at the blood–brain barrier secretion of proinflammatory cytokines such as tumor necrosis factor-α (TNF-α) and interferon-γ (IFN-γ) cause endothelial upregulation of adhesion molecules such as intercellular adhesion molecule (ICAM) and vascular cell adhesion modules (VCAM). The latter seem of special importance as demonstrated in EAE.[3] The very late activation (VLA)-4–VCAM interaction allows activated T cells to pass into the CNS parenchyma, and upon rerecognition of a specific CNS autoantigen the T cells become reactivated with secretion of proinflammatory cytokines, inducing a further inflammatory cascade with recruitment of macrophages, chemokine secretion, etc. There is also a potential role for myelin-specific autoantibodies binding to myelin with destructive potential in the form of complement-mediated lysis or opsonization for macrophage attack. Most likely a series of downmodulatory events later takes place involving an endogenous steroid response and cytokines such as tumor growth factor-β (TGF-β) and interleukin 10 (IL-10).

One of the few experimental situations we have in human disease is during clinical trials against specific targets. The most effective treatment against MS hitherto demonstrated is blockade of VLA-4 with a humanized monoclonal antibody, reducing relapses by 68% and new MRI lesions by 90%. This is striking evidence in support of the idea that inflammation causes MS pathology, and not the opposite, which occasionally has been suggested. For this reason, experimental models built on induction of a systemic immune response attacking the CNS are relevant for MS pathogenesis.

Apart from a purely immunological pathogenesis, there is a growing interest in target-related factors. Thus, the CNS structures in the form of the blood–brain barrier, glial cells, and neuron axons take part in the pathogenesis of the disease, and there may well be genetically regulated differences in the function of these structures upon an inflammatory insult potentially explaining interindividual differences in outcome. For example, certain patients seem to do well despite intense inflammation in the CNS, while others suffer from severe sequelae after less intense inflammatory insults. It is likely that there is an important neurodegenerative aspect of MS in the chronic progressive forms of the disease.

THE GENETICS OF MULTIPLE SCLEROSIS Since the genetic background is well known to profoundly affect immune responses in general and also rodent MS models, we here provide a short note on what is known about the genetics of MS. Furthermore, if the particular genes disposing for MS can be found, they will denote primary causative pathogenic pathways, which will pave the way for therapy and prevention. As discussed below, rodent MS models may assist in the finding of such genes.

There is solid evidence for genes affecting the risk and course of MS, which is particularly clear from observations in twins, with an approximate concordance rate of 30% in monozygotic twins versus 2–4% in dizygotic twins.

The human leukocyte antigen (HLA) complex is the best established and proven genome region regulating MS. It contains around 200 genes, most of which are involved in immune functions. Mapping of influences within the complex is difficult due to a strong linkage disequilibrium between genes in the complex. However, there is strong evidence that the class I and II genes are involved. Different alleles of HLA class II genes may either dispose for or protect against MS.[4] Class I genes also affect MS.[5] The basic reasons for these effects are not known, but current knowledge strongly suggests that the influence is related to the basic function of the HLA molecules, which is to bind antigenic peptides and present them to T cells. It can be argued that knowledge concerning those genes important for MS has been pivotal in the current focus on new therapies for MS directed toward interfering with T cell function.

The major histocompatibility complex (MHC; in humans equal to the HLA) molecules are among the most polymorphic genes existing in mammals, allowing a broad repertoire of peptides at the population level to be presented to the immune system; this in turn helps ensure that infectious agents do not develop strategies to escape host immunity. Thus, preferential or promiscuous binding of myelin autoantigen peptides is an explanation for the MHC regulation of MS. In this context it is worth noting that these phenomena have instrumental impact on the handling of the experimental models discussed below.

Non-MHC genes also regulate MS. However, several linkage screens have failed to demonstrate any genome regions apart from the HLA complex with major impact on MS. This in turn suggests that many genes are involved and each has a low or modest impact, which in turn has impeded their positioning. Present speculation suggests odds ratios in the range 1–2. Only a few non-MHC genes, such as the PRKCA and IL-7 receptor and a chemokine gene cluster,[6–9] have more definite evidence of support.

The outcome of human linkage screens discussed above as well as gene mappings in rodent models suggest that the number of genes with some degree of influence on MS may be around a hundred.[10] This non-MHC genetics may also have relevance for experimental modelling of MS, especially when MS genes are better defined. It also relates to the probable genetic heterogeneity of MS, that is, different genes may predispose for the same end phenotype in the form of MS, and thus potentially result in very different mechanisms. This may require different therapeutic strategies. In the future it may be possible to utilize genetic information for tailor-made therapy, in which individuals/families with particular sets of MS-disposing genes accordingly receive adapted therapies.

The non-MHC genetic background of different rodent strains is thus important to consider when evaluating the outcome of an experiment or experimental therapy. With time pharmacogenetically defined strains may be at hand (see more below).

EXPERIMENTAL AUTOIMMUNE ENCEPHALOMYELITIS: HISTORICAL NOTES AND CURRENT DEBATE EAE historically emanated from research devoted to understanding the cause of neuroparalytic accidents of rabies vaccination. Some individuals developed an acute or subacute encephalomyelitis weeks after the rabies vaccination. The virus was grown in brains and the inoculum was contaminated with CNS material. Experiments by Rivers and Schwentker tested the brain material alone, which gave rise to encephalomyelitis in a proportion of monkeys. Later, introduction of Freund's adjuvant led to a much more reproducible disease induction. This experimental disease reflects acute disseminated encephalomyelitis (ADEM) rather than MS, which has a chronic relapsing course. As described below, there are now a series of chronic relapsing models that better mimic MS. For more than 50 years there has been continuous debate on EAE as a model for MS, relating to different views on the presumed etiology, the predictability of data in the models versus validity for MS, and the "spontaneous" occurrence of MS versus the need for active induction in rodents. John Alvord, Jr., one of the pioneers in research on EAE and myelin basic protein, in 1984 presented a graph charting the ups and downs in his belief in EAE as a model for MS[11]; this varied from close to 100% when myelin basic protein (MPB) was found to be a discrete encephalitogen to 0% when antimeasles antibodies were found in MS cerebrospinal fluid, suggesting measles as the cause of MS. In 1984 he was 80% convinced that EAE was a good model for MS.

Recently at least three studies have addressed these issues. First, a negative study argued that therapies developed in EAE do not work in MS.[12] An opposing view was presented by Steinman and Zamvil,[13] who argued that at least two therapies in MS, glatiramer acetate and anti-VLA-4 blockade, are completely based on EAE experiments and are of therapeutic use in MS. An intermediate view was presented by Ransohoff.[14]

It is often argued that EAE is not a good model for MS since it needs to be induced, whereas MS is spontaneous. However, twin data and epidemiological data strongly suggest that MS is also induced, or triggered by so far unknown events (infection is a common speculation). In particular, the 70% discordance rate in monozygotic twins suggests triggering by environmental factors.

PRINCIPAL MODELS OF EXPERIMENTAL AUTOIMMUNE ENCEPHALOMYELITIS AND A NOTE ON THE SELECTION OF SPECIES

EAE can be induced in practically all species, and most commonly in the mouse or rat. For very select questions nonhuman primate models may be used, for example, when testing reagents that are specific only for primates including humans.

The mouse species has a number of advantages: it is small and therefore inexpensive to keep and breed, accessible for gene deletions, which is not yet possible in the rat, and transgenic expression of genes is easier. For genetics, the mouse displays a full genome sequence accessible on the internet, with single nucleotide polymorphism (SNP) and microsatellite maps. However, in our experience the models accessible in the mouse have been less reproducible and there is a paucity of chronic mouse models closely mimicking the course and histopathology of MS.

In this chapter we deal only with rat models, with which we have extensive experience. Here, a series of easily reproduced acute and chronic models are at hand. The rat genome sequence is now also available on the internet (ensemble.org), there is a microsatellite map, and SNP maps will appear. There is also access to a wide variety of EAE-susceptible and resistant inbred strains. Well-controlled acute monophasic as well as protracted relapsing EAE forms are at hand.

Many of these models are now routinely used by the pharmaceutical industry in evaluating new drug principles. A common problem in such experimentations is that drugs developed to fit human structures, irrespective of whether the drug is a monoclonal antibody, soluble receptor structure, or a low-molecular compound, often either have no or much lower binding affinities to the rodent target. The solution to this is either parallel development of rodent-specific compounds or transgenic expression of the human target in the rodent.

The following induction methods can be discerned in principle.

Adoptive Transfer Experimental Autoimmune Encephalomyelitis That EAE had a central T cell-mediated pathogenesis was demonstrated in the early 1980s.[15,16] Myelin antigen reactive T cells were recovered from actively immunized mice or rats, cultured *in vitro* in the presence of stimulating antigen, and after several rounds of *in vitro* stimulation monospecific T cell lines and clones were obtained that upon transfer to naive syngeneic recipients caused EAE, in most cases with an acute monophasic disease course, with widespread inflammation in the CNS, but with little or no demyelination. This methodology has had enormous impact on the understanding of certain aspects of autoimmune disease, especially on questions such as T cell trafficking and the understanding of central features of T cell biology important for neuroinflammation. This form of EAE is therefore useful in studies of similar types of questions, but perhaps not in situations in which a disease course and pathology closely mimicking MS is required. We will not describe this form of EAE further.

Cotransfer Experimental Autoimmune Encephalomyelitis For studies of the additive or synergistic role of autoantibodies, research groups invented cotransfer models, in which autoimmune T cells were transferred along with myelin-specific antibodies. For example, in this way it was demonstrated that antibodies alone did not cause disease, but required the elicitation of CNS inflammation by the T cells.[17] Consequently this type of methodology can be used to address similar questions on the role of antibodies. We will not discuss this methodology further.

Actively Induced Experimental Autoimmune Encephalomyelitis The experimental animal is immunized with a CNS autoantigen, in most cases together with an adjuvant (mineral oil + mycobacteria, and sometimes also injection of pertussis toxin). This recruits an autoimmune T and B cell response causing EAE with either an acute monophasic or chronic relapsing course, depending on the autoantigen and/or the strain of rodent used. Detailed descriptions of this methodology will be provided below.

Certain viral infections in rodents, such as Theiler's murine encephalomyelitis, are accompanied by recruitment of an autoimmune response, where there is strong evidence to support the idea that this autoimmune response may be responsible for the chronic relapses that ensue.[18]

Transgenic Induction of Experimental Autoimmune Encephalomyelitis With recent still more fine-tuned, elaborate use of transgenic techniques, "spontaneous EAE" may occur. Thus mice are constructed in which a high proportion of the T cells expresses human T cell receptors for myelin autoantigens. The mice can be further manipulated to express human MHC molecules and T cell costimulatory molecules. In addition, they can transgenically express B cells producing myelin-specific antibodies. All these mouse variants, to a variable extent, display spontaneous EAE, which interestingly also varies depending on the laboratory environment in which they are kept. Naturally, these types of mice, although highly artificial, can be used for mechanistic studies.[19–21] It is a semantic question as to whether these forms of EAE should be regarded as spontaneous or induced. We think that the genetic manipulation represents a form of active induction. Potentially, these mice could also be interesting to explore with regard to environmental triggers that may precipitate disease in extremely susceptible rodents. We will not explore these EAE models further.

ACUTE MONOPHASIC OR CHRONIC RELAPSING EXPERIMENTAL AUTOIMMUNE ENCEPHALOMYELITIS One of the oldest and best characterized EAE models is induced in LEW rats by subcutaneous immunization with MBP (derived from guinea pigs), or MBP_{gp} 63–88 peptide, emulsified in complete Freund´s adjuvant (mineral oil and heat-killed *Mycobacterium tuberculosis*). The first clinical signs of disease appear approximately day 9–10 postimmunization (pi) in the form of sudden weight loss. By approximately day 10–11 pi tail weakness is evident, progressing to hind leg weakness and eventually complete paralysis of the hind legs at day 12–13 pi A recovery phase follows, with complete (or nearly complete) gain of hind leg and tail function at day 16–17 pi. A small number (0–10%) of animals may have a second relapse around day 22–24 pi, but otherwise they stay healthy, and are resistant to further attempts to induce EAE.[22] Similar to MS, inflammatory infiltrates in the CNS are most common around small veins, but in the MBP-induced LEW rat model, lesions are found only in the spinal cord (they develop caudally to rostrally and parallel clinical signs of paresis) and not in the brain. At affected levels in the spinal cord, the whole parenchyma is infiltrated by $CD4^+$ α-β cells and activated macrophages. Some $CD8^+$ α-β cells, NK cells, γ-δ T cells, and B cells may also be present, and there are signs of activation of CNS resident cells (especially microglia).[23] Most other EAE models are "variations on the theme" described above, but may include features such as a high frequency of relapsing disease and/or demyelination.

In contrast to MBP-induced EAE in the LEW rat, rMOG (aa 1–125)-induced EAE (MOG-EAE) in DA rats displays both a high frequency of relapsing disease and focal inflammation with demyelination in the CNS.[24,25]

Clinically, relapsing-remitting signs of neurological dysfunction characterize MOG-EAE. However, there are also rats displaying chronic (nonremitting) signs of disease, acute lethal disease, or mild, very late clinical disease [occasionally DA rats have displayed no signs of clinical disease until day 50–60 postimmunization (unpublished observations)]. Therefore, MOG-EAE is similar to MS in its diversity in individual clinical courses. Rats may present with only one lesion in the optic nerves and one lesion in the spinal cord, mimicking the MS subtype Devic´s disease. Active demyelination is associated with the deposition of immunoglobulins and complement component 9 (C9), indicating antibody-dependent demyelination. Immunologically, there are signs of activation of both cellular and humoral anti-MOG-specific lymphocytes. This is again similar to MS, where both T and B cell responses to MOG and other myelin antigens are present.[26,27]

Due to the gradual release of antigen in actively induced disease, it is difficult to determine whether "chronic" EAE can be considered "truly chronic." If chronicity is defined as a "self-perpetuating" condition in which signs of disease continue indefinitely even after the removal of the initial trigger, then actively induced EAE is not proven to be chronic. Another problem with chronicity in EAE is the occurrence of irreversible neurological deficits. It is not difficult to devise a situation in which an acute inflammation causes irreversible neurological deficits due to destructive lesions in the CNS. Signs of disease may then persist indefinitely after the disappearance of the acute inflammation.

PRACTICAL CONSIDERATIONS

CENTRAL NERVE SYSTEM AUTOANTIGEN Autoantigens for EAE-inductions may be purified (myelin) proteins, recombinant proteins, synthesized peptides, or even whole spinal cord and brain homogenates. There are pros and cons with all of these approaches. Purified myelin components may contain contaminations from other myelin proteins and a mixture of antigens is obviously also the case when whole spinal cord homogenates are utilized. Recombinant proteins circumvent the problem of contamination by other myelin proteins, but introduce potential immunologically active contaminations from the production system (often bacteria). Peptide-induced EAE may be complicated by the nonphysiological tertiary structure of peptides[28] and nonphysiological epitope selections.[29]

If a relapsing/remitting disease course is desired, rMOG seems to be the most consistent antigen to use, particularly with DA and LEW.AV1 rats. rMOG here denotes recombinant MOG aa 1–125 (the N-terminal Ig-like extracellular domain), since full-length MOG is difficult to express due to its low solubility. If disease is induced by MBP peptides, monophasic EAE usually ensues.[30,31] In adoptive transfer EAE, the antigen specificity of the transferred cells determines the topography of the CNS lesions and the extent of both parenchyma inflammation and macrophage activation.[32] Thus, the choice of antigen determines both clinical and histopathological disease expression.

DOSE OF ANTIGEN Although rarely discussed in studies on EAE, the dose of antigen may dramatically affect disease expression. Usually, titration experiments with three or four doses of antigen are necessary to set up experimental conditions before the actual experiment is started. In MOG–EAE, a reduction of the amount of antigen in the immunization may result in conversion of an acute lethal disease to a chronic disease, or to no disease.[24] Suggested immunization protocols are discussed in the section on Immunization below.

ADJUVANTS The minimum adjuvant utilized in most EAE studies is incomplete Freund´s adjuvant (IFA). IFA is a mixture of mineral oils, and it has by itself a complex impact on the immune system, including triggering of oil-induced arthritis in DA rats.[33,34] IFA protects the antigen from rapid degradation and thereby prolongs its exposure to the immune system. Macrophage phagocytosis may also be facilitated by a more particulate physical state of the antigen.[35]

Heat-killed *Mycobacterium tuberculosis* may be added to the IFA, resulting in complete Freund´s adjuvant (CFA).[36] CFA is a classical adjuvant, and has been utilized extensively to boost both cell-mediated and humoral immune responses to antigens with which it has been emulsified. The downside of using CFA as adjuvant is that it adds complexity; mycobacterial antigens interact with the immune system at several levels and responses to mycobacteria are genetically regulated.[37] In most EAE induction protocols, CFA is required to induce disease, while IFA induces protective immunity. The ability to induce EAE in DA rats with rMOG/IFA alone is therefore unique and reflects both the strong sensitivity to adjuvants in rats (especially DA rats)[38,39] and the potent encephalitogenic properties of MOG.

EMULSION The relation between encephalitogenicity and the size of the water/oil emulsion droplets in the inoculum has been studied by Maatta *et al.*[40] Increased encephalitogenicity was associated with the small size of the droplets (achieved by sonication instead of extrusion). A small droplet size corresponded to a large surface area, which preferentially contained antigen. In the extruded emulsion, the antigen was buried in the droplet interior. Even so, most EAE induction protocols do not favor sonication, and a proper emulsion is acquired by mixing the water/oil emulsion in a syringe, or rather two connected syringes. Glass syringes are preferable to plastic, since plastic syringes tend to "clog" from the emulsion. At the beginning of mixing, the emulsion is easily passed through the syringe, but after a while, the texture changes and it takes more force to compress the syringe. The emulsion is then ready to be used. However, before use, the emulsion may be tested by putting a drop in a beaker filled with water. The drop should not immediately dissolve in the water, but rather remain as a drop, or dissolve very slowly.

IMMUNIZATION In our hands, subcutaneous EAE induction is less effective than intradermal immunization. However, it is not always easy to achieve a completely intradermal location, especially with a volume of 200 μl, which is a common injection volume. If CFA adjuvant is utilized, rather than IFA, the immunization must always be subcutaneous, to avoid local inflammation and formation of a wound at the site of injection. A needle diameter of 0.6 mm is most often our choice. Thinner needles are difficult to use due to the consistency of the emulsion. It is advisable to make extra emulsion (for 5–10 extra animals) to avoid having to repeat the procedure if spills or other mistakes occur during the emulsification or immunization. It is also important to avoid self-injections, since it may trigger neuroinflammatory disease in certain individuals. We are aware of one such case in which the needle penetrated the skin of the hand with a minimal deposition of inoculum, and the person developed a subacute EAE-like disease. We therefore recommend a minimal surgical excision of the injection site if self-injection indeed occurs.

STRESS An old, but seldom systematically studied observation is that nonspecific stress may profoundly suppress EAE. When designing EAE experiments, care should be taken to avoid disturbances such as noise, changes in animal caretaker staff/routines, and experimental manipulations such as blood samples. Stress effects seem to be most pronounced before the appearance of clinical signs of EAE.[41,42] It has been suggested that differences in neuroendocrine function may constitute part of the non-MHC regulation of strain susceptibility/resistance in EAE.[43]

SEASONAL EFFECTS Seasonal effects are not well recognized in the EAE literature, but there are some reports. In 2004, Teuscher *et al.* reported increased susceptibility in mice immunized during the summer.[44,45] In contrast, our own experience is that EAE susceptibility (and fertility) in rats is reduced during the summer months (unpublished observation). It is unclear if the difference is due to the different types of animals or if the local environment is more important.

AGE AND WEIGHT Very young animals with immature immune systems differ in their response to EAE induction.[46] We avoid using rats weighing less than 150 g. In adult animals, age may be a factor of importance, in turn intimately linked to weight. Weight is important mainly with "mild" induction protocols, i.e., low doses of rMOG in IFA, while induction protocols with high doses of autoantigen and adjuvant are less sensitive to differences in animal weight. Since age-dependent weight increases differ in different rat strains, it is particularly important to consider weight when comparing different strains. Among the common strains utilized in rat EAE experiments, LEW and PVG rats gain weight faster than DA and ACI rats. Also, male rats gain weight faster than female rats, so all gender comparisons need to consider this factor.

GENDER The impact of gender differs depending on the EAE model studied. In monophasic LEW rat EAE, induced by guinea pig spinal cord homogenate, males are usually more susceptible than females.[47] In SJL mice injected with PLP 139–151-specific T cells, female donor cells mediated more severe EAE than male donor cells.[48] In our own experience, monophasic MBP or MBP peptide-induced EAE is more severe and reliable in males, while MOG-EAE is more severe and reliable in females.[24,49,50] Mechanisms for gender influences in EAE may include direct hormonal effects on T cells and/or regulation by gender-specific non-MHC loci.[51,52] The Y-chromosome itself has been implicated as harboring such loci in mouse EAE.[53] In MS, the female-to-male ratio is 2 : 1.[54]

BREEDER AND SUBSTRAIN Inbred strains are, by definition, genetically homogeneous. However, if identical inbred strains are bred separately for several generations, accumulation of spontaneous mutations ("genetic drift"), residual heterozygosity, and/or accidental contamination by genetic material from other rat strains may well result in the establishment of "substrains."[55,56] Substrains are well known to differ in their susceptibility to EAE (and other autoimmune diseases).[57] Among the strains commonly utilized in EAE studies, LEW and BN rat substrains differ in their susceptibility to EAE, while there has been fewer problems regarding substrain differences in DA, PVG, and ACI rats.

MAJOR HISTOCOMPATIBILITY COMPLEX AND NON-MAJOR HISTOCOMPATIBILITY COMPLEX GENES Classically a series of different rat strains, differing in both their MHC and non-MHC genes has been divided into resistant or susceptible strains, mainly based on susceptibility to MBP or whole spinal cord-induced disease.

First, resistance or susceptibility should be used as relative terms since use of stronger immunization protocols can overcome genetic barriers.

Second, both features may reside in both the MHC complex and among non-MHC genes. As an example, the classical MBP EAE-resistant strain, the BN rat, is relatively resistant to EAE due to both non-MHC genes and an MBP nonresponder MHC haplotype.[58,59] However, the BN rat MHC (RT1N) is strikingly responsive to MOG, as demonstrated with RT1N on the EAE-susceptible Lewis non-MHC background.[24]

In the following section we discuss these matters in some detail.

WHY IS THE RAT STRAIN CRUCIAL IN SETTING UP AN EXPERIMENTAL AUTOIMMUNE ENCEPHALOMYELITIS MODEL?

MAJOR HISTOCOMPATIBILITY COMPLEX GENE IMPACT ON RAT EXPERIMENTAL AUTOIMMUNE ENCEPHALOMYELITIS The influence of the MHC gene on EAE was recognized in the early 1970s,[60] as well as the influence of the HLA gene on MS.[61–63] The full genetic composition and distribution of rat genes are known in detail.[64] There are about 200 genes in the rat species, with the region divided into classical class I (Ia), class II, class III, and nonclassical class I (Ib) genes, situated in that order on rat chromosome 20. The nomenclature for the rat MHC is RT1, in which RT1A corresponds to classical class Ia genes, which corresponds to HLA A in humans. RT1B denotes class II genes corresponding to human HLA-DQ and mouse H2-A. The rat RT1D corresponds to HLA-DR and mouse H2-E.

Rat strains can be selected according to their MHC, based on their response to particular autoantigens (Table 56–1). Thus, there are different MHC restriction patterns depending on the particular autoantigen used for immunization. It is possible to select inbred rat strains with varying MHC on different genetic backgrounds, either with a relative non-MHC resistance or susceptibility.

In addition, there are strains with recombinations within the MHC providing options for a rough positioning of MHC regulation within the MHC complex. Use of such rats has positioned the major EAE regulation to the class II genes, with additional influences from other parts of the MHC.[24,5865] Mechanistically, the influence of class II is perhaps most easy to understand, especially with regard to peptide-induced EAE.

Starting with the most classical rat EAE, induced with MBP in the Lewis rat (RT1L), immunodominant stretches of the molecule have been defined. These comprise the MBP 63–88 region and the 89–101 region. Classically heterologous MBP of guinea pig origin has been most effective in inducing EAE in the Lewis rat. Interestingly, on the RT1U haplotype the rat sequence is as effective.[66] There is only a one amino acid difference in the core region of the peptide-binding site at position 79, by a single exchange of threonine (T) to serine (S). Despite the heterologous character of this guinea pig peptide, the fact that disease ensues must mean that recruited T cells recognize endogenous rat MBP. With these peptides, the use of intra-MHC recombinant strains between responder and nonresponder MHC has resulted in the mapping of major influences to the class II genes.[59] Use of peptide-binding assays has demonstrated that RT1B molecules restrict the response to the 63–88 peptide, while RT1D restricts the MBP 89–101 response.[67] Similar studies with MOG peptides on a series of rat MHC haplotypes have shown that MOG 91–108 is dominantly encephalitogenic.[68]

Since the basic function of class II genes is to present peptides to $CD4^+$ T cells, the regulation of class II MHC is likely to depend on differences in the peptide-binding abilities of the different class II variants. Peptide-binding studies with purified class II molecules from different MHC haplotypes have shown that dominantly immunogenic peptides in the rat bind with high or moderate affinity.[67] Peptides that bind poorly tend to be nonimmunogenic. This may be a difference compared to one example in the mouse in which the encephalitogenic MBP 1–10 is a poor class II binder. In addition, only a fraction of immunogenic peptides in the rat is encephalitogenic, suggesting that there are additional factors, perhaps in the quality of a class II-regulated immune response, that determine pathogenicity of the ensuing immune response, a topic that still is unresolved in autoimmunology.

Use of intra-MHC recombinant rats has also evidenced a function of the class I-$CD8^+$ T cell pathway in EAE, in which certain alleles may recruit a protective disease downregulatory immune response.[65]

Less well studied are influences from the class III region on EAE. However, there are indications that genes in this region may influence histopathological features of MOG EAE, such as the degree of axon loss and demyelination.[69]

Table 56–1
Selected EAE models: strains, MHC haplotypes, and autoantigens[a,b]

Strain	*MHC Haplotype*	*Autoantigen*	*Type of EAE*	*Comment*
DA	AV1	rMOG(1–125)	Relapsing	MS-like model
		MBP_{rat}69–87	Protracted in some rats	
		MBP 87–101	Protracted in some rats	Rat and gp sequence identical
		PLP_{bovine}	Severe EAE	
		Spinal cord homogenate	Prolonged, relapses	Simple, robust
LEW	L	MBP_{gp}63–88	Monophasic	Classic model, robust
		MBP_{rat}63–88	Monophasic	Less severe than with the gp peptide
		MBP 89–101	Monophasic	Rat and gp sequence identical
		rMOG(1–125)	Resistant	Resistance is relative in EAE; may be overcome
		PLP_{bovine}	<50% develop clinical signs	CNS pathology includes demyelination
LEW.AV1	AV1	rMOG(1–125)	Relapsing/chronic	
LEW.N	N	rMOG(1–125)	Severe, lethal	Chronic with low doses of autoantigen
		MBP_{gp}63–88	Resistant	
		MBP 89–101	Monophasic	
		MBP 87–110	Relatively resistant	$CD8^+$ cells involved in the relative resistance
LEW.W	U	rMOG(1–125)	Semiresistant	
		MBP_{gp}63–88	Monophasic	
		MBP_{rat}63–88	Monophasic	Similar severity as with the gp peptide
PVG	C	rMOG(1–125)	Resistant	
PVG.AV1	AV1	rMOG(1–125)	Semiresistant	
ACI	AV1	rMOG(1–125)	Resistant	

[a]See references 24, 30, 31, 38, 59, 77, 78, and 79.
[b]EAE, experimental autoimmune encephalomyelitis; MHC, major histocompatibility complex; MS, multiple sclerosis; CNS, central nervous system; r, recombinant; PLP, proteolipid protein; gp, guinea pig.

Finally, with the profound influences of the MHC on EAE induced with different myelin autoantigens, the selection of strain is of utmost importance when planning an experiment.

NON-MAJOR HISTOCOMPATIBILITY COMPLEX GENE REGULATION OF EXPERIMENTAL AUTOIMMUNE ENCEPHALOMYELITIS The use of rat strains with differing non-MHC background genes but the same MHC enables demonstration of a non-MHC gene regulation of EAE. This has been demonstrated in the rat for both whole spinal cord-induced disease[70] and MOG-induced EAE.[49] Common rat strains that are relatively non-MHC gene EAE resistant include the BN, PVG ACI, and E3 strains. Susceptible strains include the LEW and DA strains. Naturally this has practical implications when selecting strains for experiments.

A long-term goal in EAE research, also extrapolated to human MS, has been to exactly position the non-MHC genes, both for the understanding of pathogenic pathways and for the selection of an appropriate model for mechanistic as well as therapeutic studies. Hypothetically, a particular experimental manipulation or drug treatment could have drastically different outcomes depending on the genetic makeup at the EAE regulating loci.

These gene mapping efforts have so far resulted in a number of quantitative trait loci (QTLs) with statistical probability for regulating disease. Typically F_2 crosses between resistant and susceptible strains are immunized and phenotype–genotype comparisons are made. In total, 19 whole-genome scans of rat or mouse crosses have been published resulting in a total of approximately 50 QTLs. Further experiments in advanced intercross lines have demonstrated that these may well be composed of several sub-QTLs, which contain more than one polymorphic gene.[71,72] Unfortunately, subsequent work to exactly position the responsible genes within the QTLs is much more laborious. At the moment, only a few examples of exact gene positioning have been achieved: in the rat the NCF1 gene regulating both arthritis[73] and EAE[74] and in the mouse IL-2.[75] Such gene positioning is achieved by creating congenic strains through selective backcrossing a selection of recombinants with phenotypic differences until very small fragments are achieved. Ideally overlapping congenics with only one gene ensues, thus providing formal proof. However, the positioning can be assisted by gene sequencing and expression analysis. The resulting congenic rats represent pharmacogenetically defined strains.

Studies of the type described represent a particular line of EAE research. For practical purposes in context with use of EAE as a model in general, it is important to know that the outcome of particular experiments can be profoundly influenced by non-MHC genes.

SCORING AND ETHICAL CONSIDERATIONS

EXPERIMENTAL AUTOIMMUNE ENCEPHALOMYELITIS PHENOTYPES The most common phenotype in EAE is the "clinical score." Most studies use a scale in which Grade 0 equals an unaffected animal; Grade 1, tail weakness or tail paralysis; Grade 2, hindleg paraparesis or hemiparesis; Grade 3, hindleg paralysis or hemiparalysis; and Grade 4, paresis in three or more extremities (tetraparesis), a moribund state, or death. There are also more detailed scales, such as Grade 0 normal, 1 tail weakness, 2 tail paralysis, 3 tail paralysis and mild waddle, 4 tail paralysis and severe waddle, 5 tail paralysis and paralysis of one limb, 6 tail paralysis and paralysis of a pair of limbs, 7 tetraparesis, and 8 premorbid or deceased.[76] Regardless of the number of grades, none of the scales is linear or parametric. Median values, percentiles, and nonparametric statistics are thus preferred. Sometimes, EAE grades are added for the entire observed disease period in a single animal, to create a "cumulative score." This is not correct since the scores are not linear; when it comes to disease severity Grade 1 for 3 days does not necessarily equal Grade 3 for 1 day. However, EAE rarely affects individual animals with higher scores for just single days, so in "real life" cumulative scores actually give a quite good overall approximation of how severe the disease has been throughout the experiment. An overall value for the clinical score for the whole experiment also avoids problems of multiple comparisons with associated loss of power, if scores are compared each day between experimental groups. Other obvious EAE phenotypes include incidence (usually Grade 1 is required for at least 2 days to score an animal as clinically EAE affected, to avoid "false positives") and maximum EAE score. Day of onset and the number of days with disease may be further phenotypes.

It is often simple to score animals with EAE during the initial bout of disease, when the disease often progresses from Grade 1 (tail paralysis) to Grade 2 (paraparesis, normally with tail paralysis), etc. In relapsing disease, it may be more complicated, because certain individuals, for instance, may recover from the tail paralysis, but still have a residual mild deficit in one hindleg. Usually such rats are scored as "2" (if the most common scoring scale is used), even if they have a normal tail.

There is, unfortunately, no common consensus on how to report EAE scores and several studies indeed report mean EAE scores, also with standard error bars! Regardless of how the EAE scores are reported, it is vital to remember that all scoring scales represent a gross simplification of the expression of the neuroimmunological disease. It is recommended that qualitative data (appearance, behavior) be added to EAE scoring sheets, when needed, to not overlook phenotypes not normally included in the classical EAE scoring scales.

Weight during the course of EAE may be a complementing phenotype. Usually, animals with EAE start to lose weight (compared to the normal slow weight increase) 1 day before the appearance of clinical signs of EAE. The weight loss during disease then roughly follows the disease course. However, weight is a complex phenotype and it may be affected by other factors besides the EAE itself. Thus, weight is not commonly reported in EAE papers, even if most centers indeed weigh animals daily during EAE experiments.

In addition to clinical scoring and weighing, there are a wide variety of histopathology, immunology, electrophysiology, and imaging techniques available to measure various aspects of EAE. It is beyond the scope of this chapter to list all available techniques.

All animal experiments should be guided by the three Rs: replace, reduce, and refine.[77] Most research animal departments and/or ethical review boards also have clear rules about when to euthanize animals, and they must of course be obeyed. As a general rule, animals reaching grade 4 EAE (tetraparesis) should be euthanized. However, it is important to also stress that reduction and refinement really need to be carefully thought through, since it is not ethical to perform underpowered experiments with too mild clinical disease expression. True replacement, refinement, and reduction, require careful planning of each experiment, to maximize the information obtained from every animal.

IMMUNIZATION PROTOCOLS

MBP_{GP}63–88 PEPTIDE-INDUCED EXPERIMENTAL AUTOIMMUNE ENCEPHALOMYELITIS IN LEW RATS For each rat use 100 μg of peptide in 100 μl of phosphate-buffered saline (PBS) and 500 μg of *Mycobacterium tuberculosis* (MT) h57 (Sigma) in 100 μl of IFA. For 20 rats this equals 100 μl of stock peptide (20 mg/ml) in 1900 μl of PBS and 25 mg MT in 2000 μl of IFA. For immunization use 200 μl per rat subcutaneously at the tail base.

RMOG (AA 1–125)-INDUCED EXPERIMENTAL AUTOIMMUNE ENCEPHALOMYELITIS IN DA RATS For each rat use 20 μg of rMOG in 100 μl of PBS and 100 μl of IFA. For 20 rats this equals 308 μl of stock rMOG (1.3 mg/ml) in 1692 μl of PBS and 2000 μl of IFA. For immunization use 200 μl per rat intradermally at the tail base.

REFERENCES

1. Trapp BD, Peterson J, Ransohoff RM, Rudick R, Mork S, Bo L. Axonal transection in the lesions of multiple sclerosis. *N Engl J Med* 1998;338(5):278–285.
2. Becker KG, Simon RM, Bailey-Wilson JE, *et al.* Clustering of non-major histocompatibility complex susceptibility candidate loci in human autoimmune diseases. *Proc Natl Acad Sci USA* 1998; 95(17):9979–9984.
3. Baron JL, Madri JA, Ruddle NH, Hashim G, Janeway CA Jr. Surface expression of alpha 4 integrin by CD4 T cells is required for their entry into brain parenchyma. *J Exp Med* 1993;177(1):57–68.
4. Dyment DA, Herrera BM, Cader MZ, *et al.* Complex interactions among MHC haplotypes in multiple sclerosis: Susceptibility and resistance. *Hum Mol Genet* 2005;14(14):2019–2026.
5. Fogdell-Hahn A, Ligers A, Gronning M, Hillert J, Olerup O. Multiple sclerosis: A modifying influence of HLA class I genes in an HLA class II associated autoimmune disease. *Tissue Antigens* 2000;55(2): 140–148.
6. Saarela J, Kallio SP, Chen D, *et al.* PRKCA and multiple sclerosis: Association in two independent populations. *PLoS Genet* 2006;2(3): e42.
7. Zhang Z, Duvefelt K, Svensson F, *et al.* Two genes encoding immune-regulatory molecules (LAG3 and IL7R) confer susceptibility to multiple sclerosis. *Genes Immun* 2005;6(2):145–152.
8. A meta-analysis of genomic screens in multiple sclerosis. The Transatlantic Multiple Sclerosis Genetics Cooperative. *Mult Scler* 2001;7(1):3–11.
9. Vyshkina T, Shugart YY, Birnbaum G, Leist TP, Kalman B. Association of haplotypes in the beta-chemokine locus with multiple sclerosis. *Eur J Hum Genet* 2005;13(2):240–247.
10. Olsson T, Jagodic M, Piehl F, Wallström E. Genetics of autoimmune neuroinflammation. *Curr Opin Immunol* 2006;18(6):643–649.
11. Alvord EC Jr. The challenge: How good a model of MS is EAE today? *Prog Clin Biol Res* 1984;146:3–5.
12. Sriram S, Steiner I. Experimental allergic encephalomyelitis: A misleading model of multiple sclerosis. *Ann Neurol* 2005;58(6): 939–945.
13. Steinman L, Zamvil SS. Virtues and pitfalls of EAE for the development of therapies for multiple sclerosis. *Trends Immunol* 2005; 26(11):565–571.
14. Ransohoff RM. EAE: Pitfalls outweigh virtues of screening potential treatments for multiple sclerosis. *Trends Immunol* 2006;27(4):167–168.
15. Pettinelli CB, McFarlin DE. Adoptive transfer of experimental allergic encephalomyelitis in SJL/J mice after in vitro activation of lymph node cells by myelin basic protein: Requirement for Lyt 1+ 2- T lymphocytes. *J Immunol* 1981;127(4):1420–1423.

16. Ben-Nun A, Wekerle H, Cohen IR. The rapid isolation of clonable antigen-specific T lymphocyte lines capable of mediating autoimmune encephalomyelitis. *Eur J Immunol* 1981;11(3):195–199.
17. Linington C, Bradl M, Lassmann H, Brunner C, Vass K. Augmentation of demyelination in rat acute allergic encephalomyelitis by circulating mouse monoclonal antibodies directed against a myelin/oligodendrocyte glycoprotein. *Am J Pathol* 1988;130(3):443–454.
18. Katz-Levy Y, Neville KL, Girvin AM, *et al.* Endogenous presentation of self myelin epitopes by CNS-resident APCs in Theiler's virus-infected mice. *J Clin Invest* 1999;104(5):599–610.
19. Gregersen JW, Holmes S, Fugger L. Humanized animal models for autoimmune diseases. *Tissue Antigens* 2004;63(5):383–394.
20. Bettelli E, Baeten D, Jager A, Sobel RA, Kuchroo VK. Myelin oligodendrocyte glycoprotein-specific T and B cells cooperate to induce a Devic-like disease in mice. *J Clin Invest* 2006;116(9):2393–2402.
21. Krishnamoorthy G, Lassmann H, Wekerle H, Holz A. Spontaneous opticospinal encephalomyelitis in a double-transgenic mouse model of autoimmune T cell/B cell cooperation. *J Clin Invest* 2006; 116(9):2385–2392.
22. Willenborg DO. Experimental allergic encephalomyelitis in the Lewis rat: Studies on the mechanism of recovery from disease and acquired resistance to reinduction. *J Immunol* 1979;123(3):1145–1150.
23. Eng LF, Ghirnikar RS, Lee YL. Inflammation in EAE—role of chemokine/cytokine expression by resident and infiltrating cells. *Neurochem Res* 1996;21(4):511–525.
24. Weissert R, Wallstrom E, Storch MK, *et al.* MHC haplotype-dependent regulation of MOG-induced EAE in rats. *J Clin Invest* 1998;102(6):1265–1273.
25. Storch MK, Stefferl A, Brehm U, *et al.* Autoimmunity to myelin oligodendrocyte glycoprotein in rats mimics the spectrum of multiple sclerosis pathology. *Brain Pathol* 1998;8(4):681–694.
26. Olsson T. Immunology of multiple sclerosis. *Curr Opin Neurol Neurosurg* 1992;5(2):195–202.
27. Steinman L. Multiple sclerosis: A coordinated immunological attack against myelin in the central nervous system. *Cell* 1996;85(3): 299–302.
28. Brehm U, Piddlesden SJ, Gardinier MV, Linington C. Epitope specificity of demyelinating monoclonal autoantibodies directed against the human myelin oligodendrocyte glycoprotein (MOG). *J Neuroimmunol* 1999;97(1–2):9–15.
29. Ma C, Whiteley PE, Cameron PM, *et al.* Role of APC in the selection of immunodominant T cell epitopes. *J Immunol* 1999;163(12): 6413–6423.
30. Stepaniak JA, Gould KE, Sun D, Swanborg RH. A comparative study of experimental autoimmune encephalomyelitis in Lewis and DA rats. *J Immunol* 1995;155(5):2762–2769.
31. Weissert R, Svenningsson A, Lobell A, Degraaf KL, Andersson R, Olsson T. Molecular and genetic requirements for preferential recruitment of TCRBV8S2(+) T cells in Lewis rat experimental autoimmune encephalomyelitis. *J Immunol* 1998;160(2):681–690.
32. Berger T, Weerth S, Kojima K, Linington C, Wekerle H, Lassmann H. Experimental autoimmune encephalomyelitis: The antigen specificity of T lymphocytes determines the topography of lesions in the central and peripheral nervous system. *Lab Invest* 1997;76(3): 355–364.
33. Kleinau S, Erlandsson H, Holmdahl R, Klareskog L. Adjuvant oils induce arthritis in the DA rat. I. Characterization of the disease and evidence for an immunological involvement. *J Autoimmun* 1991; 4(6):871–880.
34. Lorentzen JC, Glaser A, Jacobsson L, *et al.* Identification of rat susceptibility loci for adjuvant-oil-induced arthritis. *Proc Natl Acad Sci USA* 1998;95(11):6383–6387.
35. Warren HS, Vogel FR, Chedid LA. Current status of immunological adjuvants. *Annu Rev Immunol* 1986;4:369–388.
36. Freund J. Some aspects of active immunization. *Annu Rev Microbiol* 1947;1:291–309.
37. Lavebratt C, Apt AS, Nikonenko BV, Schalling M, Schurr E. Severity of tuberculosis in mice is linked to distal chromosome 3 and proximal chromosome 9. *J Infect Dis* 1999;180(1):150–155.
38. Lorentzen JC, Issazadeh S, Storch M, *et al.* Protracted, relapsing and demyelinating experimental autoimmune encephalomyelitis in DA rats immunized with syngeneic spinal cord and incomplete Freunds adjuvant. *J Neuroimmunol* 1995;63(2):193–205.
39. Lenz DC, Wolf NA, Swanborg RH. Strain variation in autoimmunity: Attempted tolerization of DA rats results in the induction of experimental autoimmune encephalomyelitis. *J Immunol* 1999;163(4): 1763–1768.
40. Maatta JA, Eralinna JP, Roytta M, Salmi AA, Hinkkanen AE. Physical state of the neuroantigen in adjuvant emulsions determines encephalitogenic status in the BALB/c mouse. *J Immunol Methods* 1996;190(1):133–141.
41. Whitacre CC, Dowdell K, Griffin AC. Neuroendocrine influences on experimental autoimmune encephalomyelitis. *Ann NY Acad Sci* 1998;840:705–716.
42. Correa SG, Rodriguez-Galan MC, Rivero VE, Riera CM. Chronic varied stress modulates experimental autoimmune encephalomyelitis in Wistar rats. *Brain Behav Immun* 1998;12(2):134–148.
43. Mason D, MacPhee I, Antoni F. The role of the neuroendocrine system in determining genetic susceptibility to experimental allergic encephalomyelitis in the rat. *Immunology* 1990;70(1):1–5.
44. Teuscher C, Bunn JY, Fillmore PD, Butterfield RJ, Zachary JF, Blankenhorn EP. Gender, age, and season at immunization uniquely influence the genetic control of susceptibility to histopathological lesions and clinical signs of experimental allergic encephalomyelitis: Implications for the genetics of multiple sclerosis. *Am J Pathol* 2004;165(5):1593–1602.
45. Teuscher C, Doerge RW, Fillmore PD, Blankenhorn EP. eae36, a locus on mouse chromosome 4, controls susceptibility to experimental allergic encephalomyelitis in older mice and mice immunized in the winter. *Genetics* 2006;172(2):1147–1153.
46. Cua DJ, Coffman RL, Stohlman SA. Exposure to T helper 2 cytokines in vivo before encounter with antigen selects for T helper subsets via alterations in antigen-presenting cell function. *J Immunol* 1996;157(7):2830–2836.
47. Trooster WJ, Teelken AW, Gerrits PO, *et al.* The effect of gonadectomy on the clinical course of chronic experimental allergic encephalomyelitis. *Clin Neurol Neurosurg* 1996;98(3):222–226.
48. Bebo BF, Schuster JC, Vandenbark AA, Offner H. Gender differences in experimental autoimmune encephalomyelitis develop during the induction of the immune response to encephalitogenic peptides. *J Neurosci Res* 1998;52(4):420–426.
49. Dahlman I, Wallstrom E, Weissert R, *et al.* Linkage analysis of myelin oligodendrocyte glycoprotein-induced experimental autoimmune encephalomyelitis in the rat identifies a locus controlling demyelination on chromosome 18. *Hum Mol Genet* 1999;8(12):2183–2190.
50. Wallstrom E, Weissert R, Lorentzen J, Olsson T. Major histocompatibility complex haplotype RT1(av1) is associated with relapsing/remitting experimental autoimmune encephalomyelitis. *Transplant Proc* 1997;29(3):1686–1689.
51. Jansson L, Holmdahl R. Estrogen-mediated immunosuppression in autoimmune diseases. *Inflamm Res* 1998;47(7):290–301.
52. Butterfield RJ, Blankenhorn EP, Roper RJ, *et al.* Genetic analysis of disease subtypes and sexual dimorphisms in mouse experimental allergic encephalomyelitis (EAE): Relapsing/remitting and monophasic remitting/nonrelapsing EAE are immunogenetically distinct. *J Immunol* 1999;162(5):3096–3102.
53. Teuscher C, Noubade R, Spach K, *et al.* Evidence that the Y chromosome influences autoimmune disease in male and female mice. *Proc Natl Acad Sci USA* 2006;103(21):8024–8029.
54. Weinshenker BG, Bass B, Rice GP, *et al.* The natural history of multiple sclerosis: A geographically based study. I. Clinical course and disability. *Brain* 1989;112(Pt. 1):133–146.
55. Hedrich HJ. *Genetic Monitoring of Inbred Strains of Rats.* New York: Gustav Fischer Verlag, 1990.
56. Olofsson P, Johansson A, Wedekind D, *et al.* Inconsistent susceptibility to autoimmunity in inbred LEW rats is due to genetic crossbreeding involving segregation of the arthritis-regulating gene Ncf1. *Genomics* 2004;83(5):765–771.

57. Gould KE, Stepaniak JA, Swanborg RH. Variable susceptibility of Lewis rats to experimental autoimmune encephalomyelitis. *J Neuroimmunol* 1994;54(1–2):145–146.
58. Stefferl A, Brehm U, Storch M, *et al.* Myelin oligodendrocyte glycoprotein induces experimental autoimmune encephalomyelitis in the "resistant" brown Norway rat: Disease susceptibility is determined by MHC and MHC-linked effects on the B cell response. *J Immunol* 1999;163(1):40–49.
59. Muhallab S, Dahlman I, Wallstrom E. Disparate MHC class II haplotypes in myelin oligodendrocyte glycoprotein- and myelin basic protein-induced experimental autoimmune encephalomyelitis. *J Neuroimmunol* 2005;161(1–2):155–161.
60. Williams RM, Moore MJ. Linkage of susceptibility to experimental allergic encephalomyelitis to the major histocompatibility locus in the rat. *J Exp Med* 1973;138(4):775–783.
61. Jersild C, Fog T, Hansen GS, Thomsen M, Svejgaard A, Dupont B. Histocompatibility determinants in multiple sclerosis, with special reference to clinical course. *Lancet* 1973;2(840):1221–1225.
62. Jersild C, Svejgaard A, Fog T. HL-A antigens and multiple sclerosis. *Lancet* 1972;1(762):1240–1241.
63. Barcellos LF, Sawcer S, Ramsay PP, *et al.* Heterogeneity at the HLA-DRB1 locus and risk for multiple sclerosis. *Hum Mol Genet* 2006;15(18):2813–2824.
64. Hurt P, Walter L, Sudbrak R, *et al.* The genomic sequence and comparative analysis of the rat major histocompatibility complex. *Genome Res* 2004;14(4):631–639.
65. Mustafa M, Vingsbo C, Olsson T, Issazadeh S, Ljungdahl A, Holmdahl R. Protective influences on experimental autoimmune encephalomyelitis by MHC class I and class II alleles. *J Immunol* 1994;153(7):3337–3344.
66. de Graaf KL, Berne GP, Herrmann MM, Hansson GK, Olsson T, Weissert R. CDR3 sequence preference of TCRBV8S2+ T cells within the CNS does not reflect single amino acid dependent avidity expansion. *J Neuroimmunol* 2005;166(1–2):47–54.
67. de Graaf KL, Weissert R, Kjellen P, Holmdahl R, Olsson T. Allelic variations in rat MHC class II binding of myelin basic protein peptides correlate with encephalitogenicity. *Int Immunol* 1999; 11(12):1981–1988.
68. Weissert R, de Graaf KL, Storch MK, *et al.* MHC class II-regulated central nervous system autoaggression and T cell responses in peripheral lymphoid tissues are dissociated in myelin oligodendrocyte glycoprotein-induced experimental autoimmune encephalomyelitis. *J Immunol* 2001;166(12):7588–7599.
69. Storch MK, Weissert R, Steffer A, *et al.* MHC gene related effects on microglia and macrophages in experimental autoimmune encephalomyelitis determine the extent of axonal injury. *Brain Pathol* 2002;12(3):287–299.
70. Dahlman I, Jacobsson L, Glaser A, *et al.* Genome-wide linkage analysis of chronic relapsing experimental autoimmune encephalomyelitis in the rat identifies a major susceptibility locus on chromosome 9. *J Immunol* 1999;162(5):2581–2588.
71. Jagodic M, Becanovic K, Sheng JR, *et al.* An advanced intercross line resolves Eae18 into two narrow quantitative trait loci syntenic to multiple sclerosis candidate loci. *J Immunol* 2004;173(2):1366–1373.
72. Jagodic M, Marta M, Becanovic K, *et al.* Resolution of a 16.8-Mb autoimmunity-regulating rat chromosome 4 region into multiple encephalomyelitis quantitative trait loci and evidence for epistasis. *J Immunol* 2005;174(2):918–924.
73. Olofsson P, Holmberg J, Tordsson J, Lu S, Akerstrom B, Holmdahl R. Positional identification of Ncf1 as a gene that regulates arthritis severity in rats. *Nat Genet* 2002;2:2.
74. Becanovic K, Jagodic M, Sheng JR, *et al.* Advanced intercross line mapping of Eae5 reveals Ncf-1 and CLDN4 as candidate genes for experimental autoimmune encephalomyelitis. *J Immunol* 2006;176(10):6055–6064.
75. Encinas JA, Wicker LS, Peterson LB, *et al.* QTL influencing autoimmune diabetes and encephalomyelitis map to a 0.15-cM region containing Il2. *Nat Genet* 1999;21(2):158–160.
76. Kjellen P, Issazadeh S, Olsson T, Holmdahl R. Genetic influence on disease course and cytokine response in relapsing experimental allergic encephalomyelitis. *Int Immunol* 1998;10(3):333–340.
77. Russell WMS, Burch RL. *The Principles of Humane Experimental Techniques.* London: Methuen, 1959.
78. Yamamura T, Namikawa T, Endoh M, Kunishita T, Tabira T. Experimental allergic encephalomyelitis induced by proteolipid apoprotein in Lewis rats. *J Neuroimmunol* 1986;12(2):143–153.
79. Issazadeh S, Kjellen P, Olsson T, Mustafa M, Holmdahl R. Major histocompatibility complex-controlled protective influences on experimental autoimmune encephalomyelitis are peptide specific. *Eur J Immunol* 1997;27(6):1584–1587.

57 Animal Models in Virology

JOSEPH L. BRYANT

ABSTRACT

The knowledge we have gained from the study of many diseases that affect humans comes from the study of disease processes in different animal species, and this has enhanced our understanding of the pathogenesis of the disease in humans. The American Medical Association says almost every advance in medical science in the twentieth century, from antibiotics and vaccines to antidepressant drugs and organ transplants, has been achieved either directly or indirectly through the use of animals as models of disease. In this chapter a brief overview of the uses of animal models for research on human viral diseases is presented.

Key Words: AIDS, Animal models, Cell biochemistry, Cell-based assays, Cervical cancer, Comparative medicine, Ebola virus, Genetic alterations, Genetic make-up, Hantavirus, Hepatitis C, Human papilloma virus, Influenza virus, Intracellular parasites, Mimicry, Nipah virus, Nonhuman primates, SARS, Transgenic mice.

INTRODUCTION

Animal models are used in almost every field of biomedical research. Almost all advances in medical knowledge and treatment, especially those in the past century, has involved work with laboratory animals. Two-thirds of the Nobel prizes awarded since 1901 have been for discoveries requiring the use of laboratory animals.

DEFINITION OF VIROLOGY Viruses are small infectious obligate intracellular molecular parasites. The most basic definition is that viruses are composed of a genome and one or more proteins coating that genome.

The simplest definition of virology is the study of viruses and viral diseases. What then is a virus? Viruses are small infectious obligate intracellular molecular parasites. Today the most basic definition is that viruses are composed of a genome and one or more proteins coating that genome. The virus genome is composed of either DNA or RNA. The genetic information for such a protein coat and other information required for replication are encoded in that genome. The HIV virus is a good example (human immunodeficiency virus/structure).[1,2] Virology is a work in progress. With much left to learn even about extensively studied viruses, we are also certain to be challenged by new, emerging viruses.[3,4] At first, virologists focused on the essential interactions of viruses with cultured cells in which they are grown. However, the strategies of viruses are often focused on interactions with systems of the host animal, notably the immune system. Present and future research in virology involves unraveling the many details of the virus–host relationship. We are likely to be frustrated over and over by the subtlety of the mechanisms that have evolved both to replicate viruses and to defend against host responses. In some cases we will see new examples of viral mimicry of cellular functions. Meanwhile, intracellular locations of viral activities and associations of viral proteins and nucleic acids with cellular counterparts will continue to illuminate details of cellular organization and regulation. Animal models of many human viral diseases are still needed to help elucidate the paths of viral dissemination and specific interactions of certain cells or organs with viruses.

SPECIFIC ROLES ANIMALS HAVE PLAYED IN RESEARCH IN HUMAN MEDICAL VIROLOGY There are several examples of specific roles animals have played in research in virology; one of the best examples is the virus that causes AIDS. The scientific literature is filled with numerous examples of the specific roles animals have played in viral research both in humans and animals.[5–8] Animal models are essential for investigating and studying diseases in humans and that includes viral diseases. In the very early 1980s, when reports of a lethal immunodeficiency disease began circulating, the news at first generated deep fears because no one knew what was causing the disease. The causative agent is a retrovirus, human immunodeficiency virus (HIV).[9] In fact, the reverse transcriptase in the viral particle provides the signal that was used to isolate the virus for the first time. Tremendous advances in our understanding of HIV have been made through the use of animal models.[9–13]

While there are several examples of specific roles animals have played in research in virology, one of the best examples is the virus that causes AIDS. However, there are limitations inherent in many of the current models, whether they are nonhuman primates or other models such as feline immunodeficiency virus (FIV). An exhausting review of the literature clearly shows that while there is no ideal model for HIV, the accumulating knowledge obtained from these models have made major contributions to understanding the pathogenesis of HIV.[14,15] As stated by Robert Gallo, "With animals, we may have a cure for AIDS, without animals we will never cure AIDS in my lifetime."[16]

From: *Sourcebook of Models for Biomedical Research*
(P. M. Conn, ed.), © 2008 Humana Press Inc., Totowa, NJ.

Scientists are also using a number of different animal models to obtain information that can have application to HIV. FIV, transgenic mice, and rats that contain part of the HIV genome or coreceptors for viral entry, and severe combined immune deficiency (SCID) mice reconstituted with human immune system cells or tissues are some of the animal models being used to study the pathogenesis of HIV.[17–19]

Other examples of specific roles of animals in viral are the coronavirus, severe acute respiratory syndrome (SARS), in which the ferret and domesticated cat are susceptible to infection by SARS.[20,21] The scientific literature is filled with numerous examples of the specific roles animals have played in research in virology.[22–24]

HISTORICAL PROSPECTIVE OF ANIMAL MODELS IN VIROLOGY

Viruses have historically provided and continue to provide the basis for much of our knowledge and understanding of modern biology, genetics, and medicine. An understanding of viral genomes and viral replication provides basic information concerning cellular processes in general.

HISTORY OF RESEARCH ON SELECTIVE VIRAL DISEASES Virology's history began over a century ago; viruses, including bacteriophage, have been the source for much of today's molecular biology. In the past 40 years of research on animal virology, we have watched the field go from just growing and identifying animal viruses to a highly sophisticated branch of science. In fact, viruses, including bacteriophage, are the source for much of today's molecular biology. This personal and, of necessity, somewhat limited record reflects how far virology has come and where it still needs to go.

The early definition of a virus was any infectious agent that passed through a filter that caught bacteria. Nonetheless, as early as 1927, visionaries such as Hermann Muller recognized that the small size of viruses had profound implications. There simply was not enough room inside a virus for much more than its genetic material. Of course, no one then knew what genetic material was.

Most studies in virology until the 1950s focused on laboratory animals such as mice, chickens, and ferrets. A pathfinding exception was the work of John Enders and his colleagues at Harvard Medical School who, in 1949, adapted poliovirus to grow in cell culture. This achievement not only revolutionized the production of vaccines, but it set the path for the biochemical analysis of this and other viruses. Virology and immunology—those twins of offense and defense—are generally in a fine balance, but the story of HIV shows that the balance can be tipped against the immune system by a particularly cunning agent. Under the leadership of Dr. Robert Gallo, the Institute of Human Virology is dedicated to the discovery, research, treatment,and prevention of chronic viral diseases, including HIV/AIDS.[16]

The study of viruses has historically provided and continues to provide the basis for much of our most fundamental understanding of modern biology, genetics, and medicine. Virology has had an impact on the study of biological marcromolecules, processes of cellular gene expression, mechanisms for generating genetic diversity, processes involved in the control of cell growth and development, aspects of molecular evolution, the mechanism of disease and response of the host to it, and the spread of disease in populations. In essence, viruses are collections of genetic information directed toward one end: their own replication. The viral genome contains the "blueprints" for virus replication enciphered in the genetic code, and must be decoded by the molecular machinery of the cell that it infects to gain this end. Viruses are obligate intracellular parasites dependent on the metabolic and genetic functions of living cells. The replication and propagation of a given virus in a population are frequently (but not always) manifest with the occurrence of an infectious disease that spreads between individuals.

The historic reason for the discovery and characterization of viruses, and a continuing major reason for their detailed study, involves the desire to understand and control the diseases and attending degrees of economic and individual distress caused by them. As science progressed, it became clear that there were many other important reasons for the study of viruses and their replication. Since viruses are parasitic on the molecular processes of gene expression and its regulation in the host cell, an understanding of viral genomes and virus replication provides basic information concerning cellular processes in general (Table 57–1).

There is archeological evidence in Egyptian mummies and medical texts of readily identifiable viral infections, including genital papillomas (warts) and poliomyelitis. There are also somewhat imperfect historical records of viral disease affecting human populations in classical and medieval times. While the recent campaign to eradicate smallpox has been successful and it no longer exists in the human population (owing to the effectiveness of vaccines against it, the genetic stability of the virus, and a well-orchestrated political and social effort to carry out the eradication), the disease periodically wreaked havoc and had profound effects on human history over thousands of years. Smallpox epidemics during the Middle Ages and later in Europe resulted in significant population losses as well as changes in the economic, religious, political, and social life of individuals. The effectiveness of vaccination strategies gradually led to the decline of the disease in other parts of the world until after World War II. Recently fears have arisen that the high virulence of the virus and its mode of spread might make it an attractive agent for bioterrorism.[24]

Depending on the infection and the focus of study, other animals have proven to be useful in infectious disease research. These animals include the rabbit, rat, guinea pig, pig, dog, and monkey. The latter, in particular, has been utilized in the study of AIDS, as primates are the genetically closest relatives to humans. The advent of molecular techniques of genetic alteration has made the development of genetically tailored animal models possible. Thus, for example, mouse models exist in which the activity of certain genes has been curtailed. These are known as transgenic animals. The involvement of the gene product in the infectious process is possible on a scale not possible without the use of the animal.[25]

Table 57–1
Human diseases caused by viruses

Acute hemorrhagic conjunctivitis
Acute hemorrhagic cystic
AIDS/acquired immune deficiency syndrome—human immunodeficiency virus
Bronchiolitis—respiratory syncytial virus
California encephalitis—California encephalitis virus
Cervical cancer—human papilloma virus
Chickenpox—varicella zoster virus
Colorado tick fever—Colorado tick fever virus
Conjunctivitis—herpes simplex virus
Cowpox—vaccinia virus
Group, infections—parainfluenza viruses
Dengue—dengue virus
"Devil's Grip"—coxsackie B
Eastern equine encephalitis—EEE virus
Ebola hemorrhagic fever—Ebola virus
Gastroenteritis—Norwalk virus
Genital HSV—herpes simplex virus
Gingivostomatitis—HSV-1
Hantavirus hemorrhagic fever/Hantaan-Korean hemorrhagic fever—Hantavirus
Hepatitis
Hepatitis A—hepatitis A virus
Hepatitis B—hepatitis B virus
Hepatitis C—hepatitis C virus
Hepatitis D—hepatitis D virus
Hepatitis E—hepatitis E virus
Herpangina—coxsackie A
Herpes, genital—HSV-2
Herpes labialis—HSV-1
Infectious myocarditis—coxsackie B1–B5
Infectious pericarditis—coxsackie B1–B5
Influenza—influenza viruses A, B, and C
Keratoconjunctivitis—adenovirus
Lass hemorrhagic fever—Marburg virus
Measles—rubella virus
Meningitis, aseptic—coxsackie A and B (enterovirus), lymphocytic choriomeningitis virus
Mononucleosis—Epstein–Barr virus
Mumps—mumps virus
Pharyngitis
Respiratory syncytial virus
Influenza virus
Parainfluenza virus
Adenovirus
Epstein–Barr virus
Pleurodynia—coxsackie B
Pneumonia, viral—respiratory syncytial virus
Polio, poliomyelitis—Poliovirus
Progressive multifocal leukoencephalopathy—JC virus
Rabies—rabies virus
Roseola—HHV-6
Rubella—rubivirus
Severe acute respiratory syndrome (SARS)—a human coronavirus
Shingles (zoster)—varicella zoster virus
Urethritis—herpes simplex virus
West equine encephalitis—WEE virus
Yellow fever—Yellow fever virus
Zoster—varicella zoster virus

ASSESSMENT OF CURRENT AND POTENTIAL ANIMAL MODELS IN VIROLOGY

A number of potential useful models exist that may facilitate and improve the understanding of the pathogenesis and treatment of viral infection. A good example is HIV-1 disease.

Animal models have provided a controlled setting for the study of human immunodeficiency virus-1 (HIV-1) disease, the preclinical testing of novel antiviral compounds, and the evaluation of vaccines. Because the animals serve as models for humans they should be closely reflective of human physiology and pathophysiology. Moreover, their use must be complementary to, and not replaceable by, experimental approaches that do not require animals. For any given experiment, the critical question is which, if any, model is most useful—and why?

ADVANCES MADE IN BIOMEDICAL RESEARCH IN VIROLOGY USING ANIMALS

Viruses are collection of genetic information directed toward replication. The viral genome contains the blueprint for virus replication encoded in the genetic code and must be decoded by molecular biology. Exploitation of viral diseases of animals may ultimately provide a useful means of understanding human viral diseases.

The literature on the history of the use of animals in human medical virology is somewhat sketchy and not well organized. This clearly indicates that the study of viruses as a pathogen in animal models is a relative new field and in most cases, due to molecular biology and genomic research studies, is now coming into its own.

Most of the major advances in modern virology during the past 25 years have been due principally to the development of refined laboratory techniques and tools and have provided an array of new knowledge and information about the nature of viral infection and pathogenesis.

Changing their role from hunters of microbes to biochemists probing the nature of life, virologists are simultaneously reflecting and leading the revolution in biomedical research. By using the post-World War II tools of tissue culture, radioactive isotopes, chromatography, density gradient centrifugation, and the electron microscope, they have acquired vast knowledge about the way viruses infect cells and cause disease. Unexpectedly, the viruses themselves have emerged as powerful probes into the nature of cellular and life processes. Because of the necessarily close relationship between viruses and their host cells, the understanding and control of viral infections depend almost wholly on knowledge of the biochemistry of cells.

Vaccines and sera have been powerful aids in the prevention of viral diseases such as polio, measles, mumps, rubella, yellow fever, and hepatitis, but they are only extensions of the fundamental principles of Jenner's smallpox vaccine and Pasteur's rabies vaccine. Compared with the modern treatment of established diseases by means of specific chemotherapy, the management of viral diseases lags behind all other forms of chemotherapy.

ACHIEVEMENTS OF ANIMAL RESEARCH IN VIROLOGY

Several animal models, primarily rodents and monkeys, have been developed that recapitulate many aspects of the disease seen in humans. Studies reviewed here demonstrate how these models have played a vital role in understanding and developing therapeutics for these viral diseases.

HUMAN PAPILLOMAVIRUS AND CERVICAL CANCER As the scientific community continues to make new discoveries about factors contributing to the development of cancer, viral pathogens have been found to play an important role. In the development of cervical cancer, one specific type of abnormal cell growth (neoplasia), DNA from human papillomavirus (HPV), has been detected in 50–95% of lesions. It is estimated that women with HPV have 10–30 times the risk of developing cervical neoplasia than those not infected. Using this system as a model, many questions on how viral infections influence the development of cancer have been explored experimentally.[26]

INFLUENZA VIRUS The animal models presently available for the study of the pathogenesis of influenza virus disease have limitations.[27,28] Influenza A virus will experimentally infect a number of Old World and New World primates. The gibbon and baboon develop clinical illness with nasal application of the virus, and the squirrel, cynomolgus, and rhesus monkeys develop illness when the virus is inoculated intratracheally. Primate models suffer from a number of disadvantages, including the limited availability of expensive animals. In addition, these animals are outbred and the models lack many of the reagents necessary to characterize the host response in detail. Mammals such as horses and pigs that are natural hosts for influenza have also been used experimentally. However, their large size and the limited number of reagents available preclude their use in the laboratory.

Small-animal models that have been used to study influenza virus pathogenesis include the ferret, in which human influenza virus was originally isolated.[29] Adult ferrets become ill after infection with unadapted influenza A viruses, exhibiting fever, lethargy, and weight loss. The ferret model has been used in recent studies of H5N1 viruses, the transmission of influenza, and the development of resistance to antiviral therapy. Unfortunately, ferrets are outbred and reagents are not available for dissecting the correlates of protective immunity.

There has been one report, published in Polish and largely overlooked, of the use of outbred cotton rats (*Sigmodon hispidus*) for pathogenesis experiments with influenza viruses. Nasal administration of virus in lightly anesthetized cotton rats resulted in virus replication, the production of pulmonary lesions, and a strong immune response. Results suggest that the cotton rat may serve as a useful model for the study of influenza pathogenesis. The animals are small, inbred, easy to handle, and relatively inexpensive to purchase and maintain.[30]

Influenza virus strains that cause worldwide outbreaks (pandemics) are classic examples of emerging viruses that are maintained in other animal hosts before transmission to humans. Influenza viruses are isolated from a variety of animals, including humans, pigs, horses, wild and domestic birds, and even sea mammals.[31] The most devastating viral infection in this century was not caused by HIV, but by Spanish influenza, which killed more than 20 million people worldwide. Genetic studies suggest that the Spanish influenza virus originally was derived from birds. Furthermore, the causative viruses from the 1957 and 1968 influenza pandemics were hybrids between human and avian influenza viruses. Because humans did not have immunity to avian influenza viruses, the hybrid viruses produced devastating consequences (70,000 and 46,500 deaths globally in the 1957 and 1968 pandemics, respectively). Thus, it is critical to understand the mechanisms by which new influenza strains capable of causing pandemics emerge. Animal models will play a critical role in understanding the mechanisms.

HANTAVIRUS Hantavirus is segmented RNA viruses belonging to the genus *Hantavirus* in the family Bunyaviridae. Hantaviruses are maintained in various rodent reservoirs, in which the hosts are persistently infected without disease symptoms. Specific hantaviruses transmitted from the contaminated urine and feces of infected rodents cause two important human diseases, hemorrhagic fever with renal syndrome (HFRS) and hantavirus pulmonary syndrome (HPS). Annually, hundreds of thousands of cases of HFRS are reported throughout Euro-Asia, whereas hundreds of cases of HPS are reported in countries in North and South America. Because rodents act as the natural reservoir for hantaviruses and human-to-human infections are rare, understanding the ecology of hantavirus within their natural reservoir is important for preventing and controlling the emergence of such diseases.[32]

The comparison of may hantavirus genomes form different rodent species has shown a clear correlation between the rodent species and the virus genotype, suggesting that hantavirus have coevolved with their natural hosts for >20 million years, since before the first humans evolved. It remains unclear how hantavirus exist within a rodent reservoir, particularly how they establish a persistent infection. In experiments with laboratory rats and mice, several groups have shown that an experimentally infected newborn animal readily develops a persistent infection, whereas an adult animal develops only a transient infection and recovers completely. On the other hand, epizootiological investigations have demonstrated that virus is transmitted between adult animals through wounds, and the adults develop a persistent infection.

EBOLA VIRUS Ebola virus is a nonsegmented RNA virus, which, together with Marburg virus, makes up the filovirus family. This now notorious group of viruses was discovered in 1967 when Marburg virus was identified as the etiological agent of a hemorrhagic fever outbreak in research facilities in Europe, which handled tissues from African green monkeys imported from Uganda. Subsequently, Ebola viruses were shown to be the cause of simultaneously occurring hemorrhagic fever outbreaks in 1976 in the Democratic Republic of Congo (DRC, formerly Zaire) and Sudan. These outbreaks were shown to be caused by two different subtypes of Ebola virus, which became known as the Zaire and Sudan subtypes. Mortality rates of up to 80% were recorded in these and more recent outbreaks in DRC and Gabon in 1995–1996. Epidemiological data from recent outbreaks indicate that close contact is necessary for efficient transmission of Ebola virus from one individual to another, and little evidence can be found for aerosol transmission of the virus. Despite considerable efforts to identify the natural reservoir for Ebola and Marburg viruses, the host species remains an enigma. Although nonhuman primates have been implicated as the source of the introduction of the virus into humans during several of the identified outbreaks, they are not considered likely to represent reservoir species because of their susceptibility to high-mortality hemorrhagic disease similar to that seen in humans.

NIPAH VIRUS Nipah virus is a newly discovered member of the paramyxovirus family of nonsegmented RNA viruses. This virus was responsible for a viral encephalitis outbreak in Malaysia that was first recognized in October 1998 and ended in midsummer 1999. This outbreak resulted in almost 300 confirmed infections, and the mortality rate for hospitalized cases was approximately 35%. Initially, Malaysian authorities thought the outbreak was caused by Japanese encephalitis (JE virus, a mosquito-borne RNA virus). However, JE vaccination and mosquito control efforts failed to halt the epidemic. The virus appeared to be first introduced into pigs, where close contact caused by intensive farming practices led to efficient pig-to-pig transmission, and subsequently pig-to-human transmission. Most human cases were in close proximity to the infected pigs. Genetic analysis showed Nipah virus to be closely related to Hendra virus, which recently was discovered in Australia as a cause of disease in horses and humans and also is maintained in *Pteropus* species fruit bats.

These examples highlight the subtle balance of environmental and genetic factors that can mold the diverse evolutionary patterns observed for RNA viruses and illustrate the complexity of these systems, which makes it difficult to predict future viral disease emergences. Much more research is needed and animal models will continue to be needed to ultimately understand and control these diseases.

RELEVANT ANIMAL MODELS

Use of animals as models for human disease has been indispensable in understanding the cause, biology, and prevention of disease. The Animal Model Division at the Institute of Human Virology has developed several relevant animal models, particularly for the study of AIDS and AIDS-associated cancers.

Animal models are required for the study of human diseases. However, the relevance of small animal models such as mice to natural human *in vivo* physiological and metabolic kinetics remains unclear. Large animal species may provide far more appropriate preclinical models that will more closely reflect human physiological characteristics and behavior. Among large animals, nonhuman primates may provide the best models because of their close phylogenetic relationship to humans It is essential to develop animal models for human diseases to reveal its mechanisms and to develop new therapeutic interventions.

Led by the author of this review at the Institute of Human Virology, the Animal Models Division is a unique feature of the Institute, enabling scientists to work with relatively inexpensive models to study AIDS and new drugs or therapies without risk to humans. Developing animal subjects for use in viral research is a science unto itself, and it is essential in taking a discovery from laboratory to clinic. The use of animals as models for human disease has been indispensable in understanding the causes, biology, and prevention of disease. "To my knowledge, almost all major scientific successes on unraveling and conquering human diseases have been with the use of animal models. AIDS and AIDS-associated diseases, will be no exception" (J.L. Bryant, personal communication).

VIRAL DISEASE IN SEARCH OF AN ANIMAL MODEL

Hepatitis viruses belong to different families and have in common a striking hepatotropism and restrictions of propagation in cell culture. Viral hepatitis represents a global public health problem. Over the past 20+ years, the chimpanzee model has served as the backbone for advancements in the hepatitis C virus (HCV) research field. Despite this remarkable progress, the chimpanzee model has some important disadvantages. Perhaps most importantly, chimpanzees are rare, expensive, and difficult to handle. Limitations to the chimpanzee model have stimulated progress toward developing alternative animal modes for HCV research. Transgenic technology, coupled with the relative ease and low cost by which mice can be reared and maintained, along with the availability of inbred mouse strains, have made the laboratory mouse an attractive animal model for HCV research.

For some human viruses [e.g., hepatitis B (HBV) and HCV], no satisfactory cell culture or animal model exists, and in these cases, inhibition of an essential viral function or activity against related viruses can be used to indicate potential activity. When no satisfactory cell culture or animal model exists for the target human virus, it is particularly important to know whether a drug's active moiety enters cells, if it has a proposed intracellular site of action, and if the intracellular concentration is consistent with biochemical studies to identify an inhibitory concentration. Cell-based assays and host cell lines for studying viruses such as HBV and HCV replication may advance and improve, but at the present time are limited. For analysis of HCV replication, a replicon system has been developed that permits studies of viral replication and can be used to assess antiviral activity of some anti-HCV drugs.

More than 200 million people worldwide are now believed to be infected with HCV, including 4 million individuals within the United States. HCV accounts for the deaths of at least 8000 to 10,000 Americans each year and is now one of the most common indicators for adult liver transplants in developed countries. Before the introduction of anti-HCV screening in mid-1990, HCV accounted for 80–90% of new cases of posttransfusion hepatitis in the United States. Currently, injection drug use is probably the most common risk factor for HCV infection, with approximately 80% of this population seropositive for HCV. A high rate of HCV infection is also seen in individuals with bleeding disorders or chronic renal failure, groups that have had frequent exposure to blood and blood products.

HCV is a global public health problem, with approximately 3% of the world population now infected. The clinical course of HCV often involves chronic infection, which can lead to liver dysfunction and hepatocellular carcinoma. Because HCV cannot be efficiently propagated in cell culture, research has relied heavily on animal models to study the physical characteristics of HCV and the course of events associated with HCV infection. The chimpanzee is the only nonhuman primate actually proven to be susceptible to HCV infection and has commonly been used to study viral hepatitis induced by HCV. Molecular cloning of the HCV genome has now allowed HCV transmission studies in chimpanzees to progress from the early work of characterizing infectious serum to a current focus of characterizing infectious HCV molecular clones. Moreover, the cloned HCV genome has paved the way for the development of alternative animal models for HCV, most notably transgenic mouse models for the study of HCV pathogenesis.[33] The expression of specific viral protein products in these animal models will provide important insights into the structure–function relation that specific HCV genome sequences impart on virus replication and pathogenesis.

Despite this remarkable progress, the chimpanzee model has some important disadvantages. Perhaps most importantly, chimpanzees are rare, expensive, and difficult to handle, and must be housed and cared for in appropriate nonhuman primate research facilities. Such research facilities must possess proper surgical support and specialized veterinary care. Moreover, the chimpanzee has been listed as an endangered species since 1988, and appropriate safeguards must be considered whenever selecting such a model for research.

These and other limitations to the chimpanzee model have stimulated progress toward developing alternative animal models for HCV research. Transgenic technology, coupled with the relative ease and low cost by which mice can be reared and maintained, along with the availability of inbred mouse strains, have made the laboratory mouse an attractive animal model for HCV research.[34] An HCV infection cannot be propagated in mouse tissues, which obviously limits the research application of mouse models. However, expression of the HCV genome or subgenomic fragments of HCV within inbred strains of mice has demonstrated the utility of mouse models for research geared toward understanding mechanisms of HCV pathogenesis.

Transgenic mice expressing HCV proteins are only beginning to provide insights into the pathobiology of HCV infection. From the studies presented, it is clear that expression of HCV proteins per se is not directly cytopathic to hepatocytes, even when expression levels are greater than those observed in HCV-infected individuals. This evidence supports the notion that the host immune response to HCV may play a significant role in HCV pathogenesis. The use of conditional transgenic expression systems, such as the Cre/LoxP system, offers exciting prospects to study the immune-mediated mechanisms of HCV-related liver injury in transgenic mice. However, the recent development of mice immunocompetent for the HCV proteins using this system awaits full characterization. It is clear that in some transgenic mouse strains, expression of the HCV genome or the core protein alone results in the development of steatosis and HCC. How does the core protein induce these pathologies, and how may the other HCV proteins contribute to this phenotype? It is unclear whether this phenotype is a result of HCV protein expression directly or whether it occurs in combination with environmental or host-derived factors leading to chronic liver cell injury. Importantly, steatosis and HCC are relevant clinical manifestations seen in HCV-infected individuals. These transgenic mice should therefore play a pivotal role in elucidating the molecular mechanisms of HCV-related liver pathology.

CONCLUSIONS

There should be no doubt that the knowledge gained form the study of viral diseases in humans in different animal species has enhanced our knowledge and understanding of the pathogenesis of the diseases in humans. One important component of the value of using animals in viral research is that the pooling of knowledge of diseases in different animal species has led to more rapid progress in understanding the pathogenesis of diseases.

It was not the intent here to discuss comprehensively the contribution of animals in viral research but to historically acknowledge that they have played and will continue to play a major role in solving many health problems, including viral diseases in humans. A review of the literature clearly shows that depending on the training and background of the scientists concerning the use of animals in viral research, there are numerous differences of animal models in viral research. It is clear that the more knowledge the scientist has of diseases in animals the better the understanding of how a model can be used. For example, the direct application of knowledge of a disease in an animal, scrapie in sheep, has led to a better understanding of a human disease, kuru. It was a veterinarian and pathologist, Dr. W. J. Hadlow, who through his experience with scrapie in sheep and goats noted the similarities of the histological lesions in the brains of kuru patients and those of sheep and goats. He suggested the inoculation of brain material from kuru patients into animals. Dr. Carlton Gajdusek, using chimpanzees, produced a fatal disease in these animals after a long inoculation period. This established a viral etiology for kuru and led to the prevention of the diseases. This is one of numerous examples of how applying the knowledge of veterinary medicine and human medicine can lead to the prevention and treatment of diseases caused by viruses.

Our understanding of emerging and reemerging viruses from an evolutionary perspective offers clear directions to follow. If we seriously plan to develop efficient and reasonable control strategies, an international concerted effort involving a thorough survey of the genetic diversity of viruses in time and space is a must. This will also help in understanding several processes that take place at the molecular level in a comparative context. More emphasis on multidisciplinary sciences is needed. Molecular biology and systematics, bioinformatics, data mining, and analysis are all becoming considerably sophisticated lately. Hepatitis C and HIV-1 are glaring proof that disease knows no national borders. The use of and development of animal models for studying viral diseases of humans will continue to be a necessary part of biomedical research in understanding the current viral diseases that affect humans and the new emerging viruses.

REFERENCES

1. Ghiara JB, Stura EA, Stanfield RL, Profy AT, Wilson IA. Crystal structure of the principal neutralization site of HIV-1. *Science* 1994;264(5155):82–85.
2. Gottfredsson M, Bohjanen PR. Human immunodeficiency virus type I as a target for gene therapy. *Front Biosci* 1997;2:619–634.
3. Drotman DP. *Emerging Infectious Diseases*. Vol. 12, No. 11, November 2006. www.cdc.gov/eid.
4. McFadden G. Smallpox: An ancient disease enters the modern era of virogenomics *Proc Natl Acad Sci USA* 2004;101(42):14994–14995.
5. Campo MS. Animal models of papillomavirus pathogenesis *Virus Res* 2002;89(2):249–261.
6. March RF. Animal model of human disease: Kuru, Creutzfeldt-Jakob disease (slow virus infections). Animal model: Transmissible mink encephalopathy, scrapie-like disease of mink. *Am J Pathol* 1972; 69(1):209–212.
7. 19th annual UCLA symposium: Animal models of human viral diseases: Relevance to developmental therapeutics. *J Cell Biochem Suppl* 1990;14D:41–75.
8. Levy JA. The value of primate models for studying human immunodeficiency virus pathogenesis. *J Med Primatol* 1996;25(3): 163–174.
9. Langley RJ, Prince GA, Ginsberg HS. HIV type-1 infection of the cotton rat (Sigmodon fulviventer and S. hispidus). *Proc Natl Acad Sci USA* 1998;95(24):14355–14360.
10. Miller CJ, Alexander NJ, Sutjipto S, Lacker AA, Gettie A, Hendrikx AG, Lowenstine LJ, Jennings M, Marx PA. Genital mucosal transmission of simian immunodeficiency virus: Animal model for heterosexual transmission of human immunodeficiency virus. *J Virol* 1989;63(10):4277–4284.
11. Lackner AA, Smith MO, Munn RJ, Martfeld DJ, Gardener MB, Marx PA, Dandekar S. Localization of simian immunodeficiency virus. *Am J Pathol* 1991;145:428–439.
12. Letvin NL, Daniel MD, Seghal PK, Desrosiers RC, Hunt RD, Waldron LM, McKey JJ, Schmidt DK, Chalifoux LV, King NW. Induction of AIDS-like disease in macaque monkeys with T-cell tropic retrovirus STLV-III. *Science* 1985;230:71–73.
13. Cohen J. Building a small-animal model for AIDS, block by block. *Science* 2001;293(5532):1034–1036.
14. Nesbit CE, Schwartz SA. In vitro and animal models of human immunodeficiency virus infection of the central nervous system. *Clin Diagn Lab Immunol* 2002;9:515–524.
15. Avgeropoulos N, Kelley B, Middaugh L, Arrigo S, Persidsky Y, Gendelman HE, Tyor WR. SCID mice with HIV encephalitis develop behavioral abnormalities. *J Acquir Immune Defic Syndr Hum Retrovirol* 1998;18:13–20.
16. Gallo R. History of The Institute of Human Virology. http://www.ihv.org/aboutihv/index.html (personal communication), 2006.
17. Bennett M, Hart CA. Feline immunodeficiency virus infection-model for HIV and AIDS? *J Med Microbiol* 1995;42:233–236.
18. McCune JM. Animal models of HIV-1 disease. *Science* 1997; 278(5346):2141–2142.
19. Reid W, Sadowska M, Denaro F, Rao S, Foulk J Jr, Hayes N, Jones O, Doodnauth D, Davis H, Sill A, O'Driscoll P, Huso D, Fouts T, Lewis G, Hill M, Kamin-Lewis R, Wei C, Ray P, Gallo R, Reitz M, Bryant J. An HIV-1 transgenic rat that develops HIV-related pathology and immunologic dysfunction *Proc Natl Acad Sci USA* 2001;98(16):9271–9276.
20. Martina BE, Haagmans BL, Kuiken T, Fouchier RA, Rimmelzwaan GF, Van Amerongen G, Peiris JS, Lim W, Osterhaus AD. Virology: SARS virus infection of cats and ferrets *Nature* 2003;425(6961): 915.
21. Navas-Martin S, Weiss SR. SARS: Lessons learned from other coronaviruses. *Viral Immunol* 2003;16(4):461–474.
22. Prince AM, Brotman B. Perspectives on hepatitis B studies with chimpanzees. *ILAR J* 2001;42(2):85–88.
23. de Harven E. Viral etiology of human cancer: A historical perspective. *Haematologica* 1999;84(5):385–389.
24. Chen Z, Zhou P, Ho DD, Landau NR, Marx PA. Genetically divergent strains of simian immunodeficiency virus use CCR5 as a coreceptor for entry. *J Virol* 1997;71:2705–2714.
25. Klotman PE, Notkins AL. Transgenic models of human immunodeficiency virus type-1. *Curr Top Microbiol Immunol* 1996 206: 198–222.
26. Murakami M, Gurski KJ, Steller MA. Human papillomavirus vaccines for cervical cancer. *J Immunother* 1999;22:212–218.
27. Govorkova EA, Rehg JE, Krauss S, *et al.* Lethality to ferrets of H5N1 influenza viruses isolated from humans and poultry in 2004. *J Virol* 2005;79:2191–2198.
28. Chotpitayasunondh T, Ungchusak K, Hanshaoworakul W, *et al.* Human disease from influenza A (H5N1), Thailand, 2004. *Emerg Infect Dis* 2005;11:201–209.
29. Zitzow LA, Rowe T, Morken T, Shieh W-J, Zaki S, Katz JM. Pathogenesis of avian influenza A (H5N1) viruses in ferrets. *J Virol* 2002;76(9):4420–4429.
30. Martin G. Ottolini JC, Blanco G, Eichelberger MC, Porter DD, Pletneva L, Richardson JY, Prince GA. The cotton rat provides a useful small-animal model for the study of influenza virus pathogenesis. *J Gen Virol* 2005;86:2823–2830.
31. CDC—Influenza http://www.cdc.gov/flue/about/fluviruses.htm.
32. Hart CA, Bennett M. Hantavirus infections: Epidemiology and pathogenesis. *Microbes Infect* 1999;1(14):1229–1237.
33. Mercer DR, Schiller DE, Elliott JF, Douglas DN, Hao C, Rinfret A, Addison WR, Fischer KP, Churchill TA, Lakey JRT, Tyrrell DLJ, Kneteman NM. Hepatitis C virus replication in mice with chimeric human livers. *Nat Med* 2001;7:927–933.
34. Barth H, Liang TJ, Baumert TF. Hepatitis C virus entry: Molecular biology and clinical implications. *Hepatology* 2006;44(3):527–535.

58 Nonhuman Primate Models for AIDS

NANCY L. HAIGWOOD, RAYMOND R. O'NEILL, AND VANESSA M. HIRSCH

ABSTRACT

Since the discovery of acquired immune deficiency syndrome (AIDS) and the human immunodeficiency viruses HIV-1 and HIV-2 in patients 25 years ago, efforts have been directed toward identifying or developing appropriate models for this disease. Several species of nonhuman primates can be productively infected with primate lentiviruses HIV-1, HIV-2, SIV, and genetic chimeras of HIV and SIV (SHIV), and the ensuing pathology has many elements in common with HIV infection of humans. This chapter summarizes the key features of the nonhuman primate models that have been developed for AIDS research as of 2006. It focuses on (1) the major contributions and limitations of these models to our understanding of HIV infection and pathogenesis and (2) uses of the models to test prophylactic and therapeutic approaches to prevent infection and/or limit disease. It is intended as a reference for investigators in the field as well as for those considering utilizing the models to address specific questions in AIDS research. In an overview of this type, comprehensive referencing of studies is not possible, and key examples are included in an effort to cite some of the many important studies performed to date. Many excellent comprehensive and focused reviews on the subject of nonhuman primate models for AIDS have been written in the past several years, and readers are encouraged to consult these reviews for details, examples, and additional references.

Key Words: HIV-1, HIV-2, SIV, SHIV, AIDS, Baboons, Chimpanzees, Macaques.

GENERAL CHARACTERISTICS

HISTORY OF AIDS MODELS Since the discovery of the etiological agent of AIDS, human immunodeficiency virus (HIV), efforts to combat the HIV/AIDS epidemic have been focused on the development of vaccines and drug treatment, as well as on behavioral interventions. Because the disease is known to be a caused by a blood-borne virus, some of the earliest efforts were directed to developing tests for screening the human blood supply. Recombinant viral antigens that were developed for diagnostic tests were also tested for their suitability as early vaccine candidates. The early vaccine work was bolstered with significant private and government-sponsored research funding commitments in the United States, Canada, the United Kingdom, and Europe, in part due to misplaced optimism about the ease of developing HIV vaccines. Much of this funding was directed toward the discovery and development of appropriate animal models to study antiviral immunity and to evaluate vaccine candidates and strategies. Although natural lentiviral infections in nonlaboratory animal species are known, none develops into an AIDS-like illness. Attempts to create transgenic mice or rats that develop AIDS have been unsuccessful to date.

Importantly, early research in academic laboratories and in industry did lead to anti-HIV, or antiretroviral (ARV) drug development in the absence of animal models. These early drugs targeted viral enzymes such as reverse transcriptase (RT) and protease (PR) with no human counterparts, and drugs could be screened for activity against the virus using *in vitro* enzymatic inhibition assays. Because this type of drug research did not require an animal model to demonstrate drug potency, nearly all of the ARVs in use at present were approved without animal model efficacy testing. With the development of PR inhibitors and their use in a cocktail of anti-RT inhibitors, patients could expect to experience significant, sustained control of virus replication and in many cases delay or prevention of opportunistic infections. These combination drug regimens were termed highly active antiretroviral treatment, or HAART.[1] Various improved formulations and combinations of HAART are the current standard of care in the developed world, and there are continuing efforts to find economic mechanisms to make these drugs available worldwide.

Despite success in developing HAART, the epidemic continues to spread worldwide through sexual transmission, primarily heterosexual, via mother-to-child transmission, and drug use via blood-contaminated needles. As with other epidemics, an important public health goal is to develop vaccines that can prevent the acquisition or limit the spread of a lethal or seriously debilitating infection. These vaccines, once discovered, can then be utilized in concert with campaigns to limit transmission via behavioral changes. AIDS vaccine research has been far more challenging than drug research due to a number of factors. First, HIV-1 is a lethal virus that persists in the host due to its integration into the genome. Therefore the use of human challenge models is prohibited, even after the development of relatively effective ARV drugs. Second, the virus targets many of the very immune cells that are needed to eliminate or control it, leading to death by

From: *Sourcebook of Models for Biomedical Research* (P. M. Conn, ed.), © 2008 Humana Press Inc., Totowa, NJ.

infection with opportunistic agents. Thus, vaccines will need to be provided prior to or early in infection to stimulate maximal immunity before the loss of key immune cells. Finally, animal models described to date are imperfect and do not recapitulate all aspects of the infection. This is a particular irony, as the viruses HIV-1 and HIV-2 certainly were endemic to nonhuman primates before bridging the gap to productive infection of humans.[2] Nonhuman primates were imported to the United States and other European countries from Africa and Asia during the 1960s continuing through the 1980s to provide animals for testing in cancer and other biomedical research. In the early 1980s, investigators observed that some of their Asian macaques that had been exposed to African sooty mangabey monkeys harboring their own lentivirus (SIVsm) were succumbing to a disease that had remarkable similarities to AIDS in humans. This disease was termed "simian AIDS." The etiological agent was a blood-borne virus that showed morphological similarities to HIV-1 in electron micrographs and was designated SIVmac to indicate its origin in macaque monkeys. This fortuitous observation was the genesis of today's models that are in wide use for vaccine,[3–5] microbicide, transmission,[6] pathogenesis and treatment,[7,8] and behavioral research.[9,10] As noted in the abstract, many fine reviews have been published to describe the models and their uses, and the reader is encouraged to pursue these for additional information.

COMPARISON OF HUMAN AND SIMIAN LENTIVIRUSES In addition to animal passage of infectious blood and tissues, molecular cloning was a key technology in the identification of virus isolates for use in nonhuman primate studies.[11,12] Fulfilling Koch's postulate, it was demonstrated directly that blood from simian immunodeficiency virus (SIV)-infected macaques[13] as well as cloned SIV was the causative agent of this nonhuman primate version of AIDS in Asian macaques.[12] Like HIV-1, this SIV is a member of the lentivirus group of retroviruses, characterized by the presence of regulatory genes in addition to the standard retroviral *gag*, *pol*, and *env* genes. These viruses cause a persistent infection and a relatively slow disease course. It was subsequently demonstrated through molecular cloning of viruses from various African species including chimpanzees, sooty mangabeys, and African green monkeys that the SIVs are endemic to many African primates[14,15] and are by definition well adapted, failing to cause disease in their natural host. While there are a variety of different SIVs that are classified based on their African primate of origin, this chapter focuses on the SIVmac/SIVsm strains that are most widely used in vaccine and pathogenesis research. This research to determine the etiology of the simian disease was performed in parallel with early work to establish an animal host for HIV-1 and HIV-2. Originally imported for research in other diseases, the macaque models arose by accidental exposure to other endemic viruses, as noted above. Some of these monkeys (sooty mangabeys) harbor an endemic nonpathogenic virus, SIVsm, which was transmitted to the Asian macaques.[16] More comprehensive searches have shown that this group of viruses is widespread in many species of African primates. Painstaking field work to obtain samples and detailed phylogenetic analyses of viruses found in these samples[17] have led to our current understanding of the sources of HIV-1 and HIV-2 in humans. HIV-1 arose by exposure of humans to a virus endemic in chimpanzees, and HIV-2 from similar exposure to macaques infected with SIV, most likely SIVsm.[18] Susceptibility of specific cell types to viral entry by the nonhuman primate lentiviruses is enabled by the envelope glycoprotein, which binds to the cellular receptor (CD4, found on monocytes, macrophages, and subpopulations of T cells) and one or more of the 7-transmembrane chemokine coreceptors (typically either CCR5 or CXCR4).[19] HIV-1 laboratory isolates that are adapted for growth in tissue culture utilize CD4 and the chemokine receptor CXCR4, while primary HIV-1 isolates utilize CCR5.[20] SIV is known to utilize CCR5 and potentially other coreceptors as yet uncharacterized.[21] Mutations in the gene encoding CCR5 can confer partial or full resistance to HIV-1 infection in humans who are heterozygous and homozygous, respectively, for a 32-base pair deletion in gene that prevents expression of functional CCR5.[22] Despite conservation of receptor and some coreceptor utilization, the primate lentivirus envelope proteins, while structurally similar, differ as much as 20% in sequence, limiting cross-protective immunity. Recent studies have identified two important restriction factors unique to different primate species that operate at a cellular level to inhibit HIV or SIV viral replication. These are APOBEC3G, which exerts a late block to retroviral replication,[23] and TRIM5α, which exerts a dominant block to infection immediately after entry.[24] Further studies of how these restriction factors limit the acquisition of the ancestors of our current human epidemic will no doubt cast light on changes that accrued during cross-species transmission and important adaptations of the current viruses to increase transmission to humans.[25]

SPECIES UTILIZED FOR AIDS RESEARCH Early work to develop an infectivity model focused on infection of chimpanzees (*Pan troglodytes*) with HIV-1,[26] since they are the closest genetic relative of *Homo sapiens*. Although this species is readily infected with HIV-1 and there is a robust antiviral immune response, there was a puzzling lack of evidence for disease progression. Evidence for disease was found only recently in captive chimpanzees infected with one or more HIV-1 isolates for 10 years or more.[27] In retrospect, this finding makes sense, as the viruses being used to infect the chimpanzees had been selected and adapted for growth and pathogenesis in humans. There was a significant effort in the 1980s to expand the number of captive chimpanzees available for AIDS research via breeding programs, since the U.S. laboratory population was severely limited in number, and additional importation from Africa was prohibited. However, even with increased breeding, the cost of performing experiments with even a small number of chimpanzees was prohibitive for all but a few groups. These studies were limited to testing approaches that could block infection, such as an early study that showed an envelope-specific monoclonal antibody directed to the third hypervariable region could block intravenous infection with HIV-IIIB.[28] Although baboons (*Papio* species) were not productively infected with HIV-1, they could be infected with HIV-2, and some efforts were directed at developing a pathogenic HIV-2 via passage *in vivo*.[29] Macaque (*Macaca nemestrina*)-passaged HIV-2 led to a model characterized by high peripheral and tissue virus and rapid, irreversible loss of $CD4^+$ T cells.[30] Species that are susceptible to infection with SIV include *Macaca fascicularis* (cynomolgus), *Macaca mulatta* (rhesus), and *M. nemestrina* (pigtailed). An excellent discussion of the merits and limitations of each of these species can be found in the review by Hu.[5]

PATHOGENESIS In humans, HIV infection with a single or small number of variants quickly expands with viral replication to a group of genetically[31] and biologically distinguishable viruses.[32] Different virus–host combinations lead to differing degrees of pathology, with some of the equation dependent upon the virus and some dependent upon the host. Typically, macaques

Table 58–1
Nonhuman primate models for human AIDS

Virus	*Animal*	*Comments*
HIV-1	Chimpanzee	Replicates but usually no disease, or disease in >10 years Chimps are limited in number, expensive, and require long-term care
	Pigtailed macaque	Very poor replication, not a useful model for disease
HIV-2	Baboon	Moderate replication, not a useful model for disease
	Pigtailed macaque	One pathogenic strain replicates to a high level and causes an acute $CD4^+$ T cell loss (HIV-2–287)
SIVsm/mac	Rhesus macaques Pigtailed macaques Cynomolgus macaques	Various strains that induce AIDS in macaques (SIVmac251, SIVmac239mc, SIVsmB670, SIVsmE660, SIVsmE543mc, SIVsmm9, SIVsmPBj) Rhesus macaques of Indian origin the most frequently used model
SHIV	Rhesus macaques	The majority of strains use the CXCR4 coreceptor expressed on naive T cells and cause an acute $CD4^+$ T cell loss (SHIV-SF33, SHIV-KU1, SHIV-89.6P, SHIV-DH12)
	Pigtailed macaques	One pathogenic strain uses the CCR5 coreceptor on memory T cells and causes more of an AIDS-like disease (SHIV-SF162)

experience high viral loads and loss of $CD4^+$ T cells within 6 months to 2 years of infection, a course that is significantly foreshortened compared with HIV-1 or HIV-2 in humans. In comparative infection studies with SIVsmE660, for example, the time to loss of $CD4^+$ T cells and disease from opportunistic infection is significantly shortened in *M. nemestrina* compared with *M. mulatta*. Experiments with a shorter duration using host and virus pairs with high pathogenicity are less expensive and can provide greater throughput at primate centers. Therefore these models are frequently favored. A number of AIDS-inducing SIV isolates and molecular clones have been derived as detailed in Table 58–1. These include the SIVmac239 and SIVmac251 viruses originally described by investigators at Harvard, SIVmne from the National Cancer Institute and University of Washington research groups, and related strains SIVsmm9, SIVsmB670, SIVsmE543, and SIVsmE660 isolated by investigators at the Yerkes Primate Center and Tulane University.[7,33] The pathogenesis of disease is similar between these viruses and is characterized by (1) an acute phase of high viremia in which rapid loss of mucosal $CD4^+$ memory T cells occurs, (2) the development of cellular and humoral immune responses and a long, relatively asymptomatic period in which a gradual decline in immune function and loss of $CD4^+$ T cells occurs, and (3) the development terminally of opportunistic infections. Most of these viruses also cause SIV encephalitis, which has provided a model for AIDS dementia.[34] Loss of gut-associated lymphoid tissue (GALT) in SIV-infected macaques is a hallmark of pathogenesis[35,36] and was shown prior to such studies in humans.[37,38] Two important findings were derived from nonhuman primate studies: (1) the pathogenic nature of live attenuated vaccines in newborn rhesus macaques, despite lack of pathogenesis in juveniles;[39,40] and (2) the reemergence of virus from vaccinated "protected" macaques after several years, underscoring the difficulty of detecting some infections[41] as well as the permanence of lentivirus infection and the potential for lethal outcomes after years of control despite years of viral control.[42–44]

TRANSMISSION One of the most useful applications of the nonhuman primate models has been to explore the routes of infection by various isolates and clones. Early research involved intravenous inoculation of chimpanzees, baboons, or macaques with blood or infected tissues, and later infectious molecular clones. Each of the known routes of transmission in humans has been demonstrated in the monkey for both SIV[45] and for SHIV[46]: oral,[47] vaginal,[48] rectal,[49] and perinatal.[6,50] The importance of hormonal changes in susceptibility of transmission could be explored in a controlled manner, demonstrating higher susceptibility during different stages of the menstrual cycle.[51] Using these models, it has been possible to quantify the difference in infectiousness of various stocks by different routes, thereby showing that SIV and SHIV are approximately 1000-fold more infectious by the intravenous route than by any mucosal route.[48] Mucosal transmission of HIV-1 in chimpanzees was shown to be relatively inefficient,[52,53] leading to loss of CD4 cells.[54] Viral divergence was seen with dual infection in chimpanzees,[55] foreshadowing findings in the clinic[56] that have shown intersubtype and even intrasubtype recombination to be an important outcome of infection. Importantly, the nonhuman primate models allow the comprehensive investigation of cells that are infected in the genital tract[57] as well as investigations as to how the virus spreads following infection.[58]

HUSBANDRY/AVAILABLE STOCKS OF NONHUMAN PRIMATES AND LENTIVIRUSES

LOCATIONS OF MAJOR RESEARCH CENTERS USING NONHUMAN PRIMATES FOR AIDS RESEARCH Scientists working in nonhuman primate models for AIDS perform this research in academic, nonprofit, industrial, and government-sponsored laboratories and centers. Numerous private concerns provide purchase, care, and infrastructure for nonhuman primate work for investigators throughout the world. In the United States, there are eight National Primate Research Centers (NPRCs) for

Table 58–2
National Primate Research Centers in the United States, 2006

Primate center name and website	*Academic affiliate*	*City and state*
California National Primate Research Center http://www.cnprc.ucdavis.edu	University of California at Davis	Davis, CA
New England National Primate Research Center http://www.hms.harvard.edu/nerprc/	Harvard University	Southborough, MA
Oregon National Primate Research Center http://onprc.ohsu.edu/	Oregon Health and Sciences University	Portland, OR
Southwest National Primate Research Center http://www.sfbr.org/pages/snprc_index.php	Southwest Foundation for Biomedical Research	San Antonio, TX
Tulane National Primate Research Center http://www.tpc.tulane.edu/	Tulane University	Covington, LA
Washington National Primate Research Center http://www.wanprc.org/WaNPRC/	University of Washington	Seattle, WA
Wisconsin National Primate Research Center http://www.primate.wisc.edu/	University of Wisconsin	Madison, WI
Yerkes National Primate Research Center http://www.yerkes.emory.edu	Emory University	Atlanta, GA

which long-term core support is provided by center grants from the National Center for Research Resources (NCRR) of the United States National Institutes of Health. Their locations, academic affiliations, and websites are listed in Table 58–2. The mission of these centers is to provide nonhuman primate resources, relevant scientific expertise, and biological samples to the research community, and each of the centers has active research in AIDS pathogenesis, vaccines, and treatments. Because of their historical involvement in AIDS research, these centers remain important resources to the research community worldwide. A current list of contact information for the NPRC, and an overview of how outside investigators may access these resources, is maintained at www.ncrr.nih.gov/ncrrprog/cmpdir/PRIMATES.asp. For specific experiments, specific pathogen-free (SPF) rhesus and pigtailed monkeys initially free of SIV, simian retrovirus D, STLV-1, and monkey herpes B virus are available from the eight NPRCs and from the Caribbean primate center in Puerto Rico (http://rcmi.rcm.upr.edu/docs/primate.html). In addition to these centers, many academic institutions, pharmaceutical and biotechnology companies, and the U.S. National Institutes of Health maintain primate colonies for in-house research. In Europe, the German Primate Center and the Biomedical Primate Research Center in the Netherlands house primates and have historically been major centers for AIDS research. This is by no means an exhaustive list; rather its objective is to acquaint the investigator seeking possible research collaborations with the national centers in the United States as a starting point.

VIRAL STOCKS The HIV-1 stocks that have been tested in chimpanzees and subsequently in macaques were derived from the early, laboratory-passaged viruses such as HIV-IIIB or HIV-BRU, HIV-LAI, HIV-MN, or HIV-SF2. These viruses replicated to high titers *in vivo* and elicited robust antiviral antibodies, including neutralizing antibodies. Although attempts were made by several investigators to infect macaques with HIV, sustained replication and pathogenesis were not observed. Today, it is strongly suspected that restriction by TRIM5α in cells from most macaque species was responsible for prevention of cross-species transmission. HIV-2 isolates were serially passaged in baboons with some efficient replication, but little or no pathogenesis.[59] After the observation that SIV infection caused pathogenesis in the Asian macaques, a number of virus isolates were derived from the blood and tissues of serially infected macaques. These three major types of SIV are based on SIVsm (naturally found in sooty mangabeys), SIVmac (from an experimental infection of *M. mulatta*), and SIVmne (from an experimental infection of *M. nemestrina*). Members of these isolates were further characterized by obtaining full-length infectious molecular clones that could recapitulate the pathogenic capacity of the isolate. Subsequently other members of the SIV family have been identified from naturally infected African green monkeys (SIVagm), mandrills (SIVmnd), colobus monkeys (SIVcol), and almost all of the various African monkeys, but not baboons. For the most part, the SIV/macaque model has utilized SIVsm or SIVmac in *M. mulatta* or *M. fascicularis*; other groups have favored the use of SIVmne in *M. fascicularis* or *M. nemestrina*.

Hayami and colleagues made a seminal contribution to the field when they constructed the first viral chimera between HIV-1 and SIV.[60–62] This virus consisted of the SIV "backbone" with the *env* gene replaced by that from HIV-1. Since this first SHIV construct, several additional SHIVs have been described.[63] However, the development of chimeric viruses has not been a trivial pursuit, and only one virus to date, SHIV-SF162P3, bears an HIV-1 envelope that utilizes the CCR5 coreceptor.[64] This virus was shown to cause depletion of $CD4^+$ T cells in the GALT, unlike CXCR4-utilizing SHIVs.[65] These chimeras have been very useful for vaccine efforts that target the development of neutralizing antibodies, since the env genes of SIV and HIV-1 are poorly cross-neutralized, if at all. SHIVs may also be useful in the pursuit of live attenuated vaccines.[66] Other important constructs have included replacement of other SIV genes, such as the RT gene, with that from HIV-1, in order to test RT inhibitors *in vivo*.[67]

Many investigators have developed stocks of SIV and SHIV that are made widely available to researchers who wish to perform challenge studies. The expense of preparing and titering these large scale (300 vials or greater) stocks *in vivo* in the relevant nonhuman primate host is quite high, certainly well above $100,000 (U.S.). Typically a full *in vivo* titration requires purchasing, infecting, and housing at least 6 and up to 10 animals for up to 1 year in BSL-2 or BSL-3 containment, in order to support statistically robust outcomes.[68] In part to defray this cost, several primate centers have prepared and titered stocks that are widely distributed, and the NIH Division of AIDS also has made several stocks available, including a recent stock of SHIV-SF162P3 grown in *M. mulatta*. Clearly the use of identical challenge stocks is a variable that can be controlled by this judicious sharing of resources. The U.S. NIH has developed a shared resource for qualified investigators that is entitled the AIDS Research Reference Reagent Repository (http://www.aidsreagent.org/ecommerce/). This resource provides a very comprehensive and well-referenced set of viral stocks, cell lines for growing virus, monoclonal antibodies, recombinant antigens, etc. to investigators who are registrants and can show that their laboratories have met any required biological and biosafety containment criteria.

ADVANTAGES/DISADVANTAGES

There are many advantages to the utilization of nonhuman primate models for AIDS research. Some examples include increased understanding of viral and host factors that affect pathogenic outcomes, infectiousness via diverse routes of entry, and the roles of specific factors in limiting infection. The use of these models allows investigators to compare outcomes of infection in multiple individuals using the identical, characterized—in many cases cloned—virus stock. By this type of study, it is apparent that there are host factors contributing to pathogenic outcomes. Conversely, viruses derived at different times of infection can themselves have differential pathogenic potential.[69] Direct inoculation of virus at various mucosal sites has allowed a clearer understanding of the differential infectivity of stocks given by different routes. Similarly, the ability to perform timed necropsies of primates infected with well-characterized and genetically identical stocks has facilitated the identification of specific cell types involved in the earliest stages of infection by SIV and SHIV, as well as the rates of mutation that accrue after infection. Information about neurovirulence has been much more approachable in the SIV model than in infected humans.[70] The contribution of regions of the genome to the pathogenic potential of a given virus has been demonstrated with viral chimeras, and more recently the study of changes in the viral envelope over time has yielded interesting information about the inherent ability of the HIV-1 envelope to tolerate mutations *in vivo*.[71] One of the key questions in vaccine research is how neutralizing antibodies are generated, particularly those with the ability to neutralize multiple viruses. Neutralizing antibodies with broad capacity can be generated by infection with the primate lentiviruses, but they are rare, as in humans.[72] These models may offer insights into the mechanisms of development of key immune responses, including innate, cellular, and humoral immunity. Laboratory nonhuman primate models for AIDS allow comprehensive studies of the events occurring in the days immediately following exposure and infection, and experiments with various combinations of competing microbes, which cannot be accomplished for exposed humans.

Another area that can be addressed much more easily in the nonhuman primate than in humans is the evaluation of vaccines and therapies to limit mother-to-child transmission. Some new models were developed in recent years to test active and passive approaches for vaccination in very young primates. These include breast milk transmission,[73] perinatal transmission,[74] and the oral inoculation of newborns with SIV[75] or SHIV.[76] Risky approaches that would be prohibited in humans, such as live attenuated vaccines[39] and passive antibody transfer,[77–79] have been important tools to understanding host immunity and protection. Even with less than fully effective vaccines, it is possible to test how well newborns and infants respond to vaccination. Newborn rhesus macaques have been vaccinated with live attenuated vaccines or recombinant viral vectors expressing SIV antigens in the first few weeks of life and they have been able to mount significant immune responses and control pathogenic viral challenge,[80] which is encouraging news for the future once more effective vaccines are developed. Important information about passive immunity[81] such as dosing and timing[82] and active vaccination of newborns[80] can be derived from these models without risking human infants.

These advantages are balanced by a number of disadvantages. Obviously, the absence of a model that uses HIV-1 as the infectious agent is the most glaring problem for the field. Although certain aspects of this problem can be overcome by the use of SHIV clones and isolates, there is as yet no substitute for clinical work to understand human immune responses to the virus and pathogenic sequelae of the resulting immune depletion. Most non-SPF adult macaques in captivity harbor a virus that is lethal to humans, herpes B. This problem has led to the development of safety regulations to prevent exposure, but there have still been rare infections of animal handlers. One approach is to develop SPF breeding colonies, and many of the centers are making great strides in this area, eliminating (as noted above) SIV, simian retrovirus (SRV), simian T-lymphotropic virus-1 (STLV-1), and herpes B. The lack of complete genotyping, specifically with a focus on MHC Class I, has made the use of nonhuman primates more challenging for both pathogenesis and vaccine studies. Certain alleles in macaques that allow responses to key epitopes in SIV Gag, for example, were very useful in understanding escape from cytotoxic T cell (CTL) control of SIV.[83] This problem is complicated by the use of multiple species of macaques by different investigators, at a time when genetic characterization of the different species is not well developed. To some degree, the use of certain species to the exclusion of others (e.g., *M. mulatta* of Indian origin) has created a shortage that can be alleviated only by expanded breeding or acceptance of other species and subspecies. Differential virus replication in rhesus macaques of Indian and Chinese origin calls into question the interchangeability of the two populations for research.[84] Certainly, another major disadvantage is the expense of purchase and maintenance of these valuable primates, which has limited the size of studies. Thus, analyses of subtle differences in immune parameters in immunogenicity studies, for examples, have been much easier to evaluate in clinical studies in humans. Some of the recent efforts from the HIV Vaccine Enterprise include infusing HIV vaccine research with funding that can support much larger nonhuman primate studies that may have sufficient statistical power to inform the choice of immunogens for clinical development. However, most of the primate facilities are not designed to handle multiple large (>100 animal) experiments at one time, which will require careful planning and shepherding of resources. Finally, research in nonhuman primates requires an appreciation of the model at many levels, and for the first time investigator there are many complex issues to consider. This last problem is one of the easiest to address, as seasoned investigators can be found to collaborate, and the NPRCs have as their mandate to serve the research community. Diverse samples of blood, tissues, and organs are available to all research investigators. Pilot studies are available by competitive application, and these can serve to match an investigator with a primate modeler, facilitating the development of preliminary data for a larger study or grant. Although the NCRR provides core support for the eight NPRCs, scores of additional grants from other Institutes of NIH that are awarded to numerous academic and research institutions provide the majority of the funding for the NPRC work. An additional resource to consider is attendance of the yearly Nonhuman Primate Models for AIDS Symposium, presently having completed its 24th Annual conference (http://www.hivatlas.org/24th_Annual_Symposium_on_Nonhuman_Primate_Models_for_AIDS) where current results and future collaborative efforts and opportunities are discussed.

PREDICTIVE VALUE OF THE MODELS As noted above, the uses of the nonhuman primate models are diverse, thus making the careful choice of the appropriate model for a particular study absolutely key to understanding the outcome(s). In addition, frequently due to availability of species and particular virus stocks, the same experimental design has been tested in different laboratories with divergent outcomes. Thus, one of the major issues confounding AIDS research has been the interpretation of results obtained utilizing different virus–primate pairs. These differences have also been compounded by the unfortunate limitations in nonhuman primate funding during the first 10 years of the epidemic, which led to use of sizes of experimental arms that were very small (three or four animals per group), insufficient to provide statistical power unless there were all-or-none outcomes. Recent studies have utilized much larger groups (6–10), allowing the interrogation of more subtle questions and comparisons between groups receiving different vaccines, for example. The current primate models for AIDS, as well as future models, would be considerably less expensive, allow greater throughput at the primate centers, and reduce the number of monkeys used yearly, if surrogate markers of vaccine success could be assessed before, or instead of, live virus challenge. Surrogate markers currently include both T cell and antibody responses in blood and at mucosal sites.

The most compelling question for the nonhuman primate models is their predictive value for human studies. In cases in which direct comparison can be made such as drug development or transmission studies, it appears that the predictive value is strong.[85] A great deal of our understanding about the biology of infection and dissemination via mucosal routes has been obtained from nonhuman primate studies, where both qualitative and quantitative data are forthcoming. Three recent reviews have touched on this point of predictive value for vaccine studies: Staprans and Feinberg,[3] Haigwood,[85] and Hu[5]. Once there is an HIV vaccine in humans with some efficacy, it will be possible to validate one or more of the nonhuman primate animal–virus pairs and to proceed with some confidence to use this model for future development. Until that time, these models will continue to serve as important adjuncts to clinical studies to understand the issues in transmission, persistence, and pathogenesis,[86] to elucidate innate and adaptive parameters of viral control,[87] and to serve as testing grounds for vaccine strategies.

REFERENCES

1. Ho DD. Time to hit HIV, early and hard. *N Engl J Med* 1995;333(7):450–451.
2. Gao F, Bailes E, Robertson DL, *et al.* Origin of HIV-1 in the chimpanzee Pan troglodytes troglodytes. *Nature* 1999;397(6718): 436–441.
3. Staprans SI, Feinberg MB. The roles of nonhuman primates in the preclinical evaluation of candidate AIDS vaccines. *Expert Rev Vaccines* 2004;3(4Suppl.):S5–32.
4. Schmidt B, Ashlock BM, Foster H, Fujimura SH, Levy JA. HIV-infected cells are major inducers of plasmacytoid dendritic cell interferon production, maturation, and migration. *Virology* 2005;343(2): 256–266.
5. Hu SL. Non-human primate models for AIDS vaccine research. *Curr Drug Targets Infect Disord* 2005;5(2):193–201.
6. Jayaraman P, Haigwood NL. Animal models for perinatal transmission of HIV-1. *Front Biosci* 2006;11:2828–2844.
7. Johnson PR, Hirsch VM. Genetic variation of simian immunodeficiency viruses in nonhuman primates. *AIDS Res Hum Retroviruses* 1992;8(3):367–372.
8. Hirsch VM, Lifson JD. Simian immunodeficiency virus infection of monkeys as a model system for the study of AIDS pathogenesis, treatment, and prevention. *Adv Pharmacol* 2000;49:437–477.
9. Weed MR, Steward DJ. Neuropsychopathology in the SIV/macaque model of AIDS. *Front Biosci* 2005;10:710–727.
10. Zink MC, Laast VA, Helke KL, *et al.* From mice to macaques–animal models of HIV nervous system disease. *Curr HIV Res* 2006; 4(3):293–305.
11. Daniel MD, Letvin NL, King NW, *et al.* Isolation of T-cell tropic HTLV-III-like retrovirus from macaques. *Science* 1985;228(4704): 1201–1204.
12. Letvin NL, Daniel MD, Sehgal PK, *et al.* Induction of AIDS-like disease in macaque monkeys with T-cell tropic retrovirus STLV-III. *Science* 1985;230(4721):71–73.
13. Gravell M, London WT, Houff SA, *et al.* Transmission of simian acquired immunodeficiency syndrome (SAIDS) with blood or filtered plasma. *Science* 1984;223(4631):74–76.
14. Sharp PM, Shaw GM, Hahn BH. Simian immunodeficiency virus infection of chimpanzees. *J Virol* 2005;79(7):3891–3902.
15. Keele BF, Van Heuverswyn F, Li Y, *et al.* Chimpanzee reservoirs of pandemic and nonpandemic HIV-1. *Science* 2006;313(5786):523–526.
16. Hirsch VM, Olmsted RA, Murphey-Corb M, Purcell RH, Johnson PR. An African primate lentivirus (SIVsm) closely related to HIV-2. *Nature* 1989;339(6223):389–392.
17. Gojobori T, Moriyama EN, Ina Y, *et al.* Evolutionary origin of human and simian immunodeficiency viruses. *Proc Natl Acad Sci USA* 1990;87(11):4108–4111.
18. Chen Z, Luckay A, Sodora DL, *et al.* Human immunodeficiency virus type 2 (HIV-2) seroprevalence and characterization of a distinct HIV-2 genetic subtype from the natural range of simian immunodeficiency virus-infected sooty mangabeys. *J Virol* 1997;71(5):3953–3960.
19. Choe H, Farzan M, Konkel M, *et al.* The orphan seven-transmembrane receptor apj supports the entry of primary T-cell-line-tropic and dualtropic human immunodeficiency virus type 1. *J Virol* 1998;72(7):6113–6118.
20. Long EM, Rainwater SM, Lavreys L, Mandaliya K, Overbaugh J. HIV type 1 variants transmitted to women in Kenya require the CCR5 coreceptor for entry, regardless of the genetic complexity of the infecting virus. *AIDS Res Hum Retroviruses* 2002;18(8):567–576.
21. Chen Z, Gettie A, Ho DD, Marx PA. Primary SIVsm isolates use the CCR5 coreceptor from sooty mangabeys naturally infected in west Africa: A comparison of coreceptor usage of primary SIVsm, HIV-2, and SIVmac. *Virology* 1998;246(1):113–124.
22. Bakshi SS, Zhang L, Ho D, Than S, Pahwa SG. Distribution of CCR5delta32 in human immunodeficiency virus-infected children and its relationship to disease course. *Clin Diagn Lab Immunol* 1998;5(1):38–40.

23. Sheehy AM, Gaddis NC, Choi JD, Malim MH. Isolation of a human gene that inhibits HIV-1 infection and is suppressed by the viral Vif protein. *Nature* 2002;418(6898):646–650.
24. Stremlau M, Owens CM, Perron MJ, Kiessling M, Autissier P, Sodroski J. The cytoplasmic body component TRIM5alpha restricts HIV-1 infection in Old World monkeys. *Nature* 2004;427(6977):848–853.
25. Heeney JL, Dalgleish AG, Weiss RA. Origins of HIV and the evolution of resistance to AIDS. *Science* 2006;313(5786):462–466.
26. Alter HJ, Eichberg JW, Masur H, *et al.* Transmission of HTLV-III infection from human plasma to chimpanzees: An animal model for AIDS. *Science* 1984;226(4674):549–552.
27. Novembre FJ, Saucier M, Anderson DC, *et al.* Development of AIDS in a chimpanzee infected with human immunodeficiency virus type 1. *J Virol* 1997;71(5):4086–4091.
28. Emini EA, Schleif WA, Nunberg JH, *et al.* Prevention of HIV-1 infection in chimpanzees by gp120 V3 domain-specific monoclonal antibody. *Nature* 1992;355(6362):728–730.
29. Castro BA, Nepomuceno M, Lerche NW, Eichberg JW, Levy JA. Persistent infection of baboons and rhesus monkeys with different strains of HIV-2. *Virology* 1991;184(1):219–226.
30. Watson A, McClure J, Ranchalis J, *et al.* Early postinfection antiviral treatment reduces viral load and prevents CD4+ cell decline in HIV type 2-infected macaques. *AIDS Res Hum Retroviruses* 1997;13(16):1375–1381.
31. Saag MS, Hahn BH, Gibbons J, *et al.* Extensive variation of human immunodeficiency virus type-1 in vivo. *Nature* 1988;334(6181):440–444.
32. Fisher AG, Ensoli B, Looney D, *et al.* Biologically diverse molecular variants within a single HIV-1 isolate. *Nature* 1988;334(6181):444–447.
33. Hirsch VM, Johnson PR. Pathogenic diversity of simian immunodeficiency viruses. *Virus Res* 1994;32(2):183–203.
34. Lackner AA, Smith MO, Munn RJ, *et al.* Localization of simian immunodeficiency virus in the central nervous system of rhesus monkeys. *Am J Pathol* 1991;139(3):609–621.
35. Heise C, Vogel P, Miller CJ, Halsted CH, Dandekar S. Simian immunodeficiency virus infection of the gastrointestinal tract of rhesus macaques. Functional, pathological, and morphological changes. *Am J Pathol* 1993;142(6):1759–1771.
36. Heise C, Miller CJ, Lackner A, Dandekar S. Primary acute simian immunodeficiency virus infection of intestinal lymphoid tissue is associated with gastrointestinal dysfunction. *J Infect Dis* 1994;169(5):1116–1120.
37. Guadalupe M, Reay E, Sankaran S, *et al.* Severe CD4+ T-cell depletion in gut lymphoid tissue during primary human immunodeficiency virus type 1 infection and substantial delay in restoration following highly active antiretroviral therapy. *J Virol* 2003;77(21):11708–11717.
38. Chase A, Zhou Y, Siliciano RF. HIV-1-induced depletion of CD4+ T cells in the gut: Mechanism and therapeutic implications. *Trends Pharmacol Sci* 2006;27(1):4–7.
39. Daniel MD, Kirchhoff F, Czajak SC, Sehgal PK, Desrosiers RC. Protective effects of a live attenuated SIV vaccine with a deletion in the nef gene. *Science* 1992;258(5090):1938–1941.
40. Ruprecht RM. Live attenuated AIDS viruses as vaccines: Promise or peril? *Immunol Rev* 1999;170:135–149.
41. McChesney MB, Collins JR, Lu D, *et al.* Occult systemic infection and persistent simian immunodeficiency virus (SIV)-specific CD4(+)-T-cell proliferative responses in rhesus macaques that were transiently viremic after intravaginal inoculation of SIV. *J Virol* 1998;72(12):10029–10035.
42. Finzi D, Hermankova M, Pierson T, *et al.* Identification of a reservoir for HIV-1 in patients on highly active antiretroviral therapy. *Science* 1997;278(5341):1295–1300.
43. Ho DD. Toward HIV eradication or remission: The tasks ahead. *Science* 1998;280(5371):1866–1867.
44. Finzi D, Blankson J, Siliciano JD, *et al.* Latent infection of CD4+ T cells provides a mechanism for lifelong persistence of HIV-1, even in patients on effective combination therapy. *Nat Med* 1999;5(5):512–517.
45. Trichel AM, Roberts ED, Wilson LA, Martin LN, Ruprecht RM, Murphey-Corb M. SIV/DeltaB670 transmission across oral, colonic, and vaginal mucosae in the macaque. *J Med Primatol* 1997;26(1–2):3–10.
46. Lu Y, Brosio P, Lafaile M, *et al.* Vaginal transmission of chimeric simian/human immunodeficiency viruses in rhesus macaques. *J Virol* 1996;70(5):3045–3050.
47. Ruprecht RM, Baba TW, Liska V, *et al.* Oral transmission of primate lentiviruses. *J Infect Dis* 1999;179(Suppl. 3):S408–412.
48. Miller CJ, Alexander NJ, Sutjipto S, *et al.* Genital mucosal transmission of simian immunodeficiency virus: Animal model for heterosexual transmission of human immunodeficiency virus. *J Virol* 1989;63(10):4277–4284.
49. Doria-Rose NA, Ohlen C, Polacino P, *et al.* Multigene DNA priming-boosting vaccines protect macaques from acute CD4+-T-cell depletion after simian-human immunodeficiency virus SHIV89.6P mucosal challenge. *J Virol* 2003;77(21):11563–11577.
50. Ruprecht RM, Fratazzi C, Sharma PL, Greene MF, Penninck D, Wyand M. Animal models for perinatal transmission of pathogenic viruses. *Ann NY Acad Sci* 1993;693:213–628.
51. Sodora DL, Gettie A, Miller CJ, Marx PA. Vaginal transmission of SIV: Assessing infectivity and hormonal influences in macaques inoculated with cell-free and cell-associated viral stocks. *AIDS Res Hum Retroviruses* 1998;14(Suppl. 1):S119–123.
52. Fultz PN, Wei Q, Yue L. Rectal transmission of human immunodeficiency virus type 1 to chimpanzees. *J Infect Dis* 1999;179(Suppl. 3):S418–421.
53. Girard M, Mahoney J, Wei Q, *et al.* Genital infection of female chimpanzees with human immunodeficiency virus type 1. *AIDS Res Hum Retroviruses* 1998;14(15):1357–1367.
54. Davis IC, Girard M, Fultz PN. Loss of CD4+ T cells in human immunodeficiency virus type 1-infected chimpanzees is associated with increased lymphocyte apoptosis. *J Virol* 1998;72(6):4623–4632.
55. Wei Q, Fultz PN. Extensive diversification of human immunodeficiency virus type 1 subtype B strains during dual infection of a chimpanzee that progressed to AIDS. *J Virol* 1998;72(4):3005–3017.
56. Zhu T, Wang N, Carr A, Wolinsky S, Ho DD. Evidence for coinfection by multiple strains of human immunodeficiency virus type 1 subtype B in an acute seroconvertor. *J Virol* 1995;69(2):1324–1327.
57. Hu J, Gardner MB, Miller CJ. Simian immunodeficiency virus rapidly penetrates the cervicovaginal mucosa after intravaginal inoculation and infects intraepithelial dendritic cells. *J Virol* 2000;74(13):6087–6095.
58. Miller CJ, Li Q, Abel K, *et al.* Propagation and dissemination of infection after vaginal transmission of simian immunodeficiency virus. *J Virol* 2005;79(14):9217–9227.
59. Barnett SW, Murthy KK, Herndier BG, Levy JA. An AIDS-like condition induced in baboons by HIV-2. *Science* 1994;266(5185):642–646.
60. Kuwata T, Igarashi T, Ido E, *et al.* Construction of human immunodeficiency virus 1/simian immunodeficiency virus strain mac chimeric viruses having vpr and/or nef of different parental origins and their in vitro and in vivo replication. *J Gen Virol* 1995;76(Pt. 9):2181–2191.
61. Kuwata T, Shioda T, Igarashi T, *et al.* Chimeric viruses between SIVmac and various HIV-1 isolates have biological properties that are similar to those of the parental HIV-1. *AIDS* 1996;10(12):1331–1337.
62. Hayami M, Igarashi T. SIV/HIV-1 chimeric viruses having HIV-1 env gene: A new animal model and a candidate for attenuated live vaccine. *Leukemia* 1997;11(Suppl.)3:95–97.
63. Karlsson GB, Halloran M, Li J, *et al.* Characterization of molecularly cloned simian-human immunodeficiency viruses causing rapid CD4+ lymphocyte depletion in rhesus monkeys. *J Virol* 1997;71(6):4218–4225.

64. Harouse JM, Gettie A, Eshetu T, *et al.* Mucosal transmission and induction of simian AIDS by CCR5-specific simian/human immunodeficiency virus SHIV(SF162P3). *J Virol* 2001;75(4):1990–1995.
65. Harouse JM, Gettie A, Tan RC, Blanchard J, Cheng-Mayer C. Distinct pathogenic sequela in rhesus macaques infected with CCR5 or CXCR4 utilizing SHIVs. *Science* 1999;284(5415):816–819.
66. Kuwata T, Miura T, Hayami M. Using SHIVs to develop an anti-HIV-1 live-attenuated vaccine. *Trends Microbiol* 2001;9(10): 475–480.
67. Isaka Y, Sato A, Kawauchi S, *et al.* Construction of the chimeric reverse transcriptase of simian immunodeficiency virus sensitive to nonnucleoside reverse transcriptase inhibitor. *Microbiol Immunol* 1998;42(3):195–202.
68. Spouge JL, Layne SP, Dembo M. Analytic results for quantifying HIV infectivity. *Bull Math Biol* 1989;51(6):715–730.
69. Kimata JT, Kuller L, Anderson DB, Dailey P, Overbaugh J. Emerging cytopathic and antigenic simian immunodeficiency virus variants influence AIDS progression. *Nat Med* 1999;5(5):535–541.
70. Mankowski JL, Flaherty MT, Spelman JP, *et al.* Pathogenesis of simian immunodeficiency virus encephalitis: Viral determinants of neurovirulence. *J Virol* 1997;71(8):6055–6060.
71. Blay WM, Gnanakaran S, Foley B, Doria-Rose NA, Korber BT, Haigwood NL. Consistent patterns of change during the divergence of human immunodeficiency virus type 1 envelope from that of the inoculated virus in simian/human immunodeficiency virus-infected macaques. *J Virol* 2006;80(2):999–1014.
72. Yuste E, Sanford HB, Carmody J, *et al.* Simian immunodeficiency virus engrafted with human immunodeficiency virus Type 1 (HIV-1)-specific epitopes: Replication, neutralization, and survey of HIV-1-positive plasma. *J Virol* 2006;80(6):3030–3041.
73. Amedee AM, Lacour N, Ratterree M. Mother-to-infant transmission of SIV via breast-feeding in rhesus macaques. *J Med Primatol* 2003;32(4–5):187–193.
74. Jayaraman P, Mohan D, Polacino P, *et al.* Perinatal transmission of SHIV-SF162P3 in Macaca nemestrina. *J Med Primatol* 2004;33(5–6):243–250.
75. Baba TW, Koch J, Mittler ES, *et al.* Mucosal infection of neonatal rhesus monkeys with cell-free SIV. *AIDS Res Hum Retroviruses* 1994;10(4):351–357.
76. Baba TW, Trichel AM, An L, *et al.* Infection and AIDS in adult macaques after nontraumatic oral exposure to cell-free SIV. *Science* 1996;272(5267):1486–1489.
77. Shibata R, Igarashi T, Haigwood N, *et al.* Neutralizing antibody directed against the HIV-1 envelope glycoprotein can completely block HIV-1/SIV chimeric virus infections of macaque monkeys. *Nat Med* 1999;5(2):204–210.
78. Mascola JR, Stiegler G, VanCott TC, *et al.* Protection of macaques against vaginal transmission of a pathogenic HIV-1/SIV chimeric virus by passive infusion of neutralizing antibodies. *Nat Med* 2000;6:207–210.
79. Haigwood NL, Montefiori DC, Sutton WF, *et al.* Passive immunotherapy in simian immunodeficiency virus-infected macaques accelerates the development of neutralizing antibodies. *J Virol* 2004; 78(11):5983–5995.
80. Van Rompay KK, Greenier JL, Cole KS, *et al.* Immunization of newborn rhesus macaques with simian immunodeficiency virus (SIV) vaccines prolongs survival after oral challenge with virulent SIVmac251. *J Virol* 2003;77(1):179–190.
81. Baba TW, Liska V, Hofmann-Lehmann R, *et al.* Human neutralizing monoclonal antibodies of the IgG1 subtype protect against mucosal simian-human immunodeficiency virus infection. *Nat Med* 2000; 6(2):200–206.
82. Nishimura Y, Igarashi T, Haigwood NL, *et al.* Transfer of neutralizing IgG to macaques 6 h but not 24 h after SHIV infection confers sterilizing protection: Implications for HIV-1 vaccine development. *Proc Natl Acad Sci USA* 2003;100(25):15131–15136.
83. Allen TM, O'Connor DH, Jing P, *et al.* Tat-specific cytotoxic T lymphocytes select for SIV escape variants during resolution of primary viraemia. *Nature* 2000;407(6802):386–390.
84. Marthas ML, Lu D, Penedo MC, Hendrickx AG, Miller CJ. Titration of an SIVmac251 stock by vaginal inoculation of Indian and Chinese origin rhesus macaques: Transmission efficiency, viral loads, and antibody responses. *AIDS Res Hum Retroviruses* 2001;17(15): 1455–1466.
85. Haigwood NL. Predictive value of primate models for AIDS. *AIDS Rev* 2004;6(4):187–198.
86. Johnson WE, Desrosiers RC. Viral persistence: HIV's strategies of immune system evasion. *Annu Rev Med* 2002;53:499–518.
87. Evans DT, Desrosiers RC. Immune evasion strategies of the primate lentiviruses. *Immunol Rev* 2001;183:141–158.

Models of Other Human Diseases

VI

59 Use of Congenic Mouse Strains for Candidate Disease Gene Identification in Complex Traits

UTE CHRISTINE ROGNER AND PHILIP AVNER

ABSTRACT

Genetic analysis aims to establish the relationship between a gene and a specific trait through the parallel study of phenotype and underlying genetic variation. Current tools for genetic analysis in the mouse extend from standard matings such as intercrosses and backcrosses to the use of more specialized resources such as recombinant inbred (RI) strains and recombinant congenic (RC) strains. Both RI and RC strains are characterized by the fixation of a genome-wide admixture of novel genetic combinations subsequent to meiotic recombination and segregation. Other resources of particular interest for the analysis of complex traits include consomic and congenic strains. In consomic or chromosome substitution strains, the genetic admixture is restricted to a single chromosome, and in congenic strains it is restricted to a single chromosomal region. Consomic and congenic mouse stocks have played a pivotal role in the genetic analysis of complex diseases that are often under the control of both environmental and multiple genetic factors and include type 1 diabetes, multiple sclerosis, obesity, and cancer. This chapter will provide an overview of the different tools available in the mouse for the genetic analysis of complex traits, with particular emphasis on different aspects of the use of congenic mouse strains.

Key Words: Consomic, Congenic, Inbred mice, Complex disease, Quantitative trait locus (QTL), Gene.

TOOLS AND SYSTEMS OF GENETIC ANALYSIS IN THE MOUSE

The mouse is the organism of choice for modeling human disease with not only many thousands of mutations already isolated and projects to inactivate all mouse genes well underway,[1] but also with 450 described inbred strains,[2] a wealth of genetic and phenotypic diversity.

It is this collection of inbred strains, rather than the collection of mutations, that has provided the raw material for studying phenotypes under complex genetic control. This has, in recent times, been facilitated by the Mouse Phenome project (see database links), which aims at garnering and curating, through a coordinated international effort, baseline phenotypic data on most of the common inbred strains.

From: *Sourcebook of Models for Biomedical Research*
(P. M. Conn, ed.), © 2008 Humana Press Inc., Totowa, NJ.

The breeding systems underlying the development of key genetic resources are shown in Figure 59–1. Each recombinant inbred (RI) strain contains a unique admixture of genetic contributions—in approximately equal proportions—from its two original progenitor inbred strains. Traditionally, such RI strains are established by crossing animals of two inbred strains, followed by 20 or more consecutive generations of brother/sister matings.[3] Recombinant congenic (RC) strains are also established by an initial crossing between two inbred strains, but this is followed by a few, usually two, backcrosses of the resulting F_1 hybrids to one of the parental strains, called the "recipient" strain, with subsequent brother/sister intercrossing as for RI lines.[4] In both RI and RC strains the result is a mosaic genetic structure with blocks of genetic material from one parent interspersed with blocks of genetic material from the other parent. RI and RC strains differ, however, in the relative contribution of the two parents to the final strains. The Complex Trait Consortium[5] represents the largest community effort to date to generate and maintain collections of RI lines having sufficient analytical power to allow for systematic quantitative trait locus (QTL) analysis. The current project aims to generate some 1000 RIs from eight different parental strains to sort out the genes behind different complex disorders.

Consomics and congenics are special types of inbred strains in which part of the genome of one mouse strain is transferred to another by backcrossing the mouse strain acting as the donor to the recipient strain followed by intercrossing in later generations to ensure fixation and homozygosity. Selection is systematically practiced to ensure retention of the desired genetic material from the donor strain. The method was first introduced by Snell who produced histoincompatible congenic strains, which he originally called "congenic resistant" strains.[6] In the case of a consomic, a whole chromosome is transferred.[7,8] In the case of a congenic strain it is a defined chromosomal segment, also termed the differential segment that is transferred[9,10] (Figure 59–1). Such congenic strains will normally carry differential regions of 10–20 Mb in size[11] unless specific efforts are made to reduce the size of the differential segment. Several hundred congenic strains currently exist. Many of these can be retrieved directly from the Jackson Laboratories. Congenic strains need to be distinguished from coisogenic strains that differ at only a single locus from their parental strain.[12] Coisogenics can be derived by both gene targeting involving homologous recombination or mutagenesis

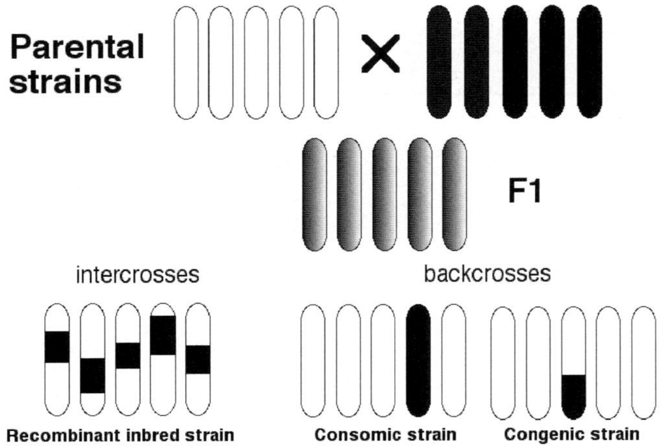

Figure 59–1. Genetic admixture in recombinant inbred strains, consomics, and congenics. Recombinant inbred strains are generated by intercrosses of F_1 mice and subsequent brother–sister interbreeding. Consomic strains are generated by repeated backcrossing of F_1 animals to the parental receiver strain, but only one chromosome is derived from the donor strain. In congenic strains only the differential chromosome segment is derived from the donor strain.

approaches. RI and RC consomic and congenic strains all have in common the advantage that they allow repeat phenotyping of large numbers of genetically homogeneous animals under defined environmental conditions.

APPROACHES TO THE IDENTIFICATION OF A GENETIC INTERVAL CONTROLLING A COMPLEX TRAIT

The starting point for the identification of a genetic locus controlling a complex trait, or QTL, is most often the generation of an F_1 intercross from two parental inbred strains differing in phenotype for the trait under study and the subsequent intercrossing of F_1 progeny to produce an F_2 generation that can be subject to genetic analysis. Alternative analytical schemes may involve the backcrossing of the F_1 progeny to either one or the other parental strains, to generate first-stage backcross generation animals (BC1) and/or eventually backcross 2 animals (BC2). The number of animals needed for the F_2, BC1, or BC2 analysis depends on the strength of the phenotypic effect conferred by each QTL under study and also on the size of the genetic interval(s) to be identified. Generally, experimental cohort sizes are in the range of several hundred animals. Most of the currently identified QTL genes are slightly atypical in presenting rather extreme phenotypes that confer above average contributions to the overall phenotype. Contributions of the average single QTL to the overall phenotype has been estimated to be 5% or less.[13] While increasing the number of F_2 or BC animals under study and using a higher marker density may increase statistical power and allow the candidate region to be better localized, the additional effort would sometimes be better directed elsewhere, for example, to fine mapping and use of congenic strains (see the discussion below).

Alternatives to the generation of *de novo* crosses involve the use of sets of RI lines and/or RC strains. The analytical power of an RI line set depends on the number of available sublines and the degree of genetic/phenotypic variation in the parental strains. While RI sets can deliver higher mapping resolution than F2 mice,[13] current RI sets will often have insufficient power to identify genes encoding small-effect QTLs. In contrast to RI sets, sets of RC strains have the property of limiting the amount of the genome that has to be searched for multiple genes involved in quantitative traits as long as they have been selected for the phenotype of interest. The genome of a standard RC strain comprises, on average, only 12.5% from the donor strain.[14]

Irrespective of the approach used in the initial QTL localization, moving on to subsequent candidate gene identification and characterization will often prove difficult with both genetic complexity and genetic background effects hampering progress. Consulting the Mouse Genome Informatics (MGI) database (see below) reveals, for example, that while an initial localization has been described for some 2000 quantitative trait loci, in less than 1% of the cases has the gene underlying the QTLs been identified. Recent advances have on occasion been linked to the use of innovative *in silico* "Hapmap" type strategies, but mostly to the exploitation of consomic and especially congenic strains, which are increasingly playing a pivotal role in the dissection of QTLs.

HAPLOTYPE MAPPING AS AN *IN SILICO* ALTERNATIVE AND COMPLEMENTATION

In silico mapping is a powerful computational-based method for predicting chromosomal regions regulating phenotypic traits.[15] The discovery based on the distribution of single nucleotide polymorphisms (SNPs) in different inbred mouse strains of an alternating mosaic pattern of relatively large, typically 1–2 Mb, genomic regions (blocks) of low or high polymorphic variation (SNPs)[16,17] is the foundation for this approach. Regions that are poor in polymorphic markers have been designated as regions of common ancestry. Haplotype blocks are conceptually defined as genomic segments harboring sets of coupled polymorphisms that reflect a common ancestral origin.[18,19] Genome-wide association studies involving the correlation of a phenotype, for instance, disease prone and disease resistant, over a wide selection of different inbred mouse strains to patterns of genetic variation in these same strains have provided powerful indications of potential candidate regions.[15] QTL genes are likely, although not necessarily, to be found in regions of different ancestry among pairs of differentially affected strains, while a given phenotype observed in common in different strains will likely be controlled by a region that is held in common between these strains.[20] The approach requires that well-standardized parental phenotype data are available for many inbred strains. The PHENOME project was established in part with this type of application in mind.

Applied to congenic strain analysis, knowledge of the haplotype block structure within a congenic interval may focus interest on a particular subregion within the congenic candidate region (Figure 59–2). There are several examples in the literature that have demonstrated the value of this type of combined approach.[21–23]

In whatever way *in silico* mapping is applied, the success of the strategy depends on the size of the underlying haplotype blocks and the panel of relevant mouse strains. The approach also depends on rates of mutation within regions of shared haplotype being sufficiently low as to not obscure the underlying patterns of haplotype variation. Recent work indicates that the definition of haplotype blocks is not that robust and that methods for QTL mapping may fail if they assume a simple block-like structure.[19]

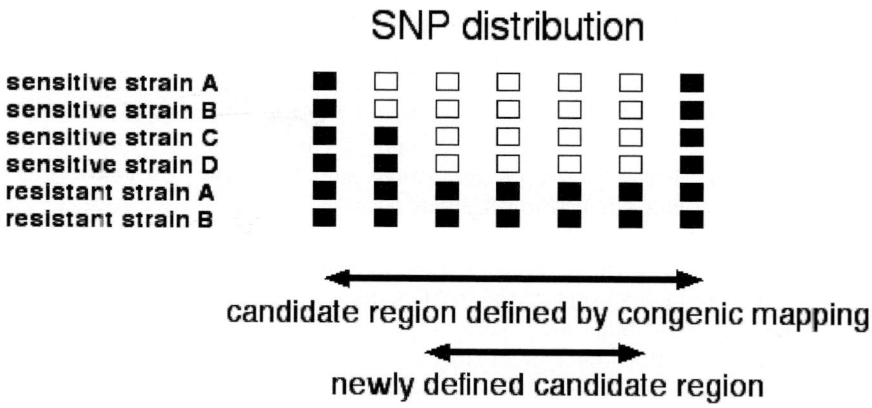

Figure 59–2. Haplotype mapping. Comparison of the SNP distribution within the candidate interval between disease-sensitive and -resistant strains may allow a candidate region to be further reduced.

USING CONSOMICS AND CONGENICS FOR DISSECTING QUANTITATIVE TRAIT LOCI INTO GENES

Consomic and especially congenic strains are proving to be key resources for the dissection of QTLs whose genetic control is very distant from the simplicity of characters showing single-gene dominant or single-gene recessive Mendelian inheritance. Understanding the role of an individual QTL is often hampered by the complexity of its genetic and phenotypic interactions with other participating QTLs. As an example, consider susceptibility to type 1 diabetes and the well-known disease model for type 1 diabetes, the nondiabetes (NOD) mouse.[24,25] It turns out that diabetes-sensitive strains such as NOD carry not only diabetes sensitivity loci but also QTL loci conferring diabetes resistance. Conversely, diabetes-resistant strains such as C57BL/6 carry loci conferring diabetes susceptibility as well as diabetes resistance genes. It is the overall balance and interactions that determine the final phenotype.[26] Some idea of the overall complexity is given by the 30 or more murine insulin-dependent diabetes loci (*Idd*) that have been genetically identified. In this complex situation, congenic strains often carrying segments derived from the diabetes-resistant C57BL/6 mouse strain have been of critical importance for breaking down the overall genetic complexity, for ensuring reproducibility of the detailed phenotyping studies, and for the comparative transcriptional profiling approaches used to narrow down the candidate region. Among non-major histocompatibility complex (MHC) *Idd* genes that have been cloned using a congenic strategy are *Idd3*, where *Il2* has been implicated,[22] *Idd5.1*, where *Ctla4* and *Icos* were suggested,[27] and *Idd5.2*, where *Nramp* is a likely candidate.[28]

The second reason to generate congenic strains is to study the effects of a gene mutation, knockout, knockin, or transgene on one or several different genetic backgrounds. This approach is of particular interest for studies on genetic modifiers.[29,30] The use of congenics, although often onerous, remains less time consuming than introducing the same modification in parallel onto different genetic backgrounds by gene targeting or transgenesis. Indeed, in many cases the latter approach may be impossible as the appropriate ES cell lines may not be available. When congenic construction is used, it should, however, be borne in mind, as already pointed out, that it is not a single gene that is being transferred but a gene and its surrounding genetic region. There have been examples showing that such genomic fragments can contribute to the phenotype or in a worst-case scenario even confer a new phenotype. Different breeding strategies to minimize such problems have been discussed.[31]

CLASSICAL AND SPEED BREEDING STRATEGIES FOR CONGENIC STRAINS

Congenic strains are derived by repeated backcrossing of the desired donor strain to the recipient strain with selection for the differential segment. This breeding is then followed by sister/brother interbreeding of the backcrossed progeny. In practice, female F_1 animals are mated with recipient strain males to establish the BC1 generation. Males heterozygous for the selected chromosome region are then repeatedly backcrossed to recipient females during congenic strain derivation. Congenic strains are then rendered homozygous for the genetic intervals under study by intercrossing heterozygous males and females of the same genotype and subsequently maintained by brother and sister mating. When repeated backcrossing is used to establish a congenic strain a minimum of nine generations of backcrossing is normally recommended to remove 99.9% of the unlinked and unwanted donor material,[32] though the exact number of required backcross generations appears somewhat arbitrary.[33] A genome scan, minimally, for other known loci relevant to the disease under study should be carried out before fixing the congenic interval so that if further backcrossing is necessary to remove a contaminating genomic fragment this can be carried out before the congenic strain is rendered homozygous.

Alternative congenic breeding schemes have been established that involve both positive selection for the desired differential segment and selection against the rest of the donor genome during early backcross generations. In such breeding schemes, called "speed congenics," the genetically "best" animals, i.e., those carrying the differential segment and minimal detectable donor strain material elsewhere in the genome, are selected. Theoretically the process can lead to the creation of a congenic strain with less than 0.5% contaminating donor genome unlinked to the differential segment within a total of five generations or four backcrosses.[34] Simulations suggest that screening between 16 and 20 male progeny per generation with markers spaced every 25 cM most efficiently reduces unlinked contaminating donor genome and represents an effective organizational strategy. Use of larger progeny cohorts and higher marker density seems of little advantage in reducing the contaminating donor genome until later backcross generations. High-density genotyping of the differential segment in later generations is, however, necessary to reduce the size of the target region below the 20–30 cM otherwise obtained at the N4 generation.[10] Experience suggests that both "best" and "second best" males should routinely be kept for breeding, in particular when intersubspecies congenic strains are being constructed and poor breeding performance may occur.

Recent simulation studies have suggested that marker-assisted breeding strategies can lead to increased background heterogeneity, or "gaps," in the recipient genetic background as compared to standard breeding procedures. This suggests that additional backcrossing may still be required in order to reduce the number and length of such gaps.[35] On the assumption of putative remaining gaps, it may be of interest to derive a given set of congenic strains from a single breeding pair and to generate at least one congenic strain carrying no differential fragment as an internal control for phenotyping.

When choosing recombinants to fix genetic intervals we recommend ensuring the highest possible density of markers within the differential fragment (1–2 cM) to avoid partial heterozygosity.

While the genetic interval conferring a particular phenotype in a given congenic strain can often be reduced and refined by identification of new recombinants during further backcrossing, it becomes increasingly difficult to obtain the necessary recombinants as the genetic distance under study is reduced and much larger breeding populations are needed. Several studies have shown the existence of sex-specific differences in recombination frequency[36–38] and it can, therefore, on occasion, be of use to change the direction of the cross and use heterozygous females instead of males or vice versa. It should be noted that recombination does not occur with random efficiency throughout the genome and is often higher at so-called hotspots. Certain theories predict that recombination will occur more often in regions in which gene density is higher and less often in "gene deserts," an expectation borne out for the human genome, where recombination rates are found to be higher in regions of the genome with higher gene density.[39] The informative polymorphic markers necessary to characterize the recombinants can be normally identified from the MGI database. Current densities of available markers ensure that only rarely will additional comparative sequencing efforts be required to identify additional SNPs.

When analyzing congenic strains, it has been observed that the phenotypic effects often get smaller as the genetic interval is reduced and subcongenics are generated. This most often occurs when the original effect was due to the combination of several genes and may reflect the relatively frequent occurrence of QTLs as haplotype blocks. In other cases, genetic interactions may lead to the suppression of phenotypes when intervals are combined and in such cases the generation of subcongenics is accompanied by an increase in the penetrance of the trait.[40] In theses cases, complexity within the genetic interval can normally still be successfully addressed, although it may require larger numbers of animals to be phenotyped and studied.

The benefit that can be obtained from a panel of congenic strains is critically dependent on the quality of the phenotyping available, which, in turn, is obviously dependent on the disease under study. The availability of subphenotypes for characterization is often critical to the fine dissection of the trait. Analysis often starts with the most robust and basic phenotype before proceeding to more subtle analysis of subphenotypes. Phenotyping methods often employed range from histology, behavioral studies, and the evaluation of physiological parameters to metabolomics and transcriptional profiling. At this stage of the project, the efforts employed in the construction of the congenic strains are usually rewarded.

CANDIDATE GENE IDENTIFICATION AND VALIDATION

Once a candidate region has been defined and characterized, the crucial difficulty of identifying and validating the causative gene(s) arises. It is important to realize that there is rarely one single approach, but rather a spectrum of complementary approaches that can be used to identify without ambiguity the gene(s) underlying a QTL.

The identification of changes in the primary nucleotide sequence, which is powerfully diagnostic in the case of mutations in monogenic disorders, is of much less certain value in the case of QTL characterization. First, because nonsense or stop codons that completely abolish gene function are much less likely to underlie QTL variation than in mutations affecting monogenic traits. Second, because the investigator cannot, a priori, know whether to expect changes in coding sequences or in noncoding regulatory sequences. And third, because extensive nucleotide variation may occur outside of the exon sequences of some genes, which renders the identification of causal polymorphisms in such regions potentially perilous. This indeterminacy may moreover be compounded if the causal gene lies within a region that shows a high degree of polymorphism between the parental strains. In such cases the polymorphism within the genomic sequence of the causal gene will likely be no more marked than that of the surrounding genes. Such regions of elevated polymorphism are to be found throughout the mouse genome and reflect the breeding history of mouse inbred strains (see above). Such caveats suggest that resequencing of entire regions in the donor and recipient strains is an approach that can help to exclude a certain proportion of candidate genes rather than lead to the unambiguous identification of the responsible gene.

The use of expression profiling for candidate gene identification is based on the idea that in many cases the QTL will reflect quantitative changes in the expression of the underlying gene(s). This approach has proven to be particularly fruitful in cases in which the phenotype of the disease under study has provided clues as to the class(es) of genes or to the tissues in which the candidate gene is likely to be expressed. Annotated lists of genes for the region under study obtained from Ensembl, MGI, or NCBI databases, including *in silico* expression profiling approaches based on exploiting data from sources such as Serial Analysis of Gene Expression (SAGE) libraries, microarray analysis-based datasets, and cumulative data on expressed sequence tags (ESTs) (see Table 59–1), allow the tissue expression profiles of the genes to be established. Where necessary these data are then validated for the most promising candidates by comparative expression profiling of the discriminatory congenic strains using techniques such as quantitative real-time polymerase chain reaction (PCR). A complementary variant of this approach uses genome-wide microarray-based expression profiling to identify possible genetic pathways.[41] The efficiency of both strategies depends on the completeness of the gene annotations and the exhaustiveness of

Table 59–1
Some useful web links

Mouse Genome Informatics (MGI) database	http://www.informatics.jax.org/
Complex Trait Consortium	http://www.complextrait.org/
Mouse Phenome Database	http://www.jax.org/phenome
Center of Rodent Genetics	http://www.niehs.nih.gov/crg/
Mouse Genome Resequencing and SNP Discovery Project	http://www.niehs.nih.gov/crg/cprc.htm
Online books on mouse genetics and human molecular genetics	http://www.informatics.jax.org/silver/index.shtml http://www.ncbi.nlm.nih.gov/books/bv.fcgi?rid=hmg
Single nucleotide polymorphisms (SNPs)	http://www.broad.mit.edu/snp/mouse/ http://mousesnp.roche.com/
Mouse sequence databases	http://www.ncbi.nlm.nih.gov/ http://www.ensembl.org/ http://mrcseq.har.mrc.ac.uk/ http://www.genome.ucsc.edu/
Gene expression data	http://www.informatics.jax.org/menus/expression_menu.shtml http://www.ncbi.nlm.nih.gov/UniGene/clust.cgi?ORG=Mm http://bodymap.ims.u-tokyo.ac.jp/
BAC mapping	http://www.bcgsc.ca/lab/mapping/mouse
The RNAi WEB	http://www.rnaiweb.com/
Transgenic animal web (links)	http://www.med.umich.edu/tamc/links.html
Knockout mouse project	http://www.nih.gov/science/models/mouse/knockout/index.html
Available mouse models	http://jaxmice.jax.org/index.html
Standard genotyping protocol	http://www.jax.org/mmr/Standard_mapping_protocol.html
Online Mendelian Inheritance in Man, OMIM	http://www.ncbi.nlm.nih.gov/entrez/query.fcgi?db=OMIM
Human/mouse relationships	http://www.ncbi.nlm.nih.gov/Omim/Homology/

the gene representation that is being exploited and is influenced by both the cellular complexity of the target tissue and relative transcript expression levels. In the case of complex tissues sensitivity may be increased by analyzing cellular subpopulations of the tissue in question.[42,43]

Neither the identification of sequence variation nor of altered expression profiles is of itself sufficient to establish causality. For this, techniques of gene inactivation, gene overexpression, and the gold standard, replacement of the allele of one strain by that of the other strain—referred to here as "allele shuffling"—are required. Ongoing programs for inactivation or mutating all mouse genes will increasingly provide "off the shelf" ES cells carrying a knockout for the candidate genes under study. Database links that provide information about existing knockouts and mutants are listed in Table 59–1. Exploiting such resources for QTL validation may in many cases be complicated by the need to cross the knockout onto the relevant genetic background. Interestingly, some of the constructs used in these global knockout programs may facilitate application of a knockin strategy, which would enable the integration of alternative functional allelic forms of the gene at the disabled locus to be undertaken.[44]

An alternative and potentially very efficient strategy for gene disabling is the use of small interfering RNA (siRNA) to inhibit/knock down gene expression. RNA interference (RNAi) is a highly evolutionary conserved process of posttranscriptional gene silencing (PTGS) in which double-stranded RNA (dsRNA), when introduced into a cell, causes sequence-specific degradation of homologous mRNA sequences.[45] Double-stranded RNAs of around 21 nucleotides in length inhibit the expression of specific genes and can work effectively both *ex vivo* and *in vivo*.[46–48] The RNAi WEB database provides a good summary of such studies and excellent practical advice. The long-term and stable inhibition of gene function normally necessary for assessing the effect of a QTL implies that delivery systems capable of supporting stable and persistent expression *in vivo* are necessary. Cloning of stable interfering short hairpin (sh) RNA molecules and use of several viral vectors and transposon-based nonviral vectors have been reported that fulfill this requirement.[49,50] Interestingly, such RNAi approaches can be used with embryos circumventing the strain restriction bottleneck associated with ES cell use. Although there are problems associated with off target non-specificity of RNAi that require stringent control, RNAi can be engineered to be specific for particular allelic forms of a given gene. Providing technical problems can be overcome, experiments aimed at targeting RNAi to alleles of a candidate gene in F_1 animals from two discriminatory congenic strains should prove particularly informative.

Overexpression studies provide an alternative if less stringent way to either further reduce the size of the candidate region or to validate formally candidate genes, when the trait under consideration is dominantly or codominantly expressed.[51,52] Such approaches are often carried out using BAC clones, which may be one to several hundred kilobases in size and therefore allow the gene to be tested along with many of the *cis*-acting sequences necessary for its regulated expression. Such studies are being facilitated by the construction of BAC libraries for many mouse strains other than C57BL/6 and 129/Sv.[53] The fingerprinting of these libraries and the ready availability of both mouse genomic draft and finished sequences[54] and of strain resequencing programs allow the DNA hybridization probes necessary for BAC isolation to be easily designed. Although BAC transgenesis is an efficient process, it should be noted that both copy number variation and variation in the site of integration in the genome leading to position effects may lead to modifications in gene expression profiles that might hamper or obscure the identification of the gene when subtle phenotypes are concerned.

COMPARATIVE GENOMICS BETWEEN MOUSE AND HUMAN

Humans and mouse have both evolved from an ancestral small mammal that split into two species toward the end of the dinosaur era. Despite 75 million years of separate evolution, only about 300 genes—1% of the 25,000–30,000 genes in the mouse genome—were found to be without an obvious counterpart in the human genome and vice versa,[55,56] leading to the idea that in the majority of cases the underlying biology and physiology of processes occurring in humans and mouse will be similar or identical. Most studies of mouse models of human disease are predicated on this. Diseases under monogenic control most often, although not necessarily always, provide support for the assumption.[57] Where the assumption fails to hold, because mutations in a given gene fail to produce the same phenotype in both species, the differences are mainly imputable to differences in physiology, to subtle differences in the gene regulation, and or/to differences in the specific mutation itself rather than to the absence of the gene. The predictive value of QTLs identified in one species for the other is generally considerably weaker. This is almost certainly due to the genetic heterogeneity underlying most complex traits, to differences in penetrance even when the same gene may be concerned, and to differences in the spectrum of naturally occurring variation at a given locus in humans and mouse. This, however, does not imply that the underlying genetic networks are highly divergent and if we take the case of type 1 diabetes this is clearly not the case. Indeed, the identification of causative genes for the autoimmune disease type 1 diabetes (T1D) in humans and candidate genes in the NOD mouse has shown that susceptibility or resistance to type 1 diabetes, involving genes and pathways contributing to autoimmune pathogenesis, are primarily held in common by the two species. To cite but one example, gene variants for the interacting molecules IL2 and CD25, members of a pathway that is essential for immune homeostasis, are present in mice and humans.[58] In this context, identifying the mouse genes involved in type1 diabetes by consolidating our knowledge of the pathway underlying the pathogenesis will both indirectly and directly identify novel targets, which can be studied for their possible role in the human pathogenic process.

Congenic strains have and will continue to play a central role in the panoply of genetic tools available to the research scientist. Indeed, their importance and pivotal role are likely to rise as the tools necessary for high-throughput systematic functional testing of candidate genes lying within the candidate regions defined by using congenic strains become more widely available.

ACKNOWLEDGMENTS

We thank Richard Harvey and Christophe Rachez for critical reading of the manuscript. This work was supported by grants from the Juvenile Diabetes Research Foundation International (1–2000–600), the Association pour la recherche sur le Diabète, and recurrent funding from the CNRS, INSERM, and the Pasteur Institute.

REFERENCES

1. Grimm D. Mouse genetics. A mouse for every gene. *Science* 2006;312(5782):1862–1866.
2. Beck JA, Lloyd S, Hafezparast M, *et al.* Genealogies of mouse inbred strains. *Nat Genet* 2000;24(1):23–25.
3. Bailey DW. Recombinant-inbred strains. An aid to finding identity, linkage, and function of histocompatibility and other genes. *Transplantation* 1971;11(3):325–327.
4. Demant P, Hart AA. Recombinant congenic strains—a new tool for analyzing genetic traits determined by more than one gene. *Immunogenetics* 1986;24(6):416–422.
5. Churchill GA, Airey DC, Allayee H, *et al.* The Collaborative Cross, a community resource for the genetic analysis of complex traits. *Nat Genet* 2004;36(11):1133–1137.
6. Snell GD. Congenic resistant strains of mice. In: Morse HC, Ed. *Origins of Inbred Mice.* New York: Academic Press, 1978:1–31.
7. Nadeau JH, Singer JB, Matin A, Lander ES. Analysing complex genetic traits with chromosome substitution strains. *Nat Genet* 2000;24(3):221–225.
8. Santos J, Montagutelli X, Acevedo A, *et al.* A new locus for resistance to gamma-radiation-induced thymic lymphoma identified using inter-specific consomic and inter-specific recombinant congenic strains of mice. *Oncogene* 2002;21(43):6680–6683.
9. Boyse EA, Bentley DR. The increasing value of congenic mice in biomedical research. *Lab Anim Sci* 1977;27(5Pt. 2):771–781.
10. Wakeland E, Morel L, Achey K, Yui M, Longmate J. Speed congenics: A classic technique in the fast lane (relatively speaking). *Immunol Today* 1997;18(10):472–477.
11. Peirce J. Looking at old tools in new ways: Using knockouts as congenics to study QTLs. *Genome Res* 2001;11(9):1469–1471.
12. Roths JB, Murphy ED, Eicher EM. A new mutation, gld, that produces lymphoproliferation and autoimmunity in C3H/HeJ mice. *J Exp Med* 1984;159(1):1–20.
13. Flint J, Valdar W, Shifman S, Mott R. Strategies for mapping and cloning quantitative trait genes in rodents. *Nat Rev Genet* 2005;6(4):271–286.
14. Stassen AP, Groot PC, Eppig JT, Demant P. Genetic composition of the recombinant congenic strains. *Mamm Genome* 1996;7(1): 55–58.
15. Grupe A, Germer S, Usuka J, *et al.* In silico mapping of complex disease-related traits in mice. *Science* 2001;292(5523):1915–1918.
16. Lindblad-Toh K, Winchester E, Daly MJ, *et al.* Large-scale discovery and genotyping of single-nucleotide polymorphisms in the mouse. *Nat Genet* 2000;24(4):381–386.
17. Wade CM, Kulbokas EJ, Kirby AW, *et al.* The mosaic structure of variation in the laboratory mouse genome. *Nature* 2002; 420(6915):574–578.
18. Frazer KA, Wade CM, Hinds DA, Patil N, Cox DR, Daly MJ. Segmental phylogenetic relationships of inbred mouse strains revealed by fine-scale analysis of sequence variation across 4.6 mb of mouse genome. *Genome Res* 2004;14(8):1493–1500.
19. Yalcin B, Fullerton J, Miller S, *et al.* Unexpected complexity in the haplotypes of commonly used inbred strains of laboratory mice. *Proc Natl Acad Sci USA* 2004;101(26):9734–9739.
20. Guo Y, Weller P, Farrell E, *et al.* In silico pharmacogenetics of warfarin metabolism. *Nat Biotechnol* 2006;24(5):531–536.
21. Hillebrandt S, Wasmuth HE, Weiskirchen R, *et al.* Complement factor 5 is a quantitative trait gene that modifies liver fibrogenesis in mice and humans. *Nat Genet* 2005;37(8):835–843.
22. Lyons PA, Armitage N, Argentina F, *et al.* Congenic mapping of the type 1 diabetes locus, Idd3, to a 780-kb region of mouse chromosome 3: Identification of a candidate segment of ancestral DNA by haplotype mapping. *Genome Res* 2000;10(4):446–453.
23. Ikegami H, Makino S, Yamato E, *et al.* Identification of a new susceptibility locus for insulin-dependent diabetes mellitus by ancestral haplotype congenic mapping. *J Clin Invest* 1995;96(4):1936–1942.
24. Makino S, Kunimoto K, Muraoka Y, Mizushima Y, Katagiri K, Tochino Y. Breeding of a non-obese, diabetic strain of mice. *Jikken Dobutsu* 1980;29(1):1–13.

25. Hattori M, Buse JB, Jackson RA, *et al.* The NOD mouse: Recessive diabetogenic gene in the major histocompatibility complex. *Science* 1986;231(4739):733–735.
26. Yang Y, Santamaria P. Lessons on autoimmune diabetes from animal models. *Clin Sci (Lond)* 2006;110(6):627–639.
27. Greve B, Vijayakrishnan L, Kubal A, *et al.* The diabetes susceptibility locus Idd5.1 on mouse chromosome 1 regulates ICOS expression and modulates murine experimental autoimmune encephalomyelitis. *J Immunol* 2004;173(1):157–163.
28. Kissler S, Stern P, Takahashi K, Hunter K, Peterson LB, Wicker LS. In vivo RNA interference demonstrates a role for Nramp1 in modifying susceptibility to type 1 diabetes. *Nat Genet* 2006;38(4):479–483.
29. Montagutelli X. Effect of the genetic background on the phenotype of mouse mutations. *J Am Soc Nephrol* 2000;11(Suppl. 1)6: S101–105.
30. Nadeau JH. Modifier genes in mice and humans. *Nat Rev Genet* 2001;2(3):165–174.
31. Wolfer DP, Crusio WE, Lipp HP. Knockout mice: Simple solutions to the problems of genetic background and flanking genes. *Trends Neurosci* 2002;25(7):336–340.
32. Silver LM. *Mouse Genetics Concepts and Applications*. Oxford, UK: Oxford University Press, 1995.
33. Festing MFW. *Inbred Strains in Biomedical Research*. London: The Macmillian Press LTD, 1979.
34. Markel P, Shu P, Ebeling C, *et al.* Theoretical and empirical issues for marker-assisted breeding of congenic mouse strains. *Nat Genet* 1997;17(3):280–284.
35. Armstrong NJ, Brodnicki TC, Speed TP. Mind the gap: Analysis of marker-assisted breeding strategies for inbred mouse strains. *Mamm Genome* 2006;17(4):273–287.
36. Lynn A, Schrump S, Cherry J, Hassold T, Hunt P. Sex, not genotype, determines recombination levels in mice. *Am J Hum Genet* 2005;77(4):670–675.
37. Morelli MA, Cohen PE. Not all germ cells are created equal: Aspects of sexual dimorphism in mammalian meiosis. *Reproduction* 2005;130(6):761–781.
38. Shiroishi T, Sagai T, Hanzawa N, Gotoh H, Moriwaki K. Genetic control of sex-dependent meiotic recombination in the major histocompatibility complex of the mouse. *EMBO J* 1991;10(3):681–686.
39. Fullerton SM, Bernardo Carvalho A, Clark AG. Local rates of recombination are positively correlated with GC content in the human genome. *Mol Biol Evol* 2001;18(6):1139–1142.
40. Rogner UC, Boitard C, Morin J, Melanitou E, Avner P. Three loci on mouse chromosome 6 influence onset and final incidence of type I diabetes in NOD.C3H congenic strains. *Genomics* 2001;74(2):163–171.
41. Eaves IA, Wicker LS, Ghandour G, *et al.* Combining mouse congenic strains and microarray gene expression analyses to study a complex trait: The NOD model of type 1 diabetes. *Genome Res* 2002;12(2): 232–243.
42. Lock C, Hermans G, Pedotti R, *et al.* Gene-microarray analysis of multiple sclerosis lesions yields new targets validated in autoimmune encephalomyelitis. *Nat Med* 2002;8(5):500–508.
43. Scearce LM, Brestelli JE, McWeeney SK, *et al.* Functional genomics of the endocrine pancreas: The pancreas clone set and PancChip, new resources for diabetes research. *Diabetes* 2002;51(7):1997–2004.
44. Garcia-Otin AL, Guillou F. Mammalian genome targeting using site-specific recombinases. *Front Biosci* 2006;11:1108–1136.
45. Fire A, Xu S, Montgomery MK, Kostas SA, Driver SE, Mello CC. Potent and specific genetic interference by double-stranded RNA in Caenorhabditis elegans. *Nature* 1998;391(6669):806–811.
46. Hasuwa H, Kaseda K, Einarsdottir T, Okabe M. Small interfering RNA and gene silencing in transgenic mice and rats. *FEBS Lett* 2002;532(1–2):227–230.
47. Xia H, Mao Q, Paulson HL, Davidson BL. siRNA-mediated gene silencing in vitro and in vivo. *Nat Biotechnol* 2002;20(10): 1006–1010.
48. Qin XF, An DS, Chen IS, Baltimore D. Inhibiting HIV-1 infection in human T cells by lentiviral-mediated delivery of small interfering RNA against CCR5. *Proc Natl Acad Sci USA* 2003;100(1): 183–188.
49. Naldini L. Lentiviruses as gene transfer agents for delivery to non-dividing cells. *Curr Opin Biotechnol* 1998;9(5):457–463.
50. Yant SR, Ehrhardt A, Mikkelsen JG, Meuse L, Pham T, Kay MA. Transposition from a gutless adeno-transposon vector stabilizes transgene expression in vivo. *Nat Biotechnol* 2002;20(10): 999–1005.
51. Symula DJ, Frazer KA, Ueda Y, *et al.* Functional screening of an asthma QTL in YAC transgenic mice. *Nat Genet* 1999;23(2): 241–244.
52. Giraldo P, Montoliu L. Size matters: Use of YACs, BACs and PACs in transgenic animals. *Transgenic Res* 2001;10(2):83–103.
53. Osoegawa K, Tateno M, Woon PY, *et al.* Bacterial artificial chromosome libraries for mouse sequencing and functional analysis. *Genome Res* 2000;10(1):116–128.
54. Waterston RH, Lindblad-Toh K, Birney E, *et al.* Initial sequencing and comparative analysis of the mouse genome. *Nature* 2002; 420(6915):520–562.
55. Okazaki Y, Furuno M, Kasukawa T, *et al.* Analysis of the mouse transcriptome based on functional annotation of 60,770 full-length cDNAs. *Nature* 2002;420(6915):563–573.
56. Marshall E. Genome sequencing. Public group completes draft of the mouse. *Science* 2002;296(5570):1005.
57. Villasenor J, Benoist C, Mathis D. AIRE and APECED: Molecular insights into an autoimmune disease. *Immunol Rev* 2005;204: 156–164.
58. Wicker LS, Clark J, Fraser HI, *et al.* Type 1 diabetes genes and pathways shared by humans and NOD mice. *J Autoimmun* 2005; 25(Suppl.):29–33.

60 Animal Models of Sudden Infant Death Syndrome

JANE BLOOD-SIEGFRIED

ABSTRACT

Sudden infant death syndrome (SIDS) is the most common cause of postneonatal infant mortality in the developed world. It is a diagnosis of exclusion, defined as the sudden and unexplained death of an infant between 1 month and 1 year of life. Examination of specific pathology may be possible only in an animal model. The three animal models most commonly used in SIDS research to examine risk factors are the piglet, neonatal rabbit, and neonatal rat. We also describe a developmental rat model of sudden unexplained death in response to viral and bacterial infections. This model offers a unique way to evaluate the role of immune development and its relationship to potential risk factors believed to be important in the etiology of SIDS.

Key Words: Animal model, Cytokine, Endotoxin, Inflammation, Influenza virus, Neonatal rabbit, Neonatal rat, Piglet, SIDS.

INTRODUCTION

Sudden infant death syndrome (SIDS) is the most common cause of postneonatal infant mortality in the developed world.[1,2] It is a diagnosis of exclusion, often defined as the sudden death of an infant not explained by a thorough autopsy, review of the clinical history, and examination of the death scene.[2,3] SIDS occurs from 1 month to 1 year with the peak incidence between 2 and 6 months of age. Although infants dying of SIDS appear normal and healthy, there is often a history of a mild viral illness that is not thought to be the cause of death.[1,2]

Epidemiological research has helped establish risk factors for SIDS including ethnicity, maternal smoking and drug use during pregnancy, environmental tobacco smoke exposure, concurrent viral illness, overheating, and sleep position; in addition, the incidence is highest in the winter months.[4] Before 1990, prone sleeping was considered the greatest risk factor for SIDS. With the decrease in SIDS due to the campaign to put infants to sleep on their backs, the most significant risk factor has become maternal smoking. Although passive smoke exposure can cause respiratory problems in infants, maternal smoking during gestation is the most important risk for SIDS.[5–7]

The etiology of SIDS is thought to be multifactorial, and it is quite likely that many different means can cause the unexplained death. Several theories of causation are prevalent in the animal model literature, in particular, pathology related to respiration. Changes, both central and systemic, that can affect oxygenation, rebreathing of carbon dioxide, and recurrent apnea are a prime focus of animal research.[8–11]

Other animal studies have focused on damage to the fetal neurological system. Autopsy evaluation of many SIDS cases reveals histological and chemical changes that indicate abnormal development in the brain.[12,13] These abnormalities are located primarily in the brainstem but are found throughout the brain as well. They reflect possible damage from anoxia, but are not limited to that particular insult. For example, prenatal nicotine exposure without hypoxia may cause structural and neurochemical changes in the brain.[14–18] An animal model offers a way to examine the developing nervous system and damage from possible insults.

A more recent area of animal research focuses on mechanisms that challenge cardiovascular responsiveness through overheating, hypoglycemia, infection, and toxins. Risk factors such as the increased incidence of SIDS in winter months parallel closely those associated with these cardiovascular challenges.[19]

Prone sleeping position is thought to play a significant role in risk because of changes in gas exchange; however, elevated body temperature may also be a factor. It is believed that the fatal event involves the inability to compensate for a vital physiological challenge to a compromised infant during sleep. Compensatory mechanisms of gasping ability, arousal, and efforts to overcome lowered blood pressure and heart rate are currently being examined.[13,20]

Some autopsy findings observed in SIDS infants have been difficult to explain. Many SIDS infants have characteristic intrathoracic petechiae and liquid blood in the heart.[21,22] These could be associated with efforts to overcome hypoxia, respiratory distress, suffocation, or resuscitation; however, they could also be due to coagulation problems associated with infection and inflammation.[23,24] Autopsy findings are varied, but there are often signs of inflammation and response to infection that are out of proportion to preexisting symptoms. Thymic involution, a sign that the infant has been responding to excessive stress, is observed in more than half of SIDS cases.[25]

A good animal model must reflect the characteristics of an uneventful history without a clear cause for death. Careful consideration must be given to associations between the human condition and the findings in the model. Insults should be plausible and compatible with daily living, and they cannot be so severe that they are obviously the cause of the death.

From: *Sourcebook of Models for Biomedical Research*
(P. M. Conn, ed.), © 2008 Humana Press Inc., Totowa, NJ.

ANIMAL MODELS OF SUDDEN INFANT DEATH SYNDROME

Many animals have been used to study sudden unexplained death in cats,[26,27] chick embryo,[28–31] dogs,[8] ferrets,[32,33] guinea pigs,[34] mice,[35–37] and several primates.[38,39] Although primates are more like humans than the other species, they present difficulties for researchers. Primates have fewer offspring and much longer gestations than other laboratory animals. In addition, many species of primates are endangered, adding moral and legal reasons not to subject them to science that involves the death of offspring. Therefore the majority of research on sudden unexplained death has been done using the piglet, neonatal rabbit, and neonatal rat (Table 60–1). This chapter therefore focuses on models developed in these three species.

PIGLET The piglet is commonly used in neonatal research, in part because it is a large animal, which allows easy manipulation of equipment and procedures. The piglet also has a tracheal morphology that is very similar to a newborn infant.[40] Many processes have been well characterized in perinatal swine models. Piglets have been used primarily for respiratory research in SIDS, but they have also been used to examine the effects of heat and heat stress, reflux, and cardiac development.

A large study published in 1994 evaluated the natural history of a group of 1921 piglets, examining all animals that died for cause of death. Eight animals died suddenly or unexpectedly; however, only one animal represented a "SIDS-like" death. It was concluded that although there was a high incidence of mortality, the incidence of sudden unexplained death in the piglet was very low, and thus it would be pointless to study the piglet in the natural state.[41]

Nevertheless, studies in piglet models have been very productive in focusing on specific systems. Galland and colleagues,[42,43] for example, in an early study, looked at the effects of hyperthermia and oxygen consumption and tested the hypothesis that a cold face and hot body might increase hyperthermia and drop O_2 consumption in a cold climate. They found that face cooling could also increase rapid eye movement (REM) sleep, a time of risk for SIDS.[42] Additionally, this study supported the idea that head covering does not increase CO_2 or decrease O_2 but has its primary effect by increasing body temperature.[44]

In another study, heated piglets developed tachycardia, hypotension, and metabolic acidosis. They also became tachypneic, with hypocapneic alkalosis and neutrophil leukocytosis. Interestingly, these animals had excessive hemorrhage in the lungs and alveolar edema, similar to that seen in SIDS cases.[45] In addition, piglets have been shown to have a delayed response to airway obstruction while they are recovering from fever.[46]

Xia and colleagues used decerebrated, vagotomized piglets to evaluate the combined effects of hyperthermia and apnea in sudden death. They found that respiratory inhibition associated with chemoreflex was prolonged when body temperature was elevated,[47] and these changes were probably mediated by the central nervous system (CNS).[48]

It has been proposed that reflux during sleep has a negative effect on the ability of the human infant to autoresuscitate.[49] The piglet model allows for manipulation of the gastrointestinal (GI) tract and artificial instillation of a mildly acidic compound to stimulate airway protective responses in naturally sleeping and sedated animals.[49,50] Richardson and Adams examined lung changes postmortem following laryngeal chemoreflux and found that these piglets had petechiae in their lungs.[51] Togari and colleagues examined the compensatory role of cerebral circulation in respiratory control in a high CO_2 environment and found that reduction of cerebral blood flow caused respiratory suppression in hypercarbic piglets.[52]

The piglet model has also been used to examine hypoxic responses to infection and smoking. Pretreatment with nicotine

Table 60–1
Animal models examining risk factors for sudden infant death syndrome in the piglet, neonatal rabbit, and neonatal rat

Risk factors	*Animal*	*Conclusion*	*Reference*
Hyperthermia	Piglet	Head covering does not decrease O_2 or increase CO_2 but has its primary effect by increasing body temperature	42–44
	Piglet	Piglets have a delayed response to airway obstruction while recovering from fever	46
	Piglet	Respiratory inhibition in response to chemoreflex was prolonged with elevated body temperature; these changes are probably mediated by the central nervous system (CNS)	47, 48
CNS	Piglet	The reduction of cerebral blood flow caused respiratory suppression in hypercarbic piglets	52
	Rabbit	Maternal cocaine exposure alters the maturation of centers in the brain important for autonomic and respiratory function	58–60
Autonomic tone	Piglet	Imbalance in autonomic tone favoring parasympathetic tone could cause a problem in the developing piglet	55–57
Nicotine	Piglet	Nicotine and infection interfere with autoresuscitation from chemoreflex or hypoxia	53, 54
	Rat	Prenatal nicotine exposure decreases autoresuscitation in response to apnea; this appears to be related to the affect on autonomic nervous system development	72–79
Toxins and infection	Rabbit	Intravenous administration of different bacterial toxins increased catecholamine levels, and induced bradycardia, hypotension, and apnea in neonatal rabbits	62, 65
	Rat	Combinations of bacterial isolates were lethal to rat pups	66, 83, 84
Development	Rat	There is a transitory period of imbalance in brainstem respiratory control in the 12-day-old rat	86–88
	Rat	Lethal response to influenza A and endotoxin challenge in the 12-day-old rat pup	83, 84

and endotoxin has been shown to interfere with autoresuscitation and produce prolonged apnea in experimental piglets exposed to a subglottic acidified saline solution.[53] Nicotine and interleukin (IL)-1β had synergistic effects, decreasing the animal's ability to respond to apnea.[53] This response was also stimulated by intermittent hypoxia down to 6% O_2, then reoxygenation. Nicotine and inflammatory mediators appear to act as cofactors in hypoxic–ischemic neurological injury.[54] Froen and colleagues did not examine the synergy between endotoxin and nicotine; however, their work suggests a relationship between either nicotine or an inflammatory response to infection and lethal apnea.[53]

SIDS infants may succumb to a fatal cardiac arrhythmia that is not detectable on autopsy, and researchers have used the piglet to examine the development of cardiac sympathetic innervation and evaluate arrhythmia.[55] These studies have established that an imbalance in the maturation of autonomic function favoring a parasympathetic response could pose a problem to the developing piglet.[56] They have also shown that the piglet can be used to evaluate prolonged QT syndrome in swine.[55,57] This syndrome can cause lethal arrhythmias in all age groups and may possibly be a mechanism of SIDS.

RABBIT The rabbit allows researchers to do larger scale manipulations than are possible in rodents. In SIDS research, rabbits have been used primarily to examine the effects of respiratory insults, infectious agents, and gastric reflux.

In an early study, a series of experiments examined the effect of maternal cocaine use in pregnancy on respiratory response in neonatal rabbits.[58,59] The researchers exposed rabbit pups to different levels of FiO_2 and found that prenatal cocaine changed the maturational response of respiratory control.[60] This supported the basic hypothesis that cocaine use alters the maturation of centers in the brain important for autonomic and respiratory function.

Gastrointestinal pathogens are known to be a leading cause of morbidity and mortality in children under the age of 5 years.[61] Siarakas and colleagues examined the effects of intravenous administration of six common bacterial toxins on the cardiorespiratory system in 1–3 kg rabbits. They observed bradycardia, hypotension, and apnea with sudden death and concluded that under the right conditions, bacteria could produce toxins that cause inflammatory responses similar to those associated with endotoxin-induced shock.[62]

The rabbit has also been used to evaluate the response to acid reflux by instilling an acid solution in the larynx of neonatal rabbits, inducing obstructive, central, and mixed apnea.[63] Similarities between the gastrointestinal tract of the young rabbit and the human infant provide a good model for evaluating potential effects of toxigenic bacteria.[64] A study using this model measured catecholamine responses to intravenous and intraluminary exposure in the gastrointestinal tract to bacterial toxins. With increasing intravenous doses, the study found a dose-related increase in catecholamine levels and sudden death. However, intraluminal doses did not increase sudden death. It appears that a healthy gut provides a good protective barrier.[65]

RAT The rat is a wonderful research animal because it is large enough to allow manipulation using standard equipment and it has the added advantage of large litters and fast turnover of progeny. A drawback to rat research is that there are fewer antibodies available for immune studies, and knockout strains are much less developed than with mice. In spite of this, the rat has often been used in SIDS research. Rat models have been developed to examine both infectious and respiratory insults, and current work is focusing on neurological changes in the brain following prenatal nicotine exposure.

Building on a body of work showing that infectious insults could cause unexplained death, Lee and colleagues showed that following subcutaneous injection of nasopharyngeal bacterial isolates cultured from human SIDS infants, 21-day-old weanling rats died rapidly without terminal signs of illness and with minimal inflammatory changes in lungs, liver, and heart.[66] When pathogen samples were paired (*Escherichia coli* and *Staphylococcus aureus*) mortality occurred much more abruptly, demonstrating the synergy between pathogens.

The identification of *Helicobacter pylori* DNA in postmortem tissues of SIDS infants led to the use of a rat model to examine fatal apnea as a response to gastroesophageal reflux in infants.[67,68] There is little evidence that *H. pylori* is the primary cause of SIDS; however, this model is consistent with the infectious challenge theories of SIDS.[69,70]

Maternal smoking is an independent risk factor for SIDS.[71] Passive exposure to environmental tobacco smoke increases risk,[5,6] but the risk of SIDS doubles with maternal smoking and is three to four times higher if the mother smokes more than 10 cigarettes per day.[7] Nicotine is currently considered a neuroteratogen and thus it is possible that maternal smoking during pregnancy is more harmful than postnatal exposure to environmental tobacco smoke.[5,72]

Prenatal exposure to nicotine has been shown to cause a decreased response to apnea and respiratory insult; in addition, it is believed that nicotine affects the development of autonomic responsiveness and delays maturity in the newborn, increasing the risk of SIDS.[73] The rat offers an excellent model in which to evaluate the effects of nicotine on development of the respiratory and central nervous systems.

Perinatal nicotine has been shown to impair respiratory function and response to hypoxia in neonatal rats during the first week of life.[73–75] More recently, investigators have used the rat model to evaluate changes in autonomic function as a result of perinatal nicotine exposure, as well as its effects on the heart,[76] brain,[77] and respiration.[78,79] The changes appear to last well into adulthood.[15,80]

A DEVELOPMENTAL MODEL OF SUDDEN UNEXPLAINED DEATH

The studies reviewed above did not always address the consistent observation that the peak incidence of SIDS occurs between 2 and 4 months in all countries for which accurate data are available.[1] The factors underlying the increased susceptibility of this particular age range need to be examined. Developmental responses to infectious challenges may be crucial in this regard.

Animal models have been important for developing and testing ideas about the physiological events leading to SIDS; however, it is difficult to show a direct connection to human immune development. We think that rat immune development may provide a useful model for human infants. The rapid development of the immune and neurological systems in the rat makes each day a significant time point for comparison to human infants.

There may be changes in the development of immune responses that turn a minor illness into a lethal event.[81] We have addressed this issue by assessing an age-dependent model of susceptibility in rats to a nonlethal strain of influenza A virus and a sublethal dose of endotoxin.

Initial experiments were done on postnatal day (PND) 10 to ascertain mortality, timing between doses, and dose response. Pups were treated with 25 µl of a rat-adapted influenza virus (RAIV) given intranasally.[82] One to 5 days post-RAIV treatment, five pups per day, per dose, were given an intraperitoneal dose of endotoxin (2 mg/kg, 0.5 mg/kg, 0.2 mg/kg, or 0.05 mg/kg). A second control group of RAIV-naive pups were given the same dose of endotoxin at each time point. RAIV controls received virus only on PND 10. Pups were examined every 4 h postendotoxin for morbidity and mortality.[83]

The animals died quietly without significant symptoms 8–10 h postendotoxin. This model produced 20% mortality on PND 10, 70–80% mortality on PND 12, but no mortality after PND 14.[83,84] This narrow window of susceptibility on PND 12 is similar to the narrow window of susceptibility to SIDS in human infants between 2 and 4 months of age.

Additional studies were conducted to establish the critical age at which the dual challenge resulted in unexplained death. Pups were treated intranasally with RAIV on day 8, 10, or 14 of life and then given endotoxin (0.5 mg/kg, 0.2 mg/kg, 0.1 mg/kg, or 0.05 mg/kg) intraperitoneally 2 days later. RAIV-naive pups were given endotoxin in the same doses and time interval as the first group. Additional controls were treated with RAIV only on PND 8, 10, or 14. Pups were examined every 4 h for morbidity and mortality, and deceased animals were removed for necropsy.[83]

Adult (8-week-old) rats were similarly challenged with the established adult dose of 100 µl of RAIV intranasally and endotoxin (2 mg/kg, 0.5 mg/kg, 0.2 mg/kg, 0.1 mg/kg) intraperitoneally 2 days post-RAIV treatment.[85] There was no morbidity or mortality in the adult animals.[83]

Interestingly, Liu and Wong-Riley showed developmental changes in the respiratory center in the brainstem of rats in the susceptible period.[86–88] On PND 12 rat pups had an abrupt switch in the expression of $GABA_A$ receptor subunits in the pre-Botzinger complex in the rat brain, the key area for respiratory control in the brainstem. The transitory period of imbalance between inhibitory and excitatory drives may leave the respiratory system vulnerable to respiratory insults, including an immune response to a significant infectious stimulus.[86–88]

These experiments have helped us to establish a model of sudden unexplained death in neonatal rats. Influenza administration on PND 10 followed by an endotoxin challenge of 0.2 mg/kg on PND 12 caused the highest mortality and the lowest morbidity in these animals. The model was used in all subsequent experiments, described below.

MATERIALS AND METHODS

ANIMALS Fischer-344 rat dams were obtained from Charles River Laboratories (Raleigh, NC).

Procedure

1. Animal cages were placed in Illinois isolation cubicles for all of the studies.
2. Dams were housed in individual flat bottom plastic cages at 72°F ± 2, humidity 50% ± 10, in with 12-h light–dark cycles.
3. All animals received food and water *ad libitum*.
4. Within 48 h of birth, rat pups were randomly cross-fostered and litters standardized to eight pups per dam.
5. A vigorous sentinel animal program was in place to evaluate viral spread. However, it has been shown that there is no between-cage contamination and little within-cage cross-infection using RAIV.[89]
6. In developmental studies each litter is considered a single group ($n = 1$); therefore for the experiments, only one pup was removed from each cage for each treatment group.

INFLUENZA We developed our model based on work done by Burleson, who showed that an influenza strain could be adapted to induce an immune response in rats without causing signs of illness.[89] RAIV was developed from Influenza A/Port-Chalmers/1/73 (H3N2) virus. It had been initially purchased from American Type Culture Collection and adapted by 10 successive passages in allantoic fluid of chicken eggs.[89,90] A detailed description of this model and methods can be found in *Methods in Immunotoxicology*.[89]

For our studies, the virus was a gift from Gary Burleson (Burleson Research Technologies, Raleigh, NC). The virus was passaged in adult male F-344 rats using isoflurane anesthesia, as described by Burleson.[89] One day after infection the lung tissues were removed and homogenized in Eagle's minimal essential medium (EMEM) solution diluted to a 10% weight/volume suspension. The virus was washed and frozen in 0.5-ml aliquots at –70°C.[89] Viral content was analyzed using the Madin–Darby canine kidney plaque assay.[89] Prior to use, the virus was deemed clear of mycoplasma, bacterial, viral, and fungal contaminants.

Procedure

1. On PND 10, stock virus was mixed with phosphate-buffered saline (PBS) in a 1:100 dilution and held on ice until used. The viral content of the 1:100 diluted solution was 1.9×10^4 plaque-forming units (PFU)/ml.
2. Pups were removed from the dam and given 25 µl of RAIV intranasally using a pipetteman. The viral solution was dripped on the nare and inspired by the pup.
3. Pups were allowed to recover before they were placed back with the dam.
4. All pups in five cages were RAIV treated and those in another five cages were treated with PBS. This was done to ensure five subjects in each treatment group.

ENDOTOXIN *E. coli* endotoxin 0127:B8 was obtained from Sigma (St Louis, MO).

Procedure

1. On PND 12, endotoxin was diluted with PBS to a concentration of 0.02 mg/ml.
2. Individual pups were weighed, and the average weight for these studies was 20 ± 2 g.
3. Endotoxin dilution was given in a single intraperitoneal dose of 0.2 ml.
4. At 2, 4, 6, or 8 h postendotoxin administration, five pups were removed from five dams in each treatment group and sedated using a 2:10 dilution of Nembutal 50 mg/ml mixed in PBS (0.2 ml intraperitoneally = 0.01 ml/g).
5. Dams were kept in cages with the pups until the end of the experiment for pup warmth and feeding.

COLLECTION OF SERUM FROM A CARDIAC PUNCTURE

Procedure

1. After establishing that the animal was adequately sedated, a 25-gauge needle and syringe were used to collect blood

from the heart. Prior to starting, the plunger of the syringe was pulled to the 0.05 ml position.

2. Placing the needle at the zyphoid process, the needle was inserted, then angled at 15–20° and advanced slowly toward the left axillae until blood appeared in the syringe.
3. Blood was slowly removed and placed in a serum separator collection tube.
4. The tubes were spun according to the package directions.
5. The serum was collected, placed in a cryotube, and frozen at −70°C.

HARVESTING OF ORGANS FROM EACH PUP

Procedure

1. The spleen was removed and placed in RPMI to be processed for cell phenotyping.
2. One kidney with adrenal gland, one lobe of lung, one lobe of liver, one half thymus, one half heart, and a small piece of spleen were placed in cassettes for histology.
3. The other half of each organ was placed in a cryotube, flash frozen in liquid nitrogen, then stored at −70°C to be used for reverse transcriptase polymerase chain reaction (RT-PCR) assay.
4. All remaining animal tissues were placed in biohazard bags and treated as infectious waste.

CHARACTERISTICS OF THE MODEL

Characteristic findings reflected increased inflammation and the release of nitric oxide, stimulated by the dual infectious challenge. At each time point following endotoxin, the RAIV-treated animals were more severely affected than the RAIV-untreated animals. At 6 h postendotoxin, the pups were mildly cyanotic and cool, and adequate cardiac blood samples were difficult to obtain. These animals have been shown to be hypotensive and bradycardic using a tail cuff measuring device (Hatteras Instruments, Inc., Raleigh, NC). Following cardiac puncture at 6 h, RAIV-treated pups continued to bleed without clotting, suggesting disorder of the clotting mechanisms, often seen with severe infection.

Gross examination revealed petechiae in the lungs, subendocardial hemorrhage, lymphoid cell necrosis of the thymic cortex, and pale kidneys in endotoxin-treated RAIV-treated animals. All pups still had milk in their stomachs, suggesting a rapid progression to death. The RAIV-treated animals died between 8 and 10 h postendotoxin. No other animals died.

HISTOLOGY Tissues (liver, spleen, thymus, heart, lung, kidney) were processed, embedded in paraffin, sectioned at 5–6 μm, and stained with hematoxylin and eosin (H&E) for microscopic examination, as described by Blood-Siegfried.[83,91] All RAIV-treated animals at 6 and 8 h postendotoxin injection had signs of inflammation and oxidative stress. The inflammatory component was minor compared to the prominent necrotic process. In addition, there were signs of altered clotting processes.[83]

CELL PHENOTYPING (FLUORESCENCE-ACTIVATED CELL SORTER) Cells were double stained for surface antigen expression, as described in Blood-Siegfried.[83,92] The macrophage cell population started migrating out of the spleen by 2 h after endotoxin administration in the RAIV-treated animals. This occurred in RAIV-untreated animals between 2 and 4 h postendotoxin administration. A similar change also occurred in T-lymphocyte populations, both CD4 and CD8. Migration out of the spleen started at 2 h postendotoxin administration in the RAIV-treated animals, which responded 2 h before the RAIV-untreated animals.[83]

BRONCHEOALVEOLAR LAVAGE FLUID Alveolar cells and fluid were harvested by lavaging the lungs of five animals per group *in situ* at least five times with 2 ml of ice-cold Ca^{2+}, Mg^{2+}-free PBS as described in Blood-Siegfried.[83,85] Cell populations identified on Wright stain from broncheoalveolar fluid (BALF) showed a predominance of alveolar macrophages with little fluctuation in cell numbers across time or between groups. RAIV treatment did not appear to increase inflammatory changes in the lungs of these animals.[83]

ENZYME-LINKED IMMUNOASSAY Commercial rat cytokine enzyme-linked immunosorbent assay (ELISA) kits (Biosource, Camarillo, CA) were used to measure tumor necrosis factor (TNF)-α, interferon (IFN)-γ, and interleukin (IL)-12 in serum, BALF samples, and liver homogenates.[83,84] The only cytokine showing a statistically significance change was IFN-γ with RAIV-untreated pups demonstrating a 2-fold greater increase at 6 h than RAIV-treated animals. There was no significant difference in TNF- α, IL-12, and IFN-γ in BALF or liver tissue cytokines across time or between groups; however, there was an increase in inflammatory cytokines in all endotoxin-exposed animals.[83,84]

REVERSE TRANSCRIPTASE POLYMERASE CHAIN REACTION ASSAYS Total RNA was isolated from flash frozen liver and lung samples according to the methods described by the manufacturer for the RNeasy kit (Quiagen, Valencia, CA).[84] Endotoxin administration stimulated a brisk upregulation of inflammatory cytokine mRNA in RAIV-treated animals and this occurred several hours earlier than in the animals not exposed to RAIV. The same pattern occurred with nitric oxide synthase mRNA. RAIV treatment thus appears to prime the immune system in these animals to respond very quickly to the endotoxin challenge.[84]

IMMUNOHISTOCHEMISTRY Tissues were fixed in 10% neutral buffered formalin, processed routinely, and embedded in paraffin. Serial sections (6 μm) of liver were stained for NF-κB, COX-2, or inducible nitric acid synthase (iNOS).[84,93] The same pattern of rapid response of iNOS was verified by immunohistochemical staining at 6 h postendotoxin; higher levels of iNOS staining were found in liver samples in animals not treated with RAIV, substantiating the PCR findings. No differences in immunostaining for NF-κB or COX-2 were noted between liver sections from RAIV-treated or untreated animals at any time points examined.[84]

CONCLUSIONS

Animal models provide a useful tool for evaluating potential mechanisms of sudden unexplained death in infants. Indeed, examination of specific pathology may be possible only in an animal model. Comparisons between species may be difficult, but some characteristics of SIDS can be evaluated in these systems.

In our model, RAIV treatment appears to prime the immune response in 12-day-old rat pups, and the response is specific to this age group. Our model thus offers a unique way to evaluate the role of immune development and its relationship to potential pathogens in the etiology of SIDS.

REFERENCES

1. Blackwell CC, Weir DM. The role of infection in sudden infant death syndrome. *FEMS Immunol Med Microbiol* 1999;25:1–6.
2. Willinger M, James LS, Catz C. Defining the sudden infant death syndrome (SIDS): Deliberations of an expert panel convened by the National Institute of Child Health and Human Development. *Pediatr Pathol* 1991;11:677–684.
3. Rognum TO. Definition and pathologic features. In: Byard RW, Krouse HF, Eds. *Sudden Infant Death Syndrome: Problems, Progress and Possibilities*. London: Hodder Arnold, 2001:4–30.
4. American Academy of Pediatrics. The changing concept of sudden infant death syndrome: Diagnostic coding shifts, controversies regarding the sleeping environment, and new variables to consider in reducing risk. *Pediatrics* 2005;116:1245–1255.
5. Alm B, Milerad J, Wennergren G, *et al.* A case-control study of smoking and sudden infant death syndrome in the Scandinavian countries, 1992 to 1995. The Nordic Epidemiological SIDS Study. *Arch Dis Child* 1998;78:329–334.
6. Klonoff-Cohen HS, Edelstein SL, Lefkowitz ES, *et al.* The effect of passive smoking and tobacco exposure through breast milk on sudden infant death syndrome. *JAMA* 1995;273:795–798.
7. MacDorman MF, Cnattingius S, Hoffman HJ, Kramer MS, Haglund B. Sudden infant death syndrome and smoking in the United States and Sweden. *Am J Epidemiol* 1997;146:249–257.
8. Duke SG, Postma GN, McGuirt WF, Ririe D, Averill DB, Koufman JA. Laryngospasm and diaphragmatic arrest in immature dogs after laryngeal acid exposure: A possible model for sudden infant death syndrome. *Ann Otol Rhinol Laryngol* 2001;110:729–733.
9. Gershan WM, Becker CG, Forster HV, Besch NS, Lowry TF. Apnea and bradycardia due to anaphylaxis to tobacco glycoprotein in the infant rabbit. *Environ Res* 2004;94:152–159.
10. Huang YH, Brown AR, Costy-Bennett S, Luo ZL, Fregosi RF. Influence of prenatal nicotine exposure on postnatal development of breathing pattern. *Respir Physiol Neurobiol* 2004;143:1–8.
11. Kahraman L, Thach BT. Inhibitory effects of hyperthermia on mechanisms involved in autoresuscitation from hypoxic apnea in mice: A model for thermal stress causing SIDS. *J Appl Physiol* 2004;97: 669–674.
12. Kinney HC, Filiano JJ, Harper RM. The neuropathology of the sudden infant death syndrome. A review. *J Neuropathol Exp Neurol* 1992;51:115–126.
13. Harper RM, Kinney HC, Fleming PJ, Thach BT. Sleep influences on homeostatic functions: Implications for sudden infant death syndrome. *Respir Physiol* 2000;119:123–132.
14. Abdel-Rahman A, Dechkovskaia A, Mehta-Simmons H, Guan X, Khan W, Abou-Donia M. Increased expression of glial fibrillary acidic protein in cerebellum and hippocampus: Differential effects on neonatal brain regional acetylcholinesterase following maternal exposure to combined chlorpyrifos and nicotine. *J Toxicol Environ Health* 2003;66:2047–2066.
15. Abdel-Rahman A, Dechkovskaia AM, Mehta-Simmons H, *et al.* Maternal exposure to nicotine and chlorpyrifos, alone and in combination, leads to persistently elevated expression of glial fibrillary acidic protein in the cerebellum of the offspring in late puberty. *Arch Toxicol* 2004;78:467–476.
16. Abdel-Rahman A, Dechkovskaia AM, Sutton JM, *et al.* Maternal exposure of rats to nicotine via infusion during gestation produces neurobehavioral deficits and elevated expression of glial fibrillary acidic protein in the cerebellum and CA1 subfield in the offspring at puberty. *Toxicology* 2005;209:245–261.
17. Huang ZG, Wang X, Evans C, Gold A, Bouairi E, Mendelowitz D. Prenatal nicotine exposure alters the types of nicotinic receptors that facilitate excitatory inputs to cardiac vagal neurons. *J Neurophysiol* 2004;92:2548–2554.
18. Lichtensteiger W, Ribary U, Schlumpf M, Odermatt B, Widmer HR. Prenatal adverse-effects of nicotine on the developing brain. *Prog Brain Res* 1988;73:137–157.
19. Blackwell CC, Moscovis SM, Gordon AE, *et al.* Cytokine responses and sudden infant death syndrome: Genetic, developmental, and environmental risk factors. *J Leukoc Biol* 2005;78:1242–1254.
20. Harper RM. Autonomic control during sleep and risk for sudden death in infancy. *Arch Ital Biol* 2001;139:185–194.
21. Berry PJ. Pathological findings in SIDS. *J Clin Pathol* 1992;45: 11–16.
22. Krous HF. Sudden infant death syndrome: Pathology and pathophysiology. *Pathol Annu* 1984;19:1–14.
23. Harrison M, Curran C, Gillan JE. Mast-cell degranulation suggests nonimmune anaphylaxis as a cause of deaths in SIDS—an electron-microscopy study. *Lab Invest* 1992;66:P5.
24. Holgate ST, Walters C, Walls AF, *et al.* The anaphylaxis hypothesis of sudden infant death syndrome (SIDS): Mast cell degranulation in cot death revealed by elevated concentrations of tryptase in serum. *Clin Exp Allergy* 1994;24:1115–1122.
25. Rambaud C, Guibert M, Briand E, Grangeot-Keros L, Coulomb-L'Hermin A, Dehan M. Microbiology in sudden infant death syndrome (SIDS) and other childhood deaths. *FEMS Immunol Med Microbiol* 1999;25:59–66.
26. Toth LA, Chaudhary MA. Developmental alterations in auditory arousal from sleep in healthy and virus-infected cats. *Sleep* 1998;21: 143–152.
27. Harper RM, Woo MA, Alger JR. Visualization of sleep influences on cerebellar and brainstem cardiac and respiratory control mechanisms. *Brain Res Bull* 2000;53:125–131.
28. Sayers NM, Drucker DB, Morris JA, Telford DR. Lethal synergy between toxins of staphylococci and enterobacteria: Implications for sudden infant death syndrome. *J Clin Pathol* 1995;48:929–932.
29. Sayers NM, Drucker DB, Morris JA, Telford DR. Significance of endotoxin in lethal synergy between bacteria associated with sudden infant death syndrome: Follow up study. *J Clin Pathol* 1996;49: 365–368.
30. Drucker DB, Aluyi HA, Morris JA, Telford DR, Oppenheim BA, Crawley BA. Possibility of separating toxins from bacteria associated with sudden infant death syndrome using anion exchange chromatography. *J Clin Pathol* 1992;45:802–805.
31. Drucker DB, Aluyi HS, Morris JA, Telford DR, Gibbs A. Lethal synergistic action of toxins of bacteria isolated from sudden infant death syndrome. *J Clin Pathol* 1992;45:799–801.
32. Jakeman KJ, Rushton DI, Smith H, Sweet C. Exacerbation of bacterial toxicity to infant ferrets by influenza virus: Possible role in sudden infant death syndrome. *J Infect Dis* 1991;163:35–40.
33. Lundemose JB, Smith H, Sweet C. Cytokine release from human peripheral blood leucocytes incubated with endotoxin with and without prior infection with influenza virus: Relevance to the sudden infant death syndrome. *Int J Exp Pathol* 1993;74:291–297.
34. Bonham AC, Chen CY, Mutoh T, Joad JP. Lung C-fiber CNS reflex: Role in the respiratory consequences of extended environmental tobacco smoke exposure in young guinea pigs. *Environ Health Perspect* 2001;109:573–578.
35. Nakamura A, Fukuda Y, Kuwaki T. Sleep apnea in mice. *FASEB J* 2003;17:A1212.
36. Nakamura A, Fukuda Y, Kuwaki T. Sleep apnea and effect of chemostimulation on breathing instability in mice. *J Appl Physiol* 2003;94:525–532.
37. Nakamura A, Fukuda Y, Kuwaki T. Mouse as an animal model of sleep apnea. *Sleep* 2003;26:A232–A233.
38. Hendrickx AG, Tarantal AF. Infant-mortality—the role of the macaque as a model for human disease. *Am J Primatol* 1994;34:35–40.
39. Emery MJ, Hlastala MP, Matsumoto AM. Depression of hypercapnic ventilatory drive by testosterone in the sleeping infant primate. *J Appl Physiol* 1994;76:1786–1793.
40. Post EJ, Wood AK, Page M, Jeffery HE. A method for simultaneous physiological and radiographic recordings from sleeping neonatal piglets. *Sleep* 1995;18:309–316.
41. Lavoue S, Dagorne M, Morvan H, Madec F, Durigon M. Is the piglet a useful animal-model of sudden-infant-death-syndrome. *Biol Neonate* 1994;65:310–316.

42. Galland BC, Peebles CM, Bolton DPG, Taylor BJ. Sleep state organization in the developing piglet during exposure to different thermal stimuli. *Sleep* 1993;16:610–619.
43. Galland BC, Peebles CM, Bolton DPG, Taylor BJ. Oxygen-consumption in the newborn piglet during combined cold face hot body exposure. *J Paediatr Child Health* 1992;28:S33–S35.
44. Galland BC, Peebles CM, Bolton DPG, Taylor BJ. The microenvironment of the sleeping newborn piglet covered by bedclothes—gas-exchange and temperature. *J Paediatr Child Health* 1994;30: 144–150.
45. Elder DE, Bolton DP, Dempster AG, Taylor BJ, Broadbent RS. Pathophysiology of overheating in a piglet model: Findings compared with sudden infant death syndrome. *J Paediatr Child Health* 1996;32:113–119.
46. Voss LJ, Bolton DPG, Galland BC, Taylor BJ. Effects of prior hypoxia exposure, endotoxin and sleep state on arousal ability to airway obstruction in piglets: Implications for sudden infant death syndrome. *Biol Neonate* 2005;88:145–155.
47. Curran AK, Xia L, Leiter JC, Bartlett D. Elevated body temperature enhances the laryngeal chemoreflex in decerebrate piglets. *J Appl Physiol* 2005;98:780–786.
48. Xia L, Leiter JC, Bartlett D. Laryngeal water receptors are insensitive to body temperature in neonatal piglets. *Respir Physiol Neurobiol* 2006;150:82–86.
49. Jeffery HE, Page M, Post EJ, Wood AKW. Physiological-studies of gastroesophageal reflux and airway protective responses in the young animal and human infant. *Clin Exp Pharmacol Physiol* 1995;22: 544–549.
50. McKelvey GM, Post EJ, Wood AK, Jeffery HE. Airway protection following simulated gastro-oesophageal reflux in sedated and sleeping neonatal piglets during active sleep. *Clin Exp Pharmacol Physiol* 2001;28:533–539.
51. Richardson MA, Adams J. Fatal apnea in piglets by way of laryngeal chemoreflex: Postmortem findings as anatomic correlates of sudden infant death syndrome in the human infant. *Laryngoscope* 2005;115: 1163–1169.
52. Togari H, Kato I, Yamaguchi N. Reduction of cerebral blood flow produces respiratory arrest in hypercarbic newborn piglets: A possible pathogenesis of sudden infant death syndrome. *Dev Brain Dysfunct* 1996;9:253–257.
53. Froen TF, Aker H, Stray-Pedersen B, Saugstad OD. Adverse effects of nicotine and interleukin-1 beta on autoresuscitation after apnea in piglets: Implications for sudden infant death syndrome. *Pediatrics* 2000;105:E52.
54. Froen JF, Amerio G, Stray-Pedersen B, Saugstad OD. Detrimental effects of nicotine and endotoxin in the newborn piglet brain during severe hypoxemia. *Biol Neonate* 2002;82:188–196.
55. Tong SW, Ingenito S, Frasier ID, Sica AL, Gootman N, Gootman PM. Effects of cardiac autonomic imbalance on postnatal changes in the electrocardiographic Q-T interval in conscious swine. *J Auton Nerv Syst* 1997;64:162–165.
56. Tong SW, Frasier ID, Ingenito S, Sica AL, Gootman N, Gootman PM. Age-related effects of cardiac sympathetic denervation on the responses to cardiopulmonary receptor stimulation in piglets. *Pediatr Res* 1997;41:72–77.
57. Tong SW, Ingenito S, Anderson JE, Gootman N, Sica AL, Gootman PM. Development of a swine animal-model for the study of sudden-infant-death-syndrome. *Lab Anim Sci* 1995;45:398–403.
58. Gingras JL, Weesemayer D. Maternal cocaine addiction. 2. An animal-model for the study of brain-stem mechanisms operative in sudden-infant-death-syndrome. *Med Hypotheses* 1990;33: 231–234.
59. Weesemayer DE, Klemkawalden LM, Gingras JL, Brouillette RT. Effects of in utero cocaine exposure on postnatal respiration in the newborn rabbit. *Pediatr Res* 1989;25:A331.
60. Weesemayer DE, Klemkawalden LM, Barkov GA, Gingras JL. Effects of prenatal cocaine on the ventilatory response to hypoxia in newborn rabbits. *Dev Pharmacol Ther* 1992;18:116–124.
61. Pickering LK, Snyder JD, Behrman RE, Kliegman RM, Jenson HB. Gastroenteritis. In: Behrman RE, Kliegman RM, Jenson HB, Eds. *Nelson Textbook of Pediatrics*, 16th ed. Philadelphia, PA: W.B. Saunders, 2000:765–768.
62. Siarakas S, Damas E, Murrell WG. Is cardiorespiratory failure induced by bacterial toxins the cause of sudden infant death syndrome? Studies with an animal model (the rabbit). *Toxicon* 1995; 33:635–649.
63. Wetmore RF. Effects of acid on the larynx of the maturing rabbit and their possible significance to the sudden-infant-death-syndrome. *Laryngoscope* 1993;103:1242–1254.
64. Kulangara AC, Schechtman AM. Passage of heterologous serum proteins from mother to foetal compartments in the rabbit. *Am J Physiol* 1962;203:1071–1080.
65. Siarakas S, Damas E, Murrell WG. The effect of enteric bacterial toxins on the catecholamine levels of the rabbit. *Pathology* 1997;29:278–285.
66. Lee S, Barson AJ, Drucker DB, Morris JA, Telford DR. Lethal challenge of gnotobiotic weanling rats with bacterial isolates from cases of sudden infant death syndrome (SIDS). *J Clin Pathol* 1987;40: 1393–1396.
67. Orienstein DM. Aspiration pneumonias and gastroesophageal reflux-related respiratory disease. In: Behrman RE, Kliegman RM, Jenson HB, Eds. *Nelson Textbook of Pediatrics*, 16th ed. Philadelphia. PA: W.B. Saunders, 2000:1288–1291.
68. Pattison CP, Marshall BJ, Scott LW, Herndon B, Willsie SK. Proposed link between Helicobacter pylori (HP) and sudden infant death syndrome (SIDS): Possible pathogenic mechanisms in an animal model. I. Effects of intratracheal urease. *Gastroenterology* 1998;114: A900.
69. Elitsur Y. Helicobacter pylori and SIDS: The jury is in at last! *Am J Gastroenterol* 2002;97:1576–1577.
70. Ho GY, Windsor HM, Snowball B, Marshall BJ. Helicobacter pylori is not the cause of sudden infant death syndrome (SIDS). *Am J Gastroenterol* 2001;96:3288–3294.
71. Blair PS, Fleming PJ, Bensley D, *et al.* Smoking and the sudden infant death syndrome: Results from 1993–1995 case-control study for confidential inquiry into stillbirths and deaths in infancy. Confidential enquiry into stillbirths and deaths regional coordinators and researchers. *BMJ* 1996;313:195–198.
72. Slotkin TA. Fetal nicotine or cocaine exposure: Which one is worse? *J Pharmacol Exp Ther* 1998;285:931–945.
73. Fewell JE, Smith FG, Ng VKY. Prenatal exposure to nicotine impairs protective responses of rat pups to hypoxia in an age-dependent manner. *Respir Physiol* 2001;127:61–73.
74. Slotkin TA, Saleh JL, McCook EC, Seidler FJ. Impaired cardiac function during postnatal hypoxia in rats exposed to nicotine prenatally: Implications for perinatal morbidity and mortality, and for sudden infant death syndrome. *Teratology* 1997;55:177–184.
75. St John WM, Leiter JC. Maternal nicotine depresses eupneic ventilation of neonatal rats. *Neurosci Lett* 1999;267:206–208.
76. Evans C, Wang J, Neff R, Mendelowitz D. Hypoxia recruits a respiratory-related excitatory pathway to brainstem premotor cardiac vagal neurons in animals exposed to prenatal nicotine. *Neuroscience* 2005;133:1073–1079.
77. Neff RA, Simmens SJ, Evans C, Mendelowitz D. Prenatal nicotine exposure alters central cardiorespiratory responses to hypoxia in rats: Implications for sudden infant death syndrome. *J Neurosci* 2004; 24:9261–9268.
78. Huang ZG, Griffioen KJS, Wang X, *et al.* Differential control of central cardiorespiratory interactions by hypercapnia and the effect of prenatal nicotine. *J Neurosci* 2006;26:21–29.
79. Huang ZG, Wang X, Dergacheva O, Mendelowitz D. Prenatal nicotine exposure recruits an excitatory pathway to brainstem parasympathetic cardioinhibitory neurons during hypoxia/hypercapnia in the rat: Implications for sudden infant death syndrome. *Pediatr Res* 2005;58:562–567.
80. Abreu-Villaca Y, Seidler FJ, Slotkin TA. Does prenatal nicotine exposure sensitize the brain to nicotine-induced neurotoxicity in adolescence? *Neuropsychopharmacology* 2004;29:1440–1450.

81. Gleeson M, Cripps AW. Development of mucosal immunity in the first year of life and relationship to sudden infant death syndrome. *FEMS Immunol Med Microbiol* 2004;42:21–33.
82. Dye JA, Morgan KT, Neldon DL, Tepper JS, Burleson GR, Costa DL. Characterization of upper respiratory disease in rats following neonatal inoculation with a rat-adapted influenza virus. *Vet Pathol* 1996;33:43–54.
83. Blood-Siegfried, Nyska A, Lieder H, *et al.* Synergistic effect of influenza A virus on endotoxin-induced mortality in rat pups: A potential model for sudden infant death syndrome. *Pediatr Res* 2002;52:481–490.
84. Blood-Siegfried, Nyska A, Geisenhoffer K, *et al.* Alteration in regulation of inflammatory response to influenza A virus and endotoxin in suckling rat pups: A potential relationship to sudden infant death syndrome. *FEMS Immunol Med Microbiol* 2004;42:85–93.
85. Ehrlich JP, Burleson GR. Enhanced and prolonged pulmonary influenza virus infection following phosgene inhalation. *J Toxicol Environ Health* 1991;34:259–273.
86. Liu QL, Wong-Riley MTT. Postnatal developmental expressions of neurotransmitters and receptors in various brain stem nuclei of rats. *J Appl Physiol* 2005;98:1442–1457.
87. Liu QL, Wong-Riley MTT. Developmental changes in the expression of GABA(A) receptor subunits alpha 1, alpha 2, and alpha 3 in the rat pre-Botzinger complex. *J Appl Physiol* 2004;96:1825–1831.
88. Wong-Riley MTT, Liu QL. Neurochemical development of brain stem nuclei involved in the control of respiration. *Respir Physiol Neurobiol* 2005;149:83–98.
89. Burleson G. Influenza virus host resistance model for assessment of imunotoxicity, immunostimulation, and antiviral compounds. In: Burleson G, Dean JH, Munson AE, Eds. *Methods in Immunotoxicology*, Vol. 2. New York: Wiley-Liss, 1995:181–202.
90. Burleson G. Pulmonary immunocompetence and pulmonary immunotoxicology. In: Smialowicz RJ, Holsapple MP, Eds. *Experimental Immunotoxicology*. Boca Raton, FL: CRC Press, 1996:119–141.
91. Lomnitski L, Nyska B, Ben-Shaul V, *et al.* Effects of antioxidants apocynin and the natural water-soluble antioxidant from spinach on cellular damage induced by lipopolysaccaride in the rat. *Toxicol Pathol* 2000;28:580–587.
92. Gehrs BC, Riddle MM, Williams WC, Smialowicz RJ. Alterations in the developing immune system of the F344 rat after perinatal exposure to 2,3,7,8-tetrachlorodibenzo-p-dioxin: II. Effects on the pup and the adult. *Toxicology* 1997;122:229.
93. Hsu SM, Raine L, Fanger H. Use of avidin-biotin-peroxidase complex (ABC) in immunoperoxidase techniques: A comparison between ABC and unlabeled antibody (PAP) procedures. *J Histochem Cytochem* 1981;29:577–580.

61 Animal Models of Posttraumatic Stress Disorder

Hagit Cohen, Michael A. Matar, and Joseph Zohar

ABSTRACT

Posttraumatic stress disorder (PTSD) affects about 20–30% of exposed individuals. Clinical studies of PTSD generally employ stringent criteria for inclusion in study populations, and yet in animal studies the data collection and analysis are generally expressed as a function of "exposed" versus "nonexposed" populations, regardless of individual variation in response. There is a need for a better parallel between the approach to understanding animal behavioral models and the contemporary understanding of the clinical conditions. We conceived an approach to understanding the consequences of stress exposure in a manner that would enable us to segregate the study animals into groups according to the degree of their response to a stressor, i.e., the degree to which their behavior is altered or disrupted. The idea was to set apart the most clearly affected, i.e., "extreme behavioral response" group from the "minimal behavioral response" at the extremes, and to establish a middle group of partial responders, groups that better reflect behavioral criteria akin to clinical symptoms already defined. This chapter presents an overview of a series of studies, examining the contribution, validity, and practicality of this animal model.

Key Words: Posttraumatic stress disorder, Animal model, Hypothalamic–pituitary–adrenal axis, Early-life stress, Juvenile stress, Corticosterone, Brain-derived neurotrophic factor.

POSTTRAUMATIC STRESS DISORDER

Posttraumatic stress disorder (PTSD) is an incapacitating chronic syndrome reflecting a disorder of cognitive, emotional, and physiological processing and/or recovery from the initial reaction to exposure to a potentially traumatic experience (PTE).[1]

In the first hours to days following the experience the majority of individuals exposed to an extreme event will demonstrate, to a varying degree, symptoms such as intense fear, helplessness, or horror followed by anxiety, depression, agitation, shock, or dissociation, and may have trouble functioning in their usual manner for a while.[2–7]

Retrospective and prospective epidemiological studies indicate that most individuals affected by a PTE will adapt within a period of 1–4 weeks following exposure,[2,6] and only a small proportion will develop long-term psychopathology.[2,6] In the United States, studies report that the rate of lifetime exposure to at least one "serious" traumatic event (excluding grief and mourning) is quite high; a conservative estimate reported 61% among men and 51% among women.[8] Other studies have found similar rates.[9–14] The lifetime prevalence of PTSD in the general population reaches about 7% overall,[15] suggesting that about 20–30% of individuals exposed to severe stressors will develop PTSD.[9] This figure varies depending on the type of trauma studied, where male rape victims suffer very high rates and populations exposed to natural disasters significantly less.[16] The discrepancy between the proportion of the general population exposed to PTE and those who eventually fulfill criteria for the disorder suggests qualitative differences in vulnerability and/or resilience.

After extensive study over the past two decades, PTSD was established as a disorder and listed in the *Diagnostic and Statistic Manual* (*DSM*). A diagnosis of PTSD is made if the required symptoms are present 1 month or more after exposure to a triggering event: (1) intrusive reexperiencing of the traumatic event in the form of nightmares and flashbacks, with an exaggerated response to trauma-related reminders/cues; (2) persistent avoidance of stimuli associated with the trauma and emotional numbing; and (3) persistent symptoms of exaggerated startle response, increased physiological arousal, and sustained preparedness for an instant alarm response.[1]

PTSD has severe effects on widespread areas of the individual's functioning, severely compromising their quality of life and affecting their workplace, family, and social life. Moreover, PTSD is also often comorbid with other disorders such as depressive and anxiety disorders, drug and alcohol abuse, cognitive and memory impairments, and sexual dysfunction.[1] The development of PTSD is often a gradual process and extends over time through a series of stages ranging from relatively contained distress to severe disability.[17] As the disorder evolves over time, pathological changes and debilitating comorbidity may become fixed and irreversible. Unlike processes in which exposure to repeated stimuli induces a process of learning or conditioning, implying an increased efficiency in processing of data to produce the required response, the psychopathology underlying PTSD produces a paradoxical vulnerability to negative sequelae upon subsequent exposure to stress.[17]

The sequelae of exposure to a traumatizing stressor are subject to extensive clinical study. Clinical research often gives rise to important questions or hypotheses as to the pathogenesis, clinical course, and outcomes of such events. Among the issues raised are

From: *Sourcebook of Models for Biomedical Research*
(P. M. Conn, ed.),

those relating to factors that may confer risk or resilience for the development of more severe stress-induced clinical outcomes, such as PTSD. By their nature, clinical studies raise issues concerning premorbid factors largely by means of extrapolating retrospectively. Prospective studies are almost impossible to conceive and would most probably be prohibitively expensive to put into practice.

An animal model can give a good approximation of certain aspects of the complex clinical disorder at best, enabling the study of questions raised in clinical research in a prospective study design and under far more controllable conditions. To maximize its validity as an extension of clinical research, the design of the animal model must strive to conform as closely as possible to the clinical entity and the contemporary conception of the disorder.

ANIMAL MODELS OF POSTTRAUMATIC STRESS DISORDER

The basis of PTSD as a clinical entity rests upon etiology and requires exposure to a severely stressful experience or traumatic event. Thus, exposure to a stressogenic/traumatic experience has formed the basis for animal models of PTSD.

There are as yet no uniformly accepted animal models for PTSD. A number of animal models have been developed in which intense stressful experiences, aversive challenges, and situational reminders of a traumatic stressor have been shown to have long-term effects on psychological/behavioral and physiological functioning, reflected in biobehavioral paradigms and biophysiological tests, e.g., hypothalamic–pituitary–adrenal (HPA) axis assays and electrophysiological studies.[18–30] Irrespective of the study design or of the stress paradigm, animal studies have generally included the entire stress-exposed population as the study population and the results discussed and conceptualized as involving this population versus "others." However, in practice there is considerable heterogeneity in the degree of behavioral disruption or response to the stressor and animals tend to demonstrate considerable individual differences, certainly in the degree to which their behavior is disrupted.

THE CUT-OFF BEHAVIORAL CRITERIA MODEL

One way to reconcile this disparity is to regard the heterogeneity of animal responses to stress paradigms as modeling the heterogeneity of human responses to traumatic stress, and to develop means by which the degree of the individual's behavioral response forms the basis for a system of classification, akin to the inclusion and exclusion criteria employed in clinical studies.

Since the scope of animal models is limited to observable and measurable parameters, the behavioral criteria for the model of stress-induced conditions developed by this study group are based on two well-established paradigms that measure anxiety-like, fearful, avoidant and hypervigilant/hyperalert behaviors, paralleling aspects of traumatic stress-induced human behaviors.[25,31] Exploratory behavior on the elevated plus maze (EPM) serves as the main platform for the assessment of these behaviors and the acoustic startle response (ASR) paradigm allows for a precise quantification of hyperalertness, in terms of magnitude of response and habituation to the stimulus.

In our model, we exposed adult rats to a predator scent (cat urine) stressor (PSS) for 10 min as previously described by Adamec *et al.*[23,25,26] and others.[28–30,32–35] Regarding the conceptual validity of the model itself, "predator exposure trauma" is a potentially life-threatening situation (criteria A) and it may represent a more ''natural'' setting than other types of stressors on a teleological level,[23,24] such as electrical tail shocks and restraint which might possibly be related to extreme conditions like torture.

Seven days after a single 10-min PSS exposure, the exposed rats showed significantly decreased time spent in the open arms of the maze and increased levels of avoidance and anxiety-like behavior in the EPM, and higher mean startle responses as compared to control rats. However, the animals' behavior was not uniformly disturbed, but rather demonstrated a broad range of variation in severity of anxiety-like behaviors (Figure 61–1).

Quantitative and qualitative assessment of individual animal behavioral responses to stress exposure provides a degree of standardization that enables a greater degree of resolution. For instance, groups of subjects can be defined by their response pattern or magnitude and selectively focused on for study or followed up over time; the prevalence of defined patterns of response may be assessed and compared, allowing the "potency" of different stressors to be compared, behaviors to be correlated to physiological parameters, and so on. Since partial responses are far more difficult to quantify and interpret, in animal models more so than in human subjects, the studies carried out according to this approach to date have preferred specificity to sensitivity, and have sought to create clearly defined groups of behaviors for comparison. By means of well-defined cut-off behavioral criteria (CBCs), the studies have focused on animals displaying an extreme behavioral response (EBR) on both the EPM and the ASR paradigms performed consecutively, compared to those displaying a minimal

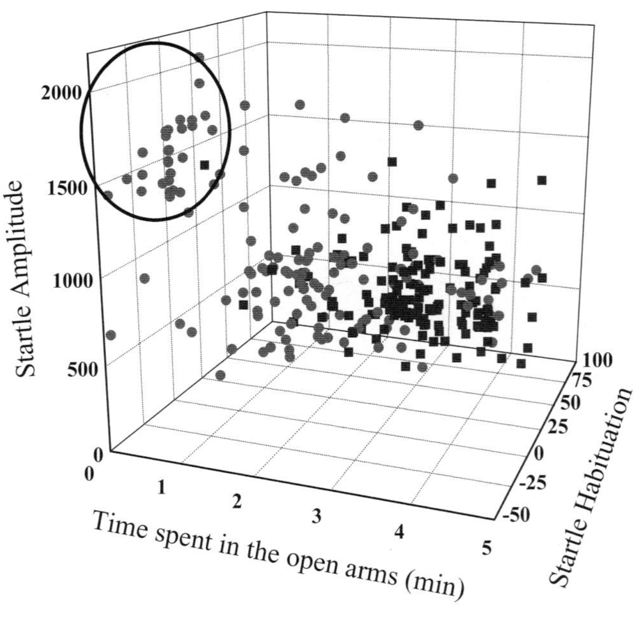

Figure 61–1. The effect of single PSS exposure versus unexposed control on rat anxiety-like behavior and acoustic startle response and habituation. The representation of the data from both paradigms (EPM and ASR) shows two obvious and rather distinct features. First, it is clear that PSS exposure alters the response of the majority of individuals to at least some degree. Second, the cluster of individuals that forms in the upper left hand corner of the graph (i.e., had the more extreme responses to exposure) is quite distinct from the majority of individuals.[36,48] (See color insert.)

behavioral response (MBR) to both, and compared to controls. "Maladaptive" responses to both of these serial CBCs were required for "inclusion" into the EBR group, whereas the reverse was required for inclusion in the MBR group.[36]

The criteria for each behavioral paradigm were as follows:

1. Extreme behavioral response: 5 min (entire session) spent in the closed arms and no entries into the open arms on the EPM; the mean amplitude of the startle response (at 110 dB) exceeds 800 units and the startle response shows no habituation over time.
2. Minimal behavioral response: 0–1 min spent in the closed arms and eight or more open arm entries on the EPM; the mean amplitude of the startle response (at 110 dB) does not exceed 700 units and habituation is demonstrated.
3. Partial behavioral response: animals neither EBR nor MBR are termed "partial behavioral responders" (PBR).

When behavioral data for the entire exposed population were examined in this way, based on the segregation of animals according to the CBCs, it was demonstrated that only about 25% develop significant extreme behavioral disruptions as the result of exposure to the stressor (Figure 61–2A), as compared to 1.3% in controls (Figure 61–2B). Similarly, the prevalence of MBR rats in the exposed groups was 24.7% (Figure 61–2A) and in the control group was 80.0% (Figure 61–2B).[36]

These results are in agreement with trends in the literature on human studies,[9] mentioned above, in which about 15–35% fulfill criteria and another 20–30% are considered as PTSD spectrum disorders.[9,10,13] It was therefore suggested that application of CBCs is both feasible and valid.

The above-mentioned approach to the analysis of animal data is conceptually equivalent to *DSM* criteria for PTSD and the criteria proposed by Yehuda and Antelman.[37]

The demonstrated reduction in the time spent by rats in the open arms of the plus maze may parallel avoidance behavior seen in human PTSD (Criteria C). Since the total number of entries into any arms of the plus maze is the same, except that the EBR rats keep to the closed arms only, the results are consistent with anxiety, and not with nonspecific impairment of locomotion. *DSM-IV* defines this symptom as a persistent avoidance of reminders of the trauma and numbing of responsiveness. Since the exposure took place in an open space it is in keeping with the described definition.

The assessment of rats after 7 days has previously been established by Adamec *et al.*,[25] and is taken to refer to PTSD and not to acute stress reaction. Our results have since shown repeatedly that as of day 7, the prevalence rate remains constant at about 25% until day 30. The *DSM-IV* defines PTSD as lasting more than a month after the exposure. So the persistence of anxiety symptoms on day 7 after the exposure fulfills the time element of PTSD.[25]

In the next step, we examined the validity of applying the CBC concept in animal studies, using dynamic unsupervised fuzzy clustering.

Unsupervised fuzzy clustering (UFC) analysis is a mathematical tool that attempts to assess the relationships among patterns of a data set by organizing the patterns into groups or clusters, such that patterns within one cluster are more similar to each other than patterns belonging to different clusters.[38–42] The main goal of most clustering methods is to supply useful information by grouping data in clusters, where within each cluster the data exhibit similarity. Similarity is defined by a distance measure, and global objective functional or regional graph theoretical criteria are optimized to find the optimal partition of data. The partition approach defines the cluster to which each data element belongs by using the elements of the membership matrix.[38–42]

The full range of the dataset was classified into three ($C = 3$) clusters. The data in the clusters that the model yielded corresponded to the data for the animals designated behaviorally by means of the CBCs as clearly affected, partially affected (including some unexposed animals), and clearly unaffected (Figure 61–3).

The UFC analysis yielded, in addition to the "anxious" and the "baseline" clusters, a third group corresponding to an extreme pathological response (see Figure 61–3). According to this analysis, about 22% develop extreme behavioral disruptions as the result of exposure to the stressor, as compared to 0.7% of the unexposed control group. This extreme response group (i.e., maladapted rats) exhibits significantly higher startle response and significantly more avoidance behavior in the EPM than other animals. Again, these proportions are in agreement with current trends in the literature on PTSD studies in human subjects.[9] Taken together, the UFC analysis supports the conceptual contention of the CBCs and thus lends validity to our previous studies. Moreover, the clustering process may help us to find and define the characteristics of the subgroup of the unexposed population that has a higher risk of developing PTSD.[43]

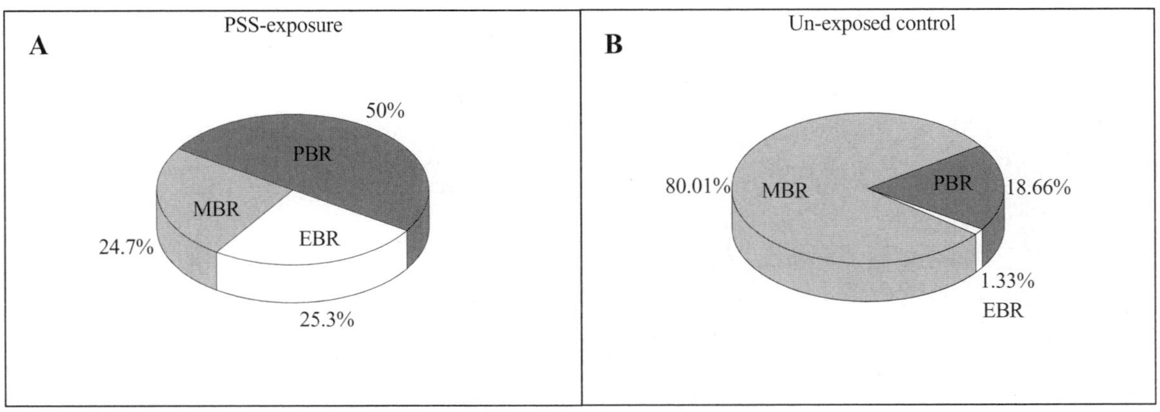

Figure 61–2. Reanalysis of data applying cut-off behavioral criteria. (A) PSS exposure. (B) Unexposed control.

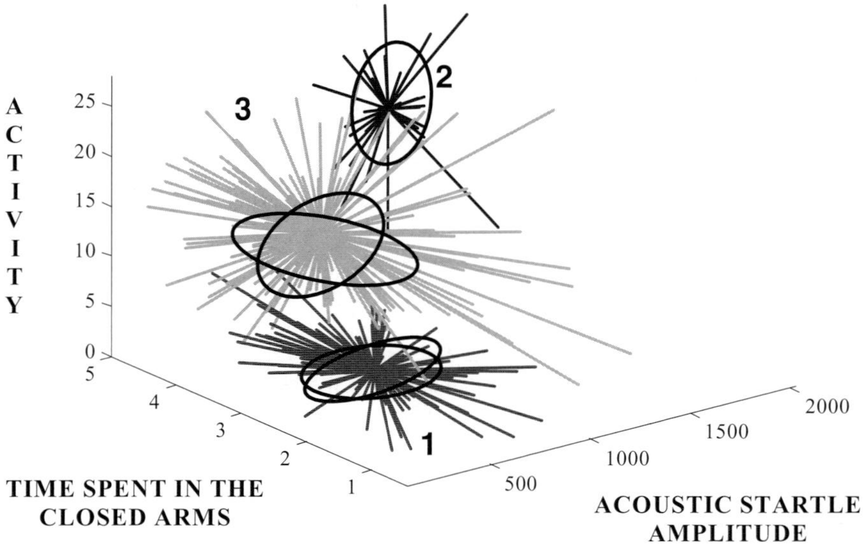

Figure 61–3. Unsupervised fuzzy clustering: Partition of four behavioral parameters into three clusters. The ellipses represent the standard error of each characteristic in each cluster. The first cluster (C1) is related to baseline behavior associated with unstressed animals and is characterized by very low levels of anxiety (MBR). The middle cluster (cluster C3) represents the subclinical or partially affected animals and is characterized by some degree of anxiety and stress levels (PBR). Cluster C2 corresponds to the exposed animals and is characterized by higher levels of anxiety and stress, namely the extremely affected animals (EBR).[43] The behavioral parameters were acoustic startle response amplitude, time spent in the closed arms on the EPM, number of entries to the closed arms, and the total number of entries to the open and the closed arms of the EMP (total activity).[43]

In the next section, we present a series of studies that were constructed to examine the prevalence of EBR and MBR responses over time, in an attempt to map the transition from the immediate reaction, concomitant with an acute stress response, to more enduring and chronic changes, representing PTSD, under different conditions.[44]

By means of singling out the clearly affected individuals labeled as EBR from among the entire population of rats exposed to the stress paradigms, and focusing on them, a time curve could be constructed to reflect the prevalence of clearly "symptomatic" behavior over time (Figure 61–4).

Immediately following the traumatic stimulus, most of the animals (90%) manifested a pattern of behaviors that fulfills criteria for EBR status for the first few hours. This can be regarded as being equivalent to the initial response observed in human subjects in the first hours or days after an extreme event. The prevalence of rats fulfilling the CBC criteria for EBR decreased rapidly, and subsequent sampling revealed that at the end of a week only 25% of the total exposed population had PTSD-like symptoms. The time curve appears to indicate that the rats responded initially with a ubiquitous, probably normative, acute response, but only about 15% went on to suffer from enduring behavioral effects.

The curve representing these responses, although gleaned from different groups of rats, very clearly resembles, both in term of time and of prevalence, the curve of human responses to exposure to a significant stressor and the proportions affected.[45] This pattern of symptoms has been demonstrated across the different kinds of traumatic events, both in human PTSD and in the different models in this study. We submit that the distinction between the acute phase of the stress response seen in most subjects immediately after exposure and the chronic enduring effects associated with stress found after several days or weeks is important in the study of animal models dealing with the behavioral and physiological consequences of stress.[44]

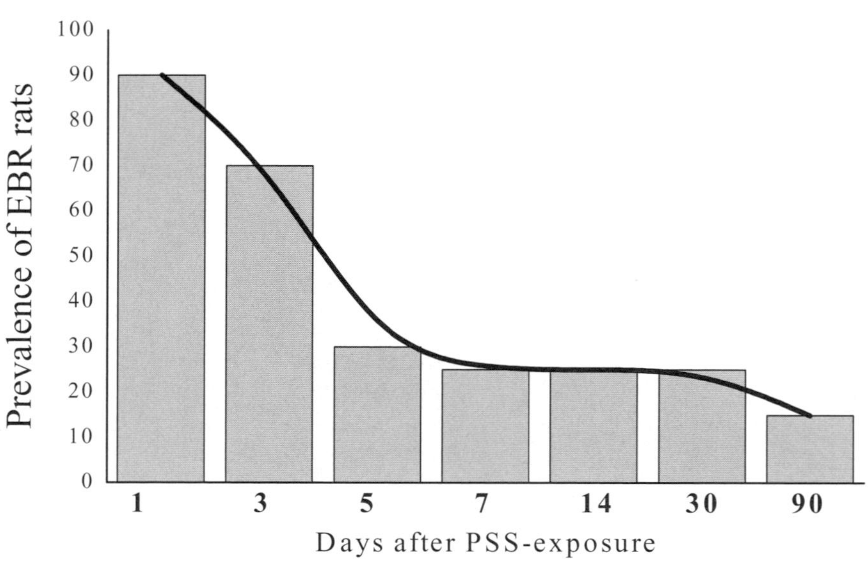

Figure 61–4. The prevalence of EBR rats after single exposure to a predator stimulus along the time axis. A steady drop in prevalence of EBR rats from 90% on day 1 to 15% by day 90 is shown. As of day 7, this rate remains constant until day 30. Our results have shown a progressive, time-dependent decline in the prevalence of EBR rats during the first few weeks that follow a trauma.[44]

BIOLOGICAL STRESS RESPONSE

It is well known that the biological stress response is key to understanding the pathophysiology of PTSD and its associated symptoms. Thus, the primary aim of the following study was to analyze separately various biobehavioral parameters of the animals that responded in a more marked manner to exposure to a stressful, potentially traumatic, event and to compare these to the parameters of rats that demonstrated no such response. We submit that this approach would better approximate the individual variance seen in posttraumatic stress responses in humans (i.e., the clinical syndrome) in a more valid manner.

In a series of studies this approach has elicited highly significant overlap between animals showing EBR and those with extreme biophysiological measures, i.e., HPA axis assays (corticosterone and neuroactive steroids), heart rate variability (reflecting autonomic nervous system activity), and BDNF mRNA and protein levels much more clearly than when the exposed group was analyzed as a whole.

HYPOTHALAMIC–PITUITARY–ADRENAL AXIS

Exposure to stress or a threat to homeostasis triggers activation of the parvicellular neurons of the paraventricular nucleus in the hypothalamus to release corticotropin-releasing factor and vasopressin. These hormones stimulate synthesis of polypeptide precursor proopiomelanocortin products, which include adrenocorticotrophic hormone (ACTH), melanocyte-stimulating hormone cells, and β-endorphin, that have been shown to interact with the immune system.[46,47] In the adrenal gland ACTH stimulates the release of glucocorticoids, typically cortisol in man and corticosterone in rodents. The responses of the HPA axis are regulated by an efficient negative feedback system, exerted by glucocorticoids. Taken together, the HPA stress hormones play a major role in mediating both adaptive and maladaptive responses.

We evaluated plasma corticosterone and ACTH levels 7 days after PSS exposure. Throughout the study data were analyzed for exposed versus unexposed populations as a whole, and then reanalyzed according to the CBC classification separately for EBR, MBR, and PBR versus unexposed controls, and finally these groups were compared to each other (Figure 61–5).

Our results have shown that a single 10-min exposure to a PSS significantly increased plasma corticosterone (Figure 61–5A) and plasma ACTH concentrations (Figure 61–5B) only in EBR rats, but not in MBR individuals or in nonexposed controls.[36,48] The fact that there is a clear-cut involvement of the HPA axis in the physiological component of the syndrome has been quite firmly established by studies in PTSD patients and animals, despite conflicting data reported by different groups.[49–52] The results of the present study are in keeping with those of certain groups of investigators.[53] Moreover, Adamec and colleagues[24–26] have found evidence of differential dysregulation of the HPA axis in certain strains of rats exposed to cats. One week after cat exposure, Hooded rats show elevations in basal plasma corticosterone, whereas Wistar rats do not. This difference in strain reactivity was proposed to be related to the higher basal anxiety in Hooded rats. We may speculate that greater basal anxiety could even stem from factors such as response to the blood sampling procedure, or possibly that different strains possess different vulnerability factors.

The fact that EBR rats show a dysfunction of the HPA axis, whereas MBR rats do not, indicates quite clearly that the population of exposed animals is not a homogeneous cohort, and that it is important to distinguish between affected and unaffected subjects in the analysis and interpretation of data from animal studies.[36,48]

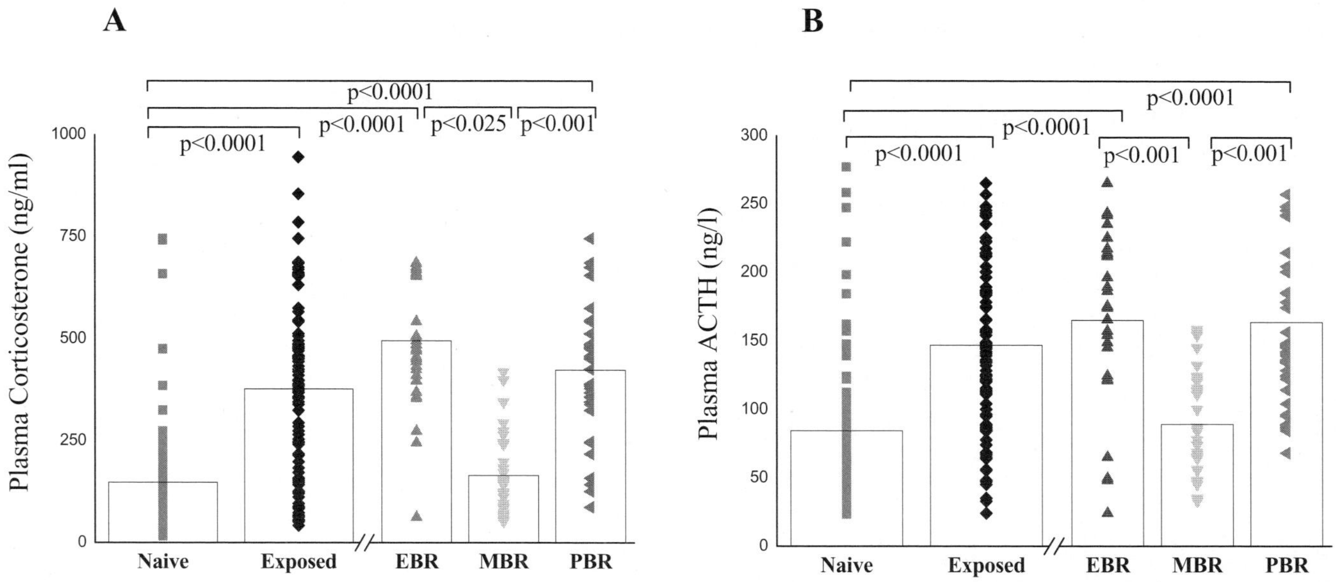

Figure 61–5. Circulating levels of (A) corticosterone and (B) ACTH at day 7 post-PSS exposure. Seven days after exposure, the PSS-exposed population demonstrated significantly increased plasma corticosterone (A) and ACTH levels (B) compared to controls [$F(1,260) = 116.2, p < 0.00001$; $F(1,260) = 83.8, p < 0.00001$, respectively].[36] Reanalysis of data according to CBCs: significant differences in plasma corticosterone and ACTH levels were demonstrated between the subgroups of exposed animals [$F(3,296) = 133.3, p < 0.0001$; $F(3,296) = 74.6, p < 0.0001$]. The post hoc Bonferroni test revealed that plasma corticosterone and ACTH levels of EBR individuals were significantly higher than in MBR individuals ($p < 0.0001$ for both) or in controls ($p < 0.0001$). The levels of PBR animals also varied significantly from those of MBR ($p < 0.001$) (and controls).[36]

Other adrenal steroids, such as the neuroactive steroids, may be involved in the development of long-term behavioral changes after exposure to a PTE, and hence in vulnerability and resilience. Dehydroepiandrosterone (DHEA) and its corresponding sulfate ester dehydroepiandrosterone sulfate (DHEA-S) are neuroactive steroids that are synthesized *in situ* within the brain and in peripheral steroidogenic organs.[54–56] DHEA and DHEA-S appear to possess a number of physiological roles and are known to act as allosteric agonists of the γ-aminobutyric acid ($GABA_A$) receptor—DHEA-S being the more potent of the two.[55] This activity may imply an effect on behavior, cognition, mood, and sleep.[57] In addition, they are reported to exert an effect on neural plasticity and to possess neuroprotective qualities. Indeed, both DHEA and DHEA-S protect rat hippocampus neurons against *N*-methyl-D-aspartate (NMDA)-induced excitotoxicity, glucocorticoid-induced neurotoxicity, and oxidative stress-induced damage induced by the $H_2O_2/FeSO_4$ complex.[58] Changes in the levels of neuroactive steroids may thus be involved in the pathophysiology of adaptive and maladaptive stress responses.

This study assessed the relationship between peripheral DHEA, DHEA-S levels, and magnitude of behavioral change, i.e., EBR and MBR individuals[59] (Figure 61–6).

Whereas both extreme and minimal behavioral response groups of exposed rats displayed decreased levels of DHEA-S compared to controls (Figure 61–6B), DHEA levels were selectively reduced to a statistically significant degree only in EBR rats, compared to both MBR rats and controls (Figure 61–6A). Assuming that grossly disrupted behavior (EBR) is maladaptive, the implication is that DHEA may be involved in mediating adaptive responses to a stressful event (resilience). Conversely, an impaired neuroactive steroid response to a stressogenic experience may be associated with an increased susceptibility to PTSD-like responses. DHEA-S levels are normally two to five times higher than those of DHEA, whereas in the exposed groups this ratio is inverted. Since DHEA-S is a more potent negative modulator of the GABAergic receptor than DHEA, this inverted ratio implies a decrease in negative modulation. This may represent an adaptive factor while stressful conditions prevail, yet may also mediate (or characterize) long-term EBR. It is of note that DHEA possesses antiglucocorticoid activity[60] and decreases corticosterone levels.[61] Thus the reduced DHEA levels in EBR rats may cause the increased levels of corticosterone.[59]

Taken together, the results suggest that concomitantly decreased circulatory levels of DHEA and elevated corticosterone levels may be associated with a pathological response to stress, while maintenance of normal levels of both steroids may be associated with resilience or a "buffering" of the negative effects of stress. It remains to be established whether changes in the synthesis or release of neurosteroids are a cause of, a risk factor for, or a consequence of stress-related disorders.[59]

The neurobiological mechanisms involved in PTSD thus presumably include changes in synaptic efficacy and plasticity in various brain areas, which, rather than improving, in fact impair the ability of the brain to respond adequately to subsequent stress responses.[62] Were it possible to correct or to induce mechanisms that would reinstate adaptive neural plasticity and remodeling, this dysfunction might be overcome.

Brain-derived neurotrophic factor (BDNF), a member of the neurotrophin (NT) family of survival-promoting molecules, plays an important role in the growth, development, maintenance, and function of several neuronal systems.[63] It is known to modulate synaptic plasticity and neurotransmitter release in a variety of neurotransmitter systems, as well as intracellular signal-transduction pathways.[63] It regulates axonal and dendritic branching and remodeling,[64–67] synaptogenesis in arborizing axon terminals, efficacy of synaptic transmission, and the functional maturation of excitatory and inhibitory synapses.[68–70]

Growing evidence suggests important roles for BDNF in the pathogenesis of mood disorders and in the mechanism of action of therapeutic agents, such as mood stabilizers and antidepressants.[71–73] BDNF may also be involved in the pathophysiology of anxiety disorders. Data derived from animal studies have

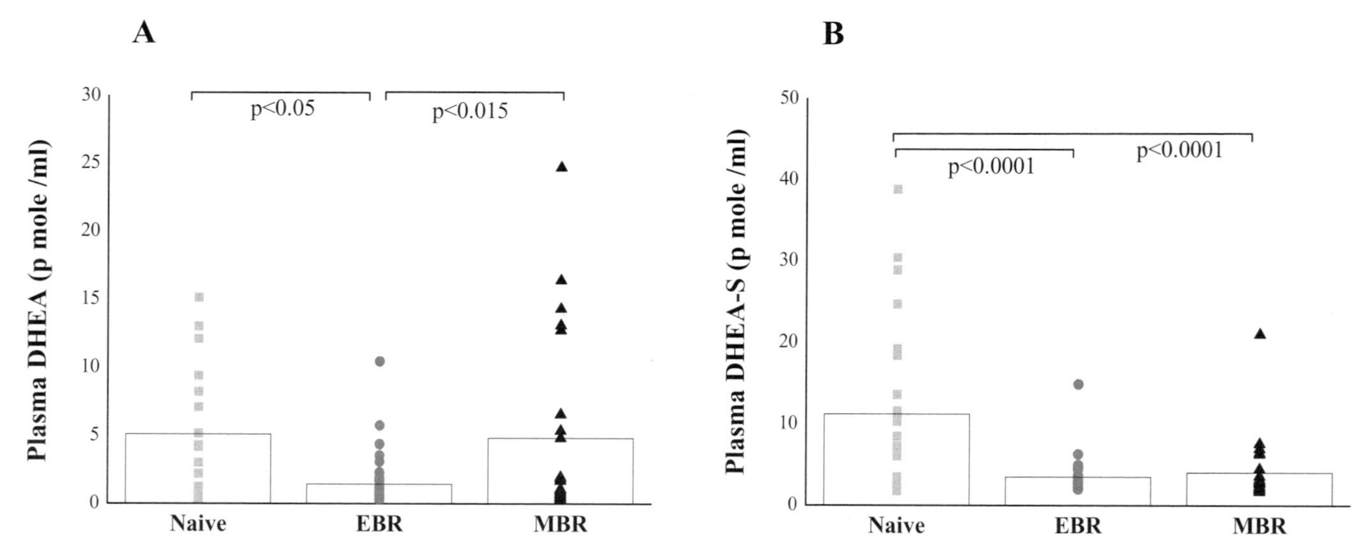

Figure 61–6. Circulating levels of (A) DHEA and (B) DHEA-S at day 7 post- PSS exposure. The plasma DHEA levels of EBR rats were significantly lower than in the MBR group (Bonferroni test $p < 0.02$) and in unexposed controls (Bonferroni test $p < 0.016$). There was no significant difference between the MBR and control groups (A). Plasma levels of DHEA-S in exposed animals, both EBR and MBR, were significantly lower than in controls (Bonferroni test $p < 0.0001$, $p < 0.0001$, respectively) (B). There was no significant difference between the DHEA-S levels of the EBR and MBR groups.

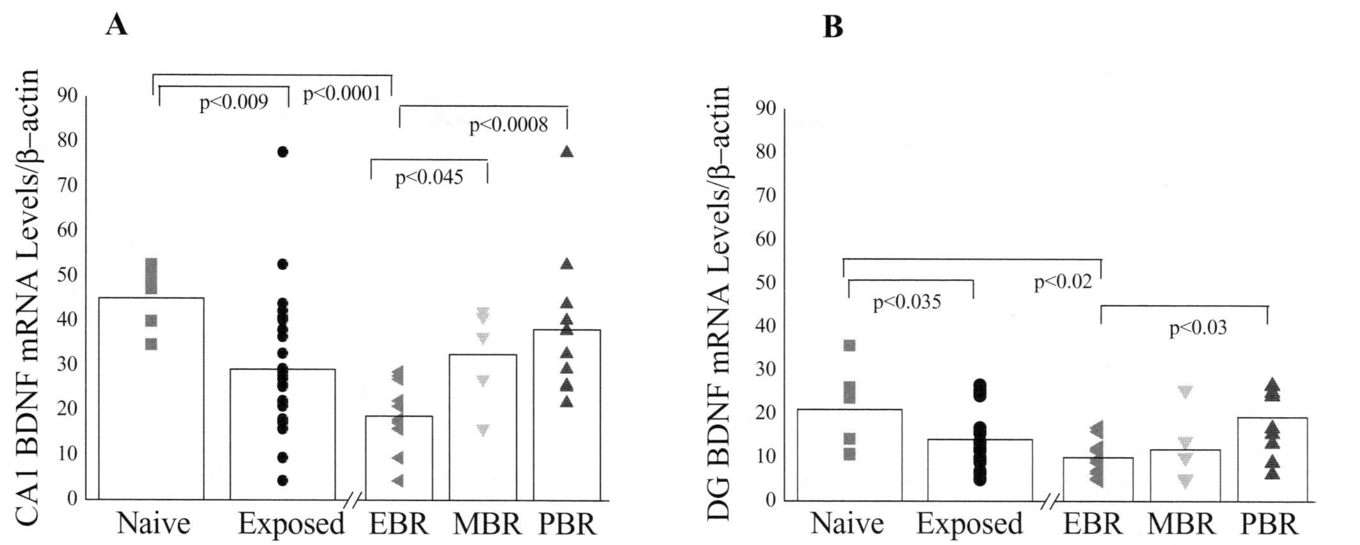

Figure 61–7. Hippocampal BDNF mRNA levels at day 7 post-PSS exposure. Exposed versus unexposed populations: the PSS-exposed population demonstrated significantly decreased CA1 (A) and DG (B) mRNA BDNF levels compared to controls [$F(1,34) = 7.8$, $p < 0.009$, and $F(1,34) = 4.1$, $p < 0.05$, respectively]. Reanalysis of data according to CBCs: in the hippocampal subregions CA1 (A) and DG (B), there were significant differences between groups [$F(3,32) = 10.35$, $p < 0.0001$, and $F(3,32) = 5.0$, $p < 0.007$, respectively]. The post hoc Bonferroni test revealed that stress exposure significantly decreased CA1 BDNF mRNA levels in EBR individuals as compared to controls ($p < 0.0001$) and as compared to MBR ($p < 0.045$) and to PBR animals ($p < 0.0008$). In the DG area the Bonferroni test revealed that PSS exposure significantly decreased BDNF mRNA levels in EBR individuals as compared to controls ($p < 0.02$) and to PBR animals ($p < 0.03$). There were no significant differences between EBR and MBR individuals in the DG subregion.[76]

demonstrated that compared with wild-type mice, BDNF conditional mutants (in which BDNF has been eliminated from the brain after birth through the use of the cre-loxP recombination system) were hyperactive after exposure to stressors and had higher anxiety levels when evaluated using the light–dark exploration paradigm.[74] On the other hand, single intrahippocampal administration of BDNF improved performance in a spatial memory task and had enduring anxiolytic effects.[75]

This experiment set out to examine the relationship between prevalence rates of distinct patterns of behavioral responses to predator stress, local levels of mRNA for BDNF, TrkB, and two other neurotrophic factors in selected brain areas (Figures 61–7 and 61–8).

The EBR animals exhibited significantly lower BDNF mRNA levels and higher TrkB mRNA in the CA1 and DG hippocampal subregions at day 7 as compared to naive unexposed rats and to the PBR group. In the CA1 subregion the EBR individuals exhibited significantly lower BDNF mRNA levels as compared to MBR or PBR animals. In other words, long-term downregulation of BDNF mRNA in the rat hippocampal CA1 subregion correlates with a PTSD-like behavioral stress response. This may represent a physiological marker characteristic of animals whose behavior is most severely affected by exposure to the stressor.[76]

Going beyond the mRNA expression, BDNF protein levels were assessed in the affected brain regions of behaviorally EBR individuals. The EBR animals exhibited significantly lower BDNF protein levels in the CA1 and DG hippocampal subregions at day 7 as compared to naive unexposed rats, concomitant with BDNF mRNA levels.[76]

Long-term BDNF deficiency in the hippocampus could have physiological consequences, inducing damage to hippocampal neurons. Neurotrophins, and particularly BDNF, are known to modulate many aspects of neuronal plasticity[77,78] and the selection of functional neuronal connections in the CNS.[79–81] Decreased expression of BDNF mRNA in the EBR group may decrease synaptic plasticity and the stabilization of synaptic connectivity, causing vulnerability to psychopathology. Interestingly, endogenous BDNF is required for the memory consolidation process through dual effects on postsynaptic gene expression (Arc) and local protein synthesis. Lower levels of BDNF after stress exposure may be involved in the anomalous traumatic memory consolidation process thought to occur in PTSD patients.[82] Because the exact mechanism of stress-induced BDNF downregulation is unclear and the expression of BDNF is regulated by various neurotransmitters, further studies would be necessary to unambiguously establish such a relationship.[76]

The consistent association between downregulation of BDNF in selected subregions of the hippocampus (CA1 and DG) and extreme behavioral responses to stress exposure may imply either a pathogenic role for BDNF and/or a physiological marker.

At 7 days the downregulation was found in both the CA1 and DG subregions and remained consistent over the entire 30-day trial period, specifically in the CA1 subregion and specifically in EBR animals. It thus extends beyond the scope of a transient change, which might be associated with maintenance of homeostasis, to represent a physiological characteristic of animals displaying severe chronic behavioral stress responses.

Taken together, a single 10-min exposure to PSS caused increased levels of "anxiety/PTSD"-related behaviors in EBR rats in the EPM and ASR paradigms 7 days later. However, the prevalence of EBR individuals in the exposed group was only 25.0%. These EBR rats exhibited significantly higher plasma corticosterone, and ACTH concentrations, lower plasma DHEA levels, lower hippocampal CA1 and DG BDNF mRNA, and higher

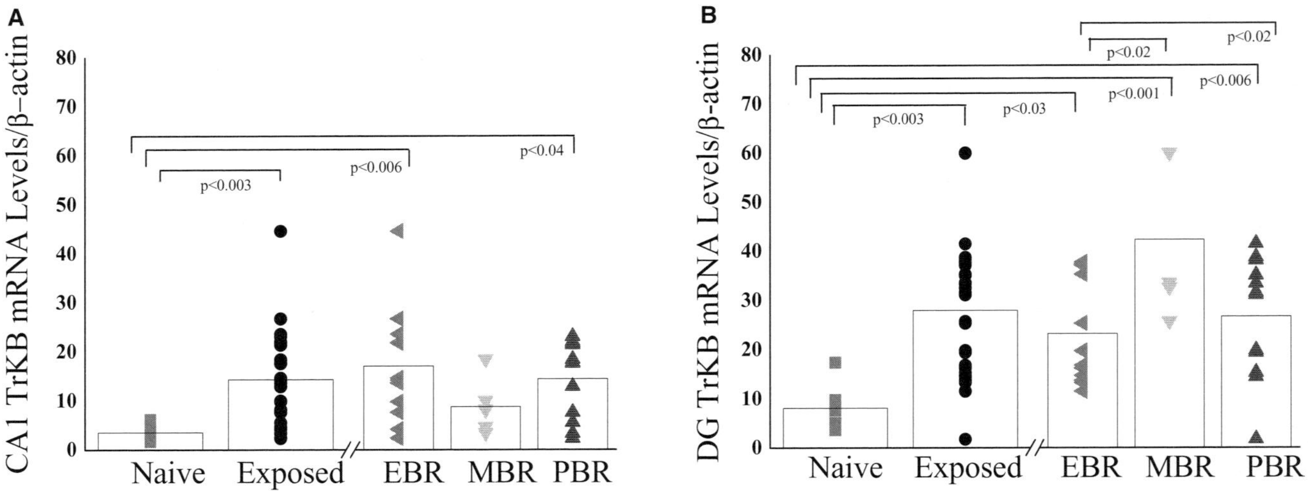

Figure 61–8. Hippocampal TrkB mRNA levels at day 7 post-PSS exposure. Exposed versus unexposed populations: the exposed population demonstrated significantly increased TrkB mRNA levels in CA1 (A) and DG (B) as compared to controls [$F(1,36) = 10.3$, $p < 0.003$, and $F(1,35) = 15.87$, $p < 0.0003$, respectively). Reanalysis of data according to CBCs: in the hippocampal CA1 (A) and DG (B) subregions, there were significant differences in TrkB mRNA levels between groups [$F(3,34) = 4.99$, $p < 0.006$, and $F(3,33) = 10.11$, $p < 0.0001$, respectively]. The post hoc Bonferroni test revealed that stress exposure significantly increased TrkB mRNA levels in the EBR groups as compared to controls ($p < 0.006$) in CA1. Moreover, TrkB mRNA levels were significantly higher in the PBR group as compared to controls ($p < 0.04$). The CA1 TrkB mRNA levels were not significantly different in the MBR group compared to controls. In the DG, the post hoc Bonferroni test revealed that stress exposure significantly increased TrkB mRNA levels in all the exposed groups as compared to controls (EBR: $p < 0.03$; MBR: $p < 0.001$; SC: $p < 0.006$). The TrkB mRNA levels were significantly higher in the MBR group as compared to the EBR group ($p < 0.02$).

hippocampal CA1 and DG TrkB mRNA. These changes were not found in MBR rats or in controls (not exposed), and therefore emphasize the importance of applying CBC (as an equivalent to the human "diagnostic criteria") in animal studies of PTSD.

The significant difference between accurate physiological parameters from EBR-exposed rats as compared to both MBR-exposed rats and controls is evidence for the construct validity of this proposed modification of animal models of stress and its sequelae.[76]

RISK FACTORS FOR DEVELOPING POSTTRAUMATIC STRESS DISORDER

Clinical studies of PTSD have elicited proposed risk factors for developing PTSD in the aftermath of stress exposure. Generally, these risk factors have arisen from retrospective analysis of premorbid characteristics of study populations. A valid animal model of PTSD can complement clinical studies and help to elucidate issues, such as the contribution of proposed risk factors, in ways that are not practical in the clinical arena.

EARLY LIFE STRESS AND SUBSEQUENT PSYCHOPATHOLOGY

Evidence suggests that early life stress is a major risk factor for the development and persistence of mental disorders such as mood disorders and PTSD.[83–87] Numerous studies in rodents have demonstrated substantial long-term biophysiological consequences of environmental manipulations in immature (preweaning) animals, in which the formation of neuroendocrine regulatory mechanisms has not been completed and associative neuronal circuits are not yet fully elaborated.[88–93] Juvenile (postweaning but prepubertal) animals have also been shown to demonstrate long-term behavioral consequences subsequent to early exposure to PTE's.[94]

Although most previous studies of early life trauma in animal models have focused on suckling pups, we focused on juvenile animals (postweaning, prepubertal), taken to represent "childhood" versus "infancy," i.e., a period in which a significant degree of plasticity in a rapidly maturing brain (and stress response system) might be found. Exposure to marked stress at this critical developmental phase in all likelihood has significant effects on later functioning, and possibly on the development of brain structures and their interactions.

ANIMAL MODEL FOR TRAUMA IN CHILDHOOD/YOUTH

Male Sprague–Dawley rats were randomly assigned to five groups: (1) unexposed control: not exposed to any adverse conditions at any time; (2) juvenile stress (JS): exposure to a predator scent (cat urine) at 28 days of age; (3) adulthood stress (AS): exposure to a predator scent in adulthood (9 weeks of age); (4) juvenile and adulthood stress (JS + AS): exposure to a predator scent both at 28 days and at 9 weeks; and (5) adulthood and adulthood stress (AS + AS): recurrent exposure to a predator scent at 9 weeks and again 30 days later.

The stress paradigm consisted of placing the rats on well-soiled cat litter for 15 min and the controls on unsoiled litter. Behavioral responses were assessed in the elevated plus maze and the acoustic startle response paradigm.[95,96]

The results demonstrate that juvenile trauma increases the vulnerability for developing subsequent long-term behavioral

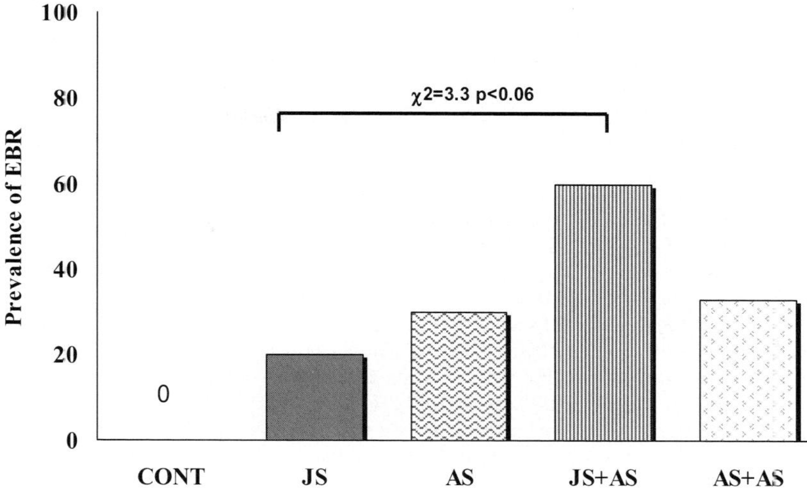

Figure 61–9. Effect of single exposure to predator scent early in life on the prevalence of extreme behavioral response. Juvenile exposure to predator scent stress significantly affected the response to reexposure in adulthood: 60% of the rats exposed to predator scent both at 28 days and at 9 weeks (JS + AS) demonstrated extreme behavioral responses, compared with 20–25% of those exposed only once either as juveniles (JS) or as adults (AS), and none in the control-unexposed group. Within the unexposed control group all rats were categorized as "unaffected" (MBR), and no rats displayed anxiety-like or depressive-like behaviors.[96]

disruptions, taken to represent posttraumatic stress symptom equivalents, after a second exposure to the same stressor in adulthood in a prospective study design (Figure 61–9). They thus serve to reassert retrospective findings from clinical studies.[84,85] This model may thus serve as a valid platform for further studies of childhood trauma.[95,96]

CONCLUSIONS

Animal models can help to elucidate certain aspects of the pathogenesis of a complex psychophysiological disorder such as PTSD in a prospective study design. Individual variability in pattern or degree of response to the stressor can be utilized to achieve greater validity of the model to the disorder it is intended to represent.

It could also help to address the debate as to whether PTSD is a disorder that reflects the upper end of a continuum of normal reactions to stress, as proposed by Ruscio *et al.*,[97] or whether it is a distinct entity intimately related to, but not on a direct continuum, with normal responses to stress.[37] So far, the distribution of the data is clustered at either end of the slope, suggesting a biphasic tendency of distribution. Results from studies of large populations will be needed to better address a qualitative or a quantitative linear approach to PTSD. However, the data presented suggest that it is important to perform a separate analysis for EBR- and MBR-exposed rats, since by means of this approach it is possible to highlight the range of individual differences in response to stress, whether they are qualitative or quantitative.

This animal model approach goes beyond accepting individual variability as an "unavoidable fact," but rather regards it with respect and focuses on individual variability as a useful source of knowledge. In this sense it may well represent an example of a more generalizable principle: individual variability in responses of participants in animal studies should be reexamined as a potential source for valuable information, and not disregarded or viewed merely as a source of unavoidable bias, throughout the scientific community.

REFERENCES

1. American Psychiatric Association. *Diagnostic and Statistical Manual of Mental Disorders*, 4th ed. Washington, DC: American Psychiatric Association, 1994.
2. Bryant RA. Recovery after the tsunami: Timeline for rehabilitation. *J Clin Psychiatry* 2006;67(Suppl. 2):50–55.
3. Connor KM, Foa EB, Davidson JR. Practical assessment and evaluation of mental health problems following a mass disaster. *J Clin Psychiatry* 2006;67(Suppl. 2):26–33.
4. Davidson JR, Stein DJ, Shalev AY, Yehuda R. Posttraumatic stress disorder: Acquisition, recognition, course, and treatment. *J Neuropsychiatry Clin Neurosci* 2004;16(2):135–147.
5. Davidson JR. Pharmacologic treatment of acute and chronic stress following trauma: 2006. *J Clin Psychiatry* 2006;67(Suppl. 2): 34–39.
6. Foa EB, Stein DJ, McFarlane AC. Symptomatology and psychopathology of mental health problems after disaster. *J Clin Psychiatry* 2006;67(Suppl 2):15–25.
7. Shalev AY. Acute stress reactions in adults. *Biol Psychiatry* 2002; 51(7):532–543.
8. Kessler RC, Sonnega A, Bromet E, Hughes M, Nelson CB. Posttraumatic stress disorder in the National Comorbidity Survey. *Arch Gen Psychiatry* 1995;52(12):1048–1060.
9. Breslau N, Davis GC, Andreski P, Peterson E. Traumatic events and posttraumatic stress disorder in an urban population of young adults. *Arch Gen Psychiatry* 1991;48(3):216–222.
10. Breslau N, Kessler RC, Chilcoat HD, Schultz LR, Davis GC, Andreski P. Trauma and posttraumatic stress disorder in the community: The 1996 Detroit Area Survey of Trauma. *Arch Gen Psychiatry* 1998;55(7):626–632.
11. Helzer JE, Robins LN, McEvoy L. Post-traumatic stress disorder in the general population. Findings of the epidemiologic catchment area survey. *N Engl J Med* 1987;317(26):1630–1634.
12. Perkonigg A, Kessler RC, Storz S, Wittchen HU. Traumatic events and post-traumatic stress disorder in the community: Prevalence, risk factors and comorbidity. *Acta Psychiatr Scand* 2000;101(1):46–59.
13. Resnick HS, Kilpatrick DG, Dansky BS, Saunders BE, Best CL. Prevalence of civilian trauma and posttraumatic stress disorder in a representative national sample of women. *J Consult Clin Psychol* 1993;61(6):984–991.
14. Shore JH, Vollmer WM, Tatum EL. Community patterns of posttraumatic stress disorders. *J Nerv Ment Dis* 1989;177(11):681–685.
15. Fairbank JA, Schlenger WE, Saigh PA, Davidson JRT. An epidemiologic profile of post-traumatic stress disorder: Prevalence, comorbidity, and risk factors. In: Charney DS, Friedman MJ, Deutch AY. Eds. *Neurobiological and Clinical Consequences of Stress: From Normal Adaptation to PTSD*. Philadelphia, PA: Lippincott-Raven, 1995:415–427.
16. Shalev AY. Measuring outcome in posttraumatic stress disorder. *J Clin Psychiatry* 2000;61(Suppl. 5):33–39; discussion 40–42.

17. Solomon Z, Shklar R, Mikulincer M. Frontline treatment of combat stress reaction: A 20-year longitudinal evaluation study. *Am J Psychiatry* 2005;162(12):2309–2314.
18. Pynoos RS, Ritzmann RF, Steinberg AM, Goenjian A, Prisecaru I. A behavioral animal model of posttraumatic stress disorder featuring repeated exposure to situational reminders. *Biol Psychiatry* 1996;39(2):129–134.
19. Servatius RJ, Ottenweller JE, Natelson BH. Delayed startle sensitization distinguishes rats exposed to one or three stress sessions: Further evidence toward an animal model of PTSD. *Biol Psychiatry* 1995;38(8):539–546.
20. Stam R, Bruijnzeel AW, Wiegant VM. Long-lasting stress sensitisation. *Eur J Pharmacol* 2000;405(1–3):217–224.
21. Richter-Levin G. Acute and long-term behavioral correlates of underwater trauma—potential relevance to stress and post-stress syndromes. *Psychiatry Res* 1998;79(1):73–83.
22. Wang J, Akirav I, Richter-Levin G. Short-term behavioral and electrophysiological consequences of underwater trauma. *Physiol Behav* 2000;70(3–4):327–332.
23. Adamec RE, Shallow T. Lasting effects on rodent anxiety of a single exposure to a cat. *Physiol Behav* 1993;54(1):101–109.
24. Adamec RE, Shallow T, Budgell J. Blockade of CCK(B) but not CCK(A) receptors before and after the stress of predator exposure prevents lasting increases in anxiety-like behavior: Implications for anxiety associated with posttraumatic stress disorder. *Behav Neurosci* 1997;111(2):435–449.
25. Adamec R. Transmitter systems involved in neural plasticity underlying increased anxiety and defense—implications for understanding anxiety following traumatic stress. *Neurosci Biobehav Rev* 1997;21(6):755–765.
26. Adamec R, Kent P, Anisman H, Shallow T, Merali Z. Neural plasticity, neuropeptides and anxiety in animals—implications for understanding and treating affective disorder following traumatic stress in humans. *Neurosci Biobehav Rev* 1998;23(2):301–318.
27. Adamec RE, Burton P, Shallow T, Budgell J. NMDA receptors mediate lasting increases in anxiety-like behavior produced by the stress of predator exposure—implications for anxiety associated with posttraumatic stress disorder. *Physiol Behav* 1999;65(4–5):723–737.
28. Cohen H, Friedberg S, Mater M, Kotler M, Zeev K. Interaction of CCK-4 induced anxiety and post-cat exposure anxiety in rats. *Depress Anxiety* 1996;4(3):144–145.
29. Cohen H, Kaplan Z, Kotler M. CCK-antagonists in a rat exposed to acute stress: Implication for anxiety associated with post-traumatic stress disorder. *Depress Anxiety* 1999;10(1):8–17.
30. Cohen H, Benjamin J, Kaplan Z, Kotler M. Administration of high-dose ketoconazole, an inhibitor of steroid synthesis, prevents post-traumatic anxiety in an animal model. *Eur Neuropsychopharmacol* 2000;10(6):429–435.
31. File SE. The interplay of learning and anxiety in the elevated plus-maze. *Behav Brain Res* 1993;58(1–2):199–202.
32. Blanchard RJ, Blanchard DC, Rodgers J, Weiss SM. The characterization and modelling of antipredator defensive behavior. *Neurosci Biobehav Rev* 1990;14(4):463–472.
33. Blanchard RJ, Blanchard DC, Weiss SM, Meyer S. The effects of ethanol and diazepam on reactions to predatory odors. *Pharmacol Biochem Behav* 1990;35(4):775–780.
34. Blanchard RJ, Shepherd JK, Rodgers RJ, Magee L, Blanchard DC. Attenuation of antipredator defensive behavior in rats following chronic treatment with imipramine. *Psychopharmacology* 1993;110(1–2):245–253.
35. Blanchard RJ, Nikulina JN, Sakai RR, McKittrick C, McEwen B, Blanchard DC. Behavioral and endocrine change following chronic predatory stress. *Physiol Behav* 1998;63(4):561–569.
36. Cohen H, Joseph Z, Matar M. The relevance of differential response to trauma in an animal model of post-traumatic stress disorder. *Biol Psychiatry* 2003;53(6):463–473.
37. Yehuda R, Antelman SM. Criteria for rationally evaluating animal models of posttraumatic stress disorder. *Biol Psychiatry* 1993;33(7):479–486.
38. Bezdek JC, Castelaz PF. Prototype classification and feature selection with fuzzy sets. *IEEE Trans SMC-7* 1971:87–92.
39. Bezdek JC, Pal NR. Some new validity indexes of cluster validity. *IEEE Trans Syst Man Cybern B Cybern* 1998;28(3):301–315.
40. Gath I, Geva AB. Fuzzy clustering for the estimation of the parameters of the components of mixtures of normal distributions. *Pattern Recogn Lett* 1989;9(3):77–86.
41. Pal NR, Bezdek JC. On cluster validity for the fuzzy C-means model. *IEEE Trans Fuzzy Syst* 1995;3(3):370–379.
42. Rezaee MR, Lelieveldt BPF, Reiber JHC. A new cluster validity for the fuzzy c-mean. *Pattern Recogn Lett* 1998;19(3–4):237–246.
43. Cohen H, Zohar J, Matar MA, Kaplan Z, Geva AB. Unsupervised fuzzy clustering analysis supports behavioral cutoff criteria in an animal model of posttraumatic stress disorder. *Biol Psychiatry* 2005;58(8):640–651.
44. Cohen H, Zohar J, Matar MA, Zeev K, Loewenthal U, Richter-Levin G. Setting apart the affected: The use of behavioral criteria in animal models of post traumatic stress disorder. *Neuropsychopharmacology* 2004;29(11):1962–1970.
45. Yehuda R, McFarlane AC, Shalev AY. Predicting the development of posttraumatic stress disorder from the acute response to a traumatic event. *Biol Psychiatry* 1998;44(12):1305–1313.
46. Lipton JM, Catania A. Mechanisms of antiinflammatory action of the neuroimmunomodulatory peptide alpha-MSH. *Ann NY Acad Sci* 1998;840:373–380.
47. Jessop DS. Beta-endorphin in the immune system—mediator of pain and stress? *Lancet* 1998;351(9119):1828–1829.
48. Cohen H, Zohar J. Animal models of post traumatic stress disorder: The use of cut off behavioral criteria. *Ann NY Acad Sci* 2004;1032:167–178.
49. Yehuda R, Giller EL, Southwick SM, Lowy MT, Mason JW. Hypothalamic-pituitary-adrenal dysfunction in posttraumatic stress disorder. *Biol Psychiatry* 1991;30(10):1031–1048.
50. Yehuda R, Kahana B, Binder-Brynes K, Southwick SM, Mason JW, Giller EL. Low urinary cortisol excretion in Holocaust survivors with posttraumatic stress disorder. *Am J Psychiatry* 1995;152(7):982–986.
51. Mason JW, Giller EL, Kosten TR, Ostroff RB, Podd L. Urinary free-cortisol levels in posttraumatic stress disorder patients. *J Nerv Ment Dis* 1986;174(3):145–149.
52. Mason JW, Giller EL, Kosten TR, Harkness L. Elevation of urinary norepinephrine/cortisol ratio in posttraumatic stress disorder. *J Nerv Ment Dis* 1988;176(8):498–502.
53. Morgan CA 3rd, Wang S, Mason J, *et al.* Hormone profiles in humans experiencing military survival training. *Biol Psychiatry* 2000;47(10):891–901.
54. Robel P, Baulieu EE. Dehydroepiandrosterone (DHEA) is a neuroactive neurosteroid. *Ann NY Acad Sci* 1995;774:82–110.
55. Baulieu EE, Robel P. Dehydroepiandrosterone (DHEA) and dehydroepiandrosterone sulfate (DHEAS) as neuroactive neurosteroids. *Proc Natl Acad Sci USA* 1998;95(8):4089–4091.
56. Baulieu EE, Robel P, Schumacher M. Neurosteroids: Beginning of the story. *Int Rev Neurobiol* 2001;46:1–32.
57. Stoffel-Wagner B. Neurosteroid biosynthesis in the human brain and its clinical implications. *Ann NY Acad Sci* 2003;1007:64–78.
58. Bastianetto S, Ramassamy C, Poirier J, Quirion R. Dehydroepiandrosterone (DHEA) protects hippocampal cells from oxidative stress-induced damage. *Brain Res Mol Brain Res* 1999;66(1–2):35–41.
59. Cohen H, Maayan R, Touati-Werner D, *et al.* Decreased circulatory levels of neuroactive steroids in behaviorally more extremely affected rats subsequent to exposure to a potentially traumatic experience. *Int J Neuropsychopharmacol* 2006;4:1–7.
60. Browne ES, Wright BE, Porter JR, Svec F. Dehydroepiandrosterone: Antiglucocorticoid action in mice. *Am J Med Sci* 1992;303(6):366–371.
61. Maayan R, Morad O, Dorfman P, Overstreet DH, Weizman A, Yadid G. The involvement of dehydroepiandrosterone (DHEA) and its sulfate ester (DHEAS) in blocking the therapeutic effect of electro-

convulsive shocks in an animal model of depression. *Eur Neuropsychopharmacol* 2005;15(3):253–262.
62. Abraham WC, Tate WP. Metaplasticity: A new vista across the field of synaptic plasticity. *Prog Neurobiol* 1997;52(4):303–323.
63. Hyman C, Hofer M, Barde YA, *et al.* BDNF is a neurotrophic factor for dopaminergic neurons of the substantia nigra. *Nature* 1991; 350(6315):230–232.
64. Lom B, Cohen-Cory S. Brain-derived neurotrophic factor differentially regulates retinal ganglion cell dendritic and axonal arborization in vivo. *J Neurosci* 1999;19(22):9928–9938.
65. McAllister AK. Subplate neurons: A missing link among neurotrophins, activity, and ocular dominance plasticity? *Proc Natl Acad Sci USA* 1999;96(24):13600–13602.
66. Shimada A, Mason CA, Morrison ME. TrkB signaling modulates spine density and morphology independent of dendrite structure in cultured neonatal Purkinje cells. *J Neurosci* 1998;18(21):8559–8570.
67. Yacoubian TA, Lo DC. Truncated and full-length TrkB receptors regulate distinct modes of dendritic growth. *Nat Neurosci* 2000;3(4): 342–349.
68. Rutherford LC, Nelson SB, Turrigiano GG. BDNF has opposite effects on the quantal amplitude of pyramidal neuron and interneuron excitatory synapses. *Neuron* 1998;21(3):521–530.
69. Vicario-Abejon C, Collin C, McKay RD, Segal M. Neurotrophins induce formation of functional excitatory and inhibitory synapses between cultured hippocampal neurons. *J Neurosci* 1998;18(18):7256–7271.
70. Seil FJ, Drake-Baumann R. TrkB receptor ligands promote activity-dependent inhibitory synaptogenesis. *J Neurosci* 2000;20(14):5367–5373.
71. Duman RS. Synaptic plasticity and mood disorders. *Mol Psychiatry* 2002;7(Suppl. 1):S29–34.
72. Duman RS. Role of neurotrophic factors in the etiology and treatment of mood disorders. *Neuromolecular Med* 2004;5(1):11–25.
73. Manji HK, Duman RS. Impairments of neuroplasticity and cellular resilience in severe mood disorders: Implications for the development of novel therapeutics. *Psychopharmacol Bull* 2001;35(2):5–49.
74. Rios M, Fan G, Fekete C, *et al.* Conditional deletion of brain-derived neurotrophic factor in the postnatal brain leads to obesity and hyperactivity. *Mol Endocrinol* 2001;15(10):1748–1757.
75. Cirulli F, Berry A, Chiarotti F, Alleva E. Intrahippocampal administration of BDNF in adult rats affects short-term behavioral plasticity in the Morris water maze and performance in the elevated plus-maze. *Hippocampus* 2004;14(7):802–807.
76. Kozlovsky N, Matar MA, Kaplan Z, Kotler M, Zohar J, Cohen H. Long-term down-regulation of BDNF mRNA in rat hippocampal CA1 subregion correlates with PTSD-like behavioural stress response. *Int J Neuropsychopharmacol* 2007;Feb 12:1–18.
77. Shieh PB, Ghosh A. Molecular mechanisms underlying activity-dependent regulation of BDNF expression. *J Neurobiol* 1999;41(1): 127–134.
78. Thoenen H. Neurotrophins and activity-dependent plasticity. *Prog Brain Res* 2000;128:183–191.
79. Poo MM. Neurotrophins as synaptic modulators. *Nat Rev Neurosci* 2001;2(1):24–32.
80. Huang EJ, Reichardt LF. Neurotrophins: Roles in neuronal development and function. *Annu Rev Neurosci* 2001;24:677–736.
81. Mamounas LA, Altar CA, Blue ME, Kaplan DR, Tessarollo L, Lyons WE. BDNF promotes the regenerative sprouting, but not survival, of injured serotonergic axons in the adult rat brain. *J Neurosci* 2000;20(2):771–782.
82. van Praag HM. The cognitive paradox in posttraumatic stress disorder: A hypothesis. *Prog Neuropsychopharmacol Biol Psychiatry* 2004;28(6):923–935.
83. Ford JD, Kidd P. Early childhood trauma and disorders of extreme stress as predictors of treatment outcome with chronic posttraumatic stress disorder. *J Trauma Stress* 1998;11(4):743–761.
84. Heim C, Nemeroff CB. The role of childhood trauma in the neurobiology of mood and anxiety disorders: Preclinical and clinical studies. *Biol Psychiatry* 2001;49(12):1023–1039.
85. Heim C, Newport DJ, Wagner D, Wilcox MM, Miller AH, Nemeroff CB. The role of early adverse experience and adulthood stress in the prediction of neuroendocrine stress reactivity in women: A multiple regression analysis. *Depress Anxiety* 2002;15(3) 117–125.
86. Breslau N. Psychiatric morbidity in adult survivors of childhood trauma. *Semin Clin Neuropsychiatry* 2002;7(2):80–88.
87. Shea A, Walsh C, Macmillan H, Steiner M. Child maltreatment and HPA axis dysregulation: Relationship to major depressive disorder and post traumatic stress disorder in females. *Psychoneuroendocrinology* 2005;30(2):162–178.
88. Francis DD, Caldji C, Champagne F, Plotsky PM, Meaney MJ. The role of corticotropin-releasing factor—norepinephrine systems in mediating the effects of early experience on the development of behavioral and endocrine responses to stress. *Biol Psychiatry* 1999;46(9):1153–1166.
89. Caldji C, Diorio J, Meaney MJ. Variations in maternal care in infancy regulate the development of stress reactivity. *Biol Psychiatry* 2000;48(12):1164–1174.
90. Heidbreder CA, Weiss IC, Domeney AM, *et al.* Behavioral, neurochemical and endocrinological characterization of the early social isolation syndrome. *Neuroscience* 2000;100(4):749–768.
91. Goldberg S, Levitan R, Leung E, *et al.* Cortisol concentrations in 12- to 18-month-old infants: Stability over time, location, and stressor. *Biol Psychiatry* 2003;54(7):719–726.
92. Maslova LN, Bulygina VV, Markel AL. Chronic stress during prepubertal development: Immediate and long-lasting effects on arterial blood pressure and anxiety-related behavior. *Psychoneuroendocrinology* 2002;27(5):549–561.
93. Maslova LN, Bulygina VV, Popova NK. Immediate and long-lasting effects of chronic stress in the prepubertal age on the startle reflex. *Physiol Behav* 2002;75(1–2):217–225.
94. Avital A, Richter-Levin G. Exposure to juvenile stress exacerbates the behavioural consequences of exposure to stress in the adult rat. *Int J Neuropsychopharmacol* 2005;8(2):163–173.
95. Tsoory M, Cohen H, Richter-Levin G. Juvenile stress induces a predisposition to either anxiety or depressive-like symptoms following stress in adulthood. *Eur Neuropsychopharmacol* 2006;17(4): 245–256.
96. Cohen H, Kaplan Z, Matar AM, Loewenthal U, Zohar J, Richter-Levin G. Long-lasting behavioral effects of juvenile trauma in an animal model of PTSD associated with a failure of the autonomic nervous system to recover. *Eur Neuropsychopharmacol* 2007; 17(6–7):464–477.
97. Ruscio AM, Ruscio J, Keane TM. The latent structure of posttraumatic stress disorder: A taxometric investigation of reactions to extreme stress. *J Abnorm Psychol* 2002;111(2):290–301.

62 Animal Models for Studying Fetal Alcohol Syndrome, Alcohol-Related Birth Defects, and Alcohol-Related Neurodevelopmental Disorder

TIMOTHY A. CUDD

ABSTRACT

Maternal alcohol abuse during gestation can result in a collection of fetal injuries, the most common of which can be broadly categorized as (1) intrauterine growth retardation, (2) craniofacial dysmorphology, and (3) central nervous system damage. These injuries are referred to by the terms (depending on specific conditions and manifestations) fetal alcohol syndrome, alcohol-related birth defects, alcohol-related neurodevelopmental disorder, and others. While the manifestations of injuries from maternal alcohol abuse during gestation have been described, there remain many unanswered questions: what are the specific mechanisms by which alcohol mediates these injuries, which women are at greatest risk, how can we identify affected children at an early age, and how might we intervene to prevent or mitigate the damage? Because of the inherent limitations of studies conducted in humans, the use of animal models is important in addressing these unanswered questions. However, no one animal species can be utilized to model all aspects of human prenatal alcohol exposure injury as no species is identical to the human with respect to development, structure, and function. Therefore, performing studies modeling human prenatal alcohol injury in animals requires careful consideration of the particular animal model and whether or not this species, studied under these specific experimental conditions, does model what occurs in humans. This chapter summarizes some of the important considerations in choosing an animal model, in general and in the context of animal models that have been utilized to study human prenatal alcohol exposure.

Key Words: Fetal alcohol syndrome, Alcohol-related neurodevelopment disorder, Alcohol-related birth defects, Teratogen, Ethanol.

INTRODUCTION

Three decades have passed since fetal alcohol syndrome (FAS) was first identified as a clinical syndrome,[1] and yet we still know surprisingly little about the range of long-term damage resulting from variations in gestational alcohol abuse and we know even less about the exact causes of or how to treat these deficits. Notwithstanding the documented clinical literature, much of the knowledge about the specific effects of gestational alcohol exposure on fetal brain, body, organ, and behavioral development has been derived from the effective use of animal models.

Human studies are handicapped in their ability to address some of the key issues related to FAS and other fetal alcohol exposure-induced damage because of variable consumption patterns of alcohol, unreliable self-estimates of alcohol intake by pregnant women, and because many women who abuse alcohol also abuse other drugs.[2] Furthermore, ethical considerations place considerable constraints on the use of human subjects to study the teratogenic impact of alcohol. The use of animal models circumvents many of the problems inherent in human studies. One major advantage of using animal models is that they make it possible to control important maternal and environmental variables such as genetic background, nutritional status of the mother, and dose and timing of the alcohol exposure, and also allow for questions to be addressed at a more mechanistic level.[3,4]

Fortunately, there is good correspondence between findings in human and animal studies,[5] which helps to validate the use of animals in FAS research. Animal studies have demonstrated that heavy alcohol exposure during development can produce deleterious effects in most of the same brain regions and organ systems that have been reported in humans with FAS. In addition, animal studies have expanded upon the gross morphological studies of humans with FAS by identifying deficits at the cellular (neuronal numbers) level in specific brain regions and alterations in the cellular morphology[6–11] and function[12] of neurons that survive the alcohol insult, as well as alterations in glia and myelin.[13–17]

ATTENTION TO DEFINITIONS

In this chapter, we will limit the discussion to *in vivo* whole animal models used to investigate the cardinal features that define FAS. The definition of FAS has remained essentially unchanged since it was originally defined in the 1970s: a characteristic pattern of facial anomalies, growth retardation, and various neurodevelopmental abnormalities.[1] A study by a special Institute of Medicine committee[18] added to that definition the criterion of a

From: *Sourcebook of Models for Biomedical Research*
(P. M. Conn, ed.), © 2008 Humana Press Inc., Totowa, NJ.

confirmed history of drinking during pregnancy. Over the years since the adaptation of the term FAS, a variety of other terms have emerged. Fortunately, an excellent review of these terms appeared recently.[19] One of the earliest was fetal alcohol effects (FAE), which was used to identify children damaged by prenatal alcohol exposure but who did not exhibit all of the FAS criteria.[20] A more encompassing term, alcohol-related birth defects (ARBD), was coined for individuals (and animals) with other less common organ anomalies who had been exposed to alcohol prenatally, but who did not have the characteristic FAS face. The FAS face, as defined by the Institute of Medicine, is a specific pattern of facial anomalies, including short palpebral fissures and abnormalities of the premaxillary zone (e.g., flat upper lip, flattened philtrum, flat midface).[18] A term that has gained popularity within the past decade to categorize behavioral and cognitive deficits, which constitute the most severe consequences of fetal alcohol exposure, is alcohol-related neurodevelopmental disorder (ARND).[18] Finally, there are terms such as "partial FAS" (which includes some but not all of the facial features and a confirmed maternal drinking history) and fetal alcohol spectrum disorder (FASD), which has been used to designate the complete spectrum of deficits attributed to *in utero* alcohol exposure.[21]

It is important to appreciate that the cardinal features of FAS (collectively) are not the result of alcohol exposure during a restricted temporal period and that it is unlikely that a single mechanism is involved. Therefore, it would be inappropriate to make an unqualified claim that an intervention prevents FAS in an animal model when the model exhibits only some of the features of FAS and when the alcohol exposure occurred over a restricted portion of gestation. Researchers using animal models must be very clear about the conclusions they draw from studies on components of FAS or parts of gestation. To be accurate, investigators should refrain from claiming to have conducted an investigation of FAS unless the particular animal model exhibits all of the features used in the definition of FAS in humans, including the facial dysmorphology. By definition, this would exclude most animal studies.

ESSENTIAL ISSUES IN THE USE OF ANIMAL MODELS

All animal models that have been used to study prenatal alcohol exposure have specific advantages (and disadvantages). The advantages of a particular animal model must result in answering the research question at hand. A desirable quality of any study using an animal model includes the ease of extrapolating the results to the human condition. On the other hand, some model systems may be chosen because they are very dissimilar to the human condition, but when used in combination with other model systems, can be used to answer important questions. However, at some point there must be some verification that findings from experiments using an animal model that is very dissimilar are relevant to humans.

Ideally, an animal model should be one in which alcohol is imbibed voluntarily and preferentially, as in humans, and where the model expresses all of the features of FAS and where development is similar to that in humans. No such model has been found. Animal models that require forcibly exposing animals to alcohol create some level of stress in the subjects. In animal models that utilize an alcohol-containing liquid diet, the subjects have only one source of nutrition and thus do not voluntarily and preferentially consume alcohol. Thus, researchers must design experiments that control for stress. When it is necessary to utilize a route of alcohol administration other than oral, it is important to appreciate that faster or slower rise time might induce different mechanistic cascades to result in the cellular damage, thus confounding the interpretation of the mechanism of induction for the developmental defect.

It is very important to consider the timing of brain development relative to birth (i.e., how "developed" are the offspring when they are born compared with human babies) and how closely the animal model brain anatomy and functional capabilities match those of the human.[22] When the object is to develop a therapy, intervention, or treatment for alcohol-induced fetal brain injury, it will be crucial to know that the animal model is a functional homologue of human brain anatomy and function.

In studying fetal damage from prenatal alcohol, the real possibility of maternal and placental roles or interactions must be considered. On the other hand, important contributions have been made by using nonmammalian animal models to study the direct actions of alcohol on fetal development that do not involve the placenta and mother as factors.[23–26] These studies suggest that the placenta and mother are not required to create the specific damage being measured other than acting as the source of alcohol to the fetus. Nonmammalian animal models can be of considerable value in addressing specific research questions that cannot be addressed in mammals and in separating out maternal–placental and nonmaternal–placental contributions to the damage. However, it is important to remember that the human condition does involve maternal–placental involvement and these components may create interactions that cannot be appreciated in simple animal models. Furthermore, the questions being addressed in these simple models are specific to certain stages of development, patterns, and concentrations of alcohol exposure and it is necessary to remember that the full spectrum of fetal alcohol damage involves many different stages of development, patterns of exposure, and mechanisms of action.

There is no single "best" animal model for addressing all FAS, ARBD, and ARND questions. However, the key issues related to the research questions at hand must be addressed. For instance, if the focus is on questions associated with drinking patterns, then a model system that enables considerable control of the administration of the parameter of alcohol exposure is crucial. Sometimes this is simply a choice of technique (e.g., intragastric gavage versus liquid diet). At other times, it may be intrinsic to the model system (e.g., intravenous for the sheep; voluntary drinking for the miniature swine).

And finally, it should be emphatically stated that some studies are best performed in human subjects. For example, several large cohorts of children have been identified around the world who can be tested directly for specific kinds of cognitive and other functional impairments resulting from moderate to heavy *in utero* alcohol exposure. It can be argued that it makes little sense to perform an animal study and then try to extrapolate the results to humans if the question can be addressed more directly in humans in the first place, since it avoids the extremely problematic issues of extrapolating results from one species to another. Therefore, we argue that the most effective use of animal studies today will be to model important questions that for various reasons still cannot be conducted efficaciously in humans.

HISTORICAL PERSPECTIVES AND CURRENT STATUS

The initial animal studies that were designed after FAS was first identified and defined understandably had goals that were broad and relatively unfocused. For instance, substantial effort was made to determine whether alcohol was actually teratogenic in the absence of other factors such as polydrug use or undernutrition, and to catalog the spectrum of effects caused by *in utero* exposure to alcohol. It is important to remember that those studies were conducted in an environment in which few children with FAS had been identified and that many investigators in the teratology field were publicly skeptical that alcohol was really teratogenic. Therefore, those early animal studies were critically important in that they provided legitimacy to the claim of alcohol teratogenicity and helped to identify the ways that alcohol could adversely influence development. Because few affected children were available for experimental purposes, and then only on a limited basis, clinicians had to be especially prudent in their selection of studies. Early animal studies played an important role in helping to drive clinical research in the field. For instance, animal studies identified problems in learning and memory[27,28] and suggested that short-term memory might be particularly vulnerable to developmental alcohol exposure,[29] which in turn stimulated clinical researchers to ask whether short-term memory is compromised in children with FAS.[30,31]

The cardinal features of FAS that have been studied using animal models are (1) intrauterine growth retardation, (2) craniofacial dysmorphology, and (3) central nervous system (CNS) deficits. Intrauterine growth retardation (IUGR) has an identifiable endpoint: low birth weight, impaired intrauterine and early postnatal growth, or stunted growth. Much of the research on alcohol-induced IUGR has been performed in rats[32] and mice.[33] Yet body growth is affected by parturition,[34] and parturition in rats and mice occurs at roughly the brain growth velocity equivalent of the end of the second trimester in humans.[22] As such, the third trimester equivalent of fetal brain growth in rats and mice occurs after parturition. For that reason, it is necessary to be cautious about using these animal models to explore the interactions of alcohol and somatic growth if the primary goal is to use such information to prevent or treat alcohol-induced IUGR in humans.

The examination of alcohol-induced craniofacial dysmorphology is a good example of progress made in one area of fetal alcohol research primarily through the use of an animal model (see Mice below).

CNS deficits are by far the most diffusely defined of the three cardinal features of FAS and range from neuropathology to behavioral alterations. While considerable progress has been made in human studies, there is still a need for animal models. Animal models have been useful in answering questions about not only what deficits are created but also how alcohol produces CNS deficits. It is still difficult to generate reliable data about the amount and timing of alcohol consumption during pregnancy from women who gave birth to affected babies. It is unreasonable to expect women to remember precise details about their alcohol consumption over a year or more prior to delivery. Furthermore, many FAS children are given up for adoption or end up in foster homes. Thus, any pertinent information about the biological mother's drinking habits prior to or during her pregnancy is no longer available. The first animal model evidence that alcohol could kill developing neurons[35] was extremely important. Additional studies showed that neurons in other brain areas and at different times of development were (third trimester,[36,37] various trimesters[38]) or were not vulnerable (locus coeruleus,[39] neocortex,[40] ventrolateral nucleus of the thalamus[38]). CNS deficits have been measured in the following ways using animal models: deficits in neuron physiology,[12] neurotransmitter levels,[41,42] neurosteroid levels,[43] and neuroendocrine function,[44,45] alterations in DNA,[46] RNA,[47] and protein expression pattern,[48] small volume of brain for body size (microencephaly[49]), holoprosencephaly,[50] hyperactivity,[51] deficits in olfactory behavior[52] as well as in spatial learning and behavior,[53] deficits in operant learning,[54] and deficits in electroencephalogram (EEG) recordings,[55] a list that is not exhaustive.

These studies have brought the field to a point at which the further documentation of cell loss under slightly different experimental conditions is providing diminishing returns. That does not mean that these types of studies lack any importance, but they may provide little additional predictive value. On the other hand, there are studies of neuron loss that could make some substantial contribution beyond identifying yet just another case of cell loss. For instance, even 30 years after the first report of alcohol-induce Purkinje cell death,[35] little progress has been made in terms of which mode of cell death is involved.[56] Furthermore, because the animal models chosen to examine alcohol-induced Purkinje cell death (loss) were mostly early postnatal (when Purkinje cells are postmitotic) or *in vitro* explant cultures, it is not clear whether the modes of alcohol-induced cell death would be the same in an *in utero* third trimester model (guinea pig or sheep, for example) or earlier in development, prior to the third trimester equivalent. An interesting and important question is whether cell death of a particular neuron type (Purkinje cells, for instance) following *in utero* alcohol exposure[11] occurs via the same mechanism as occurs with early postnatal exposure. If the answer to this question reveals differences in the modes of cell death when comparing *in utero* and postnatal exposure then it would be much more important to pursue questions of how to prevent or reduce the damage using animal models where the exposure to alcohol occurs *in utero* as in humans. If instead, the mechanisms of cell death are found to be the same, then it would be an obvious advantage to pursue the questions using animal models that are smaller, cheaper, or have other advantages that would expedite the results (e.g., neonatal rats). Again, this example highlights the importance of choosing the correct animal model to address specific relevant questions that will generate specific answers rather than simply describing yet another feature of the effect of alcohol on brain development.

In summary, there is now a relatively broad understanding of the effects of alcohol on the developing conceptus. However, there is still a need for further development of behavioral dysfunction models to serve as specific tests for intervention and treatment. It is also important to determine the specific behaviors that are spared. Better knowledge of what is damaged and what is not can help our understanding of how the damage occurs. We are now positioned to utilize animal models to address questions concerning how alcohol damages the developing conceptus at the molecular level. Knowledge of the molecular and genetic basis of alcohol damage to the CNS will be crucial in developing effective strategies for intervention, treatment, and prevention.

MAMMALIAN ANIMAL MODELS

MICE (*MUS MUSCULUS*) The usefulness of the mouse as a model for alcohol research is a function of several factors, including its relatively short gestation time (19 days), the wealth of background information available on anatomy, physiology, and pharmacokinetics, and its established use in behavioral studies where there is information about brain function and certain behaviors. In addition to these advantages, genetic manipulations are well established, including transgenic mice, mouse lineages for high and low alcohol consumption and other alcohol-related behaviors, and reverse genetics. The potential usefulness of the mouse became even greater with the completion of the sequencing of the mouse genome. And already in existence are a large variety of inbred and knockout/in strains available for specific uses.

Some of the earliest studies using the mouse to examine the effects of alcohol on brain development found deficits in fetal and offspring growth[57] and offspring behavior following gestational alcohol exposure[58] and were interpreted as consistent with those found in fetal alcohol-exposed children. One important set of studies demonstrated that the craniofacial dysmorphology observed in children diagnosed with FAS could be induced in the mouse by alcohol administration at gastrulation.[59] In these studies, Sulik's research group[50,60] and that of Webster[61,62] showed that the C57BL/6J mouse was susceptible to facial dysmorphology only if the alcohol was given at gastrulation. The critical relationship between the timing of alcohol exposure and offspring deficit was one of the important issues addressed by the experiments using the fetal C57BL/6J mouse, in addition to the notable implication that a single binge-like exposure to alcohol during a narrow window of nervous system development could result in the induction of a definitive characteristic of fetal alcohol syndrome. These findings were important in establishing the mouse as a useful model system.

However, it is also important to appreciate the shortcomings of the mouse model. Not every pregnant mouse treated with a high dose of alcohol will have severely affected fetuses or, if treated with alcohol at gastrulation, will have fetuses with any discernible craniofacial dysmorphology. Furthermore, there is no evidence that the fetuses displaying a craniofacial dysmorphology at GD14 would survive past birth and no evidence that these fetuses would continue to display this deficit as development proceeded through adulthood. In addition, the brain development of the mouse is both prenatal and postnatal, meaning that in order to study the effects of alcohol during all three trimesters (equivalent to that of humans), the mouse mother must be treated from conception to parturition and the mouse offspring must be treated from birth to about postnatal day 10–12; that much of the third trimester equivalent of brain development is postnatal and that third trimester equivalent dosing does not involve the placenta and mother make relating these data back to humans problematic.

RATS (*RATTUS NORVEGICUS*) The laboratory rat is a popular choice of animal model in fetal alcohol research. Rats were used to provide descriptive data on the effects of alcohol before pregnancy,[63] before and during pregnancy,[64,65] and during pregnancy and lactation[66] on offspring development. More recently, rats have been used to link genes with alcohol-induced behavioral changes. There is a large amount of information available on the rat including basic anatomy, physiology, pharmacokinetics, and reproduction, including the significant availability of phenotypic data. The rat is also very useful for behavioral experiments due to its large size, relatively short gestation (22 days), unique intelligence, relatively long life span, and abundant amount of historical data from behavior studies.

Some of the earliest fetal alcohol studies using the rat showed both physical[7,67] and behavioral[68] alterations in the offspring of alcohol-exposed pregnant rats. These early descriptive studies were important because they showed a very strong correlation in findings between the rat and humans with FAS, and this helped to establish the rat as a suitable animal model for many basic research questions. As the FAS field progressed, however, it became clear that the rat was not the best model system for addressing all questions. While problems with alcohol administration methods[69,70] and issues related to timing of alcohol dosing[71] were overcome, others have persisted. One of the most obvious disadvantages is that brain development (relative to that of the human fetus) occurs both *in utero* and in the early postnatal period, much like the mouse. This means that although FAS is a consequence of heavy maternal drinking throughout pregnancy, most rat studies in the literature have modeled the equivalent of alcohol exposure only during the first two trimesters in humans. Thus, studies of temporal vulnerability[49] suffer from problems of extrapolation of the results to the human condition. Furthermore, the mouse is more often a better model than the rat when genetic manipulations and the creation of transgenic and gene knockin/out lines are needed (with some exceptions). Most rat strains lack genetic homogeneity, which does not allow alcohol–gene interactions to be suitably controlled. Even though efforts to create gene knockin and gene overexpression models in rats have suffered from problems associated with embryonic stem cell stages, naturally occurring mutations, gene-specific deletions, and breeding programs to create high and low alcohol-consuming rat strains[72,73] have been explored systematically. These alcohol-preferring strains have been useful in identifying specific deficits associated with gestational alcohol exposure.[74,75]

GUINEA PIG (*CAVIA PORCELLUS*) The guinea pig has a number of important advantages including its long gestation period, larger body size compared to rats or mice, and primarily the fact that the brain development of the fetal guinea pig is more advanced at birth than that of rodents. Therefore, the brain growth spurt, which occurs *in utero* in humans, also occurs *in utero* in the guinea pig, making it a favorable choice for fetal alcohol studies of temporal vulnerability. In a series of studies, researchers have discovered several interesting changes in brain development as a result of prenatal alcohol exposure in the guinea pig model system. For example, exposure of pregnant guinea pigs to oral alcohol throughout gestation results in offspring with reduced brain and body weights,[76] increased incidence of hyperactivity,[77] altered dendritic arborization of layer V somatosensory cortex pyramidal cells,[78] decreased numbers of hippocampal CA1 pyramidal neurons,[77] altered cortical γ-aminobutyric acid [GABA(A)] receptor expression,[79] and changes in the subunit composition of *N*-methyl-D-aspartate (NMDA) and α-amino-3-hydroxy-5-methyl-4-isoxazolepropionate (AMPA) receptors in the cortex of adult offspring.[80] Of particular interest is that a recent study found that the hyperactivity induced by prenatal alcohol exposure was not related to alcohol exposure during the brain growth spurt[81] since offspring exposed to alcohol only during the brain growth spurt exhibited no behavioral changes and no changes in hippocampal CA1, CA3, or dentate granule cell or cerebellum Purkinje cell density[82] but did exhibit brain growth deficits in the cortex, hippocampus, and cerebellum. Thus, the guinea pig has been used successfully to address questions about the temporal vulnerability of the brain to alcohol-induced injury without the confounding variable of prenatal versus postnatal alcohol exposure, which is always a concern with rat and mouse models. While the size of the guinea pig and the timing of its brain development relative to parturition are clear advantages over the mouse and rat for some aspects of fetal alcohol research, the current usefulness of the guinea pig animal model is limited due to several factors, including the greater expense of housing the animals compared to rats and mice and the absence of the genomic tools available in rats and mice.

SHEEP (*OVIS ARIES*) Sheep possess a number of major advantages for the study of FAS when compared to other species both large and small. First, the adult ewe and fetus are extremely tolerant of handling, surgery, and chronic instrumentation, features that facilitate long-term treatment and blood sampling at multiple time points within the same subject. Surgically placed indwelling vascular catheters can easily be maintained for many months in adult sheep, while extraordinary measures must be employed to maintain catheters chronically in most other animal models. Because of the robustness of the sheep fetus preparation, many basic physiology questions have been addressed using this model system. As a result, a large literature exists on the normal physiology of the fetal sheep, which precludes the need for exhaustive preliminary studies simply to determine normal values and to establish the animal preparation. A second major advantage of the sheep is its large body mass. The sheep fetus weighs between 0.85 and 4.50 kg during the third trimester equivalent (this wide weight range is due to the rapid growth during this period of gestation), while the adults will weigh between 50 and 75 kg, comparable with that of a human fetus and an adult woman, respectively. When the measurement of multiple dependent variables must be performed simultaneously and at multiple time points, thus requiring the collection of relatively large volumes of blood, an animal of significant body mass is required. Such experiments simply cannot be performed in small laboratory animals. A third major advantage of sheep is that this species has a much longer length of gestation with brain development, relative to humans, occurring entirely *in utero*, which allows the investigator more opportunity to intervene or perform experiments at specific times during gestation than when using most other species. However, maintaining research sheep is expensive and requires special housing and veterinary care, and the sheep model is less well established in terms of two of the three FAS characteristics, the facial dysmorphology and central nervous system function (brain–behavior relationships). Thus the research questions that are addressed using this animal model must be sufficiently important and focused and the sheep must possess great advantages compared to some other models in order to justify the resources expended.

While sheep were used even earlier to study basic physiological issues in the fetus and the effects of ethanol on the fetal cardiovascular system pertaining to the practice during the 1970s of using alcohol as a tocolytic, Potter and co-workers[83] reported the first use of sheep in FAS research. This was one of the first to use a large animal model to demonstrate reduced organ size, including brain, in alcohol-imbibing sheep. Not long afterward, Rose and co-workers[84] utilized a sheep model system to address ethanol actions on fetal endocrine function. Cumming and co-workers[85] reported on ethanol pharmacokinetics and cardiovascular responses in fetal sheep. The most prolific investigations to date of those utilizing a sheep animal model were conducted by James Brien's group who reported on fetal brain metabolism,[86] alcohol disposition in the fetomaternal compartments,[87] actions on cardiovascular and brain activity,[88] and other issues. Falconer[89] addressed actions of alcohol on placental blood flow and glucose metabolism utilizing a sheep model system. In addition to Brien, Richardson, and co-workers,[86] Gleason and Hotchkiss[90] have also investigated alcohol actions on cerebral blood flow and metabolism in fetal sheep. Reynolds, Brien, and co-workers investigated neurotransmitter responses to alcohol in fetal sheep[91] and the actions of *in utero* alcohol exposure on behavior in neonatal lambs.[92] We have now reported the establishment of the eye blink classical condition as a functional dependent variable in this species.[93] We also have utilized this model system to address issues related to FAS and hypoxemia[94] and disturbances in the hypothalamic–pituitary–adrenal axis[95] and the hypothalamic–pituitary–thyroid axis.[96] Additionally, we have further validated the usefulness of this model system by demonstrating that fetal alcohol exposure results in Purkinje cell loss like that reported in the well-established rodent model system at similar doses and patterns of exposure.[11] While this is not an exhaustive cataloging of all FAS-related work performed in sheep, it represents the breadth of application and the efficacy of this model system for use in investigating FAS-related issues. Drawbacks of the sheep model are expense and lack of genomic tools.

MINIATURE SWINE The pig, especially the miniature swine (due to its smaller size), possesses a number of important advantages over other species for use in fetal alcohol research. A major advantage over almost all other species is that the pig will voluntarily consume alcohol, to the point of intoxication, in the presence or absence of food or water,[97] especially if it is beer (Dr. Thomas Badger, personal communication). This attribute makes the pig behaviorally more like humans than other species in that no other mammalian species will voluntarily consume such large quantities of alcohol, even with special training (e.g., rats) or conditions (e.g., physical restraint). A second advantage involves the timing of brain development relative to humans. The pig fetal brain growth velocity profile is the most like the human, in both cases peaking at the time of parturition.[22] This attribute would allow for studies of trimester-based vulnerabilities without the need of modifying the timing of alcohol administration with respect to the timing of parturition. A third advantage of the miniature swine is that alcohol metabolism and clearance are very similar to humans,[97] whereas alcohol metabolism in rats and mice is quite different. Fourth, alcohol consumption by pigs during pregnancy increases embryonic and fetal loss and reduces birth weight as in humans.[98] These similarities between the pig and humans have been utilized recently to explore the hypothesis that alcohol mediates these actions by effects on the uterine environment.[99] Others have exploited the strengths of the pig model system to investigate the actions of alcohol on the maternal and fetal liver.[100] To our knowledge, no other groups of investigators have employed the pig model system to investigate the actions of alcohol on the developing fetus, suggesting that this species may pose opportunities for other areas of investigation. However, as with other large animal species, the expense and resources needed to maintain a colony for research necessitate a clear justification of the questions addressed. Adult female miniature swine weigh over 100 pounds and they require substantial space, which restricts where they can be used. Furthermore, since the pig has been used infrequently as an animal model for FAS, the generalizability of the results for specific dependent measures must be examined carefully.

NONHUMAN PRIMATES (*MACACA MULATTA; MACACA NEMISTRINA*) Theoretically, nonhuman primates should be the best animal model of the human condition since they are most closely related to humans in terms of cognition and behavior. Additionally, nonhuman primates exhibit a similar pattern of fetal development compared to humans (with the exception that nonhuman primates are born slightly advanced of human infants with respect to motor function). Nonhuman primates possess a relatively large face and brain, and well-defined cortical regions like those of humans, in addition to intelligence that is surpassed only by humans. Much of their anatomy, physiology, pharmacokinetics, and reproductive behavior have been documented, and fetal development progresses similarly to that of the human. They exhibit advanced behaviors that have parallels to human behavior.[101] The nonhuman primate model clearly has an important role in the examination of complex human behaviors, such as executive function. These behaviors are more reliably examined in the nonhuman primate than in rodent models. Another important advantage is that not only can the same behaviors be examined in nonhuman primates as in humans, but also often the same testing apparatus and conditions of testing prevail. Consequently, the findings can be extrapolated more easily to the human condition.

Altshuler and Shippenberg[102] were the first to report FAS studies utilizing a primate model system (rhesus monkeys) instrumented with intragastric cannulas. However, their report was preliminary and contained no live births. Also around this time, Mukherjee and Hodgen[103] utilized rhesus monkeys and cynomolgus monkeys to address cardiovascular actions of alcohol on the fetus. However, these studies were performed acutely on anesthetized monkeys, an important drawback as anesthesia may have altered cardiovascular function and potentially protective reflex responses to alcohol. Fisher and co-workers[104] utilized cynomolgus monkeys to address effects of prenatal alcohol on placental nutrient transport. Hill and co-workers[105] utilized cynomolgus monkeys to address alcohol pharmacokinetics. These endeavors have in common that they each resulted in a single publication suggesting that employing a monkey model system was useful for addressing only very specific questions or that further efforts were unfruitful. Clarren and co-workers are the only group who have utilized a nonhuman primate model to produce more than a single report. Subjects (pigtailed macaques) in their studies exhibited neuroanatomical, cognitive, motor, developmental, and facial anomalies like those seen in children exposed to alcohol *in utero*.[106–110] This group has also reported neurochemical and immunological actions of prenatal alcohol exposure.[111,112]

However, this model system, like all others, possesses disadvantages as well as advantages. Among the primary disadvantages is the nearly absolute requirement of specific methods of restraint for both acute treatments and to maintain instrumentation in chronically instrumented animals; restraint thus becomes a major confounding experimental factor and a potential cause of high pregnancy losses. The high rates of reproductive losses reported in primates used in FAS research raise the distinct possibility that a selection process might confound results. A high stress level associated with restraint is a recognized problem in studies using nonhuman primates, more so than in other large (sheep, pig) animal models.[113] Schneider *et al.*[114] utilized a voluntary drinking method in a rhesus monkey animal model, which is a strength insofar as restraint for alcohol administration was avoided, but equally important, the voluntary drinking method can be used only to explore the effects of low and variable (within and between individual subjects) doses of alcohol, since primates are not fond of drinking alcohol (compared to, for example, miniature swine). As a consequence of the great expense and legitimate ethical concerns, as well as other significant practical problems (including health risks to investigators and animal care technicians), the nonhuman primate model is useful for addressing questions only when other animal models are inappropriate.

NONMAMMALIAN ANIMAL MODELS

Nonmammalian animal models have made significant contributions to many research fields, including toxicology, development, and environmental chemical, preclinical safety, and efficacy testing for new drugs under development by industry. A significant amount of information about embryonic development has been derived from nonmammalian species due to the rapid extrauterine development and embryo translucency. In addition, nonmammalian models can be used in experiments that require synchronized embryo subjects. This has been an issue in some mammalian animal models because the timing of specific developmental events within each fetus is not synchronized. For example, not all mouse embryos will be at the identical developmental stage when a treatment is administered to the pregnant mouse.[115] Synchronizing the embryos would be an exploitable advantage in order to assess the extent to which all embryos are, or are not, equally affected by the same alcohol exposure. In addition, the use of a lower organism for biomedical research is one positive step toward the goals of reduction, refinement, and replacement of the use of animals in research. Thus nonmammalian animal models may offer some distinct advantages for use in the FAS research field.

CHICKEN/CHICK EMBRYO (*GALLUS GALLUS*) The chick embryo has a long history of use in developmental research. The chick animal model offers obvious advantages such as small size, rapid embryonic development, and lack of a placenta, which may reveal the extent of maternal protective factors. This animal model has been used to study craniofacial development and its perturbation by developmental alcohol exposure. Cartwright and Smith[23] showed that increasing doses of alcohol injected into the chick embryo at the gastrulation stage resulted in increasing craniofacial and other developmental deficits (hindbrain segmentation and neural tube defects), as well as increasing mortality. While the threshold for mortality appeared to be at about 250 μl (10% ethanol solution), this dose still resulted in a significant incidence of craniofacial anomalies that correlated with extensive cell death in the neural crest cell population.[23] Further research in this area revealed that alcohol stimulates cartilage differentiation in mesenchymal cells from various facial primordia, suggesting a potential mechanistic explanation for the craniofacial dysmorphologies in the chicken with respect to the competition between bone and other tissues forming within the face.[116] In addition, the important role played by neural crest cells in the etiology of alcohol-induced craniofacial dysmorphology has been confirmed[117] and this deficit can be rescued by injection with sonic hedgehog.[118] These studies demonstrate important advances in our understanding of how neural crest cells respond to alcohol at specific stages of their dynamic development into the face. However, it is important to recognize that the alcohol-induced craniofacial dysmorphology (which is only one of the triad of FAS characteristics) in the chick, like the mouse, may not be observable in posthatching subjects, which would render this animal model suitable for only specific research questions. Thus, the utility of this model for addressing the full spectrum of FAS is limited.

One of the most serious limitations of the chick animal model is the scarcity of information on genetic mutations and the lack of knowledge about the chicken genome. At least one study has determined that various strains of chick embryo are more or less susceptible to alcohol-induced neural crest cell apoptosis,[119] and this in an important step in the FAS field. However, little work is being done on direct genetic manipulations in the chicken that would assist in determining the relative importance of alcohol-induced changes in gene expression as a risk factor for the craniofacial deficit, which is the most common utilization of the chick animal model in the FAS field. Nonetheless, experiments using the chick embryo have resulted in major advances in knowledge about many aspects of embryonic development, such as neurogenesis, axis and limb development, and somatogenesis.

FROG EMBRYOS (*XENOPUS*) The *Xenopus* embryo model has been used extensively to study basic embryological processes and the effects of environmental and other toxicants on development.[120–122] The use of the *Xenopus* embryo model in developmental toxicology stems from its external development, large size, and identifiable blastomeres, its ability to withstand extensive surgical intervention, and the fact that it can be cultured *in vitro*. The wealth of information about early embryonic patterning events has been gleaned from *Xenopus* embryos, especially the maternal contributions to embryonic development that occur prior to transcription of the embryonic genome. The *Xenopus* embryo model has the potential for exploitation in the study of alcohol-induced defects, especially resulting from perturbations in very early developmental events. The results of one such study by Nakatsuji[123] support this statement. The tadpoles (exposed to alcohol from the stages of morula through neurulation) showed alterations in physical morphology resulting from the alcohol treatment; the width of the mouth and the length of the head were consistently smaller in the alcohol-exposed compared to the control embryos. In addition, these researchers found that the changes in neural organization occurred at gastrulation and neurulation, and involved alterations in cell migration to the affected region.

Using the bullfrog, Uray and Sexton[124] examined the effect of high levels of alcohol exposure on the development of the cerebellum and reported that the timing of alcohol exposure was critical with respect to disrupting cerebellar development.[26] The usefulness of the bullfrog model for studying the effects of alcohol may be limited since the alcohol levels required were at the extreme end of the physiological relevant range for humans.[124] The *Xenopus* embryo model may be best suited for fetal alcohol research questions concerning very early (embryonic) development due to the wealth of information available about normal *Xenopus* development. Furthermore, the role of genetic manipulations and gene interactions in controlling the well-studied developmental events is not well documented compared to other nonmammalian organisms.

ZEBRAFISH (*DANIO RERIO*) The zebrafish is a powerful animal model for the examination of early embryonic development that has been utilized surprisingly little within the fetal alcohol field. The zebrafish has many advantages, especially as a genetic and developmental model, due to embryo transparency (mutant phenotypes are easily identified in whole, living embryos), fecundity (females may lay 75–120 eggs per cycle), the ease of manipulation, and the durability of the embryos to withstand various culture conditions. For the fetal alcohol field, perhaps the most important feature of the zebrafish embryo as an animal model is the ease of genetic manipulations, including forward and reverse genetics, insertional mutagenesis (morpholino knockdown/knockout), and positional cloning.[125] The zebrafish animal model has been exploited in a wide range of related fields, including development,[126,127] teratogy,[128] environmental pollution assessments,[24,129] and behavioral studies with,[130] or without, the presence of alcohol.[131] Currently, there are very few journal articles documenting zebrafish teratogenesis from exposure to alcohol per se, but the field is growing. In one study, Blader and Strahle[132] found cyclopia in alcohol zebrafish embryos exposed to 2.4% alcohol, in addition to altered migration of the prechordal plate. Some researchers have used the zebrafish to study alcohol's effects on visual development by giving alcohol (1.5%) to zebrafish embryos at various developmental time points and testing them on visual function and physiology.[133] They found that the embryos treated with alcohol during the period of eye development had significantly smaller eyes and impaired visual function relative to control embryos. Although these findings are potentially relevant for the FAS field (since both eye/facial development and function are affected by embryonic alcohol exposure), one question that remains unanswered involves the concentration of alcohol to which the embryos were exposed (1.5–2.4%). These values clearly exceed the physiological tolerance of humans and it is important to reconcile that fact with information about exactly how much alcohol the embryo receives. If the embryo absorbs only a portion of the concentration of alcohol in the media and still manifests the defect, then this animal model has great potential. On the other hand, if the embryos achieve equilibrium with the media alcohol concentration, then the utility of the model may be severely limited because the effects may be the result of toxicity rather than teratogenicity. There is some indication that the alcohol level within the chorion is the same as that within the media and zebrafish do possess the enzymes responsible for the conversion of alcohol to its primary (and toxic) metabolite, acetaldehyde,[126] but to date, no one has determined the level of alcohol *within the embryo*. In addition to its other positive attributes, the zebrafish has the potential to be a very powerful tool for the understanding of the role of genes in alcohol-induced deficits due to the recent cloning of the zebrafish genome and extensive catalog of known mutations. Some of the genes screened to date have been linked to specific human disease states.[134] The zebrafish is being actively used by one institution (Institute of Neuroscience, University of Oregon) and the publication of a guide to zebrafish[135] and an information sharing website (http://zfin.org) are available. Therefore the zebrafish is likely to be increasingly more popular in addressing basic questions associated with the relationship of gene expression and the deleterious effects of alcohol exposure during early embryonic development.

FRUIT FLY (*DROSOPHILA*) The most powerful advantage of the fruit fly for alcohol research is the abundance of genetic mutations and the potential for manipulation of the fruit fly genome to explore gene–alcohol interactions. In addition, the fruit fly is well suited to exploring multigenerational effects due to its rapid development from embryo through reproductive age adult. The effect of alcohol on multiple generations has received little attention in the fetal alcohol research field, but likely will receive further investigation as the number of children diagnosed with FAS reaching adulthood and bearing children increases. Since the entire genome of *D. melanogaster* has now been sequenced and cloned, the probability of using it to address genomic questions and multigenerational issues related to fetal alcohol-induced damage has increased. This advantage has been exploited extensively to study genetic perturbations associated with behavioral sensitivity and resistance to alcohol.[136] One other interesting attribute of the fruit fly is that it can metabolize alcohol directly to acetate, bypassing acetaldehyde formation, which has interesting implications for the study of mechanisms of alcohol teratogenicity. Although the development of the fruit fly from fertilization to adulthood has been well characterized,[137] to date little research has been performed on the teratogenesis of alcohol per se,[138] although other chemicals have been tested for teratogenicity.[139] Obviously the small size and rapid development of the fruit fly make it a suitable model for environmental and chemical teratogenesis screening, but these factors become an obstacle to some specific research questions in the fetal alcohol field, such as mechanisms of craniofacial development.

ROUNDWORM (*CAENORHABDITIS ELEGANS*) Although the roundworm, *C. elegans*, has rarely been used to attempt to study questions related to developmental alcohol exposure, it has an enduring history in the toxicology and neuroscience fields, principally due to its small size, external development, simplicity of neural organization (302 neurons and 5000 synapses), and complete genome mapping. Specifically, C. *elegans* has been used to study the molecular basis of specific behaviors such as circadian rhythms[140] and habituation.[141] Some studies have shown a dose-dependent reduction in reproductive activity[142] and the existence of an alcohol-sensitive genetic mutation.[143] This nematode has the potential to be used to address specific fetal alcohol research questions, particularly simple behavior. However, its genetic differences from humans impede its overall usefulness. Currently, there are other nonmammalian models that share genes with humans that would be more suitable to the study of the role of genes in alcohol's effects on development. Nonetheless, when dealing with behavior questions, starting with a relatively simple system and moving forward has proven effective in other areas and may provide a useful approach to stimulate ideas with respect to more complex neuronal systems. At this point, it remains to be seen whether the advantages of the *C. elegans* model can be exploited to address important, specific fetal alcohol questions.

USE OF ANIMAL MODELS IN THE FUTURE

Enduring challenges that face the fetal alcohol field include finding ways to identify the women at greatest risk of abusing alcohol during pregnancy with the hopes of preventing prenatal alcohol exposure and identifying children at birth who have been exposed to alcohol *in utero*. More elusive may be identifying effective treatments for those diagnosed with FAS. In addition to simply providing data that can be used to support the argument for the reduction or cessation of drinking during pregnancy, there are still at least three areas in which animal studies can have a significant impact on the fetal alcohol field. The first of those areas involves evidence of conditions that increase fetal risk when associated with maternal alcohol abuse. Research in this area offers to help explain much of the variation that occurs in human offspring with histories of similar amounts of alcohol exposure and may help to improve the predictability of the likelihood and extent of damage to the developing fetus.

The second area is the identification of the mechanisms of alcohol-induced fetal damage. Mechanistic studies offer the best hope for insights into developing pharmacological strategies for intervention. Research is now beginning to focus on developing therapies to prevent the induction of fetal injury. For instance, Spong and her colleagues[144] recently demonstrated that the pretreatment of pregnant mice with two novel peptides resulted in a significant reduction in fetal demise following an acute dose of alcohol early in gestation. However, it will be important in the future to better define the role played by the peptides in mediating this action and to demonstrate these protective actions in other animal models.

Another approach to therapeutic intervention that is showing considerable promise is the employment of behavioral, environmental, or cognition stimulating conditions to improve specific fetal alcohol-induced behavioral deficits. Klintsova *et al.*[145] found that 20 days of forced complex motor tasks was sufficient to partially ameliorate the effects of postnatal binge alcohol exposure in a rat model system. However, the rat is a postnatal model for what occurs prenatally in humans. And in fact, the results suggest that the complex motor training actually increases plasticity in the brain, rather than ameliorating the brain damage itself. Nonetheless, if similar results could be found using other animal models of FAS and perhaps in FAS children, it would make a compelling case for this type of intervention at a much later developmental stage.

Third, some of the best opportunities for advancing the fetal alcohol field are likely to come from behavioral studies that will promote the bidirectional flow of information between animal and human studies. Two examples illustrate this point. The first is the use of eye blink conditioning as a tool to study FAS and other alcohol-related injury. This technique is especially powerful for three important reasons. First, unlike most other tests, it can be used effectively at different stages of development and even in the same subjects over time. Second, it can be used in both animals and humans. Third, much is known about the neuronal pathways involved in this type of learning. Eye blink conditioning has been studied in humans for many years, including the assessment of cognitive development in infants.[146] Because the neural substrates of classical conditioning at different developmental ages have been mostly worked out in animals,[147] researchers have begun to realize its potential for evaluating the effects of neurotoxic/teratogenic agents,[148] including alcohol exposure in developing rats.[149] These efforts have led to the correlation of specific deficits in conditioning with altered electrical activity[150] and neuronal loss[151] in the deep cerebellar nuclei. This powerful interaction between human and animal studies will allow specific types of learning deficits to be detected in human infants and then better understood by an animal model where the relevant functional neural pathways can be dissected.

Another creative example of how studies with animals can be extended to humans is the use of the principles of the Morris water maze. This type of learning task has been shown to require an intact hippocampus, and it is an effective tool for evaluating deficits in spatial learning in rodents that have been affected by developmental alcohol exposure.[152] On the surface, the behavioral requirements of the Morris maze hardly seem applicable to humans, especially those with substantial cognitive or motor impairments. However, the wealth of neurobiological information related to place learning stimulated investigators at the University of New Mexico to develop an innovative virtual water maze that can be used to assess learning differences in adolescents with FAS.[153] Again, with this application, deficits in humans can be understood in the context of neural systems that can be manipulated experimentally in animals.

These examples illustrate how future animal and human studies can interact and drive the field forward more rapidly than would otherwise be possible. The careful choice of questions that will have the most impact on the human condition, coupled with selecting the animal models that are best suited for addressing those questions, will undoubtedly have important implications in the development of detection, intervention, and treatment strategies for FAS, FASD, and ARND.

ACKNOWLEDGMENTS

Supported by Grants AA10940 and AA15339 from the National Institute of Alcohol Abuse and Alcoholism.

REFERENCES

1. Jones KL, Smith DW. Recognition of the fetal alcohol syndrome in early infancy. *Lancet* 1973;2(7836):999–1001.
2. Weiner L, Rosett HL, Edelin KC, Alpert JJ, Zuckerman B. Alcohol consumption by pregnant women. *Obstet Gynecol* 1983;61(1): 6–12.
3. Becker HC, Randall CL, Salo AL, Saulnier JL, Weathersby RT. Animal research: Charting the course for FAS. *Alcohol Health Res* 1994;18:10–16.
4. Coles C. Critical periods for prenatal alcohol exposure. *Alcohol Health Res* 1994;18:22–29.
5. Driscoll CD, Streissguth AP, Riley EP. Prenatal alcohol exposure: Comparability of effects in humans and animal models. *Neurotoxicol Teratol* 1990;12(3):231–237.
6. Barnes DE, Walker DW. Prenatal ethanol exposure permanently reduces the number of pyramidal neurons in rat hippocampus. *Brain Res* 1981;227(3):333–340.
7. Hammer RP Jr, Scheibel AB. Morphologic evidence for a delay of neuronal maturation in fetal alcohol exposure. *Exp Neurol* 1981;74(2):587–596.
8. Hamre KM, West JR. The effects of the timing of ethanol exposure during the brain growth spurt on the number of cerebellar Purkinje and granule cell nuclear profiles. *Alcohol Clin Exp Res* 1993; 17(3):610–622.
9. Miller MW. Effects of alcohol on the generation and migration of cerebral cortical neurons. *Science* 1986;233(4770):1308–1311.
10. Smith DE, Foundas A, Canale J. Effect of perinatally administered ethanol on the development of the cerebellar granule cell. *Exp Neurol* 1986;92(3):491–501.
11. West JR, Parnell SE, Chen WJ, Cudd TA. Alcohol-mediated Purkinje cell loss in the absence of hypoxemia during the third trimester in an ovine model system. *Alcohol Clin Exp Res* 2001;25(7): 1051–1057.
12. Hsiao SH, Parrish AR, Nahm SS, Abbott LC, McCool BA, Frye GD. Effects of early postnatal ethanol intubation on GABAergic synaptic proteins. *Brain Res* 2002;138(2):177–185.
13. Goodlett CR, Leo JT, O'Callaghan JP, Mahoney JC, West JR. Transient cortical astrogliosis induced by alcohol exposure during the neonatal brain growth spurt in rats. *Brain Res* 1993;72(1): 85–97.
14. Guerri C, Pascual M, Renau-Piqueras J. Glia and fetal alcohol syndrome. *Neurotoxicology* 2001;22(5):593–599.
15. Miller MW, Potempa G. Numbers of neurons and glia in mature rat somatosensory cortex: Effects of prenatal exposure to ethanol. *J Comp Neurol* 1990;293(1):92–102.
16. Phillips DE. Effects of alcohol on glial cell development *in vivo*: Morphological studies. In: *Alcohol and Glial Cells: NIAAA Research Monograph*. Bethesda, MD: National Institute on Alcohol Abuse and Alcoholism, 1994:69–91.
17. Valles S, Sancho-Tello M, Minana R, Climent E, Renau-Piqueras J, Guerri C. Glial fibrillary acidic protein expression in rat brain and in radial glia culture is delayed by prenatal ethanol exposure. *J Neurochem* 1996;67(6):2425–2433.
18. Stratton K, Howe C, Battaglia F. *Institute of Medicine Report. Fetal Alcohol Syndrome: Diagnosis, Epidemiology, Prevention, and Treatment*. Washington, DC: National Academy Press, 1996
19. Warren KR, Foudin LL. Alcohol-related birth defects—the past, present, and future. *Alcohol Res Health* 2001;25(3):153–158.
20. Clarren SK, Smith DW. The fetal alcohol syndrome. *N Engl J Med* 1978;298(19):1063–1067.
21. Streissguth AP, O'Malley K. Neuropsychiatric implications and long-term consequences of fetal alcohol spectrum disorders. *Semin Clin Neuropsychiatry* 2000;5(3):177–190.
22. Dobbing J, Sands J. Comparative aspects of the brain growth spurt. *Early Hum Dev* 1979;3(1):79–83.
23. Cartwright MM, Smith SM. Increased cell death and reduced neural crest cell numbers in ethanol-exposed embryos: Partial basis for the fetal alcohol syndrome phenotype. *Alcohol Clin Exp Res* 1995;19(2):378–386.
24. Carvan MJ 3rd, Sonntag DM, Cmar CB, Cook RS, Curran MA, Miller GL. Oxidative stress in zebrafish cells: Potential utility of transgenic zebrafish as a deployable sentinel for site hazard ranking. *Sci Total Environ* 2001;274(1–3):183–196.
25. Su B, Debelak KA, Tessmer LL, Cartwright MM, Smith SM. Genetic influences on craniofacial outcome in an avian model of prenatal alcohol exposure. *Alcohol Clin Exp Res* 2001;25(1): 60–69.
26. Uray NJ, Sexton PS. Age-related vulnerability of cerebellar development in ethanol-treated tadpoles. *Alcohol Clin Exp Res* 1990;14:348.
27. Driscoll CD, Riley EP, Meyer LS. Delayed taste aversion learning in preweanling rats exposed to alcohol prenatally. *Alcohol* 1985; 2(2):277–280.
28. Randall CL, Becker HC, Middaugh LD. Effect of prenatal ethanol exposure on activity and shuttle avoidance behavior in adult C57 mice. *Alcohol Drug Res* 1985;6(5):351–360.
29. Goodlett CR, Bonthius DJ, Wasserman EA, West JR. An animal model of central nervous system dysfunction associated with fetal alcohol exposure: Behavioral and neuroanatomical correlates. In: Gormezano I, Wasserman EA, Eds. *Learning and Memory: Behavioral and Biological Processes*. Englewood Cliffs, NJ: Lawrence Erlbaum, 1992:183–208.
30. Streissguth AP, Barr HM, Olson HC, Sampson PD, Bookstein FL, Burgess DM. Drinking during pregnancy decreases word attack and arithmetic scores on standardized tests: Adolescent data from a population-based prospective study. *Alcohol Clin Exp Res* 1994; 18(2):248–254.
31. Uecker A, Nadel L. Spatial locations gone awry: Object and spatial memory deficits in children with fetal alcohol syndrome. *Neuropsychologia* 1996;34(3):209–223.
32. Abel EL. Effects of prenatal alcohol exposure on birth weight in rats: Is there an inverted U-shaped function? *Alcohol* 1996;13(1): 99–102.
33. Middaugh LD, Boggan WO. Postnatal growth deficits in prenatal ethanol-exposed mice: Characteristics and critical periods. *Alcohol Clin Exp Res* 1991;15(6):919–926.

34. Oppenheim RW. Ontogenetic adaptations and retrogressive processes in the development of the nervous system and behavior: A neuroembryological perspective, in clinics in developmental medicine. In: Connolly KJ, Prechtl HFR, Eds. *Maturation and Development: Biological and Psychological Perspectives*. Philadelphia, PA: Lippincott Co., 1981:73–109.
35. Bauer-Moffett C, Altman J. The effect of ethanol chronically administered to preweanling rats on cerebellar development: A morphological study. *Brain Res* 1977;119(2):249–268.
36. Goodlett CR, Eilers AT. Alcohol-induced Purkinje cell loss with a single binge exposure in neonatal rats: A stereological study of temporal windows of vulnerability. *Alcohol Clin Exp Res* 1997;21(4):738–744.
37. West JR, Goodlett CR, Bonthius DJ, Hamre KM, Marcussen BL. Cell population depletion associated with fetal alcohol brain damage: Mechanisms of BAC-dependent cell loss. *Alcohol Clin Exp Res* 1990;14(6):813–818.
38. Livy DJ, Maier SE, West JR. Fetal alcohol exposure and temporal vulnerability: Effects of binge-like alcohol exposure on the ventrolateral nucleus of the thalamus. *Alcohol Clin Exp Res* 2001; 25(5):774–780.
39. Chen WA, Parnell SE, West JR. Early postnatal alcohol exposure produced long-term deficits in brain weight, but not the number of neurons in the locus coeruleus. *Brain Res* 1999;118(1–2):33–38.
40. Mooney SM, Napper RM, West JR. Long-term effect of postnatal alcohol exposure on the number of cells in the neocortex of the rat: A stereological study. *Alcohol Clin Exp Res* 1996;20(4):615–623.
41. Druse MJ, Tajuddin N, Kuo A, Connerty M. Effects of *in utero* ethanol exposure on the developing dopaminergic system in rats. *J Neurosci Res* 1990;27(2):233–240.
42. Light KE, Serbus DC, Santiago M. Exposure of rats to ethanol from postnatal days 4 to 8: Alterations of cholinergic neurochemistry in the cerebral cortex and corpus striatum at day 20. *Alcohol Clin Exp Res* 1989;13(1):29–35.
43. Costa ET, Olivera DS, Meyer DA, *et al.* Fetal alcohol exposure alters neurosteroid modulation of hippocampal N-methyl-D-aspartate receptors. *J Biol Chem* 2000;275(49):38268–38274.
44. Taylor AN, Branch BJ, Kokka N. Neuroendocrine effects of fetal alcohol exposure. *Prog Biochem Pharmacol* 1981;18:99–110.
45. Weinberg J. Neuroendocrine effects of prenatal alcohol exposure. *Ann NY Acad Sci* 1993;697:86–96.
46. Lopez de Ochoa JA, Villa-Elizaga I, Frizell E, *et al.* Alteration in deoxyribonucleic acid and proteins in cerebral tissues from fetuses subject to alcohol in utero. *Rev Esp Fisiol* 1989;45:35–42.
47. Du X, Hamre K. Identity and neuroanatomical localization of messenger RNAs that change expression in the neural tube of mouse embryos within 1 h after ethanol exposure. *Brain Res* 2003; 144(1):9–23.
48. Yang J, Zoeller RT. Differential display identifies neuroendocrine-specific protein-A (NSP-A) and interferon-inducible protein 10 (IP-10) as ethanol-responsive genes in the fetal rat brain. *Brain Res* 2002;138(2):117–133.
49. Maier SE, Chen WJ, Miller JA, West JR. Fetal alcohol exposure and temporal vulnerability regional differences in alcohol-induced microencephaly as a function of the timing of binge-like alcohol exposure during rat brain development. *Alcohol Clin Exp Res* 1997;21(8):1418–1428.
50. Sulik KK, Johnston MC. Embryonic origin of holoprosencephaly: Interrelationship of the developing brain and face. *Scan Electron Microsc* 1982(Pt. 1):309–322.
51. Osborn JA, Kim CK, Steiger J, Weinberg J. Prenatal ethanol exposure differentially alters behavior in males and females on the elevated plus maze. *Alcohol Clin Exp Res* 1998;22(3):685–696.
52. Dominguez HD, Lopez MF, Molina JC. Neonatal responsiveness to alcohol odor and infant alcohol intake as a function of alcohol experience during late gestation. *Alcohol* 1998;16(2):109–117.
53. Goodlett CR, Johnson TB. Neonatal binge ethanol exposure using intubation: Timing and dose effects on place learning. *Neurotoxicol Teratol* 1997;19(6):435–446.
54. Gentry GD, Middaugh LD. Prenatal ethanol weakens the efficacy of reinforcers for adult mice. *Teratology* 1988;37(2):135–144.
55. Cortese BM, Krahl SE, Berman RF, Hannigan JH. Effects of prenatal ethanol exposure on hippocampal theta activity in the rat. *Alcohol* 1997;14(3):231–235.
56. Heaton MB, Moore DB, Paiva M, Madorsky I, Mayer J, Shaw G. The role of neurotrophic factors, apoptosis-related proteins, and endogenous antioxidants in the differential temporal vulnerability of neonatal cerebellum to ethanol. *Alcohol Clin Exp Res* 2003;27(4):657–669.
57. Chernoff GF. The fetal alcohol syndrome in mice: An animal model. *Teratology* 1977;15(3):223–229.
58. Randall CL, Taylor WJ. Prenatal ethanol exposure in mice: Teratogenic effects. *Teratology* 1979;19(3):305–311.
59. Sulik KK, Johnston MC, Webb MA. Fetal alcohol syndrome: Embryogenesis in a mouse model. *Science* 1981;214(4523): 936–938.
60. Sulik KK, Johnston MC. Sequence of developmental alterations following acute ethanol exposure in mice: Craniofacial features of the fetal alcohol syndrome. *Am J Anat* 1983;166(3):257–269.
61. Webster WS, Walsh DA, McEwen SE, Lipson AH. Some teratogenic properties of ethanol and acetaldehyde in C57BL/6J mice: Implications for the study of the fetal alcohol syndrome. *Teratology* 1983;27(2):231–243.
62. Webster WS, Lipson AH, Sulik KK. Interference with gastrulation during the third week of pregnancy as a cause of some facial abnormalities and CNS defects. *Am J Med Genet* 1988;31(3):505–512.
63. Ledig M, Misslin R, Kopp P, Vogel E, Tholey G, Mandel P. Alcohol exposure before pregnancy: Biochemical and behavioral effects on the offspring of rats. *Pharmacol Biochem Behav* 1990;36(2): 279–285.
64. Henderson GI, Schenker S. The effect of maternal alcohol consumption on the viability and visceral development of the newborn rat. *Res Commun Chem Pathol Pharmacol* 1977;16(1):15–32.
65. Leichter J, Lee M. Effect of maternal ethanol administration on physical growth of the offspring in rats. *Growth* 1979;43(4): 288–293.
66. Borges S, Lewis PD. A study of alcohol effects on the brain during gestation and lactation. *Teratology* 1982;25(3):283–289.
67. West JR, Hodges CA, Black AC Jr. Prenatal exposure to ethanol alters the organization of hippocampal mossy fibers in rats. *Science* 1981;211(4485):957–959.
68. Abel EL. Fetal alcohol syndrome: Behavioral teratology. *Psychol Bull* 1980;87(1):29–50.
69. Leichter J, Lee M. Method of ethanol administration as a confounding factor in studies of fetal alcohol syndrome. *Life Sci* 1982; 31(3):221–227.
70. Samson HH. Maternal ethanol consumption and fetal development in the rat: A comparison of ethanol exposure techniques. *Alcohol Clin Exp Res* 1981;5(1):67–74.
71. Diaz J, Samson HH. Impaired brain growth in neonatal rats exposed to ethanol. *Science* 1980;208(4445):751–753.
72. Li TK, Lumeng L, Doolittle DP. Selective breeding for alcohol preference and associated responses. *Behav Genet* 1993;23(2): 163–170.
73. Sinclair JD, Li TK. Long and short alcohol deprivation: Effects on AA and P alcohol-preferring rats. *Alcohol* 1989;6(6):505–509.
74. Riley EP, Barron S, Melcer T, Gonzalez D. Alterations in activity following alcohol administration during the third trimester equivalent in P and NP rats. *Alcohol Clin Exp Res* 1993;17(6):1240–1246.
75. Thomas JD, Melcer T, Weinert S, Riley EP. Neonatal alcohol exposure produces hyperactivity in high-alcohol-sensitive but not in low-alcohol-sensitive rats. *Alcohol* 1998;16(3):237–242.
76. Kimura KA, Chiu J, Reynolds JN, Brien JF. Effect of chronic prenatal ethanol exposure on nitric oxide synthase I and III proteins in the hippocampus of the near-term fetal guinea pig. *Neurotoxicol Teratol* 1999;21(3):251–259.
77. Abdollah S, Catlin MC, Brien JF. Ethanol neuro-behavioural teratogenesis in the guinea pig: Behavioural dysfunction and hippocam-

pal morphologic change. *Can J Physiol Pharmacol* 1993; 71(10–11):776–782.

78. Fabregues I, Ferrer I, Gairi JM, Cahuana A, Giner P. Effects of prenatal exposure to ethanol on the maturation of the pyramidal neurons in the cerebral cortex of the guinea-pig: A quantitative Golgi study. *Neuropathol Appl Neurobiol* 1985;11(4):291–298.
79. Bailey CD, Brien JF, Reynolds JN. Chronic prenatal ethanol exposure increases GABA(A) receptor subunit protein expression in the adult guinea pig cerebral cortex. *J Neurosci* 2001;21(12): 4381–4389.
80. Dettmer TS, Barnes A, Iqbal U, *et al.* Chronic prenatal ethanol exposure alters ionotropic glutamate receptor subunit protein levels in the adult guinea pig cerebral cortex. *Alcohol Clin Exp Res* 2003;27(4):677–681.
81. Byrnes ML, Reynolds JN, Brien JF. Effect of prenatal ethanol exposure during the brain growth spurt of the guinea pig. *Neurotoxicol Teratol* 2001;23(4):355–364.
82. Byrnes ML, Reynolds JN, Brien JF. Brain growth spurt-prenatal ethanol exposure and the guinea pig hippocampal glutamate signaling system. *Neurotoxicol Teratol* 2003;25(3):303–310.
83. Potter BJ, Belling GB, Mano MT, Hetzel BS. Experimental production of growth retardation in the sheep fetus after exposure to alcohol. *Med J Aust* 1980;2(4):191–193.
84. Rose JC, Meis PJ, Castro MI. Alcohol and fetal endocrine function. *Neurobehav Toxicol Teratol* 1981;3(2):105–110.
85. Cumming ME, Ong BY, Wade JG, Sitar DS. Maternal and fetal ethanol pharmacokinetics and cardiovascular responses in near-term pregnant sheep. *Can J Physiol Pharmacol* 1984;62(12):1435–1439.
86. Richardson BS, Patrick JE, Bousquet J, Homan J, Brien JF. Cerebral metabolism in fetal lamb after maternal infusion of ethanol. *Am J Physiol* 1985;249(5Pt. 2):R505–509.
87. Brien JF, Clarke DW, Richardson B, Patrick J. Disposition of ethanol in maternal blood, fetal blood, and amniotic fluid of third-trimester pregnant ewes. *Am J Obstet Gynecol* 1985;152(5): 583–590.
88. Patrick J, Richardson B, Hasen G, *et al.* Effects of maternal ethanol infusion on fetal cardiovascular and brain activity in lambs. *Am J Obstet Gynecol* 1985;151(7):859–867.
89. Falconer J. The effect of maternal ethanol infusion on placental blood flow and fetal glucose metabolism in sheep. *Alcohol* 1990;25(4):413–416.
90. Gleason CA, Hotchkiss KJ. Cerebral responses to acute maternal alcohol intoxication in immature fetal sheep. *Pediatric Res* 1992;31(6):645–648.
91. Reynolds JD, Penning DH, Dexter F, *et al.* Dose-dependent effects of acute in vivo ethanol exposure on extracellular glutamate concentration in the cerebral cortex of the near-term fetal sheep. *Alcohol Clin Exp Res* 1995;19(6):1447–1453.
92. Spear-Smith J, Brien JF, Grafe M, Allrich R, Reynolds JD. Chronic ethanol exposure during late gestation produces behavioral anomalies in neonatal lambs. *Neurotoxicol Teratol* 2000;22(2):205–212.
93. Johnson TB, Goodlett CR, Stanton ME, Cudd TA. Development of eyeblink classical conditioning in the lamb as a functional measure of third trimester alcohol exposure effects. *Alcohol Clin Exp Res* 2006;(Suppl. 30):31A.
94. Cudd TA, Chen WJ, Parnell SE, West JR. Third trimester binge ethanol exposure results in fetal hypercapnea and acidemia but not hypoxemia in pregnant sheep. *Alcohol Clin Exp Res* 2001;25(2): 269–276.
95. Cudd TA, Chen And WJ, West JR. Fetal and maternal sheep hypothalamus pituitary adrenal axis responses to chronic binge ethanol exposure during the third trimester equivalent. *Alcohol Clin Exp Res* 2001;25(7):1065–1071.
96. Cudd TA, Chen WJ, West JR. Fetal and maternal thyroid hormone responses to ethanol exposure during the third trimester equivalent of gestation in sheep. *Alcohol Clin Exp Res* 2002;26(1):53–58.
97. Dexter JD, Tumbleson ME, Hutcheson DP, Middleton CC. Sinclair (S-1) miniature swine as a model for the study of human alcoholism. *Ann NY Acad Sci* 1976;273:188–193.
98. Dexter JD, Tumbleson ME, Decker JD, Middleton CC. Fetal alcohol syndrome in Sinclair (S-1) miniature swine. *Alcohol Clin Exp Res* 1980;4(2):146–151.
99. Kubotsu SL, Hu J, Carnahan KG, deAvila J, Ott TL, Mirando MA. The effects of chronic ethanol consumption during early pregnancy on conceptus health and uterine function in pigs. *Alcohol Clin Exp Res* 2003;27(4):712–719.
100. Romert P, Matthiessen ME. Alcohol-induced injury of mitochondria in hepatocytes of mini-pig fetuses. *Virchows Archiv* 1983;399(3):299–305.
101. Burbacher TM, Grant KS. Methods for studying nonhuman primates in neurobehavioral toxicology and teratology. *Neurotoxicol Teratol* 2000;22(4):475–486.
102. Altshuler HL, Shippenberg TS. A subhuman primate model for fetal alcohol syndrome research. *Neurobehav Toxicol Teratol* 1981;3(2): 121–126.
103. Mukherjee AB, Hodgen GD. Maternal ethanol exposure induces transient impairment of umbilical circulation and fetal hypoxia in monkeys. *Science* 1982;218(4573):700–702.
104. Fisher SE, Atkinson M, Jacobson S, *et al.* Selective fetal malnutrition: The effect of in vivo ethanol exposure upon in vitro placental uptake of amino acids in the non-human primate. *Pediatr Res* 1983;17(9):704–707.
105. Hill DE, Slikker W Jr, Goad PT, Bailey JR, Sziszak TJ, Hendrickx AG. Maternal, fetal, and neonatal elimination of ethanol in nonhuman primates. *Dev Pharmacol Ther* 1983;6(4):259–268.
106. Bonthius DJ, Bonthius NE, Napper RM, Astley SJ, Clarren SK, West JR. Purkinje cell deficits in nonhuman primates following weekly exposure to ethanol during gestation. *Teratology* 1996;53(4): 230–236.
107. Clarren SK, Bowden DM. Fetal alcohol syndrome: A new primate model for binge drinking and its relevance to human ethanol teratogenesis. *J Pediatr* 1982;101(5):819–824.
108. Clarren SK, Astley SJ, Bowden DM. Physical anomalies and developmental delays in nonhuman primate infants exposed to weekly doses of ethanol during gestation. *Teratology* 1988;37(6):561–569.
109. Inouye RN, Kokich VG, Clarren SK, Bowden DM. Fetal alcohol syndrome: An examination of craniofacial dysmorphology in Macaca nemestrina. *J Med Primatol* 1985;14(1):35–48.
110. Miller MW, Astley SJ, Clarren SK. Number of axons in the corpus callosum of the mature Macaca nemestrina: Increases caused by prenatal exposure to ethanol. *J Comp Neurol* 1999;412(1):123–131.
111. Clarren SK, Astley SJ, Bowden DM, *et al.* Neuroanatomic and neurochemical abnormalities in nonhuman primate infants exposed to weekly doses of ethanol during gestation. *Alcohol Clin Exp Res* 1990;14(5):674–683.
112. Grossmann A, Astley SJ, Liggitt HD, *et al.* Immune function in offspring of nonhuman primates (Macaca nemestrina) exposed weekly to 1.8 g/kg ethanol during pregnancy: Preliminary observations. *Alcohol Clin Exp Res* 1993;17(4):822–827.
113. Schneider ML, Moore CF, Kraemer GW, Roberts AD, DeJesus OT. The impact of prenatal stress, fetal alcohol exposure, or both on development: Perspectives from a primate model. *Psychoneuroendocrinology* 2002;27(1–2):285–298.
114. Schneider ML, Moore CF, Becker EF. Timing of moderate alcohol exposure during pregnancy and neonatal outcome in rhesus monkeys (Macaca mulatta). *Alcohol Clin Exp Res* 2001;25(8):1238–1245.
115. Boehm SL 2nd, Lundahl KR, Caldwell J, Gilliam DM. Ethanol teratogenesis in the C57BL/6J, DBA/2J, and A/J inbred mouse strains. *Alcohol* 1997;14(4):389–395.
116. Hoffman LM, Kulyk WM. Alcohol promotes in vitro chondrogenesis in embryonic facial mesenchyme. *Int J Dev Biol* 1999;43(2): 167–174.
117. Rovasio RA, Battiato NL. Ethanol induces morphological and dynamic changes on in vivo and in vitro neural crest cells. *Alcohol Clin Exp Res* 2002;26(8):1286–1298.
118. Ahlgren SC, Thakur V, Bronner-Fraser M. Sonic hedgehog rescues cranial neural crest from cell death induced by ethanol exposure. *Proc Natl Acad Sci USA* 2002;99(16):10476–10481.

119. Debelak KA, Smith SM. Avian genetic background modulates the neural crest apoptosis induced by ethanol exposure. *Alcohol Clin Exp Res* 2000;24(3):307–314.
120. Bantle JA, Burton DT, Dawson DA, *et al.* Initial interlaboratory validation study of FETAX: Phase I testing. *J Appl Toxicol* 1994;14(3):213–223.
121. Cardellini P, Ometto L. Teratogenic and toxic effects of alcohol ethoxylate and alcohol ethoxy sulfate surfactants on Xenopus laevis embryos and tadpoles. *Ecotoxicol Environ Saf* 2001;48(2): 170–177.
122. Sabourin TD, Faulk RT, Goss LB. The efficacy of three non-mammalian test systems in the identification of chemical teratogens. *J Appl Toxicol* 1985;5(4):227–233.
123. Nakatsuji N. Craniofacial malformation in Xenopus laevis tadpoles caused by the exposure of early embryos to ethanol. *Teratology* 1983;28(2):299–305.
124. Uray NJ, Sexton PS. Morphological alterations in the cerebella of ethanol-treated larval bullfrog tadpoles. *Alcohol Clin Exp Res* 1989; 13:320.
125. Malicki JJ, Pujic Z, Thisse C, Thisse B, Wei X. Forward and reverse genetic approaches to the analysis of eye development in zebrafish. *Vision Res* 2002;42(4):527–533.
126. Dasmahapatra AK, Doucet HL, Bhattacharyya C, Carvan MJ 3rd. Developmental expression of alcohol dehydrogenase (ADH3) in zebrafish (Danio rerio). *Biochem Biophys Res Commun* 2001;286(5): 1082–1086.
127. Guo S, Brush J, Teraoka H, *et al.* Development of noradrenergic neurons in the zebrafish hindbrain requires BMP, FGF8, and the homeodomain protein soulless/Phox2a. *Neuron* 1999;24(3):555–566.
128. Van Leeuwen CJ, Grootelaar EM, Niebeek G. Fish embryos as teratogenicity screens: A comparison of embryotoxicity between fish and birds. *Ecotoxicol Environ Saf* 1990;20:42–52.
129. Carvan MJ 3rd, Dalton TP, Stuart GW, Nebert DW. Transgenic zebrafish as sentinels for aquatic pollution. *Ann NY Acad Sci* 2000;919:133–147.
130. Gerlai R, Lahav M, Guo S, Rosenthal A. Drinks like a fish: Zebra fish (Danio rerio) as a behavior genetic model to study alcohol effects. *Pharmacol Biochem Behav* 2000;67(4):773–782.
131. Hurd MW, Debruyne J, Straume M, Cahill GM. Circadian rhythms of locomotor activity in zebrafish. *Physiol Behav* 1998;65(3): 465–472.
132. Blader P, Strahle U. Ethanol impairs migration of the prechordal plate in the zebrafish embryo. *Dev Biol* 1998;201(2):185–201.
133. Bilotta J, Saszik S, Givin CM, Hardesty HR, Sutherland SE. Effects of embryonic exposure to ethanol on zebrafish visual function. *Neurotoxicol Teratol* 2002;24(6):759–766.
134. Dooley K, Zon LI. Zebrafish: A model system for the study of human disease. *Curr Opin Genet Dev* 2000;10(3):252–256.
135. Westerfield M. *The Zebrafish Book: A Guide for the Laboratory Use of Zebrafish (Danio rerio).* Eugene, OR: University of Oregon Press, 2000.
136. Heberlein U. Genetics of alcohol-induced behaviors in Drosophila. *Alcohol Res Health* 2000;24(3):185–188.
137. Hidalgo A. Interactive nervous system development: Control of cell survival in Drosophila. *Trends Neurosci* 2002;25(7):365–370.
138. Bokor K, Pecsenye K. Differences in the effect of ethanol on fertility and viability components among laboratory strains of Drosophila melanogaster. *Hereditas* 2000;132(3):215–227.
139. Lynch DW, Schuler RL, Hood RD, Davis DG. Evaluation of Drosophila for screening developmental toxicants: Test results with eighteen chemicals and presentation of a new Drosophila bioassay. *Teratog Carcinog Mutagen* 1991;11(3):147–173.
140. Saigusa T, Ishizaki S, Watabiki S, *et al.* Circadian behavioural rhythm in Caenorhabditis elegans. *Curr Biol* 2002;12(2):R46–47.
141. Rose JK, Rankin CH. Analyses of habituation in Caenorhabditis elegans. *Learn Mem* 2001;8(2):63–69.
142. Dhawan R, Dusenbery DB, Williams PL. Comparison of lethality, reproduction, and behavior as toxicological endpoints in the nematode Caenorhabditis elegans. *J Toxicol Environ Health* 1999;58(7): 451–462.
143. Morgan PG, Sedensky MM. Mutations affecting sensitivity to ethanol in the nematode, Caenorhabditis elegans. *Alcohol Clin Exp Res* 1995;19(6):1423–1429.
144. Spong CY, Abebe DT, Gozes I, Brenneman DE, Hill JM. Prevention of fetal demise and growth restriction in a mouse model of fetal alcohol syndrome. *J Pharmacol Exp Ther* 2001;297(2):774–779.
145. Klintsova AY, Goodlett CR, Greenough WT. Therapeutic motor training ameliorates cerebellar effects of postnatal binge alcohol. *Neurotoxicol Teratol* 2000;22(1):125–132.
146. Ivkovich D, Eckerman CO, Krasnegor NA, Stanton ME. Using eye blink conditioning to assess neurocognitive development in human infants. In: Woodruff DS, Steinmetz JE, Eds. *Eye Blink Classical Conditioning: Applications in Humans.* Amsterdam, The Netherlands: Kluver Academic Publishers; 2000:135–153.
147. Woodruff-Pak DS, Logan CG, Thompson RF. Neurobiological substrates of classical conditioning across the life span. In: Diamond A, Ed. *The Development and Neural Bases of Higher Cognitive Functions.* New York: The New York Academy of Sciences, 1990:150–178.
148. Freeman JH Jr, Barone S Jr, Stanton ME. Disruption of cerebellar maturation by an antimitotic agent impairs the ontogeny of eyeblink conditioning in rats. *J Neurosci* 1995;15(11):7301–7314.
149. Stanton ME, Goodlett CR. Neonatal ethanol exposure impairs eyeblink conditioning in weanling rats. *Alcohol Clin Exp Res* 1998; 22(1):270–275.
150. Green JT, Johnson TB, Goodlett CR, Steinmetz JE. Eye blink classical conditioning and interpositus nucleus activity are disrupted in adult rats exposed to ethanol as neonates. *Learning Memory* 2002;9:304–320.
151. Green JT, Tran T, Goodlett CR, Steinmetz JL. Neonatal ethanol produces cerebellar deep nuclear cell loss and correlated disruption of eye blink conditioning in adult rats. *Brain Res* 2002;956:302–311.
152. Goodlett CR. Early postnatal alcohol exposure that produces high blood alcohol levels impairs development of spatial navigation learning. *Psychobiology* 1987;15:64–74.
153. Hamilton DA, Kodituwakku P, Sutherland RJ, Savage DD. Children with fetal alcohol syndrome are impaired at place learning but not cued-navigation in a virtual Morris water task. *Behav Brain Res* 2003;143(1):85–94.

63 Modeling Drug and Alcohol Abuse

Experimental Examples of the Utility of Zebrafish

ROBERT T. GERLAI

ABSTRACT

This chapter will discuss the utility of zebrafish (*Danio rerio*) in the analysis of the biological and genetic mechanisms of acute and chronic alcohol (ethanol, ethyl alcohol, or ETOH) consumption on human brain function and it will also present examples on how zebrafish can be used as a model organism to study mechanisms of addiction to this and other drugs of abuse. The chapter's focus is biased toward behavior genetic approaches, i.e., it is not intended to provide a comprehensive overview of all possibilities with zebrafish. This bias represents a subjective view, but we also argue that genetic applications and a focus on behavioral function are perhaps particularly appropriate for modeling human alcohol and drug abuse, and we also argue that zebrafish will provide unprecedented insights into these diseases.

Key Words: Alcohol addiction, Alcoholism, Behavioral phenotyping, Drug abuse, Forward genetics, Zebrafish.

ZEBRAFISH, A FAVORITE OF DEVELOPMENTAL BIOLOGISTS AND GENETICISTS

Zebrafish has been a popular subject of developmental biology and genetics for the past three decades.[1] This popularity is mainly due to the pioneering work of Streisinger *et al.*,[2] who realized that certain characteristics of this vertebrate organism make it particularly useful for the investigation of biological processes involved in the ontogenesis of vertebrates (see Chapters 13 and 14 of this volume for a more general discussion of the use of zebrafish in biomedical research). Most importantly, zebrafish has an externally developing embryo that is transparent and thus its tissues and organs can be directly observed under the microscope. Another important advantage of this vertebrate over other model organisms, including the mouse or the rat, is that it is small (adult zebrafish reach a length of 4 cm) and is highly social, and thus can be housed in large groups in small fish tanks. Finally, a definite advantage of this species over many other laboratory vertebrate model organisms is its prolific nature. A single female can lay 200 eggs once every 2–3 days under ideal conditions.[3] After realizing these advantages, developmental biologists and geneticists started to develop genetic tools for this species[2,4–7] and as a result by now we have an impressive arsenal of genetic techniques available for zebrafish. For example, genetic markers crucial for the localization and identification of randomly induced mutations have been established. Rapid amplification of polymorphic DNA (RAPD) and amplified fragment length polymorphism (AFLP) have been utilized for positional cloning.[8–10] A genetic linkage map comprising more than 4000 polymorphic microsatellite markers [11] and radiation hybrid maps composed of both microsatellite markers and expressed sequence tags (ESTs)[12,13] are now available. Single nucleotide polymorphisms (SNPs) and oligonucleotide microarrays have been utilized for mapping of zebrafish mutations.[14] A viral infection-based insertional mutagenesis technique has also been established for the generation of insertional mutations that could be rapidly cloned due to the presence of the viral tag in the genome.[15] In addition to forward genetic approaches, reverse genetic methods have been developed. Morpholino antisense knockdown allows the inactivation of known genes in early embryos.[16] Targeted-induced local lesions in genomes (TILLING)[17] has also been successfully adapted to zebrafish.[18] Thus, now complete loss-of-function (null), gain-of-function, or partial loss-of-function (hypomorphic) mutations can be isolated and analyzed. Finally, the genome of zebrafish has been sequenced at the Sanger Center. All these tools and pieces of information are in the public domain (e.g., GenBank, Sanger Center website, and ZFIN).[19] In summary, the genetics of zebrafish is highly developed and places this species on par with the mouse and the fruit fly.[1,4,20,21] It is also important to note that genes discovered in this species have been found evolutionarily conserved and have homologs in mammals including our own species.[22] For example, such genes include genes whose products play roles in alcohol metabolism.[23,24] Furthermore, syntenic relationships between chromosomal regions of zebrafish and mammalian species have been determined.[25,26] It is thus expected that there is a good chance that genetic information discovered using zebrafish will translate well to mammalian systems. Thus could zebrafish be utilized in the mechanistic analysis of human alcoholism? Before we discuss this question, let us consider why we

From: *Sourcebook of Models for Biomedical Research* (P. M. Conn, ed.),

would like to study the biology and genetics of alcohol effects and the abuse of this substance.

ALCOHOLISM IS AN ENORMOUS PROBLEM, YET A CLEAR PICTURE OF ITS MECHANISMS HAS NOT EMERGED

Alcohol abuse and alcoholism are frequent and debilitating diseases. Their direct effects on the individual can be devastating. Alcohol abuse may lead to behavioral and psychiatric problems with severe consequences for the patient as well as for the family and society in general. The direct and indirect costs of alcohol abuse are enormous. In the United States alcohol abuse was estimated to cost more than $150 billion yearly and resulted in 40,000 deaths (National Institute on Alcohol Abuse and Alcoholism, 1990), and these figures are becoming worse.[27,28] Prevalence estimates for alcohol abuse and dependence range between 11% and 16%, amounting to a staggering number of approximately 30 million people afflicted by this disease only in the United States.[29,30]

Treatment of alcohol-related problems includes detoxification (a short-term solution) and long-term rehabilitation programs, e.g., cognitive–behavioral therapy, motivational enhancement therapy, and Alcoholics Anonymous or related 12-step programs.[31] Although detoxification may be achieved, long-term rehabilitation programs have had only limited success, as the percentage of relapse at 3 months after the start of the program approaches 50%.[32] Briefly, there is an enormous unmet medical need for pharmacological intervention. Based on clinical and preclinical findings, several pharmacotherapies have been attempted, e.g., serotonin (5-HT) receptor partial agonists, γ-aminobutyric acid A ($GABA_A$) receptor agonists, dopamine receptor agonists and antagonists, and opioid antagonists[32] have been tried. However, most drugs have failed. Notable exceptions are disulfiram, a compound converted into diethyldithiocarbamate in the body, which then chelates copper and other metals important for metalloenzymes including alcohol dehydrogenase. Efficacy has been shown[33] for naltrexone (an opioid receptor antagonist) and for acamprosate (a drug that has some structural homology to GABA itself).

Efficacious pharmacotherapies have been difficult to develop because of the fact that the mechanisms of alcohol abuse and alcoholism are not understood. Alcohol is a nonspecific and low potency drug that affects numerous molecular targets and biochemical pathways and does it in a rather complex dose-dependent manner. Alcohol is known to impair the *N*-methyl-D-aspartate receptor (NMDA-R)[34] and enhance $GABA_A$-R function[35] via allosteric modulation of these receptors. The anxiolytic and sedative effects of alcohol may be mediated by $GABA_A$-R.[36] Inhibition of NMDA-R may underlie the memory-impairing effects of alcohol.[37] The reinforcing effects of alcohol are thought to be mediated by the dopaminergic system and serotonin (e.g., $5\text{-}HT_3$-R) receptors[36] and endogenous opioids.[32] Neuropeptides, e.g., neurotensin and corticotrophin-releasing factor (CRF), have also been implicated.[38] Adenosine (an inhibitory neuromodulator) has also been shown to interact with alcohol.[39] Voltage-gated ion channels (e.g., L-type, N-type, and T-type calcium channels, and the M-type K^+ channel) are inhibited by alcohol.[40–42] Alcohol may indirectly or directly influence second messenger systems or other intracellular signaling cascades[43] leading to modulation of transcription and/or translation processes. In summary, the number of molecular targets with which alcohol interacts is large and the mode of interaction is complex. A clear picture has not emerged as to the key molecular and neurobiological mechanisms that could explain the acute and chronic effects of alcohol. Consequently, development of appropriate pharmacological intervention has been difficult. A possible research route toward the understanding of the mechanisms of the effects of alcohol is to use model organisms. In the following we discuss some notable studies in this area of research.

DEVELOPMENTAL EFFECTS OF ALCOHOL IN ZEBRAFISH

The effects of alcohol on zebrafish have been extensively studied from a developmental biology perspective. For over three decades it has been clear that alcohol is deleterious for the developing zebrafish embryo: notochord and spinal cord duplications arose in the embryo after alcohol exposure.[44] Numerous recent studies have investigated the teratogenic and toxicological properties of alcohol in zebrafish and found that exposure to this substance during embryogenesis significantly affects the visual system leading to cyclopia, delayed, and abnormal lamination of the neural retina and to abnormalities in the pattern of tectal lamination as well.[45] Hallare *et al.*[46] showed that posthatching exposure leads to significantly increased mortality at higher alcohol doses (1.5% and higher) and it also induces numerous abnormalities including edema, reduced heartbeat, diminished pigmentation, and other major developmental defects such as disrupted axial body formation. In fact, among the three solvents studied [alcohol, acetone, and dimethyl sulfoxide (DMSO)], alcohol had the most deleterious effects. Zebrafish has also been proposed as a model of fetal alcohol syndrome.[47] In this study several of the above-mentioned abnormalities were found and the degree of severity of these abnormalities correlated with both the concentration of alcohol and the age at which the embryos were exposed to the substance. Strain differences in the developmental abnormalities induced by early alcohol exposure have also been demonstrated,[48,49] suggesting the involvement of genetic factors. The behavioral effects of acute alcohol exposure on the zebrafish embryo have been studied in detail by Lockwood *et al.*[50] who found a dose-dependent alcohol effect on locomotion closely corresponding to a previously described inverted U-shaped dose–response curve in adults[51] with intermediate doses resulting in the highest increase in activity. In addition, similar to what has been described in the adult,[51] the embryo also exhibited alcohol-induced melanocyte responses, i.e., an apparent darkening of skin pigmentation in response to alcohol.

Apart from the results presented in the latter study,[51] however, little is known about the effects of alcohol on the brain function and behavior of adult zebrafish. Given that in the human society alcoholism affects a large number of adults[30] and the mechanisms of alcohol's action and of alcohol addiction are not well understood,[52] analysis and modeling of the effects of alcohol on the adult vertebrate brain continue to be an important goal.[53] Classical laboratory model species including the rat and the mouse have been utilized to study this question. Many of these studies employed genetic approaches[53] and some of these studies are reviewed below.

RODENT MODELS OF THE HUMAN DISEASE: THE ADVANTAGE OF GENETIC APPROACHES AND THEIR COMPARISON

There is no a priori reason why genetics should be more powerful than any other approaches, and thus by focusing on genetic approaches in this chapter we do not imply that factors other than genes are unimportant in the development of alcoholism and in the abuse of alcohol and of other drugs. However, genetics provides an important experimental tool with which to probe the biological mechanisms of human diseases. Even when a larger proportion of human disease cases is due to nongenetic factors, genetic methodology has significantly contributed to the understanding of disease mechanism. For example, in Alzheimer's disease the familial cases account for only less than 5% of the patient population, yet the identification of the amyloid precursor protein, and the discovery of the role of its abnormal processing leading to the accumulation of A-β peptide fragments in neurons, has been significantly facilitated by genetics[54] (also see Chapter 22 of this volume). The point of genetic approaches is thus that they provide a window to the biological mechanisms irrespective of whether the cause of the disease is primarily genetic or environmental in nature. Besides, alcoholism and alcohol addiction as well as addiction to other drugs of abuse have been found to have a significant heritable component.[52,55,56]

To address the complexities of alcohol effects and the use and abuse of this substance in the adult, numerous genetic approaches have been employed using animal models. These include quantitative genetic (e.g., inbred strains and their crosses, selective breeding, recombinant inbred and recombinant inbred congeneic lines) and molecular genetic approaches (e.g., transgenic/knockout models). Gene expression analysis, e.g., gene chips, has also been employed. One of the first genetic approaches in alcohol research was the long-sleep (LS) and short-sleep (SS) selected mouse lines. LS mice bred for increased sensitivity to the sedative effects of alcohol showed enhanced response to the $GABA_A$ agonist muscimol compared to SS mice, suggesting involvement of GABA receptors in alcoholism.[57] Bidirectional selection was also conducted for seizure susceptibility associated with alcohol withdrawal and the withdrawal-seizure-prone (WSP) and withdrawal-seizure-resistant (WSR) mouse lines were generated.[58] Analysis of these selected lines revealed that seizure proneness is associated with increased NMDA-elicited convulsions and elevated expression of Ca^{2+} channels and $GABA_A$ receptor subunits. The disadvantage of all quantitative genetic studies including selection experiments is that only genes showing variability in the studied population may be investigated.

Quantitative trait locus (or QTL) mapping is also limited to the study of genes that show variability in the experimental population. Nevertheless, QTL mapping has identified chromosomes 1, 4, and 11 as the locus of genes associated with alcohol withdrawal in the mouse,[58] and finer mapping with recombinant congeneic mouse strains has been narrowing the regions of interest[59] leading to potential identification of the actual genes. Despite the widespread use of QTL analysis, however, success has been slow[58,60] (but see Tabakoff *et al.*[37]).

The effect of overexpression and underexpression of a candidate gene may be investigated *in vivo* in transgenic mice. "Candidate" genes tested for alcohol-related questions in this manner include, e.g., those encoding $GABA_A$ receptor subunits or the 5-HT_{1B} receptor, PKCγ2, dopamine receptors (e.g., D_2 and D_4), and neuropeptide Y as well as the *fyn* intracellular tyrosine kinase.[58] The pros and cons of transgenic techniques have been extensively discussed.[61,62] Briefly, the greatest disadvantage of these approaches is that they only allow the investigation of known genes but not the discovery of the involvement of novel ones.

Discovery of novel genes may be achieved by forward genetic approaches (see, e.g., the identification of the "Clock" gene in the mouse reviewed by Pinto and Takahashi[63]). Here the mutation is induced randomly, the mutation effects are tested at the phenotypical level, and the animal carrying the mutation is identified and then bred. In case of dominant or semidominant, i.e., not completely recessive, mutation the heritable change is observed at the F_1 generation and recessive mutations are detected at the F_3 generation. Localization and subsequent identification of the mutant gene are based on genetic marker-aided linkage analysis followed by positional cloning and or gene expression analysis.[63] Forward genetic approaches are now being utilized in zebrafish. For example, Guo *et al.*[64] have been using this strategy to identify genes involved in substance abuse with particular focus on the dopaminergic system. Darland and Dowling[65] and colleagues also focus on substance abuse (cocaine)-related behavioral mutants as well as visual system-related mutants.[66] The advantage of the mutagenesis approach is several-fold. First, compared to quantitative genetic studies, forward genetics is not limited to the discovery of genes that show variability. Second, unlike transgenic approaches, forward genetics allows the discovery of novel genes. Third, several mutations may be isolated for the same gene and from these mutants it is possible to make inferences about the function of the gene product. Fourth, both loss and gain of function mutations may be identified. In the following, therefore, we mainly focus on forward genetic applications and present a few examples of how these studies contributed to, or are expected to contribute to, the understanding of the mechanisms of alcohol and other drug abuse-related human diseases with the use of zebrafish.

NOVEL GENES POTENTIALLY RELEVANT IN DRUG USE AND ABUSE: THE FIRST SUCCESS STORIES USING FORWARD GENETICS IN ZEBRAFISH

The characteristics of zebrafish make this species particularly amenable to high-throughput screening, and indeed, forward genetic approaches, which are screening intensive, have already yielded successfully results with this organism. For example, Guo *et al.*[64] have been using this strategy to identify genes involved in substance abuse with particular focus on the dopaminergic system. They implemented a mutation screen based not on behavioral phenotypes but on the histological analysis of dopaminergic neurons. Based on alterations using tyrosine hydroxylase (TH) immunohistochemistry, they detected mutants whose catecholaminergic systems were affected. For example, the "too few" mutant exhibited a reduced number of dopaminergic neurons in the hypothalamus during embryonic stages.[64] Later, Rink and Guo[67] discovered that the "too few" mutation also affected the development of serotonergic neurons, and identified that this mutation disrupted a zinc-finger-containing protein,[68] a probable transcription regulator. Another mutation they identified was "foggy."[64]

This mutant exhibited abnormalities (reduced or lack of TH staining) in diencephalic, telencephalic, and retinal DA neurons, as well as in the locus coeruleus and in peripheral sympathetic neurons. Guo *et al.*[9] identified the gene carrying the "foggy" mutation using positional cloning and found that it is a phosphorylation-dependent dual regulator of transcription elongation, a finding that provided the first key evidence that negative regulators of transcription elongation may control key aspects of neuronal development. Why is this all relevant for addiction research? It is, as Guo[69] argues, because the dopaminergic system plays crucial roles in reward and is known to be a key player in drug abuse.

Behavioral characteristics have also been utilized as criteria for the identification of mutants in forward genetic studies. Darland and Dowling have shown development of preference for cocaine in zebrafish after a single trial in a place preference paradigm[65] demonstrating that a drug abused by humans also has rewarding properties in zebrafish. In their forward genetic screen using the conditioned place preference paradigm Darland and Dowling identified three distinct mutations selectively affecting cocaine-dependent behaviors.[65]

THE BOTTLENECK: PHENOTYPICAL SCREENING PARADIGMS

Up to now, we have not discussed an important disadvantage associated with the use of zebrafish. Although this species is excellent for high-throughput screening due to the ease with which random genetic mutations may be induced and propagated in a large number of offspring, a major disadvantage of zebrafish is that appropriate phenotypical screening techniques are available mainly for the developmental biologist but not when it comes to the question of what behavioral alteration or brain dysfunction may be elicited by the mutation. Briefly, the foundation of forward genetics, the phenotypical screening methodology, is mainly missing. This critical comment, of course, does not concern the already successful forward genetic studies discussed above, but it must be noted that these studies represent the exception rather than the rule: compared to what we know about the behavior and brain function of other model organisms, zebrafish may be regarded as an uncharted species.[70,71] Given that the success of gene identification and thus the discovery of how genes may contribute to particular phenotypes rest on our ability to identify mutants and given that alterations of brain function may be best detected using behavioral testing, appropriate and sensitive behavioral screening applications are an absolute must when it comes to the question of how genes contribute to drug use and abuse. In the following, we present our first few attempts to develop behavioral screening applications for the detection of mutations affecting alcohol use and abuse.

CHARACTERIZATION OF ZEBRAFISH BEHAVIOR AND THE DEVELOPMENT OF HIGH-THROUGHPUT TEST METHODS

First let us consider an important point that is sometimes confused in the literature, the definition of an "animal model." According to our view, an animal model can be regarded as a model if it has at least some construct, face, and predictive validity. That is, if the model resembles some mechanistic aspects of the human disease, if it is similar to it in its phenotypical manifestation (appearance), and if it can reliably predict the outcome of some manipulation that is expected to influence aspects of the human disease. Why are we discussing these points here? Because we want to emphasize that the following discussion will not be about an animal model in this strict sense. It will include our latest results concerning the development of behavioral testing methodology that will be useful for the identification of mutations affecting alcohol-induced behavioral responses. That is, it will include the foundation of forward genetics.

We already mentioned that we[51] investigated the acute effects of alcohol in adult zebrafish. The tests we developed were designed to be deliberately simplistic so that automation and fast speed could be achieved in future mutagenesis studies. The results we obtained showed that alcohol-induced behavioral changes closely resembled those seen in humans. For example, locomotory activity of adult zebrafish showed a characteristic inverted U-shaped dose–response curve with intermediate doses eliciting hyperactivity and the highest applied dose eliciting a sedative effect (Figure 63–1). Similarly, intermediate doses elicited increased aggression as measured by how close a fish swam to his mirror image (Figure 63–2) or how much aggressive posture it showed toward it (Figure 63–3). The shortcoming of this study was that all behavioral responses were manually quantified, i.e., a human observer blind to the experimental conditions scored the previously recorded videotapes. While this method is very precise and is unbiased,[72] it is rather time consuming. To investigate the feasibility of automated computer-based behavioral quantification in the analysis of zebrafish behavior two manual quantification methods were

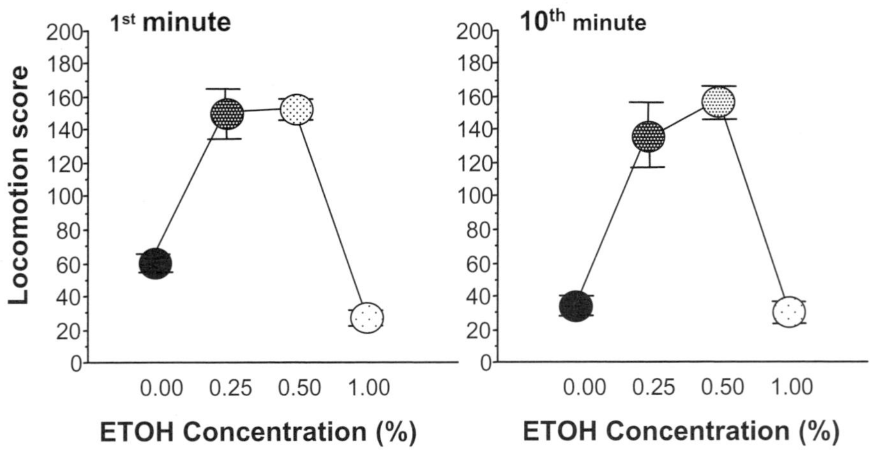

Figure 63–1. Locomotory activity of zebrafish (outbred wild-type population) after acute (1-h-long) alcohol treatment is increased by intermediate doses of alcohol. The locomotion score was manually recorded and is calculated as the number of crossings between segments of the observation tank during a 1-min observation session at the beginning (first minute) and at the end (tenth minute) of the test. Mean ± SE are shown. Sample sizes: ETOH 0.00% = 13, ETOH 0.25% = 15, ETOH 0.50% = 13, ETOH 1.00% = 16. Note that ETOH 1.00% fish exhibited depression of activity while the other two alcohol-treated groups showed a robust increase. For further details and statistical analyses see Gerlai.[70]

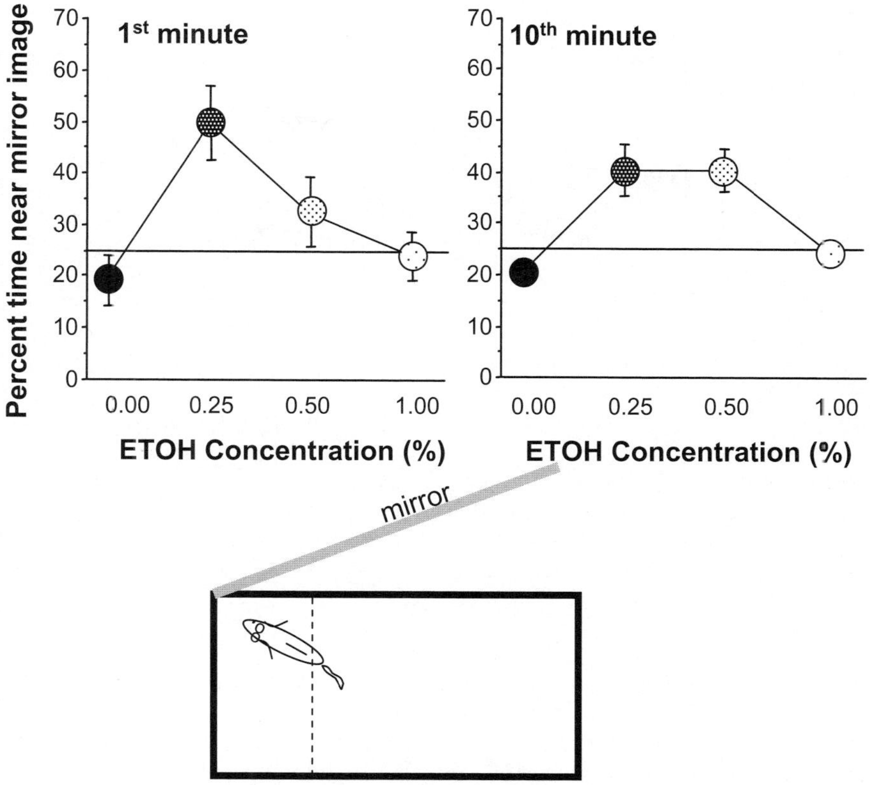

Figure 63–2. Intermediate alcohol concentrations (acute 1-h-long treatment) enhance preference for staying close to the mirror image, a "conspecific opponent," indicating enhanced aggression. Mean ± SE are shown. Sample sizes: ETOH 0.00% = 15, ETOH 0.25% = 16, ETOH 0.50% = 16, ETOH 1.00% = 18. Similar to locomotor activity, performance was quantified for the first and the tenth minute of the session. Note the inclined mirror making the mirror image appear larger when viewed by the subject from the left side of the tank. Note the inverted U-shaped dose–response curve. The solid horizontal line represents the chance level. For further details and statistical analyses see Gerlai.[70]

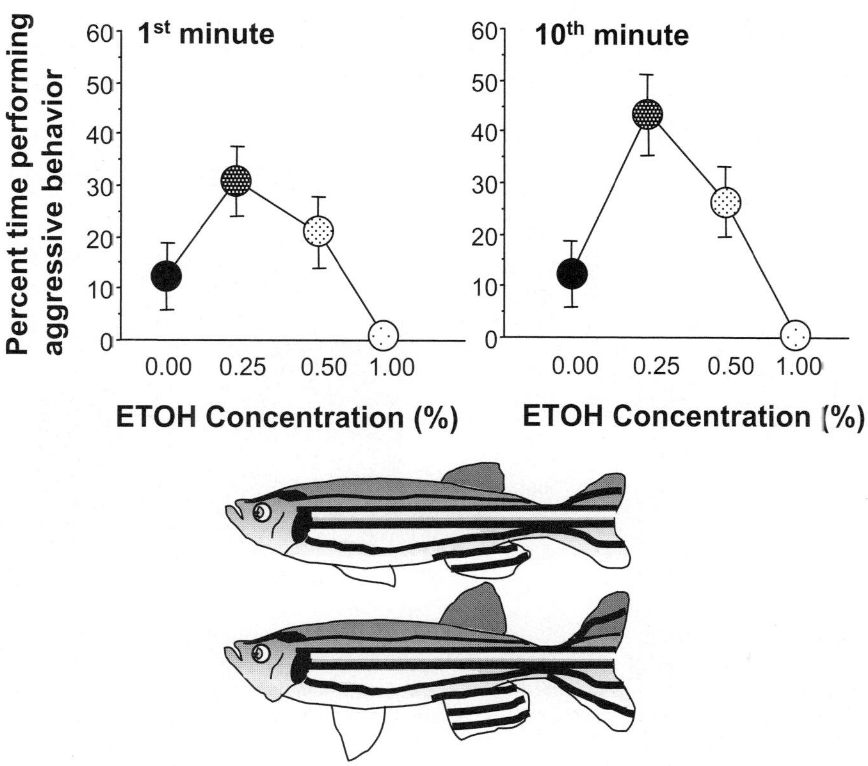

Figure 63–3. Intermediate doses of alcohol (1-h acute treatment) enhance aggressive behavioral responses elicited by the sight of an opponent (mirror image) as measured by quantifying the amount of time the fish spent with agonistic motor and posture patterns. Mean ± SE are shown. Sample sizes: ETOH 0.00% = 15, ETOH 0.25% = 16, ETOH 0.50% = 16, ETOH 1.00% = 18. Note that data are obtained by quantifying videorecordings of the same fish as in Figure 63–2 and sample sizes are identical to those shown in that figure. Also note that the lower one of the two zebrafish is displaying (showing an aggressive posture). For further details and statistical analyses see Gerlai.[70]

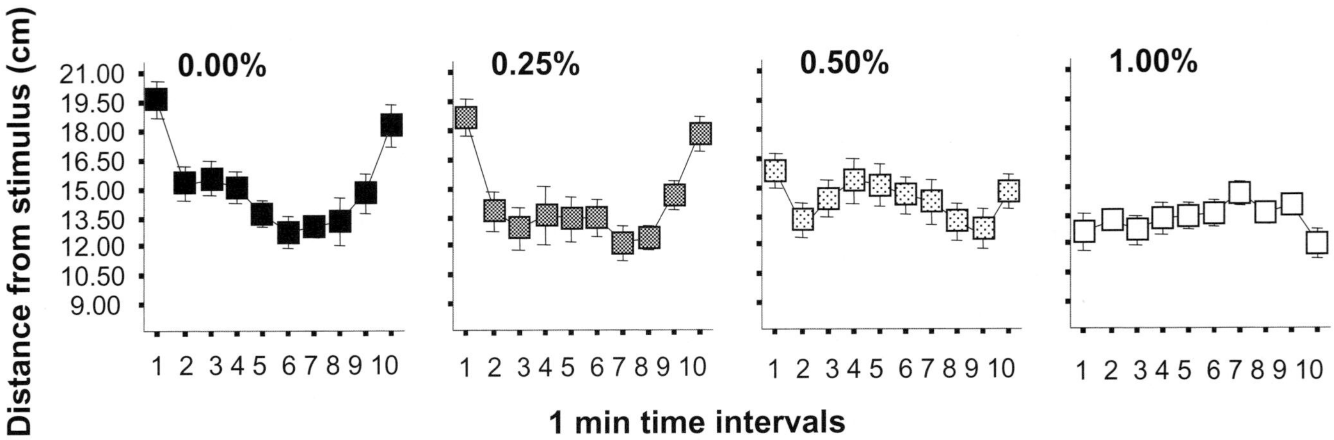

Figure 63–4. Acute (1-h-long) alcohol treatment reduces the distance of zebrafish from a predator model stimulus in a dose-dependent manner. The acute dose is indicated above the graphs. Mean ± SE are shown. Sample size (*n*) for each dose group equaled 14. Note the U-shaped time trajectories, which are due to the fact that the predator model was presented only during the first and during the last minute of the test. Also note that the predator model is very effective and elicits increased distance at 0.00% and at 0.25% alcohol doses but at higher doses its effect diminishes. For further details and statistical analyses see Gerlai *et al.*[73]

compared with a more sophisticated video-tracking-based automated technique. A very good correlation was found among the results obtained by the different methods and this suggested that automated behavioral quantification, and thus high-throughput behavioral screening, will be possible. Utilizing video-tracking and the previously developed test paradigms, we are now investigating the acute and the chronic effects of alcohol exposure in adult zebrafish. Our preliminary findings suggest that alcohol has fear-reducing properties at the doses employed (interpreted as an anxiolytic effect in the human literature) that manifest, for example, as reduced distance from a predator model (Figure 63–4). Importantly, this effect was detected without the need for a human observer to be present. Chronic alcohol exposure effects have also been successfully quantified using the automated video-tracking method. We have also found that after a 2-week-long exposure to 0.25% ethanol, fish exhibit significant adaptation to the substance and when exposed to the same concentration during a test they do not increase their activity level (Figure 63–5), i.e., we have obtained evidence for adaptation to alcohol in zebrafish.[73] Signs of withdrawal symptoms were also detected in that fish that received chronic alcohol treatment showed hyperactivity when exposed to fresh water (Figure 63–5). Briefly, these results suggest that acute and chronic exposure to alcohol has significant behavior-altering effects in zebrafish analogous in appearance to those

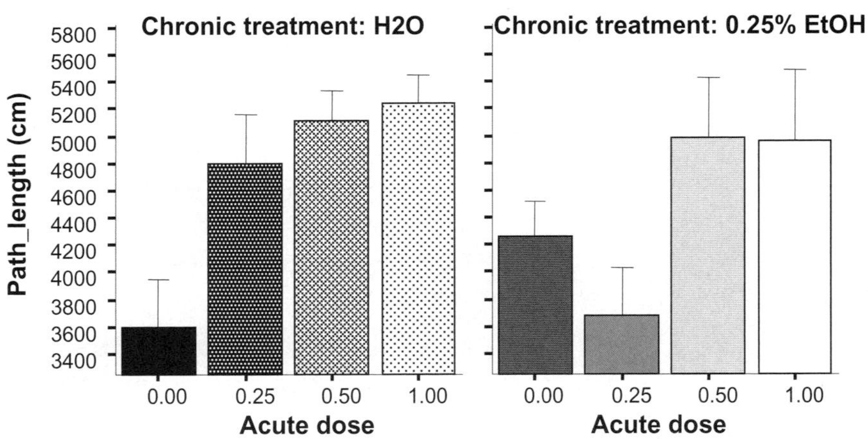

Figure 63–5. The effects of acute and chronic alcohol treatment on the locomotor activity of zebrafish: the total path length (centimeters) traveled during the 10-min recording session shows adaptation to alcohol. The left panel shows the results for fish that were not exposed to alcohol during chronic treatment (chronic treatment H_2O). The right panel shows the results for fish that received chronic ethanol (0.25% concentration) for 2 weeks (chronic treatment ETOH). Mean + SEM are shown. Sample sizes are as follows. Chronic treatment freshwater groups, $n = 14$ for each group. For chronic treatment ETOH, n = 13, 12, 12, and 13, respectively; for acute treatment ETOH, dose = 0.00%, 0.25%, 0.50%, and 1.00%, respectively. Note that the first bar on the left panel (fish that received no alcohol acutely or chronically) is not different from the second bar on the right panel (fish that received 0.25% ethanol both chronically before and acutely during the experiment), suggesting adaptation to alcohol in response to chronic treatment. Also note the somewhat elevated path length value of the first bar on the right panel (fish that received the 0.25% chronic ethanol treatment but were withdrawn from it 1 h before and during the behavioral recording session), suggesting withdrawal-induced hyperactivity. For further details and statistical analyses see Gerlai *et al.*[73]

seen in humans. Also it is important to note that these changes can be quantified using automated methods. Automated computerized behavioral quantification is an important requirement for high-throughput screening. It allows the application of a large number of test apparatuses running in parallel. The limiting factor in such a setup is no longer the amount of time an experimenter has but rather the physical space available and the cost of the equipment. Given that the test tanks are small and the equipment is also inexpensive (commercially available video equipment and computer hardware and software), high-throughput screening is now thus feasible.

In summary, the above results demonstrate that high-throughput behavioral screens are feasible with zebrafish. Given the strength of the genetics of this species and the increasing number of laboratories that utilize zebrafish as a model system, we forecast that zebrafish will be a promising animal model with which the acute and the chronic effects of drugs of abuse, including alcohol, may be investigated both in the embryo and in the adult. These investigations will undoubtedly lead to discoveries of novel genes and biochemical mechanisms involved in alcohol and drug abuse-related abnormalities and thus will allow the development of biologically sound animal models of these diseases.

REFERENCES

1. Eisen JS. Zebrafish make a big splash. *Cell* 1996;87:969–977.
2. Streisinger G, Walker C, Dower N, Knauber D, Singer F. Production of clones of homozygous diploid zebrafish (Brachydanio rerio). *Nature* 1981;291:293–296.
3. Westerfield M. *The Zebrafish Book*. Eugene, OR: University of Oregon Press, 1993.
4. Walker C, Streisinger G. Induction of mutations by r-rays in pregonial germ cells of zebrafish. *Genetics* 1983;103:109–124.
5. Driever W, Solnica-Krezel L, Schier AF, Neuhauss SCF, Malicki J, Stemple DL, Stainier DYR, Zwartkruis F, Abdelilah S, Rangini Z. A genetic screen for mutations affecting embryogenesis in zebrafish. *Development* 1996;123:37–46.
6. Haffter P, Granato M, Brand M, Mullins MC, Hammerschmidt M, Kane DA, Odenthal J, Van Eeden FJM. The identification of genes with unique and essential function in the development of the zebrafish, Danio rerio. *Development* 1996;123:1–36.
7. Granato M, Van Eeden FJM, Schach U, Trowe T, Brand M, Furutani-Seiki M, Haffter P, Hammerschmidt M, Heisenberg CP, Jiang YJ, Kane DA, Kelsh RN, Mullins MC, Odenthal J, Nusslein-Volhard C. Genes controlling and mediating locomotion behavior of the zebrafish embryo and larva. *Development* 1996;123:399–413.
8. Donovan A, Brownlie A, Zhou Y, Shepard J, Prat SJ, Moynihan J, Paw BH, Drejer A, Barut B, Zapata A. Positional cloning of zebrafish ferroportin 1 identifies a conserved vertebrate iron exporter. *Nature* 2000;403:776–781.
9. Guo S, Yamaguchi Y, Schilbach S, Wada T, Lee J, Goddard A, French D, Handa H, Rosenthal A. A regulator of transcriptional elongation controls vertebrate neuronal development. *Nature* 2000; 408:366–369.
10. Zhang J, Talbot WS, Schier AF. Positional cloning identifies zebrafish one-eyed pinhead as a permissive EGF-related ligand required during gastrulation. *Cell* 1998;92:241–251.
11. Knapik EW, Goodman A, Ekker M, Chevrette M, Delgado J, Neuhauss SCF, Shimoda N, Driever W, Fishman MC, Jacob H. A microsatellite genetic linkage map for zebrafish. *Nat Genet* 1998;18: 338–343.
12. Geisler R, Rauch GJ, Baier H, Van Bebber F, Brobeta L, Dekens MP, Finger K, Fricke C. A radiation hybrid map of the zebrafish genome. *Nat Genet* 1999;23:86–89.
13. Hukriede NA, Joly L, Tsang M, Miles J, Tellis P, Epstein JA, Barbazuk WB, Li FN, Paw B, Postlewait JH. Radiation hybrid mapping of the zebrafish genome. *Proc Natl Acad Sci USA* 1999;96: 9745–9750.
14. Stickney HL, Schmutz J, Woods IG, Holtzer CC, Dickson MC, Kelly PD, Myers RM, Talbot WS. Rapid mapping of zebrafish mutations with SNPs and oligonucleotide microarrays. *Genome Res* 2002;12: 1929–1934.
15. Amsterdam A, Burgess S, Golling G, Chen W, Sun Z, Townsend K, Farrington S, Haldi M, Hopkins N. A large-scale insertional mutagenesis screen in zebrafish. *Genes Dev* 1999;13:2713–2724.
16. Nasevicius A, Ekker SC. Effective targeted gene "knockdown" in zebrafish. *Nat Genet* 2000;26:216–220.
17. McCallum CM, Comai L, Greene EA, Henikoff S. Targeting induced local lesions in genomes (TILLING) for plant functional genomics. *Plant Physiol* 2000;123:439–442.
18. Wienholds E, Schulte-Merker S, Walderich B, Plasterk RHA. Target selected inactivation of the zebrafish rag1 gene. *Science* 2002;297: 99–102.
19. Sprague J, Doerry E, Douglas S, Westerfield M. The zebrafish information network (ZFIN): A resource for genetic, genomic and developmental research. *Nucleic Acids Res* 2001;29:87–90.
20. GranatoM, Nusslein-Volhard C. Fishing for genes controlling development. *Curr Opin Genet Dev* 1996;6:461–468.
21. Grunwald DJ. A fin-de-siècle achievement: Charting new waters in vertebrate biology. *Science* 1996;274:1634–1635.
22. Cerda J, Conrad M, Markl J, Brand M, Herrmann H. Zebrafish vimentin: Molecular characterisation, assembly properties and developmental expression. *Eur J Cell Biol* 1998;77:175–187.
23. Lassen N, Estey T, Tanguay RL, Pappa A, Reimers MJ, Vasiliou V. Molecular cloning, baculovirus expression, and tissue distribution of the zebrafish aldehyde dehydrogenase 2. *Drug Metab Dispos* 2005; 33:649–656.
24. Reimers MJ, Hahn ME, Tanguay RL. Two zebrafish alcohol dehydrogenases share common ancestry with mammalian class I, II, IV, and V alcohol dehydrogenase genes but have distinct functional characteristics. *J Biol Chem* 2004;279:38303–38312.
25. Barbazuk WB, Korf I, Kadavi C, Heyen J, Tate S, Wun E, Bedell JA, McPherson JD, Johnson SL. The syntenic relationship of the zebrafish and human genomes. *Genome Res* 2000;10:1351–1358.
26. Woods IG, Kelly PD, Chu F, Ngo-Hazelett P, Yan YL, Huang H, Postlewait JH, Talbot WS. A comparative map of the zebrafish genome. *Genome Res* 2000;10:1903–1914.
27. Rice DP. Economic costs of substance abuse, 1995. *Proc Assoc Am Physicians* 1999;111:119–125.
28. Harwood HJ, Fountain D, Livermore G. Economic costs of alcohol abuse and alcoholism. *Recent Dev Alcohol* 1998;14:307–330.
29. Robbins LN, Helzer JE, Weissman MM, Orvaschel H, Gruenberg E, Burke JD, Regier DA. Lifetime prevalence of specific psychiatric disorders in three sites. *Arch Gen Psychiatry* 1984;41:949–958.
30. Sullivan EJ, Handley SM. Alcohol and drug abuse. *Annu Rev Nurs Res* 1993;11:281–297.
31. Fuller RK, Hiller-Sturmhöfel S. Alcoholism treatment in the United States. An overview. *Alcohol Res Health* 1999;23:69–77.
32. O'Brien CP, Eckardt MJ, Linnoila MI. Pharmacotherapy of alcoholism. In: Bloom FE, Kupfer DJ, Eds. *Psychopharmacology: The Fourth Generation of Progress*. New York: Raven Press, Ltd., 1995: 1745–1755.
33. Rawson RA, McCann MJ, Hasson AJ, Ling W. Addiction pharmacotherapy 2000: New options, new challenges. *J Psychoactive Drugs* 2000;32:371–378.
34. Lovinger DM, White G, Weight FF. NMDA receptor-mediated synaptic excitation selectively inhibited by ethanol in hippocampal slice from adult rat. *J Neurosci* 1990;10:1372–1379.
35. Deitrich RA, Dunwiddie TV, Harris RA, Erwin VG. Mechanisms of action of ethanol: Initial central nervous system actions. *Pharmacol Rev* 1989;41:489–537.
36. Nestler EJ, Hyman SE, Malenka RC. *Molecular Neuropharmacology: A Foundation for Clinical Neuroscience*. New York: McGraw-Hill, 2001.

37. Tabakoff B, Bhave SV, Hoffman PL. Selective breeding, quantitative trait locus analysis, and gene arrays identify candidate genes for complex drug-related behaviors. *J Neurosci* 2003;23:4491–4498.
38. Erwin VG. Neurotensin: A potential mediator of ethanol actions. In: Deitrich RA, Erwin VG, Eds. *Pharmacological Effects of Ethanol on the Nervous System.* Boca Raton, FL: CRC Press, 1996:163–174.
39. Dunwiddie TV. Acute and chronic effects of ethanol on the brain: Interactions of ethanol with adenosine, adenosine transporters, and adenosine receptors. In: Deitrich RA, Erwin VG, Eds. *Pharmacological Effects of Ethanol on the Nervous System.* Boca Raton, FL: CRC Press, 1996:147–162.
40. Crews FT, Morrow AL, Criswell H, Breese G. Effects of ethanol on ion channels. *Int Rev Neurobiol* 1996;39:283–367.
41. Walter HJ, Messing RO. Regulation of neuronal voltage-gated calcium channels by ethanol. *Neurochem Int* 1999;35:95–101.
42. Koob GF. The neuropharmacology of ethanol's behavioral action: New data, new paradigms, new hope. In: Deitrich RA, Erwin VG, Eds. *Pharmacological Effects of Ethanol on the Nervous System.* Boca Raton, FL: CRC Press, 1996:1–12.
43. Hoek JB, Kholodenko BN. The intracellular signaling network as a target for ethanol. *Alcohol Clin Exp Res* 1998;22:224S-230S.
44. Laale HW. Ethanol induced notochord and spinal cord duplications in the embryo of the zebrafish, Brachydanio rerio. *J Exp Zool* 1971; 177:51–64.
45. Arenzana FJ, Carvan MJ 3rd, Aijon J, Sanchez-Gonzalez R, Arevalo R, Porteros A. Teratogenic effects of ethanol exposure on zebrafish visual system development. *Neurotoxicol Teratol* 2006;28:342–348.
46. Hallare A, Nagel K, Kohler HR, Triebskorn R. Comparative embryotoxicity and proteotoxicity of three carrier solvents to zebrafish (Danio rerio) embryos. *Ecotoxicol Environ Saf* 2006;63:378–388.
47. Bilotta J, Barnett JA, Hancock L, Saszik S. Ethanol exposure alters zebrafish development: A novel model of fetal alcohol syndrome. *Neurotoxicol Teratol* 2004;26:737–743.
48. Loucks E, Carvan MJ III. Strain-dependent effects of developmental ethanol exposure in zebrafish. *Neurotoxicol Teratol* 2004;26:745–755.
49. Carvan MJ III, Loucks E, Weberb ND, Williams FE. Ethanol effects on the developing zebrafish: Neurobehavior and skeletal morphogenesis. *Neurotoxicol Teratol* 2004;26:757–768.
50. Lockwood B, Bjerke S, Kobayashi K, Guo S. Acute effects of alcohol on larval zebrafish: A genetic system for large-scale screening. *Pharmacol Biochem Behav* 2004;77:647–654.
51. Gerlai R, Lahav M, Guo S, Rosenthal A. Drinks like a fish: Zebra fish (Danio rerio) as a behavior genetic model to study alcohol effects. *Pharmacol Biochem Behav* 2000;67:773–782.
52. Oroszi G, Goldman D. Alcoholism: Genes and mechanisms. *Pharmacogenomics* 2004;5:1037–1048.
53. Lovinger DM, Crabbe JC. Laboratory models of alcoholism: Treatment target identification and insight into mechanisms. *Nat Neurosci* 2005;8:1471–1480.
54. Neve RL, McPhie DL, Chen Y. Alzheimer's disease: A dysfunction of the amyloid precursor protein(1). *Brain Res* 2000;886:54–66.
55. Cloninger C. Neurogenetic adaptive mechanisms in alcoholism. *Science* 1987;236:410–416.
56. Li MD. The genetics of nicotine dependence. *Curr Psychiatry Rep* 2006;8:158–164.
57. Mihic S, Harris R. Alcohol actions at the GABAA receptor/chloride channel complex. In: Deitrich RA, Erwin VG, Eds. *Pharmacological Effects of Ethanol on the Nervous System.* Boca Raton, FL: CRC Press, 1996:51–72.
58. Browman KE, Crabbe JC. Alcohol and genetics: New animal models. *Mol Med Today* 1999;5:310–318.
59. Blizard DA, Darvasi A. Experimental strategies for mapping quantitative trait loci (QTL) analysis in laboratory animals. In: Crusio WE, Gerlai RT, Eds. *Handbook of Molecular-Genetic Techniques for Brain and Behavior Research.* Amsterdam, the Netherlands: Elsevier, 1999:61–81.
60. Crabbe JC. Alcohol and genetics: New models. *Am J Med Genet* 2002;114:969–974.
61. Gerlai R. Gene targeting studies of mammalian behavior: Is it the mutation or the background genotype? *Trends Neurosci* 1996;19: 177–181.
62. Gerlai R. Gene targeting in neuroscience: The systemic approach. *Trends Neurosci* 1996;19:188–189.
63. Pinto LH, Takahashi JS. Genetic dissection of mouse behavior using induced mutagenesis. In: Crusio WE, Gerlai RT, Eds. *Handbook of Molecular-Genetic Techniques for Brain and Behavior Research.* Amsterdam, the Netherlands: Elsevier, 1999:147–165.
64. Guo S, Wilson SW, Cooke S, Chitnis AB, Driever W, Rosenthal A. Mutations in the zebrafish unmask shared regulatory pathways controlling the development of catecholaminergic neurons. *Dev Biol* 1999;208:473–487.
65. Darland T, Dowling JE. Behavioral screening for cocaine sensitivity in mutagenized zebrafish. *Proc Natl Acad Sci USA* 2001;98:11691–11696.
66. Brockerhoff SE, Hurley JB, Janssen-Bienhold U, Neuhauss SCF, Driever W, Dowling JE. A behavioral screen for isolating zebrafish mutants with visual system defects. *Proc Natl Acad Sci USA* 1995;92:10545–10549.
67. Rink E, Guo S. The too few mutant selectively affects subgroups of monoaminergic neurons in the zebrafish forebrain. *Neuroscience* 2004;127:147–154.
68. Levkowitz G, Zeller J, Sirotkin HI, French D, Schilbach S, Hashimoto H, Hibi M, Talbot WS, Rosenthal A. Zinc finger protein too few controls the development of monoaminergic neurons. *Nat Neurosci* 2003;6:28–33.
69. Guo S. Linking genes to brain, behavior and neurological diseases: What can we learn from zebrafish? *Genes Brain Behav* 2004;3:63–74.
70. Gerlai R. Zebra fish: An uncharted behavior genetic model. *Behav Genet* 2003;33:461–468.
71. Sison M, Cawker J, Buske C, Gerlai R. Fishing for genes influencing vertebrate behavior: Zebrafish making headway. *Lab Anim* 2006;35: 33–39.
72. Blaser R, Gerlai R. Behavioral phenotyping in zebra fish: Comparison of three behavioural quantification methods. *Behav Res Methods* 2006;38:456–469.
73. Gerlai R, Lee V, Blaser R. Effects of acute and chronic ethanol exposure on the behavior of adult zebrafish (Danio rerio). *Pharmacol Biochem Behav* 2006;85(4):752–761.

64 Mouse Models for Experimental Cancer Therapy

SARA ROCKWELL AND KAREN R. ROCKWELL

ABSTRACT

Tumors in laboratory rats and mice have long been the major model systems used in experiments intended to improve the diagnosis and treatment of cancer. A wide variety of rodent models are used in experimental cancer therapy; all have strengths and weaknesses as models for human cancer and as models for developing improved therapies for cancer. The strengths and the limitations of the various models are discussed in this chapter. These should be considered carefully by researchers as they select systems for use in their projects in experimental cancer therapy, as they design and perform their experiments, and as they analyze their data and translate their studies to improve the care of cancer patients.

Key Words: Mouse, Inbred strains, Experimental cancer therapy, Tumor models, Toxicities of anticancer agents, History of laboratory mouse.

INTRODUCTION

ORIGINS OF LABORATORY MICE Popular culture associates white mice with biomedical research, but most people (including many scientists) do not realize that laboratory mice are very special creatures, with unique characteristics and pedigrees that trace back over 30 centuries.[1–4] Humans and mice have lived together since ancient times. The English word *mouse* traces back through the Latin *mus* and Greek *mys* to the Sanskrit *mush*, a name derived from the verb meaning "to steal," showing that the habits of these little rodents were familiar to Indo-Europeans at least 6000 years ago. Greek history ascribed a major victory of the Greek army to the intervention of the god Apollo, who sent mice into the enemy camp the night before the battle to chew through the leather thongs of the shields and render them useless. The Greeks celebrated this divine intervention by building temples to *Apollo Smintheus* (Apollo the god of mice) and by raising under the alters of these temples white mice, which were credited with powers of prophecy and healing. Domesticated mice with unusual coat colors, color patterns, appearance, and behavioral characteristics have been raised and studied in Asia for more that 3000 years. The Chinese character meaning "white mouse" is ancient. The symbol for "spotted mouse" was present in the first Chinese lexicon, compiled in 1100 BCE. Because mice gathered and stored food and other objects, they were considered prudent and wise. In Japan, mice were considered to be the messengers of Diakoku, the god of wealth, and the activities of mice raised in his temples were analyzed to predict the success or failure of business ventures.

People have continued to raise domesticated mice since these ancient times. Domesticated house mice and "fancy mice" with unusual characteristics appear frequently in art and literature throughout the world. By the eighteenth century, breeding of fancy mice had become a global commercial enterprise as well as a common hobby. Mouse farms, which bred and sold mice of European, Asian, and mixed origins, were common in the United States as well as Europe and Asia. Mice were bred for unusual coat colors, including white, black, yellow, and chocolate, and to have patterns that included spots, stripes, and markings like Siamese cats. Strains of dwarf mice, hairless mice, mice with curled or kinked tails, "waltzing" mice, and other unusual mice were also developed.

THE RISE OF EXPERIMENTAL BIOLOGY The emergence of biology as an experimental science in seventeenth and eighteenth century Europe saw the use of mice as laboratory models.[1,2] Harvey's studies of anatomy and circulation and Priestley's and Levoisier's studies of respiration used domesticated mice. Some historians believe that Mendel's studies of the inheritance of specific traits included studies with mice as well as pea plants. The discussions of Darwin's theory of evolution in the late 1800s and the rediscovery of Mendel's work in 1900 enhanced interest in heredity and led to experiments involving the deliberate breeding of organisms with divergent traits. Mice rapidly became important models in this research. The mouse's short gestation time (21 days), short time to sexual maturity (3 months), and relatively large litter size facilitated genetic studies. The commercially available lines of mice with readily distinguishable physical and behavioral traits provided ideal starting materials for geneticists who wished to study inheritance and assess whether specific traits could be fixed by inbreeding in animals, as they were known to be in plants. By the early 1900s, William Castel, Clarence Little, Hasely Bagg, Leonell Strong, Leo Loeb, and many other American geneticists were studying inheritance in mice.[1,2]

THE DEVELOPMENT OF INBRED MOUSE STRAINS In the early 1900s Abbie E.C. Lathrop, who owned the Grandby Mouse Farm in Grandby, CT, noticed that the different "families" of fancy mice she was raising differed dramatically in their incidences of cancer.[1,2] She sent mice from her cancer-prone mouse

From: *Sourcebook of Models for Biomedical Research*
(P. M. Conn, ed.), © 2008 Humana Press Inc., Totowa, NJ.

family to the geneticist Leo Loeb. This led to a series of collaborative studies examining the role of genetic and endocrine factors in the development of mammary tumors. About the same time, Ernest Tyzzer published a series of papers describing studies in which tumors were implanted from mice of one "race" into mice of related and unrelated lineages. Little began studying the hereditary factors that determined sensitivity or resistance to tumor transplantation, and decided that there was a need for mouse models with a "homogeneity of genetic constitution" that allowed this problem and other complex, non-Mendelian problems in genetics to be studied rigorously. By the early 1900s a number of researchers were systematically inbreeding mice, often with founding stock obtained from the mouse lines sold by Grandby Mouse Farm, to produce different strains of uniform, genetically defined animals that provided the foundation stock for modern laboratory mice.

INBRED STRAINS

The inbred mouse strains that result after ~50 generations of brother–sister mating and selection for uniformity and desired traits are unique biological models (see Chapter 20, this volume).[2–6] First, all of the animals within an inbred strain are genetically identical, except for the inevitable differences in the male and female karyotypes. Moreover, the mice are homozygous at all genetic loci; all recessive alleles are therefore expressed. The resulting animal-to-animal homogeneity in physiology, behavior, development of strain-specific disease, and response to treatments or manipulations has profound implications for the design and performance of biological experiments. In addition, because the strains have been bred to homozygosity, the first generation (F_1) hybrids resulting from a cross of two inbred strains are also uniform and genetically defined. Inbred strains of mice therefore provide powerful model systems for studying genetic diseases and for studying the effects of genetic factors on the development of injuries and diseases produced by exogenous agents, as described in detail in other chapters in this volume. The characteristics of most inbred strains of mice are well described on the mouse genome informatics website at http://www.informatics.jax.org/.[6]

The development of inbred mouse strains was intrinsically linked to their use in studying the implications of heredity factors in the development of cancer.[1,2] Their critical role in this area has continued and has expanded with the development of genetically engineered mice (GEM) that allow studies of the effects of the presence, absence, mutation, or altered expression of specific genes or combinations of genes (see Chapter 41, this volume).[6,7] These models have also found wide use in preclinical studies of the potential benefits and toxicities of agents being developed for the prevention and treatment of cancer.[3,4,8–10] They have also been used to study the mechanisms of action of these agents and to develop regimens for use in initial clinical trials of novel anticancer agents.[3]

MODEL SYSTEMS FOR STUDYING THE TOXICITIES OF ANTICANCER AGENTS

Most anticancer agents produce their beneficial effects by killing or preventing the growth of the malignant cells. Unfortunately, cancer cells develop from the normal tissues of the same patient, and although their evolution through precancerous lesions to cancer results in the acquisition of the cellular phenotypes associated with malignancy, cancer cells seldom develop absolutely unique physiological signatures that can be used to target cancer therapy with complete specificity. Instead, the targets generally appear at low levels or in slight variations on some cells in normal tissues, and the agents therefore have effects on normal tissues as well as tumor. The development and testing of new cancer therapies therefore must always involve studies of the toxic side effects of the agents, as well as studies of their antineoplastic efficacies.

Inbred strains of laboratory mice have proven invaluable for studying the response to therapy of the critical tissues that limit the intensity of cancer therapy. The value of these models results from several factors. First, the genetic homogeneity of the mice within an inbred strain minimizes animal-to-animal variations in the delivery of treatment related to factors such as size, percent body fat, or uptake from the injection site or the gastrointestinal track. Similarly, for drugs that are enzymatically activated or metabolized, the uniformity in the enzymes and enzyme levels within an inbred strain produces greater uniformity in pharmacokinetics than would be seen in outbred mice with enzyme polymorphisms. The physiological differences in different inbred strains of mice[4–6] can result in very different effects from the same agent. This is illustrated by a comparison of the duration of anesthesia produced when the same intraperitoneal dose of the same barbiturate is injected into mice of similar sex and age, but from different strains: the reported "sleeptimes" vary by a factor of about 3.[5]

Other genetic differences between inbred strains also influence the response of mice to treatment, through mechanisms that reflect both cellular and tissue level effects. Inbred strains differ in their ability to repair the DNA damage that produces the effects of many anticancer drugs. As an extreme example, *scid* mice, often used as hosts for human tumor xenografts because of their profound immunodeficiency, are immunodeficient because the *scid* mutation produces a deficiency in the repair of DNA double-strand breaks. Because of this cellular repair deficit, radiation injuries to the normal tissues of *scid* mice develop at extremely low radiation doses, one-third to one-half the dose needed to produce the same effect in normal mice.[11] Less obvious differences in DNA repair exist among inbred strains.[12] For example, studies of radiosensitive BALB/c mice, radioresistant C57BL/6 mice, and their hybrid progeny relate the different radiosensitivities of these mice to differences in DNA-PKcs expression that result in variations in nonhomologous end joining. Physiological differences at the tissue or whole animal level also affect the development of treatment-induced injuries. Susceptibility to the development of lung fibrosis after treatment with radiation or bleomycin, for example, is a heritable trait in mice. Different inbred strains and their crosses, as well as GEM, show striking variations in lung injury after irradiation that reflect the influences of multiple genetic loci.[13] The use of mouse strains with differing susceptibilities to naturally occurring diseases of specific organ systems, described elsewhere in this book, offers opportunities to determine how the development of injuries from cancer therapeutic agents mimics and how it differs from the development of the genetically induced diseases and the functional declines associated with aging.

The use of inbred strains to study the effects of anticancer agents has some significant implications for experimental design. The uniformity of the mice within a strain has statistical implications: it reduces the number of animals needed in experiments and allows detection of smaller effects than could be seen with more variable models. It has been possible, for example, to detect age- and sex-dependent differences in the toxicity of some agents that would be masked in experiments using animals from a more variable population. In fact, it should be noted that some of the toxicities commonly ascribed to therapeutic agents have been definitively identified *only* from studies of mice and other model organisms. We know from Muller's studies in fruit flies and Russell's studies in mice that irradiation of the gonads causes germ line mutations that can lead to mutations or genetic disease in the progeny.[14,15] The potential for this deleterious effect in humans is assumed and is considered when planning cancer therapy for patients who may have children in the future, but a significant increase in genetic diseases or polymorphisms above the background rate has never been demonstrated in any human populations studied, including the large, heavily irradiated populations of Hiroshima and Nagasaki.[3] Uniform, genetically defined mouse models therefore provide extremely powerful experimental systems for defining and studying the potential toxicities of anticancer drugs and allow researchers to address questions that cannot be answered either using *in vitro* systems or by studies of human populations.

There are also potential problems and pitfalls in the use of inbred mice as model systems. The breeding of inbred lines must be closely monitored to ensure that the expected genotype, phenotype, and uniformity are maintained. Breeding errors that result in the introduction of foreign genes can be disastrous for the experiments. It is important for researchers to realize that the inbred mice they purchase or acquire from different sources may not be identical, despite the fact that they have the same strain designation.[4–6] Some mouse strains were not actually fully inbred at the time the founder stock was separated to establish the different sublines now available.[2,4] Genetic drift, resulting from random spontaneous mutations, has also introduced genetic changes into the sublines.[4,6] The slow accumulation of genetic differences becomes important when comparing sublines that have been separated for many decades (i.e., dozens of generations). Moreover, some natural mutations (such as the *nude* mutation) have occurred several times and some transgenics and knockouts have been produced multiple times; these mutations are therefore available in mice of different outbred and inbred backgrounds. Investigators should consider the substrain designations, as well as strain designations, and should research the origins and characteristics of the substrains if they change vendors or sources. They should also consider the microbiological status [including the presence/absence of strain-associated viruses such as mouse mammary tumor virus (MMTV)] of the mice from different sources. The parental genotypes of some GEM, especially those produced using the genetically diverse strain 129 mice or their crosses,[6] are not well defined. Thus, although the gene of interest may be the same in the different genetically altered mice, there may be very significant differences elsewhere in the genomes.

Additional thought must be given when using hybrids between inbred strains (see Chapter 20, this volume).[5,6] Although F_1 hybrids are uniform and genetically defined, they are heterozygous, rather than homozygous, and recessive genes will therefore be hidden. Moreover, the two possible crosses will not be the same. For example, a C3H × C57BL mouse would differ from a C57BL × C3H mouse in its X and Y chromosomes, in its imprinting, in its mitochondrial DNA, and in intrauterine and postnatal exposures that reflect the mother's genetics or epigenetics (classically including exposure to MMTV in the milk).

Certain caveats must always be remembered when using any mouse model system. Disney's cartoons notwithstanding, mice are not furry little people. Researchers must always be aware that there are differences between mice and humans that can complicate or even preclude the use of mice as models for specific projects. An obvious example in cancer therapy is that mice do not vomit: they are therefore useless as models for predicting the nausea associated with cancer therapy or for developing approaches to ameliorate this toxicity. The pharmacological and pharmacokinetic differences between mice and humans must also be considered. Uptake, distribution, and elimination of drugs are often much more rapid in mice than in people. In studies of hypoxic cell radiosensitizers, such as misonidazole, where radiosensitization increased with the drug levels in the tumor at the time of irradiation and toxicity increased with the cumulative drug exposure, the mouse models showed much greater efficacy than could be obtained in patients and failed to predict the peripheral neuropathies seen in clinical trials.[16] The differences in the rates of cell proliferation in mice and people[10] must also be considered in designing studies of anticancer agents. Many antineoplastic drugs and radiation are "cycle-active" agents, which have effects that differ for cells in different phases of the cell cycle. Because mouse cells generally have much shorter cell cycle times than human cells, the effects of changing the duration of treatment or the timing of multifraction or multiagent regimens in mice will be different from those in people. Similarly, the late tissue injuries (e.g. lung fibrosis, kidney damage) that limit the delivery of many anticancer regimens will develop more rapidly in the mouse (with a lifespan of 2–3 years) than in people.

MODEL SYSTEMS FOR STUDIES OF THE EFFICACY OF ANTICANCER THERAPIES

Studies of the efficacy of potential anticancer agents require the use of model tumor systems, because even the most sophisticated cell culture systems cannot replicate the complex environments within solid tumors or the interactions between the tumors, stroma, and hosts. A wide variety of mouse tumor models exist. Each has value for addressing specific questions, but each has potential limitations, as discussed below. Researchers using any of these model tumors should also consider the strengths and limitations of mouse models discussed above, as most of those discussions are equally germane to studies of tumors.

MODEL RODENT TUMORS

Spontaneous Tumors Outbred and inbred mice develop spontaneous tumors (i.e., tumors developing from an unknown etiology) over their life spans.[4,5,8–10,17] These are seldom used in experimental cancer therapy because they develop in many organs and locations and have a wide spectrum of histologies and characteristics. Moreover, any specific type of spontaneous tumor will occur with a relatively low frequency. Because the incidence of spontaneous tumors increases with age, such tumors become common only in colonies of animals being held for life span studies: they are rarely found in experiments using young mice or even in production colonies, because breeders are culled as their production rate declines in midlife. These factors preclude the accumulation of large numbers of similar spontaneous tumors for study. Their use has therefore been limited primarily to the initiation of transplanted tumor lines, the common models described below.

Autochthonous Tumors Many inbred strains of mice develop strain-specific tumors at a high frequency or have high incidences of specific tumors if given specific treatments.[4–6] These are not true "spontaneous" tumors, because they result from the effects of specific oncogenes and/or endogenous viruses. They are called "autochthonous tumors" when they are studied in the mice in which they arose. Examples of autochthonous tumors include mammary tumors in C3H mice, which transmit the MMTV in their milk and have a genotype conferring sensitivity to the carcinogenic effects of this virus; radiation-induced lymphoma in C57BL mice, reflecting the presence and induction of the radiation-induced leukemia virus; or the characteristic tumors developing in GEM carrying activated oncogenes or harboring specific deficits in tumor suppressor genes. While most work with these tumor models has focused on studying the etiology of the cancer and the molecular pathways leading to malignancy, a few projects have examined the responses of these tumors to therapy. C3H mouse mammary tumors have been used, for example, to study the effects of hormonal manipulations, as well as effects of radiation and drugs.[16]

Autochthonous tumors are true primary tumors, growing in the site where they arose. The tumor/stroma relationship therefore bears a closer resemblance to primary human tumors than does that of tumors arising from cells transplanted into healthy mouse tissue. The limitations arise from the costs and difficulty of working with these tumors. To obtain sufficient tumors for study, large colonies of mice at risk must be maintained and monitored for tumor development over periods of months or years. Because the initial evolution of tumors from precancerous lesions is recapitulated in these models, the tumors will be quite variable. They may have different histologies, variable growth rates, and variable metastatic and invasive potentials. Tumors will develop in different locations (e.g., different mammary glands or different lymph nodes) and some may be in locations that rapidly lead to morbidity or death. Some animals will develop multiple primary tumors simultaneously or during an experiment, complicating studies of therapeutic responses. Some tumors will become detectable at an earlier stage than others. The age and health of the hosts will vary. Tumors will develop in the colony at different times spanning months or years; one or only a few tumors will be available for treatment at any one time. Each treated and control group will therefore include tumors discovered, treated, and observed separately. All of these factors introduce experimental variability and raise technical and statistical problems for the use of these model systems in studying the response of established malignancies to therapy. They are therefore used primarily in studies testing interventions that might delay the development of malignancies by preventing development of premalignant lesions, ablating such lesions, or preventing their progression to malignancy.

Transplanted Tumors

Tumor Lines Mice from the same inbred substrain are essentially identical twins, and, like identical twins, they will accept reciprocal transplants of tissues without rejection. They will also accept transplants of tumors developing in mice of the same substrain. Transplanted tumors in inbred mice are the most common mouse model systems used in experimental cancer therapy.[8–10] Use of these tumors offers notable advantages in experimental design. All tumors within an experiment can be inoculated in the same anatomical site, with the same pool of tumor cells, during the same session. Tumors within an experiment can be treated as a cohort, with treated and control tumors examined and assayed simultaneously by the same observer. The use of transplanted tumors therefore offers numerous advantages in terms of simplifying the experimental design and reducing the numbers of tumors (and therefore mice) needed to perform experiments with adequate statistical power to compare the effects of different treatments.

There are many things the researcher needs to consider before beginning a project examining the response of transplanted tumors to therapy. First is the choice of tumor model. A wide variety of experimental tumor lines are available from researchers and from institutional and commercial cell and tissue repositories. The investigator should ensure that the characteristics of the tumor line to be studied are appropriate to address the question being asked. Early passage tumors generally grow more slowly and are better differentiated than are tumor lines that have been serially transplanted in mice or passaged repeatedly in cell culture. This reflects the fact that serial passage (*in vivo* or *in vitro*) selects for the cells with the most rapid growth rate.[12] Over the first dozen serial passages the tumor cell population evolves to have shorter cell cycle times and a higher fraction of proliferating cells. Early passage tumors may offer better models for well-differentiated human tumors. However, this comes at a price: more cells or tissue must be implanted to produce a tumor and some recipient mice may not develop tumors; the tumors grow more slowly; and tumors show greater tumor-to-tumor variability in histology, behavior, and response to therapy. Late passage tumors are better characterized, are more uniform, and are rapidly growing and easy to handle. Established tumor lines vary in their invasiveness and metastatic potential. Some lines metastasize early (perhaps even before the primary tumor becomes palpable) either to specific tissues or widely; others rarely produce metastasis. Studies of metastasis will require the use of model systems different from those studies using local control as an endpoint, in which long-term observation of the site of the treated primary tumor is essential and the presence of metastasis would preclude observations long enough to use this assay meaningfully.

Selection of Hosts The host mice should be uniform in origin, age, sex, weight, and/or any other critical characteristics. Age is a critical factor. Many vendors sell very young mice (4–5 weeks of age), which are too small to carry the tumor burden needed for the studies. Moreover, the pharmacology and effects of anticancer drugs often differ in immature and adult mice, even when drug doses are given on a per gram basis. Animals should be held until they reach 2–2.5 months of age before inoculating tumors. The researcher should ensure that the substrain chosen as the host is appropriate and syngeneic with the tumor line—noting that some substrains of common inbred lines have been separate long enough that they will reject tumors (and skin grafts) from other substrains. Some long-passaged tumors (e.g., Harding-Passey melanoma, S180, and Ehrlich ascites carcinoma) arose in outbred mice or mouse strains that no longer exist, and have been carried in mice of many different strains.[17] In addition, some tumor lines exhibit tumor-specific or viral antigens that render them immunogenic even in "syngeneic" mice.[8,17] The possibility that the hosts will mount immune responses to transplanted tumors must therefore be considered when designing experiments and interpreting data.[8,17]

Care should be taken to ensure that experiments begin with mice of the best possible microbiological status. The purchase price of specific pathogen-free mice may be higher than that of "conventional" mice and the per diem charges for maintaining the mice under stringent microbiological conditions may likewise be higher. However, any loss of animals or experiments due to intercurrent infections can soon make this difference in cost disappear. Almost all cancer therapies are immunosuppressive; treated mice will therefore be at increased risk of illness and morbidity from both pathogens and "nonpathogenic" organisms. In addition, it has been shown repeatedly that both active infection and past infections alter the physiology and immune function of mice in ways that perturb tumor growth and alter the responses of tumors and normal tissues to therapeutic agents.[8] Stress from other sources (noise, handling, recaging of aggressive adult males) has also been shown to affect tumor growth, and even minor changes in husbandry can alter drug effects (e.g., changing from pine shavings to corn cob bedding alters liver microsomes and therefore changes the pharmacokinetics of some drugs).[8] Care should therefore be taken to select appropriate animals and to house and handle them appropriately throughout the experiments.

Inoculation of Tumors Many other technical factors must be considered in designing experiments with transplanted tumors. The site of tumor implantation is important. While older studies often used tumor cells injected into the peritoneal cavity, this implantation site is now used infrequently. Intraperitoneal injection produces ascites, in which tumor cells and host immune cells exist as single cell suspensions in the ascites fluid. However, ascites models mimic the growth of only a few natural tumors, such as ovarian carcinomas and leukemic infiltrates. As appreciation of the interactions between the malignant cells of solid tumors and the stroma and vasculature has improved, the limitations of ascites tumor models and the importance of studying solid tumor cells in a more realistic association with stroma have become clear. Most solid malignancies are now implanted subcutaneously, intradermally, or into tissues such as muscle, lung, brain, spleen, or mammary fat pad. Orthotopic implantation of tumors into the tissue of origin or into a common site of metastasis has theoretical advantages and in some cases provides tumors that better model invasion and/or metastasis from primary tumors of the same histology.[18] However, orthotopic implants generally require surgery to expose and visualize an internal organ for inoculation. The wound-healing responses and immunosuppression resulting from surgery can complicate interpretation of the results, and the internal location of the developing tumors precludes the sequential tumor observations needed to monitor tumor growth, to deliver treatment at a specific predetermined volume, or to follow the effects of treatment on tumor growth and cure. Single terminal observations of tumors treated at a prespecified time after treatment have often been used to assay responses with such tumors; this compromise presents many experimental design problems. Sequential imaging of tumors using X-rays, computed tomography (CT), magnetic resonance imaging (MRI), or assays of light from green fluorescent protein (GFP)-expressing tumor cells offers the potential for more rigorous delivery of treatment and sequential monitoring of tumor growth, but is labor intensive and subjects the mice to multiple anesthetizations and immobilizations for imaging. Subcutaneous, intradermal, and intramuscular injections allow better visualization and measurement of the tumors, but at the price of an artificial location.

The site of inoculation should be chosen to allow the tumor to grow to the desired experimental size without interfering with movement and without invading, metastasizing, or ulcerating to produce painful disease. For example, injecting tumors in the loose skin of the flank is preferable to injection into the very constricted skin of the foot. The study should also be designed with endpoints that allow the hosts to be euthanized at a humane point in time. One of the hallmarks of cancer is that tumors in many sites can grow to relatively large sizes without causing pain and discomfort. This, in fact, is one of the factors which makes early detection of cancer difficult and necessitates the use of screening tools such as mammography, pap smears, prostate-specific antigen (PSA), and colonoscopy to detect early asymptomatic cancer in human populations. Investigators should design experiments that use this window of opportunity as effectively as possible and thereby perform their experiments with as little pain or distress to the animals as possible.

Tumor cells for inoculation can come from several sources. Small pieces of minced tumor can be inoculated using a trocar. This technique transplants intact tumor tissue including stroma and may allow transplantation of fragile tumor cells that do not survive the procedures needed to prepare cell suspensions. Many tumors are inoculated using single-cell suspensions prepared from tumors or cell cultures.[8,9] This approach produces more uniform populations of tumor than inoculating tumor chunks, because the cell suspension can be counted and the number of cells inoculated can be precisely controlled. It is critical that any cell line inoculated into mice be prescreened for pathogenic bacteria and viruses. Cell cultures and transplanted tumors can acquire and carry rodent and human pathogens. Implantation of contaminated tumor cells can result in systemic infection of the host mice; the resulting infection may be spread via urine, feces, fomites, and/or aerosols to animals throughout the colony and, in the case of zoonoses, even to researchers and animal handlers.[8]

ASSAYS OF TUMOR RESPONSE The effects of therapy can be assayed in several different ways: biochemical, histological, and metabolic assays, tumor growth assays, tumor cell survival assays, and tumor cure assays are most common.[8] The basic design of all such experiments begins similarly: a large number of animals are inoculated with tumors and the tumors are allowed to become established. At a predetermined time, the tumors are treated with the agent of interest. The "time of treatment" may be selected in two ways. If the tumors grow relatively uniformly, an optimal experimental design may be to randomize the tumors on the day they reach a predetermined average volume, then treat all tumors on the same day. If there is considerable variability in the growth of the tumors, it may be preferable to assign each tumor to an experimental or control group using a predetermined stratification or randomization scheme when it reaches a predetermined volume, then treat it. The volume of treatment is an important consideration. It is important that this volume not be so large that the ratio of the tumor size to host size does not reflect tumors seen clinically. All experiments should include untreated control groups followed in parallel with the experimental tumors. If the treatment involves significant manipulations (e.g., anesthesia for irradiation or imaging; injection of drugs in a vehicle with possible effects) additional sham-treated controls should be included to test for any stress responses or immunological changes associated with the experimental procedures.

Biochemical assays and histological and metabolic assays of tumor response generally are of limited value in assaying the outcome of experimental cancer therapy. There is usually a very tenuous relationship between the transient biochemical and metabolic changes (e.g. DNA synthesis, glucose utilization, ATP levels) seen soon after treatment and the long-term effects of the treatment on the viability and growth of the tumor cells.[8,9,19] Similarly, early measures of histology can be misleading. For example, radiation and many alkylating agents kill tumor cells through DNA damage and chromosomal aberrations that become manifest several cell cycles after irradiation. As a result, the histology of a heavily treated tumor hours to days after treatment may look almost normal, even though most of the cells will eventually die and the tumor will regress. Measurements of apoptosis, cell proliferation, or DNA synthesis soon after treatment can likewise provide misleading underestimates of the efficacy of therapy.[8]

A more rigorous assay of the response of tumors to treatment can be obtained using a cell survival assay, in which the tumors are removed after treatment, then minced, and dispersed to form a single cell suspension. The survival (clonogenicity) of the suspended tumor cells is then measured. This can be done by injecting known numbers of cells into new hosts and quantitatively measuring their ability to form tumors in the lungs (after intravenous injection of cells from many solid tumor), in the spleen (after intravenous injection of some leukemias and lymphomas), or in subcutaneous sites (using an endpoint dilution assay).[8,9] For tumor cell lines that grow well in cell culture, the survival of the tumor cells can be assayed by plating the suspended cells at low densities in cell culture and testing individual cells for their ability to grow into macroscopic colonies.[19] The advantage of these techniques is that they directly measure the survival of the tumor cells, a critical metric, and allow the determination of dose–response curves for cell survival after graded treatments. With many tumor systems, the survival determinations are very reproducible and precise and allow accurate comparisons of the effects of different agents and regimens. The major disadvantage of the approach is that suspension of the cells after treatment removes the cells from the natural environment within the tumor, which may alter their response. Moreover, these techniques cannot be used to study protracted or fractionated regimens in which cell death or cell proliferation results in changes in cell number during treatment. The details of the techniques for measuring tumor cell survival and the appropriate use and pitfalls of these techniques have been discussed in several reviews.[8,9]

Tumor growth studies comparing the growth of treated and control tumors have many advantages for use as assays in experimental cancer therapy.[8,9] They are technically simple (albeit labor intensive). They can be used with a wide variety of murine and xenograft systems. Little prior knowledge of the efficacy of the therapeutic agents to be tested is required to design the experiments. Information on toxicity is needed to plan the experiments, because loss of mice from toxicity would preclude analysis of efficacy. Rigorous growth studies require identification of individual tumors, so that the growth of each individual tumor can be monitored before and after treatment. For subcutaneous or intradermal tumors, the three external diameters of the tumor can be measured using vernier calipers and the volume calculated from these measurements. To ensure maximal consistency and objectivity, tumors should always be measured by the same person and that observer should be blinded to the treatment. Tumors should be monitored frequently (one to three times per week, depending on growth rate) and each tumor should be followed until its volume reaches a predetermined final volume that is several times the treatment volume. The most rigorous measure of the differences in tumor growth produced by different treatments is an analysis of the difference in the time needed for tumors to grow from the treatment volume to a slightly larger predetermined assay volume, most commonly four times the treatment volume.[8,20] Animal care committees often allow maximum tumor weights no greater than 10% of the host body weight, and require euthanasia of individual animals sooner if tumor growth or the development of metastases produces signs of injury, pain, or distress; experiments should be planned to meet these requirements. Certain approaches frequently seen in the literature are to be avoided. One is measuring only one or two dimensions of the tumor and estimating the volume from these unnecessarily limited measurements. Another is the growth of control tumors to unacceptably large sizes to produce a dramatic graph. Another is ending the experiment when the control tumors become large, for convenience, to reduce costs, or to remove and weigh tumors to get "final" treated and control tumor weights. (Tumor weights can be calculated readily from tumor volumes if this metric is desired.) Such a protocol results in the loss of potentially valuable data on the regrowth of tumors that responded well to treatment. An experiment that ends with one or more groups of tumors that have not yet regrown, especially if there are groups that are still indistinguishable from one another, is a flawed experiment that has wasted time and money by providing a result that cannot be analyzed rigorously. Rigorous mathematical techniques for the analysis of tumor growth delay data are described in detail in several reviews.[8,9,20]

Tumor control studies test the ability of an experimental treatment to cause complete tumor regression that endures long enough that the tumor is not expected to recur during the lifetime of the host.[8,20] Tumor control studies measure the desired outcome of clinical cancer therapy: tumor cure. Their use is limited by the fact that they can be used to test only highly effective treatments that permanently eradicate the malignancies without producing unacceptable host toxicity and by the very large numbers of mice needed to produce precise dose–response curves for each different regimen tested. The basic approach used in these studies is similar to that outlined for tumor growth studies, except that each tumor that continues growing after treatment is followed only long enough to ensure that it is not controlled, while mice with tumors that regress are monitored for months to watch for recurrence. The tumor control dose (TCD_{50}) is then calculated from the data relating the proportion of tumors controlled to the dose of the treatment agent, using probit or logit techniques.[8,20]

In summary, several different techniques are available for assaying the response of rodent tumors to experimental cancer therapies. The assays are conceptually different: they have different biological bases, use different experimental techniques and designs, require different levels of preknowledge, measure different regions of the dose–effect curve, and are analyzed using different statistical approaches.[10,22] Researchers need to consider carefully the goals of their experiments and the agents to be tested before selecting the assay system to be used in their studies.

HUMAN TUMOR XENOGRAFTS The use of human tumor cells xenografted into immune-deficient mice raises some unique issues. These systems were developed some years ago to test cytotoxic therapies using human tumor cells in the microenvironments found within solid tumors.[8,9] Human tumor cells lines and human tumor xenografts have become increasingly important in experimental cancer therapy with the increasing development of agents aimed at molecular targets unique to human cells. Such targets include, for example, antibodies against specific human cell surface markers and SiRNA molecules that silence specific human genes. Mouse tumor lines cannot be used to test the efficacy of these agents because they lack the human target. Human tumor xenografts are therefore used to test the effects of these agents. This must be done with great care and with respect for the limitations of these model systems. First, it must be remembered that human tumor cell lines and the xenografts derived from them are established, highly selected, rapidly growing cell lines that no longer recapitulate fully the characteristics of the primary tumors from which they were derived. Furthermore, only the malignant cells of the xenografts are human: the vascular bed, the supportive stroma, and the immune cells (which together comprise about half the tumor mass) are mouse. Moreover, the host is a mouse, and the uptake, distribution, and elimination of the agent therefore reflect that of a human-targeted molecule administered to a murine host that contains only a very small proportion of human cells. In the mouse, the target will be unique to the tumor cells, but this may not be true in a human patient. The targeting of the agent to the tumor cells in a xenograft therefore may be very different from the tumor targeting that will occur in the human patient. Analogously, the toxicity to the mouse, which lacks the target, can be very different from that for human patients, who may well have critical cell populations that exhibit the same or similar targets, and therefore are susceptible to attack by the agent. Xenograft systems therefore cannot be used to assess the toxicity of these agents or to predict the potential ratio of risks and benefits.

CONCLUSIONS

Tumors and normal tissues in inbred mice and rats have long been the major model systems used in experimental cancer therapy. A wide variety of different model systems have been used in these studies. The advantages and the limitations of the different model systems should be considered carefully by researchers as they select models for use in their projects, as they design and perform their experiments, as they analyze their data, and as they translate their studies to regimens designed to improve the diagnosis or treatment of human cancers.

REFERENCES

1. Keeler CE. *The Laboratory Mouse, Its Origin, Heredity, and Culture.* Cambridge, MA: Harvard University Press, 1931.
2. Morse HC III. The laboratory mouse—A historical perspective. In: Foster HL, Small JD, Fox JG, Eds. *The Mouse in Biomedical Research,* Vol. I. New York: Academic Press, 1981:1–16.
3. Rockwell S. Experimental radiotherapy: A brief history. *Radiat Res* 1998;150(Suppl.):S157–S169.
4. Green EL, Ed. *The Biology of the Laboratory Mouse,* 2nd ed. New York: Dover Publications, Inc.,1966.
5. Foster HL, Small JD, Fox JG, Eds. *The Mouse in Biomedical Research,* Vols. I–IV. New York: Academic Press, 1981–1983.
6. Mouse Genome Informatics Website, www.informatics.jax.org.
7. Nagy A. *Manipulating the Mouse Embryo: A Laboratory Manual,* 3rd ed. Cold Spring Harbor, NY: Cold Spring Harbor Laboratory, 2003.
8. Kallman RF, Ed. *Rodent Tumor Models in Experimental Cancer Therapy.* New York: Plenum Press, 1987.
9. Teicher BA, Ed. *Tumor Models in Cancer Research.* Totowa, NJ: Humana Press, 2002.
10. Steel GG. *Growth Kinetics of Tumours.* Oxford, UK: Clarendon Press, 1977.
11. Biedermann KA, Sun J, Giaccia AJ, Tosto LM, Brown JM *scid* mutation in mice confers hypersensitivity to ionizing radiation and a deficiency in DNA double-strand break repair. *Proc Natl Acad Sci USA* 1991;88:1394–1397.
12. Okayasu R, Suetomi K, Yu Y, *et al.* A deficiency in DNA repair and DNA-PKcs expression in the radiosensitive BALB/c mouse. *Cancer Res* 2000;60:4342–4345.
13. Haston CK, Zhou X, Gumbiner-Russo L, *et al.* Universal and radiation-specific loci influence murine susceptibility to radiation-induced pulmonary fibrosis. *Cancer Res* 2002;62:3782–3788.
14. Muller HJ. Artificial transmutation of the gene. *Science* 1922;66: 84–87.
15. Russell WL. Studies in mammalian radiation genetics. *Nucleonics* 1965;25:53–62.
16. Moulder JE, Dutreix J, Rockwell S, Siemann DW. Applicability of animal tumor data to cancer therapy in humans. *Int J Radiat Oncol Biol Phys* 1988;14:913–927.
17. Hewett HB. The choice of animal tumor for experimental studies of cancer therapy. *Adv Cancer Res* 1978;27:149–200.
18. Hoffman RM. Orthotopic metastatic (MetaMouse) models for discovery and development of novel chemotherapy. *Methods Mol Med* 2005;111:297–322.
19. Elkind MM, Whitmore GF. *The Radiobiology of Cultured Mammalian Cells.* New York: Gordon and Breach, 1967.
20. Moulder JE, Rockwell S. Hypoxic fractions of solid tumors: Experimental techniques, methods of analysis, and a survey of existing data. *Int J Radiat Oncol Biol Phys* 1984;10:695–712.

65 Rat Models of Skin Wound Healing

Wanda A. Dorsett-Martin and Annette B. Wysocki

ABSTRACT

Rats provide an excellent model for skin wound healing by allowing the standardization of the type, size, shape, and depth of the wound injury, which facilitates comparison of data between studies of healing in all mammalian species. The rat is often selected for skin wound-healing models because of its ready availability, low cost, and small size, which result in a more economical and efficient use of limited laboratory space and housing facilities. Despite species differences, the availability of animals with well-defined health and genetic backgrounds along with a bounty of literature documenting biological responses and parameters for rats allows the rat to serve as a valuable research tool in the search for faster, stronger, and more anatomically correct wound healing with the ultimate goal of exact skin replacement.

Key Words: Rats, Wounds, Skin, Wound healing, Models.

INTRODUCTION

Animal models are commonly used in wound research in an attempt to replicate human wound-healing problems such as burns, dehiscence, ischemia, venous or pressure ulceration, infection, and scarring,[1] or as a way to study the molecular mechanisms of acute wound healing, which could not be determined in human studies. Wound healing is a complex process involving cell and extracellular matrix interactions that can affect various hard and soft tissues and organ systems.[2] The normal mammalian response to cutaneous injury includes inflammation, formation of granulation tissue, epithelialization, and remodeling of new tissue to restore tissue integrity.[3] The use of an *in vivo* model is inevitable in trying to obtain information on the multifactorial nature of the wound-healing process, which may be influenced by externally introduced factors[4] or by the presence of underlying pathology.

Ethics prohibit the use of humans in some wound-healing studies, especially individuals with impaired wound-healing ability who are already at risk for delayed or arrested healing resulting in further harm. One of the advantages of using animal models, such as rats, is that the wound-healing process is accelerated; this makes it possible to study the process in days rather than weeks required for humans.[3,5] Another major advantage to using animal models is the ability to standardize the type, size, shape, and depth of the wound injury, which facilitates comparison of the data between studies. Excisional wounds and analyses of tissues in quantities that could not be obtained from human subjects ensure the continuing use of animals for wound-healing studies.[3]

It is the investigator's responsibility to determine which model will yield data that are clinically relevant to humans and still feasible to use when considering the unique conditions of a specific study.[6] Chronic wounds found in humans are often a combination of impaired circulation, poor nutrition, comorbid conditions, advanced chronological age, restricted physical activity, bacterial colonization, persistent physiological imbalance, or the exposure to radiation and use of chemotherapeutic drugs or prolonged stress. Venostasis ulcers, ischemic ulcers, diabetic ulcers, pressure sores, and genodermatoses (epidermolysis bullosa) are examples of commonly occurring chronic wounds. To maintain scientific reproducibility and cost effectiveness, most researchers must limit investigations to isolated segments of human chronic wound problems with their animal models.[1] Researchers use genetic, surgical, and pharmacological approaches or a combination of approaches in animals to replicate the pathological conditions associated with chronic wounds.[3]

HEALING IN RATS

Rats have been used widely in the study of skin wound healing and the efficacy of different treatment modalities. This particular animal species is often selected for its availability, low cost, and small size. Incisional and excisional models commonly use the rat's dorsum as the wound location and have been implemented in numerous wound-healing studies. Dorsal sites tend to be more effective in keeping the animal from reaching and manipulating the wound. Rats and humans share the following skin characteristics: the presence of an epidermis, basement membrane, hair follicles, and dermis. Obviously, there are numerous anatomical and physiological differences between human and rat, some of which are listed in Table 65–1. Among the differences is the fact that rats do not form keloids or hypertrophic scars[2] but people of certain ethnic backgrounds, such as African-Americans and Asians, are predisposed to excessive scarring.[7]

CONTRACTION AND EPITHELIALIZATION Contraction is one of the major features of healing in rat wounds.[3] Contraction is primarily mediated by myofibroblasts, which are characterized by their muscle-like interaction with the extracellular matrix, especially collagen. Rats are loose-skinned animals because of their skin's elasticity or redundancy and its lack of a strong adherence to the underlying structures.[1,3,5] The skin of loose-skinned

From: *Sourcebook of Models for Biomedical Research* (P. M. Conn, ed.), © 2008 Humana Press Inc., Totowa, NJ.

Table 65–1
Comparison of skin traits

Trait	*Human*	*Rat*
Skin adherence	Tight	Loose
Panniculus carnosus	None	Present
Hair growth	Mosaic	Patches
Apocrine glands	Present	None
Eccrine glands	Present	None
Vitamin C source	Exogenous	Endogenous
Keloid/hypertrophic scar	Possible	No

rats can slide and retract over the subcutaneous fascia at the time of incision to initially produce a large gap.[1] Consequently, wound contraction, which is usually more rapid than epithelialization since new tissue is not created, causes a decrease in the overall healing time of rat wounds.[3] The tight skin found in humans makes comparison with loose-skinned animal models difficult.[1,3] Contraction can be measured by noninvasive, morphometric techniques such as video images with computer programs or by simply tracing the wound every other day. Contraction can be represented as a percent change in the wound surface area compared to the original wound's size.[3] If contraction is not desirable, the wound can be splinted by physical barriers such as retaining rings, biopolymer plugs, or other mechanical means.[1]

Epithelialization is the primary response in burns, abrasions, and partial thickness wounds when only the epithelium and superficial dermis are missing.[3] Epithelial cells move quickly out from the hair follicles and sweat glands left in the remaining dermis as well as from the leading wound edges.[3] In full-thickness open wounds, epithelialization comes only from the wound edge and occurs at the rate of 1–2 mm/day.[3] In a normal surgical incision with edge-to-edge approximation, the epithelialization process occurs in 24–48 h.[3]

PANNICULUS CARNOSUS Rats possess a subcutaneous panniculus carnosus muscle that contributes to skin healing by both contraction and collagen formation.[4] Leaving the panniculus carnosus with its rich vascular supply intact provides an anatomically constant wound bed. Thus, if the panniculus carnosus is left intact, wound healing will be facilitated.

HAIR CYCLE The hair of the rat goes through a cyclic pattern in which anagen is the active growth phase; hair follicles start to shut down during catagen, and in telogen, the resting stage, the hair may or may not fall out. All hair follicles do not proceed through the same cycle at the same time. Rats typically have an active growth phase that passes like a wave from ventral to dorsal and from rostral to caudal.[8] Rat skin thickness is affected during this hair growth cycle so that skin during the anagen phase is markedly thicker, in mice 1.5–2 μm thicker, and more vascular than areas with resting, telogen-phase skin.[9,10] Littermates can exhibit differences in skin thickness. Contraction is decreased in anagen compared to telogen skin. In contrast, humans have a mosaic pattern of hair cyclic activity, thus ensuring a more uniform skin thickness relative to body location.

Partial thickness wounds made with devices such as a dermatome may have inherent problems for the study if performed in haired animals such as the commonly used Sprague–Dawley and Wistar strains of rats. The high density of hair in the rodent model causes an exaggeration in the rate of reepithelialization.[1] Consideration of the advantages and disadvantages of the use of depilatory agents and/or shaving in haired rats needs to be made on each study.

VITAMIN C Another potential consideration in using rats to recapitulate human skin wound healing is that unlike humans, rats are not subject to scurvy and therefore do not require diets with added vitamin C. Rats possess the enzyme L-gluconolactone that converts L-gluconogammalactone to vitamin C. Primates and guinea pigs need vitamin C supplementation for collagen synthesis.

STRAIN AND SEX PREFERENCES Rats fall into two basic groups: either inbred or outbred. Inbred strains are developed through at least 20 generations of brother–sister matings.[11] Sharp and Regina describe outbred rats as those with less than 1% inbreeding per generation and maintenance in a closed colony for at least four generations.[11] Of the two most commonly used outbred stocks, the Sprague–Dawley is generally larger than the Wistar, but both are considered docile and serve well as general purpose models.[12–14] The Fisher rats are smaller than the Sprague–Dawley and the Wistar rats.[13] The Fisher and Norway rats are identified as inbred strains and the Fisher rats are also described as being a general purpose model, whereas the Norway rats are listed as appropriate for studies such as aging and kidney research.[12]

Nude rats, which are athymic, have been used in immunology-based research.[12,15] Lewis rats, an inbred strain, have been used for transplantation studies, endocrinology, multiple sclerosis, and experimentally induced autoimmune diseases.[12,13] The inbred strain ACI, a cross between the August and Copenhagen–Irish strains, has been used in research dealing with congenital genitourinary anomalies and prostatic adenocarcinomas.[13] A rationale for the selections of strains or sexes of rats is usually not included in journal articles. Currently, there is scant literature indicating the superiority of one strain over another for wound-healing studies. When determining wound contraction coefficients, Cross *et al.* found that Hooded Lister rats were more docile as compared to Sprague–Dawley rats, allowing multiple precise wound area measurements to be made daily.[5] The ability to trace wound areas with great accuracy is crucial to mathematical analysis. Personal preferences, convenience, past experiences, and expense may weigh heavily in the choices. In general, preference for the models is more likely a result of long-term experience, known breeding characteristics, cost, genetic stability, and ready accessibility.[16]

Sharp and LaRegina warn that different stocks and strains of rats have shown variability in biological parameters such as clinical chemistry, hematology, and anesthesia response, so they advise investigators not to change rat strains or their rat suppliers for the duration of the study.[11] Such variability could potentially skew data from wound-healing research. One of the reasons to produce the various strains of rats is to have diverse animal populations within the species so that they do not have the exact same behavioral characteristics and physiological parameters. Since cost-conscious reasoning usually plays a role in animal selection, it is important to note that in general, male rats consistently cost less than females of the same size.[12] A point to consider in selecting the sex of study animals would be the possible hormonal influence on wound healing, because it has been reported that estrogen deficiency is associated with impaired cutaneous wound healing.[17,18] Also, as reported by Cross *et al.*, the thinner skin found in female rats allows a faster rate of wound healing with a higher wound contraction rate, while the wounds of male rats have a greater tendency to heal by epithelialization,[5] although the effects of the hair growth cycle in this study were not addressed. Since the National Institutes of Health and the Food and Drug Administration now require the use of females in human clinical studies, it is predicted that the inclusion of female animals in studies will increase over time.

NUDE RATS Nude rats have no functioning thymus and since T-lymphocytes mature in the thymus, their rudimentary thymus renders these animals immunodeficient.[15] Immune function from killer and natural killer cells that do not depend on the thymus appears to be normal.[19] Nude rats grow at only 60–80% of the rate of normal rats.[20] Hair-deficient rats include strains such as Charles River hairless, Rowett nude, and New Zealand nude.

DIABETES Diabetes mellitus is one of the most common causes of impaired wound healing in humans.[1] Diabetic models of wound healing are either chemically induced or a result of genetics.[3] Streptozotocin sulfate at 55 mg/kg intramuscularly can be used to induce diabetes mellitus in rats.[1] Wound-healing parameters of diabetic rats include defects in biomechanical strength of incisional wounds, closure of excisional wounds, formation of granulation tissue in porous implants, and angiogenesis.[1]

Zucker rats have been developed with a diabetic phenotype. Genetically diabetic rats often have obesity levels to the point that excisional wound closure is impeded not only by physiological dysfunctions but also by excess fat, preventing skin contraction.[1] Like humans, diabetic animals show a greater tendency toward wound infection.[21] The reduced wound-healing capacity in diabetics is most likely associated with diabetic neuropathy and ischemia.[22]

KNOCKOUTS/STEROID USE AND GENETICALLY IMMUNODEFICIENT ANIMALS Glucocorticoids have an antiinflammatory activity leading to thinning of the skin and reduced healing capacity.[1] Steroids reduce the capacity of monocytes to differentiate into growth factor-expressing macrophages and then reduce the capacity of fibroblasts to synthesize collagen.[1] Steroids have been used to lengthen the normally rapid healing times seen with rats in order to study the effects of various wound-healing agents.[3] Dostal and Gamelli have reported that hydrocortisone and dexamethasone reduce wound strength.[23]

Antiproliferative agents such as doxorubicin (Adriamycin) cause immune compromise by disrupting the production of immune cells and can impair wound healing.[1] Extravasation of Adriamycin can cause severe ulceration, and the drug produces side effects such as weight loss, anemia, and cardiotoxicity that can affect the interpretation of wound-healing studies.[3,24,25] Experimental depletion of specific elements of the repair system such as antimacrophage or platelet-derived growth factor-neutralizing antibodies could provide impaired models of wound healing.[1]

The rat genome project will make it possible to create additional knockin/knockout models for wound-healing studies. Previously, most of these studies were conducted in mice. Completion of the rat genome project in 2004 with the sequencing for the Brown Norway (BN) rat led by the Bayer College of Medicine should open new areas of study for rat wound-healing research and permit analysis of genotype to ensure consistent results.

WOUND MODELS AND METHODS

Reviews[1–4] of various animal wound-healing models describe chronic wound models, ischemic conditions, and acute surgical wounds that can be categorized as incisional models and/or excisional models. Additional models include (1) dead space, (2) wound chambers, (3) burns, and/or (4) impaired. To date, most experts agree that there are no chronic wound models comparable to the chronic, nonhealing wounds found in humans.[4] Many purported chronic models are actually ischemic models and are not sufficient models for venous wounds.

INCISIONAL MODELS Incisional models can be used to investigate the wound-healing process and the influence of different systemic and local products along with various dressings, sutures, and wound-therapy protocols.[4] Incisional wounds can be analyzed for wound-breaking strength, histology, immunohistology, collagen and protein content, and *in situ* hybridization.[3] When the incisional edges are reapproximated and sutured or stapled, healing will occur primarily by reepithelialization (primary initiation healing) within 1–2 days with minimal contraction.[3]

Scalpels and scissors are the most common surgical instruments used to create a full-thickness linear wound that can be sutured closed. This is one of the most frequently used models in wound-healing research.[2] A sharp blade cutting the skin causes a rapid disruption of tissue integrity with minimal collateral damage.[1] Immediately upon incision, there is extravasation of plasma and blood cells into the new space, and dependent on the extent and timing of hemostasis, the formation of a fibrin clot that will act as a bridge between the margins of injury[1] and prime the wound for healing as soon as the wound is made.

Electrocautery or Bovie knives, while commonly used in the operating room, are not commonly reported in animal skin wound-healing studies. Incisions with laser and electrocautery may produce some collateral thermal damage but less hemorrhage than just straight edge blades.[1] The pulsed CO_2 laser may also be used as a burning device with, depending on energy input, substantial collateral damage, but computer-assisted beam control allows the creation of precise size areas of damage.[1] A newer generation of cold lasers has largely eliminated the thermal damage seen with the older class of lasers.

After performing an incision, porous materials such as the Hunt-Schilling chamber, polyvinyl alcohol, viscose sponges, and polytetrafluoroethylene are used to create a subcutaneous pocket.[3] Granulation tissue and wound fluid can be examined by implanting a chamber or sponge in a subcutaneous pocket.[4] Implant materials are chosen to be relatively inert so there is minimal inflammatory response, and biochemical assessments can be made with most implants because of the well-defined volume enclosed.[1] At early time points (3–5 days) the interstitial fluid that accumulates within the implant can be aspirated and analyzed for metabolites, cytokines, growth factors, and nonadherent cells.[1] The cavity created for an implant can be used as a reservoir for injectable or implantable material such as growth stimulants, chemoattractants, or antibodies.[1] Limitations of implants include the interference of the implant with normal scar maturation and the possibility of eventual foreign body immune responses.[1] Even polyvinyl alcohol sponges induce a giant cell reaction and lead to ectopic calcification; therefore wound data are collected for only the first 3–4 weeks with most implant models.[1]

The polyvinyl alcohol sponge was probably one of the first devices used in healing studies.[2] These sponges elicit a pronounced foreign tissue reaction with rapid infiltration of the sponge by connective tissue deposited at the wound site.[2] Connective tissue formation can be isolated from epithelialization and contraction by the use of porous subcutaneous implants.[1] These implants create an artificial tissue space into which plasma infuses, which leads to the formation of a fibrin clot and the subsequent formation of granulation tissue.[1] The polyvinyl alcohol sponge model can be used to study granulation tissue by histological and biochemical analysis.[4] Further maturation into scar tissue may occur and the implant may become surrounded by a connective tissue capsule made of several layers of collagenous fascia.[1]

Like wound chambers, polytetrafluoroethylene tubing, which is porous and allows the diffusion of gas and fluids through the device, is most often inserted in the dorsal region of an anesthetized animal.[4] These tube models are used to study the amount of collagen deposited expressed as hydroxyproline and total protein or DNA deposited per length unit of the tube.[4] Uptake of radioactively labeled thymidine can be used to determine the number of replicating cells in the tubes.[4]

In 1976 Viljanto described a combination tube/sponge device that is placed as a drain; the wound fluid can then be analyzed in the early inflammatory phase of wound healing.[4,26] The material trapped in the sponges has been studied by histology, enzyme histochemistry, biochemistry, and cytology at the light and electron microscope level.[27]

The wire mesh Hunt-Schilling chamber is placed subcutaneously and rapidly fills with wound fluid; later, as healing proceeds, it is filled with connective tissue.[2,4,28,29] Fluids, cells, and connective tissue can be collected for histological, histochemical, and biochemical analysis and the presence of various substances can be followed over time in a single wound, providing data regarding differences found over time in the wound.[2,4] In some cases, samples can be acquired for several months.[29] Equally important, the effects of different agents can be studied upon their placement in the chamber.[2]

EXCISIONAL MODELS In excisional wound models, the effect of different types of dressings, dermal substitutes, and topical agents can be investigated with the outcome measurements being an evaluation of wound size by measuring changes in area and time to complete healing and by examining the histology of the wound tissue.[4] Early healing can be documented histologically through serial/sequential biopsies and later scarring, and any deformation of the surrounding area can be observed.[4] The excision site not only provides a wound for study but also provides a sample of tissue for the analysis or quantification of substances such as collagen, RNA, DNA, glycosaminoglycans, inflammatory cells, growth factors, and other biochemical mediators involved in the healing process.[1,2] Excisional wounds can be covered with occlusive dressings, which can allow wound fluids to be collected for analysis or to assess the status of various soluble factors in the wound environment such as proteinases and cytokines.[1,30]

Excisional wounds may be enclosed with occlusive materials to permit collection of wound fluids or to concentrate topical treatments.[1] Saline-filled external vinyl chambers, "Eriksson chambers," can also be employed and are placed over the wound, maintaining a sealed wet environment to study healing in both partial and full thickness wounds.[31,32]

The split thickness injury involves the use of a sharp blade, usually a dermatome, which cuts parallel to the skin surface at a depth that includes the epidermis and upper dermis.[1,3] A substantial amount of dermis, mostly the reticular or deep dermis, remains as well as the base of most epidermal appendages. Epithelialization will occur both from the margins and the epidermal appendages.[1,3] Many animals are unsuitable for this because of the extremely high hair density, which greatly exaggerates the rate of reepithelialization; therefore hairless strains provide an advantage when using a split thickness model.[1] There is no or minimal wound contraction with partial thickness excisional wounds and the surface area can be calculated exactly.[4] Evaluation of healing is made either by separating the healing epidermis from the dermis with chaotropic or enzymatic digestion, in which case planimetric methods are conducted, or serial sections are made for histomorphometric analysis.[1] The split thickness model is useful for testing many agents and devices that promote reepithelialization, including wound dressings, topical agents, and growth factors.[1,3,4] Extravasation of lymph and blood occurs when using split thickness models, and an eschar made of the fibrin clot and expended granulocytes will form over the wound.[1] Desiccation can also be used to impair wound healing; therefore, conversely, a semiocclusive dressing or other moisture-retaining or -promoting products could be used to cover the area and promote healing.[1]

Full thickness wounds can be created by removing the skin with scissors, scalpel, biopsy punches, or a dermatome using several passes.[1,33] In full thickness models, both the epidermis and dermis are removed to the fascial planes or the subcutaneous fat. In loose-skinned animals such as the rat, the panniculus carnosus, which is firmly adherent to the base of the dermis, may or may not be removed. Healing occurs from the margins and the base of the wound by the formation of a fibrin clot that is invaded by granulation tissue and by the migration of an epidermal tongue along the interface between the granulation tissue and the clot.[1,3] Bleeding and fluid loss are more extensive in this model compared with split thickness models.[1] There is also a greater susceptibility to infection. The open defect will heal by a combination of contraction, reepithelialization, and dermal reconstitution.[4] With a full thickness model there is involvement of all the dermal components; thus, epithelialization occurs only from the wound margins. Full thickness wounds allow/permit analysis of biochemical components, histology, and cell populations found in the wound.[1] Healing rates can be monitored by measuring the filling of the space with granulation tissue, extent of reepithelialization, histological organization of connective tissue, angiogenesis, and biological content of collagen or proteoglycans.[1] Contraction, dermal reconstitution, inflammation, chemotaxis, and aesthetic and functional outcome can also be evaluated in this type of model.[4]

Sequential epidermal layers can be removed using adhesive tape applied repeatedly. This removes the stratum corneum and stratum granulosum layers to expose the basal keratinocyte skin layer.[1] If the basement membrane is left intact, this type of injury will activate the processes of epidermal repair.[1] Tape stripping is less damaging to the epidermis than the use of blister models [4] in which the basement membrane can be damaged. Abrasions of various depths have also been produced with the use a mechanically driven emery wheel. Blistering involves the rupture of the epithelial basement membrane zone, forcing detachment of the epidermis from the underlying dermis.[1] Blister models can be used for the evaluation of epidermal regeneration and the influence of different treatment options.[4] The epidermal structures will lose their nutrient supply and become necrotic unless quickly placed back in contact with the basement membrane zone.[1] If separation persists, the epidermal healing must come from the margins not from beneath.[1] Suction devices, chemical (ammonium hydroxide) or biological vesicants, heat, or a combination of suction and heat may be used to raise blisters.[1] Healing of suction blister wounds is made by reepithelialization alone.[4] Blister models can be used to study absorption of drugs,[34] dressing products, topical agents, and transepidermal water loss to study the effectiveness of the epidermis as a barrier.[35]

Partial and full thickness burns can be made by applying a heated metal object or by scalding the skin.[3] Burns have been produced by methods such as hot water (75–99°C), hot metal plates (250°C), steam, hot oil, gas torches, radiant energy, and ignited alcohol gauze.[36,37] Since burn injuries have a progressive course for up to 2 days after the inciting cause,[38,39] model standardization and consistent timing are important in reproducing burns.[3] Post-burning ice baths can also be used to control the extent of injury, but all wound depths should be determined by histology.[3]

Thermal burns create an extensive zone of necrosis that includes denatured or charred connective tissue.[1] Out of the destroyed zone there is a zone of coagulation necrosis in which denatured plasma and cellular proteins cause obstruction of blood vessels and lymphatics, which in turn prevent nutrient flow to the involved tissue.[1] General anesthesia and postoperative care including analgesics and antibiotics should be utilized with burn models.[1]

Walker and Mason, of the U.S. Army surgical Research Unit, use a scald burn produced on anesthetized rats held in a protective template that limits the area exposed to burning and the thickness of the burn is limited by timing the exposure to the water.[40] Hot water (95–100°C) or ethanol bath partial immersions can produce extensive third-degree burns without charring, but as Davidson warns, this produces a much greater physiological stress than groups of small lesions.[1]

Thermal burns can be induced on shaved skin by placing a heated conductive object such as a brass rod of uniform size heated in either a water or oil bath. The degree of tissue damage is consistent and controlled by a combination of weight, time, and temperature with the extent of damage confirmed through histological sections.[1]

The transient freezing of tissue causes the formation of internal ice crystals that rupture the cell membrane resulting in local necrosis without extensive protein denaturation or coagulation necrosis.[1] The mass of the cooled metal object and the length of exposure will determine the depth of the injury, which can be determined by histology.[1]

A number of caustic agents, especially alkalis, can produce skin lesions.[1] Dermonecrotic wounds have been made by agents such as Adriamycin and the venom of the Brown recluse (*Loxosceles*) spider.[1] Brown recluse spider venom contains sphingomyelinase, with the lesion showing hemorrhage and gradual development into a ulcer, which can be used to develop treatments to control excess activity of serine proteinases released by neutrophils.[1]

Raised flaps with compromised circulation will show necrosis and impaired healing.[1] If flaps are allowed to reattach to the dermal base, revascularization can occur.[1] To prevent reattachment, the strip of tissue can be sutured or clipped to form a tube flap that will have a perfusion gradient from each end toward the center.[1] Wounds produced along the length of the pedicle will have graded healing patterns depending on circulation proximity, which can be determined with transcutaneous pO_2 measurements or laser Doppler.[1]

Ischemia is perhaps the most significant factor contributing to the chronicity of wounds associated with diabetes, some types of peripheral vascular disease, and pressure sores.[3] Quirinia and Viidik have developed an ischemic incisional model on the dorsum of the rat in which a bipedicled H-shaped flap is created so that the cross bar of the H is the test wound.[41] The cross bar of tissue is made ischemic (up to 93% at 1 day after wounding) by transecting the branches of the central vein, which serve as the primary drainage for the arterial supply of the rat's back.[3,41] Pressure applications can prove difficult to standardize and maintain to study decubitus.[1] Peirce *et al.* report on the use of magnets and implanted steel plates to pinch the skin and cause pressure ulcer formation.[42]

Flow models with skin windows have been developed to provide direct microscopic observation of blood flow for the entire thickness of skin. A hamster model with a chronic implantation dorsal skin flap window chamber has been used to observe and quantify alterations in the cutaneous vasomotion function caused by scald burns.[43] This flow model could also be used with rats.

FACTORS THAT MAY AFFECT RESULTS

Chronic wounds commonly involve the skin, but can be very diverse in etiology and therapy. The chronic wound, though common in the human population, is not common in laboratory rats, and no model has proved to be entirely satisfactory.[1,4] Wound healing can be altered by any factor that affects the whole animal such as stress, nutrition, minor illnesses, age, weight, infection, inflammation, bleeding, necrosis, genetics, noise, environmental factors, radiation exposure, pharmacological agents, and the presence of foreign bodies.[3]

Kennedy and Cliff stated that in their studies of wound contraction, the sex of the animal, the time of day of wounding, and the size and shape of the wounds had no influence on the wound contraction curves expressed as percent of original area against time.[44]

The wound(s) can be made anywhere on the body surface, but in most cases researchers usually choose the dorsum of the animal.[4] Auberach and Auerbach reported regional differences found in the growth of normal and neoplastic cells in mice, and they reported concerns of other scientists over the need for consistency in the application of experimental procedures to cancer research.[45] Kullander and Olsson showed the tensile strength of dorsal incisional wounds of rats diminished the more caudally they were placed.[14] Kullander and Olsson suggested that when studying wound healing, it is important to compare cutaneous wounds made in exactly corresponding positions.[14] They speculated that the differences in the vascularity of different skin regions may be the reason for the differences found in tensile strength.[14]

Since an important variable in governing the rate of wound contraction is how tightly the skin is adherent to the underlying tissue, wound location is very significant. Medical personnel and their patients alike have been frustrated in their attempts to heal chronic wounds such as venous and diabetic ulcers of the lower legs and feet. In these cases, contraction is thought to make up approximately a 30% or less contribution to wound healing.[3] To ensure the accuracy of results, a standardized position for measuring the wounds must be maintained for all the animals when using "loose-skinned" species such as rats.[5]

The shape of the wound depends on the purpose of the study.[4] Size is a key consideration in the study of wounds.[5,46] Within experimental studies, no matter what the original intended shape, the resulting size and shape are influenced by the remaining skin's elastic forces with a pull along the tension lines of the particular area.[46] Montandon *et al.*[46] suggest that the decreased contraction observed in round ulcers is often a secondary phenomenon due to the lack of extensibility or to the adherence to the surrounding skin. For the study of wound healing, they recommend the use of square wounds no less than 4 cm^2 to allow the study of the effects of contraction as well as epithelialization on closure. On the other hand, Cross *et al.* explained that the metabolic and physical stress of a large wound relative to total body surface area will independently influence the results of the wound-healing research.[5] In an effort to minimize stress, Cross *et al.* chose to use a 15 × 15-mm wound on 200–350 g adult female Hooded Lister rats rather than 20 × 20 mm in size.[5] Limiting the size of wounds to avoid stress also suggests that stress could be caused by too many wounds on one animal.

Single caging is recommended once wounds are created. Animal interference in wound healing, especially by cage mates, has to be weighed against space and cost limitations.

Traditional rodent litter type bedding may become imbedded in the wound and may incite infection or a foreign body reaction. Nontraditional bedding that is softer and does not form a foreign body in the wound is recommended in light of added handling/care time, increased bedding costs, and possible increased disposal costs.

Food and water placement should be checked to ensure they are readily accessible to wounded animals that may have wound- or pain-limiting mobility. For example, rats may not wish to retrieve food from the top portions of cages if they have abdominal wounds and water conveyance tubing may have to be lowered to ease accessibility.

The nutritional status of animals that have impaired mobility or are dealing with pain should be monitored. Lack of adequate nutritional status could have detrimental effects on healing.

Wound pain, rough handling, noise levels, altered light and dark cycles, and other environmental stimuli may all influence the animal's overall state of being, which may lead to physiological changes that could effect the ability of the animals to heal. Many early studies on the effects of stress on the body were performed by Hans Selye and remain valid. Cross *et al.* used the frequent measuring of the rats' body weights as an indicator of stress in their studies.[5]

Radiation can be used to reduce local or systemic propagation of cell populations critical to tissue repair.[1,3] Localized radiation will show the importance of local cell proliferation from recruitment of peripheral, circulating populations.[1] Total body radiation can affect marrow precursors that are necessary for wound repair.[1,3] In conjunction with local therapy, local versus systemic effects of a wound treatment can be determined.[1] High-energy irradiation limited to skin cells can impair surface healing for up to 3 weeks without loss of monocytes or other hematopoietic cells.[3] Knowledge of the effects of radiation on wound healing is necessary in the use of cancer treatment radiotherapy.[3]

Most infected wounds will fail to heal if contamination is greater than 10^5 organisms per gram of tissue,[47] with the exception of β-hemolytic streptococcus in humans. The presence of infection can cause impairment due to a combination of the exaggeration and exhaustion of host defenses along with the damage caused by bacterial proteases and endotoxins.[1] Bacterial counts should ideally be confirmed by biopsy.[1] However, rats have good host immune systems and are not infection prone unless bacteria are purposefully introduced.

Rat ages and body weights can be correlated,[12] and aging has been reported to result in prolonged or poor wound healing.[5,48,49] The age and/or weight of the animal may present an additional variable in the wound-healing analysis and details about these animal characteristics should be reported.

The Harlan Laboratory Animal Company reports that weights of 200–224 g correspond to male Sprague–Dawley rats that are 49–52 days old.[12] Rats that are 3–6 months old would be comparable to young humans.[3] Mogford and Mustoe state that it is generally accepted that the elderly have less capacity to deal with environmental stress and are more apt to suffer from comorbidities, especially those that cause a decrease in skin blood flow.[3] Sensory nerve function declines during the aging process, and this is associated with decreased neurogenic inflammation and poor wound healing.[49] Experts question whether accurate comparisons can be made regarding wound healing between young rats and humans who are not all young, vigorous healers, especially in chronic wound cases that are more apt to be found among the geriatric populations.[1] Age-related healing impairment is seen in the rat between juvenile and adult animals (8 versus 16 weeks in Sprague–Dawley rats) where there is a marked decline in the rate of wound healing.[1] Age-related differences in wound healing are evident in studies using rats as well as other animal models, yet many continue to use young rapidly growing animals to study problems of impaired healing in which the overall robust nature of a young animal's healing can prove difficult to manipulate.[1]

Injectable anesthetics commonly used in the rat include sodium pentobarbital or mixtures of ketamine and xylazine. Common inhalation agents often used are ether, halothane, and isoflurane. Combinations of injectable and inhalation agents have been employed in research with rat models. Whether the use of various anesthetic agents plays a role in the rate of wound healing in regards to comparability has been questioned. Li *et al.*[50] reported that anesthetics such as ether and ketamine may influence the action of laser light treatment on wound healing. In choosing anesthetic agents, the cost, the availability, and the investigator's familiarity with the drugs must be considered. Personnel and the institution must be compliant with governmental narcotics regulations. This usually requires additional documentation and added security measures. Achieving a consistent anatomical position for measuring wounds may require the use of anesthesia, but it must be remembered that repeated exposures to anesthesia may be harmful and could potentially cause wound damage during some methods of anesthesia induction.[5] The influence of various analgesics on wound healing adds another variable in analyzing animal-derived data for relevance to humans.

OUTCOME MEASURES

Wound-healing models have been employed to determine the effectiveness of dressings, systemic and topical pharmacological agents, skin equivalents, sutures, and both viral and stem cell therapies. Investigators have used a combination of macroscopic and histological observations, biochemical and biomechanical measurements, and measurements of wound-healing markers, often in combination with analyses of cellular and immunological responses, to evaluate the progress of wound repair.[4,5] For example, histological assessments such as the progression of new epithelium, inflammation, and vascular response, thickness of the skin layers, and the formation of collagen in the wound defect are often made. Histology is used for quantitating angiogenesis.[4] Staining for nonspecific esterase activity will detect the presence of macrophages.[2] Alcian blue can be used to detect proteoglycans and Masson's trichrome will show collagen.[2] Antibodies with fluorescent tags can be used to identify specific types of collagen and growth factors.[2]

Various testing methods have been used to determine the success or failure of treatment modalities. The purpose of wound healing is to restore tissue continuity and resistance to externally applied forces.[51] In animal models, the progress of this process can be assessed by measuring the development of the mechanical strength of the wound.[51] In the majority of the studies reviewed, evaluation of the skin models was accomplished by testing both the tensile and breaking strength of the wounded area with a tensiometer. Granulation tissue, especially collagen, is the main constituent for the development of biomechanical strength of the wound.[4] The breaking strength of the closed linear wound is the force required to dissociate the healing tissue and the tensile strength is the breaking strength per unit thickness of tissue.[2] Tensile strength studies must be considered in terms of the timing of collagen deposition.[2] The incisional wound model lends itself to the evaluation of the development of breaking strength, either physiologically or when influenced by local or systemic agents.

Areas of contraction and reepithelialization were documented with tracings and area grid techniques, usually in conjunction with computer analysis.[5] Electron microscopy, dye testing of vasculature patency, immunohistochemical testings, and genetic level investigation are a few examples of the other types of testing employed to evaluate outcomes. Varying methods used in reporting results only add to the difficulties encountered in the attempt to make direct comparisons between studies[5]; testing conditions, seasonal skin variations, and wounding technique varied so widely that comparisons may not be scientifically justifiable. A separate study would have to be initiated to concentrate on the wound-healing abilities of separate strains of rats while maintaining as little variation as possible in the size and the age of the animal, the sex used, environmental conditions, animal handling, wounding technique, and all parameters used for healing analysis.

CONCLUSIONS

Characteristics such as a short gestation, short life span, docile behavior, and ready availability of animals with well-defined health and genetic backgrounds are critical in the decision to use the rat as the choice of a research animal.[13]

Developing an animal model that has all the complexity of human chronic wounds may be an unattainable goal, since nonhealing and delayed healing wounds in humans are often the result of combinations of impaired circulation, inadequate nutrition, age, limited physical activity, and/or chronic physiological imbalance.[1] Animal models cannot replace the ultimate verification of actions in human wounds due to the differences in tissue architecture and immune response.[1] Models need to be developed that demonstrate and validate these impairments to be sure they are comparable. This would permit a higher level of confidence in animal data. Results of wound contraction coefficients show that choices of strain, sexes, and ages of rats are variables that somehow must be standardized in the search for reproducibility.[5] The need for consistency is of great importance in the study of wound healing to increase the ability to compare studies and further the knowledge base.

Table 65–2
Wound-healing studies report

Parameter/characteristic
Species
Strain
Sex
Age
Weight
Stage of hair cycle
Wound type
Wound size
Wound depth
Wound shape
Wound location
Anesthetic used
Analgesia
Institutional IACUC approval
Time points of sampling
Types of samples taken
Position of the animal at the time of measuring
Method of analysis

Human clinical trials have been disappointing in comparison to the promising results seen with growth factor therapies in animal studies; thus there is a great deal of concern regarding the ability to transfer animal research data to the human clinical situation.[52] In attempting to explain impaired wound healing, most investigators try to divide the problem into components that can be shown to be reproducible while keeping costs to an acceptable level.[1] Having a model that is merely reproducible is not enough to investigate modifications to the wound-healing process.[5] The model must also have the capacity to allow the detection of the effects of potential treatments or procedures.[5] Thus, it cannot be overemphasized that models with standardization of techniques and reproducibility need to be explored so that results from different treatment modalities can be compared in a scientific manner. Accurate comparisons between studies regarding the ultimate efficacy of treatments would add to the knowledge base of wound healing. The desire to compare studies for the advancement of wound-healing knowledge is being hampered by the differences found between the studies. Standardization in reporting, as suggested in Table 65–2, could facilitate comparisons and may initiate additional research that favors the inevitable comparisons between the studies.[16] This increased knowledge base would be vital in transferring animal-derived data to human clinical situations.

REFERENCES

1. Davidson JM. Animal models for wound repair. *Arch Dermatol Res* 1998;290:S1–S11.
2. Cohen IK, Mast BA. Models of wound healing. *J Trauma* 1990; 30(12Suppl.):S149–155.
3. Mogford JE, Mustoe TA. Experimental models of wound healing. In: Falanga V, Ed. *Cutaneous Wound Healing*. London: Martin Dunitz Ltd., 2001:109–122.
4. Gottrup F, Ågren MS, Karlsmark T. Models for use in wound healing research: A survey focusing on in vitro and in vivo adult soft tissue. *Wound Repair Regen* 2000;8:83–96.

5. Cross SE, Naylor IL, Coleman RA, Teo TC. An experimental model to investigate the dynamics of wound contraction. *Br J Plast Surg* 1995;48:189–197.
6. Lindblad W. Animal models in wound healing research: Do we need more? *Wound Repair Regen* 2000;8:81–82.
7. Lepault E, Celeste C, Dore M, Martineau D, Theoret CL. Comparative study on microvascular occlusion and apoptosis in body and limb wounds in the horse. *Wound Repair Regen* 2005;13:520–529.
8. Ebling FJ, Hale PA. The composition of female rat skin in relation to region, age, hair growth cycle and hormones. *J Endocrinol* 1966; 36:177–201.
9. Zawacki BE, Jones RJ. Standard depth burns in the rat: The importance of the hair growth cycle. *Br J Plast Surg* 1967;20:347–354.
10. Delgado AV, McManus AT, Chambers JP. Exogenous administration of substance P enhances wound healing in a novel skin-injury model. *Exp Biol Med* 2005;230:271–280.
11. Sharp PE, LaRegina MC. *The Laboratory Rat*. Boca Raton, FL: CRC Press, 1998.
12. *Harlan Laboratory Animals*. Harlan Product Guide. Indianapolis, IN: Harlan World Headquarters, 2006.
13. Kohn DF, Clifford CB. Biology and diseases of rats. In: Fox JG, Anderson LC, Loew FM, Quimbly FW, Eds. *Laboratory Animal Medicine*, 2nd ed. San Diego, CA: Elsevier Science, 2002:121–165.
14. Kullander S, Olsson A. On the tensile strength of healing cutaneous wounds in pregnant rats. *Acta Endrocrinol* 1962;41:314–320.
15. Rolstad B. The athymic nude rat: An animal experimental model to reveal novel aspects of innate immune responses. *Immunol Rev* 2001; 184:136–144.
16. Dorsett-Martin WA. Rat models of skin wound healing: A review. *Wound Repair Regen* 2004;12:591–599.
17. Ashcroft GS, Dodsworth J, van Boxtel E, *et al.* Estrogen accelerates cutaneous wound healing associated with an increase in TGF-beta1 levels. *Nat Med* 1997;3:1209–1215.
18. Pirila E, Parikka M, Ramamurthy NS, *et al.* Chemically modified tetracycline (CMT-8) and estrogen promote wound healing in ovariectomized rats: Effects on matrix met alloproteinase-2, membrane type 1 matrix met alloproteinase, and laminin-5 gamma2-chain. *Wound Repair Regen* 2002;10:38–51.
19. Chassoux D, Dokhelar MC, Tursz T, Salomon JC. Antibody-dependent cellular cytotoxicity and natural killing in nude rats: Quantitative study according to age and sex. *Ann Immunol (Paris)* 1983;134D(3): 309–318.
20. Festing MFW, May D, Connors TA, Lovell D, Sparrow S. An athymic nude mutation in the rat. *Nature* 1978;274:365–366.
21. Cooperstein SJ, Watkins D. Action of toxic drugs on islet cells. In: Cooperstein SJ, Watkins D, Eds. *The Islets of Langerhans: Biochemistry, Physiology and Pathology*. New York: Academic Press, 1981:387–425.
22. Lampeter EF, Signore A, Gale EA, Pozzilli P. Lessons from the NOD mouse for the pathogenesis and immunotherapy of human type 1 (insulin-dependent) diabetes mellitus. *Diabetologia* 1989;32:703–708.
23. Dostal GH, Gamelli RL. The differential effect of corticosteroids on wound disruption strength in mice. *Arch Surg* 1990;125:636–640.
24. Tan C, Etcubanas E, Wollner N, *et al.* Adriamycin—an antitumor antibiotic in the treatment of neoplastic diseases. *Cancer* 1973; 32:9–17.
25. Rudolph R, Suzuki M, Luce JK. Experimental skin necrosis produced by adriamycin. *Cancer Treat Rep* 1979;63:529–537.
26. Viljanto J. Cellstick: A device for wound healing studies in man. Description of the method. *J Surg Res* 1976;20:115–119.
27. Viljanto JA. Assessment of wound healing speed in man. In: Barbul A, Cladwell MD, Eaglstein WH, Hunt TK, Marshall D, Pines E, Skover G, Eds. *Clinical and Experimental Approaches to Dermal and Epidermal Repair: Normal and Chronic Wounds*. New York: Wiley-Liss, 1991:279–290.
28. Schilling JA, Joel W, Shurley HM. Wound healing: A comparative study of the histochemical changes in granulation tissue contained in a stainless steel wire mesh and polyvinyl sponge cylinder. *Surgery* 1959;46:702–710.
29. Hunt TK, Twomey P, Zederfeldt B, Dunphy JE. Respiratory gas tensions and pH in healing wounds. *Am J Surg* 1967;114:302–307.
30. Wysocki AB, Staiano-Coico L, Grinnell F. Wound fluid from chronic leg ulcers contains elevated levels of metalloproteinases MMP-2 and MMP-9. *J Invest Dermatol* 1993;101:64–68.
31. Breuing K, Eriksson E, Liu P, Miller DR. Healing of partial thickness porcine skin wounds in a liquid environment. *J Surg Res* 1992;52:50–58.
32. Andree C, Swain WF, Page CP, *et al.* In vivo transfer and expression of a human epidermal growth factor gene accelerates wound repair. *Proc Natl Acad Sci USA* 1994;91:12188–12192.
33. De Vries HJ, Mekkes JR, Middelkoop E, Hinrichs WLJ, Wildevuur CH, Westerhof W. Dermal substitutes for full-thickness wounds in a one-stage grafting model. *Wound Repair Regen* 1993;1:244–254.
34. Lundin S, Svedman P, Höglunds P, Jönsson K, Melin P. Absorption of an oxytocin antagonist (antocin) and vasopressin analogue (dDAVP) through a standardized skin erosion in volunteers. *Pharm Res* 1995;12:2024–2029.
35. Pinnagoda J, Tubker RA, Agener T, Serup J. Guidelines for transepidermal water loss (TEWL) measurement. *Contact Dermatitis* 1990; 22:164–178.
36. Barrow RE, Meyer NA, Jeschke MG. Effect of varying burn sizes and ambient temperature on the hypermetabolic rate in thermally injured rats. *J Surg Res* 2001;99:253–257.
37. Johnson, TR. Selection of appropriate experimental burn/trauma models. In: Ninnemann JL, Ed. *Traumatic Injury*. Baltimore, MD: University Park Press, 1983:199.
38. Boykin JV, Eriksson E, Pittman RN. In vivo microcirculation of scald burn and the progression of postburn dermal ischemia. *Plast Reconstr Surg* 1991;66:191–198.
39. Davis SC, Mertz PM, Eaglstein WH. Second degree burn-healing: The effect of occlusive dressings and a cream. *J Surg Res* 1990;48: 245–248.
40. Walker HL, Mason AD. A standard animal burn. *J Trauma* 1968;8: 1049–1051.
41. Quirinia A, Viidik A. The influence of age on the healing of normal and ischemic incisional skin wounds. *Mech Ageing Dev* 1991;58:221–232.
42. Peirce SM, Skalak TC, Rodeheaver GT. Ischemia-reperfusion injury in chronic pressure ulcer formation: A skin model in the rat. *Wound Repair Regen* 2000;8:68–76.
43. Aggarwal SJ, Diller KR, Blake GK, Baxter CR. Burn-induced alterations in vasoactive function of the peripheral cutaneous microcirculation. *J Burn Care Rehabil* 1994;15:1–12.
44. Kennedy DF, Cliff WJ. A systematic study of wound contraction in mammalian skin. *Pathology* 1979;11:207–222.
45. Auerbach R, Auerbach W. Regional differences in the growth of normal and neoplastic cells. *Science* 1982;215:127–134.
46. Montandon D, D'Anoiran G, Gabbiani G. The mechanism of wound contraction and epithelialization: Clinical and experimental studies. *Clin Plast Surg* 1977;4:325–346.
47. Robson MC. Burn sepsis. *Crit Care Clin* 1988;4:281–298.
48. Ballas CB, Davidson JM. Delayed wound healing in aged rats is associated with increased collagen gel remodeling and contraction by skin fibroblasts, not with differences in apoptotic or myofibroblast cell populations. *Wound Repair Regen* 2001;9:223–237.
49. Khalil Z, Merhi M, Livett BG. Differential involvement of conotoxin-sensitive mechanisms in neurogenic vasodilatation responses: Effects of age. *J Gerontol A Biol Sci Med Sci* 2001;56:356–363.
50. Li Y, Liu TC, Duan R. Effects of some anesthetics on wound healing: Laser biomodulation mechanisms [published abstract]. *Lasers Surg Med* 2001;13(26Suppl.):9S.
51. Jorgensen PH. Growth hormone, skin and wound healing: Experimental studies in the rat. *APMIS Suppl* 1997;72:1–54.
52. Cohen K. An overview of wound healing biology. In: Ziegler TR, Pierce GF, Herndon DN, Eds. *Growth Factors and Wound Healing: Basic Science and Potential Clinical Applications*. New York: Springer-Verlag, Inc., 1997:3–8.

66 Animal Models of Prostate Cancer

ADEBUSOLA A. ALAGBALA AND BARBARA A. FOSTER

ABSTRACT

Prostate cancer is one of the most prevalent cancers in the United States. The current standard treatments for early stage confined disease include surgery and radiation therapy. However, advanced diseases or tumors that relapse are treated with hormone deprivation therapy. Unfortunately, most men relapse with advanced metastatic and aggressive hormone refractory disease for which there is no cure. Although prostate-specific antigen (PSA) screening and digital rectal examinations have improved detection of early stage disease, prognosis for advanced metastatic disease is still poor. Therefore, there is a need to understand the biology and genetics of the normal prostate gland and the molecular and biochemical pathways that are deregulated in prostate malignancy. Animal models of prostate cancer have been developed to delineate the key factors that contribute to the initiation, progression, and metastasis of prostate cancer. Such models are also used for preclinical screening of therapeutic regimens for the prevention and treatment of prostate cancer. Animal models of prostate cancer such as rat, canine and mouse models will be discussed. Emphasis will be placed on autochthonous genetically engineered mouse (GEM) models of prostate cancer.

Key Words: Prostate cancer, Genetically engineered mouse models, Rodent models, SV40 T antigens, Prostate gland, Canine models, Human prostate cancer xenograft models, Transgenic mice, Bigenic mice.

INTRODUCTION

Prostate cancer is one of the leading causes of cancer and cancer-related mortality in the United States.[1] It is estimated that one in six men will develop prostate cancer in their lifetime.[2] Age is one of the major risk factors for prostate cancer as most men diagnosed with prostate adenocarcinoma are over 65 years old.[3] Other risk factors for prostate cancer include race[3] and a positive family history of prostate cancer.[4,5] Early stage organ-confined prostate cancer is clinically manageable by surgery or radiation therapy. Advanced metastatic diseases or tumors that relapse are treated with hormonal therapies. However, most men progress with advanced hormone refractory prostate cancer for which there is currently no cure. The high incidence and mortality associated with prostate cancer in the United States have driven interest in the development of appropriate animal models to study the biological and genetic basis of prostate cancer progression. Such models are also useful for evaluating therapeutic interventions such as chemotherapy, immunotherapy, gene therapy, and chemoprevention.[6] Human prostate cancer is a latent, multistep process that preferentially metastasizes to the bone. It is also a heterogeneous disease as indicated by variabilities in genetics, oncogenic signaling, and tumor grade.[6] Spontaneous prostate cancer is relatively uncommon in nonhuman species with the few exceptions being some strains of rats[6,7] and dogs.[8–10] These factors create challenging obstacles in attempts to study human prostate cancer in animals. Animal models that recapitulate the key features of prostate cancer initiation, progression, and metastasis are critical for studies designed to understand prostate tumorigenesis and to screen preventive and chemotherapeutic interventions. Several animal models of prostate cancer have emerged including human prostate cancer xenograft models that are implanted in mice, canine models, and rodent models. These animal models are valuable tools for delineating the mechanisms of prostate cancer initiation, progression, and metastasis.

From: *Sourcebook of Models for Biomedical Research*
(P. M. Conn, ed.), © 2008 Humana Press Inc., Totowa, NJ.

THE PROSTATE GLAND: DEVELOPMENT, MORPHOGENESIS, AND ANATOMY

The embryonic development of the prostate gland originates from epithelial prostatic buds that arise from the urogenital sinus (UGS) in the endoderm. Prostatic buds appear during gestation at 10 weeks in humans and 17.5 days in mice. In rodents, these buds undergo androgen-dependent growth and ductal branching into the surrounding mesenchyme in the early weeks of postnatal development.[11] In contrast, ductal growth and morphogenesis in the human prostate occur prenatally and during puberty when androgen levels are higher. The development and growth of the prostate gland are dependent on androgens and the function of the androgen receptor. Surgical or chemical castration with antiandrogens or estrogen inhibits the development and growth of the prostate.[12–15]

The prostate is a secretory male accessory sex gland that surrounds the urethra and is situated beneath the urinary bladder.[15] Prostatic ducts secrete several components of semen including proteins, fructose, and zinc ions that are released into the urethra during ejaculation. The benign and pathological conditions that are associated with the prostate have generated interest in the study of its development, biology, functions, and malignancy. Interestingly, while prostate cancer occurs frequently, tumor incidence in other male accessory organs such as the seminal vesicles and bulbourethral glands are negligible or nonexistent in men.[16]

This chapter will discuss genetically engineered mouse (GEM) models mainly because of their widespread utility and the innovative research in the area. However, other animal models of prostate cancer including canine, rat, and human prostate cancer xenograft models will be briefly discussed.

PROSTATE CANCER MODELS

HUMAN PROSTATE CANCER XENOGRAFT MODELS Human prostate cancer xenograft models are generated by grafting prostate tissue samples into immunodeficient host mice. The PC-82 model was the first prostate cancer xenograft model that was successfully established by subcutaneous heterotransplantation of tissue samples in athymic nude BALB/c mice.[17] Other xenograft models were developed including androgen-dependent PC-295 and PC-310 models and hormone refractory PC-133, PC-135, PC-324, and PC-339 xenograft models.[18,19]

More details on the prostate cancer xenograft models that are available, well-characterized, and widely utilized are reviewed by Navone *et al.*[20] and van Weerden and Romijn.[19] Briefly, other human prostate cancer xenograft models include the androgen responsive CWR22[21] and hormone-refractory CWR22R[22] models, the LuCaP 23 lines that were generated from metastatic tissue samples,[23] the androgen-responsive LAPC-4 and hormone-refractory LAPC-3 models,[24] and the MDA PCa xenograft series that were established from metastatic tissue samples.[20] In addition, subrenal capsule grafting and orthotopic transplantation of human prostate cancer tissues into mice have been used to generate transplantable and metastatic prostate cancer sublines.[25] Several factors can influence the establishment of prostate cancer xenograft models. These include the strain of athymic nude mice used, transplantation in nude mice versus severe combined immunodeficiency (SCID) mice, the site of tumor implantation (subcutaneous or orthotopic), and the quality of tissue samples used.

A major limitation of human prostate cancer xenograft models is that tissues are obtained from late stage primary tumors or metastatic deposits. Therefore, these models are largely unsuitable for studying the molecular mechanisms of early prostate carcinogenesis. The use of these models for chemoprevention trials is also limited because they represent end-stage disease.

CANINE MODELS Dogs spontaneously, naturally, and frequently develop prostate carcinoma.[9] Canine prostate cancer mimics several characteristics of its human counterpart in terms of heterogeneity, prevalence in elderly populations, histopathology of the disease, and association with skeletal metastasis.[6] The median age of prostate cancer diagnosis in dogs (10 years) is equivalent to the average age of diagnosis in older men (70 years) when converted to human-years.[26] Studies of elderly pet dogs and military working dogs indicate that the frequency of high-grade prostatic intraepithelial neoplasia (HGPIN) and adenocarcinoma is higher in canines as they age.[8,9] Bone metastasis is the most common lethal clinical complication that is associated with advanced prostate cancer in men. Dogs develop osteoblastic bone lesions that mimic the human disease,[6] thus making this model relevant for preclinical testing of bone-targeted therapies. In an autopsy study that consecutively examined 129 dogs with prostate cancer, 24% presented with skeletal metastasis.[9]

Advantages: Most men die of advanced metastatic prostate cancer after failing hormone therapy. Canine models can be utilized for imaging and evaluating therapeutic modalities for prevention and treatment of bone metastasis.

Disadvantages: Clinically evident disease and skeletal metastasis occur in older dogs with low penetrance. Thus, the long latency associated with the emergence of prostate cancer presents a major obstacle for using canine models for research. Therefore experiments with canine models require a long time commitment and are rather costly.

RODENT MODELS Rodent models of prostate cancer including the laboratory rat and mouse have emerged as powerful tools for studying the biological basis of prostate cancer initiation, progression, and metastasis because they can be managed in large numbers and are easy to use.[7] There are several key similarities and differences between the rodent and human prostate that should be taken into consideration when utilizing rodent models of prostate cancer. There is a distinct anatomical difference between the rodent and human prostate. The rodent prostate is organized into four distinct lobes around the urethra (Figure 66–1A): ventral, dorsal, and lateral (dorsolateral), and anterior (coagulating gland). The individual lobular structures of the rodent prostate have unique branching duct patterns.[11] A normal adult mouse prostatic duct consists of an open lumen that is surrounded by tall columnar secretory epithelial cells and basal epithelial cells that are closely associated with the basement membrane.

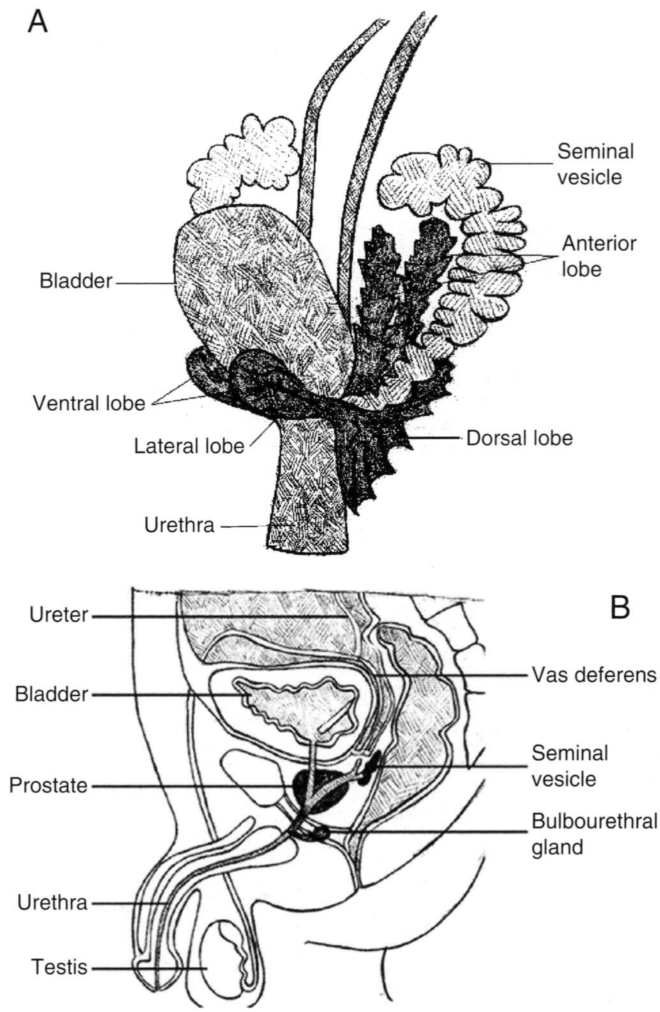

Figure 66–1. Diagrams of the male genitourinary systems of an adult (A) mouse and (B) human. (Adapted from Cunha *et al.*[15])

In contrast, the human prostate (Figure 66–1B) consists of a single gland that is located adjacent to the urethra at the base of the bladder.[15] It is classified into four anatomically distinct morphological regions, namely, the anterior fibromuscular stroma, transition zone (TZ), central zone (CZ) and peripheral zone (PZ).[27–29] These zones differ histologically and in their predisposition to benign and malignant prostatic diseases. Benign prostatic hyperplasia (BPH) is a nonneoplastic increase in proliferation of prostatic tissue and generally occurs in the TZ. In contrast, prostate carcinoma is frequently found in the PZ. There have been suggestions that the dorsolateral lobe of the rodent prostate is most similar to the human PZ where most prostate cancers occur[30]; however, this contention is still debatable.[31,32] The lobular structures of the human and rodent prostate glands are analogous during embryonic development.[32] Although similarities are found in the embryonic development of the rodent and human prostates, the anatomy of the adult prostate gland is strikingly different in each species (Figure 66–1). The rodent models that are used for prostate cancer research include rat and mouse models. Much emphasis will be placed on autochthonous GEM models, while rat models will be briefly highlighted.

Rat Models

Spontaneous Rat Models Certain strains of rats are predisposed to spontaneously developing prostate cancer. The rat models of prostatic carcinoma that are currently available were either derived from spontaneous[33,34] or induced tumors.[35] There are a number of rat models that are currently in use; however, not all models are applicable for research studies. For example, the ACI/Seg spontaneous rat model has limited usefulness because of long latency, low tumor incidence, and low incidence of invasive prostate carcinoma.[6,7]

The Dunning rat model is one of the first rat models that was established.[36] The Dunning R-3327 tumor was a spontaneous adenocarcinoma that developed in an inbred Copenhagen rat and was transplanted into syngeneic Copenhagen × Fischer F_1 hybrid rats.[33] The Dunning system currently consists of several *in vitro* and *in vivo* transplantable sublines that were derived from the original well-differentiated, slow-growing, and nonmetastatic Dunning R-3327 tumor.[37–41] These sublines represent different stages in prostate cancer progression and metastasis. R-3327-H is a well-differentiated, slow-growing, and androgen-sensitive tumor, R-3327-HI is a well-differentiated, slow-growing, and androgen-insensitive tumor, and R-3327-AT is a fast-growing, androgen-insensitive, and anaplastic tumor.[38] A Dunning R-3327 MAT LyLu line that metastasizes to the lymph nodes and lungs was derived from the R-3327-AT line.[39,40] Copenhagen rats inoculated with R-3327 MAT LyLu cells by tail vein injections develop skeletal metastasis.[42] The different sublines of the Dunning model represent a wide spectrum of stages in prostate cancer progression and metastasis. This model has been extensively characterized and is widely utilized, thus making it very attractive for research purposes. The expression of nonprostatic proteins in the Dunning H tumor has raised questions about the origin of the Dunning tumors.[43] Analysis of Dunning H tumor extracts indicated that prostate-specific secretory proteins such as SVS II and transglutaminase were not expressed; however, proteins associated with the mammary gland were detected in the tumor.[43] These findings suggest that the mammary gland or other male accessory sex glands might be the origin of the Dunning H tumor. However, these observations are tempered by the fact that the Dunning model was useful for identifying several clinically relevant markers of human prostate cancer including KAI-1, CD44, and β-thymosin.[6,7]

The Lobund–Wistar (L-W) prostate adenocarcinoma rat model originated from a colony of random bred, germ-free Lobund–Wistar rats.[34] The L-W model consists of spontaneous, induced, and transplantable tumors that metastasize.[6,34] The low tumor incidence and long latency period associated with the spontaneous L-W model have limited its use for research.[7] However, tumor incidence and metastasis are increased in the L-W model using chemical carcinogens (methylnitrosourea, MNU)[44] and hormones (testosterone propionate, TP).[45] The combination of MNU and TP further enhances tumor formation and metastasis in the L-W model.[35,44] This model is unique because it is the only rat model that spontaneously develops hormone-induced adenocarcinoma of the prostate with metastases. Most men fail androgen deprivation therapy and relapse with advanced hormone refractory prostate cancer for which there is currently no cure. The spontaneous, metastatic, and hormone-refractory L-W model[46] can be utilized to study the molecular mechanisms of progression of hormone-refractory prostate cancer and to test chemopreventive and treatment regimens for late stage and metastatic disease.

Advantages of spontaneous rat models: A broad spectrum of human disease progression ranging from PIN to metastasis is represented in the models, making them useful for studies on prevention and metastasis of prostate cancer. The long latency of spontaneous models makes them ideal for studies aimed at elucidating the mechanisms of early carcinogenesis.[7] The Dunning and L-W models are among the best-characterized and well-studied rat models of prostate cancer. Rodents have been widely used for prostate cancer research because maintenance of rodent colonies is economical, they are easily handled, and tumors can be induced or implanted.[6]

Disadvantages of spontaneous rat models: Some limitations associated with rat models include long tumor latency, lack of reproducibility, and low percentage of rats that spontaneously develop adenocarcinoma and metastasis in some models.

Transgenic Rat Models Transgenic animals carry an exogenous vector construct (transgene) inserted into the genome of all their cells, including germ cells, allowing germ line transmission to offspring.[47] Most transgenic animal models that have been developed to study the molecular basis of prostate cancer initiation, progression, and metastasis are mouse models, but transgenic rat models exist. Transgenic mouse models of prostate cancer will be discussed in detail in the next section. Rats are larger in comparison to mice, thus more sample materials (tissues, blood, etc.) can be obtained from rats. Spontaneous and carcinogen- or hormone-induced rat models sometimes have long latency periods and low tumor incidence. Asamoto *et al.* developed a transgenic rat model of prostate cancer using the androgen-regulated prostate specific minimal rat probasin (PB) gene (−426 to +33) to drive expression of oncogenic simian virus 40 (SV40) early genes (T and t antigens, Tag) in Sprague–Dawley rats.[48] The phenotype is 100% penetrant, as all PB/SV40 Tag rats develop androgen-dependent prostate adenocarcinoma in ventral, dorsolateral, and anterior lobes before 15 weeks of age.[48] Castration of 5-week-old PB/SV40 Tag rats completely inhibits prostate tumorigenesis and castration of 20-week-old tumor-bearing rats induces apoptosis, inflammation, and complete tumor involution.[48] This model may be suitable for studies designed to elucidate the mechanisms of androgen-dependent prostate cancer progression and to test therapeutic interventions targeted for early stage androgen-dependent prostate cancer. In contrast to the low tumor incidence and long latency period associated with spontaneous rat models, transgenic rats develop prostate cancer over time. This presents a window of opportunity that can be targeted for molecular, genetic, and therapeutic intervention studies.

Mouse Models Transgenic or GEM models of prostate cancer have been widely used to delineate the genetic and molecular events that contribute to prostate cancer progression and metastasis. Although there are significant anatomical differences between the mouse and human prostate, these models have been used to validate several genetic events that are clinically relevant for human prostate cancer. Several approaches can be employed to create transgenic mouse models of prostate cancer including overexpression of heterologous genes using tissue-specific promoters and genetic manipulation of oncogenes and tumor suppressor genes that are implicated in prostate carcinogenesis.

There are several GEM models of prostate cancer that develop only benign prostatic lesions such as prostatic intraepithelial neoplasia (PIN), while others develop tumors in tissues other than the prostate. Several groups have developed GEM models to mimic several aspects of human prostate carcinogenesis; however, not all these models develop invasive carcinoma in the prostate or distant metastasis. Special emphasis will be placed on models that develop autochthonous prostate cancer that mimics the relevant and lethal aspects of prostate carcinogenesis in man including progressive development of low-grade PIN, invasive carcinoma, and metastasis to distant organs. Some factors to consider when using GEM models of prostate cancer include the site of integration of the transgene, the choice of promoter used in the targeting construct, and the strain background of the mice.

The site of integration of a targeting construct into the chromosome influences the level of expression of transgenes in transgenic mice. The random insertion of a transgene into the genome produces a position effect,[47] which may result in transgenic offsprings with varying levels of transgene expression or aberrant expression of the transgene in nonspecific tissues or cell types.[49] Greenberg *et al.*[50] observed that the level of expression of a rat probasin (rPB)/chloramphenicol acetyltransferase (CAT) transgene in the mouse prostate was independent of copy number. Variabilities in the expression of the transgene in the prostate of transgenic mice were attributed to position effects that occur due to random integration of the transgene into the genome.

The use of prostate-specific promoters is critical for the expression of a target vector in the prostate. Promoter elements that have been used to generate GEM models of prostate cancer include the rat probasin (rPB),[51,52] rat C3(1) steroid-binding protein,[53,54] fetal globin γ,[55] cryptdin-2,[56] gp91-pho,[57] and prostate secretory protein of 94 amino acids (PSP94).[58] However, not all these promoter elements direct prostate-specific gene expression. In addition, the size of the promoter of choice influences the regulation of transgene expression, thereby affecting prostate cancer progression. The long 12-kb fragment of the 5′ flanking region of the rPB promoter (LPB) drives higher levels of transgene expression in prostatic epithelial cells compared to the minimal −426/+28 region of the same promoter.[59] The biological differences in prostate cancer progression in the transgenic adenocarcinoma of the mouse prostate (TRAMP) model that was generated using the minimal rPB to drive expression of SV40 large and small T antigens,[51] and the LADY transgenic model that was made with a LPB-SV40 large T antigen construct,[52] are due to the size of the promoter elements and differences in transgenes used in each model. The promoter of choice can also dictate the severity of the prostatic lesions developed in a transgenic model. For example, the ARR_2PB promoter that was constructed by linking two androgen-responsive regions (ARR) to the PB promoter strongly directs prostate-specific transgene expression in transgenic mice.[60] Compared to (−426)PB-Myc-PAI mice in which prostate-specific gene expression is driven by the weaker −426/+28 PB promoter, ARR_2PB-Myc-PAI mice develop PIN lesions and prostate carcinoma more rapidly.[61]

The strain background of mice used to create GEM models may also influence the phenotype and rate of prostatic disease progression in transgenic mice. This is exemplified in the TRAMP model, where the genetic background of the mice influences the phenotype of prostatic disease.[62] TRAMP mice in the C57BL/6 background develop tumors that invade the seminal vesicles and urethra, while mice in the C57BL/6 × FVB background develop highly vascularized solid tumors within the prostate that are less likely to invade the seminal vesicles. C57BL/6 TRAMP mice can live for up to 52 weeks, while C57BL/6 × FVB TRAMP mice are euthanized by 33 weeks due to large tumors that can obstruct the urinary tract.[62] In addition, bone metastasis has been observed only in C57BL/6 × FVB TRAMP mice.[63] These data indicate that the genotype of a mouse can influence the phenotype obtained in GEM models.

Autochthonous Transgenic Mice/Genetically Engineered Mice Several GEM models have been created to recapitulate key features of human prostate cancer. However, there is no single model that mimics the full spectrum of the human disease. Some GEM models develop tumors in nonprostatic tissues while others develop mild epithelial hyperplasia that does not progress to invasive carcinoma or metastasis. More emphasis will be placed on the discussion of mouse models that are well characterized and widely used and that recapitulate key features of human prostate cancer such as development of progressive, autochthonous prostate carcinoma and metastasis to distant organs.

The neoplastic changes in the prostate of GEM are considered to be the consequences of the genetic alterations induced in the animals by the transgene. This is based on observations that the incidence of spontaneous neoplastic prostatic diseases is very low in the mouse prostate.[32,64] Individual investigators have generally established the histopathological grading systems used to classify benign and malignant prostatic lesions of GEM models.[53,62,64–66] The Bar Harbor Classification system is a hierarchical histopathological classification system that was developed to facilitate uniformity in the descriptions of the pathological changes in the prostate of GEM.[32] Briefly, nonneoplastic changes in the prostate of mice are described as epithelial and stromal hyperplasia. Mouse prostatic intraepithelial neoplasia (mPIN) is classified as a noninvasive neoplastic proliferation of the prostate epithelium that can possess premalignant potential. Malignant proliferation of the prostate gland is described as invasive carcinoma, adenocarcinoma, and neuroendocrine carcinoma.

SV40 T ANTIGEN (TAG)-BASED MODELS Transgenic mouse models generated using vector constructs in which a promoter drives expression of SV40 large and small T antigens (T and t antigens, Tag) have been used extensively to study prostate carcinogenesis because of the transforming properties of SV40

Tag. There are currently no definitive links between SV40 infections and risk of prostate cancer; however, SV40 contaminations in polio vaccines administered from 1955 to 1963 may be associated with human mesotheliomas, ependymomas, and osteosarcomas.[67,68] The major functions of large T antigen include promotion of cellular transformation and viral replication in target cells by interacting with and blocking the functions of tumor suppressor genes including p53[69] and the retinoblastoma (Rb) family.[70] Large T also binds to transcriptional coactivators such as CBP and p300.[71] One mechanism by which small t antigen promotes uncontrolled cell proliferation is by interacting with the serine–threonine protein phosphatase 2A (PP2A) that leads to constitutive activation of the Wnt signaling pathway.[72]

TRAMP model: The TRAMP model is an autochthonous model of prostate carcinoma that was generated using the minimal rat probasin (rPB) –426/+28 regulatory element to drive expression of the SV40 early genes (T and t antigens, Tag) to the epithelium of the prostate.[51] The androgen-regulated rPB fragment directs prostate epithelium-specific expression of the transgene to the dorsolateral and ventral lobes. Transgene expression is controlled developmentally and hormonally, thus it is expressed postnatally at sexual maturity when androgen levels are high.[50,51] Besides abrogation of the tumor suppressor functions of p53 and Rb by large Tag, other genetic "hits" occur to drive tumor progression.[73] Expression of the transgene is 100% penetrant, therefore all mice eventually develop prostate cancer.[51] TRAMP mice develop progressive prostate cancer that metastasizes to distant organs.[63,64] By 12 weeks of age, TRAMP mice typically develop PIN lesions and invasive cancer.[64] By 24 weeks of age, 100% of mice develop poorly differentiated prostate cancer that is associated with metastasis to the periaortic lymph nodes and lungs and less frequently to the bone, liver, salivary glands, and kidney.[64] Based on the histological grading system established for the TRAMP model, mice develop progressive prostatic neoplasia from low grade prostatic intraepithelial neoplasia (LGPIN) to poorly differentiated cancer.[64] Another grading system for the TRAMP model is based on a unique numerical grading system that avoids terminologies associated with human prostate cancer pathology due to differences in the mouse and human prostate.[66] Transgene expression in the TRAMP model is initially androgen dependent; however, following castration of mice at 12 weeks of age, 70–80% of mice develop hormone refractory prostate cancer while ~20–30% do not present with palpable tumors by 24 weeks following androgen deprivation.[74] Neuroendocrine differentiation is associated with tumor progression, metastasis, and androgen insensitivity in human prostate cancer.[75] There is a stochastic development of the neuroendocrine phenotype that mostly occurs in poorly differentiated tumors that emerge in intact and castrated TRAMP mice.[64] The TRAMP model has been used to elucidate the role of tumor angiogenesis in prostate cancer progression. A distinct "angiogenic switch" that correlates with increased vascularity is associated with prostate cancer progression in the TRAMP model.[76] VEGFR1 is highly expressed in PIN lesions, while increased VEGFR2 expression is detected in invasive carcinoma. Changes in the temporal and spatial expression patterns of the FGF signaling axis also correlate with tumor progression in the TRAMP model.[77] The identification of distinct molecular markers of angiogenesis in the TRAMP model has been exploited to evaluate the effects of antiangiogenic therapies on tumor progression.[78] Cell lines have been established from TRAMP tumors that can either be used for *in vitro* studies or grafted into syngeneic C57BL/6 mice.[79] The TRAMP model is an autochthonous, progressive, and heterogeneous model of prostate cancer that recapitulates several key features in the natural history of human prostate cancer. It is one of the most extensively characterized and widely used transgenic models of prostate cancer and is useful for evaluating therapeutic interventions such as chemoprevention and immunotherapy.

LADY models: These models were generated using a large 12-kb fragment of the 5′-flanking sequence of the probasin gene (LPB) to drive expression of an SV40 large T antigen deletion mutant (that eliminates expression of small t antigen) in the prostate epithelium (LPB-Tag).[52,65] The decreased aggressiveness of the prostate cancer phenotype observed in LPB-Tag models is attributed to the expression of only SV40 large T antigen, thus eliminating the contributing effects of small t antigen. Expression of the transgene is 100% penetrant and mice progressively develop prostate cancer that recapitulates key aspects of human cancer. The 12T-7f line of mice develop multifocal lesions, nuclear atypia, and progressive prostatic lesions including PIN and invasive carcinoma.[52] Metastasis rarely occurs in this model; however, the prostatic lesions that develop are androgen dependent as they regress following castration. The 12T-10 line of the LADY model displays progressive high-grade PIN, poorly differentiated cancer, neuroendocrine differentiation, and metastasis to the lymph nodes, lung, spleen, and kidney.[65] These tumors also demonstrate hormone refractoriness following castration. An allograft model (NE-10) that is hormone refractory, neuroendocrine, and metastatic was derived from the prostate tumor of a 12T-10 mouse.[80] The 12T-10 model may be used to study hormone refractory, neuroendocrine prostate carcinoma progression and to test novel therapeutic modalities that inhibit development of the neuroendocrine phenotype.

C3(1)-SV40 large T/small t model: In this model, expression of SV40 early genes is controlled by the androgen-responsive rat steroid-binding protein C3(1) regulatory elements.[53] By 8 weeks of age, male mice develop PIN lesions in the ventral and dorsolateral lobes that progress to invasive carcinoma by 28 weeks of age. The model recapitulates several aspects of human prostate cancer in terms of development of PIN lesions that harbor mutations in Ha-Ras. Metastasis to distant organs occurs in $<5\%$ of mice. However, the C3(1) promoter is not prostate specific as the transgene is expressed in nonprostatic tissues such as the mammary and urethral glands. Similar observations were made in C3(1)-Polyoma middle-T (Py-MT) transgenic mice where the C3(1) regulatory sequences were used to drive expression of the Py-MT gene in the prostatic epithelium.[54] Mice develop dorsolateral and ventral prostate tumors; however, invasive tumors also develop in the vas deferens and salivary glands.

PSP94-SV40 large T/small t antigen model: The prostate secretory protein of 94 amino (PSP94) is one of the most abundant prostate secretory proteins.[81] The promoter/enhancer region of the androgen-regulated and prostate-specific mouse PSP94 gene was used to control expression of SV40 early genes in PSP94-SV40 Tag mice.[58] Histopathological changes in the dorsolateral and ventral prostate lobes correlate with the onset of puberty. Mice progressively develop low- and high-grade PIN by 12 weeks of age, poorly differentiated cancer by 16 weeks, and lymph node metastasis. Castration of PSP94-SV40 Tag mice results in involution and atrophy in the prostate, indicating that tumor progression is androgen dependent. Importantly, PSP94 is conserved across several species including humans and mice.[82] However, there is no human equivalent of the prostate-specific rat probasin (PB) gene that has been widely utilized for generating GEM models. As a result, PSP94 is being developed for generating vector constructs for gene therapy because it is a strong prostate-specific gene.[58]

Cryptdin-2-SV40 large T/small t antigen model: Cryptdins are nonprostatic, antimicrobial peptides that are secreted by the Paneth cells of the intestinal epithelium.[83] Cryptdin-SV40 Tag mice were generated using nucleotides −6500 to +34 of the mouse cryptdin-2 gene to direct expression of SV40 early genes to neuroendocrine cells in the lobes of the mouse prostate.[56] Mice rapidly develop PIN lesions and neuroendocrine carcinoma. In addition, metastasis to lymph nodes, liver, lungs, and bone occurs frequently in the model. This model may be appropriate for studies on the mechanisms of neuroendocrine prostate cancer progression.

Fetal globin γ (Gγ)-SV40 large T/small t antigen model: Targeted expression of the fetal Gγ-SV40 Tag construct in mice unexpectedly results in the development of prostate tumors consisting of mixed neuroendocrine and epithelial cells.[55] Further characterization of the model indicated that dorsolateral and ventral prostate tumors develop following castration.[84] Metastasis to the lymph nodes and other distant sites is also observed. The fetal Gγ-SV40 Tag model is a hormone-refractory and metastatic transgenic mouse model that is suitable for studies of the mechanisms of progression of aggressive and advanced prostate cancer. However, these mice also develop nonprostatic tumors including adrenocortical and brown adipose tumors.[55]

Extensive descriptions of other SV40-Tag-based models are reviewed by Kasper[85] and Abate-Shen and Shen.[31] The TRAMP and LADY models are among the first described and therefore best characterized and widely used SV40 Tag-based transgenic models. These models progressively develop prostate cancer and metastasis that recapitulates key features of the human disease. However, one limitation of these models is the rapid onset of aggressive prostate cancers with neuroendocrine features.

ONCOGENES AND TUMOR SUPPRESSOR GENES Overexpression of oncogenes or inactivation of tumor suppressor genes is associated with progression of human prostate cancer. A number of GEM models have been created to mimic the genetic perturbations that promote progression of human prostate cancer (reviewed by Huss *et al.*,[86] Abate-Shen and Shen,[31,87] and Kasper[85,88]). In contrast to most SV40 Tag models that develop PIN, invasive carcinoma, neuroendocrine carcinoma, and metastasis, significant proportions of these mice develop mild hyperplasia or noninvasive PIN lesions. Therefore, more emphasis will be placed on models that develop severe prostatic lesions. Aberrant expression of oncogenes is implicated in human prostate cancer pathogenesis. C-Myc is an oncogene that is overexpressed in invasive and metastatic human prostate cancer samples.[89] C3(1)-c-Myc transgenic mice primarily develop hyperplasia in the ventral prostate.[90] Compared to (−426) PB-Myc-PAI mice that were generated with the minimal rat probasin promoter, ARR_2PB-myc-PAI transgenic mice rapidly develop high-grade PIN with invasive carcinoma in less than 6 months.[61] Homozygous deletion of MXI1 (MXI1$^{-/-}$), a c-Myc antagonist,[91] results in prostatic dysplasia and hyperplasia.[92] Mutations in members of the Ras family of oncogenes are found in human prostate cancer specimens.[93] Overexpression of RasT24 in PB-T*ras* transgenic mice produces mild prostatic changes including epithelial and stromal hyperplasia. In PB-*RAS* mice that overexpress activated Harvey-*RAS*, PIN lesions and intestinal metaplasia develop.[94] Transgenic mice were generated to model genomic instability using the minimal probasin promoter to force expression of the c-*fos* oncogene and *ECORI* restriction enzyme.[95] While only low-grade PIN lesions develop in PB-*fos* mice, PB-*Eco*RI mice develop high-grade PIN and invasive carcinoma. Increased expression of the antiapoptotic Bcl-2 gene in C3(1)-bcl-2 mice produces epithelial and stromal hyperplasia in the prostatic epithelium.[96] However, no hyperplastic or neoplastic phenotypes were observed in the prostates of PB-bcl-2 mice.[97]

Deregulation of components of the PI3K/Akt pathway has been implicated in prostate carcinogenesis. PTEN/MMAC1 is a ubiquitously expressed protein/lipid phosphatase that functions as a tumor suppressor by inhibiting the PI3K/Akt pathway and promoting apoptosis.[98] PTEN maps to chromosome 10q23 and undergoes loss of heterozygosity in many cancers including human prostate cancer.[98] PTEN heterozygous (PTEN$^{+/-}$) mice develop epithelial hyperplasia, dysplasia, and PIN lesions in the prostate.[99,100] Homozygous PTEN (PTEN$^{-/-}$) mice are embryonic lethal[99,101]; therefore, conditional knockout mice were generated to specifically delete PTEN in the prostatic epithelium (PB-Cre4 × PTEN$^{loxp/loxp}$).[102] Probasin (PB) directs prostate epithelium-specific expression of the Cre enzyme. This leads to Cre-mediated recombination and deletion of the PTEN gene between two *loxP* sites. PB-Cre4 × PTEN$^{loxp/loxp}$ mice progressively develop PIN, prostate adenocarcinoma, and metastasis to the lymph nodes and lungs.[102] The conditional PTEN model is unique because it is the first transgenic mouse model of prostate cancer that displays multistep carcinogenesis that closely mimics the initiation, progression, and metastasis of human prostate adenocarcinoma.[102] Loss of PTEN activity promotes constitutive activation of the PI3K/Akt pathway.[103] To model the elevated Akt activity that is frequently detected in human prostate cancer,[104] a PB-myr-HA-Akt1 (MPAKT) transgenic mouse model with constitutive Akt activity was generated.[105] MPAKT mice develop PIN lesions in the ventral prostate lobe that are strikingly similar to PIN lesions in PTEN$^{+/-}$ mice.

Nkx3.1 is a prostate-specific homeobox gene that is important for normal prostate development and differentiation.[106] Although Nkx3.1 mutations have yet to be identified in prostate cancer, it is considered a candidate tumor suppressor gene because it maps to the chromosome 8p12–21 region that is frequently lost in cancer,[107,108] including prostate tumors.[109,110] Defective branching morphogenesis, epithelial hyperplasia and PIN-like prostatic lesions were observed in Nkx3.1$^{-/-}$ null mice.[106] Prostate-specific deletion of Nkx3.1 in PSA-Cre × Nkx3.1$^{f/f}$ conditional knockout mice results in the formation of PIN lesions.[111]

CELL CYCLE PATHWAYS AND GROWTH FACTOR/PEPTIDE SIGNALING Aberrant cell cycle progression and evasion of apoptosis are hallmarks of prostate cancer progression from early stage disease to late stage advanced metastatic disease.[112] p27^{Kip1} is a cyclin-dependent kinase (CDK) inhibitor that is frequently lost in human prostate cancer.[113] Loss of p27^{Kip1} expression in mice (p27$^{-/-}$) promotes only mild hyperplasia in the prostate.[113] Expression of the SKP2 complex that targets p27 for ubiquitination and proteolysis is elevated in prostate cancer.[114] Overexpression of SKP2 in ARR_2PB-SKP2 transgenic mice promotes high-grade

PIN and decreases p27 expression in the prostate.[114] Loss-of-function mouse models of negative cell cycle regulators and tumor suppressor genes such as $p53^{-/-}$ [115] and $Rb^{-/-}$ [116] null mice display normal prostate development or embryonic lethality, respectively.

Deregulation of growth factor signaling networks facilitates survival of cancer cells. Mild hyperplasia and dysplasia were observed in the fibroblast growth factor 7 (FGF7) overexpressing PB-FGF7 (PKS) mouse model.[117] Minimal probasin promoter-directed expression of dominant negative FGFR2iiib to the prostate in KDNR mice results in development of a neuroendocrine-like phenotype.[117] In ARR_2PB-FGF8b transgenic mice, increased expression of FGF8b promotes development of high-grade PIN lesions.[118] Elevated serum levels of insulin-like growth factor-1 (IGF-1) increase the risk of prostate cancer.[119] Bovine keratin 5 (BK5)-IGF-1[120] and PB-IGF-1[121] mice were generated to study the role of IGF-1 in prostate cancer progression. In contrast to PB-IGF-1 mice that did not develop hyperplastic or neoplastic changes in the prostate epithelium,[121] BK5-IGF-1 mice develop multistep prostate carcinogenesis including hyperplasia, PIN, and invasive carcinoma with neuroendocrine features.[120] Transforming growth factor-β (TGF-β) acts as a tumor suppressor via TGF-β receptors (TβR) in normal prostate cells;[122] however, it also promotes prostate cancer progression.[123] Transgenic mice were created to block endogenous TGF-β receptor type II (TβRII) using a metallothionein promoter to control dominant-negative TβRII (MT-DNIIR).[124] Crossbreeding of MT-DNIIR mice with the LADY (LPB-SV40 large Tag) model enhances metastasis in bigenic mice.[125] Activated ERBB2 (HER-2/Neu), a member of the epidermal growth factor receptor (EGFR) family, is implicated in hormone refractory prostate cancer progression.[126] Probasin promoter-driven expression of oncogenic Neu in the prostate epithelium of PB-Neu mice results in progressive PIN lesions and invasive carcinoma.[127]

STEROID RECEPTORS Overexpression of the androgen receptor (AR) promotes prostate cancer survival following androgen deprivation.[128] Minimal probasin promoter-driven expression of murine AR in the prostate epithelium of mice (PB-mAR) results in mild prostatic dysplasia, hyperplasia, and PIN after 1 year.[129] The antiproliferative effects of retinoic acid[130] and vitamin D[131] in prostate cancer are mediated by retinoic acid receptors (RARs or RXRs)[132] and the vitamin D receptor (VDR).[131] RXRα conditional knockout mice were generated by crossing PB-Cre4 mice with $RXR\alpha^{f/f}$ mice (PB-Cre4 × $RXR\alpha^{f/f}$) for Cre-mediated targeted deletion of RXRα in the prostate epithelium.[133] These mice develop hyperplasia and low-grade and high-grade PIN lesions.

BIGENIC MODELS The prevalence of mild dysplasia, hyperplasia, and noninvasive neoplastic lesions in the prostates of most GEM models with targeted loss-of-function or gain-of-function mutations of one gene suggests that prostate cancer progression and metastasis involve a complex, multistep process. Therefore, several GEM models with two or more genetic hits have been generated to better recapitulate the complex natural history of human prostate cancer.

Although no histological changes were observed in the prostates of PB-bcl-2 mice that overexpress the antiapoptotic Bcl-2 gene,[97] PB-bcl-2 × TRAMP bigenic mice rapidly develop prostate tumors.[134] This suggests that Bcl-2 decreases tumor latency in TRAMP mice. In $PTEN^{+/-}$, $p27^{-/-}$ ($Cdkn1b^{-/-}$), and $Nkx3.1^{-/-}$ knockout mice, the prostatic lesions observed include dysplasia, hyperplasia, and sometimes PIN, but not invasive carcinoma.[99,100,106,113] Bigenic $PTEN^{+/-}$ × $Cdkn1b^{-/-}$ mice develop PIN lesions that progress to invasive carcinoma in 25% of mice.[135] Crossbreeding of $PTEN^{+/-}$ and $Nkx3.1^{-/-}$ null mice is also synergistic as mice develop invasive carcinoma[136] that frequently metastasizes to the lymph nodes.[137] These studies indicate cooperativity between PTEN haploinsufficiency and loss of p27 or Nkx3.1 tumor suppressor functions. Furthermore, crossbreeding of $PTEN^{+/-}$ mice with TRAMP mice results in rapid development of invasive prostate carcinoma.[138] FGF2 levels are elevated in advanced stage metastatic human prostate cancer.[139] In TRAMP mice that were crossbred with $FGF2^{-/-}$ null mice, the incidence of metastasis and the tumor grade were decreased.[139] Hepsin is a type II transmembrane serine protease that is overexpressed in human prostate cancer.[140–142] Probasin promoter-driven overexpression of hepsin in the prostate epithelium of PB-hepsin mice causes weakened epithelial–stromal adhesion.[143] Crossbreeding of the 12T-7f line of the LADY model that develops invasive cancer without metastasis to PB-hepsin mice triggers liver, lung, and bone metastasis.[143]

In contrast to GEM models with targeted manipulation of only one gene, most bigenic mouse models develop more severe prostatic lesions that are sometimes associated with distant site metastasis. This suggests that multiple genetic events are necessary for development of prostate cancer.

ALTERNATIVE GENETICALLY ENGINEERED MOUSE MODELS The prostate gland is composed of different cellular compartments including epithelial cells, stromal cells, and endothelial cells. The epithelial cell compartment consists of secretory luminal cells, basal cells, and neuroendocrine cells. Most innovations in the field of transgenic mouse modeling involve targeted disruption or overexpression of genes in the luminal epithelial cell compartment using prostate epithelium-specific regulatory elements such as probasin and PSP94. The genetic and molecular changes that arise in other cellular compartments of the prostate must be critically evaluated to adequately recapitulate the complex, multistep changes that contribute to prostate carcinogenesis.

To examine the role of TGF-β signaling in stromal–epithelial interactions, the TGF-β type II receptor (Tgfbr2) was conditionally inactivated in mouse fibroblasts by crossing fibroblast-specific protein 1 (Fsp1)-Cre mice with $Tgfbr2^{f/f}$ mice ($Tgfbr2^{fspKO}$).[144] Fsp1 controls prostatic fibroblast-specific expression of Cre recombinase, and Cre mediates recombination at the *loxP* sites surrounding the Tgfbr2 gene in $Tgfbr2^{f/f}$ mice. Stromal hypercellularity and PIN develop in the dorsolateral, ventral, and anterior prostate lobes of $Tgfbr2^{fspKO}$ mice. The prostate stem cell antigen (PSCA) promoter directs prostate-specific gene expression to a subpopulation of epithelial cells that expresses basal and luminal epithelial cell markers.[145] The PSCA promoter may be utilized as a promoter that targets a unique subset of epithelial cells in the prostate gland. Prostate-specific promoters that can direct gene expression to prostatic endothelial cells have yet to be reported.

Some of the limitations associated with traditional GEM models such as the TRAMP and LADY models include the rapid onset of aggressive prostate cancer and a high incidence of neuroendocrine tumors. In addition, there is currently no evidence that the SV40 early genes are involved in development of human prostate cancer. Random insertion of the transgene into the genome can cause a "position effect" that leads to differences in genotype and phenotype in traditional GEM models. To circumvent some of these problems, knockin technology is being utilized to generate mouse models of prostate cancer. Knockin mice are created by targeted insertion of a heterologous gene into a specific gene locus.[146] This facilitates single copy insertion of the targeting vector into the genome, thereby resulting in uniform transgene expression and a stable phenotype. PSP94 mouse adenocarcinoma prostate model knockin mice (PSP-KIMAP) were generated by targeted insertion of SV40 early genes to the prostate-specific PSP94 promoter/enhancer gene locus.[147] Compared to PSP94-SV40 Tag transgenic mice that were generated using traditional transgenic technique to create mice with PSP94 promoter-driven prostate-specific expression of SV40 Tag,[58] prostate cancer progression was more prolonged and synchronous in PSP-KIMAP mice.[147]

CONCLUSIONS

Each spontaneous, inducible, and transgenic animal model of prostate cancer that is currently available has strengths that have been exploited to understand the genetic and molecular basis of prostate carcinogenesis. However, there is currently no single model that fully recapitulates the full spectrum of human prostate cancer initiation, progression, and metastasis. One obvious limitation of the GEM models that have been widely utilized for prostate cancer research is the lack of reproducible development of bone metastasis. Most gain- or loss-of-function models have targeted disruption of candidate genes directed to the prostate luminal epithelium. To fully model the entire continuum and natural history of prostate cancer development, it will be critical to utilize regulatory sequences that target gene expression to other prostatic cellular compartments such as basal, stromal, and endothelial cells.

REFERENCES

1. Jemal A, Siegel R, Ward E, *et al.* Cancer statistics, 2006. *CA Cancer J Clin* 2006;56:106–130.
2. Khan MA, Partin AW. Expectant management: An option for localized prostate cancer. *Prostate Cancer Prostatic Dis* 2005;8:311–315.
3. American Cancer Society (ACS) website. http://www.cancer.org.
4. Carter BS, Beaty TH, Steinberg GD, Childs B, Walsh PC. Mendelian inheritance of familial prostate cancer. *Proc Natl Acad Sci USA* 1992;89:3367–3371.
5. Steinberg GD, Carter BS, Beaty TH, Childs B, Walsh PC. Family history and the risk of prostate cancer. *Prostate* 1990;17:337–347.
6. Bostwick DG, Ramnani D, Qian J. Prostatic intraepithelial neoplasia: Animal models 2000. *Prostate* 2000;43:286–294.
7. Lucia MS, Bostwick DG, Bosland M, *et al.* Workgroup I: Rodent models of prostate cancer. *Prostate* 1998;36:49–55.
8. Waters DJ, Hayden DW, Bell FW, Klausner JS, Qian J, Bostwick DG. Prostatic intraepithelial neoplasia in dogs with spontaneous prostate cancer. *Prostate* 1997;30:92–97.
9. Waters DJ, Sakr WA, Hayden DW, *et al.* Workgroup 4: Spontaneous prostate carcinoma in dogs and nonhuman primates. *Prostate* 1998;36:64–67.
10. Aquilina JW, McKinney L, Pacelli A, *et al.* High grade prostatic intraepithelial neoplasia in military working dogs with and without prostate cancer. *Prostate* 1998;36:189–193.
11. Sugimura Y, Cunha GR, Donjacour AA. Morphogenesis of ductal networks in the mouse prostate. *Biol Reprod* 1986;34:961–971.
12. Neumann F, Elger W, Steinbeck H. Antiandrogens and reproductive development. *Philos Trans R Soc Lond B Biol Sci* 1970;259:179–184.
13. Elger W, Graf KJ, Steinbeck H, Neumann F. Hormonal control of sexual development. *Adv Biosci* 1974;13:41–69.
14. Neumann F, Graf KJ, Elger W. Hormone-induced disturbances in sexual differentiation. *Adv Biosci* 1974;13:71–101.
15. Cunha GR, Donjacour AA, Cooke PS, *et al.* The endocrinology and developmental biology of the prostate. *Endocrine Rev* 1987;8:338–362.
16. Isaacs JT. Prostatic structure and function in relation to the etiology of prostatic cancer. *Prostate* 1983;4:351–366.
17. Hoehn W, Schroeder FH, Reimann JF, Joebsis AC, Hermanek P. Human prostatic adenocarcinoma: Some characteristics of a serially transplantable line in nude mice (PC 82). *Prostate* 1980;1:95–104.
18. van Weerden WM, de Ridder CM, Verdaasdonk CL, *et al.* Development of seven new human prostate tumor xenograft models and their histopathological characterization. *Am J Pathol* 1996;149:1055–1062.
19. van Weerden WM, Romijn JC. Use of nude mouse xenograft models in prostate cancer research. *Prostate* 2000;43:263–271.
20. Navone NM, Logothetis CJ, von Eschenbach AC, Troncoso P. Model systems of prostate cancer: Uses and limitations. *Cancer Metastasis Rev* 1998;17:361–371.
21. Wainstein MA, He F, Robinson D, *et al.* CWR22: Androgen-dependent xenograft model derived from a primary human prostatic carcinoma. *Cancer Res* 1994;54:6049–6052.
22. Nagabhushan M, Miller CM, Pretlow TP, *et al.* CWR22: The first human prostate cancer xenograft with strongly androgen-dependent and relapsed strains both in vivo and in soft agar. *Cancer Res* 1996;56:3042–3046.
23. Ellis WJ, Vessella RL, Buhler KR, *et al.* Characterization of a novel androgen-sensitive, prostate-specific antigen-producing prostatic carcinoma xenograft: LuCaP 23. Clin *Cancer Res* 1996;2:1039–1048.
24. Klein KA, Reiter RE, Redula J, *et al.* Progression of metastatic human prostate cancer to androgen independence in immunodeficient SCID mice. *Nat Med* 1997;3:402–408.
25. Wang Y, Xue H, Cutz JC, *et al.* An orthotopic metastatic prostate cancer model in SCID mice via grafting of a transplantable human prostate tumor line. *Lab Invest* 2005;85:1392–1404.
26. Waters DJ, Patronek GJ, Bostwick DG, Glickman LT. Comparing the age at prostate cancer diagnosis in humans and dogs. *J Natl Cancer Inst* 1996;88:1686–1687.
27. McNeal JE. Origin and development of carcinoma in the prostate. *Cancer* 1969;23:24–34.
28. McNeal JE. The prostate gland: Morphology and pathobiology. *Monogr Urol* 1983;4:3–37.
29. McNeal JE. Normal histology of the prostate. *Am J Surg Pathol* 1988;12:619–633.
30. Price D. Comparative aspects of development and structure in the prostate. *Natl Cancer Inst Monogr* 1963;12:1–27.
31. Abate-Shen C, Shen MM. Molecular genetics of prostate cancer. *Genes Dev* 2000;14:2410–2434.
32. Shappell SB, Thomas GV, Roberts RL, *et al.* Prostate pathology of genetically engineered mice: Definitions and classification. The consensus report from the Bar Harbor meeting of the Mouse Models of Human Cancer Consortium Prostate Pathology Committee. *Cancer Res* 2004;64:2270–2305.
33. Lubaroff DM, Canfield L, Reynolds CW. The Dunning tumors. *Prog Clin Biol Res* 1980;37:243–263.
34. Pollard M, Luckert PH. Transplantable metastasizing prostate adenocarcinomas in rats. *J Natl Cancer Inst* 1975;54:643–649.
35. Pollard M, Luckert PH. Production of autochthonous prostate cancer in Lobund-Wistar rats by treatments with N-nitroso-N-methylurea and testosterone. *J Natl Cancer Inst* 1986;77:583–587.
36. Dunning WF. Prostate cancer in the rat. *Natl Cancer Inst Monogr* 1963;12:351–369.
37. Lubaroff DM, Canfield L, Rasmussen GT, Reynolds CW. An animal model for the study of prostate carcinoma. *Natl Cancer Inst Monogr* 1978;49:275–281.

38. Isaacs JT, Weissman RM, Coffey DS, Scott WW. Concepts in prostatic cancer biology: Dunning R-3327 H, HI, and AT tumors. *Prog Clin Biol Res* 1980;37:311–323.
39. Isaacs JT, Yu GW, Coffey DS. The characterization of a newly identified, highly metastatic variety of Dunning R 3327 rat prostatic adenocarcinoma system: The MAT LyLu tumor. *Invest Urol* 1981;19:20–23.
40. Isaacs JT, Isaacs WB, Feitz WF, Scheres J. Establishment and characterization of seven Dunning rat prostatic cancer cell lines and their use in developing methods for predicting metastatic abilities of prostatic cancers. *Prostate* 1986;9:261–281.
41. Isaacs JT. The R-3327 system of rat prostatic cancers. *Urol Oncol Semin Orig Invest* 1996;2:115–116.
42. Geldof AA, Rao BR. Prostatic tumor (R3327) skeletal metastasis. *Prostate* 1990;16:279–290.
43. Goebel HW, Rausch U, Steinhoff M, *et al.* Arguments against the prostatic origin of the R-3327 Dunning H tumor. *Virchows Arch B Cell Pathol Incl Mol Pathol* 1992;62:9–18.
44. Pollard M, Luckert PH. Autochthonous prostate adenocarcinomas in Lobund-Wistar rats: A model system. *Prostate* 1987;11:219–227.
45. Pollard M, Luckert PH, Schmidt MA. Induction of prostate adenocarcinomas in Lobund Wistar rats by testosterone. *Prostate* 1982;3: 563–568.
46. Pollard M, Suckow MA. Hormone-refractory prostate cancer in the Lobund-Wistar rat. *Exp Biol Med* 2005;230:520–526.
47. Sigmund C. Major approaches for generating and analyzing transgenic mice. An overview. *Hypertension* 1993;22:599–607.
48. Asamoto M, Hokaiwado N, Cho Y-M, *et al.* Prostate carcinomas developing in transgenic rats with SV40 T antigen expression under probasin promoter control are strictly androgen dependent. *Cancer Res* 2001;61:4693–4700.
49. al-Shawi R, Kinnaird J, Burke J, Bishop JO. Expression of a foreign gene in a line of transgenic mice is modulated by a chromosomal position effect. *Mol Cell Biol* 1990;10:1192–1198.
50. Greenberg NM, DeMayo FJ, Sheppard PC, *et al.* The rat probasin gene promoter directs hormonally and developmentally regulated expression of a heterologous gene specifically to the prostate in transgenic mice. *Mol Endocrinol* 1994;8:230–239.
51. Greenberg NM, DeMayo F, Finegold MJ, *et al.* Prostate cancer in a transgenic mouse. *Proc Natl Acad Sci USA* 1995;92:3439–3443.
52. Kasper S, Sheppard PC, Yan Y, *et al.* Development, progression, and androgen-dependence of prostate tumors in probasin-large T antigen transgenic mice: A model for prostate cancer. *Lab Invest* 1998;78:i–xv.
53. Shibata MA, Ward JM, Devor DE, Liu ML, Green JE. Progression of prostatic intraepithelial neoplasia to invasive carcinoma in C3(1)/SV40 large T antigen transgenic mice: Histopathological and molecular biological alterations. *Cancer Res* 1996;56:4894–4903.
54. Tehranian A, Morris DW, Min BH, Bird DJ, Cardiff RD, Barry PA. Neoplastic transformation of prostatic and urogenital epithelium by the polyoma virus middle T gene. *Am J Pathol* 1996;149:1177–1191.
55. Perez-Stable C, Altman NH, Brown J, Harbison M, Cray C, Roos BA. Prostate, adrenocortical, and brown adipose tumors in fetal globin/T antigen transgenic mice. *Lab Invest* 1996;74:363–373.
56. Garabedian EM, Humphrey PA, Gordon JI. A transgenic mouse model of metastatic prostate cancer originating from neuroendocrine cells. *Proc Natl Acad Sci USA* 1998;95:15382–15387.
57. Skalnik DG, Dorfman DM, Williams DA, Orkin SH. Restriction of neuroblastoma to the prostate gland in transgenic mice. *Mol Cell Biol* 1991;11:4518–4527.
58. Gabril MY, Onita T, Ji PG, *et al.* Prostate targeting: PSP94 gene promoter/enhancer region directed prostate tissue-specific expression in a transgenic mouse prostate cancer model. *Gene Ther* 2002;9:1589–1599.
59. Yan Y, Sheppard PC, Kasper S, *et al.* Large fragment of the probasin promoter targets high levels of transgene expression to the prostate of transgenic mice. *Prostate* 1997;32:129–139.
60. Zhang J, Thomas TZ, Kasper S, Matusik RJ. A small composite probasin promoter confers high levels of prostate-specific gene expression through regulation by androgens and glucocorticoids in vitro and in vivo. *Endocrinology* 2000;141:4698–4710.
61. Ellwood-Yen K, Graeber TG, Wongvipat J, *et al.* Myc-driven murine prostate cancer shares molecular features with human prostate tumors. *Cancer Cell* 2003;4:223–238.
62. Gingrich JR, Barrios RJ, Foster BA, Greenberg NM. Pathologic progression of autochthonous prostate cancer in the TRAMP model. *Prostate Cancer Prostatic Dis* 1999;2:70–75.
63. Gingrich JR, Barrios RJ, Morton RA, *et al.* Metastatic prostate cancer in a transgenic mouse. *Cancer Res* 1996;56:4096–4102.
64. Kaplan-Lefko PJ, Chen TM, Ittmann MM, *et al.* Pathobiology of autochthonous prostate cancer in a pre-clinical transgenic mouse model. *Prostate* 2003;55:219–237.
65. Masumori N, Thomas TZ, Chaurand P, *et al.* A probasin-large T antigen transgenic mouse line develops prostate adenocarcinoma and neuroendocrine carcinoma with metastatic potential. *Cancer Res* 2001;61:2239–2249.
66. Suttie A, Nyska A, Haseman JK, Moser GJ, Hackett TR, Goldsworthy TL. A grading scheme for the assessment of proliferative lesions of the mouse prostate in the TRAMP model. *Toxicol Pathol* 2003;31:31–38.
67. Jasani B, Cristaudo A, Emri SA, *et al.* Association of SV40 with human tumours. *Semin Cancer Biol* 2001;11:49–61.
68. Barbanti-Brodano G, Sabbioni S, Martini F, Negrini M, Corallini A, Tognon M. Simian virus 40 infection in humans and association with human diseases: Results and hypotheses. *Virology* 2004;318: 1–9.
69. Pipas JM, Levine AJ. Role of T antigen interactions with p53 in tumorigenesis. *Semin Cancer Biol* 2001;11:23–30.
70. DeCaprio JA, Ludlow JW, Figge J, *et al.* SV40 large tumor antigen forms a specific complex with the product of the retinoblastoma susceptibility gene. *Cell* 1988;54:275–283.
71. Ali SH, DeCaprio JA. Cellular transformation by SV40 large T antigen: Interaction with host proteins. *Semin Cancer Biol* 2001;11: 15–23.
72. Pallas DC, Shahrik LK, Martin BL, *et al.* Polyoma small and middle T antigens and SV40 small t antigen form stable complexes with protein phosphatase 2A. *Cell* 1990;60:167–176.
73. Ewald D, Li M, Efrat S, *et al.* Time-sensitive reversal of hyperplasia in transgenic mice expressing SV40 T antigen. *Science* 1996;273:1384–1386.
74. Gingrich JR, Barrios RJ, Kattan MW, Nahm HS, Finegold MJ, Greenberg NM. Androgen-independent prostate cancer progression in the TRAMP model. *Cancer Res* 1997;57:4687–4691.
75. Hansson J, Abrahamsson PA. Neuroendocrine pathogenesis in adenocarcinoma of the prostate. *Ann Oncol* 2001;12(Suppl. 2): S145–152.
76. Huss WJ, Hanrahan CF, Barrios RJ, Simons JW, Greenberg NM. Angiogenesis and prostate cancer: Identification of a molecular progression switch. *Cancer Res* 2001;61:2736–2743.
77. Huss WJ, Barrios RJ, Foster BA, Greenberg NM. Differential expression of specific FGF ligand and receptor isoforms during angiogenesis associated with prostate cancer progression. *Prostate* 2003;54:8–16.
78. Huss WJ, Barrios RJ, Greenberg NM. SU5416 selectively impairs angiogenesis to induce prostate cancer-specific apoptosis. *Mol Cancer Ther* 2003;2:611–616.
79. Foster BA, Gingrich JR, Kwon ED, Madias C, Greenberg NM. Characterization of prostatic epithelial cell lines derived from transgenic adenocarcinoma of the mouse prostate (TRAMP) model. *Cancer Res* 1997;57:3325–3330.
80. Masumori N, Tsuchiya K, Tu WH, *et al.* An allograft model of androgen independent prostatic neuroendocrine carcinoma derived from a large probasin promoter-T antigen transgenic mouse line. *J Urol* 2004;171:439–442.
81. Hara M, Kimura H. Two prostate-specific antigens, gamma-seminoprotein and beta-microseminoprotein. *J Lab Clin Med* 1989; 113:541–548.

82. Xuan JW, Kwong J, Chan FL, *et al.* cDNA, genomic cloning, and gene expression analysis of mouse PSP94 (prostate secretory protein of 94 amino acids). *DNA Cell Biol* 1999;18:11–26.
83. Ouellette AJ, Selsted ME. Paneth cell defensins: Endogenous peptide components of intestinal host defense. *FASEB J* 1996;10:1280–1289.
84. Perez-Stable C, Altman NH, Mehta PP, Deftos LJ, Roos BA. Prostate cancer progression, metastasis, and gene expression in transgenic mice. *Cancer Res* 1997;57:900–906.
85. Kasper S. Survey of genetically engineered mouse models for prostate cancer: Analyzing the molecular basis of prostate cancer development, progression, and metastasis. *J Cell Biochem* 2005;94: 279–297.
86. Huss WJ, Maddison LA, Greenberg NM. Autochthonous mouse models for prostate cancer: Past, present and future. *Semin Cancer Biol* 2001;11:245–260.
87. Abate-Shen C, Shen MM. Mouse models of prostate carcinogenesis. *Trends Genet* 2002;18:S1–5.
88. Kasper S, Smith JA Jr. Genetically modified mice and their use in developing therapeutic strategies for prostate cancer. *J Urol* 2004; 172:12–19.
89. Jenkins RB, Qian J, Lieber MM, Bostwick DG. Detection of c-myc oncogene amplification and chromosomal anomalies in metastatic prostatic carcinoma by fluorescence in situ hybridization. *Cancer Res* 1997;57:524–531.
90. Zhang X, Lee C, Ng PY, Rubin M, Shabsigh A, Buttyan R. Prostatic neoplasia in transgenic mice with prostate-directed overexpression of the c-myc oncoprotein. *Prostate* 2000;43:278–285.
91. Benson LQ, Coon MR, Krueger LM, Han GC, Sarnaik AA, Wechsler DS. Expression of MXI1, a Myc antagonist, is regulated by Sp1 and AP2. *J Biol Chem* 1999;274:28794–28802.
92. Schreiber-Agus N, Meng Y, Hoang T, *et al.* Role of Mxi1 in ageing organ systems and the regulation of normal and neoplastic growth. *Nature* 1998;393:483–487.
93. Konishi N, Hiasa Y, Tsuzuki T, Tao M, Enomoto T, Miller GJ. Comparison of ras activation in prostate carcinoma in Japanese and American men. *Prostate* 1997;30:53–57.
94. Scherl A, Li JF, Cardiff RD, Schreiber-Agus N. Prostatic intraepithelial neoplasia and intestinal metaplasia in prostates of probasin-RAS transgenic mice. *Prostate* 2004;59:448–459.
95. Voelkel-Johnson C, Voeks DJ, Greenberg NM, *et al.* Genomic instability-based transgenic models of prostate cancer. *Carcinogenesis* 2000;21:1623–1627.
96. Zhang X, Chen MW, Ng A, *et al.* Abnormal prostate development in C3(1)-bcl-2 transgenic mice. *Prostate* 1997;32:16–26.
97. Bruckheimer EM, Cho S, Brisbay S, *et al.* The impact of bcl-2 expression and bax deficiency on prostate homeostasis in vivo. *Oncogene* 2000;19:2404–2412.
98. Di Cristofano A, Pandolfi PP. The multiple roles of PTEN in tumor suppression. *Cell* 2000;100:387–390.
99. Di Cristofano A, Pesce B, Cordon-Cardo C, Pandolfi PP. Pten is essential for embryonic development and tumour suppression. *Nat Genet* 1998;19:348–355.
100. Podsypanina K, Ellenson LH, Nemes A, *et al.* Mutation of Pten/Mmac1 in mice causes neoplasia in multiple organ systems. *Proc Natl Acad Sci USA* 1999;96:1563–1568.
101. Suzuki A, de la Pompa JL, Stambolic V, *et al.* High cancer susceptibility and embryonic lethality associated with mutation of the PTEN tumor suppressor gene in mice. *Curr Biol* 1998;8:1169–1 178.
102. Wang S, Gao J, Lei Q, *et al.* Prostate-specific deletion of the murine Pten tumor suppressor gene leads to metastatic prostate cancer. *Cancer Cell* 2003;4:209–221.
103. Vazquez F, Sellers WR. The PTEN tumor suppressor protein: An antagonist of phosphoinositide 3-kinase signaling. *Biochim Biophys Acta Rev Cancer* 2000;1470:M21–M35.
104. Sun M, Wang G, Paciga JE, *et al.* AKT1/PKBalpha kinase is frequently elevated in human cancers and its constitutive activation is required for oncogenic transformation in NIH3T3 cells. *Am J Pathol* 2001;159:431–437.
105. Majumder PK, Yeh JJ, George DJ, *et al.* Prostate intraepithelial neoplasia induced by prostate restricted Akt activation: The MPAKT model. *Proc Natl Acad Sci USA* 2003;100:7841–7846.
106. Bhatia-Gaur R, Donjacour AA, Sciavolino PJ, *et al.* Roles for Nkx3.1 in prostate development and cancer. *Genes Dev* 1999;13: 966–977.
107. Voeller HJ, Augustus M, Madike V, Bova GS, Carter KC, Gelmann EP. Coding region of NKX3.1, a prostate-specific homeobox gene on 8p21, is not mutated in human prostate cancers. *Cancer Res* 1997;57:4455–4459.
108. He WW, Sciavolino PJ, Wing J, *et al.* A novel human prostate-specific, androgen-regulated homeobox gene (NKX3.1) that maps to 8p21, a region frequently deleted in prostate cancer. *Genomics* 1997;43:69–77.
109. Chang M, Tsuchiya K, Batchelor RH, *et al.* Deletion mapping of chromosome 8p in colorectal carcinoma and dysplasia arising in ulcerative colitis, prostatic carcinoma, and malignant fibrous histiocytomas. *Am J Pathol* 1994;144:1–6.
110. Matsuyama H, Pan Y, Skoog L, *et al.* Deletion mapping of chromosome 8p in prostate cancer by fluorescence in situ hybridization. *Oncogene* 1994;9:3071–3076.
111. Abdulkadir SA, Magee JA, Peters TJ, *et al.* Conditional loss of Nkx3.1 in adult mice induces prostatic intraepithelial neoplasia. *Mol Cell Biol* 2002;22:1495–1503.
112. Berges RR, Vukanovic J, Epstein JI, *et al.* Implication of cell kinetic changes during the progression of human prostatic cancer. *Clin Cancer Res* 1995;1:473–480.
113. Cordon-Cardo C, Koff A, Drobnjak M, *et al.* Distinct altered patterns of p27KIP1 gene expression in benign prostatic hyperplasia and prostatic carcinoma. *J Natl Cancer Inst* 1998;90:1284–1291.
114. Shim EH, Johnson L, Noh HL, *et al.* Expression of the F-box protein SKP2 induces hyperplasia, dysplasia, and low-grade carcinoma in the mouse prostate. *Cancer Res* 2003;63:1583–1588.
115. Donehower LA, Harvey M, Slagle BL, *et al.* Mice deficient for p53 are developmentally normal but susceptible to spontaneous tumours. *Nature* 1992;356:215–221.
116. Wang Y, Hayward SW, Donjacour AA, *et al.* Sex hormone-induced carcinogenesis in Rb-deficient prostate tissue. *Cancer Res* 2000;60: 6008–6017.
117. Foster BA, Evangelou A, Gingrich JR, Kaplan PJ, DeMayo F, Greenberg NM. Enforced expression of FGF-7 promotes epithelial hyperplasia whereas a dominant negative FGFR2iiib promotes the emergence of neuroendocrine phenotype in prostate glands of transgenic mice. *Differentiation* 2002;70:624–632.
118. Song Z, Wu X, Powell WC, *et al.* Fibroblast growth factor 8 isoform B overexpression in prostate epithelium: A new mouse model for prostatic intraepithelial neoplasia. *Cancer Res* 2002;62:5096–5105.
119. Chan JM, Stampfer MJ, Giovannucci E, *et al.* Plasma insulin-like growth factor-I and prostate cancer risk: A prospective study. *Science* 1998;279:563–566.
120. DiGiovanni J, Kiguchi K, Frijhoff A, *et al.* Deregulated expression of insulin-like growth factor 1 in prostate epithelium leads to neoplasia in transgenic mice. *Proc Natl Acad Sci USA* 2000;97:3455–3460.
121. Konno-Takahashi N, Takeuchi T, Shimizu T, *et al.* Engineered IGF-I expression induces glandular enlargement in the murine prostate. *J Endocrinol* 2003;177:389–398.
122. Derynck R. TGF-beta-receptor-mediated signaling. *Trends Biochem Sci* 1994;19:548–553.
123. Wikstrom P, Lindh G, Bergh A, Damber JE. Alterations of transforming growth factor beta1 (TGF-beta1) and TGFbeta receptor expressions with progression in Dunning rat prostatic adenocarcinoma sublines. *Urol Res* 1999;27:185–193.
124. Serra R, Johnson M, Filvaroff EH, *et al.* Expression of a truncated, kinase-defective TGF-beta type II receptor in mouse skeletal tissue promotes terminal chondrocyte differentiation and osteoarthritis. *J Cell Biol* 1997;139:541–552.
125. Tu WH, Thomas TZ, Masumori N, *et al.* The loss of TGF-beta signaling promotes prostate cancer metastasis. *Neoplasia* 2003;5:267–277.

126. Di Lorenzo G, Autorino R, De Laurentiis M, *et al.* HER-2/neu receptor in prostate cancer development and progression to androgen independence. *Tumori* 2004;90:163–170.
127. Li Z, Szabolcs M, Terwilliger JD, Efstratiadis A. Prostatic intraepithelial neoplasia and adenocarcinoma in mice expressing a probasin-Neu oncogenic transgene. *Carcinogenesis* 2006;27:1054–1067.
128. Feldman BJ, Feldman D. The development of androgen-independent prostate cancer. *Nat Rev Cancer* 2001;1:34–45.
129. Stanbrough M, Leav I, Kwan PW, Bubley GJ, Balk SP. Prostatic intraepithelial neoplasia in mice expressing an androgen receptor transgene in prostate epithelium. *Proc Natl Acad Sci USA* 2001;98: 10823–10828.
130. Fong CJ, Sutkowski DM, Braun EJ, *et al.* Effect of retinoic acid on the proliferation and secretory activity of androgen-responsive prostatic carcinoma cells. *J Urol* 1993;149:1190–1194.
131. Skowronski RJ, Peehl DM, Feldman D. Vitamin D and prostate cancer: 1, 25 dihydroxyvitamin D3 receptors and actions in human prostate cancer cell lines. *Endocrinology* 1993;132:1952–1960.
132. Allegretto E, McClurg M, Lazarchik S, *et al.* Transactivation properties of retinoic acid and retinoid X receptors in mammalian cells and yeast. Correlation with hormone binding and effects of metabolism. *J Biol Chem* 1993;268:26625–26633 [published erratum appears in *J Biol Chem* 1994;269(10):7834].
133. Huang J, Powell WC, Khodavirdi AC, *et al.* Prostatic intraepithelial neoplasia in mice with conditional disruption of the retinoid X receptor alpha allele in the prostate epithelium. *Cancer Res* 2002;62: 4812–4819.
134. Bruckheimer EM, Brisbay S, Johnson DJ, Gingrich JR, Greenberg N, McDonnell TJ. Bcl-2 accelerates multistep prostate carcinogenesis in vivo. *Oncogene* 2000;19:5251–5258.
135. Di Cristofano A, De Acetis M, Koff A, Cordon-Cardo C, Pandolfi PP. Pten and p27KIP1 cooperate in prostate cancer tumor suppression in the mouse. *Nat Genet* 2001;27:222–224.
136. Kim MJ, Cardiff RD, Desai N, *et al.* Cooperativity of Nkx3.1 and Pten loss of function in a mouse model of prostate carcinogenesis. *Proc Natl Acad Sci USA* 2002;99:2884–2889.
137. Abate-Shen C, Banach-Petrosky WA, Sun X, *et al.* Nkx3.1; Pten mutant mice develop invasive prostate adenocarcinoma and lymph node metastases. *Cancer Res* 2003;63:3886–3890.
138. Kwabi-Addo B, Giri D, Schmidt K, *et al.* Haploinsufficiency of the Pten tumor suppressor gene promotes prostate cancer progression. *Proc Natl Acad Sci USA* 2001;98:11563–11568.
139. Polnaszek N, Kwabi-Addo B, Peterson LE, *et al.* Fibroblast growth factor 2 promotes tumor progression in an autochthonous mouse model of prostate cancer. *Cancer Res* 2003;63:5754–5760.
140. Magee JA, Araki T, Patil S, *et al.* Expression profiling reveals hepsin overexpression in prostate cancer. *Cancer Res* 2001;61: 5692–5696.
141. Stamey TA, Warrington JA, Caldwell MC, *et al.* Molecular genetic profiling of Gleason grade 4/5 prostate cancers compared to benign prostatic hyperplasia. *J Urol* 2001;166:2171–2177.
142. Welsh JB, Sapinoso LM, Su AI, *et al.* Analysis of gene expression identifies candidate markers and pharmacological targets in prostate cancer. *Cancer Res* 2001;61:5974–5978.
143. Klezovitch O, Chevillet J, Mirosevich J, Roberts RL, Matusik RJ, Vasioukhin V. Hepsin promotes prostate cancer progression and metastasis. *Cancer Cell* 2004;6:185–195.
144. Bhowmick NA, Chytil A, Plieth D, *et al.* TGF-beta signaling in fibroblasts modulates the oncogenic potential of adjacent epithelia. *Science* 2004;303:848–851.
145. Watabe T, Lin M, Ide H, *et al.* Growth, regeneration, and tumorigenesis of the prostate activates the PSCA promoter. *Proc Natl Acad Sci USA* 2002;99:401–406.
146. Nature Cell Migration Consortium (CMC). http://www.cellmigration.org/resource/komouse/komouse_approaches.shtml.
147. Duan W, Gabril MY, Moussa M, *et al.* Knockin of SV40 Tag oncogene in a mouse adenocarcinoma of the prostate model demonstrates advantageous features over the transgenic model. *Oncogene* 2005;24:1510–1524.

67 Animal Models of Diabetes

WILLY J. MALAISSE AND ABDULLAH SENER

ABSTRACT

This review deals with animal models of diabetes mellitus. Experimentally induced diabetes, such as that caused by antiinsulin serum, pancreatectomy, glucose infusion, β-cytotoxic agents, and viruses, is first considered. Attention is also paid to diabetogenic nutritional and hormonal factors, inhibitors of insulin release, and animal models of glycosuria. Second, both non-insulin-dependent spontaneous diabetes, such as that encountered in several rodent species, and insulin-dependent spontaneous diabetes, especially in NOD mice and BB rats, are briefly discussed. Last, the more recent use of gene targeting and transgenic techniques to explore the physiopathology of diabetes and the introduction of several insulin-secreting and other pancreatic endocrine cell lines in experimental diabetology are reviewed.

Key Words: Diabetes mellitus, Animal models, Antiinsulin serum, Pancreatectomy, β-Cytotoxic agents, Spontaneous diabetes, Gene targeting and transgenic techniques, Insulin-secreting cell lines.

INTRODUCTION

Diabetes mellitus has occurred in humans at least as long as written medical reports exist, i.e., approximately 4000 years. It has been found in all human societies where any consistent search has been made. In the introduction of a Workshop on Spontaneous Diabetes in Laboratory Animals, held in Augusta (MI) in 1966, A.E. Renold and W.E. Dulin proposed that the same may be true in animals, particularly those living in association with humans, whether as domestic animals or as animals bred in the laboratory.[1] They indicated that diabetes was reported in species such as the tree shrew[2] and the dolphin.[3] They also provided a table listing, under the title "Type of inherited diabetes in mus musculus and in other small laboratory rodents," six single gene mutations and three inbred strains and F_1 hybrids in mice, and four other species, including spiny-mouse, sand rat, Chinese hamster, and the single "fatty" mutant gene in the rat. This introduction was followed by 222 pages of articles devoted to the same issue. Needless to add, in this chapter, the issue of animal models of diabetes could be reviewed in only a quite restricted and, at best, illustrative manner.

EXPERIMENTALLY INDUCED DIABETES

Several experimental procedures can be used to induce hyperglycemia in otherwise normal animals. Seven of these procedures are considered below.

From: *Sourcebook of Models for Biomedical Research*
(P. M. Conn, ed.), © 2008 Humana Press Inc., Totowa, NJ.

DIABETES INDUCED BY ANTIINSULIN SERUM A transient state of insulinopenia for several hours up to a few days can be induced by the injection, e.g., in rats or dogs, of guinea pig antiinsulin serum.[4] The restoration of normoglycemia occurs when all circulating antibodies are neutralized by endogenously secreted insulin.[5] This approach makes it possible to characterize the metabolic consequences of insulinopenia without any primary alteration of endocrine pancreatic function. It was also used to explore the consequences of a sustained increase in insulin output upon the functional behavior of pancreatic islets, e.g., after their isolation from animals previously injected with guinea pig antiinsulin serum.[6]

PANCREATECTOMY In the 1880s, von Mering and Minkowski observed that after removal of the pancreas of a dog, the animal developed polyuria and polydipsia and was eventually found to have diabetes mellitus.[7] History has also given a special place to Marjorie, one of the dogs used by Banting and Best in their seminal experiments on the isolation and purification of insulin in the 1920s.[8]

In a more recent work, partial pancreatectomy, e.g., removal of 70–90% of the β cell mass, was used to induce a mild type 2 diabetes-like state associated with a stable increase in glycemia. This approach is still widely used to explore its functional consequences upon the remaining insulin-producing cells.[9]

GLUCOSE INFUSION The hyperglycemia provoked in healthy animals by the infusion of glucose differs in several respects, e.g., by the changes in unesterified fatty acids and production of ketone bodies, from that prevailing in diabetic animals. Nevertheless, the sustained infusion of a hypertonic solution of glucose represents a suitable model to reproduce some of the hydroionic consequences of diabetic hyperglycemia. For instance, prolonged glucose infusions to dogs served as an animal model for the study of hyperosmolar nonketotic coma.[10] In exceptional cases, glucose administration for several weeks was also reported to induce permanent diabetes.[11] Moreover, chronic intravenous infusion of glucose for periods of up to 7 days has been used, inter alia, to explore the biochemical determinants of the insulin secretory defect caused by sustained hyperglycemia and often referred to as a phenomenon of B cell glucotoxicity or in-competence. For instance, it was proposed that under these experimental conditions, glycogen accumulation in pancreatic islet β cells may account for two typical features of B cell glucotoxicity, mainly a paradoxical short-lived inhibition of insulin release in response to the bolus intravenous injection of glucose and a perturbation of the anomeric specificity of glucose-induced insulin release.[12]

β-CYTOTOXIC AGENTS Alloxan and streptozotocin are the two most commonly used β-cytotoxic agents in the perspective of nonsurgical induction of experimental diabetes in animals.

The diabetogenic action of alloxan was first reported by Dunn *et al.*[13] The intravenous administration of freshly dissolved alloxan (0.3–0.7 μmol/g body weight) after 24 h starvation is recommended as the usual procedure. The selective cytotoxic action of alloxan on insulin-producing cells was attributed to the coincidence of two distinct and independent features: a rapid accumulation of the drug into islet β cells and their exquisite sensitivity to peroxides.[14]

Streptozotocin, an antibiotic isolated from *Streptomyces achromogenes*, also exerts a selective β-cytoxic action.[15] It is usually administered intravenously, at a dose close to 0.25 μmol/g body weight, to overnight starved animals. Three further modalities of streptozotocin administration merit mention. First, the combined administration of streptozotocin and nicotinamide, which protects against the diabetogenic action of the antibiotic, is currently used to develop, within about 1 year, pancreatic insulinomas.[16] Second, a multiple low-dose streptozotocin model has been extensively used to study the immunological pathways that lead to insulitis and β cell death.[17] Last, injection of neonatal rats with streptozotocin within 2 days of birth results, later in life, in a model of non-insulin-dependent diabetes, with reduced β cell mass and mild hyperglycemia.[18]

Vacor, dithizone dehydroascorbic acid, pentamidine, and 8-hydroxyquinoline were also examined for their diabetogenic effects in experimental animals.[19–21]

VIRUS-INDUCED DIABETES Viruses, such as encephalomyocarditis virus (EMC), may induce diabetes in experimental animals.[22] This occurs not by autoimmunity but by direct infection and cytolysis of pancreatic β cells.

DIABETOGENIC NUTRITIONAL AND HORMONAL FACTORS It has long been known that the long-term administration of different hormones may lead to either glucose intolerance or frank diabetes. Such is the case for growth hormone, glucocorticoids, glucagon, and thyroid hormones.[23] Likewise, obesity, such as that induced by gold thioglucose or a fat diet, may affect glucose homeostasis.[24] Attention was also drawn to the perturbation of both fuel homeostasis and islet function found in starvation,[25] protein malnutrition,[26] and pregnancy and lactation.[27] In these several instances, suitable animal experimental designs were developed to assess both the pancreatic endocrine and extrapancreatic determinants participating in the response to such diabetogenic factors. In this respect, two examples could be mentioned. First, attention has been paid to the role of diabetes in pregnancy and the intrauterine environment in the subsequent development of diabetes among offspring.[28] Second, second-generation rats deficient in long-chain polyunsaturated ω3 fatty acids were recently examined to further explore the impact of these fatty acids on metabolic variables such as liver triglyceride content, visceral obesity, insulin resistance, and pancreatic islet metabolic, ionic, and secretory parameters.[29]

INHIBITION OF INSULIN RELEASE Several chemical agents inhibit insulin secretion. For *in vivo* studies, D-mannoheptulose, adrenalin, and diazoxide merit consideration.[20]

D-Mannoheptulose is a natural ketoheptose present in several plants, e.g., avocados. By inhibiting glucose phosphorylation in islet cells, it suppresses insulin secretion. The *in vivo* use of D-mannoheptulose (1–4 μmol/g body weight given intravenously) is limited by its biological half-life, itself linked to its renal clearance.[5]

Adrenalin inhibits insulin release at the intervention of β cell α-adrenergic receptors. *In vivo* it needs to be continuously infused (about 5 pmol/g body weight/min), the insulinemia rapidly reaching high values when the infusion of the adrenergic agent is halted. Obviously, another limitation concerns the extrapancreatic effect of the hormone, e.g., in terms of glycogenolysis and lipolysis.

Diazoxide also inhibits insulin release by causing the gating of ATP-sensitive K+ channels. It has been used *in vivo* (e.g., 0.5–2.0 μmol/g body weight given orally each day), for instance, to distinguish between the respective role of hyperglycemia and sustained high insulin secretion in the functional behavior of islet cells, including granulolysis, slackening of proinsulin conversion, or anomeric specificity of glucose-stimulated insulin release.[30,31]

ANIMAL MODELS OF GLYCOSURIA Phlorizin, a naturally occurring flavinoid, inhibits the resorption of glucose in the proximal renal tubule. It can be used to restore near-normoglycemia in diabetic animals by provoking massive glycosuria and polyuria.[32] This procedure may help to distinguish between the effects of hyperglycemia and other consequences of experimental diabetes in pathological processes, such as diabetic complications.

SPONTANEOUS ANIMAL DIABETES

As already alluded to in the introduction of this chapter, an exhaustive analysis of all spontaneous animal models of diabetes would largely be beyond the limits of this review.

NON-INSULIN-DEPENDENT MODELS For instance, more than a dozen spontaneous animal models of type 2 diabetes are currently under investigation.

Chinese Hamster The incidence of diabetes is high in Chinese hamsters (*Cricetulus griseus*). In this animal, the syndrome is inherited and polygenic. Typical experiments involve comparison between aglycosuric control animals, mildly diabetic animals with slight and/or occasional glycosuria, and more severely diabetic animals with histories of severe glycolysia detected at weekly intervals for more than 2 months.[33] For instance, in such a design, insulin secretion was found to be highest in the animals with intermittent glycosuria and lowest in the more severely diabetic animals.

Goto-Kakizaki Rat The Goto-Kakizaki (GK) rat was developed by the selective breeding of Wistar rats with the highest blood glucose concentration over may generations.[34] This inherited model of type 2 diabetes presents several features that can be compared with human type 2 diabetes. Such features include inter alia, polygenic inheritance, high maternal transmission, and the occurrence of diabetic complications, such as renal lesions, structural changes in peripheral nerves, and abnormalities of the retina.

Sand Rat In its natural habitat, the sand rat (*Psammonys obesus*) has an essentially vegetarian diet. However, when fed laboratory chow, the animals become obese, insulin resistant, and hyperglycemic. Hence, the sand rat model is particularly useful when studying the effects of diet and exercise in the development of type 2 diabetes.[35] As a matter of fact, the identification of diabetes in the sand rat, which was first investigated by physiologists interested in the kidney regulation of urine volume in this desertic animal, was initially based on the observation of oral infectious lesions in the animals given free access to laboratory chow.

Spiny Mouse Environmental factors such as exercise and nutrition also play a crucial role in the development of diabetes in spiny mice (*Acomys cahirinus*), whether transferred from their normal foraging habitat or bred in the laboratory.[36] Moreover, in this model, a primary abnormality of the β cell microtubular system may participate in the impaired dynamics of insulin release.[37]

KK Mouse KK mouse were first described in 1967 as a strain bred for large body size. Typically, the mouse gradually becomes obese in adult life, this being associated with insulin resistance, compensatory islet cell hyperplasia, and hyperinsulinemia, and eventually resulting in mild hyperglycemia.[38] Some differences may now exist among the KK mouse strains maintained as colonies in different laboratories.

Nagoya-Shibata-Yasuda Mouse As with the KK mouse, the Nagoya-Shibata-Yasuda (NSY) mouse was developed by selected inbreeding. The NSY mice spontaneously develop diabetes in both an age-dependent manner and with a marked gender difference. This model is quite useful, therefore, when considering age-related phenotypic changes, such as the progressive alteration of β cell function.[39]

Otsuka Long-Evans Tokushima Fatty Rat A spontaneously diabetic rat with polyuria, polydipsia, and slight obesity was discovered in 1984 in an outbred colony of Long–Evans rats, which had been purchased from Charles River Canada in 1982 and since kept at the Tokushima Research Institute in Japan. By selective breeding, the Otsuka Long–Evans Tokushima fatty (OLETF) rat was then developed and characterized by a chronic course of disease with late onset of hyperglycemia, mild obesity, inheritance by males, hyperplastic foci of pancreatic islets, and renal complications.[40]

New Zealand Obese Mouse Diabetes in the New Zealand obese (NZO) mouse, developed in 1972 by Larkins and Martin, displays polygenic inheritance and is characterized by defective insulin secretion as a prominent feature of the syndrome.[41] Dietary restriction and normalization of body weight can partly ameliorate the β cell defect.

Fatty Zucker Rat The Zucker diabetic fatty (ZDF) rat (fa/fa) is a well-characterized model of non-insulin-dependent diabetes mellitus in which homozygous males demonstrate obesity and insulin resistance, but have normal plasma glucose levels until about 10 weeks of age when overt hyperglycemia develops. Heterozygous littermates are lean and do not develop diabetes. Since diabetes-prone ZDF rats go through a clearly defined prediabetic period, they offer the opportunity to compare selected metabolic and hormonal variables in the prediabetic and diabetic state.[42]

Diabetes Mouse The diabetic syndrome in C57BL/KsJ db/db mice is inherited as an autosomal recessive trait. Diabetic mice are hyperphagic, obese, functionally sterile with atrophic gonads and accessory structures, and hyperinsulinemic. In 1970, Coleman and Hummel proposed that the primary defect in diabetic mice may involve the hypothalamus, particularly in the regions of the satiety centers.[43] The db/db mouse, like the fa/fa rat, was eventually found to indeed harbor a mutation in a hypothalamic receptor for leptin.

Obese Mouse A single autosomal recessive gene is responsible for the obesity–diabetes syndrome in ob/ob mice. Animals homozygous for the ob gene exhibit hyperphagia and obesity as well as hyperinsulinemia and insulin resistance. The ob gene has been transferred to several strains of mice, which have been employed extensively to study various aspects of insulin secretion, obesity, and diabetes. In 1994, Friedman and colleagues reported the cloning of the mutation responsible for the massive obesity seen in the ob/ob mouse.[44] The defective ob gene coded for a protein secreted by adipocytes and termed leptin.

New Genetic Animal Models of Type 2 Diabetes New animal models of diabetes are still sought. For instance, a new genetic model of type 2 spontaneous diabetes, the Tsumura Suzuki obese diabetic (TSOD) mouse, which develops moderate degrees of obesity and diabetes that are especially apparent in animals more than 11 weeks old, displays both impairment of insulin-stimulated GLUT4 translocation in skeletal muscle and adipose tissue, and diabetic renal and neural complications.[45,46] Likewise, an outbred animal model of early onset polygenic obesity and diabetes was recently reported as the M16 mouse.[47]

INSULIN-DEPENDENT MODELS The insulin-dependent animal models of diabetes experimentally induced by alloxan, streptozotocin, or total pancreatectomy were already mentioned in this chapter. The two models of spontaneous insulin-dependent diabetes are the Bio Breeding (BB) and the nonobese diabetic (NOD) mouse. They are eventually characterized by hyperglycemia, extensive or complete degeneration of B cells, insulinopenia, weight loss, hyperphagia, polydipsia, and, if untreated, ketosis and death.

The NOD Mouse The NOD mouse was developed by selectively breeding offspring from a laboratory strain that in fact was first used in the study of cataract development.[48] Insulitis, present in mice 4–5 weeks old, is followed by subclinical β destruction, frank diabetes being present between 12 and 30 weeks of age. Genetic studies have revealed multiple susceptibility genes, the MHC region apparently playing a key role.[49] The immunological cascade that includes T-helper type 2 (Th2) cells, effector cells (CD4+, CD8+), and cytokines was also examined in the NOD mice. Immunosuppression with cyclosporine in combination with other immunomodulatory compounds such as vitamin D helps to prevent diabetes in NOD mice and also in BB rats.

The BB Rat The BB rat was first recognized in 1974 in the Bio Breeding Laboratories, a commercial breeding company based in Ottawa.[50] Current studies in this animal model of type 1 diabetes involve the comparison between diabetes-prone and control BB rats. In the diabetes-prone BB (BBdp) rats, insulinopenia develops at about 12 weeks of age, the pancreatic islets being subjected to an autoimmune attack with T cells, B cells, macrophages, and natural killer cells recruited to the insulitis.

In light of the effect of diets on the incidence of diabetes in BBdp rats, a type-1 diabetes-related protein from wheat was recently identified as the first candidate wheat protein that is not only antigenic in both diabetic rats and human patients, but is also closely linked with the autoimmune attack of the pancreas. As a matter of fact, Scott and colleagues proposed that intestinal dysfunction and dysregulation of the gut immune system play a role in the development of diabetes in BBdp rats fed a diabetes-promoting diet.[51]

Other Models Other animal models of type 1 diabetes cited by Rees and Alcolado[21] include the Long–Evans Tokushima lean

(LETL) rat, New Zealand white rabbit, Keeshond dog, Chinese hamster, and Celebes black ape (*Macacca nigra*).

GENE TARGETING AND TRANSGENIC TECHNIQUES

An explosion in the number of animal models used for the study of diabetes has come from the use of molecular biological techniques. For instance, knockout animals were produced by using a genetic construct that will disrupt a normal gene. This approach is not without limitations. To mention only one example, overexpressing or underexpressing a single gene may lead to changes in the expression of other genes as a compensatory mechanism.

Transgenic animals in diabetes research include, to cite only a few ones, knocking-out the insulin receptor gene, its substrate 1 (IRS1) or 2 (IRS2) gene, the glucokinase gene, and the GLUT4 gene.[21]

Tissue-specific knockouts can be produced so that the animals have a gene knocked out in specific tissues, such as pancreatic β cells, liver, adipose tissue, or brain. This quite powerful technique has produced much interesting data in recent years. For instance, an animal model of mitochondrial diabetes was developed by knocking out the mitochondrial transcription factor mtFA only in pancreatic β cells.[52] Likewise, the establishment of liver-specific glucokinase gene knockout mice was recently proposed as a new animal model for screening antidiabetic drugs.[53]

It is beyond the limits of this chapter to enumerate all animal models relevant to the pathophysiology of diabetes mellitus and established by gene targeting or transgenic techniques. Truly speaking, this approach conceivably covers an almost unlimited field of investigations on each factor possibly involved in the development of this disease. Only two further examples will be cited to illustrate such a diversity.

First, establishment of an NOD-Pdcd1$^{-/-}$ mice was recently proposed as an efficient model of type 1 diabetes.[54] In this study, advantage was taken of the use of NOD mice knocked out for programmed cell death (Pdcd1), an immunoreceptor belonging to the CD28/cytotoxic T-lymphocyte-associated antigen-4 family. Second, another recent study deals with the generation of mice lacking the SHIP2 gene, SHIP2 being a member of the inositol polyphosphate 5-phosphate family.[55] Loss of SHIP2 leads to increased sensitivity to insulin with severe neonatal hypoglycemia, deregulated expression of the genes involved in gluconeogenesis, and perinatal death. Adult mice that are heterozygous for the SHIP2 mutation have increased glucose tolerance and insulin sensitivity associated with an increased recruitment of the GLUT4 glucose transporter and increased glycogen synthesis in skeletal muscle.[55]

INSULIN-SECRETING AND OTHER PANCREATIC ENDOCRINE CELL LINES

INSULIN-SECRETING CELL LINES The use of primary B cells in biochemical and molecular research is limited not only by the availability of endocrine tissue, but also by the sophistication of the methods required to isolate pancreatic islets, prepare dispersed islet cells from them, and eventually separate and purify β cells from other islet cell types. Numerous investigators have attempted, therefore, to establish an insulin-secreting cell line that retains normal regulation of insulin secretion. As previously reviewed,[56,57] different approaches have been used, including induction of pancreatic tumors by irradiation or viral infection, immortalization of β cells *in vitro*, and development of transgenic mice with targeted expression of a recombinant oncogene in the β cell. The success of these attempts was limited because cell differentiation and proliferation capacities are, as a rule, mutually exclusive.

The following information refers to the most commonly used tumoral insulin-producing cell lines

RINm5F Cell Line RIN lines were derived from a transplantable islet-cell tumor, itself induced by high-dose irradiation in an inbred NEDH (New England Deaconess Hospital) rat strain.[58] Two lines (RIN-r and RIN-m) were derived from the initial tumor. An RIN-m5F subclone, derived from the RIN-m line, was selected for further studies based mainly on its relatively high insulin secretion rate. The RIN-m5F line exhibits only a modest secretory response to D-glucose and this mainly in a low range of hexose concentrations.[59]

INS-1 Cell Line Another widely used cell line for insulin secretion studies is the INS-1 cell line. It was established from the original radiation-induced tumor described above by growing the tumor material in the presence of 2-mercaptoethanol.[60] This INS-1 cell line yields a steady increase of insulin release in the 0–11.2 mM D-glucose concentration range.

HIT-T15 Cell Line The HIT-T15 cell line was obtained by subcloning one of the original cell lines derived from isolated Syrian hamster pancreatic cells that had been transformed by infection with simian virus 40.[61] This HIT-T15 cell line was selected for its high insulin content.

βTC6 Cell Line The βTC class of insulinoma cell lines was established from transgenic mice by targeted expression of the SV40 large T-antigen in β cells, under the control of the rat insulin II promoter. The first cell lines established from the insulinomas developed in these mice (βTC1, βTC2) displayed a maximal secretory response to D-glucose at a 1.25 mM control concentration of the hexose. Further cell lines (βCT3, βCT6) exhibited a half normal response to the sugar at 0.5 mM D-glucose.[62]

MIN6 Cell Line The MIN6 cell line was obtained by a procedure similar to that used for the engineering of βTC lines. It initially exhibited a secretory response to D-glucose over a range of concentrations of the hexose and with a magnitude similar to that of normal islet cells.[63] However, MIN6 cells were found to lose these attributes with time of culture.

BRIN-BD11 Cell Line The BRIN-BD11 cell line was established by electrofusion of normal pancreatic cells from NEDH rats with immortalized RINm5F cells. They are currently used in several investigations dealing, for instance, with the process of so-called β cell glucotoxicity.[64]

The present list is not exhaustive. For instance, further cell lines (e.g., NIT-1 and βHC line) and subclones of the βTC6 and INS-1 cell lines were also investigated. Attempts at preparation and stable propagation of human β cells lines have met so far with only limited success.

Although an uncommon procedure, it could be proposed that the selection of a given insulin-producing cell line may require comparison between distinct cell lines according to the very experimental device selected for the purpose of the projected investigations. To cite only one example, in a recent study dealing with the stimulus-secretion coupling of hypotonicity induced insulin release, pilot experiments were conducted in MIN6, INS-

1, and BRIN-BD11 cells. The relative magnitude of the increment in insulin output caused by extracellular hypoosmolarity was much greater in the latter cells, the only ones in which the secretory response to hypotonicity was also modulated by the extracellular concentration of D-glucose. Hence, the BRIN-BD11 cells were eventually selected for further investigations on this issue (R. Beauwens *et al.*, unpublished observations).

OTHER PANCREATIC ENDOCRINE CELL LINES Glucagon-producing α-cell-derived lines are also available, inter alia, from the Endocrine Pancreas Transcription Regulatory Regions Database (EP-TRRD). They include α-TC1 (mouse), α-Tc6 (mouse), AN 697 (rat), and MSL-GAN (rat) cell lines or glucagonoma cells (Takaki). Incidentally, glucagon storage, release, and degradation could also be investigated in tumoral insulin-producing cells of the RINm5F line.[65] Both the qualitative and quantitative aspects of these processes differed vastly, however, from those previously characterized in normal islet cells, so that the tumoral cells were eventually considered as a model of dysfunction rather than normal function of glucagon-producing cells.

Last, somatostatin-secreting cell lines were also established, e.g., from a human pancreatic islet cell carcinoma.[66] Secretion of somatostatin from these cells could be stimulated at a high extracellular K+ concentration (50 mM) and by theophylline (10 mM). They were considered, therefore, as useful for studying the regulatory mechanism of somatostatin secretion.

CONCLUSIONS

This chapter illustrated the considerable variety of animal models and animal cells presently available in the field of experimental diabetology. Such a snapshot should not obscure the tremendous contributions of scientists in this field, whether in the past, the present, and, for sure, the future.

ACKNOWLEDGMENT

We are most grateful to C. Demesmaeker for secretarial help.

REFERENCES

1. Renold AE, Dulin WE. Spontaneous diabetes in laboratory animals. Introduction by workshop editors. *Diabetologia* 1967;3:63–64.
2. Rabb GB, Getty BS, Williamson WM, Lombard IS. Spontaneous diabetes mellitus in tree shrews, Urogale everitti. *Diabetes* 1966; 15:327–330.
3. Kenney DW. Cited by Schweisshheimer W. Diabetes beim Delphin. *Der Diabetiker* 1966;16:48.
4. Armin J, Cunningham NF, Grant RT, Lloyd MK, Wright PH. Acute insulin deficiency provoked in dog, pig and sheep by single injections of anti-insulin serum. *J Physiol* 1961;157:64–73.
5. Wright PH, Rivera-Calimlim L, Malaisse WJ. Endogenous insulin secretion in the rat following injection of anti-insulin serum. *Am J Physiol* 1966;211:1089–1094.
6. Malaisse WJ, Malaisse-Lagae F, Wright PH. Insulin secretion *in vitro* by islets from insulin-deficient rats. *Proc Soc Exp Biol Med* 1967;126:474–476.
7. Minkowski O. Untersuchungen über den Diabetes Mellitus nach Extirpation des Pancreas. *Naunyn-Schmiedebergs Arch Exp Path Pharmak* 1893;31:85–189.
8. Bliss M. *The Discovery of Insulin*. Chicago, IL: The University of Chicago Press, 2000.
9. Bonner-Weir S, Trent DF, Weir GC. Partial pancreatectomy in the rat and subsequent defect in glucose-induced insulin release. *J Clin Invest* 1983;71:1544–1554.
10. Malaisse WJ. Etude expérimentale des troubles hydro-ioniques secondaires à l'hyperglycémie. In: *Journées de Diabétologie de l'Hôtel-Dieu, 1975*. Paris: Flammarion, 1975:111–127.
11. Dohan FC, Lukens FDW. Lesions of pancreatic islets produced in cats by administration of glucose. *Science* 1947;105:183.
12. Malaisse WJ. Physiology of insulin secretion and its alteration in diabetes: The concept of glucotoxicity. In: Andreani D, Gueriguian JL, Striker GE, Eds. *Diabetic Complications: Epidemiology and Pathogenetic Mechanisms*. New York: Raven Press, 1991:3–23.
13. Dunn JS, Sheehan ML, McLetchie NGB. Necrosis of islets of Langerhans produced experimentally. *Lancet* 1943;1:484–487.
14. Malaisse WJ, Malaisse-Lagae F, Sener A, Pipeleers DG. Determinants of the selective toxicity of alloxan to the pancreatic B-cell *Proc Natl Acad Sci USA* 1982;79:927–930.
15. Bono VH Jr. Review of mechanism of action studies of the nitrosoureas. *Cancer Treat Rep* 1976;60;699–702.
16. Yamagami T, Miwa A, Takasawa S, Yamamoto H, Okamoto H. Induction of rat pancreatic B-cell tumors by the combined administration of streptozotocin or alloxan and poly(adenosine diphosphate ribose) synthetase inhibitors. *Cancer Res* 1985;45:1845–1849.
17. Lukic ML, Stosic-Grujicic S, Shahin A. Effector mechanisms in low-dose streptozotocin-induced diabetes. *Dev Immunol* 1998;6:119–128.
18. Portha B, Picon L, Rosselin G. Chemical diabetes in the adult rat as the spontaneous evolution of neonatal diabetes. *Diabetologia* 1979;17:371–377.
19. Saï P, Boillot D, Boitard C, Debray-Sachs M, Reach G, Assan R. Pentamidine, a new diabetogenic drug in laboratory rodents. *Diabetologia* 1983;25:418–423.
20. Malaisse WJ, Malaisse-Lagae F. Le diabète expérimental par agents chimiques. *Nouv Pres Méd* 1972;1:473–478.
21. Rees DA, Alcolado JC. Animal models of diabetes mellitus. *Diabetic Med* 2005;22:359–370.
22. Onodera T, Toniolo A, Ray UR, Jenson AB, Knazek RA, Notkins AL. Virus-induced diabetes mellitus. *J Exp Med* 1981;153:1457–1473.
23. Malaisse W. Hormonal and environmental modification of islet activity. In Steiner DF, Freinkel N, Eds. *Handbook of Physiology: Endocrine Pancreas*, Section 7, Vol 1. Washington, DC: American Physiological Society, 1972:237–260.
24. Malaisse W, Malaisse-Lagae F. Islet function in experimental obesity. In Vague J, Ed. *Physiopathology of Adipose Tissue*. Amsterdam, the Netherlands: Excerpta Medica Foundation, 1969:70–71.
25. Malaisse WJ, Malaisse-Lagae F, Wright PH. Effect of fasting upon insulin secretion in the rat. *Am J Physiol* 1967;213:843–848.
26. Swene I, Borg LAH, Crace CJ, Schnell Lanström A. Persistent reduction of pancreatic beta-cell mass after a limited period of protein-energy malnutrition in the young rat. *Diabetologia* 1992;35:939–945.
27. Malaisse WJ, Hubinont CJ, Marynissen G. Pancreatic islet cell function in oral contraception, pregnancy and lactation—a review. *Arch Gynecol Obstet* 1989;246:125–132.
28. Aerts L, Sodoyez-Goffaux F, Sodoyez JC, Malaisse WJ, Van Assche FA. The diabetic intrauterine milieu has a long-standing effect on insulin secretion by B cells and on insulin uptake by target tissues. *Am J Obstet Gynecol* 1988;159;1287–1292.
29. Oguzhan B, Zhang Y, Louchami K, Courtois P, Portois L, Chardigny J-M, Malaisse WJ, Carpentier YA, Sener A. Pancreatic islet function in ω3 fatty acid depleted rats. Glucose metabolism and nutrient-stimulated insulin release. *Endocrine* 2006;29:467–476.
30. Valverde I, Alarcon C, Rovira A, Malaisse-Lagae F, Malaisse WJ. Diazoxide-induced long-term hyperglycemia. II. Slackening of pro-insulin conversion. *Diabetes Res* 1989;10:59–62.
31. Leclercq-Meyer V, Marchand J, Malaisse WJ. Attenuated anomeric difference of glucose-induced insulin release in the perfused pancreas of diazoxide-treated rats. *Horm Metab Res* 1991;23:257–261.
32. Jansen SW, Mertens GJ, Sweep CG, Span PN, Verhofstad AA, Hermus AR. Phlorizin treatment prevents the decrease in plasma insulin levels but not the progressive histopathological changes in the

pancreatic islets during aging of Zucker diabetic fatty rats. *J Endocrinol Invest* 2003;26:508–515.
33. Malaisse W, Malaisse-Lagae F, Gerritsen GC, Dulin WE, Wright PH. Insulin secretion in vitro by the pancreas of the Chinese hamster. *Diabetologia* 1997;3:109–114.
34. Goto Y, Kakizaki M, Mazaki N. Spontaneous diabetes produced by selective breeding of normal Wistar rats. *Proc Jpn Acad* 1975;51: 80–85.
35. Ziv E, Shafrir E, Kalman R, Galer S, Bar-On H. Changing pattern of prevalence of insulin resistance in *Psammomys obesus*, a model of nutritionally induced type 2 diabetes. *Metabolism* 1999;48;1549–1554.
36. Junod A, Letarte J, Lambert AE, Stauffacher W. Studies in Spiny mice (*Acomys cahirinus*): Metabolic state and pancreatic insulin release in vitro. *Horm Metab Res* 1969;1:45–52.
37. Malaisse-Lagae F, Ravazzola M, Amherdt M, Gutzeit A, Stauffacher W, Malaisse WJ, Orci L. An apparent abnormality of the B-cell microtubular system in Spiny mice (*Acomys cahirinus*). *Diabetologia* 1974;10:71–76.
38. Nakamura M, Yamada K. Studies on a diabetic (KK) strain of the mouse. *Diabetologia* 1967;3:212–221.
39. Ueda H, Ikegami H, Kawacuchi Y, Fujisawa T, Yamato E, Shibata M, Ogihara T. Genetic analysis of late-onset type 2 diabetes in a mouse model of human complex trait. *Diabetes* 1999;48:1168–1174.
40. Kawano K, Hirashima T, Mori S, Saitoh Y, Kurosumi M, Natori T. Spontaneous long-term hyperglycaemic rat with diabetic complications. Otsuka Long-Evans Tokushima fatty (OLETF) strain. *Diabetes* 1992;41:1422–1428.
41. Larkins RG, Martin FIR. Selective defect in insulin release in one form of spontaneous laboratory diabetes. *Nature* 1972;235:86–88.
42. MacDonald MJ, Tank J, Polonsky KS. Low mitochondrial glycerophosphate dehydrogenase and pyruvate carboxylase in pancreatic islets of Zucker diabetic fatty rats. *Diabetes* 1996;45:1626–1630.
43. Coleman DL, Hummel KP. The effects of hypothalamic lesions in genetically diabetic mice. *Diabetologia* 1970;6:263–267.
44. Zhang Y, Proenca R, Maffei M, Barone M, Leopold L, Friedman JM. Positional cloning of the mouse *obese* gene and its human homologue. *Nature* 1994;372:425–432.
45. Miura T, Suzuki W, Ishihara E, Arai I, Ishida H, Seino Y, Tanigawa K. Impairment of insulin-stimulated GLUT4 translocation in skeletal muscle and adipose tissue in the Tsumura Suzuki obese diabetic mouse: A new genetic animal model of type 2 diabetes. *Eur J Endocrinol* 2001;145:785–790.
46. Iizuka S, Suzuki W, Tabuchi M, Nagata M, Imamura S, Kobayashi Y, Kanitani M, Yanagisawa T, Kase Y, Takeda S, Aburada M, Takahashi KW. Diabetic complications in a new animal model (TSOD mouse) of spontaneous NIDDM with obesity. *Exp Anim* 2005;54: 71–83.
47. Allan MF, Eisen EJ, Pomp D. The M16 mouse: An outbred animal model of early onset polygenic obesity and diabesity. *Obesity Res* 2004;12:1397–1407.
48. Makino S, Kunimoto K, Munaoko Y, Mizushima Y, Katagiri K, Tochino Y. Breeding of a non-obese diabetic strain of mice. *Exp Anim* 1980;29:1–13.
49. Todd JA. Genetic analysis of type 1 diabetes using whole genome approaches. *Proc Natl Acad Sci USA* 1995;92:8560–8565.
50. Nakhooda AF, Like AA, Chappel CI, Murray FT, Marliss EB. The spontaneous diabetic Wistar rat. Metabolic and morphological studies. *Diabetes* 1977;26:100–112.
51. Malaisse WJ, Courtois P, Scott FW. Insulin-dependent diabetes and gut dysfunction: The BB rat model. *Horm Metab Res* 2004;36:585–594.
52. Silva JP, Köhler M, Graff C, Oldfors A, Magnuson MA, Berggren P-O, Larsson N-G. Impaired insulin secretion and β-cell loss in tissue-specific knock-out mice with mitochondrial diabetes. *Nat Genet* 2000;26:336–340.
53. Zhang Y, Tan X, Xiao M, Li H, Mao Y, Yang X, Tan H. Establishment of liver specific glucokinase gene knockout mice: A new animal model for screening anti-diabetic drugs. *Acta Pharmacol Sin* 2004;25: 1659–1665.
54. Wang J, Yoshida T, Nakaki F, Hiai H, Okazaki T, Honjo T. Establishment of NOD-*Pdcd1*$^{-/-}$ mice as an efficient model of type 1 diabetes. *Proc Natl Acad Sci USA* 2005;102:11823–11828.
55. Clement S, Krause U, Desmedt F, Tanti J-F, Behrends J, Pesesse X, Sasaki T, Penninger J, Doherty M, Malaisse W, Dumont JE, Le Marchand-Brustel Y, Erneux C, Hue L, Schurmans S. The lipid phosphatase SHIP2 controls insulin sensitivity. *Nature* 2001;409:92–97.
56. Poitout V, Olson LK, Robertson RP. Insulin-secreting cell lines: Classification, characteristics and potential applications. *Diabetes Metab* 1996;22:7–14.
57. Hohmeier HE, Newgard CB. Cell lines derived from pancreatic islets. *Mol Cell Endocrinol* 2004;228:121–128.
58. Chick WL, Warren S, Chute RN, Like AA, Lauris V, Kitchen KC. A transplantable insulinoma in the rat. *Proc Natl Acad Sci USA* 1977;74:628–632.
59. Giroix M-H, Sener A, Dufrane SP, Malaisse-Lagae F, Malaisse WJ. Glucose metabolism in insulin-producing tumoral cells. *Arch Biochem Biophys* 1985;241:561–570.
60. Asfari M, Janjic D, Meda P, Li G, Halban PA, Wollheim CB. Establishment of 2-mercaptoethanol-dependent differentiated insulin-secreting cell lines. *Endocrinology* 1992;130:167–178.
61. Santerre RF, Cook RA, Crisel RM, Sharp JD, Schmidt RJ, Williams DC, Wilson CP. Insulin synthesis in a clonal cell line of simian virus 40-transformed hamster pancreatic beta cells. *Proc Natl Acad Sci USA* 1981;78:4339–4343.
62. Poitout V, Stout LE, Armstrong MB, Walseth TF, Sorensen RL, Robertson RP. Morphological and functional characterization of βTC-6 cells—an insulin-secreting cell line derived from transgenic mice. *Diabetes* 1995;44:306–313.
63. Ishihara N, Asano T, Tsukuda K, Katagiri H, Inukai K, Anai M, Kikuchi M, Yazaki Y, Miyazaki J-I, Oka Y. Pancreatic beta cell line MIN6 exhibits characteristics of glucose metabolism and glucose-stimulated insulin secretion similar to those of normal islets. *Diabetologia* 1993;36:1139–1145.
64. McClenaghan NH, Barnett CR, Ah-Sing E, Yasser HA, Abdel-Wahab YHA, O'Harte FPM, Yoon T-W, Swanson-Flatt SK, Flatt PR. Characterization of a novel glucose-responsive insulin-secreting cell line, BRIN-BD11, produced by electrofusion. *Diabetes* 1996;45: 1132–1140.
65. Barretto M, Valverde I, Malaisse WJ. Glucagon storage, release and degradation by tumoral islet cells (RINm5F line). *Diabetes Res* 1989;10:121–123.
66. Takaki R, Ono J, Nakamura M, Yokogawa K, Kumae S, Hiraoka T, Yamaguchi K, Hamaguchi K, Uchida S. Isolation of glucagon-secreting cell lines by cloning insulinoma cells. *In vitro Cell Devel Biol* 1986;22:120–126.

68 Animal Models of Kidney Diseases

Chunsun Dai, Lawrence P. Kiss, and Youhua Liu

ABSTRACT

Kidney diseases can be devastating illnesses with a high morbidity and mortality. The incidence of end-stage renal failure is increasing at a rate of 6–8% annually worldwide. Animal models are essential for understanding the pathogenetic mechanisms governing the onset and progression of various kidney diseases and for preclinically testing novel, more effective, treatments for the patients afflicted with renal impairment. Despite the presence of several nonrodent models, rats and mice remain principal and indispensable species used by investigators in the field of nephrology. In this chapter, we provide a concise review of the most commonly used kidney disease models. Emphasis is placed on several well-established models, such as the ischemic/reperfusion acute renal failure model, remnant kidney model, and unilateral ureteral obstruction model. Undoubtedly, establishment of authentic animal models is of vital importance in delineating the mechanisms of a range of human kidney diseases and in evaluating the feasibility and efficacy of novel therapeutic interventions.

Key Words: Acute renal failure, Chronic kidney disease, Renal fibrosis, Ischemia/reperfusion, Diabetic nephropathy, Unilateral ureter obstruction, Remnant kidney.

INTRODUCTION

Despite significant improvements in patient management, kidney diseases are still devastating illnesses with a high rate of morbidity and mortality worldwide.[1–4] Understanding the pathogenetic mechanisms of kidney dysfunction is of fundamental importance in finding new and effective treatments for patients afflicted with renal disease.[5–7] For this goal, animal models have proven extremely valuable when investigating the pathogenesis of kidney diseases and in evaluating the efficacy of novel therapeutic strategies. In the past decades, many rodent and nonrodent models have been created and used by investigators in the field of nephrology. Due to easy genetic manipulations, nonrodent models, such as the worm *Caenorhabditis elegans*, the fruit fly *Drosophila melanogaster*, and the zebrafish *Danio rerio*, are very useful for studying fundamental renal biology such as embryonic and organ development.[8] However, rodent models remain essential tools that are widely employed by nephrological investigators.[9,10]

Both mice and rats have their advantages and disadvantages when creating kidney disease models. It is easier to perform surgical procedures and to collect samples from rats; additionally, rat models are usually able to faithfully recapitulate the major pathological manifestations of human kidney diseases, whereas mice are notorious for resistance to development of clinical lesions after injury.[11] However, as genetically modified mice are increasingly available, the unique transgenic or knockout mouse models provide us with an unprecedented opportunity to study the role of a particular gene in normal renal physiology as well as in the development of kidney diseases.[12] In this regard, it seems inevitable that more and more nephrological studies will be carried out in mouse models in the future. It should be noted that different strains and genetic makeup of the same species often display tremendous heterogeneity in developing renal lesions after injury.[13,14] Likewise, different genders have shown a drastic variation in susceptibility to an identical manipulation, often with males more prone to renal injury.[15,16] Therefore, several critical factors, such as animal species, strain, gender, age, and body weight, may play a decisive role in determining the outcome of a particular experiment. In this chapter, we provide a concise review of the most commonly used kidney disease models.

ACUTE RENAL FAILURE MODELS

Acute renal failure (ARF) is characterized by a rapid loss of renal function due to acute, and often severe, damage to the kidney. It is often caused by acute tubular necrosis (ATN), a term widely used in clinical practice. ARF affects as many as 20% of patients in the intensive care unit (ICU) and imposes a high mortality rate of ~50%.[17] The pathogenesis of ARF is very complicated. Ischemia, toxins, or sepsis alone or in combination could be involved in the onset and progression of ARF in clinical patients.[18] For this reason it is extremely difficult to create an ARF model that exactly stimulate the clinical findings. Based on the major causes of clinical ARF, three types of rodent models have been developed: ischemic/reperfusion, toxins, and sepsis-induced ARF.

ISCHEMIC ACUTE RENAL FAILURE MODELS Four types of ischemic models have been reported: warm ischemia–reperfusion (I/R), cold ischemia–warm reperfusion, isolated kidney with perfusion, and contrast nephropathy.[19] Here we discuss only the warm I/R model, which is the most widely utilized in both mice and rats. In rats, 1 day after I/R, serum creatinine increases significantly. Hypoxic tubular injury usually occurs in the S3 segments of the proximal tubule and in the medullary thick ascending limbs (MTAL) of the outer medulla and medullary rays. The kidneys exhibit obvious tubular atrophy and dilation, accom-

From: *Sourcebook of Models for Biomedical Research* (P. M. Conn, ed.), © 2008 Humana Press Inc., Totowa, NJ.

panied by interstitial inflammation.[20,21] The procedure of I/R is as follows. After placing the rat on a heating pad to maintain its body temperature during surgery, a laparotomy is performed and the renal pedicle is isolated by blunt dissection. Both of the renal vascular pedicles are clamped with surgical clips. The occlusion of blood flow should be confirmed by visually observing blanching of the kidneys. After 30–60 min, the clamps are released and the kidney should be reperfused in 1 min. The duration of ischemia should be optimized according to the age, gender, and genetic background. During the ischemic period, body temperature control is also very important; animals should be kept at about 37°C on a heating pad to avoid renal protective effects induced by low body temperature.

In mice the I/R procedure is similar to that in rats. Males are more susceptible to I/R injury compared to females.[15] Mouse serum creatinine and blood urea nitrogen (BUN) levels reach peak at 24–48 h after I/R.[22] The outer stripe of the medulla is the major site of injury to the kidneys during I/R. Pathological findings include loss of the proximal tubular brush border, the presence of cellular debris and casts, medullary congestion, inflammatory response, and proximal tubular cell apoptosis. Activation of caspases and endonucleases and tubular cell apoptosis have been documented in the development of ischemic ARF.[23,24] Apoptotic cells can be identified in various segments of the renal tubule.[25] Kidney function starts to recover within 2–3 days. In contrast to I/R insult, human ARF is most commonly caused by hypoperfusion. Given the obvious differences between the complete occlusion of the renal artery in animals and the complicated clinical conditions in human ARF, the I/R model appears not to replicate the human situation exactly. In this regard, care should be taken when interpreting the results obtained from this animal model.

TOXIC ACUTE RENAL FAILURE MODELS Many commonly used medications, such as gentamycin and cisplatin, can act as toxins and induce ATN-related ARF. Many toxins have been used in animal models to cause ARF.

Cisplatin-Induced Acute Renal Failure Cisplatin is a chemotherapeutic agent widely used to treat a large number of carcinomas in humans. Administration of a high dose of cisplatin often leads to renal toxicity, a severe side effect that limits its clinical use in cancer patients. In rodents, a single dose of cisplatin at 20~40 mg/kg body weight (for mice) or 7 mg/kg (for rats) by intraperitoneal injection can induce ARF.[26] In mice, cisplatin-induced injury is usually associated with tubular necrosis, predominantly in the S3 segments of the proximal tubules, manifested by cast formation, loss of tubular brush border membranes, and dilation of tubules. Kidney failure starts at day 2 and reaches a maximum at day 3 to 4, depending on the doses. Contrary to the ischemia model, females are much more susceptible to injury induced by cisplatin than their male counterparts.[15] This model is quite simple and reproducible. The tubular dysfunction and disease recovery phase is comparable to that found in humans. The mechanisms of cisplatin-induced nephrotoxicity remains largely unknown. Several mechanisms including oxidant stress, inflammation, genotoxic damage, and cell cycle arrest may account for its nephrotoxicity.

Folic Acid-Induced Acute Renal Failure An intravenous injection of high dose folic acid (FA) can induce ARF in mice or rats, which is associated with a rapid decline of renal function, tubular epithelial cell death, morphological abnormalities, and the appearance of FA crystals in renal tubules.[27,28] Mice display a rapid and marked increase in serum creatinine as early as day 1, which remains significantly elevated at least to day 3 after injection of FA (Figure 68–1). The serum creatinine level declines

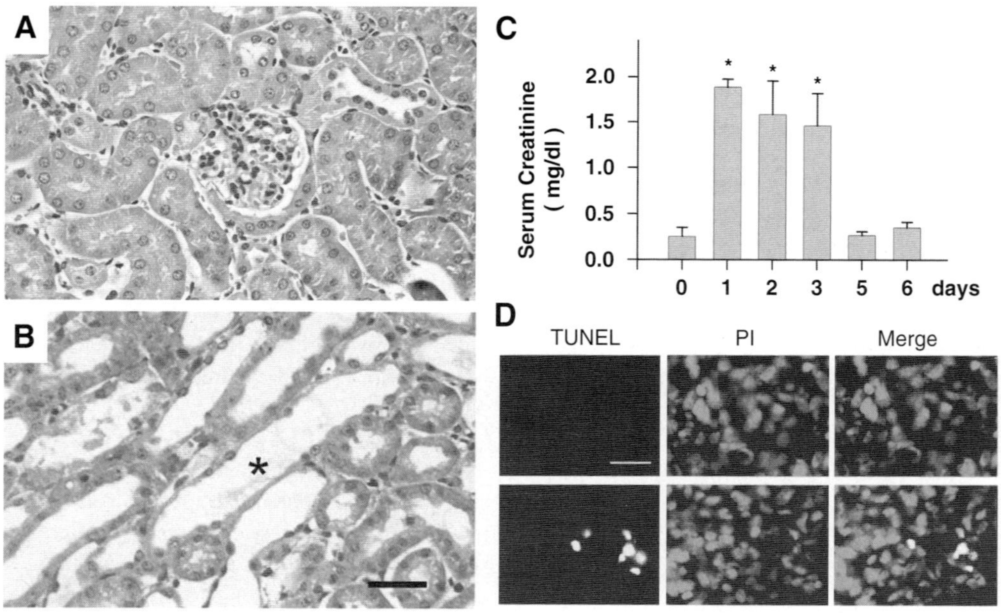

Figure 68–1. Folic acid induces acute renal failure (ARF) in mice. (A, B) Representative micrographs show kidney morphology at day 1 after folic acid injection. (A) Normal control; (B) folic acid. (C) Serum creatinine levels at different time points after folic acid injection. Data are presented as mean ± SEM from six animals per group at each time point. $*p < 0.05$. (D) Representative micrographs show apoptosis detected by *in situ* terminal deoxynucleotidyltransferase-mediated dUTP nick-end labeling (TUNEL) staining. Left panel, apoptosis-positive cells detected by use of TUNEL. Central panel, propidium iodide (PI) staining shows cell nuclei. Right panel, merge of the TUNEL and PI staining. Scale bar, 20 μm. (Adapted and modified from Dai *et al.*[29])

toward baseline thereafter. Kidney morphology typically reveals tubular cast formation, dilation, loss of brush borders, tubular epithelial cell necrosis and apoptosis, and interstitial infiltration (Figure 68–1). These pathological findings are comparable with human ARF. An increase in proliferating cell nuclear antigen (PCNA) expression, which peaks at day 2 after FA injection, can be detected. This suggests active cell cycle progression and, presumably, cell proliferation in the kidneys during the tubular repair and regeneration phase.[29] FA is administrated by a single intravenous or intraperitoneal injection at a dose of 250 mg/kg body weight in male mice or rats.[29,30] The mechanism of FA-induced ARF is believed to be associated with the upregulation of tumor necrosis factor (TNF)-α and is also related to the microcrystal formation within tubules. We have shown that FA causes a marked reduction in the expression of the prosurvival protein Bcl-xL in the kidneys, which could be an important mechanism underlying tubular epithelial cell death in this model.[29]

Gentamycin-Induced Acute Renal Failure Gentamycin, an aminoglycoside antibiotic, has been successfully used in the treatment of gram-negative bacterial infections. Due to its severe nephrotoxicity, it has become a widely used reagent to induce ARF in rodents. Gentamycin can be injected at 80–100 mg/kg by an intramuscular, subcutaneous, or intraperitoneal route daily for 6 days. After several injections, severe damage can be found in the cortex, including tubular necrosis, tubular dilation, accumulation of protein casts, and loss of the brush border.[31] In proximal tubular cells, myeloid bodies, a characteristic morphological feature of getamycin nephrotoxicity in human, can be found. Rats are more suitable for the studies of gentamycin nephrotoxicity than mice. The exact mechanisms of the gentamycin nephrotoxicity remain elusive, but they could be related to enhanced generation of reactive oxygen species in the kidney.

Pigment Nephropathy Administration of glycerol is another reliable method to induce ARF in rats or mice.[32] This disease model is very severe, accompanied by intravascular hemolysis and myoglobinemia. These pathological lesions are similar to ARF in humans induced by rhabdomyolysis and/or intravascular hemolysis.[33] Kidney dysfunction, as measured by an increased BUN, serum creatinine, and *N*-acetyl-β-glucosaminidase, can be found as early as 6 h after glycerol injection. The procedure of this model is very simple, with intramuscular injection of 50% glycerol at 8–10 ml/kg body weight. Glycerol should be administered through a deep intramuscular injection equally distributed to both hind legs. The tubular damage is caused by a combination of dehydration, toxic proximal tubular cell damage, vasoconstriction, renal inflammation, and tubular obstruction by casts.[34]

SEPSIS ACUTE RENAL FAILURE MODELS Gram-negative sepsis and septic shock-induced kidney failure is the most common cause of clinical ARF. The pathophysiological features of sepsis are high cardiac output and low systemic vascular resistance.[35] It is believed that a combination of immune, toxic, and hemodynamic conditions leads to ARF in septic patients.[36] Lipopolysaccharide (LPS), a major component of gram-negative bacteria, induces systemic upregulation of TNF-α and interleukin (IL)-1, and generates nitric oxide (NO) produced by inducible NO synthase. These in turn can directly or indirectly lead to systemic hypotension and tubular necrosis.[36] Four types of sepsis models have been reported: endotoxin infusion, intravenous bacterial infusion, cecal ligation and perforation, and intraperitoneal infusion of bacteria. Endotoxin-induced ARF is relatively simple. LPS can be given at 100 μg/kg body weight (for rats) by tail vein injection[37] or at a higher dose of 2 mg/kg (for mice) by intraperitoneal injection.[38] Kidney dysfunction and tubular epithelial cell death can be found at 24 h after LPS injection.

CHRONIC KIDNEY DISEASE MODELS

Chronic kidney disease (CKD) is a group of renal disorders with a wide variety of etiologies. The pathogenesis of CKD is a progressive process that eventually leads to end-stage renal failure (ESRF), a devastating condition that necessitates patient dependence on life-long treatments with dialysis or renal transplantation.[5] The number of patients with ESRF is growing at a rate of 6–8% annually worldwide, with the financial burden of health care costs becoming increasingly unaffordable.[2–4] In this context, it is important to characterize the mechanisms and the factors responsible for CKD progression and to identify and evaluate novel and effective renoprotective strategies. Clearly, animal models of CKD are essential in these efforts.

DIABETIC NEPHROPATHY Diabetic nephropathy (DN) is the single most important cause of CKD and accounts for 40% of the patient population with ESRF. The pathogenesis of DN is characterized by persistent albuminuria and an accumulation of extracellular matrix (ECM) that leads to glomerulosclerosis, tubulointerstitial fibrosis, and progressive loss of renal function.[39,40] Thus far, a DN animal model with typical pathological findings has not been documented.[41] All currently used DN models have some of the early DN features, including albuminuria, glomerular hypertrophy, and mild to moderate mesangial ECM deposition. However, little or no glomerular sclerosis and/or renal failure exactly mimicking human DN has been reported.

Streptozotocin (STZ) is used to induce hyperglycemia by destroying pancreatic β cells in both mice and rats, which imitates type I diabetes in humans.[42] There are two different administration methods: a high dose with a single injection or small doses with multiple intraperitoneal injections. It is believed that a single high dose of STZ can directly damage the islet β cells, leading to necrosis or apoptosis, whereas multiple small doses of STZ could induce the animal's autoimmunity to β cells. Both methods can induce islet β cell dysfunction and hyperglycemia. For the high-dose STZ protocol, rats are intravenously injected with STZ at 60 mg/kg body weight, and diabetes is typically induced at 3–5 days after injection. Insulin is sometimes used to maintain an appropriate blood glucose level. For the small-dose protocol, rats are intraperitoneally injected with STZ at 20–30 mg/kg body weight daily; elevated blood glucose is usually detectable after four or five injections. The blood glucose level may keep increasing and reach a peak 1 month after the last injection. Diabetic rats with long-term moderate hyperglycemia, however, do not develop characteristic glomerular lesions of human DN and, in fact, develop minimal glomerular injury only after 1 year of diabetes. This suggests that diabetic rats with moderate hyperglycemia can be used as an appropriate model only for early DN.[43–45] STZ can also induce diabetes in mice; however, the dose of STZ used is higher than that in rats. Usually, a single intravenous injection of STZ at 150–200 mg/kg can induce diabetes. Daily intraperitoneal injection at 40–50 mg/kg for 5 days is an alternate approach. Hyperglycemia often occurs within 2 weeks following low-dose STZ injections, and albuminuria develops in 5 weeks after diabetes is induced. Glomerular hypertrophy and mesangial expansion typically develop as the disease progresses.

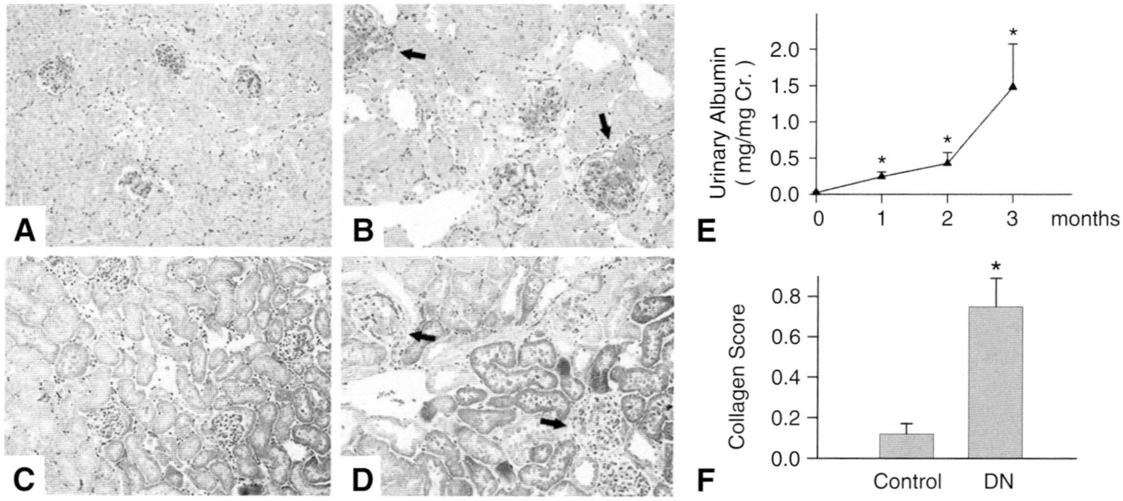

Figure 68–2. Diabetic nephropathy induced by streptozotocin (STZ) in uninephrectomized mice. One week after uninephrectomy, male CD1 mice received an intravenous injection of STZ. (A–D) Representative micrographs demonstrate glomerular enlargement, mesangial expansion, and segmental glomerulosclerosis (periodic acid–Schiff staining; A, B) and glomerular collagen deposition (Masson–Trichrome staining; C, D). (A and C) Normal control; (B and D) diabetic mice. Arrows indicate injured glomeruli. (E) Albuminuria develops in diabetic mice in a time-dependent manner. Data are presented as means ±SEM. $*p < 0.05$. (F) Glomerular collagen deposition score in diabetic and normal mice. $*p < 0.05$. (Adapted and modified from Dai *et al.*[46]) (See color insert.)

To accelerate the progression of DN, we recently reduced renal mass by unilateral nephrectomy (UNx), followed by a single intravenous injection of STZ at 150 mg/kg in male CD-1 mice.[46] Diabetic kidneys at 4 and 12 weeks after injection of STZ display a marked enlargement in size, but serum creatinine levels increase only slightly in diabetic mice at 12 weeks after STZ, when the urine albumin level reaches 1.5 mg/mg creatinine (Figure 68–2). Diabetic glomeruli exhibit an expanded mesangial area and narrowed capillary lumen, as well as an increased collagen deposition at 12 weeks (Figure 68–2). However, only three of six mice show focal and segmental glomerulosclerosis (FSGS) in 1–4% of the glomeruli (Figure 68–2).

The susceptibility of the mice to diabetes and DN is drastically different among various strains.[14,47] Males often display more robust hyperglycemia with STZ than females. It is well known that C57BL/6J is more resistant to DN than other strains. However, C57BLKS/J is more susceptible to renal lesions than C57BL/6. 129/SvJ is more susceptible to glomerulosclerosis, but this strain of mice is resistant to hyperglycemia.

Of the genetic mouse models of diabetes, *db/db* mice are the best characterized and widely used to study the progression of DN. The *db/db* mice in a C57BLKS/J background are considered a genetic DN model, with many features similar to human type II diabetes.[48] Compared to nondiabetic *db/m* mice, *db/db* mice develop hyperinsulinemia by 10 days of age and marked obesity and severe hyperglycemia by 10 weeks. The blood glucose level is ~300 and 500 mg/dl at the age of 8 and 25 weeks, respectively, in *db/db* mice, indicating an advanced diabetic state. Increased urinary albumin excretion is found in *db/db* mice and it reaches ~550 μg/24 h at 25 weeks. These mice also develop glomerular mesangial expansion and histological lesions that resemble those found in human DN. In *db/db* mice the animals display kidney hypertrophy and enlarged glomerular size, and increased fibronectin and type IV collagen accumulation in glomerular mesangium.[49,50] There is no evidence of tubular atrophy or tubulointerstitial fibrosis.

REMNANT KIDNEY MODEL Renal mass ablation is a classic approach to induce glomerular injury and tubulointerstitial lesions. The remnant kidney model after 5/6 nephrectomy, also known as subtotal nephrectomy (SNX), is the most widely used and arguably the best model of CKD that mimics many features of human diseases.[51] The kidney advances through three histological stages after SNX: phase 1, kidney hypertrophy appears at 2–4 weeks after surgery; phase 2, detectable changes of the residual renal tissue occur up to 10 weeks; and phase 3, glomerular sclerosis and interstitial fibrosis become apparent 10 weeks after SNX. The procedures of renal mass ablation are as follows.[52] After the abdominal cavity is opened with a dorsal incision, the right kidney is freed, decapsulated, ligated at the renal hilus, and finally removed. Seven days later, the cortex of the left kidney is ablated by ligation of the poles or the branches of the renal artery or by resection of the poles of the kidney while protecting the adrenal glands during the surgery. The strain (Wistar and Sprague–Dawley rats) and gender may influence the functional and morphological consequences of the remnant kidney model. After SNX, rats develop proteinuria at 4 weeks, which increases up to 10 weeks in females and males of both strains. The glomerular filtration rate (GFR) is significantly reduced after SNX, which is nearly independent of animal strain, gender, and operation protocol. At 10 weeks after surgery, rats develop moderate interstitial fibrosis with reactive lymphocytic infiltrates and progressive tubular atrophy.

In mice, a two-step procedure is also used to create uremia.[53] Briefly, cortical electrocautery is applied to the right kidney through a 2-cm flank incision and a left total nephrectomy is performed through a similar incision 2 weeks later. At 6 weeks after surgery, BUN levels increase by 225% above baseline and tend

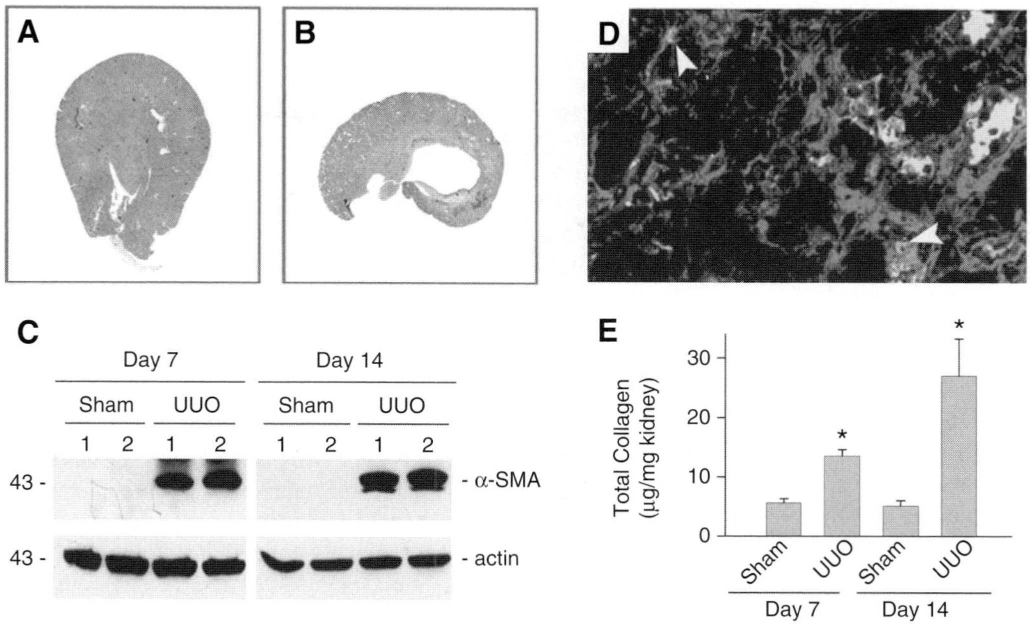

Figure 68–3. Interstitial fibrosis in the mouse model of obstructive nephropathy. (A, B) Representative micrographs show the cross sections and gross morphology of the obstructed kidneys at day 14 after ureteral ligation. (A) Sham control; (B) UUO. (C) Western blot analyses demonstrate a marked induction of α-smooth muscle actin (α-SMA), a molecular marker for myofibroblasts, in the obstructed kidney at day 14 after UUO. (D) Double immunofluorescence staining shows the α-SMA (red) and proximal tubular epithelial cell marker, fluorescein isothiocyanate (FITC)-conjugated lectin from *Tetragonolobus purpureas* (green). (E) Quantitative determination of total kidney collagen contents in sham and obstructed kidneys. Data are presented as means ±SEM. $*p < 0.01$. (Adapted and modified from Yang *et al.*[59]) (See color insert.)

to be stable, with minimal variations during the 6-week period. Uremic animals have high serum levels of calcium, phosphate, and parathyroid hormone. A model with 75% renal mass reduction in mice as a single-step procedure has also been reported.[54] Before removing the right kidney, the contralateral kidney is partially infarcted by ligating the anterior renal branch of its main artery. Histological examination of the remaining kidneys reveals progressive changes. From 1 to 12 months postsurgery, the major alterations are mesangial expansion, the presence of hyalinosis, particularly on the afferent arterioles, and the presence of some protein casts in the tubules. Segmental sclerosis is observed in some mice at 2 months after surgery. At 12 months, severe mesangial expansion, focal segmental sclerosis, hyalinosis, and protein casts in the tubules are evident.

UNILATERAL URETER OBSTRUCTION In infants and children, obstructive nephropathy is the leading cause of chronic renal failure. In older patients between 60 and 65 years old the incidence of obstructive nephrology increases again overall, particularly in men, due to the prevalence of prostate disease. Kidney histology reveals tubular atrophy and interstitial fibrosis after ureter obstruction.[55,56]

Many animal models of urinary tract obstruction have been developed, including studies of fetal sheep and neonatal rats and mice.[57] A model with complete unilateral ureteral obstruction (UUO) performed either in rats or mice has been well established and is widely used currently.[12] This model of fibrosis is characterized by renal myofibroblast activation, tubular atrophy, and interstitial fibrosis with minimal glomerular lesions. In male CD-1 mice, the obstructed kidneys display marked activation of myofibroblasts at 7 days after surgery, characterized by a dramatic increase in the abundance of α-smooth muscle actin (α-SMA) (Figure 68–3). Colocalization of α-SMA and a proximal tubular cell marker suggests a tubular epithelial-to-mesenchymal transition (EMT) in this model.[58,59] After UUO, the obstructed kidneys displayed a marked increase in transform growth factor (TGF)-β1 and its type I receptor expression at 7 days. Importantly, both TGF-β1 and its type I receptor expression are mainly upregulated in renal tubular epithelium in the diseased kidneys. One week after UUO, cellular infiltration, tubular proliferation and apoptosis, tubular EMT, myofibroblast accumulation, ECM deposition, and tubular atrophy are observable. These changes are highly predictable and reproducible, and also generally mimic the tubulointerstitial fibrosis in humans, particularly that seen in obstructive nephropathy. The disadvantage of the UUO model is that it is impossible to assess kidney dysfunction, since another kidney remains intact and appears sufficient to support normal renal function.

ADRIAMYCIN AND PUROMYCIN MODEL Puromycin, an aminonucleoside, and adriamycin (ADR), an antineoplastic antibiotic, can induce nephrotic syndrome in rats when delivered by intravenous injection.[10,60] A single dose of puromycin at 50 mg/kg body weight in rats can induce severe proteinuria within 10 days, followed by apparent resolution, and then with lower level proteinuria associated with early glomerular segmental sclerotic lesions. Rats usually develop glomerular segmental sclerosis by week 18. Multiple intraperitoneal injections of puromycin (at 10 mg/kg body weight for the first dose and additional 40 mg/kg doses every 4 weeks) will accelerate the progression of disease. Intravenous ADR injection at 2 mg/kg twice with a 3-week interval leads to early phase proteinuria after the second injection.

Kidneys often develop widespread glomerulosclerosis with tubulointerstitial fibrosis by 24 weeks. A single injection of adriamycin at 5 mg/kg body weight produces proteinuria in 1 week.

ADR is also used in mice to induce podocyte dysfunction and proteinuria.[61,62] All male BALB/c mice treated with ADR at a dosage of 10.5 mg/kg body weight develop nephrotic syndrome, and progressive renal injury. Overt proteinuria appears at day 5 and is maximal at day 7. High levels of proteinuria are maintained after 1 week. Some mice develop hematuria (35.7%) or leukocyturia (53.6%). Histological examination reveals a focal increase in resorption droplets in tubular cells and intraluminal casts at weeks 1 and 2. At week 4, tubules display severe changes, including a decrease in height of tubular epithelial cells, a loss of brush borders, and vacuolization. There is a moderate interstitial monocyte infiltration. Glomeruli are reduced in size with several vacuoles, collapse of tufts, as well as expansion of the mesangium. At week 6, in the cortex, widespread tubular atrophy and intratubular cast formation with tubulointerstitial expansion are observed. Extensive focal glomerulosclerosis and severe interstitial fibrosis and inflammation are present. Male C57BL/6J mice do not develop any significant proteinuria or histological change, even at 8 weeks after ADR injection of 15 mg/kg body weight and die within 2 weeks of ADR injection at 20 mg/kg body weight. However, in our laboratory we treat mice with mixed genetic backgrounds including C57BL/6J with ADR at 25 mg/kg body weight by intravenous infusion. All mice develop prominent albuminuria by day 3 after injection, although some mice begin to die at day 5. Immunofluorescence staining reveals that nephrin and CD2AP protein abundance are reduced after ADR administration.

CYCLOSPORINE NEPHROPATHY Cyclosporine (CsA) is an immunosuppressive agent that is used clinically in organ transplantation. However, its ability to induce acute and chronic nephrotoxicity in humans limits its usage in clinical settings. CsA was first found to induce kidney dysfunction in rodents on a low salt diet. When rats are on a salt-depleted diet, CsA administration through daily subcutaneous injection at 15 mg/kg body weight induces a striped pattern of tubulointerstitial fibrosis, tubular atrophy, and afferent arteriolar hyalinosis with renal failure at 4 weeks after injection.[63] Khan *et al.* also report that CsA given orally (50 mg/kg body weight, daily) for 14 days induces kidney dysfunction characterized by elevated serum creatinine, severe isometric vacuolization, and widening of the interstitium in male Wistar rats.[64]

PROTEIN OVERLOAD PROTEINURIA Proteinuria is the most common sign of CKD and is increased along disease progression. Filtered protein may upregulate the genes encoding vasoactive and inflammatory mediators in tubular cells, thereby exerting their detrimental effects. The protocol of protein overload is quite simple, with intraperitoneal injections of bovine serum albumin (BSA) (low endotoxin) at 2 g/day, 6 times a week, for 3 consecutive weeks in rats.[65] Kidney disease gets worse when accompanied by uninephrectomy.[66] Rats develop massive proteinuria, with mild focal glomerulosclerosis and focal areas of tubular dilation and fibrosis.

A protein overload model has also been developed in mice. Male and female C57BL/6J and 129S2/Sv mice receive increasing BSA (from 2 g/kg to 10 g/kg body weight) for 11 days. Male (but not female) C57BL/6J mice develop proteinuria. In 129S2/Sv mice, proteinuria is evident in both males and females.[67] In another protocol, mice with a mixed gene background of C57BL/6 and 129/J undergo uninephrectomy, followed by BSA injection at 10 g/kg body weight, 5 days per week, for 11 weeks. Mice develop marked glomerulosclerosis, interstitial inflammation, and tubulointerstitial injury.[67]

CONCLUSIONS

Experimental animal models are indispensable for delineating the pathogenetic mechanisms of kidney diseases and for evaluating the efficacy of various potential renoprotective strategies. Although many animal models are well established and widely used, thus far no animal model can exactly replicate the pathogenesis of various kidney diseases in humans. Therefore, faithfully modeling human renal pathogenic processes in animals continues to be a challenge in renal medicine. Novel experimental approaches, as exemplified by combining renal mass ablation and STZ-induced hyperglycemia, may generate a unique model that mimics the major pathological findings of human CKD. In addition, nonrodent models could be enormously valuable, as demonstrated by the recent observation that cisplatin also induces typical histological changes and a decline in renal function in larval zebrafish; this could be a possible unique model, amenable to both genetic manipulation and high-throughput drug screening.[68] Furthermore, as transgenic and knockout mice are increasingly available, there is an unparalleled opportunity for exploring the mechanisms of a particular gene in a range of kidney diseases. It is hoped that establishment of authentic animal models of human kidney diseases will facilitate the preclinical testing of novel therapeutic interventions, thereby ultimately benefiting patients with various kidney disorders.

REFERENCES

1. Eknoyan G, Lameire N, Barsoum R, Eckardt KU, Levin A, Levin N, Locatelli F, MacLeod A, Vanholder R, Walker R, Wang H. The burden of kidney disease: Improving global outcomes. *Kidney Int* 2004;66:1310–1314.
2. Middleton RJ, Foley RN, Hegarty J, Cheung CM, McElduff P, Gibson JM, Kalra PA, O'Donoghue DJ, New JP. The unrecognized prevalence of chronic kidney disease in diabetes. *Nephrol Dial Transplant* 2006;21:88–92.
3. White SL, Cass A, Atkins RC, Chadban SJ. Chronic kidney disease in the general population. *Adv Chronic Kidney Dis* 2005;12:5–13.
4. Schieppati A, Remuzzi G. Chronic renal diseases as a public health problem: Epidemiology, social, and economic implications. *Kidney Int Suppl* 2005;98:S7–S10.
5. Liu Y. Renal fibrosis: New insights into the pathogenesis and therapeutics. *Kidney Int* 2006;69:213–217.
6. Fogo AB. Progression versus regression of chronic kidney disease. *Nephrol Dial Transplant* 2006;21:281–284.
7. Remuzzi G, Bertani T. Pathophysiology of progressive nephropathies. *N Engl J Med* 1998;339:1448–1456.
8. Hentschel DM, Bonventre JV. Novel non-rodent models of kidney disease. *Curr Mol Med* 2005;5:537–546.
9. Zoja C, Morigi M, Benigni A, Remuzzi G. Genetics of rare diseases of the kidney: Learning from mouse models. *Cytogenet Genome Res* 2004;105:479–484.
10. Zoja C, Abbate M, Remuzzi G. Progression of chronic kidney disease: Insights from animal models. *Curr Opin Nephrol Hypertens* 2006; 15:250–257.
11. Breyer MD, Bottinger E, Brosius FC 3rd, Coffman TM, Harris RC, Heilig CW, Sharma K. Mouse models of diabetic nephropathy. *J Am Soc Nephrol* 2005;16:27–45.
12. Bascands JL, Schanstra JP. Obstructive nephropathy: Insights from genetically engineered animals. *Kidney Int* 2005;68:925–937.

13. Zheng F, Striker GE, Esposito C, Lupia E, Striker LJ. Strain differences rather than hyperglycemia determine the severity of glomerulosclerosis in mice. *Kidney Int* 1998;54:1999–2007.
14. Gurley SB, Clare SE, Snow KP, Hu A, Meyer TW, Coffman TM. Impact of genetic background on nephropathy in diabetic mice. *Am J Physiol Renal Physiol* 2006;290:F214–222.
15. Wei Q, Wang MH, Dong Z. Differential gender differences in ischemic and nephrotoxic acute renal failure. *Am J Nephrol* 2005;25: 491–499.
16. Kher A, Meldrum KK, Wang M, Tsai BM, Pitcher JM, Meldrum DR. Cellular and molecular mechanisms of sex differences in renal ischemia-reperfusion injury. *Cardiovasc Res* 2005;67:594–603.
17. Abernethy VE, Lieberthal W. Acute renal failure in the critically ill patient. *Crit Care Clin* 2002;18:203–222.
18. Lieberthal W, Nigam SK. Acute renal failure. II. Experimental models of acute renal failure: Imperfect but indispensable. *Am J Physiol Renal Physiol* 2000;278:F1–F12.
19. Heyman SN, Lieberthal W, Rogiers P, Bonventre JV. Animal models of acute tubular necrosis. *Curr Opin Crit Care* 2002;8:526–534.
20. Mohaupt M, Kramer HJ. Acute ischemic renal failure: Review of experimental studies on pathophysiology and potential protective interventions. *Renal Fail* 1989;11:177–185.
21. Lieberthal W, Nigam SK. Acute renal failure. I. Relative importance of proximal vs. distal tubular injury. *Am J Physiol Renal Physiol* 1998;275:F623–631.
22. Safirstein RL. Acute renal failure: From renal physiology to the renal transcriptome. *Kidney Int Suppl* 2004;S62–S66.
23. Bonegio R, Lieberthal W. Role of apoptosis in the pathogenesis of acute renal failure. *Curr Opin Nephrol Hypertens* 2002;11:301–308.
24. Rana A, Sathyanarayana P, Lieberthal W. Role of apoptosis of renal tubular cells in acute renal failure: Therapeutic implications. *Apoptosis* 2001;6:83–102.
25. Wei Q, Yin XM, Wang MH, Dong Z. Bid deficiency ameliorates ischemic renal failure and delays animal death in C57BL/6 mice. *Am J Physiol Renal Physiol* 2006;290:F35–F42.
26. Fillastre JP, Raguenez-Viotte G. Cisplatin nephrotoxicity. *Toxicol Lett* 1989;46:163–175.
27. Schubert GE. Folic acid-induced acute renal failure in the rat: Morphological studies. *Kidney Int Suppl* 1976;6:S46–S50.
28. Liu Y, Tolbert EM, Lin L, Thursby MA, Sun AM, Nakamura T, Dworkin LD. Up-regulation of hepatocyte growth factor receptor: An amplification and targeting mechanism for hepatocyte growth factor action in acute renal failure. *Kidney Int* 1999;55:442–453.
29. Dai C, Yang J, Liu Y. Single injection of naked plasmid encoding hepatocyte growth factor prevents cell death and ameliorates acute renal failure in mice. *J Am Soc Nephrol* 2002;13:411–422.
30. Ortega A, Ramila D, Izquierdo A, Gonzalez L, Barat A, Gazapo R, Bosch RJ, Esbrit P. Role of the renin-angiotensin system on the parathyroid hormone-related protein overexpression induced by nephrotoxic acute renal failure in the rat. *J Am Soc Nephrol* 2005;16:939–949.
31. Bledsoe G, Crickman S, Mao J, Xia CF, Murakami H, Chao L, Chao J. Kallikrein/kinin protects against gentamicin-induced nephrotoxicity by inhibition of inflammation and apoptosis. *Nephrol Dial Transplant* 2006;21:624–633.
32. Chander V, Chopra K. Protective effect of resveratrol, a polyphenolic phytoalexin on glycerol-induced acute renal failure in rat kidney. *Renal Fail* 2006:28:161–169.
33. Aydogdu N, Atmaca G, Yalcin O, Taskiran R, Tastekin E, Kaymak K. Protective effects of L-carnitine on myoglobinuric acute renal failure in rats. *Clin Exp Pharmacol Physiol* 2006;33:119–124.
34. Newaz M, Yousefipour Z, Oyekan A. Role of PPAR-gamma on the pathogenesis and vascular changes in glycerol-induced acute renal failure. *Pharmacol Res* 2006;54:234–240.
35. Langenberg C, Wan L, Egi M, May CN, Bellomo R. Renal blood flow in experimental septic acute renal failure. *Kidney Int* 2006;69: 1996–2002.
36. Wan L, Bellomo R, Di Giantomasso D, Ronco C. The pathogenesis of septic acute renal failure. *Curr Opin Crit Care* 2003;9:496–502.
37. Fu XJ, Iijima K, Nozu K, Hamahira K, Tanaka R, Oda T, Yoshikawa N, Matsuo M. Role of p38 MAP kinase pathway in a toxin-induced model of hemolytic uremic syndrome. *Pediatr Nephrol* 2004;19: 844–852.
38. Zager RA, Johnson AC, Lund S, Hanson SY, Abrass CK. Levosimendan protects against experimental endotoxemic acute renal failure. *Am J Physiol Renal Physiol* 2006;290:F1453–1462.
39. Susztak K, Bottinger EP. Diabetic nephropathy: A frontier for personalized medicine. *J Am Soc Nephrol* 2006;17:361–367.
40. Schena FP, Gesualdo L. Pathogenetic mechanisms of diabetic nephropathy. *J Am Soc Nephrol* 2005;16(Suppl. 1):S30–S33.
41. Breyer MD, Qi Z, Tchekneva E. Diabetic nephropathy: Leveraging mouse genetics. *Curr Opin Nephrol Hypertens* 2006;15:227–232.
42. Linden KC, DeHaan CL, Zhang Y, Glowacka S, Cox AJ, Kelly DJ, Rogers S. Renal expression and localization of the facilitative glucose transporters GLUT1 and GLUT12 in animal models of hypertension and diabetic nephropathy. *Am J Physiol Renal Physiol* 2006;290: F205–213.
43. O'Donnell MP, Kasiske BL, Keane WF. Glomerular hemodynamic and structural alterations in experimental diabetes mellitus. *FASEB J* 1988;2:2339–2347.
44. Matsubara T, Abe H, Arai H, Nagai K, Mima A, Kanamori H, Sumi E, Takahashi T, Matsuura M, Iehara N, Fukatsu A, Kita T, Doi T. Expression of Smad1 is directly associated with mesangial matrix expansion in rat diabetic nephropathy. *Lab Invest* 2006;86:357–368.
45. Zatz R, Meyer TW, Rennke HG, Brenner BM. Predominance of hemodynamic rather than metabolic factors in the pathogenesis of diabetic glomerulopathy. *Proc Natl Acad Sci USA* 1985;82:5963–5967.
46. Dai C, Yang J, Bastacky S, Xia J, Li Y, Liu Y. Intravenous administration of hepatocyte growth factor gene ameliorates diabetic nephropathy in mice. *J Am Soc Nephrol* 2004;15:2637–2647.
47. Breyer MD, Bottinger E, Brosius FC, Coffman TM, Fogo A Harris RC, Heilig CW, Sharma K. Diabetic nephropathy: Of mice and men. *Adv Chronic Kidney Dis* 2005;12:128–145.
48. Sharma K, McCue P, Dunn SR. Diabetic kidney disease in the db/db mouse. *Am J Physiol Renal Physiol* 2003;284:F1138–1144.
49. Sugaru E, Nakagawa T, Ono-Kishino M, Nagamine J, Tokunaga T, Kitoh M, Hume WE, Nagata R, Taiji M. SMP-534 ameliorates progression of glomerular fibrosis and urinary albumin in diabetic db/db mice. *Am J Physiol Renal Physiol* 2006;290:F813–820.
50. Park CW, Zhang Y, Zhang X, Wu J, Chen L, Cha DR, Su D, Hwang MT, Fan X, Davis L, Striker G, Zheng F, Breyer M, Guan Y. PPARalpha agonist fenofibrate improves diabetic nephropathy in db/db mice. *Kidney Int* 2006;69:1511–1517.
51. Fleck C, Appenroth D, Jonas P, Koch M, Kundt G, Nizze H, Stein G. Suitability of 5/6 nephrectomy (5/6NX) for the induction of interstitial renal fibrosis in rats—influence of sex, strain, and surgical procedure. *Exp Toxicol Pathol* 2006;57:195–205.
52. Liu ZC, Chow KM, Chang TM. Evaluation of two protocols of uremic rat model: Partial nephrectomy and infarction. *Renal Fail* 2003;25:935–943.
53. Massy ZA, Ivanovski O, Nguyen-Khoa T, Angulo J, Szumilak D, Mothu N, Phan O, Daudon M, Lacour B, Drueke TB, Muntzel MS. Uremia accelerates both atherosclerosis and arterial calcification in apolipoprotein E knockout mice. *J Am Soc Nephrol* 2005;16:109–116.
54. Al Banchaabouchi M, Marescau B, Van Marck E, D'Hooge R, De Deyn PP. Long-term effect of partial nephrectomy on biological parameters, kidney histology, and guanidino compound levels in mice. *Metabolism* 2001;50:1418–1425.
55. Chevalier RL. Pathophysiology of obstructive nephropathy in the newborn. *Semin Nephrol* 1998;18:585–593.
56. Yang J, Liu Y. Dissection of key events in tubular epithelial to myofibroblast transition and its implications in renal interstitial fibrosis. *Am J Pathol* 2001;159:1465–1475.
57. Chevalier RL. Pathogenesis of renal injury in obstructive uropathy. *Curr Opin Pediatr* 2006;18:153–160.

58. Yang J, Liu Y. Blockage of tubular epithelial to myofibroblast transition by hepatocyte growth factor prevents renal interstitial fibrosis. *J Am Soc Nephrol* 2002;13:96–107.
59. Yang J, Dai C, Liu Y. Hepatocyte growth factor gene therapy and angiotensin II blockade synergistically attenuate renal interstitial fibrosis in mice. *J Am Soc Nephrol* 2002;13:2464–2477.
60. Couser WG, Salant DJ, Stilmant MM, Arbeit LA, Darby C, Sliogeris VG. The effects of aminonucleoside of puromycin and nephrotoxic serum on subepithelial immune-deposit formation in passive Heymann nephritis. *J Lab Clin Med* 1979;94:917–932.
61. Wang Y, Wang YP, Tay YC, Harris DC. Progressive adriamycin nephropathy in mice: Sequence of histologic and immunohistochemical events. *Kidney Int* 2000;58:1797–1804.
62. Qin XH, Lee VW, Wang YP, Zheng GP, Wang Y, Alexander SI, Harris DC. A protective role for programmed death 1 in progression of murine adriamycin nephropathy. *Kidney Int* 2006;70(7):1244–1250.
63. Li C, Sun BK, Lim SW, Song JC, Kang SW, Kim YS, Kang DH, Cha JH, Kim J, Yang CW. Combined effects of losartan and pravastatin on interstitial inflammation and fibrosis in chronic cyclosporine-induced nephropathy. *Transplantation* 2005;79:1522–1529.
64. Khan M, Shobha JC, Mohan IK, Rao Naidu MU, Prayag A, Kutala VK. Spirulina attenuates cyclosporine-induced nephrotoxicity in rats. *J Appl Toxicol* 2006;26:444–451.
65. van Timmeren MM, Bakker SJ, Stegeman CA, Gans RO, van Goor H. Addition of oleic acid to delipidated bovine serum albumin aggravates renal damage in experimental protein-overload nephrosis. *Nephrol Dial Transplant* 2005;20:2349–2357.
66. van Timmeren MM, Bakker SJ, Vaidya VS, Bailly V, Schuurs TA, Damman J, Stegeman CA, Bonventre JV, van Goor H. Tubular kidney injury molecule-1 in protein-overload nephropathy. *Am J Physiol Renal Physiol* 2006;291:F456–464.
67. Ishola DA Jr, van der Giezen DM, Hahnel B, Goldschmeding R, Kriz W, Koomans HA, Joles JA. In mice, proteinuria and renal inflammatory responses to albumin overload are strain-dependent. *Nephrol Dial Transplant* 2006;21:591–597.
68. Hentschel DM, Park KM, Cilenti L, Zervos AS, Drummond I, Bonventre JV. Acute renal failure in zebrafish: A novel system to study a complex disease. *Am J Physiol Renal Physiol* 2005;288: F923–929.

69 Animal Models of Multiple Sclerosis

Colin R. Young and C. Jane Welsh

ABSTRACT

To determine whether an immunological or pharmaceutical product has potential for therapy in treating multiple sclerosis (MS), detailed animal models are required. To date many animal models for human MS have been described in mice, rats, rabbits, guinea pigs, marmosets, and rhesus monkeys. The most comprehensive studies have involved murine experimental allergic (or autoimmune) encephalomyelitis (EAE), Semliki Forest virus (SFV), mouse hepatitis virus (MHV), and Theiler's murine encephalomyelitis virus (TMEV). Here, we describe in detail multispecies animal models of human MS, namely EAE, SFV, MHV, and TMEV, in addition to chemically induced demyelination. The validity and applicability of each of these models are critically evaluated.

Key Words: Multiple sclerosis, Experimental autoimmune encephalomyelitis, Semliki Forest virus, Mouse hepatitis virus, Theiler's murine encephalomyelitis virus.

INTRODUCTION

Multiple sclerosis (MS) affects about 350,000 people in the United States and is a major cause of nervous system disability in adults between the ages of 15 and 45 years. The symptoms are diverse, ranging from tremor, nystagmus, paralysis, and disturbances in speech and vision. Extensive demyelination is seen in the neuronal lesions. The clinical heterogeneity of MS, as well as the finding of different pathological patterns, suggests that MS may be a spectrum of diseases that may represent different pathological processes.[1] This has led to the development of many different animal models, including rodents and nonhuman primates, that reflect the pathological processes and could allow for the development of therapeutic approaches. At the present time, the exact etiological mechanism in humans is not clear; however, several animal models are available providing insight into disease processes. The relative inaccessibility and sensitivity of the central nervous system (CNS) in humans preclude studies on disease pathogenesis, and so much of our understanding of infections and immune responses has been derived from experimental animal models. The experimental systems include Theiler's virus, mouse hepatitis virus, and Semliki Forest virus infections of laboratory rodents. Additional information has been obtained from studies of experimental infections of other animals that result in demyelination, notably maedi-visna virus in sheep and canine distemper virus in dogs. In humans and animals, most natural cases of demyelinating disease are rare complications of viral infections. One possible reason for the low incidence of demyelination following viral infections could be the low efficiency of neuroinvasion. However, a correlation between CNS infection and clinical disease is difficult to determine.

The role of genetics and environmental factors in MS is complex. Factors such as geographical location, ethnic background, and clustering in temperate climates all contribute to susceptibility. Individuals with a North European heritage are statistically more susceptible to MS than those from a more tropical environment and it is more common in women.[2] Epidemiological data indicate that MS is not a single-gene disorder and that additionally environmental factors contribute to the disease.[3] Data from genetic studies indicate that although MHC genes clearly contribute to disease susceptibility and/or resistance, it is probable that a combination of environmental factors may additionally contribute to disease development in genetically predisposed individuals.

To understand the initiating factors and progression of MS, researchers have turned to experimental model systems. Since this disease cannot be recreated in a tissue culture system, much effort has been directed to the use of laboratory animals. Those animal models should mirror the clinical and pathological findings observed in human MS. Ideally, the animal model should be in a species that is easy to handle, inexpensive, can be kept in large numbers, and is easily bred in laboratory conditions. The most frequently used animals are laboratory rodents, including mice, rats, guinea pigs, and hamsters. One of the most useful aspects of laboratory rodents as animal models of disease is the vast array of inbred strains of the species available, most notably in experimental mice. Additionally, very valuable information has been obtained from studies using larger animals including sheep, dogs, cats, and nonhuman primates.

Models of MS fall into two main groups: viral and nonviral. Viral models are immensely relevant since epidemiological studies suggest an environmental factor, and almost all naturally occurring CNS demyelinating diseases of humans and animals of known etiology are caused by a virus. These include in humans, subacute sclerosing panencephalitis (SSPE)—caused by measles or rubella viruses, progressive multifocal leukoencephalopathy (PML)—caused by JC virus, and human T lymphotrophic virus-1

From: *Sourcebook of Models for Biomedical Research*
(P. M. Conn, ed.), © 2008 Humana Press Inc., Totowa, NJ.

(HTLV-1)-associated myelopathy (HAM)—caused by HTLV-1; in animals, these include visna virus in sheep and canine distemper in dogs. However, no one virus has consistently been associated with human MS, although it is likely that more than one virus could trigger the disease.

Of the nonviral models of MS, experimental allergic encephalomyelitis (EAE) is the most widely studied. EAE is characterized by inflammatory infiltrates in the CNS that can be associated with demyelinating lesions. In EAE, the disease is initiated by the extraneural injection of CNS material, or purified myelin components, emulsified in an adjuvant, the most commonly employed one being complete Freund's adjuvant containing *Mycobacterium tuberculosis* H37Ra. However, no naturally occurring autoimmune correlate of this experimental disease is known, although it is extensively researched as a model of MS, with the reasoning that MS may be such a disease.

The most widely studied models of MS are the experimental infections of rodents resulting in an inflammatory demyelinating disease in the CNS, such as Theiler's virus, mouse hepatitis virus, and Semliki Forest virus.[4] Each of these infections gives rise to lesions of mononuclear cell inflammatory demyelination throughout the brain and spinal cord but not in the peripheral nervous system. As such, this histopathology correlates with human MS, although it does not preclude the fact that the viruses could gain access to the CNS via the peripheral nervous system. These viral models demonstrate how a virus can easily reproduce CNS disease, which is comparatively rare in humans, and how this can be influenced by many factors including both genetic and immunological.

Experimental studies in induced animal models have the advantage over studies in spontaneous models in that the onset and progression of the disease can be controlled. Although it has been proposed that some autoimmune diseases may have a viral etiology, virus-induced autoimmunity is a controversial subject. Epidemiological studies of MS provide strong evidence for the involvement of a viral etiology in the onset of disease. Theiler's virus-induced demyelination, a model for human MS, bears several similarities to the human disease: an immune-mediated demyelination, involvement of $CD4^+$ helper T cells and $CD8^+$ cytotoxic T cells, delayed type hypersensitivity responses to viral antigens and autoantigens, and pathology. Indeed this mouse model may provide a scenario that closely resembles chronic progressive MS.

THEILER'S VIRUS-INDUCED DEMYELINATION

Theiler's murine encephalomyelitis virus (TMEV) is a picornavirus that causes an asymptomatic gastrointestinal infection, followed by occasional paralysis. There are two main strains of TMEV, the virulent strains and persistent Theiler's original (TO) strains. The virulent GDVII strains of Theiler's virus are highly neurovirulent and when injected intracranially, cause death by encephalitis within 48h. GDVII strains also cause differing forms of paralysis depending on the route of inoculation (see Table 69–1). From these studies it appears that the GDVII virus may gain access to the CNS by retrograde axonal transport rather than by a hematogeneous route.[5]

Infection of susceptible strains of mice with the persistent TO strains BeAn, DA, WW, or Yale results in a primary demyelinating disease that closely resembles human MS.[6] Infection of resistant strains of mice with BeAn does not result in demyelinating disease, since these mice are able to clear virus from the CNS. Susceptible mice fail to clear virus from the CNS, possibly resulting from poor natural killer (NK) cell and cytotoxic T lymphocyte (CTL) responses. Persistent viral infection of the CNS is required for demyelination. Following the intracranial injection of susceptible mice with BeAn, virus replicates both in the brain and spinal cord.[7] One month postinfection, viral titers decrease, and high levels of neutralizing antibodies are detected (Figure 69–1). At this point in the disease, neurons may become infected with virus and mice may develop a nonprogressive flaccid paralysis of the forelimbs and/or hindlimbs.[8] This is sometimes referred to as a polio-like disease, but this is confusing since flaccid paralysis in mice infected with poliovirus is progressive and normally results in death. In the late phase of the disease, astrocytes, oligodendrocytes, and macrophage/microglial cells become infected with virus. Also in the demyelinating disease there is both B and T cell autoimmunity, directed against myelin and its antigenic components.

GENETICS OF PERSISTENT INFECTION AND DEMYELINATING DISEASE All inbred mouse strains inoculated intracerebrally with TMEV show early encephalomyelitis, but not all strains remain persistently infected. Resistant strains normally

Table 69–1
Specificity of paralysis due to routes of injection of Theiler's murine encephalomyelitis virus (TMEV)

Route of injection[a]	*Volume of virus*[b]	*pfu injected*[c]	*Paralysis, encephalitis, or mortality*[d]	*Days of paralysis*[e]
Intramuscular	100 µl	10^6 pfu	URLP/BRLP paralysis	2–4
Intragastric	100 µl	10^6 pfu	0%	0
Intravenous	100 µl	10^6 pfu	Fore/hindlimb paralysis 80%	6
Intraperitoneal	100 µl	10^6 pfu	Fore/hindlimb paralysis 100%	6
Intratongue	100 µl	10^6 pfu	Tongue paralysis 100%	4
Intracranial	10–20 µl	10^5 pfu	No paralysis, 100% mortality, 100% clinical encephalitis	0
Footpad	100 µl	10^6 pfu	Fore/hindlimb paralysis 100%	2–4

Source: Adapted from Villarreal *et al.*[5]
[a]Route of injection of TMEV in CBA mice.
[b]Amount of virus injected.
[c]pfu injected.
[d]Incidence of paralysis or encephalitis or mortality observed in mice.
[e]Onset of paralysis in CBA mice.

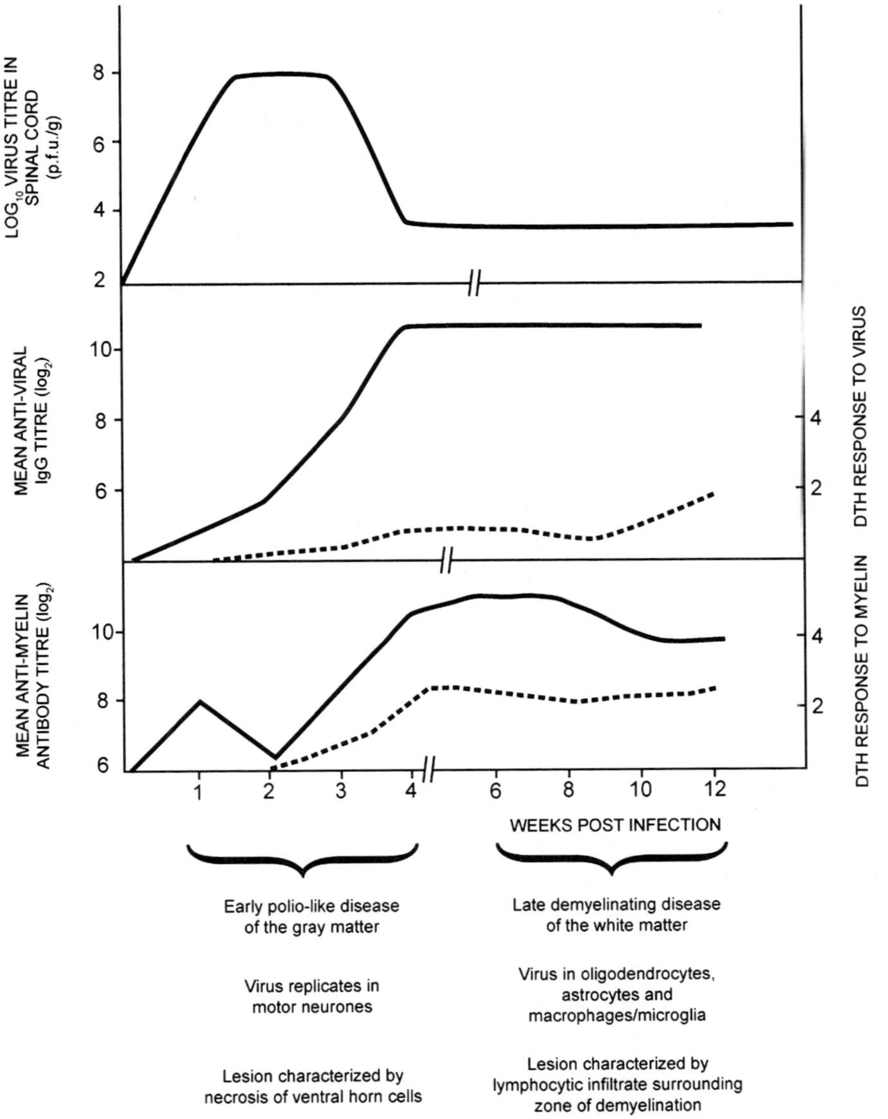

Figure 69-1. Disease course of Theiler's virus-induced demyelination in CBA mice. Intracerebral infection of CBA mice with 5×10^4 pfu of the BeAn strain of Theiler's virus results in high levels of virus in the spinal cord during the first month of infection (top panel). The CNS viral titers decrease at 4 weeks pi when neutralizing antiviral antibodies and viral T cell responses are detected (middle panel). During late disease, the virus is detected in astrocytes, oligodendrocytes, and macrophage/microglial cells and autoreactive T and B cell responses to myelin are detected in TMEV-infected mice (lower panel). (Reprinted from Welsh *et al.*[8] Copyright 2006 with permission from Springer.)

clear the virus after 14 days, whereas susceptible strains remain persistently infected for life. Thus, susceptibility/resistance refers to the demyelinating late disease and not to the encephalomyelitis. CNS viral load during persistence varies among susceptible strains, making susceptibility a quantitative trait.[9] This trait is under multigenic control, with H-2 MHC class I genes being the most prominent. Additionally, several non-H-2 quantitative trait loci (QTL) have been identified within the same H-2 haplotypes that control persistence. There is generally a good correlation in inbred strains between susceptibility to three phenotypes (viral load, pathology, and symptoms), suggesting that variations in both demyelination and clinical disease may result from how each mouse strain can control the viral load during the persistent infection.[10] Using B10 congeneic and recombinant strains of mice, susceptibility to disease has been mapped to the H-2D region.[11] Furthermore, resistant haplotypes are dominant and the same locus controls viral load during persistence and demyelination. Currently, 11 non-H-2 susceptibility loci have been identified as having an effect on susceptibility to Theiler's virus-induced disease (TVID) (see Table 69–2).

The mechanism(s) of TVID may be different for different mouse strains, but most of the information has come from studies of SJL/J mice infected with the DA or BeAn strain of virus. The virus infects oligodendrocytes, and the resulting demyelinating disease could be due, in part, to the virus killing oligodendrocytes directly or by the virus-specific $CD8^+$ CTLs present in the lesions.[7]

Table 69–2
Non-H2 loci of susceptibility to Theiler's virus-induced disease

Locus	Chr	Location (cM)	Phenotype	Viral strain
Tmevd1	6	22	Clinical signs	BeAn
Tmevd2	3	46	Clinical signs	BeAn
Tmevd3	14	12.5	Demyelination	DA
Tmevd4	14	39.5	Demyelination	DA
Tmevd5	11	60	Clinical signs	DA
Tmevd6	1	19.5	Clinical signs	BeAn
Tmevd7	5	72	Clinical signs	BeAn
Tmevd8	15	4.7	Clinical signs	BeAn
Tmevd9	1	32.8	Clinical signs	BeAn
Tmevp2	10	51.5–62	Viral load	DA
Tmevp3	10	69–70	Viral load	DA

Source: Adapted from Brahic *et al.*[9]

A series of experiments has demonstrated that demyelination correlates with the presence of a CD4⁺ T cell-mediated response against viral epitopes. These cells secrete cytokines such as interferon (IFN)-γ that activate both microglial cells and invading monocytes, which subsequently secrete factors such as tumor necrosis factor (TNF)-α and thus can cause "bystander" demyelination. Activated macrophages ingest and degrade damaged myelin. Autoantibodies[12,13] and myelin-specific $CD4^+$ T cells have been shown in SJL/J mice several months after intracranial inoculation.[14,15] Epitope spreading in these mice commences with recognition of a proteolipid protein (PLP) epitope, and then progresses to additional PLP epitopes and then to myeloid basic protein (MBP) epitopes.[14] A direct demonstration that disease can be maintained on a purely autoimmune footing, after infection has been eradicated, has not been shown.

IMMUNITY AND THEILER'S VIRUS The first response to viral infection is the production of type I interferons, which are critical for viral clearance. IFN-α/β receptor knockout mice injected with TMEV die of encephalomyelitis within 10 days of infection. NK cells are activated early in infection with certain viruses. In TMEV infection susceptible SJL mice have a 50% lower NK cell activity in comparison to the highly resistant C57BL/6 mice. This low NK activity in SJL mice is in part due to a defect in the thymus impairing the responsiveness of NK cells to stimulation by IFN-β. The pivotal role of NK cells in early TMEV clearance is demonstrated by the finding that resistant mice depleted of NK cells by monoclonal antibodies to NK 1.1 develop severe signs of gray matter disease.[16]

In the early disease, both $CD4^+$ and $CD8^+$ T cells have been shown to be important in viral clearance. In early disease $CD4^+$ T cells are required for B cells to produce antibodies for viral clearance.[12] These $CD4^+$ T cells secrete IFN-γ, which *in vitro* inhibits TMEV replication and has a protective role *in vivo*. $CD8^+$ T cells are also important in viral clearance, as demonstrated by the finding that $CD8^+$ T cell-depleted mice fail to clear virus and develop a more severe demyelinating disease.[17] $CD8^+$ T cells also provide protection against TVID when adoptively transferred to a TVID-susceptible strain, BALB/c.AnNCr. Thus, $CD8^+$ T cells are implicated in viral clearance and resistance to demyelination. Higher CTL activity has been demonstrated in TVID-resistant C57BL/6 mice as compared to resistant SJL/J mice.[18] These CTLs may play an important role in viral disease since they may recognize viral determinants and/or they may inhibit delayed type hypersensitivity responses.

The relative roles of Th1/Th2 cells in TVID are very complex, and a simple picture of a Th1 or Th2 polarization during infection may not be apparent. A pathogenic role for Th1 cells during late demyelinating disease is demonstrated by the finding that both TVID correlates with delayed type hypersensitivity responses to TMEV and that the depletion of $CD4^+$ T cells during late disease results in the amelioration of clinical signs. High levels of the proinflammatory Th1 cytokines IFN-γ and TNF-α in late disease correlate well with maximal disease activity. Evidence demonstrating the protective role of Th2 in TVID has been shown in experiments in which skewing the immune response toward Th2 immunity in TMEV infection diminishes the later demyelinating disease.[19] However, other studies have shown that the Th1/Th2 balance did not explain the difference in susceptibility to TVID. Th1 cytokines are generally pathogenic during late demyelinating disease, whereas Th2 cytokines are protective.

USE OF THEILER'S MURINE ENCEPHALOMYELITIS VIRUS MODEL TO STUDY REMYELINATION Remyelination in MS lesions was first documented in 1906[20] and is characterized by abnormally thin myelin sheaths in relation to axon diameter. To date, however, there are few reliable data on the frequency of remyelination in MS patients. Stimulation of remyelination is a potential treatment for MS. The TMEV model of MS can be used to study remyelination using remyelination-promoting antibodies. In this remyelination model, SJL/J mice, aged 4–8 weeks, are injected with a 10 μl volume containing 200,000 pfu of Daniel strain intracerebrally. All animals develop mild encephalitis, which resolves within 14 days after the injection. The infected mice then develop the chronic demyelinating disease that gradually progresses over several months. To study remyelination, mice that had been infected with TMEV for 6 months receive a single intraperitoneal injection of 0.5 mg (~0.025 g/kg body weight) of a recombinant remyelinating antibody (rHIgM22) in phosphate buffered saline (PBS).[21] In one study, 82.8% of lesions in animals treated with rHIgM22 showed retraction of varying degrees, presumably the effect of remyelination in these lesions. The direct binding of rHIgM22 to demyelinated lesions is consistent with the hypothesis that these antibodies work directly in the lesions, probably by binding to the CNS glia to induce remyelination. Thus, this murine TMEV model can also be used as a model with which to examine different modes of remyelination.

MOUSE HEPATITIS VIRUS

Mouse hepatitis virus (MHV) is a member of the Coronaviridae, a group of large positive sense enveloped RNA viruses. Depending on the strain of virus used, MHV causes a variety of diseases including enteritis, hepatitis, and demyelinating encephalomyelitis.[22] Infection of mice with the neutrotropic JHM strain of MHV causes encephalitis, followed by chronic demyelination. Virus is not cleared from the CNS, resulting in a persistent infection.

After intracerebral or intranasal infection with MHV, virus enters the brain and causes encephalitis.[23] Intranasal infection with MHV-JHM or -A59 leads to viral spread through the olfactory bulb and along the olfactory tracts, as well as along the trigeminal nerve to the mesencephalic nucleus.[23] Up to 4 days postinfection (pi) early viral spread is via specific neural pathways and neural connections. Viral titers peak at about day 5 pi in the brain and later in the spinal cord and virus is cleared by days 8–20 pi.[24] However, viral antigen is still detectable up to day 30 pi. Additionally, viral RNA is detectable in the brain as late at 10–12 months postinfection, although the amount of RNA decreases with time. Liver infection can occur after any route of infection (in, ic, ig, or ip), with viral titers peaking at day 5 pi and hepatitis developing during the first 1–2 weeks.[25]

CNS demyelination develops as active MHV infection resolves. The lesions observed are histologically very similar to those observed in MS patients. These MHV lesions are characterized by primary demyelination accompanied by naked axons,[23] and are found scattered throughout the spinal cord.[26] The peripheral nervous system is not affected. Chronic lesions are associated with lipid-laden macrophages, scattered lymphocytes, and perivascular cuffing. These chronic lesions can persist as late as day 90 pi, and demyelinating axons can be seen as late as 16 months postinfection.

Chronic diseases in MHV-infected mice are associated with ataxia, hindlimb paresis, and paralysis, followed by a recovery. This animal recovery is mediated by CNS remyelination, beginning anywhere from 14 to 70 days pi.

C57BL/6 mice (H-2^b) are susceptible to MHV infection. In this murine model adult mice (of weight 20–22 g) are anesthetized by inhalant anesthesia and receive an intracerebral injection of approximately 500 pfu of a neurotropic MHV strain in a volume of approximately 20 μl of PBS. This intracerebral injection of MHV results in a biphasic disease: an acute encephalomyelitis with myelin loss, followed 10–12 days later by an immune-mediated demyelinating encephalomyelitis with progressive destruction of the CNS.[27] There is an 80–90% survival rate of mice injected with this MHV, with animals usually succumbing during the first 2 weeks of acute infection. Animals surviving this acute stage show a 95% chance of survival.[28] Control animals injected intracerebrally with sterile PBS show no clinical signs or histological defects.

Electron micrographs of demyelinating lesions show that macrophage processes slip between layers in the myelin sheath, implying that macrophages could indeed be mediating demyelination.[29] The appearance of macrophages within the CNS also correlates with the development of lesions. Additionally, they do not appear in large numbers in the absence of lymphocytes, so it is possible that myelin damage is caused by a nonmacrophage-dependent mechanism and that macrophages may only clear up the damaged myelin. In contrast to other mouse models of demyelination, there does not appear to be a clear role for any single lymphocytic or monocytic subset mediating the demyelination. Rather, it appears that a balance of immune components may be necessary for viral clearance and that various pathways, both immune and nonimmune, may cause the ensuing demyelinating events.

Recently, progress has been made in further identifying the immune cells required for demyelination. Experimental infection of severe combined immunodeficiency (SCID) mice, lacking T cells, results in fulminate encephalitis without demyelination.[30] Adoptive transfer of splenocytes from syngeneic immunocompetent mice into infected SCID mice results in demyelination within 7–9 days posttransfer. Additional experiments indicated that either $CD4^+$ or $CD8^+$ T cell subsets are capable of initiating this process. However, mice that receive splenocytes depleted of $CD4^+$ T cells survive longer and develop more demyelination than mice receiving splenocytes depleted of $CD8^+$ T cells. Thus, experimental SCID mice demonstrate that the roles of each T cell subset in demyelinating diseases are not equal.[31]

IFN-γ is a critical mediator of homeostasis and inflammation in MS and many of its rodent models. Bone marrow chimera mice have been used to address the role of IFN-γ in bystander demyelination mediated by $CD8^+$ T cells. These chimeras as rodent models for JHM have addressed a hypothesis that IFN-γ produced by $CD8^+$ T cells, and not from other sources, was the critical component in mediating bystander demyelination. This chimeric approach did not compromise IFN-γ production by cells such as NK cells and dendritic cells, thus preserving the innate immune response to the virus.[32] The results demonstrated that IFN-γ produced by these innate cells was unable to initiate the demyelinating disease, even in the context of activated $CD8^+$ T cells lacking only the ability of produce IFN-γ. These findings highlight the role that $CD8^+$ T cells have in demyelination in JMH-infected mice.[33] It has been demonstrated that IFN-γ is critical in other animal models of demyelination and in MS.

SEMLIKI FOREST VIRUS

Semliki Forest virus (SFV) is an alphavirus of the Togaviridae. The virus has been isolated from mosquitoes, but the natural host is unknown. SFV is a single-stranded positive strand RNA virus that has been cloned and sequenced. The most commonly studied strains used in adult mice are the virulent L10 strain and the avirulent A7(74) strain. Both of these strains are avirulent in neonatal and suckling mice by all routes of infection. Experimental infection of mice with SFV is widely used as a model to study the mechanism of virus-induced CNS disease. SFV has the advantage of being neuroinvasive as well as neurotropic, thus allowing studies of viral entry into the CNS and the integrity of the blood–brain barrier (BBB). Following intraperitoneal injection with 5000 pfu SFV in 0.1 ml PBS containing 0.75% bovine serum albumin,[34] all strains replicate in muscles and other tissues, resulting in a plasma viremia. Virus then crosses the cerebral vascular endothelial cells, resulting in infection of neurons and oligodendrocytes.[35] In neonatal or adult mice, infection with virulent strains results in widespread infection that is lethal within a few days. In contrast, infection of mice with the A7(74) strain results in a CNS infection, and infectious virus is cleared from the brain by day 10. Infiltrating mononuclear cells are observed 3 days pi and peak at about day 7. Focal lesions of demyelination throughout the CNS are observed 10 days pi and peak between 14 and 21 days pi.[36]

SFV-induced demyelinating diseases have been widely studied following intraperitoneal injection of adult mice with the A7(74) strain of the virus. Following intraperitoneal injection, virus is detected in the brain by 24 h. Viral titers then rise, but rapidly decline following initiation of the immune response. Interestingly, although infectious virus can be detected only up to day 8 pi, real-time polymerase chain reaction (rt-PCR) studies detect viral RNA up to day 90 pi.[37] Thus, it is possible that there is persistence of viral antigen(s). Disturbance of the BBB occurs between 4 and 10 days pi, which corresponds to the increase in inflammatory cell infiltration and reduction in viral titer and which may be related to the influx of cells or cytokine-mediated effects. The presence of macrophages, activated microglia, and the proinflammatory cytokines TNF-α, IFN-γ, interleukin (IL)-1α, IL-2, IL-6, and granulocyte-macrophage colony-stimulating factor (GM-CSF) during SFV-induced demyelination, in addition to enhancing the inflammatory response, may also play a role in controlling viral infection since IL-6, IFN-γ, and TNF have direct antiviral activity.[38] Additionally, IFN-γ and TNF production peripherally coincides with SFV-induced encephalitis in SJL and B6 mice. Interestingly, these same cytokines predominate in MS lesions.[39] An intense inflammatory response characterized by perivascular cuffing is apparent histologically from 3 days. Demyelination, as demonstrated using luxol fast blue staining of sections, is apparent by 14 days. However, small focal lesions of demyelination can be observed using electron microscopy by day 10. A striking feature of SFV infection appears in the optic nerve, where there are demyelinating lesions and changes in visually evoked responses and axonal transport.[40] This optic neuritis also occurs in human MS.

It appears that SFV-induced demyelination in this mouse model is accompanied by neurophysiologically demonstrable visual deficits very similar to those found in MS patients. Thus, this may provide a very useful animal model for research into MS. The advantages of this model are that genetic and environmental factors can be readily controlled, while the low cost and fast reproductive rate make experimental design considerably easier.

No demyelination is observed following SFV infection of SCID mice or athymic mice. In the absence of specific immune responses, SCID mice infected with SFV A7(74) have a persistent viremia, a persistent and restricted CNS infection, and no lesions of demyelination. Comparison of the infection to that in nu/nu and BALB/c mice and studies on the transfer of immune sera show that immunoglobulin M (IgM) antibodies clear the viremia but not the brain virus and that infections of brain virus can be reduced by IgG antibodies. These IgG antibodies can abolish infectivity titers in the brain but cannot remove all viral RNA.[41] Adoptive cell transfer studies and administration of anti-CD8 antibodies demonstrate that demyelination following SFV infection is dependent on $CD8^+$ T cells.[4] This is consistent with the finding that the CNS inflammatory infiltrate is dominated by $CD8^+$ T cells. This finding is analogous to that in MS and is in contrast to that in EAE, where $CD4^+$ cells predominate.[42]

In the EAE autoimmune model of MS, studies suggest that a Th1 cytokine profile predominates. Another point of difference between the EAE model and the SFV model is shown in the Th1/Th2 profiles. Following infection with SFV, Th1 and Th2 cytokines were detected in the CNS and both were present throughout the time course studied, indicating that there was no bias of Th response in the CNS, nor were changes apparent with time.[43]

EXPERIMENTAL ALLERGIC ENCEPHALOMYELITIS

The experimental disease EAE has been investigated in many strains of animals including mice, rats, guinea pigs, rabbits, marmosets, and rhesus monkeys. EAE is an autoimmune inflammatory disease of the CNS and is characterized by perivascular and subpial inflammatory infiltrates and demyelinating lesions. The disease is usually initiated by injection of autoantigens emulsified in an adjuvant. The progression and pathology of lesions observed depend on the type of antigen used in the injection, the method of injection, and the strain of animal used. Because of its very nature, EAE as a model of MS does not address certain pertinent questions relating to MS, such as age-related onset of disease or epidemiology. A major difference between EAE and viral models of MS is that in EAE the inflammatory response is directed to autoantigens. A feature of the EAE model is that the course of the disease can be relapsing and remitting.

Studies of EAE have been used to identify antigenic determinants on components of myelin. Using bioinformatic technology these determinants have been used to search available databases of viral and bacterial proteins. Results indicate numerous viral and bacterial protein segments with probable sequence similarity to myelin basic protein determinants.[44]

EXPERIMENTAL ALLERGIC ENCEPHALOMYELITIS IN RABBITS EAE has been induced in rabbits by footpad inoculation with rabbit spinal cord homogenate, resulting in hindlimb paresis or paralysis.[45] Rabbits with 5-day paraplegia showed increased spinal cord incorporation of radioactive drugs administered in the epidural space. Thus, this demyelinating disease process may expose the spinal cord to larger amounts of substances administered neuraxially. It is therefore possible that this rabbit model could be used to investigate the incorporation of radioactive therapeutic drugs in the epidural space.

EXPERIMENTAL ALLERGIC ENCEPHALOMYELITIS IN GUINEA PIGS Guinea pigs have also been investigated to determine whether they may serve as useful EAE models of MS. The interest in guinea pigs stems from the fact that group 1 CD1 glycoprotein homologues, which in humans present foreign and self lipid and glycolipid antigens to T cells, are not found in mice and rats but are present in guinea pigs. In this guinea pig model, animals have been sensitized for EAE, and CD1 and MHC class II expression has been measured in the CNS. In normal guinea pigs low level MHC class II occurred on meningeal macrophages and microglial cells, whereas immunoreactivity for CD1 was absent. In the EAE CNS, however, the majority of infiltrating cells were MHC II^+ and microglia showed increased expression, whereas CD1 immunoreactivity was detected on astrocytes, B cells, and macrophages. Minimal CD1 and MHC II coexpression was detected on inflammatory cells or glia. Thus, in this guinea pig EAE model group 1 CD1 molecules are upregulated in the CNS on subsets of cells distinct from the majority of MHC II-bearing cells.[46] This expression of CD1 proteins in such EAE lesions broadens the potential repertoire of antigens recognized at these sites and highlights the value of this guinea pig model of human MS.

EXPERIMENTAL ALLERGIC ENCEPHALOMYELITIS IN RATS Rats were injected with spinal cord homogenate or the encephalitogen; myelin basic protein induced EAE in genetically susceptible Dark Agouti (DA) rats but not in Albino Oxford (AO) rats. Here 8- to 12-week-old rats were immunized in either or both hind footpads with 0.1 ml antigenic emulsion containing 100 µg rat spinal cord tissue in complete Freund's adjuvant (CFA).[47] Rats are monitored from day 5 after inoculation and the severity of disease was assessed by grading tail, hindlimb, and forelimb weakness, each on a sale of 0 (no disease), 1 (loss of tail tonicity), 2 (hindlimb weakness), 3 (hindlimb paralysis), to 4 (moribund or dead). Clinical disease in susceptible strains of rats is apparent in all animals, and the onset of disease occurs at day 11 postinjection. At the peak of the clinical manifestation of EAE there is a marked increase in the level of infiltration of cells accompanied by a lack of activation in susceptible DA rats, whereas it remains elevated in resistant AO rats. At the peak of clinical disease DA rat spinal cords contain high levels of $CD4^+$ T cells. DA rats also contained 10 times as many live $CD4^+$ T cells as AO rats. Astrocytosis, as an indication of CNS reaction to the presence of inflammatory cells, was clearly observed in both rat species. Microglial activation persists in resistant AO rats, whereas activation is downregulated in DA rats. In this model it is speculated that at the peak of disease, infiltrating monocytes and macrophages are the main antigen-presenting and effector cells.

Rat EAE may also be induced by the injection of xenogeneic myelin. For example, 8- to 12-week-old Lewis rats injected in both hind footpads with an emulsion containing 100 µg of guinea pig myelin basic protein and CFA develop acute EAE. Also chronic relapsing EAE (CR-EAE) may also be induced in this rat model using a regimen of intraperitoneal injections of 4 mg/kg of cyclosporin A.[48] Pathology studies indicate that in acute and CR-EAE, MCP-1 and its receptor CCR2 are significantly upregulated throughout the course of CR-EAE and that a large number of macrophages infiltrated the CR-EAE lesion. This suggests that macrophages recruited by MCP-1 and CCR2-expressing CNS cells are responsible for the development and relapse of EAE. Thus, in this rat model, in addition to T cells, macrophages are another target for immunotherapy studies for neurological autoimmune diseases.

A more recent development of a rat EAE model involves using human MBP as antigen. Here EAE was induced by the immunization of female Wistar rats with human MBP. It was found that most of the rats developed tail tone loss and hindlimb paralysis together with demyelination, infiltrative lymphocyte foci, and "neurophagia" in the cortex of cerebra and in the white matter of the spinal cord.[49] This study further demonstrated that this rat model of EAE induced by human MBP resembles many features of human MS and may promise to be a better animal model for the study of MS.[49]

The use of CFA is not a prerequisite for the development of rat EAE. For example, EAE can be induced in 10- to 16-week-old DA rats by a single hind footpad injection of an encephalitogenic emulsion consisting of rat or guinea pig spinal cord homogenate (SCH) in PBS.[50] The reason for not wanting to use CFA is that in itself it induces a strong inflammatory response and exerts numerous immunomodulatory properties. Additionally, CFA induces a strong anti-purified protein derivative (PPD) response and may induce adjuvant arthritis, another autoimmune disease. The susceptibility of DA rats to EAE induction with SCH depends upon the origin of the CNS tissue, the homologous tissue being the more efficient encephalitogen. DA rats that recovered from EAE that had been induced with homologous SCH without adjuvant and then immunized with the encephalitogenic emulsion containing CFA developed clinical signs of the disease. Neurological signs in rechallenged rats were milder, but first signs appeared earlier. The earlier onset of EAE observed in DA rats after rechallenge has been attributed to the reactivation of memory cells. Taken together, these experiments demonstrate that EAE can be efficiently and reproducibly induced in DA rats without the use of CFA. This experimental model for understanding the basic mechanisms involved in autoimmunity within the CNS, without the limitations and inherent problems imposed by the application of adjuvants, may represent one of the most reliable rodent models of MS.

The rat as an experimental model could be used to evaluate new immunotherapies of EAE. These include antigen-induced mucosal tolerance, treatment with cytokines, and dendritic cell-based immunotherapy.

The ideal treatment of diseases with an autoimmune background such as MS should specifically eliminate autoreactive T cells without affecting the integrity of the immune system. One way to achieve this would be to induce immunological tolerance to autoantigens by the oral or nasal administration of autoantigen. Several studies have shown that nasal administration of soluble antigens results in peripheral tolerance by immune deviation or the induction of other regulatory mechanisms. In the rat model this tolerance has been investigated using synthetic peptides of MBP, MBP68–86, 87–99, and 110–128. Nasal administration of the encephalitogenic MBP68–86 or 87–99 suppresses EAE. MBP68–86 and 87–99 given together had synergistic effects in suppressing EAE and reversed ongoing EAE. A problem, however, of antigen-specific therapy by the nasal route is that one antigen, or peptide, may be effective in inducing tolerance in one strain of animal but not in another. One way of treating ongoing EAE may be the use of an altered peptide ligand with high tolerogenic efficacy when administered nasally.[51]

Cytokines have been widely used in disease prevention and treatment. Cytokine immunotherapy in MS could employ one or two basic strategies: first, to administer immune response downregulatory cytokines, or second, to administer inhibitors of proinflammatory cytokines. The nasal route of administering these cytokines has been studied in the rat EAE model. Nasal administration of low doses of antiinflammatory or regulatory cytokines such as IL-4, IL-10, or tumor growth factor (TGF)-β1 inhibits development of rat EAE when given before or on the day of immunization, but by differing mechanisms. Nasally administered IL-10 reduced both peripheral immune responses and microglia activation in the CNS, whereas nasal administration of IL-4 or TGF-β1 triggered the activation of dendritic cells (DCs).

However, nasal administration of cytokines alone fails to treat ongoing Lewis rat EAE. Interestingly, nasal administration of MBP68–86 + IL-4 or MBP68–86 + IL-10 suppresses ongoing EAE in Lewis rats. The suppression of EAE by MBP68–86 + IL-10 is associated with the induction of a broad lymphocyte hyporesponsiveness. Although this combined administration of autoantigen plus cytokine may be effective in treating rat EAE, the applicability of this to human MS is severely limited by the lack of knowledge of the pathologically relevant autoantigen(s) in MS.

DCs not only activate lymphocytes, but also induce T cell tolerance to antigens.[52] Use of tolerogenic DCs is thus a possible immunotherapeutic strategy for treatment of EAE, and indeed this has been studied in some detail. However, MBP68–86-pulsed DCs only prevented the development of EAE and failed to treat ongoing EAE in Lewis rats.[53] In an attempt to treat ongoing EAE, splenic DCs have been isolated from healthy Lewis rats and modified *in vitro* with cytokines IFN-β, IL-2, IL-10, or TGF-β1. Upon subcutaneous injection into Lewis rats on day 5 pi with MBP68–86 + FCA, IFN-β or TGF-β1-modified DCs promoted immune protection from EAE.

EXPERIMENTAL ALLERGIC ENCEPHALOMYELITIS IN NONHUMAN PRIMATES

The common marmoset *Callithrix jacchus* is an outbred species characterized by a naturally occurring bone marrow chimerism. The marmoset is a primate phylogenetically close to humans, and has been studied as an animal model for MS.[54] EAE can be induced in the common marmoset by the injection of human brain white matter, dispersed in demineralized water to a concentration of 30 mg/ml and emulsified with CFA containing 0.5 mg/ml of *Mycobacterium butyricum* H37A. Monkeys are injected intracutaneously with 600 μl of emulsion into the dorsal skin at several locations. Clinical disease in this model is scored daily on a scale from 0 to 5: 0 = no clinical signs; 0.5 = apathy, loss of appetite, and an altered walking pattern without ataxia; 1 = lethargy and/or anorexia; 2 = ataxia; 2.5 = paraparesis or monoparesis and/or sensory loss and/or brainstem syndrome; 3 = paraplegia or hemiplegia; 4 = quadriplegia; and 5 = spontaneous death attributable to EAE.[55] Here the onset of disease, as measured by clinical scores, is variable among animals between 7 and 13 weeks postinoculation. Additionally, the maximal clinical scores are variable among animals and range between 2 and 4. On histopathological examination, large plaques of demyelination are observed in the white matter of cerebral hemispheres, mainly localized around the wall of lateral ventricles, in the hemispheric white matter, corpus callosum, optic nerves, and optic tracts. The demyelinated areas show a moderate or severe degree of inflammation characterized by perivascular cuffs of mononuclear cells. In the spinal cord, widespread demyelination is also observed. Areas of demyelination involve the ventral, lateral, and dorsal columns of the spinal cord, especially in the outer part of the spinal tracts. Thus, pathology in the marmoset model is characterized by inflammation, demyelination, and astrogliosis. Interestingly, this model demonstrates the presence of axonal damage in demyelinating plaques. Indeed, axonal damage and loss are well-known events in MS. In MS, axonal damage appears to be an early event, related to an acute inflammation. In the marmoset EAE, axonal damage also occurs in areas of acute and early inflammation and demyelination. This EAE in *C. jacchus* is of special interest because of the resemblance of this model to the human disease, and the similarity between the immune systems of marmosets and humans.

The type of clinical signs of EAE in marmosets depends largely on the antigens used for disease induction. Sensitization of marmosets to human myelin induces a relapsing-remitting, secondary-progressive disease course.[56] Lesions in this model represent all stages present in chronic MS. Marmosets inoculated with MBP develop only mild inflammatory disease unless *Bordetella pertussis* is used with the encephalitogen. CNS demyelination critically depends on the presence of antibodies to myelin oligodendrocyte glycoprotein (MOG), a minor CNS component. Marmosets sensitized to a chimeric protein of MBP and proteolipid protein (of myelin) develop clinical EAE only after the autoimmune reaction has spread to MOG. Marmosets immunized with recombinant human MOG 1–125 do not develop relapsing-remitting disease but only chronic-progressive disease.[57] During the asymptomatic phase of this primary progressive-like disease, which can last from 2 to 20 weeks, brain lesions are detectable using magnetic resonance imaging (MRI), but are not expressed clinically.

The induction of EAE with MBP or white matter tissue homogenate (WMH) has been well established in rhesus monkeys (*Macaca mulatta*). The rhesus monkey was the first animal species in which EAE was deliberately induced.[58] That autoimmunity to brain antigens could induce paralytic disease was confirmed by studies in rhesus monkeys given repeated inoculations of brain homogenates.[58] MOG-induced EAE has also been produced in this nonhuman primate species[59] that is a highly relevant model for the human disease. The close similarity of the human and rhesus monkey immune system is illustrated by the high degree of similarity between the polymorphic MHC and T cell receptor genes between these two primates. To produce this MOG-induced EAE, monkeys are injected, under anesthesia, with a total of 1 ml of 1:1 emulsion composed of 320 μg MOG in PBS and CFA at 10 sites into the dorsal skin. Overt clinical signs are scored daily according to the following criteria: (0) no clinical signs; (0.5) loss of appetite, apathy, and altered walking; (1) lethargy, anorexia, substantial reduction of the general condition, and loss of tail tonus; (2) ataxia, tail biting, sensory loss, and/or blindness; (2.5) incomplete paralysis of one (hemiparesis) or two sides (paraparesis); (3) complete paralysis of one (hemiplegia) or two sides (paraplegia); (4) complete paralysis (quadriplegia); and (5) death. The onset of clinical disease varies between animals, and occurs at days 15–23 after encephalitogenic challenge. All monkeys, however, develop clinical disease and all achieved a score of 4 on clinical severity.

The current available panel of nonhuman primate EAE models may reflect the spectrum of inflammatory demyelinating diseases in the human population. These EAE models can therefore be used to investigate pathogenic mechanisms and to develop more effective therapies.

EXPERIMENTAL ALLERGIC ENCEPHALOMYELITIS IN MICE

The most widely studied animal model of EAE is that of the mouse. In common with other animal models of EAE, disease induction varies depending on both the sex of the animals, the mouse strain used, as well as the origin of the spinal cord encephalitogen. In this model mice, aged 6–8 weeks, are immunized subcutaneously in four sites over the back with 200–400 μg of guinea pig MBP emulsified in equal volumes of CFA containing 200–400 μg heat-killed *Mycobacterium tuberculosis*.[60] Mice also receive 200 μg of pertussis toxin in 0.2 ml PBS intraperitoneally at the time of immunization and 48 h later. Mice are then scored daily for clinical signs of EAE for at least 35 days as follows: 0, no clinical signs; +1, limp tail or waddling gait with tail tonicity; +2, ataxia or waddling gait with tail limpness; +3, partial hindlimb paralysis; +4, total hindlimb paralysis; and 5, moribund/death. For each strain of mice there is variation in day of onset of disease, varying from day 7 to day 22 postinfection; incidence of disease, varying from 30% to 100% of animals; incidence of mortality, varying from 0% to 40% of animals; and mean clinical scores, varying from 0 to 1.6. Many mouse strains have been employed in the study of EAE, and while the SJL strain has been most frequently used to model gender differences in both disease onset and severity, the SJL model has some limitations due to its diminished $CD4^+$ T cell repertoire. Certain susceptible strains of mice, such as FVB mice, show a relapsing-remitting course of disease that bears some resemblance to MS. FVB mice therefore may serve as a mouse strain into which various transgenes may be introduced for the purpose of studying their influence on EAE and for exploring new therapeutic approaches.[61]

Since EAE is a well-studied disease in mice, mimicking many clinical and pathological features of MS, including CNS inflammation and demyelination, it is of significance that it can also be used as an appropriate model to study MS-related pain. It has been clearly demonstrated in SJL that in both "active" and "passive" EAE, there is an initial increase in tail withdrawal latency (hypoalgesia) that peaked several days prior to the peak in motor deficits during the acute disease phase. During the chronic disease phase, tail withdrawal latencies decreased and were significantly faster than control latencies for up to 38 days postimmunization. Thus, it is possible to use both murine active and passive EAE as models for MS-related pain.[62]

While specific immunotherapeutic strategies are effective in experimental model systems, translation to the human disease has genetically been poorly tolerated or has proved to be ineffective. This conflict may in part be due to the model systems used as well as the poor correlation of *in vitro* findings compared to those observed *in vivo*. In Biozzi ABH mice, which express the novel MHC class II A, EAE occurs following immunization with myelin proteins and peptide epitopes of these proteins; however, only PLP peptide 56–70, MOG peptide 8–21, or spinal cord homogenate reproducibly induces chronic relapsing EAE (CREAE) with inflammation and demyelination.[63] CREAE provides a well-characterized reproducible system to develop therapeutic strategies during established relapsing autoimmune neurological disease and is pertinent to MS. In CREAE in ABH mice, relapse and progression of disease are associated with emergence and broadening of the immune repertoire due to release of myelin antigens following myelin damage.[64] Thus, this CREAE model in Biozzi ABH mice is very well suited as a model with which to examine the effect of therapeutic strategies in a dynamic system.

Disease susceptibility in human MS is associated with three MHC class II alleles in the HLA-DR2 haplotype, DRB1*1501, DRB5*0101, and DQB1*0602.[65] An autoimmune pathogenesis has been hypothesized in which one or more of these MHC class II molecules presents CNS-derived self-antigens to autoaggressive $CD4^+$ T cells, which infiltrate the CNS initiating an inflammatory response. However, the target autoantigens in MS are unknown. Immunization of mice with myelin or other brain-associated proteins induces EAE, a disease resembling MS both clinically and pathologically. The proteins, MBP, PLP, and MOG, components of the myelin sheath, are candidate antigens.[66] Indeed, T cells that are reactive to these antigens have been demonstrated in MS patients.[67] Mice expressing the human HLA-DR2 (DRB1*1501) molecule are capable of presenting peptides from all these three MS candidate autoantigens. It is possible in MS that while T cells responding to one of these antigens may initiate the disease, epitope spreading and the recruitment of T cells with additional specificities, as the disease progresses, could lead to inflammatory responses to several proteins resulting in an escalation of the autoimmune response.[68]

Transgenic mouse models of multiple sclerosis are now well established. The following are two examples of such transgenic models. First, MS is associated with HLA class II molecules HLA-DR2, -DR3, and -DR4. In humans it is difficult to analyze the individual roles of HLA molecules in disease pathogenesis due to the heterogeneity of MHC genes, linkage disequilibrium, the influence of non-MHC genes, and the contribution of environmental factors. However, the specific roles of each of these class II molecules can be addressed using transgenic models expressing these HLA genes. This model could prove useful in deciphering the role of HLA molecules and autoantigens in MS.[69] Second, while EAE has been a valuable model for the immunopathogenesis of MS, it has sometimes been difficult to reconcile the findings and therapies in the rodent models and the cellular and molecular interactions that can be studied in human disease. Humanized transgenic mice offer a means of achieving this, through the expression of disease-implicated HLA class II molecules, coexpressed with a cognate HLA-class II-restricted, myelin-specific T cell receptor derived from a human T cell clone implicated in disease. Such transgenic mice could provide an excellent model for studying epitope spreading in a humanized immunogenetic environment and for testing of immunotherapies.[70]

APPLICABILITY OF EXPERIMENTAL ALLERGIC ENCEPHALOMYELITIS MODELS OF MULTIPLE SCLEROSIS

The majority of the current therapies being planned for phase II and III trials in MS were first examined in EAE. Thus a particularly pertinent question is whether EAE is a suitable relevant research tool for MS? Some researchers believe that while EAE is a useful model of acute human CNS demyelination, its contribution to the understanding of MS is limited.[71] EAE is an acute monophasic illness, as compared to MS, which is a chronic relapsing disease, and may be more suited as a model of acute disseminated encephalomyelitis (ADEM). Drawbacks of the EAE model include the following: (1) the nature of the inflammatory response in EAE as compared to MS; (2) Th-1-mediated disease in EAE as compared to MS; (3) differences in the pathology between EAE and MS; and (4) pitfalls in extending immunotherapies from EAE to MS (see Table 69–3).

Consequently it may be concluded that the clinical picture of EAE presented depends not only on the animal species used, but also on the route of administration of the encephalitogen and the nature of the encephalitogen, MBP, PLP, or MOG. It is thus possible that these EAE models are somewhat imprecise methods to study the pathogenesis of MS or to develop therapeutic strategies.

The nonhuman primate EAE models are of primary importance for the safety and efficacy of testing new therapeutics for

Table 69–3
A comparison of some of the characteristics of EAE and MS[a]

Characteristics	*EAE*	*MS*
Inflammatory response	$CD4^+$ MBP-reactive T cells in perivascular lesions	Predominantly $CD8^+$ T cells in lesions, whereas $CD4^+$ T cells are infrequent
Th-1 or Th-2-mediated disease	Adoptive transfer of EAE by Th-1 cells	IFN-γ (Th-1 cytokine) seen in MS lesions
	Not true in all EAE models since MBP-reactive Th-2 cell clones can also cause EAE	No clear cytokine preponderance
Fundamental differences in pathology		
Location of pathology	Perivenous sleeves of myelin loss in spinal cord and brain	Demyelination not restricted to perivenous regions of white matter. Extensive demyelination of cerebral cortex
Location of lesions	Dependent on encephalitogen used; MBP-and PLP-induced EAE shows inflammation in lumbar spinal cord, MOG-induced EAE shows inflammation in the brainstem	Periventricular, brainstem, optic nerves, upper cervical cord
CSF immunology	Antibodies to myelin antigens in CSF	Antibodies to myelin antigens in CSF infrequent

continued

Table 69–3
(continued)

Characteristics	*EAE*	*MS*
Effect of immunotherapies		
IFN-γ	Variable and depends on route of administration	Worsens the inflammatory lesions?
IFN-β	Variable and depends on the route of administration	Decreases relapse rate
Anti-TNF Ab	Reverses EAE	Worsens MS
Anti-CD4	Cures EAE	No evidence of clinical efficacy on relapse or progression

Source: Adapted from Siriam and Steiner.[70]

[a]EAE, experimental allergic encephalomyelitis; MS, multiple sclerosis; MBP, myeloid basic protein; IFN, interferon; TNF, tumor necrosis factor; PLP, proteolipid protein; MOG, myeloid oligodendrocyte glycoprotein; CSF, colony-stimulating factor.

MS that may not work sufficiently well in species distant from humans, such as rodents. Questions concerning the immunogenicity of biological therapeutics have also been addressed in nonhuman primates. Many biological therapeutics, such as anti-CD4 antibodies[72] and altered peptide ligands,[73] have been investigated in rodents. Although some of these therapeutics have been effective in treating EAE in rodents, they have proven to be partially effective, or in some cases detrimental, in MS patients.[74] This ultimately raises the question of whether rodent models are the appropriate animal models for testing new therapeutic strategies for use in human MS.

CHEMICALLY INDUCED DEMYELINATION

There are several examples, in humans and animals, of demyelinating diseases not associated with viral infections, such as demyelination associated with vitamin deficiency or toxins. Many different animal models of EAE have been studied using various MRI techniques.[75] The clinical features of such models depend greatly upon the route of inoculation of the encephalitogen as well as the species and strain of animal used. Inoculation routes such as subcutaneous, footpad, or intraperitoneal are not helpful in determining the onset or location of the lesion in the brain or spinal cord. Thus, to create demyelinating lesions of precisely known locations and time courses, stereotaxic techniques are used to inoculate animals with chemicals that induce demyelinating lesions in the brain. Several chemicals, such as ethidium bromide, cuprizone, and lysophosphatidylcholine (LPC), when injected directly into nerves or into the CNS, produce lesions of demyelination. For demyelination studies with LPC, male Wistar rats are anesthetized with sodium pentathal and fixed in a rat head restraining stereotaxic surgical table, head shaved, a burr hole created, and 0.2 μl of a 1% LPC solution in isotonic saline injected using an injector cannula. Then LPC is infused at the rate of 0.05 μl/min for the next 4–5 min. The cannula is then removed and the burr hole closed using bone wax. Rats are then observed daily and histological studies are carried out from day 3 to day 15 after LPC injection to cover the entire process of disease evolution.[76]

Using this LPC-induced demyelination it is possible to observe the complete pathological process of demyelination and remyelination in this animal model of MS. Demyelination can be observed with the maximum value occurring on day 10. After day 10, remyelination starts with a reduction in edema. This model could be particularly useful for studying remyelination. One prominent feature of all chemically induced lesions is that the demyelinating lesions, and subsequent remyelination, can be studied without the interference of immune mechanisms. This has a tremendous advantage over virally induced models of MS: since no virus was inoculated none can remain to affect the remyelination.

ACKNOWLEDGMENT

The authors would like to thank Dana Parks for expert secretarial assistance with this manuscript.

REFERENCES

1. Hart BA, Amor S. The use of animal models to investigate the pathogenesis of neuroinflammatory disorders of the central nervous system. *Curr Opin Neurol* 2003;16:375–383.
2. Dyment DA, Cader MZ, Willer CJ, Risch N, Sadovnick AD, Ebers GC. A multi-generational family with multiple sclerosis. *Brain* 2002;125:1474–1482.
3. Bulman D, Ebers G. The geography of multiple sclerosis reflects genetic susceptibility. *J Trop Geogr Neurol* 2002;2:66–72.
4. Fazakerley JK, Buchmeier MJ. Pathogenesis of virus-induced demyelination. *Adv Virus Res* 1993;42:249–254.
5. Villarreal D, Young CR, Storts R, Ting JW, Welsh CJ. A comparison of the neurotropism of Theiler's virus and poliovirus in CBA mice. *Microb Pathog* 2006;41:149–156.
6. Lipton HL. Theiler's virus infection in mice: An unusual biphasic disease leading to demyelination. *Infect Immun* 1975;11:1147–1155.
7. Welsh CJR, Blakemore WF, Tonks P, Borrow P, Nash AA. Theiler's murine encephalomyelitis virus infection in mice: A persistent viral infection of the central nervous system which induces demyelination. In: Dimmock N, Ed. *Immune Responses, Virus Infection and Disease.* Oxford, UK: Oxford University Press, 1989:125–147.
8. Welsh CJ, Mi W, Sieve A, Steelman AJ, Johnson RR, Young CR, Prentice T, Hammons A, Storts R, Welsh T, Meagher MM. The effect of restraint stress on the neuropathogenesis of Theiler's virus-induced demyelination, a murine model for multiple sclerosis. In: Welsh CJ, Meagher MW, Sternberg E, Eds. *Neural and Neuroendocrine Mechanisms in Host Defense and Autoimmunity.* New York: Springer, 2006:190–225.
9. Bureau J-F, Montagutelli X, Lefebvre S, Guenet J-L, Pla M, Brahic M. The interaction of two groups of murine genes determines the persistence of Theiler's virus in the central nervous system. *J Virol* 1992;6:4698–4704.
10. Brahic M, Bureau J-F, Michiels T. The genetics of the persistent infection and demyelinating disease caused by Theiler's virus. *Annu Rev Microbiol* 2005;59:279–298.
11. Rodriguez M, Leibowitz J, David CS. Susceptibility to Theiler's virus-induced demyelination. Mapping of the gene within the H-2D region. *J Exp Med* 1986;16:620–631.

12. Welsh CJR, Tonks P, Nash AA, Blakemore WF. The effect of L3T4 T cell depletion on the pathogenesis of Theiler's murine encephalomyelitis virus infection in CBA mice. *J Gen Virol* 1987;68:1659–1667.
13. Cash E, Bandeira A, Chirinian S, Brahic M. Characterization of B lymphocytes present in the demyelination lesions induced by Theiler's virus. *J Immunol* 1989;14:984–988.
14. Borrow P, Welsh CJR, Dean D, Tonks P, Blakemore WF, Nash AA. Investigation of the role of autoimmune responses to myelin in the pathogenesis of Theiler's virus-induced demyelinating disease. *Immunology* 1998;9:478–484.
15. Miller SD, VanDerlugt CL, Begolka WS, Pao W, Yauch RL, Neville KL, Katz-Levy Y, Carrizosa A, Kim BS. Persistent infection with Theiler's virus leads to CNS autoimmunity via epitope spreading. *Nat Med* 1997;3:1133–1136.
16. Paya CV, Patick AK, Leibson PJ, Rodriguez M. Role of natural killer cells as immune effectors in encephalitis and demyelination induced by Theiler's virus. *J Immunol* 1989;143:95–102.
17. Borrow P, Tonks P, Welsh CJR, Nash AA. The role of CD8+ T cells in the acute and chronic phases of Theiler's virus-induced disease in mice. *J Gen Virol* 1992;7:1861–1865.
18. Lyman MA, Myoung J, Mohindrum M, Kim BS. Quantitative, not qualitative, differences in CD8+ T cell responses to Theiler's murine encephalomyelitis virus between resistant C57BL/6 and susceptible SJL/J mice. *Eur J Immunol* 2004;34:2730–2739.
19. Karpus WJ, Pope JG, Peterson JD, Dal Canto MC, Miller SD. Inhibition of Theiler's virus mediated demyelination by peripheral immune tolerance induction. *J Immunol* 1995;155:947–957.
20. Deloire-Grassin MS, Brochet B, Quesson B, Delalande C, Dousset V, Canioni P, Petry KG. In vivo evaluation of remyelination in rat brain by magnetization transfer imaging. *J Neurol Sci* 2000;17: 10–16.
21. Pirko I, Ciric B, Gamez J, Bieber AJ, Warrington AE, Johnson AJ, Hanson DP, Pease LR, Macura SI, Rodriguez M. A human antibody that promotes remyelination enters the CNS and decreases lesion load as detected by T2-weighted spinal cord MRI in a virus-induced murine model of MS. *FASEB J* 2004;18:1577–1579.
22. Stohlman SA, Bergmann CC, Perlman S. In: Ahmed R, Chen I, Eds. *Persistent Viral Infections.* New York: John Wiley & Sons, Ltd., 1998:537–557.
23. Lavi E, Gilden DH, Highkin MK, Weiss SR The organ tropism of mouse hepatitis virus A59 in mice is dependent on dose and route of inoculation. *Lab Anim Sci* 1986;36:130–135.
24. Sutherland RM, Chua MM, Lavi E, Weiss SR, Paterson Y. CD4+ and CD8+ T cells are not major effectors of mouse hepatitis virus A59-induced demyelinating disease. *J Neurovirol* 1997;3:225–228.
25. Lavi E, Das Sarma J, Weiss SR. Cellular reservoirs for coronavirus infection of the brain in beta2-microglobulin knockout mice. *Pathobiology* 1989;67:75–83.
26. Lampert PW. Autoimmune and virus induced demyelinating diseases. A review. *Am J Pathol* 1978;91:176–208.
27. Lane TE, Asensio VC, Yu N, Paoletti AD, Campbell IL, Buchmeier MJ. Dynamic regulation of alpha- and beta-chemokine expression in the central nervous system during mouse hepatitis virus-induced demyelinating disease. *J Immunol* 1998;160:970–978.
28. Totoiu MO, Nistor GI, Lane TE, Keirstead HS. Remyelination, axonal sparing, and locomotor recovery following transplantation of glial-committed progenitor cells into the MHV model of multiple sclerosis. *Exp Neurol* 2004;187:254–265.
29. Powell HC, Lampert PW. Oligodendrocytes and their myelin-plasma membrane connections in JHM mouse hepatitis virus encephalomyelitis. *Lab Invest* 1975;33:440–445.
30. Wu GF, Perlman S. Macrophage infiltration, but not apoptosis, is correlated with immune-mediated demyelination, following murine infection with a neurotropic coronavirus. *J Virol* 1999;73: 8771–8780.
31. Wu GF, Dandekar AA, Pewe L, Perlman S. CD4 and CD8 T cells have redundant but not identical roles in virus-induced demyelination. *J Immunol* 2000;16:2278–2286.
32. Gandy KL, Domen J, Aguila H, Weissman IL. CD8+ T CR+ and CD8+ TcR- cells in whole bone marrow facilitate the engraftment of hematopoietic stem cells across allogeneic barriers. *Immunity* 1999; 11:579–590.
33. Dandekar AA, Anghelina D, Perlman S. Bystander CD8 T-cell-mediated demyelination is interferon-γ-dependent in a coronavirus model of multiple sclerosis. *Am J Pathol* 2004;164:363–369.
34. Smith J-P, Morris-Downes M, Brennan FR, Wallace GJ, Amor S. A role for α4-integrin in the pathology following Semliki Forest virus infection. *J Neuroimmunol* 2000;106:60–68.
35. Fazakerly JK, Pathak S, Scallan M, Amor S, Dyson H. Replication of the A7(74) strain of Semliki Forest virus is restricted in neurons. *Virology* 1993;195:627–637.
36. Kelly WR, Blakemore WF, Jagelman S, Webb HE. Demyelination induced in mice by avirulent Semliki Forest virus. II. An ultrastructural study of focal demyelination in the brain. *Neuropathol Appl Neurobiol* 1992;8:43–53.
37. Donnelly SM, Sheahan BJ, Atkins GJ. Long-term effects of Semliki Forest virus infection in the mouse central nervous system. *Neuropathol Appl Neurobiol* 1997;23:235–241.
38. Ramsay AJ, Ruby J, Ramshaw LA. A case for cytokines as effector molecules in the resolution of virus infection. *Immunol Today* 1993;14:155–157.
39. Canella B Raine CS. The adhesion molecule and cytokine profile of multiple sclerosis lesions. *Ann Neurol* 1995;37:424–435.
40. Tremain KE, Ikeda H. Physiological deficits in the visual system of mice infected with Semliki Forest virus and their correlation with those seen in patients with demyelinating disease. *Brain* 1993; 106:879–895.
41. Amor S, Scallan MF, Morris MM, Dyson H, Fazakerley JK. Role of immune responses in protection and pathogenesis during Semliki Forest virus encephalitis. *J Gen Virol* 1996;77:281–291.
42. Allen SJ, Baker D, O'Neill JK, Davison AN, Turk JL. Isolation and characterization of cells infiltrating the spinal cord during the course of chronic relapsing experimental allergic encephalomyelitis in the Biozzi AB/H mice. *Cell Immunol* 1993;146:335–350.
43. Morris MM, Dyson H, Baker, D, Harbige LS, Fazakerley JK, Amor S. Characterization of the cellular and cytokine response in the central nervous system following Semliki Forest virus infection. *J Neuroimmunol* 1997;74:185–197.
44. Klee L. Probable epitopes: Relationships between myelin basic protein antigenic determinants and viral and bacterial proteins. *Neuroinformatics* 2004;2:59–70.
45. Naidu KA, Fu ES, Prockop LP. Acute experimental allergic encephalomyelitis increases lumbar spinal cord incorporation of epidurally administered [(3)H]-D-mannitol and [(14)C]-carboxyl-inulin in rabbits. *Anesth Analg* 2002;94:208–212.
46. Cipriani B, Chen L, Hiromatsu K, Knowles H, Raine CS, Battistini L, Porcelli SA, Brosnan CF. Upregulation of group 1 CD1 antigen presenting models in guinea pigs with experimental autoimmune encephalomyelitis: An immunohistochemical study. *Brain Pathol* 2003;13:1–9.
47. Menash-Brown EPK, Shahin A, Garey LJ, Lukic ML. Neurological response after induction of experimental allergic encephalomyelitis in susceptible and resistant rat strains. *Cell Immunol* 2005;233: 140–147.
48. Jee Y, Yoon WK, Okura Y, Tanuma N, Matsumoto Y. Upregulation of monocyte chemotactic protein-1 and CC chemokine receptor 2 in the central nervous system is closely associated with relapse of autoimmune encephalomyelitis in Lewis rats. *J Neuroimmunol* 2002;128:49–57.
49. Guo L, Li Y, Lin H, Ji X, Li J, Que L, Zhang Y, Rong Y, Wang J. Evaluation of a rat model of experimental autoimmune encephalomyelitis with human MBP as antigen. *Cell Mol Immunol* 2004;5:387–391.
50. Stosic-Grujicic S, Ramic Z, Bumbasirevic V, Harhaji L, Mostarica-Stojkovic M. Induction of experimental autoimmune encephalomyelitis in Dark Agouti rats without adjuvant. *Clin Exp Immunol* 2004;136:49–55.

51. Metzler B, Wraith DC. Inhibition of experimental autoimmune encephalomyelitis by inhalation but not oral administration of the encephalitogenic peptide: Influence of MHC binding affinity. *Int Immunol* 1993;5:1159–1165.
52. Banchereau J, Steinman RM. Dendritic cells and the control of immunity. *Nature* 1998;392:245–252.
53. Link H, Xiao B-G. Rat models as tools to develop new immunotherapies. *Immunol Rev* 2001;184:117–128.
54. Uccelli A, Giunti D, Capello E, Roccatagliata L, Mancardi GL. EAE in the common marmoset Callithrix jacchus. *Int MS J* 2003;10: 6–12.
55. Mancardi G, Hart B, Roccatagliata L, Brok H, Giunti D, Bontrop R, Massacesi L, Capello E, Uccelli A. Demyelination and axonal damage in a non-human primate model of multiple sclerosis. *J Neurol Sci* 2001;184:41–49.
56. t'Hart BA, Bauer J, Muller HJ, Melchers B, Nicolay K, Brok H, Bontrop RE, Lassman H, Massacesi L. Histopathological characterization of magnetic resonance imaging-detectable brain white matter lesions in a primate model of multiple sclerosis: A correlative study in the experimental autoimmune encephalomyelitis model in common marmosets (*Callithrix jacchus*). *Am J Pathol* 1998;153: 649–663.
57. Genain CP, Hauser SL. Experimental allergic encephalomyelitis in the New World monkey, *Callithrix jacchus*. *Immunol Rev* 2001;183: 159–172.
58. Rivers TM, Sprunt DH, Berry GP. Observations on the attempts to produce acute disseminated allergic encephalomyelitis in primates. *J Exp Med* 1933;58:39–53.
59. Kerlero de Rosbo Brok HPM, Bauer J, Kaye JF, t'Hart BA, Ben-Nun A. Rhesus monkeys are highly susceptible to experimental autoimmune encephalomyelitis induced by myelin oligodendrocyte glycoprotein: Characterization of immunodominant T- and B-cell epitopes. *J Neuroimmunol* 2000;110:83–96.
60. Papenfuss TL, Rogers CJ, Gienapp I, Yurrita M, McClain M, Damico N, Valo J, Song F, Whitacre C. Sex differences in experimental autoimmune encephalomyelitis in multiple murine strains. *J Neuroimmunol* 2004;150:59–69.
61. Baker AM, Grekova M, Richert J. EAE susceptibility in FVB mice. *J Neurosci* 2000;61:140–145.
62. Aicher SA, Silverman MB, Winkler CW, Bebo BF Jr. Hyperalgesia in an animal model of multiple sclerosis. *Pain* 2004;110:560–570.
63. Heijmans N, Smith PA, Morris-Downes MM, Pryce G, Baker D, Donaldson AVJ, t'Hart B, Amor S. Encephalitogenic and tolerogenic potential of altered peptide ligands of MOG and PLP in Biozzi ABH mice. *J Neuroimmunol* 2005;167:23–33.
64. Smith PA, Heijmans N, Ouwerling B, Breji EC, Evans N, van Noort JM, Plomp A, Delarasse C, t'Hart B, Pham-Dinh D, Amor S. Native myelin oligodendrocyte glycoprotein promotes severe chronic neurological disease and demyelination in Biozzi ABH mice. *Eur J Immunol* 2005;35:1311–1319.
65. Olerup O, Hillert J. HLA class II associated genetic susceptibility in multiple sclerosis: A critical reevaluation. *Tissue Antigens* 1991;38:1–15.
66. Noseworthy JH. Progress in determining the causes and treatment of multiple sclerosis. *Nature* 1999;399:A40–47.
67. Hellings N, Baree M, Verhoeven C, D'hooghe MB, Medaer R, Bernard CC, Raus J, Stinissen P. T cell reactivity to multiple myelin antigens in multiple sclerosis patients and healthy controls. *J Neurosci Res* 2001;63:290–302.
68. Gregersen JW, Holmes S, Fugger L. Humanized animal models for autoimmune diseases. *Tissue Antigens* 2004;63:383–394.
69. Mangalam AK, Khare M, Krco C, Rodriguez M, David C. Identification of T cell epitopes on human proteolipid protein and induction of experimental autoimmune encephalomyelitis in HLA class II-transgenic mice. *Eur J Immunol* 2004;34:280–290.
70. Ellmerich S, Takacs K, Mycko M, Waldner H, Wahid F, Boyton RJ, Smith, PA, Amor S, Baker D, Hafler DA, Kuchroo VK, Altmann DM. Disease-related epitope spread in a humanized T cell receptor transgenic model of multiple sclerosis. *Eur J Immunol* 2004;34: 1839–1848.
71. Siriam S, Steiner I. Experimental allergic encephalomyelitis: A misleading model of multiple sclerosis. *Ann Neurol* 2005;58:939–945.
72. Steinman L. The use of monoclonal antibodies for treatment of autoimmune disease. *J Clin Immunol* 1990;10:30S–38S.
73. Nicholson LB, Greer JM, Sobel RA, Lees MB, Kuchroo VK. An altered peptide ligand mediates immune deviation and prevents autoimmune encephalomyelitis. *Immunity* 1995;3:397–405.
74. Hohlfeld R, Wiendl H. The ups and downs of multiple sclerosis therapeutics. *Ann Neurol* 2001;49:281–284.
75. Karlik SJ, Grant EA, Lee D, Noseworthy J H. Gadolinium enhancement in acute and chronic progressive EAE in the guinea pig. *Magn Reson Med* 1996;30:326–331.
76. Deganokar MN, Jayasundar R, Jagannathan NR. Sequential diffusion-weighted magnetic resonance image study of lysophosphatidyl choline-induced experimental demyelinating lesion: An animal model of multiple sclerosis. *J Magn Reson Imaging* 2002;16:153–159.

70 Canine and Feline Models for Cancer

Gionata De Vico and Paola Maiolino

ABSTRACT

Animal models still play a crucial role in biomedical research. Among them, spontaneous canine and feline tumors could be useful to accelerate the entry of new effective drugs into clinical practice. This is because dog and cats develop neoplastic diseases in a manner more similar to human beings than to both the *in vivo* and *in vitro* models usually employed. We will discuss the scientific foundation as well as the ethical and legal concerns related to the use of this "pet model" in comparative oncology, suggesting some "key words" as the necessary starting point for a correct approach to this. Finally, the possible use of such models for the development of effective cancer theory is also discussed.

Key Words: Spontaneous canine and feline tumors, Animal models, Ethics, Law, Cancer theory.

INTRODUCTION

In modern oncology, there is a quite evident discrepancy between the great quantity of pathogenetic knowledge accumulated and its limited translation into clinical practice and therapy of human cancers.[1–3] Usually, the time required for a new antineoplastic drug to enter into clinical oncology is estimated to be 10–15 years.[2,4] It has been suggested that this could be due in part to the complex logical sequence involved in most of the preclinical studies of new antineoplastic drugs, in which the choice of the experimental models for testing the safety and efficacy of such new therapeutics represents a critical point.[3,4]

In fact, both the *in vivo* and *in vitro* models usually employed, such as rodents and/or cell lines, display intrinsic limits related to the specific characteristics of the biological systems used, whose management is frequently very complex,[5] and whose pathology, usually induced under artificial laboratory conditions, is frequently dissimilar to the studied spontaneous human disease.[2–4]

It has therefore been suggested that clinical trials based on preclinical data obtained after a screening of animal models that develop neoplastic disease in a manner more similar to human beings could accelerate the entry of new effective drugs into clinical practice.[1,4,6]

In this regard, it is widely accepted that spontaneous canine and feline tumors could represent a unique opportunity. (1) Canine and feline genetics are closer to human genetics than that of the other animal models usually employed. (2) Dogs and cats develop tumors spontaneously after years of exposure to the same environmental oncogenic factors as the owners. (3) Canine and feline spontaneous tumors share morphological, biological, and molecular characteristics similar to human tumors (interestingly, many of the antibodies employed for human molecule identification actually cross-react with similar canine molecules). (4) The incidence of cat and dog tumors is frequently higher than human tumors, and the natural evolution of the neoplastic disease is usually more rapid. (5) The life span and the size of these animals are both sufficient to detail the safety and efficacy of single or multiple drug treatments and to perform procedures such as blood sampling, bone marrow harvests, organ biopsies, and radiation therapy in a manner similar to that in humans in subjects with no artificially altered immunity. (6) The availability of clinical and pathological material is sufficient to guarantee reliable etiopathogenetic and preclinical studies in a relatively short time.[2,5,7,8]

It has also been suggested that the involvement of pet animals in biomedical research should be accompanied by specifically developed professional, ethical, and legal arguments.[1] In this context, ethical and considerations able to unravel the conflict between the interest of the various participants involved should be implemented.[9]

In particular, in biomedical research, the concern for the concept of ethical cost in experiments involving animals has led to the general acceptance of Russell and Burch's three "Rs" (replacement, reduction, and refinement)[10]: whenever it is impossible to perform scientific tests on different nonanimal models, the number of subjects employed as well as their pain, suffering, and distress should be minimized.[8,11] However, the three Rs of Russell and Burch could be inadequate for pet trials, given the specific conditions that these involve. In fact, enrolled diseased dogs and cats should be considered primarily as animal "patients,"[1] a position rather different from that of a "mere" animal model. The management of this oncological veterinary patient could require a close and complex partnership between the owners, the veterinarians in research institutions and in public institutions, as well as selected veterinary practitioners with specific training in clinical oncology, so as to involve the greatest number of patients in the experimental trials and guarantee the adequate reliability of the results and animal well-being.[1] Therefore, different key words will have to replace the "three Rs" to address correctly the necessary social, professional, and institutional behavior. We have suggested that some of these key words could be adapted from the

From: *Sourcebook of Models for Biomedical Research*
(P. M. Conn, ed.), © 2008 Humana Press Inc., Totowa, NJ.

criteria for informed consent in veterinary practice.[1,12] Information for both the owners and veterinary physicians could play a crucial role in guaranteeing the enrollment of the greatest possible number of pets in experimental trials. Information on both the benefits and risks of the trial should be adequate, with an attempt to reach an appropriate balance between them.[13] Information should deal in particular with strict eligibility criteria to enable the physicians to identify eligible patients for a specific clinical trial and to enable the owners to identify the best trials for their pet.[1] This should lead to an awareness of the commitment and responsibility of both the owners and the veterinarians involved with regard to the animal patient's well-being during the trials. Furthermore, animal patient entry in experimental trials could frequently follow conventional treatment failure.[14] Thus, a care-based approach[15] could be a further important step in working out our responsibilities to the animal patient's welfare. Caring for these terminal veterinary patients should involve palliative or analgesic treatment to minimize the suffering and pain derived from their disease. This kind of approach based on information, responsibility, and care could, in our opinion, increase the enrollment of dogs and cats in experimental trials, by providing the owners with an ethical choice different from euthanasia.[1]

CANINE AND FELINE MODELS FOR CANCER

Several spontaneous canine and feline tumors are currently considered to be relevant models for human cancers.[1–3,7,8] Canine and feline tumors may serve either as "specific tumor type" models, that is models of the corresponding human disease, or as "molecular" models, that is models of some molecular profiles that could be targeted by therapeutics, independently of the histological tumor type.[1,2]

CANINE MODELS FOR CANCER

Canine Non-Hodgkin's Lymphoma Non-Hodgkin's lymphoma (NHL) comprises about 7–24% of all neoplasms in the dog and 83% of all hematopoietic malignancies in this species.[16] Its etiology is unknown, and middle-aged to older dogs are more frequently affected, with no sex predilection. German shepherds, boxers, basset hounds, and Saint Bernards, among other breeds, develop the disease more frequently.[17] Canine NHL equates to intermediate and high-grade NHL in humans. In the dog, NHL is generally multicentric (characterized by superficial lymphadenopathy) and has a predominantly B cell origin (70–80% of cases). Other anatomic forms of NHL, in order of decreasing prevalence, are the cranial, mediastinal, gastrointestinal, and cutaneous ones. The cutaneous form is usually epitheliotrophic and of T cell origin. Primary extranodal and epitheliotrophic forms have also been reported in intestine and bladder.[18,19] NHL of T cell origin (20–30 % of cases) generally has a poor prognosis.[20] Immune system alterations in the dog, as well as in humans, seem to be associated with a higher incidence of NHL. Like human NHL, canine NHL is very sensitive to chemotherapy and is currently used as a model for testing new chemotherapeutics and new forms of immunotherapy, as well as for the study of multiple drug resistance.[21] Furthermore, NHL has been a relevant model for (1) developing hypoxic cell markers, (2) studying the effect of whole-body hyperthermia on the pharmacokinetics of systemic chemotherapy, and (3) studying autologous bone transplantation.[3]

Soft Tissue Sarcomas Soft tissue sarcomas (STSs) are a heterogeneous group of tumors derived from a variety of mesenchymal tissues.[20] They represent about 15% of all skin and subcutaneous canine tumors, and are locally invasive (with the exception of the hemangiosarcomas) with a low-to-moderate metastatic potential (less than 20%).[21] Their etiology is unknown, but in the dog these tumors have been associated with a number of exogenous factors: parasites, radiation, trauma, and artificial orthopedic implants. Older dogs are more frequently affected, and no sex predilection is observed, except for both rhabdomyosarcomas, which occur in young dogs, and synovial sarcomas, which occur more frequently in males. STSs in dogs have a pathological appearance, clinical presentation, and behavior similar to human STSs. In humans, however, STSs seem to be more common in young subjects and metastasis would appear to be more frequent then in dogs. This is in part explained by the higher numbers of nerve sheath tumors (low metastatic rate) seen in the canine species. The relevance of this model is that canine STSs respond to radiation therapy and chemotherapy in a manner similar to human STSs.[3] Thus, canine STSs have been used to test local and whole-body hyperthermia techniques, as well as the effect of hyperthermia on the pharmacokinetics of chemotherapy, and the development of both hypoxic cell markers and imaging techniques.[2,3] Among STSs, hemangiosarcomas have a higher prevalence in dogs than in humans, but are extremely similar to the human tumor with respect to histological pattern and metastatic behavior.[2]

It is, thus, actually considered an interesting model for the treatment of metastatic disease and for developing new antiangiogenic strategies and immunotherapy.[2,3,20]

Osteosarcomas Osteosarcomas (OSAs) are the most common primary bone tumor in dogs, accounting for up to 85% of malignancies originating in the skeleton.[22] Canine OSAs develop three to four times more often in the appendicular skeleton than in the axial skeleton, being classically a cancer of large and giant breeds, and occurring in middle-aged to older animals, with a median age of 7 years.[23] OSAs are locally very aggressive and cause lysis and/or production of bone. They usually metastasize (90% of cases), mainly to the lungs, via the hematogeneous route; other bones or other soft tissues are reported sites for metastasis. Canine osteosarcoma is similar to human osteosarcoma with respect to the histology, metastatic behavior, and clinical evolution of the disease, and is considered a relevant model for immunotherapy and chemotherapy.[18] In this regard it has been demonstrated that administration of liposome-muramyl tripeptide-phosphatidylethanolamine (L-MTP-PE), a potent monocyte/macrophage activator, results in increased survival time in dogs with OSA.[3,21] These studies led to a multicenter, phase III, intergroup study involving the use of L-MTP-PE in children with OSA.[2] Furthermore, new drug delivery systems, particularly for cisplatin therapy, have been studied in dogs with OSA.[3]

Oral Malignant Melanoma Canine malignant melanoma (CMM) is a highly metastatic spontaneous tumor most commonly arising in the oral cavity, and represents about 4% of all canine tumors.[24] CMM and advanced human melanomas (HM) are diseases that are initially treated with aggressive local therapies, including surgery and/or fractionated radiation therapy, and are usually followed by a systemic metastatic disease.[25] Also, CMM and HM are chemoresistant neoplasms and share many immunological targets.[3,24] For these reasons canine oral malignant melanoma (OMM) is considered a relevant model for developing new immunotherapeutic approaches for both dogs and humans suffering from the disease.[3] In a recent study, Bergman *et al.*[25] demonstrated that human tyrosinase DNA vaccination of dogs with advanced malignant melanoma is safe and potentially active, warranting additional xenogeneic DNA vaccine investigations as Phase II studies in CMM and Phase I studies in HM.

Transitional Cell Carcinomas Transitional cell carcinomas (TCCs) are the most common urinary bladder tumor in dogs and have high metastatic potential.[26] The majority are papillary infiltrative tumors of intermediate to high grade and about 80% are aneuploid. The etiology is multifactoral. Females are more frequently affected than males and Scottish terriers develop the disease more frequently than other breeds. TCCs are localized most often in the trigone and induce partial or complete urinary tract obstruction. Canine TCCs share a histological appearance, biological behavior, and response to therapy similar to invasive human TCCs.[27] It is considered a useful model for testing new photodynamic therapy technologies and chemotherapeutic combinations, particularly for the cyclooxygenase inhibitor piroxicam.[3,28,29]

Canine Mast Cell Tumors Canine mast cell tumors (CMCTs) are one of the most common malignancies in dogs, occurring with an incidence of 16–21% in the canine population.[30] Because of their variable (frequently unforeseeable) biological behavior, most investigators regard all CMCTs as potentially malignant.[31,32] Survival time and metastasis in dogs with CMCTs have been associated with the histopathological grade of the tumor.[32] Recently, however, Maiolino *et al.*[33] and Preziosi *et al.*[34] also demonstrated that the nucleomorphometric evaluation and intratumoral vessel of CMCT tissue sections could be a useful adjunct to histopathological grading in identifying tumors at increased risk of progression, to arrive at a more objective diagnosis, and to recommend more appropriate treatment. These tumors display a molecular alteration in the c-*kit* protooncogene, whose protein product, the receptor tyrosine kinase (RTKs) KIT, plays a critical role in mast cell differentiation, proliferation, survival, and activation.[35] Mutation of exon 11 of c-*kit* occurs in 30–50% of advanced mast cell tumors (MCTs), and is related to their recurrence and metastatic potential.[35] A similar mutation in the c-*kit* protooncogene has been demonstrated in 50–90% of human gastrointestinal stromal tumors (GISTs).[35,36,37] Based on this molecular correlation, translational studies with new oncology drugs (inhibitor of RTKs) for human GISTs in canine MCTs have been proposed.[36,38]

Canine Mammary Carcinomas Canine mammary carcinomas account for 30–50% of all canine tumors. These tumors have, so far, been considered a promising specific tumor-type model, being very similar to human breast cancers because of their hormonal dependence, spontaneous development in middle-aged to older animals (higher incidence in dogs between 8 and 15 years old), metastatic behavior toward regional lymph nodes and lungs, as well as adhesion molecules and neoangiogenesis patterns.[39,40,41] Furthermore, as in human mammary carcinomas, expression of *erb-B* protooncogenes has been documented,[42] and a familiarity has been demonstrated,[43] along with the expression of BRCA genes and Rad51 proteins.[44]

Abnormalities in the nuclear DNA content have been documented both in malignant and benign canine mammary tumors, but are more frequent in humans. As in humans, nutritional factors may increase the risk of mammary tumor development. Given some molecular patterns of canine mammary tumors, it has recently been suggested that they could be used as a molecular model.[38] In this context, the dysregulation of the molecular pattern of cyclin D1 and cyclin-dependent kinases (cdks) in these tumors has recently been clarified, and the results suggest that mammary neoplasia in dogs could be a good molecular model for developing new antineoplastic strategies involving cyclin D1 and cdks.[4,45]

Canine Seminomas Spontaneous testicular tumors are common in aged dogs. The principal canine testicular tumors consist of seminomas, Sertoli cell tumors, and Leydig cell tumors, and they occur with about equal frequency.[46] Seminomas are usually solitary and unilateral, and are more frequent in the right testicle. They can coexist with Sertoli and Leydig cell tumors.[47] Despite their malignant histological appearance, in dogs, metastases occur only in a small percentage of cases. In contrast, human seminomas are the most common testicular neoplasms affecting young men and have a marked tendency to metastasize.[48] This discrepancy has been the object of two recent studies[46,49] that suggested that most canine seminomas could correspond to human spermatocytic seminomas (an uncommon histological variant in men with a low metastatic potential), thus justifying their low metastatic behavior. It will be interesting to study whether canine typical seminomas, as identified by morphological and morphometric methods in the study of Maiolino *et al.*,[46] also share a histogenetic behavior similar to that of human typical seminomas, that is, the origin from so-called *in situ* carcinoma or intratubular malignant germ cell neoplasia. If so, the canine seminomas will become a very interesting spontaneous animal model.

Canine Transmissible Venereal Tumor Canine transmissible venereal tumor (TVT) is a naturally occurring transplantable round cell tumor of histiocytic origin in dogs.[50] TVT occurs most often in young, roaming, sexually intact and active dogs, with no breed or sex predilection. It commonly affects the external genitalia and it is usually transmitted at coitus. The TVT may also involve other sites, as well as the nasal and oral cavity, and rarely subcutaneous areas, and this may also be due to autotransplantation (licking and sniffing). It has a high recurrence rate and a low metastatic rate. A viral etiology has been investigated but not verified. The TVT is antigenic in dogs and an appropriate antitumor immunological response plays an important role in inhibiting growth (can induce tumor regression) and spread of the tumor. TVT is considered a promising model for human Kaposi's sarcoma (KS). It has been suggested that analogous to TVT, KS can be transplanted by viable malignant cells during sexual intercourse.[51] Furthermore percutaneous inoculation and intraarterial transplantation of canine TVT fragments in the canine lung result in predictable patterns of tumor growth resembling the solitary pulmonary nodules and metastatic disease found in humans. Thus, this tumor is considered useful to develop a lung tumor model in a large animal to simulate human lung cancer for preclinical assessment of novel therapeutic options. A tumor model in a large animal would have a wide range of biomedical research applications, including the study of various interventional imaging techniques.[52]

FELINE MODELS

Feline Non-Hodgkin's Lymphoma Feline non-Hodgkin's lymphoma (NHL) accounts for 50–90% of all hematopoietic malignancies in the cat and for approximately one-third of all malignant neoplasms in this species.[53] The feline leukemia virus (FeLV) was the most common cause prior to the widespread use of FeLV vaccination and testing programs. Cats with a median age of approximately 10 years are predominantly affected It is primarily a B cell lymphoma (75% of cases): the mediastinal, spinal, and leukemic forms are of T cell derivation and FeLV positive; the alimentary forms are usually of B cell derivation and FeLV negative.[53] Anatomic forms traditionally associated with FeLV (mediastinal or multicentric) occur in younger cats. Male cats and Siamese/oriental breeds may be predisposed to NHL.[54] Feline immunodeficiency virus (FIV) infection can increase the incidence of lymphoma in cats (particularly the alimentary lymphoma of B cell origin) by dysregulation of the immune system or activation of oncogenic pathways. Feline NHL equates to intermediate and high-grade NHL in humans. Like human and canine NHL, feline NHL is very sensitive to chemotherapy, which is very well tolerated by cats, and it is currently used as a model for testing new alternative treatments for cancers.[53]

Feline Mammary Carcinoma Mammary neoplasia is the third most common tumor type affecting female cats.[55] The mean age of development is 10–11 years with an age-relative risk that increases up to 14 years. Domestic short-haired and Siamese cats are more frequently affected than other cats. Malignancy occurs frequently (80–96%) with high mortality and a ratio between malignant and benign neoplasms varying from 9:1 to 4:1.[56,57] Feline mammary carcinoma (FMC) shows age incidence, histopathology, and pattern of metastasis[2,3] similar to human breast cancer.[55] In particular, the lack of estrogen dependency in most tumors[4] suggests that FMC could be a suitable animal model for human hormonal unresponsive breast cancer. Furthermore, some FMC subtypes have recently been described that share common features with human inflammatory mammary carcinoma and human mammary carcinoma with osteoclast-like giant cells.[58,59] Prognostic factors are nuclear polymorphism and polymetrism, as assessed by nuclear morphometry, PCNA label index, Ag-NOR count, and mitotic index.[60–63]

From a molecular point of view, HER2 and RON overexpression qualifies FMC as homologous to the subset of HER2 overexpressing, poor prognosis human breast carcinomas and as a suitable model to test innovative approaches to therapy of aggressive tumors.[64]

Feline Vaccine-Associated Sarcoma Feline vaccine-associated sarcoma (VAS) is a biologically aggressive sarcoma that can develop in the skin of cats at sites where inactivated feline vaccines (feline leukemia virus and rabies) have been administered. The increase in the incidence of this tumor has shown a strong link between vaccination and subsequent tumor development.[65] A multistep carcinogenesis model, including genetic, iatrogenic, and local factors, seems to be the most plausible explanation for the occurrence of the tumor. Inflammation is the most accepted pathogenetic hypothesis and adjuvant, the component most commonly associated with vaccines, is considered a cause of tumor development. Among VAS, fibrosarcoma is the most common (incidence of 1–10 per 10,000 cats). The tumor displays an extremely malignant biological behavior, being both locally aggressive and metastasizing in 25–70% of the cases.

Immunohistochemical detection of p53 protein, fibroblast growth factor-β (FGF-β), tumor growth factor-α (TGF-α) and platelet-derived growth factor (PDGF) and its receptor[66] has been documented, suggesting that these growth-regulating proteins may play different roles in the development of VAS cells. Mutation in p53 has been studied in a variable number of sarcomas, but has been demonstrated in only 2 out of 10 fibrosarcomas.[67] The standard treatment is aggressive surgical removal of the tumor, but combinations of surgery, radiation, and chemotherapy [doxorubicin and decarbazine (DTIC)] are usually recommended. A significant decrease in STAT3 expression in VASs treated with doxorubicin after surgical resection has also been reported, supporting evidence for the potential role of STAT3 in oncogenesis and tumor progression.[68] The possibility of the development of fibrosarcomas at presumed sites of injection not only in cats but also in dogs has been reported.[69] It is considered a useful model of hyperthermia in cancer immunogene therapy, particularly for the cytokine genes.[70]

PET MODELS AND CANCER THEORY: COULD THEY BE USEFUL?

According to current practice, we have dealt with spontaneous canine and feline tumors as models for testing new antineoplastic strategies for human cancer. In our opinion, however, besides this "utilitarian" approach, animal tumors may also be a very useful model for searching and testing new theoretical ideas in carcinogenesis.

According to the prevailing current paradigm, the so-called somatic mutation theory (SMT), cancer derives from a single somatic cell that has accumulated multiple DNA mutations; furthermore, the default state of cell proliferation in metazoa is quiescence and cancer is considered mainly a disease caused by mutations in genes that control the cell cycle.[71,72]

However, it has been suggested that mutation alone is not sufficient to induce cancer. Accordingly, a "Darwinian" approach could represent a better theoretical framework for understanding the complex cellular, molecular, and epidemiological events leading to cancer.[73,74] According to this theory, accumulation of mutations during tumor emergence could be triggered by an adequate selective cell microenvironment, the latter being able to select mutations useful for neoplastic cells to survive and grow.

A further (and very intriguing) theoretical approach to carcinogenesis is the so-called tissue organization field theory (TOFT), which regards tumors as a disease of developmental processes.[71,72] This theory is based on premises that place cancer in a different hierarchical level of complexity from that proposed by the SMT and "Darwinian" points of view, that is (1) carcinogenesis represents a problem of tissue organization comparable to organogenesis, and (2) proliferation is the default state of all cells.[71,72]

Accordingly, neoplastic development could result from disruption of the normal dynamic interaction of neighboring cells and tissues both during early development and throughout adulthood.

From a philosophical point of view, the SMT and the "Darwinian" theories result from a reductionist approach to biology and regard tumors as a molecular or at least a cellular problem; however, according to TOFT, "development and cancer will not be reduced to complex series of protein interactions, but rather a multilevel explanation will be required. In some instances molecules will do the explanatory job, in others physical forces, but at the core they will remain a problem of tridimensional tissue organization."[71]

In this context spontaneous pet tumors could provide a novel way of thinking about these questions, which takes all levels of biological organization into consideration.

For example, canine trichoblastomas, a "benign tumor derived from or reduplicating the primitive hair germ of embryonic follicular development,"[75] seem to display all the necessary peculiarities for testing some aspects of the above cancer theories. According to their growth pattern canine trichoblastomas are subdivided into four types—ribbon type, trabecular type, granular cell type, and spindle cell type—and represent about 25% of all canine epithelial skin neoplasms.[76] These tumors are similar to human trichoblastomas, which are "not exclusively epithelial tumors, but . . . heterogeneous biological systems of epithelial–mesenchymal interaction."[77] In these neoplasms the epithelial component appears to be equivalent to the hair germ and the mesenchymal component equivalent to the dermal papilla, both essential for hair follicle development.[77] Perturbed interactions within the above tissue compartments may cause or even reflect morphofunctional alterations that eventually give rise to follicular tumors.[77,78] Interestingly, we demonstrated that the complex growth pattern of these tumors may be adequately described and quantified in mathematical terms using a morphometric approach based on fractal geometry.[79,80] Furthermore, a close correlation seems to exists between the alteration of Wnt molecular signals, cellular events occurring in early follicle development, and the emergence and growth in the above tumors (personal observations). Thus, these tumors could provide, in quantitative terms, useful information about the link between molecular, cellular, and tissue changes during tumor development. Certainly, many other spontaneous pet tumors could be useful in this context. Future efforts in this direction could provide interesting models for effective cancer theory development.

REFERENCES

1. De Vico G, Maiolino P, Restucci B, Passantino A. Spontaneous tumours of pet dog as model for human cancer: Searching for adequate guidelines. *Riv Biol* 2005;98:279–296.
2. Hansen K, Khanna C. Spontaneous and genetically engineered animal models use in preclinical cancer drug development. *Eur J Cancer* 2004;40:858–880.
3. Vail DM, Thamm DH. Spontaneously occurring tumours in companion animals as models for drug development. In: Theicher BA, Andrews PA, Eds. *Anticancer Drug Development Guide. Preclinical Screening, Clinical Trial, and Approval,* 2nd ed. Totowa, NJ: Humana Press, 2004:259–284.
4. Sfacteria A, Cervasi B, Bertani C, *et al.* Messa a punto di modelli animali per la sperimentazione di strategie antineoplastiche innovative: Il tumore mammario canino. *Chir Ital* 2001;53(6):857–868.
5. Schiffer SP. Animal welfare and colony management in cancer research. *Breast Cancer Res* 1997;46:313–331.
6. London CA, Hannah AL, Zadovoskaya R, *et al.* Phase I dose-escalating study of SU11654, a small molecule receptor tyrosine kinase inhibitor, in dogs with spontaneous malignancies. *Clin Cancer Res* 2003;9:2755–2756.
7. Kruth S. Canine models for gene therapy. *Transfus Sci* 1996;17(1): 71–77.
8. Porrello, A, Cardelli P, Spugnini P. Pet models in cancer research: General principles. *J Exp Cancer Res* 2004;23:5–17.
9. Arkow P. Application of ethics to animal welfare. *Appl Anim Behav Sci* 1998;59:193–200.
10. Russell WMS, Burch RL. *The Principles of Humane Experimental Technique*. London: Methuen, 1959.
11. Griffin G. The ethics of animal use for research purposes. *Sepsis* 1998;2:263–266.
12. Santori P, Canavacci L. Le procedure per una decisione clinica informata e responsabile. Riflessioni critiche sul cosiddetto "consenso informato" in veterinaria. In: *Documenti del Comitato Bioetico per la Veterinaria presso l'Ordine dei Medici Veterinari della Provincia di Roma*. Torino, Italy: C.G. Edizioni Medico Scientifiche, 2000.
13. Verheggen FWSM, Wijmen FCB. Informed consent in clinical trials. *Health Pol* 1996;36:131–153.
14. Winiarczyk S, Gradski Z, Kosztolich G, *et al.* A clinical protocol for treatment of canine mammary tumours using encapsulated, cytochrome P450 synthesizing cells activating cyclophosphamide: A phase I/phase II study. *J Mol Med* 2002;80:610–614.
15. Fraser D. Animal ethics and animal welfare science: Bridging the two cultures. *Appl Anim Behav Sci* 1999;65:171–189.
16. Vail DV, MacEwen EG, Young KM. Canine lymphoma and lymphoid leukemias. In: Withrow SJ, MacEwen EG, Eds. *Small Animal Clinical Oncology*, 3rd ed. Philadelphia, PA: W.B. Saunders Company, 2001:558–590.
17. MacEwen EG, Young KM. Canine lymphoma and lymphoid leukemias. In: Withrow SJ, MacEwen EG, Eds. *Small Animal Clinical Oncology*, 3rd ed. Philadelphia, PA: W.B. Saunders Company, 2001:590–611.
18. French RA, Seitz SE, Valli VE. Primary epitheliotropic alimentary T-cell lymphoma with hepatic involvement in a dog. *Vet Pathol* 1996;33(3):349–352.
19. Maiolino P, DeVico G. Primary epitheliotropic T-cell lymphoma of the urinary bladder in a dog. *Vet Pathol* 2000;37(2):184–186
20. MacEwen EG. The dog as a model for clinical trials and pharmacology in comparative oncology. In: Rossi R, Richardson R, Harshbarger J, Eds. *Spontaneous Animal Tumours: A Survey.* Proceedings of the First World Conference on Spontaneous Animal Tumours, Genoa, Italy, April 28–30, 1995:299–304.
21. Vail DM, MacEwen EG. Spontaneously occurring tumours in companion animals as models for human cancer. *Cancer Invest* 2000; 18(8):781–792.
22. Straw RC. Tumours of the skeletal system. In: Withrow SJ, MacEwen EG, Eds. *Small Animal Clinical Oncology*, 2nd ed. Philadelphia, PA: W.B. Saunders Company, 1996:287–315.
23. Slayter MV, Boosinger TR, Pool RR, *et al.* Histological classification of bone and joint tumours of domestic animals. In: *Bulletin of World Health Organization International Histological Classification of Tumours of Domestic Animals.* Washington, DC: Armed Forces Institute of Pathology, 1994.
24. MacEwen EG, Kurzman ID, Vail DM, *et al.* Adjuvant therapy for melanoma in dogs: Results of randomized clinical trials using surgery, liposome encapsulated muramyl tripeptide, and granulocyte macrophage colony-stimulating factor 1. *Clin Cancer Res* 1999;5:4249–4258.
25. Bergman PJ, McKnight J, Novosad A, *et al.* Long term survival of dogs with advanced malignant melanoma after DNA vaccination with xenogeneic human tyrosinase: A phase I trial. *Clin Cancer Res* 2003;9:1284–1290.
26. Meuten DJ. Tumors of the urinary system. In: Meuten DJ, Ed. *Tumors in Domestic Animals*, 4th ed. Ames, IA: Iowa State Press, 2002:509–546.
27. Knapp DW, Glickman NW, De Nicola DB, *et al.* Naturally-occurring canine transitional cell carcinoma of the urinary bladder. A relevant model of human invasive bladder cancer. *Urol Oncol* 2000;5:47–59.
28. Henry CJ, McCaw DL, Turnquist SE., *et al.* Clinical evaluation of mitoxantrone and piroxicam in a canine model of human invasive urinary bladder carcinoma. *Clin Cancer Res* 2003;9:906–911.
29. Mohammed, SI, Craig BA, Mutsaers AJ, *et al.* Effects of the cyclooxygenase inhibitor, piroxicam, in combination with chemotherapy on tumor response, apoptosis, and angiogenesis in a canine model of human invasive urinary bladder cancer. *Mol Cancer Ther* 2003;183(2): 183–188.
30. Thamm, DH, Vail DM. Mast cell tumors. In: Withrow SJ, MacEwen EG, Eds. *Small Animal Clinical Oncology*, 3rd ed. Philadelphia, PA: W.B. Saunders Company, 2001:261–282.
31. Bostock DE. The prognosis following surgical removal of mastocytomas in dogs. *J Small Anim Pract* 1973;14:27–41.
32. Patnaik AK, Ehler WJ, MacEwen EG. Canine cutaneous mast cell tumor: Morphologic grading and survival time in 83 dogs. *Vet Pathol* 1984;21:469–474.
33. Maiolino P, Cataldi M, Paciello O, *et al.* Nucleo-morphometric analysis of canine cutaneous mast cell tumours. *J Comp Pathol* 2005; 133(2–3):209–211.

34. Preziosi R, Sarli G, Paltrinieri M. Prognostic value of intratumoral vessel density cutaneous mast cell tumors of the dog. *J Comp Pathol* 2004;130(2–3):143–151.
35. London CA, Galli SJ, Yuuki T, *et al.* Spontaneous canine mast cell tumors express tandem duplications in the proto-oncogene c-kit. *Exp Hematol* 1999;27:689–697.
36. London, CA, Hannah AL, Zadovoskaya R, *et al.* Phase I dose-escalating study of SU11654, a small molecule receptor tyrosine kinase inhibitor, in dogs with spontaneous malignancies. *Clin Canc Res* 2003;9:2755–2756.
37. Heinrich MC, Rubin BP, Longley BJ, *et al.* Biology and genetic aspects of gastrointestinal stromal tumors: KIT activation and cytogenetic alterations. *Hum Pathol* 2002;33(5):484–495.
38. Pryer NK, Lee LB, Zadovaskaya, *et al.* Proof of target for SU11654: Inhibition of KIT phosphorylation in canine mast cell tumours. *Clin Cancer Res* 2003;9:5729–5734.
39. Sorenmo K. Canine mammary gland tumours. *Vet Clin Small Anim* 2003;33:573–596.
40. Restucci, B, De Vico G, Maiolino P. Expression of b1 integrin in normal, displastic and neoplastic canine mammary gland. *J Comp Pathol* 1995;113:165–173.
41. Restucci, B, De Vico G, Maiolino P. Evaluation of angiogenesis in canine mammary tumors by quantitative platelet endothelial cell adhesion molecule immunohistochemistry. *Vet Pathol* 2000;37(4): 297–301.
42. Matsuyama, S, Nakamura M, Yonezana K, *et al.* Expression pattern of the erb-b subfamily mRNA in canine benign and malignant mammary tumours. *J Vet Med Sci* 2001;63(9):949–954.
43. Schafer KA, Kelly G, Schrader R, *et al.* A canine model of familial mammary gland neoplasia. *Vet Pathol* 1998;35:168–177.
44. Ochiai K, Morinatsu M, Tomizawa N, *et al.* Cloning and sequencing full length of canine Brac2 and Rad51 cDNA. *J Vet Med Sci* 2001;63: 1103–1108.
45. Sfacteria A, Bertani C, Costantino G, *et al.* Cyclin D1 expression in pre-cancerous and cancerous lesions of the canine mammary gland. *J Comp Pathol* 2003;128(4):245–251.
46. Maiolino P, Restucci B, Papparella S, *et al.* Correlation of nuclear morphometric features with animal and human World Health Organization International histological classifications of canine spontaneous seminomas. *Vet Pathol* 2004;41:608–611.
47. Nielsen, SW, Kennedy PC. Tumours of the genital system. In: Moulton DJ, Ed. *Tumors of Domestic Animals*, 4th ed. Ames, IA: Iowa State University Press, 1990:479–491.
48. Kennedy PC, Cullen JM, Edwards JF, *et al. Histological Classification of the Tumours of the Genital System of Domestic Animals*, second series, Vol. IV. Washington, DC: Armed Forces Institute of Pathology American Registry of Pathology, 1998.
49. Looijenga LHJ, Olie RA, van Der Gaag I, *et al.* Seminomas of the canine testis. Counterpart of spermatocytic seminomas of men? *Lab Invest* 1994;74:490–496.
50. MacEwen EG. Canine transmissible venereal tumor. In: Withrow SJ, MacEwen EG, Eds. *Small Animal Clinical Oncology*, 3rd ed. Philadelphia, PA: W.B. Saunders Company, 2001:651–656.
51. Rechavi G, Katzir N, Ramot B. Kaposi's sarcoma among AIDS patients: Transmissible venereal tumour by cell engraftment? *Med Hypotheses* 1991;34(4):338–341.
52. Rivera B, Ahrar K, Kangasniemi MM, *et al.* Canine transmissible venereal tumor: A large-animal transplantable tumor model. *Comp Med* 2005;55(4):335–343.
53. Vail DM, MacEwen EG. Feline lymphoma and leukemias. In: Withrow SJ, MacEwen EG, Eds. *Small Animal Clinical Oncology*, 3rd ed. Philadelphia, PA: W.B. Saunders Company, 2001:590–611.
54. Gabor LJ, Malik R, Canfield PJ. Clinical and anatomical features of lymphosarcoma in 118 cats. *Aust Vet J* 1998;76(11):725–732.
55. Zappulli V, De Zan G, Cardazzo B, *et al.* Feline mammary tumours in comparative oncology. *J Dairy Res* 2005;72(Special Issue):98–106.
56. Hayes HM, Milne KL, Mandell CP. Epidemiological features of feline mammary carcinoma. *Vet Rec* 1981;108:476–479.
57. Misdorp W, Romijn A, Hart AA. Feline mammary tumors: A case-control study of hormonal factors. *Anticancer Res* 1991;11:1793–1797.
58. Pérez-Alenza MD, Jiménez A, Nieto AI, Peña L. First description of feline inflammatory mammary carcinoma: Clinicopathological and immunohistochemical characteristics of three cases. *Breast Cancer Res* 2004;6:R300–R307.
59. De Vico G, Maiolino P. "Osteoclast-like" giant cells in a feline mammary carcinoma. *J Submicrosc Cytol Pathol* 1997;29:153–155.
60. De Vico G, Maiolino P. Prognostic value of nuclear morphometry in feline mammary carcinomas. *J Comp Pathol* 1997;117:99–105.
61. Castagnaro M, Casale C, Ru G, *et al.* Argyrophilic nucleolar organiser regions (AgNORs) count as indicator of post-surgical prognosis in feline mammary carcinomas. *Res Vet Sci* 1998;64:97–100.
62. Castagnaro M, De Maria R, Bozzetta E, *et al.* Ki67 index as indicator of the post-surgical prognosis mammary carcinomas. *Res Vet Sci* 1998;65:223–226.
63. Preziosi R, Sarli G, Benazzi C, *et al.* Multiparametric survival analysis of histological stage and proliferative activity in feline mammary carcinomas. *Res Vet Sci* 2002;73:53–60.
64. De Maria R, Olivero M, Iussich S, *et al.* Spontaneous feline mammary carcinoma is a model of HER2 overexpressing poor prognosis human breast cancer. *Cancer Res* 2005;65(3):907–912.
65. Hendrick MJ, Goldschmidt MH. Do injection site reactions induce fibrosarcomas in cats? *J Am Vet Med Assoc* 1991;199(8):968.
66. Nieto A, Sanchez MA, Martinez E, *et al.* Immunohistochemical expression of p53, fibroblast growth factor-b, and transforming growth factor-alpha in feline vaccine-associated sarcomas. *Vet Pathol* 2003;40(6):651–658.
67. Mayr B, Wegscheider H, Rei¢nger M, *et al.* Cytogenetic alterations in four feline soft-tissue tumours. *Vet Res Commun* 1998;22(1):21–29.
68. Petterino M, Martano P, Cascio F, *et al.* Immunohistochemical study of STAT3 expression in feline injection-site fibrosarcomas. *J Comp Pathol* 2006;134:91–100.
69. Vascellari E, Melchiotti M, Bozza A, *et al.* Fibrosarcomas at presumed sites of injection in dogs: Characteristics and comparison with non-vaccination site fibrosarcomas and feline post-vaccinal fibrosarcomas. *J Vet Med A Physiol Pathol Clin Med* 2003;50:286–291.
70. Siddiqui F, Li CY, Zhang X, Larue SM, *et al.* Characterization of a recombinant adenovirus vector encoding heat-inducible feline interleukin-12 for use in hyperthermia-induced gene-therapy. *Int J Hyperthermia* 2006;22(2):117–134.
71. Soto AM, Sonnenschein C. Emergentism as a default: Cancer as a problem of tissue organization. *J Biosci* 2005;30:101–115.
72. Soto AM, Sonnenschein C. The somatic mutation theory of cancer: Growing problems with the paradigm? *BioEssays* 2004;26:1097–1107.
73. Vineis P, Matullo G, Manuguerra M. An evolutionary paradigm for carcinogenesis? *J Epidemiol Community Health* 2003;57:89–95.
74. Garte S. Theory in carcinogenesis and epidemiology *J Epidemiol Community Health* 2003;57:85.
75. Goldschmidt MH, Dunstan RW, Stannard AA, *et al.* Epithelial and melanocytic tumours of the skin of domestic animals. In: *World Health Organization International Histological Classification of Spontaneous Animal Tumours*, second series, Vol. III. Washington, DC: Armed Forces Institute of Pathology American Registry of Pathology, 1998:22–23.
76. Abramo F, Pratesi F, Cantile C, *et al.* Survey of canine and feline follicular tumours and tumour-like lesions in central Italy. *J Small Anim Pract* 1999;40(10):479–481.
77. Headington JT. Differentiating neoplasms of hair germs. *J Clin Pathol* 1970;23:464–471.
78. Millar S. Molecular mechanisms regulating hair follicle development. *J Invest Dermatol* 2002;118:216–225.
79. Mandelbrot BB. *The Fractal Geometry of Nature*. San Francisco, CA: Freeman, 1983.
80. De Vico G, Cataldi M, Maiolino P, *et al.* Fractal analysis of canine trichoblastoma. In: Losa GA, Merlini D, Nonnenmacher TF, Weibel RE, Eds. *Fractals in Biology and Medicine*, Vol. IV. Basel, Switzerland: Birkhauser Verlag, 2005:203–207.

71 Obese Mouse Models

DEBORAH J. GOOD

ABSTRACT

In 1902, the *lethal yellow* or *Agouti yellow* (A^y) mouse was described by French geneticist Lucien Cuenot in a paper detailing coat-color inheritance in mice. Although the Agouti mouse reportedly had been bred by European mouse fanciers since the 1800s, the phenotype of obesity was not published until 1927. In the late 1960s, the obese (*ob/ob*) and diabetic (*db/db*) mouse models were described. All three of these models have been widely used in obesity research since that time. With the advent of technologies to create transgenic and knockout strains of mice there has been a rapid increase in the number of mice displaying different defects in body weight regulation. This chapter will categorize the more than 200 mouse models of body weight disorders that now exist, more than 100 years after the *agouti yellow* mouse was first characterized. The Obesity Gene map database (http://obesitygene.pbrc.edu/) is updated yearly and is a thorough reference for all genes and loci, both rodent and human, known to modulate body weight or energy utilization. This chapter categorizes mouse models listed in that database, and includes recently published models in five groupings: juvenile-onset obese models, adult-onset obese models, obesity-resistant models, polygenic obese models, and models with altered responses to induced obesity.

Key Words: Obesity, Overweight, Mouse models, Energy homeostasis, Anorexia.

INTRODUCTION

The incidence of obese and overweight humans worldwide has skyrocketed in the past two decades. Likewise, the number of published papers describing mouse models used for the study of body weight regulation also has also increased during that time period[1–5] (Figure 71–1). With strong protein sequence homologies between mouse and human gene products, many mouse models replicate human disorders of body weight regulation, and some are identical to known human genetic disorders.[6] The large numbers of available mouse models recapitulate the different phenotypes of both obesity and anorexia, including juvenile-onset obesity, adult-onset obesity, and obesity-resistant models. Mouse models of both monogenic and polygenic obesity are being used to uncover new genetic pathways that converge to maintain energy balance homeostasis.

MODELS OF JUVENILE-ONSET OBESITY

Juvenile-onset obesity in mammals is characterized by increased body weight that occurs prior to the onset of sexual maturity. In mice, sexual maturity occurs between 7 and 8 weeks of age.[7] Therefore, models listed as having juvenile-onset obesity will be limited to those that show distinct body weight differences prior to 8 weeks of age. A total of 34 mouse models can be categorized as examples of juvenile-onset obesity (Table 71–1).

Polymorphisms in three of the four most common genes associated with human obesity, namely leptin, leptin receptor, and proopiomelanocortin (POMC),[8] have mouse model counterparts that display a juvenile-onset obesity phenotype.[9–11] These genes, along with the fourth common gene involved in monogenetic forms of obesity in humans, the melanocortin-4 receptor (MC4R) gene, constitute part of the well-characterized melanocortin pathway. The MC4R mutant mouse model displays adult-onset obesity,[12,13] while animals with mutations in both MC4R and leptin are obese at weaning—earlier than either of the single mutants.[14] These mutants of the melanocortin pathway will be useful in dissecting the role of POMC neuropeptides and their receptor on the timing of body weight gain in animals.

In addition to relatively well-characterized pathways, the list of mouse models of juvenile-onset obesity reveals some interesting gene products and signaling pathways whose specific influence on the regulation of human body weight remains unclear. For example, overexpression of a dominant negative form of N-cadherin specifically in osteoclasts leads to increased adipogenesis and body weight only in young animals.[15] The effect of the transgene on body fat is absent in adulthood with no body weight changes in age-matched adult mice. It was suggested that cell fate decisions affect the proportion of adipocytes to osteoclasts in these animals through β-catenin signaling, which leads to the differences in body composition in juveniles. This interesting role for β-catenin and N-cadherins was unknown prior to the development of the mouse model. Another interesting model is the Gnas-deficient mouse.[16] This mouse model can be used to elucidate sex-specific modes of inheritance that cause body weight gain or loss. The protein product of Gnas is a stimulatory G-protein α-subunit that binds guanine nucleotides and is involved in multiple pathways including those signaling through phospho-

From: *Sourcebook of Models for Biomedical Research*
(P. M. Conn, ed.),

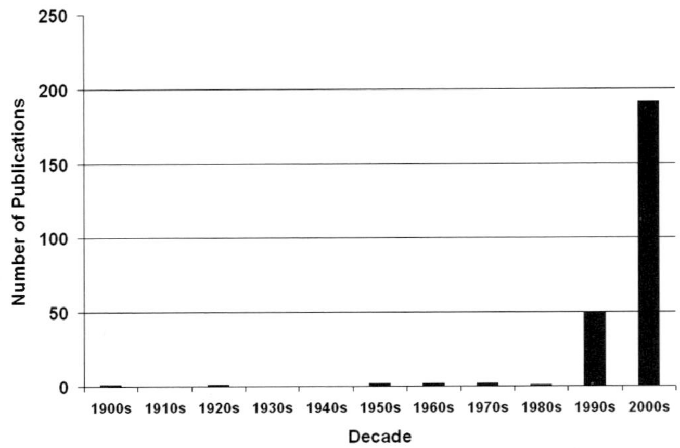

Figure 71–1. Number of publications identifying new mouse models of body weight regulation compiled by decade. References used in this chapter that describe new mouse models of obesity, anorexia, obesity resistance, or animals with altered response to diet, and genetically or chemically induced obesity were grouped by decade.

lipase C and adenylate cyclase. In this mouse model, maternal inheritance of the mutant allele leads to body weight gain by 30–40 days, while paternal inheritance has the opposite effect of reduced body weight. There is no difference in food intake between those that inherit the allele maternally or those that inherit it paternally, suggesting that the difference lies in energy utilization pathways. Humans with Albright hereditary osteodystrophy (AHO) syndrome, which is linked to mutations in GNAS,[17] display the phenotypes of obesity, skeletal defects, and short stature, similar to Gnas-deficient mice. Individuals with maternal inheritance of the mutation also display multihormone resistance.[16] The mouse model therefore resembles the human syndrome to some extent, and offers some unique insight into the role of G-proteins in sex-specific modes of inheritance that specify differences in body weight regulation and adipocyte differentiation. These are just some examples of the numerous types of models available to study aspects of juvenile-onset obesity and the possibility of extending these findings to studies of human obesity.

Table 71–1
Models of juvenile-onset obesity

Model	*Genetics*	*Phenotype*
Agouti yellow[41]	Dominant spontaneous mutation in the agouti protein	Obesity, diabetics, and some tumor formation
AGRP transgenic[42]	Transgenic overexpression of human agouti-related protein	Obesity without changes in pigmentation (as found in lethal yellow mouse) by 4–6 weeks of age
ASIP transgenic (3 types)[43–45]	Transgenic overexpression of agouti protein using three different promoters	Increased body fat and body mass
Alpha 2 transgenic-beta 3 AR deficient[46]	Transgenic containing overexpression of α_2- and deletion of β_3-adrenergic receptors in adipose tissue	Obesity due to adipocyte hyperplasia
ApoE$^{-/-}$; db/db double knockout mice[47]	Deletion of both leptin receptor and apolipoprotein E genes	Obesity similar to db/db mice with atherosclerosis
BAT ablated[48]	Transgenic expression of diphtheria toxin from the uncoupling protein 1 promoter, leading to ablation of all brown fat	Obesity by 6 weeks of age with hyperphagia
CDKI p21 deficient[49]	Deletion of the p21 subunit of cyclin-dependent kinase inhibitor (Cdkn1a) gene	Mild obesity due to adipocyte hyperplasia
CDKI p27 deficient[49]	Deletion of the p27 subunit of cyclin-dependent kinase inhibitor (Cdkn1b) gene	Mild obesity due to adipocyte hyperplasia
CDKI p21/p27 double knockout[49]	Deletion of both subunits of CDKI	Increased body fat and adipocyte numbers by 6 weeks of age
CRF transgenic[50]	Overexpression of the corticotrophin-releasing factor gene	Cushing's syndrome phenotype, including large adipose tissue deposits
Db (diabetic obese)[9,51]	Spontaneous mutation of the leptin receptor	Obesity with diabetes evident around 4 weeks of age
DRD3 deficient[52]	Deletion of the dopamine-3 receptor gene	Males show obesity on a high fat diet with normal weight on normal chow; females demonstrate no change in body weight, but increased body fat on either diet
Gal/NPY deficient[53]	Deletion of both galanin and NPY genes	Significantly increased body weight by 1 week of age and 30% increase by 24 weeks of age; hyperphagia is present
GASKO mice[54]	Deletion of the gastrin gene	Increased body fat by 2 months of age with hyperinsulinemia by 1 month of age; body weight is not visibly increased until 7 months of age; food intake is reduced

Table 71–1 (continued)

Model	*Genetics*	*Phenotype*
Gnas deficient (maternal inheritance)[16]	Deletion of the maternally inherited copy of the G-protein α-subunit; note that deletion of the paternally inherited allele has the opposite phenotype (see Table 71–3)	Mice with the maternally inherited mutant allele show increased body weight by 30–40 days of age, reduced metabolic rate, but no change in food intake relative to control animals
GSK-3βtransgenic[55]	Transgenic overexpression of human glycogen synthase kinase 3β gene by a muscle-specific promoter	Significant increase in body weight by 6 weeks of age with modest gain by 28 weeks of age; modest increase in body fat
IAP deficient[56]	Deletion of the intestinal alkaline phosphatase gene	Obesity on high fat diet with normal weight on normal chow
IDPc transgenic[57]	Transgenic overexpression of the cytosolic NADP(+)-dependent isocitrate dehydrogenase gene with liver and adipose expression	Obesity by 4–5 weeks of age in male and female transgenics with no hyperphagia
Lep/MC4R deficient[14]	Double deletion of leptin and melanocortin-4 receptor genes	Mouse models show an additive effect on obesity compared to the single mutants
LepR deficient, neuronal specific[58]	Cre-lox-mediated deletion of leptin receptor gene in neuronal cells	Increased body weight beginning by 5 weeks of age accompanied by increased food intake
bLHbeta-CTP transgenic[59]	Expression of a chimeric bovine luteinizing hormone (LH) β-subunit/human chorionic gonadotropin β-subunit COOH-terminal extension	Increased body weight by 5 weeks of age in females only with increased food intake and decreased metabolic rate; no body weight differences were found in male mice
MT deficient[60]	Deletion of both the metallothionine-I and -II genes	Increased body weight noticeable by 5–7 weeks, with significantly increased white fat by 7 weeks and hyperphagia
NMU deficient[61]	Deletion of the neuromedin U gene	Increased body weight and fat by 4 weeks of age with hyperphagia and reduced energy expenditure
Ob (obese)[11]	Spontaneous mutation of leptin	Obesity with diabetes evident around 4 weeks of age
OG2-NcadDeltaC transgenic[15]	Transgenic overexpression of a dominant negative N-cadherin	Increased juvenile adipogenesis and body weight that are absent in adulthood
POMC deficient[10]	Deletion of the proopiomelanocortin gene	Increased body weight and fat prior to 2 months of age accompanied by hyperphagia and yellow pigmentation
PSLCR1 deficient[62]	Deletion of the phospholipid scramblase 1 gene	Mildly increased adiposity prior to 2 months of age
PSLCR3 deficient[62]	Gene-trap-mediated deletion of the phospholipid scramblase 3 gene	Increased adiposity prior to 2 months of age
Resistin transgenic[63]	Adipose-specific overexpression of a dominant negative form of the resistin gene	Mild obesity by 40 days of age on normal chow or high fat diet with protection from the development of insulin resistance
SH2-B deficient[64]	Deletion of the src homology 2 domain B protein gene	Increased body weight in both males and females by 6 weeks of age with significantly increased food intake
Sim1 heterozygous[65,66]	Deletion of the mammalian single-minded 1 homologue gene using either standard knockout or floxed gene	Increased body weight before 8 weeks of age that is exacerbated by a high fat diet; increased linear growth with increased food intake and decreased energy expenditure
SLC6A1 transgenic[67]	Overexpression of the γ-aminobutyric acid transporter subtype I gene	Increased body weight and fat with reduced spontaneous activity
SREBF1 transgenic[68]	Transgenic overexpression of the sterol regulatory element-binding protein-1a in adipose tissues	Enlargement of brown fat, accompanied by fatty livers
STAT3 deficient[69]	Neuronal-specific deletion of the signal transducer and activator of transcription 3 gene	Obesity by 6–8 weeks of age with a 5-fold increase in total fat content; animals are hyperphagic, diabetic, and infertile

MODELS OF ADULT-ONSET OBESITY

In humans, adult-onset obesity is currently about twice as common as juvenile-onset obesity.[18] Mouse models of adult-onset obesity are characterized as those models displaying obesity on normal rodent chow after 9 weeks of age (Table 71–2). In all, 60 models that can be classified as having adult-onset obesity have been characterized, which is nearly twice the number of models of juvenile obesity.

An intriguing observation is that mouse models of adult-onset obesity appear to have only one side of the energy balance equation altered. Of 34 models of juvenile-onset obesity, only three models (~9%) were reported to have normal or reduced food intake, suggesting that up to 90% of mouse models of juvenile-onset obesity could be hyperphagic. For models of adult-onset obesity, 12 of 60 (~20%) display reduced or normal food intake, while only 11 of 60 (~18%) are reported to be hyperphagic. Models with juvenile-onset obesity were more likely to be both

Table 71–2
Models of adult-onset obesity

Model	*Genetics*	*Phenotype*
AqP7 deficient[70]	Deletion of the aquaporin 7 gene	Obesity after 12 weeks of age with no difference in oxygen consumption but decreased rectal temperatures after 20 weeks of age
ARKO mouse[71,72]	Deletion of the androgen receptor	Late-onset gonadal fat accumulation and increased body weight after 10 weeks of age and testicular feminization
AP2 deficient[73]	Deletion of the adipocyte fatty acid-binding protein gene	Diet-induced obesity without insulin resistance; not obese on regular chow
Ataxin-2 deficient[27]	Deletion of the ataxin-2 (Sca2) gene	Obesity on high fat diet; on normal chow increased body weight is evident after 1 year of age
Beta-less[74]	Deletion of all three β-adrenergic receptors	Slight obesity on chow by 14 weeks of age; massive obesity on high fat diet due to failure of diet-induced thermogenesis
BBS1 deficient[75]	Deletion of Bardet–Biedl syndrome 1 homolog	Obesity by 10 weeks in a subset of the animals
BBS4 deficient[75,76]	Deletion of Bardet–Biedl syndrome 4 homolog	Obesity by 12 weeks of age in males and 8 weeks of age in females
BDNF heterozygous[26,77]	Heterozygous deletion of the brain-derived neurotrophic factor gene	Approximately 50% of mice become obese by 3 months of age on normal chow; obesity can be induced earlier on a high fat diet; animals display hyperphagia with increased locomotion
BRS3 deficient[78]	Deletion of the bombesin receptor subtype-3 gene	Mild obesity by 14 weeks of age
CRF-BP transgenic[79]	Overexpression of the CRF-binding protein gene	Weight gain in female mice only between 8 and 12 months of age
Fat (cpe)[80,81]	Spontaneous mutation in the carboxypeptidase gene; there is also a Cpe knockout mouse that has the same phenotype	Obesity by 10–12 weeks of age; defects in proinsulin processing
FORKO mice[20]	Deletion of the follicle-stimulation hormone receptor gene	Mild obesity by 3–3.5 months of age in the homozygous females and obesity by 10–12 months of age in the heterozygotes; males do not become obese at any age
Gck transgenic[82]	Transgenic overexpression of glucokinase in liver	Impaired glucose tolerance by 6 months of age; obesity and insulin resistance on high fat diet
GDF-3 transgenic[83]	Adenovirus-induced growth differentiation factor-3	Adenovirus transduction occurred via tail vein at 3 weeks of age; no difference in body weight or fat on normal chow; increased body weight and fat on high fat diet
Gpr7 deficient[84]	Deletion of the receptor for neuropeptides B and W	Increased body weight by 10 weeks of age with hyperphagia, hypoactivity, and decreased metabolic rate
Gpr10 deficient[85]	Deletion of the prolactin-releasing peptide receptor gene	Obesity and hyperphagia by 16 weeks of age on a normal chow diet
Gpx1 transgenic[86]	Transgenic overexpression of the glutathione peroxidase gene	Significant body weight gain by 24 weeks of age with increased body fat and insulin resistance
GR (NR3C1) deficient[87]	Deletion of the glucocorticoid receptor (nuclear receptor 3C1) gene in neuronal cells	Reduced overall body weight with increased body fat in adulthood; reduced food intake and energy expenditure

Table 71–2 (continued)

Model	*Genetics*	*Phenotype*
H1 deficient[88]	Deletion of the histadine H1 receptor gene	Obesity by 30 weeks of age with greater food intake in obese animals and differences in nocturnal feeding in the mutants
H3 deficient[89]	Deletion of the histadine H3 receptor gene	Obesity by 28–30 weeks of age with greater food intake in obese animals and reduced locomotor activity
hGHRH transgenic[90]	Transgenic overexpression of the human growth hormone-releasing hormone gene	Modest increase in body weight by 4–6 months of age
HCRT trangenic[91]	Transgenic construct targeting ataxin-3 to orexin-containing neurons, effectively ablating them	Late-onset obesity despite eating less than normal controls; narcolepsy
HDC deficient[25]	Deletion of the histidine decarboxylase gene	Obesity by ~20 weeks of age with increased white and brown fat; there is no hyperphagia, but animals display increased metabolic efficiency and impaired glucose tolerance
11β-HSD-1 transgenic[92]	Overexpression of the 11β-hydroxysteroid dehydrogenase type 1 gene in adipose tissue	Increased body weight and body fat by 10 weeks of age on normal chow and exacerbated response to high fat diet; hyperglycemia, hyperinsulinemia, and hyperleptinemia on normal chow
5-HT(2C) deficient[93]	Deletion of the serotonin 5-HT(2C) receptor gene	Obesity after 6 months of age, with early hyperphagia accompanied by elevated physical activity
ICAM deficient[24]	Deletion of the intercellular adhesion molecule-1 gene	Increased body weight and fat by 18 weeks of age on normal diet with increased susceptibility to a high fat diet with some sex-dependent differences; no hyperphagia
IL-1R deficient[94]	Deletion of the interleukin-1 receptor gene	Increased body weight and fat by 5–6 months of age with lack of leptin responsiveness
IL-1/IL-6 double knockout[95]	Deletion of both the interleukin-1 and interleukin-6 genes	Earlier onset of obesity (by 10 weeks of age) compared to the single knockouts
IL-6 deficient[96,97]	Deletion of the interleukin-6 gene	Obesity after 6 months of age with a 20% weight gain by 9 months of age, accompanied by a 50–60% increase in total body fat; another report shows no obesity in IL-6-deficient mice
IL-18 deficient[98]	Deletion of the interleukin-18 gene	An 18.5% increase in body weight and significantly increased body fat by 6 months of age with hyperphagia and normal energy expenditure
IL-18R deficient[98]	Deletion of the interleukin-18 receptor gene	Similar phenotype to IL-18 knockout mice, with obesity developing by 6 months of age
*f*Irs2:cr^2 mice[99,100]	Conditional deletion of insulin receptor substrate 2 in β cells and hypothalamus	Adult-onset obesity by 10 weeks of age with diabetes that resolves due to β cell repopulation later in life
mIRKO mice[101,102]	Muscle-specific deletion of the insulin receptor gene	Increased adipose tissue deposits by 4 months of age without hyperinsulinemia or diabetes
Kir6.2KO[103]	Deletion of the potassium inwardly rectifying channel, Kir6.2	Adult-onset obesity following a mild impairment of glucose tolerance and failure to secrete glucose in response to insulin at younger ages
MC3R deficient[104,105]	Deletion of the melanocortin-3 receptor gene	Increased body weight and fat by 3–4 months of age with normal food intake and metabolic rate, but reduced spontaneous activity
MC4R deficient[12,13]	Deletion of the melanocotrin-4 receptor gene	Increased body weight and fat by 7–12 weeks of age with hyperphagia, reduced metabolic rate, and reduced spontaneous activity
MMP19 deficient[23]	Deletion of the matrix metalloproteinase 19 gene	On normal chow the animals show increased weight when very aged, and show diet-induced obesity due to adipocyte hypertrophy within 20 weeks of high fat diet feeding in males; animals are resistant to skin cancer

continued

Table 71–2 (continued)

Model	*Genetics*	*Phenotype*
NEIL1 deficient[22]	Deletion of the neil1 gene	Males develop significant obesity after 6 months of age with mild obesity in the females during the same time frame
Nhlh2 deficient[19,106]	Deletion of the nescient (neural) helix-loop-helix gene	Adult-onset obesity by 10–12 weeks of age in males and 7–9 weeks of age in females; animals have normal food intake but reduced spontaneous activity
NPY Y1 receptor deficient[107]	Deletion of the neuropeptide Y1 receptor gene	Increased body weight by 9 weeks of age with females showing more substantial increases than males; no hyperphagia but animals show increased feed efficiency
NPY Y5 receptor deficient[108]	Deletion of the neuropeptide Y5 receptor gene	Mild adult-onset obesity with hyperphagia
NTSR1 deficient[109]	Deletion of the neurotensin type 1 receptor gene	Obesity by 11 weeks of age accompanied by hyperphagia and a slight reduction in locomotion
P62 deficient[110]	Deletion of the signaling adapter protein p62 gene	Obesity after 4 months of age with increased food and water intake
PARP1 deficient[111]	Deletion of the poly(ADP-ribosyl)transferase gene	Increased body weight in aged animals on a mixed C57BL/6 × 129/Sv background
PC1 heterozgyous[112,113]	Deletion of the prohormone convertase I gene	Heterozygotes show increased body weight by 10–12 weeks of age; homozygous animals are growth delayed; increased proinsulin levels and reduced mature insulin levels due to lack of processing by PC1
PC1 point mutants[114]	Induced mutagenesis with identification of a point mutation in the prohormone convertase I gene	Mild obesity after 8 weeks of age that is more prominent on breeder diet than normal chow; there is no development of diabetes despite defective insulin processing
PKCtheta deficient[115]	Deletion of protein kinase C θ subunit gene	Normal body weight with increased body fat and decreased physical activity without hyperphagia on normal chow
PKCtheta transgenic[116]	Transgenic expression of a dominant negative protein kinase C θ subunit in skeletal muscle	Obesity by 4 months of age with hyperinsulinemia and reduced insulin sensitivity by 6 months of age
ProSAAS transgenic[117]	Transgenic overexpression of the PC1 inhibitor protein pro-SAAS from the β-actin promoter	Mild obesity by 9–10 weeks of age with no hyperglycemia
Prox1 deficient[118]	Deletion of the Prox1 homeobox gene	Significant increase in body weight after 9 weeks of age with large adipocytes and increased lipid deposits; mice also have defective lymphatic vasculature
PPARγ deficient[119]	Muscle-specific deletion of the peroxisome proliferators-activated receptor γ gene	No overt body weight difference by 12 months of age, but increased body fat at 4 months of age and increased susceptibility to a high fat diet; no hyperphagia
PPKO mice[120]	Deletion of the PIP3 phosphatase Pten gene in POMC neurons	Mild obesity after 16 weeks of age on normal chow in increased food intake but normal energy expenditure
PPP1R3A deficient[121]	Deletion of the glycogen targeting subunit of the protein phosphatase 1 gene	Mildly increased body weight by 4 months of age with only a 20% increase over normal animals by 1 year of age; large increases in body fat are accompanied by insulin resistance
R6/2 transgenic[122]	Transgenic for exon 1 of the human HD gene containing over 100 copies of the repeat region in Huntington disease protein	Increased body fat by 8–9 weeks of age, then wasting and death by 15 weeks of age; overall growth retardation
Rn8–12 transgenic[21]	Transgenic expression of the human renin gene	Male-specific adult-onset obesity and increased body fat by 15 weeks of age
Syndecan-1 transgenic[123]	Transgenic misexpression of syndecan-1 in the hypothalamus	Noticeably increased body weight after 8 weeks of age with a transgene dosage effect and up to a 70% increase in food intake

Table 71–2
(continued)

Model	Genetics	Phenotype
SPARC deficient[124]	Deletion of the secreted protein acidic and rich in cysteine/osteonectin/BM-40	Increased body fat (including skin) without significant increases in body weight after 10 weeks of age; increased adipocyte number and size
Tfam deficient (POMC neurons)[125]	Cre-lox-mediated deletion of the mitochondrial transcription factor Tfam in POMC neurons, resulting in progressive neurodegeneration of POMC neurons	Significant increase in body weight by 4 months of age with hyperphagia and reduced oxygen consumption
TRα mutant[126]	Insertional mutagenesis of the thyroid hormone receptor α gene with a dominant-negative form	Increased body weight (3-fold) and fat (4-fold) after 3 months of age without hyperphagia
Tub[127–129]	Spontaneous and induced (knockout) mutations in the Tubby-1 gene	Adult-onset obesity occurring after 8 weeks of age with nearly double body weight by 19 weeks of age, reduced fertility, reduced food intake, and locomotor defects; Tubby-1 mutants also display degenerative eye disease and progressive blindness

hyperphagic and have altered energy expenditure, either through metabolism or physical activity, than models with adult-onset obesity, which were more likely to be either hyperphagic or have reduced energy expenditure. While not all mice have these parameters reported, it is interesting that hyperphagia in combination with reduced metabolism or exercise appears to lead to earlier onset obesity than reduced energy expenditure alone without increased food intake. As an example of this type of model, the neuronally expressed nascient helix-loop-helix-2 (Nhlh2) gene[19] knockout mouse shows no hyperphagia or reduced body temperature, but a significant reduction in spontaneous physical activity, leading to late-onset obesity. Models of adult-onset obesity may reflect the result of having one side, but not both sides of the energy balance equation altered.

Several models in this grouping display sex differences in body weight gain. While the prevalence of obesity in humans is similar for males and females, there is a trend toward higher rates of obesity in U.S. women.[18] There are some models of adult-onset obesity that display sex-dependent differences in body weight gain. For example, in the mouse model that contains a deletion of the follicle-stimulating hormone receptor gene, only females display weight gain by 3–5 months of age, even though both males and females show the expected disruption of the reproductive axis.[20] Male-specific weight gain occurs with transgenic overexpression of the human renin gene,[21] and is more severe in males containing a deletion of the NEIL DNA endonuclease.[22] An interesting future use of these models will be in discerning the role of these proteins in sex-specific patterns of weight gain.

Several models in this category display late-onset obesity, but display an earlier onset of weight gain triggered by a high fat diet. These models are to be distinguished from some of those in Table 71–5 (p. 695) that show obesity only on high fat or modified diets. While intriguing, it is not clear whether most or even all of the adult-onset models would respond to high fat diets in this way. Those that have been identified to date range from genes involved in extracellular matrix function (MMP19,[23] ICAM,[24] and 11βHSD-1[25]), to nervous system growth and development (BDNF[26]), to proteins that may be involved in the assembly of polyribosomes (ataxin-2[27]). In each of these examples, onset of obesity can be shortened by several weeks or even several months when models are given access to a high fat diet. While there does not seem to be any obvious connection between these few genes that have been characterized thus far, more work in this area may reveal new genes and new links between genes involved in the regulation of diet-induced changes in body weight.

MODELS WITH REDUCED BODY WEIGHT

This group of models includes those with anorexia, reduced weight, or reduced body fat under normal feeding conditions (Table 71–3). There are almost as many models showing reduced body weight (74) as those with obesity (95, including both juvenile and adult-onset obesity). This suggests, as might be expected, that similar numbers of positively and negatively acting regulatory pathways exist in the control of body weight.

Many of the genes linked to reduced body weight appear to regulate adiposity and body weight through the control of food intake, as noted by the anorexic phenotype of many of the models. As an example, the M3 muscarinic receptor knockout mouse shows anorexia under normal food availability conditions and consequently displays reduced peripheral body fat and body weight.[29] While these animals also show an abnormal reduction in melanin-concentrating hormone in response to fasting,[29] it is not clear whether the effect of the receptor deletion on body weight regulation is centrally or peripherally mediated, as these receptors are found in both the nervous system and gastrointestinal smooth muscle and endocrine cells.[30]

Other models in this group exhibit no change in food intake, but a compensatory upregulation of energy utilization via either increased metabolism or increased physical activity. Examples of this include both the VGF (nonacronymic) knockout mice and the dopamine D1 receptor knockout mice. These mutant mice display increased physical activity with no change in normal food intake.[31,32] Evidence suggests that both of these factors signal through central melanocortin receptors to mediate locomotor activity.[33,34] As MC4R mutant mice show reduced locomotor activity with increased body weight,[12,13] it will be interesting in the future to determine if control of physical activity is mediated exclusively through the melanocortin system, and if agents that

Table 71–3
Models with reduced body weight

Model	*Genetics*	*Phenotype*
anx/anx[130]	Recessive mutation on chromosome 2 (spontaneous)	Anorexia, death by postnatal day 35, reduced body fat, ataxia
ACC deficient[131]	Deletion of acetyl-CoA carboxylase 2 leading to reduced malonyl-coenzyme A levels	Reduced body fat due to higher rate of fatty acid oxidation
ADAMTS1 deficient[132]	Deletion of the ADAMTS-1 gene	Reduced body weight due to adipose tissue malformation and growth retardation
ADCYAP-1 deficient[133]	Deletion of pituitary adenylate cyclase-activating polypeptide 1 (also called PACAP)	Wasting, death by postnatal day 14
AKT1 transgenic[134]	Overexpression of AKT1 in skeletal muscle	Reduced adipose tissue
AKT2 deficient[135]	Deletion of AKT2 gene	Age-dependent loss of adipose tissue
C3 deficient[136]	Deletion of complement C3, leading to loss of acylation-stimulation protein (ASP)	Reduced body weight, reduced body fat, and ability to suppress ob/ob phenotype
CAV1 deficient[137]	Deletion of caveolin 1 gene	Late-onset reduced body weight due to abnormal lipid homeostasis
CB1 deficient[138]	Deletion of the CB1 cannabinoid receptor gene	Lean with a resistance to diet-induced obesity
ChREBP deficient[139]	Deletion of the carbohydrate response element-binding protein gene	Reduced epididymal white fat and brown fat with mild insulin resistance
Cidea deficient[140]	Deletion of the cidea gene	Lean with resistance to diet-induced obesity, most likely due to higher metabolism through brown fat
Clock deficient[141,142]	Deletion of the clock gene	Impaired absorption of dietary fat with resistance to diet-induced obesity; increased body weight on normal chow with changes in diurnal feeding pattern
Contactin deficient[143]	Deletion of contactin-1 gene	Anorexia and ataxia
CPT1c deficient[144]	Deletion of the carnitine palmitoyltransferase-1c	Reduced body weight and food intake, but a unique increased susceptibility diet-induced obesity
D1R deficient[32]	Deletion of the D1 dopamine receptor	Reduced body weight, hyperactivity
DD[145,146]	Deletion of the tyrosine hydroxylase gene in all cells with restoration only in nonadrenergic cells	Anorexia, reduced spontaneous activity, death by postnatal day 28
DeltaFosB transgenic[147]	Overexpression of DeltaFosB	Reduced adipogenesis
DGAT deficient[148]	Deletion of acyl CoA:diacylglycerol acyltransferase gene	No increased weight on high fat diet, although food intake is similar, increased metabolic rate
DHCR24 deficient[149]	Deletion of the desmosterol reductase gene	Reduced cholesterol biosynthesis, reduced subcutaneous fat, infertility
DLK1 deficient[150]	Transgenic expression of the Pref-1 ectodomain fused to the human immunoglobulin constant domain driven by either the adipocyte fatty acid-binding protein promoter or the albumin promoter	Loss of adipose tissue
EIF4EBP1 deficient[151]	Deletion of the translation initiation factor 4E-binding protein gene	Reduced adipogenesis and in males, increased metabolic rate
ERRalpha deficient[152]	Deletion of the estrogen-related receptor α gene	Reduced body weight and fat, normal food intake and energy expenditure; resistance to high fat diet-induced obesity
Fatty liver dystrophy[153]	Spontaneous mutation in the lipin-1 gene	Loss of body fat and resistance to diet-induced obesity
FoxC2 transgenic[154]	Expression of human FoxC2 from adipocyte-binding protein 2 promoter	Reduced intraabdominal fat on normal diet and resistance to diet-induced obesity; increased brown fat
GAMT deficient[155]	Deletion of the guanidinoacetate *N*-methyltransferase gene	Reduced body weight and fat with no change in food intake, activity, or serum leptin levels
GDF8 deficient[156,157]	Deletion of myostatin	Reduced fat accumulation with increasing age; partial suppression of ob/ob phenotype; increased myogenesis
GFA2 deficient[158]	Deletion of glial cell-line-derived neurotrophic factor receptor $\alpha 2$ gene	Reduced fat mass and elevated metabolic rate
GHSR transgenics[159]	Overexpression of the human growth hormone secretagogue receptor 1A in GHRH neurons	Decreased adipose mass, most significant in females, without body weight change
GIPR deficient[160]	Deletion of the gastric inhibitory peptide receptor gene	Reduced metabolic rate and preferential use of fat as a substrate for energy expenditure; resistance to diet-induced and genetic obesity

Table 71–3 (continued)

Model	*Genetics*	*Phenotype*
GLUT4 deficient[161]	Deletion of the insulin-sensitive glucose transporter gene	Growth retardation and reduced fat with normal glucose regulation
GNAS deficient (paternally inherited)[16]	Deletion of the paternally inherited copy of the G-protein α subunit; note that deletion of the maternally inherited allele has the opposite phenotype (see Table 71–1)	Mice with the paternally inherited mutant allele show decreased body weight by 30–40 days of age, increased metabolic rate, but no change in food intake relative to control animals
Gng3 deficient[162]	Deletion of the G-protein γ3 subtype gene	Both males and females display reduced body weight, but in females this is more pronounced with a 25% reduction in body weight and reduced fat pads
Gpam deficient[163]	Deletion of the mitochondrial glycerol-3-phosphate acyltransferase gene	Reduced body weight and gonadal fat pads
Gpr24 deficient[36]	Deletion of the melanin-concentrating hormone-1 receptor gene	Lean, with increased metabolic rate and activity, and with increased food intake; resistant to diet-induced obesity
11β-HSD-1 deficient[164]	Deletion of the 11β-hydroxysteroid dehydrogenase type 1 gene	Reduced gonadal fat on regular chow diet and increased resistance to effects of a high fat diet in obesity-resistant MF-1 mice and C57BL/6 diabetes-susceptible mice
HSL deficient[165,166]	Deletion of the hormone-sensitive lipase gene	Reduced body fat with similar body weight on normal chow and resistance to diet-induced obesity with higher core body temperature on high fat chow
IL-1ra deficient[167,168]	Deletion of the IL-1 receptor antagonist gene	Reduced body weight by 6 weeks of age; resistant to monosodium glutamate-induced obesity
Kv1.3 deficient[169]	Deletion of the potassium voltage-gated channel subtype 3	Decreased body weight with increased basal metabolic rate and protection from diet-induced obesity
LgsKO mice[170]	Liver-specific deletion of the stimulatory G-α subunit gene	Reduced fat mass with normal body weight
Little[171]	Recessive mutation in the growth hormone-releasing hormone receptor (spontaneous)	Decreased body weight with dwarfism
LXRbeta deficient[172]	Deletion of the liver X receptor β gene	Reduced adipose mass and resistance to diet-induced obesity
M3 deficient[29]	Deletion of the M3 muscarinic receptor	Anorexia, reduced peripheral body fat
MAPK8IP1 deficient[173]	Deletion of the mitogen-activated protein kinase 8 interacting protein 1 (JNK1) gene	Reduced weight gain on normal chow and high fat diet with decreased adiposity and ability to suppress obesity in ob/ob mice
MCH deficient[174]	Deletion of the promelanin-concentrating hormone gene	Anorexia, lean, reduced metabolic rate
MCH neuron ablated[175]	Targeted ablation of melanin-concentrating hormone neurons using transgenic ataxin-3 gene expression	Reduced fat mass and body weight by 7 weeks of age with reduced food intake correlated with the reduced body weight, and increased energy expenditure, especially in male transgenics
Md[176]	Recessive mutation in the mahoganoid gene (spontaneous)	Ability to suppress obesity in A^y mice; hyperphagia with elevated metabolic rate
Mg[176]	Recessive mutation in the mahogany gene (spontaneous)	Identical to the Md mouse
Mrf-2 deficient[177]	Deletion of the Mrf-2 gene	Low weight gain in surviving homozygotes; reduced levels of white and brown adipose tissue
NPY deficient[178]	Deletion of the neuropeptide Y gene	Ability to suppress obesity in ob/ob mice
Perilipin deficient[179]	Deletion of the perilipin gene	Reduced adipose mass with increased leptin production and resistance to diet-induced obesity
PGs1 deficient[180,181]	Deletion of Gtrgeo22 using the gene trap method; later identified as polyglutamylase subunit 1	Reduced body fat and infertility in males
PI3KR1 deficient[182]	Deletion of the phosphoinositol-3-kinase regulatory subunits p50α and p55α splice variants	Increased insulin sensitivity with reduced adipose accumulation
PIP5K2B deficient[183]	Gene-trap-mediated deletion of the phosphatidylinositol-4-phosphate 5-kinase, type II, β gene	Enhanced glucose sensitivity and reduced adiposity

continued

Table 71–3
(continued)

Model	*Genetics*	*Phenotype*
PP transgenic[184]	Overexpression of the pancreatic polypeptide cDNA	Results in approximately 50% postnatal lethality due to reduced food (milk) intake; the surviving animals have reduced body weight and fat; a reduced rate of gastric emptying may be causative
PPARδ transgenic[185]	Overexpression of peroxisome proliferators activated receptor δ-VP16 fusion gene in adipose tissue	Transgenics range from reduced adipose mass to no detectable white fat; protection from both diet-induced and genetic obesity
PPARγ heterozgyous[99,185,186]	Deletion of one allele of peroxisome proliferator-activated receptor[187] or Cre-lox-mediated deletion of the gene in adipocytes[186]	Reduced fat mass and absence of insulin resistance following high fat diet
PTP-1B deficient[188,189]	Deletion of the protein tyrosine phosphatase 1B gene, or neuronal-specific deletion with similar phenotype	Reduced body weight and leanness in male mice by 7 weeks of age; resistance to high fat diet for both male and female mice
PTTG deficient[190]	Deletion of the pituitary tumor-transforming gene	Reduced body weight with no increases after 6 months of age in males and loss of epididymal fat pads; hyperglycemia and reduced insulin in response to glucose in aged males; gonadectomy of females results in hyperglycemia
RIIβ deficient[191]	Deletion of the protein kinase A RIIβ subunit	Lean with elevated metabolic rate and resistance to high fat diet-induced obesity despite normal food intake
Ras-GRF1 deficient[192]	Deletion of the Ras guanine nucleotide-releasing factor 1 gene	Reduced body weight and leanness due to lipid turnover
RSK1 deficient[193]	Deletion of the ribosomal S6 kinase 1 gene	Reduced body weight and fat with reduced adipocyte size and beta cell mass
RSK2 deficient[194]	Deletion of the ribosomal S6 kinase 2 gene	Loss of white adipose tissue and reduced body weight with age, accompanied by reduced serum leptin levels
SCD1 deficient[195]	Deletion of the steroyl-coenzyme A desaturase 1 gene	Reduced body weight and fat despite a 25% increase in food intake
SOCS3 deficient[196]	Neuron-specific deletion of the suppressor of cytokine signaling 3 using either nestin or syndecan-promoter-driven Cre recombinase	Reduced body weight due to enhanced weight loss and food intake suppression in response to leptin; resistant to diet-induced obesity
STAT5b deficient[197]	Deletion of the signal transducer and activator of transcription 5b gene	Loss of sexually dimorphic growth rates with reduced growth in males and slightly increased growth in females; overall reduced body fat in young animals with obesity occurring by 9 weeks of age in older males and females
TGFβ1 transgenic[198]	Overexpression of the transforming growth factor β1 protein in liver, kidney, and white and brown adipose tissue	Reduced body weight and 30% reduction in body fat with ability to suppress obesity in ob/ob mice; the animal is a model of lipodystrophy
TKT deficient[199]	Heterozygous deletion of the transketolase gene	While homozygous deletion is lethal, heterozygotes are growth retarded and have reduced fat
UCP1 deficient[200]	Deletion of the uncoupling protein-1 gene	Increased brown fat deposits in areas of inguinal white fat, accompanied by increased energy expenditure, reduced body temperature, and resistance to diet-induced obesity
UCP1 transgenic[201]	Overexpression of uncoupling protein-1 gene in heart and skeletal muscle	Thirty percent reduction in body weight by 5 weeks of age accompanied by similar to increased food intake and increased energy expenditure
UCP2 and UCP3 transgenic[202]	Expression of both the human uncoupling protein 2 and 3 from a bacterial artificial chromosome	Significantly reduced fat mass for one of the lines, and nearly significant reduction for a second transgenic line; both hyperphagia and reduced physical activity are present in the transgenics
Unc5c insertional transgenic[203]	Insertion of a telomerase reverse transcriptase gene in intron 1 of the Unc5c receptor gene	Up to a 30% reduction in body weight with reduced epididymal fat, ataxia, and increased spontaneous activity
Wnt10B transgenic[204]	Expression of Wnt10B from the FABP4 promoter that targets to white and brown adipose tissue	Increased body weight with a 30% reduction in body fat and resistance to diet-induced obesity
Vgf deficient[31,34]	Deletion of Vgf (no acronym) peptide	Reduced adiposity, increased metabolic rate, increased physical activity; Vgf deletion blocks development of obesity in several models
Zfp-36 deficiency[205]	Deletion of the Zfp-36 gene, encoding the putative tristetraprolin transcription factor	Reduced body weight and cachexia that may be due to aberrant tumor necrosis factor α expression

can modulate melanotonergic tone can be used for treating human body weight disorders.

MODELS TO STUDY POLYGENIC CONTRIBUTIONS TO OBESITY

Only up to 5% of all cases of human obesity can be attributed to monogenetic causes. These include single gene mutations that affect melanocortin signaling, including mutations in the leptin gene, the leptin receptor gene, the POMC gene, and the MC4R gene.[8] Most common human obesity can be attributed to multiple genes (polygenic causes), with environmental influences on the genetic penetrance or susceptibility of the resultant phenotype.[18] At present, 17 polygenic mouse models with disorders in body weight regulation exist for the study of polygenic obesity (Table 71–4) and more are being created in order to study this complex genetic disorder. There are also a number of studies not included in this listing that have demonstrated the presence of linked loci on specific mouse chromosomes that influence body weight.[5] These chromosomal regions and the genes within them may contribute to polygenic forms of obesity in humans, but this research is still ongoing. It is expected that future mouse models of polygenic obesity will be created using transgenic mutagenesis, knockout, and conditional knockout technologies to study the interaction of multiple alleles of body weight regulatory genes on this complex phenotype.

Table 71–4
Polygenic models of body weight control

Model	*Genetics*	*Phenotype*
AKR/J[35,206,207]	A substrain of AKR, which was derived by a mouse dealer named Detwiler in Norristown, PA; the AKR/J substrain has been maintained by the Jackson Laboratories	Increased body weight and fat on both normal chow and high fat diet with a strong preference for high fat foods and hyperphagia; QTLs have been identified on chromosomes 9 and 15 using these animals
Berlin fat lines[208]	Selective breeding of an outbred population of mice for increased body fat without overall increased body weight; three lines exist with different sensitivity to diet-induced obesity	Increased body weight is evident by 3 weeks of age with significant increases in body fat contents
BSB[209,210]	Strain derived from a backcross of C57BL/6J × *Mus spretus*	Body fat ranges from 1% to 50% in a backcrossed strain with no detectable sex difference in weight gain; QTLs for body fat on 6, 7, 12, and 15 have been identified using this model
C57BL/6J[35,206]	The most widely used inbred line of mice; created by mating female 57 with male 52 from Miss Abbie Lathrop's stock	Increased body fat compared to other common inbred mouse strains, with a tendency to select high fat foods
CAST/Ei	Derived from a strain of *M. musculus castaneus* from Thailand	Relatively lean with no increased body weight when given a high fat diet for up to 32 weeks; this strain is commonly used in crosses with obese lines
DBA/2J[35,206]	Very commonly used inbred line of mice; it is reportedly the oldest of all inbred strains of mice and was derived by C.C. Little by selective breeding for a dilute coat color	Moderate body fat gain with no change in body weight and a preference for high fat foods
Du6i[211]	This line is the largest selected inbred strain of mice; the line is an inbred derivative of Du6 selected for high body weight over 78 generations	Increased body weight in both males and females by 21 days of age with increases in male body weight surpassing females by 42 days of age
Fat "F" mice[212,213]	Derived in a selection strategy over 50 generations using an original three-way cross of parental lines CBA, JU, and CFLP	There is a 5- to 8-fold increase in body fat over the similarly selected lean mouse line; fat mice have a slight reduction in fertility; QTLs on chromosomes 2, 12, 15, and X have been mapped
KK mouse[214,215]	Spontaneously arising inbred line of mice selectively bred for obesity and diabetes	Maturity-onset obesity with earlier onset by high fat diet; QTLs for body weight and diabesity have been mapped to chromosomes 4, 6, and 16
M16[216,217]	Long-term selection for weight gain from an ICR parental population	Increased body weight by 6 weeks of age accompanied by increased feed efficiency, but no difference in food intake when adjusted for body weight; a QTL for fatness on chromosome 2 has been mapped using this line
NONcNZO10/LtJ[218]	Derived by crossing NZO/H1LtJ × NON/LtJ	Early-onset obesity in males with diabetes appearing in the animals with the highest level of weight gain
New Zealand obese (NZO/HlLt)[219]	Originated in a colony of inbred agouti mice and selected for coat color and obesity to F_{17}	Males show early-onset obesity and develop late-onset diabetes, while females show only obesity and appear resistant to diabetes

continued

Table 71–4
(continued)

Model	Genetics	Phenotype
SMXA-5[220,221]	Derived from SM/J and A/J parental strains	Increased body weight and fat by 10 weeks with hyperinsulinemia and impaired glucose tolerase; a QTL mapped to chromosome 2 has been characterized
SMXA-9[220]	Similar to SMXA-5	Similar to SMXA-5 with slightly higher weight gain; no QTLs have been characterized for this strain
SWR/J[206]	Inbred mouse line derived from Swiss mice in 1926	Resistant to obesity caused by a high fat diet
TallyHo[222,223]	Theiler original males identified in a colony as hyperglycemic and hyperinsulinemic were used to create an inbred line [TH(F7)] or TallyHo; this line was crossed to C57BL/6 females for QTL analysis	Obesity and diabetes with possible QTLs on chromosomes 6 and 19
TSOD[224,225]	Created from the outbred ddY strain as an inbred strain exhibiting obesity and diabetes	Both males and females exhibit juvenile onset obesity, but only male mice show a diabetic phenotype; a QTL on chromosome 5 has been linked to adiposity in this model

One interesting note on the polygenic models listed here is that the C57BL/6J inbred strain of mouse is included in this group. This inbred line of mice displays an obesity-prone phenotype when given a high fat diet.[35] Many of the knockout mouse models listed in the other tables have used the C57BL/6J line in the creation of their models. This is because blastocysts from black coat-colored mice are commonly used to distinguish chimeras containing cells from the brown coat-colored 129SvJ line. It is possible that modifying loci that have been selected for during strain creation of the C57BL/6J line have the potential to influence the phenotypes of the knockouts using this strain as a background or a mixed background of both the C57BL/6J and 129SvJ lines. One example of the effect of background strain selection on the robustness of phenotype was recently shown for mice with a mutation of the melanin-concentrating hormone (MCH). The original MCH line was created on a mixed background and displayed leanness and a trend toward hypophagia.[36] Backcrossing this line to either pure 129SvJ or C57BL/6J mice resulted in slight differences in the penetrance of the lean phenotype. MCH knockout mice on a pure 129SvJ background resisted obesity caused by the high fat diet, but not to the level of significance of those backcrossed onto a C57BL/6J line, which normally shows susceptibility to body weight gain on a high fat diet. With the MCH deletion, C57BL/6J line mice resist high fat diet-induced obesity.[37] With this example in mind, proper use of specific mouse strains and future characterization of the polygenic influences on body weight control will help in elucidating the multiple regulatory pathways in this complex phenotype.

MODELS TO STUDY ALTERED RESPONSES TO DIETARY, GENETIC, OR CHEMICALLY INDUCED OBESITY

This grouping of mouse models is separated from the others, as none of them is reported to display increased or decreased body weight, relative to their control group, on normal rodent chow and with standard housing conditions. This group of 29 mouse models reveals altered responses, relative to normal mice, to dietary, genetic, or chemically induced obesity.

In general, mouse models that show a response only to induced obesity are very interesting as they can be used to dissect pathways controlling responses to a sole stressor on the system, instead of the many facets of body weight regulation. Only seven of these mouse models show an exaggerated response to a high fat diet. An example of this phenotype is seen in mice with a deletion of the AMP-activated protein kinase (also called PRKRR2). These animals are obese only when fed a high fat diet, although they are not hyperphagic,[38] and their overall increase in body weight is greater than normal mice given the same diet. These animals can be classified as having an abnormal response to high fat diets, which is an interesting avenue to study given the changes in human diets to more processed and high fat foods in the past few decades.

The majority of the animals listed in Table 71–5 (75%) show resistance to diet-induced, genetic-induced, or chemically induced obesity. For example, the CRF2-deficient animal contains a deletion in the corticotrophin-releasing factor receptor 2 gene and displays hyperphagia when given a high fat diet but resists diet-induced obesity that ensues in normal mice given this type of diet.[28] Another interesting model in this category is the tumor necrosis factor (TNF)-α-deficient mouse model treated with gold thioglucose (GTG) injection. GTG induces obesity in mice by lesioning the ventromedial hypothalamus. The TNF-α mutation partially protects against GTG-induced obesity, particularly with respect to development of overt diabetes.[39] These animal models with resistance to obesity are vital to research that seeks to gain an understanding of the genes and protein products than can protect against body weight gain.

An understanding of the genes that control the altered responses to induced obesity is crucial to those researchers seeking pathways to target for future drug development. It is possible that these gene products may be responsible for minute adjustments in the balance of food intake and energy expenditure by the body or in the responses of this system to moderate stresses (high fat, cold, genetic modulators). These are in contrast to those listed in the other groups, which appear to have more global affects on body weight parameters, affecting energy intake and utilization under normal or standard conditions.

Table 71–5
Models with an abnormal response to diet-induced, chemically induced, or genetic obesity

Model	*Genetics*	*Phenotype*
Alpha 1bAR	Deletion of the adrenergic receptor, α_{1b}	Obesity on a high fat diet; impaired glucose metabolism
ApoC1 transgenic[226]	Transgenic overexpression of human apoC1 gene	Resistance to genetic obesity in ob/ob mouse
Beta-1 AR transgenic[227]	Transgenic overexpression of the β_1-adrenergic receptor in adipocytes	Resistance to diet-induced obesity (DIO)
CCR2 deficient[228]	Deletion of the chemokine receptor-2 gene	Resistance to DIO with reduced inflammation in adipose tissue
c-myc transgenic (liver expression)[229]	Hepatic overexpression of the myelocytomatosis oncogene, c-myc	Resistance to a high fat diet with no change in liver-specific genes normally altered with changes in diet
CRF2 deficient[28]	Deletion of the corticotrophin-releasing factor receptor 2 gene	Consume more food on a high fat diet, but are resistant to DIO
FABP4 deficient[73]	Deletion of the fatty acid-binding protein 4, or adipocyte protein 2 (aP2)	Resistance to DIO
Foxa-2 heterozygous[230]	Heterozygous deletion of the adipocyte protein Foxa-2; homozygotes are embryonic lethal	Marked obesity and induction of Foxa-2 after 7 weeks of high fat diet feeding
Ghrelin deficient[231,232]	Deletion of the ghrelin gene and replacement with a lacZ reporter	Changes in the type of metabolic substrate (fat versus carbohydrate) preferred for energy; protection from obesity
GCK[233]	Transgenic overexpression of glucokinase in skeletal muscle	Resistance to DIO accompanied by increased metabolic rate
mGPD deficient[234,235]	Deletion of the mitochondrial glycerol-3-phosphate dehydrogenase gene	Increased body weight in females by 3 weeks of age or females on a high fat diet; no effect in males, and overall thrifty phenotype
HMGIC[236,237]	Deletion of the high mobility group DNA-binding protein 1-C gene; also known as the pygmy mouse; the gene is also known as HMGA2	Resistant to both DIO and genetic obesity
Inppl1 deficient[238]	Deletion of Inppl1, which encodes SHIP2, the SH2-domain containing inositol 5-phosphatase 2	Resistant to DIO
Mac-1 mice[24]	Receptor to ICAM-1 (leukocyte integrin $\alpha M\beta 2$)	Diet-induced obesity, but it is not clear if these animals have spontaneous obesity on chow as their body weights are too variable
MPOG human transgenic × LDLR-deficient mouse[239]	Contains the human myeloperoxidase gene with a mutation in the promoter region that results in higher expression, crossed to mice containing a deletion of the LDL receptor	Increased weight gain on high fat diet only
Myostatin propeptide transgenic[240]	Muscle-specific depression of myostatin through expression of the myostatin propeptide region	Increased muscle mass with no increase in adipose tissue on a high fat diet
loxTB Mc4r restoration[241]	Floxed MC4R locus that is inactive until crossed with Sim1-Cre, which specifically activates and restores melanocortin-4 receptor to the paraventricular nucleus (PVN)	Reversal of obesity
PAI transgenic[242]	Overexpression of plasminogen activator inhibitor (PAI) 1 from the adipocyte aP2 promoter	Transgenic overexpression has a similar effect to PAI deficiency with protection from DIO in one report
Pla2g1b deficient[243]	Deletion of the pancreatic phospholipase A2, group IB gene	Resistant to DIO with reduced fat absorption (fatty stools)
PPARGC1B transgenic[244]	Transgenic overexpression of PPARγ coactivator 1 using chicken β-actin promoter	Resistant to DIO with reduced fat and body weight
PRKRR2 deficiency[38]	Deletion of the AMP-activated protein kinase gene	High fat diet-induced obesity; no difference in weight or fat when fed a normal diet
RXR gamma deficient[245]	Deletion of the retinoid X receptor gene	Resistance to DIO with lower fat mass and leptin levels
SERPINE-1[246,247]	PAI deficiency	Resistance to DIO but no difference in weight or fat on chow in two reports; rederivation of knockout line and use of wild-type littermates from the knockout line reveals an increased accumulation of gonadal fat compared to wild-type animals
Sim1 overexpression[248]	Transgenic expression of human Sim1 on a BAC clone	Resistance to DIO and genetic obesity but no change on normal chow

continued

Table 71–5 (continued)

Model	*Genetics*	*Phenotype*
Syndecan-3 deficient[249]	Deletion of the syndecan-3 gene	Resistance to DIO, but no difference in body weight or fat on chow diet for males; there is a very slight reduction in body fat for females on a low fat diet
TNF alpha deficient[39]	Deletion of the tumor necrosis factor-α gene	Reduced body fat content following gold-thioglucose injection compared to wild-type mice
TNF alpha transgenic[250]	Expression of a noncleavable transmembrane form of tumor necrosis factor-α gene	An increased adipose content on a high fat diet with no change on normal chow
UCP3 transgenic[251]	Skeletal muscle specific expression of uncoupling protein 3	A significant resistance to a high fat diet after just 4 weeks on the diet; there is a nonsignificant trend of reduced body weight on chow
Y2Y4 deficiency[252]	Double deletion of both the NPY Y2 and Y4 receptors	Resistance to a high fat diet and genetic obesity caused by Y1 receptor deletion

CONCLUSIONS

With over 200 available mouse models to study body weight regulation from a variety of aspects, including food intake, energy expenditure, genetic obesity, diet-induced obesity, and obesity resistance, the mouse has become a very powerful model system in the field of obesity research. Basic researchers are using mice to identify and connect the signaling pathways that are activated or repressed following changes in energy availability in the body. Applied researchers are using mouse models of obesity to identify potential obesity or anorexia blocking proteins and to test newly developed drugs prior to testing in humans. It is expected that more examples of these aspects of body weight regulation using the models listed in this chapter will be published in coming years.

It is also expected that additional models will be developed. As shown in Figure 71–1, the current decade of the 2000s has nearly 4-fold more published models than the decade of the 1990s, and a nearly 200-fold increase over the 1980s. The completion of the mouse genome and the push to elucidate the function of some of the newly identified gene products using resources from The Knockout Mouse Project[40] and independent research programs should yield even more unique genetic models of obesity and disturbance in weight regulation.[41–252] What should be even more prevalent in the next few years are mouse models that are refined relative to some of the global knockouts or transgenics listed here. These include models that use conditional knockout technology or targeted transgenic mice to restrict expression or confine deletion of a gene to a particular cell type or time frame during development. Finally it is expected that mouse models of polygenic obesity or obesity resistance will increase, as investigators combine models to characterize hypothesized interactions between genes, and identify new inbred or outbred strains with unique phenotypes with respect to body weight control. The coming decade promises to bring many more developments in this field using the most common laboratory rodent, *Mus musculus.*

ACKNOWLEDGMENTS

This work was supported by a grant from the National Institutes of Health R01-DK59903. The author would like to thank Dana Fox, Kristen Vella, and Franc-Eric Wiedmer for critical reading of the manuscript.

REFERENCES

1. Cuenot L. L'heredite de la pigmentation chez les souris. *Arch Zool Exp Genet* 1905;3:123–132.
2. Danforth CH. Hereditary adiposity in mice. *J Hered* 1927;18: 153–162.
3. Coleman DL. Obese and diabetes: Two mutant genes causing diabetes-obesity syndromes in mice. *Diabetologia* 1978;14(3):141–148.
4. Hummel KP, Dickie MM, Coleman DL. Diabetes, a new mutation in the mouse. *Science* 1966;153(740):1127–1128.
5. Rankinen T, Zuberi A, Chagnon YC, *et al.* The human obesity gene map: The 2005 update. *Obesity (Silver Spring)* 2006;14(4):529–644.
6. Burnside AS, Good DJ. *Genetic Diversity of Genes Involved in Body Weight Regulation.* New Delhi: Regency Publications, 2006.
7. Rugh R. *The Mouse; Its Reproduction and Development.* Minneapolis, MN: Burgess, 1968.
8. O'Rahilly S, Farooqi IS. Genetics of obesity. *Philos Trans R Soc Lond B Biol Sci* 2006;361(1471):1095–1105.
9. Chen H, Charlat O, Tartaglia LA, *et al.* Evidence that the diabetes gene encodes the leptin receptor: Identification of a mutation in the leptin receptor gene in db/db mice. *Cell* 1996;84(3):491–495.
10. Yaswen L, Diehl N, Brennan MB, Hochgeschwender U. Obesity in the mouse model of pro-opiomelanocortin deficiency responds to peripheral melanocortin. *Nat Med* 1999;5(9):1066–1070.
11. Zhang Y, Proenca R, Maffei M, Barone M, Leopold L, Friedman JM. Positional cloning of the mouse obese gene and its human homologue. *Nature* 1994;372(6505):425–432.
12. Huszar D, Lynch CA, Fairchild-Huntress V, *et al.* Targeted disruption of the melanocortin-4 receptor results in obesity in mice. *Cell* 1997;88(1):131–141.
13. Ste Marie L, Miura GI, Marsh DJ, Yagaloff K, Palmiter RD. A metabolic defect promotes obesity in mice lacking melanocortin-4 receptors. *Proc Natl Acad Sci USA* 2000;97(22):12339–12344.
14. Trevaskis JL, Butler AA. Double leptin and melanocortin-4 receptor gene mutations have an additive effect on fat mass and are associated with reduced effects of leptin on weight loss and food intake. *Endocrinology* 2005;146(10):4257–4265.
15. Castro CH, Shin CS, Stains JP, *et al.* Targeted expression of a dominant-negative N-cadherin in vivo delays peak bone mass and increases adipogenesis. *J Cell Sci* 2004;117(Pt. 13):2853–2864.
16. Yu S, Gavrilova O, Chen H, *et al.* Paternal versus maternal transmission of a stimulatory G-protein alpha subunit knockout produces opposite effects on energy metabolism. *J Clin Invest* 2000;105(5): 615–623.
17. Bastepe M, Juppner H. GNAS locus and pseudohypoparathyroidism. *Horm Res* 2005;63(2):65–74.
18. (CDC) CfDCaP. Behavioral Risk Factor Surveillance System Survey Data. In: Services USDoHaH, 2005.

19. Good DJ, Porter FD, Mahon KA, Parlow AF, Westphal H, Kirsch IR. Hypogonadism and obesity in mice with a targeted deletion of the Nhlh2 gene. *Nat Genet* 1997;15(4):397–401.
20. Danilovich N, Babu PS, Xing W, Gerdes M, Krishnamurthy H, Sairam MR. Estrogen deficiency, obesity, and skeletal abnormalities in follicle-stimulating hormone receptor knockout (FORKO) female mice. *Endocrinology* 2000;141(11):4295–4308.
21. Uehara S, Tsuchida M, Kanno T, Sasaki M, Nishikibe M, Fukamizu A. Late-onset obesity in mice transgenic for the human renin gene. *Int J Mol Med* 2003;11(6):723–727.
22. Vartanian V, Lowell B, Minko IG, *et al.* The metabolic syndrome resulting from a knockout of the NEIL1 DNA glycosylase. *Proc Natl Acad Sci USA* 2006;103(6):1864–1869.
23. Pendas AM, Folgueras AR, Llano E, *et al.* Diet-induced obesity and reduced skin cancer susceptibility in matrix metalloproteinase 19-deficient mice. *Mol Cell Biol* 2004;24(12):5304–5313.
24. Dong ZM, Gutierrez-Ramos JC, Coxon A, Mayadas TN, Wagner DD. A new class of obesity genes encodes leukocyte adhesion receptors. *Proc Natl Acad Sci USA* 1997;94(14):7526–7530.
25. Fulop AK, Foldes A, Buzas E, *et al.* Hyperleptinemia, visceral adiposity, and decreased glucose tolerance in mice with a targeted disruption of the histidine decarboxylase gene. *Endocrinology* 2003;144(10):4306–4314.
26. Fox EA, Byerly MS. A mechanism underlying mature-onset obesity: Evidence from the hyperphagic phenotype of brain-derived neurotrophic factor mutants. *Am J Physiol Regul Integr Comp Physiol* 2004;286(6):R994–1004.
27. Kiehl TR, Nechiporuk A, Figueroa KP, Keating MT, Huynh DP, Pulst SM. Generation and characterization of Sca2 (ataxin-2) knockout mice. *Biochem Biophys Res Commun* 2006;339(1):17–24.
28. Bale TL, Anderson KR, Roberts AJ, Lee KF, Nagy TR, Vale WW. Corticotropin-releasing factor receptor-2-deficient mice display abnormal homeostatic responses to challenges of increased dietary fat and cold. *Endocrinology* 2003;144(6):2580–2587.
29. Yamada M, Miyakawa T, Duttaroy A, *et al.* Mice lacking the M3 muscarinic acetylcholine receptor are hypophagic and lean. *Nature* 2001;410(6825):207–212.
30. Bymaster FP, McKinzie DL, Felder CC, Wess J. Use of M1-M5 muscarinic receptor knockout mice as novel tools to delineate the physiological roles of the muscarinic cholinergic system. *Neurochem Res* 2003;28(3–4):437–442.
31. Hahm S, Mizuno TM, Wu TJ, *et al.* Targeted deletion of the Vgf gene indicates that the encoded secretory peptide precursor plays a novel role in the regulation of energy balance. *Neuron* 1999;23(3): 537–548.
32. Xu M, Moratalla R, Gold LH, *et al.* Dopamine D1 receptor mutant mice are deficient in striatal expression of dynorphin and in dopamine-mediated behavioral responses. *Cell* 1994;79(4):729–742.
33. Alvaro JD, Taylor JR, Duman RS. Molecular and behavioral interactions between central melanocortins and cocaine. *J Pharmacol Exp Ther* 2003;304(1):391–399.
34. Watson E, Hahm S, Mizuno TM, *et al.* VGF ablation blocks the development of hyperinsulinemia and hyperglycemia in several mouse models of obesity. *Endocrinology* 2005;146(12):5151–5163.
35. Alexander J, Chang GQ, Dourmashkin JT, Leibowitz SF. Distinct phenotypes of obesity-prone AKR/J, DBA2J and C57BL/6J mice compared to control strains. *Int J Obes (Lond)* 2006;30(1):50–59.
36. Marsh DJ, Weingarth DT, Novi DE, *et al.* Melanin-concentrating hormone 1 receptor-deficient mice are lean, hyperactive, and hyperphagic and have altered metabolism. *Proc Natl Acad Sci USA* 2002;99(5):3240–3245.
37. Kokkotou E, Jeon JY, Wang X, *et al.* Mice with MCH ablation resist diet-induced obesity through strain-specific mechanisms. *Am J Physiol Regul Integr Comp Physiol* 2005;289(1):R117–124.
38. Villena JA, Viollet B, Andreelli F, Kahn A, Vaulont S, Sul HS. Induced adiposity and adipocyte hypertrophy in mice lacking the AMP-activated protein kinase-alpha2 subunit. *Diabetes* 2004;53(9): 2242–2249.
39. Ventre J, Doebber T, Wu M, *et al.* Targeted disruption of the tumor necrosis factor-alpha gene: Metabolic consequences in obese and nonobese mice. *Diabetes* 1997;46(9):1526–1531.
40. Austin CP, Battey JF, Bradley A, *et al.* The knockout mouse project. *Nat Genet* 2004;36(9):921–924.
41. Bultman SJ, Michaud EJ, Woychik RP. Molecular characterization of the mouse agouti locus. *Cell* 1992;71(7):1195–1204.
42. Ollmann MM, Wilson BD, Yang YK, *et al.* Antagonism of central melanocortin receptors in vitro and in vivo by agouti-related protein. *Science* 1997;278(5335):135–138.
43. Klebig ML, Wilkinson JE, Geisler JG, Woychik RP. Ectopic expression of the agouti gene in transgenic mice causes obesity, features of type II diabetes, and yellow fur. *Proc Natl Acad Sci USA* 1995;92(11):4728–4732.
44. Mynatt RL, Miltenberger RJ, Klebig ML, *et al.* Combined effects of insulin treatment and adipose tissue-specific agouti expression on the development of obesity. *Proc Natl Acad Sci USA* 1997;94(3): 919–922.
45. Mynatt RL, Stephens JM. Regulation of PPARgamma and obesity by agouti/melanocortin signaling in adipocytes. *Ann NY Acad Sci* 2003;994:141–146.
46. Valet P, Grujic D, Wade J, *et al.* Expression of human alpha 2-adrenergic receptors in adipose tissue of beta 3-adrenergic receptor-deficient mice promotes diet-induced obesity. *J Biol Chem* 2000; 275(44):34797–34802.
47. Wu KK, Wu TJ, Chin J, *et al.* Increased hypercholesterolemia and atherosclerosis in mice lacking both ApoE and leptin receptor. *Atherosclerosis* 2005;181(2):251–259.
48. Lowell BB, V SS, Hamann A, *et al.* Development of obesity in transgenic mice after genetic ablation of brown adipose tissue. *Nature* 1993;366(6457):740–742.
49. Naaz A, Holsberger DR, Iwamoto GA, Nelson A, Kiyokawa H, Cooke PS. Loss of cyclin-dependent kinase inhibitors produces adipocyte hyperplasia and obesity. *FASEB J* 2004;18(15):1925–1927.
50. Stenzel-Poore MP, Cameron VA, Vaughan J, Sawchenko PE, Vale W. Development of Cushing's syndrome in corticotropin-releasing factor transgenic mice. *Endocrinology* 1992;130(6):3378–3386.
51. Tartaglia LA, Dembski M, Weng X, *et al.* Identification and expression cloning of a leptin receptor, OB-R. *Cell* 1995;83(7): 1263–1271.
52. McQuade JA, Benoit SC, Xu M, Woods SC, Seeley RJ. High-fat diet induced adiposity in mice with targeted disruption of the dopamine-3 receptor gene. *Behav Brain Res* 2004;151(1–2):313–319.
53. Hohmann JG, Teklemichael DN, Weinshenker D, Wynick D, Clifton DK, Steiner RA. Obesity and endocrine dysfunction in mice with deletions of both neuropeptide Y and galanin. *Mol Cell Biol* 2004;24(7):2978–2985.
54. Cowey SL, Quast M, Belalcazar LM, *et al.* Abdominal obesity, insulin resistance, and colon carcinogenesis are increased in mutant mice lacking gastrin gene expression. *Cancer* 2005;103(12):2643–2653.
55. Pearce NJ, Arch JR, Clapham JC, *et al.* Development of glucose intolerance in male transgenic mice overexpressing human glycogen synthase kinase-3beta on a muscle-specific promoter. *Metabolism* 2004;53(10):1322–1330.
56. Narisawa S, Huang L, Iwasaki A, Hasegawa H, Alpers DH, Millan JL. Accelerated fat absorption in intestinal alkaline phosphatase knockout mice. *Mol Cell Biol* 2003;23(21):7525–7530.
57. Koh HJ, Lee SM, Son BG, *et al.* Cytosolic NADP+-dependent isocitrate dehydrogenase plays a key role in lipid metabolism. *J Biol Chem* 2004;279(38):39968–39974.
58. McMinn JE, Liu SM, Liu H, *et al.* Neuronal deletion of Lepr elicits diabesity in mice without affecting cold tolerance or fertility. *Am J Physiol Endocrinol Metab* 2005;289(3):E403–411.
59. Kero JT, Savontaus E, Mikola M, *et al.* Obesity in transgenic female mice with constitutively elevated luteinizing hormone secretion. *Am J Physiol Endocrinol Metab* 2003;285(4):E812–818.

60. Beattie JH, Wood AM, Newman AM, *et al.* Obesity and hyperleptinemia in metallothionein (-I and -II) null mice. *Proc Natl Acad Sci USA* 1998;95(1):358–363.
61. Hanada R, Teranishi H, Pearson JT, *et al.* Neuromedin U has a novel anorexigenic effect independent of the leptin signaling pathway. *Nat Med* 2004;10(10):1067–1073.
62. Wiedmer T, Zhao J, Li L, *et al.* Adiposity, dyslipidemia, and insulin resistance in mice with targeted deletion of phospholipid scramblase 3 (PLSCR3). *Proc Natl Acad Sci USA* 2004;101(36):13296–13301.
63. Kim KH, Zhao L, Moon Y, Kang C, Sul HS. Dominant inhibitory adipocyte-specific secretory factor (ADSF)/resistin enhances adipogenesis and improves insulin sensitivity. *Proc Natl Acad Sci USA* 2004;101(17):6780–6785.
64. Ren D, Li M, Duan C, Rui L. Identification of SH2-B as a key regulator of leptin sensitivity, energy balance, and body weight in mice. *Cell Metab* 2005;2(2):95–104.
65. Holder JL, Jr., Zhang L, Kublaoui BM, *et al.* Sim1 gene dosage modulates the homeostatic feeding response to increased dietary fat in mice. *Am J Physiol Endocrinol Metab* 2004;287(1):E105–113.
66. Michaud JL, Boucher F, Melnyk A, *et al.* Sim1 haploinsufficiency causes hyperphagia, obesity and reduction of the paraventricular nucleus of the hypothalamus. *Hum Mol Genet* 2001;10(14):1465–1473.
67. Ma YH, Hu JH, Zhou XG, *et al.* Transgenic mice overexpressing gamma-aminobutyric acid transporter subtype I develop obesity. *Cell Res* 2000;10(4):303–310.
68. Horton JD, Shimomura I, Ikemoto S, Bashmakov Y, Hammer RE. Overexpression of sterol regulatory element-binding protein-1a in mouse adipose tissue produces adipocyte hypertrophy, increased fatty acid secretion, and fatty liver. *J Biol Chem* 2003;278(38): 36652–36660.
69. Gao Q, Wolfgang MJ, Neschen S, *et al.* Disruption of neural signal transducer and activator of transcription 3 causes obesity, diabetes, infertility, and thermal dysregulation. *Proc Natl Acad Sci USA* 2004;101(13):4661–4666.
70. Hibuse T, Maeda N, Funahashi T, *et al.* Aquaporin 7 deficiency is associated with development of obesity through activation of adipose glycerol kinase. *Proc Natl Acad Sci USA* 2005;102(31): 10993–10998.
71. Jones ME, Thorburn AW, Britt KL, *et al.* Aromatase-deficient (ArKO) mice have a phenotype of increased adiposity. *Proc Natl Acad Sci USA* 2000;97(23):12735–12740.
72. Sato T, Matsumoto T, Yamada T, Watanabe T, Kawano H, Kato S. Late onset of obesity in male androgen receptor-deficient (AR KO) mice. *Biochem Biophys Res Commun* 2003;300(1):167–171.
73. Hotamisligil GS, Johnson RS, Distel RJ, Ellis R, Papaioannou VE, Spiegelman BM. Uncoupling of obesity from insulin resistance through a targeted mutation in aP2, the adipocyte fatty acid binding protein. *Science* 1996;274(5291):1377–1379.
74. Bachman ES, Dhillon H, Zhang CY, *et al.* BetaAR signaling required for diet-induced thermogenesis and obesity resistance. *Science* 2002;297(5582):843–845.
75. Kulaga HM, Leitch CC, Eichers ER, *et al.* Loss of BBS proteins causes anosmia in humans and defects in olfactory cilia structure and function in the mouse. *Nat Genet* 2004;36(9):994–998.
76. Mykytyn K, Mullins RF, Andrews M, *et al.* Bardet-Biedl syndrome type 4 (BBS4)-null mice implicate Bbs4 in flagella formation but not global cilia assembly. *Proc Natl Acad Sci USA* 2004;101(23): 8664–8669.
77. Kernie SG, Liebl DJ, Parada LF. BDNF regulates eating behavior and locomotor activity in mice. *EMBO J* 2000;19(6):1290–1300.
78. Ohki-Hamazaki H, Watase K, Yamamoto K, *et al.* Mice lacking bombesin receptor subtype-3 develop metabolic defects and obesity. *Nature* 1997;390(6656):165–169.
79. Lovejoy DA, Aubry JM, Turnbull A, *et al.* Ectopic expression of the CRF-binding protein: Minor impact on HPA axis regulation but induction of sexually dimorphic weight gain. *J Neuroendocrinol* 1998;10(7):483–491.
80. Cawley NX, Zhou J, Hill JM, *et al.* The carboxypeptidase E knockout mouse exhibits endocrinological and behavioral deficits. *Endocrinology* 2004;145(12):5807–5819.
81. Naggert JK, Fricker LD, Varlamov O, *et al.* Hyperproinsulinaemia in obese fat/fat mice associated with a carboxypeptidase E mutation which reduces enzyme activity. *Nat Genet* 1995;10(2):135–142.
82. Ferre T, Riu E, Franckhauser S, Agudo J, Bosch F. Long-term overexpression of glucokinase in the liver of transgenic mice leads to insulin resistance. *Diabetologia* 2003;46(12):1662–1668.
83. Wang W, Yang Y, Meng Y, Shi Y. GDF-3 is an adipogenic cytokine under high fat dietary condition. *Biochem Biophys Res Commun* 2004;321(4):1024–1031.
84. Ishii M, Fei H, Friedman JM. Targeted disruption of GPR7, the endogenous receptor for neuropeptides B and W, leads to metabolic defects and adult-onset obesity. *Proc Natl Acad Sci USA* 2003; 100(18):10540–10545.
85. Gu W, Geddes BJ, Zhang C, Foley KP, Stricker-Krongrad A. The prolactin-releasing peptide receptor (GPR10) regulates body weight homeostasis in mice. *J Mol Neurosci* 2004;22(1–2):93–103.
86. McClung JP, Roneker CA, Mu W, *et al.* Development of insulin resistance and obesity in mice overexpressing cellular glutathione peroxidase. *Proc Natl Acad Sci USA* 2004;101(24):8852–8857.
87. Kellendonk C, Eiden S, Kretz O, *et al.* Inactivation of the GR in the nervous system affects energy accumulation. *Endocrinology* 2002;143(6):2333–2340.
88. Masaki T, Chiba S, Yasuda T, *et al.* Involvement of hypothalamic histamine H1 receptor in the regulation of feeding rhythm and obesity. *Diabetes* 2004;53(9):2250–2260.
89. Takahashi K, Suwa H, Ishikawa T, Kotani H. Targeted disruption of H3 receptors results in changes in brain histamine tone leading to an obese phenotype. *J Clin Invest* 2002;110(12):1791–1799.
90. Cai A, Hyde JF. The human growth hormone-releasing hormone transgenic mouse as a model of modest obesity: Differential changes in leptin receptor (OBR) gene expression in the anterior pituitary and hypothalamus after fasting and OBR localization in somatotrophs. *Endocrinology* 1999;140(8):3609–3614.
91. Hara J, Beuckmann CT, Nambu T, *et al.* Genetic ablation of orexin neurons in mice results in narcolepsy, hypophagia, and obesity. *Neuron* 2001;30(2):345–354.
92. Paterson JM, Morton NM, Fievet C, *et al.* Metabolic syndrome without obesity: Hepatic overexpression of 11beta-hydroxysteroid dehydrogenase type 1 in transgenic mice. *Proc Natl Acad Sci USA* 2004;101(18):7088–7093.
93. Nonogaki K, Abdallah L, Goulding EH, Bonasera SJ, Tecott LH. Hyperactivity and reduced energy cost of physical activity in serotonin 5-HT(2C) receptor mutant mice. *Diabetes* 2003;52(2):315–320.
94. Garcia MC, Wernstedt I, Berndtsson A, *et al.* Mature-onset obesity in interleukin-1 receptor I knockout mice. *Diabetes* 2006;55(5): 1205–1213.
95. Chida D, Osaka T, Hashimoto O, Iwakura Y. Combined interleukin-6 and interleukin-1 deficiency causes obesity in young mice. *Diabetes* 2006;55(4):971–977.
96. Di Gregorio GB, Hensley L, Lu T, Ranganathan G, Kern PA. Lipid and carbohydrate metabolism in mice with a targeted mutation in the IL-6 gene: Absence of development of age-related obesity. *Am J Physiol Endocrinol Metab* 2004;287(1):E182–187.
97. Wallenius K, Wallenius V, Sunter D, Dickson SL, Jansson JO. Intracerebroventricular interleukin-6 treatment decreases body fat in rats. *Biochem Biophys Res Commun* 2002;293(1):560–565.
98. Netea MG, Joosten LA, Lewis E, *et al.* Deficiency of interleukin-18 in mice leads to hyperphagia, obesity and insulin resistance. *Nat Med* 2006;12(6):650–656.
99. Kubota N, Terauchi Y, Tobe K, *et al.* Insulin receptor substrate 2 plays a crucial role in beta cells and the hypothalamus. *J Clin Invest* 2004;114(7):917–927.
100. Lin X, Taguchi A, Park S, *et al.* Dysregulation of insulin receptor substrate 2 in beta cells and brain causes obesity and diabetes. *J Clin Invest* 2004;114(7):908–916.

101. Bruning JC, Michael MD, Winnay JN, *et al.* A muscle-specific insulin receptor knockout exhibits features of the metabolic syndrome of NIDDM without altering glucose tolerance. *Mol Cell* 1998;2(5):559–569.
102. Cariou B, Postic C, Boudou P, *et al.* Cellular and molecular mechanisms of adipose tissue plasticity in muscle insulin receptor knockout mice. *Endocrinology* 2004;145(4):1926–1932.
103. Winarto A, Miki T, Seino S, Iwanaga T. Morphological changes in pancreatic islets of KATP channel-deficient mice: The involvement of KATP channels in the survival of insulin cells and the maintenance of islet architecture. *Arch Histol Cytol* 2001;64(1):59–67.
104. Butler AA, Kesterson RA, Khong K, *et al.* A unique metabolic syndrome causes obesity in the melanocortin-3 receptor-deficient mouse. *Endocrinology* 2000;141(9):3518–3521.
105. Chen AS, Marsh DJ, Trumbauer ME, *et al.* Inactivation of the mouse melanocortin-3 receptor results in increased fat mass and reduced lean body mass. *Nat Genet* 2000;26(1):97–102.
106. Coyle CA, Jing E, Hosmer T, Powers JB, Wade G, Good DJ. Reduced voluntary activity precedes adult-onset obesity in Nhlh2 knockout mice. *Physiol Behav* 2002;77(2–3):387–402.
107. Kushi A, Sasai H, Koizumi H, Takeda N, Yokoyama M, Nakamura M. Obesity and mild hyperinsulinemia found in neuropeptide Y-Y1 receptor-deficient mice. *Proc Natl Acad Sci USA* 1998;95(26): 15659–15664.
108. Marsh DJ, Hollopeter G, Kafer KE, Palmiter RD. Role of the Y5 neuropeptide Y receptor in feeding and obesity. *Nat Med* 1998; 4(6):718–721.
109. Remaury A, Vita N, Gendreau S, *et al.* Targeted inactivation of the neurotensin type 1 receptor reveals its role in body temperature control and feeding behavior but not in analgesia. *Brain Res* 2002;953(1–2):63–72.
110. Rodriguez A, Duran A, Selloum M, *et al.* Mature-onset obesity and insulin resistance in mice deficient in the signaling adapter p62. *Cell Metab* 2006;3(3):211–222.
111. Wang ZQ, Auer B, Stingl L, *et al.* Mice lacking ADPRT and poly(ADP-ribosyl)ation develop normally but are susceptible to skin disease. *Genes Dev* 1995;9(5):509–520.
112. Zhu X, Orci L, Carroll R, Norrbom C, Ravazzola M, Steiner DF. Severe block in processing of proinsulin to insulin accompanied by elevation of des-64,65 proinsulin intermediates in islets of mice lacking prohormone convertase 1/3. *Proc Natl Acad Sci USA* 2002;99(16):10299–10304.
113. Zhu X, Zhou A, Dey A, *et al.* Disruption of PC1/3 expression in mice causes dwarfism and multiple neuroendocrine peptide processing defects. *Proc Natl Acad Sci USA* 2002;99(16):10293–10298.
114. Lloyd DJ, Bohan S, Gekakis N. Obesity, hyperphagia and increased metabolic efficiency in Pc1 mutant mice. *Hum Mol Genet* 2006; 15(11):1884–1893.
115. Gao Z, Wang ZQ, Zhang X, *et al.* Inactivation of PKCθ leads to increased susceptibility to obesity and dietary insulin resistance in mice. *Am J Physiol Endocrinol Metab* 2007;292(1):E84–91.
116. Serra C, Federici M, Buongiorno A, *et al.* Transgenic mice with dominant negative PKC-theta in skeletal muscle: A new model of insulin resistance and obesity. *J Cell Physiol* 2003;196(1):89–97.
117. Wei S, Feng Y, Che FY, *et al.* Obesity and diabetes in transgenic mice expressing proSAAS. *J Endocrinol* 2004;180(3):357–368.
118. Harvey NL, Srinivasan RS, Dillard ME, *et al.* Lymphatic vascular defects promoted by Prox1 haploinsufficiency cause adult-onset obesity. *Nat Genet* 2005;37(10):1072–1081.
119. Norris AW, Chen L, Fisher SJ, *et al.* Muscle-specific PPARgamma-deficient mice develop increased adiposity and insulin resistance but respond to thiazolidinediones. *J Clin Invest* 2003;112(4):608–618.
120. Plum L, Ma X, Hampel B, *et al.* Enhanced PIP3 signaling in POMC neurons causes KATP channel activation and leads to diet-sensitive obesity. *J Clin Invest* 2006;116(7):1886–1901.
121. Delibegovic M, Armstrong CG, Dobbie L, Watt PW, Smith AJ, Cohen PT. Disruption of the striated muscle glycogen targeting subunit PPP1R3A of protein phosphatase 1 leads to increased weight gain, fat deposition, and development of insulin resistance. *Diabetes* 2003;52(3):596–604.
122. Fain JN, Del Mar NA, Meade CA, Reiner A, Goldowitz D. Abnormalities in the functioning of adipocytes from R6/2 mice that are transgenic for the Huntington's disease mutation. *Hum Mol Genet* 2001;10(2):145–152.
123. Reizes O, Lincecum J, Wang Z, *et al.* Transgenic expression of syndecan-1 uncovers a physiological control of feeding behavior by syndecan-3. *Cell* 2001;106(1):105–116.
124. Bradshaw AD, Puolakkainen P, Dasgupta J, Davidson JM, Wight TN, Helene Sage E. SPARC-null mice display abnormalities in the dermis characterized by decreased collagen fibril diameter and reduced tensile strength. *J Invest Dermatol* 2003;120(6):949–955.
125. Xu AW, Kaelin CB, Morton GJ, *et al.* Effects of hypothalamic neurodegeneration on energy balance. *PLoS Biol* 2005;3(12): e415.
126. Liu YY, Schultz JJ, Brent GA. A thyroid hormone receptor alpha gene mutation (P398H) is associated with visceral adiposity and impaired catecholamine-stimulated lipolysis in mice. *J Biol Chem* 2003;278(40):38913–38920.
127. Kleyn PW, Fan W, Kovats SG, *et al.* Identification and characterization of the mouse obesity gene tubby: A member of a novel gene family. *Cell* 1996;85(2):281–290.
128. Noben-Trauth K, Naggert JK, North MA, Nishina PM. A candidate gene for the mouse mutation tubby. *Nature* 1996;380(6574): 534–538.
129. Stubdal H, Lynch CA, Moriarty A, *et al.* Targeted deletion of the tub mouse obesity gene reveals that tubby is a loss-of-function mutation. *Mol Cell Biol* 2000;20(3):878–882.
130. Maltais LJ, Lane PW, Beamer WG. Anorexia, a recessive mutation causing starvation in preweanling mice. *J Hered* 1984;75(6):468–472.
131. Abu-Elheiga L, Matzuk MM, Abo-Hashema KA, Wakil SJ. Continuous fatty acid oxidation and reduced fat storage in mice lacking acetyl-CoA carboxylase 2. *Science* 2001;291(5513):2613–2616.
132. Shindo T, Kurihara H, Kuno K, *et al.* ADAMTS-1: A metalloproteinase-disintegrin essential for normal growth, fertility, and organ morphology and function. *J Clin Invest* 2000;105(10):1345–1352.
133. Gray SL, Cummings KJ, Jirik FR, Sherwood NM. Targeted disruption of the pituitary adenylate cyclase-activating polypeptide gene results in early postnatal death associated with dysfunction of lipid and carbohydrate metabolism. *Mol Endocrinol* 2001;15(10):1739–1747.
134. Lai KM, Gonzalez M, Poueymirou WT, *et al.* Conditional activation of akt in adult skeletal muscle induces rapid hypertrophy. *Mol Cell Biol* 2004;24(21):9295–9304.
135. Garofalo RS, Orena SJ, Rafidi K, *et al.* Severe diabetes, age-dependent loss of adipose tissue, and mild growth deficiency in mice lacking Akt2/PKB beta. *J Clin Invest* 2003;112(2):197–208.
136. Xia Z, Sniderman AD, Cianflone K. Acylation-stimulating protein (ASP) deficiency induces obesity resistance and increased energy expenditure in ob/ob mice. *J Biol Chem* 2002;277(48):45874–45879.
137. Razani B, Combs TP, Wang XB, *et al.* Caveolin-1-deficient mice are lean, resistant to diet-induced obesity, and show hypertriglyceridemia with adipocyte abnormalities. *J Biol Chem* 2002;277(10): 8635–8647.
138. Ravinet Trillou C, Delgorge C, Menet C, Arnone M, Soubrie P. CB1 cannabinoid receptor knockout in mice leads to leanness, resistance to diet-induced obesity and enhanced leptin sensitivity. *Int J Obes Relat Metab Disord* 2004;28(4):640–648.
139. Iizuka K, Bruick RK, Liang G, Horton JD, Uyeda K. Deficiency of carbohydrate response element-binding protein (ChREBP) reduces lipogenesis as well as glycolysis. *Proc Natl Acad Sci USA* 2004; 101(19):7281–7286.
140. Zhou Z, Yon Toh S, Chen Z, *et al.* Cidea-deficient mice have lean phenotype and are resistant to obesity. *Nat Genet* 2003;35(1): 49–56.

141. Oishi K, Atsumi G, Sugiyama S, *et al.* Disrupted fat absorption attenuates obesity induced by a high-fat diet in Clock mutant mice. *FEBS Lett* 2006;580(1):127–130.
142. Turek FW, Joshu C, Kohsaka A, *et al.* Obesity and metabolic syndrome in circadian Clock mutant mice. *Science* 2005;308(5724): 1043–1045.
143. Fetissov SO, Bergstrom U, Johansen JE, Hokfelt T, Schalling M, Ranscht B. Alterations of arcuate nucleus neuropeptidergic development in contactin-deficient mice: Comparison with anorexia and food-deprived mice. *Eur J Neurosci* 2005;22(12):3217–3228.
144. Wolfgang MJ, Kurama T, Dai Y, *et al.* The brain-specific carnitine palmitoyltransferase-1c regulates energy homeostasis. *Proc Natl Acad Sci USA* 2006;103(19):7282–7287.
145. Szczypka MS, Rainey MA, Kim DS, *et al.* Feeding behavior in dopamine-deficient mice. *Proc Natl Acad Sci USA* 1999;96(21): 12138–12143.
146. Zhou QY, Palmiter RD. Dopamine-deficient mice are severely hypoactive, adipsic, and aphagic. *Cell* 1995;83(7):1197–1209.
147. Sabatakos G, Sims NA, Chen J, *et al.* Overexpression of DeltaFosB transcription factor(s) increases bone formation and inhibits adipogenesis. *Nat Med* 2000;6(9):985–990.
148. Smith SJ, Cases S, Jensen DR, *et al.* Obesity resistance and multiple mechanisms of triglyceride synthesis in mice lacking Dgat. *Nat Genet* 2000;25(1):87–90.
149. Wechsler A, Brafman A, Shafir M, *et al.* Generation of viable cholesterol-free mice. *Science* 2003;302(5653):2087.
150. Lee K, Villena JA, Moon YS, *et al.* Inhibition of adipogenesis and development of glucose intolerance by soluble preadipocyte factor-1 (Pref-1). *J Clin Invest* 2003;111(4):453–461.
151. Tsukiyama-Kohara K, Poulin F, Kohara M, *et al.* Adipose tissue reduction in mice lacking the translational inhibitor 4E-BP1. *Nat Med* 2001;7(10):1128–1132.
152. Luo J, Sladek R, Carrier J, Bader JA, Richard D, Giguere V. Reduced fat mass in mice lacking orphan nuclear receptor estrogen-related receptor alpha. *Mol Cell Biol* 2003;23(22):7947–7956.
153. Peterfy M, Phan J, Xu P, Reue K. Lipodystrophy in the fld mouse results from mutation of a new gene encoding a nuclear protein, lipin. *Nat Genet* 2001;27(1):121–124.
154. Cederberg A, Gronning LM, Ahren B, Tasken K, Carlsson P, Enerback S. FOXC2 is a winged helix gene that counteracts obesity, hypertriglyceridemia, and diet-induced insulin resistance. *Cell* 2001;106(5):563–573.
155. Schmidt A, Marescau B, Boehm EA, *et al.* Severely altered guanidino compound levels, disturbed body weight homeostasis and impaired fertility in a mouse model of guanidinoacetate N-methyltransferase (GAMT) deficiency. *Hum Mol Genet* 2004;13(9): 905–921.
156. Lin J, Arnold HB, Della-Fera MA, Azain MJ, Hartzell DL, Baile CA. Myostatin knockout in mice increases myogenesis and decreases adipogenesis. *Biochem Biophys Res Commun* 2002;291(3):701–706.
157. McPherron AC, Lee SJ. Suppression of body fat accumulation in myostatin-deficient mice. *J Clin Invest* 2002;109(5):595–601.
158. Rossi J, Herzig KH, Voikar V, Hiltunen PH, Segerstrale M, Airaksinen MS. Alimentary tract innervation deficits and dysfunction in mice lacking GDNF family receptor alpha2. *J Clin Invest* 2003; 112(5):707–716.
159. Lall S, Balthasar N, Carmignac D, *et al.* Physiological studies of transgenic mice overexpressing growth hormone (GH) secretagogue receptor 1A in GH-releasing hormone neurons. *Endocrinology* 2004;145(4):1602–1611.
160. Miyawaki K, Yamada Y, Ban N, *et al.* Inhibition of gastric inhibitory polypeptide signaling prevents obesity. *Nat Med* 2002;8(7):738–742.
161. Katz EB, Stenbit AE, Hatton K, DePinho R, Charron MJ. Cardiac and adipose tissue abnormalities but not diabetes in mice deficient in GLUT4. *Nature* 1995;377(6545):151–155.
162. Schwindinger WF, Giger KE, Betz KS, *et al.* Mice with deficiency of G protein gamma3 are lean and have seizures. *Mol Cell Biol* 2004;24(17):7758–7768.
163. Hammond LE, Gallagher PA, Wang S, *et al.* Mitochondrial glycerol-3-phosphate acyltransferase-deficient mice have reduced weight and liver triacylglycerol content and altered glycerolipid fatty acid composition. *Mol Cell Biol* 2002;22(23):8204–8214.
164. Morton NM, Paterson JM, Masuzaki H, *et al.* Novel adipose tissue-mediated resistance to diet-induced visceral obesity in 11 beta-hydroxysteroid dehydrogenase type 1-deficient mice. *Diabetes* 2004;53(4):931–938.
165. Harada K, Shen WJ, Patel S, *et al.* Resistance to high-fat diet-induced obesity and altered expression of adipose-specific genes in HSL-deficient mice. *Am J Physiol Endocrinol Metab* 2003;285(6): E1182–1195.
166. Wang SP, Laurin N, Himms-Hagen J, *et al.* The adipose tissue phenotype of hormone-sensitive lipase deficiency in mice. *Obes Res* 2001;9(2):119–128.
167. Horai R, Asano M, Sudo K, *et al.* Production of mice deficient in genes for interleukin (IL)-1alpha, IL-1beta, IL-1alpha/beta, and IL-1 receptor antagonist shows that IL-1beta is crucial in turpentine-induced fever development and glucocorticoid secretion. *J Exp Med* 1998;187(9):1463–1475.
168. Matsuki T, Horai R, Sudo K, Iwakura Y. IL-1 plays an important role in lipid metabolism by regulating insulin levels under physiological conditions. *J Exp Med* 2003;198(6):877–888.
169. Xu J, Koni PA, Wang P, *et al.* The voltage-gated potassium channel Kv1.3 regulates energy homeostasis and body weight. *Hum Mol Genet* 2003;12(5):551–559.
170. Chen M, Gavrilova O, Zhao WQ, *et al.* Increased glucose tolerance and reduced adiposity in the absence of fasting hypoglycemia in mice with liver-specific Gs alpha deficiency. *J Clin Invest* 2005; 115(11):3217–3227.
171. Eicher EM, Beamer WG. Inherited ateliotic dwarfism in mice. Characteristics of the mutation, little, on chromosome 6. *J Hered* 1976; 67(2):87–91.
172. Gerin I, Dolinsky VW, Shackman JG, *et al.* LXRbeta is required for adipocyte growth, glucose homeostasis, and beta cell function. *J Biol Chem* 2005;280(24):23024–23031.
173. Hirosumi J, Tuncman G, Chang L, *et al.* A central role for JNK in obesity and insulin resistance. *Nature* 2002;420(6913):333–336.
174. Shimada M, Tritos NA, Lowell BB, Flier JS, Maratos-Flier E. Mice lacking melanin-concentrating hormone are hypophagic and lean. *Nature* 1998;396(6712):670–674.
175. Alon T, Friedman JM. Late-onset leanness in mice with targeted ablation of melanin concentrating hormone neurons. *J Neurosci* 2006;26(2):389–397.
176. Miller KA, Gunn TM, Carrasquillo MM, Lamoreux ML, Galbraith DB, Barsh GS. Genetic studies of the mouse mutations mahogany and mahoganoid. *Genetics* 1997;146(4):1407–1415.
177. Whitson RH, Tsark W, Huang TH, Itakura K. Neonatal mortality and leanness in mice lacking the ARID transcription factor Mrf-2. *Biochem Biophys Res Commun* 2003;312(4):997–1004.
178. Erickson JC, Hollopeter G, Palmiter RD. Attenuation of the obesity syndrome of ob/ob mice by the loss of neuropeptide Y. *Science* 1996;274(5293):1704–1707.
179. Tansey JT, Sztalryd C, Gruia-Gray J, *et al.* Perilipin ablation results in a lean mouse with aberrant adipocyte lipolysis, enhanced leptin production, and resistance to diet-induced obesity. *Proc Natl Acad Sci USA* 2001;98(11):6494–6499.
180. Campbell PK, Waymire KG, Heier RL, *et al.* Mutation of a novel gene results in abnormal development of spermatid flagella, loss of intermale aggression and reduced body fat in mice. *Genetics* 2002;162(1):307–320.
181. Regnard C, Fesquet D, Janke C, *et al.* Characterisation of PGs1, a subunit of a protein complex co-purifying with tubulin polyglutamylase. *J Cell Sci* 2003;116(Pt. 20):4181–4190.
182. Chen D, Mauvais-Jarvis F, Bluher M, *et al.* p50alpha/p55alpha phosphoinositide 3-kinase knockout mice exhibit enhanced insulin sensitivity. *Mol Cell Biol* 2004;24(1):320–329.
183. Lamia KA, Peroni OD, Kim YB, Rameh LE, Kahn BB, Cantley LC. Increased insulin sensitivity and reduced adiposity in phosphati-

dylinositol 5-phosphate 4-kinase beta-/- mice. *Mol Cell Biol* 2004; 24(11):5080–5087.

184. Ueno N, Inui A, Iwamoto M, *et al.* Decreased food intake and body weight in pancreatic polypeptide-overexpressing mice. *Gastroenterology* 1999;117(6):1427–1432.

185. Wang YX, Lee CH, Tiep S, *et al.* Peroxisome-proliferator-activated receptor delta activates fat metabolism to prevent obesity. *Cell* 2003;113(2):159–170.

186. Jones JR, Barrick C, Kim KA, *et al.* Deletion of PPARgamma in adipose tissues of mice protects against high fat diet-induced obesity and insulin resistance. *Proc Natl Acad Sci USA* 2005;102(17): 6207–6212.

187. Kubota N, Terauchi Y, Miki H, *et al.* PPAR gamma mediates high-fat diet-induced adipocyte hypertrophy and insulin resistance. *Mol Cell* 1999;4(4):597–609.

188. Bence KK, Delibegovic M, Xue B, *et al.* Neuronal PTP1B regulates body weight, adiposity and leptin action. *Nat Med* 2006;12(8): 917–924.

189. Klaman LD, Boss O, Peroni OD, *et al.* Increased energy expenditure, decreased adiposity, and tissue-specific insulin sensitivity in protein-tyrosine phosphatase 1B-deficient mice. *Mol Cell Biol* 2000;20(15):5479–5489.

190. Wang Z, Moro E, Kovacs K, Yu R, Melmed S. Pituitary tumor transforming gene-null male mice exhibit impaired pancreatic beta cell proliferation and diabetes. *Proc Natl Acad Sci USA* 2003;100(6): 3428–3432.

191. Cummings DE, Brandon EP, Planas JV, Motamed K, Idzerda RL, McKnight GS. Genetically lean mice result from targeted disruption of the RII beta subunit of protein kinase A. *Nature* 1996;382(6592): 622–626.

192. Font de Mora J, Esteban LM, Burks DJ, *et al.* Ras-GRF1 signaling is required for normal beta-cell development and glucose homeostasis. *EMBO J* 2003;22(12):3039–3049.

193. Um SH, Frigerio F, Watanabe M, *et al.* Absence of S6K1 protects against age- and diet-induced obesity while enhancing insulin sensitivity. *Nature* 2004;431(7005):200–205.

194. El-Haschimi K, Dufresne SD, Hirshman MF, Flier JS, Goodyear LJ, Bjorbaek C. Insulin resistance and lipodystrophy in mice lacking ribosomal S6 kinase 2. *Diabetes* 2003;52(6):1340–1346.

195. Ntambi JM, Miyazaki M, Stoehr JP, *et al.* Loss of stearoyl-CoA desaturase-1 function protects mice against adiposity. *Proc Natl Acad Sci USA* 2002;99(17):11482–11486.

196. Mori H, Hanada R, Hanada T, *et al.* Socs3 deficiency in the brain elevates leptin sensitivity and confers resistance to diet-induced obesity. *Nat Med* 2004;10(7):739–743.

197. Udy GB, Towers RP, Snell RG, *et al.* Requirement of STAT5b for sexual dimorphism of body growth rates and liver gene expression. *Proc Natl Acad Sci USA* 1997;94(14):7239–7244.

198. Clouthier DE, Comerford SA, Hammer RE. Hepatic fibrosis, glomerulosclerosis, and a lipodystrophy-like syndrome in PEPCK-TGF-beta1 transgenic mice. *J Clin Invest* 1997;100(11):2697–2713.

199. Xu ZP, Wawrousek EF, Piatigorsky J. Transketolase haploinsufficiency reduces adipose tissue and female fertility in mice. *Mol Cell Biol* 2002;22(17):6142–6147.

200. Liu X, Rossmeisl M, McClaine J, Riachi M, Harper ME, Kozak LP. Paradoxical resistance to diet-induced obesity in UCP1-deficient mice. *J Clin Invest* 2003;111(3):399–407.

201. Couplan E, Gelly C, Goubern M, *et al.* High level of uncoupling protein 1 expression in muscle of transgenic mice selectively affects muscles at rest and decreases their IIb fiber content. *J Biol Chem* 2002;277(45):43079–43088.

202. Horvath TL, Diano S, Miyamoto S, *et al.* Uncoupling proteins-2 and 3 influence obesity and inflammation in transgenic mice. *Int J Obes Relat Metab Disord* 2003;27(4):433–442.

203. Choi YS, Hong SB, Jeon HK, *et al.* Insertional mutation in the intron 1 of Unc5h3 gene induces ataxic, lean and hyperactive phenotype in mice. *Exp Anim* 2003;52(4):273–283.

204. Longo KA, Wright WS, Kang S, *et al.* Wnt10b inhibits development of white and brown adipose tissues. *J Biol Chem* 2004;279(34): 35503–35509.

205. Taylor GA, Carballo E, Lee DM, *et al.* A pathogenetic role for TNF alpha in the syndrome of cachexia, arthritis, and autoimmunity resulting from tristetraprolin (TTP) deficiency. *Immunity* 1996;4(5): 445–454.

206. West DB, Boozer CN, Moody DL, Atkinson RL. Dietary obesity in nine inbred mouse strains. *Am J Physiol Regul Integr Comp Physiol* 1992;262(6Pt. 2):R1025–1032.

207. West DB, Goudey-Lefevre J, York B, Truett GE. Dietary obesity linked to genetic loci on chromosomes 9 and 15 in a polygenic mouse model. *J Clin Invest* 1994;94(4):1410–1416.

208. Wagener A, Schmitt AO, Aksu S, Schlote W, Neuschl C, Brockmann GA. Genetic, sex, and diet effects on body weight and obesity in the Berlin Fat Mouse inbred lines. *Physiol Genomics* 2006;27(3): 264–270.

209. Fisler JS, Warden CH, Pace MJ, Lusis AJ. BSB: A new mouse model of multigenic obesity. *Obes Res* 1993;1(4):271–280.

210. Warden CH, Fisler JS, Shoemaker SM, *et al.* Identification of four chromosomal loci determining obesity in a multifactorial mouse model. *J Clin Invest* 1995;95(4):1545–1552.

211. Bevova MR, Aulchenko YS, Aksu S, Renne U, Brockmann GA. Chromosome-wise dissection of the genome of the extremely big mouse line DU6i. *Genetics* 2006;172(1):401–410.

212. Bunger L, Hill WG. Inbred lines of mice derived from long-term divergent selection on fat content and body weight. *Mamm Genome* 1999;10(6):645–648.

213. Horvat S, Bunger L, Falconer VM, *et al.* Mapping of obesity QTLs in a cross between mouse lines divergently selected on fat content. *Mamm Genome* 2000;11(1):2–7.

214. Ikeda H. KK mouse. *Diabetes Res Clin Pract* 1994;24(Suppl.): S313–316.

215. Suto J, Matsuura S, Imamura K, Yamanaka H, Sekikawa K. Genetics of obesity in KK mouse and effects of A(y) allele on quantitative regulation. *Mamm Genome* 1998;9(7):506–510.

216. Allan MF, Eisen EJ, Pomp D. The M16 mouse: An outbred animal model of early onset polygenic obesity and diabesity. *Obes Res* 2004;12(9):1397–1407.

217. Jerez-Timaure NC, Eisen EJ, Pomp D. Fine mapping of a QTL region with large effects on growth and fatness on mouse chromosome 2. *Physiol Genomics* 2005;21(3):411–422.

218. Leiter EH, Reifsnyder PC. Differential levels of diabetogenic stress in two new mouse models of obesity and type 2 diabetes. *Diabetes* 2004;53(Suppl. 1):S4–11.

219. Bielschowsky F, Bielschowsky M. The New Zealand strain of obese mice; their response to stilboestrol and to insulin. *Aust J Exp Biol Med Sci* 1956;34(3):181–198.

220. Anunciado RV, Horio F, Ohno T, Tanaka S, Nishimura M, Namikawa T. Characterization of hyperinsulinemic recombinant inbred (RI) strains (SMXA-5 and SMXA-9) derived from normoinsulinemic SM/J and A/J mice. *Exp Anim* 2000;49(2):83–90.

221. Kobayashi M, Io F, Kawai T, *et al.* Major quantitative trait locus on chromosome 2 for glucose tolerance in diabetic SMXA-5 mouse established from non-diabetic SM/J and A/J strains. *Diabetologia* 2006;49(3):486–495.

222. Kim JH, Sen S, Avery CS, *et al.* Genetic analysis of a new mouse model for non-insulin-dependent diabetes. *Genomics* 2001;74(3): 273–286.

223. Kim JH, Stewart TP, Zhang W, Kim HY, Nishina PM, Naggert JK. Type 2 diabetes mouse model TallyHo carries an obesity gene on chromosome 6 that exaggerates dietary obesity. *Physiol Genomics* 2005;22(2):171–181.

224. Hirayama I, Yi Z, Izumi S, *et al.* Genetic analysis of obese diabetes in the TSOD mouse. *Diabetes* 1999;48(5):1183–1191.

225. Mizutani S, Gomi H, Hirayama I, Izumi T. Chromosome 2 locus Nidd5 has a potent effect on adiposity in the TSOD mouse. *Mamm Genome* 2006;17(5):375–384.

226. Jong MC, Voshol PJ, Muurling M, *et al.* Protection from obesity and insulin resistance in mice overexpressing human apolipoprotein C1. *Diabetes* 2001;50(12):2779–2785.

227. Soloveva V, Graves RA, Rasenick MM, Spiegelman BM, Ross SR. Transgenic mice overexpressing the beta 1-adrenergic receptor in

adipose tissue are resistant to obesity. *Mol Endocrinol* 1997;11(1):27–38.

228. Chen A, Mumick S, Zhang C, *et al.* Diet induction of monocyte chemoattractant protein-1 and its impact on obesity. *Obes Res* 2005;13(8):1311–1320.
229. Riu E, Ferre T, Hidalgo A, *et al.* Overexpression of c-myc in the liver prevents obesity and insulin resistance. *FASEB J* 2003;17(12):1715–1717.
230. Wolfrum C, Shih DQ, Kuwajima S, Norris AW, Kahn CR, Stoffel M. Role of Foxa-2 in adipocyte metabolism and differentiation. *J Clin Invest* 2003;112(3):345–356.
231. Wortley KE, Anderson KD, Garcia K, *et al.* Genetic deletion of ghrelin does not decrease food intake but influences metabolic fuel preference. *Proc Natl Acad Sci USA* 2004;101(21):8227–8232.
232. Wortley KE, del Rincon JP, Murray JD, *et al.* Absence of ghrelin protects against early-onset obesity. *J Clin Invest* 2005;115(12):3573–3578.
233. Otaegui PJ, Ferre T, Riu E, Bosch F. Prevention of obesity and insulin resistance by glucokinase expression in skeletal muscle of transgenic mice. *FASEB J* 2003;17(14):2097–2099.
234. Alfadda A, DosSantos RA, Stepanyan Z, Marrif H, Silva JE. Mice with deletion of the mitochondrial glycerol-3-phosphate dehydrogenase gene exhibit a thrifty phenotype: Effect of gender. *Am J Physiol Regul Integr Comp Physiol* 2004;287(1):R147–156.
235. Brown LJ, Koza RA, Everett C, *et al.* Normal thyroid thermogenesis but reduced viability and adiposity in mice lacking the mitochondrial glycerol phosphate dehydrogenase. *J Biol Chem* 2002;277(36):32892–32898.
236. Anand A, Chada K. In vivo modulation of Hmgic reduces obesity. *Nat Genet* 2000;24(4):377–380.
237. Berg JP. Pygmy mouse gene mutation protects against obesity. *Eur J Endocrinol* 2000;143(3):317–318.
238. Sleeman MW, Wortley KE, Lai KM, *et al.* Absence of the lipid phosphatase SHIP2 confers resistance to dietary obesity. *Nat Med* 2005;11(2):199–205.
239. Castellani LW, Chang JJ, Wang X, Lusis AJ, Reynolds WF. Transgenic mice express human MPO -463G/A alleles at atherosclerotic lesions, developing hyperlipidemia and obesity in -463G males. *J Lipid Res* 2006;47(7):1366–1377.
240. Zhao B, Wall RJ, Yang J. Transgenic expression of myostatin propeptide prevents diet-induced obesity and insulin resistance. *Biochem Biophys Res Commun* 2005;337(1):248–255.
241. Balthasar N, Dalgaard LT, Lee CE, *et al.* Divergence of melanocortin pathways in the control of food intake and energy expenditure. *Cell* 2005;123(3):493–505.
242. Lijnen HR, Maquoi E, Morange P, *et al.* Nutritionally induced obesity is attenuated in transgenic mice overexpressing plasminogen activator inhibitor-1. *Arterioscler Thromb Vasc Biol* 2003;23(1):78–84.
243. Huggins KW, Camarota LM, Howles PN, Hui DY. Pancreatic triglyceride lipase deficiency minimally affects dietary fat absorption but dramatically decreases dietary cholesterol absorption in mice. *J Biol Chem* 2003;278(44):42899–42905.
244. Kamei Y, Ohizumi H, Fujitani Y, *et al.* PPARgamma coactivator 1beta/ERR ligand 1 is an ERR protein ligand, whose expression induces a high-energy expenditure and antagonizes obesity. *Proc Natl Acad Sci USA* 2003;100(21):12378–12383.
245. Haugen BR, Jensen DR, Sharma V, *et al.* Retinoid X receptor gamma-deficient mice have increased skeletal muscle lipoprotein lipase activity and less weight gain when fed a high-fat diet. *Endocrinology* 2004;145(8):3679–3685.
246. Lijnen HR. Effect of plasminogen activator inhibitor-1 deficiency on nutritionally-induced obesity in mice. *Thromb Haemost* 2005;93(5):816–819.
247. Ma LJ, Mao SL, Taylor KL, *et al.* Prevention of obesity and insulin resistance in mice lacking plasminogen activator inhibitor 1. *Diabetes* 2004;53(2):336–346.
248. Kublaoui BM, Holder JL, Jr., Tolson KP, Gemelli T, Zinn AR. SIM1 overexpression partially rescues agouti yellow and diet-induced obesity by normalizing food intake. *Endocrinology* 2006;147(10):4542–4549.
249. Strader AD, Reizes O, Woods SC, Benoit SC, Seeley RJ. Mice lacking the syndecan-3 gene are resistant to diet-induced obesity. *J Clin Invest* 2004;114(9):1354–1360.
250. Voros G, Maquoi E, Collen D, Lijnen HR. Influence of membrane-bound tumor necrosis factor (TNF)-alpha on obesity and glucose metabolism. *J Thromb Haemost* 2004;2(3):507–513.
251. Son C, Hosoda K, Ishihara K, *et al.* Reduction of diet-induced obesity in transgenic mice overexpressing uncoupling protein 3 in skeletal muscle. *Diabetologia* 2004;47(1):47–54.
252. Sainsbury A, Bergen HT, Boey D, *et al.* Y2Y4 receptor double knockout protects against obesity due to a high-fat diet or Y1 receptor deficiency in mice. *Diabetes* 2006;55(1):19–26.

72 Study of Polycystic Kidney Disease in the Nematode *Caenorhabditis elegans*

MATTHEW J. BUECHNER

ABSTRACT

Polycystic kidney disease (PKD) is characterized by the uncontrolled slow swelling of tubules in the kidneys and other tissues. PKDs are some of the most common genetic diseases, and are caused by defects in the function of either of two polycystin ion channels, or of their transport to the primary cilium, one of the locations where these proteins function to detect mechanical stress on the cell. Homologues of the polycystins are found in the nematode *Caenorhabditis elegans*, where these proteins function in male-specific neurons to detect the location of the hermaphrodite vulva. Mutations in nematode polycystin genes cause defects in male mating behavior, which can be measured via various behavioral and electrophysiological assays. Mutations that affect formation of sensory cilia, or transport of materials into the cilium, also cause defects in male mating behavior, as well as defects in various tactic behaviors mediated through other neurons. The affected genes identify many homologues of other human disease genes, including nephronophthisis and Bardet–Biedl syndrome. The nematode mutations can be detected through multiple behavioral and cell biological techniques, including measurement of fluorescently-labeled polycystin into the cilium.

Key Words: *Caenorhabditis elegans*, Polycystin, Polycystic kidney disease, Nephronophthisis, Bardet–Biedl, TRP channels, Behavioral assay, Chemotaxis assay, Electrophysiology.

INTRODUCTION

Caenorhabditis elegans does not have the advanced metanephric kidney of mammals; nonetheless, this little roundworm has been instrumental in greatly increasing our understanding of polycystic kidney disease (PKD). A brief review of the discoveries found in these studies provides a good example of the strengths and limitations of invertebrate animal models in the study of human disease, and the importance of genomic information for determining the function of a protein family in cell structure. The ability of evolution to adapt conserved biochemical mechanisms to new tissues and biological roles in different organisms has dictated that study of PKD in *C. elegans* requires the ability to develop and perform behavioral and electrophysiological assays, rather than assays used to examine mammalian renal function and development.

From: *Sourcebook of Models for Biomedical Research*
(P. M. Conn, ed.), © 2008 Humana Press Inc., Totowa, NJ.

POLYCYSTIC KIDNEY DISEASES

PKDs are a series of genetic diseases that affects the diameter of tubules in many tissues.[1] In the kidney, a subset of the numerous small nephrons and collecting tubules swells over the course of years into large fluid-filled cysts that block flow through the kidney and lead to end-stage renal disease. In addition to the kidney, PKD affects other tubular tissues, especially the liver and pancreas, and death is often caused by aneurysms of the cerebral vasculature.

AUTOSOMAL DOMINANT POLYCYSTIC KIDNEY DISEASE Autosomal dominant PKD (ADPKD) is the most prevalent lethal genetic disease, with estimates as high as 1 out of 200 people affected.[1,2] The high incidence of the disease is the result of the disease being caused by dominant mutation, and by the late age of onset of disease symptoms, often in the 30s, so that patients have passed the affected gene on to the next generation before symptoms occur. PKD is caused predominantly by mutations in one of two genes: PKD1 on chromosome 16, accounting for 85% of cases; and PKD2 on chromosome 4. Polycystic kidneys can also occur as a symptom of other genetic diseases, including the recessive nephronophthises,[3] von Hippel–Lindau disease,[4] Bardet–Biedl syndrome,[5] oral-facial-digital syndrome,[5] and Meckel–Gruber syndrome.[7]

Major advances in our understanding of polycystic kidneys have come from the cloning of the affected genes for PKD in the 1990s,[8,9] and of the nephronophthises and Bardet–Biedl syndrome genes[10–12] in the past decade. Studies in model organisms[13] have provided crucial information toward understanding the function of the encoded proteins in normal nephrons, and how the loss of that function in PKD patients causes unregulated growth of tubule cysts.

PKD1 and PKD2 both encode large transmembrane proteins, called "polycystins," of the TRP family of divalent ion channels.[14–16] TRP channels are a widespread and ancient family, with members found in all animals, and are used as receptors for a wide range of stimuli, including chemical odors, taste, heat, and osmolarity. Human PKD2 contains six transmembrane domains, and a cytoplasmic EF-hand to bind to calcium,[9] an endoplasmic reticulum-targeting signal, and a coiled coil to interact with other proteins, including PKD1 and CD2-associated protein.[17–19] At the cytoplasmic N-terminus was found a site to target the protein to the cilium, and a GSK3 phosphorylation site. Between the last two transmembrane domains is a loop found to interact with HAX-1[20] (Figure 72–1). *C. elegans* PKD-2 has a similar structure,

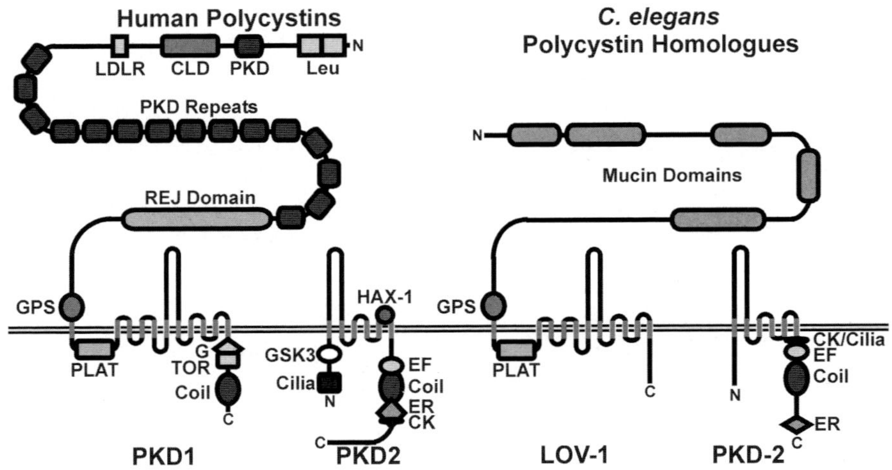

Figure 72–1. Diagram of the relative size and domains of human and *C. elegans* polycystins. Human polycystin-like proteins (e.g., PKDREJ, PKD2L) are not shown. Diagrams are not to scale, but overall protein lengths are approximately accurate. Termini of proteins are designated by N and C. Transmembrane domains are shown as gray lines crossing the membrane. For cilia and plasma membranes, the extracellular domain is on the top half and the cytoplasmic below. Other domains indicated: PKD, PKD repeats (Ig-like folds); Leu, leucine-rich domains; CLD, C-lectin domain; LDLR, low-density lipoprotein-like region; REJ, sea urchin receptor for egg jelly domain; GPS, G-protein-coupled receptor site of proteolysis; PLAT, polycystin/lipoxygenase/α-toxin domain; G, G-protein-binding and -activating site; TOR, region of interaction with the target of rapamycin pathway; Coil, coiled-coil domain; Cilia, domain necessary to target protein to cilium; GSK3, glycogen synthase kinase 3 phosphorylation site; HAX-1, binding site for HAX-1; EF, EF-hand calcium-binding motif; ER, endoplasmic reticulum retention site; CK, casein kinase 2 phosphorylation site; Mucin, numerous serine/threonine O-glycosylation sites.

with regions homologous to most of these domains, although binding at all of these has not been proven. While *C. elegans* PKD-2 does not have a canonical EF-hand, it does have a region that is theoretically similar enough to function as an EF-hand.[21] A phosphorylation site for casein kinase II is present in both mammalian and *C. elegans* PKD2 that modulates activity, and in *C. elegans*, this site is required to target the protein to the cilium.[22,23]

PKD1 is a much larger protein. It includes the same six transmembrane domains as does PKD2, plus an additional five transmembrane domains upstream of these, and a huge extracellular domain.[8] It is believed that PKD1 acts as a mechanosensor,[24] but it is not known if the extracellular domain binds to a specific target in order to become active, or if the protein has intrinsic ability to detect membrane stretch as do other pressure and stretch receptors.[25] Both the mammalian protein and its *C. elegans* homologue have a receptor for egg jelly and a GPS domain just upstream from the first transmembrane domain, which allow the N-terminal domain to be cleaved from the rest of the protein after binding to its extracellular ligand.[26] The cytoplasmic domain has several signaling domains in addition to a coiled coil used to bind to PKD2. Mammalian PKD1 has been linked to signaling through the mTOR, phosphatidylinositol 3-kinase/Akt, JAK/STAT, protein kinase C, AP-1, Wnt, mitogen-activated protein (MAP) kinase, and heterotrimeric G-protein pathways.[27–32] The *C. elegans* location of vulva (LOV)-1 protein C-terminus shares a conserved tyrosine phosphorylation site with human PKD1, but the homology to other domains is not as clear.

C. elegans was of course the first animal whose genome was completely sequenced.[33] This effort revealed the presence of homologues of both polycystin genes. Initial efforts focused on a reverse genetics approach to create mutants deleted for one or both of the polycystin homologues. Surprisingly, these mutants showed no structural or functional defects in the primitive nematode excretory system, nor did these mutants exhibit formation of cysts in any other tubular structures in the nematode. At the same time, incisive forward genetic studies on a seemingly completely unrelated aspect of nematode biology, the neurobiology of male mating behavior, found that the function of the PKD1 homologue, location of vulva (LOV)-1, is necessary to allow male nematodes to detect the location of the hermaphrodite vulva.[34]

Expression studies showed that both LOV-1 and the PKD2 homologue PKD-2 are expressed only in male-specific head and tail nematode sensory neurons that use a cilium as the sensory receptor.[34] LOV-1 is expressed in the cilium, while PKD-2 is strongly expressed in the neuronal cell body as well as in the cilium. Similar cellular positions were then confirmed for the mammalian PKD proteins, and emphasized the importance of the primary cilium in multicellular tubules as a sensory receptor for detecting mechanical stress on the cell. Exciting findings from many laboratories suggest that the primary cilium acts as a rigid antenna sticking up into the lumen of nephrons and other tubules to act as a flow sensor.[35] Physically bending the cilium causes calcium to enter the cell, but only if PKD1 and PKD2 are present in the cilium.[36]

The mammalian polycystins appear additionally to be present at cell–cell junctions[37] and at the focal adhesions[38,39] that anchor epithelial tubules to the underlying basement membrane. Both of these sites are also excellent locations for detecting mechanical pressure attempting to bend, move, or deform cells, and underscore a physiological role of polycystins as mechanosensors.

A question as to the usefulness of *C. elegans* as a model for PKD arises in considering why mutations in polycystins cause recessive mutations in nematodes, but a dominant disease in humans. The late onset of autosomal dominant PKD is consistent with the "two-hit hypothesis,"[40] which proposes that the onset of ADPKD requires inactivation of both copies of the polycystin gene. Individuals heterozygous for a mutated polycystin gene will suffer a random mutation in the other copy of that gene in some renal cells during their lifetime, which allows a cyst to begin forming. Individuals homozygous for a mutation in one of the nonpolycystin genes (see below), however, will already have both copies of that gene inactivated, and so cyst formation will begin early, even in the embryo for some genes. The two-hit hypothesis also explains why many dominant mutations in polycystin genes are large deletions or nonsense mutations early in the gene. There are no known dominant mutations of the *lov-1* or *pkd-2* genes in *C. elegans*, but since dominance in mammalian PKD seems to reflect complete loss of function of polycystin genes in a few cells rather than a partial loss of function in the whole organism, the study of the nematode genes is not invalidated as a model for understanding the function of the mammalian proteins.

OTHER POLYCYSTIC DISEASES Rare mutations in several other genes can cause recessive forms of PKD. These recessive diseases tend to have an earlier onset than does ADPKD. One series of recessive genetic diseases that causes juvenile onset of polycystic kidneys are the nephronophthises.[41] Some of the extrarenal manifestations of these diseases include defects in the retina, a neural tissue. Six mammalian nephronophthisis genes have been cloned.[11,12] The encoded "nephrocystins" do not show homology with each other, but most of them are located in the primary cilium, the same location as PKD1 and PKD2. Some types of nephrocystins have also been found at the cell–cell junctions and focal adhesions, again places where polycystins are located; and several nephrocystins are additionally located at the centrosome. Overall, the positions of the nephrocystins and polycystins suggest that both of these sets of proteins are essential for measuring and transducing physical stretch, caused either by bending of the cilium or tension at other sites in the cell that causes it to pull on neighboring cells or on the underlying basement membrane.

Clear homologues to nephrocystin-1 and nephrocystin-4 are found in *C. elegans*.[42] These proteins associate with PKD-2 in male-specific neurons and are required for proper mating behavior. Other *C. elegans* proteins show weaker homology to segments of nephrocystin-2 (also called "inversin"), nephrocystin-3, and nephrocystin-6, but it is not clear if there is functional homology here between worms and mammals.

Mutations in another series of genes cause the rare Bardet–Biedl syndrome,[43] which is characterized by any of a large host of symptoms, some of which clearly involve the primary cilium, including polycystic kidneys, *situs inversus*, and retinal degradation. At least 11 such genes have been cloned so far, and 9 of these have clear homologues in *C. elegans*. One of the first indications of the function of these proteins came from the observation that the *C. elegans* homologues are expressed solely in ciliated sensory neurons[44] (in both sexes, not just males). Further study in vertebrates and the alga *Chlamydomonas*, as well as in *C. elegans*, found that the encoded proteins are present predominantly in the primary cilium, and in the ciliary basal body, a modified centrosome. The *C. elegans* BBS genes include *osm-12*, homologous to human BBS7. The *osm* mutants were originally isolated as being unable to undergo osmotaxis, the avoidance of areas of high osmolarity in behavioral assays.[45] This behavior is mediated primarily through ciliated sensory neurons in the head of the animal.

BBS protein function is necessary for the synthesis of the primary cilium, while it has no effect on the synthesis of motile cilia. In *C. elegans*, mutations in the BBS genes alter the transport of material both into and out of the cilium, a process called intraflagellar transport (IFT), first described for *Chlamydomonas*.[46] Lack of this process prevents the transport of receptors located in the cilium that mediate many types of behaviors. Knockdown of IFT gene function in *C. elegans* via RNAi of several of the BBS genes causes defects in osmotaxis and chemotaxis.

Mutations in several other genes have been found that alter the behavior of *C. elegans* as well as intraflagellar transport. One example is the mouse *polaris* gene, which was discovered as a mutant that caused polycystic kidneys and generated left-right asymmetry of internal organs.[47] This asymmetry was found to be caused by primary cilium defects in the node in early mammalian development. Close homologues are found in humans and *Chlamydomonas* IFT particles as well as the *osm-5* gene of *C. elegans*,[48] another gene required for normal ciliogenesis, and detected via defects in osmotactic behavior.

Other proteins related to polycystic kidney diseases are still being discovered, and *C. elegans* will undoubtedly be a useful model for some of these conditions. For example, recently the human gene defective in Meckel–Gruber syndrome was cloned.[7] Clone F35D2.4 is a close homologue in *C. elegans* and was independently identified as a likely ciliary gene.[49]

USE OF *CAENORHABDITIS ELEGANS* TO STUDY POLYCYSTIN FUNCTION

It is necessary to take advantage of the strength of the organism as an object of study and to investigate the proteins of interest in the tissues where they are expressed. To examine PKD function in nematodes, it is essential to take advantage of genetics, and to assay neural function through behavioral assays, electrophysiological assays, and cell biological observations.

Major genetic methods for working with *C. elegans* are described in more detail elsewhere (Brenner,[50] www.WormBook.org, and Chapter 12 in this volume), but time spent becoming familiar with the organism is invaluable. Subtle differences in movement or cell structure may be obvious to one observer, but undetectable to another. A compound microscope with DIC (Nomarski) optics is essential for observing the placement and development of individual cells within the organism. Since the creature is clear, tagging the gene of interest by linking it to a fluorescent gene, such as *gfp* from jellyfish or *dsRed* from coral, will allow visualization of the synthesized chimeric protein inside the organism, to allow determination of the time and place of expression including subcellular location.

Do take advantage of the tremendous online resources for working with this creature. The online resource WormBase.org includes the complete genome sequence, information on every gene in the creature, and references to every *C. elegans* paper published, as well as unpublished abstracts and experiments reported only in meetings and in the early *Worm Breeder's Gazette* newsletter (some of the initial data of Barr and Sternberg on the identity of LOV-1 as a PKD1 homologue were reported here long before publication of their initial report in *Nature*). WormAtlas.org has detailed ultrastructural information on all the cells and tissues of the animal, including clear diagrams of the ciliated neurons of the male reproductive system, where the polycystins are expressed. And the *C. elegans* Stock Center (cgsc.org) has available many of the mutant stocks needed to study these diseases, including deletion mutants in both *lov-1* and *pkd-2*. Two *C. elegans* knockout projects, one in Japan and one in Oklahoma

City and Vancouver, Canada are also attempting to isolate strains carrying deletion mutations of each of the 19,000 genes of *C. elegans*. The Oklahoma/Vancouver project routinely send their creations to the *C. elegans* stock center.

BEHAVIORAL ASSAYS The initial technique used to isolate nematode polycystin mutants was a forward genetic screen to identify mutagenized male worms that were no longer capable of performing one of the many steps required to complete mating.[34] Normally, male worms appear to sense the presence of a hermaphrodite through chemical receptors. The cilia of many of the neural endings that contain polycystins are contained in a series of epithelial open-ended tubes arranged around the male copulatory bursa. The rays are used by the male to detect chemical and mechanical cues to place the bursa over the hermaphrodite vulva for copulation. To do this, the male approaches the hermaphrodite and engages in a series of stereotypical behaviors.[51] These include placing his copulatory bursa alongside the hermaphrodite, and backing up along the hermaphrodite. If the male does not find the hermaphrodite vulva by the time its tail reaches the end of the hermaphrodite, the male flexes his tail sharply and continues moving backward along the opposite side of the hermaphrodite. Once the male passes his tail alongside the hermaphrodite vulva, the animal stops (it is not known if the signal is mechanical, chemical, or both), and inserts rigid cuticular structures called spicules into the vulva. The male then ejaculates motile sperm into the vulva, which then travel to the hermaphrodite spermatheca. In some strains of nematode, the male finally secretes a copulatory plug to prevent other males from mating with the hermaphrodite.

To study male mating behavior, it is very helpful to use the *him-5* mutant instead of wild-type worms. Since *C. elegans* is a self-fertilizing hermaphrodite, it does not need males to give rise to a new generation; in a typical well-fed colony, males make up only about 0.5% of the population. Mutations in the *him-5* gene allow a higher incidence of spontaneous disjunction of the X chromosome to occur,[52] which gives rise to a higher frequency of male (haploid for the X chromosome) animals instead of hermaphrodite (XX) animals. As much as one-third of a *him-5* colony will be male, which allows much easier observation of male mating behavior.

A series of studies primarily from the Sternberg laboratory isolated mutants in each of the steps described above. Arduous work by Barr and Sternberg[34] detected rare mutants in which the males were capable of detecting the hermaphrodite, moving backward along the hermaphrodite, and flexing the tail in order to search the other side of the animal, but had great difficulty finding the vulva (*lov*, for defective in *l*ocation *o*f *v*ulva). Mutant males continue to circle the hermaphrodite repeatedly, until they stumble upon the hermaphrodite vulva. This behavior can be observed in individual animals either by counting the number of times that a male circles the hermaphrodite before attempting to insert its spicules in the vulva, or by measuring the amount of time needed for a male to attempt to insert its spicules after brushing up against a hermaphrodite.

Genetic mapping of the *lov-1* mutation was accomplished through standard techniques for *C. elegans*, including measuring linkage to known mutants, and rescue of the mutant through microinjection of wild-type DNA. The complete genome sequence of *C. elegans* is freely available from the *C. elegans* Sequencing Project[33] as cosmid fragments, each containing approximately 40,000 base pairs of genomic DNA. The progeny of *lov-1* mutants could be cured through injection into the parent of DNA containing the nematode homologue of human PKD1.[34]

Later studies showed that mutants carrying a deletion of the nematode homologue of *pkd-2* are also incapable of locating the hermaphrodite vulva, and that a mutant deficient in both *lov-1* and *pkd-2* was no more defective in vulva location than a mutant for only one of these genes, consistent with the idea that these two proteins function together in a signal transduction pathway.[53]

Further searches for other mutations that prevent male nematodes from locating the hermaphrodite vulva revealed the *klp-6* gene,[54] which encodes a kinesin motor protein. Mutation in *klp-6* prevents LOV-1 from reaching the cilium (as visualized through the use of a *lov-1::gfp* transgene), a result that reinforces both the importance of ciliary function and the importance of proper trafficking of proteins to the cilium.

Another method of observing differences in wild-type nematodes and nematodes defective in polycystin function is to look at behavior in groups.[21] In addition to the male-specific neurons located in the male tail, four ciliated neurons (CEM neurons) are located in the head of males.[55] The function of the CEM neurons is not known, but it is reasonable to assume that they are involved in male detection of hermaphrodite nematodes. Male nematodes deleted for either *lov-1* or *pkd-2* exhibit less social behavior. By placing a large number of male nematodes together with hermaphrodites on a plate of food bacteria, it is possible to observe easily and immediately a large difference in behavior. While the hermaphrodites scatter about the plate foraging for food, most of the wild-type males will attempt to mate with hermaphrodites, and many will group in "clumps" of several males attempting to mate with a single hermaphrodite. In contrast, *lov-1* or *pkd-2* mutant males will mostly ignore the hermaphrodites, and predominantly spend their time foraging for food as well. It is important to use plates in which the lawn of food bacteria is spread evenly across the entire plate up to the edge, since thicker growth of bacteria at lawn edges attracts worms and can affect the rate of "clumping" by restricting the worms to specific areas of the plate. This assay has been useful in identifying suppressor strains that are "interested" in mating even though they are deleted for polycystin-2 function.

The strengths and disadvantages to the social behavior and LOV assays are complementary. The LOV assay is time and energy intensive, requiring placing an individual male with a hermaphrodite, and watching that individual's attempts to locate the hermaphrodite vulva. On the other hand, this assay clearly lets the observer know the individual male animal's phenotype. The social behavior assay, on the other hand, is relatively quick and easy to perform, and the result can be determined by an observer without extensive training or time at the microscope. The assay, however, cannot be performed on a single worm, but must be performed on a group. The test animals can all be the progeny of a single *him-5* hermaphrodite, but you must wait a generation until the animal to be tested has enough progeny to allow the performance of this assay.

Since polycystins must be expressed in the cilium in order to perform at least some of their functions in both worms and humans, mutants that affect the formation of cilia, or transport of proteins into the cilium (IFT), will show defects in polycystin function. In humans, this is the cause of polycystic kidneys in Bardet–Biedl syndrome and nephronophthisis. In *C. elegans*, the male-specific neurons that express the polycystins are a subset of the sensory neurons. Many ciliated sensory neurons, especially in the head of the animal, detect other chemical, thermal, and osmotic stimuli. Defects in ciliary formation or IFT can be detected via effects on osmotactic and chemotactic behaviors. Assays for these behaviors are described in great detail elsewhere.[56] The standard method for assaying osmotaxis is to place worms on agar surrounded by a ring of either 8 M glycerol, 4 M NaCl, or 4 M fructose (plus dye to make the ring visible) gently pressed onto the surface of the agar and allowed to soak in for a few minutes.[45] The worms are observed as soon as they are placed in the center of the ring. Wild-type animals will avoid the glycerol and will generally stay within the ring for at least 10 min. Severely osmotaxis-defective animals will not detect the ring and most will attempt to cross it fairly quickly. *C. elegans* cannot survive osmolarities greater than about 600 mOsM, much lower than the concentration of glycerol imprinted on the agar surface, and so many of the animals will die in attempting to cross the barrier. If attempting to enrich mutagenized animals for those defective in osmotaxis, it is important to select potential mutants before they are dehydrated.

Chemotaxis has been examined from the very beginning of modern studies on *C. elegans*,[57,58] and multiple methods for assaying chemotactic responses to various compounds have been developed. Mutations in *che-3*, which encodes a dynein heavy chain component of the IFT machinery, were isolated through use of an assay with Petri plates divided into four sections via plastic separators.[59] Agar containing attractant is placed in two opposite quadrants, and agar with no attractant (or a lower concentration, or a mild repellent) is placed in the remaining two quadrants. After solidifying, agar (with no attractant) "bridges" are placed over the plastic dividers to connect the quadrants. A population of animals is then placed in the center of the plate, and allowed to swarm. Wild-type nematodes swim to quadrants containing high levels of attractant, while animals that move to sections of the plate containing no attractant are enriched for chemotaxis mutants.

An alternative method to gauge chemotaxis has been used to isolate mutants defective in detecting volatile odorants.[60] A population of worms is placed in the center of a single large plate that has been spotted at one point near the edge with the attractant and with an anesthetic (sodium azide), and with anesthetic alone on the opposite side of the plate. Most worms will smell the attractant and move toward it. Chemotaxis-defective mutants will crawl in many directions, and eventually be incapacitated above a spot of anesthetic. Worms trapped over the spot that contains no attractant will be greatly enriched for chemotaxis-defective mutations.

CELL BIOLOGICAL STUDIES Polycystin 1 (LOV-1) appears to be located predominantly in the primary cilium of the nematode neurons where it is active.[34] PKD-2 is also expressed in the cilium, although it is expressed at high levels in other compartments of the cell as well. If the polycystins are not brought to the cilium, then the animal is unable to mate properly. While polycystins are found only in male-specific cilia, other ciliated neurons in the head and tail are used as sensory receptors for a wide range of stimuli in both sexes.[61] The ciliated endings are open to the environment to allow detection of attractant and repellent chemicals as well as osmosensation. A classic study of ciliary structure was reported by Perkins *et al.*,[62] who stained the animals with the mucin-soluble dye 5-fluorescein isothiocyanate (FITC). Mucin is a highly glycosylated protein that forms a mucous layer to trap small chemicals to be sensed by the cilia, analogous to the function of mucus around mammalian olfactory receptors. Not all ciliated neurons stain with FITC, and in fact none of the male-specific ciliated neurons normally is stained by FITC. But isolation of mutants in which FITC staining is abnormal (*dyf* for *dye-filling* defective) has been used to identify mutants defective in ciliary construction and transport of proteins into the cilium. Many of these mutants turned out to be defective in osmotaxis, chemotaxis, and detection of a specific pheromone that enables nematodes to suspend growth under harsh conditions (the "dauer" form[63]). The researchers also isolated rare mutants defective in osmotaxis in which the male-specific ray neurons of the tail are stained by FITC. Mutants in this gene, *osm-1*, were later found to be defective in male mating as well as in osmotaxis.[64]

The availability of polycystin constructs linked to the *gfp* gene is very useful for examining transport of polycystins to the cilium, as well as IFT. The Barr and Scholey laboratories have demonstrated the transport of these proteins into sensory neuronal cilia.[54] As noted above, defective kinesin KLP-6 prevents GFP-tagged LOV-1 from reaching its proper ciliary position. Several other kinesin proteins, including OSM-3 and KAP-1, function together and sequentially to transport material into sensory cilia.[65] By examining mutagenized nematodes and isolating rare animals in which these fluorescent proteins do not travel into the cilium, these laboratories have also discovered mutants in genes required for polycystin activity in both *C. elegans* and humans. Such mutants have identified proteins required in many cell types for polycystin to be brought up to and into the cilium, as well as cell-type-specific factors that are required only in nematode male sensory neurons.

To observe transport, the nematodes must be transformed via microinjections of plasmids expressing the gene of interest fused to the *gfp* or *dsRed* marker gene. The methods for producing and observing transgenic nematodes are fairly standard in every *C. elegans* laboratory (see Chapter 12, this volume). A series of expression vectors was originally developed by the Fire laboratory (ftp://ftp.wormbase.org/pub/wormbase/datasets/fire_vectors/), and is now commercially available (www.addgene.org), that contains a multicloning site upstream of the marker gene, which contains introns to increase expression in *C. elegans*. Once constructed, the plasmid must be microinjected into the syncytial gonad of adult hermaphrodite worms, where the DNA is unstably incorporated in the oocytes as a linear array of multiple copies of the plasmid maintained separately from the chromosome.[66] It is generally straightforward to use a dissecting fluorescence microscope to search for progeny that express the array. Since the DNA array is not linked to a chromosome, it is unstably maintained at cell division, and is therefore not transmitted in subsequent generations. To prevent loss of the transgene, animals carrying the array are treated with DNA-breaking agents such as trimethylpsoralen plus ultraviolet light, in order to incorporate the array into a chromosome. A stable line that passes the transgene reliably to subsequent generations can thereby be generated.

Examining subcellular location of green fluorescent proteins (GFP)-labeled proteins is difficult in the small neurons of *C. elegans*. The animals must be anesthetized with sodium azide, levamisole, or phenoxypropanol, or contain a muscle mutation such as *unc-54* in order to prevent them from moving under the microscope. A good fluorescence DIC or confocal microscope is required, with a 63× or 100× objective, to see smaller processes, and experience with identifying neurons is essential. Fortunately, the position of most neuronal cell bodies and axons in *C. elegans* is invariant, and has been well characterized.[55] Detailed descriptions of neuron structure is available online (http://www.wormatlas.org). With the well-characterized ciliated sensory neurons of the head, it has even been possible via time-lapse confocal microscopy to measure the rate of movement of labeled proteins in axons and on IFT particles into the cilium, and to distinguish differences in movement of polycystins into the cilium in various mutants.[65–69]

ELECTROPHYSIOLOGICAL STUDIES The polycystins are medium-conductance relatively nonselective divalent cation channels.[16] Direct study of polycystin function within the narrow cilia is not possible with current electrophysiological techniques. There are several ways to study the activity of ion channels in other areas of *C. elegans* neurons, but direct electrophysiological studies are difficult, both because *C. elegans* neurons are small (the cell bodies are only 2–3 μm in diameter), and because the nematode body plan features a flexible collagenous cuticle surrounding the animal, which acts as both skeleton and skin. Since muscles pull on the cuticle to move the animal, the cuticle must be maintained in a relatively rigid state. This is accomplished by osmotically maintaining a strong turgor pressure in the body fluid relative to liquid outside the animal's body.[70] Any cutting of the cuticle therefore causes the internal material to explode out of the animal. Indeed, mutants have been isolated in which the hermaphrodite uterus and gonads spontaneously pop out of the vulva, and mishandling of male worms often causes the spicules to protrude explosively, rendering the male incapable of mating. Attempts to isolate neurons for electrophysiological studies yield the same result, with the identity of neurons being unrecognizable after they have exploded from out of an incision.

Fortunately, there are a large number of molecular biology constructs linked to the *gfp* gene that can be used to identify specific neurons. By microinjecting these constructs into nematodes and selecting strains that pass the construct reliably to the next generation, researchers have generated a large series of strains that expresses GFP only in specific cells of the animal. Many of these strains are available from the *Caenorhabditis* Genetics Center, and many such constructs for a given gene are identified on WormBase.org. By opening the animal near the nerve ring, the major ganglion of the worm, and letting neurons or muscle cells pop out of the worm, on-cell electrophysiological recordings have been made from single identified neurons.[71–74] Animals can be opened either directly via cutting, or by use of collagenase to weaken the cuticle. The released cells can also be placed in primary tissue culture,[75] and electrical properties measured directly. To date, direct electrophysiological methods of recording from cells *in vivo* have not been reported for recording activity of male-specific neurons in the tail, however.

Another method that has been used to study other *C. elegans* ion channels is to express the gene in a cell to which electrodes can be attached easily, such as a *Xenopus laevis* oocyte. An excellent primer on the use of this technique for studying *C. elegans* channels is available from the Driscoll laboratory.[76] Using the gene predictions of WormBase.org, a cDNA clone can be synthesized from mRNA. It is important to make sure that the gene predictions are correct, especially for the 5′ and 3′ ends of the gene, since the GeneFinder computer algorithm used to predict exons is less accurate at the ends of genes. For example, the original computer prediction of the *lov-1* polycystin gene sequence showed two genes (clones ZK945.9 and ZK945.10), with only one of these predicted genes showing homology to human PKD1. Later work proved that these two predicted genes are in actuality two halves of a single transcript, and that the piece with low homology to human PKD1 is actually the N-terminal ligand-binding domain of LOV-1. Our work (unpublished) similarly found that a predicted SK3 potassium channel gene is actually twice as large as predicted, with a large unpredicted N-terminal domain.

Once a cDNA clone has been synthesized, it is linearized and a kit can be used to create cRNA, an artificially produced mRNA to be injected into oocytes. The oocytes can be prepared from commercially available sources of *Xenopus* ovaries, a great advantage since many *C. elegans* laboratories do not maintain the paperwork necessary for working with vertebrate animals. The oocytes are prepared, injected, and observed via a dissection microscope. Measurement of channel activity is either through standard two-electrode voltage clamp or via single-electrode patch clamp on oocytes whose vitelline membrane has been removed to allow formation of a clean gigaOhm seal to the pipette. These techniques have been used successfully with a large number of *C. elegans* channels, including the degenerin channels that mediate touch sensitivity.[73] They have not been reported for use on *C. elegans* polycystins as yet. In part this may reflect the fact that the *lov-1* gene is quite large for a *C. elegans* gene (though small by human gene standards), and reports of difficulties in functional expression of PKD1 have made this a less attractive target (although expression of the C-terminal half of the protein yields currents[77]). Human PKD2 has been studied more thoroughly in *Xenopus* oocytes, and such experiments confirmed that PKD2 is a calcium channel and interacts with the IP_3 receptor at the endoplasmic reticulum (ER) membrane.[78,79]

Both human and *C. elegans* polycystin electrical activity have been measured using planar bilayer membranes (BLM).[21,80,81] Several plates of animals are collected, and their cuticles weakened and lysed via treatment with collagenase, to release cells from inside of the animal. The ER vesicles are then purified via standard centrifugal techniques. It is essential to have antibodies to the ion channel of interest to identify the protein in ER vesicles. A great advantage in using *C. elegans* is the ability to compare vesicles from wild-type animals with those from mutants deleted for the *pkd-2* gene. The vesicles are placed in a BLM bath, and electrical activity is measured as the vesicles fuse with the membrane in the center. *C. elegans* PKD-2 was found to have electrical properties similar to those of mammalian PKD2. PKD-2 is strongly expressed in the ER, while LOV-1 is not, so it was not possible to determine if LOV-1 forms channel activities by itself or in combination with PKD-2. Purification of ciliary membranes has not been attempted in *C. elegans*, due to the small number of ciliated neurons in the animal. If ciliary membrane vesicles could be purified successfully from wild-type and mutant animals, the BLM technique would be very useful for determining the influence of many IFT and ciliary proteins on polycystin activity.

A new method of examining *C. elegans* polycystin electrical activity has been reported.[21] Nematodes (*him-5* mutants, to increase the number of males and therefore the percentage of cells that are ciliated neurons) were treated with collagenase as before, and triturated in order to separate cells from each other, and loaded with Ca^{2+} and the calcium-sensitive dye Fluo3-AM. The cells were depolarized by means of extracellular application of KCl. The excitable cells fluoresce as calcium is released from extracellular stores, which can be measured via fluorescence microscopy. Again, if ciliated neuron cells were to be marked with a dsRed marker, it might be possible to use this technique to explore the ability of neuronal stimulation to release calcium from intracellular stores in various mutants that potentially affect polycystin function.

LIMITATIONS ON *CAENORHABDITIS ELEGANS* AS A MODEL As noted above, *C. elegans* provides a strong model for studying polycystin function in ciliated cells. But there are limitations to the usefulness of the nematode as a model for understanding polycystic kidney disease. First, not all of the proteins that are necessary for maintenance of proper nephron shape and prevention of cyst formation are present in *C. elegans*. It is not possible to study the function of, e.g., nephrocystin-5 in nematodes if the gene does not exist there. On the other hand, this indicates that nephrocystin-5 is not an essential protein for calcium signaling via ciliary polycystin function, and that the likely function of nephrocystin-5 in human nephrons involves some other aspect of polycystin's function in epithelial cells.

In addition, some of the proteins present in both worms and humans have significant differences. For example, the polycystin-1 homologue LOV-1 has a completely different extracellular N-terminus from that of mammalian PKD1 (Figure 72–1). This almost certainly indicates that these two proteins bind to different chemical substances in order to signal inside the cell. In fact, the N-terminal domain of PKD1/LOV-1 is very different in insects from that of nematodes and mammals as well,[82] which suggests that the PKD1 signaling domain regulates a common process (presumably chemo/mechanotransduction) that has been adapted to multiple purposes over evolution.

In nematodes, the polycystins are found only in male-specific neurons, and have a single task: When the cilium is stretched and/or while LOV-1 binds to the mechanical/chemical stimulus presented at the vulva, the neuron must rapidly admit enough calcium to depolarize sufficiently to cause an action potential. In the mammalian tubular epithelial cells, polycystin has a more difficult function: The nephritic cell's primary cilium acts as a flow sensor that is presumably bent repeatedly as liquid passes through the nephron.[35] Each tubule cell encounters other conditions that stretch the junctions to other cells and to the basement membrane as well. Indeed, liquid passing through the tubule should exert pressure on the apical membrane that should be transmitted to the cytoskeleton and exert tension on these points. Hydrostatic pressure without fluid flow can cause tubule expansion.[83] And without polycystin function, tubules slowly enlarge to form a cyst; therefore, polycystin must act as an inhibitor for this process. Signaling from polycystins at these points must be integrated to provide information to the epithelial cell as to whether it should reform the basement membrane, change shape, or divide in order to make the tubule larger. Accordingly, mammalian PKD1 has a larger C-terminal cytoplasmic domain than does the nematode LOV-1 protein (Figure 72–1), with several functional homologies to signaling modules that are not obviously present in *C. elegans* LOV-1, such as to the mTOR, Wnt, and JAK/STAT signal transduction pathways. *C. elegans* has these signaling pathways, and regulation of epithelial shape and tubule size is seen in all metazoan organisms,[84] so these aspects of epithelial cell structure and regulation can be studied in *C. elegans*, but polycystin signaling to these processes would have to be proved before they could be well studied in this organism.

CONCLUSIONS

C. elegans provides an ideal model for understanding many, but not all, aspects of polycystic kidney disease. The use (and NIH support) of forward genetic studies, many times on seemingly unrelated biological phenomena such as nematode male mating or osmotaxis, can nevertheless prove to be critically important in understanding human disease. In particular, LOV-1 and PKD-2 provide an excellent model for mammalian primary cilium synthesis, polycystin synthesis and transport, and polycystin function in calcium signaling in the cilium and endoplasmic reticulum (Figure 72–2), but vertebrate models (e.g., zebrafish and mouse)

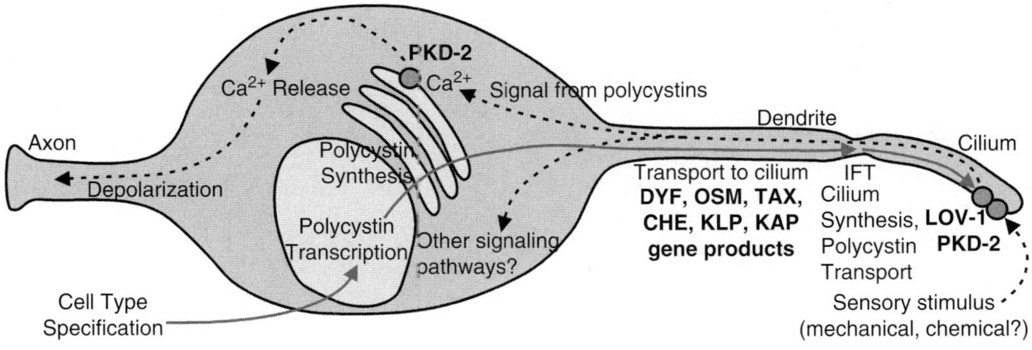

Figure 72–2. Diagram of synthesis, transport, and possible mechanism of signaling of polycystins in a *C. elegans* male-specific neuron. Gray solid lines indicate synthesis and transport of polycystins and of cell components necessary for polycystin function. Black dotted lines indicate signaling from polycystins in neurons. The sensory dendrite is on the right and the axon on the left. Constriction in the dendrite indicates the transition zone at the base of the sensory cilium. Cell type is specified by transcription factors. Components necessary to construct a sensory cilium are transported to the end of the dendrite. IFT (intraflagellar transport) is necessary to construct flagellum. Polycystins are transcribed and transported through the endoplasmic reticulum, where much of the PKD-2 protein is retained. Transport of LOV-1 and PKD-2 polycystins requires various kinesins (KLP proteins and KAP, kinesin-associated proteins) as well as other components discovered as defective in DYF (DYe-Filling), OSMotaxis, TAXis, and CHEmotaxis. IFT is required to transport LOV-1 and PKD-2 to sites of activity in the cilium. Mechanical and/or chemical stimulation of the LOV-1 protein at the vulval opening signals polycystin activity. LOV-1 and PKD-2 are presumed to act together as an ion channel to admit divalent cations into the cell. This signal partially depolarizes the cell, to stimulate PKD-2 at the endoplasmic reticulum to open to release calcium from intracellular stores. The neuron is thereby depolarized to elicit a behavioral response. Other signaling pathways after stimulation of ciliary polycystins are possible but unproven in *C. elegans* neurons.

may be more useful for studying aspects of polycystin function that may not be present in *C. elegans*. Most aspects of ciliary formation, and transport of receptor polycystins and nephrocystins into the cilium, can also be studied in *C. elegans* via a wide variety of behavioral, cell biological, and electrophysiological methods, and provide a promising source of future results to deepen our understanding of the normal physiology of the kidney tubules and the prevention of polycystic kidney disease.

ACKNOWLEDGMENTS

I am greatly indebted to the support of the entire *C. elegans* research community, and from NIH R01 DK55526 and the Kansas Interdisciplinary Center for PKD Disease Research.

REFERENCES

1. Grantham JJ. The etiology, pathogenesis, and treatment of autosomal dominant polycystic kidney disease: Recent advances. *Am J Kidney Dis* 1996;28:788–803.
2. Gabow PA. Cystic diseases of the kidney. In: Morgan SH, Grünfeld J-P, Eds. *Inherited Disorders of the Kidney*. Oxford, UK: Oxford University Press, 1998:132–162.
3. Hildebrandt F, Waldherr R, Kutt R, Brandis M. The nephronophthisis complex: Clinical and genetic aspects. *Clin Invest* 1992;70:802–808.
4. Lonser RR, Glenn GM, Walther M, *et al.* von Hippel-Lindau disease. *Lancet* 2003;361:2059–2067.
5. Green JS, Parfrey PS, Harnett JD, *et al.* The cardinal manifestations of Bardet-Biedl syndrome, a form of Laurence-Moon-Biedl syndrome. *N Engl J Med* 1989;321:1002–1009.
6. Curry NS, Milutinovic J, Grossnickle M, Munden M. Renal cystic disease associated with orofaciodigital syndrome. *Urol Radiol* 1992;13:153–157.
7. Smith UM, Consugar M, Tee LJ, *et al.* The transmembrane protein meckelin (MKS3) is mutated in Meckel-Gruber syndrome and the *wpk* rat. *Nat Genet* 2006;38:191–196.
8. Ward CJ, Peral B, Hughes J, *et al.* The polycystic kidney disease 1 gene encodes a 14 kb transcript and lies within a duplicated region on chromosome 16. *Cell* 1994;77:881–894.
9. Mochizuki T, Wu G, Hayashi T, *et al. PKD2*, a gene for polycystic kidney disease that encodes an integral membrane protein. *Science* 1996;272:1339–1342.
10. Beales PL. Lifting the lid on Pandora's box: The Bardet-Biedl syndrome. *Curr Opin Genet Dev* 2005;15:315–323.
11. Otto EA, Loeys B, Khanna H, *et al.* Nephrocystin-5, a ciliary IQ domain protein, is mutated in Senior-Loken syndrome and interacts with RPGR and calmodulin. *Nat Genet* 2005;37:282–288.
12. Sayer JA, Otto EA, O'Toole JF, *et al.* The centrosomal protein nephrocystin-6 is mutated in Joubert syndrome and activates transcription factor ATF4. *Nat Genet* 2006;38:674–681.
13. Hentschel DM, Bonventre JV. Novel non-rodent models of kidney disease. *Curr Mol Med* 2005;5:537–546.
14. Kahn-Kirby AH, Bargmann CI. TRP channels in *C. elegans*. *Annu Rev Physiol* 2006;68:719–736.
15. Montell C, Rubin GM. Molecular characterization of the *Drosophila trp* locus: A putative integral membrane protein required for phototransduction. *Neuron* 1989;2:1313–1323.
16. Clapham DE. TRP channels as cellular sensors. *Nature* 2003;426:517–524.
17. Tsiokas L, Kim E, Arnould T, Sukhatme VP, Walz G. Homo- and heterodimeric interactions between the gene products of *PKD1* and *PKD2*. *Proc Natl Acad Sci USA* 1997;94:6965–6970.
18. Qian F, Germino FJ, Cai Y, Zhang X, Somlo S, Germino GG. PKD1 interacts with PKD2 through a probable coiled-coil domain. *Nat Genet* 1997;16:179–183.
19. Lehtonen S, Ora A, Olkkonen VM, *et al. In vivo* interaction of the adapter protein CD2-associated protein with the type 2 polycystic kidney disease protein, polycystin-2. *J Biol Chem* 2000;275:32888–32893.
20. Gallagher AR, Cedzich A, Gretz N, Somlo S, Witzgall R. The polycystic kidney disease protein PKD2 interacts with Hax-1, a protein associated with the actin cytoskeleton. *Proc Natl Acad Sci USA* 2000;97:4017–4022.
21. Koulen P, Duncan RS, Liu J, *et al.* Polycystin-2 accelerates Ca^{2+} release from intracellular stores in *Caenorhabditis elegans*. *Cell Calcium* 2005;37:593–601.
22. Cai Y, Anyatonwu G, Okuhara D, *et al.* Calcium dependence of polycystin-2 channel activity is modulated by phosphorylation at Ser812. *J Biol Chem* 2004;279:19987–19995.
23. Hu J, Bae YK, Knobel KM, Barr MM. Casein kinase II and calcineurin modulate TRPP function and ciliary localization. *Mol Biol Cell* 2006;17:2200–2211.
24. Qian F, Wei W, Germino G, Oberhauser A. The nanomechanics of polycystin-1 extracellular region. *J Biol Chem* 2005;280:40723–40730.
25. Hamill OP. Twenty odd years of stretch-sensitive channels. *Pflugers Arch* 2006;453(3):333–351.
26. Qian F, Boletta A, Bhunia AK, *et al.* Cleavage of polycystin-1 requires the receptor for egg jelly domain and is disrupted by human autosomal-dominant polycystic kidney disease 1-associated mutations. *Proc Natl Acad Sci USA* 2002;99:16981–16986.
27. Shillingford JM, Murcia NS, Larson CH, *et al.* The mTOR pathway is regulated by polycystin-1, and its inhibition reverses renal cystogenesis in polycystic kidney disease. *Proc Natl Acad Sci USA* 2006;103:5466–5471.
28. Boca M, Distefano G, Qian F, Bhunia AK, Germino GG, Boletta A. Polycystin-1 induces resistance to apoptosis through the phosphatidylinositol 3-kinase/Akt signaling pathway. *J Am Soc Nephrol* 2006;17:637–647.
29. Bhunia AK, Piontek K, Boletta A, *et al.* PKD1 induces p21(waf1) and regulation of the cell cycle via direct activation of the JAK-STAT signaling pathway in a process requiring PKD2. *Cell* 2002;109:157–168.
30. Arnould T, Kim E, Tsiokas L, *et al.* The polycystic kidney disease 1 gene product mediates protein kinase C alpha-dependent and c-Jun N-terminal kinase-dependent activation of the transcription factor AP-1. *J Biol Chem* 1998;273:6013–6018.
31. Kim E, Arnould T, Sellin LK, *et al.* The polycystic kidney disease 1 gene product modulates Wnt signaling. *J Biol Chem* 1999;274:4947–4953.
32. Yamaguchi T, Wallace DP, Magenheimer BS, Hempson SJ, Grantham JJ, Calvet JP. Calcium restriction allows cAMP activation of the B-Raf/ERK pathway, switching cells to a cAMP-dependent growth-stimulated phenotype. *J Biol Chem* 2004;279:40419–40430.
33. *C. elegans* Sequencing Consortium. Genome sequence of the nematode *C. elegans*: A platform for investigating biology. *Science* 1998;282:2012–2018.
34. Barr MM, Sternberg PW. A polycystic kidney-disease gene homologue required for male mating behaviour in *C. elegans*. *Nature* 1999;401:386–389.
35. Praetorius HA, Spring KR. The renal cell primary cilium functions as a flow sensor. *Curr Opin Nephrol Hypertens* 2003;12:517–520.
36. Nauli SM, Alenghat FJ, Luo Y, *et al.* Polycystins 1 and 2 mediate mechanosensation in the primary cilium of kidney cells. *Nat Genet* 2003;33:129–137.
37. Silberberg M, Charron AJ, Bacallao R, Wandinger-Ness A. Mispolarization of desmosomal proteins and altered intercellular adhesion in autosomal dominant polycystic kidney disease. *Am J Physiol Renal Physiol* 2005;288:F1153–1163.
38. Wilson PD, Geng L, Li X, Burrow CR. The *PKD1* gene product, "polycystin-1," is a tyrosine-phosphorylated protein that colocalizes with alpha2beta1-integrin in focal clusters in adherent renal epithelia. *Lab Invest* 1999;79:1311–1323.
39. Kaletta T, Van der Craen M, Van Geel A, *et al.* Towards understanding the polycystins. *Nephron Exp Nephrol* 2003;93:e9–17.
40. Watnick T, Germino GG. Molecular basis of autosomal dominant polycystic kidney disease. *Semin Nephrol* 1999;19:327–343.

41. Saunier S, Salomon R, Antignac C. Nephronophthisis. *Curr Opin Genet Dev* 2005;15:324–331.
42. Jauregui AR, Barr MM. Functional characterization of the *C. elegans* nephrocystins NPHP-1 and NPHP-4 and their role in cilia and male sensory behaviors. *Exp Cell Res* 2005;305:333–342.
43. Blacque OE, Leroux MR. Bardet-Biedl syndrome: An emerging pathomechanism of intracellular transport. *Cell Mol Life Sci* 2006;63(18):2145–2161.
44. Ansley SJ, Badano JL, Blacque OE, *et al.* Basal body dysfunction is a likely cause of pleiotropic Bardet-Biedl syndrome. *Nature* 2003;425:628–633.
45. Culotti JG, Russell RL. Osmotic avoidance defective mutants of the nematode *Cænorhabditis elegans*. *Genetics* 1978;90:243–256.
46. Kozminski KG, Johnson KA, Forscher P, Rosenbaum JL. A motility in the eukaryotic flagellum unrelated to flagellar beating. *Proc Natl Acad Sci USA* 1993;90:5519–5523.
47. Murcia NS, Richards WG, Yoder BK, Mucenski ML, Dunlap JR, Woychik RP. The Oak Ridge Polycystic Kidney (*orpk*) disease gene is required for left-right axis determination. *Development* 2000;127:2347–2355.
48. Qin H, Rosenbaum JL, Barr MM. An autosomal recessive polycystic kidney disease gene homolog is involved in intraflagellar transport in *C. elegans* ciliated sensory neurons. *Curr Biol* 2001;11:457–461.
49. Blacque OE, Perens EA, Boroevich KA, *et al.* Functional genomics of the cilium, a sensory organelle. *Curr Biol* 2005;15:935–941.
50. Brenner S. The genetics of *Caenorhabditis elegans*. *Genetics* 1974;77:71–94.
51. Liu KS, Sternberg PW. Sensory regulation of male mating behavior in *Caenorhabditis elegans*. *Neuron* 1995;14:79–89.
52. Goldstein P. The synaptonemal complexes of *Caenorhabditis elegans*: Pachytene karyotype analysis of hermaphrodites from the recessive him-5 and him-7 mutants. *J Cell Sci* 1986;82:119–127.
53. Barr MM, DeModena J, Braun D, Nguyen CQ, Hall DH, Sternberg PW. The *Caenorhabditis elegans* autosomal dominant polycystic kidney disease gene homologs *lov-1* and *pkd-2* act in the same pathway. *Curr Biol* 2001;11:1341–1346.
54. Peden EM, Barr MM. The KLP-6 kinesin is required for male mating behaviors and polycystin localization in *Caenorhabditis elegans*. *Curr Biol* 2005;15:394–404.
55. White JG, Southgate E, Thomson JN, Brenner S. The structure of the nervous system of the nematode *Caenorhabditis elegans*. *Philos Trans R Soc Lond B Biol Sci* 1986;314:1–340.
56. Hart AC. Behavior. In: The Community of *C. elegans* Researchers, Eds. *WormBook*, 2006.
57. Ward S. Chemotaxis by the nematode *Caenorhabditis elegans*: Identification of attractants and analysis of the response by use of mutants. *Proc Natl Acad Sci USA* 1973;70:817–821.
58. Dusenbery DB, Sheridan RE, Russell RL. Chemotaxis-defective mutants of the nematode *Caenorhabditis elegans*. *Genetics* 1975;80:297–309.
59. Wicks SR, de Vries CJ, van Luenen HG, Plasterk RH. CHE-3, a cytosolic dynein heavy chain, is required for sensory cilia structure and function in *Caenorhabditis elegans*. *Dev Biol* 2000;221: 295–307.
60. Bargmann CI, Hartwieg E, Horvitz HR. Odorant-selective genes and neurons mediate olfaction in *C. elegans*. *Cell* 1993;74:515–527.
61. Ward S, Thomson N, White JG, Brenner S. Electron microscopical reconstruction of the anterior sensory anatomy of the nematode *Caenorhabditis elegans*. *J Comp Neurol* 1975;160:313–337.
62. Perkins LA, Hedgecock EM, Thomson JN, Culotti JG. Mutant sensory cilia in the nematode *Caenorhabditis elegans*. *Dev Biol* 1986;117:456–487.
63. Riddle DL, Albert PA. Genetic and environmental regulation of dauer larva development. In: Riddle DL, Meyer BJ, Priess JR, Blumenthal T, Eds. *C. elegans II*. Cold Spring Harbor, NY: Cold Spring Harbor Press, 1997:739–768.
64. Bell LR, Stone S, Yochem J, Shaw JE, Herman RK. The molecular identities of the *Caenorhabditis elegans* intraflagellar transport genes *dyf-6*, *daf-10* and *osm-1*. *Genetics* 2006;173:1275–1286.
65. Snow JJ, Ou G, Gunnarson AL, *et al.* Two anterograde intraflagellar transport motors cooperate to build sensory cilia on *C. elegans* neurons. *Nat Cell Biol* 2004;6:1109–1113.
66. Stinchcomb DT, Shaw JE, Carr SH, Hirsh D. Extrachromosomal DNA transformation of *Cænorhabditis elegans*. *Mol Cell Biol* 1985;5:3484–3496.
67. Zhou HM, Brust-Mascher I, Scholey JM. Direct visualization of the movement of the monomeric axonal transport motor UNC-104 along neuronal processes in living *Caenorhabditis elegans*. *J Neurosci* 2001;21:3749–3755.
68. Bae YK, Qin H, Knobel KM, Hu J, Rosenbaum JL, Barr MM. General and cell-type specific mechanisms target TRPP2/PKD-2 to cilia. *Development* 2006;133:3859–3870.
69. Qin H, Burnette DT, Bae YK, Forscher P, Barr MM, Rosenbaum JL. Intraflagellar transport is required for the vectorial movement of TRPV channels in the ciliary membrane. *Curr Biol* 2005;15:1695–1699.
70. Lamitina ST, Morrison R, Moeckel GW, Strange K. Adaptation of the nematode *Caenorhabditis elegans* to extreme osmotic stress. *Am J Physiol Cell Physiol* 2004;286:C785–791.
71. Brockie PJ, Mellem JE, Hills T, Madsen DM, Maricq AV. The *C. elegans* glutamate receptor subunit NMR-1 is required for slow NMDA-activated currents that regulate reversal frequency during locomotion. *Neuron* 2001;31:617–630.
72. Richmond JE, Jorgensen EM. One GABA and two acetylcholine receptors function at the *C. elegans* neuromuscular junction. *Nat Neurosci* 1999;2:791–797.
73. O'Hagan R, Chalfie M, Goodman MB. The MEC-4 DEG/ENaC channel of *Caenorhabditis elegans* touch receptor neurons transduces mechanical signals. *Nat Neurosci* 2005;8:43–50.
74. Lockery SR, Goodman MB. Tight-seal whole-cell patch clamping of *Caenorhabditis elegans* neurons. *Methods Enzymol* 1998;293:201–217.
75. Christensen M, Estevez A, Yin X, *et al.* A primary culture system for functional analysis of *C. elegans* neurons and muscle cells. *Neuron* 2002;33:503–514.
76. Bianchi L, Driscoll M. Heterologous expression of *C. elegans* ion channels in *Xenopus* oocytes. In: The Community of *C. elegans* Researchers, Ed. WormBook, 2006.
77. Chernova MN, Vandorpe DH, Clark JS, Alper SL. Expression of the polycystin-1 C-terminal cytoplasmic tail increases Cl channel activity in *Xenopus* oocytes. *Kidney Int* 2005;68:632–641.
78. Luo Y, Vassilev PM, Li X, Kawanabe Y, Zhou J. Native polycystin 2 functions as a plasma membrane Ca^{2+}-permeable cation channel in renal epithelia. *Mol Cell Biol* 2003;23:2600–2607.
79. Li Y, Wright JM, Qian F, Germino GG, Guggino WB. Polycystin 2 interacts with type I inositol 1,4,5-trisphosphate receptor to modulate intracellular Ca^{2+} signaling. *J Biol Chem* 2005;280:41298–41306.
80. Koulen P, Cai Y, Geng L, *et al.* Polycystin-2 is an intracellular calcium release channel. *Nat Cell Biol* 2002;4:191–197.
81. Gonzalez-Perrett S, Kim K, Ibarra C, *et al.* Polycystin-2, the protein mutated in autosomal dominant polycystic kidney disease (ADPKD), is a Ca^{2+}-permeable nonselective cation channel. *Proc Natl Acad Sci USA* 2001;98:1182–1187.
82. Montell C. Drosophila TRP channels. *Pflugers Arch* 2005;451: 19–28.
83. Tanner GA, McQuillan PF, Maxwell MR, Keck JK, McAteer JA. An *in vitro* test of the cell stretch-proliferation hypothesis of renal cyst enlargement. *J Am Soc Nephrol* 1995;6:1230–1241.
84. Lubarsky B, Krasnow MA. Tube morphogenesis: Making and shaping biological tubes. *Cell* 2003;112:19–28.

73 Animal Models of Myelofibrosis

Alessandro M. Vannucchi, Jean-Luc Villeval,
Orianne Wagner-Ballon, Paola Guglielmelli,
and Anna Rita Migliaccio

ABSTRACT

The pathogenesis of idiopathic myelofibrosis has not yet been completely elucidated, but the availability of animal models of the disease has made it possible, in the past few years, to focus on the cells (megakaryocytes) and cytokines (transforming growth factor-β, among others) potentially involved. This review will deal with the two best characterized animal models (the TPO-overexpressing and the GATA-1^{low} mouse) as well as with the newest ones that have been developed following the identification of the *JAK2*^{V617F} and *MPL*$^{W515K/L}$ mutations in the human form of the disease. Each of these unique animal models conceivably represents an invaluable tool for further characterizing the disease mechanisms as well as for testing novel therapeutic agents.

Key Words: Myelofibrosis, Mpl, GATA-1, Animal models, Transforming growth factor-β, Megakaryocyte, Osteosclerosis.

INTRODUCTION

Idiopathic myelofibrosis (IM), also called myelofibrosis with myeloid metaplasia (MMM),[1] is a malignant hematological disorders that together with polycythemia vera (PV), essential thrombocythemia (ET), and other less frequent entities belongs to the Philadelphia-negative chronic myeloproliferative disorders (MPD), according to the World Health Organization (WHO) classification of myeloid neoplasms.[2] All these are clonal disorders that originate at the level of a pluripotent hematopoietic stem/progenitor cell that, while retaining a relatively normal differentiation program unlike that in acute leukemia, has lost some of the regulatory mechanisms that control proliferation; indeed, one of the best characterized abnormalities of hematopoietic cells in these disorders is their substantial independence from regulatory cytokines, in particular erythropoietin (EPO)[3] or thrombopoietin (TPO),[4] the two main cytokines for the erythroid and megakaryocytic (Mk) lineage, respectively. MMM is characterized by anemia, extramedullary hematopoiesis, bone marrow (BM) structural changes that include fibrosis, neoangiogenesis, and osteosclerosis, and by constitutive circulation of stem/progenitor cells with dacryocytes, nucleated erythroid cells, and immature myeloid cells in the peripheral blood (PB); in some cases, MMM represents the disease evolution of a preexisting PV or ET (so called, post-polycythemic/thrombocythemic MMM), and has an intrinsic tendency to evolve into acute leukemia.[1] Accumulated evidence indicates that the changes in the structure of the BM that are characteristic of MMM represent a typical reactive process (the cells of the stroma do not belong to the neoplastic clone), likely mediated by cytokines abnormally released by clonal hematopoietic cells. On the other hand, the primary lesion occurring in the hematopoietic stem/progenitor cell of MMM is still not well understood, although recently an acquired point mutation in the *JAK2* gene, consisting of a valine-to-phenylalanine substitution (*JAK2*^{V617F}) in the autoinhibitory JH2 (JAK homology) domain of the molecule, has been found to occur in most PV patients and in about half of those with ET or MMM.[5–8] An even more infrequent point mutation of the TPO receptor, *MPL*^{W515L}, has been detected in less than 10% of MMM patients.[9,10] Although both these mutations when expressed in murine models recapitulate several aspects of MMM (see below), their exact positioning in the sequence of a multistep series of events forming the basis of the pathogenesis of the disease remains to be defined.

A frequent observation in MMM patients has been the occurrence of the discrete involvement of the Mk lineage in BM biopsies[11]; actually, the morphological appearance of the BM Mk population, beyond its quantitative expansion, is one of the key points in the WHO classification, allowing differential diagnosis of MMM from other MPDs, as well as the identification of a "prefibrotic" form of IM.[12] This critical role of Mks has been corroborated by the study of animal models of the disease that, either incidentally discovered or not (Table 73–1), have been consistently characterized by the accumulation of Mks eventually associated with thrombocytosis. This chapter will review the production and characterization of two extensively studied models of MMM, represented by TPO^{high} and GATA-1^{low} mice, together with the newest findings concerning the development of IM in mice carrying hematopoietic cells mutated at the *JAK2* or *MPL* locus.

From: *Sourcebook of Models for Biomedical Research*
(P. M. Conn, ed.), © 2008 Humana Press Inc., Totowa, NJ.

Table 73–1
Additional murine models presenting myelofibrosis[a]

Model	*Main features*
MPSV-induced mice	MPSV; splenic and medullary fibrosis; splenic myeloid hyperplasia; myelemia; anemia; thrombocytopenia; possibly due to high levels of IL-1a, IL-6, GM-CSF, G-CSF, CSF-1, and TNF-α[62–64]
TEL/PDGFRβR transgenic mice	Overproduction of mature neutrophils and dysplastic megakaryocytes in the bone marrow; splenomegaly with extramedullary hematopoiesis; high levels of IL-3 and GM-CSF[65–67]
TEL/JAK2 and *TEL/TRKC* transgenic mice	Extramedullary hematopoiesis, intensive myeloproliferation in the bone marrow and the spleen with splenomegaly, granulocytic leukocytosis; high levels of IL-3 and GM-CSF[66,68]
Bach1 transgenic mice	Transgenic mice bearing human BACH1 cDNA under the control of the GATA1 locus; significant thrombocytopenia associated with impaired maturation of megakaryocytes and development of myelofibrosis[69]
VEGFR-1 transgenic mice (op/op$Flt1^{TK-/-}$)	Extensive osteoclast deficiency, resulting in bone marrow fibrosis and extramedullary hematopoiesis[70,71]
LIF overexpressing mice	Developed in DBA/2 mice engrafted with FDC-P1 cells overexpressing LIF; myelosclerosis, splenomegaly, extramedullary hematopoiesis in the liver, neutrophil leukocytosis[72]
OPG/OCIF transgenic mice	Early and progressive osteopetrosis, splenomegaly, and extramedullary hematopoiesis[73]
BUBR1 mutant mice ($BUBR^{+/-}$)	Splenomegaly, extramedullary megakaryocytopoiesis; increased megakaryocytic, and decreased erythroid, progenitor content in the bone marrow[74]
FLT3-ITD mutant mice	Myelofibrosis, splenomegaly, extramedullary hematopoiesis, leukocytosis with a few immature myeloid cells; myeloproliferative disease has a latency of 40–60 days and does not evolve into AML[75]
Lnk mutant mice ($Lnk^{-/-}$)	Splenomegaly with fibrosis and extramedullary hematopoiesis, hyperplasia of the megakaryocytic lineage, increased number of erythroid cells in the spleen but decreased in bone marrow; defects of B lymphopoiesis; leukocytosis and thrombocytosis; increased proliferative responses of hematopoietic progenitors to cytokines such as SCF, IL-3, and IL-7[76]

[a]MPSV, myeloproliferative sarcoma virus; IL, interleukin; GM-CSF, granulocyte-macrophage colony-stimulating factor; TNF, tumor necrosis factor; LIF, leukemia inhibitory factor; AML, acute myeloid leukemia; SCF, stem cell factor.

THE TPO^{high} MODELS

TPO, also known as MPL ligand or megakaryocyte growth and development factor (MGDF), is the major physiological humoral regulator of megakaryopoiesis and platelet production.[13,14] The TPO^{high} model of myelofibrosis was discovered during investigations aimed to determine pathologies and side effects associated with long-term exposure to high doses of the recently discovered TPO, an obvious drug to increase platelet levels in patients. Several models are described, including the simplest approach by systemic repeated administration of a truncated version of recombinant TPO associated with a polyethylene glycol tail (PEG-rh-MDGF)[15] and more sophisticated approaches by which TPO overexpression was achieved by adenovirus vectors,[16–19] retrovirus vectors,[20,21] or by transgenesis.[22,23] Except for one model of transgenic mice,[22] all the other models described the occurrence of BM fibrosis in mice. We will mainly focus on the retroviral models and adenoviral models that have been studied extensively.

TPO^{high} MODEL BY SYSTEMIC REPEATED ADMINISTRATION OF PEG-RH-MDGF Immunocompetent BALB/c mice were injected daily for 14 days with increasing doses of PEG-rh-MDGF from 10 to 500 μg/kg.[15] All animals developed dose-dependent thrombocytosis, moderate leukocytosis, anemia, splenomegaly, and extramedullary hematopoiesis. A BM fibrosis was observed at the highest dose injected, which reversed rapidly when the injections were stopped. Transient myelofibrosis was also observed in rats receiving PEG-rh-MDGF.[24] The effects of systemic administration of TPO in the GATA-1^{low} model of myelofibrosis will be described below.[25]

TPO^{high} MODELS BY ADENOVIRAL VECTORS An alternative approach was developed using replication-deficient adenovirus vectors. All the studies were performed using injections of adenovirus carrying a human TPO cDNA. However, the promoter driving the expression of TPO was either a cytomegalovirus (CMV) or a Rous sarcoma virus (RSV) promoter. Furthermore, several routes of injection were used (subcutaneous,[16] intratracheal,[17] intraperitoneal,[19] and intravenous[18]). Finally, different stains of mice were treated: immunocompetent BALB/c mice and immunodeficient nonobese diabetic/severe combined immunodeficient (NOD/SCID) and SCID mice. Immunocompetent mice did not develop fibrosis because an immune thrombocytopenia occurred after a brief thrombocytosis that peaked at day 7, accompanied by Mk hyperplasia in the spleen. This immune reaction was due to the generation of neutralizing antibodies directed against the transgene human TPO that cross-reacted with murine TPO.[26] To avoid this reaction, models were developed in immunodeficient mice.[16,17,19,26] SCID (severe combined immunodeficient; T and B cell deficient) mice developed thrombocytosis, leukocytosis, and marked Mk hyperplasia in the spleen and BM with a progressive myelofibrosis and osteosclerosis as early as 6 weeks after injection. Surprisingly, no fibrotic or sclerotic changes were seen in NOD/SCID [T and B cell deficient as well as presenting multiple other defects in innate immunity including natural

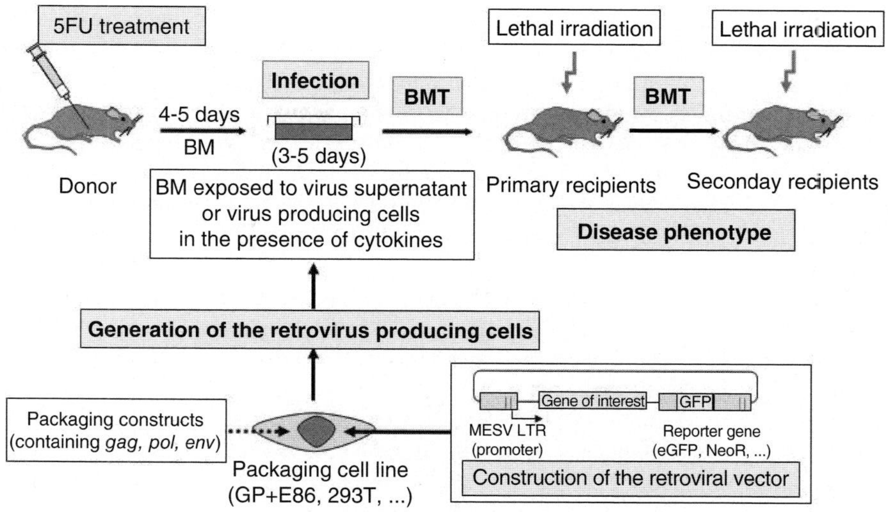

Figure 73–1. A representation of the procedure used for the adoptive transfer of bone marrow cells infected by a retrovirus expressing the gene of interest (TPO, $JAK2^{V617F}$). BMT, bone marrow transplantation; 5FU, 5-fluorouracil.

killer (NK) and monocyte–macrophage function] mice.[26] Because NOD/SCID mice differ from SCID mice by impaired mononuclear phagocyte functions, it was postulated that monocytes and macrophages are required for promotion of myelofibrosis in TPO^{high} mice. However, this theory is not supported by another recent study demonstrating that NOD/SCID mice exposed to high and sustained TPO levels develop BM fibrosis.[27]

TPO^{high} MODELS BY RETROVIRAL VECTORS Two models were described, one by Yan *et al.*[20] and one by our group.[21] In both investigations, the murine TPO cDNA was cloned into a retroviral vector and infectious nonreplicative recombinant retroviral particles were produced from the transfected GP + E86 packaging cell line. A high virus titer-producing clone was selected. In our study, this clone, called GP122, also produced very high TPO level (6000 U/ml).[21] The principle consists of transducing hematopoietic stem cells (HSC) of a mouse with the virus in order to sustain high levels of TPO production by the hematopoietic system throughout the life of a mouse. The procedure is based on BM transplantation (BMT). BM cells, collected from donor mice, are infected with the retrovirus and injected into recipient mice depleted of their own hematopoietic system as a result of lethal irradiation (9.5 Gy for the C57BL/6 strain). To transduce HSC efficiently, which will ensure the lifetime maintenance of a transduced progeny, donor mice are treated with 5-fluorouracil (5FU, a single intraperitoneal or intravenous injection at 150 mg/kg for 4 days). The duration of treatment allows the maximal number of HSC to be infected by the retrovirus, corresponding to the maximal number of cycling HSC. The infection is carried out by a coculture of the 5FU-treated cells (two femurs) with the virus-producing cells (10^5 cells in a 10-cm tissue culture Petri dish) for 4–5 days. The number of adherent virus-producing cells is calculated in order to obtain a confluent culture at the day of BM cell collection. The infection is carried out in the presence of cytokines able to induce HSC proliferation and survival. We used a spleen cell-conditioned medium, but this can be replaced with a mixture of early-acting cytokines such as interleukin 3 (IL-3), stem cell factor (SCF), and IL-6 (TPO being secreted by the virus-producing cells).[20] The infected BM cells are collected as the nonadherent cells, and care should be taken not to detach too many virus-producing cells, which will be detrimental to the mice during the injection (these cells, derived from Swiss mice, will be rejected by the immunocompetent C57BL/6J recipient mice). In the study by Yan *et al.*, the infection was not carried by coculture but by serial exposures of the 5FU-treated cells to a virus-producing cell supernatant.[20] Lethally irradiated syngeneic mice (2 months old) are finally injected with 10^6 BM cells (Figure 73–1). The mice are not studied before full blood reconstitution (5 weeks for myeloid and B cells and 7 weeks for T cells). The result of this procedure is the genomic integration of the virus in HSC and their progeny that could be analyzed by Southern blot to evaluate the numbers of both provirus copies, compared to the endogenous gene, and of individual transduced HSC marked by the sites of provirus integration. Provirus, as efficient vectors of expression, produced high level of TPO under the control of the viral long terminal repeat (LTR) promoter.

Using this strategy, we obtained a two-step disease that mimicked the evolution of human MMM and resulted in the death of 100% of the animals within 10 months after transplantation.[21] The first phase of the disease (2 months after BMT) is characterized by thrombocytosis (a 4-fold increase), leukocytosis (up to a 10-fold increase), and decrease of the hematocrit. Giant platelets were observed in PB smears. A striking elevation in the PB, BM and spleen progenitor cell numbers, except the colony-forming unit (CFU)-E, was noticed. Massive hyperplasia of Mks, often organized in clusters, and of granulocytes was observed in BM and spleen with a hypoplasia of erythroblasts in the BM. From the third month post-BMT, platelets, leukocytes, and red cell numbers dropped dramatically and mice developed a severe pancytopenia. Progenitor cell numbers dropped in BM and spleen but increased in the PB and peritoneal cavity. Extramedullary hematopoiesis was observed in liver, kidney, lymph nodes, and lung. Histological examination revealed severe fibrosis in the BM and spleen and osteosclerosis in BM. Notably, 2 mice out of 32 primary recipients and 1 mouse out of 30 secondary recipients died prematurely from undifferentiated acute leukemia. Thus, this fatal myeloproliferative disorder mimicked most of the features observed in the human disease.

Yan *et al.* reported a less dramatic phenotype.[20] The platelet counts increased 5-fold and remained at a steady level for up to 73 weeks. A moderate anemia was observed and a slight increase of the number of PB nucleated cells was described. Mice developed extramedullary hematopoiesis and splenomegaly with hyperplasia of Mks and granulocytes. BM was also obturated with a dense network of reticulin fibers and new bone trabeculae. However, mice did not display the second phase of our disease model characterized by severe thrombocytopenia, anemia, and leukopenia. They remained alive and active over a year and no leukemic transformation was described. The difference may be related to the plasma level of TPO; in the study by Yan *et al.*,[20] plasma TPO reached 30 ng/ml compared to up to 1000 ng/ml in our study.[21]

TPO^{high} MODELS BY TRANSGENESIS Two models of transgenic mice with contradictory results were developed. In the first one,[22] transgenic mice overexpressing human TPO under the control of a liver-specific apolipoprotein E enhancer/promoter did not display fibrosis despite elevated plasma TPO levels (3 ng/ml), and a 4-fold increase in platelet counts and in tissue Mk content for more than 16 months. In contrast, in another recent transgenic model,[23] expressing full-length murine TPO under the control of the mouse H chain enhancer/promoter, similar mild leukocytosis (2-fold), anemia (a 10% decrease in hematocrit), thrombocytosis (3-fold), and increase of plasma TPO level (3 ng/ml) were associated with BM fibrosis starting at 9 months with further progression at 12 months of age. Local production of TPO in hematopoietic tissues in the second model, in contrast to the first one, might have been the reason for the occurrence of fibrosis.[23]

All these TPO^{high} models showed that TPO overexpression in mice leads to a myeloproliferative disorder featuring most aspects of MMM: leukocytosis, anemia, thrombocytosis, and in some cases pancytopenia, splenomegaly, extramedullary hematopoiesis, fibrosis, and osteosclerosis. Depending on the approach used, the severity of the disease (pancytopenia, leukemia, massive splenomegaly, fibrosis, and osteosclerosis) seems to be linked to the levels of TPO, the duration of exposure, as well as to the site of TPO production.[28]

THE GATA-1^{low} MODEL

GATA-1 is a transcription factor, encoded by a gene located on the X chromosome, that plays essential roles in the development of the erythroid,[29] megakaryocytic,[30] eosinophilic,[31] and mast cell[32] lineages, and that has been clarified by the phenotype of mice harboring experimentally induced alterations of its expression. Animal knockouts for the gene die *in utero* at day E10.5 because of severe anemia,[33] while those expressing ectopic levels of *Gata1* showed an increased rate of erythroid differentiation.[34] GATA-1 is member of the family of GATA transcription factors that controls the expression of target genes by binding to the consensus WGATAR sequence; since they share a highly conserved structure, the cellular functions of GATA-1 are strictly dependent on the appropriate lineage and on the differentiation stage of activation.[35] As a result of studies performed in the laboratory of Dr. Stuart Horkin in Boston, concerning the role of the region upstream to *Gata1* in regulating gene expression in erythroid and Mk differentiation, McDevitt *et al.* generated a strain of mice (*Gata1*$^{neo\delta HS}$) characterized by the targeted deletion of an ≅8-kb upstream region that included a DNase hypersensitive (HS) site (HSI) replaced by a neomycin-resistance cassette by homologous recombination.[36] A targeting vector, containing a *loxP* site-flanked neomycin resistance gene driven by the mouse phosphoglycerate kinase promoter and a herpes simplex virus thymidine kinase gene, was used to disrupt the region of interest. The construct was electroporated into 129S4/SvJae-derived J1 embryonic stem cells and correctly targeted embryonic stem cells were injected into C57BL/6 blastocysts[36] (Figure 73–2). The resulting *Gata1*$^{neo\delta HS}$ mice were anemic and demonstrated marked impairment in erythroid cell maturation[36] because of an increased apoptotic rate due to a 4- to5-fold decrease in GATA-1 expression. However, the most striking finding was that these mutants were severely thrombocytopenic because of the almost complete lack of *Gata1* expression in Mks that remain immature and release a few abnormally large platelets.[37,38] The proliferative rate of GATA-1^{low} Mks is increased, while the expression of lineage-specific genes, including the *mpl*, glycoproteins Ibα and Ibβ, and platelet factor 4, is reduced.[37,38] At the ultrastructural level, the Mks are blocked between stage I and II of maturation, and show poor organization of the α-granules due to a reduced expression of some α-granule-specific proteins, such as von Willebrand factor, or an atypical localization of other proteins, such as P-selectin, that, although being expressed at normal levels, are found on the demarcation membrane systems (DMS) rather than in the granules themselves[39] (see also below).

In the original C57BL/6 mutant strain, the mortality was >90% of the mutants around birth; however, when the mutation was introduced in the CD1 background in our laboratory by backcross-

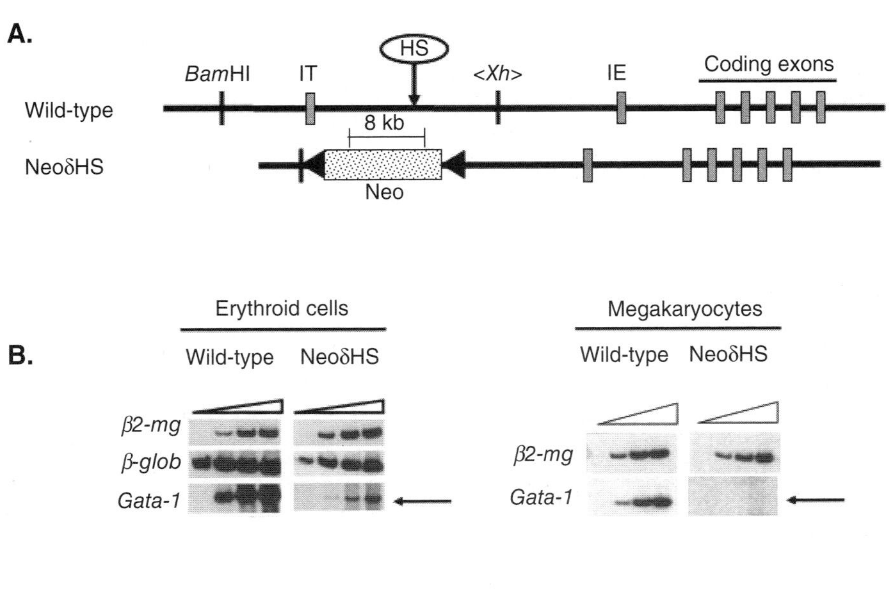

Figure 73–2. Generation and characterization of the GATA-1^{low} mice. (A) The procedure employed to generate GATA-1^{low} mice in the original C57BL/6 strain is described.[34] HS, DNase hypersensitive site I. IT and IE, "testis" and "erythroid" alternative first *GATA1* exons, respectively. (B) The expression levels (by semi-quantitative RT-PCR) of *GATA-1* in erythroid and megakaryocytic cells, to demonstrate the selective abrogation of *GATA-1* in the latter, are presented.

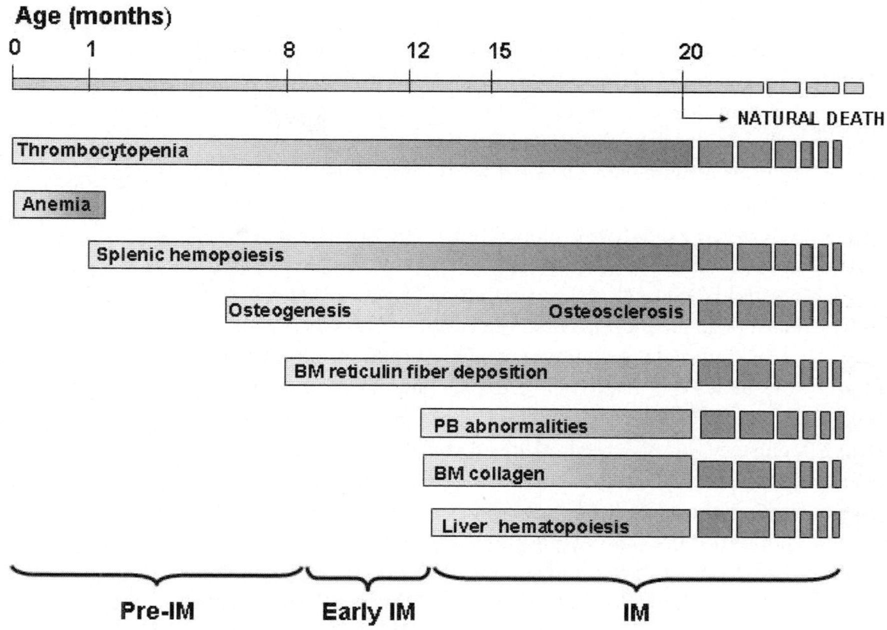

Figure 73–3. The age-dependent development of myelofibrosis in GATA-1^{low} mice.

ing to CD1 mice, starting from one breeder pair of B6 mice provided by Dr. S. Orkin, most of the mutants survived to reach adulthood, and, although severely anemic at birth, recovered from their anemia within the first month. On the other hand, marked thrombocytopenia (with counts less than 10% of the normal value) is maintained throughout their life. Thrombocytopenia is due to impaired Mk maturation, since the number of Mks in the BM and spleen is markedly increased; these cells are of varying size, with a population of large, almost mature Mk together with smaller forms, which are often found in clusters, especially in the spleen.[40] These morphological features, which resembled those observed in IM biopsies and in hematopoietic tissues from the TPO^{high} mice, prompted us to investigate whether these mice would develop myelofibrosis.

By analyzing a large cohort of GATA-1^{low} mice, we have been able to describe the natural history of myelofibrosis in these mice (Figure 73–3).[40,41] During their first year, the only overt hematological abnormality of these mice is thrombocytopenia and splenic enlargement due to the expansion of the Mks lineage; after the first month, they are no longer anemic, although a greater than normal degree of erythroid progenitor cell apoptosis persists, and they are able to mount a supraadequate response to anemia induction or EPO treatment.[42] From the first year of life, the first signs of MMM appear, with a progressive reduction (up to one-fifth the normal value) of the nucleated cell content in the BM cavity, the appearance of BM reticulin fibrosis that proceeds to collagen fiber deposition, and eventually osteosclerosis. The size of the spleen increases further (more than 5- to 10-fold that of the wild-type animals), and extramedullary hematopoiesis in the liver and less frequently in other organs can be detected. A moderate leukocytosis, with a few immature myeloid cells but no blasts, and dacryocytes with nucleated erythroblasts can be found in the PB. Thus, at 15–18 months of age, a full myelofibrotic phenotype is presented by the mice. However, unlike the mice overexpressing TPO, there is no leukemic terminal phase in the GATA-1^{low} animals, and the survival of the mice is not significantly reduced, since most die of old age at 20–24 months.[40] It is of interest that these genetically modified animal models have been made possible by their development in a CD1 background, since the high mortality rate in the original C57BL/6 strain would have likely precluded their recognition. Furthermore, when the mutation was introduced in the DBA/2 background, these mice showed severe anemia but did not express a full myelofibrotic phenotype, including extramedullary hematopoiesis and osteosclerosis. These observations point to the possible role of gene modifiers in the different strains, and might also configure a model to study the variability of the clinical phenotype of the disease in humans.[43] The GATA-1^{low} mice in the CD1 background are available from the Jackson Laboratories (at http://jaxmice.jax.org/, Strain Name: STOCK *Gata1^{tm2Sho}*, Stock Number 004655).

THE *JAK2*^{V617F} MODEL

The *JAK2*^{V617F} model of myelofibrosis was established during investigations trying to determine the exact role of the *JAK2*^{V617F} mutation recently identified in MPDs.[5–8] To analyze the consequences of this mutation for normal hematopoiesis, two groups studied the *in vivo* effects of *JAK2*^{V617F} expression in hematopoietic cells using an adoptive transfer in recipient mice of marrow cells transduced by a retrovirus expressing *JAK2*^{V617F}.[44,45] Both studies demonstrated that mice transplanted with HSC expressing *JAK2*^{V617F} develop a disease mimicking human PV, including its evolution toward the "spent" phase characterized by myelofibrosis.

In our study, we used a procedure similar to the one previously described for the TPO overexpressing model except that cocultures were carried out in the presence of IL-3, IL-6, SCF, and TPO.[44] Furthermore, virus-producing clones were generated from two infections of the GP + E86 cells with vesicular stomatitis virus (VSV)-G pseudotyped virus. Wernig *et al.* used transient transfections of 293T cells with the retroviral vector and the packaging plasmids to produce viral supernatants that were used to transduce 5FU-treated BM cells.[45] The infection procedure was carried out in the presence of IL-3, IL-6, and SCF and consisted of 1 day of prestimulation followed by 2 successive days of brief exposure to the virus supernatant (90 min for 1800 × *g* centrifugations). Recipients were treated with two serial sets of irradiation (5.5 Gy each), instead of one (9.5 Gy) in our study, before injection of the transduced BM cells.[44] In both studies, bicistronic viruses carrying an IRESeGFP cassette were used with similar promoters [murine stem cell virus (MSCV) LTR derived].

During the 2–4 months of transplantation, using both strategies, C57BL/6 mice rapidly developed polycythemia (hematocrit: 60–90%) with reticulocytosis. Wernig *et al.* reported low EPO levels.[45] A mild leukocytosis was observed with an excess of polynuclear elements and rare immature forms. The platelet size was increased but platelet counts remained normal except in one of our low *JAK2*^{V617F} expresser groups, where transient mild thrombocytosis was observed. Myeloid hyperplasia of the marrow with Mk clusters was observed. Mild splenomegaly (2- to 3-fold) developed with expansion of the red pulp and hyperplasia of myeloid and erythroid cells (4- to 5-fold) and clusters of dysplastic and apoptotic Mks with neutrophil emperipolesis. Wernig *et al.* described decreased Mk ploidy and extramedulary hematopoiesis in the liver with myeloid, erythroid, and Mk cells.[45] The total number of myeloid and erythroid progenitor cells increased due to the expansion of splenic hematopoiesis. Endogenous erythroid colonies (growing without addition of EPO) developed from BM and spleen cells (around 30% of the total CFU-E) and constitutive activation of JAK2 downstream effectors (Stat5, ERK1/2) was detected in spleen and BM cells. No fibrosis was observed at this stage. In our study, mice were followed for 4 months after BMT.[44] During this late phase, the disease evolved to a very different phenotype. We observed abatement of polycythemia, abnormal red blood cell morphology, giant platelets as a sign of dysmegakaryopoiesis, and fibrosis in the BM and spleen. Ultimately, development of severe fibrosis (with reticulin and collagen deposition) in the BM and spleen was associated with anemia, thrombocytopenia, severe granulocytosis, marrow hypocellularity, and significant splenomegaly (6-fold). Mild osteosclerosis was observed in femurs. C57BL/6 mice usually survived over the 6-month survey; however, 3 out of 18 mice in Wernig's study[45] and 3 out of 65 mice in our study[44] prematurely died with very elevated white blood cell counts (58–102 × 10^9 cell/ml) with no sign of acute leukemia. The disease phenotype was not transplantable in sublethally irradiated mice but was transferable in lethally irradiated mice. Southern blots revealed oligoclonal retrovirally marked hematopoiesis early or late after transplantation.

Wernig *et al.* described a more dramatic phenotype in BALB/c than in C57B6 mice.[45] These mice develop marked leukocytosis (up to 17-fold), huge splenomegaly (10-fold), and fibrosis at early stages (1 month post-BMT), and had a short survival (from 50 to 95 days), suggesting that genetic modifiers are important in disease phenotype.[45]

In summary, these models demonstrated that high levels of *JAK2*^{V617F} expression induce a disease with features similar to those observed in PV patients, including evolution into a myelofibrosis-like disease. However, how a unique mutation can be involved in numerous diverse human disease phenotypes, such as in the MPD, is not explained by these models, and further work will be needed to establish the mechanisms leading to *JAK2*^{V617F}-positive pathologies that are not associated with erythrocytosis, especially as concerns MMM. Finally, this model offers a unique opportunity to assess novel therapeutic approaches for *JAK2*^{V617F}-positive PV.

THE *MPL*^{W515L} MODEL

Searching for novel mutations involving other components of the JAK-STAT signal transduction pathway in *JAK2*^{V617F}-negative MPD, a novel somatic mutation in the transmembrane region of MPL was identified, resulting in a tryptophan-to-leucine substitution at codon 515 (*MPL*^{W515L}).[9] This mutation is present in approximately 5–10% of *JAK2*^{V617F}-negative IM patients,[9,10] and results in constitutive activation of JAK-STAT signaling. Hematopoietic cells expressing *MPL*^{W515L} showed a selective proliferative advantage and a cytokine-independent growth when compared to wild-type cells.[9] With the aim to evaluate the relevance of this mutation in the pathogenesis of IM, a murine transplant assay was developed. Bone marrow cells were transduced with murine stem cell virus–internal ribosome entry site–enhanced green fluorescent protein (MSCV-IRES-EGFP) vectors containing *MPL*^{WT} or *MPL*^{W515L} and transplanted into lethally irradiated BALB/c mice. Only mice transplanted with *MPL*^{W515L}-transduced HSC developed, with a median latency of 18 days, a fully penetrant lethal myeloproliferative disease characterized by hepatomegaly and splenomegaly, leukocytosis, marked thrombocytosis, extramedullary hematopoiesis, and myelofibrosis.[9] Compared to *MPL*^{WT} mice, the BM of *MPL*^{W515L} animals showed a predominance of maturing myeloid cells and atypical and dysplastic Mks, sometime collected in clusters, that frequently exhibited the phenomenon of emperipolesis of neutrophils in their cytoplasm. Similar findings were reported in the spleen and the liver. When cultured *in vitro*, spleen cells derived from *MPL*^{W515L}-transduced mice demonstrated a marked increase of MK-CFUs that originated larger colonies as compared to those derived from BM cells.[9] In summary, this model offered a novel mechanism for activation of downstream signal transduction pathways in *JAK2*^{V617F}-negative MPD patients that results in myelofibrosis, and may represent a useful model for the study of drugs targeted to *MPL*^{W515L}-harboring cells.

INSIGHTS FROM MURINE MODELS INTO THE PATHOGENESIS OF THE STROMAL REACTIONS IN MYELOFIBROSIS

The stromal reaction observed in the BM microenvironment in MMM is characterized by fibrosis (excessive deposits of extracellular matrix proteins), osteosclerosis (new bone formation), and neoangiogenesis. This aberrant stromal reaction is known to be secondary to the stem cell disorder and is supposedly mediated by cytokines released by cells belonging to the abnormal clone, and especially from the Mks. Both the TPO^{high} and the GATA-1^{low} mouse model have provided new insight into the pathogenesis of the stromal reaction. The role of transforming growth factor-β1 (TGF-β1) and osteoprotegerin (OPG) in the promotion of BM fibrosis and osteosclerosis, respectively, has been elucidated. Additionally, a possible mechanism for release of cytokines in the BM microenvironment by abnormal Mks has been identified in the process of emperipolesis. Finally, the role of the monocyte/macrophage lineage in the promotion of the stromal reaction has been addressed in the TPO^{high} model. Thus, both TPO^{high} and GATA-1^{low} mouse models have proved to be useful tools with which to investigate both the cytokines and the cells involved in the modification of the BM microenvironment that occurs in MMM.

CYTOKINES INVOLVED IN THE STROMAL REACTION

Role of Transforming Growth Factor-β1 *In vivo* and *in vitro* studies have involved several cytokines in the development of myelofibrosis, such as TGF-β1, platelet-derived growth factor (PDGF), or basic fibroblast growth factor (bFGF). Among these, TGF-β1 has been the most likely candidate based on indirect observations. First, it potently stimulates the production of extracellular matrix components, by inducing fibroblasts to proliferate and secrete extracellular matrix and cell adhesion proteins, and prevents matrix degradation through the enhanced expression of protease that inhibits enzymes involved in the degradation of the extracellular matrix.[46] Second, TGF-β1 promotes fibrosis of a variety of nonhematopoietic tissues.[46] More recently, it has been shown to be directly involved in the pathogenesis of BM fibrosis in hairy cell leukemia.[49] Third, increased plasma levels of TGF-β1 have been reported both in patients with MMM and in TPO^{high} mice, and increased levels of TGF-β1 were found in extracellular fluids of BM and spleen in TPO^{high} and GATA-1^{low} mice.[25]

The prominent role of TGF-β1 in the promotion of BM fibrosis has been confirmed using the TPO^{high} model.[50] Lethally irradiated wild-type hosts (used because *TGF-β1*$^{-/-}$ mice die at weaning age from a multifocal inflammatory syndrome) were transplanted for long-term reconstitution with HSC from mice genetically deficient for *TGF-β1* (*TGF-β1*$^{-/-}$) or from wild-type (*TGF-β1*$^{+/+}$) mice infected with a retrovirus encoding the mouse TPO. Over the 4 months of follow-up, TPO levels in plasma were markedly elevated in both groups of mice and all animals typically developed a myeloproliferative syndrome characterized by thrombocytosis, leukocytosis, splenomegaly, increased numbers of progenitors in the PB, and extramedullary hematopoiesis. However, whereas prominent fibrosis was observed in BM and spleen from all mice engrafted with *TGF-β1*$^{+/+}$ cells, none of the mice repopulated with *TGF-β1*$^{-/-}$ cells showed deposition of reticulin or collagen fibers. In accordance with the development of fibrosis, latent TGF-β1 levels in plasma and extracellular fluids of the spleen from mice engrafted with *TGF-β1*$^{+/+}$ cells were increased 4- to 8-fold over baseline levels, whereas no increase was detected in hosts engrafted with *TGF-β1*$^{-/-}$ hematopoietic cells. Taken together, these data strongly supported the prominent role of TGF-β1 produced by hematopoietic cells in the pathogenesis of myelofibrosis induced by high TPO levels. Significantly, increased levels of TGF-β1 are contained in the extracellular fluids of BM and spleen in GATA-1^{low} mice, but they are 2- to 6-fold lower than those reported in the TPO^{high} mice; it has been suggested that the different disease evolution and morbidity in these two animal models might be correlated with the levels of TGF-β1 in the BM microenvironment.[25] However, other growth factors, such as PDGF or bFGF, might also contribute to the development of fibrosis, and would require investigation. Finally, another mechanism remains unknown; this concerns the fact that TGF-β1 has to be activated to exert its biological effect. In fact, even if increased plasma levels of latent TGF-β1 were detected in TPO^{high} mice, no increase in the active form of TGF-β1 was measured in the circulation.[50] Accordingly, no fibrotic lesions were identified in distant organs known to be targets of the fibrogenic effect of TGF-β1, such as lung or kidney. This indicates that critical regulatory mechanisms controlling TGF-β1 activation must take place within the hematopoietic environment at secretion sites. These mechanisms responsible for the local activation of TGF-β1 are not yet understood, but might involve potent *in vivo* activators of TGF-β1, such as thrombospondin-1 (TSP-1), integrins $\alpha_v\beta_6$, $\alpha_{IIb}\beta_3$, and $\alpha_v\beta_3$, or proteases. MKs and platelets synthesize and store TSP-1 in the same organelles as TGF-β1, and, moreover, they express integrins $\alpha_{IIb}\beta_3$ and $\alpha_v\beta_3$. Thus, Mks may participate in the local activation of TGF-β1; additionally, monocytes/macrophages are sources of cytokines that can activate TGF-β1 *in vivo* as well.

Role of Osteoprotegerin The rapid development in TPO^{high} mice of osteosclerosis also offered a model to investigate the underlying causes of abnormal bone growth. Bone remodeling depends on the tightly integrated activity of two distinct cell types, the osteoblasts (the bone-forming cells) and the osteoclasts (the bone-resorbing cells). Osteoblasts derive from mesenchymal progenitors through the regulatory action of cell–cell and cell–matrix interactions and the actions of several growth factors such as TGF-β1, which could exert both stimulatory and inhibitory effects. Osteoclasts are monocyte-derived multinucleated cells They require macrophage-colony-stimulating factor (M-CSF) for proliferation and receptor activation of nuclear factor-Kappa-ligand (RANK-L), via its binding to the receptor RANK expressed at the surface of osteoclast progenitors, for differentiation and maturation. OPG is a decoy soluble receptor that binds RANK-L and inhibits osteoclastogenesis.

The role of OPG in the development of osteosclerosis in TPO^{high} mice was demonstrated using OPG-deficient mice (*OPG*$^{-/-}$).[51] Because these animals are viable and develop normally, the four combinations of graft/host were performed, allowing the discrimination of the potential role of hematopoietic OPG (from donor cells) or stromal OPG (host cells). Regardless of the combination, all animals developed the expected myeloproliferative syndrome with BM fibrosis associated with increased TGF-β1 plasma levels. Osteosclerosis occurred only in *OPG*$^{+/+}$ hosts regardless of donor cell genotype with a marked elevation of OPG plasma levels. In contrast, only rare bone growth was observed in *OPG*$^{-/-}$ hosts with no OPG detectable in the circulation. These findings strongly suggest the role of the stromal OPG secreted by the host microenvironment in the promotion of osteosclerosis. The mechanism leading to stromal OPG upregulation remains unknown, but it is not mediated by increased TGF-β1 in TPO^{high} mice. Indeed, retrospective analysis of TPO^{high} mice engrafted with *TGF-β1*$^{-/-}$ donor cells showed no correlation between TGF-β1 levels and OPG levels.[51] All mice displayed similarly increased OPG plasma levels regardless of donor cell genotype (*TGF-β1*$^{-/-}$ or *TGF-β1*$^{+/+}$). However, TPO^{high} mice engrafted with *TGF-β1*$^{-/-}$ donor cells developed a delayed osteosclerosis compared to the control group, suggesting the involvement of TGF-β1 in this process, likely by stimulating osteoblasts proliferation. Furthermore, a recent study showed that TPO^{high} mice also displayed also high plasma levels of IL-1, which was demonstrated to originate in part from platelets.[27] This cytokine may be involved in stromal OPG upregulation as well. Osteosclerosis and increased bone mRNA levels for osteocalcin are also developed by GATA-1^{low} mice, starting around the sixth month of age. By crossing experiments, it has been observed that the extent of osteosclerosis is dependent on the strain employed, since bone marrow formation is maximal in C57BL/6 mice, very limited in DB/2 mice, and intermediate in the standard CD1 strain;[43] thus these models offer the opportunity to study novel factors implied in the process of osteosclerosis. Although GATA-1^{low} Mks are able to promote osteoblast proliferation *in vitro* by direct cell contact,[52] we failed to demonstrate an *in vivo* correlation between the number of Mks in the BM of different mutated strains and the degree of osteosclerosis.[43]

CELLS INVOLVED IN THE STROMAL REACTION

Prominent Role of Megakaryocytes and the Process of Emperipolesis Several lines of evidence, obtained both from studies of patients with MMM and from murine models as described above, favor a crucial role of Mks in the development of the stromal reaction (Figure 73–4). First, Mks and platelets are the main source of TGF-β1. Second, hyperplasia and disturbed differentiation of Mks are observed in MMM patients and in both TPO^{high} and GATA-1^{low} mice. In addition, on BM biopsies, Mks are often found in close proximity to or surrounded by reticulin fibers, both in humans and mice. Third, BM fibrosis is associated with other diseases involving Mks, such as megakaryoblastic leukemia or gray platelets syndrome, and is also presented by other genetically modified animals (Table 73–1). TPO^{high} mice display Mk hyperplasia with dysmorphic Mks, suggesting that a long-lasting high production of TPO could affect Mk maturation. However, since GATA-1^{low} mice display normal levels of plasma TPO, notwithstanding their severe lifelong thrombocytopenia (likely because of TPO sequestration by the enlarged Mk mass), it seems unlikely that myelofibrosis in these mice is linked to TPO overstimulation. However, interestingly enough, Mks from mice treated with high TPO doses expressed low GATA-1 levels, positioning TPO/MPL and GATA-1 in a downstream relationship in the pathogenesis of the development of abnormal megakaryocytopoiesis and myelofibrosis.[25] How a high level of TPO can induce a dysmegakaryopoiesis remains unknown. One hypothesis is that a long-lasting TPO stimulation leads to MPL traffic defects and altered signaling (S. Constantinescu, unpublished data).

One possible mechanism for inappropriate cytokine release in MMM could be a pathological form of emperipolesis. This rare physiological process is defined by the random passage of different types of BM cells through the Mk cytoplasm without any important consequences for either the host or the invading cells. An increased frequency of neutrophil polymorphonuclear (PMN) cell emperipolesis within Mks is a common feature shared by the majority of patients suffering from extreme thrombocytosis (either reactive or myeloproliferative), but is not necessarily associated with BM fibrosis.[53] On the other hand, both in patients with MMM and in TPO^{high} and GATA-1^{low} mice, increased neutrophil and eosinophil PMN cell emperipolesis related to abnormal P-selectin localization on the DMS has been described.[54] The Mks from both these mice contain very few normal, mature α-granules; rather they have the appearance of large empty vacuoles derived from the Golgi apparatus and are almost devoid of normal constituents. In particular, P-selectin (GMP-140, CD62), a leukocyte receptor and a normal constituent of α-granules from which it is released during platelet activation, is found in small vacuolar structures scattered in the cytoplasm and on the membranes of DMS.[39] A model has been proposed by which the increased PMN emperipolesis in Mks, favored by the abnormal P-selectin localization on the DMS, would result in paraapoptosis of Mks due to the release of neutrophil enzymes in the cytoplasm; in fact, the cytoplasm of engulfed PMN is almost devoid of myeloperoxidase-positive granules, which conversely can be found in the Mk cytoplasm. This pathological and specific interaction between Mks and PMN cells could lead progressively to PMN-mediated Mk destruction, especially by causing α-granule lysis and the discharge of growth factors, including TGF-β1 normally contained in the granules, outside the Mk itself in the BM microenvironment through the DMS. This would finally result in the stimulation of reticulin and collagen synthesis and deposition by fibroblasts.[37]

Eventually, this same sequence of events might play a role in the constitutive mobilization of HSC in the circulation, which is

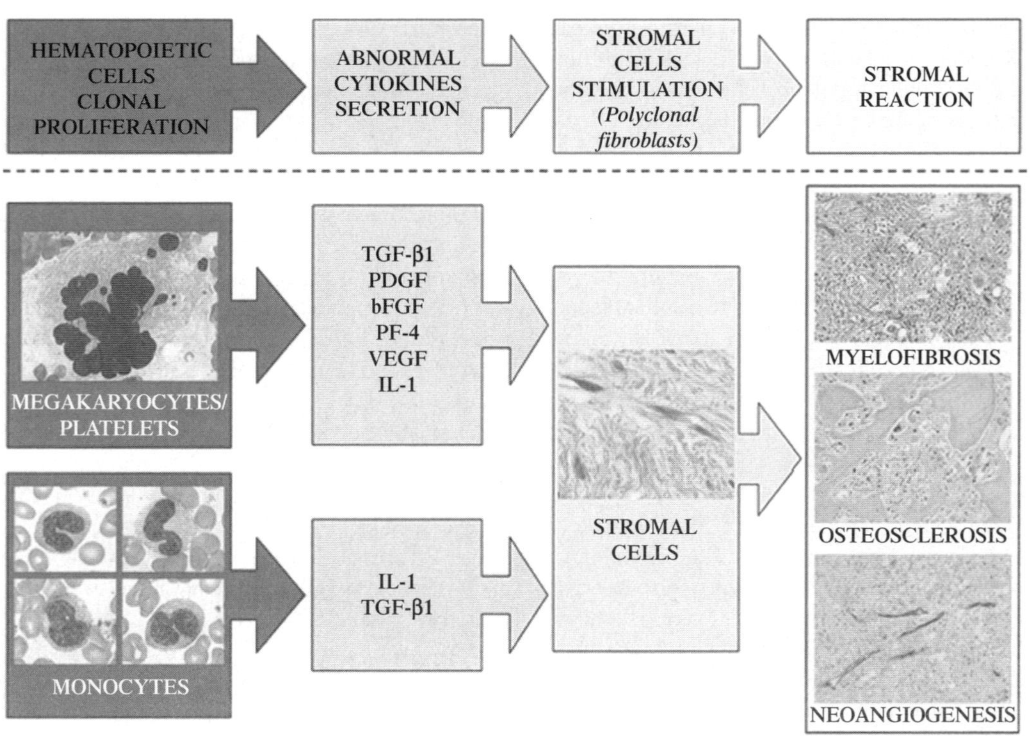

Figure 73–4. A pathogenetic model for the cytokine-mediated stromal reaction observed in MMM. (See color insert.)

characteristic of both the murine models and patients with MMM.[55–57] The PMN proteases released through the process of emperipolesis might result in the cleavage of adhesion receptors from the surface of the cells constituting the hematopoietic niche in the BM, as well as from the membrane of HSC, similar to what happens in the HSC mobilization induced by granulocyte-colony-stimulating factor (G-CSF).

Minor Role of Monocyte/Macrophage A central role for the changes occurring in the monocyte/macrophage lineage in the BM microenvironment in MMM has been supported by two different observations. First, Rameshwar *et al* reported that monocytes from patients with MMM were spontaneously activated and abnormally secreted TGF-β1.[58] Second, Frey *et al.* reported that TPO overexpression in SCID mice, but not in NOD/SCID mice, led to BM fibrosis, indicating a prominent role for the monocyte/macrophage lineage in the promotion of the stromal reaction.[19] However, a recent study demonstrated that NOD/SCID mice exposed to high TPO levels developed marrow fibrosis, strongly suggesting that fully functional monocytes are not required for the stromal reaction to occur, and pointing again to Mk as the key cell responsible for the pathogenesis of the changes in the BM microenvironment.[27]

RELEVANCE OF MURINE MODELS FOR HUMAN MYELOFIBROSIS

Since TPO^{high} models were first described, the role played by the TPO/MPL pathway in MMM has been elucidated. Recently two activating mutations of *MPL*, MPL^{W515L} and MPL^{W515K}, have been described in 10% of MMM patients and have been shown to induce fibrosis in mice.[9,10] Furthermore, the activating $JAK2^{V617F}$ mutation, found in 50% of MMM, also induces fibrosis in mice[59] and is directly linked to deregulated MPL signaling. On the other hand, although mutations in *GATA1* have not been reported to occur in MMM, the levels of GATA-1 in the MKs from MMM patients are significantly reduced, independent of *JAK2* mutational status, which points to a defect in the GATA-1-dependent pathway(s).[60] Thus, there is enough to suggest that the downstream relationship existing between the TPO^{high} and the $GATA\text{-}1^{low}$ animal models would mirror the situation in human disease. Furthermore, it is likely that the prominent role of TGF-β1 in fibrosis and of OPG in the promotion of osteosclerosis, shown with these models, will also be confirmed in the new animal models of MMM; of interest, abnormalities in OPG levels have already been described in human MMM.[61]

In conclusion, even though the newly discovered molecular lesions as a basis for MMM in humans appear to be different from those found in the animal models, these models have been shown to recapitulate the same cascade of events that culminate in the specific phenotypic abnormalities of the human disorder. Thus, these animal models are expected to further improve our understanding of the pathogenesis of MMM, with potential clinical implications, not to mention their usefulness for testing novel therapeutic agents to treat this incurable disease.[62–76]

ACKNOWLEDGMENTS

Supported by a grant from Associazione Italiana per la Ricerca sul Cancro (Milano, I) to A.M.V., La Ligue Nationale contre le Cancer (équipe labellisée 2004) to J.L.V. and O.W.B., and NIH to A.R.M.

REFERENCES

1. Barosi G, Hoffman R. Idiopathic myelofibrosis. *Semin Hematol* 2005;42(4):248–258.
2. Vardiman JW, Harris NL, Brunning RD. The World Health Organization (WHO) classification of the myeloid neoplasms. *Blood* 2002;100(7):2292–2302.
3. Prchal JF, Axelrad AA. Letter: Bone-marrow responses in polycythemia vera. *N Engl J Med* 1974;290(24):1382.
4. Taksin AL, Couedic JP, Dusanter-Fourt I, *et al.* Autonomous megakaryocyte growth in essential thrombocythemia and idiopathic myelofibrosis is not related to a c-mpl mutation or to an autocrine stimulation by Mpl-L. *Blood* 1999;93(1):125–139.
5. James C, Ugo V, Le Couedic JP, *et al.* A unique clonal JAK2 mutation leading to constitutive signalling causes polycythaemia vera. *Nature* 2005;434(7037):1144–1148.
6. Baxter EJ, Scott LM, Campbell PJ, *et al.* Acquired mutation of the tyrosine kinase JAK2 in human myeloproliferative disorders. *Lancet* 2005;365(9464):1054–1061.
7. Levine RL, Wadleigh M, Cools J, *et al.* Activating mutation in the tyrosine kinase JAK2 in polycythemia vera, essential thrombocythemia, and myeloid metaplasia with myelofibrosis. *Cancer Cell* 2005;7(4):387–397.
8. Kralovics R, Passamonti F, Buser AS, *et al.* A gain-of-function mutation of JAK2 in myeloproliferative disorders. *N Engl J Med* 2005;352(17):1779–1790.
9. Pikman Y, Lee BH, Mercher T, *et al.* MPLW515L is a novel somatic activating mutation in myelofibrosis with myeloid metaplasia. *PLoS Med* 2006;3(7):e270.
10. Pardanani AD, Levine RL, Lasho T, *et al.* MPL515 mutations in myeloproliferative and other myeloid disorders: A study of 1182 patients. *Blood* 2006;108(10):3472–3476.
11. Thiele J, Kuemmel T, Sander C, Fischer R. Ultrastructure of bone marrow tissue in so-called primary (idiopathic) myelofibrosis-osteomyelosclerosis (agnogenic myeloid metaplasia). I. Abnormalities of megakaryopoiesis and thrombocytes. *J Submicrosc Cytol Pathol* 1991;23(1):93–107.
12. Thiele JPR, Imbert M, Vardiman JW, Brunning RD, Flandrin G. Chronic idiopathic myelofibrosis. In: Jaffe ES, Harris NL, Stein H, Vardiman JW, Eds. *World Health Organization–Tumors of Hematopoietic and Lymphoid Tissues.* Washington, DC: IARC Press, 2001:35–38.
13. Kaushansky K. Thrombopoietin: The primary regulator of platelet production. *Blood* 1995;86(2):419–431.
14. de Sauvage FJ, Carver-Moore K, Luoh SM, *et al.* Physiological regulation of early and late stages of megakaryocytopoiesis by thrombopoietin. *J Exp Med* 1996;183(2):651–656.
15. Ulich TR, del Castillo J, Senaldi G, *et al.* Systemic hematologic effects of PEG-rHuMGDF-induced megakaryocyte hyperplasia in mice. *Blood* 1996;87(12):5006–5015.
16. Ohwada A, Rafii S, Moore MA, Crystal RG. In vivo adenovirus vector-mediated transfer of the human thrombopoietin cDNA maintains platelet levels during radiation-and chemotherapy-induced bone marrow suppression. *Blood* 1996;88(3):778–784.
17. Cannizzo SJ, Frey BM, Rafii S, *et al.* Augmentation of blood platelet levels by intratracheal administration of an adenovirus vector encoding human thrombopoietin cDNA. *Nat Biotechnol* 1997;15(6): 570–573.
18. Abina MA, Tulliez M, Duffour MT, *et al.* Thrombopoietin (TPO) knockout phenotype induced by cross-reactive antibodies against TPO following injection of mice with recombinant adenovirus encoding human TPO. *J Immunol* 1998;160(9):4481–4489.
19. Frey BM, Rafii S, Teterson M, Eaton D, Crystal RG, Moore MA. Adenovector-mediated expression of human thrombopoietin cDNA in immune-compromised mice: Insights into the pathophysiology of osteomyelofibrosis. *J Immunol* 1998;160(2):691–699.
20. Yan XQ, Lacey D, Fletcher F, *et al.* Chronic exposure to retroviral vector encoded MGDF (mpl-ligand) induces lineage-specific growth and differentiation of megakaryocytes in mice. *Blood* 1995;86(11):4025–4033.

21. Villeval JL, Cohen-Solal K, Tulliez M, *et al.* High thrombopoietin production by hematopoietic cells induces a fatal myeloproliferative syndrome in mice. *Blood* 1997;90(11):4369–4383.
22. Zhou W, Toombs CF, Zou T, Guo J, Robinson MO. Transgenic mice overexpressing human c-mpl ligand exhibit chronic thrombocytosis and display enhanced recovery from 5-fluorouracil or antiplatelet serum treatment. *Blood* 1997;89(5):1551–1559.
23. Kakumitsu H, Kamezaki K, Shimoda K, *et al.* Transgenic mice overexpressing murine thrombopoietin develop myelofibrosis and osteosclerosis. *Leuk Res* 2005;29(7):761–769.
24. Yanagida M, Ide Y, Imai A, *et al.* The role of transforming growth factor-beta in PEG-rHuMGDF-induced reversible myelofibrosis in rats. *Br J Haematol* 1997;99(4):739–745.
25. Vannucchi AM, Bianchi L, Paoletti F, *et al.* A pathobiologic pathway linking thrombopoietin, GATA-1, and TGF-beta1 in the development of myelofibrosis. *Blood* 2005;105(9):3493–3501.
26. Abina MA, Tulliez M, Lacout C, *et al.* Major effects of TPO delivered by a single injection of a recombinant adenovirus on prevention of septicemia and anemia associated with myelosuppression in mice: Risk of sustained expression inducing myelofibrosis due to immunosuppression. *Gene Ther* 1998;5(4):497–506.
27. Wagner-Ballon O, Chagraoui H, Prina E, *et al.* Monocyte/macrophage dysfunctions do not impair the promotion of myelofibrosis by high levels of thrombopoietin. *J Immunol* 2006;176(11): 6425–6433.
28. Tefferi A. Experimental myelofibrosis in mice and the implications to human disease. *Leuk Res* 2005;29(7):723–726.
29. Tsai SF, Martin DI, Zon LI, D'Andrea AD, Wong GG, Orkin SH. Cloning of cDNA for the major DNA-binding protein of the erythroid lineage through expression in mammalian cells. *Nature* 1989; 339(6224):446–451.
30. Romeo PH, Prandini MH, Joulin V, *et al.* Megakaryocytic and erythrocytic lineages share specific transcription factors. *Nature* 1990;344(6265):447–449.
31. Yu C, Cantor AB, Yang H, *et al.* Targeted deletion of a high-affinity GATA-binding site in the GATA-1 promoter leads to selective loss of the eosinophil lineage in vivo. *J Exp Med* 2002;195(11): 1387–1395.
32. Migliaccio AR, Rana RA, Sanchez M, *et al.* GATA-1 as a regulator of mast cell differentiation revealed by the phenotype of the GATA-1low mouse mutant. *J Exp Med* 2003;197(3):281–296.
33. Pevny L, Simon MC, Robertson E, *et al.* Erythroid differentiation in chimaeric mice blocked by a targeted mutation in the gene for transcription factor GATA-1. *Nature* 1991;349(6306):257–260.
34. Farina SF, Girard LJ, Vanin EF, Nienhuis AW, Bodine DM. Dysregulated expression of GATA-1 following retrovirus-mediated gene transfer into murine hematopoietic stem cells increases erythropoiesis. *Blood* 1995;86(11):4124–4133.
35. Cantor AB, Orkin SH. Hematopoietic development: A balancing act. *Curr Opin Genet Dev* 2001;11(5):513–519.
36. McDevitt MA, Shivdasani RA, Fujiwara Y, Yang H, Orkin SH. A "knockdown" mutation created by cis-element gene targeting reveals the dependence of erythroid cell maturation on the level of transcription factor GATA-1. *Proc Natl Acad Sci USA* 1997;94(13): 6781–6785.
37. Shivdasani RA, Fujiwara Y, McDevitt MA, Orkin SH. A lineage-selective knockout establishes the critical role of transcription factor GATA-1 in megakaryocyte growth and platelet development. *EMBO J* 1997;16(13):3965–3973.
38. Vyas P, Ault K, Jackson CW, Orkin SH, Shivdasani RA. Consequences of GATA-1 deficiency in megakaryocytes and platelets. *Blood* 1999;93(9):2867–2875.
39. Centurione L, Di Baldassarre A, Zingariello M, *et al.* Increased and pathologic emperipolesis of neutrophils within megakaryocytes associated with marrow fibrosis in GATA-1(low) mice. *Blood* 2004;104(12):3573–3580.
40. Vannucchi AM, Bianchi L, Cellai C, *et al.* Development of myelofibrosis in mice genetically impaired for GATA-1 expression (GATA-1(low) mice). *Blood* 2002;100(4):1123–1132.
41. Vannucchi AM, Migliaccio AR, Paoletti F, Chagraoui H, Wendling F. Pathogenesis of myelofibrosis with myeloid metaplasia: Lessons from mouse models of the disease. *Semin Oncol* 2005;32:365–372.
42. Vannucchi AM, Bianchi L, Cellai C, Paoletti F, Carrai V, Calzolari A, Centurione L, Lorenzini L, Carta C, Alfani E, Sanchez M, Migliaccio G., Miagliaccio AR. Accentuated response to PHZ and EPO in mice genetically impaired for their GATA-1 expression. *Blood* 2001;97:3040–3050.
43. Martelli F, Ghinassi B, Panetta B, *et al.* Variegation of the phenotype induced by the GATA-1low mutation in mice of different genetic background. *Blood* 2005;105(13):4102–4113.
44. Lacout C, Pisani DF, Tulliez M, Moreau Gachelin F, Vainchenker W, Villeval JL. JAK2V617F expression in murine hematopoietic cells leads to MPD mimicking human PV with secondary myelofibrosis. *Blood* 2006;108(5):1652–1660.
45. Wernig G, Mercher T, Okabe R, Levine RL, Lee BH, Gilliland DG. Expression of Jak2V617F causes a polycythemia vera-like disease with associated myelofibrosis in a murine bone marrow transplant model. *Blood* 2006;107(11):4274–4281.
46. Roberts AB, Sporn MB, Assoian RK, *et al.* Transforming growth factor type beta: Rapid induction of fibrosis and angiogenesis in vivo and stimulation of collagen formation in vitro. *Proc Natl Acad Sci USA* 1986;83(12):4167–4171.
47. Kimura A, Katoh O, Hyodo H, Kuramoto A. Transforming growth factor-beta regulates growth as well as collagen and fibronectin synthesis of human marrow fibroblasts. *Br J Haematol* 1989;72(4): 486–491.
48. Border WA, Noble NA. Transforming growth factor beta in tissue fibrosis. *N Engl J Med* 1994;331(19):1286–1292.
49. Shehata M, Schwarzmeier JD, Hilgarth M, Hubmann R, Duechler M, Gisslinger H. TGF-beta1 induces bone marrow reticulin fibrosis in hairy cell leukemia. *J Clin Invest* 2004;113(5):676–685.
50. Chagraoui H, Komura E, Tulliez M, Giraudier S, Vainchenker W, Wendling F. Prominent role of TGF-beta 1 in thrombopoietin-induced myelofibrosis in mice. *Blood* 2002;100(10):3495–3503.
51. Chagraoui H, Tulliez M, Smayra T, *et al.* Stimulation of osteoprotegerin production is responsible for osteosclerosis in mice overexpressing TPO. *Blood* 2003;101(8):2983–2989.
52. Kacena MA, Shivdasani RA, Wilson K, *et al.* Megakaryocyte-osteoblast interaction revealed in mice deficient in transcription factors GATA-1 and NF-E2. *J Bone Miner Res* 2004;19(4):652–660.
53. Cashell AW, Buss DH. The frequency and significance of megakaryocytic emperipolesis in myeloproliferative and reactive states. *Ann Hematol* 1992;64(6):273–276.
54. Schmitt A, Jouault H, Guichard J, Wendling F, Drouin A, Cramer EM. Pathologic interaction between megakaryocytes and polymorphonuclear leukocytes in myelofibrosis. *Blood* 2000;96(4):1342–1347.
55. Barosi G, Viarengo G, Pecci A, *et al.* Diagnostic and clinical relevance of the number of circulating CD34(+) cells in myelofibrosis with myeloid metaplasia. *Blood* 2001;98(12):3249–3255.
56. Xu M, Bruno E, Chao J, *et al.* The constitutive mobilization of CD34+ cells into the peripheral blood in idiopathic myelofibrosis may be due to the action of a number of proteases. *Blood* 2005;105:4508–4515.
57. Xu M, Bruno E, Chao J, *et al.* The constitutive mobilization of bone marrow-repopulating cells into the peripheral blood in idiopathic myelofibrosis. *Blood* 2005;105(4):1699–1705.
58. Rameshwar P, Narayanan R, Qian J, Denny TN, Colon C, Gascon P. NF-kappa B as a central mediator in the induction of TGF-beta in monocytes from patients with idiopathic myelofibrosis: An inflammatory response beyond the realm of homeostasis. *J Immunol* 2000;165(4):2271–2277.
59. Villeval JL, James C, Pisani DF, Casadevall N, Vainchenker W. New insights into the pathogenesis of JAK2 V617F-positive myeloproliferative disorders and consequences for the management of patients. *Semin Thromb Hemost* 2006;32(4Pt. 2):341–351.
60. Vannucchi AM, Pancrazzi A, Guglielmelli P, *et al.* Abnormalities of GATA-1 in megakaryocytes from patients with idiopathic myelofibrosis. *Am J Pathol* 2005;167(3):849–858.

61. Wang JC, Hemavathy K, Charles W, *et al.* Osteosclerosis in idiopathic myelofibrosis is related to the overproduction of osteoprotegerin (OPG). *Exp Hematol* 2004;32(10):905–910.
62. Le Bousse-Kerdiles MC, Smadja-Joffe F, Klein B, Caillou B, Jasmin C. Study of a virus-induced myeloproliferative syndrome associated with tumor formation in mice. *Eur J Cancer* 1980;16(1):43–51.
63. Le Bousse-Kerdiles MC, Souyri M, Smadja-Joffe F, Praloran V, Jasmin C, Ziltener HJ. Enhanced hematopoietic growth factor production in an experimental myeloproliferative syndrome. *Blood* 1992;79(12):3179–3187.
64. Ostertag W, Vehmeyer K, Fagg B, *et al.* Myeloproliferative virus, a cloned murine sarcoma virus with spleen focus-forming properties in adult mice. *J Virol* 1980;33(2):573–582.
65. Tomasson MH, Sternberg DW, Williams IR, *et al.* Fatal myeloproliferation, induced in mice by TEL/PDGFbetaR expression, depends on PDGFbetaR tyrosines 579/581. *J Clin Invest* 2000;105(4):423–432.
66. Tomasson MH, Williams IR, Li S, *et al.* Induction of myeloproliferative disease in mice by tyrosine kinase fusion oncogenes does not require granulocyte-macrophage colony-stimulating factor or interleukin-3. *Blood* 2001;97(5):1435–1441.
67. Ritchie KA, Aprikyan AA, Bowen-Pope DF, *et al.* The Tel-PDGFR-beta fusion gene produces a chronic myeloproliferative syndrome in transgenic mice. *Leukemia* 1999;13(11):1790–1803.
68. Schwaller J, Frantsve J, Aster J, *et al.* Transformation of hematopoietic cell lines to growth-factor independence and induction of a fatal myelo- and lymphoproliferative disease in mice by retrovirally transduced TEL/JAK2 fusion genes. *Embo J* 1998;17(18):5321–5333.
69. Toki T, Katsuoka F, Kanezaki R, *et al.* Transgenic expression of BACH1 transcription factor results in megakaryocytic impairment. *Blood* 2005;105(8):3100–3108.
70. Lyden D, Hattori K, Dias S, *et al.* Impaired recruitment of bone-marrow-derived endothelial and hematopoietic precursor cells blocks tumor angiogenesis and growth. *Nat Med* 2001;7(11):1194–1201.
71. Gerber HP, Malik AK, Solar GP, *et al.* VEGF regulates haematopoietic stem cell survival by an internal autocrine loop mechanism. *Nature* 2002;417(6892):954–958.
72. Metcalf D, Gearing DP. A myelosclerotic syndrome in mice engrafted with cells producing high levels of leukemia inhibitory factor (LIF). *Leukemia* 1989;3(12):847–852.
73. Hofbauer LC. Osteoprotegerin ligand and osteoprotegerin: Novel implications for osteoclast biology and bone metabolism. *Eur J Endocrinol* 1999;141(3):195–210.
74. Wang Q, Liu T, Fang Y, *et al.* BUBR1 deficiency results in abnormal megakaryopoiesis. *Blood* 2004;103(4):1278–1285.
75. Kelly LM, Liu Q, Kutok JL, Williams IR, Boulton CL, Gilliland DG. FLT3 internal tandem duplication mutations associated with human acute myeloid leukemias induce myeloproliferative disease in a murine bone marrow transplant model. *Blood* 2002;99(1):310–318.
76. Velazquez L, Cheng AM, Fleming HE, *et al.* Cytokine signaling and hematopoietic homeostasis are disrupted in Lnk-deficient mice. *J Exp Med* 2002;195(12):1599–1611.

74 Animal Models for Bone Tissue Engineering Purposes

Véronique Viateau, Delphine Logeart-Avramoglou, Geneviève Guillemin, and Hervé Petite

ABSTRACT

To assess the efficacy of engineered tissues, it is necessary to have (1) appropriate large animal models that mimic the clinical setting and (2) relevant methods of monitoring the biofuntionality of these tissues. However, developing these tissue constructs is a step-by-step process in which numerous variables such as scaffold design, source of stem cells and mode of growth factor application have to be optimized. After an *in vitro* optimization phase, the use of small animal models to optimize these various parameters and sort out any teething problems is recommended before launching into large animal models. Depending on the experimental aims, engineered tissues can be transplanted into either ectopic sites (subcutaneously or intramuscularly) or orthotopic sites. In all these experimental studies, non invasive imaging methods (X-ray, magnetic resonance, *in vivo* fluorescence, ultrasound imaging methods, etc.) as well as detailed quantitative molecular and histological analyses have been used to monitor the *in vivo* behavior of the engineered constructs. In this chapter we take stock of the present state of the art in this field.

Key Words: Bone, Bone healing, Tissue engineering, Stem cells, Growth factors, Biomechanics.

GENERAL CONSIDERATIONS

Tissue engineering and regenerative medicine are new research areas dealing with how to repair and regenerate organs and tissues using the natural signaling pathways and components of the organism. Basically, the three primary constituents of a tissue-engineering approach are (1) cells, (2) a scaffold for delivering and/or retaining cells, and (3) bioactive factors.[1]

The development of bioengineered bone constructs is a step-by-step process that begins with a preliminary but essential *in vitro* optimization step in which parameters such as scaffold design, cell proliferation, cell behavior on the material surface, cell osteogenic capability, and the kinetics of bioactive factor release are assessed.

After this necessary *in vitro* optimization phase, and because the ultimate goal of this approach is the reconstruction of a functional tissue that requires a complex biological and biomechanical *in vivo* environment, the use of animal models usually constitutes the next step of the bone construct developmental process before eventually performing human clinical trials. Animal models have therefore been consistently used to study and test the primary constituents separately, in various combinations, and lastly, within a defective bone site. Animal tests are a long and cumbersome stage in the bone construct developmental process because of the large number of biological and biomechanical variables that have to be assessed before it is possible to apply this engineering approach in the setting of human clinical practice.

CRITERIA FOR APPROPRIATE CHOICE OF ANIMAL MODELS IN BONE REPAIR

As previously stated by Einhorn,[2] "appropriate use of animal models in tissue-engineering research begins with careful consideration of the question asked."[2] Since many questions can potentially be asked in the field of tissue engineering, the answers are usually obtained using different animal models. The problem to be solved should therefore first be clearly identified in order to define the most appropriate experimental model. The complexity of the experimental design will depend on the criteria to be assessed (biochemical, molecular, mechanical testing, histology, functionality, etc.). It is worth noting that however carefully the experimental design and the equipment required are planned and however suitable the animal model selected may be, the results of studies involving no negative and positive control data that can be used to make comparisons will be useless, irrelevant, or worse still, misleading. Thus, determining the appropriateness of an animal model can be the most difficult and time-consuming aspect of the study design.

The focus of the study, especially whether it is biological, mechanical, or both, is a significant point. If biological issues are at stake, small, skeletally mature animals may be useful, and since these animals are inexpensive and give fast healing times, they can be used to set up large pools of data. Many biological tissue engineering questions can be answered using relatively simple small animal models. For instance, experiments in which construct is implanted into ectopic (subcutaneous or intramuscular) sites in small animals make it possible to easily determine the biocompatibility, biodegradability, osteointegration, osteoconductivity, and osteoinductive and osteogenic potential of the bone engineered construct *in vivo*. Larger, more expensive animals can be more suitable for studies that involve biomechanical issues because they simulate more closely the size and structure of human bones. Large animal models simulating the clinical situa-

From: *Sourcebook of Models for Biomedical Research*
(P. M. Conn, ed.), © 2008 Humana Press Inc., Totowa, NJ.

tion as closely as possible are also usually preferred for preclinical trials.

The size of the defect in which the biomaterial will be tested, whether it is critical or noncritical (i.e., whether or not the defect will heal spontaneously during the animal's lifetime), its location (see below for various orthotopic implantations), as well as the mechanical environment (nonloaded or loaded models) are also critical factors. In addition, while the relevance of the model is of utmost importance, other parameters such as surgical feasibility and reproducibility, animal morbidity, and technical and financial constraints also have to be taken into account.

As far as bone tissue engineering is concerned, the main questions that arise tend to focus on the selection of the cell source, the scaffold material(s), the appropriate growth factor(s), the growth factor concentration required, and the mode of application (carrier systems and/or release mechanisms, for example).

CELL-RELATED CRITERIA Osteoprogenitor cells for the development of bioengineered bone have been isolated from a variety of tissues including periosteum, bone marrow, and adipose tissue, which are the most popular sources. The yield obtained in terms of the skeletal cell isolation rates and the osteogenic potential of material obtained from animal species shows some variability. For instance, harvested murine bone-marrow-derived stromal cells (BMSCs) are usually contaminated by the hematopoietic fraction,[3] isolation of rabbit BMSCs is weakly reproducible,[4] and sheep BMSCs show very weak levels of expression of alkaline phosphatase, an early marker of osteoblast differentiation. Cells can also be defined depending on their donor, which can determine the choice of animal model. Skeletal stem cells can be harvested from the animal into which they will be reimplanted (this is known as an autologous source), for instance. Although this autologous approach ensures the immunocompatibility of the cellular component, it has several drawbacks: (1) it is time consuming and cumbersome, since many animals are required to be able to perform statistically significant data analysis, (2) it is difficult to obtain cells showing a reproducible and consistent osteogenic potential from different animals, and (3) when using small animals, additional animals (from genetically identical animals, such as twins, clones, or highly inbred research animal models) have to be sacrificed to harvest a sufficiently large cellular sample. Last, for the sake of greater clinical relevance, skeletal stem cells are often harvested from human patients (xenogeneic source). However, the use of human cells involves implanting a hybrid construct into immunocompromised nude or severely combined immunodeficient (SCID) animals in which the humoral and cellular immune system is immature.[5] Only nude or SCID mice and nude rats are currently available for this purpose on the market, which drastically reduces the range of possible animal models when using human cellular components; in addition, the cost of these immunocompromised animals is usually 5- to 6-fold higher than that of their wild-type homologs. In specific cases in which it is necessary to assess the osteogenicity of xenogeneic cells grown *in vitro*, the diffusion chamber assay can be used. Cells are inoculated into a small chamber made of semipermeable membranes allowing the diffusion of gases and nutrients, while shielding xenogeneic cells from invading host cells.[6,7]

Some new cellular models have been recently developed using novel genetic methods developed for tissue-engineering applications. Genetically engineered cells that express (either constitutively or under the control of specific promoters) protein reporters such as enhanced green fluorescent protein (GFP) or luciferase were designed to track cells implanted *in vivo*.[8] Fluorescent and bioluminescent imaging methods provide a new noninvasive means of imaging the spatial patterns of marker gene expression. Several aspects of bone formation can thus be followed in individual animals from the time of implantation up to the moment when newly formed bone is remodeled and fully integrated.[9,10] While techniques of this kind open promising new perspectives, their use is nevertheless restricted to small-sized animal models.

SCAFFOLD-RELATED CRITERIA Scaffolds are mainly used for the delivery and retention of osteogenic cells in bone tissue engineering. Apart from serving as a mechanical substrate, scaffolds provide a favorable environment in which bone cells can migrate, proliferate, differentiate, and deposit bone matrix (i.e., they promote osteoconduction). Subcutaneous or intramuscular implantation is usually performed to test the biocompatibility and biodegradation/bioresorption of the scaffold, mostly in rats and lagomorphs for the sake of economy and technical simplicity. The size, shape (massive, granular, or even injectable), hardness, and chemical composition are the main parameters of the scaffold that have to be taken into consideration when choosing the animal species (small vs. large animals), the anatomical site (an intramembranous vs. endochondral bone site; a load-bearing vs. a non-load-bearing site), and the shape of the bone defect (a closed cavity vs. an open defect that will require additional fixing devices).

GROWTH FACTOR-RELATED CRITERIA Growth factors could be used to enhance the proliferation and differentiation of host or implanted osteoprogenitors as well as to promote local processes of angiogenesis. Strategies involving the use of bone morphogenetic proteins (BMPs), which induce differentiation of uncommitted progenitors into osteoblasts,[11] or angioinductive growth factors such as vascular endothelial growth factor (VEGF) or platelet-derived growth factor (PDGF)[12] seem to be particularly promising. Bioactive molecules are usually incorporated onto or into a three-dimensional scaffold. Simple tests can be performed on ectopic sites to assess the bioactivity of molecules. For instance, the osteoinductivity of BMPs is classically assessed by quantifying the rates of new bone formation after implanting BMPs at intramuscular or subcutaneous sites in rodents 3–4 weeks previously. When methods of this kind are used at orthopic sites, it is the features of the carrier material (see the previous section) rather than the type of growth factor applied that will determine the choice of animal model to be used.

The kinetics of growth factor release are of crucial importance, since the cellular response to biological stimuli will depend on the amount of biomolecule released and the time lapse of cell exposure. It is therefore essential to determine the *in vivo* pharmacokinetic profiles of the growth factors used in innovative delivery systems by tagging the protein before its implantation (for instance, using radioactive, fluorescent, biotinylated labels or Flag fusion protein production). Experiments along these lines could easily be performed ectopically or orthopically on small-sized animals.

Based on these considerations relating to the constituents of engineered bone constructs, in many cases the variables of interest can be tested using relatively simple models. However, most of the studies carried out so far on the monitoring of biological cues in the context of the whole organism have been restricted to the use of rodent models. Small-sized animal models can therefore be used to optimize many of the steps involved, although the results obtained will have to be applicable to medium- and large-sized animal models in order to simulate human clinical settings more closely.

DECISION MAKING: ANIMAL MODELS CURRENTLY USED IN TISSUE ENGINEERED BONE CONSTRUCT ASSESSMENT

Whereas evaluation of tissue engineered bone construct (TEBC) biofunctionality is independent of its future field of application and is always performed in the same animal models, preclinical evaluations are performed in an animal model that reflects a specific human clinical setting.

PRELIMINARY ASSESSMENT OF THE BIOFUNCTIONALITY OF NEW TISSUE ENGINEERED BONE CONSTRUCTS

Heterotopic Implantation Subcutaneous or intramuscular implantation is the gold standard for testing the biocompatibility as well as the osteogenic and osteoinductive potential of bone constructs *in vivo*. These simple surgical procedures are mostly performed on rats and rabbits for financial reasons and because of their technical simplicity, but they have also been tested on larger species such as sheep[13] and goats[14] in cases involving larger, clinically sized implants where cell survival is a major concern. The TEBCs are usually directly implanted into the recipient bed. Alternatively, the fate of *in vitro* expanded cells can be monitored by inoculating them into diffusion chambers prior to their implantation. The nucleopore filters of the diffusion chambers prevent the invasion of implanted material by host cells, while allowing the diffusion of nutrients and cell contacts through the pores. Intraperitoneal implantation of these chambers is also being currently performed.[15–17]

Orthotopic Implantation Orthotopic implantation methods make it possible to assess the osteoconductivity and osteointegration of the implanted material as well as its biodegradability in a bone environment. Rat calvaria is the gold standard model used for orthotopic implantation purposes. Calvaria, which develops from a membrane precursor, has a poor blood supply and relatively little bone marrow. Calvarial defects therefore create a hostile environment for bone healing. Single midsagittal 8-mm circular lesions[18] and bilateral 5-mm parietal lesions[19] can be created in rats: both result in critical size defects (CSDs) as they give rise to a fibrous nonunion when bone loss is not replaced (Figure 74–1A). Craniotomies are easy to perform, have low morbidity rates, and are highly reproducible if performed with a circular trephine. Material of all kinds can be tested in that location, which is particularly well suited for assaying granular or paste-like materials. Moreover, a large number of animals can be operated on allowing significant statistical analysis. Laterally performed craniotomies have the advantage of allowing paired design and minimized morbidity while avoiding accidental damage to the midsagittal sinus. Yet the close vicinity of adjacent defects may allow substance diffusion and impair model relevance.[20]

Bone chamber models implanted in the diaphysis of long bones can be used to test some scaffold material in a reproducible loading or unloading environment. Bone chambers have been occasionally used at the femoral level in the rabbit and goat (1) to compare scaffolds loaded with various growth factor concentrations, (2) to study differences in scaffold processing, and (3) to assess the effects of loads on bone healing. The material to be tested is implanted in a titanium chamber where tissue ingrowth can occur. Depending on the type of chamber used, either single or repeated sampling procedures can be carried out. Repeated sampling is useful as it significantly reduces the number of animals to be operated on.

PRECLINICAL EVALUATIONS OF AUGMENTING BONE-HEALING PROPERTIES IN BONE REPLACEMENT PROCEDURES

A relevant animal model for the preclinical evaluation of a TEBC will (1) mimic the surgical technique that is utilized

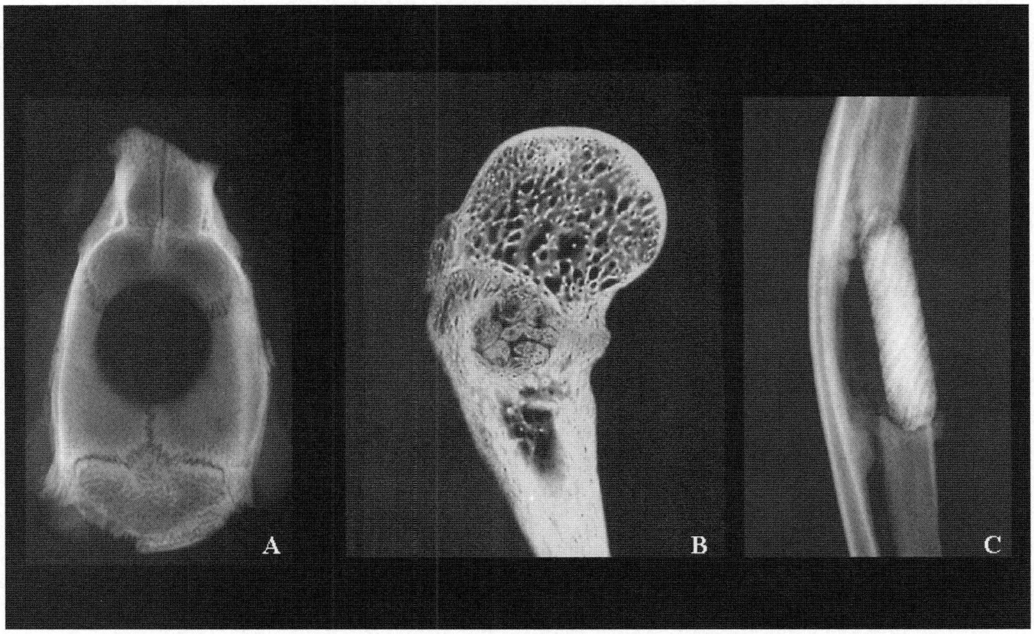

Figure 74–1. (A) Four-week postoperative rat calvaria bone defects left empty. (B) One-month postoperative microradiography of metaphyseal defects created in rabbit distal femoral condyles and filled with a TEBC. (C) Four-week postoperative rabbit ulnar defect filled with a TEBC.

Table 74–1
Animal models currently used in preclinical trials in the field of cranial and maxillofacial surgery[a]

Animal species	*References*	*Bone*	*Defect length (mm)*	*BFT*	*Observation period (weeks)*	*Negative controls*	*CSD*	*Morbidity*
	Bosch *et al.*[19]	Calvaria	8	–	34–52	+	+	+
Rat	Lindsey *et al.*[26]	Nasal bone	20 × 8	–	24	+	+	0
Rabbit	Frame[29]	Calvaria	15	–	36	+	+	0
Dog	Prolo *et al.*[30]	Calvaria	20	–	24	+	+	0
	Hjorting-Hansen and Andreasen[31]	Mandible	8	–	16	+		0
	Leake and Rappoport[32]	Mandible	30	APT	24	+		++
Pig	Lindholm *et al.*[33]	Skull		–				
Sheep	Viljanen *et al.*[20]	Calvaria	22	–		+	+	0
	Salmon and Duncan[25]							
	Schliepake *et al.*[27]	Mandible	9	–	6–12			0
	Hanson *et al.*[28]	Mandible	Unicortical	–		+	+	0
		Orbital wall	–					
				–	24	–		0
			15 × 20					
NHP	Hollinger and Kleinschmidt[21]	Calvaria	15	–	8	+		0

[a]BFT, bone fixation technique; CSD, critical size defect; APT, alloplastic trays; NHP, nonhuman primate.

in humans, (2) yield rates of nonunion similar to those observed in humans, (3) mimic the biomechanical environment present in humans, and (4) include animals receiving autografts (positive controls) as well as untreated animals (negative controls). The difficulty of inducing new bone formation depends partly on the anatomical location at which the construct is tested (metaphyseal or diaphyseal bone or spine, for example). The success of a TEBC performed at one anatomical site is therefore not predictive of its performances at another location.

Cranial Surgery Studies on calvarial, parietal, and skull defects have been performed on rabbit, dog, sheep, minipig, and nonhuman primate (NHP) (Table 74–1).[18,20–22] These lesions are easy to create, are highly reproducible, and involve low morbidity rates. The experimental designs used in these studies accurately reflect the clinical settings in which human cranioplasty is performed, although the calvarial regenerative capacity of animals is greater than that of humans, in whom the calvaria is devoid of muscle insertions. Several defects can be created in the same animal in order to compare different biomaterials and to obtain negative and positive control data on the same animal. Up to 12 22-mm lesions per animal have been created, for example, by Viljanen on sheep.[20]

MAXILLOFACIAL SURGERY

Filling Defects Filling defects in maxillofacial surgery often result from alveolar bone resorption at tooth extraction sites (Table 74–1). Suitable models have been developed in the dog by performing mandibular and maxillary premolar extractions.[23] Bone fillers have also recently been tested on unicortical or bicortical mandibular filling defects induced in dogs,[21] miniature pigs,[24] and sheep.[25] Mandibular defects have several advantages over periodontal wounds[25]: (1) they can be standardized in terms of their size and shape, (2) they involve a closed wound rather than an open wound, as occurs in the mouth, and (3) the only tissue regenerated is bone. Calvarial defects such as those described in rabbits, dogs, and NHPs are also currently used because of the anatomical similarities existing with the human mandible (two cortical plates with intervening cancellous bone).[18] Critical-sized nasal defects have been recently described in large Sprague–Dawley rats.[26] The advantage is that these defects involve an endochondral type of bone, whereas calvarial defects are induced in a membranous type of bone.

Segmental Bone Defects Mandibular discontinuity defects have been induced for experimental purposes to mimic the segmental bone resections that are currently performed in oncological maxillofacial surgery. These models partly reproduce the adverse clinical settings in which bone replacement usually takes place during maxillofacial surgery: (1) bone loading requires preventing segmental motion using bone fixation procedures when performing extensive bone resection, and (2) bone replacement is performed in the vicinity of a potentially highly contaminated site. Masticatory stresses differ, however, between humans and animals and vary from one species to another, depending on nutritional habits. Models have been described in the rat, rabbit, dog, sheep, and NHP.[21,27] As previously stated by Hollinger and Kleinschmidt,[21] mandibular bone is very thin in rodents and rabbits and ensuring the retention of implants is a challenging problem. We do not recommend the use of mandibular resection on these species: it is preferable to induce mandibular discontinuity defects in dogs, sheep, or NHPs, where they are easier to perform. Although there is little objective information available on the subject, Schmitz and Hollinger[18] have suggested that mandibular CSD can be as long as 40 mm in adult foxhounds and up to 25% of the total length of the mandible in NHPs. Mandibular resections are prone to postoperative infections, and when they are extensive, they are difficult to stabilize. Complications can be minimized by (1) performing preventive extraction of the teeth bordering the defect (including the maxillary teeth) 2 weeks before bone resection,[21] (2) using submandibular approaches to prevent intraoral perforation, and (3) ensuring mandibular bone fixation with bone plates.

Orbital wall defects have been performed in sheep.[28] The reproducibility of this model may be greater than in the case of mandibular defects and the morbidity may be lower, but as these defects are located in nonloaded bones, this model should be used only to test material to be used with unloaded bones.

ORTHOPEDIC SURGERY

Filling Defects Many TEBCs are used as "bone fillers" to deal with defects resulting from either bone sampling for grafting procedures, metaphyseal traumas, or surgical removal of benign or malignant tumors. Bone fillers are engineered as paste-like, granula, or massive (preset blocks) materials and experimental defects must be compatible with these specific presentations.[29–33]

Standardized metaphyseal defects have been induced in rabbit[34] and sheep[35] distal femoral condyles (Figure 74–1B) and in the dog proximal tibia.[36] These defects do not require any special additional immobilization and since they are highly reproducible, statistical analysis can be carried out on the histological data obtained. Defective lumens are sealed either with a bone flap or with methylmethacrylate cement to prevent the leakage of material. As they are located in loaded bones, these defects are more useful than the critical size ilial defect induced in the goat.[37]

TEBCs are also used as bone fillers in vertebroplasties to maintain or augment vertebral body volume in patients with bone loss resulting from osteoporosis or tumor removal. A sheep model for vertebral bone loss has been described[38] in which a defect is created surgically in the lumbar vertebral body, and either left intact (negative controls) or filled with coral (using an innovative technique) (Figure 74–2A) or with an autologous corticocancellous graft (positive controls).

Segmental Bone Defects According to Keys's hypothesis, under experimental conditions, a segmental long bone defect 1.5 times the size of the diaphyseal diameter will be beyond the regenerative capacities of bone in skeletally mature dogs and will result in nonunion when the missing bone is not replaced.[39] This hypothesis was also found to hold true in the case of feline tibia.[40] The length of the bone resections at which nonunion occurs is species and bone dependent and must therefore be established whenever a new model is developed (negative controls). Key's findings nevertheless roughly apply to many species (dogs, cats, pigs, sheep, NHPs) and can serve as guidelines for developing new models (Table 74–2). Rodents and rabbits are exceptions to this rule, however, as only much longer resections are associated with nonunion in these species.

As mentioned above, segmental defects must be kept in a perfectly stable biomechanical environment while healing occurs. Weight bearing is more difficult to prevent in animals than in humans; however. Achilles tenotomy is not a reliable adjunctive technique as it prevents weight bearing for only a short time.[41] Models in which bone fixation techniques (such as locked intramedullary nailing, bone plating) are used appropriately to neutralize any forces generated are therefore preferable to unlocked intramedullary nailing or external devices such as plaster casts, which do not neutralize the compression, traction, or rotation forces. Although external fixation methods may be satisfactory from the biomechanical point of view, they have several limitations: (1) the weight bearing pattern is irregular, and (2) bone lysis occurs around the pins with time, thus causing instability and reducing the reproducibility and relevance of the model.[42,43]

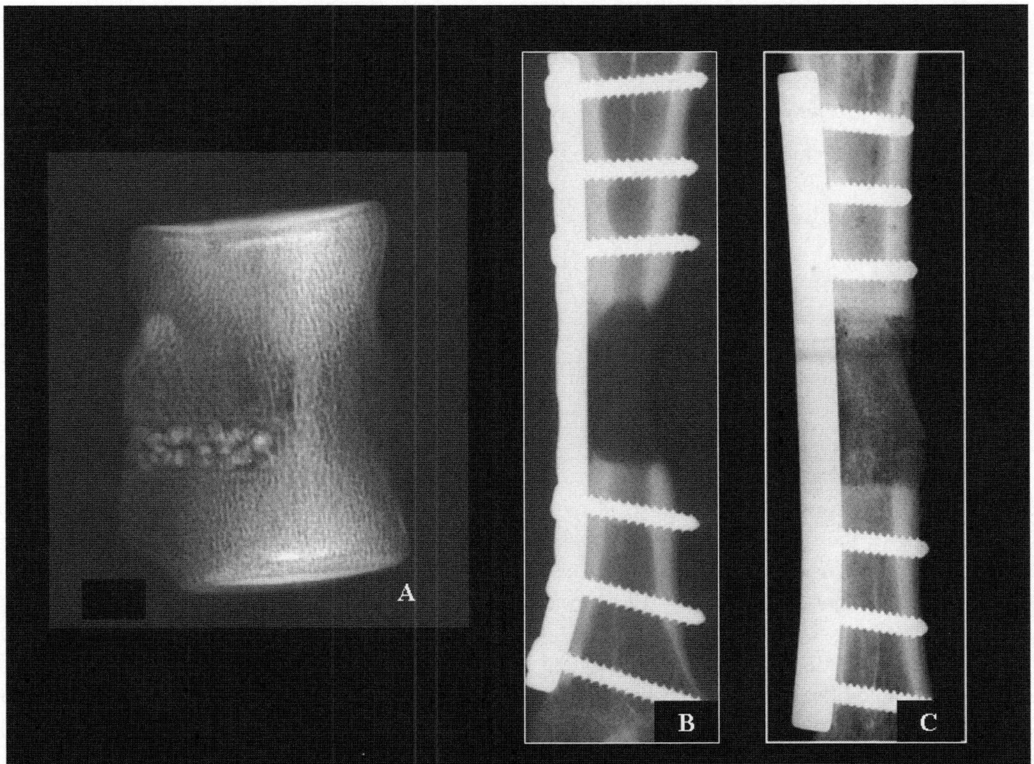

Figure 74–2. (A) Immediate postoperative radiograph of a surgically created vertebral defect filled with coral granules. Radiograph of a 16-week postoperative sheep metatarsal defects left empty (B) or filled with an autologous bone graft (C).

Table 74–2
Animal models used in preclinical trials involving long bone segmental resection[a]

Animal species	*References*	*Bone*	*Resection length (mm)*	*BFT*	*Observation period (weeks)*	*Negative controls*	*CSD*	*Positive controls*	*Morbidity*
Rat	Werntz *et al.*[64]	Femur	5	Bone plate	12	+	+	–	ND
	Einhorn *et al.*[45]	Femur	6	External fixator	12	+	+	–	ND
	Wolff and Stevenson[44]	Femur	8	Bone plate	8–16	–	ND	–	ND
Rabbit	Dahners[65]	Ulna	10	–	6	+	–	+	ND
	Gaullier[66]	Ulna	20	IM pin	8	+	–	–	–
	Bolander and Balian[46]	Ulna	20	–	12	+	+	+	–
	Wittbjer *et al.*[50]	Radius	12	–	4	–	ND	–	–
	Kitsugi *et al.*[51]	Tibia	16	IM pin	25	ND	–	–	+++
Dog	Johnson *et al.*[43]	Radius	25	External fixator	24	+	–	+	+
	Johnson *et al.*[55]	Ulna	35	Bone plate	14	–	ND	+	ND
	Bruder *et al.*[53]	Femur	20	Bone plate	16	+	+	–	–
	Cong *et al.*[52]	Femur	15	Bone plate	8–24	–	ND	–	ND
	Johnson *et al.*[54]	Tibia	15	External fixator	12	+	–	+	+
Cat	Toombs *et al.*[40]	Tibia	10	Bone plate	12	+	+	+	–
	Tooms and Wallace[67]								
Pig	Sendowski *et al.*[60]	Femur	25	External fixator	24	–	ND	–	+++
	Sendowski *et al.*[60]			or bone plate					
Minipig	Meinig *et al.*[59]	Femur	25	–	24	–	ND	–	+++
Minipig	Ehrnberg *et al.*[68]	Radius	30	Bone plate	12	+	–	–	–
	Gerhart *et al.*[69]								
Sheep	Wippermann *et al.*[41]	Femur	40	External fixator	16	+	+	–	+
	DenBoer *et al.*[70]	Femur	25	Bone plate	12	+	+	+	
	Muir *et al.*[57]	Tibia	20		12	+	–	+	+
	Gao[56]	Tibia	30	Bone plate	12	+	–	+	+
	Mastrogiacomo *et al.*[71]	Tibia	50	Locked IM pin	14	–	ND	+	–
				Locked IM pin					–
	Viateau *et al.*[58]	Tibia	25	Two bone plates	16	–	ND	–	
		Tibia	48	Bone plate	12–24–48–	–	ND	–	–
		Metatarsus	25	Bone plate	96	+	+	+	+++
					16				
NHP	Andersson *et al.*[62,63]	Humerus	50	Bone plate	24–240	–	ND	–	–
	Andersson *et al.*[62,63]	Femur	76	Bone plate	24–240	–	ND	–	–
	Andersson *et al.*[62,63]	Tibia	63	Bone plate	24–240	–	ND	–	–
	Cook *et al.*[61]	Ulna	20	External coaptation	20	–	ND	–	–
	Cook *et al.*[61]	Tibia	20	IM pin + external coaptation	20	–	ND	–	–

[a]BFT, bone fixation technique; CSD, critical size defects; IM, intramuscular; NHP, nonhuman primate.

Femoral resection has been performed on rats: 5- to 8-mm-long bone resections (20–25% of the femoral bone length) stabilized by custom-made bone plates[44] or external fixation devices[45] were found to result in nonunion after approximately 12 weeks.

The rabbit ulna is one of the most commonly used segmental long bone resection models.[46,47] In this case, remedial surgical procedures can be performed bilaterally,[48] they are easy to perform, and no internal fixation is normally required, as radioulnar synostosis makes for ulnar stability. The ulna is a weight-bearing bone with which implant loading is possible and the morbidity rate is low. This method has several limitations, however: (1) spontaneous bone healing occurs in 30% of the animals after 10-mm-long resections; since bone healing with 20-mm-long ostectomies has been observed, this is more of a model for delayed bone healing than for CSDs[49]; (2) bone curvature and synostosis make implant positioning and stabilization difficult to achieve when solid materials are used; and (3) individual variations in bone curvatures reduce the reproducibility of the procedures used to sample bone for histological analysis and biomechanical assays. Radial resections performed on rabbits suffer from the same disadvantages.[50] Tibial resections stabilized using unlocked intramedullary nailing methods have been reported to be associated with rotational instability and high rates of morbidity, which preclude their use.[51]

A highly reproducible CSD model has been developed based on the cat tibia.[40] Femoral,[46,52,53] tibial,[54] radial,[43] and ulnar[55] resections performed on dogs were stabilized using either bone plating[52,53] or external fixation methods.[43,54] Some of these models have shown spontaneous bone healing and are therefore not appropriate models for CSDs.[43] It was found that 25-mm-long ulnar resections performed on dogs have the same advantages and disadvantages as those described in the rabbit. Radial synostosis is nevertheless more strictly localized in dogs than in rabbits, which simplifies the positioning and stabilization of the implants.

Femoral resections stabilized by means of bone plates and tibial resections stabilized using either bone plates[56] or an intramedullary interlocking nail[57] have been performed in sheep. The recently described metatarsal resection method[58] has many advantages: (1) it is a true CSD, as fibrous nonunion was observed in 25-mm-long bone defects after a 4-month observation period, (2) it is performed in a straight bone lending itself to reproducible histomorphometric and biomechanical assays, and (3) it is associated with a low rate of morbidity (Figure 74–2B and C).

Postoperative handling of pigs is often difficult, and so is orthopedic surgery. A radial defect induced in the miniature pig was associated with very low rates of morbidity.[59] Femoral defects stabilized using either bone plate or external fixation methods have been described by Sendowski in Pitmann Moore minipigs and Large White pigs. High morbidity rates have been recorded with these models: external fixation devices have been unsuccessful in animals of both sizes and the use of bone plates has been satisfactory only with minipigs.[60]

Femoral, tibial, ulnar, and humeral resections have been performed on NHPs.[61–63] Although these models were developed in a species similar to humans as far as bone healing and biomechanical loading processes are concerned, they have some major drawbacks: (1) lack of negative controls[61–63] and (2) the excellent mechanical strength of the constructs used, which were solid metal implants.[62–71]

Spinal Arthrodesis (Fusion) As mentioned by Kruyt *et al.*,[14] spinal arthrodesis is one of the most challenging applications for TEBCs since even the use of autologous bone, the gold standard, is associated with relatively high failure rates. The objectives of spinal fusion differ from those of bone loss replacement: fusion is performed in order to create a mechanically optimum solution in a pathological spinal unit, and not to restore normal anatomy.[72] Regional differences in the spinal column can also significantly affect patients' healing potential.[73] The anterior spine is mainly cancellous and presents a large surface area for bone-healing purposes, thus providing greater opportunities for successful interbody fusion, whereas the posterior column has a greater proportion of cortical bone and a sublumbar healing environment, which have resulted in high failure rates among the attempts at fusion made at this location. Lastly, the fusion rates depend on (1) the number of vertebral units treated (the larger their numbers, the higher the risk of failure will be) and (2) the degree of mobility (left–right side comparisons can be misleading if appropriate instrumentation is lacking). TEBC methods must therefore be tested in an anatomical location similar to where it is to be applied, and the procedure used should be tested on a suitably large number of vertebral units.

Extensive critical reviews of the animal models used in preclinical spinal fusion trials[72–90] have been published by Schimandle and Boden,[91] Khan and Lane,[72] and Kruyt *et al.*[14] Interbody fusions obtained using either anterior or posterior surgical approaches as well as laminar, facet, and posterolateral intertransverse process fusion (PLF) techniques have been described in the rat,[90,92] rabbit,[93] dog,[74,94] pig,[9,95] sheep,[83,86,96] goat,[77,81,97,98] and NHP.[88–90] However, positive and negative controls have not always been included in these studies (Tables 74–3, 74–4, and 74–5).

Lumbar PLF performed on rabbits, which are the most extensively used animal models for spinal fusion, gives overall nonunion rates of 30–40%. These models can be used to make cost-effective comparisons between different osteoinductive constructs.[93] Dog models have also been extensively used. However, models developed on the basis of these small to medium sized species have several anatomical and biomechanical limitations: (1) the experimental fusion techniques used on animals can differ significantly from those performed on humans (the pedicle screw fixation method, which is currently used on humans to optimize biomechanical stability cannot be performed on rats, rabbits, or dogs), (2) implantable bone constructs designed for animals are necessarily smaller than those that can be used on the human spine, and (3) the loads exerted on the spine differ significantly between animals and humans.

Table 74–3
Animal models used in cervical anterior interbody fusion (AIF) tests

Animal species	*References*	*Fusion level*	*Bone fixation technique*	*Observation period (weeks)*	*Negative controls*	*Positive controls*	*Complications*
Dog	Cook *et al.*[74]	C3–C4, C4–C5	–	6, 12, 26	–	+	Implant extrusion
Sheep	Kandziora *et al.*[75]	C3–C4	–	12	–	+	–
Goat	Zdeblick *et al.*[76]	C2–C3, C3–C4, C4–C5	+/– Plating	12	+	+	–
	Zdeblick *et al.*[77]	C2–C3, C3–C4, C4–C5	Cage	12	–	+	–
	Toth *et al.*[78]	C2–C3, C5–C6	–	12, 24	+	+	Implant extrusion

Table 74–4
Animal models used in lumbar posterolateral fusion (PLF) tests

Animal species	*References*	*Fusion level*	*Bone fixation technique*	*Observation Period (weeks)*	*Negative controls*	*Positive controls*	Morbidity
Rat	Huang *et al.*[79]	L4–L5	–	8	–	+	–
Rabbit	Boden *et al.*[80]	L5–L6	–	2; 3; 4; 5; 6; 10	+	+	–
Dog	Sandhu *et al.*[81]	L4–L5	–	12	+	+	–
Pig	Xue *et al.*[82]	L2–L3, L5–L6	Pedicle screw-rod system	12	–	+	–
Sheep	Baramki *et al.*[83]	L3–L4, L4–L5	Pedicle screw-rod system	20	+	+	–
NHP[a]	Barnes *et al.*[84]	L4–L5	–	24	–	–	–

[a]NHP, nonhuman primate.

Although models for spinal fusion developed in large animals do not simulate the graft-healing environment of humans any better than those involving smaller animals such as rabbits,[14] they do overcome some of the drawbacks encountered with small animal models: (1) implants such as cages and pedicle screws designed for human applications can also be used in the case of large animals, (2) some anatomical similarities with the human spine do exist [goat cervical anterior interbody fusion (AIF) models have gained popularity because this species has an erect cervical spine],[76] and (3) biomechanical trials can be more reliably performed on large-sized specimens.

NHP models undeniably recreate biological and biomechanical environments most similar to those described in humans. Failure to achieve posterolateral intertransverse fusion often occurs with NHPs, as with humans, despite decortication and the implantation of an autograft in the host bed/tissue/bone. Yet because of the low availability of animals and for ethical reasons, the use of NHP models is limited and only a few animals can usually be included in trials.

QUANTITATIVE ASSESSMENT OF BONE HEALING

Since bone exists in a variety of shapes and sizes, there are no established standards for assessing bone-healing processes. We will provide a brief overview of the techniques that are most commonly used to assess bone healing.

Table 74–5
Selected animal models for lumbar interbody fusion

Animal species	*References*	*Localization of fusion*	*Bone fixation technique*	*Observation Period (weeks)*	*Negative controls*	*Positive controls*	*Morbidity*
Rabbit	Kai *et al.*[85]	L5–L6	Bone plate	12	+	+	–
Pig	Li *et al.*[9]	L3–L4, L4–L5	Pedicle screw-rod system + cage	12	–	+	–
	Zou *et al.*[82]	L2–L3, L4–L5, L6–L7	Pedicle screw-rod system or staple + cage or ring	12	–	+	–
Sheep	Sandhu *et al.*[86]	L4–L5	Cage	24	+	+	Cage misplacement
	Sandhu and Kahn[87]	L4–L5	Cage	24	–	+	–
NHP	Boden *et al.*[88]	L7S1	Cage	24	–	–	–
	Hecht *et al.*[89]	L7-S1	–	12, 24	–	+	–
	Wang *et al.*[90]	L3–L4; L5–L6	–	12	–	+	–

[a]NHP, nonhuman primate.

CLINICAL ASSESSMENT OF UNION The limb should be examined with no cast for any persistent edema and tenderness at the operated site, which may suggest incomplete union. Persistent mobility at the operated site is a definite sign of incomplete union.

RADIOGRAPHIC ASSESSMENT

Plain X-Rays Since very little scientific information is available about the evolution of the bone repair process on the basis of radiographic evidence, radiographic assessments of bone repair depend largely on radiologists' experience. However, a number of characteristic features of bone repair (fracture gap widening, sclerosis at the fracture margin, periosteal reaction, the presence of callus, a callus density greater than cortex, bridging, periosteal reaction incorporation, remodeling) can be objectively observed on X-rays and some of them can be quantified. In addition, radiographic scales have been described for objectively assessing bone repair processes.

Dual X-Ray Absorptiometry and Computed Tomography Bone densitometric and morphological parameters can be measured quantitatively by dual X-ray absorptiometry (DXA) and volumetric quantitative computed tomography (vQCT).[99] vQCT is based on the scanning of the entire region of interest (i.e., the entire bone defect area) and a three-dimensional reconstruction of these data into anatomically relevant projections, using anatomical landmarks to define a system of coordinates. Techniques of this kind provide an accurate volumetric picture, making it possible to determine the bone mineral content (BMC) and bone mineral density (BMD) either of the entire region of interest or, by separate measurements, of the trabecular or cortical components. More sophisticated apparatuses [such as micro-computed tomography (CT)] are now being developed for three-dimensional *in vitro* analysis of small bone pieces. These methods yield images of a sufficiently good quality to be able to determine standard histomorphometric parameters such as trabecular thickness and separation. This information is useful as bone strength is only partly explained by BMD. The latest generation of micro-CTs collects three-dimensional data sets that provide a basis for finite-element modeling, which are then used for performing virtual biomechanical tests to predict mechanical properties.

HISTOMORPHOMETRY

Quantitative analysis of histological samples (histomorphometry) is based on the following measurements: (1) static bone development indices, (2) dynamic rates of bone formation and mineralization, and (3) microarchitectural properties of bone that contribute to ultimate bone strength. These measurements of skeletal parameters in tissue sections are performed with a bone-specific image analysis system and a microscope on stained sections. All the skeletal parameters used should be in compliance with and calculated according to the recommendations of the histomorphometry nomenclature committee of the American Society of Bone and Mineral Research.[100] In a recent review by Gerstenfeld *et al.*, it was recommended that the following parameters should be determined to study bone repair: callus diameter, total callus area, area of the cartilage, area of the fibrous tissue, area of the void, area of the total osseous tissues, and osteoclast and osteoblast volume density.[101] More specialized parameters that are also relevant to ongoing bone repair processes include the number of vessels per callus area, the number of apoptotic cells per cartilage area, and the total number of apoptotic cells per callus area. One rather critical issue arising in the quantitative analysis of histological samples is how to accurately sample such a heterogeneous tissue in order to obtain a set of statistically meaningful area measurements reflecting changes in the biomechanical and radiological properties. To address this issue, Gerstenfeld *et al.*[101] made the following specific technical recommendations: (1) adopt a transverse sectioning strategy rather than a longitudinal one, and (2) sections should be collected at fixed increments along the long axes of bone to ensure complete sampling across the entire volume of the bone defect and to obtain statistically valid measurements representative of all possible individual measurements of the total bone defect volume.

BIOMECHANICAL ANALYSIS

All treatments involving the use of an engineered bone are designed to regenerate a new bone that will withstand physiological loads in exactly the same way as a normal bone. It is therefore essential to be able to assess the mechanical properties of the newly formed bone in order to determine whether the new bone has been successfully or unsuccessfully engineered. Biomechanical tests can be performed on specimens machined from the bone piece to determine the intrinsic properties of the repaired tissue. Such measurements reflect the quality of newly formed bone but do not yield any information about the ability of the segment of newly formed bone to withstand physiological loads. Alternatively, tests can be carried out on whole segments of bone to determine the extrinsic structural properties of the bone segment. Tests of this kind can be used to predict the functional performances of the bone segment, but are rather prone to variations, as the mechanical properties of bone segments depend on parameters such as their size and their geometry.

Biomechanical studies are best performed on large animals that mimic the size and structure of human bone more closely than small animals. Tension, compression, torsion, and bending tests are routinely performed to assess the mechanical properties of bone. The choice of a specific test will depend on the question addressed and should, whenever possible, mimic the specific physiological loading conditions encountered in humans. Although no standardized procedures are available so far for assessing the mechanical properties of bone, detailed descriptions of biomechanical tests and recommendations on specimen preparation and preservation have been given.[102,103] In their tutorial, Turner and Burr[103] recommend (1) the use of sound testing methods, (2) validation of the testing procedure using material standards, and (3) the use of proper specimen preservation and preparation procedures.

One of the main drawbacks of biomechanical tests is that they are destructive and require a sufficiently large number of specimens for statistical analysis to be possible. Alternatively, ultrasonic methods involving sound velocity measurements through the bone structure can be used.[104] These methods are not destructive, but the bone strength is not measured directly and has to be inferred from the sound velocity data.

CONCLUSIONS

When testing a new TEBC, preliminary studies on biofunctionality (i.e., proof of the concept)[87] can be performed on mice and rats by carrying out subcutaneous or intramuscular implantation procedures (at ectopic sites) followed by rat calvarial CSD (at orthotopic sites).

Preclinical studies should ideally be performed first on a small animal model to assess the feasibility of the approach tested, and then on a large animal model to assess the efficacy. These models should be suited to the future field of application: (1) dogs or NHP mandibular CSDs if the construct is to be used for maxillofacial surgery, (2) rabbit ulnar CSDs followed by sheep metatarsal CSDs if it is to be used in orthopedic surgery for segmental long bone replacement purposes, and (3) rabbit then sheep lumbar PLF, or rabbit then goat or sheep cervical AIF, if the material is to be used to enhance spinal fusion in the lumbar and cervical spine, respectively.

The appropriate choice of experimental model is essential to obtaining relevant results. A complete review of the literature and a sharp critical analysis of previously validated models are essential preliminary steps. Ethical considerations must also be taken into consideration, and the animal model associated with the lowest morbidity rates and the least painful procedures should always be chosen. In some instances, custom-made experimental designs are required. Preliminary trials must first be performed on a few animals to assess the feasibility of the method tested and the resulting morbidity rates. Whichever model is chosen, all the results obtained must be interpreted strictly in the context of the experimental model used and great care should be taken about extrapolating these data to humans.

REFERENCES

1. Logeart-Avramoglou D, Anagnostou F, Bizios R, Petite H. Engineering bone: Challenges and obstacles. *J Cell Mol Med* 2005;9:72–84.
2. Einhorn TA. Clinically applied models of bone regeneration in tissue engineering research. *Clin Orthop Relat Res* 1999;(367Suppl.): S59–67.
3. Phinney DG, Kopen G, Isaacson RL, Prockop DJ. Plastic adherent stromal cells from the bone marrow of commonly used strains of inbred mice: Variations in yield, growth, and differentiation. *J Cell Biochem* 1999;72:570–585.
4. Solchaga LA, Johnstone B, Yoo JU, Goldberg VM, Caplan AI. High variability in rabbit bone marrow-derived mesenchymal cell preparations. *Cell Transplant* 1999;8:511–519.
5. Krebsbach PH, *et al.* Bone formation in vivo: Comparison of osteogenesis by transplanted mouse and human marrow stromal fibroblasts. *Transplantation* 1997;63:1059–1069.
6. Bab I, *et al.* Osteogenesis in in vivo diffusion chamber cultures of human marrow cells. *Bone Miner* 1988;4:373–386.
7. Friedenstein AJ, Chailakhyan RK, Gerasimov UV. Bone marrow osteogenic stem cells: In vitro cultivation and transplantation in diffusion chambers. *Cell Tissue Kinet* 1987;20:263–272.
8. Brazelton TR, Blau HM. Optimizing techniques for tracking transplanted stem cells in vivo. *Stem Cells* 2005;23:1251–1265.
9. de Boer J, van Blitterswijk C, Lowik C. Bioluminescent imaging: Emerging technology for non-invasive imaging of bone tissue engineering. *Biomaterials* 2006;27:1851–1858.
10. Yang M, Baranov E, Moossa AR, Penman S, Hoffman RM. Visualizing gene expression by whole-body fluorescence imaging. *Proc Natl Acad Sci USA* 2000;97:12278–12282.
11. Wozney JM, Rosen V. Bone morphogenetic protein and bone morphogenetic protein gene family in bone formation and repair. *Clin Orthop Relat Res* 1998;346:26–37.
12. Bouhadir KH, Mooney DJ. Promoting angiogenesis in engineered tissues. *J Drug Target* 2001;9:397–406.
13. Gosain AK, *et al.* A 1-year study of osteoinduction in hydroxyapatite-derived biomaterials in an adult sheep model: Part I. *Plast Reconstr Surg* 2002;109:619–630.
14. Kruyt MC, *et al.* Bone tissue engineering and spinal fusion: The potential of hybrid constructs by combining osteoprogenitor cells and scaffolds. *Biomaterials* 2004;25:1463–1473.
15. Budenz R, Bernard GW. Osteogenesis and leukopoiesis within diffusion chamber implants of isolated bone marrow subpopulation. *Am J Anat* 1980;159:455–474.
16. Ashton BA, Allen TD, Howlet CR, Eaglesom CC, Hattori A, Owen M. Formation of bone and cartilage by marrow stromal cells in diffusion chambers in vivo. *Clin Orthop Related Res* 1980;151: 294–307.
17. Haynesworth SE, Goshima J, Goldberg VM, Caplan AI. Characterization of cells with osteogenic potential from human marrow. *Bone* 1992;13:81–88.
18. Schmitz JP, Hollinger JO. The critical size defect as an experimental model for craniomandibulofacial nonunions. *Clin Orthop Related Res* 1986;205:299–308.
19. Bosch C, Melsen B, Vargervik K. Importance of the critical-size bone defect in testing bone-regenerating materials. *J Craniofac Surg* 1998;9:310–316.
20. Viljanen VV, Lindholm TC, Gao TJ, Lindholm TS. Low dosage of native allogeneic bone morphogenetic protein in repair of sheep calvaria defects. *Int J Oral Maxillofac Surg* 1997;26:389–393.
21. Hollinger JO, Kleinschmidt JC. The critical size defect as an experimental model to test bone repair materials. *J Craniofac Surg* 1990;1:60–68.
22. Schmitz JP, Schwartz Z, Hollinger JO, Boyan B. Characterization of rat calvarial nonunion defects. *Acta Anat* 1990;138:185–192.
23. Indovina A, Block MS. Comparison of three bone substitutes in canine extraction sites. *J Oral Maxillofac Surg* 2002;60:53–58.
24. Buser D, Hoffmann B, Bernard JP, Lussi A, Mettler D. Evaluation of filling materials in membrane-protected bone defects. A comparative histomorphometric study in the mandible of miniature pigs. *Clin Oral Implants Res* 1998;9:137–150.
25. Salmon R, Duncan W. Determination of the critical size for nonhealing defects in the mandibular bone of sheep. Part 1: A pilot study. *J NZ Soc Periodontol* 1997;81:6–15.
26. Lindsey WH, Franz DA, Toung JS, London SD, Ogle RO. A nasal critical size defect. *Arch Otolaryngol Head Neck Surg* 1998;124:912–915.
27. Schliephake H, Knebel JW, Aufderheide M, Tauscher M. Use of cultivated osteoprogenitor cells to increase bone formation in segmental mandibular defects: An experimental pilot study in sheep. *Int J Oral Maxillofac Surg* 2001;30:531–537.
28. Hanson LJ, Donovan MG, Hellstein JW, Dickerson NC. Experimental evaluation of expanded polytetrafluoroethylene for reconstruction of orbital floor defects. *Int J Oral Maxillofac Surg* 1994;52:1050–1055.
29. Frame JW. A convenient animal model for testing bone substitute materials. *J Oral Surg* 1980;38:176–180.
30. Prolo D, Pedrotti PW, Burres KP, Oklund S. Superior osteogenesis in transplanted allogeneic canine skull following chemical sterilization. *Clin Orthop* 1980;108:230.
31. Hjorting-Hansen E, Andreasen JO. Incomplete bone healing of experimental cavities in dog mandibles. *Br J Oral Surg* 1971;9:33.
32. Leake DL, Rappoport M. Mandibular reconstruction: Bone induction in an alloplastic tray. *Surgery* 1972;72:332.
33. Lindholm TC, Lindholm TS, Marttinen A, Urist M. Bovine bone morphogenetic protein (bBMP/NCP)-induced repair of skull trephine defects in pigs. *Clin Orthop Relat Res* 1994;301:263–270.
34. Pasquier G, Flautre B, Blary MC, Anselme K, Hardouin P. Injectable percutaneous bone biomaterials: An experimental study in a rabbit model. *J Mater Sci Mater Med* 1996;7:683–690.
35. Flautre B, Delecourt C, Blary MC, Van Landuyt P, Lemaitre J, Hardouin P. Volume effect on biological properties of a calcium phosphate hydraulic cement: An experimental study in the sheep. *Bone* 1999;25:35S–39S.
36. Frankenburg EP, Goldstein SA, Bauer TW, Harris SA, Poser RD. Biomechanical and histological evaluation of a calcium phosphate cement. *J Bone Joint Surg Am* 1998;80-A:1112–1124.

37. Anderson ML, Dhert WJ, de Bruijn JD, Dalmeijer RA, Leenders H, van Blitterswijk Verbout AJ. Critical size defect in the goat's ilium. *Clin Orthop Relat Res* 1999;364:231–239.
38. Cunin G, Boissonnet H, Petite H, Blanchat C, Guillemenin G. Experimental vertebroplasty using osteoconductive granular material. *Spine* 2000;25:1070–1076.
39. Key J. The effect of local calcium depot on osteogenesis and healing of fractures. *J Bone Joint Surg Am* 1934;16-A:176–184.
40. Toombs JP, Wallace LJ, Bjorling DE, Rowland G. Evaluation of Key's hypothesis in the feline tibia: An experimental model for augmented bone healing studies. *Am J Vet Res* 1985;46:513–518.
41. Wippermann BW, Zwipp H, Saemann T, Tscherne H. 40th Annual Meeting, Orthopaedic Research Society, New Orleans, LA, February 21–24, 1994:545.
42. Aron DN, Toombs JP, Hollingsworth SC. Primary treatment of severe fractures by external skeletal fixation: Threaded pins compared with smooth pins. *J Am Anim Hosp Assoc* 1986;22;659–670.
43. Johnson KD, August A, Sciadini MF, Smith C. Evaluation of ground cortical autograft as a bone graft material in a new canine bilateral segmental long bone defect model. *J Orthop Trauma* 1996;10: 28–36.
44. Wolff D, Goldberg VM, Stevenson S. Histomorphometric analysis of the repair of a segmental diaphyseal defect with ceramic and titanium fibermetal implants: Effects of bone marrow. *J Orthop Res* 1994;12:439–446.
45. Einhorn TA, Lane JM, Burstein AH, Kopman CR, Vigorita VJ. The healing of segmental bone defects induced by demineralized bone matrix. *J Bone Joint Surg Am* 1984;66-A:274–279.
46. Bolander ME, Balian G. The use of demineralized bone matrix in repair of segmental defects: Augmentation with extracted matrix proteins and a comparison with autologous grafts. *J Bone Joint Surg Am* 1986;68-A:1264–1274.
47. Tuli SM, Singh AD. The osteoinductive property of decalcified bone matrix. *J Bone Joint Surg Br* 1978;60-B:116–123.
48. Perka C, Schultz O, Spitzer RS, Lindenhayn K, Burmester GR. Segmental bone repair by tissue-engineered periosteal transplants with bioresorbable fleece and fibrin scaffolds in rabbits. *Biomaterials* 2000;21:1145–1153.
49. Louisia S. 40 (Paris VII, Paris, 1994).
50. Wittbjer J, Palmer B, Rohlin M, Thorngren KG. Osteogenic activity in composite grafts of demineralized compact bone and marrow. *Clin Orthop Relat Res* 1983;173:229–238.
51. Kitsugi T, Yamamuro T, Kokubo A. Bonding behaviour of a glass ceramic containing apatite and wallatsonite in segmental replacement of the rabbit tibia under load-bearing condition. *J Bone Joint Surg Am* 1989;71-A:264–272.
52. Cong Z, Jianxin W, Huaizhi F, Bing L, Xingdong Z. Repairing segmental bone defects with living porous ceramic cylinders: An experimental study in the dog femora. *J Biomed Mater Res* 2000;55:28–32.
53. Bruder SP, Kraus KH, Golberg VM. The effect of implants loaded with autologous mesenchymal stem cells on the healing of canine segmental bone defects. *J Bone Joint Surg Am* 1998;80-A: 985–996.
54. Johnson AL, Stein LE, Roe SC. Evaluation of collagen as a retainer for autogenous cancellous bone used in repair of full thickness cortical bone defects. *Vet Surg* 1987;16:146–150.
55. Johnson EE, Urist MR, Schmalzried TP, Chotivichit A, Huang HK, Finerman AM. Autogeneic cancellous bone grafts in extensive segmental ulnar defects in dogs. *Clin Orthop Relat Res* 1989;243: 254–265.
56. Gao TJ. In: Wilson J, Hench L, Greenspan D, Eds. *Bioceramics*, Vol. 8. New York: Elsevier Science, 1995:199–204.
57. Muir P, Johnson KA. Tibial intercalary allograft incorporation: Comparison of fixation with locked intramedullary nail and dynamic compression plate. *J Orthop Res* 1995;13:132–137.
58. Viateau VT, Reviron T, Meunier A, Guillemin N, Petite H, Sedel L. 8th Annual Scientific Meeting of the European College of Veterinary Surgeons, Brugge, Belgium, 1999:118.
59. Meinig RP, Buesing CM, Helm J, Gogolewski S. Regeneration of diaphyseal bone defects using resorbable poly(L/DL-lactide) and poly(D-lactide) membranes in the Yucatan pig model. *J Orthop Trauma* 1997;11:551–558.
60. Sendowski JCP, Bazan JF, Grimaldi JF, Merrien YL, Martin DF. 21 Services Vétérinaires des Armées, D.M., No. 307, DEF, DCSSA, AST, 1986.
61. Cook SD, Wolfe MW, Salked SL, Rueger DC. Recombinant human osteogenic protein 1 heals segmental defects in non-human primates. *J Bone Joint Surg Am* 1995;77-A:734–735.
62. Andersson G, Gaechter A, Galante J, Rostoker W. Segmental replacement of long bones in baboons using a fiber titanium implant. *J Bone Joint Surg Am* 1978;60-A:31–40.
63. Andersson G, Lereim P, Galante J, Rostoker W. Segmental replacement of long bones in baboons with fiber metal implants and autologous bone grafts of different particle size. *Acta Orthop Scand* 1982;53:349–354.
64. Werntz JR, Lane J, Klein C, Decarlo E. The osteogenic potential of bone marrow to heal segmental long bone defects. *Proc 33rd Annu Meeting Orthop Res Soc* 1987:441.
65. Dahners LE. *Proc 34th Annu Meeting Orthop Res Soc* 1988;390:3901988.
66. OG. 27 (Paris VII, 1996).
67. Toombs JP, Wallace LJ. Evaluation of autogeneic and allogeneic cortical chip grafting in a feline tibial nonunion model. *Am J Vet Res* 1985;46:519–528.
68. Ehrnberg A, De Pablos J, Martinez-Lotti G, Kreicbergs A. Comparison of demineralized allogeneic bone matrix grafting (the Urist procedure) and the Ilizarov procedure in large diaphyseal defects in sheep. *J Orthop Res* 1993;11:438–447.
69. Gehrart TN, Kirker-Head CA, Kriz MJ, Holtrop ME, Hennig GE, Hipp J, Schielling SH, Wang E. Healing segmental femoral defects in sheep using recombinant human bone morphogenetic protein. *Clin Orthop Relat Res* 1993;293:317–326.
70. Den Boer FC, Patka P, Bakker FC, Wippermann BW, Van Lingen A, Vink GQ, Boshuizen K, Haarman HJ. New segmental long bone defect model in sheep: Quantitative analysis of healing with dual energy x-ray absorptiometry. *J Orthop Res* 1999;17:654–660.
71. Mastrogiacomo M, *et al.* Reconstruction of extensive long bone defects in sheep using resorbable bioceramics based on silicon stabilized tricalcium phosphate. *Tissue Eng* 2006;12:1261–1273.
72. Khan SN, Lane JM. Spinal fusion surgery: Animal models for tissue-engineered bone constructs. *Biomaterials* 2004;25:1475–1485.
73. Louis-Ugbo J, Boden S. In: Bono CM, Garfin SR, Tornetta P, Einhorn TA, Eds. *Spine*. Philadelphia, PA: Lippincott Williams & Wilkins, 2004:343.
74. Cook SD, *et al.* In vivo evaluation of anterior cervical fusions with hydroxylapatite graft material. *Spine* 1994;19:1856–1866.
75. Kandziora F, *et al.* [Experimental fusion of the sheep cervical spine. Part I: Effect of cage design on interbody fusion]. *Chirurg* 2002;73:909–917.
76. Zdeblick TA, Cooke ME, Wilson D, Kunz DN, McCabe R. Anterior cervical discectomy, fusion, and plating. A comparative animal study. *Spine* 1993;18:1974–1983.
77. Zdeblick TA, *et al.* Cervical interbody fusion cages. An animal model with and without bone morphogenetic protein. *Spine* 1998;23:758–765; discussion 766.
78. Toth JM, *et al.* Evaluation of porous biphasic calcium phosphate ceramics for anterior cervical interbody fusion in a caprine model. *Spine* 1995;20:2203–2210.
79. Huang RC, *et al.* Alendronate inhibits spine fusion in a rat model. *Spine* 2005;30:2516–2522.
80. Boden SD, Schimandle JH, Hutton WC. An experimental lumbar intertransverse process spinal fusion model. Radiographic, histologic, and biomechanical healing characteristics *Spine* 1995;20:412–420.
81. Sandhu HS, *et al.* Evaluation of rhBMP-2 with an OPLA carrier in a canine posterolateral (transverse process) spinal fusion model. *Spine* 1995;20:2669–2682.

82. Xue Q, *et al.* The influence of alendronate treatment and bone graft volume on posterior lateral spine fusion in a porcine model. *Spine* 2005;30:1116–1121.
83. Baramki HG, Steffen T, Lander P, Chang M, Marchesi D. The efficacy of interconnected porous hydroxyapatite in achieving posterolateral lumbar fusion in sheep. *Spine* 2000;25:1053–1060.
84. Barnes B, *et al.* Lower dose of rhBMP-2 achieves spine fusion when combined with an osteoconductive bulking agent in non-human primates. *Spine* 2005;30:1127–1133.
85. Kai T, Shao-qing G, Geng-ting D. In vivo evaluation of bone marrow stromal-derived osteoblasts-porous calcium phosphate ceramic composites as bone graft substitute for lumbar intervertebral spinal fusion. *Spine* 2003;28:1653–1658.
86. Sandhu HS, *et al.* Distractive properties of a threaded interbody fusion device. An in vivo model. *Spine* 1996;21:1201–1210.
87. Sandhu HS, Khan SN. Animal models for preclinical assessment of bone morphogenetic proteins in the spine. *Spine* 2002;27:S32–38.
88. Boden SD, Martin GJ Jr, Horton WC, Truss TL, Sandhu HS. Laparoscopic anterior spinal arthrodesis with rhBMP-2 in a titanium interbody threaded cage. *J Spinal Disord* 1998;11:95–101.
89. Hecht BP, *et al.* The use of recombinant human bone morphogenetic protein 2 (rhBMP-2) to promote spinal fusion in a nonhuman primate anterior interbody fusion model. *Spine* 1999;24:629–636.
90. Wang T, Dang G, Guo Z, Yang M. Evaluation of autologous bone marrow mesenchymal stem cell-calcium phosphate ceramic composite for lumbar fusion in rhesus monkey interbody fusion model. *Tissue Eng* 2005;11:1159–1167.
91. Schimandle JH, Boden SD. The use of animal models to study spinal fusion. *Spine* 1994;19:1998–2006.
92. Alden TD, *et al.* Percutaneous spinal fusion using bone morphogenetic protein-2 gene therapy. *J Neurosurg* 1999;90:109–114.
93. Boden SD. Biology of lumbar spine fusion and use of bone graft substitutes: Present, future, and next generation. *Tissue Eng* 2000;6:383–399.
94. Frenkel SR, Moskovich R, Spivak J, Zhang ZH, Prewett AB. Demineralized bone matrix. Enhancement of spinal fusion. *Spine* 1993;18:1634–1639.
95. Rubino F, *et al.* Minimally invasive spine surgery: An animal model for endoscopic approach to the anterior cervical and upper thoracic spine. *J Laparoendosc Adv Surg Tech A* 2000;10:309–313.
96. Cunningham BW, *et al.* Osteogenic protein versus autologous interbody arthrodesis in the sheep thoracic spine. A comparative endoscopic study using the Bagby and Kuslich interbody fusion device. *Spine* 1999;24:509–518.
97. Pintar FA, *et al.* Fusion rate and biomechanical stiffness of hydroxylapatite versus autogenous bone grafts for anterior discectomy. An in vivo animal study. *Spine* 1994;19:2524–2528.
98. Sidhu KS, *et al.* Anterior cervical interbody fusion with rhBMP-2 and tantalum in a goat model. *Spine J* 2001;1:331–340.
99. Genant HK, Jiang Y. Advanced imaging assessment of bone quality. *Ann NY Acad Sci* 2006;1068:410–428.
100. Parfitt AM, *et al.* Bone histomorphometry: Standardization of nomenclature, symbols, and units. Report of the ASBMR Histomorphometry Nomenclature Committee. *J Bone Miner Res* 1987;2: 595–610.
101. Gerstenfeld LC, Wronski TJ, Hollinger JO, Einhorn TA. Application of histomorphometric methods to the study of bone repair. *J Bone Miner Res* 2005;20:1715–1722.
102. Athanasiou KA, Zhu C, Lanctot DR, Agrawal CM, Wang X. Fundamentals of biomechanics in tissue engineering of bone. *Tissue Eng* 2000;6:361–381.
103. Turner CH, Burr DB. Basic biomechanical measurements of bone: A tutorial. *Bone* 1993;14:595–608.
104. Weiss S, Zimmerman MC, Harten RD, Alberta FG, Meunier A. The acoustic and structural properties of the human femur. *J Biomech Eng* 1998;120:71–76.

OTHER TOOLS VII

75 Markov Processes for Biomedical Data Analysis

RICARDO OCAÑA-RIOLA

ABSTRACT

Changes over time in a random variable make up a sequence of temporal values known as a stochastic process. Neuronal activity, depression status, and viral load are just a few examples of stochastic processes of interest in the field of biomedical research. Understanding the evolution and predicting the future status of the process under certain conditions both provide key information that is extremely useful in animal experimentation and in clinical and epidemiological studies in humans. A Markov process is a particular kind of stochastic process in which its future status will depend only on its present status, not on its past history. Since the Russian mathematician Andrei Andreyevich Markov (1856–1922) first defined this kind of process in 1906, there have been numerous applications in biomedical research. This chapter shows some of the Markov models that are currently the most widely used in the fields of biomedicine and the health sciences. The content has been drafted from an eminently practical standpoint, with particular emphasis on the main techniques for analysis, interpretation of results, the usefulness of the models, and the software available for applications. Interesting references have been included in the bibliography cited at the end of the chapter for any readers keen to learn more about the more theoretical aspects of this topic.

Key Words: Markov, Stochastic process, Probability, Autoregressive models, Biomedical research, Statistical analysis.

STOCHASTIC PROCESSES

Many of the variables used in biomedical research exhibit values that change over time. In rat breast cancer models, response to the carcinogen can be evaluated by recording the number of animals that have developed cancer on a monthly basis or by measuring tumour growth every week until the end of the experimental period.[1] In humans, blood pressure, the number of CD4 cells, or the status of patients during the course of their disease are all variables that show changing values over time giving rise to a time sequence of observations for each of them.

Generally speaking, the value of a variable X at time t is usually expressed as $X(t)$, such that $X(t_1), \ldots, X(t_k)$ form a sequence of values over time called a *stochastic process*.[2] The values that the variable may show are known as states and the process, as it develops over time, will be subject to changes in state and transitions between different states. In the usual terminology, *state space* refers to all possible values that the variable defining the process may take.

An analysis of stochastic processes will enable us to predict the state of a given process in the future, based on information available on its past states or values. The complexity of such predictions depends largely on the kind of process involved. However, the characteristics of some of these make them easier to analyze and interpret. The Markov process is a particularly well studied case, i.e., a stochastic process in which future status will depend solely on present status but not on past history.[3,4]

Since the Russian mathematician Andrei Andreyevich Markov (1856–1922) first defined this kind of process in 1906, there have been numerous applications in biomedical research. Both in animal experimentation[5–8] and in human studies,[9–12] the Markov models still prove to be effective today and remain the most widely used methods to study transitions between different states of a process and how these develop over time.

The methodology used for data analysis will differ according to the particular conditions set for each study. *Markov chains* are referred to when the values that the process may take are discrete, while the term *Markov process* is generally reserved for processes with a continuous state space. Although there is no standardized terminology, the general classification of Markov models can be outlined as shown in Table 75–1.

Many researchers today continue to explore new applications for stochastic processes in the field of biomedicine and the health sciences, giving rise to a major area of theoretical and applied research. Of all these processes, it is the Markov chains, both in discrete and continuous time, that are most widely used in biomedical research.[13] For this reason, these will be studied in greater detail in this chapter. Markov processes entail a greater mathematical complexity and their analysis falls beyond the scope of this chapter. However, the bibliography cited at the end of the chapter may be of interest to the keen reader.[2,14,15]

From: *Sourcebook of Models for Biomedical Research* (P. M. Conn, ed.), © 2008 Humana Press Inc., Totowa, NJ.

Table 75–1
Classification of Markov models

		State space	
		Discrete	*Continuous*
Observation time	Discrete	Discrete Markov chain	Discrete Markov process
	Continuous	Continuous Markov chain	Diffusion process

MARKOV CHAINS IN DISCRETE TIME

When the study variable can take only discrete values and observation is performed discontinuously over time, at discrete moments in time, the stochastic process can be represented by $X(1), X(2), X(3)$, where $X(n)$ is the value for the variable X in time n, where $n = 1, 2, 3. \ldots$

If this is a Markov chain, the future state will depend only on the present state and not on its past states or values. In this case, the probabilities of transition from one state to another are defined as

$$p_{ij}(n) = P[X(n+1) = j|X(n) = i]\ i,j = 1, \ldots, m$$

where $p_{ij}(n)$ is the probability that variable X will take value j at observation time $n + 1$ if its value currently, at time n, is i. In other words, $p_{ij}(n)$ is the probability that the process will change from state i to state j at the time n.

The evolution of the Markov chain can be determined by ascertaining these probabilities that form the transition matrix $P(n)$:

$$P(n) = \begin{pmatrix} p_{11}(n) & p_{12}(n) & \cdots & p_{1m}(n) \\ p_{21}(n) & p_{22}(n) & \cdots & p_{2m}(n) \\ \vdots & \vdots & & \vdots \\ p_{m1}(n) & p_{m2}(n) & \cdots & p_{mm}(n) \end{pmatrix}$$

Since we are dealing with probabilities, all the components in the transition matrix must lie between 0 and 1, while the sum of the values in each row must equal 1. This is known as a *stochastic matrix*.

In the process described, the transition probabilities change over time, i.e., there is a distinct transition matrix for each observation time. This is what happens in viral epidemics such as chickenpox outbreaks where the probability of changing from being ill to being healthy will be almost zero on the first day of the infection and will approach 1 between 7 and 10 days after the onset of symptoms.[16] When this occurs, the discrete Markov chain will be *nonhomogeneous*.

When the transition probabilities do not change over time, the probability of passing from state i to state j will be the same at all times during the observation of the process. In this case, there will be a single transition probability p_{ij} that is constant over time such that $p_{ij} = p_{ij}(1) = p_{ij}(2) = \ldots = p_{ij}(n)$. Then the process will be a *homogeneous* or *stationary* Markov chain in discrete time, with a single transition matrix expressed as follows:

$$P = \begin{pmatrix} p_{11} & p_{12} & \cdots & p_{1m} \\ p_{21} & p_{22} & \cdots & p_{2m} \\ \vdots & \vdots & & \vdots \\ p_{m1} & p_{m2} & \cdots & p_{mm} \end{pmatrix}$$

ANALYSIS OF HOMOGENEOUS MARKOV CHAINS IN DISCRETE TIME Research related to the dynamics of populations exposed to infectious diseases is intended to shed light on the pattern of the epidemic when an infected subject enters the population, to study the rate of viral propagation, and to estimate the number of individuals who will be infected during the epidemic.

Since the early 1980s, AIDS has become one of the greatest pandemics of our time. Spain is currently one of the most severely affected Western European countries.[17] As a result, understanding the future developments of the epidemic has become a priority that should enable appropriate health policies to be implemented.

The following sections will describe how the HIV/AIDS epidemic in Spain can be analyzed based on Markov chains with yearly observation intervals.

Defining the States of the Process and Transition Mechanisms between Them One of the simplest models to study the AIDS epidemic establishes that each subject in the population may be in only one of the following states: susceptible (S), HIV-infected (HIV), with AIDS (AIDS), or died as a result of the disease (D).

The susceptible subject is not infected (state 1, S). After contact with an infected person, the subject will develop the infection and may infect other individuals in the population (state 2, HIV). An infected subject who develops the disease will become an AIDS case (state 3, AIDS), and may die as a result of the disease (state 4, D). Figure 75–1 shows the possible transitions between states together with the theoretical transition probabilities.

States "S," "HIV," and "AIDS" are called *transitory*, given that individuals will never go back to these states once they have gone on to another stage in the disease. State "D," however, is an *absorbent state,* as the subjects reaching this state remain in it and have no possibility of changing to another state.

This epidemic model is a variant of the SIR and Reed-Frost models that have been widely used to study dynamics in infectious processes in both human and nonhuman populations.[18,19]

Selection of the Most Suitable Markov Model to Study the Process Subjects are observed on a yearly basis so that both the states and the moments in time of the process are discrete. Moreover, the future state individuals may be in will depend only on their current state, not on their past history. As a result, the stochastic process is a discrete Markov chain.

If no changes are made to population health habits, HIV prevention policies, and treatment guidelines, both the incidence of HIV-AIDS and the mortality rate for AIDS cases will remain constant. As a result, there will no be change over time in

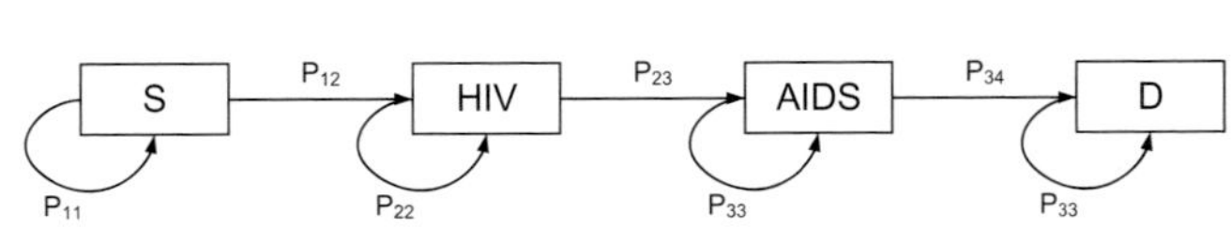

Figure 75–1. States in the HIV-AIDS epidemic.

transition probabilities. According to this idealistic hypothesis, the discrete Markov chain will also be homogeneous. In this case, this is the most appropriate Markov model to study how the epidemic will evolve.

Determining Transition Probabilities between States In human epidemics, the transition matrix is usually derived either from the data published on the population, incidence, prevalence, and mortality, or by means of specific purpose-designed research efforts, usually based on longitudinal studies. The most recent figures available for the HIV-AIDS epidemic in Spain date to 2004.[17] This information will be used to calculate the transition probabilities, starting at the first row of the matrix.

According to data published in Spain, the estimated number of new cases in 2004 was 2750 in a population of 43,197,684 inhabitants, i.e., a contagion rate of 64 cases per million inhabitants.[20] Consequently, $p_{12} = 0.000064$. Since a subject in a "susceptible" state cannot directly pass on to the "AIDS" or the "D" states, then $p_{13} = p_{14} = 0$. The sum of probabilities in a row of the transition matrix must be 1, giving $p_{11} = 1 - p_{12} - p_{13} - p_{14} = 0.999936$.

The second row of the matrix refers to the transitions that may occur from the HIV state. Thus, an infected individual will not revert to being susceptible, so $p_{21} = 0$. The incidence for AIDS in 2004 was 804 new cases[17] in a prevalent population estimated at 125,000 HIV-infected individuals,[21] giving $p_{23} = 0.006432$. An HIV^+ subject cannot go directly to state "D," so $p_{24} = 0$. Moreover, given that the values for the row must add up to 1, then $p_{22} = 1 - p_{21} - p_{23} - p_{24} = 0.993568$.

Those subjects who have developed AIDS will not return to either the "S" or "HIV" states, therefore $p_{31} = p_{32} = 0$. In 2004, there were 542 deaths from a total of 6609 AIDS cases,[17] which gives $p_{34} = 0.0082$. Finally,

$$p_{33} = 1 - p_{31} - p_{32} - p_{34} = 0.917991.$$

Since "D" is an absorbent state and the subjects remain in this state, then $p_{44} = 1$ and $p_{41} = p_{42} = p_{43} = 0$.

As a result, the transition matrix for the HIV-AIDS epidemic in Spain will be

$$P = \begin{matrix} \\ S \\ HIV \\ AIDS \\ D \end{matrix} \begin{pmatrix} S & HIV & AIDS & D \\ 0.999936 & 0.000064 & 0 & 0 \\ 0 & 0.993568 & 0.006432 & 0 \\ 0 & 0 & 0.917991 & 0.0082 \\ 0 & 0 & 0 & 1 \end{pmatrix}$$

Analyzing the Evolution of the Epidemic over Time The transition matrix contains the probabilities for passing from one state to another in a single year, i.e., the probability that a susceptible subject is infected by the next year is 0.000064. However, to understand how the epidemic will evolve, it is necessary to calculate the probability that a subject who is susceptible today may be infected within 2 years. The Capman–Kolmogorov equations[3] enable these probabilities to be calculated by multiplying the transition matrix by itself, giving rise to a two-step transition matrix, i.e.,

$$P^{(2)} = P \times P \begin{pmatrix} 0.999872 & 0.000128 & 0.0000004 & 0 \\ 0 & 0.987177 & 0.012295 & 0.000053 \\ 0 & 0 & 0.842708 & 0.015728 \\ 0 & 0 & 0 & 1 \end{pmatrix}$$

Thus, the probability that a subject who is susceptible today may be infected within 2 years is 0.000128 or, in other words, there will be 128 *de novo* infections in 2 years. Similarly, the probability that individuals can change from their current state to another in 3 years can be calculated from the three-step transition matrix $P^{(3)} = P^{(2)} \times P$. Generally speaking, the transition matrix in n steps is merely the product of successive matrices, with n being a whole number greater than or equal to 2.

The course of developments in the epidemic over time will depend on the initial conditions present for the population under study vis-à-vis the virus, i.e., the current distribution of individuals in each of the disease states. In 2004 there were 43,065,533 susceptible subjects in Spain, 125,000 HIV infected and alive, 6609 AIDS cases, and 542 AIDS-related deaths. These figures make up the vector of initial prevalent cases that mark the starting point for the study of the future developments in the epidemic. This vector is represented by the sum of all its values, and must coincide with the overall population under study, taken as 43,197,684 inhabitants. The number of subjects that will be in each of the disease states within n years is derived from the product of the vector for initial prevalences and the transition matrix in n steps, i.e., $v \times P^{(n)}$.[2,13]

In 2006, 2 years after the latest recorded information, the number of subjects in each of the states will be $v \times P^{(2)}$ = (43 060 021, 128 892, 7124, 653). Thus, the number of infected cases in 2006 will have increased to 128,892, the number of AIDS cases will be 7124 and the AIDS-related deaths will amount to 653. The number of people susceptible to contagion will be 43,060,021.

The vector of initial prevalences may also be expressed in terms of probabilities or relative frequencies, dividing each component among the overall population. In this case, we would have v = (0.99694, 0.002894, 0.000153, 0.000013) and $v \times P^{(2)}$ would result in the probability of being in one of the states 2 years after commencement of the study.

When studying population epidemics, the vector for initial prevalences is usually expressed as a number of cases per one million inhabitants. In this case, assuming an initial population of 1,000,000 people, there would be v = (996940, 2894, 153, 13). Figure 75–2 shows the development of the epidemic using this vector as the starting point and calculating rates per million for each of the disease states up to 2024, i.e., 20 years after commencement of the study. For each year, the rate is derived from $v \times P^{(n)}$ with $n = 1, \ldots 20$.

If there is no change in the incidence and mortality for HIV-AIDS, the trend for HIV and AIDS cases will be on the increase. In 2024, the number of HIV infections will have risen to 3744 per million inhabitants, the number of people living with AIDS will amount to 248 per million inhabitants, while AIDS-related deaths will rise to 46 per million inhabitants. These figures mean that in 20 years, the number of *de novo* infections will have increased by 29%, AIDS cases will rise to 62%, and AIDS-related deaths will be 154% higher.

Different simulations for the hypothetical evolution of the epidemic in Spain can be derived by altering the vector for initial prevalences. Understanding what might happen in the future if the current situation were different provides a major source of information for decision-making purposes. This is one of the greatest contributions to biomedical research made by Markov models.

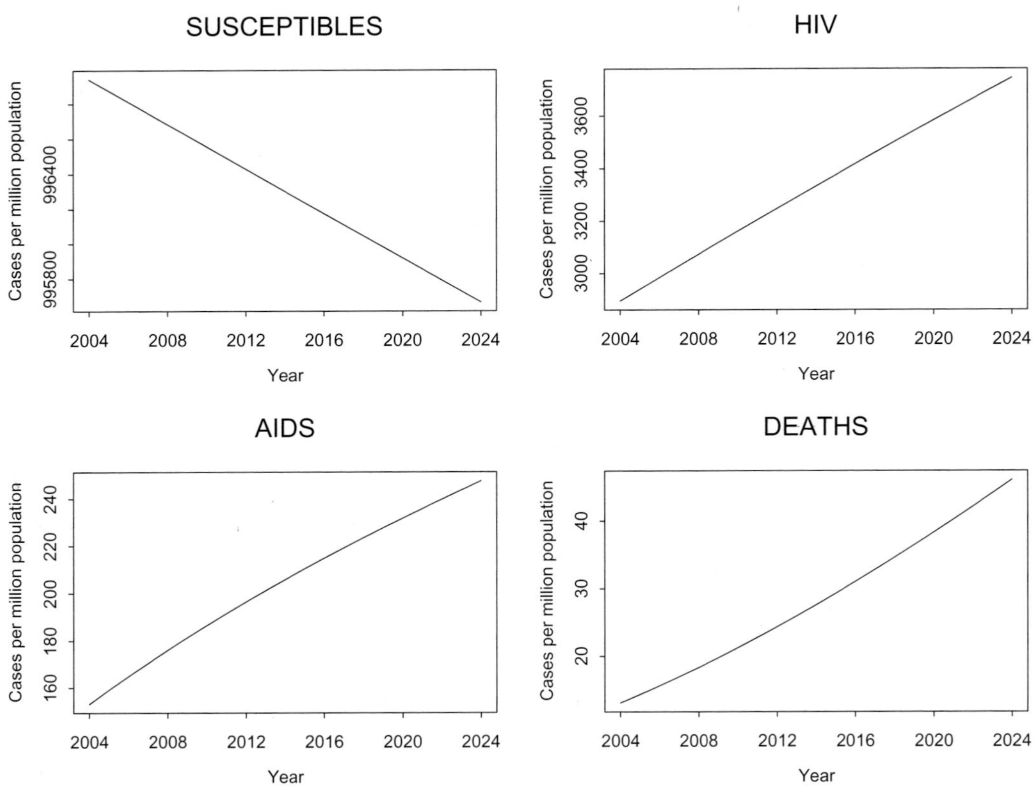

Figure 75–2. Predictions on the evolution of the HIV-AIDS epidemic in Spain.

ANALYSIS OF NONHOMOGENEOUS MARKOV CHAINS IN DISCRETE TIME Certain animal models for spatial learning in rats use a T-shaped maze with a single starting point and two exits placed on either side of the maze. Food or water is placed at one of these exits, such that the rat receives positive reinforcement every time it goes to this particular side of the maze.[22,23]

Let us suppose that the following experiment was repeated on five consecutive occasions with 20 adult rats. The side of the maze chosen by each rat was noted for each of the repetitions of the experiment, and recorded as correct (C) if this led to positive reinforcement or as incorrect (I) otherwise. The aim was to check whether the rats learned to go toward the side where the food was placed as well as to study the developments in the learning curve over time.

Definition of the States in the Process and Transition Mechanisms between States In the animal learning model, the space of states comprises two alternatives: choosing the right (state 1, C) or the wrong (state 2, I) side. Every time the experiment is repeated, the rat can choose either of the two options and can also repeat the same option several times consecutively. Figure 75–3 shows the possible transitions between both states called *recurrent*, since the rat, after selecting any of the states, could choose the same state again in later repetitions of the experiment.

Selection of the Most Suitable Markov Model to Study the Process The experiment is repeated five times consecutively, taking each repetition as an observation time. In this case, both the states and moments in time in the process are discrete. In addition, the side of the maze chosen by the rat in the following repetition will depend only on the current choice and not on past decisions. As a result, the stochastic process comprises a discrete Markov chain.

If the repetition of the experiment makes the animal learn to choose the right route, the probability of guessing correctly on the next repetition if the preceding choice had been wrong will be different at the beginning and at the end of the experiment. In line with this hypothesis, the probabilities of a transition between states will not be constant over time. As a result, the discrete Markov chain will also be nonhomogeneous. Consequently, there will be a transition matrix for each of the repetitions in the experiment.

Determination of the Probabilities of Transition between States In models of this kind, transition probabilities are usually estimated through longitudinal studies designed for this purpose. In this case, the experiment was conducted on 20 rats, 10 (50%) of which chose the right side of the maze at the first attempt (Table 75–2). All 10 received positive reinforcement.

When the experiment was repeated for the first time, 6 of the 10 rats that had initially chosen the right side of the maze chose this option again, thus $p_{11}(1) = 0.60$ $p_{11}(1) = 0.60$. Despite having chosen the right route initially, the remaining rats took the wrong route on the first repetition, such that $p_{12}(1) = 0.40$ $p_{12}(1) = 0.40$ (Table 75–2).

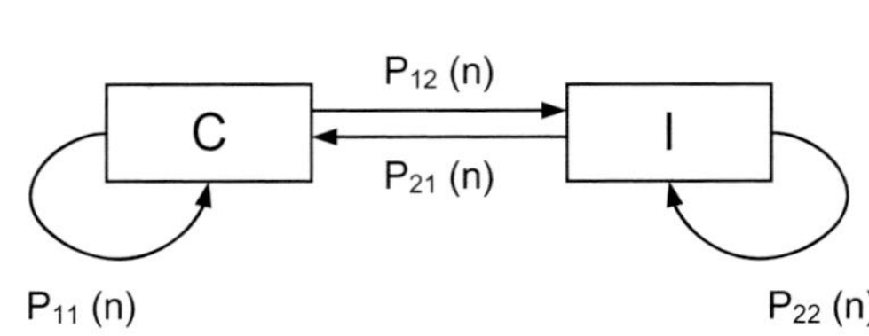

Figure 75–3. States in an animal learning model in rats.

Table 75–2
Transition matrices in an animal model for spatial learning in rats.[a]

Initial probabilities				*First repetition*				*Second repetition*				*Third repetition*				*Fourth repetition*		
				Time 1 P(1)				*Time 2 P(2)*				*Time 3 P(3)*				*Time 4 P(4)*		
C	*I*			*C*	*I*			*C*	*I*			*C*	*I*			*C*	*I*	
			C	6 (0.60)	4 (0.40)	10	C	6 (0.75)	2 (0.25)	8	C	9 (0.82)	2 (0.18)	11	C	14 (0.93)	1 (0.07)	15
			I	2 (0.20)	8 (0.80)	10	I	5 (0.42)	7 (0.58)	12	I	6 (0.67)	3 (0.33)	9	I	4 (0.80)	1 (0.20)	5
10 (0.50)	10 (0.50)	20		8	12	20		11	9	20		15	5	20		18	2	20

[a]Simulated data for didactic purposes. The cells of each matrix contain the number of rats that has gone from one state to another. In parentheses are the transition probabilities.

Of the 10 rats that originally took the wrong route, 2 changed their option in the repetition, thus $p_{21}(1) = 0.20$ $p_{21}(1) = 0.20$. As these are complementary, then $p_{22}(1) = 0.80$ $p_{22}(1) = 0.80$ (Table 75–2).

Analysis of Developments in Learning As in the case of the homogeneous Markov chains, the Chapman–Kolmogorov equations[3] enable the transition matrix in n times to be calculated, expressed by $P^{(n)}$, so that we can ascertain the probability of a rat choosing the right side of the maze on the nth repetition (time n) if it chose the wrong route on the first repetition. This transition matrix in n times is derived from the product of all the preceding transition matrices in time n, i.e., $P^{(n)} = P(1) \times P(2) \times \ldots \times P(n)$. Thus, the two-step transition matrix is expressed as

$$P^{(2)} = P(1) \times P(2) = \begin{pmatrix} 0.60 & 0.40 \\ 0.20 & 0.80 \end{pmatrix} \times \begin{pmatrix} 0.75 & 0.25 \\ 0.42 & 0.58 \end{pmatrix} = \begin{pmatrix} 0.62 & 0.38 \\ 0.49 & 0.51 \end{pmatrix}$$

such that if a rat chose the wrong route in the maze on the first repetition, the probability of choosing correctly on the second repetition is $p_{21}^{(2)} = 0.49$.

The three-step and four-step transition matrices will be $P^{(3)} = P^{(2)} \times P(3)$ and $P^{(4)} = P^{(3)} \times P(4)$, respectively. This concept is similar to that seen for homogeneous Markov chains. The only difference is that previously, as a homogeneous process, the transition matrices multiplied were identical given that they do not change over time.

The initial conditions for the process will set the learning curve for the rats, such that if 50% of the animals initially choose the right route in the maze and 50% the wrong route, the vector for initial probabilities is $v = (0.50, 0.50)$.

The probability of choosing the right or wrong route when the experiment has been repeated n times is the result of the product of the vector of initial probabilities and the transition matrix in n steps, i.e., $v \times P^{(n)}$ where $n = 1,2,3,4$.

Figure 75–4 shows the learning curve for the rats, where we can see the growing trend toward choosing the right route as the experiment is repeated again and again. After four repetitions, the probability that a rat will choose the route in the maze leading to the positive reinforcement is 0.90 and only 10% of the animals will choose the opposite option.

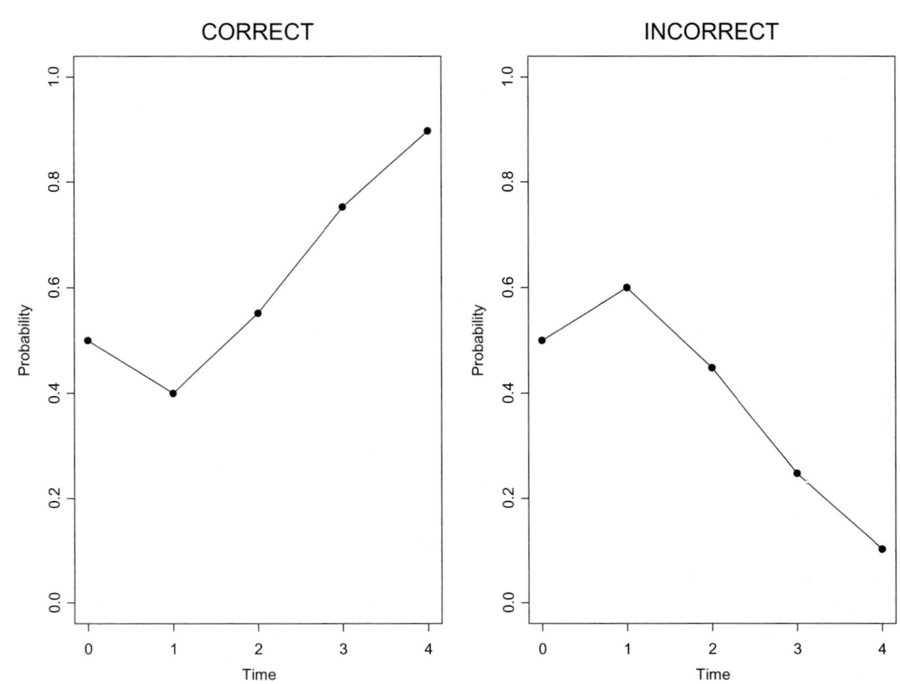

Figure 75–4. Learning curve in an experimental spatial learning model in rats.

MARKOV CHAINS IN CONTINUOUS TIME

Occasionally, changes in the status of a variable may arise at any time, and not only in discrete times. As a result, the stochastic process is represented by a succession of values $X(t)$ that represent the value for variable X at any moment in time t, where $t \geq 0$.

Similar to discrete Markov chains, if the process is Markovian, the probabilities of transition from one state to another are now defined as

$$p_{ij}(t,t+h) = P[X(t+h) = j|X(t) = i] \quad i,j = 1,2,\ldots,m$$

Thus, if in time t the variable is in state i, $p_{ij}(t,t+h)$ is the probability that a change in state j may occur between t and $t + h$, where h is a very small increase in time approaching zero. Indeed, these are *instant probabilities* of transition.

In the process described, the value of these transition probabilities depends on moment t when the change arises, i.e., the instant probability of going from state i to state j is different at time $t = 0$, $t = 1.72$, $t = 4.21$, etc. As a result, we will have a *nonhomogeneous* continuous Markov chain.

However, if the probabilities of transition between states do not depend on time, then we will have a *homogeneous* or *stationary* continuous Markov chain. In this case,

$$p_{ij}(h) = P[X(t+h) = j|X(t) = i] \quad i,j = 1,2,\ldots,m$$

Both homogeneous and nonhomogeneous types of model have been applied in genetic studies,[24,25] in research on viral load in HIV-positive patients,[26] or in variations in the status of asthmatic patients,[27,28] among other fields. However, it is not as straightforward to estimate the probabilities of transitions and to plot the graphs for evolution of the process over time as in discrete Markov chains, especially for nonhomogeneous models.

Over the past few decades, different techniques to analyze this kind of stochastic process have been proposed. Among these are the resolution of differential equations,[3,4] algorithms based on panel data,[29,30] parametric measurements, and nonparametric approaches.[31,32] Many of these methods are based on complex mathematical equations that are not included in the computer programs generally used to analyze biomedical data. For this reason, specific macros have been developed over the past few years enabling some of these models to be applied with less effort.[33]

Currently, both continuous time Markov chains and Markov processes with a continuous state space comprise major areas for research. Of particular interest is the development of new statistical methods and the implementation of specific software that can facilitate data analysis.

ADDITIONAL FEATURES OF MARKOV CHAIN ANALYSIS

Before the above-mentioned methods can be applied, we must first check that the process fulfils the Markov property. With the right parameterizations, any stochastic process can be expressed as an autoregressive generalized linear model in which past values act as variables accounting for the current state of the process.[34] To check whether the chain is Markovian, we need only to check that all the regression coefficients in the model are zero except the coefficient for the state of the process at the preceding moment in time. In this case, we will have an autoregressive model of the first order, equivalent to the Markov property. In general terms, stationary Markov chains are easier to model than nonhomogeneous ones. However, it is not good practice to assume the process is homogeneous without a prior check to guarantee the appropriate use of stationary models. The test proposed by Kalbfleisch and Lawless[29] is generally the most widely used to check this hypothesis. In this test, the number of transitions observed at each time interval is compared with the number of expected transitions according to the hypothesis of homogeneity. This gives a likelihood ratio statistic based on a chi-square distribution.

Any software that enables macros to be programmed can be used to analyze any kind of Markov chain. *SAS* (www.sas.com), *Stata* (www.stata.com), *SPLUS* (www.insightful.com), or *R* (www.r-project.org) are the most widely used programs for this purpose. The fact that specific routines have already been devised for some of these provides a further advantage for researchers.[2,33] In any case, basic computer programming skills are required to handle these programs, especially if we intend to implement new macros that have not yet been devised by other authors.

Other specific software, such as *Relex Markov* (www.relex.com), *DecisionPro* (www.vanguardsw.com), or *TreeAge Pro* (www.treeage.com) have been devised exclusively to analyze Markov chains. These are easy-to-use programs, although some are intended only for the analysis of Markov chains in discrete time and tend to pose certain limitations on the estimation of transition probabilities from a database with information from a longitudinal study.

Whether one software or another is used, be it specific for Markov chains analysis or for general statistics, will depend largely on the complexity of the study itself and the researcher's requirements.[35,36]

ACKNOWLEDGMENTS

This work has been supported in part by the Network of Public Health and Epidemiology Research Centres (RCESP). The author thanks Maria Moya, Alberto Fernández, and Gracia Maroto for very helpful comments on earlier drafts of this chapter.

REFERENCES

1. Thompson H, Sporn MB. Mammary cancer in rats. In: Teicher BA, Ed. *Tumor Models in Cancer Research.* Totowa, NJ: Humana Press, 2002.
2. Lindsey JK. *Statistical Analysis of Stochastic Processes in Time.* Cambridge, UK: Cambridge University Press, 2004.
3. Feller WF. *An Introduction to Probability Theory and Its Applications*, 3rd ed., Vol. I. New York: John Wiley & Sons, 1968.
4. Drake AW. *Fundamentals of Applied Probability Theory.* New York: McGraw-Hill, 1988.
5. Uchida G, Fukuda M, Tanifuji M. Correlated transition between two activity state of neurons. *Phys Rev E Stat Nonlin Soft Matter Phys* 2006;73(3Pt. 1):031910.
6. Yang HC, Chao A. Modelling animal's behavioural response by Markov chain models for capture-recapture experiments. *Biometrics* 2005;61:1010–1017.
7. Zhai J, Morris RW. A Markov chain model for animal estrus cycling data. *Biometrics* 2005;61:141–150.
8. Carey JR, Liedo P, Muller HG, Wang JL, Zhang Y, Harshman L. Stochastic dietary restriction using a Markov chain feeding protocol elicits complex, life history response in medflies. *Aging Cell* 2005;4:31–39.
9. Kopycka-Kedzierawski DT, Billings RJ. Application of nonhomogeneous Markov models for analysing longitudinal caries risk. *Community Dent Oral Epidemiol* 2006;34:123–129.

10. Berg AT, Lin J, Ebrahimi N, Testa FM, Levy SR, Shinnar S. Modelling remission and relapse in pediatric epilepsy: Application of a Markov process. *Epilepsy Res* 2004;60:31–40.
11. Alexandersson M, Cawley S, Pachter L. SLAM: Cross-species gene finding and alignment with a generalized pair hidden Markov model. *Genome Res* 2003;13:496–502.
12. Trajstman AC. A Markov chain model for Newcastle disease and its relevance to the intracerebral pathogenicity index. *Biom J* 2002;44: 43–57.
13. Norris JR. *Markov Chains*. Cambridge, UK: Cambridge University Press, 1998.
14. Tuerlinckx F, Maris E, Ratcliff R, De Boeck P. A comparison of four methods for simulating the diffusion process. *Behav Res Methods Instrum Comput* 2001;33:443–456.
15. Karatzas I, Schreve SE. *Brownian Motion and Stochastic Calculus*, 2nd ed. New York: Springer-Verlag, 1991.
16. Arvin AM, Gershon AA. *Varicella-Zoster Virus: Virology and Clinical Management*. Cambridge, UK: Cambridge University Press, 2001.
17. EuroHIV. HIV/AIDS Surveillance in Europe (End-year report 2004). Saint-Maurice: Institut de Veille Sanitaire, 2005.
18. Chen WY, Bokka S. Stochastic modelling of nonlinear epidemiology. *J Theor Biol* 2005;234:455–470.
19. Hurd HS, Kaneene JB. The application of simulation models and systems analysis in epidemiology: A review. *Prev Vet Med* 1993;15: 81–99.
20. Centro Nacional de Epidemiología. *Valoración de la epidemia de VIH en España a partir de los sistemas de notificación de casos de las Comunidades Autónomas*. Madrid: CNE, 2005.
21. Cañas EJ, García-León FJ, Andérica-Frías G. Epidemiología de la infección VIH y SIDA. In: Pachón J, Pujol E, Rivero A, Eds. *La infección por el VIH: Guía práctica,* 2nd ed. Seville: Sociedad Andaluza de Enfermedades Infecciosas, 2003.
22. Dickinson A. *Contemporary Animal Learning Theory*. Cambridge, UK: Cambridge University Press, 1980.
23. Domjan M, Burkhard B. *The Principles of Learning and Behavior*. Monterey, CA: Brooks-Cole, 1986.
24. Tyvand PA, Thorvaldsen S. Markov model of haploid random mating with given distribution of population size. *Bull Math Biol* 2006;68: 807–819.
25. De Gruttola V, Foulkes AS. Validation and discovery in Markov models of genetics data. *Stat Appl Genet Mol Biol* 2004;3:Article 38.
26. Mathieu E, Loup P, Dellamonica P, Daures JP. Markov modelling of immunological and virological states in HIV-1 infected patients. *Biom J* 2005;47:834–846.
27. Saint-Pierre P, Bourdin A, Chanez P, Daures JP, Godard P. Are overweight asthmatics more difficult to control? *Allergy* 2006;61: 79–84.
28. Combescure C, Chanez P, Saint-Pierre P, Daures JP, Proudhon H, Godard P. Assessment of variations in control of asthma over time. *Eur Respir J* 2003;22:298–304.
29. Kalbfleisch JD, Lawless JF. The analysis of panel data under a Markov assumption. *J Am Stat Assoc* 1985;80863–80871.
30. Ocaña-Riola R. Non-homogeneous Markov processes for biomedical data analysis. *Biom J* 2005;47:369–376.
31. Commenges D. Inference for multi-state models from interval-censored data. *Stat Methods Med Res* 2002;11:167–182.
32. Frydman H. A nonparametric estimation procedure for a periodically observed three-state Markov process with application to AIDS. *J Roy Stat Soc B* 1992;54:853–866.
33. Hui-Min W, Ming-Fang Y, Chen TH. SAS macro program for non-homogeneous Markov process in modelling multi-state disease progression. *Comput Methods Programs Biomed* 2004;75:95–105.
34. Hardin J, Hilbe J. *Generalized Linear Models and Extensions*, 2nd ed. College Station, TX: Stata Press, 2006.
35. Hazen GB. Stochastic trees and the StoTree modeling environment: Models and software for medical decision analysis. *J Med Syst* 2002;26:399–413.
36. Adalsteinsson D, McMillen D, Elston TC. Biochemical Network Stochastic Simulator (BioNetS): Software for stochastic modeling of biochemical networks. *BMC Bioinformatics* 2004;5:24.

76 Software Tools for Modeling Biomedical Systems

KARL THOMASETH

ABSTRACT

Biomedical applications of computer modeling and simulation are manifold and range from virtual reality for training purposes to codification of knowledge of complex physiological systems to formulation and testing of hypotheses about biological phenomena. While development of complex simulation tools and comprehensive modeling requires considerable effort, the use of simple models can, even with minor investments, be useful to experimenters to organize ideas, analyze data efficiently, and plan experiments. Even for this restricted modeling framework, the variety of possible applications with different needs and the number of available software tools are very large. This chapter will therefore draw attention only to a few methodological issues and available software tools specific to biomedical modeling, in particular, model representation paradigms and statistical identification approaches. The intent is not to endorse the use of any particular modeling approach or software tool, but rather to delineate, with a few reference points, a broader modeling perspective that may be tailored to the needs of particular modeling studies on biomedical systems.

Key Words: Mathematical models, Physiologically based models, Minimal models, Parameter estimation, Statistical models, Hierarchical models.

INTRODUCTION

Computer simulation is used in all fields of scientific research to mimic the dynamic behavior of real systems under various operating conditions. This is accomplished by means of models that formally embed knowledge and hypotheses about the functioning of the systems. Simulation thus consists of reproducing real world behavior by operating on a model in a virtual environment, making it possible to test and to better understand the implications of beliefs about the actual system even in situations that would not be feasible in practice. Although simulation models are synthetic conceptual representations of reality, typically described in the form of mathematical equations, if they capture either explicitly or implicitly basic physical characteristics of actual systems they can produce valid predictions of their behavior even in broad ranges of testing situations. Because systems are in general entities that can be controlled and observed from the environment only through special inputs and outputs, the external validity of simulation models can be judged only in terms of the correspondence between simulated predictions and actual system responses observed under comparable input conditions. Nevertheless, confidence in simulation outcomes that cannot be readily verified in practice can still rely on the intrinsic validity of the assumptions embedded in the model. The formulation of simulation models is therefore simplified if the behavior of the system is governed by general physical or empirical laws that allow an accurate prediction of the system's response by means of mathematical equations. This approach is usually attainable and actually implemented with man-made systems, e.g., from simple products up to large industrial plants, enabling designers and users to predict the dynamic behavior of the assembly before its first use or even during the planning phase to identify and correct limitations and drawbacks. It is for this reason that early developments of computer simulation were made within space flight programs.

VIRTUAL REALITY Simulation is also useful to gain insight into the characteristics and limitations of physical systems that have to be manipulated or controlled by human operators. Simulators in biomedicine, e.g., for virtual surgery and clinical treatments,[1] may become tools that are as normal in the education of surgeons and medical staff as flight simulators currently are in the training of pilots. In these simulators, the analogy to reality is not only related to the accuracy of the model responses, but also to the mechanisms provided for interaction with the simulation system. Thus, the development of realistic human interfaces plays a major role in the design of such simulation systems.

KNOWLEDGE REPRESENTATION A different application of computerized models consists of the systematic collection and representation of knowledge about physical systems without the immediate goal of numerical simulation. An example is the bioinformatics Biological Systems Database resource (KEGG),[2] which provides access to a large amount of information, for instance, on the molecular structures of chemical compounds and drugs, and on the interactions of substrates, enzymes, and genes in metabolic pathways in different species. Available information in KEGG already makes it possible, for instance, to automatically generate by computer software mathematical equations of biochemical enzyme reactions, in particular metabolic pathways. However, to effectively simulate such reactions, numerical values should be assigned to substrates and to enzyme concentrations as well as to dissociation and rate constants appearing in the dynamic

From: *Sourcebook of Models for Biomedical Research*
(P. M. Conn, ed.), © 2008 Humana Press Inc., Totowa, NJ.

equations. These are, however, mostly unavailable. Therefore, qualitative knowledge about the structure of a biological system, e.g., metabolic pathways, combined with general laws, e.g., mass action and enzyme kinetics, is sufficient to formulate a theoretical model and to write down its mathematical equations. However, additional quantitative information is required, e.g., on model parameters, to predict through computer simulation real life situations and to possibly validate the theoretical model against experimental data.

The most extensive undertaking for the integration of models of human physiology is the so-called Physiome Project, promoted by the International Union of Physiological Sciences.[3,4] The aim is the comprehensive codification, within a multiscale modeling framework, of current knowledge on biological structures and their functions at various levels of aggregation, ranging from molecules and cells, through tissues, organs, and body regions, up to the whole organism. The challenging computational modeling problems include the linking of properties of biological structures described in a hierarchical way at different levels of detail, and the integration of multidomain characteristics of biological material, e.g., mechanical, electrical, and biochemical, within time-varying, nonhomogeneous spatially distributed representations of substructures.

STRUCTURAL MODELING A more heuristic approach to describing the dynamic properties of generic systems is the so-called lumped-parameter modeling method, in which microscopic and regional aspects of materials and processes are not taken into consideration in favor of minimalistic representations. Typical biomedical examples are compartmental models that lump, in a finite number of state variables, substances distributed in anatomical regions and/or chemical binding states assuming that concentrations are homogeneous within the compartment's distribution volumes and that mixing, i.e., the equilibration of concentration after the supply or removal of the substance, occurs instantaneously. The seemingly oversimplification of the reality of lumped-parameter models is, however, fully compensated by their practical usefulness due to a higher level of abstraction that leads to simpler, more tractable mathematical model equations. This makes lumped-parameter models suitable for model identification, i.e., determination of a model that best fits experimental data, which can be a valuable complement to scientific research. In particular, if a model is structured consistently with knowledge or belief about the functioning of the system being studied, then estimated values of parameters provide quantitative information about particular functions of the system. This makes it possible, for instance, to assess and interpret the effects of external interventions, e.g., drug administration, by planning, carrying out, and model-based analysis of repeated experiments under different treatment conditions. More generally, if a particular structured model describes experimental data more accurately than similarly credible competing models, then even reasonable hypotheses can be put aside in favor of others. Mathematical models can therefore contribute to the advancement of knowledge in biomedical research insofar as they lead to the formulation and testing of hypotheses in conjunction with experimental observations.

In relation to this latter modeling framework, the following sections will first stress the importance of using structured models with physiological interpretation, then recall some graphic and textual conventions used in various software tools for specifying models of dynamic biomedical systems, and finally discuss principal statistical methods that should be provided by modeling tools designed for biomedical applications.

BIOMEDICAL SYSTEMS REPRESENTATIONS

EQUIVALENCE AND DISTINCTION OF MODELS Experiments consist in provoking with external inputs a response of a system to reveal internal characteristics that cannot be accessed or observed directly. Observations collected in dynamic, i.e., time-varying, conditions are clearly more informative about the system's structure than those collected at steady state, but require model-based calculations to extract quantitative information from the data. For this purpose a mathematical model of a given system-experiment pair can in general be expressed as a function of time

$$y(t) = f[t, u(t), p] \qquad (76\text{–}1)$$

where, at the generic time instant t, $y(t)$ represents the array of all measurable quantities, $u(t)$ the array of all external, possibly time-varying, inputs and events that characterize that particular experiment, and p represents the array of model parameters. Actual measurements are in general noisy and possibly taken at discrete time points

$$z(t_s) = y(t_s) + e(t_s) \qquad (76\text{–}2)$$

where $e(t_s)$ is measurement noise and the generic observation time point t_s belongs to the sampling set SS = $\{t_1, t_2, \ldots\}$.

Let us initially assume that model Eqs. (76–1) and (76–2) may correctly predict the outcome of experiments for any inputs $u(t)$ and sampling SS with particular values of parameters p, but without disclosing anything about the biological nature of the observed phenomenon. In this case the model would not be useful to interpret observed variations in the parameters p, e.g., between different subjects or following drug administration, because these would be meaningless without a physiological basis of the model. Mathematical models should therefore not only be able to fit experimental data but also be useful in explaining differences in responses observed under different experimental conditions. In this regard the structure of a model, from which the equations are derived, represents the main factor influencing the interpretation of results. For instance, Figure 76–1 shows two-compartmental models, which have clearly different meanings but that are equivalent from a dynamic viewpoint. That is, for any different system structure there exists a different mathematical expression of the type of Eq. (76–1), which can produce, with some restrictions on parameters, the same model output. Furthermore, from a mathematical point of view the dynamics of all models in Figure 76–1 can be summarized by the so-called impulse response function given by the sum of two exponential terms

$$h(t) = Ae^{-at} + Be^{-bt}, \quad t \geq 0;\ 0, \quad t < 0 \qquad (76\text{–}3)$$

with $p = [A, a, B, b]$ in Eq. (76–1). The model output is then given as a convolution integral,

$$y(t) = \int_{-\infty}^{t} h(t-s)u(s)ds \qquad (76\text{–}4)$$

The equivalence of the various models of Figure 76–1 relies on the fact that for admissible values of p there exist, for each model, parameter values $p_1, p_2, \ldots$ and nonlinear functions $g_1, g_2, \ldots$ that produce the same system output as in Eq. (76–4),

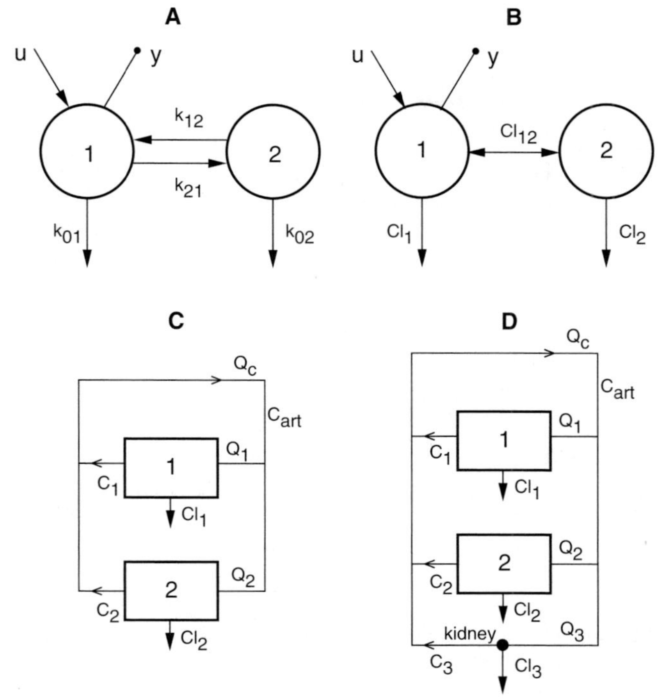

Figure 76–1. Different representations of two-compartment models exhibiting the same dynamic behavior under parameter constraints. (A) Traditional model with k_{ij} representing fractional clearances (time^{-1}), u input of material, and y concentration measurement. The distribution volume V_1 of compartment 1 must be assigned to relate concentration to mass. The model is not uniquely identifiable, unless either k_{01} or k_{02} is zero. (B) Physiological parameterization with Cls representing clearances (volume per time). The exchange of material between compartments depends, as in diffusion, on concentration differences, thus both compartmental volumes V_1 and V_2 must be assigned, giving a total number of parameters as in (A). (C) Physiologically based circulatory model with compartments interpreted typically as tissues with high and low perfusion and Qs and Cs representing blood flows and concentrations, respectively. Additional parameters (not shown) are distribution volumes and blood-tissue partition coefficients. The total number of parameters is thus considerably larger than for the minimal representations in (A) and (B). (D) The same as in (C) with additional elimination from the kidneys assumed having negligible blood volume.

i.e., with $p = g_1(p_1) = g_2(p_2)$. . . . Should the primary biological significance in parameter variations be provided by, say, p_1 then analysis of experimental data based on model Eq. (76–3) would show variability of p rather than p_1, i.e., useful information would be confused by the nonlinear function g_1. This can be extrapolated to the problem of comparing two competing models having both plausible physiological structures. One model may be preferred to another because of more realistic variations in parameters, e.g., in relationship to a disease or drug effect. The claim that the two-exponential function [Eq. (76–3)] is sufficient to describe a linear two-compartmental model is therefore at least misleading as regards the possible interpretations that can be given to between-subject variability of parameters.

Another advantage of adopting physiologically based structured modeling is the straightforward reutilization of already tested system structures to define new models, e.g., by adding compartments or introducing control actions on the transfer rates between compartments.

The concept of equivalence between models can be generalized to the so-called identifiability problem, which is often understated because it is difficult to solve. Let us assume, for example, that a real system's response may be accurately described through Eq. (76–4), i.e., by means of the four parameters in p, but that the experimenter may try to fit the data to a more complex model, i.e., with more than four parameters. In this case the equivalence $p = g_1(p_1)$ would be satisfied by multiple values of p_1 (maybe an infinite number), causing uncertainty in the results and making analyses on parameter variations meaningless. A more subtle situation is given with the same number of parameters but with multiple solutions for p_1 that satisfy the equation $p = g_1(p_1) = g_1(q_1)$, with p_1 different from q_1. To address these methodological issues at least one software tool has been developed.[5]

MODELING SOFTWARE TOOLS

A multitude of general purpose and specialized software tools exist for solving specific problems encountered in modeling and identification of biomedical systems. However, rarely can one software satisfy all the needs in a certain study. Fortunately, modern operating systems of desktop computers greatly facilitate exchange of data and information between different applications. Personal working environments are commonly tailored to individual attitudes and needs, e.g., from data entry through data analysis to final reporting. One should therefore consider also the possibility of utilizing different tools for organizing personalized modeling environments. In this way one could benefit from state-of-the-art, often free open-source, software developed in different disciplines for modeling, numerical simulation and statistical analysis, with added scientific, besides economic, value compared to all-in-one, usually niche commercial, software. Accessibility to a particular software is however often limited mainly by lack of human resources (trainee or expert user) or by the learning curves, that are usually the steeper the more flexible a particular software is. The choice of a software for pursuing a particular modeling project is therefore often suboptimal, and a weakness in the design of a chosen software can become a constraint in the study implementation. Thus, time investment in getting acquainted with more sophisticated and flexible software brings certainly advantages on the long run.

MODELING PARADIGMS Models of biomedical systems should be formulated as explicitly as possible to reveal and ensure coherence in the conceptualization of the actual systems. For instance, mass balance equations are often written implicitly in terms of dynamics of concentrations rather than masses to eliminate unknown distribution volumes viewed as "nuisance" parameters. This can cause trivial mistakes, such as implicitly assuming that a substance and its metabolites have the same distribution volume when they do not, e.g., extracellular versus total body water. Therefore, explicitly declaring even unknown parameters by setting them to conventional values may help to avoid oversights.

The mathematical equations that ultimately characterize the model itself are usually inadequate for a clear representation of systems, especially if documentation is scarce. The equations are typically described in terms of so-called ordinary differential equations (ODE) that can be efficiently solved numerically by computer algorithms. However, as a rule, ODE lose their clear relation to the underlying modeling assumptions and are commonly not self-explanatory except for modeling experts. The feature of modeling software that mainly influences clarity and productivity is, therefore, in our view, the capability for domain-specific rather than equation-oriented modeling. In the traditional equation-oriented approach, mathematical equations that describe the dynamics of a system must be declared explicitly, such as in ACSL (Advanced Continuous Simulation Language) (e.g., www.aegistg.com), or coded in some programming language, such as Fortran or C, that requires specific mathematical modeling and programming skills. Moreover, deriving mathematical equations[6] from a system's concept can be an uncircumventable hurdle for most nonmathematicians. In this regard, to facilitate dissemination and reuse of models, a standard representation for coding mathematical models of biomedical systems, the CellML markup language, was devised at the University of Auckland (www.cellml.org).

In contrast to equation-based modeling, domain-specific modeling consists of declaring a system's structure by means of graphic or textual formalisms conventionally used in a particular discipline, e.g., representation of biochemical reactions or electrical circuit networks, leaving to the software the burden of generating the mathematical model equations or simulation code. Domain-specific modeling thus makes it possible to concentrate on the system's structure and modeling assumptions, reducing the risk of misspecifying model equations due to automated code generation. In the following a few domain-specific modeling paradigms of interest in biomedical research are discussed.

System Dynamics Modeling This modeling paradigm was introduced in the 1960s by J.W. Forrester at MIT to analyze the evolution of economical phenomena and is currently being promoted by the System Dynamics Society (www.systemdynamics.org). The modeling approach consists of a graphic specification of system structure in terms of so-called stock-and-flow diagrams, e.g., reservoirs and transfer of material, also depicting the relationship between variables, in particular, how flows between stocks are controlled by other variables. Thus, this kind of representation naturally fulfills conservation of mass. System dynamics modeling has been widely used in education and has been applied to social as well as physical and biological sciences.[7] Among various software that implement Forrester's System Dynamics paradigm the mostly used in biomedical research is STELLA (ISEE Systems, www.iseesystems.com). Another software historically associated with STELLA is Berkeley Madonna (www.berkeleymadonna.com), which used to lack a graphic model specification interface and relied on model files produced with STELLA. Now Madonna implements a simplified graphic interface based on the System Dynamics paradigm, as well as a textual interface for defining chemical reactions. A distinctive feature of Madonna is the implementation of tools to easily evaluate dependencies of simulation outcomes to parameters through sensitivity analysis and parameter estimation.

Compartmental Models Although the term "compartment" is often used to denote a distinct entity in a lumped-parameter representation of a distributed system, here it is understood in the more restricted sense of a pooled representation of a substance uniformly dissolved in a well-stirred volume of solvent (Figure 76–1). Compartmental models are completely specified by the initial quantity of material and distribution volume of each compartment and by the time-varying flow of material exchanged between compartments. This system representation is thus closely related to system dynamics modeling, but with a succinct graphic

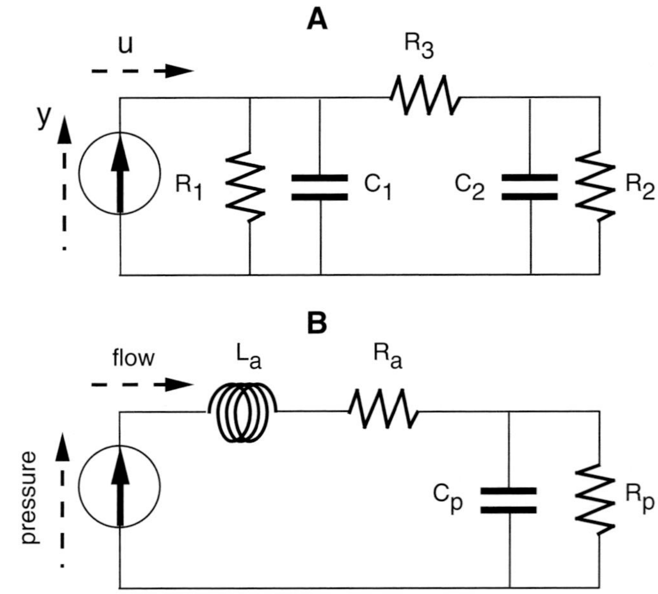

Figure 76–2. Simple electrical circuits with *R*, *C*, and *L* representing resistance, compliance, and inertia, respectively. (A) Dynamically equivalent circuit of the physiological compartment model of Figure 76–1B, if corresponding resistances are the inverses of clearances, i.e., $R = 1/\mathrm{Cl}$, and capacitances are equal to distribution volumes. (B) Electrical analog of a respiratory model with forced ventilation. Subscripts a and p indicate airways and periphery, respectively. L_a accounts for inertia of air.

representation that evidences only compartments and direction of flows. A mathematically well-defined subclass, especially useful for modeling the kinetics of tracers, is represented by linear compartmental models, in which the flow rate between two compartments is proportional to the concentration in the source compartment (Figure 76–1A). Modern software that explicitly tackles this type of models is, for instance, SAAM II (http://depts.washington.edu/saam2).

Electrical/Hydraulic Models Lumped-parameter models of the pressure–flow relationship in airways and blood vessels are conceptually equivalent to electrical circuit networks used, for instance, to model the membrane potential along the dendritic tree of a neuron. It is sufficient to equate pressure with potential and flow with current. Constitutive elements of such models are compliances (capacitors), inertial components (inductances), resistances, and sources of pressure (potential) or flow (current) (Figure 76–2). Conservation principles of mass (charge) and energy yield the so-called Kirchhoff's laws for electric circuits, which are useful for determining the pressures (potentials) at all junctions (nodes) and flows (currents) through each branch in a hydraulic (electric) network. It is worth noting that there are also analogies between electrical circuits and thermodynamic network models. The pioneering software used to simulate this kind of models is SPICE (Simulation Program with Integrated Circuits Emphasis) originally developed by L. Nagel and D. Pederson in the early 1970s at the University of California, Berkeley (http://bwrc.eecs.berkeley.edu/Classes/IcBook/SPICE/).

Signal Flow Diagrams Signal flow diagrams are mainly used in control engineering with blocks representing subcomponents of a system and with directed paths that link the outputs of components with the inputs of other ones (Figure 76–3). Addi-

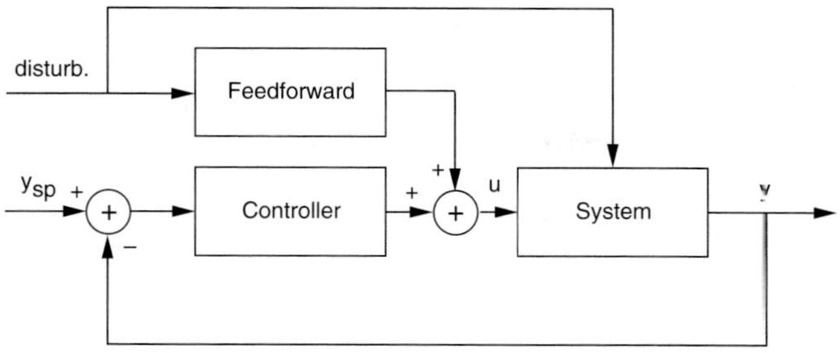

Figure 76–3. Example of a "homeostatic" control system with set point y_{sp}; y and u are controlled and control variable, respectively. The main controller produces its output dependent on the difference $y_{sp} - y$, and the so-called feedforward mechanism anticipates control actions in response to external disturbances of the system. The diagram can be viewed as a sketch of glucose regulation, where the pancreas is the main controller and gastrointestinal hormones are feedforward signals during meal ingestion.

tional graphic elements include summation and other mathematical operator blocks, which are important to specify closed control loops in the system. A fundamental concept of signal flow diagrams is that outputs of subsystems are only measures of some internal quantity and do not represent, e.g., a loss of material. Only inputs to subsystems affect the internal dynamics through their signaling action. This system representation well serves to clarify the control loops and signals in a system, but the implementation of conservation principles, e.g., of mass, can yield quite intricate representations. The software mostly used in implementing this modeling paradigm is Simulink, a graphic package that extends the numerical software Matlab (The MathWorks, www.mathworks.com). Incidentally, a recently developed Matlab extension of interest in biomedical research is named SimBiology, which provides tools for simulating and analyzing biochemical pathways. Nevertheless, other software tools exist for modeling biochemical reactions, such as the free program Gepasi (www.gepasi.org).

Bond Graphs Bond graphs were proposed in 1959 by H.M. Paynter to represent physical systems in terms of elements (ports) that store, dissipate, supply, and transform energy in various physical forms, and bonds that represent the exchange of power between these elements. Special components, so-called junctions, define the way components are connected, i.e., in series or in parallel. The power exchange paradigm naturally implements energy conservation across multiple energetic domains, e.g., mechanical, hydraulic, electrical, chemical, and thermodynamic. This makes bond graphs suited for specifying models of multidomain systems. The two factors whose product defines power, i.e., effort and flow in bond graph terminology, have different interpretations in different energetic domains, e.g., as described previously for hydraulic and electrical systems, but are treated consistently throughout the bond graph theory. The formalism is useful even if the product of effort and flow does not represent a power; in these cases we speak of pseudobond graphs. An important characteristic of bond graphs is that mathematical model equations can be derived symbolically from the graphic representation in a systematic, algorithmic way that can be implemented in computer software. This makes bond graphs intrinsically a multidomain-specific modeling approach suitable for generating model equations that can be easily exported for simulation to other software, e.g., Matlab or ACSL. The first bond graph modeling software developed in the early 1970s by R.C. Rosenberg, ENPORT, has been followed by many others, mainly oriented toward control engineering and industrial applications. The distinctive feature of bond graphs of algorithmic and symbolic derivation of mathematical model equations has pioneered the development of other multidisciplinary modeling languages, of which MODELICA (www.modelica.org) appears to be the most promising. The feasibility of multidisciplinary modeling software for biomedical systems has also been investigated.[8]

STATISTICAL SOFTWARE TOOLS

Statistical evaluation of the agreement between model predictions and experimental data for model identification and validation is a widespread and general issue in all experimental sciences. Biomedical systems modeling, as considered here, can largely take advantage of statistical methods implemented in general purpose statistical software. Only recently emerging modeling issues in postgenomic studies pose new challenging problems to statistical research that require new theoretical solutions and ad hoc software development, e.g., BioConductor (www.bioconductor.org). The integration of software specifically designed for modeling biomedical systems with general purpose statistical tools would therefore be a means to maximize quality and cost effectiveness in software design.

With regard to the more traditional statistical methods, of particular relevance for modeling dynamic biomedical systems are the so-called population pharmacokinetic methods that were pioneered in the early 1980s by L.B. Sheiner and others, which led to the development of established software such as NONMEM (www.globomaxservice.com). In the classical two-stage approach a model is first fitted to dynamic experimental data of each individual, and, in the second stage, model parameter distributions of the studied group and possible effects of covariates are analyzed. Population approaches aim instead at estimating simultaneously all parameters of a hierarchical model with different modeling levels. Typically, the dynamic system response of a single individual represents the first modeling level, and the statistical properties of model parameters and causal relationships between parameters and covariates are defined at the second or higher modeling levels.

The two mainstream population approaches are the Bayesian and the nonlinear mixed-effects modeling methods. Bayesian methods intrinsically use prior information to infer from experimental data the so-called posterior distribution of parameters in a single individual or in a group of subjects. That is, prior information available from previous studies or from the literature is updated, even with sparse experimental data, to obtain model parameter estimates in an individual or a group. Normally these estimates would not be obtainable from the same data without using prior information. For instance, in the extreme case of no availability of data, the estimates returned are simply the prior distributions. Bayesian methods thus have the substantial advantage of providing in any case model parameter estimates, which are more or less biased toward the a priori mean depending on the available information and data. This has clear implications for the design of complex studies in which one part of the experiment can provide prior information from a few individuals, and in the other part, sparse data are collected from many individuals to provide complementary information on between-subject variability. General purpose Bayesian software tools are not very common because, on the one hand, the solution of inference problems typically requires ad hoc implementations, which are, on the other hand, easily put together by experienced statisticians due to powerful yet simple computational methods. Of note is the widespread software BUGS (Bayesian inference Using Gibbs Sampling) (www.mrc-bsu.cam.ac.uk/bugs) and the derived pharmacokinetic-oriented tool PKBUGS.

Nonlinear mixed effects (NLME) modeling methods are, in short, population approaches that view group data as one ensemble and distinguish between parameters that characterize the whole group, so-called fixed effects such as average value of model parameters and effect of treatment, and individual variations from these reference values, so-called random effects. NLME methods make efficient use of experimental data because fixed effects are estimated from the entire data set, making it possible not only to collect sparse data in individual experiments, but also to increase the success rate of model fitting over individual data fitting. Moreover, NLME are more robust than two-step approaches in disentangling between-individual variability of model parameters from statistical uncertainty related to residual errors. NLME estimation procedures are available, in addition to the aforementioned NONMEM, in most advanced statistical packages and data analysis environments. An extensively used software is the NLME package[9] provided with the free open-source statistical software R (www.r-project.org).

CONCLUSIONS

The study of biomedical systems, in a broad sense, involves different scientific disciplines that have historically evolved independently from each other with distinct conventional representations and heuristic mathematical laws used to describe specific physical phenomena. Developing mathematical models within a multidisciplinary context is, therefore, more difficult to accomplish than using software designed for a specific domain, especially if interacting phenomena belong to different domains and the software lacks a common specification dictionary. Unfortunately, no prevailing model description language has yet emerged to allow the exchange and reutilization of models across different simulation software—-not even bond graph modeling, which has been around for decades and presents unique characteristics as to the unifying representation of systems and the algorithmic approach to derive mathematical model equations. Perhaps modern object-oriented modeling languages adopted by different software developers and vendors, such as MODELICA, will fill this gap. An additional limitation of general purpose modeling software is usually the unavailability of robust statistical methods needed for model identification and testing of hypotheses that are indispensable in experimental research. At present, researchers interested in biomedical systems modeling therefore ought to select software tools that best meet their particular needs, either from biomedically oriented niche software that may exhibit adequate modeling and sufficient data analysis features, or from general purpose software tools that generally require integration and interface with other software to create advanced multipurpose modeling and simulation environments. To accomplish this, in addition to some programming skills, access to crucial information on data exchange formats and the like may be required, which is usually available only with nonproprietary software.

REFERENCES

1. Clinical Simulation Center. University of Michigan Health System, Ann Arbor, MI. www.med.umich.edu/umcsc/.
2. Kyoto Encyclopedia of Genes and Genomes (KEGG). Bioinformatics Center and Institute for Chemical Research, Kyoto University, Kyoto, Japan. www.genome.jp/kegg/.
3. The Physiome Project, National Simulation Resource, University of Washington, Seattle, WA. www.physiome.org/.
4. Hunter P, Nielsen P. A strategy for integrative computational physiology. *Physiology* 2005;20:316–325.
5. Audoly S, D'Angio L, Saccomani MP, Cobelli C. Global identifiability of linear compartmental models. A computer algebra algorithm. *IEEE Trans Biomed Eng* 1998;45:36–47.
6. Keener J, Sneyd J. *Mathematical Physiology*. New York: Springer-Verlag, 1998.
7. Hargrove JL. *Dynamic Modeling in the Health Sciences*. New York: Springer-Verlag, 1998.
8. Thomaseth K. Multidisciplinary modelling of biomedical systems. *Comput Methods Programs Biomed* 2003;71:189–201.
9. Pinheiro JC, Bates DM. *Mixed-Effects Models in S and S-PLUS*. New York: Springer-Verlag, 2000.

77 Developing Websites for Biomedical Research and Training

KRISHNA B. SINGH

ABSTRACT

The Internet and web technologies are changing rapidly, bringing a steady stream of exciting new applications and abilities to both the developer and the end user. Web-based applications have unique attributes to provide new and updated teaching materials in an interactive format. Websites designed for biomedical research and training provide rich contents with multiple hyperlinks and multimedia objects. Web-based tools for information exchange include forums, bulletin boards and discussion lists, chat rooms, videoconferencing, and webcasts using streaming technology. Regardless of the delivery format, the design of websites for biomedical research or training raises many issues such as selection of hardware and software authoring tools, collaborative design efforts, pedagogy, copyright or intellectual property issues, and concerns regarding security, privacy, and confidentiality. This chapter briefly reviews the client–server technology, and describes commercial tools for developing an academic website from personal experience. The challenges and motivators for creating websites for biomedical education and research are discussed.

Key Words: Information technology, Internet, World Wide Web, Website, Online education, Medical research, Biomedicine.

INTRODUCTION

Modern information technology (IT) and telecommunications through the Internet play a significant role in biomedical research and training across the globe. The Internet provides a repository of websites and databases on a variety of topics in biomedicine, and many more are constantly added and updated.[1,2] The web technology is changing rapidly, bringing a steady stream of exciting new applications and abilities to both the developer and the end user. Web-based applications have unique attributes to provide new and updated teaching materials in an interactive format. Websites designed for biomedical research and training provide rich contents with multiple hyperlinks and multimedia objects.

Several articles offer comprehensive reviews of website designs and implementation of web-based information.[3–5] Website designs involve distance delivery of print materials, graphics, animations, and audiovisual materials. Web-based tools for information exchange include forums, bulletin boards and discussion lists, chat rooms, videoconferencing, and webcasts using streaming technology.

This chapter briefly reviews the client–server technology, and describes commercial tools for developing an academic website from personal experience. The challenges and motivators for creating websites for biomedical education and research are discussed.

CLIENT–SERVER TECHNOLOGY

The Internet uses standard TCP/IP (transmission control protocol/Internet protocol) stack, which is a set of protocols for data communication over various computer networks. The protocols work together to allow computers using different operating systems and protocols to communicate with one another with a common language.

An intranet is a "private" network of connected computers used to share information within an institution, organization, company, and their affiliates and clients. The principal difference between an intranet and the Internet is that the intranet is not intended for public access. With privacy and security features installed, both Internet and intranets can be used to share a wide variety of media including text documents, graphics, animations, streaming audio and videos, and custom-designed applications.[6–9]

The methodology of website development involves composing web pages in standard hypertext markup language (HTML) and sending them over the Internet or intranet using a protocol called HTTP, which is an acronym for hypertext transfer protocol. However, other web authoring syntax and a variety of applications could also be used as needed. Software applications called a web browser such as Internet Explorer find information from websites and display them on computers connected to the Internet any time and from anywhere in the world. The web content resides on computers called web servers, and the client machine is any computer accessing that server's data.

HARDWARE REQUIREMENTS

In a typical institution or company, the IT personnel install and manage many computers using different operating systems, and hardware and software. The client workstations may include computers running Microsoft Windows, Apple Macintosh, and other

From: *Sourcebook of Models for Biomedical Research*
(P. M. Conn, ed.), © 2008 Humana Press Inc., Totowa, NJ.

platforms with different operating systems and applications. The servers and workstations are connected with an assortment of hardware and software, which might include Ethernet (10 megabits/second, Mbps), fast Ethernet (100 Mbps), and much higher speeds. In such environments, a local area network (LAN) and wide area network (WAN) of interconnected computers can be used for website development. Several routers and hubs are used for interconnectivity and traffic control.

The essential hardware and software for developing websites consist of a web server connected to a network of computers. In our department, the web server runs under a Windows 2003 operating system (Dell Computer Corporation, Round Rock, TX). The ancillary units consist of an automated tape backup, an uninterruptible power supply, and an antivirus program. Technicians from the Microcomputer Services of our institution installed and configured the hardware and software systems.

Based on our experience, the following routines are suggested to maximize the performance of a departmental network using the client–server technology:

1. The central processing unit of the server should be as fast as possible; multiple high-speed microprocessors may be needed for database and image intensive tasks.
2. Random access memory of 1–2 gigabytes or higher should be used.
3. Large hard drives with a capacity of 50–100 gigabytes or more are needed for storage of data and should be as fast as possible; redundant hard disk arrays will provide data backup and security.
4. Network adapters are required for networking. Most enterprise LANs and WANs are Ethernet based, where computers are connected by one of the network interface cards. The 10/100 Base-T card communicates at 10 or 100 Mbps, and higher speeds are becoming common at many academic institutions.
5. The network should offer domain name service, which converts easily remembered alphabetic addresses to the numeric IP address to find other computers over the Internet.

SOFTWARE REQUIREMENTS

Over the past few years, building web-based services and websites has become a popular way of sharing information for biomedical research projects.[1,2] The open source is a trustworthy mechanism for website deployment and distribution of information. Open source technology has become popular because it decreases duplicate efforts and produces rapid development of websites at participating institutions. Several custom software projects have been described that run using an open source model.[7,8] However, many universities are only beginning to draft policies regarding copyright issues for developing open source software.

The basic software needed for creating websites consists of one or more authoring tools and editing applications for texts, images, graphics, animations, simulations, and audio and video streams. Until recently, authoring web pages needed a working knowledge of HTML syntax, which, although not difficult to learn, can be tedious. Web pages can be authored using any text editor, and there are currently a variety of high-quality WYSIWYG (what you see is what you get) editors on the market that do not need any knowledge of HTML syntax. With these editors, anyone who can use a word processor can create static web pages. For a summary of website development tools, visit http://horton.com/tools.

For using database technologies, presenting live content and customization of websites, advanced computer skills are needed. These include a working knowledge of a scripting language such as JavaScript, and sophisticated programming languages such as Java and Perl. The ASP (Active Server Pages) technology was created by Microsoft to allow an easy combination of HTML, scripts (such as JavaScript and Microsoft's VBScript), and ActiveX. Another programming language called Extensible Markup Language or XML provides the foundation for many web services. Much like HTML, XML uses programming tags that define the properties of data on web pages. However, XML tags are far more versatile and can define precisely how data are displayed as well as contents.

Web-based data mining technology presents particular challenges to the standard HTTP protocol and website developers need to learn advanced skills. Webmasters should also have a working knowledge of at least one basic web-enabled database solution such as Microsoft Access. For database intensive tasks, SQL (Structured Query Language) is an example of a database management application that has become popular for querying and managing databases on websites. For skilled webmasters, other client and server-side programming and database creation solutions are available on the market.

CONTENT DEVELOPMENT

This section describes how the author used off-the-shelf commercial tools and created an educational website in the department of obstetrics and gynecology at Louisiana State University Health Sciences Center in Shreveport, Louisiana. The website was developed according to the guidelines of the American Medical Association.[10,11] The departmental faculty provided teaching materials and the authors independently edited the content for consistency and style before postings were done. The departmental website is located at http://obg.lsuhsc-s.edu/obgyn/index.html.

The website development tool automatically creates the web pages with standard HTML syntax (NetObjects Fusion, Website Pros, Inc., Jacksonville, FL). Slideshows were produced using Microsoft PowerPoint, and the training modules were created using Trainersoft software (San Diego, CA). Multiple-choice questions were written using Questionmark software (Stamford, CT). The training videos were edited using Camtasia Studio software (Okemos, MI). CodeCharge software (Utica, MI) was used to create backend databases for case studies on the intranet using Microsoft Access application. To encourage private discussions among the participants a bulletin board was created and hosted on the intranet, which is password protected.

Currently the departmental site includes 238 pages with 3771 links, including those found on external servers providing women's healthcare education. The website provides access to interactive lectures, computer-generated tests, quizzes, searchable databases, and a bulletin board for approved users. Various sections of the website provide links that include administrative and teaching materials. The Lessons and the Visiting Professors sections contain modular topics in reproductive medicine such as menstrual disorders, polycystic ovary syndrome, endometriosis, infertility, and laparoscopy. The Library section is an online source for educational materials in reproductive medicine (Figure 77–1). Table 77–1 provides the main goals and website addresses. Table 77–2 provides information on visitor profiles at the departmental sites from 1998 to 2004.

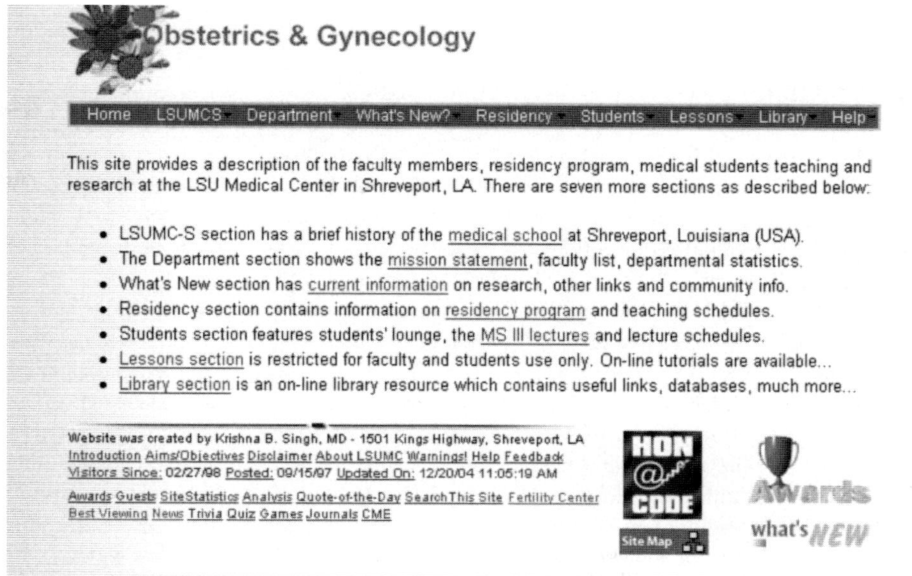

Figure 77–1. Screenshot of the home page at the departmental website showing the navigation bar and links for various sections.

Table 77–1
Websites of the obstetrics and gynecology department (1997–2006)

Website address	Purpose
http://obg.lsuhsc-s.edu/obgyn/index.html	Home page of the department
http://obg.lsuhsc-s.edu/art/index.html	Home page of the Fertility Center
http://obg.lsuhsc-s.edu/movie/index.html	Website resources/history
http://obg.lsuhsc-s.edu/sites/index.html	Academic departments links
http://obg.lsuhsc-s.edu/education/index.html	Educational links; references
http://obg.lsuhsc-s.edu/trainer/quick_start_.html	Resident orientation lectures
http://obg.lsuhsc-s.edu/pco.index.html	Animal PCO[a] models/archives
http://obg.lsuhsc-s.edu/gallery/index.html	Photo archives on PCOS[a]
http://obg.lsuhsc-s.edu/video/index.html	Laparoscopy training videos
http://obg.lsuhsc-s.edu/library/index.html	Healthcare resources/links
http://obg.lsuhsc-s.edu/test/index.html	Announcements, new pages
http://obg.lsuhsc-s.edu/data/index.html	Intranet for case studies[b]

[a]PCO, polycystic ovary; PCOS, polycystic ovary syndrome.
[b]Password-protected intranet site for authenticated users.

Table 77–2
Visitor data on departmental website: 1998–2004[a]

Visitor data	Number
Hits	
Entire site (successful)	3,719.073
Average per day	1,522
Home page	38,447
Page views	
Page views (impressions)	1,010,159
Average per day	413
Document views	1,004,701
Visits	
Total visits	283,300
Average per day	115
Average visit length	00:14:02
Median visit length	NA[b]
International visits	8.28%
Visits of unknown origin	17.33%
Visits from the United States	74.38%
Visitors	
Unique visitors	79,207
Visitors who visited once	66,108
Visitors who visited more than once	13,099

[a]The website is located at http://obg.lsuhsc-s.edu/obgyn/index.html.
[b]NA, not applicable.

DISCUSSION

Web-based virtual learning (also called electronic learning or e-learning) can be defined broadly as any use of the Internet and web technologies to create learning experiences. Web-based training (WBT) is individualized or group tutoring or mentoring delivered over public or private computer networks and displayed by a web browser. WBT is an example of on-demand interactive training stored in a server and accessed online across a network of computers.

Today many software tools for website development are available on the market, mostly at a low cost to individual users for professional website designs and publication. The Internet provides massive amounts of information including interactive tutorials on website development. Popular search engines provide addresses of websites for authoring as well as other website development tools. Online bookstores provide listings for hundreds of books on website development.

During the past decade, the virtual learning knowledge base has exploded, and increasing numbers of institutions and organizations are using innovative methods for online teaching and research.[12–17] To see some examples, visit the Association of Professors of Gynecology and Obstetrics website at http://www.apgo.org/members/index.cfm and review the reproductive medicine sites at several health sciences centers in the United States and Canada. The academic department's websites can also be searched using a tool provided at our departmental website (see Table 77–1).

Developing quality websites symbolizes an evolution that needs experienced website designers, surveys of the target audience, focus groups, and analysis of server data or a combination of these methods. However, posting of information on the web does not always undergo peer reviews or need some standard for publication. The design and implementation of websites need readily available technical support personnel and a variety of tools and techniques.[18,19] The collaboration of IT experts with content experts allows interactive websites to be developed that are cost effective and geared to end users needs. Our website is international in scope, collaborative, asynchronous in delivery, flexible, and responsive to learner needs.[16,17]

The websites designed for biomedical education and research can provide rich contents with many hyperlinks and multimedia objects. Web-based tools for information exchange include forums, bulletin boards and discussion lists, chat rooms, videoconferencing, and webcasting using streaming technology. Webcasting is a new instructional technology used to deliver audio and video presentations via the Internet, enabling learners to participate in a live class via a personal computer. Increasingly, scientific conferences are being webcast to extend the benefits of online education.[20,21]

Syndicated content delivery, known as Really Simple Syndication (RSS), is a popular way to deliver information from websites. Using RSS, a news site or a blog can automatically feed articles to a group of subscribers. Because RSS is still new, not all websites can publish using this medium. However, RSS feeds can be integrated on websites with relative ease to receive the latest information on a given subject.[22] This technology provides a speedy way of updating contents of academic websites.

Web logs, also known as "blogs," are an emerging tool that is easy to use, is web based, and can improve communication, collaboration, and information-gathering skills. Blog postings, typically updated daily, can include images, photos, links, video, audio, or simple text. The postings are archived by date and sometimes by category or by author. Another communication and reference tool called wiki is a hypertext-based collaborative software that enables authoring of documents collectively using a web browser.[23,24]

The blogs and wikis are web-based applications for collaborative knowledge warehouse. These tools have the potential to offer all the features of complex and expensive IT solutions for sharing and exchange of information (see wikipedia.org). An emerging technology called Web 2.0 refers to a second generation of services available on the web that lets people collaborate and share information online. In contrast to the first generation, Web 2.0 gives users an experience closer to desktop applications than the traditional static web pages. These collaboration tools can enable researchers to share knowledge and experience quickly using the available technologies.

Regardless of the delivery format, the design of websites for biomedical education and research raises many issues such as selection of hardware and software authoring tools, collaborative design efforts, pedagogy, and copyright or intellectual property issues. Significant concerns regarding the security, privacy, and confidentiality of consumers have also been raised. Motivating reasons for effective website development include team attributes such as strong leadership and judicious use of available resources. Barriers to development include administrative environments that do not yet fully integrate IT into educational vision and framework, lack of academic incentives for the faculty, and limited technical support staff.[25]

Recently many virtual communities for providing health education as well as support groups for patients have emerged. These cover a wide range of clinical specialties, technologies, and stakeholders. However, more research is needed to evaluate under which conditions and for whom electronic support groups are effective and how the effectiveness for delivering social support electronically can be assessed.[12,13] Chapter 78 (this volume) describes a variety of tools and techniques to build virtual communities for biomedical education and research.

CONCLUSIONS

Web-based applications have unique attributes that provide new and updated teaching materials in an interactive format. Websites designed for biomedical research and training provide rich contents with multiple hyperlinks and multimedia objects. Today many software packages for website development are available on the market, mostly at a low cost to individual users for professional website designs. The Internet and web technologies can promote online collaboration using forums, discussion groups, and content sharing. Newer technologies enable real-time collaboration in various forms. Properly designed websites can motivate, teach, and quickly update learners and teachers involved in biomedical education and research.

REFERENCES

1. Fox JA, Butland SL, McMillan S, Campbell G, Ouellette BF. The Bioinformatics Links Directory: A compilation of molecular biology web servers. *Nucleic Acids Res* 2005;33:W3–24.
2. Neerincx PB, Leunissen JA. Evolution of web services in bioinformatics. *Brief Bioinform* 2005;6:178–188.

3. Liaskos J, Diomidus M. Multimedia technologies in education. *Stud Health Technol Inform* 2002;65:359–372.
4. Huang C. Designing high-quality interactive multimedia learning modules. *Comput Med Imaging Graph* 2005;29:223–233.
5. Cook DA, Dupras DM. A practical guide to developing effective web-based learning. *J Gen Intern Med* 2004;19:698–707.
6. Schleyer TK, Teasley SD, Bhatnagar R. Comparative case study of two biomedical research collaboratories. *J Med Internet Res* 2005;7(5):e53. Available at http://www.jmir.org/2005/5/e53/. Accessed May 31, 2006.
7. McDonald CJ, Schadow G, Barnes M, Dexter P, Overhage JM, Mamlin B, *et al.* Open Source software in medical informatics—why, how and what. *Int J Med Inform* 2003;69:175–184.
8. Jakobovits RM, Rosse C, Brinkley JF. WIRM: An open source toolkit for building biomedical web applications. *J Am Med Inform Assoc* 2002;9:557–570.
9. Frank MS, Dreyer K. Empowering radiologic education on the Internet: A new virtual website technology for hosting interactive educational content on the World Wide Web. *J Digit Imaging* 2001;14:113–116.
10. Silberg WM, Lundberg GD, Musacchio RA. Assessing, controlling, and assuring the quality of medical information on the Internet: Caveant lector et viewor—-Let the reader and viewer beware. *JAMA* 1997;277:1244–1245.
11. Winker MA, Flanagin A, Chi-Lum B, White J, Andrews K, Kennett RL, *et al.* Guidelines for medical and health information sites on the Internet: Principles governing AMA web sites. *JAMA* 2000;283: 1600–1606.
12. Jadad AR, Enkin M. The new alchemy: Transmuting information to knowledge in the electronic age. *CMAJ* 2000;162:1826–1828.
13. Jadad AR, Enkin MW, Glouberman S, Groff P, Stern A. Are virtual communities good for our health? *BMJ* 2006;332:925–926.
14. Jenkins JM. The Internet, intranets and reproductive medicine. *Hum Reprod* 1999;14:586–589.
15. Zondervan K, Cardon L, Kennedy S. Development of a web site for the genetic epidemiology of endometriosis. *Fertil Steril* 2002;78: 777–781.
16. Singh KB. How one academic web site was created, and content standards used (letter). *Am Med News*, October 21, 2002. Available at http://www.ama-assn.org/amednews/2002/ltr02.htm. Accessed May 31, 2006.
17. Singh KB. An academic website in reproductive medicine. *Acad Exchange Q* 2005;9:303–306.
18. Grannell MS, Singh RR, Tang R, Mansoor S, Walsh TN. Designing a medical web site. *Ir J Med Sci* 2001;170:123–125.
19. Ryan AG, Louis LJ, Yee WC. Informatics in radiology (infoRAD): HTML and web site design for the radiologist: A primer. *Radiographics* 2005;25:1101–1118.
20. Locatis C, Fontelo P, Sneiderman C, Ackerman M, Uijtdehaage S, Candler C, *et al.* Webcasting videoconferences over IP: A synchronous communication experiment. *J Am Med Inform Assoc* 2003;10:150–153.
21. Yagi Y, Ahmed I, Gross W, Becich MJ, Demetris AJ, Wells A, *et al.* Webcasting pathology department conferences in a geographically distributed medical center. *Hum Pathol* 2004;35:790–797.
22. Cunningham K. Health and medical news on the web: Comparing the results of news-providing web resources. *Med Ref Serv Q* 2005;24:17–40.
23. Sauer IM, Bialek D, Efimova E, Schwartlander R, Pless G, Neuhaus P. "Blogs" and "wikis" are valuable software tools for communication within research groups. *Artif Organs* 2005;29:82–83.
24. Altmann U. Representation of medical informatics in the wikipedia and its perspectives. *Stud Health Technol Inform* 2005;116:755–760.
25. Simmons C, Nyhof-Young J, Bradley J. Shoestring budgets, band-aids, and team work: Challenges and motivators in the development of a web-based resource for undergraduate clinical skills teaching. *J Med Internet Res* 2005;7:e14. Available at http://www.jmir.org/2005/2/e14/. Accessed May 31, 2006.

78 Building Virtual Research Communities Using Web Technology

Krishna B. Singh

ABSTRACT

Exchanging information and building communication channels are critical ingredients of biomedical education and research. Collaboration tools can help researchers work in harmony and learn together at a distance. This category spans a wide variety of applications from simple text-based e-mail clients to complex online meeting tools. E-mail is the oldest, most widely used, and effective collaboration tool. Online discussions go by various formats and names such as discussion groups, bulletin boards, and discussion forums. The Internet and web technologies have the potential to increase the productivity of biomedical research. The Internet collaboratories can support expensive equipment to address complex problems, which can speed up discovery and innovation in research. Having the right tools and technology is a necessary foundation and building a community needs conscious effort among website designers, community promoters, and leaders. Integrating electronic collaborative tools into routine scientific practice can be successful but requires further research.

Key Words: Internet, World Wide Web, Online education, Biomedical research, Groupware, Collaboratory.

INTRODUCTION

Advances in Internet and web technologies over the past decade have resulted in increased opportunities for consumers and the medical community to become more innovative in biomedical research. These innovations have allowed instantaneous access to medical information at the point of need. Web-based applications have unique attributes to provide new and updated teaching materials in an interactive format. By capitalizing on technologies such as laptop or notebook computers, personal digital assistants, and web-based applications people have portable and remote access to biomedical information almost instantaneously. In the literature, several types of virtual collaboration tools have been described.[1–5] The use of electronic tools such as e-mail, bulletin boards, and web messaging for collaboration is also emerging in the medical practice and allied fields.[6–9]

This chapter explains the terminology and reviews potential uses of a wide range of Internet and web-based collaboration tools. Information on how some institutions have used the technology to develop large-scale collaborative research initiatives is also presented. Finally, how integrating information technology might harmonize biomedical education and research and its potential significance are discussed.

TERMINOLOGY

Many collaboration tools use the server–client architecture or its variants on a network of computers. In technical terms, a client is a computer system that accesses the service on another computer called a server using network connections. The client–server model is still used today on the Internet, where a user may connect to a service operating on a remote system through the Internet protocol suites. Web browsers are clients that connect to web servers and retrieve web pages for display. Most people use e-mail clients to retrieve their e-mail from their Internet service provider's mail storage servers. E-mail list servers are a popular Internet tool for online discussions. Instant messaging consists of instant communications between two or more people on the Internet. Instant messaging requires the use of a client program that hooks up an instant messaging service. An instant message differs from e-mail in that conversations happen in real time. Popular instant messaging services include Microsoft's MSN Messenger, AOL Instant Messenger, Yahoo! Messenger, Google Talk, and iChat for Apple computers. An older and still popular online chat medium is known as Internet Relay Chat or IRC. Technically, instant messaging typically boosts communication and allows easy collaboration among participants. Online chat can refer to any kind of communication over the Internet, but generally refers to a direct one-on-one chat in chat rooms, or using tools such as instant messenger applications (visit http://en.wikipedia.org).

Web-based virtual learning (also called electronic learning or e-Learning) can be defined broadly as any use of the Internet and web technologies to create teaching and learning experiences. Web-based training (WBT) is individualized or group tutoring or mentoring delivered over public or private computer networks and displayed by a web browser. WBT is an example of on-demand interactive online training stored in a server and accessed across a network of computers. WBT can be updated very rapidly, and the training provider administers access to the training. The phrase virtual community or online community is often hyped in different ways (visit http://en.wikipedia.org). The digital divide is a term that describes the differences between those who have access to the Internet and those who do not because of economic reasons, lack of computer competency or self-efficacy, or different knowledge management styles.

From: *Sourcebook of Models for Biomedical Research* (P. M. Conn, ed.), © 2008 Humana Press Inc., Totowa, NJ.

Table 78–1
Synchronous collaboration tools for real-time virtual meetings

Tool	*What it does*
Instant messaging/chat	Real-time one-to-one text messaging delivered over the Internet
Internet relay chat	Real-time many-to-many text messaging system delivered over the Internet
Whiteboarding/annotating	Real-time display screen that is shared and viewed by multiple users over the Internet and on which users can write or draw
Application and desktop sharing	Allows multiple users to view and share a computer desktop or application over the Internet
Audio conferencing	Allows multiple users to talk with each other in real time over the Internet
Video conferencing/video chat	Allows multiple users to see and hear each other in real time using streaming video and audio over the Internet
Interactive web collaboration tools	These tools combine many of the applications listed above and enable users to interact in a number of ways

Source: Roberts S. Choosing tools for real-time virtual meetings. Available at http://coe.sdsu.edu/eet/articles/sitools/index.htm.

Communication technologies can be broken down into two basic categories: (1) those that allow asynchronous collaboration in which communication does not occur in real time, and (2) those that allow synchronous collaboration in which communication takes place over networks in real time. Examples of asynchronous technologies include traditional correspondence courses, e-mail, database, and message boards. There are advantages to asynchronous learning when using the web-based technologies because hypermedia and hyperlinks can provide large amounts of information in distant locations. Examples of synchronous technologies include instant messaging and chat applications, Internet relay chat, whiteboarding, application sharing, and audioconferencing and videoconferencing using the streaming technology (Table 78–1). These media are expensive to buy or build but permit more collaborative work, brainstorming, learning, or tutoring in real time, and may also be more productive than asynchronous communications.[2–5]

Exchanging information and building communication channels are critical ingredients of biomedical education and research. Biomedical researchers face hard constraints for managing the vast amount of information produced during research projects. Collaboration tools can help researchers work in harmony and learn together at a distance. Most researchers desirous of creating a virtual space to share common experiences with colleagues can use the tools described below.

COLLABORATION TOOLS

This category spans a wide variety of applications from simple text-based e-mail clients to complex online meeting tools. The collaboration tools let participants share their ideas, and are essential for collaborative online learning and knowledge management initiatives. Providing a collaborative environment may require several separate tools using the Internet and web technologies.

E-mail is the oldest, most widely used, and effective collaborative tool preferred by patients and doctors alike.[6,7] The literature contains several articles in which the uses of bulletin boards and instant messaging have been described.[8,9] Online discussions go by various formats and names such as discussion groups, bulletin boards, and discussion forums. These applications are provided in a general-purpose server such as Microsoft Exchange Server. Alternatively, they can be part of an institution-wide virtual learning system such as Blackboard (blackboard.com). Simple bulletin boards and forums can also be integrated at websites using site development tools such as NetObjects Fusion (see Chapter 77, this volume).

With tools known as groupware, researchers can communicate easily with one another to perform difficult tasks even though they are remotely located or rarely overlap in time. There are several groupware applications such as Lotus Notes (IBM, Armonk, NY), eRoom (EMC Corporation, Pleasanton, CA), and Groove (Microsoft, Inc., Redmond, WA). Groove is a secure, peer-to-peer collaborative tool that integrates with a wide variety of applications such as file sharing, threaded discussions, web links, document review, and calendar into a single workspace. The product also integrates well with Microsoft Office suite. Because of its peer-to-peer architecture, the administrative overhead is much lower than that of some server-based applications such as Lotus Notes. For a list of groupware tools, visit http://www.grantbow.com/groupware.html.

With modern technology, research collaborations have advanced beyond simply providing shared access to documents on the Internet. Researchers engaged in complex initiatives have realized that running collaborative projects with great efficiency needs far more structure and organization than a basic shared network can manage. This problem is further compounded because research projects often span departments, divisions, and institutional boundaries on a global scale. Examples of newer and promising collaboration tools used in the medical and allied field include wikipedia, blogs, syndicated media, podcasting, and webcasting using streaming technologies.[10–15] Interactive tutorials on several tools and technologies for building virtual communities are available at http://www.imarkgroup.org/moduledescrC_en.asp. The next section describes custom-built collaborating systems, called collaboratories, and discusses what they are and what research domains they serve.

COLLABORATORY COMMUNITIES

The term collaboratory was first coined in the late 1980s to describe large-scale collaborations among researchers using the Internet infrastructure and tools. A collaboratory can be defined as an information technology large-scale infrastructure that supports cooperation among individuals, groups, or organizations in pursuit of a shared goal by simplifying communication and knowledge sharing.[16–18] Recently several collaboratories have been developed for online education and biomedical research. Using collaboratories, researchers can share access to large datasets and shared environments, support expensive equipment to address complex problems, and speed up discovery and innovation. They can also use an emerging class of advanced network-based applications referred to as "grid" computing. This approach

Table 78–2
Collaboratories in the United States and their domains of expertise

Name	*Purpose*	*Web reference*
Biomedical Informatics Research Network	Data sharing across neuroimaging databases in the United States	http://www.nbirn.net
Biological Collaborative Environment	Molecular modeling and simulation	http://www.ks.uiuc.edu/research/biocore
Molecular Modeling Collaboratory	Extensible, interactive molecular modeling software	http://www.cgl.ucsf.edu/research/collaboratory
National Laboratory for the Study of Rural Telemedicine	Established the first virtual hospital	http://www.vh.org
Visible Human Project	Three-dimensional representations of the normal male and female bodies	http://vhp.med.umich.edu
Environmental Molecular Sciences Laboratory	New communications technologies	http://collaboratory.emsl.pnl.gov

is flexible, secure, and provides coordinated resource sharing among many individuals and resources.

The Environmental Molecular Sciences Laboratory (EMSL) is a new national user facility housing various expensive scientific instruments. Development of the EMSL collaboratory is a symmetric collaboration between computer scientists, domain scientists (physical and biological sciences), and sociologists. The collaboratory relies on developing new communications technologies—shared computer displays, electronic notebooks, and virtual reality collaboration spaces with an integration of these technologies with videoconferencing and e-mail abilities. Table 78–2 provides a list of unique research projects incorporating collaboratory technologies in the United States.

DISCUSSION

The number of virtual communities that focus on healthcare topics has recently exploded. For example, Yahoo Groups (http://groups.yahoo.com) shows thousands of virtual communities related to healthcare and alternative medicine. Explosive growth of the Internet and web technologies at affordable costs plus consumer satisfaction with communities partly explain this phenomenon. The implications of rapidly growing health-related virtual communities need further research.[19–21]

Conducting online Internet-based research raises several ethical questions, especially about privacy and informed consent.[22–24] Researchers and peers from institutional review boards primarily decide whether research is intrusive and has the potential for harm, how confidentiality can be protected, and whether and how informed consent should be obtained.[21] Developing private intranet communities and websites internally provides administrative control and choices for leasing or buying hardware and software. However, it may be more proper to use the services of a community aggregator or application service provider depending on the aims of the project and available finances.

Application service providers can offer a rich collection of templates and tools to those who want to run their own online community. Internet service providers and major search engines also offer a wide range of collaborative tools at no extra cost to end users. For example, MSN provides different types of web-based e-mail servers, MSN messenger, chat rooms, and an application called spaces for creating weblogs or blogs.

The benefits of having an online community are real, but building a virtual community is not always straightforward. Having the right tools and technology is a necessary foundation, but does not guarantee that a community will work as expected. Building a community needs conscious effort among website designers, community promoters, and leaders. The builder of a virtual education and research community needs to determine what will hold it together as it expands with partnerships in the future.

Interactive websites for patient communication called patient portals may improve communication between patients and their clinics and physicians. In a randomized controlled trial, portal group patients demonstrated increased satisfaction with communication and overall care. Patients in the portal group particularly valued the portal's convenience, reduced communication barriers, and direct physician responses. More online messages from patients contained informational and psychosocial content compared to telephone calls, which may enhance the patient–physician relationship.[25] Institutions and organizations must promote cost-effective methods for researchers to communicate and collaborate efficiently with peers.

The major barriers to virtual networking include lack of time and motivation and negative attitudes to information technology among participants. Other factors for implementing virtual networks are related to technical and logistic issues. Integrating electronic collaborative tools into routine scientific practice can be successful, but requires further research on the technical, social, and behavioral factors influencing the adoption and use of collaboration tools.[18] Personal digital assistants, mobile telephones, fast services by direct mail, fax machines, and audioconferencing and videoconferencing are common tools for modern communication. For more demanding needs, researchers now have many choices for combining interactive tools with the traditional systems.

CONCLUSIONS

Web-based applications have unique attributes that provide new and updated teaching materials in an interactive format. Today many off-the-shelf communication tools for information gathering and biomedical research can be integrated at websites or used as stand-alone applications. With rapidly advancing information technology, the biomedical research community continues to harness trends in an emerging cyberspace infrastructure. These trends include collaborative scientific experiments and integration of resources from diversified fields. Developing a collaboratory empowers teamwork and collaborative approaches that support large-scale and information-rich biomedical investigations. Scientific research at regional and national collaboratories provides new insight into the biology of health as well as the prevention and cure of diseases.

REFERENCES

1. Bowes J. Building online communities for professional networks. Available at http://globalsummit.educationau.edu.au/globalsummit/papers/jbowes.htm. Accessed May 30, 2006.
2. Boettcher JV. Designing for the virtual interactive classroom. Available at http://www.campus-technology.com/article.asp?id=11046. Accessed May 30, 2006.
3. Roberts S. Choosing tools for real-time virtual meetings. Available at http://coe.sdsu.edu/eet/articles/sitools/index.htm. Accessed May 30, 2006.
4. Kirwin A. Technology and collaborative environments in distance education. Available at http://coe.sdsu.edu/eet/articles/collabenv/index.htm. Accessed May 30, 2006.
5. Horton W, Horton K. *E-Learning Tools and Technologies: A Consumer's Guide for Trainers, Teachers, Educators, and Instructional Designers*, 1st ed. Indianapolis, IN: Wiley Publishing Inc., 2003.
6. Brooks RG, Menachemi N. Physician use of e-mail with patients: Factors influencing electronic communication and adherence to best practices. *J Med Internet Res* 2006;8:e2. Available at http://www.jmir.org/2006/1/e2/. Accessed May 30, 2006.
7. Moyer CA, Stern DT, Dobias KS, Cox DT, Katz SJ. Bridging the electronic divide: Patient and provider perspectives on e-mail communication in primary care. *Am J Manag Care* 2002;8:427–433.
8. Culver JD, Gerr F, Frumkin H. Medical information on the Internet: A study of an electronic bulletin board. *J Gen Intern Med* 1997;12:466–470.
9. Liederman EM, Lee JC, Baquero VH, Seites PG. Patient-physician web messaging. The impact on message volume and satisfaction. *J Gen Intern Med* 2005;20:52–57.
10. Altmann U. Representation of Medical Informatics in the wikipedia and its perspectives. *Stud Health Technol Inform* 2005;116:755–760.
11. Sauer IM, Bialek D, Efimova E, Schwartlander R, Pless G, Neuhaus P. "Blogs" and "wikis" are valuable software tools for communication within research groups. *Artif Organs* 2005;29:82–83.
12. Maag M. The potential use of "blogs" in nursing education. *Comput Inform Nurs* 2005;23:16–24.
13. Cunningham K. Health and medical news on the web: Comparing the results of news-providing web resources. *Med Ref Serv Q* 2005;24:17–40.
14. Maag M. Podcasting and MP3 players: Emerging education technologies. *Comput Inform Nurs* 2006;24:9–13.
15. DiMaria-Ghalili RA, Ostrow L, Rodney K. Webcasting: A new instructional technology in distance graduate nursing education. *J Nurs Educ* 2005;44:11–18.
16. Gantenbein RE. Designing an Internet-based collaboratory for biomedical research. *Biomed Sci Instrum* 2002;38:399–404.
17. Schleyer TK. Collaboratories: Leveraging information technology for cooperative research. *J Dent Res* 2001;80:1508–1512.
18. Schleyer TKL, Teasley SD, Bhatnagar R. Comparative case study of two biomedical research collaboratories. *J Med Internet Res* 2005;7:e53. Available at http://www.jmir.org/2005/5/e53/. Accessed May 30, 2006.
19. Jadad AR, Enkin MW, Glouberman S, Groff P, Stern A. Are virtual communities good for our health? *BMJ* 2006;332:925–926.
20. Eysenbach G, Powell J, Englesakis M, Rizo C, Stern A. Health related virtual communities and electronic support groups: Systematic review of the effects of online peer to peer interactions. *BMJ* 2004;328:1166.
21. Eysenbach G, Till JE. Ethical issues in qualitative research on Internet communities. *BMJ* 2001;323:1103–1105.
22. The Department of Health and Human Services. Standards for privacy of individually identifiable health information. Available at http://www.hipaadvisory.com/regs/regs_in_PDF/finalprivmod.pdf. Accessed May 30, 2006.
23. Houghton G, Singh S, Fraser J. Using email as a research tool in general practice: Starting to implement the National Service Framework for Mental Health. *Inform Prim Care* 2003;11:27–31.
24. Kralik D, Warren J, Price K, Koch T, Pignone G. The ethics of research using electronic mail discussion groups. *J Adv Nurs* 2005;52:537–545.
25. Lin CT, Wittevrongel L, Moore L, Beaty BL, Ross SE. An Internet-based patient-provider communication system: Randomized controlled trial. *J Med Internet Res* 2005;7:e47. Available at http://www.jmir.org/2005/4/e47. Accessed May 30, 2006.

Index

D

E

N

O

P

T